Commercial Renovation Costs with RSMeans data

Thomas Lane, Senior Editor

2018
39th annual edition

WITHDRAWN

D0911487

Chief Data Officer
Noam Reininger

Engineering Director
Bob Mewis, CCP

Contributing Editors
Christopher Babbitt
Sam Babbitt
Michelle Curran
Matthew Doheny (8)
Cheryl Elsmore
Linval Gentles
John Gomes (13, 41)
Derrick Hale, PE (2, 31, 32, 33, 34, 35, 44, 46)
Wafaa Hamitou (11, 12)

Joseph Kelble (14, 21, 22, 23, 25, 26, 27, 28, 48)
Charles Kibbee (1, 4)
Gerard Lafond, PE
Thomas Lane (6, 7)
Genevieve Medeiros
Elisa Mello
Ken Monty
Marilyn Phelan, AIA (9, 10)
Stephen C. Plotner (3, 5)
Callum Riley
Stephen Rosenberg
Jeff Sessions
Gabe Sirota

Matthew Sorrentino
Kevin Souza
Keegan Spraker
Tim Tonello
Jen Walsh
David Yazbek

Product Manager
Andrea Sillah

Production Manager
Debbie Panarelli

Production
Jonathan Forgit
Mary Lou Geary

Sharon Larsen
Sheryl Rose

Technical Support
Judy Abbruzzese
Gary L. Hoitt

Cover Design
Blaire Collins

Data Analytics
Tim Duggan
Todd Glowac
Matthew Kelliher-Gibson

Numbers in italics are the divisional responsibilities for each editor. Please contact the designated editor directly with any questions.

Gordian RSMeans data
Construction Publishers & Consultants
1099 Hingham Street, Suite 201
Rockland, MA 02370
United States of America
1-800-448-8182
www.RSMeans.com

Copyright 2017 by The Gordian Group Inc.
All rights reserved.
Cover photo © iStock.com/Pulvret80

Printed in the United States of America
ISSN 2161-4628
ISBN 978-1-946872-02-9

0048 $266.99 per copy (in United States)
Price is subject to change without prior notice.

Related Data and Services

2018 Commercial Renovation Costs with RSMeans data has been tirelessly researched and carefully compiled to provide construction cost data for estimating small jobs with small quantities of work and large jobs that encompass major renovations. As with all RSMeans cost data, with reasonable exercise of judgment, this data can be used for estimating jobs with a wide range of costs.

Our engineers recommend the following products and services to complement *Commercial Renovation Costs with RSMeans data:*

Annual Cost Data Books
2018 Facilities Construction Costs with RSMeans data
2018 Facilities Maintenance & Repair Costs with RSMeans data

Reference Books
Estimating Building Costs
RSMeans Estimating Handbook
Green Building: Project Planning & Estimating
How to Estimate with RSMeans data
Plan Reading & Material Takeoff
Project Scheduling & Management for Construction
Universal Design for Style, Comfort & Safety

Seminars and In-House Training
Facilities Construction Estimating
Training for our online estimating solution
Practical Project Management for Construction Professionals
Scheduling with MSProject for Construction Professionals
Mechanical & Electrical Estimating

RSMeans data Online
For access to the latest cost data, an intuitive search, and an easy-to-use estimate builder, take advantage of the time savings available from our online application. To learn more visit: www.RSMeans.com/2018online.

Enterprise Solutions
Building owners, facility managers, building product manufacturers, and attorneys across the public and private sectors engage with RSMeans data Enterprise to solve unique challenges where trusted construction cost data is critical. To learn more visit: www.RSMeans.com/Enterprise.

Custom Built Data Sets
Building and Space Models: Quickly plan construction costs across multiple locations based on geography, project size, building system component, product options, and other variables for precise budgeting and cost control.

Predictive Analytics: Accurately plan future builds with custom graphical interactive dashboards, negotiate future costs of tenant build-outs, and identify and compare national account pricing.

Consulting
Building Product Manufacturing Analytics: Validate your claims and assist with new product launches.

Third-Party Legal Resources: Used in cases of construction cost or estimate disputes, construction product failure vs. installation failure, eminent domain, class action construction product liability, and more.

API
For resellers or internal application integration, RSMeans data is offered via API. Deliver Unit, Assembly, and Square Foot Model data within your interface. To learn more about how you can provide your customers with the latest in localized construction cost data visit: www.RSMeans.com/API.

Table of Contents

Foreword

The Value of RSMeans data from Gordian

Since 1942, RSMeans data has been the industry-standard materials, labor, and equipment cost information database for contractors, facility owners and managers, architects, engineers, and anyone else that requires the latest localized construction cost information. Over 75 years later, the objective remains the same: to provide facility and construction professionals with the most current and comprehensive construction cost database possible.

With the constant influx of new construction methods and materials, in addition to ever-changing labor and material costs, last year's cost data is not reliable for today's designs, estimates, or budgets. The RSMeans data engineers invest over 22,000 hours in cost research annually and apply real-world construction experience to identify and quantify new building products and methodologies, adjust productivity rates, and adjust costs to local market conditions across the nation. This unparalleled construction cost expertise is why so many facility and construction professionals rely on RSMeans data year over year.

About Gordian

Gordian originated in the spirit of innovation and a strong commitment to helping clients reach and exceed their construction goals. In 1982, Gordian's Chairman and Founder, Harry H. Mellon, created Job Order Contracting while serving as Chief Engineer at the Supreme Headquarters Allied Powers Europe. Job Order Contracting is a unique indefinite delivery/indefinite quantity (IDIQ) process, which enables facility owners to complete a substantial number of repair, maintenance, and construction projects with a single, competitively awarded contract. Realizing facility and infrastructure owners across various industries could greatly benefit from the time and cost saving advantages of this innovative construction procurement solution, he established Gordian in 1990.

Continuing the commitment to providing the most relevant and accurate facility and construction data, software, and expertise in the industry, Gordian enhanced the fortitude of its data with the acquisition of RSMeans in 2014. And in an effort to expand its facility management capabilities, Gordian acquired Sightlines, the leading provider of facilities benchmarking data and analysis, in 2015.

Our Offerings

Gordian is the leader in facility and construction cost data, software, and expertise for all phases of the building life cycle. From planning to design, procurement, construction, and operations, Gordian's solutions help clients maximize efficiency, optimize cost savings, and increase building quality with its highly specialized data engineers, software, and unique proprietary data sets.

Our Commitment

At Gordian, we do more than talk about the quality of our data and the usefulness of its application. We stand behind all of our RSMeans data — from historical cost indexes to construction materials and techniques — to craft current costs and predict future trends. If you have any questions about our products or services, please call us toll-free at 800-448-8182 or visit our website at www.gordian.com.

MasterFormat® 2014/ MasterFormat® 2016 Comparison Table

This table compares the 2014 edition of the Construction Specifications Institute's MasterFormat® to the expanded 2016 edition. For your convenience, all revised 2014 numbers and titles are listed along with the corresponding 2016 numbers and titles. In some cases, a designation of RSMeans is used to identify sections of data numbered exclusively in RSMeans products.

CSI 2014 MF ID	CSI 2014 MF Description	2014 Designation	CSI 2016 MF ID	CSI 2016 MF Description	2016 Designation
35 20 23.23	Hydraulic Dredging	CSI	35 24 13.13	Cutter Suction Dredging	RSMeans
35 20 23.13	Mechanical Dredging	CSI	35 24 23.13	Mechanical Dredging	CSI
35 20 23	Dredging	CSI	35 24 23	Clamshell Dredging	RSMeans
35 20 16.73	Slide Gates	RSMeans	35 22 73.16	Slide Gates	CSI
35 20 16.69	Knife Gates	RSMeans	35 22 69.16	Knife Gates	RSMeans
35 20 16.66	Flap Gates	RSMeans	35 22 66.16	Flap Gates	RSMeans
35 20 16.63	Canal Gates	RSMeans	35 22 63.16	Canal Gates	RSMeans
35 20 16.26	Hydraulic Sluice Gates	CSI	35 22 26.16	Hydraulic Sluice Gates	RSMeans
35 20 16	Hydraulic Gates	CSI	35 22 26	Sluice Gates	RSMeans
35 20	Waterway and Marine Construction and Equipment	CSI	35 22	Hydraulic Gates	RSMeans
33 72 33.46	Substation Converter Stations	CSI	33 78 33.46	Substation Converter Stations	RSMeans
33 72 33.36	Cable Trays For Utility Substations	CSI	26 05 36.36	Cable Trays For Utility Substations	CSI
33 72 33.33	Raceway/Boxes For Utility Substations	CSI	26 05 33.33	Raceway/Boxes For Utility Substations	CSI
33 52 43.13	Aviation Fuel Piping	CSI	33 52 13.43	Aviation Fuel Piping	RSMeans
33 52 16.13	Gasoline Piping	CSI	33 52 13.16	Gasoline Piping	RSMeans
33 52 13.14	Petroleum Products	RSMeans	33 52 13.19	Petroleum Products	RSMeans
33 51 33.10	Piping, Valves & Meters, Gas Distribution	RSMeans	33 59 33.10	Piping, Valves & Meters, Gas Distribution	CSI
33 51 33	Natural-Gas Metering	CSI	33 59 33	Natural-Gas Metering	CSI
33 51 13.30	Piping, Gas Service & Distribution, Steel	RSMeans	33 52 16.16	Piping, Gas Service & Distribution, Steel	RSMeans
33 51 13.20	Piping, Gas Service & Distribution, Steel	RSMeans	33 52 16.13	Steel Natural Gas Piping	RSMeans
33 51 13.10	Piping, Gas Service and Distribution, Polyethylene	RSMeans	33 52 16.20	Piping, Gas Service And Distribution, Polyethylene	CSI
33 49 23	Storm Drainage Water Retention Structures	CSI	33 46 23	Modular Buried Stormwater Storage Units	CSI
33 49 13	Storm Drainage Manholes, Frames, and Covers	CSI	33 05 61	Concrete Manholes	RSMeans
33 47 19	Water Ponds and Reservoirs	CSI	33 46 11	Stormwater Ponds	CSI
33 47 13.54	Garden Ponds	RSMeans	13 12 13.54	Garden Ponds	CSI
33 47 13.53	Reservoir Liners HDPE	CSI	31 05 19.53	Reservoir Liners HDPE	RSMeans
33 47 13	Pond and Reservoir Liners	CSI	31 05 19	Geosynthetics for Earthwork	RSMeans
33 46 26.10	Geotextiles For Subsurface Drainage	RSMeans	33 41 23.19	Geosynthetic Drainage Layers	RSMeans
33 46 26	Geotextile Subsurface Drainage Filtration	CSI	33 41 23	Drainage Layers	CSI
33 46 16	Subdrainage Piping	CSI	33 41 16	Subdrainage Piping	CSI
33 46	Subdrainage	CSI	33 41	Subdrainage	CSI
33 44 16	Utility Trench Drains	CSI	33 42 36	Stormwater Trench Drains	CSI
33 44 13	Utility Area Drains	CSI	33 42 33	Stormwater Curbside Drains and Inlets	RSMeans
33 42 16.15	Oval Arch Culverts	RSMeans	33 42 13.15	Oval Arch Culverts	RSMeans
33 42 16.13	Culverts & Box Trench Sections	CSI	33 42 13.14	Culverts & Box Trench Sections	RSMeans
33 41 13	Public Storm Utility Drainage Piping	CSI	33 42 11	Stormwater Gravity Piping	CSI
33 41	Storm Utility Drainage Piping	CSI	33 42	Stormwater Conveyance	CSI
33 36 50	Drainage Field Systems	RSMeans	33 34 51	Drainage Field System	CSI
33 36 33.13	Utility Septic Tank Tile Drainage Field	CSI	33 34 51.13	Utility Septic Tank Tile Drainage Field	RSMeans
33 36 19	Utility Septic Tank Effluent Filter	CSI	33 34 16	Septic Tank Effluent Filters	CSI
33 36 13.19	Polyethylene Utility Septic Tank	CSI	33 34 13.33	Polyethylene Septic Tanks	CSI
33 36 13.13	Concrete Utility Septic Tank	CSI	33 34 13.13	Concrete Septic Tanks	RSMeans
33 36 13	Utility Septic Tank and Effluent Wet Wells	CSI	33 34 13	Septic Tanks	CSI
33 36	Utility Septic Tanks	CSI	33 34	Onsite Wastewater Disposal	CSI
33 31 13	Public Sanitary Utility Sewerage Piping	CSI	33 31 11	Public Sanitary Sewerage Gravity Piping	CSI
33 21 13	Public Water Supply Wells	CSI	33 11 13	Potable Water Supply Wells	RSMeans
33 21	Water Supply Wells	CSI	33 11	Groundwater Sources	CSI
33 16 19.50	Elevated Water Storage Tanks	RSMeans	33 16 11.50	Elevated Water Storage Tanks	CSI
33 16 19	Elevated Water Utility Storage Tanks	CSI	33 16 11	Elevated Composite Water Storage Tanks	CSI
33 16 13.29	Wood Water Storage Tanks	RSMeans	33 16 59.29	Wood Water Storage Tanks	CSI
33 16 13.23	Plastic-Coated Fabric Pillow Water Tanks	RSMeans	33 16 56.23	Plastic-Coated Fabric Pillow Water Tanks	CSI
33 16 13.19	Horizontal Plastic Water Tanks	CSI	33 16 56.19	Horizontal Plastic Water Tanks	CSI
33 16 13.16	Prestressed Conc. Water Storage Tanks	CSI	33 16 36.16	Prestressed Conc. Water Storage Tanks	RSMeans
33 16 13.13	Steel Water Storage Tanks	CSI	33 16 23.13	Steel Water Storage Tanks	RSMeans
33 16 13	Aboveground Water Utility Storage Tanks	CSI	33 16 23	Ground-Level Steel Water Storage Tanks	CSI
33 12 19.10	Fire Hydrants	RSMeans	33 14 19.30	Fire Hydrants	CSI
33 12 16.20	Valves	RSMeans	33 14 19.20	Valves	CSI

33 12 16.10	Valves	RSMeans	33 14 19.10	Valves	CSI
33 12 16	Water Utility Distribution Valves	CSI	33 14 19	Valves and Hydrants for Water Utility Service	CSI
33 12 13.15	Tapping, Crosses and Sleeves	RSMeans	33 14 17.15	Tapping, Crosses and Sleeves	CSI
33 12 13	Water Service Connections	CSI	33 14 17	Site Water Utility Service Laterals	RSMeans
33 11 13	Public Water Utility Distribution Piping	CSI	33 14 13	Public Water Utility Distribution Piping	RSMeans
33 11	Water Utility Distribution Piping	CSI	33 14	Water Utility Transmission and Distribution	CSI
33 05 26	Utility Identification	CSI	33 05 97	Identification and Signage for Utilities	RSMeans
33 05 23.22	Directional Drilling	RSMeans	33 05 07.13	Utility Directional Drilling	RSMeans
33 05 23.20	Horizontal Boring	RSMeans	33 05 07.23	Utility Boring and Jacking	CSI
33 05 23.19	Microtunneling	CSI	33 05 07.36	Microtunneling	CSI
33 05 23	Trenchless Utility Installation	CSI	33 05 07	Trenchless Installation of Utility Piping	RSMeans
33 05 16	Utility Structures	CSI	33 05 63	Concrete Vaults and Chambers	RSMeans
33 01 30.71	Rehabilitation of Sewer Utilities	CSI	33 01 30.23	Pipe Bursting	RSMeans
33 01 30.16	TV Inspection of Sewer Pipelines	CSI	33 01 30.11	Television Inspection of Sewers	CSI
32 31 13.30	Fence, Chain Link, Gates & Posts	RSMeans	32 31 11.10	Gate Operators	CSI
31 71 21.10	Cut and Cover Tunnels	RSMeans	31 71 23.10	Cut and Cover Tunnels	CSI
31 71 21	Tunnel Excavation by Cut and Cover	RSMeans	31 71 23	Tunneling by Cut and Cover	CSI
28 46 13	Hard-Wired Detention Monitoring & Control Systems	CSI	28 52 11	Detention Monitoring and Control Systems	CSI
28 46	Electronic Detention Monitoring & Control Systems	CSI	28 52	Detention Security Systems	CSI
28 41 13	Building Systems	RSMeans	28 33 11	Electronic Structural Monitoring Systems	RSMeans
28 39 10	Notification Systems	RSMeans	28 47 12	Notification Systems	CSI
28 39	Mass Notification Systems	CSI	28 47	Mass Notification	RSMeans
28 33 33	Gas Detection Sensors	CSI	28 42 15	Gas Detection Sensors	CSI
28 33	Gas Detection and Alarm	CSI	28 42	Gas Detection and Alarm	RSMeans
28 32 33	Radiation Detection Sensors	CSI	28 41 15	Radiation Detection Sensors	CSI
28 31 49.50	Carbon-Monoxide Detectors	RSMeans	28 46 11.21	Carbon-Monoxide Detection Sensors	CSI
28 31 46.50	Smoke Detectors	RSMeans	28 46 11.27	Other Sensors	CSI
28 31 43	Fire Detection Sensors	CSI	28 46 11	Fire Sensors and Detectors	CSI
28 31 23	Fire Det. & Alarm Annunciation Panels & Fire Station	CSI	28 46 21	Fire Alarm	CSI
28 23 19.10	Digital Video Recorder (DVR)	RSMeans	28 05 19.11	Digital Video Recorders	RSMeans
28 23 19	Digital Video Recorders & Analog Recording Devices	CSI	28 05 19	Storage Appliances for Electronic Safety & Security	RSMeans
28 16 16	Intrusion Detection Systems Infrastructure	CSI	28 31 16	Intrusion Detection Systems Infrastructure	RSMeans
28 13 53.39	Security Access Full Body Imaging Machine	RSMeans	28 18 15.39	Security Access Full Body Imaging Machine	CSI
28 13 53.36	Security Access Debugging Kit	RSMeans	28 18 53.36	Security Access Debugging Kit	CSI
28 13 53.33	Security Access Counterfeit Money Detector	RSMeans	28 18 53.33	Security Access Counterfeit Money Detector	CSI
28 13 53.23	Security Access Explosive Detection Equipment	CSI	28 18 15.23	Security Access Explosive Detection Equipment	CSI
28 13 53.16	Security Access X-Ray Equipment	CSI	28 18 13.16	Security Access X-Ray Equipment	CSI
28 13 53.13	Security Access Metal Detectors	CSI	28 18 11.13	Security Access Metal Detectors	RSMeans
28 13 53	Security Access Detection	CSI	28 18 11	Security Access Metal Detectors	CSI
28 13 23.50	Vehicle Barriers	RSMeans	28 19 15.50	Vehicle Barriers	CSI
28 13 23	Access Control Remote Devices	RSMeans	28 19 15	Perimeter Vehicle Access Management Systems	RSMeans
28 13	Access Control	CSI	28 19	Access Control Vehicle Identification Systems	CSI
28 05 13.23	Fire Alarm Communications Conductors and Cables	CSI	27 15 01.19	Fire Alarm Communications Conductors and Cables	RSMeans
28 05 13.10	Alarm & Communications Cable	RSMeans	27 15 01.11	Conductors & Cables For Electronic Safety & Security	CSI
28 01 30.51	Maint. and Admin. of Elec. Detection and Alarm	CSI	28 01 80.51	Maint. & Administration of Fire Detection & Alarm	RSMeans
28 01 30	Operation and Maint. of Elec. Detection and Alarm	CSI	28 01 80	Operation and Maint. of Fire Detection and Alarm	RSMeans
26 56 19.20	Roadway Luminaire	RSMeans	26 56 21.20	Roadway Luminaire	RSMeans
26 56 19.10	Roadway Lighting Fixtures	RSMeans	26 56 21.10	LED Exterior Lighting	RSMeans
26 56 16.55	Parking LED Lighting	RSMeans	26 56 19.60	Parking LED Lighting	RSMeans
26 53 13.10	Exit Lighting Fixtures	RSMeans	26 52 13.16	Exit Signs	CSI
26 26 13.10	Power Distribution Unit	RSMeans	26 27 33.20	Power Distribution Unit	RSMeans
13 34 23.35	Geodesic Domes	RSMeans	13 33 13.35	Geodesic Domes	CSI
13 34 23.15	Domes	RSMeans	13 34 56.15	Domes	CSI
11 14 13.13	Portable Posts and Railings	CSI	11 14 19.13	Portable Posts and Railings	CSI
10 21 13.20	Plastic Toilet Compartment Components	RSMeans	10 21 14.19	Plastic Toilet Compartment Components	CSI
10 21 13.17	Plastic-Laminate Clad Toilet Compartment Components	RSMeans	10 21 14.16	Plastic-Laminate Clad Toilet Compartment Components	CSI
10 21 13.14	Metal Toilet Compartment Components	RSMeans	10 21 14.13	Metal Toilet Compartment Components	CSI
08 74 23.50	Security Access Control Accessories	RSMeans	28 15 11.19	Security Access Control Accessories	CSI
08 74 19.50	Biometric Identity Access	RSMeans	28 15 11.15	Biometric Identity Devices	RSMeans
08 74 16.50	Keypad Access	RSMeans	28 15 11.13	Keypads	RSMeans
08 74 13	Card Key Access Control Hardware	CSI	28 15 11	Integrated Credential Readers & Field Entry Mgmt	RSMeans
08 74	Access Control Hardware	CSI	28 15	Access Control Hardware Devices	CSI
08 56 63	Detention Windows	CSI	11 98 21	Detention Windows	RSMeans
08 34 63	Detention Doors and Frames	CSI	11 98 12	Detention Doors and Frames	RSMeans
07 72 73.10	Pitch Pockets, Variable Sizes	RSMeans	07 71 16.20	Pitch Pockets, Variable Sizes	CSI
05 53 13.70	Expanded Steel Grating, at Ground	RSMeans	05 53 19.20	Expanded Grating, Steel	CSI
02 85 33	Removal and Disposal of Materials with Mold	CSI	02 87 13.33	Removal and Disposal of Materials with Mold	RSMeans
02 85 16	Mold Remediation Preparation and Containment	CSI	02 87 13	Mold Remediation	RSMeans
02 85	Mold Remediation	CSI	02 87	Biohazard Remediation	CSI

For additional tools that help with the utilization of the Construction Specifications Institute's 2016 edition of MasterFormat® please visit the following website: http://www.masterformat.com/revisions/

How the Cost Data Is Built: An Overview

Unit Prices*

All cost data have been divided into 50 divisions according to the MasterFormat® system of classification and numbering.

Assemblies*

The cost data in this section have been organized in an "Assemblies" format. These assemblies are the functional elements of a building and are arranged according to the 7 elements of the UNIFORMAT II classification system. For a complete explanation of a typical "Assembly", see "RSMeans data: Assemblies—How They Work."

Residential Models*

Model buildings for four classes of construction—economy, average, custom, and luxury—are developed and shown with complete costs per square foot.

Commercial/Industrial/ Institutional Models*

This section contains complete costs for 77 typical model buildings expressed as costs per square foot.

Green Commercial/Industrial/ Institutional Models*

This section contains complete costs for 25 green model buildings expressed as costs per square foot.

References*

This section includes information on Equipment Rental Costs, Crew Listings, Historical Cost Indexes, City Cost Indexes, Location Factors, Reference Tables, and Change Orders, as well as a listing of abbreviations.

- **Equipment Rental Costs:** Included are the average costs to rent and operate hundreds of pieces of construction equipment.
- **Crew Listings:** This section lists all the crews referenced in the cost data. A crew is composed of more than one trade classification and/or the addition of power equipment to any trade classification. Power equipment is included in the cost of the crew. Costs are shown both with bare labor rates and with the installing contractor's overhead and profit added. For each, the total crew cost per eight-hour day and the composite cost per labor-hour are listed.

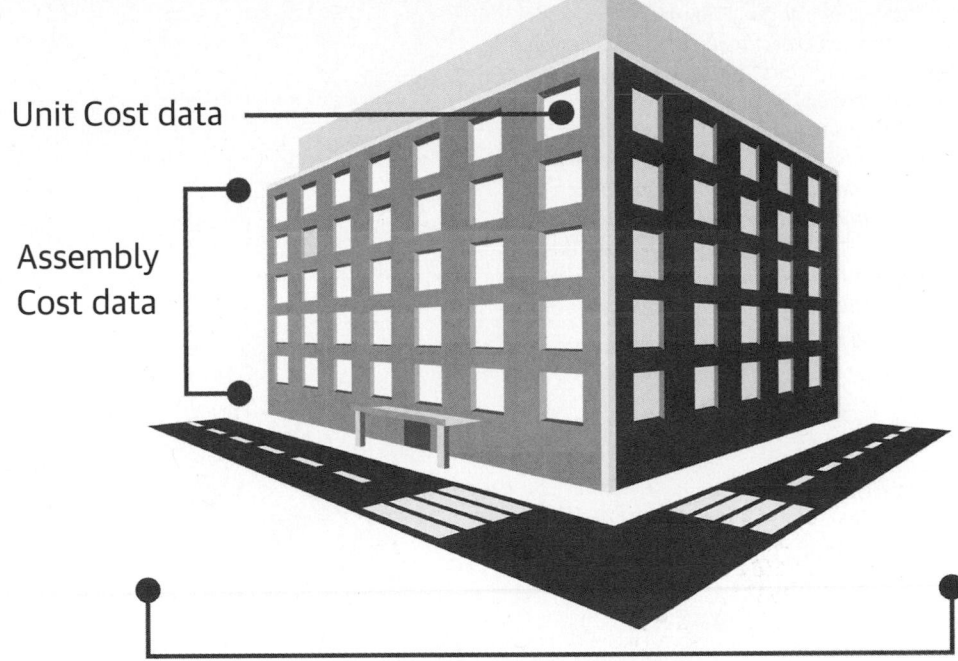

Unit Cost data

Assembly Cost data

Square Foot Models

- **Historical Cost Indexes:** These indexes provide you with data to adjust construction costs over time.
- **City Cost Indexes:** All costs in this data set are U.S. national averages. Costs vary by region. You can adjust for this by CSI Division to over 730 cities in 900+ 3-digit zip codes throughout the U.S. and Canada by using this data.
- **Location Factors:** You can adjust total project costs to over 730 cities in 900+ 3-digit zip codes throughout the U.S. and Canada by using the weighted number, which applies across all divisions.
- **Reference Tables:** At the beginning of selected major classifications in the Unit Prices are reference numbers indicators. These numbers refer you to related information in the Reference Section. In this section, you'll find reference tables, explanations, and estimating information that support how we develop the unit price data, technical data, and estimating procedures.
- **Change Orders:** This section includes information on the factors that influence the pricing of change orders.

- **Abbreviations:** A listing of abbreviations used throughout this information, along with the terms they represent, is included.

Index (printed versions only)

A comprehensive listing of all terms and subjects will help you quickly find what you need when you are not sure where it occurs in MasterFormat®.

Conclusion

This information is designed to be as comprehensive and easy to use as possible.

> The Construction Specifications Institute (CSI) and Construction Specifications Canada (CSC) have produced the 2016 edition of MasterFormat®, a system of titles and numbers used extensively to organize construction information.
>
> All unit prices in the RSMeans cost data are now arranged in the 50-division MasterFormat® 2016 system.

---

* Not all information is available in all data sets

Note: The material prices in RSMeans cost data are "contractor's prices." They are the prices that contractors can expect to pay at the lumberyards, suppliers'/distributors' warehouses, etc. Small orders of specialty items would be higher than the costs shown, while very large orders, such as truckload lots, would be less. The variation would depend on the size, timing, and negotiating power of the contractor. The labor costs are primarily for new construction or major renovation rather than repairs or minor alterations. With reasonable exercise of judgment, the figures can be used for any building work.

Estimating with RSMeans data: Unit Prices

Following these steps will allow you to complete an accurate estimate using RSMeans data Unit Prices.

1. Scope Out the Project

- Think through the project and identify the CSI divisions needed in your estimate.
- Identify the individual work tasks that will need to be covered in your estimate.
- The Unit Price data have been divided into 50 divisions according to CSI MasterFormat® 2016.
- In printed versions, the Unit Price Section Table of Contents on page 1 may also be helpful when scoping out your project.
- Experienced estimators find it helpful to begin with Division 2 and continue through completion. Division 1 can be estimated after the full project scope is known.

2. Quantify

- Determine the number of units required for each work task that you identified.
- Experienced estimators include an allowance for waste in their quantities. (Waste is not included in our Unit Price line items unless otherwise stated.)

3. Price the Quantities

- Use the search tools available to locate individual Unit Price line items for your estimate.
- Reference Numbers indicated within a Unit Price section refer to additional information that you may find useful.
- The crew indicates who is performing the work for that task. Crew codes are expanded in the Crew Listings in the Reference Section to include all trades and equipment that comprise the crew.
- The Daily Output is the amount of work the crew is expected to complete in one day.
- The Labor-Hours value is the amount of time it will take for the crew to install one unit of work.
- The abbreviated Unit designation indicates the unit of measure upon which the crew, productivity, and prices are based.
- Bare Costs are shown for materials, labor, and equipment needed to complete the Unit Price line item. Bare costs do not include waste, project overhead, payroll insurance, payroll taxes, main office overhead, or profit.
- The Total Incl O&P cost is the billing rate or invoice amount of the installing contractor or subcontractor who performs the work for the Unit Price line item.

4. Multiply

- Multiply the total number of units needed for your project by the Total Incl O&P cost for each Unit Price line item.
- Be careful that your take off unit of measure matches the unit of measure in the Unit column.
- The price you calculate is an estimate for a completed item of work.
- Keep scoping individual tasks, determining the number of units required for those tasks, matching each task with individual Unit Price line items, and multiplying quantities by Total Incl O&P costs.
- An estimate completed in this manner is priced as if a subcontractor, or set of subcontractors, is performing the work. The estimate does not yet include Project Overhead or Estimate Summary components such as general contractor markups on subcontracted work, general contractor office overhead and profit, contingency, and location factors.

5. Project Overhead

- Include project overhead items from Division 1-General Requirements.
- These items are needed to make the job run. They are typically, but not always, provided by the general contractor. Items include, but are not limited to, field personnel, insurance, performance bond, permits, testing, temporary utilities, field office and storage facilities, temporary scaffolding and platforms, equipment mobilization and demobilization, temporary roads and sidewalks, winter protection, temporary barricades and fencing, temporary security, temporary signs, field engineering and layout, final cleaning, and commissioning.
- Each item should be quantified and matched to individual Unit Price line items in Division 1, then priced and added to your estimate.
- An alternate method of estimating project overhead costs is to apply a percentage of the total project cost, usually 5% to 15% with an average of 10% (see General Conditions).
- Include other project related expenses in your estimate such as:
  - Rented equipment not itemized in the Crew Listings
  - Rubbish handling throughout the project (see section 02 41 19.19)

6. Estimate Summary

- Include sales tax as required by laws of your state or county.
- Include the general contractor's markup on self-performed work, usually 5% to 15% with an average of 10%.
- Include the general contractor's markup on subcontracted work, usually 5% to 15% with an average of 10%.
- Include the general contractor's main office overhead and profit:
  - RSMeans data provides general guidelines on the general contractor's main office overhead (see section 01 31 13.60 and Reference Number R013113-50).
  - Markups will depend on the size of the general contractor's operations, projected annual revenue, the level of risk, and the level of competition in the local area and for this project in particular.
- Include a contingency, usually 3% to 5%, if appropriate.
- Adjust your estimate to the project's location by using the City Cost Indexes or the Location Factors in the Reference Section:
  - Look at the rules in "How to Use the City Cost Indexes" to see how to apply the Indexes for your location.
  - When the proper Index or Factor has been identified for the project's location, convert it to a multiplier by dividing it by 100, then multiply that multiplier by your estimated total cost. The original estimated total cost will now be adjusted up or down from the national average to a total that is appropriate for your location.

Editors' Note:
We urge you to spend time reading and
understanding the supporting material. An
accurate estimate requires experience, knowledge,
and careful calculation. The more you know about
how we at RSMeans developed the data, the more
accurate your estimate will be. In addition, it is
important to take into consideration the reference
material such as Equipment Listings, Crew
Listings, City Cost Indexes, Location Factors, and
Reference Tables.

How to Use the Cost Data: The Details

What's Behind the Numbers? The Development of Cost data

RSMeans data engineers continually monitor developments in the construction industry in order to ensure reliable, thorough, and up-to-date cost information. While overall construction costs may vary relative to general economic conditions, price fluctuations within the industry are dependent upon many factors. Individual price variations may, in fact, be opposite to overall economic trends. Therefore, costs are constantly tracked, and complete updates are performed yearly. Also, new items are frequently added in response to changes in materials and methods.

Costs in U.S. Dollars

All costs represent U.S. national averages and are given in U.S. dollars. The City Cost Index (CCI) with RSMeans data can be used to adjust costs to a particular location. The CCI for Canada can be used to adjust U.S. national averages to local costs in Canadian dollars. No exchange rate conversion is necessary because it has already been factored in.

G The processes or products identified by the green symbol in our publications have been determined to be environmentally responsible and/or resource-efficient solely by RSMeans data engineering staff. The inclusion of the green symbol does not represent compliance with any specific industry association or standard.

Material Costs

RSMeans data engineers contact manufacturers, dealers, distributors, and contractors all across the U.S. and Canada to determine national average material costs. If you have access to current material costs for your specific location, you may wish to make adjustments to reflect differences from the national average. Included within material costs are fasteners for a normal installation. RSMeans data engineers use manufacturers' recommendations, written specifications, and/or standard construction practices for the sizing and spacing of fasteners. Adjustments to material costs may be required for your specific application or location. The manufacturer's warranty is assumed. Extended warranties are not included in the material costs. **Material costs do not include sales tax.**

Labor Costs

Labor costs are based upon a mathematical average of trade-specific wages in 30 major U.S. cities. The type of wage (union, open shop, or residential) is identified on the inside back cover of printed publications or is selected by the estimator when using the electronic products. Markups for the wages can also be found on the inside back cover of printed publications and/or under the labor references found in the electronic products.

- If wage rates in your area vary from those used, or if rate increases are expected within a given year, labor costs should be adjusted accordingly.

Labor costs reflect productivity based on actual working conditions. In addition to actual installation, these figures include time spent during a normal weekday on tasks, such as material receiving and handling, mobilization at site, site movement, breaks, and cleanup.

Productivity data is developed over an extended period so as not to be influenced by abnormal variations and reflects a typical average.

Equipment Costs

Equipment costs include not only rental but also operating costs for equipment under normal use. The operating costs include parts and labor for routine servicing, such as the repair and replacement of pumps, filters, and worn lines. Normal operating expendables, such as fuel, lubricants, tires, and electricity (where applicable), are also included. Extraordinary operating expendables with highly variable wear patterns, such as diamond bits and blades, are excluded. These costs are included under materials. Equipment rental rates are obtained from industry sources throughout North America—contractors, suppliers, dealers, manufacturers, and distributors.

Rental rates can also be treated as reimbursement costs for contractor-owned equipment. Owned equipment costs include depreciation, loan payments, interest, taxes, insurance, storage, and major repairs.

Equipment costs do not include operators' wages.

Equipment Cost/Day—The cost of equipment required for each crew is included in the Crew Listings in the Reference Section (small tools that are considered essential everyday tools are not listed out separately). The Crew Listings itemize specialized tools and heavy equipment along with labor trades. The daily cost of itemized equipment included in a crew is based on dividing the weekly bare rental rate by 5 (number of working days per week), then adding the hourly operating cost times 8 (the number of hours per day). This Equipment Cost/Day is shown in the last column of the Equipment Rental Costs in the Reference Section.

Mobilization, Demobilization—The cost to move construction equipment from an equipment yard or rental company to the job site and back again is not included in equipment costs. Mobilization (to the site) and demobilization (from the site) costs can be found in the Unit Price Section. If a piece of equipment is already at the job site, it is not appropriate to utilize mobilization or demobilization costs again in an estimate.

Overhead and Profit

Total Cost including O&P for the installing contractor is shown in the last column of the Unit Price and/or Assemblies. This figure is the sum of the bare material cost plus 10% for profit, the bare labor cost plus total overhead and profit, and the bare equipment cost plus 10% for profit. Details for the calculation of overhead and profit on labor are shown on the inside back cover of the printed product and in the Reference Section of the electronic product.

General Conditions

Cost data in this data set are presented in two ways: Bare Costs and Total Cost including O&P (Overhead and Profit). General Conditions, or General Requirements, of the contract should also be added to the Total Cost including O&P when applicable. Costs for General Conditions are listed in Division 1 of the Unit Price Section and in the Reference Section.

General Conditions for the installing contractor may range from 0% to 10% of the Total Cost including O&P. For the general or prime contractor, costs for General Conditions may range from 5% to 15% of the Total Cost including O&P, with a figure of 10% as the most typical allowance. If applicable, the Assemblies and Models sections use costs that include the installing contractor's overhead and profit (O&P).

Factors Affecting Costs

Costs can vary depending upon a number of variables. Here's a listing of some factors that affect costs and points to consider.

Quality—The prices for materials and the workmanship upon which productivity is based represent sound construction work. They are also in line with industry standard and manufacturer specifications and are frequently used by federal, state, and local governments.

Overtime—We have made no allowance for overtime. If you anticipate premium time or work beyond normal working hours, be sure to make an appropriate adjustment to your labor costs.

Productivity—The productivity, daily output, and labor-hour figures for each line item are based on an eight-hour work day in daylight hours in moderate temperatures, and up to a 14' working height unless otherwise indicated. For work that extends beyond normal work hours or is performed under adverse conditions, productivity may decrease.

Size of Project—The size, scope of work, and type of construction project will have a significant impact on cost. Economies of scale can reduce costs for large projects. Unit costs can often run higher for small projects.

Location—Material prices are for metropolitan areas. However, in dense urban areas, traffic and site storage limitations may increase costs. Beyond a 20-mile radius of metropolitan areas, extra trucking or transportation charges may also increase the material costs slightly. On the other hand, lower wage rates may be in effect. Be sure to consider both of these factors when preparing an estimate, particularly if the job site is located in a central city or remote rural location. In addition, highly specialized subcontract items may require travel and per-diem expenses for mechanics.

Other Factors—

- season of year
- contractor management
- weather conditions
- local union restrictions
- building code requirements
- availability of:
  - adequate energy
  - skilled labor
  - building materials
- owner's special requirements/restrictions
- safety requirements
- environmental considerations
- access

Unpredictable Factors—General business conditions influence "in-place" costs of all items. Substitute materials and construction methods may have to be employed. These may affect the installed cost and/or life cycle costs. Such factors may be difficult to evaluate and cannot necessarily be predicted on the basis of the job's location in a particular section of the country. Thus, where these factors apply, you may find significant but unavoidable cost variations for which you will have to apply a measure of judgment to your estimate.

Rounding of Costs

In printed publications only, all unit prices in excess of $5.00 have been rounded to make them easier to use and still maintain adequate precision of the results.

How Subcontracted Items Affect Costs

A considerable portion of all large construction jobs is usually subcontracted. In fact, the percentage done by subcontractors is constantly increasing and may run over 90%. Since the workers employed by these companies do nothing else but install their particular product, they soon become experts in that line. The result is, installation by these firms is accomplished so efficiently that the total in-place cost, even with the general contractor's overhead and profit, is no more, and often less, than if the principal contractor had handled the installation. Companies that deal with construction specialties are anxious to have their product perform well and, consequently, the installation will be the best possible.

Contingencies

The allowance for contingencies generally provides for unforeseen construction difficulties. On alterations or repair jobs, 20% is not too much. If drawings are final and only field contingencies are being considered, 2% or 3% is probably sufficient, and often nothing needs to be added. Contractually, changes in plans will be covered by extras. The contractor should consider inflationary price trends and possible material shortages during the course of the job. These escalation factors are dependent upon both economic conditions and the anticipated time between the estimate and actual construction. If drawings are not complete or approved, or a budget cost is wanted, it is wise to add 5% to 10%. Contingencies, then, are a matter of judgment.

Important Estimating Considerations

The productivity, or daily output, of each craftsman or crew assumes a well-managed job where tradesmen with the proper tools and equipment, along with the appropriate construction materials, are present. Included are daily set-up and cleanup time, break time, and plan layout time. Unless otherwise indicated, time for material movement on site (for items

that can be transported by hand) of up to 200' into the building and to the first or second floor is also included. If material has to be transported by other means, over greater distances, or to higher floors, an additional allowance should be considered by the estimator.

While horizontal movement is typically a sole function of distances, vertical transport introduces other variables that can significantly impact productivity. In an occupied building, the use of elevators (assuming access, size, and required protective measures are acceptable) must be understood at the time of the estimate. For new construction, hoist wait and cycle times can easily be 15 minutes and may result in scheduled access extending beyond the normal work day. Finally, all vertical transport will impose strict weight limits likely to preclude the use of any motorized material handling.

The productivity, or daily output, also assumes installation that meets manufacturer/designer/ standard specifications. A time allowance for quality control checks, minor adjustments, and any task required to ensure the proper function or operation is also included. For items that require connections to services, time is included for positioning, leveling, securing the unit, and for making all the necessary connections (and start up where applicable), ensuring a complete installation. Estimating of the services themselves (electrical, plumbing, water, steam, hydraulics, dust collection, etc.) is separate.

In some cases, the estimator must consider the use of a crane and an appropriate crew for the installation of large or heavy items. For those situations where a crane is not included in the assigned crew and as part of the line item cost, then equipment rental costs, mobilization and demobilization costs, and operator and support personnel costs must be considered.

Labor-Hours

The labor-hours expressed in this publication are derived by dividing the total daily labor-hours for the crew by the daily output. Based on average installation time and the assumptions listed above, the labor-hours include: direct labor, indirect labor, and nonproductive time. A typical day for a craftsman might include, but is not limited to:

- Direct Work
  - ☐ Measuring and layout
  - ☐ Preparing materials
  - ☐ Actual installation
  - ☐ Quality assurance/quality control
- Indirect Work
  - ☐ Reading plans or specifications
  - ☐ Preparing space
  - ☐ Receiving materials
  - ☐ Material movement
  - ☐ Giving or receiving instruction
  - ☐ Miscellaneous
- Non-Work
  - ☐ Chatting
  - ☐ Personal issues
  - ☐ Breaks
  - ☐ Interruptions (i.e., sickness, weather, material or equipment shortages, etc.)

If any of the items for a typical day do not apply to the particular work or project situation, the estimator should make any necessary adjustments.

Final Checklist

Estimating can be a straightforward process provided you remember the basics. Here's a checklist of some of the steps you should remember to complete before finalizing your estimate.

Did you remember to:

- factor in the City Cost Index for your locale?
- take into consideration which items have been marked up and by how much?
- mark up the entire estimate sufficiently for your purposes?
- read the background information on techniques and technical matters that could impact your project time span and cost?
- include all components of your project in the final estimate?
- double check your figures for accuracy?
- call RSMeans data engineers if you have any questions about your estimate or the data you've used? Remember, Gordian stands behind all of our products, including our extensive RSMeans data solutions. If you have any questions about your estimate, about the costs you've used from our data, or even about the technical aspects of the job that may affect your estimate, feel free to call the Gordian RSMeans editors at 1-800-448-8182.

Access Quarterly Data Updates

rsmeans.com/2018books

Using Minimum Labor/Equipment Charges for Small Quantities

Estimating small construction or repair tasks often creates situations in which the quantity of work to be performed is very small. When this occurs, the labor and/or equipment costs to perform the work may be too low to allow for the crew to get to the job, receive instructions, find materials, get set up, perform the work, clean up, and get to the next job. In these situations, the estimator should compare the developed labor and/or equipment costs for performing the work (e.g., quantity × labor and/or equipment costs) with the *minimum labor/equipment charge* within the Unit Price section of the data set. (These minimum labor/equipment charge line items appear only in *Facilities Construction Costs with RSMeans data* and *Commercial Renovation Costs with RSMeans data*.)

If the labor and/or equipment costs developed by the estimator are LOWER THAN the *minimum labor/equipment charge* listed at the bottom of specific sections of Unit Price costs, the estimator should adjust the developed costs upward to the *minimum labor/equipment charge*. The proper use of a *minimum labor/equipment charge* results in having enough money in the estimate to cover the contractor's higher cost of performing a very small amount of work during a partial workday.

A *minimum labor/equipment charge* should be used only when the task being estimated is the only task the crew will perform at the job site that day. **If, however, the crew will be able to perform other tasks at the job site that day, the use of a *minimum labor/equipment charge* is not appropriate.**

08 52 Wood Windows

08 52 10 – Plain Wood Windows

08 52 10.40 Casement Window		Crew	Daily Output	Labor-Hours	Unit	Material	2018 Bare Costs Labor	Equipment	Total	Total Incl O&P
0010	**CASEMENT WINDOW**, including frame, screen and grilles									
0100	2'-0" x 3'-0" H, double insulated glass	G 1 Carp	10	.800	Ea.	282	40.50		322.50	375
0150	Low E glass	G	10	.800		275	40.50		315.50	365
0200	2'-0" x 4'-6" high, double insulated glass	G	9	.889		385	45		430	500
0250	Low E glass	G	9	.889		385	45		430	500
9000	Minimum labor/equipment charge	1 Carp	3	2.667	Job		135		135	219

Example:

Establish the bid price to install two casement windows. Assume installation of 2' × 4'-6" metal clad windows with insulating glass [Unit Price line number 08 52 10.40 0250] and that this is the only task this crew will perform at the job site that day.

Solution:

Step One — Develop the Bare Labor Cost for this task:
Bare Labor Cost = 2 windows @ $45.00/each = $90.00

Step Two — Evaluate the *minimum labor/equipment charge* for this Unit Price section against the developed Bare Labor Cost for this task:
Minimum labor/equipment charge = $135.00 (compare with $90.00)

Step Three — Choose to adjust the developed labor cost upward to the *minimum labor/equipment charge.*

Step Four — Develop the bid price for this task (including O&P):
Add together the marked-up Bare Material Cost for this task and the marked-up *minimum labor/equipment charge* for this Unit Price section.

2 × ($385.00 + 10%) + ($135.00 + 63.9%)

=   2 × ($385.00 + $38.50) + ($135.00 + $82.27)

=   2 × $423.50 + $217.27

= $847.00 + $217.27

= $1,064.27

ANSWER: $1,064.27 is the correct bid price to use. This sum takes into consideration the material cost (with 10% for profit) for these two windows, plus the *minimum labor/equipment charge* (with O&P included) for this section of the Unit Price.

Unit Price Section

Table of Contents

1

Table of Contents (cont.)

Table of Contents (cont.)

3

RSMeans data: Unit Prices— How They Work

All RSMeans data: Unit Prices are organized in the same way.

03 30 Cast-In-Place Concrete

03 30 53 – Miscellaneous Cast-In-Place Concrete

03 30 53.40 Concrete In Place		Crew	Daily Output	Labor-Hours	Unit	Material	2018 Bare Costs Labor	Equipment	Total	Total Incl O&P
0010 **CONCRETE IN PLACE**	R033105-10									
0020 Including forms (4 uses), Grade 60 rebar, concrete (Portland cement	R033105-20									
0050 Type I), placement and finishing unless otherwise indicated	R033105-70									
0300 Beams (3500 psi), 5 kip/L.F., 10' span	R033105-85	C-14A	15.62	12.804	C.Y.	340	645	58	1,043	1,475
0350 25' span		"	18.55	10.782		355	545	49	949	1,325
0500 Chimney foundations (5000 psi), over 5 C.Y.		C-14C	32.22	3.476		171	167	.80	338.80	460
0510 (3500 psi), under 5 C.Y.		"	23.71	4.724		198	227	1.09	426.09	585
0700 Columns, square (4000 psi), 12" x 12", up to 1% reinforcing by area		C-14A	11.96	16.722		380	845	76	1,301	1,875
3540 Equipment pad (3000 psi), 3' x 3' x 6" thick		C-14H	45	1.067	Ea.	44.50	52.50	.57	97.57	135
3550 4' x 4' x 6" thick			30	1.600		69	79	.85	148.85	204
3560 5' x 5' x 8" thick			18	2.667		126	132	1.41	259.41	355
3570 6' x 6' x 8" thick			14	3.429		173	169	1.82	343.82	465
3580 8' x 8' x 10" thick			8	6		370	296	3.18	669.18	885
3590 10' x 10' x 12" thick			5	9.600		645	475	5.10	1,125.10	1,475

It is important to understand the structure of RSMeans data: Unit Prices, so that you can find information easily and use it correctly.

---

1 Line Numbers

Line Numbers consist of 12 characters, which identify a unique location in the database for each task. The first 6 or 8 digits conform to the Construction Specifications Institute MasterFormat® 2016. The remainder of the digits are a further breakdown in order to arrange items in understandable groups of similar tasks. Line numbers are consistent across all of our publications, so a line number in any of our products will always refer to the same item of work.

2 Descriptions

Descriptions are shown in a hierarchical structure to make them readable. In order to read a complete description, read up through the indents to the top of the section. Include everything that is above and to the left that is not contradicted by information below. For instance, the complete description for line 03 30 53.40 3550 is "Concrete in place, including forms (4 uses), Grade 60 rebar, concrete (Portland cement Type 1), placement and finishing unless otherwise indicated; Equipment pad (3000 psi), 4' × 4' × 6" thick."

3 RSMeans data

When using **RSMeans data**, it is important to read through an entire section to ensure that you use the data that most closely matches your work. Note that sometimes there is additional information shown in the section that may improve your price. There are frequently lines that further describe, add to, or adjust data for specific situations.

4 Reference Information

Gordian's RSMeans engineers have created **reference** information to assist you in your estimate. **If** there is information that applies to a section, it will be indicated at the start of the section. The Reference Section is located in the back of the data set.

5 Crews

Crews include labor and/or equipment necessary to accomplish each task. In this case, Crew C-14H is used. Gordian's RSMeans staff selects a crew to represent the workers and equipment that are

typically used for that task. In this case, Crew C-14H consists of one carpenter foreman (outside), two carpenters, one rodman, one laborer, one cement finisher, and one gas engine vibrator. Details of all crews can be found in the Reference Section.

Crews - Renovation

Crew No.	Bare Costs		Incl. Subs O&P		Cost Per Labor-Hour	
Crew C-14H	Hr.	Daily	Hr.	Daily	Bare Costs	Incl. O&P
1 Carpenter Foreman (outside)	$52.70	$421.60	$85.55	$684.40	$49.36	$79.56
2 Carpenters	50.70	811.20	82.30	1316.80		
1 Rodman (reinf.)	54.65	437.20	87.30	698.40		
1 Laborer	39.85	318.80	64.70	517.60		
1 Cement Finisher	47.55	380.40	75.20	601.60		
1 Gas Engine Vibrator		25.60		28.16	.53	.59
48 L.H., Daily Totals		$2394.80		$3846.96	$49.89	$80.14

6 Daily Output

The **Daily Output** is the amount of work that the crew can do in a normal 8-hour workday, including mobilization, layout, movement of materials, and cleanup. In this case, crew C-14H can install thirty 4' × 4' × 6" thick concrete pads in a day. Daily output is variable and based on many factors, including the size of the job, location, and environmental conditions. RSMeans data represents work done in daylight (or adequate lighting) and temperate conditions.

7 Labor-Hours

The figure in the **Labor-Hours** column is the amount of labor required to perform one unit of work—in this case the amount of labor required to construct one 4' × 4' equipment pad. This figure is calculated by dividing the number of hours of labor in the crew by the daily output (48 labor-hours divided by 30 pads = 1.6 hours of labor per pad). Multiply 1.6 times 60 to see the value in minutes: 60 × 1.6 = 96

minutes. Note: the labor-hour figure is not dependent on the crew size. A change in crew size will result in a corresponding change in daily output, but the labor-hours per unit of work will not change.

8 Unit of Measure

All RSMeans data: Unit Prices include the typical **Unit of Measure** used for estimating that item. For concrete-in-place the typical unit is cubic yards (C.Y.) or each (Ea.). For installing broadloom carpet it is square yard, and for gypsum board it is square foot. The estimator needs to take special care that the unit in the data matches the unit in the take-off. Unit conversions may be found in the Reference Section.

9 Bare Costs

Bare Costs are the costs of materials, labor, and equipment that the installing contractor pays. They represent the cost, in U.S. dollars, for one unit of work. They do not include any markups for profit or labor burden.

10 Bare Total

The **Total column** represents the total bare cost for the installing contractor in U.S. dollars. In this case, the sum of $69 for material + $79 for labor + $.85 for equipment is $148.85

11 Total Incl O&P

The **Total Incl O&P column** is the total cost, including overhead and profit, that the installing contractor will charge the customer. This represents the cost of materials plus 10% profit, the cost of labor plus labor burden and 10% profit, and the cost of equipment plus 10% profit. It does not include the general contractor's overhead and profit. Note: See the inside back cover of the printed product or the Reference Section of the electronic product for details of how the labor burden is calculated.

National Average

*The RSMeans data in our print publications represents a "national average" cost. This data should be modified to the project location using the **City Cost Indexes** or **Location Factors** tables found in the Reference Section. Use the Location Factors to adjust estimate totals if the project covers multiple trades. Use the City Cost Indexes (CCI) for single trade*

projects or projects where a more detailed analysis is required. All figures in the two tables are derived from the same research. The last row of data in the CCI—the weighted average—is the same as the numbers reported for each location in the location factor table.

RSMeans data: Unit Prices—
How They Work (Continued)

Project Name:	Pre-Engineered Steel Building		Architect: As Shown					
Location:	Anywhere, USA						01/01/18	R+R
Line Number	Description	Qty	Unit	Material	Labor	Equipment	SubContract	Estimate Total
03 30 53.40 3940	Strip footing, 12" x 24", reinforced	34	C.Y.	$5,406.00	$3,808.00	$18.36	$0.00	
03 30 53.40 3950	Strip footing, 12" x 36", reinforced	15	C.Y.	$2,295.00	$1,350.00	$6.45	$0.00	
03 11 13.65 3000	Concrete slab edge forms	500	L.F.	$145.00	$1,280.00	$0.00	$0.00	
03 22 11.10 0200	Welded wire fabric reinforcing	150	C.S.F.	$2,940.00	$4,200.00	$0.00	$0.00	
03 31 13.35 0300	Ready mix concrete, 4000 psi for slab on grade	278	C.Y.	$35,584.00	$0.00	$0.00	$0.00	
03 31 13.70 4300	Place, strike off & consolidate concrete slab	278	C.Y.	$0.00	$5,031.80	$130.66	$0.00	
03 35 13.30 0250	Machine float & trowel concrete slab	15,000	S.F.	$0.00	$9,450.00	$300.00	$0.00	
03 15 16.20 0140	Cut control joints in concrete slab	950	L.F.	$47.50	$399.00	$57.00	$0.00	
03 39 23.13 0300	Sprayed concrete curing membrane	150	C.S.F.	$1,815.00	$1,005.00	$0.00	$0.00	
Division 03	**Subtotal**			**$48,232.50**	**$26,523.80**	**$512.47**	**$0.00**	**$75,268.77**
08 36 13.10 2650	Manual 10' x 10' steel sectional overhead door	8	Ea.	$10,400.00	$3,600.00	$0.00	$0.00	
08 36 13.10 2860	Insulation and steel back panel for OH door	800	S.F.	$4,000.00	$0.00	$0.00	$0.00	
Division 08	**Subtotal**			**$14,400.00**	**$3,600.00**	**$0.00**	**$0.00**	**$18,000.00**
13 34 19.50 1100	Pre-Engineered Steel Building, 100' x 150' x 24'	15,000	SF Flr.	$0.00	$0.00	$0.00	$375,000.00	
13 34 19.50 6050	Framing for PESB door opening, 3' x 7'	4	Opng.	$0.00	$0.00	$0.00	$2,320.00	
13 34 19.50 6100	Framing for PESB door opening, 10' x 10'	8	Opng.	$0.00	$0.00	$0.00	$9,400.00	
13 34 19.50 6200	Framing for PESB window opening, 4' x 3'	6	Opng.	$0.00	$0.00	$0.00	$3,390.00	
13 34 19.50 5750	PESB door, 3' x 7', single leaf	4	Opng.	$2,620.00	$700.00	$0.00	$0.00	
13 34 19.50 7750	PESB sliding window, 4' x 3' with screen	6	Opng.	$2,550.00	$600.00	$45.30	$0.00	
13 34 19.50 6550	PESB gutter, eave type, 26 ga., painted	300	L.F.	$2,220.00	$819.00	$0.00	$0.00	
13 34 19.50 8650	PESB roof vent, 12" wide x 10' long	15	Ea.	$555.00	$3,285.00	$0.00	$0.00	
13 34 19.50 6900	PESB insulation, vinyl faced, 4" thick	27,400	S.F.	$13,152.00	$9,590.00	$0.00	$0.00	
Division 13	**Subtotal**			**$21,097.00**	**$14,994.00**	**$45.30**	**$390,110.00**	**$426,246.30**
			Subtotal	$83,729.50	$45,117.80	$557.77	$390,110.00	$519,515.07
Division 01	**General Requirements @ 7%**			5,861.07	3,158.25	39.04	27,307.70	
			Estimate Subtotal	$89,590.57	$48,276.05	$596.81	$417,417.70	$519,515.07
			Sales Tax @ 5%	4,479.53		29.84	10,435.44	
			Subtotal A	94,070.09	48,276.05	626.65	427,853.14	
			GC O & P	9,407.01	29,496.66	62.67	42,785.31	
			Subtotal B	103,477.10	77,772.71	689.32	470,638.46	$652,577.59
			Contingency @ 5%					32,628.88
			Subtotal C					$685,206.47
			Bond @ $12/1000 +10% O&P					9,044.73
			Subtotal D					$694,251.19
			Location Adjustment Factor		102.30			15,967.78
			Grand Total					**$710,218.97**

This estimate is based on an interactive spreadsheet. You are free to download it and adjust it to your methodology.
A copy of this spreadsheet is available at **www.RSMeans.com/2018books.**

Sample Estimate

This sample demonstrates the elements of an estimate, including a tally of the RSMeans data lines and a summary of the markups on a contractor's work to arrive at a total cost to the owner. The Location Factor with RSMeans data is added at the bottom of the estimate to adjust the cost of the work to a specific location.

---

1 Work Performed

The body of the estimate shows the RSMeans data selected, including the line number, a brief description of each item, its take-off unit and quantity, and the bare costs of materials, labor, and equipment. This estimate also includes a column titled "SubContract." This data is taken from the column "Total Incl O&P" and represents the total that a subcontractor would charge a general contractor for the work, including the sub's markup for overhead and profit.

2 Division 1, General Requirements

This is the first division numerically but the last division estimated. Division 1 includes project-wide needs provided by the general contractor. These requirements vary by project but may include temporary facilities and utilities, security, testing, project cleanup, etc. For small projects a percentage can be used—typically between 5% and 15% of project cost. For large projects the costs may be itemized and priced individually.

3 Sales Tax

If the work is subject to state or local sales taxes, the amount must be added to the estimate. Sales tax may be added to material costs, equipment costs, and subcontracted work. In this case, sales tax was added in all three categories. It was assumed that approximately half the subcontracted work would be material cost, so the tax was applied to 50% of the subcontract total.

4 GC O&P

This entry represents the general contractor's markup on material, labor, equipment, and subcontractor costs. Our standard markup on materials, equipment, and subcontracted work is 10%. In this estimate, the markup on the labor performed by the GC's workers uses "Skilled Workers Average" shown in Column F on the table "Installing Contractor's Overhead & Profit," which can be found on the inside back cover of the printed product or in the Reference Section of the electronic product.

5 Contingency

A factor for contingency may be added to any estimate to represent the cost of unknowns that may occur between the time that the estimate is performed and the time the project is constructed. The amount of the allowance will depend on the stage of design at which the estimate is done and the contractor's assessment of the risk involved. Refer to section 01 21 16.50 for contingency allowances.

6 Bonds

Bond costs should be added to the estimate. The figures here represent a typical performance bond, ensuring the owner that if the general contractor does not complete the obligations in the construction contract the bonding company will pay the cost for completion of the work.

7 Location Adjustment

Published prices are based on national average costs. If necessary, adjust the total cost of the project using a location factor from the "Location Factor" table or the "City Cost Index" table. Use location factors if the work is general, covering multiple trades. If the work is by a single trade (e.g., masonry) use the more specific data found in the "City Cost Indexes."

Estimating Tips
01 20 00 Price and Payment Procedures

- Allowances that should be added to estimates to cover contingencies and job conditions that are not included in the national average material and labor costs are shown in Section 01 21.

- When estimating historic preservation projects (depending on the condition of the existing structure and the owner's requirements), a 15%–20% contingency or allowance is recommended, regardless of the stage of the drawings.

01 30 00 Administrative Requirements

- Before determining a final cost estimate, it is a good practice to review all the items listed in Subdivisions 01 31 and 01 32 to make final adjustments for items that may need customizing to specific job conditions.

- Requirements for initial and periodic submittals can represent a significant cost to the General Requirements of a job. Thoroughly check the submittal specifications when estimating a project to determine any costs that should be included.

01 40 00 Quality Requirements

- All projects will require some degree of quality control. This cost is not included in the unit cost of construction listed in each division. Depending upon the terms of the contract, the various costs of inspection and testing can be the responsibility of either the owner or the contractor. Be sure to include the required costs in your estimate.

01 50 00 Temporary Facilities and Controls

- Barricades, access roads, safety nets, scaffolding, security, and many more requirements for the execution of a safe project are elements of direct cost. These costs can easily be overlooked when preparing an estimate. When looking through the major classifications of this subdivision, determine which items apply to each division in your estimate.

- Construction equipment rental costs can be found in the Reference Section in Section 01 54 33. Operators' wages are not included in equipment rental costs.

- Equipment mobilization and demobilization costs are not included in equipment rental costs and must be considered separately.

- The cost of small tools provided by the installing contractor for his workers is covered in the "Overhead" column on the "Installing Contractor's Overhead and Profit" table that lists labor trades, base rates, and markups and, therefore, is included in the "Total Incl. O&P" cost of any unit price line item.

01 70 00 Execution and Closeout Requirements

- When preparing an estimate, thoroughly read the specifications to determine the requirements for Contract Closeout. Final cleaning, record documentation, operation and maintenance data, warranties and bonds, and spare parts and maintenance materials can all be elements of cost for the completion of a contract. Do not overlook these in your estimate.

Reference Numbers
Reference numbers are shown at the beginning of some major classifications. These numbers refer to related items in the Reference Section. The reference information may be an estimating procedure, an alternate pricing method, or technical information.

Note: Not all subdivisions listed here necessarily appear. ∎

Did you know?

RSMeans data is available through our online application with 24/7 access:

- Search for unit prices by keyword
- Leverage the most up-to-date data
- Build and export estimates

Try it free for 30 days!
www.rsmeans.com/2018freetrial

01 11 Summary of Work

01 11 31 – Professional Consultants

	Crew	Daily Output	Labor-Hours	Unit	Material	2018 Bare Costs Labor	Equipment	Total	Total Incl O&P
01 11 31.10 Architectural Fees									
0010 **ARCHITECTURAL FEES** R011110-10									
0020 For new construction									
0060 Minimum				Project				4.90%	4.90%
0090 Maximum								16%	16%
0100 For alteration work, to $500,000, add to new construction fee								50%	50%
0150 Over $500,000, add to new construction fee								25%	25%
01 11 31.20 Construction Management Fees									
0010 **CONSTRUCTION MANAGEMENT FEES**									
0060 For work to $100,000				Project				10%	10%
0070 To $250,000								9%	9%
0090 To $1,000,000								6%	6%
0100 To $5,000,000								5%	5%
0110 To $10,000,000								4%	4%
01 11 31.30 Engineering Fees									
0010 **ENGINEERING FEES** R011110-30									
0020 Educational planning consultant, minimum				Project				.50%	.50%
0100 Maximum				"				2.50%	2.50%
0400 Elevator & conveying systems, minimum				Contrct				2.50%	2.50%
0500 Maximum								5%	5%
1000 Mechanical (plumbing & HVAC), minimum								4.10%	4.10%
1100 Maximum								10.10%	10.10%
1200 Structural, minimum				Project				1%	1%
1300 Maximum				"				2.50%	2.50%
01 11 31.75 Renderings									
0010 **RENDERINGS** Color, matted, 20" x 30", eye level,									
0020 1 building, minimum				Ea.	2,225			2,225	2,450
0050 Average					3,125			3,125	3,450
0100 Maximum					5,050			5,050	5,550
1000 5 buildings, minimum					4,150			4,150	4,550
1100 Maximum					8,175			8,175	9,000

01 21 Allowances

01 21 16 – Contingency Allowances

01 21 16.50 Contingencies

	Crew	Daily Output	Labor-Hours	Unit	Material	2018 Bare Costs Labor	Equipment	Total	Total Incl O&P
0010 **CONTINGENCIES**, Add to estimate									
0020 Conceptual stage				Project				20%	20%
0050 Schematic stage								15%	15%
0100 Preliminary working drawing stage (Design Dev.)								10%	10%
0150 Final working drawing stage								3%	3%

01 21 53 – Factors Allowance

01 21 53.50 Factors

	Crew	Daily Output	Labor-Hours	Unit	Material	2018 Bare Costs Labor	Equipment	Total	Total Incl O&P
0010 **FACTORS** Cost adjustments R012153-10									
0100 Add to construction costs for particular job requirements									
0500 Cut & patch to match existing construction, add, minimum				Costs	2%	3%			
0550 Maximum					5%	9%			
0800 Dust protection, add, minimum					1%	2%			
0850 Maximum					4%	11%			
1100 Equipment usage curtailment, add, minimum					1%	1%			
1150 Maximum					3%	10%			
1400 Material handling & storage limitation, add, minimum					1%	1%			

01 21 Allowances

01 21 53 – Factors Allowance

01 21 53.50 Factors		Crew	Daily Output	Labor-Hours	Unit	Material	2018 Bare Costs Labor	Equipment	Total	Total Incl O&P
1450	Maximum				Costs	6%	7%			
1700	Protection of existing work, add, minimum					2%	2%			
1750	Maximum					5%	7%			
2000	Shift work requirements, add, minimum						5%			
2050	Maximum						30%			
2300	Temporary shoring and bracing, add, minimum					2%	5%			
2350	Maximum					5%	12%			

01 21 53.60 Security Factors

		Crew	Daily Output	Labor-Hours	Unit	Material	2018 Bare Costs Labor	Equipment	Total	Total Incl O&P
0010	**SECURITY FACTORS** R012153-60									
0100	Additional costs due to security requirements									
0110	Daily search of personnel, supplies, equipment and vehicles									
0120	Physical search, inventory and doc of assets, at entry				Costs		30%			
0130	At entry and exit						50%			
0140	Physical search, at entry						6.25%			
0150	At entry and exit						12.50%			
0160	Electronic scan search, at entry						2%			
0170	At entry and exit						4%			
0180	Visual inspection only, at entry						.25%			
0190	At entry and exit						.50%			
0200	ID card or display sticker only, at entry						.12%			
0210	At entry and exit						.25%			
0220	Day 1 as described below, then visual only for up to 5 day job duration									
0230	Physical search, inventory and doc of assets, at entry				Costs		5%			
0240	At entry and exit						10%			
0250	Physical search, at entry						1.25%			
0260	At entry and exit						2.50%			
0270	Electronic scan search, at entry						.42%			
0280	At entry and exit						.83%			
0290	Day 1 as described below, then visual only for 6-10 day job duration									
0300	Physical search, inventory and doc of assets, at entry				Costs		2.50%			
0310	At entry and exit						5%			
0320	Physical search, at entry						.63%			
0330	At entry and exit						1.25%			
0340	Electronic scan search, at entry						.21%			
0350	At entry and exit						.42%			
0360	Day 1 as described below, then visual only for 11-20 day job duration									
0370	Physical search, inventory and doc of assets, at entry				Costs		1.25%			
0380	At entry and exit						2.50%			
0390	Physical search, at entry						.31%			
0400	At entry and exit						.63%			
0410	Electronic scan search, at entry						.10%			
0420	At entry and exit						.21%			
0430	Beyond 20 days, costs are negligible									
0440	Escort required to be with tradesperson during work effort				Costs		6.25%			

01 21 55 – Job Conditions Allowance

01 21 55.50 Job Conditions		Crew	Daily Output	Labor-Hours	Unit	Material	2018 Bare Costs Labor	Equipment	Total	Total Incl O&P
0010	**JOB CONDITIONS** Modifications to applicable									
0020	cost summaries									
0100	Economic conditions, favorable, deduct				Project				2%	2%
0200	Unfavorable, add								5%	5%
0300	Hoisting conditions, favorable, deduct								2%	2%
0400	Unfavorable, add								5%	5%

01 21 Allowances

01 21 55 – Job Conditions Allowance

01 21 55.50 Job Conditions	Crew	Daily Output	Labor-Hours	Unit	Material	2018 Bare Costs Labor	Equipment	Total	Total Incl O&P
0500 Gonoral contractor management, experienced, deduct				Project				2%	2%
0600 Inexperienced, add								10%	10%
0700 Labor availability, surplus, deduct								1%	1%
0800 Shortage, add								10%	10%
0900 Material storage area, available, deduct								1%	1%
1000 Not available, add								2%	2%
1100 Subcontractor availability, surplus, deduct								5%	5%
1200 Shortage, add								12%	12%
1300 Work space, available, deduct								2%	2%
1400 Not available, add								5%	5%

01 21 63 – Taxes

01 21 63.10 Taxes				Unit	Material	Labor	Equipment	Total	Total Incl O&P
0010 **TAXES** R012909-80									
0020 Sales tax, State, average				%	5.08%				
0050 Maximum R012909-85					7.50%				
0200 Social Security, on first $118,500 of wages						7.65%			
0300 Unemployment, combined Federal and State, minimum R012909-86						.60%			
0350 Average						9.60%			
0400 Maximum						12%			

01 31 Project Management and Coordination

01 31 13 – Project Coordination

01 31 13.20 Field Personnel

01 31 13.20 Field Personnel				Unit	Material	Labor	Equipment	Total	Total Incl O&P
0010 **FIELD PERSONNEL**									
0020 Clerk, average				Week		485		485	780
0160 General purpose laborer, average						1,600		1,600	2,575
0180 Project manager, minimum						2,125		2,125	3,450
0200 Average						2,450		2,450	3,950
0220 Maximum						2,800		2,800	4,500
0240 Superintendent, minimum						2,075		2,075	3,350
0260 Average						2,275		2,275	3,675
0280 Maximum						2,600		2,600	4,175

01 31 13.30 Insurance

01 31 13.30 Insurance				Unit	Material	Labor	Equipment	Total	Total Incl O&P
0010 **INSURANCE** R013113-40									
0020 Builders risk, standard, minimum				Job				.24%	.24%
0050 Maximum R013113-50								.64%	.64%
0200 All-risk type, minimum								.25%	.25%
0250 Maximum R013113-60								.62%	.62%
0400 Contractor's equipment floater, minimum				Value				.50%	.50%
0450 Maximum				"				1.50%	1.50%
0600 Public liability, average				Job				2.02%	2.02%
0810 Workers' compensation & employer's liability									
2000 Range of 35 trades in 50 states, excl. wrecking & clerical, min.				Payroll		1.30%			
2100 Average						11.53%			
2200 Maximum						120.29%			

01 31 13.40 Main Office Expense

01 31 13.40 Main Office Expense				Unit	Material	Labor	Equipment	Total	Total Incl O&P
0010 **MAIN OFFICE EXPENSE** Average for General Contractors									
0020 As a percentage of their annual volume									
0030 Annual volume to $300,000, minimum				% Vol.				20%	
0040 Maximum								30%	

01 31 Project Management and Coordination

01 31 13 – Project Coordination

01 31 13.40 Main Office Expense

		Crew	Daily Output	Labor-Hours	Unit	Material	2018 Bare Costs Labor	Equipment	Total	Total Incl O&P
0060	To $500,000, minimum				% Vol.				17%	
0070	Maximum								22%	
0080	To $1,000,000, minimum								16%	
0090	Maximum								19%	
0110	To $3,000,000, minimum								14%	
0120	Maximum								16%	
0130	To $5,000,000, minimum								8%	
0140	Maximum								10%	

01 31 13.50 General Contractor's Mark-Up

									Total	Total Incl O&P
0010	**GENERAL CONTRACTOR'S MARK-UP** on Change Orders									
0200	Extra work, by subcontractors, add				%				10%	10%
0250	By General Contractor, add								15%	15%
0400	Omitted work, by subcontractors, deduct all but								5%	5%
0450	By General Contractor, deduct all but								7.50%	7.50%
0600	Overtime work, by subcontractors, add								15%	15%
0650	By General Contractor, add								10%	10%

01 31 13.80 Overhead and Profit

									Total	Total Incl O&P
0010	**OVERHEAD & PROFIT** Allowance to add to items in this	R013113-50								
0020	book that do not include Subs O&P, average				%				25%	
0100	Allowance to add to items in this book that	R013113-55								
0110	do include Subs O&P, minimum				%				5%	5%
0150	Average								10%	10%
0200	Maximum								15%	15%
0290	Typical, by size of project, under $50,000								40%	
0310	$50,000 to $100,000								35%	
0320	$100,000 to $500,000								25%	
0330	$500,000 to $1,000,000								20%	

01 31 13.90 Performance Bond

									Total	Total Incl O&P
0010	**PERFORMANCE BOND**									
0020	For buildings, minimum				Job				.60%	.60%
0100	Maximum				"				2.50%	2.50%

01 41 Regulatory Requirements

01 41 26 – Permit Requirements

01 41 26.50 Permits

									Total	Total Incl O&P
0010	**PERMITS**									
0020	Rule of thumb, most cities, minimum				Job				.50%	.50%
0100	Maximum				"				2%	2%

For customer support on your Commercial Renovation Costs with RSMeans data, call 800.448.8182.

13

01 45 Quality Control

01 45 23 – Testing and Inspecting Services

01 45 23.50 Testing	Crew	Daily Output	Labor-Hours	Unit	Material	2018 Bare Costs Labor	Equipment	Total	Total Incl O&P
0010 **TESTING** and Inspecting Services									
1800 Compressive test, cylinder, delivered to lab, ASTM C 39				Ea.				12	13
1900 Picked up by lab, minimum								14	15
1950 Average								18	20
2000 Maximum								27	30
2200 Compressive strength, cores (not incl. drilling), ASTM C 42								36	40
2300 Patching core holes				↓				22	24
4730 Soil testing									
4735 Soil density, nuclear method, ASTM D 2922				Ea.				35	38.50
4740 Sand cone method, ASTM D 1556								27	30
4750 Moisture content, ASTM D 2216								9	10
4780 Permeability test, double ring infiltrometer								500	550
4800 Permeability, var. or constant head, undist., ASTM D 2434								227	250
4850 Recompacted								250	275
4900 Proctor compaction, 4" standard mold, ASTM D 698								123	135
4950 6" modified mold								68	75
5100 Shear tests, triaxial, minimum								410	450
5150 Maximum								545	600
5550 Technician for inspection, per day, earthwork								320	350
5650 Bolting								400	440
5750 Roofing								480	530
5790 Welding				↓				480	530
5820 Non-destructive metal testing, dye penetrant				Day				310	340
5840 Magnetic particle								310	340
5860 Radiography								450	495
5880 Ultrasonic				↓				310	340
6000 Welding certification, minimum				Ea.				91	100
6100 Maximum				"				250	275
7000 Underground storage tank									
7500 Volumetric tightness test, <=12,000 gal.				Ea.				435	480
7510 <=30,000 gal.				"				615	675
7600 Vadose zone (soil gas) sampling, 10-40 samples, min.				Day				1,375	1,500
7610 Maximum				"				2,275	2,500
7700 Ground water monitoring incl. drilling 3 wells, min.				Total				4,550	5,000
7710 Maximum				"				6,375	7,000
8000 X-ray concrete slabs				Ea.				182	200

01 51 Temporary Utilities

01 51 13 – Temporary Electricity

01 51 13.80 Temporary Utilities	Crew	Daily Output	Labor-Hours	Unit	Material	2018 Bare Costs Labor	Equipment	Total	Total Incl O&P
0010 **TEMPORARY UTILITIES**									
0350 Lighting, lamps, wiring, outlets, 40,000 S.F. building, 8 strings	1 Elec	34	.235	CSF Flr	5.60	13.70		19.30	27
0360 16 strings	"	17	.471		11.20	27.50		38.70	55
0400 Power for temp lighting only, 6.6 KWH, per month								.92	1.01
0430 11.8 KWH, per month								1.65	1.82
0450 23.6 KWH, per month								3.30	3.63
0600 Power for job duration incl. elevator, etc., minimum								47	51.50
0650 Maximum				↓				110	121

01 52 Construction Facilities

01 52 13 – Field Offices and Sheds

01 52 13.20 Office and Storage Space

01 52 13.20 Office and Storage Space	Crew	Daily Output	Labor-Hours	Unit	Material	2018 Bare Costs Labor	2018 Bare Costs Equipment	Total	Total Incl O&P
0010 **OFFICE AND STORAGE SPACE**									
0020 Office trailer, furnished, no hookups, 20' x 8', buy	2 Skwk	1	16	Ea.	8,900	840		9,740	11,100
0250 Rent per month					198			198	218
0300 32' x 8', buy	2 Skwk	.70	22.857		14,200	1,200		15,400	17,500
0350 Rent per month					247			247	272
0400 50' x 10', buy	2 Skwk	.60	26.667		29,300	1,400		30,700	34,600
0450 Rent per month					355			355	395
0500 50' x 12', buy	2 Skwk	.50	32		25,900	1,675		27,575	31,200
0550 Rent per month					450			450	495
0700 For air conditioning, rent per month, add					50			50	55
0800 For delivery, add per mile				Mile	12			12	13.20
0890 Delivery each way				Ea.	2,725			2,725	3,000
1200 Storage boxes, 20' x 8', buy	2 Skwk	1.80	8.889		3,325	465		3,790	4,425
1250 Rent per month					84.50			84.50	93
1300 40' x 8', buy	2 Skwk	1.40	11.429		3,875	600		4,475	5,250
1350 Rent per month					110			110	121

01 54 Construction Aids

01 54 16 – Temporary Hoists

01 54 16.50 Weekly Forklift Crew

	Crew	Daily Output	Labor-Hours	Unit	Material	Labor	Equipment	Total	Total Incl O&P
0010 **WEEKLY FORKLIFT CREW**									
0100 All-terrain forklift, 45' lift, 35' reach, 9000 lb. capacity	A-3P	.20	40	Week		2,050	2,425	4,475	5,925

01 54 19 – Temporary Cranes

01 54 19.50 Daily Crane Crews

	Crew	Daily Output	Labor-Hours	Unit	Material	Labor	Equipment	Total	Total Incl O&P
0010 **DAILY CRANE CREWS** for small jobs, portal to portal									
0100 12-ton truck-mounted hydraulic crane	A-3H	1	8	Day		450	630	1,080	1,400
0200 25-ton	A-3I	1	8			450	760	1,210	1,550
0300 40-ton	A-3J	1	8			450	1,325	1,775	2,150
0400 55-ton	A-3K	1	16			840	1,500	2,340	2,975
0500 80-ton	A-3L	1	16			840	2,250	3,090	3,800
0900 If crane is needed on a Saturday, Sunday or Holiday									
0910 At time-and-a-half, add				Day		50%			
0920 At double time, add				"		100%			

01 54 19.60 Monthly Tower Crane Crew

	Crew	Daily Output	Labor-Hours	Unit	Material	Labor	Equipment	Total	Total Incl O&P
0010 **MONTHLY TOWER CRANE CREW**, excludes concrete footing									
0100 Static tower crane, 130' high, 106' jib, 6200 lb. capacity	A-3N	.05	176	Month		9,875	26,000	35,875	44,200

01 54 23 – Temporary Scaffolding and Platforms

01 54 23.60 Pump Staging

		Crew	Daily Output	Labor-Hours	Unit	Material	Labor	Equipment	Total	Total Incl O&P
0010 **PUMP STAGING**, Aluminum	R015423-20									
0200 24' long pole section, buy					Ea.	425			425	465
0300 18' long pole section, buy						330			330	360
0400 12' long pole section, buy						222			222	244
0500 6' long pole section, buy						117			117	128
0600 6' long splice joint section, buy						86.50			86.50	95.50
0700 Pump jack, buy						174			174	192
0900 Foldable brace, buy						69.50			69.50	76
1000 Workbench/back safety rail support, buy						93.50			93.50	103
1100 Scaffolding planks/workbench, 14" wide x 24' long, buy						775			775	850
1200 Plank end safety rail, buy						355			355	390
1250 Safety net, 22' long, buy						425			425	465

01 54 23.60 Pump Staging	Crew	Daily Output	Labor-Hours	Unit	Material	2018 Bare Costs Labor	Equipment	Total	Total Incl O&P
1300 System in place, 50' working height, per use based on 50 uses	2 Carp	84.80	.189	C.S.F.	7.05	9.55		16.60	23.50
1400 100 uses		84.80	.189		3.53	9.55		13.08	19.45
1500 150 uses	↓	84.80	.189	↓	2.36	9.55		11.91	18.15

01 54 23.70 Scaffolding	Crew	Daily Output	Labor-Hours	Unit	Material	2018 Bare Costs Labor	Equipment	Total	Total Incl O&P
0010 **SCAFFOLDING** R015423-10									
0015 Steel tube, regular, no plank, labor only to erect & dismantle									
0090 Building exterior, wall face, 1 to 5 stories, 6'-4" x 5' frames	3 Carp	8	3	C.S.F.		152		152	247
0200 6 to 12 stories	4 Carp	8	4			203		203	330
0301 13 to 20 stories	5 Clab	8	5			199		199	325
0460 Building interior, wall face area, up to 16' high	3 Carp	12	2	↓		101		101	165
0560 16' to 40' high		10	2.400			122		122	198
0800 Building interior floor area, up to 30' high	↓	150	.160	C.C.F.		8.10		8.10	13.15
0900 Over 30' high	4 Carp	160	.200	"		10.15		10.15	16.45
0906 Complete system for face of walls, no plank, material only rent/mo				C.S.F.	33			33	36
0908 Interior spaces, no plank, material only rent/mo				C.C.F.	3.77			3.77	4.14
0910 Steel tubular, heavy duty shoring, buy									
0920 Frames 5' high 2' wide				Ea.	99.50			99.50	109
0925 5' high 4' wide					114			114	125
0930 6' high 2' wide					115			115	127
0935 6' high 4' wide				↓	126			126	139
0940 Accessories									
0945 Cross braces				Ea.	20			20	22
0950 U-head, 8" x 8"					22			22	24
0955 J-head, 4" x 8"					16			16	17.60
0960 Base plate, 8" x 8"					17.70			17.70	19.45
0965 Leveling jack				↓	39			39	42.50
1000 Steel tubular, regular, buy									
1100 Frames 3' high 5' wide				Ea.	91			91	100
1150 5' high 5' wide					107			107	117
1200 6'-4" high 5' wide					99.50			99.50	109
1350 7'-6" high 6' wide					170			170	187
1500 Accessories, cross braces					16			16	17.60
1550 Guardrail post					20			20	22
1600 Guardrail 7' section					8.05			8.05	8.85
1650 Screw jacks & plates					25.50			25.50	28
1700 Sidearm brackets					22.50			22.50	24.50
1750 8" casters					37			37	40.50
1800 Plank 2" x 10" x 16'-0"					66			66	72.50
1900 Stairway section					283			283	310
1910 Stairway starter bar					32			32	35
1920 Stairway inside handrail					53			53	58
1930 Stairway outside handrail					84			84	92.50
1940 Walk-thru frame guardrail				↓	41.50			41.50	46
2000 Steel tubular, regular, rent/mo.									
2100 Frames 3' high 5' wide				Ea.	4.38			4.38	4.82
2150 5' high 5' wide					4.38			4.38	4.82
2200 6'-4" high 5' wide					5.25			5.25	5.80
2250 7'-6" high 6' wide					9.75			9.75	10.75
2500 Accessories, cross braces					.88			.88	.97
2550 Guardrail post					.88			.88	.97
2600 Guardrail 7' section					.88			.88	.97
2650 Screw jacks & plates				↓	1.75			1.75	1.93

01 54 Construction Aids

01 54 23 – Temporary Scaffolding and Platforms

01 54 23.70 Scaffolding

		Crew	Daily Output	Labor-Hours	Unit	Material	2018 Bare Costs Labor	Equipment	Total	Total Incl O&P
2700	Sidearm brackets				Ea.	1.75			1.75	1.93
2750	8" casters					7			7	7.70
2800	Outrigger for rolling tower					2.63			2.63	2.89
2850	Plank 2" x 10" x 16'-0"					9.75			9.75	10.75
2900	Stairway section					32.50			32.50	36
2940	Walk-thru frame guardrail				↓	2.19			2.19	2.41
3000	Steel tubular, heavy duty shoring, rent/mo.									
3250	5' high 2' & 4' wide				Ea.	8.30			8.30	9.15
3300	6' high 2' & 4' wide					8.30			8.30	9.15
3500	Accessories, cross braces					.88			.88	.97
3600	U-head, 8" x 8"					2.44			2.44	2.68
3650	J-head, 4" x 8"					2.44			2.44	2.68
3700	Base plate, 8" x 8"					.88			.88	.97
3750	Leveling jack					2.44			2.44	2.68
5700	Planks, 2" x 10" x 16'-0", labor only to erect & remove to 50' H	3 Carp	72	.333			16.90		16.90	27.50
5800	Over 50' high	4 Carp	80	.400	↓		20.50		20.50	33

01 54 23.75 Scaffolding Specialties

		Crew	Daily Output	Labor-Hours	Unit	Material	2018 Bare Costs Labor	Equipment	Total	Total Incl O&P
0010	**SCAFFOLDING SPECIALTIES**									
1200	Sidewalk bridge, heavy duty steel posts & beams, including									
1210	parapet protection & waterproofing (material cost is rent/month)									
1220	8' to 10' wide, 2 posts	3 Carp	15	1.600	L.F.	45	81		126	181
1230	3 posts	"	10	2.400	"	69	122		191	274
1500	Sidewalk bridge using tubular steel scaffold frames including									
1510	planking (material cost is rent/month)	3 Carp	45	.533	L.F.	8.25	27		35.25	53
1600	For 2 uses per month, deduct from all above					50%				
1700	For 1 use every 2 months, add to all above					100%				
1900	Catwalks, 20" wide, no guardrails, 7' span, buy				Ea.	150			150	165
2000	10' span, buy					211			211	232
3720	Putlog, standard, 8' span, with hangers, buy					75.50			75.50	83
3730	Rent per month					16			16	17.60
3750	12' span, buy					99.50			99.50	109
3755	Rent per month					20			20	22
3760	Trussed type, 16' span, buy					255			255	281
3770	Rent per month					24			24	26.50
3790	22' span, buy					275			275	300
3795	Rent per month					32			32	35
4000	7 step					815			815	900
4100	Rolling towers, buy, 5' wide, 7' long, 10' high					1,300			1,300	1,425
4200	For additional 5' high sections, to buy				↓	245			245	270
4300	Complete incl. wheels, railings, outriggers,									
4350	21' high, to buy				Ea.	2,225			2,225	2,450
4400	Rent/month = 5% of purchase cost				"	196			196	216
5000	Motorized work platform, mast climber									
5050	Base unit, 50' W, less than 100' tall, rent/mo				Ea.	3,150			3,150	3,450
5100	Less than 200' tall, rent/mo					3,750			3,750	4,125
5150	Less than 300' tall, rent/mo					4,400			4,400	4,850
5200	Less than 400' tall, rent/mo				↓	4,975			4,975	5,475
5250	Set up and demob, per unit, less than 100' tall	B-68F	16.60	1.446	C.S.F.		76.50	16.80	93.30	143
5300	Less than 200' tall		25	.960			51	11.15	62.15	94.50
5350	Less than 300' tall		30	.800			42.50	9.30	51.80	79
5400	Less than 400' tall	↓	33	.727	↓		38.50	8.45	46.95	71.50
5500	Mobilization (price includes freight in and out) per unit				Ea.				1,100	1,225

For customer support on your Commercial Renovation Costs with RSMeans data, call 800.448.8182.

17

01 54 23.80 Staging Aids	Crew	Daily Output	Labor-Hours	Unit	Material	2018 Bare Costs Labor	Equipment	Total	Total Incl O&P
0010 **STAGING AIDS** and fall protection equipment									
0100 Sidewall staging bracket, tubular, buy				Ea.	57.50			57.50	63.50
0110 Cost each per day, based on 250 days use				Day	.23			.23	.25
0200 Guard post, buy				Ea.	54.50			54.50	60
0210 Cost each per day, based on 250 days use				Day	.22			.22	.24
0300 End guard chains, buy per pair				Pair	42			42	46
0310 Cost per set per day, based on 250 days use				Day	.23			.23	.25
1010 Cost each per day, based on 250 days use				"	.04			.04	.05
1100 Wood bracket, buy				Ea.	24.50			24.50	27
1110 Cost each per day, based on 250 days use				Day	.10			.10	.11
2010 Cost per pair per day, based on 250 days use					.51			.51	.56
3010 Cost each per day, based on 250 days use					.23			.23	.25
3100 Aluminum scaffolding plank, 20" wide x 24' long, buy				Ea.	815			815	900
3110 Cost each per day, based on 250 days use				Day	3.27			3.27	3.59
4000 Nylon full body harness, lanyard and rope grab				Ea.	166			166	182
4010 Cost each per day, based on 250 days use				Day	.66			.66	.73
4100 Rope for safety line, 5/8" x 100' nylon, buy				Ea.	56			56	61.50
4110 Cost each per day, based on 250 days use				Day	.22			.22	.25
4200 Permanent U-Bolt roof anchor, buy				Ea.	30			30	33
4300 Temporary (one use) roof ridge anchor, buy				"	6.40			6.40	7
5000 Installation (setup and removal) of staging aids									
5010 Sidewall staging bracket	2 Carp	64	.250	Ea.		12.70		12.70	20.50
5020 Guard post with 2 wood rails	"	64	.250			12.70		12.70	20.50
5030 End guard chains, set	1 Carp	64	.125			6.35		6.35	10.30
5100 Roof shingling bracket		96	.083			4.22		4.22	6.85
5200 Ladder jack		64	.125			6.35		6.35	10.30
5300 Wood plank, 2" x 10" x 16'	2 Carp	80	.200			10.15		10.15	16.45
5310 Aluminum scaffold plank, 20" x 24'	"	40	.400			20.50		20.50	33
5410 Safety rope	1 Carp	40	.200			10.15		10.15	16.45
5420 Permanent U-Bolt roof anchor (install only)	2 Carp	40	.400			20.50		20.50	33
5430 Temporary roof ridge anchor (install only)	1 Carp	64	.125			6.35		6.35	10.30

01 54 26 – Temporary Swing Staging

01 54 26.50 Swing Staging

01 54 26.50 Swing Staging	Crew	Daily Output	Labor-Hours	Unit	Material	2018 Bare Costs Labor	Equipment	Total	Total Incl O&P
0010 **SWING STAGING**, 500 lb. cap., 2' wide to 24' long, hand operated									
0020 steel cable type, with 60' cables, buy				Ea.	5,725			5,725	6,300
0030 Rent per month				"	570			570	630
0600 Lightweight (not for masons) 24' long for 150' height,									
0610 manual type, buy				Ea.	11,900			11,900	13,100
0620 Rent per month					1,200			1,200	1,325
0700 Powered, electric or air, to 150' high, buy					30,000			30,000	33,000
0710 Rent per month					2,100			2,100	2,300
0780 To 300' high, buy					30,500			30,500	33,500
0800 Rent per month					2,125			2,125	2,350
1000 Bosun's chair or work basket 3' x 3.5', to 300' high, electric, buy					12,200			12,200	13,400
1010 Rent per month					850			850	935
2200 Move swing staging (setup and remove)	E-4	2	16	Move		880	49.50	929.50	1,550

01 54 Construction Aids

01 54 36 – Equipment Mobilization

01 54 36.50 Mobilization	Crew	Daily Output	Labor-Hours	Unit	Material	2018 Bare Costs Labor	Equipment	Total	Total Incl O&P
0010 **MOBILIZATION** (Use line item again for demobilization) R015436-50									
0015 Up to 25 mi. haul dist. (50 mi. RT for mob/demob crew)									
1200 Small equipment, placed in rear of, or towed by pickup truck	A-3A	4	2	Ea.		103	31.50	134.50	197
1300 Equipment hauled on 3-ton capacity towed trailer	A-3Q	2.67	3			154	57	211	305
1400 20-ton capacity	B-34U	2	8			390	212	602	855
1500 40-ton capacity	B-34N	2	8			400	330	730	995
1600 50-ton capacity	B-34V	1	24			1,225	910	2,135	2,950
1700 Crane, truck-mounted, up to 75 ton (driver only)	1 Eqhv	4	2			112		112	177
1800 Over 75 ton (with chase vehicle)	A-3E	2.50	6.400			325	50.50	375.50	575
2400 Crane, large lattice boom, requiring assembly	B-34W	.50	144	▼		7,025	6,725	13,750	18,600
2500 For each additional 5 miles haul distance, add						10%	10%		
3000 For large pieces of equipment, allow for assembly/knockdown									
3100 For mob/demob of micro-tunneling equip, see Section 33 05 23.19									
3200 For mob/demob of pile driving equip, see Section 31 62 19.10									

01 55 Vehicular Access and Parking

01 55 23 – Temporary Roads

01 55 23.50 Roads and Sidewalks

	Crew	Daily Output	Labor-Hours	Unit	Material	Labor	Equipment	Total	Total Incl O&P
0010 **ROADS AND SIDEWALKS** Temporary									
0050 Roads, gravel fill, no surfacing, 4" gravel depth	B-14	715	.067	S.Y.	3.36	2.83	.44	6.63	8.75
0100 8" gravel depth	"	615	.078	"	6.70	3.29	.51	10.50	13.25

01 56 Temporary Barriers and Enclosures

01 56 13 – Temporary Air Barriers

01 56 13.60 Tarpaulins

	Crew	Daily Output	Labor-Hours	Unit	Material	Labor	Equipment	Total	Total Incl O&P
0010 **TARPAULINS**									
0020 Cotton duck, 10-13.13 oz./S.Y., 6' x 8'				S.F.	.85			.85	.94
0050 30' x 30'					.59			.59	.65
0100 Polyvinyl coated nylon, 14-18 oz., minimum					1.44			1.44	1.58
0150 Maximum					1.44			1.44	1.58
0200 Reinforced polyethylene 3 mils thick, white					.04			.04	.04
0300 4 mils thick, white, clear or black					.13			.13	.14
0400 5.5 mils thick, clear					.19			.19	.21
0500 White, fire retardant					.61			.61	.67
0600 12 mils, oil resistant, fire retardant					.49			.49	.54
0700 8.5 mils, black					.18			.18	.20
0720 Steel reinforced polyethylene, 20 mils thick					.54			.54	.59
0730 Polyester reinforced w/integral fastening system, 11 mils thick					.19			.19	.21
0740 Polyethylene, reflective, 23 mils thick				▼	1.35			1.35	1.49

01 56 13.90 Winter Protection

	Crew	Daily Output	Labor-Hours	Unit	Material	Labor	Equipment	Total	Total Incl O&P
0010 **WINTER PROTECTION**									
0100 Framing to close openings	2 Clab	500	.032	S.F.	.45	1.28		1.73	2.56
0200 Tarpaulins hung over scaffolding, 8 uses, not incl. scaffolding		1500	.011		.25	.43		.68	.97
0250 Tarpaulin polyester reinf. w/integral fastening system, 11 mils thick		1600	.010		.21	.40		.61	.88
0300 Prefab fiberglass panels, steel frame, 8 uses	▼	1200	.013	▼	2.52	.53		3.05	3.63

For customer support on your Commercial Renovation Costs with RSMeans data, call 800.448.8182.

19

01 56 Temporary Barriers and Enclosures

01 56 16 – Temporary Dust Barriers

01 56 16.10 Dust Barriers, Temporary

		Crew	Daily Output	Labor-Hours	Unit	Material	2018 Bare Costs Labor	Equipment	Total	Total Incl O&P
0010	**DUST BARRIERS, TEMPORARY**									
0020	Spring loaded telescoping pole & head, to 12', erect and dismantle	1 Clab	240	.033	Ea.		1.33		1.33	2.16
0025	Cost per day (based upon 250 days)				Day	.29			.29	.31
0030	To 21', erect and dismantle	1 Clab	240	.033	Ea.		1.33		1.33	2.16
0035	Cost per day (based upon 250 days)				Day	.58			.58	.63
0040	Accessories, caution tape reel, erect and dismantle	1 Clab	480	.017	Ea.		.66		.66	1.08
0045	Cost per day (based upon 250 days)				Day	.34			.34	.37
0060	Foam rail and connector, erect and dismantle	1 Clab	240	.033	Ea.		1.33		1.33	2.16
0065	Cost per day (based upon 250 days)				Day	.12			.12	.13
0070	Caution tape	1 Clab	384	.021	C.L.F.	3.13	.83		3.96	4.80
0080	Zipper, standard duty		60	.133	Ea.	7.30	5.30		12.60	16.65
0090	Heavy duty		48	.167	"	9.75	6.65		16.40	21.50
0100	Polyethylene sheet, 4 mil		37	.216	Sq.	2.58	8.60		11.18	16.85
0110	6 mil		37	.216	"	3.73	8.60		12.33	18.10
1000	Dust partition, 6 mil polyethylene, 1" x 3" frame	2 Carp	2000	.008	S.F.	.32	.41		.73	1.01
1080	2" x 4" frame	"	2000	.008	"	.35	.41		.76	1.04
1085	Negative air machine, 1800 CFM				Ea.	860			860	950
1090	Adhesive strip application, 2" width	1 Clab	192	.042	C.L.F.	6.55	1.66		8.21	9.90

01 56 23 – Temporary Barricades

01 56 23.10 Barricades

		Crew	Daily Output	Labor-Hours	Unit	Material	2018 Bare Costs Labor	Equipment	Total	Total Incl O&P
0010	**BARRICADES**									
0020	5' high, 3 rail @ 2" x 8", fixed	2 Carp	20	.800	L.F.	6.05	40.50		46.55	72.50
0150	Movable	"	30	.533	"	5	27		32	49.50
0300	Stock units, 58' high, 8' wide, reflective, buy				Ea.	211			211	232
0350	With reflective tape, buy				"	360			360	395
0400	Break-a-way 3" PVC pipe barricade									
0410	with 3 ea. 1' x 4' reflectorized panels, buy				Ea.	125			125	137
0500	Barricades, plastic, 8" x 24" wide, foldable					59			59	65
0800	Traffic cones, PVC, 18" high					11			11	12.10
0850	28" high					17.75			17.75	19.55
0900	Barrels, 55 gal., with flasher	1 Clab	96	.083		111	3.32		114.32	127
1000	Guardrail, wooden, 3' high, 1" x 6" on 2" x 4" posts	2 Carp	200	.080	L.F.	1.32	4.06		5.38	8.05
1100	2" x 6" on 4" x 4" posts	"	165	.097		2.54	4.92		7.46	10.80
1200	Portable metal with base pads, buy					14.25			14.25	15.65
1250	Typical installation, assume 10 reuses	2 Carp	600	.027		2.30	1.35		3.65	4.72
1300	Barricade tape, polyethylene, 7 mil, 3" wide x 500' long roll				Ea.	25			25	27.50
3000	Detour signs, set up and remove									
3010	Reflective aluminum, MUTCD, 24" x 24", post mounted	1 Clab	20	.400	Ea.	2.54	15.95		18.49	29
4000	Roof edge portable barrier stands and warning flags, 50 uses	1 Rohe	9100	.001	L.F.	.06	.03		.09	.12
4010	100 uses	"	9100	.001	"	.03	.03		.06	.08

01 56 26 – Temporary Fencing

01 56 26.50 Temporary Fencing

		Crew	Daily Output	Labor-Hours	Unit	Material	2018 Bare Costs Labor	Equipment	Total	Total Incl O&P
0010	**TEMPORARY FENCING**									
0020	Chain link, 11 ga., 4' high	2 Clab	400	.040	L.F.	1.63	1.59		3.22	4.38
0100	6' high		300	.053		4.55	2.13		6.68	8.45
0200	Rented chain link, 6' high, to 1000' (up to 12 mo.)		400	.040		2.99	1.59		4.58	5.90
0250	Over 1000' (up to 12 mo.)		300	.053		3.19	2.13		5.32	6.95
0350	Plywood, painted, 2" x 4" frame, 4' high	A-4	135	.178		6.35	8.55		14.90	21
0400	4" x 4" frame, 8' high	"	110	.218		12.10	10.45		22.55	30.50
0500	Wire mesh on 4" x 4" posts, 4' high	2 Carp	100	.160		10.35	8.10		18.45	24.50
0550	8' high	"	80	.200		15.60	10.15		25.75	33.50

01 56 Temporary Barriers and Enclosures

01 56 29 – Temporary Protective Walkways

01 56 29.50 Protection

		Crew	Daily Output	Labor-Hours	Unit	Material	2018 Bare Costs Labor	2018 Bare Costs Equipment	Total	Total Incl O&P
0010	**PROTECTION**									
0020	Stair tread, 2" x 12" planks, 1 use	1 Carp	75	.107	Tread	5.35	5.40		10.75	14.65
0100	Exterior plywood, 1/2" thick, 1 use		65	.123		1.85	6.25		8.10	12.20
0200	3/4" thick, 1 use		60	.133		2.77	6.75		9.52	14
2200	Sidewalks, 2" x 12" planks, 2 uses		350	.023	S.F.	.89	1.16		2.05	2.86
2300	Exterior plywood, 2 uses, 1/2" thick		750	.011		.31	.54		.85	1.22
2400	5/8" thick		650	.012		.38	.62		1	1.43
2500	3/4" thick		600	.013		.46	.68		1.14	1.61

01 56 32 – Temporary Security

01 56 32.50 Watchman

		Crew	Daily Output	Labor-Hours	Unit	Material	Labor	Equipment	Total	Total Incl O&P
0010	**WATCHMAN**									
0020	Service, monthly basis, uniformed person, minimum				Hr.				25	27.50
0100	Maximum								45.50	50
0200	Person and command dog, minimum								31	34
0300	Maximum								54.50	60
0500	Sentry dog, leased, with job patrol (yard dog), 1 dog				Week				290	320
0600	2 dogs				"				390	430
0800	Purchase, trained sentry dog, minimum				Ea.				1,375	1,500
0900	Maximum				"				2,725	3,000

01 58 Project Identification

01 58 13 – Temporary Project Signage

01 58 13.50 Signs

		Crew	Daily Output	Labor-Hours	Unit	Material	Labor	Equipment	Total	Total Incl O&P
0010	**SIGNS**									
0020	High intensity reflectorized, no posts, buy				Ea.	25			25	27.50

01 66 Product Storage and Handling Requirements

01 66 19 – Material Handling

01 66 19.10 Material Handling

		Crew	Daily Output	Labor-Hours	Unit	Material	Labor	Equipment	Total	Total Incl O&P
0010	**MATERIAL HANDLING**									
0020	Above 2nd story, via stairs, per C.Y. of material per floor	2 Clab	145	.110	C.Y.		4.40		4.40	7.15
0030	Via elevator, per C.Y. of material		240	.067			2.66		2.66	4.31
0050	Distances greater than 200', per C.Y. of material per each addl 200'		300	.053			2.13		2.13	3.45

01 71 Examination and Preparation

01 71 23 – Field Engineering

01 71 23.13 Construction Layout

		Crew	Daily Output	Labor-Hours	Unit	Material	Labor	Equipment	Total	Total Incl O&P
0010	**CONSTRUCTION LAYOUT**									
1100	Crew for layout of building, trenching or pipe laying, 2 person crew	A-6	1	16	Day		805	50.50	855.50	1,350
1200	3 person crew	A-7	1	24			1,325	50.50	1,375.50	2,150
1400	Crew for roadway layout, 4 person crew	A-8	1	32			1,700	50.50	1,750.50	2,775

01 74 Cleaning and Waste Management

01 74 13 – Progress Cleaning

01 74 13.20 Cleaning Up		Crew	Daily Output	Labor-Hours	Unit	Material	2018 Bare Costs Labor	Equipment	Total	Total Incl O&P
0010	**CLEANING UP**									
0020	After job completion, allow, minimum				Job				.30%	.30%
0040	Maximum				"				1%	1%
0052	Cleanup of floor area, continuous, per day, during const.	A-5	16	1.125	M.S.F.	2.41	45.50	2.95	50.86	79.50
0100	Final by GC at end of job	"	11.50	1.565	"	2.54	63	4.10	69.64	109

01 76 Protecting Installed Construction

01 76 13 – Temporary Protection of Installed Construction

01 76 13.20 Temporary Protection		Crew	Daily Output	Labor-Hours	Unit	Material	Labor	Equipment	Total	Total Incl O&P
0010	**TEMPORARY PROTECTION**									
0020	Flooring, 1/8" tempered hardboard, taped seams	2 Carp	1500	.011	S.F.	.45	.54		.99	1.38
0030	Peel away carpet protection	1 Clab	3200	.003	"	.13	.10		.23	.30

01 91 Commissioning

01 91 13 – General Commissioning Requirements

01 91 13.50 Building Commissioning		Crew	Daily Output	Labor-Hours	Unit	Material	Labor	Equipment	Total	Total Incl O&P
0010	**BUILDING COMMISSIONING**									
0100	Systems operation and verification during turnover				%				.25%	.25%
0150	Including all systems subcontractors								.50%	.50%
0200	Systems design assistance, operation, verification and training								.50%	.50%
0250	Including all systems subcontractors								1%	1%

01 93 Facility Maintenance

01 93 09 – Facility Equipment

01 93 09.50 Moving Equipment		Crew	Daily Output	Labor-Hours	Unit	Material	Labor	Equipment	Total	Total Incl O&P
0010	**MOVING EQUIPMENT**, Remove and reset, 100' distance,									
0020	No obstructions, no assembly or leveling unless noted									
0100	Annealing furnace, 24' overall	B-67	4	4	Ea.		208	67	275	400
0200	Annealing oven, small		14	1.143			59.50	19.10	78.60	115
0240	Very large		1	16			830	268	1,098	1,600
0400	Band saw, small		12	1.333			69.50	22.50	92	134
0440	Large		8	2			104	33.50	137.50	200
0500	Blue print copy machine		7	2.286			119	38	157	229
0600	Bonding mill, 6"		7	2.286			119	38	157	229
0620	12"		6	2.667			139	44.50	183.50	267
0640	18"		4	4			208	67	275	400
0660	24"		2	8			415	134	549	800
0700	Boring machine (jig)	B-68	7	3.429			179	38	217	325
0800	Bridgeport mill, standard	B-67	14	1.143			59.50	19.10	78.60	115
1000	Calibrator, 6 unit	"	14	1.143			59.50	19.10	78.60	115
1100	Comparitor, bench top	2 Clab	14	1.143			45.50		45.50	74
1140	Floor mounted	B-67	7	2.286			119	38	157	229
1200	Computer, desk top	2 Clab	25	.640			25.50		25.50	41.50
1300	Copy machine	"	25	.640			25.50		25.50	41.50
1500	Deflasher	B-67	14	1.143			59.50	19.10	78.60	115
1600	Degreaser, small		14	1.143			59.50	19.10	78.60	115
1640	Large 24' overall		1	16			830	268	1,098	1,600

01 93 Facility Maintenance

01 93 09 – Facility Equipment

01 93 09.50 Moving Equipment		Crew	Daily Output	Labor-Hours	Unit	Material	2018 Bare Costs		Total	Total Incl O&P
							Labor	Equipment		
1700	Desk with chair	2 Clab	25	.640	Ea.		25.50		25.50	41.50
1725	Desk only	"	22	.727			29		29	47
1750	Chair only	1 Clab	300	.027			1.06		1.06	1.73
1800	Dial press	B-67	7	2.286			119	38	157	229
1900	Drafting table	2 Clab	14	1.143			45.50		45.50	74
2000	Drill press, bench top	"	14	1.143			45.50		45.50	74
2040	Floor mounted	B-67	14	1.143			59.50	19.10	78.60	115
2080	Industrial radial	"	7	2.286			119	38	157	229
2100	Dust collector, portable	2 Clab	25	.640			25.50		25.50	41.50
2140	Stationary, small	B-67	7	2.286			119	38	157	229
2180	Stationary, large	"	2	8			415	134	549	800
2300	Electric discharge machine	B-68	7	3.429			179	38	217	325
2400	Environmental chamber walls, including assembly	4 Clab	18	1.778	L.F.		71		71	115
2600	File cabinet	2 Clab	25	.640	Ea.		25.50		25.50	41.50
2800	Grinder/sander, pedestal mount	B-67	14	1.143			59.50	19.10	78.60	115
3000	Hack saw, power	2 Clab	24	.667			26.50		26.50	43
3100	Hydraulic press	B-67	14	1.143			59.50	19.10	78.60	115
3500	Laminar flow tables	"	14	1.143			59.50	19.10	78.60	115
3600	Lathe, bench	2 Clab	14	1.143			45.50		45.50	74
3640	6"	B-67	14	1.143			59.50	19.10	78.60	115
3680	10"		13	1.231			64	20.50	84.50	123
3720	12"		12	1.333			69.50	22.50	92	134
4000	Milling machine		8	2			104	33.50	137.50	200
4100	Molding press, 25 ton		5	3.200			167	53.50	220.50	320
4140	60 ton		4	4			208	67	275	400
4180	100 ton		2	8			415	134	549	800
4220	150 ton		1.50	10.667			555	178	733	1,075
4260	200 ton		1	16			830	268	1,098	1,600
4300	300 ton		.75	21.333			1,100	355	1,455	2,150
4700	Oil pot stand		14	1.143			59.50	19.10	78.60	115
5000	Press, 10 ton		14	1.143			59.50	19.10	78.60	115
5040	15 ton		12	1.333			69.50	22.50	92	134
5080	20 ton		10	1.600			83.50	27	110.50	161
5120	30 ton		8	2			104	33.50	137.50	200
5160	45 ton		6	2.667			139	44.50	183.50	267
5200	60 ton		4	4			208	67	275	400
5240	75 ton		2.50	6.400			335	107	442	640
5280	100 ton		2	8			415	134	549	800
5500	Raised floor, including assembly	2 Carp	250	.064	S.F.		3.24		3.24	5.25
5600	Rolling mill, 6"	B-67	7	2.286	Ea.		119	38	157	229
5640	9"		6	2.667			139	44.50	183.50	267
5680	12"		4	4			208	67	275	400
5720	13"		3.50	4.571			238	76.50	314.50	460
5760	18"		2	8			415	134	549	800
5800	25"		1	16			830	268	1,098	1,600
6000	Sander, floor stand		14	1.143			59.50	19.10	78.60	115
6100	Screw machine		7	2.286			119	38	157	229
6200	Shaper, 16"		14	1.143			59.50	19.10	78.60	115
6300	Shear, power assist		4	4			208	67	275	400
6400	Slitter, 6"		14	1.143			59.50	19.10	78.60	115
6440	8"		13	1.231			64	20.50	84.50	123
6480	10"		12	1.333			69.50	22.50	92	134
6520	12"		11	1.455			75.50	24.50	100	146

01 93 09.50 Moving Equipment		Crew	Daily Output	Labor-Hours	Unit	Material	2018 Bare Costs		Total	Total Incl O&P
							Labor	Equipment		
6560	16"	B-67	10	1.600	Ea.		83.50	27	110.50	161
6600	20"		8	2			104	33.50	137.50	200
6640	24"		6	2.667			139	44.50	183.50	267
6800	Snag and tap machine		7	2.286			119	38	157	229
6900	Solder machine (auto)		7	2.286			119	38	157	229
7000	Storage cabinet metal, small	2 Clab	36	.444			17.70		17.70	29
7040	Large		25	.640			25.50		25.50	41.50
7100	Storage rack open, small		14	1.143			45.50		45.50	74
7140	Large		7	2.286			91		91	148
7200	Surface bench, small	B-67	14	1.143			59.50	19.10	78.60	115
7240	Large	"	5	3.200			167	53.50	220.50	320
7300	Surface grinder, large wet	B-68	5	4.800			251	53.50	304.50	455
7500	Time check machine	2 Clab	14	1.143			45.50		45.50	74
8000	Welder, 30 kVA (bench)		14	1.143			45.50		45.50	74
8100	Work bench with chair		25	.640			25.50		25.50	41.50
8900	For moving distance to 200', add						25%			

Estimating Tips
02 30 00 Subsurface Investigation
In preparing estimates on structures involving earthwork or foundations, all information concerning soil characteristics should be obtained. Look particularly for hazardous waste, evidence of prior dumping of debris, and previous stream beds.

02 40 00 Demolition and Structure Moving
The costs shown for selective demolition do not include rubbish handling or disposal. These items should be estimated separately using RSMeans data or other sources.

- Historic preservation often requires that the contractor remove materials from the existing structure, rehab them, and replace them. The estimator must be aware of any related measures and precautions that must be taken when doing selective demolition and cutting and patching. Requirements may include special handling and storage, as well as security.

- In addition to Subdivision 02 41 00, you can find selective demolition items in each division. Example: Roofing demolition is in Division 7.
- Absent of any other specific reference, an approximate demolish-in-place cost can be obtained by halving the new-install labor cost. To remove for reuse, allow the entire new-install labor figure.

02 40 00 Building Deconstruction
This section provides costs for the careful dismantling and recycling of most low-rise building materials.

02 50 00 Containment of Hazardous Waste
This section addresses on-site hazardous waste disposal costs.

02 80 00 Hazardous Material Disposal/Remediation
This subdivision includes information on hazardous waste handling, asbestos remediation, lead remediation, and mold remediation. See reference numbers

RO28213-20 and RO28319-60 for further guidance in using these unit price lines.

02 90 00 Monitoring Chemical Sampling, Testing Analysis
This section provides costs for on-site sampling and testing hazardous waste.

Reference Numbers
Reference numbers are shown at the beginning of some major classifications. These numbers refer to related items in the Reference Section. The reference information may be an estimating procedure, an alternate pricing method, or technical information.

Note: Not all subdivisions listed here necessarily appear. ■

Did you know?

RSMeans data is available through our online application with 24/7 access:
- Search for unit prices by keyword
- Leverage the most up-to-date data
- Build and export estimates

Try it free for 30 days!
www.rsmeans.com/2018freetrial

02 21 13.09 Topographical Surveys	Crew	Daily Output	Labor-Hours	Unit	Material	2018 Bare Costs Labor	Equipment	Total	Total Incl O&P
0010 **TOPOGRAPHICAL SURVEYS**									
0020 Topographical surveying, conventional, minimum	A-7	3.30	7.273	Acre	21.50	400	15.25	436.75	675
0100 Maximum	A-8	.60	53.333	"	58.50	2,850	84.50	2,993	4,700

02 21 13.13 Boundary and Survey Markers

	Crew	Daily Output	Labor-Hours	Unit	Material	Labor	Equipment	Total	Total Incl O&P
0010 **BOUNDARY AND SURVEY MARKERS**									
0300 Lot location and lines, large quantities, minimum	A-7	2	12	Acre	34.50	660	25	719.50	1,125
0320 Average	"	1.25	19.200		58.50	1,050	40.50	1,149	1,775
0400 Small quantities, maximum	A-8	1	32	↓	73	1,700	50.50	1,823.50	2,850

02 32 13.10 Boring and Exploratory Drilling

	Crew	Daily Output	Labor-Hours	Unit	Material	Labor	Equipment	Total	Total Incl O&P
0010 **BORING AND EXPLORATORY DRILLING**									
0020 Borings, initial field stake out & determination of elevations	A-6	1	16	Day		805	50.50	855.50	1,350
0100 Drawings showing boring details				Total		335		335	425
0200 Report and recommendations from P.E.						775		775	970
0300 Mobilization and demobilization	B-55	4	6	↓		248	249	497	675
0350 For over 100 miles, per added mile		450	.053	Mile		2.21	2.21	4.42	6
0600 Auger holes in earth, no samples, 2-1/2" diameter		78.60	.305	L.F.		12.65	12.70	25.35	34.50
0650 4" diameter		67.50	.356			14.70	14.75	29.45	40.50
0800 Cased borings in earth, with samples, 2-1/2" diameter		55.50	.432		16.65	17.90	17.95	52.50	67
0850 4" diameter	↓	32.60	.736		28	30.50	30.50	89	114
1000 Drilling in rock, "BX" core, no sampling	B-56	34.90	.458			21	41.50	62.50	79.50
1050 With casing & sampling		31.70	.505		16.65	23	46	85.65	106
1200 "NX" core, no sampling		25.92	.617			28	56	84	107
1250 With casing and sampling	↓	25	.640	↓	19.75	29	58	106.75	133
1400 Borings, earth, drill rig and crew with truck mounted auger	B-55	1	24	Day		995	995	1,990	2,700
1450 Rock using crawler type drill	B-56	1	16	"		730	1,450	2,180	2,775
1500 For inner city borings add, minimum								10%	10%
1510 Maximum								20%	20%

02 41 13.15 Hydrodemolition

	Crew	Daily Output	Labor-Hours	Unit	Material	Labor	Equipment	Total	Total Incl O&P
0010 **HYDRODEMOLITION**									
0015 Hydrodemolition, concrete pavement									
0120 20,000 psi, Crew to include loader/vacuum truck as required									
0130 2" depth	B-5	1000	.056	S.F.		2.47	1.44	3.91	5.55
0410 4" depth		800	.070			3.09	1.80	4.89	6.95
0420 6" depth	↓	600	.093	↓		4.12	2.40	6.52	9.25

02 41 13.17 Demolish, Remove Pavement and Curb

	Crew	Daily Output	Labor-Hours	Unit	Material	Labor	Equipment	Total	Total Incl O&P
0010 **DEMOLISH, REMOVE PAVEMENT AND CURB** R024119-10									
5010 Pavement removal, bituminous roads, up to 3" thick	B-38	690	.058	S.Y.		2.63	1.63	4.26	6
5050 4"-6" thick		420	.095			4.32	2.68	7	9.85
5100 Bituminous driveways		640	.063			2.83	1.76	4.59	6.45
5200 Concrete to 6" thick, hydraulic hammer, mesh reinforced		255	.157			7.10	4.41	11.51	16.25
5300 Rod reinforced		200	.200	↓		9.05	5.65	14.70	20.50
5400 Concrete, 7"-24" thick, plain		33	1.212	C.Y.		55	34	89	126
5500 Reinforced	↓	24	1.667	"		75.50	47	122.50	173

26

For customer support on your Commercial Renovation Costs with RSMeans data, call 800.448.8182.

02 41 Demolition

02 41 13 – Selective Site Demolition

02 41 13.17 Demolish, Remove Pavement and Curb

		Crew	Daily Output	Labor-Hours	Unit	Material	2018 Bare Costs Labor	2018 Bare Costs Equipment	Total	Total Incl O&P
5590	Minimum labor/equipment charge	B-38	6	6.667	Job		300	188	488	690
5600	With hand held air equipment, bituminous, to 6" thick	B-39	1900	.025	S.F.		1.06	.12	1.18	1.86
5700	Concrete to 6" thick, no reinforcing		1600	.030			1.26	.15	1.41	2.20
5800	Mesh reinforced		1400	.034			1.44	.17	1.61	2.51
5900	Rod reinforced	↓	765	.063	↓		2.64	.31	2.95	4.60
5990	Minimum labor/equipment charge	B-38	6	6.667	Job		300	188	488	690
6000	Curbs, concrete, plain	B-6	360	.067	L.F.		2.91	.87	3.78	5.60
6100	Reinforced		275	.087			3.81	1.14	4.95	7.35
6200	Granite		360	.067			2.91	.87	3.78	5.60
6300	Bituminous		528	.045	↓		1.98	.59	2.57	3.84
6390	Minimum labor/equipment charge	↓	6	4	Job		175	52	227	340

02 41 13.23 Utility Line Removal

		Crew	Daily Output	Labor-Hours	Unit	Material	2018 Bare Costs Labor	2018 Bare Costs Equipment	Total	Total Incl O&P
0010	**UTILITY LINE REMOVAL**									
0015	No hauling, abandon catch basin or manhole	B-6	7	3.429	Ea.		150	44.50	194.50	289
0020	Remove existing catch basin or manhole, masonry		4	6			262	78	340	505
0030	Catch basin or manhole frames and covers, stored		13	1.846			80.50	24	104.50	156
0040	Remove and reset	↓	7	3.429			150	44.50	194.50	289
0900	Hydrants, fire, remove only	B-21A	5	8			400	72.50	472.50	710
0950	Remove and reset	"	2	20	↓		1,000	182	1,182	1,775
0990	Minimum labor/equipment charge	2 Plum	2	8	Job		495		495	775
2900	Pipe removal, sewer/water, no excavation, 12" diameter	B-6	175	.137	L.F.		6	1.79	7.79	11.55
2930	15"-18" diameter	B-12Z	150	.160			7.25	9.40	16.65	22
2960	21"-24" diameter		120	.200			9.05	11.75	20.80	27.50
3000	27"-36" diameter	↓	90	.267			12.05	15.70	27.75	36.50
3200	Steel, welded connections, 4" diameter	B-6	160	.150			6.55	1.95	8.50	12.65
3300	10" diameter		80	.300	↓		13.10	3.91	17.01	25.50
3390	Minimum labor/equipment charge	↓	3	8	Job		350	104	454	675

02 41 13.30 Minor Site Demolition

		Crew	Daily Output	Labor-Hours	Unit	Material	2018 Bare Costs Labor	2018 Bare Costs Equipment	Total	Total Incl O&P
0010	**MINOR SITE DEMOLITION** R024119-10									
0100	Roadside delineators, remove only	B-80	175	.183	Ea.		8.10	3.50	11.60	16.90
0110	Remove and reset	"	100	.320	"		14.20	6.10	20.30	30
0400	Minimum labor/equipment charge	B-6	4	6	Job		262	78	340	505
0800	Guiderail, corrugated steel, remove only	B-80A	100	.240	L.F.		9.55	2.38	11.93	18.15
0850	Remove and reset	"	40	.600	"		24	5.95	29.95	45.50
0860	Guide posts, remove only	B-80B	120	.267	Ea.		11.40	1.97	13.37	20.50
0870	Remove and reset	B-55	50	.480	"		19.85	19.95	39.80	54
0890	Minimum labor/equipment charge	2 Clab	4	4	Job		159		159	259
1000	Masonry walls, block, solid	B-5	1800	.031	C.F.		1.37	.80	2.17	3.09
1200	Brick, solid		900	.062			2.74	1.60	4.34	6.15
1400	Stone, with mortar		900	.062			2.74	1.60	4.34	6.15
1500	Dry set	↓	1500	.037	↓		1.65	.96	2.61	3.70
1600	Median barrier, precast concrete, remove and store	B-3	430	.112	L.F.		4.97	5.30	10.27	13.85
1610	Remove and reset	"	390	.123	"		5.50	5.85	11.35	15.25
1650	Minimum labor/equipment charge	A-1	4	2	Job		79.50	17.80	97.30	149
4000	Sidewalk removal, bituminous, 2" thick	B-6	350	.069	S.Y.		2.99	.89	3.88	5.80
4010	2-1/2" thick		325	.074			3.23	.96	4.19	6.25
4050	Brick, set in mortar		185	.130			5.65	1.69	7.34	10.95
4100	Concrete, plain, 4"		160	.150			6.55	1.95	8.50	12.65
4110	Plain, 5"		140	.171			7.50	2.23	9.73	14.45
4120	Plain, 6"		120	.200			8.75	2.60	11.35	16.85
4200	Mesh reinforced, concrete, 4"		150	.160			7	2.08	9.08	13.50
4210	5" thick	↓	131	.183	↓		8	2.39	10.39	15.45

27

For customer support on your Commercial Renovation Costs with RSMeans data, call 800.448.8182.

02 41 13 – Selective Site Demolition

02 41 13.30 Minor Site Demolition	Crew	Daily Output	Labor-Hours	Unit	Material	2018 Bare Costs Labor	2018 Bare Costs Equipment	Total	Total Incl O&P
4220	6" thick	B-6	112	.214	S.Y.		9.35	2.79	12.14	18.10
4290	Minimum labor/equipment charge	B-39	12	4	Job		168	19.50	187.50	294

02 41 13.33 Railtrack Removal

		Crew	Daily Output	Labor-Hours	Unit	Material	Labor	Equipment	Total	Total Incl O&P
0010	**RAILTRACK REMOVAL**									
3500	Railroad track removal, ties and track	B-13	330	.170	L.F.		7.40	1.79	9.19	13.90
3600	Ballast	B-14	500	.096	C.Y.		4.04	.63	4.67	7.20
3700	Remove and re-install, ties & track using new bolts & spikes		50	.960	L.F.		40.50	6.25	46.75	72
3800	Turnouts using new bolts and spikes		1	48	Ea.		2,025	310	2,335	3,600
3890	Minimum labor/equipment charge		5	9.600	Job		405	62.50	467.50	720

02 41 13.60 Selective Demolition Fencing

		Crew	Daily Output	Labor-Hours	Unit	Material	Labor	Equipment	Total	Total Incl O&P
0010	**SELECTIVE DEMOLITION FENCING** R024119-10									
1600	Fencing, barbed wire, 3 strand	2 Clab	430	.037	L.F.		1.48		1.48	2.41
1650	5 strand	"	280	.057			2.28		2.28	3.70
1700	Chain link, posts & fabric, 8'-10' high, remove only	B-6	445	.054			2.36	.70	3.06	4.55

02 41 16 – Structure Demolition

02 41 16.13 Building Demolition

		Crew	Daily Output	Labor-Hours	Unit	Material	Labor	Equipment	Total	Total Incl O&P
0010	**BUILDING DEMOLITION** Large urban projects, incl. 20 mi. haul R024119-10									
0011	No foundation or dump fees, C.F. is vol. of building standing									
0020	Steel	B-8	21500	.003	C.F.		.14	.13	.27	.37
0050	Concrete		15300	.004			.19	.19	.38	.52
0080	Masonry		20100	.003			.15	.14	.29	.40
0100	Mixture of types		20100	.003			.15	.14	.29	.40
0500	Small bldgs, or single bldgs, no salvage included, steel	B-3	14800	.003			.14	.15	.29	.40
0600	Concrete		11300	.004			.19	.20	.39	.52
0650	Masonry		14800	.003			.14	.15	.29	.40
0700	Wood		14800	.003			.14	.15	.29	.40
0750	For buildings with no interior walls, deduct								30%	30%
1000	Demolition single family house, one story, wood 1600 S.F.	B-3	1	48	Ea.		2,150	2,300	4,450	5,975
1020	3200 S.F.		.50	96			4,275	4,575	8,850	11,900
1200	Demolition two family house, two story, wood 2400 S.F.		.67	71.964			3,200	3,425	6,625	8,925
1220	4200 S.F.		.38	128			5,700	6,100	11,800	15,900
1300	Demolition three family house, three story, wood 3200 S.F.		.50	96			4,275	4,575	8,850	11,900
1320	5400 S.F.		.30	160			7,125	7,625	14,750	19,900
5000	For buildings with no interior walls, deduct								30%	30%

02 41 16.17 Building Demolition Footings and Foundations

		Crew	Daily Output	Labor-Hours	Unit	Material	Labor	Equipment	Total	Total Incl O&P
0010	**BUILDING DEMOLITION FOOTINGS AND FOUNDATIONS** R024119-10									
0200	Floors, concrete slab on grade,									
0240	4" thick, plain concrete	B-13L	5000	.003	S.F.		.18	.39	.57	.71
0280	Reinforced, wire mesh		4000	.004			.22	.49	.71	.89
0300	Rods		4500	.004			.20	.44	.64	.80
0400	6" thick, plain concrete		4000	.004			.22	.49	.71	.89
0420	Reinforced, wire mesh		3200	.005			.28	.61	.89	1.11
0440	Rods		3600	.004			.25	.54	.79	.99
1000	Footings, concrete, 1' thick, 2' wide		300	.053	L.F.		2.99	6.50	9.49	11.90
1080	1'-6" thick, 2' wide		250	.064			3.59	7.85	11.44	14.25
1120	3' wide		200	.080			4.49	9.80	14.29	17.85
1140	2' thick, 3' wide		175	.091			5.15	11.20	16.35	20.50
1200	Average reinforcing, add								10%	10%
1220	Heavy reinforcing, add								20%	20%
2000	Walls, block, 4" thick	B-13L	8000	.002	S.F.		.11	.24	.35	.45
2040	6" thick		6000	.003			.15	.33	.48	.60

02 41 16 – Structure Demolition

02 41 16.17 Building Demolition Footings and Foundations

		Crew	Daily Output	Labor-Hours	Unit	Material	2018 Bare Costs Labor	Equipment	Total	Total Incl O&P
2080	8" thick	B-13L	4000	.004	S.F.		.22	.49	.71	.89
2100	12" thick	↓	3000	.005			.30	.65	.95	1.19
2200	For horizontal reinforcing, add								10%	10%
2220	For vertical reinforcing, add								20%	20%
2400	Concrete, plain concrete, 6" thick	B-13L	4000	.004			.22	.49	.71	.89
2420	8" thick		3500	.005			.26	.56	.82	1.01
2440	10" thick		3000	.005			.30	.65	.95	1.19
2500	12" thick	↓	2500	.006			.36	.78	1.14	1.43
2600	For average reinforcing, add								10%	10%
2620	For heavy reinforcing, add				↓				20%	20%
9000	Minimum labor/equipment charge	A-1	2	4	Job		159	35.50	194.50	298

02 41 16.19 Minor Building Deconstruction

			Crew	Daily Output	Labor-Hours	Unit	Material	Labor	Equipment	Total	Total Incl O&P
0010	**MINOR BUILDING DECONSTRUCTION**	R024119-10									
0011	For salvage, avg house										
0225	2 story, pre 1970 house, 1400 S.F., labor for all salv mat, min	G	6 Clab	25	1.920	SF Flr.		76.50		76.50	124
0230	Maximum	G	"	15	3.200	"		128		128	207
0235	Salvage of carpet, tackless	G	2 Clab	2400	.007	S.F.		.27		.27	.43
0240	Wood floors, incl denailing and packaging	G	3 Clab	270	.089	"		3.54		3.54	5.75
0260	Wood doors and trim, standard	G	1 Clab	16	.500	Ea.		19.95		19.95	32.50
0280	Base or cove mouldings, incl denailing and packaging	G		500	.016	L.F.		.64		.64	1.04
0320	Closet shelving and trim, incl denailing and packaging	G	↓	18	.444	Set		17.70		17.70	29
0340	Kitchen cabinets, uppers and lowers, prefab type	G	2 Clab	24	.667	L.F.		26.50		26.50	43
0360	Kitchen cabinets, uppers and lowers, built-in	G		12	1.333	"		53		53	86.50
0370	Bath fixt, incl toilet, tub, vanity, and med cabinet	G	↓	3	5.333	Set		213		213	345

02 41 19 – Selective Demolition

02 41 19.13 Selective Building Demolition

0010	**SELECTIVE BUILDING DEMOLITION**										
0020	Costs related to selective demolition of specific building components										
0025	are included under Common Work Results (XX 05)										
0030	in the component's appropriate division.										

02 41 19.16 Selective Demolition, Cutout

			Crew	Daily Output	Labor-Hours	Unit	Material	Labor	Equipment	Total	Total Incl O&P
0010	**SELECTIVE DEMOLITION, CUTOUT**	R024119-10									
0020	Concrete, elev. slab, light reinforcement, under 6 C.F.		B-9	65	.615	C.F.		25	3.60	28.60	44
0050	Light reinforcing, over 6 C.F.			75	.533	"		21.50	3.12	24.62	38.50
0200	Slab on grade to 6" thick, not reinforced, under 8 S.F.			85	.471	S.F.		18.95	2.75	21.70	34
0250	8-16 S.F.		↓	175	.229	"		9.20	1.34	10.54	16.40
0255	For over 16 S.F. see Line 02 41 16.17 0400										
0600	Walls, not reinforced, under 6 C.F.		B-9	60	.667	C.F.		27	3.90	30.90	48
0650	6-12 C.F.		"	80	.500	"		20	2.93	22.93	35.50
0655	For over 12 C.F. see Line 02 41 16.17 2500										
1000	Concrete, elevated slab, bar reinforced, under 6 C.F.		B-9	45	.889	C.F.		36	5.20	41.20	63.50
1050	Bar reinforced, over 6 C.F.			50	.800	"		32	4.68	36.68	57.50
1200	Slab on grade to 6" thick, bar reinforced, under 8 S.F.			75	.533	S.F.		21.50	3.12	24.62	38.50
1250	8-16 S.F.		↓	150	.267	"		10.75	1.56	12.31	19.10
1255	For over 16 S.F. see Line 02 41 16.17 0440										
1400	Walls, bar reinforced, under 6 C.F.		B-9	50	.800	C.F.		32	4.68	36.68	57.50
1450	6-12 C.F.		"	70	.571	"		23	3.34	26.34	41
1455	For over 12 C.F. see Lines 02 41 16.17 2500 and 2600										
2000	Brick, to 4 S.F. opening, not including toothing										
2040	4" thick		B-9	30	1.333	Ea.		53.50	7.80	61.30	95.50
2060	8" thick			18	2.222			89.50	13	102.50	159
2080	12" thick		↓	10	4	↓		161	23.50	184.50	287

02 41 Demolition

02 41 19 – Selective Demolition

02 41 19.16 Selective Demolition, Cutout

		Crew	Daily Output	Labor-Hours	Unit	Material	2018 Bare Costs Labor	Equipment	Total	Total Incl O&P
2400	Concrete block, to 4 S.F. opening, 2" thick	B-9	35	1.143	Ea.		46	6.70	52.70	82
2420	4" thick		30	1.333			53.50	7.80	61.30	95.50
2440	8" thick		27	1.481			59.50	8.65	68.15	107
2460	12" thick		24	1.667			67	9.75	76.75	120
2600	Gypsum block, to 4 S.F. opening, 2" thick		80	.500			20	2.93	22.93	35.50
2620	4" thick		70	.571			23	3.34	26.34	41
2640	8" thick		55	.727			29.50	4.25	33.75	52
2800	Terra cotta, to 4 S.F. opening, 4" thick		70	.571			23	3.34	26.34	41
2840	8" thick		65	.615			25	3.60	28.60	44
2880	12" thick	↓	50	.800	↓		32	4.68	36.68	57.50
4000	For toothing masonry, see Section 04 01 20.50									
6000	Walls, interior, not including re-framing,									
6010	openings to 5 S.F.									
6100	Drywall to 5/8" thick	1 Clab	24	.333	Ea.		13.30		13.30	21.50
6200	Paneling to 3/4" thick		20	.400			15.95		15.95	26
6300	Plaster, on gypsum lath		20	.400			15.95		15.95	26
6340	On wire lath	↓	14	.571	↓		23		23	37
7000	Wood frame, not including re-framing, openings to 5 S.F.									
7200	Floors, sheathing and flooring to 2" thick	1 Clab	5	1.600	Ea.		64		64	104
7310	Roofs, sheathing to 1" thick, not including roofing		6	1.333			53		53	86.50
7410	Walls, sheathing to 1" thick, not including siding		7	1.143			45.50		45.50	74
8500	Minimum labor/equipment charge	↓	4	2	Job		79.50		79.50	129

02 41 19.18 Selective Demolition, Disposal Only

		Crew	Daily Output	Labor-Hours	Unit	Material	2018 Bare Costs Labor	Equipment	Total	Total Incl O&P
0010	**SELECTIVE DEMOLITION, DISPOSAL ONLY** R024119-10									
0015	Urban bldg w/salvage value allowed									
0020	Including loading and 5 mile haul to dump									
0200	Steel frame	B-3	430	.112	C.Y.		4.97	5.30	10.27	13.85
0300	Concrete frame		365	.132			5.85	6.25	12.10	16.30
0400	Masonry construction		445	.108			4.81	5.15	9.96	13.40
0500	Wood frame	↓	247	.194	↓		8.65	9.25	17.90	24

02 41 19.19 Selective Demolition

		Crew	Daily Output	Labor-Hours	Unit	Material	2018 Bare Costs Labor	Equipment	Total	Total Incl O&P
0010	**SELECTIVE DEMOLITION**, Rubbish Handling R024119-10									
0020	The following are to be added to the demolition prices									
0050	The following are components for a complete chute system									
0100	Top chute circular steel, 4' long, 18" diameter	B-1C	15	1.600	Ea.	272	65	30.50	367.50	440
0102	23" diameter		15	1.600		295	65	30.50	390.50	465
0104	27" diameter		15	1.600		320	65	30.50	415.50	490
0106	30" diameter		15	1.600		340	65	30.50	435.50	515
0108	33" diameter		15	1.600		365	65	30.50	460.50	540
0110	36" diameter		15	1.600		385	65	30.50	480.50	565
0112	Regular chute, 18" diameter		15	1.600		204	65	30.50	299.50	365
0114	23" diameter		15	1.600		227	65	30.50	322.50	390
0116	27" diameter		15	1.600		249	65	30.50	344.50	415
0118	30" diameter		15	1.600		261	65	30.50	356.50	425
0120	33" diameter		15	1.600		295	65	30.50	390.50	465
0122	36" diameter		15	1.600		320	65	30.50	415.50	490
0124	Control door chute, 18" diameter		15	1.600		385	65	30.50	480.50	565
0126	23" diameter		15	1.600		410	65	30.50	505.50	590
0128	27" diameter		15	1.600		430	65	30.50	525.50	615
0130	30" diameter		15	1.600		455	65	30.50	550.50	640
0132	33" diameter		15	1.600		475	65	30.50	570.50	665
0134	36" diameter	↓	15	1.600		500	65	30.50	595.50	690

02 41 19 – Selective Demolition

02 41 19.19 Selective Demolition	Crew	Daily Output	Labor-Hours	Unit	Material	2018 Bare Costs Labor	2018 Bare Costs Equipment	Total	Total Incl O&P
0136	Chute liners, 14 ga., 18"-30" diameter	B-1C	15	1.600	Ea.	209	65	30.50	304.50	370
0138	33"-36" diameter		15	1.600		261	65	30.50	356.50	425
0140	17% thinner chute, 30" diameter		15	1.600		220	65	30.50	315.50	380
0142	33% thinner chute, 30" diameter	↓	15	1.600		166	65	30.50	261.50	320
0144	Top chute cover	1 Clab	24	.333		143	13.30		156.30	179
0146	Door chute cover	"	24	.333		143	13.30		156.30	179
0148	Top chute trough	2 Clab	12	1.333		475	53		528	610
0150	Bolt down frame & counter weights, 250 lb.	B-1	4	6		4,175	243		4,418	5,000
0152	500 lb.		4	6		6,175	243		6,418	7,200
0154	750 lb.		4	6		8,925	243		9,168	10,200
0156	1000 lb.		2.67	8.989		9,800	365		10,165	11,400
0158	1500 lb.	↓	2.67	8.989		12,600	365		12,965	14,400
0160	Chute warning light system, 5 stories	B-1C	4	6		8,900	243	114	9,257	10,300
0162	10 stories	"	2	12		14,200	485	227	14,912	16,600
0164	Dust control device for dumpsters	1 Clab	8	1		133	40		173	211
0166	Install or replace breakaway cord		8	1		25	40		65	92
0168	Install or replace warning sign	↓	16	.500	↓	10	19.95		29.95	43.50
0600	Dumpster, weekly rental, 1 dump/week, 6 C.Y. capacity (2 tons)				Week	415			415	455
0700	10 C.Y. capacity (3 tons)					480			480	530
0725	20 C.Y. capacity (5 tons)					565			565	625
0800	30 C.Y. capacity (7 tons)					730			730	800
0840	40 C.Y. capacity (10 tons)				↓	775			775	850
0900	Alternate pricing for dumpsters									
0910	Delivery, average for all sizes				Ea.	75			75	82.50
0920	Haul, average for all sizes					235			235	259
0930	Rent per day, average for all sizes					20			20	22
0940	Rent per month, average for all sizes				↓	80			80	88
0950	Disposal fee per ton, average for all sizes				Ton	88			88	97
2000	Load, haul, dump and return, 0'-50' haul, hand carried	2 Clab	24	.667	C.Y.		26.50		26.50	43
2005	Wheeled		37	.432			17.25		17.25	28
2040	0'-100' haul, hand carried		16.50	.970			38.50		38.50	62.50
2045	Wheeled	↓	25	.640			25.50		25.50	41.50
2050	Forklift	A-3R	25	.320			16.40	5.95	22.35	32.50
2080	Haul and return, add per each extra 100' haul, hand carried	2 Clab	35.50	.451			17.95		17.95	29
2085	Wheeled		54	.296			11.80		11.80	19.15
2120	For travel in elevators, up to 10 floors, add		140	.114			4.55		4.55	7.40
2130	0'-50' haul, incl. up to 5 riser stairs, hand carried		23	.696			27.50		27.50	45
2135	Wheeled		35	.457			18.20		18.20	29.50
2140	6-10 riser stairs, hand carried		22	.727			29		29	47
2145	Wheeled		34	.471			18.75		18.75	30.50
2150	11-20 riser stairs, hand carried		20	.800			32		32	52
2155	Wheeled		31	.516			20.50		20.50	33.50
2160	21-40 riser stairs, hand carried		16	1			40		40	64.50
2165	Wheeled		24	.667			26.50		26.50	43
2170	0-100' haul, incl. 5 riser stairs, hand carried		15	1.067			42.50		42.50	69
2175	Wheeled		23	.696			27.50		27.50	45
2180	6-10 riser stairs, hand carried		14	1.143			45.50		45.50	74
2185	Wheeled		21	.762			30.50		30.50	49.50
2190	11-20 riser stairs, hand carried		12	1.333			53		53	86.50
2195	Wheeled		18	.889			35.50		35.50	57.50
2200	21-40 riser stairs, hand carried		8	2			79.50		79.50	129
2205	Wheeled		12	1.333			53		53	86.50
2210	Haul and return, add per each extra 100' haul, hand carried		35.50	.451	↓		17.95		17.95	29

R024119-20

02 41 Demolition

02 41 19 – Selective Demolition

02 41 19.19 Selective Demolition	Crew	Daily Output	Labor-Hours	Unit	Material	2018 Bare Costs Labor	2018 Bare Costs Equipment	Total	Total Incl O&P
2215	Wheeled	2 Clab	54	.296	C.Y.		11.80		11.80	19.15
2220	For each additional flight of stairs, up to 5 risers, add		550	.029	Flight		1.16		1.16	1.88
2225	6-10 risers, add		275	.058			2.32		2.32	3.76
2230	11-20 risers, add		138	.116			4.62		4.62	7.50
2235	21-40 risers, add		69	.232			9.25		9.25	15
3000	Loading & trucking, including 2 mile haul, chute loaded	B-16	45	.711	C.Y.		30	12.05	42.05	61.50
3040	Hand loading truck, 50' haul	"	48	.667			28	11.30	39.30	57.50
3080	Machine loading truck	B-17	120	.267			11.80	5.40	17.20	25
3120	Wheeled 50' and ramp dump loaded	2 Clab	24	.667			26.50		26.50	43
5000	Haul, per mile, up to 8 C.Y. truck	B-34B	1165	.007			.32	.47	.79	1.02
5100	Over 8 C.Y. truck	"	1550	.005			.24	.35	.59	.77

02 41 19.20 Selective Demolition, Dump Charges

		Crew	Daily Output	Labor-Hours	Unit	Material	2018 Bare Costs Labor	2018 Bare Costs Equipment	Total	Total Incl O&P
0010	**SELECTIVE DEMOLITION, DUMP CHARGES** R024119-10									
0020	Dump charges, typical urban city, tipping fees only									
0100	Building construction materials				Ton	74			74	81
0200	Trees, brush, lumber					63			63	69.50
0300	Rubbish only					63			63	69.50
0500	Reclamation station, usual charge					74			74	81

02 41 19.21 Selective Demolition, Gutting

		Crew	Daily Output	Labor-Hours	Unit	Material	2018 Bare Costs Labor	2018 Bare Costs Equipment	Total	Total Incl O&P
0010	**SELECTIVE DEMOLITION, GUTTING** R024119-10									
0020	Building interior, including disposal, dumpster fees not included									
0500	Residential building									
0560	Minimum	B-16	400	.080	SF Flr.		3.35	1.36	4.71	6.90
0580	Maximum	"	360	.089	"		3.72	1.51	5.23	7.70
0900	Commercial building									
1000	Minimum	B-16	350	.091	SF Flr.		3.83	1.55	5.38	7.90
1020	Maximum		250	.128	"		5.35	2.17	7.52	11.10
3000	Minimum labor/equipment charge		4	8	Job		335	136	471	690

02 41 19.25 Selective Demolition, Saw Cutting

		Crew	Daily Output	Labor-Hours	Unit	Material	2018 Bare Costs Labor	2018 Bare Costs Equipment	Total	Total Incl O&P
0010	**SELECTIVE DEMOLITION, SAW CUTTING** R024119-10									
0015	Asphalt, up to 3" deep	B-89	1050	.015	L.F.	.19	.73	.40	1.32	1.81
0020	Each additional inch of depth	"	1800	.009		.06	.43	.23	.72	1.01
1200	Masonry walls, hydraulic saw, brick, per inch of depth	B-89B	300	.053		.06	2.55	2.24	4.85	6.60
1220	Block walls, solid, per inch of depth	"	250	.064		.06	3.07	2.69	5.82	7.90
2000	Brick or masonry w/hand held saw, per inch of depth	A-1	125	.064		.04	2.55	.57	3.16	4.82
5000	Wood sheathing to 1" thick, on walls	1 Carp	200	.040			2.03		2.03	3.29
5020	On roof	"	250	.032			1.62		1.62	2.63
9000	Minimum labor/equipment charge	A-1	2	4	Job		159	35.50	194.50	298

02 41 19.27 Selective Demolition, Torch Cutting

		Crew	Daily Output	Labor-Hours	Unit	Material	2018 Bare Costs Labor	2018 Bare Costs Equipment	Total	Total Incl O&P
0010	**SELECTIVE DEMOLITION, TORCH CUTTING** R024119-10									
0020	Steel, 1" thick plate	E-25	333	.024	L.F.	.87	1.36	.04	2.27	3.32
0040	1" diameter bar	"	600	.013	Ea.	.15	.76	.02	.93	1.47
1000	Oxygen lance cutting, reinforced concrete walls									
1040	12"-16" thick walls	1 Clab	10	.800	L.F.		32		32	52
1080	24" thick walls	"	6	1.333	"		53		53	86.50
1090	Minimum labor/equipment charge	E-25	2	4	Job		227	6.30	233.30	390
1100	See Section 05 05 21.10									

02 42 10 – Building Deconstruction

02 42 10.10 Estimated Salvage Value or Savings

		Crew	Daily Output	Labor-Hours	Unit	Material	2018 Bare Costs Labor	2018 Bare Costs Equipment	Total	Total Incl O&P
0010	**ESTIMATED SALVAGE VALUE OR SAVINGS**									
0015	Excludes material handling, packaging, container costs and									
0020	transportation for salvage or disposal									
0050	All items in Section 02 42 10.10 are credit deducts and not costs									
0100	Copper wire salvage value	G			Lb.				1.60	1.60
0110	Disposal savings	G							.04	.04
0200	Copper pipe salvage value	G							2.50	2.50
0210	Disposal savings	G							.05	.05
0300	Steel pipe salvage value	G							.06	.06
0310	Disposal savings	G							.03	.03
0400	Cast iron pipe salvage value	G							.03	.03
0410	Disposal savings	G							.01	.01
0500	Steel doors or windows salvage value	G							.06	.06
0510	Aluminum	G							.55	.55
0520	Disposal savings	G							.03	.03
0600	Aluminum siding salvage value	G							.49	.49
0630	Disposal savings	G							.03	.03
0640	Wood siding (no lead or asbestos)	G			C.Y.				12	12
0800	Clean concrete disposal savings	G			Ton				62	62
0850	Asphalt shingles disposal savings	G			"				60	60
1000	Wood wall framing clean salvage value	G			M.B.F.				55	55
1010	Painted	G							44	44
1020	Floor framing	G							55	55
1030	Painted	G							44	44
1050	Roof framing	G							55	55
1060	Painted	G							44	44
1100	Wood beams salvage value	G							55	55
1200	Wood framing and beams disposal savings	G			Ton				66	66
1220	Wood sheathing and sub-base flooring	G							72.50	72.50
1230	Wood wall paneling (1/4" thick)	G							66	66
1300	Wood panel 3/4"-1" thick low salvage value	G			S.F.				.55	.55
1350	High salvage value	G			"				2.20	2.20
1400	Disposal savings	G			Ton				66	66
1500	Flooring tongue and groove 25/32" thick low salvage value	G			S.F.				.55	.55
1530	High salvage value	G			"				1.10	1.10
1560	Disposal savings	G			Ton				66	66
1600	Drywall or sheet rock salvage value	G							22	22
1650	Disposal savings	G							66	66

02 42 10.20 Deconstruction of Building Components

		Crew	Daily Output	Labor-Hours	Unit	Material	2018 Bare Costs Labor	2018 Bare Costs Equipment	Total	Total Incl O&P
0010	**DECONSTRUCTION OF BUILDING COMPONENTS**									
0012	Buildings one or two stories only									
0015	Excludes material handling, packaging, container costs and									
0020	transportation for salvage or disposal									
0050	Deconstruction of plumbing fixtures									
0100	Wall hung or countertop lavatory	G	2 Clab	16	1	Ea.		40		40	64.50
0110	Single or double compartment kitchen sink	G		14	1.143			45.50		45.50	74
0120	Wall hung urinal	G		14	1.143			45.50		45.50	74
0130	Floor mounted	G		8	2			79.50		79.50	129
0140	Floor mounted water closet	G		16	1			40		40	64.50
0150	Wall hung	G		14	1.143			45.50		45.50	74
0160	Water fountain, free standing	G		16	1			40		40	64.50
0170	Wall hung or deck mounted	G		12	1.333			53		53	86.50

02 42 Removal and Salvage of Construction Materials

02 42 10 – Building Deconstruction

02 42 10.20 Deconstruction of Building Components

		Crew	Daily Output	Labor-Hours	Unit	Material	2018 Bare Costs Labor	2018 Bare Costs Equipment	Total	Total Incl O&P
0180	Bathtub, steel or fiberglass	2 Clab	10	1.600	Ea.		64		64	104
0190	Cast iron	G	8	2			79.50		79.50	129
0200	Shower, single	G	6	2.667			106		106	173
0210	Group	G	7	2.286			91		91	148
0300	Deconstruction of electrical fixtures									
0310	Surface mounted incandescent fixtures	2 Clab	48	.333	Ea.		13.30		13.30	21.50
0320	Fluorescent, 2 lamp	G	32	.500			19.95		19.95	32.50
0330	4 lamp	G	24	.667			26.50		26.50	43
0340	Strip fluorescent, 1 lamp	G	40	.400			15.95		15.95	26
0350	2 lamp	G	32	.500			19.95		19.95	32.50
0400	Recessed drop-in fluorescent fixture, 2 lamp	G	27	.593			23.50		23.50	38.50
0410	4 lamp	G	18	.889			35.50		35.50	57.50
0500	Deconstruction of appliances									
0510	Cooking stoves	2 Clab	26	.615	Ea.		24.50		24.50	40
0520	Dishwashers	"	26	.615	"		24.50		24.50	40
0600	Deconstruction of millwork and trim									
0610	Cabinets, wood	2 Carp	40	.400	L.F.		20.50		20.50	33
0620	Countertops	G	100	.160	"		8.10		8.10	13.15
0630	Wall paneling, 1" thick	G	500	.032	S.F.		1.62		1.62	2.63
0640	Ceiling trim	G	500	.032	L.F.		1.62		1.62	2.63
0650	Wainscoting	G	500	.032	S.F.		1.62		1.62	2.63
0660	Base, 3/4"-1" thick	G	600	.027	L.F.		1.35		1.35	2.19
0700	Deconstruction of doors and windows									
0710	Doors, wrap, interior, wood, single, no closers	2 Carp	21	.762	Ea.	4.75	38.50		43.25	68
0720	Double	G	13	1.231		9.50	62.50		72	111
0730	Solid core, single, exterior or interior	G	10	1.600		4.75	81		85.75	137
0740	Double	G	8	2		9.50	101		110.50	175
0810	Windows, wrap, wood, single									
0812	with no casement or cladding	2 Carp	21	.762	Ea.	4.75	38.50		43.25	68
0820	with casement and/or cladding	"	18	.889	"	4.75	45		49.75	78.50
0900	Deconstruction of interior finishes									
0910	Drywall for recycling	2 Clab	1775	.009	S.F.		.36		.36	.58
0920	Plaster wall, first floor	G	1775	.009			.36		.36	.58
0930	Second floor	G	1330	.012			.48		.48	.78
1000	Deconstruction of roofing and accessories									
1010	Built-up roofs	2 Clab	570	.028	S.F.		1.12		1.12	1.82
1020	Gutters, fascia and rakes	"	1140	.014	L.F.		.56		.56	.91
2000	Deconstruction of wood components									
2010	Roof sheeting	2 Clab	570	.028	S.F.		1.12		1.12	1.82
2020	Main roof framing	G	760	.021	L.F.		.84		.84	1.36
2030	Porch roof framing	G	445	.036			1.43		1.43	2.33
2040	Beams 4" x 8"	B-1	375	.064			2.59		2.59	4.21
2050	4" x 10"	G	300	.080			3.24		3.24	5.25
2055	4" x 12"	G	250	.096			3.89		3.89	6.30
2060	6" x 8"	G	250	.096			3.89		3.89	6.30
2065	6" x 10"	G	200	.120			4.86		4.86	7.90
2070	6" x 12"	G	170	.141			5.70		5.70	9.30
2075	8" x 12"	G	126	.190			7.70		7.70	12.55
2080	10" x 12"	G	100	.240			9.70		9.70	15.80
2100	Ceiling joists	2 Clab	800	.020			.80		.80	1.29
2150	Wall framing, interior	G	1230	.013			.52		.52	.84
2160	Sub-floor	G	2000	.008	S.F.		.32		.32	.52
2170	Floor joists	G	2000	.008	L.F.		.32		.32	.52

02 42 Removal and Salvage of Construction Materials

02 42 10 – Building Deconstruction

02 42 10.20 Deconstruction of Building Components		Crew	Daily Output	Labor-Hours	Unit	Material	2018 Bare Costs Labor	2018 Bare Costs Equipment	Total	Total Incl O&P
2200	Wood siding (no lead or asbestos)	G 2 Clab	1300	.012	S.F.		.49		.49	.80
2300	Wall framing, exterior	G	1600	.010	L.F.		.40		.40	.65
2400	Stair risers	G	53	.302	Ea.		12.05		12.05	19.55
2500	Posts	G	800	.020	L.F.		.80		.80	1.29
3000	Deconstruction of exterior brick walls									
3010	Exterior brick walls, first floor	G 2 Clab	200	.080	S.F.		3.19		3.19	5.20
3020	Second floor	G	64	.250	"		9.95		9.95	16.20
3030	Brick chimney	G	100	.160	C.F.		6.40		6.40	10.35
4000	Deconstruction of concrete									
4010	Slab on grade, 4" thick, plain concrete	G B-9	500	.080	S.F.		3.22	.47	3.69	5.75
4020	Wire mesh reinforced	G	470	.085			3.43	.50	3.93	6.10
4030	Rod reinforced	G	400	.100			4.03	.59	4.62	7.20
4110	Foundation wall, 6" thick, plain concrete	G	160	.250			10.05	1.46	11.51	17.95
4120	8" thick	G	140	.286			11.50	1.67	13.17	20.50
4130	10" thick	G	120	.333			13.40	1.95	15.35	24
9000	Deconstruction process, support equipment as needed									
9010	Daily use, portal to portal, 12-ton truck-mounted hydraulic crane crew	G A-3H	1	8	Day		450	630	1,080	1,400
9020	Daily use, skid steer and operator	G A-3C	1	8			410	365	775	1,050
9030	Daily use, backhoe 48 HP, operator and labor	G "	1	8			410	365	775	1,050

02 42 10.30 Deconstruction Material Handling

		Crew	Daily Output	Labor-Hours	Unit	Material	2018 Bare Costs Labor	2018 Bare Costs Equipment	Total	Total Incl O&P
0010	**DECONSTRUCTION MATERIAL HANDLING**									
0012	Buildings one or two stories only									
0100	Clean and stack brick on pallet	G 2 Clab	1200	.013	Ea.		.53		.53	.86
0200	Haul 50' and load rough lumber up to 2" x 8" size	G	2000	.008	"		.32		.32	.52
0210	Lumber larger than 2" x 8"	G	3200	.005	B.F.		.20		.20	.32
0300	Finish wood for recycling stack and wrap per pallet	G	8	2	Ea.	38	79.50		117.50	171
0350	Light fixtures		6	2.667		68.50	106		174.50	248
0375	Windows		6	2.667		64.50	106		170.50	244
0400	Miscellaneous materials		8	2		19	79.50		98.50	150
1000	See Section 02 41 19.19 for bulk material handling									

02 43 Structure Moving

02 43 13 – Structure Relocation

02 43 13.13 Building Relocation

		Crew	Daily Output	Labor-Hours	Unit	Material	2018 Bare Costs Labor	2018 Bare Costs Equipment	Total	Total Incl O&P
0010	**BUILDING RELOCATION**									
0011	One day move, up to 24' wide									
0020	Reset on existing foundation					Total			11,500	11,500
0040	Wood or steel frame bldg., based on ground floor area	G B-4	185	.259	S.F.		10.70	2.54	13.24	20
0060	Masonry bldg., based on ground floor area	G "	137	.350			14.45	3.43	17.88	27.50
0200	For 24'-42' wide, add								15%	15%

For customer support on your Commercial Renovation Costs with RSMeans data, call 800.448.8182.

35

02 58 Snow Control

02 58 13 – Snow Fencing

02 58 13.10 Snow Fencing System		Crew	Daily Output	Labor-Hours	Unit	Material	2018 Bare Costs Labor	Equipment	Total	Total Incl O&P
0010	**SNOW FENCING SYSTEM**									
7001	Snow fence on steel posts 10' OC, 4' high	B-1	500	.048	L.F.	.46	1.95		2.41	3.67

02 65 Underground Storage Tank Removal

02 65 10 – Underground Tank and Contaminated Soil Removal

02 65 10.30 Removal of Underground Storage Tanks

			Crew	Daily Output	Labor-Hours	Unit	Material	2018 Bare Costs Labor	Equipment	Total	Total Incl O&P
0010	**REMOVAL OF UNDERGROUND STORAGE TANKS**	R026510-20									
0011	Petroleum storage tanks, non-leaking										
0100	Excavate & load onto trailer										
0110	3,000 gal. to 5,000 gal. tank	G	B-14	4	12	Ea.		505	78	583	900
0120	6,000 gal. to 8,000 gal. tank	G	B-3A	3	13.333			570	298	868	1,250
0130	9,000 gal. to 12,000 gal. tank	G	"	2	20			855	445	1,300	1,875
0190	Known leaking tank, add					%				100%	100%
0200	Remove sludge, water and remaining product from bottom										
0201	of tank with vacuum truck										
0300	3,000 gal. to 5,000 gal. tank	G	A-13	5	1.600	Ea.		82	144	226	288
0310	6,000 gal. to 8,000 gal. tank	G		4	2			103	180	283	360
0320	9,000 gal. to 12,000 gal. tank	G		3	2.667			137	240	377	480
0390	Dispose of sludge off-site, average					Gal.				6.25	6.80
0400	Insert inert solid CO_2 "dry ice" into tank										
0401	For cleaning/transporting tanks (1.5 lb./100 gal. cap)	G	1 Clab	500	.016	Lb.	1.19	.64		1.83	2.35
0403	Insert solid carbon dioxide, 1.5 lb./100 gal.	G	"	400	.020	"	1.19	.80		1.99	2.60
0503	Disconnect and remove piping	G	1 Plum	160	.050	L.F.		3.11		3.11	4.84
0603	Transfer liquids, 10% of volume	G	"	1600	.005	Gal.		.31		.31	.48
0703	Cut accessway into underground storage tank	G	1 Clab	5.33	1.501	Ea.		60		60	97
0813	Remove sludge, wash and wipe tank, 500 gal.	G	1 Plum	8	1			62		62	96.50
0823	3,000 gal.	G		6.67	1.199			74.50		74.50	116
0833	5,000 gal.	G		6.15	1.301			81		81	126
0843	8,000 gal.	G		5.33	1.501			93.50		93.50	145
0853	10,000 gal.	G		4.57	1.751			109		109	169
0863	12,000 gal.	G		4.21	1.900			118		118	184
1020	Haul tank to certified salvage dump, 100 miles round trip										
1023	3,000 gal. to 5,000 gal. tank					Ea.				760	830
1026	6,000 gal. to 8,000 gal. tank									880	960
1029	9,000 gal. to 12,000 gal. tank									1,050	1,150
1100	Disposal of contaminated soil to landfill										
1110	Minimum					C.Y.				145	160
1111	Maximum					"				400	440
1120	Disposal of contaminated soil to										
1121	bituminous concrete batch plant										
1130	Minimum					C.Y.				80	88
1131	Maximum					"				115	125
1203	Excavate, pull, & load tank, backfill hole, 8,000 gal. +	G	B-12C	.50	32	Ea.		1,525	2,100	3,625	4,775
1213	Haul tank to certified dump, 100 miles rt, 8,000 gal. +	G	B-34K	1	8			370	835	1,205	1,500
1223	Excavate, pull, & load tank, backfill hole, 500 gal.	G	B-11C	1	16			750	310	1,060	1,550
1233	Excavate, pull, & load tank, backfill hole, 3,000-5,000 gal.	G	B-11M	.50	32			1,500	770	2,270	3,250
1243	Haul tank to certified dump, 100 miles rt, 500 gal.	G	B-34L	1	8			410	188	598	855
1253	Haul tank to certified dump, 100 miles rt, 3,000-5,000 gal.	G	B-34M	1	8			410	238	648	910
2010	Decontamination of soil on site incl poly tarp on top/bottom										
2011	Soil containment berm and chemical treatment										
2020	Minimum	G	B-11C	100	.160	C.Y.	7.45	7.50	3.12	18.07	23.50

For customer support on your Commercial Renovation Costs with RSMeans data, call 800.448.8182.

02 65 Underground Storage Tank Removal

02 65 10 - Underground Tank and Contaminated Soil Removal

02 65 10.30 Removal of Underground Storage Tanks		Crew	Daily Output	Labor-Hours	Unit	Material	2018 Bare Costs Labor	Equipment	Total	Total Incl O&P
2021	Maximum	G B-11C	100	.160	C.Y.	9.65	7.50	3.12	20.27	26
2050	Disposal of decontaminated soil, minimum								135	150
2055	Maximum								400	440

02 81 Transportation and Disposal of Hazardous Materials

02 81 20 - Hazardous Waste Handling

02 81 20.10 Hazardous Waste Cleanup/Pickup/Disposal

		Crew	Daily Output	Labor-Hours	Unit	Material	Labor	Equipment	Total	Total Incl O&P
0010	**HAZARDOUS WASTE CLEANUP/PICKUP/DISPOSAL**									
0100	For contractor rental equipment, i.e., dozer,									
0110	Front end loader, dump truck, etc., see 01 54 33 Reference Section									
1000	Solid pickup									
1100	55 gal. drums				Ea.				240	265
1120	Bulk material, minimum				Ton				190	210
1130	Maximum				"				595	655
1200	Transportation to disposal site									
1220	Truckload = 80 drums or 25 C.Y. or 18 tons									
1260	Minimum				Mile				3.95	4.45
1270	Maximum				"				7.25	7.35
3000	Liquid pickup, vacuum truck, stainless steel tank									
3100	Minimum charge, 4 hours									
3110	1 compartment, 2200 gallon				Hr.				140	155
3120	2 compartment, 5000 gallon				"				200	225
3400	Transportation in 6900 gallon bulk truck				Mile				7.95	8.75
3410	In teflon lined truck				"				10.20	11.25
5000	Heavy sludge or dry vacuumable material				Hr.				140	160
6000	Dumpsite disposal charge, minimum				Ton				140	155
6020	Maximum				"				415	455

02 82 Asbestos Remediation

02 82 13 - Asbestos Abatement

02 82 13.39 Asbestos Remediation Plans and Methods

		Crew	Daily Output	Labor-Hours	Unit	Material	Labor	Equipment	Total	Total Incl O&P
0010	**ASBESTOS REMEDIATION PLANS AND METHODS**									
0100	Building Survey-Commercial Building				Ea.				2,200	2,400
0200	Asbestos Abatement Remediation Plan				"				1,350	1,475

02 82 13.41 Asbestos Abatement Equipment

		Crew	Daily Output	Labor-Hours	Unit	Material	Labor	Equipment	Total	Total Incl O&P
0010	**ASBESTOS ABATEMENT EQUIPMENT** R028213-20									
0011	Equipment and supplies, buy									
0200	Air filtration device, 2000 CFM				Ea.	820			820	900
0250	Large volume air sampling pump, minimum					315			315	345
0260	Maximum					360			360	395
0300	Airless sprayer unit, 2 gun					2,175			2,175	2,375
0350	Light stand, 500 watt					38			38	42
0400	Personal respirators									
0410	Negative pressure, 1/2 face, dual operation, minimum				Ea.	27			27	29.50
0420	Maximum					30			30	33
0450	P.A.P.R., full face, minimum					675			675	740
0460	Maximum					1,400			1,400	1,550
0470	Supplied air, full face, including air line, minimum					380			380	420
0480	Maximum					525			525	580

For customer support on your Commercial Renovation Costs with RSMeans data, call 800.448.8182.

37

02 82 Asbestos Remediation

02 82 13 – Asbestos Abatement

02 82 13.41 Asbestos Abatement Equipment

		Crew	Daily Output	Labor-Hours	Unit	Material	2018 Bare Costs Labor	Equipment	Total	Total Incl O&P
0500	Personnel sampling pump				Ea.	245			245	270
1500	Power panel, 20 unit, including GFI					445			445	490
1600	Shower unit, including pump and filters					1,075			1,075	1,175
1700	Supplied air system (type C)					3,500			3,500	3,850
1750	Vacuum cleaner, HEPA, 16 gal., stainless steel, wet/dry					1,100			1,100	1,200
1760	55 gallon					1,300			1,300	1,425
1800	Vacuum loader, 9-18 ton/hr.					97,000			97,000	107,000
1900	Water atomizer unit, including 55 gal. drum					289			289	320
2000	Worker protection, whole body, foot, head cover & gloves, plastic					7.95			7.95	8.75
2500	Respirator, single use					26			26	28.50
2550	Cartridge for respirator					4.40			4.40	4.84
2570	Glove bag, 7 mil, 50" x 64"					7.60			7.60	8.35
2580	10 mil, 44" x 60"					8.90			8.90	9.75
3000	HEPA vacuum for work area, minimum					330			330	360
3050	Maximum					725			725	795
6000	Disposable polyethylene bags, 6 mil, 3 C.F.					.86			.86	.95
6300	Disposable fiber drums, 3 C.F.					18.70			18.70	20.50
6400	Pressure sensitive caution labels, 3" x 5"					2.42			2.42	2.66
6450	11" x 17"					7.60			7.60	8.35
6500	Negative air machine, 1800 CFM					860			860	950

02 82 13.42 Preparation of Asbestos Containment Area

		Crew	Daily Output	Labor-Hours	Unit	Material	2018 Bare Costs Labor	Equipment	Total	Total Incl O&P
0010	**PREPARATION OF ASBESTOS CONTAINMENT AREA**									
0100	Pre-cleaning, HEPA vacuum and wet wipe, flat surfaces	A-9	12000	.005	S.F.	.01	.30		.31	.49
0200	Protect carpeted area, 2 layers 6 mil poly on 3/4" plywood	"	1000	.064		2.02	3.61		5.63	8
0300	Separation barrier, 2" x 4" @ 16", 1/2" plywood ea. side, 8' high	2 Carp	400	.040		2.81	2.03		4.84	6.40
0310	12' high		320	.050		2.75	2.54		5.29	7.15
0320	16' high		200	.080		2.73	4.06		6.79	9.60
0400	Personnel decontam. chamber, 2" x 4" @ 16", 3/4" ply ea. side		280	.057		3.36	2.90		6.26	8.40
0450	Waste decontam. chamber, 2" x 4" studs @ 16", 3/4" ply ea. side		360	.044		3.36	2.25		5.61	7.35
0500	Cover surfaces with polyethylene sheeting									
0501	Including glue and tape									
0550	Floors, each layer, 6 mil	A-9	8000	.008	S.F.	.04	.45		.49	.76
0551	4 mil		9000	.007		.03	.40		.43	.67
0560	Walls, each layer, 6 mil		6000	.011		.04	.60		.64	1
0561	4 mil		7000	.009		.03	.52		.55	.86
0570	For heights above 14', add						20%			
0575	For heights above 20', add						30%			
0580	For fire retardant poly, add					100%				
0590	For large open areas, deduct					10%	20%			
0600	Seal floor penetrations with foam firestop to 36 sq. in.	2 Carp	200	.080	Ea.	11.35	4.06		15.41	19.10
0610	36 sq. in. to 72 sq. in.		125	.128		22.50	6.50		29	35.50
0615	72 sq. in. to 144 sq. in.		80	.200		45.50	10.15		55.65	66.50
0620	Wall penetrations, to 36 sq. in.		180	.089		11.35	4.51		15.86	19.80
0630	36 sq. in. to 72 sq. in.		100	.160		22.50	8.10		30.60	38
0640	72 sq. in. to 144 sq. in.		60	.267		45.50	13.50		59	72
0800	Caulk seams with latex	1 Carp	230	.035	L.F.	.17	1.76		1.93	3.05
0900	Set up neg. air machine, 1-2k CFM/25 M.C.F. volume	1 Asbe	4.30	1.860	Ea.		105		105	168
0950	Set up and remove portable shower unit	2 Asbe	4	4	"		225		225	360

02 82 13 – Asbestos Abatement

02 82 13.43 Bulk Asbestos Removal	Crew	Daily Output	Labor-Hours	Unit	Material	2018 Bare Costs Labor	Equipment	Total	Total Incl O&P
0010 **BULK ASBESTOS REMOVAL**									
0020 Includes disposable tools and 2 suits and 1 respirator filter/day/worker									
0100 Beams, W 10 x 19	A-9	235	.272	L.F.	.69	15.35		16.04	25.50
0110 W 12 x 22		210	.305		.77	17.20		17.97	28.50
0120 W 14 x 26		180	.356		.90	20		20.90	33
0130 W 16 x 31		160	.400		1.02	22.50		23.52	37
0140 W 18 x 40		140	.457		1.16	26		27.16	43
0150 W 24 x 55		110	.582		1.48	33		34.48	54
0160 W 30 x 108		85	.753		1.91	42.50		44.41	70
0170 W 36 x 150		72	.889	▼	2.26	50		52.26	83
0200 Boiler insulation	▼	480	.133	S.F.	.41	7.50		7.91	12.50
0210 With metal lath, add				%				50%	50%
0300 Boiler breeching or flue insulation	A-9	520	.123	S.F.	.31	6.95		7.26	11.45
0310 For active boiler, add				%				100%	100%
0400 Duct or AHU insulation	A-10B	440	.073	S.F.	.18	4.10		4.28	6.75
0500 Duct vibration isolation joints, up to 24 sq. in. duct	A-9	56	1.143	Ea.	2.90	64.50		67.40	106
0520 25 sq. in. to 48 sq. in. duct		48	1.333		3.39	75		78.39	124
0530 49 sq. in. to 76 sq. in. duct		40	1.600	▼	4.06	90		94.06	148
0600 Pipe insulation, air cell type, up to 4" diameter pipe		900	.071	L.F.	.18	4.01		4.19	6.60
0610 4" to 8" diameter pipe		800	.080		.20	4.51		4.71	7.40
0620 10" to 12" diameter pipe		700	.091		.23	5.15		5.38	8.50
0630 14" to 16" diameter pipe		550	.116	▼	.30	6.55		6.85	10.85
0650 Over 16" diameter pipe		650	.098	S.F.	.25	5.55		5.80	9.20
0700 With glove bag up to 3" diameter pipe		200	.320	L.F.	9.50	18.05		27.55	39.50
1000 Pipe fitting insulation up to 4" diameter pipe		320	.200	Ea.	.51	11.25		11.76	18.60
1100 6" to 8" diameter pipe		304	.211		.53	11.85		12.30	19.60
1110 10" to 12" diameter pipe		192	.333		.85	18.80		19.65	31
1120 14" to 16" diameter pipe		128	.500	▼	1.27	28		29.27	46.50
1130 Over 16" diameter pipe		176	.364	S.F.	.92	20.50		21.42	34
1200 With glove bag, up to 8" diameter pipe		75	.853	L.F.	6.45	48		54.45	84
2000 Scrape foam fireproofing from flat surface		2400	.027	S.F.	.07	1.50		1.57	2.48
2100 Irregular surfaces		1200	.053		.14	3.01		3.15	4.97
3000 Remove cementitious material from flat surface		1800	.036		.09	2		2.09	3.31
3100 Irregular surface		1000	.064		.12	3.61		3.73	5.95
4000 Scrape acoustical coating/fireproofing, from ceiling		3200	.020		.05	1.13		1.18	1.87
5000 Remove VAT and mastic from floor by hand	▼	2400	.027		.07	1.50		1.57	2.48
5100 By machine	A-11	4800	.013	▼	.03	.75	.01	.79	1.25
5150 For 2 layers, add				%				50%	50%
6000 Remove contaminated soil from crawl space by hand	A-9	400	.160	C.F.	.41	9		9.41	14.90
6100 With large production vacuum loader	A-12	700	.091	"	.23	5.15	1.03	6.41	9.65
7000 Radiator backing, not including radiator removal	A-9	1200	.053	S.F.	.14	3.01		3.15	4.97
8000 Cement-asbestos transite board and cement wall board	2 Asbe	1000	.016		.15	.90		1.05	1.61
8100 Transite shingle siding	A-10B	750	.043		.21	2.41		2.62	4.09
8200 Shingle roofing	"	2000	.016		.08	.90		.98	1.54
8250 Built-up, no gravel, non-friable	B-2	1400	.029		.08	1.15		1.23	1.96
8260 Bituminous flashing	1 Rofc	300	.027		.08	1.17		1.25	2.18
8300 Asbestos millboard, flat board and VAT contaminated plywood	2 Asbe	1000	.016	▼	.08	.90		.98	1.53
9000 For type B (supplied air) respirator equipment, add				%				10%	10%

02 82 13.44 Demolition In Asbestos Contaminated Area

	Crew	Daily Output	Labor-Hours	Unit	Material	Labor	Equipment	Total	Total Incl O&P
0010 **DEMOLITION IN ASBESTOS CONTAMINATED AREA**									
0200 Ceiling, including suspension system, plaster and lath	A-9	2100	.030	S.F.	.08	1.72		1.80	2.84
0210 Finished plaster, leaving wire lath	▼	585	.109	▼	.28	6.15		6.43	10.20

02 82 13 – Asbestos Abatement

02 82 13.44 Demolition In Asbestos Contaminated Area

		Crew	Daily Output	Labor-Hours	Unit	Material	2018 Bare Costs Labor	Equipment	Total	Total Incl O&P
0220	Suspended acoustical tile	A-9	3500	.018	S.F.	.05	1.03		1.08	1.70
0230	Concealed tile grid system		3000	.021		.05	1.20		1.25	1.99
0240	Metal pan grid system		1500	.043		.11	2.40		2.51	3.97
0250	Gypsum board		2500	.026		.07	1.44		1.51	2.38
0260	Lighting fixtures up to 2' x 4'		72	.889	Ea.	2.26	50		52.26	83
0400	Partitions, non load bearing									
0410	Plaster, lath, and studs	A-9	690	.093	S.F.	.88	5.25		6.13	9.35
0450	Gypsum board and studs	"	1390	.046	"	.12	2.59		2.71	4.29
9000	For type B (supplied air) respirator equipment, add				%				10%	10%

02 82 13.45 OSHA Testing

		Crew	Daily Output	Labor-Hours	Unit	Material	2018 Bare Costs Labor	Equipment	Total	Total Incl O&P
0010	**OSHA TESTING**									
0100	Certified technician, minimum				Day				200	220
0110	Maximum								300	330
0121	Industrial hygienist, minimum	1 Asbe	1.75	4.571			257		257	410
0130	Maximum								400	440
0200	Asbestos sampling and PCM analysis, NIOSH 7400, minimum	1 Asbe	8	1	Ea.	16.05	56.50		72.55	108
0210	Maximum		4	2		27.50	113		140.50	211
1000	Cleaned area samples		8	1		202	56.50		258.50	310
1100	PCM air sample analysis, NIOSH 7400, minimum		8	1		15.10	56.50		71.60	107
1110	Maximum		4	2		2.22	113		115.22	182
1200	TEM air sample analysis, NIOSH 7402, minimum								80	106
1210	Maximum								360	450

02 82 13.46 Decontamination of Asbestos Containment Area

		Crew	Daily Output	Labor-Hours	Unit	Material	2018 Bare Costs Labor	Equipment	Total	Total Incl O&P
0010	**DECONTAMINATION OF ASBESTOS CONTAINMENT AREA**									
0100	Spray exposed substrate with surfactant (bridging)									
0200	Flat surfaces	A-9	6000	.011	S.F.	.40	.60		1	1.40
0250	Irregular surfaces		4000	.016	"	.40	.90		1.30	1.88
0300	Pipes, beams, and columns		2000	.032	L.F.	.60	1.80		2.40	3.55
1000	Spray encapsulate polyethylene sheeting		8000	.008	S.F.	.40	.45		.85	1.16
1100	Roll down polyethylene sheeting		8000	.008	"		.45		.45	.72
1500	Bag polyethylene sheeting		400	.160	Ea.	1	9		10	15.55
2000	Fine clean exposed substrate, with nylon brush		2400	.027	S.F.		1.50		1.50	2.41
2500	Wet wipe substrate		4800	.013			.75		.75	1.20
2600	Vacuum surfaces, fine brush		6400	.010			.56		.56	.90
3000	Structural demolition									
3100	Wood stud walls	A-9	2800	.023	S.F.		1.29		1.29	2.06
3500	Window manifolds, not incl. window replacement		4200	.015			.86		.86	1.38
3600	Plywood carpet protection		2000	.032			1.80		1.80	2.89
4000	Remove custom decontamination facility	A-10A	8	3	Ea.	15.15	169		184.15	288
4100	Remove portable decontamination facility	3 Asbe	12	2	"	14.15	113		127.15	196
5000	HEPA vacuum, shampoo carpeting	A-9	4800	.013	S.F.	.10	.75		.85	1.31
9000	Final cleaning of protected surfaces	A-10A	8000	.003	"		.17		.17	.27

02 82 13.47 Asbestos Waste Pkg., Handling, and Disp.

		Crew	Daily Output	Labor-Hours	Unit	Material	2018 Bare Costs Labor	Equipment	Total	Total Incl O&P
0010	**ASBESTOS WASTE PACKAGING, HANDLING, AND DISPOSAL**									
0100	Collect and bag bulk material, 3 C.F. bags, by hand	A-9	400	.160	Ea.	.86	9		9.86	15.40
0200	Large production vacuum loader	A-12	880	.073		1	4.10	.82	5.92	8.55
1000	Double bag and decontaminate	A-9	960	.067		.86	3.76		4.62	6.95
2000	Containerize bagged material in drums, per 3 C.F. drum	"	800	.080		18.70	4.51		23.21	27.50
3000	Cart bags 50' to dumpster	2 Asbe	400	.040			2.25		2.25	3.61
5000	Disposal charges, not including haul, minimum				C.Y.				61	67
5020	Maximum				"				355	395
9000	For type B (supplied air) respirator equipment, add				%				10%	10%

40

For customer support on your Commercial Renovation Costs with RSMeans data, call 800.448.8182.

02 82 Asbestos Remediation

02 82 13 – Asbestos Abatement

02 82 13.48 Asbestos Encapsulation With Sealants

		Crew	Daily Output	Labor-Hours	Unit	Material	2018 Bare Costs Labor	Equipment	Total	Total Incl O&P
0010	**ASBESTOS ENCAPSULATION WITH SEALANTS**									
0100	Ceilings and walls, minimum	A-9	21000	.003	S.F.	.40	.17		.57	.72
0110	Maximum		10600	.006		.51	.34		.85	1.11
0200	Columns and beams, minimum		13300	.005		.40	.27		.67	.87
0210	Maximum		5325	.012		.61	.68		1.29	1.76
0300	Pipes to 12" diameter including minor repairs, minimum		800	.080	L.F.	.51	4.51		5.02	7.75
0310	Maximum		400	.160	"	1.21	9		10.21	15.80

02 83 Lead Remediation

02 83 19 – Lead-Based Paint Remediation

02 83 19.21 Lead Paint Remediation Plans and Methods

		Crew	Daily Output	Labor-Hours	Unit	Material	2018 Bare Costs Labor	Equipment	Total	Total Incl O&P
0010	**LEAD PAINT REMEDIATION PLANS AND METHODS**									
0100	Building Survey-Commercial Building				Ea.				2,050	2,250
0200	Lead Abatement Remediation Plan								1,225	1,350
0300	Lead Paint Testing, AAS Analysis								51	56
0400	Lead Paint Testing, X-Ray Fluorescence								51	56

02 83 19.22 Preparation of Lead Containment Area

		Crew	Daily Output	Labor-Hours	Unit	Material	2018 Bare Costs Labor	Equipment	Total	Total Incl O&P
0010	**PREPARATION OF LEAD CONTAINMENT AREA**									
0020	Lead abatement work area, test kit, per swab	1 Skwk	16	.500	Ea.	3.12	26		29.12	45.50
0025	For dust barriers see Section 01 56 16.10									
0050	Caution sign	1 Skwk	48	.167	Ea.	8.30	8.75		17.05	23
0100	Pre-cleaning, HEPA vacuum and wet wipe, floor and wall surfaces	3 Skwk	5000	.005	S.F.	.01	.25		.26	.41
0105	Ceiling, 6'-11' high		4100	.006		.07	.31		.38	.57
0108	12'-15' high		3550	.007		.07	.35		.42	.65
0115	Over 15' high		3000	.008		.09	.42		.51	.77
0500	Cover surfaces with polyethylene sheeting									
0550	Floors, each layer, 6 mil	3 Skwk	8000	.003	S.F.	.04	.16		.20	.29
0560	Walls, each layer, 6 mil	"	6000	.004	"	.04	.21		.25	.38
0570	For heights above 14', add						20%			
2400	Post abatement cleaning of protective sheeting, HEPA vacuum & wet wipe	3 Skwk	5000	.005	S.F.	.01	.25		.26	.41
2450	Doff, bag and seal protective sheeting		12000	.002		.01	.10		.11	.18
2500	Post abatement cleaning, HEPA vacuum & wet wipe		5000	.005		.01	.25		.26	.41

02 83 19.23 Encapsulation of Lead-Based Paint

		Crew	Daily Output	Labor-Hours	Unit	Material	2018 Bare Costs Labor	Equipment	Total	Total Incl O&P
0010	**ENCAPSULATION OF LEAD-BASED PAINT**									
0020	Interior, brushwork, trim, under 6"	1 Pord	240	.033	L.F.	2.32	1.42		3.74	4.83
0030	6"-12" wide		180	.044		3.03	1.89		4.92	6.35
0040	Balustrades		300	.027		2.02	1.13		3.15	4.04
0050	Pipe to 4" diameter		500	.016		1.21	.68		1.89	2.42
0060	To 8" diameter		375	.021		1.52	.91		2.43	3.13
0070	To 12" diameter		250	.032		2.22	1.36		3.58	4.63
0080	To 16" diameter		170	.047		3.33	2		5.33	6.85
0090	Cabinets, ornate design		200	.040	S.F.	3.03	1.70		4.73	6.05
0100	Simple design		250	.032	"	2.32	1.36		3.68	4.74
0110	Doors, 3' x 7', both sides, incl. frame & trim									
0120	Flush	1 Pord	6	1.333	Ea.	28.50	56.50		85	122
0130	French, 10-15 lite		3	2.667		6.05	113		119.05	189
0140	Panel		4	2		35.50	85		120.50	176
0150	Louvered		2.75	2.909		32.50	124		156.50	235
0160	Windows, per interior side, per 15 S.F.									
0170	1-6 lite	1 Pord	14	.571	Ea.	20	24.50		44.50	61

For customer support on your Commercial Renovation Costs with RSMeans data, call 800.448.8182.

41

02 83 19 – Lead-Based Paint Remediation

02 83 19.23 Encapsulation of Lead-Based Paint

		Crew	Daily Output	Labor-Hours	Unit	Material	2018 Bare Costs Labor	Equipment	Total	Total Incl O&P
0180	7-10 lite	1 Pord	7.50	1.067	Ea.	22	45.50		67.50	97.50
0190	12 lite		5.75	1.391		30.50	59		89.50	129
0200	Radiators		8	1		70.50	42.50		113	147
0210	Grilles, vents		275	.029	S.F.	2.02	1.24		3.26	4.21
0220	Walls, roller, drywall or plaster		1000	.008		.61	.34		.95	1.22
0230	With spunbonded reinforcing fabric		720	.011		.71	.47		1.18	1.54
0240	Wood		800	.010		.71	.43		1.14	1.46
0250	Ceilings, roller, drywall or plaster		900	.009		.71	.38		1.09	1.39
0260	Wood		700	.011		.81	.49		1.30	1.67
0270	Exterior, brushwork, gutters and downspouts		300	.027	L.F.	1.92	1.13		3.05	3.93
0280	Columns		400	.020	S.F.	1.42	.85		2.27	2.93
0290	Spray, siding		600	.013	"	1.01	.57		1.58	2.02
0300	Miscellaneous									
0310	Electrical conduit, brushwork, to 2" diameter	1 Pord	500	.016	L.F.	1.21	.68		1.89	2.42
0320	Brick, block or concrete, spray		500	.016	S.F.	1.21	.68		1.89	2.42
0330	Steel, flat surfaces and tanks to 12"		500	.016		1.21	.68		1.89	2.42
0340	Beams, brushwork		400	.020		1.41	.85		2.26	2.92
0350	Trusses		400	.020		1.41	.85		2.26	2.92

02 83 19.26 Removal of Lead-Based Paint

		Crew	Daily Output	Labor-Hours	Unit	Material	2018 Bare Costs Labor	Equipment	Total	Total Incl O&P
0010	**REMOVAL OF LEAD-BASED PAINT** R028319-60									
0011	By chemicals, per application									
0050	Baseboard, to 6" wide	1 Pord	64	.125	L.F.	.80	5.30		6.10	9.45
0070	To 12" wide		32	.250	"	1.47	10.65		12.12	18.70
0200	Balustrades, one side		28	.286	S.F.	1.50	12.15		13.65	21
1400	Cabinets, simple design		32	.250		1.37	10.65		12.02	18.60
1420	Ornate design		25	.320		1.60	13.60		15.20	24
1600	Cornice, simple design		60	.133		1.60	5.65		7.25	10.85
1620	Ornate design		20	.400		5.55	17		22.55	33.50
2800	Doors, one side, flush		84	.095		1.73	4.05		5.78	8.40
2820	Two panel		80	.100		1.37	4.26		5.63	8.35
2840	Four panel		45	.178		1.43	7.55		8.98	13.70
2880	For trim, one side, add		64	.125	L.F.	.77	5.30		6.07	9.40
3000	Fence, picket, one side		30	.267	S.F.	1.37	11.35		12.72	19.70
3200	Grilles, one side, simple design		30	.267		1.37	11.35		12.72	19.70
3220	Ornate design		25	.320		1.47	13.60		15.07	23.50
3240	Handrails		90	.089	L.F.	1.37	3.78		5.15	7.55
4400	Pipes, to 4" diameter		90	.089		1.73	3.78		5.51	7.95
4420	To 8" diameter		50	.160		3.47	6.80		10.27	14.75
4440	To 12" diameter		36	.222		5.20	9.45		14.65	21
4460	To 16" diameter		20	.400		6.95	17		23.95	35
4500	For hangers, add		40	.200	Ea.	2.53	8.50		11.03	16.45
4800	Siding		90	.089	S.F.	1.32	3.78		5.10	7.50
5000	Trusses, open		55	.145	SF Face	2.02	6.20		8.22	12.15
6200	Windows, one side only, double-hung, 1/1 light, 24" x 48" high		4	2	Ea.	24	85		109	164
6220	30" x 60" high		3	2.667		32.50	113		145.50	218
6240	36" x 72" high		2.50	3.200		38.50	136		174.50	261
6280	40" x 80" high		2	4		47.50	170		217.50	325
6400	Colonial window, 6/6 light, 24" x 48" high		2	4		47.50	170		217.50	325
6420	30" x 60" high		1.50	5.333		62.50	227		289.50	435
6440	36" x 72" high		1	8		94	340		434	650
6480	40" x 80" high		1	8		94	340		434	650
6600	8/8 light, 24" x 48" high		2	4		47.50	170		217.50	325

42

For customer support on your Commercial Renovation Costs with RSMeans data, call 800.448.8182.

02 83 Lead Remediation

02 83 19 – Lead-Based Paint Remediation

02 83 19.26 Removal of Lead-Based Paint		Crew	Daily Output	Labor-Hours	Unit	Material	2018 Bare Costs Labor	Equipment	Total	Total Incl O&P
6620	40" x 80" high	1 Pord	1	8	Ea.	94	340		434	650
6800	12/12 light, 24" x 48" high		1	8		94	340		434	650
6820	40" x 80" high		.75	10.667		126	455		581	870
6840	Window frame & trim items, included in pricing above									
7000	Hand scraping and HEPA vacuum, less than 4 S.F.	1 Pord	8	1	Ea.	.86	42.50		43.36	69.50
8000	Collect and bag bulk material, 3 C.F. bags, by hand		30	.267	"	.86	11.35		12.21	19.15
9000	Minimum labor/equipment charge		3	2.667	Job		113		113	182

02 87 Biohazard Remediation

02 87 13 – Mold Remediation

02 87 13.16 Mold Remediation Preparation and Containment

		Crew	Daily Output	Labor-Hours	Unit	Material	2018 Bare Costs Labor	Equipment	Total	Total Incl O&P
0010	**MOLD REMEDIATION PREPARATION AND CONTAINMENT**									
0015	Mold remediation plans and methods									
0020	Initial inspection, areas to 2500 S.F.				Total				268	295
0030	Areas to 5000 S.F.								440	485
0032	Areas to 10000 S.F.								440	485
0040	Testing, air sample each								126	139
0050	Swab sample								115	127
0060	Tape sample								125	140
0070	Post remediation air test								126	139
0080	Mold abatement plan, area to 2500 S.F.								1,225	1,325
0090	Areas to 5000 S.F.								1,625	1,775
0095	Areas to 10000 S.F.								2,550	2,800
0100	Packup & removal of contents, average 3 bedroom home, excl storage								8,150	8,975
0110	Average 5 bedroom home, excl storage								15,300	16,800
0600	For demolition in mold contaminated areas, see Section 02 85 33.50									
0610	For personal protection equipment, see Section 02 82 13.41									
6010	Preparation of mold containment area									
6100	Pre-cleaning, HEPA vacuum and wet wipe, flat surfaces	A-9	12000	.005	S.F.	.01	.30		.31	.49
6300	Separation barrier, 2" x 4" @ 16", 1/2" plywood ea. side, 8' high	2 Carp	400	.040		3.43	2.03		5.46	7.05
6310	12' high		320	.050		3.43	2.54		5.97	7.90
6320	16' high		200	.080		2.32	4.06		6.38	9.15
6400	Personnel decontam. chamber, 2" x 4" @ 16", 3/4" ply ea. side		280	.057		4.44	2.90		7.34	9.60
6450	Waste decontam. chamber, 2" x 4" studs @ 16", 3/4" ply each side		360	.044		4.44	2.25		6.69	8.55
6500	Cover surfaces with polyethylene sheeting									
6501	Including glue and tape									
6550	Floors, each layer, 6 mil	A-9	8000	.008	S.F.	.04	.45		.49	.76
6551	4 mil		9000	.007		.03	.40		.43	.67
6560	Walls, each layer, 6 mil		6000	.011		.04	.60		.64	1
6561	4 mil		7000	.009		.03	.52		.55	.86
6570	For heights above 14', add						20%			
6575	For heights above 20', add						30%			
6580	For fire retardant poly, add					100%				
6590	For large open areas, deduct					10%	20%			
6600	Seal floor penetrations with foam firestop to 36 sq. in.	2 Carp	200	.080	Ea.	11.35	4.06		15.41	19.10
6610	36 sq. in. to 72 sq. in.		125	.128		22.50	6.50		29	35.50
6615	72 sq. in. to 144 sq. in.		80	.200		45.50	10.15		55.65	66.50
6620	Wall penetrations, to 36 sq. in.		180	.089		11.35	4.51		15.86	19.80
6630	36 sq. in. to 72 sq. in.		100	.160		22.50	8.10		30.60	38
6640	72 sq. in. to 144 sq. in.		60	.267		45.50	13.50		59	72
6800	Caulk seams with latex caulk	1 Carp	230	.035	L.F.	.17	1.76		1.93	3.05

For customer support on your Commercial Renovation Costs with RSMeans data, call 800.448.8182.

43

02 87 Biohazard Remediation

02 87 13 – Mold Remediation

02 87 13.16 Mold Remediation Preparation and Containment	Crew	Daily Output	Labor-Hours	Unit	Material	2018 Bare Costs Labor	Equipment	Total	Total Incl O&P
6900 Set up neg. air machine, 1-2k CFM/25 M.C.F. volume	1 Asbe	4.30	1.860	Ea.		105		105	168

02 87 13.33 Removal and Disposal of Materials With Mold

		Crew	Daily Output	Labor-Hours	Unit	Material	2018 Bare Costs Labor	Equipment	Total	Total Incl O&P
0010	**REMOVAL AND DISPOSAL OF MATERIALS WITH MOLD**									
0015	Demolition in mold contaminated area									
0200	Ceiling, including suspension system, plaster and lath	A-9	2100	.030	S.F.	.08	1.72		1.80	2.84
0210	Finished plaster, leaving wire lath		585	.109		.28	6.15		6.43	10.20
0220	Suspended acoustical tile		3500	.018		.05	1.03		1.08	1.70
0230	Concealed tile grid system		3000	.021		.05	1.20		1.25	1.99
0240	Metal pan grid system		1500	.043		.11	2.40		2.51	3.97
0250	Gypsum board		2500	.026		.07	1.44		1.51	2.38
0255	Plywood		2500	.026		.07	1.44		1.51	2.38
0260	Lighting fixtures up to 2' x 4'		72	.889	Ea.	2.26	50		52.26	83
0400	Partitions, non load bearing									
0410	Plaster, lath, and studs	A-9	690	.093	S.F.	.88	5.25		6.13	9.35
0450	Gypsum board and studs		1390	.046		.12	2.59		2.71	4.29
0465	Carpet & pad		1390	.046		.12	2.59		2.71	4.29
0600	Pipe insulation, air cell type, up to 4" diameter pipe		900	.071	L.F.	.18	4.01		4.19	6.60
0610	4" to 8" diameter pipe		800	.080		.20	4.51		4.71	7.40
0620	10" to 12" diameter pipe		700	.091		.23	5.15		5.38	8.50
0630	14" to 16" diameter pipe		550	.116		.30	6.55		6.85	10.85
0650	Over 16" diameter pipe		650	.098	S.F.	.25	5.55		5.80	9.20
9000	For type B (supplied air) respirator equipment, add				%				10%	10%

Estimating Tips
General

- Carefully check all the plans and specifications. Concrete often appears on drawings other than structural drawings, including mechanical and electrical drawings for equipment pads. The cost of cutting and patching is often difficult to estimate. See Subdivision 03 81 for Concrete Cutting, Subdivision 02 41 19.16 for Cutout Demolition, Subdivision 03 05 05.10 for Concrete Demolition, and Subdivision 02 41 19.19 for Rubbish Handling (handling, loading, and hauling of debris).

- Always obtain concrete prices from suppliers near the job site. A volume discount can often be negotiated, depending upon competition in the area. Remember to add for waste, particularly for slabs and footings on grade.

03 10 00 Concrete Forming and Accessories

- A primary cost for concrete construction is forming. Most jobs today are constructed with prefabricated forms. The selection of the forms best suited for the job and the total square feet of forms required for efficient concrete forming and placing are key elements in estimating concrete construction. Enough forms must be available for erection to make efficient use of the concrete placing equipment and crew.

- Concrete accessories for forming and placing depend upon the systems used. Study the plans and specifications to ensure that all special accessory requirements have been included in the cost estimate, such as anchor bolts, inserts, and hangers.

- Included within costs for forms-in-place are all necessary bracing and shoring.

03 20 00 Concrete Reinforcing

- Ascertain that the reinforcing steel supplier has included all accessories, cutting, bending, and an allowance for lapping, splicing, and waste. A good rule of thumb is 10% for lapping, splicing, and waste. Also, 10% waste should be allowed for welded wire fabric.

- The unit price items in the subdivisions for Reinforcing In Place, Glass Fiber Reinforcing, and Welded Wire Fabric include the labor to install accessories such as beam and slab bolsters, high chairs, and bar ties and tie wire. The material cost for these accessories is not included; they may be obtained from the Accessories Subdivisions.

03 30 00 Cast-In-Place Concrete

- When estimating structural concrete, pay particular attention to requirements for concrete additives, curing methods, and surface treatments. Special consideration for climate, hot or cold, must be included in your estimate. Be sure to include requirements for concrete placing equipment and concrete finishing.

- For accurate concrete estimating, the estimator must consider each of the following major components individually: forms, reinforcing steel, ready-mix concrete, placement of the concrete, and finishing of the top surface. For faster estimating, Subdivision 03 30 53.40 for Concrete-In-Place can be used; here, various items of concrete work are presented that include the costs of all five major components (unless specifically stated otherwise).

03 40 00 Precast Concrete
03 50 00 Cast Decks and Underlayment

- The cost of hauling precast concrete structural members is often an important factor. For this reason, it is important to get a quote from the nearest supplier. It may become economically feasible to set up precasting beds on the site if the hauling costs are prohibitive.

Reference Numbers

Reference numbers are shown at the beginning of some major classifications. These numbers refer to related items in the Reference Section. The reference information may be an estimating procedure, an alternate pricing method, or technical information.

Note: Not all subdivisions listed here necessarily appear. ■

Did you know?

RSMeans data is available through our online application with 24/7 access:

- Search for unit prices by keyword
- Leverage the most up-to-date data
- Build and export estimates

Try it free for 30 days!
www.rsmeans.com/2018freetrial

03 01 30.62 Concrete Patching

		Crew	Daily Output	Labor-Hours	Unit	Material	2018 Bare Costs Labor	2018 Bare Costs Equipment	Total	Total Incl O&P
0010	**CONCRETE PATCHING**									
0100	Floors, 1/4" thick, small areas, regular grout	1 Cefi	170	.047	S.F.	1.54	2.24		3.78	5.25
0150	Epoxy grout	"	100	.080	"	8.65	3.80		12.45	15.55
2000	Walls, including chipping, cleaning and epoxy grout									
2100	1/4" deep	1 Cefi	65	.123	S.F.	7.45	5.85		13.30	17.40
2150	1/2" deep		50	.160		14.85	7.60		22.45	28.50
2200	3/4" deep		40	.200		22.50	9.50		32	39.50
9000	Minimum labor/equipment charge		4.50	1.778	Job		84.50		84.50	134

03 01 30.71 Concrete Crack Repair

		Crew	Daily Output	Labor-Hours	Unit	Material	2018 Bare Costs Labor	2018 Bare Costs Equipment	Total	Total Incl O&P
0010	**CONCRETE CRACK REPAIR**									
1000	Structural repair of concrete cracks by epoxy injection (ACI RAP-1)									
1001	suitable for horizontal, vertical and overhead repairs									
1010	Clean/grind concrete surface(s) free of contaminants	1 Cefi	400	.020	L.F.		.95		.95	1.50
1015	Rout crack with v-notch crack chaser, if needed	C-32	600	.027		.03	1.17	.15	1.35	2.08
1020	Blow out crack with oil-free dry compressed air (1 pass)	C-28	3000	.003			.13	.01	.14	.21
1030	Install surface-mounted entry ports (spacing = concrete depth)	1 Cefi	400	.020	Ea.	1.62	.95		2.57	3.28
1040	Cap crack at surface with epoxy gel (per side/face)		400	.020	L.F.	.36	.95		1.31	1.90
1050	Snap off ports, grind off epoxy cap residue after injection		200	.040	"		1.90		1.90	3.01
1100	Manual injection with 2-part epoxy cartridge, excludes prep									
1110	Up to 1/32" (0.03125") wide x 4" deep	1 Cefi	160	.050	L.F.	.18	2.38		2.56	3.96
1120	6" deep		120	.067		.27	3.17		3.44	5.30
1130	8" deep		107	.075		.36	3.56		3.92	6
1140	10" deep		100	.080		.45	3.80		4.25	6.50
1150	12" deep		96	.083		.54	3.96		4.50	6.85
1210	Up to 1/16" (0.0625") wide x 4" deep		160	.050		.36	2.38		2.74	4.16
1220	6" deep		120	.067		.54	3.17		3.71	5.60
1230	8" deep		107	.075		.72	3.56		4.28	6.40
1240	10" deep		100	.080		.90	3.80		4.70	7
1250	12" deep		96	.083		1.08	3.96		5.04	7.45
1310	Up to 3/32" (0.09375") wide x 4" deep		160	.050		.54	2.38		2.92	4.35
1320	6" deep		120	.067		.81	3.17		3.98	5.90
1330	8" deep		107	.075		1.08	3.56		4.64	6.80
1340	10" deep		100	.080		1.35	3.80		5.15	7.50
1350	12" deep		96	.083		1.62	3.96		5.58	8.05
1410	Up to 1/8" (0.125") wide x 4" deep		160	.050		.72	2.38		3.10	4.55
1420	6" deep		120	.067		1.08	3.17		4.25	6.20
1430	8" deep		107	.075		1.44	3.56		5	7.20
1440	10" deep		100	.080		1.80	3.80		5.60	8
1450	12" deep		96	.083		2.16	3.96		6.12	8.60
1500	Pneumatic injection with 2-part bulk epoxy, excludes prep									
1510	Up to 5/32" (0.15625") wide x 4" deep	C-31	240	.033	L.F.	.25	1.58	1.34	3.17	4.26
1520	6" deep		180	.044		.38	2.11	1.78	4.27	5.70
1530	8" deep		160	.050		.51	2.38	2.01	4.90	6.55
1540	10" deep		150	.053		.64	2.54	2.14	5.32	7.05
1550	12" deep		144	.056		.76	2.64	2.23	5.63	7.45
1610	Up to 3/16" (0.1875") wide x 4" deep		240	.033		.31	1.58	1.34	3.23	4.32
1620	6" deep		180	.044		.46	2.11	1.78	4.35	5.80
1630	8" deep		160	.050		.61	2.38	2.01	5	6.65
1640	10" deep		150	.053		.76	2.54	2.14	5.44	7.20
1650	12" deep		144	.056		.92	2.64	2.23	5.79	7.65
1710	Up to 1/4" (0.1875") wide x 4" deep		240	.033		.41	1.58	1.34	3.33	4.43
1720	6" deep		180	.044		.61	2.11	1.78	4.50	5.95

03 01 30 – Maintenance of Cast-In-Place Concrete

03 01 30.71 Concrete Crack Repair	Crew	Daily Output	Labor-Hours	Unit	Material	2018 Bare Costs Labor	Equipment	Total	Total Incl O&P
1730	8" deep	C-31	160	.050	L.F.	.82	2.38	2.01	5.21	6.85
1740	10" deep		150	.053		1.02	2.54	2.14	5.70	7.50
1750	12" deep	↓	144	.056	↓	1.22	2.64	2.23	6.09	8
2000	Non-structural filling of concrete cracks by gravity-fed resin (ACI RAP-2)									
2001	suitable for individual cracks in stable horizontal surfaces only									
2010	Clean/grind concrete surface(s) free of contaminants	1 Cefi	400	.020	L.F.		.95		.95	1.50
2020	Rout crack with v-notch crack chaser, if needed	C-32	600	.027		.03	1.17	.15	1.35	2.08
2030	Blow out crack with oil-free dry compressed air (1 pass)	C-28	3000	.003			.13	.01	.14	.21
2040	Cap crack with epoxy gel at underside of elevated slabs, if needed	1 Cefi	400	.020		.36	.95		1.31	1.90
2050	Insert backer rod into crack, if needed		400	.020		.02	.95		.97	1.53
2060	Partially fill crack with fine dry sand, if needed		800	.010		.02	.48		.50	.77
2070	Apply two beads of sealant alongside crack to form reservoir, if needed	↓	400	.020	↓	.26	.95		1.21	1.79
2100	Manual filling with squeeze bottle of 2-part epoxy resin									
2110	Full depth crack up to 1/16" (0.0625") wide x 4" deep	1 Cefi	300	.027	L.F.	.09	1.27		1.36	2.11
2120	6" deep		240	.033		.14	1.58		1.72	2.66
2130	8" deep		200	.040		.19	1.90		2.09	3.21
2140	10" deep		185	.043		.23	2.06		2.29	3.51
2150	12" deep		175	.046		.28	2.17		2.45	3.75
2210	Full depth crack up to 1/8" (0.125") wide x 4" deep		270	.030		.19	1.41		1.60	2.43
2220	6" deep		215	.037		.28	1.77		2.05	3.11
2230	8" deep		180	.044		.37	2.11		2.48	3.75
2240	10" deep		165	.048		.46	2.31		2.77	4.16
2250	12" deep		155	.052		.56	2.45		3.01	4.49
2310	Partial depth crack up to 3/16" (0.1875") wide x 1" deep		300	.027		.07	1.27		1.34	2.09
2410	Up to 1/4" (0.250") wide x 1" deep		270	.030		.09	1.41		1.50	2.33
2510	Up to 5/16" (0.3125") wide x 1" deep		250	.032		.12	1.52		1.64	2.54
2610	Up to 3/8" (0.375") wide x 1" deep	↓	240	.033	↓	.14	1.58		1.72	2.66

03 01 30.72 Concrete Surface Repairs

		Crew	Daily Output	Labor-Hours	Unit	Material	Labor	Equipment	Total	Total Incl O&P
0010	**CONCRETE SURFACE REPAIRS**									
9000	Surface repair by Methacrylate flood coat (ACI RAP-13)									
9010	Suitable for healing and sealing horizontal surfaces only									
9020	Large cracks must previously have been repaired or filled									
9100	Shotblast entire surface to remove contaminants	A-1A	4000	.002	S.F.		.10	.05	.15	.23
9200	Blow off dust and debris with oil-free dry compressed air	C-28	16000	.001			.02		.02	.04
9300	Flood coat surface w/Methacrylate, distribute w/broom/squeegee, no prep.	3 Cefi	8000	.003		1	.14		1.14	1.33
9400	Lightly broadcast even coat of dry silica sand while sealer coat is tacky	1 Cefi	8000	.001	↓	.01	.05		.06	.09

03 05 05 – Selective Demolition for Concrete

03 05 05.10 Selective Demolition, Concrete

		Crew	Daily Output	Labor-Hours	Unit	Material	Labor	Equipment	Total	Total Incl O&P
0010	**SELECTIVE DEMOLITION, CONCRETE**	R024119-10								
0012	Excludes saw cutting, torch cutting, loading or hauling									
0050	Break into small pieces, reinf. less than 1% of cross-sectional area	B-9	24	1.667	C.Y.		67	9.75	76.75	120
0060	Reinforcing 1% to 2% of cross-sectional area		16	2.500			101	14.65	115.65	179
0070	Reinforcing more than 2% of cross-sectional area	↓	8	5	↓		201	29.50	230.50	355
0150	Remove whole pieces, up to 2 tons per piece	E-18	36	1.111	Ea.		61	30	91	135
0160	2-5 tons per piece		30	1.333			73	36	109	163
0170	5-10 tons per piece		24	1.667			91.50	45	136.50	204
0180	10-15 tons per piece		18	2.222			122	60	182	271
0250	Precast unit embedded in masonry, up to 1 C.F.	D-1	16	1			45		45	73
0260	1-2 C.F.	↓	12	1.333	↓		59.50		59.50	97.50

03 05 05 – Selective Demolition for Concrete

03 05 05.10 Selective Demolition, Concrete	Crew	Daily Output	Labor-Hours	Unit	Material	2018 Bare Costs Labor	Equipment	Total	Total Incl O&P
0270	2-5 C.F.	D-1	10	1.600	Ea.		71.50		71.50	117
0280	5-10 C.F.	↓	8	2	↓		89.50		89.50	146
0990	For hydrodemolition see Section 02 41 13.15									
1910	Minimum labor/equipment charge	B-9	2	20	Job		805	117	922	1,425

03 05 13 – Basic Concrete Materials

03 05 13.20 Concrete Admixtures and Surface Treatments

		Crew	Daily Output	Labor-Hours	Unit	Material	Labor	Equipment	Total	Total Incl O&P
0010	**CONCRETE ADMIXTURES AND SURFACE TREATMENTS**									
0040	Abrasives, aluminum oxide, over 20 tons				Lb.	2.53			2.53	2.78
0050	1 to 20 tons					2.63			2.63	2.89
0070	Under 1 ton					2.73			2.73	3.01
0100	Silicon carbide, black, over 20 tons					4.32			4.32	4.75
0110	1 to 20 tons					4.49			4.49	4.94
0120	Under 1 ton				↓	4.67			4.67	5.15
0200	Air entraining agent, .7 to 1.5 oz. per bag, 55 gallon drum				Gal.	21			21	23
0220	5 gallon pail					29			29	32
0300	Bonding agent, acrylic latex, 250 S.F./gallon, 5 gallon pail					24.50			24.50	27
0320	Epoxy resin, 80 S.F./gallon, 4 gallon case				↓	65.50			65.50	72
0400	Calcium chloride, 50 lb. bags, T.L. lots				Ton	1,225			1,225	1,325
0420	Less than truckload lots				Bag	22.50			22.50	25
0500	Carbon black, liquid, 2 to 8 lb. per bag of cement				Lb.	13.70			13.70	15.05
0600	Concrete admixture, integral colors, dry pigment, 5 lb. bag				Ea.	28			28	31
0610	10 lb. bag					36.50			36.50	40
0620	25 lb. bag				↓	74			74	81.50
0920	Dustproofing compound, 250 S.F./gal., 5 gallon pail				Gal.	7.10			7.10	7.80
1010	Epoxy based, 125 S.F./gal., 5 gallon pail				"	59			59	65
1100	Hardeners, metallic, 55 lb. bags, natural (grey)				Lb.	.86			.86	.95
1200	Colors					1.48			1.48	1.63
1300	Non-metallic, 55 lb. bags, natural grey					.63			.63	.70
1320	Colors				↓	.80			.80	.88
1550	Release agent, for tilt slabs, 5 gallon pail				Gal.	18.25			18.25	20
1570	For forms, 5 gallon pail					11.85			11.85	13
1590	Concrete release agent for forms, 100% biodegradable, zero VOC, 5 gal. pail Ⓖ					21.50			21.50	23.50
1595	55 gallon drum Ⓖ					18.15			18.15	19.95
1600	Sealer, hardener and dustproofer, epoxy-based, 125 S.F./gal., 5 gallon unit					59			59	65
1620	3 gallon unit					65			65	71.50
1630	Sealer, solvent-based, 250 S.F./gal., 55 gallon drum					21.50			21.50	23.50
1640	5 gallon pail					32			32	35
1650	Sealer, water based, 350 S.F., 55 gallon drum					21.50			21.50	23.50
1660	5 gallon pail					24			24	26.50
1900	Set retarder, 100 S.F./gal., 1 gallon pail				↓	6.20			6.20	6.80
2000	Waterproofing, integral 1 lb. per bag of cement				Lb.	3.19			3.19	3.51
2100	Powdered metallic, 40 lbs. per 100 S.F., standard colors					3.79			3.79	4.17
2120	Premium colors				↓	5.30			5.30	5.85
3000	For colored ready-mix concrete, add to prices in section 03 31 13.35									
3100	Subtle shades, 5 lb. dry pigment per C.Y., add				C.Y.	28			28	31
3400	Medium shades, 10 lb. dry pigment per C.Y., add					36.50			36.50	40
3700	Deep shades, 25 lb. dry pigment per C.Y., add				↓	74			74	81.50

03 05 13.85 Winter Protection

		Crew	Daily Output	Labor-Hours	Unit	Material	Labor	Equipment	Total	Total Incl O&P
0010	**WINTER PROTECTION**									
0012	For heated ready mix, add				C.Y.	5.35			5.35	5.90
0100	Temporary heat to protect concrete, 24 hours	2 Clab	50	.320	M.S.F.	200	12.75		212.75	241
0200	Temporary shelter for slab on grade, wood frame/polyethylene sheeting									

48

For customer support on your Commercial Renovation Costs with RSMeans data, call 800.448.8182.

03 05 Common Work Results for Concrete

03 05 13 – Basic Concrete Materials

03 05 13.85 Winter Protection

		Crew	Daily Output	Labor-Hours	Unit	Material	2018 Bare Costs Labor	Equipment	Total	Total Incl O&P
0201	Build or remove, light framing for short spans	2 Carp	10	1.600	M.S.F.	305	81		386	465
0210	Large framing for long spans	"	3	5.333	"	410	270		680	890
0710	Electrically heated pads, 15 watts/S.F., 20 uses				S.F.	.56			.56	.61

03 11 Concrete Forming

03 11 13 – Structural Cast-In-Place Concrete Forming

03 11 13.20 Forms In Place, Beams and Girders

			Crew	Daily Output	Labor-Hours	Unit	Material	2018 Bare Costs Labor	Equipment	Total	Total Incl O&P
0010	**FORMS IN PLACE, BEAMS AND GIRDERS**	R031113-40									
0500	Exterior spandrel, job-built plywood, 12" wide, 1 use	R031113-60	C-2	225	.213	SFCA	3.05	10.50		13.55	20.50
0650	4 use			310	.155		.99	7.60		8.59	13.45
1000	18" wide, 1 use			250	.192		2.70	9.45		12.15	18.30
1150	4 use			315	.152		.88	7.50		8.38	13.15
1500	24" wide, 1 use			265	.181		2.46	8.90		11.36	17.15
1650	4 use			325	.148		.80	7.25		8.05	12.70
2000	Interior beam, job-built plywood, 12" wide, 1 use			300	.160		3.71	7.90		11.61	16.90
2150	4 use			377	.127		1.21	6.25		7.46	11.50
2500	24" wide, 1 use			320	.150		2.51	7.40		9.91	14.75
2650	4 use			395	.122		.81	6		6.81	10.60
3000	Encasing steel beam, hung, job-built plywood, 1 use			325	.148		3.15	7.25		10.40	15.25
3150	4 use			430	.112		1.02	5.50		6.52	10.05
9000	Minimum labor/equipment charge		2 Carp	2	8	Job		405		405	660

03 11 13.25 Forms In Place, Columns

			Crew	Daily Output	Labor-Hours	Unit	Material	2018 Bare Costs Labor	Equipment	Total	Total Incl O&P
0010	**FORMS IN PLACE, COLUMNS**	R031113-40									
0500	Round fiberglass, 4 use per mo., rent, 12" diameter		C-1	160	.200	L.F.	8.15	9.60		17.75	24.50
0550	16" diameter	R031113-60		150	.213		9.70	10.25		19.95	27.50
0600	18" diameter			140	.229		10.85	10.95		21.80	30
0650	24" diameter			135	.237		13.55	11.40		24.95	33.50
0700	28" diameter			130	.246		15.15	11.80		26.95	36
0800	30" diameter			125	.256		15.80	12.30		28.10	37.50
0850	36" diameter			120	.267		21	12.80		33.80	44
1500	Round fiber tube, recycled paper, 1 use, 8" diameter	G		155	.206		2.88	9.90		12.78	19.25
1550	10" diameter	G		155	.206		3.61	9.90		13.51	20
1600	12" diameter	G		150	.213		4.05	10.25		14.30	21
1650	14" diameter	G		145	.221		4.29	10.60		14.89	22
1700	16" diameter	G		140	.229		5.30	10.95		16.25	23.50
1720	18" diameter	G		140	.229		6.20	10.95		17.15	24.50
1750	20" diameter	G		135	.237		7.75	11.40		19.15	27
1800	24" diameter	G		130	.246		9.65	11.80		21.45	30
1850	30" diameter	G		125	.256		14.50	12.30		26.80	36
1900	36" diameter	G		115	.278		19.60	13.35		32.95	43
1950	42" diameter	G		100	.320		43.50	15.35		58.85	73
2000	48" diameter	G		85	.376		54.50	18.05		72.55	89.50
2200	For seamless type, add						15%				
3000	Round, steel, 4 use per mo., rent, regular duty, 14" diameter	G	C-1	145	.221	L.F.	18.10	10.60		28.70	37
3050	16" diameter	G		125	.256		18.45	12.30		30.75	40.50
3100	Heavy duty, 20" diameter	G		105	.305		20	14.65		34.65	45.50
3150	24" diameter	G		85	.376		22	18.05		40.05	53.50
3200	30" diameter	G		70	.457		25	22		47	63
3250	36" diameter	G		60	.533		27	25.50		52.50	71
3300	48" diameter	G		50	.640		40	30.50		70.50	94
3350	60" diameter	G		45	.711		49.50	34		83.50	110

For customer support on your Commercial Renovation Costs with RSMeans data, call 800.448.8182.

49

03 11 Concrete Forming

03 11 13 – Structural Cast-In-Place Concrete Forming

03 11 13.25 Forms In Place, Columns		Crew	Daily Output	Labor-Hours	Unit	Material	2018 Bare Costs Labor	Equipment	Total	Total Incl O&P
4500	For second and succeeding months, deduct					50%				
5000	Job-built plywood, 8" x 8" columns, 1 use	C-1	165	.194	SFCA	2.66	9.30		11.96	18.05
5050	2 use		195	.164		1.52	7.90		9.42	14.50
5100	3 use		210	.152		1.06	7.30		8.36	13
5150	4 use		215	.149		.88	7.15		8.03	12.55
5500	12" x 12" columns, 1 use		180	.178		2.58	8.55		11.13	16.70
5550	2 use		210	.152		1.42	7.30		8.72	13.40
5600	3 use		220	.145		1.03	7		8.03	12.50
5650	4 use		225	.142		.84	6.85		7.69	12
6000	16" x 16" columns, 1 use		185	.173		2.59	8.30		10.89	16.30
6050	2 use		215	.149		1.38	7.15		8.53	13.10
6100	3 use		230	.139		1.04	6.70		7.74	12
6150	4 use		235	.136		.85	6.55		7.40	11.55
6500	24" x 24" columns, 1 use		190	.168		2.94	8.10		11.04	16.35
6550	2 use		216	.148		1.61	7.10		8.71	13.35
6600	3 use		230	.139		1.17	6.70		7.87	12.15
6650	4 use		238	.134		.95	6.45		7.40	11.50
7000	36" x 36" columns, 1 use		200	.160		2.16	7.70		9.86	14.85
7050	2 use		230	.139		1.22	6.70		7.92	12.20
7100	3 use		245	.131		.86	6.25		7.11	11.10
7150	4 use		250	.128		.70	6.15		6.85	10.70
7400	Steel framed plywood, based on 50 uses of purchased									
7420	forms, and 4 uses of bracing lumber									
7500	8" x 8" column	C-1	340	.094	SFCA	2.18	4.52		6.70	9.75
7550	10" x 10"		350	.091		1.89	4.39		6.28	9.20
7600	12" x 12"		370	.086		1.61	4.15		5.76	8.50
7650	16" x 16"		400	.080		1.25	3.84		5.09	7.65
7700	20" x 20"		420	.076		1.11	3.66		4.77	7.15
7750	24" x 24"		440	.073		.80	3.49		4.29	6.55
7755	30" x 30"		440	.073		1.02	3.49		4.51	6.75
7760	36" x 36"		460	.070		.89	3.34		4.23	6.40
9000	Minimum labor/equipment charge	2 Carp	2	8	Job		405		405	660

03 11 13.35 Forms In Place, Elevated Slabs

03 11 13.35 Forms In Place, Elevated Slabs		Crew	Daily Output	Labor-Hours	Unit	Material	2018 Bare Costs Labor	Equipment	Total	Total Incl O&P
0010	**FORMS IN PLACE, ELEVATED SLABS** R031113-40									
1000	Flat plate, job-built plywood, to 15' high, 1 use	C-2	470	.102	S.F.	3.87	5.05		8.92	12.40
1150	4 use R031113-60		560	.086		1.26	4.22		5.48	8.25
1500	15' to 20' high ceilings, 4 use		495	.097		1.26	4.77		6.03	9.15
2000	Flat slab, drop panels, job-built plywood, to 15' high, 1 use		449	.107		4.45	5.25		9.70	13.45
2150	4 use		544	.088		1.45	4.34		5.79	8.65
2250	15' to 20' high ceilings, 4 use		480	.100		2.55	4.92		7.47	10.80
3500	Floor slab, with 1-way joist pans, 1 use		415	.116		7.95	5.70		13.65	18
3650	4 use		500	.096		4.95	4.73		9.68	13.10
4500	With 2-way waffle domes, 1 use		405	.119		8.20	5.85		14.05	18.50
4550	4 use		470	.102		5.20	5.05		10.25	13.90
5000	Box out for slab openings, over 16" deep, 1 use		190	.253	SFCA	3.48	12.45		15.93	24
5050	2 use		240	.200	"	1.92	9.85		11.77	18.10
5500	Shallow slab box outs, to 10 S.F.		42	1.143	Ea.	11.60	56.50		68.10	104
5550	Over 10 S.F. (use perimeter)		600	.080	L.F.	1.55	3.94		5.49	8.10
6000	Bulkhead forms for slab, with keyway, 1 use, 2 piece		500	.096		1.72	4.73		6.45	9.55
6100	3 piece (see also edge forms)		460	.104		1.91	5.15		7.06	10.45
6200	Slab bulkhead form, 4-1/2" high, exp metal, w/keyway & stakes G	C-1	1200	.027		.91	1.28		2.19	3.08
6210	5-1/2" high G		1100	.029		1.14	1.40		2.54	3.52

03 11 Concrete Forming

03 11 13 – Structural Cast-In-Place Concrete Forming

03 11 13.35 Forms In Place, Elevated Slabs

		Crew	Daily Output	Labor-Hours	Unit	Material	2018 Bare Costs Labor	Equipment	Total	Total Incl O&P
6215	7-1/2" high G	C-1	960	.033	L.F.	1.37	1.60		2.97	4.11
6220	9-1/2" high G		840	.038		1.49	1.83		3.32	4.61
6500	Curb forms, wood, 6" to 12" high, on elevated slabs, 1 use		180	.178	SFCA	1.71	8.55		10.26	15.75
6550	2 use		205	.156		.94	7.50		8.44	13.20
6600	3 use		220	.145		.69	7		7.69	12.10
6650	4 use		225	.142		.56	6.85		7.41	11.70
7000	Edge forms to 6" high, on elevated slab, 4 use		500	.064	L.F.	.21	3.07		3.28	5.20
7070	7" to 12" high, 1 use		162	.198	SFCA	1.30	9.50		10.80	16.85
7080	2 use		198	.162		.72	7.75		8.47	13.40
7090	3 use		222	.144		.52	6.90		7.42	11.80
7101	4 use		350	.091		.21	4.39		4.60	7.35
7500	Depressed area forms to 12" high, 4 use		300	.107	L.F.	.77	5.10		5.87	9.15
7550	12" to 24" high, 4 use		175	.183		1.05	8.80		9.85	15.40
8000	Perimeter deck and rail for elevated slabs, straight		90	.356		11.95	17.05		29	40.50
8050	Curved		65	.492		16.45	23.50		39.95	56.50
8500	Void forms, round plastic, 8" high x 3" diameter G		450	.071	Ea.	2.02	3.41		5.43	7.75
8550	4" diameter G		425	.075		2.29	3.61		5.90	8.35
8600	6" diameter G		400	.080		4.09	3.84		7.93	10.75
8650	8" diameter G		375	.085		6.75	4.10		10.85	14.10
9000	Minimum labor/equipment charge	2 Carp	2	8	Job		405		405	660

03 11 13.40 Forms In Place, Equipment Foundations

		Crew	Daily Output	Labor-Hours	Unit	Material	2018 Bare Costs Labor	Equipment	Total	Total Incl O&P
0010	**FORMS IN PLACE, EQUIPMENT FOUNDATIONS** R031113-40									
0020	1 use	C-2	160	.300	SFCA	2.74	14.75		17.49	27
0050	2 use R031113-60		190	.253		1.51	12.45		13.96	21.50
0100	3 use		200	.240		1.10	11.80		12.90	20.50
0150	4 use		205	.234		.90	11.55		12.45	19.70
9000	Minimum labor/equipment charge	1 Carp	3	2.667	Job		135		135	219

03 11 13.45 Forms In Place, Footings

		Crew	Daily Output	Labor-Hours	Unit	Material	2018 Bare Costs Labor	Equipment	Total	Total Incl O&P
0010	**FORMS IN PLACE, FOOTINGS** R031113-40									
0020	Continuous wall, plywood, 1 use	C-1	375	.085	SFCA	6.60	4.10		10.70	13.90
0150	4 use R031113-60	"	485	.066	"	2.15	3.17		5.32	7.50
1500	Keyway, 4 use, tapered wood, 2" x 4"	1 Carp	530	.015	L.F.	.22	.77		.99	1.48
1550	2" x 6"	"	500	.016	"	.32	.81		1.13	1.67
3000	Pile cap, square or rectangular, job-built plywood, 1 use	C-1	290	.110	SFCA	2.92	5.30		8.22	11.80
3150	4 use		383	.084		.95	4.01		4.96	7.55
5000	Spread footings, job-built lumber, 1 use		305	.105		2.10	5.05		7.15	10.45
5150	4 use		414	.077		.68	3.71		4.39	6.75
9000	Minimum labor/equipment charge	1 Carp	3	2.667	Job		135		135	219

03 11 13.47 Forms In Place, Gas Station Forms

		Crew	Daily Output	Labor-Hours	Unit	Material	2018 Bare Costs Labor	Equipment	Total	Total Incl O&P
0010	**FORMS IN PLACE, GAS STATION FORMS**									
0050	Curb fascia, with template, 12 ga. steel, left in place, 9" high G	1 Carp	50	.160	L.F.	14.10	8.10		22.20	28.50
1000	Sign or light bases, 18" diameter, 9" high G		9	.889	Ea.	.89	45		134	171
1050	30" diameter, 13" high G		8	1	"	141	50.50		191.50	238
1990	Minimum labor/equipment charge		2	4	Job		203		203	330
2000	Island forms, 10' long, 9" high, 3'-6" wide G	C-1	10	3.200	Ea.	395	154		549	685
2050	4' wide G		9	3.556		405	171		576	725
2500	20' long, 9" high, 4' wide G		6	5.333		655	256		911	1,125
2550	5' wide G		5	6.400		680	305		985	1,250
9000	Minimum labor/equipment charge	1 Carp	3	2.667	Job		135		135	219

For customer support on your Commercial Renovation Costs with RSMeans data, call 800.448.8182.

51

03 11 13 – Structural Cast-In-Place Concrete Forming

03 11 13.50 Forms In Place, Grade Beam

		Crew	Daily Output	Labor-Hours	Unit	Material	2018 Bare Costs Labor	Equipment	Total	Total Incl O&P
0010	**FORMS IN PLACE, GRADE BEAM**	R031113-40								
0020	Job-built plywood, 1 use	C-2	530	.091	SFCA	2.97	4.46		7.43	10.50
0150	4 use	"	605	.079	"	.96	3.91		4.87	7.40
9000	Minimum labor/equipment charge	R031113-60	2 Carp	2	8	Job		405		405	660

03 11 13.65 Forms In Place, Slab On Grade

		Crew	Daily Output	Labor-Hours	Unit	Material	2018 Bare Costs Labor	Equipment	Total	Total Incl O&P
0010	**FORMS IN PLACE, SLAB ON GRADE**	R031113-40								
1000	Bulkhead forms w/keyway, wood, 6" high, 1 use	R031113-60	C-1	510	.063	L.F.	1.05	3.01		4.06	6.05
1400	Bulkhead form for slab, 4-1/2" high, exp metal, incl keyway & stakes	G		1200	.027		.91	1.28		2.19	3.08
1410	5-1/2" high	G		1100	.029		1.14	1.40		2.54	3.52
1420	7-1/2" high	G		960	.033		1.37	1.60		2.97	4.11
1430	9-1/2" high	G		840	.038		1.49	1.83		3.32	4.61
2000	Curb forms, wood, 6" to 12" high, on grade, 1 use			215	.149	SFCA	1.90	7.15		9.05	13.70
2150	4 use			275	.116	"	.62	5.60		6.22	9.75
3000	Edge forms, wood, 4 use, on grade, to 6" high			600	.053	L.F.	.29	2.56		2.85	4.47
3050	7" to 12" high			435	.074	SFCA	.69	3.53		4.22	6.50
3060	Over 12"			350	.091	"	.93	4.39		5.32	8.10
3500	For depressed slabs, 4 use, to 12" high			300	.107	L.F.	.76	5.10		5.86	9.15
3550	To 24" high			175	.183		1.01	8.80		9.81	15.35
4000	For slab blockouts, to 12" high, 1 use			200	.160		.82	7.70		8.52	13.35
4050	To 24" high, 1 use			120	.267		1.04	12.80		13.84	22
4100	Plastic (extruded), to 6" high, multiple use, on grade			800	.040		6.65	1.92		8.57	10.45
5000	Screed, 24 ga. metal key joint, see Section 03 15 16.30										
5020	Wood, incl. wood stakes, 1" x 3"		C-1	900	.036	L.F.	.85	1.71		2.56	3.71
5050	2" x 4"			900	.036	"	.84	1.71		2.55	3.69
6000	Trench forms in floor, wood, 1 use			160	.200	SFCA	1.89	9.60		11.49	17.70
6150	4 use			185	.173	"	.61	8.30		8.91	14.10
8760	Void form, corrugated fiberboard, 4" x 12", 4' long	G		3000	.011	S.F.	3.43	.51		3.94	4.60
8770	6" x 12", 4' long			3000	.011		4.10	.51		4.61	5.35
8780	1/4" thick hardboard protective cover for void form		2 Carp	1500	.011		.68	.54		1.22	1.63
9000	Minimum labor/equipment charge		1 Carp	2	4	Job		203		203	330

03 11 13.85 Forms In Place, Walls

		Crew	Daily Output	Labor-Hours	Unit	Material	2018 Bare Costs Labor	Equipment	Total	Total Incl O&P
0010	**FORMS IN PLACE, WALLS**	R031113-10								
0100	Box out for wall openings, to 16" thick, to 10 S.F.		C-2	24	2	Ea.	26.50	98.50		125	189
0150	Over 10 S.F. (use perimeter)	R031113-40	"	280	.171	L.F.	2.28	8.45		10.73	16.20
0250	Brick shelf, 4" w, add to wall forms, use wall area above shelf										
0260	1 use	R031113-60	C-2	240	.200	SFCA	2.47	9.85		12.32	18.70
0350	4 use			300	.160	"	.99	7.90		8.89	13.90
0500	Bulkhead, wood with keyway, 1 use, 2 piece			265	.181	L.F.	2.19	8.90		11.09	16.85
0600	Bulkhead forms with keyway, 1 piece expanded metal, 8" wall	G	C-1	1000	.032		1.37	1.54		2.91	4
0610	10" wall	G		800	.040		1.49	1.92		3.41	4.76
0620	12" wall	G		525	.061		1.79	2.93		4.72	6.70
0700	Buttress, to 8' high, 1 use		C-2	350	.137	SFCA	4.23	6.75		10.98	15.60
0850	4 use			480	.100	"	1.40	4.92		6.32	9.55
1000	Corbel or haunch, to 12" wide, add to wall forms, 1 use			150	.320	L.F.	2.38	15.75		18.13	28
1150	4 use			180	.267	"	.77	13.15		13.92	22.50
2000	Wall, job-built plywood, to 8' high, 1 use			370	.130	SFCA	2.76	6.40		9.16	13.40
2050	2 use			435	.110		1.75	5.45		7.20	10.75
2100	3 use			495	.097		1.28	4.77		6.05	9.15
2150	4 use			505	.095		1.04	4.68		5.72	8.75
2400	Over 8' to 16' high, 1 use			280	.171		3.05	8.45		11.50	17.05
2450	2 use			345	.139		1.33	6.85		8.18	12.55
2500	3 use			375	.128		.95	6.30		7.25	11.30

03 11 Concrete Forming

03 11 13 – Structural Cast-In-Place Concrete Forming

03 11 13.85 Forms In Place, Walls

		Crew	Daily Output	Labor-Hours	Unit	Material	2018 Bare Costs Labor	2018 Bare Costs Equipment	Total	Total Incl O&P
2550	4 use	C-2	395	.122	SFCA	.78	6		6.78	10.55
2700	Over 16' high, 1 use		235	.204		2.71	10.05		12.76	19.30
2750	2 use		290	.166		1.49	8.15		9.64	14.90
2800	3 use		315	.152		1.08	7.50		8.58	13.40
2850	4 use		330	.145		.88	7.15		8.03	12.55
4000	Radial, smooth curved, job-built plywood, 1 use		245	.196		2.55	9.65		12.20	18.45
4150	4 use		335	.143		.83	7.05		7.88	12.35
4200	Below grade, job-built plywood, 1 use		225	.213		2.71	10.50		13.21	20
4210	2 use		225	.213		1.50	10.50		12	18.70
4220	3 use		225	.213		1.24	10.50		11.74	18.40
4230	4 use		225	.213		.89	10.50		11.39	18
4600	Retaining wall, battered, job-built plyw'd, to 8' high, 1 use		300	.160		2.03	7.90		9.93	15.05
4750	4 use		390	.123		.66	6.05		6.71	10.55
4900	Over 8' to 16' high, 1 use		240	.200		2.21	9.85		12.06	18.45
5050	4 use		320	.150		.72	7.40		8.12	12.80
7500	Lintel or sill forms, 1 use	1 Carp	30	.267		3.25	13.50		16.75	25.50
7560	4 use	"	37	.216		1.05	10.95		12	18.95
7800	Modular prefabricated plywood, based on 20 uses of purchased									
7820	forms, and 4 uses of bracing lumber									
7860	To 8' high	C-2	800	.060	SFCA	1.09	2.95		4.04	6
8060	Over 8' to 16' high		600	.080		1.15	3.94		5.09	7.65
8600	Pilasters, 1 use		270	.178		3.20	8.75		11.95	17.70
8660	4 use		385	.125		1.04	6.15		7.19	11.10
9010	Steel framed plywood, based on 50 uses of purchased									
9020	forms, and 4 uses of bracing lumber									
9060	To 8' high	C-2	600	.080	SFCA	.73	3.94		4.67	7.20
9260	Over 8' to 16' high		450	.107		.73	5.25		5.98	9.30
9460	Over 16' to 20' high		400	.120		.73	5.90		6.63	10.40
9475	For elevated walls, add						10%			
9480	For battered walls, 1 side battered, add					10%	10%			
9485	For battered walls, 2 sides battered, add					15%	15%			
9900	Minimum labor/equipment charge	2 Carp	2	8	Job		405		405	660

03 11 16 – Architectural Cast-in-Place Concrete Forming

03 11 16.13 Concrete Form Liners

		Crew	Daily Output	Labor-Hours	Unit	Material	2018 Bare Costs Labor	2018 Bare Costs Equipment	Total	Total Incl O&P
0010	**CONCRETE FORM LINERS**									
5750	Liners for forms (add to wall forms), ABS plastic									
5800	Aged wood, 4" wide, 1 use	1 Carp	256	.031	SFCA	3.47	1.58		5.05	6.40
5820	2 use		256	.031		1.91	1.58		3.49	4.67
5830	3 use		256	.031		1.39	1.58		2.97	4.10
5840	4 use		256	.031		1.13	1.58		2.71	3.81
5900	Fractured rope rib, 1 use		192	.042		5.05	2.11		7.16	9
5925	2 use		192	.042		2.77	2.11		4.88	6.45
5950	3 use		192	.042		2.01	2.11		4.12	5.65
6000	4 use		192	.042		1.63	2.11		3.74	5.25
6100	Ribbed, 3/4" deep x 1-1/2" OC, 1 use		224	.036		5.05	1.81		6.86	8.50
6125	2 use		224	.036		2.77	1.81		4.58	6
6150	3 use		224	.036		2.01	1.81		3.82	5.15
6200	4 use		224	.036		1.63	1.81		3.44	4.74
6300	Rustic brick pattern, 1 use		224	.036		3.47	1.81		5.28	6.75
6325	2 use		224	.036		1.91	1.81		3.72	5.05
6350	3 use		224	.036		1.39	1.81		3.20	4.47
6400	4 use		224	.036		1.13	1.81		2.94	4.18

For customer support on your Commercial Renovation Costs with RSMeans data, call 800.448.8182.

03 11 Concrete Forming

03 11 16 – Architectural Cast-in-Place Concrete Forming

03 11 16.13 Concrete Form Liners		Crew	Daily Output	Labor-Hours	Unit	Material	2018 Bare Costs Labor	Equipment	Total	Total Incl O&P
6500	3/8" striated, random, 1 use	1 Carp	224	.036	SFCA	3.47	1.81		5.28	6.75
6525	2 use		224	.036		1.91	1.81		3.72	5.05
6550	3 use		224	.036		1.39	1.81		3.20	4.47
6600	4 use		224	.036		1.13	1.81		2.94	4.18

03 15 Concrete Accessories

03 15 05 – Concrete Forming Accessories

03 15 05.70 Shores

			Crew	Daily Output	Labor-Hours	Unit	Material	2018 Bare Costs Labor	Equipment	Total	Total Incl O&P
0010	**SHORES**										
0020	Erect and strip, by hand, horizontal members										
0500	Aluminum joists and stringers	G	2 Carp	60	.267	Ea.		13.50		13.50	22
0600	Steel, adjustable beams	G		45	.356			18.05		18.05	29.50
0700	Wood joists			50	.320			16.20		16.20	26.50
0800	Wood stringers			30	.533			27		27	44
1000	Vertical members to 10' high	G		55	.291			14.75		14.75	24
1050	To 13' high	G		50	.320			16.20		16.20	26.50
1100	To 16' high	G		45	.356			18.05		18.05	29.50
1500	Reshoring	G		1400	.011	S.F.	.61	.58		1.19	1.61
1600	Flying truss system	G	C-17D	9600	.009	SFCA		.46	.08	.54	.84
1760	Horizontal, aluminum joists, 6-1/4" high x 5' to 21' span, buy	G				L.F.	16.25			16.25	17.85
1770	Beams, 7-1/4" high x 4' to 30' span	G				"	19			19	21
1810	Horizontal, steel beam, W8x10, 7' span, buy	G				Ea.	61			61	67.50
1830	10' span	G					71			71	78
1920	15' span	G					122			122	135
1940	20' span	G					172			172	189
1970	Steel stringer, W8x10, 4' to 16' span, buy	G				L.F.	7.10			7.10	7.85
3000	Rent for job duration, aluminum joist @ 2' OC, per mo.	G				SF Flr.	.41			.41	.45
3050	Steel W8x10	G					.18			.18	.20
3060	Steel adjustable	G					.18			.18	.20
3500	#1 post shore, steel, 5'-7" to 9'-6" high, 10,000# cap., buy	G				Ea.	155			155	171
3550	#2 post shore, 7'-3" to 12'-10" high, 7800# capacity	G					180			180	198
3600	#3 post shore, 8'-10" to 16'-1" high, 3800# capacity	G					196			196	216
5010	Frame shoring systems, steel, 12,000#/leg, buy										
5040	Frame, 2' wide x 6' high	G				Ea.	115			115	127
5250	X-brace	G					20			20	22
5550	Base plate	G					17.70			17.70	19.45
5600	Screw jack	G					39			39	42.50
5650	U-head, 8" x 8"	G					22			22	24

03 15 05.85 Stair Tread Inserts

			Crew	Daily Output	Labor-Hours	Unit	Material	2018 Bare Costs Labor	Equipment	Total	Total Incl O&P
0010	**STAIR TREAD INSERTS**										
0105	Cast nosing insert, abrasive surface, pre-drilled, includes screws										
0110	Aluminum, 3" wide x 3' long		1 Cefi	32	.250	Ea.	52	11.90		63.90	76.50
0120	4' long			31	.258		69.50	12.25		81.75	96
0130	5' long			30	.267		86.50	12.70		99.20	116
0135	Extruded nosing insert, black abrasive strips, continuous anchor										
0140	Aluminum, 3" wide x 3' long		1 Cefi	64	.125	Ea.	33.50	5.95		39.45	46.50
0150	4' long			60	.133		45	6.35		51.35	59.50
0160	5' long			56	.143		56	6.80		62.80	72.50
0165	Extruded nosing insert, black abrasive strips, pre-drilled, incl. screws										
0170	Aluminum, 3" wide x 3' long		1 Cefi	32	.250	Ea.	42	11.90		53.90	65
0180	4' long			31	.258		56	12.25		68.25	81

03 15 Concrete Accessories

03 15 05 – Concrete Forming Accessories

03 15 05.85 Stair Tread Inserts		Crew	Daily Output	Labor-Hours	Unit	Material	2018 Bare Costs Labor	Equipment	Total	Total Incl O&P
0190	5' long	1 Cefi	30	.267	Ea.	70	12.70		82.70	97
9000	Minimum labor/equipment charge	1 Carp	4	2	Job		101		101	165

03 15 13 – Waterstops

03 15 13.50 Waterstops		Crew	Daily Output	Labor-Hours	Unit	Material	Labor	Equipment	Total	Total Incl O&P
0010	**WATERSTOPS**, PVC and Rubber									
0020	PVC, ribbed 3/16" thick, 4" wide	1 Carp	155	.052	L.F.	1.43	2.62		4.05	5.80
0050	6" wide		145	.055		2.41	2.80		5.21	7.20
0500	With center bulb, 6" wide, 3/16" thick		135	.059		2.43	3		5.43	7.55
0550	3/8" thick		130	.062		4.43	3.12		7.55	9.90
0600	9" wide x 3/8" thick		125	.064		7.20	3.24		10.44	13.20

03 15 16 – Concrete Construction Joints

03 15 16.20 Control Joints, Saw Cut		Crew	Daily Output	Labor-Hours	Unit	Material	Labor	Equipment	Total	Total Incl O&P
0010	**CONTROL JOINTS, SAW CUT**									
0100	Sawcut control joints in green concrete									
0120	1" depth	C-27	2000	.008	L.F.	.03	.38	.05	.46	.70
0140	1-1/2" depth		1800	.009		.05	.42	.06	.53	.79
0160	2" depth		1600	.010		.07	.48	.06	.61	.90
0180	Sawcut joint reservoir in cured concrete									
0182	3/8" wide x 3/4" deep, with single saw blade	C-27	1000	.016	L.F.	.05	.76	.10	.91	1.37
0184	1/2" wide x 1" deep, with double saw blades		900	.018		.10	.85	.11	1.06	1.58
0186	3/4" wide x 1-1/2" deep, with double saw blades		800	.020		.21	.95	.13	1.29	1.87
0190	Water blast joint to wash away laitance, 2 passes	C-29	2500	.003			.13	.03	.16	.24
0200	Air blast joint to blow out debris and air dry, 2 passes	C-28	2000	.004			.19	.01	.20	.31
0300	For backer rod, see Section 07 91 23.10									
0340	For joint sealant, see Section 03 15 16.30 or 07 92 13.20									
0900	For replacement of joint sealant, see Section 07 01 90.01									

03 15 16.30 Expansion Joints			Crew	Daily Output	Labor-Hours	Unit	Material	Labor	Equipment	Total	Total Incl O&P
0010	**EXPANSION JOINTS**										
0020	Keyed, cold, 24 ga., incl. stakes, 3-1/2" high	G	1 Carp	200	.040	L.F.	.85	2.03		2.88	4.23
0050	4-1/2" high	G		200	.040		.91	2.03		2.94	4.29
0100	5-1/2" high	G		195	.041		1.14	2.08		3.22	4.63
0150	7-1/2" high	G		190	.042		1.37	2.14		3.51	4.98
0160	9-1/2" high	G		185	.043		1.49	2.19		3.68	5.20
0300	Poured asphalt, plain, 1/2" x 1"		1 Clab	450	.018		.24	.71		.95	1.41
0350	1" x 2"			400	.020		.95	.80		1.75	2.34
0500	Neoprene, liquid, cold applied, 1/2" x 1"			450	.018		2.28	.71		2.99	3.66
0550	1" x 2"			400	.020		9.10	.80		9.90	11.35
0700	Polyurethane, poured, 2 part, 1/2" x 1"			400	.020		1.39	.80		2.19	2.82
0750	1" x 2"			350	.023		5.55	.91		6.46	7.60
0900	Rubberized asphalt, hot or cold applied, 1/2" x 1"			450	.018		.30	.71		1.01	1.48
0950	1" x 2"			400	.020		1.19	.80		1.99	2.60
1100	Hot applied, fuel resistant, 1/2" x 1"			450	.018		.45	.71		1.16	1.65
1150	1" x 2"			400	.020		1.79	.80		2.59	3.25
2000	Premolded, bituminous fiber, 1/2" x 6"		1 Carp	375	.021		.42	1.08		1.50	2.22
2050	1" x 12"			300	.027		1.97	1.35		3.32	4.36
2140	Concrete expansion joint, recycled paper and fiber, 1/2" x 6"	G		390	.021		.42	1.04		1.46	2.15
2150	1/2" x 12"	G		360	.022		.84	1.13		1.97	2.76
2250	Cork with resin binder, 1/2" x 6"			375	.021		1.17	1.08		2.25	3.05
2300	1" x 12"			300	.027		3.20	1.35		4.55	5.70
2500	Neoprene sponge, closed cell, 1/2" x 6"			375	.021		2.31	1.08		3.39	4.30
2550	1" x 12"			300	.027		8.85	1.35		10.20	11.90

03 15 16 – Concrete Construction Joints

03 15 16.30 Expansion Joints		Crew	Daily Output	Labor-Hours	Unit	Material	2018 Bare Costs Labor	Equipment	Total	Total Incl O&P
2750	Polyethylene foam, 1/2" x 6"	1 Carp	375	.021	L.F.	.52	1.08		1.60	2.33
2800	1" x 12"		300	.027		2.78	1.35		4.13	5.25
3000	Polyethylene backer rod, 3/8" diameter		460	.017		.03	.88		.91	1.47
3050	3/4" diameter		460	.017		.07	.88		.95	1.50
3100	1" diameter		460	.017		.12	.88		1	1.56
3500	Polyurethane foam, with polybutylene, 1/2" x 1/2"		475	.017		1.19	.85		2.04	2.70
3550	1" x 1"		450	.018		3.03	.90		3.93	4.79
3750	Polyurethane foam, regular, closed cell, 1/2" x 6"		375	.021		.89	1.08		1.97	2.74
3800	1" x 12"		300	.027		3.15	1.35		4.50	5.65
4000	Polyvinyl chloride foam, closed cell, 1/2" x 6"		375	.021		2.40	1.08		3.48	4.40
4050	1" x 12"		300	.027		8.25	1.35		9.60	11.30
4250	Rubber, gray sponge, 1/2" x 6"		375	.021		2.02	1.08		3.10	3.98
4300	1" x 12"		300	.027		7.25	1.35		8.60	10.20
4400	Redwood heartwood, 1" x 4"		400	.020		1.16	1.01		2.17	2.93
4450	1" x 6"	▼	375	.021	▼	1.75	1.08		2.83	3.69
5000	For installation in walls, add						75%			
5250	For installation in boxouts, add						25%			

03 15 19 – Cast-In Concrete Anchors

03 15 19.05 Anchor Bolt Accessories

0010	**ANCHOR BOLT ACCESSORIES**									
0015	For anchor bolts set in fresh concrete, see Section 03 15 19.10									
8150	Anchor bolt sleeve, plastic, 1" diameter bolts	1 Carp	60	.133	Ea.	13.40	6.75		20.15	25.50
8500	1-1/2" diameter		28	.286		19.75	14.50		34.25	45
8600	2" diameter		24	.333		19.45	16.90		36.35	49
8650	3" diameter	▼	20	.400	▼	35.50	20.50		56	72

03 15 19.10 Anchor Bolts

0010	**ANCHOR BOLTS**									
0015	Made from recycled materials									
0025	Single bolts installed in fresh concrete, no templates									
0030	Hooked w/nut and washer, 1/2" diameter, 8" long Ⓖ	1 Carp	132	.061	Ea.	1.42	3.07		4.49	6.55
0040	12" long Ⓖ		131	.061		1.58	3.10		4.68	6.80
0050	5/8" diameter, 8" long Ⓖ		129	.062		3.80	3.14		6.94	9.30
0060	12" long Ⓖ		127	.063		4.67	3.19		7.86	10.35
0070	3/4" diameter, 8" long Ⓖ		127	.063		4.67	3.19		7.86	10.35
0080	12" long Ⓖ	▼	125	.064	▼	5.85	3.24		9.09	11.65
0090	2-bolt pattern, including job-built 2-hole template, per set									
0100	J-type, incl. hex nut & washer, 1/2" diameter x 6" long Ⓖ	1 Carp	21	.381	Set	5.15	19.30		24.45	37
0110	12" long Ⓖ		21	.381		5.75	19.30		25.05	38
0120	18" long Ⓖ		21	.381		6.70	19.30		26	39
0130	3/4" diameter x 8" long Ⓖ		20	.400		11.95	20.50		32.45	46
0140	12" long Ⓖ		20	.400		14.30	20.50		34.80	48.50
0150	18" long Ⓖ		20	.400		17.80	20.50		38.30	52.50
0160	1" diameter x 12" long Ⓖ		19	.421		22.50	21.50		44	59.50
0170	18" long Ⓖ		19	.421		27	21.50		48.50	64
0180	24" long Ⓖ		19	.421		32	21.50		53.50	70
0190	36" long Ⓖ		18	.444		43	22.50		65.50	84
0200	1-1/2" diameter x 18" long Ⓖ		17	.471		41.50	24		65.50	84
0210	24" long Ⓖ		16	.500		48.50	25.50		74	94.50
0300	L-type, incl. hex nut & washer, 3/4" diameter x 12" long Ⓖ		20	.400		14.10	20.50		34.60	48.50
0310	18" long Ⓖ		20	.400		17.35	20.50		37.85	52
0320	24" long Ⓖ		20	.400		20.50	20.50		41	55.50
0330	30" long Ⓖ	▼	20	.400		25.50	20.50		46	61

03 15 19 – Cast-In Concrete Anchors

03 15 19.10 Anchor Bolts

			Crew	Daily Output	Labor-Hours	Unit	Material	2018 Bare Costs Labor	Equipment	Total	Total Incl O&P
0340	36" long	G	1 Carp	20	.400	Set	28.50	20.50		49	64.50
0350	1" diameter x 12" long	G		19	.421		22	21.50		43.50	58.50
0360	18" long	G		19	.421		26.50	21.50		48	63.50
0370	24" long	G		19	.421		32	21.50		53.50	70
0380	30" long	G		19	.421		37.50	21.50		59	75.50
0390	36" long	G		18	.444		42.50	22.50		65	83
0400	42" long	G		18	.444		51	22.50		73.50	92.50
0410	48" long	G		18	.444		57	22.50		79.50	99
0420	1-1/4" diameter x 18" long	G		18	.444		33.50	22.50		56	73.50
0430	24" long	G		18	.444		39	22.50		61.50	79.50
0440	30" long	G		17	.471		45	24		69	88
0450	36" long	G		17	.471		50.50	24		74.50	94.50
1000	4-bolt pattern, including job-built 4-hole template, per set										
1100	J-type, incl. hex nut & washer, 1/2" diameter x 6" long	G	1 Carp	19	.421	Set	7.65	21.50		29.15	43
1110	12" long	G		19	.421		8.95	21.50		30.45	44.50
1120	18" long	G		18	.444		10.85	22.50		33.35	48.50
1130	3/4" diameter x 8" long	G		17	.471		21.50	24		45.50	62
1140	12" long	G		17	.471		26	24		50	67
1150	18" long	G		17	.471		33	24		57	75
1160	1" diameter x 12" long	G		16	.500		42.50	25.50		68	88
1170	18" long	G		15	.533		51	27		78	100
1180	24" long	G		15	.533		61.50	27		88.50	112
1190	36" long	G		15	.533		83.50	27		110.50	136
1200	1-1/2" diameter x 18" long	G		13	.615		80	31		111	139
1210	24" long	G		12	.667		94.50	34		128.50	159
1300	L-type, incl. hex nut & washer, 3/4" diameter x 12" long	G		17	.471		25.50	24		49.50	66.50
1310	18" long	G		17	.471		32	24		56	74
1320	24" long	G		17	.471		38.50	24		62.50	81
1330	30" long	G		16	.500		48	25.50		73.50	94
1340	36" long	G		16	.500		54.50	25.50		80	101
1350	1" diameter x 12" long	G		16	.500		41	25.50		66.50	86
1360	18" long	G		15	.533		50.50	27		77.50	99.50
1370	24" long	G		15	.533		61.50	27		88.50	112
1380	30" long	G		15	.533		72	27		99	124
1390	36" long	G		15	.533		82	27		109	135
1400	42" long	G		14	.571		99.50	29		128.50	156
1410	48" long	G		14	.571		111	29		140	169
1420	1-1/4" diameter x 18" long	G		14	.571		64	29		93	118
1430	24" long	G		14	.571		76	29		105	131
1440	30" long	G		13	.615		87.50	31		118.50	147
1450	36" long	G		13	.615		99	31		130	160

03 15 19.45 Machinery Anchors

			Crew	Daily Output	Labor-Hours	Unit	Material	2018 Bare Costs Labor	Equipment	Total	Total Incl O&P
0010	**MACHINERY ANCHORS**, heavy duty, incl. sleeve, floating base nut,										
0020	lower stud & coupling nut, fiber plug, connecting stud, washer & nut.										
0030	For flush mounted embedment in poured concrete heavy equip. pads.										
0200	Stud & bolt, 1/2" diameter	G	E-16	40	.400	Ea.	52.50	22.50	2.46	77.46	98.50
0300	5/8" diameter	G		35	.457		61	25.50	2.81	89.31	114
0500	3/4" diameter	G		30	.533		72.50	29.50	3.28	105.28	134
0600	7/8" diameter	G		25	.640		82.50	35.50	3.94	121.94	155
0800	1" diameter	G		20	.800		88.50	44.50	4.92	137.92	179
0900	1-1/4" diameter	G		15	1.067		118	59.50	6.55	184.05	238

03 21 11 – Plain Steel Reinforcement Bars

03 21 11.60 Reinforcing In Place

		Crew	Daily Output	Labor-Hours	Unit	Material	2018 Bare Costs Labor	Equipment	Total	Total Incl O&P
0010	**REINFORCING IN PLACE**, 50-60 ton lots, A615 Grade 60	R032110-10								
0020	Includes labor, but not material cost, to install accessories									
0030	Made from recycled materials									
0102	Beams & Girders, #3 to #7	G	4 Rodm	3200	.010	Lb.	.48	.55		1.03	1.40
0152	#8 to #18	G		5400	.006		.48	.32		.80	1.05
0202	Columns, #3 to #7	G		3000	.011		.48	.58		1.06	1.46
0252	#8 to #18	G		4600	.007		.48	.38		.86	1.14
0402	Elevated slabs, #4 to #7	G		5800	.006		.48	.30		.78	1.01
0502	Footings, #4 to #7	G		4200	.008		.48	.42		.90	1.20
0552	#8 to #18	G		7200	.004		.48	.24		.72	.92
0602	Slab on grade, #3 to #7	G		4200	.008		.48	.42		.90	1.20
0702	Walls, #3 to #7	G		6000	.005		.48	.29		.77	1
0752	#8 to #18	G	↓	8000	.004	↓	.48	.22		.70	.88
0900	For other than 50-60 ton lots										
1000	Under 10 ton job, #3 to #7, add						25%	10%			
1010	#8 to #18, add						20%	10%			
1050	10-50 ton job, #3 to #7, add						10%				
1060	#8 to #18, add						5%				
1100	60-100 ton job, #3 to #7, deduct						5%				
1110	#8 to #18, deduct						10%				
1150	Over 100 ton job, #3 to #7, deduct						10%				
1160	#8 to #18, deduct						15%				
2000	Unloading & sorting, add to above		C-5	100	.560	Ton		30.50	5.90	36.40	55
2200	Crane cost for handling, 90 picks/day, up to 1.5 tons/bundle, add to above			135	.415			22.50	4.38	26.88	41
2210	1.0 ton/bundle			92	.609			33	6.40	39.40	59.50
2220	0.5 ton/bundle		↓	35	1.600	↓		87	16.90	103.90	157
2400	Dowels, 2 feet long, deformed, #3	G	2 Rodm	520	.031	Ea.	.40	1.68		2.08	3.13
2410	#4	G		480	.033		.70	1.82		2.52	3.68
2420	#5	G		435	.037		1.10	2.01		3.11	4.42
2430	#6	G		360	.044	↓	1.58	2.43		4.01	5.60
2450	Longer and heavier dowels, add	G		725	.022	Lb.	.53	1.21		1.74	2.51
2500	Smooth dowels, 12" long, 1/4" or 3/8" diameter	G		140	.114	Ea.	.73	6.25		6.98	10.80
2520	5/8" diameter	G		125	.128		1.28	7		8.28	12.55
2530	3/4" diameter	G	↓	110	.145	↓	1.58	7.95		9.53	14.45
2600	Dowel sleeves for CIP concrete, 2-part system										
2610	Sleeve base, plastic, for 5/8" smooth dowel sleeve, fasten to edge form		1 Rodm	200	.040	Ea.	.53	2.19		2.72	4.07
2615	Sleeve, plastic, 12" long, for 5/8" smooth dowel, snap onto base			400	.020		1.33	1.09		2.42	3.21
2620	Sleeve base, for 3/4" smooth dowel sleeve			175	.046		.53	2.50		3.03	4.57
2625	Sleeve, 12" long, for 3/4" smooth dowel			350	.023		1.21	1.25		2.46	3.33
2630	Sleeve base, for 1" smooth dowel sleeve			150	.053		.68	2.91		3.59	5.40
2635	Sleeve, 12" long, for 1" smooth dowel		↓	300	.027		1.42	1.46		2.88	3.89
2700	Dowel caps, visual warning only, plastic, #3 to #8		2 Rodm	800	.020		.30	1.09		1.39	2.08
2720	#8 to #18			750	.021		.74	1.17		1.91	2.68
2750	Impalement protective, plastic, #4 to #9		↓	800	.020	↓	1.08	1.09		2.17	2.94
9000	Minimum labor/equipment charge		1 Rodm	4	2	Job		109		109	175

03 21 21 – Composite Reinforcement Bars

03 21 21.11 Glass Fiber-Reinforced Polymer Reinf. Bars

		Crew	Daily Output	Labor-Hours	Unit	Material	2018 Bare Costs Labor	Equipment	Total	Total Incl O&P
0010	**GLASS FIBER-REINFORCED POLYMER REINFORCEMENT BARS**									
0020	Includes labor, but not material cost, to install accessories									
0050	#2 bar, .043 lb./L.F.	4 Rodm	9500	.003	L.F.	.40	.18		.58	.73
0100	#3 bar, .092 lb./L.F.		9300	.003		.63	.19		.82	.99
0150	#4 bar, .160 lb./L.F.	↓	9100	.004		.91	.19		1.10	1.31

03 21 Reinforcement Bars

03 21 21 – Composite Reinforcement Bars

03 21 21.11 Glass Fiber-Reinforced Polymer Reinf. Bars		Crew	Daily Output	Labor-Hours	Unit	Material	2018 Bare Costs Labor	Equipment	Total	Total Incl O&P
0200	#5 bar, .258 lb./L.F.	4 Rodm	8700	.004	L.F.	1.09	.20		1.29	1.52
0250	#6 bar, .372 lb./L.F.		8300	.004		1.81	.21		2.02	2.33
0300	#7 bar, .497 lb./L.F.		7900	.004		2.37	.22		2.59	2.96
0350	#8 bar, .620 lb./L.F.		7400	.004		2.38	.24		2.62	3
0400	#9 bar, .800 lb./L.F.		6800	.005		3.99	.26		4.25	4.80
0450	#10 bar, 1.08 lb./L.F.		5800	.006		3.75	.30		4.05	4.61
0500	For bends, add per bend				Ea.	1.54			1.54	1.69

03 22 Fabric and Grid Reinforcing

03 22 11 – Plain Welded Wire Fabric Reinforcing

03 22 11.10 Plain Welded Wire Fabric

			Crew	Daily Output	Labor-Hours	Unit	Material	2018 Bare Costs Labor	Equipment	Total	Total Incl O&P
0010	**PLAIN WELDED WIRE FABRIC** ASTM A185	R032205-30									
0020	Includes labor, but not material cost, to install accessories										
0030	Made from recycled materials										
0050	Sheets										
0100	6 x 6 - W1.4 x W1.4 (10 x 10) 21 lb./C.S.F.	G	2 Rodm	35	.457	C.S.F.	15.10	25		40.10	56.50
0200	6 x 6 - W2.1 x W2.1 (8 x 8) 30 lb./C.S.F.	G		31	.516		19.60	28		47.60	66.50
0300	6 x 6 - W2.9 x W2.9 (6 x 6) 42 lb./C.S.F.	G		29	.552		25.50	30		55.50	76
0400	6 x 6 - W4 x W4 (4 x 4) 58 lb./C.S.F.	G		27	.593		35	32.50		67.50	90
0500	4 x 4 - W1.4 x W1.4 (10 x 10) 31 lb./C.S.F.	G		31	.516		22.50	28		50.50	70
0600	4 x 4 - W2.1 x W2.1 (8 x 8) 44 lb./C.S.F.	G		29	.552		28	30		58	79
0650	4 x 4 - W2.9 x W2.9 (6 x 6) 61 lb./C.S.F.	G		27	.593		45	32.50		77.50	101
0700	4 x 4 - W4 x W4 (4 x 4) 85 lb./C.S.F.	G		25	.640		56.50	35		91.50	118
0750	Rolls										
0800	2 x 2 - #14 galv., 21 lb./C.S.F., beam & column wrap	G	2 Rodm	6.50	2.462	C.S.F.	46.50	135		181.50	266
0900	2 x 2 - #12 galv. for gunite reinforcing	G	"	6.50	2.462	"	70	135		205	292
9000	Minimum labor/equipment charge		1 Rodm	4	2	Job		109		109	175

03 22 13 – Galvanized Welded Wire Fabric Reinforcing

03 22 13.10 Galvanized Welded Wire Fabric

						Unit	Material			Total	Total Incl O&P
0010	**GALVANIZED WELDED WIRE FABRIC**										
0100	Add to plain welded wire pricing for galvanized welded wire					Lb.	.23			.23	.25

03 22 16 – Epoxy-Coated Welded Wire Fabric Reinforcing

03 22 16.10 Epoxy-Coated Welded Wire Fabric

						Unit	Material			Total	Total Incl O&P
0010	**EPOXY-COATED WELDED WIRE FABRIC**										
0100	Add to plain welded wire pricing for epoxy-coated welded wire					Lb.	.39			.39	.43

03 24 Fibrous Reinforcing

03 24 05 – Reinforcing Fibers

03 24 05.30 Synthetic Fibers

						Unit	Material			Total	Total Incl O&P
0010	**SYNTHETIC FIBERS**										
0100	Synthetic fibers, add to concrete					Lb.	4.71			4.71	5.20
0110	1-1/2 lb./C.Y.					C.Y.	7.30			7.30	8

03 24 05.70 Steel Fibers

						Unit	Material			Total	Total Incl O&P
0010	**STEEL FIBERS**										
0140	ASTM A850, Type V, continuously deformed, 1-1/2" long x 0.045" diam.										
0150	Add to price of ready mix concrete	G				Lb.	1.22			1.22	1.34
0205	Alternate pricing, dosing at 5 lb./C.Y., add to price of RMC	G				C.Y.	6.10			6.10	6.70
0210	10 lb./C.Y.	G					12.20			12.20	13.40

03 24 Fibrous Reinforcing

03 24 05 – Reinforcing Fibers

03 24 05.70 Steel Fibers		Crew	Daily Output	Labor-Hours	Unit	Material	2018 Bare Costs Labor	Equipment	Total	Total Incl O&P
0215	15 lb./C.Y.	G			C.Y.	18.30			18.30	20
0220	20 lb./C.Y.	G				24.50			24.50	27
0225	25 lb./C.Y.	G				30.50			30.50	33.50
0230	30 lb./C.Y.	G				36.50			36.50	40.50
0235	35 lb./C.Y.	G				42.50			42.50	47
0240	40 lb./C.Y.	G				49			49	53.50
0250	50 lb./C.Y.	G				61			61	67
0275	75 lb./C.Y.	G				91.50			91.50	101
0300	100 lb./C.Y.	G				122			122	134

03 30 Cast-In-Place Concrete

03 30 53 – Miscellaneous Cast-In-Place Concrete

03 30 53.40 Concrete In Place		Crew	Daily Output	Labor-Hours	Unit	Material	2018 Bare Costs Labor	Equipment	Total	Total Incl O&P
0010	**CONCRETE IN PLACE**									
0020	Including forms (4 uses), Grade 60 rebar, concrete (Portland cement	R033105-10 R033105-20								
0050	Type I), placement and finishing unless otherwise indicated	R033105-70								
0300	Beams (3500 psi), 5 kip/L.F., 10' span	R033105-85	C-14A	15.62	12.804	C.Y.	340	645	58	1,043	1,475
0350	25' span		"	18.55	10.782		355	545	49	949	1,325
0500	Chimney foundations (5000 psi), over 5 C.Y.		C-14C	32.22	3.476		171	167	.80	338.80	460
0510	(3500 psi), under 5 C.Y.		"	23.71	4.724		198	227	1.09	426.09	585
0700	Columns, square (4000 psi), 12" x 12", up to 1% reinforcing by area		C-14A	11.96	16.722		380	845	76	1,301	1,875
0720	Up to 2% reinforcing by area			10.13	19.743		580	1,000	89.50	1,669.50	2,325
0740	Up to 3% reinforcing by area			9.03	22.148		850	1,125	101	2,076	2,850
0800	16" x 16", up to 1% reinforcing by area			16.22	12.330		305	625	56	986	1,400
0820	Up to 2% reinforcing by area			12.57	15.911		495	805	72	1,372	1,925
0840	Up to 3% reinforcing by area			10.25	19.512		750	985	88.50	1,823.50	2,525
0900	24" x 24", up to 1% reinforcing by area			23.66	8.453		261	425	38.50	724.50	1,025
0920	Up to 2% reinforcing by area			17.71	11.293		440	570	51.50	1,061.50	1,450
0940	Up to 3% reinforcing by area			14.15	14.134		685	715	64	1,464	1,975
1000	36" x 36", up to 1% reinforcing by area			33.69	5.936		232	300	27	559	770
1020	Up to 2% reinforcing by area			23.32	8.576		390	435	39	864	1,175
1040	Up to 3% reinforcing by area			17.82	11.223		640	565	51	1,256	1,675
1100	Columns, round (4000 psi), tied, 12" diameter, up to 1% reinforcing by area			20.97	9.537		380	480	43.50	903.50	1,250
1120	Up to 2% reinforcing by area			15.27	13.098		575	660	59.50	1,294.50	1,775
1140	Up to 3% reinforcing by area			12.11	16.515		835	835	75	1,745	2,350
1200	16" diameter, up to 1% reinforcing by area			31.49	6.351		320	320	29	669	900
1220	Up to 2% reinforcing by area			19.12	10.460		520	530	47.50	1,097.50	1,475
1240	Up to 3% reinforcing by area			13.77	14.524		755	735	66	1,556	2,075
1300	20" diameter, up to 1% reinforcing by area			41.04	4.873		310	246	22	578	765
1320	Up to 2% reinforcing by area			24.05	8.316		495	420	38	953	1,250
1340	Up to 3% reinforcing by area			17.01	11.758		750	595	53.50	1,398.50	1,850
1400	24" diameter, up to 1% reinforcing by area			51.85	3.857		290	195	17.50	502.50	655
1420	Up to 2% reinforcing by area			27.06	7.391		490	375	33.50	898.50	1,175
1440	Up to 3% reinforcing by area			18.29	10.935		730	555	49.50	1,334.50	1,750
1500	36" diameter, up to 1% reinforcing by area			75.04	2.665		296	135	12.10	443.10	555
1520	Up to 2% reinforcing by area			37.49	5.335		475	270	24	769	980
1540	Up to 3% reinforcing by area			22.84	8.757		720	445	40	1,205	1,550
1900	Elevated slab (4000 psi), flat slab with drops, 125 psf Sup. Load, 20' span		C-14B	38.45	5.410		284	273	23.50	580.50	775
1950	30' span			50.99	4.079		298	206	17.80	521.80	675
2100	Flat plate, 125 psf Sup. Load, 15' span			30.24	6.878		261	345	30	636	880
2150	25' span			49.60	4.194		270	211	18.30	499.30	655

03 30 53 – Miscellaneous Cast-In-Place Concrete

03 30 53.40 Concrete In Place	Crew	Daily Output	Labor-Hours	Unit	Material	2018 Bare Costs Labor	Equipment	Total	Total Incl O&P
2300	Waffle const., 30" domes, 125 psf Sup. Load, 20' span	C-14B	37.07	5.611	C.Y.	297	283	24.50	604.50	805
2350	30' span		44.07	4.720		275	238	20.50	533.50	715
2500	One way joists, 30" pans, 125 psf Sup. Load, 15' span		27.38	7.597		400	385	33	818	1,100
2550	25' span		31.15	6.677		365	335	29	729	980
2700	One way beam & slab, 125 psf Sup. Load, 15' span		20.59	10.102		279	510	44	833	1,175
2750	25' span		28.36	7.334		265	370	32	667	920
2900	Two way beam & slab, 125 psf Sup. Load, 15' span		24.04	8.652		270	435	37.50	742.50	1,050
2950	25' span		35.87	5.799		236	292	25.50	553.50	755
3100	Elevated slabs, flat plate, including finish, not									
3110	including forms or reinforcing									
3150	Regular concrete (4000 psi), 4" slab	C-8	2613	.021	S.F.	1.70	.95	.34	2.99	3.76
3200	6" slab		2585	.022		2.48	.96	.34	3.78	4.65
3250	2-1/2" thick floor fill		2685	.021		1.11	.92	.33	2.36	3.06
3300	Lightweight, 110 #/C.F., 2-1/2" thick floor fill		2585	.022		1.14	.96	.34	2.44	3.18
3400	Cellular concrete, 1-5/8" fill, under 5000 S.F.		2000	.028		.79	1.24	.44	2.47	3.34
3450	Over 10,000 S.F.		2200	.025		.75	1.13	.40	2.28	3.08
3500	Add per floor for 3 to 6 stories high		31800	.002			.08	.03	.11	.16
3520	For 7 to 20 stories high		21200	.003			.12	.04	.16	.24
3540	Equipment pad (3000 psi), 3' x 3' x 6" thick	C-14H	45	1.067	Ea.	44.50	52.50	.57	97.57	135
3550	4' x 4' x 6" thick		30	1.600		69	79	.85	148.85	204
3560	5' x 5' x 8" thick		18	2.667		126	132	1.41	259.41	355
3570	6' x 6' x 8" thick		14	3.429		173	169	1.82	343.82	465
3580	8' x 8' x 10" thick		8	6		370	296	3.18	669.18	885
3590	10' x 10' x 12" thick		5	9.600		645	475	5.10	1,125.10	1,475
3800	Footings (3000 psi), spread under 1 C.Y.	C-14C	28	4	C.Y.	185	192	.92	377.92	515
3825	1 C.Y. to 5 C.Y.		43	2.605		219	125	.60	344.60	445
3850	Over 5 C.Y.		75	1.493		203	72	.34	275.34	340
3900	Footings, strip (3000 psi), 18" x 9", unreinforced	C-14L	40	2.400		146	113	.65	259.65	345
3920	18" x 9", reinforced	C-14C	35	3.200		169	154	74	323.74	435
3925	20" x 10", unreinforced	C-14L	45	2.133		142	100	.58	242.58	320
3930	20" x 10", reinforced	C-14C	40	2.800		161	135	.64	296.64	395
3935	24" x 12", unreinforced	C-14L	55	1.745		140	82	.47	222.47	288
3940	24" x 12", reinforced	C-14C	48	2.333		159	112	.54	271.54	355
3945	36" x 12", unreinforced	C-14L	70	1.371		136	64.50	.37	200.87	254
3950	36" x 12", reinforced	C-14C	60	1.867		153	90	.43	243.43	315
4000	Foundation mat (3000 psi), under 10 C.Y.		38.67	2.896		221	139	.67	360.67	470
4050	Over 20 C.Y.		56.40	1.986		196	95.50	.46	291.96	370
4200	Wall, free-standing (3000 psi), 8" thick, 8' high	C-14D	45.83	4.364		182	219	19.80	420.80	575
4250	14' high		27.26	7.337		211	370	33.50	614.50	865
4260	12" thick, 8' high		64.32	3.109		165	156	14.10	335.10	450
4270	14' high		40.01	4.999		175	251	22.50	448.50	620
4300	15" thick, 8' high		80.02	2.499		160	126	11.35	297.35	390
4350	12' high		51.26	3.902		160	196	17.70	373.70	510
4500	18' high		48.85	4.094		178	206	18.60	402.60	550
4520	Handicap access ramp (4000 psi), railing both sides, 3' wide	C-14H	14.58	3.292	L.F.	360	163	1.74	524.74	660
4525	5' wide		12.22	3.928		370	194	2.08	566.08	725
4530	With 6" curb and rails both sides, 3' wide		8.55	5.614		370	277	2.98	649.98	860
4535	5' wide		7.31	6.566		375	325	3.48	703.48	940
4650	Slab on grade (3500 psi), not including finish, 4" thick	C-14E	60.75	1.449	C.Y.	145	71	.42	216.42	274
4700	6" thick	"	92	.957	"	140	47	.28	187.28	230
4701	Thickened slab edge (3500 psi), for slab on grade poured									
4702	monolithically with slab; depth is in addition to slab thickness;									
4703	formed vertical outside edge, earthen bottom and inside slope									

03 30 53.40 Concrete In Place	Crew	Daily Output	Labor-Hours	Unit	Material	2018 Bare Costs Labor	Equipment	Total	Total Incl O&P
4705	8" deep x 8" wide bottom, unreinforced	C-14L	2190	.044	L.F.	3.97	2.06	.01	6.04	7.70
4710	8" x 8", reinforced	C-14C	1670	.067		6.20	3.22	.02	9.44	12.05
4715	12" deep x 12" wide bottom, unreinforced	C-14L	1800	.053		8.15	2.51	.01	10.67	13.05
4720	12" x 12", reinforced	C-14C	1310	.086		12.25	4.11	.02	16.38	20
4725	16" deep x 16" wide bottom, unreinforced	C-14L	1440	.067		13.80	3.13	.02	16.95	20.50
4730	16" x 16", reinforced	C-14C	1120	.100		18.70	4.81	.02	23.53	28.50
4735	20" deep x 20" wide bottom, unreinforced	C-14L	1150	.083		21	3.92	.02	24.94	29.50
4740	20" x 20", reinforced	C-14C	920	.122		27	5.85	.03	32.88	39.50
4745	24" deep x 24" wide bottom, unreinforced	C-14L	930	.103		29.50	4.85	.03	34.38	40.50
4750	24" x 24", reinforced	C-14C	740	.151	↓	38	7.30	.03	45.33	53.50
4751	Slab on grade (3500 psi), incl. troweled finish, not incl. forms									
4760	or reinforcing, over 10,000 S.F., 4" thick	C-14F	3425	.021	S.F.	1.61	.95	.01	2.57	3.30
4820	6" thick		3350	.021		2.36	.97	.01	3.34	4.16
4840	8" thick		3184	.023		3.23	1.02	.01	4.26	5.20
4900	12" thick		2734	.026		4.84	1.19	.01	6.04	7.25
4950	15" thick	↓	2505	.029	↓	6.10	1.30	.01	7.41	8.80
5000	Slab on grade (3000 psi), incl. broom finish, not incl. forms									
5001	or reinforcing, 4" thick	C-14G	2873	.019	S.F.	1.57	.87	.01	2.45	3.13
5010	6" thick		2590	.022		2.46	.96	.01	3.43	4.25
5020	8" thick	↓	2320	.024	↓	3.20	1.08	.01	4.29	5.25
5200	Lift slab in place above the foundation, incl. forms, reinforcing,									
5210	concrete (4000 psi) and columns, over 20,000 S.F./floor	C-14B	2113	.098	S.F.	7.45	4.96	.43	12.84	16.65
5250	10,000 S.F. to 20,000 S.F./floor		1650	.126		8.10	6.35	.55	15	19.75
5300	Under 10,000 S.F./floor	↓	1500	.139	↓	8.75	7	.60	16.35	21.50
5500	Lightweight, ready mix, including screed finish only,									
5510	not including forms or reinforcing									
5550	1:4 (2500 psi) for structural roof decks	C-14B	260	.800	C.Y.	116	40.50	3.49	159.99	196
5600	1:6 (3000 psi) for ground slab with radiant heat	C-14F	92	.783		127	35.50	.28	162.78	196
5650	1:3:2 (2000 psi) with sand aggregate, roof deck	C-14B	260	.800		115	40.50	3.49	158.99	195
5700	Ground slab (2000 psi)	C-14F	107	.673		115	30.50	.24	145.74	175
5900	Pile caps (3000 psi), incl. forms and reinf., sq. or rect., under 10 C.Y.	C-14C	54.14	2.069		185	99.50	.48	284.98	365
5950	Over 10 C.Y.		75	1.493		175	72	.34	247.34	310
6000	Triangular or hexagonal, under 10 C.Y.		53	2.113		142	102	.49	244.49	320
6050	Over 10 C.Y.	↓	85	1.318		158	63.50	.30	221.80	275
6200	Retaining walls (3000 psi), gravity, 4' high see Section 32 32	C-14D	66.20	3.021		160	152	13.70	325.70	435
6250	10' high		125	1.600		154	80.50	7.25	241.75	305
6300	Cantilever, level backfill loading, 8' high		70	2.857		171	143	12.95	326.95	435
6350	16' high	↓	91	2.198	↓	165	110	10	285	370
6800	Stairs (3500 psi), not including safety treads, free standing, 3'-6" wide	C-14H	83	.578	LF Nose	6.05	28.50	.31	34.86	53
6850	Cast on ground		125	.384	"	5.15	18.95	.20	24.30	36.50
7000	Stair landings, free standing		200	.240	S.F.	4.83	11.85	.13	16.81	24.50
7050	Cast on ground	↓	475	.101	"	3.93	4.99	.05	8.97	12.45
9000	Minimum labor/equipment charge	2 Carp	1	16	Job		810		810	1,325

03 31 Structural Concrete

03 31 13 – Heavyweight Structural Concrete

03 31 13.25 Concrete, Hand Mix

		Crew	Daily Output	Labor-Hours	Unit	Material	2018 Bare Costs Labor	Equipment	Total	Total Incl O&P
0010	**CONCRETE, HAND MIX** for small quantities or remote areas									
0050	Includes bulk local aggregate, bulk sand, bagged Portland									
0060	cement (Type I) and water, using gas powered cement mixer									
0125	2500 psi	C-30	135	.059	C.F.	3.80	2.36	1.19	7.35	9.30
0130	3000 psi		135	.059		4.11	2.36	1.19	7.66	9.65
0135	3500 psi		135	.059		4.29	2.36	1.19	7.84	9.85
0140	4000 psi		135	.059		4.50	2.36	1.19	8.05	10.10
0145	4500 psi		135	.059		4.75	2.36	1.19	8.30	10.35
0150	5000 psi	↓	135	.059	↓	5.05	2.36	1.19	8.60	10.75
0300	Using pre-bagged dry mix and wheelbarrow (80-lb. bag = 0.6 C.F.)									
0340	4000 psi	1 Clab	48	.167	C.F.	6.85	6.65		13.50	18.35

03 31 13.30 Concrete, Volumetric Site-Mixed

		Crew	Daily Output	Labor-Hours	Unit	Material	2018 Bare Costs Labor	Equipment	Total	Total Incl O&P
0010	**CONCRETE, VOLUMETRIC SITE-MIXED**									
0015	Mixed on-site in volumetric truck									
0020	Includes local aggregate, sand, Portland cement (Type I) and water									
0025	Excludes all additives and treatments									
0100	3000 psi, 1 C.Y. mixed and discharged				C.Y.	214			214	235
0110	2 C.Y.					158			158	174
0120	3 C.Y.					138			138	152
0130	4 C.Y.					125			125	137
0140	5 C.Y.				↓	112			112	123
0200	For truck holding/waiting time past first 2 on-site hours, add				Hr.	93.50			93.50	103
0210	For trip charge beyond first 20 miles, each way, add				Mile	3.64			3.64	4
0220	For each additional increase of 500 psi, add				Ea.	4.17			4.17	4.59

03 31 13.35 Heavyweight Concrete, Ready Mix

		Crew	Daily Output	Labor-Hours	Unit	Material	2018 Bare Costs Labor	Equipment	Total	Total Incl O&P
0010	**HEAVYWEIGHT CONCRETE, READY MIX**, delivered									
0012	Includes local aggregate, sand, Portland cement (Type I) and water									
0015	Excludes all additives and treatments R033105-20									
0020	2000 psi				C.Y.	115			115	127
0100	2500 psi					119			119	130
0150	3000 psi					121			121	133
0200	3500 psi					124			124	137
0300	4000 psi					128			128	140
0350	4500 psi					131			131	144
0400	5000 psi					134			134	148
0411	6000 psi					138			138	152
0412	8000 psi					145			145	160
0413	10,000 psi					153			153	168
0414	12,000 psi					160			160	176
1000	For high early strength (Portland cement Type III), add					10%				
1300	For winter concrete (hot water), add					5.35			5.35	5.90
1410	For mid-range water reducer, add					3.59			3.59	3.95
1420	For high-range water reducer/superplasticizer, add					6.30			6.30	6.90
1430	For retarder, add					3.23			3.23	3.55
1440	For non-Chloride accelerator, add					6.40			6.40	7.05
1450	For Chloride accelerator, per 1%, add					3.86			3.86	4.25
1460	For fiber reinforcing, synthetic (1 lb./C.Y.), add					8			8	8.80
1500	For Saturday delivery, add					8.50			8.50	9.35
1510	For truck holding/waiting time past 1st hour per load, add				Hr.	105			105	116
1520	For short load (less than 4 C.Y.), add per load				Ea.	86			86	94.50
2000	For all lightweight aggregate, add				C.Y.	45%				

03 31 Structural Concrete

03 31 13 – Heavyweight Structural Concrete

03 31 13.70 Placing Concrete

		Crew	Daily Output	Labor-Hours	Unit	Material	2018 Bare Costs Labor	2018 Bare Costs Equipment	Total	Total Incl O&P
0010	**PLACING CONCRETE** R033105-70									
0020	Includes labor and equipment to place, level (strike off) and consolidate									
0050	Beams, elevated, small beams, pumped	C-20	60	1.067	C.Y.		45.50	15.55	61.05	90.50
0100	With crane and bucket	C-7	45	1.600			69.50	24	93.50	138
0200	Large beams, pumped	C-20	90	.711			30.50	10.35	40.85	60.50
0250	With crane and bucket	C-7	65	1.108			48	16.45	64.45	95.50
0400	Columns, square or round, 12" thick, pumped	C-20	60	1.067			45.50	15.55	61.05	90.50
0450	With crane and bucket	C-7	40	1.800			78	27	105	156
0600	18" thick, pumped	C-20	90	.711			30.50	10.35	40.85	60.50
0650	With crane and bucket	C-7	55	1.309			57	19.45	76.45	113
0800	24" thick, pumped	C-20	92	.696			30	10.15	40.15	59
0850	With crane and bucket	C-7	70	1.029			44.50	15.30	59.80	89
1000	36" thick, pumped	C-20	140	.457			19.55	6.65	26.20	39
1050	With crane and bucket	C-7	100	.720			31.50	10.70	42.20	62.50
1400	Elevated slabs, less than 6" thick, pumped	C-20	140	.457			19.55	6.65	26.20	39
1450	With crane and bucket	C-7	95	.758			33	11.25	44.25	65.50
1500	6" to 10" thick, pumped	C-20	160	.400			17.10	5.85	22.95	34
1550	With crane and bucket	C-7	110	.655			28.50	9.75	38.25	56
1600	Slabs over 10" thick, pumped	C-20	180	.356			15.20	5.20	20.40	30
1650	With crane and bucket	C-7	130	.554			24	8.25	32.25	47.50
1900	Footings, continuous, shallow, direct chute	C-6	120	.400			16.60	.43	17.03	27.50
1950	Pumped	C-20	150	.427			18.25	6.20	24.45	36.50
2000	With crane and bucket	C-7	90	.800			35	11.90	46.90	69
2100	Footings, continuous, deep, direct chute	C-6	140	.343			14.20	.37	14.57	23.50
2150	Pumped	C-20	160	.400			17.10	5.85	22.95	34
2200	With crane and bucket	C-7	110	.655			28.50	9.75	38.25	56
2400	Footings, spread, under 1 C.Y., direct chute	C-6	55	.873			36	.93	36.93	59.50
2450	Pumped	C-20	65	.985			42	14.35	56.35	84
2500	With crane and bucket	C-7	45	1.600			69.50	24	93.50	138
2600	Over 5 C.Y., direct chute	C-6	120	.400			16.60	.43	17.03	27.50
2650	Pumped	C-20	150	.427			18.25	6.20	24.45	36.50
2700	With crane and bucket	C-7	100	.720			31.50	10.70	42.20	62.50
2900	Foundation mats, over 20 C.Y., direct chute	C-6	350	.137			5.70	.15	5.85	9.35
2950	Pumped	C-20	400	.160			6.85	2.33	9.18	13.60
3000	With crane and bucket	C-7	300	.240			10.45	3.57	14.02	20.50
3200	Grade beams, direct chute	C-6	150	.320			13.25	.34	13.59	22
3250	Pumped	C-20	180	.356			15.20	5.20	20.40	30
3300	With crane and bucket	C-7	120	.600			26	8.90	34.90	52
3500	High rise, for more than 5 stories, pumped, add per story	C-20	2100	.030			1.30	.44	1.74	2.59
3510	With crane and bucket, add per story	C-7	2100	.034			1.49	.51	2	2.95
3700	Pile caps, under 5 C.Y., direct chute	C-6	90	.533			22	.57	22.57	36
3750	Pumped	C-20	110	.582			25	8.50	33.50	49.50
3800	With crane and bucket	C-7	80	.900			39	13.35	52.35	77.50
3850	Pile cap, 5 C.Y. to 10 C.Y., direct chute	C-6	175	.274			11.35	.29	11.64	18.65
3900	Pumped	C-20	200	.320			13.70	4.67	18.37	27
3950	With crane and bucket	C-7	150	.480			21	7.15	28.15	41.50
4000	Over 10 C.Y., direct chute	C-6	215	.223			9.25	.24	9.49	15.20
4050	Pumped	C-20	240	.267			11.40	3.89	15.29	22.50
4100	With crane and bucket	C-7	185	.389			16.90	5.80	22.70	33.50
4300	Slab on grade, up to 6" thick, direct chute	C-6	110	.436			18.10	.47	18.57	29.50
4350	Pumped	C-20	130	.492			21	7.20	28.20	42
4400	With crane and bucket	C-7	110	.655			28.50	9.75	38.25	56

03 31 Structural Concrete

03 31 13 – Heavyweight Structural Concrete

03 31 13.70 Placing Concrete

		Crew	Daily Output	Labor-Hours	Unit	Material	2018 Bare Costs Labor	Equipment	Total	Total Incl O&P
4600	Over 6" thick, direct chute	C-6	165	.291	C.Y.		12.05	.31	12.36	19.85
4650	Pumped	C-20	185	.346			14.80	5.05	19.85	29.50
4700	With crane and bucket	C-7	145	.497			21.50	7.40	28.90	42.50
4900	Walls, 8" thick, direct chute	C-6	90	.533			22	.57	22.57	36
4950	Pumped	C-20	100	.640			27.50	9.35	36.85	54.50
5000	With crane and bucket	C-7	80	.900			39	13.35	52.35	77.50
5050	12" thick, direct chute	C-6	100	.480			19.90	.51	20.41	32.50
5100	Pumped	C-20	110	.582			25	8.50	33.50	49.50
5200	With crane and bucket	C-7	90	.800			35	11.90	46.90	69
5300	15" thick, direct chute	C-6	105	.457			18.95	.49	19.44	31
5350	Pumped	C-20	120	.533			23	7.80	30.80	45.50
5400	With crane and bucket	C-7	95	.758			33	11.25	44.25	65.50
5600	Wheeled concrete dumping, add to placing costs above									
5610	Walking cart, 50' haul, add	C-18	32	.281	C.Y.		11.25	1.81	13.06	20.50
5620	150' haul, add		24	.375			15.05	2.41	17.46	27
5700	250' haul, add		18	.500			20	3.21	23.21	36
5800	Riding cart, 50' haul, add	C-19	80	.113			4.51	1.22	5.73	8.65
5810	150' haul, add		60	.150			6	1.62	7.62	11.55
5900	250' haul, add		45	.200			8	2.16	10.16	15.40
6000	Concrete in-fill for pan-type metal stairs and landings. Manual placement									
6010	includes up to 50' horizontal haul from point of concrete discharge.									
6100	Stair pan treads, 2" deep									
6110	Flights in 1st floor level up/down from discharge point	C-8A	3200	.015	S.F.		.64		.64	1.03
6120	2nd floor level		2500	.019			.82		.82	1.32
6130	3rd floor level		2000	.024			1.03		1.03	1.65
6140	4th floor level		1800	.027			1.14		1.14	1.83
6200	Intermediate stair landings, pan-type 4" deep									
6210	Flights in 1st floor level up/down from discharge point	C-8A	2000	.024	S.F.		1.03		1.03	1.65
6220	2nd floor level		1500	.032			1.37		1.37	2.20
6230	3rd floor level		1200	.040			1.71		1.71	2.75
6240	4th floor level		1000	.048			2.05		2.05	3.30
9000	Minimum labor/equipment charge	C-6	2	24	Job		995	25.50	1,020.50	1,625

03 35 Concrete Finishing

03 35 13 – High-Tolerance Concrete Floor Finishing

03 35 13.30 Finishing Floors, High Tolerance

		Crew	Daily Output	Labor-Hours	Unit	Material	2018 Bare Costs Labor	Equipment	Total	Total Incl O&P
0010	**FINISHING FLOORS, HIGH TOLERANCE**									
0012	Finishing of fresh concrete flatwork requires that concrete									
0013	first be placed, struck off & consolidated									
0015	Basic finishing for various unspecified flatwork									
0100	Bull float only	C-10	4000	.006	S.F.		.27		.27	.43
0125	Bull float & manual float		2000	.012			.54		.54	.86
0150	Bull float, manual float & broom finish, w/edging & joints		1850	.013			.58		.58	.93
0200	Bull float, manual float & manual steel trowel		1265	.019			.85		.85	1.36
0210	For specified Random Access Floors in ACI Classes 1, 2, 3 and 4 to achieve									
0215	Composite Overall Floor Flatness and Levelness values up to FF35/FL25									
0250	Bull float, machine float & machine trowel (walk-behind)	C-10C	1715	.014	S.F.		.63	.02	.65	1.03
0300	Power screed, bull float, machine float & trowel (walk-behind)	C-10D	2400	.010			.45	.05	.50	.77
0350	Power screed, bull float, machine float & trowel (ride-on)	C-10E	4000	.006			.27	.06	.33	.50
0352	For specified Random Access Floors in ACI Classes 5, 6, 7 and 8 to achieve									
0354	Composite Overall Floor Flatness and Levelness values up to FF50/FL50									

For customer support on your Commercial Renovation Costs with RSMeans data, call 800.448.8182.

65

03 35 Concrete Finishing

03 35 13 – High-Tolerance Concrete Floor Finishing

03 35 13.30 Finishing Floors, High Tolerance

		Crew	Daily Output	Labor-Hours	Unit	Material	2018 Bare Costs Labor	Equipment	Total	Total Incl O&P
0356	Add for two-dimensional restraightening after power float	C-10	6000	.004	S.F.		.18		.18	.29
0358	For specified Random or Defined Access Floors in ACI Class 9 to achieve									
0360	Composite Overall Floor Flatness and Levelness values up to FF100/FL100									
0362	Add for two-dimensional restraightening after bull float & power float	C-10	3000	.008	S.F.		.36		.36	.57
0364	For specified Superflat Defined Access Floors in ACI Class 9 to achieve									
0366	Minimum Floor Flatness and Levelness values of FF100/FL100									
0368	Add for 2-dim'l restraightening after bull float, power float, power trowel	C-10	2000	.012	S.F.		.54		.54	.86
9100	Minimum labor/equipment charge	"	2	12	Job		540		540	860

03 35 29 – Tooled Concrete Finishing

03 35 29.60 Finishing Walls

		Crew	Daily Output	Labor-Hours	Unit	Material	2018 Bare Costs Labor	Equipment	Total	Total Incl O&P
0010	**FINISHING WALLS**									
0020	Break ties and patch voids	1 Cefi	540	.015	S.F.	.04	.70		.74	1.16
0050	Burlap rub with grout	"	450	.018		.04	.85		.89	1.39
0300	Bush hammer, green concrete	B-39	1000	.048			2.02	.23	2.25	3.52
0350	Cured concrete	"	650	.074			3.11	.36	3.47	5.40
0500	Acid etch	1 Cefi	575	.014		.13	.66		.79	1.20
0700	Sandblast, light penetration	E-11	1100	.029		.55	1.30	.18	2.03	2.98
0750	Heavy penetration	"	375	.085		1.09	3.80	.54	5.43	8.20
0850	Grind form fins flush	1 Clab	700	.011	L.F.		.46		.46	.74
9000	Minimum labor/equipment charge	C-10	2	12	Job		540		540	860

03 35 33 – Stamped Concrete Finishing

03 35 33.50 Slab Texture Stamping

		Crew	Daily Output	Labor-Hours	Unit	Material	2018 Bare Costs Labor	Equipment	Total	Total Incl O&P
0010	**SLAB TEXTURE STAMPING**									
0050	Stamping requires that concrete first be placed, struck off, consolidated,									
0060	bull floated and free of bleed water. Decorative stamping tasks include:									
0100	Step 1 - first application of dry shake colored hardener	1 Cefi	6400	.001	S.F.	.43	.06		.49	.56
0110	Step 2 - bull float		6400	.001			.06		.06	.09
0130	Step 3 - second application of dry shake colored hardener		6400	.001		.21	.06		.27	.32
0140	Step 4 - bull float, manual float & steel trowel	3 Cefi	1280	.019			.89		.89	1.41
0150	Step 5 - application of dry shake colored release agent	1 Cefi	6400	.001		.10	.06		.16	.20
0160	Step 6 - place, tamp & remove mats	3 Cefi	2400	.010		.81	.48		1.29	1.64
0170	Step 7 - touch up edges, mat joints & simulated grout lines	1 Cefi	1280	.006			.30		.30	.47
0300	Alternate stamping estimating method includes all tasks above	4 Cefi	800	.040		1.55	1.90		3.45	4.71
0400	Step 8 - pressure wash @ 3000 psi after 24 hours	1 Cefi	1600	.005			.24		.24	.38
0500	Step 9 - roll 2 coats cure/seal compound when dry	"	800	.010		.63	.48		1.11	1.44

03 35 43 – Polished Concrete Finishing

03 35 43.10 Polished Concrete Floors

		Crew	Daily Output	Labor-Hours	Unit	Material	2018 Bare Costs Labor	Equipment	Total	Total Incl O&P
0010	**POLISHED CONCRETE FLOORS**									
0015	Processing of cured concrete to include grinding, honing,									
0020	and polishing of interior floors with 22" segmented diamond									
0025	planetary floor grinder (2 passes in different directions per grit)									
0100	Removal of pre-existing coatings, dry, with carbide discs using									
0105	dry vacuum pick-up system, final hand sweeping									
0110	Glue, adhesive or tar	J-4	1.60	15	M.S.F.	21	675	152	848	1,275
0120	Paint, epoxy, 1 coat		3.60	6.667		21	300	67.50	388.50	580
0130	2 coats		1.80	13.333		21	600	135	756	1,125
0200	Grinding and edging, wet, including wet vac pick-up and auto									
0205	scrubbing between grit changes									
0210	40-grit diamond/metal matrix	J-4A	1.60	20	M.S.F.	41.50	875	310	1,226.50	1,775
0220	80-grit diamond/metal matrix		2	16		41.50	700	249	990.50	1,450
0230	120-grit diamond/metal matrix		2.40	13.333		41.50	585	207	833.50	1,200

03 35 Concrete Finishing

03 35 43 – Polished Concrete Finishing

03 35 43.10 Polished Concrete Floors		Crew	Daily Output	Labor-Hours	Unit	Material	2018 Bare Costs Labor	Equipment	Total	Total Incl O&P
0240	200-grit diamond/metal matrix	J-4A	2.80	11.429	M.S.F.	41.50	500	178	719.50	1,050
0300	Spray on dye or stain (1 coat)	1 Cefi	16	.500		222	24		246	282
0400	Spray on densifier/hardener (2 coats)	"	8	1		365	47.50		412.50	475
0410	Auto scrubbing after 2nd coat, when dry	J-4B	16	.500			19.95	15.90	35.85	50
0500	Honing and edging, wet, including wet vac pick-up and auto									
0505	scrubbing between grit changes									
0510	100-grit diamond/resin matrix	J-4A	2.80	11.429	M.S.F.	41.50	500	178	719.50	1,050
0520	200-grit diamond/resin matrix	"	2.80	11.429	"	41.50	500	178	719.50	1,050
0530	Dry, including dry vacuum pick-up system, final hand sweeping									
0540	400-grit diamond/resin matrix	J-4A	2.80	11.429	M.S.F.	41.50	500	178	719.50	1,050
0600	Polishing and edging, dry, including dry vac pick-up and hand									
0605	sweeping between grit changes									
0610	800-grit diamond/resin matrix	J-4A	2.80	11.429	M.S.F.	41.50	500	178	719.50	1,050
0620	1500-grit diamond/resin matrix		2.80	11.429		41.50	500	178	719.50	1,050
0630	3000-grit diamond/resin matrix		2.80	11.429		41.50	500	178	719.50	1,050
0700	Auto scrubbing after final polishing step	J-4B	16	.500			19.95	15.90	35.85	50

03 37 Specialty Placed Concrete

03 37 13 – Shotcrete

03 37 13.30 Gunite (Dry-Mix)

		Crew	Daily Output	Labor-Hours	Unit	Material	2018 Bare Costs Labor	Equipment	Total	Total Incl O&P
0010	**GUNITE (DRY-MIX)**									
0020	Typical in place, 1" layers, no mesh included	C-16	2000	.028	S.F.	.36	1.24	.19	1.79	2.60
0300	Typical in place, including mesh, 2" thick, flat surfaces		1000	.056		1.43	2.48	.39	4.30	6
0350	Curved surfaces		500	.112		1.43	4.96	.78	7.17	10.35
0500	4" thick, flat surfaces		750	.075		2.16	3.31	.52	5.99	8.25
0550	Curved surfaces		350	.160		2.16	7.10	1.11	10.37	14.95
0900	Prepare old walls, no scaffolding, good condition	C-10	1000	.024			1.08		1.08	1.72
0950	Poor condition	"	275	.087			3.93		3.93	6.25
1100	For high finish requirement or close tolerance, add						50%			
1150	Very high						110%			
9000	Minimum labor/equipment charge	C-10	1	24	Job		1,075		1,075	1,725

03 37 13.60 Shotcrete (Wet-Mix)

		Crew	Daily Output	Labor-Hours	Unit	Material	2018 Bare Costs Labor	Equipment	Total	Total Incl O&P
0010	**SHOTCRETE (WET-MIX)**									
0020	Wet mix, placed @ up to 12 C.Y./hour, 3000 psi	C-8C	80	.600	C.Y.	118	26.50	6.25	150.75	179
0100	Up to 35 C.Y./hour	C-8E	240	.200	"	106	8.70	2.40	117.10	134
1010	Fiber reinforced, 1" thick	C-8C	1740	.028	S.F.	.87	1.21	.29	2.37	3.21
1020	2" thick		900	.053		1.74	2.33	.56	4.63	6.25
1030	3" thick		825	.058		2.60	2.55	.61	5.76	7.60
1040	4" thick		750	.064		3.47	2.80	.67	6.94	9.05

For customer support on your Commercial Renovation Costs with RSMeans data, call 800.448.8182.

67

03 39 Concrete Curing

03 39 13 – Water Concrete Curing

03 39 13.50 Water Curing

		Crew	Daily Output	Labor-Hours	Unit	Material	2018 Bare Costs Labor	Equipment	Total	Total Incl O&P
0010	**WATER CURING**									
0015	With burlap, 4 uses assumed, 7.5 oz.	2 Clab	55	.291	C.S.F.	14.75	11.60		26.35	35
0100	10 oz.	"	55	.291	"	26.50	11.60		38.10	48
0400	Curing blankets, 1" to 2" thick, buy				S.F.	.26			.26	.29
9000	Minimum labor/equipment charge	1 Clab	5	1.600	Job		64		64	104

03 39 23 – Membrane Concrete Curing

03 39 23.13 Chemical Compound Membrane Concrete Curing

		Crew	Daily Output	Labor-Hours	Unit	Material	2018 Bare Costs Labor	Equipment	Total	Total Incl O&P
0010	**CHEMICAL COMPOUND MEMBRANE CONCRETE CURING**									
0300	Sprayed membrane curing compound	2 Clab	95	.168	C.S.F.	12.10	6.70		18.80	24.50
0700	Curing compound, solvent based, 400 S.F./gal., 55 gallon lots				Gal.	21.50			21.50	23.50
0720	5 gallon lots					32			32	35
0800	Curing compound, water based, 250 S.F./gal., 55 gallon lots					21.50			21.50	23.50
0820	5 gallon lots					24			24	26.50

03 39 23.23 Sheet Membrane Concrete Curing

		Crew	Daily Output	Labor-Hours	Unit	Material	2018 Bare Costs Labor	Equipment	Total	Total Incl O&P
0010	**SHEET MEMBRANE CONCRETE CURING**									
0200	Curing blanket, burlap/poly, 2-ply	2 Clab	70	.229	C.S.F.	25	9.10		34.10	42.50

03 41 Precast Structural Concrete

03 41 13 – Precast Concrete Hollow Core Planks

03 41 13.50 Precast Slab Planks

		Crew	Daily Output	Labor-Hours	Unit	Material	2018 Bare Costs Labor	Equipment	Total	Total Incl O&P
0010	**PRECAST SLAB PLANKS**	R034105-30								
0020	Prestressed roof/floor members, grouted, solid, 4" thick	C-11	2400	.030	S.F.	8	1.63	.84	10.47	12.45
0050	6" thick		2800	.026		8.95	1.40	.72	11.07	13
0100	Hollow, 8" thick		3200	.023		10	1.22	.63	11.85	13.75
0150	10" thick		3600	.020		10.35	1.09	.56	12	13.85
0200	12" thick		4000	.018		10.75	.98	.50	12.23	14

03 41 23 – Precast Concrete Stairs

03 41 23.50 Precast Stairs

		Crew	Daily Output	Labor-Hours	Unit	Material	2018 Bare Costs Labor	Equipment	Total	Total Incl O&P
0010	**PRECAST STAIRS**									
0020	Precast concrete treads on steel stringers, 3' wide	C-12	75	.640	Riser	153	32	6.60	191.60	227
0300	Front entrance, 5' wide with 48" platform, 2 risers		16	3	Flight	625	150	31	806	965
0350	5 risers		12	4		995	201	41.50	1,237.50	1,475
0500	6' wide, 2 risers		15	3.200		695	160	33	888	1,050
0550	5 risers		11	4.364		1,100	219	45	1,364	1,600
1200	Basement entrance stairwell, 6 steps, incl. steel bulkhead door	B-51	22	2.182		1,875	89.50	8.55	1,973.05	2,200
1250	14 steps	"	11	4.364		3,125	179	17.15	3,321.15	3,725

03 41 33 – Precast Structural Pretensioned Concrete

03 41 33.10 Precast Beams

		Crew	Daily Output	Labor-Hours	Unit	Material	2018 Bare Costs Labor	Equipment	Total	Total Incl O&P
0010	**PRECAST BEAMS**	R034105-30								
0011	L-shaped, 20' span, 12" x 20"	C-11	32	2.250	Ea.	4,100	122	63	4,285	4,800
0060	18" x 36"		24	3		5,600	163	84	5,847	6,550
0100	24" x 44"		22	3.273		6,725	178	91.50	6,994.50	7,800
0150	30' span, 12" x 36"		24	3		7,575	163	84	7,822	8,700
0200	18" x 44"		20	3.600		9,225	196	101	9,522	10,600
0250	24" x 52"		16	4.500		11,100	245	126	11,471	12,700
0400	40' span, 12" x 52"		20	3.600		11,800	196	101	12,097	13,300
0450	18" x 52"		16	4.500		13,100	245	126	13,471	14,900
0500	24" x 52"		12	6		14,800	325	168	15,293	16,900
1200	Rectangular, 20' span, 12" x 20"		32	2.250		3,975	122	63	4,160	4,650

03 41 Precast Structural Concrete

03 41 33 – Precast Structural Pretensioned Concrete

03 41 33.10 Precast Beams

		Crew	Daily Output	Labor-Hours	Unit	Material	2018 Bare Costs Labor	Equipment	Total	Total Incl O&P
1250	18" x 36"	C-11	24	3	Ea.	4,925	163	84	5,172	5,800
1300	24" x 44"		22	3.273		5,875	178	91.50	6,144.50	6,875
1400	30' span, 12" x 36"		24	3		6,625	163	84	6,872	7,675
1450	18" x 44"		20	3.600		8,000	196	101	8,297	9,250
1500	24" x 52"		16	4.500		9,625	245	126	9,996	11,100
1600	40' span, 12" x 52"		20	3.600		9,850	196	101	10,147	11,200
1650	18" x 52"		16	4.500		11,200	245	126	11,571	12,800
1700	24" x 52"		12	6		12,800	325	168	13,293	14,800
2000	"T" shaped, 20' span, 12" x 20"		32	2.250		4,675	122	63	4,860	5,400
2050	18" x 36"		24	3		6,275	163	84	6,522	7,275
2100	24" x 44"		22	3.273		7,525	178	91.50	7,794.50	8,675
2200	30' span, 12" x 36"		24	3		8,625	163	84	8,872	9,850
2250	18" x 44"		20	3.600		10,500	196	101	10,797	11,900
2300	24" x 52"		16	4.500		12,500	245	126	12,871	14,300
2500	40' span, 12" x 52"		20	3.600		14,200	196	101	14,497	16,000
2550	18" x 52"		16	4.500		15,100	245	126	15,471	17,100
2600	24" x 52"		12	6		16,700	325	168	17,193	19,000

03 41 33.15 Precast Columns

			Crew	Daily Output	Labor-Hours	Unit	Material	2018 Bare Costs Labor	Equipment	Total	Total Incl O&P
0010	**PRECAST COLUMNS**	R034105-30									
0020	Rectangular to 12' high, 16" x 16"		C-11	120	.600	L.F.	234	32.50	16.80	283.30	330
0050	24" x 24"		"	96	.750	"	315	41	21	377	435

03 45 Precast Architectural Concrete

03 45 13 – Faced Architectural Precast Concrete

03 45 13.50 Precast Wall Panels

		Crew	Daily Output	Labor-Hours	Unit	Material	2018 Bare Costs Labor	Equipment	Total	Total Incl O&P
0010	**PRECAST WALL PANELS**									
2200	Fiberglass reinforced cement with urethane core									
2210	R20, 8' x 8', 5" plain finish	E-2	750	.075	S.F.	28	4.05	2.25	34.30	40
2220	Exposed aggregate or brick finish	"	600	.093	"	42.50	5.05	2.81	50.36	58

03 48 Precast Concrete Specialties

03 48 43 – Precast Concrete Trim

03 48 43.40 Precast Lintels

		Crew	Daily Output	Labor-Hours	Unit	Material	2018 Bare Costs Labor	Equipment	Total	Total Incl O&P
0010	**PRECAST LINTELS**, smooth gray, prestressed, stock units only									
0800	4" wide, 8" high, x 4' long	D-10	28	1.143	Ea.	32.50	56.50	13	102	142
0850	8' long		24	1.333		82	66	15.15	163.15	214
1000	6" wide, 8" high, x 4' long		26	1.231		48	61	14	123	167
1050	10' long		22	1.455		120	72	16.50	208.50	266
1200	8" wide, 8" high, x 4' long		24	1.333		56.50	66	15.15	137.65	186
1250	12' long		20	1.600		183	79	18.20	280.20	350
1275	For custom sizes, types, colors, or finishes of precast lintels, add					150%				

03 48 43.90 Precast Window Sills

		Crew	Daily Output	Labor-Hours	Unit	Material	2018 Bare Costs Labor	Equipment	Total	Total Incl O&P
0010	**PRECAST WINDOW SILLS**									
0600	Precast concrete, 4" tapers to 3", 9" wide	D-1	70	.229	L.F.	19	10.25		29.25	37.50
0650	11" wide	"	60	.267	"	31	11.95		42.95	53.50

03 51 Cast Roof Decks

03 51 13 – Cementitious Wood Fiber Decks

03 51 13.50 Cementitious/Wood Fiber Planks		Crew	Daily Output	Labor-Hours	Unit	Material	2018 Bare Costs Labor	Equipment	Total	Total Incl O&P
0010	**CEMENTITIOUS/WOOD FIBER PLANKS**									
0050	Plank, beveled edge, 1" thick	2 Carp	1000	.016	S.F.	3.41	.81		4.22	5.05
0100	1-1/2" thick		975	.016		4.42	.83		5.25	6.20
0150	T&G, 2" thick		950	.017		3.75	.85		4.60	5.50
0200	2-1/2" thick	↓	925	.017		4.09	.88		4.97	5.90
1000	Bulb tee, sub-purlin and grout, 6' span, add	E-1	5000	.005		2.16	.26	.02	2.44	2.83
1100	8' span	"	4200	.006	↓	2.16	.31	.02	2.49	2.93

03 52 Lightweight Concrete Roof Insulation

03 52 16 – Lightweight Insulating Concrete

03 52 16.13 Lightweight Cellular Insulating Concrete

			Crew	Daily Output	Labor-Hours	Unit	Material	Labor	Equipment	Total	Total Incl O&P
0010	**LIGHTWEIGHT CELLULAR INSULATING CONCRETE**	R035216-10									
0020	Portland cement and foaming agent	**G**	C-8	50	1.120	C.Y.	128	49.50	17.65	195.15	239

03 52 16.16 Lightweight Aggregate Insulating Concrete

			Crew	Daily Output	Labor-Hours	Unit	Material	Labor	Equipment	Total	Total Incl O&P
0010	**LIGHTWEIGHT AGGREGATE INSULATING CONCRETE**	R035216-10									
0100	Poured vermiculite or perlite, field mix,										
0110	1:6 field mix	**G**	C-8	50	1.120	C.Y.	255	49.50	17.65	322.15	380
0200	Ready mix, 1:6 mix, roof fill, 2" thick	**G**		10000	.006	S.F.	1.42	.25	.09	1.76	2.06
0250	3" thick	**G**	↓	7700	.007	"	2.13	.32	.11	2.56	2.99

03 53 Concrete Topping

03 53 16 – Iron-Aggregate Concrete Topping

03 53 16.50 Floor Topping

		Crew	Daily Output	Labor-Hours	Unit	Material	Labor	Equipment	Total	Total Incl O&P
0010	**FLOOR TOPPING**									
0400	Integral topping/finish, on fresh concrete, using 1:1:2 mix, 3/16" thick	C-10B	1000	.040	S.F.	.12	1.72	.24	2.08	3.17
0450	1/2" thick		950	.042		.33	1.81	.26	2.40	3.54
0500	3/4" thick		850	.047		.49	2.02	.29	2.80	4.09
0600	1" thick		750	.053		.66	2.29	.32	3.27	4.75
0800	Granolithic topping, on fresh or cured concrete, 1:1:1-1/2 mix, 1/2" thick		590	.068		.36	2.91	.41	3.68	5.50
0820	3/4" thick		580	.069		.55	2.96	.42	3.93	5.80
0850	1" thick		575	.070		.73	2.99	.42	4.14	6.05
0950	2" thick	↓	500	.080	↓	1.46	3.43	.49	5.38	7.65

03 54 Cast Underlayment

03 54 13 – Gypsum Cement Underlayment

03 54 13.50 Poured Gypsum Underlayment

		Crew	Daily Output	Labor-Hours	Unit	Material	Labor	Equipment	Total	Total Incl O&P
0010	**POURED GYPSUM UNDERLAYMENT**									
0400	Underlayment, gypsum based, self-leveling 2500 psi, pumped, 1/2" thick	C-8	24000	.002	S.F.	.43	.10	.04	.57	.69
0500	3/4" thick		20000	.003		.65	.12	.04	.81	.97
0600	1" thick	↓	16000	.004		.87	.16	.06	1.09	1.27
1400	Hand placed, 1/2" thick	C-18	450	.020		.43	.80	.13	1.36	1.92
1500	3/4" thick	"	300	.030	↓	.65	1.20	.19	2.04	2.88

03 54 16 – Hydraulic Cement Underlayment

03 54 16.50 Cement Underlayment

		Crew	Daily Output	Labor-Hours	Unit	Material	Labor	Equipment	Total	Total Incl O&P
0010	**CEMENT UNDERLAYMENT**									
2510	Underlayment, P.C. based, self-leveling, 4100 psi, pumped, 1/4" thick	C-8	20000	.003	S.F.	1.68	.12	.04	1.84	2.10
2520	1/2" thick	↓	19000	.003		3.36	.13	.05	3.54	3.95

03 54 Cast Underlayment

03 54 16 – Hydraulic Cement Underlayment

03 54 16.50 Cement Underlayment		Crew	Daily Output	Labor-Hours	Unit	Material	2018 Bare Costs Labor	Equipment	Total	Total Incl O&P
2530	3/4" thick	C-8	18000	.003	S.F.	5.05	.14	.05	5.24	5.80
2540	1" thick		17000	.003		6.70	.15	.05	6.90	7.70
2550	1-1/2" thick	↓	15000	.004		10.10	.17	.06	10.33	11.40
2560	Hand placed, 1/2" thick	C-18	450	.020		3.36	.80	.13	4.29	5.15
2610	Topping, P.C. based, self-leveling, 6100 psi, pumped, 1/4" thick	C-8	20000	.003		2.47	.12	.04	2.63	2.96
2620	1/2" thick		19000	.003		4.93	.13	.05	5.11	5.70
2630	3/4" thick		18000	.003		7.40	.14	.05	7.59	8.40
2660	1" thick		17000	.003		9.85	.15	.05	10.05	11.15
2670	1-1/2" thick	↓	15000	.004		14.80	.17	.06	15.03	16.60
2680	Hand placed, 1/2" thick	C-18	450	.020	↓	4.93	.80	.13	5.86	6.90

03 63 Epoxy Grouting

03 63 05 – Grouting of Dowels and Fasteners

03 63 05.10 Epoxy Only

		Crew	Daily Output	Labor-Hours	Unit	Material	2018 Bare Costs Labor	Equipment	Total	Total Incl O&P
0010	**EPOXY ONLY**									
1500	Chemical anchoring, epoxy cartridge, excludes layout, drilling, fastener									
1530	For fastener 3/4" diam. x 6" embedment	2 Skwk	72	.222	Ea.	5.20	11.65		16.85	24.50
1535	1" diam. x 8" embedment		66	.242		7.80	12.70		20.50	29
1540	1-1/4" diam. x 10" embedment		60	.267		15.65	13.95		29.60	39.50
1545	1-3/4" diam. x 12" embedment		54	.296		26	15.50		41.50	53.50
1550	14" embedment		48	.333		31.50	17.45		48.95	62.50
1555	2" diam. x 12" embedment		42	.381		41.50	19.95		61.45	78
1560	18" embedment	↓	32	.500	↓	52	26		78	99.50

03 81 Concrete Cutting

03 81 13 – Flat Concrete Sawing

03 81 13.50 Concrete Floor/Slab Cutting

		Crew	Daily Output	Labor-Hours	Unit	Material	2018 Bare Costs Labor	Equipment	Total	Total Incl O&P
0010	**CONCRETE FLOOR/SLAB CUTTING**									
0050	Includes blade cost, layout and set-up time									
0300	Saw cut concrete slabs, plain, up to 3" deep	B-89	1060	.015	L.F.	.12	.72	.40	1.24	1.73
0320	Each additional inch of depth		3180	.005		.04	.24	.13	.41	.58
0400	Mesh reinforced, up to 3" deep		980	.016		.14	.78	.43	1.35	1.86
0420	Each additional inch of depth		2940	.005		.05	.26	.14	.45	.62
0500	Rod reinforced, up to 3" deep		800	.020		.17	.96	.53	1.66	2.29
0520	Each additional inch of depth		2400	.007	↓	.06	.32	.18	.56	.76
0590	Minimum labor/equipment charge	↓	2	8	Job		385	210	595	840

03 81 13.75 Concrete Saw Blades

		Crew	Daily Output	Labor-Hours	Unit	Material	2018 Bare Costs Labor	Equipment	Total	Total Incl O&P
0010	**CONCRETE SAW BLADES**									
3000	Blades for saw cutting, included in cutting line items									
3020	Diamond, 12" diameter				Ea.	220			220	242
3040	18" diameter					415			415	460
3080	24" diameter					695			695	765
3120	30" diameter					980			980	1,075
3160	36" diameter					1,300			1,300	1,425
3200	42" diameter				↓	2,550			2,550	2,800

03 81 Concrete Cutting

03 81 16 – Track Mounted Concrete Wall Sawing

03 81 16.50 Concrete Wall Cutting	Crew	Daily Output	Labor-Hours	Unit	Material	2018 Bare Costs Labor	Equipment	Total	Total Incl O&P
0010 **CONCRETE WALL CUTTING**									
0750 Includes blade cost, layout and set-up time									
0800 Concrete walls, hydraulic saw, plain, per inch of depth	B-89B	250	.064	L.F.	.04	3.07	2.69	5.80	7.90
0820 Rod reinforcing, per inch of depth		150	.107	"	.06	5.10	4.48	9.64	13.15
0890 Minimum labor/equipment charge		2	8	Job		385	335	720	980

03 82 Concrete Boring

03 82 13 – Concrete Core Drilling

03 82 13.10 Core Drilling

	Crew	Daily Output	Labor-Hours	Unit	Material	2018 Bare Costs Labor	Equipment	Total	Total Incl O&P
0010 **CORE DRILLING**									
0015 Includes bit cost, layout and set-up time									
0020 Reinforced concrete slab, up to 6" thick									
0100 1" diameter core	B-89A	17	.941	Ea.	.18	43.50	6.60	50.28	77.50
0150 For each additional inch of slab thickness in same hole, add		1440	.011		.03	.51	.08	.62	.95
0200 2" diameter core		16.50	.970		.29	44.50	6.80	51.59	80.50
0250 For each additional inch of slab thickness in same hole, add		1080	.015		.05	.68	.10	.83	1.26
0300 3" diameter core		16	1		.39	46	7	53.39	82.50
0350 For each additional inch of slab thickness in same hole, add		720	.022		.07	1.02	.16	1.25	1.90
0500 4" diameter core		15	1.067		.50	49	7.50	57	88.50
0550 For each additional inch of slab thickness in same hole, add		480	.033		.08	1.54	.23	1.85	2.83
0700 6" diameter core		14	1.143		.77	52.50	8	61.27	94.50
0750 For each additional inch of slab thickness in same hole, add		360	.044		.13	2.05	.31	2.49	3.79
0900 8" diameter core		13	1.231		1.18	56.50	8.65	66.33	102
0950 For each additional inch of slab thickness in same hole, add		288	.056		.20	2.56	.39	3.15	4.79
1100 10" diameter core		12	1.333		1.49	61.50	9.35	72.34	111
1150 For each additional inch of slab thickness in same hole, add		240	.067		.25	3.07	.47	3.79	5.75
1300 12" diameter core		11	1.455		1.83	67	10.20	79.03	121
1350 For each additional inch of slab thickness in same hole, add		206	.078		.30	3.58	.54	4.42	6.75
1500 14" diameter core		10	1.600		2.12	74	11.20	87.32	134
1550 For each additional inch of slab thickness in same hole, add		180	.089		.35	4.10	.62	5.07	7.75
1700 18" diameter core		9	1.778		2.92	82	12.45	97.37	150
1750 For each additional inch of slab thickness in same hole, add		144	.111		.49	5.10	.78	6.37	9.70
1754 24" diameter core		8	2		4.14	92	14	110.14	169
1756 For each additional inch of slab thickness in same hole, add		120	.133		.69	6.15	.93	7.77	11.75
1760 For horizontal holes, add to above						20%	20%		
1770 Prestressed hollow core plank, 8" thick									
1780 1" diameter core	B-89A	17.50	.914	Ea.	.24	42	6.40	48.64	75.50
1790 For each additional inch of plank thickness in same hole, add		3840	.004		.03	.19	.03	.25	.37
1794 2" diameter core		17.25	.928		.39	43	6.50	49.89	76.50
1796 For each additional inch of plank thickness in same hole, add		2880	.006		.05	.26	.04	.35	.50
1800 3" diameter core		17	.941		.52	43.50	6.60	50.62	78
1810 For each additional inch of plank thickness in same hole, add		1920	.008		.07	.38	.06	.51	.75
1820 4" diameter core		16.50	.970		.67	44.50	6.80	51.97	80.50
1830 For each additional inch of plank thickness in same hole, add		1280	.013		.08	.58	.09	.75	1.12
1840 6" diameter core		15.50	1.032		1.03	47.50	7.25	55.78	86
1850 For each additional inch of plank thickness in same hole, add		960	.017		.13	.77	.12	1.02	1.51
1860 8" diameter core		15	1.067		1.57	49	7.50	58.07	89.50
1870 For each additional inch of plank thickness in same hole, add		768	.021		.20	.96	.15	1.31	1.93
1880 10" diameter core		14	1.143		1.99	52.50	8	62.49	96
1890 For each additional inch of plank thickness in same hole, add		640	.025		.25	1.15	.18	1.58	2.32

03 82 Concrete Boring

03 82 13 – Concrete Core Drilling

03 82 13.10 Core Drilling

		Crew	Daily Output	Labor-Hours	Unit	Material	2018 Bare Costs Labor	Equipment	Total	Total Incl O&P
1900	12" diameter core	B-89A	13.50	1.185	Ea.	2.44	54.50	8.30	65.24	100
1910	For each additional inch of plank thickness in same hole, add		548	.029		.30	1.35	.20	1.85	2.75
1999	Drilling, core, minimum labor/equipment charge		2	8	Job		370	56	426	655
3000	Bits for core drilling, included in drilling line items									
3010	Diamond, premium, 1" diameter				Ea.	73			73	80.50
3020	2" diameter					116			116	127
3030	3" diameter					156			156	172
3040	4" diameter					201			201	222
3060	6" diameter					310			310	340
3080	8" diameter					470			470	520
3110	10" diameter					600			600	660
3120	12" diameter					730			730	805
3140	14" diameter					850			850	935
3180	18" diameter					1,175			1,175	1,275
3240	24" diameter					1,650			1,650	1,825

03 82 16 – Concrete Drilling

03 82 16.10 Concrete Impact Drilling

		Crew	Daily Output	Labor-Hours	Unit	Material	2018 Bare Costs Labor	Equipment	Total	Total Incl O&P
0010	**CONCRETE IMPACT DRILLING**									
0020	Includes bit cost, layout and set-up time, no anchors									
0050	Up to 4" deep in concrete/brick floors/walls									
0100	Holes, 1/4" diameter	1 Carp	75	.107	Ea.	.07	5.40		5.47	8.90
0150	For each additional inch of depth in same hole, add		430	.019		.02	.94		.96	1.55
0200	3/8" diameter		63	.127		.06	6.45		6.51	10.50
0250	For each additional inch of depth in same hole, add		340	.024		.01	1.19		1.20	1.96
0300	1/2" diameter		50	.160		.06	8.10		8.16	13.20
0350	For each additional inch of depth in same hole, add		250	.032		.01	1.62		1.63	2.65
0400	5/8" diameter		48	.167		.10	8.45		8.55	13.80
0450	For each additional inch of depth in same hole, add		240	.033		.03	1.69		1.72	2.77
0500	3/4" diameter		45	.178		.13	9		9.13	14.80
0550	For each additional inch of depth in same hole, add		220	.036		.03	1.84		1.87	3.03
0600	7/8" diameter		43	.186		.18	9.45		9.63	15.50
0650	For each additional inch of depth in same hole, add		210	.038		.05	1.93		1.98	3.19
0700	1" diameter		40	.200		.16	10.15		10.31	16.60
0750	For each additional inch of depth in same hole, add		190	.042		.04	2.14		2.18	3.51
0800	1-1/4" diameter		38	.211		.29	10.65		10.94	17.65
0850	For each additional inch of depth in same hole, add		180	.044		.07	2.25		2.32	3.74
0900	1-1/2" diameter		35	.229		.50	11.60		12.10	19.35
0950	For each additional inch of depth in same hole, add		165	.048		.12	2.46		2.58	4.13
1000	For ceiling installations, add						40%			

For customer support on your Commercial Renovation Costs with RSMeans data, call 800.448.8182.

73

Division Notes

	CREW	DAILY OUTPUT	LABOR-HOURS	UNIT	BARE COSTS				TOTAL INCL O&P
					MAT.	LABOR	EQUIP.	TOTAL	

Estimating Tips
04 05 00 Common Work Results for Masonry

- The terms mortar and grout are often used interchangeably—and incorrectly. Mortar is used to bed masonry units, seal the entry of air and moisture, provide architectural appearance, and allow for size variations in the units. Grout is used primarily in reinforced masonry construction and to bond the masonry to the reinforcing steel. Common mortar types are M (2500 psi), S (1800 psi), N (750 psi), and O (350 psi), and they conform to ASTM C270. Grout is either fine or coarse and conforms to ASTM C476, and in-place strengths generally exceed 2500 psi. Mortar and grout are different components of masonry construction and are placed by entirely different methods. An estimator should be aware of their unique uses and costs.

- Mortar is included in all assembled masonry line items. The mortar cost, part of the assembled masonry material cost, includes all ingredients, all labor, and all equipment required. Please see reference number RO40513-10.

- Waste, specifically the loss/droppings of mortar and the breakage of brick and block, Is Included in all unit cost lines that include mortar and masonry units in this division. A factor of 25% is added for mortar and 3% for brick and concrete masonry units.

- Scaffolding or staging is not included in any of the Division 4 costs. Refer to Subdivision 01 54 23 for scaffolding and staging costs.

04 20 00 Unit Masonry

- The most common types of unit masonry are brick and concrete masonry. The major classifications of brick are building brick (ASTM C62), facing brick (ASTM C216), glazed brick, fire brick, and pavers. Many varieties of texture and appearance can exist within these classifications, and the estimator would be wise to check local custom and availability within the project area. For repair and remodeling jobs, matching the existing brick may be the most important criteria.

- Brick and concrete block are priced by the piece and then converted into a price per square foot of wall. Openings less than two square feet are generally ignored by the estimator because any savings in units used are offset by the cutting and trimming required.

- It is often difficult and expensive to find and purchase small lots of historic brick. Costs can vary widely. Many design issues affect costs, selection of mortar mix, and repairs or replacement of masonry materials. Cleaning techniques must be reflected in the estimate.

- All masonry walls, whether interior or exterior, require bracing. The cost of bracing walls during construction should be included by the estimator, and this bracing must remain in place until permanent bracing is complete. Permanent bracing of masonry walls is accomplished by masonry itself, in the form of pilasters or abutting wall corners, or by anchoring the walls to the structural frame. Accessories in the form of anchors, anchor slots, and ties are used, but their supply and installation can be by different trades. For instance, anchor slots on spandrel beams and columns are supplied and welded in place by the steel fabricator, but the ties from the slots into the masonry are installed by the bricklayer. Regardless of the installation method, the estimator must be certain that these accessories are accounted for in pricing.

Reference Numbers
Reference numbers are shown at the beginning of some major classifications. These numbers refer to related items in the Reference Section. The reference information may be an estimating procedure, an alternate pricing method, or technical information.

Note: Not all subdivisions listed here necessarily appear. ∎

---

04 01 20.20 Pointing Masonry

		Crew	Daily Output	Labor-Hours	Unit	Material	2018 Bare Costs Labor	2018 Bare Costs Equipment	Total	Total Incl O&P
0010	**POINTING MASONRY**									
0300	Cut and repoint brick, hard mortar, running bond	1 Bric	80	.100	S.F.	.56	5.05		5.61	8.80
0320	Common bond		77	.104		.56	5.25		5.81	9.15
0360	Flemish bond		70	.114		.59	5.75		6.34	10.05
0400	English bond		65	.123		.59	6.20		6.79	10.75
0600	Soft old mortar, running bond		100	.080		.56	4.02		4.58	7.15
0620	Common bond		96	.083		.56	4.19		4.75	7.45
0640	Flemish bond		90	.089		.59	4.47		5.06	7.95
0680	English bond		82	.098		.59	4.91		5.50	8.65
0700	Stonework, hard mortar		140	.057	L.F.	.74	2.87		3.61	5.50
0720	Soft old mortar		160	.050	"	.74	2.52		3.26	4.92
1000	Repoint, mask and grout method, running bond		95	.084	S.F.	.74	4.24		4.98	7.70
1020	Common bond		90	.089		.74	4.47		5.21	8.10
1040	Flemish bond		86	.093		.78	4.68		5.46	8.50
1060	English bond		77	.104		.78	5.25		6.03	9.40
2000	Scrub coat, sand grout on walls, thin mix, brushed		120	.067		3.36	3.35		6.71	9.15
2020	Troweled		98	.082		4.68	4.11		8.79	11.85
9000	Minimum labor/equipment charge		3	2.667	Job		134		134	219

04 01 20.30 Pointing CMU

		Crew	Daily Output	Labor-Hours	Unit	Material	2018 Bare Costs Labor	2018 Bare Costs Equipment	Total	Total Incl O&P
0010	**POINTING CMU**									
0300	Cut and repoint block, hard mortar, running bond	1 Bric	190	.042	S.F.	.23	2.12		2.35	3.71
0310	Stacked bond		200	.040		.23	2.01		2.24	3.53
0600	Soft old mortar, running bond		230	.035		.23	1.75		1.98	3.10
0610	Stacked bond		245	.033		.23	1.64		1.87	2.93

04 01 20.41 Unit Masonry Stabilization

		Crew	Daily Output	Labor-Hours	Unit	Material	2018 Bare Costs Labor	2018 Bare Costs Equipment	Total	Total Incl O&P
0010	**UNIT MASONRY STABILIZATION**									
0100	Structural repointing method									
0110	Cut/grind mortar joint	1 Bric	240	.033	L.F.		1.68		1.68	2.73
0120	Clean and mask joint		2500	.003		.11	.16		.27	.38
0130	Epoxy paste and 1/4" FRP rod		240	.033		2.06	1.68		3.74	5
0132	3/8" FRP rod		160	.050		2.92	2.52		5.44	7.30
0140	Remove masking		14400	.001			.03		.03	.05
0300	Structural fabric method									
0310	Primer	1 Bric	600	.013	S.F.	.85	.67		1.52	2.03
0320	Apply filling/leveling paste		720	.011		.87	.56		1.43	1.87
0330	Epoxy, glass fiber fabric		720	.011		9.25	.56		9.81	11.05
0340	Carbon fiber fabric		720	.011		22.50	.56		23.06	26

04 01 20.50 Toothing Masonry

		Crew	Daily Output	Labor-Hours	Unit	Material	2018 Bare Costs Labor	2018 Bare Costs Equipment	Total	Total Incl O&P
0010	**TOOTHING MASONRY**									
0500	Brickwork, soft old mortar	1 Clab	40	.200	V.L.F.		7.95		7.95	12.95
0520	Hard mortar		30	.267			10.65		10.65	17.25
0700	Blockwork, soft old mortar		70	.114			4.55		4.55	7.40
0720	Hard mortar		50	.160			6.40		6.40	10.35
9000	Minimum labor/equipment charge		4	2	Job		79.50		79.50	129

04 01 30 – Unit Masonry Cleaning

04 01 30.20 Cleaning Masonry

		Crew	Daily Output	Labor-Hours	Unit	Material	2018 Bare Costs Labor	2018 Bare Costs Equipment	Total	Total Incl O&P
0010	**CLEANING MASONRY**									
0200	By chemical, brush and rinse, new work, light construction dust	D-1	1000	.016	S.F.	.06	.72		.78	1.24
0220	Medium construction dust		800	.020		.09	.90		.99	1.56
0240	Heavy construction dust, drips or stains		600	.027		.12	1.19		1.31	2.08
0260	Low pressure wash and rinse, light restoration, light soil		800	.020		.14	.90		1.04	1.62

04 01 30 – Unit Masonry Cleaning

04 01 30.20 Cleaning Masonry

		Crew	Daily Output	Labor-Hours	Unit	Material	2018 Bare Costs Labor	Equipment	Total	Total Incl O&P
0270	Average soil, biological staining	D-1	400	.040	S.F.	.22	1.79		2.01	3.16
0280	Heavy soil, biological and mineral staining, paint		330	.048		.29	2.17		2.46	3.86
0300	High pressure wash and rinse, heavy restoration, light soil		600	.027		.10	1.19		1.29	2.06
0310	Average soil, biological staining		400	.040		.16	1.79		1.95	3.09
0320	Heavy soil, biological and mineral staining, paint		250	.064		.21	2.87		3.08	4.91
0400	High pressure wash, water only, light soil	C-29	500	.016			.64	.13	.77	1.18
0420	Average soil, biological staining		375	.021			.85	.17	1.02	1.57
0440	Heavy soil, biological and mineral staining, paint		250	.032			1.28	.25	1.53	2.35
0800	High pressure water and chemical, light soil		450	.018		.17	.71	.14	1.02	1.50
0820	Average soil, biological staining		300	.027		.26	1.06	.21	1.53	2.24
0840	Heavy soil, biological and mineral staining, paint		200	.040		.34	1.59	.32	2.25	3.32
1200	Sandblast, wet system, light soil	J-6	1750	.018		.36	.81	.12	1.29	1.82
1220	Average soil, biological staining		1100	.029		.55	1.28	.18	2.01	2.85
1240	Heavy soil, biological and mineral staining, paint		700	.046		.73	2.01	.29	3.03	4.35
1400	Dry system, light soil		2500	.013		.36	.56	.08	1	1.39
1420	Average soil, biological staining		1750	.018		.55	.81	.12	1.48	2.02
1440	Heavy soil, biological and mineral staining, paint		1000	.032		.73	1.41	.20	2.34	3.28
1800	For walnut shells, add					.80			.80	.88
1820	For corn chips, add					.80			.80	.88
2000	Steam cleaning, light soil	A-1H	750	.011			.43	.10	.53	.80
2020	Average soil, biological staining		625	.013			.51	.12	.63	.96
2040	Heavy soil, biological and mineral staining		375	.021			.85	.20	1.05	1.60
4000	Add for masking doors and windows	1 Clab	800	.010		.07	.40		.47	.73
4200	Add for pedestrian protection				Job				10%	10%
9000	Minimum labor/equipment charge	D-4	2	16	"		720	64.50	784.50	1,250

04 01 30.60 Brick Washing

		Crew	Daily Output	Labor-Hours	Unit	Material	2018 Bare Costs Labor	Equipment	Total	Total Incl O&P
0010	**BRICK WASHING**									
0012	Acid cleanser, smooth brick surface	1 Bric	560	.014	S.F.	.05	.72		.77	1.23
0050	Rough brick		400	.020		.07	1.01		1.08	1.71
0060	Stone, acid wash		600	.013		.08	.67		.75	1.18
1000	Muriatic acid, price per gallon in 5 gallon lots				Gal.	10.10			10.10	11.10

04 05 Common Work Results for Masonry

04 05 05 – Selective Demolition for Masonry

04 05 05.10 Selective Demolition

		Crew	Daily Output	Labor-Hours	Unit	Material	2018 Bare Costs Labor	Equipment	Total	Total Incl O&P
0010	**SELECTIVE DEMOLITION** R024119-10									
0200	Bond beams, 8" block with #4 bar	2 Clab	32	.500	L.F.		19.95		19.95	32.50
0300	Concrete block walls, unreinforced, 2" thick		1200	.013	S.F.		.53		.53	.86
0310	4" thick		1150	.014			.55		.55	.90
0320	6" thick		1100	.015			.58		.58	.94
0330	8" thick		1050	.015			.61		.61	.99
0340	10" thick		1000	.016			.64		.64	1.04
0360	12" thick		950	.017			.67		.67	1.09
0380	Reinforced alternate courses, 2" thick		1130	.014			.56		.56	.92
0390	4" thick		1080	.015			.59		.59	.96
0400	6" thick		1035	.015			.62		.62	1
0410	8" thick		990	.016			.64		.64	1.05
0420	10" thick		940	.017			.68		.68	1.10
0430	12" thick		890	.018			.72		.72	1.16
0440	Reinforced alternate courses & vertically 48" OC, 4" thick		900	.018			.71		.71	1.15
0450	6" thick		850	.019			.75		.75	1.22

04 05 05.10 Selective Demolition	Crew	Daily Output	Labor-Hours	Unit	Material	2018 Bare Costs Labor	Equipment	Total	Total Incl O&P
0460	8" thick	2 Clab	800	.020	S.F.		.80		.80	1.29
0480	10" thick		750	.021			.85		.85	1.38
0490	12" thick	▼	700	.023	▼		.91		.91	1.48
1000	Chimney, 16" x 16", soft old mortar	1 Clab	55	.145	C.F.		5.80		5.80	9.40
1020	Hard mortar		40	.200			7.95		7.95	12.95
1030	16" x 20", soft old mortar		55	.145			5.80		5.80	9.40
1040	Hard mortar		40	.200			7.95		7.95	12.95
1050	16" x 24", soft old mortar		55	.145			5.80		5.80	9.40
1060	Hard mortar		40	.200			7.95		7.95	12.95
1080	20" x 20", soft old mortar		55	.145			5.80		5.80	9.40
1100	Hard mortar		40	.200			7.95		7.95	12.95
1110	20" x 24", soft old mortar		55	.145			5.80		5.80	9.40
1120	Hard mortar		40	.200			7.95		7.95	12.95
1140	20" x 32", soft old mortar		55	.145			5.80		5.80	9.40
1160	Hard mortar		40	.200			7.95		7.95	12.95
1200	48" x 48", soft old mortar		55	.145			5.80		5.80	9.40
1220	Hard mortar	▼	40	.200	▼		7.95		7.95	12.95
1250	Metal, high temp steel jacket, 24" diameter	E-2	130	.431	V.L.F.		23.50	13	36.50	53.50
1260	60" diameter	"	60	.933			50.50	28	78.50	116
1280	Flue lining, up to 12" x 12"	1 Clab	200	.040			1.59		1.59	2.59
1282	Up to 24" x 24"		150	.053			2.13		2.13	3.45
2000	Columns, 8" x 8", soft old mortar		48	.167			6.65		6.65	10.80
2020	Hard mortar		40	.200			7.95		7.95	12.95
2060	16" x 16", soft old mortar		16	.500			19.95		19.95	32.50
2100	Hard mortar		14	.571			23		23	37
2140	24" x 24", soft old mortar		8	1			40		40	64.50
2160	Hard mortar		6	1.333			53		53	86.50
2200	36" x 36", soft old mortar		4	2			79.50		79.50	129
2220	Hard mortar		3	2.667	▼		106		106	173
2230	Alternate pricing method, soft old mortar		30	.267	C.F.		10.65		10.65	17.25
2240	Hard mortar	▼	23	.348	"		13.85		13.85	22.50
3000	Copings, precast or masonry, to 8" wide									
3020	Soft old mortar	1 Clab	180	.044	L.F.		1.77		1.77	2.88
3040	Hard mortar	"	160	.050	"		1.99		1.99	3.24
3100	To 12" wide									
3120	Soft old mortar	1 Clab	160	.050	L.F.		1.99		1.99	3.24
3140	Hard mortar	"	140	.057	"		2.28		2.28	3.70
4000	Fireplace, brick, 30" x 24" opening									
4020	Soft old mortar	1 Clab	2	4	Ea.		159		159	259
4040	Hard mortar		1.25	6.400			255		255	415
4100	Stone, soft old mortar		1.50	5.333			213		213	345
4120	Hard mortar		1	8	▼		320		320	520
5000	Veneers, brick, soft old mortar		140	.057	S.F.		2.28		2.28	3.70
5020	Hard mortar		125	.064			2.55		2.55	4.14
5050	Glass block, up to 4" thick		500	.016			.64		.64	1.04
5100	Granite and marble, 2" thick		180	.044			1.77		1.77	2.88
5120	4" thick		170	.047			1.88		1.88	3.04
5140	Stone, 4" thick		180	.044			1.77		1.77	2.88
5160	8" thick		175	.046	▼		1.82		1.82	2.96
5400	Alternate pricing method, stone, 4" thick		60	.133	C.F.		5.30		5.30	8.65
5420	8" thick		85	.094	"		3.75		3.75	6.10
9000	Minimum labor/equipment charge	▼	2	4	Job		159		159	259

78

For customer support on your Commercial Renovation Costs with RSMeans data, call 800.448.8182.

04 05 13.10 Cement	Crew	Daily Output	Labor-Hours	Unit	Material	2018 Bare Costs Labor	Equipment	Total	Total Incl O&P
0010 **CEMENT**									
0100 Masonry, 70 lb. bag, T.L. lots				Bag	13.80			13.80	15.20
0150 L.T.L. lots					14.65			14.65	16.10
0200 White, 70 lb. bag, T.L. lots					18.05			18.05	19.85
0250 L.T.L. lots				↓	19.95			19.95	22

04 05 13.20 Lime

	Crew	Daily Output	Labor-Hours	Unit	Material	Labor	Equipment	Total	Total Incl O&P
0010 **LIME**									
0020 Masons, hydrated, 50 lb. bag, T.L. lots				Bag	11.20			11.20	12.30
0050 L.T.L. lots					12.30			12.30	13.55
0200 Finish, double hydrated, 50 lb. bag, T.L. lots					10.40			10.40	11.45
0250 L.T.L. lots				↓	11.45			11.45	12.60

04 05 13.23 Surface Bonding Masonry Mortaring

	Crew	Daily Output	Labor-Hours	Unit	Material	Labor	Equipment	Total	Total Incl O&P
0010 **SURFACE BONDING MASONRY MORTARING**									
0020 Gray or white colors, not incl. block work	1 Bric	540	.015	S.F.	.15	.74		.89	1.39

04 05 13.30 Mortar

	Crew	Daily Output	Labor-Hours	Unit	Material	Labor	Equipment	Total	Total Incl O&P
0010 **MORTAR**									
0020 With masonry cement									
0100 Type M, 1:1:6 mix	1 Brhe	143	.056	C.F.	6.10	2.20		8.30	10.35
0200 Type N, 1:3 mix		143	.056		6.10	2.20		8.30	10.30
0300 Type O, 1:3 mix		143	.056		4.90	2.20		7.10	9
0400 Type PM, 1:1:6 mix, 2500 psi		143	.056		6	2.20		8.20	10.20
0500 Type S, 1/2:1:4 mix	↓	143	.056	↓	6.40	2.20		8.60	10.65
2000 With Portland cement and lime									
2100 Type M, 1:1/4:3 mix	1 Brhe	143	.056	C.F.	9.15	2.20		11.35	13.70
2200 Type N, 1:1:6 mix, 750 psi		143	.056		7.15	2.20		9.35	11.45
2300 Type O, 1:2:9 mix (Pointing Mortar)		143	.056		8.55	2.20		10.75	13
2400 Type PL, 1:1/2:4 mix, 2500 psi		143	.056		6	2.20		8.20	10.20
2600 Type S, 1:1/2:4 mix, 1800 psi	↓	143	.056		8.40	2.20		10.60	12.85
2650 Pre-mixed, type S or N					6.30			6.30	6.95
2700 Mortar for glass block	1 Brhe	143	.056	↓	12.60	2.20		14.80	17.45
2900 Mortar for fire brick, dry mix, 10 lb. pail				Ea.	22.50			22.50	25

04 05 13.91 Masonry Restoration Mortaring

	Crew	Daily Output	Labor-Hours	Unit	Material	Labor	Equipment	Total	Total Incl O&P
0010 **MASONRY RESTORATION MORTARING**									
0020 Masonry restoration mix				Lb.	.71			.71	.78
0050 White				"	1.18			1.18	1.30

04 05 13.93 Mortar Pigments

	Crew	Daily Output	Labor-Hours	Unit	Material	Labor	Equipment	Total	Total Incl O&P
0010 **MORTAR PIGMENTS**, 50 lb. bags (2 bags per M bricks)									
0020 Color admixture, range 2 to 10 lb. per bag of cement, light colors				Lb.	5.55			5.55	6.10
0050 Medium colors					7.40			7.40	8.15
0100 Dark colors				↓	15.50			15.50	17.05

04 05 13.95 Sand

	Crew	Daily Output	Labor-Hours	Unit	Material	Labor	Equipment	Total	Total Incl O&P
0010 **SAND**, screened and washed at pit									
0020 For mortar, per ton				Ton	21			21	23
0050 With 10 mile haul					41.50			41.50	45.50
0100 With 30 mile haul				↓	62.50			62.50	69
0200 Screened and washed, at the pit				C.Y.	29.50			29.50	32.50
0250 With 10 mile haul					57.50			57.50	63.50
0300 With 30 mile haul				↓	87			87	95.50

For customer support on your Commercial Renovation Costs with RSMeans data, call 800.448.8182.

79

04 05 13 – Masonry Mortaring

04 05 13.98 Mortar Admixtures

		Crew	Daily Output	Labor-Hours	Unit	Material	2018 Bare Costs Labor	Equipment	Total	Total Incl O&P
0010	**MORTAR ADMIXTURES**									
0020	Waterproofing admixture, per quart (1 qt. to 2 bags of masonry cement)				Qt.	4.98			4.98	5.50

04 05 16 – Masonry Grouting

04 05 16.30 Grouting

		Crew	Daily Output	Labor-Hours	Unit	Material	2018 Bare Costs Labor	Equipment	Total	Total Incl O&P
0010	**GROUTING**									
0011	Bond beams & lintels, 8" deep, 6" thick, 0.15 C.F./L.F.	D-4	1480	.022	L.F.	.75	.97	.09	1.81	2.50
0020	8" thick, 0.2 C.F./L.F.		1400	.023		1.22	1.03	.09	2.34	3.10
0050	10" thick, 0.25 C.F./L.F.		1200	.027		1.26	1.20	.11	2.57	3.44
0060	12" thick, 0.3 C.F./L.F.		1040	.031		1.51	1.39	.12	3.02	4.04
0200	Concrete block cores, solid, 4" thk., by hand, 0.067 C.F./S.F. of wall	D-8	1100	.036	S.F.	.34	1.67		2.01	3.09
0210	6" thick, pumped, 0.175 C.F./S.F.	D-4	720	.044		.88	2	.18	3.06	4.41
0250	8" thick, pumped, 0.258 C.F./S.F.		680	.047		1.30	2.12	.19	3.61	5.05
0300	10" thick, pumped, 0.340 C.F./S.F.		660	.048		1.71	2.18	.19	4.08	5.60
0350	12" thick, pumped, 0.422 C.F./S.F.		640	.050		2.12	2.25	.20	4.57	6.20
0500	Cavity walls, 2" space, pumped, 0.167 C.F./S.F. of wall		1700	.019		.84	.85	.08	1.77	2.37
0550	3" space, 0.250 C.F./S.F.		1200	.027		1.26	1.20	.11	2.57	3.44
0600	4" space, 0.333 C.F./S.F.		1150	.028		1.67	1.25	.11	3.03	3.99
0700	6" space, 0.500 C.F./S.F.		800	.040		2.51	1.80	.16	4.47	5.85
0800	Door frames, 3' x 7' opening, 2.5 C.F. per opening		60	.533	Opng.	12.55	24	2.14	38.69	55
0850	6' x 7' opening, 3.5 C.F. per opening		45	.711	"	17.60	32	2.86	52.46	74.50
2000	Grout, C476, for bond beams, lintels and CMU cores		350	.091	C.F.	5.05	4.12	.37	9.54	12.60
9000	Minimum labor/equipment charge	1 Bric	2	4	Job		201		201	330

04 05 19 – Masonry Anchorage and Reinforcing

04 05 19.05 Anchor Bolts

		Crew	Daily Output	Labor-Hours	Unit	Material	2018 Bare Costs Labor	Equipment	Total	Total Incl O&P
0010	**ANCHOR BOLTS**									
0015	Installed in fresh grout in CMU bond beams or filled cores, no templates									
0020	Hooked, with nut and washer, 1/2" diameter, 8" long	1 Bric	132	.061	Ea.	1.42	3.05		4.47	6.55
0030	12" long		131	.061		1.58	3.07		4.65	6.75
0040	5/8" diameter, 8" long		129	.062		3.80	3.12		6.92	9.30
0050	12" long		127	.063		4.67	3.17		7.84	10.30
0060	3/4" diameter, 8" long		127	.063		4.67	3.17		7.84	10.30
0070	12" long		125	.064		5.85	3.22		9.07	11.65

04 05 19.16 Masonry Anchors

		Crew	Daily Output	Labor-Hours	Unit	Material	2018 Bare Costs Labor	Equipment	Total	Total Incl O&P
0010	**MASONRY ANCHORS**									
0020	For brick veneer, galv., corrugated, 7/8" x 7", 22 ga.	1 Bric	10.50	.762	C	16.05	38.50		54.55	80
0100	24 ga.		10.50	.762		10.15	38.50		48.65	73.50
0150	16 ga.		10.50	.762		30	38.50		68.50	95.50
0200	Buck anchors, galv., corrugated, 16 ga., 2" bend, 8" x 2"		10.50	.762		65	38.50		103.50	134
0250	8" x 3"		10.50	.762		67.50	38.50		106	137
0300	Adjustable, rectangular, 4-1/8" wide									
0350	Anchor and tie, 3/16" wire, mill galv.									
0400	2-3/4" eye, 3-1/4" tie	1 Bric	1.05	7.619	M	475	385		860	1,150
0500	4-3/4" tie		1.05	7.619		510	385		895	1,200
0520	5-1/2" tie		1.05	7.619		555	385		940	1,225
0550	4-3/4" eye, 3-1/4" tie		1.05	7.619		525	385		910	1,200
0570	4-3/4" tie		1.05	7.619		565	385		950	1,250
0580	5-1/2" tie		1.05	7.619		615	385		1,000	1,300
0660	Cavity wall, Z-type, galvanized, 6" long, 1/8" diam.		10.50	.762	C	24	38.50		62.50	88.50
0670	3/16" diameter		10.50	.762		33	38.50		71.50	98.50
0680	1/4" diameter		10.50	.762		40.50	38.50		79	108
0850	8" long, 3/16" diameter		10.50	.762		26	38.50		64.50	91

04 05 19 – Masonry Anchorage and Reinforcing

04 05 19.16 Masonry Anchors	Crew	Daily Output	Labor-Hours	Unit	Material	2018 Bare Costs Labor	Equipment	Total	Total Incl O&P
0855	1/4" diameter	1 Bric	10.50	.762	C	47.50	38.50		86	115
1000	Rectangular type, galvanized, 1/4" diameter, 2" x 6"		10.50	.762		75.50	38.50		114	146
1050	4" x 6"		10.50	.762		91	38.50		129.50	163
1100	3/16" diameter, 2" x 6"		10.50	.762		48	38.50		86.50	116
1150	4" x 6"		10.50	.762		55	38.50		93.50	123
1200	Mesh wall tie, 1/2" mesh, hot dip galvanized									
1400	16 ga., 12" long, 3" wide	1 Bric	9	.889	C	92.50	44.50		137	174
1420	6" wide		9	.889		135	44.50		179.50	222
1440	12" wide		8.50	.941		215	47.50		262.50	315
1500	Rigid partition anchors, plain, 8" long, 1" x 1/8"		10.50	.762		237	38.50		275.50	325
1550	1" x 1/4"		10.50	.762		279	38.50		317.50	370
1580	1-1/2" x 1/8"		10.50	.762		261	38.50		299.50	350
1600	1-1/2" x 1/4"		10.50	.762		325	38.50		363.50	425
1650	2" x 1/8"		10.50	.762		310	38.50		348.50	405
1700	2" x 1/4"		10.50	.762		405	38.50		443.50	510
2000	Column flange ties, wire, galvanized									
2300	3/16" diameter, up to 3" wide	1 Bric	10.50	.762	C	88.50	38.50		127	160
2350	To 5" wide		10.50	.762		96.50	38.50		135	169
2400	To 7" wide		10.50	.762		104	38.50		142.50	177
2600	To 9" wide		10.50	.762		111	38.50		149.50	185
2650	1/4" diameter, up to 3" wide		10.50	.762		111	38.50		149.50	185
2700	To 5" wide		10.50	.762		120	38.50		158.50	195
2800	To 7" wide		10.50	.762		135	38.50		173.50	211
2850	To 9" wide		10.50	.762		145	38.50		183.50	223
2900	For hot dip galvanized, add					35%				
4000	Channel slots, 1-3/8" x 1/2" x 8"									
4100	12 ga., plain	1 Bric	10.50	.762	C	218	38.50		256.50	305
4150	16 ga., galvanized	"	10.50	.762	"	149	38.50		187.50	227
4200	Channel slot anchors									
4300	16 ga., galvanized, 1-1/4" x 3-1/2"				C	58			58	64
4350	1-1/4" x 5-1/2"					68.50			68.50	75
4400	1-1/4" x 7-1/2"					77.50			77.50	85.50
4500	1/8" plain, 1-1/4" x 3-1/2"					144			144	158
4550	1-1/4" x 5-1/2"					153			153	169
4600	1-1/4" x 7-1/2"					167			167	184
4700	For corrugation, add					77			77	85
4750	For hot dip galvanized, add					35%				
5000	Dowels									
5100	Plain, 1/4" diameter, 3" long				C	56			56	61.50
5150	4" long					62			62	68
5200	6" long					75.50			75.50	83.50
5300	3/8" diameter, 3" long					73.50			73.50	81
5350	4" long					91.50			91.50	101
5400	6" long					106			106	116
5500	1/2" diameter, 3" long					108			108	118
5550	4" long					128			128	140
5600	6" long					165			165	182
5700	5/8" diameter, 3" long					147			147	162
5750	4" long					181			181	199
5800	6" long					249			249	274
6000	3/4" diameter, 3" long					183			183	202
6100	4" long					232			232	256
6150	6" long					335			335	365

04 05 19 – Masonry Anchorage and Reinforcing

04 05 19.16 Masonry Anchors	Crew	Daily Output	Labor-Hours	Unit	Material	2018 Bare Costs Labor	Equipment	Total	Total Incl O&P
6300	For hot dip galvanized, add				C	35%				

04 05 19.26 Masonry Reinforcing Bars

	04 05 19.26 Masonry Reinforcing Bars	Crew	Daily Output	Labor-Hours	Unit	Material	2018 Bare Costs Labor	Equipment	Total	Total Incl O&P
0010	**MASONRY REINFORCING BARS**									
0015	Steel bars A615, placed horiz., #3 & #4 bars	1 Bric	450	.018	Lb.	.48	.89		1.37	1.99
0020	#5 & #6 bars		800	.010		.48	.50		.98	1.35
0050	Placed vertical, #3 & #4 bars		350	.023		.48	1.15		1.63	2.41
0060	#5 & #6 bars		650	.012		.48	.62		1.10	1.54
0200	Joint reinforcing, regular truss, to 6" wide, mill std galvanized		30	.267	C.L.F.	24	13.40		37.40	48.50
0250	12" wide		20	.400		28	20		48	64
0400	Cavity truss with drip section, to 6" wide		30	.267		22.50	13.40		35.90	47
0450	12" wide		20	.400		26	20		46	61.50
0500	Joint reinforcing, ladder type, mill std galvanized									
0600	9 ga. sides, 9 ga. ties, 4" wall	1 Bric	30	.267	C.L.F.	23	13.40		36.40	47.50
0650	6" wall		30	.267		20.50	13.40		33.90	44.50
0700	8" wall		25	.320		21.50	16.10		37.60	50.50
0750	10" wall		20	.400		24.50	20		44.50	59.50
0800	12" wall		20	.400		25.50	20		45.50	61
1000	Truss type									
1100	9 ga. sides, 9 ga. ties, 4" wall	1 Bric	30	.267	C.L.F.	24	13.40		37.40	48
1150	6" wall		30	.267		24	13.40		37.40	48.50
1200	8" wall		25	.320		29	16.10		45.10	58.50
1250	10" wall		20	.400		23.50	20		43.50	59
1300	12" wall		20	.400		24.50	20		44.50	60
1500	3/16" sides, 9 ga. ties, 4" wall		30	.267		27.50	13.40		40.90	52
1550	6" wall		30	.267		34	13.40		47.40	59.50
1600	8" wall		25	.320		35.50	16.10		51.60	65.50
1650	10" wall		20	.400		36.50	20		56.50	73
1700	12" wall		20	.400		37	20		57	74
2000	3/16" sides, 3/16" ties, 4" wall		30	.267		38.50	13.40		51.90	64
2050	6" wall		30	.267		39.50	13.40		52.90	65
2100	8" wall		25	.320		40.50	16.10		56.60	71.50
2150	10" wall		20	.400		42.50	20		62.50	80
2200	12" wall		20	.400		45	20		65	82.50
2500	Cavity truss type, galvanized									
2600	9 ga. sides, 9 ga. ties, 4" wall	1 Bric	25	.320	C.L.F.	43	16.10		59.10	74
2650	6" wall		25	.320		40.50	16.10		56.60	71
2700	8" wall		20	.400		45	20		65	82.50
2750	10" wall		15	.533		47.50	27		74.50	96
2800	12" wall		15	.533		43.50	27		70.50	92
3000	3/16" sides, 9 ga. ties, 4" wall		25	.320		41	16.10		57.10	71.50
3050	6" wall		25	.320		38	16.10		54.10	68
3100	8" wall		20	.400		54	20		74	92.50
3150	10" wall		15	.533		39	27		66	87
3200	12" wall		15	.533		43.50	27		70.50	91.50
3500	For hot dip galvanizing, add				Ton	460			460	505

04 05 23 – Masonry Accessories

04 05 23.13 Masonry Control and Expansion Joints

	04 05 23.13 Masonry Control and Expansion Joints	Crew	Daily Output	Labor-Hours	Unit	Material	2018 Bare Costs Labor	Equipment	Total	Total Incl O&P
0010	**MASONRY CONTROL AND EXPANSION JOINTS**									
0020	Rubber, for double wythe 8" minimum wall (Brick/CMU)	1 Bric	400	.020	L.F.	2.24	1.01		3.25	4.10
0025	"T" shaped		320	.025		1.25	1.26		2.51	3.43
0030	Cross-shaped for CMU units		280	.029		1.66	1.44		3.10	4.17
0050	PVC, for double wythe 8" minimum wall (Brick/CMU)		400	.020		1.66	1.01		2.67	3.47

04 05 Common Work Results for Masonry

04 05 23 – Masonry Accessories

04 05 23.13 Masonry Control and Expansion Joints

		Crew	Daily Output	Labor-Hours	Unit	Material	2018 Bare Costs Labor	Equipment	Total	Total Incl O&P
0120	"T" shaped	1 Bric	320	.025	L.F.	.84	1.26		2.10	2.97
0160	Cross-shaped for CMU units	↓	280	.029	↓	1.16	1.44		2.60	3.62

04 05 23.19 Masonry Cavity Drainage, Weepholes, and Vents

		Crew	Daily Output	Labor-Hours	Unit	Material	2018 Bare Costs Labor	Equipment	Total	Total Incl O&P
0010	**MASONRY CAVITY DRAINAGE, WEEPHOLES, AND VENTS**									
0020	Extruded aluminum, 4" deep, 2-3/8" x 8-1/8"	1 Bric	30	.267	Ea.	38.50	13.40		51.90	64.50
0050	5" x 8-1/8"		25	.320		50	16.10		66.10	82
0100	2-1/4" x 25"		25	.320		87	16.10		103.10	122
0150	5" x 16-1/2"		22	.364		60	18.30		78.30	96
0175	5" x 24"		22	.364		90.50	18.30		108.80	130
0200	6" x 16-1/2"		22	.364		97	18.30		115.30	137
0250	7-3/4" x 16-1/2"	↓	20	.400		88	20		108	130
0400	For baked enamel finish, add					35%				
0500	For cast aluminum, painted, add					60%				
1000	Stainless steel ventilators, 6" x 6"	1 Bric	25	.320		222	16.10		238.10	272
1050	8" x 8"		24	.333		248	16.75		264.75	300
1100	12" x 12"		23	.348		285	17.50		302.50	345
1150	12" x 6"		24	.333		269	16.75		285.75	325
1200	Foundation block vent, galv., 1-1/4" thk, 8" high, 16" long, no damper	↓	30	.267		16.20	13.40		29.60	40
1250	For damper, add				↓	3.28			3.28	3.61

04 05 23.95 Wall Plugs

		Crew	Daily Output	Labor-Hours	Unit	Material	2018 Bare Costs Labor	Equipment	Total	Total Incl O&P
0010	**WALL PLUGS** (for nailing to brickwork)									
0020	25 ga., galvanized, plain	1 Bric	10.50	.762	C	26.50	38.50		65	91.50
0050	Wood filled	"	10.50	.762	"	67.50	38.50		106	137

04 21 Clay Unit Masonry

04 21 13 – Brick Masonry

04 21 13.13 Brick Veneer Masonry

			Crew	Daily Output	Labor-Hours	Unit	Material	2018 Bare Costs Labor	Equipment	Total	Total Incl O&P
0010	**BRICK VENEER MASONRY**, T.L. lots, excl. scaff., grout & reinforcing	R042110-20									
0015	Material costs incl. 3% brick and 25% mortar waste										
2000	Standard, sel. common, 4" x 2-2/3" x 8" (6.75/S.F.)	R042110-50	D-8	230	.174	S.F.	4.32	8		12.32	17.75
2020	Red, 4" x 2-2/3" x 8", running bond			220	.182		4.06	8.35		12.41	18.05
2050	Full header every 6th course (7.88/S.F.)			185	.216		4.73	9.90		14.63	21.50
2100	English, full header every 2nd course (10.13/S.F.)			140	.286		6.05	13.10		19.15	28
2150	Flemish, alternate header every course (9.00/S.F.)			150	.267		5.40	12.25		17.65	26
2200	Flemish, alt. header every 6th course (7.13/S.F.)			205	.195		4.29	8.95		13.24	19.30
2250	Full headers throughout (13.50/S.F.)			105	.381		8.05	17.50		25.55	37.50
2300	Rowlock course (13.50/S.F.)			100	.400		8.05	18.35		26.40	39
2350	Rowlock stretcher (4.50/S.F.)			310	.129		2.74	5.90		8.64	12.65
2400	Soldier course (6.75/S.F.)			200	.200		4.06	9.20		13.26	19.40
2450	Sailor course (4.50/S.F.)			290	.138		2.74	6.35		9.09	13.35
2600	Buff or gray face, running bond (6.75/S.F.)			220	.182		4.30	8.35		12.65	18.35
2700	Glazed face brick, running bond			210	.190		13	8.75		21.75	28.50
2750	Full header every 6th course (7.88/S.F.)			170	.235		15.15	10.80		25.95	34.50
3000	Jumbo, 6" x 4" x 12" running bond (3.00/S.F.)			435	.092		5.45	4.22		9.67	12.90
3050	Norman, 4" x 2-2/3" x 12" running bond (4.5/S.F.)			320	.125		6.40	5.75		12.15	16.40
3100	Norwegian, 4" x 3-1/5" x 12" (3.75/S.F.)			375	.107		5.65	4.90		10.55	14.25
3150	Economy, 4" x 4" x 8" (4.50/S.F.)			310	.129		4.46	5.90		10.36	14.55
3200	Engineer, 4" x 3-1/5" x 8" (5.63/S.F.)			260	.154		3.92	7.05		10.97	15.80
3250	Roman, 4" x 2" x 12" (6.00/S.F.)			250	.160		7.30	7.35		14.65	20
3300	S.C.R., 6" x 2-2/3" x 12" (4.50/S.F.)			310	.129	↓	6.30	5.90		12.20	16.60

For customer support on your Commercial Renovation Costs with RSMeans data, call 800.448.8182.

83

04 21 Clay Unit Masonry

04 21 13 – Brick Masonry

04 21 13.13 Brick Veneer Masonry	Crew	Daily Output	Labor-Hours	Unit	Material	2018 Bare Costs Labor	Equipment	Total	Total Incl O&P
3350	Utility, 4" x 4" x 12" (3.00/S.F.)	D-8	360	.111	S.F.	5.10	5.10		10.20	13.90
3360	For less than truck load lots, add					.10%				
3400	For cavity wall construction, add						15%			
3450	For stacked bond, add						10%			
3500	For interior veneer construction, add						15%			
3550	For curved walls, add						30%			
9000	Minimum labor/equipment charge	D-1	2	8	Job		360		360	585

04 21 13.14 Thin Brick Veneer

		Crew	Daily Output	Labor-Hours	Unit	Material	Labor	Equipment	Total	Total Incl O&P
0010	**THIN BRICK VENEER**									
0015	Material costs incl. 3% brick and 25% mortar waste									
0020	On & incl. metal panel support sys, modular, 2-2/3" x 5/8" x 8", red	D-7	92	.174	S.F.	9.60	7.30		16.90	22
0100	Closure, 4" x 5/8" x 8"		110	.145		9.35	6.10		15.45	19.95
0110	Norman, 2-2/3" x 5/8" x 12"		110	.145		9.40	6.10		15.50	20
0120	Utility, 4" x 5/8" x 12"		125	.128		9.15	5.40		14.55	18.55
0130	Emperor, 4" x 3/4" x 16"		175	.091		10.30	3.84		14.14	17.40
0140	Super emperor, 8" x 3/4" x 16"		195	.082		10.60	3.45		14.05	17.15
0150	For L shaped corners with 4" return, add				L.F.	9.25			9.25	10.20
0200	On masonry/plaster back-up, modular, 2-2/3" x 5/8" x 8", red	D-7	137	.117	S.F.	4.49	4.91		9.40	12.70
0210	Closure, 4" x 5/8" x 8"		165	.097		4.24	4.08		8.32	11.10
0220	Norman, 2-2/3" x 5/8" x 12"		165	.097		4.28	4.08		8.36	11.15
0230	Utility, 4" x 5/8" x 12"		185	.086		4	3.64		7.64	10.15
0240	Emperor, 4" x 3/4" x 16"		260	.062		5.15	2.59		7.74	9.80
0250	Super emperor, 8" x 3/4" x 16"		285	.056		5.50	2.36		7.86	9.80
0260	For L shaped corners with 4" return, add				L.F.	9.25			9.25	10.20
0270	For embedment into pre-cast concrete panels, add				S.F.	14.40			14.40	15.85

04 21 13.15 Chimney

		Crew	Daily Output	Labor-Hours	Unit	Material	Labor	Equipment	Total	Total Incl O&P
0010	**CHIMNEY**, excludes foundation, scaffolding, grout and reinforcing									
0100	Brick, 16" x 16", 8" flue	D-1	18.20	.879	V.L.F.	25.50	39.50		65	92.50
0150	16" x 20" with one 8" x 12" flue		16	1		40	45		85	117
0200	16" x 24" with two 8" x 8" flues		14	1.143		58.50	51		109.50	148
0250	20" x 20" with one 12" x 12" flue		13.70	1.168		48	52.50		100.50	138
0300	20" x 24" with two 8" x 12" flues		12	1.333		66.50	59.50		126	171
0350	20" x 32" with two 12" x 12" flues		10	1.600		84.50	71.50		156	210

04 21 13.18 Columns

		Crew	Daily Output	Labor-Hours	Unit	Material	Labor	Equipment	Total	Total Incl O&P
0010	**COLUMNS**, solid, excludes scaffolding, grout and reinforcing									
0050	Brick, 8" x 8", 9 brick/V.L.F.	D-1	56	.286	V.L.F.	5.30	12.80		18.10	27
0100	12" x 8", 13.5 brick/V.L.F.		37	.432		7.95	19.35		27.30	40.50
0200	12" x 12", 20 brick/V.L.F.		25	.640		11.80	28.50		40.30	60
0300	16" x 12", 27 brick/V.L.F.		19	.842		15.95	37.50		53.45	79
0400	16" x 16", 36 brick/V.L.F.		14	1.143		21.50	51		72.50	107
0500	20" x 16", 45 brick/V.L.F.		11	1.455		26.50	65		91.50	136
0600	20" x 20", 56 brick/V.L.F.		9	1.778		33	79.50		112.50	167
0700	24" x 20", 68 brick/V.L.F.		7	2.286		40	102		142	211
0800	24" x 24", 81 brick/V.L.F.		6	2.667		48	119		167	248
1000	36" x 36", 182 brick/V.L.F.		3	5.333		108	239		347	510
9000	Minimum labor/equipment charge		2	8	Job		360		360	585

04 21 13.30 Oversized Brick

		Crew	Daily Output	Labor-Hours	Unit	Material	Labor	Equipment	Total	Total Incl O&P
0010	**OVERSIZED BRICK**, excludes scaffolding, grout and reinforcing									
0100	Veneer, 4" x 2.25" x 16"	D-8	387	.103	S.F.	5.40	4.74		10.14	13.70
0102	8" x 2.25" x 16", multicell		265	.151		17.10	6.95		24.05	30
0105	4" x 2.75" x 16"		412	.097		5.60	4.46		10.06	13.40
0107	8" x 2.75" x 16", multicell		295	.136		17	6.20		23.20	29

04 21 Clay Unit Masonry

04 21 13 – Brick Masonry

04 21 13.30 Oversized Brick

		Crew	Daily Output	Labor-Hours	Unit	Material	2018 Bare Costs Labor	Equipment	Total	Total Incl O&P
0110	4" x 4" x 16"	D-8	460	.087	S.F.	3.61	3.99		7.60	10.50
0120	4" x 8" x 16"		533	.075		4.26	3.44		7.70	10.30
0122	4" x 8" x 16" multicell		327	.122		16.05	5.60		21.65	27
0125	Loadbearing, 6" x 4" x 16", grouted and reinforced		387	.103		11.15	4.74		15.89	20
0130	8" x 4" x 16", grouted and reinforced		327	.122		12.25	5.60		17.85	22.50
0132	10" x 4" x 16", grouted and reinforced		327	.122		25	5.60		30.60	36.50
0135	6" x 8" x 16", grouted and reinforced		440	.091		14.40	4.17		18.57	22.50
0140	8" x 8" x 16", grouted and reinforced		400	.100		15.45	4.59		20.04	24.50
0145	Curtainwall/reinforced veneer, 6" x 4" x 16"		387	.103		16.15	4.74		20.89	25.50
0150	8" x 4" x 16"		327	.122		19.65	5.60		25.25	30.50
0152	10" x 4" x 16"		327	.122		27.50	5.60		33.10	39.50
0155	6" x 8" x 16"		440	.091		19.90	4.17		24.07	29
0160	8" x 8" x 16"		400	.100		28	4.59		32.59	38
0200	For 1 to 3 slots in face, add					15%				
0210	For 4 to 7 slots in face, add					25%				
0220	For bond beams, add					20%				
0230	For bullnose shapes, add					20%				
0240	For open end knockout, add					10%				
0250	For white or gray color group, add					10%				
0260	For 135 degree corner, add					250%				

04 21 13.35 Common Building Brick

		Crew	Daily Output	Labor-Hours	Unit	Material	2018 Bare Costs Labor	Equipment	Total	Total Incl O&P
0010	COMMON BUILDING BRICK, C62, T.L. lots, material only R042110-20									
0020	Standard				M	570			570	625
0050	Select				"	535			535	585

04 21 13.40 Structural Brick

		Crew	Daily Output	Labor-Hours	Unit	Material	2018 Bare Costs Labor	Equipment	Total	Total Incl O&P
0010	STRUCTURAL BRICK C652, Grade SW, incl. mortar, scaffolding not incl.									
0100	Standard unit, 4-5/8" x 2-3/4" x 9-5/8"	D-8	245	.163	S.F.	4.27	7.50		11.77	16.90
0120	Bond beam		225	.178		4.27	8.15		12.42	18
0140	V cut bond beam		225	.178		4.27	8.15		12.42	18
0160	Stretcher quoin, 5-5/8" x 2-3/4" x 9-5/8"		245	.163		7.90	7.50		15.40	21
0180	Corner quoin		245	.163		7.90	7.50		15.40	21
0200	Corner, 45 degree, 4-5/8" x 2-3/4" x 10-7/16"		235	.170		7.85	7.80		15.65	21.50

04 21 13.45 Face Brick

		Crew	Daily Output	Labor-Hours	Unit	Material	2018 Bare Costs Labor	Equipment	Total	Total Incl O&P
0010	FACE BRICK Material Only, C216, T.L. lots R042110-20									
0300	Standard modular, 4" x 2-2/3" x 8"				M	495			495	545
0450	Economy, 4" x 4" x 8"					870			870	955
0510	Economy, 4" x 4" x 12"					1,375			1,375	1,525
0550	Jumbo, 6" x 4" x 12"					1,650			1,650	1,825
0610	Jumbo, 8" x 4" x 12"					1,650			1,650	1,825
0650	Norwegian, 4" x 3-1/5" x 12"					1,375			1,375	1,500
0710	Norwegian, 6" x 3-1/5" x 12"					1,700			1,700	1,875
0850	Standard glazed, plain colors, 4" x 2-2/3" x 8"					1,775			1,775	1,950
1000	Deep trim shades, 4" x 2-2/3" x 8"					2,275			2,275	2,500
1080	Jumbo utility, 4" x 4" x 12"					1,550			1,550	1,700
1120	4" x 8" x 8"					1,975			1,975	2,175
1140	4" x 8" x 16"					5,850			5,850	6,425
1260	Engineer, 4" x 3-1/5" x 8"					585			585	645
1350	King, 4" x 2-3/4" x 10"					540			540	595
1770	Standard modular, double glazed, 4" x 2-2/3" x 8"					2,700			2,700	2,950
1850	Jumbo, colored glazed ceramic, 6" x 4" x 12"					2,825			2,825	3,125
2050	Jumbo utility, glazed, 4" x 4" x 12"					5,325			5,325	5,850
2100	4" x 8" x 8"					6,250			6,250	6,875

04 21 13 – Brick Masonry

04 21 13.45 Face Brick	Crew	Daily Output	Labor-Hours	Unit	Material	2018 Bare Costs Labor	Equipment	Total	Total Incl O&P
2150	4" x 16" x 8"				M	7,325			7,325	8,050
2170	For less than truck load lots, add					15			15	16.50
2180	For buff or gray brick, add					16			16	17.60
3050	Used brick					440			440	480
3150	Add for brick to match existing work, minimum					5%				
3200	Maximum					50%				

04 21 26 – Glazed Structural Clay Tile Masonry

04 21 26.10 Structural Facing Tile

		Crew	Daily Output	Labor-Hours	Unit	Material	Labor	Equipment	Total	Total Incl O&P
0010	**STRUCTURAL FACING TILE**, std. colors, excl. scaffolding, grout, reinforcing									
0020	6T series, 5-1/3" x 12", 2.3 pieces per S.F., glazed 1 side, 2" thick	D-8	225	.178	S.F.	8.65	8.15		16.80	23
0100	4" thick		220	.182		13.10	8.35		21.45	28
0150	Glazed 2 sides		195	.205		17.40	9.40		26.80	34.50
0250	6" thick		210	.190		17.60	8.75		26.35	33.50
0300	Glazed 2 sides		185	.216		21	9.90		30.90	39
0400	8" thick		180	.222		23	10.20		33.20	42
0500	Special shapes, group 1		400	.100	Ea.	8.35	4.59		12.94	16.65
0550	Group 2		375	.107		13.50	4.90		18.40	23
0600	Group 3		350	.114		17.40	5.25		22.65	27.50
0650	Group 4		325	.123		36	5.65		41.65	48.50
0700	Group 5		300	.133		43	6.10		49.10	57.50
0750	Group 6		275	.145		58	6.70		64.70	75
1000	Fire rated, 4" thick, 1 hr. rating		210	.190	S.F.	18.65	8.75		27.40	35
1050	2 hr. rating		170	.235		26.50	10.80		37.30	46.50
1300	Acoustic, 4" thick		210	.190		39.50	8.75		48.25	57.50
1310	6" thick		170	.235		40.50	10.80		51.30	62
1320	8" thick		145	.276		43	12.65		55.65	67.50
1350	8T acoustic, 4" thick		265	.151		18.05	6.95		25	31
1360	6" thick		256	.156		28.50	7.15		35.65	43
1370	8" thick		208	.192		31	8.85		39.85	48.50
2000	8W series, 8" x 16", 1.125 pieces per S.F.									
2050	2" thick, glazed 1 side	D-8	360	.111	S.F.	10.10	5.10		15.20	19.45
2100	4" thick, glazed 1 side		345	.116		13.85	5.30		19.15	24
2150	Glazed 2 sides		325	.123		16.50	5.65		22.15	27.50
2200	6" thick, glazed 1 side		330	.121		28	5.55		33.55	40
2210	Glazed 2 sides		250	.160		27	7.35		34.35	42
2250	8" thick, glazed 1 side		310	.129		29	5.90		34.90	41.50
2260	Glazed 2 sides		220	.182		34.50	8.35		42.85	51.50
2500	Special shapes, group 1		300	.133	Ea.	17	6.10		23.10	28.50
2550	Group 2		280	.143		22.50	6.55		29.05	35.50
2600	Group 3		260	.154		24	7.05		31.05	38
2650	Group 4		250	.160		50.50	7.35		57.85	67.50
2700	Group 5		240	.167		43.50	7.65		51.15	60.50
2750	Group 6		230	.174		94.50	8		102.50	117
2900	Fire rated, 2" thick, glazed 1 side		360	.111	S.F.	13.40	5.10		18.50	23
3000	4" thick, glazed 1 side		345	.116		15.90	5.30		21.20	26
3020	Glazed 2 sides		345	.116		21.50	5.30		26.80	32
3040	6" thick, glazed 1 side		330	.121		25	5.55		30.55	36.50
3045	Glazed 2 sides		250	.160		22.50	7.35		29.85	37
3047	8" thick, glazed 1 side		310	.129		30	5.90		35.90	42.50
3048	Glazed 2 sides		220	.182		31.50	8.35		39.85	48
3050	Special shapes, group 1, fire rated		300	.133		21.50	6.10		27.60	33.50
3060	Group 2		280	.143		29	6.55		35.55	42

04 21 Clay Unit Masonry

04 21 26 – Glazed Structural Clay Tile Masonry

04 21 26.10 Structural Facing Tile		Crew	Daily Output	Labor-Hours	Unit	Material	2018 Bare Costs Labor	Equipment	Total	Total Incl O&P
3070	Group 3	D-8	260	.154	S.F.	26.50	7.05		33.55	41
3080	Group 4		250	.160		47.50	7.35		54.85	64.50
3090	Group 5		240	.167		55.50	7.65		63.15	73.50
3100	Acoustic, 4" thick	↓	345	.116	↓	19.10	5.30		24.40	29.50
3120	4W series, 8" x 8", 2.25 pieces per S.F.									
3125	2" thick, glazed 1 side	D-8	360	.111	S.F.	9.90	5.10		15	19.20
3130	4" thick, glazed 1 side		345	.116		11.65	5.30		16.95	21.50
3135	Glazed 2 sides		325	.123		16.05	5.65		21.70	27
3140	6" thick, glazed 1 side		330	.121		16.60	5.55		22.15	27.50
3145	Glazed 2 sides		250	.160		20	7.35		27.35	34
3150	8" thick, glazed 1 side		310	.129	↓	24	5.90		29.90	36
3155	Special shapes, group 1		300	.133	Ea.	7.75	6.10		13.85	18.55
3160	Group 2	↓	280	.143	"	8.95	6.55		15.50	20.50
3200	For designer colors, add					25%				
3300	For epoxy mortar joints, add				S.F.	1.79			1.79	1.97
9000	Minimum labor/equipment charge	D-1	2	8	Job		360		360	585

04 21 29 – Terra Cotta Masonry

04 21 29.10 Terra Cotta Masonry Components

		Crew	Daily Output	Labor-Hours	Unit	Material	2018 Bare Costs Labor	Equipment	Total	Total Incl O&P
0010	**TERRA COTTA MASONRY COMPONENTS**									
0020	Coping, split type, not glazed, 9" wide	D-1	90	.178	L.F.	10.50	7.95		18.45	24.50
0100	13" wide		80	.200		15.35	8.95		24.30	31.50
0200	Coping, split type, glazed, 9" wide		90	.178		17.85	7.95		25.80	32.50
0250	13" wide	↓	80	.200	↓	23.50	8.95		32.45	40.50
0500	Partition or back-up blocks, scored, in C.L. lots									
0700	Non-load bearing 12" x 12", 3" thick, special order	D-8	550	.073	S.F.	19.50	3.34		22.84	27
0750	4" thick, standard		500	.080		6.10	3.67		9.77	12.70
0800	6" thick		450	.089		8.20	4.08		12.28	15.70
0850	8" thick		400	.100		10.30	4.59		14.89	18.85
1000	Load bearing, 12" x 12", 4" thick, in walls		500	.080		5.75	3.67		9.42	12.30
1050	In floors		750	.053		5.75	2.45		8.20	10.30
1200	6" thick, in walls		450	.089		9.30	4.08		13.38	16.90
1250	In floors		675	.059		9.30	2.72		12.02	14.70
1400	8" thick, in walls		400	.100		11.55	4.59		16.14	20.50
1450	In floors		575	.070		11.55	3.19		14.74	17.95
1600	10" thick, in walls, special order		350	.114		27	5.25		32.25	38.50
1650	In floors, special order		500	.080		27	3.67		30.67	36
1800	12" thick, in walls, special order		300	.133		26.50	6.10		32.60	39
1850	In floors, special order	↓	450	.089		26.50	4.08		30.58	35.50
2000	For reinforcing with steel rods, add to above				↓	15%	5%			
2025	Placed vertical, #3 & #4 bars	1 Bric	350	.023	Lb.	.48	1.15		1.63	2.41
2100	For smooth tile instead of scored, add				S.F.	2.83			2.83	3.11
2200	For L.C.L. quantities, add				"	10%	10%			
9000	Minimum labor/equipment charge	D-1	2	8	Job		360		360	585

04 21 29.20 Terra Cotta Tile

		Crew	Daily Output	Labor-Hours	Unit	Material	2018 Bare Costs Labor	Equipment	Total	Total Incl O&P
0010	**TERRA COTTA TILE**, on walls, dry set, 1/2" thick									
0100	Square, hexagonal or lattice shapes, unglazed	1 Tilf	135	.059	S.F.	4.74	2.78		7.52	9.60
0300	Glazed, plain colors		130	.062		7.25	2.88		10.13	12.50
0400	Intense colors	↓	125	.064	↓	8.50	3		11.50	14.10

04 22 10.11 Autoclave Aerated Concrete Block

		Crew	Daily Output	Labor-Hours	Unit	Material	2018 Bare Costs Labor	Equipment	Total	Total Incl O&P
0010	**AUTOCLAVE AERATED CONCRETE BLOCK**, excl. scaffolding, grout & reinforcing									
0050	Solid, 4" x 8" x 24", incl. mortar [G]	D-8	600	.067	S.F.	1.50	3.06		4.56	6.65
0060	6" x 8" x 24" [G]		600	.067		2.25	3.06		5.31	7.45
0070	8" x 8" x 24" [G]		575	.070		3	3.19		6.19	8.50
0080	10" x 8" x 24" [G]		575	.070		3.66	3.19		6.85	9.20
0090	12" x 8" x 24" [G]		550	.073		4.50	3.34		7.84	10.40

04 22 10.14 Concrete Block, Back-Up

		Crew	Daily Output	Labor-Hours	Unit	Material	2018 Bare Costs Labor	Equipment	Total	Total Incl O&P
0010	**CONCRETE BLOCK, BACK-UP**, C90, 2000 psi									
0020	Normal weight, 8" x 16" units, tooled joint 1 side									
0050	Not-reinforced, 2000 psi, 2" thick	D-8	475	.084	S.F.	1.60	3.87		5.47	8.05
0200	4" thick		460	.087		1.92	3.99		5.91	8.60
0300	6" thick		440	.091		2.50	4.17		6.67	9.55
0350	8" thick		400	.100		2.66	4.59		7.25	10.45
0400	10" thick		330	.121		3.15	5.55		8.70	12.55
0450	12" thick	D-9	310	.155		4.36	6.95		11.31	16.10
1000	Reinforced, alternate courses, 4" thick	D-8	450	.089		2.10	4.08		6.18	8.95
1100	6" thick		430	.093		2.68	4.27		6.95	9.90
1150	8" thick		395	.101		2.88	4.65		7.53	10.75
1200	10" thick		320	.125		3.32	5.75		9.07	13
1250	12" thick	D-9	300	.160		4.54	7.15		11.69	16.70
9000	Minimum labor/equipment charge	D-1	2	8	Job		360		360	585

04 22 10.16 Concrete Block, Bond Beam

		Crew	Daily Output	Labor-Hours	Unit	Material	2018 Bare Costs Labor	Equipment	Total	Total Incl O&P
0010	**CONCRETE BLOCK, BOND BEAM**, C90, 2000 psi									
0020	Not including grout or reinforcing									
0125	Regular block, 6" thick	D-8	584	.068	L.F.	2.75	3.14		5.89	8.20
0130	8" high, 8" thick	"	565	.071		2.86	3.25		6.11	8.45
0150	12" thick	D-9	510	.094		4.05	4.22		8.27	11.35
0525	Lightweight, 6" thick	D-8	592	.068		2.98	3.10		6.08	8.30
0530	8" high, 8" thick	"	575	.070		3.59	3.19		6.78	9.15
0550	12" thick	D-9	520	.092		4.84	4.14		8.98	12.05
2000	Including grout and 2 #5 bars									
2100	Regular block, 8" high, 8" thick	D-8	300	.133	L.F.	5.05	6.10		11.15	15.55
2150	12" thick	D-9	250	.192		6.90	8.60		15.50	21.50
2500	Lightweight, 8" high, 8" thick	D-8	305	.131		5.80	6		11.80	16.20
2550	12" thick	D-9	255	.188		7.70	8.45		16.15	22
9000	Minimum labor/equipment charge	D-1	2	8	Job		360		360	585

04 22 10.19 Concrete Block, Insulation Inserts

		Crew	Daily Output	Labor-Hours	Unit	Material	2018 Bare Costs Labor	Equipment	Total	Total Incl O&P
0010	**CONCRETE BLOCK, INSULATION INSERTS**									
0100	Styrofoam, plant installed, add to block prices									
0200	8" x 16" units, 6" thick				S.F.	1.22			1.22	1.34
0250	8" thick					1.37			1.37	1.51
0300	10" thick					1.42			1.42	1.56
0350	12" thick					1.57			1.57	1.73
0500	8" x 8" units, 8" thick					1.22			1.22	1.34
0550	12" thick					1.42			1.42	1.56

04 22 10.23 Concrete Block, Decorative

		Crew	Daily Output	Labor-Hours	Unit	Material	2018 Bare Costs Labor	Equipment	Total	Total Incl O&P
0010	**CONCRETE BLOCK, DECORATIVE**, C90, 2000 psi									
0020	Embossed, simulated brick face									
0100	8" x 16" units, 4" thick	D-8	400	.100	S.F.	2.88	4.59		7.47	10.65
0200	8" thick		340	.118		3.13	5.40		8.53	12.25
0250	12" thick		300	.133		5.30	6.10		11.40	15.80
0400	Embossed both sides									

88

For customer support on your Commercial Renovation Costs with RSMeans data, call 800.448.8182.

04 22 10 – Concrete Masonry Units

04 22 10.23 Concrete Block, Decorative	Crew	Daily Output	Labor-Hours	Unit	Material	2018 Bare Costs Labor	Equipment	Total	Total Incl O&P
0500	8" thick	D-8	300	.133	S.F.	4.23	6.10		10.33	14.65
0550	12" thick	"	275	.145	"	5.55	6.70		12.25	17
1000	Fluted high strength									
1100	8" x 16" x 4" thick, flutes 1 side,	D-8	345	.116	S.F.	4.09	5.30		9.39	13.20
1150	Flutes 2 sides		335	.119		4.69	5.50		10.19	14.10
1200	8" thick	↓	300	.133		5.50	6.10		11.60	16.05
1250	For special colors, add				↓	.67			.67	.73
1400	Deep grooved, smooth face									
1450	8" x 16" x 4" thick	D-8	345	.116	S.F.	2.63	5.30		7.93	11.60
1500	8" thick	"	300	.133	"	4.14	6.10		10.24	14.55
2000	Formblock, incl. inserts & reinforcing									
2100	8" x 16" x 8" thick	D-8	345	.116	S.F.	3.65	5.30		8.95	12.70
2150	12" thick	"	310	.129	"	4.83	5.90		10.73	14.95
2500	Ground face									
2600	8" x 16" x 4" thick	D-8	345	.116	S.F.	3.97	5.30		9.27	13.05
2650	6" thick		325	.123		4.74	5.65		10.39	14.40
2700	8" thick		300	.133		5.20	6.10		11.30	15.75
2725	10" thick	↓	280	.143		6.05	6.55		12.60	17.35
2750	12" thick	D-9	265	.181	↓	5.70	8.10		13.80	19.55
2900	For special colors, add, minimum					15%				
2950	For special colors, add, maximum					45%				
4000	Slump block									
4100	4" face height x 16" x 4" thick	D-1	165	.097	S.F.	4.37	4.34		8.71	11.90
4150	6" thick		160	.100		6.25	4.48		10.73	14.20
4200	8" thick		155	.103		6.20	4.62		10.82	14.40
4250	10" thick		140	.114		12.95	5.10		18.05	22.50
4300	12" thick		130	.123		13.10	5.50		18.60	23.50
4400	6" face height x 16" x 6" thick		155	.103		5.80	4.62		10.42	13.90
4450	8" thick		150	.107		8.75	4.78		13.53	17.45
4500	10" thick		130	.123		13.45	5.50		18.95	24
4550	12" thick	↓	120	.133	↓	13.95	5.95		19.90	25
5000	Split rib profile units, 1" deep ribs, 8 ribs									
5100	8" x 16" x 4" thick	D-8	345	.116	S.F.	4.06	5.30		9.36	13.15
5150	6" thick		325	.123		4.61	5.65		10.26	14.25
5200	8" thick		300	.133		5.20	6.10		11.30	15.70
5225	10" thick	↓	300	.133		5.70	6.10		11.80	16.30
5250	12" thick	D-9	275	.175		6.05	7.80		13.85	19.40
5400	For special deeper colors, 4" thick, add					1.37			1.37	1.50
5450	12" thick, add					1.42			1.42	1.57
5600	For white, 4" thick, add					1.37			1.37	1.50
5650	6" thick, add					1.39			1.39	1.53
5700	8" thick, add					1.44			1.44	1.58
5750	12" thick, add				↓	1.48			1.48	1.63
6000	Split face									
6100	8" x 16" x 4" thick	D-8	350	.114	S.F.	3.77	5.25		9.02	12.70
6150	6" thick		325	.123		4.28	5.65		9.93	13.90
6200	8" thick		300	.133		4.82	6.10		10.92	15.30
6225	10" thick	↓	275	.145		5.45	6.70		12.15	16.90
6250	12" thick	D-9	270	.178		5.85	7.95		13.80	19.40
6300	For scored, add					.40			.40	.44
6400	For special deeper colors, 4" thick, add					.67			.67	.73
6450	6" thick, add					.79			.79	.87
6500	8" thick, add					.81			.81	.89

04 22 Concrete Unit Masonry

04 22 10 – Concrete Masonry Units

04 22 10.23 Concrete Block, Decorative

		Crew	Daily Output	Labor-Hours	Unit	Material	2018 Bare Costs Labor	Equipment	Total	Total Incl O&P
6550	12" thick, add				S.F.	.84			.84	.92
6650	For white, 4" thick, add					1.36			1.36	1.49
6700	6" thick, add					1.37			1.37	1.50
6750	8" thick, add					1.38			1.38	1.52
6800	12" thick, add					1.42			1.42	1.57
7000	Scored ground face, 2 to 5 scores									
7100	8" x 16" x 4" thick	D-8	340	.118	S.F.	8.80	5.40		14.20	18.45
7150	6" thick		310	.129		9.70	5.90		15.60	20.50
7200	8" thick		290	.138		10.90	6.35		17.25	22.50
7250	12" thick	D-9	265	.181		14.70	8.10		22.80	29.50
8000	Hexagonal face profile units, 8" x 16" units									
8100	4" thick, hollow	D-8	340	.118	S.F.	3.83	5.40		9.23	13
8200	Solid		340	.118		4.90	5.40		10.30	14.20
8300	6" thick, hollow		310	.129		4.07	5.90		9.97	14.10
8350	8" thick, hollow		290	.138		4.64	6.35		10.99	15.45
8500	For stacked bond, add						26%			
8550	For high rise construction, add per story	D-8	67.80	.590	M.S.F.		27		27	44
8600	For scored block, add					10%				
8650	For honed or ground face, per face, add				Ea.	1.21			1.21	1.33
8700	For honed or ground end, per end, add				"	1.21			1.21	1.33
8750	For bullnose block, add					10%				
8800	For special color, add					13%				
9000	Minimum labor/equipment charge	D-1	2	8	Job		360		360	585

04 22 10.24 Concrete Block, Exterior

		Crew	Daily Output	Labor-Hours	Unit	Material	2018 Bare Costs Labor	Equipment	Total	Total Incl O&P
0010	**CONCRETE BLOCK, EXTERIOR**, C90, 2000 psi									
0020	Reinforced alt courses, tooled joints 2 sides									
0100	Normal weight, 8" x 16" x 6" thick	D-8	395	.101	S.F.	2.46	4.65		7.11	10.30
0200	8" thick		360	.111		3.80	5.10		8.90	12.50
0250	10" thick		290	.138		4.41	6.35		10.76	15.20
0300	12" thick	D-9	250	.192		5.15	8.60		13.75	19.70
9000	Minimum labor/equipment charge	D-1	2	8	Job		360		360	585

04 22 10.26 Concrete Block Foundation Wall

		Crew	Daily Output	Labor-Hours	Unit	Material	2018 Bare Costs Labor	Equipment	Total	Total Incl O&P
0010	**CONCRETE BLOCK FOUNDATION WALL**, C90/C145									
0050	Normal-weight, cut joints, horiz joint reinf, no vert reinf.									
0200	Hollow, 8" x 16" x 6" thick	D-8	455	.088	S.F.	2.93	4.04		6.97	9.80
0250	8" thick		425	.094		3.11	4.32		7.43	10.45
0300	10" thick		350	.114		3.57	5.25		8.82	12.45
0350	12" thick	D-9	300	.160		4.78	7.15		11.93	16.95
0500	Solid, 8" x 16" block, 6" thick	D-8	440	.091		2.99	4.17		7.16	10.10
0550	8" thick	"	415	.096		4.23	4.42		8.65	11.85
0600	12" thick	D-9	350	.137		6.10	6.15		12.25	16.70
1000	Reinforced, #4 vert @ 48"									
1100	Hollow, 8" x 16" block, 4" thick	D-8	455	.088	S.F.	2.95	4.04		6.99	9.85
1125	6" thick		445	.090		3.90	4.13		8.03	11.05
1150	8" thick		415	.096		4.51	4.42		8.93	12.15
1200	10" thick		340	.118		5.35	5.40		10.75	14.70
1250	12" thick	D-9	290	.166		7	7.40		14.40	19.80
1500	Solid, 8" x 16" block, 6" thick	D-8	430	.093		3.02	4.27		7.29	10.25
1600	8" thick	"	405	.099		4.23	4.53		8.76	12.05
1650	12" thick	D-9	340	.141		6.10	6.30		12.40	17
9000	Minimum labor/equipment charge	D-1	2	8	Job		360		360	585

For customer support on your Commercial Renovation Costs with RSMeans data, call 800.448.8182.

04 22 Concrete Unit Masonry

04 22 10 – Concrete Masonry Units

04 22 10.30 Concrete Block, Interlocking

	04 22 10.30 Concrete Block, Interlocking	Crew	Daily Output	Labor-Hours	Unit	Material	2018 Bare Costs Labor	Equipment	Total	Total Incl O&P
0010	**CONCRETE BLOCK, INTERLOCKING**									
0100	Not including grout or reinforcing									
0200	8" x 16" units, 2,000 psi, 8" thick	D-1	245	.065	S.F.	3.15	2.93		6.08	8.25
0300	12" thick	"	220	.073		4.69	3.26		7.95	10.45
0400	Including grout & reinforcing, 8" thick	D-4	245	.131		8.60	5.90	.53	15.03	19.55
0450	12" thick		220	.145		10.30	6.55	.58	17.43	22.50
0500	16" thick		185	.173		12.85	7.80	.70	21.35	27.50
9000	Minimum labor/equipment charge	D-1	2	8	Job		360		360	585

04 22 10.32 Concrete Block, Lintels

	04 22 10.32 Concrete Block, Lintels	Crew	Daily Output	Labor-Hours	Unit	Material	2018 Bare Costs Labor	Equipment	Total	Total Incl O&P
0010	**CONCRETE BLOCK, LINTELS**, C90, normal weight									
0100	Including grout and horizontal reinforcing									
0200	8" x 8" x 8", 1 #4 bar	D-4	300	.107	L.F.	4.03	4.81	.43	9.27	12.65
0250	2 #4 bars		295	.108		4.24	4.89	.44	9.57	13.05
0400	8" x 16" x 8", 1 #4 bar		275	.116		3.69	5.25	.47	9.41	13
0450	2 #4 bars		270	.119		3.91	5.35	.48	9.74	13.45
1000	12" x 8" x 8", 1 #4 bar		275	.116		5.50	5.25	.47	11.22	15
1100	2 #4 bars		270	.119		5.75	5.35	.48	11.58	15.45
1150	2 #5 bars		270	.119		6	5.35	.48	11.83	15.70
1200	2 #6 bars		265	.121		6.25	5.45	.49	12.19	16.25
1500	12" x 16" x 8", 1 #4 bar		250	.128		6.60	5.75	.51	12.86	17.10
1600	2 #3 bars		245	.131		6.60	5.90	.53	13.03	17.35
1650	2 #4 bars		245	.131		6.80	5.90	.53	13.23	17.55
1700	2 #5 bars		240	.133		7.05	6	.54	13.59	18.05
9000	Minimum labor/equipment charge	D-1	2	8	Job		360		360	585

04 22 10.33 Lintel Block

	04 22 10.33 Lintel Block	Crew	Daily Output	Labor-Hours	Unit	Material	2018 Bare Costs Labor	Equipment	Total	Total Incl O&P
0010	**LINTEL BLOCK**									
3481	Lintel block 6" x 8" x 8"	D-1	300	.053	Ea.	1.36	2.39		3.75	5.40
3501	6" x 16" x 8"		275	.058		2.10	2.61		4.71	6.55
3521	8" x 8" x 8"		275	.058		1.21	2.61		3.82	5.60
3561	8" x 16" x 8"		250	.064		1.90	2.87		4.77	6.75

04 22 10.34 Concrete Block, Partitions

	04 22 10.34 Concrete Block, Partitions	Crew	Daily Output	Labor-Hours	Unit	Material	2018 Bare Costs Labor	Equipment	Total	Total Incl O&P
0010	**CONCRETE BLOCK, PARTITIONS**, excludes scaffolding									
1000	Lightweight block, tooled joints, 2 sides, hollow									
1100	Not reinforced, 8" x 16" x 4" thick	D-8	440	.091	S.F.	1.95	4.17		6.12	8.95
1150	6" thick		410	.098		2.75	4.48		7.23	10.35
1200	8" thick		385	.104		3.36	4.77		8.13	11.50
1250	10" thick		370	.108		4.07	4.96		9.03	12.60
1300	12" thick	D-9	350	.137		4.27	6.15		10.42	14.70
1500	Reinforced alternate courses, 4" thick	D-8	435	.092		2.12	4.22		6.34	9.25
1600	6" thick		405	.099		2.90	4.53		7.43	10.60
1650	8" thick		380	.105		3.52	4.83		8.35	11.75
1700	10" thick		365	.110		4.25	5.05		9.30	12.90
1750	12" thick	D-9	345	.139		4.46	6.25		10.71	15.05
4000	Regular block, tooled joints, 2 sides, hollow									
4100	Not reinforced, 8" x 16" x 4" thick	D-8	430	.093	S.F.	1.82	4.27		6.09	8.95
4150	6" thick		400	.100		2.40	4.59		6.99	10.15
4200	8" thick		375	.107		2.57	4.90		7.47	10.80
4250	10" thick		360	.111		3.05	5.10		8.15	11.65
4300	12" thick	D-9	340	.141		4.26	6.30		10.56	15
4500	Reinforced alternate courses, 8" x 16" x 4" thick	D-8	425	.094		2	4.32		6.32	9.25
4550	6" thick		395	.101		2.58	4.65		7.23	10.45

04 22 Concrete Unit Masonry

04 22 10 – Concrete Masonry Units

04 22 10.34 Concrete Block, Partitions		Crew	Daily Output	Labor-Hours	Unit	Material	2018 Bare Costs Labor	Equipment	Total	Total Incl O&P
4600	8" thick	D-8	370	.108	S.F.	2.79	4.96		7.75	11.15
4650	10" thick	↓	355	.113		4.30	5.15		9.45	13.20
4700	12" thick	D-9	335	.143	↓	4.45	6.40		10.85	15.35
9000	Minimum labor/equipment charge	D-1	2	8	Job		360		360	585

04 22 10.38 Concrete Brick

		Crew	Daily Output	Labor-Hours	Unit	Material	2018 Bare Costs Labor	Equipment	Total	Total Incl O&P
0010	**CONCRETE BRICK**, C55, grade N, type 1									
0100	Regular, 4" x 2-1/4" x 8"	D-8	660	.061	Ea.	.49	2.78		3.27	5.10
0125	Rusticated, 4" x 2-1/4" x 8"		660	.061		.55	2.78		3.33	5.15
0150	Frog, 4" x 2-1/4" x 8"		660	.061		.53	2.78		3.31	5.10
0200	Double, 4" x 4-7/8" x 8"	↓	535	.075	↓	.86	3.43		4.29	6.55

04 22 10.44 Glazed Concrete Block

		Crew	Daily Output	Labor-Hours	Unit	Material	2018 Bare Costs Labor	Equipment	Total	Total Incl O&P
0010	**GLAZED CONCRETE BLOCK** C744									
0100	Single face, 8" x 16" units, 2" thick	D-8	360	.111	S.F.	10.80	5.10		15.90	20
0200	4" thick		345	.116		11.10	5.30		16.40	21
0250	6" thick		330	.121		12	5.55		17.55	22.50
0300	8" thick		310	.129		12.60	5.90		18.50	23.50
0350	10" thick	↓	295	.136		14.65	6.20		20.85	26.50
0400	12" thick	D-9	280	.171		15.60	7.70		23.30	29.50
0700	Double face, 8" x 16" units, 4" thick	D-8	340	.118		15.95	5.40		21.35	26.50
0750	6" thick		320	.125		19.05	5.75		24.80	30.50
0800	8" thick		300	.133	↓	20	6.10		26.10	32
1000	Jambs, bullnose or square, single face, 8" x 16", 2" thick		315	.127	Ea.	20.50	5.85		26.35	32
1050	4" thick		285	.140	"	21	6.45		27.45	33.50
1200	Caps, bullnose or square, 8" x 16", 2" thick		420	.095	L.F.	19.15	4.37		23.52	28
1250	4" thick		380	.105	"	21.50	4.83		26.33	31.50
1256	Corner, bullnose or square, 2" thick		280	.143	Ea.	22.50	6.55		29.05	35
1258	4" thick		270	.148		25	6.80		31.80	38.50
1260	6" thick		260	.154		30	7.05		37.05	44.50
1270	8" thick		250	.160		36.50	7.35		43.85	52.50
1280	10" thick		240	.167		37	7.65		44.65	53.50
1290	12" thick	↓	230	.174	↓	39	8		47	56
1500	Cove base, 8" x 16", 2" thick		315	.127	L.F.	9.90	5.85		15.75	20.50
1550	4" thick		285	.140		10	6.45		16.45	21.50
1600	6" thick		265	.151		10.70	6.95		17.65	23
1650	8" thick	↓	245	.163	↓	11.20	7.50		18.70	24.50
9000	Minimum labor/equipment charge	D-1	2	8	Job		360		360	585

04 23 Glass Unit Masonry

04 23 13 – Vertical Glass Unit Masonry

04 23 13.10 Glass Block

		Crew	Daily Output	Labor-Hours	Unit	Material	2018 Bare Costs Labor	Equipment	Total	Total Incl O&P
0010	**GLASS BLOCK**									
0100	Plain, 4" thick, under 1,000 S.F., 6" x 6"	D-8	115	.348	S.F.	26	15.95		41.95	55
0150	8" x 8"		160	.250		16.40	11.50		27.90	37
0160	end block		160	.250		60.50	11.50		72	85
0170	90 degree corner		160	.250		63.50	11.50		75	88
0180	45 degree corner		160	.250		54	11.50		65.50	77.50
0200	12" x 12"		175	.229		24	10.50		34.50	43.50
0210	4" x 8"		160	.250		30.50	11.50		42	52
0220	6" x 8"		160	.250		19.90	11.50		31.40	40.50
0300	1,000 to 5,000 S.F., 6" x 6"	↓	135	.296		25.50	13.60		39.10	50

04 23 Glass Unit Masonry

04 23 13 – Vertical Glass Unit Masonry

04 23 13.10 Glass Block

		Crew	Daily Output	Labor-Hours	Unit	Material	2018 Bare Costs Labor	Equipment	Total	Total Incl O&P
0350	8" x 8"	D-8	190	.211	S.F.	16.05	9.65		25.70	33.50
0400	12" x 12"		215	.186		24	8.55		32.55	40
0410	4" x 8"		215	.186		29.50	8.55		38.05	46.50
0420	6" x 8"		215	.186		19.50	8.55		28.05	35.50
0500	Over 5,000 S.F., 6" x 6"	↓	145	.276	↓	25	12.65		37.65	48
0700	For solar reflective blocks, add					100%				
1000	Thinline, plain, 3-1/8" thick, under 1,000 S.F., 6" x 6"	D-8	115	.348	S.F.	23.50	15.95		39.45	52
1050	8" x 8"		160	.250		13.15	11.50		24.65	33
1250	8" x 8"		215	.186		12.60	8.55		21.15	28
1400	For cleaning block after installation (both sides), add	↓	1000	.040	↓	.16	1.84		2	3.17
9000	Minimum labor/equipment charge	D-1	2	8	Job		360		360	585

04 27 Multiple-Wythe Unit Masonry

04 27 10 – Multiple-Wythe Masonry

04 27 10.10 Cornices

		Crew	Daily Output	Labor-Hours	Unit	Material	2018 Bare Costs Labor	Equipment	Total	Total Incl O&P
0010	**CORNICES**									
0110	Face bricks, 12 brick/S.F.	D-1	30	.533	SF Face	5.85	24		29.85	45.50
0150	15 brick/S.F.		23	.696	"	7	31		38	58.50
9000	Minimum labor/equipment charge	↓	1.50	10.667	Job		480		480	780

04 27 10.30 Brick Walls

		Crew	Daily Output	Labor-Hours	Unit	Material	2018 Bare Costs Labor	Equipment	Total	Total Incl O&P
0010	**BRICK WALLS**, including mortar, excludes scaffolding									
0020	Estimating by number of brick									
0140	Face brick, 4" thick wall, 6.75 brick/S.F.	D-8	1.45	27.586	M	590	1,275		1,865	2,725
0150	Common brick, 4" thick wall, 6.75 brick/S.F.		1.60	25		665	1,150		1,815	2,600
0204	8" thick, 13.50 brick/S.F.		1.80	22.222		685	1,025		1,710	2,425
0250	12" thick, 20.25 brick/S.F.		1.90	21.053		690	965		1,655	2,325
0304	16" thick, 27.00 brick/S.F.		2	20		695	920		1,615	2,275
0500	Reinforced, face brick, 4" thick wall, 6.75 brick/S.F.		1.40	28.571		615	1,300		1,915	2,825
0520	Common brick, 4" thick wall, 6.75 brick/S.F.		1.55	25.806		690	1,175		1,865	2,675
0550	8" thick, 13.50 brick/S.F.		1.75	22.857		710	1,050		1,760	2,475
0600	12" thick, 20.25 brick/S.F.		1.85	21.622		710	990		1,700	2,400
0650	16" thick, 27.00 brick/S.F.	↓	1.95	20.513	↓	720	940		1,660	2,325
0790	Alternate method of figuring by square foot									
0800	Face brick, 4" thick wall, 6.75 brick/S.F.	D-8	215	.186	S.F.	3.99	8.55		12.54	18.35
0850	Common brick, 4" thick wall, 6.75 brick/S.F. R042110-50		240	.167		4.50	7.65		12.15	17.45
0900	8" thick, 13.50 brick/S.F.		135	.296		9.25	13.60		22.85	32
1000	12" thick, 20.25 brick/S.F.		95	.421		13.90	19.35		33.25	47
1050	16" thick, 27.00 brick/S.F.		75	.533		18.75	24.50		43.25	60.50
1200	Reinforced, face brick, 4" thick wall, 6.75 brick/S.F.		210	.190		4.15	8.75		12.90	18.80
1220	Common brick, 4" thick wall, 6.75 brick/S.F.		235	.170		4.66	7.80		12.46	17.90
1250	8" thick, 13.50 brick/S.F.		130	.308		9.55	14.10		23.65	33.50
1300	12" thick, 20.25 brick/S.F.		90	.444		14.40	20.50		34.90	49.50
1350	16" thick, 27.00 brick/S.F.	↓	70	.571	↓	19.40	26		45.40	64.50
9000	Minimum labor/equipment charge	D-1	2	8	Job		360		360	585

04 27 10.40 Steps

		Crew	Daily Output	Labor-Hours	Unit	Material	2018 Bare Costs Labor	Equipment	Total	Total Incl O&P
0010	**STEPS**									
0012	Entry steps, select common brick	D-1	.30	53.333	M	535	2,400		2,935	4,475

For customer support on your Commercial Renovation Costs with RSMeans data, call 800.448.8182.

93

04 41 Dry-Placed Stone

04 41 10 – Dry Placed Stone

04 41 10.10 Rough Stone Wall

		Crew	Daily Output	Labor-Hours	Unit	Material	2018 Bare Costs Labor	2018 Bare Costs Equipment	Total	Total Incl O&P
0011	**ROUGH STONE WALL**, Dry									
0012	Dry laid (no mortar), under 18" thick Ⓖ	D-1	60	.267	C.F.	13.50	11.95		25.45	34.50
0100	Random fieldstone, under 18" thick Ⓖ	D-12	60	.533		13.50	24		37.50	54.50
0150	Over 18" thick Ⓖ	"	63	.508	↓	16.20	23		39.20	55.50
0500	Field stone veneer Ⓖ	D-8	120	.333	S.F.	12.65	15.30		27.95	39
0510	Valley stone veneer Ⓖ	↓	120	.333		12.65	15.30		27.95	39
0520	River stone veneer Ⓖ		120	.333	↓	12.65	15.30		27.95	39
0600	Rubble stone walls, in mortar bed, up to 18" thick Ⓖ	D-11	75	.320	C.F.	16.30	15.15		31.45	42.50
9000	Minimum labor/equipment charge	D-1	2	8	Job		360		360	585

04 43 Stone Masonry

04 43 10 – Masonry with Natural and Processed Stone

04 43 10.05 Ashlar Veneer

		Crew	Daily Output	Labor-Hours	Unit	Material	2018 Bare Costs Labor	2018 Bare Costs Equipment	Total	Total Incl O&P
0011	**ASHLAR VENEER** +/- 4" thk, random or random rectangular									
0150	Sawn face, split joints, low priced stone	D-8	140	.286	S.F.	12.20	13.10		25.30	35
0200	Medium priced stone		130	.308		13.85	14.10		27.95	38
0300	High priced stone		120	.333		18.40	15.30		33.70	45
0600	Seam face, split joints, medium price stone		125	.320		19.35	14.70		34.05	45.50
0700	High price stone		120	.333		19.10	15.30		34.40	46
1000	Split or rock face, split joints, medium price stone		125	.320		11.55	14.70		26.25	36.50
1100	High price stone	↓	120	.333	↓	17.55	15.30		32.85	44.50

04 43 10.10 Bluestone

		Crew	Daily Output	Labor-Hours	Unit	Material	2018 Bare Costs Labor	2018 Bare Costs Equipment	Total	Total Incl O&P
0010	**BLUESTONE**, cut to size									
0500	Sills, natural cleft, 10" wide to 6' long, 1-1/2" thick	D-11	70	.343	L.F.	13.75	16.20		29.95	41.50
0550	2" thick		63	.381		15.05	18		33.05	46
0600	Smooth finish, 1-1/2" thick		70	.343		12.85	16.20		29.05	40.50
0650	2" thick		63	.381		13.90	18		31.90	45
0800	Thermal finish, 1-1/2" thick		70	.343		12.85	16.20		29.05	40.50
0850	2" thick	↓	63	.381		13.90	18		31.90	45
1000	Stair treads, natural cleft, 12" wide, 6' long, 1-1/2" thick	D-10	115	.278		12.85	13.75	3.16	29.76	40
1050	2" thick		105	.305		13.90	15.10	3.46	32.46	43.50
1100	Smooth finish, 1-1/2" thick		115	.278		12.85	13.75	3.16	29.76	40
1150	2" thick		105	.305		13.90	15.10	3.46	32.46	43.50
1300	Thermal finish, 1-1/2" thick		115	.278		12.85	13.75	3.16	29.76	40
1350	2" thick	↓	105	.305	↓	13.90	15.10	3.46	32.46	43.50
9000	Minimum labor/equipment charge	D-1	2.50	6.400	Job		287		287	470

04 43 10.45 Granite

		Crew	Daily Output	Labor-Hours	Unit	Material	2018 Bare Costs Labor	2018 Bare Costs Equipment	Total	Total Incl O&P
0010	**GRANITE**, cut to size									
0050	Veneer, polished face, 3/4" to 1-1/2" thick									
0150	Low price, gray, light gray, etc.	D-10	130	.246	S.F.	27.50	12.20	2.80	42.50	53.50
0180	Medium price, pink, brown, etc.		130	.246		30.50	12.20	2.80	45.50	56.50
0220	High price, red, black, etc.	↓	130	.246	↓	43	12.20	2.80	58	70
0300	1-1/2" to 2-1/2" thick, veneer									
0350	Low price, gray, light gray, etc.	D-10	130	.246	S.F.	29	12.20	2.80	44	55
0500	Medium price, pink, brown, etc.	"	130	.246	"	34	12.20	2.80	49	60.50
2500	Steps, copings, etc., finished on more than one surface									
2550	Low price, gray, light gray, etc.	D-10	50	.640	C.F.	94	31.50	7.25	132.75	162
2575	Medium price, pink, brown, etc.		50	.640		122	31.50	7.25	160.75	193
2600	High price, red, black, etc.	↓	50	.640	↓	150	31.50	7.25	188.75	224
9000	Minimum labor/equipment charge	D-1	2	8	Job		360		360	585

04 43 Stone Masonry

04 43 10 – Masonry with Natural and Processed Stone

04 43 10.55 Limestone

	04 43 10.55 Limestone	Crew	Daily Output	Labor-Hours	Unit	Material	2018 Bare Costs Labor	Equipment	Total	Total Incl O&P
0010	**LIMESTONE**, cut to size									
0020	Veneer facing panels									
0500	Texture finish, light stick, 4-1/2" thick, 5' x 12'	D-4	300	.107	S.F.	20.50	4.81	.43	25.74	30.50
0750	5" thick, 5' x 14' panels	D-10	275	.116		21.50	5.75	1.32	28.57	35
1000	Sugarcube finish, 2" thick, 3' x 5' panels		275	.116		29	5.75	1.32	36.07	43
1050	3" thick, 4' x 9' panels		275	.116		25.50	5.75	1.32	32.57	39.50
1200	4" thick, 5' x 11' panels		275	.116		31.50	5.75	1.32	38.57	45.50
1400	Sugarcube, textured finish, 4-1/2" thick, 5' x 12'		275	.116		32.50	5.75	1.32	39.57	47
1450	5" thick, 5' x 14' panels		275	.116		34	5.75	1.32	41.07	48
2000	Coping, sugarcube finish, top & 2 sides		30	1.067	C.F.	68.50	53	12.10	133.60	174
2100	Sills, lintels, jambs, trim, stops, sugarcube finish, simple		20	1.600		68.50	79	18.20	165.70	224
2150	Detailed		20	1.600		68.50	79	18.20	165.70	224
2300	Steps, extra hard, 14" wide, 6" rise		50	.640	L.F.	25.50	31.50	7.25	64.25	87
3000	Quoins, plain finish, 6" x 12" x 12"	D-12	25	1.280	Ea.	41	58		99	140
3050	6" x 16" x 24"	"	25	1.280	"	55	58		113	155
9000	Minimum labor/equipment charge	D-1	2	8	Job		360		360	585

04 43 10.60 Marble

	04 43 10.60 Marble	Crew	Daily Output	Labor-Hours	Unit	Material	2018 Bare Costs Labor	Equipment	Total	Total Incl O&P
0011	**MARBLE**, ashlar, split face, +/- 4" thick, random									
0040	Lengths 1' to 4' & heights 2" to 7-1/2", average	D-8	175	.229	S.F.	18.30	10.50		28.80	37
0100	Base, polished, 3/4" or 7/8" thick, polished, 6" high	D-10	65	.492	L.F.	12	24.50	5.60	42.10	59
0300	Carvings or bas-relief, from templates, simple design		80	.400	S.F.	150	19.80	4.54	174.34	202
0350	Intricate design		80	.400	"	350	19.80	4.54	374.34	415
1000	Facing, polished finish, cut to size, 3/4" to 7/8" thick									
1050	Carrara or equal	D-10	130	.246	S.F.	22.50	12.20	2.80	37.50	48
1100	Arabescato or equal		130	.246		39.50	12.20	2.80	54.50	66.50
1300	1-1/4" thick, Botticino Classico or equal		125	.256		24	12.65	2.91	39.56	50
1350	Statuarietto or equal		125	.256		41.50	12.65	2.91	57.06	69
1500	2" thick, Crema Marfil or equal		120	.267		48.50	13.20	3.03	64.73	78.50
1550	Cafe Pinta or equal		120	.267		70	13.20	3.03	86.23	102
1700	Rubbed finish, cut to size, 4" thick									
1740	Average	D-10	100	.320	S.F.	41.50	15.85	3.64	60.99	75
1780	Maximum	"	100	.320	"	71.50	15.85	3.64	90.99	109
2200	Window sills, 6" x 3/4" thick	D-1	85	.188	L.F.	11.10	8.45		19.55	26
2500	Flooring, polished tiles, 12" x 12" x 3/8" thick									
2510	Thin set, Giallo Solare or equal	D-11	90	.267	S.F.	17.50	12.60		30.10	40
2600	Sky Blue or equal		90	.267		16	12.60		28.60	38
2700	Mortar bed, Giallo Solare or equal		65	.369		17.50	17.45		34.95	48
2740	Sky Blue or equal		65	.369		16	17.45		33.45	46
2780	Travertine, 3/8" thick, Sierra or equal	D-10	130	.246		9.40	12.20	2.80	24.40	33
2790	Silver or equal	"	130	.246		26	12.20	2.80	41	51.50
2800	Patio tile, non-slip, 1/2" thick, flame finish	D-11	75	.320		11.15	15.15		26.30	37
2900	Shower or toilet partitions, 7/8" thick partitions									
3050	3/4" or 1-1/4" thick stiles, polished 2 sides, average	D-11	75	.320	S.F.	46.50	15.15		61.65	75.50
3201	Soffits, add to above prices				"	20%	100%			
3210	Stairs, risers, 7/8" thick x 6" high	D-10	115	.278	L.F.	15.60	13.75	3.16	32.51	43
3360	Treads, 12" wide x 1-1/4" thick	"	115	.278	"	44	13.75	3.16	60.91	74.50
3500	Thresholds, 3' long, 7/8" thick, 4" to 5" wide, plain	D-12	24	1.333	Ea.	35.50	60.50		96	138
3550	Beveled		24	1.333	"	71.50	60.50		132	178
3700	Window stools, polished, 7/8" thick, 5" wide		85	.376	L.F.	21.50	17.05		38.55	52
9000	Minimum labor/equipment charge	D-1	2	8	Job		360		360	585

04 43 Stone Masonry

04 43 10 – Masonry with Natural and Processed Stone

04 43 10.75 Sandstone or Brownstone

		Crew	Daily Output	Labor-Hours	Unit	Material	2018 Bare Costs Labor	Equipment	Total	Total Incl O&P
0011	**SANDSTONE OR BROWNSTONE**									
0100	Sawed face veneer, 2-1/2" thick, to 2' x 4' panels	D-10	130	.246	S.F.	21.50	12.20	2.80	36.50	46.50
0150	4" thick, to 3'-6" x 8' panels		100	.320		21.50	15.85	3.64	40.99	53
0300	Split face, random sizes	▼	100	.320	▼	13.40	15.85	3.64	32.89	44.50
9000	Minimum labor/equipment charge	D-1	2.50	6.400	Job		287		287	470

04 43 10.80 Slate

		Crew	Daily Output	Labor-Hours	Unit	Material	2018 Bare Costs Labor	Equipment	Total	Total Incl O&P
0010	**SLATE**									
0040	Pennsylvania - blue gray to black									
0050	Vermont - unfading green, mottled green & purple, gray & purple									
0100	Virginia - blue black									
3100	Stair landings, 1" thick, black, clear	D-1	65	.246	S.F.	21	11.05		32.05	41.50
3200	Ribbon	"	65	.246	"	23.50	11.05		34.55	43.50
3500	Stair treads, sand finish, 1" thick x 12" wide									
3550	Under 3 L.F.	D-10	85	.376	L.F.	23.50	18.65	4.28	46.43	60
3600	3 L.F. to 6 L.F.	"	120	.267	"	25	13.20	3.03	41.23	52.50
3700	Ribbon, sand finish, 1" thick x 12" wide									
3750	To 6 L.F.	D-10	120	.267	L.F.	21	13.20	3.03	37.23	48.50
4000	Stools or sills, sand finish, 1" thick, 6" wide	D-12	160	.200		12.20	9.05		21.25	28
4100	Honed finish		160	.200		11.65	9.05		20.70	27.50
4200	10" wide		90	.356		18.80	16.10		34.90	47
4250	Honed finish		90	.356		17.50	16.10		33.60	46
4400	2" thick, 6" wide		140	.229		19.60	10.35		29.95	38.50
4450	Honed finish		140	.229		18.65	10.35		29	37.50
4600	10" wide		90	.356		30.50	16.10		46.60	60.50
4650	Honed finish	▼	90	.356	▼	29	16.10		45.10	58.50
4800	For lengths over 3', add					25%				
9000	Minimum labor/equipment charge	D-1	2.50	6.400	Job		287		287	470

04 43 10.85 Window Sill

		Crew	Daily Output	Labor-Hours	Unit	Material	2018 Bare Costs Labor	Equipment	Total	Total Incl O&P
0010	**WINDOW SILL**									
0020	Bluestone, thermal top, 10" wide, 1-1/2" thick	D-1	85	.188	S.F.	9.60	8.45		18.05	24.50
0050	2" thick		75	.213	"	9.60	9.55		19.15	26
0100	Cut stone, 5" x 8" plain		48	.333	L.F.	12.55	14.95		27.50	38.50
0200	Face brick on edge, brick, 8" wide		80	.200		5.50	8.95		14.45	20.50
0400	Marble, 9" wide, 1" thick		85	.188		8.95	8.45		17.40	23.50
0900	Slate, colored, unfading, honed, 12" wide, 1" thick		85	.188		8.95	8.45		17.40	23.50
0950	2" thick	▼	70	.229	▼	8.95	10.25		19.20	26.50
9000	Minimum labor/equipment charge	1 Bric	2	4	Job		201		201	330

04 54 Refractory Brick Masonry

04 54 10 – Refractory Brick Work

04 54 10.10 Fire Brick

		Crew	Daily Output	Labor-Hours	Unit	Material	2018 Bare Costs Labor	Equipment	Total	Total Incl O&P
0010	**FIRE BRICK**									
0012	Low duty, 2000°F, 9" x 2-1/2" x 4-1/2"	D-1	.60	26.667	M	1,650	1,200		2,850	3,750
0050	High duty, 3000°F	"	.60	26.667	"	2,800	1,200		4,000	5,025

04 54 10.20 Fire Clay

		Crew	Daily Output	Labor-Hours	Unit	Material	2018 Bare Costs Labor	Equipment	Total	Total Incl O&P
0010	**FIRE CLAY**									
0020	Gray, high duty, 100 lb. bag				Bag	38			38	42
0050	100 lb. drum, premixed (400 brick per drum)				Drum	43			43	47.50

04 57 Masonry Fireplaces

04 57 10 – Brick or Stone Fireplaces

04 57 10.10 Fireplace

	Crew	Daily Output	Labor-Hours	Unit	Material	2018 Bare Costs Labor	2018 Bare Costs Equipment	Total	Total Incl O&P
0010 **FIREPLACE**									
0100 Brick fireplace, not incl. foundations or chimneys									
0110 30" x 29" opening, incl. chamber, plain brickwork	D-1	.40	40	Ea.	590	1,800		2,390	3,575
0200 Fireplace box only (110 brick)	"	2	8	"	162	360		522	765
0300 For elaborate brickwork and details, add					35%	35%			
0400 For hearth, brick & stone, add	D-1	2	8	Ea.	213	360		573	820
0410 For steel, damper, cleanouts, add		4	4		17.55	179		196.55	310
0600 Plain brickwork, incl. metal circulator		.50	32		900	1,425		2,325	3,350
0800 Face brick only, standard size, 8" x 2-2/3" x 4"		.30	53.333	M	620	2,400		3,020	4,575
0900 Stone fireplace, fieldstone, add				SF Face	8.95			8.95	9.80
1000 Cut stone, add				"	8.40			8.40	9.25
9000 Minimum labor/equipment charge	D-1	2	8	Job		360		360	585

04 71 Manufactured Brick Masonry

04 71 10 – Simulated or Manufactured Brick

04 71 10.10 Simulated Brick

	Crew	Daily Output	Labor-Hours	Unit	Material	2018 Bare Costs Labor	2018 Bare Costs Equipment	Total	Total Incl O&P
0010 **SIMULATED BRICK**									
0020 Aluminum, baked on colors	1 Carp	200	.040	S.F.	4.50	2.03		6.53	8.25
0050 Fiberglass panels		200	.040		10	2.03		12.03	14.30
0100 Urethane pieces cemented in mastic		150	.053		8.60	2.70		11.30	13.85
0150 Vinyl siding panels		200	.040		10.90	2.03		12.93	15.30

04 72 Cast Stone Masonry

04 72 10 – Cast Stone Masonry Features

04 72 10.10 Coping

	Crew	Daily Output	Labor-Hours	Unit	Material	2018 Bare Costs Labor	2018 Bare Costs Equipment	Total	Total Incl O&P
0010 **COPING**, stock units									
0050 Precast concrete, 10" wide, 4" tapers to 3-1/2", 8" wall	D-1	75	.213	L.F.	22.50	9.55		32.05	40.50
0100 12" wide, 3-1/2" tapers to 3", 10" wall		70	.229		24.50	10.25		34.75	43.50
0110 14" wide, 4" tapers to 3-1/2", 12" wall		65	.246		28	11.05		39.05	48.50
0150 16" wide, 4" tapers to 3-1/2", 14" wall		60	.267		30	11.95		41.95	52.50
0300 Limestone for 12" wall, 4" thick		90	.178		14.80	7.95		22.75	29.50
0350 6" thick		80	.200		22	8.95		30.95	39
0500 Marble, to 4" thick, no wash, 9" wide		90	.178		12.90	7.95		20.85	27
0550 12" wide		80	.200		18	8.95		26.95	34.50
0700 Terra cotta, 9" wide		90	.178		6.45	7.95		14.40	20
0750 12" wide		80	.200		8.80	8.95		17.75	24.50
0800 Aluminum, for 12" wall		80	.200		9.20	8.95		18.15	24.50
9000 Minimum labor/equipment charge		2	8	Job		360		360	585

04 72 20 – Cultured Stone Veneer

04 72 20.10 Cultured Stone Veneer Components

	Crew	Daily Output	Labor-Hours	Unit	Material	2018 Bare Costs Labor	2018 Bare Costs Equipment	Total	Total Incl O&P
0010 **CULTURED STONE VENEER COMPONENTS**									
0110 On wood frame and sheathing substrate, random sized cobbles, corner stones	D-8	70	.571	V.L.F.	10.40	26		36.40	54.50
0120 Field stones		140	.286	S.F.	7.55	13.10		20.65	30
0130 Random sized flats, corner stones		70	.571	V.L.F.	10.70	26		36.70	55
0140 Field stones		140	.286	S.F.	8.90	13.10		22	31.50
0150 Horizontal lined ledgestones, corner stones		75	.533	V.L.F.	10.40	24.50		34.90	51.50
0160 Field stones		150	.267	S.F.	7.55	12.25		19.80	28.50
0170 Random shaped flats, corner stones		65	.615	V.L.F.	10.40	28.50		38.90	57.50
0180 Field stones		150	.267	S.F.	7.55	12.25		19.80	28.50

For customer support on your Commercial Renovation Costs with RSMeans data, call 800.448.8182.

04 72 Cast Stone Masonry

04 72 20 – Cultured Stone Veneer

04 72 20.10 Cultured Stone Veneer Components	Crew	Daily Output	Labor-Hours	Unit	Material	2018 Bare Costs Labor	Equipment	Total	Total Incl O&P
0190	Random shaped/textured face, corner stones	D-8	65	.615	V.L.F.	10.40	28.50		38.90	57.50
0200	Field stones		130	.308	S.F.	7.55	14.10		21.65	31.50
0210	Random shaped river rock, corner stones		65	.615	V.L.F.	10.40	28.50		38.90	57.50
0220	Field stones		130	.308	S.F.	7.55	14.10		21.65	31.50
0240	On concrete or CMU substrate, random sized cobbles, corner stones		70	.571	V.L.F.	9.70	26		35.70	53.50
0250	Field stones		140	.286	S.F.	7.20	13.10		20.30	29.50
0260	Random sized flats, corner stones		70	.571	V.L.F.	10	26		36	54
0270	Field stones		140	.286	S.F.	8.55	13.10		21.65	31
0280	Horizontal lined ledgestones, corner stones		75	.533	V.L.F.	9.70	24.50		34.20	50.50
0290	Field stones		150	.267	S.F.	7.20	12.25		19.45	28
0300	Random shaped flats, corner stones		70	.571	V.L.F.	9.70	26		35.70	53.50
0310	Field stones		140	.286	S.F.	7.20	13.10		20.30	29.50
0320	Random shaped/textured face, corner stones		65	.615	V.L.F.	9.70	28.50		38.20	56.50
0330	Field stones		130	.308	S.F.	7.20	14.10		21.30	31
0340	Random shaped river rock, corner stones		65	.615	V.L.F.	9.70	28.50		38.20	56.50
0350	Field stones		130	.308	S.F.	7.20	14.10		21.30	31
0360	Cultured stone veneer, #15 felt weather resistant barrier	1 Clab	3700	.002	Sq.	5.35	.09		5.44	6.05
0370	Expanded metal lath, diamond, 2.5 lb./S.Y., galvanized	1 Lath	85	.094	S.Y.	3.10	4.63		7.73	10.70
0390	Water table or window sill, 18" long	1 Bric	80	.100	Ea.	10	5.05		15.05	19.20

04 73 Manufactured Stone Masonry

04 73 20 – Simulated or Manufactured Stone

04 73 20.10 Simulated Stone

0010	SIMULATED STONE									
0100	Insulated fiberglass panels, 5/8" ply backer	L-4	200	.120	S.F.	10	5.70		15.70	20.50

Estimating Tips
05 05 00 Common Work Results for Metals

- Nuts, bolts, washers, connection angles, and plates can add a significant amount to both the tonnage of a structural steel job and the estimated cost. As a rule of thumb, add 10% to the total weight to account for these accessories.

- Type 2 steel construction, commonly referred to as "simple construction," consists generally of field-bolted connections with lateral bracing supplied by other elements of the building, such as masonry walls or x-bracing. The estimator should be aware, however, that shop connections may be accomplished by welding or bolting. The method may be particular to the fabrication shop and may have an impact on the estimated cost.

05 10 00 Structural Steel

- Steel items can be obtained from two sources: a fabrication shop or a metals service center. Fabrication shops can fabricate items under more controlled conditions than crews in the field can. They are also more efficient and can produce items more economically. Metal service centers serve as a source of long mill shapes to both fabrication shops and contractors.

- Most line items in this structural steel subdivision, and most items in 05 50 00 Metal Fabrications, are indicated as being shop fabricated. The bare material cost for these shop fabricated items is the "Invoice Cost" from the shop and includes the mill base price of steel plus mill extras, transportation to the shop, shop drawings and detailing where warranted, shop fabrication and handling, sandblasting and a shop coat of primer paint, all necessary structural bolts, and delivery to the job site. The bare labor cost and bare equipment cost for these shop fabricated items are for field installation or erection.

- Line items in Subdivision 05 12 23.40 Lightweight Framing, and other items scattered in Division 5, are indicated as being field fabricated. The bare material cost for these field fabricated items is the "Invoice Cost" from the metals service center and includes the mill base price of steel plus mill extras, transportation to the metals service center, material handling, and delivery of long lengths of mill shapes to the job site. Material costs for structural bolts and welding rods should be added to the estimate. The bare labor cost and bare equipment cost for these items are for both field fabrication and field installation or erection, and include time for cutting, welding, and drilling in the fabricated metal items. Drilling into concrete and fasteners to fasten field fabricated items to other work are not included and should be added to the estimate.

05 20 00 Steel Joist Framing

- In any given project the total weight of open web steel joists is determined by the loads to be supported and the design. However, economies can be realized in minimizing the amount of labor used to place the joists. This is done by maximizing the joist spacing, and therefore minimizing the number of joists required to be installed on the job. Certain spacings and locations may be required by the design, but in other cases maximizing the spacing and keeping it as uniform as possible will keep the costs down.

05 30 00 Steel Decking

- The takeoff and estimating of a metal deck involves more than the area of the floor or roof and the type of deck specified or shown on the drawings. Many different sizes and types of openings may exist. Small openings for individual pipes or conduits may be drilled after the floor/roof is installed, but larger openings may require special deck lengths as well as reinforcing or structural support. The estimator should determine who will be supplying this reinforcing. Additionally, some deck terminations are part of the deck package, such as screed angles and pour stops, and others will be part of the steel contract, such as angles attached to structural members and cast-in-place angles and plates. The estimator must ensure that all pieces are accounted for in the complete estimate.

05 50 00 Metal Fabrications

- The most economical steel stairs are those that use common materials, standard details, and most importantly, a uniform and relatively simple method of field assembly. Commonly available A36/A992 channels and plates are very good choices for the main stringers of the stairs, as are angles and tees for the carrier members. Risers and treads are usually made by specialty shops, and it is most economical to use a typical detail in as many places as possible. The stairs should be pre-assembled and shipped directly to the site. The field connections should be simple and straightforward enough to be accomplished efficiently, and with minimum equipment and labor.

Reference Numbers

Reference numbers are shown at the beginning of some major classifications. These numbers refer to related items in the Reference Section. The reference information may be an estimating procedure, an alternate pricing method, or technical information.

Note: Not all subdivisions listed here necessarily appear. ■

Did you know?

RSMeans data is available through our online application with 24/7 access:

- Search for unit prices by keyword
- Leverage the most up-to-date data
- Build and export estimates

Try it free for 30 days!
www.rsmeans.com/2018freetrial

05 05 05 – Selective Demolition for Metals

05 05 05.10 Selective Demolition, Metals

		Daily Output	Labor-Hours	Unit	Material	2018 Bare Costs Labor	Equipment	Total	Total Incl O&P
						Crew			
0010	**SELECTIVE DEMOLITION, METALS** R024119-10								
0015	Excludes shores, bracing, cutting, loading, hauling, dumping								
0020	Remove nuts only up to 3/4" diameter	1 Sswk	480	.017	Ea.		.91		.91	1.55
0030	7/8" to 1-1/4" diameter		240	.033			1.82		1.82	3.11
0040	1-3/8" to 2" diameter		160	.050			2.73		2.73	4.66
0060	Unbolt and remove structural bolts up to 3/4" diameter		240	.033			1.82		1.82	3.11
0070	7/8" to 2" diameter		160	.050			2.73		2.73	4.66
0140	Light weight framing members, remove whole or cut up, up to 20 lb.		240	.033			1.82		1.82	3.11
0150	21-40 lb.	2 Sswk	210	.076			4.16		4.16	7.10
0160	41-80 lb.	3 Sswk	180	.133			7.30		7.30	12.45
0170	81-120 lb.	4 Sswk	150	.213			11.65		11.65	19.90
0230	Structural members, remove whole or cut up, up to 500 lb.	E-19	48	.500			27	22.50	49.50	70
0240	1/4-2 tons	E-18	36	1.111			61	30	91	135
0250	2-5 tons	E-24	30	1.067			58	19.70	77.70	119
0260	5-10 tons	E-20	24	2.667			145	55	200	305
0270	10-15 tons	E-2	18	3.111			169	93.50	262.50	385
0340	Fabricated item, remove whole or cut up, up to 20 lb.	1 Sswk	96	.083			4.55		4.55	7.75
0350	21-40 lb.	2 Sswk	84	.190			10.40		10.40	17.75
0360	41-80 lb.	3 Sswk	72	.333			18.20		18.20	31
0370	81-120 lb.	4 Sswk	60	.533			29		29	49.50
0380	121-500 lb.	E-19	48	.500			27	22.50	49.50	70
0390	501-1000 lb.	"	36	.667			36	30	66	93
0500	Steel roof decking, uncovered, bare	B-2	5000	.008	S.F.		.32		.32	.52
2950	Minimum labor/equipment charge	E-19	2	12	Job		650	540	1,190	1,675

05 05 19 – Post-Installed Concrete Anchors

05 05 19.10 Chemical Anchors

		Crew	Daily Output	Labor-Hours	Unit	Material	Labor	Equipment	Total	Total Incl O&P
0010	**CHEMICAL ANCHORS**									
0020	Includes layout & drilling									
1430	Chemical anchor, w/rod & epoxy cartridge, 3/4" diameter x 9-1/2" long	B-89A	27	.593	Ea.	10.85	27.50	4.15	42.50	60.50
1435	1" diameter x 11-3/4" long		24	.667		20	30.50	4.67	55.17	77
1440	1-1/4" diameter x 14" long		21	.762		38	35	5.35	78.35	104
1445	1-3/4" diameter x 15" long		20	.800		69.50	37	5.60	112.10	142
1450	18" long		17	.941		83	43.50	6.60	133.10	169
1455	2" diameter x 18" long		16	1		113	46	7	166	206
1460	24" long		15	1.067		147	49	7.50	203.50	249

05 05 19.20 Expansion Anchors

			Crew	Daily Output	Labor-Hours	Unit	Material	Labor	Equipment	Total	Total Incl O&P
0010	**EXPANSION ANCHORS**										
0100	Anchors for concrete, brick or stone, no layout and drilling										
0200	Expansion shields, zinc, 1/4" diameter, 1-5/16" long, single	G	1 Carp	90	.089	Ea.	.45	4.51		4.96	7.80
0300	1-3/8" long, double	G		85	.094		.58	4.77		5.35	8.40
0400	3/8" diameter, 1-1/2" long, single	G		85	.094		.69	4.77		5.46	8.50
0500	2" long, double	G		80	.100		1.15	5.05		6.20	9.50
0600	1/2" diameter, 2-1/16" long, single	G		80	.100		1.26	5.05		6.31	9.65
0700	2-1/2" long, double	G		75	.107		2.13	5.40		7.53	11.15
0800	5/8" diameter, 2-5/8" long, single	G		75	.107		2.18	5.40		7.58	11.20
0900	2-3/4" long, double	G		70	.114		2.92	5.80		8.72	12.60
1000	3/4" diameter, 2-3/4" long, single	G		70	.114		3.29	5.80		9.09	13
1100	3-15/16" long, double	G		65	.123		5.45	6.25		11.70	16.15
2100	Hollow wall anchors for gypsum wall board, plaster or tile										
2300	1/8" diameter, short	G	1 Carp	160	.050	Ea.	.33	2.54		2.87	4.48
2400	Long	G		150	.053		.32	2.70		3.02	4.74
2500	3/16" diameter, short	G		150	.053		.51	2.70		3.21	4.95

05 05 Common Work Results for Metals

05 05 19 – Post-Installed Concrete Anchors

05 05 19.20 Expansion Anchors

		Crew	Daily Output	Labor-Hours	Unit	Material	2018 Bare Costs Labor	2018 Bare Costs Equipment	Total	Total Incl O&P
2600	Long	1 Carp [G]	140	.057	Ea.	.67	2.90		3.57	5.45
2700	1/4" diameter, short	[G]	140	.057		.73	2.90		3.63	5.50
2800	Long	[G]	130	.062		.80	3.12		3.92	5.95
3000	Toggle bolts, bright steel, 1/8" diameter, 2" long	[G]	85	.094		.25	4.77		5.02	8.05
3100	4" long	[G]	80	.100		.29	5.05		5.34	8.55
3200	3/16" diameter, 3" long	[G]	80	.100		.33	5.05		5.38	8.60
3300	6" long	[G]	75	.107		.45	5.40		5.85	9.30
3400	1/4" diameter, 3" long	[G]	75	.107		.41	5.40		5.81	9.25
3500	6" long	[G]	70	.114		.59	5.80		6.39	10.05
3600	3/8" diameter, 3" long	[G]	70	.114		.86	5.80		6.66	10.35
3700	6" long	[G]	60	.133		1.50	6.75		8.25	12.60
3800	1/2" diameter, 4" long	[G]	60	.133		1.65	6.75		8.40	12.75
3900	6" long	[G]	50	.160		2.08	8.10		10.18	15.45
4000	Nailing anchors									
4100	Nylon nailing anchor, 1/4" diameter, 1" long	1 Carp	3.20	2.500	C	23	127		150	231
4200	1-1/2" long		2.80	2.857		25.50	145		170.50	264
4300	2" long		2.40	3.333		27.50	169		196.50	305
4400	Metal nailing anchor, 1/4" diameter, 1" long	[G]	3.20	2.500		22.50	127		149.50	231
4500	1-1/2" long	[G]	2.80	2.857		29	145		174	267
4600	2" long	[G]	2.40	3.333		35.50	169		204.50	315
8000	Wedge anchors, not including layout or drilling									
8050	Carbon steel, 1/4" diameter, 1-3/4" long	1 Carp [G]	150	.053	Ea.	.51	2.70		3.21	4.95
8100	3-1/4" long	[G]	140	.057		.67	2.90		3.57	5.45
8150	3/8" diameter, 2-1/4" long	[G]	145	.055		.54	2.80		3.34	5.15
8200	5" long	[G]	140	.057		.95	2.90		3.85	5.75
8250	1/2" diameter, 2-3/4" long	[G]	140	.057		1.07	2.90		3.97	5.85
8300	7" long	[G]	125	.064		1.82	3.24		5.06	7.25
8350	5/8" diameter, 3-1/2" long	[G]	130	.062		1.81	3.12		4.93	7.05
8400	8-1/2" long	[G]	115	.070		3.86	3.53		7.39	10
8450	3/4" diameter, 4-1/4" long	[G]	115	.070		2.75	3.53		6.28	8.75
8500	10" long	[G]	95	.084		6.25	4.27		10.52	13.80
8550	1" diameter, 6" long	[G]	100	.080		8.70	4.06		12.76	16.20
8575	9" long	[G]	85	.094		11.35	4.77		16.12	20
8600	12" long	[G]	75	.107		12.25	5.40		17.65	22.50
8650	1-1/4" diameter, 9" long	[G]	70	.114		22	5.80		27.80	33.50
8700	12" long	[G]	60	.133		28	6.75		34.75	41.50
8750	For type 303 stainless steel, add					350%				
8800	For type 316 stainless steel, add					450%				
8950	Self-drilling concrete screw, hex washer head, 3/16" diam. x 1-3/4" long	1 Carp [G]	300	.027	Ea.	.19	1.35		1.54	2.40
8960	2-1/4" long	[G]	250	.032		.21	1.62		1.83	2.86
8970	Phillips flat head, 3/16" diam. x 1-3/4" long	[G]	300	.027		.19	1.35		1.54	2.40
8980	2-1/4" long	[G]	250	.032		.21	1.62		1.83	2.86
9000	Minimum labor/equipment charge		4	2	Job		101		101	165

05 05 21 – Fastening Methods for Metal

05 05 21.10 Cutting Steel

		Crew	Daily Output	Labor-Hours	Unit	Material	2018 Bare Costs Labor	2018 Bare Costs Equipment	Total	Total Incl O&P
0010	**CUTTING STEEL**									
0020	Hand burning, incl. preparation, torch cutting & grinding, no staging									
0050	Steel to 1/4" thick	E-25	400	.020	L.F.	.20	1.13	.03	1.36	2.18
0100	1/2" thick		320	.025		.37	1.42	.04	1.83	2.87
0150	3/4" thick		260	.031		.61	1.74	.05	2.40	3.69
0200	1" thick		200	.040		.87	2.27	.06	3.20	4.89
9000	Minimum labor/equipment charge		2	4	Job		227	6.30	233.30	390

For customer support on your Commercial Renovation Costs with RSMeans data, call 800.448.8182.

101

05 05 21 – Fastening Methods for Metal

05 05 21.15 Drilling Steel		Crew	Daily Output	Labor-Hours	Unit	Material	2018 Bare Costs Labor	2018 Bare Costs Equipment	Total	Total Incl O&P
0010	**DRILLING STEEL**									
1910	Drilling & layout for steel, up to 1/4" deep, no anchor									
1920	Holes, 1/4" diameter	1 Sswk	112	.071	Ea.	.10	3.90		4	6.75
1925	For each additional 1/4" depth, add		336	.024		.10	1.30		1.40	2.33
1930	3/8" diameter		104	.077		.09	4.20		4.29	7.25
1935	For each additional 1/4" depth, add		312	.026		.09	1.40		1.49	2.49
1940	1/2" diameter		96	.083		.11	4.55		4.66	7.85
1945	For each additional 1/4" depth, add		288	.028		.11	1.52		1.63	2.71
1950	5/8" diameter		88	.091		.16	4.97		5.13	8.65
1955	For each additional 1/4" depth, add		264	.030		.16	1.66		1.82	3
1960	3/4" diameter		80	.100		.20	5.45		5.65	9.50
1965	For each additional 1/4" depth, add		240	.033		.20	1.82		2.02	3.33
1970	7/8" diameter		72	.111		.27	6.05		6.32	10.65
1975	For each additional 1/4" depth, add		216	.037		.27	2.02		2.29	3.74
1980	1" diameter		64	.125		.23	6.85		7.08	11.90
1985	For each additional 1/4" depth, add		192	.042		.23	2.28		2.51	4.13
1990	For drilling up, add						40%			
2000	Minimum labor/equipment charge	1 Carp	4	2	Job		101		101	165

05 05 21.90 Welding Steel

		Crew	Daily Output	Labor-Hours	Unit	Material	2018 Bare Costs Labor	2018 Bare Costs Equipment	Total	Total Incl O&P
0010	**WELDING STEEL**, Structural R050521-20									
0020	Field welding, 1/8" E6011, cost per welder, no operating engineer	E-14	8	1	Hr.	4.71	56.50	12.30	73.51	115
0200	With 1/2 operating engineer	E-13	8	1.500		4.71	82.50	12.30	99.51	156
0300	With 1 operating engineer	E-12	8	2		4.71	108	12.30	125.01	197
0500	With no operating engineer, 2# weld rod per ton	E-14	8	1	Ton	4.71	56.50	12.30	73.51	115
0600	8# E6011 per ton	"	2	4		18.85	227	49	294.85	460
0800	With one operating engineer per welder, 2# E6011 per ton	E-12	8	2		4.71	108	12.30	125.01	197
0900	8# E6011 per ton	"	2	8		18.85	430	49	497.85	785
1200	Continuous fillet, down welding									
1300	Single pass, 1/8" thick, 0.1#/L.F.	E-14	150	.053	L.F.	.24	3.02	.66	3.92	6.15
1400	3/16" thick, 0.2#/L.F.		75	.107		.47	6.05	1.31	7.83	12.25
1500	1/4" thick, 0.3#/L.F.		50	.160		.71	9.05	1.97	11.73	18.40
1610	5/16" thick, 0.4#/L.F.		38	.211		.94	11.95	2.59	15.48	24.50
1800	3 passes, 3/8" thick, 0.5#/L.F.		30	.267		1.18	15.10	3.28	19.56	31
2010	4 passes, 1/2" thick, 0.7#/L.F.		22	.364		1.65	20.50	4.47	26.62	41.50
2200	5 to 6 passes, 3/4" thick, 1.3#/L.F.		12	.667		3.06	38	8.20	49.26	77
2400	8 to 11 passes, 1" thick, 2.4#/L.F.		6	1.333		5.65	75.50	16.40	97.55	153
2600	For vertical joint welding, add						20%			
2700	Overhead joint welding, add						300%			
2900	For semi-automatic welding, obstructed joints, deduct						5%			
3000	Exposed joints, deduct						15%			
4000	Cleaning and welding plates, bars, or rods									
4010	to existing beams, columns, or trusses	E-14	12	.667	L.F.	1.18	38	8.20	47.38	75
9000	Minimum labor/equipment charge	"	4	2	Job		113	24.50	137.50	220

05 05 23 – Metal Fastenings

05 05 23.30 Lag Screws

			Crew	Daily Output	Labor-Hours	Unit	Material	2018 Bare Costs Labor	2018 Bare Costs Equipment	Total	Total Incl O&P
0010	**LAG SCREWS**										
0020	Steel, 1/4" diameter, 2" long	G	1 Carp	200	.040	Ea.	.10	2.03		2.13	3.40
0200	1/2" diameter, 3" long	G		130	.062		.66	3.12		3.78	5.80
0300	5/8" diameter, 3" long	G		120	.067		1.29	3.38		4.67	6.90

05 05 Common Work Results for Metals

05 05 23 – Metal Fastenings

05 05 23.50 Powder Actuated Tools and Fasteners		Crew	Daily Output	Labor-Hours	Unit	Material	2018 Bare Costs Labor	Equipment	Total	Total Incl O&P
0010	**POWDER ACTUATED TOOLS & FASTENERS**									
0020	Stud driver, .22 caliber, single shot				Ea.	152			152	167
0100	.27 caliber, semi automatic, strip				"	460			460	505
0300	Powder load, single shot, .22 cal, power level 2, brown				C	5.65			5.65	6.25
0400	Strip, .27 cal, power level 4, red					8.15			8.15	9
0600	Drive pin, .300 x 3/4" long	G	1 Carp	4.80	1.667		4.39	84.50		88.89	142
0700	.300 x 3" long with washer	G	"	4	2		13.40	101		114.40	180

05 05 23.70 Structural Blind Bolts

		Crew	Daily Output	Labor-Hours	Unit	Material	Labor	Equipment	Total	Total Incl O&P
0010	**STRUCTURAL BLIND BOLTS**									
0100	1/4" diameter x 1/4" grip	G	1 Sswk	240	.033	Ea.	1.73	1.82		3.55	5
0150	1/2" grip	G		216	.037		1.86	2.02		3.88	5.50
0200	3/8" diameter x 1/2" grip	G		232	.034		3.12	1.88		5	6.65
0250	3/4" grip	G		208	.038		3.27	2.10		5.37	7.20
0300	1/2" diameter x 1/2" grip	G		224	.036		6.05	1.95		8	10
0350	3/4" grip	G		200	.040		8.20	2.19		10.39	12.75
0400	5/8" diameter x 3/4" grip	G		216	.037		10.90	2.02		12.92	15.45
0450	1" grip	G		192	.042		14.80	2.28		17.08	20

05 05 23.87 Weld Studs

		Crew	Daily Output	Labor-Hours	Unit	Material	Labor	Equipment	Total	Total Incl O&P
0010	**WELD STUDS**									
0020	1/4" diameter, 2-11/16" long	G	E-10	1120	.014	Ea.	.37	.80	.30	1.47	2.10
0100	4-1/8" long	G		1080	.015		.35	.82	.31	1.48	2.14
0200	3/8" diameter, 4-1/8" long	G		1080	.015		.40	.82	.31	1.53	2.19
0300	6-1/8" long	G		1040	.015		.52	.86	.32	1.70	2.39
9000	Minimum labor/equipment charge		1 Sswk	2	4	Job		219		219	375

05 12 Structural Steel Framing

05 12 23 – Structural Steel for Buildings

05 12 23.05 Canopy Framing

		Crew	Daily Output	Labor-Hours	Unit	Material	Labor	Equipment	Total	Total Incl O&P
0010	**CANOPY FRAMING**									
0020	6" and 8" members, shop fabricated	G	E-4	3000	.011	Lb.	1.62	.59	.03	2.24	2.82
9000	Minimum labor/equipment charge		1 Sswk	1	8	Job		435		435	745

05 12 23.10 Ceiling Supports

		Crew	Daily Output	Labor-Hours	Unit	Material	Labor	Equipment	Total	Total Incl O&P
0010	**CEILING SUPPORTS**									
1000	Entrance door/folding partition supports, shop fabricated	G	E-4	60	.533	L.F.	27	29.50	1.64	58.14	81.50
1100	Linear accelerator door supports	G		14	2.286		123	126	7.05	256.05	360
1200	Lintels or shelf angles, hung, exterior hot dipped galv.	G		267	.120		18.40	6.60	.37	25.37	31.50
1250	Two coats primer paint instead of galv.	G		267	.120		15.95	6.60	.37	22.92	29
1400	Monitor support, ceiling hung, expansion bolted	G		4	8	Ea.	425	440	24.50	889.50	1,250
1450	Hung from pre-set inserts	G		6	5.333		460	294	16.45	770.45	1,025
1600	Motor supports for overhead doors	G		4	8		217	440	24.50	681.50	1,025
1700	Partition support for heavy folding partitions, without pocket	G		24	1.333	L.F.	61.50	73.50	4.11	139.11	197
1750	Supports at pocket only	G		12	2.667		123	147	8.20	278.20	395
2000	Rolling grilles & fire door supports	G		34	.941		52.50	52	2.90	107.40	150
2100	Spider-leg light supports, expansion bolted to ceiling slab	G		8	4	Ea.	175	221	12.30	408.30	580
2150	Hung from pre-set inserts	G		12	2.667	"	188	147	8.20	343.20	465
2400	Toilet partition support	G		36	.889	L.F.	61.50	49	2.74	113.24	154
2500	X-ray travel gantry support	G		12	2.667	"	210	147	8.20	365.20	490

For customer support on your Commercial Renovation Costs with RSMeans data, call 800.448.8182.

103

05 12 23.17 Columns, Structural

		Crew	Daily Output	Labor-Hours	Unit	Material	2018 Bare Costs Labor	2018 Bare Costs Equipment	Total	Total Incl O&P
0010	**COLUMNS, STRUCTURAL**	R051223-10								
0015	Made from recycled materials									
0020	Shop fab'd for 100-ton, 1-2 story project, bolted connections									
0800	Steel, concrete filled, extra strong pipe, 3-1/2" diameter	E-2	660	.085	L.F.	44.50	4.61	2.56	51.67	59.50
0830	4" diameter		780	.072		49.50	3.90	2.16	55.56	63.50
0890	5" diameter		1020	.055		59	2.98	1.65	63.63	72
0930	6" diameter		1200	.047		78	2.53	1.41	81.94	92
0940	8" diameter		1100	.051	↓	78	2.76	1.53	82.29	92.50
1500	Steel pipe, extra strong, no concrete, 3" to 5" diameter	Ⓖ	16000	.004	Lb.	1.35	.19	.11	1.65	1.92
3300	Structural tubing, square, A500GrB, 4" to 6" square, light section	Ⓖ	11270	.005	"	1.35	.27	.15	1.77	2.09
8090	For projects 75 to 99 tons, add				%	10%				
8092	50 to 74 tons, add					20%				
8094	25 to 49 tons, add					30%	10%			
8096	10 to 24 tons, add					50%	25%			
8098	2 to 9 tons, add					75%	50%			
8099	Less than 2 tons, add				↓	100%	100%			
9000	Minimum labor/equipment charge	1 Sswk	1	8	Job		435		435	745

05 12 23.18 Corner Guards

		Crew	Daily Output	Labor-Hours	Unit	Material	Labor	Equipment	Total	Total Incl O&P
0010	**CORNER GUARDS**									
0020	Steel angle w/anchors, 1" x 1" x 1/4", 1.5#/L.F.	2 Carp	160	.100	L.F.	8.25	5.05		13.30	17.35
0100	2" x 2" x 1/4" angles, 3.2#/L.F.	"	150	.107	"	10.70	5.40		16.10	20.50
9000	Minimum labor/equipment charge	1 Carp	2	4	Job		203		203	330

05 12 23.20 Curb Edging

		Crew	Daily Output	Labor-Hours	Unit	Material	Labor	Equipment	Total	Total Incl O&P
0010	**CURB EDGING**									
0020	Steel angle w/anchors, shop fabricated, on forms, 1" x 1", 0.8#/L.F.	Ⓖ E-4	350	.091	L.F.	1.67	5.05	.28	7	10.75
0300	4" x 4" angles, 8.2#/L.F.	Ⓖ	275	.116	"	14.25	6.40	.36	21.01	27
9000	Minimum labor/equipment charge	↓	4	8	Job		440	24.50	464.50	775

05 12 23.40 Lightweight Framing

		Crew	Daily Output	Labor-Hours	Unit	Material	Labor	Equipment	Total	Total Incl O&P
0010	**LIGHTWEIGHT FRAMING**									
0015	Made from recycled materials									
0200	For load-bearing steel studs see Section 05 41 13.30									
0400	Angle framing, field fabricated, 4" and larger	Ⓖ E-3	440	.055	Lb.	.78	3.02	.22	4.02	6.25
0450	Less than 4" angles	Ⓖ	265	.091		.81	5	.37	6.18	9.85
0600	Channel framing, field fabricated, 8" and larger	Ⓖ	500	.048		.81	2.66	.20	3.67	5.65
0650	Less than 8" channels	Ⓖ	335	.072		.81	3.96	.29	5.06	7.95
1000	Continuous slotted channel framing system, shop fab, simple framing	Ⓖ 2 Sswk	2400	.007		4.17	.36		4.53	5.20
1200	Complex framing	Ⓖ "	1600	.010		4.71	.55		5.26	6.15
1250	Plate & bar stock for reinforcing beams and trusses	Ⓖ				1.48			1.48	1.63
1300	Cross bracing, rods, shop fabricated, 3/4" diameter	Ⓖ E-3	700	.034		1.62	1.90	.14	3.66	5.15
1310	7/8" diameter	Ⓖ	850	.028		1.62	1.56	.12	3.30	4.57
1320	1" diameter	Ⓖ	1000	.024		1.62	1.33	.10	3.05	4.15
1330	Angle, 5" x 5" x 3/8"	Ⓖ	2800	.009		1.62	.47	.04	2.13	2.63
1350	Hanging lintels, shop fabricated	Ⓖ	850	.028		1.62	1.56	.12	3.30	4.57
1380	Roof frames, shop fabricated, 3'-0" square, 5' span	Ⓖ E-2	4200	.013		1.62	.72	.40	2.74	3.43
1400	Tie rod, not upset, 1-1/2" to 4" diameter, with turnbuckle	Ⓖ 2 Sswk	800	.020		1.75	1.09		2.84	3.79
1420	No turnbuckle	Ⓖ	700	.023		1.68	1.25		2.93	3.98
1500	Upset, 1-3/4" to 4" diameter, with turnbuckle	Ⓖ	800	.020		1.75	1.09		2.84	3.79
1520	No turnbuckle	Ⓖ	700	.023	↓	1.68	1.25		2.93	3.98
9000	Minimum labor/equipment charge	↓	2	8	Job		435		435	745

05 12 23.45 Lintels

05 12 23.45 Lintels		Crew	Daily Output	Labor-Hours	Unit	Material	2018 Bare Costs Labor	Equipment	Total	Total Incl O&P
0010	**LINTELS**									
0015	Made from recycled materials									
0020	Plain steel angles, shop fabricated, under 500 lb.	G	1 Bric	550	.015	Lb.	1.04	.73		1.77	2.33
0100	500 to 1,000 lb.	G		640	.013		1.01	.63		1.64	2.14
0200	1,000 to 2,000 lb.	G		640	.013		.98	.63		1.61	2.11
0300	2,000 to 4,000 lb.	G		640	.013		.96	.63		1.59	2.08
0500	For built-up angles and plates, add to above	G					1.35			1.35	1.48
0700	For engineering, add to above						.13			.13	.15
0900	For galvanizing, add to above, under 500 lb.						.30			.30	.33
0950	500 to 2,000 lb.						.27			.27	.30
1000	Over 2,000 lb.						.25			.25	.28
2000	Steel angles, 3-1/2" x 3", 1/4" thick, 2'-6" long	G	1 Bric	47	.170	Ea.	14.55	8.55		23.10	30
2100	4'-6" long	G		26	.308		26	15.50		41.50	54.50
2500	3-1/2" x 3-1/2" x 5/16", 5'-0" long	G		18	.444		39	22.50		61.50	79
2600	4" x 3-1/2", 1/4" thick, 5'-0" long	G		21	.381		33.50	19.15		52.65	68
2700	9'-0" long	G		12	.667		60	33.50		93.50	121
2800	4" x 3-1/2" x 5/16", 7'-0" long	G		12	.667		58	33.50		91.50	119
2900	5" x 3-1/2" x 5/16", 10'-0" long	G		8	1		93.50	50.50		144	185
9000	Minimum labor/equipment charge			4	2	Job		101		101	164

05 12 23.60 Pipe Support Framing

05 12 23.60 Pipe Support Framing		Crew	Daily Output	Labor-Hours	Unit	Material	Labor	Equipment	Total	Total Incl O&P
0010	**PIPE SUPPORT FRAMING**										
0020	Under 10#/L.F., shop fabricated	G	E-4	3900	.008	Lb.	1.80	.45	.03	2.28	2.78
0200	10.1 to 15#/L.F.	G		4300	.007		1.78	.41	.02	2.21	2.68
0400	15.1 to 20#/L.F.	G		4800	.007		1.75	.37	.02	2.14	2.58
0600	Over 20#/L.F.	G		5400	.006		1.72	.33	.02	2.07	2.48

05 12 23.65 Plates

05 12 23.65 Plates		Crew	Daily Output	Labor-Hours	Unit	Material	Labor	Equipment	Total	Total Incl O&P
0010	**PLATES**										
0015	Made from recycled materials										
0020	For connections & stiffener plates, shop fabricated										
0050	1/8" thick (5.1 lb./S.F.)	G				S.F.	6.85			6.85	7.55
0100	1/4" thick (10.2 lb./S.F.)	G					13.75			13.75	15.10
0300	3/8" thick (15.3 lb./S.F.)	G					20.50			20.50	22.50
0400	1/2" thick (20.4 lb./S.F.)	G					27.50			27.50	30
0450	3/4" thick (30.6 lb./S.F.)	G					41			41	45.50
0500	1" thick (40.8 lb./S.F.)	G					55			55	60.50
2000	Steel plate, warehouse prices, no shop fabrication										
2100	1/4" thick (10.2 lb./S.F.)	G				S.F.	7.30			7.30	8

05 12 23.75 Structural Steel Members

05 12 23.75 Structural Steel Members		Crew	Daily Output	Labor-Hours	Unit	Material	Labor	Equipment	Total	Total Incl O&P
0010	**STRUCTURAL STEEL MEMBERS** R051223-10										
0015	Made from recycled materials										
0020	Shop fab'd for 100-ton, 1-2 story project, bolted connections										
0102	Beam or girder, W 6 x 9 R051223-15	G	E-2	600	.093	L.F.	13.35	5.05	2.81	21.21	26
0302	W 8 x 10	G		600	.093		14.80	5.05	2.81	22.66	28
0502	x 31	G		550	.102		46	5.55	3.07	54.62	63
0702	W 10 x 22	G		600	.093		32.50	5.05	2.81	40.36	47.50
0902	x 49	G		550	.102		72.50	5.55	3.07	81.12	92.50
1102	W 12 x 16	G		880	.064		23.50	3.45	1.92	28.87	34
1302	x 22	G		880	.064		32.50	3.45	1.92	37.87	44
1502	x 26	G		880	.064		38.50	3.45	1.92	43.87	50.50
1902	W 14 x 26	G		990	.057		38.50	3.07	1.70	43.27	49.50
2102	x 30	G		900	.062		44.50	3.38	1.87	49.75	56.50

05 12 23.75 Structural Steel Members

		Crew	Daily Output	Labor-Hours	Unit	Material	2018 Bare Costs Labor	Equipment	Total	Total Incl O&P
2702	W 16 x 26 [G]	E-2	1000	.056	L.F.	38.50	3.04	1.69	43.23	49.50
2902	x 31 [G]	↓	900	.062		46	3.38	1.87	51.25	58
8490	For projects 75 to 99 tons, add					10%				
8492	50 to 74 tons, add					20%				
8494	25 to 49 tons, add					30%	10%			
8496	10 to 24 tons, add					50%	25%			
8498	2 to 9 tons, add					75%	50%			
8499	Less than 2 tons, add				↓	100%	100%			
9000	Minimum labor/equipment charge	E-2	2	28	Job		1,525	845	2,370	3,475

05 12 23.77 Structural Steel Projects

		Crew	Daily Output	Labor-Hours	Unit	Material	2018 Bare Costs Labor	Equipment	Total	Total Incl O&P
0010	**STRUCTURAL STEEL PROJECTS**									
0015	Made from recycled materials									
0020	Shop fab'd for 100-ton, 1-2 story project, bolted connections									
0700	Offices, hospitals, etc., steel bearing, 1 to 2 stories R050521-20 [G]	E-5	10.30	7.767	Ton	2,700	425	173	3,298	3,850
1300	Industrial bldgs., 1 story, beams & girders, steel bearing [G]		12.90	6.202		2,700	340	138	3,178	3,675
1400	Masonry bearing R051223-10 [G]		10	8		2,700	435	179	3,314	3,875
1600	1 story with roof trusses, steel bearing [G]		10.60	7.547		3,175	410	168	3,753	4,375
1700	Masonry bearing R051223-15 [G]	↓	8.30	9.639		3,175	525	215	3,915	4,625
5390	For projects 75 to 99 tons, add					10%				
5392	50 to 74 tons, add					20%				
5394	25 to 49 tons, add					30%	10%			
5396	10 to 24 tons, add					50%	25%			
5398	2 to 9 tons, add					75%	50%			
5399	Less than 2 tons, add				↓	100%	100%			

05 12 23.78 Structural Steel Secondary Members

		Crew	Daily Output	Labor-Hours	Unit	Material	2018 Bare Costs Labor	Equipment	Total	Total Incl O&P
0010	**STRUCTURAL STEEL SECONDARY MEMBERS**									
0015	Made from recycled materials									
0020	Shop fabricated for 20-ton girt/purlin framing package, materials only									
0100	Girts/purlins, C/Z-shapes, includes clips and bolts									
0110	6" x 2-1/2" x 2-1/2", 16 ga., 3.0 lb./L.F.				L.F.	3.63			3.63	4
0115	14 ga., 3.5 lb./L.F.					4.24			4.24	4.66
0120	8" x 2-3/4" x 2-3/4", 16 ga., 3.4 lb./L.F.					4.12			4.12	4.53
0125	14 ga., 4.1 lb./L.F.					4.97			4.97	5.45
0130	12 ga., 5.6 lb./L.F.					6.80			6.80	7.45
0135	10" x 3-1/2" x 3-1/2", 14 ga., 4.7 lb./L.F.					5.70			5.70	6.25
0140	12 ga., 6.7 lb./L.F.					8.10			8.10	8.95
0145	12" x 3-1/2" x 3-1/2", 14 ga., 5.3 lb./L.F.					6.40			6.40	7.05
0150	12 ga., 7.4 lb./L.F.				↓	8.95			8.95	9.85
0200	Eave struts, C-shape, includes clips and bolts									
0210	6" x 4" x 3", 16 ga., 3.1 lb./L.F.				L.F.	3.76			3.76	4.13
0215	14 ga., 3.9 lb./L.F.					4.73			4.73	5.20
0220	8" x 4" x 3", 16 ga., 3.5 lb./L.F.					4.24			4.24	4.66
0225	14 ga., 4.4 lb./L.F.					5.35			5.35	5.85
0230	12 ga., 6.2 lb./L.F.					7.50			7.50	8.25
0235	10" x 5" x 3", 14 ga., 5.2 lb./L.F.					6.30			6.30	6.95
0240	12 ga., 7.3 lb./L.F.					8.85			8.85	9.75
0245	12" x 5" x 4", 14 ga., 6.0 lb./L.F.					7.25			7.25	8
0250	12 ga., 8.4 lb./L.F.				↓	10.20			10.20	11.20
0300	Rake/base angle, excludes concrete drilling and expansion anchors									
0310	2" x 2", 14 ga., 1.0 lb./L.F.	2 Sswk	640	.025	L.F.	1.21	1.37		2.58	3.66
0315	3" x 2", 14 ga., 1.3 lb./L.F.		535	.030		1.58	1.63		3.21	4.52
0320	3" x 3", 14 ga., 1.6 lb./L.F.	↓	500	.032		1.94	1.75		3.69	5.10

05 12 Structural Steel Framing

05 12 23 – Structural Steel for Buildings

05 12 23.78 Structural Steel Secondary Members		Crew	Daily Output	Labor-Hours	Unit	Material	2018 Bare Costs Labor	Equipment	Total	Total Incl O&P
0325	4" x 3", 14 ga., 1.8 lb./L.F.	2 Sswk	480	.033	L.F.	2.18	1.82		4	5.50
0600	Installation of secondary members, erection only									
0610	Girts, purlins, eave struts, 16 ga., 6" deep	E-18	100	.400	Ea.		22	10.85	32.85	49
0615	8" deep		80	.500			27.50	13.55	41.05	61
0620	14 ga., 6" deep		80	.500			27.50	13.55	41.05	61
0625	8" deep		65	.615			34	16.65	50.65	75
0630	10" deep		55	.727			40	19.70	59.70	88.50
0635	12" deep		50	.800			44	21.50	65.50	98
0640	12 ga., 8" deep		50	.800			44	21.50	65.50	98
0645	10" deep		45	.889			49	24	73	109
0650	12" deep	↓	40	1	↓		55	27	82	122
0900	For less than 20-ton job lots									
0905	For 15 to 19 tons, add				%	10%				
0910	For 10 to 14 tons, add					25%				
0915	For 5 to 9 tons, add					50%	50%	50%		
0920	For 1 to 4 tons, add					75%	75%	75%		
0925	For less than 1 ton, add				↓	100%	100%	100%		
0990	Minimum labor/equipment charge	E-18	1	40	Job		2,200	1,075	3,275	4,900

05 15 Wire Rope Assemblies

05 15 16 – Steel Wire Rope Assemblies

05 15 16.70 Temporary Cable Safety Railing

			Crew	Daily Output	Labor-Hours	Unit	Material	2018 Bare Costs Labor	Equipment	Total	Total Incl O&P
0010	**TEMPORARY CABLE SAFETY RAILING**, Each 100' strand incl.										
0020	2 eyebolts, 1 turnbuckle, 100' cable, 2 thimbles, 6 clips										
0025	Made from recycled materials										
0100	One strand using 1/4" cable & accessories	G	2 Sswk	4	4	C.L.F.	153	219		372	545
0200	1/2" cable & accessories	G	"	2	8	"	325	435		760	1,100

05 21 Steel Joist Framing

05 21 13 – Deep Longspan Steel Joist Framing

05 21 13.50 Deep Longspan Joists

			Crew	Daily Output	Labor-Hours	Unit	Material	2018 Bare Costs Labor	Equipment	Total	Total Incl O&P
0010	**DEEP LONGSPAN JOISTS**										
3010	DLH series, 40-ton job lots, bolted cross bridging, shop primer										
3015	Made from recycled materials										
3040	Spans to 144' (shipped in 2 pieces)	G	E-7	13	6.154	Ton	2,125	335	145	2,605	3,050
3500	For less than 40-ton job lots										
3502	For 30 to 39 tons, add					%	10%				
3504	20 to 29 tons, add						20%				
3506	10 to 19 tons, add						30%				
3507	5 to 9 tons, add						50%	25%			
3508	1 to 4 tons, add						75%	50%			
3509	Less than 1 ton, add					↓	100%	100%			
4010	SLH series, 40-ton job lots, bolted cross bridging, shop primer										
4040	Spans to 200' (shipped in 3 pieces)	G	E-7	13	6.154	Ton	2,200	335	145	2,680	3,150
6100	For less than 40-ton job lots										
6102	For 30 to 39 tons, add					%	10%				
6104	20 to 29 tons, add						20%				
6106	10 to 19 tons, add						30%				
6107	5 to 9 tons, add					↓	50%	25%			

05 21 Steel Joist Framing

05 21 13 – Deep Longspan Steel Joist Framing

05 21 13.50 Deep Longspan Joists		Crew	Daily Output	Labor-Hours	Unit	Material	2018 Bare Costs Labor	Equipment	Total	Total Incl O&P
6108	1 to 4 tons, add				%	75%	50%			
6109	Less than 1 ton, add					100%	100%			

05 21 16 – Longspan Steel Joist Framing

05 21 16.50 Longspan Joists

			Crew	Daily Output	Labor-Hours	Unit	Material	2018 Bare Costs Labor	Equipment	Total	Total Incl O&P
0010	**LONGSPAN JOISTS**										
2000	LH series, 40-ton job lots, bolted cross bridging, shop primer										
2015	Made from recycled materials										
2040	Longspan joists, LH series, up to 96'	G	E-7	13	6.154	Ton	2,025	335	145	2,505	2,975
2600	For less than 40-ton job lots										
2602	For 30 to 39 tons, add					%	10%				
2604	20 to 29 tons, add						20%				
2606	10 to 19 tons, add						30%				
2607	5 to 9 tons, add						50%	25%			
2608	1 to 4 tons, add						75%	50%			
2609	Less than 1 ton, add						100%	100%			

05 21 19 – Open Web Steel Joist Framing

05 21 19.10 Open Web Joists

			Crew	Daily Output	Labor-Hours	Unit	Material	2018 Bare Costs Labor	Equipment	Total	Total Incl O&P
0010	**OPEN WEB JOISTS**										
0015	Made from recycled materials										
0050	K series, 40-ton lots, horiz. bridging, spans to 30', shop primer	G	E-7	12	6.667	Ton	1,800	365	157	2,322	2,750
0440	K series, 30' to 50' spans	G	"	17	4.706	"	1,775	257	111	2,143	2,500
0800	For less than 40-ton job lots										
0802	For 30 to 39 tons, add					%	10%				
0804	20 to 29 tons, add						20%				
0806	10 to 19 tons, add						30%				
0807	5 to 9 tons, add						50%	25%			
0808	1 to 4 tons, add						75%	50%			
0809	Less than 1 ton, add						100%	100%			
1010	CS series, 40-ton job lots, horizontal bridging, shop primer										
1040	Spans to 30'	G	E-7	12	6.667	Ton	1,850	365	157	2,372	2,825
1500	For less than 40-ton job lots										
1502	For 30 to 39 tons, add					%	10%				
1504	20 to 29 tons, add						20%				
1506	10 to 19 tons, add						30%				
1507	5 to 9 tons, add						50%	25%			
1508	1 to 4 tons, add						75%	50%			
1509	Less than 1 ton, add						100%	100%			
6400	Individual steel bearing plate, 6" x 6" x 1/4" with J-hook	G	1 Bric	160	.050	Ea.	8.10	2.52		10.62	13
9000	Minimum labor and equipment, bar joists		F-6	1	40	Job		1,900	495	2,395	3,600

For customer support on your Commercial Renovation Costs with RSMeans data, call 800.448.8182.

05 31 Steel Decking

05 31 13 – Steel Floor Decking

05 31 13.50 Floor Decking

		Crew	Daily Output	Labor-Hours	Unit	Material	2018 Bare Costs Labor	Equipment	Total	Total Incl O&P
0010	**FLOOR DECKING** R053100-10									
0015	Made from recycled materials									
5100	Non-cellular composite decking, galvanized, 1-1/2" deep, 16 ga. G	E-4	3500	.009	S.F.	4.01	.50	.03	4.54	5.30
5120	18 ga. G		3650	.009		3.24	.48	.03	3.75	4.41
5140	20 ga. G		3800	.008		2.58	.46	.03	3.07	3.66
5200	2" deep, 22 ga. G		3860	.008		2.22	.46	.03	2.71	3.25
5300	20 ga. G		3600	.009		2.48	.49	.03	3	3.60
5400	18 ga. G		3380	.009		3.17	.52	.03	3.72	4.40
5500	16 ga. G		3200	.010		3.96	.55	.03	4.54	5.35
5700	3" deep, 22 ga. G		3200	.010		2.42	.55	.03	3	3.63
5800	20 ga.. G		3000	.011		2.73	.59	.03	3.35	4.04
5900	18 ga. G		2850	.011		3.37	.62	.03	4.02	4.81
6000	16 ga. G		2700	.012		4.50	.65	.04	5.19	6.10
9000	Minimum labor/equipment charge	1 Sswk	1	8	Job		435		435	745

05 31 23 – Steel Roof Decking

05 31 23.50 Roof Decking

		Crew	Daily Output	Labor-Hours	Unit	Material	2018 Bare Costs Labor	Equipment	Total	Total Incl O&P
0010	**ROOF DECKING**									
0015	Made from recycled materials									
2100	Open type, 1-1/2" deep, Type B, wide rib, galv., 22 ga., under 50 sq. G	E-4	4500	.007	S.F.	2.38	.39	.02	2.79	3.31
2400	Over 500 squares G	"	5100	.006	"	1.71	.35	.02	2.08	2.50

05 31 33 – Steel Form Decking

05 31 33.50 Form Decking

		Crew	Daily Output	Labor-Hours	Unit	Material	2018 Bare Costs Labor	Equipment	Total	Total Incl O&P
0010	**FORM DECKING**									
0015	Made from recycled materials									
6100	Slab form, steel, 28 ga., 9/16" deep, Type UFS, uncoated G	E-4	4000	.008	S.F.	1.80	.44	.02	2.26	2.76
6200	Galvanized G	"	4000	.008	"	1.59	.44	.02	2.05	2.53

05 35 Raceway Decking Assemblies

05 35 13 – Steel Cellular Decking

05 35 13.50 Cellular Decking

		Crew	Daily Output	Labor-Hours	Unit	Material	2018 Bare Costs Labor	Equipment	Total	Total Incl O&P
0010	**CELLULAR DECKING**									
0015	Made from recycled materials									
0200	Cellular units, galv, 1-1/2" deep, Type BC, 20-20 ga., over 15 squares G	E-4	1460	.022	S.F.	10.10	1.21	.07	11.38	13.25
0400	3" deep, Type NC, galvanized, 20-20 ga. G		1375	.023		11.10	1.28	.07	12.45	14.45
1000	4-1/2" deep, Type JC, galvanized, 18-20 ga. G		1100	.029		15.45	1.60	.09	17.14	19.85
1500	For acoustical deck, add					15%				
1900	For multi-story or congested site, add						50%			

05 41 Structural Metal Stud Framing

05 41 13 – Load-Bearing Metal Stud Framing

05 41 13.05 Bracing

		Crew	Daily Output	Labor-Hours	Unit	Material	2018 Bare Costs Labor	Equipment	Total	Total Incl O&P
0010	**BRACING**, shear wall X-bracing, per 10' x 10' bay, one face									
0015	Made of recycled materials									
0120	Metal strap, 20 ga. x 4" wide G	2 Carp	18	.889	Ea.	18.90	45		63.90	94
0130	6" wide G		18	.889		31.50	45		76.50	108
0160	18 ga. x 4" wide G		16	1		33	50.50		83.50	119
0170	6" wide G		16	1		48.50	50.50		99	136
0410	Continuous strap bracing, per horizontal row on both faces									

For customer support on your Commercial Renovation Costs with RSMeans data, call 800.448.8182.

109

05 41 Structural Metal Stud Framing

05 41 13 – Load-Bearing Metal Stud Framing

05 41 13.05 Bracing

		Crew	Daily Output	Labor-Hours	Unit	Material	2018 Bare Costs Labor	2018 Bare Costs Equipment	Total	Total Incl O&P
0420	Metal strap, 20 ga. x 2" wide, studs 12" OC	G 1 Carp	7	1.143	C.L.F.	57	58		115	157
0430	16" OC	G	8	1		57	50.50		107.50	146
0440	24" OC	G	10	.800		57	40.50		97.50	129
0450	18 ga. x 2" wide, studs 12" OC	G	6	1.333		79	67.50		146.50	197
0460	16" OC	G	7	1.143		79	58		137	181
0470	24" OC	G	8	1		79	50.50		129.50	170

05 41 13.10 Bridging

		Crew	Daily Output	Labor-Hours	Unit	Material	2018 Bare Costs Labor	2018 Bare Costs Equipment	Total	Total Incl O&P
0010	**BRIDGING**, solid between studs w/1-1/4" leg track, per stud bay									
0015	Made from recycled materials									
0200	Studs 12" OC, 18 ga. x 2-1/2" wide	G 1 Carp	125	.064	Ea.	.95	3.24		4.19	6.30
0210	3-5/8" wide	G	120	.067		1.16	3.38		4.54	6.75
0220	4" wide	G	120	.067		1.22	3.38		4.60	6.85
0230	6" wide	G	115	.070		1.60	3.53		5.13	7.50
0240	8" wide	G	110	.073		1.97	3.69		5.66	8.15
0300	16 ga. x 2-1/2" wide	G	115	.070		1.21	3.53		4.74	7.10
0310	3-5/8" wide	G	110	.073		1.47	3.69		5.16	7.60
0320	4" wide	G	110	.073		1.57	3.69		5.26	7.75
0330	6" wide	G	105	.076		2.01	3.86		5.87	8.45
0340	8" wide	G	100	.080		2.51	4.06		6.57	9.35
1200	Studs 16" OC, 18 ga. x 2-1/2" wide	G	125	.064		1.22	3.24		4.46	6.60
1210	3-5/8" wide	G	120	.067		1.49	3.38		4.87	7.15
1220	4" wide	G	120	.067		1.57	3.38		4.95	7.20
1230	6" wide	G	115	.070		2.05	3.53		5.58	8
1240	8" wide	G	110	.073		2.52	3.69		6.21	8.80
1300	16 ga. x 2-1/2" wide	G	115	.070		1.55	3.53		5.08	7.45
1310	3-5/8" wide	G	110	.073		1.88	3.69		5.57	8.05
1320	4" wide	G	110	.073		2.01	3.69		5.70	8.20
1330	6" wide	G	105	.076		2.57	3.86		6.43	9.10
1340	8" wide	G	100	.080		3.22	4.06		7.28	10.15
2200	Studs 24" OC, 18 ga. x 2-1/2" wide	G	125	.064		1.77	3.24		5.01	7.20
2210	3-5/8" wide	G	120	.067		2.15	3.38		5.53	7.85
2220	4" wide	G	120	.067		2.27	3.38		5.65	8
2230	6" wide	G	115	.070		2.96	3.53		6.49	9
2240	8" wide	G	110	.073		3.65	3.69		7.34	10
2300	16 ga. x 2-1/2" wide	G	115	.070		2.24	3.53		5.77	8.20
2310	3-5/8" wide	G	110	.073		2.72	3.69		6.41	9
2320	4" wide	G	110	.073		2.91	3.69		6.60	9.20
2330	6" wide	G	105	.076		3.72	3.86		7.58	10.35
2340	8" wide	G	100	.080		4.65	4.06		8.71	11.70
3000	Continuous bridging, per row									
3100	16 ga. x 1-1/2" channel thru studs 12" OC	G 1 Carp	6	1.333	C.L.F.	51.50	67.50		119	167
3110	16" OC	G	7	1.143		51.50	58		109.50	151
3120	24" OC	G	8.80	.909		51.50	46		97.50	132
4100	2" x 2" angle x 18 ga., studs 12" OC	G	7	1.143		80.50	58		138.50	183
4110	16" OC	G	9	.889		80.50	45		125.50	162
4120	24" OC	G	12	.667		80.50	34		114.50	144
4200	16 ga., studs 12" OC	G	5	1.600		101	81		182	243
4210	16" OC	G	7	1.143		101	58		159	205
4220	24" OC	G	10	.800		101	40.50		141.50	177

05 41 13 – Load-Bearing Metal Stud Framing

05 41 13.25 Framing, Boxed Headers/Beams		Crew	Daily Output	Labor-Hours	Unit	Material	2018 Bare Costs Labor	Equipment	Total	Total Incl O&P
0010	**FRAMING, BOXED HEADERS/BEAMS**									
0015	Made from recycled materials									
0200	Double, 18 ga. x 6" deep	G	2 Carp	220	.073	L.F.	5.50	3.69		9.19	12.05
0210	8" deep	G		210	.076		6.05	3.86		9.91	12.90
0220	10" deep	G		200	.080		7.40	4.06		11.46	14.75
0230	12" deep	G		190	.084		8.05	4.27		12.32	15.80
0300	16 ga. x 8" deep	G		180	.089		7	4.51		11.51	15
0310	10" deep	G		170	.094		8.45	4.77		13.22	17
0320	12" deep	G		160	.100		9.15	5.05		14.20	18.35
0400	14 ga. x 10" deep	G		140	.114		9.75	5.80		15.55	20
0410	12" deep	G		130	.123		10.65	6.25		16.90	22
1210	Triple, 18 ga. x 8" deep	G		170	.094		8.75	4.77		13.52	17.40
1220	10" deep	G		165	.097		10.60	4.92		15.52	19.65
1230	12" deep	G		160	.100		11.60	5.05		16.65	21
1300	16 ga. x 8" deep	G		145	.110		10.15	5.60		15.75	20.50
1310	10" deep	G		140	.114		12.15	5.80		17.95	23
1320	12" deep	G		135	.119		13.25	6		19.25	24.50
1400	14 ga. x 10" deep	G		115	.139		13.25	7.05		20.30	26
1410	12" deep	G		110	.145		14.65	7.35		22	28

05 41 13.30 Framing, Stud Walls		Crew	Daily Output	Labor-Hours	Unit	Material	2018 Bare Costs Labor	Equipment	Total	Total Incl O&P
0010	**FRAMING, STUD WALLS** w/top & bottom track, no openings,									
0020	Headers, beams, bridging or bracing									
0025	Made from recycled materials									
4100	8' high walls, 18 ga. x 2-1/2" wide, studs 12" OC	G	2 Carp	54	.296	L.F.	9.15	15		24.15	34.50
4110	16" OC	G		77	.208		7.35	10.55		17.90	25
4120	24" OC	G		107	.150		5.50	7.60		13.10	18.35
4130	3-5/8" wide, studs 12" OC	G		53	.302		10.85	15.30		26.15	37
4140	16" OC	G		76	.211		8.70	10.65		19.35	27
4150	24" OC	G		105	.152		6.55	7.75		14.30	19.75
4160	4" wide, studs 12" OC	G		52	.308		11.30	15.60		26.90	38
4170	16" OC	G		74	.216		9.05	10.95		20	28
4180	24" OC	G		103	.155		6.80	7.90		14.70	20.50
4190	6" wide, studs 12" OC	G		51	.314		14.40	15.90		30.30	42
4200	16" OC	G		73	.219		11.55	11.10		22.65	31
4210	24" OC	G		101	.158		8.70	8.05		16.75	22.50
4220	8" wide, studs 12" OC	G		50	.320		17.45	16.20		33.65	45.50
4230	16" OC	G		72	.222		14	11.25		25.25	33.50
4240	24" OC	G		100	.160		10.60	8.10		18.70	25
4300	16 ga. x 2-1/2" wide, studs 12" OC	G		47	.340		10.85	17.25		28.10	40
4310	16" OC	G		68	.235		8.60	11.95		20.55	29
4320	24" OC	G		94	.170		6.35	8.65		15	21
4330	3-5/8" wide, studs 12" OC	G		46	.348		12.95	17.65		30.60	42.50
4340	16" OC	G		66	.242		10.25	12.30		22.55	31
4350	24" OC	G		92	.174		7.55	8.80		16.35	22.50
4360	4" wide, studs 12" OC	G		45	.356		13.55	18.05		31.60	44.50
4370	16" OC	G		65	.246		10.75	12.50		23.25	32.50
4380	24" OC	G		90	.178		7.95	9		16.95	23.50
4390	6" wide, studs 12" OC	G		44	.364		17.05	18.45		35.50	49
4400	16" OC	G		64	.250		13.55	12.70		26.25	35.50
4410	24" OC	G		88	.182		10.05	9.20		19.25	26
4420	8" wide, studs 12" OC	G		43	.372		21	18.85		39.85	53.50
4430	16" OC	G		63	.254		16.60	12.90		29.50	39.50

05 41 Structural Metal Stud Framing

05 41 13 – Load-Bearing Metal Stud Framing

05 41 13.30 Framing, Stud Walls		Crew	Daily Output	Labor-Hours	Unit	Material	2018 Bare Costs Labor	Equipment	Total	Total Incl O&P
4440	24" OC	G	2 Carp	86	.186	L.F.	12.30	9.45		21.75	29
5100	10' high walls, 18 ga. x 2-1/2" wide, studs 12" OC	G		54	.296		11	15		26	36.50
5110	16" OC	G		77	.208		8.70	10.55		19.25	26.50
5120	24" OC	G		107	.150		6.40	7.60		14	19.35
5130	3-5/8" wide, studs 12" OC	G		53	.302		13	15.30		28.30	39.50
5140	16" OC	G		76	.211		10.30	10.65		20.95	28.50
5150	24" OC	G		105	.152		7.60	7.75		15.35	21
5160	4" wide, studs 12" OC	G		52	.308		13.55	15.60		29.15	40.50
5170	16" OC	G		74	.216		10.75	10.95		21.70	29.50
5180	24" OC	G		103	.155		7.95	7.90		15.85	21.50
5190	6" wide, studs 12" OC	G		51	.314		17.25	15.90		33.15	45
5200	16" OC	G		73	.219		13.70	11.10		24.80	33
5210	24" OC	G		101	.158		10.15	8.05		18.20	24
5220	8" wide, studs 12" OC	G		50	.320		21	16.20		37.20	49.50
5230	16" OC	G		72	.222		16.55	11.25		27.80	36.50
5240	24" OC	G		100	.160		12.30	8.10		20.40	26.50
5300	16 ga. x 2-1/2" wide, studs 12" OC	G		47	.340		13.10	17.25		30.35	42.50
5310	16" OC	G		68	.235		10.30	11.95		22.25	30.50
5320	24" OC	G		94	.170		7.45	8.65		16.10	22
5330	3-5/8" wide, studs 12" OC	G		46	.348		15.60	17.65		33.25	45.50
5340	16" OC	G		66	.242		12.25	12.30		24.55	33.50
5350	24" OC	G		92	.174		8.90	8.80		17.70	24
5360	4" wide, studs 12" OC	G		45	.356		16.35	18.05		34.40	47.50
5370	16" OC	G		65	.246		12.85	12.50		25.35	34.50
5380	24" OC	G		90	.178		9.35	9		18.35	25
5390	6" wide, studs 12" OC	G		44	.364		20.50	18.45		38.95	52.50
5400	16" OC	G		64	.250		16.15	12.70		28.85	38.50
5410	24" OC	G		88	.182		11.80	9.20		21	28
5420	8" wide, studs 12" OC	G		43	.372		25	18.85		43.85	58
5430	16" OC	G		63	.254		19.80	12.90		32.70	43
5440	24" OC	G		86	.186		14.45	9.45		23.90	31
6190	12' high walls, 18 ga. x 6" wide, studs 12" OC	G		41	.390		20	19.80		39.80	54
6200	16" OC	G		58	.276		15.80	14		29.80	40
6210	24" OC	G		81	.198		11.55	10		21.55	29
6220	8" wide, studs 12" OC	G		40	.400		24.50	20.50		45	59.50
6230	16" OC	G		57	.281		19.15	14.25		33.40	44
6240	24" OC	G		80	.200		14	10.15		24.15	32
6390	16 ga. x 6" wide, studs 12" OC	G		35	.457		24	23		47	64
6400	16" OC	G		51	.314		18.80	15.90		34.70	46.50
6410	24" OC	G		70	.229		13.55	11.60		25.15	33.50
6420	8" wide, studs 12" OC	G		34	.471		29.50	24		53.50	71
6430	16" OC	G		50	.320		23	16.20		39.20	52
6440	24" OC	G		69	.232		16.60	11.75		28.35	37.50
6530	14 ga. x 3-5/8" wide, studs 12" OC	G		34	.471		22.50	24		46.50	63.50
6540	16" OC	G		48	.333		17.75	16.90		34.65	47
6550	24" OC	G		65	.246		12.75	12.50		25.25	34.50
6560	4" wide, studs 12" OC	G		33	.485		24	24.50		48.50	66.50
6570	16" OC	G		47	.340		18.75	17.25		36	48.50
6580	24" OC	G		64	.250		13.50	12.70		26.20	35.50
6730	12 ga. x 3-5/8" wide, studs 12" OC	G		31	.516		31.50	26		57.50	77
6740	16" OC	G		43	.372		24.50	18.85		43.35	57
6750	24" OC	G		59	.271		17.15	13.75		30.90	41.50
6760	4" wide, studs 12" OC	G		30	.533		33.50	27		60.50	81

05 41 Structural Metal Stud Framing

05 41 13 – Load-Bearing Metal Stud Framing

05 41 13.30 Framing, Stud Walls

			Crew	Daily Output	Labor-Hours	Unit	Material	2018 Bare Costs Labor	Equipment	Total	Total Incl O&P
6770	16" OC	G	2 Carp	42	.381	L.F.	26	19.30		45.30	60
6780	24" OC	G		58	.276		18.30	14		32.30	42.50
7390	16' high walls, 16 ga. x 6" wide, studs 12" OC	G		33	.485		31	24.50		55.50	74
7400	16" OC	G		48	.333		24	16.90		40.90	54
7410	24" OC	G		67	.239		17.05	12.10		29.15	38.50
7420	8" wide, studs 12" OC	G		32	.500		38	25.50		63.50	83
7430	16" OC	G		47	.340		29.50	17.25		46.75	60.50
7440	24" OC	G		66	.242		21	12.30		33.30	43
7560	14 ga. x 4" wide, studs 12" OC	G		31	.516		31	26		57	76.50
7570	16" OC	G		45	.356		24	18.05		42.05	56
7580	24" OC	G		61	.262		17	13.30		30.30	40
7590	6" wide, studs 12" OC	G		30	.533		39	27		66	87
7600	16" OC	G		44	.364		30	18.45		48.45	63
7610	24" OC	G		60	.267		21.50	13.50		35	45.50
7760	12 ga. x 4" wide, studs 12" OC	G		29	.552		44	28		72	93.50
7770	16" OC	G		40	.400		33.50	20.50		54	70
7780	24" OC	G		55	.291		23.50	14.75		38.25	49.50
7790	6" wide, studs 12" OC	G		28	.571		55.50	29		84.50	108
7800	16" OC	G		39	.410		42.50	21		63.50	80.50
7810	24" OC	G		54	.296		29.50	15		44.50	57
8590	20' high walls, 14 ga. x 6" wide, studs 12" OC	G		29	.552		48	28		76	98
8600	16" OC	G		42	.381		37	19.30		56.30	72
8610	24" OC	G		57	.281		26	14.25		40.25	51.50
8620	8" wide, studs 12" OC	G		28	.571		52	29		81	104
8630	16" OC	G		41	.390		40	19.80		59.80	76
8640	24" OC	G		56	.286		28.50	14.50		43	54.50
8790	12 ga. x 6" wide, studs 12" OC	G		27	.593		68	30		98	124
8800	16" OC	G		37	.432		52	22		74	93
8810	24" OC	G		51	.314		36	15.90		51.90	65.50
8820	8" wide, studs 12" OC	G		26	.615		82.50	31		113.50	142
8830	16" OC	G		36	.444		63	22.50		85.50	106
8840	24" OC	G		50	.320		43.50	16.20		59.70	74.50
9000	Minimum labor/equipment charge			4	4	Job		203		203	330

05 42 Cold-Formed Metal Joist Framing

05 42 13 – Cold-Formed Metal Floor Joist Framing

05 42 13.05 Bracing

			Crew	Daily Output	Labor-Hours	Unit	Material	2018 Bare Costs Labor	Equipment	Total	Total Incl O&P
0010	**BRACING**, continuous, per row, top & bottom										
0015	Made from recycled materials										
0120	Flat strap, 20 ga. x 2" wide, joists at 12" OC	G	1 Carp	4.67	1.713	C.L.F.	60	87		147	207
0130	16" OC	G		5.33	1.501		57.50	76		133.50	188
0140	24" OC	G		6.66	1.201		55.50	61		116.50	160
0150	18 ga. x 2" wide, joists at 12" OC	G		4	2		78.50	101		179.50	252
0160	16" OC	G		4.67	1.713		77	87		164	226
0170	24" OC	G		5.33	1.501		75.50	76		151.50	207

05 42 13.10 Bridging

			Crew	Daily Output	Labor-Hours	Unit	Material	2018 Bare Costs Labor	Equipment	Total	Total Incl O&P
0010	**BRIDGING**, solid between joists w/1-1/4" leg track, per joist bay										
0015	Made from recycled materials										
0230	Joists 12" OC, 18 ga. track x 6" wide	G	1 Carp	80	.100	Ea.	1.60	5.05		6.65	10
0240	8" wide	G		75	.107		1.97	5.40		7.37	10.95
0250	10" wide	G		70	.114		2.46	5.80		8.26	12.10

05 42 13 – Cold-Formed Metal Floor Joist Framing

05 42 13.10 Bridging		Crew	Daily Output	Labor-Hours	Unit	Material	2018 Bare Costs Labor	2018 Bare Costs Equipment	Total	Total Incl O&P
0260	12" wide	G 1 Carp	65	.123	Ea.	2.78	6.25		9.03	13.20
0330	16 ga. track x 6" wide	G	70	.114		2.01	5.80		7.81	11.60
0340	8" wide	G	65	.123		2.51	6.25		8.76	12.90
0350	10" wide	G	60	.133		3.10	6.75		9.85	14.35
0360	12" wide	G	55	.145		3.58	7.35		10.93	15.90
0440	14 ga. track x 8" wide	G	60	.133		3.13	6.75		9.88	14.40
0450	10" wide	G	55	.145		3.89	7.35		11.24	16.25
0460	12" wide	G	50	.160		4.49	8.10		12.59	18.10
0550	12 ga. track x 10" wide	G	45	.178		5.70	9		14.70	21
0560	12" wide	G	40	.200		5.80	10.15		15.95	23
1230	16" OC, 18 ga. track x 6" wide	G	80	.100		2.05	5.05		7.10	10.50
1240	8" wide	G	75	.107		2.52	5.40		7.92	11.60
1250	10" wide	G	70	.114		3.15	5.80		8.95	12.85
1260	12" wide	G	65	.123		3.56	6.25		9.81	14.05
1330	16 ga. track x 6" wide	G	70	.114		2.57	5.80		8.37	12.25
1340	8" wide	G	65	.123		3.22	6.25		9.47	13.70
1350	10" wide	G	60	.133		3.98	6.75		10.73	15.30
1360	12" wide	G	55	.145		4.59	7.35		11.94	17
1440	14 ga. track x 8" wide	G	60	.133		4.01	6.75		10.76	15.35
1450	10" wide	G	55	.145		4.98	7.35		12.33	17.45
1460	12" wide	G	50	.160		5.75	8.10		13.85	19.50
1550	12 ga. track x 10" wide	G	45	.178		7.30	9		16.30	22.50
1560	12" wide	G	40	.200		7.45	10.15		17.60	24.50
2230	24" OC, 18 ga. track x 6" wide	G	80	.100		2.96	5.05		8.01	11.50
2240	8" wide	G	75	.107		3.65	5.40		9.05	12.80
2250	10" wide	G	70	.114		4.56	5.80		10.36	14.40
2260	12" wide	G	65	.123		5.15	6.25		11.40	15.80
2330	16 ga. track x 6" wide	G	70	.114		3.72	5.80		9.52	13.50
2340	8" wide	G	65	.123		4.65	6.25		10.90	15.25
2350	10" wide	G	60	.133		5.75	6.75		12.50	17.30
2360	12" wide	G	55	.145		6.65	7.35		14	19.25
2440	14 ga. track x 8" wide	G	60	.133		5.80	6.75		12.55	17.35
2450	10" wide	G	55	.145		7.20	7.35		14.55	19.90
2460	12" wide	G	50	.160		8.35	8.10		16.45	22.50
2550	12 ga. track x 10" wide	G	45	.178		10.55	9		19.55	26.50
2560	12" wide	G	40	.200		10.80	10.15		20.95	28.50

05 42 13.25 Framing, Band Joist

		Crew	Daily Output	Labor-Hours	Unit	Material	2018 Bare Costs Labor	2018 Bare Costs Equipment	Total	Total Incl O&P
0010	**FRAMING, BAND JOIST** (track) fastened to bearing wall									
0015	Made from recycled materials									
0220	18 ga. track x 6" deep	G 2 Carp	1000	.016	L.F.	1.30	.81		2.11	2.75
0230	8" deep	G	920	.017		1.61	.88		2.49	3.20
0240	10" deep	G	860	.019		2.01	.94		2.95	3.74
0320	16 ga. track x 6" deep	G	900	.018		1.64	.90		2.54	3.26
0330	8" deep	G	840	.019		2.05	.97		3.02	3.82
0340	10" deep	G	780	.021		2.53	1.04		3.57	4.47
0350	12" deep	G	740	.022		2.92	1.10		4.02	4.99
0430	14 ga. track x 8" deep	G	750	.021		2.55	1.08		3.63	4.57
0440	10" deep	G	720	.022		3.17	1.13		4.30	5.30
0450	12" deep	G	700	.023		3.66	1.16		4.82	5.90
0540	12 ga. track x 10" deep	G	670	.024		4.64	1.21		5.85	7.05
0550	12" deep	G	650	.025		4.75	1.25		6	7.25

05 42 Cold-Formed Metal Joist Framing

05 42 13 – Cold-Formed Metal Floor Joist Framing

05 42 13.30 Framing, Boxed Headers/Beams

		Crew	Daily Output	Labor-Hours	Unit	Material	2018 Bare Costs Labor	Equipment	Total	Total Incl O&P
0010	**FRAMING, BOXED HEADERS/BEAMS**									
0015	Made from recycled materials									
0200	Double, 18 ga. x 6" deep	G	2 Carp	220	.073	L.F.	5.50	3.69		9.19	12.05
0210	8" deep	G		210	.076		6.05	3.86		9.91	12.90
0220	10" deep	G		200	.080		7.40	4.06		11.46	14.75
0230	12" deep	G		190	.084		8.05	4.27		12.32	15.80
0300	16 ga. x 8" deep	G		180	.089		7	4.51		11.51	15
0310	10" deep	G		170	.094		8.45	4.77		13.22	17
0320	12" deep	G		160	.100		9.15	5.05		14.20	18.35
0400	14 ga. x 10" deep	G		140	.114		9.75	5.80		15.55	20
0410	12" deep	G		130	.123		10.65	6.25		16.90	22
0500	12 ga. x 10" deep	G		110	.145		12.85	7.35		20.20	26
0510	12" deep	G		100	.160		14.15	8.10		22.25	29
1210	Triple, 18 ga. x 8" deep	G		170	.094		8.75	4.77		13.52	17.40
1220	10" deep	G		165	.097		10.60	4.92		15.52	19.65
1230	12" deep	G		160	.100		11.60	5.05		16.65	21
1300	16 ga. x 8" deep	G		145	.110		10.15	5.60		15.75	20.50
1310	10" deep	G		140	.114		12.15	5.80		17.95	23
1320	12" deep	G		135	.119		13.25	6		19.25	24.50
1400	14 ga. x 10" deep	G		115	.139		14.10	7.05		21.15	27
1410	12" deep	G		110	.145		15.50	7.35		22.85	29
1500	12 ga. x 10" deep	G		90	.178		18.75	9		27.75	35
1510	12" deep	G		85	.188		21	9.55		30.55	38.50

05 42 13.40 Framing, Joists

		Crew	Daily Output	Labor-Hours	Unit	Material	2018 Bare Costs Labor	Equipment	Total	Total Incl O&P
0010	**FRAMING, JOISTS**, no band joists (track), web stiffeners, headers,									
0020	Beams, bridging or bracing									
0025	Made from recycled materials									
0030	Joists (2" flange) and fasteners, materials only									
0220	18 ga. x 6" deep	G				L.F.	1.69			1.69	1.86
0230	8" deep	G					1.98			1.98	2.18
0240	10" deep	G					2.34			2.34	2.58
0320	16 ga. x 6" deep	G					2.07			2.07	2.28
0330	8" deep	G					2.47			2.47	2.71
0340	10" deep	G					2.89			2.89	3.18
0350	12" deep	G					3.28			3.28	3.60
0430	14 ga. x 8" deep	G					3.10			3.10	3.41
0440	10" deep	G					3.57			3.57	3.93
0450	12" deep	G					4.06			4.06	4.47
0540	12 ga. x 10" deep	G					5.20			5.20	5.70
0550	12" deep	G					5.90			5.90	6.50
1010	Installation of joists to band joists, beams & headers, labor only									
1220	18 ga. x 6" deep		2 Carp	110	.145	Ea.		7.35		7.35	11.95
1230	8" deep			90	.178			9		9	14.65
1240	10" deep			80	.200			10.15		10.15	16.45
1320	16 ga. x 6" deep			95	.168			8.55		8.55	13.85
1330	8" deep			70	.229			11.60		11.60	18.80
1340	10" deep			60	.267			13.50		13.50	22
1350	12" deep			55	.291			14.75		14.75	24
1430	14 ga. x 8" deep			65	.246			12.50		12.50	20.50
1440	10" deep			45	.356			18.05		18.05	29.50
1450	12" deep			35	.457			23		23	37.50
1540	12 ga. x 10" deep			40	.400			20.50		20.50	33

05 42 13 – Cold-Formed Metal Floor Joist Framing

05 42 13.40 Framing, Joists		Crew	Daily Output	Labor-Hours	Unit	Material	2018 Bare Costs Labor	2018 Bare Costs Equipment	Total	Total Incl O&P
1550	12" deep	2 Carp	30	.533	Ea.		27		27	44
9000	Minimum labor/equipment charge		4	4	Job		203		203	330

05 42 13.45 Framing, Web Stiffeners

	05 42 13.45 Framing, Web Stiffeners		Crew	Daily Output	Labor-Hours	Unit	Material	Labor	Equipment	Total	Total Incl O&P
0010	**FRAMING, WEB STIFFENERS** at joist bearing, fabricated from										
0020	Stud piece (1-5/8" flange) to stiffen joist (2" flange)										
0025	Made from recycled materials										
2120	For 6" deep joist, with 18 ga. x 2-1/2" stud	G	1 Carp	120	.067	Ea.	.92	3.38		4.30	6.50
2130	3-5/8" stud	G		110	.073		1.08	3.69		4.77	7.20
2140	4" stud	G		105	.076		1.12	3.86		4.98	7.50
2150	6" stud	G		100	.080		1.42	4.06		5.48	8.15
2160	8" stud	G		95	.084		1.71	4.27		5.98	8.85
2220	8" deep joist, with 2-1/2" stud	G		120	.067		1.23	3.38		4.61	6.85
2230	3-5/8" stud	G		110	.073		1.45	3.69		5.14	7.60
2240	4" stud	G		105	.076		1.50	3.86		5.36	7.90
2250	6" stud	G		100	.080		1.90	4.06		5.96	8.70
2260	8" stud	G		95	.084		2.29	4.27		6.56	9.45
2320	10" deep joist, with 2-1/2" stud	G		110	.073		1.53	3.69		5.22	7.70
2330	3-5/8" stud	G		100	.080		1.79	4.06		5.85	8.55
2340	4" stud	G		95	.084		1.86	4.27		6.13	9
2350	6" stud	G		90	.089		2.36	4.51		6.87	9.90
2360	8" stud	G		85	.094		2.84	4.77		7.61	10.85
2420	12" deep joist, with 2-1/2" stud	G		110	.073		1.84	3.69		5.53	8
2430	3-5/8" stud	G		100	.080		2.16	4.06		6.22	9
2440	4" stud	G		95	.084		2.24	4.27		6.51	9.40
2450	6" stud	G		90	.089		2.84	4.51		7.35	10.40
2460	8" stud	G		85	.094		3.42	4.77		8.19	11.50
3130	For 6" deep joist, with 16 ga. x 3-5/8" stud	G		100	.080		1.34	4.06		5.40	8.05
3140	4" stud	G		95	.084		1.40	4.27		5.67	8.50
3150	6" stud	G		90	.089		1.75	4.51		6.26	9.25
3160	8" stud	G		85	.094		2.14	4.77		6.91	10.10
3230	8" deep joist, with 3-5/8" stud	G		100	.080		1.80	4.06		5.86	8.60
3240	4" stud	G		95	.084		1.88	4.27		6.15	9
3250	6" stud	G		90	.089		2.35	4.51		6.86	9.90
3260	8" stud	G		85	.094		2.87	4.77		7.64	10.90
3330	10" deep joist, with 3-5/8" stud	G		85	.094		2.22	4.77		6.99	10.20
3340	4" stud	G		80	.100		2.32	5.05		7.37	10.80
3350	6" stud	G		75	.107		2.91	5.40		8.31	12
3360	8" stud	G		70	.114		3.55	5.80		9.35	13.30
3430	12" deep joist, with 3-5/8" stud	G		85	.094		2.68	4.77		7.45	10.70
3440	4" stud	G		80	.100		2.80	5.05		7.85	11.35
3450	6" stud	G		75	.107		3.50	5.40		8.90	12.65
3460	8" stud	G		70	.114		4.28	5.80		10.08	14.10
4230	For 8" deep joist, with 14 ga. x 3-5/8" stud	G		90	.089		2.22	4.51		6.73	9.75
4240	4" stud	G		85	.094		2.35	4.77		7.12	10.35
4250	6" stud	G		80	.100		2.95	5.05		8	11.50
4260	8" stud	G		75	.107		3.15	5.40		8.55	12.25
4330	10" deep joist, with 3-5/8" stud	G		75	.107		2.76	5.40		8.16	11.85
4340	4" stud	G		70	.114		2.91	5.80		8.71	12.60
4350	6" stud	G		65	.123		3.65	6.25		9.90	14.15
4360	8" stud	G		60	.133		3.90	6.75		10.65	15.25
4430	12" deep joist, with 3-5/8" stud	G		75	.107		3.32	5.40		8.72	12.45
4440	4" stud	G		70	.114		3.50	5.80		9.30	13.25

05 42 Cold-Formed Metal Joist Framing

05 42 13 – Cold-Formed Metal Floor Joist Framing

05 42 13.45 Framing, Web Stiffeners

			Crew	Daily Output	Labor-Hours	Unit	Material	2018 Bare Costs Labor	Equipment	Total	Total Incl O&P
4450	6" stud	G	1 Carp	65	.123	Ea.	4.40	6.25		10.65	15
4460	8" stud	G		60	.133		4.70	6.75		11.45	16.10
5330	For 10" deep joist, with 12 ga. x 3-5/8" stud	G		65	.123		3.97	6.25		10.22	14.50
5340	4" stud	G		60	.133		4.23	6.75		10.98	15.60
5350	6" stud	G		55	.145		5.35	7.35		12.70	17.85
5360	8" stud	G		50	.160		6.45	8.10		14.55	20.50
5430	12" deep joist, with 3-5/8" stud	G		65	.123		4.78	6.25		11.03	15.40
5440	4" stud	G		60	.133		5.10	6.75		11.85	16.55
5450	6" stud	G		55	.145		6.45	7.35		13.80	19.05
5460	8" stud	G		50	.160		7.80	8.10		15.90	21.50

05 42 23 – Cold-Formed Metal Roof Joist Framing

05 42 23.05 Framing, Bracing

			Crew	Daily Output	Labor-Hours	Unit	Material	2018 Bare Costs Labor	Equipment	Total	Total Incl O&P
0010	**FRAMING, BRACING**										
0015	Made from recycled materials										
0020	Continuous bracing, per row										
0100	16 ga. x 1-1/2" channel thru rafters/trusses @ 16" OC	G	1 Carp	4.50	1.778	C.L.F.	51.50	90		141.50	203
0120	24" OC	G		6	1.333		51.50	67.50		119	167
0300	2" x 2" angle x 18 ga., rafters/trusses @ 16" OC	G		6	1.333		80.50	67.50		148	199
0320	24" OC	G		8	1		80.50	50.50		131	171
0400	16 ga., rafters/trusses @ 16" OC	G		4.50	1.778		101	90		191	257
0420	24" OC	G		6.50	1.231		101	62.50		163.50	212

05 42 23.10 Framing, Bridging

			Crew	Daily Output	Labor-Hours	Unit	Material	2018 Bare Costs Labor	Equipment	Total	Total Incl O&P
0010	**FRAMING, BRIDGING**										
0015	Made from recycled materials										
0020	Solid, between rafters w/1-1/4" leg track, per rafter bay										
1200	Rafters 16" OC, 18 ga. x 4" deep	G	1 Carp	60	.133	Ea.	1.57	6.75		8.32	12.65
1210	6" deep	G		57	.140		2.05	7.10		9.15	13.80
1220	8" deep	G		55	.145		2.52	7.35		9.87	14.75
1230	10" deep	G		52	.154		3.15	7.80		10.95	16.10
1240	12" deep	G		50	.160		3.56	8.10		11.66	17.05
2200	24" OC, 18 ga. x 4" deep	G		60	.133		2.27	6.75		9.02	13.45
2210	6" deep	G		57	.140		2.96	7.10		10.06	14.80
2220	8" deep	G		55	.145		3.65	7.35		11	15.95
2230	10" deep	G		52	.154		4.56	7.80		12.36	17.65
2240	12" deep	G		50	.160		5.15	8.10		13.25	18.80

05 42 23.50 Framing, Parapets

			Crew	Daily Output	Labor-Hours	Unit	Material	2018 Bare Costs Labor	Equipment	Total	Total Incl O&P
0010	**FRAMING, PARAPETS**										
0015	Made from recycled materials										
0100	3' high installed on 1st story, 18 ga. x 4" wide studs, 12" OC	G	2 Carp	100	.160	L.F.	5.70	8.10		13.80	19.40
0110	16" OC	G		150	.107		4.85	5.40		10.25	14.15
0120	24" OC	G		200	.080		4.01	4.06		8.07	11
0200	6" wide studs, 12" OC	G		100	.160		7.30	8.10		15.40	21
0210	16" OC	G		150	.107		6.25	5.40		11.65	15.65
0220	24" OC	G		200	.080		5.15	4.06		9.21	12.30
1100	Installed on 2nd story, 18 ga. x 4" wide studs, 12" OC	G		95	.168		5.70	8.55		14.25	20
1110	16" OC	G		145	.110		4.85	5.60		10.45	14.45
1120	24" OC	G		190	.084		4.01	4.27		8.28	11.35
1200	6" wide studs, 12" OC	G		95	.168		7.30	8.55		15.85	22
1210	16" OC	G		145	.110		6.25	5.60		11.85	15.95
1220	24" OC	G		190	.084		5.15	4.27		9.42	12.65
2100	Installed on gable, 18 ga. x 4" wide studs, 12" OC	G		85	.188		5.70	9.55		15.25	22
2110	16" OC	G		130	.123		4.85	6.25		11.10	15.50

117

05 42 23 – Cold-Formed Metal Roof Joist Framing

05 42 23.50 Framing, Parapets		Crew	Daily Output	Labor-Hours	Unit	Material	2018 Bare Costs Labor	2018 Bare Costs Equipment	Total	Total Incl O&P
2120	24" OC	G 2 Carp	170	.094	L.F.	4.01	4.77		8.78	12.15
2200	6" wide studs, 12" OC	G	85	.188		7.30	9.55		16.85	23.50
2210	16" OC	G	130	.123		6.25	6.25		12.50	17
2220	24" OC	G	170	.094		5.15	4.77		9.92	13.45

05 42 23.60 Framing, Roof Rafters

		Crew	Daily Output	Labor-Hours	Unit	Material	Labor	Equipment	Total	Total Incl O&P
0010	**FRAMING, ROOF RAFTERS**									
0015	Made from recycled materials									
0100	Boxed ridge beam, double, 18 ga. x 6" deep	G 2 Carp	160	.100	L.F.	5.50	5.05		10.55	14.30
0110	8" deep	G	150	.107		6.05	5.40		11.45	15.45
0120	10" deep	G	140	.114		7.40	5.80		13.20	17.55
0130	12" deep	G	130	.123		8.05	6.25		14.30	19
0200	16 ga. x 6" deep	G	150	.107		6.20	5.40		11.60	15.65
0210	8" deep	G	140	.114		7	5.80		12.80	17.10
0220	10" deep	G	130	.123		8.45	6.25		14.70	19.40
0230	12" deep	G	120	.133		9.15	6.75		15.90	21
1100	Rafters, 2" flange, material only, 18 ga. x 6" deep	G				1.69			1.69	1.86
1110	8" deep	G				1.98			1.98	2.18
1120	10" deep	G				2.34			2.34	2.58
1130	12" deep	G				2.70			2.70	2.97
1200	16 ga. x 6" deep	G				2.07			2.07	2.28
1210	8" deep	G				2.47			2.47	2.71
1220	10" deep	G				2.89			2.89	3.18
1230	12" deep	G				3.28			3.28	3.60
2100	Installation only, ordinary rafter to 4:12 pitch, 18 ga. x 6" deep	2 Carp	35	.457	Ea.		23		23	37.50
2110	8" deep		30	.533			27		27	44
2120	10" deep		25	.640			32.50		32.50	52.50
2130	12" deep		20	.800			40.50		40.50	66
2200	16 ga. x 6" deep		30	.533			27		27	44
2210	8" deep		25	.640			32.50		32.50	52.50
2220	10" deep		20	.800			40.50		40.50	66
2230	12" deep		15	1.067			54		54	88
8100	Add to labor, ordinary rafters on steep roofs						25%			
8110	Dormers & complex roofs						50%			
8200	Hip & valley rafters to 4:12 pitch						25%			
8210	Steep roofs						50%			
8220	Dormers & complex roofs						75%			
8300	Hip & valley jack rafters to 4:12 pitch						50%			
8310	Steep roofs						75%			
8320	Dormers & complex roofs						100%			
9000	Minimum labor/equipment charge	2 Carp	4	4	Job		203		203	330

05 42 23.70 Framing, Soffits and Canopies

		Crew	Daily Output	Labor-Hours	Unit	Material	Labor	Equipment	Total	Total Incl O&P
0010	**FRAMING, SOFFITS & CANOPIES**									
0015	Made from recycled materials									
0130	Continuous ledger track @ wall, studs @ 16" OC, 18 ga. x 4" wide	G 2 Carp	535	.030	L.F.	1.05	1.52		2.57	3.61
0140	6" wide	G	500	.032		1.36	1.62		2.98	4.13
0150	8" wide	G	465	.034		1.68	1.74		3.42	4.68
0160	10" wide	G	430	.037		2.10	1.89		3.99	5.35
0230	Studs @ 24" OC, 18 ga. x 4" wide	G	800	.020		1	1.01		2.01	2.75
0240	6" wide	G	750	.021		1.30	1.08		2.38	3.19
0250	8" wide	G	700	.023		1.61	1.16		2.77	3.65
0260	10" wide	G	650	.025		2.01	1.25		3.26	4.24
1000	Horizontal soffit and canopy members, material only									

05 42 Cold-Formed Metal Joist Framing

05 42 23 – Cold-Formed Metal Roof Joist Framing

05 42 23.70 Framing, Soffits and Canopies		Crew	Daily Output	Labor-Hours	Unit	Material	2018 Bare Costs Labor	Equipment	Total	Total Incl O&P
1030	1-5/8" flange studs, 18 ga. x 4" deep	G			L.F.	1.34			1.34	1.48
1040	6" deep	G				1.70			1.70	1.87
1050	8" deep	G				2.05			2.05	2.26
1140	2" flange joists, 18 ga. x 6" deep	G				1.93			1.93	2.13
1150	8" deep	G				2.27			2.27	2.49
1160	10" deep	G				2.68			2.68	2.94
4030	Installation only, 18 ga., 1-5/8" flange x 4" deep		2 Carp	130	.123	Ea.		6.25		6.25	10.15
4040	6" deep			110	.145			7.35		7.35	11.95
4050	8" deep			90	.178			9		9	14.65
4140	2" flange, 18 ga. x 6" deep			110	.145			7.35		7.35	11.95
4150	8" deep			90	.178			9		9	14.65
4160	10" deep			80	.200			10.15		10.15	16.45
6010	Clips to attach fascia to rafter tails, 2" x 2" x 18 ga. angle	G	1 Carp	120	.067		.95	3.38		4.33	6.55
6020	16 ga. angle	G	"	100	.080		1.20	4.06		5.26	7.90
9000	Minimum labor/equipment charge		2 Carp	4	4	Job		203		203	330

05 44 Cold-Formed Metal Trusses

05 44 13 – Cold-Formed Metal Roof Trusses

05 44 13.60 Framing, Roof Trusses		Crew	Daily Output	Labor-Hours	Unit	Material	2018 Bare Costs Labor	Equipment	Total	Total Incl O&P
0010	**FRAMING, ROOF TRUSSES**									
0015	Made from recycled materials									
0020	Fabrication of trusses on ground, Fink (W) or King Post, to 4:12 pitch									
0120	18 ga. x 4" chords, 16' span	G	2 Carp	12	1.333	Ea.	62.50	67.50		130	179
0130	20' span	G		11	1.455		78.50	74		152.50	206
0140	24' span	G		11	1.455		94	74		168	223
0150	28' span	G		10	1.600		110	81		191	253
0160	32' span	G		10	1.600		125	81		206	270
0250	6" chords, 28' span	G		9	1.778		139	90		229	299
0260	32' span	G		9	1.778		159	90		249	320
0270	36' span	G		8	2		179	101		280	360
0280	40' span	G		8	2		199	101		300	385
1120	5:12 to 8:12 pitch, 18 ga. x 4" chords, 16' span	G		10	1.600		71.50	81		152.50	211
1130	20' span	G		9	1.778		89.50	90		179.50	245
1140	24' span	G		9	1.778		108	90		198	264
1150	28' span	G		8	2		125	101		226	305
1160	32' span	G		8	2		143	101		244	325
1250	6" chords, 28' span	G		7	2.286		159	116		275	365
1260	32' span	G		7	2.286		182	116		298	390
1270	36' span	G		6	2.667		204	135		339	445
1280	40' span	G		6	2.667		227	135		362	470
2120	9:12 to 12:12 pitch, 18 ga. x 4" chords, 16' span	G		8	2		89.50	101		190.50	264
2130	20' span	G		7	2.286		112	116		228	310
2140	24' span	G		7	2.286		134	116		250	335
2150	28' span	G		6	2.667		157	135		292	390
2160	32' span	G		6	2.667		179	135		314	415
2250	6" chords, 28' span	G		5	3.200		199	162		361	480
2260	32' span	G		5	3.200		227	162		389	515
2270	36' span	G		4	4		256	203		459	610
2280	40' span	G		4	4		284	203		487	640
4900	Minimum labor/equipment charge			4	4	Job		203		203	330
5120	Erection only of roof trusses, to 4:12 pitch, 16' span		F-6	48	.833	Ea.		39.50	10.35	49.85	75.50

05 44 Cold-Formed Metal Trusses

05 44 13 – Cold-Formed Metal Roof Trusses

05 44 13.60 Framing, Roof Trusses

		Crew	Daily Output	Labor-Hours	Unit	Material	2018 Bare Costs Labor	2018 Bare Costs Equipment	Total	Total Incl O&P
5130	20' span	F-6	46	.870	Ea.		41.50	10.80	52.30	78.50
5140	24' span		44	.909			43	11.25	54.25	82
5150	28' span		42	.952			45	11.80	56.80	86
5160	32' span		40	1			47.50	12.40	59.90	90
5170	36' span		38	1.053			50	13.05	63.05	95
5180	40' span		36	1.111			52.50	13.80	66.30	100
5220	5:12 to 8:12 pitch, 16' span		42	.952			45	11.80	56.80	86
5230	20' span		40	1			47.50	12.40	59.90	90
5240	24' span		38	1.053			50	13.05	63.05	95
5250	28' span		36	1.111			52.50	13.80	66.30	100
5260	32' span		34	1.176			56	14.60	70.60	106
5270	36' span		32	1.250			59.50	15.50	75	113
5280	40' span		30	1.333			63.50	16.55	80.05	120
5320	9:12 to 12:12 pitch, 16' span		36	1.111			52.50	13.80	66.30	100
5330	20' span		34	1.176			56	14.60	70.60	106
5340	24' span		32	1.250			59.50	15.50	75	113
5350	28' span		30	1.333			63.50	16.55	80.05	120
5360	32' span		28	1.429			68	17.70	85.70	129
5370	36' span		26	1.538			73	19.10	92.10	139
5380	40' span		24	1.667			79	20.50	99.50	151
9000	Minimum labor/equipment charge		2	20	Job		950	248	1,198	1,800

05 51 Metal Stairs

05 51 13 – Metal Pan Stairs

05 51 13.50 Pan Stairs

			Crew	Daily Output	Labor-Hours	Unit	Material	2018 Bare Costs Labor	2018 Bare Costs Equipment	Total	Total Incl O&P
0010	**PAN STAIRS**, shop fabricated, steel stringers										
0015	Made from recycled materials										
0200	Metal pan tread for concrete in-fill, picket rail, 3'-6" wide	G	E-4	35	.914	Riser	560	50.50	2.82	613.32	705
0300	4'-0" wide	G		30	1.067		625	59	3.29	687.29	790
0350	Wall rail, both sides, 3'-6" wide	G		53	.604		425	33.50	1.86	460.36	530
1500	Landing, steel pan, conventional	G		160	.200	S.F.	74.50	11.05	.62	86.17	101

05 51 16 – Metal Floor Plate Stairs

05 51 16.50 Floor Plate Stairs

			Crew	Daily Output	Labor-Hours	Unit	Material	2018 Bare Costs Labor	2018 Bare Costs Equipment	Total	Total Incl O&P
0010	**FLOOR PLATE STAIRS**, shop fabricated, steel stringers										
0015	Made from recycled materials										
0400	Cast iron tread and pipe rail, 3'-6" wide	G	E-4	35	.914	Riser	600	50.50	2.82	653.32	750
0450	4'-0" wide	G		30	1.067		630	59	3.29	692.29	795
0475	5'-0" wide	G		28	1.143		695	63	3.52	761.52	870
0500	Checkered plate tread, industrial, 3'-6" wide	G		28	1.143		365	63	3.52	431.52	510
0550	Circular, for tanks, 3'-0" wide	G		33	.970		415	53.50	2.99	471.49	555
0600	For isolated stairs, add							100%			
0800	Custom steel stairs, 3'-6" wide, economy	G	E-4	35	.914		560	50.50	2.82	613.32	705
0810	Medium priced	G		30	1.067		740	59	3.29	802.29	920
0900	Deluxe	G		20	1.600		920	88	4.93	1,012.93	1,175
1100	For 4' wide stairs, add							5%	5%		
1300	For 5' wide stairs, add							10%	10%		

05 51 Metal Stairs

05 51 19 – Metal Grating Stairs

05 51 19.50 Grating Stairs

05 51 19.50 Grating Stairs		Crew	Daily Output	Labor-Hours	Unit	Material	2018 Bare Costs Labor	Equipment	Total	Total Incl O&P
0010	**GRATING STAIRS**, shop fabricated, steel stringers, safety nosing on treads									
0015	Made from recycled materials									
0020	Grating tread and pipe railing, 3'-6" wide	G	E-4	35	.914	Riser	365	50.50	2.82	418.32	490
0100	4'-0" wide	G		30	1.067	"	420	59	3.29	482.29	570
9000	Minimum labor/equipment charge	↓		2	16	Job		880	49.50	929.50	1,550

05 51 23 – Metal Fire Escapes

05 51 23.25 Fire Escapes

05 51 23.25 Fire Escapes		Crew	Daily Output	Labor-Hours	Unit	Material	Labor	Equipment	Total	Total Incl O&P
0010	**FIRE ESCAPES**, shop fabricated									
0200	2' wide balcony, 1" x 1/4" bars 1-1/2" OC, with railing	G	2 Sswk	10	1.600	L.F.	60	87.50		147.50	215
0400	1st story cantilevered stair, standard, with railing	G		.50	32	Ea.	2,500	1,750		4,250	5,725
0500	Cable counterweighted, with railing	G		.40	40	"	2,325	2,175		4,500	6,275
0700	36" x 40" platform & fixed stair, with railing	G	↓	.40	40	Flight	1,100	2,175		3,275	4,950
0900	For 3'-6" wide escapes, add to above						100%	150%			

05 51 23.50 Fire Escape Stairs

05 51 23.50 Fire Escape Stairs		Crew	Daily Output	Labor-Hours	Unit	Material	Labor	Equipment	Total	Total Incl O&P
0010	**FIRE ESCAPE STAIRS**, portable										
0100	Portable ladder					Ea.	128			128	141

05 51 33 – Metal Ladders

05 51 33.13 Vertical Metal Ladders

05 51 33.13 Vertical Metal Ladders		Crew	Daily Output	Labor-Hours	Unit	Material	Labor	Equipment	Total	Total Incl O&P
0010	**VERTICAL METAL LADDERS**, shop fabricated									
0015	Made from recycled materials									
0020	Steel, 20" wide, bolted to concrete, with cage	G	E-4	50	.640	V.L.F.	67	35.50	1.97	104.47	136
0100	Without cage	G		85	.376		36.50	21	1.16	58.66	77
0300	Aluminum, bolted to concrete, with cage	G		50	.640		133	35.50	1.97	170.47	208
0400	Without cage	G		85	.376	↓	52	21	1.16	74.16	94
9000	Minimum labor/equipment charge		↓	2	16	Job		880	49.50	929.50	1,550

05 51 33.16 Inclined Metal Ladders

05 51 33.16 Inclined Metal Ladders		Crew	Daily Output	Labor-Hours	Unit	Material	Labor	Equipment	Total	Total Incl O&P
0010	**INCLINED METAL LADDERS**, shop fabricated									
0015	Made from recycled materials									
3900	Industrial ships ladder, steel, 24" W, grating treads, 2 line pipe rail	G	E-4	30	1.067	Riser	214	59	3.29	276.29	340
4000	Aluminum	G	"	30	1.067	"	299	59	3.29	361.29	435

05 51 33.23 Alternating Tread Ladders

05 51 33.23 Alternating Tread Ladders		Crew	Daily Output	Labor-Hours	Unit	Material	Labor	Equipment	Total	Total Incl O&P
0010	**ALTERNATING TREAD LADDERS**, shop fabricated									
0015	Made from recycled materials									
0800	Alternating tread ladders, 68-degree angle of incline									
0810	8' vertical rise, steel, 149 lb., standard paint color		B-68G	3	5.333	Ea.	2,400	291	93	2,784	3,250
0820	Non-standard paint color			3	5.333		2,850	291	93	3,234	3,725
0830	Galvanized			3	5.333		2,825	291	93	3,209	3,700
0840	Stainless			3	5.333		4,075	291	93	4,459	5,100
0850	Aluminum, 87 lb.			3	5.333		2,975	291	93	3,359	3,875
1010	10' vertical rise, steel, 181 lb., standard paint color			2.75	5.818		2,925	320	101	3,346	3,850
1020	Non-standard paint color			2.75	5.818		3,400	320	101	3,821	4,375
1030	Galvanized			2.75	5.818		3,400	320	101	3,821	4,400
1040	Stainless			2.75	5.818		4,925	320	101	5,346	6,050
1050	Aluminum, 103 lb.			2.75	5.818		3,600	320	101	4,021	4,600
1210	12' vertical rise, steel, 245 lb., standard paint color			2.50	6.400		3,425	350	112	3,887	4,500
1220	Non-standard paint color			2.50	6.400		3,925	350	112	4,387	5,050
1230	Galvanized			2.50	6.400		4,000	350	112	4,462	5,125
1240	Stainless			2.50	6.400		5,750	350	112	6,212	7,050
1250	Aluminum, 103 lb.			2.50	6.400		4,200	350	112	4,662	5,350
1410	14' vertical rise, steel, 281 lb., standard paint color			2.25	7.111	↓	3,925	390	124	4,439	5,125

05 51 Metal Stairs

05 51 33 – Metal Ladders

05 51 33.23 Alternating Tread Ladders		Crew	Daily Output	Labor-Hours	Unit	Material	2018 Bare Costs Labor	Equipment	Total	Total Incl O&P
1420	Non-standard paint color	B-68G	2.25	7.111	Ea.	4,475	390	124	4,989	5,725
1430	Galvanized		2.25	7.111		4,600	390	124	5,114	5,850
1440	Stainless		2.25	7.111		6,575	390	124	7,089	8,050
1450	Aluminum, 136 lb.		2.25	7.111		4,825	390	124	5,339	6,100
1610	16' vertical rise, steel, 317 lb., standard paint color		2	8		4,450	435	140	5,025	5,775
1620	Non-standard paint color		2	8		5,025	435	140	5,600	6,425
1630	Galvanized		2	8		5,175	435	140	5,750	6,600
1640	Stainless		2	8		7,425	435	140	8,000	9,050
1650	Aluminum, 153 lb.		2	8		5,425	435	140	6,000	6,875

05 52 Metal Railings

05 52 13 – Pipe and Tube Railings

05 52 13.50 Railings, Pipe			Crew	Daily Output	Labor-Hours	Unit	Material	2018 Bare Costs Labor	Equipment	Total	Total Incl O&P
0010	**RAILINGS, PIPE**, shop fab'd, 3'-6" high, posts @ 5' OC										
0015	Made from recycled materials										
0020	Aluminum, 2 rail, satin finish, 1-1/4" diameter	G	E-4	160	.200	L.F.	52.50	11.05	.62	64.17	77.50
0030	Clear anodized	G		160	.200		64.50	11.05	.62	76.17	90.50
0040	Dark anodized	G		160	.200		71.50	11.05	.62	83.17	98
0080	1-1/2" diameter, satin finish	G		160	.200		61.50	11.05	.62	73.17	87.50
0090	Clear anodized	G		160	.200		69.50	11.05	.62	81.17	95.50
0100	Dark anodized	G		160	.200		76.50	11.05	.62	88.17	103
0140	Aluminum, 3 rail, 1-1/4" diam., satin finish	G		137	.234		68	12.90	.72	81.62	97.50
0150	Clear anodized	G		137	.234		84.50	12.90	.72	98.12	116
0160	Dark anodized	G		137	.234		93	12.90	.72	106.62	125
0200	1-1/2" diameter, satin finish	G		137	.234		80.50	12.90	.72	94.12	111
0210	Clear anodized	G		137	.234		91.50	12.90	.72	105.12	124
0220	Dark anodized	G		137	.234		101	12.90	.72	114.62	134
0500	Steel, 2 rail, on stairs, primed, 1-1/4" diameter	G		160	.200		29	11.05	.62	40.67	51.50
0520	1-1/2" diameter	G		160	.200		31.50	11.05	.62	43.17	54
0540	Galvanized, 1-1/4" diameter	G		160	.200		39	11.05	.62	50.67	62.50
0560	1-1/2" diameter	G		160	.200		44	11.05	.62	55.67	68
0580	Steel, 3 rail, primed, 1-1/4" diameter	G		137	.234		43	12.90	.72	56.62	70.50
0600	1-1/2" diameter	G		137	.234		45.50	12.90	.72	59.12	73
0620	Galvanized, 1-1/4" diameter	G		137	.234		60.50	12.90	.72	74.12	89.50
0640	1-1/2" diameter	G		137	.234		70	12.90	.72	83.62	100
0700	Stainless steel, 2 rail, 1-1/4" diam., #4 finish	G		137	.234		128	12.90	.72	141.62	163
0720	High polish	G		137	.234		206	12.90	.72	219.62	250
0740	Mirror polish	G		137	.234		258	12.90	.72	271.62	305
0760	Stainless steel, 3 rail, 1-1/2" diam., #4 finish	G		120	.267		192	14.70	.82	207.52	237
0770	High polish	G		120	.267		320	14.70	.82	335.52	375
0780	Mirror finish	G		120	.267		390	14.70	.82	405.52	450
0900	Wall rail, alum. pipe, 1-1/4" diam., satin finish	G		213	.150		25	8.30	.46	33.76	42
0905	Clear anodized	G		213	.150		31.50	8.30	.46	40.26	49
0910	Dark anodized	G		213	.150		37	8.30	.46	45.76	55.50
0915	1-1/2" diameter, satin finish	G		213	.150		28	8.30	.46	36.76	45.50
0920	Clear anodized	G		213	.150		35	8.30	.46	43.76	53
0925	Dark anodized	G		213	.150		44	8.30	.46	52.76	63
0930	Steel pipe, 1-1/4" diameter, primed	G		213	.150		17.35	8.30	.46	26.11	34
0935	Galvanized	G		213	.150		25	8.30	.46	33.76	42
0940	1-1/2" diameter	G		176	.182		17.90	10.05	.56	28.51	37.50
0945	Galvanized	G		213	.150		25	8.30	.46	33.76	42

05 52 Metal Railings

05 52 13 – Pipe and Tube Railings

05 52 13.50 Railings, Pipe

		Crew	Daily Output	Labor-Hours	Unit	Material	2018 Bare Costs Labor	Equipment	Total	Total Incl O&P
0955	Stainless steel pipe, 1-1/2" diam., #4 finish	G E-4	107	.299	L.F.	102	16.50	.92	119.42	141
0960	High polish	G	107	.299		207	16.50	.92	224.42	257
0965	Mirror polish	G ↓	107	.299	↓	245	16.50	.92	262.42	299
2000	2-line pipe rail (1-1/2" T&B) with 1/2" pickets @ 4-1/2" OC,									
2005	attached handrail on brackets									
2010	42" high aluminum, satin finish, straight & level	G E-4	120	.267	L.F.	268	14.70	.82	283.52	320
2050	42" high steel, primed, straight & level	G "	120	.267		147	14.70	.82	162.52	187
4000	For curved and level rails, add						10%	10%		
4100	For sloped rails for stairs, add				↓	30%	30%			
9000	Minimum labor/equipment charge	1 Sswk	2	4	Job		219		219	375

05 52 16 – Industrial Railings

05 52 16.50 Railings, Industrial

		Crew	Daily Output	Labor-Hours	Unit	Material	2018 Bare Costs Labor	Equipment	Total	Total Incl O&P
0010	**RAILINGS, INDUSTRIAL**, shop fab'd, 3'-6" high, posts @ 5' OC									
0020	2 rail, 3'-6" high, 1-1/2" pipe	G E-4	255	.125	L.F.	41.50	6.90	.39	48.79	57.50
0200	For 4" high kick plate, 10 ga., add	G				5.90			5.90	6.45
0500	For curved level rails, add					10%	10%			
0550	For sloped rails for stairs, add				↓	30%	30%			
9000	Minimum labor/equipment charge	1 Sswk	2	4	Job		219		219	375

05 55 Metal Stair Treads and Nosings

05 55 13 – Metal Stair Treads

05 55 13.50 Stair Treads

		Crew	Daily Output	Labor-Hours	Unit	Material	2018 Bare Costs Labor	Equipment	Total	Total Incl O&P
0010	**STAIR TREADS**, stringers and bolts not included									
3000	Diamond plate treads, steel, 1/8" thick									
3005	Open riser, black enamel									
3010	9" deep x 36" long	G 2 Sswk	48	.333	Ea.	91.50	18.20		109.70	131
3020	42" long	G	48	.333		96.50	18.20		114.70	137
3030	48" long	G	48	.333		101	18.20		119.20	142
3040	11" deep x 36" long	G	44	.364		96.50	19.85		116.35	140
3050	42" long	G	44	.364		102	19.85		121.85	146
3060	48" long	G	44	.364		108	19.85		127.85	153
3110	Galvanized, 9" deep x 36" long	G	48	.333		144	18.20		162.20	190
3120	42" long	G	48	.333		153	18.20		171.20	200
3130	48" long	G	48	.333		163	18.20		181.20	210
3140	11" deep x 36" long	G	44	.364		149	19.85		168.85	198
3150	42" long	G	44	.364		163	19.85		182.85	213
3160	48" long	G ↓	44	.364	↓	169	19.85		188.85	220
3200	Closed riser, black enamel									
3210	12" deep x 36" long	G 2 Sswk	40	.400	Ea.	110	22		132	159
3220	42" long	G	40	.400		119	22		141	169
3230	48" long	G	40	.400		126	22		148	176
3240	Galvanized, 12" deep x 36" long	G	40	.400		173	22		195	229
3250	42" long	G	40	.400		189	22		211	246
3260	48" long	G ↓	40	.400	↓	198	22		220	256
4000	Bar grating treads									
4005	Steel, 1-1/4" x 3/16" bars, anti-skid nosing, black enamel									
4010	8-5/8" deep x 30" long	G 2 Sswk	48	.333	Ea.	54	18.20		72.20	90.50
4020	36" long	G	48	.333		63.50	18.20		81.70	101
4030	48" long	G	48	.333		97.50	18.20		115.70	138
4040	10-15/16" deep x 36" long	G ↓	44	.364		70	19.85		89.85	111

For customer support on your Commercial Renovation Costs with RSMeans data, call 800.448.8182.

123

05 55 Metal Stair Treads and Nosings

05 55 13 – Metal Stair Treads

05 55 13.50 Stair Treads		Crew	Daily Output	Labor-Hours	Unit	Material	2018 Bare Costs Labor	Equipment	Total	Total Incl O&P
4050	48" long	2 Sswk	44	.364	Ea.	100	19.85		119.85	144
4060	Galvanized, 8-5/8" deep x 30" long	G	48	.333		62	18.20		80.20	99.50
4070	36" long	G	48	.333		73.50	18.20		91.70	112
4080	48" long	G	48	.333		108	18.20		126.20	150
4090	10-15/16" deep x 36" long	G	44	.364		86.50	19.85		106.35	129
4100	48" long	G	44	.364		112	19.85		131.85	157
4200	Aluminum, 1-1/4" x 3/16" bars, serrated, with nosing									
4210	7-5/8" deep x 18" long	G	2 Sswk	52	.308	Ea.	50	16.80		66.80	83.50
4220	24" long	G	52	.308		59	16.80		75.80	93.50
4230	30" long	G	52	.308		68.50	16.80		85.30	104
4240	36" long	G	52	.308		177	16.80		193.80	224
4250	8-13/16" deep x 18" long	G	48	.333		67	18.20		85.20	105
4260	24" long	G	48	.333		99.50	18.20		117.70	140
4270	30" long	G	48	.333		115	18.20		133.20	157
4280	36" long	G	48	.333		194	18.20		212.20	245
4290	10" deep x 18" long	G	44	.364		130	19.85		149.85	177
4300	30" long	G	44	.364		173	19.85		192.85	224
4310	36" long	G	44	.364		210	19.85		229.85	265
5000	Channel grating treads									
5005	Steel, 14 ga., 2-1/2" thick, galvanized									
5010	9" deep x 36" long	G	2 Sswk	48	.333	Ea.	103	18.20		121.20	144
5020	48" long	G	48	.333	"	139	18.20		157.20	183
9000	Minimum labor/equipment charge		4	4	Job		219		219	375

05 55 19 – Metal Stair Tread Covers

05 55 19.50 Stair Tread Covers for Renovation

		Crew	Daily Output	Labor-Hours	Unit	Material	2018 Bare Costs Labor	Equipment	Total	Total Incl O&P
0010	**STAIR TREAD COVERS FOR RENOVATION**									
0205	Extruded tread cover with nosing, pre-drilled, includes screws									
0210	Aluminum with black abrasive strips, 9" wide x 3' long	1 Carp	24	.333	Ea.	104	16.90		120.90	143
0220	4' long		22	.364		139	18.45		157.45	183
0230	5' long		20	.400		179	20.50		199.50	230
0240	11" wide x 3' long		24	.333		135	16.90		151.90	176
0250	4' long		22	.364		180	18.45		198.45	228
0260	5' long		20	.400		221	20.50		241.50	276
0305	Black abrasive strips with yellow front strips									
0310	Aluminum, 9" wide x 3' long	1 Carp	24	.333	Ea.	117	16.90		133.90	157
0320	4' long		22	.364		156	18.45		174.45	202
0330	5' long		20	.400		195	20.50		215.50	248
0340	11" wide x 3' long		24	.333		141	16.90		157.90	183
0350	4' long		22	.364		186	18.45		204.45	234
0360	5' long		20	.400		236	20.50		256.50	292
0405	Black abrasive strips with photoluminescent front strips									
0410	Aluminum, 9" wide x 3' long	1 Carp	24	.333	Ea.	150	16.90		166.90	193
0420	4' long		22	.364		183	18.45		201.45	232
0430	5' long		20	.400		229	20.50		249.50	285
0440	11" wide x 3' long		24	.333		163	16.90		179.90	207
0450	4' long		22	.364		218	18.45		236.45	269
0460	5' long		20	.400		272	20.50		292.50	330
9000	Minimum labor/equipment charge		4	2	Job		101		101	165

05 56 Metal Castings

05 56 13 – Metal Construction Castings

05 56 13.50 Construction Castings		Crew	Daily Output	Labor-Hours	Unit	Material	2018 Bare Costs Labor	Equipment	Total	Total Incl O&P
0010	**CONSTRUCTION CASTINGS**									
0020	Manhole covers and frames, see Section 33 44 13.13									
0100	Column bases, cast iron, 16" x 16", approx. 65 lb.	G	E-4	46	.696	Ea.	142	38.50	2.14	182.64	224
0200	32" x 32", approx. 256 lb.	G		23	1.391	"	520	76.50	4.29	600.79	710
0600	Miscellaneous C.I. castings, light sections, less than 150 lb.	G		3200	.010	Lb.	8.95	.55	.03	9.53	10.80
1300	Special low volume items	G		3200	.010	"	11.25	.55	.03	11.83	13.30

05 58 Formed Metal Fabrications

05 58 13 – Column Covers

05 58 13.05 Column Covers

			Crew	Daily Output	Labor-Hours	Unit	Material	2018 Bare Costs Labor	Equipment	Total	Total Incl O&P
0010	**COLUMN COVERS**										
0015	Made from recycled materials										
0020	Excludes structural steel, light ga. metal framing, misc. metals, sealants										
0100	Round covers, 2 halves with 2 vertical joints for backer rod and sealant										
0110	Up to 12' high, no horizontal joints										
0120	12" diameter, 0.125" aluminum, anodized/painted finish	G	2 Sswk	32	.500	V.L.F.	31	27.50		58.50	80.50
0130	Type 304 stainless steel, 16 gauge, #4 brushed finish	G		32	.500		44	27.50		71.50	95
0140	Type 316 stainless steel, 16 gauge, #4 brushed finish	G		32	.500		55	27.50		82.50	107
0150	18" diameter, aluminum	G		32	.500		46.50	27.50		74	98
0160	Type 304 stainless steel	G		32	.500		66	27.50		93.50	119
0170	Type 316 stainless steel	G		32	.500		82.50	27.50		110	138
0180	24" diameter, aluminum	G		32	.500		62	27.50		89.50	115
0190	Type 304 stainless steel	G		32	.500		88	27.50		115.50	144
0200	Type 316 stainless steel	G		32	.500		110	27.50		137.50	168
0210	30" diameter, aluminum	G		30	.533		78	29		107	135
0220	Type 304 stainless steel	G		30	.533		110	29		139	171
0230	Type 316 stainless steel	G		30	.533		138	29		167	201
0240	36" diameter, aluminum	G		30	.533		93.50	29		122.50	153
0250	Type 304 stainless steel	G		30	.533		132	29		161	195
0260	Type 316 stainless steel	G		30	.533		165	29		194	232
0400	Up to 24' high, 2 stacked sections with 1 horizontal joint										
0410	18" diameter, aluminum	G	2 Sswk	28	.571	V.L.F.	49	31		80	108
0450	Type 304 stainless steel	G		28	.571		69.50	31		100.50	130
0460	Type 316 stainless steel	G		28	.571		86.50	31		117.50	149
0470	24" diameter, aluminum	G		28	.571		65.50	31		96.50	126
0480	Type 304 stainless steel	G		28	.571		92.50	31		123.50	156
0490	Type 316 stainless steel	G		28	.571		116	31		147	181
0500	30" diameter, aluminum	G		24	.667		81.50	36.50		118	152
0510	Type 304 stainless steel	G		24	.667		116	36.50		152.50	189
0520	Type 316 stainless steel	G		24	.667		145	36.50		181.50	221
0530	36" diameter, aluminum	G		24	.667		98	36.50		134.50	170
0540	Type 304 stainless steel	G		24	.667		139	36.50		175.50	215
0550	Type 316 stainless steel	G		24	.667		173	36.50		209.50	253

05 58 23 – Formed Metal Guards

05 58 23.90 Window Guards

			Crew	Daily Output	Labor-Hours	Unit	Material	2018 Bare Costs Labor	Equipment	Total	Total Incl O&P
0010	**WINDOW GUARDS**, shop fabricated										
0015	Expanded metal, steel angle frame, permanent	G	E-4	350	.091	S.F.	23.50	5.05	.28	28.83	34.50
0025	Steel bars, 1/2" x 1/2", spaced 5" OC	G	"	290	.110	"	16.25	6.10	.34	22.69	28.50
0030	Hinge mounted, add	G				Opng.	47			47	52
0040	Removable type, add	G				"	29.50			29.50	32.50

125

For customer support on your Commercial Renovation Costs with RSMeans data, call 800.448.8182.

05 58 Formed Metal Fabrications

05 58 23 – Formed Metal Guards

05 58 23.90 Window Guards		Crew	Daily Output	Labor-Hours	Unit	Material	2018 Bare Costs Labor	Equipment	Total	Total Incl O&P
0050	For galvanized guards, add				S.F.	35%				
0070	For pivoted or projected type, add					105%	40%			
0100	Mild steel, stock units, economy	G	E-4	405	.079		6.40	4.36	.24	11	14.70
0200	Deluxe	G		405	.079		12.95	4.36	.24	17.55	22
0400	Woven wire, stock units, 3/8" channel frame, 3' x 5' opening	G		40	.800	Opng.	172	44	2.46	218.46	268
0500	4' x 6' opening	G		38	.842		276	46.50	2.59	325.09	385
0800	Basket guards for above, add	G					238			238	262
1000	Swinging guards for above, add	G					81			81	89
9000	Minimum labor/equipment charge		1 Sswk	2	4	Job		219		219	375

05 58 25 – Formed Lamp Posts

05 58 25.40 Lamp Posts		Crew	Daily Output	Labor-Hours	Unit	Material	2018 Bare Costs Labor	Equipment	Total	Total Incl O&P
0010	**LAMP POSTS**										
0020	Aluminum, 7' high, stock units, post only	G	1 Carp	16	.500	Ea.	85.50	25.50		111	135
0100	Mild steel, plain	G		16	.500	"	63.50	25.50		89	111
9000	Minimum labor/equipment charge			4	2	Job		101		101	165

05 71 Decorative Metal Stairs

05 71 13 – Fabricated Metal Spiral Stairs

05 71 13.50 Spiral Stairs		Crew	Daily Output	Labor-Hours	Unit	Material	2018 Bare Costs Labor	Equipment	Total	Total Incl O&P
0010	**SPIRAL STAIRS**										
1805	Shop fabricated, custom ordered										
1810	Aluminum, 5'-0" diameter, plain units	G	E-4	45	.711	Riser	580	39	2.19	621.19	705
1820	Fancy units	G		45	.711		1,000	39	2.19	1,041.19	1,175
1900	Cast iron, 4'-0" diameter, plain units	G		45	.711		745	39	2.19	786.19	890
1920	Fancy units	G		25	1.280		1,300	70.50	3.94	1,374.44	1,550
2000	Steel, industrial checkered plate, 4' diameter	G		45	.711		490	39	2.19	531.19	610
2200	6' diameter	G		40	.800		695	44	2.46	741.46	840
3100	Spiral stair kits, 12 stacking risers to fit exact floor height										
3110	Steel, flat metal treads, primed, 3'-6" diameter	G	2 Carp	1.60	10	Flight	1,350	505		1,855	2,300
3120	4'-0" diameter	G		1.45	11.034		1,525	560		2,085	2,575
3130	4'-6" diameter	G		1.35	11.852		1,700	600		2,300	2,825
3140	5'-0" diameter	G		1.25	12.800		1,850	650		2,500	3,100
3210	Galvanized, 3'-6" diameter	G		1.60	10		1,850	505		2,355	2,875
3220	4'-0" diameter	G		1.45	11.034		2,200	560		2,760	3,325
3230	4'-6" diameter	G		1.35	11.852		2,425	600		3,025	3,650
3240	5'-0" diameter	G		1.25	12.800		2,650	650		3,300	3,975
3310	Checkered plate tread, primed, 3'-6" diameter			1.45	11.034		1,600	560		2,160	2,650
3320	4'-0" diameter	G		1.35	11.852		1,800	600		2,400	2,950
3330	4'-6" diameter	G		1.25	12.800		1,975	650		2,625	3,200
3340	5'-0" diameter	G		1.15	13.913		2,150	705		2,855	3,525
3410	Galvanized, 3'-6" diameter			1.45	11.034		2,200	560		2,760	3,300
3420	4'-0" diameter	G		1.35	11.852		2,525	600		3,125	3,750
3430	4'-6" diameter	G		1.25	12.800		2,775	650		3,425	4,100
3440	5'-0" diameter	G		1.15	13.913		3,000	705		3,705	4,450
3510	Red oak covers on flat metal treads, 3'-6" diameter			1.35	11.852		2,450	600		3,050	3,650
3520	4'-0" diameter			1.25	12.800		2,925	650		3,575	4,250
3530	4'-6" diameter			1.15	13.913		3,150	705		3,855	4,625
3540	5'-0" diameter			1.05	15.238		3,400	775		4,175	4,975

05 73 Decorative Metal Railings

05 73 16 – Wire Rope Decorative Metal Railings

05 73 16.10 Cable Railings		Crew	Daily Output	Labor-Hours	Unit	Material	2018 Bare Costs Labor	Equipment	Total	Total Incl O&P
0010	**CABLE RAILINGS**, with 316 stainless steel 1 x 19 cable, 3/16" diameter									
0015	Made from recycled materials									
0100	1-3/4" diameter stainless steel posts x 42" high, cables 4" OC	G	2 Sswk	25	.640	L.F.	45.50	35		80.50	110

05 73 23 – Ornamental Railings

05 73 23.50 Railings, Ornamental		Crew	Daily Output	Labor-Hours	Unit	Material	2018 Bare Costs Labor	Equipment	Total	Total Incl O&P
0010	**RAILINGS, ORNAMENTAL**, 3'-6" high, posts @ 6' OC									
0020	Bronze or stainless, hand forged, plain	G	2 Sswk	24	.667	L.F.	260	36.50		296.50	350
0100	Fancy	G		18	.889		520	48.50		568.50	655
0200	Aluminum, panelized, plain	G		24	.667		12.20	36.50		48.70	75.50
0300	Fancy	G		18	.889		25.50	48.50		74	111
0400	Wrought iron, hand forged, plain	G		24	.667		94	36.50		130.50	166
0500	Fancy	G		18	.889		231	48.50		279.50	340
0550	Steel, panelized, plain	G		24	.667		20.50	36.50		57	84.50
0560	Fancy	G		18	.889		30	48.50		78.50	116
0600	Composite metal/wood/glass, plain			18	.889		142	48.50		190.50	239
0700	Fancy	▼		12	1.333	▼	284	73		357	440
9000	Minimum labor/equipment charge		1 Sswk	2	4	Job		219		219	375

05 75 Decorative Formed Metal

05 75 13 – Columns

05 75 13.10 Aluminum Columns		Crew	Daily Output	Labor-Hours	Unit	Material	2018 Bare Costs Labor	Equipment	Total	Total Incl O&P
0010	**ALUMINUM COLUMNS**									
0015	Made from recycled materials									
0020	Aluminum, extruded, stock units, no cap or base, 6" diameter	G	E-4	240	.133	L.F.	12.85	7.35	.41	20.61	27
0100	8" diameter	G	"	170	.188	"	16.75	10.40	.58	27.73	37
0460	Caps, ornamental, plain	G				Set	310			310	340
0470	Fancy	G				"	1,550			1,550	1,725
0500	For square columns, add to column prices above					L.F.	50%				

Division Notes

	CREW	DAILY OUTPUT	LABOR-HOURS	UNIT	BARE COSTS				TOTAL INCL O&P
					MAT.	LABOR	EQUIP.	TOTAL	

Estimating Tips

06 05 00 Common Work Results for Wood, Plastics, and Composites

- Common to any wood-framed structure are the accessory connector items such as screws, nails, adhesives, hangers, connector plates, straps, angles, and hold-downs. For typical wood-framed buildings, such as residential projects, the aggregate total for these items can be significant, especially in areas where seismic loading is a concern. For floor and wall framing, the material cost is based on 10 to 25 lbs. of accessory connectors per MBF. Hold-downs, hangers, and other connectors should be taken off by the piece.

  Included with material costs are fasteners for a normal installation. Gordian's RSMeans engineers use manufacturers' recommendations, written specifications, and/or standard construction practice for the sizing and spacing of fasteners. Prices for various fasteners are shown for informational purposes only. Adjustments should be made if unusual fastening conditions exist.

06 10 00 Carpentry

- Lumber is a traded commodity and therefore sensitive to supply and demand in the marketplace. Even with "budgetary" estimating of wood-framed projects, it is advisable to call local suppliers for the latest market pricing.

- The common quantity unit for wood-framed projects is "thousand board feet" (MBF). A board foot is a volume of wood—1" x 1" x 1' or 144 cubic inches. Board-foot quantities are generally calculated using nominal material dimensions— dressed sizes are ignored. Board foot per lineal foot of any stick of lumber can be calculated by dividing the nominal cross-sectional area by 12. As an example, 2,000 lineal feet of 2 x 12 equates to 4 MBF by dividing the nominal area, 2 x 12, by 12, which equals 2, and multiplying by 2,000 to give 4,000 board feet. This simple rule applies to all nominal dimensioned lumber.

- Waste is an issue of concern at the quantity takeoff for any area of construction. Framing lumber is sold in even foot lengths, i.e., 8', 10', 12', 14', 16' and depending on spans, wall heights, and the grade of lumber, waste is inevitable. A rule of thumb for lumber waste is 5%–10% depending on material quality and the complexity of the framing.

- Wood in various forms and shapes is used in many projects, even where the main structural framing is steel, concrete, or masonry. Plywood as a back-up partition material and 2x boards used as blocking and cant strips around roof edges are two common examples. The estimator should ensure that the costs of all wood materials are included in the final estimate.

06 20 00 Finish Carpentry

- It is necessary to consider the grade of workmanship when estimating labor costs for erecting millwork and an interior finish. In practice, there are three grades: premium, custom, and economy. The RSMeans daily output for base and case moldings is in the range of 200 to 250 L.F. per carpenter per day. This is appropriate for most average custom-grade projects. For premium projects, an adjustment to productivity of 25%–50% should be made, depending on the complexity of the job.

Reference Numbers

Reference numbers are shown at the beginning of some major classifications. These numbers refer to related items in the Reference Section. The reference information may be an estimating procedure, an alternate pricing method, or technical information.

Note: Not all subdivisions listed here necessarily appear. ■

Did you know?

RSMeans data is available through our online application with 24/7 access:

- Search for unit prices by keyword
- Leverage the most up-to-date data
- Build and export estimates

Try it free for 30 days!
www.rsmeans.com/2018freetrial

06 05 05.10 Selective Demolition Wood Framing	Crew	Daily Output	Labor-Hours	Unit	Material	2018 Bare Costs Labor	Equipment	Total	Total Incl O&P
0010 **SELECTIVE DEMOLITION WOOD FRAMING** R024119-10									
0100 Timber connector, nailed, small	1 Clab	96	.083	Ea.		3.32		3.32	5.40
0110 Medium		60	.133			5.30		5.30	8.65
0120 Large		48	.167			6.65		6.65	10.80
0130 Bolted, small		48	.167			6.65		6.65	10.80
0140 Medium		32	.250			9.95		9.95	16.20
0150 Large		24	.333			13.30		13.30	21.50
2958 Beams, 2" x 6"	2 Clab	1100	.015	L.F.		.58		.58	.94
2960 2" x 8"		825	.019			.77		.77	1.25
2965 2" x 10"		665	.024			.96		.96	1.56
2970 2" x 12"		550	.029			1.16		1.16	1.88
2972 2" x 14"		470	.034			1.36		1.36	2.20
2975 4" x 8"	B-1	413	.058			2.35		2.35	3.82
2980 4" x 10"		330	.073			2.95		2.95	4.78
2985 4" x 12"		275	.087			3.54		3.54	5.75
3000 6" x 8"		275	.087			3.54		3.54	5.75
3040 6" x 10"		220	.109			4.42		4.42	7.15
3080 6" x 12"		185	.130			5.25		5.25	8.55
3120 8" x 12"		140	.171			6.95		6.95	11.30
3160 10" x 12"		110	.218			8.85		8.85	14.35
3162 Alternate pricing method		1.10	21.818	M.B.F.		885		885	1,425
3170 Blocking, in 16" OC wall framing, 2" x 4"	1 Clab	600	.013	L.F.		.53		.53	.86
3172 2" x 6"		400	.020			.80		.80	1.29
3174 In 24" OC wall framing, 2" x 4"		600	.013			.53		.53	.86
3176 2" x 6"		400	.020			.80		.80	1.29
3178 Alt method, wood blocking removal from wood framing		.40	20	M.B.F.		795		795	1,300
3179 Wood blocking removal from steel framing		.36	22.222	"		885		885	1,450
3180 Bracing, let in, 1" x 3", studs 16" OC		1050	.008	L.F.		.30		.30	.49
3181 Studs 24" OC		1080	.007			.30		.30	.48
3182 1" x 4", studs 16" OC		1050	.008			.30		.30	.49
3183 Studs 24" OC		1080	.007			.30		.30	.48
3184 1" x 6", studs 16" OC		1050	.008			.30		.30	.49
3185 Studs 24" OC		1080	.007			.30		.30	.48
3186 2" x 3", studs 16" OC		800	.010			.40		.40	.65
3187 Studs 24" OC		830	.010			.38		.38	.62
3188 2" x 4", studs 16" OC		800	.010			.40		.40	.65
3189 Studs 24" OC		830	.010			.38		.38	.62
3190 2" x 6", studs 16" OC		800	.010			.40		.40	.65
3191 Studs 24" OC		830	.010			.38		.38	.62
3192 2" x 8", studs 16" OC		800	.010			.40		.40	.65
3193 Studs 24" OC		830	.010			.38		.38	.62
3194 "T" shaped metal bracing, studs at 16" OC		1060	.008			.30		.30	.49
3195 Studs at 24" OC		1200	.007			.27		.27	.43
3196 Metal straps, studs at 16" OC		1200	.007			.27		.27	.43
3197 Studs at 24" OC		1240	.006			.26		.26	.42
3200 Columns, round, 8' to 14' tall		40	.200	Ea.		7.95		7.95	12.95
3202 Dimensional lumber sizes	2 Clab	1.10	14.545	M.B.F.		580		580	940
3250 Blocking, between joists	1 Clab	320	.025	Ea.		1		1	1.62
3252 Bridging, metal strap, between joists		320	.025	Pr.		1		1	1.62
3254 Wood, between joists		320	.025	"		1		1	1.62
3260 Door buck, studs, header & access., 8' high 2" x 4" wall, 3' wide		32	.250	Ea.		9.95		9.95	16.20
3261 4' wide		32	.250			9.95		9.95	16.20
3262 5' wide		32	.250			9.95		9.95	16.20

06 05 05 – Selective Demolition for Wood, Plastics, and Composites

06 05 05.10 Selective Demolition Wood Framing	Crew	Daily Output	Labor-Hours	Unit	Material	2018 Bare Costs Labor	Equipment	Total	Total Incl O&P
3263	6' wide	1 Clab	32	.250	Ea.		9.95		9.95	16.20
3264	8' wide		30	.267			10.65		10.65	17.25
3265	10' wide		30	.267			10.65		10.65	17.25
3266	12' wide		30	.267			10.65		10.65	17.25
3267	2" x 6" wall, 3' wide		32	.250			9.95		9.95	16.20
3268	4' wide		32	.250			9.95		9.95	16.20
3269	5' wide		32	.250			9.95		9.95	16.20
3270	6' wide		32	.250			9.95		9.95	16.20
3271	8' wide		30	.267			10.65		10.65	17.25
3272	10' wide		30	.267			10.65		10.65	17.25
3273	12' wide		30	.267			10.65		10.65	17.25
3274	Window buck, studs, header & access., 8' high 2" x 4" wall, 2' wide		24	.333			13.30		13.30	21.50
3275	3' wide		24	.333			13.30		13.30	21.50
3276	4' wide		24	.333			13.30		13.30	21.50
3277	5' wide		24	.333			13.30		13.30	21.50
3278	6' wide		24	.333			13.30		13.30	21.50
3279	7' wide		24	.333			13.30		13.30	21.50
3280	8' wide		22	.364			14.50		14.50	23.50
3281	10' wide		22	.364			14.50		14.50	23.50
3282	12' wide		22	.364			14.50		14.50	23.50
3283	2" x 6" wall, 2' wide		24	.333			13.30		13.30	21.50
3284	3' wide		24	.333			13.30		13.30	21.50
3285	4' wide		24	.333			13.30		13.30	21.50
3286	5' wide		24	.333			13.30		13.30	21.50
3287	6' wide		24	.333			13.30		13.30	21.50
3288	7' wide		24	.333			13.30		13.30	21.50
3289	8' wide		22	.364			14.50		14.50	23.50
3290	10' wide		22	.364			14.50		14.50	23.50
3291	12' wide		22	.364			14.50		14.50	23.50
3360	Deck or porch decking		825	.010	L.F.		.39		.39	.63
3400	Fascia boards, 1" x 6"		500	.016			.64		.64	1.04
3440	1" x 8"		450	.018			.71		.71	1.15
3480	1" x 10"		400	.020			.80		.80	1.29
3490	2" x 6"		450	.018			.71		.71	1.15
3500	2" x 8"		400	.020			.80		.80	1.29
3510	2" x 10"		350	.023			.91		.91	1.48
3610	Furring, on wood walls or ceiling		4000	.002	S.F.		.08		.08	.13
3620	On masonry or concrete walls or ceiling		1200	.007	"		.27		.27	.43
3800	Headers over openings, 2 @ 2" x 6"		110	.073	L.F.		2.90		2.90	4.71
3840	2 @ 2" x 8"		100	.080			3.19		3.19	5.20
3880	2 @ 2" x 10"		90	.089			3.54		3.54	5.75
3885	Alternate pricing method		.26	30.651	M.B.F.		1,225		1,225	1,975
3920	Joists, 1" x 4"		1250	.006	L.F.		.26		.26	.41
3930	1" x 6"		1135	.007			.28		.28	.46
3940	1" x 8"		1000	.008			.32		.32	.52
3950	1" x 10"		895	.009			.36		.36	.58
3960	1" x 12"		765	.010			.42		.42	.68
4200	2" x 4"	2 Clab	1000	.016			.64		.64	1.04
4230	2" x 6"		970	.016			.66		.66	1.07
4240	2" x 8"		940	.017			.68		.68	1.10
4250	2" x 10"		910	.018			.70		.70	1.14
4280	2" x 12"		880	.018			.72		.72	1.18
4281	2" x 14"		850	.019			.75		.75	1.22

06 05 05 – Selective Demolition for Wood, Plastics, and Composites

06 05 05.10 Selective Demolition Wood Framing	Crew	Daily Output	Labor-Hours	Unit	Material	2018 Bare Costs Labor	Equipment	Total	Total Incl O&P
4282	Composite joists, 9-1/2"	2 Clab	960	.017	L.F.		.66		.66	1.08
4283	11-7/8"		930	.017			.69		.69	1.11
4284	14"		897	.018			.71		.71	1.15
4285	16"		865	.019			.74		.74	1.20
4290	Wood joists, alternate pricing method		1.50	10.667	M.B.F.		425		425	690
4500	Open web joist, 12" deep		500	.032	L.F.		1.28		1.28	2.07
4505	14" deep		475	.034			1.34		1.34	2.18
4510	16" deep		450	.036			1.42		1.42	2.30
4520	18" deep		425	.038			1.50		1.50	2.44
4530	24" deep		400	.040			1.59		1.59	2.59
4550	Ledger strips, 1" x 2"	1 Clab	1200	.007			.27		.27	.43
4560	1" x 3"		1200	.007			.27		.27	.43
4570	1" x 4"		1200	.007			.27		.27	.43
4580	2" x 2"		1100	.007			.29		.29	.47
4590	2" x 4"		1000	.008			.32		.32	.52
4600	2" x 6"		1000	.008			.32		.32	.52
4601	2" x 8 or 2" x 10"		800	.010			.40		.40	.65
4602	4" x 6"		600	.013			.53		.53	.86
4604	4" x 8"		450	.018			.71		.71	1.15
5400	Posts, 4" x 4"	2 Clab	800	.020			.80		.80	1.29
5405	4" x 6"		550	.029			1.16		1.16	1.88
5410	4" x 8"		440	.036			1.45		1.45	2.35
5425	4" x 10"		390	.041			1.64		1.64	2.65
5430	4" x 12"		350	.046			1.82		1.82	2.96
5440	6" x 6"		400	.040			1.59		1.59	2.59
5445	6" x 8"		350	.046			1.82		1.82	2.96
5450	6" x 10"		320	.050			1.99		1.99	3.24
5455	6" x 12"		290	.055			2.20		2.20	3.57
5480	8" x 8"		300	.053			2.13		2.13	3.45
5500	10" x 10"		240	.067			2.66		2.66	4.31
5660	T&G floor planks		2	8	M.B.F.		320		320	520
5682	Rafters, ordinary, 16" OC, 2" x 4"		880	.018	S.F.		.72		.72	1.18
5683	2" x 6"		840	.019			.76		.76	1.23
5684	2" x 8"		820	.020			.78		.78	1.26
5685	2" x 10"		820	.020			.78		.78	1.26
5686	2" x 12"		810	.020			.79		.79	1.28
5687	24" OC, 2" x 4"		1170	.014			.55		.55	.89
5688	2" x 6"		1117	.014			.57		.57	.93
5689	2" x 8"		1091	.015			.58		.58	.95
5690	2" x 10"		1091	.015			.58		.58	.95
5691	2" x 12"		1077	.015			.59		.59	.96
5795	Rafters, ordinary, 2" x 4" (alternate method)		862	.019	L.F.		.74		.74	1.20
5800	2" x 6" (alternate method)		850	.019			.75		.75	1.22
5840	2" x 8" (alternate method)		837	.019			.76		.76	1.24
5855	2" x 10" (alternate method)		825	.019			.77		.77	1.25
5865	2" x 12" (alternate method)		812	.020			.79		.79	1.27
5870	Sill plate, 2" x 4"	1 Clab	1170	.007			.27		.27	.44
5871	2" x 6"		780	.010			.41		.41	.66
5872	2" x 8"		586	.014			.54		.54	.88
5873	Alternate pricing method		.78	10.256	M.B.F.		410		410	665
5885	Ridge board, 1" x 4"	2 Clab	900	.018	L.F.		.71		.71	1.15
5886	1" x 6"		875	.018			.73		.73	1.18
5887	1" x 8"		850	.019			.75		.75	1.22

06 05 05 – Selective Demolition for Wood, Plastics, and Composites

06 05 05.10 Selective Demolition Wood Framing	Crew	Daily Output	Labor-Hours	Unit	Material	2018 Bare Costs Labor	Equipment	Total	Total Incl O&P
5888	1" x 10"	2 Clab	825	.019	L.F.		.77		.77	1.25
5889	1" x 12"		800	.020			.80		.80	1.29
5890	2" x 4"		900	.018			.71		.71	1.15
5892	2" x 6"		875	.018			.73		.73	1.18
5894	2" x 8"		850	.019			.75		.75	1.22
5896	2" x 10"		825	.019			.77		.77	1.25
5898	2" x 12"		800	.020			.80		.80	1.29
5900	Hip & valley rafters, 2" x 6"		500	.032			1.28		1.28	2.07
5940	2" x 8"		420	.038			1.52		1.52	2.47
6050	Rafter tie, 1" x 4"		1250	.013			.51		.51	.83
6052	1" x 6"		1135	.014			.56		.56	.91
6054	2" x 4"		1000	.016			.64		.64	1.04
6056	2" x 6"		970	.016			.66		.66	1.07
6070	Sleepers, on concrete, 1" x 2"	1 Clab	4700	.002			.07		.07	.11
6075	1" x 3"		4000	.002			.08		.08	.13
6080	2" x 4"		3000	.003			.11		.11	.17
6085	2" x 6"		2600	.003			.12		.12	.20
6086	Sheathing from roof, 5/16"	2 Clab	1600	.010	S.F.		.40		.40	.65
6088	3/8"		1525	.010			.42		.42	.68
6090	1/2"		1400	.011			.46		.46	.74
6092	5/8"		1300	.012			.49		.49	.80
6094	3/4"		1200	.013			.53		.53	.86
6096	Board sheathing from roof		1400	.011			.46		.46	.74
6100	Sheathing, from walls, 1/4"		1200	.013			.53		.53	.86
6110	5/16"		1175	.014			.54		.54	.88
6120	3/8"		1150	.014			.55		.55	.90
6130	1/2"		1125	.014			.57		.57	.92
6140	5/8"		1100	.015			.58		.58	.94
6150	3/4"		1075	.015			.59		.59	.96
6152	Board sheathing from walls		1500	.011			.43		.43	.69
6158	Subfloor/roof deck, with boards		2200	.007			.29		.29	.47
6159	Subfloor/roof deck, with tongue & groove boards		2000	.008			.32		.32	.52
6160	Plywood, 1/2" thick		768	.021			.83		.83	1.35
6162	5/8" thick		760	.021			.84		.84	1.36
6164	3/4" thick		750	.021			.85		.85	1.38
6165	1-1/8" thick		720	.022			.89		.89	1.44
6166	Underlayment, particle board, 3/8" thick	1 Clab	780	.010			.41		.41	.66
6168	1/2" thick		768	.010			.42		.42	.67
6170	5/8" thick		760	.011			.42		.42	.68
6172	3/4" thick		750	.011			.43		.43	.69
6200	Stairs and stringers, straight run	2 Clab	40	.400	Riser		15.95		15.95	26
6240	With platforms, winders or curves	"	26	.615	"		24.50		24.50	40
6300	Components, tread	1 Clab	110	.073	Ea.		2.90		2.90	4.71
6320	Riser		80	.100	"		3.99		3.99	6.45
6390	Stringer, 2" x 10"		260	.031	L.F.		1.23		1.23	1.99
6400	2" x 12"		260	.031			1.23		1.23	1.99
6410	3" x 10"		250	.032			1.28		1.28	2.07
6420	3" x 12"		250	.032			1.28		1.28	2.07
6590	Wood studs, 2" x 3"	2 Clab	3076	.005			.21		.21	.34
6600	2" x 4"		2000	.008			.32		.32	.52
6640	2" x 6"		1600	.010			.40		.40	.65
6720	Wall framing, including studs, plates and blocking, 2" x 4"	1 Clab	600	.013	S.F.		.53		.53	.86
6740	2" x 6"		480	.017	"		.66		.66	1.08

06 05 05.10 Selective Demolition Wood Framing		Crew	Daily Output	Labor-Hours	Unit	Material	2018 Bare Costs Labor	Equipment	Total	Total Incl O&P
6750	Headers, 2" x 4"	1 Clab	1125	.007	L.F.		.28		.28	.46
6755	2" x 6"		1125	.007			.28		.28	.46
6760	2" x 8"		1050	.008			.30		.30	.49
6765	2" x 10"		1050	.008			.30		.30	.49
6770	2" x 12"		1000	.008			.32		.32	.52
6780	4" x 10"		525	.015			.61		.61	.99
6785	4" x 12"		500	.016			.64		.64	1.04
6790	6" x 8"		560	.014			.57		.57	.92
6795	6" x 10"		525	.015			.61		.61	.99
6797	6" x 12"		500	.016			.64		.64	1.04
7000	Trusses									
7050	12' span	2 Clab	74	.216	Ea.		8.60		8.60	14
7150	24' span	F-3	66	.606			31.50	7.50	39	59
7200	26' span		64	.625			32.50	7.75	40.25	60.50
7250	28' span		62	.645			33.50	8	41.50	63
7300	30' span		58	.690			35.50	8.55	44.05	67
7350	32' span		56	.714			37	8.85	45.85	69.50
7400	34' span		54	.741			38.50	9.20	47.70	72
7450	36' span		52	.769			40	9.55	49.55	75
8000	Soffit, T&G wood	1 Clab	520	.015	S.F.		.61		.61	1
8010	Hardboard, vinyl or aluminum	"	640	.013			.50		.50	.81
8030	Plywood	2 Carp	315	.051			2.58		2.58	4.18
9000	Minimum labor/equipment charge	1 Clab	4	2	Job		79.50		79.50	129
9500	See Section 02 41 19.19 for rubbish handling									

06 05 05.20 Selective Demolition Millwork and Trim

06 05 05.20 Selective Demolition Millwork and Trim		Crew	Daily Output	Labor-Hours	Unit	Material	2018 Bare Costs Labor	Equipment	Total	Total Incl O&P
0010	**SELECTIVE DEMOLITION MILLWORK AND TRIM** R024119-10									
1000	Cabinets, wood, base cabinets, per L.F.	2 Clab	80	.200	L.F.		7.95		7.95	12.95
1020	Wall cabinets, per L.F.	"	80	.200	"		7.95		7.95	12.95
1060	Remove and reset, base cabinets	2 Carp	18	.889	Ea.		45		45	73
1070	Wall cabinets		20	.800			40.50		40.50	66
1072	Oven cabinet, 7' high		11	1.455			74		74	120
1074	Cabinet door, up to 2' high	1 Clab	66	.121			4.83		4.83	7.85
1076	2' - 4' high	"	46	.174			6.95		6.95	11.25
1100	Steel, painted, base cabinets	2 Clab	60	.267	L.F.		10.65		10.65	17.25
1120	Wall cabinets		60	.267	"		10.65		10.65	17.25
1200	Casework, large area		320	.050	S.F.		1.99		1.99	3.24
1220	Selective		200	.080	"		3.19		3.19	5.20
1500	Counter top, straight runs		200	.080	L.F.		3.19		3.19	5.20
1510	L, U or C shapes		120	.133			5.30		5.30	8.65
1550	Remove and reset, straight runs	2 Carp	50	.320			16.20		16.20	26.50
1560	L, U or C shape	"	40	.400			20.50		20.50	33
2000	Paneling, 4' x 8' sheets	2 Clab	2000	.008	S.F.		.32		.32	.52
2100	Boards, 1" x 4"		700	.023			.91		.91	1.48
2120	1" x 6"		750	.021			.85		.85	1.38
2140	1" x 8"		800	.020			.80		.80	1.29
3000	Trim, baseboard, to 6" wide		1200	.013	L.F.		.53		.53	.86
3040	Greater than 6" and up to 12" wide		1000	.016			.64		.64	1.04
3080	Remove and reset, minimum	2 Carp	400	.040			2.03		2.03	3.29
3090	Maximum	"	300	.053			2.70		2.70	4.39
3100	Ceiling trim	2 Clab	1000	.016			.64		.64	1.04
3120	Chair rail		1200	.013			.53		.53	.86
3140	Railings with balusters		240	.067			2.66		2.66	4.31

06 05 Common Work Results for Wood, Plastics, and Composites

06 05 05 – Selective Demolition for Wood, Plastics, and Composites

06 05 05.20 Selective Demolition Millwork and Trim	Crew	Daily Output	Labor-Hours	Unit	Material	2018 Bare Costs Labor	Equipment	Total	Total Incl O&P
3160	Wainscoting	2 Clab	700	.023	S.F.		.91		.91	1.48
9000	Minimum labor/equipment charge	1 Clab	4	2	Job		79.50		79.50	129

06 05 23 – Wood, Plastic, and Composite Fastenings

06 05 23.10 Nails

		Crew	Daily Output	Labor-Hours	Unit	Material	2018 Bare Costs Labor	Equipment	Total	Total Incl O&P
0010	**NAILS**, material only, based upon 50# box purchase									
0020	Copper nails, plain				Lb.	11.15			11.15	12.30
0400	Stainless steel, plain					8.55			8.55	9.40
0500	Box, 3d to 20d, bright					1.43			1.43	1.57
0520	Galvanized					2.34			2.34	2.57
0600	Common, 3d to 60d, plain					1.15			1.15	1.27
0700	Galvanized					2.19			2.19	2.41
0800	Aluminum					10.80			10.80	11.90
1000	Annular or spiral thread, 4d to 60d, plain					3.19			3.19	3.51
1200	Galvanized					3.02			3.02	3.32
1400	Drywall nails, plain					1.78			1.78	1.96
1600	Galvanized					1.84			1.84	2.02
1800	Finish nails, 4d to 10d, plain					1.24			1.24	1.36
2000	Galvanized					1.81			1.81	1.99
2100	Aluminum					7.95			7.95	8.75
2300	Flooring nails, hardened steel, 2d to 10d, plain					3.54			3.54	3.89
2400	Galvanized					3.75			3.75	4.13
2500	Gypsum lath nails, 1-1/8", 13 ga. flathead, blued					3.32			3.32	3.65
2600	Masonry nails, hardened steel, 3/4" to 3" long, plain					2.34			2.34	2.57
2700	Galvanized					3.90			3.90	4.29
2900	Roofing nails, threaded, galvanized					2.76			2.76	3.04
3100	Aluminum					7.25			7.25	8
3300	Compressed lead head, threaded, galvanized					2.97			2.97	3.27
3600	Siding nails, plain shank, galvanized					2.53			2.53	2.78
3800	Aluminum					5.75			5.75	6.35
5000	Add to prices above for cement coating					.15			.15	.17
5200	Zinc or tin plating					.25			.25	.28
5500	Vinyl coated sinkers, 8d to 16d					2.56			2.56	2.82

06 05 23.50 Wood Screws

		Crew	Daily Output	Labor-Hours	Unit	Material	2018 Bare Costs Labor	Equipment	Total	Total Incl O&P
0010	**WOOD SCREWS**									
0020	#8, 1" long, steel				C	4.60			4.60	5.05
0100	Brass					12.20			12.20	13.40
0200	#8, 2" long, steel					8			8	8.80
0300	Brass					24			24	26.50
0400	#10, 1" long, steel					3.30			3.30	3.63
0500	Brass					15.60			15.60	17.15
0600	#10, 2" long, steel					5.30			5.30	5.85
0700	Brass					27			27	29.50
0800	#10, 3" long, steel					8.85			8.85	9.75
1000	#12, 2" long, steel					7.30			7.30	8.05
1100	Brass					34.50			34.50	38
1500	#12, 3" long, steel					11.15			11.15	12.30
2000	#12, 4" long, steel					24.50			24.50	27

06 05 23.60 Timber Connectors

		Crew	Daily Output	Labor-Hours	Unit	Material	2018 Bare Costs Labor	Equipment	Total	Total Incl O&P
0010	**TIMBER CONNECTORS**									
0020	Add up cost of each part for total cost of connection									
0100	Connector plates, steel, with bolts, straight	2 Carp	75	.213	Ea.	27.50	10.80		38.30	48
0110	Tee, 7 ga.		50	.320		34	16.20		50.20	64

For customer support on your Commercial Renovation Costs with RSMeans data, call 800.448.8182.

135

06 05 23.60 Timber Connectors	Crew	Daily Output	Labor-Hours	Unit	Material	2018 Bare Costs Labor	Equipment	Total	Total Incl O&P
0120 T- Strap, 14 ga., 12" x 8" x 2"	2 Carp	50	.320	Ea.	34	16.20		50.20	64
0150 Anchor plates, 7 ga., 9" x 7"	↓	75	.213		27.50	10.80		38.30	48
0200 Bolts, machine, sq. hd. with nut & washer, 1/2" diameter, 4" long	1 Carp	140	.057		.75	2.90		3.65	5.55
0300 7-1/2" long		130	.062		1.37	3.12		4.49	6.55
0500 3/4" diameter, 7-1/2" long		130	.062		3.20	3.12		6.32	8.55
0610 Machine bolts, w/nut, washer, 3/4" diameter, 15" L, HD's & beam hangers		95	.084	↓	5.95	4.27		10.22	13.50
0800 Drilling bolt holes in timber, 1/2" diameter		450	.018	Inch		.90		.90	1.46
0900 1" diameter		350	.023	"		1.16		1.16	1.88
1100 Framing anchor, angle, 3" x 3" x 1-1/2", 12 ga.		175	.046	Ea.	2.43	2.32		4.75	6.45
1150 Framing anchors, 18 ga., 4-1/2" x 2-3/4"		175	.046		2.43	2.32		4.75	6.45
1160 Framing anchors, 18 ga., 4-1/2" x 3"		175	.046		2.43	2.32		4.75	6.45
1170 Clip anchors plates, 18 ga., 12" x 1-1/8"		175	.046		2.43	2.32		4.75	6.45
1250 Holdowns, 3 ga. base, 10 ga. body		8	1		38	50.50		88.50	125
1260 Holdowns, 7 ga. 11-1/16" x 3-1/4"		8	1		38	50.50		88.50	125
1270 Holdowns, 7 ga. 14-3/8" x 3-1/8"		8	1		38	50.50		88.50	125
1275 Holdowns, 12 ga. 8" x 2-1/2"		8	1		38	50.50		88.50	125
1300 Joist and beam hangers, 18 ga. galv., for 2" x 4" joist		175	.046		.75	2.32		3.07	4.59
1400 2" x 6" to 2" x 10" joist		165	.048		1.27	2.46		3.73	5.40
1600 16 ga. galv., 3" x 6" to 3" x 10" joist		160	.050		2.92	2.54		5.46	7.35
1700 3" x 10" to 3" x 14" joist		160	.050		4.32	2.54		6.86	8.85
1800 4" x 6" to 4" x 10" joist		155	.052		2.89	2.62		5.51	7.45
1900 4" x 10" to 4" x 14" joist		155	.052		4.89	2.62		7.51	9.65
2000 Two-2" x 6" to two-2" x 10" joists		150	.053		3.83	2.70		6.53	8.60
2100 Two-2" x 10" to two-2" x 14" joists		150	.053		4.29	2.70		6.99	9.10
2300 3/16" thick, 6" x 8" joist		145	.055		59.50	2.80		62.30	70
2400 6" x 10" joist		140	.057		62	2.90		64.90	72.50
2500 6" x 12" joist		135	.059		64.50	3		67.50	76
2700 1/4" thick, 6" x 14" joist	↓	130	.062		67	3.12		70.12	78.50
2900 Plywood clips, extruded aluminum H clip, for 3/4" panels					.22			.22	.24
3000 Galvanized 18 ga. back-up clip					.18			.18	.20
3200 Post framing, 16 ga. galv. for 4" x 4" base, 2 piece	1 Carp	130	.062		15.15	3.12		18.27	22
3300 Cap		130	.062		21	3.12		24.12	28
3500 Rafter anchors, 18 ga. galv., 1-1/2" wide, 5-1/4" long		145	.055		.45	2.80		3.25	5.05
3600 10-3/4" long		145	.055		1.37	2.80		4.17	6.05
3800 Shear plates, 2-5/8" diameter		120	.067		2.47	3.38		5.85	8.20
3900 4" diameter		115	.070		2.40	3.53		5.93	8.40
4000 Sill anchors, embedded in concrete or block, 25-1/2" long		115	.070		11.60	3.53		15.13	18.50
4100 Spike grids, 3" x 6"		120	.067		1.09	3.38		4.47	6.70
4400 Split rings, 2-1/2" diameter		120	.067		2.44	3.38		5.82	8.20
4500 4" diameter		110	.073		2.78	3.69		6.47	9.05
4550 Tie plate, 20 ga., 7" x 3-1/8"		110	.073		2.78	3.69		6.47	9.05
4560 5" x 4-1/8"		110	.073		2.78	3.69		6.47	9.05
4575 Twist straps, 18 ga., 12" x 1-1/4"		110	.073		2.78	3.69		6.47	9.05
4580 16" x 1-1/4"		110	.073		2.78	3.69		6.47	9.05
4600 Strap ties, 20 ga., 2-1/16" wide, 12-13/16" long		180	.044		.87	2.25		3.12	4.62
4700 16 ga., 1-3/8" wide, 12" long		180	.044		.87	2.25		3.12	4.62
4800 1-1/4" wide, 21-5/8" long		160	.050		2.70	2.54		5.24	7.10
5000 Toothed rings, 2-5/8" or 4" diameter		90	.089	↓	2.05	4.51		6.56	9.55
5200 Truss plates, nailed, 20 ga., up to 32' span	↓	17	.471	Truss	12.85	24		36.85	52.50
5400 Washers, 2" x 2" x 1/8"				Ea.	.46			.46	.51
5500 3" x 3" x 3/16"				"	1.32			1.32	1.45
9000 Minimum labor/equipment charge	1 Carp	4	2	Job		101		101	165

06 05 Common Work Results for Wood, Plastics, and Composites

06 05 23 – Wood, Plastic, and Composite Fastenings

06 05 23.80 Metal Bracing		Crew	Daily Output	Labor-Hours	Unit	Material	2018 Bare Costs Labor	Equipment	Total	Total Incl O&P
0010	**METAL BRACING**									
0302	Let-in, "T" shaped, 22 ga. galv. steel, studs at 16" OC	1 Carp	580	.014	L.F.	.81	.70		1.51	2.02
0402	Studs at 24" OC		600	.013		.81	.68		1.49	1.99
0502	Steel straps, 16 ga. galv. steel, studs at 16" OC		600	.013		1.05	.68		1.73	2.26
0602	Studs at 24" OC	↓	620	.013	↓	1.05	.65		1.70	2.22

06 11 Wood Framing

06 11 10 – Framing with Dimensional, Engineered or Composite Lumber

06 11 10.01 Forest Stewardship Council Certification

0010	**FOREST STEWARDSHIP COUNCIL CERTIFICATION**									
0020	For Forest Stewardship Council (FSC) cert dimension lumber, add [G]					65%				

06 11 10.02 Blocking

0010	**BLOCKING**									
2600	Miscellaneous, to wood construction									
2620	2" x 4"	1 Carp	.17	47.059	M.B.F.	630	2,375		3,005	4,575
2625	Pneumatic nailed		.21	38.095		635	1,925		2,560	3,825
2660	2" x 8"		.27	29.630		670	1,500		2,170	3,175
2665	Pneumatic nailed	↓	.33	24.242	↓	680	1,225		1,905	2,750
2720	To steel construction									
2740	2" x 4"	1 Carp	.14	57.143	M.B.F.	630	2,900		3,530	5,400
2780	2" x 8"		.21	38.095	"	670	1,925		2,595	3,875
9000	Minimum labor/equipment charge	↓	4	2	Job		101		101	165

06 11 10.04 Wood Bracing

0010	**WOOD BRACING**									
0012	Let-in, with 1" x 6" boards, studs @ 16" OC	1 Carp	150	.053	L.F.	.78	2.70		3.48	5.25
0202	Studs @ 24" OC	"	230	.035	"	.78	1.76		2.54	3.72

06 11 10.06 Bridging

0010	**BRIDGING**									
0012	Wood, for joists 16" OC, 1" x 3"	1 Carp	130	.062	Pr.	.69	3.12		3.81	5.80
0017	Pneumatic nailed		170	.047		.76	2.39		3.15	4.71
0102	2" x 3" bridging		130	.062		.72	3.12		3.84	5.85
0107	Pneumatic nailed		170	.047		.75	2.39		3.14	4.70
0302	Steel, galvanized, 18 ga., for 2" x 10" joists at 12" OC		130	.062		1.71	3.12		4.83	6.95
0352	16" OC		135	.059		1.71	3		4.71	6.75
0402	24" OC		140	.057		2.57	2.90		5.47	7.55
0602	For 2" x 14" joists at 16" OC		130	.062		1.86	3.12		4.98	7.10
0902	Compression type, 16" OC, 2" x 8" joists		200	.040		1.26	2.03		3.29	4.67
1002	2" x 12" joists		200	.040		1.25	2.03		3.28	4.67

06 11 10.10 Beam and Girder Framing

0010	**BEAM AND GIRDER FRAMING** R061110-30									
1000	Single, 2" x 6"	2 Carp	700	.023	L.F.	.64	1.16		1.80	2.59
1005	Pneumatic nailed		812	.020		.65	1		1.65	2.34
1020	2" x 8"		650	.025		.89	1.25		2.14	3.01
1025	Pneumatic nailed		754	.021		.90	1.08		1.98	2.74
1040	2" x 10"		600	.027		1.41	1.35		2.76	3.74
1045	Pneumatic nailed		696	.023		1.42	1.17		2.59	3.45
1060	2" x 12"		550	.029		1.79	1.47		3.26	4.36
1065	Pneumatic nailed		638	.025		1.81	1.27		3.08	4.05
1080	2" x 14"	↓	500	.032	↓	2.27	1.62		3.89	5.10

For customer support on your Commercial Renovation Costs with RSMeans data, call 800.448.8182.

137

06 11 Wood Framing

06 11 10 – Framing with Dimensional, Engineered or Composite Lumber

06 11 10.10 Beam and Girder Framing		Crew	Daily Output	Labor-Hours	Unit	Material	2018 Bare Costs Labor	Equipment	Total	Total Incl O&P
1085	Pneumatic nailed	2 Carp	580	.028	L.F.	2.28	1.40		3.68	4.78
1100	3" x 8"		550	.029		2.87	1.47		4.34	5.55
1120	3" x 10"		500	.032		3.62	1.62		5.24	6.60
1140	3" x 12"		450	.036		4.34	1.80		6.14	7.70
1160	3" x 14"		400	.040		5.05	2.03		7.08	8.85
1170	4" x 6"	F-3	1100	.036		3.06	1.88	.45	5.39	6.90
1180	4" x 8"		1000	.040		4.15	2.07	.50	6.72	8.45
1200	4" x 10"		950	.042		4.96	2.18	.52	7.66	9.55
1220	4" x 12"		900	.044		5.50	2.30	.55	8.35	10.35
1240	4" x 14"		850	.047		6.40	2.44	.58	9.42	11.60
1250	6" x 8"		525	.076		7.90	3.95	.94	12.79	16.10
1260	6" x 10"		500	.080		6.90	4.14	.99	12.03	15.40
1290	8" x 12"		300	.133		16.55	6.90	1.65	25.10	31
2000	Double, 2" x 6"	2 Carp	625	.026		1.29	1.30		2.59	3.53
2005	Pneumatic nailed		725	.022		1.30	1.12		2.42	3.25
2020	2" x 8"		575	.028		1.79	1.41		3.20	4.26
2025	Pneumatic nailed		667	.024		1.81	1.22		3.03	3.96
2040	2" x 10"		550	.029		2.82	1.47		4.29	5.50
2045	Pneumatic nailed		638	.025		2.85	1.27		4.12	5.20
2060	2" x 12"		525	.030		3.59	1.55		5.14	6.45
2065	Pneumatic nailed		610	.026		3.62	1.33		4.95	6.15
2080	2" x 14"		475	.034		4.53	1.71		6.24	7.75
2085	Pneumatic nailed		551	.029		4.56	1.47		6.03	7.40
3000	Triple, 2" x 6"		550	.029		1.93	1.47		3.40	4.51
3005	Pneumatic nailed		638	.025		1.95	1.27		3.22	4.21
3020	2" x 8"		525	.030		2.68	1.55		4.23	5.45
3025	Pneumatic nailed		609	.026		2.71	1.33		4.04	5.15
3040	2" x 10"		500	.032		4.23	1.62		5.85	7.30
3045	Pneumatic nailed		580	.028		4.27	1.40		5.67	6.95
3060	2" x 12"		475	.034		5.40	1.71		7.11	8.65
3065	Pneumatic nailed		551	.029		5.45	1.47		6.92	8.35
3080	2" x 14"		450	.036		7.40	1.80		9.20	11.05
3085	Pneumatic nailed		522	.031		6.85	1.55		8.40	10.05
9000	Minimum labor/equipment charge	1 Carp	2	4	Job		203		203	330

06 11 10.12 Ceiling Framing

		Crew	Daily Output	Labor-Hours	Unit	Material	2018 Bare Costs Labor	Equipment	Total	Total Incl O&P
0010	**CEILING FRAMING**									
6000	Suspended, 2" x 3"	2 Carp	1000	.016	L.F.	.42	.81		1.23	1.79
6050	2" x 4"		900	.018		.42	.90		1.32	1.92
6100	2" x 6"		800	.020		.64	1.01		1.65	2.36
6150	2" x 8"		650	.025		.89	1.25		2.14	3.01
9000	Minimum labor/equipment charge	1 Carp	4	2	Job		101		101	165

06 11 10.14 Posts and Columns

		Crew	Daily Output	Labor-Hours	Unit	Material	2018 Bare Costs Labor	Equipment	Total	Total Incl O&P
0010	**POSTS AND COLUMNS**									
0100	4" x 4"	2 Carp	390	.041	L.F.	1.83	2.08		3.91	5.40
0150	4" x 6"		275	.058		3.06	2.95		6.01	8.15
0200	4" x 8"		220	.073		4.15	3.69		7.84	10.55
0250	6" x 6"		215	.074		5.35	3.77		9.12	12
0300	6" x 8"		175	.091		7.90	4.64		12.54	16.20
0350	6" x 10"		150	.107		6.90	5.40		12.30	16.40
9000	Minimum labor/equipment charge	1 Carp	2	4	Job		203		203	330

06 11 Wood Framing

06 11 10 – Framing with Dimensional, Engineered or Composite Lumber

06 11 10.18 Joist Framing	Crew	Daily Output	Labor-Hours	Unit	Material	2018 Bare Costs Labor	Equipment	Total	Total Incl O&P
0010 **JOIST FRAMING** R061110-30									
2000　Joists, 2" x 4"	2 Carp	1250	.013	L.F.	.42	.65		1.07	1.51
2005　　Pneumatic nailed		1438	.011		.42	.56		.98	1.39
2100　　2" x 6"		1250	.013		.64	.65		1.29	1.76
2105　　Pneumatic nailed		1438	.011		.65	.56		1.21	1.64
2150　　2" x 8"		1100	.015		.89	.74		1.63	2.18
2155　　Pneumatic nailed		1265	.013		.90	.64		1.54	2.03
2200　　2" x 10"		900	.018		1.41	.90		2.31	3.01
2205　　Pneumatic nailed		1035	.015		1.42	.78		2.20	2.83
2250　　2" x 12"		875	.018		1.79	.93		2.72	3.48
2255　　Pneumatic nailed		1006	.016		1.81	.81		2.62	3.30
2300　　2" x 14"		770	.021		2.27	1.05		3.32	4.20
2305　　Pneumatic nailed		886	.018		2.28	.92		3.20	4
2350　　3" x 6"		925	.017		2	.88		2.88	3.62
2400　　3" x 10"		780	.021		3.62	1.04		4.66	5.65
2450　　3" x 12"		600	.027		4.34	1.35		5.69	6.95
2500　　4" x 6"		800	.020		3.06	1.01		4.07	5
2550　　4" x 10"		600	.027		4.96	1.35		6.31	7.65
2600　　4" x 12"		450	.036		5.50	1.80		7.30	9
2605　Sister joist, 2" x 6"		800	.020		.64	1.01		1.65	2.36
2606　　Pneumatic nailed		960	.017		.65	.85		1.50	2.09
2610　　2" x 8"		640	.025		.89	1.27		2.16	3.04
2611　　Pneumatic nailed		768	.021		.90	1.06		1.96	2.70
2615　　2" x 10"		535	.030		1.41	1.52		2.93	4.01
2616　　Pneumatic nailed		642	.025		1.42	1.26		2.68	3.61
2620　　2" x 12"		455	.035		1.79	1.78		3.57	4.86
2625　　Pneumatic nailed		546	.029		1.81	1.49		3.30	4.40
3000　Composite wood joist 9-1/2" deep		.90	17.778	M.L.F.	1,775	900		2,675	3,425
3010　　11-1/2" deep		.00	10.102		2,050	920		2,970	3,750
3020　　14" deep		.82	19.512		2,550	990		3,540	4,425
3030　　16" deep		.78	20.513		4,075	1,050		5,125	6,175
4000　Open web joist 12" deep		.88	18.182		3,600	920		4,520	5,475
4002　　Per linear foot		880	.018	L.F.	3.60	.92		4.52	5.45
4004　　Treated, per linear foot		880	.018	"	4.53	.92		5.45	6.50
4010　　14" deep		.82	19.512	M.L.F.	3,900	990		4,890	5,900
4012　　Per linear foot		820	.020	L.F.	3.90	.99		4.89	5.90
4014　　Treated, per linear foot		820	.020	"	4.98	.99		5.97	7.10
4020　　16" deep		.78	20.513	M.L.F.	3,850	1,050		4,900	5,950
4022　　Per linear foot		780	.021	L.F.	3.86	1.04		4.90	5.95
4024　　Treated, per linear foot		780	.021	"	5.10	1.04		6.14	7.30
4030　　18" deep		.74	21.622	M.L.F.	4,225	1,100		5,325	6,425
4032　　Per linear foot		740	.022	L.F.	4.22	1.10		5.32	6.40
4034　　Treated, per linear foot		740	.022	"	5.60	1.10		6.70	8
6000　Composite rim joist, 1-1/4" x 9-1/2"		.90	17.778	M.L.F.	2,075	900		2,975	3,750
6010　　1-1/4" x 11-1/2"		.88	18.182		2,275	920		3,195	4,000
6020　　1-1/4" x 14-1/2"		.82	19.512		3,000	990		3,990	4,900
6030　　1-1/4" x 16-1/2"		.78	20.513		2,750	1,050		3,800	4,725
9000　Minimum labor/equipment charge	1 Carp	4	2	Job		101		101	165

06 11 10.24 Miscellaneous Framing

	Crew	Daily Output	Labor-Hours	Unit	Material	Labor	Equipment	Total	Total Incl O&P
0010 **MISCELLANEOUS FRAMING**									
2000　Firestops, 2" x 4"	2 Carp	780	.021	L.F.	.42	1.04		1.46	2.15
2005　　Pneumatic nailed		952	.017		.42	.85		1.27	1.85

For customer support on your Commercial Renovation Costs with RSMeans data, call 800.448.8182.

139

06 11 Wood Framing

06 11 10 – Framing with Dimensional, Engineered or Composite Lumber

06 11 10.24 Miscellaneous Framing

		Crew	Daily Output	Labor-Hours	Unit	Material	2018 Bare Costs Labor	Equipment	Total	Total Incl O&P
2100	2" x 6"	2 Carp	600	.027	L.F.	.64	1.35		1.99	2.90
2105	Pneumatic nailed		732	.022		.65	1.11		1.76	2.52
5000	Nailers, treated, wood construction, 2" x 4"		800	.020		.59	1.01		1.60	2.30
5005	Pneumatic nailed		960	.017		.59	.85		1.44	2.02
5100	2" x 6"		750	.021		.74	1.08		1.82	2.57
5105	Pneumatic nailed		900	.018		.74	.90		1.64	2.28
5120	2" x 8"		700	.023		1.12	1.16		2.28	3.11
5125	Pneumatic nailed		840	.019		1.13	.97		2.10	2.81
5200	Steel construction, 2" x 4"		750	.021		.59	1.08		1.67	2.41
5220	2" x 6"		700	.023		.74	1.16		1.90	2.69
5240	2" x 8"		650	.025		1.12	1.25		2.37	3.26
7000	Rough bucks, treated, for doors or windows, 2" x 6"		400	.040		.74	2.03		2.77	4.10
7005	Pneumatic nailed		480	.033		.74	1.69		2.43	3.56
7100	2" x 8"		380	.042		1.12	2.14		3.26	4.70
7105	Pneumatic nailed		456	.035		1.13	1.78		2.91	4.13
8000	Stair stringers, 2" x 10"		130	.123		1.41	6.25		7.66	11.70
8100	2" x 12"		130	.123		1.79	6.25		8.04	12.10
8150	3" x 10"		125	.128		3.62	6.50		10.12	14.55
8200	3" x 12"		125	.128		4.34	6.50		10.84	15.35
8870	Laminated structural lumber, 1-1/4" x 11-1/2"		130	.123		2.28	6.25		8.53	12.65
8880	1-1/4" x 14-1/2"		130	.123		2.97	6.25		9.22	13.40
9000	Minimum labor/equipment charge	1 Carp	4	2	Job		101		101	165

06 11 10.26 Partitions

		Crew	Daily Output	Labor-Hours	Unit	Material	2018 Bare Costs Labor	Equipment	Total	Total Incl O&P
0010	**PARTITIONS**									
0020	Single bottom and double top plate, no waste, std. & better lumber									
0180	2" x 4" studs, 8' high, studs 12" OC	2 Carp	80	.200	L.F.	4.60	10.15		14.75	21.50
0185	12" OC, pneumatic nailed		96	.167		4.66	8.45		13.11	18.80
0200	16" OC		100	.160		3.77	8.10		11.87	17.30
0205	16" OC, pneumatic nailed		120	.133		3.81	6.75		10.56	15.15
0300	24" OC		125	.128		2.93	6.50		9.43	13.75
0305	24" OC, pneumatic nailed		150	.107		2.96	5.40		8.36	12.05
0380	10' high, studs 12" OC		80	.200		5.45	10.15		15.60	22.50
0385	12" OC, pneumatic nailed		96	.167		5.50	8.45		13.95	19.75
0400	16" OC		100	.160		4.39	8.10		12.49	18
0405	16" OC, pneumatic nailed		120	.133		4.44	6.75		11.19	15.85
0500	24" OC		125	.128		3.35	6.50		9.85	14.25
0505	24" OC, pneumatic nailed		150	.107		3.39	5.40		8.79	12.50
0580	12' high, studs 12" OC		65	.246		6.30	12.50		18.80	27.50
0585	12" OC, pneumatic nailed		78	.205		6.35	10.40		16.75	24
0600	16" OC		80	.200		5	10.15		15.15	22
0605	16" OC, pneumatic nailed		96	.167		5.10	8.45		13.55	19.30
0700	24" OC		100	.160		3.77	8.10		11.87	17.30
0705	24" OC, pneumatic nailed		120	.133		3.81	6.75		10.56	15.15
0780	2" x 6" studs, 8' high, studs 12" OC		70	.229		7.10	11.60		18.70	26.50
0785	12" OC, pneumatic nailed		84	.190		7.15	9.65		16.80	23.50
0800	16" OC		90	.178		5.80	9		14.80	21
0805	16" OC, pneumatic nailed		108	.148		5.85	7.50		13.35	18.65
0900	24" OC		115	.139		4.50	7.05		11.55	16.40
0905	24" OC, pneumatic nailed		138	.116		4.55	5.90		10.45	14.55
0980	10' high, studs 12" OC		70	.229		8.35	11.60		19.95	28
0985	12" OC, pneumatic nailed		84	.190		8.45	9.65		18.10	25
1000	16" OC		90	.178		6.75	9		15.75	22

06 11 Wood Framing

06 11 10 – Framing with Dimensional, Engineered or Composite Lumber

06 11 10.26 Partitions		Crew	Daily Output	Labor-Hours	Unit	Material	2018 Bare Costs Labor	Equipment	Total	Total Incl O&P
1005	16" OC, pneumatic nailed	2 Carp	108	.148	L.F.	6.85	7.50		14.35	19.70
1100	24" OC		115	.139		5.15	7.05		12.20	17.10
1105	24" OC, pneumatic nailed		138	.116		5.20	5.90		11.10	15.25
1180	12' high, studs 12" OC		55	.291		9.65	14.75		24.40	34.50
1185	12" OC, pneumatic nailed		66	.242		9.75	12.30		22.05	30.50
1200	16" OC		70	.229		7.70	11.60		19.30	27.50
1205	16" OC, pneumatic nailed		84	.190		7.80	9.65		17.45	24.50
1300	24" OC		90	.178		5.80	9		14.80	21
1305	24" OC, pneumatic nailed		108	.148		5.85	7.50		13.35	18.65
1400	For horizontal blocking, 2" x 4", add		600	.027		.42	1.35		1.77	2.65
1500	2" x 6", add		600	.027		.64	1.35		1.99	2.90
1600	For openings, add	▼	250	.064	▼		3.24		3.24	5.25
1702	Headers for above openings, material only, add				B.F.	.74			.74	.81
9000	Minimum labor/equipment charge	1 Carp	4	2	Job		101		101	165

06 11 10.28 Porch or Deck Framing

		Crew	Daily Output	Labor-Hours	Unit	Material	2018 Bare Costs Labor	Equipment	Total	Total Incl O&P
0010	**PORCH OR DECK FRAMING**									
0100	Treated lumber, posts or columns, 4" x 4"	2 Carp	390	.041	L.F.	1.20	2.08		3.28	4.70
0110	4" x 6"		275	.058		1.94	2.95		4.89	6.95
0120	4" x 8"		220	.073		3.96	3.69		7.65	10.35
0130	Girder, single, 4" x 4"		675	.024		1.20	1.20		2.40	3.27
0140	4" x 6"		600	.027		1.94	1.35		3.29	4.33
0150	4" x 8"		525	.030		3.96	1.55		5.51	6.85
0160	Double, 2" x 4"		625	.026		1.21	1.30		2.51	3.44
0170	2" x 6"		600	.027		1.52	1.35		2.87	3.86
0180	2" x 8"		575	.028		2.30	1.41		3.71	4.82
0190	2" x 10"		550	.029		2.88	1.47		4.35	5.55
0200	2" x 12"		525	.030		4.15	1.55		5.70	7.05
0210	Triple, 2" x 4"		575	.028		1.81	1.41		3.22	4.28
0220	2" x 6"		550	.029		2.28	1.47		3.75	4.89
0230	2" x 8"		525	.030		3.45	1.55		5	6.30
0240	2" x 10"		500	.032		4.32	1.62		5.94	7.40
0250	2" x 12"		475	.034		6.20	1.71		7.91	9.60
0260	Ledger, bolted 4' OC, 2" x 4"		400	.040		.75	2.03		2.78	4.11
0270	2" x 6"		395	.041		.89	2.05		2.94	4.31
0280	2" x 8"		390	.041		1.27	2.08		3.35	4.78
0290	2" x 10"		385	.042		1.55	2.11		3.66	5.10
0300	2" x 12"		380	.042		2.17	2.14		4.31	5.85
0310	Joists, 2" x 4"		1250	.013		.60	.65		1.25	1.71
0320	2" x 6"		1250	.013		.76	.65		1.41	1.89
0330	2" x 8"		1100	.015		1.15	.74		1.89	2.47
0340	2" x 10"		900	.018		1.44	.90		2.34	3.05
0350	2" x 12"	▼	875	.018		1.73	.93		2.66	3.41
0360	Railings and trim, 1" x 4"	1 Carp	300	.027		.51	1.35		1.86	2.75
0370	2" x 2"		300	.027		.43	1.35		1.78	2.66
0380	2" x 4"		300	.027		.59	1.35		1.94	2.84
0390	2" x 6"		300	.027	▼	.74	1.35		2.09	3.01
0400	Decking, 1" x 4"		275	.029	S.F.	2.76	1.47		4.23	5.40
0410	2" x 4"		300	.027		2	1.35		3.35	4.39
0420	2" x 6"		320	.025		1.60	1.27		2.87	3.82
0430	5/4" x 6"	▼	320	.025	▼	2.15	1.27		3.42	4.42
0440	Balusters, square, 2" x 2"	2 Carp	660	.024	L.F.	.43	1.23		1.66	2.47
0450	Turned, 2" x 2"	▼	420	.038	▼	.57	1.93		2.50	3.77

For customer support on your Commercial Renovation Costs with RSMeans data, call 800.448.8182.

141

06 11 10.28 Porch or Deck Framing

		Crew	Daily Output	Labor-Hours	Unit	Material	2018 Bare Costs Labor	Equipment	Total	Total Incl O&P
0460	Stair stringer, 2" x 10"	2 Carp	130	.123	L.F.	1.44	6.25		7.69	11.75
0470	2" x 12"		130	.123		1.73	6.25		7.98	12.05
0480	Stair treads, 1" x 4"		140	.114		2.76	5.80		8.56	12.45
0490	2" x 4"		140	.114		.60	5.80		6.40	10.05
0500	2" x 6"		160	.100		.91	5.05		5.96	9.25
0510	5/4" x 6"		160	.100		1	5.05		6.05	9.35
0520	Turned handrail post, 4" x 4"		64	.250	Ea.	34	12.70		46.70	58
0530	Lattice panel, 4' x 8', 1/2"		1600	.010	S.F.	.72	.51		1.23	1.61
0535	3/4"		1600	.010	"	1.07	.51		1.58	1.99
0540	Cedar, posts or columns, 4" x 4"		390	.041	L.F.	3.69	2.08		5.77	7.45
0550	4" x 6"		275	.058		6.90	2.95		9.85	12.35
0560	4" x 8"		220	.073		11	3.69		14.69	18.10
0800	Decking, 1" x 4"		550	.029		2.75	1.47		4.22	5.40
0810	2" x 4"		600	.027		5.60	1.35		6.95	8.35
0820	2" x 6"		640	.025		10.15	1.27		11.42	13.20
0830	5/4" x 6"		640	.025		6.30	1.27		7.57	8.95
0840	Railings and trim, 1" x 4"		600	.027		2.75	1.35		4.10	5.20
0860	2" x 4"		600	.027		5.60	1.35		6.95	8.35
0870	2" x 6"		600	.027		10.15	1.35		11.50	13.35
0920	Stair treads, 1" x 4"		140	.114		2.75	5.80		8.55	12.40
0930	2" x 4"		140	.114		5.60	5.80		11.40	15.55
0940	2" x 6"		160	.100		10.15	5.05		15.20	19.40
0950	5/4" x 6"		160	.100		6.30	5.05		11.35	15.15
0980	Redwood, posts or columns, 4" x 4"		390	.041		6.45	2.08		8.53	10.50
0990	4" x 6"		275	.058		12.60	2.95		15.55	18.65
1000	4" x 8"		220	.073		23.50	3.69		27.19	32
1240	Decking, 1" x 4"	1 Carp	275	.029	S.F.	3.97	1.47		5.44	6.75
1260	2" x 6"		340	.024		7.50	1.19		8.69	10.20
1270	5/4" x 6"		320	.025		4.77	1.27		6.04	7.30
1280	Railings and trim, 1" x 4"	2 Carp	600	.027	L.F.	1.17	1.35		2.52	3.48
1310	2" x 6"		600	.027		7.50	1.35		8.85	10.45
1420	Alternative decking, wood/plastic composite, 5/4" x 6" [G]		640	.025		3.27	1.27		4.54	5.65
1440	1" x 4" square edge fir		550	.029		2.77	1.47		4.24	5.45
1450	1" x 4" tongue and groove fir		450	.036		1.51	1.80		3.31	4.59
1460	1" x 4" mahogany		550	.029		2.07	1.47		3.54	4.66
1462	5/4" x 6" PVC		550	.029		3.36	1.47		4.83	6.10
1465	Framing, porch or deck, alt deck fastening, screws, add	1 Carp	240	.033	S.F.		1.69		1.69	2.74
1470	Accessories, joist hangers, 2" x 4"		160	.050	Ea.	.75	2.54		3.29	4.95
1480	2" x 6" through 2" x 12"		150	.053		1.27	2.70		3.97	5.80
1530	Post footing, incl excav, backfill, tube form & concrete, 4' deep, 8" diam.	F-7	12	2.667		17.55	121		138.55	215
1540	10" diameter		11	2.909		24	132		156	241
1550	12" diameter		10	3.200		30.50	145		175.50	269

06 11 10.30 Roof Framing

		Crew	Daily Output	Labor-Hours	Unit	Material	2018 Bare Costs Labor	Equipment	Total	Total Incl O&P
0010	**ROOF FRAMING** R061110-30									
1900	Rough fascia, 2" x 6"	2 Carp	250	.064	L.F.	.64	3.24		3.88	5.95
2000	2" x 8"		225	.071		.89	3.61		4.50	6.85
2100	2" x 10"		180	.089		1.41	4.51		5.92	8.85
5000	Rafters, to 4 in 12 pitch, 2" x 6", ordinary		1000	.016		.64	.81		1.45	2.03
5060	2" x 8", ordinary		950	.017		.89	.85		1.74	2.37
5120	2" x 10", ordinary		630	.025		1.41	1.29		2.70	3.64
5180	2" x 12", ordinary		575	.028		1.79	1.41		3.20	4.26
5250	Composite rafter, 9-1/2" deep		575	.028		1.76	1.41		3.17	4.23

06 11 Wood Framing

06 11 10 – Framing with Dimensional, Engineered or Composite Lumber

06 11 10.30 Roof Framing		Crew	Daily Output	Labor-Hours	Unit	Material	2018 Bare Costs Labor	Equipment	Total	Total Incl O&P
5260	11-1/2" deep	2 Carp	575	.028	L.F.	2.05	1.41		3.46	4.54
5300	Hip and valley rafters, 2" x 6", ordinary		760	.021		.64	1.07		1.71	2.44
5360	2" x 8", ordinary		720	.022		.89	1.13		2.02	2.81
5420	2" x 10", ordinary		570	.028		1.41	1.42		2.83	3.86
5480	2" x 12", ordinary		525	.030		1.79	1.55		3.34	4.48
5540	Hip and valley jacks, 2" x 6", ordinary		600	.027		.64	1.35		1.99	2.90
5600	2" x 8", ordinary		490	.033		.89	1.66		2.55	3.67
5660	2" x 10", ordinary		450	.036		1.41	1.80		3.21	4.48
5720	2" x 12", ordinary		375	.043		1.79	2.16		3.95	5.50
5761	For slopes steeper than 4 in 12, add						30%			
5780	Rafter tie, 1" x 4", #3	2 Carp	800	.020	L.F.	.51	1.01		1.52	2.21
5790	2" x 4", #3		800	.020		.42	1.01		1.43	2.11
5800	Ridge board, #2 or better, 1" x 6"		600	.027		.78	1.35		2.13	3.05
5820	1" x 8"		550	.029		1.30	1.47		2.77	3.82
5840	1" x 10"		500	.032		1.69	1.62		3.31	4.49
5860	2" x 6"		500	.032		.64	1.62		2.26	3.34
5880	2" x 8"		450	.036		.89	1.80		2.69	3.91
5900	2" x 10"		400	.040		1.41	2.03		3.44	4.84
5920	Roof cants, split, 4" x 4"		650	.025		1.83	1.25		3.08	4.04
5940	6" x 6"		600	.027		5.35	1.35		6.70	8.10
5960	Roof curbs, untreated, 2" x 6"		520	.031		.64	1.56		2.20	3.24
5980	2" x 12"		400	.040		1.79	2.03		3.82	5.25
6000	Sister rafters, 2" x 6"		800	.020		.64	1.01		1.65	2.36
6020	2" x 8"		640	.025		.89	1.27		2.16	3.04
6040	2" x 10"		535	.030		1.41	1.52		2.93	4.01
6060	2" x 12"		455	.035		1.79	1.78		3.57	4.86
9000	Minimum labor/equipment charge	1 Carp	4	2	Job		101		101	165

06 11 10.32 Sill and Ledger Framing

06 11 10.32 Sill and Ledger Framing		Crew	Daily Output	Labor-Hours	Unit	Material	2018 Bare Costs Labor	Equipment	Total	Total Incl O&P
0010	**SILL AND LEDGER FRAMING**									
0020	Extruded polystyrene sill sealer, 5-1/2" wide	1 Carp	1600	.005	L.F.	.16	.25		.41	.59
2002	Ledgers, nailed, 2" x 4"	2 Carp	755	.021		.42	1.07		1.49	2.20
2052	2" x 6"		600	.027		.64	1.35		1.99	2.90
2102	Bolted, not including bolts, 3" x 6"		325	.049		1.98	2.50		4.48	6.25
2152	3" x 12"		233	.069		4.31	3.48		7.79	10.40
2602	Mud sills, redwood, construction grade, 2" x 4"		895	.018		2.26	.91		3.17	3.96
2622	2" x 6"		780	.021		3.40	1.04		4.44	5.45
4002	Sills, 2" x 4"		600	.027		.41	1.35		1.76	2.64
4052	2" x 6"		550	.029		.63	1.47		2.10	3.09
4082	2" x 8"		500	.032		.88	1.62		2.50	3.60
4101	2" x 10"		450	.036		1.39	1.80		3.19	4.46
4121	2" x 12"		400	.040		1.78	2.03		3.81	5.25
4202	Treated, 2" x 4"		550	.029		.58	1.47		2.05	3.03
4222	2" x 6"		500	.032		.73	1.62		2.35	3.43
4242	2" x 8"		450	.036		1.11	1.80		2.91	4.15
4261	2" x 10"		400	.040		1.38	2.03		3.41	4.81
4281	2" x 12"		350	.046		2.01	2.32		4.33	5.95
4402	4" x 4"		450	.036		1.16	1.80		2.96	4.20
4422	4" x 6"		350	.046		1.88	2.32		4.20	5.85
4462	4" x 8"		300	.053		3.88	2.70		6.58	8.65
4481	4" x 10"		260	.062		5.60	3.12		8.72	11.20
9000	Minimum labor/equipment charge	1 Carp	4	2	Job		101		101	165

06 11 10 – Framing with Dimensional, Engineered or Composite Lumber

06 11 10.34 Sleepers

		Crew	Daily Output	Labor-Hours	Unit	Material	2018 Bare Costs Labor	Equipment	Total	Total Incl O&P
0010	**SLEEPERS**									
0100	On concrete, treated, 1" x 2"	2 Carp	2350	.007	L.F.	.30	.35		.65	.89
0150	1" x 3"		2000	.008		.49	.41		.90	1.20
0200	2" x 4"		1500	.011		.63	.54		1.17	1.57
0250	2" x 6"		1300	.012		.82	.62		1.44	1.91
9000	Minimum labor/equipment charge	1 Carp	4	2	Job		101		101	165

06 11 10.36 Soffit and Canopy Framing

		Crew	Daily Output	Labor-Hours	Unit	Material	2018 Bare Costs Labor	Equipment	Total	Total Incl O&P
0010	**SOFFIT AND CANOPY FRAMING**									
1002	Canopy or soffit framing, 1" x 4"	2 Carp	900	.018	L.F.	.51	.90		1.41	2.02
1021	1" x 6"		850	.019		.78	.95		1.73	2.41
1042	1" x 8"		750	.021		1.30	1.08		2.38	3.19
1102	2" x 4"		620	.026		.42	1.31		1.73	2.58
1121	2" x 6"		560	.029		.64	1.45		2.09	3.06
1142	2" x 8"		500	.032		.89	1.62		2.51	3.61
1202	3" x 4"		500	.032		1.20	1.62		2.82	3.95
1221	3" x 6"		400	.040		2	2.03		4.03	5.50
1242	3" x 10"		300	.053		3.62	2.70		6.32	8.35
9000	Minimum labor/equipment charge	1 Carp	4	2	Job		101		101	165

06 11 10.38 Treated Lumber Framing Material

		Crew	Daily Output	Labor-Hours	Unit	Material	2018 Bare Costs Labor	Equipment	Total	Total Incl O&P
0010	**TREATED LUMBER FRAMING MATERIAL**									
0100	2" x 4"				M.B.F.	870			870	960
0110	2" x 6"					725			725	800
0120	2" x 8"					830			830	915
0130	2" x 10"					830			830	915
0140	2" x 12"					1,000			1,000	1,100
0200	4" x 4"					870			870	955
0210	4" x 6"					940			940	1,025
0220	4" x 8"					1,450			1,450	1,600

06 11 10.40 Wall Framing

		Crew	Daily Output	Labor-Hours	Unit	Material	2018 Bare Costs Labor	Equipment	Total	Total Incl O&P
0010	**WALL FRAMING** R061110-30									
0100	Door buck, studs, header, access, 8' high, 2" x 4" wall, 3' wide	1 Carp	32	.250	Ea.	17.55	12.70		30.25	40
0110	4' wide		32	.250		18.85	12.70		31.55	41
0120	5' wide		32	.250		22.50	12.70		35.20	45.50
0130	6' wide		32	.250		24.50	12.70		37.20	47.50
0140	8' wide		30	.267		36.50	13.50		50	62
0150	10' wide		30	.267		50	13.50		63.50	77
0160	12' wide		30	.267		68.50	13.50		82	97.50
0170	2" x 6" wall, 3' wide		32	.250		25	12.70		37.70	47.50
0180	4' wide		32	.250		26	12.70		38.70	49
0190	5' wide		32	.250		30	12.70		42.70	53.50
0200	6' wide		32	.250		31.50	12.70		44.20	55.50
0210	8' wide		30	.267		43.50	13.50		57	70
0220	10' wide		30	.267		57	13.50		70.50	85
0230	12' wide		30	.267		76	13.50		89.50	106
0240	Window buck, studs, header & access, 8' high 2" x 4" wall, 2' wide		24	.333		18.60	16.90		35.50	48
0250	3' wide		24	.333		21.50	16.90		38.40	51.50
0260	4' wide		24	.333		24	16.90		40.90	53.50
0270	5' wide		24	.333		27.50	16.90		44.40	58
0280	6' wide		24	.333		31	16.90		47.90	61.50
0290	7' wide		24	.333		41	16.90		57.90	72.50
0300	8' wide		22	.364		45.50	18.45		63.95	80

144

For customer support on your Commercial Renovation Costs with RSMeans data, call 800.448.8182.

06 11 10 – Framing with Dimensional, Engineered or Composite Lumber

06 11 10.40 Wall Framing		Crew	Daily Output	Labor-Hours	Unit	Material	2018 Bare Costs Labor	2018 Bare Costs Equipment	Total	Total Incl O&P
0310	10' wide	1 Carp	22	.364	Ea.	60	18.45		78.45	96
0320	12' wide		22	.364		81	18.45		99.45	119
0330	2" x 6" wall, 2' wide		24	.333		27.50	16.90		44.40	58
0340	3' wide		24	.333		31	16.90		47.90	61.50
0350	4' wide		24	.333		33.50	16.90		50.40	64.50
0360	5' wide		24	.333		37.50	16.90		54.40	69
0370	6' wide		24	.333		41.50	16.90		58.40	73
0380	7' wide		24	.333		52.50	16.90		69.40	85
0390	8' wide		22	.364		57	18.45		75.45	93
0400	10' wide		22	.364		72.50	18.45		90.95	110
0410	12' wide		22	.364		95	18.45		113.45	135
2002	Headers over openings, 2" x 6"	2 Carp	360	.044	L.F.	.64	2.25		2.89	4.37
2007	2" x 6", pneumatic nailed		432	.037		.65	1.88		2.53	3.77
2052	2" x 8"		340	.047		.89	2.39		3.28	4.85
2057	2" x 8", pneumatic nailed		408	.039		.90	1.99		2.89	4.22
2101	2" x 10"		320	.050		1.41	2.54		3.95	5.65
2106	2" x 10", pneumatic nailed		384	.042		1.42	2.11		3.53	4.99
2152	2" x 12"		300	.053		1.79	2.70		4.49	6.35
2157	2" x 12", pneumatic nailed		360	.044		1.81	2.25		4.06	5.65
2180	4" x 8"		260	.062		4.15	3.12		7.27	9.60
2185	4" x 8", pneumatic nailed		312	.051		4.17	2.60		6.77	8.80
2191	4" x 10"		240	.067		4.96	3.38		8.34	10.95
2196	4" x 10", pneumatic nailed		288	.056		4.99	2.82		7.81	10.05
2202	4" x 12"		190	.084		5.50	4.27		9.77	13
2207	4" x 12", pneumatic nailed		228	.070		5.50	3.56		9.06	11.85
2241	6" x 10"		165	.097		6.90	4.92		11.82	15.60
2246	6" x 10", pneumatic nailed		198	.081		6.95	4.10		11.05	14.30
2251	6" x 12"		140	.114		8.70	5.80		14.50	18.95
2256	6" x 12", pneumatic nailed		168	.095		8.75	4.83		13.58	17.45
5002	Plates, untreated, 2" x 3"		850	.019		.42	.95		1.37	2.02
5007	2" x 3", pneumatic nailed		1020	.016		.43	.80		1.23	1.76
5022	2" x 4"		800	.020		.42	1.01		1.43	2.11
5027	2" x 4", pneumatic nailed		960	.017		.42	.85		1.27	1.84
5041	2" x 6"		750	.021		.64	1.08		1.72	2.47
5046	2" x 6", pneumatic nailed		900	.018		.65	.90		1.55	2.18
5122	Studs, 8' high wall, 2" x 3"		1200	.013		.42	.68		1.10	1.57
5127	2" x 3", pneumatic nailed		1440	.011		.43	.56		.99	1.38
5142	2" x 4"		1100	.015		.41	.74		1.15	1.65
5147	2" x 4", pneumatic nailed		1320	.012		.42	.61		1.03	1.47
5162	2" x 6"		1000	.016		.64	.81		1.45	2.03
5167	2" x 6", pneumatic nailed		1200	.013		.65	.68		1.33	1.82
5182	3" x 4"		800	.020		1.20	1.01		2.21	2.97
5187	3" x 4", pneumatic nailed		960	.017		1.21	.85		2.06	2.70
8200	For 12' high walls, deduct						5%			
8220	For stub wall, 6' high, add						20%			
8240	3' high, add						40%			
8250	For second story & above, add						5%			
8300	For dormer & gable, add						15%			
9000	Minimum labor/equipment charge	1 Carp	4	2	Job		101		101	165

06 11 10.42 Furring

		Crew	Daily Output	Labor-Hours	Unit	Material	2018 Bare Costs Labor	Equipment	Total	Total Incl O&P
0010	**FURRING**									
0012	Wood strips, 1" x 2", on walls, on wood	1 Carp	550	.015	L.F.	.27	.74		1.01	1.49
0015	On wood, pneumatic nailed		710	.011		.27	.57		.84	1.22
0300	On masonry		495	.016		.29	.82		1.11	1.65
0400	On concrete		260	.031		.29	1.56		1.85	2.85
0600	1" x 3", on walls, on wood		550	.015		.43	.74		1.17	1.67
0605	On wood, pneumatic nailed		710	.011		.43	.57		1	1.40
0700	On masonry		495	.016		.46	.82		1.28	1.84
0800	On concrete		260	.031		.46	1.56		2.02	3.04
0850	On ceilings, on wood		350	.023		.43	1.16		1.59	2.35
0855	On wood, pneumatic nailed		450	.018		.43	.90		1.33	1.93
0900	On masonry		320	.025		.46	1.27		1.73	2.57
0950	On concrete		210	.038	↓	.46	1.93		2.39	3.65
9000	Minimum labor/equipment charge	↓	4	2	Job		101		101	165

06 11 10.44 Grounds

		Crew	Daily Output	Labor-Hours	Unit	Material	2018 Bare Costs Labor	Equipment	Total	Total Incl O&P
0010	**GROUNDS**									
0020	For casework, 1" x 2" wood strips, on wood	1 Carp	330	.024	L.F.	.27	1.23		1.50	2.29
0100	On masonry		285	.028		.29	1.42		1.71	2.63
0200	On concrete		250	.032		.29	1.62		1.91	2.95
0400	For plaster, 3/4" deep, on wood		450	.018		.27	.90		1.17	1.75
0500	On masonry		225	.036		.29	1.80		2.09	3.25
0600	On concrete		175	.046		.29	2.32		2.61	4.08
0700	On metal lath		200	.040	↓	.29	2.03	·	2.32	3.61
9000	Minimum labor/equipment charge	↓	4	2	Job		101		101	165

06 12 Structural Panels
06 12 10 – Structural Insulated Panels
06 12 10.10 OSB Faced Panels

			Crew	Daily Output	Labor-Hours	Unit	Material	2018 Bare Costs Labor	Equipment	Total	Total Incl O&P
0010	**OSB FACED PANELS**										
0100	Structural insul. panels, 7/16" OSB both faces, EPS insul., 3-5/8" T	G	F-3	2075	.019	S.F.	3.65	1	.24	4.89	5.90
0110	5-5/8" thick	G		1725	.023		4.10	1.20	.29	5.59	6.75
0120	7-3/8" thick	G		1425	.028		4.45	1.45	.35	6.25	7.65
0130	9-3/8" thick	G		1125	.036		4.75	1.84	.44	7.03	8.70
0140	7/16" OSB one face, EPS insul., 3-5/8" thick	G		2175	.018		3.75	.95	.23	4.93	5.90
0150	5-5/8" thick	G		1825	.022		4.35	1.14	.27	5.76	6.90
0160	7-3/8" thick	G		1525	.026		4.85	1.36	.33	6.54	7.90
0170	9-3/8" thick	G		1225	.033		5.35	1.69	.40	7.44	9.10
0190	7/16" OSB - 1/2" GWB faces, EPS insul., 3-5/8" T	G		2075	.019		3.45	1	.24	4.69	5.65
0200	5-5/8" thick	G		1725	.023		4.10	1.20	.29	5.59	6.75
0210	7-3/8" thick	G		1425	.028		4.65	1.45	.35	6.45	7.85
0220	9-3/8" thick	G		1125	.036		5.25	1.84	.44	7.53	9.25
0240	7/16" OSB - 1/2" MRGWB faces, EPS insul., 3-5/8" T	G		2075	.019		3.55	1	.24	4.79	5.80
0250	5-5/8" thick	G		1725	.023		4.25	1.20	.29	5.74	6.95
0260	7-3/8" thick	G		1425	.028		4.65	1.45	.35	6.45	7.85
0270	9-3/8" thick	G	↓	1125	.036		5.35	1.84	.44	7.63	9.35
0300	For 1/2" GWB added to OSB skin, add	G					1.40			1.40	1.54
0310	For 1/2" MRGWB added to OSB skin, add	G					1.40			1.40	1.54
0320	For one T1-11 skin, add to OSB-OSB	G					1.95			1.95	2.15
0330	For one 19/32" CDX skin, add to OSB-OSB	G				↓	1.50			1.50	1.65
0500	Structural insulated panel, 7/16" OSB both sides, straw core										

06 12 Structural Panels

06 12 10 – Structural Insulated Panels

06 12 10.10 OSB Faced Panels

		Crew	Daily Output	Labor-Hours	Unit	Material	2018 Bare Costs Labor	Equipment	Total	Total Incl O&P
0510	4-3/8" T, walls (w/sill, splines, plates)	G F-6	2400	.017	S.F.	7.55	.79	.21	8.55	9.80
0520	Floors (w/splines)	G	2400	.017		7.55	.79	.21	8.55	9.80
0530	Roof (w/splines)	G	2400	.017		7.55	.79	.21	8.55	9.80
0550	7-7/8" T, walls (w/sill, splines, plates)	G	2400	.017		11.40	.79	.21	12.40	14.05
0560	Floors (w/splines)	G	2400	.017		11.40	.79	.21	12.40	14.05
0570	Roof (w/splines)	G	2400	.017		11.40	.79	.21	12.40	14.05

06 12 19 – Composite Shearwall Panels

06 12 19.10 Steel and Wood Composite Shearwall Panels

		Crew	Daily Output	Labor-Hours	Unit	Material	2018 Bare Costs Labor	Equipment	Total	Total Incl O&P
0010	**STEEL & WOOD COMPOSITE SHEARWALL PANELS**									
0020	Anchor bolts, 36" long (must be placed in wet concrete)	1 Carp	150	.053	Ea.	37	2.70		39.70	45
0030	On concrete, 2" x 4" & 2" x 6" walls, 7'-10' high, 360 lb. shear, 12" wide	2 Carp	8	2		460	101		561	670
0040	715 lb. shear, 15" wide		8	2		510	101		611	725
0050	1860 lb. shear, 18" wide		8	2		525	101		626	745
0060	2780 lb. shear, 21" wide		8	2		560	101		661	780
0070	3790 lb. shear, 24" wide		8	2		640	101		741	865
0080	2" x 6" walls, 11'-13' high, 1180 lb. shear, 18" wide		6	2.667		645	135		780	930
0090	1555 lb. shear, 21" wide		6	2.667		720	135		855	1,000
0100	2280 lb. shear, 24" wide		6	2.667		805	135		940	1,100
0110	For installing above on wood floor frame, add									
0120	Coupler nuts, threaded rods, bolts, shear transfer plate kit	1 Carp	16	.500	Ea.	65	25.50		90.50	113
0130	Framing anchors, angle (2 required)	"	96	.083	"	2.43	4.22		6.65	9.50
0140	For blocking see Section 06 11 10.02									
0150	For installing above, first floor to second floor, wood floor frame, add									
0160	Add stack option to first floor wall panel				Ea.	71.50			71.50	78.50
0170	Threaded rods, bolts, shear transfer plate kit	1 Carp	16	.500		75	25.50		100.50	124
0180	Framing anchors, angle (2 required)	"	96	.083		2.43	4.22		6.65	9.50
0190	For blocking see section 06 11 10.02									
0200	For installing stacked panels, balloon framing									
0210	Add stack option to first floor wall panel				Ea.	71.50			71.50	78.50
0220	Threaded rods, bolts kit	1 Carp	16	.500	"	45	25.50		70.50	90.50

06 13 Heavy Timber Construction

06 13 23 – Heavy Timber Framing

06 13 23.10 Heavy Framing

		Crew	Daily Output	Labor-Hours	Unit	Material	2018 Bare Costs Labor	Equipment	Total	Total Incl O&P
0010	**HEAVY FRAMING**									
0020	Beams, single 6" x 10"	2 Carp	1.10	14.545	M.B.F.	1,550	735		2,285	2,925
0100	Single 8" x 16"		1.20	13.333	"	1,925	675		2,600	3,225
0202	Built from 2" lumber, multiple 2" x 14"		900	.018	B.F.	.96	.90		1.86	2.52
0212	Built from 3" lumber, multiple 3" x 6"		700	.023		1.32	1.16		2.48	3.33
0222	Multiple 3" x 8"		800	.020		1.43	1.01		2.44	3.22
0232	Multiple 3" x 10"		900	.018		1.44	.90		2.34	3.04
0242	Multiple 3" x 12"		1000	.016		1.44	.81		2.25	2.90
0252	Built from 4" lumber, multiple 4" x 6"		800	.020		1.52	1.01		2.53	3.33
0262	Multiple 4" x 8"		900	.018		1.55	.90		2.45	3.16
0272	Multiple 4" x 10"		1000	.016		1.48	.81		2.29	2.95
0282	Multiple 4" x 12"		1100	.015		1.36	.74		2.10	2.70
0292	Columns, structural grade, 1500f, 4" x 4"		450	.036	L.F.	1.78	1.80		3.58	4.89
0302	6" x 6"		225	.071		4.17	3.61		7.78	10.45
0402	8" x 8"		240	.067		8	3.38		11.38	14.30
0502	10" x 10"		90	.178		13.50	9		22.50	29.50

06 13 Heavy Timber Construction

06 13 23 – Heavy Timber Framing

06 13 23.10 Heavy Framing

06 13 23.10 Heavy Framing	Crew	Daily Output	Labor-Hours	Unit	Material	2018 Bare Costs Labor	Equipment	Total	Total Incl O&P
0602	12" x 12"	2 Carp	70	.229	L.F.	18.55	11.60		30.15	39.50
0802	Floor planks, 2" thick, T&G, 2" x 6"		1050	.015	B.F.	1.61	.77		2.38	3.02
0902	2" x 10"		1100	.015		1.63	.74		2.37	2.99
1102	3" thick, 3" x 6"		1050	.015		1.62	.77		2.39	3.03
1202	3" x 10"		1100	.015		1.65	.74		2.39	3.02
1402	Girders, structural grade, 12" x 12"		800	.020		1.55	1.01		2.56	3.35
1502	10" x 16"		1000	.016		2.55	.81		3.36	4.13
2302	Roof purlins, 4" thick, structural grade		1050	.015		1.55	.77		2.32	2.95
9000	Minimum labor/equipment charge	1 Carp	2	4	Job		203		203	330

06 15 Wood Decking

06 15 16 – Wood Roof Decking

06 15 16.10 Solid Wood Roof Decking

06 15 16.10 Solid Wood Roof Decking	Crew	Daily Output	Labor-Hours	Unit	Material	2018 Bare Costs Labor	Equipment	Total	Total Incl O&P
0010	**SOLID WOOD ROOF DECKING**									
0350	Cedar planks, 2" thick	2 Carp	350	.046	S.F.	6.65	2.32		8.97	11.10
0400	3" thick		320	.050		10	2.54		12.54	15.10
0500	4" thick		250	.064		13.35	3.24		16.59	19.90
0550	6" thick		200	.080		20	4.06		24.06	28.50
0650	Douglas fir, 2" thick		350	.046		2.78	2.32		5.10	6.80
0700	3" thick		320	.050		4.17	2.54		6.71	8.70
0800	4" thick		250	.064		5.55	3.24		8.79	11.35
0850	6" thick		200	.080		8.35	4.06		12.41	15.75
0950	Hemlock, 2" thick		350	.046		2.83	2.32		5.15	6.85
1000	3" thick		320	.050		4.25	2.54		6.79	8.80
1100	4" thick		250	.064		5.65	3.24		8.89	11.50
1150	6" thick		200	.080		8.50	4.06		12.56	15.95
1250	Western white spruce, 2" thick		350	.046		1.81	2.32		4.13	5.75
1300	3" thick		320	.050		2.71	2.54		5.25	7.10
1400	4" thick		250	.064		3.61	3.24		6.85	9.25
1450	6" thick		200	.080		5.40	4.06		9.46	12.55
9000	Minimum labor/equipment charge	1 Carp	2	4	Job		203		203	330

06 15 23 – Laminated Wood Decking

06 15 23.10 Laminated Roof Deck

06 15 23.10 Laminated Roof Deck	Crew	Daily Output	Labor-Hours	Unit	Material	2018 Bare Costs Labor	Equipment	Total	Total Incl O&P
0010	**LAMINATED ROOF DECK**									
0020	Pine or hemlock, 3" thick	2 Carp	425	.038	S.F.	5.75	1.91		7.66	9.45
0100	4" thick		325	.049		7.50	2.50		10	12.30
0300	Cedar, 3" thick		425	.038		7.05	1.91		8.96	10.85
0400	4" thick		325	.049		9.45	2.50		11.95	14.40
0600	Fir, 3" thick		425	.038		6	1.91		7.91	9.70
0700	4" thick		325	.049		7.55	2.50		10.05	12.35
9000	Minimum labor/equipment charge	1 Carp	3	2.667	Job		135		135	219

06 16 13 – Insulating Sheathing

06 16 13.10 Insulating Sheathing

06 16 13.10 Insulating Sheathing		Crew	Daily Output	Labor-Hours	Unit	Material	2018 Bare Costs Labor	Equipment	Total	Total Incl O&P
0010	**INSULATING SHEATHING**									
0020	Expanded polystyrene, 1#/C.F. density, 3/4" thick, R2.89	G	2 Carp	1400	.011	S.F.	.36	.58		.94	1.34
0030	1" thick, R3.85	G		1300	.012		.43	.62		1.05	1.48
0040	2" thick, R7.69	G		1200	.013		.70	.68		1.38	1.87
0050	Extruded polystyrene, 15 psi compressive strength, 1" thick, R5	G		1300	.012		.71	.62		1.33	1.79
0060	2" thick, R10	G		1200	.013		.87	.68		1.55	2.06
0070	Polyisocyanurate, 2#/C.F. density, 3/4" thick	G		1400	.011		.58	.58		1.16	1.58
0080	1" thick	G		1300	.012		.59	.62		1.21	1.66
0090	1-1/2" thick	G		1250	.013		.74	.65		1.39	1.86
0100	2" thick	G		1200	.013		.90	.68		1.58	2.09

06 16 23 – Subflooring

06 16 23.10 Subfloor

06 16 23.10 Subfloor			Crew	Daily Output	Labor-Hours	Unit	Material	2018 Bare Costs Labor	Equipment	Total	Total Incl O&P
0010	**SUBFLOOR**	R061636-20									
0011	Plywood, CDX, 1/2" thick		2 Carp	1500	.011	SF Flr.	.62	.54		1.16	1.56
0015	Pneumatic nailed			1860	.009		.62	.44		1.06	1.39
0102	5/8" thick			1350	.012		.77	.60		1.37	1.82
0107	Pneumatic nailed			1674	.010		.77	.48		1.25	1.63
0202	3/4" thick			1250	.013		.92	.65		1.57	2.07
0207	Pneumatic nailed			1550	.010		.92	.52		1.44	1.87
0302	1-1/8" thick, 2-4-1 including underlayment			1050	.015		2.07	.77		2.84	3.53
0440	With boards, 1" x 6", S4S, laid regular			900	.018		1.70	.90		2.60	3.32
0452	1" x 8", laid regular			1000	.016		2.07	.81		2.88	3.60
0462	Laid diagonal			850	.019		2.07	.95		3.02	3.83
0502	1" x 10", laid regular			1100	.015		2.12	.74		2.86	3.53
0602	Laid diagonal			900	.018		2.12	.90		3.02	3.79
8990	Subfloor adhesive, 3/8" bead		1 Carp	2300	.003	L.F.	.11	.18		.29	.41
9000	Minimum labor/equipment charge		"	4	2	Job		101		101	165

06 16 26 – Underlayment

06 16 26.10 Wood Product Underlayment

06 16 26.10 Wood Product Underlayment			Crew	Daily Output	Labor-Hours	Unit	Material	2018 Bare Costs Labor	Equipment	Total	Total Incl O&P
0010	**WOOD PRODUCT UNDERLAYMENT**	R061636-20									
0015	Plywood, underlayment grade, 1/4" thick		2 Carp	1500	.011	S.F.	.94	.54		1.48	1.91
0018	Pneumatic nailed			1860	.009		.94	.44		1.38	1.74
0030	3/8" thick			1500	.011		1.04	.54		1.58	2.02
0070	Pneumatic nailed			1860	.009		1.04	.44		1.48	1.85
0102	1/2" thick			1450	.011		1.23	.56		1.79	2.26
0107	Pneumatic nailed			1798	.009		1.23	.45		1.68	2.08
0202	5/8" thick			1400	.011		1.36	.58		1.94	2.44
0207	Pneumatic nailed			1736	.009		1.36	.47		1.83	2.26
0302	3/4" thick			1300	.012		1.47	.62		2.09	2.63
0306	Pneumatic nailed			1612	.010		1.47	.50		1.97	2.44
0502	Particle board, 3/8" thick			1500	.011		.41	.54		.95	1.33
0507	Pneumatic nailed			1860	.009		.41	.44		.85	1.16
0602	1/2" thick			1450	.011		.43	.56		.99	1.38
0607	Pneumatic nailed			1798	.009		.43	.45		.88	1.20
0802	5/8" thick			1400	.011		.55	.58		1.13	1.55
0807	Pneumatic nailed			1736	.009		.55	.47		1.02	1.37
0902	3/4" thick			1300	.012		.67	.62		1.29	1.75
0907	Pneumatic nailed			1612	.010		.67	.50		1.17	1.56
1100	Hardboard, underlayment grade, 4' x 4', .215" thick	G		1500	.011		.68	.54		1.22	1.63
9000	Minimum labor/equipment charge		1 Carp	4	2	Job		101		101	165

For customer support on your Commercial Renovation Costs with RSMeans data, call 800.448.8182.

149

06 16 Sheathing

06 16 33 – Wood Board Sheathing

06 16 33.10 Board Sheathing		Crew	Daily Output	Labor-Hours	Unit	Material	2018 Bare Costs Labor	Equipment	Total	Total Incl O&P
0009	**BOARD SHEATHING**									
0010	Roof, 1" x 6" boards, laid horizontal	2 Carp	725	.022	S.F.	1.70	1.12		2.82	3.68
0020	On steep roof		520	.031		1.70	1.56		3.26	4.39
0040	On dormers, hips, & valleys		480	.033		1.70	1.69		3.39	4.60
0050	Laid diagonal		650	.025		1.70	1.25		2.95	3.89
0070	1" x 8" boards, laid horizontal		875	.018		2.07	.93		3	3.79
0080	On steep roof		635	.025		2.07	1.28		3.35	4.35
0090	On dormers, hips, & valleys		580	.028		2.12	1.40		3.52	4.60
0100	Laid diagonal		725	.022		2.07	1.12		3.19	4.10
0110	Skip sheathing, 1" x 4", 7" OC	1 Carp	1200	.007		.63	.34		.97	1.25
0120	1" x 6", 9" OC		1450	.006		.78	.28		1.06	1.31
0180	T&G sheathing/decking, 1" x 6"		1000	.008		1.80	.41		2.21	2.64
0190	2" x 6"		1000	.008		3.86	.41		4.27	4.91
0200	Walls, 1" x 6" boards, laid regular	2 Carp	650	.025		1.70	1.25		2.95	3.89
0210	Laid diagonal		585	.027		1.70	1.39		3.09	4.11
0220	1" x 8" boards, laid regular		765	.021		2.07	1.06		3.13	4
0230	Laid diagonal		650	.025		2.07	1.25		3.32	4.31

06 16 36 – Wood Panel Product Sheathing

06 16 36.10 Sheathing

06 16 36.10 Sheathing		Crew	Daily Output	Labor-Hours	Unit	Material	2018 Bare Costs Labor	Equipment	Total	Total Incl O&P
0010	**SHEATHING** R061110-30									
0012	Plywood on roofs, CDX									
0032	5/16" thick	2 Carp	1600	.010	S.F.	.59	.51		1.10	1.46
0037	Pneumatic nailed R061636-20		1952	.008		.59	.42		1.01	1.31
0052	3/8" thick		1525	.010		.60	.53		1.13	1.52
0057	Pneumatic nailed		1860	.009		.60	.44		1.04	1.37
0102	1/2" thick		1400	.011		.62	.58		1.20	1.62
0103	Pneumatic nailed		1708	.009		.62	.48		1.10	1.45
0202	5/8" thick		1300	.012		.77	.62		1.39	1.85
0207	Pneumatic nailed		1586	.010		.77	.51		1.28	1.67
0302	3/4" thick		1200	.013		.92	.68		1.60	2.12
0307	Pneumatic nailed		1464	.011		.92	.55		1.47	1.92
0502	Plywood on walls, with exterior CDX, 3/8" thick		1200	.013		.60	.68		1.28	1.76
0507	Pneumatic nailed		1488	.011		.60	.55		1.15	1.54
0602	1/2" thick		1125	.014		.62	.72		1.34	1.85
0607	Pneumatic nailed		1395	.011		.62	.58		1.20	1.62
0702	5/8" thick		1050	.015		.77	.77		1.54	2.09
0707	Pneumatic nailed		1302	.012		.77	.62		1.39	1.85
0802	3/4" thick		975	.016		.92	.83		1.75	2.37
0807	Pneumatic nailed		1209	.013		.92	.67		1.59	2.11
1000	For shear wall construction, add						20%			
1200	For structural 1 exterior plywood, add				S.F.	10%				
3000	Wood fiber, regular, no vapor barrier, 1/2" thick	2 Carp	1200	.013		.64	.68		1.32	1.80
3100	5/8" thick		1200	.013		.70	.68		1.38	1.87
3300	No vapor barrier, in colors, 1/2" thick		1200	.013		.81	.68		1.49	1.99
3400	5/8" thick		1200	.013		.85	.68		1.53	2.04
3600	With vapor barrier one side, white, 1/2" thick		1200	.013		.63	.68		1.31	1.79
3700	Vapor barrier 2 sides, 1/2" thick		1200	.013		.84	.68		1.52	2.02
3800	Asphalt impregnated, 25/32" thick		1200	.013		.32	.68		1	1.45
3850	Intermediate, 1/2" thick		1200	.013		.32	.68		1	1.45
4500	Oriented strand board, on roof, 7/16" thick G		1460	.011		.48	.56		1.04	1.43
4505	Pneumatic nailed G		1780	.009		.48	.46		.94	1.27
4550	1/2" thick G		1400	.011		.48	.58		1.06	1.47

06 16 Sheathing

06 16 36 – Wood Panel Product Sheathing

06 16 36.10 Sheathing

		Crew	Daily Output	Labor-Hours	Unit	Material	2018 Bare Costs Labor	2018 Bare Costs Equipment	Total	Total Incl O&P
4555	Pneumatic nailed	G	2 Carp	1736	.009	S.F.	.48	.47		.95	1.29
4600	5/8" thick	G		1300	.012		.54	.62		1.16	1.60
4605	Pneumatic nailed	G		1586	.010		.54	.51		1.05	1.42
4610	On walls, 7/16" thick	G		1200	.013		.48	.68		1.16	1.63
4615	Pneumatic nailed	G		1488	.011		.48	.55		1.03	1.41
4620	1/2" thick	G		1195	.013		.48	.68		1.16	1.63
4625	Pneumatic nailed	G		1325	.012		.48	.61		1.09	1.52
4630	5/8" thick	G		1050	.015		.54	.77		1.31	1.84
4635	Pneumatic nailed	G		1302	.012		.54	.62		1.16	1.60
4700	Oriented strand board, factory laminated W.R. barrier, on roof, 1/2" thick	G		1400	.011		.75	.58		1.33	1.77
4705	Pneumatic nailed	G		1736	.009		.75	.47		1.22	1.59
4720	5/8" thick	G		1300	.012		.91	.62		1.53	2.01
4725	Pneumatic nailed	G		1586	.010		.91	.51		1.42	1.83
4730	5/8" thick, T&G	G		1150	.014		1.09	.71		1.80	2.34
4735	Pneumatic nailed, T&G	G		1400	.011		1.09	.58		1.67	2.14
4740	On walls, 7/16" thick	G		1200	.013		.69	.68		1.37	1.86
4745	Pneumatic nailed	G		1488	.011		.69	.55		1.24	1.64
4750	1/2" thick	G		1195	.013		.75	.68		1.43	1.93
4755	Pneumatic nailed	G		1325	.012		.75	.61		1.36	1.82
4800	Joint sealant tape, 3-1/2"			7600	.002	L.F.	.28	.11		.39	.48
4810	Joint sealant tape, 6"			7600	.002	"	.41	.11		.52	.62
9000	Minimum labor/equipment charge		1 Carp	2	4	Job		203		203	330

06 16 43 – Gypsum Sheathing

06 16 43.10 Gypsum Sheathing

		Crew	Daily Output	Labor-Hours	Unit	Material	2018 Bare Costs Labor	2018 Bare Costs Equipment	Total	Total Incl O&P
0010	**GYPSUM SHEATHING**										
0020	Gypsum, weatherproof, 1/2" thick		2 Carp	1125	.014	S.F.	.46	.72		1.18	1.68
0040	With embedded glass mats		"	1100	.015	"	.71	.74		1.45	1.98

06 17 Shop-Fabricated Structural Wood

06 17 33 – Wood I-Joists

06 17 33.10 Wood and Composite I-Joists

		Crew	Daily Output	Labor-Hours	Unit	Material	2018 Bare Costs Labor	2018 Bare Costs Equipment	Total	Total Incl O&P
0010	**WOOD AND COMPOSITE I-JOISTS**										
0100	Plywood webs, incl. bridging & blocking, panels 24" OC										
1200	15' to 24' span, 50 psf live load	F-5	2400	.013	SF Flr.	1.99	.68		2.67	3.30
1300	55 psf live load		2250	.014		2.31	.73		3.04	3.72
1400	24' to 30' span, 45 psf live load		2600	.012		2.90	.63		3.53	4.21
1500	55 psf live load		2400	.013		4.62	.68		5.30	6.20

06 17 53 – Shop-Fabricated Wood Trusses

06 17 53.10 Roof Trusses

		Crew	Daily Output	Labor-Hours	Unit	Material	2018 Bare Costs Labor	2018 Bare Costs Equipment	Total	Total Incl O&P
0010	**ROOF TRUSSES**										
0100	Fink (W) or King post type, 2'-0" OC										
0200	Metal plate connected, 4 in 12 slope										
0210	24' to 29' span	F-3	3000	.013	SF Flr.	1.68	.69	.17	2.54	3.14
0300	30' to 43' span		3000	.013		2.39	.69	.17	3.25	3.91
0400	44' to 60' span		3000	.013		2.39	.69	.17	3.25	3.92
0700	Glued and nailed, add						50%				

For customer support on your Commercial Renovation Costs with RSMeans data, call 800.448.8182.

151

06 18 13 – Glued-Laminated Beams

06 18 13.20 Laminated Framing	Crew	Daily Output	Labor-Hours	Unit	Material	2018 Bare Costs Labor	Equipment	Total	Total Incl O&P
0010 **LAMINATED FRAMING**									
0020 30 lb., short term live load, 15 lb. dead load									
0200 Straight roof beams, 20' clear span, beams 8' OC	F-3	2560	.016	SF Flr.	2.26	.81	.19	3.26	4.01
0300 Beams 16' OC		3200	.013		1.65	.65	.16	2.46	3.03
0500 40' clear span, beams 8' OC		3200	.013		4.29	.65	.16	5.10	5.95
0600 Beams 16' OC	↓	3840	.010		3.54	.54	.13	4.21	4.90
0800 60' clear span, beams 8' OC	F-4	2880	.017		7.35	.85	.35	8.55	9.85
0900 Beams 16' OC	"	3840	.013		5.50	.64	.26	6.40	7.35
1100 Tudor arches, 30' to 40' clear span, frames 8' OC	F-3	1680	.024		9.60	1.23	.30	11.13	12.85
1200 Frames 16' OC	"	2240	.018		7.50	.92	.22	8.64	10
1400 50' to 60' clear span, frames 8' OC	F-4	2200	.022		10.30	1.12	.45	11.87	13.65
1500 Frames 16' OC		2640	.018		8.80	.93	.38	10.11	11.60
1700 Radial arches, 60' clear span, frames 8' OC		1920	.025		9.65	1.28	.52	11.45	13.30
1800 Frames 16' OC		2880	.017		7.65	.85	.35	8.85	10.15
2000 100' clear span, frames 8' OC		1600	.030		9.95	1.54	.62	12.11	14.10
2100 Frames 16' OC		2400	.020		8.80	1.03	.41	10.24	11.75
2300 120' clear span, frames 8' OC		1440	.033		13.25	1.71	.69	15.65	18.05
2400 Frames 16' OC	↓	1920	.025		12.10	1.28	.52	13.90	15.95
2600 Bowstring trusses, 20' OC, 40' clear span	F-3	2400	.017		6	.86	.21	7.07	8.20
2700 60' clear span	F-4	3600	.013		5.40	.68	.28	6.36	7.35
2800 100' clear span		4000	.012		7.65	.62	.25	8.52	9.65
2900 120' clear span	↓	3600	.013		8.10	.68	.28	9.06	10.30
3000 For less than 1000 B.F., add					20%				
3100 For premium appearance, add to S.F. prices					5%				
3300 For industrial type, deduct					15%				
3500 For stain and varnish, add					5%				
3900 For 3/4" laminations, add to straight					25%				
4100 Add to curved					15%				
4300 Alternate pricing method: (use nominal footage of									
4310 components). Straight beams, camber less than 6"	F-3	3.50	11.429	M.B.F.	3,175	590	142	3,907	4,600
4400 Columns, including hardware		2	20		3,425	1,025	248	4,698	5,700
4600 Curved members, radius over 32'		2.50	16		3,500	830	198	4,528	5,400
4700 Radius 10' to 32'	↓	3	13.333		3,475	690	165	4,330	5,100
4900 For complicated shapes, add maximum					100%				
5100 For pressure treating, add to straight					35%				
5200 Add to curved					45%				
6000 Laminated veneer members, southern pine or western species									
6050 1-3/4" wide x 5-1/2" deep	2 Carp	480	.033	L.F.	3.47	1.69		5.16	6.55
6100 9-1/2" deep		480	.033		4.25	1.69		5.94	7.40
6150 14" deep		450	.036		7.35	1.80		9.15	11.05
6200 18" deep	↓	450	.036	↓	10.30	1.80		12.10	14.25
6300 Parallel strand members, southern pine or western species									
6350 1-3/4" wide x 9-1/4" deep	2 Carp	480	.033	L.F.	4.77	1.69		6.46	8
6400 11-1/4" deep		450	.036		4.94	1.80		6.74	8.40
6450 14" deep		400	.040		7.60	2.03		9.63	11.65
6500 3-1/2" wide x 9-1/4" deep		480	.033		16.10	1.69		17.79	20.50
6550 11-1/4" deep		450	.036		19.90	1.80		21.70	25
6600 14" deep		400	.040		22.50	2.03		24.53	28
6650 7" wide x 9-1/4" deep		450	.036		33.50	1.80		35.30	40
6700 11-1/4" deep		420	.038		42	1.93		43.93	49.50
6750 14" deep	↓	400	.040	↓	50	2.03		52.03	58.50
9000 Minimum labor/equipment charge	F-3	2.50	16	Job		830	198	1,028	1,550

06 22 13 – Standard Pattern Wood Trim

06 22 13.15 Moldings, Base	Crew	Daily Output	Labor-Hours	Unit	Material	Labor	2018 Bare Costs Equipment	Total	Total Incl O&P
0010 MOLDINGS, BASE									
5100 Classic profile, 5/8" x 5-1/2", finger jointed and primed	1 Carp	250	.032	L.F.	1.59	1.62		3.21	4.38
5105 Poplar		240	.033		1.62	1.69		3.31	4.52
5110 Red oak		220	.036		2.49	1.84		4.33	5.75
5115 Maple		220	.036		3.75	1.84		5.59	7.10
5120 Cherry		220	.036		4.45	1.84		6.29	7.90
5125 3/4" x 7-1/2", finger jointed and primed		250	.032		2	1.62		3.62	4.83
5130 Poplar		240	.033		2.59	1.69		4.28	5.60
5135 Red oak		220	.036		3.61	1.84		5.45	6.95
5140 Maple		220	.036		4.99	1.84		6.83	8.50
5145 Cherry		220	.036		5.95	1.84		7.79	9.55
5150 Modern profile, 5/8" x 3-1/2", finger jointed and primed		250	.032		.96	1.62		2.58	3.68
5155 Poplar		240	.033		1.03	1.69		2.72	3.87
5160 Red oak		220	.036		1.70	1.84		3.54	4.86
5165 Maple		220	.036		2.63	1.84		4.47	5.90
5170 Cherry		220	.036		2.84	1.84		4.68	6.10
5175 Ogee profile, 7/16" x 3", finger jointed and primed		250	.032		.66	1.62		2.28	3.35
5180 Poplar		240	.033		.67	1.69		2.36	3.48
5185 Red oak		220	.036		.95	1.84		2.79	4.03
5200 9/16" x 3-1/2", finger jointed and primed		250	.032		.66	1.62		2.28	3.35
5205 Pine		240	.033		1.17	1.69		2.86	4.03
5210 Red oak		220	.036		2.80	1.84		4.64	6.05
5215 9/16" x 4-1/2", red oak		220	.036		4.19	1.84		6.03	7.60
5220 5/8" x 3-1/2", finger jointed and primed		250	.032		.98	1.62		2.60	3.71
5225 Poplar		240	.033		1.03	1.69		2.72	3.87
5230 Red oak		220	.036		1.70	1.84		3.54	4.86
5235 Maple		220	.036		2.63	1.84		4.47	5.90
5240 Cherry		220	.036		2.84	1.84		4.68	6.10
5245 5/8" x 4", finger jointed and primed		250	.032		1.25	1.62		2.87	4
5250 Poplar		240	.033		1.33	1.69		3.02	4.20
5255 Red oak		220	.036		1.91	1.84		3.75	5.10
5260 Maple		220	.036		2.94	1.84		4.78	6.20
5265 Cherry		220	.036		3.08	1.84		4.92	6.40
5270 Rectangular profile, oak, 3/8" x 1-1/4"		260	.031		1.24	1.56		2.80	3.90
5275 1/2" x 2-1/2"		255	.031		2.35	1.59		3.94	5.15
5280 1/2" x 3-1/2"		250	.032		2.78	1.62		4.40	5.70
5285 1" x 6"		240	.033		4.14	1.69		5.83	7.30
5290 1" x 8"		240	.033		5.15	1.69		6.84	8.40
5295 Pine, 3/8" x 1-3/4"		260	.031		.50	1.56		2.06	3.08
5300 7/16" x 2-1/2"		255	.031		.78	1.59		2.37	3.44
5305 1" x 6"		240	.033		.72	1.69		2.41	3.53
5310 1" x 8"		240	.033		.93	1.69		2.62	3.76
5315 Shoe, 1/2" x 3/4", primed		260	.031		.51	1.56		2.07	3.09
5320 Pine		240	.033		.34	1.69		2.03	3.11
5325 Poplar		240	.033		.41	1.69		2.10	3.19
5330 Red oak		220	.036		.55	1.84		2.39	3.59
5335 Maple		220	.036		.73	1.84		2.57	3.79
5340 Cherry		220	.036		.83	1.84		2.67	3.90
5345 11/16" x 1-1/2", pine		240	.033		.73	1.69		2.42	3.54
5350 Caps, 11/16" x 1-3/8", pine		240	.033		.58	1.69		2.27	3.38
5355 3/4" x 1-3/4", finger jointed and primed		260	.031		.77	1.56		2.33	3.38
5360 Poplar		240	.033		1.04	1.69		2.73	3.88
5365 Red oak		220	.036		1.25	1.84		3.09	4.36

06 22 13 – Standard Pattern Wood Trim

06 22 13.15 Moldings, Base

		Crew	Daily Output	Labor-Hours	Unit	Material	2018 Bare Costs Labor	Equipment	Total	Total Incl O&P
5370	Maple	1 Carp	220	.036	L.F.	1.63	1.84		3.47	4.78
5375	Cherry		220	.036		3.30	1.84		5.14	6.60
5380	Combination base & shoe, 9/16" x 3-1/2" & 1/2" x 3/4", pine		125	.064		1.51	3.24		4.75	6.90
5385	Three piece oak, 6" high		80	.100		5.95	5.05		11	14.80
5390	Including 3/4" x 1" base shoe		70	.114		6.45	5.80		12.25	16.50
5395	Flooring cant strip, 3/4" x 3/4", pre-finished pine		260	.031		.44	1.56		2	3.01
5400	For pre-finished, stain and clear coat, add					.55			.55	.61
5405	Clear coat only, add					.45			.45	.50
9000	Minimum labor/equipment charge	1 Carp	4	2	Job		101		101	165

06 22 13.30 Moldings, Casings

		Crew	Daily Output	Labor-Hours	Unit	Material	2018 Bare Costs Labor	Equipment	Total	Total Incl O&P
0010	**MOLDINGS, CASINGS**									
0085	Apron, 9/16" x 2-1/2", pine	1 Carp	250	.032	L.F.	1.43	1.62		3.05	4.20
0090	5/8" x 2-1/2", pine		250	.032		1.71	1.62		3.33	4.51
0110	5/8" x 3-1/2", pine		220	.036		1.94	1.84		3.78	5.10
0300	Band, 11/16" x 1-1/8", pine		270	.030		.75	1.50		2.25	3.26
0310	11/16" x 1-1/2", finger jointed and primed		270	.030		.76	1.50		2.26	3.27
0320	Pine		270	.030		1	1.50		2.50	3.54
0330	11/16" x 1-3/4", finger jointed and primed		270	.030		.95	1.50		2.45	3.48
0350	Pine		270	.030		.95	1.50		2.45	3.48
0355	Beaded, 3/4" x 3-1/2", finger jointed and primed		220	.036		.98	1.84		2.82	4.07
0360	Poplar		220	.036		1.15	1.84		2.99	4.25
0365	Red oak		220	.036		1.70	1.84		3.54	4.86
0370	Maple		220	.036		2.63	1.84		4.47	5.90
0375	Cherry		220	.036		3.18	1.84		5.02	6.50
0380	3/4" x 4", finger jointed and primed		220	.036		1.22	1.84		3.06	4.33
0385	Poplar		220	.036		1.44	1.84		3.28	4.57
0390	Red oak		220	.036		2.02	1.84		3.86	5.20
0395	Maple		220	.036		2.65	1.84		4.49	5.90
0400	Cherry		220	.036		3.36	1.84		5.20	6.70
0405	3/4" x 5-1/2", finger jointed and primed		200	.040		1.35	2.03		3.38	4.77
0410	Poplar		200	.040		1.99	2.03		4.02	5.50
0415	Red oak		200	.040		2.93	2.03		4.96	6.50
0420	Maple		200	.040		3.85	2.03		5.88	7.50
0425	Cherry		200	.040		4.43	2.03		6.46	8.15
0430	Classic profile, 3/4" x 2-3/4", finger jointed and primed		250	.032		.86	1.62		2.48	3.57
0435	Poplar		250	.032		1.03	1.62		2.65	3.76
0440	Red oak		250	.032		1.51	1.62		3.13	4.29
0445	Maple		250	.032		2.16	1.62		3.78	5
0450	Cherry		250	.032		2.45	1.62		4.07	5.30
0455	Fluted, 3/4" x 3-1/2", poplar		220	.036		1.15	1.84		2.99	4.25
0460	Red oak		220	.036		1.70	1.84		3.54	4.86
0465	Maple		220	.036		2.63	1.84		4.47	5.90
0470	Cherry		220	.036		3.18	1.84		5.02	6.50
0475	3/4" x 4", poplar		220	.036		1.44	1.84		3.28	4.57
0480	Red oak		220	.036		2.02	1.84		3.86	5.20
0485	Maple		220	.036		2.65	1.84		4.49	5.90
0490	Cherry		220	.036		3.36	1.84		5.20	6.70
0495	3/4" x 5-1/2", poplar		200	.040		1.99	2.03		4.02	5.50
0500	Red oak		200	.040		2.93	2.03		4.96	6.50
0505	Maple		200	.040		3.85	2.03		5.88	7.50
0510	Cherry		200	.040		4.43	2.03		6.46	8.15
0515	3/4" x 7-1/2", poplar		190	.042		1.26	2.14		3.40	4.85

06 22 13 – Standard Pattern Wood Trim

06 22 13.30 Moldings, Casings

		Crew	Daily Output	Labor-Hours	Unit	Material	2018 Bare Costs Labor	Equipment	Total	Total Incl O&P
0520	Red oak	1 Carp	190	.042	L.F.	3.93	2.14		6.07	7.80
0525	Maple		190	.042		5.65	2.14		7.79	9.65
0530	Cherry		190	.042		6.75	2.14		8.89	10.90
0535	3/4" x 9-1/2", poplar		180	.044		4.12	2.25		6.37	8.20
0540	Red oak		180	.044		6.55	2.25		8.80	10.85
0545	Maple		180	.044		9	2.25		11.25	13.55
0550	Cherry		180	.044		9.80	2.25		12.05	14.45
0555	Modern profile, 9/16" x 2-1/4", poplar		250	.032		.78	1.62		2.40	3.49
0560	Red oak		250	.032		.81	1.62		2.43	3.52
0565	11/16" x 2-1/2", finger jointed & primed		250	.032		.84	1.62		2.46	3.55
0570	Pine		250	.032		1.34	1.62		2.96	4.10
0575	3/4" x 2-1/2", poplar		250	.032		.91	1.62		2.53	3.63
0580	Red oak		250	.032		1.23	1.62		2.85	3.98
0585	Maple		250	.032		1.86	1.62		3.48	4.67
0590	Cherry		250	.032		2.56	1.62		4.18	5.45
0595	Mullion, 5/16" x 2", pine		270	.030		.90	1.50		2.40	3.43
0600	9/16" x 2-1/2", finger jointed and primed		250	.032		.98	1.62		2.60	3.71
0605	Pine		250	.032		1.34	1.62		2.96	4.10
0610	Red oak		250	.032		3.28	1.62		4.90	6.25
0615	1-1/16" x 3-3/4", red oak		220	.036		6.95	1.84		8.79	10.60
0620	Ogee, 7/16" x 2-1/2", poplar		250	.032		.74	1.62		2.36	3.44
0625	Red oak		250	.032		.94	1.62		2.56	3.66
0630	9/16" x 2-1/4", finger jointed and primed		250	.032		.55	1.62		2.17	3.23
0635	Poplar		250	.032		.59	1.62		2.21	3.28
0640	Red oak		250	.032		.81	1.62		2.43	3.52
0645	11/16" x 2-1/2", finger jointed and primed		250	.032		.72	1.62		2.34	3.42
0700	Pine		250	.032		1.42	1.62		3.04	4.19
0701	Red oak		250	.032		3.25	1.62		4.87	6.20
0730	11/16" x 3-1/2", finger jointed and primed		220	.036		1.37	1.84		3.21	4.50
0750	Pine		220	.036		1.79	1.84		3.63	4.96
0755	3/4" x 2-1/2", finger jointed and primed		250	.032		.77	1.62		2.39	3.48
0760	Poplar		250	.032		.91	1.62		2.53	3.63
0765	Red oak		250	.032		1.23	1.62		2.85	3.98
0770	Maple		250	.032		1.86	1.62		3.48	4.67
0775	Cherry		250	.032		2.34	1.62		3.96	5.20
0780	3/4" x 3-1/2", finger jointed and primed		220	.036		.98	1.84		2.82	4.07
0785	Poplar		220	.036		1.15	1.84		2.99	4.25
0790	Red oak		220	.036		1.70	1.84		3.54	4.86
0795	Maple		220	.036		2.63	1.84		4.47	5.90
0800	Cherry		220	.036		3.18	1.84		5.02	6.50
4700	Square profile, 1" x 1", teak		215	.037		2.24	1.89		4.13	5.50
4800	Rectangular profile, 1" x 3", teak		200	.040		6.70	2.03		8.73	10.65
9000	Minimum labor/equipment charge		4	2	Job		101		101	165

06 22 13.35 Moldings, Ceilings

		Crew	Daily Output	Labor-Hours	Unit	Material	2018 Bare Costs Labor	Equipment	Total	Total Incl O&P
0010	**MOLDINGS, CEILINGS**									
0600	Bed, 9/16" x 1-3/4", pine	1 Carp	270	.030	L.F.	1.20	1.50		2.70	3.76
0650	9/16" x 2", pine		270	.030		1.19	1.50		2.69	3.75
0710	9/16" x 1-3/4", oak		270	.030		2.35	1.50		3.85	5
1200	Cornice, 9/16" x 1-3/4", pine		270	.030		1	1.50		2.50	3.54
1300	9/16" x 2-1/4", pine		265	.030		1.31	1.53		2.84	3.92
1350	Cove, 1/2" x 2-1/4", poplar		265	.030		1.07	1.53		2.60	3.65
1360	Red oak		265	.030		1.37	1.53		2.90	3.98

06 22 13.35 Moldings, Ceilings	Crew	Daily Output	Labor-Hours	Unit	Material	2018 Bare Costs Labor	Equipment	Total	Total Incl O&P
1370	Hard maple	1 Carp	265	.030	L.F.	1.64	1.53		3.17	4.28
1380	Cherry		265	.030		2.27	1.53		3.80	4.97
2400	9/16" x 1-3/4", pine		270	.030		1	1.50		2.50	3.54
2500	11/16" x 2-3/4", pine		265	.030		1.87	1.53		3.40	4.53
2510	Crown, 5/8" x 5/8", poplar		300	.027		.46	1.35		1.81	2.69
2520	Red oak		300	.027		.54	1.35		1.89	2.78
2530	Hard maple		300	.027		.73	1.35		2.08	2.99
2540	Cherry		300	.027		.77	1.35		2.12	3.04
2600	9/16" x 3-5/8", pine		250	.032		2.10	1.62		3.72	4.94
2700	11/16" x 4-1/4", pine		250	.032		2.97	1.62		4.59	5.90
2705	Oak		250	.032		6.30	1.62		7.92	9.60
2710	3/4" x 1-3/4", poplar		270	.030		.74	1.50		2.24	3.25
2720	Red oak		270	.030		1.09	1.50		2.59	3.64
2730	Hard maple		270	.030		1.44	1.50		2.94	4.02
2740	Cherry		270	.030		1.77	1.50		3.27	4.39
2750	3/4" x 2", poplar		270	.030		.97	1.50		2.47	3.51
2760	Red oak		270	.030		1.30	1.50		2.80	3.87
2770	Hard maple		270	.030		1.74	1.50		3.24	4.35
2780	Cherry		270	.030		1.94	1.50		3.44	4.57
2790	3/4" x 2-3/4", poplar		265	.030		1.05	1.53		2.58	3.63
2800	Red oak		265	.030		1.62	1.53		3.15	4.26
2810	Hard maple		265	.030		2.16	1.53		3.69	4.85
2820	Cherry		265	.030		2.48	1.53		4.01	5.20
2830	3/4" x 3-1/2", poplar		250	.032		1.34	1.62		2.96	4.10
2840	Red oak		250	.032		2.02	1.62		3.64	4.85
2850	Hard maple		250	.032		2.64	1.62		4.26	5.55
2860	Cherry		250	.032		3.09	1.62		4.71	6.05
2870	FJP poplar		250	.032		.99	1.62		2.61	3.72
2880	3/4" x 5", poplar		245	.033		1.99	1.66		3.65	4.88
2890	Red oak		245	.033		2.94	1.66		4.60	5.90
2900	Hard maple		245	.033		3.86	1.66		5.52	6.95
2910	Cherry		245	.033		4.45	1.66		6.11	7.60
2920	FJP poplar		245	.033		1.41	1.66		3.07	4.24
2930	3/4" x 6-1/4", poplar		240	.033		2.36	1.69		4.05	5.35
2940	Red oak		240	.033		3.57	1.69		5.26	6.65
2950	Hard maple		240	.033		4.66	1.69		6.35	7.90
2960	Cherry		240	.033		5.55	1.69		7.24	8.85
2970	7/8" x 8-3/4", poplar		220	.036		3.89	1.84		5.73	7.25
2980	Red oak		220	.036		6.05	1.84		7.89	9.65
2990	Hard maple		220	.036		8.40	1.84		10.24	12.20
3000	Cherry		220	.036		9.80	1.84		11.64	13.75
3010	1" x 7-1/4", poplar		220	.036		3.89	1.84		5.73	7.25
3020	Red oak		220	.036		6.05	1.84		7.89	9.65
3030	Hard maple		220	.036		8.40	1.84		10.24	12.20
3040	Cherry		220	.036		9.80	1.84		11.64	13.75
3050	1-1/16" x 4-1/4", poplar		250	.032		2.36	1.62		3.98	5.20
3060	Red oak		250	.032		2.78	1.62		4.40	5.70
3070	Hard maple		250	.032		3.67	1.62		5.29	6.65
3080	Cherry		250	.032		5.25	1.62		6.87	8.40
3090	Dentil crown, 3/4" x 5", poplar		250	.032		1.99	1.62		3.61	4.82
3100	Red oak		250	.032		2.94	1.62		4.56	5.85
3110	Hard maple		250	.032		3.86	1.62		5.48	6.85
3120	Cherry		250	.032		4.45	1.62		6.07	7.50

06 22 13 – Standard Pattern Wood Trim

06 22 13.35 Moldings, Ceilings

		Crew	Daily Output	Labor-Hours	Unit	Material	2018 Bare Costs Labor	Equipment	Total	Total Incl O&P
3130	Dentil piece for above, 1/2" x 1/2", poplar	1 Carp	300	.027	L.F.	2.82	1.35		4.17	5.30
3140	Red oak		300	.027		3.24	1.35		4.59	5.75
3150	Hard maple		300	.027		3.68	1.35		5.03	6.25
3160	Cherry		300	.027		4.01	1.35		5.36	6.60
9000	Minimum labor/equipment charge		4	2	Job		101		101	165

06 22 13.40 Moldings, Exterior

		Crew	Daily Output	Labor-Hours	Unit	Material	2018 Bare Costs Labor	Equipment	Total	Total Incl O&P
0010	**MOLDINGS, EXTERIOR**									
0100	Band board, cedar, rough sawn, 1" x 2"	1 Carp	300	.027	L.F.	.69	1.35		2.04	2.95
0110	1" x 3"		300	.027		1.02	1.35		2.37	3.32
0120	1" x 4"		250	.032		1.36	1.62		2.98	4.13
0130	1" x 6"		250	.032		2.05	1.62		3.67	4.88
0140	1" x 8"		225	.036		2.73	1.80		4.53	5.95
0150	1" x 10"		225	.036		3.39	1.80		5.19	6.65
0160	1" x 12"		200	.040		4.08	2.03		6.11	7.75
0240	STK, 1" x 2"		300	.027		.45	1.35		1.80	2.68
0250	1" x 3"		300	.027		.49	1.35		1.84	2.73
0260	1" x 4"		250	.032		.81	1.62		2.43	3.52
0270	1" x 6"		250	.032		1.32	1.62		2.94	4.08
0280	1" x 8"		225	.036		2.01	1.80		3.81	5.15
0290	1" x 10"		225	.036		2.77	1.80		4.57	5.95
0300	1" x 12"		200	.040		4.29	2.03		6.32	8
0310	Pine, #2, 1" x 2"		300	.027		.28	1.35		1.63	2.50
0320	1" x 3"		300	.027		.45	1.35		1.80	2.68
0330	1" x 4"		250	.032		.54	1.62		2.16	3.23
0340	1" x 6"		250	.032		.82	1.62		2.44	3.53
0350	1" x 8"		225	.036		1.33	1.80		3.13	4.39
0360	1" x 10"		225	.036		1.74	1.80		3.54	4.84
0370	1" x 12"		200	.040		2.13	2.03		4.16	5.65
0380	D & better, 1" x 2"		300	.027		.50	1.35		1.85	2.74
0390	1" x 3"		300	.027		.64	1.35		1.99	2.89
0400	1" x 4"		250	.032		.82	1.62		2.44	3.53
0410	1" x 6"		250	.032		1.07	1.62		2.69	3.80
0420	1" x 8"		225	.036		1.62	1.80		3.42	4.71
0430	1" x 10"		225	.036		2.24	1.80		4.04	5.40
0440	1" x 12"		200	.040		2.69	2.03		4.72	6.25
0450	Redwood, clear all heart, 1" x 2"		300	.027		.64	1.35		1.99	2.89
0460	1" x 3"		300	.027		.95	1.35		2.30	3.23
0470	1" x 4"		250	.032		1.20	1.62		2.82	3.95
0480	1" x 6"		252	.032		1.79	1.61		3.40	4.58
0490	1" x 8"		225	.036		2.37	1.80		4.17	5.55
0500	1" x 10"		225	.036		3.98	1.80		5.78	7.30
0510	1" x 12"		200	.040		4.77	2.03		6.80	8.55
0530	Corner board, cedar, rough sawn, 1" x 2"		225	.036		.69	1.80		2.49	3.69
0540	1" x 3"		225	.036		1.02	1.80		2.82	4.06
0550	1" x 4"		200	.040		1.36	2.03		3.39	4.79
0560	1" x 6"		200	.040		2.05	2.03		4.08	5.55
0570	1" x 8"		200	.040		2.73	2.03		4.76	6.30
0580	1" x 10"		175	.046		3.39	2.32		5.71	7.50
0590	1" x 12"		175	.046		4.08	2.32		6.40	8.25
0670	STK, 1" x 2"		225	.036		.45	1.80		2.25	3.42
0680	1" x 3"		225	.036		.49	1.80		2.29	3.47
0690	1" x 4"		200	.040		.79	2.03		2.82	4.16

For customer support on your Commercial Renovation Costs with RSMeans data, call 800.448.8182.

06 22 13.40 Moldings, Exterior

		Crew	Daily Output	Labor-Hours	Unit	Material	2018 Bare Costs Labor	Equipment	Total	Total Incl O&P
0700	1" x 6"	1 Carp	200	.040	L.F.	1.32	2.03		3.35	4.74
0710	1" x 8"		200	.040		2.01	2.03		4.04	5.50
0720	1" x 10"		175	.046		2.77	2.32		5.09	6.80
0730	1" x 12"		175	.046		4.29	2.32		6.61	8.45
0740	Pine, #2, 1" x 2"		225	.036		.28	1.80		2.08	3.24
0750	1" x 3"		225	.036		.45	1.80		2.25	3.42
0760	1" x 4"		200	.040		.54	2.03		2.57	3.89
0770	1" x 6"		200	.040		.82	2.03		2.85	4.19
0780	1" x 8"		200	.040		1.33	2.03		3.36	4.75
0790	1" x 10"		175	.046		1.74	2.32		4.06	5.65
0800	1" x 12"		175	.046		2.13	2.32		4.45	6.10
0810	D & better, 1" x 2"		225	.036		.50	1.80		2.30	3.48
0820	1" x 3"		225	.036		.64	1.80		2.44	3.63
0830	1" x 4"		200	.040		.82	2.03		2.85	4.19
0840	1" x 6"		200	.040		1.07	2.03		3.10	4.46
0850	1" x 8"		200	.040		1.62	2.03		3.65	5.05
0860	1" x 10"		175	.046		2.24	2.32		4.56	6.20
0870	1" x 12"		175	.046		2.69	2.32		5.01	6.70
0880	Redwood, clear all heart, 1" x 2"		225	.036		.64	1.80		2.44	3.63
0890	1" x 3"		225	.036		.95	1.80		2.75	3.97
0900	1" x 4"		200	.040		1.20	2.03		3.23	4.61
0910	1" x 6"		200	.040		1.79	2.03		3.82	5.25
0920	1" x 8"		200	.040		2.37	2.03		4.40	5.90
0930	1" x 10"		175	.046		3.98	2.32		6.30	8.15
0940	1" x 12"		175	.046		4.77	2.32		7.09	9
0950	Cornice board, cedar, rough sawn, 1" x 2"		330	.024		.69	1.23		1.92	2.76
0960	1" x 3"		290	.028		1.02	1.40		2.42	3.40
0970	1" x 4"		250	.032		1.36	1.62		2.98	4.13
0980	1" x 6"		250	.032		2.05	1.62		3.67	4.88
0990	1" x 8"		200	.040		2.73	2.03		4.76	6.30
1000	1" x 10"		180	.044		3.39	2.25		5.64	7.40
1010	1" x 12"		180	.044		4.08	2.25		6.33	8.15
1020	STK, 1" x 2"		330	.024		.45	1.23		1.68	2.49
1030	1" x 3"		290	.028		.49	1.40		1.89	2.81
1040	1" x 4"		250	.032		.81	1.62		2.43	3.52
1050	1" x 6"		250	.032		1.32	1.62		2.94	4.08
1060	1" x 8"		200	.040		2.01	2.03		4.04	5.50
1070	1" x 10"		180	.044		2.77	2.25		5.02	6.70
1080	1" x 12"		180	.044		4.29	2.25		6.54	8.35
1500	Pine, #2, 1" x 2"		330	.024		.28	1.23		1.51	2.31
1510	1" x 3"		290	.028		.30	1.40		1.70	2.60
1600	1" x 4"		250	.032		.54	1.62		2.16	3.23
1700	1" x 6"		250	.032		.82	1.62		2.44	3.53
1800	1" x 8"		200	.040		1.33	2.03		3.36	4.75
1900	1" x 10"		180	.044		1.74	2.25		3.99	5.55
2000	1" x 12"		180	.044		2.13	2.25		4.38	6
2020	D & better, 1" x 2"		330	.024		.50	1.23		1.73	2.55
2030	1" x 3"		290	.028		.64	1.40		2.04	2.97
2040	1" x 4"		250	.032		.82	1.62		2.44	3.53
2050	1" x 6"		250	.032		1.07	1.62		2.69	3.80
2060	1" x 8"		200	.040		1.62	2.03		3.65	5.05
2070	1" x 10"		180	.044		2.24	2.25		4.49	6.10
2080	1" x 12"		180	.044		2.69	2.25		4.94	6.60

06 22 13 – Standard Pattern Wood Trim

06 22 13.40 Moldings, Exterior		Crew	Daily Output	Labor-Hours	Unit	Material	2018 Bare Costs Labor	2018 Bare Costs Equipment	Total	Total Incl O&P
2090	Redwood, clear all heart, 1" x 2"	1 Carp	330	.024	L.F.	.64	1.23		1.87	2.70
2100	1" x 3"		290	.028		.95	1.40		2.35	3.31
2110	1" x 4"		250	.032		1.20	1.62		2.82	3.95
2120	1" x 6"		250	.032		1.79	1.62		3.41	4.60
2130	1" x 8"		200	.040		2.37	2.03		4.40	5.90
2140	1" x 10"		180	.044		3.98	2.25		6.23	8.05
2150	1" x 12"		180	.044		4.77	2.25		7.02	8.90
2160	3 piece, 1" x 2", 1" x 4", 1" x 6", rough sawn cedar		80	.100		4.11	5.05		9.16	12.75
2180	STK cedar		80	.100		2.57	5.05		7.62	11.10
2200	#2 pine		80	.100		1.64	5.05		6.69	10.05
2210	D & better pine		80	.100		2.38	5.05		7.43	10.85
2220	Clear all heart redwood		80	.100		3.62	5.05		8.67	12.25
2230	1" x 8", 1" x 10", 1" x 12", rough sawn cedar		65	.123		10.15	6.25		16.40	21.50
2240	STK cedar		65	.123		9	6.25		15.25	20
2300	#2 pine		65	.123		5.15	6.25		11.40	15.85
2320	D & better pine		65	.123		6.50	6.25		12.75	17.30
2330	Clear all heart redwood		65	.123		11.05	6.25		17.30	22.50
2340	Door/window casing, cedar, rough sawn, 1" x 2"		275	.029		.69	1.47		2.16	3.15
2350	1" x 3"		275	.029		1.02	1.47		2.49	3.52
2360	1" x 4"		250	.032		1.36	1.62		2.98	4.13
2370	1" x 6"		250	.032		2.05	1.62		3.67	4.88
2380	1" x 8"		230	.035		2.73	1.76		4.49	5.85
2390	1" x 10"		230	.035		3.39	1.76		5.15	6.60
2395	1" x 12"		210	.038		4.08	1.93		6.01	7.60
2410	STK, 1" x 2"		275	.029		.45	1.47		1.92	2.88
2420	1" x 3"		275	.029		.49	1.47		1.96	2.93
2430	1" x 4"		250	.032		.81	1.62		2.43	3.52
2440	1" x 6"		250	.032		1.32	1.62		2.94	4.08
2450	1" x 8"		230	.035		2.01	1.76		3.77	5.05
2460	1" x 10"		230	.035		2.77	1.76		4.53	5.90
2470	1" x 12"		210	.038		4.29	1.93		6.22	7.85
2550	Pine, #2, 1" x 2"		275	.029		.28	1.47		1.75	2.70
2560	1" x 3"		275	.029		.45	1.47		1.92	2.88
2570	1" x 4"		250	.032		.54	1.62		2.16	3.23
2580	1" x 6"		250	.032		.82	1.62		2.44	3.53
2590	1" x 8"		230	.035		1.33	1.76		3.09	4.32
2600	1" x 10"		230	.035		1.74	1.76		3.50	4.77
2610	1" x 12"		210	.038		2.13	1.93		4.06	5.50
2620	Pine, D & better, 1" x 2"		275	.029		.50	1.47		1.97	2.94
2630	1" x 3"		275	.029		.64	1.47		2.11	3.09
2640	1" x 4"		250	.032		.82	1.62		2.44	3.53
2650	1" x 6"		250	.032		1.07	1.62		2.69	3.80
2660	1" x 8"		230	.035		1.62	1.76		3.38	4.64
2670	1" x 10"		230	.035		2.24	1.76		4	5.30
2680	1" x 12"		210	.038		2.69	1.93		4.62	6.10
2690	Redwood, clear all heart, 1" x 2"		275	.029		.64	1.47		2.11	3.09
2695	1" x 3"		275	.029		.95	1.47		2.42	3.43
2710	1" x 4"		250	.032		1.20	1.62		2.82	3.95
2715	1" x 6"		250	.032		1.79	1.62		3.41	4.60
2730	1" x 8"		230	.035		2.37	1.76		4.13	5.45
2740	1" x 10"		230	.035		3.98	1.76		5.74	7.25
2750	1" x 12"		210	.038		4.77	1.93		6.70	8.40
3500	Bellyband, pine, 11/16" x 4-1/4"		250	.032		2.82	1.62		4.44	5.75

06 22 13.40 Moldings, Exterior	Crew	Daily Output	Labor-Hours	Unit	Material	2018 Bare Costs Labor	Equipment	Total	Total Incl O&P
3610	Brickmold, pine, 1-1/4" x 2"	1 Carp	200	.040	L.F.	2.10	2.03		4.13	5.60
3620	FJP, 1-1/4" x 2"		200	.040		.97	2.03		3	4.36
5100	Fascia, cedar, rough sawn, 1" x 2"		275	.029		.69	1.47		2.16	3.15
5110	1" x 3"		275	.029		1.02	1.47		2.49	3.52
5120	1" x 4"		250	.032		1.36	1.62		2.98	4.13
5200	1" x 6"		250	.032		2.05	1.62		3.67	4.88
5300	1" x 8"		230	.035		2.73	1.76		4.49	5.85
5310	1" x 10"		230	.035		3.39	1.76		5.15	6.60
5320	1" x 12"		210	.038		4.08	1.93		6.01	7.60
5400	2" x 4"		220	.036		1.07	1.84		2.91	4.16
5500	2" x 6"		220	.036		1.59	1.84		3.43	4.74
5600	2" x 8"		200	.040		2.12	2.03		4.15	5.65
5700	2" x 10"		180	.044		2.64	2.25		4.89	6.55
5800	2" x 12"		170	.047		6.30	2.39		8.69	10.75
6120	STK, 1" x 2"		275	.029		.45	1.47		1.92	2.88
6130	1" x 3"		275	.029		.49	1.47		1.96	2.93
6140	1" x 4"		250	.032		.81	1.62		2.43	3.52
6150	1" x 6"		250	.032		1.32	1.62		2.94	4.08
6160	1" x 8"		230	.035		2.01	1.76		3.77	5.05
6170	1" x 10"		230	.035		2.77	1.76		4.53	5.90
6180	1" x 12"		210	.038		4.29	1.93		6.22	7.85
6185	2" x 2"		260	.031		.74	1.56		2.30	3.34
6190	Pine, #2, 1" x 2"		275	.029		.28	1.47		1.75	2.70
6200	1" x 3"		275	.029		.45	1.47		1.92	2.88
6210	1" x 4"		250	.032		.54	1.62		2.16	3.23
6220	1" x 6"		250	.032		.82	1.62		2.44	3.53
6230	1" x 8"		230	.035		1.33	1.76		3.09	4.32
6240	1" x 10"		230	.035		1.74	1.76		3.50	4.77
6250	1" x 12"		210	.038		2.13	1.93		4.06	5.50
6260	D & better, 1" x 2"		275	.029		.50	1.47		1.97	2.94
6270	1" x 3"		275	.029		.64	1.47		2.11	3.09
6280	1" x 4"		250	.032		.82	1.62		2.44	3.53
6290	1" x 6"		250	.032		1.07	1.62		2.69	3.80
6300	1" x 8"		230	.035		1.62	1.76		3.38	4.64
6310	1" x 10"		230	.035		2.24	1.76		4	5.30
6312	1" x 12"		210	.038		2.69	1.93		4.62	6.10
6330	Southern yellow, 1-1/4" x 5"		240	.033		2.88	1.69		4.57	5.90
6340	1-1/4" x 6"		240	.033		2.57	1.69		4.26	5.55
6350	1-1/4" x 8"		215	.037		3.65	1.89		5.54	7.05
6360	1-1/4" x 12"		190	.042		5.35	2.14		7.49	9.35
6370	Redwood, clear all heart, 1" x 2"		275	.029		.64	1.47		2.11	3.09
6380	1" x 3"		275	.029		1.20	1.47		2.67	3.71
6390	1" x 4"		250	.032		1.20	1.62		2.82	3.95
6400	1" x 6"		250	.032		1.79	1.62		3.41	4.60
6410	1" x 8"		230	.035		2.37	1.76		4.13	5.45
6420	1" x 10"		230	.035		3.98	1.76		5.74	7.25
6430	1" x 12"		210	.038		4.77	1.93		6.70	8.40
6440	1-1/4" x 5"		240	.033		1.87	1.69		3.56	4.80
6450	1-1/4" x 6"		240	.033		2.23	1.69		3.92	5.20
6460	1-1/4" x 8"		215	.037		3.54	1.89		5.43	6.95
6470	1-1/4" x 12"		190	.042		7.15	2.14		9.29	11.30
6580	Frieze, cedar, rough sawn, 1" x 2"		275	.029		.69	1.47		2.16	3.15
6590	1" x 3"		275	.029		1.02	1.47		2.49	3.52

06 22 13 – Standard Pattern Wood Trim

06 22 13.40 Moldings, Exterior	Crew	Daily Output	Labor-Hours	Unit	Material	2018 Bare Costs Labor	Equipment	Total	Total Incl O&P
6600	1" x 4"	1 Carp	250	.032	L.F.	1.36	1.62		2.98	4.13
6610	1" x 6"		250	.032		2.05	1.62		3.67	4.88
6620	1" x 8"		250	.032		2.73	1.62		4.35	5.65
6630	1" x 10"		225	.036		3.39	1.80		5.19	6.65
6640	1" x 12"		200	.040		4.04	2.03		6.07	7.75
6650	STK, 1" x 2"		275	.029		.45	1.47		1.92	2.88
6660	1" x 3"		275	.029		.49	1.47		1.96	2.93
6670	1" x 4"		250	.032		.81	1.62		2.43	3.52
6680	1" x 6"		250	.032		1.32	1.62		2.94	4.08
6690	1" x 8"		250	.032		2.01	1.62		3.63	4.84
6700	1" x 10"		225	.036		2.77	1.80		4.57	5.95
6710	1" x 12"		200	.040		4.29	2.03		6.32	8
6790	Pine, #2, 1" x 2"		275	.029		.28	1.47		1.75	2.70
6800	1" x 3"		275	.029		.45	1.47		1.92	2.88
6810	1" x 4"		250	.032		.54	1.62		2.16	3.23
6820	1" x 6"		250	.032		.82	1.62		2.44	3.53
6830	1" x 8"		250	.032		1.33	1.62		2.95	4.09
6840	1" x 10"		225	.036		1.74	1.80		3.54	4.84
6850	1" x 12"		200	.040		2.13	2.03		4.16	5.65
6860	D & better, 1" x 2"		275	.029		.50	1.47		1.97	2.94
6870	1" x 3"		275	.029		.64	1.47		2.11	3.09
6880	1" x 4"		250	.032		.82	1.62		2.44	3.53
6890	1" x 6"		250	.032		1.07	1.62		2.69	3.80
6900	1" x 8"		250	.032		1.62	1.62		3.24	4.41
6910	1" x 10"		225	.036		2.24	1.80		4.04	5.40
6920	1" x 12"		200	.040		2.69	2.03		4.72	6.25
6930	Redwood, clear all heart, 1" x 2"		275	.029		.64	1.47		2.11	3.09
6940	1" x 3"		275	.029		.95	1.47		2.42	3.43
6950	1" x 4"		250	.032		1.20	1.62		2.82	3.95
6960	1" x 6"		250	.032		1.79	1.62		3.41	4.60
6970	1" x 8"		250	.032		2.37	1.62		3.99	5.25
6980	1" x 10"		225	.036		3.98	1.80		5.78	7.30
6990	1" x 12"		200	.040		4.77	2.03		6.80	8.55
7000	Grounds, 1" x 1", cedar, rough sawn		300	.027		.35	1.35		1.70	2.58
7010	STK		300	.027		.28	1.35		1.63	2.49
7020	Pine, #2		300	.027		.18	1.35		1.53	2.39
7030	D & better		300	.027		.31	1.35		1.66	2.53
7050	Redwood		300	.027		.39	1.35		1.74	2.62
7060	Rake/verge board, cedar, rough sawn, 1" x 2"		225	.036		.69	1.80		2.49	3.69
7070	1" x 3"		225	.036		1.02	1.80		2.82	4.06
7080	1" x 4"		200	.040		1.36	2.03		3.39	4.79
7090	1" x 6"		200	.040		2.05	2.03		4.08	5.55
7100	1" x 8"		190	.042		2.73	2.14		4.87	6.45
7110	1" x 10"		190	.042		3.39	2.14		5.53	7.20
7120	1" x 12"		180	.044		4.08	2.25		6.33	8.15
7130	STK, 1" x 2"		225	.036		.45	1.80		2.25	3.42
7140	1" x 3"		225	.036		.49	1.80		2.29	3.47
7150	1" x 4"		200	.040		.81	2.03		2.84	4.18
7160	1" x 6"		200	.040		1.32	2.03		3.35	4.74
7170	1" x 8"		190	.042		2.01	2.14		4.15	5.70
7180	1" x 10"		190	.042		2.77	2.14		4.91	6.50
7190	1" x 12"		180	.044		4.29	2.25		6.54	8.35
7200	Pine, #2, 1" x 2"		225	.036		.28	1.80		2.08	3.24

06 22 13.40 Moldings, Exterior		Crew	Daily Output	Labor-Hours	Unit	Material	2018 Bare Costs Labor	Equipment	Total	Total Incl O&P
7210	1" x 3"	1 Carp	225	.036	L.F.	.45	1.80		2.25	3.42
7220	1" x 4"		200	.040		.54	2.03		2.57	3.89
7230	1" x 6"		200	.040		.82	2.03		2.85	4.19
7240	1" x 8"		190	.042		1.33	2.14		3.47	4.93
7250	1" x 10"		190	.042		1.74	2.14		3.88	5.40
7260	1" x 12"		180	.044		2.13	2.25		4.38	6
7340	D & better, 1" x 2"		225	.036		.50	1.80		2.30	3.48
7350	1" x 3"		225	.036		.64	1.80		2.44	3.63
7360	1" x 4"		200	.040		.82	2.03		2.85	4.19
7370	1" x 6"		200	.040		1.07	2.03		3.10	4.46
7380	1" x 8"		190	.042		1.62	2.14		3.76	5.25
7390	1" x 10"		190	.042		2.24	2.14		4.38	5.95
7400	1" x 12"		180	.044		2.69	2.25		4.94	6.60
7410	Redwood, clear all heart, 1" x 2"		225	.036		.64	1.80		2.44	3.63
7420	1" x 3"		225	.036		.95	1.80		2.75	3.97
7430	1" x 4"		200	.040		1.20	2.03		3.23	4.61
7440	1" x 6"		200	.040		1.79	2.03		3.82	5.25
7450	1" x 8"		190	.042		2.37	2.14		4.51	6.05
7460	1" x 10"		190	.042		3.98	2.14		6.12	7.85
7470	1" x 12"		180	.044		4.77	2.25		7.02	8.90
7480	2" x 4"		200	.040		2.30	2.03		4.33	5.80
7490	2" x 6"		182	.044		3.44	2.23		5.67	7.40
7500	2" x 8"		165	.048		4.57	2.46		7.03	9
7630	Soffit, cedar, rough sawn, 1" x 2"	2 Carp	440	.036		.69	1.84		2.53	3.75
7640	1" x 3"		440	.036		1.02	1.84		2.86	4.12
7650	1" x 4"		420	.038		1.36	1.93		3.29	4.64
7660	1" x 6"		420	.038		2.05	1.93		3.98	5.40
7670	1" x 8"		420	.038		2.73	1.93		4.66	6.15
7680	1" x 10"		400	.040		3.39	2.03		5.42	7
7690	1" x 12"		400	.040		4.08	2.03		6.11	7.75
7700	STK, 1" x 2"		440	.036		.45	1.84		2.29	3.48
7710	1" x 3"		440	.036		.49	1.84		2.33	3.53
7720	1" x 4"		420	.038		.81	1.93		2.74	4.03
7730	1" x 6"		420	.038		1.32	1.93		3.25	4.59
7740	1" x 8"		420	.038		2.01	1.93		3.94	5.35
7750	1" x 10"		400	.040		2.77	2.03		4.80	6.35
7760	1" x 12"		400	.040		4.29	2.03		6.32	8
7770	Pine, #2, 1" x 2"		440	.036		.28	1.84		2.12	3.30
7780	1" x 3"		440	.036		.45	1.84		2.29	3.48
7790	1" x 4"		420	.038		.54	1.93		2.47	3.74
7800	1" x 6"		420	.038		.82	1.93		2.75	4.04
7810	1" x 8"		420	.038		1.33	1.93		3.26	4.60
7820	1" x 10"		400	.040		1.74	2.03		3.77	5.20
7830	1" x 12"		400	.040		2.13	2.03		4.16	5.65
7840	D & better, 1" x 2"		440	.036		.50	1.84		2.34	3.54
7850	1" x 3"		440	.036		.64	1.84		2.48	3.69
7860	1" x 4"		420	.038		.82	1.93		2.75	4.04
7870	1" x 6"		420	.038		1.07	1.93		3	4.31
7880	1" x 8"		420	.038		1.62	1.93		3.55	4.92
7890	1" x 10"		400	.040		2.24	2.03		4.27	5.75
7900	1" x 12"		400	.040		2.69	2.03		4.72	6.25
7910	Redwood, clear all heart, 1" x 2"		440	.036		.64	1.84		2.48	3.69
7920	1" x 3"		440	.036		.95	1.84		2.79	4.03

06 22 13 – Standard Pattern Wood Trim

06 22 13.40 Moldings, Exterior

		Crew	Daily Output	Labor-Hours	Unit	Material	2018 Bare Costs Labor	Equipment	Total	Total Incl O&P
7930	1" x 4"	2 Carp	420	.038	L.F.	1.20	1.93		3.13	4.46
7940	1" x 6"		420	.038		1.79	1.93		3.72	5.10
7950	1" x 8"		420	.038		2.37	1.93		4.30	5.75
7960	1" x 10"		400	.040		3.98	2.03		6.01	7.65
7970	1" x 12"		400	.040		4.77	2.03		6.80	8.55
8050	Trim, crown molding, pine, 11/16" x 4-1/4"	1 Carp	250	.032		3.93	1.62		5.55	6.95
8060	Back band, 11/16" x 1-1/16"		250	.032		.90	1.62		2.52	3.62
8070	Insect screen frame stock, 1-1/16" x 1-3/4"		395	.020		2.23	1.03		3.26	4.12
8080	Dentils, 2-1/2" x 2-1/2" x 4", 6" OC		30	.267		1.22	13.50		14.72	23.50
8100	Fluted, 5-1/2"		165	.048		5.05	2.46		7.51	9.55
8110	Stucco bead, 1-3/8" x 1-5/8"		250	.032		2.33	1.62		3.95	5.20
9000	Minimum labor/equipment charge		4	2	Job		101		101	165

06 22 13.45 Moldings, Trim

		Crew	Daily Output	Labor-Hours	Unit	Material	2018 Bare Costs Labor	Equipment	Total	Total Incl O&P
0010	**MOLDINGS, TRIM**									
0200	Astragal, stock pine, 11/16" x 1-3/4"	1 Carp	255	.031	L.F.	1.24	1.59		2.83	3.94
0250	1-5/16" x 2-3/16"		240	.033		2.62	1.69		4.31	5.60
0800	Chair rail, stock pine, 5/8" x 2-1/2"		270	.030		1.54	1.50		3.04	4.13
0900	5/8" x 3-1/2"		240	.033		2.33	1.69		4.02	5.30
1000	Closet pole, stock pine, 1-1/8" diameter		200	.040		1.16	2.03		3.19	4.57
1100	Fir, 1-5/8" diameter		200	.040		2.05	2.03		4.08	5.55
3300	Half round, stock pine, 1/4" x 1/2"		270	.030		.25	1.50		1.75	2.72
3350	1/2" x 1"		255	.031		.64	1.59		2.23	3.28
3400	Handrail, fir, single piece, stock, hardware not included									
3450	1-1/2" x 1-3/4"	1 Carp	80	.100	L.F.	2.10	5.05		7.15	10.55
3470	Pine, 1-1/2" x 1-3/4"		80	.100		1.97	5.05		7.02	10.40
3500	1-1/2" x 2-1/2"		76	.105		2.57	5.35		7.92	11.50
3600	Lattice, stock pine, 1/4" x 1-1/8"		270	.030		.36	1.50		1.86	2.84
3700	1/4" x 1-3/4"		250	.032		.61	1.62		2.23	3.30
3800	Miscellaneous, custom, pine, 1" x 1"		270	.030		.44	1.50		1.94	2.92
3850	1" x 2"		265	.030		.88	1.53		2.41	3.45
3900	1" x 3"		240	.033		1.32	1.69		3.01	4.19
4100	Birch or oak, nominal 1" x 1"		240	.033		.39	1.69		2.08	3.16
4200	Nominal 1" x 3"		215	.037		1.16	1.89		3.05	4.33
4400	Walnut, nominal 1" x 1"		215	.037		.60	1.89		2.49	3.72
4500	Nominal 1" x 3"		200	.040		1.79	2.03		3.82	5.25
4700	Teak, nominal 1" x 1"		215	.037		2.79	1.89		4.68	6.10
4800	Nominal 1" x 3"		200	.040		8.35	2.03		10.38	12.50
4900	Quarter round, stock pine, 1/4" x 1/4"		275	.029		.25	1.47		1.72	2.67
4950	3/4" x 3/4"		255	.031		.52	1.59		2.11	3.15
5600	Wainscot moldings, 1-1/8" x 9/16", 2' high, minimum		76	.105	S.F.	12.35	5.35		17.70	22
5700	Maximum		65	.123	"	16.25	6.25		22.50	28
9000	Minimum labor/equipment charge		4	2	Job		101		101	165

06 22 13.50 Moldings, Window and Door

		Crew	Daily Output	Labor-Hours	Unit	Material	2018 Bare Costs Labor	Equipment	Total	Total Incl O&P
0010	**MOLDINGS, WINDOW AND DOOR**									
2800	Door moldings, stock, decorative, 1-1/8" wide, plain	1 Carp	17	.471	Set	48	24		72	91
2900	Detailed		17	.471	"	110	24		134	160
2960	Clear pine door jamb, no stops, 11/16" x 4-9/16"		240	.033	L.F.	4.93	1.69		6.62	8.15
3150	Door trim set, 1 head and 2 sides, pine, 2-1/2" wide		12	.667	Opng.	24	34		58	81.50
3170	3-1/2" wide		11	.727	"	30.50	37		67.50	93.50
3250	Glass beads, stock pine, 3/8" x 1/2"		275	.029	L.F.	.31	1.47		1.78	2.73
3270	3/8" x 7/8"		270	.030		.40	1.50		1.90	2.88
4850	Parting bead, stock pine, 3/8" x 3/4"		275	.029		.47	1.47		1.94	2.91

06 22 13 – Standard Pattern Wood Trim

06 22 13.50 Moldings, Window and Door

		Crew	Daily Output	Labor-Hours	Unit	Material	2018 Bare Costs Labor	Equipment	Total	Total Incl O&P
4870	1/2" x 3/4"	1 Carp	255	.031	L.F.	.41	1.59		2	3.03
5000	Stool caps, stock pine, 11/16" x 3-1/2"		200	.040		2.17	2.03		4.20	5.70
5100	1-1/16" x 3-1/4"		150	.053		3.40	2.70		6.10	8.15
5300	Threshold, oak, 3' long, inside, 5/8" x 3-5/8"		32	.250	Ea.	11.20	12.70		23.90	33
5400	Outside, 1-1/2" x 7-5/8"		16	.500	"	47	25.50		72.50	92.50
5900	Window trim sets, including casings, header, stops,									
5910	stool and apron, 2-1/2" wide, FJP	1 Carp	13	.615	Opng.	32	31		63	86
5950	Pine		10	.800		38	40.50		78.50	108
6000	Oak		6	1.333		77	67.50		144.50	195
9000	Minimum labor/equipment charge		4	2	Job		101		101	165

06 22 13.60 Moldings, Soffits

		Crew	Daily Output	Labor-Hours	Unit	Material	2018 Bare Costs Labor	Equipment	Total	Total Incl O&P
0010	**MOLDINGS, SOFFITS** R061110-30									
0200	Soffits, pine, 1" x 4"	2 Carp	420	.038	L.F.	.51	1.93		2.44	3.70
0210	1" x 6"		420	.038		.78	1.93		2.71	4
0220	1" x 8"		420	.038		1.29	1.93		3.22	4.56
0230	1" x 10"		400	.040		1.68	2.03		3.71	5.15
0240	1" x 12"		400	.040		2.07	2.03		4.10	5.55
0250	STK cedar, 1" x 4"		420	.038		.77	1.93		2.70	3.99
0260	1" x 6"		420	.038		1.28	1.93		3.21	4.55
0270	1" x 8"		420	.038		1.97	1.93		3.90	5.30
0280	1" x 10"		400	.040		2.71	2.03		4.74	6.25
0290	1" x 12"		400	.040		4.23	2.03		6.26	7.95
1000	Exterior AC plywood, 1/4" thick		400	.040	S.F.	1	2.03		3.03	4.39
1050	3/8" thick		400	.040		1.04	2.03		3.07	4.43
1100	1/2" thick		400	.040		1.23	2.03		3.26	4.64
1150	Polyvinyl chloride, white, solid	1 Carp	230	.035		2.17	1.76		3.93	5.25
1160	Perforated	"	230	.035		2.17	1.76		3.93	5.25
1170	Accessories, "J" channel 5/8"	2 Carp	700	.023	L.F.	.50	1.16		1.66	2.43
9000	Minimum labor/equipment charge	"	5	3.200	Job		162		162	263

06 25 Prefinished Paneling

06 25 13 – Prefinished Hardboard Paneling

06 25 13.10 Paneling, Hardboard

			Crew	Daily Output	Labor-Hours	Unit	Material	2018 Bare Costs Labor	Equipment	Total	Total Incl O&P
0010	**PANELING, HARDBOARD**										
0050	Not incl. furring or trim, hardboard, tempered, 1/8" thick	G	2 Carp	500	.032	S.F.	.45	1.62		2.07	3.13
0100	1/4" thick	G		500	.032		.64	1.62		2.26	3.33
0300	Tempered pegboard, 1/8" thick	G		500	.032		.44	1.62		2.06	3.11
0400	1/4" thick	G		500	.032		.70	1.62		2.32	3.40
0600	Untempered hardboard, natural finish, 1/8" thick	G		500	.032		.44	1.62		2.06	3.11
0700	1/4" thick	G		500	.032		.53	1.62		2.15	3.21
0900	Untempered pegboard, 1/8" thick	G		500	.032		.48	1.62		2.10	3.16
1000	1/4" thick	G		500	.032		.48	1.62		2.10	3.16
1200	Plastic faced hardboard, 1/8" thick	G		500	.032		.62	1.62		2.24	3.31
1300	1/4" thick	G		500	.032		.84	1.62		2.46	3.55
1500	Plastic faced pegboard, 1/8" thick	G		500	.032		.62	1.62		2.24	3.31
1600	1/4" thick	G		500	.032		.80	1.62		2.42	3.51
1800	Wood grained, plain or grooved, 1/8" thick	G		500	.032		.65	1.62		2.27	3.35
1900	1/4" thick	G		425	.038		1.42	1.91		3.33	4.66
2100	Moldings, wood grained MDF			500	.032	L.F.	.41	1.62		2.03	3.08
2200	Pine			425	.038	"	1.41	1.91		3.32	4.65
9000	Minimum labor/equipment charge		1 Carp	2	4	Job		203		203	330

06 25 Prefinished Paneling

06 25 16 – Prefinished Plywood Paneling

06 25 16.10 Paneling, Plywood

		Crew	Daily Output	Labor-Hours	Unit	Material	2018 Bare Costs Labor	Equipment	Total	Total Incl O&P
0010	**PANELING, PLYWOOD** R061636-20									
2400	Plywood, prefinished, 1/4" thick, 4' x 8' sheets									
2410	with vertical grooves. Birch faced, economy	2 Carp	500	.032	S.F.	1.47	1.62		3.09	4.25
2420	Average		420	.038		1.20	1.93		3.13	4.46
2430	Custom		350	.046		1.10	2.32		3.42	4.97
2600	Mahogany, African		400	.040		2.45	2.03		4.48	6
2700	Philippine (Lauan)		500	.032		.65	1.62		2.27	3.35
2900	Oak		500	.032		1.34	1.62		2.96	4.10
3000	Cherry		400	.040		2	2.03		4.03	5.50
3200	Rosewood		320	.050		3.10	2.54		5.64	7.55
3400	Teak		400	.040		3.12	2.03		5.15	6.70
3600	Chestnut		375	.043		5.30	2.16		7.46	9.30
3800	Pecan		400	.040		2.50	2.03		4.53	6.05
3900	Walnut, average		500	.032		2.47	1.62		4.09	5.35
3950	Custom		400	.040		2.46	2.03		4.49	6
4000	Plywood, prefinished, 3/4" thick, stock grades, economy		320	.050		1.44	2.54		3.98	5.70
4100	Average		224	.071		4.93	3.62		8.55	11.30
4300	Architectural grade, custom		224	.071		5.40	3.62		9.02	11.80
4400	Luxury		160	.100		5.30	5.05		10.35	14.05
4600	Plywood, "A" face, birch, VC, 1/2" thick, natural		450	.036		1.38	1.80		3.18	4.45
4700	Select		450	.036		1.95	1.80		3.75	5.10
4900	Veneer core, 3/4" thick, natural		320	.050		2.25	2.54		4.79	6.60
5000	Select		320	.050		2.53	2.54		5.07	6.90
5200	Lumber core, 3/4" thick, natural		320	.050		3.14	2.54		5.68	7.55
5500	Plywood, knotty pine, 1/4" thick, A2 grade		450	.036		1.53	1.80		3.33	4.61
5600	A3 grade		450	.036		2.09	1.80		3.89	5.25
5800	3/4" thick, veneer core, A2 grade		320	.050		2.31	2.54		4.85	6.65
5900	A3 grade		320	.050		2.39	2.54		4.93	6.75
6100	Aromatic cedar, 1/4" thick, plywood		400	.040		2.25	2.03		4.28	5.75
6200	1/4" thick, particle board	↓	400	.040	↓	1.15	2.03		3.18	4.56
9000	Minimum labor/equipment charge	1 Carp	2	4	Job		203		203	330

06 25 26 – Panel System

06 25 26.10 Panel Systems

		Crew	Daily Output	Labor-Hours	Unit	Material	2018 Bare Costs Labor	Equipment	Total	Total Incl O&P
0010	**PANEL SYSTEMS**									
0100	Raised panel, eng. wood core w/wood veneer, std., paint grade	2 Carp	300	.053	S.F.	12.15	2.70		14.85	17.75
0110	Oak veneer		300	.053		24	2.70		26.70	31
0120	Maple veneer		300	.053		30	2.70		32.70	37.50
0130	Cherry veneer		300	.053		37.50	2.70		40.20	46
0300	Class I fire rated, paint grade		300	.053		13.50	2.70		16.20	19.25
0310	Oak veneer		300	.053		30	2.70		32.70	37.50
0320	Maple veneer		300	.053		41	2.70		43.70	49.50
0330	Cherry veneer		300	.053		49	2.70		51.70	58.50
0510	Beadboard, 5/8" MDF, standard, primed		300	.053		9.35	2.70		12.05	14.70
0520	Oak veneer, unfinished		300	.053		14.70	2.70		17.40	20.50
0530	Maple veneer, unfinished		300	.053		15.80	2.70		18.50	22
0610	Rustic paneling, 5/8" MDF, standard, maple veneer, unfinished	↓	300	.053	↓	18.50	2.70		21.20	25

06 26 Board Paneling

06 26 13 – Profile Board Paneling

06 26 13.10 Paneling, Boards	Crew	Daily Output	Labor-Hours	Unit	Material	2018 Bare Costs Labor	Equipment	Total	Total Incl O&P
0010 **PANELING, BOARDS**									
6400 Wood board paneling, 3/4" thick, knotty pine	2 Carp	300	.053	S.F.	2.05	2.70		4.75	6.65
6500 Rough sawn cedar		300	.053		3.30	2.70		6	8
6700 Redwood, clear, 1" x 4" boards		300	.053		4.96	2.70		7.66	9.85
6900 Aromatic cedar, closet lining, boards		275	.058		2.42	2.95		5.37	7.45
9000 Minimum labor/equipment charge	1 Carp	2	4	Job		203		203	330

06 43 Wood Stairs and Railings

06 43 13 – Wood Stairs

06 43 13.20 Prefabricated Wood Stairs

	Crew	Daily Output	Labor-Hours	Unit	Material	2018 Bare Costs Labor	Equipment	Total	Total Incl O&P
0010 **PREFABRICATED WOOD STAIRS**									
0100 Box stairs, prefabricated, 3'-0" wide									
0110 Oak treads, up to 14 risers	2 Carp	39	.410	Riser	99.50	21		120.50	143
0600 With pine treads for carpet, up to 14 risers	"	39	.410	"	64	21		85	105
1100 For 4' wide stairs, add				Flight	25%				
1550 Stairs, prefabricated stair handrail with balusters	1 Carp	30	.267	L.F.	82	13.50		95.50	113
1700 Basement stairs, prefabricated, pine treads									
1710 Pine risers, 3' wide, up to 14 risers	2 Carp	52	.308	Riser	64	15.60		79.60	96
4000 Residential, wood, oak treads, prefabricated		1.50	10.667	Flight	1,300	540		1,840	2,300
4200 Built in place		.44	36.364	"	2,300	1,850		4,150	5,525
4400 Spiral, oak, 4'-6" diameter, unfinished, prefabricated,									
4500 incl. railing, 9' high	2 Carp	1.50	10.667	Flight	2,750	540		3,290	3,900
9000 Minimum labor/equipment charge	"	3	5.333	Job		270		270	440

06 43 13.40 Wood Stair Parts

	Crew	Daily Output	Labor-Hours	Unit	Material	2018 Bare Costs Labor	Equipment	Total	Total Incl O&P
0010 **WOOD STAIR PARTS**									
0020 Pin top balusters, 1-1/4", oak, 34"	1 Carp	96	.083	Ea.	5.20	4.22		9.42	12.55
0030 38"		96	.083		5.85	4.22		10.07	13.30
0040 42"		96	.083		6.35	4.22		10.57	13.85
0050 Poplar, 34"		96	.083		3.37	4.22		7.59	10.55
0060 38"		96	.083		3.99	4.22		8.21	11.25
0070 42"		96	.083		8.80	4.22		13.02	16.50
0080 Maple, 34"		96	.083		4.90	4.22		9.12	12.25
0090 38"		96	.083		5.60	4.22		9.82	13.05
0100 42"		96	.083		6.50	4.22		10.72	14
0130 Primed, 34"		96	.083		3.70	4.22		7.92	10.90
0140 38"		96	.083		4.25	4.22		8.47	11.55
0150 42"		96	.083		4.89	4.22		9.11	12.25
0180 Box top balusters, 1-1/4", oak, 34"		60	.133		8.75	6.75		15.50	20.50
0190 38"		60	.133		11.85	6.75		18.60	24
0200 42"		60	.133		13	6.75		19.75	25.50
0210 Poplar, 34"		60	.133		7.50	6.75		14.25	19.20
0220 38"		60	.133		8.10	6.75		14.85	19.85
0230 42"		60	.133		8.95	6.75		15.70	21
0240 Maple, 34"		60	.133		9.80	6.75		16.55	22
0250 38"		60	.133		10.75	6.75		17.50	23
0260 42"		60	.133		11.85	6.75		18.60	24
0290 Primed, 34"		60	.133		8.25	6.75		15	20
0300 38"		60	.133		9.50	6.75		16.25	21.50
0310 42"		60	.133		9.85	6.75		16.60	22
0340 Square balusters, cut from lineal stock, pine, 1-1/16" x 1-1/16"		180	.044	L.F.	1.31	2.25		3.56	5.10
0350 1-5/16" x 1-5/16"		180	.044		1.98	2.25		4.23	5.85

06 43 13.40 Wood Stair Parts	Crew	Daily Output	Labor-Hours	Unit	Material	2018 Bare Costs Labor	Equipment	Total	Total Incl O&P
0360	1-5/8" x 1-5/8"	1 Carp	180	.044	L.F.	2.68	2.25		4.93	6.60
0370	Turned newel, oak, 3-1/2" square, 48" high		8	1	Ea.	105	50.50		155.50	199
0380	62" high		8	1		101	50.50		151.50	194
0390	Poplar, 3-1/2" square, 48" high		8	1		46.50	50.50		97	134
0400	62" high		8	1		58	50.50		108.50	147
0410	Maple, 3-1/2" square, 48" high		8	1		65	50.50		115.50	154
0420	62" high		8	1		80	50.50		130.50	171
0430	Square newel, oak, 3-1/2" square, 48" high		8	1		55	50.50		105.50	143
0440	58" high		8	1		68.50	50.50		119	158
0450	Poplar, 3-1/2" square, 48" high		8	1		35.50	50.50		86	122
0460	58" high		8	1		43.50	50.50		94	131
0470	Maple, 3" square, 48" high		8	1		54	50.50		104.50	142
0480	58" high		8	1		68	50.50		118.50	158
0490	Railings, oak, economy		96	.083	L.F.	9.45	4.22		13.67	17.25
0500	Average		96	.083		13.50	4.22		17.72	21.50
0510	Custom		96	.083		17.95	4.22		22.17	26.50
0520	Maple, economy		96	.083		11.90	4.22		16.12	19.90
0530	Average		96	.083		13.50	4.22		17.72	21.50
0540	Custom		96	.083		20.50	4.22		24.72	29.50
0550	Oak, for bending rail, economy		48	.167		26	8.45		34.45	42
0560	Average		48	.167		29.50	8.45		37.95	46
0570	Custom		48	.167		33	8.45		41.45	50
0580	Maple, for bending rail, economy		48	.167		29	8.45		37.45	45
0590	Average		48	.167		31.50	8.45		39.95	48
0600	Custom		48	.167		35	8.45		43.45	51.50
0610	Risers, oak, 3/4" x 8", 36" long		80	.100	Ea.	13	5.05		18.05	22.50
0620	42" long		70	.114		15.15	5.80		20.95	26
0630	48" long		63	.127		17.30	6.45		23.75	29.50
0640	54" long		56	.143		19.50	7.25		26.75	33.50
0650	60" long		50	.160		21.50	8.10		29.60	37
0660	72" long		42	.190		26	9.65		35.65	44
0670	Poplar, 3/4" x 8", 36" long		80	.100		12.75	5.05		17.80	22.50
0680	42" long		71	.113		14.90	5.70		20.60	25.50
0690	48" long		63	.127		17	6.45		23.45	29
0700	54" long		56	.143		19.15	7.25		26.40	33
0710	60" long		50	.160		21.50	8.10		29.60	36.50
0720	72" long		42	.190		25.50	9.65		35.15	43.50
0730	Pine, 1" x 8", 36" long		80	.100		3.88	5.05		8.93	12.50
0740	42" long		70	.114		4.53	5.80		10.33	14.40
0750	48" long		63	.127		5.20	6.45		11.65	16.15
0760	54" long		56	.143		5.80	7.25		13.05	18.15
0770	60" long		50	.160		6.45	8.10		14.55	20.50
0780	72" long		42	.190		7.75	9.65		17.40	24.50
0790	Treads, oak, no returns, 1-1/32" x 11-1/2" x 36" long		32	.250		29.50	12.70		42.20	53
0800	42" long		32	.250		34.50	12.70		47.20	58.50
0810	48" long		32	.250		39.50	12.70		52.20	63.50
0820	54" long		32	.250		44	12.70		56.70	69
0830	60" long		32	.250		49	12.70		61.70	74.50
0840	72" long		32	.250		59	12.70		71.70	85.50
0850	Mitred return one end, 1-1/32" x 11-1/2" x 36" long		24	.333		39	16.90		55.90	70.50
0860	42" long		24	.333		45.50	16.90		62.40	77.50
0870	48" long		24	.333		52	16.90		68.90	84.50
0880	54" long		24	.333		58.50	16.90		75.40	92

06 43 Wood Stairs and Railings

06 43 13 – Wood Stairs

06 43 13.40 Wood Stair Parts

		Crew	Daily Output	Labor-Hours	Unit	Material	2018 Bare Costs Labor	Equipment	Total	Total Incl O&P
0890	60" long	1 Carp	24	.333	Ea.	65	16.90		81.90	99
0900	72" long		24	.333		78	16.90		94.90	113
0910	Mitred return two ends, 1-1/32" x 11-1/2" x 36" long		12	.667		49	34		83	109
0920	42" long		12	.667		57	34		91	118
0930	48" long		12	.667		65.50	34		99.50	127
0940	54" long		12	.667		73.50	34		107.50	136
0950	60" long		12	.667		81.50	34		115.50	145
0960	72" long		12	.667		98	34		132	163
0970	Starting step, oak, 48", bullnose		8	1		172	50.50		222.50	272
0980	Double end bullnose		8	1		254	50.50		304.50	360
1030	Skirt board, pine, 1" x 10"		55	.145	L.F.	1.68	7.35		9.03	13.80
1040	1" x 12"		52	.154	"	2.07	7.80		9.87	14.95
1050	Oak landing tread, 1-1/16" thick		54	.148	S.F.	7.95	7.50		15.45	21
1060	Oak cove molding		96	.083	L.F.	1	4.22		5.22	7.95
1070	Oak stringer molding		96	.083	"	4	4.22		8.22	11.25
1090	Rail bolt, 5/16" x 3-1/2"		48	.167	Ea.	2.75	8.45		11.20	16.75
1100	5/16" x 4-1/2"		48	.167		2.75	8.45		11.20	16.75
1120	Newel post anchor		16	.500		13	25.50		38.50	55.50
1130	Tapered plug, 1/2"		240	.033		1	1.69		2.69	3.84
1140	1"		240	.033		.99	1.69		2.68	3.83
9000	Minimum labor/equipment charge		3	2.667	Job		135		135	219

06 43 16 – Wood Railings

06 43 16.10 Wood Handrails and Railings

		Crew	Daily Output	Labor-Hours	Unit	Material	2018 Bare Costs Labor	Equipment	Total	Total Incl O&P
0010	**WOOD HANDRAILS AND RAILINGS**									
0020	Custom design, architectural grade, hardwood, plain	1 Carp	38	.211	L.F.	12.65	10.65		23.30	31.50
0100	Shaped		30	.267		53	13.50		66.50	80.50
0300	Stock interior railing with spindles 4" OC, 4' long		40	.200		46.50	10.15		56.65	68
0400	8' long		48	.167		46.50	8.45		54.95	65
9000	Minimum labor/equipment charge		3	2.667	Job		135		135	219

06 44 Ornamental Woodwork

06 44 19 – Wood Grilles

06 44 19.10 Grilles

		Crew	Daily Output	Labor-Hours	Unit	Material	2018 Bare Costs Labor	Equipment	Total	Total Incl O&P
0010	**GRILLES** and panels, hardwood, sanded									
0020	2' x 4' to 4' x 8', custom designs, unfinished, economy	1 Carp	38	.211	S.F.	58	10.65		68.65	81.50
0050	Average		30	.267		68.50	13.50		82	97.50
0100	Custom		19	.421		72	21.50		93.50	114
9000	Minimum labor/equipment charge		2	4	Job		203		203	330

06 44 33 – Wood Mantels

06 44 33.10 Fireplace Mantels

		Crew	Daily Output	Labor-Hours	Unit	Material	2018 Bare Costs Labor	Equipment	Total	Total Incl O&P
0010	**FIREPLACE MANTELS**									
0015	6" molding, 6' x 3'-6" opening, plain, paint grade	1 Carp	5	1.600	Opng.	450	81		531	625
0100	Ornate, oak		5	1.600		600	81		681	785
0300	Prefabricated pine, colonial type, stock, deluxe		2	4		1,825	203		2,028	2,325
0400	Economy		3	2.667		790	135		925	1,100
9000	Minimum labor/equipment charge		3	2.667	Job		135		135	219

06 44 33.20 Fireplace Mantel Beam

		Crew	Daily Output	Labor-Hours	Unit	Material	2018 Bare Costs Labor	Equipment	Total	Total Incl O&P
0010	**FIREPLACE MANTEL BEAM**									
0020	Rough texture wood, 4" x 8"	1 Carp	36	.222	L.F.	8.05	11.25		19.30	27
0100	4" x 10"		35	.229	"	10.80	11.60		22.40	30.50

06 44 Ornamental Woodwork

06 44 33 – Wood Mantels

06 44 33.20 Fireplace Mantel Beam

		Crew	Daily Output	Labor-Hours	Unit	Material	2018 Bare Costs Labor	Equipment	Total	Total Incl O&P
0300	Laminated hardwood, 2-1/4" x 10-1/2" wide, 6' long	1 Carp	5	1.600	Ea.	105	81		186	247
0400	8' long		5	1.600	"	146	81		227	292
0600	Brackets for above, rough sawn		12	.667	Pr.	10.10	34		44.10	66
0700	Laminated		12	.667	"	19.50	34		53.50	76.50
9000	Minimum labor/equipment charge		4	2	Job		101		101	165

06 44 39 – Wood Posts and Columns

06 44 39.10 Decorative Beams

		Crew	Daily Output	Labor-Hours	Unit	Material	2018 Bare Costs Labor	Equipment	Total	Total Incl O&P
0010	**DECORATIVE BEAMS**									
0020	Rough sawn cedar, non-load bearing, 4" x 4"	2 Carp	180	.089	L.F.	1.62	4.51		6.13	9.10
0100	4" x 6"		170	.094		1.77	4.77		6.54	9.70
0200	4" x 8"		160	.100		2.32	5.05		7.37	10.80
0300	4" x 10"		150	.107		4.04	5.40		9.44	13.25
0400	4" x 12"		140	.114		4.66	5.80		10.46	14.55
0500	8" x 8"		130	.123		4.77	6.25		11.02	15.40
9000	Minimum labor/equipment charge	1 Carp	3	2.667	Job		135		135	219

06 44 39.20 Columns

		Crew	Daily Output	Labor-Hours	Unit	Material	2018 Bare Costs Labor	Equipment	Total	Total Incl O&P
0010	**COLUMNS**									
0050	Aluminum, round colonial, 6" diameter	2 Carp	80	.200	V.L.F.	17.75	10.15		27.90	36
0100	8" diameter		62.25	.257		18.55	13.05		31.60	41.50
0200	10" diameter		55	.291		21	14.75		35.75	47.50
0250	Fir, stock units, hollow round, 6" diameter		80	.200		26	10.15		36.15	45
0300	8" diameter		80	.200		33.50	10.15		43.65	53
0350	10" diameter		70	.229		41	11.60		52.60	64
0360	12" diameter		65	.246		59	12.50		71.50	85.50
0400	Solid turned, to 8' high, 3-1/2" diameter		80	.200		10.15	10.15		20.30	27.50
0500	4-1/2" diameter		75	.213		12.10	10.80		22.90	31
0600	5-1/2" diameter		70	.229		16.35	11.60		27.95	37
0800	Square columns, built-up, 5" x 5"		65	.246		33.50	12.50		46	57.50
0900	Solid, 3-1/2" x 3-1/2"		130	.123		10.15	6.25		16.40	21.50
1600	Hemlock, tapered, T&G, 12" diam., 10' high		100	.160		47.50	8.10		55.60	65
1700	16' high		65	.246		79	12.50		91.50	108
1900	14" diameter, 10' high		100	.160		96	8.10		104.10	118
2000	18' high		65	.246		106	12.50		118.50	137
2200	18" diameter, 12' high		65	.246		173	12.50		185.50	212
2300	20' high		50	.320		124	16.20		140.20	163
2500	20" diameter, 14' high		40	.400		187	20.50		207.50	239
2600	20' high		35	.457		177	23		200	233
2800	For flat pilasters, deduct					33%				
3000	For splitting into halves, add				Ea.	110			110	121
4000	Rough sawn cedar posts, 4" x 4"	2 Carp	250	.064	V.L.F.	4.24	3.24		7.48	9.90
4100	4" x 6"		235	.068		8.30	3.45		11.75	14.70
4200	6" x 6"		220	.073		14	3.69		17.69	21.50
4300	8" x 8"		200	.080		18.70	4.06		22.76	27
9000	Minimum labor/equipment charge	1 Carp	3	2.667	Job		135		135	219

06 48 13 – Exterior Wood Door Frames

06 48 13.10 Exterior Wood Door Frames and Accessories	Crew	Daily Output	Labor-Hours	Unit	Material	2018 Bare Costs Labor	Equipment	Total	Total Incl O&P
0010 **EXTERIOR WOOD DOOR FRAMES AND ACCESSORIES**									
0400 Exterior frame, incl. ext. trim, pine, 5/4 x 4-9/16" deep	2 Carp	375	.043	L.F.	6.75	2.16		8.91	10.95
0420 5-3/16" deep		375	.043		8.30	2.16		10.46	12.60
0440 6-9/16" deep		375	.043		9.75	2.16		11.91	14.25
0600 Oak, 5/4 x 4-9/16" deep		350	.046		12.55	2.32		14.87	17.55
0620 5-3/16" deep		350	.046		13.80	2.32		16.12	18.90
0640 6-9/16" deep		350	.046		18.55	2.32		20.87	24.50
1000 Sills, 8/4 x 8" deep, oak, no horns		100	.160		7.15	8.10		15.25	21
1020 2" horns		100	.160		21.50	8.10		29.60	36.50
1040 3" horns		100	.160		21.50	8.10		29.60	36.50
1100 8/4 x 10" deep, oak, no horns		90	.178		6.55	9		15.55	22
1120 2" horns		90	.178		27	9		36	44
1140 3" horns		90	.178		27	9		36	44
2000 Wood frame & trim, ext., colonial, 3' opng., fluted pilasters, flat head		22	.727	Ea.	505	37		542	615
2010 Dentil head		21	.762		600	38.50		638.50	725
2020 Ram's head		20	.800		650	40.50		690.50	780
2100 5'-4" opening, in-swing, fluted pilasters, flat head		17	.941		445	47.50		492.50	570
2120 Ram's head		15	1.067		1,325	54		1,379	1,550
2140 Out-swing, fluted pilasters, flat head		17	.941		480	47.50		527.50	605
2160 Ram's head		15	1.067		1,500	54		1,554	1,750
2400 6'-0" opening, in-swing, fluted pilasters, flat head		16	1		480	50.50		530.50	610
2420 Ram's head		10	1.600		1,500	81		1,581	1,775
2460 Out-swing, fluted pilasters, flat head		16	1		480	50.50		530.50	615
2480 Ram's head		10	1.600		1,500	81		1,581	1,775
2600 For two sidelights, flat head, add		30	.533	Opng.	277	27		304	350
2620 Ram's head, add		20	.800	"	920	40.50		960.50	1,075
2700 Custom birch frame, 3'-0" opening		16	1	Ea.	232	50.50		282.50	340
2750 6'-0" opng.		16	1		365	50.50		415.50	485
2900 Exterior, modern, plain trim, 3' opng., in-swing, FJP		26	.615		48	31		79	104
2920 Fir		24	.667		56.50	34		90.50	117
2940 Oak		22	.727		66	37		103	133

06 48 16 – Interior Wood Door Frames

06 48 16.10 Interior Wood Door Jamb and Frames

	Crew	Daily Output	Labor-Hours	Unit	Material	2018 Bare Costs Labor	Equipment	Total	Total Incl O&P
0010 **INTERIOR WOOD DOOR JAMB AND FRAMES**									
3000 Interior frame, pine, 11/16" x 3-5/8" deep	2 Carp	375	.043	L.F.	4.15	2.16		6.31	8.10
3020 4-9/16" deep		375	.043		4.80	2.16		6.96	8.80
3200 Oak, 11/16" x 3-5/8" deep		350	.046		2.76	2.32		5.08	6.80
3220 4-9/16" deep		350	.046		11	2.32		13.32	15.85
3240 5-3/16" deep		350	.046		17.85	2.32		20.17	23.50
3400 Walnut, 11/16" x 3-5/8" deep		350	.046		9.10	2.32		11.42	13.75
3420 4-9/16" deep		350	.046		10.80	2.32		13.12	15.60
3440 5-3/16" deep		350	.046		9.65	2.32		11.97	14.40
3600 Pocket door frame		16	1	Ea.	86	50.50		136.50	177
3800 Threshold, oak, 5/8" x 3-5/8" deep		200	.080	L.F.	3.57	4.06		7.63	10.55
3820 4-5/8" deep		190	.084		3.95	4.27		8.22	11.30
3840 5-5/8" deep		180	.089		6.65	4.51		11.16	14.60
9000 Minimum labor/equipment charge	1 Carp	4	2	Job		101		101	165

06 49 19 – Exterior Wood Shutters

06 49 19.10 Shutters, Exterior	Crew	Daily Output	Labor-Hours	Unit	Material	2018 Bare Costs Labor	Equipment	Total	Total Incl O&P
0010 **SHUTTERS, EXTERIOR**									
0012 Aluminum, louvered, 1'-4" wide, 3'-0" long	1 Carp	10	.800	Pr.	192	40.50		232.50	277
0200 4'-0" long		10	.800		232	40.50		272.50	320
0300 5'-4" long		10	.800		274	40.50		314.50	365
0400 6'-8" long		9	.889		340	45		385	450
1000 Pine, louvered, primed, each 1'-2" wide, 3'-3" long		10	.800		235	40.50		275.50	325
1100 4'-7" long		10	.800		270	40.50		310.50	365
1250 Each 1'-4" wide, 3'-0" long		10	.800		263	40.50		303.50	355
1350 5'-3" long		10	.800		320	40.50		360.50	420
1500 Each 1'-6" wide, 3'-3" long		10	.800		255	40.50		295.50	345
1600 4'-7" long		10	.800		315	40.50		355.50	415
1620 Cedar, louvered, 1'-2" wide, 5'-7" long		10	.800		320	40.50		360.50	415
1630 Each 1'-4" wide, 2'-2" long		10	.800		171	40.50		211.50	254
1640 3'-0" long		10	.800		219	40.50		259.50	305
1650 3'-3" long		10	.800		229	40.50		269.50	320
1660 3'-11" long		10	.800		265	40.50		305.50	355
1670 4'-3" long		10	.800		294	40.50		334.50	390
1680 5'-3" long		10	.800		335	40.50		375.50	435
1690 5'-11" long		10	.800		350	40.50		390.50	450
1700 Door blinds, 6'-9" long, each 1'-3" wide		9	.889		385	45		430	500
1710 1'-6" wide		9	.889		415	45		460	530
1720 Cedar, solid raised panel, each 1'-4" wide, 3'-3" long		10	.800		325	40.50		365.50	420
1730 3'-11" long		10	.800		325	40.50		365.50	425
1740 4'-3" long		10	.800		335	40.50		375.50	435
1750 4'-7" long		10	.800		370	40.50		410.50	475
1760 4'-11" long		10	.800		400	40.50		440.50	505
1770 5'-11" long		10	.800		515	40.50		555.50	635
1800 Door blinds, 6'-9" long, each 1'-3" wide		9	.889		490	45		535	615
1900 1'-6" wide		9	.889		580	45		625	715
2500 Polystyrene, solid raised panel, each 1'-4" wide, 3'-3" long		10	.800		82	40.50		122.50	156
2600 3'-11" long		10	.800		90	40.50		130.50	165
2700 4'-7" long		10	.800		102	40.50		142.50	178
2800 5'-3" long		10	.800		116	40.50		156.50	193
2900 6'-8" long		9	.889		144	45		189	231
4500 Polystyrene, louvered, each 1'-2" wide, 3'-3" long		10	.800		38.50	40.50		79	108
4600 4'-7" long		10	.800		45	40.50		85.50	116
4750 5'-3" long		10	.800		53.50	40.50		94	125
4850 6'-8" long		9	.889		69	45		114	149
6000 Vinyl, louvered, each 1'-2" x 4'-7" long		10	.800		58	40.50		98.50	130
6200 Each 1'-4" x 6'-8" long		9	.889		81.50	45		126.50	163
9000 Minimum labor/equipment charge		4	2	Job		101		101	165

For customer support on your Commercial Renovation Costs with RSMeans data, call 800.448.8182.

171

06 51 Structural Plastic Shapes and Plates

06 51 13 – Plastic Lumber

06 51 13.10 Recycled Plastic Lumber

			Crew	Daily Output	Labor-Hours	Unit	Material	2018 Bare Costs Labor	Equipment	Total	Total Incl O&P
0010	**RECYCLED PLASTIC LUMBER**										
4000	Sheeting, recycled plastic, black or white, 4' x 8' x 1/8"	G	2 Carp	1100	.015	S.F.	1.09	.74		1.83	2.40
4010	4' x 8' x 3/16"	G		1100	.015		1.51	.74		2.25	2.86
4020	4' x 8' x 1/4"	G		950	.017		1.81	.85		2.66	3.38
4030	4' x 8' x 3/8"	G		950	.017		3.02	.85		3.87	4.71
4040	4' x 8' x 1/2"	G		900	.018		4.03	.90		4.93	5.90
4050	4' x 8' x 5/8"	G		900	.018		6	.90		6.90	8.05
4060	4' x 8' x 3/4"	G		850	.019		7.50	.95		8.45	9.80
4070	Add for colors	G				Ea.	5%				
8500	100% recycled plastic, var colors, NLB, 2" x 2"	G				L.F.	1.90			1.90	2.09
8510	2" x 4"	G					3.90			3.90	4.29
8520	2" x 6"	G					6.10			6.10	6.75
8530	2" x 8"	G					8.60			8.60	9.50
8540	2" x 10"	G					12.25			12.25	13.50
8550	5/4" x 4"	G					4.58			4.58	5.05
8560	5/4" x 6"	G					5.70			5.70	6.25
8570	1" x 6"	G					2.83			2.83	3.11
8580	1/2" x 8"	G					3.32			3.32	3.65
8590	2" x 10" T&G	G					12.65			12.65	13.90
8600	3" x 10" T&G	G					18.95			18.95	21
8610	Add for premium colors	G					20%				

06 51 13.12 Structural Plastic Lumber

			Crew	Daily Output	Labor-Hours	Unit	Material	2018 Bare Costs Labor	Equipment	Total	Total Incl O&P
0010	**STRUCTURAL PLASTIC LUMBER**										
1320	Plastic lumber, posts or columns, 4" x 4"		2 Carp	390	.041	L.F.	11.10	2.08		13.18	15.65
1325	4" x 6"			275	.058		13.30	2.95		16.25	19.45
1330	4" x 8"			220	.073		19.55	3.69		23.24	27.50
1340	Girder, single, 4" x 4"			675	.024		11.10	1.20		12.30	14.20
1345	4" x 6"			600	.027		13.30	1.35		14.65	16.85
1350	4" x 8"			525	.030		19.55	1.55		21.10	24
1352	Double, 2" x 4"			625	.026		7.95	1.30		9.25	10.85
1354	2" x 6"			600	.027		9.65	1.35		11	12.80
1356	2" x 8"			575	.028		16.85	1.41		18.26	21
1358	2" x 10"			550	.029		23.50	1.47		24.97	28.50
1360	2" x 12"			525	.030		25	1.55		26.55	30
1362	Triple, 2" x 4"			575	.028		11.90	1.41		13.31	15.40
1364	2" x 6"			550	.029		14.45	1.47		15.92	18.30
1366	2" x 8"			525	.030		25	1.55		26.55	30.50
1368	2" x 10"			500	.032		35.50	1.62		37.12	41.50
1370	2" x 12"			475	.034		37.50	1.71		39.21	44
1372	Ledger, bolted 4' OC, 2" x 4"			400	.040		4.12	2.03		6.15	7.80
1374	2" x 6"			550	.029		4.89	1.47		6.36	7.80
1376	2" x 8"			390	.041		8.55	2.08		10.63	12.80
1378	2" x 10"			385	.042		11.90	2.11		14.01	16.50
1380	2" x 12"			380	.042		12.60	2.14		14.74	17.30
1382	Joists, 2" x 4"			1250	.013		3.97	.65		4.62	5.40
1384	2" x 6"			1250	.013		4.83	.65		5.48	6.35
1386	2" x 8"			1100	.015		8.40	.74		9.14	10.45
1388	2" x 10"			500	.032		11.90	1.62		13.52	15.75
1390	2" x 12"			875	.018		12.50	.93		13.43	15.25
1392	Railings and trim, 5/4" x 4"		1 Carp	300	.027		4.35	1.35		5.70	7
1394	2" x 2"			300	.027		1.91	1.35		3.26	4.29
1396	2" x 4"			300	.027		3.95	1.35		5.30	6.55

06 51 Structural Plastic Shapes and Plates

06 51 13 – Plastic Lumber

06 51 13.12 Structural Plastic Lumber	Crew	Daily Output	Labor-Hours	Unit	Material	2018 Bare Costs Labor	Equipment	Total	Total Incl O&P
1398	2" x 6"	1 Carp	300	.027	L.F.	4.79	1.35		6.14	7.45

06 52 Plastic Structural Assemblies

06 52 10 – Fiberglass Structural Assemblies

06 52 10.30 Fiberglass Grating

		Crew	Daily Output	Labor-Hours	Unit	Material	2018 Bare Costs Labor	Equipment	Total	Total Incl O&P
0010	**FIBERGLASS GRATING**									
0100	Molded, green (for mod. corrosive environment)									
0140	1" x 4" mesh, 1" thick	2 Sswk	400	.040	S.F.	7.40	2.19		9.59	11.90
0180	1-1/2" square mesh, 1" thick		400	.040		15.75	2.19		17.94	21
0220	1-1/4" thick		400	.040		9.05	2.19		11.24	13.70
0260	1-1/2" thick		400	.040		26.50	2.19		28.69	32.50
0300	2" square mesh, 2" thick		320	.050		27.50	2.73		30.23	34.50
1000	Orange (for highly corrosive environment)									
1040	1" x 4" mesh, 1" thick	2 Sswk	400	.040	S.F.	16.95	2.19		19.14	22.50
1080	1-1/2" square mesh, 1" thick		400	.040		18.95	2.19		21.14	24.50
1120	1-1/4" thick		400	.040		19.95	2.19		22.14	25.50
1160	1-1/2" thick		400	.040		21	2.19		23.19	26.50
1200	2" square mesh, 2" thick		320	.050		21	2.73		23.73	27.50
3000	Pultruded, green (for mod. corrosive environment)									
3040	1" OC bar spacing, 1" thick	2 Sswk	400	.040	S.F.	13.95	2.19		16.14	19.10
3080	1-1/2" thick		320	.050		14.65	2.73		17.38	21
3120	1-1/2" OC bar spacing, 1" thick		400	.040		14	2.19		16.19	19.15
3160	1-1/2" thick		400	.040		20.50	2.19		22.69	26.50
4000	Grating support legs, fixed height, no base				Ea.	46			46	50.50
4040	With base					40.50			40.50	45
4080	Adjustable to 60"					61.50			61.50	67.50

06 52 10.40 Fiberglass Floor Grating

		Crew	Daily Output	Labor-Hours	Unit	Material	2018 Bare Costs Labor	Equipment	Total	Total Incl O&P
0010	**FIBERGLASS FLOOR GRATING**									
0100	Reinforced polyester, fire retardant, 1" x 4" grid, 1" thick	E-4	510	.063	S.F.	14.50	3.46	.19	18.15	22
0200	1-1/2" x 6" mesh, 1-1/2" thick		500	.064		16.85	3.53	.20	20.58	24.50
0300	With grit surface, 1-1/2" x 6" grid, 1-1/2" thick		500	.064		17.20	3.53	.20	20.93	25

06 63 Plastic Railings

06 63 10 – Plastic (PVC) Railings

06 63 10.10 Plastic Railings

		Crew	Daily Output	Labor-Hours	Unit	Material	2018 Bare Costs Labor	Equipment	Total	Total Incl O&P
0010	**PLASTIC RAILINGS**									
0100	Horizontal PVC handrail with balusters, 3-1/2" wide, 36" high	1 Carp	96	.083	L.F.	29	4.22		33.22	38.50
0150	42" high		96	.083		28.50	4.22		32.72	38.50
0200	Angled PVC handrail with balusters, 3-1/2" wide, 36" high		72	.111		18.85	5.65		24.50	30
0250	42" high		72	.111		28.50	5.65		34.15	40.50
0300	Post sleeve for 4 x 4 post		96	.083		14.70	4.22		18.92	23
0400	Post cap for 4 x 4 post, flat profile		48	.167	Ea.	13	8.45		21.45	28
0450	Newel post style profile		48	.167		25.50	8.45		33.95	41.50
0500	Raised corbeled profile		48	.167		36.50	8.45		44.95	54
0550	Post base trim for 4 x 4 post		96	.083		20.50	4.22		24.72	29.50

06 65 10.10 PVC Trim, Exterior		Crew	Daily Output	Labor-Hours	Unit	Material	2018 Bare Costs Labor	Equipment	Total	Total Incl O&P
0010	**PVC TRIM, EXTERIOR**									
0100	Cornerboards, 5/4" x 6" x 6"	1 Carp	240	.033	L.F.	7.20	1.69		8.89	10.65
0110	Door/window casing, 1" x 4"		200	.040		1.39	2.03		3.42	4.82
0120	1" x 6"		200	.040		2.06	2.03		4.09	5.55
0130	1" x 8"		195	.041		2.72	2.08		4.80	6.35
0140	1" x 10"		195	.041		3.44	2.08		5.52	7.15
0150	1" x 12"		190	.042		4.22	2.14		6.36	8.10
0160	5/4" x 4"		195	.041		1.69	2.08		3.77	5.25
0170	5/4" x 6"		195	.041		2.67	2.08		4.75	6.30
0180	5/4" x 8"		190	.042		3.50	2.14		5.64	7.30
0190	5/4" x 10"		190	.042		4.50	2.14		6.64	8.40
0200	5/4" x 12"		185	.043		5.35	2.19		7.54	9.45
0210	Fascia, 1" x 4"		250	.032		1.39	1.62		3.01	4.16
0220	1" x 6"		250	.032		2.06	1.62		3.68	4.90
0230	1" x 8"		225	.036		2.72	1.80		4.52	5.90
0240	1" x 10"		225	.036		3.44	1.80		5.24	6.70
0250	1" x 12"		200	.040		4.22	2.03		6.25	7.95
0260	5/4" x 4"		240	.033		1.69	1.69		3.38	4.60
0270	5/4" x 6"		240	.033		2.67	1.69		4.36	5.70
0280	5/4" x 8"		215	.037		3.50	1.89		5.39	6.90
0290	5/4" x 10"		215	.037		4.50	1.89		6.39	8
0300	5/4" x 12"		190	.042		5.35	2.14		7.49	9.35
0310	Frieze, 1" x 4"		250	.032		1.39	1.62		3.01	4.16
0320	1" x 6"		250	.032		2.06	1.62		3.68	4.90
0330	1" x 8"		225	.036		2.72	1.80		4.52	5.90
0340	1" x 10"		225	.036		3.44	1.80		5.24	6.70
0350	1" x 12"		200	.040		4.22	2.03		6.25	7.95
0360	5/4" x 4"		240	.033		1.69	1.69		3.38	4.60
0370	5/4" x 6"		240	.033		2.67	1.69		4.36	5.70
0380	5/4" x 8"		215	.037		3.50	1.89		5.39	6.90
0390	5/4" x 10"		215	.037		4.50	1.89		6.39	8
0400	5/4" x 12"		190	.042		5.35	2.14		7.49	9.35
0410	Rake, 1" x 4"		200	.040		1.39	2.03		3.42	4.82
0420	1" x 6"		200	.040		2.06	2.03		4.09	5.55
0430	1" x 8"		190	.042		2.72	2.14		4.86	6.45
0440	1" x 10"		190	.042		3.44	2.14		5.58	7.25
0450	1" x 12"		180	.044		4.22	2.25		6.47	8.30
0460	5/4" x 4"		195	.041		1.69	2.08		3.77	5.25
0470	5/4" x 6"		195	.041		2.67	2.08		4.75	6.30
0480	5/4" x 8"		185	.043		3.50	2.19		5.69	7.40
0490	5/4" x 10"		185	.043		4.50	2.19		6.69	8.50
0500	5/4" x 12"		175	.046		5.35	2.32		7.67	9.65
0510	Rake trim, 1" x 4"		225	.036		1.39	1.80		3.19	4.46
0520	1" x 6"		225	.036		2.06	1.80		3.86	5.20
0560	5/4" x 4"		220	.036		1.69	1.84		3.53	4.85
0570	5/4" x 6"		220	.036		2.67	1.84		4.51	5.95
0610	Soffit, 1" x 4"	2 Carp	420	.038		1.39	1.93		3.32	4.67
0620	1" x 6"		420	.038		2.06	1.93		3.99	5.40
0630	1" x 8"		420	.038		2.72	1.93		4.65	6.15
0640	1" x 10"		400	.040		3.44	2.03		5.47	7.05
0650	1" x 12"		400	.040		4.22	2.03		6.25	7.95
0660	5/4" x 4"		410	.039		1.69	1.98		3.67	5.05
0670	5/4" x 6"		410	.039		2.67	1.98		4.65	6.15

06 65 Plastic Trim

06 65 10 – PVC Trim

06 65 10.10 PVC Trim, Exterior

		Crew	Daily Output	Labor-Hours	Unit	Material	2018 Bare Costs Labor	2018 Bare Costs Equipment	Total	Total Incl O&P
0680	5/4" x 8"	2 Carp	410	.039	L.F.	3.50	1.98		5.48	7.05
0690	5/4" x 10"		390	.041		4.50	2.08		6.58	8.35
0700	5/4" x 12"		390	.041		5.35	2.08		7.43	9.30

06 80 Composite Fabrications

06 80 10 – Composite Decking

06 80 10.10 Woodgrained Composite Decking

		Crew	Daily Output	Labor-Hours	Unit	Material	2018 Bare Costs Labor	2018 Bare Costs Equipment	Total	Total Incl O&P
0010	**WOODGRAINED COMPOSITE DECKING**									
0100	Woodgrained composite decking, 1" x 6"	2 Carp	640	.025	L.F.	3.72	1.27		4.99	6.15
0110	Grooved edge		660	.024		3.87	1.23		5.10	6.25
0120	2" x 6"		640	.025		3.81	1.27		5.08	6.25
0130	Encased, 1" x 6"		640	.025		3.81	1.27		5.08	6.25
0140	Grooved edge		660	.024		3.96	1.23		5.19	6.35
0150	2" x 6"		640	.025		5.35	1.27		6.62	7.90

06 81 Composite Railings

06 81 10 – Encased Railings

06 81 10.10 Encased Composite Railings

		Crew	Daily Output	Labor-Hours	Unit	Material	2018 Bare Costs Labor	2018 Bare Costs Equipment	Total	Total Incl O&P
0010	**ENCASED COMPOSITE RAILINGS**									
0100	Encased composite railing, 6' long, 36" high, incl. balusters	1 Carp	16	.500	Ea.	143	25.50		168.50	199
0110	42" high, incl. balusters		16	.500		207	25.50		232.50	269
0120	8' long, 36" high, incl. balusters		12	.667		165	34		199	237
0130	42" high, incl. balusters		12	.667		156	34		190	226
0140	Accessories, post sleeve, 4" x 4", 39" long		32	.250		25	12.70		37.70	48
0150	96" long		24	.333		77	16.90		93.90	112
0160	6" x 6", 39" long		32	.250		54	12.70		66.70	80
0170	96" long		24	.333		155	16.90		171.90	198
0180	Accessories, post skirt, 4" x 4"		96	.083		3.99	4.22		8.21	11.25
0190	6" x 6"		96	.083		4.72	4.22		8.94	12.05
0200	Post cap, 4" x 4", flat		48	.167		6.65	8.45		15.10	21
0210	Pyramid		48	.167		5.70	8.45		14.15	19.95
0220	Post cap, 6" x 6", flat		48	.167		11.30	8.45		19.75	26
0230	Pyramid		48	.167		10.80	8.45		19.25	25.50

For customer support on your Commercial Renovation Costs with RSMeans data, call 800.448.8182.

175

Division Notes

	CREW	DAILY OUTPUT	LABOR-HOURS	UNIT	BARE COSTS				TOTAL INCL O&P
					MAT.	LABOR	EQUIP.	TOTAL	

Estimating Tips

07 10 00 Dampproofing and Waterproofing

- Be sure of the job specifications before pricing this subdivision. The difference in cost between waterproofing and dampproofing can be great. Waterproofing will hold back standing water. Dampproofing prevents the transmission of water vapor. Also included in this section are vapor retarding membranes.

07 20 00 Thermal Protection

- Insulation and fireproofing products are measured by area, thickness, volume or R-value. Specifications may give only what the specific R-value should be in a certain situation. The estimator may need to choose the type of insulation to meet that R-value.

07 30 00 Steep Slope Roofing
07 40 00 Roofing and Siding Panels

- Many roofing and siding products are bought and sold by the square. One square is equal to an area that measures 100 square feet.

  This simple change in unit of measure could create a large error if the estimator is not observant. Accessories necessary for a complete Installation must be figured into any calculations for both material and labor.

07 50 00 Membrane Roofing
07 60 00 Flashing and Sheet Metal
07 70 00 Roofing and Wall Specialties and Accessories

- The items in these subdivisions compose a roofing system. No one component completes the installation, and all must be estimated. Built-up or single-ply membrane roofing systems are made up of many products and installation trades. Wood blocking at roof perimeters or penetrations, parapet coverings, reglets, roof drains, gutters, downspouts, sheet metal flashing, skylights, smoke vents, and roof hatches all need to be considered along with the roofing material. Several different installation trades will need to work together on the roofing system. Inherent difficulties in the scheduling and coordination of various trades must be accounted for when estimating labor costs.

07 90 00 Joint Protection

- To complete the weather-tight shell, the sealants and caulkings must be estimated. Where different materials meet—at expansion joints, at flashing penetrations, and at hundreds of other locations throughout a construction project—caulking and sealants provide another line of defense against water penetration. Often, an entire system is based on the proper location and placement of caulking or sealants. The detailed drawings that are included as part of a set of architectural plans show typical locations for these materials. When caulking or sealants are shown at typical locations, this means the estimator must include them for all the locations where this detail is applicable. Be careful to keep different types of sealants separate, and remember to consider backer rods and primers if necessary.

Reference Numbers

Reference numbers are shown at the beginning of some major classifications. These numbers refer to related items in the Reference Section. The reference information may be an estimating procedure, an alternate pricing method, or technical information.

Note: Not all subdivisions listed here necessarily appear. ■

Did you know?

RSMeans data is available through our online application with 24/7 access:

- Search for unit prices by keyword
- Leverage the most up-to-date data
- Build and export estimates

Try it free for 30 days!
www.rsmeans.com/2018freetrial

07 01 50 – Maintenance of Membrane Roofing

07 01 50.10 Roof Coatings		Crew	Daily Output	Labor-Hours	Unit	Material	2018 Bare Costs Labor	Equipment	Total	Total Incl O&P
0010	**ROOF COATINGS**									
0012	Asphalt, brush grade, material only				Gal.	8.85			8.85	9.75
0200	Asphalt base, fibered aluminum coating [G]					8.40			8.40	9.25
0300	Asphalt primer, 5 gal.					7.50			7.50	8.25
0600	Coal tar pitch, 200 lb. barrels				Ton	1,250			1,250	1,375
0700	Tar roof cement, 5 gal. lots				Gal.	14.25			14.25	15.65
0800	Glass fibered roof & patching cement, 5 gal.				"	8.40			8.40	9.25
0900	Reinforcing glass membrane, 450 S.F./roll				Ea.	59.50			59.50	65.50
1000	Neoprene roof coating, 5 gal., 2 gal./sq.				Gal.	30.50			30.50	33.50
1100	Roof patch & flashing cement, 5 gal.					8.90			8.90	9.80
1200	Roof resaturant, glass fibered, 3 gal./sq.					9.65			9.65	10.60
1600	Reflective roof coating, white, elastomeric, approx. 50 S.F./gal. [G]					17.20			17.20	18.90

07 01 90 – Maintenance of Joint Protection

07 01 90.81 Joint Sealant Replacement		Crew	Daily Output	Labor-Hours	Unit	Material	Labor	Equipment	Total	Total Incl O&P
0010	**JOINT SEALANT REPLACEMENT**									
0050	Control joints in concrete floors/slabs									
0100	Option 1 for joints with hard dry sealant									
0110	Step 1: Sawcut to remove 95% of old sealant									
0112	1/4" wide x 1/2" deep, with single saw blade	C-27	4800	.003	L.F.	.02	.16	.02	.20	.29
0114	3/8" wide x 3/4" deep, with single saw blade		4000	.004		.03	.19	.03	.25	.36
0116	1/2" wide x 1" deep, with double saw blades		3600	.004		.06	.21	.03	.30	.42
0118	3/4" wide x 1-1/2" deep, with double saw blades		3200	.005		.12	.24	.03	.39	.55
0120	Step 2: Water blast joint faces and edges	C-29	2500	.003			.13	.03	.16	.24
0130	Step 3: Air blast joint faces and edges	C-28	2000	.004			.19	.01	.20	.31
0140	Step 4: Sand blast joint faces and edges	E-11	2000	.016			.71	.10	.81	1.31
0150	Step 5: Air blast joint faces and edges	C-28	2000	.004			.19	.01	.20	.31
0200	Option 2 for joints with soft pliable sealant									
0210	Step 1: Plow joint with rectangular blade	B-62	2600	.009	L.F.		.40	.07	.47	.72
0220	Step 2: Sawcut to re-face joint faces									
0222	1/4" wide x 1/2" deep, with single saw blade	C-27	2400	.007	L.F.	.02	.32	.04	.38	.57
0224	3/8" wide x 3/4" deep, with single saw blade		2000	.008		.04	.38	.05	.47	.70
0226	1/2" wide x 1" deep, with double saw blades		1800	.009		.08	.42	.06	.56	.81
0228	3/4" wide x 1-1/2" deep, with double saw blades		1600	.010		.16	.48	.06	.70	.99
0230	Step 3: Water blast joint faces and edges	C-29	2500	.003			.13	.03	.16	.24
0240	Step 4: Air blast joint faces and edges	C-28	2000	.004			.19	.01	.20	.31
0250	Step 5: Sand blast joint faces and edges	E-11	2000	.016			.71	.10	.81	1.31
0260	Step 6: Air blast joint faces and edges	C-28	2000	.004			.19	.01	.20	.31
0290	For saw cutting new control joints, see Section 03 15 16.20									
8910	For backer rod, see Section 07 91 23.10									
8920	For joint sealant, see Section 03 15 16.30 or 07 92 13.20									

07 05 Common Work Results for Thermal and Moisture Protection

07 05 05 – Selective Demolition for Thermal and Moisture Protection

07 05 05.10 Selective Demo., Thermal and Moist. Protection	Crew	Daily Output	Labor-Hours	Unit	Material	2018 Bare Costs Labor	Equipment	Total	Total Incl O&P
0010	**SELECTIVE DEMO., THERMAL AND MOISTURE PROTECTION**									
0020	Caulking/sealant, to 1" x 1" joint R024119-10	1 Clab	600	.013	L.F.		.53		.53	.86
0120	Downspouts, including hangers		350	.023	"		.91		.91	1.48
0220	Flashing, sheet metal		290	.028	S.F.		1.10		1.10	1.79
0420	Gutters, aluminum or wood, edge hung		240	.033	L.F.		1.33		1.33	2.16
0520	Built-in		100	.080	"		3.19		3.19	5.20
0620	Insulation, air/vapor barrier		3500	.002	S.F.		.09		.09	.15
0670	Batts or blankets		1400	.006	C.F.		.23		.23	.37
0720	Foamed or sprayed in place	2 Clab	1000	.016	B.F.		.64		.64	1.04
0770	Loose fitting	1 Clab	3000	.003	C.F.		.11		.11	.17
0870	Rigid board		3450	.002	B.F.		.09		.09	.15
1120	Roll roofing, cold adhesive		12	.667	Sq.		26.50		26.50	43
1170	Roof accessories, adjustable metal chimney flashing		9	.889	Ea.		35.50		35.50	57.50
1325	Plumbing vent flashing		32	.250	"		9.95		9.95	16.20
1375	Ridge vent strip, aluminum		310	.026	L.F.		1.03		1.03	1.67
1620	Skylight to 10 S.F.		8	1	Ea.		40		40	64.50
2120	Roof edge, aluminum soffit and fascia		570	.014	L.F.		.56		.56	.91
2170	Concrete coping, up to 12" wide	2 Clab	160	.100			3.99		3.99	6.45
2220	Drip edge	1 Clab	1000	.008			.32		.32	.52
2270	Gravel stop		950	.008			.34		.34	.54
2370	Sheet metal coping, up to 12" wide		240	.033			1.33		1.33	2.16
2470	Roof insulation board, over 2" thick	B-2	7800	.005	B.F.		.21		.21	.34
2520	Up to 2" thick	"	3900	.010	S.F.		.41		.41	.67
2620	Roof ventilation, louvered gable vent	1 Clab	16	.500	Ea.		19.95		19.95	32.50
2670	Remove, roof hatch	G-3	15	2.133			106		106	170
2675	Rafter vents	1 Clab	960	.008			.33		.33	.54
2720	Soffit vent and/or fascia vent		575	.014	L.F.		.55		.55	.90
2775	Soffit vent strip, aluminum, 3" to 4" wide		160	.050			1.99		1.99	3.24
2820	Roofing accessories, shingle moulding, to 1" x 4"		1600	.005			.20		.20	.32
2870	Cant strip	B-2	2000	.020			.81		.81	1.31
2920	Concrete block walkway	1 Clab	230	.035			1.39		1.39	2.25
3070	Roofing, felt paper, #15		70	.114	Sq.		4.55		4.55	7.40
3125	#30 felt		30	.267	"		10.65		10.65	17.25
3170	Asphalt shingles, 1 layer	B-2	3500	.011	S.F.		.46		.46	.75
3180	2 layers		1750	.023	"		.92		.92	1.49
3370	Modified bitumen		26	1.538	Sq.		62		62	101
3420	Built-up, no gravel, 3 ply		25	1.600			64.50		64.50	105
3470	4 ply		21	1.905			76.50		76.50	124
3620	5 ply		1600	.025	S.F.		1.01		1.01	1.63
3720	5 ply, with gravel		890	.045			1.81		1.81	2.94
3725	Loose gravel removal		5000	.008			.32		.32	.52
3730	Embedded gravel removal		2000	.020			.81		.81	1.31
3870	Fiberglass sheet		1200	.033			1.34		1.34	2.18
4120	Slate shingles		1900	.021			.85		.85	1.38
4170	Ridge shingles, clay or slate		2000	.020	L.F.		.81		.81	1.31
4320	Single ply membrane, attached at seams		52	.769	Sq.		31		31	50.50
4370	Ballasted		75	.533			21.50		21.50	35
4420	Fully adhered		39	1.026			41.50		41.50	67
4550	Roof hatch, 2'-6" x 3'-0"	1 Clab	10	.800	Ea.		32		32	52
4670	Wood shingles	B-2	2200	.018	S.F.		.73		.73	1.19
4820	Sheet metal roofing	"	2150	.019			.75		.75	1.22
4970	Siding, horizontal wood clapboards	1 Clab	380	.021			.84		.84	1.36
5025	Exterior insulation finish system	"	120	.067			2.66		2.66	4.31

07 05 05 – Selective Demolition for Thermal and Moisture Protection

07 05 05.10 Selective Demo., Thermal and Moist. Protection	Crew	Daily Output	Labor-Hours	Unit	Material	2018 Bare Costs Labor	Equipment	Total	Total Incl O&P
5070	Tempered hardboard, remove and reset	1 Carp	380	.021	S.F.		1.07		1.07	1.73
5120	Tempered hardboard sheet siding	"	375	.021	↓		1.08		1.08	1.76
5170	Metal, corner strips	1 Clab	850	.009	L.F.		.38		.38	.61
5225	Horizontal strips		444	.018	S.F.		.72		.72	1.17
5320	Vertical strips		400	.020			.80		.80	1.29
5520	Wood shingles		350	.023			.91		.91	1.48
5620	Stucco siding		360	.022			.89		.89	1.44
5670	Textured plywood		725	.011			.44		.44	.71
5720	Vinyl siding		510	.016	↓		.63		.63	1.02
5770	Corner strips		900	.009	L.F.		.35		.35	.58
5870	Wood, boards, vertical		400	.020	S.F.		.80		.80	1.29
5880	Steel siding, corrugated/ribbed	↓	402.50	.020	"		.79		.79	1.29
5920	Waterproofing, protection/drain board	2 Clab	3900	.004	B.F.		.16		.16	.27
5970	Over 1/2" thick		1750	.009	S.F.		.36		.36	.59
6020	To 1/2" thick	↓	2000	.008	"		.32		.32	.52
9000	Minimum labor/equipment charge	1 Clab	2	4	Job		159		159	259

07 11 Dampproofing

07 11 13 – Bituminous Dampproofing

07 11 13.10 Bituminous Asphalt Coating

		Crew	Daily Output	Labor-Hours	Unit	Material	Labor	Equipment	Total	Total Incl O&P
0010	**BITUMINOUS ASPHALT COATING**									
0030	Brushed on, below grade, 1 coat	1 Rofc	665	.012	S.F.	.22	.53		.75	1.18
0100	2 coat		500	.016		.44	.70		1.14	1.74
0300	Sprayed on, below grade, 1 coat		830	.010		.22	.42		.64	1
0400	2 coat	↓	500	.016	↓	.43	.70		1.13	1.73
0500	Asphalt coating, with fibers				Gal.	8.40			8.40	9.25
0600	Troweled on, asphalt with fibers, 1/16" thick	1 Rofc	500	.016	S.F.	.37	.70		1.07	1.65
0700	1/8" thick		400	.020		.65	.88		1.53	2.28
1000	1/2" thick		350	.023	↓	2.11	1		3.11	4.11
9000	Minimum labor/equipment charge	↓	3	2.667	Job		117		117	209

07 11 16 – Cementitious Dampproofing

07 11 16.20 Cementitious Parging

		Crew	Daily Output	Labor-Hours	Unit	Material	Labor	Equipment	Total	Total Incl O&P
0010	**CEMENTITIOUS PARGING**									
0020	Portland cement, 2 coats, 1/2" thick	D-1	250	.064	S.F.	.37	2.87		3.24	5.10
0100	Waterproofed Portland cement, 1/2" thick, 2 coats	"	250	.064	"	5.80	2.87		8.67	11.05

07 12 Built-up Bituminous Waterproofing

07 12 13 – Built-Up Asphalt Waterproofing

07 12 13.20 Membrane Waterproofing

		Crew	Daily Output	Labor-Hours	Unit	Material	Labor	Equipment	Total	Total Incl O&P
0010	**MEMBRANE WATERPROOFING**									
0012	On slabs, 1 ply, felt, mopped	G-1	3000	.019	S.F.	.43	.77	.17	1.37	2.04
0015	On walls, 1 ply, felt, mopped		3000	.019		.43	.77	.17	1.37	2.04
0100	On slabs, 1 ply, glass fiber fabric, mopped		2100	.027		.47	1.10	.24	1.81	2.74
0105	On walls, 1 ply, glass fiber fabric, mopped		2100	.027		.47	1.10	.24	1.81	2.74
0300	On slabs, 2 ply, felt, mopped		2500	.022		.86	.92	.21	1.99	2.82
0305	On walls, 2 ply, felt, mopped		2500	.022		.86	.92	.21	1.99	2.82
0400	On slabs, 2 ply, glass fiber fabric, mopped		1650	.034		1.04	1.39	.31	2.74	3.96
0405	On walls, 2 ply, glass fiber fabric, mopped		1650	.034		1.04	1.39	.31	2.74	3.96
0600	On slabs, 3 ply, felt, mopped		2100	.027		1.30	1.10	.24	2.64	3.65

07 12 Built-up Bituminous Waterproofing

07 12 13 – Built-Up Asphalt Waterproofing

07 12 13.20 Membrane Waterproofing

	07 12 13.20 Membrane Waterproofing	Crew	Daily Output	Labor-Hours	Unit	Material	2018 Bare Costs Labor	Equipment	Total	Total Incl O&P
0605	On walls, 3 ply, felt, mopped	G-1	2100	.027	S.F.	1.30	1.10	.24	2.64	3.65
0700	On slabs, 3 ply, glass fiber fabric, mopped		1550	.036		1.42	1.48	.33	3.23	4.57
0705	On walls, 3 ply, glass fiber fabric, mopped	↓	1550	.036		1.42	1.48	.33	3.23	4.57
0710	Asphaltic hardboard protection board, 1/8" thick	2 Rofc	500	.032		.67	1.41		2.08	3.25
0715	1/4" thick		450	.036		1.56	1.56		3.12	4.51
1000	EPS membrane protection board, 1/4" thick		3500	.005		.34	.20		.54	.73
1050	3/8" thick		3500	.005		.37	.20		.57	.77
1060	1/2" thick		3500	.005		.41	.20		.61	.81
1070	Fiberglass fabric, black, 20/10 mesh		116	.138	Sq.	14.70	6.05		20.75	27
9000	Minimum labor/equipment charge	↓	2	8	Job		350		350	625

07 13 Sheet Waterproofing

07 13 53 – Elastomeric Sheet Waterproofing

07 13 53.10 Elastomeric Sheet Waterproofing and Access.

		Crew	Daily Output	Labor-Hours	Unit	Material	2018 Bare Costs Labor	Equipment	Total	Total Incl O&P
0010	**ELASTOMERIC SHEET WATERPROOFING AND ACCESS.**									
0090	EPDM, plain, 45 mils thick	2 Rofc	580	.028	S.F.	1.47	1.21		2.68	3.78
0100	60 mils thick		570	.028		1.58	1.23		2.81	3.93
0300	Nylon reinforced sheets, 45 mils thick		580	.028		1.66	1.21		2.87	3.98
0400	60 mils thick	↓	570	.028	↓	1.69	1.23		2.92	4.05
0600	Vulcanizing splicing tape for above, 2" wide				C.L.F.	63			63	69
0700	4" wide				"	121			121	134
0900	Adhesive, bonding, 60 S.F./gal.				Gal.	26.50			26.50	29
1000	Splicing, 75 S.F./gal.				"	38			38	42
1200	Neoprene sheets, plain, 45 mils thick	2 Rofc	580	.028	S.F.	2.11	1.21		3.32	4.48
1300	60 mils thick		570	.028		2.23	1.23		3.46	4.65
1500	Nylon reinforced, 45 mils thick		580	.028		2.15	1.21		3.36	4.53
1600	60 mils thick		570	.028		2.78	1.23		4.01	5.25
1800	120 mils thick	↓	500	.032	↓	5.65	1.41		7.06	8.70
1900	Adhesive, splicing, 150 S.F./gal. per coat				Gal.	38			38	42
2100	Fiberglass reinforced, fluid applied, 1/8" thick	2 Rofc	500	.032	S.F.	1.63	1.41		3.04	4.30
2200	Polyethylene and rubberized asphalt sheets, 60 mils thick		550	.029		.94	1.28		2.22	3.31
2400	Polyvinyl chloride sheets, plain, 10 mils thick		580	.028		.15	1.21		1.36	2.33
2500	20 mils thick		570	.028		.20	1.23		1.43	2.42
2700	30 mils thick	↓	560	.029	↓	.25	1.26		1.51	2.52
3000	Adhesives, trowel grade, 40-100 S.F./gal.				Gal.	24			24	26.50
3100	Brush grade, 100-250 S.F./gal.				"	21.50			21.50	24
3300	Bitumen modified polyurethane, fluid applied, 55 mils thick	2 Rofc	665	.024	S.F.	.99	1.06		2.05	2.98
9000	Minimum labor/equipment charge	"	2	8	Job		350		350	625

07 16 Cementitious and Reactive Waterproofing

07 16 16 – Crystalline Waterproofing

07 16 16.20 Cementitious Waterproofing

		Crew	Daily Output	Labor-Hours	Unit	Material	2018 Bare Costs Labor	Equipment	Total	Total Incl O&P
0010	**CEMENTITIOUS WATERPROOFING**									
0020	1/8" application, sprayed on	G-2A	1000	.024	S.F.	.68	.93	.61	2.22	3.03
0050	4 coat cementitious metallic slurry	1 Cefi	1.20	6.667	C.S.F.	39	315		354	545

For customer support on your Commercial Renovation Costs with RSMeans data, call 800.448.8182.

181

07 19 Water Repellents

07 19 19 – Silicone Water Repellents

07 19 19.10 Silicone Based Water Repellents		Crew	Daily Output	Labor-Hours	Unit	Material	2010 Bare Costs Labor	Equipment	Total	Total Incl O&P
0010	**SILICONE BASED WATER REPELLENTS**									
0020	Water base liquid, roller applied	2 Rofc	7000	.002	S.F.	.42	.10		.52	.64
0200	Silicone or stearate, sprayed on CMU, 1 coat	1 Rofc	4000	.002		.39	.09		.48	.59
0300	2 coats		3000	.003		.79	.12		.91	1.08
9000	Minimum labor/equipment charge		3	2.667	Job		117		117	209

07 21 Thermal Insulation

07 21 13 – Board Insulation

07 21 13.10 Rigid Insulation

07 21 13.10 Rigid Insulation			Crew	Daily Output	Labor-Hours	Unit	Material	2010 Bare Costs Labor	Equipment	Total	Total Incl O&P
0010	**RIGID INSULATION**, for walls										
0040	Fiberglass, 1.5#/C.F., unfaced, 1" thick, R4.1	G	1 Carp	1000	.008	S.F.	.34	.41		.75	1.03
0060	1-1/2" thick, R6.2	G		1000	.008		.40	.41		.81	1.10
0080	2" thick, R8.3	G		1000	.008		.50	.41		.91	1.21
0120	3" thick, R12.4	G		800	.010		.60	.51		1.11	1.48
0370	3#/C.F., unfaced, 1" thick, R4.3	G		1000	.008		.54	.41		.95	1.25
0390	1-1/2" thick, R6.5	G		1000	.008		.79	.41		1.20	1.53
0400	2" thick, R8.7	G		890	.009		1.07	.46		1.53	1.92
0420	2-1/2" thick, R10.9	G		800	.010		1.11	.51		1.62	2.04
0440	3" thick, R13	G		800	.010		1.62	.51		2.13	2.60
0520	Foil faced, 1" thick, R4.3	G		1000	.008		.84	.41		1.25	1.58
0540	1-1/2" thick, R6.5	G		1000	.008		1.25	.41		1.66	2.04
0560	2" thick, R8.7	G		890	.009		1.58	.46		2.04	2.48
0580	2-1/2" thick, R10.9	G		800	.010		1.86	.51		2.37	2.87
0600	3" thick, R13	G		800	.010		2.09	.51		2.60	3.12
1600	Isocyanurate, 4' x 8' sheet, foil faced, both sides										
1610	1/2" thick	G	1 Carp	800	.010	S.F.	.31	.51		.82	1.16
1620	5/8" thick	G		800	.010		.48	.51		.99	1.35
1630	3/4" thick	G		800	.010		.45	.51		.96	1.32
1640	1" thick	G		800	.010		.61	.51		1.12	1.49
1650	1-1/2" thick	G		730	.011		.67	.56		1.23	1.64
1660	2" thick	G		730	.011		.90	.56		1.46	1.89
1670	3" thick	G		730	.011		2.76	.56		3.32	3.94
1680	4" thick	G		730	.011		2.52	.56		3.08	3.67
1700	Perlite, 1" thick, R2.77	G		800	.010		.47	.51		.98	1.34
1750	2" thick, R5.55	G		730	.011		.78	.56		1.34	1.76
1900	Extruded polystyrene, 25 psi compressive strength, 1" thick, R5	G		800	.010		.57	.51		1.08	1.45
1940	2" thick, R10	G		730	.011		1.14	.56		1.70	2.15
1960	3" thick, R15	G		730	.011		1.59	.56		2.15	2.65
2100	Expanded polystyrene, 1" thick, R3.85	G		800	.010		.27	.51		.78	1.12
2120	2" thick, R7.69	G		730	.011		.54	.56		1.10	1.49
2140	3" thick, R11.49	G		730	.011		.81	.56		1.37	1.79
9000	Minimum labor/equipment charge			4	2	Job		101		101	165

07 21 13.13 Foam Board Insulation

07 21 13.13 Foam Board Insulation			Crew	Daily Output	Labor-Hours	Unit	Material	2010 Bare Costs Labor	Equipment	Total	Total Incl O&P
0010	**FOAM BOARD INSULATION**										
0600	Polystyrene, expanded, 1" thick, R4	G	1 Carp	680	.012	S.F.	.27	.60		.87	1.27
0700	2" thick, R8	G		675	.012	"	.54	.60		1.14	1.57
9000	Minimum labor/equipment charge			4	2	Job		101		101	165

07 21 16.10 Blanket Insulation for Floors/Ceilings

	07 21 16.10 Blanket Insulation for Floors/Ceilings		Crew	Daily Output	Labor-Hours	Unit	Material	2018 Bare Costs Labor	Equipment	Total	Total Incl O&P
0010	**BLANKET INSULATION FOR FLOORS/CEILINGS**										
0020	Including spring type wire fasteners										
2000	Fiberglass, blankets or batts, paper or foil backing										
2100	3-1/2" thick, R13	G	1 Carp	700	.011	S.F.	.41	.58		.99	1.39
2150	6-1/4" thick, R19	G		600	.013		.51	.68		1.19	1.66
2210	9-1/2" thick, R30	G		500	.016		.73	.81		1.54	2.12
2220	12" thick, R38	G		475	.017		1.05	.85		1.90	2.55
3000	Unfaced, 3-1/2" thick, R13	G		600	.013		.33	.68		1.01	1.46
3010	6-1/4" thick, R19	G		500	.016		.39	.81		1.20	1.75
3020	9-1/2" thick, R30	G		450	.018		.60	.90		1.50	2.12
3030	12" thick, R38	G		425	.019		.75	.95		1.70	2.38
9000	Minimum labor/equipment charge			4	2	Job		101		101	165

07 21 16.20 Blanket Insulation for Walls

	07 21 16.20 Blanket Insulation for Walls		Crew	Daily Output	Labor-Hours	Unit	Material	2018 Bare Costs Labor	Equipment	Total	Total Incl O&P
0010	**BLANKET INSULATION FOR WALLS**										
0020	Kraft faced fiberglass, 3-1/2" thick, R11, 15" wide	G	1 Carp	1350	.006	S.F.	.32	.30		.62	.84
0030	23" wide	G		1600	.005		.32	.25		.57	.76
0060	R13, 11" wide	G		1150	.007		.34	.35		.69	.94
0080	15" wide	G		1350	.006		.34	.30		.64	.86
0100	23" wide	G		1600	.005		.34	.25		.59	.78
0110	R15, 11" wide	G		1150	.007		.49	.35		.84	1.11
0120	15" wide	G		1350	.006		.49	.30		.79	1.03
0130	23" wide	G		1600	.005		.49	.25		.74	.95
0140	6" thick, R19, 11" wide	G		1150	.007		.44	.35		.79	1.05
0160	15" wide	G		1350	.006		.44	.30		.74	.97
0180	23" wide	G		1600	.005		.44	.25		.69	.89
0182	R21, 11" wide	G		1150	.007		.66	.35		1.01	1.30
0184	15" wide	G		1350	.006		.66	.30		.96	1.22
0186	23" wide	G		1600	.005		.66	.75		91	1 14
0188	9" thick, R30, 11" wide	G		985	.008		.73	.41		1.14	1.47
0200	15" wide	G		1150	.007		.73	.35		1.08	1.37
0220	23" wide	G		1350	.006		.73	.30		1.03	1.29
0230	12" thick, R38, 11" wide	G		985	.008		1.05	.41		1.46	1.83
0240	15" wide	G		1150	.007		1.05	.35		1.40	1.73
0260	23" wide	G		1350	.006		1.05	.30		1.35	1.65
0410	Foil faced fiberglass, 3-1/2" thick, R13, 11" wide	G		1150	.007		.47	.35		.82	1.09
0420	15" wide	G		1350	.006		.47	.30		.77	1.01
0440	23" wide	G		1600	.005		.47	.25		.72	.93
0442	R15, 11" wide	G		1150	.007		.48	.35		.83	1.10
0444	15" wide	G		1350	.006		.48	.30		.78	1.02
0446	23" wide	G		1600	.005		.48	.25		.73	.94
0448	6" thick, R19, 11" wide	G		1150	.007		.62	.35		.97	1.25
0460	15" wide	G		1350	.006		.62	.30		.92	1.17
0480	23" wide	G		1600	.005		.62	.25		.87	1.09
0482	R21, 11" wide	G		1150	.007		.64	.35		.99	1.27
0484	15" wide	G		1350	.006		.64	.30		.94	1.19
0486	23" wide	G		1600	.005		.64	.25		.89	1.11
0488	9" thick, R30, 11" wide	G		985	.008		.94	.41		1.35	1.70
0500	15" wide	G		1150	.007		.94	.35		1.29	1.60
0550	23" wide	G		1350	.006		.94	.30		1.24	1.52
0560	12" thick, R38, 11" wide	G		985	.008		1.15	.41		1.56	1.94
0570	15" wide	G		1150	.007		1.15	.35		1.50	1.84
0580	23" wide	G		1350	.006		1.15	.30		1.45	1.76

07 21 16 – Blanket Insulation

07 21 16.20 Blanket Insulation for Walls		Crew	Daily Output	Labor-Hours	Unit	Material	2018 Bare Costs Labor	Equipment	Total	Total Incl O&P
0620	Unfaced fiberglass, 3-1/2" thick, R13, 11" wide	G 1 Carp	1150	.007	S.F.	.33	.35		.68	.93
0820	15" wide	G	1350	.006		.33	.30		.63	.85
0830	23" wide	G	1600	.005		.33	.25		.58	.77
0832	R15, 11" wide	G	1150	.007		.46	.35		.81	1.08
0836	23" wide	G	1600	.005		.46	.25		.71	.92
0838	6" thick, R19, 11" wide	G	1150	.007		.39	.35		.74	1
0860	15" wide	G	1150	.007		.39	.35		.74	1
0880	23" wide	G	1350	.006		.39	.30		.69	.92
0882	R21, 11" wide	G	1150	.007		.69	.35		1.04	1.33
0886	15" wide	G	1350	.006		.69	.30		.99	1.25
0888	23" wide	G	1600	.005		.69	.25		.94	1.17
0890	9" thick, R30, 11" wide	G	985	.008		.60	.41		1.01	1.33
0900	15" wide	G	1150	.007		.60	.35		.95	1.23
0920	23" wide	G	1350	.006		.60	.30		.90	1.15
0930	12" thick, R38, 11" wide	G	985	.008		.75	.41		1.16	1.50
0940	15" wide	G	1000	.008		.75	.41		1.16	1.49
0960	23" wide	G	1150	.007		.75	.35		1.10	1.40
1300	Wall or ceiling insulation, mineral wool batts									
1320	3-1/2" thick, R15	G 1 Carp	1600	.005	S.F.	.77	.25		1.02	1.26
1340	5-1/2" thick, R23	G	1600	.005		1.21	.25		1.46	1.74
1380	7-1/4" thick, R30	G	1350	.006		1.60	.30		1.90	2.24
1700	Non-rigid insul., recycled blue cotton fiber, unfaced batts, R13, 16" wide	G	1600	.005		.99	.25		1.24	1.50
1710	R19, 16" wide	G	1600	.005		1.35	.25		1.60	1.90
1850	Friction fit wire insulation supports, 16" OC		960	.008	Ea.	.07	.42		.49	.77
9000	Minimum labor/equipment charge		4	2	Job		101		101	165

07 21 19 – Foamed In Place Insulation

07 21 19.10 Masonry Foamed In Place Insulation

0010	**MASONRY FOAMED IN PLACE INSULATION**									
0100	Amino-plast foam, injected into block core, 6" block	G G-2A	6000	.004	Ea.	.17	.16	.10	.43	.57
0110	8" block	G	5000	.005		.20	.19	.12	.51	.68
0120	10" block	G	4000	.006		.25	.23	.15	.63	.85
0130	12" block	G	3000	.008		.34	.31	.20	.85	1.15
0140	Injected into cavity wall	G	13000	.002	B.F.	.06	.07	.05	.18	.24
0150	Preparation, drill holes into mortar joint every 4 V.L.F., 5/8" diameter	1 Clab	960	.008	Ea.		.33		.33	.54
0160	7/8" diameter		680	.012			.47		.47	.76
0170	Patch drilled holes, 5/8" diameter		1800	.004		.04	.18		.22	.33
0180	7/8" diameter		1200	.007		.05	.27		.32	.49
9000	Minimum labor/equipment charge	G-2A	2	12	Job		465	305	770	1,150
9010	Minimum labor/equipment charge	1 Clab	4	2	"		79.50		79.50	129

07 21 23 – Loose-Fill Insulation

07 21 23.10 Poured Loose-Fill Insulation

0010	**POURED LOOSE-FILL INSULATION**									
0020	Cellulose fiber, R3.8 per inch	G 1 Carp	200	.040	C.F.	.69	2.03		2.72	4.05
0021	4" thick	G	1000	.008	S.F.	.17	.41		.58	.84
0022	6" thick	G	800	.010	"	.28	.51		.79	1.13
0080	Fiberglass wool, R4 per inch	G	200	.040	C.F.	.62	2.03		2.65	3.97
0081	4" thick	G	600	.013	S.F.	.21	.68		.89	1.33
0082	6" thick	G	400	.020	"	.30	1.01		1.31	1.98
0100	Mineral wool, R3 per inch	G	200	.040	C.F.	.49	2.03		2.52	3.83
0101	4" thick	G	600	.013	S.F.	.16	.68		.84	1.28
0102	6" thick	G	400	.020	"	.25	1.01		1.26	1.92
0300	Polystyrene, R4 per inch	G	200	.040	C.F.	1.42	2.03		3.45	4.85

07 21 Thermal Insulation

07 21 23 – Loose-Fill Insulation

07 21 23.10 Poured Loose-Fill Insulation

		Crew	Daily Output	Labor-Hours	Unit	Material	2018 Bare Costs Labor	Equipment	Total	Total Incl O&P
0301	4" thick	G	1 Carp	600	.013	S.F.	.47	.68		1.15	1.62
0302	6" thick	G		400	.020	"	.71	1.01		1.72	2.43
0400	Perlite, R2.78 per inch	G		200	.040	C.F.	5.30	2.03		7.33	9.10
0401	4" thick	G		1000	.008	S.F.	1.76	.41		2.17	2.60
0402	6" thick	G		800	.010	"	2.65	.51		3.16	3.73
9000	Minimum labor/equipment charge		▼	4	2	Job		101		101	165

07 21 23.20 Masonry Loose-Fill Insulation

		Crew	Daily Output	Labor-Hours	Unit	Material	2018 Bare Costs Labor	Equipment	Total	Total Incl O&P
0010	**MASONRY LOOSE-FILL INSULATION**, vermiculite or perlite										
0100	In cores of concrete block, 4" thick wall, .115 C.F./S.F.	G	D-1	4800	.003	S.F.	.61	.15		.76	.91
0200	6" thick wall, .175 C.F./S.F.	G		3000	.005		.93	.24		1.17	1.41
0300	8" thick wall, .258 C.F./S.F.	G		2400	.007		1.36	.30		1.66	1.99
0400	10" thick wall, .340 C.F./S.F.	G		1850	.009		1.80	.39		2.19	2.61
0500	12" thick wall, .422 C.F./S.F.	G		1200	.013	▼	2.23	.60		2.83	3.43
0600	Poured cavity wall, vermiculite or perlite, water repellent	G	▼	250	.064	C.F.	5.30	2.87		8.17	10.50
0700	Foamed in place, urethane in 2-5/8" cavity	G	G-2A	1035	.023	S.F.	1.35	.90	.59	2.84	3.69
0800	For each 1" added thickness, add	G	"	2372	.010	"	.51	.39	.26	1.16	1.53

07 21 26 – Blown Insulation

07 21 26.10 Blown Insulation

		Crew	Daily Output	Labor-Hours	Unit	Material	2018 Bare Costs Labor	Equipment	Total	Total Incl O&P
0010	**BLOWN INSULATION** Ceilings, with open access										
0020	Cellulose, 3-1/2" thick, R13	G	G-4	5000	.005	S.F.	.24	.19	.06	.49	.65
0030	5-3/16" thick, R19	G		3800	.006		.35	.26	.08	.69	.90
0050	6-1/2" thick, R22	G		3000	.008		.45	.32	.10	.87	1.14
0100	8-11/16" thick, R30	G		2600	.009		.61	.37	.12	1.10	1.41
0120	10-7/8" thick, R38	G		1800	.013		.78	.54	.17	1.49	1.92
1000	Fiberglass, 5.5" thick, R11	G		3800	.006		.21	.26	.08	.55	.74
1050	6" thick, R12	G		3000	.008		.30	.32	.10	.72	.97
1100	8.8" thick, R19	G		2200	.011		.37	.44	.14	.95	1.28
1200	10" thick, R22	G		1800	.013		.43	.54	.17	1.14	1.55
1300	11.5" thick, R26	G		1500	.016		.52	.65	.20	1.37	1.84
1350	13" thick, R30	G		1400	.017		.60	.69	.22	1.51	2.03
1450	16" thick, R38	G		1145	.021		.77	.85	.27	1.89	2.51
1500	20" thick, R49	G		920	.026	▼	1.01	1.06	.33	2.40	3.20
9000	Minimum labor/equipment charge		▼	4	6	Job		243	76.50	319.50	480

07 21 27 – Reflective Insulation

07 21 27.10 Reflective Insulation Options

		Crew	Daily Output	Labor-Hours	Unit	Material	2018 Bare Costs Labor	Equipment	Total	Total Incl O&P
0010	**REFLECTIVE INSULATION OPTIONS**										
0020	Aluminum foil on reinforced scrim	G	1 Carp	19	.421	C.S.F.	15	21.50		36.50	51
0100	Reinforced with woven polyolefin	G		19	.421		22	21.50		43.50	59
0500	With single bubble air space, R8.8	G		15	.533		26	27		53	73
0600	With double bubble air space, R9.8	G		15	.533	▼	31.50	27		58.50	78.50
9000	Minimum labor/equipment charge		▼	4	2	Job		101		101	165

07 21 29 – Sprayed Insulation

07 21 29.10 Sprayed-On Insulation

		Crew	Daily Output	Labor-Hours	Unit	Material	2018 Bare Costs Labor	Equipment	Total	Total Incl O&P
0010	**SPRAYED-ON INSULATION**										
0020	Fibrous/cementitious, finished wall, 1" thick, R3.7	G	G-2	2050	.012	S.F.	.36	.49	.06	.91	1.26
0100	Attic, 5.2" thick, R19	G		1550	.015	"	.42	.65	.08	1.15	1.60
0200	Fiberglass, R4 per inch, vertical	G		1600	.015	B.F.	.19	.63	.08	.90	1.31
0210	Horizontal	G	▼	1200	.020	"	.19	.84	.11	1.14	1.68
0300	Closed cell, spray polyurethane foam, 2 lb./C.F. density										
0310	1" thick	G	G-2A	6000	.004	S.F.	.51	.16	.10	.77	.94
0320	2" thick	G	▼	3000	.008	▼	1.03	.31	.20	1.54	1.90

07 21 Thermal Insulation

07 21 29 – Sprayed Insulation

07 21 29.10 Sprayed-On Insulation		Crew	Daily Output	Labor-Hours	Unit	Material	2018 Bare Costs Labor	2018 Bare Costs Equipment	Total	Total Incl O&P
0330	3" thick	G	G-2A	2000	.012	S.F.	1.54	.47	.31	2.32	2.84
0335	3-1/2" thick	G		1715	.014		1.80	.54	.36	2.70	3.31
0340	4" thick	G		1500	.016		2.05	.62	.41	3.08	3.79
0350	5" thick	G		1200	.020		2.57	.78	.51	3.86	4.72
0355	5-1/2" thick	G		1090	.022		2.82	.86	.56	4.24	5.20
0360	6" thick	G		1000	.024		3.08	.93	.61	4.62	5.70
9000	Minimum labor/equipment charge		G-2	2	12	Job		505	64.50	569.50	880

07 22 Roof and Deck Insulation

07 22 16 – Roof Board Insulation

07 22 16.10 Roof Deck Insulation

07 22 16.10 Roof Deck Insulation			Crew	Daily Output	Labor-Hours	Unit	Material	2018 Bare Costs Labor	2018 Bare Costs Equipment	Total	Total Incl O&P
0010	**ROOF DECK INSULATION**, fastening excluded										
0016	Asphaltic cover board, fiberglass lined, 1/8" thick		1 Rofc	1400	.006	S.F.	.48	.25		.73	.98
0018	1/4" thick			1400	.006		.96	.25		1.21	1.51
0020	Fiberboard low density, 1/2" thick, R1.39	G		1300	.006		.35	.27		.62	.87
0030	1" thick, R2.78	G		1040	.008		.58	.34		.92	1.24
0080	1-1/2" thick, R4.17	G		1040	.008		.89	.34		1.23	1.58
0100	2" thick, R5.56	G		1040	.008		1.15	.34		1.49	1.87
0110	Fiberboard high density, 1/2" thick, R1.3	G		1300	.006		.27	.27		.54	.78
0120	1" thick, R2.5	G		1040	.008		.56	.34		.90	1.22
0130	1-1/2" thick, R3.8	G		1040	.008		.83	.34		1.17	1.51
0200	Fiberglass, 3/4" thick, R2.78	G		1300	.006		.62	.27		.89	1.16
0400	15/16" thick, R3.70	G		1300	.006		.82	.27		1.09	1.38
0460	1-1/16" thick, R4.17	G		1300	.006		1.07	.27		1.34	1.66
0600	1-5/16" thick, R5.26	G		1300	.006		1.42	.27		1.69	2.04
0650	2-1/16" thick, R8.33	G		1040	.008		1.52	.34		1.86	2.27
0700	2-7/16" thick, R10	G		1040	.008		1.72	.34		2.06	2.49
0800	Gypsum cover board, fiberglass mat facer, 1/4" thick			1400	.006		.48	.25		.73	.98
0810	1/2" thick			1300	.006		.59	.27		.86	1.13
0820	5/8" thick			1200	.007		.62	.29		.91	1.20
0830	Primed fiberglass mat facer, 1/4" thick			1400	.006		.50	.25		.75	1
0840	1/2" thick			1300	.006		.58	.27		.85	1.12
0850	5/8" thick			1200	.007		.61	.29		.90	1.19
1650	Perlite, 1/2" thick, R1.32	G		1365	.006		.27	.26		.53	.76
1655	3/4" thick, R2.08	G		1040	.008		.38	.34		.72	1.02
1660	1" thick, R2.78	G		1040	.008		.54	.34		.88	1.19
1670	1-1/2" thick, R4.17	G		1040	.008		.80	.34		1.14	1.48
1680	2" thick, R5.56	G		910	.009		1.08	.39		1.47	1.88
1685	2-1/2" thick, R6.67	G		910	.009		1.45	.39		1.84	2.29
1690	Tapered for drainage	G		1040	.008	B.F.	1.05	.34		1.39	1.76
1700	Polyisocyanurate, 2#/C.F. density, 3/4" thick	G		1950	.004	S.F.	.42	.18		.60	.78
1705	1" thick	G		1820	.004		.43	.19		.62	.81
1715	1-1/2" thick	G		1625	.005		.58	.22		.80	1.03
1725	2" thick	G		1430	.006		.74	.25		.99	1.25
1735	2-1/2" thick	G		1365	.006		.99	.26		1.25	1.55
1745	3" thick	G		1300	.006		1.10	.27		1.37	1.69
1755	3-1/2" thick	G		1300	.006		1.71	.27		1.98	2.36
1765	Tapered for drainage	G		1820	.004	B.F.	.52	.19		.71	.91
1900	Extruded polystyrene										
1910	15 psi compressive strength, 1" thick, R5	G	1 Rofc	1950	.004	S.F.	.55	.18		.73	.93
1920	2" thick, R10	G		1625	.005		.72	.22		.94	1.18

07 22 16 – Roof Board Insulation

07 22 16.10 Roof Deck Insulation		Crew	Daily Output	Labor-Hours	Unit	Material	2018 Bare Costs Labor	Equipment	Total	Total Incl O&P
1930	3" thick, R15	G	1 Rofc	1300	.006	S.F.	1.43	.27		1.70	2.05
1932	4" thick, R20	G		1300	.006		1.93	.27		2.20	2.60
1934	Tapered for drainage	G		1950	.004	B.F.	.51	.18		.69	.88
1940	25 psi compressive strength, 1" thick, R5	G		1950	.004	S.F.	1.11	.18		1.29	1.54
1942	2" thick, R10	G		1625	.005		2.11	.22		2.33	2.71
1944	3" thick, R15	G		1300	.006		3.22	.27		3.49	4.02
1946	4" thick, R20	G		1300	.006		4.44	.27		4.71	5.35
1948	Tapered for drainage	G		1950	.004	B.F.	.56	.18		.74	.94
1950	40 psi compressive strength, 1" thick, R5	G		1950	.004	S.F.	.87	.18		1.05	1.28
1952	2" thick, R10	G		1625	.005		1.65	.22		1.87	2.21
1954	3" thick, R15	G		1300	.006		2.39	.27		2.66	3.11
1956	4" thick, R20	G		1300	.006		3.13	.27		3.40	3.93
1958	Tapered for drainage	G		1820	.004	B.F.	.87	.19		1.06	1.30
1960	60 psi compressive strength, 1" thick, R5	G		1885	.004	S.F.	1.05	.19		1.24	1.49
1962	2" thick, R10	G		1560	.005		2	.23		2.23	2.59
1964	3" thick, R15	G		1270	.006		3.26	.28		3.54	4.07
1966	4" thick, R20	G		1235	.006		4.04	.28		4.32	4.96
1968	Tapered for drainage	G		1820	.004	B.F.	1	.19		1.19	1.44
2010	Expanded polystyrene, 1#/C.F. density, 3/4" thick, R2.89	G		1950	.004	S.F.	.20	.18		.38	.54
2020	1" thick, R3.85	G		1950	.004		.27	.18		.45	.62
2100	2" thick, R7.69	G		1625	.005		.54	.22		.76	.98
2110	3" thick, R11.49	G		1625	.005		.81	.22		1.03	1.28
2120	4" thick, R15.38	G		1625	.005		1.08	.22		1.30	1.58
2130	5" thick, R19.23	G		1495	.005		1.35	.24		1.59	1.91
2140	6" thick, R23.26	G		1495	.005		1.62	.24		1.86	2.20
2150	Tapered for drainage	G		1950	.004	B.F.	.53	.18		.71	.90
2400	Composites with 2" EPS										
2410	1" fiberboard	G	1 Rofc	1325	.006	S.F.	1.45	.27		1.72	2.07
2420	7/16" oriented strand board	G		1040	.008		1.15	.34		1.49	1.87
2430	1/2" plywood	G		1040	.008		1.44	.34		1.78	2.18
2440	1" perlite	G		1040	.008		1.17	.34		1.51	1.89
2450	Composites with 1-1/2" polyisocyanurate										
2460	1" fiberboard	G	1 Rofc	1040	.008	S.F.	1.07	.34		1.41	1.78
2470	1" perlite	G		1105	.007		.96	.32		1.28	1.63
2480	7/16" oriented strand board	G		1040	.008		.84	.34		1.18	1.52
3000	Fastening alternatives, coated screws, 2" long			3744	.002	Ea.	.06	.09		.15	.24
3010	4" long			3120	.003		.11	.11		.22	.32
3020	6" long			2675	.003		.19	.13		.32	.44
3030	8" long			2340	.003		.28	.15		.43	.58
3040	10" long			1872	.004		.47	.19		.66	.85
3050	Pre-drill and drive wedge spike, 2-1/2"			1248	.006		.40	.28		.68	.94
3060	3-1/2"			1101	.007		.58	.32		.90	1.21
3070	4-1/2"			936	.009		.65	.38		1.03	1.39
3075	3" galvanized deck plates			7488	.001		.08	.05		.13	.17
3080	Spot mop asphalt		G-1	295	.190	Sq.	5.80	7.80	1.74	15.34	22
3090	Full mop asphalt		"	192	.292		11.60	12	2.68	26.28	37
3110	Low-rise polyurethane adhesive, from 5 gallon kit, 12" OC beads		1 Rofc	45	.178		33	7.80		40.80	50.50
3120	6" OC beads			32	.250		66	11		77	92
3130	4" OC beads			30	.267		99	11.70		110.70	130
9000	Minimum labor/equipment charge			3.25	2.462	Job		108		108	193

187

07 24 Exterior Insulation and Finish Systems

07 24 13 – Polymer-Based Exterior Insulation and Finish System

07 24 13.10 Exterior Insulation and Finish Systems		Crew	Daily Output	Labor-Hours	Unit	Material	2018 Bare Costs Labor	Equipment	Total	Total Incl O&P
0010	**EXTERIOR INSULATION AND FINISH SYSTEMS**									
0095	Field applied, 1" EPS insulation	G	J-1	390	.103	S.F.	1.80	4.49	.34	6.63	9.55
0100	With 1/2" cement board sheathing	G		268	.149		2.59	6.55	.50	9.64	13.85
0105	2" EPS insulation	G		390	.103		2.07	4.49	.34	6.90	9.85
0110	With 1/2" cement board sheathing	G		268	.149		2.86	6.55	.50	9.91	14.15
0115	3" EPS insulation	G		390	.103		2.34	4.49	.34	7.17	10.15
0120	With 1/2" cement board sheathing	G		268	.149		3.13	6.55	.50	10.18	14.45
0125	4" EPS insulation	G		390	.103		2.61	4.49	.34	7.44	10.45
0130	With 1/2" cement board sheathing	G		268	.149		4.18	6.55	.50	11.23	15.60
0140	Premium finish add			1265	.032		.36	1.38	.11	1.85	2.73
0150	Heavy duty reinforcement add			914	.044		.60	1.92	.15	2.67	3.88
0160	2.5#/S.Y. metal lath substrate add		1 Lath	75	.107	S.Y.	3.10	5.25		8.35	11.65
0170	3.4#/S.Y. metal lath substrate add		"	75	.107	"	4.44	5.25		9.69	13.15
0180	Color or texture change		J-1	1265	.032	S.F.	.82	1.38	.11	2.31	3.23
0190	With substrate leveling base coat		1 Plas	530	.015		.83	.70		1.53	2.03
0210	With substrate sealing base coat		1 Pord	1224	.007		.12	.28		.40	.58
0370	V groove shape in panel face					L.F.	.68			.68	.75
0380	U groove shape in panel face					"	.85			.85	.94
0433	Crack repair, acrylic rubber, fluid applied, 20 mils thick		1 Plas	350	.023	S.F.	1.51	1.06		2.57	3.36
0437	50 mils thick, reinforced		"	200	.040	"	2.71	1.86		4.57	5.95
0440	For higher than one story, add							25%			

07 25 Weather Barriers

07 25 10 – Weather Barriers or Wraps

07 25 10.10 Weather Barriers

		Crew	Daily Output	Labor-Hours	Unit	Material	2018 Bare Costs Labor	Equipment	Total	Total Incl O&P
0010	**WEATHER BARRIERS**									
0400	Asphalt felt paper, #15	1 Carp	37	.216	Sq.	5.35	10.95		16.30	23.50
0401	Per square foot	"	3700	.002	S.F.	.05	.11		.16	.24
0450	Housewrap, exterior, spun bonded polypropylene									
0470	Small roll	1 Carp	3800	.002	S.F.	.15	.11		.26	.34
0480	Large roll	"	4000	.002	"	.14	.10		.24	.31
2100	Asphalt felt roof deck vapor barrier, class 1 metal decks	1 Rofc	37	.216	Sq.	22.50	9.50		32	41.50
2200	For all other decks	"	37	.216		16.95	9.50		26.45	35.50
2800	Asphalt felt, 50% recycled content, 15 lb., 4 sq./roll	1 Carp	36	.222		5.65	11.25		16.90	24.50
2810	30 lb., 2 sq./roll	"	36	.222		9	11.25		20.25	28
3000	Building wrap, spun bonded polyethylene	2 Carp	8000	.002	S.F.	.18	.10		.28	.36
9960	Minimum labor/equipment charge	1 Carp	2	4	Job		203		203	330

07 26 Vapor Retarders

07 26 10 – Above-Grade Vapor Retarders

07 26 10.10 Vapor Retarders

			Crew	Daily Output	Labor-Hours	Unit	Material	2018 Bare Costs Labor	Equipment	Total	Total Incl O&P
0010	**VAPOR RETARDERS**										
0020	Aluminum and kraft laminated, foil 1 side	G	1 Carp	37	.216	Sq.	13.65	10.95		24.60	33
0100	Foil 2 sides	G		37	.216		14.15	10.95		25.10	33.50
0600	Polyethylene vapor barrier, standard, 2 mil	G		37	.216		1.50	10.95		12.45	19.45
0700	4 mil	G		37	.216		2.58	10.95		13.53	20.50
0900	6 mil	G		37	.216		3.73	10.95		14.68	22
1200	10 mil	G		37	.216		8.75	10.95		19.70	27.50
1300	Clear reinforced, fire retardant, 8 mil	G		37	.216		26.50	10.95		37.45	47

07 26 Vapor Retarders

07 26 10 – Above-Grade Vapor Retarders

07 26 10.10 Vapor Retarders		Crew	Daily Output	Labor-Hours	Unit	Material	2018 Bare Costs Labor	Equipment	Total	Total Incl O&P
1350	Cross laminated type, 3 mil	[G] 1 Carp	37	.216	Sq.	12.35	10.95		23.30	31.50
1400	4 mil	[G]	37	.216		13.10	10.95		24.05	32.50
1800	Reinf. waterproof, 2 mil polyethylene backing, 1 side		37	.216		9.95	10.95		20.90	29
1900	2 sides		37	.216		12.90	10.95		23.85	32
2400	Waterproofed kraft with sisal or fiberglass fibers		37	.216		21	10.95		31.95	41
9950	Minimum labor/equipment charge		4	2	Job		101		101	165

07 27 Air Barriers

07 27 13 – Modified Bituminous Sheet Air Barriers

07 27 13.10 Modified Bituminous Sheet Air Barrier

		Crew	Daily Output	Labor-Hours	Unit	Material	2018 Bare Costs Labor	Equipment	Total	Total Incl O&P
0010	**MODIFIED BITUMINOUS SHEET AIR BARRIER**									
0100	SBS modified sheet laminated to polyethylene sheet, 40 mils, 4" wide	1 Carp	1200	.007	L.F.	.33	.34		.67	.91
0120	6" wide		1100	.007		.45	.37		.82	1.09
0140	9" wide		1000	.008		.62	.41		1.03	1.35
0160	12" wide		900	.009		.80	.45		1.25	1.61
0180	18" wide	2 Carp	1700	.009	S.F.	.75	.48		1.23	1.59
0200	36" wide	"	1800	.009		.73	.45		1.18	1.53
0220	Adhesive for above	1 Carp	1400	.006		.31	.29		.60	.82

07 27 26 – Fluid-Applied Membrane Air Barriers

07 27 26.10 Fluid Applied Membrane Air Barrier

		Crew	Daily Output	Labor-Hours	Unit	Material	2018 Bare Costs Labor	Equipment	Total	Total Incl O&P
0010	**FLUID APPLIED MEMBRANE AIR BARRIER**									
0100	Spray applied vapor barrier, 25 S.F./gallon	1 Pord	1375	.006	S.F.	.01	.25		.26	.42

07 31 Shingles and Shakes

07 31 13 – Asphalt Shingles

07 31 13.10 Asphalt Roof Shingles

		Crew	Daily Output	Labor-Hours	Unit	Material	2018 Bare Costs Labor	Equipment	Total	Total Incl O&P
0010	**ASPHALT ROOF SHINGLES**									
0100	Standard strip shingles									
0150	Inorganic, class A, 25 year	1 Rofc	5.50	1.455	Sq.	75	64		139	197
0155	Pneumatic nailed		7	1.143		75	50		125	172
0200	30 year		5	1.600		95.50	70.50		166	230
0205	Pneumatic nailed		6.25	1.280		95.50	56.50		152	205
0250	Standard laminated multi-layered shingles									
0300	Class A, 240-260 lb./square	1 Rofc	4.50	1.778	Sq.	110	78		188	260
0305	Pneumatic nailed		5.63	1.422		110	62.50		172.50	232
0350	Class A, 250-270 lb./square		4	2		110	88		198	278
0355	Pneumatic nailed		5	1.600		110	70.50		180.50	246
0400	Premium, laminated multi-layered shingles									
0450	Class A, 260-300 lb./square	1 Rofc	3.50	2.286	Sq.	144	100		244	335
0455	Pneumatic nailed		4.37	1.831		144	80.50		224.50	300
0500	Class A, 300-385 lb./square		3	2.667		260	117		377	495
0505	Pneumatic nailed		3.75	2.133		260	94		354	455
0800	#15 felt underlayment		64	.125		5.35	5.50		10.85	15.70
0825	#30 felt underlayment		58	.138		10.30	6.05		16.35	22
0850	Self adhering polyethylene and rubberized asphalt underlayment		22	.364		78	16		94	114
0900	Ridge shingles		330	.024	L.F.	2.27	1.07		3.34	4.40
0905	Pneumatic nailed		412.50	.019	"	2.27	.85		3.12	4.02
1000	For steep roofs (7 to 12 pitch or greater), add						50%			
9000	Minimum labor/equipment charge	1 Rofc	3	2.667	Job		117		117	209

For customer support on your Commercial Renovation Costs with RSMeans data, call 800.448.8182.

189

07 31 Shingles and Shakes

07 31 16 – Metal Shingles

07 31 16.10 Aluminum Shingles

07 31 16.10 Aluminum Shingles		Crew	Daily Output	Labor-Hours	Unit	Material	2018 Bare Costs Labor	Equipment	Total	Total Incl O&P
0010	**ALUMINUM SHINGLES**									
0020	Mill finish, .019" thick	1 Carp	5	1.600	Sq.	228	81		309	385
0100	.020" thick	"	5	1.600		257	81		338	415
0300	For colors, add					21.50			21.50	24
0600	Ridge cap, .024" thick	1 Carp	170	.047	L.F.	4	2.39		6.39	8.25
0700	End wall flashing, .024" thick		170	.047		2.27	2.39		4.66	6.35
0900	Valley section, .024" thick		170	.047		3.97	2.39		6.36	8.25
1000	Starter strip, .024" thick		400	.020		1.82	1.01		2.83	3.65
1200	Side wall flashing, .024" thick		170	.047		2.22	2.39		4.61	6.30
1500	Gable flashing, .024" thick		400	.020		1.77	1.01		2.78	3.60
9000	Minimum labor/equipment charge		3	2.667	Job		135		135	219

07 31 16.20 Steel Shingles

07 31 16.20 Steel Shingles		Crew	Daily Output	Labor-Hours	Unit	Material	Labor	Equipment	Total	Total Incl O&P
0010	**STEEL SHINGLES**									
0012	Galvanized, 26 ga.	1 Rots	2.20	3.636	Sq.	360	161		521	680
0200	24 ga.	"	2.20	3.636		355	161		516	675
0300	For colored galvanized shingles, add					58			58	64
9000	Minimum labor/equipment charge	1 Rots	3	2.667	Job		118		118	210

07 31 26 – Slate Shingles

07 31 26.10 Slate Roof Shingles

07 31 26.10 Slate Roof Shingles			Crew	Daily Output	Labor-Hours	Unit	Material	Labor	Equipment	Total	Total Incl O&P
0010	**SLATE ROOF SHINGLES**										
0100	Buckingham Virginia black, 3/16" - 1/4" thick	G	1 Rots	1.75	4.571	Sq.	555	202		757	970
0200	1/4" thick	G		1.75	4.571		555	202		757	970
0900	Pennsylvania black, Bangor, #1 clear	G		1.75	4.571		495	202		697	905
1200	Vermont, unfading, green, mottled green	G		1.75	4.571		505	202		707	915
1300	Semi-weathering green & gray	G		1.75	4.571		360	202		562	755
1400	Purple	G		1.75	4.571		435	202		637	840
1500	Black or gray	G		1.75	4.571		485	202		687	895
2500	Slate roof repair, extensive replacement			1	8		600	355		955	1,300
2600	Repair individual pieces, scattered			19	.421	Ea.	7.05	18.60		25.65	41
2700	Ridge shingles, slate			200	.040	L.F.	10.10	1.77		11.87	14.25
9000	Minimum labor/equipment charge			3	2.667	Job		118		118	210

07 31 29 – Wood Shingles and Shakes

07 31 29.13 Wood Shingles

07 31 29.13 Wood Shingles		Crew	Daily Output	Labor-Hours	Unit	Material	Labor	Equipment	Total	Total Incl O&P
0010	**WOOD SHINGLES** R061110-30									
0012	16" No. 1 red cedar shingles, 5" exposure, on roof	1 Carp	2.50	3.200	Sq.	295	162		457	590
0015	Pneumatic nailed		3.25	2.462		295	125		420	530
0200	7-1/2" exposure, on walls		2.05	3.902		196	198		394	535
0205	Pneumatic nailed		2.67	2.996		196	152		348	465
0300	18" No. 1 red cedar perfections, 5-1/2" exposure, on roof		2.75	2.909		253	147		400	515
0305	Pneumatic nailed		3.57	2.241		253	114		367	460
0500	7-1/2" exposure, on walls		2.25	3.556		186	180		366	495
0505	Pneumatic nailed		2.92	2.740		186	139		325	430
0600	Resquared and rebutted, 5-1/2" exposure, on roof		3	2.667		290	135		425	540
0605	Pneumatic nailed		3.90	2.051		290	104		394	490
0900	7-1/2" exposure, on walls		2.45	3.265		214	166		380	505
0905	Pneumatic nailed		3.18	2.516		214	128		342	440
1000	Add to above for fire retardant shingles					55			55	60.50
1060	Preformed ridge shingles	1 Carp	400	.020	L.F.	3.89	1.01		4.90	5.95
2000	White cedar shingles, 16" long, extras, 5" exposure, on roof		2.40	3.333	Sq.	197	169		366	490
2005	Pneumatic nailed		3.12	2.564		197	130		327	430
2050	5" exposure on walls		2	4		197	203		400	545

07 31 Shingles and Shakes

07 31 29 – Wood Shingles and Shakes

07 31 29.13 Wood Shingles

		Crew	Daily Output	Labor-Hours	Unit	Material	2018 Bare Costs Labor	Equipment	Total	Total Incl O&P
2055	Pneumatic nailed	1 Carp	2.60	3.077	Sq.	197	156		353	470
2100	7-1/2" exposure, on walls		2	4		141	203		344	485
2105	Pneumatic nailed		2.60	3.077		141	156		297	410
2150	"B" grade, 5" exposure on walls		2	4		172	203		375	520
2155	Pneumatic nailed		2.60	3.077		172	156		328	440
2300	For #15 organic felt underlayment on roof, 1 layer, add		64	.125		5.35	6.35		11.70	16.20
2400	2 layers, add	▼	32	.250		10.75	12.70		23.45	32.50
2600	For steep roofs (7/12 pitch or greater), add to above				▼		50%			
2700	Panelized systems, No.1 cedar shingles on 5/16" CDX plywood									
2800	On walls, 8' strips, 7" or 14" exposure	2 Carp	700	.023	S.F.	6.40	1.16		7.56	8.95
3500	On roofs, 8' strips, 7" or 14" exposure	1 Carp	3	2.667	Sq.	665	135		800	950
3505	Pneumatic nailed		4	2	"	665	101		766	895
9000	Minimum labor/equipment charge	▼	3	2.667	Job		135		135	219

07 31 29.16 Wood Shakes

		Crew	Daily Output	Labor-Hours	Unit	Material	2018 Bare Costs Labor	Equipment	Total	Total Incl O&P
0010	**WOOD SHAKES**									
1100	Hand-split red cedar shakes, 1/2" thick x 24" long, 10" exp. on roof	1 Carp	2.50	3.200	Sq.	300	162		462	595
1105	Pneumatic nailed		3.25	2.462		300	125		425	535
1110	3/4" thick x 24" long, 10" exp. on roof		2.25	3.556		300	180		480	625
1115	Pneumatic nailed		2.92	2.740		300	139		439	555
1200	1/2" thick, 18" long, 8-1/2" exp. on roof		2	4		275	203		478	630
1205	Pneumatic nailed		2.60	3.077		275	156		431	555
1210	3/4" thick x 18" long, 8-1/2" exp. on roof		1.80	4.444		275	225		500	665
1215	Pneumatic nailed		2.34	3.419		275	173		448	580
1255	10" exposure on walls		2	4		265	203		468	620
1260	10" exposure on walls, pneumatic nailed	▼	2.60	3.077		265	156		421	545
1700	Add to above for fire retardant shakes, 24" long					55			55	60.50
1800	18" long				▼	55			55	60.50
1810	Ridge shakes	1 Carp	350	.023	L.F.	5.75	1.16		6.91	8.25

07 32 Roof Tiles

07 32 13 – Clay Roof Tiles

07 32 13.10 Clay Tiles

		Crew	Daily Output	Labor-Hours	Unit	Material	2018 Bare Costs Labor	Equipment	Total	Total Incl O&P
0010	**CLAY TILES**, including accessories									
0300	Flat shingle, interlocking, 15", 166 pcs./sq., fireflashed blend	3 Rots	6	4	Sq.	470	177		647	830
0500	Terra cotta red		6	4		520	177		697	885
0600	Roman pan and top, 18", 102 pcs./sq., fireflashed blend	▼	5.50	4.364		505	193		698	900
0640	Terra cotta red	1 Rots	2.40	3.333		570	147		717	885
1100	Barrel mission tile, 18", 166 pcs./sq., fireflashed blend	3 Rots	5.50	4.364		415	193		608	800
1140	Terra cotta red		5.50	4.364		420	193		613	805
1700	Scalloped edge flat shingle, 14", 145 pcs./sq., fireflashed blend		6	4		1,150	177		1,327	1,600
1800	Terra cotta red	▼	6	4		1,050	177		1,227	1,475
3010	#15 felt underlayment	1 Rofc	64	.125		5.35	5.50		10.85	15.70
3020	#30 felt underlayment		58	.138		10.30	6.05		16.35	22
3040	Polyethylene and rubberized asph. underlayment	▼	22	.364		78	16		94	114
9010	Minimum labor/equipment charge	1 Rots	2	4	Job		177		177	315

07 32 Roof Tiles

07 32 16 – Concrete Roof Tiles

07 32 16.10 Concrete Tiles

07 32 16.10 Concrete Tiles	Crew	Daily Output	Labor-Hours	Unit	Material	2018 Bare Costs Labor	Equipment	Total	Total Incl O&P
0010 **CONCRETE TILES**									
0020 Corrugated, 13" x 16-1/2", 90 per sq., 950 lb./sq.									
0050 Earthtone colors, nailed to wood deck	1 Rots	1.35	5.926	Sq.	105	262		367	580
0150 Blues		1.35	5.926		105	262		367	580
0200 Greens		1.35	5.926		106	262		368	580
0250 Premium colors		1.35	5.926		106	262		368	580
0500 Shakes, 13" x 16-1/2", 90 per sq., 950 lb./sq.									
0600 All colors, nailed to wood deck	1 Rots	1.50	5.333	Sq.	126	235		361	560
1500 Accessory pieces, ridge & hip, 10" x 16-1/2", 8 lb. each	"	120	.067	Ea.	3.80	2.94		6.74	9.45
1700 Rake, 6-1/2" x 16-3/4", 9 lb. each					3.80			3.80	4.18
1800 Mansard hip, 10" x 16-1/2", 9.2 lb. each					3.80			3.80	4.18
1900 Hip starter, 10" x 16-1/2", 10.5 lb. each					10.50			10.50	11.55
2000 3 or 4 way apex, 10" each side, 11.5 lb. each					12			12	13.20
9000 Minimum labor/equipment charge	1 Rots	3	2.667	Job		118		118	210

07 32 19 – Metal Roof Tiles

07 32 19.10 Metal Roof Tiles

07 32 19.10 Metal Roof Tiles	Crew	Daily Output	Labor-Hours	Unit	Material	2018 Bare Costs Labor	Equipment	Total	Total Incl O&P
0010 **METAL ROOF TILES**									
0020 Accessories included, .032" thick aluminum, mission tile	1 Carp	2.50	3.200	Sq.	830	162		992	1,175
0200 Spanish tiles		3	2.667	"	550	135		685	825
9000 Minimum labor/equipment charge		3	2.667	Job		135		135	219

07 33 Natural Roof Coverings

07 33 63 – Vegetated Roofing

07 33 63.10 Green Roof Systems

07 33 63.10 Green Roof Systems		Crew	Daily Output	Labor-Hours	Unit	Material	2018 Bare Costs Labor	Equipment	Total	Total Incl O&P
0010 **GREEN ROOF SYSTEMS**										
0020 Soil mixture for green roof 30% sand, 55% gravel, 15% soil										
0100 Hoist and spread soil mixture 4" depth up to 5 stories tall roof	G	B-13B	4000	.014	S.F.	.23	.61	.25	1.09	1.50
0150 6" depth	G		2667	.021		.35	.92	.37	1.64	2.27
0200 8" depth	G		2000	.028		.46	1.22	.50	2.18	3.03
0250 10" depth	G		1600	.035		.58	1.53	.62	2.73	3.77
0300 12" depth	G		1335	.042		.69	1.83	.74	3.26	4.53
0310 Alt. man-made soil mix, hoist & spread, 4" deep up to 5 stories tall roof	G		4000	.014		1.92	.61	.25	2.78	3.36
0350 Mobilization 55 ton crane to site	G	1 Eqhv	3.60	2.222	Ea.		125		125	197
0355 Hoisting cost to 5 stories per day (Avg. 28 picks per day)	G	B-13B	1	56	Day		2,450	995	3,445	5,025
0360 Mobilization or demobilization, 100 ton crane to site driver & escort	G	A-3E	2.50	6.400	Ea.		325	50.50	375.50	575
0365 Hoisting cost 6-10 stories per day (Avg. 21 picks per day)	G	B-13C	1	56	Day		2,450	1,875	4,325	5,975
0370 Hoist and spread soil mixture 4" depth 6-10 stories tall roof	G		4000	.014	S.F.	.23	.61	.47	1.31	1.74
0375 6" depth	G		2667	.021		.35	.92	.70	1.97	2.63
0380 8" depth	G		2000	.028		.46	1.22	.94	2.62	3.51
0385 10" depth	G		1600	.035		.58	1.53	1.17	3.28	4.38
0390 12" depth	G		1335	.042		.69	1.83	1.40	3.92	5.25
0400 Green roof edging treated lumber 4" x 4", no hoisting included	G	2 Carp	400	.040	L.F.	1.20	2.03		3.23	4.61
0410 4" x 6"	G		400	.040		1.94	2.03		3.97	5.45
0420 4" x 8"	G		360	.044		3.96	2.25		6.21	8
0430 4" x 6" double stacked	G		300	.053		3.89	2.70		6.59	8.65
0500 Green roof edging redwood lumber 4" x 4", no hoisting included	G		400	.040		6.45	2.03		8.48	10.40
0510 4" x 6"	G		400	.040		12.60	2.03		14.63	17.15
0520 4" x 8"	G		360	.044		23.50	2.25		25.75	29.50
0530 4" x 6" double stacked	G		300	.053		25	2.70		27.70	32
0550 Components, not including membrane or insulation:										

07 33 Natural Roof Coverings

07 33 63 – Vegetated Roofing

07 33 63.10 Green Roof Systems

		Crew	Daily Output	Labor-Hours	Unit	Material	2018 Bare Costs Labor	2018 Bare Costs Equipment	Total	Total Incl O&P
0560	Fluid applied rubber membrane, reinforced, 215 mil thick	G	G-5	350	.114	S.F.	.30	4.56	.52	5.38	9.05
0570	Root barrier	G	2 Rofc	775	.021		.70	.91		1.61	2.39
0580	Moisture retention barrier and reservoir	G	"	900	.018		2.66	.78		3.44	4.32
0600	Planting sedum, light soil, potted, 2-1/4" diameter, 2 per S.F.	G	1 Clab	420	.019		6.10	.76		6.86	7.95
0610	1 per S.F.	G	"	840	.010		3.05	.38		3.43	3.98
0630	Planting sedum mat per S.F. including shipping (4000 S.F. min)	G	4 Clab	4000	.008		7.30	.32		7.62	8.55
0640	Installation sedum mat system (no soil required) per S.F. (4000 S.F. min)	G	"	4000	.008		10.20	.32		10.52	11.70
0645	Note: pricing of sedum mats shipped in full truck loads (4000-5000 S.F.)										

07 41 Roof Panels

07 41 13 – Metal Roof Panels

07 41 13.10 Aluminum Roof Panels

		Crew	Daily Output	Labor-Hours	Unit	Material	2018 Bare Costs Labor	2018 Bare Costs Equipment	Total	Total Incl O&P
0010	**ALUMINUM ROOF PANELS**									
0020	Corrugated or ribbed, .0155" thick, natural	G-3	1200	.027	S.F.	1	1.33		2.33	3.22
0300	Painted		1200	.027		1.45	1.33		2.78	3.72
0400	Corrugated, .018" thick, on steel frame, natural finish		1200	.027		1.24	1.33		2.57	3.48
0600	Painted		1200	.027		1.55	1.33		2.88	3.83
0700	Corrugated, on steel frame, natural, .024" thick		1200	.027		1.80	1.33		3.13	4.10
0800	Painted		1200	.027		2.20	1.33		3.53	4.54
0900	.032" thick, natural		1200	.027		3.04	1.33		4.37	5.45
1200	Painted		1200	.027		3.28	1.33		4.61	5.75
9000	Minimum labor/equipment charge	1 Rofc	3	2.667	Job		117		117	209

07 41 13.20 Steel Roofing Panels

		Crew	Daily Output	Labor-Hours	Unit	Material	2018 Bare Costs Labor	2018 Bare Costs Equipment	Total	Total Incl O&P
0010	**STEEL ROOFING PANELS**									
0012	Corrugated or ribbed, on steel framing, 30 ga. galv	G-3	1100	.029	S.F.	1.67	1.45		3.12	4.15
0100	28 ga.		1050	.030		1.44	1.52		2.96	4
0300	26 ga.		1000	.032		2.02	1.59		3.61	4.76
0400	24 ga.		950	.034		3	1.68		4.68	6
0600	Colored, 28 ga.		1050	.030		1.81	1.52		3.33	4.41
0700	26 ga.		1000	.032		2.12	1.59		3.71	4.87
0710	Flat profile, 1-3/4" standing seams, 10" wide, standard finish, 26 ga.		1000	.032		3.93	1.59		5.52	6.85
0715	24 ga.		950	.034		4.60	1.68		6.28	7.75
0720	22 ga.		900	.036		5.70	1.77		7.47	9.10
0725	Zinc aluminum alloy finish, 26 ga.		1000	.032		3.20	1.59		4.79	6.05
0730	24 ga.		950	.034		3.72	1.68		5.40	6.75
0735	22 ga.		900	.036		4.24	1.77		6.01	7.50
0740	12" wide, standard finish, 26 ga.		1000	.032		3.98	1.59		5.57	6.90
0745	24 ga.		950	.034		5.15	1.68		6.83	8.35
0750	Zinc aluminum alloy finish, 26 ga.		1000	.032		4.44	1.59		6.03	7.40
0755	24 ga.		950	.034		3.72	1.68		5.40	6.75
0840	Flat profile, 1" x 3/8" batten, 12" wide, standard finish, 26 ga.		1000	.032		3.46	1.59		5.05	6.35
0845	24 ga.		950	.034		4.08	1.68		5.76	7.15
0850	22 ga.		900	.036		4.91	1.77		6.68	8.25
0855	Zinc aluminum alloy finish, 26 ga.		1000	.032		3.36	1.59		4.95	6.25
0860	24 ga.		950	.034		3.82	1.68		5.50	6.90
0865	22 ga.		900	.036		4.34	1.77		6.11	7.60
0870	16-1/2" wide, standard finish, 24 ga.		950	.034		4.03	1.68		5.71	7.10
0875	22 ga.		900	.036		4.55	1.77		6.32	7.85
0880	Zinc aluminum alloy finish, 24 ga.		950	.034		3.51	1.68		5.19	6.55
0885	22 ga.		900	.036		3.98	1.77		5.75	7.20
0890	Flat profile, 2" x 2" batten, 12" wide, standard finish, 26 ga.		1000	.032		4.03	1.59		5.62	6.95

07 41 Roof Panels

07 41 13 – Metal Roof Panels

07 41 13.20 Steel Roofing Panels		Crew	Daily Output	Labor-Hours	Unit	Material	2018 Bare Costs Labor	Equipment	Total	Total Incl O&P
0895	24 ga.	G-3	950	.034	S.F.	4.86	1.68		6.54	8.05
0900	22 ga.		900	.036		5.90	1.77		7.67	9.35
0905	Zinc aluminum alloy finish, 26 ga.		1000	.032		3.82	1.59		5.41	6.75
0910	24 ga.		950	.034		4.29	1.68		5.97	7.40
0915	22 ga.		900	.036		4.96	1.77		6.73	8.30
0920	16-1/2" wide, standard finish, 24 ga.		950	.034		4.44	1.68		6.12	7.55
0925	22 ga.		900	.036		5.10	1.77		6.87	8.50
0930	Zinc aluminum alloy finish, 24 ga.		950	.034		3.98	1.68		5.66	7.05
0935	22 ga.	↓	900	.036	↓	4.55	1.77		6.32	7.85
9000	Minimum labor/equipment charge	1 Rofc	2	4	Job		176		176	315

07 41 33 – Plastic Roof Panels

07 41 33.10 Fiberglass Panels

		Crew	Daily Output	Labor-Hours	Unit	Material	2018 Bare Costs Labor	Equipment	Total	Total Incl O&P
0010	**FIBERGLASS PANELS**									
0012	Corrugated panels, roofing, 8 oz./S.F.	G-3	1000	.032	S.F.	2.47	1.59		4.06	5.25
0100	12 oz./S.F.		1000	.032		4.52	1.59		6.11	7.50
0300	Corrugated siding, 6 oz./S.F.		880	.036		1.99	1.81		3.80	5.10
0400	8 oz./S.F.		880	.036		2.47	1.81		4.28	5.60
0500	Fire retardant		880	.036		3.98	1.81		5.79	7.25
0600	12 oz. siding, textured		880	.036		3.87	1.81		5.68	7.15
0700	Fire retardant		880	.036		4.48	1.81		6.29	7.80
0900	Flat panels, 6 oz./S.F., clear or colors		880	.036		2.52	1.81		4.33	5.65
1100	Fire retardant, class A		880	.036		3.56	1.81		5.37	6.80
1300	8 oz./S.F., clear or colors	↓	880	.036	↓	2.47	1.81		4.28	5.60
9000	Minimum labor/equipment charge	1 Rofc	2	4	Job		176		176	315

07 42 Wall Panels

07 42 13 – Metal Wall Panels

07 42 13.10 Mansard Panels

		Crew	Daily Output	Labor-Hours	Unit	Material	2018 Bare Costs Labor	Equipment	Total	Total Incl O&P
0010	**MANSARD PANELS**									
0600	Aluminum, stock units, straight surfaces	1 Shee	115	.070	S.F.	4.17	4.16		8.33	11.15
0700	Concave or convex surfaces, add		75	.107	"	2.20	6.40		8.60	12.45
0800	For framing, to 5' high, add		115	.070	L.F.	4	4.16		8.16	10.95
0900	Soffits, to 1' wide		125	.064	S.F.	2.75	3.83		6.58	9.10
9000	Minimum labor/equipment charge	↓	2.50	3.200	Job		191		191	300

07 42 13.20 Aluminum Siding Panels

		Crew	Daily Output	Labor-Hours	Unit	Material	2018 Bare Costs Labor	Equipment	Total	Total Incl O&P
0010	**ALUMINUM SIDING PANELS**									
0012	Corrugated, on steel framing, .019" thick, natural finish	G-3	775	.041	S.F.	1.66	2.06		3.72	5.10
0100	Painted		775	.041		1.82	2.06		3.88	5.30
0400	Farm type, .021" thick on steel frame, natural		775	.041		1.72	2.06		3.78	5.15
0600	Painted		775	.041		1.82	2.06		3.88	5.30
0700	Industrial type, corrugated, on steel, .024" thick, mill		775	.041		2.37	2.06		4.43	5.90
0900	Painted		775	.041		2.52	2.06		4.58	6.05
1000	.032" thick, mill		775	.041		2.64	2.06		4.70	6.20
1200	Painted		775	.041		3.17	2.06		5.23	6.75
1300	V-Beam, on steel frame, .032" thick, mill		775	.041		2.99	2.06		5.05	6.55
1500	Painted		775	.041		3.33	2.06		5.39	6.95
1600	.040" thick, mill		775	.041		3.59	2.06		5.65	7.25
1800	Painted	↓	775	.041		4.17	2.06		6.23	7.85
3800	Horizontal, colored clapboard, 8" wide, plain	2 Carp	515	.031		2.50	1.58		4.08	5.30
3810	Insulated	↓	515	.031	↓	2.88	1.58		4.46	5.75

07 42 Wall Panels

07 42 13 – Metal Wall Panels

07 42 13.20 Aluminum Siding Panels

		Crew	Daily Output	Labor-Hours	Unit	Material	2018 Bare Costs Labor	Equipment	Total	Total Incl O&P
3830	8" embossed, painted	2 Carp	515	.031	S.F.	2.85	1.58		4.43	5.70
3840	Insulated		515	.031		3.13	1.58		4.71	6
3860	12" painted, smooth		600	.027		2.70	1.35		4.05	5.15
3870	Insulated		600	.027		2.93	1.35		4.28	5.40
3890	12" embossed, painted		600	.027		2.88	1.35		4.23	5.35
3900	Insulated		515	.031		2.78	1.58		4.36	5.60
4000	Vertical board & batten, colored, non-insulated	↓	515	.031		2.10	1.58		3.68	4.87
4200	For simulated wood design, add				↓	.15			.15	.17
4300	Corners for above, outside	2 Carp	515	.031	V.L.F.	3.64	1.58		5.22	6.55
4500	Inside corners	"	515	.031	"	1.77	1.58		3.35	4.51
9000	Minimum labor/equipment charge	1 Carp	3	2.667	Job		135		135	219

07 42 13.30 Steel Siding

		Crew	Daily Output	Labor-Hours	Unit	Material	2018 Bare Costs Labor	Equipment	Total	Total Incl O&P
0010	**STEEL SIDING**									
0020	Beveled, vinyl coated, 8" wide	1 Carp	265	.030	S.F.	1.80	1.53		3.33	4.46
0050	10" wide	"	275	.029		1.95	1.47		3.42	4.54
0080	Galv, corrugated or ribbed, on steel frame, 30 ga.	G-3	800	.040		1.25	1.99		3.24	4.56
0100	28 ga.		795	.040		1.35	2.01		3.36	4.69
0300	26 ga.		790	.041		1.78	2.02		3.80	5.20
0400	24 ga.		785	.041		2.25	2.03		4.28	5.70
0600	22 ga.		770	.042		2.35	2.07		4.42	5.90
0700	Colored, corrugated/ribbed, on steel frame, 10 yr. finish, 28 ga.		800	.040		2.05	1.99		4.04	5.45
0900	26 ga.		795	.040		2.11	2.01		4.12	5.50
1000	24 ga.		790	.041		2.36	2.02		4.38	5.80
1020	20 ga.	↓	785	.041	↓	3	2.03		5.03	6.55
9000	Minimum labor/equipment charge	1 Carp	3	2.667	Job		135		135	219

07 46 Siding

07 46 23 – Wood Siding

07 46 23.10 Wood Board Siding

		Crew	Daily Output	Labor-Hours	Unit	Material	2018 Bare Costs Labor	Equipment	Total	Total Incl O&P
0010	**WOOD BOARD SIDING**									
3200	Wood, cedar bevel, A grade, 1/2" x 6"	1 Carp	295	.027	S.F.	4.42	1.38		5.80	7.10
3300	1/2" x 8"		330	.024		7.50	1.23		8.73	10.25
3500	3/4" x 10", clear grade		375	.021		7.30	1.08		8.38	9.75
3600	"B" grade		375	.021		3.99	1.08		5.07	6.15
3800	Cedar, rough sawn, 1" x 4", A grade, natural		220	.036		7.30	1.84		9.14	11.05
3900	Stained		220	.036		7.45	1.84		9.29	11.20
4100	1" x 12", board & batten, #3 & Btr., natural		420	.019		4.76	.97		5.73	6.80
4200	Stained		420	.019		5.10	.97		6.07	7.15
4400	1" x 8" channel siding, #3 & Btr., natural		330	.024		4.74	1.23		5.97	7.20
4500	Stained		330	.024		5	1.23		6.23	7.50
4700	Redwood, clear, beveled, vertical grain, 1/2" x 4"		220	.036		4.81	1.84		6.65	8.30
4750	1/2" x 6"		295	.027		4.80	1.38		6.18	7.55
4800	1/2" x 8"		330	.024		5.20	1.23		6.43	7.70
5000	3/4" x 10"		375	.021		4.95	1.08		6.03	7.20
5200	Channel siding, 1" x 10", B grade		375	.021		4.60	1.08		5.68	6.80
5250	Redwood, T&G boards, B grade, 1" x 4"		220	.036		7.50	1.84		9.34	11.25
5270	1" x 8"		330	.024		7.90	1.23		9.13	10.70
5400	White pine, rough sawn, 1" x 8", natural		330	.024		2.48	1.23		3.71	4.73
5500	Stained		330	.024	↓	2.38	1.23		3.61	4.62
9000	Minimum labor/equipment charge	↓	2	4	Job		203		203	330

For customer support on your Commercial Renovation Costs with RSMeans data, call 800.448.8182.

195

07 46 29 – Plywood Siding

07 46 29.10 Plywood Siding Options	Crew	Daily Output	Labor-Hours	Unit	Material	2018 Bare Costs Labor	Equipment	Total	Total Incl O&P
0010 **PLYWOOD SIDING OPTIONS**									
0900 Plywood, medium density overlaid, 3/8" thick	2 Carp	750	.021	S.F.	1.35	1.08		2.43	3.25
1000 1/2" thick		700	.023		1.55	1.16		2.71	3.59
1100 3/4" thick		650	.025		1.90	1.25		3.15	4.12
1600 Texture 1-11, cedar, 5/8" thick, natural		675	.024		2.60	1.20		3.80	4.81
1700 Factory stained		675	.024		2.87	1.20		4.07	5.10
1900 Texture 1-11, fir, 5/8" thick, natural		675	.024		1.37	1.20		2.57	3.46
2000 Factory stained		675	.024		1.91	1.20		3.11	4.05
2050 Texture 1-11, S.Y.P., 5/8" thick, natural		675	.024		1.44	1.20		2.64	3.53
2100 Factory stained		675	.024		1.51	1.20		2.71	3.61
2200 Rough sawn cedar, 3/8" thick, natural		675	.024		1.27	1.20		2.47	3.35
2300 Factory stained		675	.024		1.57	1.20		2.77	3.68
2500 Rough sawn fir, 3/8" thick, natural		675	.024		.92	1.20		2.12	2.96
2600 Factory stained		675	.024		1.09	1.20		2.29	3.15
2800 Redwood, textured siding, 5/8" thick		675	.024		2.01	1.20		3.21	4.16
3000 Polyvinyl chloride coated, 3/8" thick		750	.021		1.16	1.08		2.24	3.04
9000 Minimum labor/equipment charge	1 Carp	2	4	Job		203		203	330

07 46 33 – Plastic Siding

07 46 33.10 Vinyl Siding

07 46 33.10 Vinyl Siding	Crew	Daily Output	Labor-Hours	Unit	Material	2018 Bare Costs Labor	Equipment	Total	Total Incl O&P
0010 **VINYL SIDING**									
3995 Clapboard profile, woodgrain texture, .048 thick, double 4	2 Carp	495	.032	S.F.	1.07	1.64		2.71	3.84
4000 Double 5		550	.029		1.07	1.47		2.54	3.57
4005 Single 8		495	.032		1.37	1.64		3.01	4.17
4010 Single 10		550	.029		1.65	1.47		3.12	4.20
4015 .044 thick, double 4		495	.032		1.05	1.64		2.69	3.82
4020 Double 5		550	.029		1.07	1.47		2.54	3.57
4025 .042 thick, double 4		495	.032		1.07	1.64		2.71	3.84
4030 Double 5		550	.029		1.07	1.47		2.54	3.57
4035 Cross sawn texture, .040 thick, double 4		495	.032		.73	1.64		2.37	3.47
4040 Double 5		550	.029		.67	1.47		2.14	3.13
4045 Smooth texture, .042 thick, double 4		495	.032		.80	1.64		2.44	3.54
4050 Double 5		550	.029		.80	1.47		2.27	3.27
4055 Single 8		495	.032		.80	1.64		2.44	3.54
4060 Cedar texture, .044 thick, double 4		495	.032		1.14	1.64		2.78	3.92
4065 Double 6		600	.027		1.15	1.35		2.50	3.46
4070 Dutch lap profile, woodgrain texture, .048 thick, double 5		550	.029		1.09	1.47		2.56	3.59
4075 .044 thick, double 4.5		525	.030		1.09	1.55		2.64	3.71
4080 .042 thick, double 4.5		525	.030		.92	1.55		2.47	3.53
4085 .040 thick, double 4.5		525	.030		.73	1.55		2.28	3.32
4100 Shake profile, 10" wide		400	.040		3.71	2.03		5.74	7.40
4105 Vertical pattern, .046 thick, double 5		550	.029		1.56	1.47		3.03	4.11
4110 .044 thick, triple 3		550	.029		1.78	1.47		3.25	4.35
4115 .040 thick, triple 4		550	.029		1.68	1.47		3.15	4.24
4120 .040 thick, triple 2.66		550	.029		1.78	1.47		3.25	4.35
4125 Insulation, fan folded extruded polystyrene, 1/4"		2000	.008		.29	.41		.70	.98
4130 3/8"		2000	.008		.32	.41		.73	1.01
4135 Accessories, J channel, 5/8" pocket		700	.023	L.F.	.51	1.16		1.67	2.44
4140 3/4" pocket		695	.023		.56	1.17		1.73	2.50
4145 1-1/4" pocket		680	.024		.86	1.19		2.05	2.88
4150 Flexible, 3/4" pocket		600	.027		2.39	1.35		3.74	4.82
4155 Under sill finish trim		500	.032		.56	1.62		2.18	3.24
4160 Vinyl starter strip		700	.023		.66	1.16		1.82	2.60

07 46 Siding

07 46 33 – Plastic Siding

07 46 33.10 Vinyl Siding

		Crew	Daily Output	Labor-Hours	Unit	Material	2018 Bare Costs Labor	Equipment	Total	Total Incl O&P
4165	Aluminum starter strip	2 Carp	700	.023	L.F.	.29	1.16		1.45	2.20
4170	Window casing, 2-1/2" wide, 3/4" pocket		510	.031		1.71	1.59		3.30	4.46
4175	Outside corner, woodgrain finish, 4" face, 3/4" pocket		700	.023		2.15	1.16		3.31	4.25
4180	5/8" pocket		700	.023		2.13	1.16		3.29	4.23
4185	Smooth finish, 4" face, 3/4" pocket		700	.023		2.13	1.16		3.29	4.23
4190	7/8" pocket		690	.023		2.03	1.18		3.21	4.15
4195	1-1/4" pocket		700	.023		1.42	1.16		2.58	3.45
4200	Soffit and fascia, 1' overhang, solid		120	.133		4.72	6.75		11.47	16.15
4205	Vented		120	.133		4.72	6.75		11.47	16.15
4207	18" overhang, solid		110	.145		5.50	7.35		12.85	18
4208	Vented		110	.145		5.50	7.35		12.85	18
4210	2' overhang, solid		100	.160		6.30	8.10		14.40	20
4215	Vented		100	.160		6.30	8.10		14.40	20
4217	3' overhang, solid		100	.160		7.85	8.10		15.95	22
4218	Vented		100	.160		7.85	8.10		15.95	22
4220	Colors for siding and soffits, add				S.F.	.15			.15	.17
4225	Colors for accessories and trim, add				L.F.	.31			.31	.34
9000	Minimum labor/equipment charge	1 Carp	3	2.667	Job		135		135	219

07 46 33.20 Polypropylene Siding

		Crew	Daily Output	Labor-Hours	Unit	Material	2018 Bare Costs Labor	Equipment	Total	Total Incl O&P
0010	**POLYPROPYLENE SIDING**									
4090	Shingle profile, random grooves, double 7	2 Carp	400	.040	S.F.	3.24	2.03		5.27	6.85
4092	Cornerpost for above	1 Carp	365	.022	L.F.	13.10	1.11		14.21	16.20
4095	Triple 5	2 Carp	400	.040	S.F.	3.24	2.03		5.27	6.85
4097	Cornerpost for above	1 Carp	365	.022	L.F.	12.30	1.11		13.41	15.35
5000	Staggered butt, double 7"	2 Carp	400	.040	S.F.	3.70	2.03		5.73	7.35
5002	Cornerpost for above	1 Carp	365	.022	L.F.	13.10	1.11		14.21	16.20
5010	Half round, double 6-1/4"	2 Carp	360	.044	S.F.	3.70	2.25		5.95	7.75
5020	Shake profile, staggered butt, double 9"	"	510	.031	"	3.70	1.59		5.29	6.65
5022	Cornerpost for above	1 Carp	365	.022	L.F.	9.80	1.11		10.91	12.55
5030	Straight butt, double 7"	2 Carp	400	.040	S.F.	3.70	2.03		5.73	7.35
5032	Cornerpost for above	1 Carp	365	.022	L.F.	13	1.11		14.11	16.10
6000	Accessories, J channel, 5/8" pocket	2 Carp	700	.023		.51	1.16		1.67	2.44
6010	3/4" pocket		695	.023		.56	1.17		1.73	2.50
6020	1-1/4" pocket		680	.024		.86	1.19		2.05	2.88
6030	Aluminum starter strip		700	.023		.29	1.16		1.45	2.20

07 46 46 – Fiber Cement Siding

07 46 46.10 Fiber Cement Siding

		Crew	Daily Output	Labor-Hours	Unit	Material	2018 Bare Costs Labor	Equipment	Total	Total Incl O&P
0010	**FIBER CEMENT SIDING**									
0020	Lap siding, 5/16" thick, 6" wide, 4-3/4" exposure, smooth texture	2 Carp	415	.039	S.F.	1.31	1.95		3.26	4.61
0025	Woodgrain texture		415	.039		1.31	1.95		3.26	4.61
0030	7-1/2" wide, 6-1/4" exposure, smooth texture		425	.038		1.63	1.91		3.54	4.90
0035	Woodgrain texture		425	.038		1.63	1.91		3.54	4.90
0040	8" wide, 6-3/4" exposure, smooth texture		425	.038		1.23	1.91		3.14	4.45
0045	Rough sawn texture		425	.038		1.23	1.91		3.14	4.45
0050	9-1/2" wide, 8-1/4" exposure, smooth texture		440	.036		1.29	1.84		3.13	4.41
0055	Woodgrain texture		440	.036		1.29	1.84		3.13	4.41
0060	12" wide, 10-3/8" exposure, smooth texture		455	.035		2.09	1.78		3.87	5.20
0065	Woodgrain texture		455	.035		2.09	1.78		3.87	5.20
0070	Panel siding, 5/16" thick, smooth texture		750	.021		1.33	1.08		2.41	3.22
0075	Stucco texture		750	.021		1.33	1.08		2.41	3.22
0080	Grooved woodgrain texture		750	.021		1.33	1.08		2.41	3.22
0085	V - grooved woodgrain texture		750	.021		1.33	1.08		2.41	3.22

07 46 Siding

07 46 46 – Fiber Cement Siding

07 46 46.10 Fiber Cement Siding	Crew	Daily Output	Labor-Hours	Unit	Material	2018 Bare Costs Labor	Equipment	Total	Total Incl O&P
0088	Shingle siding, 48" x 15-1/4" panels, 7" exposure	2 Carp	700	.023	S.F.	4.19	1.16		5.35	6.50
0090	Wood starter strip	↓	400	.040	L.F.	.45	2.03		2.48	3.79

07 46 73 – Soffit

07 46 73.10 Soffit Options

		Crew	Daily Output	Labor-Hours	Unit	Material	Labor	Equipment	Total	Total Incl O&P
0010	**SOFFIT OPTIONS**									
0012	Aluminum, residential, .020" thick	1 Carp	210	.038	S.F.	2.07	1.93		4	5.40
0100	Baked enamel on steel, 16 or 18 ga.		105	.076		6.05	3.86		9.91	12.90
0300	Polyvinyl chloride, white, solid		230	.035		2.17	1.76		3.93	5.25
0400	Perforated	↓	230	.035		2.17	1.76		3.93	5.25
0500	For colors, add				↓	.15			.15	.17
9000	Minimum labor/equipment charge	1 Carp	3	2.667	Job		135		135	219

07 51 Built-Up Bituminous Roofing

07 51 13 – Built-Up Asphalt Roofing

07 51 13.10 Built-Up Roofing Components

		Crew	Daily Output	Labor-Hours	Unit	Material	Labor	Equipment	Total	Total Incl O&P
0010	**BUILT-UP ROOFING COMPONENTS**									
0012	Asphalt saturated felt, #30, 2 sq./roll	1 Rofc	58	.138	Sq.	10.30	6.05		16.35	22
0200	#15, 4 sq./roll, plain or perforated, not mopped		58	.138		5.35	6.05		11.40	16.70
0300	Roll roofing, smooth, #65		15	.533		10.40	23.50		33.90	53.50
0500	#90		12	.667		37	29.50		66.50	92.50
0520	Mineralized		12	.667		36	29.50		65.50	91.50
0540	D.C. (double coverage), 19" selvage edge	↓	10	.800	↓	48	35		83	115
0580	Adhesive (lap cement)				Gal.	8.45			8.45	9.30
0800	Steep, flat or dead level asphalt, 10 ton lots, packaged				Ton	965			965	1,075
9000	Minimum labor/equipment charge	1 Rofc	4	2	Job		88		88	157

07 51 13.13 Cold-Applied Built-Up Asphalt Roofing

		Crew	Daily Output	Labor-Hours	Unit	Material	Labor	Equipment	Total	Total Incl O&P
0010	**COLD-APPLIED BUILT-UP ASPHALT ROOFING**									
0020	3 ply system, installation only (components listed below)	G-5	50	.800	Sq.		32	3.62	35.62	61
0100	Spunbond poly. fabric, 1.35 oz./S.Y., 36"W, 10.8 sq./roll				Ea.	133			133	147
0500	Base & finish coat, 3 gal./sq., 5 gal./can				Gal.	8			8	8.80
0600	Coating, ceramic granules, 1/2 sq./bag				Ea.	23.50			23.50	26
0700	Aluminum, 2 gal./sq.				Gal.	11.80			11.80	13
0800	Emulsion, fibered or non-fibered, 4 gal./sq.				"	6.30			6.30	6.95

07 51 13.20 Built-Up Roofing Systems

		Crew	Daily Output	Labor-Hours	Unit	Material	Labor	Equipment	Total	Total Incl O&P
0010	**BUILT-UP ROOFING SYSTEMS**									
0120	Asphalt flood coat with gravel/slag surfacing, not including									
0140	Insulation, flashing or wood nailers									
0200	Asphalt base sheet, 3 plies #15 asphalt felt, mopped	G-1	22	2.545	Sq.	104	105	23.50	232.50	325
0350	On nailable decks		21	2.667		107	109	24.50	240.50	340
0500	4 plies #15 asphalt felt, mopped		20	2.800		142	115	25.50	282.50	390
0550	On nailable decks		19	2.947		126	121	27	274	385
0700	Coated glass base sheet, 2 plies glass (type IV), mopped		22	2.545		111	105	23.50	239.50	335
0850	3 plies glass, mopped		20	2.800		134	115	25.50	274.50	380
0950	On nailable decks		19	2.947		126	121	27	274	385
1100	4 plies glass fiber felt (type IV), mopped		20	2.800		165	115	25.50	305.50	415
1150	On nailable decks		19	2.947		149	121	27	297	410
1200	Coated & saturated base sheet, 3 plies #15 asph. felt, mopped		20	2.800		116	115	25.50	256.50	360
1250	On nailable decks		19	2.947		108	121	27	256	365
1300	4 plies #15 asphalt felt, mopped	↓	22	2.545	↓	135	105	23.50	263.50	360
2000	Asphalt flood coat, smooth surface									

07 51 13.20 Built-Up Roofing Systems

		Crew	Daily Output	Labor-Hours	Unit	Material	2018 Bare Costs Labor	Equipment	Total	Total Incl O&P
2200	Asphalt base sheet & 3 plies #15 asphalt felt, mopped	G-1	24	2.333	Sq.	110	96	21.50	227.50	315
2400	On nailable decks		23	2.435		102	100	22.50	224.50	315
2600	4 plies #15 asphalt felt, mopped		24	2.333		129	96	21.50	246.50	335
2700	On nailable decks		23	2.435		121	100	22.50	243.50	335
2900	Coated glass fiber base sheet, mopped, and 2 plies of									
2910	glass fiber felt (type IV)	G-1	25	2.240	Sq.	106	92	20.50	218.50	305
3100	On nailable decks		24	2.333		100	96	21.50	217.50	305
3200	3 plies, mopped		23	2.435		129	100	22.50	251.50	345
3300	On nailable decks		22	2.545		121	105	23.50	249.50	345
3800	4 plies glass fiber felt (type IV), mopped		23	2.435		152	100	22.50	274.50	370
3900	On nailable decks		22	2.545		144	105	23.50	272.50	370
4000	Coated & saturated base sheet, 3 plies #15 asph. felt, mopped		24	2.333		111	96	21.50	228.50	315
4200	On nailable decks		23	2.435		103	100	22.50	225.50	315
4300	4 plies #15 organic felt, mopped		22	2.545		130	105	23.50	258.50	355
4500	Coal tar pitch with gravel/slag surfacing									
4600	4 plies #15 tarred felt, mopped	G-1	21	2.667	Sq.	199	109	24.50	332.50	440
4800	3 plies glass fiber felt (type IV), mopped	"	19	2.947	"	164	121	27	312	425
5000	Coated glass fiber base sheet, and 2 plies of									
5010	glass fiber felt (type IV), mopped	G-1	19	2.947	Sq.	170	121	27	318	435
5300	On nailable decks		18	3.111		148	128	28.50	304.50	425
5600	4 plies glass fiber felt (type IV), mopped		21	2.667		228	109	24.50	361.50	475
5800	On nailable decks		20	2.800		207	115	25.50	347.50	460

07 51 13.30 Cants

		Crew	Daily Output	Labor-Hours	Unit	Material	Labor	Equipment	Total	Total Incl O&P
0010	**CANTS**									
0012	Lumber, treated, 4" x 4" cut diagonally	1 Rofc	325	.025	L.F.	1.91	1.08		2.99	4.03
0300	Mineral or fiber, trapezoidal, 1" x 4" x 48"		325	.025		.30	1.08		1.38	2.26
0400	1-1/2" x 5-5/8" x 48"		325	.025		.48	1.08		1.56	2.46
9000	Minimum labor/equipment charge		4	2	Job		88		88	157

07 51 13.40 Felts

		Crew	Daily Output	Labor-Hours	Unit	Material	Labor	Equipment	Total	Total Incl O&P
0010	**FELTS**									
0012	Glass fibered roofing felt, #15, not mopped	1 Rofc	58	.138	Sq.	9.40	6.05		15.45	21
0300	Base sheet, #80, channel vented		58	.138		43.50	6.05		49.55	59
0400	#70, coated		58	.138		18.20	6.05		24.25	31
0500	Cap, #87, mineral surfaced		58	.138		83	6.05		89.05	102
0600	Flashing membrane, #65		16	.500		10.40	22		32.40	50.50
0800	Coal tar fibered, #15, no mopping		58	.138		17.75	6.05		23.80	30.50
0900	Asphalt felt, #15, 4 sq./roll, no mopping		58	.138		5.35	6.05		11.40	16.70
1100	#30, 2 sq./roll		58	.138		10.30	6.05		16.35	22
1200	Double coated, #33		58	.138		11.20	6.05		17.25	23
1400	#40, base sheet		58	.138		11.35	6.05		17.40	23.50
1450	Coated and saturated		58	.138		12.35	6.05		18.40	24.50
1500	Tarred felt, organic, #15, 4 sq. rolls		58	.138		12.90	6.05		18.95	25
1550	#30, 2 sq. roll		58	.138		25.50	6.05		31.55	39
1700	Add for mopping above felts, per ply, asphalt, 24 lb./sq.	G-1	192	.292		11.60	12	2.68	26.28	37
1800	Coal tar mopping, 30 lb./sq.		186	.301		18.90	12.35	2.76	34.01	46
1900	Flood coat, with asphalt, 60 lb./sq.		60	.933		29	38.50	8.55	76.05	110
2000	With coal tar, 75 lb./sq.		56	1		47.50	41	9.20	97.70	135
9000	Minimum labor/equipment charge	1 Rofc	4	2	Job		88		88	157

07 51 13.50 Walkways for Built-Up Roofs

		Crew	Daily Output	Labor-Hours	Unit	Material	Labor	Equipment	Total	Total Incl O&P
0010	**WALKWAYS FOR BUILT-UP ROOFS**									
0020	Asphalt impregnated, 3' x 6' x 1/2" thick	1 Rofc	400	.020	S.F.	1.87	.88		2.75	3.62
0100	3' x 3' x 3/4" thick		400	.020	"	5.35	.88		6.23	7.45

07 51 Built-Up Bituminous Roofing

07 51 13 – Built-Up Asphalt Roofing

07 51 13.50 Walkways for Built-Up Roofs		Crew	Daily Output	Labor-Hours	Unit	Material	2018 Bare Costs Labor	Equipment	Total	Total Incl O&P
0600	100% recycled rubber, 3' x 4' x 3/8"	1 Rofc ⒼＧ	400	.020	L.F.	6.50	.88		7.38	8.70
0610	3' x 4' x 1/2"	ⒼＧ	400	.020		6.85	.88		7.73	9.10
0620	3' x 4' x 3/4"	ⒼＧ	400	.020	↓	8.50	.88		9.38	10.90
9000	Minimum labor/equipment charge	↓	2.75	2.909	Job		128		128	228

07 52 Modified Bituminous Membrane Roofing

07 52 13 – Atactic-Polypropylene-Modified Bituminous Membrane Roofing

07 52 13.10 APP Modified Bituminous Membrane

07 52 13.10 APP Modified Bituminous Membrane		Crew	Daily Output	Labor-Hours	Unit	Material	2018 Bare Costs Labor	Equipment	Total	Total Incl O&P
0010	**APP MODIFIED BITUMINOUS MEMBRANE** R075213-30									
0020	Base sheet, #15 glass fiber felt, nailed to deck	1 Rofc	58	.138	Sq.	10.95	6.05		17	23
0030	Spot mopped to deck	G-1	295	.190		15.20	7.80	1.74	24.74	32.50
0040	Fully mopped to deck	"	192	.292		21	12	2.68	35.68	47.50
0050	#15 organic felt, nailed to deck	1 Rofc	58	.138		6.90	6.05		12.95	18.40
0060	Spot mopped to deck	G-1	295	.190		11.15	7.80	1.74	20.69	28
0070	Fully mopped to deck	"	192	.292	↓	16.95	12	2.68	31.63	43
2100	APP mod., smooth surf. cap sheet, poly. reinf., torched, 160 mils	G-5	2100	.019	S.F.	.76	.76	.09	1.61	2.29
2150	170 mils		2100	.019		.77	.76	.09	1.62	2.30
2200	Granule surface cap sheet, poly. reinf., torched, 180 mils		2000	.020		.96	.80	.09	1.85	2.58
2250	Smooth surface flashing, torched, 160 mils		1260	.032		.76	1.27	.14	2.17	3.26
2300	170 mils		1260	.032		.77	1.27	.14	2.18	3.27
2350	Granule surface flashing, torched, 180 mils	↓	1260	.032		.96	1.27	.14	2.37	3.48
2400	Fibrated aluminum coating	1 Rofc	3800	.002	↓	.08	.09		.17	.26
2450	Seam heat welding	"	205	.039	L.F.	.08	1.71		1.79	3.15

07 52 16 – Styrene-Butadiene-Styrene Modified Bituminous Membrane Roofing

07 52 16.10 SBS Modified Bituminous Membrane

07 52 16.10 SBS Modified Bituminous Membrane		Crew	Daily Output	Labor-Hours	Unit	Material	2018 Bare Costs Labor	Equipment	Total	Total Incl O&P
0010	**SBS MODIFIED BITUMINOUS MEMBRANE**									
0080	Mod. bit. rfng., SBS mod, gran surf. cap sheet, poly. reinf.									
1150	For reflective granules, add				S.F.	.71	1.02	.28	2.01	2.93
1600	Smooth surface cap sheet, mopped, 145 mils	G-1	2100	.027		.82	1.10	.24	2.16	3.12
1620	Lightweight base sheet, fiberglass reinforced, 35 to 47 mil		2100	.027		.29	1.10	.24	1.63	2.54
1625	Heavyweight base/ply sheet, reinforced, 87 to 120 mil thick	↓	2100	.027		.93	1.10	.24	2.27	3.24
1650	Granulated walkpad, 180 to 220 mils	1 Rofc	400	.020		1.88	.88		2.76	3.64
1700	Smooth surface flashing, 145 mils	G-1	1260	.044		.82	1.82	.41	3.05	4.60
1800	150 mils		1260	.044		.51	1.82	.41	2.74	4.26
1900	Granular surface flashing, 150 mils		1260	.044		.72	1.82	.41	2.95	4.49
2000	160 mils	↓	1260	.044		.75	1.82	.41	2.98	4.53
2010	Elastomeric asphalt primer	1 Rofc	2600	.003		.17	.14		.31	.43
2015	Roofing asphalt, 30 lb./square	G-1	19000	.003		.15	.12	.03	.30	.41
2020	Cold process adhesive, 20 to 30 mils thick	1 Rofc	750	.011		.25	.47		.72	1.12
2025	Self adhering vapor retarder, 30 to 45 mils thick	G-5	2150	.019	↓	1.07	.74	.08	1.89	2.59
2050	Seam heat welding	1 Rofc	205	.039	L.F.	.08	1.71		1.79	3.15

For customer support on your Commercial Renovation Costs with RSMeans data, call 800.448.8182.

07 53 Elastomeric Membrane Roofing

07 53 16 – Chlorosulfonate-Polyethylene Roofing

07 53 16.10 Chlorosulfonated Polyethylene Roofing

07 53 16.10 Chlorosulfonated Polyethylene Roofing	Crew	Daily Output	Labor-Hours	Unit	Material	2018 Bare Costs Labor	Equipment	Total	Total Incl O&P
0010 **CHLOROSULFONATED POLYETHYLENE ROOFING**									
0800 Chlorosulfonated polyethylene (CSPE)									
0900 45 mils, heat welded seams, plate attachment	G-5	35	1.143	Sq.	241	45.50	5.20	291.70	355
1100 Heat welded seams, plate attachment and ballasted		26	1.538		252	61.50	6.95	320.45	395
1200 60 mils, heat welded seams, plate attachment		35	1.143		310	45.50	5.20	360.70	430
1300 Heat welded seams, plate attachment and ballasted		26	1.538		320	61.50	6.95	388.45	470

07 53 23 – Ethylene-Propylene-Diene-Monomer Roofing

07 53 23.20 Ethylene-Propylene-Diene-Monomer Roofing

07 53 23.20 Ethylene-Propylene-Diene-Monomer Roofing	Crew	Daily Output	Labor-Hours	Unit	Material	2018 Bare Costs Labor	Equipment	Total	Total Incl O&P
0010 **ETHYLENE-PROPYLENE-DIENE-MONOMER ROOFING (EPDM)**									
3500 Ethylene-propylene-diene-monomer (EPDM), 45 mils, 0.28 psf									
3600 Loose-laid & ballasted with stone (10 psf)	G-5	51	.784	Sq.	85	31.50	3.55	120.05	153
3700 Mechanically attached		35	1.143		79	45.50	5.20	129.70	174
3800 Fully adhered with adhesive		26	1.538		112	61.50	6.95	180.45	241
4500 60 mils, 0.40 psf									
4600 Loose-laid & ballasted with stone (10 psf)	G-5	51	.784	Sq.	101	31.50	3.55	136.05	171
4700 Mechanically attached		35	1.143		94	45.50	5.20	144.70	190
4800 Fully adhered with adhesive		26	1.538		127	61.50	6.95	195.45	257
4810 45 mil, 0.28 psf, membrane only					49			49	54
4820 60 mil, 0.40 psf, membrane only					62.50			62.50	68.50
4850 Seam tape for membrane, 3" x 100' roll				Ea.	44.50			44.50	49
4900 Batten strips, 10' sections					3.94			3.94	4.33
4910 Cover tape for batten strips, 6" x 100' roll					187			187	206
4930 Plate anchors				M	81			81	89
4970 Adhesive for fully adhered systems, 60 S.F./gal.				Gal.	20.50			20.50	22.50

07 53 29 – Polyisobutylene Roofing

07 53 29.10 Polyisobutylene Roofing

07 53 29.10 Polyisobutylene Roofing	Crew	Daily Output	Labor-Hours	Unit	Material	2018 Bare Costs Labor	Equipment	Total	Total Incl O&P
0010 **POLYISOBUTYLENE ROOFING**									
7500 Polyisobutylene (PIB), 100 mils, 0.57 psf									
7600 Loose-laid & ballasted with stone/gravel (10 psf)	G-5	51	.784	Sq.	211	31.50	3.55	246.05	292
7700 Partially adhered with adhesive		35	1.143		251	45.50	5.20	301.70	365
7800 Hot asphalt attachment		35	1.143		241	45.50	5.20	291.70	350
7900 Fully adhered with contact cement		26	1.538		262	61.50	6.95	330.45	405

07 54 Thermoplastic Membrane Roofing

07 54 19 – Polyvinyl-Chloride Roofing

07 54 19.10 Polyvinyl-Chloride Roofing (PVC)

07 54 19.10 Polyvinyl-Chloride Roofing (PVC)	Crew	Daily Output	Labor-Hours	Unit	Material	2018 Bare Costs Labor	Equipment	Total	Total Incl O&P
0010 **POLYVINYL-CHLORIDE ROOFING (PVC)**									
8200 Heat welded seams									
8700 Reinforced, 48 mils, 0.33 psf									
8750 Loose-laid & ballasted with stone/gravel (12 psf)	G-5	51	.784	Sq.	113	31.50	3.55	148.05	184
8800 Mechanically attached		35	1.143		106	45.50	5.20	156.70	204
8850 Fully adhered with adhesive		26	1.538		153	61.50	6.95	221.45	285
8860 Reinforced, 60 mils, 0.40 psf									
8870 Loose-laid & ballasted with stone/gravel (12 psf)	G-5	51	.784	Sq.	114	31.50	3.55	149.05	185
8880 Mechanically attached		35	1.143		107	45.50	5.20	157.70	205
8890 Fully adhered with adhesive		26	1.538		154	61.50	6.95	222.45	286

For customer support on your Commercial Renovation Costs with RSMeans data, call 800.448.8182.

201

07 54 Thermoplastic Membrane Roofing

07 54 23 – Thermoplastic-Polyolefin Roofing

07 54 23.10 Thermoplastic Polyolefin Roofing (T.P.O.)	Crew	Daily Output	Labor-Hours	Unit	Material	2018 Bare Costs Labor	Equipment	Total	Total Incl O&P
0010 **THERMOPLASTIC POLYOLEFIN ROOFING (T.P.O.)**									
0100 45 mil, loose laid & ballasted with stone (1/2 ton/sq.)	G-5	51	.784	Sq.	88	31.50	3.55	123.05	157
0120 Fully adhered		25	1.600		78.50	64	7.25	149.75	208
0140 Mechanically attached		34	1.176		78	47	5.35	130.35	175
0160 Self adhered		35	1.143		78.50	45.50	5.20	129.20	174
0180 60 mil membrane, heat welded seams, ballasted		50	.800		100	32	3.62	135.62	171
0200 Fully adhered		25	1.600		90	64	7.25	161.25	221
0220 Mechanically attached		34	1.176		94	47	5.35	146.35	193
0240 Self adhered		35	1.143		106	45.50	5.20	156.70	204

07 54 30 – Ketone Ethylene Ester Roofing

07 54 30.10 Ketone Ethylene Ester Roofing

	Crew	Daily Output	Labor-Hours	Unit	Material	Labor	Equipment	Total	Total Incl O&P
0010 **KETONE ETHYLENE ESTER ROOFING**									
0100 Ketone ethylene ester roofing, 50 mil, fully adhered	G-5	26	1.538	Sq.	202	61.50	6.95	270.45	340
0120 Mechanically attached		35	1.143		131	45.50	5.20	181.70	231
0140 Ballasted with stone		51	.784		138	31.50	3.55	173.05	212
0160 50 mil, fleece backed, adhered w/hot asphalt	G-1	26	2.154		162	88.50	19.75	270.25	360
0180 Accessories, pipe boot	1 Rofc	32	.250	Ea.	25.50	11		36.50	48
0200 Pre-formed corners		32	.250	"	8.90	11		19.90	29.50
0220 Ketone clad metal, including up to 4 bends		330	.024	S.F.	4.12	1.07		5.19	6.45
0240 Walkway pad	2 Rofc	800	.020	"	4.33	.88		5.21	6.35
0260 Stripping material	1 Rofc	310	.026	L.F.	1	1.13		2.13	3.12

07 55 Protected Membrane Roofing

07 55 10 – Protected Membrane Roofing Components

07 55 10.10 Protected Membrane Roofing Components

	Crew	Daily Output	Labor-Hours	Unit	Material	Labor	Equipment	Total	Total Incl O&P
0010 **PROTECTED MEMBRANE ROOFING COMPONENTS**									
0100 Choose roofing membrane from 07 50									
0120 Then choose roof deck insulation from 07 22									
0130 Filter fabric	2 Rofc	10000	.002	S.F.	.09	.07		.16	.23
0140 Ballast, 3/8" - 1/2" in place	G-1	36	1.556	Ton	20.50	64	14.30	98.80	152
0150 3/4" - 1-1/2" in place	"	36	1.556	"	20.50	64	14.30	98.80	152
0200 2" concrete blocks, natural	1 Clab	115	.070	S.F.	3.51	2.77		6.28	8.35
0210 Colors	"	115	.070	"	3.77	2.77		6.54	8.65

07 56 Fluid-Applied Roofing

07 56 10 – Fluid-Applied Roofing Elastomers

07 56 10.10 Elastomeric Roofing

	Crew	Daily Output	Labor-Hours	Unit	Material	Labor	Equipment	Total	Total Incl O&P
0010 **ELASTOMERIC ROOFING**									
0020 Acrylic, 44% solids, 2 coats, on corrugated metal	2 Rofc	2400	.007	S.F.	.56	.29		.85	1.14
0025 On smooth metal		3000	.005		.45	.23		.68	.92
0030 On foam or modified bitumen		1500	.011		.90	.47		1.37	1.83
0035 On concrete		1500	.011		.90	.47		1.37	1.83
0040 On tar and gravel		1500	.011		.90	.47		1.37	1.83
0045 36% solids, 2 coats, on corrugated metal		2400	.007		.52	.29		.81	1.10
0050 On smooth metal		3000	.005		.42	.23		.65	.88
0055 On foam or modified bitumen		1500	.011		.84	.47		1.31	1.76
0060 On concrete		1500	.011		.84	.47		1.31	1.76
0065 On tar and gravel		1500	.011		.84	.47		1.31	1.76

07 56 Fluid-Applied Roofing

07 56 10 – Fluid-Applied Roofing Elastomers

07 56 10.10 Elastomeric Roofing	Crew	Daily Output	Labor-Hours	Unit	Material	2018 Bare Costs Labor	Equipment	Total	Total Incl O&P
0070	Primer if required, 2 coats on corrugated metal	2 Rofc	2400	.007	S.F.	.51	.29		.80	1.08
0075	On smooth metal		3000	.005		.51	.23		.74	.98
0080	On foam or modified bitumen		1500	.011		.74	.47		1.21	1.66
0085	On concrete		1500	.011		.56	.47		1.03	1.45
0090	On tar & gravel/rolled roof		1500	.011		.74	.47		1.21	1.66
0110	Acrylic rubber, fluid applied, 20 mils thick	G-5	2000	.020		2.22	.80	.09	3.11	3.96
0120	50 mils, reinforced		1200	.033		3.28	1.33	.15	4.76	6.15
0130	For walking surface, add		900	.044		1.22	1.77	.20	3.19	4.72
0300	Neoprene, fluid applied, 20 mil thick, not reinforced	G-1	1135	.049		1.44	2.03	.45	3.92	5.70
0600	Non-woven polyester, reinforced		960	.058		1.57	2.40	.54	4.51	6.60
0700	5 coat neoprene deck, 60 mil thick, under 10,000 S.F.		325	.172		4.81	7.10	1.58	13.49	19.65
0900	Over 10,000 S.F.		625	.090		4.81	3.68	.82	9.31	12.75
9000	Minimum labor/equipment charge	1 Rofc	2	4	Job		176		176	315

07 57 Coated Foamed Roofing

07 57 13 – Sprayed Polyurethane Foam Roofing

07 57 13.10 Sprayed Polyurethane Foam Roofing (S.P.F.)	Crew	Daily Output	Labor-Hours	Unit	Material	2018 Bare Costs Labor	Equipment	Total	Total Incl O&P
0010	**SPRAYED POLYURETHANE FOAM ROOFING (S.P.F.)**									
0100	Primer for metal substrate (when required)	G-2A	3000	.008	S.F.	.49	.31	.20	1	1.31
0200	Primer for non-metal substrate (when required)		3000	.008		.19	.31	.20	.70	.98
0300	Closed cell spray, polyurethane foam, 3 lb./C.F. density, 1", R6.7		15000	.002		.64	.06	.04	.74	.87
0400	2", R13.4		13125	.002		1.29	.07	.05	1.41	1.58
0500	3", R18.6		11485	.002		1.93	.08	.05	2.06	2.32
0550	4", R24.8		10080	.002		2.57	.09	.06	2.72	3.06
0700	Spray-on silicone coating		2500	.010		1.24	.37	.25	1.06	2.28
0800	Warranty 5-20 year manufacturer's								.15	.15
0900	Warranty 20 year, no dollar limit								.20	.20

07 58 Roll Roofing

07 58 10 – Asphalt Roll Roofing

07 58 10.10 Roll Roofing	Crew	Daily Output	Labor-Hours	Unit	Material	2018 Bare Costs Labor	Equipment	Total	Total Incl O&P
0010	**ROLL ROOFING**									
0100	Asphalt, mineral surface									
0200	1 ply #15 organic felt, 1 ply mineral surfaced									
0300	Selvage roofing, lap 19", nailed & mopped	G-1	27	2.074	Sq.	71	85	19.05	175.05	251
0400	3 plies glass fiber felt (type IV), 1 ply mineral surfaced									
0500	Selvage roofing, lapped 19", mopped	G-1	25	2.240	Sq.	123	92	20.50	235.50	320
0600	Coated glass fiber base sheet, 2 plies of glass fiber									
0700	Felt (type IV), 1 ply mineral surfaced selvage									
0800	Roofing, lapped 19", mopped	G-1	25	2.240	Sq.	131	92	20.50	243.50	330
0900	On nailable decks	"	24	2.333	"	120	96	21.50	237.50	325
1000	3 plies glass fiber felt (type III), 1 ply mineral surfaced									
1100	Selvage roofing, lapped 19", mopped	G-1	25	2.240	Sq.	123	92	20.50	235.50	320

For customer support on your Commercial Renovation Costs with RSMeans data, call 800.448.8182.

203

07 61 Sheet Metal Roofing

07 61 13 – Standing Seam Sheet Metal Roofing

07 61 13.10 Standing Seam Sheet Metal Roofing, Field Fab.	Crew	Daily Output	Labor-Hours	Unit	Material	2018 Bare Costs Labor	Equipment	Total	Total Incl O&P
0010 **STANDING SEAM SHEET METAL ROOFING, FIELD FABRICATED**									
0400 Copper standing seam roofing, over 10 squares, 16 oz., 125 lb./sq.	1 Shee	1.30	6.154	Sq.	1,025	370		1,395	1,700
0600 18 oz., 140 lb./sq.		1.20	6.667		1,125	400		1,525	1,875
0700 20 oz., 150 lb./sq.		1.10	7.273		1,350	435		1,785	2,150
1200 For abnormal conditions or small areas, add					25%	100%			
1300 For lead-coated copper, add					25%				
1400 Lead standing seam roofing, 5 lb./S.F.	1 Shee	1.30	6.154		1,575	370		1,945	2,300
1500 Zinc standing seam roofing, .020" thick		1.20	6.667		1,275	400		1,675	2,025
1510 .027" thick		1.15	6.957		1,600	415		2,015	2,400
1520 .032" thick		1.10	7.273		1,825	435		2,260	2,675
1530 .040" thick		1.05	7.619		2,350	455		2,805	3,300
9000 Minimum labor/equipment charge		2	4	Job		239		239	375

07 61 16 – Batten Seam Sheet Metal Roofing

07 61 16.10 Batten Seam Sheet Metal Roofing, Field Fab.

	Crew	Daily Output	Labor-Hours	Unit	Material	2018 Bare Costs Labor	Equipment	Total	Total Incl O&P
0010 **BATTEN SEAM SHEET METAL ROOFING, FIELD FABRICATED**									
0012 Copper batten seam roofing, over 10 sq., 16 oz., 130 lb./sq.	1 Shee	1.10	7.273	Sq.	1,275	435		1,710	2,075
0020 Lead batten seam roofing, 5 lb./S.F.		1.20	6.667		1,650	400		2,050	2,425
0100 Zinc/copper alloy batten seam roofing, .020" thick		1.20	6.667		1,325	400		1,725	2,075
0200 Copper roofing, batten seam, over 10 sq., 18 oz., 145 lb./sq.		1	8		1,425	480		1,905	2,325
0300 20 oz., 160 lb./sq.		1	8		1,700	480		2,180	2,625
0500 Stainless steel batten seam roofing, type 304, 28 ga.		1.20	6.667		655	400		1,055	1,350
0600 26 ga.		1.15	6.957		755	415		1,170	1,475
0800 Zinc, copper alloy roofing, batten seam, .027" thick		1.15	6.957		1,675	415		2,090	2,475
0900 .032" thick		1.10	7.273		1,900	435		2,335	2,775
1000 .040" thick		1.05	7.619		2,450	455		2,905	3,425
9000 Minimum labor/equipment charge		2	4	Job		239		239	375

07 61 19 – Flat Seam Sheet Metal Roofing

07 61 19.10 Flat Seam Sheet Metal Roofing, Field Fabricated

	Crew	Daily Output	Labor-Hours	Unit	Material	2018 Bare Costs Labor	Equipment	Total	Total Incl O&P
0010 **FLAT SEAM SHEET METAL ROOFING, FIELD FABRICATED**									
0900 Copper flat seam roofing, over 10 squares, 16 oz., 115 lb./sq.	1 Shee	1.20	6.667	Sq.	940	400		1,340	1,650
0950 18 oz., 130 lb./sq.		1.15	6.957		1,050	415		1,465	1,800
1000 20 oz., 145 lb./sq.		1.10	7.273		1,250	435		1,685	2,050
1008 Zinc flat seam roofing, .020" thick		1.20	6.667		1,125	400		1,525	1,875
1010 .027" thick		1.15	6.957		1,425	415		1,840	2,225
1020 .032" thick		1.12	7.143		1,625	425		2,050	2,450
1030 .040" thick		1.05	7.619		2,100	455		2,555	3,025
1100 Lead flat seam roofing, 5 lb./S.F.		1.30	6.154		1,400	370		1,770	2,125
9000 Minimum labor/equipment charge		2.75	2.909	Job		174		174	274

07 62 Sheet Metal Flashing and Trim

07 62 10 – Sheet Metal Trim

07 62 10.10 Sheet Metal Cladding

	Crew	Daily Output	Labor-Hours	Unit	Material	2018 Bare Costs Labor	Equipment	Total	Total Incl O&P
0010 **SHEET METAL CLADDING**									
0100 Aluminum, up to 6 bends, .032" thick, window casing	1 Carp	180	.044	S.F.	1.80	2.25		4.05	5.65
0200 Window sill		72	.111	L.F.	1.80	5.65		7.45	11.15
0300 Door casing		180	.044	S.F.	1.80	2.25		4.05	5.65
0400 Fascia		250	.032		1.80	1.62		3.42	4.61
0500 Rake trim		225	.036		1.80	1.80		3.60	4.91
0700 .024" thick, window casing		180	.044		1.42	2.25		3.67	5.20
0800 Window sill		72	.111	L.F.	1.42	5.65		7.07	10.70

07 62 Sheet Metal Flashing and Trim

07 62 10 - Sheet Metal Trim

07 62 10.10 Sheet Metal Cladding		Crew	Daily Output	Labor-Hours	Unit	Material	2018 Bare Costs Labor	Equipment	Total	Total Incl O&P
0900	Door casing	1 Carp	180	.044	S.F.	1.42	2.25		3.67	5.20
1000	Fascia		250	.032		1.42	1.62		3.04	4.19
1100	Rake trim		225	.036		1.42	1.80		3.22	4.49
1200	Vinyl coated aluminum, up to 6 bends, window casing		180	.044		1.83	2.25		4.08	5.65
1300	Window sill		72	.111	L.F.	1.83	5.65		7.48	11.15
1400	Door casing		180	.044	S.F.	1.83	2.25		4.08	5.65
1500	Fascia		250	.032		1.83	1.62		3.45	4.64
1600	Rake trim		225	.036		1.83	1.80		3.63	4.94

07 65 Flexible Flashing

07 65 10 - Sheet Metal Flashing

07 65 10.10 Sheet Metal Flashing and Counter Flashing

		Crew	Daily Output	Labor-Hours	Unit	Material	2018 Bare Costs Labor	Equipment	Total	Total Incl O&P
0010	**SHEET METAL FLASHING AND COUNTER FLASHING**									
0011	Including up to 4 bends									
0020	Aluminum, mill finish, .013" thick	1 Rofc	145	.055	S.F.	.83	2.42		3.25	5.25
0030	.016" thick		145	.055		.98	2.42		3.40	5.40
0060	.019" thick		145	.055		1.40	2.42		3.82	5.85
0100	.032" thick		145	.055		1.35	2.42		3.77	5.80
0200	.040" thick		145	.055		2.29	2.42		4.71	6.85
0300	.050" thick		145	.055		2.75	2.42		5.17	7.35
0325	Mill finish 5" x 7" step flashing, .016" thick		1920	.004	Ea.	.15	.18		.33	.50
0350	Mill finish 12" x 12" step flashing, .016" thick		1600	.005	"	.55	.22		.77	1
0400	Painted finish, add				S.F.	.33			.33	.36
1000	Mastic-coated 2 sides, .005" thick	1 Rofc	330	.024		1.81	1.07		2.88	3.89
1100	.016" thick		330	.024		2	1.07		3.07	4.10
1600	Copper, 16 oz. sheets, under 1000 lb.		115	.070		8.10	3.06		11.16	14.35
1700	Over 4000 lb.		155	.052		8.10	2.27		10.37	12.95
1900	20 oz. sheets, under 1000 lb.		110	.073		10.70	3.20		13.90	17.50
2000	Over 4000 lb.		145	.055		10.15	2.42		12.57	15.50
2200	24 oz. sheets, under 1000 lb.		105	.076		14.75	3.35		18.10	22
2300	Over 4000 lb.		135	.059		14	2.60		16.60	20
2500	32 oz. sheets, under 1000 lb.		100	.080		19	3.52		22.52	27.50
2600	Over 4000 lb.		130	.062		18.05	2.70		20.75	24.50
2700	W shape for valleys, 16 oz., 24" wide		100	.080	L.F.	16.40	3.52		19.92	24.50
5800	Lead, 2.5 lb./S.F., up to 12" wide		135	.059	S.F.	6.15	2.60		8.75	11.40
5900	Over 12" wide		135	.059		4.04	2.60		6.64	9.10
8900	Stainless steel sheets, 32 ga.		155	.052		3.35	2.27		5.62	7.75
9000	28 ga.		155	.052		4.66	2.27		6.93	9.20
9100	26 ga.		155	.052		4.50	2.27		6.77	9
9200	24 ga.		155	.052		5	2.27		7.27	9.55
9290	For mechanically keyed flashing, add					40%				
9320	Steel sheets, galvanized, 20 ga.	1 Rofc	130	.062	S.F.	1.27	2.70		3.97	6.20
9322	22 ga.		135	.059		1.25	2.60		3.85	6
9324	24 ga.		140	.057		.95	2.51		3.46	5.55
9326	26 ga.		148	.054		.83	2.38		3.21	5.15
9328	28 ga.		155	.052		.72	2.27		2.99	4.83
9340	30 ga.		160	.050		.60	2.20		2.80	4.58
9400	Terne coated stainless steel, .015" thick, 28 ga.		155	.052		8.10	2.27		10.37	12.95
9500	.018" thick, 26 ga.		155	.052		9.05	2.27		11.32	14
9600	Zinc and copper alloy (brass), .020" thick		155	.052		10.25	2.27		12.52	15.35
9700	.027" thick		155	.052		12.25	2.27		14.52	17.55

07 65 Flexible Flashing

07 65 10 – Sheet Metal Flashing

07 65 10.10 Sheet Metal Flashing and Counter Flashing	Crew	Daily Output	Labor-Hours	Unit	Material	2018 Bare Costs Labor	Equipment	Total	Total Incl O&P
9800	.032" thick	1 Rofc	155	.052	S.F.	15.50	2.27		17.77	21
9900	.040" thick		155	.052		20.50	2.27		22.77	26.50
9950	Minimum labor/equipment charge		3	2.667	Job		117		117	209

07 65 12 – Fabric and Mastic Flashings

07 65 12.10 Fabric and Mastic Flashing and Counter Flashing

		Crew	Daily Output	Labor-Hours	Unit	Material	Labor	Equipment	Total	Total Incl O&P
0010	**FABRIC AND MASTIC FLASHING AND COUNTER FLASHING**									
1300	Asphalt flashing cement, 5 gallon				Gal.	9.65			9.65	10.60
4900	Fabric, asphalt-saturated cotton, specification grade	1 Rofc	35	.229	S.Y.	3.20	10.05		13.25	21.50
5000	Utility grade		35	.229		1.50	10.05		11.55	19.55
5300	Close-mesh fabric, saturated, 17 oz./S.Y.		35	.229		2.17	10.05		12.22	20.50
5500	Fiberglass, resin-coated		35	.229		1.07	10.05		11.12	19.10
8500	Shower pan, bituminous membrane, 7 oz.		155	.052	S.F.	1.75	2.27		4.02	5.95

07 65 13 – Laminated Sheet Flashing

07 65 13.10 Laminated Sheet Flashing

		Crew	Daily Output	Labor-Hours	Unit	Material	Labor	Equipment	Total	Total Incl O&P
0010	**LAMINATED SHEET FLASHING**, Including up to 4 bends									
0500	Aluminum, fabric-backed 2 sides, mill finish, .004" thick	1 Rofc	330	.024	S.F.	1.60	1.07		2.67	3.66
0700	.005" thick		330	.024		1.80	1.07		2.87	3.88
0750	Mastic-backed, self adhesive		460	.017		3.35	.76		4.11	5.05
0800	Mastic-coated 2 sides, .004" thick		330	.024		1.60	1.07		2.67	3.66
2800	Copper, paperbacked 1 side, 2 oz.		330	.024		2.18	1.07		3.25	4.30
2900	3 oz.		330	.024		3.12	1.07		4.19	5.35
3100	Paperbacked 2 sides, 2 oz.		330	.024		2.35	1.07		3.42	4.49
3150	3 oz.		330	.024		2.20	1.07		3.27	4.32
3200	5 oz.		330	.024		3.38	1.07		4.45	5.60
3250	7 oz.		330	.024		6.70	1.07		7.77	9.25
3400	Mastic-backed 2 sides, copper, 2 oz.		330	.024		2	1.07		3.07	4.10
3500	3 oz.		330	.024		2.50	1.07		3.57	4.65
3700	5 oz.		330	.024		3.80	1.07		4.87	6.10
3800	Fabric-backed 2 sides, copper, 2 oz.		330	.024		2.01	1.07		3.08	4.11
4000	3 oz.		330	.024		2.80	1.07		3.87	4.98
4100	5 oz.		330	.024		3.90	1.07		4.97	6.20
4300	Copper-clad stainless steel, .015" thick, under 500 lb.		115	.070		6.65	3.06		9.71	12.75
4400	Over 2000 lb.		155	.052		6.75	2.27		9.02	11.50
4600	.018" thick, under 500 lb.		100	.080		7.95	3.52		11.47	15
4700	Over 2000 lb.		145	.055		7.75	2.42		10.17	12.85
8550	Shower pan, 3 ply copper and fabric, 3 oz.		155	.052		4	2.27		6.27	8.45
8600	7 oz.		155	.052		4.73	2.27		7	9.25
9300	Stainless steel, paperbacked 2 sides, .005" thick		330	.024		3.92	1.07		4.99	6.20

07 65 19 – Plastic Sheet Flashing

07 65 19.10 Plastic Sheet Flashing and Counter Flashing

		Crew	Daily Output	Labor-Hours	Unit	Material	Labor	Equipment	Total	Total Incl O&P
0010	**PLASTIC SHEET FLASHING AND COUNTER FLASHING**									
7300	Polyvinyl chloride, black, 10 mil	1 Rofc	285	.028	S.F.	.26	1.23		1.49	2.49
7400	20 mil		285	.028		.26	1.23		1.49	2.49
7600	30 mil		285	.028		.33	1.23		1.56	2.56
7700	60 mil		285	.028		.85	1.23		2.08	3.14
7900	Black or white for exposed roofs, 60 mil		285	.028		1.22	1.23		2.45	3.54
8060	PVC tape, 5" x 45 mils, for joint covers, 100 L.F./roll				Ea.	177			177	194
8850	Polyvinyl chloride, 30 mil	1 Rofc	160	.050	S.F.	1.50	2.20		3.70	5.55

07 65 Flexible Flashing

07 65 23 – Rubber Sheet Flashing

07 65 23.10 Rubber Sheet Flashing and Counter Flashing	Crew	Daily Output	Labor-Hours	Unit	Material	2018 Bare Costs Labor	Equipment	Total	Total Incl O&P
0010 **RUBBER SHEET FLASHING AND COUNTER FLASHING**									
4810 EPDM 90 mils, 1" diameter pipe flashing	1 Rofc	32	.250	Ea.	19.95	11		30.95	41.50
4820 2" diameter		30	.267		19.90	11.70		31.60	43
4830 3" diameter		28	.286		21	12.55		33.55	46
4840 4" diameter		24	.333		28.50	14.65		43.15	57.50
4850 6" diameter		22	.364		28.50	16		44.50	60
8100 Rubber, butyl, 1/32" thick		285	.028	S.F.	2.22	1.23		3.45	4.64
8200 1/16" thick		285	.028		3.27	1.23		4.50	5.80
8300 Neoprene, cured, 1/16" thick		285	.028		2.60	1.23		3.83	5.05
8400 1/8" thick		285	.028		6.10	1.23		7.33	8.90

07 65 26 – Self-Adhering Sheet Flashing

07 65 26.10 Self-Adhering Sheet or Roll Flashing	Crew	Daily Output	Labor-Hours	Unit	Material	2018 Bare Costs Labor	Equipment	Total	Total Incl O&P
0010 **SELF-ADHERING SHEET OR ROLL FLASHING**									
0020 Self-adhered flashing, 25 mil cross laminated HDPE, 4" wide	1 Rofc	960	.008	L.F.	.20	.37		.57	.87
0040 6" wide		896	.009		.30	.39		.69	1.03
0060 9" wide		832	.010		.45	.42		.87	1.25
0080 12" wide		768	.010		.60	.46		1.06	1.48

07 71 Roof Specialties

07 71 16 – Manufactured Counterflashing Systems

07 71 16.10 Roof Drain Boot

	Crew	Daily Output	Labor-Hours	Unit	Material	2018 Bare Costs Labor	Equipment	Total	Total Incl O&P
0010 **ROOF DRAIN BOOT**									
0100 Cast iron, 4" diameter	1 Shee	125	.064	L.F.	45	3.83		48.83	55.50
0300 4" x 3"		125	.064		63.50	3.83		67.33	75.50
0400 5" x 4"		125	.064		123	3.83		126.83	142

07 71 16.20 Pitch Pockets, Variable Sizes

	Crew	Daily Output	Labor-Hours	Unit	Material	2018 Bare Costs Labor	Equipment	Total	Total Incl O&P
0010 **PITCH POCKETS, VARIABLE SIZES**									
0100 Adjustable, 4" to 7", welded corners, 4" deep	1 Rofc	48	.167	Ea.	16.65	7.35		24	31.50
0200 Side extenders, 6"	"	240	.033	"	2.78	1.46		4.24	5.65

07 71 19 – Manufactured Gravel Stops and Fasciae

07 71 19.10 Gravel Stop

	Crew	Daily Output	Labor-Hours	Unit	Material	2018 Bare Costs Labor	Equipment	Total	Total Incl O&P
0010 **GRAVEL STOP**									
0020 Aluminum, .050" thick, 4" face height, mill finish	1 Shee	145	.055	L.F.	6.55	3.30		9.85	12.40
0080 Duranodic finish		145	.055		7.40	3.30		10.70	13.35
0100 Painted		145	.055		7.50	3.30		10.80	13.45
0300 6" face height		135	.059		6.85	3.54		10.39	13.10
0350 Duranodic finish		135	.059		7.95	3.54		11.49	14.35
0400 Painted		135	.059		8.75	3.54		12.29	15.25
0600 8" face height		125	.064		7.75	3.83		11.58	14.60
0650 Duranodic finish		125	.064		9.15	3.83		12.98	16.10
0700 Painted		125	.064		9.75	3.83		13.58	16.80
0900 12" face height, .080" thick, 2 piece		100	.080		11.05	4.78		15.83	19.70
0950 Duranodic finish		100	.080		11	4.78		15.78	19.65
1000 Painted		100	.080		13.50	4.78		18.28	22.50
1200 Copper, 16 oz., 3" face height		145	.055		24.50	3.30		27.80	32
1300 6" face height		135	.059		33.50	3.54		37.04	42.50
1350 Galv steel, 24 ga., 4" leg, plain, with continuous cleat, 4" face		145	.055		6.50	3.30		9.80	12.35
1360 6" face height		145	.055		6.55	3.30		9.85	12.40
1500 Polyvinyl chloride, 6" face height		135	.059		5.75	3.54		9.29	11.95

207

07 71 Roof Specialties

07 71 19 – Manufactured Gravel Stops and Fasciae

07 71 19.10 Gravel Stop

	07 71 19.10 Gravel Stop	Crew	Daily Output	Labor-Hours	Unit	Material	2018 Bare Costs Labor	Equipment	Total	Total Incl O&P
1600	9" face height	1 Shee	125	.064	L.F.	6.90	3.83		10.73	13.65
1800	Stainless steel, 24 ga., 6" face height		135	.059		15.80	3.54		19.34	23
1900	12" face height		100	.080		23	4.78		27.78	33
2100	20 ga., 6" face height		135	.059		19.50	3.54		23.04	27
2200	12" face height		100	.080		28	4.78		32.78	38
9000	Minimum labor/equipment charge		3.50	2.286	Job		137		137	215

07 71 19.30 Fascia

		Crew	Daily Output	Labor-Hours	Unit	Material	Labor	Equipment	Total	Total Incl O&P
0010	**FASCIA**									
0100	Aluminum, reverse board and batten, .032" thick, colored, no furring incl.	1 Shee	145	.055	S.F.	7	3.30		10.30	12.90
0200	Residential type, aluminum	1 Carp	200	.040	L.F.	2.02	2.03		4.05	5.50
0220	Vinyl	"	200	.040	"	2.17	2.03		4.20	5.70
0300	Steel, galv and enameled, stock, no furring, long panels	1 Shee	145	.055	S.F.	5.20	3.30		8.50	10.90
0600	Short panels		115	.070	"	5.20	4.16		9.36	12.25
9000	Minimum labor/equipment charge		4	2	Job		120		120	189

07 71 23 – Manufactured Gutters and Downspouts

07 71 23.10 Downspouts

		Crew	Daily Output	Labor-Hours	Unit	Material	Labor	Equipment	Total	Total Incl O&P
0010	**DOWNSPOUTS**									
0020	Aluminum, embossed, .020" thick, 2" x 3"	1 Shee	190	.042	L.F.	.91	2.52		3.43	4.97
0100	Enameled		190	.042		1.36	2.52		3.88	5.45
0300	.024" thick, 2" x 3"		180	.044		2.12	2.66		4.78	6.50
0400	3" x 4"		140	.057		1.95	3.42		5.37	7.55
0600	Round, corrugated aluminum, 3" diameter, .020" thick		190	.042		2.08	2.52		4.60	6.25
0700	4" diameter, .025" thick		140	.057		3.19	3.42		6.61	8.90
0900	Wire strainer, round, 2" diameter		155	.052	Ea.	1.74	3.09		4.83	6.75
1000	4" diameter		155	.052		2.38	3.09		5.47	7.50
1200	Rectangular, perforated, 2" x 3"		145	.055		2.32	3.30		5.62	7.75
1300	3" x 4"		145	.055		3.33	3.30		6.63	8.85
1500	Copper, round, 16 oz., stock, 2" diameter		190	.042	L.F.	8.35	2.52		10.87	13.10
1600	3" diameter		190	.042		8.60	2.52		11.12	13.40
1800	4" diameter		145	.055		9.80	3.30		13.10	16
1900	5" diameter		130	.062		14.65	3.68		18.33	22
2100	Rectangular, corrugated copper, stock, 2" x 3"		190	.042		8.30	2.52		10.82	13.10
2200	3" x 4"		145	.055		9.35	3.30		12.65	15.50
2400	Rectangular, plain copper, stock, 2" x 3"		190	.042		11.25	2.52		13.77	16.35
2500	3" x 4"		145	.055		14	3.30		17.30	20.50
2700	Wire strainers, rectangular, 2" x 3"		145	.055	Ea.	17.20	3.30		20.50	24
2800	3" x 4"		145	.055		17.95	3.30		21.25	25
3000	Round, 2" diameter		145	.055		6.50	3.30		9.80	12.35
3100	3" diameter		145	.055		7.25	3.30		10.55	13.20
3300	4" diameter		145	.055		12.25	3.30		15.55	18.70
3400	5" diameter		115	.070		22.50	4.16		26.66	31.50
3600	Lead-coated copper, round, stock, 2" diameter		190	.042	L.F.	22.50	2.52		25.02	29
3700	3" diameter		190	.042		23	2.52		25.52	29.50
3900	4" diameter		145	.055		24	3.30		27.30	31.50
4000	5" diameter, corrugated		130	.062		24	3.68		27.68	32
4200	6" diameter, corrugated		105	.076		31.50	4.56		36.06	41.50
4300	Rectangular, corrugated, stock, 2" x 3"		190	.042		15.15	2.52		17.67	20.50
4500	Plain, stock, 2" x 3"		190	.042		24.50	2.52		27.02	31
4600	3" x 4"		145	.055		33	3.30		36.30	41.50
4800	Steel, galvanized, round, corrugated, 2" or 3" diameter, 28 ga.		190	.042		2.07	2.52		4.59	6.25
4900	4" diameter, 28 ga.		145	.055		2.08	3.30		5.38	7.50
5100	5" diameter, 26 ga.		130	.062		3.53	3.68		7.21	9.70

07 71 23 – Manufactured Gutters and Downspouts

07 71 23.10 Downspouts

		Crew	Daily Output	Labor-Hours	Unit	Material	2018 Bare Costs Labor	Equipment	Total	Total Incl O&P
5400	6" diameter, 28 ga.	1 Shee	105	.076	L.F.	3.48	4.56		8.04	11.05
5500	26 ga.		105	.076		3.78	4.56		8.34	11.35
5700	Rectangular, corrugated, 28 ga., 2" x 3"		190	.042		2.04	2.52		4.56	6.20
5800	3" x 4"		145	.055		2.20	3.30		5.50	7.60
6000	Rectangular, plain, 28 ga., galvanized, 2" x 3"		190	.042		3.83	2.52		6.35	8.20
6100	3" x 4"		145	.055		4.32	3.30		7.62	9.95
6300	Epoxy painted, 24 ga., corrugated, 2" x 3"		190	.042		2.40	2.52		4.92	6.60
6400	3" x 4"		145	.055		2.91	3.30		6.21	8.40
6600	Wire strainers, rectangular, 2" x 3"		145	.055	Ea.	17.30	3.30		20.60	24.50
6700	3" x 4"		145	.055		19	3.30		22.30	26
6900	Round strainers, 2" or 3" diameter		145	.055		4.23	3.30		7.53	9.85
7000	4" diameter		145	.055		6.25	3.30		9.55	12.10
9000	Minimum labor/equipment charge		4	2	Job		120		120	189

07 71 23.20 Downspout Elbows

		Crew	Daily Output	Labor-Hours	Unit	Material	2018 Bare Costs Labor	Equipment	Total	Total Incl O&P
0010	**DOWNSPOUT ELBOWS**									
0020	Aluminum, embossed, 2" x 3", .020" thick	1 Shee	100	.080	Ea.	.93	4.78		5.71	8.55
0100	Enameled		100	.080		1.78	4.78		6.56	9.50
0200	Embossed, 3" x 4", .025" thick		100	.080		3.10	4.78		7.88	10.95
0300	Enameled		100	.080		3.93	4.78		8.71	11.85
0400	Embossed, corrugated, 3" diameter, .020" thick		100	.080		3.10	4.78		7.88	10.95
0500	4" diameter, .025" thick		100	.080		6.55	4.78		11.33	14.75
0600	Copper, 16 oz., 2" diameter		100	.080		9.45	4.78		14.23	17.95
0700	3" diameter		100	.080		9.10	4.78		13.88	17.55
0800	4" diameter		100	.080		13.95	4.78		18.73	23
1000	Rectangular, 2" x 3" corrugated		100	.080		9.25	4.78		14.03	17.70
1100	3" x 4" corrugated		100	.080		15.10	4.78		19.88	24
1300	Vinyl, 2-1/2" diameter, 45 or 75 degree bend		100	.080		3.98	4.78		8.76	11.95
1400	Tee Y junction		75	.107		12.95	6.40		19.35	24.50
9000	Minimum labor/equipment charge		4	2	Job		120		120	109

07 71 23.30 Gutters

		Crew	Daily Output	Labor-Hours	Unit	Material	2018 Bare Costs Labor	Equipment	Total	Total Incl O&P
0010	**GUTTERS**									
0012	Aluminum, stock units, 5" K type, .027" thick, plain	1 Shee	125	.064	L.F.	2.80	3.83		6.63	9.15
0100	Enameled		125	.064		2.89	3.83		6.72	9.25
0300	5" K type, .032" thick, plain		125	.064		3.53	3.83		7.36	9.95
0400	Enameled		125	.064		3.52	3.83		7.35	9.90
0700	Copper, half round, 16 oz., stock units, 4" wide		125	.064		8.50	3.83		12.33	15.40
0900	5" wide		125	.064		7.10	3.83		10.93	13.85
1000	6" wide		118	.068		11.25	4.05		15.30	18.80
1200	K type, 16 oz., stock, 5" wide		125	.064		8.25	3.83		12.08	15.15
1300	6" wide		125	.064		9.05	3.83		12.88	16
1500	Lead coated copper, 16 oz., half round, stock, 4" wide		125	.064		15.55	3.83		19.38	23
1600	6" wide		118	.068		18.50	4.05		22.55	27
1800	K type, stock, 5" wide		125	.064		18.50	3.83		22.33	26.50
1900	6" wide		125	.064		18.50	3.83		22.33	26.50
2100	Copper clad stainless steel, K type, 5" wide		125	.064		7.75	3.83		11.58	14.60
2200	6" wide		125	.064		9.80	3.83		13.63	16.85
2400	Steel, galv, half round or box, 28 ga., 5" wide, plain		125	.064		2.18	3.83		6.01	8.45
2500	Enameled		125	.064		2.28	3.83		6.11	8.55
2700	26 ga., stock, 5" wide		125	.064		2.47	3.83		6.30	8.75
2800	6" wide		125	.064		2.50	3.83		6.33	8.80
3000	Vinyl, O.G., 4" wide	1 Carp	115	.070		1.34	3.53		4.87	7.20
3100	5" wide		115	.070		1.64	3.53		5.17	7.55

For customer support on your Commercial Renovation Costs with RSMeans data, call 800.448.8182.

209

07 71 23.30 Gutters		Crew	Daily Output	Labor-Hours	Unit	Material	2018 Bare Costs Labor	Equipment	Total	Total Incl O&P
3200	4" half round, stock units	1 Carp	115	.070	L.F.	1.39	3.53		4.92	7.30
3250	Joint connectors				Ea.	3.03			3.03	3.33
3300	Wood, clear treated cedar, fir or hemlock, 3" x 4"	1 Carp	100	.080	L.F.	11	4.06		15.06	18.70
3400	4" x 5"	"	100	.080	"	22	4.06		26.06	30.50
5000	Accessories, end cap, K type, aluminum 5"	1 Shee	625	.013	Ea.	.72	.77		1.49	2
5010	6"		625	.013		1.56	.77		2.33	2.93
5020	Copper, 5"		625	.013		3.49	.77		4.26	5.05
5030	6"		625	.013		3.54	.77		4.31	5.10
5040	Lead coated copper, 5"		625	.013		13	.77		13.77	15.50
5050	6"		625	.013		13.90	.77		14.67	16.45
5060	Copper clad stainless steel, 5"		625	.013		3.52	.77		4.29	5.10
5070	6"		625	.013		3.52	.77		4.29	5.10
5080	Galvanized steel, 5"		625	.013		1.36	.77		2.13	2.71
5090	6"		625	.013		2.34	.77		3.11	3.78
5100	Vinyl, 4"	1 Carp	625	.013		6.20	.65		6.85	7.85
5110	5"	"	625	.013		6.55	.65		7.20	8.25
5120	Half round, copper, 4"	1 Shee	625	.013		4.40	.77		5.17	6.05
5130	5"		625	.013		4.72	.77		5.49	6.40
5140	6"		625	.013		7.65	.77		8.42	9.65
5150	Lead coated copper, 5"		625	.013		14.65	.77		15.42	17.30
5160	6"		625	.013		22	.77		22.77	25
5170	Copper clad stainless steel, 5"		625	.013		4.53	.77		5.30	6.20
5180	6"		625	.013		4.53	.77		5.30	6.20
5190	Galvanized steel, 5"		625	.013		2.40	.77		3.17	3.85
5200	6"		625	.013		3	.77		3.77	4.51
5210	Outlet, aluminum, 2" x 3"		420	.019		.62	1.14		1.76	2.48
5220	3" x 4"		420	.019		1.05	1.14		2.19	2.96
5230	2-3/8" round		420	.019		.56	1.14		1.70	2.42
5240	Copper, 2" x 3"		420	.019		7.20	1.14		8.34	9.70
5250	3" x 4"		420	.019		8.15	1.14		9.29	10.80
5260	2-3/8" round		420	.019		4.55	1.14		5.69	6.80
5270	Lead coated copper, 2" x 3"		420	.019		26	1.14		27.14	30.50
5280	3" x 4"		420	.019		29.50	1.14		30.64	34
5290	2-3/8" round		420	.019		26	1.14		27.14	30.50
5300	Copper clad stainless steel, 2" x 3"		420	.019		7.20	1.14		8.34	9.70
5310	3" x 4"		420	.019		8.15	1.14		9.29	10.80
5320	2-3/8" round		420	.019		4.55	1.14		5.69	6.80
5330	Galvanized steel, 2" x 3"		420	.019		3.40	1.14		4.54	5.55
5340	3" x 4"		420	.019		5.25	1.14		6.39	7.55
5350	2-3/8" round		420	.019		4.25	1.14		5.39	6.50
5360	K type mitres, aluminum		65	.123		3.38	7.35		10.73	15.30
5370	Copper		65	.123		13.15	7.35		20.50	26
5380	Lead coated copper		65	.123		55	7.35		62.35	72
5390	Copper clad stainless steel		65	.123		27.50	7.35		34.85	41.50
5400	Galvanized steel		65	.123		24	7.35		31.35	38
5420	Half round mitres, copper		65	.123		63.50	7.35		70.85	81.50
5430	Lead coated copper		65	.123		90	7.35		97.35	111
5440	Copper clad stainless steel		65	.123		56.50	7.35		63.85	73.50
5450	Galvanized steel		65	.123		29	7.35		36.35	43.50
5460	Vinyl mitres and outlets		65	.123		10.45	7.35		17.80	23
5470	Sealant		940	.009	L.F.	.01	.51		.52	.81
5480	Soldering		96	.083	"	.26	4.98		5.24	8.15
9000	Minimum labor/equipment charge		3.75	2.133	Job		128		128	201

For customer support on your Commercial Renovation Costs with RSMeans data, call 800.448.8182.

07 71 Roof Specialties

07 71 23 – Manufactured Gutters and Downspouts

07 71 23.35 Gutter Guard

		Crew	Daily Output	Labor-Hours	Unit	Material	2018 Bare Costs Labor	2018 Bare Costs Equipment	Total	Total Incl O&P
0010	**GUTTER GUARD**									
0020	6" wide strip, aluminum mesh	1 Carp	500	.016	L.F.	2.58	.81		3.39	4.16
0100	Vinyl mesh		500	.016	"	2.78	.81		3.59	4.38
9000	Minimum labor/equipment charge		4	2	Job		101		101	165

07 71 26 – Reglets

07 71 26.10 Reglets and Accessories

		Crew	Daily Output	Labor-Hours	Unit	Material	2018 Bare Costs Labor	2018 Bare Costs Equipment	Total	Total Incl O&P
0010	**REGLETS AND ACCESSORIES**									
0020	Reglet, aluminum, .025" thick, in parapet	1 Carp	225	.036	L.F.	1.57	1.80		3.37	4.66
0300	16 oz. copper		225	.036		6.50	1.80		8.30	10.10
0400	Galvanized steel, 24 ga.		225	.036		1.23	1.80		3.03	4.28
0600	Stainless steel, .020" thick		225	.036		3.85	1.80		5.65	7.15
0900	Counter flashing for above, 12" wide, .032" aluminum	1 Shee	150	.053		2.16	3.19		5.35	7.45
1200	16 oz. copper		150	.053		6.25	3.19		9.44	11.95
1300	Galvanized steel, 26 ga.		150	.053		1.25	3.19		4.44	6.45
1500	Stainless steel, .020" thick		150	.053		6.30	3.19		9.49	12
9000	Minimum labor/equipment charge	1 Carp	3	2.667	Job		135		135	219

07 71 29 – Manufactured Roof Expansion Joints

07 71 29.10 Expansion Joints

		Crew	Daily Output	Labor-Hours	Unit	Material	2018 Bare Costs Labor	2018 Bare Costs Equipment	Total	Total Incl O&P
0010	**EXPANSION JOINTS**									
0300	Butyl or neoprene center with foam insulation, metal flanges									
0400	Aluminum, .032" thick for openings to 2-1/2"	1 Rofc	165	.048	L.F.	12.10	2.13		14.23	17.10
0600	For joint openings to 3-1/2"		165	.048		12.10	2.13		14.23	17.10
0610	For joint openings to 5"		165	.048		14.15	2.13		16.28	19.40
0620	For joint openings to 8"		165	.048		17	2.13		19.13	22.50
0700	Copper, 16 oz. for openings to 2-1/2"		165	.048		19.05	2.13		21.18	25
0900	For joint openings to 3-1/2"		165	.048		19.70	2.13		21.83	25.50
0910	For joint openings to 5"		165	.048		22	2.13		24.13	28.50
0920	For joint openings to 8"		165	.048		25	2.13		27.13	32
1000	Galvanized steel, 26 ga. for openings to 2-1/2"		165	.048		10.30	2.13		12.43	15.15
1200	For joint openings to 3-1/2"		165	.048		10.30	2.13		12.43	15.15
1210	For joint openings to 5"		165	.048		11.85	2.13		13.98	16.85
1220	For joint openings to 8"		165	.048		15.45	2.13		17.58	21
1300	Lead-coated copper, 16 oz. for openings to 2-1/2"		165	.048		34.50	2.13		36.63	42
1500	For joint openings to 3-1/2"		165	.048		34.50	2.13		36.63	42
1600	Stainless steel, .018", for openings to 2-1/2"		165	.048		13.90	2.13		16.03	19.10
1800	For joint openings to 3-1/2"		165	.048		13.90	2.13		16.03	19.10
1810	For joint openings to 5"		165	.048		15.20	2.13		17.33	20.50
1820	For joint openings to 8"		165	.048		21.50	2.13		23.63	28
1900	Neoprene, double-seal type with thick center, 4-1/2" wide		125	.064		15.15	2.81		17.96	21.50
1950	Polyethylene bellows, with galv steel flat flanges		100	.080		6.55	3.52		10.07	13.50
1960	With galvanized angle flanges		100	.080		7.05	3.52		10.57	14.05
2000	Roof joint with extruded aluminum cover, 2"	1 Shee	115	.070		29	4.16		33.16	38
2100	Roof joint, plastic curbs, foam center, standard	1 Rofc	100	.080		13.65	3.52		17.17	21.50
2200	Large	"	100	.080		17.70	3.52		21.22	25.50
2500	Roof to wall joint with extruded aluminum cover	1 Shee	115	.070		29	4.16		33.16	38
2700	Wall joint, closed cell foam on PVC cover, 9" wide	1 Rofc	125	.064		5.55	2.81		8.36	11.10
2800	12" wide	"	115	.070		6.60	3.06		9.66	12.75
9000	Minimum labor/equipment charge	1 Shee	3	2.667	Job		159		159	251

For customer support on your Commercial Renovation Costs with RSMeans data, call 800.448.8182.

07 71 Roof Specialties

07 71 43 – Drip Edge

07 71 43.10 Drip Edge, Rake Edge, Ice Belts	Crew	Daily Output	Labor-Hours	Unit	Material	2018 Bare Costs Labor	Equipment	Total	Total Incl O&P
0010 **DRIP EDGE, RAKE EDGE, ICE BELTS**									
0020 Aluminum, .016" thick, 5" wide, mill finish	1 Carp	400	.020	L.F.	.58	1.01		1.59	2.29
0100 White finish		400	.020		.64	1.01		1.65	2.35
0200 8" wide, mill finish		400	.020		1.51	1.01		2.52	3.31
0300 Ice belt, 28" wide, mill finish		100	.080		7.85	4.06		11.91	15.20
0310 Vented, mill finish		400	.020		2.26	1.01		3.27	4.14
0320 Painted finish		400	.020		2.54	1.01		3.55	4.44
0400 Galvanized, 5" wide		400	.020		.61	1.01		1.62	2.32
0500 8" wide, mill finish		400	.020		.83	1.01		1.84	2.56
0510 Rake edge, aluminum, 1-1/2" x 1-1/2"		400	.020		.33	1.01		1.34	2.01
0520 3-1/2" x 1-1/2"		400	.020		.45	1.01		1.46	2.15
9000 Minimum labor/equipment charge		4	2	Job		101		101	165

07 72 Roof Accessories

07 72 23 – Relief Vents

07 72 23.10 Roof Vents

	Crew	Daily Output	Labor-Hours	Unit	Material	2018 Bare Costs Labor	Equipment	Total	Total Incl O&P
0010 **ROOF VENTS**									
0020 Mushroom shape, for built-up roofs, aluminum	1 Rofc	30	.267	Ea.	65	11.70		76.70	92.50
0100 PVC, 6" high		30	.267	"	21.50	11.70		33.20	45
9000 Minimum labor/equipment charge		2.75	2.909	Job		128		128	228

07 72 23.20 Vents

	Crew	Daily Output	Labor-Hours	Unit	Material	2018 Bare Costs Labor	Equipment	Total	Total Incl O&P
0010 **VENTS**									
0100 Soffit or eave, aluminum, mill finish, strips, 2-1/2" wide	1 Carp	200	.040	L.F.	.44	2.03		2.47	3.77
0200 3" wide		200	.040		.47	2.03		2.50	3.81
0300 Enamel finish, 3" wide		200	.040		.53	2.03		2.56	3.87
0400 Mill finish, rectangular, 4" x 16"		72	.111	Ea.	1.58	5.65		7.23	10.90
0500 8" x 16"		72	.111		2.50	5.65		8.15	11.90
2420 Roof ventilator	Q-9	16	1		47	54		101	137
2500 Vent, roof vent	1 Rofc	24	.333		23.50	14.65		38.15	52

07 72 26 – Ridge Vents

07 72 26.10 Ridge Vents and Accessories

	Crew	Daily Output	Labor-Hours	Unit	Material	2018 Bare Costs Labor	Equipment	Total	Total Incl O&P
0010 **RIDGE VENTS AND ACCESSORIES**									
0100 Aluminum strips, mill finish	1 Rofc	160	.050	L.F.	2.40	2.20		4.60	6.55
0150 Painted finish		160	.050	"	4.17	2.20		6.37	8.50
0200 Connectors		48	.167	Ea.	4.95	7.35		12.30	18.50
0300 End caps		48	.167	"	2.25	7.35		9.60	15.55
0400 Galvanized strips		160	.050	L.F.	3.71	2.20		5.91	8
0430 Molded polyethylene, shingles not included		160	.050	"	2.78	2.20		4.98	7
0440 End plugs		48	.167	Ea.	2.25	7.35		9.60	15.55
0450 Flexible roll, shingles not included		160	.050	L.F.	2.37	2.20		4.57	6.55
2300 Ridge vent strip, mill finish	1 Shee	155	.052	"	3.93	3.09		7.02	9.20

07 72 33 – Roof Hatches

07 72 33.10 Roof Hatch Options

	Crew	Daily Output	Labor-Hours	Unit	Material	2018 Bare Costs Labor	Equipment	Total	Total Incl O&P
0010 **ROOF HATCH OPTIONS**									
0500 2'-6" x 3', aluminum curb and cover	G-3	10	3.200	Ea.	885	159		1,044	1,225
0520 Galvanized steel curb and aluminum cover		10	3.200		845	159		1,004	1,175
0540 Galvanized steel curb and cover		10	3.200		615	159		774	930
0600 2'-6" x 4'-6", aluminum curb and cover		9	3.556		1,075	177		1,252	1,475
0800 Galvanized steel curb and aluminum cover		9	3.556		950	177		1,127	1,325

212

07 72 Roof Accessories

07 72 33 – Roof Hatches

07 72 33.10 Roof Hatch Options

07 72 33.10 Roof Hatch Options	Crew	Daily Output	Labor-Hours	Unit	Material	2018 Bare Costs Labor	2018 Bare Costs Equipment	Total	Total Incl O&P
0900 Galvanized steel curb and cover	G-3	9	3.556	Ea.	995	177		1,172	1,375
1100 4' x 4' aluminum curb and cover		8	4		1,775	199		1,974	2,275
1120 Galvanized steel curb and aluminum cover		8	4		1,700	199		1,899	2,200
1140 Galvanized steel curb and cover		8	4		1,100	199		1,299	1,525
1200 2'-6" x 8'-0", aluminum curb and cover		6.60	4.848		2,050	242		2,292	2,650
1400 Galvanized steel curb and aluminum cover		6.60	4.848		1,900	242		2,142	2,450
1500 Galvanized steel curb and cover	▼	6.60	4.848		1,200	242		1,442	1,700
1800 For plexiglass panels, 2'-6" x 3'-0", add to above				▼	475			475	525
9000 Minimum labor/equipment charge	2 Carp	2	8	Job		405		405	660

07 72 36 – Smoke Vents

07 72 36.10 Smoke Hatches

				Unit	Material	Labor			
0010 **SMOKE HATCHES**									
0200 For 3'-0" long, add to roof hatches from Section 07 72 33.10				Ea.	25%	5%			
0250 For 4'-0" long, add to roof hatches from Section 07 72 33.10					20%	5%			
0300 For 8'-0" long, add to roof hatches from Section 07 72 33.10				▼	10%	5%			

07 72 36.20 Smoke Vent Options

	Crew	Daily Output	Labor-Hours	Unit	Material	Labor	Equipment	Total	Total Incl O&P
0010 **SMOKE VENT OPTIONS**									
0100 4' x 4' aluminum cover and frame	G-3	13	2.462	Ea.	2,200	123		2,323	2,600
0200 Galvanized steel cover and frame		13	2.462		1,650	123		1,773	2,025
0300 4' x 8' aluminum cover and frame		8	4		2,825	199		3,024	3,425
0400 Galvanized steel cover and frame	▼	8	4	▼	2,350	199		2,549	2,900
9000 Minimum labor/equipment charge	2 Carp	2	8	Job		405		405	660

07 72 53 – Snow Guards

07 72 53.10 Snow Guard Options

	Crew	Daily Output	Labor-Hours	Unit	Material	Labor	Equipment	Total	Total Incl O&P
0010 **SNOW GUARD OPTIONS**									
0100 Slate & asphalt shingle roofs, fastened with nails	1 Rofc	160	.050	Ea.	12.35	2.20		14.55	17.50
0200 Standing seam metal roofs, fastened with set screws		48	.167		17.55	7.35		24.90	32.50
0300 Surface mount for metal roofs, fastened with solder		48	.167	▼	7.45	7.35		14.80	21.50
0400 Double rail pipe type, including pipe	▼	130	.062	L.F.	34	2.70		36.70	42.50

07 72 80 – Vents

07 72 80.30 Vent Options

	Crew	Daily Output	Labor-Hours	Unit	Material	Labor	Equipment	Total	Total Incl O&P
0010 **VENT OPTIONS**									
0020 Plastic, for insulated decks, 1 per M.S.F.	1 Rofc	40	.200	Ea.	19.85	8.80		28.65	37.50
0100 Heavy duty		20	.400		50	17.60		67.60	86.50
0300 Aluminum		30	.267		21	11.70		32.70	44.50
0800 Polystyrene baffles, 12" wide for 16" OC rafter spacing	1 Carp	90	.089		.44	4.51		4.95	7.80
0900 For 24" OC rafter spacing		110	.073	▼	.79	3.69		4.48	6.85
9000 Minimum labor/equipment charge	▼	3	2.667	Job		135		135	219

07 76 Roof Pavers

07 76 16 – Roof Decking Pavers

07 76 16.10 Roof Pavers and Supports	Crew	Daily Output	Labor-Hours	Unit	Material	2018 Bare Costs Labor	Equipment	Total	Total Incl O&P
0010 **ROOF PAVERS AND SUPPORTS**									
1000 Roof decking pavers, concrete blocks, 2" thick, natural	1 Clab	115	.070	S.F.	3.51	2.77		6.28	8.35
1100 Colors		115	.070	"	3.77	2.77		6.54	8.65
1200 Support pedestal, bottom cap		960	.008	Ea.	3	.33		3.33	3.84
1300 Top cap		960	.008		4.80	.33		5.13	5.85
1400 Leveling shims, 1/16"		1920	.004		1.20	.17		1.37	1.59
1500 1/8"		1920	.004		1.20	.17		1.37	1.59
1600 Buffer pad		960	.008		2.50	.33		2.83	3.29
1700 PVC legs (4" SDR 35)		2880	.003	Inch	.14	.11		.25	.33
2000 Alternate pricing method, system in place		101	.079	S.F.	7.15	3.16		10.31	13

07 81 Applied Fireproofing

07 81 16 – Cementitious Fireproofing

07 81 16.10 Sprayed Cementitious Fireproofing

	Crew	Daily Output	Labor-Hours	Unit	Material	2018 Bare Costs Labor	Equipment	Total	Total Incl O&P
0010 **SPRAYED CEMENTITIOUS FIREPROOFING**									
0050 Not including canvas protection, normal density									
0100 Per 1" thick, on flat plate steel	G-2	3000	.008	S.F.	.57	.34	.04	.95	1.21
0200 Flat decking		2400	.010		.57	.42	.05	1.04	1.36
0400 Beams		1500	.016		.57	.67	.09	1.33	1.79
0500 Corrugated or fluted decks		1250	.019		.85	.81	.10	1.76	2.35
0700 Columns, 1-1/8" thick		1100	.022		.64	.92	.12	1.68	2.30
0800 2-3/16" thick		700	.034		1.37	1.44	.18	2.99	4.03
0900 For canvas protection, add		5000	.005		.10	.20	.03	.33	.46
1000 Not including canvas protection, high density									
1100 Per 1" thick, on flat plate steel	G-2	3000	.008	S.F.	2.23	.34	.04	2.61	3.04
1110 On flat decking		2400	.010		2.23	.42	.05	2.70	3.19
1120 On beams		1500	.016		2.23	.67	.09	2.99	3.62
1130 Corrugated or fluted decks		1250	.019		2.23	.81	.10	3.14	3.86
1140 Columns, 1-1/8" thick		1100	.022		2.51	.92	.12	3.55	4.36
1150 2-3/16" thick		1100	.022		5	.92	.12	6.04	7.10
1170 For canvas protection, add		5000	.005		.10	.20	.03	.33	.46
1200 Not including canvas protection, retrofitting									
1210 Per 1" thick, on flat plate steel	G-2	1500	.016	S.F.	.50	.67	.09	1.26	1.72
1220 On flat decking		1200	.020		.50	.84	.11	1.45	2.02
1230 On beams		750	.032		.50	1.34	.17	2.01	2.90
1240 Corrugated or fluted decks		625	.038		.75	1.61	.21	2.57	3.65
1250 Columns, 1-1/8" thick		550	.044		.56	1.83	.23	2.62	3.83
1260 2-3/16" thick		500	.048		1.13	2.02	.26	3.41	4.76
1400 Accessories, preliminary spattered texture coat		4500	.005		.05	.22	.03	.30	.44
1410 Bonding agent	1 Plas	1000	.008		.10	.37		.47	.70
9000 Minimum labor/equipment charge	G-2	3	8	Job		335	43	378	585

07 84 13 – Penetration Firestopping

07 84 13.10 Firestopping		Crew	Daily Output	Labor-Hours	Unit	Material	2018 Bare Costs Labor	Equipment	Total	Total Incl O&P
0010	**FIRESTOPPING** R078413-30									
0100	Metallic piping, non insulated									
0110	Through walls, 2" diameter	1 Carp	16	.500	Ea.	3.63	25.50		29.13	45
0120	4" diameter		14	.571		6.35	29		35.35	54
0130	6" diameter		12	.667		9.30	34		43.30	65.50
0140	12" diameter		10	.800		18.60	40.50		59.10	86.50
0150	Through floors, 2" diameter		32	.250		1.85	12.70		14.55	22.50
0160	4" diameter		28	.286		3.25	14.50		17.75	27
0170	6" diameter		24	.333		4.74	16.90		21.64	32.50
0180	12" diameter		20	.400		9.55	20.50		30.05	43.50
0190	Metallic piping, insulated									
0200	Through walls, 2" diameter	1 Carp	16	.500	Ea.	6.35	25.50		31.85	48
0210	4" diameter		14	.571		9.30	29		38.30	57.50
0220	6" diameter		12	.667		12.50	34		46.50	69
0230	12" diameter		10	.800		21.50	40.50		62	89.50
0240	Through floors, 2" diameter		32	.250		3.25	12.70		15.95	24
0250	4" diameter		28	.286		4.74	14.50		19.24	28.50
0260	6" diameter		24	.333		6.40	16.90		23.30	34.50
0270	12" diameter		20	.400		9.75	20.50		30.25	44
0280	Non metallic piping, non insulated									
0290	Through walls, 2" diameter	1 Carp	12	.667	Ea.	46	34		80	106
0300	4" diameter		10	.800		89.50	40.50		130	164
0310	6" diameter		8	1		172	50.50		222.50	272
0330	Through floors, 2" diameter		16	.500		30	25.50		55.50	74
0340	4" diameter		6	1.333		57.50	67.50		125	173
0350	6" diameter		6	1.333		111	67.50		178.50	232
0370	Ductwork, insulated & non insulated, round									
0380	Through walls, 6" diameter	1 Carp	12	.667	Ea.	12.50	34		46.50	69
0390	12" diameter		10	.800		21.50	40.50		62	89.50
0400	18" diameter		8	1		26	50.50		76.50	111
0410	Through floors, 6" diameter		16	.500		6.40	25.50		31.90	48
0420	12" diameter		14	.571		10.90	29		39.90	59
0430	18" diameter		12	.667		13.25	34		47.25	69.50
0440	Ductwork, insulated & non insulated, rectangular									
0450	With stiffener/closure angle, through walls, 6" x 12"	1 Carp	8	1	Ea.	16.70	50.50		67.20	101
0460	12" x 24"		6	1.333		33.50	67.50		101	147
0470	24" x 48"		4	2		67	101		168	239
0480	With stiffener/closure angle, through floors, 6" x 12"		10	.800		8.35	40.50		48.85	75
0490	12" x 24"		8	1		16.70	50.50		67.20	101
0500	24" x 48"		6	1.333		33.50	67.50		101	147
0510	Multi trade openings									
0520	Through walls, 6" x 12"	1 Carp	2	4	Ea.	46.50	203		249.50	380
0530	12" x 24"	"	1	8		152	405		557	825
0540	24" x 48"	2 Carp	1	16		540	810		1,350	1,925
0550	48" x 96"	"	.75	21.333		2,025	1,075		3,100	3,975
0560	Through floors, 6" x 12"	1 Carp	2	4		42	203		245	375
0570	12" x 24"	"	1	8		58	405		463	725
0580	24" x 48"	2 Carp	.75	21.333		139	1,075		1,214	1,900
0590	48" x 96"	"	.50	32		330	1,625		1,955	3,000
0600	Structural penetrations, through walls									
0610	Steel beams, W8 x 10	1 Carp	8	1	Ea.	70.50	50.50		121	160
0620	W12 x 14		6	1.333		102	67.50		169.50	222
0630	W21 x 44		5	1.600		146	81		227	292

07 84 13.10 Firestopping		Crew	Daily Output	Labor-Hours	Unit	Material	2018 Bare Costs Labor	Equipment	Total	Total Incl O&P
0640	W36 x 135	1 Carp	3	2.667	Ea.	226	135		361	465
0650	Bar joists, 18" deep		6	1.333		41.50	67.50		109	156
0660	24" deep		6	1.333		55.50	67.50		123	171
0670	36" deep		5	1.600		83	81		164	224
0680	48" deep		4	2		111	101		212	287
0690	Construction joints, floor slab at exterior wall									
0700	Precast, brick, block or drywall exterior									
0710	2" wide joint	1 Carp	125	.064	L.F.	9.80	3.24		13.04	16
0720	4" wide joint	"	75	.107	"	15.30	5.40		20.70	25.50
0730	Metal panel, glass or curtain wall exterior									
0740	2" wide joint	1 Carp	40	.200	L.F.	9.80	10.15		19.95	27
0750	4" wide joint	"	25	.320	"	15.30	16.20		31.50	43.50
0760	Floor slab to drywall partition									
0770	Flat joint	1 Carp	100	.080	L.F.	13.60	4.06		17.66	21.50
0780	Fluted joint		50	.160		15.15	8.10		23.25	30
0790	Etched fluted joint		75	.107		19.05	5.40		24.45	30
0800	Floor slab to concrete/masonry partition									
0810	Flat joint	1 Carp	75	.107	L.F.	13.60	5.40		19	24
0820	Fluted joint	"	50	.160	"	15.50	8.10		23.60	30
0830	Concrete/CMU wall joints									
0840	1" wide	1 Carp	100	.080	L.F.	19.40	4.06		23.46	28
0850	2" wide		75	.107		24	5.40		29.40	35.50
0860	4" wide		50	.160		33.50	8.10		41.60	50
0870	Concrete/CMU floor joints									
0880	1" wide	1 Carp	200	.040	L.F.	19.40	2.03		21.43	25
0890	2" wide		150	.053		24	2.70		26.70	31
0900	4" wide		100	.080		34.50	4.06		38.56	44

07 91 Preformed Joint Seals

07 91 13 – Compression Seals

07 91 13.10 Compression Seals

		Crew	Daily Output	Labor-Hours	Unit	Material	Labor	Equipment	Total	Total Incl O&P
0010	**COMPRESSION SEALS**									
4900	O-ring type cord, 1/4"	1 Bric	472	.017	L.F.	.44	.85		1.29	1.87
4910	1/2"		440	.018		1.20	.91		2.11	2.81
4920	3/4"		424	.019		2.35	.95		3.30	4.14
4930	1"		408	.020		4.07	.99		5.06	6.10
4940	1-1/4"		384	.021		7.75	1.05		8.80	10.20
4950	1-1/2"		368	.022		9.50	1.09		10.59	12.25
4960	1-3/4"		352	.023		16.20	1.14		17.34	19.70
4970	2"		344	.023		22	1.17		23.17	26

07 91 16 – Joint Gaskets

07 91 16.10 Joint Gaskets

		Crew	Daily Output	Labor-Hours	Unit	Material	Labor	Equipment	Total	Total Incl O&P
0010	**JOINT GASKETS**									
4400	Joint gaskets, neoprene, closed cell w/adh, 1/8" x 3/8"	1 Bric	240	.033	L.F.	.33	1.68		2.01	3.09
4500	1/4" x 3/4"		215	.037		.63	1.87		2.50	3.74
4700	1/2" x 1"		200	.040		1.50	2.01		3.51	4.93
4800	3/4" x 1-1/2"		165	.048		1.63	2.44		4.07	5.75

07 91 23 – Backer Rods

07 91 23.10 Backer Rods		Crew	Daily Output	Labor-Hours	Unit	Material	2018 Bare Costs Labor	Equipment	Total	Total Incl O&P
0010	**BACKER RODS**									
0032	Backer rod, polyethylene, 1/4" diameter	1 Bric	460	.017	L.F.	.02	.87		.89	1.46
0052	1/2" diameter		460	.017		.04	.87		.91	1.48
0072	3/4" diameter		460	.017		.07	.87		.94	1.50
0092	1" diameter		460	.017		.12	.87		.99	1.56

07 91 26 – Joint Fillers

07 91 26.10 Joint Fillers		Crew	Daily Output	Labor-Hours	Unit	Material	2018 Bare Costs Labor	Equipment	Total	Total Incl O&P
0010	**JOINT FILLERS**									
4360	Butyl rubber filler, 1/4" x 1/4"	1 Bric	290	.028	L.F.	.22	1.39		1.61	2.51
4365	1/2" x 1/2"		250	.032		.89	1.61		2.50	3.61
4370	1/2" x 3/4"		210	.038		1.34	1.92		3.26	4.60
4375	3/4" x 3/4"		230	.035		2.01	1.75		3.76	5.05
4380	1" x 1"		180	.044		2.68	2.24		4.92	6.60
4390	For coloring, add					12%				
4980	Polyethylene joint backing, 1/4" x 2"	1 Bric	2.08	3.846	C.L.F.	13.50	193		206.50	330
4990	1/4" x 6"		1.28	6.250	"	29	315		344	545
5600	Silicone, room temp vulcanizing foam seal, 1/4" x 1/2"		1312	.006	L.F.	.45	.31		.76	1
5610	1/2" x 1/2"		656	.012		.90	.61		1.51	1.99
5620	1/2" x 3/4"		442	.018		1.35	.91		2.26	2.98
5630	3/4" x 3/4"		328	.024		2.03	1.23		3.26	4.24
5640	1/8" x 1"		1312	.006		.45	.31		.76	1
5650	1/8" x 3"		442	.018		1.35	.91		2.26	2.98
5670	1/4" x 3"		295	.027		2.71	1.36		4.07	5.20
5680	1/4" x 6"		148	.054		5.40	2.72		8.12	10.40
5690	1/2" x 6"		82	.098		10.85	4.91		15.76	19.90
5700	1/2" x 9"		52.50	.152		16.25	7.65		23.90	30.50
5710	1/2" x 12"		33	.242		21.50	12.20		33.70	44

07 92 13 – Elastomeric Joint Sealants

07 92 13.10 Masonry Joint Sealants		Crew	Daily Output	Labor-Hours	Unit	Material	2018 Bare Costs Labor	Equipment	Total	Total Incl O&P
0010	**MASONRY JOINT SEALANTS**, 1/2" x 1/2" joint									
0050	Re-caulk only, oil base	1 Bric	225	.036	L.F.	.54	1.79		2.33	3.51
0100	Acrylic latex		205	.039		.23	1.96		2.19	3.46
0200	Polyurethane		200	.040		.71	2.01		2.72	4.06
0300	Silicone		195	.041		.70	2.06		2.76	4.14
1000	Cut out and re-caulk, oil base		145	.055		.54	2.78		3.32	5.10
1050	Acrylic latex		130	.062		.23	3.10		3.33	5.30
1100	Polyurethane		125	.064		.71	3.22		3.93	6.05
1150	Silicone		120	.067		.70	3.35		4.05	6.20
9000	Minimum labor/equipment charge		4	2	Job		101		101	164

07 92 13.20 Caulking and Sealant Options

07 92 13.20 Caulking and Sealant Options		Crew	Daily Output	Labor-Hours	Unit	Material	2018 Bare Costs Labor	Equipment	Total	Total Incl O&P
0010	**CAULKING AND SEALANT OPTIONS**									
0050	Latex acrylic based, bulk				Gal.	29.50			29.50	32.50
0055	Bulk in place 1/4" x 1/4" bead	1 Bric	300	.027	L.F.	.09	1.34		1.43	2.29
0060	1/4" x 3/8"		294	.027		.16	1.37		1.53	2.40
0065	1/4" x 1/2"		288	.028		.21	1.40		1.61	2.51
0075	3/8" x 3/8"		284	.028		.23	1.42		1.65	2.57
0080	3/8" x 1/2"		280	.029		.31	1.44		1.75	2.68
0085	3/8" x 5/8"		276	.029		.39	1.46		1.85	2.81

07 92 13 – Elastomeric Joint Sealants

07 92 13.20 Caulking and Sealant Options	Crew	Daily Output	Labor-Hours	Unit	Material	2018 Bare Costs Labor	Equipment	Total	Total Incl O&P
0095	3/8" x 3/4"	1 Bric	272	.029	L.F.	.47	1.48		1.95	2.92
0100	1/2" x 1/2"		275	.029		.42	1.46		1.88	2.85
0105	1/2" x 5/8"		269	.030		.52	1.50		2.02	3.01
0110	1/2" x 3/4"		263	.030		.62	1.53		2.15	3.19
0115	1/2" x 7/8"		256	.031		.73	1.57		2.30	3.36
0120	1/2" x 1"		250	.032		.83	1.61		2.44	3.54
0125	3/4" x 3/4"		244	.033		.94	1.65		2.59	3.72
0130	3/4" x 1"		225	.036		1.25	1.79		3.04	4.29
0135	1" x 1"	▼	200	.040	▼	1.66	2.01		3.67	5.10
0190	Cartridges				Gal.	32.50			32.50	36
0200	11 fl. oz. cartridge				Ea.	2.81			2.81	3.09
0500	1/4" x 1/2"	1 Bric	288	.028	L.F.	.23	1.40		1.63	2.53
0600	1/2" x 1/2"		275	.029		.46	1.46		1.92	2.89
0800	3/4" x 3/4"		244	.033		1.03	1.65		2.68	3.83
0900	3/4" x 1"		225	.036		1.38	1.79		3.17	4.44
1000	1" x 1"	▼	200	.040	▼	1.72	2.01		3.73	5.20
1400	Butyl based, bulk				Gal.	37			37	40.50
1500	Cartridges				"	40.50			40.50	44.50
1700	1/4" x 1/2", 154 L.F./gal.	1 Bric	288	.028	L.F.	.24	1.40		1.64	2.54
1800	1/2" x 1/2", 77 L.F./gal.	"	275	.029	"	.48	1.46		1.94	2.92
2300	Polysulfide compounds, 1 component, bulk				Gal.	83			83	91
2400	Cartridges				"	128			128	141
2600	1 or 2 component, in place, 1/4" x 1/4", 308 L.F./gal.	1 Bric	300	.027	L.F.	.27	1.34		1.61	2.49
2700	1/2" x 1/4", 154 L.F./gal.		288	.028		.54	1.40		1.94	2.87
2900	3/4" x 3/8", 68 L.F./gal.		272	.029		1.22	1.48		2.70	3.75
3000	1" x 1/2", 38 L.F./gal.	▼	250	.032	▼	2.18	1.61		3.79	5.05
3200	Polyurethane, 1 or 2 component				Gal.	53.50			53.50	58.50
3300	Cartridges				"	67			67	73.50
3500	Bulk, in place, 1/4" x 1/4"	1 Bric	300	.027	L.F.	.17	1.34		1.51	2.38
3655	1/2" x 1/4"		288	.028		.35	1.40		1.75	2.66
3800	3/4" x 3/8"		272	.029		.79	1.48		2.27	3.27
3900	1" x 1/2"	▼	250	.032	▼	1.39	1.61		3	4.16
4100	Silicone rubber, bulk				Gal.	57			57	62.50
4200	Cartridges				"	52.50			52.50	57.50
4300	Bulk in place, 1/4" x 1/2", 154 L.F./gal.	1 Bric	235	.034	L.F.	.37	1.71		2.08	3.20

07 92 16 – Rigid Joint Sealants

07 92 16.10 Rigid Joint Sealants

		Crew	Daily Output	Labor-Hours	Unit	Material	Labor	Equipment	Total	Total Incl O&P
0010	**RIGID JOINT SEALANTS**									
5802	Tapes, sealant, PVC foam adhesive, 1/16" x 1/4"				L.F.	.09			.09	.10
5902	1/16" x 1/2"					.09			.09	.10
5952	1/16" x 1"					.16			.16	.18
6002	1/8" x 1/2"				▼	.09			.09	.10

07 92 19 – Acoustical Joint Sealants

07 92 19.10 Acoustical Sealant

		Crew	Daily Output	Labor-Hours	Unit	Material	Labor	Equipment	Total	Total Incl O&P
0010	**ACOUSTICAL SEALANT**									
0020	Acoustical sealant, elastomeric, cartridges				Ea.	8.45			8.45	9.25
0025	In place, 1/4" x 1/4"	1 Bric	300	.027	L.F.	.34	1.34		1.68	2.57
0030	1/4" x 1/2"		288	.028		.69	1.40		2.09	3.04
0035	1/2" x 1/2"		275	.029		1.38	1.46		2.84	3.90
0040	1/2" x 3/4"		263	.030		2.07	1.53		3.60	4.77
0045	3/4" x 3/4"		244	.033		3.10	1.65		4.75	6.10
0050	1" x 1"	▼	200	.040	▼	5.50	2.01		7.51	9.35

For customer support on your Commercial Renovation Costs with RSMeans data, call 800.448.8182.

07 95 Expansion Control

07 95 13 – Expansion Joint Cover Assemblies

07 95 13.50 Expansion Joint Assemblies	Crew	Daily Output	Labor-Hours	Unit	Material	2018 Bare Costs Labor	Equipment	Total	Total Incl O&P
0010 **EXPANSION JOINT ASSEMBLIES**									
0200 Floor cover assemblies, 1" space, aluminum	1 Sswk	38	.211	L.F.	19	11.50		30.50	40.50
0300 Bronze		38	.211		60.50	11.50		72	86
0500 2" space, aluminum		38	.211		18.50	11.50		30	40
0600 Bronze		38	.211		60.50	11.50		72	86
0800 Wall and ceiling assemblies, 1" space, aluminum		38	.211		16.75	11.50		28.25	38
0900 Bronze		38	.211		53.50	11.50		65	78.50
1100 2" space, aluminum		38	.211		17.10	11.50		28.60	38.50
1200 Bronze		38	.211		52.50	11.50		64	77
1400 Floor to wall assemblies, 1" space, aluminum		38	.211		19.50	11.50		31	41
1500 Bronze or stainless		38	.211		64.50	11.50		76	90.50
1700 Gym floor angle covers, aluminum, 3" x 3" angle		46	.174		19.50	9.50		29	37.50
1800 3" x 4" angle		46	.174		23	9.50		32.50	41.50
2000 Roof closures, aluminum, flat roof, low profile, 1" space		57	.140		24	7.65		31.65	39.50
2100 High profile		57	.140		26	7.65		33.65	41.50
2300 Roof to wall, low profile, 1" space		57	.140		23.50	7.65		31.15	38.50
2400 High profile		57	.140		26	7.65		33.65	41.50
9000 Minimum labor/equipment charge		2	4	Job		219		219	375

Division Notes

	CREW	DAILY OUTPUT	LABOR-HOURS	UNIT	BARE COSTS				TOTAL INCL O&P
					MAT.	LABOR	EQUIP.	TOTAL	

Estimating Tips
08 10 00 Doors and Frames
All exterior doors should be addressed for their energy conservation (insulation and seals).

- Most metal doors and frames look alike, but there may be significant differences among them. When estimating these items, be sure to choose the line item that most closely compares to the specification or door schedule requirements regarding:
  - □ type of metal
  - □ metal gauge
  - □ door core material
  - □ fire rating
  - □ finish
- Wood and plastic doors vary considerably in price. The primary determinant is the veneer material. Lauan, birch, and oak are the most common veneers. Other variables include the following:
  - □ hollow or solid core
  - □ fire rating
  - □ flush or raised panel
  - □ finish
- Door pricing includes bore for cylindrical locksets and mortise for hinges.

08 30 00 Specialty Doors and Frames
- There are many varieties of special doors, and they are usually priced per each. Add frames, hardware, or operators required for a complete installation.

08 40 00 Entrances, Storefronts, and Curtain Walls
- Glazed curtain walls consist of the metal tube framing and the glazing material. The cost data in this subdivision is presented for the metal tube framing alone or the composite wall. If your estimate requires a detailed takeoff of the framing, be sure to add the glazing cost and any tints.

08 50 00 Windows
- Steel windows are unglazed and aluminum can be glazed or unglazed. Some metal windows are priced without glass. Refer to 08 80 00 Glazing for glass pricing. The grade C indicates commercial grade windows, usually ASTM C-35.
- All wood windows and vinyl are priced preglazed. The glazing is insulating glass. Add the cost of screens and grills if required and not already included.

08 70 00 Hardware
- Hardware costs add considerably to the cost of a door. The most efficient method to determine the hardware requirements for a project is to review the door and hardware schedule together. One type of door may have different hardware, depending on the door usage.
- Door hinges are priced by the pair, with most doors requiring 1-1/2 pairs per door. The hinge prices do not include installation labor, because it is included in door installation.

Hinges are classified according to the frequency of use, base material, and finish.

08 80 00 Glazing
- Different openings require different types of glass. The most common types are:
  - □ float
  - □ tempered
  - □ insulating
  - □ impact-resistant
  - □ ballistic-resistant
- Most exterior windows are glazed with insulating glass. Entrance doors and window walls, where the glass is less than 18" from the floor, are generally glazed with tempered glass. Interior windows and some residential windows are glazed with float glass.
- Coastal communities require the use of impact-resistant glass, dependent on wind speed.
- The insulation or 'u' value is a strong consideration, along with solar heat gain, to determine total energy efficiency.

Reference Numbers
Reference numbers are shown at the beginning of some major classifications. These numbers refer to related items in the Reference Section. The reference information may be an estimating procedure, an alternate pricing method, or technical information.

Note: Not all subdivisions listed here necessarily appear. ■

08 01 Operation and Maintenance of Openings

08 01 11 – Operation and Maintenance of Metal Doors and Frames

08 01 11.10 Door and Window Maintenance

		Crew	Daily Output	Labor-Hours	Unit	Material	2018 Bare Costs Labor	Equipment	Total	Total Incl O&P
0010	**DOOR & WINDOW MAINTENANCE**									
1130	Install door	2 Carp	13.07	1.224	Ea.		62		62	101
1140	Remove deadbolt	1 Carp	13.85	.578			29.50		29.50	47.50
1150	Remove panic bar		7.69	1.040			52.50		52.50	85.50
1420	Remove door, plane to fit, rail, reinstall door	↓	9.60	.833	↓		42.50		42.50	68.50

08 01 14 – Operation and Maintenance of Wood Doors

08 01 14.15 Door and Window Maintenance

		Crew	Daily Output	Labor-Hours	Unit	Material	2018 Bare Costs Labor	Equipment	Total	Total Incl O&P
0010	**DOOR & WINDOW MAINTENANCE**									
0050	Rehang single door	1 Carp	32	.250	Ea.		12.70		12.70	20.50
0100	Rehang double door		16	.500	Pr.		25.50		25.50	41
0350	Install window sash cord	↓	16	.500	Ea.	4.54	25.50		30.04	46

08 01 53 – Operation and Maintenance of Plastic Windows

08 01 53.81 Solid Vinyl Replacement Windows

			Crew	Daily Output	Labor-Hours	Unit	Material	2018 Bare Costs Labor	Equipment	Total	Total Incl O&P
0010	**SOLID VINYL REPLACEMENT WINDOWS**	R085313-20									
0020	Double-hung, insulated glass, up to 83 united inches	G	2 Carp	8	2	Ea.	173	101		274	355
0040	84 to 93	G		8	2		199	101		300	385
0060	94 to 101	G		6	2.667		199	135		334	440
0080	102 to 111	G		6	2.667		208	135		343	450
0100	112 to 120	G		6	2.667	↓	222	135		357	465
0120	For each united inch over 120, add	G		800	.020	Inch	2.79	1.01		3.80	4.72
0140	Casement windows, one operating sash, 42 to 60 united inches	G		8	2	Ea.	228	101		329	415
0160	61 to 70	G		8	2		259	101		360	450
0180	71 to 80	G		8	2		283	101		384	475
0200	81 to 96	G		8	2		298	101		399	495
0220	Two operating sash, 58 to 78 united inches	G		8	2		455	101		556	665
0240	79 to 88	G		8	2		490	101		591	700
0260	89 to 98	G		8	2		525	101		626	745
0280	99 to 108	G		6	2.667		555	135		690	830
0300	109 to 121	G		6	2.667		595	135		730	875
0320	Two operating, one fixed sash, 73 to 108 united inches	G		8	2		715	101		816	955
0340	109 to 118	G		8	2		755	101		856	995
0360	119 to 128	G		6	2.667		775	135		910	1,075
0380	129 to 138	G		6	2.667		820	135		955	1,125
0400	139 to 156	G		6	2.667		860	135		995	1,175
0420	Four operating sash, 98 to 118 united inches	G		8	2		1,025	101		1,126	1,300
0440	119 to 128	G		8	2		1,100	101		1,201	1,375
0460	129 to 138	G		6	2.667		1,175	135		1,310	1,500
0480	139 to 148	G		6	2.667		1,225	135		1,360	1,575
0500	149 to 168	G		6	2.667		1,300	135		1,435	1,675
0520	169 to 178	G		6	2.667		1,425	135		1,560	1,775
0560	Fixed picture window, up to 63 united inches	G		8	2		170	101		271	350
0580	64 to 83	G		8	2		200	101		301	385
0600	84 to 101	G		8	2	↓	240	101		341	430
0620	For each united inch over 101, add	G	↓	900	.018	Inch	2.85	.90		3.75	4.60
0800	Cellulose fiber insulation, poured into sash balance cavity	G	1 Carp	36	.222	C.F.	.69	11.25		11.94	19.05
0820	Silicone caulking at perimeter	G	"	800	.010	L.F.	.17	.51		.68	1.01
2000	Impact resistant replacement windows										
2005	Laminated glass, 120 MPH rating, measure in united inches										
2010	Installation labor does not cover any rework of the window opening										
2020	Double-hung, insulated glass, up to 101 united inches		2 Carp	8	2	Ea.	520	101		621	735
2025	For each united inch over 101, add			80	.200	Inch	3.30	10.15		13.45	20
2100	Casement windows, impact resistant, up to 60 united inches		↓	8	2	Ea.	480	101		581	690

222

08 01 Operation and Maintenance of Openings

08 01 53 – Operation and Maintenance of Plastic Windows

08 01 53.81 Solid Vinyl Replacement Windows

		Crew	Daily Output	Labor-Hours	Unit	Material	2018 Bare Costs Labor	Equipment	Total	Total Incl O&P
2120	61 to 70	2 Carp	8	2	Ea.	505	101		606	720
2130	71 to 80		8	2		535	101		636	750
2140	81 to 100		8	2		550	101		651	765
2150	For each united inch over 100, add		80	.200	Inch	4.65	10.15		14.80	21.50
2200	Awning windows, impact resistant, up to 60 united inches		8	2	Ea.	490	101		591	705
2220	61 to 70		8	2		515	101		616	730
2230	71 to 80		8	2		525	101		626	740
2240	For each united inch over 80, add		80	.200	Inch	4.65	10.15		14.80	21.50
2300	Picture windows, impact resistant, up to 63 united inches		8	2	Ea.	355	101		456	555
2320	63 to 83		8	2		385	101		486	585
2330	84 to 101		8	2		420	101		521	625
2340	For each united inch over 101, add		80	.200	Inch	3.30	10.15		13.45	20

08 05 Common Work Results for Openings

08 05 05 – Selective Demolition for Openings

08 05 05.10 Selective Demolition Doors

		Crew	Daily Output	Labor-Hours	Unit	Material	2018 Bare Costs Labor	Equipment	Total	Total Incl O&P
0010	**SELECTIVE DEMOLITION DOORS** R024119-10									
0200	Doors, exterior, 1-3/4" thick, single, 3' x 7' high	1 Clab	16	.500	Ea.		19.95		19.95	32.50
0202	Doors, exterior, 3' - 6' wide x 7' high		12	.667			26.50		26.50	43
0210	3' x 8' high		10	.800			32		32	52
0215	Double, 3' x 8' high		6	1.333			53		53	86.50
0220	Double, 6' x 7' high		12	.667			26.50		26.50	43
0500	Interior, 1-3/8" thick, single, 3' x 7' high		20	.400			15.95		15.95	26
0520	Double, 6' x 7' high		16	.500			19.95		19.95	32.50
0700	Bi folding, 3' x 6'-8" high		20	.400			15.95		15.95	26
0720	6' x 6'-8" high		18	.444			17.70		17.70	29
0900	Bi-passing, 3' x 6'-8" high		16	.500			19.95		19.95	32.50
0940	6' x 6'-8" high		14	.571			23		23	37
0960	Interior metal door 1-3/4" thick, 3'-0" x 6'-8"		18	.444			17.70		17.70	29
0980	Interior metal door 1-3/4" thick, 3'-0" x 7'-0"		18	.444			17.70		17.70	29
1000	Interior wood door 1-3/4" thick, 3'-0" x 6'-8"		20	.400			15.95		15.95	26
1020	Interior wood door 1-3/4" thick, 3'-0" x 7'-0"		20	.400			15.95		15.95	26
1100	Door demo, floor door	2 Sswk	5	3.200			175		175	298
1500	Remove and reset, hollow core	1 Carp	8	1			50.50		50.50	82.50
1520	Solid		6	1.333			67.50		67.50	110
2000	Frames, including trim, metal		8	1			50.50		50.50	82.50
2200	Wood	2 Carp	32	.500			25.50		25.50	41
2201	Alternate pricing method	1 Carp	200	.040	L.F.		2.03		2.03	3.29
2205	Remove door hardware	"	45.70	.175	Ea.		8.90		8.90	14.40
3000	Special doors, counter doors	2 Carp	6	2.667			135		135	219
3100	Double acting		10	1.600			81		81	132
3200	Floor door (trap type), or access type		8	2			101		101	165
3300	Glass, sliding, including frames		12	1.333			67.50		67.50	110
3400	Overhead, commercial, 12' x 12' high		4	4			203		203	330
3440	up to 20' x 16' high		3	5.333			270		270	440
3445	up to 35' x 30' high		1	16			810		810	1,325
3500	Residential, 9' x 7' high		8	2			101		101	165
3540	16' x 7' high		7	2.286			116		116	188
3600	Remove and reset, small		4	4			203		203	330
3620	Large		2.50	6.400			325		325	525
3700	Roll-up grille		5	3.200			162		162	263

For customer support on your Commercial Renovation Costs with RSMeans data, call 800.448.8182.

223

08 05 05 – Selective Demolition for Openings

08 05 05.10 Selective Demolition Doors

		Crew	Daily Output	Labor-Hours	Unit	Material	2018 Bare Costs Labor	Equipment	Total	Total Incl O&P
3800	Revolving door, no wire or connections	2 Carp	2	8	Ea.		405		405	660
3900	Storefront swing door	↓	3	5.333			270		270	440
3902	Cafe/bar swing door	2 Clab	8	2			79.50		79.50	129
4000	Residential lockset, exterior	1 Carp	28	.286			14.50		14.50	23.50
4224	Pocket door, no frame		8	1			50.50		50.50	82.50
4300	Remove and reset deadbolt	↓	8	1			50.50		50.50	82.50
5500	Remove door closer	1 Clab	10	.800			32		32	52
5520	Remove push/pull plate		20	.400			15.95		15.95	26
5540	Remove door bolt		30	.267			10.65		10.65	17.25
5560	Remove kick/mop plate		20	.400			15.95		15.95	26
5585	Remove panic device		10	.800			32		32	52
5590	Remove mail slot		45	.178			7.10		7.10	11.50
5595	Remove peep hole	↓	45	.178			7.10		7.10	11.50
5604	Remove 9' x 7' garage door	2 Carp	9.50	1.684			85.50		85.50	139
5624	16' x 7' garage door		8	2			101		101	165
5626	9' x 7' swing up garage door		7	2.286			116		116	188
5628	16' x 7' swing up garage door	↓	7	2.286			116		116	188
5630	Garage door demolition, remove garage door track	1 Carp	6	1.333			67.50		67.50	110
5644	Remove overhead door opener	1 Clab	5	1.600			64		64	104
5774	Remove heavy gage sectional door, 20 ga., 8' x 8'	2 Carp	7	2.286			116		116	188
5784	10' x 10'		6	2.667			135		135	219
5794	12' x 12'		5	3.200			162		162	263
5805	14' x 14'	↓	4	4			203		203	330
6384	Remove French door unit	1 Clab	8	1			40		40	64.50
7100	Remove double swing pneumatic doors, openers and sensors	2 Skwk	.50	32	Opng.		1,675		1,675	2,700
7570	Remove shock absorbing door	2 Sswk	1.90	8.421	"		460		460	785
9000	Minimum labor/equipment charge	1 Carp	4	2	Job		101		101	165

08 05 05.20 Selective Demolition of Windows

		Crew	Daily Output	Labor-Hours	Unit	Material	2018 Bare Costs Labor	Equipment	Total	Total Incl O&P
0010	**SELECTIVE DEMOLITION OF WINDOWS** R024119-10									
0200	Aluminum, including trim, to 12 S.F.	1 Clab	16	.500	Ea.		19.95		19.95	32.50
0240	To 25 S.F.		11	.727			29		29	47
0280	To 50 S.F.		5	1.600			64		64	104
0320	Storm windows/screens, to 12 S.F.		27	.296			11.80		11.80	19.15
0360	To 25 S.F.		21	.381			15.20		15.20	24.50
0400	To 50 S.F.	↓	16	.500			19.95		19.95	32.50
0600	Glass, up to 10 S.F./window		200	.040	S.F.		1.59		1.59	2.59
0620	Over 10 S.F./window		150	.053	"		2.13		2.13	3.45
1000	Steel, including trim, to 12 S.F.		13	.615	Ea.		24.50		24.50	40
1020	To 25 S.F.		9	.889			35.50		35.50	57.50
1040	To 50 S.F.		4	2			79.50		79.50	129
2000	Wood, including trim, to 12 S.F.		22	.364			14.50		14.50	23.50
2020	To 25 S.F.		18	.444			17.70		17.70	29
2060	To 50 S.F.		13	.615			24.50		24.50	40
2065	To 180 S.F.	↓	8	1	↓		40		40	64.50
4400	Remove skylight, prefabricated glass block with metal frame	G-3	180	.178	S.F.		8.85		8.85	14.15
4410	Remove skylight, plstc domes, flush/curb mtd	2 Carp	395	.041	"		2.05		2.05	3.33
4420	Remove skylight, plstc/glass up to 2' x 3'	1 Carp	15	.533	Ea.		27		27	44
4440	Remove skylight, plstc/glass up to 4' x 6'	2 Carp	10	1.600			81		81	132
4480	Remove roof window up to 3' x 4'	1 Carp	8	1			50.50		50.50	82.50
4500	Remove roof window up to 4' x 6'	2 Carp	6	2.667			135		135	219
5020	Remove and reset window, up to a 2' x 2' window	1 Carp	6	1.333			67.50		67.50	110
5040	Up to a 3' x 3' window	↓	4	2			101		101	165

08 05 Common Work Results for Openings

08 05 05 – Selective Demolition for Openings

08 05 05.20 Selective Demolition of Windows	Crew	Daily Output	Labor-Hours	Unit	Material	2018 Bare Costs Labor	2018 Bare Costs Equipment	Total	Total Incl O&P
5080	Up to a 4' x 5' window	1 Carp	2	4	Ea.		203		203	330
6000	Screening only	1 Clab	4000	.002	S.F.		.08		.08	.13
9000	Minimum labor/equipment charge	"	4	2	Job		79.50		79.50	129

08 11 Metal Doors and Frames

08 11 16 – Aluminum Doors and Frames

08 11 16.10 Entrance Doors

	08 11 16.10 Entrance Doors	Crew	Daily Output	Labor-Hours	Unit	Material	2018 Bare Costs Labor	2018 Bare Costs Equipment	Total	Total Incl O&P
0010	**ENTRANCE DOORS** and frame, aluminum, narrow stile									
0011	Including standard hardware, clear finish, no glass									
0012	Top and bottom offset pivots, 1/4" beveled glass stops, threshold									
0013	Dead bolt lock with inside thumb screw, standard push pull									
0020	3'-0" x 7'-0" opening	2 Sswk	2	8	Ea.	945	435		1,380	1,800
0025	Anodizing aluminum entr. door & frame, add					113			113	124
0030	3'-6" x 7'-0" opening	2 Sswk	2	8		870	435		1,305	1,700
0100	3'-0" x 10'-0" opening, 3' high transom		1.80	8.889		1,350	485		1,835	2,300
0200	3'-6" x 10'-0" opening, 3' high transom		1.80	8.889		1,375	485		1,860	2,350
0280	5'-0" x 7'-0" opening		2	8		1,600	435		2,035	2,500
0300	6'-0" x 7'-0" opening		1.30	12.308		1,150	675		1,825	2,425
0301	6'-0" x 7'-0" opening		1.30	12.308	Pr.	1,150	675		1,825	2,425
0400	6'-0" x 10'-0" opening, 3' high transom		1.10	14.545	"	1,475	795		2,270	2,975
0520	3'-0" x 7'-0" opening, wide stile		2	8	Ea.	925	435		1,360	1,775
0540	3'-6" x 7'-0" opening		2	8		1,000	435		1,435	1,850
0560	5'-0" x 7'-0" opening		2	8		1,350	435		1,785	2,250
0580	6'-0" x 7'-0" opening		1.30	12.308	Pr.	1,550	675		2,225	2,850
0600	7'-0" x 7'-0" opening		1	16	"	2,275	875		3,150	4,000
1200	For non-standard size, add				Leaf	80%				
1250	For installation of non-standard size, add						20%			
1300	Light bronze finish, add				Leaf	36%				
1400	Dark bronze finish, add					25%				
1500	For black finish, add					40%				
1600	Concealed panic device, add					970			970	1,075
1700	Electric striker release, add				Opng.	320			320	350
1800	Floor check, add				Leaf	670			670	740
1900	Concealed closer, add				"	550			550	600
9000	Minimum labor/equipment charge	2 Carp	4	4	Job		203		203	330

08 11 63 – Metal Screen and Storm Doors and Frames

08 11 63.23 Aluminum Screen and Storm Doors and Frames

	08 11 63.23 Aluminum Screen and Storm Doors and Frames	Crew	Daily Output	Labor-Hours	Unit	Material	2018 Bare Costs Labor	2018 Bare Costs Equipment	Total	Total Incl O&P
0010	**ALUMINUM SCREEN AND STORM DOORS AND FRAMES**									
0020	Combination storm and screen									
0420	Clear anodic coating, 2'-8" wide	2 Carp	14	1.143	Ea.	220	58		278	335
0440	3'-0" wide	"	14	1.143	"	191	58		249	305
0500	For 7'-0" door height, add					8%				
1020	Mill finish, 2'-8" wide	2 Carp	14	1.143	Ea.	250	58		308	370
1040	3'-0" wide	"	14	1.143		270	58		328	390
1100	For 7'-0" door, add					8%				
1520	White painted, 2'-8" wide	2 Carp	14	1.143		239	58		297	355
1540	3'-0" wide	"	14	1.143		315	58		373	440
1600	For 7'-0" door, add					8%				
2000	Wood door & screen, see Section 08 14 33.20									
9000	Minimum labor/equipment charge	1 Carp	4	2	Job		101		101	165

225

08 12 13 – Hollow Metal Frames

08 12 13.13 Standard Hollow Metal Frames		Crew	Daily Output	Labor-Hours	Unit	Material	2018 Bare Costs Labor	Equipment	Total	Total Incl O&P
0010	**STANDARD HOLLOW METAL FRAMES**										
0020	16 ga., up to 5-3/4" jamb depth									
0025	3'-0" x 6'-8" single	G	2 Carp	16	1	Ea.	121	50.50		171.50	216
0028	3'-6" wide, single	G		16	1		232	50.50		282.50	340
0030	4'-0" wide, single	G		16	1		268	50.50		318.50	380
0040	6'-0" wide, double	G		14	1.143		227	58		285	345
0045	8'-0" wide, double	G		14	1.143		227	58		285	345
0100	3'-0" x 7'-0" single	G		16	1		192	50.50		242.50	294
0110	3'-6" wide, single			16	1		194	50.50		244.50	296
0112	4'-0" wide, single			16	1		227	50.50		277.50	335
0140	6'-0" wide, double			14	1.143		227	58		285	345
0145	8'-0" wide, double			14	1.143		237	58		295	355
1000	16 ga., up to 4-7/8" deep, 3'-0" x 7'-0" single	G		16	1		181	50.50		231.50	282
1140	6'-0" wide, double	G		14	1.143		204	58		262	320
1200	16 ga., 8-3/4" deep, 3'-0" x 7'-0" single	G		16	1		192	50.50		242.50	294
1240	6'-0" wide, double	G		14	1.143		256	58		314	375
2800	14 ga., up to 3-7/8" deep, 3'-0" x 7'-0" single	G		16	1		247	50.50		297.50	355
2840	6'-0" wide, double	G		14	1.143		278	58		336	400
3000	14 ga., up to 5-3/4" deep, 3'-0" x 6'-8" single	G		16	1		154	50.50		204.50	252
3002	3'-6" wide, single	G		16	1		239	50.50		289.50	345
3005	4'-0" wide, single	G		16	1		155	50.50		205.50	254
3600	up to 5-3/4" jamb depth, 4'-0" x 7'-0" single			15	1.067		237	54		291	350
3620	6'-0" wide, double			12	1.333		177	67.50		244.50	305
3640	8'-0" wide, double			12	1.333		282	67.50		349.50	420
3700	8'-0" high, 4'-0" wide, single			15	1.067		270	54		324	385
3740	8'-0" wide, double			12	1.333		320	67.50		387.50	460
4000	6-3/4" deep, 4'-0" x 7'-0" single	G		15	1.067		266	54		320	380
4020	6'-0" wide, double	G		12	1.333		300	67.50		367.50	440
4040	8'-0" wide, double	G		12	1.333		224	67.50		291.50	355
4100	8'-0" high, 4'-0" wide, single	G		15	1.067		176	54		230	281
4140	8'-0" wide, double	G		12	1.333		415	67.50		482.50	565
4400	8-3/4" deep, 4'-0" x 7'-0", single			15	1.067		385	54		439	510
4440	8'-0" wide, double			12	1.333		405	67.50		472.50	560
4500	4'-0" x 8'-0", single			15	1.067		440	54		494	575
4540	8'-0" wide, double			12	1.333		450	67.50		517.50	605
4900	For welded frames, add						51			51	56
5380	Steel frames, KD, 14 ga., "B" label, to 5-3/4" throat, to 3'-0" x 7'-0"		2 Carp	15	1.067		225	54		279	335
5400	14 ga., "B" label, up to 5-3/4" deep, 4'-0" x 7'-0" single	G		15	1.067		167	54		221	272
5440	8'-0" wide, double	G		12	1.333		219	67.50		286.50	350
5800	6-3/4" deep, 7'-0" high, 4'-0" wide, single	G		15	1.067		173	54		227	279
5840	8'-0" wide, double	G		12	1.333		305	67.50		372.50	450
6200	8-3/4" deep, 4'-0" x 7'-0" single	G		15	1.067		247	54		301	360
6240	8'-0" wide, double	G		12	1.333		380	67.50		447.50	530
6300	For "A" label use same price as "B" label										
6400	For baked enamel finish, add						30%	15%			
6500	For galvanizing, add						20%				
6600	For hospital stop, add					Ea.	235			235	259
6620	For hospital stop, stainless steel, add					"	109			109	120
7900	Transom lite frames, fixed, add		2 Carp	155	.103	S.F.	43	5.25		48.25	56
8000	Movable, add		"	130	.123	"	55	6.25		61.25	70.50
9000	Minimum labor/equipment charge		1 Carp	4	2	Job		101		101	165

08 12 13.15 Borrowed Lites	Crew	Daily Output	Labor-Hours	Unit	Material	2018 Bare Costs Labor	Equipment	Total	Total Incl O&P
0010 **BORROWED LITES**									
0100	Hollow metal section 20 ga., 3-1/2" with glass stop	2 Carp	100	.160	L.F.	17.90	8.10		26	33
0110	3-3/4" with glass stop		100	.160		8.25	8.10		16.35	22
0120	4" with glass stop		100	.160		8.35	8.10		16.45	22.50
0130	4-5/8" with glass stop		100	.160		8.65	8.10		16.75	22.50
0140	4-7/8" with glass stop		100	.160		8.80	8.10		16.90	23
0150	5" with glass stop		100	.160		9.60	8.10		17.70	23.50
0160	5-3/8" with glass stop		100	.160		11.30	8.10		19.40	25.50
0300	Hollow metal section 18 ga., 3-1/2" with glass stop		80	.200		8.90	10.15		19.05	26.50
0310	3-3/4" with glass stop		80	.200		9.10	10.15		19.25	26.50
0320	4" with glass stop		80	.200		12.60	10.15		22.75	30.50
0330	4-5/8" with glass stop		80	.200		9.65	10.15		19.80	27
0340	4-7/8" with glass stop		80	.200		9.90	10.15		20.05	27.50
0350	5" with glass stop		80	.200		10	10.15		20.15	27.50
0360	5-3/8" with glass stop		80	.200		10	10.15		20.15	27.50
0370	6-5/8" with glass stop		80	.200		11.15	10.15		21.30	28.50
0380	6-7/8" with glass stop		80	.200		11.30	10.15		21.45	29
0390	7-1/4" with glass stop		80	.200		11.80	10.15		21.95	29.50
0500	Mullion section 18 ga., 3-1/2" with double stop		50	.320		20.50	16.20		36.70	49.50
0510	3-3/4" with double stop		50	.320		21	16.20		37.20	49.50
0530	4-5/8" with double stop		50	.320		22.50	16.20		38.70	51.50
0540	4-7/8" with double stop		50	.320		23	16.20		39.20	52
0550	5" with double stop		50	.320		23	16.20		39.20	52
0560	5-3/8" with double stop		50	.320		24	16.20		40.20	53
0570	6-5/8" with double stop		50	.320		26	16.20		42.20	55
0580	6-7/8" with double stop		50	.320		26.50	16.20		42.70	55.50
0590	7-1/4" with double stop		50	.320		27	16.20		43.20	56
1000	Assembled frame 2'-0" x 2'-0" 16 ga., 3-3/4" with glass stop		10	1.600	Ea.	202	81		283	355
1010	3'-0" x 2'-0"		10	1.600		202	81		283	355
1020	4'-0" x 2'-0"		10	1.600		202	81		283	355
1030	5'-0" x 2'-0"		10	1.600		202	81		283	355
1040	6'-0" x 2'-0"		8	2		300	101		401	495
1050	7'-0" x 2'-0"		8	2		300	101		401	495
1060	8'-0" x 2'-0"		8	2		300	101		401	495
1100	2'-0" x 7'-0"		8	2		300	101		401	495
1110	3'-0" x 7'-0"		8	2		300	101		401	495
1120	4'-0" x 7'-0"		8	2		300	101		401	495
1130	5'-0" x 7'-0"		8	2		300	101		401	495
1140	6'-0" x 7'-0"		6	2.667		325	135		460	580
1150	7'-0" x 7'-0"		6	2.667		325	135		460	580
1160	8'-0" x 7'-0"		6	2.667		325	135		460	580
1200	Assembled frame 2'-0" x 2'-0" 16 ga., 4-7/8" with glass stop		10	1.600		202	81		283	355
1210	3'-0" x 2'-0"		10	1.600		202	81		283	355
1220	4'-0" x 2'-0"		8	2		202	101		303	385
1230	5'-0" x 2'-0"		8	2		300	101		401	495
1240	6'-0" x 2'-0"		8	2		300	101		401	495
1250	7'-0" x 2'-0"		8	2		300	101		401	495
1260	8'-0" x 2'-0"		8	2		300	101		401	495
1300	2'-0" x 7'-0"		8	2		300	101		401	495
1310	3'-0" x 7'-0"		8	2		300	101		401	495
1320	4'-0" x 7'-0"		8	2		300	101		401	495
1330	5'-0" x 7'-0"		6	2.667		300	135		435	550

For customer support on your Commercial Renovation Costs with RSMeans data, call 800.448.8182.

227

08 12 13 – Hollow Metal Frames

08 12 13.15 Borrowed Lites

		Crew	Daily Output	Labor-Hours	Unit	Material	2018 Bare Costs Labor	Equipment	Total	Total Incl O&P
1340	6'-0" x 7'-0"	2 Carp	6	2.667	Ea.	325	135		460	580
1350	7'-0" x 7'-0"		6	2.667		325	135		460	580
1360	8'-0" x 7'-0"		6	2.667		325	135		460	580
1400	Assembled frame 2'-0" x 3'-0" 16 ga., 6-1/8" with glass stop		6	2.667		202	135		337	440
1410	3'-0" x 3'-0"		8	2		202	101		303	385
1420	4'-0" x 3'-0"		8	2		202	101		303	385
1430	5'-0" x 3'-0"		6	2.667		300	135		435	550
1440	6'-0" x 3'-0"		6	2.667		300	135		435	550
1450	7'-0" x 3'-0"		6	2.667		300	135		435	550
1460	8'-0" x 3'-0"		10	1.600		300	81		381	460

08 12 13.25 Channel Metal Frames

			Crew	Daily Output	Labor-Hours	Unit	Material	2018 Bare Costs Labor	Equipment	Total	Total Incl O&P
0010	**CHANNEL METAL FRAMES**										
0020	Steel channels with anchors and bar stops										
0100	6" channel @ 8.2#/L.F., 3' x 7' door, weighs 150#	G	E-4	13	2.462	Ea.	242	136	7.60	385.60	505
0200	8" channel @ 11.5#/L.F., 6' x 8' door, weighs 275#	G		9	3.556		445	196	10.95	651.95	835
0300	8' x 12' door, weighs 400#	G		6.50	4.923		645	272	15.15	932.15	1,200
0400	10" channel @ 15.3#/L.F., 10' x 10' door, weighs 500#	G		6	5.333		810	294	16.45	1,120.45	1,400
0500	12' x 12' door, weighs 600#	G		5.50	5.818		970	320	17.90	1,307.90	1,650
0600	12" channel @ 20.7#/L.F., 12' x 12' door, weighs 825#	G		4.50	7.111		1,325	390	22	1,737	2,175
0700	12' x 16' door, weighs 1000#	G		4	8		1,625	440	24.50	2,089.50	2,550
0800	For frames without bar stops, light sections, deduct						15%				
0900	Heavy sections, deduct						10%				
9000	Minimum labor/equipment charge		E-4	4	8	Job		440	24.50	464.50	775

08 13 Metal Doors

08 13 13 – Hollow Metal Doors

08 13 13.13 Standard Hollow Metal Doors

			Crew	Daily Output	Labor-Hours	Unit	Material	2018 Bare Costs Labor	Equipment	Total	Total Incl O&P
0010	**STANDARD HOLLOW METAL DOORS**	R081313-20									
0015	Flush, full panel, hollow core										
0017	When noted doors are prepared but do not include glass or louvers										
0020	1-3/8" thick, 20 ga., 2'-0" x 6'-8"	G	2 Carp	20	.800	Ea.	340	40.50		380.50	440
0040	2'-8" x 6'-8"	G		18	.889		350	45		395	460
0060	3'-0" x 6'-8"	G		17	.941		355	47.50		402.50	470
0100	3'-0" x 7'-0"			17	.941		365	47.50		412.50	485
0120	For vision lite, add						108			108	119
0140	For narrow lite, add						116			116	128
0320	Half glass, 20 ga., 2'-0" x 6'-8"	G	2 Carp	20	.800		600	40.50		640.50	725
0340	2'-8" x 6'-8"	G		18	.889		600	45		645	735
0360	3'-0" x 6'-8"	G		17	.941		600	47.50		647.50	740
0400	3'-0" x 7'-0"	G		17	.941		400	47.50		447.50	520
0410	1-3/8" thick, 18 ga., 2'-0" x 6'-8"	G		20	.800		405	40.50		445.50	510
0420	3'-0" x 6'-8"	G		17	.941		410	47.50		457.50	530
0425	3'-0" x 7'-0"	G		17	.941		430	47.50		477.50	555
0450	For vision lite, add						108			108	119
0452	For narrow lite, add						116			116	128
0460	Half glass, 18 ga., 2'-0" x 6'-8"	G	2 Carp	20	.800		560	40.50		600.50	685
0465	2'-8" x 6'-8"	G		18	.889		580	45		625	715
0470	3'-0" x 6'-8"	G		17	.941		570	47.50		617.50	705
0475	3'-0" x 7'-0"	G		17	.941		585	47.50		632.50	725
0500	Hollow core, 1-3/4" thick, full panel, 20 ga., 2'-8" x 6'-8"	G		18	.889		435	45		480	555
0520	3'-0" x 6'-8"	G		17	.941		435	47.50		482.50	560

For customer support on your Commercial Renovation Costs with RSMeans data, call 800.448.8182.

08 13 13 – Hollow Metal Doors

08 13 13.13 Standard Hollow Metal Doors		Crew	Daily Output	Labor-Hours	Unit	Material	2018 Bare Costs Labor	Equipment	Total	Total Incl O&P
0640	3'-0" x 7'-0"	G	2 Carp	17	.941	Ea.	465	47.50		512.50	590
0680	4'-0" x 7'-0"	G		15	1.067		635	54		689	785
0700	4'-0" x 8'-0"	G		13	1.231		750	62.50		812.50	920
1000	18 ga., 2'-8" x 6'-8"	G		17	.941		485	47.50		532.50	610
1020	3'-0" x 6'-8"	G		16	1		485	50.50		535.50	615
1120	3'-0" x 7'-0"	G		17	.941		495	47.50		542.50	625
1150	3'-6" x 7'-0"		1 Carp	14	.571		580	29		609	680
1180	4'-0" x 7'-0"	G	2 Carp	14	1.143		695	58		753	860
1200	4'-0" x 8'-0"	G	"	17	.941		850	47.50		897.50	1,025
1212	For vision lite, add						108			108	119
1214	For narrow lite, add						116			116	128
1230	Half glass, 20 ga., 2'-8" x 6'-8"	G	2 Carp	20	.800		580	40.50		620.50	705
1240	3'-0" x 6'-8"	G		18	.889		580	45		625	710
1260	3'-0" x 7'-0"	G		18	.889		595	45		640	730
1280	Embossed panel, 1-3/4" thick, poly core, 20 ga., 3'-0" x 7'-0"	G		18	.889		420	45		465	535
1290	Half glass, 1-3/4" thick, poly core, 20 ga., 3'-0" x 7'-0"	G		18	.889		615	45		660	750
1320	18 ga., 2'-8" x 6'-8"	G		18	.889		640	45		685	780
1340	3'-0" x 6'-8"	G		17	.941		635	47.50		682.50	775
1360	3'-0" x 7'-0"	G		17	.941		645	47.50		692.50	790
1380	4'-0" x 7'-0"	G		15	1.067		795	54		849	965
1400	4'-0" x 8'-0"	G		14	1.143		895	58		953	1,075
1500	Flush full panel, 16 ga., steel hollow core										
1520	2'-0" x 6'-8"	G	2 Carp	20	.800	Ea.	570	40.50		610.50	690
1530	2'-8" x 6'-8"	G		20	.800		570	40.50		610.50	690
1540	3'-0" x 6'-8"	G		20	.800		555	40.50		595.50	675
1560	2'-8" x 7'-0"	G		18	.889		580	45		625	715
1570	3'-0" x 7'-0"	G		18	.889		560	45		605	695
1580	3'-6" x 7'-0"	G		18	.889		665	45		710	810
1590	4'-0" x 7'-0"	G		18	.889		730	45		775	000
1600	2'-8" x 8'-0"	G		18	.889		715	45		760	860
1620	3'-0" x 8'-0"	G		18	.889		730	45		775	875
1630	3'-6" x 8'-0"	G		18	.889		810	45		855	965
1640	4'-0" x 8'-0"	G		18	.889		850	45		895	1,000
1650	1-13/16", 14 ga., 2'-8" x 7'-0"	G		10	1.600		1,150	81		1,231	1,375
1670	3'-0" x 7'-0"	G		10	1.600		1,100	81		1,181	1,350
1690	3'-6" x 7'-0"	G		10	1.600		1,225	81		1,306	1,450
1700	4'-0" x 7'-0"	G		10	1.600		1,275	81		1,356	1,525
1720	Insulated, 1-3/4" thick, full panel, 18 ga., 3'-0" x 6'-8"	G		15	1.067		485	54		539	625
1740	2'-8" x 7'-0"	G		16	1		505	50.50		555.50	640
1760	3'-0" x 7'-0"	G		15	1.067		505	54		559	645
1800	4'-0" x 8'-0"	G		13	1.231		780	62.50		842.50	960
1820	Half glass, 18 ga., 3'-0" x 6'-8"	G		16	1		635	50.50		685.50	780
1840	2'-8" x 7'-0"	G		17	.941		665	47.50		712.50	810
1860	3'-0" x 7'-0"	G		16	1		690	50.50		740.50	845
1900	4'-0" x 8'-0"	G		14	1.143		555	58		613	705
2000	For vision lite, add						108			108	119
2010	For narrow lite, add						116			116	128
8100	For bottom louver, add						224			224	246
8110	For baked enamel finish, add						30%	15%			
8120	For galvanizing, add						20%				
8190	Commercial door, flush steel, fire door see section 08 13 13 15										
8270	For dutch door with shelf, add to standard door					Ea.	300			300	330
9000	Minimum labor/equipment charge		1 Carp	4	2	Job		101		101	165

08 13 13.15 Metal Fire Doors

		Crew	Daily Output	Labor-Hours	Unit	Material	2018 Bare Costs Labor	Equipment	Total	Total Incl O&P
0010	**METAL FIRE DOORS** R081313-20									
0015	Steel, flush, "B" label, 90 minutes									
0020	Full panel, 20 ga., 2'-0" x 6'-8"	2 Carp	20	.800	Ea.	425	40.50		465.50	530
0040	2'-8" x 6'-8"		18	.889		440	45		485	560
0060	3'-0" x 6'-8"		17	.941		440	47.50		487.50	565
0080	3'-0" x 7'-0"		17	.941		455	47.50		502.50	580
0140	18 ga., 3'-0" x 6'-8"		16	1		505	50.50		555.50	640
0160	2'-8" x 7'-0"		17	.941		535	47.50		582.50	665
0180	3'-0" x 7'-0"		16	1		515	50.50		565.50	655
0200	4'-0" x 7'-0"		15	1.067		650	54		704	805
0220	For "A" label, 3 hour, 18 ga., use same price as "B" label									
0240	For vision lite, add				Ea.	163			163	179
0300	Full panel, 16 ga., 2'-0" x 6'-8"	2 Carp	20	.800		315	40.50		355.50	410
0310	2'-8" x 6'-8"		18	.889		550	45		595	680
0320	3'-0" x 6'-8"		17	.941		440	47.50		487.50	560
0350	2'-8" x 7'-0"		17	.941		570	47.50		617.50	705
0360	3'-0" x 7'-0"		16	1		555	50.50		605.50	695
0370	4'-0" x 7'-0"		15	1.067		715	54		769	880
0520	Flush, "B" label, 90 minutes, egress core, 20 ga., 2'-0" x 6'-8"		18	.889		665	45		710	810
0540	2'-8" x 6'-8"		17	.941		670	47.50		717.50	815
0560	3'-0" x 6'-8"		16	1		675	50.50		725.50	825
0580	3'-0" x 7'-0"		16	1		700	50.50		750.50	855
0640	Flush, "A" label, 3 hour, egress core, 18 ga., 3'-0" x 6'-8"		15	1.067		730	54		784	890
0660	2'-8" x 7'-0"		16	1		765	50.50		815.50	925
0680	3'-0" x 7'-0"		15	1.067		750	54		804	910
0700	4'-0" x 7'-0"		14	1.143		910	58		968	1,100
9000	Minimum labor/equipment charge	1 Carp	4	2	Job		101		101	165

08 13 13.20 Residential Steel Doors

			Crew	Daily Output	Labor-Hours	Unit	Material	2018 Bare Costs Labor	Equipment	Total	Total Incl O&P
0010	**RESIDENTIAL STEEL DOORS**										
0020	Prehung, insulated, exterior										
0030	Embossed, full panel, 2'-8" x 6'-8"	G	2 Carp	17	.941	Ea.	405	47.50		452.50	530
0040	3'-0" x 6'-8"	G		15	1.067		290	54		344	410
0060	3'-0" x 7'-0"	G		15	1.067		350	54		404	475
0070	5'-4" x 6'-8", double	G		8	2		575	101		676	800
0220	Half glass, 2'-8" x 6'-8"	G		17	.941		330	47.50		377.50	445
0240	3'-0" x 6'-8"	G		16	1		330	50.50		380.50	450
0260	3'-0" x 7'-0"	G		16	1		400	50.50		450.50	525
0270	5'-4" x 6'-8", double	G		8	2		990	101		1,091	1,250
1320	Flush face, full panel, 2'-8" x 6'-8"	G		16	1		291	50.50		341.50	405
1340	3'-0" x 6'-8"	G		15	1.067		291	54		345	410
1360	3'-0" x 7'-0"	G		15	1.067		310	54		364	430
1380	5'-4" x 6'-8", double	G		8	2		720	101		821	955
1420	Half glass, 2'-8" x 6'-8"	G		17	.941		350	47.50		397.50	465
1440	3'-0" x 6'-8"	G		16	1		350	50.50		400.50	470
1460	3'-0" x 7'-0"	G		16	1		435	50.50		485.50	565
1480	5'-4" x 6'-8", double	G		8	2		665	101		766	895
1500	Sidelight, full lite, 1'-0" x 6'-8" with grille	G					345			345	375
1510	1'-0" x 6'-8", low E	G					365			365	405
1520	1'-0" x 6'-8", half lite	G					305			305	335
1530	1'-0" x 6'-8", half lite, low E	G					296			296	325
2300	Interior, residential, closet, bi-fold, 2'-0" x 6'-8"	G	2 Carp	16	1		222	50.50		272.50	325
2330	3'-0" wide	G		16	1		260	50.50		310.50	370

For customer support on your Commercial Renovation Costs with RSMeans data, call 800.448.8182.

08 13 Metal Doors

08 13 13 – Hollow Metal Doors

08 13 13.20 Residential Steel Doors

		Crew	Daily Output	Labor-Hours	Unit	Material	2018 Bare Costs Labor	Equipment	Total	Total Incl O&P
2360	4'-0" wide	[G] 2 Carp	15	1.067	Ea.	280	54		334	400
2400	5'-0" wide	[G]	14	1.143		340	58		398	470
2420	6'-0" wide	[G] ↓	13	1.231	↓	300	62.50		362.50	430
9000	Minimum labor/equipment charge	1 Carp	4	2	Job		101		101	165

08 13 13.25 Doors Hollow Metal

		Crew	Daily Output	Labor-Hours	Unit	Material	2018 Bare Costs Labor	Equipment	Total	Total Incl O&P
0010	**DOORS HOLLOW METAL**									
0500	Exterior, commercial, flush, 20 ga., 1-3/4" x 7'-0" x 2'-6" wide	[G] 2 Carp	15	1.067	Ea.	291	54		345	410
0530	2'-8" wide	[G]	15	1.067		291	54		345	410
0560	3'-0" wide	[G]	14	1.143		292	58		350	415
0590	3'-6" wide	[G]	14	1.143		370	58		428	500
1000	18 ga., 1-3/4" x 7'-0" x 2'-6" wide	[G]	15	1.067		320	54		374	440
1030	2'-8" wide	[G]	15	1.067		320	54		374	445
1060	3'-0" wide	[G]	14	1.143		315	58		373	440
1500	16 ga., 1-3/4" x 7'-0" x 2'-6" wide	[G]	15	1.067		370	54		424	500
1530	2'-8" wide	[G]	15	1.067		370	54		424	500
1560	3'-0" wide	[G]	14	1.143		370	58		428	505
1590	3'-6" wide	[G]	14	1.143		690	58		748	855
2900	Fire door, "A" label, 18 gauge, 1-3/4" x 2'-6" x 7'-0"	[G]	15	1.067		795	54		849	965
2930	2'-8" wide	[G]	15	1.067		690	54		744	850
2960	3'-0" wide	[G]	14	1.143		690	58		748	855
2990	3'-6" wide	[G]	14	1.143		470	58		528	615
3100	"B" label, 2'-6" wide	[G]	15	1.067		585	54		639	735
3130	2'-8" wide	[G]	15	1.067		585	54		639	735
3160	3'-0" wide	[G] ↓	14	1.143	↓	585	58		643	740

08 13 16 – Aluminum Doors

08 13 16.10 Commercial Aluminum Doors

		Crew	Daily Output	Labor-Hours	Unit	Material	2018 Bare Costs Labor	Equipment	Total	Total Incl O&P
0010	**COMMERCIAL ALUMINUM DOORS**, flush, no glazing									
5000	Flush panel doors, pair of 2'-6" x 7'-0"	2 Sswk	2	8	Pr.	1,300	435		1,735	2,175
5050	3'-0" x 7'-0", single		2.50	6.400	Ea.	790	350		1,140	1,450
5100	Pair of 3'-0" x 7'-0"		2	8	Pr.	1,400	435		1,835	2,275
5150	3'-6" x 7'-0", single	↓	2.50	6.400	Ea.	915	350		1,265	1,600

08 14 Wood Doors

08 14 13 – Carved Wood Doors

08 14 13.10 Types of Wood Doors, Carved

		Crew	Daily Output	Labor-Hours	Unit	Material	2018 Bare Costs Labor	Equipment	Total	Total Incl O&P
0010	**TYPES OF WOOD DOORS, CARVED**									
3000	Solid wood, 1-3/4" thick stile and rail									
3020	Mahogany, 3'-0" x 7'-0", six panel	2 Carp	14	1.143	Ea.	1,200	58		1,258	1,400
3030	With two lites		10	1.600		3,050	81		3,131	3,475
3040	3'-6" x 8'-0", six panel		10	1.600		1,625	81		1,706	1,925
3050	With two lites		8	2		2,850	101		2,951	3,325
3100	Pine, 3'-0" x 7'-0", six panel		14	1.143		510	58		568	655
3110	With two lites		10	1.600		805	81		886	1,025
3120	3'-6" x 8'-0", six panel		10	1.600		1,025	81		1,106	1,250
3130	With two lites		8	2		1,950	101		2,051	2,325
3200	Red oak, 3'-0" x 7'-0", six panel		14	1.143		1,325	58		1,383	1,550
3210	With two lites		10	1.600		2,525	81		2,606	2,900
3220	3'-6" x 8'-0", six panel		10	1.600		2,700	81		2,781	3,100
3230	With two lites	↓	8	2	↓	3,500	101		3,601	4,025
4000	Hand carved door, mahogany									

08 14 13 – Carved Wood Doors

08 14 13.10 Types of Wood Doors, Carved		Crew	Daily Output	Labor-Hours	Unit	Material	2018 Bare Costs Labor	Equipment	Total	Total Incl O&P
4020	3'-0" x 7'-0", simple design	2 Carp	14	1.143	Ea.	1,750	58		1,808	2,025
4030	Intricate design		11	1.455		3,675	74		3,749	4,150
4040	3'-6" x 8'-0", simple design		10	1.600		2,775	81		2,856	3,175
4050	Intricate design	▼	8	2		3,675	101		3,776	4,200
4400	For custom finish, add					575			575	635
4600	Side light, mahogany, 7'-0" x 1'-6" wide, 4 lites	2 Carp	18	.889		1,100	45		1,145	1,275
4610	6 lites		14	1.143		2,525	58		2,583	2,875
4620	8'-0" x 1'-6" wide, 4 lites		14	1.143		2,075	58		2,133	2,375
4630	6 lites		10	1.600		2,075	81		2,156	2,400
4640	Side light, oak, 7'-0" x 1'-6" wide, 4 lites		18	.889		1,200	45		1,245	1,400
4650	6 lites		14	1.143		2,100	58		2,158	2,400
4660	8'-0" x 1'-6" wide, 4 lites		14	1.143		1,250	58		1,308	1,475
4670	6 lites	▼	10	1.600	▼	2,100	81		2,181	2,425

08 14 16 – Flush Wood Doors

08 14 16.09 Smooth Wood Doors

08 14 16.09 Smooth Wood Doors		Crew	Daily Output	Labor-Hours	Unit	Material	2018 Bare Costs Labor	Equipment	Total	Total Incl O&P
0010	**SMOOTH WOOD DOORS**									
0015	Flush, interior, hollow core									
0025	Lauan face, 1-3/8", 3'-0" x 6'-8"	2 Carp	17	.941	Ea.	55	47.50		102.50	138
0030	4'-0" x 6'-8"		16	1		126	50.50		176.50	222
0080	1-3/4", 2'-0" x 6'-8"	▼	17	.941		55	47.50		102.50	138
0085	3'-0" x 6'-8"	1 Carp	16	.500		55	25.50		80.50	102
0108	3'-0" x 7'-0"	2 Carp	16	1	▼	206	50.50		256.50	310
0112	Pair of 3'-0" x 7'-0"		9	1.778	Pr.	132	90		222	291
0140	Birch face, 1-3/8", 2'-6" x 6'-8"		17	.941	Ea.	113	47.50		160.50	203
0180	3'-0" x 6'-8"		17	.941		97	47.50		144.50	185
0200	4'-0" x 6'-8"		16	1		158	50.50		208.50	257
0202	1-3/4", 2'-0" x 6'-8"		17	.941		57.50	47.50		105	141
0204	2'-4" x 7'-0"		16	1		120	50.50		170.50	215
0206	2'-6" x 7'-0"		16	1		124	50.50		174.50	219
0210	3'-0" x 7'-0"		16	1	▼	158	50.50		208.50	256
0214	Pair of 3'-0" x 7'-0"		9	1.778	Pr.	259	90		349	430
0220	Oak face, 1-3/8", 2'-0" x 6'-8"		17	.941	Ea.	110	47.50		157.50	199
0280	3'-0" x 6'-8"		17	.941		115	47.50		162.50	204
0300	4'-0" x 6'-8"		16	1		141	50.50		191.50	238
0305	1-3/4", 2'-6" x 6'-8"		17	.941		125	47.50		172.50	215
0310	3'-0" x 7'-0"		16	1		250	50.50		300.50	360
0320	Walnut face, 1-3/8", 2'-0" x 6'-8"		17	.941		185	47.50		232.50	282
0340	2'-6" x 6'-8"		17	.941		192	47.50		239.50	289
0380	3'-0" x 6'-8"		17	.941		200	47.50		247.50	298
0400	4'-0" x 6'-8"	▼	16	1		219	50.50		269.50	325
0430	For 7'-0" high, add					27			27	30
0440	For 8'-0" high, add					39.50			39.50	43.50
0480	For prefinishing, clear, add					48.50			48.50	53
0500	For prefinishing, stain, add					60.50			60.50	66.50
1320	M.D. overlay on hardboard, 1-3/8", 2'-0" x 6'-8"	2 Carp	17	.941		123	47.50		170.50	213
1340	2'-6" x 6'-8"		17	.941		124	47.50		171.50	214
1380	3'-0" x 6'-8"		17	.941		130	47.50		177.50	221
1400	4'-0" x 6'-8"	▼	16	1		185	50.50		235.50	287
1420	For 7'-0" high, add					17.90			17.90	19.65
1440	For 8'-0" high, add					35.50			35.50	39
1720	H.P. plastic laminate, 1-3/8", 2'-0" x 6'-8"	2 Carp	16	1		275	50.50		325.50	390
1740	2'-6" x 6'-8"	▼	16	1	▼	270	50.50		320.50	380

232

For customer support on your Commercial Renovation Costs with RSMeans data, call 800.448.8182.

08 14 16.09 Smooth Wood Doors		Crew	Daily Output	Labor-Hours	Unit	Material	2018 Bare Costs Labor	Equipment	Total	Total Incl O&P
1780	3'-0" x 6'-8"	2 Carp	15	1.067	Ea.	299	54		353	420
1800	4'-0" x 6'-8"	↓	14	1.143		395	58		453	530
1820	For 7'-0" high, add					17.90			17.90	19.70
1840	For 8'-0" high, add					35			35	39
2020	Particle core, lauan face, 1-3/8", 2'-6" x 6'-8"	2 Carp	15	1.067		97	54		151	195
2040	3'-0" x 6'-8"		14	1.143		101	58		159	205
2080	3'-0" x 7'-0"		13	1.231		120	62.50		182.50	233
2085	4'-0" x 7'-0"		12	1.333		129	67.50		196.50	252
2110	1-3/4", 3'-0" x 7'-0"		13	1.231		206	62.50		268.50	330
2120	Birch face, 1-3/8", 2'-6" x 6'-8"		15	1.067		110	54		164	209
2140	3'-0" x 6'-8"		14	1.143		121	58		179	227
2180	3'-0" x 7'-0"		13	1.231		132	62.50		194.50	246
2200	4'-0" x 7'-0"		12	1.333		148	67.50		215.50	273
2205	1-3/4", 3'-0" x 7'-0"		13	1.231		195	62.50		257.50	315
2220	Oak face, 1-3/8", 2'-6" x 6'-8"		15	1.067		122	54		176	222
2240	3'-0" x 6'-8"		14	1.143		134	58		192	241
2280	3'-0" x 7'-0"		13	1.231		140	62.50		202.50	255
2300	4'-0" x 7'-0"		12	1.333		166	67.50		233.50	293
2305	1-3/4", 3'-0" x 7'-0"		13	1.231		236	62.50		298.50	360
2320	Walnut face, 1-3/8", 2'-0" x 6'-8"		15	1.067		131	54		185	232
2340	2'-6" x 6'-8"		14	1.143		148	58		206	257
2380	3'-0" x 6'-8"		13	1.231		166	62.50		228.50	284
2400	4'-0" x 6'-8"	↓	12	1.333		219	67.50		286.50	350
2440	For 8'-0" high, add					44.50			44.50	49
2460	For 8'-0" high walnut, add					40.50			40.50	44.50
2720	For prefinishing, clear, add					39.50			39.50	43.50
2740	For prefinishing, stain, add					57			57	62.50
3320	M.D. overlay on hardboard, 1-3/8", 2'-6" x 6'-8"	2 Carp	14	1.143		178	58		236	290
3340	3'-0" x 6'-8"		13	1.231		193	62.50		255.50	315
3380	3'-0" x 7'-0"		12	1.333		196	67.50		263.50	325
3400	4'-0" x 7'-0"	↓	10	1.600		265	81		346	425
3440	For 8'-0" height, add					39.50			39.50	43.50
3460	For solid wood core, add					70.50			70.50	77.50
3720	H.P. plastic laminate, 1-3/8", 2'-6" x 6'-8"	2 Carp	13	1.231		160	62.50		222.50	277
3740	3'-0" x 6'-8"		12	1.333		188	67.50		255.50	315
3780	3'-0" x 7'-0"		11	1.455		193	74		267	330
3800	4'-0" x 7'-0"	↓	8	2		380	101		481	580
3840	For 8'-0" height, add					39.50			39.50	43.50
3860	For solid wood core, add					43			43	47.50
4000	Exterior, flush, solid core, birch, 1-3/4" x 2'-6" x 7'-0"	2 Carp	15	1.067		166	54		220	271
4020	2'-8" wide		15	1.067		247	54		301	360
4040	3'-0" wide	↓	14	1.143		217	58		275	335
4045	3'-0" x 8'-0"	1 Carp	8	1		475	50.50		525.50	610
4100	Oak faced 1-3/4" x 2'-6" x 7'-0"	2 Carp	15	1.067		222	54		276	330
4120	2'-8" wide		15	1.067		227	54		281	340
4140	3'-0" wide	↓	14	1.143		215	58		273	330
4160	Walnut faced, 1-3/4" x 3'-0" x 6'-8"	1 Carp	17	.471		510	24		534	600
4180	3'-6" wide	"	17	.471		615	24		639	715
4200	Walnut faced, 1-3/4" x 2'-6" x 7'-0"	2 Carp	15	1.067		325	54		379	450
4220	2'-8" wide		15	1.067		300	54		354	420
4240	3'-0" wide	↓	14	1.143		291	58		349	415
4300	For 6'-8" high door, deduct from 7'-0" door					18.05			18.05	19.90
5000	Wood doors, for vision lite, add					108			108	119

08 14 16 – Flush Wood Doors

08 14 16.09 Smooth Wood Doors	Crew	Daily Output	Labor-Hours	Unit	Material	2018 Bare Costs Labor	2018 Bare Costs Equipment	Total	Total Incl O&P
5010	Wood doors, for narrow lite, add				Ea.	116			116	128
5015	Wood doors, for bottom (or top) louver, add					224			224	246
9000	Minimum labor/equipment charge	1 Carp	4	2	Job		101		101	165

08 14 16.10 Wood Doors Decorator	Crew	Daily Output	Labor-Hours	Unit	Material	Labor	Equipment	Total	Total Incl O&P
0010	**WOOD DOORS DECORATOR**									
1800	Exterior, flush, solid wood core, birch 1-3/4" x 2'-6" x 7'-0"	2 Carp	15	1.067	Ea.	315	54		369	435
1820	2'-8" wide		15	1.067		350	54		404	475
1840	3'-0" wide		14	1.143		355	58		413	485
1900	Oak faced, 1-3/4" x 2'-6" x 7'-0"		15	1.067		204	54		258	310
1920	2'-8" wide		15	1.067		440	54		494	575
1940	3'-0" wide		14	1.143		455	58		513	595
2100	Walnut faced, 1-3/4" x 2'-6" x 7'-0"		15	1.067		395	54		449	525
2120	2'-8" wide		15	1.067		410	54		464	545
2140	3'-0" wide		14	1.143		435	58		493	575

08 14 16.20 Wood Fire Doors	Crew	Daily Output	Labor-Hours	Unit	Material	Labor	Equipment	Total	Total Incl O&P
0010	**WOOD FIRE DOORS**									
0020	Particle core, 7 face plys, "B" label,									
0040	1 hour, birch face, 1-3/4" x 2'-6" x 6'-8"	2 Carp	14	1.143	Ea.	435	58		493	570
0080	3'-0" x 6'-8"		13	1.231		490	62.50		552.50	635
0090	3'-0" x 7'-0"		12	1.333		455	67.50		522.50	610
0100	4'-0" x 7'-0"		12	1.333		655	67.50		722.50	830
0140	Oak face, 2'-6" x 6'-8"		14	1.143		490	58		548	635
0180	3'-0" x 6'-8"		13	1.231		500	62.50		562.50	650
0190	3'-0" x 7'-0"		12	1.333		500	67.50		567.50	660
0200	4'-0" x 7'-0"		12	1.333		635	67.50		702.50	810
0240	Walnut face, 2'-6" x 6'-8"		14	1.143		500	58		558	645
0280	3'-0" x 6'-8"		13	1.231		515	62.50		577.50	670
0290	3'-0" x 7'-0"		12	1.333		550	67.50		617.50	720
0300	4'-0" x 7'-0"		12	1.333		820	67.50		887.50	1,025
0440	M.D. overlay on hardboard, 2'-6" x 6'-8"		15	1.067		340	54		394	460
0480	3'-0" x 6'-8"		14	1.143		400	58		458	535
0490	3'-0" x 7'-0"		13	1.231		425	62.50		487.50	570
0500	4'-0" x 7'-0"		12	1.333		450	67.50		517.50	605
0740	90 minutes, birch face, 1-3/4" x 2'-6" x 6'-8"		14	1.143		335	58		393	460
0780	3'-0" x 6'-8"		13	1.231		380	62.50		442.50	520
0790	3'-0" x 7'-0"		12	1.333		385	67.50		452.50	535
0800	4'-0" x 7'-0"		12	1.333		520	67.50		587.50	685
0840	Oak face, 2'-6" x 6'-8"		14	1.143		495	58		553	640
0880	3'-0" x 6'-8"		13	1.231		505	62.50		567.50	655
0890	3'-0" x 7'-0"		12	1.333		560	67.50		627.50	725
0900	4'-0" x 7'-0"		12	1.333		660	67.50		727.50	835
0940	Walnut face, 2'-6" x 6'-8"		14	1.143		500	58		558	645
0980	3'-0" x 6'-8"		13	1.231		440	62.50		502.50	585
0990	3'-0" x 7'-0"		12	1.333		505	67.50		572.50	670
1000	4'-0" x 7'-0"		12	1.333		655	67.50		722.50	835
1140	M.D. overlay on hardboard, 2'-6" x 6'-8"		15	1.067		405	54		459	535
1180	3'-0" x 6'-8"		14	1.143		415	58		473	550
1190	3'-0" x 7'-0"		13	1.231		495	62.50		557.50	645
1200	4'-0" x 7'-0"		12	1.333		560	67.50		627.50	725
1240	For 8'-0" height, add					82			82	90
1260	For 8'-0" height walnut, add					96.50			96.50	106
2200	Custom architectural "B" label, flush, 1-3/4" thick, birch,									

08 14 Wood Doors

08 14 16 – Flush Wood Doors

08 14 16.20 Wood Fire Doors		Crew	Daily Output	Labor-Hours	Unit	Material	2018 Bare Costs Labor	Equipment	Total	Total Incl O&P
2210	Solid core									
2220	2'-6" x 7'-0"	2 Carp	15	1.067	Ea.	395	54		449	520
2260	3'-0" x 7'-0"		14	1.143		340	58		398	470
2300	4'-0" x 7'-0"		13	1.231		520	62.50		582.50	670
2420	4'-0" x 8'-0"		11	1.455		525	74		599	700
2480	For oak veneer, add					50%				
2500	For walnut veneer, add					75%				
9000	Minimum labor/equipment charge	1 Carp	4	2	Job		101		101	165

08 14 33 – Stile and Rail Wood Doors

08 14 33.10 Wood Doors Paneled

		Crew	Daily Output	Labor-Hours	Unit	Material	2018 Bare Costs Labor	Equipment	Total	Total Incl O&P
0010	**WOOD DOORS PANELED**									
0020	Interior, six panel, hollow core, 1-3/8" thick									
0040	Molded hardboard, 2'-0" x 6'-8"	2 Carp	17	.941	Ea.	62	47.50		109.50	146
0060	2'-6" x 6'-8"		17	.941		64	47.50		111.50	148
0070	2'-8" x 6'-8"		17	.941		67	47.50		114.50	151
0080	3'-0" x 6'-8"		17	.941		73.50	47.50		121	159
0140	Embossed print, molded hardboard, 2'-0" x 6'-8"		17	.941		64	47.50		111.50	148
0160	2'-6" x 6'-8"		17	.941		64	47.50		111.50	148
0180	3'-0" x 6'-8"		17	.941		73.50	47.50		121	159
0540	Six panel, solid, 1-3/8" thick, pine, 2'-0" x 6'-8"		15	1.067		155	54		209	259
0560	2'-6" x 6'-8"		14	1.143		154	58		212	264
0580	3'-0" x 6'-8"		13	1.231		148	62.50		210.50	264
1020	Two panel, bored rail, solid, 1-3/8" thick, pine, 1'-6" x 6'-8"		16	1		245	50.50		295.50	350
1040	2'-0" x 6'-8"		15	1.067		320	54		374	445
1060	2'-6" x 6'-8"		14	1.143		365	58		423	495
1340	Two panel, solid, 1-3/8" thick, fir, 2'-0" x 6'-8"		15	1.067		160	54		214	264
1360	2'-6" x 6'-8"		14	1.143		215	58		273	330
1380	3'-0" x 6'-8"		13	1.231		375	62.50		437.50	515
1740	Five panel, solid, 1-3/8" thick, fir, 2'-0" x 6'-8"		15	1.067		286	54		340	405
1760	2'-6" x 6'-8"		14	1.143		380	58		438	515
1780	3'-0" x 6'-8"		13	1.231		380	62.50		442.50	520
4190	Exterior, Knotty pine, paneled, 1-3/4", 3'-0" x 6'-8"		16	1		795	50.50		845.50	960
4195	Double 1-3/4", 3'-0" x 6'-8"		16	1		1,600	50.50		1,650.50	1,825
4200	Ash, paneled, 1-3/4", 3'-0" x 6'-8"		16	1		895	50.50		945.50	1,075
4205	Double 1-3/4", 3'-0" x 6'-8"		16	1		1,800	50.50		1,850.50	2,050
4210	Cherry, paneled, 1-3/4", 3'-0" x 6'-8"		16	1		1,025	50.50		1,075.50	1,225
4215	Double 1-3/4", 3'-0" x 6'-8"		16	1		2,075	50.50		2,125.50	2,350
4230	Ash, paneled, 1-3/4", 3'-0" x 8'-0"		16	1		1,075	50.50		1,125.50	1,275
4235	Double 1-3/4", 3'-0" x 8'-0"		16	1		2,175	50.50		2,225.50	2,450
4240	Hard maple, paneled, 1-3/4", 3'-0" x 8'-0"		16	1		1,175	50.50		1,225.50	1,375
4245	Double 1-3/4", 3'-0" x 8'-0"		16	1		2,375	50.50		2,425.50	2,675
4250	Cherry, paneled, 1-3/4", 3'-0" x 8'-0"		16	1		1,175	50.50		1,225.50	1,375
4255	Double 1-3/4", 3'-0" x 8'-0"		16	1		2,375	50.50		2,425.50	2,675
9000	Minimum labor/equipment charge	1 Carp	4	2	Job		101		101	165

08 14 33.20 Wood Doors Residential

		Crew	Daily Output	Labor-Hours	Unit	Material	2018 Bare Costs Labor	Equipment	Total	Total Incl O&P
0010	**WOOD DOORS RESIDENTIAL**									
0200	Exterior, combination storm & screen, pine									
0260	2'-8" wide	2 Carp	10	1.600	Ea.	310	81		391	475
0280	3'-0" wide		9	1.778		325	90		415	505
0300	7'-1" x 3'-0" wide		9	1.778		355	90		445	535
0400	Full lite, 6'-9" x 2'-6" wide		11	1.455		325	74		399	480
0420	2'-8" wide		10	1.600		330	81		411	495

For customer support on your Commercial Renovation Costs with RSMeans data, call 800.448.8182.

235

08 14 33.20 Wood Doors Residential	Crew	Daily Output	Labor-Hours	Unit	Material	2018 Bare Costs Labor	2018 Bare Costs Equipment	Total	Total Incl O&P
0440	3'-0" wide	2 Carp	9	1.778	Ea.	335	90		425	510
0500	7'-1" x 3'-0" wide		9	1.778		365	90		455	545
0700	Dutch door, pine, 1-3/4" x 2'-8" x 6'-8", 6 panel		12	1.333		695	67.50		762.50	875
0720	Half glass		10	1.600		985	81		1,066	1,200
0800	3'-0" wide, 6 panel		12	1.333		590	67.50		657.50	760
0820	Half glass		10	1.600		1,050	81		1,131	1,275
1000	Entrance door, colonial, 1-3/4" x 6'-8" x 2'-8" wide		16	1		570	50.50		620.50	710
1020	6 panel pine, 3'-0" wide		15	1.067		560	54		614	705
1100	8 panel pine, 2'-8" wide		16	1		685	50.50		735.50	835
1120	3'-0" wide	▼	15	1.067		635	54		689	790
1200	For tempered safety glass lites (min. of 2), add					85.50			85.50	94
1300	Flush, birch, solid core, 1-3/4" x 6'-8" x 2'-8" wide	2 Carp	16	1		142	50.50		192.50	240
1320	3'-0" wide		15	1.067		152	54		206	255
1350	7'-0" x 2'-8" wide		16	1		133	50.50		183.50	230
1360	3'-0" wide	▼	15	1.067		155	54		209	259
1380	For tempered safety glass lites, add					114			114	126
1420	6'-8" x 3'-0" wide, fir	2 Carp	16	1		505	50.50		555.50	645
1720	Carved mahogany 3'-0" x 6'-8"		15	1.067		1,400	54		1,454	1,650
1800	Lauan, solid core, 1-3/4" x 7'-0" x 2'-4" wide		16	1		137	50.50		187.50	234
1810	2'-6" wide		15	1.067		144	54		198	246
1820	2'-8" wide		9	1.778		149	90		239	310
1830	3'-0" wide		16	1		164	50.50		214.50	263
1840	3'-4" wide		16	1	▼	254	50.50		304.50	365
1850	Pair of 3'-0" wide	▼	15	1.067	Pr.	330	54		384	450
2700	Interior, closet, bi-fold, w/hardware, no frame or trim incl.									
2720	Flush, birch, 2'-6" x 6'-8"	2 Carp	13	1.231	Ea.	73	62.50		135.50	181
2740	3'-0" wide		13	1.231		74.50	62.50		137	183
2760	4'-0" wide		12	1.333		115	67.50		182.50	237
2780	5'-0" wide		11	1.455		114	74		188	246
2800	6'-0" wide		10	1.600		135	81		216	280
2820	Flush, hardboard, primed, 6'-8" x 2'-6" wide		13	1.231		73.50	62.50		136	182
2840	3'-0" wide		13	1.231		85	62.50		147.50	195
2860	4'-0" wide		12	1.333		148	67.50		215.50	273
2880	5'-0" wide		11	1.455		177	74		251	315
2900	6'-0" wide		10	1.600		179	81		260	330
3000	Raised panel pine, 6'-6" or 6'-8" x 2'-6" wide		13	1.231		179	62.50		241.50	298
3020	3'-0" wide		13	1.231		289	62.50		351.50	420
3040	4'-0" wide		12	1.333		315	67.50		382.50	455
3060	5'-0" wide		11	1.455		405	74		479	565
3080	6'-0" wide		10	1.600		445	81		526	620
3200	Louvered, pine, 6'-6" or 6'-8" x 2'-6" wide		13	1.231		154	62.50		216.50	270
3220	3'-0" wide		13	1.231		239	62.50		301.50	365
3240	4'-0" wide		12	1.333		251	67.50		318.50	385
3260	5'-0" wide		11	1.455		278	74		352	425
3280	6'-0" wide	▼	10	1.600	▼	310	81		391	470
4400	Bi-passing closet, incl. hardware and frame, no trim incl.									
4420	Flush, lauan, 6'-8" x 4'-0" wide	2 Carp	12	1.333	Opng.	167	67.50		234.50	294
4440	5'-0" wide		11	1.455		183	74		257	320
4460	6'-0" wide		10	1.600		169	81		250	320
4600	Flush, birch, 6'-8" x 4'-0" wide		12	1.333		254	67.50		321.50	390
4620	5'-0" wide		11	1.455		223	74		297	365
4640	6'-0" wide		10	1.600		335	81		416	500
4800	Louvered, pine, 6'-8" x 4'-0" wide	▼	12	1.333		485	67.50		552.50	640

08 14 33 – Stile and Rail Wood Doors

08 14 33.20 Wood Doors Residential	Crew	Daily Output	Labor-Hours	Unit	Material	2018 Bare Costs Labor	Equipment	Total	Total Incl O&P
4820	5'-0" wide	2 Carp	11	1.455	Opng.	610	74		684	790
4840	6'-0" wide		10	1.600		720	81		801	920
5000	Paneled, pine, 6'-8" x 4'-0" wide		12	1.333		490	67.50		557.50	650
5020	5'-0" wide		11	1.455		630	74		704	810
5040	6'-0" wide		10	1.600		840	81		921	1,050
5042	8'-0" wide		12	1.333		980	67.50		1,047.50	1,175
6100	Folding accordion, closet, including track and frame									
6120	Vinyl, 2 layer, stock	2 Carp	10	1.600	Ea.	68	81		149	207
6200	Rigid PVC	"	10	1.600	"	108	81		189	251
7310	Passage doors, flush, no frame included									
7320	Hardboard, hollow core, 1-3/8" x 6'-8" x 1'-6" wide	2 Carp	18	.889	Ea.	41.50	45		86.50	119
7330	2'-0" wide		18	.889		45	45		90	123
7340	2'-6" wide		18	.889		52.50	45		97.50	131
7350	2'-8" wide		18	.889		53	45		98	132
7360	3'-0" wide		17	.941		55	47.50		102.50	138
7420	Lauan, hollow core, 1-3/8" x 6'-8" x 1'-6" wide		18	.889		53	45		98	132
7440	2'-0" wide		18	.889		60	45		105	139
7450	2'-4" wide		18	.889		68	45		113	148
7460	2'-6" wide		18	.889		68	45		113	148
7480	2'-8" wide		18	.889		70	45		115	150
7500	3'-0" wide		17	.941		75	47.50		122.50	160
7700	Birch, hollow core, 1-3/8" x 6'-8" x 1'-6" wide		18	.889		62	45		107	141
7720	2'-0" wide		18	.889		69	45		114	149
7740	2'-6" wide		18	.889		79	45		124	160
7760	2'-8" wide		18	.889		82	45		127	163
7780	3'-0" wide		17	.941		88	47.50		135.50	175
8000	Pine louvered, 1-3/8" x 6'-8" x 1'-6" wide		19	.842		148	42.50		190.50	233
8020	2'-0" wide		18	.889		164	45		209	253
8040	2'-6" wide		18	.889		199	45		244	292
8060	2'-8" wide		18	.889		209	45		254	305
8080	3'-0" wide		17	.941		219	47.50		266.50	320
8300	Pine paneled, 1-3/8" x 6'-8" x 1'-6" wide		19	.842		191	42.50		233.50	280
8320	2'-0" wide		18	.889		223	45		268	320
8330	2'-4" wide		18	.889		241	45		286	340
8340	2'-6" wide		18	.889		250	45		295	350
8360	2'-8" wide		18	.889		258	45		303	355
8380	3'-0" wide		17	.941		271	47.50		318.50	375
8804	Pocket door, 6 panel pine, 2'-6" x 6'-8" with frame		10.50	1.524		370	77.50		447.50	530
8814	2'-8" x 6'-8"		10.50	1.524		375	77.50		452.50	535
8824	3'-0" x 6'-8"		10.50	1.524		385	77.50		462.50	545
9000	Passage doors, flush, no frame, birch, solid core, 1-3/8" x 2'-4" x 7'-0"		16	1		125	50.50		175.50	221
9020	2'-8" wide		16	1		133	50.50		183.50	229
9040	3'-0" wide		16	1		146	50.50		196.50	244
9060	3'-4" wide		15	1.067		271	54		325	385
9080	Pair of 3'-0" wide		9	1.778	Pr.	291	90		381	465
9100	Lauan, solid core, 1-3/8" x 7'-0" x 2'-4" wide		16	1	Ea.	163	50.50		213.50	262
9120	2'-8" wide		16	1		147	50.50		197.50	245
9140	3'-0" wide		16	1		192	50.50		242.50	294
9160	3'-4" wide		15	1.067		206	54		260	315
9180	Pair of 3'-0" wide		9	1.778	Pr.	385	90		475	565
9200	Hardboard, solid core, 1-3/8" x 7'-0" x 2'-4" wide		16	1	Ea.	167	50.50		217.50	266
9220	2'-8" wide		16	1		174	50.50		224.50	274
9240	3'-0" wide		16	1		179	50.50		229.50	280

08 14 33 – Stile and Rail Wood Doors

08 14 33.20 Wood Doors Residential	Crew	Daily Output	Labor-Hours	Unit	Material	2018 Bare Costs Labor	Equipment	Total	Total Incl O&P
9260	3'-4" wide	2 Carp	15	1.067	Ea.	340	54		394	465
9900	Minimum labor/equipment charge	1 Carp	4	2	Job		101		101	165

08 14 35 – Torrified Doors

08 14 35.10 Torrified Exterior Doors

		Crew	Daily Output	Labor-Hours	Unit	Material	Labor	Equipment	Total	Total Incl O&P
0010	**TORRIFIED EXTERIOR DOORS**									
0020	Wood doors made from torrified wood, exterior									
0030	All doors require a finish be applied, all glass is insulated									
0040	All doors require pilot holes for all fasteners									
0100	6 panel, paint grade poplar, 1-3/4" x 3'-0" x 6'-8"	2 Carp	12	1.333	Ea.	1,325	67.50		1,392.50	1,550
0120	Half glass 3'-0" x 6'-8"	"	12	1.333		1,425	67.50		1,492.50	1,650
0200	Side lite, full glass, 1-3/4" x 1'-2" x 6'-8"					930			930	1,025
0220	Side lite, half glass, 1-3/4" x 1'-2" x 6'-8"					910			910	1,000
0300	Raised face, 2 panel, paint grade poplar, 1-3/4" x 3'-0" x 7'-0"	2 Carp	12	1.333		1,325	67.50		1,392.50	1,550
0320	Side lite, raised face, half glass, 1-3/4" x 1'-2" x 7'-0"					1,050			1,050	1,175
0500	6 panel, Fir, 1-3/4" x 3'-0" x 6'-8"	2 Carp	12	1.333		1,875	67.50		1,942.50	2,150
0520	Half glass 3'-0" x 6'-8"	"	12	1.333		1,850	67.50		1,917.50	2,150
0600	Side lite, full glass, 1-3/4" x 1'-2" x 6'-8"					940			940	1,025
0620	Side lite, half glass, 1-3/4" x 1'-2" x 6'-8"					970			970	1,075
0700	6 panel, Mahogany, 1-3/4" x 3'-0" x 6'-8"	2 Carp	12	1.333		1,975	67.50		2,042.50	2,275
0800	Side lite, full glass, 1-3/4" x 1'-2" x 6'-8"					1,050			1,050	1,150
0820	Side lite, half glass, 1-3/4" x 1'-2" x 6'-8"					1,050			1,050	1,150

08 14 40 – Interior Cafe Doors

08 14 40.10 Cafe Style Doors

		Crew	Daily Output	Labor-Hours	Unit	Material	Labor	Equipment	Total	Total Incl O&P
0010	**CAFE STYLE DOORS**									
6520	Interior cafe doors, 2'-6" opening, stock, panel pine	2 Carp	16	1	Ea.	400	50.50		450.50	525
6540	3'-0" opening	"	16	1	"	445	50.50		495.50	570
6550	Louvered pine									
6560	2'-6" opening	2 Carp	16	1	Ea.	320	50.50		370.50	435
8000	3'-0" opening		16	1		335	50.50		385.50	455
8010	2'-6" opening, hardwood		16	1		360	50.50		410.50	480
8020	3'-0" opening		16	1		395	50.50		445.50	520
9000	Minimum labor/equipment charge	1 Carp	4	2	Job		101		101	165

08 16 Composite Doors

08 16 13 – Fiberglass Doors

08 16 13.10 Entrance Doors, Fibrous Glass

			Crew	Daily Output	Labor-Hours	Unit	Material	Labor	Equipment	Total	Total Incl O&P
0010	**ENTRANCE DOORS, FIBROUS GLASS**										
0020	Exterior, fiberglass, door, 2'-8" wide x 6'-8" high	G	2 Carp	15	1.067	Ea.	281	54		335	400
0040	3'-0" wide x 6'-8" high	G		15	1.067		274	54		328	390
0060	3'-0" wide x 7'-0" high	G		15	1.067		490	54		544	625
0080	3'-0" wide x 6'-8" high, with two lites	G		15	1.067		340	54		394	465
0100	3'-0" wide x 8'-0" high, with two lites	G		15	1.067		525	54		579	670
0110	Half glass, 3'-0" wide x 6'-8" high	G		15	1.067		450	54		504	585
0120	3'-0" wide x 6'-8" high, low E	G		15	1.067		490	54		544	625
0130	3'-0" wide x 8'-0" high	G		15	1.067		585	54		639	735
0140	3'-0" wide x 8'-0" high, low E	G		15	1.067		675	54		729	835
0150	Side lights, 1'-0" wide x 6'-8" high	G					281			281	310
0160	1'-0" wide x 6'-8" high, low E	G					289			289	320
0180	1'-0" wide x 6'-8" high, full glass	G					340			340	370
0190	1'-0" wide x 6'-8" high, low E	G					360			360	395

08 16 Composite Doors

08 16 14 – French Doors

08 16 14.10 Exterior Doors With Glass Lites

		Crew	Daily Output	Labor-Hours	Unit	Material	2018 Bare Costs Labor	Equipment	Total	Total Incl O&P
0010	**EXTERIOR DOORS WITH GLASS LITES**									
0020	French, Fir, 1-3/4", 3'-0" wide x 6'-8" high	2 Carp	12	1.333	Ea.	620	67.50		687.50	790
0025	Double		12	1.333		1,250	67.50		1,317.50	1,450
0030	Maple, 1-3/4", 3'-0" wide x 6'-8" high		12	1.333		695	67.50		762.50	875
0035	Double		12	1.333		1,400	67.50		1,467.50	1,625
0040	Cherry, 1-3/4", 3'-0" wide x 6'-8" high		12	1.333		810	67.50		877.50	1,000
0045	Double		12	1.333		1,625	67.50		1,692.50	1,875
0100	Mahogany, 1-3/4", 3'-0" wide x 8'-0" high		10	1.600		825	81		906	1,050
0105	Double		10	1.600		1,650	81		1,731	1,950
0110	Fir, 1-3/4", 3'-0" wide x 8'-0" high		10	1.600		1,225	81		1,306	1,475
0115	Double		10	1.600		2,475	81		2,556	2,850
0120	Oak, 1-3/4", 3'-0" wide x 8'-0" high		10	1.600		1,875	81		1,956	2,200
0125	Double	↓	10	1.600	↓	3,750	81		3,831	4,250

08 17 Integrated Door Opening Assemblies

08 17 13 – Integrated Metal Door Opening Assemblies

08 17 13.10 Hollow Metal Doors and Frames

		Crew	Daily Output	Labor-Hours	Unit	Material	2018 Bare Costs Labor	Equipment	Total	Total Incl O&P
0010	**HOLLOW METAL DOORS AND FRAMES**									
0100	Prehung, flush, 18 ga., 1-3/4" x 6'-8" x 2'-8" wide	1 Carp	16	.500	Ea.	640	25.50		665.50	745
0120	3'-0" wide		15	.533		640	27		667	750
0130	3'-6" wide		13	.615		705	31		736	825
0140	4'-0" wide		10	.800		770	40.50		810.50	910
0300	Double, 1-3/4" x 3'-0" x 6'-8"		5	1.600		1,325	81		1,406	1,575
0350	3'-0" x 8'-0"	↓	4	2	↓	1,325	101		1,426	1,650

08 17 13.20 Stainless Steel Doors and Frames

			Crew	Daily Output	Labor-Hours	Unit	Material	2018 Bare Costs Labor	Equipment	Total	Total Incl O&P
0010	**STAINLESS STEEL DOORS AND FRAMES**										
0020	Stainless steel (304) prehung 24 ga. 2'-6" x 6'-8" door w/16 g frame	G	2 Carp	6	2.667	Ea.	1,775	135		1,910	2,200
0025	2'-8" x 6'-8"	G		6	2.667		1,700	135		1,835	2,100
0030	3'-0" x 6'-8"	G		6	2.667		1,700	135		1,835	2,075
0040	3'-0" x 7'-0"	G		6	2.667		1,725	135		1,860	2,125
0050	4'-0" x 7'-0"	G		5	3.200		2,225	162		2,387	2,725
0100	Stainless steel (316) prehung 24 ga. 2'-6" x 6'-8" door w/16 g frame	G		6	2.667		2,525	135		2,660	3,000
0110	2'-8" x 6'-8"	G		6	2.667		2,375	135		2,510	2,825
0120	3'-0" x 6'-8"	G		6	2.667		2,425	135		2,560	2,900
0150	3'-0" x 7'-0"	G		6	2.667		2,575	135		2,710	3,050
0160	4'-0" x 7'-0"	G		6	2.667		3,200	135		3,335	3,750
0300	Stainless steel (304) prehung 18 ga. 2'-6" x 6'-8" door w/16 g frame	G		6	2.667		2,875	135		3,010	3,375
0310	2'-8" x 6'-8"	G		6	2.667		2,925	135		3,060	3,450
0320	3'-0" x 6'-8"	G		6	2.667		2,975	135		3,110	3,500
0350	3'-0" x 7'-0"	G		6	2.667		3,025	135		3,160	3,550
0360	4'-0" x 7'-0"	G		5	3.200		3,150	162		3,312	3,750
0500	Stainless steel, prehung door, foam core, 14 ga., 3'-0" x 7'-0"	G		5	3.200		3,150	162		3,312	3,750
0600	Stainless steel, prehung double door, foam core, 14 ga., 3'-0" x 7'-0"	G	↓	4	4	↓	6,200	203		6,403	7,150

08 17 23 – Integrated Wood Door Opening Assemblies

08 17 23.10 Pre-Hung Doors

		Crew	Daily Output	Labor-Hours	Unit	Material	2018 Bare Costs Labor	Equipment	Total	Total Incl O&P
0010	**PRE-HUNG DOORS**									
0300	Exterior, wood, comb. storm & screen, 6'-9" x 2'-6" wide	2 Carp	15	1.067	Ea.	248	54		302	360
0320	2'-8" wide		15	1.067		330	54		384	455
0340	3'-0" wide		15	1.067		310	54		364	430
0360	For 7'-0" high door, add	↓			↓	40			40	44

08 17 23 – Integrated Wood Door Opening Assemblies

08 17 23.10 Pre-Hung Doors		Crew	Daily Output	Labor-Hours	Unit	Material	2018 Bare Costs Labor	Equipment	Total	Total Incl O&P
1600	Entrance door, flush, birch, solid core									
1620	4-5/8" solid jamb, 1-3/4" x 6'-8" x 2'-8" wide	2 Carp	16	1	Ea.	300	50.50		350.50	415
1640	3'-0" wide		16	1		405	50.50		455.50	530
1642	5-5/8" jamb		16	1		345	50.50		395.50	465
1680	For 7'-0" high door, add					25.50			25.50	28
2000	Entrance door, colonial, 6 panel pine									
2020	4-5/8" solid jamb, 1-3/4" x 6'-8" x 2'-8" wide	2 Carp	16	1	Ea.	660	50.50		710.50	810
2040	3'-0" wide	"	16	1		690	50.50		740.50	845
2060	For 7'-0" high door, add					56.50			56.50	62
2200	For 5-5/8" solid jamb, add					44.50			44.50	49
2230	French style, exterior, 1 lite, 1-3/4" x 3'-0" x 6'-8"	1 Carp	14	.571		655	29		684	765
2235	9 lites	"	14	.571		705	29		734	820
2245	15 lites	2 Carp	14	1.143		735	58		793	905
2250	Double, 15 lites, 2'-0" x 6'-8", 4'-0" opening		7	2.286	Pr.	1,275	116		1,391	1,600
2260	2'-6" x 6'-8", 5'-0" opening		7	2.286		1,425	116		1,541	1,750
2280	3'-0" x 6'-8", 6'-0" opening		7	2.286		1,550	116		1,666	1,900
2430	3'-0" x 7'-0", 15 lites		14	1.143	Ea.	1,000	58		1,058	1,200
2432	Two 3'-0" x 7'-0"		7	2.286	Pr.	2,075	116		2,191	2,475
2435	3'-0" x 8'-0"		14	1.143	Ea.	1,075	58		1,133	1,275
2437	Two, 3'-0" x 8'-0"		7	2.286	Pr.	2,200	116		2,316	2,625
4000	Interior, passage door, 4-5/8" solid jamb									
4350	Paneled, primed, hollow core, 2'-8" wide	2 Carp	17	.941	Ea.	125	47.50		172.50	216
4360	3'-0" wide		17	.941		193	47.50		240.50	291
4400	Lauan, flush, solid core, 1-3/8" x 6'-8" x 2'-6" wide		17	.941		194	47.50		241.50	291
4420	2'-8" wide		17	.941		194	47.50		241.50	291
4440	3'-0" wide		16	1		211	50.50		261.50	315
4600	Hollow core, 1-3/8" x 6'-8" x 2'-6" wide		17	.941		136	47.50		183.50	228
4620	2'-8" wide		17	.941		139	47.50		186.50	230
4640	3'-0" wide		16	1		142	50.50		192.50	240
4700	For 7'-0" high door, add					41			41	45.50
5000	Birch, flush, solid core, 1-3/8" x 6'-8" x 2'-6" wide	2 Carp	17	.941		294	47.50		341.50	405
5020	2'-8" wide		17	.941		209	47.50		256.50	310
5040	3'-0" wide		16	1		320	50.50		370.50	440
5200	Hollow core, 1-3/8" x 6'-8" x 2'-6" wide		17	.941		242	47.50		289.50	345
5220	2'-8" wide		17	.941		273	47.50		320.50	380
5240	3'-0" wide		16	1		264	50.50		314.50	375
5280	For 7'-0" high door, add					34.50			34.50	38
5500	Hardboard paneled, 1-3/8" x 6'-8" x 2'-6" wide	2 Carp	17	.941		152	47.50		199.50	245
5520	2'-8" wide		17	.941		163	47.50		210.50	257
5540	3'-0" wide		16	1		161	50.50		211.50	261
6000	Pine paneled, 1-3/8" x 6'-8" x 2'-6" wide		17	.941		276	47.50		323.50	385
6020	2'-8" wide		17	.941		293	47.50		340.50	405
6040	3'-0" wide		16	1		300	50.50		350.50	415
8200	Birch, flush, solid core, 1-3/4" x 6'-8" x 2'-4" wide	1 Carp	17	.471		237	24		261	300
8220	2'-6" wide		17	.471		229	24		253	291
8240	2'-8" wide		17	.471		236	24		260	298
8260	3'-0" wide		16	.500		240	25.50		265.50	305
8280	3'-6" wide		15	.533		375	27		402	460
8500	Pocket door frame with lauan, flush, hollow core, 1-3/8" x 3'-0" x 6'-8"		17	.471		271	24		295	340
9000	Minimum labor/equipment charge		4	2	Job		101		101	165

08 31 Access Doors and Panels

08 31 13 – Access Doors and Frames

08 31 13.10 Types of Framed Access Doors

		Crew	Daily Output	Labor-Hours	Unit	Material	2018 Bare Costs Labor	2018 Bare Costs Equipment	Total	Total Incl O&P
0010	**TYPES OF FRAMED ACCESS DOORS**									
1000	Fire rated door with lock									
1100	Metal, 12" x 12"	1 Carp	10	.800	Ea.	162	40.50		202.50	244
1150	18" x 18"		9	.889		203	45		248	296
1200	24" x 24"		9	.889		350	45		395	455
1250	24" x 36"		8	1		350	50.50		400.50	470
1300	24" x 48"		8	1		400	50.50		450.50	520
1350	36" x 36"		7.50	1.067		520	54		574	660
1400	48" x 48"		7.50	1.067		670	54		724	830
1600	Stainless steel, 12" x 12"		10	.800		345	40.50		385.50	440
1650	18" x 18"		9	.889		375	45		420	490
1700	24" x 24"		9	.889		550	45		595	685
1750	24" x 36"	↓	8	1	↓	705	50.50		755.50	860
2000	Flush door for finishing									
2100	Metal 8" x 8"	1 Carp	10	.800	Ea.	36	40.50		76.50	106
2150	12" x 12"	"	10	.800	"	37.50	40.50		78	107
3000	Recessed door for acoustic tile									
3100	Metal, 12" x 12"	1 Carp	4.50	1.778	Ea.	52	90		142	203
3150	12" x 24"		4.50	1.778		107	90		197	264
3200	24" x 24"		4	2		99	101		200	274
3250	24" x 36"	↓	4	2	↓	164	101		265	345
4000	Recessed door for drywall									
4100	Metal 12" x 12"	1 Carp	6	1.333	Ea.	74.50	67.50		142	192
4150	12" x 24"		5.50	1.455		119	74		193	251
4200	24" x 36"	↓	5	1.600	↓	168	81		249	315
6000	Standard door									
6100	Metal, 8" x 8"	1 Carp	10	.800	Ea.	28.50	40.50		69	97.50
6150	12" x 12"		10	.800		31	40.50		71.50	100
6200	18" x 18"		9	.889		37	45		82	114
6250	24" x 24"		9	.889		51	45		96	129
6300	24" x 36"		8	1		88	50.50		138.50	179
6350	36" x 36"		8	1		100	50.50		150.50	193
6500	Stainless steel, 8" x 8"		10	.800		80	40.50		120.50	154
6550	12" x 12"		10	.800		99	40.50		139.50	175
6600	18" x 18"		9	.889		190	45		235	281
6650	24" x 24"		9	.889	↓	281	45		326	385
9000	Minimum labor/equipment charge	↓	4	2	Job		101		101	165

08 31 13.20 Bulkhead/Cellar Doors

		Crew	Daily Output	Labor-Hours	Unit	Material	2018 Bare Costs Labor	2018 Bare Costs Equipment	Total	Total Incl O&P
0010	**BULKHEAD/CELLAR DOORS**									
0020	Steel, not incl. sides, 44" x 62"	1 Carp	5.50	1.455	Ea.	655	74		729	840
0100	52" x 73"		5.10	1.569		825	79.50		904.50	1,025
0500	With sides and foundation plates, 57" x 45" x 24"		4.70	1.702		870	86.50		956.50	1,100
0600	42" x 49" x 51"		4.30	1.860	↓	595	94.50		689.50	810
9000	Minimum labor/equipment charge	↓	2	4	Job		203		203	330

08 31 13.30 Commercial Floor Doors

		Crew	Daily Output	Labor-Hours	Unit	Material	2018 Bare Costs Labor	2018 Bare Costs Equipment	Total	Total Incl O&P
0010	**COMMERCIAL FLOOR DOORS**									
0020	Aluminum tile, steel frame, one leaf, 2' x 2' opng.	2 Sswk	3.50	4.571	Opng.	630	250		880	1,125
0050	3'-6" x 3'-6" opening		3.50	4.571		1,150	250		1,400	1,675
0500	Double leaf, 4' x 4' opening		3	5.333		1,625	291		1,916	2,300
0550	5' x 5' opening		3	5.333	↓	2,875	291		3,166	3,650
9000	Minimum labor/equipment charge	↓	2	8	Job		435		435	745

For customer support on your Commercial Renovation Costs with RSMeans data, call 800.448.8182.

241

08 31 Access Doors and Panels

08 31 13 – Access Doors and Frames

08 31 13.35 Industrial Floor Doors	Crew	Daily Output	Labor-Hours	Unit	Material	2018 Bare Costs Labor	Equipment	Total	Total Incl O&P
0010 **INDUSTRIAL FLOOR DOORS**									
0020 Steel 300 psf L.L., single leaf, 2' x 2', 175#	2 Sswk	6	2.667	Opng.	750	146		896	1,075
0050 3' x 3' opening, 300#		5.50	2.909		1,100	159		1,259	1,475
0300 Double leaf, 4' x 4' opening, 455#		5	3.200		2,500	175		2,675	3,050
0350 5' x 5' opening, 645#		4.50	3.556		2,875	194		3,069	3,475
1000 Aluminum, 300 psf L.L., single leaf, 2' x 2', 60#		6	2.667		750	146		896	1,075
1050 3' x 3' opening, 100#		5.50	2.909		935	159		1,094	1,300
1500 Double leaf, 4' x 4' opening, 160#		5	3.200		2,150	175		2,325	2,650
1550 5' x 5' opening, 235#	▼	4.50	3.556	▼	2,875	194		3,069	3,475
9000 Minimum labor/equipment charge	▼	2	8	Job		435		435	745

08 32 Sliding Glass Doors

08 32 13 – Sliding Aluminum-Framed Glass Doors

08 32 13.10 Sliding Aluminum Doors

08 32 13.10 Sliding Aluminum Doors	Crew	Daily Output	Labor-Hours	Unit	Material	2018 Bare Costs Labor	Equipment	Total	Total Incl O&P
0010 **SLIDING ALUMINUM DOORS**									
0350 Aluminum, 5/8" tempered insulated glass, 6' wide									
0400 Premium	2 Carp	4	4	Ea.	1,550	203		1,753	2,025
0450 Economy		4	4		820	203		1,023	1,225
0500 8' wide, premium		3	5.333		1,700	270		1,970	2,325
0550 Economy		3	5.333		1,475	270		1,745	2,075
0600 12' wide, premium		2.50	6.400		3,000	325		3,325	3,825
0650 Economy		2.50	6.400		1,575	325		1,900	2,250
4000 Aluminum, baked on enamel, temp glass, 6'-8" x 10'-0" wide		4	4		1,100	203		1,303	1,525
4020 Insulating glass, 6'-8" x 6'-0" wide		4	4		955	203		1,158	1,375
4040 8'-0" wide		3	5.333		1,125	270		1,395	1,675
4060 10'-0" wide		2	8		1,400	405		1,805	2,200
4080 Anodized, temp glass, 6'-8" x 6'-0" wide		4	4		455	203		658	830
4100 8'-0" wide		3	5.333		575	270		845	1,075
4120 10'-0" wide	▼	2	8	▼	660	405		1,065	1,375
5000 Aluminum sliding glass door system									
5010 Sliding door 4' wide opening single side	2 Carp	2	8	Ea.	6,000	405		6,405	7,250
5015 8' wide opening single side		2	8		9,000	405		9,405	10,600
5020 Telescoping glass door system, 4' wide opening biparting		2	8		5,100	405		5,505	6,250
5025 8' wide opening biparting		2	8		6,000	405		6,405	7,250
5030 Folding glass door, 4' wide opening biparting		2	8		8,000	405		8,405	9,450
5035 8' wide opening biparting		2	8		10,000	405		10,405	11,700
5040 ICU-CCU sliding telescoping glass door, 4' x 7', single side opening		2	8		3,050	405		3,455	4,000
5045 8' x 7', single side opening	▼	2	8		4,625	405		5,030	5,750
7000 Electric swing door operator and control, single door w/sensors	1 Carp	4	2		2,825	101		2,926	3,275
7005 Double door w/sensors		2	4		5,650	203		5,853	6,525
7010 Electric folding door operator and control, single door w/sensors		4	2		6,600	101		6,701	7,425
7015 Bi-folding door		4	2		7,800	101		7,901	8,750
7020 Electric swing door operator and control, single door	▼	4	2	▼	2,825	101		2,926	3,275

08 32 19 – Sliding Wood-Framed Glass Doors

08 32 19.10 Sliding Wood Doors

08 32 19.10 Sliding Wood Doors	Crew	Daily Output	Labor-Hours	Unit	Material	2018 Bare Costs Labor	Equipment	Total	Total Incl O&P
0010 **SLIDING WOOD DOORS**									
0020 Wood, tempered insul. glass, 6' wide, premium	2 Carp	4	4	Ea.	1,450	203		1,653	1,925
0100 Economy		4	4		1,200	203		1,403	1,650
0150 8' wide, wood, premium		3	5.333		1,850	270		2,120	2,475
0200 Economy	▼	3	5.333	▼	1,525	270		1,795	2,125

08 32 Sliding Glass Doors

08 32 19 – Sliding Wood-Framed Glass Doors

08 32 19.10 Sliding Wood Doors

		Crew	Daily Output	Labor-Hours	Unit	Material	2018 Bare Costs Labor	Equipment	Total	Total Incl O&P
0235	10' wide, wood, premium	2 Carp	2.50	6.400	Ea.	2,675	325		3,000	3,450
0240	Economy		2.50	6.400		2,275	325		2,600	3,025
0250	12' wide, wood, premium		2.50	6.400		3,100	325		3,425	3,925
0300	Economy	↓	2.50	6.400	↓	2,475	325		2,800	3,250
9000	Minimum labor/equipment charge	1 Carp	2	4	Job		203		203	330

08 32 19.15 Sliding Glass Vinyl-Clad Wood Doors

			Crew	Daily Output	Labor-Hours	Unit	Material	2018 Bare Costs Labor	Equipment	Total	Total Incl O&P
0010	**SLIDING GLASS VINYL-CLAD WOOD DOORS**										
0020	Glass, sliding vinyl-clad, insul. glass, 6'-0" x 6'-8"	G	2 Carp	4	4	Opng.	1,525	203		1,728	2,000
0025	6'-0" x 6'-10" high	G		4	4		1,700	203		1,903	2,175
0030	6'-0" x 8'-0" high	G		4	4		2,050	203		2,253	2,575
0050	5'-0" x 6'-8" high	G		4	4		1,575	203		1,778	2,075
0100	8'-0" x 6'-10" high	G		4	4		2,025	203		2,228	2,550
0150	8'-0" x 8'-0" high	G		4	4		2,350	203		2,553	2,925
0500	4 leaf, 9'-0" x 6'-10" high	G		3	5.333		3,350	270		3,620	4,150
0550	9'-0" x 8'-0" high	G		3	5.333		3,925	270		4,195	4,750
0600	12'-0" x 6'-10" high	G		3	5.333		4,025	270		4,295	4,875
0650	12'-0" x 8'-0" high	G	↓	3	5.333	↓	4,025	270		4,295	4,875
9000	Minimum labor/equipment charge		1 Carp	4	2	Job		101		101	165

08 33 Coiling Doors and Grilles

08 33 13 – Coiling Counter Doors

08 33 13.10 Counter Doors, Coiling Type

		Crew	Daily Output	Labor-Hours	Unit	Material	2018 Bare Costs Labor	Equipment	Total	Total Incl O&P
0010	**COUNTER DOORS, COILING TYPE**									
0020	Manual, incl. frame and hardware, galv. stl., 4' roll-up, 6' long	2 Carp	2	8	Opng.	1,300	405		1,705	2,075
0300	Galvanized steel, UL label		1.80	8.889		1,325	450		1,775	2,200
0600	Stainless steel, 4' high roll-up, 6' long		2	8		2,225	405		2,630	3,075
0700	10' long		1.80	8.889		2,650	450		3,100	3,650
2000	Aluminum, 4' high, 4' long		2.20	7.273		1,675	370		2,045	2,425
2020	6' long		2	8		2,000	405		2,405	2,850
2040	8' long		1.90	8.421		2,275	425		2,700	3,200
2060	10' long		1.80	8.889		2,275	450		2,725	3,250
2080	14' long		1.40	11.429		2,925	580		3,505	4,175
2100	6' high, 4' long		2	8		2,025	405		2,430	2,875
2120	6' long		1.60	10		1,800	505		2,305	2,800
2140	10' long	↓	1.40	11.429	↓	2,375	580		2,955	3,550
9000	Minimum labor/equipment charge	1 Carp	2	4	Job		203		203	330

08 33 16 – Coiling Counter Grilles

08 33 16.10 Coiling Grilles

		Crew	Daily Output	Labor-Hours	Unit	Material	2018 Bare Costs Labor	Equipment	Total	Total Incl O&P
0010	**COILING GRILLES**									
2020	Aluminum, manual operated, mill finish	2 Sswk	82	.195	S.F.	30	10.65		40.65	51
2040	Bronze anodized		82	.195	"	46.50	10.65		57.15	69
2060	Steel, manual operated, 10' x 10' high		1	16	Opng.	2,675	875		3,550	4,450
2080	15' x 8' high	↓	.80	20	"	3,000	1,100		4,100	5,175
3000	For safety edge bottom bar, electric, add				L.F.	50.50			50.50	55.50
8000	For motor operation, add	2 Sswk	5	3.200	Opng.	1,300	175		1,475	1,725
9000	Minimum labor/equipment charge	"	1	16	Job		875		875	1,500

08 33 Coiling Doors and Grilles

08 33 23 – Overhead Coiling Doors

08 33 23.10 Coiling Service Doors	Crew	Daily Output	Labor-Hours	Unit	Material	2018 Bare Costs Labor	Equipment	Total	Total Incl O&P
0010 **COILING SERVICE DOORS** Steel, manual, 20 ga., incl. hardware									
0050　8' x 8' high	2 Sswk	1.60	10	Ea.	725	545		1,270	1,725
0130　12' x 12' high, standard		1.20	13.333		2,100	730		2,830	3,550
0160　10' x 20' high, standard		.50	32		2,575	1,750		4,325	5,825
2000　Class A fire doors, manual, 20 ga., 8' x 8' high		1.40	11.429		1,575	625		2,200	2,800
2100　10' x 10' high		1.10	14.545		2,150	795		2,945	3,700
2200　20' x 10' high		.80	20		4,400	1,100		5,500	6,700
2300　12' x 12' high		1	16		3,375	875		4,250	5,200
2400　20' x 12' high		.80	20		4,750	1,100		5,850	7,100
2500　14' x 14' high		.60	26.667		3,700	1,450		5,150	6,550
2600　20' x 16' high		.50	32		5,775	1,750		7,525	9,325
2700　10' x 20' high		.40	40		4,575	2,175		6,750	8,750
3000　For 18 ga. doors, add				S.F.	1.10			1.10	1.21
3300　For enamel finish, add				"	2.60			2.60	2.86
3600　For safety edge bottom bar, pneumatic, add				L.F.	22			22	24.50
4000　For weatherstripping, extruded rubber, jambs, add					18			18	19.80
4100　Hood, add					9			9	9.90
4200　Sill, add					9			9	9.90
4500　Motor operators, to 14' x 14' opening	2 Sswk	5	3.200	Ea.	1,175	175		1,350	1,600
4700　For fire door, additional fusible link, add				"	75			75	82.50
9000　Minimum labor/equipment charge	2 Sswk	1	16	Job		875		875	1,500

08 34 Special Function Doors

08 34 13 – Cold Storage Doors

08 34 13.10 Doors for Cold Area Storage

	Crew	Daily Output	Labor-Hours	Unit	Material	2018 Bare Costs Labor	Equipment	Total	Total Incl O&P
0010 **DOORS FOR COLD AREA STORAGE**									
0020　Single, 20 ga. galvanized steel									
0300　Horizontal sliding, 5' x 7', manual operation, 3.5" thick	2 Carp	2	8	Ea.	3,050	405		3,455	4,000
0400　4" thick		2	8		3,350	405		3,755	4,350
0500　6" thick		2	8		3,425	405		3,830	4,425
0800　5' x 7', power operation, 2" thick		1.90	8.421		5,600	425		6,025	6,850
0900　4" thick		1.90	8.421		5,700	425		6,125	6,975
1000　6" thick		1.90	8.421		6,475	425		6,900	7,825
1300　9' x 10', manual operation, 2" insulation		1.70	9.412		4,450	475		4,925	5,675
1400　4" insulation		1.70	9.412		4,675	475		5,150	5,925
1500　6" insulation		1.70	9.412		5,600	475		6,075	6,925
1800　Power operation, 2" insulation		1.60	10		7,700	505		8,205	9,300
1900　4" insulation		1.60	10		7,850	505		8,355	9,475
2000　6" insulation		1.70	9.412		8,250	475		8,725	9,850
2300　For stainless steel face, add					25%				
3000　Hinged, lightweight, 3' x 7'-0", 2" thick	2 Carp	2	8	Ea.	1,325	405		1,730	2,100
3050　4" thick		1.90	8.421		1,650	425		2,075	2,525
3300　Polymer doors, 3' x 7'-0"		1.90	8.421		1,350	425		1,775	2,175
3350　6" thick		1.40	11.429		2,350	580		2,930	3,525
3600　Stainless steel, 3' x 7'-0", 4" thick		1.90	8.421		1,725	425		2,150	2,600
3650　6" thick		1.40	11.429		2,750	580		3,330	3,975
3900　Painted, 3' x 7'-0", 4" thick		1.90	8.421		1,325	425		1,750	2,150
3950　6" thick		1.40	11.429		2,350	580		2,930	3,525
5000　Bi-parting, electric operated									
5010　6' x 8' opening, galv. faces, 4" thick for cooler	2 Carp	.80	20	Opng.	7,225	1,025		8,250	9,600

08 34 Special Function Doors

08 34 13 – Cold Storage Doors

08 34 13.10 Doors for Cold Area Storage

		Crew	Daily Output	Labor-Hours	Unit	Material	2018 Bare Costs Labor	2018 Bare Costs Equipment	Total	Total Incl O&P
5050	For freezer, 4" thick	2 Carp	.80	20	Opng.	7,950	1,025		8,975	10,400
5300	For door buck framing and door protection, add		2.50	6.400		670	325		995	1,250
6000	Galvanized batten door, galvanized hinges, 4' x 7'		2	8		1,825	405		2,230	2,675
6050	6' x 8'		1.80	8.889		2,475	450		2,925	3,450
6500	Fire door, 3 hr., 6' x 8', single slide		.80	20		8,250	1,025		9,275	10,700
6550	Double, bi-parting		.70	22.857		11,400	1,150		12,550	14,400
9000	Minimum labor/equipment charge	1 Carp	2	4	Job		203		203	330

08 34 36 – Darkroom Doors

08 34 36.10 Various Types of Darkroom Doors

		Crew	Daily Output	Labor-Hours	Unit	Material	2018 Bare Costs Labor	2018 Bare Costs Equipment	Total	Total Incl O&P
0010	**VARIOUS TYPES OF DARKROOM DOORS**									
0015	Revolving, standard, 2 way, 36" diameter	2 Carp	3.10	5.161	Opng.	3,025	262		3,287	3,750
0020	41" diameter		3.10	5.161		3,150	262		3,412	3,900
0050	3 way, 51" diameter		1.40	11.429		4,000	580		4,580	5,350
1000	4 way, 49" diameter		1.40	11.429		4,125	580		4,705	5,500
2000	Hinged safety, 2 way, 41" diameter		2.30	6.957		3,925	355		4,280	4,875
2500	3 way, 51" diameter		1.40	11.429		4,225	580		4,805	5,600
3000	Pop out safety, 2 way, 41" diameter		3.10	5.161		4,225	262		4,487	5,075
4000	3 way, 51" diameter		1.40	11.429		5,200	580		5,780	6,650
5000	Wheelchair-type, pop out, 51" diameter		1.40	11.429		5,800	580		6,380	7,350
5020	72" diameter		.90	17.778		9,400	900		10,300	11,800

08 34 53 – Security Doors and Frames

08 34 53.20 Steel Door

		Crew	Daily Output	Labor-Hours	Unit	Material	2018 Bare Costs Labor	2018 Bare Costs Equipment	Total	Total Incl O&P
0010	**STEEL DOOR** flush with ballistic core and welded frame both 14 ga.									
0120	UL 752 Level 3, 1-3/4", 3'-0" x 7'-0"	2 Carp	1.50	10.667	Opng.	2,600	540		3,140	3,725
0125	1-3/4", 3'-6" x 7'-0"		1.50	10.667		2,850	540		3,390	4,000
0130	1-3/4", 4'-0" x 7'-0"		1.20	13.333		3,050	675		3,725	4,450
0150	UL 752 Level 8, 1-3/4", 3'-0" x 7'-0"		1.50	10.667		13,500	540		14,040	15,700
0155	1-3/4", 3'-6" x 7'-0"		1.50	10.667		14,700	540		15,240	17,100
0160	1-3/4", 4'-0" x 7'-0"		1.20	13.333		14,700	675		15,375	17,300
1000	Safe room sliding door and hardware, 1-3/4", 3'-0" x 7'-0" UL 752 Level 3		.50	32		23,700	1,625		25,325	28,700
1050	Safe room swinging door and hardware, 1-3/4", 3'-0" x 7'-0" UL 752 Level 3		.50	32		28,400	1,625		30,025	33,900

08 34 53.30 Wood Ballistic Doors

		Crew	Daily Output	Labor-Hours	Unit	Material	2018 Bare Costs Labor	2018 Bare Costs Equipment	Total	Total Incl O&P
0010	**WOOD BALLISTIC DOORS** with frames and hardware									
0050	Wood, 1-3/4", 3'-0" x 7'-0" UL 752 Level 3	2 Carp	1.50	10.667	Opng.	2,575	540		3,115	3,725

08 34 56 – Security Gates

08 34 56.10 Gates

		Crew	Daily Output	Labor-Hours	Unit	Material	2018 Bare Costs Labor	2018 Bare Costs Equipment	Total	Total Incl O&P
0010	**GATES**									
0015	Driveway gates include mounting hardware									
0500	Wood, security gate, driveway, dual, 10' wide	H-4	.80	25	Opng.	3,775	1,175		4,950	6,050
0505	12' wide		.80	25		3,650	1,175		4,825	5,900
0510	15' wide		.80	25		4,325	1,175		5,500	6,675
0600	Steel, security gate, driveway, single, 10' wide		.80	25		1,700	1,175		2,875	3,775
0605	12' wide		.80	25		1,900	1,175		3,075	4,000
0620	Steel, security gate, driveway, dual, 12' wide		.80	25		2,050	1,175		3,225	4,175
0625	14' wide		.80	25		2,250	1,175		3,425	4,375
0630	16' wide		.80	25		2,450	1,175		3,625	4,600
0700	Aluminum, security gate, driveway, dual, 10' wide		.80	25		3,200	1,175		4,375	5,425
0705	12' wide		.80	25		3,850	1,175		5,025	6,125
0710	16' wide		.80	25		4,375	1,175		5,550	6,725
1000	Security gate, driveway, opener 12 VDC				Ea.	830			830	910
1010	Wireless					1,900			1,900	2,075

08 34 Special Function Doors

08 34 56 – Security Gates

08 34 56.10 Gates

		Crew	Daily Output	Labor-Hours	Unit	Material	2018 Bare Costs Labor	Equipment	Total	Total Incl O&P
1020	Security gate, driveway, opener 24 VDC				Ea.	1,425			1,425	1,550
1030	Wireless					1,575			1,575	1,750
1040	Security gate, driveway, opener 12 VDC, solar panel 10 watt	1 Elec	2	4		445	233		678	850
1050	20 watt	"	2	4		585	233		818	1,000

08 34 73 – Sound Control Door Assemblies

08 34 73.10 Acoustical Doors

		Crew	Daily Output	Labor-Hours	Unit	Material	2018 Bare Costs Labor	Equipment	Total	Total Incl O&P
0010	**ACOUSTICAL DOORS**									
0020	Including framed seals, 3' x 7', wood, 40 STC rating	2 Carp	1.50	10.667	Ea.	3,350	540		3,890	4,550
0100	Steel, 41 STC rating		1.50	10.667		3,600	540		4,140	4,825
0200	45 STC rating		1.50	10.667		3,975	540		4,515	5,250
0300	48 STC rating		1.50	10.667		4,650	540		5,190	6,000
0400	52 STC rating		1.50	10.667		5,200	540		5,740	6,575
9000	Minimum labor/equipment charge	1 Carp	4	2	Job		101		101	165

08 36 Panel Doors

08 36 13 – Sectional Doors

08 36 13.10 Overhead Commercial Doors

		Crew	Daily Output	Labor-Hours	Unit	Material	2018 Bare Costs Labor	Equipment	Total	Total Incl O&P
0010	**OVERHEAD COMMERCIAL DOORS**									
1000	Stock, sectional, heavy duty, wood, 1-3/4" thick, 8' x 8' high	2 Carp	2	8	Ea.	1,225	405		1,630	2,000
1200	12' x 12' high		1.50	10.667		2,350	540		2,890	3,450
1300	Chain hoist, 14' x 14' high		1.30	12.308		3,650	625		4,275	5,025
1600	20' x 16' high		.65	24.615		5,975	1,250		7,225	8,600
2100	For medium duty custom door, deduct					5%	5%			
2150	For medium duty stock doors, deduct					10%	5%			
2300	Fiberglass and aluminum, heavy duty, sectional, 12' x 12' high	2 Carp	1.50	10.667	Ea.	3,350	540		3,890	4,550
2450	Chain hoist, 20' x 20' high	"	.50	32		7,450	1,625		9,075	10,800
2900	For electric trolley operator, 1/3 HP, to 12' x 12', add	1 Carp	2	4		1,100	203		1,303	1,550
2950	Over 12' x 12', 1/2 HP, add		1	8		1,200	405		1,605	1,975
2980	Overhead, for row of clear lites, add		1	8		164	405		569	840
9000	Minimum labor/equipment charge	2 Carp	1.50	10.667	Job		540		540	880

08 36 13.20 Residential Garage Doors

		Crew	Daily Output	Labor-Hours	Unit	Material	2018 Bare Costs Labor	Equipment	Total	Total Incl O&P
0010	**RESIDENTIAL GARAGE DOORS**									
0050	Hinged, wood, custom, double door, 9' x 7'	2 Carp	4	4	Ea.	875	203		1,078	1,300
0070	16' x 7'		3	5.333		1,300	270		1,570	1,875
0200	Overhead, sectional, incl. hardware, fiberglass, 9' x 7', standard		5	3.200		1,025	162		1,187	1,400
0220	Deluxe		5	3.200		1,200	162		1,362	1,600
0300	16' x 7', standard		6	2.667		1,625	135		1,760	2,025
0320	Deluxe		6	2.667		2,250	135		2,385	2,700
0500	Hardboard, 9' x 7', standard		8	2		715	101		816	950
0520	Deluxe		8	2		860	101		961	1,100
0600	16' x 7', standard		6	2.667		1,300	135		1,435	1,650
0620	Deluxe		6	2.667		1,500	135		1,635	1,875
0700	Metal, 9' x 7', standard		8	2		885	101		986	1,150
0720	Deluxe		6	2.667		995	135		1,130	1,325
0800	16' x 7', standard		6	2.667		1,075	135		1,210	1,400
0820	Deluxe		5	3.200		1,450	162		1,612	1,875
0900	Wood, 9' x 7', standard		8	2		1,025	101		1,126	1,300
0920	Deluxe		8	2		2,250	101		2,351	2,650
1000	16' x 7', standard		6	2.667		1,675	135		1,810	2,075
1020	Deluxe		6	2.667		3,125	135		3,260	3,675

08 36 Panel Doors

08 36 13 – Sectional Doors

08 36 13.20 Residential Garage Doors

		Crew	Daily Output	Labor-Hours	Unit	Material	2018 Bare Costs Labor	Equipment	Total	Total Incl O&P
1800	Door hardware, sectional	1 Carp	4	2	Ea.	360	101		461	565
1810	Door tracks only		4	2		170	101		271	350
1820	One side only		7	1.143		128	58		186	234
4000	For electric operator, economy, add		8	1		440	50.50		490.50	570
4100	Deluxe, including remote control	▼	8	1	▼	635	50.50		685.50	780
4500	For transmitter/receiver control, add to operator				Total	113			113	124
4600	Transmitters, additional				"	63.50			63.50	70
6000	Replace section, on sectional door, fiberglass, 9' x 7'	1 Carp	4	2	Ea.	635	101		736	860
6020	16' x 7'		3.50	2.286		760	116		876	1,025
6200	Hardboard, 9' x 7'		4	2		176	101		277	360
6220	16' x 7'		3.50	2.286		238	116		354	450
6300	Metal, 9' x 7'		4	2		225	101		326	410
6320	16' x 7'		3.50	2.286		330	116		446	555
6500	Wood, 9' x 7'		4	2		125	101		226	300
6520	16' x 7'	▼	3.50	2.286		250	116		366	465
7010	Garage doors, row of lites				▼	126			126	138
9000	Minimum labor/equipment charge	1 Carp	2.50	3.200	Job		162		162	263

08 36 19 – Multi-Leaf Vertical Lift Doors

08 36 19.10 Sectional Vertical Lift Doors

		Crew	Daily Output	Labor-Hours	Unit	Material	2018 Bare Costs Labor	Equipment	Total	Total Incl O&P
0010	**SECTIONAL VERTICAL LIFT DOORS**									
0020	Motorized, 14 ga. steel, incl. frame and control panel									
0050	16' x 16' high	L-10	.50	48	Ea.	21,800	2,675	990	25,465	29,600
0100	10' x 20' high		1.30	18.462		36,600	1,025	380	38,005	42,400
0120	15' x 20' high		1.30	18.462		44,800	1,025	380	46,205	51,500
0140	20' x 20' high		1	24		52,500	1,350	495	54,345	60,500
0160	25' x 20' high		1	24		59,000	1,350	495	60,845	67,500
0170	32' x 24' high		.75	32		50,000	1,775	660	52,435	58,500
0180	20' x 25' high		1	24		60,500	1,350	495	62,345	69,500
0200	25' x 25' high		.70	34.286		73,500	1,925	710	76,135	85,000
0220	25' x 30' high		.70	34.286		79,500	1,925	710	82,135	91,000
0240	30' x 30' high		.70	34.286		90,500	1,925	710	93,135	103,500
0260	35' x 30' high	▼	.70	34.286	▼	95,500	1,925	710	98,135	109,000

08 38 Traffic Doors

08 38 19 – Rigid Traffic Doors

08 38 19.20 Double Acting Swing Doors

		Crew	Daily Output	Labor-Hours	Unit	Material	2018 Bare Costs Labor	Equipment	Total	Total Incl O&P
0010	**DOUBLE ACTING SWING DOORS**									
0020	Including frame, closer, hardware and vision panel									
1000	Polymer, 7'-0" high, 4'-0" wide	2 Carp	4.20	3.810	Pr.	2,100	193		2,293	2,625
1025	6'-0" wide		4	4		2,200	203		2,403	2,750
1050	6'-8" wide	▼	4	4	▼	2,400	203		2,603	2,975
2000	3/4" thick, stainless steel									
2010	Stainless steel, 7' high opening, 4' wide	2 Carp	4	4	Pr.	2,600	203		2,803	3,175
2050	7' wide		3.80	4.211	"	2,800	213		3,013	3,425
9000	Minimum labor/equipment charge	▼	2	8	Job		405		405	660

08 38 19.30 Shock Absorbing Doors

		Crew	Daily Output	Labor-Hours	Unit	Material	2018 Bare Costs Labor	Equipment	Total	Total Incl O&P
0010	**SHOCK ABSORBING DOORS**									
0020	Rigid, no frame, 1-1/2" thick, 5' x 7'	2 Sswk	1.90	8.421	Opng.	1,525	460		1,985	2,450
0100	8' x 8'		1.80	8.889		2,000	485		2,485	3,025
0500	Flexible, no frame, insulated, .16" thick, economy, 5' x 7'		2	8	▼	1,750	435		2,185	2,675

08 38 Traffic Doors

08 38 19 – Rigid Traffic Doors

08 38 19.30 Shock Absorbing Doors

		Crew	Daily Output	Labor-Hours	Unit	Material	2018 Bare Costs Labor	Equipment	Total	Total Incl O&P
0600	Deluxe	2 Sswk	1.90	8.421	Opng.	2,625	460		3,085	3,650
1000	8' x 8' opening, economy		2	8		2,750	435		3,185	3,775
1100	Deluxe		1.90	8.421		3,500	460		3,960	4,625
9000	Minimum labor/equipment charge		2	8	Job		435		435	745

08 41 Entrances and Storefronts

08 41 13 – Aluminum-Framed Entrances and Storefronts

08 41 13.20 Tube Framing

		Crew	Daily Output	Labor-Hours	Unit	Material	2018 Bare Costs Labor	Equipment	Total	Total Incl O&P
0010	**TUBE FRAMING**, For window walls and storefronts, aluminum stock									
0050	Plain tube frame, mill finish, 1-3/4" x 1-3/4"	2 Glaz	103	.155	L.F.	10.50	7.55		18.05	23.50
0150	1-3/4" x 4"		98	.163		14.05	7.90		21.95	28.50
0200	1-3/4" x 4-1/2"		95	.168		16.90	8.15		25.05	32
0250	2" x 6"		89	.180		25	8.70		33.70	41.50
0350	4" x 4"		87	.184		28.50	8.90		37.40	45.50
0400	4-1/2" x 4-1/2"		85	.188		30	9.15		39.15	48
0450	Glass bead		240	.067		3.27	3.23		6.50	8.80
1000	Flush tube frame, mill finish, 1/4" glass, 1-3/4" x 4", open header		80	.200		14.05	9.70		23.75	31
1050	Open sill		82	.195		11.45	9.45		20.90	28
1100	Closed back header		83	.193		20	9.35		29.35	37
1150	Closed back sill		85	.188		19.20	9.15		28.35	36
1160	Tube fmg., spandrel cover both sides, alum 1" wide	1 Sswk	85	.094	S.F.	96	5.15		101.15	115
1170	Tube fmg., spandrel cover both sides, alum 2" wide	"	85	.094	"	38	5.15		43.15	51
1200	Vertical mullion, one piece	2 Glaz	75	.213	L.F.	20.50	10.35		30.85	39
1250	Two piece		73	.219		21.50	10.65		32.15	40.50
1300	90° or 180° vertical corner post		75	.213		33	10.35		43.35	53
1400	1-3/4" x 4-1/2", open header		80	.200		16.75	9.70		26.45	34
1450	Open sill		82	.195		13.90	9.45		23.35	30.50
1500	Closed back header		83	.193		20.50	9.35		29.85	37.50
1550	Closed back sill		85	.188		19.50	9.15		28.65	36.50
1600	Vertical mullion, one piece		75	.213		22	10.35		32.35	40.50
1650	Two piece		73	.219		23	10.65		33.65	42.50
1700	90° or 180° vertical corner post		75	.213		23.50	10.35		33.85	42.50
2000	Flush tube frame, mil fin.,ins. glass w/thml brk, 2" x 4-1/2", open header		75	.213		16	10.35		26.35	34.50
2050	Open sill		77	.208		13.45	10.10		23.55	31
2100	Closed back header		78	.205		15.05	9.95		25	32.50
2150	Closed back sill		80	.200		15.70	9.70		25.40	33
2200	Vertical mullion, one piece		70	.229		16.95	11.10		28.05	36.50
2250	Two piece		68	.235		18.30	11.40		29.70	38.50
2300	90° or 180° vertical corner post		70	.229		18.65	11.10		29.75	38.50
5000	Flush tube frame, mill fin., thermal brk., 2-1/4" x 4-1/2", open header		74	.216		16.80	10.50		27.30	35.50
5050	Open sill		75	.213		14.85	10.35		25.20	33
5100	Vertical mullion, one piece		69	.232		18.75	11.25		30	38.50
5150	Two piece		67	.239		21	11.60		32.60	41.50
5200	90° or 180° vertical corner post		69	.232		18.80	11.25		30.05	38.50
5295	Flush tube frame, mill finish, 4-1/2" x 4-1/2"		85	.188		30	9.15		39.15	48
5300	Plain tube frame, mill finish, 2" x 3" jamb		115	.139		11.70	6.75		18.45	24
5310	2" x 4"		115	.139		16.55	6.75		23.30	29
5320	3" x 5-1/2"		107	.150		28.50	7.25		35.75	42.50
5330	Head, 2" x 3"		200	.080		19.95	3.88		23.83	28.50
5340	Mullion, 2" x 3"		200	.080		20.50	3.88		24.38	29
5350	2" x 4"		115	.139		26	6.75		32.75	39.50

08 41 13 – Aluminum-Framed Entrances and Storefronts

08 41 13.20 Tube Framing		Crew	Daily Output	Labor-Hours	Unit	Material	2018 Bare Costs Labor	2018 Bare Costs Equipment	Total	Total Incl O&P
5360	3" x 5-1/2"	2 Glaz	107	.150	L.F.	34.50	7.25		41.75	49.50
5370	4" corner mullion		90	.178		29.50	8.60		38.10	46.50
5380	Horizontal, 2" x 3"		115	.139		23.50	6.75		30.25	37
5390	3" x 5-1/2"		107	.150		29	7.25		36.25	43.50
5430	Sill section, 1/8" x 6"		133	.120		36	5.85		41.85	49.50
5440	1/8" x 7"		133	.120		43.50	5.85		49.35	57
5450	1/8" x 8-1/2"		133	.120		50.50	5.85		56.35	65
5460	Column covers, aluminum, 1/8" x 26"		57	.281		65.50	13.60		79.10	94
5470	1/8" x 34"		53	.302		84	14.65		98.65	116
5480	1/8" x 38"		53	.302		93.50	14.65		108.15	127
5700	Vertical mullions, clear finish, 1/4" thick glass		528	.030		13.45	1.47		14.92	17.15
5720	3/8" thick		445	.036		13.45	1.74		15.19	17.55
5740	1/2" thick		400	.040		13.90	1.94		15.84	18.45
5760	3/4" thick		344	.047		14.40	2.26		16.66	19.50
5780	1" thick		304	.053		14.60	2.55		17.15	20
6980	Door stop (snap in)	↓	380	.042	↓	3.66	2.04		5.70	7.35
7000	For joints, 90°, clip type, add				Ea.	25.50			25.50	28
7050	Screw spline joint, add					23.50			23.50	26
7100	For joint other than 90°, add				↓	49.50			49.50	54.50
8000	For bronze anodized aluminum, add					15%				
8020	For black finish, add					30%				
8050	For stainless steel materials, add					350%				
8100	For monumental grade, add					53%				
8150	For steel stiffener, add	2 Glaz	200	.080	L.F.	11.30	3.88		15.18	18.65
8200	For 2 to 5 stories, add per story				Story		8%			
9000	Minimum labor/equipment charge	2 Glaz	2	8	Job		390		390	625

08 41 19 – Stainless-Steel-Framed Entrances and Storefronts

08 41 19.10 Stainless-Steel and Glass Entrance Unit

		Crew	Daily Output	Labor-Hours	Unit	Material	2018 Bare Costs Labor	2018 Bare Costs Equipment	Total	Total Incl O&P
0010	**STAINLESS-STEEL AND GLASS ENTRANCE UNIT**, narrow stiles									
0020	3' x 7' opening, including hardware, minimum	2 Sswk	1.60	10	Opng.	7,450	545		7,995	9,125
0050	Average		1.40	11.429		7,950	625		8,575	9,825
0100	Maximum	↓	1.20	13.333		8,475	730		9,205	10,600
1000	For solid bronze entrance units, statuary finish, add				↓	64%				
1100	Without statuary finish, add					45%				
2000	Balanced doors, 3' x 7', economy	2 Sswk	.90	17.778	Ea.	10,100	970		11,070	12,800
2100	Premium		.70	22.857	"	16,500	1,250		17,750	20,300
9000	Minimum labor/equipment charge	↓	2	8	Job		435		435	745

08 41 26 – All-Glass Entrances and Storefronts

08 41 26.10 Window Walls Aluminum, Stock

		Crew	Daily Output	Labor-Hours	Unit	Material	2018 Bare Costs Labor	2018 Bare Costs Equipment	Total	Total Incl O&P
0010	**WINDOW WALLS ALUMINUM, STOCK**, including glazing									
0020	Minimum	H-2	160	.150	S.F.	50	6.85		56.85	66
0050	Average		140	.171		69	7.80		76.80	88
0100	Maximum	↓	110	.218	↓	183	9.95		192.95	217
0500	For translucent sandwich wall systems, see Section 07 41 33.10									
0850	Cost of the above walls depends on material,									
0860	finish, repetition, and size of units.									
0870	The larger the opening, the lower the S.F. cost									

For customer support on your Commercial Renovation Costs with RSMeans data, call 800.448.8182.

249

08 42 Entrances

08 42 26 – All-Glass Entrances

08 42 26.10 Swinging Glass Doors	Crew	Daily Output	Labor-Hours	Unit	Material	2018 Bare Costs Labor	Equipment	Total	Total Incl O&P
0010 **SWINGING GLASS DOORS**									
0020 Including hardware, 1/2" thick, tempered, 3' x 7' opening	2 Glaz	2	8	Opng.	2,325	390		2,715	3,200
0100 6' x 7' opening		1.40	11.429	"	4,575	555		5,130	5,950
9000 Minimum labor/equipment charge		2	8	Job		390		390	625

08 42 36 – Balanced Door Entrances

08 42 36.10 Balanced Entrance Doors	Crew	Daily Output	Labor-Hours	Unit	Material	2018 Bare Costs Labor	Equipment	Total	Total Incl O&P
0010 **BALANCED ENTRANCE DOORS**									
0020 Hardware & frame, alum. & glass, 3' x 7', econ.	2 Sswk	.90	17.778	Ea.	7,025	970		7,995	9,375
0150 Premium		.70	22.857	"	8,425	1,250		9,675	11,400
9000 Minimum labor/equipment charge		1	16	Job		875		875	1,500

08 43 Storefronts

08 43 13 – Aluminum-Framed Storefronts

08 43 13.10 Aluminum-Framed Entrance Doors and Frames	Crew	Daily Output	Labor-Hours	Unit	Material	2018 Bare Costs Labor	Equipment	Total	Total Incl O&P
0010 **ALUMINUM-FRAMED ENTRANCE DOORS AND FRAMES**									
0015 Standard hardware and glass stops but no glass									
0020 Entrance door, 3' x 7' opening, clear anodized finish	2 Sswk	7	2.286	Opng.	625	125		750	900
0040 Bronze finish		7	2.286		675	125		800	960
0060 Black finish		7	2.286		690	125		815	975
0200 3'-6" x 7'-0", mill finish		7	2.286		730	125		855	1,025
0220 Bronze finish		7	2.286		865	125		990	1,175
0240 Black finish		7	2.286		915	125		1,040	1,225
0500 6' x 7' opening, clear finish		6	2.667		980	146		1,126	1,325
0520 Bronze finish		6	2.667		1,075	146		1,221	1,425
0600 Door frame for above doors 3'-0" x 7'-0", mill finish		6	2.667		445	146		591	735
0620 Bronze finish		6	2.667		500	146		646	800
0640 Black finish		6	2.667		550	146		696	855
0700 3'-6" x 7'-0", mill finish		6	2.667		365	146		511	650
0720 Bronze finish		6	2.667		365	146		511	650
0740 Black finish		6	2.667		365	146		511	650
0800 6'-0" x 7'-0", mill finish		6	2.667		365	146		511	650
0820 Bronze finish		6	2.667		370	146		516	655
0840 Black finish		6	2.667		400	146		546	690
1000 With 3' high transom above, 3' x 7' opening, clear finish		5.50	2.909		540	159		699	865
1050 Bronze finish		5.50	2.909		550	159		709	875
1100 Black finish		5.50	2.909		630	159		789	960
1300 3'-6" x 7'-0" opening, clear finish		5.50	2.909		375	159		534	685
1320 Bronze finish		5.50	2.909		395	159		554	705
1340 Black finish		5.50	2.909		405	159		564	715
1500 6' x 7' opening, clear finish		5.50	2.909		645	159		804	980
1550 Bronze finish		5.50	2.909		670	159		829	1,000
1600 Black finish		5.50	2.909		740	159		899	1,075
8000 For 8' high doors, add				Ea.	200			200	220
9000 Minimum labor/equipment charge	2 Sswk	4	4	Job		219		219	375

08 43 13.20 Storefront Systems

	Crew	Daily Output	Labor-Hours	Unit	Material	2018 Bare Costs Labor	Equipment	Total	Total Incl O&P
0010 **STOREFRONT SYSTEMS**, aluminum frame clear 3/8" plate glass									
0020 incl. 3' x 7' door with hardware (400 sq. ft. max. wall)									
0500 Wall height to 12' high, commercial grade	2 Glaz	150	.107	S.F.	24	5.15		29.15	35
0600 Institutional grade		130	.123		29.50	5.95		35.45	42
0700 Monumental grade		115	.139		82	6.75		88.75	101

08 43 Storefronts

08 43 13 – Aluminum-Framed Storefronts

08 43 13.20 Storefront Systems	Crew	Daily Output	Labor-Hours	Unit	Material	2018 Bare Costs Labor	Equipment	Total	Total Incl O&P
1000	6' x 7' door with hardware, commercial grade	2 Glaz	135	.119	S.F.	75	5.75		80.75	92
1100	Institutional grade		115	.139		58.50	6.75		65.25	75.50
1200	Monumental grade		100	.160		113	7.75		120.75	137
1500	For bronze anodized finish, add					15%				
1600	For black anodized finish, add					36%				
1700	For stainless steel framing, add to monumental					78%				
9000	Minimum labor/equipment charge	2 Glaz	1	16	Job		775		775	1,250

08 43 29 – Sliding Storefronts

08 43 29.10 Sliding Panels

		Crew	Daily Output	Labor-Hours	Unit	Material	2018 Bare Costs Labor	Equipment	Total	Total Incl O&P
0010	**SLIDING PANELS**									
0020	Mall fronts, aluminum & glass, 15' x 9' high	2 Glaz	1.30	12.308	Opng.	3,825	595		4,420	5,200
0100	24' x 9' high		.70	22.857		5,500	1,100		6,600	7,850
0200	48' x 9' high, with fixed panels		.90	17.778		9,950	860		10,810	12,300
0500	For bronze finish, add					17%				
9000	Minimum labor/equipment charge	2 Glaz	1	16	Job		775		775	1,250

08 45 Translucent Wall and Roof Assemblies

08 45 10 – Translucent Roof Assemblies

08 45 10.10 Skyroofs

		Crew	Daily Output	Labor-Hours	Unit	Material	2018 Bare Costs Labor	Equipment	Total	Total Incl O&P
0010	**SKYROOFS**									
1200	Skylights, circular, clear, double glazed acrylic									
1230	30" diameter	2 Carp	3	5.333	Ea.	3,000	270		3,270	3,750
1250	60" diameter		3	5.333		4,000	270		4,270	4,850
1290	96" diameter		2	8		5,000	405		5,405	6,150
1300	Skylight Barrel Vault, clear, double glazed, acrylic									
1330	3'-0" x 12'-0"	G-3	3	10.667	Ea.	5,000	530		5,530	6,350
1350	4'-0" x 12'-0"		3	10.667		5,500	530		6,030	6,900
1390	5'-0" x 12'-0"		2	16		6,000	795		6,795	7,875
1400	Skylight Pyramid, aluminum frame, clear low E laminated glass									
1410	The glass is installed in the frame except where noted									
1430	Square, 3' x 3'	G-3	3	10.667	Ea.	6,000	530		6,530	7,450
1440	4' x 4'		3	10.667		7,000	530		7,530	8,550
1450	5' x 5', glass must be field installed		3	10.667		8,000	530		8,530	9,650
1460	6' x 6', glass must be field installed		2	16		10,000	795		10,795	12,300
1550	Install pre-cut laminated glass in aluminum frame on a flat roof	2 Glaz	55	.291	SF Surf		14.10		14.10	23
1560	Install pre-cut laminated glass in aluminum frame on a sloped roof	"	40	.400	"		19.40		19.40	31.50
9000	Minimum labor/equipment charge	2 Carp	8	2	Job		101		101	165

08 51 Metal Windows

08 51 13 – Aluminum Windows

08 51 13.10 Aluminum Sash

		Crew	Daily Output	Labor-Hours	Unit	Material	2018 Bare Costs Labor	Equipment	Total	Total Incl O&P
0010	**ALUMINUM SASH**									
0020	Stock, grade C, glaze & trim not incl., casement	2 Sswk	200	.080	S.F.	41	4.37		45.37	53
0050	Double-hung		200	.080		41.50	4.37		45.87	53
0100	Fixed casement		200	.080		18.20	4.37		22.57	27.50
0150	Picture window		200	.080		19.55	4.37		23.92	29
0200	Projected window		200	.080		37.50	4.37		41.87	48.50
0250	Single-hung		200	.080		17.20	4.37		21.57	26.50
0300	Sliding		200	.080		22.50	4.37		26.87	32

251

08 51 Metal Windows

08 51 13 - Aluminum Windows

08 51 13.10 Aluminum Sash

		Crew	Daily Output	Labor-Hours	Unit	Material	2018 Bare Costs Labor	Equipment	Total	Total Incl O&P
1000	Mullions for above, tubular	2 Sswk	240	.067	L.F.	6.50	3.64		10.14	13.35
9000	Minimum labor/equipment charge	1 Sswk	2	4	Job		219		219	375

08 51 13.20 Aluminum Windows

		Crew	Daily Output	Labor-Hours	Unit	Material	2018 Bare Costs Labor	Equipment	Total	Total Incl O&P
0010	**ALUMINUM WINDOWS**, incl. frame and glazing, commercial grade									
1000	Stock units, casement, 3'-1" x 3'-2" opening	2 Sswk	10	1.600	Ea.	380	87.50		467.50	570
1050	Add for storms					122			122	134
1600	Projected, with screen, 3'-1" x 3'-2" opening	2 Sswk	10	1.600		360	87.50		447.50	545
1700	Add for storms					119			119	131
2000	4'-5" x 5'-3" opening	2 Sswk	8	2		410	109		519	635
2100	Add for storms					128			128	141
2500	Enamel finish windows, 3'-1" x 3'-2"	2 Sswk	10	1.600		365	87.50		452.50	555
2600	4'-5" x 5'-3"		8	2		415	109		524	640
3000	Single-hung, 2' x 3' opening, enameled, standard glazed		10	1.600		212	87.50		299.50	380
3100	Insulating glass		10	1.600		257	87.50		344.50	430
3300	2'-8" x 6'-8" opening, standard glazed		8	2		370	109		479	590
3400	Insulating glass		8	2		480	109		589	715
3700	3'-4" x 5'-0" opening, standard glazed		9	1.778		305	97		402	500
3800	Insulating glass		9	1.778		340	97		437	540
4000	Sliding aluminum, 3' x 2' opening, standard glazed		10	1.600		220	87.50		307.50	390
4100	Insulating glass		10	1.600		236	87.50		323.50	410
4300	5' x 3' opening, standard glazed		9	1.778		335	97		432	535
4400	Insulating glass		9	1.778		390	97		487	595
4600	8' x 4' opening, standard glazed		6	2.667		355	146		501	645
4700	Insulating glass		6	2.667		575	146		721	885
5000	9' x 5' opening, standard glazed		4	4		540	219		759	970
5100	Insulating glass		4	4		850	219		1,069	1,300
5500	Sliding, with thermal barrier and screen, 6' x 4', 2 track		8	2		725	109		834	980
5700	4 track		8	2		910	109		1,019	1,175
6000	For above units with bronze finish, add					15%				
6200	For installation in concrete openings, add					8%				
7000	Double-hung, insulating glass, 2'-0" x 2'-0"	2 Sswk	11	1.455		117	79.50		196.50	265
7020	2'-0" x 2'-6"		11	1.455		131	79.50		210.50	281
7040	3'-0" x 1'-6"		10	1.600		270	87.50		357.50	445
7060	3'-0" x 2'-0"		10	1.600		315	87.50		402.50	495
7080	3'-0" x 2'-6"		10	1.600		380	87.50		467.50	570
7100	3'-0" x 3'-0"		10	1.600		425	87.50		512.50	615
7120	3'-0" x 3'-6"		10	1.600		445	87.50		532.50	640
7140	3'-0" x 4'-0"		10	1.600		480	87.50		567.50	680
7160	3'-0" x 5'-0"		10	1.600		530	87.50		617.50	730
7180	3'-0" x 6'-0"		9	1.778		560	97		657	780

08 51 13.30 Impact Resistant Aluminum Windows

		Crew	Daily Output	Labor-Hours	Unit	Material	2018 Bare Costs Labor	Equipment	Total	Total Incl O&P
0010	**IMPACT RESISTANT ALUMINUM WINDOWS**, incl. frame and glazing									
0100	Single-hung, impact resistant, 2'-8" x 5'-0"	2 Carp	9	1.778	Ea.	1,250	90		1,340	1,525
0120	3'-0" x 5'-0"		9	1.778		1,350	90		1,440	1,650
0130	4'-0" x 5'-0"		9	1.778		1,450	90		1,540	1,750
0250	Horizontal slider, impact resistant, 5'-5" x 5'-2"		9	1.778		1,625	90		1,715	1,925

08 51 23 - Steel Windows

08 51 23.10 Steel Sash

		Crew	Daily Output	Labor-Hours	Unit	Material	2018 Bare Costs Labor	Equipment	Total	Total Incl O&P
0010	**STEEL SASH** Custom units, glazing and trim not included									
0100	Casement, 100% vented	2 Sswk	200	.080	S.F.	67.50	4.37		71.87	81.50
0200	50% vented		200	.080		55.50	4.37		59.87	68.50
0300	Fixed		200	.080		29.50	4.37		33.87	40

For customer support on your Commercial Renovation Costs with RSMeans data, call 800.448.8182.

08 51 Metal Windows

08 51 23 – Steel Windows

08 51 23.10 Steel Sash

		Crew	Daily Output	Labor-Hours	Unit	Material	2018 Bare Costs Labor	Equipment	Total	Total Incl O&P
1000	Projected, commercial, 40% vented	2 Sswk	200	.080	S.F.	52.50	4.37		56.87	65
1100	Intermediate, 50% vented		200	.080		60	4.37		64.37	73.50
1500	Industrial, horizontally pivoted		200	.080		55.50	4.37		59.87	68.50
1600	Fixed		200	.080		32	4.37		36.37	43
2000	Industrial security sash, 50% vented		200	.080		60	4.37		64.37	73.50
2100	Fixed		200	.080		48.50	4.37		52.87	61
2500	Picture window		200	.080		31	4.37		35.37	41.50
3000	Double-hung		200	.080		60.50	4.37		64.87	74
5000	Mullions for above, open interior face		240	.067	L.F.	10.65	3.64		14.29	17.90
5100	With interior cover		240	.067	"	17.60	3.64		21.24	25.50
5200	Single glazing for above, add	2 Glaz	200	.080	S.F.	7	3.88		10.88	13.95
6000	Double glazing for above, add		200	.080		13.30	3.88		17.18	21
6100	Triple glazing for above, add		85	.188		12.50	9.15		21.65	28.50
9000	Minimum labor/equipment charge	1 Sswk	2	4	Job		219		219	375

08 51 23.20 Steel Windows

		Crew	Daily Output	Labor-Hours	Unit	Material	2018 Bare Costs Labor	Equipment	Total	Total Incl O&P
0010	STEEL WINDOWS Stock, including frame, trim and insul. glass R085123-10									
1000	Custom units, double-hung, 2'-8" x 4'-6" opening	2 Sswk	12	1.333	Ea.	725	73		798	920
1100	2'-4" x 3'-9" opening		12	1.333		595	73		668	780
1500	Commercial projected, 3'-9" x 5'-5" opening		10	1.600		1,250	87.50		1,337.50	1,525
1600	6'-9" x 4'-1" opening		7	2.286		1,650	125		1,775	2,050
2000	Intermediate projected, 2'-9" x 4'-1" opening		12	1.333		700	73		773	895
2100	4'-1" x 5'-5" opening		10	1.600		1,425	87.50		1,512.50	1,725
9000	Minimum labor/equipment charge	1 Sswk	3	2.667	Job		146		146	249

08 51 66 – Metal Window Screens

08 51 66.10 Screens

		Crew	Daily Output	Labor-Hours	Unit	Material	2018 Bare Costs Labor	Equipment	Total	Total Incl O&P
0010	SCREENS									
0020	For metal sash, aluminum or bronze mesh, flat screen	2 Sswk	1200	.013	S.F.	4.50	.73		5.23	6.20
0500	Wicket screen, inside window		1000	.016		6.85	.87		7.72	9.05
0800	Security screen, aluminum frame with stainless steel cloth		1200	.013		24.50	.73		25.23	28
0900	Steel grate, painted, on steel frame		1600	.010		13.50	.55		14.05	15.80
1000	Screens for solar louvers		160	.100		25.50	5.45		30.95	37.50
4000	See Section 05 58 23.90 for window guards									

08 52 Wood Windows

08 52 10 – Plain Wood Windows

08 52 10.20 Awning Window

		Crew	Daily Output	Labor-Hours	Unit	Material	2018 Bare Costs Labor	Equipment	Total	Total Incl O&P
0010	AWNING WINDOW, Including frame, screens and grilles									
0100	34" x 22", insulated glass	1 Carp	10	.800	Ea.	276	40.50		316.50	370
0200	Low E glass		10	.800		298	40.50		338.50	395
0300	40" x 28", insulated glass		9	.889		315	45		360	420
0400	Low E glass		9	.889		345	45		390	455
0500	48" x 36", insulated glass		8	1		470	50.50		520.50	600
0600	Low E glass		8	1		495	50.50		545.50	630
4000	Impact windows, minimum, add					60%				
4010	Impact windows, maximum, add					160%				
9000	Minimum labor/equipment charge	1 Carp	4	2	Job		101		101	165

08 52 10.30 Wood Windows	Crew	Daily Output	Labor-Hours	Unit	Material	2018 Bare Costs Labor	Equipment	Total	Total Incl O&P
0010 **WOOD WINDOWS**, double-hung									
0020 Including frame, double insulated glass, screens and grilles									
0040 Double-hung, 2'-2" x 3'-4" high	2 Carp	15	1.067	Ea.	215	54		269	325
0060 2'-2" x 4'-4"		14	1.143		234	58		292	350
0080 2'-6" x 3'-4"		13	1.231		225	62.50		287.50	350
0100 2'-6" x 4'-0"		12	1.333		233	67.50		300.50	365
0120 2'-6" x 4'-8"		12	1.333		253	67.50		320.50	390
0140 2'-10" x 3'-4"		10	1.600		227	81		308	380
0160 2'-10" x 4'-0"		10	1.600		252	81		333	410
0180 3'-7" x 3'-4"		9	1.778		259	90		349	430
0200 3'-7" x 5'-4"		9	1.778		294	90		384	470
0220 3'-10" x 5'-4"	↓	8	2		520	101		621	735
3800 Triple glazing for above, add				↓	25%				

08 52 10.40 Casement Window

	Crew	Daily Output	Labor-Hours	Unit	Material	2018 Bare Costs Labor	Equipment	Total	Total Incl O&P
0010 **CASEMENT WINDOW**, including frame, screen and grilles									
0100 2'-0" x 3'-0" H, double insulated glass G	1 Carp	10	.800	Ea.	282	40.50		322.50	375
0150 Low E glass G		10	.800		275	40.50		315.50	365
0200 2'-0" x 4'-6" high, double insulated glass G		9	.889		385	45		430	500
0250 Low E glass G		9	.889		385	45		430	500
0260 Casement 4'-2" x 4'-2" double insulated glass G		11	.727		920	37		957	1,050
0270 4'-0" x 4'-0" Low E glass G		11	.727		550	37		587	660
0290 6'-4" x 5'-7" Low E glass G		9	.889		1,175	45		1,220	1,375
0300 2'-4" x 6'-0" high, double insulated glass G		8	1		445	50.50		495.50	575
0350 Low E glass G		8	1		440	50.50		490.50	570
0522 Vinyl-clad, premium, double insulated glass, 2'-0" x 3'-0" G		10	.800		285	40.50		325.50	380
0524 2'-0" x 4'-0" G		9	.889		330	45		375	440
0525 2'-0" x 5'-0" G		8	1		380	50.50		430.50	505
0528 2'-0" x 6'-0" G		8	1		400	50.50		450.50	530
0600 3'-0" x 5'-0" G		8	1		700	50.50		750.50	850
0700 4'-0" x 3'-0" G		8	1		765	50.50		815.50	925
0710 4'-0" x 4'-0" G		8	1		655	50.50		705.50	805
0720 4'-8" x 4'-0" G		8	1		720	50.50		770.50	880
0730 4'-8" x 5'-0" G		6	1.333		825	67.50		892.50	1,025
0740 4'-8" x 6'-0" G		6	1.333		930	67.50		997.50	1,125
0750 6'-0" x 4'-0" G		6	1.333		845	67.50		912.50	1,050
0800 6'-0" x 5'-0" G	↓	6	1.333		930	67.50		997.50	1,125
0900 5'-6" x 5'-6" G	2 Carp	15	1.067		1,475	54		1,529	1,725
8190 For installation, add per leaf				↓		15%			
8200 For multiple leaf units, deduct for stationary sash									
8220 2' high				Ea.	24			24	26.50
8240 4'-6" high					27			27	30
8260 6' high					36.50			36.50	40
8300 Impact windows, minimum, add					60%				
8310 Impact windows, maximum, add				↓	160%				
9000 Minimum labor/equipment charge	1 Carp	3	2.667	Job		135		135	219

08 52 10.50 Double-Hung

	Crew	Daily Output	Labor-Hours	Unit	Material	2018 Bare Costs Labor	Equipment	Total	Total Incl O&P
0010 **DOUBLE-HUNG**, Including frame, screens and grilles									
0100 2'-0" x 3'-0" high, low E insul. glass G	1 Carp	10	.800	Ea.	193	40.50		233.50	278
0200 3'-0" x 4'-0" high, double insulated glass G		9	.889		288	45		333	390
0300 4'-0" x 4'-6" high, low E insulated glass G	↓	8	1		335	50.50		385.50	455
8000 Impact windows, minimum, add					60%				
8010 Impact windows, maximum, add				↓	160%				

08 52 Wood Windows

08 52 10 – Plain Wood Windows

08 52 10.50 Double-Hung

		Crew	Daily Output	Labor-Hours	Unit	Material	2018 Bare Costs Labor	2018 Bare Costs Equipment	Total	Total Incl O&P
9000	Minimum labor/equipment charge	1 Carp	3	2.667	Job		135		135	219

08 52 10.55 Picture Window

		Crew	Daily Output	Labor-Hours	Unit	Material	Labor	Equipment	Total	Total Incl O&P
0010	**PICTURE WINDOW**, Including frame and grilles									
0100	3'-6" x 4'-0" high, double insulated glass	2 Carp	12	1.333	Ea.	435	67.50		502.50	590
0150	Low E glass		12	1.333		445	67.50		512.50	600
0200	4'-0" x 4'-6" high, double insulated glass		11	1.455		550	74		624	725
0250	Low E glass		11	1.455		530	74		604	705
0300	5'-0" x 4'-0" high, double insulated glass		11	1.455		580	74		654	760
0350	Low E glass		11	1.455		605	74		679	785
0400	6'-0" x 4'-6" high, double insulated glass		10	1.600		640	81		721	830
0450	Low E glass		10	1.600		640	81		721	835

08 52 10.65 Wood Sash

		Crew	Daily Output	Labor-Hours	Unit	Material	Labor	Equipment	Total	Total Incl O&P
0010	**WOOD SASH**, Including glazing but not trim									
0050	Custom, 5'-0" x 4'-0", 1" double glazed, 3/16" thick lites	2 Carp	3.20	5	Ea.	233	254		487	665
0100	1/4" thick lites		5	3.200		263	162		425	550
0200	1" thick, triple glazed		5	3.200		430	162		592	740
0300	7'-0" x 4'-6" high, 1" double glazed, 3/16" thick lites		4.30	3.721		430	189		619	775
0400	1/4" thick lites		4.30	3.721		490	189		679	840
0500	1" thick, triple glazed		4.30	3.721		560	189		749	920
0600	8'-6" x 5'-0" high, 1" double glazed, 3/16" thick lites		3.50	4.571		580	232		812	1,025
0700	1/4" thick lites		3.50	4.571		640	232		872	1,075
0800	1" thick, triple glazed		3.50	4.571		675	232		907	1,125
0900	Window frames only, based on perimeter length				L.F.	4.13			4.13	4.54
1200	Window sill, stock, per lineal foot					8.65			8.65	9.50
1250	Casing, stock					3.38			3.38	3.72

08 52 10.70 Sliding Windows

			Crew	Daily Output	Labor-Hours	Unit	Material	Labor	Equipment	Total	Total Incl O&P
0010	**SLIDING WINDOWS**										
0100	3'-0" x 3'-0" high, double insulated	G	1 Carp	10	.800	Ea.	293	40.50		333.50	385
0120	Low E glass	G		10	.800		320	40.50		360.50	415
0200	4'-0" x 3'-6" high, double insulated	G		9	.889		380	45		425	495
0220	Low E glass	G		9	.889		385	45		430	495
0300	6'-0" x 5'-0" high, double insulated	G		8	1		505	50.50		555.50	640
0320	Low E glass	G		8	1		550	50.50		600.50	690
9000	Minimum labor/equipment charge			3	2.667	Job		135		135	219

08 52 13 – Metal-Clad Wood Windows

08 52 13.10 Awning Windows, Metal-Clad

		Crew	Daily Output	Labor-Hours	Unit	Material	Labor	Equipment	Total	Total Incl O&P
0010	**AWNING WINDOWS, METAL-CLAD**									
2000	Metal-clad, awning deluxe, double insulated glass, 34" x 22"	1 Carp	9	.889	Ea.	255	45		300	355
2050	36" x 25"		9	.889		276	45		321	380
2100	40" x 22"		9	.889		295	45		340	400
2150	40" x 30"		9	.889		340	45		385	450
2200	48" x 28"		8	1		350	50.50		400.50	470
2250	60" x 36"		8	1		375	50.50		425.50	495

08 52 13.20 Casement Windows, Metal-Clad

			Crew	Daily Output	Labor-Hours	Unit	Material	Labor	Equipment	Total	Total Incl O&P
0010	**CASEMENT WINDOWS, METAL-CLAD**										
0100	Metal-clad, deluxe, dbl. insul. glass, 2'-0" x 3'-0" high	G	1 Carp	10	.800	Ea.	294	40.50		334.50	390
0120	2'-0" x 4'-0" high	G		9	.889		320	45		365	425
0130	2'-0" x 5'-0" high	G		8	1		340	50.50		390.50	460
0140	2'-0" x 6'-0" high	G		8	1		380	50.50		430.50	505
0300	Metal-clad, casement, bldrs mdl, 6'-0" x 4'-0", dbl. insul. glass, 3 panels		2 Carp	10	1.600		1,200	81		1,281	1,450
0310	9'-0" x 4'-0", 4 panels			8	2		1,550	101		1,651	1,875

255

08 52 Wood Windows

08 52 13 – Metal-Clad Wood Windows

08 52 13.20 Casement Windows, Metal-Clad		Crew	Daily Output	Labor-Hours	Unit	Material	2018 Bare Costs Labor	Equipment	Total	Total Incl O&P
0320	10'-0" x 5'-0", 5 panels	2 Carp	7	2.286	Ea.	2,100	116		2,216	2,525
0330	12'-0" x 6'-0", 6 panels	↓	6	2.667	↓	2,700	135		2,835	3,175

08 52 13.30 Double-Hung Windows, Metal-Clad

			Crew	Daily Output	Labor-Hours	Unit	Material	Labor	Equipment	Total	Total Incl O&P
0010	**DOUBLE-HUNG WINDOWS, METAL-CLAD**										
0100	Metal-clad, deluxe, dbl. insul. glass, 2'-6" x 3'-0" high	G	1 Carp	10	.800	Ea.	285	40.50		325.50	380
0120	3'-0" x 3'-6" high	G		10	.800		325	40.50		365.50	425
0140	3'-0" x 4'-0" high	G		9	.889		340	45		385	450
0160	3'-0" x 4'-6" high	G		9	.889		355	45		400	465
0180	3'-0" x 5'-0" high	G		8	1		385	50.50		435.50	510
0200	3'-6" x 6'-0" high	G	↓	8	1	↓	465	50.50		515.50	595

08 52 13.35 Picture and Sliding Windows Metal-Clad

			Crew	Daily Output	Labor-Hours	Unit	Material	Labor	Equipment	Total	Total Incl O&P
0010	**PICTURE AND SLIDING WINDOWS METAL-CLAD**										
2000	Metal-clad, dlx picture, dbl. insul. glass, 4'-0" x 4'-0" high		2 Carp	12	1.333	Ea.	380	67.50		447.50	530
2100	4'-0" x 6'-0" high			11	1.455		560	74		634	735
2200	5'-0" x 6'-0" high			10	1.600		620	81		701	810
2300	6'-0" x 6'-0" high		↓	10	1.600		710	81		791	910
2400	Metal-clad, dlx sliding, dbl. insul. glass, 3'-0" x 3'-0" high	G	1 Carp	10	.800		330	40.50		370.50	425
2420	4'-0" x 3'-6" high	G		9	.889		400	45		445	515
2440	5'-0" x 4'-0" high	G		9	.889		480	45		525	600
2460	6'-0" x 5'-0" high	G	↓	8	1	↓	745	50.50		795.50	905
9000	Minimum labor/equipment charge		2 Carp	2.75	5.818	Job		295		295	480

08 52 13.40 Bow and Bay Windows, Metal-Clad

			Crew	Daily Output	Labor-Hours	Unit	Material	Labor	Equipment	Total	Total Incl O&P
0010	**BOW AND BAY WINDOWS, METAL-CLAD**										
0100	Metal-clad, deluxe, dbl. insul. glass, 8'-0" x 5'-0" high, 4 panels		2 Carp	10	1.600	Ea.	1,700	81		1,781	2,000
0120	10'-0" x 5'-0" high, 5 panels			8	2		1,825	101		1,926	2,175
0140	10'-0" x 6'-0" high, 5 panels			7	2.286		2,150	116		2,266	2,575
0160	12'-0" x 6'-0" high, 6 panels			6	2.667		2,925	135		3,060	3,450
0400	Double-hung, bldrs. model, bay, 8' x 4' high, dbl. insul. glass			10	1.600		1,350	81		1,431	1,600
0440	Low E glass			10	1.600		1,450	81		1,531	1,725
0460	9'-0" x 5'-0" high, dbl. insul. glass			6	2.667		1,450	135		1,585	1,825
0480	Low E glass			6	2.667		1,525	135		1,660	1,900
0500	Metal-clad, deluxe, dbl. insul. glass, 7'-0" x 4'-0" high			10	1.600		1,300	81		1,381	1,550
0520	8'-0" x 4'-0" high			8	2		1,350	101		1,451	1,650
0540	8'-0" x 5'-0" high			7	2.286		1,375	116		1,491	1,725
0560	9'-0" x 5'-0" high		↓	6	2.667	↓	1,475	135		1,610	1,850

08 52 16 – Plastic-Clad Wood Windows

08 52 16.10 Bow Window

			Crew	Daily Output	Labor-Hours	Unit	Material	Labor	Equipment	Total	Total Incl O&P
0010	**BOW WINDOW** including frames, screens, and grilles										
0020	End panels operable										
1000	Bow type, casement, wood, bldrs. mdl., 8' x 5' dbl. insul. glass, 4 panel		2 Carp	10	1.600	Ea.	1,575	81		1,656	1,850
1050	Low E glass			10	1.600		1,325	81		1,406	1,575
1100	10'-0" x 5'-0", dbl. insul. glass, 6 panels			6	2.667		1,350	135		1,485	1,725
1200	Low E glass, 6 panels			6	2.667		1,450	135		1,585	1,825
1300	Vinyl-clad, bldrs. model, dbl. insul. glass, 6'-0" x 4'-0", 3 panel			10	1.600		1,050	81		1,131	1,275
1340	9'-0" x 4'-0", 4 panel			8	2		1,425	101		1,526	1,750
1380	10'-0" x 6'-0", 5 panels			7	2.286		2,350	116		2,466	2,775
1420	12'-0" x 6'-0", 6 panels			6	2.667		3,075	135		3,210	3,600
2000	Bay window, 8' x 5', dbl. insul. glass			10	1.600		1,925	81		2,006	2,225
2050	Low E glass			10	1.600		2,325	81		2,406	2,675
2100	12'-0" x 6'-0", dbl. insul. glass, 6 panels			6	2.667		2,400	135		2,535	2,850
2200	Low E glass		↓	6	2.667	↓	3,250	135		3,385	3,800

08 52 Wood Windows

08 52 16 – Plastic-Clad Wood Windows

08 52 16.10 Bow Window

		Crew	Daily Output	Labor-Hours	Unit	Material	2018 Bare Costs Labor	Equipment	Total	Total Incl O&P
2300	Vinyl-clad, premium, dbl. insul. glass, 8'-0" x 5'-0"	2 Carp	10	1.600	Ea.	1,800	81		1,881	2,100
2340	10'-0" x 5'-0"		8	2		2,400	101		2,501	2,825
2380	10'-0" x 6'-0"		7	2.286		2,800	116		2,916	3,275
2420	12'-0" x 6'-0"		6	2.667		3,350	135		3,485	3,925
3300	Vinyl-clad, premium, dbl. insul. glass, 7'-0" x 4'-6"		10	1.600		1,400	81		1,481	1,650
3340	8'-0" x 4'-6"		8	2		1,425	101		1,526	1,725
3380	8'-0" x 5'-0"		7	2.286		1,500	116		1,616	1,850
3420	9'-0" x 5'-0"		6	2.667		1,525	135		1,660	1,900
9000	Minimum labor/equipment charge		2.50	6.400	Job		325		325	525

08 52 16.15 Awning Window Vinyl-Clad

		Crew	Daily Output	Labor-Hours	Unit	Material	2018 Bare Costs Labor	Equipment	Total	Total Incl O&P
0010	**AWNING WINDOW VINYL-CLAD** including frames, screens, and grilles									
0240	Vinyl-clad, 34" x 22"	1 Carp	10	.800	Ea.	267	40.50		307.50	360
0280	36" x 28"		9	.889		305	45		350	415
0300	36" x 36"		9	.889		350	45		395	460
0340	40" x 22"		10	.800		295	40.50		335.50	390
0360	48" x 28"		8	1		370	50.50		420.50	495
0380	60" x 36"		8	1		520	50.50		570.50	655

08 52 16.20 Half Round, Vinyl-Clad

		Crew	Daily Output	Labor-Hours	Unit	Material	2018 Bare Costs Labor	Equipment	Total	Total Incl O&P
0010	**HALF ROUND, VINYL-CLAD**, double insulated glass, incl. grille									
0800	14" height x 24" base	2 Carp	9	1.778	Ea.	420	90		510	605
1040	15" height x 25" base		8	2		405	101		506	610
1060	16" height x 28" base		7	2.286		430	116		546	660
1080	17" height x 29" base		7	2.286		445	116		561	675
2000	19" height x 33" base	1 Carp	6	1.333		490	67.50		557.50	645
2100	20" height x 35" base		6	1.333		485	67.50		552.50	640
2200	21" height x 37" base		6	1.333		495	67.50		562.50	655
2250	23" height x 41" base	2 Carp	6	2.667		535	135		670	805
2300	26" height x 48" base		6	2.667		555	135		690	830
2350	30" height x 56" base		6	2.667		645	135		780	930
3000	36" height x 67" base	1 Carp	4	2		1,150	101		1,251	1,425
3040	38" height x 71" base	2 Carp	5	3.200		1,000	162		1,162	1,375
3050	40" height x 75" base	"	5	3.200		1,325	162		1,487	1,725
5000	Elliptical, 71" x 16"	1 Carp	11	.727		995	37		1,032	1,150
5100	95" x 21"	"	10	.800		1,425	40.50		1,465.50	1,650

08 52 16.35 Double-Hung Window

			Crew	Daily Output	Labor-Hours	Unit	Material	2018 Bare Costs Labor	Equipment	Total	Total Incl O&P
0010	**DOUBLE-HUNG WINDOW** including frames, screens, and grilles										
0300	Vinyl-clad, premium, double insulated glass, 2'-6" x 3'-0"	G	1 Carp	10	.800	Ea.	340	40.50		380.50	435
0305	2'-6" x 4'-0"	G		10	.800		375	40.50		415.50	480
0400	3'-0" x 3'-6"	G		10	.800		345	40.50		385.50	440
0500	3'-0" x 4'-0"	G		9	.889		400	45		445	515
0600	3'-0" x 4'-6"	G		9	.889		425	45		470	545
0700	3'-0" x 5'-0"	G		8	1		460	50.50		510.50	590
0790	3'-4" x 5'-0"	G		8	1		450	50.50		500.50	580
0800	3'-6" x 6'-0"	G		8	1		515	50.50		565.50	650
0820	4'-0" x 5'-0"	G		7	1.143		570	58		628	725
0830	4'-0" x 6'-0"	G		7	1.143		725	58		783	890

08 52 16.40 Transom Windows

		Crew	Daily Output	Labor-Hours	Unit	Material	2018 Bare Costs Labor	Equipment	Total	Total Incl O&P
0010	**TRANSOM WINDOWS**									
0050	Vinyl-clad, premium, dbl. insul. glass, 32" x 8"	1 Carp	16	.500	Ea.	191	25.50		216.50	251
0100	36" x 8"		16	.500		206	25.50		231.50	268
0110	36" x 12"		16	.500		221	25.50		246.50	284
2000	Custom sizes, up to 350 sq. in.		12	.667		280	34		314	365

08 52 16 – Plastic-Clad Wood Windows

08 52 16.40 Transom Windows

		Daily Output	Labor-Hours	Unit	Material	2018 Bare Costs Labor	2018 Bare Costs Equipment	Total	Total Incl O&P
						Labor	Equipment		
2100	351 to 750 sq. in.	1 Carp	12	.667	Ea.	330	34		364	415
2200	751 to 1150 sq. in.		11	.727		420	37		457	520
2300	1151 to 1450 sq. in.		11	.727		480	37		517	590
2400	1451 to 1850 sq. in.	2 Carp	12	1.333		540	67.50		607.50	705
2500	1851 to 2250 sq. in.		12	1.333		640	67.50		707.50	815
2600	2251 to 2650 sq. in.		11	1.455		700	74		774	890
2700	2651 to 3050 sq. in.		11	1.455		715	74		789	910
2800	3051 to 3450 sq. in.		11	1.455		780	74		854	980
2900	3451 to 3850 sq. in.		10	1.600		860	81		941	1,075
3000	3851 to 4250 sq. in.		10	1.600		880	81		961	1,100
3100	4251 to 4650 sq. in.		10	1.600		895	81		976	1,125
3200	4651 to 5050 sq. in.		10	1.600		960	81		1,041	1,175
3300	5051 to 5450 sq. in.		9	1.778		1,100	90		1,190	1,350
3400	5451 to 5850 sq. in.		9	1.778		1,000	90		1,090	1,250
3600	6251 to 6650 sq. in.		8	2		1,200	101		1,301	1,500
3700	6651 to 7050 sq. in.		8	2		1,275	101		1,376	1,575

08 52 16.45 Trapezoid Windows

		Daily Output	Labor-Hours	Unit	Material	Labor	Equipment	Total	Total Incl O&P
0010	**TRAPEZOID WINDOWS**								
0900	20" base x 44" leg x 53" leg	2 Carp	13	1.231	Ea.	400	62.50		462.50	540
1000	24" base x 90" leg x 102" leg		8	2		700	101		801	935
3000	36" base x 40" leg x 22" leg		12	1.333		440	67.50		507.50	595
3010	36" base x 44" leg x 25" leg		13	1.231		465	62.50		527.50	610
3050	36" base x 26" leg x 48" leg		9	1.778		470	90		560	660
3100	36" base x 42" legs, 50" peak		9	1.778		555	90		645	755
3200	36" base x 60" leg x 81" leg		11	1.455		730	74		804	920
4320	44" base x 23" leg x 56" leg		11	1.455		570	74		644	745
4350	44" base x 59" leg x 92" leg		10	1.600		880	81		961	1,100
4500	46" base x 15" leg x 46" leg		8	2		440	101		541	650
4550	46" base x 16" leg x 48" leg		8	2		465	101		566	675
4600	46" base x 50" leg x 80" leg		7	2.286		700	116		816	960
6600	66" base x 12" leg x 42" leg		8	2		600	101		701	825
6650	66" base x 12" legs, 28" peak		9	1.778		480	90		570	675
6700	68" base x 23" legs, 31" peak		8	2		600	101		701	825

08 52 16.70 Vinyl-Clad, Premium, DBL. Insul. Glass

			Daily Output	Labor-Hours	Unit	Material	Labor	Equipment	Total	Total Incl O&P
0010	**VINYL-CLAD, PREMIUM, DBL. INSUL. GLASS**									
1000	Sliding, 3'-0" x 3'-0"	G	1 Carp	10	.800	Ea.	625	40.50		665.50	750
1050	4'-0" x 3'-6"	G		9	.889		690	45		735	835
1100	5'-0" x 4'-0"	G		9	.889		920	45		965	1,075
1150	6'-0" x 5'-0"	G		8	1		1,175	50.50		1,225.50	1,350

08 52 50 – Window Accessories

08 52 50.10 Window Grille or Muntin

		Daily Output	Labor-Hours	Unit	Material	Labor	Equipment	Total	Total Incl O&P
0010	**WINDOW GRILLE OR MUNTIN**, snap in type								
0020	Standard pattern interior grilles								
2000	Wood, awning window, glass size, 28" x 16" high	1 Carp	30	.267	Ea.	30.50	13.50		44	55.50
2060	44" x 24" high		32	.250		44	12.70		56.70	69
2100	Casement, glass size, 20" x 36" high		30	.267		33.50	13.50		47	59
2180	20" x 56" high		32	.250		46.50	12.70		59.20	71.50
2200	Double-hung, glass size, 16" x 24" high		24	.333	Set	56	16.90		72.90	89
2280	32" x 32" high		34	.235	"	140	11.95		151.95	173
2500	Picture, glass size, 48" x 48" high		30	.267	Ea.	128	13.50		141.50	163
2580	60" x 68" high		28	.286	"	196	14.50		210.50	240
2600	Sliding, glass size, 14" x 36" high		24	.333	Set	37.50	16.90		54.40	69

08 52 Wood Windows

08 52 50 – Window Accessories

08 52 50.10 Window Grille or Muntin		Crew	Daily Output	Labor-Hours	Unit	Material	2018 Bare Costs Labor	Equipment	Total	Total Incl O&P
2680	36" x 36" high	1 Carp	22	.364	Set	45	18.45		63.45	79.50
9000	Minimum labor/equipment charge	↓	5	1.600	Job		81		81	132

08 52 66 – Wood Window Screens

08 52 66.10 Wood Screens

		Crew	Daily Output	Labor-Hours	Unit	Material	Labor	Equipment	Total	Total Incl O&P
0010	**WOOD SCREENS**									
0020	Over 3 S.F., 3/4" frames	2 Carp	375	.043	S.F.	4.75	2.16		6.91	8.75
0100	1-1/8" frames		375	.043		8.45	2.16		10.61	12.80
0200	Rescreen wood frame	↓	500	.032	↓	1.66	1.62		3.28	4.46
9000	Minimum labor/equipment charge	1 Carp	4	2	Job		101		101	165

08 52 69 – Wood Storm Windows

08 52 69.10 Storm Windows

			Crew	Daily Output	Labor-Hours	Unit	Material	Labor	Equipment	Total	Total Incl O&P
0010	**STORM WINDOWS**, aluminum residential										
0300	Basement, mill finish, incl. fiberglass screen										
0320	1'-10" x 1'-0" high	G	2 Carp	30	.533	Ea.	36	27		63	83.50
0340	2'-9" x 1'-6" high	G		30	.533		39	27		66	87
0360	3'-4" x 2'-0" high	G	↓	30	.533	↓	46	27		73	94.50
1600	Double-hung, combination, storm & screen										
1700	Custom, clear anodic coating, 2'-0" x 3'-5" high		2 Carp	30	.533	Ea.	99	27		126	153
1720	2'-6" x 5'-0" high			28	.571		120	29		149	179
1740	4'-0" x 6'-0" high			25	.640		235	32.50		267.50	310
1800	White painted, 2'-0" x 3'-5" high			30	.533		115	27		142	171
1820	2'-6" x 5'-0" high			28	.571		175	29		204	240
1840	4'-0" x 6'-0" high			25	.640		290	32.50		322.50	375
2000	Clear anodic coating, 2'-0" x 3'-5" high	G		30	.533		97	27		124	151
2020	2'-6" x 5'-0" high	G		28	.571		124	29		153	183
2040	4'-0" x 6'-0" high	G		25	.640		135	32.50		167.50	202
2400	White painted, 2'-0" x 3'-5" high	G		30	.533		95	27		122	149
2420	2'-6" x 5'-0" high	G		28	.571		100	29		179	157
2440	4'-0" x 6'-0" high	G		25	.640		115	32.50		147.50	180
2600	Mill finish, 2'-0" x 3'-5" high	G		30	.533		90	27		117	143
2620	2'-6" x 5'-0" high	G		28	.571		95	29		124	152
2640	4'-0" x 6'-8" high	G	↓	25	.640	↓	115	32.50		147.50	180
4000	Picture window, storm, 1 lite, white or bronze finish										
4020	4'-6" x 4'-6" high		2 Carp	25	.640	Ea.	140	32.50		172.50	207
4040	5'-8" x 4'-6" high			20	.800		155	40.50		195.50	237
4400	Mill finish, 4'-6" x 4'-6" high			25	.640		140	32.50		172.50	207
4420	5'-8" x 4'-6" high		↓	20	.800	↓	160	40.50		200.50	242
4600	3 lite, white or bronze finish										
4620	4'-6" x 4'-6" high		2 Carp	25	.640	Ea.	165	32.50		197.50	235
4640	5'-8" x 4'-6" high			20	.800		175	40.50		215.50	259
4800	Mill finish, 4'-6" x 4'-6" high			25	.640		160	32.50		192.50	229
4820	5'-8" x 4'-6" high		↓	20	.800	↓	165	40.50		205.50	248
6000	Sliding window, storm, 2 lite, white or bronze finish										
6020	3'-4" x 2'-7" high		2 Carp	28	.571	Ea.	143	29		172	204
6040	4'-4" x 3'-3" high			25	.640		160	32.50		192.50	229
6060	5'-4" x 6'-0" high		↓	20	.800	↓	216	40.50		256.50	305
8000	PVC framed										
8100	Double-hung, combination, storm & screen										
8120	2'-6" x 3'-5"		2 Carp	15	1.067	Ea.	135	54		189	237
8140	4' x 6'		"	12.50	1.280	"	171	65		236	293
8600	Single lite picture storm										
8620	4'-6" x 4'-6"		2 Carp	12.50	1.280	Ea.	171	65		236	293

259

08 52 Wood Windows

08 52 69 – Wood Storm Windows

08 52 69.10 Storm Windows		Crew	Daily Output	Labor-Hours	Unit	Material	2018 Bare Costs Labor	2018 Bare Costs Equipment	Total	Total Incl O&P
8640	5'-8" x 4'-6"	2 Carp	10	1.600	Ea.	187	81		268	340
9000	Interior storm window									
9100	Storm window interior glass	1 Glaz	107	.075	S.F.	6.20	3.63		9.83	12.70
9410	Minimum labor/equipment charge	1 Carp	4	2	Job		101		101	165

08 53 Plastic Windows

08 53 13 – Vinyl Windows

08 53 13.20 Vinyl Single-Hung Windows

			Crew	Daily Output	Labor-Hours	Unit	Material	2018 Bare Costs Labor	2018 Bare Costs Equipment	Total	Total Incl O&P
0010	**VINYL SINGLE-HUNG WINDOWS**, insulated glass										
0020	Grids, low E, J fin, extension jambs										
0130	25" x 41"	G	2 Carp	20	.800	Ea.	210	40.50		250.50	297
0140	25" x 49"	G		18	.889		220	45		265	315
0150	25" x 57"	G		17	.941		230	47.50		277.50	330
0160	25" x 65"	G		16	1		250	50.50		300.50	360
0170	29" x 41"	G		18	.889		210	45		255	305
0180	29" x 53"	G		18	.889		230	45		275	325
0190	29" x 57"	G		17	.941		240	47.50		287.50	340
0200	29" x 65"	G		16	1		250	50.50		300.50	360
0210	33" x 41"	G		20	.800		225	40.50		265.50	315
0220	33" x 53"	G		18	.889		245	45		290	345
0230	33" x 57"	G		17	.941		250	47.50		297.50	355
0240	33" x 65"	G		16	1		260	50.50		310.50	370
0250	37" x 41"	G		20	.800		250	40.50		290.50	340
0260	37" x 53"	G		18	.889		275	45		320	380
0270	37" x 57"	G		17	.941		285	47.50		332.50	395
0280	37" x 65"	G		16	1		300	50.50		350.50	415

08 53 13.30 Vinyl Double-Hung Windows

			Crew	Daily Output	Labor-Hours	Unit	Material	2018 Bare Costs Labor	2018 Bare Costs Equipment	Total	Total Incl O&P
0010	**VINYL DOUBLE-HUNG WINDOWS**, insulated glass										
0100	Grids, low E, J fin, ext. jambs, 21" x 53"	G	2 Carp	18	.889	Ea.	220	45		265	315
0102	21" x 37"	G		18	.889		230	45		275	325
0104	21" x 41"	G		18	.889		240	45		285	335
0106	21" x 49"	G		18	.889		250	45		295	350
0110	21" x 57"	G		17	.941		270	47.50		317.50	375
0120	21" x 65"	G		16	1		285	50.50		335.50	400
0128	25" x 37"	G		20	.800		240	40.50		280.50	330
0130	25" x 41"	G		20	.800		250	40.50		290.50	340
0140	25" x 49"	G		18	.889		260	45		305	360
0145	25" x 53"	G		18	.889		275	45		320	380
0150	25" x 57"	G		17	.941		280	47.50		327.50	390
0160	25" x 65"	G		16	1		295	50.50		345.50	410
0162	25" x 69"	G		16	1		300	50.50		350.50	415
0164	25" x 77"	G		16	1		325	50.50		375.50	445
0168	29" x 37"	G		18	.889		250	45		295	350
0170	29" x 41"	G		18	.889		260	45		305	360
0172	29" x 49"	G		18	.889		270	45		315	370
0180	29" x 53"	G		18	.889		280	45		325	385
0190	29" x 57"	G		17	.941		290	47.50		337.50	400
0200	29" x 65"	G		16	1		310	50.50		360.50	425
0202	29" x 69"	G		16	1		315	50.50		365.50	430
0205	29" x 77"	G		16	1		340	50.50		390.50	460
0208	33" x 37"	G		20	.800		265	40.50		305.50	360

08 53 Plastic Windows

08 53 13 – Vinyl Windows

08 53 13.30 Vinyl Double-Hung Windows

			Crew	Daily Output	Labor-Hours	Unit	Material	2018 Bare Costs Labor	Equipment	Total	Total Incl O&P
0210	33" x 41"	G	2 Carp	20	.800	Ea.	270	40.50		310.50	365
0215	33" x 49"	G		20	.800		285	40.50		325.50	380
0220	33" x 53"	G		18	.889		290	45		335	395
0230	33" x 57"	G		17	.941		300	47.50		347.50	410
0240	33" x 65"	G		16	1		320	50.50		370.50	435
0242	33" x 69"	G		16	1		330	50.50		380.50	450
0246	33" x 77"	G		16	1		350	50.50		400.50	470
0250	37" x 41"	G		20	.800		295	40.50		335.50	390
0255	37" x 49"	G		20	.800		297	40.50		337.50	390
0260	37" x 53"	G		18	.889		320	45		365	425
0270	37" x 57"	G		17	.941		340	47.50		387.50	455
0280	37" x 65"	G		16	1		360	50.50		410.50	480
0282	37" x 69"	G		16	1		365	50.50		415.50	485
0286	37" x 77"	G		16	1		380	50.50		430.50	505
0300	Solid vinyl, average quality, double insulated glass, 2'-0" x 3'-0"	G	1 Carp	10	.800		291	40.50		331.50	385
0310	3'-0" x 4'-0"	G		9	.889		214	45		259	310
0330	Premium, double insulated glass, 2'-6" x 3'-0"	G		10	.800		276	40.50		316.50	370
0340	3'-0" x 3'-6"	G		9	.889		310	45		355	415
0350	3'-0" x 4'-0"	G		9	.889		330	45		375	440
0360	3'-0" x 4'-6"	G		9	.889		335	45		380	445
0370	3'-0" x 5'-0"	G		8	1		360	50.50		410.50	480
0380	3'-6" x 6'-0"	G		8	1		395	50.50		445.50	520

08 53 13.40 Vinyl Casement Windows

			Crew	Daily Output	Labor-Hours	Unit	Material	2018 Bare Costs Labor	Equipment	Total	Total Incl O&P
0010	**VINYL CASEMENT WINDOWS**, insulated glass										
0015	Grids, low E, J fin, extension jambs, screens										
0100	One lite, 21" x 41"	G	2 Carp	20	.800	Ea.	315	40.50		355.50	410
0110	21" x 47"	G		20	.800		330	40.50		370.50	430
0120	21" x 53"	G		20	.800		365	40.50		405.50	465
0128	24" x 35"	G		19	.842		297	42.50		339.50	395
0130	24" x 41"	G		19	.842		320	42.50		362.50	420
0140	24" x 47"	G		19	.842		355	42.50		397.50	460
0150	24" x 53"	G		19	.842		380	42.50		422.50	490
0158	28" x 35"	G		19	.842		310	42.50		352.50	410
0160	28" x 41"	G		19	.842		340	42.50		382.50	445
0170	28" x 47"	G		19	.842		375	42.50		417.50	485
.0180	28" x 53"	G		19	.842		405	42.50		447.50	515
0184	28" x 59"	G		19	.842		415	42.50		457.50	530
0188	Two lites, 33" x 35"	G		18	.889		495	45		540	620
0190	33" x 41"	G		18	.889		530	45		575	660
0200	33" x 47"	G		18	.889		565	45		610	695
0210	33" x 53"	G		18	.889		580	45		625	715
0212	33" x 59"	G		18	.889		635	45		680	775
0215	33" x 72"	G		18	.889		665	45		710	810
0220	41" x 41"	G		18	.889		550	45		595	680
0230	41" x 47"	G		18	.889		610	45		655	745
0240	41" x 53"	G		17	.941		650	47.50		697.50	795
0242	41" x 59"	G		17	.941		690	47.50		737.50	835
0246	41" x 72"	G		17	.941		720	47.50		767.50	870
0250	47" x 41"	G		17	.941		585	47.50		632.50	720
0260	47" x 47"	G		17	.941		625	47.50		672.50	770
0270	47" x 53"	G		17	.941		665	47.50		712.50	815
0272	47" x 59"	G		17	.941		725	47.50		772.50	875

08 53 Plastic Windows

08 53 13 – Vinyl Windows

08 53 13.40 Vinyl Casement Windows

			Crew	Daily Output	Labor-Hours	Unit	Material	2018 Bare Costs Labor	Equipment	Total	Total Incl O&P
0280	56" x 41"	G	2 Carp	15	1.067	Ea.	625	54		679	780
0290	56" x 47"	G		15	1.067		665	54		719	825
0300	56" x 53"	G		15	1.067		725	54		779	885
0302	56" x 59"	G		15	1.067		750	54		804	915
0310	56" x 72"	G		15	1.067		810	54		864	980
0340	Solid vinyl, premium, double insulated glass, 2'-0" x 3'-0" high	G	1 Carp	10	.800		270	40.50		310.50	365
0360	2'-0" x 4'-0" high	G		9	.889		299	45		344	405
0380	2'-0" x 5'-0" high	G		8	1		335	50.50		385.50	455

08 53 13.50 Vinyl Picture Windows

		Crew	Daily Output	Labor-Hours	Unit	Material	Labor	Equipment	Total	Total Incl O&P
0010	**VINYL PICTURE WINDOWS**, insulated glass									
0120	Grids, low E, J fin, ext. jambs, 47" x 35"	2 Carp	12	1.333	Ea.	305	67.50		372.50	445
0130	47" x 41"		12	1.333		400	67.50		467.50	550
0140	47" x 47"		12	1.333		350	67.50		417.50	495
0150	47" x 53"		11	1.455		375	74		449	535
0160	71" x 35"		11	1.455		400	74		474	560
0170	71" x 41"		11	1.455		420	74		494	580
0180	71" x 47"		11	1.455		450	74		524	615

08 53 13.60 Vinyl Half Round Windows

		Crew	Daily Output	Labor-Hours	Unit	Material	Labor	Equipment	Total	Total Incl O&P
0010	**VINYL HALF ROUND WINDOWS**, Including grille, J fin, low E, ext. jambs									
0100	10" height x 20" base	2 Carp	8	2	Ea.	475	101		576	690
0110	15" height x 30" base		8	2		480	101		581	695
0120	17" height x 34" base		7	2.286		395	116		511	625
0130	19" height x 38" base		7	2.286		440	116		556	675
0140	19" height x 33" base		7	2.286		575	116		691	825
0150	24" height x 48" base	1 Carp	6	1.333		500	67.50		567.50	660
0160	25" height x 50" base	"	6	1.333		775	67.50		842.50	965
0170	30" height x 60" base	2 Carp	6	2.667		795	135		930	1,100

08 54 Composite Windows

08 54 13 – Fiberglass Windows

08 54 13.10 Fiberglass Single-Hung Windows

			Crew	Daily Output	Labor-Hours	Unit	Material	Labor	Equipment	Total	Total Incl O&P
0010	**FIBERGLASS SINGLE-HUNG WINDOWS**										
0100	Grids, low E, 18" x 24"	G	2 Carp	18	.889	Ea.	335	45		380	445
0110	18" x 40"	G		17	.941		340	47.50		387.50	455
0130	24" x 40"	G		20	.800		360	40.50		400.50	460
0230	36" x 36"	G		17	.941		370	47.50		417.50	485
0250	36" x 48"	G		20	.800		405	40.50		445.50	510
0260	36" x 60"	G		18	.889		445	45		490	565
0280	36" x 72"	G		16	1		470	50.50		520.50	600
0290	48" x 40"	G		16	1		470	50.50		520.50	600

08 54 13.30 Fiberglass Slider Windows

			Crew	Daily Output	Labor-Hours	Unit	Material	Labor	Equipment	Total	Total Incl O&P
0010	**FIBERGLASS SLIDER WINDOWS**										
0100	Grids, low E, 36" x 24"	G	2 Carp	20	.800	Ea.	340	40.50		380.50	440
0110	36" x 36"	G	"	20	.800	"	390	40.50		430.50	495

08 54 13.50 Fiberglass Bay Windows

			Crew	Daily Output	Labor-Hours	Unit	Material	Labor	Equipment	Total	Total Incl O&P
0010	**FIBERGLASS BAY WINDOWS**										
0150	48" x 36"	G	2 Carp	11	1.455	Ea.	1,100	74		1,174	1,325

08 56 Special Function Windows

08 56 46 – Radio-Frequency-Interference Shielding Windows

08 56 46.10 Radio-Frequency-Interference Mesh	Crew	Daily Output	Labor-Hours	Unit	Material	2018 Bare Costs Labor	Equipment	Total	Total Incl O&P
0010 **RADIO-FREQUENCY-INTERFERENCE MESH**									
0100 16 mesh copper 0.011" wire				S.F.	4.94			4.94	5.45
0150 22 mesh copper 0.015" wire					6.10			6.10	6.70
0200 100 mesh copper 0.022" wire					8.75			8.75	9.60
0250 100 mesh stainless steel 0.0012" wire					14.80			14.80	16.30

08 61 Roof Windows

08 61 13 – Metal Roof Windows

08 61 13.10 Roof Windows

08 61 13.10 Roof Windows	Crew	Daily Output	Labor-Hours	Unit	Material	2018 Bare Costs Labor	Equipment	Total	Total Incl O&P
0010 **ROOF WINDOWS**, fixed high perf tmpd glazing, metallic framed									
0020 46" x 21-1/2", Flashed for shingled roof	1 Carp	8	1	Ea.	281	50.50		331.50	395
0100 46" x 28"		8	1		305	50.50		355.50	420
0125 57" x 44"		6	1.333		375	67.50		442.50	525
0130 72" x 28"		7	1.143		375	58		433	510
0150 Fixed, laminated tempered glazing, 46" x 21-1/2"		8	1		465	50.50		515.50	595
0175 46" x 28"		8	1		515	50.50		565.50	650
0200 57" x 44"		6	1.333		485	67.50		552.50	640
0500 Vented flashing set for shingled roof, 46" x 21-1/2"		7	1.143		465	58		523	605
0525 46" x 28"		6	1.333		515	67.50		582.50	675
0550 57" x 44"		5	1.600		645	81		726	840
0560 72" x 28"		5	1.600		645	81		726	840
0575 Flashing set for low pitched roof, 46" x 21-1/2"		7	1.143		535	58		593	685
0600 46" x 28"		7	1.143		585	58		643	740
0625 57" x 44"		5	1.600		730	81		811	930
0650 Flashing set for curb, 46" x 21 1/2"		7	1.143		620	58		678	775
0675 46" x 28"		7	1.143		670	58		728	835
0700 57" x 44"		5	1.600		825	81		906	1,050

08 62 Unit Skylights

08 62 13 – Domed Unit Skylights

08 62 13.10 Domed Skylights

08 62 13.10 Domed Skylights		Crew	Daily Output	Labor-Hours	Unit	Material	2018 Bare Costs Labor	Equipment	Total	Total Incl O&P
0010 **DOMED SKYLIGHTS**										
0020 Skylight, fixed dome type, 22" x 22"	G	G-3	12	2.667	Ea.	216	133		349	450
0030 22" x 46"	G		10	3.200		269	159		428	550
0040 30" x 30"	G		12	2.667		286	133		419	525
0050 30" x 46"	G		10	3.200		380	159		539	675
0110 Fixed, double glazed, 22" x 27"	G		12	2.667		275	133		408	515
0120 22" x 46"	G		10	3.200		293	159		452	575
0130 44" x 46"	G		10	3.200		430	159		589	730
0210 Operable, double glazed, 22" x 27"	G		12	2.667		405	133		538	655
0220 22" x 46"	G		10	3.200		465	159		624	770
0230 44" x 46"	G		10	3.200		925	159		1,084	1,275
9000 Minimum labor/equipment charge			2	16	Job		795		795	1,275

08 62 13.20 Skylights

08 62 13.20 Skylights		Crew	Daily Output	Labor-Hours	Unit	Material	2018 Bare Costs Labor	Equipment	Total	Total Incl O&P
0010 **SKYLIGHTS**, flush or curb mounted										
2120 Ventilating insulated plexiglass dome with										
2130 curb mounting, 36" x 36"	G	G-3	12	2.667	Ea.	480	133		613	740
2150 52" x 52"	G		12	2.667		670	133		803	945
2160 28" x 52"	G		10	3.200		495	159		654	800

For customer support on your Commercial Renovation Costs with RSMeans data, call 800.448.8182.

263

08 62 Unit Skylights

08 62 13 – Domed Unit Skylights

08 62 13.20 Skylights

		Crew	Daily Output	Labor-Hours	Unit	Material	2018 Bare Costs Labor	Equipment	Total	Total Incl O&P
2170	36" x 52"	G	G-3	10	3.200	Ea.	545	159		704	850
2180	For electric opening system, add	G					315			315	350
2210	Operating skylight, with thermopane glass, 24" x 48"	G	G-3	10	3.200		595	159		754	910
2220	32" x 48"	G		9	3.556		620	177		797	970
2300	Insulated safety glass with aluminum frame	G		160	.200	S.F.	95	9.95		104.95	121
4000	Skylight, solar tube kit, incl. dome, flashing, diffuser, 1 pipe, 10" diam.	G	1 Carp	2	4	Ea.	300	203		503	660
4010	14" diam.	G		2	4		400	203		603	770
4020	21" diam.	G		2	4		490	203		693	870
4030	Accessories for, 1' long x 9" diam. pipe	G		24	.333		50	16.90		66.90	82.50
4040	2' long x 9" diam. pipe	G		24	.333		43	16.90		59.90	75
4050	4' long x 9" diam. pipe	G		20	.400		75	20.50		95.50	116
4060	1' long x 13" diam. pipe	G		24	.333		70	16.90		86.90	105
4070	2' long x 13" diam. pipe	G		24	.333		55	16.90		71.90	88
4080	4' long x 13" diam. pipe	G		20	.400		102	20.50		122.50	145
4090	6.5" turret ext. for 21" diam. pipe	G		16	.500		110	25.50		135.50	162
4100	12' long x 21" diam. flexible pipe	G		12	.667		90	34		124	154
4110	45 degree elbow, 10"	G		16	.500		155	25.50		180.50	212
4120	14"	G		16	.500		85	25.50		110.50	135
4130	Interior decorative ring, 9"	G		20	.400		45	20.50		65.50	82.50
4140	13"	G		20	.400		60	20.50		80.50	99

08 63 Metal-Framed Skylights

08 63 13 – Domed Metal-Framed Skylights

08 63 13.20 Skylight Rigid Metal-Framed

		Crew	Daily Output	Labor-Hours	Unit	Material	2018 Bare Costs Labor	Equipment	Total	Total Incl O&P
0010	**SKYLIGHT RIGID METAL-FRAMED** skylight framing is aluminum										
0050	Fixed acrylic double domes, curb mount, 25-1/2" x 25-1/2"		G-3	10	3.200	Ea.	210	159		369	485
0060	25-1/2" x 33-1/2"			10	3.200		242	159		401	520
0070	25-1/2" x 49-1/2"			6	5.333		279	266		545	730
0080	33-1/2" x 33-1/2"			8	4		305	199		504	655
0090	37-1/2" x 25-1/2"			8	4		250	199		449	595
0100	37-1/2" x 37-1/2"			8	4		245	199		444	590
0110	37-1/2" x 49-1/2"			6	5.333		420	266		686	885
0120	49-1/2" x 33-1/2"			6	5.333		375	266		641	835
0130	49-1/2" x 49-1/2"			6	5.333		470	266		736	940
1000	Fixed tempered glass, curb mount, 17-1/2" x 33-1/2"			10	3.200		170	159		329	440
1020	17-1/2" x 49-1/2"			6	5.333		190	266		456	635
1030	25-1/2" x 25-1/2"			10	3.200		170	159		329	440
1040	25-1/2" x 33-1/2"			10	3.200		200	159		359	475
1050	25-1/2" x 37-1/2"			8	4		210	199		409	550
1060	25-1/2" x 49-1/2"			6	5.333		222	266		488	670
1070	25-1/2" x 73-1/2"			6	5.333		375	266		641	835
1080	33-1/2" x 33-1/2"			8	4		235	199		434	580
2000	Manual vent tempered glass & screen, curb, 25-1/2" x 25-1/2"			10	3.200		410	159		569	705
2020	25-1/2" x 37-1/2"			10	3.200		465	159		624	765
2030	25-1/2" x 49-1/2"			8	4		505	199		704	875
2040	33-1/2" x 33-1/2"			8	4		535	199		734	910
2050	33-1/2" x 49-1/2"			6	5.333		675	266		941	1,175
2060	37-1/2" x 37-1/2"			6	5.333		615	266		881	1,100
2070	49-1/2" x 49-1/2"			6	5.333		790	266		1,056	1,300
3000	Electric vent tempered glass, curb mount, 25-1/2" x 25-1/2"			10	3.200		960	159		1,119	1,300
3020	25-1/2" x 37-1/2"			10	3.200		1,050	159		1,209	1,400

08 63 Metal-Framed Skylights

08 63 13 – Domed Metal-Framed Skylights

08 63 13.20 Skylight Rigid Metal-Framed		Crew	Daily Output	Labor-Hours	Unit	Material	2018 Bare Costs Labor	Equipment	Total	Total Incl O&P
3030	25-1/2" x 49-1/2"	G-3	8	4	Ea.	1,125	199		1,324	1,550
3040	33-1/2" x 33-1/2"		8	4		1,125	199		1,324	1,575
3050	33-1/2" x 49-1/2"		6	5.333		1,225	266		1,491	1,750
3060	37-1/2" x 37-1/2"		6	5.333		1,200	266		1,466	1,725
3070	49-1/2" x 49-1/2"		6	5.333		1,325	266		1,591	1,875

08 71 Door Hardware

08 71 13 – Automatic Door Operators

08 71 13.10 Automatic Openers Commercial

		Crew	Daily Output	Labor-Hours	Unit	Material	2018 Bare Costs Labor	Equipment	Total	Total Incl O&P
0010	**AUTOMATIC OPENERS COMMERCIAL**									
0020	Pneumatic door opener incl. motion sens, control box, tubing, compressor									
0050	For single swing door, per opening	2 Skwk	.80	20	Ea.	4,625	1,050		5,675	6,750
0100	Pair, per opening		.50	32	Opng.	7,500	1,675		9,175	11,000
1000	For single sliding door, per opening		.60	26.667		5,075	1,400		6,475	7,825
1300	Bi-parting pair		.50	32		7,600	1,675		9,275	11,100
1420	Electronic door opener incl. motion sens, 12 V control box, motor									
1450	For single swing door, per opening	2 Skwk	.80	20	Opng.	3,750	1,050		4,800	5,800
1500	Pair, per opening		.50	32		6,925	1,675		8,600	10,300
1600	For single sliding door, per opening		.60	26.667		4,575	1,400		5,975	7,275
1700	Bi-parting pair		.50	32		5,475	1,675		7,150	8,725
1750	Handicap actuator buttons, 2, incl. 12 V DC wiring, add	1 Carp	1.50	5.333	Pr.	480	270		750	965
2000	Electric panic button for door	2 Skwk	.50	32	Opng.	202	1,675		1,877	2,925

08 71 20 – Hardware

08 71 20.15 Hardware

		Crew	Daily Output	Labor-Hours	Unit	Material	2018 Bare Costs Labor	Equipment	Total	Total Incl O&P
0010	**HARDWARE**									
0020	Average hardware cost									
1000	Door hardware, apartment, Interior	1 Carp	4	2	Door	505	101		606	720
1300	Average, door hardware, motel/hotel interior, with access card		4	2		660	101		761	890
1500	Hospital bedroom, average quality		4	2		685	101		786	915
2000	High quality		3	2.667		670	135		805	960
2250	School, single exterior, incl. lever, incl. panic device		3	2.667		1,450	135		1,585	1,825
2500	Single interior, regular use, lever included		3	2.667		550	135		685	825
2600	Average, door hdwe. set, school, classroom, ANSI F88, incl. lever		3	2.667		850	135		985	1,150
2850	Stairway, single interior		3	2.667		620	135		755	905
3100	Double exterior, with panic device		2	4	Pr.	2,725	203		2,928	3,325
6020	Add for fire alarm door holder, electro-magnetic	1 Elec	4	2	Ea.	104	116		220	294

08 71 20.30 Door Closers

		Crew	Daily Output	Labor-Hours	Unit	Material	2018 Bare Costs Labor	Equipment	Total	Total Incl O&P
0010	**DOOR CLOSERS** adjustable backcheck, multiple mounting									
0015	and rack and pinion									
0020	Standard Regular Arm	1 Carp	6	1.333	Ea.	201	67.50		268.50	330
0040	Hold open arm		6	1.333		96.50	67.50		164	216
0100	Fusible link		6.50	1.231		172	62.50		234.50	290
1500	Concealed closers, normal use, head, pivot hung, interior		5.50	1.455		380	74		454	540
1510	Exterior		5.50	1.455		445	74		519	610
1520	Overhead concealed, all sizes, regular arm		5.50	1.455		210	74		284	350
1525	Concealed arm		5	1.600		335	81		416	495
1530	Concealed in door, all sizes, regular arm		5.50	1.455		350	74		424	505
1535	Concealed arm		5	1.600		274	81		355	430
1560	Floor concealed, all sizes, single acting		2.20	3.636		555	184		739	910
1565	Double acting		2.20	3.636		535	184		719	885

265

08 71 20.30 Door Closers

		Crew	Daily Output	Labor-Hours	Unit	Material	2018 Bare Costs Labor	2018 Bare Costs Equipment	Total	Total Incl O&P
1570	Interior, floor, offset pivot, single acting	1 Carp	3.50	2.286	Ea.	880	116		996	1,150
1590	Exterior		3.50	2.286		965	116		1,081	1,275
1610	Hold open arm		6	1.333		450	67.50		517.50	605
1620	Double acting, standard arm		6	1.333		830	67.50		897.50	1,025
1630	Hold open arm		6	1.333		840	67.50		907.50	1,025
1640	Floor, center hung, single acting, bottom arm		6	1.333		440	67.50		507.50	595
1650	Double acting		6	1.333		500	67.50		567.50	665
1660	Offset hung, single acting, bottom arm		6	1.333		600	67.50		667.50	770
2000	Backcheck and adjustable power, hinge face mount									
5000	For cast aluminum cylinder, deduct				Ea.	37			37	40.50
5040	For delayed action, add					52			52	57.50
5080	For fusible link arm, add					43.50			43.50	48
5120	For shock absorbing arm, add					51.50			51.50	57
5160	For spring power adjustment, add					41.50			41.50	45.50
6000	Closer-holder, hinge face mount, all sizes, exposed arm	1 Carp	6.50	1.231		219	62.50		281.50	340
6500	Electro magnetic closer/holder									
6510	Single point, no detector	1 Carp	4	2	Ea.	550	101		651	770
6515	Including detector		4	2		705	101		806	940
6520	Multi-point, no detector		4	2		965	101		1,066	1,225
6524	Including detector		4	2		1,450	101		1,551	1,750
6550	Electric automatic operators									
6555	Operator	1 Carp	4	2	Ea.	2,050	101		2,151	2,425
6570	Wall plate actuator		4	2		228	101		329	415
7000	Electronic closer-holder, hinge facemount, concealed arm		5	1.600		445	81		526	620
7400	With built-in detector		5	1.600		615	81		696	805
9000	Minimum labor/equipment charge		4	2	Job		101		101	165

08 71 20.36 Panic Devices

		Crew	Daily Output	Labor-Hours	Unit	Material	2018 Bare Costs Labor	2018 Bare Costs Equipment	Total	Total Incl O&P
0010	**PANIC DEVICES**	R087110-10								
0015	Touch bars various styles									
0040	Single door exit only, rim device, wide stile									
0050	Economy US28	1 Carp	5	1.600	Ea.	279	81		360	435
0060	Standard duty US28		5	1.600		715	81		796	920
0065	US26D		5	1.600		805	81		886	1,025
0070	US10		5	1.600		775	81		856	985
0075	US3		5	1.600		885	81		966	1,100
0080	Night latch, economy US28		5	1.600		380	81		461	550
0085	Standard duty, night latch US28		5	1.600		815	81		896	1,025
0090	US26D		5	1.600		1,125	81		1,206	1,375
0095	US10		5	1.600		1,075	81		1,156	1,325
0100	US3		5	1.600		1,225	81		1,306	1,475
0500	Single door exit only, rim device, narrow stile									
0540	Economy US28	1 Carp	5	1.600	Ea.	465	81		546	645
0550	Standard duty US28		5	1.600		875	81		956	1,100
0565	US26D		5	1.600		780	81		861	985
0570	US10		5	1.600		795	81		876	1,000
0575	US3		5	1.600		1,000	81		1,081	1,225
0580	Economy with night latch US28		5	1.600		335	81		416	500
0585	Standard duty US28		5	1.600		940	81		1,021	1,150
0590	US26D		5	1.600		1,125	81		1,206	1,375
0595	US10		5	1.600		945	81		1,026	1,175
0600	US32D		5	1.600		1,250	81		1,331	1,500
1040	Single door exit only, surface vertical rod									

266

For customer support on your Commercial Renovation Costs with RSMeans data, call 800.448.8182.

08 71 20 – Hardware

08 71 20.36 Panic Devices	Crew	Daily Output	Labor-Hours	Unit	Material	2018 Bare Costs Labor	Equipment	Total	Total Incl O&P
1050	Surface vertical rod, economy US28	1 Carp	4	2	Ea.	490	101		591	705
1060	Surface vertical rod, standard duty US28		4	2		1,125	101		1,226	1,425
1065	US26D		4	2		1,250	101		1,351	1,525
1070	US10		4	2		1,150	101		1,251	1,450
1075	US3		4	2		1,200	101		1,301	1,500
1080	US32D		4	2		1,250	101		1,351	1,575
1500	Single door exit, rim, narrow stile, economy, surface vertical rod									
1540	Surface vertical rod, narrow, economy US28	1 Carp	4	2	Ea.	500	101		601	715
2040	Single door exit only, concealed vertical rod									
2042	These devices will require separate and extra door preparation									
2060	Concealed rod, standard duty US28	1 Carp	4	2	Ea.	1,050	101		1,151	1,325
2065	US26D		4	2		1,175	101		1,276	1,450
2070	US10		4	2		1,275	101		1,376	1,575
2560	Concealed rod, narrow, standard duty US28		4	2		1,100	101		1,201	1,400
2565	US26D		4	2		1,325	101		1,426	1,625
2567	US26		4	2		1,400	101		1,501	1,725
2570	US10		4	2		1,150	101		1,251	1,450
2575	US3		4	2		1,300	101		1,401	1,625
3040	Single door exit only, mortise device, wide stile									
3042	These devices must be combined with a mortise lock									
3060	Mortise, standard duty US28	1 Carp	4	2	Ea.	1,125	101		1,226	1,400
3065	US26D		4	2		1,075	101		1,176	1,375
3067	US26		4	2		1,250	101		1,351	1,550
3070	US10		4	2		1,150	101		1,251	1,450
3075	US3		4	2		1,425	101		1,526	1,750
3080	US32D		4	2		1,125	101		1,226	1,400
3500	Single door exit only, mortise device, narrow stile									
3510	These devices must be combined with a mortise lock									
3540	Mortise, economy US28	1 Carp	4	2	Ea.	510	101		611	725

08 71 20.40 Lockset	Crew	Daily Output	Labor-Hours	Unit	Material	2018 Bare Costs Labor	Equipment	Total	Total Incl O&P
0010	**LOCKSET**, Standard duty									
0020	Non-keyed, passage, w/sect. trim	1 Carp	12	.667	Ea.	78	34		112	141
0100	Privacy		12	.667		80	34		114	143
0400	Keyed, single cylinder function		10	.800		151	40.50		191.50	233
0420	Hotel (see also Section 08 71 20.15)		8	1		210	50.50		260.50	315
0500	Lever handled, keyed, single cylinder function		10	.800		127	40.50		167.50	206
1000	Heavy duty with sectional trim, non-keyed, passages		12	.667		127	34		161	194
1100	Privacy		12	.667		157	34		191	228
1400	Keyed, single cylinder function		10	.800		188	40.50		228.50	272
1420	Hotel		8	1		525	50.50		575.50	665
1600	Communicating		10	.800		310	40.50		350.50	405
1690	For re-core cylinder, add					52			52	57
3800	Cipher lockset w/key pad (security item)	1 Carp	13	.615		1,000	31		1,031	1,150
3900	Cipher lockset with dial for swinging doors (security item)		13	.615		1,850	31		1,881	2,075
3920	with dial for swinging doors & drill resistant plate (security item)		12	.667		2,250	34		2,284	2,525
3950	Cipher lockset with dial for safe/vault door (security item)		12	.667		1,650	34		1,684	1,850
3980	Keyless, pushbutton type									
4000	Residential/light commercial, deadbolt, standard	1 Carp	9	.889	Ea.	144	45		189	232
4010	Heavy duty		9	.889		241	45		286	340
4020	Industrial, heavy duty, with deadbolt		9	.889		425	45		470	540
4030	Key override		9	.889		425	45		470	545
4040	Lever activated handle		9	.889		435	45		480	550

267

08 71 Door Hardware

08 71 20 – Hardware

08 71 20.40 Lockset		Crew	Daily Output	Labor-Hours	Unit	Material	2018 Bare Costs Labor	2018 Bare Costs Equipment	Total	Total Incl O&P
4050	Key override	1 Carp	9	.889	Ea.	445	45		490	565
4060	Double sided pushbutton type		8	1		800	50.50		850.50	965
4070	Key override		8	1		820	50.50		870.50	990
9000	Minimum labor/equipment charge		6	1.333	Job		67.50		67.50	110

08 71 20.41 Dead Locks

		Crew	Daily Output	Labor-Hours	Unit	Material	Labor	Equipment	Total	Total Incl O&P
0010	**DEAD LOCKS**									
0011	Mortise heavy duty outside key (security item)	1 Carp	9	.889	Ea.	182	45		227	273
0020	Double cylinder		9	.889		182	45		227	273
0100	Medium duty, outside key		10	.800		112	40.50		152.50	189
0110	Double cylinder		10	.800		121	40.50		161.50	199
1000	Tubular, standard duty, outside key		10	.800		50.50	40.50		91	122
1010	Double cylinder		10	.800		65.50	40.50		106	139
1200	Night latch, outside key		10	.800		50	40.50		90.50	121
1203	Deadlock night latch		7.70	1.039		50	52.50		102.50	141
1420	Deadbolt lock, single cylinder		10	.800		46.50	40.50		87	117
1440	Double cylinder		10	.800		64	40.50		104.50	136

08 71 20.42 Mortise Locksets

		Crew	Daily Output	Labor-Hours	Unit	Material	Labor	Equipment	Total	Total Incl O&P
0010	**MORTISE LOCKSETS**, Comm., wrought knobs & full escutcheon trim									
0015	Assumes mortise is cut									
0020	Non-keyed, passage, Grade 3	1 Carp	9	.889	Ea.	157	45		202	246
0030	Grade 1		8	1		425	50.50		475.50	550
0040	Privacy set, Grade 3		9	.889		172	45		217	263
0050	Grade 1		8	1		465	50.50		515.50	595
0100	Keyed, office/entrance/apartment, Grade 2		8	1		201	50.50		251.50	305
0110	Grade 1		7	1.143		530	58		588	675
0120	Single cylinder, typical, Grade 3		8	1		193	50.50		243.50	295
0130	Grade 1		7	1.143		525	58		583	675
0200	Hotel, room, Grade 3		7	1.143		192	58		250	305
0210	Grade 1 (see also Section 08 71 20.15)		6	1.333		530	67.50		597.50	690
0300	Double cylinder, Grade 3		8	1		229	50.50		279.50	335
0310	Grade 1		7	1.143		545	58		603	695
1000	Wrought knobs and sectional trim, non-keyed, passage, Grade 3		10	.800		133	40.50		173.50	212
1010	Grade 1		9	.889		425	45		470	545
1040	Privacy, Grade 3		10	.800		153	40.50		193.50	235
1050	Grade 1		9	.889		480	45		525	605
1100	Keyed, entrance, office/apartment, Grade 3		9	.889		225	45		270	320
1103	Install lockset		6.92	1.156		225	58.50		283.50	340
1110	Grade 1		8	1		550	50.50		600.50	685
1120	Single cylinder, Grade 3		9	.889		230	45		275	325
1130	Grade 1		8	1		510	50.50		560.50	645
2000	Cast knobs and full escutcheon trim									
2010	Non-keyed, passage, Grade 3	1 Carp	9	.889	Ea.	280	45		325	385
2020	Grade 1		8	1		385	50.50		435.50	510
2040	Privacy, Grade 3		9	.889		325	45		370	435
2050	Grade 1		8	1		450	50.50		500.50	580
2120	Keyed, single cylinder, Grade 3		8	1		335 *	50.50		385.50	455
2123	Mortise lock		6.15	1.301		335	66		401	475
2130	Grade 1		7	1.143		535	58		593	685
3000	Cast knob and sectional trim, non-keyed, passage, Grade 3		10	.800		215	40.50		255.50	300
3010	Grade 1		10	.800		395	40.50		435.50	495
3040	Privacy, Grade 3		10	.800		229	40.50		269.50	320
3050	Grade 1		10	.800		450	40.50		490.50	560

08 71 Door Hardware

08 71 20 – Hardware

		Crew	Daily Output	Labor-Hours	Unit	Material	2018 Bare Costs Labor	Equipment	Total	Total Incl O&P
08 71 20.42 Mortise Locksets										
3100	Keyed, office/entrance/apartment, Grade 3	1 Carp	9	.889	Ea.	260	45		305	360
3110	Grade 1		9	.889		590	45		635	720
3120	Single cylinder, Grade 3		9	.889		265	45		310	365
3130	Grade 1		9	.889		535	45		580	665
3190	For re-core cylinder, add					86			86	94.50
08 71 20.44 Anti-Ligature Locksets Grade 1										
0010	**ANTI-LIGATURE LOCKSETS GRADE 1**									
0100	Anti-ligature cylindrical locksets									
0500	Arch handle passage set, US32D	1 Carp	8	1	Ea.	500	50.50		550.50	635
0510	privacy set		8	1		600	50.50		650.50	745
0520	classroom set		8	1		615	50.50		665.50	765
0530	storeroom set		8	1		615	50.50		665.50	765
0550	Lever handle passage set US32D		8	1		475	50.50		525.50	610
0560	Hospital/privacy set		8	1		495	50.50		545.50	630
0570	Office set		8	1		445	50.50		495.50	575
0580	Classroom set		8	1		500	50.50		550.50	635
0590	Storeroom set		8	1		445	50.50		495.50	575
0600	Exit set		8	1		485	50.50		535.50	620
0610	Entry set		8	1		490	50.50		540.50	625
0620	Asylum set		8	1		460	50.50		510.50	590
0630	Back to back dummy set		8	1		295	50.50		345.50	410
0690	Anti-ligature mortise locksets									
0700	Knob handle privacy set, US32D	1 Carp	8	1	Ea.	620	50.50		670.50	770
0710	Office set		8	1		725	50.50		775.50	885
0720	Classroom set		8	1		725	50.50		775.50	885
0730	Hotel set		8	1		740	50.50		790.50	900
0740	Apartment set		8	1		740	50.50		790.50	900
0750	Institutional privacy set		8	1		785	50.50		835.50	950
0760	Lever handle office set US32D		8	1		600	50.50		650.50	745
0770	Classroom set sectional trim		8	1		600	50.50		650.50	745
0780	Hotel set		8	1		615	50.50		665.50	765
0790	Apartment set		8	1		615	50.50		665.50	765
0800	Institutional privacy set		8	1		665	50.50		715.50	820
0810	Office set		8	1		765	50.50		815.50	930
0820	Classroom set escutcheon trim		8	1		765	50.50		815.50	930
0830	Hotel set escutcheon trim		8	1		785	50.50		835.50	945
0840	Apartment set escutcheon trim		8	1		785	50.50		835.50	945
0850	Institutional privacy set escutcheon trim		8	1		785	50.50		835.50	945
08 71 20.45 Peepholes										
0010	**PEEPHOLES**									
2010	Peephole	1 Carp	32	.250	Ea.	8.95	12.70		21.65	30.50
2020	Peephole, wide view	"	32	.250	"	12.55	12.70		25.25	34.50
08 71 20.50 Door Stops										
0010	**DOOR STOPS**									
0020	Holder & bumper, floor or wall	1 Carp	32	.250	Ea.	36	12.70		48.70	60
1300	Wall bumper, 4" diameter, with rubber pad, aluminum		32	.250		12.70	12.70		25.40	34.50
1600	Door bumper, floor type, aluminum		32	.250		3.32	12.70		16.02	24
1620	Brass		32	.250		11.15	12.70		23.85	33
1630	Bronze		32	.250		19.20	12.70		31.90	41.50
1900	Plunger type, door mounted		32	.250		30.50	12.70		43.20	54.50
2500	Holder, floor type, aluminum		32	.250		20.50	12.70		33.20	43
2520	Wall type, aluminum		32	.250		32	12.70		44.70	55.50

08 71 Door Hardware

08 71 20 – Hardware

08 71 20.50 Door Stops

		Crew	Daily Output	Labor-Hours	Unit	Material	2018 Bare Costs Labor	2018 Bare Costs Equipment	Total	Total Incl O&P
2530	Overhead type, bronze	1 Carp	32	.250	Ea.	110	12.70		122.70	142
2540	Plunger type, aluminum		32	.250		30.50	12.70		43.20	54.50
2560	Brass		32	.250		46	12.70		58.70	71
3000	Electromagnetic, wall mounted, US3		3	2.667		282	135		417	530
3020	Floor mounted, US3		3	2.667		335	135		470	590
4000	Doorstop, ceiling mounted		3	2.667		18.65	135		153.65	240
4030	Doorstop, header		3	2.667		45	135		180	269
9000	Minimum labor/equipment charge		6	1.333	Job		67.50		67.50	110

08 71 20.55 Push-Pull Plates

		Crew	Daily Output	Labor-Hours	Unit	Material	2018 Bare Costs Labor	2018 Bare Costs Equipment	Total	Total Incl O&P
0010	**PUSH-PULL PLATES**									
0090	Push plate, 0.050 thick, 3" x 12", aluminum	1 Carp	12	.667	Ea.	5.95	34		39.95	61.50
0100	4" x 16"		12	.667		14.50	34		48.50	71
0110	6" x 16"		12	.667		8	34		42	64
0120	8" x 16"		12	.667		9.40	34		43.40	65.50
0200	Push plate, 0.050 thick, 3" x 12", brass		12	.667		13.70	34		47.70	70
0210	4" x 16"		12	.667		17.10	34		51.10	74
0220	6" x 16"		12	.667		26.50	34		60.50	84
0230	8" x 16"		12	.667		34.50	34		68.50	93
0250	Push plate, 0.050 thick, 3" x 12", satin brass		12	.667		13.90	34		47.90	70.50
0260	4" x 16"		12	.667		17.30	34		51.30	74
0270	6" x 16"		12	.667		26.50	34		60.50	84
0280	8" x 16"		12	.667		34.50	34		68.50	93
0490	Push plate, 0.050 thick, 3" x 12", bronze		12	.667		16.50	34		50.50	73
0500	4" x 16"		12	.667		26	34		60	84
0510	6" x 16"		12	.667		31.50	34		65.50	89.50
0520	8" x 16"		12	.667		38	34		72	96.50
0600	Push plate, antimicrobial copper alloy finish, 3.5" x 15"		13	.615		19.75	31		50.75	72
0610	4" x 16"		13	.615		24	31		55	76.50
0620	6" x 16"		13	.615		28.50	31		59.50	81.50
0630	6" x 20"		13	.615		34.50	31		65.50	88.50
0740	Push plate, 0.050 thick, 3" x 12", stainless steel		12	.667		11.65	34		45.65	68
0750	4" x 16"		12	.667		28.50	34		62.50	86.50
0760	6" x 16"		12	.667		18.30	34		52.30	75
0780	8" x 16"		12	.667		23.50	34		57.50	80.50
0790	Push plate, 0.050 thick, 3" x 12", satin stainless steel		12	.667		6.60	34		40.60	62.50
0810	4" x 16"		12	.667		8.10	34		42.10	64
0820	6" x 16"		12	.667		11.30	34		45.30	67.50
0830	8" x 16"		12	.667		15.60	34		49.60	72
0980	Pull plate, 0.050 thick, 3" x 12", aluminum		12	.667		25.50	34		59.50	83
1000	4" x 16"		12	.667		26.50	34		60.50	84.50
1050	Pull plate, 0.050 thick, 3" x 12", brass		12	.667		39.50	34		73.50	98.50
1060	4" x 16"		12	.667		34.50	34		68.50	93
1080	Pull plate, 0.050 thick, 3" x 12", bronze		12	.667		50	34		84	110
1100	4" x 16"		12	.667		58	34		92	119
1180	Pull plate, 0.050 thick, 3" x 12", stainless steel		12	.667		45.50	34		79.50	105
1200	4" x 16"		12	.667		60.50	34		94.50	122
1250	Pull plate, 0.050 thick, 3" x 12", chrome		12	.667		45.50	34		79.50	105
1270	4" x 16"		12	.667		47.50	34		81.50	107
1500	Pull handle and push bar, aluminum		11	.727		123	37		160	195
2000	Bronze		10	.800		166	40.50		206.50	249
9800	Minimum labor/equipment charge		5	1.600	Job		81		81	132

08 71 20 – Hardware

08 71 20.60 Entrance Locks	Crew	Daily Output	Labor-Hours	Unit	Material	2018 Bare Costs Labor	Equipment	Total	Total Incl O&P
0010 **ENTRANCE LOCKS**									
0015 Cylinder, grip handle deadlocking latch	1 Carp	9	.889	Ea.	181	45		226	272
0020 Deadbolt		8	1		195	50.50		245.50	297
0100 Push and pull plate, dead bolt		8	1		234	50.50		284.50	340
0900 For handicapped lever, add					152			152	167

08 71 20.65 Thresholds

	Crew	Daily Output	Labor-Hours	Unit	Material	2018 Bare Costs Labor	Equipment	Total	Total Incl O&P
0010 **THRESHOLDS**									
0011 Threshold 3' long saddles aluminum	1 Carp	48	.167	L.F.	10.40	8.45		18.85	25
0100 Aluminum, 8" wide, 1/2" thick		12	.667	Ea.	50	34		84	110
0500 Bronze		60	.133	L.F.	45.50	6.75		52.25	61
0600 Bronze, panic threshold, 5" wide, 1/2" thick		12	.667	Ea.	162	34		196	233
0700 Rubber, 1/2" thick, 5-1/2" wide		20	.400		41	20.50		61.50	78
0800 2-3/4" wide		20	.400		49.50	20.50		70	87
1950 ADA compliant thresholds									
2000 Threshold, wood oak 3-1/2" wide x 24" long	1 Carp	12	.667	Ea.	11.90	34		45.90	68
2010 3-1/2" wide x 36" long		12	.667		16	34		50	72.50
2020 3-1/2" wide x 48" long		12	.667		21.50	34		55.50	78.50
2030 4-1/2" wide x 24" long		12	.667		13.75	34		47.75	70
2040 4-1/2" wide x 36" long		12	.667		20	34		54	77
2050 4-1/2" wide x 48" long		12	.667		26.50	34		60.50	84
2060 6-1/2" wide x 24" long		12	.667		19.10	34		53.10	76
2070 6-1/2" wide x 36" long		12	.667		28.50	34		62.50	86.50
2080 6-1/2" wide x 48" long		12	.667		38	34		72	97
2090 Threshold, wood cherry 3-1/2" wide x 24" long		12	.667		14.50	34		48.50	71
2100 3-1/2" wide x 36" long		12	.667		28.50	34		62.50	86.50
2110 3-1/2" wide x 48" long		12	.667		35.50	34		69.50	94
2120 4-1/2" wide x 24" long		12	.667		18.50	34		52.50	75.50
2130 4-1/2" wide x 36" long		12	.667		34.50	34		68.50	93
2140 4-1/2" wide x 48" long		12	.667		43.50	34		77.50	103
2150 6-1/2" wide x 24" long		12	.667		26.50	34		60.50	84
2160 6-1/2" wide x 36" long		12	.667		46.50	34		80.50	106
2170 6-1/2" wide x 48" long		12	.667		59.50	34		93.50	121
2180 Threshold, wood walnut 3-1/2" wide x 24" long		12	.667		17.50	34		51.50	74.50
2190 3-1/2" wide x 36" long		12	.667		33.50	34		67.50	92
2200 3-1/2" wide x 48" long		12	.667		42	34		76	101
2210 4-1/2" wide x 24" long		12	.667		25	34		59	82.50
2220 4-1/2" wide x 36" long		12	.667		44	34		78	104
2230 4-1/2" wide x 48" long		12	.667		56	34		90	117
2240 6-1/2" wide x 24" long		12	.667		42	34		76	101
2250 6-1/2" wide x 36" long		12	.667		69	34		103	131
2260 6-1/2" wide x 48" long		12	.667		90	34		124	154
2300 Threshold, aluminum 4" wide x 36" long		12	.667		25	34		59	82.50
2310 4" wide x 48" long		12	.667		30	34		64	88
2320 4" wide x 72" long		12	.667		39	34		73	98
2330 5" wide x 36" long		12	.667		33	34		67	91.50
2340 5" wide x 48" long		12	.667		40	34		74	99
2350 5" wide x 72" long		12	.667		67	34		101	129
2360 6" wide x 36" long		12	.667		39	34		73	98
2370 6" wide x 48" long		12	.667		48	34		82	108
2380 6" wide x 72" long		12	.667		67	34		101	129
2390 7" wide x 36" long		12	.667		49	34		83	109
2400 7" wide x 48" long		12	.667		60	34		94	121

271

For customer support on your Commercial Renovation Costs with RSMeans data, call 800.448.8182.

08 71 20.65 Thresholds

		Crew	Daily Output	Labor-Hours	Unit	Material	2018 Bare Costs Labor	2018 Bare Costs Equipment	Total	Total Incl O&P
2410	7" wide x 72" long	1 Carp	12	.667	Ea.	91	34		125	155
2500	Threshold, ramp, aluminum or rubber 24" x 24"		12	.667	↓	181	34		215	254
9000	Minimum labor/equipment charge	↓	4	2	Job		101		101	165

08 71 20.75 Door Hardware Accessories

		Crew	Daily Output	Labor-Hours	Unit	Material	2018 Bare Costs Labor	2018 Bare Costs Equipment	Total	Total Incl O&P
0010	**DOOR HARDWARE ACCESSORIES**									
0050	Door closing coordinator, 36" (for paired openings up to 56")	1 Carp	8	1	Ea.	113	50.50		163.50	208
0060	48" (for paired openings up to 84")		8	1		120	50.50		170.50	215
0070	56" (for paired openings up to 96")		8	1		129	50.50		179.50	225
1000	Knockers, brass, standard		16	.500		64	25.50		89.50	112
1100	Deluxe	↓	10	.800	↓	90.50	40.50		131	166
2000	Torsion springs for overhead doors									
2050	1-3/4" diam., 32" long, 0.243" spring	1 Carp	8	1	Ea.	49	50.50		99.50	137
2060	1-3/4" diam., 32" long, 0.250" spring		8	1		49	50.50		99.50	136
2070	2" diam., 32" long, 0.243" spring		8	1		42	50.50		92.50	129
2080	2" diam., 32" long, 0.253" spring		8	1		50	50.50		100.50	138
4500	Rubber door silencers		540	.015		.39	.75		1.14	1.65
9000	Minimum labor/equipment charge	↓	6	1.333	Job		67.50		67.50	110

08 71 20.80 Hasps

		Crew	Daily Output	Labor-Hours	Unit	Material	2018 Bare Costs Labor	2018 Bare Costs Equipment	Total	Total Incl O&P
0010	**HASPS**, steel assembly									
0015	3"	1 Carp	26	.308	Ea.	5.45	15.60		21.05	31.50
0020	4-1/2"		13	.615		6.70	31		37.70	58
0040	6"	↓	12.50	.640	↓	9.95	32.50		42.45	63.50

08 71 20.90 Hinges

		Crew	Daily Output	Labor-Hours	Unit	Material	2018 Bare Costs Labor	2018 Bare Costs Equipment	Total	Total Incl O&P
0010	**HINGES**									
0012	Full mortise, avg. freq., steel base, USP, 4-1/2" x 4-1/2"				Pr.	39.50			39.50	43.50
0100	5" x 5", USP					59			59	65
0200	6" x 6", USP					124			124	137
0400	Brass base, 4-1/2" x 4-1/2", US10					64			64	70
0500	5" x 5", US10					71.50			71.50	78.50
0600	6" x 6", US10					164			164	180
0800	Stainless steel base, 4-1/2" x 4-1/2", US32				↓	77.50			77.50	85
0900	For non removable pin, add (security item)				Ea.	5.55			5.55	6.10
0910	For floating pin, driven tips, add					3.40			3.40	3.74
0930	For hospital type tip on pin, add					14.20			14.20	15.60
0940	For steeple type tip on pin, add				↓	19.30			19.30	21
0950	Full mortise, high frequency, steel base, 3-1/2" x 3-1/2", US26D				Pr.	31			31	34
1000	4-1/2" x 4-1/2", USP					66.50			66.50	73.50
1100	5" x 5", USP					52.50			52.50	58
1200	6" x 6", USP					139			139	153
1400	Brass base, 3-1/2" x 3-1/2", US4					52.50			52.50	57.50
1430	4-1/2" x 4-1/2", US10					76			76	83.50
1500	5" x 5", US10					127			127	139
1600	6" x 6", US10					171			171	188
1800	Stainless steel base, 4-1/2" x 4-1/2", US32					104			104	115
1810	5" x 4-1/2", US32				↓	140			140	154
1930	For hospital type tip on pin, add				Ea.	14.60			14.60	16.05
1950	Full mortise, low frequency, steel base, 3-1/2" x 3-1/2", US26D				Pr.	28			28	31
2000	4-1/2" x 4-1/2", USP					23.50			23.50	25.50
2100	5" x 5", USP					49			49	54
2200	6" x 6", USP					94			94	104
2300	4-1/2" x 4-1/2", US3					17.70			17.70	19.45
2310	5" x 5", US3				↓	42.50			42.50	46.50

08 71 Door Hardware

08 71 20 – Hardware

08 71 20.90 Hinges

		Crew	Daily Output	Labor-Hours	Unit	Material	2018 Bare Costs Labor	Equipment	Total	Total Incl O&P
2400	Brass bass, 4-1/2" x 4-1/2", US10				Pr.	55			55	60.50
2500	5" x 5", US10					79			79	86.50
2800	Stainless steel base, 4-1/2" x 4-1/2", US32					76.50			76.50	84.50
8000	Install hinge	1 Carp	34	.235	▼		11.95		11.95	19.35

08 71 20.91 Special Hinges

		Crew	Daily Output	Labor-Hours	Unit	Material	2018 Bare Costs Labor	Equipment	Total	Total Incl O&P
0010	**SPECIAL HINGES**									
0015	Paumelle, high frequency									
0020	Steel base, 6" x 4-1/2", US10				Pr.	138			138	152
0100	Brass base, 5" x 4-1/2", US10				Ea.	248			248	273
0200	Paumelle, average frequency, steel base, 4-1/2" x 3-1/2", US10				Pr.	93.50			93.50	103
0400	Olive knuckle, low frequency, brass base, 6" x 4-1/2", US10				Ea.	143			143	158
1000	Electric hinge with concealed conductor, average frequency									
1010	Steel base, 4-1/2" x 4-1/2", US26D				Pr.	315			315	345
1100	Bronze base, 4-1/2" x 4-1/2", US26D				"	335			335	365
1200	Electric hinge with concealed conductor, high frequency									
1210	Steel base, 4-1/2" x 4-1/2", US26D				Pr.	252			252	277
1600	Double weight, 800 lb., steel base, removable pin, 5" x 6", USP					485			485	535
1700	Steel base-welded pin, 5" x 6", USP					170			170	187
1800	Triple weight, 2000 lb., steel base, welded pin, 5" x 6", USP					630			630	690
2000	Pivot reinf., high frequency, steel base, 7-3/4" door plate, USP					151			151	167
2200	Bronze base, 7-3/4" door plate, US10				▼	238			238	262
3000	Swing clear, full mortise, full or half surface, high frequency,									
3010	Steel base, 5" high, USP				Pr.	144			144	159
3200	Swing clear, full mortise, average frequency									
3210	Steel base, 4-1/2" high, USP				Pr.	130			130	143
4000	Wide throw, average frequency, steel base, 4-1/2" x 6", USP					90.50			90.50	99.50
4200	High frequency, steel base, 4-1/2" x 6", USP					113			113	125
4440	US26D				▼	90			90	99
4600	Spring hinge, single acting, 6" flange, steel				Ea.	53			53	58
4700	Brass					92.50			92.50	102
4900	Double acting, 6" flange, steel					84			84	92.50
4950	Brass				▼	130			130	143
5000	T-strap, galvanized, 4"				Pr.	21.50			21.50	23.50
5010	6"					29			29	32
5020	8"				▼	54.50			54.50	60
8000	Continuous hinges									
8010	Steel, piano, 2" x 72"	1 Carp	20	.400	Ea.	23	20.50		43.50	58.50
8020	Brass, piano, 1-1/16" x 30"		30	.267		9	13.50		22.50	32
8030	Acrylic, piano, 1-3/4" x 12"		40	.200		16	10.15		26.15	34
8040	Aluminum, door, standard duty, 7'		3	2.667		135	135		270	370
8050	Heavy duty, 7'		3	2.667		142	135		277	375
8060	8'		3	2.667		160	135		295	395
8070	Steel, door, heavy duty, 7'		3	2.667		200	135		335	440
8080	8'		3	2.667		250	135		385	495
8090	Stainless steel, door, heavy duty, 7'		3	2.667		260	135		395	505
8100	8'		3	2.667		280	135		415	530
9000	Continuous hinge, steel, full mortise, heavy duty, 96"		2	4		380	203		583	750
9200	Continuous geared hinge, aluminum, full mortise, standard duty, 83"		3	2.667		148	135		283	380
9250	Continuous geared hinge, aluminum, full mortise, heavy duty, 83"		3	2.667	▼	242	135		377	485

08 71 20.95 Kick Plates

		Crew	Daily Output	Labor-Hours	Unit	Material	2018 Bare Costs Labor	Equipment	Total	Total Incl O&P
0010	**KICK PLATES**									
0020	Stainless steel, .050", 16 ga., 8" x 28", US32	1 Carp	15	.533	Ea.	41	27		68	89

08 71 20.95 Kick Plates		Crew	Daily Output	Labor-Hours	Unit	Material	2018 Bare Costs Labor	Equipment	Total	Total Incl O&P
0030	8" x 30"	1 Carp	15	.533	Ea.	44.50	27		71.50	92.50
0040	8" x 34"		15	.533		50	27		77	99
0050	10" x 28"		15	.533		82	27		109	134
0060	10" x 30"		15	.533		88.50	27		115.50	142
0070	10" x 34"		15	.533		99	27		126	153
0080	Mop/Kick, 4" x 28"		15	.533		37	27		64	84.50
0090	4" x 30"		15	.533		39	27		66	87
0100	4" x 34"		15	.533		44	27		71	92.50
0110	6" x 28"		15	.533		42	27		69	90
0120	6" x 30"		15	.533		51	27		78	100
0130	6" x 34"		15	.533		57	27		84	107
0500	Bronze, .050", 8" x 28"		15	.533		72	27		99	124
0510	8" x 30"		15	.533		69	27		96	120
0520	8" x 34"		15	.533		77.50	27		104.50	130
0530	10" x 28"		15	.533		80.50	27		107.50	133
0540	10" x 30"		15	.533		81.50	27		108.50	134
0550	10" x 34"		15	.533		98	27		125	152
0560	Mop/Kick, 4" x 28"		15	.533		35	27		62	83
0570	4" x 30"		15	.533		38.50	27		65.50	86.50
0580	4" x 34"		15	.533		39.50	27		66.50	87.50
0590	6" x 28"		15	.533		49	27		76	98
0600	6" x 30"		15	.533		50.50	27		77.50	99.50
0610	6" x 34"		15	.533		60	27		87	110
1000	Acrylic, .125", 8" x 26"		15	.533		27	27		54	74
1010	8" x 36"		15	.533		37.50	27		64.50	85
1020	8" x 42"		15	.533		44	27		71	92.50
1030	10" x 26"		15	.533		34	27		61	81.50
1040	10" x 36"		15	.533		47.50	27		74.50	96
1050	10" x 42"		15	.533		55	27		82	105
1060	Mop/Kick, 4" x 26"		15	.533		16.40	27		43.40	62
1070	4" x 36"		15	.533		22.50	27		49.50	68.50
1080	4" x 42"		15	.533		26	27		53	72.50
1090	6" x 26"		15	.533		24	27		51	70.50
1100	6" x 36"		15	.533		33	27		60	80.50
1110	6" x 42"		15	.533		38.50	27		65.50	86.50
1220	Brass, .050", 8" x 26"		15	.533		69.50	27		96.50	120
1230	8" x 36"		15	.533		73.50	27		100.50	125
1240	8" x 42"		15	.533		84.50	27		111.50	137
1250	10" x 26"		15	.533		70	27		97	121
1260	10" x 36"		15	.533		89.50	27		116.50	143
1270	10" x 42"		15	.533		103	27		130	158
1320	Mop/Kick, 4" x 26"		15	.533		23	27		50	69.50
1330	4" x 36"		15	.533		32	27		59	79
1340	4" x 42"		15	.533		37	27		64	85
1350	6" x 26"		15	.533		34.50	27		61.50	82
1360	6" x 36"		15	.533		48	27		75	96.50
1370	6" x 42"		15	.533		55	27		82	105
1800	Aluminum, .050", 8" x 26"		15	.533		38	27		65	85.50
1810	8" x 36"		15	.533		36	27		63	83.50
1820	8" x 42"		15	.533		42.50	27		69.50	90.50
1830	10" x 26"		15	.533		32.50	27		59.50	80
1840	10" x 36"		15	.533		45	27		72	93.50
1850	10" x 42"		15	.533		62	27		89	113

08 71 Door Hardware

08 71 20 – Hardware

08 71 20.95 Kick Plates

		Crew	Daily Output	Labor-Hours	Unit	Material	2018 Bare Costs Labor	Equipment	Total	Total Incl O&P
1860	Mop/Kick, 4" x 26"	1 Carp	15	.533	Ea.	15.85	27		42.85	61.50
1870	4" x 36"		15	.533		21	27		48	67
1880	4" x 42"		15	.533		25.50	27		52.50	72
1890	6" x 26"		15	.533		23	27		50	69.50
1900	6" x 36"		15	.533		31.50	27		58.50	78.50
1910	6" x 42"		15	.533		36.50	27		63.50	84.50
9000	Minimum labor/equipment charge		6	1.333	Job		67.50		67.50	110

08 71 21 – Astragals

08 71 21.10 Exterior Mouldings, Astragals

		Crew	Daily Output	Labor-Hours	Unit	Material	2018 Bare Costs Labor	Equipment	Total	Total Incl O&P
0010	**EXTERIOR MOULDINGS, ASTRAGALS**									
0400	One piece, overlapping cadmium plated steel, flat, 3/16" x 2"	1 Carp	90	.089	L.F.	3.88	4.51		8.39	11.55
0600	Prime coated steel, flat, 1/8" x 3"		90	.089		5.90	4.51		10.41	13.80
0800	Stainless steel, flat, 3/32" x 1-5/8"		90	.089		15.55	4.51		20.06	24.50
1000	Aluminum, flat, 1/8" x 2"		90	.089		3.98	4.51		8.49	11.70
1200	Nail on, "T" extrusion		120	.067		2.05	3.38		5.43	7.75
1300	Vinyl bulb insert		105	.076		2.66	3.86		6.52	9.20
1600	Screw on, "T" extrusion		90	.089		3.79	4.51		8.30	11.45
1700	Vinyl insert		75	.107		4.37	5.40		9.77	13.60
2000	"L" extrusion, neoprene bulbs		75	.107		4.38	5.40		9.78	13.60
2100	Neoprene sponge insert		75	.107		6.70	5.40		12.10	16.15
2200	Magnetic		75	.107		10.30	5.40		15.70	20
2400	Spring hinged security seal, with cam		75	.107		6.60	5.40		12	16.05
2600	Spring loaded locking bolt, vinyl insert		45	.178		8.95	9		17.95	24.50
2800	Neoprene sponge strip, "Z" shaped, aluminum		60	.133		8.05	6.75		14.80	19.80
2900	Solid neoprene strip, nail on aluminum strip		90	.089		3.93	4.51		8.44	11.60
3000	One piece stile protection									
3020	Neoprene fabric loop, nail on aluminum strips	1 Carp	60	.133	L.F.	1.15	6.75		7.90	12.20
3110	Flush mounted aluminum extrusion, 1/2" x 1-1/4"		60	.133		6.70	6.75		13.45	18.30
3140	3/4" x 1-3/8"		60	.133		3.98	6.75		10.73	15.35
3160	1-1/8" x 1-3/4"		60	.133		4.56	6.75		11.31	15.95
3300	Mortise, 9/16" x 3/4"		60	.133		3.98	6.75		10.73	15.35
3320	13/16" x 1-3/8"		60	.133		4.17	6.75		10.92	15.55
3600	Spring bronze strip, nail on type		105	.076		1.90	3.86		5.76	8.35
3620	Screw on, with retainer		75	.107		2.62	5.40		8.02	11.70
3800	Flexible stainless steel housing, pile insert, 1/2" door		105	.076		7.05	3.86		10.91	14
3820	3/4" door		105	.076		7.85	3.86		11.71	14.90
4000	Extruded aluminum retainer, flush mount, pile insert		105	.076		2.18	3.86		6.04	8.65
4080	Mortise, felt insert		90	.089		4.42	4.51		8.93	12.15
4160	Mortise with spring, pile insert		90	.089		3.30	4.51		7.81	10.95
4400	Rigid vinyl retainer, mortise, pile insert		105	.076		2.62	3.86		6.48	9.15
4600	Wool pile filler strip, aluminum backing		105	.076		2.77	3.86		6.63	9.30
5000	Two piece overlapping astragal, extruded aluminum retainer									
5010	Pile insert	1 Carp	60	.133	L.F.	3.25	6.75		10	14.55
5020	Vinyl bulb insert		60	.133		1.88	6.75		8.63	13
5040	Vinyl flap insert		60	.133		3.54	6.75		10.29	14.85
5060	Solid neoprene flap insert		60	.133		6.35	6.75		13.10	17.95
5080	Hypalon rubber flap insert		60	.133		6.45	6.75		13.20	18.05
5090	Snap on cover, pile insert		60	.133		9.15	6.75		15.90	21
5400	Magnetic aluminum, surface mounted		60	.133		23.50	6.75		30.25	36.50
5500	Interlocking aluminum, 5/8" x 1" neoprene bulb insert		45	.178		5.55	9		14.55	21
5600	Adjustable aluminum, 9/16" x 21/32", pile insert		45	.178		17.05	9		26.05	33.50
5800	Magnetic, adjustable, 9/16" x 21/32"		45	.178		22	9		31	39

For customer support on your Commercial Renovation Costs with RSMeans data, call 800.448.8182.

275

08 71 Door Hardware

08 71 21 – Astragals

08 71 21.10 Exterior Mouldings, Astragals

08 71 21.10 Exterior Mouldings, Astragals	Crew	Daily Output	Labor-Hours	Unit	Material	2018 Bare Costs Labor	Equipment	Total	Total Incl O&P
6000	Two piece stile protection								
6010	Cloth backed rubber loop, 1" gap, nail on aluminum strips	1 Carp	45	.178	L.F.	4.22	9		13.22	19.30
6040	Screw on aluminum strips		45	.178		6.35	9		15.35	21.50
6100	1-1/2" gap, screw on aluminum extrusion		45	.178		5.70	9		14.70	21
6240	Vinyl fabric loop, slotted aluminum extrusion, 1" gap		45	.178		2.14	9		11.14	17
6300	1-1/4" gap		45	.178		6	9		15	21.50

08 71 25 – Weatherstripping

08 71 25.10 Mechanical Seals, Weatherstripping

08 71 25.10 Mechanical Seals, Weatherstripping	Crew	Daily Output	Labor-Hours	Unit	Material	Labor	Equipment	Total	Total Incl O&P
0010	**MECHANICAL SEALS, WEATHERSTRIPPING**									
1000	Doors, wood frame, interlocking, for 3' x 7' door, zinc	1 Carp	3	2.667	Opng.	47.50	135		182.50	272
1100	Bronze		3	2.667		60.50	135		195.50	286
1300	6' x 7' opening, zinc		2	4		58.50	203		261.50	395
1400	Bronze		2	4		69.50	203		272.50	405
1700	Wood frame, spring type, bronze									
1800	3' x 7' door	1 Carp	7.60	1.053	Opng.	25	53.50		78.50	114
1900	6' x 7' door	"	7	1.143	"	31	58		89	128
2200	Metal frame, spring type, bronze									
2300	3' x 7' door	1 Carp	3	2.667	Opng.	49.50	135		184.50	273
2400	6' x 7' door	"	2.50	3.200	"	53	162		215	320
2500	For stainless steel, spring type, add					133%				
2700	Metal frame, extruded sections, 3' x 7' door, aluminum	1 Carp	3	2.667	Opng.	28	135		163	250
2800	Bronze		3	2.667		82.50	135		217.50	310
3100	6' x 7' door, aluminum		1.50	5.333		35	270		305	480
3200	Bronze		1.50	5.333		140	270		410	595
3500	Threshold weatherstripping									
3650	Door sweep, flush mounted, aluminum	1 Carp	25	.320	Ea.	20.50	16.20		36.70	49
3700	Vinyl		25	.320		18.35	16.20		34.55	46.50
5000	Garage door bottom weatherstrip, 12' aluminum, clear		14	.571		25.50	29		54.50	75
5010	Bronze		14	.571		91.50	29		120.50	148
5050	Bottom protection, rubber		14	.571		41.50	29		70.50	92.50
5100	Threshold		14	.571		74.50	29		103.50	129
9000	Minimum labor/equipment charge		3	2.667	Job		135		135	219

08 75 Window Hardware

08 75 10 – Window Handles and Latches

08 75 10.10 Handles and Latches

08 75 10.10 Handles and Latches	Crew	Daily Output	Labor-Hours	Unit	Material	Labor	Equipment	Total	Total Incl O&P
0010	**HANDLES AND LATCHES**									
1000	Handles, surface mounted, aluminum	1 Carp	24	.333	Ea.	5.45	16.90		22.35	33.50
1020	Brass		24	.333		5.75	16.90		22.65	34
1040	Chrome		24	.333		8.20	16.90		25.10	36.50
1200	Window handles window crank ADA		24	.333		16	16.90		32.90	45
1500	Recessed, aluminum		12	.667		3.39	34		37.39	58.50
1520	Brass		12	.667		5.10	34		39.10	60.50
1540	Chrome		12	.667		4.24	34		38.24	59.50
2000	Latches, aluminum		20	.400		4.05	20.50		24.55	37.50
2020	Brass		20	.400		5.35	20.50		25.85	39
2040	Chrome		20	.400		3.95	20.50		24.45	37.50
9000	Minimum labor/equipment charge		6	1.333	Job		67.50		67.50	110

08 75 10.15 Window Opening Control

0010	**WINDOW OPENING CONTROL**									

For customer support on your Commercial Renovation Costs with RSMeans data, call 800.448.8182.

08 75 Window Hardware

08 75 10 – Window Handles and Latches

08 75 10.15 Window Opening Control

		Crew	Daily Output	Labor-Hours	Unit	Material	2018 Bare Costs Labor	2018 Bare Costs Equipment	Total	Total Incl O&P
0015	Window stops									
0020	For double-hung window	1 Carp	35	.229	Ea.	38.50	11.60		50.10	61.50
0030	Cam action for sliding window		35	.229		4.92	11.60		16.52	24
0040	Thumb screw for sliding window	↓	35	.229	↓	4.65	11.60		16.25	24
0100	Window guards, child safety bars for single or double-hung									
0110	14" to 17" wide max vert opening 26"	1 Carp	16	.500	Ea.	70	25.50		95.50	118
0120	17" to 23" wide max vert opening 26"		16	.500		74	25.50		99.50	123
0130	23" to 36" wide max vert opening 26"		16	.500		79	25.50		104.50	128
0140	35" to 58" wide max vert opening 26"		16	.500		94	25.50		119.50	144
0150	58" to 90" wide max vert opening 26"		16	.500		125	25.50		150.50	179
0160	73" to 120" wide max vert opening 26"	↓	16	.500	↓	227	25.50		252.50	291

08 75 30 – Weatherstripping

08 75 30.10 Mechanical Weather Seals

		Crew	Daily Output	Labor-Hours	Unit	Material	2018 Bare Costs Labor	2018 Bare Costs Equipment	Total	Total Incl O&P
0010	**MECHANICAL WEATHER SEALS**, Window, double-hung, 3' x 5'									
0020	Zinc	1 Carp	7.20	1.111	Opng.	22	56.50		78.50	116
0100	Bronze		7.20	1.111		42	56.50		98.50	138
0200	Vinyl V strip		7	1.143		10.65	58		68.65	106
0500	As above but heavy duty, zinc		4.60	1.739		20.50	88		108.50	166
0600	Bronze	↓	4.60	1.739	↓	75.50	88		163.50	226
9000	Minimum labor/equipment charge	1 Clab	4.60	1.739	Job		69.50		69.50	113

08 79 Hardware Accessories

08 79 20 – Door Accessories

08 79 20.10 Door Hardware Accessories

		Crew	Daily Output	Labor-Hours	Unit	Material	2018 Bare Costs Labor	2018 Bare Costs Equipment	Total	Total Incl O&P
0010	**DOOR HARDWARE ACCESSORIES**									
0140	Door bolt, surface, 4"	1 Carp	32	.250	Ea.	15.55	12.70		28.25	37.50
0160	Door latch	"	12	.667	"	8.75	34		42.75	64.50
0200	Sliding closet door									
0220	Track and hanger, single	1 Carp	10	.800	Ea.	67	40.50		107.50	140
0240	Double		8	1		81.50	50.50		132	173
0260	Door guide, single		48	.167		31	8.45		39.45	47.50
0280	Double		48	.167		41	8.45		49.45	58.50
0600	Deadbolt and lock cover plate, brass or stainless steel		30	.267		30	13.50		43.50	55
0620	Hole cover plate, brass or chrome		35	.229		8.20	11.60		19.80	28
2240	Mortise lockset, passage, lever handle		9	.889		161	45		206	250
4000	Security chain, standard		18	.444		12.95	22.50		35.45	51
4100	Deluxe	↓	18	.444	↓	45.50	22.50		68	86.50

08 81 Glass Glazing

08 81 10 – Float Glass

08 81 10.10 Various Types and Thickness of Float Glass

		Crew	Daily Output	Labor-Hours	Unit	Material	2018 Bare Costs Labor	2018 Bare Costs Equipment	Total	Total Incl O&P
0010	**VARIOUS TYPES AND THICKNESS OF FLOAT GLASS**									
0020	3/16" plain	2 Glaz	130	.123	S.F.	6.80	5.95		12.75	17.15
0200	Tempered, clear		130	.123		7.25	5.95		13.20	17.65
0300	Tinted		130	.123		6.80	5.95		12.75	17.15
0600	1/4" thick, clear, plain		120	.133		9.90	6.45		16.35	21.50
0700	Tinted		120	.133		9.45	6.45		15.90	21
0800	Tempered, clear		120	.133		6.80	6.45		13.25	17.95
0900	Tinted		120	.133	↓	11.70	6.45		18.15	23.50

08 81 Glass Glazing

08 81 10 – Float Glass

08 81 10.10 Various Types and Thickness of Float Glass	Crew	Daily Output	Labor-Hours	Unit	Material	2018 Bare Costs Labor	Equipment	Total	Total Incl O&P
1600	3/8" thick, clear, plain	2 Glaz	75	.213	S.F.	11.20	10.35		21.55	29
1700	Tinted		75	.213		16.55	10.35		26.90	35
1800	Tempered, clear		75	.213		17.70	10.35		28.05	36
1900	Tinted		75	.213		20	10.35		30.35	38.50
2200	1/2" thick, clear, plain		55	.291		19.05	14.10		33.15	44
2300	Tinted		55	.291		29	14.10		43.10	55
2400	Tempered, clear		55	.291		26.50	14.10		40.60	52
2500	Tinted		55	.291		28	14.10		42.10	53.50
2800	5/8" thick, clear, plain		45	.356		29	17.25		46.25	60
2900	Tempered, clear		45	.356		33	17.25		50.25	64.50
3200	3/4" thick, clear, plain		35	.457		37.50	22		59.50	77
3300	Tempered, clear		35	.457		43.50	22		65.50	84
3600	1" thick, clear, plain	↓	30	.533		62	26		88	111
8900	For low emissivity coating for 3/16" & 1/4" only, add to above				↓	18%				
9000	Minimum labor/equipment charge	1 Glaz	2	4	Job		194		194	315

08 81 17 – Fire Glass

08 81 17.10 Fire Resistant Glass

		Crew	Daily Output	Labor-Hours	Unit	Material	2018 Bare Costs Labor	Equipment	Total	Total Incl O&P
0010	**FIRE RESISTANT GLASS**									
0020	Fire glass minimum	2 Glaz	40	.400	S.F.	40.50	19.40		59.90	76
0030	Mid range		40	.400		82.50	19.40		101.90	122
0050	High end	↓	40	.400	↓	365	19.40		384.40	430

08 81 20 – Vision Panels

08 81 20.10 Full Vision

		Crew	Daily Output	Labor-Hours	Unit	Material	2018 Bare Costs Labor	Equipment	Total	Total Incl O&P
0010	**FULL VISION**, window system with 3/4" glass mullions									
0020	Up to 10' high	H-2	130	.185	S.F.	67.50	8.40		75.90	88
0100	10' to 20' high, minimum		110	.218		72	9.95		81.95	95
0150	Average		100	.240		77.50	10.95		88.45	103
0200	Maximum	↓	80	.300	↓	86.50	13.70		100.20	118
9000	Minimum labor/equipment charge	1 Glaz	2	4	Job		194		194	315

08 81 25 – Glazing Variables

08 81 25.10 Applications of Glazing

		Crew	Daily Output	Labor-Hours	Unit	Material	2018 Bare Costs Labor	Equipment	Total	Total Incl O&P
0010	**APPLICATIONS OF GLAZING** R088110-10									
0600	For glass replacement, add				S.F.		100%			
0700	For gasket settings, add				L.F.	6.10			6.10	6.75
0900	For sloped glazing, add				S.F.		26%			
2000	Fabrication, polished edges, 1/4" thick				Inch	.59			.59	.65
2100	1/2" thick					1.38			1.38	1.52
2500	Mitered edges, 1/4" thick					1.38			1.38	1.52
2600	1/2" thick				↓	2.29			2.29	2.52

08 81 30 – Insulating Glass

08 81 30.10 Reduce Heat Transfer Glass

			Crew	Daily Output	Labor-Hours	Unit	Material	2018 Bare Costs Labor	Equipment	Total	Total Incl O&P
0010	**REDUCE HEAT TRANSFER GLASS**										
0015	2 lites 1/8" float, 1/2" thk under 15 S.F.										
0020	Clear R088110-10	G	2 Glaz	95	.168	S.F.	10.35	8.15		18.50	24.50
0100	Tinted	G		95	.168		14.50	8.15		22.65	29
0200	2 lites 3/16" float, for 5/8" thk unit, 15 to 30 S.F., clear	G		90	.178		13.80	8.60		22.40	29
0400	1" thk, dbl. glazed, 1/4" float, 30 to 70 S.F., clear	G		75	.213		17.35	10.35		27.70	36
0500	Tinted	G		75	.213		23.50	10.35		33.85	42.50
2000	Both lites, light & heat reflective	G		85	.188		31.50	9.15		40.65	50
2500	Heat reflective, film inside, 1" thick unit, clear	G	↓	85	.188		27.50	9.15		36.65	45.50

278

08 81 Glass Glazing

08 81 30 – Insulating Glass

08 81 30.10 Reduce Heat Transfer Glass

			Crew	Daily Output	Labor-Hours	Unit	Material	2018 Bare Costs Labor	Equipment	Total	Total Incl O&P
2600	Tinted	G	2 Glaz	85	.188	S.F.	30	9.15		39.15	48.50
3000	Film on weatherside, clear, 1/2" thick unit	G		95	.168		19.75	8.15		27.90	34.50
3100	5/8" thick unit	G		90	.178		20.50	8.60		29.10	36.50
3200	1" thick unit	G	↓	85	.188		27.50	9.15		36.65	45
3350	Clear heat reflective film on inside	G	1 Glaz	50	.160		13.35	7.75		21.10	27
3360	Heat reflective film on inside, tinted	G		25	.320		14.15	15.50		29.65	40.50
3370	Heat reflective film on the inside, metalized	G	↓	20	.400		15.35	19.40		34.75	48.50
5000	Spectrally selective film, on ext., blocks solar gain/allows 70% of light	G	2 Glaz	95	.168	↓	15.20	8.15		23.35	30
9000	Minimum labor/equipment charge		1 Glaz	2	4	Job		194		194	315

08 81 40 – Plate Glass

08 81 40.10 Plate Glass

		Crew	Daily Output	Labor-Hours	Unit	Material	Labor	Equipment	Total	Total Incl O&P
0010	PLATE GLASS Twin ground, polished,									
0020	3/16" thick	2 Glaz	100	.160	S.F.	5.70	7.75		13.45	18.80
0100	1/4" thick		94	.170		7.80	8.25		16.05	22
0200	3/8" thick		60	.267		13.35	12.95		26.30	35.50
0300	1/2" thick	↓	40	.400	↓	26	19.40		45.40	60

08 81 55 – Window Glass

08 81 55.10 Sheet Glass

		Crew	Daily Output	Labor-Hours	Unit	Material	Labor	Equipment	Total	Total Incl O&P
0010	SHEET GLASS (window), clear float, stops, putty bed									
0015	1/8" thick, clear float	2 Glaz	480	.033	S.F.	3.92	1.62		5.54	6.90
0500	3/16" thick, clear		480	.033		6.30	1.62		7.92	9.55
0600	Tinted		480	.033		7.85	1.62		9.47	11.25
0700	Tempered		480	.033	↓	9.75	1.62		11.37	13.30
9000	Minimum labor/equipment charge	↓	5	3.200	Job		155		155	251

08 81 65 – Wire Glass

08 81 65.10 Glass Reinforced With Wire

		Crew	Daily Output	Labor-Hours	Unit	Material	Labor	Equipment	Total	Total Incl O&P
0010	GLASS REINFORCED WITH WIRE									
0012	1/4" thick rough obscure	2 Glaz	135	.119	S.F.	25.50	5.75		31.25	37.50
1000	Polished wire, 1/4" thick, diamond, clear		135	.119		29.50	5.75		35.25	42
1500	Pinstripe, obscure	↓	135	.119	↓	54	5.75		59.75	69

08 83 Mirrors

08 83 13 – Mirrored Glass Glazing

08 83 13.10 Mirrors

		Crew	Daily Output	Labor-Hours	Unit	Material	Labor	Equipment	Total	Total Incl O&P
0010	MIRRORS, No frames, wall type, 1/4" plate glass, polished edge									
0100	Up to 5 S.F.	2 Glaz	125	.128	S.F.	9.80	6.20		16	21
0200	Over 5 S.F.		160	.100		9.55	4.85		14.40	18.35
0500	Door type, 1/4" plate glass, up to 12 S.F.		160	.100		9.15	4.85		14	17.90
1000	Float glass, up to 10 S.F., 1/8" thick		160	.100		6.10	4.85		10.95	14.55
1100	3/16" thick		150	.107		7.50	5.15		12.65	16.60
1500	12" x 12" wall tiles, square edge, clear		195	.082		2.61	3.98		6.59	9.25
1600	Veined		195	.082		6.55	3.98		10.53	13.60
2000	1/4" thick, stock sizes, one way transparent		125	.128		20.50	6.20		26.70	32.50
2010	Bathroom, unframed, laminated		160	.100		14.75	4.85		19.60	24
2500	Tempered	↓	160	.100	↓	18.50	4.85		23.35	28.50

08 83 13.15 Reflective Glass

			Crew	Daily Output	Labor-Hours	Unit	Material	Labor	Equipment	Total	Total Incl O&P
0010	REFLECTIVE GLASS										
0100	1/4" float with fused metallic oxide fixed	G	2 Glaz	115	.139	S.F.	17.90	6.75		24.65	30.50
0500	1/4" float glass with reflective applied coating	G	"	115	.139	"	14.50	6.75		21.25	27

For customer support on your Commercial Renovation Costs with RSMeans data, call 800.448.8182.

279

08 84 Plastic Glazing

08 84 10 – Plexiglass Glazing

08 84 10.10 Plexiglass Acrylic

		Crew	Daily Output	Labor-Hours	Unit	Material	2018 Bare Costs Labor	2018 Bare Costs Equipment	Total	Total Incl O&P
0010	**PLEXIGLASS ACRYLIC**, clear, masked,									
0020	1/8" thick, cut sheets	2 Glaz	170	.094	S.F.	11.35	4.56		15.91	19.85
0200	Full sheets		195	.082		5.75	3.98		9.73	12.70
0500	1/4" thick, cut sheets		165	.097		13.35	4.70		18.05	22.50
0600	Full sheets		185	.086		9.30	4.19		13.49	17
0900	3/8" thick, cut sheets		155	.103		21	5		26	31
1000	Full sheets		180	.089		16.75	4.31		21.06	25.50
1300	1/2" thick, cut sheets		135	.119		31.50	5.75		37.25	44
1400	Full sheets		150	.107		18.85	5.15		24	29
1700	3/4" thick, cut sheets		115	.139		31	6.75		37.75	45.50
1800	Full sheets		130	.123		18	5.95		23.95	29.50
2100	1" thick, cut sheets		105	.152		35	7.40		42.40	50.50
2200	Full sheets		125	.128		22	6.20		28.20	34
3000	Colored, 1/8" thick, cut sheets		170	.094		20	4.56		24.56	29.50
3200	Full sheets		195	.082		11.40	3.98		15.38	18.95
3500	1/4" thick, cut sheets		165	.097		22	4.70		26.70	32
3600	Full sheets		185	.086		14.80	4.19		18.99	23
4000	Mirrors, untinted, cut sheets, 1/8" thick		185	.086		13.30	4.19		17.49	21.50
4200	1/4" thick		180	.089		17.15	4.31		21.46	26

08 84 20 – Polycarbonate

08 84 20.10 Thermoplastic

		Crew	Daily Output	Labor-Hours	Unit	Material	2018 Bare Costs Labor	2018 Bare Costs Equipment	Total	Total Incl O&P
0010	**THERMOPLASTIC**, clear, masked, cut sheets									
0020	1/8" thick	2 Glaz	170	.094	S.F.	15.20	4.56		19.76	24
0500	3/16" thick		165	.097		18.60	4.70		23.30	28
1000	1/4" thick		155	.103		18.25	5		23.25	28
1500	3/8" thick		150	.107		28.50	5.15		33.65	40
9000	Minimum labor/equipment charge	1 Glaz	2	4	Job		194		194	315

08 85 Glazing Accessories

08 85 10 – Miscellaneous Glazing Accessories

08 85 10.10 Glazing Gaskets

		Crew	Daily Output	Labor-Hours	Unit	Material	2018 Bare Costs Labor	2018 Bare Costs Equipment	Total	Total Incl O&P
0010	**GLAZING GASKETS**, Neoprene for glass tongued mullion									
0015	1/4" glass	2 Glaz	200	.080	L.F.	.89	3.88		4.77	7.25
0020	3/8"		200	.080		1.12	3.88		5	7.50
0040	1/2"		200	.080		1.26	3.88		5.14	7.65
0060	3/4"		180	.089		1.82	4.31		6.13	8.95
0080	1"		180	.089		1.49	4.31		5.80	8.60
1000	Glazing compound, wood	1 Glaz	58	.138		1.20	6.70		7.90	12.10
1005	Glazing compound, per window, up to 30 L.F.					1.20			1.20	1.32
1006	Glazing compound, per window, up to 30 L.F.				Ea.	36			36	39.50
1020	Metal	1 Glaz	58	.138	L.F.	1.20	6.70		7.90	12.10

08 85 20 – Structural Glazing Sealants

08 85 20.10 Structural Glazing Adhesives

		Crew	Daily Output	Labor-Hours	Unit	Material	2018 Bare Costs Labor	2018 Bare Costs Equipment	Total	Total Incl O&P
0010	**STRUCTURAL GLAZING ADHESIVES**									
0050	Structural glazing adhesive, 1/4" x 1/4" joint seal	1 Glaz	200	.040	L.F.	13.95	1.94		15.89	18.45

08 87 Glazing Surface Films

08 87 13 – Solar Control Films

08 87 13.10 Solar Films On Glass

08 87 13.10 Solar Films On Glass	Crew	Daily Output	Labor-Hours	Unit	Material	2018 Bare Costs Labor	Equipment	Total	Total Incl O&P
0010 **SOLAR FILMS ON GLASS** (glass not included)									
1000 Bronze, 20% VLT	H-2	180	.133	S.F.	1.60	6.10		7.70	11.60
1020 50% VLT		180	.133		1.80	6.10		7.90	11.85
1050 Neutral, 20% VLT		180	.133		1.80	6.10		7.90	11.85
1100 Silver, 15% VLT		180	.133		1.04	6.10		7.14	11
1120 35% VLT		180	.133		3.02	6.10		9.12	13.15
1150 68% VLT		180	.133		.40	6.10		6.50	10.30
3000 One way mirror with night vision 5% in and 15% out VLT		180	.133		3.03	6.10		9.13	13.20

08 87 16 – Glass Safety Films

08 87 16.10 Safety Films On Glass

08 87 16.10 Safety Films On Glass	Crew	Daily Output	Labor-Hours	Unit	Material	2018 Bare Costs Labor	Equipment	Total	Total Incl O&P
0010 **SAFETY FILMS ON GLASS** (glass not included)									
0015 Safety film helps hold glass together when broken									
0050 Safety film, clear, 2 mil	H-2	180	.133	S.F.	1.21	6.10		7.31	11.20
0100 Safety film, clear, 4 mil		180	.133		1.81	6.10		7.91	11.85
0110 Safety film, tinted, 4 mil		180	.133		3.46	6.10		9.56	13.65
0150 Safety film, black tint, 8 mil		180	.133		2.23	6.10		8.33	12.30

08 87 26 – Bird Control Film

08 87 26.10 Bird Control Film

08 87 26.10 Bird Control Film	Crew	Daily Output	Labor-Hours	Unit	Material	2018 Bare Costs Labor	Equipment	Total	Total Incl O&P
0010 **BIRD CONTROL FILM**									
0050 Patterned, adhered to glass	2 Glaz	180	.089	S.F.	6.75	4.31		11.06	14.40
0200 Decals small, adhered to glass	1 Glaz	50	.160	Ea.	2.48	7.75		10.23	15.30
0250 Decals large, adhered to glass		50	.160	"	6.95	7.75		14.70	20
0300 Decals set of 4, adhered to glass		25	.320	Set	6.95	15.50		22.45	32.50
0400 Bird control film tape		10	.800	Roll	19.95	39		58.95	84.50

08 87 33 – Electrically Tinted Window Film

08 87 33.20 Window Film

08 87 33.20 Window Film	Crew	Daily Output	Labor-Hours	Unit	Material	2018 Bare Costs Labor	Equipment	Total	Total Incl O&P
0010 **WINDOW FILM** adhered on glass (glass not included)									
0015 Film is pre-wired and can be trimmed									
0100 Window film 4 S.F. adhered on glass	1 Glaz	200	.040	S.F.	224	1.94		225.94	250
0120 6 S.F.		180	.044		335	2.16		337.16	375
0160 9 S.F.		170	.047		505	2.28		507.28	560
0180 10 S.F.		150	.053		560	2.59		562.59	620
0200 12 S.F.		144	.056		670	2.69		672.69	740
0220 14 S.F.		140	.057		780	2.77		782.77	860
0240 16 S.F.		128	.063		890	3.03		893.03	985
0260 18 S.F.		126	.063		1,000	3.08		1,003.08	1,100
0280 20 S.F.		120	.067		1,125	3.23		1,128.23	1,225

08 87 53 – Security Films On Glass

08 87 53.10 Security Film Adhered On Glass

08 87 53.10 Security Film Adhered On Glass	Crew	Daily Output	Labor-Hours	Unit	Material	2018 Bare Costs Labor	Equipment	Total	Total Incl O&P
0010 **SECURITY FILM ADHERED ON GLASS** (glass not included)									
0020 Security film, clear, 7 mil	1 Glaz	200	.040	S.F.	1.09	1.94		3.03	4.33
0030 8 mil		200	.040		1.97	1.94		3.91	5.30
0040 9 mil		200	.040		1.08	1.94		3.02	4.32
0050 10 mil		200	.040		1.87	1.94		3.81	5.20
0060 12 mil		200	.040		1.84	1.94		3.78	5.15
0075 15 mil		200	.040		2.32	1.94		4.26	5.70
0100 Security film, sealed with structural adhesive, 7 mil		180	.044		5.55	2.16		7.71	9.60
0110 8 mil		180	.044		6.45	2.16		8.61	10.55
0140 14 mil		180	.044		7.05	2.16		9.21	11.25

08 88 Special Function Glazing

08 88 52 – Prefabricated Glass Block Windows

08 88 52.10 Prefabricated Glass Block Windows

		Crew	Daily Output	Labor-Hours	Unit	Material	2018 Bare Costs Labor	Equipment	Total	Total Incl O&P
0010	**PREFABRICATED GLASS BLOCK WINDOWS**									
0015	Includes frame and silicone seal									
0020	Glass Block Window 16" x 8"	2 Glaz	6	2.667	Ea.	164	129		293	390
0025	16" x 16"		6	2.667		188	129		317	415
0030	16" x 24"		6	2.667		218	129		347	450
0050	16" x 48"		4	4		355	194		549	705
0100	Glass Block Window 24" x 8"		6	2.667		181	129		310	410
0110	24" x 16"		6	2.667		218	129		347	450
0120	24" x 24"		6	2.667		286	129		415	525
0150	24" x 48"		6	2.667		380	129		509	625
0200	Glass Block Window 32" x 8"		6	2.667		181	129		310	410
0210	32" x 16"		6	2.667		243	129		372	475
0220	32" x 24"		4	4		292	194		486	635
0250	32" x 48"		4	4		485	194		679	850
0300	Glass Block Window 40" x 8"		6	2.667		240	129		369	475
0310	40" x 16"		6	2.667		273	129		402	510
0320	40" x 24"		6	2.667		335	129		464	575
0350	40" x 48"		6	2.667		560	129		689	830
0400	Glass Block Window 48" x 8"		6	2.667		269	129		398	505
0410	48" x 16"		6	2.667		355	129		484	600
0420	48" x 24"		6	2.667		380	129		509	625
0450	48" x 48"		6	2.667		635	129		764	905
0500	Glass Block Window 56" x 8"		6	2.667		252	129		381	485
0510	56" x 16"		4	4		390	194		584	745
0520	56" x 24"		4	4		450	194		644	810
0550	56" x 48"		4	4		865	194		1,059	1,275

08 88 52.20 Prefabricated Glass Block Windows Hurricane Resistant

		Crew	Daily Output	Labor-Hours	Unit	Material	2018 Bare Costs Labor	Equipment	Total	Total Incl O&P
0010	**PREFABRICATED GLASS BLOCK WINDOWS HURRICANE RESISTANT**									
0015	Includes frame and silicone seal									
3020	Glass Block Window 16" x 16"	2 Glaz	6	2.667	Ea.	425	129		554	680
3030	16" x 24"		6	2.667		505	129		634	765
3040	16" x 32"		4	4		585	194		779	960
3050	16" x 40"		4	4		630	194		824	1,000
3060	16" x 48"		4	4		665	194		859	1,050
3070	16" x 56"		4	4		825	194		1,019	1,225
3080	Glass Block Window 24" x 24"		6	2.667		585	129		714	855
3090	24" x 32"		4	4		665	194		859	1,050
3100	24" x 40"		4	4		825	194		1,019	1,225
3110	24" x 48"		4	4		870	194		1,064	1,275
3120	24" x 56"		4	4		905	194		1,099	1,300
3130	Glass Block Window 32" x 32"		4	4		745	194		939	1,125
3140	32" x 40"		4	4		850	194		1,044	1,250
3150	32" x 48"		4	4		985	194		1,179	1,400
3160	32" x 56"		4	4		1,150	194		1,344	1,575
3200	Glass Block Window 40" x 40"		4	4		985	194		1,179	1,400
3210	40" x 48"		3	5.333		1,200	259		1,459	1,725
3220	40" x 56"		3	5.333		1,325	259		1,584	1,875
3260	Glass Block Window 48" x 48"		3	5.333		1,375	259		1,634	1,950
3270	48" x 56"		3	5.333		1,550	259		1,809	2,125
3310	Glass Block Window 56" x 56"		3	5.333		1,700	259		1,959	2,300

08 88 52 – Prefabricated Glass Block Windows

08 88 52.30 Prefabricated Acrylic Block Casement Windows	Crew	Daily Output	Labor-Hours	Unit	Material	2018 Bare Costs Labor	Equipment	Total	Total Incl O&P
0010 **PREFABRICATED ACRYLIC BLOCK CASEMENT WINDOWS**									
0015 Windows include frame and nail fin.									
3050 6" x 6" x 1-3/4" Block Casement Window 2' x 2'	2 Glaz	6	2.667	Ea.	425	129		554	675
3055 2' x 3'		6	2.667		460	129		589	720
3060 2' x 4'		6	2.667		460	129		589	720
3065 2' x 5'		6	2.667		565	129		694	830
3070 2' x 6'		6	2.667		620	129		749	895
3075 2' x 7'		4	4		650	194		844	1,025
3080 2' x 8'		4	4		700	194		894	1,075
3085 2' x 9'		4	4		750	194		944	1,150
3090 3' x 3'		4	4		505	194		699	875
3095 3' x 4'		4	4		555	194		749	925
3100 3' x 5'		4	4		640	194		834	1,025
3110 3' x 6'		4	4		695	194		889	1,075
3115 3' x 7'		4	4		765	194		959	1,150
3120 3' x 8'		4	4		825	194		1,019	1,225
3125 3' x 9'		3	5.333		890	259		1,149	1,400
3130 4' x 4'		3	5.333		665	259		924	1,150
3135 4' x 5'		3	5.333		750	259		1,009	1,250
3140 4' x 6'		3	5.333		790	259		1,049	1,275
3145 4' x 7'		3	5.333		690	259		949	1,175
3150 4' x 8'		3	5.333		790	259		1,049	1,275
3500 8" x 8" x 1-3/4" Block Casement Window 2' x 2'		6	2.667		390	129		519	635
3510 2' x 3'		6	2.667		390	129		519	640
3515 2' x 4'		6	2.667		480	129		609	735
3520 2' x 5'		6	2.667		560	129		689	825
3525 2' x 6'		4	4		620	194		814	1,000
3530 2' x 7'		4	4		650	194		844	1,025
3535 2' x 8'		3	5.333		705	259		964	1,200
3540 2' x 9'		3	5.333		735	259		994	1,225
3545 2' x 10'		3	5.333		785	259		1,044	1,275
3550 2' x 11'		3	5.333		925	259		1,184	1,450
3555 3' x 3'		4	4		475	194		669	840
3560 3' x 4'		4	4		525	194		719	890
3565 3' x 5'		4	4		640	194		834	1,025
3570 3' x 6'		4	4		700	194		894	1,075
3575 3' x 7'		4	4		775	194		969	1,175
3580 3' x 8'		3	5.333		835	259		1,094	1,350
3585 3' x 9'		3	5.333		870	259		1,129	1,375
3590 4' x 4'		3	5.333		640	259		899	1,125
3595 4' x 5'		3	5.333		755	259		1,014	1,250
3600 4' x 6'		3	5.333		810	259		1,069	1,325

08 88 56 – Ballistics-Resistant Glazing

08 88 56.10 Laminated Glass

	Crew	Daily Output	Labor-Hours	Unit	Material	2018 Bare Costs Labor	Equipment	Total	Total Incl O&P
0010 **LAMINATED GLASS**									
0020 Clear float .03" vinyl 1/4" thick	2 Glaz	90	.178	S.F.	12.75	8.60		21.35	28
0100 3/8" thick		78	.205		27.50	9.95		37.45	46.50
0200 .06" vinyl, 1/2" thick		65	.246		28	11.95		39.95	50.50
1000 5/8" thick		90	.178		30.50	8.60		39.10	47.50
2000 Bullet-resisting, 1-3/16" thick, to 15 S.F.		16	1		117	48.50		165.50	207
2100 Over 15 S.F.		16	1		150	48.50		198.50	244
2200 2" thick, to 15 S.F.		10	1.600		149	77.50		226.50	289

For customer support on your Commercial Renovation Costs with RSMeans data, call 800.448.8182.

283

08 88 Special Function Glazing

08 88 56 – Ballistics-Resistant Glazing

08 88 56.10 Laminated Glass		Crew	Daily Output	Labor-Hours	Unit	Material	2018 Bare Costs Labor	Equipment	Total	Total Incl O&P
2300	Over 15 S.F.	2 Glaz	10	1.600	S.F.	127	77.50		204.50	265
2500	2-1/4" thick, to 15 S.F.		12	1.333		207	64.50		271.50	330
2600	Over 15 S.F.		12	1.333		196	64.50		260.50	320
2700	Level 2 (.357 magnum), NIJ and UL		12	1.333		58	64.50		122.50	168
2750	Level 3A (.44 magnum) NIJ, UL 3		12	1.333		63	64.50		127.50	174
2800	Level 4 (AK-47) NIJ, UL 7 & 8		12	1.333		83	64.50		147.50	196
2850	Level 5 (M-16) UL		12	1.333		93	64.50		157.50	206
2900	Level 3 (7.62 Armor Piercing) NIJ, UL 4 & 5		12	1.333		108	64.50		172.50	223

08 91 Louvers

08 91 16 – Operable Wall Louvers

08 91 16.10 Movable Blade Louvers

		Crew	Daily Output	Labor-Hours	Unit	Material	2018 Bare Costs Labor	Equipment	Total	Total Incl O&P
0010	**MOVABLE BLADE LOUVERS**									
0100	PVC, commercial grade, 12" x 12"	1 Shee	20	.400	Ea.	80.50	24		104.50	126
0110	16" x 16"		14	.571		94	34		128	157
0120	18" x 18"		14	.571		97	34		131	160
0130	24" x 24"		14	.571		138	34		172	206
0140	30" x 30"		12	.667		166	40		206	246
0150	36" x 36"		12	.667		196	40		236	278
0300	Stainless steel, commercial grade, 12" x 12"		20	.400		218	24		242	278
0310	16" x 16"		14	.571		264	34		298	345
0320	18" x 18"		14	.571		279	34		313	360
0330	20" x 20"		14	.571		320	34		354	410
0340	24" x 24"		14	.571		385	34		419	475
0350	30" x 30"		12	.667		430	40		470	535
0360	36" x 36"		12	.667		520	40		560	635

08 91 19 – Fixed Louvers

08 91 19.10 Aluminum Louvers

		Crew	Daily Output	Labor-Hours	Unit	Material	2018 Bare Costs Labor	Equipment	Total	Total Incl O&P
0010	**ALUMINUM LOUVERS**									
0020	Aluminum with screen, residential, 8" x 8"	1 Carp	38	.211	Ea.	20	10.65		30.65	39.50
0100	12" x 12"		38	.211		16	10.65		26.65	35
0200	12" x 18"		35	.229		21	11.60		32.60	42.50
0250	14" x 24"		30	.267		29	13.50		42.50	54
0300	18" x 24"		27	.296		32	15		47	59.50
0500	24" x 30"		24	.333		61.50	16.90		78.40	95
0700	Triangle, adjustable, small		20	.400		58.50	20.50		79	97.50
0800	Large		15	.533		80.50	27		107.50	133
2100	Midget, aluminum, 3/4" deep, 1" diameter		85	.094		.86	4.77		5.63	8.70
2150	3" diameter		60	.133		2.94	6.75		9.69	14.20
2200	4" diameter		50	.160		5.60	8.10		13.70	19.30
2250	6" diameter		30	.267		3.78	13.50		17.28	26
3000	PVC, commercial grade, 12" x 12"	1 Shee	20	.400		95	24		119	143
3010	12" x 18"		20	.400		107	24		131	156
3020	12" x 24"		20	.400		144	24		168	196
3030	14" x 24"		20	.400		150	24		174	203
3100	Aluminum, commercial grade, 12" x 12"		20	.400		169	24		193	224
3110	24" x 24"		16	.500		247	30		277	320
3120	24" x 36"		16	.500		299	30		329	375
3130	36" x 24"		16	.500		310	30		340	385
3140	36" x 36"		14	.571		385	34		419	480

08 91 Louvers

08 91 19 – Fixed Louvers

08 91 19.10 Aluminum Louvers

		Crew	Daily Output	Labor-Hours	Unit	Material	2018 Bare Costs Labor	Equipment	Total	Total Incl O&P
3150	36" x 48"	1 Shee	10	.800	Ea.	400	48		448	515
3160	48" x 36"		10	.800		390	48		438	500
3170	48" x 48"		10	.800		430	48		478	550
3180	60" x 48"		8	1		490	60		550	635
3190	60" x 60"		8	1		590	60		650	745

08 91 19.20 Steel Louvers

		Crew	Daily Output	Labor-Hours	Unit	Material	2018 Bare Costs Labor	Equipment	Total	Total Incl O&P
0010	**STEEL LOUVERS**									
3300	Galvanized steel, fixed blades, commercial grade, 18" x 18"	1 Shee	20	.400	Ea.	194	24		218	252
3310	24" x 24"		20	.400		229	24		253	289
3320	24" x 36"		16	.500		296	30		326	370
3330	36" x 24"		16	.500		295	30		325	370
3340	36" x 36"		14	.571		380	34		414	470
3350	36" x 48"		10	.800		385	48		433	495
3360	48" x 36"		10	.800		385	48		433	495
3370	48" x 48"		10	.800		415	48		463	535
3380	60" x 48"		10	.800		490	48		538	615
3390	60" x 60"		10	.800		585	48		633	720

08 91 26 – Door Louvers

08 91 26.10 Steel Louvers, 18 Gauge, Fixed Blade

		Crew	Daily Output	Labor-Hours	Unit	Material	2018 Bare Costs Labor	Equipment	Total	Total Incl O&P
0010	**STEEL LOUVERS, 18 GAUGE, FIXED BLADE**									
0050	12" x 12", with powder coat	1 Carp	20	.400	Ea.	85	20.50		105.50	127
0055	18" x 12"		20	.400		87.50	20.50		108	129
0060	18" x 18"		20	.400		91.50	20.50		112	134
0065	24" x 12"		20	.400		113	20.50		133.50	158
0070	24" x 18"		20	.400		124	20.50		144.50	169
0075	24" x 24"		20	.400		152	20.50		172.50	200
0100	12" x 12", galvanized		20	.400		77	20.50		97.50	118
0105	18" x 12"		20	.400		79.50	20.50		100	121
0115	24" x 12"		20	.400		103	20.50		123.50	147
0125	24" x 24"		20	.400		143	20.50		163.50	191
0300	6" x 12", painted		20	.400		51	20.50		71.50	89
0320	10" x 16"		20	.400		79	20.50		99.50	120
0340	12" x 22"		20	.400		93.50	20.50		114	136
0360	14" x 14"		20	.400		91	20.50		111.50	133
0400	18" x 18"		20	.400		80	20.50		100.50	121
0420	20" x 26"		20	.400		139	20.50		159.50	186
0440	22" x 22"		20	.400		133	20.50		153.50	179
0460	26" x 26"		20	.400		208	20.50		228.50	262
3000	Fire rated									
3010	12" x 12"	1 Carp	10	.800	Ea.	159	40.50		199.50	241
3050	18" x 18"		10	.800		227	40.50		267.50	315
3100	24" x 12"		10	.800		242	40.50		282.50	330
3120	24" x 18"		10	.800		255	40.50		295.50	345
3130	24" x 24"		10	.800		283	40.50		323.50	375

08 91 26.20 Aluminum Louvers

		Crew	Daily Output	Labor-Hours	Unit	Material	2018 Bare Costs Labor	Equipment	Total	Total Incl O&P
0010	**ALUMINUM LOUVERS**, 18 ga., fixed blade, clear anodized									
0050	6" x 12"	1 Carp	20	.400	Ea.	62	20.50		82.50	101
0060	10" x 10"		20	.400		73	20.50		93.50	114
0065	10" x 14"		20	.400		87.50	20.50		108	129
0080	12" x 22"		20	.400		116	20.50		136.50	160
0090	14" x 14"		20	.400		104	20.50		124.50	148
0100	18" x 10"		20	.400		86.50	20.50		107	128

08 91 Louvers

08 91 26 – Door Louvers

08 91 26.20 Aluminum Louvers		Crew	Daily Output	Labor-Hours	Unit	Material	2018 Bare Costs Labor	Equipment	Total	Total Incl O&P
0110	18" x 18"	1 Carp	20	.400	Ea.	134	20.50		154.50	181
0120	22" x 22"		20	.400		190	20.50		210.50	242
0130	26" x 26"		20	.400		243	20.50		263.50	300

08 95 Vents

08 95 13 – Soffit Vents

08 95 13.10 Wall Louvers

		Crew	Daily Output	Labor-Hours	Unit	Material	2018 Bare Costs Labor	Equipment	Total	Total Incl O&P
0010	**WALL LOUVERS**									
2340	Baked enamel finish	1 Carp	200	.040	L.F.	5.70	2.03		7.73	9.60
2400	Under eaves vent, aluminum, mill finish, 16" x 4"		48	.167	Ea.	1.89	8.45		10.34	15.80
2500	16" x 8"		48	.167	"	2.61	8.45		11.06	16.55

08 95 16 – Wall Vents

08 95 16.10 Louvers

		Crew	Daily Output	Labor-Hours	Unit	Material	2018 Bare Costs Labor	Equipment	Total	Total Incl O&P
0010	**LOUVERS**									
0020	Redwood, 2'-0" diameter, full circle	1 Carp	16	.500	Ea.	275	25.50		300.50	345
0100	Half circle		16	.500		214	25.50		239.50	276
0200	Octagonal		16	.500		194	25.50		219.50	254
0300	Triangular, 5/12 pitch, 5'-0" at base		16	.500		555	25.50		580.50	650
1000	Rectangular, 1'-4" x 1'-3"		16	.500		25.50	25.50		51	69
1100	Rectangular, 1'-4" x 1'-8"		16	.500		38.50	25.50		64	83
1200	1'-4" x 2'-2"		15	.533		44	27		71	92.50
1300	1'-9" x 2'-2"		15	.533		52.50	27		79.50	102
1400	2'-3" x 2'-2"		14	.571		70	29		99	124
1700	2'-4" x 2'-11"		13	.615		70	31		101	128
2000	Aluminum, 12" x 16"		25	.320		22.50	16.20		38.70	51
2010	16" x 20"		25	.320		30	16.20		46.20	59.50
2020	24" x 30"		25	.320		61.50	16.20		77.70	94
2100	6' triangle		12	.667		181	34		215	254
3100	Round, 2'-2" diameter		16	.500		179	25.50		204.50	238
7000	Vinyl gable vent, 8" x 8"		38	.211		17	10.65		27.65	36
7020	12" x 12"		38	.211		30	10.65		40.65	50.50
7080	12" x 18"		35	.229		38.50	11.60		50.10	61.50
7200	18" x 24"		30	.267		55.50	13.50		69	83
9000	Minimum labor/equipment charge		3.50	2.286	Job		116		116	188

Estimating Tips
General

- Room Finish Schedule: A complete set of plans should contain a room finish schedule. If one is not available, it would be well worth the time and effort to obtain one.

09 20 00 Plaster and Gypsum Board

- Lath is estimated by the square yard plus a 5% allowance for waste. Furring, channels, and accessories are measured by the linear foot. An extra foot should be allowed for each accessory miter or stop.

- Plaster is also estimated by the square yard. Deductions for openings vary by preference, from zero deduction to 50% of all openings over 2 feet in width. The estimator should allow one extra square foot for each linear foot of horizontal interior or exterior angle located below the ceiling level. Also, double the areas of small radius work.

- Drywall accessories, studs, track, and acoustical caulking are all measured by the linear foot. Drywall taping is figured by the square foot. Gypsum wallboard is estimated by the square foot. No material deductions should be made for door or window openings under 32 S.F.

09 60 00 Flooring

- Tile and terrazzo areas are taken off on a square foot basis. Trim and base materials are measured by the linear foot. Accent tiles are listed per each. Two basic methods of installation are used. Mud set is approximately 30% more expensive than thin set. The cost of grout is included with tile unit price lines unless otherwise noted. In terrazzo work, be sure to include the linear footage of embedded decorative strips, grounds, machine rubbing, and power cleanup.

- Wood flooring is available in strip, parquet, or block configuration. The latter two types are set in adhesives with quantities estimated by the square foot. The laying pattern will influence labor costs and material waste. In addition to the material and labor for laying wood floors, the estimator must make allowances for sanding and finishing these areas, unless the flooring is prefinished.

- Sheet flooring is measured by the square yard. Roll widths vary, so consideration should be given to use the most economical width, as waste must be figured into the total quantity. Consider also the installation methods available—direct glue down or stretched. Direct glue-down installation is assumed with sheet carpet unit price lines unless otherwise noted.

09 70 00 Wall Finishes

- Wall coverings are estimated by the square foot. The area to be covered is measured—length by height of the wall above the baseboards—to calculate the square footage of each wall. This figure is divided by the number of square feet in the single roll which is being used. Deduct, in full, the areas of openings such as doors and windows. Where a pattern match is required allow 25%–30% waste.

09 80 00 Acoustic Treatment

- Acoustical systems fall into several categories. The takeoff of these materials should be by the square foot of area with a 5% allowance for waste. Do not forget about scaffolding, if applicable, when estimating these systems.

09 90 00 Painting and Coating

- A major portion of the work in painting involves surface preparation. Be sure to include cleaning, sanding, filling, and masking costs in the estimate.

- Protection of adjacent surfaces is not included in painting costs. When considering the method of paint application, an important factor is the amount of protection and masking required. These must be estimated separately and may be the determining factor in choosing the method of application.

Reference Numbers

Reference numbers are shown at the beginning of some major classifications. These numbers refer to related items in the Reference Section. The reference information may be an estimating procedure, an alternate pricing method, or technical information.

Note: Not all subdivisions listed here necessarily appear. ■

09 01 Maintenance of Finishes

09 01 70 – Maintenance of Wall Finishes

09 01 70.10 Gypsum Wallboard Repairs

		Crew	Daily Output	Labor-Hours	Unit	Material	2018 Bare Costs Labor	Equipment	Total	Total Incl O&P
0010	**GYPSUM WALLBOARD REPAIRS**									
0100	Fill and sand, pin/nail holes	1 Carp	960	.008	Ea.		.42		.42	.69
0110	Screw head pops		480	.017			.85		.85	1.37
0120	Dents, up to 2" square		48	.167		.01	8.45		8.46	13.70
0130	2" to 4" square		24	.333		.03	16.90		16.93	27.50
0140	Cut square, patch, sand and finish, holes, up to 2" square		12	.667		.03	34		34.03	55
0150	2" to 4" square		11	.727		.09	37		37.09	60
0160	4" to 8" square		10	.800		.24	40.50		40.74	66.50
0170	8" to 12" square		8	1		.48	50.50		50.98	83
0180	12" to 32" square		6	1.333		1.60	67.50		69.10	112
0210	16" by 48"		5	1.600		2.76	81		83.76	135
0220	32" by 48"		4	2		4.52	101		105.52	170
0230	48" square		3.50	2.286		6.40	116		122.40	195
0240	60" square		3.20	2.500		9.95	127		136.95	217
0500	Skim coat surface with joint compound		1600	.005	S.F.	.03	.25		.28	.44
0510	Prepare, retape and refinish joints		60	.133	L.F.	.63	6.75		7.38	11.65
9000	Minimum labor/equipment charge		2	4	Job		203		203	330

09 05 Common Work Results for Finishes

09 05 05 – Selective Demolition for Finishes

09 05 05.10 Selective Demolition, Ceilings

		Crew	Daily Output	Labor-Hours	Unit	Material	2018 Bare Costs Labor	Equipment	Total	Total Incl O&P
0010	**SELECTIVE DEMOLITION, CEILINGS** R024119-10									
0200	Ceiling, gypsum wall board, furred and nailed or screwed	2 Clab	800	.020	S.F.		.80		.80	1.29
0220	On metal frame		760	.021			.84		.84	1.36
0240	On suspension system, including system		720	.022			.89		.89	1.44
1000	Plaster, lime and horse hair, on wood lath, incl. lath		700	.023			.91		.91	1.48
1020	On metal lath		570	.028			1.12		1.12	1.82
1100	Gypsum, on gypsum lath		720	.022			.89		.89	1.44
1120	On metal lath		500	.032			1.28		1.28	2.07
1200	Suspended ceiling, mineral fiber, 2' x 2' or 2' x 4'		1500	.011			.43		.43	.69
1250	On suspension system, incl. system		1200	.013			.53		.53	.86
1500	Tile, wood fiber, 12" x 12", glued		900	.018			.71		.71	1.15
1540	Stapled		1500	.011			.43		.43	.69
1580	On suspension system, incl. system		760	.021			.84		.84	1.36
2000	Wood, tongue and groove, 1" x 4"		1000	.016			.64		.64	1.04
2040	1" x 8"		1100	.015			.58		.58	.94
2400	Plywood or wood fiberboard, 4' x 8' sheets		1200	.013			.53		.53	.86
2500	Remove & refinish textured ceiling	1 Plas	222	.036		.03	1.67		1.70	2.71
9000	Minimum labor/equipment charge	1 Clab	2	4	Job		159		159	259

09 05 05.20 Selective Demolition, Flooring

		Crew	Daily Output	Labor-Hours	Unit	Material	2018 Bare Costs Labor	Equipment	Total	Total Incl O&P
0010	**SELECTIVE DEMOLITION, FLOORING** R024119-10									
0200	Brick with mortar	2 Clab	475	.034	S.F.		1.34		1.34	2.18
0400	Carpet, bonded, including surface scraping		2000	.008			.32		.32	.52
0480	Tackless		9000	.002			.07		.07	.12
0550	Carpet tile, releasable adhesive		5000	.003			.13		.13	.21
0560	Permanent adhesive		1850	.009			.34		.34	.56
0600	Composition, acrylic or epoxy		400	.040			1.59		1.59	2.59
0700	Concrete, scarify skin	A-1A	225	.036			1.86	.95	2.81	4.05
0800	Resilient, sheet goods	2 Clab	1400	.011			.46		.46	.74
0820	For gym floors	"	900	.018			.71		.71	1.15
0850	Vinyl or rubber cove base	1 Clab	1000	.008	L.F.		.32		.32	.52

288

09 05 05 – Selective Demolition for Finishes

09 05 05.20 Selective Demolition, Flooring

		Crew	Daily Output	Labor-Hours	Unit	Material	2018 Bare Costs Labor	Equipment	Total	Total Incl O&P
0860	Vinyl or rubber cove base, molded corner	1 Clab	1000	.008	Ea.		.32		.32	.52
0870	For glued and caulked installation, add to labor						50%			
0900	Vinyl composition tile, 12" x 12"	2 Clab	1000	.016	S.F.		.64		.64	1.04
2000	Tile, ceramic, thin set		675	.024			.94		.94	1.53
2020	Mud set		625	.026			1.02		1.02	1.66
2200	Marble, slate, thin set		675	.024			.94		.94	1.53
2220	Mud set		625	.026			1.02		1.02	1.66
2600	Terrazzo, thin set		450	.036			1.42		1.42	2.30
2620	Mud set		425	.038			1.50		1.50	2.44
2640	Terrazzo, cast in place	▼	300	.053			2.13		2.13	3.45
3000	Wood, block, on end	1 Carp	400	.020			1.01		1.01	1.65
3200	Parquet		450	.018			.90		.90	1.46
3400	Strip flooring, interior, 2-1/4" x 25/32" thick		325	.025			1.25		1.25	2.03
3500	Exterior, porch flooring, 1" x 4"		220	.036			1.84		1.84	2.99
3800	Subfloor, tongue and groove, 1" x 6"		325	.025			1.25		1.25	2.03
3820	1" x 8"		430	.019			.94		.94	1.53
3840	1" x 10"		520	.015			.78		.78	1.27
4000	Plywood, nailed		600	.013			.68		.68	1.10
4100	Glued and nailed		400	.020			1.01		1.01	1.65
4200	Hardboard, 1/4" thick	▼	760	.011			.53		.53	.87
8000	Remove flooring, bead blast, simple floor plan	A-1A	1000	.008		.05	.42	.21	.68	.97
8100	Complex floor plan		400	.020		.05	1.05	.54	1.64	2.34
8150	Mastic only	▼	1500	.005	▼	.05	.28	.14	.47	.67
9000	Minimum labor/equipment charge	1 Clab	4	2	Job		79.50		79.50	129
9050	For grinding concrete floors, see Section 03 35 43.10									

09 05 05.30 Selective Demolition, Walls and Partitions

		Crew	Daily Output	Labor-Hours	Unit	Material	2018 Bare Costs Labor	Equipment	Total	Total Incl O&P
0010	**SELECTIVE DEMOLITION, WALLS AND PARTITIONS** R024119-10									
0020	Walls, concrete, reinforced	B-39	120	.400	C.F.		16.85	1.95	18.80	29
0025	Plain	"	160	.300			12.65	1.46	14.11	22
0100	Brick, 4" to 12" thick	B-9	220	.182	▼		7.30	1.06	8.36	13.05
0200	Concrete block, 4" thick		1150	.035	S.F.		1.40	.20	1.60	2.49
0280	8" thick		1050	.038			1.53	.22	1.75	2.74
0300	Exterior stucco 1" thick over mesh	▼	3200	.013			.50	.07	.57	.90
1000	Gypsum wallboard, nailed or screwed	1 Clab	1000	.008			.32		.32	.52
1010	2 layers		400	.020			.80		.80	1.29
1020	Glued and nailed		900	.009			.35		.35	.58
1500	Fiberboard, nailed		900	.009			.35		.35	.58
1520	Glued and nailed		800	.010			.40		.40	.65
1568	Plenum barrier, sheet lead		300	.027			1.06		1.06	1.73
1600	Glass block		65	.123			4.90		4.90	7.95
2000	Movable walls, metal, 5' high		300	.027			1.06		1.06	1.73
2020	8' high	▼	400	.020			.80		.80	1.29
2200	Metal or wood studs, finish 2 sides, fiberboard	B-1	520	.046			1.87		1.87	3.04
2250	Lath and plaster		260	.092			3.74		3.74	6.05
2300	Gypsum wallboard		520	.046			1.87		1.87	3.04
2350	Plywood	▼	450	.053			2.16		2.16	3.51
2800	Paneling, 4' x 8' sheets	1 Clab	475	.017			.67		.67	1.09
3000	Plaster, lime and horsehair, on wood lath		400	.020			.80		.80	1.29
3020	On metal lath		335	.024			.95		.95	1.55
3400	Gypsum or perlite, on gypsum lath		410	.020			.78		.78	1.26
3420	On metal lath	▼	300	.027	▼		1.06		1.06	1.73
3450	Plaster, interior gypsum, acoustic, or cement		60	.133	S.Y.		5.30		5.30	8.65

For customer support on your Commercial Renovation Costs with RSMeans data, call 800.448.8182.

289

09 05 05 – Selective Demolition for Finishes

09 05 05.30 Selective Demolition, Walls and Partitions	Crew	Daily Output	Labor-Hours	Unit	Material	2018 Bare Costs Labor	Equipment	Total	Total Incl O&P
3500	Stucco, on masonry	1 Clab	145	.055	S.Y.		2.20		2.20	3.57
3510	Commercial 3-coat		80	.100			3.99		3.99	6.45
3520	Interior stucco		25	.320			12.75		12.75	20.50
3753	Remove damaged glass block	B-1	134.62	.178	S.F.		7.20		7.20	11.75
3760	Tile, ceramic, on walls, thin set	1 Clab	300	.027			1.06		1.06	1.73
3765	Mud set		250	.032			1.28		1.28	2.07
3800	Toilet partitions, slate or marble		5	1.600	Ea.		64		64	104
3820	Metal or plastic		8	1	"		40		40	64.50
5000	Wallcovering, vinyl	1 Pape	700	.011	S.F.		.49		.49	.78
5010	With release agent		1500	.005			.23		.23	.36
5025	Wallpaper, 2 layers or less, by hand		250	.032			1.36		1.36	2.19
5035	3 layers or more		165	.048			2.07		2.07	3.31
9000	Minimum labor/equipment charge	1 Clab	4	2	Job		79.50		79.50	129

09 05 71 – Acoustic Underlayment

09 05 71.10 Acoustical Underlayment

		Crew	Daily Output	Labor-Hours	Unit	Material	2018 Bare Costs Labor	Equipment	Total	Total Incl O&P
0010	**ACOUSTICAL UNDERLAYMENT**									
0100	Rubber underlayment, 5/64"	1 Tilf	275	.029	S.F.	.32	1.36		1.68	2.50
0110	1/8" thk.		275	.029		.51	1.36		1.87	2.71
0120	13/64" thk.		270	.030		.69	1.39		2.08	2.95
0130	15/64" thk.		270	.030		.92	1.39		2.31	3.20
0140	23/64" thk.		265	.030		1.26	1.41		2.67	3.62
0150	15/32" thk.		265	.030		1.60	1.41		3.01	3.99
0400	Cork underlayment, 3/32"		275	.029		.39	1.36		1.75	2.58
0410	1/8" thk.		275	.029		.40	1.36		1.76	2.59
0420	5/32" thk.		275	.029		.47	1.36		1.83	2.67
0430	15/64" thk.		270	.030		.61	1.39		2	2.86
0440	15/32" thk.		265	.030		1.27	1.41		2.68	3.63
0600	Rubber cork underlayment, 3/64"		275	.029		.31	1.36		1.67	2.49
0610	13/64" thk.		270	.030		1.27	1.39		2.66	3.59
0620	3/8" thk.		270	.030		2.42	1.39		3.81	4.85
0630	25/64" thk.		270	.030		2.55	1.39		3.94	5
0800	Foam underlayment, 5/64"		275	.029		.54	1.36		1.90	2.74
0810	1/8" thk.		275	.029		.59	1.36		1.95	2.80
0820	15/32" thk.		265	.030		.49	1.41		1.90	2.77
4000	Nylon matting 0.4" thick, with carbon black spinerette									
4010	plus polyester fabric, on floor	D-7	1600	.010	S.F.	1.41	.42		1.83	2.21

09 21 Plaster and Gypsum Board Assemblies

09 21 16 – Gypsum Board Assemblies

09 21 16.23 Gypsum Board Shaft Wall Assemblies

		Crew	Daily Output	Labor-Hours	Unit	Material	2018 Bare Costs Labor	Equipment	Total	Total Incl O&P
0010	**GYPSUM BOARD SHAFT WALL ASSEMBLIES**									
0020	Cavity type on 25 ga. J-track & C-H studs, 24" OC									
0030	1" thick coreboard wall liner on shaft side									
0040	2-hour assembly with double layer									
0060	5/8" f.r. gyp bd on rm side, 2-1/2" J-track & C-H studs	2 Carp	220	.073	S.F.	2.95	3.69		6.64	9.25
0065	4" J-track & C-H studs		220	.073		3.11	3.69		6.80	9.40
0070	6" J-track & C-H studs		220	.073		3.33	3.69		7.02	9.65
0100	3-hour assembly with triple layer									
0300	5/8" f.r. gyp bd on rm side, 2-1/2" J-track & C-H studs	2 Carp	180	.089	S.F.	2.60	4.51		7.11	10.15
0305	4" J-track & C-H studs		180	.089		2.76	4.51		7.27	10.35

09 21 Plaster and Gypsum Board Assemblies

09 21 16 – Gypsum Board Assemblies

09 21 16.23 Gypsum Board Shaft Wall Assemblies		Crew	Daily Output	Labor-Hours	Unit	Material	2018 Bare Costs Labor	Equipment	Total	Total Incl O&P
0310	6" J-track & C-H studs	2 Carp	180	.089	S.F.	2.98	4.51		7.49	10.55
0400	4-hour assembly, 1" coreboard, 5/8" fire rated gypsum board									
0600	and 3/4" galv. metal furring channels, 24" OC, with									
0700	Dbl. layer 5/8" f.r. gyp bd on rm side, 2-1/2" trk. & C-H studs	2 Carp	110	.145	S.F.	3.33	7.35		10.68	15.60
0705	4" J-track & C-H studs		110	.145		3.11	7.35		10.46	15.35
0710	6" J-track & C-H studs		110	.145		3.33	7.35		10.68	15.60
0900	For taping & finishing, add per side	1 Carp	1050	.008		.05	.39		.44	.68
1000	For insulation, see Section 07 21									
5200	For work over 8' high, add	2 Carp	3060	.005	S.F.		.27		.27	.43
5300	For distribution cost 3 stories and above, add per story	"	6100	.003	"		.13		.13	.22

09 21 16.33 Partition Wall

		Crew	Daily Output	Labor-Hours	Unit	Material	2018 Bare Costs Labor	Equipment	Total	Total Incl O&P
0010	**PARTITION WALL** Stud wall, 8' to 12' high									
0050	1/2", interior, gypsum board, std, tape & finish 2 sides									
0500	Installed on and incl., 2" x 4" wood studs, 16" OC	2 Carp	310	.052	S.F.	1.24	2.62		3.86	5.60
1000	Metal studs, NLB, 25 ga., 16" OC, 3-5/8" wide		350	.046		1.15	2.32		3.47	5.05
1200	6" wide		330	.048		1.28	2.46		3.74	5.40
1400	Water resistant, on 2" x 4" wood studs, 16" OC		310	.052		1.40	2.62		4.02	5.80
1600	Metal studs, NLB, 25 ga., 16" OC, 3-5/8" wide		350	.046		1.31	2.32		3.63	5.20
1800	6" wide		330	.048		1.44	2.46		3.90	5.55
2000	Fire res., 2 layers, 1-1/2 hr., on 2" x 4" wood studs, 16" OC		210	.076		2.04	3.86		5.90	8.50
2200	Metal studs, NLB, 25 ga., 16" OC, 3-5/8" wide		250	.064		1.95	3.24		5.19	7.40
2400	6" wide		230	.070		2.08	3.53		5.61	8.05
2600	Fire & water res., 2 layers, 1-1/2 hr., 2" x 4" studs, 16" OC		210	.076		2.04	3.86		5.90	8.50
2800	Metal studs, NLB, 25 ga., 16" OC, 3-5/8" wide		250	.064		1.95	3.24		5.19	7.40
3000	6" wide		230	.070		2.08	3.53		5.61	8.05
3200	5/8", interior, gypsum board, standard, tape & finish 2 sides									
3400	Installed on and including 2" x 4" wood studs, 16" OC	2 Carp	300	.053	S.F.	1.26	2.70		3.96	5.75
3600	24" OC		330	.048		1.15	2.46		3.61	5.25
3800	Metal studs, NLB, 25 ga., 16" OC, 3-5/8" wide		340	.047		1.17	2.39		3.56	5.15
4000	6" wide		320	.050		1.30	2.54		3.84	5.55
4200	24" OC, 3-5/8" wide		360	.044		1.07	2.25		3.32	4.84
4400	6" wide		340	.047		1.16	2.39		3.55	5.15
4800	Water resistant, on 2" x 4" wood studs, 16" OC		300	.053		1.44	2.70		4.14	5.95
5000	24" OC		330	.048		1.33	2.46		3.79	5.45
5200	Metal studs, NLB, 25 ga. 16" OC, 3-5/8" wide		340	.047		1.35	2.39		3.74	5.35
5400	6" wide		320	.050		1.48	2.54		4.02	5.75
5600	24" OC, 3-5/8" wide		360	.044		1.25	2.25		3.50	5.05
5800	6" wide		340	.047		1.34	2.39		3.73	5.35
6000	Fire resistant, 2 layers, 2 hr., on 2" x 4" wood studs, 16" OC		205	.078		2	3.96		5.96	8.60
6200	24" OC		235	.068		1.89	3.45		5.34	7.70
6400	Metal studs, NLB, 25 ga., 16" OC, 3-5/8" wide		245	.065		1.95	3.31		5.26	7.55
6600	6" wide		225	.071		2.04	3.61		5.65	8.10
6800	24" OC, 3-5/8" wide		265	.060		1.81	3.06		4.87	6.95
7000	6" wide		245	.065		1.90	3.31		5.21	7.50
7200	Fire & water resistant, 2 layers, 2 hr., 2" x 4" studs, 16" OC		205	.078		2	3.96		5.96	8.60
7400	24" OC		235	.068		1.89	3.45		5.34	7.70
7600	Metal studs, NLB, 25 ga., 16" OC, 3-5/8" wide		245	.065		1.91	3.31		5.22	7.50
7800	6" wide		225	.071		2.04	3.61		5.65	8.10
8000	24" OC, 3-5/8" wide		265	.060		1.81	3.06		4.87	6.95
8200	6" wide		245	.065		1.90	3.31		5.21	7.50
8600	1/2" blueboard, mesh tape both sides									
8620	Installed on and including 2" x 4" wood studs, 16" OC	2 Carp	300	.053	S.F.	1.40	2.70		4.10	5.95

09 21 Plaster and Gypsum Board Assemblies

09 21 16 – Gypsum Board Assemblies

09 21 16.33 Partition Wall

		Crew	Daily Output	Labor-Hours	Unit	Material	2018 Bare Costs Labor	Equipment	Total	Total Incl O&P
8640	Metal studs, NLB, 25 ga., 16" OC, 3-5/8" wide	2 Carp	340	.047	S.F.	1.31	2.39		3.70	5.30
8660	6" wide	↓	320	.050	↓	1.44	2.54		3.98	5.70
8800	Hospital security partition, 5/8" fiber reinf. high abuse gyp. bd.									
8810	Mtl. studs, NLB, 20 ga., 16" OC, 3-5/8" wide, w/sec. mesh, gyp. bd.	2 Carp	208	.077	S.F.	4.25	3.90		8.15	11.05
9000	Exterior, 1/2" gypsum sheathing, 1/2" gypsum finished, interior,									
9100	including foil faced insulation, metal studs, 20 ga.									
9200	16" OC, 3-5/8" wide	2 Carp	290	.055	S.F.	1.79	2.80		4.59	6.50
9400	6" wide	↓	270	.059		1.99	3		4.99	7.05
9600	Partitions, for work over 8' high, add	↓	1530	.010	↓		.53		.53	.86

09 22 Supports for Plaster and Gypsum Board

09 22 03 – Fastening Methods for Finishes

09 22 03.20 Drilling Plaster/Drywall

		Crew	Daily Output	Labor-Hours	Unit	Material	2018 Bare Costs Labor	Equipment	Total	Total Incl O&P
0010	**DRILLING PLASTER/DRYWALL**									
1100	Drilling & layout for drywall/plaster walls, up to 1" deep, no anchor									
1200	Holes, 1/4" diameter	1 Carp	150	.053	Ea.	.01	2.70		2.71	4.40
1300	3/8" diameter		140	.057		.01	2.90		2.91	4.71
1400	1/2" diameter		130	.062		.01	3.12		3.13	5.05
1500	3/4" diameter		120	.067		.02	3.38		3.40	5.50
1600	1" diameter		110	.073		.02	3.69		3.71	6
1700	1-1/4" diameter		100	.080		.04	4.06		4.10	6.65
1800	1-1/2" diameter	↓	90	.089		.06	4.51		4.57	7.35
1900	For ceiling installations, add				↓		40%			

09 22 13 – Metal Furring

09 22 13.13 Metal Channel Furring

		Crew	Daily Output	Labor-Hours	Unit	Material	2018 Bare Costs Labor	Equipment	Total	Total Incl O&P
0010	**METAL CHANNEL FURRING**									
0030	Beams and columns, 7/8" hat channels, galvanized, 12" OC	1 Lath	155	.052	S.F.	.44	2.54		2.98	4.48
0050	16" OC		170	.047		.36	2.31		2.67	4.04
0070	24" OC		185	.043		.24	2.13		2.37	3.61
0100	Ceilings, on steel, 7/8" hat channels, galvanized, 12" OC		210	.038		.40	1.87		2.27	3.39
0300	16" OC		290	.028		.36	1.36		1.72	2.53
0400	24" OC		420	.019		.24	.94		1.18	1.74
0600	1-5/8" hat channels, galvanized, 12" OC		190	.042		.53	2.07		2.60	3.85
0700	16" OC		260	.031		.47	1.51		1.98	2.91
0900	24" OC		390	.021		.32	1.01		1.33	1.94
0930	7/8" hat channels with sound isolation clips, 12" OC		120	.067		1.72	3.28		5	7.05
0940	16" OC		100	.080		1.29	3.93		5.22	7.60
0950	24" OC		165	.048		.86	2.38		3.24	4.70
0960	1-5/8" hat channels, galvanized, 12" OC		110	.073		1.85	3.57		5.42	7.70
0970	16" OC		100	.080		1.38	3.93		5.31	7.70
0980	24" OC		155	.052		.92	2.54		3.46	5
1000	Walls, 7/8" hat channels, galvanized, 12" OC		235	.034		.40	1.67		2.07	3.08
1200	16" OC		265	.030		.36	1.48		1.84	2.73
1300	24" OC		350	.023		.24	1.12		1.36	2.03
1500	1-5/8" hat channels, galvanized, 12" OC		210	.038		.53	1.87		2.40	3.53
1600	16" OC		240	.033		.47	1.64		2.11	3.10
1800	24" OC		305	.026		.32	1.29		1.61	2.38
1920	7/8" hat channels with sound isolation clips, 12" OC		125	.064		1.72	3.15		4.87	6.85
1940	16" OC		100	.080		1.29	3.93		5.22	7.60
1950	24" OC		150	.053		.86	2.62		3.48	5.10

292

09 22 Supports for Plaster and Gypsum Board

09 22 13 – Metal Furring

09 22 13.13 Metal Channel Furring		Crew	Daily Output	Labor-Hours	Unit	Material	2018 Bare Costs Labor	Equipment	Total	Total Incl O&P
1960	1-5/8" hat channels, galvanized, 12" OC	1 Lath	115	.070	S.F.	1.85	3.42		5.27	7.45
1970	16" OC		95	.084		1.38	4.14		5.52	8.05
1980	24" OC		140	.057		.92	2.81		3.73	5.45
3000	Z Furring, walls, 1" deep, 25 ga., 24" OC		350	.023		1.38	1.12		2.50	3.28
3010	48" OC		700	.011		.69	.56		1.25	1.65
3020	1-1/2" deep, 24" OC		345	.023		1.61	1.14		2.75	3.57
3030	48" OC		695	.012		.80	.57		1.37	1.77
3040	2" deep, 24" OC		340	.024		1.93	1.16		3.09	3.94
3050	48" OC		690	.012		.97	.57		1.54	1.96
3060	1" deep, 20 ga., 24" OC		350	.023		2.28	1.12		3.40	4.27
3070	48" OC		700	.011		1.14	.56		1.70	2.14
3080	1-1/2" deep, 24" OC		345	.023		2.62	1.14		3.76	4.68
3090	48" OC		695	.012		1.31	.57		1.88	2.33
4000	2" deep, 24" OC		340	.024		3.22	1.16		4.38	5.35
4010	48" OC		690	.012		1.61	.57		2.18	2.67
9000	Minimum labor/equipment charge		4	2	Job		98.50		98.50	155

09 22 16 – Non-Structural Metal Framing

09 22 16.13 Non-Structural Metal Stud Framing		Crew	Daily Output	Labor-Hours	Unit	Material	2018 Bare Costs Labor	Equipment	Total	Total Incl O&P
0010	**NON-STRUCTURAL METAL STUD FRAMING**									
1600	Non-load bearing, galv., 8' high, 25 ga. 1-5/8" wide, 16" OC	1 Carp	619	.013	S.F.	.27	.66		.93	1.36
1610	24" OC		950	.008		.20	.43		.63	.91
1620	2-1/2" wide, 16" OC		613	.013		.36	.66		1.02	1.46
1630	24" OC		938	.009		.27	.43		.70	.99
1640	3-5/8" wide, 16" OC		600	.013		.40	.68		1.08	1.54
1650	24" OC		925	.009		.30	.44		.74	1.04
1660	4" wide, 16" OC		594	.013		.45	.68		1.13	1.60
1670	24" OC		925	.009		.33	.44		.77	1.08
1680	6" wide, 16" OC		588	.014		.53	.69		1.22	1.70
1690	24" OC		906	.009		.40	.45		.85	1.17
1700	20 ga. studs, 1-5/8" wide, 16" OC		494	.016		.34	.82		1.16	1.71
1710	24" OC		763	.010		.26	.53		.79	1.14
1720	2-1/2" wide, 16" OC		488	.016		.44	.83		1.27	1.83
1730	24" OC		750	.011		.33	.54		.87	1.24
1740	3-5/8" wide, 16" OC		481	.017		.50	.84		1.34	1.92
1750	24" OC		738	.011		.37	.55		.92	1.30
1760	4" wide, 16" OC		475	.017		.60	.85		1.45	2.05
1770	24" OC		738	.011		.45	.55		1	1.39
1780	6" wide, 16" OC		469	.017		.71	.86		1.57	2.19
1790	24" OC		725	.011		.54	.56		1.10	1.50
2000	Non-load bearing, galv., 10' high, 25 ga. 1-5/8" wide, 16" OC		495	.016		.25	.82		1.07	1.61
2100	24" OC		760	.011		.19	.53		.72	1.08
2200	2-1/2" wide, 16" OC		490	.016		.34	.83		1.17	1.71
2250	24" OC		750	.011		.25	.54		.79	1.15
2300	3-5/8" wide, 16" OC		480	.017		.38	.85		1.23	1.79
2350	24" OC		740	.011		.28	.55		.83	1.20
2400	4" wide, 16" OC		475	.017		.42	.85		1.27	1.86
2450	24" OC		740	.011		.31	.55		.86	1.23
2500	6" wide, 16" OC		470	.017		.50	.86		1.36	1.95
2550	24" OC		725	.011		.37	.56		.93	1.32
2600	20 ga. studs, 1-5/8" wide, 16" OC		395	.020		.32	1.03		1.35	2.03
2650	24" OC		610	.013		.24	.66		.90	1.34
2700	2-1/2" wide, 16" OC		390	.021		.41	1.04		1.45	2.14

09 22 16.13 Non-Structural Metal Stud Framing	Crew	Daily Output	Labor-Hours	Unit	Material	2018 Bare Costs Labor	Equipment	Total	Total Incl O&P
2750 24" OC	1 Carp	600	.013	S.F.	.30	.68		.98	1.43
2800 3-5/8" wide, 16" OC		385	.021		.47	1.05		1.52	2.23
2850 24" OC		590	.014		.35	.69		1.04	1.50
2900 4" wide, 16" OC		380	.021		.57	1.07		1.64	2.36
2950 24" OC		590	.014		.42	.69		1.11	1.58
3000 6" wide, 16" OC		375	.021		.68	1.08		1.76	2.50
3050 24" OC		580	.014		.50	.70		1.20	1.68
3060 Non-load bearing, galv., 12' high, 25 ga. 1-5/8" wide, 16" OC		413	.019		.24	.98		1.22	1.86
3070 24" OC		633	.013		.18	.64		.82	1.24
3080 2-1/2" wide, 16" OC		408	.020		.32	.99		1.31	1.97
3090 24" OC		625	.013		.24	.65		.89	1.31
3100 3-5/8" wide, 16" OC		400	.020		.36	1.01		1.37	2.05
3110 24" OC		617	.013		.26	.66		.92	1.36
3120 4" wide, 16" OC		396	.020		.40	1.02		1.42	2.10
3130 24" OC		617	.013		.30	.66		.96	1.39
3140 6" wide, 16" OC		392	.020		.48	1.03		1.51	2.21
3150 24" OC		604	.013		.35	.67		1.02	1.48
3160 20 ga. studs, 1-5/8" wide, 16" OC		329	.024		.31	1.23		1.54	2.34
3170 24" OC		508	.016		.23	.80		1.03	1.55
3180 2-1/2" wide, 16" OC		325	.025		.39	1.25		1.64	2.46
3190 24" OC		500	.016		.29	.81		1.10	1.64
3200 3-5/8" wide, 16" OC		321	.025		.45	1.26		1.71	2.55
3210 24" OC		492	.016		.33	.82		1.15	1.70
3220 4" wide, 16" OC		317	.025		.55	1.28		1.83	2.68
3230 24" OC		492	.016		.40	.82		1.22	1.78
3240 6" wide, 16" OC		313	.026		.65	1.30		1.95	2.81
3250 24" OC		483	.017		.47	.84		1.31	1.88
3260 Non-load bearing, galv., 16' high, 25 ga. 4" wide, 12" OC		195	.041		.52	2.08		2.60	3.95
3270 16" OC		275	.029		.41	1.47		1.88	2.84
3280 24" OC		400	.020		.30	1.01		1.31	1.98
3290 6" wide, 12" OC		190	.042		.62	2.14		2.76	4.15
3300 16" OC		280	.029		.49	1.45		1.94	2.88
3310 24" OC		400	.020		.35	1.01		1.36	2.04
3320 20 ga. studs, 2-1/2" wide, 12" OC		180	.044		.50	2.25		2.75	4.21
3330 16" OC		254	.032		.40	1.60		2	3.03
3340 24" OC		390	.021		.29	1.04		1.33	2.01
3350 3-5/8" wide, 12" OC		170	.047		.58	2.39		2.97	4.50
3360 16" OC		251	.032		.45	1.62		2.07	3.12
3370 24" OC		384	.021		.33	1.06		1.39	2.07
3380 4" wide, 12" OC		170	.047		.70	2.39		3.09	4.64
3390 16" OC		247	.032		.55	1.64		2.19	3.28
3400 24" OC		384	.021		.40	1.06		1.46	2.15
3410 6" wide, 12" OC		175	.046		.83	2.32		3.15	4.67
3420 16" OC		245	.033		.65	1.66		2.31	3.41
3430 24" OC		400	.020		.48	1.01		1.49	2.18
3440 Non-load bearing, galv., 20' high, 25 ga. 6" wide, 12" OC		125	.064		.60	3.24		3.84	5.90
3450 16" OC		220	.036		.47	1.84		2.31	3.51
3460 24" OC		360	.022		.34	1.13		1.47	2.21
3470 20 ga. studs, 4" wide, 12" OC		120	.067		.69	3.38		4.07	6.25
3480 16" OC		215	.037		.54	1.89		2.43	3.65
3490 6" wide, 12" OC		115	.070		.81	3.53		4.34	6.65
3500 16" OC		215	.037		.64	1.89		2.53	3.76
3510 24" OC		331	.024		.46	1.23		1.69	2.50

For customer support on your Commercial Renovation Costs with RSMeans data, call 800.448.8182.

09 22 Supports for Plaster and Gypsum Board

09 22 16 – Non-Structural Metal Framing

09 22 16.13 Non-Structural Metal Stud Framing	Crew	Daily Output	Labor-Hours	Unit	Material	2018 Bare Costs Labor	Equipment	Total	Total Incl O&P
5000	For load bearing studs, see Section 05 41 13.30									
9000	Minimum labor/equipment charge	1 Carp	4	2	Job		101		101	165

09 22 26 – Suspension Systems

09 22 26.13 Ceiling Suspension Systems

		Crew	Daily Output	Labor-Hours	Unit	Material	Labor	Equipment	Total	Total Incl O&P
0010	CEILING SUSPENSION SYSTEMS for gypsum board or plaster									
8000	Suspended ceilings, including carriers									
8200	1-1/2" carriers, 24" OC with:									
8300	7/8" channels, 16" OC	1 Lath	275	.029	S.F.	.58	1.43		2.01	2.89
8320	24" OC		310	.026		.46	1.27		1.73	2.50
8400	1-5/8" channels, 16" OC		205	.039		.69	1.92		2.61	3.79
8420	24" OC		250	.032		.53	1.57		2.10	3.07
8600	2" carriers, 24" OC with:									
8700	7/8" channels, 16" OC	1 Lath	250	.032	S.F.	.66	1.57		2.23	3.20
8720	24" OC		285	.028		.54	1.38		1.92	2.77
8800	1-5/8" channels, 16" OC		190	.042		.77	2.07		2.84	4.12
8820	24" OC		225	.036		.62	1.75		2.37	3.44

09 22 36 – Lath

09 22 36.13 Gypsum Lath

			Crew	Daily Output	Labor-Hours	Unit	Material	Labor	Equipment	Total	Total Incl O&P
0010	GYPSUM LATH	R092000-50									
0020	Plain or perforated, nailed, 3/8" thick		1 Lath	85	.094	S.Y.	3.15	4.63		7.78	10.75
0100	1/2" thick			80	.100		2.52	4.92		7.44	10.50
0300	Clipped to steel studs, 3/8" thick			75	.107		3.15	5.25		8.40	11.70
0400	1/2" thick			70	.114		2.52	5.60		8.12	11.60
1500	For ceiling installations, add			216	.037			1.82		1.82	2.87
1600	For columns and beams, add			170	.047			2.31		2.31	3.65
9000	Minimum labor/equipment charge			4.25	1.882	Job		92.50		92.50	146

09 22 36.23 Metal Lath

			Crew	Daily Output	Labor-Hours	Unit	Material	Labor	Equipment	Total	Total Incl O&P
0010	METAL LATH	R092000-50									
0020	Diamond, expanded, 2.5 lb./S.Y., painted					S.Y.	4.15			4.15	4.57
0100	Galvanized						3.10			3.10	3.41
0300	3.4 lb./S.Y., painted						4.33			4.33	4.76
0400	Galvanized						4.44			4.44	4.88
0600	For #15 asphalt sheathing paper, add						.48			.48	.53
0900	Flat rib, 1/8" high, 2.75 lb., painted						3.60			3.60	3.96
1000	Foil backed						3.81			3.81	4.19
1200	3.4 lb./S.Y., painted						4.54			4.54	4.99
1300	Galvanized						4.62			4.62	5.10
1500	For #15 asphalt sheathing paper, add						.48			.48	.53
1800	High rib, 3/8" high, 3.4 lb./S.Y., painted						4.72			4.72	5.20
1900	Galvanized						4.28			4.28	4.71
2400	3/4" high, painted, .60 lb./S.F.					S.F.	.73			.73	.80
2500	.75 lb./S.F.					"	1.58			1.58	1.74
2800	Stucco mesh, painted, 3.6 lb.					S.Y.	4.29			4.29	4.72
3000	K-lath, perforated, absorbent paper, regular						4.42			4.42	4.86
3100	Heavy duty						5.20			5.20	5.75
3300	Waterproof, heavy duty, grade B backing						5.10			5.10	5.60
3400	Fire resistant backing						5.65			5.65	6.20
3600	2.5 lb. diamond painted, on wood framing, on walls		1 Lath	85	.094		4.15	4.63		8.78	11.85
3700	On ceilings			75	.107		4.15	5.25		9.40	12.80
3900	3.4 lb. diamond painted, on wood framing, on walls			80	.100		4.54	4.92		9.46	12.75
4000	On ceilings			70	.114		4.54	5.60		10.14	13.85

For customer support on your Commercial Renovation Costs with RSMeans data, call 800.448.8182.

295

09 22 36.23 Metal Lath

09 22 36.23 Metal Lath		Crew	Daily Output	Labor-Hours	Unit	Material	2018 Bare Costs Labor	Equipment	Total	Total Incl O&P
4200	3.4 lb. diamond painted, wired to steel framing	1 Lath	75	.107	S.Y.	4.54	5.25		9.79	13.25
4300	On ceilings		60	.133		4.54	6.55		11.09	15.35
4600	Cornices, wired to steel		35	.229		4.54	11.25		15.79	22.50
4800	Screwed to steel studs, 2.5 lb.		80	.100		4.15	4.92		9.07	12.30
4900	3.4 lb.		75	.107		4.33	5.25		9.58	13
5100	Rib lath, painted, wired to steel, on walls, 2.5 lb.		75	.107		3.60	5.25		8.85	12.20
5200	3.4 lb.		70	.114		4.72	5.60		10.32	14.05
5400	4.0 lb.		65	.123		5.75	6.05		11.80	15.85
5500	For self-furring lath, add					.12			.12	.13
5700	Suspended ceiling system, incl. 3.4 lb. diamond lath, painted	1 Lath	15	.533		4.36	26		30.36	46.50
5800	Galvanized	"	15	.533		4.54	26		30.54	46.50
6000	Hollow metal stud partitions, 3.4 lb. painted lath both sides									
6010	Non-load bearing, 25 ga., w/rib lath, 2-1/2" studs, 12" OC	1 Lath	20.30	.394	S.Y.	13.30	19.35		32.65	45
6300	16" OC		21.10	.379		12.45	18.65		31.10	43
6350	24" OC		22.70	.352		11.65	17.30		28.95	40.50
6400	3-5/8" studs, 16" OC		19.50	.410		12.85	20		32.85	46
6600	24" OC		20.40	.392		11.95	19.25		31.20	43.50
6700	4" studs, 16" OC		20.40	.392		13.25	19.25		32.50	45
6900	24" OC		21.60	.370		12.25	18.20		30.45	42
7000	6" studs, 16" OC		19.50	.410		13.95	20		33.95	47.50
7100	24" OC		21.10	.379		12.75	18.65		31.40	43.50
7200	L.B. partitions, 16 ga., w/rib lath, 2-1/2" studs, 16" OC		20	.400		13.70	19.65		33.35	46
7300	3-5/8" studs, 16 ga.		19.70	.406		15.45	19.95		35.40	48.50
7500	4" studs, 16 ga.		19.50	.410		16	20		36	49.50
7600	6" studs, 16 ga.		18.70	.428		19	21		40	54
9000	Minimum labor/equipment charge		4.25	1.882	Job		92.50		92.50	146

09 23 Gypsum Plastering
09 23 13 – Acoustical Gypsum Plastering
09 23 13.10 Perlite or Vermiculite Plaster

		Crew	Daily Output	Labor-Hours	Unit	Material	Labor	Equipment	Total	Total Incl O&P
0010	**PERLITE OR VERMICULITE PLASTER** R092000-50									
0020	In 100 lb. bags, under 200 bags				Bag	18.20			18.20	20
0300	2 coats, no lath included, on walls	J-1	92	.435	S.Y.	5.95	19.05	1.45	26.45	38.50
0400	On ceilings		79	.506		5.95	22	1.69	29.64	44
0900	3 coats, no lath included, on walls		74	.541		6.40	23.50	1.80	31.70	47
1000	On ceilings		63	.635		6.40	28	2.11	36.51	54
1700	For irregular or curved surfaces, add to above						30%			
1800	For columns and beams, add to above						50%			
1900	For soffits, add to ceiling prices						40%			
9000	Minimum labor/equipment charge	1 Plas	1	8	Job		370		370	595

09 23 20 – Gypsum Plaster
09 23 20.10 Gypsum Plaster On Walls and Ceilings

		Crew	Daily Output	Labor-Hours	Unit	Material	Labor	Equipment	Total	Total Incl O&P
0010	**GYPSUM PLASTER ON WALLS AND CEILINGS** R092000-50									
0020	80# bag, less than 1 ton				Bag	15.80			15.80	17.40
0300	2 coats, no lath included, on walls	J-1	105	.381	S.Y.	3.76	16.65	1.27	21.68	32
0400	On ceilings		92	.435		3.76	19.05	1.45	24.26	36
0900	3 coats, no lath included, on walls		87	.460		5.40	20	1.53	26.93	39.50
1000	On ceilings		78	.513		5.40	22.50	1.71	29.61	44
1600	For irregular or curved surfaces, add						30%			
1800	For columns & beams, add						50%			

09 23 Gypsum Plastering

09 23 20 – Gypsum Plaster

09 23 20.10 Gypsum Plaster On Walls and Ceilings	Crew	Daily Output	Labor-Hours	Unit	Material	2018 Bare Costs Labor	Equipment	Total	Total Incl O&P
9000 Minimum labor/equipment charge	1 Plas	1	8	Job		370		370	595

09 24 Cement Plastering

09 24 23 – Cement Stucco

09 24 23.40 Stucco

		Crew	Daily Output	Labor-Hours	Unit	Material	2018 Bare Costs Labor	Equipment	Total	Total Incl O&P
0010	**STUCCO** R092000-50									
0015	3 coats 7/8" thick, float finish, with mesh, on wood frame	J-2	63	.762	S.Y.	8.35	34	2.12	44.47	66
0100	On masonry construction, no mesh incl.	J-1	67	.597		3.43	26	1.99	31.42	48
0300	For trowel finish, add	1 Plas	170	.047			2.19		2.19	3.50
0600	For coloring, add	J-1	685	.058		.42	2.56	.19	3.17	4.76
0700	For special texture, add	"	200	.200		1.46	8.75	.67	10.88	16.35
0900	For soffits, add	J-2	155	.310		2.25	13.85	.86	16.96	25.50
1000	Stucco, with bonding agent, 3 coats, on walls, no mesh incl.	J-1	200	.200		4.36	8.75	.67	13.78	19.55
1200	Ceilings		180	.222		3.77	9.75	.74	14.26	20.50
1300	Beams		80	.500		3.77	22	1.67	27.44	41
1500	Columns		100	.400		3.77	17.50	1.33	22.60	33.50
1550	Minimum labor/equipment charge	1 Plas	1	8	Job		370		370	595
1600	Mesh, galvanized, nailed to wood, 1.8 lb.	1 Lath	60	.133	S.Y.	7.35	6.55		13.90	18.45
1800	3.6 lb.		55	.145		4.29	7.15		11.44	16
1900	Wired to steel, galvanized, 1.8 lb.		53	.151		7.35	7.40		14.75	19.80
2100	3.6 lb.		50	.160		4.29	7.85		12.14	17.10
9000	Minimum labor/equipment charge		4	2	Job		98.50		98.50	155

09 25 Other Plastering

09 25 23 – Lime Based Plastering

09 25 23.10 Venetian Plaster

		Crew	Daily Output	Labor-Hours	Unit	Material	2018 Bare Costs Labor	Equipment	Total	Total Incl O&P
0010	**VENETIAN PLASTER**									
0100	Walls, 1 coat primer, roller applied	1 Plas	950	.008	S.F.	.17	.39		.56	.82
0200	Plaster, 3 coats, incl. sanding	2 Plas	700	.023		.42	1.06		1.48	2.16
0210	For pigment, light colors add per S.F. plaster					.02			.02	.02
0220	For pigment, dark colors add per S.F. plaster					.04			.04	.04
0300	For sealer/wax coat incl. burnishing, add	1 Plas	300	.027		.38	1.24		1.62	2.39

09 26 Veneer Plastering

09 26 13 – Gypsum Veneer Plastering

09 26 13.20 Blueboard

		Crew	Daily Output	Labor-Hours	Unit	Material	2018 Bare Costs Labor	Equipment	Total	Total Incl O&P
0010	**BLUEBOARD** For use with thin coat									
0100	plaster application see Section 09 26 13.80									
1000	3/8" thick, on walls or ceilings, standard, no finish included	2 Carp	1900	.008	S.F.	.34	.43		.77	1.06
1100	With thin coat plaster finish		875	.018		.45	.93		1.38	2.01
1400	On beams, columns, or soffits, standard, no finish included		675	.024		.39	1.20		1.59	2.38
1450	With thin coat plaster finish		475	.034		.50	1.71		2.21	3.32
3000	1/2" thick, on walls or ceilings, standard, no finish included		1900	.008		.35	.43		.78	1.08
3100	With thin coat plaster finish		875	.018		.46	.93		1.39	2.02
3300	Fire resistant, no finish included		1900	.008		.35	.43		.78	1.08
3400	With thin coat plaster finish		875	.018		.46	.93		1.39	2.02
3450	On beams, columns, or soffits, standard, no finish included		675	.024		.40	1.20		1.60	2.39

09 26 Veneer Plastering

09 26 13 – Gypsum Veneer Plastering

09 26 13.20 Blueboard

		Crew	Daily Output	Labor-Hours	Unit	Material	2018 Bare Costs Labor	Equipment	Total	Total Incl O&P
3500	With thin coat plaster finish	2 Carp	475	.034	S.F.	.52	1.71		2.23	3.34
3700	Fire resistant, no finish included		675	.024		.40	1.20		1.60	2.39
3800	With thin coat plaster finish		475	.034		.52	1.71		2.23	3.34
5000	5/8" thick, on walls or ceilings, fire resistant, no finish included		1900	.008		.36	.43		.79	1.09
5100	With thin coat plaster finish		875	.018		.47	.93		1.40	2.03
5500	On beams, columns, or soffits, no finish included		675	.024		.41	1.20		1.61	2.41
5600	With thin coat plaster finish		475	.034		.53	1.71		2.24	3.35
6000	For high ceilings, over 8' high, add		3060	.005			.27		.27	.43
6500	For distribution costs 3 stories and above, add per story	↓	6100	.003	↓		.13		.13	.22
9000	Minimum labor/equipment charge	1 Carp	2	4	Job		203		203	330

09 26 13.80 Thin Coat Plaster

		Crew	Daily Output	Labor-Hours	Unit	Material	2018 Bare Costs Labor	Equipment	Total	Total Incl O&P
0010	**THIN COAT PLASTER** R092000-50									
0012	1 coat veneer, not incl. lath	J-1	3600	.011	S.F.	.11	.49	.04	.64	.94
1000	In 50 lb. bags				Bag	15.20			15.20	16.75

09 28 Backing Boards and Underlayments

09 28 13 – Cementitious Backing Boards

09 28 13.10 Cementitious Backerboard

		Crew	Daily Output	Labor-Hours	Unit	Material	2018 Bare Costs Labor	Equipment	Total	Total Incl O&P
0010	**CEMENTITIOUS BACKERBOARD**									
0070	Cementitious backerboard, on floor, 3' x 4' x 1/2" sheets	2 Carp	525	.030	S.F.	.86	1.55		2.41	3.46
0080	3' x 5' x 1/2" sheets		525	.030		.79	1.55		2.34	3.38
0090	3' x 6' x 1/2" sheets		525	.030		.78	1.55		2.33	3.37
0100	3' x 4' x 5/8" sheets		525	.030		.95	1.55		2.50	3.56
0110	3' x 5' x 5/8" sheets		525	.030		.95	1.55		2.50	3.55
0120	3' x 6' x 5/8" sheets		525	.030		.95	1.55		2.50	3.56
0150	On wall, 3' x 4' x 1/2" sheets		350	.046		.86	2.32		3.18	4.71
0160	3' x 5' x 1/2" sheets		350	.046		.79	2.32		3.11	4.63
0170	3' x 6' x 1/2" sheets		350	.046		.78	2.32		3.10	4.62
0180	3' x 4' x 5/8" sheets		350	.046		.95	2.32		3.27	4.81
0190	3' x 5' x 5/8" sheets		350	.046		.95	2.32		3.27	4.80
0200	3' x 6' x 5/8" sheets		350	.046		.95	2.32		3.27	4.81
0250	On counter, 3' x 4' x 1/2" sheets		180	.089		.86	4.51		5.37	8.25
0260	3' x 5' x 1/2" sheets		180	.089		.79	4.51		5.30	8.15
0270	3' x 6' x 1/2" sheets		180	.089		.78	4.51		5.29	8.15
0300	3' x 4' x 5/8" sheets		180	.089		.95	4.51		5.46	8.35
0310	3' x 5' x 5/8" sheets		180	.089		.95	4.51		5.46	8.35
0320	3' x 6' x 5/8" sheets	↓	180	.089	↓	.95	4.51		5.46	8.35

09 29 Gypsum Board

09 29 10 – Gypsum Board Panels

09 29 10.30 Gypsum Board

		Crew	Daily Output	Labor-Hours	Unit	Material	2018 Bare Costs Labor	Equipment	Total	Total Incl O&P
0010	**GYPSUM BOARD** on walls & ceilings R092910-10									
0100	Nailed or screwed to studs unless otherwise noted									
0110	1/4" thick, on walls or ceilings, standard, no finish included	2 Carp	1330	.012	S.F.	.37	.61		.98	1.40
0115	1/4" thick, on walls or ceilings, flexible, no finish included		1050	.015		.53	.77		1.30	1.83
0117	1/4" thick, on columns or soffits, flexible, no finish included		1050	.015		.53	.77		1.30	1.83
0130	1/4" thick, standard, no finish included, less than 800 S.F.		510	.031		.37	1.59		1.96	2.99
0150	3/8" thick, on walls, standard, no finish included		2000	.008		.36	.41		.77	1.06
0200	On ceilings, standard, no finish included		1800	.009		.36	.45		.81	1.13

09 29 Gypsum Board

09 29 10 – Gypsum Board Panels

09 29 10.30 Gypsum Board		Crew	Daily Output	Labor-Hours	Unit	Material	2018 Bare Costs Labor	Equipment	Total	Total Incl O&P
0250	On beams, columns, or soffits, no finish included	2 Carp	675	.024	S.F.	.36	1.20		1.56	2.35
0300	1/2" thick, on walls, standard, no finish included		2000	.008		.34	.41		.75	1.03
0350	Taped and finished (level 4 finish)		965	.017		.39	.84		1.23	1.79
0390	With compound skim coat (level 5 finish)		775	.021		.44	1.05		1.49	2.18
0400	Fire resistant, no finish included		2000	.008		.37	.41		.78	1.07
0450	Taped and finished (level 4 finish)		965	.017		.42	.84		1.26	1.82
0490	With compound skim coat (level 5 finish)		775	.021		.47	1.05		1.52	2.22
0500	Water resistant, no finish included		2000	.008		.42	.41		.83	1.12
0550	Taped and finished (level 4 finish)		965	.017		.47	.84		1.31	1.87
0590	With compound skim coat (level 5 finish)		775	.021		.52	1.05		1.57	2.27
0600	Prefinished, vinyl, clipped to studs		900	.018		.51	.90		1.41	2.02
0700	Mold resistant, no finish included		2000	.008		.44	.41		.85	1.14
0710	Taped and finished (level 4 finish)		965	.017		.49	.84		1.33	1.90
0720	With compound skim coat (level 5 finish)		775	.021		.54	1.05		1.59	2.29
1000	On ceilings, standard, no finish included		1800	.009		.34	.45		.79	1.10
1050	Taped and finished (level 4 finish)		765	.021		.39	1.06		1.45	2.15
1090	With compound skim coat (level 5 finish)		610	.026		.44	1.33		1.77	2.64
1100	Fire resistant, no finish included		1800	.009		.37	.45		.82	1.14
1150	Taped and finished (level 4 finish)		765	.021		.42	1.06		1.48	2.18
1195	With compound skim coat (level 5 finish)		610	.026		.47	1.33		1.80	2.68
1200	Water resistant, no finish included		1800	.009		.42	.45		.87	1.19
1250	Taped and finished (level 4 finish)		765	.021		.47	1.06		1.53	2.23
1290	With compound skim coat (level 5 finish)		610	.026		.52	1.33		1.85	2.73
1310	Mold resistant, no finish included		1800	.009		.44	.45		.89	1.21
1320	Taped and finished (level 4 finish)		765	.021		.49	1.06		1.55	2.26
1330	With compound skim coat (level 5 finish)		610	.026		.54	1.33		1.87	2.75
1350	Sag resistant, no finish included		1600	.010		.36	.51		.87	1.22
1360	Taped and finished (level 4 finish)		765	.021		.41	1.06		1.47	2.17
1370	With compound skim coat (level 5 finish)		610	.026		.46	1.33		1.79	2.66
1500	On beams, columns, or soffits, standard, no finish included		675	.024		.39	1.20		1.59	2.38
1550	Taped and finished (level 4 finish)		540	.030		.39	1.50		1.89	2.87
1590	With compound skim coat (level 5 finish)		475	.034		.44	1.71		2.15	3.25
1600	Fire resistant, no finish included		675	.024		.37	1.20		1.57	2.36
1650	Taped and finished (level 4 finish)		540	.030		.42	1.50		1.92	2.90
1690	With compound skim coat (level 5 finish)		475	.034		.47	1.71		2.18	3.29
1700	Water resistant, no finish included		675	.024		.48	1.20		1.68	2.48
1750	Taped and finished (level 4 finish)		540	.030		.47	1.50		1.97	2.95
1790	With compound skim coat (level 5 finish)		475	.034		.52	1.71		2.23	3.34
1800	Mold resistant, no finish included		675	.024		.51	1.20		1.71	2.51
1810	Taped and finished (level 4 finish)		540	.030		.49	1.50		1.99	2.98
1820	With compound skim coat (level 5 finish)		475	.034		.54	1.71		2.25	3.36
1850	Sag resistant, no finish included		675	.024		.41	1.20		1.61	2.41
1860	Taped and finished (level 4 finish)		540	.030		.41	1.50		1.91	2.89
1870	With compound skim coat (level 5 finish)		475	.034		.46	1.71		2.17	3.27
2000	5/8" thick, on walls, standard, no finish included		2000	.008		.35	.41		.76	1.05
2050	Taped and finished (level 4 finish)		965	.017		.40	.84		1.24	1.80
2090	With compound skim coat (level 5 finish)		775	.021		.45	1.05		1.50	2.19
2100	Fire resistant, no finish included		2000	.008		.36	.41		.77	1.06
2150	Taped and finished (level 4 finish)		965	.017		.41	.84		1.25	1.81
2195	With compound skim coat (level 5 finish)		775	.021		.46	1.05		1.51	2.20
2200	Water resistant, no finish included		2000	.008		.44	.41		.85	1.14
2250	Taped and finished (level 4 finish)		965	.017		.49	.84		1.33	1.90
2290	With compound skim coat (level 5 finish)		775	.021		.54	1.05		1.59	2.29

09 29 10.30 Gypsum Board	Crew	Daily Output	Labor-Hours	Unit	Material	2018 Bare Costs Labor	Equipment	Total	Total Incl O&P
2300	Prefinished, vinyl, clipped to studs	2 Carp	900	.018	S.F.	.79	.90		1.69	2.33
2510	Mold resistant, no finish included		2000	.008		.49	.41		.90	1.20
2520	Taped and finished (level 4 finish)		965	.017		.54	.84		1.38	1.95
2530	With compound skim coat (level 5 finish)		775	.021		.59	1.05		1.64	2.35
3000	On ceilings, standard, no finish included		1800	.009		.35	.45		.80	1.12
3050	Taped and finished (level 4 finish)		765	.021		.40	1.06		1.46	2.16
3090	With compound skim coat (level 5 finish)		615	.026		.45	1.32		1.77	2.63
3100	Fire resistant, no finish included		1800	.009		.36	.45		.81	1.13
3150	Taped and finished (level 4 finish)		765	.021		.41	1.06		1.47	2.17
3190	With compound skim coat (level 5 finish)		615	.026		.46	1.32		1.78	2.64
3200	Water resistant, no finish included		1800	.009		.44	.45		.89	1.21
3250	Taped and finished (level 4 finish)		765	.021		.49	1.06		1.55	2.26
3290	With compound skim coat (level 5 finish)		615	.026		.54	1.32		1.86	2.73
3300	Mold resistant, no finish included		1800	.009		.49	.45		.94	1.27
3310	Taped and finished (level 4 finish)		765	.021		.54	1.06		1.60	2.31
3320	With compound skim coat (level 5 finish)		615	.026		.59	1.32		1.91	2.79
3500	On beams, columns, or soffits, no finish included		675	.024		.40	1.20		1.60	2.39
3550	Taped and finished (level 4 finish)		475	.034		.46	1.71		2.17	3.27
3590	With compound skim coat (level 5 finish)		380	.042		.52	2.14		2.66	4.04
3600	Fire resistant, no finish included		675	.024		.41	1.20		1.61	2.41
3650	Taped and finished (level 4 finish)		475	.034		.47	1.71		2.18	3.28
3690	With compound skim coat (level 5 finish)		380	.042		.46	2.14		2.60	3.97
3700	Water resistant, no finish included		675	.024		.51	1.20		1.71	2.51
3750	Taped and finished (level 4 finish)		475	.034		.54	1.71		2.25	3.36
3790	With compound skim coat (level 5 finish)		380	.042		.56	2.14		2.70	4.09
3800	Mold resistant, no finish included		675	.024		.56	1.20		1.76	2.57
3810	Taped and finished (level 4 finish)		475	.034		.59	1.71		2.30	3.42
3820	With compound skim coat (level 5 finish)		380	.042		.62	2.14		2.76	4.15
4000	Fireproofing, beams or columns, 2 layers, 1/2" thick, incl finish		330	.048		.83	2.46		3.29	4.91
4010	Mold resistant		330	.048		.97	2.46		3.43	5.05
4050	5/8" thick		300	.053		.81	2.70		3.51	5.30
4060	Mold resistant		300	.053		1.07	2.70		3.77	5.55
4100	3 layers, 1/2" thick		225	.071		1.25	3.61		4.86	7.20
4110	Mold resistant		225	.071		1.46	3.61		5.07	7.45
4150	5/8" thick		210	.076		1.22	3.86		5.08	7.60
4160	Mold resistant		210	.076		1.61	3.86		5.47	8
5050	For 1" thick coreboard on columns		480	.033		.77	1.69		2.46	3.59
5100	For foil-backed board, add					.18			.18	.20
5200	For work over 8' high, add	2 Carp	3060	.005			.27		.27	.43
5270	For textured spray, add	2 Lath	1600	.010		.04	.49		.53	.82
5300	For distribution cost 3 stories and above, add per story	2 Carp	6100	.003			.13		.13	.22
5350	For finishing inner corners, add		950	.017	L.F.	.10	.85		.95	1.50
5355	For finishing outer corners, add		1250	.013		.23	.65		.88	1.30
5500	For acoustical sealant, add per bead	1 Carp	500	.016		.04	.81		.85	1.37
5550	Sealant, 1 quart tube				Ea.	7.05			7.05	7.80
6000	Gypsum sound dampening panels									
6010	1/2" thick on walls, multi-layer, lightweight, no finish included	2 Carp	1500	.011	S.F.	2.16	.54		2.70	3.26
6015	Taped and finished (level 4 finish)		725	.022		2.21	1.12		3.33	4.25
6020	With compound skim coat (level 5 finish)		580	.028		2.26	1.40		3.66	4.75
6025	5/8" thick on walls, for wood studs, no finish included		1500	.011		2.21	.54		2.75	3.31
6030	Taped and finished (level 4 finish)		725	.022		2.26	1.12		3.38	4.30
6035	With compound skim coat (level 5 finish)		580	.028		2.31	1.40		3.71	4.81
6040	For metal stud, no finish included		1500	.011		2.28	.54		2.82	3.39

09 29 Gypsum Board

09 29 10 – Gypsum Board Panels

09 29 10.30 Gypsum Board

		Crew	Daily Output	Labor-Hours	Unit	Material	2018 Bare Costs Labor	Equipment	Total	Total Incl O&P
6045	Taped and finished (level 4 finish)	2 Carp	725	.022	S.F.	2.33	1.12		3.45	4.38
6050	With compound skim coat (level 5 finish)		580	.028		2.38	1.40		3.78	4.89
6055	Abuse resist, no finish included		1500	.011		4.10	.54		4.64	5.40
6060	Taped and finished (level 4 finish)		725	.022		4.15	1.12		5.27	6.40
6065	With compound skim coat (level 5 finish)		580	.028		4.20	1.40		5.60	6.90
6070	Shear rated, no finish included		1500	.011		5.15	.54		5.69	6.55
6075	Taped and finished (level 4 finish)		725	.022		5.20	1.12		6.32	7.50
6080	With compound skim coat (level 5 finish)		580	.028		5.25	1.40		6.65	8
6085	For SCIF applications, no finish included		1500	.011		5.15	.54		5.69	6.55
6090	Taped and finished (level 4 finish)		725	.022		5.20	1.12		6.32	7.50
6095	With compound skim coat (level 5 finish)		580	.028		5.25	1.40		6.65	8
6100	1-3/8" thick on walls, THX certified, no finish included		1500	.011		9.15	.54		9.69	10.95
6105	Taped and finished (level 4 finish)		725	.022		9.20	1.12		10.32	11.90
6110	With compound skim coat (level 5 finish)		580	.028		9.25	1.40		10.65	12.40
6115	5/8" thick on walls, score & snap installation, no finish included		2000	.008		1.96	.41		2.37	2.82
6120	Taped and finished (level 4 finish)		965	.017		2.01	.84		2.85	3.57
6125	With compound skim coat (level 5 finish)		775	.021		2.06	1.05		3.11	3.96
7020	5/8" thick on ceilings, for wood joists, no finish included		1200	.013		2.21	.68		2.89	3.53
7025	Taped and finished (level 4 finish)		510	.031		2.26	1.59		3.85	5.05
7030	With compound skim coat (level 5 finish)		410	.039		2.31	1.98		4.29	5.75
7035	For metal joists, no finish included		1200	.013		2.28	.68		2.96	3.61
7040	Taped and finished (level 4 finish)		510	.031		2.33	1.59		3.92	5.15
7045	With compound skim coat (level 5 finish)		410	.039		2.38	1.98		4.36	5.85
7050	Abuse resist, no finish included		1200	.013		4.10	.68		4.78	5.60
7055	Taped and finished (level 4 finish)		510	.031		4.15	1.59		5.74	7.15
7060	With compound skim coat (level 5 finish)		410	.039		4.20	1.98		6.18	7.85
7065	Shear rated, no finish included		1200	.013		5.15	.68		5.83	6.75
7070	Taped and finished (level 4 finish)		510	.031		5.20	1.59		6.79	8.30
7075	With compound skim coat (level 5 finish)		410	.039		5.25	1.98		7.23	8.95
7080	For SCIF applications, no finish included		1200	.013		5.15	.68		5.83	6.75
7085	Taped and finished (level 4 finish)		510	.031		5.20	1.59		6.79	8.30
7090	With compound skim coat (level 5 finish)		410	.039		5.25	1.98		7.23	8.95
8010	5/8" thick on ceilings, score & snap installation, no finish included		1600	.010		1.96	.51		2.47	2.98
8015	Taped and finished (level 4 finish)		680	.024		2.01	1.19		3.20	4.15
8020	With compound skim coat (level 5 finish)		545	.029	S.F.	2.06	1.49		3.55	4.68
9000	Minimum labor/equipment charge	1 Carp	2	4	Job		203		203	330

09 29 10.50 High Abuse Gypsum Board

		Crew	Daily Output	Labor-Hours	Unit	Material	2018 Bare Costs Labor	Equipment	Total	Total Incl O&P
0010	**HIGH ABUSE GYPSUM BOARD**, fiber reinforced, nailed or									
0100	screwed to studs unless otherwise noted									
0110	1/2" thick, on walls, no finish included	2 Carp	1800	.009	S.F.	.78	.45		1.23	1.59
0120	Taped and finished (level 4 finish)		870	.018		.83	.93		1.76	2.42
0130	With compound skim coat (level 5 finish)		700	.023		.88	1.16		2.04	2.85
0150	On ceilings, no finish included		1620	.010		.78	.50		1.28	1.67
0160	Taped and finished (level 4 finish)		690	.023		.83	1.18		2.01	2.82
0170	With compound skim coat (level 5 finish)		550	.029		.88	1.47		2.35	3.36
0210	5/8" thick, on walls, no finish included		1800	.009		.81	.45		1.26	1.62
0220	Taped and finished (level 4 finish)		870	.018		.86	.93		1.79	2.45
0230	With compound skim coat (level 5 finish)		700	.023		.91	1.16		2.07	2.88
0250	On ceilings, no finish included		1620	.010		.81	.50		1.31	1.70
0260	Taped and finished (level 4 finish)		690	.023		.86	1.18		2.04	2.85
0270	With compound skim coat (level 5 finish)		550	.029		.91	1.47		2.38	3.39
0272	5/8" thick, on roof sheathing for 1-hour rating		1620	.010		.81	.50		1.31	1.70

09 29 10 – Gypsum Board Panels

09 29 10.50 High Abuse Gypsum Board

		Crew	Daily Output	Labor-Hours	Unit	Material	2018 Bare Costs Labor	2018 Bare Costs Equipment	Total	Total Incl O&P
0310	5/8" thick, on walls, very high impact, no finish included	2 Carp	1800	.009	S.F.	.90	.45		1.35	1.72
0320	Taped and finished (level 4 finish)		870	.018		.95	.93		1.88	2.55
0330	With compound skim coat (level 5 finish)		700	.023		1	1.16		2.16	2.98
0350	On ceilings, no finish included		1620	.010		.90	.50		1.40	1.80
0360	Taped and finished (level 4 finish)		690	.023		.95	1.18		2.13	2.95
0370	With compound skim coat (level 5 finish)	↓	550	.029	↓	1	1.47		2.47	3.49
0400	High abuse, gypsum core, paper face									
0410	1/2" thick, on walls, no finish included	2 Carp	1800	.009	S.F.	.64	.45		1.09	1.43
0420	Taped and finished (level 4 finish)		870	.018		.69	.93		1.62	2.27
0430	With compound skim coat (level 5 finish)		700	.023		.74	1.16		1.90	2.69
0450	On ceilings, no finish included		1620	.010		.64	.50		1.14	1.51
0460	Taped and finished (level 4 finish)		690	.023		.69	1.18		1.87	2.67
0470	With compound skim coat (level 5 finish)		550	.029		.74	1.47		2.21	3.20
0510	5/8" thick, on walls, no finish included		1800	.009		.69	.45		1.14	1.49
0520	Taped and finished (level 4 finish)		870	.018		.74	.93		1.67	2.32
0530	With compound skim coat (level 5 finish)		700	.023		.79	1.16		1.95	2.75
0550	On ceilings, no finish included		1620	.010		.69	.50		1.19	1.57
0560	Taped and finished (level 4 finish)		690	.023		.74	1.18		1.92	2.72
0570	With compound skim coat (level 5 finish)		550	.029		.79	1.47		2.26	3.26
1000	For high ceilings, over 8' high, add		2750	.006			.30		.30	.48
1010	For distribution cost 3 stories and above, add per story	↓	5500	.003			.15		.15	.24

09 29 15 – Gypsum Board Accessories

09 29 15.10 Accessories, Gypsum Board

		Crew	Daily Output	Labor-Hours	Unit	Material	2018 Bare Costs Labor	2018 Bare Costs Equipment	Total	Total Incl O&P
0010	**ACCESSORIES, GYPSUM BOARD**									
0020	Casing bead, galvanized steel	1 Carp	2.90	2.759	C.L.F.	24	140		164	254
0100	Vinyl		3	2.667		23	135		158	245
0300	Corner bead, galvanized steel, 1" x 1"		4	2		15.50	101		116.50	182
0400	1-1/4" x 1-1/4"		3.50	2.286		16.95	116		132.95	207
0600	Vinyl		4	2		20.50	101		121.50	188
0900	Furring channel, galv. steel, 7/8" deep, standard		2.60	3.077		34.50	156		190.50	291
1000	Resilient		2.55	3.137		26	159		185	287
1100	J trim, galvanized steel, 1/2" wide		3	2.667		22.50	135		157.50	244
1120	5/8" wide		2.95	2.712		32.50	137		169.50	259
1500	Z stud, galvanized steel, 1-1/2" wide		2.60	3.077	↓	41.50	156		197.50	299
9000	Minimum labor/equipment charge	↓	3	2.667	Job		135		135	219

09 30 Tiling

09 30 13 – Ceramic Tiling

09 30 13.20 Ceramic Tile Repairs

		Crew	Daily Output	Labor-Hours	Unit	Material	2018 Bare Costs Labor	2018 Bare Costs Equipment	Total	Total Incl O&P
0010	**CERAMIC TILE REPAIRS**									
1000	Grout removal, carbide tipped, rotary grinder	1 Clab	240	.033	L.F.		1.33		1.33	2.16

09 30 13.45 Ceramic Tile Accessories

		Crew	Daily Output	Labor-Hours	Unit	Material	2018 Bare Costs Labor	2018 Bare Costs Equipment	Total	Total Incl O&P
0010	**CERAMIC TILE ACCESSORIES**									
0100	Spacers, 1/8"				C	1.98			1.98	2.18
1310	Sealer for natural stone tile, installed	1 Tilf	650	.012	S.F.	.05	.58		.63	.97

09 30 29 – Metal Tiling

09 30 29.10 Metal Tile

		Crew	Daily Output	Labor-Hours	Unit	Material	2018 Bare Costs Labor	2018 Bare Costs Equipment	Total	Total Incl O&P
0010	**METAL TILE** 4' x 4' sheet, 24 ga., tile pattern, nailed									
0200	Stainless steel	2 Carp	512	.031	S.F.	28	1.58		29.58	33.50
0400	Aluminized steel	"	512	.031	"	19	1.58		20.58	23.50

For customer support on your Commercial Renovation Costs with RSMeans data, call 800.448.8182.

09 30 Tiling

09 30 29 – Metal Tiling

09 30 29.10 Metal Tile

	Crew	Daily Output	Labor-Hours	Unit	Material	2018 Bare Costs Labor	Equipment	Total	Total Incl O&P
9000	Minimum labor/equipment charge	1 Carp	4	2	Job		101		101	165

09 30 95 – Tile & Stone Setting Materials and Specialties

09 30 95.10 Moisture Resistant, Anti-Fracture Membrane

	Crew	Daily Output	Labor-Hours	Unit	Material	2018 Bare Costs Labor	Equipment	Total	Total Incl O&P
0010	**MOISTURE RESISTANT, ANTI-FRACTURE MEMBRANE**									
0200	Elastomeric membrane, 1/16" thick	D-7	275	.058	S.F.	1.12	2.45		3.57	5.10

09 31 Thin-Set Tiling

09 31 13 – Thin-Set Ceramic Tiling

09 31 13.10 Thin-Set Ceramic Tile

		Crew	Daily Output	Labor-Hours	Unit	Material	2018 Bare Costs Labor	Equipment	Total	Total Incl O&P
0010	**THIN-SET CERAMIC TILE**									
0700	Cove base, 4-1/4" x 4-1/4"	D-7	128	.125	L.F.	4.19	5.25		9.44	12.90
1000	6" x 4-1/4" high		137	.117		4.32	4.91		9.23	12.50
1300	Sanitary cove base, 6" x 4-1/4" high		124	.129		4.64	5.40		10.04	13.65
1600	6" x 6" high		117	.137		5.30	5.75		11.05	14.90
2500	Bullnose trim, 4-1/4" x 4-1/4"		128	.125		4.17	5.25		9.42	12.90
2800	2" x 6"		124	.129		4.22	5.40		9.62	13.20
3300	Ceramic tile, porcelain type, 1 color, color group 2, 1" x 1"		183	.087	S.F.	6.40	3.67		10.07	12.85
3310	2" x 2" or 2" x 1"		190	.084		6.25	3.54		9.79	12.50
3350	For random blend, 2 colors, add					1			1	1.10
3360	4 colors, add					1.50			1.50	1.65
4300	Specialty tile, 4-1/4" x 4-1/4" x 1/2", decorator finish	D-7	183	.087		12.70	3.67		16.37	19.80
4500	Add for epoxy grout, 1/16" joint, 1" x 1" tile		800	.020		.68	.84		1.52	2.08
4600	2" x 2" tile		820	.020		.62	.82		1.44	1.98
4610	Add for epoxy grout, 1/8" joint, 8" x 8" x 3/8" tile, add		900	.018		1.42	.75		2.17	2.74
4800	Pregrouted sheets, walls, 4-1/4" x 4-1/4", 6" x 4-1/4"									
4810	and 8-1/2" x 4-1/4", 4 S.F. sheets, silicone grout	D-7	240	.067	S.F.	5.60	2.80		8.40	10.65
5100	Floors, unglazed, 2 S.F. sheets,									
5110	urethane adhesive	D-7	180	.089	S.F.	2.03	3.74		5.77	8.15
5400	Walls, interior, 4-1/4" x 4-1/4" tile		190	.084		2.61	3.54		6.15	8.45
5500	6" x 4-1/4" tile		190	.084		3.15	3.54		6.69	9.05
5700	8-1/2" x 4-1/4" tile		190	.084		5.65	3.54		9.19	11.85
5800	6" x 6" tile		175	.091		3.57	3.84		7.41	10
5810	8" x 8" tile		170	.094		5.30	3.96		9.26	12.10
5820	12" x 12" tile		160	.100		4.66	4.20		8.86	11.80
5830	16" x 16" tile		150	.107		5.15	4.48		9.63	12.75
6000	Decorated wall tile, 4-1/4" x 4-1/4", color group 1		270	.059		3.68	2.49		6.17	8
6100	Color group 4		180	.089		52.50	3.74		56.24	63.50
9300	Ceramic tiles, recycled glass, standard colors, 2" x 2" thru 6" x 6" **G**		190	.084		22	3.54		25.54	30
9310	6" x 6" **G**		175	.091		22	3.84		25.84	30.50
9320	8" x 8" **G**		170	.094		23.50	3.96		27.46	32.50
9330	12" x 12" **G**		160	.100		23.50	4.20		27.70	32.50
9340	Earthtones, 2" x 2" to 4" x 8" **G**		190	.084		26	3.54		29.54	34
9350	6" x 6" **G**		175	.091		26	3.84		29.84	34.50
9360	8" x 8" **G**		170	.094		27	3.96		30.96	36
9370	12" x 12" **G**		160	.100		27	4.20		31.20	36
9380	Deep colors, 2" x 2" to 4" x 8" **G**		190	.084		30.50	3.54		34.04	39.50
9390	6" x 6" **G**		175	.091		30.50	3.84		34.34	40
9400	8" x 8" **G**		170	.094		32	3.96		35.96	42
9410	12" x 12" **G**		160	.100		32	4.20		36.20	42

09 31 Thin-Set Tiling

09 31 33 – Thin-Set Stone Tiling

09 31 33.10 Tiling, Thin-Set Stone	Crew	Daily Output	Labor-Hours	Unit	Material	2018 Bare Costs Labor	Equipment	Total	Total Incl O&P
0010 **TILING, THIN-SET STONE**									
3000 Floors, natural clay, random or uniform, color group 1	D-7	183	.087	S.F.	4.31	3.67		7.98	10.55
3100 Color group 2	"	183	.087	"	5.95	3.67		9.62	12.35

09 32 Mortar-Bed Tiling

09 32 13 – Mortar-Bed Ceramic Tiling

09 32 13.10 Ceramic Tile

	Crew	Daily Output	Labor-Hours	Unit	Material	2018 Bare Costs Labor	Equipment	Total	Total Incl O&P
0010 **CERAMIC TILE**									
0600 Cove base, 4-1/4" x 4-1/4" high	D-7	91	.176	L.F.	4.33	7.40		11.73	16.40
0900 6" x 4-1/4" high		100	.160		4.46	6.70		11.16	15.50
1200 Sanitary cove base, 6" x 4-1/4" high		93	.172		4.78	7.25		12.03	16.65
1500 6" x 6" high		84	.190		5.45	8		13.45	18.60
2400 Bullnose trim, 4-1/4" x 4-1/4"		82	.195		4.27	8.20		12.47	17.65
2700 2" x 6" bullnose trim		84	.190		4.29	8		12.29	17.35
6210 Wall tile, 4-1/4" x 4-1/4", better grade	1 Tilf	50	.160	S.F.	9.50	7.50		17	22.50
6240 2" x 2"		50	.160		6.50	7.50		14	19
6250 6" x 6"		55	.145		10.15	6.80		16.95	22
6260 8" x 8"		60	.133		9.40	6.25		15.65	20
6600 Crystalline glazed, 4-1/4" x 4-1/4", plain	D-7	100	.160		4.52	6.70		11.22	15.55
6700 4-1/4" x 4-1/4", scored tile		100	.160		6.50	6.70		13.20	17.75
6900 6" x 6" plain		93	.172		5.75	7.25		13	17.75
7000 For epoxy grout, 1/16" joints, 4-1/4" tile, add		800	.020		.41	.84		1.25	1.78
7200 For tile set in dry mortar, add		1735	.009			.39		.39	.61
7300 For tile set in Portland cement mortar, add		290	.055		.18	2.32		2.50	3.86
9500 Minimum labor/equipment charge		3.25	4.923	Job		207		207	325

09 32 16 – Mortar-Bed Quarry Tiling

09 32 16.10 Quarry Tile

	Crew	Daily Output	Labor-Hours	Unit	Material	2018 Bare Costs Labor	Equipment	Total	Total Incl O&P
0010 **QUARRY TILE**									
0100 Base, cove or sanitary, to 5" high, 1/2" thick	D-7	110	.145	L.F.	6.40	6.10		12.50	16.70
0300 Bullnose trim, red, 6" x 6" x 1/2" thick		120	.133		5.40	5.60		11	14.75
0400 4" x 4" x 1/2" thick		110	.145		5	6.10		11.10	15.15
0600 4" x 8" x 1/2" thick, using 8" as edge		130	.123		5.35	5.15		10.50	14.05
0700 Floors, 1,000 S.F. lots, red, 4" x 4" x 1/2" thick		120	.133	S.F.	8.70	5.60		14.30	18.45
0900 6" x 6" x 1/2" thick		140	.114		8.25	4.80		13.05	16.70
1000 4" x 8" x 1/2" thick		130	.123		6.60	5.15		11.75	15.45
1300 For waxed coating, add					.76			.76	.84
1500 For non-standard colors, add					.46			.46	.51
1600 For abrasive surface, add					.52			.52	.57
1800 Brown tile, imported, 6" x 6" x 3/4"	D-7	120	.133		7.20	5.60		12.80	16.75
1900 8" x 8" x 1"		110	.145		9.55	6.10		15.65	20
2100 For thin set mortar application, deduct		700	.023			.96		.96	1.52
2700 Stair tread, 6" x 6" x 3/4", plain		50	.320		7.20	13.45		20.65	29.50
2800 Abrasive		47	.340		8.55	14.30		22.85	32
3000 Wainscot, 6" x 6" x 1/2", thin set, red		105	.152		6.30	6.40		12.70	17
3100 Non-standard colors		105	.152		6.40	6.40		12.80	17.10
3300 Window sill, 6" wide, 3/4" thick		90	.178	L.F.	8.65	7.45		16.10	21.50
3400 Corners		80	.200	Ea.	6	8.40		14.40	19.90
9000 Minimum labor/equipment charge		3.25	4.923	Job		207		207	325

09 32 Mortar-Bed Tiling

09 32 23 – Mortar-Bed Glass Mosaic Tiling

09 32 23.10 Glass Mosaics		Crew	Daily Output	Labor-Hours	Unit	Material	2018 Bare Costs Labor	Equipment	Total	Total Incl O&P
0010	**GLASS MOSAICS** 3/4" tile on 12" sheets, standard grout									
1020	1" tile on 12" sheets, opalescent finish	D-7	73	.219	S.F.	18.40	9.20		27.60	35
1040	1" x 2" tile on 12" sheet, blend		73	.219		21	9.20		30.20	37.50
1060	2" tile on 12" sheet, blend		73	.219		17.60	9.20		26.80	34
1080	5/8" x random tile, linear, on 12" sheet, blend		73	.219		25	9.20		34.20	41.50
1600	Dots on 12" sheet		73	.219		25	9.20		34.20	42
1700	For glass mosaic tiles set in dry mortar, add		290	.055		.45	2.32		2.77	4.16
1720	For glass mosaic tiles set in Portland cement mortar, add		290	.055	▼	.01	2.32		2.33	3.67
1730	For polyblend sanded tile grout	▼	96.15	.166	Lb.	2.19	7		9.19	13.45

09 34 Waterproofing-Membrane Tiling

09 34 13 – Waterproofing-Membrane Ceramic Tiling

09 34 13.10 Ceramic Tile Waterproofing Membrane

		Crew	Daily Output	Labor-Hours	Unit	Material	Labor	Equipment	Total	Total Incl O&P
0010	**CERAMIC TILE WATERPROOFING MEMBRANE**									
0020	On floors, including thinset									
0030	Fleece laminated polyethylene grid, 1/8" thick	D-7	250	.064	S.F.	2.28	2.69		4.97	6.75
0040	5/16" thick	"	250	.064	"	2.60	2.69		5.29	7.10
0050	On walls, including thinset									
0060	Fleece laminated polyethylene sheet, 8 mil thick	D-7	480	.033	S.F.	2.28	1.40		3.68	4.71
0070	Accessories, including thinset									
0080	Joint and corner sheet, 4 mils thick, 5" wide	1 Tilf	240	.033	L.F.	1.35	1.56		2.91	3.95
0090	7-1/4" wide		180	.044		1.71	2.08		3.79	5.15
0100	10" wide		120	.067	▼	2.08	3.12		5.20	7.20
0110	Pre-formed corners, inside		32	.250	Ea.	7.85	11.70		19.55	27
0120	Outside		32	.250		7.65	11.70		19.35	27
0130	2" flanged floor drain with 6" stainless steel grate		16	.500	▼	370	23.50		393.50	445
0140	EPS, sloped shower floor		480	.017	S.F.	5.55	.78		6.33	7.35
0150	Curb	▼	32	.250	L.F.	14.05	11.70		25.75	34

09 35 Chemical-Resistant Tiling

09 35 13 – Chemical-Resistant Ceramic Tiling

09 35 13.10 Chemical-Resistant Ceramic Tiling

		Crew	Daily Output	Labor-Hours	Unit	Material	Labor	Equipment	Total	Total Incl O&P
0010	**CHEMICAL-RESISTANT CERAMIC TILING**									
0100	4-1/4" x 4-1/4" x 1/4", 1/8" joint	D-7	130	.123	S.F.	12.05	5.15		17.20	21.50
0200	6"x 6" x 1/2" thick		120	.133		9.75	5.60		15.35	19.55
0300	8"x 8" x 1/2" thick		110	.145		10.90	6.10		17	21.50
0400	4-1/4" x 4-1/4" x 1/4", 1/4" joint		130	.123		12.80	5.15		17.95	22.50
0500	6"x 6" x 1/2" thick		120	.133		10.85	5.60		16.45	21
0600	8"x 8" x 1/2" thick		110	.145		11.55	6.10		17.65	22.50
0700	4-1/4" x 4-1/4" x 1/4", 3/8" joint		130	.123		13.50	5.15		18.65	23
0800	6"x 6" x 1/2" thick		120	.133		11.80	5.60		17.40	22
0900	8"x 8" x 1/2" thick	▼	110	.145	▼	12.70	6.10		18.80	23.50

09 35 16 – Chemical-Resistant Quarry Tiling

09 35 16.10 Chemical-Resistant Quarry Tiling

		Crew	Daily Output	Labor-Hours	Unit	Material	Labor	Equipment	Total	Total Incl O&P
0010	**CHEMICAL-RESISTANT QUARRY TILING**									
0100	4"x 8" x 1/2" thick, 1/8" joint	D-7	130	.123	S.F.	11.20	5.15		16.35	20.50
0200	6"x 6" x 1/2" thick		120	.133		11.25	5.60		16.85	21
0300	8"x 8" x 1/2" thick	▼	110	.145	▼	10.35	6.10		16.45	21

09 35 16 – Chemical-Resistant Quarry Tiling

09 35 16.10 Chemical-Resistant Quarry Tiling	Crew	Daily Output	Labor-Hours	Unit	Material	2018 Bare Costs Labor	2018 Bare Costs Equipment	Total	Total Incl O&P
0400	4" x 8" x 1/2" thick, 1/4" joint	D-7	130	.123	S.F.	12.40	5.15		17.55	22
0500	6" x 6" x 1/2" thick		120	.133		12.35	5.60		17.95	22.50
0600	8" x 8" x 1/2" thick		110	.145		11	6.10		17.10	22
0700	4" x 8" x 1/2" thick, 3/8" joint		130	.123		13.50	5.15		18.65	23
0800	6" x 6" x 1/2" thick		120	.133		13.30	5.60		18.90	23.50
0900	8" x 8" x 1/2" thick		110	.145		12.15	6.10		18.25	23

09 51 Acoustical Ceilings

09 51 13 – Acoustical Panel Ceilings

09 51 13.10 Ceiling, Acoustical Panel

		Crew	Daily Output	Labor-Hours	Unit	Material	2018 Bare Costs Labor	2018 Bare Costs Equipment	Total	Total Incl O&P
0010	**CEILING, ACOUSTICAL PANEL**									
0100	Fiberglass boards, film faced, 2' x 2' or 2' x 4', 5/8" thick	1 Carp	625	.013	S.F.	1.26	.65		1.91	2.44
0120	3/4" thick		600	.013		3	.68		3.68	4.40
0130	3" thick, thermal, R11		450	.018		3.50	.90		4.40	5.30

09 51 14 – Acoustical Fabric-Faced Panel Ceilings

09 51 14.10 Ceiling, Acoustical Fabric-Faced Panel

		Crew	Daily Output	Labor-Hours	Unit	Material	2018 Bare Costs Labor	2018 Bare Costs Equipment	Total	Total Incl O&P
0010	**CEILING, ACOUSTICAL FABRIC-FACED PANEL**									
0100	Glass cloth faced fiberglass, 3/4" thick	1 Carp	500	.016	S.F.	2.99	.81		3.80	4.61
0120	1" thick		485	.016		3.61	.84		4.45	5.35
0130	1-1/2" thick, nubby face		475	.017		2.73	.85		3.58	4.39

09 51 23 – Acoustical Tile Ceilings

09 51 23.10 Suspended Acoustic Ceiling Tiles

		Crew	Daily Output	Labor-Hours	Unit	Material	2018 Bare Costs Labor	2018 Bare Costs Equipment	Total	Total Incl O&P
0010	**SUSPENDED ACOUSTIC CEILING TILES**, not including									
0100	suspension system									
1110	Mineral fiber tile, lay-in, 2' x 2' or 2' x 4', 5/8" thick, fine texture	1 Carp	625	.013	S.F.	.79	.65		1.44	1.92
1115	Rough textured		625	.013		.75	.65		1.40	1.88
1125	3/4" thick, fine textured		600	.013		2.11	.68		2.79	3.42
1130	Rough textured		600	.013		1.73	.68		2.41	3
1135	Fissured		600	.013		2.17	.68		2.85	3.49
1150	Tegular, 5/8" thick, fine textured		470	.017		1.12	.86		1.98	2.63
1155	Rough textured		470	.017		1.34	.86		2.20	2.87
1165	3/4" thick, fine textured		450	.018		2.33	.90		3.23	4.02
1170	Rough textured		450	.018		1.53	.90		2.43	3.14
1175	Fissured		450	.018		2.21	.90		3.11	3.89
1185	For plastic film face, add					.81			.81	.89
1190	For fire rating, add					.50			.50	.55
3720	Mineral fiber, 24" x 24" or 48", reveal edge, painted, 5/8" thick	1 Carp	600	.013		1.36	.68		2.04	2.60
3740	3/4" thick		575	.014		1.67	.71		2.38	2.98
5020	66-78% recycled content, 3/4" thick G		600	.013		2.08	.68		2.76	3.39
5040	Mylar, 42% recycled content, 3/4" thick G		600	.013		5	.68		5.68	6.60
6000	Remove & install new ceiling tiles, min fiber, 2' x 2' or 2' x 4', 5/8"thk.		335	.024		.79	1.21		2	2.84
9000	Minimum labor/equipment charge		4	2	Job		101		101	165

09 51 23.30 Suspended Ceilings, Complete

		Crew	Daily Output	Labor-Hours	Unit	Material	2018 Bare Costs Labor	2018 Bare Costs Equipment	Total	Total Incl O&P
0010	**SUSPENDED CEILINGS, COMPLETE**, incl. standard									
0100	suspension system but not incl. 1-1/2" carrier channels									
0600	Fiberglass ceiling board, 2' x 4' x 5/8", plain faced	1 Carp	500	.016	S.F.	2.05	.81		2.86	3.58
0700	Offices, 2' x 4' x 3/4"		380	.021		3.79	1.07		4.86	5.90
0800	Mineral fiber, on 15/16" T bar susp. 2' x 2' x 3/4" lay-in board		345	.023		3.13	1.18		4.31	5.35
0810	2' x 4' x 5/8" tile		380	.021		2.21	1.07		3.28	4.16
0820	Tegular, 2' x 2' x 5/8" tile on 9/16" grid		250	.032		2.58	1.62		4.20	5.45

09 51 Acoustical Ceilings

09 51 23 – Acoustical Tile Ceilings

09 51 23.30 Suspended Ceilings, Complete

		Crew	Daily Output	Labor-Hours	Unit	Material	2018 Bare Costs Labor	Equipment	Total	Total Incl O&P
0830	2' x 4' x 3/4" tile	1 Carp	275	.029	S.F.	2.82	1.47		4.29	5.50
0900	Luminous panels, prismatic, acrylic		255	.031		3.72	1.59		5.31	6.65
1200	Metal pan with acoustic pad, steel		75	.107		4.93	5.40		10.33	14.20
1300	Painted aluminum		75	.107		3.49	5.40		8.89	12.65
1500	Aluminum, degreased finish		75	.107		4.94	5.40		10.34	14.25
1600	Stainless steel		75	.107		9.30	5.40		14.70	19.05
1800	Tile, Z bar suspension, 5/8" mineral fiber tile		150	.053		2.14	2.70		4.84	6.75
1900	3/4" mineral fiber tile	↓	150	.053	↓	2.42	2.70		5.12	7.05
2402	For strip lighting, see Section 26 51 13.50									
2500	For rooms under 500 S.F., add				S.F.		25%			
9000	Minimum labor/equipment charge	1 Carp	2	4	Job		203		203	330

09 51 33 – Acoustical Metal Pan Ceilings

09 51 33.10 Ceiling, Acoustical Metal Pan

		Crew	Daily Output	Labor-Hours	Unit	Material	2018 Bare Costs Labor	Equipment	Total	Total Incl O&P
0010	**CEILING, ACOUSTICAL METAL PAN**									
0100	Metal panel, lay-in, 2' x 2', sq. edge	1 Carp	500	.016	S.F.	10.90	.81		11.71	13.30
0110	Tegular edge		500	.016		14	.81		14.81	16.70
0120	2' x 4', sq. edge		500	.016		13.75	.81		14.56	16.45
0130	Tegular edge		500	.016		14	.81		14.81	16.70
0140	Perforated alum. clip-in, 2' x 2'		500	.016		14	.81		14.81	16.70
0150	2' x 4'	↓	500	.016	↓	11.40	.81		12.21	13.85

09 51 53 – Direct-Applied Acoustical Ceilings

09 51 53.10 Ceiling Tile

		Crew	Daily Output	Labor-Hours	Unit	Material	2018 Bare Costs Labor	Equipment	Total	Total Incl O&P
0010	**CEILING TILE**, stapled or cemented									
0100	12" x 12" or 12" x 24", not including furring									
0600	Mineral fiber, vinyl coated, 5/8" thick	1 Carp	300	.027	S.F.	2.34	1.35		3.69	4.76
0700	3/4" thick		300	.027		3.05	1.35		4.40	5.55
0900	Fire rated, 3/4" thick, plain faced		300	.027		1.42	1.35		2.77	3.75
1000	Plastic coated face		300	.027		2.10	1.35		3.45	4.50
1200	Aluminum faced, 5/8" thick, plain		300	.027		1.83	1.35		3.18	4.20
3000	Wood fiber tile, 1/2" thick		400	.020		1.20	1.01		2.21	2.97
3100	3/4" thick	↓	400	.020		1.16	1.01		2.17	2.93
3300	For flameproofing, add					.07			.07	.08
3400	For sculptured 3 dimensional, add					.32			.32	.35
3900	For ceiling primer, add					.12			.12	.13
4000	For ceiling cement, add				↓	.40			.40	.44
9000	Minimum labor/equipment charge	1 Carp	4	2	Job		101		101	165

09 53 Acoustical Ceiling Suspension Assemblies

09 53 23 – Metal Acoustical Ceiling Suspension Assemblies

09 53 23.30 Ceiling Suspension Systems

		Crew	Daily Output	Labor-Hours	Unit	Material	2018 Bare Costs Labor	Equipment	Total	Total Incl O&P
0010	**CEILING SUSPENSION SYSTEMS** for boards and tile									
0050	Class A suspension system, 15/16" T bar, 2' x 4' grid	1 Carp	800	.010	S.F.	.79	.51		1.30	1.69
0300	2' x 2' grid		650	.012		1.02	.62		1.64	2.13
0310	25% recycled steel, 2' x 4' grid ⒼG		800	.010		.83	.51		1.34	1.73
0320	2' x 2' grid ⒼG	↓	650	.012		1.04	.62		1.66	2.15
0350	For 9/16" grid, add					.16			.16	.18
0360	For fire rated grid, add					.09			.09	.10
0370	For colored grid, add					.21			.21	.23
0400	Concealed Z bar suspension system, 12" module	1 Carp	520	.015		.93	.78		1.71	2.29
0600	1-1/2" carrier channels, 4' OC, add	"	470	.017	↓	.12	.86		.98	1.53

For customer support on your Commercial Renovation Costs with RSMeans data, call 800.448.8182.

307

09 53 23 – Metal Acoustical Ceiling Suspension Assemblies

09 53 23.30 Ceiling Suspension Systems	Crew	Daily Output	Labor-Hours	Unit	Material	2018 Bare Costs Labor	Equipment	Total	Total Incl O&P
0700 Carrier channels for ceilings with									
0900 recessed lighting fixtures, add	1 Carp	460	.017	S.F.	.22	.88		1.10	1.67
3000 Seismic ceiling bracing, IBC Site Class D, Occupancy Category II									
3050 For ceilings less than 2500 S.F.									
3060 Seismic clips at attached walls	1 Carp	180	.044	Ea.	1.13	2.25		3.38	4.90
3100 For ceilings greater than 2500 S.F., add									
3120 Seismic clips, joints at cross tees	1 Carp	120	.067	Ea.	1.85	3.38		5.23	7.55
3140 At cross tees and mains, mains field cut	"	60	.133	"	1.85	6.75		8.60	13
3200 Compression posts, telescopic, attached to structure above									
3210 To 30" high	1 Carp	26	.308	Ea.	38.50	15.60		54.10	68
3220 30" to 48" high		25.50	.314		42	15.90		57.90	72.50
3230 48" to 84" high		25	.320		51	16.20		67.20	82.50
3240 84" to 102" high		24.50	.327		58	16.55		74.55	91
3250 102" to 120" high		24	.333		83	16.90		99.90	119
3260 120" to 144" high		24	.333		92	16.90		108.90	129
3300 Stabilizer bars									
3310 12" long	1 Carp	240	.033	Ea.	1.05	1.69		2.74	3.89
3320 24" long		235	.034		.99	1.73		2.72	3.89
3330 36" long		230	.035		.96	1.76		2.72	3.91
3340 48" long		220	.036		.79	1.84		2.63	3.86
3400 Wire support for light fixtures, per L.F. height to structure above									
3410 Less than 10 lb.	1 Carp	400	.020	L.F.	.29	1.01		1.30	1.97
3420 10 lb. to 56 lb.	"	240	.033	"	.58	1.69		2.27	3.38
3500 Retrofit existing suspended ceiling grid to current code									
3510 Less than 2500 S.F. (using clips @ perimeter)	1 Carp	1455	.006	S.F.	.07	.28		.35	.52
3520 Greater than 2500 S.F. (using compression posts and clips)		550	.015		.43	.74		1.17	1.67
4000 Remove and replace ceiling grid, class A, 15/16" T bar, 2' x 2' grid		435	.018		1.02	.93		1.95	2.63
4010 2' x 4' grid		535	.015		.79	.76		1.55	2.10

09 54 Specialty Ceilings

09 54 16 – Luminous Ceilings

09 54 16.10 Ceiling, Luminous

	Crew	Daily Output	Labor-Hours	Unit	Material	2018 Bare Costs Labor	Equipment	Total	Total Incl O&P
0010 **CEILING, LUMINOUS**									
0020 Translucent lay-in panels, 2' x 2'	1 Carp	500	.016	S.F.	23	.81		23.81	26.50
0030 2' x 6'	"	500	.016	"	17.80	.81		18.61	21

09 54 23 – Linear Metal Ceilings

09 54 23.10 Metal Ceilings

	Crew	Daily Output	Labor-Hours	Unit	Material	2018 Bare Costs Labor	Equipment	Total	Total Incl O&P
0010 **METAL CEILINGS**									
0015 Solid alum. planks, 3-1/4" x 12', open reveal	1 Carp	500	.016	S.F.	2.35	.81		3.16	3.91
0020 Closed reveal		500	.016		3	.81		3.81	4.62
0030 7-1/4" x 12', open reveal		500	.016		4	.81		4.81	5.70
0040 Closed reveal		500	.016		5.10	.81		5.91	6.90
0050 Metal, open cell, 2' x 2', 6" cell		500	.016		8.70	.81		9.51	10.85
0060 8" cell		500	.016		9.60	.81		10.41	11.85
0070 2' x 4', 6" cell		500	.016		5.70	.81		6.51	7.55
0080 8" cell		500	.016		5.70	.81		6.51	7.55

09 54 26 – Suspended Wood Ceilings

09 54 26.10 Wood Ceilings

	Crew	Daily Output	Labor-Hours	Unit	Material	2018 Bare Costs Labor	Equipment	Total	Total Incl O&P
0010 **WOOD CEILINGS**									
1000 4"-6" wood slats on heavy duty 15/16" T-bar grid	2 Carp	250	.064	S.F.	26	3.24		29.24	34.50

308

For customer support on your Commercial Renovation Costs with RSMeans data, call 800.448.8182.

09 54 Specialty Ceilings

09 54 33 – Decorative Panel Ceilings

09 54 33.20 Metal Panel Ceilings

		Crew	Daily Output	Labor-Hours	Unit	Material	2018 Bare Costs Labor	Equipment	Total	Total Incl O&P
0010	**METAL PANEL CEILINGS**									
0020	Lay-in or screwed to furring, not including grid									
0100	Tin ceilings, 2' x 2' or 2' x 4', bare steel finish	2 Carp	300	.053	S.F.	2.14	2.70		4.84	6.75
0120	Painted white finish		300	.053	"	3.56	2.70		6.26	8.30
0140	Copper, chrome or brass finish		300	.053	L.F.	6.30	2.70		9	11.35
0200	Cornice molding, 2-1/2" to 3-1/2" wide, 4' long, bare steel finish		200	.080	S.F.	2.21	4.06		6.27	9.05
0220	Painted white finish		200	.080		2.81	4.06		6.87	9.70
0240	Copper, chrome or brass finish		200	.080		4.01	4.06		8.07	11
0320	5" to 6-1/2" wide, 4' long, bare steel finish		150	.107		3.40	5.40		8.80	12.55
0340	Painted white finish		150	.107		3.87	5.40		9.27	13.05
0360	Copper, chrome or brass finish		150	.107		5.60	5.40		11	14.95
0420	Flat molding, 3-1/2" to 5" wide, 4' long, bare steel finish		250	.064		3.79	3.24		7.03	9.40
0440	Painted white finish		250	.064		3.96	3.24		7.20	9.60
0460	Copper, chrome or brass finish		250	.064		7.20	3.24		10.44	13.15

09 61 Flooring Treatment

09 61 19 – Concrete Floor Staining

09 61 19.40 Floors, Interior

		Crew	Daily Output	Labor-Hours	Unit	Material	2018 Bare Costs Labor	Equipment	Total	Total Incl O&P
0010	**FLOORS, INTERIOR**									
0300	Acid stain and sealer									
0310	Stain, one coat	1 Pord	650	.012	S.F.	.14	.52		.66	.99
0320	Two coats		570	.014		.28	.60		.88	1.27
0330	Acrylic sealer, one coat		2600	.003		.22	.13		.35	.46
0340	Two coats		1400	.006		.45	.24		.69	.88

09 62 Specialty Flooring

09 62 19 – Laminate Flooring

09 62 19.10 Floating Floor

		Crew	Daily Output	Labor-Hours	Unit	Material	2018 Bare Costs Labor	Equipment	Total	Total Incl O&P
0010	**FLOATING FLOOR**									
8300	Floating floor, laminate, wood pattern strip, complete	1 Clab	133	.060	S.F.	4.32	2.40		6.72	8.65
8310	Components, T&G wood composite strips					3.83			3.83	4.22
8320	Film					.17			.17	.18
8330	Foam					.26			.26	.29
8340	Adhesive					.43			.43	.47
8350	Installation kit					.19			.19	.21
8360	Trim, 2" wide x 3' long				L.F.	4.30			4.30	4.73
8370	Reducer moulding				"	5.70			5.70	6.25

09 62 23 – Bamboo Flooring

09 62 23.10 Flooring, Bamboo

			Crew	Daily Output	Labor-Hours	Unit	Material	2018 Bare Costs Labor	Equipment	Total	Total Incl O&P
0010	**FLOORING, BAMBOO**										
8600	Flooring, wood, bamboo strips, unfinished, 5/8" x 4" x 3'	G	1 Carp	255	.031	S.F.	5.70	1.59		7.29	8.85
8610	5/8" x 4" x 4'	G		275	.029		5.90	1.47		7.37	8.90
8620	5/8" x 4" x 6'	G		295	.027		6.45	1.38		7.83	9.35
8630	Finished, 5/8" x 4" x 3'	G		255	.031		6.25	1.59		7.84	9.50
8640	5/8" x 4" x 4'	G		275	.029		6.55	1.47		8.02	9.60
8650	5/8" x 4" x 6'	G		295	.027		4.99	1.38		6.37	7.75
8660	Stair treads, unfinished, 1-1/16" x 11-1/2" x 4'	G		18	.444	Ea.	54.50	22.50		77	96.50
8670	Finished, 1-1/16" x 11-1/2" x 4'	G		18	.444		83	22.50		105.50	128

09 62 Specialty Flooring

09 62 23 – Bamboo Flooring

09 62 23.10 Flooring, Bamboo

		Crew	Daily Output	Labor-Hours	Unit	Material	2018 Bare Costs Labor	Equipment	Total	Total Incl O&P
8680	Stair risers, unfinished, 5/8" x 7-1/2" x 4'	G	1 Carp	18	.444	Ea.	20	22.50		42.50	58.50
8690	Finished, 5/8" x 7-1/2" x 4'	G		18	.444		38	22.50		60.50	78.50
8700	Stair nosing, unfinished, 6' long	G		16	.500		44.50	25.50		70	90
8710	Finished, 6' long	G		16	.500		42	25.50		67.50	87.50

09 62 29 – Cork Flooring

09 62 29.10 Cork Tile Flooring

		Crew	Daily Output	Labor-Hours	Unit	Material	2018 Bare Costs Labor	Equipment	Total	Total Incl O&P
0010	**CORK TILE FLOORING**										
2200	Cork tile, standard finish, 1/8" thick	G	1 Tilf	315	.025	S.F.	5.25	1.19		6.44	7.65
2250	3/16" thick	G		315	.025		5.80	1.19		6.99	8.25
2300	5/16" thick	G		315	.025		6.45	1.19		7.64	9
2350	1/2" thick	G		315	.025		7	1.19		8.19	9.60
2500	Urethane finish, 1/8" thick	G		315	.025		5.25	1.19		6.44	7.70
2550	3/16" thick	G		315	.025		7.30	1.19		8.49	9.95
2600	5/16" thick	G		315	.025		7.25	1.19		8.44	9.85
2650	1/2" thick	G		315	.025		7.20	1.19		8.39	9.80

09 63 Masonry Flooring

09 63 13 – Brick Flooring

09 63 13.10 Miscellaneous Brick Flooring

		Crew	Daily Output	Labor-Hours	Unit	Material	2018 Bare Costs Labor	Equipment	Total	Total Incl O&P
0010	**MISCELLANEOUS BRICK FLOORING**										
0020	Acid-proof shales, red, 8" x 3-3/4" x 1-1/4" thick		D-7	.43	37.209	M	715	1,575		2,290	3,250
0050	2-1/4" thick		D-1	.40	40		1,050	1,800		2,850	4,075
0200	Acid-proof clay brick, 8" x 3-3/4" x 2-1/4" thick	G		.40	40		1,000	1,800		2,800	4,025
0250	9" x 4-1/2" x 3"	G		95	.168	S.F.	4.41	7.55		11.96	17.15
0260	Cast ceramic, pressed, 4" x 8" x 1/2", unglazed		D-7	100	.160		6.85	6.70		13.55	18.15
0270	Glazed			100	.160		9.15	6.70		15.85	20.50
0280	Hand molded flooring, 4" x 8" x 3/4", unglazed			95	.168		9.05	7.10		16.15	21
0290	Glazed			95	.168		11.40	7.10		18.50	24
0300	8" hexagonal, 3/4" thick, unglazed			85	.188		9.95	7.90		17.85	23.50
0310	Glazed			85	.188		17.95	7.90		25.85	32.50
0400	Heavy duty industrial, cement mortar bed, 2" thick, not incl. brick		D-1	80	.200		1.13	8.95		10.08	15.85
0450	Acid-proof joints, 1/4" wide		"	65	.246		1.58	11.05		12.63	19.75
0500	Pavers, 8" x 4", 1" to 1-1/4" thick, red		D-7	95	.168		3.99	7.10		11.09	15.60
0510	Ironspot		"	95	.168		5.65	7.10		12.75	17.40
0540	1-3/8" to 1-3/4" thick, red		D-1	95	.168		3.85	7.55		11.40	16.55
0560	Ironspot			95	.168		5.60	7.55		13.15	18.45
0580	2-1/4" thick, red			90	.178		3.92	7.95		11.87	17.30
0590	Ironspot			90	.178		6.10	7.95		14.05	19.70
0700	Paver, adobe brick, 6" x 12", 1/2" joint	G		42	.381		1.44	17.05		18.49	29.50
0710	Mexican red, 12" x 12"	G	1 Tilf	48	.167		1.85	7.80		9.65	14.40
0720	Saltillo, 12" x 12"	G	"	48	.167		1.49	7.80		9.29	14
0800	For sidewalks and patios with pavers, see Section 32 14 16.10										
0870	For epoxy joints, add		D-1	600	.027	S.F.	3.05	1.19		4.24	5.30
0880	For Furan underlayment, add		"	600	.027		2.53	1.19		3.72	4.73
0890	For waxed surface, steam cleaned, add		A-1H	1000	.008		.21	.32	.07	.60	.83
9000	Minimum labor/equipment charge		1 Bric	2	4	Job		201		201	330

09 63 Masonry Flooring

09 63 40 – Stone Flooring

09 63 40.10 Marble

		Crew	Daily Output	Labor-Hours	Unit	Material	2018 Bare Costs Labor	2018 Bare Costs Equipment	Total	Total Incl O&P
0010	**MARBLE**									
0020	Thin gauge tile, 12" x 6", 3/8", white Carara	D-7	60	.267	S.F.	17	11.20		28.20	36.50
0100	Travertine		60	.267		8.80	11.20		20	27.50
0200	12" x 12" x 3/8", thin set, floors		60	.267		11.10	11.20		22.30	30
0300	On walls		52	.308		9.85	12.95		22.80	31.50
1000	Marble threshold, 4" wide x 36" long x 5/8" thick, white		60	.267	Ea.	11.10	11.20		22.30	30
9000	Minimum labor/equipment charge		3	5.333	Job		224		224	355

09 63 40.20 Slate Tile

		Crew	Daily Output	Labor-Hours	Unit	Material	2018 Bare Costs Labor	2018 Bare Costs Equipment	Total	Total Incl O&P
0010	**SLATE TILE**									
0020	Vermont, 6" x 6" x 1/4" thick, thin set	D-7	180	.089	S.F.	7.70	3.74		11.44	14.35
9000	Minimum labor/equipment charge	"	3	5.333	Job		224		224	355

09 64 Wood Flooring

09 64 16 – Wood Block Flooring

09 64 16.10 End Grain Block Flooring

		Crew	Daily Output	Labor-Hours	Unit	Material	2018 Bare Costs Labor	2018 Bare Costs Equipment	Total	Total Incl O&P
0010	**END GRAIN BLOCK FLOORING**									
0020	End grain flooring, coated, 2" thick	1 Carp	295	.027	S.F.	3.65	1.38		5.03	6.25
0400	Natural finish, 1" thick, fir		125	.064		3.77	3.24		7.01	9.40
0600	1-1/2" thick, pine		125	.064		3.70	3.24		6.94	9.30
0700	2" thick, pine		125	.064		4.91	3.24		8.15	10.65
9000	Minimum labor/equipment charge		2	4	Job		203		203	330

09 64 23 – Wood Parquet Flooring

09 64 23.10 Wood Parquet

		Crew	Daily Output	Labor-Hours	Unit	Material	2018 Bare Costs Labor	2018 Bare Costs Equipment	Total	Total Incl O&P
0010	**WOOD PARQUET** flooring									
5200	Parquetry, 5/16" thk, no finish, oak, plain pattern	1 Carp	160	.050	S.F.	5.45	2.54		7.99	10.10
5300	Intricate pattern		100	.080		9.90	4.06		13.96	17.45
5500	Teak, plain pattern		160	.050		6.35	2.54		8.89	11.05
5600	Intricate pattern		100	.080		10.75	4.06		14.81	18.45
5650	13/16" thick, select grade oak, plain pattern		160	.050		11.85	2.54		14.39	17.10
5700	Intricate pattern		100	.080		17.35	4.06		21.41	25.50
5800	Custom parquetry, including finish, plain pattern		100	.080		17.10	4.06		21.16	25.50
5900	Intricate pattern		50	.160		25	8.10		33.10	40.50
6700	Parquetry, prefinished white oak, 5/16" thick, plain pattern		160	.050		8	2.54		10.54	12.90
6800	Intricate pattern		100	.080		8.65	4.06		12.71	16.10
7000	Walnut or teak, parquetry, plain pattern		160	.050		8.50	2.54		11.04	13.45
7100	Intricate pattern		100	.080		13.65	4.06		17.71	21.50
7200	Acrylic wood parquet blocks, 12" x 12" x 5/16",									
7210	Irradiated, set in epoxy	1 Carp	160	.050	S.F.	10.45	2.54		12.99	15.60

09 64 29 – Wood Strip and Plank Flooring

09 64 29.10 Wood

		Crew	Daily Output	Labor-Hours	Unit	Material	2018 Bare Costs Labor	2018 Bare Costs Equipment	Total	Total Incl O&P
0010	**WOOD** R061110-30									
0020	Fir, vertical grain, 1" x 4", not incl. finish, grade B & better	1 Carp	255	.031	S.F.	3.45	1.59		5.04	6.40
0100	Grade C & better		255	.031		3.25	1.59		4.84	6.15
0300	Flat grain, 1" x 4", not incl. finish, grade B & better		255	.031		3.94	1.59		5.53	6.90
0400	Grade C & better		255	.031		3.79	1.59		5.38	6.75
4000	Maple, strip, 25/32" x 2-1/4", not incl. finish, select		170	.047		4.93	2.39		7.32	9.25
4100	#2 & better		170	.047		4.84	2.39		7.23	9.15
4300	33/32" x 3-1/4", not incl. finish, #1 grade		170	.047		5.65	2.39		8.04	10.05
4400	#2 & better		170	.047		5	2.39		7.39	9.35

09 64 Wood Flooring

09 64 29 – Wood Strip and Plank Flooring

09 64 29.10 Wood	Crew	Daily Output	Labor-Hours	Unit	Material	2018 Bare Costs Labor	Equipment	Total	Total Incl O&P
4600	Oak, white or red, 25/32" x 2-1/4", not incl. finish									
4700	#1 common	1 Carp	170	.047	S.F.	3.42	2.39		5.81	7.65
4900	Select quartered, 2-1/4" wide		170	.047		4.29	2.39		6.68	8.60
5000	Clear		170	.047		4.23	2.39		6.62	8.50
6100	Prefinished, white oak, prime grade, 2-1/4" wide		170	.047		5.25	2.39		7.64	9.60
6200	3-1/4" wide		185	.043		5.70	2.19		7.89	9.85
6400	Ranch plank		145	.055		7.30	2.80		10.10	12.55
6500	Hardwood blocks, 9" x 9", 25/32" thick		160	.050		7.45	2.54		9.99	12.25
7400	Yellow pine, 3/4" x 3-1/8", T&G, C & better, not incl. finish		200	.040		1.67	2.03		3.70	5.15
7500	Refinish wood floor, sand, 2 coats poly, wax, soft wood	1 Clab	400	.020		.21	.80		1.01	1.52
7600	Hardwood		130	.062		.21	2.45		2.66	4.21
7800	Sanding and finishing, 2 coats polyurethane		295	.027		.21	1.08		1.29	1.98
7900	Subfloor and underlayment, see Section 06 16									
8015	Transition molding, 2-1/4" wide, 5' long	1 Carp	19.20	.417	Ea.	18.65	21		39.65	55
9000	Minimum labor/equipment charge	"	2	4	Job		203		203	330

09 65 Resilient Flooring

09 65 10 – Resilient Tile Underlayment

09 65 10.10 Latex Underlayment

		Crew	Daily Output	Labor-Hours	Unit	Material	Labor	Equipment	Total	Total Incl O&P
0010	**LATEX UNDERLAYMENT**									
3600	Latex underlayment, 1/8" thk., cementitious for resilient flooring	1 Tilf	160	.050	S.F.	1.25	2.34		3.59	5.10
4000	Liquid, fortified				Gal.	32			32	35

09 65 13 – Resilient Base and Accessories

09 65 13.13 Resilient Base

		Crew	Daily Output	Labor-Hours	Unit	Material	Labor	Equipment	Total	Total Incl O&P
0010	**RESILIENT BASE**									
0690	1/8" vinyl base, 2-1/2" H, straight or cove, standard colors	1 Tilf	315	.025	L.F.	.70	1.19		1.89	2.65
0700	4" high		315	.025		1.19	1.19		2.38	3.19
0710	6" high		315	.025		1.47	1.19		2.66	3.50
0720	Corners, 2-1/2" high		315	.025	Ea.	2.21	1.19		3.40	4.31
0730	4" high		315	.025		2.52	1.19		3.71	4.65
0740	6" high		315	.025		2.84	1.19		4.03	5
0800	1/8" rubber base, 2-1/2" H, straight or cove, standard colors		315	.025	L.F.	1.10	1.19		2.29	3.09
1100	4" high		315	.025		1.28	1.19		2.47	3.29
1110	6" high		315	.025		1.90	1.19		3.09	3.97
1150	Corners, 2-1/2" high		315	.025	Ea.	2.46	1.19		3.65	4.59
1153	4" high		315	.025		2.54	1.19		3.73	4.67
1155	6" high		315	.025		3.11	1.19		4.30	5.30
1450	For premium color/finish add					50%				
1500	Millwork profile	1 Tilf	315	.025	L.F.	6.25	1.19		7.44	8.80

09 65 13.23 Resilient Stair Treads and Risers

		Crew	Daily Output	Labor-Hours	Unit	Material	Labor	Equipment	Total	Total Incl O&P
0010	**RESILIENT STAIR TREADS AND RISERS**									
0300	Rubber, molded tread, 12" wide, 5/16" thick, black	1 Tilf	115	.070	L.F.	14.30	3.26		17.56	21
0400	Colors		115	.070		15.70	3.26		18.96	22.50
0600	1/4" thick, black		115	.070		13.70	3.26		16.96	20
0700	Colors		115	.070		15.35	3.26		18.61	22
0900	Grip strip safety tread, colors, 5/16" thick		115	.070		20	3.26		23.26	27
1000	3/16" thick		120	.067		14.75	3.12		17.87	21
1200	Landings, smooth sheet rubber, 1/8" thick		120	.067	S.F.	8.20	3.12		11.32	13.95
1300	3/16" thick		120	.067	"	9.05	3.12		12.17	14.90
1500	Nosings, 3" wide, 3/16" thick, black		140	.057	L.F.	4.61	2.68		7.29	9.30

09 65 13.23 Resilient Stair Treads and Risers		Crew	Daily Output	Labor-Hours	Unit	Material	2018 Bare Costs Labor	Equipment	Total	Total Incl O&P
1600	Colors	1 Tilf	140	.057	L.F.	6.05	2.68		8.73	10.90
1800	Risers, 7" high, 1/8" thick, flat		250	.032		8.55	1.50		10.05	11.75
1900	Coved		250	.032		9.35	1.50		10.85	12.65
2100	Vinyl, molded tread, 12" wide, colors, 1/8" thick		115	.070		5.95	3.26		9.21	11.70
2200	1/4" thick		115	.070		8.05	3.26		11.31	14
2300	Landing material, 1/8" thick		200	.040	S.F.	6.15	1.87		8.02	9.75
2400	Riser, 7" high, 1/8" thick, coved		175	.046	L.F.	2.94	2.14		5.08	6.60
2500	Tread and riser combined, 1/8" thick		80	.100	"	10.05	4.69		14.74	18.45
9000	Minimum labor/equipment charge		3	2.667	Job		125		125	197

09 65 13.37 Vinyl Transition Strips

		Crew	Daily Output	Labor-Hours	Unit	Material	2018 Bare Costs Labor	Equipment	Total	Total Incl O&P
0010	**VINYL TRANSITION STRIPS**									
0100	Various mats. to various mats., adhesive applied, 1/4" to 1/8"	1 Tilf	315	.025	L.F.	1.47	1.19		2.66	3.50
0105	0.08" to 1/8"		315	.025		1.33	1.19		2.52	3.34
0110	0.08" to 1/4"		315	.025		1.45	1.19		2.64	3.48
0115	1/4" to 3/8"		315	.025		1.35	1.19		2.54	3.37
0120	1/4" to 1/2"		315	.025		1.35	1.19		2.54	3.37
0125	1/4" to 0.08"		315	.025		1.45	1.19		2.64	3.48
0200	Vinyl wheeled trans. strips, carpet to var. mats., 1/4" to 1/8" x 2-1/2"		315	.025		4.71	1.19		5.90	7.10
0205	1/4" to 1/8" x 4"		315	.025		6.10	1.19		7.29	8.60
0210	Various mats. to various mats. 1/4" to 0.08" x 2-1/2"		315	.025		5.05	1.19		6.24	7.45
0215	Carpet to various materials, 1/4" to flush x 2-1/2"		315	.025		3.93	1.19		5.12	6.20
0220	1/4" to flush x 4"		315	.025		6.05	1.19		7.24	8.55
0225	Various materials to resilient, 3/8" to 1/8" x 2-1/2"		315	.025		3.93	1.19		5.12	6.20
0230	Carpet to various materials, 3/8" to 1/4" x 2-1/2"		315	.025		5.40	1.19		6.59	7.85
0235	1/4" to 1/4" x 2-1/2"		315	.025		6.05	1.19		7.24	8.55
0240	Various materials to resilient, 1/8" to 1/8" x 2-1/2"		315	.025		4.57	1.19		5.76	6.95
0245	Various materials to var. mats., 1/8" to flush x 2-1/2"		315	.025		3.06	1.19		4.25	5.25
0250	3/8" to flush x 4"		315	.025		6.15	1.19		7.34	8.70
0255	1/2" to flush x 4"		315	.025		8.25	1.19		9.44	11
0260	Various materials to resilient, 1/8" to 0.08" x 2-1/2"		315	.025		3.55	1.19		4.74	5.80
0265	0.08" to 0.08" x 2-1/2"		315	.025		3.37	1.19		4.56	5.60
0270	3/8" to 0.08" x 2-1/2"		315	.025		3.37	1.19		4.56	5.60

09 65 16.10 Rubber and Vinyl Sheet Flooring

			Crew	Daily Output	Labor-Hours	Unit	Material	2018 Bare Costs Labor	Equipment	Total	Total Incl O&P
0010	**RUBBER AND VINYL SHEET FLOORING**										
5500	Linoleum, sheet goods	G	1 Tilf	360	.022	S.F.	3.59	1.04		4.63	5.60
5900	Rubber, sheet goods, 36" wide, 1/8" thick			120	.067		8.25	3.12		11.37	14.05
5950	3/16" thick			100	.080		10.20	3.75		13.95	17.15
6000	1/4" thick			90	.089		12.05	4.16		16.21	19.85
8000	Vinyl sheet goods, backed, .065" thick, plain pattern/colors			250	.032		4.30	1.50		5.80	7.10
8050	Intricate pattern/colors			200	.040		3.94	1.87		5.81	7.30
8100	.080" thick, plain pattern/colors			230	.035		4.22	1.63		5.85	7.20
8150	Intricate pattern/colors			200	.040		6.45	1.87		8.32	10.05
8200	.125" thick, plain pattern/colors			230	.035		3.87	1.63		5.50	6.85
8250	Intricate pattern/colors			200	.040		7.50	1.87		9.37	11.20
8400	For welding seams, add			100	.080	L.F.	.24	3.75		3.99	6.15
8450	For integral cove base, add			175	.046	"	.81	2.14		2.95	4.27
8700	Adhesive cement, 1 gallon per 200 to 300 S.F.					Gal.	31.50			31.50	34.50
8800	Asphalt primer, 1 gallon per 300 S.F.						14.95			14.95	16.45
8900	Emulsion, 1 gallon per 140 S.F.						18.95			18.95	21

09 65 19 – Resilient Tile Flooring

09 65 19.19 Vinyl Composition Tile Flooring		Crew	Daily Output	Labor-Hours	Unit	Material	2018 Bare Costs Labor	Equipment	Total	Total Incl O&P
0010	**VINYL COMPOSITION TILE FLOORING**									
7000	Vinyl composition tile, 12" x 12", 1/16" thick	1 Tilf	500	.016	S.F.	1.22	.75		1.97	2.52
7050	Embossed		500	.016		2.66	.75		3.41	4.11
7100	Marbleized		500	.016		2.66	.75		3.41	4.11
7150	Solid		500	.016		3.44	.75		4.19	4.96
7200	3/32" thick, embossed		500	.016		1.55	.75		2.30	2.89
7250	Marbleized		500	.016		3.06	.75		3.81	4.55
7300	Solid		500	.016		2.85	.75		3.60	4.32
7350	1/8" thick, marbleized		500	.016		2.44	.75		3.19	3.86
7400	Solid		500	.016		1.77	.75		2.52	3.13
7450	Conductive		500	.016		5.90	.75		6.65	7.70

09 65 19.23 Vinyl Tile Flooring

		Crew	Daily Output	Labor-Hours	Unit	Material	Labor	Equipment	Total	Total Incl O&P
0010	**VINYL TILE FLOORING**									
7500	Vinyl tile, 12" x 12", 3/32" thick, standard colors/patterns	1 Tilf	500	.016	S.F.	3.71	.75		4.46	5.25
7550	1/8" thick, standard colors/patterns		500	.016		5.20	.75		5.95	6.90
7600	1/8" thick, premium colors/patterns		500	.016		7	.75		7.75	8.90
7650	Solid colors		500	.016		3.23	.75		3.98	4.73
7700	Marbleized or Travertine pattern		500	.016		6.20	.75		6.95	8
7750	Florentine pattern		500	.016		6.60	.75		7.35	8.45
7800	Premium colors/patterns		500	.016		6.25	.75		7	8.05
9500	Minimum labor/equipment charge		4	2	Job		93.50		93.50	148

09 65 19.33 Rubber Tile Flooring

		Crew	Daily Output	Labor-Hours	Unit	Material	Labor	Equipment	Total	Total Incl O&P
0010	**RUBBER TILE FLOORING**									
6050	Rubber tile, marbleized colors, 12" x 12", 1/8" thick	1 Tilf	400	.020	S.F.	5.80	.94		6.74	7.85
6100	3/16" thick		400	.020		8	.94		8.94	10.30
6300	Special tile, plain colors, 1/8" thick		400	.020		8.15	.94		9.09	10.45
6350	3/16" thick		400	.020		9.75	.94		10.69	12.20

09 65 33 – Conductive Resilient Flooring

09 65 33.10 Conductive Rubber and Vinyl Flooring

		Crew	Daily Output	Labor-Hours	Unit	Material	Labor	Equipment	Total	Total Incl O&P
0010	**CONDUCTIVE RUBBER AND VINYL FLOORING**									
1700	Conductive flooring, rubber tile, 1/8" thick	1 Tilf	315	.025	S.F.	7.05	1.19		8.24	9.65
1800	Homogeneous vinyl tile, 1/8" thick	"	315	.025	"	6.80	1.19		7.99	9.40

09 65 66 – Resilient Athletic Flooring

09 65 66.10 Resilient Athletic Flooring

		Crew	Daily Output	Labor-Hours	Unit	Material	Labor	Equipment	Total	Total Incl O&P
0010	**RESILIENT ATHLETIC FLOORING**									
1000	Recycled rubber rolled goods, for weight rooms, 3/8" thk.	1 Tilf	315	.025	S.F.	2.66	1.19		3.85	4.81
1050	Interlocking 2'x2' squares, rubber, 1/4" thk.		310	.026		1.89	1.21		3.10	3.99
1055	5/16" thk.		310	.026		2.29	1.21		3.50	4.43
1060	3/8" thk.		300	.027		2.98	1.25		4.23	5.25
1065	1/2" thk.		310	.026		3.08	1.21		4.29	5.30
2000	Vinyl sheet flooring, 1/4" thk.		315	.025		4.28	1.19		5.47	6.60

09 66 Terrazzo Flooring

09 66 13 – Portland Cement Terrazzo Flooring

09 66 13.10 Portland Cement Terrazzo

		Crew	Daily Output	Labor-Hours	Unit	Material	2018 Bare Costs Labor	Equipment	Total	Total Incl O&P
0010	**PORTLAND CEMENT TERRAZZO**, cast-in-place									
0020	Cove base, 6" high, 16 ga. zinc	1 Mstz	20	.400	L.F.	3.30	18.80		22.10	33
0100	Curb, 6" high and 6" wide		6	1.333		6.25	62.50		68.75	106
0300	Divider strip for floors, 14 ga., 1-1/4" deep, zinc		375	.021		1.47	1		2.47	3.20
0400	Brass		375	.021		2.49	1		3.49	4.32
0600	Heavy top strip 1/4" thick, 1-1/4" deep, zinc		300	.027		2.16	1.25		3.41	4.36
1200	For thin set floors, 16 ga., 1/2" x 1/2", zinc		350	.023		1.29	1.07		2.36	3.12
1500	Floor, bonded to concrete, 1-3/4" thick, gray cement	J-3	75	.213	S.F.	3.57	9.20	4.15	16.92	23
1600	White cement, mud set		75	.213		4.16	9.20	4.15	17.51	23.50
1800	Not bonded, 3" total thickness, gray cement		70	.229		4.40	9.85	4.45	18.70	25.50
1900	White cement, mud set		70	.229		5.40	9.85	4.45	19.70	26.50
9000	Minimum labor/equipment charge	1 Mstz	1	8	Job		375		375	595

09 66 16 – Terrazzo Floor Tile

09 66 16.10 Tile or Terrazzo Base

		Crew	Daily Output	Labor-Hours	Unit	Material	2018 Bare Costs Labor	Equipment	Total	Total Incl O&P
0010	**TILE OR TERRAZZO BASE**									
0020	Scratch coat only	1 Mstz	150	.053	S.F.	.48	2.51		2.99	4.49
0500	Scratch and brown coat only		75	.107	"	.96	5		5.96	8.95
9000	Minimum labor/equipment charge		1	8	Job		375		375	595

09 66 16.13 Portland Cement Terrazzo Floor Tile

		Crew	Daily Output	Labor-Hours	Unit	Material	2018 Bare Costs Labor	Equipment	Total	Total Incl O&P
0010	**PORTLAND CEMENT TERRAZZO FLOOR TILE**									
1200	Floor tiles, non-slip, 1" thick, 12" x 12"	D-1	60	.267	S.F.	24.50	11.95		36.45	46.50
1300	1-1/4" thick, 12" x 12"		60	.267		25.50	11.95		37.45	47.50
1500	16" x 16"		50	.320		27.50	14.35		41.85	54
1600	1-1/2" thick, 16" x 16"		45	.356		25	15.95		40.95	54
1800	For Venetian terrazzo, add					7.55			7.55	8.35
1900	For white cement, add					.71			.71	.78

09 66 16.30 Terrazzo, Precast

		Crew	Daily Output	Labor-Hours	Unit	Material	2018 Bare Costs Labor	Equipment	Total	Total Incl O&P
0010	**TERRAZZO, PRECAST**									
0020	Base, 6" high, straight	1 Mstz	70	.114	L.F.	12.35	5.35		17.70	22
0100	Cove		60	.133		16.50	6.25		22.75	28
0300	8" high, straight		60	.133		14.75	6.25		21	26
0400	Cove		50	.160		21.50	7.50		29	36
0600	For white cement, add					.55			.55	.61
0700	For 16 ga. zinc toe strip, add					2.18			2.18	2.40
0900	Curbs, 4" x 4" high	1 Mstz	40	.200		40	9.40		49.40	59
1000	8" x 8" high	"	30	.267		46.50	12.55		59.05	71
2400	Stair treads, 1-1/2" thick, non-slip, three line pattern	2 Mstz	70	.229		50	10.75		60.75	72.50
2500	Nosing and two lines		70	.229		50	10.75		60.75	72.50
2700	2" thick treads, straight		60	.267		58	12.55		70.55	83.50
2800	Curved		50	.320		75.50	15.05		90.55	107
3000	Stair risers, 1" thick, to 6" high, straight sections		60	.267		13.75	12.55		26.30	35
3100	Cove		50	.320		19.15	15.05		34.20	45
3300	Curved, 1" thick, to 6" high, vertical		48	.333		26	15.65		41.65	54
3400	Cove		38	.421		42	19.80		61.80	78
3600	Stair tread and riser, single piece, straight, smooth surface		60	.267		64.50	12.55		77.05	91
3700	Non skid surface		40	.400		83.50	18.80		102.30	122
3900	Curved tread and riser, smooth surface		40	.400		90.50	18.80		109.30	129
4000	Non skid surface		32	.500		114	23.50		137.50	163
4200	Stair stringers, notched, 1" thick		25	.640		37	30		67	88
4300	2" thick		22	.727		43.50	34		77.50	102
4500	Stair landings, structural, non-slip, 1-1/2" thick		85	.188	S.F.	40.50	8.85		49.35	58.50
4600	3" thick		75	.213		57	10.05		67.05	79

315

09 66 Terrazzo Flooring

09 66 16 – Terrazzo Floor Tile

09 66 16.30 Terrazzo, Precast	Crew	Daily Output	Labor-Hours	Unit	Material	2018 Bare Costs Labor	Equipment	Total	Total Incl O&P
4800 Wainscot, 12" x 12" x 1" tiles	1 Mstz	12	.667	S.F.	8.75	31.50		40.25	59
4900 16" x 16" x 1-1/2" tiles	"	8	1	↓	17.45	47		64.45	93.50
9500 Minimum labor/equipment charge	1 Tilf	2	4	Job		187		187	296

09 66 23 – Resinous Matrix Terrazzo Flooring

09 66 23.16 Epoxy-Resin Terrazzo Flooring

	Crew	Daily Output	Labor-Hours	Unit	Material	Labor	Equipment	Total	Total Incl O&P
0010 **EPOXY-RESIN TERRAZZO FLOORING**									
1800 Epoxy terrazzo, 1/4" thick, chemical resistant, granite chips	J-3	200	.080	S.F.	6.35	3.44	1.56	11.35	14.15
1900 Recycled porcelain	"	150	.107	"	9.70	4.59	2.08	16.37	20

09 67 Fluid-Applied Flooring

09 67 13 – Elastomeric Liquid Flooring

09 67 13.13 Elastomeric Liquid Flooring

	Crew	Daily Output	Labor-Hours	Unit	Material	Labor	Equipment	Total	Total Incl O&P
0010 **ELASTOMERIC LIQUID FLOORING**									
0020 Cementitious acrylic, 1/4" thick	C-6	520	.092	S.F.	1.75	3.83	.10	5.68	8.25
0200 Methyl methachrylate, 1/4" thick	C-8A	3000	.016	"	7.15	.68		7.83	8.95

09 67 23 – Resinous Flooring

09 67 23.23 Resinous Flooring

	Crew	Daily Output	Labor-Hours	Unit	Material	Labor	Equipment	Total	Total Incl O&P
0010 **RESINOUS FLOORING**									
1200 Heavy duty epoxy topping, 1/4" thick,									
1300 500 to 1,000 S.F.	C-6	420	.114	S.F.	5.95	4.74	.12	10.81	14.35
1500 1,000 to 2,000 S.F.		450	.107		5.15	4.42	.11	9.68	12.90
1600 Over 10,000 S.F.	↓	480	.100	↓	4.75	4.15	.11	9.01	12.05

09 67 26 – Quartz Flooring

09 67 26.26 Quartz Flooring

	Crew	Daily Output	Labor-Hours	Unit	Material	Labor	Equipment	Total	Total Incl O&P
0010 **QUARTZ FLOORING**									
0600 Epoxy, with colored quartz chips, broadcast, 3/8" thick	C-6	675	.071	S.F.	3.08	2.95	.08	6.11	8.25
0700 1/2" thick		490	.098		4.44	4.06	.10	8.60	11.55
0900 Troweled, minimum		560	.086		3.57	3.55	.09	7.21	9.80
1000 Maximum	↓	480	.100	↓	5.90	4.15	.11	10.16	13.30

09 68 Carpeting

09 68 05 – Carpet Accessories

09 68 05.11 Flooring Transition Strip

	Crew	Daily Output	Labor-Hours	Unit	Material	Labor	Equipment	Total	Total Incl O&P
0010 **FLOORING TRANSITION STRIP**									
0107 Clamp down brass divider, 12' strip, vinyl to carpet	1 Tilf	31.25	.256	Ea.	14.65	12		26.65	35
0117 Vinyl to hard surface	"	31.25	.256	"	14.65	12		26.65	35

09 68 10 – Carpet Pad

09 68 10.10 Commercial Grade Carpet Pad

	Crew	Daily Output	Labor-Hours	Unit	Material	Labor	Equipment	Total	Total Incl O&P
0010 **COMMERCIAL GRADE CARPET PAD**									
9000 Sponge rubber pad, 20 oz./sq. yd.	1 Tilf	150	.053	S.Y.	4.69	2.50		7.19	9.10
9100 40 to 62 oz./sq. yd.		150	.053		8.75	2.50		11.25	13.60
9200 Felt pad, 20 oz./sq. yd.		150	.053		5.80	2.50		8.30	10.30
9300 32 to 56 oz./sq. yd.		150	.053		11.10	2.50		13.60	16.20
9400 Bonded urethane pad, 2.7 density		150	.053		5.95	2.50		8.45	10.50
9500 13.0 density		150	.053		8	2.50		10.50	12.75
9600 Prime urethane pad, 2.7 density		150	.053		3.51	2.50		6.01	7.80
9700 13.0 density	↓	150	.053	↓	6.55	2.50		9.05	11.20

09 68 Carpeting

09 68 13 – Tile Carpeting

09 68 13.10 Carpet Tile

		Crew	Daily Output	Labor-Hours	Unit	Material	2018 Bare Costs Labor	Equipment	Total	Total Incl O&P
0010	**CARPET TILE**									
0100	Tufted nylon, 18" x 18", hard back, 20 oz.	1 Tilf	80	.100	S.Y.	27	4.69		31.69	37
0110	26 oz.		80	.100		25.50	4.69		30.19	35.50
0200	Cushion back, 20 oz.		80	.100		25	4.69		29.69	35
0210	26 oz.		80	.100		30	4.69		34.69	41
1100	Tufted, 24" x 24", hard back, 24 oz. nylon		80	.100		31	4.69		35.69	41.50
1180	35 oz.		80	.100		36	4.69		40.69	47.50
5060	42 oz.		80	.100		47	4.69		51.69	59.50
6000	Electrostatic dissapative carpet tile, 24" x 24", 24 oz.		80	.100		37.50	4.69		42.19	48.50
6100	Electrostatic dissapative carpet tile for access floors, 24" x 24", 24 oz.		80	.100		47	4.69		51.69	59

09 68 16 – Sheet Carpeting

09 68 16.10 Sheet Carpet

		Crew	Daily Output	Labor-Hours	Unit	Material	2018 Bare Costs Labor	Equipment	Total	Total Incl O&P
0010	**SHEET CARPET**									
0700	Nylon, level loop, 26 oz., light to medium traffic	1 Tilf	75	.107	S.Y.	22	5		27	32
0720	28 oz., light to medium traffic		75	.107		30	5		35	41
0900	32 oz., medium traffic		75	.107		39.50	5		44.50	51.50
1100	40 oz., medium to heavy traffic		75	.107		49	5		54	61.50
2920	Nylon plush, 30 oz., medium traffic		75	.107		28.50	5		33.50	39.50
3000	36 oz., medium traffic		75	.107		36	5		41	48
3100	42 oz., medium to heavy traffic		70	.114		45.50	5.35		50.85	58.50
3200	46 oz., medium to heavy traffic		70	.114		52.50	5.35		57.85	66.50
3300	54 oz., heavy traffic		70	.114		59.50	5.35		64.85	73.50
4110	Wool, level loop, 40 oz., medium traffic		70	.114		114	5.35		119.35	133
4500	50 oz., medium to heavy traffic		70	.114		109	5.35		114.35	128
4700	Patterned, 32 oz., medium to heavy traffic		70	.114		98.50	5.35		103.85	116
4900	48 oz., heavy traffic		70	.114		109	5.35		114.35	128
5000	For less than full roll (approx. 1500 S.F.), add					25%				
5100	For small rooms, less than 12' wide, add						25%			
5200	For large open areas (no cuts), deduct						25%			
5600	For bound carpet baseboard, add	1 Tilf	300	.027	L.F.	1.81	1.25		3.06	3.96
5610	For stairs, not incl. price of carpet, add	"	30	.267	Riser		12.50		12.50	19.75
5620	For borders and patterns, add to labor						18%			
8950	For tackless, stretched installation, add padding from 09 68 10.10 to above									
9850	For brand-named specific fiber, add				S.Y.	25%				
9900	Carpet cleaning machine, rent				Day				38	42
9910	Minimum labor/equipment charge	1 Tilf	3	2.667	Job		125		125	197

09 68 20 – Athletic Carpet

09 68 20.10 Indoor Athletic Carpet

		Crew	Daily Output	Labor-Hours	Unit	Material	2018 Bare Costs Labor	Equipment	Total	Total Incl O&P
0010	**INDOOR ATHLETIC CARPET**									
3700	Polyethylene, in rolls, no base incl., landscape surfaces	1 Tilf	275	.029	S.F.	4.08	1.36		5.44	6.65
3800	Nylon action surface, 1/8" thick		275	.029		3.94	1.36		5.30	6.50
3900	1/4" thick		275	.029		5.70	1.36		7.06	8.40
4000	3/8" thick		275	.029		7.15	1.36		8.51	10
5500	Polyvinyl chloride, sheet goods for gyms, 1/4" thick		80	.100		6.85	4.69		11.54	14.95
5600	3/8" thick		60	.133		10.05	6.25		16.30	21

09 69 Access Flooring

09 69 13 – Rigid-Grid Access Flooring

09 69 13.10 Access Floors	Crew	Daily Output	Labor-Hours	Unit	Material	2018 Bare Costs Labor	Equipment	Total	Total Incl O&P
0010 **ACCESS FLOORS**									
0015 Access floor pkg. including panel, pedestal, & stringers 1500 lbs. load									
0100 Package pricing, conc. fill panels, no fin., 6" ht.	4 Carp	750	.043	S.F.	16.70	2.16		18.86	22
0105 12" ht.		750	.043		17.10	2.16		19.26	22.50
0110 18" ht.		750	.043		17.15	2.16		19.31	22.50
0115 24" ht.		750	.043		17.55	2.16		19.71	23
0120 Package pricing, steel panels, no fin., 6" ht.		750	.043		13.35	2.16		15.51	18.15
0125 12" ht.		750	.043		16.75	2.16		18.91	22
0130 18" ht.		750	.043		16.85	2.16		19.01	22
0135 24" ht.		750	.043		16.95	2.16		19.11	22
0140 Package pricing, wood core panels, no fin., 6" ht.		750	.043		13.50	2.16		15.66	18.35
0145 12" ht.		750	.043		13.75	2.16		15.91	18.60
0150 18" ht.		750	.043		13.75	2.16		15.91	18.60
0155 24" ht.		750	.043		14	2.16		16.16	18.90
0160 Pkg. pricing, conc. fill pnls, no stringers, no fin., 6" ht, 1250 lbs. load		700	.046		10.80	2.32		13.12	15.60
0165 12" ht.		700	.046		10	2.32		12.32	14.75
0170 18" ht.		700	.046		10.95	2.32		13.27	15.75
0175 24" ht.		700	.046		11.90	2.32		14.22	16.80
0250 Panels, 2' x 2' conc. fill, no fin.	2 Carp	500	.032		23	1.62		24.62	27.50
0255 With 1/8" high pressure laminate		500	.032		56	1.62		57.62	64.50
0260 Metal panels with 1/8" high pressure laminate		500	.032		66.50	1.62		68.12	75.50
0265 Wood core panels with 1/8" high pressure laminate		500	.032		43.50	1.62		45.12	50.50
0400 Aluminum panels, no fin.		500	.032		33.50	1.62		35.12	39
0600 For carpet covering, add					9.05			9.05	10
0700 For vinyl floor covering, add					9.35			9.35	10.30
0900 For high pressure laminate covering, add					7.70			7.70	8.45
0910 For snap on stringer system, add	2 Carp	1000	.016		1.59	.81		2.40	3.07
1000 Machine cutouts after initial installation	1 Carp	50	.160	Ea.	20	8.10		28.10	35
1050 Pedestals, 6" to 12"	2 Carp	85	.188		8.65	9.55		18.20	25
1100 Air conditioning grilles, 4" x 12"	1 Carp	17	.471		72.50	24		96.50	119
1150 4" x 18"	"	14	.571		99.50	29		128.50	157
1200 Approach ramps, steel	2 Carp	60	.267	S.F.	26.50	13.50		40	51
1300 Aluminum	"	40	.400	"	34	20.50		54.50	70.50
1500 Handrail, 2 rail, aluminum	1 Carp	15	.533	L.F.	117	27		144	173

09 72 Wall Coverings

09 72 13 – Cork Wall Coverings

09 72 13.10 Covering, Cork Wall

	Crew	Daily Output	Labor-Hours	Unit	Material	2018 Bare Costs Labor	Equipment	Total	Total Incl O&P
0010 **COVERING, CORK WALL**									
0600 Cork tiles, light or dark, 12" x 12" x 3/16"	1 Pape	240	.033	S.F.	4.26	1.42		5.68	6.95
0700 5/16" thick		235	.034		3.33	1.45		4.78	6
0900 1/4" basket weave		240	.033		3.39	1.42		4.81	6
1000 1/2" natural, non-directional pattern		240	.033		6.80	1.42		8.22	9.75
1100 3/4" natural, non-directional pattern		240	.033		11.60	1.42		13.02	15.05
1200 Granular surface, 12" x 36", 1/2" thick		385	.021		1.30	.89		2.19	2.85
1300 1" thick		370	.022		1.66	.92		2.58	3.31
1500 Polyurethane coated, 12" x 12" x 3/16" thick		240	.033		4.03	1.42		5.45	6.70
1600 5/16" thick		235	.034		6	1.45		7.45	8.95
1800 Cork wallpaper, paperbacked, natural		480	.017		1.60	.71		2.31	2.90
1900 Colors		480	.017		2.84	.71		3.55	4.26

09 72 Wall Coverings

09 72 16 – Vinyl-Coated Fabric Wall Coverings

09 72 16.13 Flexible Vinyl Wall Coverings

		Crew	Daily Output	Labor-Hours	Unit	Material	2018 Bare Costs Labor	Equipment	Total	Total Incl O&P
0010	**FLEXIBLE VINYL WALL COVERINGS**									
3000	Vinyl wall covering, fabric-backed, lightweight, type 1 (12-15 oz./S.Y.)	1 Pape	640	.013	S.F.	1.07	.53		1.60	2.03
3300	Medium weight, type 2 (20-24 oz./S.Y.)		480	.017		.91	.71		1.62	2.14
3400	Heavy weight, type 3 (28 oz./S.Y.)		435	.018		1.40	.78		2.18	2.80
3600	Adhesive, 5 gal. lots (18 S.Y./gal.)				Gal.	12.05			12.05	13.25

09 72 16.16 Rigid-Sheet Vinyl Wall Coverings

		Crew	Daily Output	Labor-Hours	Unit	Material	2018 Bare Costs Labor	Equipment	Total	Total Incl O&P
0010	**RIGID-SHEET VINYL WALL COVERINGS**									
0100	Acrylic, modified, semi-rigid PVC, .028" thick	2 Carp	330	.048	S.F.	1.30	2.46		3.76	5.40
0110	.040" thick	"	320	.050	"	1.85	2.54		4.39	6.15

09 72 19 – Textile Wall Coverings

09 72 19.10 Textile Wall Covering

		Crew	Daily Output	Labor-Hours	Unit	Material	2018 Bare Costs Labor	Equipment	Total	Total Incl O&P
0010	**TEXTILE WALL COVERING**, including sizing; add 10-30% waste @ takeoff									
0020	Silk	1 Pape	640	.013	S.F.	4.72	.53		5.25	6.05
0030	Cotton		640	.013		7.05	.53		7.58	8.65
0040	Linen		640	.013		1.92	.53		2.45	2.96
0050	Blend		640	.013		3.26	.53		3.79	4.44
0090	Grass cloth, natural fabric G		400	.020		1.79	.85		2.64	3.34
0100	Grass cloths with lining paper G		400	.020		1.38	.85		2.23	2.89
0110	Premium texture/color G		350	.023		3.15	.97		4.12	5.05

09 72 20 – Natural Fiber Wall Covering

09 72 20.10 Natural Fiber Wall Covering

		Crew	Daily Output	Labor-Hours	Unit	Material	2018 Bare Costs Labor	Equipment	Total	Total Incl O&P
0010	**NATURAL FIBER WALL COVERING**, including sizing; add 10-30% waste @ takeoff									
0015	Bamboo	1 Pape	640	.013	S.F.	2.22	.53		2.75	3.29
0030	Burlap		640	.013		1.88	.53		2.41	2.92
0045	Jute		640	.013		1.33	.53		1.86	2.31
0060	Sisal		640	.013		1.53	.53		2.06	2.53

09 72 23 – Wallpapering

09 72 23.10 Wallpaper

		Crew	Daily Output	Labor-Hours	Unit	Material	2018 Bare Costs Labor	Equipment	Total	Total Incl O&P
0010	**WALLPAPER** including sizing; add 10-30% waste @ takeoff									
0050	Aluminum foil	1 Pape	275	.029	S.F.	1.03	1.24		2.27	3.12
0100	Copper sheets, .025" thick, vinyl backing		240	.033		5.50	1.42		6.92	8.35
0300	Phenolic backing		240	.033		7.10	1.42		8.52	10.15
2400	Gypsum-based, fabric-backed, fire resistant									
2500	for masonry walls, 21 oz./S.Y.	1 Pape	800	.010	S.F.	.77	.43		1.20	1.53
2600	Average		720	.011		1.26	.47		1.73	2.15
2700	Small quantities		640	.013		.68	.53		1.21	1.60
3700	Wallpaper, average workmanship, solid pattern, low cost paper		640	.013		.61	.53		1.14	1.52
3900	Basic patterns (matching required), avg. cost paper		535	.015		1.19	.64		1.83	2.33
4000	Paper at $85 per double roll, quality workmanship		435	.018		2.11	.78		2.89	3.58
5990	Wallpaper removal, 1 layer		800	.010		.02	.43		.45	.70
6000	Wallpaper removal, 3 layer		400	.020		.07	.85		.92	1.44
9000	Minimum labor/equipment charge		2	4	Job		170		170	273

09 74 Flexible Wood Sheets

09 74 16 – Flexible Wood Veneers

09 74 16.10 Veneer, Flexible Wood	Crew	Daily Output	Labor-Hours	Unit	Material	2018 Bare Costs Labor	Equipment	Total	Total Incl O&P
0010 **VENEER, FLEXIBLE WOOD**									
0100 Flexible wood veneer, 1/32" thick, plain woods	1 Pape	100	.080	S.F.	2.41	3.41		5.82	8.10
0110 Exotic woods	"	95	.084	"	3.64	3.59		7.23	9.75

09 77 Special Wall Surfacing

09 77 30 – Fiberglass Reinforced Panels

09 77 30.10 Fiberglass Reinforced Plastic Panels

		Crew	Daily Output	Labor-Hours	Unit	Material	Labor	Equipment	Total	Total Incl O&P
0010	**FIBERGLASS REINFORCED PLASTIC PANELS**, .090" thick									
0020	On walls, adhesive mounted, embossed surface	2 Carp	640	.025	S.F.	1.20	1.27		2.47	3.38
0030	Smooth surface		640	.025		1.48	1.27		2.75	3.69
0040	Fire rated, embossed surface		640	.025		2.15	1.27		3.42	4.43
0050	Nylon rivet mounted, on drywall, embossed surface		480	.033		1.20	1.69		2.89	4.06
0060	Smooth surface		480	.033		1.48	1.69		3.17	4.37
0070	Fire rated, embossed surface		480	.033		2.15	1.69		3.84	5.10
0080	On masonry, embossed surface		320	.050		1.20	2.54		3.74	5.45
0090	Smooth surface		320	.050		1.48	2.54		4.02	5.75
0100	Fire rated, embossed surface		320	.050		2.15	2.54		4.69	6.50
0110	Nylon rivet and adhesive mounted, on drywall, embossed surface		240	.067		1.36	3.38		4.74	7
0120	Smooth surface		240	.067		1.36	3.38		4.74	7
0130	Fire rated, embossed surface		240	.067		2.37	3.38		5.75	8.10
0140	On masonry, embossed surface		190	.084		1.36	4.27		5.63	8.45
0150	Smooth surface		190	.084		1.36	4.27		5.63	8.45
0160	Fire rated, embossed surface		190	.084		2.37	4.27		6.64	9.55
0170	For moldings, add	1 Carp	250	.032	L.F.	.28	1.62		1.90	2.94
0180	On ceilings, for lay in grid system, embossed surface		400	.020	S.F.	1.20	1.01		2.21	2.97
0190	Smooth surface		400	.020		1.48	1.01		2.49	3.28
0200	Fire rated, embossed surface		400	.020		2.15	1.01		3.16	4.02

09 77 43 – Panel Systems

09 77 43.20 Slatwall Panels and Accessories

		Crew	Daily Output	Labor-Hours	Unit	Material	Labor	Equipment	Total	Total Incl O&P
0010	**SLATWALL PANELS AND ACCESSORIES**									
0100	Slatwall panel, 4' x 8' x 3/4" T, MDF, paint grade	1 Carp	500	.016	S.F.	1.56	.81		2.37	3.04
0110	Melamine finish		500	.016		2.04	.81		2.85	3.56
0120	High pressure plastic laminate finish		500	.016		3.41	.81		4.22	5.05
0125	Wood veneer		500	.016		3.80	.81		4.61	5.50
0130	Aluminum channel inserts, add					2.64			2.64	2.90
0200	Accessories, corner forms, 8' L				L.F.	5.30			5.30	5.85
0210	T-connector, 8' L					7.50			7.50	8.25
0220	J-mold, 8' L					1.28			1.28	1.41
0230	Edge cap, 8' L					1.72			1.72	1.89
0240	Finish end cap, 8' L					3.24			3.24	3.56
0300	Display hook, metal, 4" L				Ea.	.42			.42	.46
0310	6" L					.48			.48	.53
0320	8" L					.51			.51	.56
0330	10" L					.57			.57	.63
0340	12" L					.64			.64	.70
0350	Acrylic, 4" L					1.23			1.23	1.35
0360	6" L					.99			.99	1.09
0400	Waterfall hanger, metal, 12"-16"					3.75			3.75	4.13
0410	Acrylic					11.50			11.50	12.65
0500	Shelf bracket, metal, 8"					1.90			1.90	2.09

09 77 Special Wall Surfacing

09 77 43 – Panel Systems

09 77 43.20 Slatwall Panels and Accessories	Crew	Daily Output	Labor-Hours	Unit	Material	2018 Bare Costs Labor	Equipment	Total	Total Incl O&P
0510	10"				Ea.	2.06			2.06	2.27
0520	12"					2.28			2.28	2.51
0530	14"					2.50			2.50	2.75
0540	16"					2.85			2.85	3.14
0550	Acrylic, 8"					3.82			3.82	4.20
0560	10"					4.11			4.11	4.52
0570	12"					4.61			4.61	5.05
0580	14"					6.05			6.05	6.65
0600	Shelf, acrylic, 12" x 16" x 1/4"					22			22	24.50
0610	12" x 24" x 1/4"					25			25	27.50

09 81 Acoustic Insulation

09 81 13 – Acoustic Board Insulation

09 81 13.10 Acoustic Board Insulation

		Crew	Daily Output	Labor-Hours	Unit	Material	Labor	Equipment	Total	Total Incl O&P
0010	**ACOUSTIC BOARD INSULATION**									
0020	Cellulose fiber board, 1/2" thk.	1 Carp	800	.010	S.F.	.88	.51		1.39	1.79

09 81 16 – Acoustic Blanket Insulation

09 81 16.10 Sound Attenuation Blanket

		Crew	Daily Output	Labor-Hours	Unit	Material	Labor	Equipment	Total	Total Incl O&P
0010	**SOUND ATTENUATION BLANKET**									
0020	Blanket, 1" thick	1 Carp	925	.009	S.F.	.26	.44		.70	1
0500	1-1/2" thick		920	.009		.30	.44		.74	1.05
1000	2" thick		915	.009		.42	.44		.86	1.18
1500	3" thick		910	.009		.61	.45		1.06	1.39
2000	Wall hung, STC 18-21, 1" thick, 4' x 20'	2 Carp	22	.727	Ea.	525	37		562	640
2010	10' x 20'	"	19	.842		1,325	42.50		1,367.50	1,525
2020	Wall hung, STC 27-28, 3" thick, 4' x 20'	3 Carp	12	2		575	101		676	795
2030	10' x 20'	"	9	2.667		1,425	135		1,560	1,800
3400	Urethane plastic foam, open cell, on wall, 2" thick	2 Carp	2050	.008	S.F.	3.32	.40		3.72	4.29
3500	3" thick		1550	.010		4.41	.52		4.93	5.70
3600	4" thick		1050	.015		6.20	.77		6.97	8.05
3700	On ceiling, 2" thick		1700	.009		3.31	.48		3.79	4.41
3800	3" thick		1300	.012		4.41	.62		5.03	5.85
3900	4" thick		900	.018		6.20	.90		7.10	8.25
9000	Minimum labor/equipment charge	1 Carp	5	1.600	Job		81		81	132

09 84 Acoustic Room Components

09 84 13 – Fixed Sound-Absorptive Panels

09 84 13.10 Fixed Panels

		Crew	Daily Output	Labor-Hours	Unit	Material	Labor	Equipment	Total	Total Incl O&P
0010	**FIXED PANELS** Perforated steel facing, painted with									
0100	Fiberglass or mineral filler, no backs, 2-1/4" thick, modular									
0200	space units, ceiling or wall hung, white or colored	1 Carp	100	.080	S.F.	9.45	4.06		13.51	17
0300	Fiberboard sound deadening panels, 1/2" thick	"	600	.013	"	.34	.68		1.02	1.47
0500	Fiberglass panels, 4' x 8' x 1" thick, with									
0600	glass cloth face for walls, cemented	1 Carp	155	.052	S.F.	9.30	2.62		11.92	14.50
0700	1-1/2" thick, dacron covered, inner aluminum frame,									
0710	wall mounted	1 Carp	300	.027	S.F.	9.10	1.35		10.45	12.20
0900	Mineral fiberboard panels, fabric covered, 30" x 108",									
1000	3/4" thick, concealed spline, wall mounted	1 Carp	150	.053	S.F.	6.55	2.70		9.25	11.60
9000	Minimum labor/equipment charge	"	4	2	Job		101		101	165

09 91 Painting

09 91 03 – Paint Restoration

09 91 03.20 Sanding

		Crew	Daily Output	Labor-Hours	Unit	Material	2018 Bare Costs Labor	Equipment	Total	Total Incl O&P
0010	**SANDING** and puttying interior trim, compared to									
0100	Painting 1 coat, on quality work				L.F.		100%			
0300	Medium work						50%			
0400	Industrial grade				↓		25%			
0500	Surface protection, placement and removal									
0510	Surface protection, placement and removal, basic drop cloths	1 Pord	6400	.001	S.F.		.05		.05	.09
0520	Masking with paper		800	.010		.07	.43		.50	.76
0530	Volume cover up (using plastic sheathing or building paper)	↓	16000	.001	↓		.02		.02	.03

09 91 03.30 Exterior Surface Preparation

		Crew	Daily Output	Labor-Hours	Unit	Material	2018 Bare Costs Labor	Equipment	Total	Total Incl O&P
0010	**EXTERIOR SURFACE PREPARATION**									
0015	Doors, per side, not incl. frames or trim									
0020	Scrape & sand									
0030	Wood, flush	1 Pord	616	.013	S.F.		.55		.55	.89
0040	Wood, detail		496	.016			.69		.69	1.10
0050	Wood, louvered		280	.029			1.22		1.22	1.95
0060	Wood, overhead	↓	616	.013	↓		.55		.55	.89
0070	Wire brush									
0080	Metal, flush	1 Pord	640	.013	S.F.		.53		.53	.85
0090	Metal, detail		520	.015			.65		.65	1.05
0100	Metal, louvered		360	.022			.95		.95	1.52
0110	Metal or fibr., overhead		640	.013			.53		.53	.85
0120	Metal, roll up		560	.014			.61		.61	.98
0130	Metal, bulkhead	↓	640	.013	↓		.53		.53	.85
0140	Power wash, based on 2500 lb. operating pressure									
0150	Metal, flush	A-1H	2240	.004	S.F.		.14	.03	.17	.27
0160	Metal, detail		2120	.004			.15	.04	.19	.28
0170	Metal, louvered		2000	.004			.16	.04	.20	.30
0180	Metal or fibr., overhead		2400	.003			.13	.03	.16	.25
0190	Metal, roll up		2400	.003			.13	.03	.16	.25
0200	Metal, bulkhead	↓	2200	.004	↓		.15	.03	.18	.28
0400	Windows, per side, not incl. trim									
0410	Scrape & sand									
0420	Wood, 1-2 lite	1 Pord	320	.025	S.F.		1.06		1.06	1.71
0430	Wood, 3-6 lite		280	.029			1.22		1.22	1.95
0440	Wood, 7-10 lite		240	.033			1.42		1.42	2.28
0450	Wood, 12 lite		200	.040			1.70		1.70	2.73
0460	Wood, Bay/Bow	↓	320	.025	↓		1.06		1.06	1.71
0470	Wire brush									
0480	Metal, 1-2 lite	1 Pord	480	.017	S.F.		.71		.71	1.14
0490	Metal, 3-6 lite		400	.020			.85		.85	1.37
0500	Metal, Bay/Bow	↓	480	.017	↓		.71		.71	1.14
0510	Power wash, based on 2500 lb. operating pressure									
0520	1-2 lite	A-1H	4400	.002	S.F.		.07	.02	.09	.14
0530	3-6 lite		4320	.002			.07	.02	.09	.14
0540	7-10 lite		4240	.002			.08	.02	.10	.14
0550	12 lite		4160	.002			.08	.02	.10	.14
0560	Bay/Bow	↓	4400	.002	↓		.07	.02	.09	.14
0600	Siding, scrape and sand, light=10-30%, med.=30-70%									
0610	Heavy=70-100% of surface to sand									
0650	Texture 1-11, light	1 Pord	480	.017	S.F.		.71		.71	1.14
0660	Med.		440	.018			.77		.77	1.24
0670	Heavy		360	.022	↓		.95		.95	1.52

09 91 03 – Paint Restoration

09 91 03.30 Exterior Surface Preparation

		Crew	Daily Output	Labor-Hours	Unit	Material	2018 Bare Costs Labor	2018 Bare Costs Equipment	Total	Total Incl O&P
0680	Wood shingles, shakes, light	1 Pord	440	.018	S.F.		.77		.77	1.24
0690	Med.		360	.022			.95		.95	1.52
0700	Heavy		280	.029			1.22		1.22	1.95
0710	Clapboard, light		520	.015			.65		.65	1.05
0720	Med.		480	.017			.71		.71	1.14
0730	Heavy	▼	400	.020	▼		.85		.85	1.37
0740	Wire brush									
0750	Aluminum, light	1 Pord	600	.013	S.F.		.57		.57	.91
0760	Med.		520	.015			.65		.65	1.05
0770	Heavy	▼	440	.018	▼		.77		.77	1.24
0780	Pressure wash, based on 2500 lb. operating pressure									
0790	Stucco	A-1H	3080	.003	S.F.		.10	.02	.12	.20
0800	Aluminum or vinyl		3200	.003			.10	.02	.12	.19
0810	Siding, masonry, brick & block	▼	2400	.003	▼		.13	.03	.16	.25
1300	Miscellaneous, wire brush									
1310	Metal, pedestrian gate	1 Pord	100	.080	S.F.		3.40		3.40	5.45
1320	Aluminum chain link, both sides		250	.032			1.36		1.36	2.19
1400	Existing galvanized surface, clean and prime, prep for painting	▼	380	.021	▼	.13	.90		1.03	1.58
8000	For chemical washing, see Section 04 01 30									
8010	For steam cleaning, see Section 04 01 30.20									

09 91 03.40 Interior Surface Preparation

		Crew	Daily Output	Labor-Hours	Unit	Material	2018 Bare Costs Labor	2018 Bare Costs Equipment	Total	Total Incl O&P
0010	**INTERIOR SURFACE PREPARATION**									
0020	Doors, per side, not incl. frames or trim									
0030	Scrape & sand									
0040	Wood, flush	1 Pord	616	.013	S.F.		.55		.55	.89
0050	Wood, detail		496	.016			69		.69	1.10
0060	Wood, louvered	▼	280	.029	▼		1.22		1.22	1.95
0070	Wire brush									
0080	Metal, flush	1 Pord	640	.013	S.F.		.53		.53	.85
0090	Metal, detail		520	.015			.65		.65	1.05
0100	Metal, louvered	▼	360	.022	▼		.95		.95	1.52
0110	Hand wash									
0120	Wood, flush	1 Pord	2160	.004	S.F.		.16		.16	.25
0130	Wood, detail		2000	.004			.17		.17	.27
0140	Wood, louvered		1360	.006			.25		.25	.40
0150	Metal, flush		2160	.004			.16		.16	.25
0160	Metal, detail		2000	.004			.17		.17	.27
0170	Metal, louvered	▼	1360	.006	▼		.25		.25	.40
0400	Windows, per side, not incl. trim									
0410	Scrape & sand									
0420	Wood, 1-2 lite	1 Pord	360	.022	S.F.		.95		.95	1.52
0430	Wood, 3-6 lite		320	.025			1.06		1.06	1.71
0440	Wood, 7-10 lite		280	.029			1.22		1.22	1.95
0450	Wood, 12 lite		240	.033			1.42		1.42	2.28
0460	Wood, Bay/Bow	▼	360	.022	▼		.95		.95	1.52
0470	Wire brush									
0480	Metal, 1-2 lite	1 Pord	520	.015	S.F.		.65		.65	1.05
0490	Metal, 3-6 lite		440	.018			.77		.77	1.24
0500	Metal, Bay/Bow	▼	520	.015	▼		.65		.65	1.05
0600	Walls, sanding, light=10-30%, medium=30-70%,									
0610	heavy=70-100% of surface to sand									
0650	Walls, sand									

09 91 03 – Paint Restoration

09 91 03.40 Interior Surface Preparation

09 91 03.40 Interior Surface Preparation		Crew	Daily Output	Labor-Hours	Unit	Material	2018 Bare Costs Labor	2018 Bare Costs Equipment	Total	Total Incl O&P
0660	Gypsum board or plaster, light	1 Pord	3077	.003	S.F.		.11		.11	.18
0670	Gypsum board or plaster, medium		2160	.004			.16		.16	.25
0680	Gypsum board or plaster, heavy		923	.009			.37		.37	.59
0690	Wood, T&G, light		2400	.003			.14		.14	.23
0700	Wood, T&G, medium		1600	.005			.21		.21	.34
0710	Wood, T&G, heavy	↓	800	.010	↓		.43		.43	.68
0720	Walls, wash									
0730	Gypsum board or plaster	1 Pord	3200	.003	S.F.		.11		.11	.17
0740	Wood, T&G		3200	.003			.11		.11	.17
0750	Masonry, brick & block, smooth		2800	.003			.12		.12	.20
0760	Masonry, brick & block, coarse	↓	2000	.004	↓		.17		.17	.27
8000	For chemical washing, see Section 04 01 30									
8010	For steam cleaning, see Section 04 01 30.20									
9010	Minimum labor/equipment charge	1 Pord	3	2.667	Job		113		113	182

09 91 03.41 Scrape After Fire Damage

09 91 03.41 Scrape After Fire Damage		Crew	Daily Output	Labor-Hours	Unit	Material	2018 Bare Costs Labor	2018 Bare Costs Equipment	Total	Total Incl O&P
0010	**SCRAPE AFTER FIRE DAMAGE**									
0050	Boards, 1" x 4"	1 Pord	336	.024	L.F.		1.01		1.01	1.63
0060	1" x 6"		260	.031			1.31		1.31	2.10
0070	1" x 8"		207	.039			1.64		1.64	2.64
0080	1" x 10"		174	.046			1.96		1.96	3.14
0500	Framing, 2" x 4"		265	.030			1.28		1.28	2.06
0510	2" x 6"		221	.036			1.54		1.54	2.47
0520	2" x 8"		190	.042			1.79		1.79	2.88
0530	2" x 10"		165	.048			2.06		2.06	3.31
0540	2" x 12"		144	.056			2.36		2.36	3.79
1000	Heavy framing, 3" x 4"		226	.035			1.51		1.51	2.42
1010	4" x 4"		210	.038			1.62		1.62	2.60
1020	4" x 6"		191	.042			1.78		1.78	2.86
1030	4" x 8"		165	.048			2.06		2.06	3.31
1040	4" x 10"		144	.056			2.36		2.36	3.79
1060	4" x 12"		131	.061	↓		2.60		2.60	4.17
2900	For sealing, light damage		825	.010	S.F.	.15	.41		.56	.83
2920	Heavy damage	↓	460	.017	"	.33	.74		1.07	1.55
3000	For sandblasting, see Section 04 01 30.20									
9000	Minimum labor/equipment charge	1 Pord	3	2.667	Job		113		113	182

09 91 13 – Exterior Painting

09 91 13.30 Fences

09 91 13.30 Fences		Crew	Daily Output	Labor-Hours	Unit	Material	2018 Bare Costs Labor	2018 Bare Costs Equipment	Total	Total Incl O&P
0010	**FENCES**									
0100	Chain link or wire metal, one side, water base									
0110	Roll & brush, first coat	1 Pord	960	.008	S.F.	.07	.35		.42	.65
0120	Second coat		1280	.006		.07	.27		.34	.50
0130	Spray, first coat		2275	.004	↓	.07	.15		.22	.32
0140	Second coat	↓	2600	.003	↓	.07	.13		.20	.29
0150	Picket, water base									
0160	Roll & brush, first coat	1 Pord	865	.009	S.F.	.08	.39		.47	.71
0170	Second coat		1050	.008		.08	.32		.40	.60
0180	Spray, first coat		2275	.004		.08	.15		.23	.32
0190	Second coat	↓	2600	.003	↓	.08	.13		.21	.29
0200	Stockade, water base									
0210	Roll & brush, first coat	1 Pord	1040	.008	S.F.	.08	.33		.41	.61
0220	Second coat		1200	.007		.08	.28		.36	.54
0230	Spray, first coat	↓	2275	.004		.08	.15		.23	.32

324

For customer support on your Commercial Renovation Costs with RSMeans data, call 800.448.8182.

09 91 Painting

09 91 13 – Exterior Painting

09 91 13.30 Fences

	Crew	Daily Output	Labor-Hours	Unit	Material	2018 Bare Costs Labor	2018 Bare Costs Equipment	Total	Total Incl O&P
0240 Second coat	1 Pord	2600	.003	S.F.	.08	.13		.21	.29
9000 Minimum labor/equipment charge	↓	2	4	Job		170		170	273

09 91 13.42 Miscellaneous, Exterior

	Crew	Daily Output	Labor-Hours	Unit	Material	2018 Bare Costs Labor	2018 Bare Costs Equipment	Total	Total Incl O&P
0010 **MISCELLANEOUS, EXTERIOR**									
0100 Railing, ext., decorative wood, incl. cap & baluster									
0110 Newels & spindles @ 12" OC									
0120 Brushwork, stain, sand, seal & varnish									
0130 First coat	1 Pord	90	.089	L.F.	.88	3.78		4.66	7
0140 Second coat	"	120	.067	"	.88	2.84		3.72	5.50
0150 Rough sawn wood, 42" high, 2" x 2" verticals, 6" OC									
0160 Brushwork, stain, each coat	1 Pord	90	.089	L.F.	.31	3.78		4.09	6.40
0170 Wrought iron, 1" rail, 1/2" sq. verticals									
0180 Brushwork, zinc chromate, 60" high, bars 6" OC									
0190 Primer	1 Pord	130	.062	L.F.	.88	2.62		3.50	5.15
0200 Finish coat		130	.062		1.13	2.62		3.75	5.45
0210 Additional coat	↓	190	.042	↓	1.32	1.79		3.11	4.33
0220 Shutters or blinds, single panel, 2' x 4', paint all sides									
0230 Brushwork, primer	1 Pord	20	.400	Ea.	.69	17		17.69	28.50
0240 Finish coat, exterior latex		20	.400		.57	17		17.57	28
0250 Primer & 1 coat, exterior latex		13	.615		1.11	26		27.11	43
0260 Spray, primer		35	.229		1.01	9.75		10.76	16.70
0270 Finish coat, exterior latex		35	.229		1.21	9.75		10.96	16.95
0280 Primer & 1 coat, exterior latex	↓	20	.400	↓	1.09	17		18.09	28.50
0290 For louvered shutters, add				S.F.	10%				
0300 Stair stringers, exterior, metal									
0310 Roll & brush, zinc chromate, to 14", each coat	1 Pord	320	.025	L.F.	.38	1.06		1.44	2.12
0320 Rough sawn wood, 4" x 12"									
0330 Roll & brush, exterior latex, each coat	1 Pord	215	.037	L.F.	.08	1.58		1.66	2.63
0340 Trellis/lattice, 2" x 2" @ 3" OC with 2" x 8" supports									
0350 Spray, latex, per side, each coat	1 Pord	475	.017	S.F.	.08	.72		.80	1.24
0450 Decking, ext., sealer, alkyd, brushwork, sealer coat		1140	.007		.10	.30		.40	.59
0460 1st coat		1140	.007		.12	.30		.42	.61
0470 2nd coat		1300	.006		.09	.26		.35	.51
0500 Paint, alkyd, brushwork, primer coat		1140	.007		.11	.30		.41	.60
0510 1st coat		1140	.007		.14	.30		.44	.63
0520 2nd coat		1300	.006		.10	.26		.36	.53
0600 Sand paint, alkyd, brushwork, 1 coat		150	.053	↓	.14	2.27		2.41	3.79
9000 Minimum labor/equipment charge	↓	2	4	Job		170		170	273

09 91 13.60 Siding Exterior

	Crew	Daily Output	Labor-Hours	Unit	Material	2018 Bare Costs Labor	2018 Bare Costs Equipment	Total	Total Incl O&P
0010 **SIDING EXTERIOR**, Alkyd (oil base)									
0450 Steel siding, oil base, paint 1 coat, brushwork	2 Pord	2015	.008	S.F.	.11	.34		.45	.66
0500 Spray		4550	.004		.17	.15		.32	.43
0800 Paint 2 coats, brushwork		1300	.012		.23	.52		.75	1.09
1000 Spray		2750	.006		.15	.25		.40	.57
1200 Stucco, rough, oil base, paint 2 coats, brushwork		1300	.012		.23	.52		.75	1.09
1400 Roller		1625	.010		.24	.42		.66	.93
1600 Spray		2925	.005		.25	.23		.48	.65
1800 Texture 1-11 or clapboard, oil base, primer coat, brushwork		1300	.012		.14	.52		.66	1
2000 Spray		4550	.004		.14	.15		.29	.40
2400 Paint 2 coats, brushwork		810	.020		.33	.84		1.17	1.71
2600 Spray		2600	.006		.36	.26		.62	.82
3400 Stain 2 coats, brushwork	↓	950	.017	↓	.20	.72		.92	1.38

09 91 13.60 Siding Exterior	Crew	Daily Output	Labor-Hours	Unit	Material	2018 Bare Costs Labor	2018 Bare Costs Equipment	Total	Total Incl O&P
4000	Spray	2 Pord	3050	.005	S.F.	.23	.22		.45	.61
4200	Wood shingles, oil base primer coat, brushwork		1300	.012		.13	.52		.65	.99
4400	Spray		3900	.004		.12	.17		.29	.42
5000	Paint 2 coats, brushwork		810	.020		.27	.84		1.11	1.65
5200	Spray		2275	.007		.26	.30		.56	.77
6500	Stain 2 coats, brushwork		950	.017		.20	.72		.92	1.38
7000	Spray		2660	.006		.28	.26		.54	.72
8000	For latex paint, deduct					10%				
8100	For work over 12' H, from pipe scaffolding, add						15%			
8200	For work over 12' H, from extension ladder, add						25%			
8300	For work over 12' H, from swing staging, add						35%			
9000	Minimum labor/equipment charge	1 Pord	2	4	Job		170		170	273

09 91 13.62 Siding, Misc.	Crew	Daily Output	Labor-Hours	Unit	Material	2018 Bare Costs Labor	2018 Bare Costs Equipment	Total	Total Incl O&P
0010	**SIDING, MISC.**, latex paint									
0100	Aluminum siding									
0110	Brushwork, primer	2 Pord	2275	.007	S.F.	.06	.30		.36	.55
0120	Finish coat, exterior latex		2275	.007		.06	.30		.36	.54
0130	Primer & 1 coat exterior latex		1300	.012		.13	.52		.65	.99
0140	Primer & 2 coats exterior latex		975	.016		.19	.70		.89	1.33
0150	Mineral fiber shingles									
0160	Brushwork, primer	2 Pord	1495	.011	S.F.	.14	.46		.60	.89
0170	Finish coat, industrial enamel		1495	.011		.18	.46		.64	.93
0180	Primer & 1 coat enamel		810	.020		.32	.84		1.16	1.70
0190	Primer & 2 coats enamel		540	.030		.50	1.26		1.76	2.57
0200	Roll, primer		1625	.010		.16	.42		.58	.84
0210	Finish coat, industrial enamel		1625	.010		.19	.42		.61	.88
0220	Primer & 1 coat enamel		975	.016		.35	.70		1.05	1.51
0230	Primer & 2 coats enamel		650	.025		.55	1.05		1.60	2.28
0240	Spray, primer		3900	.004		.12	.17		.29	.42
0250	Finish coat, industrial enamel		3900	.004		.16	.17		.33	.46
0260	Primer & 1 coat enamel		2275	.007		.28	.30		.58	.79
0270	Primer & 2 coats enamel		1625	.010		.45	.42		.87	1.16
0280	Waterproof sealer, first coat		4485	.004		.12	.15		.27	.37
0290	Second coat		5235	.003		.11	.13		.24	.33
0300	Rough wood incl. shingles, shakes or rough sawn siding									
0310	Brushwork, primer	2 Pord	1280	.013	S.F.	.14	.53		.67	1
0320	Finish coat, exterior latex		1280	.013		.10	.53		.63	.96
0330	Primer & 1 coat exterior latex		960	.017		.24	.71		.95	1.41
0340	Primer & 2 coats exterior latex		700	.023		.34	.97		1.31	1.94
0350	Roll, primer		2925	.005		.19	.23		.42	.58
0360	Finish coat, exterior latex		2925	.005		.12	.23		.35	.50
0370	Primer & 1 coat exterior latex		1790	.009		.31	.38		.69	.95
0380	Primer & 2 coats exterior latex		1300	.012		.43	.52		.95	1.32
0390	Spray, primer		3900	.004		.16	.17		.33	.45
0400	Finish coat, exterior latex		3900	.004		.09	.17		.26	.38
0410	Primer & 1 coat exterior latex		2600	.006		.25	.26		.51	.70
0420	Primer & 2 coats exterior latex		2080	.008		.35	.33		.68	.91
0430	Waterproof sealer, first coat		4485	.004		.21	.15		.36	.47
0440	Second coat		4485	.004		.12	.15		.27	.37
0450	Smooth wood incl. butt, T&G, beveled, drop or B&B siding									
0460	Brushwork, primer	2 Pord	2325	.007	S.F.	.10	.29		.39	.58
0470	Finish coat, exterior latex		1280	.013		.10	.53		.63	.96

09 91 13 – Exterior Painting

09 91 13.62 Siding, Misc.

		Crew	Daily Output	Labor-Hours	Unit	Material	2018 Bare Costs Labor	Equipment	Total	Total Incl O&P
0480	Primer & 1 coat exterior latex	2 Pord	800	.020	S.F.	.20	.85		1.05	1.59
0490	Primer & 2 coats exterior latex		630	.025		.31	1.08		1.39	2.07
0500	Roll, primer		2275	.007		.11	.30		.41	.60
0510	Finish coat, exterior latex		2275	.007		.11	.30		.41	.60
0520	Primer & 1 coat exterior latex		1300	.012		.22	.52		.74	1.08
0530	Primer & 2 coats exterior latex		975	.016		.33	.70		1.03	1.49
0540	Spray, primer		4550	.004		.09	.15		.24	.34
0550	Finish coat, exterior latex		4550	.004		.09	.15		.24	.34
0560	Primer & 1 coat exterior latex		2600	.006		.18	.26		.44	.62
0570	Primer & 2 coats exterior latex		1950	.008		.28	.35		.63	.86
0580	Waterproof sealer, first coat		5230	.003		.12	.13		.25	.34
0590	Second coat	▼	5980	.003		.12	.11		.23	.31
0600	For oil base paint, add					10%				
9000	Minimum labor/equipment charge	1 Pord	2	4	Job		170		170	273

09 91 13.70 Doors and Windows, Exterior

		Crew	Daily Output	Labor-Hours	Unit	Material	2018 Bare Costs Labor	Equipment	Total	Total Incl O&P
0010	**DOORS AND WINDOWS, EXTERIOR**									
0100	Door frames & trim, only									
0110	Brushwork, primer	1 Pord	512	.016	L.F.	.06	.67		.73	1.14
0120	Finish coat, exterior latex		512	.016		.07	.67		.74	1.15
0130	Primer & 1 coat, exterior latex		300	.027	▼	.13	1.13		1.26	1.97
0135	2 coats, exterior latex, both sides		15	.533	Ea.	6.35	22.50		28.85	43.50
0140	Primer & 2 coats, exterior latex	▼	265	.030	L.F.	.21	1.28		1.49	2.29
0150	Doors, flush, both sides, incl. frame & trim									
0160	Roll & brush, primer	1 Pord	10	.800	Ea.	4.80	34		38.80	60
0170	Finish coat, exterior latex		10	.800		5.40	34		39.40	60.50
0180	Primer & 1 coat, exterior latex		7	1.143		10.20	48.50		58.70	89.50
0190	Primer & 2 coats, exterior latex		5	1.600		15.65	68		83.65	126
0200	Brushwork, stain, sealer & 2 coats polyurethane	▼	4	2	▼	30	85		115	170
0210	Doors, French, both sides, 10-15 lite, incl. frame & trim									
0220	Brushwork, primer	1 Pord	6	1.333	Ea.	2.40	56.50		58.90	93.50
0230	Finish coat, exterior latex		6	1.333		2.71	56.50		59.21	94
0240	Primer & 1 coat, exterior latex		3	2.667		5.10	113		118.10	188
0250	Primer & 2 coats, exterior latex		2	4		7.65	170		177.65	281
0260	Brushwork, stain, sealer & 2 coats polyurethane	▼	2.50	3.200	▼	11.10	136		147.10	231
0270	Doors, louvered, both sides, incl. frame & trim									
0280	Brushwork, primer	1 Pord	7	1.143	Ea.	4.80	48.50		53.30	83.50
0290	Finish coat, exterior latex		7	1.143		5.40	48.50		53.90	84
0300	Primer & 1 coat, exterior latex		4	2		10.20	85		95.20	148
0310	Primer & 2 coats, exterior latex		3	2.667		15.35	113		128.35	199
0320	Brushwork, stain, sealer & 2 coats polyurethane	▼	4.50	1.778	▼	30	75.50		105.50	154
0330	Doors, panel, both sides, incl. frame & trim									
0340	Roll & brush, primer	1 Pord	6	1.333	Ea.	4.80	56.50		61.30	96.50
0350	Finish coat, exterior latex		6	1.333		5.40	56.50		61.90	97
0360	Primer & 1 coat, exterior latex		3	2.667		10.20	113		123.20	193
0370	Primer & 2 coats, exterior latex		2.50	3.200		15.35	136		151.35	236
0380	Brushwork, stain, sealer & 2 coats polyurethane	▼	3	2.667	▼	30	113		143	215
0400	Windows, per ext. side, based on 15 S.F.									
0410	1 to 6 lite									
0420	Brushwork, primer	1 Pord	13	.615	Ea.	.95	26		26.95	43
0430	Finish coat, exterior latex		13	.615		1.07	26		27.07	43
0440	Primer & 1 coat, exterior latex		8	1		2.02	42.50		44.52	70.50
0450	Primer & 2 coats, exterior latex	▼	6	1.333	▼	3.03	56.50		59.53	94.50

327

09 91 13.70 Doors and Windows, Exterior

		Crew	Daily Output	Labor-Hours	Unit	Material	2018 Bare Costs Labor	2018 Bare Costs Equipment	Total	Total Incl O&P
0460	Stain, sealer & 1 coat varnish	1 Pord	7	1.143	Ea.	4.38	48.50		52.88	83
0470	7 to 10 lite									
0480	Brushwork, primer	1 Pord	11	.727	Ea.	.95	31		31.95	50.50
0490	Finish coat, exterior latex		11	.727		1.07	31		32.07	50.50
0500	Primer & 1 coat, exterior latex		7	1.143		2.02	48.50		50.52	80
0510	Primer & 2 coats, exterior latex		5	1.600		3.03	68		71.03	112
0520	Stain, sealer & 1 coat varnish		6	1.333		4.38	56.50		60.88	96
0530	12 lite									
0540	Brushwork, primer	1 Pord	10	.800	Ea.	.95	34		34.95	55.50
0550	Finish coat, exterior latex		10	.800		1.07	34		35.07	55.50
0560	Primer & 1 coat, exterior latex		6	1.333		2.02	56.50		58.52	93
0570	Primer & 2 coats, exterior latex		5	1.600		3.03	68		71.03	112
0580	Stain, sealer & 1 coat varnish		6	1.333		4.28	56.50		60.78	95.50
0590	For oil base paint, add					10%				
9000	Minimum labor/equipment charge	1 Pord	2	4	Job		170		170	273

09 91 13.80 Trim, Exterior

		Crew	Daily Output	Labor-Hours	Unit	Material	2018 Bare Costs Labor	2018 Bare Costs Equipment	Total	Total Incl O&P
0010	**TRIM, EXTERIOR**									
0100	Door frames & trim (see Doors, interior or exterior)									
0110	Fascia, latex paint, one coat coverage									
0120	1" x 4", brushwork	1 Pord	640	.013	L.F.	.02	.53		.55	.87
0130	Roll		1280	.006		.02	.27		.29	.46
0140	Spray		2080	.004		.02	.16		.18	.28
0150	1" x 6" to 1" x 10", brushwork		640	.013		.08	.53		.61	.93
0160	Roll		1230	.007		.08	.28		.36	.53
0170	Spray		2100	.004		.06	.16		.22	.33
0180	1" x 12", brushwork		640	.013		.08	.53		.61	.93
0190	Roll		1050	.008		.08	.32		.40	.61
0200	Spray		2200	.004		.06	.15		.21	.32
0210	Gutters & downspouts, metal, zinc chromate paint									
0220	Brushwork, gutters, 5", first coat	1 Pord	640	.013	L.F.	.40	.53		.93	1.29
0230	Second coat		960	.008		.38	.35		.73	.98
0240	Third coat		1280	.006		.30	.27		.57	.77
0250	Downspouts, 4", first coat		640	.013		.40	.53		.93	1.29
0260	Second coat		960	.008		.38	.35		.73	.98
0270	Third coat		1280	.006		.30	.27		.57	.77
0280	Gutters & downspouts, wood									
0290	Brushwork, gutters, 5", primer	1 Pord	640	.013	L.F.	.06	.53		.59	.92
0300	Finish coat, exterior latex		640	.013		.06	.53		.59	.92
0310	Primer & 1 coat exterior latex		400	.020		.13	.85		.98	1.52
0320	Primer & 2 coats exterior latex		325	.025		.21	1.05		1.26	1.91
0330	Downspouts, 4", primer		640	.013		.06	.53		.59	.92
0340	Finish coat, exterior latex		640	.013		.06	.53		.59	.92
0350	Primer & 1 coat exterior latex		400	.020		.13	.85		.98	1.52
0360	Primer & 2 coats exterior latex		325	.025		.10	1.05		1.15	1.79
0370	Molding, exterior, up to 14" wide									
0380	Brushwork, primer	1 Pord	640	.013	L.F.	.08	.53		.61	.93
0390	Finish coat, exterior latex		640	.013		.08	.53		.61	.93
0400	Primer & 1 coat exterior latex		400	.020		.16	.85		1.01	1.55
0410	Primer & 2 coats exterior latex		315	.025		.16	1.08		1.24	1.91
0420	Stain & fill		1050	.008		.12	.32		.44	.66
0430	Shellac		1850	.004		.15	.18		.33	.47
0440	Varnish		1275	.006		.11	.27		.38	.55

328

For customer support on your Commercial Renovation Costs with RSMeans data, call 800.448.8182.

09 91 Painting

09 91 13 – Exterior Painting

09 91 13.80 Trim, Exterior

		Crew	Daily Output	Labor-Hours	Unit	Material	2018 Bare Costs Labor	2018 Bare Costs Equipment	Total	Total Incl O&P
9000	Minimum labor/equipment charge	1 Pord	2	4	Job		170		170	273

09 91 13.90 Walls, Masonry (CMU), Exterior

		Crew	Daily Output	Labor-Hours	Unit	Material	2018 Bare Costs Labor	2018 Bare Costs Equipment	Total	Total Incl O&P
0010	**WALLS, MASONRY (CMU), EXTERIOR**									
0360	Concrete masonry units (CMU), smooth surface									
0370	Brushwork, latex, first coat	1 Pord	640	.013	S.F.	.06	.53		.59	.92
0380	Second coat		960	.008		.05	.35		.40	.62
0390	Waterproof sealer, first coat		736	.011		.27	.46		.73	1.04
0400	Second coat		1104	.007		.27	.31		.58	.80
0410	Roll, latex, paint, first coat		1465	.005		.07	.23		.30	.45
0420	Second coat		1790	.004		.06	.19		.25	.37
0430	Waterproof sealer, first coat		1680	.005		.27	.20		.47	.63
0440	Second coat		2060	.004		.27	.17		.44	.57
0450	Spray, latex, paint, first coat		1950	.004		.06	.17		.23	.34
0460	Second coat		2600	.003		.05	.13		.18	.26
0470	Waterproof sealer, first coat		2245	.004		.27	.15		.42	.54
0480	Second coat	▼	2990	.003	▼	.27	.11		.38	.48
0490	Concrete masonry unit (CMU), porous									
0500	Brushwork, latex, first coat	1 Pord	640	.013	S.F.	.12	.53		.65	.98
0510	Second coat		960	.008		.06	.35		.41	.64
0520	Waterproof sealer, first coat		736	.011		.27	.46		.73	1.04
0530	Second coat		1104	.007		.27	.31		.58	.80
0540	Roll latex, first coat		1465	.005		.09	.23		.32	.47
0550	Second coat		1790	.004		.06	.19		.25	.37
0560	Waterproof sealer, first coat		1680	.005		.27	.20		.47	.63
0570	Second coat		2060	.004		.27	.17		.44	.57
0580	Spray latex, first coat		1950	.004		.07	.17		.24	.35
0590	Second coat		2600	.003		.05	.13		.18	.26
0600	Waterproof sealer, first coat		2245	.004		.27	.15		.42	.54
0610	Second coat		2990	.003	▼	.27	.11		.38	.40
9000	Minimum labor/equipment charge	▼	2	4	Job		170		170	273

09 91 23 – Interior Painting

09 91 23.20 Cabinets and Casework

		Crew	Daily Output	Labor-Hours	Unit	Material	2018 Bare Costs Labor	2018 Bare Costs Equipment	Total	Total Incl O&P
0010	**CABINETS AND CASEWORK**									
1000	Primer coat, oil base, brushwork	1 Pord	650	.012	S.F.	.07	.52		.59	.91
2000	Paint, oil base, brushwork, 1 coat		650	.012		.12	.52		.64	.97
3000	Stain, brushwork, wipe off		650	.012		.10	.52		.62	.95
4000	Shellac, 1 coat, brushwork		650	.012		.13	.52		.65	.98
4500	Varnish, 3 coats, brushwork, sand after 1st coat	▼	325	.025		.27	1.05		1.32	1.98
5000	For latex paint, deduct				▼	10%				
6300	Strip, prep and refinish wood furniture									
6310	Remove paint using chemicals, wood furniture	1 Pord	28	.286	S.F.	1.60	12.15		13.75	21.50
6320	Prep for painting, sanding		75	.107		.24	4.54		4.78	7.55
6350	Stain and wipe, brushwork		600	.013		.10	.57		.67	1.02
6355	Spray applied		900	.009		.09	.38		.47	.71
6360	Sealer or varnish, brushwork		1080	.007		.09	.32		.41	.61
6365	Spray applied		2100	.004		.09	.16		.25	.36
6370	Paint, primer, brushwork		720	.011		.07	.47		.54	.83
6375	Spray applied		2100	.004		.06	.16		.22	.33
6380	Finish coat, brushwork		810	.010		.12	.42		.54	.80
6385	Spray applied		2100	.004	▼	.11	.16		.27	.38
9010	Minimum labor/equipment charge	▼	2	4	Job		170		170	273

09 91 23.33 Doors and Windows, Interior Alkyd (Oil Base)

09 91 23.33 Doors and Windows, Interior Alkyd (Oil Base)	Crew	Daily Output	Labor-Hours	Unit	Material	2018 Bare Costs Labor	Equipment	Total	Total Incl O&P
0010 **DOORS AND WINDOWS, INTERIOR ALKYD (OIL BASE)**									
0500 Flush door & frame, 3' x 7', oil, primer, brushwork	1 Pord	10	.800	Ea.	3.98	34		37.98	59
1000 Paint, 1 coat		10	.800		4.16	34		38.16	59
1400 Stain, brushwork, wipe off		18	.444		2.15	18.90		21.05	33
1600 Shellac, 1 coat, brushwork		25	.320		2.64	13.60		16.24	25
1800 Varnish, 3 coats, brushwork, sand after 1st coat		9	.889		5.75	38		43.75	67
2000 Panel door & frame, 3' x 7', oil, primer, brushwork		6	1.333		2.50	56.50		59	94
2200 Paint, 1 coat		6	1.333		4.16	56.50		60.66	95.50
2600 Stain, brushwork, panel door, 3' x 7', not incl. frame		16	.500		2.15	21.50		23.65	36.50
2800 Shellac, 1 coat, brushwork		22	.364		2.64	15.45		18.09	28
3000 Varnish, 3 coats, brushwork, sand after 1st coat		7.50	1.067		5.75	45.50		51.25	79.50
4400 Windows, including frame and trim, per side									
4600 Colonial type, 6/6 lites, 2' x 3', oil, primer, brushwork	1 Pord	14	.571	Ea.	.39	24.50		24.89	39.50
5800 Paint, 1 coat		14	.571		.66	24.50		25.16	39.50
6200 3' x 5' opening, 6/6 lites, primer coat, brushwork		12	.667		.99	28.50		29.49	46.50
6400 Paint, 1 coat		12	.667		1.64	28.50		30.14	47.50
6800 4' x 8' opening, 6/6 lites, primer coat, brushwork		8	1		2.11	42.50		44.61	71
7000 Paint, 1 coat		8	1		3.50	42.50		46	72.50
8000 Single lite type, 2' x 3', oil base, primer coat, brushwork		33	.242		.39	10.30		10.69	17
8200 Paint, 1 coat		33	.242		.66	10.30		10.96	17.25
8600 3' x 5' opening, primer coat, brushwork		20	.400		.99	17		17.99	28.50
8800 Paint, 1 coat		20	.400		1.64	17		18.64	29.50
9200 4' x 8' opening, primer coat, brushwork		14	.571		2.11	24.50		26.61	41.50
9400 Paint, 1 coat		14	.571		3.50	24.50		28	43
9900 Minimum labor/equipment charge		2	4	Job		170		170	273

09 91 23.35 Doors and Windows, Interior Latex

09 91 23.35 Doors and Windows, Interior Latex	Crew	Daily Output	Labor-Hours	Unit	Material	2018 Bare Costs Labor	Equipment	Total	Total Incl O&P
0010 **DOORS & WINDOWS, INTERIOR LATEX**									
0100 Doors, flush, both sides, incl. frame & trim									
0110 Roll & brush, primer	1 Pord	10	.800	Ea.	4.08	34		38.08	59
0120 Finish coat, latex		10	.800		5.55	34		39.55	60.50
0130 Primer & 1 coat latex		7	1.143		9.65	48.50		58.15	88.50
0140 Primer & 2 coats latex		5	1.600		14.90	68		82.90	125
0160 Spray, both sides, primer		20	.400		4.30	17		21.30	32
0170 Finish coat, latex		20	.400		5.85	17		22.85	34
0180 Primer & 1 coat latex		11	.727		10.20	31		41.20	60.50
0190 Primer & 2 coats latex		8	1		15.75	42.50		58.25	86
0200 Doors, French, both sides, 10-15 lite, incl. frame & trim									
0210 Roll & brush, primer	1 Pord	6	1.333	Ea.	2.04	56.50		58.54	93.50
0220 Finish coat, latex		6	1.333		2.78	56.50		59.28	94
0230 Primer & 1 coat latex		3	2.667		4.82	113		117.82	187
0240 Primer & 2 coats latex		2	4		7.45	170		177.45	281
0260 Doors, louvered, both sides, incl. frame & trim									
0270 Roll & brush, primer	1 Pord	7	1.143	Ea.	4.08	48.50		52.58	82.50
0280 Finish coat, latex		7	1.143		5.55	·48.50		54.05	84
0290 Primer & 1 coat, latex		4	2		9.40	85		94.40	147
0300 Primer & 2 coats, latex		3	2.667		15.20	113		128.20	199
0320 Spray, both sides, primer		20	.400		4.30	17		21.30	32
0330 Finish coat, latex		20	.400		5.85	17		22.85	34
0340 Primer & 1 coat, latex		11	.727		10.20	31		41.20	60.50
0350 Primer & 2 coats, latex		8	1		16.10	42.50		58.60	86
0360 Doors, panel, both sides, incl. frame & trim									
0370 Roll & brush, primer	1 Pord	6	1.333	Ea.	4.30	56.50		60.80	95.50

09 91 Painting

09 91 23 – Interior Painting

09 91 23.35 Doors and Windows, Interior Latex

		Crew	Daily Output	Labor-Hours	Unit	Material	2018 Bare Costs Labor	Equipment	Total	Total Incl O&P
0380	Finish coat, latex	1 Pord	6	1.333	Ea.	5.55	56.50		62.05	97
0390	Primer & 1 coat, latex		3	2.667		9.65	113		122.65	193
0400	Primer & 2 coats, latex		2.50	3.200		15.20	136		151.20	236
0420	Spray, both sides, primer		10	.800		4.30	34		38.30	59
0430	Finish coat, latex		10	.800		5.85	34		39.85	61
0440	Primer & 1 coat, latex		5	1.600		10.20	68		78.20	120
0450	Primer & 2 coats, latex	▼	4	2	▼	16.10	85		101.10	155
0460	Windows, per interior side, based on 15 S.F.									
0470	1 to 6 lite									
0480	Brushwork, primer	1 Pord	13	.615	Ea.	.81	26		26.81	43
0490	Finish coat, enamel		13	.615		1.10	26		27.10	43
0500	Primer & 1 coat enamel		8	1		1.90	42.50		44.40	70.50
0510	Primer & 2 coats enamel	▼	6	1.333	▼	3	56.50		59.50	94.50
0530	7 to 10 lite									
0540	Brushwork, primer	1 Pord	11	.727	Ea.	.81	31		31.81	50.50
0550	Finish coat, enamel		11	.727		1.10	31		32.10	50.50
0560	Primer & 1 coat enamel		7	1.143		1.90	48.50		50.40	80
0570	Primer & 2 coats enamel	▼	5	1.600	▼	3	68		71	112
0590	12 lite									
0600	Brushwork, primer	1 Pord	10	.800	Ea.	.81	34		34.81	55.50
0610	Finish coat, enamel		10	.800		1.10	34		35.10	55.50
0620	Primer & 1 coat enamel		6	1.333		1.90	56.50		58.40	93
0630	Primer & 2 coats enamel	▼	5	1.600		3	68		71	112
0650	For oil base paint, add				▼	10%				
9000	Minimum labor/equipment charge	1 Pord	2	4	Job		170		170	273

09 91 23.39 Doors and Windows, Interior Latex, Zero Voc

			Crew	Daily Output	Labor-Hours	Unit	Material	2018 Bare Costs Labor	Equipment	Total	Total Incl O&P
0010	**DOORS & WINDOWS, INTERIOR LATEX, ZERO VOC**										
0100	Doors flush, both sides, incl. frame & trim										
0110	Roll & brush, primer	G	1 Pord	10	.800	Ea.	6.30	34		40.30	61.50
0120	Finish coat, latex	G		10	.800		6.95	34		40.95	62
0130	Primer & 1 coat latex	G		7	1.143		13.25	48.50		61.75	92.50
0140	Primer & 2 coats latex	G		5	1.600		19.80	68		87.80	131
0160	Spray, both sides, primer	G		20	.400		6.65	17		23.65	35
0170	Finish coat, latex	G		20	.400		7.30	17		24.30	35.50
0180	Primer & 1 coat latex	G		11	.727		14.05	31		45.05	65
0190	Primer & 2 coats latex	G	▼	8	1	▼	21	42.50		63.50	91.50
0200	Doors, French, both sides, 10-15 lite, incl. frame & trim										
0210	Roll & brush, primer	G	1 Pord	6	1.333	Ea.	3.15	56.50		59.65	94.50
0220	Finish coat, latex	G		6	1.333		3.48	56.50		59.98	95
0230	Primer & 1 coat latex	G		3	2.667		6.65	113		119.65	189
0240	Primer & 2 coats latex	G	▼	2	4	▼	9.90	170		179.90	284
0360	Doors, panel, both sides, incl. frame & trim										
0370	Roll & brush, primer	G	1 Pord	6	1.333	Ea.	6.65	56.50		63.15	98.50
0380	Finish coat, latex	G		6	1.333		6.95	56.50		63.45	98.50
0390	Primer & 1 coat, latex	G		3	2.667		13.25	113		126.25	197
0400	Primer & 2 coats, latex	G		2.50	3.200		20	136		156	241
0420	Spray, both sides, primer	G		10	.800		6.65	34		40.65	62
0430	Finish coat, latex	G		10	.800		7.30	34		41.30	62.50
0440	Primer & 1 coat, latex	G		5	1.600		14.05	68		82.05	124
0450	Primer & 2 coats, latex	G	▼	4	2	▼	21.50	85		106.50	161
0460	Windows, per interior side, based on 15 S.F.										
0470	1 to 6 lite										

09 91 23.39 Doors and Windows, Interior Latex, Zero Voc

		Crew	Daily Output	Labor-Hours	Unit	Material	2018 Bare Costs Labor	Equipment	Total	Total Incl O&P
0480	Brushwork, primer [G]	1 Pord	13	.615	Ea.	1.24	26		27.24	43.50
0490	Finish coat, enamel [G]		13	.615		1.37	26		27.37	43.50
0500	Primer & 1 coat enamel [G]		8	1		2.62	42.50		45.12	71.50
0510	Primer & 2 coats enamel [G]		6	1.333		3.99	56.50		60.49	95.50
9000	Minimum labor/equipment charge		2	4	Job		170		170	273

09 91 23.40 Floors, Interior

		Crew	Daily Output	Labor-Hours	Unit	Material	2018 Bare Costs Labor	Equipment	Total	Total Incl O&P
0010	**FLOORS, INTERIOR**									
0100	Concrete paint, latex									
0110	Brushwork									
0120	1st coat	1 Pord	975	.008	S.F.	.16	.35		.51	.73
0130	2nd coat		1150	.007		.10	.30		.40	.59
0140	3rd coat		1300	.006		.08	.26		.34	.51
0150	Roll									
0160	1st coat	1 Pord	2600	.003	S.F.	.21	.13		.34	.44
0170	2nd coat		3250	.002		.12	.10		.22	.31
0180	3rd coat		3900	.002		.09	.09		.18	.24
0190	Spray									
0200	1st coat	1 Pord	2600	.003	S.F.	.18	.13		.31	.41
0210	2nd coat		3250	.002		.10	.10		.20	.28
0220	3rd coat		3900	.002		.08	.09		.17	.23

09 91 23.44 Anti-Slip Floor Treatments

		Crew	Daily Output	Labor-Hours	Unit	Material	2018 Bare Costs Labor	Equipment	Total	Total Incl O&P
0010	**ANTI-SLIP FLOOR TREATMENTS**									
1000	Walking surface treatment, ADA compliant, mop on and rinse									
1100	For tile, terrazzo, stone or smooth concrete	1 Pord	4000	.002	S.F.	.33	.09		.42	.51
1110	For marble		4000	.002		.24	.09		.33	.40
1120	For wood		4000	.002		.22	.09		.31	.39
1130	For baths and showers		500	.016		.38	.68		1.06	1.51
2000	Granular additive for paint or sealer, add to paint cost					.02			.02	.02

09 91 23.52 Miscellaneous, Interior

		Crew	Daily Output	Labor-Hours	Unit	Material	2018 Bare Costs Labor	Equipment	Total	Total Incl O&P
0010	**MISCELLANEOUS, INTERIOR**									
2400	Floors, conc./wood, oil base, primer/sealer coat, brushwork	2 Pord	1950	.008	S.F.	.09	.35		.44	.66
2450	Roller		5200	.003		.09	.13		.22	.31
2600	Spray		6000	.003		.09	.11		.20	.28
2650	Paint 1 coat, brushwork		1950	.008		.11	.35		.46	.68
2800	Roller		5200	.003		.11	.13		.24	.33
2850	Spray		6000	.003		.12	.11		.23	.31
3000	Stain, wood floor, brushwork, 1 coat		4550	.004		.10	.15		.25	.35
3200	Roller		5200	.003		.11	.13		.24	.33
3250	Spray		6000	.003		.11	.11		.22	.30
3400	Varnish, wood floor, brushwork		4550	.004		.09	.15		.24	.34
3450	Roller		5200	.003		.10	.13		.23	.32
3600	Spray		6000	.003		.10	.11		.21	.29
3800	Grilles, per side, oil base, primer coat, brushwork	1 Pord	520	.015		.13	.65		.78	1.19
3850	Spray		1140	.007		.14	.30		.44	.63
3920	Paint 2 coats, brushwork		325	.025		.43	1.05		1.48	2.15
3940	Spray		650	.012		.49	.52		1.01	1.38
4600	Miscellaneous surfaces, metallic paint, spray applied									
4610	Water based, non-tintable, warm silver	1 Pord	1140	.007	S.F.	.78	.30		1.08	1.34
4620	Rusted iron		1140	.007		1.01	.30		1.31	1.59
4630	Low VOC, tintable		1140	.007		.35	.30		.65	.86
5000	Pipe, 1"-4" diameter, primer or sealer coat, oil base, brushwork	2 Pord	1250	.013	L.F.	.09	.54		.63	.97
5100	Spray		2165	.007		.09	.31		.40	.60

09 91 23.52 Miscellaneous, Interior

	Crew	Daily Output	Labor-Hours	Unit	Material	2018 Bare Costs Labor	Equipment	Total	Total Incl O&P
5350 Paint 2 coats, brushwork	2 Pord	775	.021	L.F.	.20	.88		1.08	1.63
5400 Spray		1240	.013		.23	.55		.78	1.13
6300 13"-16" diameter, primer or sealer coat, brushwork		310	.052		.37	2.20		2.57	3.94
6450 Spray		540	.030		.47	1.26		1.73	2.53
6500 Paint 2 coats, brushwork		195	.082		.81	3.49		4.30	6.50
6550 Spray	↓	310	.052	↓	.90	2.20		3.10	4.53
7000 Trim, wood, incl. puttying, under 6" wide									
7200 Primer coat, oil base, brushwork	1 Pord	650	.012	L.F.	.03	.52		.55	.88
7250 Paint, 1 coat, brushwork		650	.012		.05	.52		.57	.90
7450 3 coats		325	.025		.16	1.05		1.21	1.85
7500 Over 6" wide, primer coat, brushwork		650	.012		.07	.52		.59	.91
7550 Paint, 1 coat, brushwork		650	.012		.11	.52		.63	.96
7650 3 coats		325	.025	↓	.32	1.05		1.37	2.03
8000 Cornice, simple design, primer coat, oil base, brushwork		650	.012	S.F.	.07	.52		.59	.91
8250 Paint, 1 coat		650	.012		.11	.52		.63	.96
8350 Ornate design, primer coat		350	.023		.07	.97		1.04	1.63
8400 Paint, 1 coat		350	.023		.11	.97		1.08	1.68
8600 Balustrades, primer coat, oil base, brushwork		520	.015		.07	.65		.72	1.12
8900 Trusses and wood frames, primer coat, oil base, brushwork		800	.010		.07	.43		.50	.75
8950 Spray		1200	.007		.07	.28		.35	.54
9220 Paint 2 coats, brushwork		500	.016		.21	.68		.89	1.32
9240 Spray		600	.013		.24	.57		.81	1.17
9260 Stain, brushwork, wipe off		600	.013		.10	.57		.67	1.02
9280 Varnish, 3 coats, brushwork	↓	275	.029		.27	1.24		1.51	2.29
9350 For latex paint, deduct				↓	10%				
9900 Minimum labor/equipment charge	1 Pord	2	4	Job		170		170	273

09 91 23.62 Electrostatic Painting

	Crew	Daily Output	Labor-Hours	Unit	Material	2018 Bare Costs Labor	Equipment	Total	Total Incl O&P
0010 **ELECTROSTATIC PAINTING**									
0100 In shop									
0200 Flat surfaces (lockers, casework, elevator doors, etc.)									
0300 One coat	1 Pord	200	.040	S.F.	.63	1.70		2.33	3.42
0400 Two coats	"	120	.067	"	.91	2.84		3.75	5.55
0500 Irregular surfaces (furniture, door frames, etc.)									
0600 One coat	1 Pord	150	.053	S.F.	.63	2.27		2.90	4.33
0700 Two coats	"	100	.080	"	.91	3.40		4.31	6.45
0800 On site									
0900 Flat surfaces (lockers, casework, elevator doors, etc.)									
1000 One coat	1 Pord	150	.053	S.F.	.63	2.27		2.90	4.33
1100 Two coats	"	100	.080	"	.91	3.40		4.31	6.45
1200 Irregular surfaces (furniture, door frames, etc.)									
1300 One coat	1 Pord	115	.070	S.F.	.63	2.96		3.59	5.45
1400 Two coats	↓	70	.114	↓	.91	4.86		5.77	8.80
2000 Anti-microbial coating, hospital application		150	.053		.05	2.27		2.32	3.70

09 91 23.72 Walls and Ceilings, Interior

	Crew	Daily Output	Labor-Hours	Unit	Material	2018 Bare Costs Labor	Equipment	Total	Total Incl O&P
0010 **WALLS AND CEILINGS, INTERIOR**									
0100 Concrete, drywall or plaster, latex, primer or sealer coat									
0200 Smooth finish, brushwork	1 Pord	1150	.007	S.F.	.06	.30		.36	.55
0240 Roller		1350	.006		.06	.25		.31	.48
0280 Spray		2750	.003		.05	.12		.17	.26
0300 Sand finish, brushwork		975	.008		.06	.35		.41	.63
0340 Roller		1150	.007		.06	.30		.36	.55
0380 Spray	↓	2275	.004	↓	.05	.15		.20	.30

For customer support on your Commercial Renovation Costs with RSMeans data, call 800.448.8182.

333

09 91 23.72 Walls and Ceilings, Interior

		Crew	Daily Output	Labor-Hours	Unit	Material	2018 Bare Costs Labor	Equipment	Total	Total Incl O&P
0800	Paint 2 coats, smooth finish, brushwork	1 Pord	680	.012	S.F.	.14	.50		.64	.96
0840	Roller		800	.010		.14	.43		.57	.84
0880	Spray		1625	.005		.13	.21		.34	.49
0900	Sand finish, brushwork		605	.013		.14	.56		.70	1.06
0940	Roller		1020	.008		.14	.33		.47	.70
0980	Spray		1700	.005		.13	.20		.33	.47
1600	Glaze coating, 2 coats, spray, clear		1200	.007		.56	.28		.84	1.08
1640	Multicolor		1200	.007		.85	.28		1.13	1.40
1660	Painting walls, complete, including surface prep, primer &									
1670	2 coats finish, on drywall or plaster, with roller	1 Pord	325	.025	S.F.	.21	1.05		1.26	1.91
1700	For oil base paint, add					10%				
1800	For ceiling installations, add						25%			
2000	Masonry or concrete block, primer/sealer, latex paint									
2100	Primer, smooth finish, brushwork	1 Pord	1000	.008	S.F.	.15	.34		.49	.72
2110	Roller		1150	.007		.11	.30		.41	.60
2180	Spray		2400	.003		.10	.14		.24	.34
2200	Sand finish, brushwork		850	.009		.11	.40		.51	.76
2210	Roller		975	.008		.11	.35		.46	.68
2280	Spray		2050	.004		.10	.17		.27	.38
2800	Primer plus one finish coat, smooth brush		525	.015		.30	.65		.95	1.37
2810	Roller		615	.013		.19	.55		.74	1.10
2880	Spray		1200	.007		.17	.28		.45	.65
2900	Sand finish, brushwork		450	.018		.19	.76		.95	1.42
2910	Roller		515	.016		.19	.66		.85	1.27
2980	Spray		1025	.008		.17	.33		.50	.72
3600	Glaze coating, 3 coats, spray, clear		900	.009		.80	.38		1.18	1.49
3620	Multicolor		900	.009		1.05	.38		1.43	1.76
4000	Block filler, 1 coat, brushwork		425	.019		.13	.80		.93	1.43
4100	Silicone, water repellent, 2 coats, spray		2000	.004		.45	.17		.62	.76
4120	For oil base paint, add					10%				
8200	For work 8'-15' H, add						10%			
8300	For work over 15' H, add						20%			
8400	For light textured surfaces, add						10%			
8410	Heavy textured, add						25%			
9900	Minimum labor/equipment charge	1 Pord	2	4	Job		170		170	273

09 91 23.74 Walls and Ceilings, Interior, Zero VOC Latex

			Crew	Daily Output	Labor-Hours	Unit	Material	2018 Bare Costs Labor	Equipment	Total	Total Incl O&P
0010	**WALLS AND CEILINGS, INTERIOR, ZERO VOC LATEX**										
0100	Concrete, dry wall or plaster, latex, primer or sealer coat										
0200	Smooth finish, brushwork	G	1 Pord	1150	.007	S.F.	.08	.30		.38	.57
0240	Roller	G		1350	.006		.08	.25		.33	.50
0280	Spray	G		2750	.003		.06	.12		.18	.27
0300	Sand finish, brushwork	G		975	.008		.08	.35		.43	.65
0340	Roller	G		1150	.007		.09	.30		.39	.58
0380	Spray	G		2275	.004		.07	.15		.22	.31
0800	Paint 2 coats, smooth finish, brushwork	G		680	.012		.18	.50		.68	1
0840	Roller	G		800	.010		.19	.43		.62	.88
0880	Spray	G		1625	.005		.16	.21		.37	.52
0900	Sand finish, brushwork	G		605	.013		.18	.56		.74	1.09
0940	Roller	G		1020	.008		.19	.33		.52	.74
0980	Spray	G		1700	.005		.16	.20		.36	.50
1800	For ceiling installations, add	G						25%			
8200	For work 8'-15' H, add							10%			

09 91 Painting

09 91 23 – Interior Painting

09 91 23.74 Walls and Ceilings, Interior, Zero VOC Latex	Crew	Daily Output	Labor-Hours	Unit	Material	2018 Bare Costs Labor	Equipment	Total	Total Incl O&P
8300	For work over 15' H, add				S.F.		20%			
9900	Minimum labor/equipment charge	1 Pord	2	4	Job		170		170	273

09 91 23.75 Dry Fall Painting

		Crew	Daily Output	Labor-Hours	Unit	Material	Labor	Equipment	Total	Total Incl O&P
0010	**DRY FALL PAINTING**									
0100	Sprayed on walls, gypsum board or plaster									
0220	One coat	1 Pord	2600	.003	S.F.	.09	.13		.22	.30
0250	Two coats		1560	.005		.17	.22		.39	.54
0280	Concrete or textured plaster, one coat		1560	.005		.09	.22		.31	.44
0310	Two coats		1300	.006		.17	.26		.43	.61
0340	Concrete block, one coat		1560	.005		.09	.22		.31	.44
0370	Two coats		1300	.006		.17	.26		.43	.61
0400	Wood, one coat		877	.009		.09	.39		.48	.71
0430	Two coats		650	.012		.17	.52		.69	1.03
0440	On ceilings, gypsum board or plaster									
0470	One coat	1 Pord	1560	.005	S.F.	.09	.22		.31	.44
0500	Two coats		1300	.006		.17	.26		.43	.61
0530	Concrete or textured plaster, one coat		1560	.005		.09	.22		.31	.44
0560	Two coats		1300	.006		.17	.26		.43	.61
0570	Structural steel, bar joists or metal deck, one coat		1560	.005		.09	.22		.31	.44
0580	Two coats		1040	.008		.17	.33		.50	.72
9900	Minimum labor/equipment charge		2	4	Job		170		170	273

09 93 Staining and Transparent Finishing

09 93 23 – Interior Staining and Finishing

09 93 23.10 Varnish

		Crew	Daily Output	Labor-Hours	Unit	Material	Labor	Equipment	Total	Total Incl O&P
0010	**VARNISH**									
0012	1 coat + sealer, on wood trim, brush, no sanding included	1 Pord	400	.020	S.F.	.07	.85		.92	1.45
0020	1 coat + sealer, on wood trim, brush, no sanding included, no VOC		400	.020		.22	.85		1.07	1.61
0100	Hardwood floors, 2 coats, no sanding included, roller		1890	.004		.15	.18		.33	.46
9000	Minimum labor/equipment charge		4	2	Job		85		85	137

09 96 High-Performance Coatings

09 96 23 – Graffiti-Resistant Coatings

09 96 23.10 Graffiti-Resistant Treatments

		Crew	Daily Output	Labor-Hours	Unit	Material	Labor	Equipment	Total	Total Incl O&P
0010	**GRAFFITI-RESISTANT TREATMENTS**, sprayed on walls									
0100	Non-sacrificial, permanent non-stick coating, clear, on metals	1 Pord	2000	.004	S.F.	2.06	.17		2.23	2.54
0200	Concrete		2000	.004		2.35	.17		2.52	2.85
0300	Concrete block		2000	.004		3.03	.17		3.20	3.61
0400	Brick		2000	.004		3.44	.17		3.61	4.05
0500	Stone		2000	.004		3.44	.17		3.61	4.05
0600	Unpainted wood		2000	.004		3.97	.17		4.14	4.64
2000	Semi-permanent cross linking polymer primer, on metals		2000	.004		.65	.17		.82	.98
2100	Concrete		2000	.004		.78	.17		.95	1.13
2200	Concrete block		2000	.004		.97	.17		1.14	1.34
2300	Brick		2000	.004		.78	.17		.95	1.13
2400	Stone		2000	.004		.78	.17		.95	1.13
2500	Unpainted wood		2000	.004		1.08	.17		1.25	1.46
3000	Top coat, on metals		2000	.004		.55	.17		.72	.87
3100	Concrete		2000	.004		.62	.17		.79	.96

09 96 High-Performance Coatings

09 96 23 – Graffiti-Resistant Coatings

09 96 23.10 Graffiti-Resistant Treatments	Crew	Daily Output	Labor-Hours	Unit	Material	2018 Bare Costs Labor	2018 Bare Costs Equipment	Total	Total Incl O&P
3200	Concrete block	1 Pord	2000	.004	S.F.	.87	.17		1.04	1.23
3300	Brick		2000	.004		.73	.17		.90	1.07
3400	Stone		2000	.004		.73	.17		.90	1.07
3500	Unpainted wood		2000	.004		.87	.17		1.04	1.23
5000	Sacrificial, water based, on metal		2000	.004		.32	.17		.49	.63
5100	Concrete		2000	.004		.32	.17		.49	.63
5200	Concrete block		2000	.004		.32	.17		.49	.63
5300	Brick		2000	.004		.32	.17		.49	.63
5400	Stone		2000	.004		.32	.17		.49	.63
5500	Unpainted wood		2000	.004		.32	.17		.49	.63
8000	Cleaner for use after treatment									
8100	Towels or wipes, per package of 30				Ea.	.63			.63	.70
8200	Aerosol spray, 24 oz. can				"	18			18	19.80

09 96 46 – Intumescent Painting

09 96 46.10 Coatings, Intumescent

		Crew	Daily Output	Labor-Hours	Unit	Material	Labor	Equipment	Total	Total Incl O&P
0010	**COATINGS, INTUMESCENT**, spray applied									
0100	On exterior structural steel, 0.25" d.f.t.	1 Pord	475	.017	S.F.	.47	.72		1.19	1.66
0150	0.51" d.f.t.		350	.023		.47	.97		1.44	2.07
0200	0.98" d.f.t.		280	.029		.47	1.22		1.69	2.46
0300	On interior structural steel, 0.108" d.f.t.		300	.027		.46	1.13		1.59	2.32
0350	0.310" d.f.t.		150	.053		.46	2.27		2.73	4.14
0400	0.670" d.f.t.		100	.080		.46	3.40		3.86	5.95

09 96 53 – Elastomeric Coatings

09 96 53.10 Coatings, Elastomeric

		Crew	Daily Output	Labor-Hours	Unit	Material	Labor	Equipment	Total	Total Incl O&P
0010	**COATINGS, ELASTOMERIC**									
0020	High build, water proof, one coat system									
0100	Concrete, brush	1 Pord	650	.012	S.F.	.29	.52		.81	1.16

09 96 56 – Epoxy Coatings

09 96 56.20 Wall Coatings

		Crew	Daily Output	Labor-Hours	Unit	Material	Labor	Equipment	Total	Total Incl O&P
0010	**WALL COATINGS**									
0100	Acrylic glazed coatings, matte	1 Pord	525	.015	S.F.	.35	.65		1	1.43
0200	Gloss		305	.026		.73	1.12		1.85	2.59
0300	Epoxy coatings, solvent based		525	.015		.45	.65		1.10	1.54
0400	Water based		170	.047		.32	2		2.32	3.56
2400	Sprayed perlite or vermiculite, 1/16" thick, solvent based		2935	.003		.28	.12		.40	.50
2500	Water based		640	.013		.83	.53		1.36	1.76
2700	Vinyl plastic wall coating, solvent based		735	.011		.38	.46		.84	1.16
2800	Water based		240	.033		.93	1.42		2.35	3.30
3000	Urethane on smooth surface, 2 coats, solvent based		1135	.007		.34	.30		.64	.85
3100	Water based		665	.012		.60	.51		1.11	1.48

09 97 Special Coatings

09 97 35 – Dry Erase Coatings

09 97 35.10 Dry Erase Coatings	Crew	Daily Output	Labor-Hours	Unit	Material	2018 Bare Costs Labor	Equipment	Total	Total Incl O&P
0010 **DRY ERASE COATINGS**									
0020 Dry erase coatings, clear, roller applied	1 Pord	1325	.006	S.F.	2.08	.26		2.34	2.70

For customer support on your Commercial Renovation Costs with RSMeans data, call 800.448.8182.

337

Division Notes

	CREW	DAILY OUTPUT	LABOR-HOURS	UNIT	BARE COSTS				TOTAL INCL O&P
					MAT.	LABOR	EQUIP.	TOTAL	

Estimating Tips
General

- The items in this division are usually priced per square foot or each.

- Many items in Division 10 require some type of support system or special anchors that are not usually furnished with the item. The required anchors must be added to the estimate in the appropriate division.

- Some items in Division 10, such as lockers, may require assembly before installation. Verify the amount of assembly required. Assembly can often exceed installation time.

10 20 00 Interior Specialties

- Support angles and blocking are not included in the installation of toilet compartments, shower/dressing compartments, or cubicles. Appropriate line items from Division 5 or 6 may need to be added to support the installations.

- Toilet partitions are priced by the stall. A stall consists of a side wall, pilaster, and door with hardware. Toilet tissue holders and grab bars are extra.

- The required acoustical rating of a folding partition can have a significant impact on costs. Verify the sound transmission coefficient rating of the panel priced against the specification requirements.

- Grab bar installation does not include supplemental blocking or backing to support the required load. When grab bars are installed at an existing facility, provisions must be made to attach the grab bars to a solid structure.

Reference Numbers

Reference numbers are shown at the beginning of some major classifications. These numbers refer to related items in the Reference Section. The reference information may be an estimating procedure, an alternate pricing method, or technical information.

Note: Not all subdivisions listed here necessarily appear. ■

Did you know?

RSMeans data is available through our online application with 24/7 access:

- Search for unit prices by keyword
- Leverage the most up-to-date data
- Build and export estimates

Try it free for 30 days!
www.rsmeans.com/2018freetrial

10 11 13.13 Fixed Chalkboards

		Crew	Daily Output	Labor-Hours	Unit	Material	2018 Bare Costs Labor	2018 Bare Costs Equipment	Total	Total Incl O&P
0010	**FIXED CHALKBOARDS** Porcelain enamel steel									
3900	Wall hung									
4000	Aluminum frame and chalktrough									
4300	3' x 5'	2 Carp	15	1.067	Ea.	310	54		364	430
4600	4' x 12'	"	13	1.231	"	560	62.50		622.50	715
4700	Wood frame and chalktrough									
4800	3' x 4'	2 Carp	16	1	Ea.	212	50.50		262.50	315
5300	4' x 8'	"	13	1.231	"	365	62.50		427.50	500
5400	Liquid chalk, white porcelain enamel, wall hung									
5420	Deluxe units, aluminum trim and chalktrough									
5450	4' x 4'	2 Carp	16	1	Ea.	300	50.50		350.50	415
5550	4' x 12'	"	12	1.333	"	620	67.50		687.50	790
5700	Wood trim and chalktrough									
5900	4' x 4'	2 Carp	16	1	Ea.	750	50.50		800.50	910
6200	4' x 8'	↓	14	1.143	"	1,025	58		1,083	1,225
9000	Minimum labor/equipment charge		3	5.333	Job		270		270	440

10 11 13.43 Portable Chalkboards

		Crew	Daily Output	Labor-Hours	Unit	Material	2018 Bare Costs Labor	2018 Bare Costs Equipment	Total	Total Incl O&P
0010	**PORTABLE CHALKBOARDS**									
0100	Freestanding, reversible									
0120	Economy, wood frame, 4' x 6'									
0140	Chalkboard both sides				Ea.	710			710	780
0160	Chalkboard one side, cork other side				"	615			615	680
0200	Standard, lightweight satin finished aluminum, 4' x 6'									
0220	Chalkboard both sides				Ea.	700			700	770
0240	Chalkboard one side, cork other side				"	620			620	685
0300	Deluxe, heavy duty extruded aluminum, 4' x 6'									
0320	Chalkboard both sides				Ea.	1,150			1,150	1,250
0340	Chalkboard one side, cork other side				"	1,150			1,150	1,250

10 11 16 – Markerboards

10 11 16.53 Electronic Markerboards

		Crew	Daily Output	Labor-Hours	Unit	Material	2018 Bare Costs Labor	2018 Bare Costs Equipment	Total	Total Incl O&P
0010	**ELECTRONIC MARKERBOARDS**									
0100	Wall hung or free standing, 3' x 4' to 4' x 6'	2 Carp	8	2	S.F.	85.50	101		186.50	259
0150	5' x 6' to 4' x 8'		8	2	"	61	101		162	232
0500	Interactive projection module for existing whiteboards	↓	8	2	Ea.	1,300	101		1,401	1,600

10 11 23 – Tackboards

10 11 23.10 Fixed Tackboards

		Crew	Daily Output	Labor-Hours	Unit	Material	2018 Bare Costs Labor	2018 Bare Costs Equipment	Total	Total Incl O&P
0010	**FIXED TACKBOARDS**									
2120	Prefabricated, 1/4" cork, 3' x 5' with aluminum frame	2 Carp	16	1	Ea.	140	50.50		190.50	237
2140	Wood frame	"	16	1	"	164	50.50		214.50	264
2300	Glass enclosed cabinets, alum., cork panel, hinged doors									
2600	4' x 7', 3 door	2 Carp	10	1.600	Ea.	1,875	81		1,956	2,175
9000	Minimum labor/equipment charge	"	4	4	Job		203		203	330

10 11 23.20 Control Boards

		Crew	Daily Output	Labor-Hours	Unit	Material	2018 Bare Costs Labor	2018 Bare Costs Equipment	Total	Total Incl O&P
0010	**CONTROL BOARDS**									
1000	Hospital patient display board, 4-color custom design									
1010	Porcelain steel dry erase board, 36" x 24"	2 Carp	7.50	2.133	Ea.	235	108		343	435

For customer support on your Commercial Renovation Costs with RSMeans data, call 800.448.8182.

10 13 Directories

10 13 10 – Building Directories

10 13 10.10 Directory Boards

		Crew	Daily Output	Labor-Hours	Unit	Material	2018 Bare Costs Labor	2018 Bare Costs Equipment	Total	Total Incl O&P
0010	**DIRECTORY BOARDS**									
0050	Plastic, glass covered, 30" x 20"	2 Carp	3	5.333	Ea.	207	270		477	670
0100	36" x 48"		2	8		915	405		1,320	1,650
0900	Outdoor, weatherproof, black plastic, 36" x 24"		2	8		700	405		1,105	1,425
1000	36" x 36"	↓	1.50	10.667	↓	810	540		1,350	1,775
9000	Minimum labor/equipment charge	1 Carp	1	8	Job		405		405	660

10 14 Signage

10 14 19 – Dimensional Letter Signage

10 14 19.10 Exterior Signs

		Crew	Daily Output	Labor-Hours	Unit	Material	2018 Bare Costs Labor	2018 Bare Costs Equipment	Total	Total Incl O&P
0010	**EXTERIOR SIGNS**									
3900	Plaques, custom, 20" x 30", for up to 450 letters, cast aluminum	2 Carp	4	4	Ea.	1,875	203		2,078	2,375
4000	Cast bronze		4	4		1,900	203		2,103	2,400
4200	30" x 36", up to 900 letters, cast aluminum		3	5.333		2,775	270		3,045	3,500
4300	Cast bronze	↓	3	5.333	↓	4,325	270		4,595	5,225
5100	Exit signs, 24 ga. alum., 14" x 12" surface mounted	1 Carp	30	.267		51.50	13.50		65	79
5200	10" x 7"	"	20	.400		30.50	20.50		51	66.50
6400	Replacement sign faces, 6" or 8"	1 Clab	50	.160	↓	66	6.40		72.40	83
9000	Minimum labor/equipment charge	1 Carp	4	2	Job		101		101	165

10 14 23 – Panel Signage

10 14 23.13 Engraved Panel Signage

		Crew	Daily Output	Labor-Hours	Unit	Material	2018 Bare Costs Labor	2018 Bare Costs Equipment	Total	Total Incl O&P
0010	**ENGRAVED PANEL SIGNAGE**, interior									
1010	Flexible door sign, adhesive back, w/Braille, 5/8" letters, 4" x 4"	1 Clab	32	.250	Ea.	34	9.95		43.95	53
1050	6" x 6"		32	.250		49.50	9.95		59.45	70.50
1100	8" x 2"		32	.250		35	9.95		44.95	54.50
1150	8" x 4"		32	.250		44	9.95		53.95	64.50
1200	8" x 8"		32	.250		69.50	9.95		79.45	92.50
1250	12" x 2"		32	.250		35	9.95		44.95	54.50
1300	12" x 6"		32	.250		41	9.95		50.95	61
1350	12" x 12"		32	.250		155	9.95		164.95	186
1500	Graphic symbols, 2" x 2"		32	.250		12	9.95		21.95	29.50
1550	6" x 6"		32	.250		31	9.95		40.95	50
1600	8" x 8"		32	.250		39	9.95		48.95	58.50
2010	Corridor, stock acrylic, 2-sided, with mounting bracket, 2" x 8"	1 Carp	24	.333		29.50	16.90		46.40	60
2020	2" x 10"		24	.333		35	16.90		51.90	66
2050	3" x 8"		24	.333		27	16.90		43.90	57
2060	3" x 10"		24	.333		36.50	16.90		53.40	67.50
2070	3" x 12"		24	.333		34.50	16.90		51.40	65.50
2100	4" x 8"		24	.333		22	16.90		38.90	51.50
2110	4" x 10"		24	.333		40	16.90		56.90	71.50
2120	4" x 12"		24	.333	↓	50	16.90		66.90	82.50

10 14 26 – Post and Panel/Pylon Signage

10 14 26.10 Post and Panel Signage

		Crew	Daily Output	Labor-Hours	Unit	Material	2018 Bare Costs Labor	2018 Bare Costs Equipment	Total	Total Incl O&P
0010	**POST AND PANEL SIGNAGE**, Alum., incl. 2 posts,									
0011	2-sided panel (blank), concrete footings per post									
0025	24" W x 18" H	B-1	25	.960	Ea.	905	39		944	1,050
0050	36" W x 24" H		25	.960		970	39		1,009	1,150
0100	57" W x 36" H		25	.960		1,825	39		1,864	2,075
0150	81" W x 46" H	↓	25	.960	↓	2,225	39		2,264	2,525

341

10 14 Signage

10 14 43 – Photoluminescent Signage

10 14 43.10 Photoluminescent Signage	Crew	Daily Output	Labor-Hours	Unit	Material	2018 Bare Costs Labor	2018 Bare Costs Equipment	Total	Total Incl O&P
0010 **PHOTOLUMINESCENT SIGNAGE**									
0011 Photoluminescent exit sign, plastic, 7"x10"	1 Carp	32	.250	Ea.	27.50	12.70		40.20	51
0022 Polyester		32	.250		14.80	12.70		27.50	37
0033 Aluminum		32	.250		52.50	12.70		65.20	78
0044 Plastic, 10"x14"		32	.250		58.50	12.70		71.20	85
0055 Polyester		32	.250		35	12.70		47.70	59
0066 Aluminum		32	.250		69.50	12.70		82.20	97
0111 Directional exit sign, plastic, 10"x14"		32	.250		57.50	12.70		70.20	83.50
0122 Polyester		32	.250		45	12.70		57.70	70
0133 Aluminum		32	.250		67	12.70		79.70	94.50
0153 Kick plate exit sign, 10"x34"		28	.286		258	14.50		272.50	310
0173 Photoluminescent tape, 1"x60'		300	.027	L.F.	.55	1.35		1.90	2.79
0183 2"x60'		290	.028	"	1.16	1.40		2.56	3.55
0211 Photoluminescent directional arrows		42	.190	Ea.	1.28	9.65		10.93	17.10

10 14 53 – Traffic Signage

10 14 53.20 Traffic Signs

	Crew	Daily Output	Labor-Hours	Unit	Material	Labor	Equipment	Total	Total Incl O&P
0010 **TRAFFIC SIGNS**									
0012 Stock, 24" x 24", no posts, .080" alum. reflectorized	B-80	70	.457	Ea.	88	20.50	8.75	117.25	139
0100 High intensity		70	.457		101	20.50	8.75	130.25	153
0300 30" x 30", reflectorized		70	.457		122	20.50	8.75	151.25	176
0400 High intensity		70	.457		132	20.50	8.75	161.25	187
0600 Guide and directional signs, 12" x 18", reflectorized		70	.457		41	20.50	8.75	70.25	87
0700 High intensity		70	.457		54.50	20.50	8.75	83.75	102
0900 18" x 24", stock signs, reflectorized		70	.457		48	20.50	8.75	77.25	95
1000 High intensity		70	.457		53.50	20.50	8.75	82.75	101
1200 24" x 24", stock signs, reflectorized		70	.457		58.50	20.50	8.75	87.75	106
1300 High intensity		70	.457		63	20.50	8.75	92.25	111
1500 Add to above for steel posts, galvanized, 10'-0" upright, bolted		200	.160		32.50	7.10	3.06	42.66	51
1600 12'-0" upright, bolted		140	.229		39	10.15	4.37	53.52	64
1800 Highway road signs, aluminum, over 20 S.F., reflectorized		350	.091	S.F.	16.50	4.06	1.75	22.31	26.50
2000 High intensity		350	.091		16.50	4.06	1.75	22.31	26.50
2200 Highway, suspended over road, 80 S.F. min., reflectorized		165	.194		16.50	8.60	3.71	28.81	36
2300 High intensity		165	.194		18.50	8.60	3.71	30.81	38.50
9000 Replace directional sign		6	5.333	Ea.	132	237	102	471	635

10 17 Telephone Specialties

10 17 16 – Telephone Enclosures

10 17 16.10 Commercial Telephone Enclosures

	Crew	Daily Output	Labor-Hours	Unit	Material	Labor	Equipment	Total	Total Incl O&P
0010 **COMMERCIAL TELEPHONE ENCLOSURES**									
0300 Shelf type, wall hung, recessed	2 Carp	5	3.200	Ea.	810	162		972	1,150
0400 Surface mount		5	3.200	"	1,675	162		1,837	2,125
9000 Minimum labor/equipment charge		4	4	Job		203		203	330

10 21 Compartments and Cubicles

10 21 13 – Toilet Compartments

10 21 13.13 Metal Toilet Compartments

	10 21 13.13 Metal Toilet Compartments	Crew	Daily Output	Labor-Hours	Unit	Material	2018 Bare Costs Labor	Equipment	Total	Total Incl O&P
0010	**METAL TOILET COMPARTMENTS**									
0110	Cubicles, ceiling hung									
0200	Powder coated steel	2 Carp	4	4	Ea.	555	203		758	940
0500	Stainless steel	"	4	4		1,150	203		1,353	1,575
0600	For handicap units, add ♿				▼	465			465	515
0900	Floor and ceiling anchored									
1000	Powder coated steel	2 Carp	5	3.200	Ea.	600	162		762	925
1300	Stainless steel	"	5	3.200		1,250	162		1,412	1,650
1400	For handicap units, add ♿				▼	330			330	360
1610	Floor anchored									
1700	Powder coated steel	2 Carp	7	2.286	Ea.	625	116		741	875
2000	Stainless steel	"	7	2.286		1,225	116		1,341	1,550
2100	For handicap units, add ♿					330			330	360
2200	For juvenile units, deduct				▼	43.50			43.50	48
2450	Floor anchored, headrail braced									
2500	Powder coated steel	2 Carp	6	2.667	Ea.	390	135		525	650
2804	Stainless steel	"	6	2.667		1,075	135		1,210	1,400
2900	For handicap units, add ♿					330			330	360
3000	Wall hung partitions, powder coated steel	2 Carp	7	2.286		670	116		786	925
3300	Stainless steel	"	7	2.286		1,725	116		1,841	2,100
3400	For handicap units, add				▼	330			330	360
4000	Screens, entrance, floor mounted, 58" high, 48" wide									
4200	Powder coated steel	2 Carp	15	1.067	Ea.	242	54		296	355
4500	Stainless steel	"	15	1.067	"	960	54		1,014	1,150
4650	Urinal screen, 18" wide									
4704	Powder coated steel	2 Carp	6.15	2.602	Ea.	230	132		362	465
5004	Stainless steel	"	6.15	2.602	"	535	132		667	805
5100	Floor mounted, headrail braced									
5300	Powder coated steel	2 Carp	8	2	Ea.	230	101		331	420
5600	Stainless steel	"	8	2	"	590	101		691	810
5750	Pilaster, flush									
5800	Powder coated steel	2 Carp	10	1.600	Ea.	276	81		357	435
6100	Stainless steel		10	1.600		610	81		691	800
6300	Post braced, powder coated steel		10	1.600		169	81		250	320
6800	Powder coated steel		10	1.600		170	81		251	320
7800	Wedge type, powder coated steel		10	1.600		139	81		220	285
8100	Stainless steel	▼	10	1.600	▼	610	81		691	800
9000	Minimum labor/equipment charge	1 Carp	2.50	3.200	Job		162		162	263

10 21 13.16 Plastic-Laminate-Clad Toilet Compartments

		Crew	Daily Output	Labor-Hours	Unit	Material	2018 Bare Costs Labor	Equipment	Total	Total Incl O&P
0010	**PLASTIC-LAMINATE-CLAD TOILET COMPARTMENTS**									
0110	Cubicles, ceiling hung									
0300	Plastic laminate on particle board ♿	2 Carp	4	4	Ea.	550	203		753	935
0600	For handicap units, add				"	465			465	515
0900	Floor and ceiling anchored									
1100	Plastic laminate on particle board	2 Carp	5	3.200	Ea.	765	162		927	1,100
1400	For handicap units, add ♿				"	330			330	360
1610	Floor mounted									
1800	Plastic laminate on particle board	2 Carp	7	2.286	Ea.	550	116		666	795
2450	Floor mounted, headrail braced									
2600	Plastic laminate on particle board	2 Carp	6	2.667	Ea.	835	135		970	1,125
3400	For handicap units, add ♿					330			330	360
4300	Entrance screen, floor mtd., plas. lam., 58" high, 48" wide ♿	2 Carp	15	1.067	▼	515	54		569	655

10 21 13.16 Plastic-Laminate-Clad Toilet Compartments

		Crew	Daily Output	Labor-Hours	Unit	Material	2018 Bare Costs Labor	Equipment	Total	Total Incl O&P
4800	Urinal screen, 18" wide, ceiling braced, plastic laminate	2 Carp	8	2	Ea.	211	101		312	395
5400	Floor mounted, headrail braced		8	2		211	101		312	395
5900	Pilaster, flush, plastic laminate		10	1.600		425	81		506	600
6400	Post braced, plastic laminate		10	1.600		256	81		337	415
6700	Wall hung, bracket supported									
6900	Plastic laminate on particle board	2 Carp	10	1.600	Ea.	97.50	81		178.50	239
7450	Flange supported									
7500	Plastic laminate on particle board	2 Carp	10	1.600	Ea.	245	81		326	400
9000	Minimum labor/equipment charge	1 Carp	2.50	3.200	Job		162		162	263

10 21 13.19 Plastic Toilet Compartments

		Crew	Daily Output	Labor-Hours	Unit	Material	2018 Bare Costs Labor	Equipment	Total	Total Incl O&P
0010	**PLASTIC TOILET COMPARTMENTS**									
0110	Cubicles, ceiling hung									
0250	Phenolic	2 Carp	4	4	Ea.	910	203		1,113	1,325
0260	Polymer plastic	"	4	4		995	203		1,198	1,425
0600	For handicap units, add					465			465	515
0900	Floor and ceiling anchored									
1050	Phenolic	2 Carp	5	3.200	Ea.	860	162		1,022	1,225
1060	Polymer plastic	"	5	3.200		995	162		1,157	1,375
1400	For handicap units, add					330			330	360
1610	Floor mounted									
1750	Phenolic	2 Carp	7	2.286	Ea.	750	116		866	1,025
1760	Polymer plastic	"	7	2.286		885	116		1,001	1,175
2100	For handicap units, add					330			330	360
2200	For juvenile units, deduct					43.50			43.50	48
2450	Floor mounted, headrail braced									
2550	Phenolic	2 Carp	6	2.667	Ea.	750	135		885	1,050
3600	Polymer plastic		6	2.667		860	135		995	1,175
3810	Entrance screen, polymer plastic, flr. mtd., 48"x58"		6	2.667		505	135		640	775
3820	Entrance screen, polymer plastic, flr. to clg pilaster, 48"x58"		6	2.667		600	135		735	880
6110	Urinal screen, polymer plastic, pilaster flush, 18" w		6	2.667		435	135		570	700
7110	Wall hung		6	2.667		745	135		880	1,050
7710	Flange mounted		6	2.667		930	135		1,065	1,250
9000	Minimum labor/equipment charge	1 Carp	2.50	3.200	Job		162		162	263

10 21 13.40 Stone Toilet Compartments

		Crew	Daily Output	Labor-Hours	Unit	Material	2018 Bare Costs Labor	Equipment	Total	Total Incl O&P
0010	**STONE TOILET COMPARTMENTS**									
0100	Cubicles, ceiling hung, marble	2 Marb	2	8	Ea.	1,650	390		2,040	2,475
0600	For handicap units, add					465			465	515
0800	Floor & ceiling anchored, marble	2 Marb	2.50	6.400		1,750	315		2,065	2,425
1400	For handicap units, add					330			330	360
1600	Floor mounted, marble	2 Marb	3	5.333		1,075	262		1,337	1,625
2400	Floor mounted, headrail braced, marble	"	3	5.333		1,200	262		1,462	1,750
2900	For handicap units, add					330			330	360
4100	Entrance screen, floor mounted marble, 58" high, 48" wide	2 Marb	9	1.778		700	87		787	910
4600	Urinal screen, 18" wide, ceiling braced, marble	D-1	6	2.667		795	119		914	1,075
5100	Floor mounted, headrail braced									
5200	Marble	D-1	6	2.667	Ea.	675	119		794	940
5700	Pilaster, flush, marble		9	1.778		885	79.50		964.50	1,100
6200	Post braced, marble		9	1.778		865	79.50		944.50	1,075
9000	Minimum labor/equipment charge	1 Carp	2.50	3.200	Job		162		162	263

10 21 Compartments and Cubicles

10 21 14 – Toilet Compartment Components

10 21 14.13 Metal Toilet Compartment Components

		Crew	Daily Output	Labor-Hours	Unit	Material	2018 Bare Costs Labor	Equipment	Total	Total Incl O&P
0010	**METAL TOILET COMPARTMENT COMPONENTS**									
0100	Pilasters									
0110	Overhead braced, powder coated steel, 7" wide x 82" high	2 Carp	22.20	.721	Ea.	66.50	36.50		103	133
0120	Stainless steel		22.20	.721		132	36.50		168.50	205
0130	Floor braced, powder coated steel, 7" wide x 70" high		23.30	.687		104	35		139	171
0140	Stainless steel		23.30	.687		218	35		253	297
0150	Ceiling hung, powder coated steel, 7" wide x 83" high		13.30	1.203		139	61		200	252
0160	Stainless steel		13.30	1.203		285	61		346	415
0170	Wall hung, powder coated steel, 3" wide x 58" high		18.90	.847		133	43		176	216
0180	Stainless steel	↓	18.90	.847	↓	169	43		212	256
0200	Panels									
0210	Powder coated steel, 31" wide x 58" high	2 Carp	18.90	.847	Ea.	123	43		166	206
0220	Stainless steel		18.90	.847		330	43		373	430
0230	Powder coated steel, 53" wide x 58" high		18.90	.847		151	43		194	236
0240	Stainless steel		18.90	.847		400	43		443	510
0250	Powder coated steel, 63" wide x 58" high		18.90	.847		180	43		223	268
0260	Stainless steel	↓	18.90	.847	↓	495	43		538	615
0300	Doors									
0310	Powder coated steel, 24" wide x 58" high	2 Carp	14.10	1.135	Ea.	127	57.50		184.50	234
0320	Stainless steel		14.10	1.135		310	57.50		367.50	435
0330	Powder coated steel, 26" wide x 58" high		14.10	1.135		151	57.50		208.50	260
0340	Stainless steel		14.10	1.135		320	57.50		377.50	445
0350	Powder coated steel, 28" wide x 58" high		14.10	1.135		171	57.50		228.50	282
0360	Stainless steel		14.10	1.135		335	57.50		392.50	465
0370	Powder coated steel, 36" wide x 58" high		14.10	1.135		185	57.50		242.50	298
0380	Stainless steel	↓	14.10	1.135	↓	395	57.50		452.50	530
0400	Headrails									
0410	For powder coated steel, 62" long	2 Carp	65	.246	Ea.	21	12.50		33.50	43.50
0420	Stainless steel		65	.246		21	12.50		33.50	43.50
0430	For powder coated steel, 84" long		50	.320		30	16.20		46.20	59.50
0440	Stainless steel		50	.320		30	16.20		46.20	59.50
0450	For powder coated steel, 120" long		30	.533		40	27		67	87.50
0460	Stainless steel	↓	30	.533	↓	40	27		67	87.50
9000	Minimum labor/equipment charge	1 Carp	4	2	Job		101		101	165

10 21 14.16 Plastic-Lam. Clad Toilet Compart. Components

		Crew	Daily Output	Labor-Hours	Unit	Material	2018 Bare Costs Labor	Equipment	Total	Total Incl O&P
0010	**PLASTIC-LAMINATE CLAD TOILET COMPARTMENT COMPONENTS**									
0100	Pilasters									
0110	Overhead braced, 7" wide x 82" high	2 Carp	22.20	.721	Ea.	103	36.50		139.50	173
0130	Floor anchored, 7" wide x 70" high		23.30	.687		103	35		138	170
0150	Ceiling hung, 7" wide x 83" high	↓	13.30	1.203		107	61		168	217
0180	Wall hung, 3" wide x 58" high	↓	18.90	.847	↓	96	43		139	175
0200	Panels									
0210	31" wide x 58" high	2 Carp	18.90	.847	Ea.	153	43		196	238
0230	51" wide x 58" high		18.90	.847		209	43		252	299
0250	63" wide x 58" high	↓	18.90	.847	↓	242	43		285	335
0300	Doors									
0310	24" wide x 58" high	2 Carp	14.10	1.135	Ea.	151	57.50		208.50	260
0330	26" wide x 58" high		14.10	1.135		156	57.50		213.50	265
0350	28" wide x 58" high		14.10	1.135		161	57.50		218.50	271
0370	36" wide x 58" high	↓	14.10	1.135	↓	185	57.50		242.50	297
0400	Headrails									
0410	62" long	2 Carp	65	.246	Ea.	22.50	12.50		35	45.50

For customer support on your Commercial Renovation Costs with RSMeans data, call 800.448.8182.

345

10 21 14 – Toilet Compartment Components

10 21 14.16 Plastic-Lam. Clad Toilet Compart. Components	Crew	Daily Output	Labor-Hours	Unit	Material	2018 Bare Costs Labor	Equipment	Total	Total Incl O&P
0430	84" long	2 Carp	60	.267	Ea.	29.50	13.50		43	54.50
0450	120" long	↓	30	.533	↓	39.50	27		66.50	87.50
9000	Minimum labor/equipment charge	1 Carp	4	2	Job		101		101	165

10 21 14.19 Plastic Toilet Compartment Components

		Crew	Daily Output	Labor-Hours	Unit	Material	2018 Bare Costs Labor	Equipment	Total	Total Incl O&P
0010	**PLASTIC TOILET COMPARTMENT COMPONENTS**									
0100	Pilasters									
0110	Overhead braced, polymer plastic, 7" wide x 82" high	2 Carp	22.20	.721	Ea.	114	36.50		150.50	186
0120	Phenolic		22.20	.721		141	36.50		177.50	215
0130	Floor braced, polymer plastic, 7" wide x 70" high		23.30	.687		183	35		218	258
0140	Phenolic		23.30	.687		135	35		170	205
0150	Ceiling hung, polymer plastic, 7" wide x 83" high		13.30	1.203		171	61		232	287
0160	Phenolic		13.30	1.203		164	61		225	280
0180	Wall hung, phenolic, 3" wide x 58" high	↓	18.90	.847	↓	105	43		148	186
0200	Panels									
0203	Polymer plastic, 18" wide x 55" high	2 Carp	18.90	.847	Ea.	290	43		333	390
0206	Phenolic, 18" wide x 58" high		18.90	.847		251	43		294	345
0210	Polymer plastic, 31" wide x 55" high		18.90	.847		297	43		340	395
0220	Phenolic, 31" wide x 58" high		18.90	.847		278	43		321	375
0223	Polymer plastic, 48" wide x 55" high		18.90	.847		535	43		578	660
0226	Phenolic, 48" wide x 58" high		18.90	.847		470	43		513	585
0230	Polymer plastic, 51" wide x 55" high		18.90	.847		450	43		493	565
0240	Phenolic, 51" wide x 58" high		18.90	.847		400	43		443	510
0250	Polymer plastic, 63" wide x 55" high		18.90	.847		545	43		588	670
0260	Phenolic, 63" wide x 58" high	↓	18.90	.847	↓	420	43		463	530
0300	Doors									
0310	Polymer plastic, 24" wide x 55" high	2 Carp	14.10	1.135	Ea.	218	57.50		275.50	335
0320	Phenolic, 24" wide x 58" high		14.10	1.135		310	57.50		367.50	435
0330	Polymer plastic, 26" wide x 55" high		14.10	1.135		242	57.50		299.50	360
0340	Phenolic, 26" wide x 58" high		14.10	1.135		330	57.50		387.50	455
0350	Polymer plastic, 28" wide x 55" high		14.10	1.135		260	57.50		317.50	380
0360	Phenolic, 28" wide x 58" high		14.10	1.135		345	57.50		402.50	475
0370	Polymer plastic, 36" wide x 55" high		14.10	1.135		310	57.50		367.50	440
0380	Phenolic, 36" wide x 58" high	↓	14.10	1.135	↓	430	57.50		487.50	570
0400	Headrails									
0410	For polymer plastic, 62" long	2 Carp	65	.246	Ea.	23	12.50		35.50	46
0420	Phenolic		65	.246		22.50	12.50		35	45.50
0430	For polymer plastic, 84" long		50	.320		31.50	16.20		47.70	61
0440	Phenolic		50	.320		34.50	16.20		50.70	64.50
0450	For polymer plastic, 120" long		30	.533		41	27		68	89.50
0460	Phenolic	↓	30	.533		41	27		68	89
9000	Minimum labor/equipment charge	1 Carp	4	2	Job		101		101	165

10 21 23 – Cubicle Curtains and Track

10 21 23.16 Cubicle Track and Hardware

		Crew	Daily Output	Labor-Hours	Unit	Material	2018 Bare Costs Labor	Equipment	Total	Total Incl O&P
0010	**CUBICLE TRACK AND HARDWARE**									
0020	Curtain track, box channel, ceiling mounted	1 Carp	135	.059	L.F.	6.35	3		9.35	11.90
0100	Suspended		100	.080		8.55	4.06		12.61	16
0550	Polyester, antimicrobial, 9' ceiling height		425	.019		19.65	.95		20.60	23
0560	8' ceiling height	↓	425	.019	↓	17.50	.95		18.45	21

346

For customer support on your Commercial Renovation Costs with RSMeans data, call 800.448.8182.

10 22 16 – Folding Gates

10 22 16.10 Security Gates

	Crew	Daily Output	Labor-Hours	Unit	Material	2018 Bare Costs Labor	Equipment	Total	Total Incl O&P
0010 **SECURITY GATES**									
0015 For roll up type, see Section 08 33 13.10									
0300 Scissors type folding gate, ptd. steel, single, 6-1/2' high, 5-1/2' wide	2 Sswk	4	4	Opng.	225	219		444	625
0350 6-1/2' wide		4	4		238	219		457	635
0400 7-1/2' wide		4	4		237	219		456	635
0600 Double gate, 8' high, 8' wide		2.50	6.400		390	350		740	1,025
0650 10' wide		2.50	6.400		435	350		785	1,075
0700 12' wide		2	8		590	435		1,025	1,400
0750 14' wide		2	8		620	435		1,055	1,425
0900 Door gate, folding steel, 4' wide, 61" high		4	4		134	219		353	520
1000 71" high		4	4		176	219		395	570
1200 81" high		4	4		182	219		401	575
1300 Window gates, 2' to 4' wide, 31" high		4	4		85	219		304	470
1500 55" high		3.75	4.267		122	233		355	535
1600 79" high		3.50	4.571		144	250		394	585

10 22 19 – Demountable Partitions

10 22 19.43 Demountable Composite Partitions

	Crew	Daily Output	Labor-Hours	Unit	Material	2018 Bare Costs Labor	Equipment	Total	Total Incl O&P
0010 **DEMOUNTABLE COMPOSITE PARTITIONS**, add for doors									
0100 Do not deduct door openings from total L.F.									
0900 Demountable gypsum system on 2" to 2-1/2"									
1000 Steel studs, 9' high, 3" to 3-3/4" thick									
1200 Vinyl-clad gypsum	2 Carp	48	.333	L.F.	60	16.90		76.90	93.50
1300 Fabric clad gypsum		44	.364		150	18.45		168.45	195
1500 Steel clad gypsum		40	.400		167	20.50		187.50	217
1600 1.75 system, aluminum framing, vinyl-clad hardboard,									
1800 Paper honeycomb core panel, 1 3/4" to 2 1/2" thick									
1900 9' high	2 Carp	48	.333	L.F.	101	16.90		117.90	139
2100 7' high		60	.267		90.50	13.50		104	122
2200 5' high		80	.200		76.50	10.15		86.65	100
2250 Unitized gypsum system									
2300 Unitized panel, 9' high, 2" to 2-1/2" thick									
2350 Vinyl-clad gypsum	2 Carp	48	.333	L.F.	130	16.90		146.90	171
2400 Fabric clad gypsum	"	44	.364	"	214	18.45		232.45	265
2500 Unitized mineral fiber system									
2510 Unitized panel, 9' high, 2-1/4" thick, aluminum frame									
2550 Vinyl-clad mineral fiber	2 Carp	48	.333	L.F.	129	16.90		145.90	170
2600 Fabric clad mineral fiber	"	44	.364	"	193	18.45		211.45	242
2800 Movable steel walls, modular system									
2900 Unitized panels, 9' high, 48" wide									
3100 Baked enamel, pre-finished	2 Carp	60	.267	L.F.	146	13.50		159.50	183
3200 Fabric clad steel	"	56	.286	"	212	14.50		226.50	257
5500 For acoustical partitions, add, unsealed				S.F.	2.36			2.36	2.60
5550 Sealed				"	11			11	12.10
5700 For doors, see Section 08 16									
5800 For door hardware, see Section 08 71									
6100 In-plant modular office system, w/prehung hollow core door									
6200 3" thick polystyrene core panels									
6250 12' x 12', 2 wall	2 Clab	3.80	4.211	Ea.	4,825	168		4,993	5,575
6300 4 wall		1.90	8.421		7,550	335		7,885	8,850
6350 16' x 16', 2 wall		3.60	4.444		7,850	177		8,027	8,925
6400 4 wall		1.80	8.889		10,000	355		10,355	11,600
9000 Minimum labor/equipment charge	2 Carp	3	5.333	Job		270		270	440

347

10 22 Partitions

10 22 23 – Portable Partitions, Screens, and Panels

10 22 23.13 Wall Screens

		Crew	Daily Output	Labor-Hours	Unit	Material	2018 Bare Costs Labor	Equipment	Total	Total Incl O&P
0010	**WALL SCREENS**, divider panels, free standing, fiber core									
0020	Fabric face straight									
0100	3'-0" long, 4'-0" high	2 Carp	100	.160	L.F.	134	8.10		142.10	161
0200	5'-0" high		90	.178		117	9		126	144
0500	6'-0" high		75	.213		122	10.80		132.80	152
0900	5'-0" long, 4'-0" high		175	.091		65.50	4.64		70.14	79.50
1000	5'-0" high		150	.107		88	5.40		93.40	106
1500	6'-0" high		125	.128		110	6.50		116.50	132
1600	6'-0" long, 5'-0" high		162	.099		110	5		115	129
3200	Economical panels, fabric face, 4'-0" long, 5'-0" high		132	.121		59	6.15		65.15	75
3250	6'-0" high		112	.143		66.50	7.25		73.75	85
3300	5'-0" long, 5'-0" high		150	.107		60	5.40		65.40	75
3350	6'-0" high		125	.128		54	6.50		60.50	69.50
3450	Acoustical panels, 60 to 90 NRC, 3'-0" long, 5'-0" high		90	.178		78	9		87	101
3550	6'-0" high		75	.213		79.50	10.80		90.30	105
3600	5'-0" long, 5'-0" high		150	.107		64.50	5.40		69.90	80
3650	6'-0" high		125	.128		69	6.50		75.50	86.50
3700	6'-0" long, 5'-0" high		162	.099		54.50	5		59.50	68
3750	6'-0" high		138	.116		82.50	5.90		88.40	101
3800	Economy acoustical panels, 40 NRC, 4'-0" long, 5'-0" high		132	.121		59	6.15		65.15	75
3850	6'-0" high		112	.143		66.50	7.25		73.75	85
3900	5'-0" long, 6'-0" high		125	.128		54	6.50		60.50	69.50
3950	6'-0" long, 5'-0" high		162	.099		49.50	5		54.50	62.50
9000	Minimum labor/equipment charge		3	5.333	Job		270		270	440

10 22 33 – Accordion Folding Partitions

10 22 33.10 Partitions, Accordion Folding

		Crew	Daily Output	Labor-Hours	Unit	Material	2018 Bare Costs Labor	Equipment	Total	Total Incl O&P
0010	**PARTITIONS, ACCORDION FOLDING**									
0100	Vinyl covered, over 150 S.F., frame not included									
0300	Residential, 1.25 lb./S.F., 8' maximum height	2 Carp	300	.053	S.F.	22.50	2.70		25.20	29.50
0400	Commercial, 1.75 lb./S.F., 8' maximum height		225	.071		26	3.61		29.61	34.50
0900	Acoustical, 3 lb./S.F., 17' maximum height		100	.160		27.50	8.10		35.60	43
1200	5 lb./S.F., 20' maximum height		95	.168		38	8.55		46.55	56
1500	Vinyl-clad wood or steel, electric operation, 5.0 psf		160	.100		56.50	5.05		61.55	71
1900	Wood, non-acoustic, birch or mahogany, to 10' high		300	.053		30.50	2.70		33.20	38
9000	Minimum labor/equipment charge		4	4	Job		203		203	330

10 22 39 – Folding Panel Partitions

10 22 39.10 Partitions, Folding Panel

		Crew	Daily Output	Labor-Hours	Unit	Material	2018 Bare Costs Labor	Equipment	Total	Total Incl O&P
0010	**PARTITIONS, FOLDING PANEL**, acoustic, wood									
0100	Vinyl faced, to 18' high, 6 psf, economy trim	2 Carp	60	.267	S.F.	58.50	13.50		72	86.50
0150	Standard trim		45	.356		70	18.05		88.05	107
0200	Premium trim		30	.533		90	27		117	144
0400	Plastic laminate or hardwood finish, standard trim		60	.267		60.50	13.50		74	88.50
0500	Premium trim		30	.533		64.50	27		91.50	115
0600	Wood, low acoustical type, 4.5 psf, to 14' high		50	.320		44	16.20		60.20	74.50
9000	Minimum labor/equipment charge		4	4	Job		203		203	330

10 26 Wall and Door Protection

10 26 23 – Protective Wall Covering

10 26 23.10 Wall Covering, Protective

		Crew	Daily Output	Labor-Hours	Unit	Material	2018 Bare Costs Labor	Equipment	Total	Total Incl O&P
0010	**WALL COVERING, PROTECTIVE**									
1775	Wall end guard, stainless steel, 16 ga., 36" tall, screwed to studs	1 Skwk	30	.267	Ea.	25.50	13.95		39.45	51

10 26 23.13 Impact Resistant Wall Protection

		Crew	Daily Output	Labor-Hours	Unit	Material	2018 Bare Costs Labor	Equipment	Total	Total Incl O&P
0010	**IMPACT RESISTANT WALL PROTECTION**									
0100	Vinyl wall protection, complete instl. incl. panels, trim, adhesive									
0110	.040" thk., std. colors	2 Carp	320	.050	S.F.	16.05	2.54		18.59	22
0120	.040" thk., element patterns		300	.053		17.25	2.70		19.95	23.50
0130	.060" thk., std. colors		300	.053		17.25	2.70		19.95	23.50
0140	.060" thk., element patterns		280	.057		19.05	2.90		21.95	25.50
1775	Wall protection, stainless steel, 16 ga., 48" x 36" tall, screwed to studs	2 Skwk	500	.032		7.95	1.68		9.63	11.45

10 26 33 – Door and Frame Protection

10 26 33.10 Protection, Door and Frame

		Crew	Daily Output	Labor-Hours	Unit	Material	2018 Bare Costs Labor	Equipment	Total	Total Incl O&P
0010	**PROTECTION, DOOR AND FRAME**									
0100	Door frame guard, vinyl, 3" x 3" x 4'	1 Carp	30	.267	Ea.	83	13.50		96.50	114
0110	Door frame guard, vinyl, 3" x 3" x 8'		26	.308		129	15.60		144.60	168
0120	Door frame guard, stainless steel, 3" x 3" x 4'		30	.267		420	13.50		433.50	485
0500	Steel door track/wheel guards, 4'-0" high	E-4	22	1.455		109	80	4.48	193.48	262

10 28 Toilet, Bath, and Laundry Accessories

10 28 13 – Toilet Accessories

10 28 13.13 Commercial Toilet Accessories

		Crew	Daily Output	Labor-Hours	Unit	Material	2018 Bare Costs Labor	Equipment	Total	Total Incl O&P
0010	**COMMERCIAL TOILET ACCESSORIES**									
0200	Curtain rod, stainless steel, 5' long, 1" diameter	1 Carp	13	.615	Ea.	27.50	31		58.50	80.50
0300	1-1/4" diameter		13	.615		30	31		61	83.50
0350	Chrome, 1" diameter		13	.615		28	31		59	81.50
0360	For vinyl curtain, add		1950	.004	S.F.	.95	.21		1.16	1.39
0400	Diaper changing station, horizontal, wall mounted, plastic		10	.800	Ea.	248	40.50		288.50	340
0420	Vertical		10	.800		222	40.50		262.50	310
0430	Oval shaped		10	.800		222	40.50		262.50	310
0440	Recessed, with stainless steel flange		6	1.333		510	67.50		577.50	670
0500	Dispenser units, combined soap & towel dispensers,									
0510	Mirror and shelf, flush mounted	1 Carp	10	.800	Ea.	365	40.50		405.50	465
0600	Towel dispenser and waste receptacle,									
0610	18 gallon capacity	1 Carp	10	.800	Ea.	330	40.50		370.50	430
0800	Grab bar, straight, 1-1/4" diameter, stainless steel, 18" long		24	.333		30	16.90		46.90	60.50
0900	24" long		23	.348		29.50	17.65		47.15	61
1000	30" long		22	.364		32.50	18.45		50.95	66
1100	36" long		20	.400		35	20.50		55.50	71.50
1105	42" long		20	.400		36.50	20.50		57	73
1120	Corner, 36" long		20	.400		92.50	20.50		113	135
1200	1-1/2" diameter, 24" long		23	.348		31.50	17.65		49.15	63
1300	36" long		20	.400		34	20.50		54.50	70.50
1310	42" long		18	.444		37.50	22.50		60	78
1500	Tub bar, 1-1/4" diameter, 24" x 36"		14	.571		88.50	29		117.50	145
1600	Plus vertical arm		12	.667		93.50	34		127.50	158
1900	End tub bar, 1" diameter, 90° angle, 16" x 32"		12	.667		104	34		138	169
2010	Tub/shower/toilet, 2-wall, 36" x 24"		12	.667		94	34		128	159
2300	Hand dryer, surface mounted, electric, 115 volt, 20 amp		4	2		490	101		591	705
2400	230 volt, 10 amp		4	2		655	101		756	885
2450	Hand dryer, touch free, 1400 watt, 81,000 rpm		4	2		1,375	101		1,476	1,675

For customer support on your Commercial Renovation Costs with RSMeans data, call 800.448.8182.

349

10 28 13.13 Commercial Toilet Accessories	Crew	Daily Output	Labor-Hours	Unit	Material	2018 Bare Costs Labor	Equipment	Total	Total Incl O&P
2600 Hat and coat strip, stainless steel, 4 hook, 36" long	1 Carp	24	.333	Ea.	68	16.90		84.90	103
2700 6 hook, 60" long		20	.400		124	20.50		144.50	170
3000 Mirror, with stainless steel 3/4" square frame, 18" x 24"		20	.400		47	20.50		67.50	85
3100 36" x 24"		15	.533		123	27		150	179
3200 48" x 24"		10	.800		179	40.50		219.50	263
3300 72" x 24"		6	1.333		282	67.50		349.50	420
3500 With 5" stainless steel shelf, 18" x 24"		20	.400		189	20.50		209.50	241
3600 36" x 24"		15	.533		238	27		265	305
3700 48" x 24"		10	.800		257	40.50		297.50	350
3800 72" x 24"		6	1.333		259	67.50		326.50	395
4100 Mop holder strip, stainless steel, 5 holders, 48" long		20	.400		80	20.50		100.50	121
4200 Napkin/tampon dispenser, recessed		15	.533		620	27		647	725
4220 Semi-recessed		6.50	1.231		370	62.50		432.50	505
4250 Napkin receptacle, recessed		6.50	1.231		170	62.50		232.50	288
4300 Robe hook, single, regular		96	.083		19.45	4.22		23.67	28.50
4400 Heavy duty, concealed mounting		56	.143		22.50	7.25		29.75	37
4600 Soap dispenser, chrome, surface mounted, liquid		20	.400		51	20.50		71.50	89
5000 Recessed stainless steel, liquid		10	.800		161	40.50		201.50	243
5600 Shelf, stainless steel, 5" wide, 18 ga., 24" long		24	.333		82	16.90		98.90	118
5700 48" long		16	.500		157	25.50		182.50	213
5800 8" wide shelf, 18 ga., 24" long		22	.364		69.50	18.45		87.95	107
5900 48" long		14	.571		145	29		174	206
6000 Toilet seat cover dispenser, stainless steel, recessed		20	.400		173	20.50		193.50	224
6050 Surface mounted		15	.533		30	27		57	77
6100 Toilet tissue dispenser, surface mounted, SS, single roll		30	.267		18.05	13.50		31.55	42
6200 Double roll		24	.333		22	16.90		38.90	52
6240 Plastic, twin/jumbo dbl. roll		24	.333		28.50	16.90		45.40	59
6290 Toilet seat	1 Plum	40	.200		30	12.45		42.45	52.50
6400 Towel bar, stainless steel, 18" long	1 Carp	23	.348		42.50	17.65		60.15	75.50
6500 30" long		21	.381		52	19.30		71.30	88.50
6700 Towel dispenser, stainless steel, surface mounted		16	.500		43	25.50		68.50	88
6800 Flush mounted, recessed		10	.800		112	40.50		152.50	190
6900 Plastic, touchless, battery operated		16	.500		91	25.50		116.50	141
7000 Towel holder, hotel type, 2 guest size		20	.400		56	20.50		76.50	94.50
7200 Towel shelf, stainless steel, 24" long, 8" wide		20	.400		64.50	20.50		85	104
7400 Tumbler holder, for tumbler only		30	.267		25	13.50		38.50	49.50
7410 Tumbler holder, recessed		20	.400		8	20.50		28.50	42
7500 Soap, tumbler & toothbrush		30	.267		19.60	13.50		33.10	43.50
7510 Tumbler & toothbrush holder		20	.400		13.45	20.50		33.95	48
8000 Waste receptacles, stainless steel, with top, 13 gallon		10	.800		310	40.50		350.50	405
8100 36 gallon		8	1		410	50.50		460.50	535
9000 Minimum labor/equipment charge		5	1.600	Job		81		81	132
9996 Bathroom access., grab bar, straight, 1-1/2" diam., SS, 42" L install only		18	.444	Ea.		22.50		22.50	36.50

10 28 16 – Bath Accessories

10 28 16.20 Medicine Cabinets

	Crew	Daily Output	Labor-Hours	Unit	Material	2018 Bare Costs Labor	Equipment	Total	Total Incl O&P
0010 **MEDICINE CABINETS**									
0020 With mirror, sst frame, 16" x 22", unlighted	1 Carp	14	.571	Ea.	99	29		128	156
0100 Wood frame		14	.571		136	29		165	197
0300 Sliding mirror doors, 20" x 16" x 4-3/4", unlighted		7	1.143		124	58		182	231
0400 24" x 19" x 8-1/2", lighted		5	1.600		206	81		287	360
0600 Triple door, 30" x 32", unlighted, plywood body		7	1.143		350	58		408	480
0700 Steel body		7	1.143		375	58		433	505

10 28 Toilet, Bath, and Laundry Accessories

10 28 16 – Bath Accessories

10 28 16.20 Medicine Cabinets

	10 28 16.20 Medicine Cabinets	Crew	Daily Output	Labor-Hours	Unit	Material	2018 Bare Costs Labor	Equipment	Total	Total Incl O&P
0900	Oak door, wood body, beveled mirror, single door	1 Carp	7	1.143	Ea.	175	58		233	286
1000	Double door		6	1.333		370	67.50		437.50	515
1200	Hotel cabinets, stainless, with lower shelf, unlighted		10	.800		218	40.50		258.50	305
1300	Lighted		5	1.600		340	81		421	505
9000	Minimum labor/equipment charge		4	2	Job		101		101	165

10 28 19 – Tub and Shower Enclosures

10 28 19.10 Partitions, Shower

	10 28 19.10 Partitions, Shower	Crew	Daily Output	Labor-Hours	Unit	Material	2018 Bare Costs Labor	Equipment	Total	Total Incl O&P
0010	**PARTITIONS, SHOWER** floor mounted, no plumbing									
0400	Cabinet, one piece, fiberglass, 32" x 32"	2 Carp	5	3.200	Ea.	625	162		787	955
0420	36" x 36"		5	3.200		695	162		857	1,025
0440	36" x 48"		5	3.200		1,400	162		1,562	1,800
0460	Acrylic, 32" x 32"		5	3.200		360	162		522	660
0480	36" x 36"		5	3.200		1,075	162		1,237	1,450
0500	36" x 48"		5	3.200		1,300	162		1,462	1,700
0520	Shower door for above, clear plastic, 24" wide	1 Carp	8	1		182	50.50		232.50	283
0540	28" wide		8	1		249	50.50		299.50	355
0560	Tempered glass, 24" wide		8	1		239	50.50		289.50	345
0580	28" wide		8	1		281	50.50		331.50	395
4100	Shower doors, economy plastic, 24" wide	1 Shee	9	.889		137	53		190	235
4200	Tempered glass door, economy		8	1		293	60		353	415
4700	Deluxe, tempered glass, chrome on brass frame, 42" to 44"		8	1		440	60		500	575
4800	39" to 48" wide		1	8		670	480		1,150	1,500
9990	Minimum labor/equipment charge		2.50	3.200	Job		191		191	300

10 31 Manufactured Fireplaces

10 31 13 – Manufactured Fireplace Chimneys

10 31 13.10 Fireplace Chimneys

	10 31 13.10 Fireplace Chimneys	Crew	Daily Output	Labor-Hours	Unit	Material	2018 Bare Costs Labor	Equipment	Total	Total Incl O&P
0010	**FIREPLACE CHIMNEYS**									
0500	Chimney dbl. wall, all stainless, over 8'-6", 7" diam., add to fireplace	1 Carp	33	.242	V.L.F.	87.50	12.30		99.80	116
0600	10" diameter, add to fireplace		32	.250		112	12.70		124.70	145
0700	12" diameter, add to fireplace		31	.258		172	13.10		185.10	211
0800	14" diameter, add to fireplace		30	.267		227	13.50		240.50	272
1000	Simulated brick chimney top, 4' high, 16" x 16"		10	.800	Ea.	455	40.50		495.50	570
1100	24" x 24"		7	1.143	"	560	58		618	710

10 31 13.20 Chimney Accessories

	10 31 13.20 Chimney Accessories	Crew	Daily Output	Labor-Hours	Unit	Material	2018 Bare Costs Labor	Equipment	Total	Total Incl O&P
0010	**CHIMNEY ACCESSORIES**									
0020	Chimney screens, galv., 13" x 13" flue	1 Bric	8	1	Ea.	58.50	50.50		109	146
0050	24" x 24" flue		5	1.600		124	80.50		204.50	268
0200	Stainless steel, 13" x 13" flue		8	1		97.50	50.50		148	189
0250	20" x 20" flue		5	1.600		152	80.50		232.50	298
2400	Squirrel and bird screens, galvanized, 8" x 8" flue		16	.500		56	25		81	103
2450	13" x 13" flue		12	.667		66.50	33.50		100	128
9000	Minimum labor/equipment charge		3.50	2.286	Job		115		115	188

10 31 16 – Manufactured Fireplace Forms

10 31 16.10 Fireplace Forms

	10 31 16.10 Fireplace Forms	Crew	Daily Output	Labor-Hours	Unit	Material	2018 Bare Costs Labor	Equipment	Total	Total Incl O&P
0010	**FIREPLACE FORMS**									
1800	Fireplace forms, no accessories, 32" opening	1 Bric	3	2.667	Ea.	745	134		879	1,050
1900	36" opening		2.50	3.200		950	161		1,111	1,325
2000	40" opening		2	4		1,275	201		1,476	1,725
2100	78" opening		1.50	5.333		1,850	268		2,118	2,475

351

10 31 Manufactured Fireplaces

10 31 23 – Prefabricated Fireplaces

10 31 23.10 Fireplace, Prefabricated	Crew	Daily Output	Labor-Hours	Unit	Material	2018 Bare Costs Labor	Equipment	Total	Total Incl O&P
0010 **FIREPLACE, PREFABRICATED**, free standing or wall hung									
0100 With hood & screen, painted	1 Carp	1.30	6.154	Ea.	1,625	310		1,935	2,275
0150 Average		1	8		1,825	405		2,230	2,650
0200 Stainless steel		.90	8.889		3,250	450		3,700	4,300
1500 Simulated logs, gas fired, 40,000 BTU, 2' long, manual safety pilot		7	1.143	Set	545	58		603	695
1600 Adjustable flame remote pilot		6	1.333		1,275	67.50		1,342.50	1,500
1700 Electric, 1,500 BTU, 1'-6" long, incandescent flame		7	1.143		260	58		318	380
1800 1,500 BTU, LED flame		6	1.333		345	67.50		412.50	490
2000 Fireplace, built-in, 36" hearth, radiant		1.30	6.154	Ea.	720	310		1,030	1,300
2100 Recirculating, small fan		1	8		925	405		1,330	1,675
2150 Large fan		.90	8.889		2,075	450		2,525	3,025
2200 42" hearth, radiant		1.20	6.667		970	340		1,310	1,625
2300 Recirculating, small fan		.90	8.889		1,250	450		1,700	2,125
2350 Large fan		.80	10		1,450	505		1,955	2,425
2400 48" hearth, radiant		1.10	7.273		2,350	370		2,720	3,200
2500 Recirculating, small fan		.80	10		2,475	505		2,980	3,550
2550 Large fan		.70	11.429		2,575	580		3,155	3,775
3000 See through, including doors		.80	10		2,375	505		2,880	3,425
3200 Corner (2 wall)		1	8		3,150	405		3,555	4,125

10 32 Fireplace Specialties

10 32 13 – Fireplace Dampers

10 32 13.10 Dampers	Crew	Daily Output	Labor-Hours	Unit	Material	2018 Bare Costs Labor	Equipment	Total	Total Incl O&P
0010 **DAMPERS**									
0800 Damper, rotary control, steel, 30" opening	1 Bric	6	1.333	Ea.	119	67		186	240
0850 Cast iron, 30" opening		6	1.333		125	67		192	246
0880 36" opening		6	1.333		127	67		194	249
0900 48" opening		6	1.333		167	67		234	293
0920 60" opening		6	1.333		355	67		422	500
1000 84" opening, special order		5	1.600		910	80.50		990.50	1,125
1050 96" opening, special order		4	2		925	101		1,026	1,200
1200 Steel plate, poker control, 60" opening		8	1		320	50.50		370.50	435
1250 84" opening, special order		5	1.600		585	80.50		665.50	775
1400 "Universal" type, chain operated, 32" x 20" opening		8	1		250	50.50		300.50	360
1450 48" x 24" opening		5	1.600		375	80.50		455.50	540

10 32 23 – Fireplace Doors

10 32 23.10 Doors	Crew	Daily Output	Labor-Hours	Unit	Material	2018 Bare Costs Labor	Equipment	Total	Total Incl O&P
0010 **DOORS**									
0400 Cleanout doors and frames, cast iron, 8" x 8"	1 Bric	12	.667	Ea.	54.50	33.50		88	115
0450 12" x 12"		10	.800		89	40		129	164
0500 18" x 24"		8	1		150	50.50		200.50	247
0550 Cast iron frame, steel door, 24" x 30"		5	1.600		315	80.50		395.50	475
1600 Dutch oven door and frame, cast iron, 12" x 15" opening		13	.615		131	31		162	196
1650 Copper plated, 12" x 15" opening		13	.615		257	31		288	335

10 35 Stoves

10 35 13 – Heating Stoves

10 35 13.10 Wood Burning Stoves	Crew	Daily Output	Labor-Hours	Unit	Material	2018 Bare Costs Labor	Equipment	Total	Total Incl O&P
0010 **WOOD BURNING STOVES**									
0015 Cast iron, less than 1,500 S.F.	2 Carp	1.30	12.308	Ea.	1,350	625		1,975	2,500
0020 1,500 to 2,000 S.F.		1	16		2,075	810		2,885	3,600
0030 greater than 2,000 S.F.		.80	20		2,800	1,025		3,825	4,725
0050 For gas log lighter, add					47			47	52

10 43 Emergency Aid Specialties

10 43 13 – Defibrillator Cabinets

10 43 13.05 Defibrillator Cabinets

	Crew	Daily Output	Labor-Hours	Unit	Material	Labor	Equipment	Total	Total Incl O&P
0010 **DEFIBRILLATOR CABINETS**, not equipped, stainless steel									
0050 Defibrillator cabinet, stainless steel with strobe & alarm 12" x 27"	1 Carp	10	.800	Ea.	455	40.50		495.50	565
0100 Automatic External Defibrillator	"	30	.267	"	1,375	13.50		1,388.50	1,550

10 44 Fire Protection Specialties

10 44 13 – Fire Protection Cabinets

10 44 13.53 Fire Equipment Cabinets

	Crew	Daily Output	Labor-Hours	Unit	Material	Labor	Equipment	Total	Total Incl O&P
0010 **FIRE EQUIPMENT CABINETS**, not equipped, 20 ga. steel box									
0040 Recessed, D.S. glass in door, box size given									
1000 Portable extinguisher, single, 8" x 12" x 27", alum. door & frame	Q-12	8	2	Ea.	134	108		242	315
1100 Steel door and frame	"	8	2	"	128	108		236	310
3000 Hose rack assy., 1-1/2" valve & 100' hose, 24" x 40" x 5-1/2"									
3200 Steel door and frame	Q-12	6	2.667	Ea.	271	144		415	520
4000 Hose rack assy., 2-1/2" x 1-1/2" valve, 100' hose, 24" x 40" x 8"									
4200 Steel door and frame	Q-12	6	2.667	Ea.	335	144		479	590
5000 Hose rack assy., 2-1/2" x 1-1/2" valve, 100' hose									
5010 and extinguisher, 30" x 40" x 8"									
5200 Steel door and frame	Q-12	5	3.200	Ea.	291	172		463	590

10 44 16 – Fire Extinguishers

10 44 16.13 Portable Fire Extinguishers

	Crew	Daily Output	Labor-Hours	Unit	Material	Labor	Equipment	Total	Total Incl O&P
0010 **PORTABLE FIRE EXTINGUISHERS**									
0140 CO_2, with hose and "H" horn, 10 lb.				Ea.	300			300	330
1000 Dry chemical, pressurized									
1040 Standard type, portable, painted, 2-1/2 lb.				Ea.	41			41	45
1080 10 lb.					85.50			85.50	94
1100 20 lb.					138			138	152
1120 30 lb.					425			425	465
2000 ABC all purpose type, portable, 2-1/2 lb.					23.50			23.50	26
2080 9-1/2 lb.					51.50			51.50	56.50
3500 Halotron 1, 2-1/2 lb.					132			132	145
3600 5 lb.					223			223	246
3700 11 lb.					445			445	490

10 51 13.10 Lockers	Crew	Daily Output	Labor-Hours	Unit	Material	2018 Bare Costs Labor	Equipment	Total	Total Incl O&P
0011 **LOCKERS** steel, baked enamel, knock down construction									
0012 Body- 24ga, door- 16 or 18ga									
0110 1-tier locker, 12" x 15" x 72"	1 Shee	20	.400	Ea.	271	24		295	335
0120 18" x 15" x 72"		20	.400		233	24		257	294
0130 12" x 18" x 72"		20	.400		197	24		221	255
0140 18" x 18" x 72"		20	.400		270	24		294	335
0410 2- tier, 12" x 15" x 36"		30	.267		252	15.95		267.95	300
0420 18" x 15" x 36"		30	.267		235	15.95		250.95	284
0430 12" x 18" x 36"		30	.267		300	15.95		315.95	355
0440 18" x 18" x 36"		30	.267		256	15.95		271.95	305
0450 3- tier 12" x 15" x 24"		30	.267		285	15.95		300.95	340
0455 12" x 18" x 24"		30	.267		294	15.95		309.95	350
0480 4- tier 12" x 15" x 18"		30	.267		365	15.95		380.95	425
0485 12" x 18" x 18"		30	.267		380	15.95		395.95	445
0500 Two person, 18" x 15" x 72"		20	.400		335	24		359	405
0510 18" x 18" x 72"		20	.400		300	24		324	370
0520 Duplex, 15" x 15" x 72"		20	.400		335	24		359	410
0530 15" x 21" x 72"		20	.400		375	24		399	455
1098 All welded, ventilated athletic lockers, body- 16 ga., door- 14 ga.									
1100 1-tier, 12" x 18" x 72"	1 Shee	7.50	1.067	Ea.	335	64		399	470
1110 18" x 15" x 72"		7.50	1.067		545	64		609	695
1115 18" x 18" x 72"		7.50	1.067		555	64		619	710
1130 2-tier, 12" x 15" x 36"		7.50	1.067		565	64		629	725
1135 12" x 18" x 36"		7.50	1.067		605	64		669	765
1140 18" x 15" x 36"		7.50	1.067		660	64		724	825
1145 18" x 18" x 36"		7.50	1.067		670	64		734	835
1160 3-tier, 12" x 18" x 24"		7.50	1.067		775	64		839	950
1180 4-tier, 12" x 18" x 18"		7.50	1.067		590	64		654	745
1190 5H box, 12" x 18" x 14.4"		7.50	1.067		605	64		669	765
1195 18" x 18" x 14.4"		7.50	1.067		775	64		839	950
1196 6H box, 12" x 15" x 12"		7.50	1.067		485	64		549	635
1197 12" x 18" x 12"		7.50	1.067		595	64		659	755
1198 18" x 18" x 12"		7.50	1.067		735	64		799	905
2399 Standard duty lockers, body- 24 ga., doors- 16-18 ga.									
2400 16-person locker unit with clothing rack									
2500 72" wide x 15" deep x 72" high	1 Shee	15	.533	Ea.	500	32		532	600
2550 18" deep		15	.533	"	660	32		692	780
3250 Rack w/24 wire mesh baskets		1.50	5.333	Set	400	320		720	945
3260 30 baskets		1.25	6.400		390	385		775	1,025
3270 36 baskets		.95	8.421		445	505		950	1,275
3280 42 baskets		.80	10		455	600		1,055	1,450
3600 For hanger rods, add				Ea.	2.48			2.48	2.73
3650 For number plate kit, 100 plates #1 - #100, add	1 Shee	4	2		82	120		202	279
3700 For locker base, closed front panel		90	.089		5.10	5.30		10.40	14
3710 End panel, bolted		36	.222		8.40	13.30		21.70	30.50
3800 For sloping top, 12" wide		24	.333		31.50	19.95		51.45	66
3810 15" wide		24	.333		31	19.95		50.95	65.50
3820 18" wide		24	.333		34	19.95		53.95	69
3850 Sloping top end panel, 12" deep		72	.111		16.05	6.65		22.70	28
3860 15" deep		72	.111		16.60	6.65		23.25	28.50
3870 18" deep		72	.111		17.10	6.65		23.75	29.50
3900 For finish end panels, steel, 60" high, 15" deep		12	.667		30	40		70	96
3910 72" high, 12" deep		12	.667		27	40		67	92.50

10 51 Lockers

10 51 13 – Metal Lockers

10 51 13.10 Lockers

		Crew	Daily Output	Labor-Hours	Unit	Material	2018 Bare Costs Labor	Equipment	Total	Total Incl O&P
3920	18" deep	1 Shee	12	.667	Ea.	41.50	40		81.50	109
5000	For "ready to assemble" lockers,									
5010	Add to labor						75%			
5020	Deduct from material					20%				
6000	Heavy duty for detention facility, tamper proof, 14 ga. welded steel, solid									
6100	24" W x 24" D x 74" H, single tier	1 Shee	18	.444	Ea.	540	26.50		566.50	635
6110	Double tier		18	.444		535	26.50		561.50	630
6120	Triple tier		18	.444		555	26.50		581.50	650
9000	Minimum labor/equipment charge		2.50	3.200	Job		191		191	300

10 55 Postal Specialties

10 55 23 – Mail Boxes

10 55 23.10 Commercial Mail Boxes

		Crew	Daily Output	Labor-Hours	Unit	Material	2018 Bare Costs Labor	Equipment	Total	Total Incl O&P
0010	**COMMERCIAL MAIL BOXES**									
0020	Horiz., key lock, 5"H x 6"W x 15"D, alum., rear load	1 Carp	34	.235	Ea.	42.50	11.95		54.45	66.50
0100	Front loading		34	.235		42.50	11.95		54.45	66.50
0200	Double, 5"H x 12"W x 15"D, rear loading		26	.308		66.50	15.60		82.10	98.50
0300	Front loading		26	.308		74.50	15.60		90.10	108
0500	Quadruple, 10"H x 12"W x 15"D, rear loading		20	.400		115	20.50		135.50	160
0600	Front loading		20	.400		95.50	20.50		116	138
1600	Vault type, horizontal, for apartments, 4" x 5"		34	.235		26.50	11.95		38.45	48.50
1700	Alphabetical directories, 120 names		10	.800		104	40.50		144.50	180
1900	Letter slot, residential		20	.400		80.50	20.50		101	122
2000	Post office type		8	1		120	50.50		170.50	215
9000	Minimum labor/equipment charge		5	1.600	Job		81		81	132

10 56 Storage Assemblies

10 56 13 – Metal Storage Shelving

10 56 13.10 Shelving

		Crew	Daily Output	Labor-Hours	Unit	Material	2018 Bare Costs Labor	Equipment	Total	Total Incl O&P
0010	**SHELVING**									
0020	Metal, industrial, cross-braced, 3' W, 12" D	1 Sswk	175	.046	SF Shlf	8.10	2.50		10.60	13.15
0100	24" D		330	.024		6.10	1.32		7.42	8.95
2200	Wide span, 1600 lb. capacity per shelf, 6' W, 24" D		380	.021		7.15	1.15		8.30	9.85
2400	36" D		440	.018		6.10	.99		7.09	8.40
9000	Minimum labor/equipment charge	1 Carp	4	2	Job		101		101	165

10 57 Wardrobe and Closet Specialties

10 57 23 – Closet and Utility Shelving

10 57 23.19 Wood Closet and Utility Shelving

		Crew	Daily Output	Labor-Hours	Unit	Material	2018 Bare Costs Labor	Equipment	Total	Total Incl O&P
0010	**WOOD CLOSET AND UTILITY SHELVING**									
0020	Pine, clear grade, no edge band, 1" x 8"	1 Carp	115	.070	L.F.	3.52	3.53		7.05	9.60
0100	1" x 10"		110	.073		4.38	3.69		8.07	10.80
0200	1" x 12"		105	.076		5.30	3.86		9.16	12.05
0600	Plywood, 3/4" thick with lumber edge, 12" wide		75	.107		1.91	5.40		7.31	10.90
0700	24" wide		70	.114		3.38	5.80		9.18	13.10
0900	Bookcase, clear grade pine, shelves 12" OC, 8" deep, per S.F. shelf		70	.114	S.F.	11.45	5.80		17.25	22
1000	12" deep shelves		65	.123	"	17.15	6.25		23.40	29

10 57 Wardrobe and Closet Specialties

10 57 23 – Closet and Utility Shelving

10 57 23.19 Wood Closet and Utility Shelving	Crew	Daily Output	Labor-Hours	Unit	Material	2018 Bare Costs Labor	Equipment	Total	Total Incl O&P
1200	Adjustable closet rod and shelf, 12" wide, 3' long	1 Carp	20	.400	Ea.	12.60	20.50		33.10	47
1300	8' long		15	.533	"	24.50	27		51.50	71
1500	Prefinished shelves with supports, stock, 8" wide		75	.107	L.F.	5.60	5.40		11	14.95
1600	10" wide		70	.114	"	5.40	5.80		11.20	15.35
9000	Minimum labor/equipment charge		4	2	Job		101		101	165

10 73 Protective Covers

10 73 13 – Awnings

10 73 13.10 Awnings, Fabric

10 73 13.10 Awnings, Fabric		Crew	Daily Output	Labor-Hours	Unit	Material	2018 Bare Costs Labor	Equipment	Total	Total Incl O&P
0010	**AWNINGS, FABRIC**									
0020	Including acrylic canvas and frame, standard design									
0100	Door and window, slope, 3' high, 4' wide	1 Carp	4.50	1.778	Ea.	755	90		845	975
0110	6' wide		3.50	2.286		970	116		1,086	1,275
0120	8' wide		3	2.667		1,200	135		1,335	1,525
0200	Quarter round convex, 4' wide		3	2.667		1,175	135		1,310	1,525
0210	6' wide		2.25	3.556		1,525	180		1,705	1,975
0220	8' wide		1.80	4.444		1,875	225		2,100	2,425
0300	Dome, 4' wide		7.50	1.067		455	54		509	590
0310	6' wide		3.50	2.286		1,025	116		1,141	1,325
0320	8' wide		2	4		1,800	203		2,003	2,325
0350	Elongated dome, 4' wide		1.33	6.015		1,700	305		2,005	2,375
0360	6' wide		1.11	7.207		2,025	365		2,390	2,825
0370	8' wide		1	8		2,375	405		2,780	3,275
1000	Entry or walkway, peak, 12' long, 4' wide	2 Carp	.90	17.778		5,475	900		6,375	7,500
1010	6' wide		.60	26.667		8,450	1,350		9,800	11,500
1020	8' wide		.40	40		11,700	2,025		13,725	16,100
1100	Radius with dome end, 4' wide		1.10	14.545		4,150	735		4,885	5,775
1110	6' wide		.70	22.857		6,675	1,150		7,825	9,225
1120	8' wide		.50	32		9,500	1,625		11,125	13,000
2000	Retractable lateral arm awning, manual									
2010	To 12' wide, 8'-6" projection	2 Carp	1.70	9.412	Ea.	1,250	475		1,725	2,125
2020	To 14' wide, 8'-6" projection		1.10	14.545		1,450	735		2,185	2,800
2030	To 19' wide, 8'-6" projection		.85	18.824		1,950	955		2,905	3,700
2040	To 24' wide, 8'-6" projection		.67	23.881		2,475	1,200		3,675	4,700
2050	Motor for above, add	1 Carp	2.67	3		1,050	152		1,202	1,425
3000	Patio/deck canopy with frame									
3010	12' wide, 12' projection	2 Carp	2	8	Ea.	1,750	405		2,155	2,575
3020	16' wide, 14' projection	"	1.20	13.333		2,725	675		3,400	4,100
9000	For fire retardant canvas, add					7%				
9010	For lettering or graphics, add					35%				
9020	For painted or coated acrylic canvas, deduct					8%				
9030	For translucent or opaque vinyl canvas, add					10%				
9040	For 6 or more units, deduct					20%	15%			

10 73 16 – Canopies

10 73 16.10 Canopies, Residential

10 73 16.10 Canopies, Residential		Crew	Daily Output	Labor-Hours	Unit	Material	2018 Bare Costs Labor	Equipment	Total	Total Incl O&P
0010	**CANOPIES, RESIDENTIAL** Prefabricated									
0500	Carport, free standing, baked enamel, alum., .032", 40 psf									
0520	16' x 8', 4 posts	2 Carp	3	5.333	Ea.	4,850	270		5,120	5,775
0600	20' x 10', 6 posts		2	8		4,975	405		5,380	6,125
0605	30' x 10', 8 posts		2	8		7,450	405		7,855	8,850

10 73 Protective Covers

10 73 16 – Canopies

10 73 16.10 Canopies, Residential

		Crew	Daily Output	Labor-Hours	Unit	Material	2018 Bare Costs Labor	Equipment	Total	Total Incl O&P
1000	Door canopies, extruded alum., .032", 42" projection, 4' wide	1 Carp	8	1	Ea.	207	50.50		257.50	310
1020	6' wide	"	6	1.333		291	67.50		358.50	430
1040	8' wide	2 Carp	9	1.778		435	90		525	625
1060	10' wide		7	2.286		510	116		626	755
1080	12' wide		5	3.200		585	162		747	910
1200	54" projection, 4' wide	1 Carp	8	1		232	50.50		282.50	340
1220	6' wide	"	6	1.333		296	67.50		363.50	435
1240	8' wide	2 Carp	9	1.778		355	90		445	540
1260	10' wide		7	2.286		815	116		931	1,100
1280	12' wide		5	3.200		880	162		1,042	1,225
1300	Painted, add					20%				
1310	Bronze anodized, add					50%				
3000	Window awnings, aluminum, window 3' high, 4' wide	1 Carp	10	.800		160	40.50		200.50	242
3020	6' wide	"	8	1		192	50.50		242.50	294
3040	9' wide	2 Carp	9	1.778		495	90		585	690
3060	12' wide	"	5	3.200		495	162		657	810
3100	Window, 4' high, 4' wide	1 Carp	10	.800		289	40.50		329.50	385
3120	6' wide	"	8	1		355	50.50		405.50	475
3140	9' wide	2 Carp	9	1.778		625	90		715	835
3160	12' wide	"	5	3.200		710	162		872	1,050
3200	Window, 6' high, 4' wide	1 Carp	10	.800		340	40.50		380.50	440
3220	6' wide	"	8	1		390	50.50		440.50	515
3240	9' wide	2 Carp	9	1.778		955	90		1,045	1,200
3260	12' wide	"	5	3.200		1,450	162		1,612	1,875
3400	Roll-up aluminum, 2'-6" wide	1 Carp	14	.571		208	29		237	275
3420	3' wide		12	.667		213	34		247	289
3440	4' wide		10	.800		267	40.50		307.50	360
3460	6' wide		8	1		286	50.50		336.50	400
3480	9' wide	2 Carp	9	1.778		410	90		500	595
3500	12' wide	"	5	3.200		520	162		682	835
3600	Window awnings, canvas, 24" drop, 3' wide	1 Carp	30	.267	L.F.	58	13.50		71.50	86
3620	4' wide		40	.200		46	10.15		56.15	67
3700	30" drop, 3' wide		30	.267		60	13.50		73.50	88
3720	4' wide		40	.200		50	10.15		60.15	71.50
3740	5' wide		45	.178		46	9		55	65.50
3760	6' wide		48	.167		42.50	8.45		50.95	60.50
3780	8' wide		48	.167		39.50	8.45		47.95	57
3800	10' wide		50	.160		41.50	8.10		49.60	58.50
3900	Repair canvas, minimum		8	1	Ea.		50.50		50.50	82.50
3920	Maximum		4	2	"		101		101	165
9000	Minimum labor/equipment charge		3	2.667	Job		135		135	219

10 73 16.20 Metal Canopies

		Crew	Daily Output	Labor-Hours	Unit	Material	2018 Bare Costs Labor	Equipment	Total	Total Incl O&P
0010	**METAL CANOPIES**									
0020	Wall hung, .032", aluminum, prefinished, 8' x 10'	K-2	1.30	18.462	Ea.	2,500	960	183	3,643	4,550
0300	8' x 20'		1.10	21.818		4,150	1,125	216	5,491	6,700
1000	12' x 20'		1	24		5,625	1,250	238	7,113	8,550
2300	Aluminum entrance canopies, flat soffit, .032"									
2500	3'-6" x 4'-0", clear anodized	2 Carp	4	4	Ea.	1,050	203		1,253	1,475
9000	Minimum labor/equipment charge	"	2	8	Job		405		405	660

10 74 Manufactured Exterior Specialties

10 74 23 – Cupolas

10 74 23.10 Wood Cupolas

10 74 23.10 Wood Cupolas	Crew	Daily Output	Labor-Hours	Unit	Material	2018 Bare Costs Labor	Equipment	Total	Total Incl O&P
0010 **WOOD CUPOLAS**									
0020 Stock units, pine, painted, 18" sq., 28" high, alum. roof	1 Carp	4.10	1.951	Ea.	285	99		384	475
0100 Copper roof		3.80	2.105		291	107		398	495
0300 23" square, 33" high, aluminum roof		3.70	2.162		440	110		550	665
0400 Copper roof		3.30	2.424		590	123		713	850
0600 30" square, 37" high, aluminum roof		3.70	2.162		600	110		710	840
0700 Copper roof		3.30	2.424		715	123		838	990
0900 Hexagonal, 31" wide, 46" high, copper roof		4	2		945	101		1,046	1,225
1000 36" wide, 50" high, copper roof		3.50	2.286		1,675	116		1,791	2,025
1200 For deluxe stock units, add to above					25%				
1400 For custom built units, add to above					50%	50%			
9000 Minimum labor/equipment charge	1 Carp	2.75	2.909	Job		147		147	239

10 74 29 – Steeples

10 74 29.10 Prefabricated Steeples

10 74 29.10 Prefabricated Steeples	Crew	Daily Output	Labor-Hours	Unit	Material	2018 Bare Costs Labor	Equipment	Total	Total Incl O&P
0010 **PREFABRICATED STEEPLES**									
4000 Steeples, translucent fiberglass, 30" square, 15' high	F-3	2	20	Ea.	9,300	1,025	248	10,573	12,100
4150 25' high	"	1.80	22.222		10,500	1,150	276	11,926	13,800
4600 Aluminum, baked finish, 16" square, 14' high					6,150			6,150	6,750
4640 35' high, 8' base					38,400			38,400	42,200
4680 152' high, custom					628,500			628,500	691,500

10 74 33 – Weathervanes

10 74 33.10 Residential Weathervanes

10 74 33.10 Residential Weathervanes	Crew	Daily Output	Labor-Hours	Unit	Material	2018 Bare Costs Labor	Equipment	Total	Total Incl O&P
0010 **RESIDENTIAL WEATHERVANES**									
0020 Residential types, 18" to 24"	1 Carp	8	1	Ea.	134	50.50		184.50	231
0100 24" to 48"		2	4	"	1,825	203		2,028	2,325
9000 Minimum labor/equipment charge		4	2	Job		101		101	165

10 74 46 – Window Wells

10 74 46.10 Area Window Wells

10 74 46.10 Area Window Wells	Crew	Daily Output	Labor-Hours	Unit	Material	2018 Bare Costs Labor	Equipment	Total	Total Incl O&P
0010 **AREA WINDOW WELLS**, Galvanized steel									
0020 20 ga., 3'-2" wide, 1' deep	1 Sswk	29	.276	Ea.	16.95	15.10		32.05	44
0100 2' deep		23	.348		31	19		50	67
0300 16 ga., 3'-2" wide, 1' deep		29	.276		23	15.10		38.10	51
0400 3' deep		23	.348		47	19		66	84
0600 Welded grating for above, 15 lb., painted		45	.178		91.50	9.70		101.20	118
0700 Galvanized		45	.178		124	9.70		133.70	154
0900 Translucent plastic cap for above		60	.133		20.50	7.30		27.80	35

10 86 Security Mirrors and Domes

10 86 10 – Security Mirrors

10 86 10.10 Exterior Traffic Control Mirrors

10 86 10.10 Exterior Traffic Control Mirrors	Crew	Daily Output	Labor-Hours	Unit	Material	2018 Bare Costs Labor	Equipment	Total	Total Incl O&P
0010 **EXTERIOR TRAFFIC CONTROL MIRRORS**									
0100 Convex, stainless steel, 20 ga., 26" diameter	1 Carp	12	.667	Ea.	172	34		206	244

10 86 20 – Security Domes

10 86 20.10 Domes

10 86 20.10 Domes	Crew	Daily Output	Labor-Hours	Unit	Material	2018 Bare Costs Labor	Equipment	Total	Total Incl O&P
0010 **DOMES** for security cameras (CCTV)									
0100 Ceiling mounted, 10" diameter	1 Carp	30	.267	Ea.	12.50	13.50		26	36
0110 12" diameter	"	30	.267	"	9	13.50		22.50	32

Estimating Tips
General

- The items in this division are usually priced per square foot or each. Many of these items are purchased by the owner for installation by the contractor. Check the specifications for responsibilities and include time for receiving, storage, installation, and mechanical and electrical hookups in the appropriate divisions.

- Many items in Division 11 require some type of support system that is not usually furnished with the item. Examples of these systems include blocking for the attachment of casework and support angles for ceiling-hung projection screens. The required blocking or supports must be added to the estimate in the appropriate division.

- Some items in Division 11 may require assembly or electrical hookups. Verify the amount of assembly required or the need for a hard electrical connection and add the appropriate costs.

Reference Numbers

Reference numbers are shown at the beginning of some major classifications. These numbers refer to related items in the Reference Section. The reference information may be an estimating procedure, an alternate pricing method, or technical information.

Note: Not all subdivisions listed here necessarily appear. ■

Did you know?

RSMeans data is available through our online application with 24/7 access:

- Search for unit prices by keyword
- Leverage the most up-to-date data
- Build and export estimates

Try it free for 30 days!
www.rsmeans.com/2018freetrial

11 11 Vehicle Service Equipment

11 11 13 – Compressed-Air Vehicle Service Equipment

11 11 13.10 Compressed Air Equipment

11 11 13.10 Compressed Air Equipment	Crew	Daily Output	Labor-Hours	Unit	Material	2018 Bare Costs Labor	Equipment	Total	Total Incl O&P
0010 **COMPRESSED AIR EQUIPMENT**									
0030 Compressors, electric, 1-1/2 HP, standard controls	L-4	1.50	16	Ea.	485	760		1,245	1,750
0550 Dual controls		1.50	16		1,025	760		1,785	2,350
0600 5 HP, 115/230 volt, standard controls		1	24		2,025	1,150		3,175	4,100
0650 Dual controls		1	24		3,225	1,150		4,375	5,400

11 11 19 – Vehicle Lubrication Equipment

11 11 19.10 Lubrication Equipment

11 11 19.10 Lubrication Equipment	Crew	Daily Output	Labor-Hours	Unit	Material	2018 Bare Costs Labor	Equipment	Total	Total Incl O&P
0010 **LUBRICATION EQUIPMENT**									
3000 Lube equipment, 3 reel type, with pumps, not including piping	L-4	.50	48	Set	10,200	2,275		12,475	14,900
3100 Hose reel, including hose, oil/lube, 1000 psi	2 Sswk	2	8	Ea.	465	435		900	1,250
3200 Grease, 5000 psi		2	8		510	435		945	1,300
3300 Air, 50', 160 psi		2	8		610	435		1,045	1,425
3350 25', 160 psi		2	8		365	435		800	1,150

11 11 33 – Vehicle Spray Painting Equipment

11 11 33.10 Spray Painting Equipment

11 11 33.10 Spray Painting Equipment	Crew	Daily Output	Labor-Hours	Unit	Material	2018 Bare Costs Labor	Equipment	Total	Total Incl O&P
0010 **SPRAY PAINTING EQUIPMENT**									
4000 Spray painting booth, 26' long, complete	L-4	.40	60	Ea.	10,500	2,850		13,350	16,100

11 12 Parking Control Equipment

11 12 13 – Parking Key and Card Control Units

11 12 13.10 Parking Control Units

11 12 13.10 Parking Control Units	Crew	Daily Output	Labor-Hours	Unit	Material	2018 Bare Costs Labor	Equipment	Total	Total Incl O&P
0010 **PARKING CONTROL UNITS**									
5100 Card reader	1 Elec	2	4	Ea.	1,400	233		1,633	1,900
5120 Proximity with customer display	2 Elec	1	16		6,000	930		6,930	8,050
6000 Parking control software, basic functionality	1 Elec	.50	16		24,000	930		24,930	27,900
6020 Multi-function	"	.20	40		98,000	2,325		100,325	111,500

11 12 16 – Parking Ticket Dispensers

11 12 16.10 Ticket Dispensers

11 12 16.10 Ticket Dispensers	Crew	Daily Output	Labor-Hours	Unit	Material	2018 Bare Costs Labor	Equipment	Total	Total Incl O&P
0010 **TICKET DISPENSERS**									
5900 Ticket spitter with time/date stamp, standard	2 Elec	2	8	Ea.	6,675	465		7,140	8,050
5920 Mag stripe encoding	"	2	8	"	19,000	465		19,465	21,500

11 12 26 – Parking Fee Collection Equipment

11 12 26.13 Parking Fee Coin Collection Equipment

11 12 26.13 Parking Fee Coin Collection Equipment	Crew	Daily Output	Labor-Hours	Unit	Material	2018 Bare Costs Labor	Equipment	Total	Total Incl O&P
0010 **PARKING FEE COIN COLLECTION EQUIPMENT**									
5200 Cashier booth, average	B-22	1	30	Ea.	10,600	1,400	194	12,194	14,200
5300 Collector station, pay on foot	2 Elec	.20	80		112,000	4,650		116,650	130,000
5320 Credit card only	"	.50	32		20,700	1,850		22,550	25,600

11 12 26.23 Fee Equipment

11 12 26.23 Fee Equipment	Crew	Daily Output	Labor-Hours	Unit	Material	2018 Bare Costs Labor	Equipment	Total	Total Incl O&P
0010 **FEE EQUIPMENT**									
5600 Fee computer	1 Elec	1.50	5.333	Ea.	12,200	310		12,510	13,900

11 12 33 – Parking Gates

11 12 33.13 Lift Arm Parking Gates

11 12 33.13 Lift Arm Parking Gates	Crew	Daily Output	Labor-Hours	Unit	Material	2018 Bare Costs Labor	Equipment	Total	Total Incl O&P
0010 **LIFT ARM PARKING GATES**									
5000 Barrier gate with programmable controller	2 Elec	3	5.333	Ea.	3,050	310		3,360	3,825
5020 Industrial		3	5.333		5,250	310		5,560	6,250
5050 Non-programmable, with reader and 12' arm		3	5.333		1,925	310		2,235	2,600
5500 Exit verifier		1	16		19,500	930		20,430	22,900
5700 Full sign, 4" letters	1 Elec	2	4		1,250	233		1,483	1,725

11 12 Parking Control Equipment

11 12 33 – Parking Gates

11 12 33.13 Lift Arm Parking Gates

		Crew	Daily Output	Labor-Hours	Unit	Material	2018 Bare Costs Labor	Equipment	Total	Total Incl O&P
5800	Inductive loop	2 Elec	4	4	Ea.	210	233		443	590
5950	Vehicle detector, microprocessor based	1 Elec	3	2.667		440	155		595	725
7100	Traffic spike unit, flush mount, spring loaded, 72" L	B-89	4	4		1,275	192	105	1,572	1,825
7200	Surface mount, 72" L	2 Skwk	10	1.600	↓	2,375	84		2,459	2,725

11 13 Loading Dock Equipment

11 13 13 – Loading Dock Bumpers

11 13 13.10 Dock Bumpers

		Crew	Daily Output	Labor-Hours	Unit	Material	2018 Bare Costs Labor	Equipment	Total	Total Incl O&P
0010	**DOCK BUMPERS** Bolts not included									
0012	2" x 6" to 4" x 8", average	1 Carp	300	.027	B.F.	.72	1.35		2.07	2.98
0050	Bumpers, lam. rubber blocks 4-1/2" thick, 10" high, 14" long		26	.308	Ea.	54.50	15.60		70.10	85.50
0200	24" long		22	.364		98.50	18.45		116.95	138
0300	36" long		17	.471		271	24		295	335
0500	12" high, 14" long	↓	25	.320	↓	74.50	16.20		90.70	108

11 13 16 – Loading Dock Seals and Shelters

11 13 16.10 Dock Seals and Shelters

		Crew	Daily Output	Labor-Hours	Unit	Material	2018 Bare Costs Labor	Equipment	Total	Total Incl O&P
0010	**DOCK SEALS AND SHELTERS**									
6200	Shelters, fabric, for truck or train, scissor arms, minimum	1 Carp	1	8	Ea.	2,125	405		2,530	2,975
6300	Maximum	"	.50	16	"	2,600	810		3,410	4,200

11 13 19 – Stationary Loading Dock Equipment

11 13 19.10 Dock Equipment

		Crew	Daily Output	Labor-Hours	Unit	Material	2018 Bare Costs Labor	Equipment	Total	Total Incl O&P
0010	**DOCK EQUIPMENT**									
2200	Dock boards, heavy duty, 60" x 60", aluminum, 5,000 lb. capacity				Ea.	1,400			1,400	1,550
4200	Platform lifter, 6' x 6', portable, 3,000 lb. capacity				"	9,550			9,550	10,500
6000	Dock leveler, 15 ton capacity									
6100	I-beam construction, mechanical, 6' x 8'	E-16	.50	32	Ea.	5,100	1,775	197	7,072	8,875
6150	Hydraulic		.50	32		6,000	1,775	197	7,972	9,850
6200	Formed beam deck construction, mechanical, 6' x 8'		.50	32		3,825	1,775	197	5,797	7,450
6250	Hydraulic		.50	32		5,225	1,775	197	7,197	9,000
6300	Edge of dock leveler, mechanical, 15 ton capacity		2	8		1,150	445	49	1,644	2,075
6301	Edge of dock leveler, mechanical, 10 ton capacity		2	8		1,150	445	49	1,644	2,075
7000	22.5 ton capacity	↓		↓						
7100	Vertical storing dock leveler, hydraulic, 6' x 6'	E-16	.40	40	Ea.	8,125	2,225	246	10,596	13,000

11 21 Retail and Service Equipment

11 21 13 – Cash Registers and Checking Equipment

11 21 13.10 Checkout Counter

		Crew	Daily Output	Labor-Hours	Unit	Material	2018 Bare Costs Labor	Equipment	Total	Total Incl O&P
0010	**CHECKOUT COUNTER**									
0020	Supermarket conveyor, single belt	2 Clab	10	1.600	Ea.	3,400	64		3,464	3,850
0100	Double belt, power take-away		9	1.778		4,900	71		4,971	5,500
0400	Double belt, power take-away, incl. side scanning		7	2.286		5,750	91		5,841	6,475
0800	Warehouse or bulk type	↓	6	2.667	↓	6,800	106		6,906	7,650
1000	Scanning system, 2 lanes, w/registers, scan gun & memory				System	17,800			17,800	19,600
1100	10 lanes, single processor, full scan, with scales				"	169,000			169,000	186,000
2000	Register, restaurant, minimum				Ea.	790			790	870
2100	Maximum					3,350			3,350	3,675
2150	Store, minimum					790			790	870
2200	Maximum				↓	3,350			3,350	3,675

361

For customer support on your Commercial Renovation Costs with RSMeans data, call 800.448.8182.

11 21 33 – Checkroom Equipment

11 21 33.10 Clothes Check Equipment

		Crew	Daily Output	Labor-Hours	Unit	Material	2018 Bare Costs Labor	2018 Bare Costs Equipment	Total	Total Incl O&P
0010	**CLOTHES CHECK EQUIPMENT**									
0030	Clothes check rack, free standing, st. stl., 2-tier, 90 bag capacity				Ea.	1,550			1,550	1,700
0050	Wall mounted, 45 bag capacity	L-2	8	2		895	88.50		983.50	1,125
0100	Garment checking bag, green mesh fabric, 21" H x 17" W with 4.5" hook					27			27	30

11 21 53 – Barber and Beauty Shop Equipment

11 21 53.10 Barber Equipment

		Crew	Daily Output	Labor-Hours	Unit	Material	2018 Bare Costs Labor	2018 Bare Costs Equipment	Total	Total Incl O&P
0010	**BARBER EQUIPMENT**									
0020	Chair, hydraulic, movable, minimum	1 Carp	24	.333	Ea.	595	16.90		611.90	685
0050	Maximum	"	16	.500		3,775	25.50		3,800.50	4,200
0200	Wall hung styling station with mirrors, minimum	L-2	8	2		575	88.50		663.50	780
0300	Maximum	"	4	4		2,525	177		2,702	3,075
0500	Sink, hair washing basin, rough plumbing not incl.	1 Plum	8	1		495	62		557	640
1000	Sterilizer, liquid solution for tools					161			161	177
1100	Total equipment, rule of thumb, per chair, minimum	L-8	1	20		2,100	1,050		3,150	4,000
1150	Maximum	"	1	20		5,600	1,050		6,650	7,850

11 21 73 – Commercial Laundry and Dry Cleaning Equipment

11 21 73.13 Dry Cleaning Equipment

		Crew	Daily Output	Labor-Hours	Unit	Material	2018 Bare Costs Labor	2018 Bare Costs Equipment	Total	Total Incl O&P
0010	**DRY CLEANING EQUIPMENT**									
2000	Dry cleaners, electric, 20 lb. capacity, not incl. rough-in	L-1	.20	80	Ea.	33,700	4,825		38,525	44,600
2050	25 lb. capacity		.17	94.118		51,500	5,675		57,175	65,500
2100	30 lb. capacity		.15	107		54,000	6,425		60,425	69,500
2150	60 lb. capacity		.09	178		78,500	10,700		89,200	102,500

11 21 73.16 Drying and Conditioning Equipment

		Crew	Daily Output	Labor-Hours	Unit	Material	2018 Bare Costs Labor	2018 Bare Costs Equipment	Total	Total Incl O&P
0010	**DRYING AND CONDITIONING EQUIPMENT**									
0100	Dryers, not including rough-in									
1500	Industrial, 30 lb. capacity	1 Plum	2	4	Ea.	3,375	249		3,624	4,075
1600	50 lb. capacity	"	1.70	4.706		3,825	292		4,117	4,650
4700	Lint collector, ductwork not included, 8,000 to 10,000 CFM	Q-10	.30	80		9,750	4,475		14,225	17,800

11 21 73.19 Finishing Equipment

		Crew	Daily Output	Labor-Hours	Unit	Material	2018 Bare Costs Labor	2018 Bare Costs Equipment	Total	Total Incl O&P
0010	**FINISHING EQUIPMENT**									
3500	Folders, blankets & sheets, minimum	1 Elec	.17	47.059	Ea.	35,100	2,750		37,850	42,800
3700	King size with automatic stacker		.10	80		63,500	4,650		68,150	77,000
3800	For conveyor delivery, add		.45	17.778		16,100	1,025		17,125	19,400

11 21 73.23 Commercial Ironing Equipment

		Crew	Daily Output	Labor-Hours	Unit	Material	2018 Bare Costs Labor	2018 Bare Costs Equipment	Total	Total Incl O&P
0010	**COMMERCIAL IRONING EQUIPMENT**									
4500	Ironers, institutional, 110", single roll	1 Elec	.20	40	Ea.	34,000	2,325		36,325	41,000

11 21 73.26 Commercial Washers and Extractors

		Crew	Daily Output	Labor-Hours	Unit	Material	2018 Bare Costs Labor	2018 Bare Costs Equipment	Total	Total Incl O&P
0010	**COMMERCIAL WASHERS AND EXTRACTORS**, not including rough-in									
6000	Combination washer/extractor, 20 lb. capacity	L-6	1.50	8	Ea.	6,575	485		7,060	7,975
6100	30 lb. capacity		.80	15		10,000	910		10,910	12,400
6200	50 lb. capacity		.68	17.647		12,700	1,075		13,775	15,700
6300	75 lb. capacity		.30	40		18,700	2,425		21,125	24,400
6350	125 lb. capacity		.16	75		29,100	4,550		33,650	39,200

11 21 73.33 Coin-Operated Laundry Equipment

		Crew	Daily Output	Labor-Hours	Unit	Material	2018 Bare Costs Labor	2018 Bare Costs Equipment	Total	Total Incl O&P
0010	**COIN-OPERATED LAUNDRY EQUIPMENT**									
0990	Dryer, gas fired									
1000	Commercial, 30 lb. capacity, coin operated, single	1 Plum	3	2.667	Ea.	3,500	166		3,666	4,100
1100	Double stacked	"	2	4	"	8,200	249		8,449	9,400
5290	Clothes washer									
5300	Commercial, coin operated, average	1 Plum	3	2.667	Ea.	1,325	166		1,491	1,700

11 21 Retail and Service Equipment

11 21 83 – Photo Processing Equipment

11 21 83.13 Darkroom Equipment

		Crew	Daily Output	Labor-Hours	Unit	Material	2018 Bare Costs Labor	2018 Bare Costs Equipment	Total	Total Incl O&P
0010	**DARKROOM EQUIPMENT**									
0020	Developing sink, 5" deep, 24" x 48"	Q-1	2	8	Ea.	580	445		1,025	1,325
0050	48" x 52"		1.70	9.412		1,400	525		1,925	2,350
0200	10" deep, 24" x 48"		1.70	9.412		1,700	525		2,225	2,675
0250	24" x 108"		1.50	10.667		3,550	595		4,145	4,825
0500	Dryers, dehumidified filtered air, 36" x 25" x 68" high	L-7	6	4.667		2,325	227		2,552	2,925
3000	Viewing lights, 20" x 24"		6	4.667		286	227		513	680
3100	20" x 24" with color correction		6	4.667		415	227		642	820

11 22 Banking Equipment

11 22 13 – Vault Equipment

11 22 13.16 Safes

		Crew	Daily Output	Labor-Hours	Unit	Material	2018 Bare Costs Labor	2018 Bare Costs Equipment	Total	Total Incl O&P
0010	**SAFES**									
0800	Money, "B" label, 9" x 14" x 14"				Ea.	565			565	620
0900	Tool resistive, 24" x 24" x 20"					4,125			4,125	4,550
1050	Tool and torch resistive, 24" x 24" x 20"					8,350			8,350	9,200

11 22 16 – Teller and Service Equipment

11 22 16.13 Teller Equipment Systems

		Crew	Daily Output	Labor-Hours	Unit	Material	2018 Bare Costs Labor	2018 Bare Costs Equipment	Total	Total Incl O&P
0010	**TELLER EQUIPMENT SYSTEMS**									
0020	Alarm system, police	2 Elec	1.60	10	Ea.	4,975	580		5,555	6,375
0400	Bullet resistant teller window, 44" x 60"	1 Glaz	.60	13.333		4,425	645		5,070	5,925
0500	48" x 60"	"	.60	13.333		6,100	645		6,745	7,775
3000	Counters for banks, frontal only	2 Carp	1	16	Station	1,950	810		2,760	3,450
3100	Complete with steel undercounter		.50	32	"	3,625	1,625		5,250	6,625
5400	Partitions, bullet-resistant, 1-3/16" glass, 8' high		10	1.600	L.F.	201	81		282	355
5600	Pass thru, bullet-res. window, painted steel, 24" x 36"	2 Sswk	1.60	10	Ea.	2,925	545		3,470	4,125

11 30 Residential Equipment

11 30 13 – Residential Appliances

11 30 13.15 Cooking Equipment

		Crew	Daily Output	Labor-Hours	Unit	Material	2018 Bare Costs Labor	2018 Bare Costs Equipment	Total	Total Incl O&P
0010	**COOKING EQUIPMENT**									
0020	Cooking range, 30" free standing, 1 oven, minimum	2 Clab	10	1.600	Ea.	470	64		534	620
0050	Maximum	"	4	4		2,300	159		2,459	2,775
0350	Built-in, 30" wide, 1 oven, minimum	1 Elec	6	1.333		800	77.50		877.50	1,000
0400	Maximum	2 Carp	2	8		1,750	405		2,155	2,575
0900	Countertop cooktops, 4 burner, standard, minimum	1 Elec	6	1.333		325	77.50		402.50	480
0950	Maximum		3	2.667		1,750	155		1,905	2,175
1250	Microwave oven, minimum		4	2		97	116		213	287
1300	Maximum		2	4		460	233		693	865
5380	Oven, built-in, standard		4	2		900	116		1,016	1,175
5390	Deluxe		2	4		3,050	233		3,283	3,700

11 30 13.16 Refrigeration Equipment

		Crew	Daily Output	Labor-Hours	Unit	Material	2018 Bare Costs Labor	2018 Bare Costs Equipment	Total	Total Incl O&P
0010	**REFRIGERATION EQUIPMENT**									
5450	Refrigerator, no frost, 6 C.F.	2 Clab	15	1.067	Ea.	370	42.50		412.50	480
5500	Refrigerator, no frost, 10 C.F. to 12 C.F., minimum		10	1.600		440	64		504	590
5600	Maximum		6	2.667		535	106		641	765
6790	Energy-star qualified, 18 C.F., minimum [G]	2 Carp	4	4		535	203		738	920
6795	Maximum [G]		2	8		1,600	405		2,005	2,400

11 30 13 – Residential Appliances

11 30 13.16 Refrigeration Equipment		Crew	Daily Output	Labor-Hours	Unit	Material	2018 Bare Costs Labor	Equipment	Total	Total Incl O&P
6797	21.7 C.F., minimum	G	2 Carp	4	4	Ea.	1,400	203		1,603	1,875
6799	Maximum	G	↓	4	4	↓	1,875	203		2,078	2,400

11 30 13.17 Kitchen Cleaning Equipment

		Crew	Daily Output	Labor-Hours	Unit	Material	Labor	Equipment	Total	Total Incl O&P
0010	**KITCHEN CLEANING EQUIPMENT**									
2750	Dishwasher, built-in, 2 cycles, minimum	L-1	4	4	Ea.	296	241		537	700
2800	Maximum		2	8		455	480		935	1,250
3100	Energy-star qualified, minimum	G	4	4		380	241		621	795
3110	Maximum	G	↓	2	8		1,700	480		2,180	2,625

11 30 13.18 Waste Disposal Equipment

0010	**WASTE DISPOSAL EQUIPMENT**									
1750	Compactor, residential size, 4 to 1 compaction, minimum	1 Carp	5	1.600	Ea.	690	81		771	890
1800	Maximum	"	3	2.667		1,150	135		1,285	1,500
3300	Garbage disposal, sink type, minimum	L-1	10	1.600	.	106	96.50		202.50	265
3350	Maximum	"	10	1.600	↓	204	96.50		300.50	375

11 30 13.19 Kitchen Ventilation Equipment

0010	**KITCHEN VENTILATION EQUIPMENT**									
4150	Hood for range, 2 speed, vented, 30" wide, minimum	L-3	5	3.200	Ea.	93	176		269	380
4200	Maximum		3	5.333		910	293		1,203	1,475
4300	42" wide, minimum		5	3.200		170	176		346	465
4330	Custom		5	3.200		1,650	176		1,826	2,075
4350	Maximum	↓	3	5.333		2,000	293		2,293	2,675
4400	Ventless hood, 2 speed, 30" wide, minimum	1 Elec	8	1		108	58		166	208
4450	Maximum		7	1.143		620	66.50		686.50	785
4510	36" wide, minimum		7	1.143		78	66.50		144.50	189
4550	Maximum		6	1.333		680	77.50		757.50	870
4580	42" wide, minimum		6	1.333		168	77.50		245.50	305
4600	Maximum	↓	5	1.600		625	93		718	835
4650	For vented 1 speed, deduct from maximum				↓	63			63	69

11 30 13.24 Washers

0010	**WASHERS**									
5000	Residential, 4 cycle, average	1 Plum	3	2.667	Ea.	975	166		1,141	1,325
6750	Energy star qualified, front loading, minimum	G	3	2.667		700	166		866	1,025
6760	Maximum	G	1	8		1,800	495		2,295	2,750
6764	Top loading, minimum	G	3	2.667		640	166		806	965
6766	Maximum	G	↓	3	2.667	↓	1,000	166		1,166	1,350

11 30 13.25 Dryers

0010	**DRYERS**									
0500	Gas fired residential, 16 lb. capacity, average	1 Plum	3	2.667	Ea.	715	166		881	1,050
6770	Electric, front loading, energy-star qualified, minimum	G	L-2	3	5.333		405	236		641	830
6780	Maximum	G	"	2	8	↓	1,475	355		1,830	2,175

11 30 15 – Miscellaneous Residential Appliances

11 30 15.13 Sump Pumps

0010	**SUMP PUMPS**									
6400	Cellar drainer, pedestal, 1/3 HP, molded PVC base	1 Plum	3	2.667	Ea.	140	166		306	410
6450	Solid brass	"	2	4	"	240	249		489	650
6460	Sump pump, see also Section 22 14 29.16									

364

For customer support on your Commercial Renovation Costs with RSMeans data, call 800.448.8182.

11 30 Residential Equipment

11 30 33 – Retractable Stairs

11 30 33.10 Disappearing Stairway

		Crew	Daily Output	Labor-Hours	Unit	Material	2018 Bare Costs Labor	2018 Bare Costs Equipment	Total	Total Incl O&P
0010	**DISAPPEARING STAIRWAY** No trim included									
0020	One piece, yellow pine, 8'-0" ceiling	2 Carp	4	4	Ea.	227	203		430	580
0030	9'-0" ceiling		4	4		269	203		472	625
0040	10'-0" ceiling		3	5.333		251	270		521	715
0050	11'-0" ceiling		3	5.333		325	270		595	795
0060	12'-0" ceiling		3	5.333		335	270		605	810
0100	Custom grade, pine, 8'-6" ceiling, minimum	1 Carp	4	2		219	101		320	405
0150	Average		3.50	2.286		243	116		359	455
0200	Maximum		3	2.667		291	135		426	540
0500	Heavy duty, pivoted, from 7'-7" to 12'-10" floor to floor		3	2.667		1,375	135		1,510	1,750
0600	16'-0" ceiling		2	4		1,575	203		1,778	2,050
0800	Economy folding, pine, 8'-6" ceiling		4	2		177	101		278	360
0900	9'-6" ceiling		4	2		205	101		306	390
1100	Automatic electric, aluminum, floor to floor height, 8' to 9'	2 Carp	1	16		9,350	810		10,160	11,600
1400	11' to 12'		.90	17.778		9,925	900		10,825	12,400
1700	14' to 15'		.70	22.857		10,600	1,150		11,750	13,600
9000	Minimum labor/equipment charge	1 Carp	2	4	Job		203		203	330

11 32 Unit Kitchens

11 32 13 – Metal Unit Kitchens

11 32 13.10 Commercial Unit Kitchens

		Crew	Daily Output	Labor-Hours	Unit	Material	2018 Bare Costs Labor	2018 Bare Costs Equipment	Total	Total Incl O&P
0010	**COMMERCIAL UNIT KITCHENS**									
1500	Combination range, refrigerator and sink, 30" wide, minimum	L-1	2	8	Ea.	1,225	480		1,705	2,100
1550	Maximum	"	1	16	"	1,075	965		2,040	2,675
1640	Combination range, refrigerator, sink, microwave									
1660	Oven and ice maker	L-1	.80	20	Ea.	4,825	1,200		6,025	7,175

11 41 Foodservice Storage Equipment

11 41 13 – Refrigerated Food Storage Cases

11 41 13.10 Refrigerated Food Cases

		Crew	Daily Output	Labor-Hours	Unit	Material	2018 Bare Costs Labor	2018 Bare Costs Equipment	Total	Total Incl O&P
0010	**REFRIGERATED FOOD CASES**									
0030	Dairy, multi-deck, 12' long	Q-5	3	5.333	Ea.	10,700	300		11,000	12,300
0100	For rear sliding doors, add					1,925			1,925	2,125
0200	Delicatessen case, service deli, 12' long, single deck	Q-5	3.90	4.103		7,975	233		8,208	9,125
0300	Multi-deck, 18 S.F. shelf display		3	5.333		7,950	300		8,250	9,225
0400	Freezer, self-contained, chest-type, 30 C.F.		3.90	4.103		4,375	233		4,608	5,150
0500	Glass door, upright, 78 C.F.		3.30	4.848		9,250	275		9,525	10,600
0600	Frozen food, chest type, 12' long		3.30	4.848		7,825	275		8,100	9,025
0700	Glass door, reach-in, 5 door		3	5.333		9,800	300		10,100	11,300
0800	Island case, 12' long, single deck		3.30	4.848		7,650	275		7,925	8,825
0900	Multi-deck		3	5.333		8,925	300		9,225	10,300
1000	Meat case, 12' long, single deck		3.30	4.848		7,650	275		7,925	8,825
1050	Multi-deck		3.10	5.161		9,875	293		10,168	11,400
1100	Produce, 12' long, single deck		3.30	4.848		6,550	275		6,825	7,625
1200	Multi-deck		3.10	5.161		8,300	293		8,593	9,575

For customer support on your Commercial Renovation Costs with RSMeans data, call 800.448.8182.

365

11 41 13 – Refrigerated Food Storage Cases

11 41 13.20 Refrigerated Food Storage Equipment	Crew	Daily Output	Labor-Hours	Unit	Material	2018 Bare Costs Labor	Equipment	Total	Total Incl O&P
0010 **REFRIGERATED FOOD STORAGE EQUIPMENT**									
2350 Cooler, reach-in, beverage, 6' long	Q-1	6	2.667	Ea.	3,200	149		3,349	3,750
4300 Freezers, reach-in, 44 C.F.		4	4		4,725	224		4,949	5,550
4500 68 C.F.		3	5.333		5,800	298		6,098	6,850
4600 Freezer, pre-fab, 8' x 8' w/refrigeration	2 Carp	.45	35.556		10,500	1,800		12,300	14,500
4620 8' x 12'		.35	45.714		11,400	2,325		13,725	16,300
4640 8' x 16'		.25	64		13,800	3,250		17,050	20,500
4660 8' x 20'		.17	94.118		19,700	4,775		24,475	29,500
4680 Reach-in, 1 compartment	Q-1	4	4		3,225	224		3,449	3,900
4685 Energy star rated [G]	R-18	7.80	3.333		2,900	170		3,070	3,475
4700 2 compartment	Q-1	3	5.333		4,450	298		4,748	5,350
4705 Energy star rated [G]	R-18	6.20	4.194		3,575	214		3,789	4,275
4710 3 compartment	Q-1	3	5.333		6,150	298		6,448	7,225
4715 Energy star rated [G]	R-18	5.60	4.643		4,925	237		5,162	5,775
8320 Refrigerator, reach-in, 1 compartment		7.80	3.333		2,925	170		3,095	3,475
8325 Energy star rated [G]		7.80	3.333		2,900	170		3,070	3,475
8330 2 compartment		6.20	4.194		4,175	214		4,389	4,925
8335 Energy star rated [G]		6.20	4.194		3,575	214		3,789	4,275
8340 3 compartment		5.60	4.643		5,650	237		5,887	6,575
8345 Energy star rated [G]		5.60	4.643		4,925	237		5,162	5,775
8350 Pre-fab, with refrigeration, 8' x 8'	2 Carp	.45	35.556		6,900	1,800		8,700	10,500
8360 8' x 12'		.35	45.714		8,350	2,325		10,675	13,000
8370 8' x 16'		.25	64		10,400	3,250		13,650	16,800
8380 8' x 20'		.17	94.118		13,200	4,775		17,975	22,300
8390 Pass-thru/roll-in, 1 compartment	R-18	7.80	3.333		3,775	170		3,945	4,425
8400 2 compartment		6.24	4.167		5,325	213		5,538	6,200
8410 3 compartment		5.60	4.643		7,950	237		8,187	9,125
8420 Walk-in, alum, door & floor only, no refrig, 6' x 6' x 7'-6"	2 Carp	1.40	11.429		6,600	580		7,180	8,200
8430 10' x 6' x 7'-6"		.55	29.091		9,725	1,475		11,200	13,100
8440 12' x 14' x 7'-6"		.25	64		13,200	3,250		16,450	19,900
8450 12' x 20' x 7'-6"		.17	94.118		12,200	4,775		16,975	21,200
8460 Refrigerated cabinets, mobile					4,350			4,350	4,775
8470 Refrigerator/freezer, reach-in, 1 compartment	R-18	5.60	4.643		6,450	237		6,687	7,450
8480 2 compartment	"	4.80	5.417		7,175	277		7,452	8,325

11 41 13.30 Wine Cellar

	Crew	Daily Output	Labor-Hours	Unit	Material	Labor	Equipment	Total	Total Incl O&P
0010 **WINE CELLAR**, refrigerated, Redwood interior, carpeted, walk-in type									
0020 6'-8" high, including racks									
0200 80" W x 48" D for 900 bottles	2 Carp	1.50	10.667	Ea.	4,200	540		4,740	5,500
0250 80" W x 72" D for 1300 bottles		1.33	12.030		5,125	610		5,735	6,625
0300 80" W x 94" D for 1900 bottles		1.17	13.675		6,200	695		6,895	7,950
0400 80" W x 124" D for 2500 bottles		1	16		7,300	810		8,110	9,350

11 41 33 – Foodservice Shelving

11 41 33.20 Metal Food Storage Shelving

	Crew	Daily Output	Labor-Hours	Unit	Material	Labor	Equipment	Total	Total Incl O&P
0010 **METAL FOOD STORAGE SHELVING**									
8600 Stainless steel shelving, louvered 4-tier, 20" x 3'	1 Clab	6	1.333	Ea.	1,450	53		1,503	1,675
8605 20" x 4'		6	1.333		1,650	53		1,703	1,875
8610 20" x 6'		6	1.333		1,725	53		1,778	1,975
8615 24" x 3'		6	1.333		2,050	53		2,103	2,350
8620 24" x 4'		6	1.333		2,450	53		2,503	2,775
8625 24" x 6'		6	1.333		3,400	53		3,453	3,825
8630 Flat 4-tier, 20" x 3'		6	1.333		1,175	53		1,228	1,375

11 41 Foodservice Storage Equipment

11 41 33 – Foodservice Shelving

11 41 33.20 Metal Food Storage Shelving

		Crew	Daily Output	Labor-Hours	Unit	Material	2018 Bare Costs Labor	Equipment	Total	Total Incl O&P
8635	20" x 4'	1 Clab	6	1.333	Ea.	1,400	53		1,453	1,625
8640	20" x 5'		6	1.333		1,600	53		1,653	1,825
8645	24" x 3'		6	1.333		905	53		958	1,075
8650	24" x 4'		6	1.333		2,300	53		2,353	2,600
8655	24" x 6'		6	1.333		2,750	53		2,803	3,100
8700	Galvanized shelving, louvered 4-tier, 20" x 3'		6	1.333		790	53		843	955
8705	20" x 4'		6	1.333		895	53		948	1,075
8710	20" x 6'		6	1.333		950	53		1,003	1,125
8715	24" x 3'		6	1.333		725	53		778	880
8720	24" x 4'		6	1.333		955	53		1,008	1,125
8725	24" x 6'		6	1.333		1,375	53		1,428	1,575
8730	Flat 4-tier, 20" x 3'		6	1.333		730	53		783	890
8735	20" x 4'		6	1.333		640	53		693	790
8740	20" x 6'		6	1.333		950	53		1,003	1,125
8745	24" x 3'		6	1.333		715	53		768	870
8750	24" x 4'		6	1.333		550	53		603	690
8755	24" x 6'		6	1.333		850	53		903	1,025
8760	Stainless steel dunnage rack, 24" x 3'		8	1		350	40		390	450
8765	24" x 4'		8	1		445	40		485	555
8770	Galvanized dunnage rack, 24" x 3'		8	1		131	40		171	209
8775	24" x 4'		8	1		192	40		232	276

11 42 Food Preparation Equipment

11 42 10 – Commercial Food Preparation Equipment

11 42 10.10 Choppers, Mixers and Misc. Equipment

		Crew	Daily Output	Labor-Hours	Unit	Material	2018 Bare Costs Labor	Equipment	Total	Total Incl O&P
0010	**CHOPPERS, MIXERS AND MISC. EQUIPMENT**									
1700	Choppers, 5 pounds	R-18	7	3.714	Ea.	2,725	190		2,915	3,300
1720	16 pounds		5	5.200		2,525	266		2,791	3,175
1740	35 to 40 pounds		4	6.500		3,575	330		3,905	4,450
1840	Coffee brewer, 5 burners	1 Plum	3	2.667		915	166		1,081	1,250
1850	Coffee urn, twin 6 gallon urns		2	4		2,300	249		2,549	2,900
1860	Single, 3 gallon		3	2.667		1,600	166		1,766	2,025
3000	Fast food equipment, total package, minimum	6 Skwk	.08	600		211,500	31,400		242,900	283,500
3100	Maximum	"	.07	686		288,500	35,900		324,400	375,000
3800	Food mixers, bench type, 20 quarts	L-7	7	4		3,175	195		3,370	3,825
3850	40 quarts		5.40	5.185		9,450	252		9,702	10,800
3900	60 quarts		5	5.600		12,000	273		12,273	13,600
4040	80 quarts		3.90	7.179		18,900	350		19,250	21,300
4100	Floor type, 20 quarts		15	1.867		3,625	91		3,716	4,150
4120	60 quarts		14	2		10,700	97.50		10,797.50	12,000
4140	80 quarts		12	2.333		17,400	114		17,514	19,400
4160	140 quarts		8.60	3.256		26,700	158		26,858	29,700
6700	Peelers, small	R-18	8	3.250		1,825	166		1,991	2,250
6720	Large	"	6	4.333		3,150	221		3,371	3,825
6800	Pulper/extractor, close coupled, 5 HP	1 Plum	1.90	4.211		2,475	262		2,737	3,125
8580	Slicer with table	R-18	9	2.889		3,925	148		4,073	4,550

11 43 13 – Food Delivery Carts

11 43 13.10 Mobile Carts, Racks and Trays	Crew	Daily Output	Labor-Hours	Unit	Material	2018 Bare Costs Labor	Equipment	Total	Total Incl O&P
0010 **MOBILE CARTS, RACKS AND TRAYS**									
1650 Cabinet, heated, 1 compartment, reach-in	R-18	5.60	4.643	Ea.	2,850	237		3,087	3,525
1655 Pass-thru roll-in		5.60	4.643		3,575	237		3,812	4,300
1660 2 compartment, reach-in		4.80	5.417		6,250	277		6,527	7,300
1670 Mobile					3,725			3,725	4,100
2000 Hospital food cart, hot and cold service, 20 tray capacity					15,900			15,900	17,500
6850 Mobile rack w/pan slide					1,225			1,225	1,350
9180 Tray and silver dispenser, mobile	1 Clab	16	.500		650	19.95		669.95	750

11 44 Food Cooking Equipment

11 44 13 – Commercial Ranges

11 44 13.10 Cooking Equipment	Crew	Daily Output	Labor-Hours	Unit	Material	2018 Bare Costs Labor	Equipment	Total	Total Incl O&P
0010 **COOKING EQUIPMENT**									
0020 Bake oven, gas, one section	Q-1	8	2	Ea.	5,725	112		5,837	6,475
0300 Two sections		7	2.286		9,425	128		9,553	10,600
0600 Three sections		6	2.667		12,000	149		12,149	13,400
0900 Electric convection, single deck	L-7	4	7		5,525	340		5,865	6,625
1300 Broiler, without oven, standard	Q-1	8	2		3,875	112		3,987	4,425
1550 Infrared	L-7	4	7		7,425	340		7,765	8,725
4750 Fryer, with twin baskets, modular model	Q-1	7	2.286		1,375	128		1,503	1,725
5000 Floor model, on 6" legs	"	5	3.200		2,575	179		2,754	3,100
5100 Extra single basket, large					50			50	55
5170 Energy star rated, 50 lb. capacity G	R-18	4	6.500		4,750	330		5,080	5,750
5175 85 lb. capacity G	"	4	6.500		9,900	330		10,230	11,400
5300 Griddle, SS, 24" plate, w/4" legs, elec, 208 V, 3 phase, 3' long	Q-1	7	2.286		2,200	128		2,328	2,600
5550 4' long	"	6	2.667		2,025	149		2,174	2,450
6200 Iced tea brewer	1 Plum	3.44	2.326		630	145		775	915
6350 Kettle, w/steam jacket, tilting, w/positive lock, SS, 20 gallons	L-7	7	4		8,700	195		8,895	9,900
6600 60 gallons	"	6	4.667		20,000	227		20,227	22,400
6900 Range, restaurant type, 6 burners and 1 standard oven, 36" wide	Q-1	7	2.286		3,075	128		3,203	3,575
6950 Convection		7	2.286		4,550	128		4,678	5,225
7150 2 standard ovens, 24" griddle, 60" wide		6	2.667		5,325	149		5,474	6,100
7200 1 standard, 1 convection oven		6	2.667		9,500	149		9,649	10,600
7450 Heavy duty, single 34" standard oven, open top		5	3.200		5,550	179		5,729	6,375
7500 Convection oven		5	3.200		6,000	179		6,179	6,875
7700 Griddle top		6	2.667		2,175	149		2,324	2,625
7750 Convection oven		6	2.667		3,175	149		3,324	3,725
7760 Induction cooker, electric	L-7	7	4		1,825	195		2,020	2,325
8850 Steamer, electric 27 KW		7	4		11,800	195		11,995	13,200
9100 Electric, 10 KW or gas 100,000 BTU		5	5.600		7,200	273		7,473	8,350
9150 Toaster, conveyor type, 16-22 slices/minute					1,250			1,250	1,375
9160 Pop-up, 2 slot					645			645	705
9200 For deluxe models of above equipment, add					75%				
9400 Rule of thumb: Equipment cost based									
9410 on kitchen work area									
9420 Office buildings, minimum	L-7	77	.364	S.F.	94	17.70		111.70	133
9450 Maximum		58	.483		159	23.50		182.50	213
9550 Public eating facilities, minimum		77	.364		123	17.70		140.70	165
9600 Maximum		46	.609		201	29.50		230.50	269
9750 Hospitals, minimum		58	.483		127	23.50		150.50	178
9800 Maximum		39	.718		234	35		269	315

11 46 Food Dispensing Equipment

11 46 16 – Service Line Equipment

11 46 16.10 Commercial Food Dispensing Equipment

		Crew	Daily Output	Labor-Hours	Unit	Material	2018 Bare Costs Labor	Equipment	Total	Total Incl O&P
0010	**COMMERCIAL FOOD DISPENSING EQUIPMENT**									
1050	Butter pat dispenser	1 Clab	13	.615	Ea.	835	24.50		859.50	955
1100	Bread dispenser, counter top		13	.615		750	24.50		774.50	865
1900	Cup and glass dispenser, drop in		4	2		565	79.50		644.50	755
1920	Disposable cup, drop in		16	.500		655	19.95		674.95	760
2650	Dish dispenser, drop in, 12"		11	.727		2,575	29		2,604	2,875
2660	Mobile		10	.800		2,250	32		2,282	2,525
3300	Food warmer, counter, 1.2 KW					720			720	790
3550	1.6 KW					2,200			2,200	2,425
3600	Well, hot food, built-in, rectangular, 12" x 20"	R-30	10	2.600		740	122		862	1,000
3610	Circular, 7 qt.		10	2.600		405	122		527	640
3620	Refrigerated, 2 compartments		10	2.600		3,200	122		3,322	3,725
3630	3 compartments		9	2.889		3,950	136		4,086	4,575
3640	4 compartments		8	3.250		4,600	153		4,753	5,300
4720	Frost cold plate		9	2.889		21,400	136		21,536	23,700
5700	Hot chocolate dispenser	1 Plum	4	2		1,150	124		1,274	1,475
5750	Ice dispenser 567 pound	Q-1	6	2.667		5,450	149		5,599	6,225
6250	Jet spray dispenser	R-18	4.50	5.778		1,850	295		2,145	2,475
6300	Juice dispenser, concentrate	"	4.50	5.778		1,925	295		2,220	2,575
6690	Milk dispenser, bulk, 2 flavor	R-30	8	3.250		1,850	153		2,003	2,300
6695	3 flavor	"	8	3.250		2,375	153		2,528	2,875
8800	Serving counter, straight	1 Carp	40	.200	L.F.	1,425	10.15		1,435.15	1,575
8820	Curved section	"	30	.267	"	2,100	13.50		2,113.50	2,350
8825	Solid surface, see Section 12 36 61.16									
8860	Sneeze guard with lights, 60" L	1 Clab	16	.500	Ea.	345	19.95		364.95	415
8900	Sneeze guard, stainless steel and glass, single sided									
8910	Portable, 48" W				Ea.	435			435	480
8920	Portable, 72" W					415			415	460
8930	Adjustable, 36" W	1 Carp	24	.333		250	16.90		266.90	305
8940	Adjustable, 48" W	"	20	.400		315	20.50		335.50	385
9100	Soft serve ice cream machine, medium	R-18	11	2.364		7,700	121		7,821	8,650
9110	Large	"	9	2.889		21,900	148		22,048	24,200

11 46 83 – Ice Machines

11 46 83.10 Commercial Ice Equipment

		Crew	Daily Output	Labor-Hours	Unit	Material	2018 Bare Costs Labor	Equipment	Total	Total Incl O&P
0010	**COMMERCIAL ICE EQUIPMENT**									
5800	Ice cube maker, 50 lbs./day	Q-1	6	2.667	Ea.	1,700	149		1,849	2,100
5810	65 lbs./day, energy star rated		6	2.667		1,625	149		1,774	2,000
5900	250 lbs./day		1.20	13.333		2,525	745		3,270	3,925
5950	300 lbs./day, remote condensing		1.20	13.333		2,400	745		3,145	3,775
6050	500 lbs./day		4	4		2,850	224		3,074	3,475
6060	With bin		1.20	13.333		3,975	745		4,720	5,525
6070	Modular, with bin and condenser		1.20	13.333		4,025	745		4,770	5,575
6090	1000 lbs./day, with bin		1	16		5,150	895		6,045	7,075
6100	Ice flakers, 300 lbs./day		1.60	10		3,300	560		3,860	4,500
6120	600 lbs./day		.95	16.842		3,975	940		4,915	5,850
6130	1000 lbs./day		.75	21.333		4,875	1,200		6,075	7,225
6140	2000 lbs./day		.65	24.615		22,400	1,375		23,775	26,900
6160	Ice storage bin, 500 pound capacity	Q-5	1	16		1,125	905		2,030	2,650
6180	1000 pound	"	.56	28.571		2,900	1,625		4,525	5,725

For customer support on your Commercial Renovation Costs with RSMeans data, call 800.448.8182.

11 48 Foodservice Cleaning and Disposal Equipment

11 48 13 – Commercial Dishwashers

11 48 13.10 Dishwashers	Crew	Daily Output	Labor-Hours	Unit	Material	2018 Bare Costs Labor	Equipment	Total	Total Incl O&P
0010 **DISHWASHERS**									
2700 Dishwasher, commercial, rack type									
2720 10 to 12 racks/hour	Q-1	3.20	5	Ea.	3,400	280		3,680	4,150
2730 Energy star rated, 35 to 40 racks/hour [G]		1.30	12.308		4,250	690		4,940	5,750
2740 50 to 60 racks/hour [G]	↓	1.30	12.308		11,100	690		11,790	13,300
2800 Automatic, 190 to 230 racks/hour	L-6	.35	34.286		12,600	2,075		14,675	17,100
2820 235 to 275 racks/hour		.25	48		26,200	2,925		29,125	33,300
2840 8,750 to 12,500 dishes/hour	↓	.10	120	↓	47,200	7,300		54,500	63,500
2950 Dishwasher hood, canopy type	L-3A	10	1.200	L.F.	1,025	66		1,091	1,250
2960 Pant leg type	"	2.50	4.800	Ea.	7,925	264		8,189	9,150
5200 Garbage disposal 1.5 HP, 100 GPH	L-1	4.80	3.333		1,500	201		1,701	1,950
5210 3 HP, 120 GPH		4.60	3.478		2,125	209		2,334	2,675
5220 5 HP, 250 GPH	↓	4.50	3.556	↓	2,825	214		3,039	3,425
6750 Pot sink, 3 compartment	1 Plum	7.25	1.103	L.F.	970	68.50		1,038.50	1,175
6760 Pot washer, low temp wash/rinse		1.60	5	Ea.	5,475	310		5,785	6,500
6770 High pressure wash, high temperature rinse		1.20	6.667		38,000	415		38,415	42,400
9170 Trash compactor, small, up to 125 lb. compacted weight	L-4	4	6		24,300	285		24,585	27,300
9175 Large, up to 175 lb. compacted weight	"	3	8	↓	28,900	380		29,280	32,400

11 52 Audio-Visual Equipment

11 52 13 – Projection Screens

11 52 13.10 Projection Screens, Wall or Ceiling Hung

	Crew	Daily Output	Labor-Hours	Unit	Material	2018 Bare Costs Labor	Equipment	Total	Total Incl O&P
0010 **PROJECTION SCREENS, WALL OR CEILING HUNG**, matte white									
0100 Manually operated, economy	2 Carp	500	.032	S.F.	6.45	1.62		8.07	9.70
0300 Intermediate		450	.036		7.85	1.80		9.65	11.60
0400 Deluxe		400	.040	↓	9.75	2.03		11.78	14.05
9000 Minimum labor/equipment charge	↓	3	5.333	Job		270		270	440

11 52 16 – Projectors

11 52 16.10 Movie Equipment

	Crew	Daily Output	Labor-Hours	Unit	Material	2018 Bare Costs Labor	Equipment	Total	Total Incl O&P
0011 **MOVIE EQUIPMENT**									
0020 Changeover, minimum				Ea.	515			515	565
3000 Projection screens, rigid, in wall, acrylic, 1/4" thick	2 Glaz	195	.082	S.F.	46.50	3.98		50.48	57.50
3100 1/2" thick	"	130	.123	"	54	5.95		59.95	69
3700 Sound systems, incl. amplifier, mono, minimum	1 Elec	.90	8.889	Ea.	3,725	515		4,240	4,900
3800 Dolby/Super Sound, maximum		.40	20		20,100	1,175		21,275	23,900
4100 Dual system, 2 channel, front surround, minimum		.70	11.429		5,150	665		5,815	6,700
4200 Dolby/Super Sound, 4 channel, maximum	↓	.40	20	↓	18,800	1,175		19,975	22,400
5700 Seating, painted steel, upholstered, minimum	2 Carp	35	.457		152	23		175	205
5800 Maximum	"	28	.571	↓	500	29		529	595

11 53 Laboratory Equipment

11 53 33 – Emergency Safety Appliances

11 53 33.13 Emergency Equipment

		Crew	Daily Output	Labor-Hours	Unit	Material	2018 Bare Costs Labor	Equipment	Total	Total Incl O&P
0010	**EMERGENCY EQUIPMENT**									
1400	Safety equipment, eye wash, hand held				Ea.	415			415	455
1450	Deluge shower				"	820			820	900

11 53 43 – Service Fittings and Accessories

11 53 43.13 Fittings

		Crew	Daily Output	Labor-Hours	Unit	Material	2018 Bare Costs Labor	Equipment	Total	Total Incl O&P
0010	**FITTINGS**									
1600	Sink, one piece plastic, flask wash, hose, free standing	1 Plum	1.60	5	Ea.	2,025	310		2,335	2,700
1610	Epoxy resin sink, 25" x 16" x 10"	"	2	4	"	229	249		478	635
1630	Steel table, open underneath				L.F.	545			545	600
1640	Doors underneath				"	875			875	960
1900	Utility cabinet, stainless steel				Ea.	1,600			1,600	1,775
1950	Utility table, acid resistant top with drawers	2 Carp	30	.533	L.F.	171	27		198	232
8000	Alternate pricing method: as percent of lab furniture									
8050	Installation, not incl. plumbing & duct work				% Furn.				22%	22%
8100	Plumbing, final connections, simple system								10%	10%
8110	Moderately complex system								15%	15%
8120	Complex system								20%	20%
8150	Electrical, simple system								10%	10%
8160	Moderately complex system								20%	20%
8170	Complex system								35%	35%

11 53 53 – Biological Safety Cabinets

11 53 53.10 Pharmacy Cabinets

		Crew	Daily Output	Labor-Hours	Unit	Material	2018 Bare Costs Labor	Equipment	Total	Total Incl O&P
0010	**PHARMACY CABINETS**, vertical flow									
0100	Class II, type B2, 6' L	2 Carp	1.50	10.667	Ea.	14,700	540		15,240	17,100

11 57 Vocational Shop Equipment

11 57 10 – Shop Equipment

11 57 10.10 Vocational School Shop Equipment

		Crew	Daily Output	Labor-Hours	Unit	Material	2018 Bare Costs Labor	Equipment	Total	Total Incl O&P
0010	**VOCATIONAL SCHOOL SHOP EQUIPMENT**									
0020	Benches, work, wood, average	2 Carp	5	3.200	Ea.	715	162		877	1,050
0100	Metal, average		5	3.200		450	162		612	760
0400	Combination belt & disc sander, 6"		4	4		1,700	203		1,903	2,200
0700	Drill press, floor mounted, 12", 1/2 HP		4	4		400	203		603	770
0800	Dust collector, not incl. ductwork, 6" diameter	1 Shee	1.10	7.273		5,250	435		5,685	6,450
0810	Dust collector bag, 20" diameter	"	5	1.600		585	95.50		680.50	790
1000	Grinders, double wheel, 1/2 HP	2 Carp	5	3.200		215	162		377	500
1300	Jointer, 4", 3/4 HP		4	4		1,325	203		1,528	1,800
1600	Kilns, 16 C.F., to 2000°		4	4		1,575	203		1,778	2,050
1900	Lathe, woodworking, 10", 1/2 HP		4	4		575	203		778	960
2200	Planer, 13" x 6"		4	4		1,175	203		1,378	1,625
2500	Potter's wheel, motorized		4	4		1,200	203		1,403	1,625
2800	Saws, band, 14", 3/4 HP		4	4		1,100	203		1,303	1,550
3100	Metal cutting band saw, 14"		4	4		2,475	203		2,678	3,050
3400	Radial arm saw, 10", 2 HP		4	4		1,275	203		1,478	1,750
3700	Scroll saw, 24"		4	4		625	203		828	1,025
4000	Table saw, 10", 3 HP		4	4		2,900	203		3,103	3,525
4300	Welder AC arc, 30 amp capacity		4	4		3,175	203		3,378	3,825

For customer support on your Commercial Renovation Costs with RSMeans data, call 800.448.8182.

371

11 61 Broadcast, Theater, and Stage Equipment

11 61 23 – Folding and Portable Stages

11 61 23.10 Portable Stages	Crew	Daily Output	Labor-Hours	Unit	Material	2018 Bare Costs Labor	Equipment	Total	Total Incl O&P
0010 **PORTABLE STAGES**									
5000 Stages, portable with steps, folding legs, stock, 8" high				SF Stg.	41			41	45.50
5100 16" high					59.50			59.50	65.50
5200 32" high					61			61	67
5300 40" high					62.50			62.50	69

11 61 33 – Rigging Systems and Controls

11 61 33.10 Controls

	Crew	Daily Output	Labor-Hours	Unit	Material	Labor	Equipment	Total	Total Incl O&P
0010 **CONTROLS**									
0050 Control boards with dimmers and breakers, minimum	1 Elec	1	8	Ea.	14,600	465		15,065	16,800
0150 Maximum	"	.20	40	"	140,000	2,325		142,325	157,500

11 61 43 – Stage Curtains

11 61 43.10 Curtains

	Crew	Daily Output	Labor-Hours	Unit	Material	Labor	Equipment	Total	Total Incl O&P
0010 **CURTAINS**									
0500 Curtain track, straight, light duty	2 Carp	20	.800	L.F.	30.50	40.50		71	99.50
0700 Curved sections		12	1.333	"	196	67.50		263.50	325
1000 Curtains, velour, medium weight		600	.027	S.F.	8.60	1.35		9.95	11.65

11 66 Athletic Equipment

11 66 13 – Exercise Equipment

11 66 13.10 Physical Training Equipment

	Crew	Daily Output	Labor-Hours	Unit	Material	Labor	Equipment	Total	Total Incl O&P
0010 **PHYSICAL TRAINING EQUIPMENT**									
0020 Abdominal rack, 2 board capacity				Ea.	520			520	575
0050 Abdominal board, upholstered					735			735	805
0200 Bicycle trainer, minimum					520			520	570
0300 Deluxe, electric					4,875			4,875	5,350
0400 Barbell set, chrome plated steel, 25 lb.					291			291	320
0420 100 lb.					585			585	645
0450 200 lb.					775			775	850
0500 Weight plates, cast iron, per lb.				Lb.	3.62			3.62	3.98
0520 Storage rack, 10 station				Ea.	1,025			1,025	1,125
0600 Circuit training apparatus, 12 machines minimum	2 Clab	1.25	12.800	Set	32,500	510		33,010	36,500
0700 Average		1	16		40,500	640		41,140	45,600
0800 Maximum		.75	21.333		48,600	850		49,450	55,000
0820 Dumbbell set, cast iron, with rack and 5 pair					445			445	490
0900 Squat racks	2 Clab	5	3.200	Ea.	900	128		1,028	1,200
4150 Exercise equipment, bicycle trainer					965			965	1,075
4180 Chinning bar, adjustable, wall mounted	1 Carp	5	1.600		218	81		299	370
4200 Exercise ladder, 16' x 1'-7", suspended	L-2	3	5.333		1,425	236		1,661	1,950
4210 High bar, floor plate attached	1 Carp	4	2		2,525	101		2,626	2,950
4240 Parallel bars, adjustable		4	2		1,650	101		1,751	2,000
4270 Uneven parallel bars, adjustable		4	2		3,725	101		3,826	4,275
4280 Wall mounted, adjustable	L-2	1.50	10.667	Set	975	470		1,445	1,850
4300 Rope, ceiling mounted, 18' long	1 Carp	3.66	2.186	Ea.	201	111		312	400
4330 Side horse, vaulting		5	1.600		1,475	81		1,556	1,725
4360 Treadmill, motorized, deluxe, training type		5	1.600		3,925	81		4,006	4,425
4390 Weight lifting multi-station, minimum	2 Clab	1	16		280	640		920	1,325

11 66 Athletic Equipment

11 66 23 – Gymnasium Equipment

11 66 23.13 Basketball Equipment	Crew	Daily Output	Labor-Hours	Unit	Material	2018 Bare Costs Labor	Equipment	Total	Total Incl O&P
0010 **BASKETBALL EQUIPMENT**									
1000 Backstops, wall mtd., 6' extended, fixed, minimum	L-2	1	16	Ea.	1,600	710		2,310	2,900
1100 Maximum		1	16		1,875	710		2,585	3,225
1200 Swing up, minimum		1	16		1,750	710		2,460	3,075
1250 Maximum		1	16		3,050	710		3,760	4,500
1300 Portable, manual, heavy duty, spring operated		1.90	8.421		13,100	375		13,475	15,100
1400 Ceiling suspended, stationary, minimum		.78	20.513		4,275	910		5,185	6,175
1450 Fold up, with accessories, maximum		.40	40		6,350	1,775		8,125	9,900
1600 For electrically operated, add	1 Elec	1	8		2,550	465		3,015	3,550
5800 Wall pads, 1-1/2" thick, standard (not fire rated)	2 Carp	640	.025	S.F.	6.25	1.27		7.52	8.95

11 66 23.19 Boxing Ring

	Crew	Daily Output	Labor-Hours	Unit	Material	Labor	Equipment	Total	Total Incl O&P
0010 **BOXING RING**									
4100 Elevated, 22' x 22'	L-4	.10	240	Ea.	5,725	11,400		17,125	24,800
4110 For cellular plastic foam padding, add		.10	240		1,225	11,400		12,625	19,900
4120 Floor level, including posts and ropes only, 20' x 20'		.80	30		4,175	1,425		5,600	6,900
4130 Canvas, 30' x 30'		5	4.800		1,525	228		1,753	2,050

11 66 23.47 Gym Mats

	Crew	Daily Output	Labor-Hours	Unit	Material	Labor	Equipment	Total	Total Incl O&P
0010 **GYM MATS**									
5500 2" thick, naugahyde covered				S.F.	5.20			5.20	5.70
5600 Vinyl/nylon covered					8.30			8.30	9.10
6000 Wrestling mats, 1" thick, heavy duty					5			5	5.50

11 66 43 – Interior Scoreboards

11 66 43.10 Scoreboards

	Crew	Daily Output	Labor-Hours	Unit	Material	Labor	Equipment	Total	Total Incl O&P
0010 **SCOREBOARDS**									
7000 Baseball, minimum	R-3	1.30	15.385	Ea.	3,700	890	99.50	4,689.50	5,550
7200 Maximum		.05	400		20,700	23,200	2,600	46,500	61,500
7300 Football, minimum		.86	23.256		5,125	1,350	151	6,626	7,925
7400 Maximum		.20	100		19,000	5,800	650	25,450	30,600
7500 Basketball (one side), minimum		2.07	9.662		2,475	560	62.50	3,097.50	3,675
7600 Maximum		.30	66.667		7,550	3,875	430	11,855	14,800
7700 Hockey-basketball (four sides), minimum		.25	80		9,300	4,650	520	14,470	18,000
7800 Maximum		.15	133		16,200	7,725	865	24,790	30,800

11 66 53 – Gymnasium Dividers

11 66 53.10 Divider Curtains

	Crew	Daily Output	Labor-Hours	Unit	Material	Labor	Equipment	Total	Total Incl O&P
0010 **DIVIDER CURTAINS**									
4500 Gym divider curtain, mesh top, vinyl bottom, manual	L-4	500	.048	S.F.	10.25	2.28		12.53	15
4700 Electric roll up	L-7	400	.070	"	13.60	3.41		17.01	20.50

11 71 Medical Sterilizing Equipment

11 71 10 – Medical Sterilizers & Distillers

11 71 10.10 Sterilizers and Distillers

	Crew	Daily Output	Labor-Hours	Unit	Material	Labor	Equipment	Total	Total Incl O&P
0010 **STERILIZERS AND DISTILLERS**									
3010 Portable, top loading, 105-135 degree C, 3 to 30 psi, 50 L chamber				Ea.	9,625			9,625	10,600
3020 Stainless steel basket, 10.7" diam. x 11.8" H					212			212	233
3025 Stainless steel pail, 10.7" diam. x 10.7" H					315			315	345
3050 85 L chamber					17,300			17,300	19,000
3060 Stainless steel basket, 15.3" diam. x 11.5" H					445			445	485
3065 Stainless steel pail, 15.3" diam. x 11" H					665			665	730

For customer support on your Commercial Renovation Costs with RSMeans data, call 800.448.8182.

373

11 76 Operating Room Equipment

11 76 10 – Operating Room Equipment

11 76 10.10 Surgical Equipment		Crew	Daily Output	Labor-Hours	Unit	Material	2018 Bare Costs Labor	Equipment	Total	Total Incl O&P
0010	**SURGICAL EQUIPMENT**									
6550	Major surgery table, minimum	1 Sswk	.50	16	Ea.	25,100	875		25,975	29,100
6570	Maximum		.50	16		35,200	875		36,075	40,200
6600	Hydraulic, hand-held control, general surgery		.60	13.333		37,900	730		38,630	43,000
6650	Stationary, universal		.50	16		49,300	875		50,175	56,000
6800	Surgical lights, major operating room, dual head, minimum	2 Elec	1	16		5,050	930		5,980	7,025
6850	Maximum		1	16		37,500	930		38,430	42,700
6900	Ceiling mount articulation, single arm		1	16		4,650	930		5,580	6,575

11 81 Facility Maintenance Equipment

11 81 19 – Vacuum Cleaning Systems

11 81 19.10 Vacuum Cleaning		Crew	Daily Output	Labor-Hours	Unit	Material	2018 Bare Costs Labor	Equipment	Total	Total Incl O&P
0010	**VACUUM CLEANING**									
0020	Central, 3 inlet, residential	1 Skwk	.90	8.889	Total	1,150	465		1,615	2,025
0200	Commercial		.70	11.429		1,350	600		1,950	2,475
0400	5 inlet system, residential		.50	16		1,675	840		2,515	3,175
0600	7 inlet system, commercial		.40	20		1,875	1,050		2,925	3,750
4010	Rule of thumb: First 1200 S.F., installed								1,425	1,575
4020	For each additional S.F., add				S.F.				.26	.26

11 82 Facility Solid Waste Handling Equipment

11 82 26 – Facility Waste Compactors

11 82 26.10 Compactors		Crew	Daily Output	Labor-Hours	Unit	Material	2018 Bare Costs Labor	Equipment	Total	Total Incl O&P
0010	**COMPACTORS**									
0020	Compactors, 115 volt, 250 lbs./hr., chute fed	L-4	1	24	Ea.	12,500	1,150		13,650	15,600
0100	Hand fed		2.40	10		16,000	475		16,475	18,400
1000	Heavy duty industrial compactor, 0.5 C.Y. capacity		1	24		10,700	1,150		11,850	13,700
1050	1.0 C.Y. capacity		1	24		16,400	1,150		17,550	19,900
1400	For handling hazardous waste materials, 55 gallon drum packer, std.					20,900			20,900	23,000
1410	55 gallon drum packer w/HEPA filter					26,100			26,100	28,700
1420	55 gallon drum packer w/charcoal & HEPA filter					34,800			34,800	38,200
1430	All of the above made explosion proof, add					1,525			1,525	1,675

11 82 39 – Medical Waste Disposal Systems

11 82 39.10 Off-Site Disposal		Crew	Daily Output	Labor-Hours	Unit	Material	2018 Bare Costs Labor	Equipment	Total	Total Incl O&P
0010	**OFF-SITE DISPOSAL**									
0100	Medical waste disposal, Red Bag system, pick up & treat, 200 lbs./week				Week	190			190	209
0110	Per month				Month	730			730	800
0150	Red bags, 7-10 gal., 1.2 mil, pkg of 500				Ea.	70			70	77
0200	15 gal., package of 250					66			66	72.50
0250	33 gal., package of 250					69			69	76
0300	45 gal., package of 100					58			58	64

11 82 39.20 Disposal Carts		Crew	Daily Output	Labor-Hours	Unit	Material	2018 Bare Costs Labor	Equipment	Total	Total Incl O&P
0010	**DISPOSAL CARTS**									
2010	Medical waste disposal cart, HDPE, w/lid, 28 gal. capacity				Ea.	278			278	305
2020	96 gal. capacity					320			320	355
2030	150 gal. capacity, low profile					500			500	550
2040	200 gal. capacity					965			965	1,050

11 82 Facility Solid Waste Handling Equipment

11 82 39 – Medical Waste Disposal Systems

11 82 39.30 Medical Waste Sanitizers	Crew	Daily Output	Labor-Hours	Unit	Material	2018 Bare Costs Labor	Equipment	Total	Total Incl O&P
0010 **MEDICAL WASTE SANITIZERS**									
2010 Small, hand loaded, 1.5 C.Y., 225 lb. capacity				Ea.	83,000			83,000	91,000
2020 Medium, cart loaded, 6.25 C.Y., 938 lb. capacity					110,500			110,500	121,500
2030 Large, cart loaded, 15 C.Y., 2250 lb. capacity					131,500			131,500	144,500
3010 Cart, aluminum, 75 lb. capacity					2,175			2,175	2,400
3020 95 lb. capacity					2,375			2,375	2,600
4010 Stainless steel, 173 lb. capacity					3,100			3,100	3,425
4020 232 lb. capacity					3,475			3,475	3,825
4030 Cart lift, hydraulic scissor type					6,150			6,150	6,775
4040 Portable aluminum ramp					1,725			1,725	1,900
4050 Fold-down steel tracks					1,400			1,400	1,550
4060 Pull-out drawer, small					6,625			6,625	7,300
4070 Medium					9,575			9,575	10,500
4080 Large					13,500			13,500	14,900
5000 Medical waste treatment, sanitize, on-site									
5010 Less than 15,000 lbs./month				Lb.	.20			.20	.22
5020 Over 15,000 lbs./month				"	.16			.16	.18

11 91 Religious Equipment

11 91 13 – Baptisteries

11 91 13.10 Baptistry

	Crew	Daily Output	Labor-Hours	Unit	Material	2018 Bare Costs Labor	Equipment	Total	Total Incl O&P
0010 **BAPTISTRY**									
0150 Fiberglass, 3'-6" deep, x 13'-7" long,									
0160 steps at both ends, incl. plumbing, minimum	L-8	1	20	Ea.	5,550	1,050		6,600	7,800
0200 Maximum	"	.70	28.571		9,325	1,525		10,850	12,700
0250 Add for filter, heater and lights					1,850			1,850	2,025

11 91 23 – Sanctuary Equipment

11 91 23.10 Sanctuary Furnishings

	Crew	Daily Output	Labor-Hours	Unit	Material	2018 Bare Costs Labor	Equipment	Total	Total Incl O&P
0010 **SANCTUARY FURNISHINGS**									
0020 Altar, wood, custom design, plain	1 Carp	1.40	5.714	Ea.	2,575	290		2,865	3,325
0050 Deluxe		.20	40	"	12,400	2,025		14,425	17,000
5000 Wall cross, aluminum, extruded, 2" x 2" section		34	.235	L.F.	242	11.95		253.95	285
5150 4" x 4" section		29	.276		350	14		364	410
5300 Bronze, extruded, 1" x 2" section		31	.258		475	13.10		488.10	545
5350 2-1/2" x 2-1/2" section		34	.235		720	11.95		731.95	815

11 97 Security Equipment

11 97 30 – Security Drawers

11 97 30.10 Pass Through Drawer

	Crew	Daily Output	Labor-Hours	Unit	Material	2018 Bare Costs Labor	Equipment	Total	Total Incl O&P
0010 **PASS THROUGH DRAWER**									
0100 Pass-thru drawer for personal items, 18" x 15" x 24"	1 Skwk	2	4	Ea.	2,900	209		3,109	3,500
0110 Including speakers	"	1.50	5.333	"	3,250	279		3,529	4,025

11 98 Detention Equipment

11 98 21 – Detention Windows

11 98 21.13 Visitor Cubicle Windows	Crew	Daily Output	Labor-Hours	Unit	Material	2018 Bare Costs		Total	Total Incl O&P
						Labor	Equipment		
0010 **VISITOR CUBICLE WINDOWS**									
4000 Visitor cubicle, vision panel, no intercom	E-4	2	16	Ea.	3,350	880	49.50	4,279.50	5,250

11 98 30 – Detention Cell Equipment

11 98 30.10 Cell Equipment

	Crew	Daily Output	Labor-Hours	Unit	Material	Labor	Equipment	Total	Total Incl O&P
0010 **CELL EQUIPMENT**									
3000 Toilet apparatus including wash basin, average	L-8	1.50	13.333	Ea.	3,400	705		4,105	4,875

For customer support on your Commercial Renovation Costs with RSMeans data, call 800.448.8182.

Estimating Tips

General

- The items in this division are usually priced per square foot or each. Most of these items are purchased by the owner and installed by the contractor. Do not assume the items in Division 12 will be purchased and installed by the contractor. Check the specifications for responsibilities and include receiving, storage, installation, and mechanical and electrical hookups in the appropriate divisions.

- Some items in this division require some type of support system that is not usually furnished with the item. Examples of these systems include blocking for the attachment of casework and heavy drapery rods. The required blocking must be added to the estimate in the appropriate division.

Reference Numbers

Reference numbers are shown at the beginning of some major classifications. These numbers refer to related items in the Reference Section. The reference information may be an estimating procedure, an alternate pricing method, or technical information.

Note: Not all subdivisions listed here necessarily appear. ■

Did you know?

RSMeans data is available through our online application with 24/7 access:

- Search for unit prices by keyword
- Leverage the most up-to-date data
- Build and export estimates

Try it free for 30 days!
www.rsmeans.com/2018freetrial

12 21 Window Blinds

12 21 13 – Horizontal Louver Blinds

12 21 13.13 Metal Horizontal Louver Blinds

12 21 13.13 Metal Horizontal Louver Blinds	Crew	Daily Output	Labor-Hours	Unit	Material	2018 Bare Costs Labor	Equipment	Total	Total Incl O&P
0010 **METAL HORIZONTAL LOUVER BLINDS**									
0020 Horizontal, 1" aluminum slats, solid color, stock	1 Carp	590	.014	S.F.	5.90	.69		6.59	7.60

12 21 13.33 Vinyl Horizontal Louver Blinds

12 21 13.33 Vinyl Horizontal Louver Blinds	Crew	Daily Output	Labor-Hours	Unit	Material	2018 Bare Costs Labor	Equipment	Total	Total Incl O&P
0010 **VINYL HORIZONTAL LOUVER BLINDS**									
0100 2" composite, 48" wide, 48" high	1 Carp	30	.267	Ea.	97	13.50		110.50	129
0120 72" high		29	.276		131	14		145	167
0140 96" high		28	.286		179	14.50		193.50	221
0200 60" wide, 60" high		27	.296		135	15		150	174
0220 72" high		25	.320		156	16.20		172.20	198
0240 96" high		24	.333		233	16.90		249.90	284
0300 72" wide, 72" high		25	.320		209	16.20		225.20	256
0320 96" high		23	.348		288	17.65		305.65	345
0400 96" wide, 96" high		20	.400		425	20.50		445.50	500
1000 2" faux wood, 48" wide, 48" high		30	.267		76	13.50		89.50	106
1020 72" high		29	.276		99	14		113	132
1040 96" high		28	.286		118	14.50		132.50	154
1300 72" wide, 72" high		25	.320		128	16.20		144.20	168
1320 96" high		23	.348		192	17.65		209.65	240
1400 96" wide, 96" high		20	.400		256	20.50		276.50	315

12 23 Interior Shutters

12 23 10 – Wood Interior Shutters

12 23 10.10 Wood Interior Shutters

12 23 10.10 Wood Interior Shutters	Crew	Daily Output	Labor-Hours	Unit	Material	2018 Bare Costs Labor	Equipment	Total	Total Incl O&P
0010 **WOOD INTERIOR SHUTTERS**, louvered									
0200 Two panel, 27" wide, 36" high	1 Carp	5	1.600	Set	160	81		241	310
0300 33" wide, 36" high		5	1.600		206	81		287	360
0500 47" wide, 36" high		5	1.600		276	81		357	435
1000 Four panel, 27" wide, 36" high		5	1.600		158	81		239	305
1100 33" wide, 36" high		5	1.600		203	81		284	355
1300 47" wide, 36" high		5	1.600		271	81		352	430
1400 Plantation shutters, 16" x 48"		5	1.600	Ea.	181	81		262	330
1450 16" x 96"		4	2		297	101		398	490
1460 36" x 96"		3	2.667		655	135		790	940

12 23 10.13 Wood Panels

12 23 10.13 Wood Panels	Crew	Daily Output	Labor-Hours	Unit	Material	2018 Bare Costs Labor	Equipment	Total	Total Incl O&P
0010 **WOOD PANELS**									
3000 Wood folding panels with movable louvers, 7" x 20" each	1 Carp	17	.471	Pr.	92	24		116	140
4000 Fixed louver type, stock units, 8" x 20" each		17	.471		96.50	24		120.50	145
4450 18" x 40" each		17	.471		138	24		162	191

12 24 Window Shades

12 24 13 – Roller Window Shades

12 24 13.10 Shades

		Crew	Daily Output	Labor-Hours	Unit	Material	2018 Bare Costs Labor	Equipment	Total	Total Incl O&P
0010	**SHADES**									
0020	Basswood, roll-up, stain finish, 3/8" slats	1 Carp	300	.027	S.F.	19.75	1.35		21.10	23.50
0030	Double layered, heat reflective		685	.012		13.30	.59		13.89	15.55
0200	7/8" slats		300	.027		19.60	1.35		20.95	23.50
0900	Mylar, single layer, non-heat reflective		685	.012		3	.59		3.59	4.26
0910	Mylar, single layer, heat reflective		685	.012		2.54	.59		3.13	3.76
1000	Double layered, heat reflective		685	.012		6.05	.59		6.64	7.60
1100	Triple layered, heat reflective	▼	685	.012	▼	7.60	.59		8.19	9.30
2000	Polyester, room darkening, with continuous cord, GEI									
2010	36" x 72"	1 Carp	38	.211	Ea.	142	10.65		152.65	173
2020	48" x 72"		28	.286		164	14.50		178.50	204
2030	60" x 72"		23	.348		192	17.65		209.65	240
2040	72" x 72"		19	.421	▼	237	21.50		258.50	295
5000	Thermal, roll up, R4		44	.182	S.F.	15.95	9.20		25.15	32.50
5030	R10.7	▼	44	.182	"	18.35	9.20		27.55	35
5050	Magnetic clips, set of 20				Set	34.50			34.50	38
6011	Solar screening, fiberglass [G]	1 Carp	85	.094	S.F.	7.65	4.77		12.42	16.15

12 32 Manufactured Wood Casework

12 32 23 – Hardwood Casework

12 32 23.10 Manufactured Wood Casework, Stock Units

		Crew	Daily Output	Labor-Hours	Unit	Material	2018 Bare Costs Labor	Equipment	Total	Total Incl O&P
0010	**MANUFACTURED WOOD CASEWORK, STOCK UNITS**									
0300	Built-in drawer units, pine, 18" deep, 32" high, unfinished									
0400	Minimum	2 Carp	53	.302	L.F.	137	15.30		152.30	176
0500	Maximum	"	40	.400	"	161	20.50		181.50	211
0700	Kitchen base cabinets, hardwood, not incl. counter tops,									
0710	24" deep, 35" high, prefinished									
0800	One top drawer, one door below, 12" wide	2 Carp	24.80	.645	Ea.	300	32.50		332.50	385
0820	15" wide		24	.667		315	34		349	400
0840	18" wide		23.30	.687		340	35		375	430
0860	21" wide		22.70	.705		350	35.50		385.50	450
0880	24" wide		22.30	.717		415	36.50		451.50	515
1000	Four drawers, 12" wide		24.80	.645		315	32.50		347.50	405
1020	15" wide		24	.667		320	34		354	410
1040	18" wide		23.30	.687		355	35		390	445
1060	24" wide		22.30	.717		390	36.50		426.50	490
1200	Two top drawers, two doors below, 27" wide		22	.727		445	37		482	550
1220	30" wide		21.40	.748		485	38		523	595
1240	33" wide		20.90	.766		505	39		544	620
1260	36" wide		20.30	.788		525	40		565	640
1280	42" wide		19.80	.808		550	41		591	675
1300	48" wide		18.90	.847		590	43		633	720
1500	Range or sink base, two doors below, 30" wide		21.40	.748		405	38		443	505
1520	33" wide		20.90	.766		430	39		469	540
1540	36" wide		20.30	.788		455	40		495	565
1560	42" wide		19.80	.808		475	41		516	590
1580	48" wide	▼	18.90	.847		500	43		543	620
1800	For sink front units, deduct					179			179	197
2000	Corner base cabinets, 36" wide, standard	2 Carp	18	.889		745	45		790	895
2100	Lazy Susan with revolving door	"	16.50	.970	▼	955	49		1,004	1,125
4000	Kitchen wall cabinets, hardwood, 12" deep with two doors									

12 32 23.10 Manufactured Wood Casework, Stock Units	Crew	Daily Output	Labor-Hours	Unit	Material	2018 Bare Costs Labor	Equipment	Total	Total Incl O&P
4050	12" high, 30" wide	2 Carp	24.80	.645	Ea.	272	32.50		304.50	355
4100	36" wide		24	.667		75.50	34		109.50	139
4400	15" high, 30" wide		24	.667		277	34		311	360
4420	33" wide		23.30	.687		340	35		375	430
4440	36" wide		22.70	.705		330	35.50		365.50	425
4450	42" wide		22.70	.705		375	35.50		410.50	470
4700	24" high, 30" wide		23.30	.687		370	35		405	465
4720	36" wide		22.70	.705		405	35.50		440.50	510
4740	42" wide		22.30	.717		108	36.50		144.50	178
5000	30" high, one door, 12" wide		22	.727		261	37		298	345
5020	15" wide		21.40	.748		273	38		311	360
5040	18" wide		20.90	.766		300	39		339	395
5060	24" wide		20.30	.788		350	40		390	450
5300	Two doors, 27" wide		19.80	.808		390	41		431	490
5320	30" wide		19.30	.829		410	42		452	520
5340	36" wide		18.80	.851		465	43		508	585
5360	42" wide		18.50	.865		505	44		549	625
5380	48" wide		18.40	.870		570	44		614	695
6000	Corner wall, 30" high, 24" wide		18	.889		400	45		445	515
6050	30" wide		17.20	.930		425	47		472	540
6100	36" wide		16.50	.970		485	49		534	615
6500	Revolving Lazy Susan		15.20	1.053		127	53.50		180.50	227
7000	Broom cabinet, 84" high, 24" deep, 18" wide		10	1.600		745	81		826	950
7500	Oven cabinets, 84" high, 24" deep, 27" wide		8	2		1,125	101		1,226	1,425
7750	Valance board trim		396	.040	L.F.	18.05	2.05		20.10	23
7780	Toe kick trim	1 Carp	256	.031	"	3.45	1.58		5.03	6.35
7790	Base cabinet corner filler		16	.500	Ea.	48.50	25.50		74	94
7800	Cabinet filler, 3" x 24"		20	.400		19.05	20.50		39.55	54
7810	3" x 30"		20	.400		24	20.50		44.50	59
7820	3" x 42"		18	.444		33.50	22.50		56	73
7830	3" x 80"		16	.500		63.50	25.50		89	111
7850	Cabinet panel		50	.160	S.F.	10.35	8.10		18.45	24.50
9000	For deluxe models of all cabinets, add					40%				
9500	For custom built in place, add					25%	10%			
9558	Rule of thumb, kitchen cabinets not including									
9560	appliances & counter top, minimum	2 Carp	30	.533	L.F.	202	27		229	266
9600	Maximum	"	25	.640	"	440	32.50		472.50	540
9700	Minimum labor/equipment charge	1 Carp	3	2.667	Job		135		135	219

12 32 23.30 Manufactured Wood Casework Vanities

		Crew	Daily Output	Labor-Hours	Unit	Material	Labor	Equipment	Total	Total Incl O&P
0010	**MANUFACTURED WOOD CASEWORK VANITIES**									
8000	Vanity bases, 2 doors, 30" high, 21" deep, 24" wide	2 Carp	20	.800	Ea.	350	40.50		390.50	450
8050	30" wide		16	1		420	50.50		470.50	545
8100	36" wide		13.33	1.200		405	61		466	545
8150	48" wide		11.43	1.400		530	71		601	700
9000	For deluxe models of all vanities, add to above					40%				
9500	For custom built in place, add to above					25%	10%			

12 32 23.35 Manufactured Wood Casework Hardware

		Crew	Daily Output	Labor-Hours	Unit	Material	Labor	Equipment	Total	Total Incl O&P
0010	**MANUFACTURED WOOD CASEWORK HARDWARE**									
1000	Catches, minimum	1 Carp	235	.034	Ea.	1.40	1.73		3.13	4.34
1040	Maximum	"	80	.100	"	8.05	5.05		13.10	17.10
2000	Door/drawer pulls, handles									
2200	Handles and pulls, projecting, metal, minimum	1 Carp	48	.167	Ea.	5.20	8.45		13.65	19.40

12 32 Manufactured Wood Casework

12 32 23 – Hardwood Casework

12 32 23.35 Manufactured Wood Casework Hardware	Crew	Daily Output	Labor-Hours	Unit	Material	2018 Bare Costs Labor	Equipment	Total	Total Incl O&P
2240	Maximum	1 Carp	36	.222	Ea.	10.65	11.25		21.90	30
2300	Wood, minimum		48	.167		5.30	8.45		13.75	19.55
2340	Maximum		36	.222		9.80	11.25		21.05	29
2400	Drawer pulls, antimicrobial copper alloy finish		50	.160		19.55	8.10		27.65	34.50
2600	Flush, metal, minimum		48	.167		5.30	8.45		13.75	19.55
2640	Maximum		36	.222		9.90	11.25		21.15	29
2900	Drawer knobs, antimicrobial copper alloy finish		50	.160		12.95	8.10		21.05	27.50
3000	Drawer tracks/glides, minimum		48	.167	Pr.	8.95	8.45		17.40	23.50
3040	Maximum		24	.333		26	16.90		42.90	56
4000	Cabinet hinges, minimum		160	.050		3.10	2.54		5.64	7.55
4040	Maximum		68	.118		14.10	5.95		20.05	25
7000	Appliance pulls, antimicrobial copper alloy finish		50	.160	L.F.	70	8.10		78.10	90

12 34 Manufactured Plastic Casework

12 34 16 – Manufactured Solid-Plastic Casework

12 34 16.10 Outdoor Casework

		Crew	Daily Output	Labor-Hours	Unit	Material	2018 Bare Costs Labor	Equipment	Total	Total Incl O&P
0010	**OUTDOOR CASEWORK**									
0020	Cabinet, base, sink/range, 36"	2 Carp	20.30	.788	Ea.	480	40		520	590
0100	Base, 36"		20.30	.788		4,600	40		4,640	5,125
0200	Filler strip, 1" x 30"		158	.101		43.50	5.15		48.65	56.50
0210	Filler strip, 2" x 30"		158	.101		44.50	5.15		49.65	57.50

12 35 Specialty Casework

12 35 53 – Laboratory Casework

12 35 53.13 Metal Laboratory Casework

		Crew	Daily Output	Labor-Hours	Unit	Material	2018 Bare Costs Labor	Equipment	Total	Total Incl O&P
0010	**METAL LABORATORY CASEWORK**									
0020	Cabinets, base, door units, metal	2 Carp	18	.889	L.F.	253	45		298	350
0300	Drawer units		18	.889		560	45		605	695
0700	Tall storage cabinets, open, 7' high		20	.800		525	40.50		565.50	645
0900	With glazed doors		20	.800		870	40.50		910.50	1,025
1300	Wall cabinets, metal, 12-1/2" deep, open		20	.800		211	40.50		251.50	298
1500	With doors		20	.800		415	40.50		455.50	525

12 35 70 – Healthcare Casework

12 35 70.13 Hospital Casework

		Crew	Daily Output	Labor-Hours	Unit	Material	2018 Bare Costs Labor	Equipment	Total	Total Incl O&P
0010	**HOSPITAL CASEWORK**									
3000	Hospital cabinets, stainless steel with glass door(s), lockable									
3010	One door, 24" W x 18" D x 60" H	2 Clab	18	.889	Ea.	2,900	35.50		2,935.50	3,250
3020	Two doors, 36" W x 18" D x 60" H		15	1.067		4,175	42.50		4,217.50	4,675
3030	36" W x 24" D x 67" H		15	1.067		5,925	42.50		5,967.50	6,600
3040	48" W x 24" D x 66" H		12	1.333		4,850	53		4,903	5,400
3050	48" W x 24" D x 72" H		12	1.333		6,025	53		6,078	6,700
3060	60" W x 24" D x 72" H		9	1.778		6,050	71		6,121	6,775

12 36 Countertops

12 36 16 – Metal Countertops

12 36 16.10 Stainless Steel Countertops		Crew	Daily Output	Labor-Hours	Unit	Material	2018 Bare Costs Labor	Equipment	Total	Total Incl O&P
0010	**STAINLESS STEEL COUNTERTOPS**									
3200	Stainless steel, custom	1 Carp	24	.333	S.F.	179	16.90		195.90	225

12 36 19 – Wood Countertops

12 36 19.10 Maple Countertops

		Crew	Daily Output	Labor-Hours	Unit	Material	Labor	Equipment	Total	Total Incl O&P
0010	**MAPLE COUNTERTOPS**									
2900	Solid, laminated, 1-1/2" thick, no splash	1 Carp	28	.286	L.F.	88.50	14.50		103	121
3000	With square splash		28	.286	"	105	14.50		119.50	140
3400	Recessed cutting block with trim, 16" x 20" x 1"		8	1	Ea.	108	50.50		158.50	201

12 36 23 – Plastic Countertops

12 36 23.13 Plastic-Laminate-Clad Countertops

		Crew	Daily Output	Labor-Hours	Unit	Material	Labor	Equipment	Total	Total Incl O&P
0010	**PLASTIC-LAMINATE-CLAD COUNTERTOPS**									
0020	Stock, 24" wide w/backsplash, minimum	1 Carp	30	.267	L.F.	17.25	13.50		30.75	41
0100	Maximum		25	.320		39	16.20		55.20	69.50
0300	Custom plastic, 7/8" thick, aluminum molding, no splash		30	.267		34.50	13.50		48	60
0400	Cove splash		30	.267		41	13.50		54.50	67
0600	1-1/4" thick, no splash		28	.286		39	14.50		53.50	66
0700	Square splash		28	.286		43.50	14.50		58	71
0900	Square edge, plastic face, 7/8" thick, no splash		30	.267		34	13.50		47.50	59.50
1000	With splash		30	.267		41	13.50		54.50	67.50
1200	For stainless channel edge, 7/8" thick, add					3.66			3.66	4.03
1300	1-1/4" thick, add					4.36			4.36	4.80
1500	For solid color suede finish, add					5.80			5.80	6.40
1700	For end splash, add				Ea.	21			21	23
1900	For cut outs, standard, add, minimum	1 Carp	32	.250		17.40	12.70		30.10	39.50
2000	Maximum		8	1		7	50.50		57.50	90
2100	Postformed, including backsplash and front edge		30	.267	L.F.	12.60	13.50		26.10	36
2110	Mitred, add		12	.667	Ea.		34		34	55
2200	Built-in place, 25" wide, plastic laminate		25	.320	L.F.	60	16.20		76.20	92.50
9000	Minimum labor/equipment charge		3.75	2.133	Job		108		108	176

12 36 33 – Tile Countertops

12 36 33.10 Ceramic Tile Countertops

		Crew	Daily Output	Labor-Hours	Unit	Material	Labor	Equipment	Total	Total Incl O&P
0010	**CERAMIC TILE COUNTERTOPS**									
2300	Ceramic tile mosaic	1 Carp	25	.320	L.F.	39.50	16.20		55.70	70

12 36 40 – Stone Countertops

12 36 40.10 Natural Stone Countertops

		Crew	Daily Output	Labor-Hours	Unit	Material	Labor	Equipment	Total	Total Incl O&P
0010	**NATURAL STONE COUNTERTOPS**									
2500	Marble, stock, with splash, 1/2" thick, minimum	1 Bric	17	.471	L.F.	49	23.50		72.50	92.50
2700	3/4" thick, maximum	"	13	.615	"	123	31		154	186

12 36 53 – Laboratory Countertops

12 36 53.10 Laboratory Countertops and Sinks

		Crew	Daily Output	Labor-Hours	Unit	Material	Labor	Equipment	Total	Total Incl O&P
0010	**LABORATORY COUNTERTOPS AND SINKS**									
0020	Countertops, epoxy resin, not incl. base cabinets, acid-proof, minimum	2 Carp	82	.195	S.F.	47.50	9.90		57.40	68
0030	Maximum		70	.229		47.50	11.60		59.10	71.50
0040	Stainless steel		82	.195		165	9.90		174.90	197

12 36 61 – Simulated Stone Countertops

12 36 61.16 Solid Surface Countertops

		Crew	Daily Output	Labor-Hours	Unit	Material	Labor	Equipment	Total	Total Incl O&P
0010	**SOLID SURFACE COUNTERTOPS**, Acrylic polymer									
0020	Pricing for orders of 100 L.F. or greater									
0100	25" wide, solid colors	2 Carp	28	.571	L.F.	52.50	29		81.50	105

12 36 Countertops

12 36 61 – Simulated Stone Countertops

12 36 61.16 Solid Surface Countertops

		Crew	Daily Output	Labor-Hours	Unit	Material	2018 Bare Costs Labor	2018 Bare Costs Equipment	Total	Total Incl O&P
0200	Patterned colors	2 Carp	28	.571	L.F.	66.50	29		95.50	120
0300	Premium patterned colors		28	.571		88.50	29		117.50	145
0400	With silicone attached 4" backsplash, solid colors		27	.593		61	30		91	117
0500	Patterned colors		27	.593		77.50	30		107.50	135
0600	Premium patterned colors		27	.593		96.50	30		126.50	155
0700	With hard seam attached 4" backsplash, solid colors		23	.696		61	35.50		96.50	125
0800	Patterned colors		23	.696		77.50	35.50		113	143
0900	Premium patterned colors		23	.696		96.50	35.50		132	164
1000	Pricing for order of 51-99 L.F.									
1100	25" wide, solid colors	2 Carp	24	.667	L.F.	60	34		94	121
1200	Patterned colors		24	.667		76.50	34		110.50	139
1300	Premium patterned colors		24	.667		102	34		136	167
1400	With silicone attached 4" backsplash, solid colors		23	.696		70.50	35.50		106	135
1500	Patterned colors		23	.696		89	35.50		124.50	156
1600	Premium patterned colors		23	.696		111	35.50		146.50	180
1700	With hard seam attached 4" backsplash, solid colors		20	.800		70.50	40.50		111	144
1800	Patterned colors		20	.800		89	40.50		129.50	164
1900	Premium patterned colors		20	.800		111	40.50		151.50	188
2000	Pricing for order of 1-50 L.F.									
2100	25" wide, solid colors	2 Carp	20	.800	L.F.	70.50	40.50		111	144
2200	Patterned colors		20	.800		90	40.50		130.50	165
2300	Premium patterned colors		20	.800		119	40.50		159.50	197
2400	With silicone attached 4" backsplash, solid colors		19	.842		82.50	42.50		125	161
2500	Patterned colors		19	.842		105	42.50		147.50	185
2600	Premium patterned colors		19	.842		130	42.50		172.50	213
2700	With hard seam attached 4" backsplash, solid colors		15	1.067		82.50	54		136.50	179
2800	Patterned colors		15	1.067		105	54		159	203
2900	Premium patterned colors		15	1.067		130	54		184	231
3000	Sinks, pricing for order of 100 or greater units									
3100	Single bowl, hard seamed, solid colors, 13" x 17"	1 Carp	3	2.667	Ea.	375	135		510	635
3200	10" x 15"		7	1.143		174	58		232	286
3300	Cutouts for sinks		8	1			50.50		50.50	82.50
3400	Sinks, pricing for order of 51-99 units									
3500	Single bowl, hard seamed, solid colors, 13" x 17"	1 Carp	2.55	3.137	Ea.	435	159		594	735
3600	10" x 15"		6	1.333		200	67.50		267.50	330
3700	Cutouts for sinks		7	1.143			58		58	94
3800	Sinks, pricing for order of 1-50 units									
3900	Single bowl, hard seamed, solid colors, 13" x 17"	1 Carp	2	4	Ea.	510	203		713	890
4000	10" x 15"		4.55	1.758		235	89		324	405
4100	Cutouts for sinks		5.25	1.524			77.50		77.50	125
4200	Cooktop cutouts, pricing for 100 or greater units		4	2		25.50	101		126.50	193
4300	51-99 units		3.40	2.353		29.50	119		148.50	227
4400	1-50 units		3	2.667		34.50	135		169.50	257

12 36 61.17 Solid Surface Vanity Tops

		Crew	Daily Output	Labor-Hours	Unit	Material	2018 Bare Costs Labor	2018 Bare Costs Equipment	Total	Total Incl O&P
0010	**SOLID SURFACE VANITY TOPS**									
0015	Solid surface, center bowl, 17" x 19"	1 Carp	12	.667	Ea.	181	34		215	254
0020	19" x 25"		12	.667		185	34		219	258
0030	19" x 31"		12	.667		216	34		250	293
0040	19" x 37"		12	.667		252	34		286	330
0050	22" x 25"		10	.800		330	40.50		370.50	430
0060	22" x 31"		10	.800		385	40.50		425.50	490
0070	22" x 37"		10	.800		450	40.50		490.50	560

12 36 Countertops

12 36 61 - Simulated Stone Countertops

12 36 61.17 Solid Surface Vanity Tops

		Crew	Daily Output	Labor-Hours	Unit	Material	2018 Bare Costs Labor	Equipment	Total	Total Incl O&P
0080	22" x 43"	1 Carp	10	.800	Ea.	520	40.50		560.50	635
0090	22" x 49"		10	.800		560	40.50		600.50	685
0110	22" x 55"		8	1		405	50.50		455.50	530
0120	22" x 61"		8	1		460	50.50		510.50	590
0220	Double bowl, 22" x 61"		8	1		515	50.50		565.50	655
0230	Double bowl, 22" x 73"		8	1		980	50.50		1,030.50	1,150
0240	For aggregate colors, add					35%				
0250	For faucets and fittings, see Section 22 41 39.10									

12 36 61.19 Quartz Agglomerate Countertops

		Crew	Daily Output	Labor-Hours	Unit	Material	2018 Bare Costs Labor	Equipment	Total	Total Incl O&P
0010	**QUARTZ AGGLOMERATE COUNTERTOPS**									
0100	25" wide, 4" backsplash, color group A, minimum	2 Carp	15	1.067	L.F.	67.50	54		121.50	162
0110	Maximum		15	1.067		94	54		148	192
0120	Color group B, minimum		15	1.067		72	54		126	167
0130	Maximum		15	1.067		98.50	54		152.50	196
0140	Color group C, minimum		15	1.067		81	54		135	177
0150	Maximum		15	1.067		115	54		169	215
0160	Color group D, minimum		15	1.067		88	54		142	185
0170	Maximum		15	1.067		119	54		173	219

12 48 Rugs and Mats

12 48 13 - Entrance Floor Mats and Frames

12 48 13.13 Entrance Floor Mats

			Crew	Daily Output	Labor-Hours	Unit	Material	2018 Bare Costs Labor	Equipment	Total	Total Incl O&P
0010	**ENTRANCE FLOOR MATS**										
2000	Recycled rubber tire tile, 12" x 12" x 3/8" thick	G	1 Clab	125	.064	S.F.	9.95	2.55		12.50	15.10
2510	Natural cocoa fiber, 1/2" thick	G		125	.064		7.10	2.55		9.65	11.95
2520	3/4" thick	G		125	.064		6.45	2.55		9	11.20
2530	1" thick	G		125	.064		12.90	2.55		15.45	18.35
3000	Hospital tacky mats, package of 30 with frame					Ea.	52			52	57.50
3010	4 packages of 30					"	85			85	93.50

12 51 Office Furniture

12 51 16 - Case Goods

12 51 16.13 Metal Case Goods

			Unit	Material	Total	Total Incl O&P
0010	**METAL CASE GOODS**					
0020	Desks, 29" high, double pedestal, 30" x 60", metal, minimum		Ea.	575	575	630
0030	Maximum		"	895	895	985

12 54 Hospitality Furniture

12 54 13 - Hotel and Motel Furniture

12 54 13.10 Hotel Furniture

			Unit	Material	Total	Total Incl O&P
0010	**HOTEL FURNITURE**					
0020	Standard quality set, minimum		Room	2,550	2,550	2,800
0200	Maximum		"	8,525	8,525	9,375

12 54 Hospitality Furniture

12 54 16 – Restaurant Furniture

12 54 16.20 Furniture, Restaurant	Crew	Daily Output	Labor-Hours	Unit	Material	2018 Bare Costs Labor	Equipment	Total	Total Incl O&P
0010 **FURNITURE, RESTAURANT**									
0020 Bars, built-in, front bar	1 Carp	5	1.600	L.F.	297	81		378	455
0200 Back bar		5	1.600	"	216	81		297	370
0500 Booth unit, molded plastic, stub wall and 2 seats, minimum		2	4	Set	340	203		543	705
0600 Maximum		1.50	5.333	"	1,450	270		1,720	2,050
0800 Booth seat, upholstered, foursome, single (end) minimum		5	1.600	Ea.	800	81		881	1,000
0900 Maximum		4	2		1,025	101		1,126	1,300
1000 Foursome, double, minimum		4	2		1,175	101		1,276	1,475
1100 Maximum		3	2.667		1,975	135		2,110	2,400
1300 Circle booth, upholstered, 1/4 circle, minimum		3	2.667		1,375	135		1,510	1,750
1400 Maximum		2	4		1,750	203		1,953	2,250
1500 3/4 circle, minimum		1.50	5.333		6,625	270		6,895	7,725
1600 Maximum	↓	1	8	↓	7,000	405		7,405	8,375

12 55 Detention Furniture

12 55 13 – Detention Bunks

12 55 13.13 Cots

	Crew	Daily Output	Labor-Hours	Unit	Material	2018 Bare Costs Labor	Equipment	Total	Total Incl O&P
0010 **COTS**									
2500 Bolted, single, painted steel	E-4	20	1.600	Ea.	340	88	4.93	432.93	525
2700 Stainless steel	"	20	1.600	"	975	88	4.93	1,067.93	1,225

12 56 Institutional Furniture

12 56 43 – Dormitory Furniture

12 56 43.10 Dormitory Furnishings

	Crew	Daily Output	Labor-Hours	Unit	Material	2018 Bare Costs Labor	Equipment	Total	Total Incl O&P
0010 **DORMITORY FURNISHINGS**									
0300 Bunkable bed, twin, minimum				Ea.	395			395	435
0320 Maximum				"	650			650	715

12 56 51 – Library Furniture

12 56 51.10 Library Furnishings

	Crew	Daily Output	Labor-Hours	Unit	Material	2018 Bare Costs Labor	Equipment	Total	Total Incl O&P
0010 **LIBRARY FURNISHINGS**									
1710 Carrels, hardwood, 36" x 24", minimum	1 Carp	5	1.600	Ea.	560	81		641	745
1720 Maximum		4	2		1,625	101		1,726	1,950
1730 Metal, minimum		5	1.600		268	81		349	425
1740 Maximum		4	2	↓	420	101		521	625
6010 Bookshelf, metal, 90" high, 10" shelf, double face		11.50	.696	L.F.	187	35.50		222.50	264
6020 Single face	↓	12	.667	"	117	34		151	184
6050 For 8" shelving, subtract from above					10%				
6070 For 42" high with countertop, subtract from above					20%				

12 56 70 – Healthcare Furniture

12 56 70.10 Furniture, Hospital

	Crew	Daily Output	Labor-Hours	Unit	Material	2018 Bare Costs Labor	Equipment	Total	Total Incl O&P
0010 **FURNITURE, HOSPITAL**									
0020 Beds, manual, minimum				Ea.	855			855	940
0100 Maximum				"	2,825			2,825	3,100
1100 Patient wall systems, not incl. plumbing, minimum				Room	1,400			1,400	1,550
1200 Maximum				"	2,000			2,000	2,200

For customer support on your Commercial Renovation Costs with RSMeans data, call 800.448.8182.

385

12 63 Stadium and Arena Seating

12 63 13 – Stadium and Arena Bench Seating

12 63 13.13 Bleachers

		Crew	Daily Output	Labor-Hours	Unit	Material	2018 Bare Costs Labor	Equipment	Total	Total Incl O&P
0010	**BLEACHERS**									
3000	Telescoping, manual to 15 tier, minimum	F-5	65	.492	Seat	114	25		139	166
3100	Maximum		60	.533		163	27.50		190.50	225
3300	16 to 20 tier, minimum		60	.533		274	27.50		301.50	345
3400	Maximum		55	.582		325	30		355	410
3600	21 to 30 tier, minimum		50	.640		262	33		295	340
3700	Maximum		40	.800		360	41		401	460
3900	For integral power operation, add, minimum	2 Elec	300	.053		53	3.10		56.10	63.50
4000	Maximum	"	250	.064		86.50	3.72		90.22	101
5000	Benches, folding, in wall, 14' table, 2 benches	L-4	2	12	Set	965	570		1,535	1,975

12 67 Pews and Benches

12 67 13 – Pews

12 67 13.13 Sanctuary Pews

		Crew	Daily Output	Labor-Hours	Unit	Material	2018 Bare Costs Labor	Equipment	Total	Total Incl O&P
0010	**SANCTUARY PEWS**									
1500	Bench type, hardwood, minimum	1 Carp	20	.400	L.F.	100	20.50		120.50	143
1550	Maximum	"	15	.533		175	27		202	236
1570	For kneeler, add					22.50			22.50	24.50

12 93 Interior Public Space Furnishings

12 93 13 – Bicycle Racks

12 93 13.10 Bicycle Racks

		Crew	Daily Output	Labor-Hours	Unit	Material	2018 Bare Costs Labor	Equipment	Total	Total Incl O&P
0010	**BICYCLE RACKS**									
0020	Single side, grid, 1-5/8" OD stl. pipe, w/1/2" bars, galv, 5 bike cap	2 Clab	10	1.600	Ea.	315	64		379	450
0025	Powder coat finish		10	1.600		320	64		384	455
0030	Single side, grid, 1-5/8" OD stl. pipe, w/1/2" bars, galv, 9 bike cap		8	2		400	79.50		479.50	570
0035	Powder coat finish		8	2		415	79.50		494.50	585
0040	Single side, grid, 1-5/8" OD stl. pipe, w/1/2" bars, galv, 18 bike cap		4	4		465	159		624	770
0045	Powder coat finish		4	4		525	159		684	835
0050	S curve, 1-7/8" OD stl. pipe, 11 ga., galv, 5 bike cap		10	1.600		157	64		221	277
0055	Powder coat finish		10	1.600		157	64		221	277
0060	S curve, 1-7/8" OD stl. pipe, 11 ga., galv, 7 bike cap		9	1.778		167	71		238	299
0065	Powder coat finish		9	1.778		212	71		283	350
0070	S curve, 1-7/8" OD stl. pipe, 11 ga., galv, 9 bike cap		8	2		320	79.50		399.50	480
0075	Powder coat finish		8	2		310	79.50		389.50	470
0080	S curve, 1-7/8" OD stl. pipe, 11 ga., galv, 11 bike cap		6	2.667		430	106		536	650
0085	Powder coat finish		6	2.667		415	106		521	630

12 93 23 – Trash and Litter Receptacles

12 93 23.10 Trash Receptacles

			Crew	Daily Output	Labor-Hours	Unit	Material	2018 Bare Costs Labor	Equipment	Total	Total Incl O&P
0010	**TRASH RECEPTACLES**										
0500	Recycled plastic, var colors, round, 32 gal., 28" x 38" high	G	2 Clab	5	3.200	Ea.	660	128		788	930
0510	32 gal., 31" x 32" high	G		5	3.200		770	128		898	1,050
9110	Plastic, with dome lid, 32 gal. capacity			35	.457		62.50	18.20		80.70	98
9120	Recycled plastic slats, plastic dome lid, 32 gal. capacity			35	.457		297	18.20		315.20	355

Estimating Tips

General

- The items and systems in this division are usually estimated, purchased, supplied, and installed as a unit by one or more subcontractors. The estimator must ensure that all parties are operating from the same set of specifications and assumptions, and that all necessary items are estimated and will be provided. Many times the complex items and systems are covered, but the more common ones, such as excavation or a crane, are overlooked for the very reason that everyone assumes nobody could miss them. The estimator should be the central focus and be able to ensure that all systems are complete.

- Another area where problems can develop in this division is at the interface between systems. The estimator must ensure, for instance, that anchor bolts, nuts, and washers are estimated and included for the air-supported structures and pre-engineered buildings to be bolted to their foundations. Utility supply is a common area where essential items or pieces of equipment can be missed or overlooked, because each subcontractor may feel it is another's responsibility. The estimator should also be aware of certain items which may be supplied as part of a package but installed by others, and ensure that the installing contractor's estimate includes the cost of installation. Conversely, the estimator must also ensure that items are not costed by two different subcontractors, resulting in an inflated overall estimate.

13 30 00 Special Structures

- The foundations and floor slab, as well as rough mechanical and electrical, should be estimated, as this work is required for the assembly and erection of the structure. Generally, as noted in the data set, the pre-engineered building comes as a shell. Pricing is based on the size and structural design parameters stated in the reference section. Additional features, such as windows and doors with their related structural framing, must also be included by the estimator. Here again, the estimator must have a clear understanding of the scope of each portion of the work and all the necessary interfaces.

Reference Numbers

Reference numbers are shown at the beginning of some major classifications. These numbers refer to related items in the Reference Section. The reference information may be an estimating procedure, an alternate pricing method, or technical information.

Note: Not all subdivisions listed here necessarily appear. ■

Did you know?

RSMeans data is available through our online application with 24/7 access:

- Search for unit prices by keyword
- Leverage the most up-to-date data
- Build and export estimates

Try it free for 30 days!
www.rsmeans.com/2018freetrial

13 05 05.10 Selective Demolition, Air Supported Structures	Crew	Daily Output	Labor-Hours	Unit	Material	2018 Bare Costs Labor	Equipment	Total	Total Incl O&P
0010 **SELECTIVE DEMOLITION, AIR SUPPORTED STRUCTURES**									
0020 Tank covers, scrim, dbl. layer, vinyl poly w/hdwe., blower & controls									
0050 Round and rectangular · R024119-10	B-2	9000	.004	S.F.		.18		.18	.29
0100 Warehouse structures									
0120 Poly/vinyl fabric, 28 oz., incl. tension cables & inflation system	4 Clab	9000	.004	SF Flr.		.14		.14	.23
0150 Reinforced vinyl, 12 oz., 3,000 S.F.	"	5000	.006			.26		.26	.41
0200 12,000 to 24,000 S.F.	8 Clab	20000	.003			.13		.13	.21
0250 Tedlar vinyl fabric, 28 oz. w/liner, to 3,000 S.F.	4 Clab	5000	.006			.26		.26	.41
0300 12,000 to 24,000 S.F.	8 Clab	20000	.003	↓		.13		.13	.21
0350 Greenhouse/shelter, woven polyethylene with liner									
0400 3,000 S.F.	4 Clab	5000	.006	SF Flr.		.26		.26	.41
0450 12,000 to 24,000 S.F.	8 Clab	20000	.003			.13		.13	.21
0500 Tennis/gymnasium, poly/vinyl fabric, 28 oz., incl. thermal liner	4 Clab	9000	.004			.14		.14	.23
0600 Stadium/convention center, teflon coated fiberglass, incl. thermal liner	9 Clab	40000	.002	↓		.07		.07	.12
0700 Doors, air lock, 15' long, 10' x 10'	2 Carp	1.50	10.667	Ea.		540		540	880
0720 15' x 15'		.80	20			1,025		1,025	1,650
0750 Revolving personnel door, 6' diam. x 6'-6" high	↓	1.50	10.667	↓		540		540	880

13 05 05.20 Selective Demolition, Garden Houses

	Crew	Daily Output	Labor-Hours	Unit	Material	Labor	Equipment	Total	Total Incl O&P
0010 **SELECTIVE DEMOLITION, GARDEN HOUSES**									
0020 Prefab, wood, excl. foundation, average	2 Clab	400	.040	SF Flr.		1.59		1.59	2.59

13 05 05.25 Selective Demolition, Geodesic Domes

	Crew	Daily Output	Labor-Hours	Unit	Material	Labor	Equipment	Total	Total Incl O&P
0010 **SELECTIVE DEMOLITION, GEODESIC DOMES**									
0050 Shell only, interlocking plywood panels, 30' diameter	F-5	3.20	10	Ea.		510		510	830
0060 34' diameter		2.30	13.913			710		710	1,150
0070 39' diameter	↓	2	16			820		820	1,325
0080 45' diameter	F-3	2.20	18.182			940	225	1,165	1,775
0090 55' diameter		2	20			1,025	248	1,273	1,950
0100 60' diameter		2	20			1,025	248	1,273	1,950
0110 65' diameter	↓	1.60	25	↓		1,300	310	1,610	2,450

13 05 05.30 Selective Demolition, Greenhouses

	Crew	Daily Output	Labor-Hours	Unit	Material	Labor	Equipment	Total	Total Incl O&P
0010 **SELECTIVE DEMOLITION, GREENHOUSES**									
0020 Resi-type, free standing, excl. foundations, 9' long x 8' wide	2 Clab	160	.100	SF Flr.		3.99		3.99	6.45
0030 9' long x 11' wide		170	.094			3.75		3.75	6.10
0040 9' long x 14' wide		220	.073			2.90		2.90	4.71
0050 9' long x 17' wide		320	.050			1.99		1.99	3.24
0060 Lean-to type, 4' wide		64	.250			9.95		9.95	16.20
0070 7' wide		120	.133	↓		5.30		5.30	8.65
0080 Geodesic hemisphere, 1/8" plexiglass glazing, 8' diam.		4	4	Ea.		159		159	259
0090 24' diam.		.80	20			795		795	1,300
0100 48' diam.	↓	.40	40	↓		1,600		1,600	2,600

13 05 05.35 Selective Demolition, Hangars

	Crew	Daily Output	Labor-Hours	Unit	Material	Labor	Equipment	Total	Total Incl O&P
0010 **SELECTIVE DEMOLITION, HANGARS**									
0020 T type hangars, prefab, steel, galv roof & walls, incl doors, excl fndtn	E-2	2550	.022	SF Flr.		1.19	.66	1.85	2.72
0030 Circular type, prefab, steel frame, plastic skin, incl foundation, 80' diam	"	.50	112	Total		6,075	3,375	9,450	13,900

13 05 05.45 Selective Demolition, Lightning Protection

	Crew	Daily Output	Labor-Hours	Unit	Material	Labor	Equipment	Total	Total Incl O&P
0010 **SELECTIVE DEMOLITION, LIGHTNING PROTECTION**									
0020 Air terminal & base, copper, 3/8" diam. x 10", to 75' H	1 Clab	16	.500	Ea.		19.95		19.95	32.50
0030 1/2" diam. x 12", over 75' H		16	.500			19.95		19.95	32.50
0050 Aluminum, 1/2" diam. x 12", to 75' H		16	.500			19.95		19.95	32.50
0060 5/8" diam. x 12", over 75' H		16	.500	↓		19.95		19.95	32.50
0070 Cable, copper, 220 lb. per thousand feet, to 75' H	↓	640	.013	L.F.		.50		.50	.81

13 05 05 – Selective Demolition for Special Construction

13 05 05.45 Selective Demolition, Lightning Protection	Crew	Daily Output	Labor-Hours	Unit	Material	2018 Bare Costs Labor	Equipment	Total	Total Incl O&P
0080	375 lb. per thousand feet, over 75' H	1 Clab	460	.017	L.F.		.69		.69	1.13
0090	Aluminum, 101 lb. per thousand feet, to 75' H		560	.014			.57		.57	.92
0100	199 lb. per thousand feet, over 75' H		480	.017			.66		.66	1.08
0110	Arrester, 175 V AC, to ground		16	.500	Ea.		19.95		19.95	32.50
0120	650 V AC, to ground		13	.615	"		24.50		24.50	40

13 05 05.50 Selective Demolition, Pre-Engineered Steel Buildings

13 05 05.50	Crew	Daily Output	Labor-Hours	Unit	Material	2018 Bare Costs Labor	Equipment	Total	Total Incl O&P
0010	**SELECTIVE DEMOLITION, PRE-ENGINEERED STEEL BUILDINGS**									
0500	Pre-engd. steel bldgs., rigid frame, clear span & multi post, excl. salvage									
0550	3,500 to 7,500 S.F.	L-10	1000	.024	SF Flr.		1.34	.50	1.84	2.78
0600	7,501 to 12,500 S.F.		1500	.016			.89	.33	1.22	1.84
0650	12,501 S.F. or greater		1650	.015			.81	.30	1.11	1.68
0700	Pre-engd. steel building components									
0710	Entrance canopy, including frame 4' x 4'	E-24	8	4	Ea.		218	74	292	445
0720	4' x 8'	"	7	4.571			249	84.50	333.50	510
0730	HM doors, self framing, single leaf	2 Skwk	8	2			105		105	169
0740	Double leaf		5	3.200			168		168	270
0760	Gutter, eave type		600	.027	L.F.		1.40		1.40	2.25
0770	Sash, single slide, double slide or fixed		24	.667	Ea.		35		35	56
0780	Skylight, fiberglass, to 30 S.F.		16	1			52.50		52.50	84.50
0785	Roof vents, circular, 12" to 24" diameter		12	1.333			70		70	112
0790	Continuous, 10' long		8	2			105		105	169
0900	Shelters, aluminum frame									
0910	Acrylic glazing, 3' x 9' x 8' high	2 Skwk	2	8	Ea.		420		420	675
0920	9' x 12' x 8' high	"	1.50	10.667	"		560		560	900

13 05 05.60 Selective Demolition, Silos

13 05 05.60	Crew	Daily Output	Labor-Hours	Unit	Material	2018 Bare Costs Labor	Equipment	Total	Total Incl O&P
0010	**SELECTIVE DEMOLITION, SILOS**									
0020	Conc stave, indstrl, conical/sloping bott, excl fndtn, 12' diam., 35' H	E-24	.18	178	Ea.		9,675	3,275	12,950	19,800
0030	16' diam., 45' H		.12	267			14,500	4,925	19,425	29,700
0040	25' diam., 75' H		.08	400			21,800	7,375	29,175	44,500
0050	Steel, factory fabricated, 30,000 gal. cap, painted or epoxy lined	L-5	2	28			1,550	295	1,845	2,925

13 05 05.65 Selective Demolition, Sound Control

13 05 05.65	Crew	Daily Output	Labor-Hours	Unit	Material	2018 Bare Costs Labor	Equipment	Total	Total Incl O&P
0010	**SELECTIVE DEMOLITION, SOUND CONTROL**									
0120	Acoustical enclosure, 4" thick walls & ceiling panels, 8 lb./S.F.	3 Carp	144	.167	SF Surf		8.45		8.45	13.70
0130	10.5 lb./S.F.		128	.188			9.50		9.50	15.45
0140	Reverb chamber, parallel walls, 4" thick		120	.200			10.15		10.15	16.45
0150	Skewed walls, parallel roof, 4" thick		110	.218			11.05		11.05	17.95
0160	Skewed walls/roof, 4" layer/air space		96	.250			12.70		12.70	20.50
0170	Sound-absorbing panels, painted metal, 2'-6" x 8', under 1,000 S.F.		430	.056			2.83		2.83	4.59
0180	Over 1,000 S.F.		480	.050			2.54		2.54	4.12
0190	Flexible transparent curtain, clear	3 Shee	430	.056			3.34		3.34	5.25
0192	50% clear, 50% foam		430	.056			3.34		3.34	5.25
0194	25% clear, 75% foam		430	.056			3.34		3.34	5.25
0196	100% foam		430	.056			3.34		3.34	5.25
0200	Audio-masking sys., incl. speakers, amplfr., signal gnrtr.									
0205	Ceiling mounted, 5,000 S.F.	2 Elec	4800	.003	S.F.		.19		.19	.30
0210	10,000 S.F.		5600	.003			.17		.17	.26
0220	Plenum mounted, 5,000 S.F.		7600	.002			.12		.12	.19
0230	10,000 S.F.		8800	.002			.11		.11	.16

13 05 05.70 Selective Demolition, Special Purpose Rooms

13 05 05.70	Crew	Daily Output	Labor-Hours	Unit	Material	2018 Bare Costs Labor	Equipment	Total	Total Incl O&P
0010	**SELECTIVE DEMOLITION, SPECIAL PURPOSE ROOMS**									
0100	Audiometric rooms, under 500 S.F. surface	4 Carp	200	.160	SF Surf		8.10		8.10	13.15
0110	Over 500 S.F. surface	"	240	.133	"		6.75		6.75	10.95

13 05 05.70 Selective Demolition, Special Purpose Rooms

		Crew	Daily Output	Labor-Hours	Unit	Material	2018 Bare Costs Labor	Equipment	Total	Total Incl O&P
0200	Clean rooms, 12' x 12' soft wall, class 100	1 Carp	.30	26.667	Ea.		1,350		1,350	2,200
0210	Class 1,000		.30	26.667			1,350		1,350	2,200
0220	Class 10,000		.35	22.857			1,150		1,150	1,875
0230	Class 100,000		.35	22.857			1,150		1,150	1,875
0300	Darkrooms, shell complete, 8' high	2 Carp	220	.073	SF Flr.		3.69		3.69	6
0310	12' high		110	.145	"		7.35		7.35	11.95
0350	Darkrooms doors, mini-cylindrical, revolving		4	4	Ea.		203		203	330
0400	Music room, practice modular		140	.114	SF Surf		5.80		5.80	9.40
0500	Refrigeration structures and finishes									
0510	Wall finish, 2 coat Portland cement plaster, 1/2" thick	1 Clab	200	.040	S.F.		1.59		1.59	2.59
0520	Fiberglass panels, 1/8" thick		400	.020			.80		.80	1.29
0530	Ceiling finish, polystyrene plastic, 1" to 2" thick		500	.016			.64		.64	1.04
0540	4" thick		450	.018			.71		.71	1.15
0550	Refrigerator, prefab aluminum walk-in, 7'-6" high, 6' x 6' OD	2 Carp	100	.160	SF Flr.		8.10		8.10	13.15
0560	10' x 10' OD		160	.100			5.05		5.05	8.25
0570	Over 150 S.F.		200	.080			4.06		4.06	6.60
0600	Sauna, prefabricated, including heater & controls, 7' high, to 30 S.F.		120	.133			6.75		6.75	10.95
0610	To 40 S.F.		140	.114			5.80		5.80	9.40
0620	To 60 S.F.		175	.091			4.64		4.64	7.50
0630	To 100 S.F.		220	.073			3.69		3.69	6
0640	To 130 S.F.		250	.064			3.24		3.24	5.25
0650	Steam bath, heater, timer, head, single, to 140 C.F.	1 Plum	2.20	3.636	Ea.		226		226	350
0660	To 300 C.F.		2.20	3.636			226		226	350
0670	Steam bath, comm. size, w/blow-down assembly, to 800 C.F.		1.80	4.444			276		276	430
0680	To 2500 C.F.		1.60	5			310		310	485
0690	Steam bath, comm. size, multiple, for motels, apts, 500 C.F., 2 baths		2	4			249		249	385
0700	1,000 C.F., 4 baths		1.40	5.714			355		355	555

13 05 05.75 Selective Demolition, Storage Tanks

		Crew	Daily Output	Labor-Hours	Unit	Material	2018 Bare Costs Labor	Equipment	Total	Total Incl O&P
0010	**SELECTIVE DEMOLITION, STORAGE TANKS**									
0500	Steel tank, single wall, above ground, not incl. fdn., pumps or piping									
0510	Single wall, 275 gallon	Q-1	3	5.333	Ea.		298		298	465
0520	550 thru 2,000 gallon	B-34P	2	12			645	275	920	1,325
0530	5,000 thru 10,000 gallon	B-34Q	2	12			655	530	1,185	1,600
0540	15,000 thru 30,000 gallon	B-34S	2	16			910	1,625	2,535	3,200
0600	Steel tank, double wall, above ground not incl. fdn., pumps & piping									
0620	500 thru 2,000 gallon	B-34P	2	12	Ea.		645	275	920	1,325

13 05 05.85 Selective Demolition, Swimming Pool Equip

		Crew	Daily Output	Labor-Hours	Unit	Material	2018 Bare Costs Labor	Equipment	Total	Total Incl O&P
0010	**SELECTIVE DEMOLITION, SWIMMING POOL EQUIP**									
0020	Diving stand, stainless steel, 3 meter	2 Clab	3	5.333	Ea.		213		213	345
0030	1 meter		5	3.200			128		128	207
0040	Diving board, 16' long, aluminum		5.40	2.963			118		118	192
0050	Fiberglass		5.40	2.963			118		118	192
0070	Ladders, heavy duty, stainless steel, 2 tread		14	1.143			45.50		45.50	74
0080	4 tread		12	1.333			53		53	86.50
0090	Lifeguard chair, stainless steel, fixed		5	3.200			128		128	207
0100	Slide, tubular, fiberglass, aluminum handrails & ladder, 5', straight		4	4			159		159	259
0110	8', curved		6	2.667			106		106	173
0120	10', curved		3	5.333			213		213	345
0130	12' straight, with platform		2.50	6.400			255		255	415
0140	Removable access ramp, stainless steel		4	4			159		159	259
0150	Removable stairs, stainless steel, collapsible		4	4			159		159	259

13 05 05 – Selective Demolition for Special Construction

13 05 05.90 Selective Demolition, Tension Structures	Crew	Daily Output	Labor-Hours	Unit	Material	2018 Bare Costs Labor	Equipment	Total	Total Incl O&P
0010 **SELECTIVE DEMOLITION, TENSION STRUCTURES**									
0020 Steel/alum. frame, fabric shell, 60' clear span, 6,000 S.F.	B-41	2000	.022	SF Flr.		.91	.17	1.08	1.66
0030 12,000 S.F.		2200	.020			.83	.16	.99	1.51
0040 80' clear span, 20,800 S.F.		2440	.018			.75	.14	.89	1.36
0050 100' clear span, 10,000 S.F.	L-5	4350	.013			.71	.14	.85	1.35
0060 26,000 S.F.		4600	.012			.67	.13	.80	1.27
0070 36,000 S.F.		5000	.011			.62	.12	.74	1.17

13 05 05.95 Selective Demo, X-Ray/Radio Freq Protection

	Crew	Daily Output	Labor-Hours	Unit	Material	Labor	Equipment	Total	Total Incl O&P
0010 **SELECTIVE DEMO, X-RAY/RADIO FREQ PROTECTION**									
0020 Shielding lead, lined door frame, excl. hdwe., 1/16" thick	1 Clab	4.80	1.667	Ea.		66.50		66.50	108
0030 Lead sheets, 1/16" thick	2 Clab	270	.059	S.F.		2.36		2.36	3.83
0040 1/8" thick		240	.067			2.66		2.66	4.31
0050 Lead shielding, 1/4" thick		270	.059			2.36		2.36	3.83
0060 1/2" thick		240	.067			2.66		2.66	4.31
0070 Lead glass, 1/4" thick, 2.0 mm LE, 12" x 16"	2 Glaz	16	1	Ea.		48.50		48.50	78.50
0080 24" x 36"		8	2			97		97	157
0090 36" x 60"		4	4			194		194	315
0100 Lead glass window frame, with 1/16" lead & voice passage, 36" x 60"		4	4			194		194	315
0110 Lead glass window frame, 24" x 36"		8	2			97		97	157
0120 Lead gypsum board, 5/8" thick with 1/16" lead	2 Clab	320	.050	S.F.		1.99		1.99	3.24
0130 1/8" lead		280	.057			2.28		2.28	3.70
0140 1/32" lead		400	.040			1.59		1.59	2.59
0150 Butt joints, 1/8" lead or thicker, 2" x 7' long batten strip		480	.033	Ea.		1.33		1.33	2.16
0160 X-ray protection, average radiography room, up to 300 S.F., 1/16" lead, min		.50	32	Total		1,275		1,275	2,075
0170 Maximum		.30	53.333			2,125		2,125	3,450
0180 Deep therapy X ray room, 250 kV cap, up to 300 S.F., 1/4" load, min		.20	80			3,200		3,200	5,175
0190 Maximum		.12	133			5,325		5,325	8,625
0880 Radio frequency shielding, prefab or screen-type copper or steel, minimum		360	.044	SF Surf		1.77		1.77	2.88
0890 Average		310	.052			2.06		2.06	3.34
0895 Maximum		290	.055			2.20		2.20	3.57

13 12 Fountains

13 12 13 – Exterior Fountains

13 12 13.10 Outdoor Fountains

	Crew	Daily Output	Labor-Hours	Unit	Material	Labor	Equipment	Total	Total Incl O&P
0010 **OUTDOOR FOUNTAINS**									
0100 Outdoor fountain, 48" high with bowl and figures	2 Clab	2	8	Ea.	570	320		890	1,150
0200 Commercial, concrete or cast stone, 40-60" H, simple		2	8		775	320		1,095	1,375
0220 Average		2	8		1,350	320		1,670	2,000
0240 Ornate		2	8		2,625	320		2,945	3,400
0260 Metal, 72" high		2	8		1,200	320		1,520	1,850
0280 90" high		2	8		1,875	320		2,195	2,600
0300 120" high		2	8		5,000	320		5,320	6,025
0320 Resin or fiberglass, 40-60" H, wall type		2	8		480	320		800	1,050
0340 Waterfall type		2	8		1,000	320		1,320	1,625

13 12 23 – Interior Fountains

13 12 23.10 Indoor Fountains

	Crew	Daily Output	Labor-Hours	Unit	Material	Labor	Equipment	Total	Total Incl O&P
0010 **INDOOR FOUNTAINS**									
0100 Commercial, floor type, resin or fiberglass, lighted, cascade type	2 Clab	2	8	Ea.	286	320		606	835
0120 Tiered type		2	8		285	320		605	835
0140 Waterfall type		2	8		275	320		595	820

13 17 Tubs and Pools

13 17 13 – Hot Tubs

13 17 13.10 Redwood Hot Tub System	Crew	Daily Output	Labor-Hours	Unit	Material	2018 Bare Costs Labor	Equipment	Total	Total Incl O&P
0010 **REDWOOD HOT TUB SYSTEM**									
7050 4' diameter x 4' deep	Q-1	1	16	Ea.	3,225	895		4,120	4,950
7150 6' diameter x 4' deep		.80	20		4,950	1,125		6,075	7,175
7200 8' diameter x 4' deep		.80	20		7,250	1,125		8,375	9,725

13 17 33 – Whirlpool Tubs

13 17 33.10 Whirlpool Bath	Crew	Daily Output	Labor-Hours	Unit	Material	2018 Bare Costs Labor	Equipment	Total	Total Incl O&P
0010 **WHIRLPOOL BATH**									
6000 Whirlpool, bath with vented overflow, molded fiberglass									
6100 66" x 36" x 24"	Q-1	1	16	Ea.	1,025	895		1,920	2,525
6400 72" x 36" x 21"		1	16		1,250	895		2,145	2,775
6500 60" x 34" x 21"		1	16		1,175	895		2,070	2,700
6600 72" x 42" x 23"		1	16		1,275	895		2,170	2,800
6710 For color, add					10%				
6711 For designer colors and trim, add					25%				

13 21 Controlled Environment Rooms

13 21 26 – Cold Storage Rooms

13 21 26.50 Refrigeration	Crew	Daily Output	Labor-Hours	Unit	Material	2018 Bare Costs Labor	Equipment	Total	Total Incl O&P
0010 **REFRIGERATION**									
0020 Curbs, 12" high, 4" thick, concrete	2 Carp	58	.276	L.F.	3.48	14		17.48	26.50
6300 Rule of thumb for complete units, w/o doors & refrigeration, cooler		146	.110	SF Flr.	165	5.55		170.55	190
6400 Freezer		109.60	.146	"	119	7.40		126.40	143

13 24 Special Activity Rooms

13 24 16 – Saunas

13 24 16.50 Saunas and Heaters	Crew	Daily Output	Labor-Hours	Unit	Material	2018 Bare Costs Labor	Equipment	Total	Total Incl O&P
0010 **SAUNAS AND HEATERS**									
0020 Prefabricated, incl. heater & controls, 7' high, 6' x 4', C/C	L-7	2.20	12.727	Ea.	5,125	620		5,745	6,650
1700 Door only, cedar, 2'x6', with 1' x 4' tempered insulated glass window	2 Carp	3.40	4.706		620	239		859	1,075
1800 Prehung, incl. jambs, pulls & hardware	"	12	1.333		680	67.50		747.50	860
2500 Heaters only (incl. above), wall mounted, to 200 C.F.					775			775	855
4480 For additional equipment, see Section 11 66 13.10									

13 24 26 – Steam Baths

13 24 26.50 Steam Baths and Components	Crew	Daily Output	Labor-Hours	Unit	Material	2018 Bare Costs Labor	Equipment	Total	Total Incl O&P
0010 **STEAM BATHS AND COMPONENTS**									
0020 Heater, timer & head, single, to 140 C.F.	1 Plum	1.20	6.667	Ea.	2,325	415		2,740	3,200
0500 To 300 C.F.	"	1.10	7.273		2,600	450		3,050	3,550
2000 Multiple, motels, apts., 2 baths, w/blow-down assm., 500 C.F.	Q-1	1.30	12.308		6,675	690		7,365	8,425
2500 4 baths	"	.70	22.857		10,700	1,275		11,975	13,700

13 28 Athletic and Recreational Special Construction

13 28 33 – Athletic and Recreational Court Walls

13 28 33.50 Sport Court	Crew	Daily Output	Labor-Hours	Unit	Material	2018 Bare Costs Labor	Equipment	Total	Total Incl O&P
0010 **SPORT COURT**									
0020 Floors, No. 2 & better maple, 25/32" thick				SF Flr.				6.05	6.65
0300 Squash, regulation court in existing building, minimum				Court	40,300			40,300	44,400
0400 Maximum				"	44,900			44,900	49,400
0450 Rule of thumb for components:									
0470 Walls	3 Carp	.15	160	Court	12,100	8,100		20,200	26,500
0500 Floor	"	.25	96		9,575	4,875		14,450	18,400
0550 Lighting	2 Elec	.60	26.667	↓	2,300	1,550		3,850	4,925

13 34 Fabricated Engineered Structures

13 34 13 – Glazed Structures

13 34 13.13 Greenhouses

	Crew	Daily Output	Labor-Hours	Unit	Material	2018 Bare Costs Labor	Equipment	Total	Total Incl O&P
0010 **GREENHOUSES**, Shell only, stock units, not incl. 2' stub walls,									
0020 foundation, floors, heat or compartments									
0300 Residential type, free standing, 8'-6" long x 7'-6" wide	2 Carp	59	.271	SF Flr.	24	13.75		37.75	49
0400 10'-6" wide		85	.188		44.50	9.55		54.05	64.50
0600 13'-6" wide		108	.148		46	7.50		53.50	62.50
0700 17'-0" wide		160	.100		46	5.05		51.05	59
0900 Lean-to type, 3'-10" wide		34	.471		44	24		68	86.50
1000 6'-10" wide		58	.276		29	14		43	54.50
1050 8'-0" wide	↓	60	.267	↓	64	13.50		77.50	92.50
1100 Wall mounted to existing window, 3' x 3'	1 Carp	4	2	Ea.	2,125	101		2,226	2,500
1120 4' x 5'	"	3	2.667		2,325	135		2,460	2,775
3900 Cooling, 1,200 CFM exhaust fan, add					330			330	360
4000 7,850 CFM					1,050			1,050	1,150
4200 For heaters, 10 MBH, add					230			230	253
4300 60 MBH, add					800			800	880
4500 For benches, 2' x 8', add					175			175	192
4600 4' x 10', add				↓	213			213	234
4800 For ventilation & humidity control w/4 integrated outlets, add				Total	262			262	288
4900 For environmental controls and automation, 8 outputs, 9 stages, add				"	1,050			1,050	1,150
5100 For humidification equipment, add				Ea.	325			325	360
5200 For vinyl shading, add				S.F.	.42			.42	.46

13 34 13.19 Swimming Pool Enclosures

	Crew	Daily Output	Labor-Hours	Unit	Material	2018 Bare Costs Labor	Equipment	Total	Total Incl O&P
0010 **SWIMMING POOL ENCLOSURES** Translucent, free standing									
0020 not including foundations, heat or light									
0200 Economy	2 Carp	200	.080	SF Hor.	51	4.06		55.06	62.50
0600 Deluxe	"	70	.229	"	93	11.60		104.60	121

13 34 23 – Fabricated Structures

13 34 23.16 Fabricated Control Booths

	Crew	Daily Output	Labor-Hours	Unit	Material	2018 Bare Costs Labor	Equipment	Total	Total Incl O&P
0010 **FABRICATED CONTROL BOOTHS**									
0100 Guard House, prefab conc. w/bullet resistant doors & windows, roof & wiring									
0110 8' x 8', Level III	L-10	1	24	Ea.	54,000	1,350	495	55,845	62,500
0120 8' x 8', Level IV	"	1	24	"	61,500	1,350	495	63,345	70,500

13 34 23.45 Kiosks

	Crew	Daily Output	Labor-Hours	Unit	Material	2018 Bare Costs Labor	Equipment	Total	Total Incl O&P
0010 **KIOSKS**									
0020 Round, advertising type, 5' diameter, 7' high, aluminum wall, illuminated				Ea.	23,600			23,600	26,000
0100 Aluminum wall, non-illuminated					22,600			22,600	24,900
0500 Rectangular, 5' x 9', 7'-6" high, aluminum wall, illuminated					25,600			25,600	28,200
0600 Aluminum wall, non-illuminated				↓	24,100			24,100	26,500

For customer support on your Commercial Renovation Costs with RSMeans data, call 800.448.8182.

393

13 34 Fabricated Engineered Structures

13 34 63 – Natural Fiber Construction

13 34 63.50 Straw Bale Construction	Crew	Daily Output	Labor-Hours	Unit	Material	2018 Bare Costs Labor	2018 Bare Costs Equipment	Total	Total Incl O&P
0010 **STRAW BALE CONSTRUCTION**									
2020 Straw bales in walls w/modified post and beam frame [G]	2 Carp	320	.050	S.F.	6.35	2.54		8.89	11.05

13 42 Building Modules

13 42 63 – Detention Cell Modules

13 42 63.16 Steel Detention Cell Modules

	Crew	Daily Output	Labor-Hours	Unit	Material	Labor	Equipment	Total	Total Incl O&P
0010 **STEEL DETENTION CELL MODULES**									
2000 Cells, prefab., 5' to 6' wide, 7' to 8' high, 7' to 8' deep,									
2010 bar front, cot, not incl. plumbing	E-4	1.50	21.333	Ea.	9,700	1,175	65.50	10,940.50	12,800

13 48 Sound, Vibration, and Seismic Control

13 48 13 – Manufactured Sound and Vibration Control Components

13 48 13.50 Audio Masking

	Crew	Daily Output	Labor-Hours	Unit	Material	Labor	Equipment	Total	Total Incl O&P
0010 **AUDIO MASKING**, acoustical enclosure, 4" thick wall and ceiling									
0020 8 lb./S.F., up to 12' span	3 Carp	72	.333	SF Surf	31.50	16.90		48.40	62
0300 Better quality panels, 10.5 #/S.F		64	.375		36	19		55	70.50
0400 Reverb-chamber, 4" thick, parallel walls		60	.400		44.50	20.50		65	82

13 49 Radiation Protection

13 49 13 – Integrated X-Ray Shielding Assemblies

13 49 13.50 Lead Sheets

	Crew	Daily Output	Labor-Hours	Unit	Material	Labor	Equipment	Total	Total Incl O&P
0010 **LEAD SHEETS**									
0300 Lead sheets, 1/16" thick	2 Lath	135	.119	S.F.	10.90	5.85		16.75	21
0400 1/8" thick		120	.133		22	6.55		28.55	35
0500 Lead shielding, 1/4" thick		135	.119		42.50	5.85		48.35	55.50
0550 1/2" thick		120	.133		79.50	6.55		86.05	98
1200 X-ray protection, average radiography or fluoroscopy									
1210 room, up to 300 S.F. floor, 1/16" lead, economy	2 Lath	.25	64	Total	11,100	3,150		14,250	17,300
1500 7'-0" walls, deluxe	"	.15	107	"	13,400	5,250		18,650	23,000
1600 Deep therapy X-ray room, 250 kV capacity,									
1800 up to 300 S.F. floor, 1/4" lead, economy	2 Lath	.08	200	Total	31,100	9,825		40,925	49,700
1900 7'-0" walls, deluxe		.06	267	"	38,400	13,100		51,500	63,000
1999 Minimum labor/equipment charge		4.50	3.556	Job		175		175	276

13 49 21 – Lead Glazing

13 49 21.50 Lead Glazing

	Crew	Daily Output	Labor-Hours	Unit	Material	Labor	Equipment	Total	Total Incl O&P
0010 **LEAD GLAZING**									
0600 Lead glass, 1/4" thick, 2.0 mm LE, 12" x 16"	2 Glaz	13	1.231	Ea.	390	59.50		449.50	525
0700 24" x 36"	"	8	2	"	1,375	97		1,472	1,650
2000 X-ray viewing panels, clear lead plastic									
2010 7 mm thick, 0.3 mm LE, 2.3 lb./S.F.	H-3	139	.115	S.F.	271	4.97		275.97	305
2020 12 mm thick, 0.5 mm LE, 3.9 lb./S.F.		82	.195		385	8.40		393.40	435
2030 18 mm thick, 0.8 mm LE, 5.9 lb./S.F.		54	.296		440	12.80		452.80	505
2040 22 mm thick, 1.0 mm LE, 7.2 lb./S.F.		44	.364		575	15.70		590.70	655
2050 35 mm thick, 1.5 mm LE, 11.5 lb./S.F.		28	.571		880	24.50		904.50	1,000
2060 46 mm thick, 2.0 mm LE, 15.0 lb./S.F.		21	.762		1,150	33		1,183	1,300
2090 For panels 12 S.F. to 48 S.F., add crating charge				Ea.				50	50

13 49 23 – Integrated RFI/EMI Shielding Assemblies

13 49 23.50 Modular Shielding Partitions	Crew	Daily Output	Labor-Hours	Unit	Material	2018 Bare Costs Labor	Equipment	Total	Total Incl O&P
0010 MODULAR SHIELDING PARTITIONS									
4000 X-ray barriers, modular, panels mounted within framework for									
4002 attaching to floor, wall or ceiling, upper portion is clear lead									
4005 plastic window panels 48"H, lower portion is opaque leaded									
4008 steel panels 36"H, structural supports not incl.									
4010 1-section barrier, 36"W x 84"H overall									
4020 0.5 mm LE panels	H-3	6.40	2.500	Ea.	8,800	108		8,908	9,850
4030 0.8 mm LE panels		6.40	2.500		9,500	108		9,608	10,600
4040 1.0 mm LE panels		5.33	3.002		11,100	130		11,230	12,400
4050 1.5 mm LE panels		5.33	3.002		14,800	130		14,930	16,500
4060 2-section barrier, 72"W x 84"H overall									
4070 0.5 mm LE panels	H-3	4	4	Ea.	12,800	173		12,973	14,300
4080 0.8 mm LE panels		4	4		14,100	173		14,273	15,800
4090 1.0 mm LE panels		3.56	4.494		17,300	194		17,494	19,400
5000 1.5 mm LE panels		3.20	5		24,700	216		24,916	27,500
5010 3-section barrier, 108"W x 84"H overall									
5020 0.5 mm LE panels	H-3	3.20	5	Ea.	19,100	216		19,316	21,400
5030 0.8 mm LE panels		3.20	5		21,200	216		21,416	23,700
5040 1.0 mm LE panels		2.67	5.993		26,000	259		26,259	29,000
5050 1.5 mm LE panels		2.46	6.504		37,000	281		37,281	41,200
7000 X-ray barriers, mobile, mounted within framework w/casters on									
7005 bottom, clear lead plastic window panels on upper portion,									
7010 opaque on lower, 30"W x 75"H overall, incl. framework									
7020 24"H upper w/0.5 mm LE, 48"H lower w/0.8 mm LE	1 Carp	16	.500	Ea.	4,225	25.50		4,250.50	4,700
7030 48"W x 75"H overall, incl. framework									
7040 36"H upper w/0.5 mm LE, 36"H lower w/0.8 mm LE	1 Carp	16	.500	Ea.	6,675	25.50		6,700.50	7,400
7050 36"H upper w/1.0 mm LE, 36"H lower w/1.5 mm LE	"	16	.500	"	7,900	25.50		7,925.50	8,750
7060 72"W x 75"H overall, incl. framework									
7070 36"H upper w/0.5 mm LE, 36"H lower w/0.8 mm LE	1 Carp	16	.500	Ea.	7,900	25.50		7,925.50	8,750
7080 36"H upper w/1.0 mm LE, 36"H lower w/1.5 mm LE	"	16	.500	"	9,900	25.50		9,925.50	10,900

For customer support on your Commercial Renovation Costs with RSMeans data, call 800.448.8182.

395

Division Notes

	CREW	DAILY OUTPUT	LABOR-HOURS	UNIT	BARE COSTS				TOTAL INCL O&P
					MAT.	LABOR	EQUIP.	TOTAL	

Estimating Tips
General

- Many products in Division 14 will require some type of support or blocking for installation not included with the item itself. Examples are supports for conveyors or tube systems, attachment points for lifts, and footings for hoists or cranes. Add these supports in the appropriate division.

14 10 00 Dumbwaiters
14 20 00 Elevators

- Dumbwaiters and elevators are estimated and purchased in a method similar to buying a car. The manufacturer has a base unit with standard features. Added to this base unit price will be whatever options the owner or specifications require. Increased load capacity, additional vertical travel, additional stops, higher speed, and cab finish options are items to be considered. When developing an estimate for dumbwaiters and elevators, remember that some items needed by the installers may have to be included as part of the general contract.

Examples are:

- ☐ shaftway
- ☐ rail support brackets
- ☐ machine room
- ☐ electrical supply
- ☐ sill angles
- ☐ electrical connections
- ☐ pits
- ☐ roof penthouses
- ☐ pit ladders

Check the job specifications and drawings before pricing.

- Installation of elevators and handicapped lifts in historic structures can require significant additional costs. The associated structural requirements may involve cutting into and repairing finishes, moldings, flooring, etc. The estimator must account for these special conditions.

14 30 00 Escalators and Moving Walks

- Escalators and moving walks are specialty items installed by specialty contractors. There are numerous options associated with these items. For specific options, contact a manufacturer or contractor. In a method similar to estimating dumbwaiters and elevators, you should verify the extent of general contract work and add items as necessary.

14 40 00 Lifts
14 90 00 Other Conveying Equipment

- Products such as correspondence lifts, chutes, and pneumatic tube systems, as well as other items specified in this subdivision, may require trained installers. The general contractor might not have any choice as to who will perform the installation or when it will be performed. Long lead times are often required for these products, making early decisions in scheduling necessary.

Reference Numbers

Reference numbers are shown at the beginning of some major classifications. These numbers refer to related items in the Reference Section. The reference information may be an estimating procedure, an alternate pricing method, or technical information.

Note: Not all subdivisions listed here necessarily appear. ■

---

14 11 Manual Dumbwaiters

14 11 10 – Hand Operated Dumbwaiters

14 11 10.20 Manual Dumbwaiters		Crew	Daily Output	Labor-Hours	Unit	Material	2018 Bare Costs Labor	Equipment	Total	Total Incl O&P
0010	**MANUAL DUMBWAITERS**									
0020	2 stop, hand powered, up to 75 lb. capacity	2 Elev	.75	21.333	Ea.	3,100	1,750		4,850	6,100
0100	76 lb. capacity and up		.50	32	"	6,900	2,625		9,525	11,700
0300	For each additional stop, add		.75	21.333	Stop	1,125	1,750		2,875	3,950

14 12 Electric Dumbwaiters

14 12 10 – Dumbwaiters

14 12 10.10 Electric Dumbwaiters

		Crew	Daily Output	Labor-Hours	Unit	Material	2018 Bare Costs Labor	Equipment	Total	Total Incl O&P
0010	**ELECTRIC DUMBWAITERS**									
0020	2 stop, up to 75 lb. capacity	2 Elev	.13	123	Ea.	7,625	10,100		17,725	24,000
0100	76 lb. capacity and up		.11	145	"	23,000	12,000		35,000	43,700
0600	For each additional stop, add		.54	29.630	Stop	3,400	2,425		5,825	7,500

14 21 Electric Traction Elevators

14 21 13 – Electric Traction Freight Elevators

14 21 13.10 Electric Traction Freight Elevators and Options

			Crew	Daily Output	Labor-Hours	Unit	Material	2018 Bare Costs Labor	Equipment	Total	Total Incl O&P
0010	**ELECTRIC TRACTION FREIGHT ELEVATORS AND OPTIONS**	R142000-10									
0425	Electric freight, base unit, 4000 lb., 200 fpm, 4 stop, std. fin.	R142000-20	2 Elev	.05	320	Ea.	122,000	26,300		148,300	175,000
0450	For 5000 lb. capacity, add	R142000-30					6,375			6,375	7,000
0500	For 6000 lb. capacity, add	R142000-40					15,000			15,000	16,500
0525	For 7000 lb. capacity, add						20,100			20,100	22,100
0550	For 8000 lb. capacity, add						24,300			24,300	26,800
0575	For 10000 lb. capacity, add						33,700			33,700	37,000
0600	For 12000 lb. capacity, add						40,900			40,900	45,000
0625	For 16000 lb. capacity, add						49,600			49,600	54,500
0650	For 20000 lb. capacity, add						55,000			55,000	61,000
0675	For increased speed, 250 fpm, add						14,600			14,600	16,000
0700	300 fpm, geared electric, add						18,200			18,200	20,000
0725	350 fpm, geared electric, add						22,200			22,200	24,400
0750	400 fpm, geared electric, add						26,200			26,200	28,800
0775	500 fpm, gearless electric, add						36,100			36,100	39,700
0800	600 fpm, gearless electric, add						40,400			40,400	44,400
0825	700 fpm, gearless electric, add						47,700			47,700	52,500
0850	800 fpm, gearless electric, add						53,000			53,000	58,000
0875	For class "B" loading, add						2,900			2,900	3,175
0900	For class "C-1" loading, add						7,250			7,250	7,975
0925	For class "C-2" loading, add						8,650			8,650	9,500
0950	For class "C-3" loading, add						12,000			12,000	13,200
0975	For travel over 40 V.L.F., add		2 Elev	7.25	2.207	V.L.F.	640	181		821	985
1000	For number of stops over 4, add		"	.27	59.259	Stop	4,675	4,875		9,550	12,700

14 21 23 – Electric Traction Passenger Elevators

14 21 23.10 Electric Traction Passenger Elevators and Options

			Crew	Daily Output	Labor-Hours	Unit	Material	2018 Bare Costs Labor	Equipment	Total	Total Incl O&P
0010	**ELECTRIC TRACTION PASSENGER ELEVATORS AND OPTIONS**	R142000-10									
1625	Electric pass., base unit, 2,000 lb., 200 fpm, 4 stop, std. fin.	R142000-20	2 Elev	.05	320	Ea.	105,000	26,300		131,300	156,000
1650	For 2,500 lb. capacity, add	R142000-30					4,200			4,200	4,600
1675	For 3,000 lb. capacity, add	R142000-40					4,675			4,675	5,150
1700	For 3,500 lb. capacity, add						6,250			6,250	6,900
1725	For 4,000 lb. capacity, add						7,275			7,275	8,000
1750	For 4,500 lb. capacity, add						10,100			10,100	11,100

14 21 23 – Electric Traction Passenger Elevators

14 21 23.10 Electric Traction Passenger Elevators and Options

		Crew	Daily Output	Labor-Hours	Unit	Material	2018 Bare Costs Labor	Equipment	Total	Total Incl O&P
1775	For 5000 lb. capacity, add				Ea.	12,900			12,900	14,200
1800	For increased speed, 250 fpm, geared electric, add					3,475			3,475	3,825
1825	300 fpm, geared electric, add					6,900			6,900	7,600
1850	350 fpm, geared electric, add					8,600			8,600	9,475
1875	400 fpm, geared electric, add					12,700			12,700	14,000
1900	500 fpm, gearless electric, add					31,100			31,100	34,200
1925	600 fpm, gearless electric, add					49,400			49,400	54,500
1950	700 fpm, gearless electric, add					55,500			55,500	61,500
1975	800 fpm, gearless electric, add					61,500			61,500	67,500
2000	For travel over 40 V.L.F., add	2 Elev	7.25	2.207	V.L.F.	770	181		951	1,125
2025	For number of stops over 4, add		.27	59.259	Stop	3,275	4,875		8,150	11,100
2400	Electric hospital, base unit, 4,000 lb., 200 fpm, 4 stop, std fin.		.05	320	Ea.	90,000	26,300		116,300	139,500
2425	For 4,500 lb. capacity, add					6,225			6,225	6,825
2450	For 5,000 lb. capacity, add					8,150			8,150	8,950
2475	For increased speed, 250 fpm, geared electric, add					3,900			3,900	4,275
2500	300 fpm, geared electric, add					7,400			7,400	8,150
2525	350 fpm, geared electric, add					8,950			8,950	9,825
2550	400 fpm, geared electric, add					13,100			13,100	14,400
2575	500 fpm, gearless electric, add					36,600			36,600	40,300
2600	600 fpm, gearless electric, add					53,500			53,500	59,000
2625	700 fpm, gearless electric, add					59,500			59,500	65,500
2650	800 fpm, gearless electric, add					66,500			66,500	73,500
2675	For travel over 40 V.L.F., add	2 Elev	7.25	2.207	V.L.F.	185	181		366	480
2700	For number of stops over 4, add	"	.27	59.259	Stop	4,850	4,875		9,725	12,900

14 21 33 – Electric Traction Residential Elevators

14 21 33.20 Residential Elevators

		Crew	Daily Output	Labor-Hours	Unit	Material	2018 Bare Costs Labor	Equipment	Total	Total Incl O&P
0010	**RESIDENTIAL ELEVATORS**									
7000	Residential, cab type, 1 floor, 2 stop, economy model	2 Elev	.20	80	Ea.	10,800	6,575		17,375	21,900
7100	Custom model		.10	160		18,200	13,200		31,400	40,300
7200	2 floor, 3 stop, economy model		.12	133		16,000	11,000		27,000	34,500
7300	Custom model		.06	267		26,100	21,900		48,000	62,500

14 24 Hydraulic Elevators

14 24 13 – Hydraulic Freight Elevators

14 24 13.10 Hydraulic Freight Elevators and Options

			Crew	Daily Output	Labor-Hours	Unit	Material	2018 Bare Costs Labor	Equipment	Total	Total Incl O&P
0010	**HYDRAULIC FREIGHT ELEVATORS AND OPTIONS**										
1025	Hydraulic freight, base unit, 2,000 lb., 50 fpm, 2 stop, std. fin.	R142000-10 R142000-20	2 Elev	.10	160	Ea.	84,500	13,200		97,700	113,500
1050	For 2,500 lb. capacity, add	R142000-30					4,225			4,225	4,625
1075	For 3,000 lb. capacity, add	R142000-40					5,575			5,575	6,125
1100	For 3,500 lb. capacity, add						9,125			9,125	10,000
1125	For 4,000 lb. capacity, add						10,400			10,400	11,400
1150	For 4,500 lb. capacity, add						12,100			12,100	13,300
1175	For 5,000 lb. capacity, add						16,400			16,400	18,100
1200	For 6,000 lb. capacity, add						16,800			16,800	18,400
1225	For 7,000 lb. capacity, add						25,700			25,700	28,300
1250	For 8,000 lb. capacity, add						30,500			30,500	33,500
1275	For 10,000 lb. capacity, add						32,900			32,900	36,200
1300	For 12,000 lb. capacity, add						40,000			40,000	44,000
1325	For 16,000 lb. capacity, add						53,000			53,000	58,000
1350	For 20,000 lb. capacity, add						59,000			59,000	65,000

14 24 13 – Hydraulic Freight Elevators

14 24 13.10 Hydraulic Freight Elevators and Options	Crew	Daily Output	Labor-Hours	Unit	Material	2018 Bare Costs Labor	Equipment	Total	Total Incl O&P
1375	For increased speed, 100 fpm, add				Ea.	1,200			1,200	1,325
1400	125 fpm, add					3,100			3,100	3,425
1425	150 fpm, add					4,950			4,950	5,450
1450	175 fpm, add					6,950			6,950	7,650
1475	For class "B" loading, add					2,775			2,775	3,050
1500	For class "C-1" loading, add					7,100			7,100	7,800
1525	For class "C-2" loading, add					8,500			8,500	9,350
1550	For class "C-3" loading, add					11,700			11,700	12,900
1575	For travel over 20 V.L.F., add	2 Elev	7.25	2.207	V.L.F.	900	181		1,081	1,275
1600	For number of stops over 2, add	"	.27	59.259	Stop	2,250	4,875		7,125	10,000

14 24 23 – Hydraulic Passenger Elevators

14 24 23.10 Hydraulic Passenger Elevators and Options

		Crew	Daily Output	Labor-Hours	Unit	Material	2018 Bare Costs Labor	Equipment	Total	Total Incl O&P
0010	**HYDRAULIC PASSENGER ELEVATORS AND OPTIONS** R142000-10									
2050	Hyd. pass., base unit, 1,500 lb., 100 fpm, 2 stop, std. fin. R142000-20	2 Elev	.10	160	Ea.	42,900	13,200		56,100	67,500
2075	For 2,000 lb. capacity, add R142000-30					865			865	955
2100	For 2,500 lb. capacity, add R142000-40					3,200			3,200	3,500
2125	For 3,000 lb. capacity, add					4,525			4,525	4,975
2150	For 3,500 lb. capacity, add					7,625			7,625	8,400
2175	For 4,000 lb. capacity, add					9,200			9,200	10,100
2200	For 4,500 lb. capacity, add					12,100			12,100	13,300
2225	For 5,000 lb. capacity, add					17,100			17,100	18,800
2250	For increased speed, 125 fpm, add					1,200			1,200	1,325
2275	150 fpm, add					2,750			2,750	3,025
2300	175 fpm, add					5,450			5,450	6,000
2325	200 fpm, add					10,300			10,300	11,300
2350	For travel over 12 V.L.F., add	2 Elev	7.25	2.207	V.L.F.	730	181		911	1,075
2375	For number of stops over 2, add		.27	59.259	Stop	1,025	4,875		5,900	8,625
2725	Hydraulic hospital, base unit, 4,000 lb., 100 fpm, 2 stop, std. fin.		.10	160	Ea.	66,500	13,200		79,700	94,000
2775	For 4,500 lb. capacity, add					7,500			7,500	8,250
2800	For 5,000 lb. capacity, add					11,000			11,000	12,100
2825	For increased speed, 125 fpm, add					2,250			2,250	2,475
2850	150 fpm, add					3,775			3,775	4,175
2875	175 fpm, add					6,350			6,350	6,975
2900	200 fpm, add					9,300			9,300	10,200
2925	For travel over 12 V.L.F., add	2 Elev	7.25	2.207	V.L.F.	405	181		586	725
2950	For number of stops over 2, add	"	.27	59.259	Stop	4,650	4,875		9,525	12,600

14 27 Custom Elevator Cabs and Doors

14 27 13 – Custom Elevator Cab Finishes

14 27 13.10 Cab Finishes

		Crew	Daily Output	Labor-Hours	Unit	Material	2018 Bare Costs Labor	Equipment	Total	Total Incl O&P
0010	**CAB FINISHES**									
3325	Passenger elevator cab finishes (based on 3,500 lb. cab size)									
3350	Acrylic panel ceiling				Ea.	800			800	880
3375	Aluminum eggcrate ceiling					720			720	795
3400	Stainless steel doors					4,150			4,150	4,550
3425	Carpet flooring					635			635	695
3450	Epoxy flooring					480			480	530
3475	Quarry tile flooring					920			920	1,000
3500	Slate flooring					1,675			1,675	1,850
3525	Textured rubber flooring					670			670	735

14 27 Custom Elevator Cabs and Doors

14 27 13 – Custom Elevator Cab Finishes

14 27 13.10 Cab Finishes	Crew	Daily Output	Labor-Hours	Unit	Material	2018 Bare Costs Labor	Equipment	Total	Total Incl O&P
3550	Stainless steel walls				Ea.	4,275			4,275	4,700
3575	Stainless steel returns at door					1,225			1,225	1,350
4450	Hospital elevator cab finishes (based on 3,500 lb. cab size)									
4475	Aluminum eggcrate ceiling				Ea.	715			715	785
4500	Stainless steel doors					4,150			4,150	4,550
4525	Epoxy flooring					480			480	530
4550	Quarry tile flooring					920			920	1,000
4575	Textured rubber flooring					670			670	735
4600	Stainless steel walls					4,775			4,775	5,250
4625	Stainless steel returns at door					930			930	1,025

14 28 Elevator Equipment and Controls

14 28 10 – Elevator Equipment and Control Options

14 28 10.10 Elevator Controls and Doors

		Crew	Daily Output	Labor-Hours	Unit	Material	2018 Bare Costs Labor	Equipment	Total	Total Incl O&P
0010	**ELEVATOR CONTROLS AND DOORS**									
2975	Passenger elevator options									
3000	2 car group automatic controls	2 Elev	.66	24.242	Ea.	3,575	2,000		5,575	7,000
3025	3 car group automatic controls		.44	36.364		5,425	3,000		8,425	10,600
3050	4 car group automatic controls		.33	48.485		9,250	3,975		13,225	16,400
3075	5 car group automatic controls		.26	61.538		13,700	5,050		18,750	22,900
3100	6 car group automatic controls		.22	72.727		20,900	5,975		26,875	32,200
3125	Intercom service		3	5.333		570	440		1,010	1,300
3150	Duplex car selective collective		.66	24.242		4,000	2,000		6,000	7,475
3175	Center opening 1 speed doors		2	8		2,075	660		2,735	3,325
3200	Center opening 2 speed doors		2	8		2,925	660		3,585	4,225
3225	Rear opening doors (opposite front)		2	8		4,500	660		5,160	5,975
3250	Side opening 2 speed doors		2	8		7,625	660		8,285	9,425
3275	Automatic emergency power switching		.66	24.242		1,275	2,000		3,275	4,475
3300	Manual emergency power switching		8	2		540	164		704	850
3625	Hall finishes, stainless steel doors					1,500			1,500	1,625
3650	Stainless steel frames					1,525			1,525	1,675
3675	12 month maintenance contract								3,750	4,125
3700	Signal devices, hall lanterns	2 Elev	8	2		545	164		709	855
3725	Position indicators, up to 3		9.40	1.702		345	140		485	595
3750	Position indicators, per each over 3		32	.500		96.50	41		137.50	170
3775	High speed heavy duty door opener					2,725			2,725	3,000
3800	Variable voltage, O.H. gearless machine, min.	2 Elev	.16	100		35,200	8,225		43,425	51,500
3815	Maximum		.07	229		77,500	18,800		96,300	114,000
3825	Basement installed geared machine		.33	48.485		13,800	3,975		17,775	21,400
3850	Freight elevator options									
3875	Doors, bi-parting	2 Elev	.66	24.242	Ea.	6,175	2,000		8,175	9,850
3900	Power operated door and gate	"	.66	24.242		25,300	2,000		27,300	30,900
3925	Finishes, steel plate floor					1,125			1,125	1,225
3950	14 ga. 1/4" x 4' steel plate walls					2,075			2,075	2,275
3975	12 month maintenance contract								3,750	4,125
4000	Signal devices, hall lanterns	2 Elev	8	2		530	164		694	840
4025	Position indicators, up to 3		9.40	1.702		370	140		510	625
4050	Position indicators, per each over 3		32	.500		105	41		146	179
4075	Variable voltage basement installed geared machine		.66	24.242		20,900	2,000		22,900	26,100
4100	Hospital elevator options									
4125	2 car group automatic controls	2 Elev	.66	24.242	Ea.	3,575	2,000		5,575	7,000

14 28 10 – Elevator Equipment and Control Options

14 28 10.10 Elevator Controls and Doors	Crew	Daily Output	Labor-Hours	Unit	Material	2018 Bare Costs Labor	Equipment	Total	Total Incl O&P
4150	3 car group automatic controls	2 Elev	.44	36.364	Ea.	5,375	3,000		8,375	10,500
4175	4 car group automatic controls		.33	48.485		13,400	3,975		17,375	20,900
4200	5 car group automatic controls		.26	61.538		13,200	5,050		18,250	22,400
4225	6 car group automatic controls		.22	72.727		20,300	5,975		26,275	31,600
4250	Intercom service		3	5.333		540	440		980	1,275
4275	Duplex car selective collective		.66	24.242		3,900	2,000		5,900	7,350
4300	Center opening 1 speed doors		2	8		2,075	660		2,735	3,300
4325	Center opening 2 speed doors		2	8		2,725	660		3,385	4,025
4350	Rear opening doors (opposite front)		2	8		4,475	660		5,135	5,950
4375	Side opening 2 speed doors		2	8		6,725	660		7,385	8,425
4400	Automatic emergency power switching		.66	24.242		1,225	2,000		3,225	4,425
4425	Manual emergency power switching		8	2		525	164		689	830
4675	Hall finishes, stainless steel doors					1,550			1,550	1,700
4700	Stainless steel frames					1,550			1,550	1,700
4725	12 month maintenance contract								3,750	4,125
4750	Signal devices, hall lanterns	2 Elev	8	2		520	164		684	825
4775	Position indicators, up to 3		9.40	1.702		350	140		490	600
4800	Position indicators, per each over 3		32	.500		96.50	41		137.50	170
4825	High speed heavy duty door opener					2,775			2,775	3,050
4850	Variable voltage, O.H. gearless machine, min.	2 Elev	.16	100		35,800	8,225		44,025	52,000
4865	Maximum		.07	229		78,000	18,800		96,800	115,000
4875	Basement installed geared machine		.33	48.485		18,000	3,975		21,975	26,000
5000	Drilling for piston, casing included, 18" diameter	B-48	80	.700	V.L.F.	64.50	31.50	34.50	130.50	160

14 42 Wheelchair Lifts

14 42 13 – Inclined Wheelchair Lifts

14 42 13.10 Inclined Wheelchair Lifts and Starclimbers

0010	INCLINED WHEELCHAIR LIFTS AND STAIRCLIMBERS									
7700	Stair climber (chair lift), single seat, minimum	2 Elev	1	16	Ea.	5,025	1,325		6,350	7,550
7800	Maximum		.20	80		6,875	6,575		13,450	17,700
8700	Stair lift, minimum		1	16		13,600	1,325		14,925	17,000
8900	Maximum		.20	80		21,600	6,575		28,175	33,800

14 42 16 – Vertical Wheelchair Lifts

14 42 16.10 Wheelchair Lifts

0010	WHEELCHAIR LIFTS									
8000	Wheelchair lift, minimum	2 Elev	1	16	Ea.	6,875	1,325		8,200	9,575
8500	Maximum	"	.50	32	"	16,300	2,625		18,925	22,000

14 45 Vehicle Lifts

14 45 10 – Hydraulic Vehicle Lifts

14 45 10.10 Hydraulic Lifts

0010	HYDRAULIC LIFTS									
2200	Single post, 8,000 lb. capacity	L-4	.40	60	Ea.	6,625	2,850		9,475	11,900
2810	Double post, 6,000 lb. capacity		2.67	8.989		9,350	425		9,775	11,000
2815	9,000 lb. capacity		2.29	10.480		22,200	500		22,700	25,200
2820	15,000 lb. capacity		2	12		25,100	570		25,670	28,500
2822	Four post, 26,000 lb. capacity		1.80	13.333		13,900	635		14,535	16,300
2825	30,000 lb. capacity		1.60	15		55,500	715		56,215	62,000
2830	Ramp style, 4 post, 25,000 lb. capacity		2	12		21,700	570		22,270	24,700

14 45 Vehicle Lifts

14 45 10 – Hydraulic Vehicle Lifts

14 45 10.10 Hydraulic Lifts		Crew	Daily Output	Labor-Hours	Unit	Material	2018 Bare Costs Labor	Equipment	Total	Total Incl O&P
2835	35,000 lb. capacity	L-4	1	24	Ea.	101,000	1,150		102,150	113,000
2840	50,000 lb. capacity		1	24		113,000	1,150		114,150	126,500
2845	75,000 lb. capacity	↓	1	24		131,000	1,150		132,150	146,500
2850	For drive thru tracks, add, minimum					1,125			1,125	1,225
2855	Maximum					1,925			1,925	2,125
2860	Ramp extensions, 3′ (set of 2)					1,150			1,150	1,250
2865	Rolling jack platform					3,975			3,975	4,375
2870	Electric/hydraulic jacking beam					10,600			10,600	11,700
2880	Scissor lift, portable, 6,000 lb. capacity				↓	10,400			10,400	11,500

14 91 Facility Chutes

14 91 33 – Laundry and Linen Chutes

14 91 33.10 Chutes

		Crew	Daily Output	Labor-Hours	Unit	Material	2018 Bare Costs Labor	Equipment	Total	Total Incl O&P
0011	**CHUTES**, linen, trash or refuse									
0050	Aluminized steel, 16 ga., 18″ diameter	2 Shee	3.50	4.571	Floor	1,700	273		1,973	2,300
0100	24″ diameter		3.20	5		1,875	299		2,174	2,550
0400	Galvanized steel, 16 ga., 18″ diameter		3.50	4.571		1,025	273		1,298	1,550
0500	24″ diameter		3.20	5		1,150	299		1,449	1,725
0800	Stainless steel, 18″ diameter		3.50	4.571		3,000	273		3,273	3,725
0900	24″ diameter	↓	3.20	5	↓	3,150	299		3,449	3,925

14 91 82 – Trash Chutes

14 91 82.10 Trash Chutes and Accessories

		Crew	Daily Output	Labor-Hours	Unit	Material	2018 Bare Costs Labor	Equipment	Total	Total Incl O&P
0010	**TRASH CHUTES AND ACCESSORIES**									
2900	Package chutes, spiral type, minimum	2 Shee	4.50	3.556	Floor	2,425	213		2,638	3,000
3000	Maximum	"	1.50	10.667	"	6,325	640		6,965	7,975
9000	Minimum labor/equipment charge	1 Shee	1	8	Job		480		480	755

14 92 Pneumatic Tube Systems

14 92 10 – Conventional, Automatic and Computer Controlled Pneumatic Tube Systems

14 92 10.10 Pneumatic Tube Systems

		Crew	Daily Output	Labor-Hours	Unit	Material	2018 Bare Costs Labor	Equipment	Total	Total Incl O&P
0010	**PNEUMATIC TUBE SYSTEMS**									
0020	100′ long, single tube, 2 stations, stock									
0100	3″ diameter	2 Stpi	.12	133	Total	3,375	8,400		11,775	16,800
0300	4″ diameter	"	.09	178	"	4,275	11,200		15,475	22,100
0400	Twin tube, two stations or more, conventional system									
0700	3″ round	2 Stpi	46	.348	L.F.	38	22		60	76
1050	Add for blower		2	8	System	5,200	505		5,705	6,500
1110	Plus for each round station, add		7.50	2.133	Ea.	1,350	134		1,484	1,700
1200	Alternate pricing method: base cost, economy model		.75	21.333	Total	5,750	1,350		7,100	8,425
1300	Custom model		.25	64	"	11,500	4,025		15,525	19,000
1500	Plus total system length, add, for economy model		93.40	.171	L.F.	8.25	10.80		19.05	26
1600	For custom model	↓	37.60	.426	"	25	27		52	69

For customer support on your Commercial Renovation Costs with RSMeans data, call 800.448.8182.

403

Division Notes

	CREW	DAILY OUTPUT	LABOR-HOURS	UNIT	BARE COSTS				TOTAL INCL O&P
					MAT.	LABOR	EQUIP.	TOTAL	

Estimating Tips

Pipe for fire protection and all uses is located in Subdivisions 21 11 13 and 22 11 13.

The labor adjustment factors listed in Subdivision 22 01 02.20 also apply to Division 21.

Many, but not all, areas in the U.S. require backflow protection in the fire system. Insurance underwriters may have specific requirements for the type of materials to be installed or design requirements based on the hazard to be protected. Local jurisdictions may have requirements not covered by code. It is advisable to be aware of any special conditions.

For your reference, the following is a list of the most applicable Fire Codes and Standards which may be purchased from the NFPA, 1 Batterymarch Park, Quincy, MA 02169-7471.

- NFPA 1: Uniform Fire Code
- NFPA 10: Portable Fire Extinguishers
- NFPA 11: Low-, Medium-, and High-Expansion Foam
- NFPA 12: Carbon Dioxide Extinguishing Systems (Also companion 12A)
- NFPA 13: Installation of Sprinkler Systems (Also companion 13D, 13E, and 13R)
- NFPA 14: Installation of Standpipe and Hose Systems
- NFPA 15: Water Spray Fixed Systems for Fire Protection
- NFPA 16: Installation of Foam-Water Sprinkler and Foam-Water Spray Systems
- NFPA 17: Dry Chemical Extinguishing Systems (Also companion 17A)
- NFPA 18: Wetting Agents
- NFPA 20: Installation of Stationary Pumps for Fire Protection
- NFPA 22: Water Tanks for Private Fire Protection
- NFPA 24: Installation of Private Fire Service Mains and their Appurtenances
- NFPA 25: Inspection, Testing and Maintenance of Water-Based Fire Protection

Reference Numbers

Reference numbers are shown at the beginning of some major classifications. These numbers refer to related items in the Reference Section. The reference information may be an estimating procedure, an alternate pricing method, or technical information.

Note: Not all subdivisions listed here necessarily appear. ■

Did you know?

RSMeans data is available through our online application with 24/7 access:

- Search for unit prices by keyword
- Leverage the most up-to-date data
- Build and export estimates

Try it free for 30 days!
www.rsmeans.com/2018freetrial

21 05 Common Work Results for Fire Suppression

21 05 23 – General-Duty Valves for Water-Based Fire-Suppression Piping

21 05 23.50 General-Duty Valves	Crew	Daily Output	Labor-Hours	Unit	Material	2018 Bare Costs Labor	Equipment	Total	Total Incl O&P
0010 **GENERAL-DUTY VALVES**, for water-based fire suppression									
6500 Check, swing, C.I. body, brass fittings, auto. ball drip									
6520 4" size	Q-12	3	5.333	Ea.	380	287		667	865
9990 Minimum labor/equipment charge	1 Spri	3	2.667	Job		160		160	249

21 11 Facility Fire-Suppression Water-Service Piping

21 11 16 – Facility Fire Hydrants

21 11 16.50 Fire Hydrants for Buildings	Crew	Daily Output	Labor-Hours	Unit	Material	2018 Bare Costs Labor	Equipment	Total	Total Incl O&P
0010 **FIRE HYDRANTS FOR BUILDINGS**									
3750 Hydrants, wall, w/caps, single, flush, polished brass									
3800 2-1/2" x 2-1/2"	Q-12	5	3.200	Ea.	248	172		420	540
4350 Double, projecting, polished brass									
4400 2-1/2" x 2-1/2" x 4"	Q-12	5	3.200	Ea.	296	172		468	595

21 11 19 – Fire-Department Connections

21 11 19.50 Connections for the Fire-Department	Crew	Daily Output	Labor-Hours	Unit	Material	2018 Bare Costs Labor	Equipment	Total	Total Incl O&P
0010 **CONNECTIONS FOR THE FIRE-DEPARTMENT**									
7140 Standpipe connections, wall, w/plugs & chains									
7280 Double, flush, polished brass									
7300 2-1/2" x 2-1/2" x 4"	Q-12	5	3.200	Ea.	760	172		932	1,100
9990 Minimum labor/equipment charge	1 Spri	4	2	Job		120		120	187

21 12 Fire-Suppression Standpipes

21 12 13 – Fire-Suppression Hoses and Nozzles

21 12 13.50 Fire Hoses and Nozzles	Crew	Daily Output	Labor-Hours	Unit	Material	2018 Bare Costs Labor	Equipment	Total	Total Incl O&P
0010 **FIRE HOSES AND NOZZLES**									
0200 Adapters, rough brass, straight hose threads									
0220 One piece, female to male, rocker lugs									
0240 1" x 1"				Ea.	58			58	63.50
0320 2" x 2"					42			42	46
0380 2-1/2" x 2-1/2"					19			19	21
2200 Hose, less couplings									
2260 Synthetic jacket, lined, 300 lb. test, 1-1/2" diameter	Q-12	2600	.006	L.F.	3	.33		3.33	3.82
2270 2" diameter		2200	.007		2.45	.39		2.84	3.31
2280 2-1/2" diameter		2200	.007		5.55	.39		5.94	6.70
2290 3" diameter		2200	.007		3.10	.39		3.49	4.02
2360 High strength, 500 lb. test, 1-1/2" diameter		2600	.006		2.40	.33		2.73	3.16
2380 2-1/2" diameter		2200	.007		5.35	.39		5.74	6.50
5600 Nozzles, brass									
5620 Adjustable fog, 3/4" booster line				Ea.	140			140	154
5640 1-1/2" leader line					103			103	113
5660 2-1/2" direct connection					184			184	202
5680 2-1/2" playpipe nozzle					230			230	253
5780 For chrome plated, add					8%				
9900 Minimum labor/equipment charge	1 Plum	2	4	Job		249		249	385

21 12 Fire-Suppression Standpipes

21 12 19 – Fire-Suppression Hose Racks

21 12 19.50 Fire Hose Racks	Crew	Daily Output	Labor-Hours	Unit	Material	2018 Bare Costs Labor	Equipment	Total	Total Incl O&P
0010 FIRE HOSE RACKS									
2600 Hose rack, swinging, for 1-1/2" diameter hose,									
2620 Enameled steel, 50' and 75' lengths of hose	Q-12	20	.800	Ea.	69	43		112	144

21 12 23 – Fire-Suppression Hose Valves

21 12 23.70 Fire Hose Valves

	Crew	Daily Output	Labor-Hours	Unit	Material	2018 Bare Costs Labor	Equipment	Total	Total Incl O&P
0010 FIRE HOSE VALVES									
3000 Gate, hose, wheel handle, N.R.S., rough brass, 1-1/2"	1 Spri	12	.667	Ea.	154	40		194	232
3040 2-1/2", 300 lb.	"	7	1.143		222	68.50		290.50	350
3080 For polished brass, add					40%				
3090 For polished chrome, add			↓		50%				

21 13 Fire-Suppression Sprinkler Systems

21 13 13 – Wet-Pipe Sprinkler Systems

21 13 13.50 Wet-Pipe Sprinkler System Components

	Crew	Daily Output	Labor-Hours	Unit	Material	2018 Bare Costs Labor	Equipment	Total	Total Incl O&P
0010 WET-PIPE SPRINKLER SYSTEM COMPONENTS									
1220 Water motor, complete with gong	1 Spri	4	2	Ea.	445	120		565	675
1800 Firecycle system, controls, includes panel,									
1900 Flexible sprinkler head connectors									
1910 Braided stainless steel hose with mounting bracket									
1920 1/2" and 3/4" outlet size									
1940 40" length	1 Spri	30	.267	Ea.	76	15.95		91.95	109
1960 60" length	"	22	.364	"	87.50	22		109.50	131
1982 May replace hard-pipe armovers									
1984 for wet and pre-action systems.									
2600 Sprinkler heads, not including supply piping									
3700 Standard spray, pendent or upright, brass, 135°F to 286°F									
3740 1/2" NPT, 1/2" orifice	1 Spri	16	.500	Ea.	10.40	30		40.40	58
3790 For riser & feeder piping, see Section 22 11 13.00									

21 13 16 – Dry-Pipe Sprinkler Systems

21 13 16.50 Dry-Pipe Sprinkler System Components

	Crew	Daily Output	Labor-Hours	Unit	Material	2018 Bare Costs Labor	Equipment	Total	Total Incl O&P
0010 DRY-PIPE SPRINKLER SYSTEM COMPONENTS									
0600 Accelerator	1 Spri	8	1	Ea.	820	60		880	995
0800 Air compressor for dry pipe system, automatic, complete									
0820 30 gal. system capacity, 3/4 HP	1 Spri	1.30	6.154	Ea.	1,150	370		1,520	1,825
2600 Sprinkler heads, not including supply piping									
2640 Dry, pendent, 1/2" orifice, 3/4" or 1" NPT									
2660 1/2" to 6" length	1 Spri	14	.571	Ea.	132	34		166	199
2700 15-1/4" to 18" length	"	14	.571	"	154	34		188	223
6330 Valves and components									
8200 Dry pipe valve, incl. trim and gauges, 3" size	Q-12	2	8	Ea.	2,800	430		3,230	3,750
8220 4" size	"	1	16	"	3,075	860		3,935	4,750

21 13 39 – Foam-Water Systems

21 13 39.50 Foam-Water System Components

	Crew	Daily Output	Labor-Hours	Unit	Material	2018 Bare Costs Labor	Equipment	Total	Total Incl O&P
0010 FOAM-WATER SYSTEM COMPONENTS									
2600 Sprinkler heads, not including supply piping									
3600 Foam-water, pendent or upright, 1/2" NPT	1 Spri	12	.667	Ea.	216	40		256	300

407

For customer support on your Commercial Renovation Costs with RSMeans data, call 800.448.8182.

Division Notes

	CREW	DAILY OUTPUT	LABOR-HOURS	UNIT	BARE COSTS				TOTAL INCL O&P
					MAT.	LABOR	EQUIP.	TOTAL	

Estimating Tips
22 10 00 Plumbing Piping and Pumps

This subdivision is primarily basic pipe and related materials. The pipe may be used by any of the mechanical disciplines, i.e., plumbing, fire protection, heating, and air conditioning.

Note: CPVC plastic piping approved for fire protection is located in 21 11 13.

- The labor adjustment factors listed in Subdivision 22 01 02.20 apply throughout Divisions 21, 22, and 23. CAUTION: the correct percentage may vary for the same items. For example, the percentage add for the basic pipe installation should be based on the maximum height that the installer must install for that particular section. If the pipe is to be located 14' above the floor but it is suspended on threaded rod from beams, the bottom flange of which is 18' high (4' rods), then the height is actually 18' and the add is 20%. The pipe cover, however, does not have to go above the 14', and so the add should be 10%.

- Most pipe is priced first as straight pipe with a joint (coupling, weld, etc.) every 10' and a hanger usually every 10'. There are exceptions with hanger spacing such as for cast iron pipe (5')

and plastic pipe (3 per 10'). Following each type of pipe there are several lines listing sizes and the amount to be subtracted to delete couplings and hangers. This is for pipe that is to be buried or supported together on trapeze hangers. The reason that the couplings are deleted is that these runs are usually long, and frequently longer lengths of pipe are used. By deleting the couplings, the estimator is expected to look up and add back the correct reduced number of couplings.

- When preparing an estimate, it may be necessary to approximate the fittings. Fittings usually run between 25% and 50% of the cost of the pipe. The lower percentage is for simpler runs, and the higher number is for complex areas, such as mechanical rooms.

- For historic restoration projects, the systems must be as invisible as possible, and pathways must be sought for pipes, conduit, and ductwork. While installations in accessible spaces (such as basements and attics) are relatively straightforward to estimate, labor costs may be more difficult to determine when delivery systems must be concealed.

22 40 00 Plumbing Fixtures

- Plumbing fixture costs usually require two lines: the fixture itself and its "rough-in, supply, and waste."

- In the Assemblies Section (Plumbing D2010) for the desired fixture, the System Components Group at the center of the page shows the fixture on the first line. The rest of the list (fittings, pipe, tubing, etc.) will total up to what we refer to in the Unit Price section as "Rough-in, supply, waste, and vent." Note that for most fixtures we allow a nominal 5' of tubing to reach from the fixture to a main or riser.

- Remember that gas- and oil-fired units need venting.

Reference Numbers

Reference numbers are shown at the beginning of some major classifications. These numbers refer to related items in the Reference Section. The reference information may be an estimating procedure, an alternate pricing method, or technical information.

Note: Not all subdivisions listed here necessarily appear. ■

Did you know?

RSMeans data is available through our online application with 24/7 access:

- Search for unit prices by keyword
- Leverage the most up-to-date data
- Build and export estimates

Try it free for 30 days!
www.rsmeans.com/2018freetrial

22 01 02.20 Labor Adjustment Factors	Crew	Daily Output	Labor-Hours	Unit	Material	2018 Bare Costs Labor	Equipment	Total	Total Incl O&P
0010 **LABOR ADJUSTMENT FACTORS** (For Div. 21, 22 and 23) R220102-20									
0100 Labor factors, The below are reasonable suggestions, however									
0110 each project must be evaluated for its own peculiarities, and									
0120 the adjustments be increased or decreased depending on the									
0130 severity of the special conditions.									
1000 Add to labor for elevated installation (Above floor level)									
1080 10' to 14.5' high						10%			
1100 15' to 19.5' high						20%			
1120 20' to 24.5' high						25%			
1140 25' to 29.5' high						35%			
1160 30' to 34.5' high						40%			
1180 35' to 39.5' high						50%			
1200 40' and higher						55%			
2000 Add to labor for crawl space									
2100 3' high						40%			
2140 4' high						30%			
3000 Add to labor for multi-story building									
3010 For new construction (No elevator available)									
3100 Add for floors 3 thru 10						5%			
3110 Add for floors 11 thru 15						10%			
3120 Add for floors 16 thru 20						15%			
3130 Add for floors 21 thru 30						20%			
3140 Add for floors 31 and up						30%			
3170 For existing structure (Elevator available)									
3180 Add for work on floor 3 and above						2%			
4000 Add to labor for working in existing occupied buildings									
4100 Hospital						35%			
4140 Office building						25%			
4180 School						20%			
4220 Factory or warehouse						15%			
4260 Multi dwelling						15%			
5000 Add to labor, miscellaneous									
5100 Cramped shaft						35%			
5140 Congested area						15%			
5180 Excessive heat or cold						30%			
9000 Labor factors, The above are reasonable suggestions, however									
9010 each project should be evaluated for its own peculiarities.									
9100 Other factors to be considered are:									
9140 Movement of material and equipment through finished areas									
9180 Equipment room									
9220 Attic space									
9260 No service road									
9300 Poor unloading/storage area									
9340 Congested site area/heavy traffic									

22 05 05 – Selective Demolition for Plumbing

22 05 05.10 Plumbing Demolition

	Crew	Daily Output	Labor-Hours	Unit	Material	2018 Bare Costs Labor	Equipment	Total	Total Incl O&P
0010 **PLUMBING DEMOLITION** R220105-10									
1020 Fixtures, including 10' piping									
1100 Bathtubs, cast iron R024119-10	1 Plum	4	2	Ea.		124		124	193
1120 Fiberglass		6	1.333			83		83	129
1140 Steel		5	1.600			99.50		99.50	155
1200 Lavatory, wall hung		10	.800			49.50		49.50	77.50
1220 Counter top		8	1			62		62	96.50
1300 Sink, single compartment		8	1			62		62	96.50
1320 Double compartment		7	1.143			71		71	111
1400 Water closet, floor mounted		8	1			62		62	96.50
1420 Wall mounted		7	1.143			71		71	111
1500 Urinal, floor mounted		4	2			124		124	193
1520 Wall mounted		7	1.143			71		71	111
1600 Water fountains, free standing		8	1			62		62	96.50
1620 Wall or deck mounted		6	1.333			83		83	129
2000 Piping, metal, up thru 1-1/2" diameter		200	.040	L.F.		2.49		2.49	3.87
2050 2" thru 3-1/2" diameter		150	.053			3.31		3.31	5.15
2100 4" thru 6" diameter	2 Plum	100	.160			9.95		9.95	15.45
2150 8" thru 14" diameter	"	60	.267			16.55		16.55	26
2153 16" thru 20" diameter	Q-18	70	.343			20	.85	20.85	32.50
2155 24" thru 26" diameter		55	.436			25.50	1.08	26.58	41
2156 30" thru 36" diameter		40	.600			35.50	1.48	36.98	56.50
2160 Plastic pipe with fittings, up thru 1-1/2" diameter	1 Plum	250	.032			1.99		1.99	3.09
2162 2" thru 3" diameter	"	200	.040			2.49		2.49	3.87
2164 4" thru 6" diameter	Q-1	200	.080			4.47		4.47	6.95
2166 8" thru 14" diameter		150	.107			5.95		5.95	9.30
2168 16" diameter		100	.160			8.95		8.95	13.90
2250 Water heater, 40 gal.	1 Plum	6	1.333	Ea.		83		83	129
6000 Remove and reset fixtures, easy access		6	1.333			83		83	129
6100 Difficult access		4	2			124		124	193
9000 Minimum labor/equipment charge		2	4	Job		249		249	385

22 05 23 – General-Duty Valves for Plumbing Piping

22 05 23.20 Valves, Bronze

	Crew	Daily Output	Labor-Hours	Unit	Material	2018 Bare Costs Labor	Equipment	Total	Total Incl O&P
0010 **VALVES, BRONZE**									
1020 Angle, 150 lb., rising stem, threaded									
1070 3/4"	1 Plum	20	.400	Ea.	211	25		236	271
1080 1"		19	.421		300	26		326	375
1100 1-1/2"		13	.615		450	38.50		488.50	555
1102 Soldered same price as threaded									
1110 2"	1 Plum	11	.727	Ea.	725	45		770	870
1300 Ball									
1398 Threaded, 150 psi									
1460 3/4"	1 Plum	20	.400	Ea.	24	25		49	65
1522 Solder the same price as threaded									
1750 Check, swing, class 150, regrinding disc, threaded									
1850 1/2"	1 Plum	24	.333	Ea.	66.50	20.50		87	105
1860 3/4"		20	.400		86.50	25		111.50	134
1870 1"		19	.421		141	26		167	196
1890 1-1/2"		13	.615		263	38.50		301.50	350
1900 2"		11	.727		305	45		350	405
2850 Gate, N.R.S., soldered, 125 psi									
2920 1/2"	1 Plum	24	.333	Ea.	56	20.50		76.50	93.50

22 05 23.20 Valves, Bronze

22 05 23.20 Valves, Bronze		Crew	Daily Output	Labor-Hours	Unit	Material	2018 Bare Costs Labor	Equipment	Total	Total Incl O&P
2940	3/4"	1 Plum	20	.400	Ea.	64	25		89	109
2950	1"		19	.421		74.50	26		100.50	123
2970	1-1/2"		13	.615		154	38.50		192.50	230
2980	2"		11	.727		181	45		226	270
4850	Globe, class 150, rising stem, threaded									
4950	1/2"	1 Plum	24	.333	Ea.	102	20.50		122.50	145
4960	3/4"	"	20	.400	"	137	25		162	189
5600	Relief, pressure & temperature, self-closing, ASME, threaded									
5640	3/4"	1 Plum	28	.286	Ea.	209	17.75		226.75	258
5650	1"		24	.333		335	20.50		355.50	400
5660	1-1/4"		20	.400		685	25		710	795
5670	1-1/2"		18	.444		1,325	27.50		1,352.50	1,500
5680	2"		16	.500		1,425	31		1,456	1,625
6400	Pressure, water, ASME, threaded									
6440	3/4"	1 Plum	28	.286	Ea.	144	17.75		161.75	186
6450	1"		24	.333		296	20.50		316.50	355
6470	1-1/2"		18	.444		650	27.50		677.50	755
6900	Reducing, water pressure									
6920	300 psi to 25-75 psi, threaded or sweat									
6940	1/2"	1 Plum	24	.333	Ea.	440	20.50		460.50	515
6950	3/4"		20	.400		450	25		475	535
6960	1"		19	.421		700	26		726	810
8350	Tempering, water, sweat connections									
8400	1/2"	1 Plum	24	.333	Ea.	106	20.50		126.50	148
8440	3/4"	"	20	.400	"	139	25		164	192
8650	Threaded connections									
8700	1/2"	1 Plum	24	.333	Ea.	148	20.50		168.50	195
8740	3/4"	"	20	.400	"	875	25		900	1,000
8800	Water heater water & gas safety shut off									
8810	Protection against a leaking water heater									
8814	Shut off valve	1 Plum	16	.500	Ea.	185	31		216	253
8818	Water heater dam		32	.250		31.50	15.55		47.05	58.50
8822	Gas control wiring harness		32	.250		19.75	15.55		35.30	45.50
8830	Whole house flood safety shut off									
8834	Connections									
8838	3/4" NPT	1 Plum	12	.667	Ea.	935	41.50		976.50	1,100
8842	1" NPT		11	.727		960	45		1,005	1,125
8846	1-1/4" NPT		10	.800		995	49.50		1,044.50	1,175
9000	Minimum labor/equipment charge		4	2	Job		124		124	193

22 05 23.60 Valves, Plastic

22 05 23.60 Valves, Plastic		Crew	Daily Output	Labor-Hours	Unit	Material	2018 Bare Costs Labor	Equipment	Total	Total Incl O&P
0010	**VALVES, PLASTIC**									
1150	Ball, PVC, socket or threaded, true union									
1230	1/2"	1 Plum	26	.308	Ea.	42.50	19.10		61.60	77
1240	3/4"		25	.320		42.50	19.90		62.40	77.50
1250	1"		23	.348		50.50	21.50		72	89
1260	1-1/4"		21	.381		80	23.50		103.50	126
1270	1-1/2"		20	.400		80	25		105	127
1280	2"		17	.471		127	29.50		156.50	185
1650	CPVC, socket or threaded, single union									
1700	1/2"	1 Plum	26	.308	Ea.	55.50	19.10		74.60	91
1720	3/4"		25	.320		69	19.90		88.90	107
1730	1"		23	.348		84	21.50		105.50	126

22 05 Common Work Results for Plumbing

22 05 23 – General-Duty Valves for Plumbing Piping

	22 05 23.60 Valves, Plastic	Crew	Daily Output	Labor-Hours	Unit	Material	2018 Bare Costs Labor	Equipment	Total	Total Incl O&P
1750	1-1/4"	1 Plum	21	.381	Ea.	146	23.50		169.50	197
1760	1-1/2"		20	.400		134	25		159	186
1770	2"	↓	17	.471	↓	185	29.50		214.50	249
3150	Ball check, PVC, socket or threaded									
3290	2"	1 Plum	17	.471	Ea.	140	29.50		169.50	200
9000	Minimum labor/equipment charge	"	3.50	2.286	Job		142		142	221

22 05 76 – Facility Drainage Piping Cleanouts

22 05 76.10 Cleanouts

	22 05 76.10 Cleanouts	Crew	Daily Output	Labor-Hours	Unit	Material	2018 Bare Costs Labor	Equipment	Total	Total Incl O&P
0010	**CLEANOUTS**									
0060	Floor type									
0080	Round or square, scoriated nickel bronze top									
0100	2" pipe size	1 Plum	10	.800	Ea.	184	49.50		233.50	281
0120	3" pipe size		8	1		249	62		311	370
0140	4" pipe size	↓	6	1.333	↓	277	83		360	435
0980	Round top, recessed for terrazzo									
1000	2" pipe size	1 Plum	9	.889	Ea.	515	55		570	650
1080	3" pipe size		6	1.333		530	83		613	715
1100	4" pipe size		4	2		650	124		774	910
1120	5" pipe size	Q-1	6	2.667	↓	940	149		1,089	1,250
9000	Minimum labor/equipment charge	1 Plum	3	2.667	Job		166		166	258

22 05 76.20 Cleanout Tees

	22 05 76.20 Cleanout Tees	Crew	Daily Output	Labor-Hours	Unit	Material	2018 Bare Costs Labor	Equipment	Total	Total Incl O&P
0010	**CLEANOUT TEES**									
0100	Cast iron, B&S, with countersunk plug									
0220	3" pipe size	1 Plum	3.60	2.222	Ea.	150	138		288	380
0240	4" pipe size	"	3.30	2.424	"	232	151		383	490
0500	For round smooth access cover, same price									
4000	Plastic, tees and adapters. Add plugs									
4010	ABS, DWV									
4020	Cleanout tee, 1-1/2" pipe size	1 Plum	15	.533	Ea.	12.20	33		45.20	65
9000	Minimum labor/equipment charge	"	2.75	2.909	Job		181		181	281

22 07 Plumbing Insulation

22 07 16 – Plumbing Equipment Insulation

22 07 16.10 Insulation for Plumbing Equipment

	22 07 16.10 Insulation for Plumbing Equipment		Crew	Daily Output	Labor-Hours	Unit	Material	2018 Bare Costs Labor	Equipment	Total	Total Incl O&P
0010	**INSULATION FOR PLUMBING EQUIPMENT**										
2900	Domestic water heater wrap kit										
2920	1-1/2" with vinyl jacket, 20 to 60 gal.	G	1 Plum	8	1	Ea.	16.50	62		78.50	115
2925	50 to 80 gal.	G	"	8	1	"	27.50	62		89.50	127

22 07 19 – Plumbing Piping Insulation

22 07 19.10 Piping Insulation

	22 07 19.10 Piping Insulation	Crew	Daily Output	Labor-Hours	Unit	Material	2018 Bare Costs Labor	Equipment	Total	Total Incl O&P
0010	**PIPING INSULATION**									
0110	Insulation req'd. is based on the surface size/area to be covered									
0230	Insulated protectors (ADA)									
0235	For exposed piping under sinks or lavatories ♿									
0240	Vinyl coated foam, velcro tabs									
0245	P Trap, 1-1/4" or 1-1/2"	1 Plum	32	.250	Ea.	15.90	15.55		31.45	41.50
0260	Valve and supply cover									
0265	1/2", 3/8", and 7/16" pipe size	1 Plum	32	.250	Ea.	15.60	15.55		31.15	41
0280	Tailpiece offset (wheelchair) ♿									
0285	1-1/4" pipe size	1 Plum	32	.250	Ea.	12.90	15.55		28.45	38

413

22 07 Plumbing Insulation

22 07 19 – Plumbing Piping Insulation

22 07 19.10 Piping Insulation		Crew	Daily Output	Labor-Hours	Unit	Material	2018 Bare Costs Labor	Equipment	Total	Total Incl O&P
0600	Pipe covering (price copper tube one size less than IPS)									
6120	2" wall, 1/2" iron pipe size	G	Q-14	145	.110	L.F.	6.10	5.60		11.70	15.70
6210	4" iron pipe size		"	125	.128	"	11.15	6.50		17.65	22.50
6600	Fiberglass, with all service jacket									
6840	1" wall, 1/2" iron pipe size	G	Q-14	240	.067	L.F.	.85	3.38		4.23	6.35
6860	3/4" iron pipe size	G		230	.070		.92	3.53		4.45	6.65
6870	1" iron pipe size	G		220	.073		.99	3.69		4.68	7
6880	1-1/4" iron pipe size	G		210	.076		1.07	3.86		4.93	7.40
6890	1-1/2" iron pipe size	G		210	.076		1.15	3.86		5.01	7.45
6900	2" iron pipe size	G		200	.080		1.53	4.05		5.58	8.20
7080	1-1/2" wall, 1/2" iron pipe size	G		230	.070		1.98	3.53		5.51	7.85
7140	2" iron pipe size	G		190	.084		2.67	4.27		6.94	9.80
7160	3" iron pipe size	G		170	.094		2.99	4.77		7.76	10.95
7879	Rubber tubing, flexible closed cell foam									
8100	1/2" wall, 1/4" iron pipe size	G	1 Asbe	90	.089	L.F.	.72	5		5.72	8.80
8120	3/8" iron pipe size	G		90	.089		.77	5		5.77	8.85
8130	1/2" iron pipe size	G		89	.090		.84	5.05		5.89	9
8140	3/4" iron pipe size	G		89	.090		.95	5.05		6	9.15
8150	1" iron pipe size	G		88	.091		.82	5.10		5.92	9.10
8170	1-1/2" iron pipe size	G		87	.092		1.49	5.20		6.69	9.95
8180	2" iron pipe size	G		86	.093		1.91	5.25		7.16	10.50
8200	3" iron pipe size	G		85	.094		2.37	5.30		7.67	11.10
8220	4" iron pipe size	G		80	.100		3.57	5.65		9.22	12.95
8300	3/4" wall, 1/4" iron pipe size	G		90	.089		.89	5		5.89	9
8330	1/2" iron pipe size	G		89	.090		1.09	5.05		6.14	9.30
8340	3/4" iron pipe size	G		89	.090		1.54	5.05		6.59	9.80
8350	1" iron pipe size	G		88	.091		1.77	5.10		6.87	10.15
8360	1-1/4" iron pipe size	G		87	.092		2.39	5.20		7.59	10.95
8370	1-1/2" iron pipe size	G		87	.092		2.62	5.20		7.82	11.20
8380	2" iron pipe size	G		86	.093		3.04	5.25		8.29	11.75
8444	1" wall, 1/2" iron pipe size	G		86	.093		2.32	5.25		7.57	10.95
8445	3/4" iron pipe size	G		84	.095		2.81	5.35		8.16	11.70
8446	1" iron pipe size	G		84	.095		3.39	5.35		8.74	12.35
8447	1-1/4" iron pipe size	G		82	.098		3.75	5.50		9.25	12.95
8448	1-1/2" iron pipe size	G		82	.098		4.88	5.50		10.38	14.15
8449	2" iron pipe size	G		80	.100		6.10	5.65		11.75	15.70
8450	2-1/2" iron pipe size	G		80	.100		7.45	5.65		13.10	17.20
8456	Rubber insulation tape, 1/8" x 2" x 30'	G				Ea.	21			21	23
9600	Minimum labor/equipment charge		1 Plum	4	2	Job		124		124	193

22 11 Facility Water Distribution

22 11 13 – Facility Water Distribution Piping

22 11 13.14 Pipe, Brass

		Crew	Daily Output	Labor-Hours	Unit	Material	2018 Bare Costs Labor	Equipment	Total	Total Incl O&P
0010	**PIPE, BRASS**, Plain end									
0900	Field threaded, coupling & clevis hanger assembly 10' OC									
0920	Regular weight									
1120	1/2" diameter	1 Plum	48	.167	L.F.	7.65	10.35		18	24.50
1140	3/4" diameter		46	.174		10.05	10.80		20.85	28
1160	1" diameter		43	.186		6.90	11.55		18.45	25.50
1180	1-1/4" diameter	Q-1	72	.222		22.50	12.45		34.95	44.50
1200	1-1/2" diameter		65	.246		27	13.75		40.75	51

22 11 13 – Facility Water Distribution Piping

22 11 13.14 Pipe, Brass	Crew	Daily Output	Labor-Hours	Unit	Material	2018 Bare Costs Labor	Equipment	Total	Total Incl O&P
1220	2" diameter	Q-1	53	.302	L.F.	18.45	16.90		35.35	47
9000	Minimum labor/equipment charge	1 Plum	4	2	Job		124		124	193

22 11 13.16 Pipe Fittings, Brass

		Crew	Daily Output	Labor-Hours	Unit	Material	2018 Bare Costs Labor	Equipment	Total	Total Incl O&P
0010	**PIPE FITTINGS, BRASS**, Rough bronze, threaded, lead free.									
1000	Standard wt., 90° elbow									
1100	1/2"	1 Plum	12	.667	Ea.	22	41.50		63.50	88.50
1120	3/4"		11	.727		29.50	45		74.50	103
1140	1"	↓	10	.800		48	49.50		97.50	131
1160	1-1/4"	Q-1	17	.941		77.50	52.50		130	167
1180	1-1/2"	"	16	1		96	56		152	192
1500	45° elbow, 1/8"	1 Plum	13	.615		27	38.50		65.50	89.50
1580	1/2"		12	.667		27	41.50		68.50	94.50
1600	3/4"		11	.727		38.50	45		83.50	113
1620	1"	↓	10	.800		65.50	49.50		115	150
1640	1-1/4"	Q-1	17	.941		104	52.50		156.50	197
1660	1-1/2"	"	16	1		131	56		187	231
2000	Tee, 1/8"	1 Plum	9	.889		26	55		81	115
2080	1/2"		8	1		26	62		88	125
2100	3/4"		7	1.143		37	71		108	152
2120	1"	↓	6	1.333		66.50	83		149.50	202
2140	1-1/4"	Q-1	10	1.600		114	89.50		203.50	265
2160	1-1/2"	"	9	1.778		129	99.50		228.50	297
2500	Coupling, 1/8"	1 Plum	26	.308		18.45	19.10		37.55	50.50
2580	1/2"		15	.533		18.45	33		51.45	72
2600	3/4"		14	.571		26	35.50		61.50	84
2620	1"	↓	13	.615		44	38.50		82.50	108
2640	1-1/4"	Q-1	22	.727		73.50	40.50		114	145
2660	1-1/2"		20	.800		96	44.50		140.50	175
2680	2"	↓	18	.889	↓	159	49.50		208.50	252
9000	Minimum labor/equipment charge	1 Plum	4	2	Job		124		124	193

22 11 13.23 Pipe/Tube, Copper

		Crew	Daily Output	Labor-Hours	Unit	Material	2018 Bare Costs Labor	Equipment	Total	Total Incl O&P
0010	**PIPE/TUBE, COPPER**, Solder joints									
1000	Type K tubing, couplings & clevis hanger assemblies 10' OC									
1100	1/4" diameter	1 Plum	84	.095	L.F.	4.11	5.90		10.01	13.70
1200	1" diameter		66	.121		12.30	7.55		19.85	25
1260	2" diameter	↓	40	.200	↓	26	12.45		38.45	48
2000	Type L tubing, couplings & clevis hanger assemblies 10' OC									
2100	1/4" diameter	1 Plum	88	.091	L.F.	2.64	5.65		8.29	11.70
2120	3/8" diameter		84	.095		3.25	5.90		9.15	12.75
2140	1/2" diameter		81	.099		3.34	6.15		9.49	13.25
2160	5/8" diameter		79	.101		5.75	6.30		12.05	16.15
2180	3/4" diameter		76	.105		4.49	6.55		11.04	15.15
2200	1" diameter		68	.118		6.90	7.30		14.20	19
2220	1-1/4" diameter		58	.138		11.45	8.55		20	26
2240	1-1/2" diameter		52	.154		10.30	9.55		19.85	26.50
2260	2" diameter	↓	42	.190		17.90	11.85		29.75	38
2280	2-1/2" diameter	Q-1	62	.258		25.50	14.45		39.95	50.50
2300	3" diameter		56	.286		43	16		59	72
2320	3-1/2" diameter		43	.372		50.50	21		71.50	88
2340	4" diameter		39	.410		62	23		85	104
2360	5" diameter	↓	34	.471		143	26.50		169.50	199
2380	6" diameter	Q-2	40	.600	↓	145	35		180	213

For customer support on your Commercial Renovation Costs with RSMeans data, call 800.448.8182.

415

22 11 13.23 Pipe/Tube, Copper

	Crew	Daily Output	Labor-Hours	Unit	Material	2018 Bare Costs Labor	Equipment	Total	Total Incl O&P
2410	For other than full hard temper, add				L.F.	21%				
2590	For silver solder, add						15%			
3000	Type M tubing, couplings & clevis hanger assemblies 10' OC									
3140	1/2" diameter	1 Plum	84	.095	L.F.	3.70	5.90		9.60	13.25
3180	3/4" diameter		78	.103		5.20	6.35		11.55	15.65
3200	1" diameter	↓	70	.114	↓	8.20	7.10		15.30	20
4000	Type DWV tubing, couplings & clevis hanger assemblies 10' OC									
4100	1-1/4" diameter	1 Plum	60	.133	L.F.	11.60	8.30		19.90	25.50
4120	1-1/2" diameter		54	.148		11.25	9.20		20.45	26.50
4140	2" diameter	↓	44	.182		18.05	11.30		29.35	37.50
4160	3" diameter	Q-1	58	.276		24.50	15.45		39.95	51
4180	4" diameter		40	.400		55	22.50		77.50	95.50
4200	5" diameter	↓	36	.444		97.50	25		122.50	146
4220	6" diameter	Q-2	42	.571	↓	142	33		175	209
9000	Minimum labor/equipment charge	1 Plum	4	2	Job		124		124	193

22 11 13.25 Pipe/Tube Fittings, Copper

	Crew	Daily Output	Labor-Hours	Unit	Material	2018 Bare Costs Labor	Equipment	Total	Total Incl O&P
0010	**PIPE/TUBE FITTINGS, COPPER**, Wrought unless otherwise noted									
0020	For silver solder, add						15%			
0040	Solder joints, copper x copper									
0070	90° elbow, 1/4"	1 Plum	22	.364	Ea.	3.15	22.50		25.65	38.50
0100	1/2"		20	.400		1.13	25		26.13	39.50
0120	3/4"		19	.421		2.52	26		28.52	43.50
0130	1"		16	.500		6.20	31		37.20	55.50
0250	45° elbow, 1/4"		22	.364		8	22.50		30.50	44
0270	3/8"		22	.364		6.85	22.50		29.35	42.50
0280	1/2"		20	.400		2.83	25		27.83	41.50
0290	5/8"		19	.421		12	26		38	53.50
0300	3/4"		19	.421		3.56	26		29.56	44.50
0310	1"		16	.500		8.95	31		39.95	58.50
0320	1-1/4"		15	.533		16.20	33		49.20	69.50
0330	1-1/2"		13	.615		15	38.50		53.50	76
0340	2"	↓	11	.727		25	45		70	98
0350	2-1/2"	Q-1	13	1.231		51.50	69		120.50	164
0360	3"		13	1.231		90	69		159	206
0370	3-1/2"		10	1.600		128	89.50		217.50	279
0380	4"		9	1.778		156	99.50		255.50	325
0390	5"	↓	6	2.667		590	149		739	880
0400	6"	Q-2	9	2.667		930	155		1,085	1,275
0450	Tee, 1/4"	1 Plum	14	.571		7.70	35.50		43.20	64
0470	3/8"		14	.571		6.10	35.50		41.60	62
0480	1/2"		13	.615		1.92	38.50		40.42	61.50
0490	5/8"		12	.667		16.45	41.50		57.95	82.50
0500	3/4"		12	.667		6.35	41.50		47.85	71.50
0510	1"		10	.800		14.10	49.50		63.60	93
0520	1-1/4"		9	.889		22	55		77	110
0530	1-1/2"		8	1		33.50	62		95.50	133
0540	2"	↓	7	1.143		48	71		119	164
0550	2-1/2"	Q-1	8	2		104	112		216	288
0560	3"		7	2.286		145	128		273	360
0570	3-1/2"		6	2.667		485	149		634	765
0580	4"		5	3.200		330	179		509	645
0590	5"	↓	4	4		850	224		1,074	1,275

22 11 13 – Facility Water Distribution Piping

22 11 13.25 Pipe/Tube Fittings, Copper		Crew	Daily Output	Labor-Hours	Unit	Material	2018 Bare Costs Labor	Equipment	Total	Total Incl O&P
0600	6"	Q-2	6	4	Ea.	1,175	232		1,407	1,625
0612	Tee, reducing on the outlet, 1/4"	1 Plum	15	.533		14.70	33		47.70	67.50
0613	3/8"		15	.533		14.25	33		47.25	67
0614	1/2"		14	.571		13.15	35.50		48.65	70
0615	5/8"		13	.615		26.50	38.50		65	88.50
0616	3/4"		12	.667		8.50	41.50		50	74
0617	1"		11	.727		23.50	45		68.50	96.50
0618	1-1/4"		10	.800		28.50	49.50		78	109
0619	1-1/2"		9	.889		30	55		85	119
0620	2"		8	1		49	62		111	151
0621	2-1/2"	Q-1	9	1.778		136	99.50		235.50	305
0622	3"		8	2		140	112		252	330
0623	4"		6	2.667		283	149		432	540
0624	5"		5	3.200		1,550	179		1,729	1,975
0625	6"	Q-2	7	3.429		2,000	199		2,199	2,500
0626	8"	"	6	4		8,125	232		8,357	9,300
0630	Tee, reducing on the run, 1/4"	1 Plum	15	.533		19	33		52	72.50
0631	3/8"		15	.533		26.50	33		59.50	80.50
0632	1/2"		14	.571		18.10	35.50		53.60	75.50
0633	5/8"		13	.615		28.50	38.50		67	91
0634	3/4"		12	.667		19.80	41.50		61.30	86.50
0635	1"		11	.727		23	45		68	96
0636	1-1/4"		10	.800		36.50	49.50		86	118
0637	1-1/2"		9	.889		62	55		117	154
0638	2"		8	1		82	62		144	187
0639	2-1/2"	Q-1	9	1.778		175	99.50		274.50	350
0640	3"		8	2		256	112		368	455
0641	4"		6	2.667		520	149		669	805
0642	5"		5	3.200		1,175	179		1,654	1,875
0643	6"	Q-2	7	3.429		2,225	199		2,424	2,750
0644	8"	"	6	4		8,875	232		9,107	10,100
0650	Coupling, 1/4"	1 Plum	24	.333		.95	20.50		21.45	33
0670	3/8"		24	.333		1.24	20.50		21.74	33.50
0680	1/2"		22	.364		.86	22.50		23.36	36
0690	5/8"		21	.381		3.43	23.50		26.93	41
0700	3/4"		21	.381		2.35	23.50		25.85	39.50
0710	1"		18	.444		4.66	27.50		32.16	48
0715	1-1/4"		17	.471		6.10	29.50		35.60	52
0716	1-1/2"		15	.533		8.15	33		41.15	60.50
0718	2"		13	.615		13.60	38.50		52.10	74.50
0721	2-1/2"	Q-1	15	1.067		35	59.50		94.50	132
0722	3"		13	1.231		40.50	69		109.50	152
0724	3-1/2"		8	2		94.50	112		206.50	278
0726	4"		7	2.286		84.50	128		212.50	292
0728	5"		6	2.667		203	149		352	455
0731	6"	Q-2	8	3		335	174		509	640
2000	DWV, solder joints, copper x copper									
2030	90° elbow, 1-1/4"	1 Plum	13	.615	Ea.	16.50	38.50		55	77.50
2050	1-1/2"		12	.667		22	41.50		63.50	88.50
2070	2"		10	.800		35	49.50		84.50	116
2090	3"	Q-1	10	1.600		84.50	89.50		174	232
2100	4"	"	9	1.778		410	99.50		509.50	605
2250	Tee, sanitary, 1-1/4"	1 Plum	9	.889		25.50	55		80.50	114

22 11 13.25 Pipe/Tube Fittings, Copper

		Crew	Daily Output	Labor-Hours	Unit	Material	2018 Bare Costs Labor	Equipment	Total	Total Incl O&P
2270	1-1/2"	1 Plum	8	1	Ea.	31.50	62		93.50	131
2290	2"	▼	7	1.143		49.50	71		120.50	165
2310	3"	Q-1	7	2.286		188	128		316	405
2330	4"	"	6	2.667		455	149		604	730
2400	Coupling, 1-1/4"	1 Plum	14	.571		6.85	35.50		42.35	63
2420	1-1/2"		13	.615		8.50	38.50		47	69
2440	2"	▼	11	.727		11.80	45		56.80	83.50
2460	3"	Q-1	11	1.455		27.50	81.50		109	157
2480	4"	"	10	1.600	▼	60.50	89.50		150	206
9000	Minimum labor/equipment charge	1 Plum	4	2	Job		124		124	193

22 11 13.44 Pipe, Steel

		Crew	Daily Output	Labor-Hours	Unit	Material	2018 Bare Costs Labor	Equipment	Total	Total Incl O&P
0010	**PIPE, STEEL** R221113-50									
0020	All pipe sizes are to Spec. A-53 unless noted otherwise									
0050	Schedule 40, threaded, with couplings, and clevis hanger									
0060	assemblies sized for covering, 10' OC									
0540	Black, 1/4" diameter	1 Plum	66	.121	L.F.	5.55	7.55		13.10	17.80
0560	1/2" diameter		63	.127		3.82	7.90		11.72	16.50
0570	3/4" diameter		61	.131		4.18	8.15		12.33	17.30
0580	1" diameter	▼	53	.151		4.38	9.40		13.78	19.40
0590	1-1/4" diameter	Q-1	89	.180		5.15	10.05		15.20	21.50
0600	1-1/2" diameter		80	.200		5.70	11.20		16.90	23.50
0610	2" diameter		64	.250		11.95	14		25.95	35
0620	2-1/2" diameter		50	.320		16.05	17.90		33.95	45.50
0630	3" diameter		43	.372		18.80	21		39.80	53
0640	3-1/2" diameter		40	.400		24.50	22.50		47	62
0650	4" diameter	▼	36	.444	▼	21.50	25		46.50	62
1280	All pipe sizes are to Spec. A-53 unless noted otherwise									
1281	Schedule 40, threaded, with couplings and clevis hanger									
1282	assemblies sized for covering, 10' OC									
1290	Galvanized, 1/4" diameter	1 Plum	66	.121	L.F.	7.25	7.55		14.80	19.70
1310	1/2" diameter		63	.127		3.86	7.90		11.76	16.55
1320	3/4" diameter		61	.131		4.43	8.15		12.58	17.55
1330	1" diameter	▼	53	.151		4.87	9.40		14.27	19.95
1350	1-1/2" diameter	Q-1	80	.200		6.40	11.20		17.60	24.50
1360	2" diameter		64	.250		12.95	14		26.95	36.50
1370	2-1/2" diameter		50	.320		17.80	17.90		35.70	47.50
1380	3" diameter		43	.372		21	21		42	55.50
1400	4" diameter	▼	36	.444	▼	24	25		49	65
2000	Welded, sch. 40, on yoke & roll hanger assy's, sized for covering, 10' OC									
2040	Black, 1" diameter	Q-15	93	.172	L.F.	4.18	9.60	.64	14.42	20.50
2070	2" diameter		61	.262		11.50	14.65	.97	27.12	37
2090	3" diameter		43	.372		17	21	1.38	39.38	52.50
2110	4" diameter		37	.432		16.85	24	1.60	42.45	58
2120	5" diameter	▼	32	.500		34.50	28	1.85	64.35	83.50
2130	6" diameter	Q-16	36	.667	▼	42.50	38.50	1.65	82.65	108
9990	Minimum labor/equipment charge	1 Plum	3	2.667	Job		166		166	258

22 11 13.45 Pipe Fittings, Steel, Threaded

		Crew	Daily Output	Labor-Hours	Unit	Material	2018 Bare Costs Labor	Equipment	Total	Total Incl O&P
0010	**PIPE FITTINGS, STEEL, THREADED** R221113-50									
0020	Cast iron									
0040	Standard weight, black									
0060	90° elbow, straight									
0070	1/4"	1 Plum	16	.500	Ea.	10.75	31		41.75	60.50

22 11 Facility Water Distribution

22 11 13 – Facility Water Distribution Piping

22 11 13.45 Pipe Fittings, Steel, Threaded	Crew	Daily Output	Labor-Hours	Unit	Material	2018 Bare Costs Labor	Equipment	Total	Total Incl O&P
0080	3/8"	1 Plum	16	.500	Ea.	15.55	31		46.55	65.50
0090	1/2"		15	.533		6.80	33		39.80	59
0100	3/4"		14	.571		7.10	35.50		42.60	63.50
0110	1"		13	.615		8.40	38.50		46.90	69
0130	1-1/2"	Q-1	20	.800		16.50	44.50		61	87.50
0140	2"		18	.889		26	49.50		75.50	106
0150	2-1/2"		14	1.143		62	64		126	168
0160	3"		10	1.600		101	89.50		190.50	250
0180	4"		6	2.667		187	149		336	440
0500	Tee, straight									
0510	1/4"	1 Plum	10	.800	Ea.	16.80	49.50		66.30	96
0520	3/8"		10	.800		16.35	49.50		65.85	95.50
0540	3/4"		9	.889		12.35	55		67.35	99.50
0550	1"		8	1		11	62		73	109
0570	1-1/2"	Q-1	13	1.231		26	69		95	136
0580	2"		11	1.455		36.50	81.50		118	167
0590	2-1/2"		9	1.778		94.50	99.50		194	259
0600	3"		6	2.667		145	149		294	390
0620	4"		4	4		283	224		507	660
0700	Standard weight, galvanized cast iron									
0720	90° elbow, straight									
0730	1/4"	1 Plum	16	.500	Ea.	18	31		49	68.50
0740	3/8"		16	.500		18	31		49	68.50
0750	1/2"		15	.533		20.50	33		53.50	74
0760	3/4"		14	.571		20	35.50		55.50	77.50
0770	1"		13	.615		23.50	38.50		62	85
0790	1-1/2"	Q-1	20	.800		49.50	44.50		94	124
0800	2"		18	.889		73.50	49.50		123	159
0810	2-1/2"		14	1.143		151	64		215	266
0820	3"		10	1.600		229	89.50		318.50	390
0840	4"		6	2.667		420	149		569	690
1100	Tee, straight									
1110	1/4"	1 Plum	10	.800	Ea.	18.90	49.50		68.40	98.50
1120	3/8"		10	.800		21.50	49.50		71	102
1140	3/4"		9	.889		29	55		84	118
1150	1"		8	1		31.50	62		93.50	131
1170	1-1/2"	Q-1	13	1.231		73	69		142	187
1180	2"		11	1.455		91	81.50		172.50	227
1190	2-1/2"		9	1.778		185	99.50		284.50	360
1200	3"		6	2.667		415	149		564	685
1220	4"		4	4		580	224		804	990
5000	Malleable iron, 150 lb.									
5020	Black									
5040	90° elbow, straight									
5060	1/4"	1 Plum	16	.500	Ea.	5.30	31		36.30	54.50
5070	3/8"		16	.500		5.30	31		36.30	54.50
5090	3/4"		14	.571		4.41	35.50		39.91	60.50
5100	1"		13	.615		7.70	38.50		46.20	68
5120	1-1/2"	Q-1	20	.800		16.65	44.50		61.15	88
5130	2"		18	.889		28.50	49.50		78	109
5140	2-1/2"		14	1.143		64	64		128	170
5150	3"		10	1.600		93.50	89.50		183	242
5170	4"		6	2.667		201	149		350	455

For customer support on your Commercial Renovation Costs with RSMeans data, call 800.448.8182.

419

22 11 13.45 Pipe Fittings, Steel, Threaded

		Crew	Daily Output	Labor-Hours	Unit	Material	2018 Bare Costs Labor	Equipment	Total	Total Incl O&P
5450	Tee, straight									
5470	1/4"	1 Plum	10	.800	Ea.	7.70	49.50		57.20	86
5480	3/8"		10	.800		7.70	49.50		57.20	86
5500	3/4"		9	.889		7.05	55		62.05	94
5510	1"		8	1		12.05	62		74.05	110
5520	1-1/4"	Q-1	14	1.143		19.50	64		83.50	121
5530	1-1/2"		13	1.231		24	69		93	134
5540	2"		11	1.455		41.50	81.50		123	173
5550	2-1/2"		9	1.778		89	99.50		188.50	253
5560	3"		6	2.667		131	149		280	375
5570	3-1/2"		5	3.200		305	179		484	615
5580	4"		4	4		315	224		539	700
5650	Coupling									
5670	1/4"	1 Plum	19	.421	Ea.	6.55	26		32.55	47.50
5680	3/8"		19	.421		6.55	26		32.55	47.50
5690	1/2"		19	.421		5.05	26		31.05	46
5700	3/4"		18	.444		5.95	27.50		33.45	49.50
5710	1"		15	.533		8.90	33		41.90	61.50
5720	1-1/4"	Q-1	26	.615		11.50	34.50		46	66
5730	1-1/2"		24	.667		15.55	37.50		53.05	75
5740	2"		21	.762		23	42.50		65.50	92
5750	2-1/2"		18	.889		63.50	49.50		113	148
5760	3"		14	1.143		86	64		150	194
5780	4"		10	1.600		173	89.50		262.50	330
6000	For galvanized elbows, tees, and couplings, add					20%				
9990	Minimum labor/equipment charge	1 Plum	4	2	Job		124		124	193

22 11 13.47 Pipe Fittings, Steel

		Crew	Daily Output	Labor-Hours	Unit	Material	2018 Bare Costs Labor	Equipment	Total	Total Incl O&P
0010	**PIPE FITTINGS, STEEL**, flanged, welded & special									
3000	Weld joint, butt, carbon steel, standard weight									
3040	90° elbow, long radius									
3050	1/2" pipe size	Q-15	16	1	Ea.	45	56	3.70	104.70	141
3060	3/4" pipe size		16	1		45	56	3.70	104.70	141
3070	1" pipe size		16	1		21	56	3.70	80.70	114
3080	1-1/4" pipe size		14	1.143		21	64	4.23	89.23	127
3090	1-1/2" pipe size		13	1.231		21	69	4.55	94.55	135
3100	2" pipe size		10	1.600		22.50	89.50	5.90	117.90	171
3110	2-1/2" pipe size		8	2		27.50	112	7.40	146.90	213
3120	3" pipe size		7	2.286		29	128	8.45	165.45	240
3130	4" pipe size		5	3.200		48	179	11.85	238.85	345
3136	5" pipe size		4	4		103	224	14.80	341.80	480
3140	6" pipe size	Q-16	5	4.800		106	278	11.85	395.85	565
3350	Tee, straight									
3360	1/2" pipe size	Q-15	10	1.600	Ea.	110	89.50	5.90	205.40	267
3370	3/4" pipe size		10	1.600		110	89.50	5.90	205.40	267
3380	1" pipe size		10	1.600		54.50	89.50	5.90	149.90	206
3390	1-1/4" pipe size		9	1.778		68.50	99.50	6.60	174.60	238
3400	1-1/2" pipe size		8	2		68.50	112	7.40	187.90	258
3410	2" pipe size		6	2.667		54.50	149	9.85	213.35	305
3420	2-1/2" pipe size		5	3.200		75.50	179	11.85	266.35	375
3430	3" pipe size		4	4		84	224	14.80	322.80	460
3440	4" pipe size		3	5.333		118	298	19.75	435.75	615
3446	5" pipe size		2.50	6.400		195	360	23.50	578.50	795

22 11 13 – Facility Water Distribution Piping

22 11 13.47 Pipe Fittings, Steel

		Crew	Daily Output	Labor-Hours	Unit	Material	2018 Bare Costs Labor	Equipment	Total	Total Incl O&P
3450	6" pipe size	Q-16	3	8	Ea.	203	465	19.75	687.75	965
9805	2" x 1"	1 Plum	50	.160		98	9.95		107.95	123
9810	2" x 1-1/4"		50	.160		112	9.95		121.95	138
9815	2" x 1-1/2"	↓	50	.160		112	9.95		121.95	138
9820	2-1/2" x 1"	Q-1	80	.200		98	11.20		109.20	125
9825	2-1/2" x 1-1/4"		80	.200		134	11.20		145.20	164
9830	2-1/2" x 1-1/2"		80	.200		134	11.20		145.20	164
9835	2-1/2" x 2"		80	.200		98	11.20		109.20	125
9840	3" x 1"		80	.200		110	11.20		121.20	138
9845	3" x 1-1/4"		80	.200		134	11.20		145.20	164
9850	3" x 1-1/2"		80	.200		134	11.20		145.20	164
9852	3" x 2"		80	.200		155	11.20		166.20	188
9856	4" x 1"		67	.239		134	13.35		147.35	168
9858	4" x 1-1/4"		67	.239		163	13.35		176.35	200
9860	4" x 1-1/2"		67	.239		163	13.35		176.35	200
9862	4" x 2"		67	.239		174	13.35		187.35	213
9864	4" x 2-1/2"		67	.239		174	13.35		187.35	213
9866	4" x 3"	↓	67	.239		181	13.35		194.35	220
9870	6" x 1-1/2"	Q-2	50	.480		204	28		232	268
9872	6" x 2"		50	.480		212	28		240	278
9874	6" x 2-1/2"		50	.480		212	28		240	277
9876	6" x 3"		50	.480		234	28		262	300
9878	6" x 4"		50	.480		258	28		286	330
9880	8" x 2"		42	.571		355	33		388	445
9882	8" x 2-1/2"		42	.571		355	33		388	445
9884	8" x 3"		42	.571		375	33		408	465
9886	8" x 4"	↓	42	.571	↓	765	33		798	890
9990	Minimum labor/equipment charge	Q-15	3	5.333	Job		298	19.75	317.75	485

22 11 13.48 Pipe, Fittings and Valves, Steel, Grooved-Joint

		Crew	Daily Output	Labor-Hours	Unit	Material	2018 Bare Costs Labor	Equipment	Total	Total Incl O&P
0010	**PIPE, FITTINGS AND VALVES, STEEL, GROOVED-JOINT**									
0012	Fittings are ductile iron. Steel fittings noted.									
0020	Pipe includes coupling & clevis type hanger assemblies, 10' OC									
1000	Schedule 40, black									
1040	3/4" diameter	1 Plum	71	.113	L.F.	6.05	7		13.05	17.55
1050	1" diameter		63	.127		5.85	7.90		13.75	18.75
1060	1-1/4" diameter		58	.138		6.95	8.55		15.50	21
1070	1-1/2" diameter		51	.157		7.55	9.75		17.30	23.50
1080	2" diameter	↓	40	.200		8.75	12.45		21.20	29
1090	2-1/2" diameter	Q-1	57	.281		11.45	15.70		27.15	37
1100	3" diameter		50	.320		13.85	17.90		31.75	43.50
1110	4" diameter		45	.356		24	19.90		43.90	57.50
1120	5" diameter	↓	37	.432	↓	43.50	24		67.50	85
3990	Fittings: coupling material required at joints not incl. in fitting price.									
3994	Add 1 selected coupling, material only, per joint for installed price.									
4000	Elbow, 90° or 45°, painted									
4030	3/4" diameter	1 Plum	50	.160	Ea.	72	9.95		81.95	94.50
4040	1" diameter		50	.160		38.50	9.95		48.45	57.50
4050	1-1/4" diameter		40	.200		38.50	12.45		50.95	61.50
4060	1-1/2" diameter		33	.242		38.50	15.05		53.55	65.50
4070	2" diameter	↓	25	.320		38.50	19.90		58.40	73
4080	2-1/2" diameter	Q-1	40	.400		38.50	22.50		61	77
4090	3" diameter	↓	33	.485	↓	67.50	27		94.50	117

22 11 13.48 Pipe, Fittings and Valves, Steel, Grooved-Joint	Crew	Daily Output	Labor-Hours	Unit	Material	2018 Bare Costs Labor	Equipment	Total	Total Incl O&P
4100	4" diameter	Q-1	25	.640	Ea.	73.50	36		109.50	136
4110	5" diameter	↓	20	.800		175	44.50		219.50	262
4250	For galvanized elbows, add				↓	26%				
4690	Tee, painted									
4700	3/4" diameter	1 Plum	38	.211	Ea.	77	13.10		90.10	105
4740	1" diameter		33	.242		59.50	15.05		74.55	89
4750	1-1/4" diameter		27	.296		59.50	18.40		77.90	94
4760	1-1/2" diameter		22	.364		59.50	22.50		82	101
4770	2" diameter	↓	17	.471		59.50	29.50		89	111
4780	2-1/2" diameter	Q-1	27	.593		59.50	33		92.50	117
4790	3" diameter		22	.727		81	40.50		121.50	153
4800	4" diameter		17	.941		123	52.50		175.50	218
4810	5" diameter	↓	13	1.231		287	69		356	420
4900	For galvanized tees, add					24%				
4933	20" diameter	Q-2	16	1.500	↓	630	87		717	830
4940	Flexible, standard, painted									
4950	3/4" diameter	1 Plum	100	.080	Ea.	21.50	4.97		26.47	32
4960	1" diameter		100	.080		21.50	4.97		26.47	32
4970	1-1/4" diameter		80	.100		28	6.20		34.20	40.50
4980	1-1/2" diameter		67	.119		30.50	7.40		37.90	45
4990	2" diameter		50	.160		33	9.95		42.95	51.50
5000	2-1/2" diameter	Q-1	80	.200		38	11.20		49.20	59.50
5010	3" diameter		67	.239		42	13.35		55.35	67.50
5020	3-1/2" diameter		57	.281		60.50	15.70		76.20	91
5030	4" diameter		50	.320		61	17.90		78.90	95
5040	5" diameter	↓	40	.400		91.50	22.50		114	136
5050	6" diameter	Q-2	50	.480	↓	108	28		136	163
5200	For galvanized couplings, add					33%				
7790	Valves: coupling material required at joints not incl. in valve price.									
7794	Add 1 selected coupling, material only, per joint for installed price.									
9990	Minimum labor/equipment charge	1 Plum	4	2	Job		124		124	193

22 11 13.74 Pipe, Plastic

		Crew	Daily Output	Labor-Hours	Unit	Material	2018 Bare Costs Labor	Equipment	Total	Total Incl O&P
0010	**PIPE, PLASTIC**									
1800	PVC, couplings 10' OC, clevis hanger assemblies, 3 per 10'									
1820	Schedule 40									
1860	1/2" diameter	1 Plum	54	.148	L.F.	5.10	9.20		14.30	19.95
1870	3/4" diameter		51	.157		5.60	9.75		15.35	21.50
1880	1" diameter		46	.174		8.75	10.80		19.55	26.50
1890	1-1/4" diameter		42	.190		9.50	11.85		21.35	29
1900	1-1/2" diameter	↓	36	.222		9.60	13.80		23.40	32
1910	2" diameter	Q-1	59	.271		11.10	15.15		26.25	35.50
1920	2-1/2" diameter		56	.286		10.60	16		26.60	36.50
1930	3" diameter		53	.302		12.90	16.90		29.80	40.50
1940	4" diameter		48	.333		30.50	18.65		49.15	62.50
1950	5" diameter		43	.372		39	21		60	75
1960	6" diameter	↓	39	.410	↓	27	23		50	65.50
4100	DWV type, schedule 40, couplings 10' OC, clevis hanger assy's, 3 per 10'									
4210	ABS, schedule 40, foam core type									
4212	Plain end black									
4214	1-1/2" diameter	1 Plum	39	.205	L.F.	8	12.75		20.75	28.50
4216	2" diameter	Q-1	62	.258		8.55	14.45		23	32
4218	3" diameter	↓	56	.286	↓	8.35	16		24.35	34

22 11 13 – Facility Water Distribution Piping

22 11 13.74 Pipe, Plastic

		Crew	Daily Output	Labor-Hours	Unit	Material	2018 Bare Costs Labor	2018 Bare Costs Equipment	Total	Total Incl O&P
4220	4" diameter	Q-1	51	.314	L.F.	25	17.55		42.55	55
4222	6" diameter	↓	42	.381	↓	18.95	21.50		40.45	54
4240	To delete coupling & hangers, subtract									
4244	1-1/2" diam. to 6" diam.					43%	48%			
4400	PVC									
4410	1-1/4" diameter	1 Plum	42	.190	L.F.	8.50	11.85		20.35	28
4460	2" diameter	Q-1	59	.271		8.65	15.15		23.80	33
4470	3" diameter		53	.302		8.55	16.90		25.45	36
4480	4" diameter		48	.333		10.20	18.65		28.85	40
4490	6" diameter	↓	39	.410	↓	17.35	23		40.35	54.50
5300	CPVC, socket joint, couplings 10' OC, clevis hanger assemblies, 3 per 10'									
5302	Schedule 40									
5304	1/2" diameter	1 Plum	54	.148	L.F.	5.85	9.20		15.05	21
5305	3/4" diameter		51	.157		6.90	9.75		16.65	23
5306	1" diameter		46	.174		10.50	10.80		21.30	28.50
5307	1-1/4" diameter		42	.190		11.85	11.85		23.70	31.50
5308	1-1/2" diameter	↓	36	.222		11.75	13.80		25.55	34.50
5309	2" diameter	Q-1	59	.271		14.80	15.15		29.95	40
5310	2-1/2" diameter		56	.286		18.75	16		34.75	45.50
5311	3" diameter	↓	53	.302	↓	21.50	16.90		38.40	50
5360	CPVC, threaded, couplings 10' OC, clevis hanger assemblies, 3 per 10'									
5380	Schedule 40									
5460	1/2" diameter	1 Plum	54	.148	L.F.	6.70	9.20		15.90	21.50
5470	3/4" diameter		51	.157		8.35	9.75		18.10	24.50
5480	1" diameter		46	.174		12	10.80		22.80	30
5490	1-1/4" diameter		42	.190		13	11.85		24.85	32.50
5500	1-1/2" diameter	↓	36	.222		12.75	13.80		26.55	35.50
5510	2" diameter	Q-1	59	.271		16.05	15.15		31.20	41
5520	2-1/2" diameter		56	.286		20	16		36	47
5530	3" diameter	↓	53	.302	↓	23	16.90		39.90	52
9900	Minimum labor/equipment charge	1 Plum	4	2	Job		124		124	193

22 11 13.76 Pipe Fittings, Plastic

		Crew	Daily Output	Labor-Hours	Unit	Material	2018 Bare Costs Labor	2018 Bare Costs Equipment	Total	Total Incl O&P
0010	**PIPE FITTINGS, PLASTIC**									
2700	PVC (white), schedule 40, socket joints									
2760	90° elbow, 1/2"	1 Plum	33.30	.240	Ea.	.50	14.95		15.45	23.50
2810	2"	Q-1	36.40	.440	"	3.04	24.50		27.54	42
4500	DWV, ABS, non pressure, socket joints									
4540	1/4 bend, 1-1/4"	1 Plum	20.20	.396	Ea.	4.46	24.50		28.96	43.50
4560	1-1/2"	"	18.20	.440		3.44	27.50		30.94	46.50
4570	2"	Q-1	33.10	.483	↓	5.30	27		32.30	48
4650	1/8 bend, same as 1/4 bend									
4800	Tee, sanitary									
4820	1-1/4"	1 Plum	13.50	.593	Ea.	5.90	37		42.90	64
4830	1-1/2"	"	12.10	.661		5.10	41		46.10	69.50
4840	2"	Q-1	20	.800	↓	7.85	44.50		52.35	78
5000	DWV, PVC, schedule 40, socket joints									
5040	1/4 bend, 1-1/4"	1 Plum	20.20	.396	Ea.	8.80	24.50		33.30	48
5060	1-1/2"	"	18.20	.440		2.51	27.50		30.01	45.50
5070	2"	Q-1	33.10	.483		3.95	27		30.95	46.50
5080	3"		20.80	.769		11.60	43		54.60	80
5090	4"		16.50	.970		23	54		77	110
5100	6"	↓	10.10	1.584	↓	81	88.50		169.50	227

For customer support on your Commercial Renovation Costs with RSMeans data, call 800.448.8182.

423

22 11 13 – Facility Water Distribution Piping

22 11 13.76 Pipe Fittings, Plastic		Crew	Daily Output	Labor-Hours	Unit	Material	2018 Bare Costs Labor	Equipment	Total	Total Incl O&P
5105	8"	Q-2	9.30	2.581	Ea.	105	150		255	350
5110	1/4 bend, long sweep, 1-1/2"	1 Plum	18.20	.440		5.80	27.50		33.30	49
5112	2"	Q-1	33.10	.483		6.50	27		33.50	49
5114	3"		20.80	.769		14.90	43		57.90	83.50
5116	4"	↓	16.50	.970		28.50	54		82.50	116
5150	1/8 bend, 1-1/4"	1 Plum	20.20	.396		6	24.50		30.50	45
5215	8"	Q-2	9.30	2.581		116	150		266	360
5250	Tee, sanitary 1-1/4"	1 Plum	13.50	.593		9.40	37		46.40	68
5254	1-1/2"	"	12.10	.661		4.37	41		45.37	69
5255	2"	Q-1	20	.800		6.45	44.50		50.95	76.50
5256	3"		13.90	1.151		16.95	64.50		81.45	119
5257	4"		11	1.455		31	81.50		112.50	162
5259	6"	↓	6.70	2.388		125	134		259	345
5261	8"	Q-2	6.20	3.871		277	225		502	655
5264	2" x 1-1/2"	Q-1	22	.727		5.70	40.50		46.20	70
5266	3" x 1-1/2"		15.50	1.032		12.40	57.50		69.90	104
5268	4" x 3"		12.10	1.322		36.50	74		110.50	155
5271	6" x 4"	↓	6.90	2.319		121	130		251	335
5314	Combination Y & 1/8 bend, 1-1/2"	1 Plum	12.10	.661		10.65	41		51.65	76
5315	2"	Q-1	20	.800		13.30	44.50		57.80	84
5317	3"		13.90	1.151		29	64.50		93.50	132
5318	4"	↓	11	1.455	↓	58	81.50		139.50	191
5324	Combination Y & 1/8 bend, reducing									
5325	2" x 2" x 1-1/2"	Q-1	22	.727	Ea.	15	40.50		55.50	80
5327	3" x 3" x 1-1/2"		15.50	1.032		27	57.50		84.50	120
5328	3" x 3" x 2"		15.30	1.046		19.95	58.50		78.45	113
5329	4" x 4" x 2"	↓	12.20	1.311		30	73.50		103.50	147
5331	Wye, 1-1/4"	1 Plum	13.50	.593		12.05	37		49.05	71
5332	1-1/2"	"	12.10	.661		8	41		49	73
5333	2"	Q-1	20	.800		7.85	44.50		52.35	78
5334	3"		13.90	1.151		21	64.50		85.50	124
5335	4"		11	1.455		38.50	81.50		120	170
5336	6"	↓	6.70	2.388		112	134		246	330
5337	8"	Q-2	6.20	3.871		197	225		422	565
5341	2" x 1-1/2"	Q-1	22	.727		9.65	40.50		50.15	74
5342	3" x 1-1/2"		15.50	1.032		14.30	57.50		71.80	106
5343	4" x 3"		12.10	1.322		31.50	74		105.50	150
5344	6" x 4"	↓	6.90	2.319		85	130		215	296
5345	8" x 6"	Q-2	6.40	3.750		186	218		404	545
5347	Double wye, 1-1/2"	1 Plum	9.10	.879		18.05	54.50		72.55	105
5348	2"	Q-1	16.60	.964		20	54		74	106
5349	3"		10.40	1.538		42	86		128	180
5350	4"	↓	8.25	1.939	↓	85	108		193	263
5353	Double wye, reducing									
5354	2" x 2" x 1-1/2" x 1-1/2"	Q-1	16.80	.952	Ea.	18.40	53.50		71.90	103
5355	3" x 3" x 2" x 2"		10.60	1.509		31	84.50		115.50	166
5356	4" x 4" x 3" x 3"		8.45	1.893		67.50	106		173.50	239
5357	6" x 6" x 4" x 4"	↓	7.25	2.207		236	123		359	450
5374	Coupling, 1-1/4"	1 Plum	20.20	.396		5.70	24.50		30.20	45
5376	1-1/2"	"	18.20	.440		1.17	27.50		28.67	44
5378	2"	Q-1	33.10	.483		1.61	27		28.61	44
5380	3"		20.80	.769		5.65	43		48.65	73
5390	4"	↓	16.50	.970		9.55	54		63.55	95

22 11 13.76 Pipe Fittings, Plastic	Crew	Daily Output	Labor-Hours	Unit	Material	2018 Bare Costs Labor	Equipment	Total	Total Incl O&P
5410	Reducer bushing, 2" x 1-1/4"	Q-1	36.50	.438	Ea.	3.35	24.50		27.85	41.50
5412	3" x 1-1/2"		27.30	.586		9.90	33		42.90	62
5414	4" x 2"		18.20	.879		17.20	49		66.20	95.50
5416	6" x 4"	↓	11.10	1.441		44	80.50		124.50	174
5418	8" x 6"	Q-2	10.20	2.353	↓	86.50	136		222.50	305
5500	CPVC, Schedule 80, threaded joints									
5540	90° elbow, 1/4"	1 Plum	32	.250	Ea.	12.80	15.55		28.35	38
5560	1/2"		30.30	.264		7.40	16.40		23.80	33.50
5570	3/4"		26	.308		11.10	19.10		30.20	42
5580	1"		22.70	.352		15.60	22		37.60	51
5590	1-1/4"		20.20	.396		30	24.50		54.50	71.50
5600	1-1/2"	↓	18.20	.440		32.50	27.50		60	78
5610	2"	Q-1	33.10	.483		43.50	27		70.50	89.50
5620	2-1/2"		24.20	.661		134	37		171	206
5630	3"	↓	20.80	.769		144	43		187	225
5730	Coupling, 1/4"	1 Plum	32	.250		16.30	15.55		31.85	42
5732	1/2"		30.30	.264		13.40	16.40		29.80	40.50
5734	3/4"		26	.308		21.50	19.10		40.60	54
5736	1"		22.70	.352		24.50	22		46.50	61
5738	1-1/4"		20.20	.396		26	24.50		50.50	67
5740	1-1/2"	↓	18.20	.440		28	27.50		55.50	73.50
5742	2"	Q-1	33.10	.483		33	27		60	78.50
5744	2-1/2"		24.20	.661		59	37		96	123
5746	3"	↓	20.80	.769	↓	69	43		112	143
5900	CPVC, Schedule 80, socket joints									
5904	90° elbow, 1/4"	1 Plum	32	.250	Ea.	12.15	15.55		27.70	37.50
5906	1/2"		30.30	.264		4.77	16.40		21.17	31
5908	3/4"		26	.308		6.10	19.10		25.20	36.50
5910	1"		22.70	.352		9.65	22		31.65	44.50
5912	1-1/4"		20.20	.396		21	24.50		45.50	61.50
5914	1-1/2"	↓	18.20	.440		23.50	27.50		51	68
5916	2"	Q-1	33.10	.483		28	27		55	73
5918	2-1/2"		24.20	.661		65	37		102	129
5920	3"	↓	20.80	.769		73.50	43		116.50	148
5930	45° elbow, 1/4"	1 Plum	32	.250		18.10	15.55		33.65	44
5932	1/2"		30.30	.264		5.85	16.40		22.25	32
5934	3/4"		26	.308		8.40	19.10		27.50	39.50
5936	1"		22.70	.352		13.40	22		35.40	49
5938	1-1/4"		20.20	.396		26.50	24.50		51	67.50
5940	1-1/2"	↓	18.20	.440		27	27.50		54.50	72
5942	2"	Q-1	33.10	.483		30.50	27		57.50	75.50
5944	2-1/2"		24.20	.661		62	37		99	126
5946	3"	↓	20.80	.769		79.50	43		122.50	155
5990	Coupling, 1/4"	1 Plum	32	.250		12.95	15.55		28.50	38.50
5992	1/2"		30.30	.264		5.05	16.40		21.45	31
5994	3/4"		26	.308		7.05	19.10		26.15	38
5996	1"		22.70	.352		9.50	22		31.50	44.50
5998	1-1/4"		20.20	.396		14.20	24.50		38.70	54
6000	1-1/2"	↓	18.20	.440		17.85	27.50		45.35	62
6002	2"	Q-1	33.10	.483		21	27		48	65
6004	2-1/2"		24.20	.661		46	37		83	109
6006	3"	↓	20.80	.769	↓	50	43		93	123
9990	Minimum labor/equipment charge	1 Plum	4	2	Job		124		124	193

22 11 Facility Water Distribution

22 11 19 – Domestic Water Piping Specialties

22 11 19.10 Flexible Connectors

		Crew	Daily Output	Labor-Hours	Unit	Material	2018 Bare Costs Labor	2018 Bare Costs Equipment	Total	Total Incl O&P
0010	**FLEXIBLE CONNECTORS**, Corrugated, 7/8" OD, 1/2" ID									
0050	Gas, seamless brass, steel fittings									
0200	12" long	1 Plum	36	.222	Ea.	6.40	13.80		20.20	28.50
0220	18" long		36	.222		7.95	13.80		21.75	30.50
0240	24" long		34	.235		9.40	14.60		24	33.50
0260	30" long		34	.235		10.15	14.60		24.75	34
9000	Minimum labor/equipment charge		4	2	Job		124		124	193

22 11 19.18 Mixing Valve

		Crew	Daily Output	Labor-Hours	Unit	Material	2018 Bare Costs Labor	2018 Bare Costs Equipment	Total	Total Incl O&P
0010	**MIXING VALVE**, Automatic, water tempering.									
0040	1/2" size	1 Stpi	19	.421	Ea.	580	26.50		606.50	675
0050	3/4" size		18	.444	"	580	28		608	680
9000	Minimum labor/equipment charge		5	1.600	Job		101		101	157

22 11 19.38 Water Supply Meters

		Crew	Daily Output	Labor-Hours	Unit	Material	2018 Bare Costs Labor	2018 Bare Costs Equipment	Total	Total Incl O&P
0010	**WATER SUPPLY METERS**									
2000	Domestic/commercial, bronze									
2020	Threaded									
2060	5/8" diameter, to 20 GPM	1 Plum	16	.500	Ea.	50.50	31		81.50	104
2080	3/4" diameter, to 30 GPM		14	.571		92	35.50		127.50	157
2100	1" diameter, to 50 GPM		12	.667		140	41.50		181.50	219
2300	Threaded/flanged									
2340	1-1/2" diameter, to 100 GPM	1 Plum	8	1	Ea.	340	62		402	470
9000	Minimum labor/equipment charge	"	3.25	2.462	Job		153		153	238

22 11 19.42 Backflow Preventers

		Crew	Daily Output	Labor-Hours	Unit	Material	2018 Bare Costs Labor	2018 Bare Costs Equipment	Total	Total Incl O&P
0010	**BACKFLOW PREVENTERS**, Includes valves									
0020	and four test cocks, corrosion resistant, automatic operation									
1000	Double check principle									
1010	Threaded, with ball valves									
1020	3/4" pipe size	1 Plum	16	.500	Ea.	231	31		262	305
1030	1" pipe size		14	.571		265	35.50		300.50	345
1040	1-1/2" pipe size		10	.800		565	49.50		614.50	700
1050	2" pipe size		7	1.143		655	71		726	830
1080	Threaded, with gate valves									
1100	3/4" pipe size	1 Plum	16	.500	Ea.	1,025	31		1,056	1,175
1120	1" pipe size		14	.571		1,025	35.50		1,060.50	1,175
1140	1-1/2" pipe size		10	.800		1,175	49.50		1,224.50	1,375
1160	2" pipe size		7	1.143		1,625	71		1,696	1,875
4000	Reduced pressure principle									
4100	Threaded, bronze, valves are ball									
4120	3/4" pipe size	1 Plum	16	.500	Ea.	455	31		486	550
4140	1" pipe size		14	.571		480	35.50		515.50	585
4150	1-1/4" pipe size		12	.667		895	41.50		936.50	1,050
4160	1-1/2" pipe size		10	.800		985	49.50		1,034.50	1,150
4180	2" pipe size		7	1.143		1,100	71		1,171	1,325
5000	Flanged, bronze, valves are OS&Y									
5060	2-1/2" pipe size	Q-1	5	3.200	Ea.	5,325	179		5,504	6,125
5080	3" pipe size		4.50	3.556		6,050	199		6,249	6,950
5100	4" pipe size		3	5.333		7,550	298		7,848	8,800
5120	6" pipe size	Q-2	3	8		11,000	465		11,465	12,800
5600	Flanged, iron, valves are OS&Y									
5660	2-1/2" pipe size	Q-1	5	3.200	Ea.	3,025	179		3,204	3,600
5680	3" pipe size		4.50	3.556		3,175	199		3,374	3,800

22 11 19 – Domestic Water Piping Specialties

22 11 19.42 Backflow Preventers	Crew	Daily Output	Labor-Hours	Unit	Material	2018 Bare Costs Labor	Equipment	Total	Total Incl O&P
5700	4" pipe size	Q-1	3	5.333	Ea.	4,000	298		4,298	4,875
5720	6" pipe size	Q-2	3	8		5,800	465		6,265	7,100
5800	Rebuild 4" diameter reduced pressure	1 Plum	4	2		360	124		484	590
5810	6" diameter		2.66	3.008		380	187		567	710
5820	8" diameter		2	4		470	249		719	905
5830	10" diameter		1.60	5		550	310		860	1,100
9010	Minimum labor/equipment charge		2	4	Job		249		249	385

22 11 19.50 Vacuum Breakers

		Crew	Daily Output	Labor-Hours	Unit	Material	Labor	Equipment	Total	Total Incl O&P
0010	**VACUUM BREAKERS**									
0013	See also backflow preventers Section 22 11 19.42									
1000	Anti-siphon continuous pressure type									
1010	Max. 150 psi - 210°F									
1020	Bronze body									
1030	1/2" size	1 Stpi	24	.333	Ea.	171	21		192	221
1040	3/4" size		20	.400		171	25		196	227
1050	1" size		19	.421		176	26.50		202.50	236
1060	1-1/4" size		15	.533		345	33.50		378.50	435
1070	1-1/2" size		13	.615		425	39		464	530
1080	2" size		11	.727		440	46		486	555
1200	Max. 125 psi with atmospheric vent									
1210	Brass, in-line construction									
1220	1/4" size	1 Stpi	24	.333	Ea.	133	21		154	179
1230	3/8" size	"	24	.333		133	21		154	179
1260	For polished chrome finish, add					13%				
2000	Anti-siphon, non-continuous pressure type									
2010	Hot or cold water 125 psi - 210°F									
2020	Bronze body									
2030	1/4" size	1 Stpi	24	.333	Ea.	73	21		94	113
2040	3/8" size		24	.333		73	21		94	113
2050	1/2" size		24	.333		82.50	21		103.50	123
2060	3/4" size		20	.400		98.50	25		123.50	147
2070	1" size		19	.421		152	26.50		178.50	210
2080	1-1/4" size		15	.533		267	33.50		300.50	345
2090	1-1/2" size		13	.615		315	39		354	405
2100	2" size		11	.727		485	46		531	605
2110	2-1/2" size		8	1		1,400	63		1,463	1,650
2120	3" size		6	1.333		1,850	84		1,934	2,175
2150	For polished chrome finish, add					50%				
3000	Air gap fitting									
3020	1/2" NPT size	1 Plum	19	.421	Ea.	57	26		83	103
3030	1" NPT size	"	15	.533		65.50	33		98.50	124
3040	2" NPT size	Q-1	21	.762		128	42.50		170.50	208
3050	3" NPT size		14	1.143		253	64		317	380
3060	4" NPT size		10	1.600		253	89.50		342.50	415

22 11 19.54 Water Hammer Arresters/Shock Absorbers

		Crew	Daily Output	Labor-Hours	Unit	Material	Labor	Equipment	Total	Total Incl O&P
0010	**WATER HAMMER ARRESTERS/SHOCK ABSORBERS**									
0490	Copper									
0500	3/4" male IPS For 1 to 11 fixtures	1 Plum	12	.667	Ea.	28	41.50		69.50	95.50
0600	1" male IPS For 12 to 32 fixtures		8	1		47.50	62		109.50	149
0700	1-1/4" male IPS For 33 to 60 fixtures		8	1		48	62		110	150
0800	1-1/2" male IPS For 61 to 113 fixtures		8	1		69	62		131	173
0900	2" male IPS For 114 to 154 fixtures		8	1		101	62		163	208

22 11 Facility Water Distribution

22 11 19 – Domestic Water Piping Specialties

22 11 19.54 Water Hammer Arresters/Shock Absorbers		Crew	Daily Output	Labor-Hours	Unit	Material	2018 Bare Costs Labor	Equipment	Total	Total Incl O&P
1000	2-1/2" male IPS For 155 to 330 fixtures	1 Plum	4	2	Ea.	305	124		429	530
9000	Minimum labor/equipment charge	↓	3.50	2.286	Job		142		142	221

22 11 19.64 Hydrants

		Crew	Daily Output	Labor-Hours	Unit	Material	2018 Bare Costs Labor	Equipment	Total	Total Incl O&P
0010	**HYDRANTS**									
0050	Wall type, moderate climate, bronze, encased									
0200	3/4" IPS connection	1 Plum	16	.500	Ea.	850	31		881	985
1000	Non-freeze, bronze, exposed									
1100	3/4" IPS connection, 4" to 9" thick wall	1 Plum	14	.571	Ea.	380	35.50		415.50	475
1120	10" to 14" thick wall	"	12	.667		410	41.50		451.50	515
1280	For anti-siphon type, add				↓	153			153	168
9000	Minimum labor/equipment charge	1 Plum	3	2.667	Job		166		166	258

22 13 Facility Sanitary Sewerage

22 13 16 – Sanitary Waste and Vent Piping

22 13 16.20 Pipe, Cast Iron

		Crew	Daily Output	Labor-Hours	Unit	Material	2018 Bare Costs Labor	Equipment	Total	Total Incl O&P
0010	**PIPE, CAST IRON**, Soil, on clevis hanger assemblies, 5' OC R221113-50									
0020	Single hub, service wt., lead & oakum joints 10' OC									
2120	2" diameter	Q-1	63	.254	L.F.	16.90	14.20		31.10	40.50
2140	3" diameter		60	.267		20	14.90		34.90	45
2160	4" diameter	↓	55	.291		35	16.25		51.25	64
2180	5" diameter	Q-2	76	.316		43.50	18.30		61.80	76
2200	6" diameter	"	73	.329		41.50	19.05		60.55	75.50
2220	8" diameter	Q-3	59	.542	↓	63.50	32		95.50	120
4000	No hub, couplings 10' OC									
4100	1-1/2" diameter	Q-1	71	.225	L.F.	16	12.60		28.60	37
4120	2" diameter		67	.239		17.10	13.35		30.45	40
4140	3" diameter		64	.250		19.60	14		33.60	43.50
4160	4" diameter	↓	58	.276		34.50	15.45		49.95	61.50
4180	5" diameter	Q-2	83	.289	↓	43.50	16.75		60.25	73.50
9000	Minimum labor/equipment charge	1 Plum	4	2	Job		124		124	193

22 13 16.30 Pipe Fittings, Cast Iron

		Crew	Daily Output	Labor-Hours	Unit	Material	2018 Bare Costs Labor	Equipment	Total	Total Incl O&P
0010	**PIPE FITTINGS, CAST IRON**, Soil R221113-50									
0040	Hub and spigot, service weight, lead & oakum joints									
0080	1/4 bend, 2"	Q-1	16	1	Ea.	20.50	56		76.50	110
0120	3"		14	1.143		27.50	64		91.50	130
0140	4"	↓	13	1.231		43	69		112	155
0160	5"	Q-2	18	1.333		60	77.50		137.50	186
0180	6"	"	17	1.412		75	82		157	209
0200	8"	Q-3	11	2.909		225	172		397	515
0340	1/8 bend, 2"	Q-1	16	1		14.65	56		70.65	103
0350	3"		14	1.143		23	64		87	125
0360	4"	↓	13	1.231		33.50	69		102.50	144
0380	5"	Q-2	18	1.333		47.50	77.50		125	172
0400	6"	"	17	1.412		57	82		139	190
0420	8"	Q-3	11	2.909		170	172		342	455
0500	Sanitary tee, 2"	Q-1	10	1.600		28.50	89.50		118	171
0540	3"		9	1.778		46.50	99.50		146	206
0620	4"	↓	8	2		57	112		169	237
0700	5"	Q-2	12	2		113	116		229	305
0800	6"	"	11	2.182		128	127		255	340

22 13 16.30 Pipe Fittings, Cast Iron		Crew	Daily Output	Labor-Hours	Unit	Material	2018 Bare Costs Labor	Equipment	Total	Total Incl O&P
0880	8"	Q-3	7	4.571	Ea.	340	270		610	795
5990	No hub									
6000	Cplg. & labor required at joints not incl. in fitting									
6010	price. Add 1 coupling per joint for installed price									
6020	1/4 bend, 1-1/2"				Ea.	10.55			10.55	11.65
6060	2"					11.55			11.55	12.70
6080	3"					16.10			16.10	17.70
6120	4"					24			24	26
6184	1/4 bend, long sweep, 1-1/2"					27			27	29.50
6186	2"					25			25	27.50
6188	3"					30.50			30.50	33.50
6189	4"					48.50			48.50	53.50
6190	5"					94			94	103
6191	6"					107			107	118
6192	8"					291			291	320
6193	10"					585			585	645
6200	1/8 bend, 1-1/2"					8.90			8.90	9.80
6210	2"					9.95			9.95	10.95
6212	3"					13.30			13.30	14.65
6214	4"					17.45			17.45	19.15
6380	Sanitary tee, tapped, 1-1/2"					21			21	23
6382	2" x 1-1/2"					18.60			18.60	20.50
6384	2"					19.95			19.95	22
6386	3" x 2"					29.50			29.50	32.50
6388	3"					51			51	56.50
6390	4" x 1-1/2"					26.50			26.50	29
6392	4" x 2"					30			30	33
6393	4"					30			30	33
6394	6" x 1-1/2"					68.50			68.50	75.50
6396	6" x 2"					70			70	77
6459	Sanitary tee, 1-1/2"					14.85			14.85	16.30
6460	2"					15.90			15.90	17.50
6470	3"					19.60			19.60	21.50
6472	4"				▼	37			37	41
8000	Coupling, standard (by CISPI Mfrs.)									
8020	1-1/2"	Q-1	48	.333	Ea.	16.10	18.65		34.75	46.50
8040	2"		44	.364		16.10	20.50		36.60	49
8080	3"		38	.421		19.25	23.50		42.75	57.50
8120	4"	▼	33	.485	▼	22.50	27		49.50	67
8300	Coupling, cast iron clamp & neoprene gasket (by MG)									
8310	1-1/2"	Q-1	48	.333	Ea.	8.30	18.65		26.95	38
8320	2"		44	.364		11.15	20.50		31.65	44
8330	3"		38	.421		11.10	23.50		34.60	48.50
8340	4"	▼	33	.485	▼	15.30	27		42.30	59
8600	Coupling, stainless steel, heavy duty									
8620	1-1/2"	Q-1	48	.333	Ea.	5.90	18.65		24.55	35.50
8630	2"		44	.364		6.10	20.50		26.60	38
8640	2" x 1-1/2"		44	.364		10.15	20.50		30.65	42.50
8650	3"		38	.421		6.65	23.50		30.15	44
8660	4"	▼	33	.485	▼	7.50	27		34.50	50.50
9000	Minimum labor/equipment charge	1 Plum	4	2	Job		124		124	193

22 13 16.60 Traps

	Crew	Daily Output	Labor-Hours	Unit	Material	2018 Bare Costs Labor	Equipment	Total	Total Incl O&P
0010 TRAPS									
0030 Cast iron, service weight									
0050 Running P trap, without vent									
1100 2"	Q-1	16	1	Ea.	148	56		204	250
1150 4"	"	13	1.231		148	69		217	270
1160 6"	Q-2	17	1.412		720	82		802	915
3000 P trap, B&S, 2" pipe size	Q-1	16	1		38.50	56		94.50	130
3040 3" pipe size	"	14	1.143		57.50	64		121.50	163
4700 Copper, drainage, drum trap									
4840 3" x 6" swivel, 1-1/2" pipe size	1 Plum	16	.500	Ea.	275	31		306	350
5100 P trap, standard pattern									
5200 1-1/4" pipe size	1 Plum	18	.444	Ea.	88.50	27.50		116	140
5240 1-1/2" pipe size		17	.471		98.50	29.50		128	154
5260 2" pipe size		15	.533		152	33		185	219
5280 3" pipe size		11	.727		475	45		520	595
9000 Minimum labor/equipment charge		3	2.667	Job		166		166	258

22 13 16.80 Vent Flashing and Caps

	Crew	Daily Output	Labor-Hours	Unit	Material	2018 Bare Costs Labor	Equipment	Total	Total Incl O&P
0010 VENT FLASHING AND CAPS									
0120 Vent caps									
0140 Cast iron									
0180 2-1/2" to 3-5/8" pipe	1 Plum	21	.381	Ea.	49.50	23.50		73	91.50
0190 4" to 4-1/8" pipe	"	19	.421	"	69	26		95	117
0900 Vent flashing									
1000 Aluminum with lead ring									
1050 3" pipe	1 Plum	17	.471	Ea.	5.50	29.50		35	51.50
1060 4" pipe	"	16	.500	"	6.65	31		37.65	56
1350 Copper with neoprene ring									
1440 2" pipe	1 Plum	18	.444	Ea.	70.50	27.50		98	121
1450 3" pipe		17	.471		85.50	29.50		115	140
1460 4" pipe		16	.500		85.50	31		116.50	143
9000 Minimum labor/equipment charge		4	2	Job		124		124	193

22 13 19 – Sanitary Waste Piping Specialties

22 13 19.13 Sanitary Drains

	Crew	Daily Output	Labor-Hours	Unit	Material	2018 Bare Costs Labor	Equipment	Total	Total Incl O&P
0010 SANITARY DRAINS									
0400 Deck, auto park, CI, 13" top									
0440 3", 4", 5", and 6" pipe size	Q-1	8	2	Ea.	1,650	112		1,762	1,975
0480 For galvanized body, add				"	890			890	980
2000 Floor, medium duty, CI, deep flange, 7" diam. top									
2040 2" and 3" pipe size	Q-1	12	1.333	Ea.	230	74.50		304.50	370
2080 For galvanized body, add					111			111	122
2120 With polished bronze top					360			360	395
2500 Heavy duty, cleanout & trap w/bucket, CI, 15" top									
2540 2", 3", and 4" pipe size	Q-1	6	2.667	Ea.	7,525	149		7,674	8,500
2560 For galvanized body, add					1,925			1,925	2,125
2580 With polished bronze top					8,350			8,350	9,200

22 13 19.14 Floor Receptors

	Crew	Daily Output	Labor-Hours	Unit	Material	2018 Bare Costs Labor	Equipment	Total	Total Incl O&P
0010 FLOOR RECEPTORS, For connection to 2", 3" & 4" diameter pipe									
0200 12-1/2" square top, 25 sq. in. open area	Q-1	10	1.600	Ea.	1,100	89.50		1,189.50	1,375
0300 For grate with 4" diameter x 3-3/4" high funnel, add					315			315	345
0400 For grate with 6" diameter x 6" high funnel, add					256			256	282
0700 For acid-resisting bucket, add					315			315	345

22 13 Facility Sanitary Sewerage

22 13 19 – Sanitary Waste Piping Specialties

22 13 19.14 Floor Receptors

		Crew	Daily Output	Labor-Hours	Unit	Material	2018 Bare Costs Labor	Equipment	Total	Total Incl O&P
0900	For stainless steel mesh bucket liner, add				Ea.	249			249	273
2000	12-5/8" diameter top, 40 sq. in. open area	Q-1	10	1.600	↓	880	89.50		969.50	1,100
2100	For options, add same prices as square top									
3000	8" x 4" rectangular top, 7.5 sq. in. open area	Q-1	14	1.143	Ea.	1,025	64		1,089	1,225
3100	For trap primer connections, add				"	102			102	112
9000	Minimum labor/equipment charge	Q-1	3	5.333	Job		298		298	465

22 13 19.15 Sink Waste Treatment

		Crew	Daily Output	Labor-Hours	Unit	Material	2018 Bare Costs Labor	Equipment	Total	Total Incl O&P
0010	**SINK WASTE TREATMENT**, System for commercial kitchens									
0100	includes clock timer & fittings									
0200	System less chemical, wall mounted cabinet	1 Plum	16	.500	Ea.	530	31		561	635
2000	Chemical, 1 gallon, add					34.50			34.50	38
2100	6 gallons, add					157			157	172
2200	15 gallons, add					425			425	470
2300	30 gallons, add					800			800	880
2400	55 gallons, add				↓	1,350			1,350	1,500

22 13 23 – Sanitary Waste Interceptors

22 13 23.10 Interceptors

		Crew	Daily Output	Labor-Hours	Unit	Material	2018 Bare Costs Labor	Equipment	Total	Total Incl O&P
0010	**INTERCEPTORS**									
0150	Grease, fabricated steel, 4 GPM, 8 lb. fat capacity	1 Plum	4	2	Ea.	1,325	124		1,449	1,650
0200	7 GPM, 14 lb. fat capacity		4	2		1,825	124		1,949	2,225
1000	10 GPM, 20 lb. fat capacity	↓	4	2		2,150	124		2,274	2,575
1120	50 GPM, 100 lb. fat capacity	Q-1	2	8		7,200	445		7,645	8,625
1160	100 GPM, 200 lb. fat capacity		2	8		16,300	445		16,745	18,700
1240	300 GPM, 600 lb. fat capacity	↓	1	16	↓	36,100	895		36,995	41,100
9000	Minimum labor/equipment charge	1 Plum	3	2.667	Job		166		166	258

22 13 29 – Sanitary Sewerage Pumps

22 13 29.14 Sewage Ejector Pumps

		Crew	Daily Output	Labor-Hours	Unit	Material	2018 Bare Costs Labor	Equipment	Total	Total Incl O&P
0010	**SEWAGE EJECTOR PUMPS**, With operating and level controls									
0100	Simplex system incl. tank, cover, pump 15' head									
0500	37 gal. PE tank, 12 GPM, 1/2 HP, 2" discharge	Q-1	3.20	5	Ea.	490	280		770	975
0510	3" discharge		3.10	5.161		535	289		824	1,050
0530	87 GPM, .7 HP, 2" discharge		3.20	5		755	280		1,035	1,275
0540	3" discharge		3.10	5.161		820	289		1,109	1,350
0600	45 gal. coated stl. tank, 12 GPM, 1/2 HP, 2" discharge		3	5.333		880	298		1,178	1,425
0610	3" discharge		2.90	5.517		915	310		1,225	1,475
0630	87 GPM, .7 HP, 2" discharge		3	5.333		1,125	298		1,423	1,725
0640	3" discharge		2.90	5.517		1,200	310		1,510	1,775
0660	134 GPM, 1 HP, 2" discharge		2.80	5.714		1,225	320		1,545	1,850
0680	3" discharge		2.70	5.926		1,275	330		1,605	1,950
0700	70 gal. PE tank, 12 GPM, 1/2 HP, 2" discharge		2.60	6.154		950	345		1,295	1,575
0710	3" discharge		2.40	6.667		1,000	375		1,375	1,675
0730	87 GPM, .7 HP, 2" discharge		2.50	6.400		1,225	360		1,585	1,900
0740	3" discharge		2.30	6.957		1,300	390		1,690	2,025
0760	134 GPM, 1 HP, 2" discharge		2.20	7.273		1,325	405		1,730	2,100
0770	3" discharge		2	8	↓	1,425	445		1,870	2,275
9000	Minimum labor/equipment charge	↓	2.50	6.400	Job		360		360	555

For customer support on your Commercial Renovation Costs with RSMeans data, call 800.448.8182.

431

22 14 Facility Storm Drainage

22 14 26 – Facility Storm Drains

22 14 26.13 Roof Drains

	Crew	Daily Output	Labor-Hours	Unit	Material	2018 Bare Costs Labor	Equipment	Total	Total Incl O&P
0010 **ROOF DRAINS**									
0140 Cornice, CI, 45° or 90° outlet									
0200 3" and 4" pipe size	Q-1	12	1.333	Ea.	380	74.50		454.50	530
0260 For galvanized body, add					87.50			87.50	96.50
0280 For polished bronze dome, add				↓	108			108	119
3860 Roof, flat metal deck, CI body, 12" CI dome									
3890 3" pipe size	Q-1	14	1.143	Ea.	330	64		394	460
3900 4" pipe size	"	13	1.231	"	450	69		519	600
4620 Main, all aluminum, 12" low profile dome									
4640 2", 3" and 4" pipe size	Q-1	14	1.143	Ea.	500	64		564	650
9000 Minimum labor/equipment charge	1 Plum	4	2	Job		124		124	193

22 14 29 – Sump Pumps

22 14 29.16 Submersible Sump Pumps

	Crew	Daily Output	Labor-Hours	Unit	Material	2018 Bare Costs Labor	Equipment	Total	Total Incl O&P
0010 **SUBMERSIBLE SUMP PUMPS**									
1000 Elevator sump pumps, automatic									
1010 Complete systems, pump, oil detector, controls and alarm									
1020 1-1/2" discharge, does not include the sump pit/tank									
1040 1/3 HP, 115 V	1 Plum	4.40	1.818	Ea.	1,900	113		2,013	2,275
1050 1/2 HP, 115 V		4	2		1,975	124		2,099	2,375
1060 1/2 HP, 230 V		4	2		2,025	124		2,149	2,425
1070 3/4 HP, 115 V		3.60	2.222		2,050	138		2,188	2,500
1080 3/4 HP, 230 V	↓	3.60	2.222	↓	2,100	138		2,238	2,550
1100 Sump pump only									
1110 1/3 HP, 115 V	1 Plum	6.40	1.250	Ea.	174	77.50		251.50	310
1120 1/2 HP, 115 V		5.80	1.379		248	85.50		333.50	405
1130 1/2 HP, 230 V		5.80	1.379		293	85.50		378.50	455
1140 3/4 HP, 115 V		5.40	1.481		330	92		422	510
1150 3/4 HP, 230 V	↓	5.40	1.481	↓	380	92		472	560
1200 Oil detector, control and alarm only									
1210 115 V	1 Plum	8	1	Ea.	1,725	62		1,787	2,000
1220 230 V	"	8	1	"	1,725	62		1,787	2,000
7000 Sump pump, automatic									
7100 Plastic, 1-1/4" discharge, 1/4 HP	1 Plum	6.40	1.250	Ea.	150	77.50		227.50	286
7140 1/3 HP		6	1.333		217	83		300	370
7180 1-1/2" discharge, 1/2 HP		5.20	1.538		305	95.50		400.50	485
7500 Cast iron, 1-1/4" discharge, 1/4 HP		6	1.333		211	83		294	360
7540 1/3 HP		6	1.333		248	83		331	400
7560 1/2 HP		5	1.600	↓	300	99.50		399.50	485
9000 Minimum labor/equipment charge	↓	4	2	Job		124		124	193

22 15 General Service Compressed-Air Systems

22 15 19 – General Service Packaged Air Compressors and Receivers

22 15 19.10 Air Compressors

	Crew	Daily Output	Labor-Hours	Unit	Material	2018 Bare Costs Labor	Equipment	Total	Total Incl O&P
0010 **AIR COMPRESSORS**									
5250 Air, reciprocating air cooled, splash lubricated, tank mounted									
5300 Single stage, 1 phase, 140 psi									
5303 1/2 HP, 17 gal. tank	1 Stpi	3	2.667	Ea.	1,875	168		2,043	2,325
5305 3/4 HP, 30 gal. tank	↓	2.60	3.077		1,825	194		2,019	2,300
5307 1 HP, 30 gal. tank	↓	2.20	3.636	↓	1,975	229		2,204	2,525

22 31 Domestic Water Softeners

22 31 13 – Residential Domestic Water Softeners

22 31 13.10 Residential Water Softeners	Crew	Daily Output	Labor-Hours	Unit	Material	2018 Bare Costs Labor	Equipment	Total	Total Incl O&P
0010 **RESIDENTIAL WATER SOFTENERS**									
7350 Water softener, automatic, to 30 grains per gallon	2 Plum	5	3.200	Ea.	370	199		569	715
7400 To 100 grains per gallon	"	4	4	"	840	249		1,089	1,300

22 33 Electric Domestic Water Heaters

22 33 13 – Instantaneous Electric Domestic Water Heaters

22 33 13.20 Instantaneous Elec. Point-Of-Use Water Heaters

		Crew	Daily Output	Labor-Hours	Unit	Material	2018 Bare Costs Labor	Equipment	Total	Total Incl O&P
0010 **INSTANTANEOUS ELECTRIC POINT-OF-USE WATER HEATERS**										
8965 Point of use, electric, glass lined										
8969 Energy saver										
8970 2.5 gal. single element	G	1 Plum	2.80	2.857	Ea.	320	178		498	630
8971 4 gal. single element	G		2.80	2.857		355	178		533	665
8974 6 gal. single element	G		2.50	3.200		390	199		589	740
8975 10 gal. single element	G		2.50	3.200		430	199		629	780
8976 15 gal. single element	G		2.40	3.333		500	207		707	870
8977 20 gal. single element	G		2.40	3.333		570	207		777	945
8978 30 gal. single element	G		2.30	3.478		640	216		856	1,050
8979 40 gal. single element	G		2.20	3.636		1,075	226		1,301	1,525
8988 Commercial (ASHRAE energy std. 90)										
8989 6 gallon	G	1 Plum	2.50	3.200	Ea.	925	199		1,124	1,325
8990 10 gallon	G		2.50	3.200		1,000	199		1,199	1,400
8991 15 gallon	G		2.40	3.333		1,075	207		1,282	1,500
8992 20 gallon	G		2.40	3.333		1,150	207		1,357	1,575
8993 30 gallon	G		2.30	3.478		2,425	216		2,641	3,000
8995 Under the sink, copper, w/bracket										
8996 2.5 gallon	G	1 Plum	4	2	Ea.	365	124		489	600
9000 Minimum labor/equipment charge		"	1.75	4.571	Job		284		284	440

22 33 30 – Residential, Electric Domestic Water Heaters

22 33 30.13 Residential, Small-Capacity Elec. Water Heaters

	Crew	Daily Output	Labor-Hours	Unit	Material	2018 Bare Costs Labor	Equipment	Total	Total Incl O&P
0010 **RESIDENTIAL, SMALL-CAPACITY ELECTRIC DOMESTIC WATER HEATERS**									
1000 Residential, electric, glass lined tank, 5 yr., 10 gal., single element	1 Plum	2.30	3.478	Ea.	430	216		646	805
1060 30 gallon, double element		2.20	3.636		1,000	226		1,226	1,450
1080 40 gallon, double element		2	4		1,075	249		1,324	1,550
1100 52 gallon, double element		2	4		1,200	249		1,449	1,700
1120 66 gallon, double element		1.80	4.444		1,650	276		1,926	2,225
1140 80 gallon, double element		1.60	5		1,850	310		2,160	2,525

22 33 33 – Light-Commercial Electric Domestic Water Heaters

22 33 33.10 Commercial Electric Water Heaters

	Crew	Daily Output	Labor-Hours	Unit	Material	2018 Bare Costs Labor	Equipment	Total	Total Incl O&P
0010 **COMMERCIAL ELECTRIC WATER HEATERS**									
4000 Commercial, 100° rise. NOTE: for each size tank, a range of									
4010 heaters between the ones shown are available									
4020 Electric									
4100 5 gal., 3 kW, 12 GPH, 208 volt	1 Plum	2	4	Ea.	4,775	249		5,024	5,625
4160 50 gal., 36 kW, 148 GPH, 208 volt	"	1.80	4.444		11,200	276		11,476	12,700
4480 400 gal., 210 kW, 860 GPH, 480 volt	Q-1	1	16		67,000	895		67,895	75,000

For customer support on your Commercial Renovation Costs with RSMeans data, call 800.448.8182.

433

22 34 Fuel-Fired Domestic Water Heaters

22 34 13 – Instantaneous, Tankless, Gas Domestic Water Heaters

22 34 13.10 Instantaneous, Tankless, Gas Water Heaters

22 34 13.10 Instantaneous, Tankless, Gas Water Heaters		Crew	Daily Output	Labor-Hours	Unit	Material	2018 Bare Costs Labor	Equipment	Total	Total Incl O&P
0010	**INSTANTANEOUS, TANKLESS, GAS WATER HEATERS**									
9410	Natural gas/propane, 3.2 GPM [G]	1 Plum	2	4	Ea.	500	249		749	935
9420	6.4 GPM [G]		1.90	4.211		715	262		977	1,200
9430	8.4 GPM [G]		1.80	4.444		855	276		1,131	1,375
9440	9.5 GPM [G]		1.60	5		1,000	310		1,310	1,575

22 34 30 – Residential Gas Domestic Water Heaters

22 34 30.13 Residential, Atmos, Gas Domestic Wtr Heaters

22 34 30.13 Residential, Atmos, Gas Domestic Wtr Heaters		Crew	Daily Output	Labor-Hours	Unit	Material	Labor	Equipment	Total	Total Incl O&P
0010	**RESIDENTIAL, ATMOSPHERIC, GAS DOMESTIC WATER HEATERS**									
2000	Gas fired, foam lined tank, 10 yr., vent not incl.									
2040	30 gallon	1 Plum	2	4	Ea.	1,775	249		2,024	2,325
2060	40 gallon		1.90	4.211		1,775	262		2,037	2,350
2100	75 gallon		1.50	5.333		2,700	330		3,030	3,500
3000	Tank leak safety, water & gas shut off see 22 05 23.20 8800									

22 34 36 – Commercial Gas Domestic Water Heaters

22 34 36.13 Commercial, Atmos., Gas Domestic Water Htrs.

22 34 36.13 Commercial, Atmos., Gas Domestic Water Htrs.		Crew	Daily Output	Labor-Hours	Unit	Material	Labor	Equipment	Total	Total Incl O&P
0010	**COMMERCIAL, ATMOSPHERIC, GAS DOMESTIC WATER HEATERS**									
6000	Gas fired, flush jacket, std. controls, vent not incl.									
6040	75 MBH input, 73 GPH	1 Plum	1.40	5.714	Ea.	3,775	355		4,130	4,700
6060	98 MBH input, 95 GPH		1.40	5.714		7,475	355		7,830	8,775
6180	200 MBH input, 192 GPH		.60	13.333		10,200	830		11,030	12,500
6900	For low water cutoff, add		8	1		365	62		427	495
6960	For bronze body hot water circulator, add		4	2		2,150	124		2,274	2,575

22 34 46 – Oil-Fired Domestic Water Heaters

22 34 46.10 Residential Oil-Fired Water Heaters

22 34 46.10 Residential Oil-Fired Water Heaters		Crew	Daily Output	Labor-Hours	Unit	Material	Labor	Equipment	Total	Total Incl O&P
0010	**RESIDENTIAL OIL-FIRED WATER HEATERS**									
3000	Oil fired, glass lined tank, 5 yr., vent not included, 30 gallon	1 Plum	2	4	Ea.	1,275	249		1,524	1,775
3040	50 gallon	"	1.80	4.444	"	1,575	276		1,851	2,150

22 34 46.20 Commercial Oil-Fired Water Heaters

22 34 46.20 Commercial Oil-Fired Water Heaters		Crew	Daily Output	Labor-Hours	Unit	Material	Labor	Equipment	Total	Total Incl O&P
0010	**COMMERCIAL OIL-FIRED WATER HEATERS**									
8000	Oil fired, glass lined, UL listed, std. controls, vent not incl.									
8060	140 gal., 140 MBH input, 134 GPH	Q-1	2.13	7.512	Ea.	26,000	420		26,420	29,300
8080	140 gal., 199 MBH input, 191 GPH		2	8		26,900	445		27,345	30,300
8160	140 gal., 540 MBH input, 519 GPH		.96	16.667		36,700	930		37,630	41,900
8280	201 gal., 1,250 MBH input, 1,200 GPH	Q-2	1.22	19.672		58,000	1,150		59,150	65,500

22 41 Residential Plumbing Fixtures

22 41 13 – Residential Water Closets, Urinals, and Bidets

22 41 13.13 Water Closets

22 41 13.13 Water Closets		Crew	Daily Output	Labor-Hours	Unit	Material	Labor	Equipment	Total	Total Incl O&P
0010	**WATER CLOSETS**									
0032	For automatic flush, see Line 22 42 39.10 0972									
0150	Tank type, vitreous china, incl. seat, supply pipe w/stop, 1.6 gpf or noted									
0200	Wall hung									
0400	Two piece, close coupled	Q-1	5.30	3.019	Ea.	400	169		569	705
0960	For rough-in, supply, waste, vent and carrier	"	2.73	5.861	"	1,150	330		1,480	1,775
0999	Floor mounted									
1100	Two piece, close coupled	Q-1	5.30	3.019	Ea.	230	169		399	515
1102	Economy		5.30	3.019		126	169		295	400
1110	Two piece, close coupled, dual flush		5.30	3.019		320	169		489	620
1140	Two piece, close coupled, 1.28 gpf, ADA [G]		5.30	3.019		310	169		479	605

22 41 Residential Plumbing Fixtures

22 41 13 – Residential Water Closets, Urinals, and Bidets

22 41 13.13 Water Closets

22 41 13.13 Water Closets	Crew	Daily Output	Labor-Hours	Unit	Material	2018 Bare Costs Labor	Equipment	Total	Total Incl O&P
1960	For color, add					30%				
1980	For rough-in, supply, waste and vent	Q-1	3.05	5.246	Ea.	360	293		653	850

22 41 16 – Residential Lavatories and Sinks

22 41 16.13 Lavatories

22 41 16.13 Lavatories	Crew	Daily Output	Labor-Hours	Unit	Material	2018 Bare Costs Labor	Equipment	Total	Total Incl O&P
0010	**LAVATORIES**, With trim, white unless noted otherwise									
0500	Vanity top, porcelain enamel on cast iron									
0600	20" x 18"	Q-1	6.40	2.500	Ea.	300	140		440	550
0640	33" x 19" oval	"	6.40	2.500	"	520	140		660	795
0860	For color, add					25%				
1000	Cultured marble, 19" x 17", single bowl	Q-1	6.40	2.500	Ea.	127	140		267	355
1120	25" x 22", single bowl	"	6.40	2.500	"	166	140		306	400
1580	For color, same price									
1900	Stainless steel, self-rimming, 25" x 22", single bowl, ledge	Q-1	6.40	2.500	Ea.	320	140		460	570
1960	17" x 22", single bowl		6.40	2.500		305	140		445	560
2600	Steel, enameled, 20" x 17", single bowl		5.80	2.759		129	154		283	380
2900	Vitreous china, 20" x 16", single bowl		5.40	2.963		213	166		379	490
2960	20" x 17", single bowl		5.40	2.963		116	166		282	385
3580	Rough-in, supply, waste and vent for all above lavatories	↓	2.30	6.957	↓	238	390		628	865
4000	Wall hung									
4040	Porcelain enamel on cast iron, 16" x 14", single bowl	Q-1	8	2	Ea.	445	112		557	665
4180	20" x 18", single bowl	"	8	2	"	246	112		358	445
4580	For color, add					30%				
6000	Vitreous china, 18" x 15", single bowl with backsplash	Q-1	7	2.286	Ea.	161	128		289	375
6960	Rough-in, supply, waste and vent for above lavatories	"	1.66	9.639	"	465	540		1,005	1,350
7000	Pedestal type									
7600	Vitreous china, 27" x 21", white	Q-1	6.60	2.424	Ea.	670	136		806	950
7610	27" x 21", colored		6.60	2.424		855	136		991	1,150
7620	27" x 21", premium color		6.60	2.424		975	136		1,111	1,275
7660	26" x 20", white		6.60	2.424		660	136		796	935
7670	26" x 20", colored		6.60	2.424		840	136		976	1,125
7680	26" x 20", premium color		6.60	2.424		955	136		1,091	1,250
7700	24" x 20", white		6.60	2.424		470	136		606	730
7710	24" x 20", colored		6.60	2.424		595	136		731	865
7720	24" x 20", premium color		6.60	2.424		675	136		811	955
7760	21" x 18", white		6.60	2.424		258	136		394	495
7770	21" x 18", colored		6.60	2.424		278	136		414	515
7990	Rough-in, supply, waste and vent for pedestal lavatories	↓	1.66	9.639	↓	465	540		1,005	1,350
9000	Minimum labor/equipment charge	1 Plum	3	2.667	Job		166		166	258

22 41 16.16 Sinks

22 41 16.16 Sinks	Crew	Daily Output	Labor-Hours	Unit	Material	2018 Bare Costs Labor	Equipment	Total	Total Incl O&P
0010	**SINKS**, With faucets and drain									
2000	Kitchen, counter top style, PE on CI, 24" x 21" single bowl	Q-1	5.60	2.857	Ea.	305	160		465	585
2100	31" x 22" single bowl		5.60	2.857		660	160		820	975
2200	32" x 21" double bowl		4.80	3.333		375	186		561	705
3000	Stainless steel, self rimming, 19" x 18" single bowl		5.60	2.857		615	160		775	925
3100	25" x 22" single bowl		5.60	2.857		680	160		840	1,000
3200	33" x 22" double bowl		4.80	3.333		985	186		1,171	1,375
3300	43" x 22" double bowl		4.80	3.333		1,150	186		1,336	1,550
4000	Steel, enameled, with ledge, 24" x 21" single bowl		5.60	2.857		525	160		685	830
4100	32" x 21" double bowl	↓	4.80	3.333		540	186		726	885
4960	For color sinks except stainless steel, add					10%				
4980	For rough-in, supply, waste and vent, counter top sinks	Q-1	2.14	7.477	↓	275	420		695	955
5790	For rough-in, supply, waste & vent, sinks	"	1.85	8.649	↓	275	485		760	1,050

For customer support on your Commercial Renovation Costs with RSMeans data, call 800.448.8182.

435

22 41 Residential Plumbing Fixtures

22 41 19 – Residential Bathtubs

22 41 19.10 Baths

22 41 19.10 Baths	Crew	Daily Output	Labor-Hours	Unit	Material	2018 Bare Costs Labor	Equipment	Total	Total Incl O&P
0010 **BATHS**									
0100 Tubs, recessed porcelain enamel on cast iron, with trim									
0180 48" x 42"	Q-1	4	4	Ea.	2,800	224		3,024	3,425
0220 72" x 36"	"	3	5.333	"	2,850	298		3,148	3,625
0300 Mat bottom									
0380 5' long	Q-1	4.40	3.636	Ea.	1,175	203		1,378	1,625
0560 Corner 48" x 44"		4.40	3.636		2,800	203		3,003	3,400
2000 Enameled formed steel, 4'-6" long		5.80	2.759		490	154		644	780
4600 Module tub & showerwall surround, molded fiberglass									
4610 5' long x 34" wide x 76" high	Q-1	4	4	Ea.	760	224		984	1,175
4750 ADA compliant with 1-1/2" OD grab bar, antiskid bottom									
4760 60" x 32-3/4" x 72" high	Q-1	4	4	Ea.	580	224		804	990
4770 60" x 30" x 71" high with molded seat		3.50	4.571		745	256		1,001	1,225
9600 Rough-in, supply, waste and vent, for all above tubs, add		2.07	7.729		430	430		860	1,150
9900 Minimum labor/equipment charge		3	5.333	Job		298		298	465

22 41 23 – Residential Showers

22 41 23.20 Showers

22 41 23.20 Showers	Crew	Daily Output	Labor-Hours	Unit	Material	2018 Bare Costs Labor	Equipment	Total	Total Incl O&P
0011 **SHOWERS**, Stall, with drain only									
1520 32" square	Q-1	5	3.200	Ea.	1,175	179		1,354	1,550
1530 36" square		4.80	3.333		2,925	186		3,111	3,500
1540 Terrazzo receptor, 32" square		5	3.200		1,350	179		1,529	1,750
1580 36" corner angle		4.80	3.333		1,750	186		1,936	2,225
3000 Fiberglass, one piece, with 3 walls, 32" x 32" square		5.50	2.909		345	163		508	635
3100 36" x 36" square		5.50	2.909		400	163		563	695
4200 Rough-in, supply, waste and vent for above showers		2.05	7.805		370	435		805	1,100

22 41 23.40 Shower System Components

22 41 23.40 Shower System Components		Crew	Daily Output	Labor-Hours	Unit	Material	2018 Bare Costs Labor	Equipment	Total	Total Incl O&P
0010 **SHOWER SYSTEM COMPONENTS**										
5500 Head, water economizer, 1.6 GPM	G	1 Plum	24	.333	Ea.	49.50	20.50		70	86.50

22 41 36 – Residential Laundry Trays

22 41 36.10 Laundry Sinks

22 41 36.10 Laundry Sinks	Crew	Daily Output	Labor-Hours	Unit	Material	2018 Bare Costs Labor	Equipment	Total	Total Incl O&P
0010 **LAUNDRY SINKS**, With trim									
0020 Porcelain enamel on cast iron, black iron frame									
0050 24" x 21", single compartment	Q-1	6	2.667	Ea.	595	149		744	880
0100 26" x 21", single compartment	"	6	2.667	"	620	149		769	915
2000 Molded stone, on wall hanger or legs									
2020 22" x 23", single compartment	Q-1	6	2.667	Ea.	185	149		334	435
2100 45" x 21", double compartment	"	5	3.200	"	310	179		489	625
3000 Plastic, on wall hanger or legs									
3020 18" x 23", single compartment	Q-1	6.50	2.462	Ea.	145	138		283	375
3100 20" x 24", single compartment		6.50	2.462		165	138		303	395
3200 36" x 23", double compartment		5.50	2.909		213	163		376	485
3300 40" x 24", double compartment		5.50	2.909		287	163		450	570
5000 Stainless steel, counter top, 22" x 17" single compartment		6	2.667		76.50	149		225.50	315
5200 33" x 22", double compartment		5	3.200		91.50	179		270.50	380
9600 Rough-in, supply, waste and vent, for all laundry sinks		2.14	7.477		275	420		695	955
9810 Minimum labor/equipment charge	1 Plum	3	2.667	Job		166		166	258

22 41 39 – Residential Faucets, Supplies and Trim

22 41 39.10 Faucets and Fittings

22 41 39.10 Faucets and Fittings	Crew	Daily Output	Labor-Hours	Unit	Material	2018 Bare Costs Labor	Equipment	Total	Total Incl O&P
0010 **FAUCETS AND FITTINGS**									
0150 Bath, faucets, diverter spout combination, sweat	1 Plum	8	1	Ea.	86	62		148	191

22 41 39 – Residential Faucets, Supplies and Trim

22 41 39.10 Faucets and Fittings	Crew	Daily Output	Labor-Hours	Unit	Material	2018 Bare Costs Labor	Equipment	Total	Total Incl O&P
0200 For integral stops, IPS unions, add				Ea.	111			111	122
0420 Bath, press-bal mix valve w/diverter, spout, shower head, arm/flange	1 Plum	8	1		175	62		237	289
0500 Drain, central lift, 1-1/2" IPS male		20	.400		49	25		74	92.50
0600 Trip lever, 1-1/2" IPS male	↓	20	.400	↓	60	25		85	105
0810 Bidet									
0812 Fitting, over the rim, swivel spray/pop-up drain	1 Plum	8	1	Ea.	260	62		322	385
1000 Kitchen sink faucets, top mount, cast spout		10	.800		83.50	49.50		133	169
1100 For spray, add		24	.333		17.10	20.50		37.60	51
1200 Wall type, swing tube spout	↓	10	.800	↓	73.50	49.50		123	159
1300 Single control lever handle									
1310 With pull out spray									
1320 Polished chrome	1 Plum	10	.800	Ea.	196	49.50		245.50	294
2000 Laundry faucets, shelf type, IPS or copper unions		12	.667		61.50	41.50		103	132
2100 Lavatory faucet, centerset, without drain		10	.800		67	49.50		116.50	152
2120 With pop-up drain	↓	6.66	1.201	↓	58.50	74.50		133	180
2210 Porcelain cross handles and pop-up drain									
2220 Polished chrome	1 Plum	6.66	1.201	Ea.	207	74.50		281.50	345
2230 Polished brass	"	6.66	1.201	"	310	74.50		384.50	460
2260 Single lever handle and pop-up drain									
2280 Satin nickel	1 Plum	6.66	1.201	Ea.	273	74.50		347.50	415
2290 Polished chrome		6.66	1.201		197	74.50		271.50	335
2800 Self-closing, center set		10	.800		149	49.50		198.50	242
2810 Automatic sensor and operator, with faucet head [G]		6.15	1.301		480	81		561	655
4000 Shower by-pass valve with union		18	.444		57	27.50		84.50	106
4100 Shower arm with flange and head		22	.364		20.50	22.50		43	57.50
4140 Shower, hand held, pin mount, massage action, chrome		22	.364		83.50	22.50		106	127
4142 Polished brass		22	.364		156	22.50		178.50	207
4144 Shower, hand held, wall mtd, adj. spray, 2 wall mounts, chrome		20	.400		112	25		137	163
4146 Polished brass		20	.400		212	25		237	273
4148 Shower, hand held head, bar mounted 24", adj. spray, chrome		20	.400		175	25		200	232
4150 Polished brass		20	.400		360	25		385	440
4200 Shower thermostatic mixing valve, concealed, with shower head trim kit	↓	8	1	↓	350	62		412	480
4220 Shower pressure balancing mixing valve									
4230 With shower head, arm, flange and diverter tub spout									
4240 Chrome	1 Plum	6.14	1.303	Ea.	410	81		491	580
4250 Satin nickel		6.14	1.303		555	81		636	735
4260 Polished graphite		6.14	1.303		555	81		636	735
5000 Sillcock, compact, brass, IPS or copper to hose	↓	24	.333	↓	10.55	20.50		31.05	43.50
6000 Stop and waste valves, bronze									
6100 Angle, solder end 1/2"	1 Plum	24	.333	Ea.	23	20.50		43.50	57
6110 3/4"		20	.400		29	25		54	70.50
6300 Straightway, solder end 3/8"		24	.333		18.80	20.50		39.30	52.50
6310 1/2"		24	.333		18.05	20.50		38.55	52
6410 Straightway, threaded 1/2"		24	.333		21.50	20.50		42	55.50
6420 3/4"		20	.400		24	25		49	65
6430 1"		19	.421		27	26		53	70.50
7800 Water closet, wax gasket		96	.083		1.73	5.20		6.93	9.95
7820 Gasket toilet tank to bowl		32	.250		2.89	15.55		18.44	27
7830 Replacement diaphragm washer assy for ballcock valve		12	.667		2.89	41.50		44.39	67.50
7850 Dual flush valve	↓	12	.667		127	41.50		168.50	205
8000 Water supply stops, polished chrome plate									
8200 Angle, 3/8"	1 Plum	24	.333	Ea.	8.85	20.50		29.35	42
8300 1/2"	↓	22	.364	↓	8.85	22.50		31.35	45

22 41 Residential Plumbing Fixtures

22 41 39 – Residential Faucets, Supplies and Trim

22 41 39.10 Faucets and Fittings	Crew	Daily Output	Labor-Hours	Unit	Material	2018 Bare Costs Labor	Equipment	Total	Total Incl O&P
8400	Straight, 3/8"	1 Plum	26	.308	Ea.	9.30	19.10		28.40	40
8500	1/2"		24	.333		9.30	20.50		29.80	42
8600	Water closet, angle, w/flex riser, 3/8"		24	.333		30	20.50		50.50	65
9000	Minimum labor/equipment charge		4	2	Job		124		124	193

22 42 Commercial Plumbing Fixtures

22 42 13 – Commercial Water Closets, Urinals, and Bidets

22 42 13.13 Water Closets

22 42 13.13 Water Closets		Crew	Daily Output	Labor-Hours	Unit	Material	2018 Bare Costs Labor	Equipment	Total	Total Incl O&P
0010	WATER CLOSETS									
3000	Bowl only, with flush valve, seat, 1.6 gpf unless noted									
3100	Wall hung	Q-1	5.80	2.759	Ea.	970	154		1,124	1,325
3200	For rough-in, supply, waste and vent, single WC		2.56	6.250		1,200	350		1,550	1,850
3360	With floor outlet, 1.28 gpf [G]		5.80	2.759		550	154		704	845
3362	With floor outlet, 1.28 gpf, ADA [G]		5.80	2.759		575	154		729	870
3370	For rough-in, supply, waste and vent, single WC		2.84	5.634		395	315		710	925
3390	Floor mounted children's size, 10-3/4" high									
3392	With automatic flush sensor, 1.6 gpf	Q-1	6.20	2.581	Ea.	620	144		764	910
3396	With automatic flush sensor, 1.28 gpf		6.20	2.581	"	620	144		764	910
9000	Minimum labor/equipment charge		4	4	Job		224		224	350

22 42 13.16 Urinals

22 42 13.16 Urinals		Crew	Daily Output	Labor-Hours	Unit	Material	2018 Bare Costs Labor	Equipment	Total	Total Incl O&P
0010	URINALS									
3000	Wall hung, vitreous china, with self-closing valve									
3100	Siphon jet type	Q-1	3	5.333	Ea.	281	298		579	775
3120	Blowout type		3	5.333		470	298		768	980
3140	Water saving .5 gpf [G]		3	5.333		555	298		853	1,075
3300	Rough-in, supply, waste & vent		2.83	5.654		670	315		985	1,225
5000	Stall type, vitreous china, includes valve		2.50	6.400		770	360		1,130	1,400
6980	Rough-in, supply, waste and vent		1.99	8.040		430	450		880	1,175
9000	Minimum labor/equipment charge		4	4	Job		224		224	350

22 42 16 – Commercial Lavatories and Sinks

22 42 16.13 Lavatories

22 42 16.13 Lavatories		Crew	Daily Output	Labor-Hours	Unit	Material	2018 Bare Costs Labor	Equipment	Total	Total Incl O&P
0010	LAVATORIES, With trim, white unless noted otherwise									
0020	Commercial lavatories same as residential. See Section 22 41 16									

22 42 16.40 Service Sinks

22 42 16.40 Service Sinks		Crew	Daily Output	Labor-Hours	Unit	Material	2018 Bare Costs Labor	Equipment	Total	Total Incl O&P
0010	SERVICE SINKS									
6650	Service, floor, corner, PE on CI, 28" x 28"	Q-1	4.40	3.636	Ea.	1,025	203		1,228	1,450
6790	For rough-in, supply, waste & vent, floor service sinks		1.64	9.756	"	960	545		1,505	1,900
9000	Minimum labor/equipment charge		4	4	Job		224		224	350

22 42 23 – Commercial Showers

22 42 23.30 Group Showers

22 42 23.30 Group Showers		Crew	Daily Output	Labor-Hours	Unit	Material	2018 Bare Costs Labor	Equipment	Total	Total Incl O&P
0010	GROUP SHOWERS									
6000	Group, w/pressure balancing valve, rough-in and rigging not included									
6800	Column, 6 heads, no receptors, less partitions	Q-1	3	5.333	Ea.	9,400	298		9,698	10,800
6900	With stainless steel partitions		1	16		12,100	895		12,995	14,800
7600	5 heads, no receptors, less partitions		3	5.333		6,475	298		6,773	7,600
7620	4 heads (1 ADA compliant) no receptors, less partitions		3	5.333		5,750	298		6,048	6,800
7700	With stainless steel partitions		1	16		6,725	895		7,620	8,800
9000	Minimum labor/equipment charge		4	4	Job		224		224	350

22 42 33 – Wash Fountains

22 42 33.20 Commercial Wash Fountains	Crew	Daily Output	Labor-Hours	Unit	Material	2018 Bare Costs Labor	Equipment	Total	Total Incl O&P
0010 **COMMERCIAL WASH FOUNTAINS**									
1900 Group, foot control									
2000 Precast terrazzo, circular, 36" diam., 5 or 6 persons	Q-2	3	8	Ea.	7,425	465		7,890	8,875
2100 54" diam. for 8 or 10 persons		2.50	9.600		9,250	555		9,805	11,100
2400 Semi-circular, 36" diam. for 3 persons		3	8		6,500	465		6,965	7,875
2500 54" diam. for 4 or 5 persons	↓	2.50	9.600	↓	8,750	555		9,305	10,500
5610 Group, infrared control, barrier free									
5614 Precast terrazzo									
5620 Semi-circular 36" diam. for 3 persons	Q-2	3	8	Ea.	7,400	465		7,865	8,850
5630 46" diam. for 4 persons		2.80	8.571		7,975	495		8,470	9,550
5640 Circular, 54" diam. for 8 persons, button control	↓	2.50	9.600		9,675	555		10,230	11,500
5700 Rough-in, supply, waste and vent for above wash fountains	Q-1	1.82	8.791	↓	455	490		945	1,275
9000 Minimum labor/equipment charge	Q-2	3	8	Job		465		465	720

22 42 39 – Commercial Faucets, Supplies, and Trim

22 42 39.10 Faucets and Fittings

22 42 39.10 Faucets and Fittings	Crew	Daily Output	Labor-Hours	Unit	Material	2018 Bare Costs Labor	Equipment	Total	Total Incl O&P
0010 **FAUCETS AND FITTINGS**									
0840 Flush valves, with vacuum breaker									
0850 Water closet									
0860 Exposed, rear spud	1 Plum	8	1	Ea.	156	62		218	269
0870 Top spud		8	1		182	62		244	297
0880 Concealed, rear spud		8	1		210	62		272	330
0890 Top spud		8	1		171	62		233	286
0900 Wall hung		8	1		191	62		253	310
0910 Dual flush flushometer		12	.667		223	41.50		264.50	310
0912 Flushometer retrofit kit	↓	18	.444	↓	17.60	27.50		45.10	62.50
0920 Urinal									
0930 Exposed, stall	1 Plum	8	1	Ea.	182	62		244	297
0940 Wall (washout)		8	1		144	62		206	255
0950 Pedestal, top spud		8	1		146	62		208	258
0960 Concealed, stall		8	1		168	62		230	282
0970 Wall (washout)	↓	8	1	↓	181	62		243	297
0971 Automatic flush sensor and operator for									
0972 urinals or water closets, standard [G]	1 Plum	8	1	Ea.	475	62		537	615
0980 High efficiency water saving									
0984 Water closets, 1.28 gpf [G]	1 Plum	8	1	Ea.	415	62		477	555
0988 Urinals, .5 gpf [G]	"	8	1	"	415	62		477	555
2790 Faucets for lavatories									
2800 Self-closing, center set	1 Plum	10	.800	Ea.	149	49.50		198.50	242
2810 Automatic sensor and operator, with faucet head		6.15	1.301		480	81		561	655
3000 Service sink faucet, cast spout, pail hook, hose end	↓	14	.571	↓	76	35.50		111.50	139

22 42 39.30 Carriers and Supports

22 42 39.30 Carriers and Supports	Crew	Daily Output	Labor-Hours	Unit	Material	2018 Bare Costs Labor	Equipment	Total	Total Incl O&P
0010 **CARRIERS AND SUPPORTS**, For plumbing fixtures									
3000 Lavatory, concealed arm									
3050 Floor mounted, single									
3100 High back fixture	1 Plum	6	1.333	Ea.	605	83		688	795
3200 Flat slab fixture		6	1.333		520	83		603	700
3220 ADA compliant	↓	6	1.333	↓	555	83		638	740
6980 Water closet, siphon jet									
7000 Horizontal, adjustable, caulk									
7040 Single, 4" pipe size	1 Plum	5.33	1.501	Ea.	965	93.50		1,058.50	1,225
7060 5" pipe size	↓	5.33	1.501	↓	1,100	93.50		1,193.50	1,375

22 42 Commercial Plumbing Fixtures

22 42 39 – Commercial Faucets, Supplies, and Trim

22 42 39.30 Carriers and Supports	Crew	Daily Output	Labor-Hours	Unit	Material	2018 Bare Costs Labor	Equipment	Total	Total Incl O&P
7100	Double, 4" pipe size	1 Plum	5	1.600	Ea.	1,600	99.50		1,699.50	1,925
7120	5" pipe size	↓	5	1.600	↓	1,950	99.50		2,049.50	2,275
8200	Water closet, residential									
8220	Vertical centerline, floor mount									
8240	Single, 3" caulk, 2" or 3" vent	1 Plum	6	1.333	Ea.	695	83		778	895
8260	4" caulk, 2" or 4" vent	↓	6	1.333	"	895	83		978	1,125
9990	Minimum labor/equipment charge	↓	3.50	2.286	Job		142		142	221

22 43 Healthcare Plumbing Fixtures

22 43 39 – Healthcare Faucets

22 43 39.10 Faucets and Fittings

		Crew	Daily Output	Labor-Hours	Unit	Material	2018 Bare Costs Labor	Equipment	Total	Total Incl O&P
0010	**FAUCETS AND FITTINGS**									
2850	Medical, bedpan cleanser, with pedal valve,	1 Plum	12	.667	Ea.	805	41.50		846.50	950
2860	With screwdriver stop valve		12	.667		415	41.50		456.50	525
2870	With self-closing spray valve		12	.667		257	41.50		298.50	350
2900	Faucet, gooseneck spout, wrist handles, grid drain		10	.800		209	49.50		258.50	310
2940	Mixing valve, knee action, screwdriver stops	↓	4	2	↓	425	124		549	665

22 45 Emergency Plumbing Fixtures

22 45 13 – Emergency Showers

22 45 13.10 Emergency Showers

		Crew	Daily Output	Labor-Hours	Unit	Material	2018 Bare Costs Labor	Equipment	Total	Total Incl O&P
0010	**EMERGENCY SHOWERS**, Rough-in not included									
5000	Shower, single head, drench, ball valve, pull, freestanding	Q-1	4	4	Ea.	385	224		609	770
5200	Horizontal or vertical supply		4	4		560	224		784	965
6000	Multi-nozzle, eye/face wash combination		4	4		690	224		914	1,100
6400	Multi-nozzle, 12 spray, shower only		4	4		2,050	224		2,274	2,600
6600	For freeze-proof, add		6	2.667	↓	485	149		634	765
9000	Minimum labor/equipment charge	↓	3	5.333	Job		298		298	465

22 45 16 – Eyewash Equipment

22 45 16.10 Eyewash Safety Equipment

		Crew	Daily Output	Labor-Hours	Unit	Material	2018 Bare Costs Labor	Equipment	Total	Total Incl O&P
0010	**EYEWASH SAFETY EQUIPMENT**, Rough-in not included									
1000	Eye wash fountain									
1400	Plastic bowl, pedestal mounted	Q-1	4	4	Ea.	294	224		518	675

22 47 Drinking Fountains and Water Coolers

22 47 13 – Drinking Fountains

22 47 13.10 Drinking Water Fountains

		Crew	Daily Output	Labor-Hours	Unit	Material	2018 Bare Costs Labor	Equipment	Total	Total Incl O&P
0010	**DRINKING WATER FOUNTAINS**, For connection to cold water supply									
1000	Wall mounted, non-recessed									
2700	Stainless steel, single bubbler, no back	1 Plum	4	2	Ea.	1,000	124		1,124	1,300
2740	With back		4	2		1,025	124		1,149	1,325
2780	Dual handle, ADA compliant		4	2		720	124		844	985
2820	Dual level, ADA compliant		3.20	2.500		1,575	155		1,730	1,975
3980	For rough-in, supply and waste, add	↓	2.21	3.620	↓	205	225		430	575
4000	Wall mounted, semi-recessed									
4200	Poly-marble, single bubbler	1 Plum	4	2	Ea.	955	124		1,079	1,250
4600	Stainless steel, satin finish, single bubbler	"	4	2	"	1,300	124		1,424	1,650

22 47 Drinking Fountains and Water Coolers

22 47 13 – Drinking Fountains

22 47 13.10 Drinking Water Fountains

22 47 13.10 Drinking Water Fountains	Crew	Daily Output	Labor-Hours	Unit	Material	2018 Bare Costs Labor	Equipment	Total	Total Incl O&P
6000	Wall mounted, fully recessed									
6400	Poly-marble, single bubbler	1 Plum	4	2	Ea.	1,675	124		1,799	2,025
6800	Stainless steel, single bubbler		4	2		1,525	124		1,649	1,875
7580	For rough-in, supply and waste, add		1.83	4.372		205	272		477	650
7600	Floor mounted, pedestal type									
8600	Enameled iron, heavy duty service, 2 bubblers	1 Plum	2	4	Ea.	2,850	249		3,099	3,500
8880	For freeze-proof valve system, add		2	4		705	249		954	1,150
8900	For rough-in, supply and waste, add		1.83	4.372		205	272		477	650
9000	Minimum labor/equipment charge		2	4	Job		249		249	385

22 47 16 – Pressure Water Coolers

22 47 16.10 Electric Water Coolers

22 47 16.10 Electric Water Coolers	Crew	Daily Output	Labor-Hours	Unit	Material	2018 Bare Costs Labor	Equipment	Total	Total Incl O&P
0010	ELECTRIC WATER COOLERS									
0100	Wall mounted, non-recessed									
0140	4 GPH	Q-1	4	4	Ea.	680	224		904	1,100
0160	8 GPH, barrier free, sensor operated		4	4		1,075	224		1,299	1,525
1000	Dual height, 8.2 GPH		3.80	4.211		1,975	235		2,210	2,550
1040	14.3 GPH		3.80	4.211		1,475	235		1,710	2,000
3300	Semi-recessed, 8.1 GPH		4	4		825	224		1,049	1,250
4600	Floor mounted, flush-to-wall									
4640	4 GPH	1 Plum	3	2.667	Ea.	800	166		966	1,150
4980	For stainless steel cabinet, add					138			138	152
5000	Dual height, 8.2 GPH	1 Plum	2	4		1,200	249		1,449	1,700
9000	Minimum labor/equipment charge	"	2	4	Job		249		249	385

22 66 Chemical-Waste Systems for Lab. and Healthcare Facilities

22 66 53 – Laboratory Chemical-Waste and Vent Piping

22 66 53.30 Glass Pipe

22 66 53.30 Glass Pipe	Crew	Daily Output	Labor-Hours	Unit	Material	2018 Bare Costs Labor	Equipment	Total	Total Incl O&P
0010	GLASS PIPE, Borosilicate, couplings & clevis hanger assemblies, 10' OC									
0020	Drainage									
1100	1-1/2" diameter	Q-1	52	.308	L.F.	13.70	17.20		30.90	42
1120	2" diameter		44	.364		17.35	20.50		37.85	50.50
1140	3" diameter		39	.410		20.50	23		43.50	58
1160	4" diameter		30	.533		43.50	30		73.50	94.50
1180	6" diameter		26	.615		71	34.50		105.50	132
9000	Minimum labor/equipment charge	1 Plum	4	2	Job		124		124	193

22 66 53.40 Pipe Fittings, Glass

22 66 53.40 Pipe Fittings, Glass	Crew	Daily Output	Labor-Hours	Unit	Material	2018 Bare Costs Labor	Equipment	Total	Total Incl O&P
0010	PIPE FITTINGS, GLASS									
0020	Drainage, beaded ends									
0040	Coupling & labor required at joints not incl. in fitting									
0050	price. Add 1 per joint for installed price									
0070	90° bend or sweep, 1-1/2"				Ea.	35.50			35.50	39
0090	2"					45			45	49.50
0100	3"					74			74	81.50
0110	4"					119			119	130
0120	6" (sweep only)					360			360	395
0350	Tee, single sanitary, 1-1/2"					57.50			57.50	63.50
0370	2"					57.50			57.50	63.50
0380	3"					86			86	94.50
0390	4"					154			154	169
0400	6"					410			410	455

For customer support on your Commercial Renovation Costs with RSMeans data, call 800.448.8182.

441

22 66 53.40 Pipe Fittings, Glass

		Crew	Daily Output	Labor-Hours	Unit	Material	2018 Bare Costs Labor	Equipment	Total	Total Incl O&P
0500	Coupling, stainless steel, TFE seal ring									
0520	1-1/2"	Q-1	32	.500	Ea.	26.50	28		54.50	72.50
0530	2"		30	.533		33.50	30		63.50	83
0540	3"		25	.640		31	36		67	90
0550	4"		23	.696		77.50	39		116.50	146
0560	6"		20	.800		174	44.50		218.50	261
9000	Minimum labor/equipment charge	1 Plum	4	2	Job		124		124	193

22 66 53.60 Corrosion Resistant Pipe

		Crew	Daily Output	Labor-Hours	Unit	Material	2018 Bare Costs Labor	Equipment	Total	Total Incl O&P
0010	**CORROSION RESISTANT PIPE**, No couplings or hangers									
0020	Iron alloy, drain, mechanical joint									
1000	1-1/2" diameter	Q-1	70	.229	L.F.	78	12.80		90.80	106
1100	2" diameter		66	.242		80.50	13.55		94.05	110
1120	3" diameter		60	.267		87	14.90		101.90	119
1140	4" diameter		52	.308		112	17.20		129.20	150
2980	Plastic, epoxy, fiberglass filament wound, B&S joint									
3000	2" diameter	Q-1	62	.258	L.F.	12.10	14.45		26.55	36
3100	3" diameter		51	.314		14.15	17.55		31.70	43
3120	4" diameter		45	.356		20	19.90		39.90	53
3160	8" diameter	Q-2	38	.632		44.50	36.50		81	106
3200	12" diameter	"	28	.857		73	49.50		122.50	158
9800	Minimum labor/equipment charge	1 Plum	4	2	Job		124		124	193

22 66 53.70 Pipe Fittings, Corrosion Resistant

		Crew	Daily Output	Labor-Hours	Unit	Material	2018 Bare Costs Labor	Equipment	Total	Total Incl O&P
0010	**PIPE FITTINGS, CORROSION RESISTANT**									
0030	Iron alloy									
0050	Mechanical joint									
0060	1/4 bend, 1-1/2"	Q-1	12	1.333	Ea.	111	74.50		185.50	238
0080	2"		10	1.600		182	89.50		271.50	340
0090	3"		9	1.778		218	99.50		317.50	395
0100	4"		8	2		250	112		362	450
0160	Tee and Y, sanitary, straight									
0170	1-1/2"	Q-1	8	2	Ea.	121	112		233	305
0180	2"		7	2.286		162	128		290	375
0190	3"		6	2.667		250	149		399	510
0200	4"		5	3.200		460	179		639	785
0360	Coupling, 1-1/2"		14	1.143		68.50	64		132.50	175
0380	2"		12	1.333		78	74.50		152.50	202
0390	3"		11	1.455		82	81.50		163.50	217
0400	4"		10	1.600		92.50	89.50		182	241
3000	Epoxy, filament wound									
3030	Quick-lock joint									
3040	90° elbow, 2"	Q-1	28	.571	Ea.	98.50	32		130.50	158
3060	3"		16	1		113	56		169	212
3070	4"		13	1.231		154	69		223	277
3190	Tee, 2"		19	.842		235	47		282	335
3200	3"		11	1.455		283	81.50		364.50	435
3210	4"		9	1.778		340	99.50		439.50	530
9000	Minimum labor/equipment charge	1 Plum	4	2	Job		124		124	193

Estimating Tips

The labor adjustment factors listed in Subdivision 22 01 02.20 also apply to Division 23.

23 10 00 Facility Fuel Systems

- The prices in this subdivision for above- and below-ground storage tanks do not include foundations or hold-down slabs, unless noted. The estimator should refer to Divisions 3 and 31 for foundation system pricing. In addition to the foundations, required tank accessories, such as tank gauges, leak detection devices, and additional manholes and piping, must be added to the tank prices.

23 50 00 Central Heating Equipment

- When estimating the cost of an HVAC system, check to see who is responsible for providing and installing the temperature control system. It is possible to overlook controls, assuming that they would be included in the electrical estimate.

- When looking up a boiler, be careful on specified capacity. Some

manufacturers rate their products on output while others use input.

- Include HVAC insulation for pipe, boiler, and duct (wrap and liner).

- Be careful when looking up mechanical items to get the correct pressure rating and connection type (thread, weld, flange).

23 70 00 Central HVAC Equipment

- Combination heating and cooling units are sized by the air conditioning requirements. (See Reference No. R236000-20 for the preliminary sizing guide.)

- A ton of air conditioning is nominally 400 CFM.

- Rectangular duct is taken off by the linear foot for each size, but its cost is usually estimated by the pound. Remember that SMACNA standards now base duct on internal pressure.

- Prefabricated duct is estimated and purchased like pipe: straight sections and fittings.

- Note that cranes or other lifting equipment are not included on any

lines in Division 23. For example, if a crane is required to lift a heavy piece of pipe into place high above a gym floor, or to put a rooftop unit on the roof of a four-story building, etc., it must be added. Due to the potential for extreme variation—from nothing additional required to a major crane or helicopter—we feel that including a nominal amount for "lifting contingency" would be useless and detract from the accuracy of the estimate. When using equipment rental cost data from RSMeans, do not forget to include the cost of the operator(s).

Reference Numbers

Reference numbers are shown at the beginning of some major classifications. These numbers refer to related items in the Reference Section. The reference information may be an estimating procedure, an alternate pricing method, or technical information.

Note: Not all subdivisions listed here necessarily appear. ■

23 05 05.10 HVAC Demolition	Crew	Daily Output	Labor-Hours	Unit	Material	2018 Bare Costs Labor	Equipment	Total	Total Incl O&P
0010	**HVAC DEMOLITION** R220105-10								
0100	Air conditioner, split unit, 3 ton	Q-5	2	8	Ea.		455		455	705
0150	Package unit, 3 ton R024119-10	Q-6	3	8	"		470		470	730
0298	Boilers									
0300	Electric, up thru 148 kW	Q-19	2	12	Ea.		685		685	1,075
0310	150 thru 518 kW	"	1	24			1,375		1,375	2,125
0320	550 thru 2,000 kW	Q-21	.40	80			4,700		4,700	7,300
0330	2,070 kW and up	"	.30	107			6,250		6,250	9,725
0340	Gas and/or oil, up thru 150 MBH	Q-7	2.20	14.545			870		870	1,350
0350	160 thru 2,000 MBH		.80	40			2,400		2,400	3,725
0360	2,100 thru 4,500 MBH		.50	64			3,850		3,850	5,975
0370	4,600 thru 7,000 MBH		.30	107			6,400		6,400	9,950
0380	7,100 thru 12,000 MBH		.16	200			12,000		12,000	18,700
0390	12,200 thru 25,000 MBH	▼	.12	267	▼		16,000		16,000	24,900
1000	Ductwork, 4" high, 8" wide	1 Clab	200	.040	L.F.		1.59		1.59	2.59
1020	10" wide		190	.042			1.68		1.68	2.72
1040	14" wide		180	.044			1.77		1.77	2.88
1100	6" high, 8" wide		165	.048			1.93		1.93	3.14
1120	12" wide		150	.053			2.13		2.13	3.45
1140	18" wide		135	.059			2.36		2.36	3.83
1200	10" high, 12" wide		125	.064			2.55		2.55	4.14
1220	18" wide		115	.070			2.77		2.77	4.50
1240	24" wide		110	.073			2.90		2.90	4.71
1300	12"-14" high, 16"-18" wide		85	.094			3.75		3.75	6.10
1320	24" wide		75	.107			4.25		4.25	6.90
1340	48" wide		71	.113			4.49		4.49	7.30
1400	18" high, 24" wide		67	.119			4.76		4.76	7.75
1420	36" wide		63	.127			5.05		5.05	8.20
1440	48" wide		59	.136			5.40		5.40	8.75
1500	30" high, 36" wide		56	.143			5.70		5.70	9.25
1520	48" wide		53	.151			6		6	9.75
1540	72" wide	▼	50	.160	▼		6.40		6.40	10.35
1550	Duct heater, electric strip	1 Elec	8	1	Ea.		58		58	90
1850	Minimum labor/equipment charge	1 Clab	3	2.667	Job		106		106	173
2200	Furnace, electric	Q-20	2	10	Ea.		545		545	860
2300	Gas or oil, under 120 MBH	Q-9	4	4			215		215	340
2340	Over 120 MBH	"	3	5.333			287		287	450
2800	Heat pump, package unit, 3 ton	Q-5	2.40	6.667			380		380	590
2840	Split unit, 3 ton		2	8	▼		455		455	705
3000	Mechanical equipment, light items. Unit is weight, not cooling.		.90	17.778	Ton		1,000		1,000	1,575
3600	Heavy items	▼	1.10	14.545	"		825		825	1,275
5090	Remove refrigerant from system	1 Stpi	40	.200	Lb.		12.60		12.60	19.60
9000	Minimum labor/equipment charge	Q-6	3	8	Job		470		470	730

23 05 23.30 Valves, Iron Body

		Crew	Daily Output	Labor-Hours	Unit	Material	Labor	Equipment	Total	Total Incl O&P
0010	**VALVES, IRON BODY**									
1020	Butterfly, wafer type, gear actuator, 200 lb.									
1030	2"	1 Plum	14	.571	Ea.	105	35.50		140.50	171
1060	4"	Q-1	5	3.200	"	123	179		302	415
1650	Gate, 125 lb., N.R.S.									
2150	Flanged									
2240	2-1/2"	Q-1	5	3.200	Ea.	605	179		784	945

444

For customer support on your Commercial Renovation Costs with RSMeans data, call 800.448.8182.

23 05 Common Work Results for HVAC

23 05 23 – General-Duty Valves for HVAC Piping

23 05 23.30 Valves, Iron Body	Crew	Daily Output	Labor-Hours	Unit	Material	2018 Bare Costs Labor	Equipment	Total	Total Incl O&P
2260	3"	Q-1	4.50	3.556	Ea.	680	199		879	1,050
2280	4"	↓	3	5.333	↓	975	298		1,273	1,550
3550	OS&Y, 125 lb., flanged									
3680	4"	Q-1	3	5.333	Ea.	665	298		963	1,200
3690	5"	Q-2	3.40	7.059		1,050	410		1,460	1,775
3700	6"	"	3	8	↓	1,050	465		1,515	1,875
9000	Minimum labor/equipment charge	1 Plum	3	2.667	Job		166		166	258

23 07 HVAC Insulation

23 07 13 – Duct Insulation

23 07 13.10 Duct Thermal Insulation

		Crew	Daily Output	Labor-Hours	Unit	Material	2018 Bare Costs Labor	Equipment	Total	Total Incl O&P
0010	**DUCT THERMAL INSULATION**									
0110	Insulation req'd. is based on the surface size/area to be covered									
3000	Ductwork									
3020	Blanket type, fiberglass, flexible									
3030	Fire rated for grease and hazardous exhaust ducts									
3060	1-1/2" thick	Q-14	84	.190	S.F.	4.54	9.65		14.19	20.50
3090	Fire rated for plenums									
3100	1/2" x 24" x 25'	Q-14	1.94	8.247	Roll	167	420		587	855
3110	1/2" x 24" x 25'		98	.163	S.F.	3.35	8.25		11.60	16.95
3120	1/2" x 48" x 25'		1.04	15.385	Roll	335	780		1,115	1,625
3126	1/2" x 48" x 25'	↓	104	.154	S.F.	3.35	7.80		11.15	16.20
3140	FSK vapor barrier wrap, .75 lb. density									
3160	1" thick [G]	Q-14	350	.046	S.F.	.22	2.32		2.54	3.95
3170	1-1/2" thick [G]		320	.050		.27	2.53		2.80	4.36
3180	2" thick [G]		300	.053		.32	2.70		3.02	4.68
3190	3" thick [G]		260	.062		.45	3.12		3.57	5.50
3200	4" thick [G]	↓	242	.066	↓	.64	3.35		3.99	6.05
3210	Vinyl jacket, same as FSK									
9600	Minimum labor/equipment charge	1 Stpi	4	2	Job		126		126	196

23 07 16 – HVAC Equipment Insulation

23 07 16.10 HVAC Equipment Thermal Insulation

		Crew	Daily Output	Labor-Hours	Unit	Material	2018 Bare Costs Labor	Equipment	Total	Total Incl O&P
0010	**HVAC EQUIPMENT THERMAL INSULATION**									
0110	Insulation req'd. is based on the surface size/area to be covered									
9610	Minimum labor/equipment charge	1 Stpi	4	2	Job		126		126	196

23 09 Instrumentation and Control for HVAC

23 09 53 – Pneumatic and Electric Control System for HVAC

23 09 53.10 Control Components

		Crew	Daily Output	Labor-Hours	Unit	Material	2018 Bare Costs Labor	Equipment	Total	Total Incl O&P
0010	**CONTROL COMPONENTS**									
5000	Thermostats									
5030	Manual	1 Stpi	8	1	Ea.	43	63		106	145
5040	1 set back, electric, timed [G]		8	1		76.50	63		139.50	183
5050	2 set back, electric, timed [G]	↓	8	1		228	63		291	350
5200	24 hour, automatic, clock [G]	1 Shee	8	1	↓	235	60		295	355
6000	Valves, motorized zone									
6100	Sweat connections, 1/2" C x C	1 Stpi	20	.400	Ea.	156	25		181	211
6110	3/4" C x C		20	.400		151	25		176	205
6120	1" C x C	↓	19	.421	↓	204	26.50		230.50	266

23 09 53.10 Control Components	Crew	Daily Output	Labor-Hours	Unit	Material	2018 Bare Costs Labor	2018 Bare Costs Equipment	Total	Total Incl O&P
9000	Minimum labor/equipment charge	1 Plum	4	2	Job		124		124	193

23 21 Hydronic Piping and Pumps
23 21 20 – Hydronic HVAC Piping Specialties

23 21 20.34 Dielectric Unions

		Crew	Daily Output	Labor-Hours	Unit	Material	Labor	Equipment	Total	Total Incl O&P
0010	**DIELECTRIC UNIONS**, Standard gaskets for water and air									
0020	250 psi maximum pressure									
0280	Female IPT to sweat, straight									
0340	3/4" pipe size	1 Plum	20	.400	Ea.	7.85	25		32.85	47
0780	Female IPT to female IPT, straight									
0800	1/2" pipe size	1 Plum	24	.333	Ea.	14.10	20.50		34.60	47.50
0840	3/4" pipe size		20	.400	"	16.25	25		41.25	56.50
9000	Minimum labor/equipment charge		4	2	Job		124		124	193

23 21 20.46 Expansion Tanks

		Crew	Daily Output	Labor-Hours	Unit	Material	Labor	Equipment	Total	Total Incl O&P
0010	**EXPANSION TANKS**									
2000	Steel, liquid expansion, ASME, painted, 15 gallon capacity	Q-5	17	.941	Ea.	720	53.50		773.50	875
2040	30 gallon capacity		12	1.333		830	75.50		905.50	1,025
2080	60 gallon capacity		8	2		1,175	113		1,288	1,450
2120	100 gallon capacity		6	2.667		1,700	151		1,851	2,075
3000	Steel ASME expansion, rubber diaphragm, 19 gal. cap. accept.		12	1.333		2,725	75.50		2,800.50	3,125
3020	31 gallon capacity		8	2		3,025	113		3,138	3,525
3040	61 gallon capacity		6	2.667		4,425	151		4,576	5,100
3080	119 gallon capacity		4	4		4,600	227		4,827	5,400
9000	Minimum labor/equipment charge		4	4	Job		227		227	355

23 21 20.70 Steam Traps

		Crew	Daily Output	Labor-Hours	Unit	Material	Labor	Equipment	Total	Total Incl O&P
0010	**STEAM TRAPS**									
0030	Cast iron body, threaded									
0040	Inverted bucket									
0050	1/2" pipe size	1 Stpi	12	.667	Ea.	157	42		199	239
0100	1" pipe size		9	.889		350	56		406	470
0120	1-1/4" pipe size		8	1		525	63		588	680
1000	Float & thermostatic, 15 psi									
1010	3/4" pipe size	1 Stpi	16	.500	Ea.	149	31.50		180.50	213
1020	1" pipe size		15	.533		180	33.50		213.50	251
1040	1-1/2" pipe size		9	.889		315	56		371	435
1060	2" pipe size		6	1.333		905	84		989	1,125
9000	Minimum labor/equipment charge		4	2	Job		126		126	196

23 21 20.76 Strainers, Y Type, Bronze Body

		Crew	Daily Output	Labor-Hours	Unit	Material	Labor	Equipment	Total	Total Incl O&P
0010	**STRAINERS, Y TYPE, BRONZE BODY**									
0050	Screwed, 125 lb., 1/4" pipe size	1 Stpi	24	.333	Ea.	33.50	21		54.50	69
0100	1/2" pipe size		20	.400		36.50	25		61.50	79.50
0120	3/4" pipe size		19	.421		45	26.50		71.50	91
0140	1" pipe size		17	.471		60.50	29.50		90	113
0160	1-1/2" pipe size		14	.571		131	36		167	200
0180	2" pipe size		13	.615		173	39		212	251
0182	3" pipe size		12	.667		1,100	42		1,142	1,275
0200	300 lb., 2-1/2" pipe size	Q-5	17	.941		720	53.50		773.50	880
0220	3" pipe size		16	1		1,275	56.50		1,331.50	1,500
0240	4" pipe size		15	1.067		2,325	60.50		2,385.50	2,650
1000	Flanged, 150 lb., 1-1/2" pipe size	1 Stpi	11	.727		475	46		521	595

23 21 Hydronic Piping and Pumps

23 21 20 – Hydronic HVAC Piping Specialties

23 21 20.76 Strainers, Y Type, Bronze Body

		Crew	Daily Output	Labor-Hours	Unit	Material	2018 Bare Costs Labor	Equipment	Total	Total Incl O&P
1020	2" pipe size	1 Stpi	8	1	Ea.	645	63		708	810
1030	2-1/2" pipe size	Q-5	5	3.200		945	181		1,126	1,325
1040	3" pipe size		4.50	3.556		1,175	202		1,377	1,625
1060	4" pipe size	↓	3	5.333		1,800	300		2,100	2,450
1100	6" pipe size	Q-6	3	8		3,475	470		3,945	4,550
1106	8" pipe size	"	2.60	9.231	↓	3,775	545		4,320	5,000
1500	For 300 lb. rating, add					40%				
9000	Minimum labor/equipment charge	1 Stpi	3.75	2.133	Job		134		134	209

23 21 23 – Hydronic Pumps

23 21 23.13 In-Line Centrifugal Hydronic Pumps

		Crew	Daily Output	Labor-Hours	Unit	Material	2018 Bare Costs Labor	Equipment	Total	Total Incl O&P
0010	**IN-LINE CENTRIFUGAL HYDRONIC PUMPS**									
0600	Bronze, sweat connections, 1/40 HP, in line									
0640	3/4" size	Q-1	16	1	Ea.	246	56		302	360
1000	Flange connection, 3/4" to 1-1/2" size									
1040	1/12 HP	Q-1	6	2.667	Ea.	645	149		794	940
1060	1/8 HP		6	2.667		1,100	149		1,249	1,450
1100	1/3 HP		6	2.667		1,250	149		1,399	1,600
1140	2" size, 1/6 HP		5	3.200		1,600	179		1,779	2,025
1180	2-1/2" size, 1/4 HP		5	3.200		2,025	179		2,204	2,500
1220	3" size, 1/4 HP		4	4	↓	2,175	224		2,399	2,725
9000	Minimum labor/equipment charge	↓	3.25	4.923	Job		275		275	430

23 23 Refrigerant Piping

23 23 16 – Refrigerant Piping Specialties

23 23 16.16 Refrigerant Line Sets

		Crew	Daily Output	Labor-Hours	Unit	Material	2018 Bare Costs Labor	Equipment	Total	Total Incl O&P
0010	**REFRIGERANT LINE SETS**, Standard									
0100	Copper tube									
0110	1/2" insulation, both tubes									
0120	Combination 1/4" and 1/2" tubes									
0130	10' set	Q-5	42	.381	Ea.	47.50	21.50		69	85.50
0135	15' set		42	.381		76.50	21.50		98	118
0140	20' set		40	.400		74	22.50		96.50	117
0150	30' set		37	.432		105	24.50		129.50	153
0160	40' set		35	.457		130	26		156	185
0170	50' set		32	.500		162	28.50		190.50	222
0180	100' set	↓	22	.727	↓	350	41		391	450
0300	Combination 1/4" and 3/4" tubes									
0310	10' set	Q-5	40	.400	Ea.	53	22.50		75.50	94
0320	20' set		38	.421		94	24		118	140
0330	30' set		35	.457		140	26		166	195
0340	40' set		33	.485		184	27.50		211.50	245
0350	50' set		30	.533		231	30		261	300
0380	100' set	↓	20	.800	↓	525	45.50		570.50	650
0500	Combination 3/8" & 3/4" tubes									
0510	10' set	Q-5	28	.571	Ea.	61	32.50		93.50	118
0520	20' set		36	.444		95	25		120	143
0530	30' set		34	.471		129	26.50		155.50	184
0540	40' set		31	.516		170	29.50		199.50	233
0550	50' set		28	.571		199	32.50		231.50	270
0580	100' set	↓	18	.889		590	50.50		640.50	725

23 23 Refrigerant Piping

23 23 16 – Refrigerant Piping Specialties

23 23 16.16 Refrigerant Line Sets	Crew	Daily Output	Labor-Hours	Unit	Material	2018 Bare Costs Labor	Equipment	Total	Total Incl O&P
0700	Combination 3/8" & 1-1/8" tubes								
0710	10' set	Q-5	36	.444	Ea.	107	25		132	156
0720	20' set		33	.485		178	27.50		205.50	239
0730	30' set		31	.516		240	29.50		269.50	310
0740	40' set		28	.571		360	32.50		392.50	445
0750	50' set		26	.615		350	35		385	440
0900	Combination 1/2" & 3/4" tubes								
0910	10' set	Q-5	37	.432	Ea.	64.50	24.50		89	109
0920	20' set		35	.457		114	26		140	167
0930	30' set		33	.485		172	27.50		199.50	232
0940	40' set		30	.533		226	30		256	296
0950	50' set		27	.593		284	33.50		317.50	370
0980	100' set		17	.941		640	53.50		693.50	790
2100	Combination 1/2" & 1-1/8" tubes								
2110	10' set	Q-5	35	.457	Ea.	110	26		136	162
2120	20' set		31	.516		188	29.50		217.50	253
2130	30' set		29	.552		283	31.50		314.50	360
2140	40' set		25	.640		380	36.50		416.50	475
2150	50' set		14	1.143		465	65		530	610
2300	For 1" thick insulation add					30%	15%			
3000	Refrigerant line sets, min-split, flared								
3100	Combination 1/4" & 3/8" tubes								
3120	15' set	Q-5	41	.390	Ea.	78	22		100	120
3140	25' set		38.50	.416		111	23.50		134.50	159
3160	35' set		37	.432		138	24.50		162.50	190
3180	50' set		30	.533		182	30		212	247
3200	Combination 1/4" & 1/2" tubes								
3220	15' set	Q-5	41	.390	Ea.	85	22		107	128
3240	25' set		38.50	.416		113	23.50		136.50	161
3260	35' set		37	.432		144	24.50		168.50	196
3280	50' set		30	.533		190	30		220	256

23 31 HVAC Ducts and Casings

23 31 13 – Metal Ducts

23 31 13.13 Rectangular Metal Ducts

23 31 13.13 Rectangular Metal Ducts	Crew	Daily Output	Labor-Hours	Unit	Material	2018 Bare Costs Labor	Equipment	Total	Total Incl O&P
0010	**RECTANGULAR METAL DUCTS**								
0020	Fabricated rectangular, includes fittings, joints, supports,								
0021	allowance for flexible connections and field sketches.								
0030	Does not include "as-built dwgs." or insulation.								
0031	NOTE: Fabrication and installation are combined								
0040	as LABOR cost. Approx. 25% fittings assumed.								
0042	Fabrication/Inst. is to commercial quality standards								
0043	(SMACNA or equiv.) for structure, sealing, leak testing, etc.								
0100	Aluminum, alloy 3003-H14, under 100 lb.	Q-10	75	.320	Lb.	3.19	17.85		21.04	31.50
0110	100 to 500 lb.		80	.300		1.88	16.75		18.63	28.50
0120	500 to 1,000 lb.		95	.253		1.81	14.10		15.91	24
0160	Over 5,000 lb.		145	.166		1.77	9.25		11.02	16.50
0500	Galvanized steel, under 200 lb.		235	.102		.57	5.70		6.27	9.65
0520	200 to 500 lb.		245	.098		.56	5.45		6.01	9.20
0540	500 to 1,000 lb.		255	.094		.55	5.25		5.80	8.90
0560	1,000 to 2,000 lb.		265	.091		.54	5.05		5.59	8.55

23 31 HVAC Ducts and Casings

23 31 13 – Metal Ducts

23 31 13.13 Rectangular Metal Ducts

23 31 13.13 Rectangular Metal Ducts	Crew	Daily Output	Labor-Hours	Unit	Material	2018 Bare Costs Labor	Equipment	Total	Total Incl O&P
0580	Over 5,000 lb.	Q-10	285	.084	Lb.	.53	4.70		5.23	8
1000	Stainless steel, type 304, under 100 lb.		165	.145		4.71	8.10		12.81	18
1020	100 to 500 lb.		175	.137		3.87	7.65		11.52	16.30
1030	500 to 1,000 lb.		190	.126		2.58	7.05		9.63	13.95

23 31 13.16 Round and Flat-Oval Spiral Ducts

		Crew	Daily Output	Labor-Hours	Unit	Material	Labor	Equipment	Total	Total Incl O&P
0010	**ROUND AND FLAT-OVAL SPIRAL DUCTS**									
0020	Fabricated round and flat oval spiral, includes hangers,									
0021	supports and field sketches.									
5400	Spiral preformed, steel, galv., straight lengths, max 10" spwg.									
5410	4" diameter, 26 ga.	Q-9	360	.044	L.F.	1.86	2.39		4.25	5.80
5416	5" diameter, 26 ga.		320	.050		1.86	2.69		4.55	6.30
5420	6" diameter, 26 ga.		280	.057		1.86	3.08		4.94	6.90
5425	7" diameter, 26 ga.		240	.067		2.17	3.59		5.76	8.05
5430	8" diameter, 26 ga.		200	.080		2.48	4.31		6.79	9.55
5440	10" diameter, 26 ga.		160	.100		3.10	5.40		8.50	11.90
9990	Minimum labor/equipment charge	1 Shee	3	2.667	Job		159		159	251

23 31 16 – Nonmetal Ducts

23 31 16.13 Fibrous-Glass Ducts

		Crew	Daily Output	Labor-Hours	Unit	Material	Labor	Equipment	Total	Total Incl O&P
0010	**FIBROUS-GLASS DUCTS**									
3490	Rigid fiberglass duct board, foil reinf. kraft facing									
3500	Rectangular, 1" thick, alum. faced, (FRK), std. weight	Q-10	350	.069	SF Surf	.90	3.83		4.73	7.05
9990	Minimum labor/equipment charge	1 Shee	3	2.667	Job		159		159	251

23 31 16.19 PVC Ducts

		Crew	Daily Output	Labor-Hours	Unit	Material	Labor	Equipment	Total	Total Incl O&P
0010	**PVC DUCTS**									
4000	Rigid plastic, corrosive fume resistant PVC									
4020	Straight, 6" diameter	Q-9	220	.073	L.F.	13.45	3.92		17.37	21
4040	8" diameter		160	.100		19.50	5.40		24.90	30
4060	10" diameter		120	.133		24.50	7.20		31.70	38.50
4070	12" diameter		100	.160		28.50	8.60		37.10	45
4080	14" diameter		70	.229		35.50	12.30		47.80	59
4090	16" diameter		62	.258		41.50	13.90		55.40	67.50
4100	18" diameter		58	.276		56.50	14.85		71.35	86
4110	20" diameter	Q-10	75	.320		53	17.85		70.85	86
4130	24" diameter	"	55	.436		72	24.50		96.50	118
4250	Coupling, 6" diameter	Q-9	88	.182	Ea.	27.50	9.80		37.30	46
4270	8" diameter		78	.205		36	11.05		47.05	57
4290	10" diameter		70	.229		41	12.30		53.30	65
4300	12" diameter		55	.291		45	15.65		60.65	74
4310	14" diameter		44	.364		58	19.55		77.55	94.50
4320	16" diameter		40	.400		61	21.50		82.50	102
4330	18" diameter		38	.421		79	22.50		101.50	123
4340	20" diameter	Q-10	55	.436		95	24.50		119.50	143
4360	24" diameter	"	45	.533		119	30		149	178
4470	Elbow, 90°, 6" diameter	Q-9	44	.364		125	19.55		144.55	168
4490	8" diameter		28	.571		140	31		171	203
4510	10" diameter		18	.889		170	48		218	263
4520	12" diameter		15	1.067		194	57.50		251.50	305
4540	16" diameter		10	1.600		292	86		378	455
4550	18" diameter	Q-10	15	1.600		405	89.50		494.50	585
4560	20" diameter		14	1.714		605	95.50		700.50	815
4580	24" diameter		12	2		740	112		852	990
4750	Elbow 45°, use 90° and deduct					25%				

23 31 HVAC Ducts and Casings

23 31 16 – Nonmetal Ducts

23 31 16.19 PVC Ducts	Crew	Daily Output	Labor-Hours	Unit	Material	2018 Bare Costs Labor	Equipment	Total	Total Incl O&P
9990	Minimum labor/equipment charge	1 Shee	3	2.667	Job		159		159	251

23 33 Air Duct Accessories

23 33 13 – Dampers

23 33 13.13 Volume-Control Dampers

	23 33 13.13 Volume-Control Dampers	Crew	Daily Output	Labor-Hours	Unit	Material	2018 Bare Costs Labor	Equipment	Total	Total Incl O&P
0010	**VOLUME-CONTROL DAMPERS**									
5990	Multi-blade dampers, opposed blade, 8" x 6"	1 Shee	24	.333	Ea.	29	19.95		48.95	63.50
5994	8" x 8"		22	.364		30	22		52	67.50
5996	10" x 10"		21	.381		35.50	23		58.50	75
6000	12" x 12"		21	.381		39.50	23		62.50	79.50
6020	12" x 18"		18	.444		53	26.50		79.50	101
6030	14" x 10"		20	.400		38.50	24		62.50	80
6031	14" x 14"		17	.471		47	28		75	96.50
6033	16" x 12"		17	.471		47	28		75	96.50
6035	16" x 16"		16	.500		58.50	30		88.50	112
6037	18" x 16".		15	.533		64.50	32		96.50	122
6038	18" x 18"		15	.533		69.50	32		101.50	127
6070	20" x 16"		14	.571		69.50	34		103.50	130
6072	20" x 20"		13	.615		84	37		121	150
6074	22" x 18"		14	.571		84	34		118	146
6076	24" x 16"		11	.727		82	43.50		125.50	159
6078	24" x 20"		8	1		97	60		157	202
6080	24" x 24"		8	1		114	60		174	220
6110	26" x 26"		6	1.333		127	79.50		206.50	266
6133	30" x 30"	Q-9	6.60	2.424		183	131		314	405
6135	32" x 32"		6.40	2.500		206	135		341	440
6180	48" x 36"		5.60	2.857		340	154		494	615
7500	Variable volume modulating motorized damper, incl. elect. mtr.									
7504	8" x 6"	1 Shee	15	.533	Ea.	158	32		190	225
7506	10" x 6"		14	.571		158	34		192	228
7510	10" x 10"		13	.615		163	37		200	237
7520	12" x 12"		12	.667		169	40		209	249
7522	12" x 16"		11	.727		170	43.50		213.50	256
7524	16" x 10"		12	.667		167	40		207	246
7526	16" x 14"		10	.800		173	48		221	267
7528	16" x 18"		9	.889		180	53		233	282
7542	18" x 18"		8	1		184	60		244	297
7544	20" x 14"		8	1		189	60		249	305
7546	20" x 18"		7	1.143		196	68.50		264.50	325
7560	24" x 12"		8	1		191	60		251	305
7562	24" x 18"		7	1.143		206	68.50		274.50	335
7564	24" x 24"		6	1.333		214	79.50		293.50	360
7568	28" x 10"		7	1.143		191	68.50		259.50	320
7590	30" x 14"		5	1.600		202	95.50		297.50	375
7600	30" x 18"		4	2		265	120		385	480
7610	30" x 24"		3.80	2.105		350	126		476	590
7700	For thermostat, add		8	1		41.50	60		101.50	140
8000	Multi-blade dampers, parallel blade									
8100	8" x 8"	1 Shee	24	.333	Ea.	110	19.95		129.95	153
8140	16" x 10"		20	.400		138	24		162	190
8160	18" x 12"		18	.444		151	26.50		177.50	208

23 33 Air Duct Accessories

23 33 13 – Dampers

23 33 13.13 Volume-Control Dampers

		Crew	Daily Output	Labor-Hours	Unit	Material	2018 Bare Costs Labor	Equipment	Total	Total Incl O&P
8220	28" x 16"	1 Shee	10	.800	Ea.	201	48		249	297

23 33 13.16 Fire Dampers

		Crew	Daily Output	Labor-Hours	Unit	Material	2018 Bare Costs Labor	Equipment	Total	Total Incl O&P
0010	**FIRE DAMPERS**									
3000	Fire damper, curtain type, 1-1/2 hr. rated, vertical, 6" x 6"	1 Shee	24	.333	Ea.	32.50	19.95		52.45	67
3020	8" x 6"		22	.364		32.50	22		54.50	70
3240	16" x 14"		18	.444		59	26.50		85.50	107
3400	24" x 20"	↓	8	1	↓	58.50	60		118.50	159

23 33 19 – Duct Silencers

23 33 19.10 Duct Silencers

		Crew	Daily Output	Labor-Hours	Unit	Material	2018 Bare Costs Labor	Equipment	Total	Total Incl O&P
0010	**DUCT SILENCERS**									
9000	Silencers, noise control for air flow, duct				MCFM	76.50			76.50	84

23 33 23 – Turning Vanes

23 33 23.13 Air Turning Vanes

		Crew	Daily Output	Labor-Hours	Unit	Material	2018 Bare Costs Labor	Equipment	Total	Total Incl O&P
0010	**AIR TURNING VANES**									
9400	Turning vane components									
9410	Turning vane rail	1 Shee	160	.050	L.F.	.76	2.99		3.75	5.55
9420	Double thick, factory fab. vane		300	.027		1.18	1.59		2.77	3.81
9428	12" high set		170	.047		1.68	2.81		4.49	6.30
9432	14" high set		160	.050	↓	1.96	2.99		4.95	6.85
9900	Minimum labor/equipment charge	↓	4	2	Job		120		120	189

23 33 46 – Flexible Ducts

23 33 46.10 Flexible Air Ducts

		Crew	Daily Output	Labor-Hours	Unit	Material	2018 Bare Costs Labor	Equipment	Total	Total Incl O&P
0010	**FLEXIBLE AIR DUCTS**									
1280	Add to labor for elevated installation									
1282	of prefabricated (purchased) ductwork									
1283	10' to 15' high						10%			
1204	15' to 20' high						20%			
1285	20' to 25' high						25%			
1286	25' to 30' high						35%			
1287	30' to 35' high						40%			
1288	35' to 40' high						50%			
1289	Over 40' high						55%			
1300	Flexible, coated fiberglass fabric on corr. resist. metal helix									
1400	pressure to 12" (WG) UL-181									
1500	Noninsulated, 3" diameter	Q-9	400	.040	L.F.	1.20	2.15		3.35	4.71
1540	5" diameter		320	.050		1.38	2.69		4.07	5.75
1560	6" diameter		280	.057		1.60	3.08		4.68	6.60
1580	7" diameter		240	.067		1.78	3.59		5.37	7.60
1600	8" diameter		200	.080		2.02	4.31		6.33	9
1640	10" diameter		160	.100		2.61	5.40		8.01	11.35
1660	12" diameter		120	.133		3.10	7.20		10.30	14.70
1900	Insulated, 1" thick, PE jacket, 3" diameter G		380	.042		2.62	2.27		4.89	6.45
1910	4" diameter G		340	.047		2.80	2.53		5.33	7.05
1920	5" diameter G		300	.053		2.93	2.87		5.80	7.75
1940	6" diameter G		260	.062		3.11	3.31		6.42	8.60
1960	7" diameter G		220	.073		3.60	3.92		7.52	10.10
1980	8" diameter G		180	.089		3.80	4.78		8.58	11.75
2020	10" diameter G		140	.114		4.55	6.15		10.70	14.70
2040	12" diameter G		100	.160		5.25	8.60		13.85	19.35
2060	14" diameter G	↓	80	.200	↓	6.40	10.75		17.15	24
9990	Minimum labor/equipment charge	1 Shee	3	2.667	Job		159		159	251

451

23 33 53.10 Duct Liner Board

		Crew	Daily Output	Labor-Hours	Unit	Material	2018 Bare Costs Labor	2018 Bare Costs Equipment	Total	Total Incl O&P
0010	**DUCT LINER BOARD**									
3490	Board type, fiberglass liner, 3 lb. density									
3680	No finish									
3700	1" thick	G	Q-14	170	.094	S.F.	.69	4.77		5.46	8.40
3710	1-1/2" thick	G		140	.114		1.03	5.80		6.83	10.45
3720	2" thick	G		130	.123		1.38	6.25		7.63	11.50
3940	Board type, non-fibrous foam									
3950	Temperature, bacteria and fungi resistant									
3960	1" thick	G	Q-14	150	.107	S.F.	2.57	5.40		7.97	11.50
3970	1-1/2" thick	G		130	.123		4.03	6.25		10.28	14.45
3980	2" thick	G		120	.133		4.02	6.75		10.77	15.20

23 34 HVAC Fans
23 34 13 – Axial HVAC Fans

23 34 13.10 Axial Flow HVAC Fans

		Crew	Daily Output	Labor-Hours	Unit	Material	Labor	Equipment	Total	Total Incl O&P
0010	**AXIAL FLOW HVAC FANS**									
0020	Air conditioning and process air handling									

23 34 14 – Blower HVAC Fans

23 34 14.10 Blower Type HVAC Fans

		Crew	Daily Output	Labor-Hours	Unit	Material	Labor	Equipment	Total	Total Incl O&P
0010	**BLOWER TYPE HVAC FANS**									
2500	Ceiling fan, right angle, extra quiet, 0.10" S.P.									
2540	210 CFM	Q-20	19	1.053	Ea.	345	57.50		402.50	470
2580	885 CFM		16	1.250		860	68.50		928.50	1,050
2620	2,960 CFM		11	1.818		1,575	99.50		1,674.50	1,900
2640	For wall or roof cap, add	1 Shee	16	.500		291	30		321	365

23 34 16 – Centrifugal HVAC Fans

23 34 16.10 Centrifugal Type HVAC Fans

		Crew	Daily Output	Labor-Hours	Unit	Material	Labor	Equipment	Total	Total Incl O&P
0010	**CENTRIFUGAL TYPE HVAC FANS**									
0200	In-line centrifugal, supply/exhaust booster									
0220	aluminum wheel/hub, disconnect switch, 1/4" S.P.									
0240	500 CFM, 10" diameter connection	Q-20	3	6.667	Ea.	1,575	365		1,940	2,300
0280	1,520 CFM, 16" diameter connection		2	10		1,650	545		2,195	2,650
0320	3,480 CFM, 20" diameter connection		.80	25		2,075	1,375		3,450	4,450
0326	5,080 CFM, 20" diameter connection		.75	26.667		2,175	1,450		3,625	4,700
5000	Utility set, centrifugal, V belt drive, motor									
5020	1/4" S.P., 1,200 CFM, 1/4 HP	Q-20	6	3.333	Ea.	1,875	182		2,057	2,325
5040	1,520 CFM, 1/3 HP		5	4		2,450	219		2,669	3,050
5060	1,850 CFM, 1/2 HP		4	5		2,425	274		2,699	3,075
5080	2,180 CFM, 3/4 HP		3	6.667		2,875	365		3,240	3,750
7000	Roof exhauster, centrifugal, aluminum housing, 12" galvanized									
7020	curb, bird screen, back draft damper, 1/4" S.P.									
7100	Direct drive, 320 CFM, 11" sq. damper	Q-20	7	2.857	Ea.	770	156		926	1,100
7120	600 CFM, 11" sq. damper	"	6	3.333	"	985	182		1,167	1,350
9900	Minimum labor/equipment charge	1 Elec	4	2	Job		116		116	180

23 34 23 – HVAC Power Ventilators

23 34 23.10 HVAC Power Circulators and Ventilators	Crew	Daily Output	Labor-Hours	Unit	Material	2018 Bare Costs Labor	Equipment	Total	Total Incl O&P
0010	**HVAC POWER CIRCULATORS AND VENTILATORS**									
6650	Residential, bath exhaust, grille, back draft damper									
6660	50 CFM	Q-20	24	.833	Ea.	52.50	45.50		98	129
6670	110 CFM		22	.909		104	49.50		153.50	192
6680	Light combination, squirrel cage, 100 watt, 70 CFM	↓	24	.833	↓	114	45.50		159.50	197
6700	Light/heater combination, ceiling mounted									
6710	70 CFM, 1,450 watt	Q-20	24	.833	Ea.	140	45.50		185.50	226
6800	Heater combination, recessed, 70 CFM		24	.833		69.50	45.50		115	148
6820	With 2 infrared bulbs		23	.870		97	47.50		144.50	182
6900	Kitchen exhaust, grille, complete, 160 CFM		22	.909		105	49.50		154.50	194
6910	180 CFM		20	1		106	54.50		160.50	203
6920	270 CFM	↓	18	1.111	↓	230	61		291	350
6940	Residential roof jacks and wall caps									
6944	Wall cap with back draft damper									
6946	3" & 4" diam. round duct	1 Shee	11	.727	Ea.	25	43.50		68.50	96
6948	6" diam. round duct	"	11	.727	"	72.50	43.50		116	148
6958	Roof jack with bird screen and back draft damper									
6960	3" & 4" diam. round duct	1 Shee	11	.727	Ea.	24.50	43.50		68	95.50
6962	3-1/4" x 10" rectangular duct	"	10	.800	"	44	48		92	124
6980	Transition									
6982	3-1/4" x 10" to 6" diam. round	1 Shee	20	.400	Ea.	32	24		56	72.50
8020	Attic, roof type									
8030	Aluminum dome, damper & curb									
8080	12" diameter, 1,000 CFM (gravity)	1 Elec	10	.800	Ea.	600	46.50		646.50	730
8090	16" diameter, 1,500 CFM (gravity)		9	.889		725	51.50		776.50	880
8100	20" diameter, 2,500 CFM (gravity)		8	1		890	58		948	1,075
8110	26" diameter, 4,000 CFM (gravity)		7	1.143		1,075	66.50		1,141.50	1,275
8120	32" diameter, 6,500 CFM (gravity)		6	1.333		1,475	77.50		1,552.50	1,750
8130	38" diameter, 8,000 CFM (gravity)		5	1.600		2,200	93		2,293	2,575
8140	50" diameter, 13,000 CFM (gravity)	↓	4	2	↓	3,175	116		3,291	3,675
8160	Plastic, ABS dome									
8180	1,050 CFM	1 Elec	14	.571	Ea.	177	33.50		210.50	247
8200	1,600 CFM	"	12	.667	"	265	39		304	350
8240	Attic, wall type, with shutter, one speed									
8250	12" diameter, 1,000 CFM	1 Elec	14	.571	Ea.	380	33.50		413.50	470
8260	14" diameter, 1,500 CFM		12	.667		415	39		454	515
8270	16" diameter, 2,000 CFM	↓	9	.889	↓	465	51.50		516.50	595
8290	Whole house, wall type, with shutter, one speed									
8300	30" diameter, 4,800 CFM	1 Elec	7	1.143	Ea.	1,000	66.50		1,066.50	1,200
8310	36" diameter, 7,000 CFM		6	1.333		1,075	77.50		1,152.50	1,325
8320	42" diameter, 10,000 CFM		5	1.600		1,225	93		1,318	1,500
8330	48" diameter, 16,000 CFM	↓	4	2		1,525	116		1,641	1,850
8340	For two speed, add				↓	91			91	100
8350	Whole house, lay-down type, with shutter, one speed									
8360	30" diameter, 4,500 CFM	1 Elec	8	1	Ea.	1,075	58		1,133	1,275
8370	36" diameter, 6,500 CFM		7	1.143		1,150	66.50		1,216.50	1,350
8380	42" diameter, 9,000 CFM		6	1.333		1,250	77.50		1,327.50	1,500
8390	48" diameter, 12,000 CFM	↓	5	1.600		1,425	93		1,518	1,725
8440	For two speed, add					68.50			68.50	75.50
8450	For 12 hour timer switch, add	1 Elec	32	.250	↓	68.50	14.55		83.05	98

23 36 Air Terminal Units

23 36 13 – Constant-Air-Volume Units

23 36 13.10 Constant Volume Mixing Boxes	Crew	Daily Output	Labor-Hours	Unit	Material	2018 Bare Costs Labor	Equipment	Total	Total Incl O&P
0010 **CONSTANT VOLUME MIXING BOXES**									
5180 Mixing box, includes electric or pneumatic motor									
5200 Constant volume, 150 to 270 CFM	Q-9	12	1.333	Ea.	955	72		1,027	1,175
5210 270 to 600 CFM	"	11	1.455	"	985	78.50		1,063.50	1,200

23 36 16 – Variable-Air-Volume Units

23 36 16.10 Variable Volume Mixing Boxes

	Crew	Daily Output	Labor-Hours	Unit	Material	2018 Bare Costs Labor	Equipment	Total	Total Incl O&P
0010 **VARIABLE VOLUME MIXING BOXES**									
5180 Mixing box, includes electric or pneumatic motor									
5500 VAV cool only, pneumatic, pressure independent 300 to 600 CFM	Q-9	11	1.455	Ea.	765	78.50		843.50	965
5510 500 to 1,000 CFM		9	1.778		785	95.50		880.50	1,025
5520 800 to 1,600 CFM		9	1.778		810	95.50		905.50	1,050
5530 1,100 to 2,000 CFM		8	2		830	108		938	1,075
5540 1,500 to 3,000 CFM		7	2.286		885	123		1,008	1,175
5550 2,000 to 4,000 CFM		6	2.667		895	144		1,039	1,200
5560 For electric, w/thermostat, pressure dependent, add					23			23	25.50

23 37 Air Outlets and Inlets

23 37 13 – Diffusers, Registers, and Grilles

23 37 13.10 Diffusers

	Crew	Daily Output	Labor-Hours	Unit	Material	2018 Bare Costs Labor	Equipment	Total	Total Incl O&P
0010 **DIFFUSERS**, Aluminum, opposed blade damper unless noted									
0100 Ceiling, linear, also for sidewall									
0120 2" wide	1 Shee	32	.250	L.F.	17.45	14.95		32.40	42.50
0160 4" wide		26	.308		23.50	18.40		41.90	54.50
0180 6" wide		24	.333		29	19.95		48.95	63
0200 8" wide		22	.364		33.50	22		55.50	71.50
0500 Perforated, 24" x 24" lay-in panel size, 6" x 6"		16	.500	Ea.	158	30		188	221
0520 8" x 8"		15	.533		166	32		198	234
0530 9" x 9"		14	.571		169	34		203	240
0540 10" x 10"		14	.571		169	34		203	240
0560 12" x 12"		12	.667		176	40		216	257
0590 16" x 16"		11	.727		197	43.50		240.50	286
0600 18" x 18"		10	.800		211	48		259	310
0610 20" x 20"		10	.800		228	48		276	325
0620 24" x 24"		9	.889		250	53		303	360
1000 Rectangular, 1 to 4 way blow, 6" x 6"		16	.500		42	30		72	93
1010 8" x 8"		15	.533		58.50	32		90.50	115
1014 9" x 9"		15	.533		52.50	32		84.50	108
1016 10" x 10"		15	.533		80.50	32		112.50	139
1020 12" x 6"		15	.533		72	32		104	130
1040 12" x 9"		14	.571		76.50	34		110.50	138
1060 12" x 12"		12	.667		70.50	40		110.50	141
1070 14" x 6"		13	.615		78.50	37		115.50	144
1074 14" x 14"		12	.667		129	40		169	205
1150 18" x 18"		9	.889		114	53		167	209
1170 24" x 12"		10	.800		165	48		213	257
2000 T-bar mounting, 24" x 24" lay-in frame, 6" x 6"		16	.500		69.50	30		99.50	124
2020 8" x 8"		14	.571		69.50	34		103.50	131
2040 12" x 12"		12	.667		85.50	40		125.50	157
2060 16" x 16"		11	.727		106	43.50		149.50	185
2080 18" x 18"		10	.800		117	48		165	205

454

23 37 Air Outlets and Inlets

23 37 13 – Diffusers, Registers, and Grilles

23 37 13.10 Diffusers

	23 37 13.10 Diffusers	Crew	Daily Output	Labor-Hours	Unit	Material	2018 Bare Costs Labor	Equipment	Total	Total Incl O&P
6000	For steel diffusers instead of aluminum, deduct				Ea.	10%				
9000	Minimum labor/equipment charge	1 Shee	4	2	Job		120		120	189

23 37 13.30 Grilles

	23 37 13.30 Grilles	Crew	Daily Output	Labor-Hours	Unit	Material	2018 Bare Costs Labor	Equipment	Total	Total Incl O&P
0010	**GRILLES**									
0020	Aluminum, unless noted otherwise									
1000	Air return, steel, 6" x 6"	1 Shee	26	.308	Ea.	19.80	18.40		38.20	51
1020	10" x 6"		24	.333		19.80	19.95		39.75	53.50
1080	16" x 8"		22	.364		28	22		50	65.50
1100	12" x 12"		22	.364		28	22		50	65.50
1120	24" x 12"		18	.444		37	26.50		63.50	82.50
1300	48" x 24"		12	.667		133	40		173	209
9000	Minimum labor/equipment charge		4	2	Job		120		120	189

23 37 13.60 Registers

	23 37 13.60 Registers	Crew	Daily Output	Labor-Hours	Unit	Material	2018 Bare Costs Labor	Equipment	Total	Total Incl O&P
0010	**REGISTERS**									
0980	Air supply									
1000	Ceiling/wall, O.B. damper, anodized aluminum									
1010	One or two way deflection, adj. curved face bars									
1120	12" x 12"	1 Shee	18	.444	Ea.	29	26.50		55.50	74
1140	14" x 8"		17	.471		19.10	28		47.10	65.50
1240	20" x 6"		18	.444		26.50	26.50		53	71
1340	24" x 8"		13	.615		35.50	37		72.50	97
1350	24" x 18"		12	.667		63.50	40		103.50	133
2700	Above registers in steel instead of aluminum, deduct					10%				
3000	Baseboard, hand adj. damper, enameled steel									
3012	8" x 6"	1 Shee	26	.308	Ea.	6.85	18.40		25.25	36.50
3020	10" x 6"		24	.333		7.45	19.95		27.40	39.50
3040	12" x 5"		23	.348		6.05	21		27.05	39.50
3060	12" x 6"		23	.348		8.10	21		29.10	42
3080	12" x 8"		22	.364		11.75	22		33.75	47.50
3100	14" x 6"		20	.400		8.80	24		32.80	47
9000	Minimum labor/equipment charge		4	2	Job		120		120	189

23 37 23 – HVAC Gravity Ventilators

23 37 23.10 HVAC Gravity Air Ventilators

	23 37 23.10 HVAC Gravity Air Ventilators	Crew	Daily Output	Labor-Hours	Unit	Material	2018 Bare Costs Labor	Equipment	Total	Total Incl O&P
0010	**HVAC GRAVITY AIR VENTILATORS**, Includes base									
1280	Rotary ventilators, wind driven, galvanized									
1300	4" neck diameter	Q-9	20	.800	Ea.	47.50	43		90.50	121
1400	12" neck diameter		10	1.600		67	86		153	210
1500	24" neck diameter		8	2		299	108		407	500
1600	For aluminum, add					30%				
9000	Minimum labor/equipment charge	1 Plum	2	4	Job		249		249	385

23 38 Ventilation Hoods

23 38 13 – Commercial-Kitchen Hoods

23 38 13.10 Hood and Ventilation Equipment	Crew	Daily Output	Labor-Hours	Unit	Material	2018 Bare Costs Labor	Equipment	Total	Total Incl O&P
0010 **HOOD AND VENTILATION EQUIPMENT**									
2970 Exhaust hood, sst, gutter on all sides, 4' x 4' x 2'	1 Carp	1.80	4.444	Ea.	4,300	225		4,525	5,100
2980 4' x 4' x 7'	"	1.60	5	"	6,850	254		7,104	7,925

23 41 Particulate Air Filtration

23 41 13 – Panel Air Filters

23 41 13.10 Panel Type Air Filters

0010 **PANEL TYPE AIR FILTERS**									
2950 Mechanical media filtration units									
3000 High efficiency type, with frame, non-supported Ⓖ				MCFM	35			35	38.50
3100 Supported type Ⓖ				"	55			55	60.50
5500 Throwaway glass or paper media type, 12" x 36" x 1"				Ea.	2.39			2.39	2.63

23 41 16 – Renewable-Media Air Filters

23 41 16.10 Disposable Media Air Filters

0010 **DISPOSABLE MEDIA AIR FILTERS**									
5000 Renewable disposable roll				C.S.F.	5.25			5.25	5.75

23 41 19 – Washable Air Filters

23 41 19.10 Permanent Air Filters

0010 **PERMANENT AIR FILTERS**									
4500 Permanent washable Ⓖ				MCFM	25			25	27.50

23 41 23 – Extended Surface Filters

23 41 23.10 Expanded Surface Filters

0010 **EXPANDED SURFACE FILTERS**									
4000 Medium efficiency, extended surface Ⓖ				MCFM	6			6	6.60

23 42 Gas-Phase Air Filtration

23 42 13 – Activated-Carbon Air Filtration

23 42 13.10 Charcoal Type Air Filtration

0010 **CHARCOAL TYPE AIR FILTRATION**									
0050 Activated charcoal type, full flow				MCFM	650			650	715
0060 Full flow, impregnated media 12" deep					225			225	248
0070 HEPA filter & frame for field erection					450			450	495
0080 HEPA filter-diffuser, ceiling install.				↓	350			350	385

23 43 Electronic Air Cleaners

23 43 13 – Washable Electronic Air Cleaners

23 43 13.10 Electronic Air Cleaners

0010 **ELECTRONIC AIR CLEANERS**									
2000 Electronic air cleaner, duct mounted									
2150 1,000 CFM	1 Shee	4	2	Ea.	430	120		550	665
2200 1,200 CFM		3.80	2.105		505	126		631	755
2250 1,400 CFM	↓	3.60	2.222	↓	530	133		663	795

23 51 Breechings, Chimneys, and Stacks

23 51 13 – Draft Control Devices

23 51 13.16 Vent Dampers

		Crew	Daily Output	Labor-Hours	Unit	Material	2018 Bare Costs Labor	Equipment	Total	Total Incl O&P
0010	**VENT DAMPERS**									
5000	Vent damper, bi-metal, gas, 3" diameter	Q-9	24	.667	Ea.	57	36		93	119
5010	4" diameter		24	.667		57	36		93	119
5020	5" diameter		23	.696		57	37.50		94.50	122
5030	6" diameter		22	.727		57	39		96	124
5040	7" diameter		21	.762		60.50	41		101.50	131
5050	8" diameter		20	.800		63	43		106	137
9000	Minimum labor/equipment charge	1 Shee	4	2	Job		120		120	189

23 51 13.19 Barometric Dampers

		Crew	Daily Output	Labor-Hours	Unit	Material	2018 Bare Costs Labor	Equipment	Total	Total Incl O&P
0010	**BAROMETRIC DAMPERS**									
1000	Barometric, gas fired system only, 6" size for 5" and 6" pipes	1 Shee	20	.400	Ea.	104	24		128	153
1040	8" size, for 7" and 8" pipes	"	18	.444	"	146	26.50		172.50	203
2000	All fuel, oil, oil/gas, coal									
2020	10" for 9" and 10" pipes	1 Shee	15	.533	Ea.	253	32		285	330
3260	For thermal switch for above, add	"	24	.333	"	104	19.95		123.95	147

23 51 23 – Gas Vents

23 51 23.10 Gas Chimney Vents

		Crew	Daily Output	Labor-Hours	Unit	Material	2018 Bare Costs Labor	Equipment	Total	Total Incl O&P
0010	**GAS CHIMNEY VENTS**, Prefab metal, UL listed									
0020	Gas, double wall, galvanized steel									
0080	3" diameter	Q-9	72	.222	V.L.F.	6.60	11.95		18.55	26
0100	4" diameter		68	.235		8.15	12.65		20.80	29
0120	5" diameter		64	.250		9.25	13.45		22.70	31
0140	6" diameter		60	.267		11.05	14.35		25.40	34.50
0160	7" diameter		56	.286		21	15.40		36.40	47
0180	8" diameter		52	.308		22.50	16.55		39.05	51
0200	10" diameter		48	.333		43	17.95		60.95	76
0220	12" diameter		44	.364		51.50	19.55		71.05	87.50
0260	16" diameter		40	.400		123	21.50		144.50	169
0300	20" diameter	Q-10	36	.667		179	37		216	255
0340	24" diameter	"	32	.750		280	42		322	375

23 51 26 – All-Fuel Vent Chimneys

23 51 26.30 All-Fuel Vent Chimneys, Double Wall, St. Stl.

		Crew	Daily Output	Labor-Hours	Unit	Material	2018 Bare Costs Labor	Equipment	Total	Total Incl O&P
0010	**ALL-FUEL VENT CHIMNEYS, DOUBLE WALL, STAINLESS STEEL**									
7780	All fuel, pressure tight, double wall, 4" insulation, UL listed, 1,400°F.									
7790	304 stainless steel liner, aluminized steel outer jacket									
7800	6" diameter	Q-9	60	.267	V.L.F.	65	14.35		79.35	94
7804	8" diameter		52	.308		75	16.55		91.55	108
7806	10" diameter		48	.333		83.50	17.95		101.45	120
7808	12" diameter		44	.364		95.50	19.55		115.05	136
7810	14" diameter		42	.381		107	20.50		127.50	151
7880	For 316 stainless steel liner, add				L.F.	30%				
8000	All fuel, double wall, stainless steel fittings									
8010	Roof support, 6" diameter	Q-9	30	.533	Ea.	115	28.50		143.50	172
8030	8" diameter		26	.615		134	33		167	200
8040	10" diameter		24	.667		142	36		178	213
8050	12" diameter		22	.727		151	39		190	228
8060	14" diameter		21	.762		158	41		199	239
8100	Elbow 45°, 6" diameter		30	.533		246	28.50		274.50	315
8140	8" diameter		26	.615		276	33		309	355
8160	10" diameter		24	.667		315	36		351	400
8180	12" diameter		22	.727		360	39		399	455

23 51 Breechings, Chimneys, and Stacks

23 51 26 – All-Fuel Vent Chimneys

23 51 26.30 All-Fuel Vent Chimneys, Double Wall, St. Stl.

		Crew	Daily Output	Labor-Hours	Unit	Material	2018 Bare Costs Labor	2018 Bare Costs Equipment	Total	Total Incl O&P
8200	14" diameter	Q-9	21	.762	Ea.	400	41		441	505
8300	Insulated tee, 6" diameter		30	.533		289	28.50		317.50	365
8360	8" diameter		26	.615		320	33		353	400
8380	10" diameter		24	.667		355	36		391	445
8400	12" diameter		22	.727		405	39		444	505
8420	14" diameter		20	.800		470	43		513	590
8500	Boot tee, 6" diameter		28	.571		550	31		581	655
8520	8" diameter		24	.667		600	36		636	715
8530	10" diameter		22	.727		700	39		739	830
8540	12" diameter		20	.800		830	43		873	980
8550	14" diameter		18	.889		945	48		993	1,100
8600	Rain cap with bird screen, 6" diameter		30	.533		296	28.50		324.50	370
8640	8" diameter		26	.615		345	33		378	425
8660	10" diameter		24	.667		400	36		436	495
8680	12" diameter		22	.727		460	39		499	570
8700	14" diameter		21	.762		525	41		566	640
8800	Flat roof flashing, 6" diameter		30	.533		108	28.50		136.50	164
8840	8" diameter		26	.615		118	33		151	182
8860	10" diameter		24	.667		127	36		163	196
8880	12" diameter		22	.727		139	39		178	215
8900	14" diameter		21	.762		142	41		183	221

23 52 Heating Boilers

23 52 13 – Electric Boilers

23 52 13.10 Electric Boilers, ASME

		Crew	Daily Output	Labor-Hours	Unit	Material	2018 Bare Costs Labor	2018 Bare Costs Equipment	Total	Total Incl O&P
0010	**ELECTRIC BOILERS, ASME**, Standard controls and trim									
1000	Steam, 6 KW, 20.5 MBH	Q-19	1.20	20	Ea.	4,250	1,150		5,400	6,450
1160	60 KW, 205 MBH		1	24		6,925	1,375		8,300	9,725
1300	296 KW, 1,010 MBH		.45	53.333		25,000	3,050		28,050	32,300
1400	592 KW, 2,020 MBH	Q-21	.34	94.118		35,400	5,525		40,925	47,500
1540	1,110 KW, 3,788 MBH	"	.19	168		49,600	9,875		59,475	70,000
2000	Hot water, 7.5 KW, 25.6 MBH	Q-19	1.30	18.462		5,125	1,050		6,175	7,300
2040	30 KW, 102 MBH		1.20	20		5,525	1,150		6,675	7,850
2060	45 KW, 164 MBH		1.20	20		5,625	1,150		6,775	7,950
2070	60 KW, 205 MBH		1.20	20		5,750	1,150		6,900	8,075
2080	75 KW, 256 MBH		1.10	21.818		6,125	1,250		7,375	8,675
2100	90 KW, 307 MBH		1.10	21.818		6,100	1,250		7,350	8,650
9000	Minimum labor/equipment charge	Q-20	1	20	Job		1,100		1,100	1,725

23 52 23 – Cast-Iron Boilers

23 52 23.20 Gas-Fired Boilers

		Crew	Daily Output	Labor-Hours	Unit	Material	2018 Bare Costs Labor	2018 Bare Costs Equipment	Total	Total Incl O&P
0010	**GAS-FIRED BOILERS**, Natural or propane, standard controls, packaged									
1000	Cast iron, with insulated jacket									
2000	Steam, gross output, 81 MBH	Q-7	1.40	22.857	Ea.	2,300	1,375		3,675	4,675
2060	163 MBH		.90	35.556		3,175	2,125		5,300	6,825
2200	440 MBH		.51	62.500		6,150	3,750		9,900	12,600
2320	1,875 MBH		.30	107		24,200	6,400		30,600	36,700
2400	3,570 MBH		.18	182		34,500	10,900		45,400	55,000
2540	6,970 MBH		.10	320		95,000	19,200		114,200	134,500
3000	Hot water, gross output, 80 MBH		1.46	21.918		1,925	1,325		3,250	4,175
3020	100 MBH		1.35	23.704		2,450	1,425		3,875	4,900

23 52 Heating Boilers

23 52 23 – Cast-Iron Boilers

23 52 23.20 Gas-Fired Boilers

		Crew	Daily Output	Labor-Hours	Unit	Material	2018 Bare Costs Labor	Equipment	Total	Total Incl O&P
3040	122 MBH	Q-7	1.10	29.091	Ea.	2,500	1,750		4,250	5,475
3060	163 MBH		1	32		3,025	1,925		4,950	6,300
3200	440 MBH		.58	54.983		5,725	3,300		9,025	11,400
3320	2,000 MBH		.26	125		20,600	7,500		28,100	34,300
3540	6,970 MBH		.09	360		126,500	21,600		148,100	173,000
7000	For tankless water heater, add					10%				
7050	For additional zone valves up to 312 MBH, add					196			196	216
9900	Minimum labor/equipment charge	Q-6	1	24	Job		1,400		1,400	2,200

23 52 23.30 Gas/Oil Fired Boilers

		Crew	Daily Output	Labor-Hours	Unit	Material	2018 Bare Costs Labor	Equipment	Total	Total Incl O&P
0010	**GAS/OIL FIRED BOILERS**, Combination with burners and controls, packaged									
1000	Cast iron with insulated jacket									
2000	Steam, gross output, 720 MBH	Q-7	.43	74.074	Ea.	14,200	4,450		18,650	22,500
2140	2,700 MBH		.19	166		27,000	9,950		36,950	45,200
2380	6,970 MBH		.09	372		98,500	22,300		120,800	143,000
2900	Hot water, gross output									
2910	200 MBH	Q-6	.62	39.024	Ea.	10,200	2,300		12,500	14,900
2920	300 MBH		.49	49.080		10,200	2,875		13,075	15,800
2930	400 MBH		.41	57.971		12,000	3,400		15,400	18,500
2940	500 MBH		.36	67.039		12,900	3,950		16,850	20,300
3000	584 MBH	Q-7	.44	72.072		14,300	4,325		18,625	22,400
3060	1,460 MBH		.28	113		35,400	6,775		42,175	49,500
3300	13,500 MBH, 403.3 BHP		.04	727		185,000	43,600		228,600	271,500

23 52 23.40 Oil-Fired Boilers

		Crew	Daily Output	Labor-Hours	Unit	Material	2018 Bare Costs Labor	Equipment	Total	Total Incl O&P
0010	**OIL-FIRED BOILERS**, Standard controls, flame retention burner, packaged									
1000	Cast iron, with insulated flush jacket									
2000	Steam, gross output, 109 MBH	Q-7	1.20	26.667	Ea.	2,225	1,600		3,825	4,950
2060	207 MBH		.90	35.556		3,025	2,125		5,150	6,650
2180	1,084 MBH		.38	85.106		10,500	5,100		15,600	19,600
2200	1,360 MBH		.33	98.160		12,000	5,900		17,900	22,400
2240	2,175 MBH		.24	134		17,400	8,025		25,425	31,700
2340	4,360 MBH		.15	215		41,200	12,900		54,100	65,500
2460	6,970 MBH		.09	364		97,500	21,800		119,300	141,000
3000	Hot water, same price as steam									
4000	For tankless coil in smaller sizes, add				Ea.	15%				

23 52 26 – Steel Boilers

23 52 26.40 Oil-Fired Boilers

		Crew	Daily Output	Labor-Hours	Unit	Material	2018 Bare Costs Labor	Equipment	Total	Total Incl O&P
0010	**OIL-FIRED BOILERS**, Standard controls, flame retention burner									
5000	Steel, with insulated flush jacket									
7000	Hot water, gross output, 103 MBH	Q-6	1.60	15	Ea.	1,875	880		2,755	3,450
7020	122 MBH		1.45	16.506		2,000	970		2,970	3,700
7040	137 MBH		1.36	17.595		2,350	1,025		3,375	4,175
7060	168 MBH		1.30	18.405		2,250	1,075		3,325	4,150
7080	225 MBH		1.22	19.704		3,250	1,150		4,400	5,375
7100	315 MBH		.96	25.105		3,325	1,475		4,800	5,950
7120	420 MBH		.70	34.483		5,775	2,025		7,800	9,500
9000	Minimum labor/equipment charge		1.75	13.714	Job		805		805	1,250

23 52 Heating Boilers

23 52 88 – Burners

23 52 88.10 Replacement Type Burners

23 52 88.10 Replacement Type Burners	Crew	Daily Output	Labor-Hours	Unit	Material	2018 Bare Costs Labor	Equipment	Total	Total Incl O&P
0010 **REPLACEMENT TYPE BURNERS**									
0990 Residential, conversion, gas fired, LP or natural									
1000 Gun type, atmospheric input 50 to 225 MBH	Q-1	2.50	6.400	Ea.	1,000	360		1,360	1,650
1020 100 to 400 MBH	2	8		1,675	445		2,120	2,550
1040 300 to 1,000 MBH	↓	1.70	9.412	↓	5,100	525		5,625	6,425
3000 Flame retention oil fired assembly, input									
3020 0.50 to 2.25 GPH	Q-1	2.40	6.667	Ea.	350	375		725	965
3040 2.0 to 5.0 GPH	2	8		340	445		785	1,075
3060 3.0 to 7.0 GPH	1.80	8.889		580	495		1,075	1,400
3080 6.0 to 12.0 GPH	↓	1.60	10		930	560		1,490	1,900

23 54 Furnaces

23 54 13 – Electric-Resistance Furnaces

23 54 13.10 Electric Furnaces

23 54 13.10 Electric Furnaces	Crew	Daily Output	Labor-Hours	Unit	Material	2018 Bare Costs Labor	Equipment	Total	Total Incl O&P
0010 **ELECTRIC FURNACES**, Hot air, blowers, std. controls									
0011 not including gas, oil or flue piping									
1000 Electric, UL listed									
1070 10.2 MBH	Q-20	4.40	4.545	Ea.	335	249		584	760
1080 17.1 MBH	4.60	4.348		415	238		653	830
1100 34.1 MBH	↓	4.40	4.545	↓	520	249		769	960

23 54 16 – Fuel-Fired Furnaces

23 54 16.13 Gas-Fired Furnaces

23 54 16.13 Gas-Fired Furnaces	Crew	Daily Output	Labor-Hours	Unit	Material	2018 Bare Costs Labor	Equipment	Total	Total Incl O&P
0010 **GAS-FIRED FURNACES**									
3000 Gas, AGA certified, upflow, direct drive models									
3020 45 MBH input	Q-9	4	4	Ea.	645	215		860	1,050
3040 60 MBH input	3.80	4.211		640	227		867	1,050
3060 75 MBH input	3.60	4.444		710	239		949	1,150
3100 100 MBH input	3.20	5		740	269		1,009	1,250
3120 125 MBH input	3	5.333		770	287		1,057	1,300
3130 150 MBH input	2.80	5.714		1,550	310		1,860	2,175
3140 200 MBH input	↓	2.60	6.154	↓	3,175	330		3,505	4,025

23 54 16.16 Oil-Fired Furnaces

23 54 16.16 Oil-Fired Furnaces	Crew	Daily Output	Labor-Hours	Unit	Material	2018 Bare Costs Labor	Equipment	Total	Total Incl O&P
0010 **OIL-FIRED FURNACES**									
6000 Oil, UL listed, atomizing gun type burner									
6020 56 MBH output	Q-9	3.60	4.444	Ea.	2,825	239		3,064	3,475
6040 95 MBH output	3.40	4.706		2,875	253		3,128	3,550
6060 134 MBH output	3.20	5		2,825	269		3,094	3,525
6080 151 MBH output	3	5.333		3,100	287		3,387	3,875
6100 200 MBH input	2.60	6.154	↓	3,550	330		3,880	4,425
9000 Minimum labor/equipment charge	↓	2.75	5.818	Job		315		315	495

23 54 24 – Furnace Components for Cooling

23 54 24.10 Furnace Components and Combinations

23 54 24.10 Furnace Components and Combinations	Crew	Daily Output	Labor-Hours	Unit	Material	2018 Bare Costs Labor	Equipment	Total	Total Incl O&P
0010 **FURNACE COMPONENTS AND COMBINATIONS**									
0080 Coils, A.C. evaporator, for gas or oil furnaces									
0090 Add-on, with holding charge									
0100 Upflow									
0120 1-1/2 ton cooling	Q-5	4	4	Ea.	209	227		436	585
0130 2 ton cooling	↓	3.70	4.324	↓	263	245		508	670
0140 3 ton cooling		3.30	4.848		400	275		675	870

23 54 Furnaces

23 54 24 – Furnace Components for Cooling

23 54 24.10 Furnace Components and Combinations	Crew	Daily Output	Labor-Hours	Unit	Material	2018 Bare Costs Labor	Equipment	Total	Total Incl O&P
0150	4 ton cooling	Q-5	3	5.333	Ea.	550	300		850	1,075
0160	5 ton cooling	↓	2.70	5.926	↓	590	335		925	1,175
0300	Downflow									
0330	2-1/2 ton cooling	Q-5	3	5.333	Ea.	315	300		615	815
0340	3-1/2 ton cooling		2.60	6.154		465	350		815	1,050
0350	5 ton cooling	↓	2.20	7.273	↓	590	410		1,000	1,300
0600	Horizontal									
0630	2 ton cooling	Q-5	3.90	4.103	Ea.	385	233		618	785
0640	3 ton cooling		3.50	4.571		430	259		689	880
0650	4 ton cooling	↓	3.20	5		480	284		764	965
0660	5 ton cooling	↓	2.90	5.517	↓	480	315		795	1,000
2000	Cased evaporator coils for air handlers									
2100	1-1/2 ton cooling	Q-5	4.40	3.636	Ea.	310	206		516	660
2110	2 ton cooling		4.10	3.902		355	221		576	735
2120	2-1/2 ton cooling		3.90	4.103		375	233		608	775
2130	3 ton cooling		3.70	4.324		420	245		665	840
2140	3-1/2 ton cooling		3.50	4.571		520	259		779	980
2150	4 ton cooling		3.20	5		610	284		894	1,100
2160	5 ton cooling	↓	2.90	5.517	↓	600	315		915	1,150
3010	Air handler, modular									
3100	With cased evaporator cooling coil									
3120	1-1/2 ton cooling	Q-5	3.80	4.211	Ea.	965	239		1,204	1,450
3130	2 ton cooling		3.50	4.571		965	259		1,224	1,475
3140	2-1/2 ton cooling		3.30	4.848		1,025	275		1,300	1,550
3150	3 ton cooling		3.10	5.161		1,175	293		1,468	1,750
3160	3-1/2 ton cooling		2.90	5.517		1,450	315		1,765	2,075
3170	4 ton cooling		2.50	6.400		1,450	365		1,815	2,175
3180	5 ton cooling	↓	2.10	7.619	↓	1,750	430		2,180	2,600
3500	With no cooling coil									
3520	1-1/2 ton coil size	Q-5	12	1.333	Ea.	1,100	75.50		1,175.50	1,325
3530	2 ton coil size		10	1.600		1,175	90.50		1,265.50	1,450
3540	2-1/2 ton coil size		10	1.600		1,225	90.50		1,315.50	1,475
3554	3 ton coil size		9	1.778		1,200	101		1,301	1,475
3560	3-1/2 ton coil size		9	1.778		1,250	101		1,351	1,525
3570	4 ton coil size		8.50	1.882		1,325	107		1,432	1,625
3580	5 ton coil size	↓	8	2	↓	1,425	113		1,538	1,725
4000	With heater									
4120	5 kW, 17.1 MBH	Q-5	16	1	Ea.	455	56.50		511.50	590
4130	7.5 kW, 25.6 MBH		15.60	1.026		1,100	58		1,158	1,300
4140	10 kW, 34.2 MBH		15.20	1.053		1,175	59.50		1,234.50	1,400
4150	12.5 kW, 42.7 MBH	↓	14.80	1.081	↓	1,550	61.50		1,611.50	1,800

23 55 Fuel-Fired Heaters

23 55 13 – Fuel-Fired Duct Heaters

23 55 13.16 Gas-Fired Duct Heaters

		Crew	Daily Output	Labor-Hours	Unit	Material	2018 Bare Costs Labor	2018 Bare Costs Equipment	Total	Total Incl O&P
0010	**GAS-FIRED DUCT HEATERS**, Includes burner, controls, stainless steel									
0020	heat exchanger. Gas fired, electric ignition									
1000	Outdoor installation, with power venter									
1120	225 MBH output	Q-5	2.50	6.400	Ea.	5,250	365		5,615	6,350
1160	375 MBH output		1.60	10		8,400	565		8,965	10,100
1180	450 MBH output		1.40	11.429		9,450	650		10,100	11,400

23 55 33 – Fuel-Fired Unit Heaters

23 55 33.16 Gas-Fired Unit Heaters

		Crew	Daily Output	Labor-Hours	Unit	Material	2018 Bare Costs Labor	2018 Bare Costs Equipment	Total	Total Incl O&P
0010	**GAS-FIRED UNIT HEATERS**, Cabinet, grilles, fan, ctrls., burner, no piping									
0022	thermostat, no piping. For flue see Section 23 51 23.10									
1000	Gas fired, floor mounted									
1100	60 MBH output	Q-5	10	1.600	Ea.	725	90.50		815.50	940
1180	180 MBH output		6	2.667		1,150	151		1,301	1,475
2000	Suspension mounted, propeller fan, 20 MBH output		8.50	1.882		1,575	107		1,682	1,900
2040	60 MBH output		7	2.286		1,800	130		1,930	2,200
2100	130 MBH output		5	3.200		2,350	181		2,531	2,850
2240	320 MBH output		2	8		4,350	455		4,805	5,500
2500	For powered venter and adapter, add					405			405	445
5000	Wall furnace, 17.5 MBH output	Q-5	6	2.667		635	151		786	935
5020	24 MBH output		5	3.200		635	181		816	980
5040	35 MBH output		4	4		980	227		1,207	1,425
9000	Minimum labor/equipment charge		3.50	4.571	Job		259		259	405

23 56 Solar Energy Heating Equipment

23 56 16 – Packaged Solar Heating Equipment

23 56 16.40 Solar Heating Systems

			Crew	Daily Output	Labor-Hours	Unit	Material	2018 Bare Costs Labor	2018 Bare Costs Equipment	Total	Total Incl O&P
0010	**SOLAR HEATING SYSTEMS**										
0020	System/package prices, not including connecting										
0030	pipe, insulation, or special heating/plumbing fixtures										
0500	Hot water, standard package, low temperature										
0580	2 collectors, circulator, fittings, 120 gal. tank	G	Q-1	.40	40	Ea.	5,400	2,225		7,625	9,425

23 56 19 – Solar Heating Components

23 56 19.50 Solar Heating Ancillary

			Crew	Daily Output	Labor-Hours	Unit	Material	2018 Bare Costs Labor	2018 Bare Costs Equipment	Total	Total Incl O&P
0010	**SOLAR HEATING ANCILLARY**										
2300	Circulators, air										
2310	Blowers										
2550	Booster fan 6" diameter, 120 CFM	G	Q-9	16	1	Ea.	37	54		91	126
2580	8" diameter, 150 CFM	G		16	1		41.50	54		95.50	131
2660	Shutter/damper	G		12	1.333		57	72		129	176
2670	Shutter motor	G		16	1		141	54		195	240
2800	Circulators, liquid, 1/25 HP, 5.3 GPM	G	Q-1	14	1.143		193	64		257	310
2820	1/20 HP, 17 GPM	G	"	12	1.333		267	74.50		341.50	410
3000	Collector panels, air with aluminum absorber plate										
3010	Wall or roof mount										
3040	Flat black, plastic glazing										
3080	4' x 8'	G	Q-9	6	2.667	Ea.	670	144		814	965
3300	Collector panels, liquid with copper absorber plate										
3320	Black chrome, tempered glass glazing										
3330	Alum. frame, 4' x 8', 5/32" single glazing	G	Q-1	9.50	1.684	Ea.	1,025	94		1,119	1,275
3450	Flat black, alum. frame, 3.5' x 7.5'	G		9	1.778		880	99.50		979.50	1,125

23 56 19.50 Solar Heating Ancillary		Crew	Daily Output	Labor-Hours	Unit	Material	2018 Bare Costs Labor	Equipment	Total	Total Incl O&P
3500	4' x 8'	Q-1	5.50	2.909	Ea.	1,050	163		1,213	1,400
3600	Liquid, full wetted, plastic, alum. frame, 4' x 10'	G	5	3.200		330	179		509	640
3650	Collector panel mounting, flat roof or ground rack	G	7	2.286	▼	248	128		376	470
3670	Roof clamps	G	70	.229	Set	2.87	12.80		15.67	23
3700	Roof strap, teflon	1 Plum	205	.039	L.F.	25	2.43		27.43	31.50
3900	Differential controller with two sensors									
3930	Thermostat, hard wired	1 Plum	8	1	Ea.	103	62		165	210
4050	Pool valve system	G	2.50	3.200		163	199		362	490
4070	With 12 VAC actuator	G	2	4	▼	330	249		579	745
4150	Sensors									
4220	Freeze prevention	1 Plum	32	.250	Ea.	23	15.55		38.55	49.50
4300	Heat exchanger									
4315	includes coil, blower, circulator									
4316	and controller for DHW and space hot air									
4330	Fluid to air coil, up flow, 45 MBH	Q-1	4	4	Ea.	320	224		544	705
4380	70 MBH	G	3.50	4.571		360	256		616	800
4400	80 MBH	G	3	5.333	▼	485	298		783	995
4580	Fluid to fluid package includes two circulating pumps									
4590	expansion tank, check valve, relief valve									
4600	controller, high temperature cutoff and sensors	Q-1	2.50	6.400	Ea.	810	360		1,170	1,450
4650	Heat transfer fluid									
4700	Propylene glycol, inhibited anti-freeze	1 Plum	28	.286	Gal.	23.50	17.75		41.25	53.50
4800	Solar storage tanks, knocked down									
5190	6'-3" high, 7' x 7' = 306 C.F./2,000 gallons	Q-10	1.20	20	Ea.	14,400	1,125		15,525	17,700
5210	7' x 14' = 613 C.F./4,000 gallons	G	.60	40		21,600	2,225		23,825	27,300
5230	10'-6" x 14' = 919 C.F./6,000 gallons	G	.40	60		25,300	3,350		28,650	33,100
5250	14' x 17'-6" = 1,531 C.F./10,000 gallons	Q-11	.30	107	▼	32,500	6,075		38,575	45,400
7000	Solar control valves and vents									
7070	Air eliminator, automatic 3/4" size	1 Plum	32	.250	Ea.	30	15.55		45.55	57
7090	Air vent, automatic, 1/8" fitting	G	32	.250		18.55	15.55		34.10	44.50
7100	Manual, 1/8" NPT	G	32	.250		2.63	15.55		18.18	27
7120	Backflow preventer, 1/2" pipe size	G	16	.500		83	31		114	140
7130	3/4" pipe size	G	16	.500		85	31		116	142
7150	Balancing valve, 3/4" pipe size	G	20	.400		63	25		88	108
7180	Draindown valve, 1/2" copper tube	G	9	.889		215	55		270	325
7200	Flow control valve, 1/2" pipe size	G	22	.364		138	22.50		160.50	187
7220	Expansion tank, up to 5 gal.	G	32	.250		69	15.55		84.55	100
7400	Pressure gauge, 2" dial	G	32	.250		24.50	15.55		40.05	50.50
7450	Relief valve, temp. and pressure 3/4" pipe size	G	30	.267	▼	24	16.55		40.55	52.50
7500	Solenoid valve, normally closed									
7520	Brass, 3/4" NPT, 24V	1 Plum	9	.889	Ea.	136	55		191	236
7530	1" NPT, 24V	G	9	.889		221	55		276	330
7750	Vacuum relief valve, 3/4" pipe size	G	32	.250	▼	30	15.55		45.55	57
7800	Thermometers									
7820	Digital temperature monitoring, 4 locations	1 Plum	2.50	3.200	Ea.	139	199		338	465
7900	Upright, 1/2" NPT	"	8	1	"	39.50	62		101.50	140
8250	Water storage tank with heat exchanger and electric element									
8270	66 gal. with 2" x 2 lb. density insulation	1 Plum	1.60	5	Ea.	1,675	310		1,985	2,325
8300	80 gal. with 2" x 2 lb. density insulation	"	1.60	5	"	1,675	310		1,985	2,325
8500	Water storage module, plastic									
8600	Tubular, 12" diameter, 4' high	1 Carp	48	.167	Ea.	89	8.45		97.45	112
8610	12" diameter, 8' high	"	40	.200		143	10.15		153.15	173
8650	Cap, 12" diameter	G			▼	19.80			19.80	22

23 56 Solar Energy Heating Equipment

23 56 19 – Solar Heating Components

23 56 19.50 Solar Heating Ancillary	Crew	Daily Output	Labor-Hours	Unit	Material	2018 Bare Costs Labor	Equipment	Total	Total Incl O&P
9000 Minimum labor/equipment charge	1 Plum	2	4	Job		249		249	385

23 61 Refrigerant Compressors

23 61 16 – Reciprocating Refrigerant Compressors

23 61 16.10 Reciprocating Compressors

		Crew	Daily Output	Labor-Hours	Unit	Material	2018 Bare Costs Labor	Equipment	Total	Total Incl O&P
0010	**RECIPROCATING COMPRESSORS**									
0990	Refrigeration, recip. hermetic, switches & protective devices									
1000	10 ton	Q-5	1	16	Ea.	15,800	905		16,705	18,700
1100	20 ton	Q-6	.72	33.333		23,300	1,950		25,250	28,700
1400	50 ton	"	.20	120		33,700	7,050		40,750	48,100
1600	130 ton	Q-7	.21	152	↓	42,600	9,150		51,750	61,000

23 62 Packaged Compressor and Condenser Units

23 62 13 – Packaged Air-Cooled Refrigerant Compressor and Condenser Units

23 62 13.10 Packaged Air-Cooled Refrig. Condensing Units

		Crew	Daily Output	Labor-Hours	Unit	Material	2018 Bare Costs Labor	Equipment	Total	Total Incl O&P
0010	**PACKAGED AIR-COOLED REFRIGERANT CONDENSING UNITS**									
0020	Condensing unit									
0030	Air cooled, compressor, standard controls									
0050	1.5 ton	Q-5	2.50	6.400	Ea.	1,025	365		1,390	1,700
0200	2.5 ton		1.70	9.412		1,350	535		1,885	2,325
0300	3 ton		1.30	12.308		1,350	700		2,050	2,550
0400	4 ton		.90	17.778		1,775	1,000		2,775	3,525
0500	5 ton	↓	.60	26.667	↓	2,150	1,500		3,650	4,725

23 64 Packaged Water Chillers

23 64 19 – Reciprocating Water Chillers

23 64 19.10 Reciprocating Type Water Chillers

		Crew	Daily Output	Labor-Hours	Unit	Material	2018 Bare Costs Labor	Equipment	Total	Total Incl O&P
0010	**RECIPROCATING TYPE WATER CHILLERS**, With standard controls									
0494	Water chillers, integral air cooled condenser									
0980	Water cooled, multiple compressor, semi-hermetic, tower not incl.									
1090	45 ton cooling	Q-7	.29	111	Ea.	22,700	6,675		29,375	35,400
1160	100 ton cooling		.18	180		50,500	10,800		61,300	72,500
1200	145 ton cooling	↓	.16	203	↓	59,500	12,100		71,600	84,500

23 64 23 – Scroll Water Chillers

23 64 23.10 Scroll Water Chillers

		Crew	Daily Output	Labor-Hours	Unit	Material	2018 Bare Costs Labor	Equipment	Total	Total Incl O&P
0010	**SCROLL WATER CHILLERS**, With standard controls									
0480	Packaged w/integral air cooled condenser									
0482	10 ton cooling	Q-7	.34	94.118	Ea.	15,000	5,650		20,650	25,200
0490	15 ton cooling		.37	86.486		18,000	5,175		23,175	27,900
0500	20 ton cooling		.34	94.118		21,800	5,650		27,450	32,800
0520	40 ton cooling	↓	.30	108	↓	32,800	6,475		39,275	46,200
0680	Scroll water cooled, single compressor, hermetic, tower not incl.									
0760	10 ton cooling	Q-6	.36	67.039	Ea.	6,700	3,950		10,650	13,500
0900	Scroll water cooled, multiple compressor, hermetic, tower not incl.									

23 64 Packaged Water Chillers

23 64 26 - Rotary-Screw Water Chillers

23 64 26.10 Rotary-Screw Type Water Chillers

		Crew	Daily Output	Labor-Hours	Unit	Material	2018 Bare Costs Labor	Equipment	Total	Total Incl O&P
0010	**ROTARY-SCREW TYPE WATER CHILLERS**, With standard controls									
0110	Screw, liquid chiller, air cooled, insulated evaporator									
0120	130 ton	Q-7	.14	229	Ea.	94,500	13,700		108,200	125,500
0124	160 ton	"	.13	246	"	120,500	14,800		135,300	155,500

23 64 33 - Modular Water Chillers

23 64 33.10 Direct Expansion Type Water Chillers

		Crew	Daily Output	Labor-Hours	Unit	Material	2018 Bare Costs Labor	Equipment	Total	Total Incl O&P
0010	**DIRECT EXPANSION TYPE WATER CHILLERS**, With standard controls									
8000	Direct expansion, shell and tube type, for built up systems									
8020	1 ton	Q-5	2	8	Ea.	7,825	455		8,280	9,325
8030	5 ton		1.90	8.421		13,000	475		13,475	15,000
8040	10 ton	↓	1.70	9.412	↓	16,100	535		16,635	18,500
9000	Minimum labor/equipment charge	Q-6	1	24	Job		1,400		1,400	2,200

23 65 Cooling Towers

23 65 13 - Forced-Draft Cooling Towers

23 65 13.10 Forced-Draft Type Cooling Towers

		Crew	Daily Output	Labor-Hours	Unit	Material	2018 Bare Costs Labor	Equipment	Total	Total Incl O&P
0010	**FORCED-DRAFT TYPE COOLING TOWERS**, Packaged units									
0070	Galvanized steel									
0080	Induced draft, crossflow									
0100	Vertical, belt drive, 61 tons	Q-6	90	.267	TonAC	192	15.70		207.70	236
0150	100 ton		100	.240		185	14.10		199.10	226
0200	115 ton		109	.220		161	12.95		173.95	197
0250	131 ton		120	.200		193	11.75		204.75	231
0260	162 ton	↓	132	.182	↓	156	10.70		166.70	189
1000	For higher capacities, use multiples									
1500	Induced air, double flow									
1900	Vertical, gear drive, 167 ton	Q-6	126	.190	TonAC	152	11.20		163.20	184
2000	297 ton		129	.186		93.50	10.95		104.45	120
2100	582 ton		132	.182		52	10.70		62.70	73.50
2200	1,016 ton	↓	150	.160	↓	70.50	9.40		79.90	92
3000	For higher capacities, use multiples									
3500	For pumps and piping, add	Q-6	38	.632	TonAC	96	37		133	164
4000	For absorption systems, add				"	75%	75%			
4100	Cooling water chemical feeder	Q-5	3	5.333	Ea.	410	300		710	920
5000	Fiberglass tower on galvanized steel support structure									
5010	Draw thru									
5100	100 ton	Q-6	1.40	17.143	Ea.	13,900	1,000		14,900	16,900
5120	120 ton		1.20	20		16,200	1,175		17,375	19,600
5140	140 ton		1	24		17,500	1,400		18,900	21,400
5160	160 ton		.80	30		19,400	1,775		21,175	24,200
5180	180 ton		.65	36.923		22,100	2,175		24,275	27,700
5251	600 ton		.16	150		75,000	8,825		83,825	96,000
5252	1,000 ton	↓	.10	250	↓	125,000	14,700		139,700	160,500
5300	For stainless steel support structure, add					30%				
5360	For higher capacities, use multiples of each size									
6000	Stainless steel									
6010	Induced draft, crossflow, horizontal, belt drive									
6100	57 ton	Q-6	1.50	16	Ea.	29,600	940		30,540	34,000
6120	91 ton		.99	24.242		35,800	1,425		37,225	41,500
6140	111 ton	↓	.43	55.814		46,800	3,275		50,075	56,500

23 65 Cooling Towers

23 65 13 – Forced-Draft Cooling Towers

23 65 13.10 Forced-Draft Type Cooling Towers	Crew	Daily Output	Labor-Hours	Unit	Material	2018 Bare Costs Labor	Equipment	Total	Total Incl O&P
6160	126 ton	Q-6	.22	109	Ea.	48,600	6,425		55,025	63,500
6170	Induced draft, crossflow, vertical, gear drive									
6172	167 ton	Q-6	.75	32	Ea.	56,500	1,875		58,375	65,000
6174	297 ton		.43	55.814		62,500	3,275		65,775	74,000
6176	582 ton		.23	104		101,500	6,125		107,625	121,500
6180	1,016 ton		.15	160		173,500	9,400		182,900	205,500
9000	Minimum labor/equipment charge		1	24	Job		1,400		1,400	2,200

23 72 Air-to-Air Energy Recovery Equipment

23 72 13 – Heat-Wheel Air-to-Air Energy-Recovery Equipment

23 72 13.10 Heat-Wheel Air-to-Air Energy Recovery Equip.

		Crew	Daily Output	Labor-Hours	Unit	Material	Labor	Equipment	Total	Total Incl O&P
0010	**HEAT-WHEEL AIR-TO-AIR ENERGY RECOVERY EQUIPMENT**									
0100	Air to air									
9000	Minimum labor/equipment charge	1 Shee	4	2	Job		120		120	189

23 73 Indoor Central-Station Air-Handling Units

23 73 39 – Indoor, Direct Gas-Fired Heating and Ventilating Units

23 73 39.10 Make-Up Air Unit

		Crew	Daily Output	Labor-Hours	Unit	Material	Labor	Equipment	Total	Total Incl O&P
0010	**MAKE-UP AIR UNIT**									
0020	Indoor suspension, natural/LP gas, direct fired,									
0032	standard control. For flue see Section 23 51 23.10									
0040	70°F temperature rise, MBH is input									
0100	75 MBH input	Q-6	3.60	6.667	Ea.	5,150	390		5,540	6,275
0220	225 MBH input		2.40	10		7,000	590		7,590	8,625
0300	400 MBH input		1.60	15		16,500	880		17,380	19,600

23 74 Packaged Outdoor HVAC Equipment

23 74 23 – Packaged, Outdoor, Heating-Only Makeup-Air Units

23 74 23.16 Packaged, Ind.-Fired, In/Outdoor

		Crew	Daily Output	Labor-Hours	Unit	Material	Labor	Equipment	Total	Total Incl O&P
0010	**PACKAGED, IND.-FIRED, IN/OUTDOOR**, Heat-Only Makeup-Air Units									
1000	Rooftop unit, natural gas, gravity vent, S.S. exchanger									
1010	70°F temperature rise, MBH is input									
1020	250 MBH	Q-6	4	6	Ea.	15,900	355		16,255	18,100
1040	400 MBH		3.60	6.667		18,400	390		18,790	20,900
1060	550 MBH		3.30	7.273		20,800	430		21,230	23,600
1600	For cleanable filters, add					5%				
1700	For electric modulating gas control, add					10%				
9000	Minimum labor/equipment charge	Q-6	2.75	8.727	Job		515		515	800

23 74 33 – Dedicated Outdoor-Air Units

23 74 33.10 Rooftop Air Conditioners

		Crew	Daily Output	Labor-Hours	Unit	Material	Labor	Equipment	Total	Total Incl O&P
0010	**ROOFTOP AIR CONDITIONERS**, Standard controls, curb, economizer									
1000	Single zone, electric cool, gas heat									
1100	3 ton cooling, 60 MBH heating	Q-5	.70	22.857	Ea.	2,800	1,300		4,100	5,100
1120	4 ton cooling, 95 MBH heating		.61	26.403		3,325	1,500		4,825	5,975
1150	7.5 ton cooling, 170 MBH heating		.50	32.258		5,825	1,825		7,650	9,250
1156	8.5 ton cooling, 170 MBH heating		.46	34.783		7,075	1,975		9,050	10,900
1160	10 ton cooling, 200 MBH heating	Q-6	.67	35.982		9,150	2,125		11,275	13,400

466

23 74 Packaged Outdoor HVAC Equipment

23 74 33 – Dedicated Outdoor-Air Units

23 74 33.10 Rooftop Air Conditioners	Crew	Daily Output	Labor-Hours	Unit	Material	2018 Bare Costs Labor	Equipment	Total	Total Incl O&P
1220	30 ton cooling, 540 MBH heating	Q-7	.47	68.376	Ea.	35,700	4,100		39,800	45,700
1240	40 ton cooling, 675 MBH heating	"	.35	91.168		43,900	5,475		49,375	56,500
2000	Multizone, electric cool, gas heat, economizer									
2120	20 ton cooling, 360 MBH heating	Q-7	.53	60.038	Ea.	70,500	3,600		74,100	83,000
2180	30 ton cooling, 540 MBH heating		.37	85.562		109,000	5,125		114,125	128,000
2210	50 ton cooling, 540 MBH heating		.23	142		156,000	8,525		164,525	185,000
2220	70 ton cooling, 1,500 MBH heating		.16	199		167,500	11,900		179,400	202,500
2240	80 ton cooling, 1,500 MBH heating		.14	229		191,000	13,700		204,700	232,000
2280	105 ton cooling, 1,500 MBH heating		.11	291		218,000	17,400		235,400	267,000
2400	For hot water heat coil, deduct					5%				
2500	For steam heat coil, deduct					2%				
2600	For electric heat, deduct					3%	5%			
9000	Minimum labor/equipment charge	Q-5	1.50	10.667	Job		605		605	940

23 81 Decentralized Unitary HVAC Equipment

23 81 19 – Self-Contained Air-Conditioners

23 81 19.10 Window Unit Air Conditioners

		Crew	Daily Output	Labor-Hours	Unit	Material	2018 Bare Costs Labor	Equipment	Total	Total Incl O&P
0010	**WINDOW UNIT AIR CONDITIONERS**									
4000	Portable/window, 15 amp, 125 V grounded receptacle required									
4500	10,000 BTUH	1 Carp	6	1.333	Ea.	680	67.50		747.50	855
4520	12,000 BTUH	L-2	8	2	"	1,925	88.50		2,013.50	2,250
9000	Minimum labor/equipment charge	1 Carp	2	4	Job		203		203	330

23 81 19.20 Self-Contained Single Package

		Crew	Daily Output	Labor-Hours	Unit	Material	2018 Bare Costs Labor	Equipment	Total	Total Incl O&P
0010	**SELF-CONTAINED SINGLE PACKAGE**									
0100	Air cooled, for free blow or duct, not incl. remote condenser									
0110	Constant volume									
0200	3 ton cooling	Q-5	1	16	Ea.	3,750	905		4,655	5,525
0210	4 ton cooling	"	.80	20		4,100	1,125		5,225	6,275
0240	10 ton cooling	Q-7	1	32		7,675	1,925		9,600	11,400
0280	30 ton cooling	"	.80	40		25,500	2,400		27,900	31,700
1000	Water cooled for free blow or duct, not including tower									
1010	Constant volume									
1100	3 ton cooling	Q-6	1	24	Ea.	3,700	1,400		5,100	6,275
1120	5 ton cooling	"	1	24		4,525	1,400		5,925	7,175
1140	10 ton cooling	Q-7	.90	35.556		8,550	2,125		10,675	12,700
1180	30 ton cooling	"	.70	45.714		36,700	2,750		39,450	44,700
1240	60 ton cooling	Q-8	.30	107		59,500	6,400	197	66,097	75,000
1300	For hot water or steam heat coils, add					12%	10%			
9000	Minimum labor/equipment charge	Q-5	1	16	Job		905		905	1,400

23 81 23 – Computer-Room Air-Conditioners

23 81 23.10 Computer Room Units

		Crew	Daily Output	Labor-Hours	Unit	Material	2018 Bare Costs Labor	Equipment	Total	Total Incl O&P
0010	**COMPUTER ROOM UNITS**									
1000	Air cooled, includes remote condenser but not									
1020	interconnecting tubing or refrigerant									
1160	6 ton	Q-5	.30	53.333	Ea.	48,600	3,025		51,625	58,000
1240	10 ton		.25	64		51,500	3,625		55,125	62,000
1260	12 ton		.24	66.667		53,000	3,775		56,775	64,500
1300	20 ton	Q-6	.26	92.308		67,500	5,425		72,925	83,000
1320	22 ton	"	.24	100		68,500	5,875		74,375	84,500

23 81 Decentralized Unitary HVAC Equipment

23 81 26 – Split-System Air-Conditioners

23 81 26.10 Split Ductless Systems	Crew	Daily Output	Labor-Hours	Unit	Material	2018 Bare Costs Labor	Equipment	Total	Total Incl O&P
0010 **SPLIT DUCTLESS SYSTEMS**									
0100 Cooling only, single zone									
0110 Wall mount									
0120 3/4 ton cooling	Q-5	2	8	Ea.	1,100	455		1,555	1,925
0130 1 ton cooling		1.80	8.889		1,225	505		1,730	2,125
0140 1-1/2 ton cooling		1.60	10		1,975	565		2,540	3,050
0150 2 ton cooling	↓	1.40	11.429	↓	2,250	650		2,900	3,475
1000 Ceiling mount									
1020 2 ton cooling	Q-5	1.40	11.429	Ea.	2,000	650		2,650	3,200
1030 3 ton cooling	"	1.20	13.333	"	2,550	755		3,305	3,975
3000 Multizone									
3010 Wall mount									
3020 2 @ 3/4 ton cooling	Q-5	1.80	8.889	Ea.	3,375	505		3,880	4,475
5000 Cooling/Heating									
5010 Wall mount									
5110 1 ton cooling	Q-5	1.70	9.412	Ea.	1,300	535		1,835	2,250
5120 1-1/2 ton cooling	"	1.50	10.667	"	1,975	605		2,580	3,125
7000 Accessories for all split ductless systems									
7010 Add for ambient frost control	Q-5	8	2	Ea.	126	113		239	315
7020 Add for tube/wiring kit (line sets)									
7030 15' kit	Q-5	32	.500	Ea.	88	28.50		116.50	141
7036 25' kit		28	.571		133	32.50		165.50	197
7040 35' kit		24	.667		200	38		238	279
7050 50' kit	↓	20	.800	↓	202	45.50		247.50	293

23 81 29 – Variable Refrigerant Flow HVAC Systems

23 81 29.10 Heat Pump, Gas Driven

	Crew	Daily Output	Labor-Hours	Unit	Material	Labor	Equipment	Total	Total Incl O&P
0010 **HEAT PUMP, GAS DRIVEN**, Variable refrigerant volume (VRV) type									
0020 Not including interconnecting tubing or multi-zone controls									
1000 For indoor fan VRV type AHU see 23 82 19.40									
1010 Multi-zone split									
1020 Outdoor unit									
1100 8 tons cooling, for up to 17 zones	Q-5	1.30	12.308	Ea.	29,100	700		29,800	33,100
1110 Isolation rails		2.60	6.154	Pair	1,025	350		1,375	1,675
1160 15 tons cooling, for up to 33 zones		1	16	Ea.	34,800	905		35,705	39,600
1170 Isolation rails	↓	2.60	6.154	Pair	1,200	350		1,550	1,875
2000 Packaged unit									
2020 Outdoor unit									
2200 11 tons cooling	Q-5	1.30	12.308	Ea.	37,500	700		38,200	42,400
2210 Roof curb adapter	"	2.60	6.154	↓	650	350		1,000	1,250
2220 Thermostat	1 Stpi	1.30	6.154	↓	390	390		780	1,025

23 81 43 – Air-Source Unitary Heat Pumps

23 81 43.10 Air-Source Heat Pumps

	Crew	Daily Output	Labor-Hours	Unit	Material	Labor	Equipment	Total	Total Incl O&P
0010 **AIR-SOURCE HEAT PUMPS**, Not including interconnecting tubing									
1000 Air to air, split system, not including curbs, pads, fan coil and ductwork									
1012 Outside condensing unit only, for fan coil see Section 23 82 19.10									
1020 2 ton cooling, 8.5 MBH heat @ 0°F	Q-5	2	8	Ea.	1,650	455		2,105	2,500
1040 3 ton cooling, 13 MBH heat @ 0°F		1.20	13.333		2,100	755		2,855	3,500
1080 7.5 ton cooling, 33 MBH heat @ 0°F	↓	.45	35.556		3,875	2,025		5,900	7,375
1120 15 ton cooling, 64 MBH heat @ 0°F	Q-6	.50	48		8,725	2,825		11,550	14,000
1130 20 ton cooling, 85 MBH heat @ 0°F		.35	68.571		17,900	4,025		21,925	26,000
1140 25 ton cooling, 119 MBH heat @ 0°F	↓	.25	96		21,100	5,650		26,750	32,000

23 81 Decentralized Unitary HVAC Equipment

23 81 43 – Air-Source Unitary Heat Pumps

23 81 43.10 Air-Source Heat Pumps

		Crew	Daily Output	Labor-Hours	Unit	Material	2018 Bare Costs Labor	Equipment	Total	Total Incl O&P
1500	Single package, not including curbs, pads, or plenums									
1520	2 ton cooling, 6.5 MBH heat @ 0°F	Q-5	1.50	10.667	Ea.	3,200	605		3,805	4,475
1560	3 ton cooling, 10 MBH heat @ 0°F	"	1.20	13.333	"	3,675	755		4,430	5,200
6000	Air to water, single package, excluding storage tank and ductwork									
6010	Includes circulating water pump, air duct connections, digital temperature									
6020	controller with remote tank temp. probe and sensor for storage tank.									
6040	Water heating - air cooling capacity									
6110	35.5 MBH heat water, 2.3 ton cool air	Q-5	1.60	10	Ea.	17,100	565		17,665	19,700
6120	58 MBH heat water, 3.8 ton cool air		1.10	14.545		19,500	825		20,325	22,800
6130	76 MBH heat water, 4.9 ton cool air		.87	18.391		23,300	1,050		24,350	27,200
6140	98 MBH heat water, 6.5 ton cool air		.62	25.806		30,100	1,475		31,575	35,400
6150	113 MBH heat water, 7.4 ton cool air		.59	27.119		33,000	1,550		34,550	38,700
6160	142 MBH heat water, 9.2 ton cool air		.52	30.769		40,000	1,750		41,750	46,700
6170	171 MBH heat water, 11.1 ton cool air		.49	32.653		47,100	1,850		48,950	55,000

23 81 46 – Water-Source Unitary Heat Pumps

23 81 46.10 Water Source Heat Pumps

		Crew	Daily Output	Labor-Hours	Unit	Material	2018 Bare Costs Labor	Equipment	Total	Total Incl O&P
0010	**WATER SOURCE HEAT PUMPS**, Not incl. connecting tubing or water source									
2000	Water source to air, single package									
2100	1 ton cooling, 13 MBH heat @ 75°F	Q-5	2	8	Ea.	1,900	455		2,355	2,775
2140	2 ton cooling, 19 MBH heat @ 75°F		1.70	9.412		2,225	535		2,760	3,275
2220	5 ton cooling, 29 MBH heat @ 75°F		.90	17.778		3,300	1,000		4,300	5,225
9000	Minimum labor/equipment charge		1.75	9.143	Job		520		520	805

23 82 Convection Heating and Cooling Units

23 82 16 – Air Coils

23 82 16.10 Flanged Coils

		Crew	Daily Output	Labor-Hours	Unit	Material	2018 Bare Costs Labor	Equipment	Total	Total Incl O&P
0010	**FLANGED COILS**									
1500	Hot water heating, 1 row, 24" x 48"	Q-5	4	4	Ea.	1,750	227		1,977	2,275
9000	Minimum labor/equipment charge	1 Plum	1	8	Job		495		495	775

23 82 16.20 Duct Heaters

		Crew	Daily Output	Labor-Hours	Unit	Material	2018 Bare Costs Labor	Equipment	Total	Total Incl O&P
0010	**DUCT HEATERS**, Electric, 480 V, 3 Ph.									
0020	Finned tubular insert, 500°F									
0100	8" wide x 6" high, 4.0 kW	Q-20	16	1.250	Ea.	795	68.50		863.50	980
0120	12" high, 8.0 kW		15	1.333		1,325	73		1,398	1,575
0140	18" high, 12.0 kW		14	1.429		1,850	78		1,928	2,150
0160	24" high, 16.0 kW		13	1.538		2,375	84		2,459	2,750
0180	30" high, 20.0 kW		12	1.667		2,900	91		2,991	3,350
0300	12" wide x 6" high, 6.7 kW		15	1.333		845	73		918	1,050
0320	12" high, 13.3 kW		14	1.429		1,375	78		1,453	1,625
0340	18" high, 20.0 kW		13	1.538		1,925	84		2,009	2,225
8000	To obtain BTU multiply kW by 3413									
9900	Minimum labor/equipment charge	1 Shee	4	2	Job		120		120	189

23 82 19 – Fan Coil Units

23 82 19.10 Fan Coil Air Conditioning

		Crew	Daily Output	Labor-Hours	Unit	Material	2018 Bare Costs Labor	Equipment	Total	Total Incl O&P
0010	**FAN COIL AIR CONDITIONING**									
0030	Fan coil AC, cabinet mounted, filters and controls									
0100	Chilled water, 1/2 ton cooling	Q-5	8	2	Ea.	555	113		668	785
0120	1 ton cooling		6	2.667		830	151		981	1,150
0160	2-1/2 ton cooling		5	3.200		1,750	181		1,931	2,225
0180	3 ton cooling		4	4		1,950	227		2,177	2,500

23 82 19.10 Fan Coil Air Conditioning

		Crew	Daily Output	Labor-Hours	Unit	Material	2018 Bare Costs Labor	Equipment	Total	Total Incl O&P
0262	For hot water coil, add					40%	10%			
0320	1 ton cooling	Q-5	6	2.667	Ea.	1,625	151		1,776	2,000
0940	Direct expansion, for use w/air cooled condensing unit, 1-1/2 ton cooling		5	3.200		610	181		791	955
1000	5 ton cooling		3	5.333		1,075	300		1,375	1,650
1040	10 ton cooling	Q-6	2.60	9.231		2,000	545		2,545	3,050
1060	20 ton cooling	"	.70	34.286		3,850	2,025		5,875	7,350
1500	For hot water coil, add					40%	10%			
1512	For condensing unit add see Section 23 62									
3000	Chilled water, horizontal unit, housing, 2 pipe, fan control, no valves									
3100	1/2 ton cooling	Q-5	8	2	Ea.	1,325	113		1,438	1,625
3110	1 ton cooling		6	2.667		1,550	151		1,701	1,925
3120	1-1/2 ton cooling		5.50	2.909		1,775	165		1,940	2,225
3130	2 ton cooling		5.25	3.048		2,125	173		2,298	2,625
3140	3 ton cooling		4	4		2,575	227		2,802	3,175
3150	3-1/2 ton cooling		3.80	4.211		2,825	239		3,064	3,500
3160	4 ton cooling		3.80	4.211		2,825	239		3,064	3,500
4000	With electric heat, 2 pipe									
4100	1/2 ton cooling	Q-5	8	2	Ea.	1,550	113		1,663	1,900
4105	3/4 ton cooling		7	2.286		1,725	130		1,855	2,100
4110	1 ton cooling		6	2.667		1,900	151		2,051	2,300
4120	1-1/2 ton cooling		5.50	2.909		2,100	165		2,265	2,575
4130	2 ton cooling		5.25	3.048		2,650	173		2,823	3,200
4135	2-1/2 ton cooling		4.80	3.333		3,675	189		3,864	4,325
4140	3 ton cooling		4	4		4,950	227		5,177	5,800
4150	3-1/2 ton cooling		3.80	4.211		5,900	239		6,139	6,875
4160	4 ton cooling		3.60	4.444		6,825	252		7,077	7,900
9000	Minimum labor/equipment charge		3.75	4.267	Job		242		242	375

23 82 19.40 Fan Coil Air Conditioning

			Crew	Daily Output	Labor-Hours	Unit	Material	2018 Bare Costs Labor	Equipment	Total	Total Incl O&P
0010	**FAN COIL AIR CONDITIONING**, Variable refrigerant volume (VRV) type										
0020	Not including interconnecting tubing or multi-zone controls										
0030	For VRV condensing unit see section 23 81 29.10										
0050	Indoor type, ducted										
0100	Vertical concealed										
0130	1 ton cooling	G	Q-5	2.97	5.387	Ea.	2,475	305		2,780	3,200
0140	1.5 ton cooling	G		2.77	5.776		2,525	330		2,855	3,275
0150	2 ton cooling	G		2.70	5.926		2,725	335		3,060	3,525
0160	2.5 ton cooling	G		2.60	6.154		2,875	350		3,225	3,700
0170	3 ton cooling	G		2.50	6.400		2,925	365		3,290	3,775
0180	3.5 ton cooling	G		2.30	6.957		3,075	395		3,470	4,000
0190	4 ton cooling	G		2.19	7.306		3,075	415		3,490	4,050
0200	4.5 ton cooling	G		2.08	7.692		3,400	435		3,835	4,400
0230	Outside air connection possible										
1100	Ceiling concealed										
1130	0.6 ton cooling	G	Q-5	2.60	6.154	Ea.	1,625	350		1,975	2,325
1140	0.75 ton cooling	G		2.60	6.154		1,675	350		2,025	2,400
1150	1 ton cooling	G		2.60	6.154		1,775	350		2,125	2,500
1152	Screening door	G	1 Stpi	5.20	1.538		99	97		196	260
1160	1.5 ton cooling	G	Q-5	2.60	6.154		1,825	350		2,175	2,575
1162	Screening door	G	1 Stpi	5.20	1.538		114	97		211	277
1170	2 ton cooling	G	Q-5	2.60	6.154		2,175	350		2,525	2,925
1172	Screening door	G	1 Stpi	5.20	1.538		99	97		196	260
1180	2.5 ton cooling	G	Q-5	2.60	6.154		2,500	350		2,850	3,300

23 82 Convection Heating and Cooling Units

23 82 19 – Fan Coil Units

23 82 19.40 Fan Coil Air Conditioning		Crew	Daily Output	Labor-Hours	Unit	Material	2018 Bare Costs Labor	Equipment	Total	Total Incl O&P
1182	Screening door	G 1 Stpi	5.20	1.538	Ea.	172	97		269	340
1190	3 ton cooling	G Q-5	2.60	6.154		2,650	350		3,000	3,475
1192	Screening door	G 1 Stpi	5.20	1.538		172	97		269	340
1200	4 ton cooling	G Q-5	2.30	6.957		2,725	395		3,120	3,600
1202	Screening door	G 1 Stpi	5.20	1.538		172	97		269	340
1220	6 ton cooling	G Q-5	2.08	7.692		4,625	435		5,060	5,750
1230	8 ton cooling	G "	1.89	8.466		5,200	480		5,680	6,475
1260	Outside air connection possible									
4050	Indoor type, duct-free									
4100	Ceiling mounted cassette									
4130	0.75 ton cooling	G Q-5	3.46	4.624	Ea.	1,875	262		2,137	2,475
4140	1 ton cooling	G	2.97	5.387		2,000	305		2,305	2,675
4150	1.5 ton cooling	G	2.77	5.776		2,075	330		2,405	2,775
4160	2 ton cooling	G	2.60	6.154		2,200	350		2,550	2,975
4170	2.5 ton cooling	G	2.50	6.400		2,250	365		2,615	3,050
4180	3 ton cooling	G	2.30	6.957		2,425	395		2,820	3,275
4190	4 ton cooling	G	2.08	7.692		2,700	435		3,135	3,650
4198	For ceiling cassette decoration panel, add	G 1 Stpi	5.20	1.538		296	97		393	475
4230	Outside air connection possible									
4500	Wall mounted									
4530	0.6 ton cooling	G Q-5	5.20	3.077	Ea.	995	174		1,169	1,375
4540	0.75 ton cooling	G	5.20	3.077		1,050	174		1,224	1,425
4550	1 ton cooling	G	4.16	3.846		1,200	218		1,418	1,650
4560	1.5 ton cooling	G	3.77	4.244		1,300	241		1,541	1,825
4570	2 ton cooling	G	3.46	4.624		1,400	262		1,662	1,950
4590	Condensate pump for the above air handlers	G 1 Stpi	6.46	1.238		213	78		291	355
4700	Floor standing unit									
4730	1 ton cooling	G Q-5	2.60	6.154	Ea.	1,675	350		2,025	2,400
4740	1.5 ton cooling	G	2.60	6.154		1,900	350		2,250	2,650
4750	2 ton cooling	G	2.60	6.154		2,050	350		2,400	2,800
4800	Floor standing unit, concealed									
4830	1 ton cooling	G Q-5	2.60	6.154	Ea.	1,600	350		1,950	2,325
4840	1.5 ton cooling	G	2.60	6.154		1,825	350		2,175	2,550
4850	2 ton cooling	G	2.60	6.154		1,950	350		2,300	2,700
4880	Outside air connection possible									
8000	Accessories									
8100	Branch divergence pipe fitting									
8110	Capacity under 76 MBH	1 Stpi	2.50	3.200	Ea.	151	202		353	480
8120	76 MBH to 112 MBH		2.50	3.200		178	202		380	510
8130	112 MBH to 234 MBH		2.50	3.200		505	202		707	870
8200	Header pipe fitting									
8210	Max 4 branches									
8220	Capacity under 76 MBH	1 Stpi	1.70	4.706	Ea.	221	296		517	705
8260	Max 8 branches									
8270	76 MBH to 112 MBH	1 Stpi	1.70	4.706	Ea.	450	296		746	955
8280	112 MBH to 234 MBH	"	1.30	6.154	"	720	390		1,110	1,400

For customer support on your Commercial Renovation Costs with RSMeans data, call 800.448.8182.

471

23 82 Convection Heating and Cooling Units

23 82 29 – Radiators

23 82 29.10 Hydronic Heating

23 82 29.10 Hydronic Heating	Crew	Daily Output	Labor-Hours	Unit	Material	2018 Bare Costs Labor	Equipment	Total	Total Incl O&P
0010 **HYDRONIC HEATING**, Terminal units, not incl. main supply pipe									
1000 Radiation									
1100 Panel, baseboard, C.I., including supports, no covers	Q-5	46	.348	L.F.	44.50	19.70		64.20	79.50
3000 Radiators, cast iron									
3100 Free standing or wall hung, 6 tube, 25" high	Q-5	96	.167	Section	62	9.45		71.45	82.50
9000 Minimum labor/equipment charge	"	3.20	5	Job		284		284	440

23 82 36 – Finned-Tube Radiation Heaters

23 82 36.10 Finned Tube Radiation

23 82 36.10 Finned Tube Radiation	Crew	Daily Output	Labor-Hours	Unit	Material	2018 Bare Costs Labor	Equipment	Total	Total Incl O&P
0010 **FINNED TUBE RADIATION**, Terminal units, not incl. main supply pipe									
1150 Fin tube, wall hung, 14" slope top cover, with damper									
1200 1-1/4" copper tube, 4-1/4" alum. fin	Q-5	38	.421	L.F.	48	24		72	90
1250 1-1/4" steel tube, 4-1/4" steel fin		36	.444		41	25		66	84
1255 2" steel tube, 4-1/4" steel fin		32	.500		44.50	28.50		73	93
1310 Baseboard, pkgd, 1/2" copper tube, alum. fin, 7" high		60	.267		11.60	15.10		26.70	36.50
1320 3/4" copper tube, alum. fin, 7" high		58	.276		7.85	15.65		23.50	33
1340 1" copper tube, alum. fin, 8-7/8" high		56	.286		21	16.20		37.20	48
1360 1-1/4" copper tube, alum. fin, 8-7/8" high		54	.296		31.50	16.80		48.30	60.50
1500 Note: fin tube may also require corners, caps, etc.									

23 82 39 – Unit Heaters

23 82 39.16 Propeller Unit Heaters

23 82 39.16 Propeller Unit Heaters	Crew	Daily Output	Labor-Hours	Unit	Material	2018 Bare Costs Labor	Equipment	Total	Total Incl O&P
0010 **PROPELLER UNIT HEATERS**									
3950 Unit heaters, propeller, 115 V 2 psi steam, 60°F entering air									
4000 Horizontal, 12 MBH	Q-5	12	1.333	Ea.	380	75.50		455.50	540
4060 43.9 MBH		8	2		580	113		693	810
4240 286.9 MBH		2	8		1,750	455		2,205	2,625
4260 364 MBH		1.80	8.889		2,175	505		2,680	3,175
4270 404 MBH		1.60	10		2,275	565		2,840	3,375
4300 Vertical diffuser same price									
4310 Vertical flow, 40 MBH	Q-5	11	1.455	Ea.	570	82.50		652.50	760
4326 131 MBH	"	4	4		890	227		1,117	1,325
4354 420 MBH (460 V)	Q-6	1.80	13.333		2,275	785		3,060	3,725
4358 500 MBH (460 V)		1.71	14.035		3,025	825		3,850	4,600
4362 570 MBH (460 V)		1.40	17.143		4,150	1,000		5,150	6,125
4366 620 MBH (460 V)		1.30	18.462		4,150	1,075		5,225	6,250
4370 960 MBH (460 V)		1.10	21.818		7,875	1,275		9,150	10,700

23 83 Radiant Heating Units

23 83 33 – Electric Radiant Heaters

23 83 33.10 Electric Heating

23 83 33.10 Electric Heating	Crew	Daily Output	Labor-Hours	Unit	Material	2018 Bare Costs Labor	Equipment	Total	Total Incl O&P
0010 **ELECTRIC HEATING**, not incl. conduit or feed wiring									
1100 Rule of thumb: Baseboard units, including control	1 Elec	4.40	1.818	kW	109	106		215	282
1300 Baseboard heaters, 2' long, 350 watt		8	1	Ea.	27.50	58		85.50	121
1400 3' long, 750 watt		8	1		31.50	58		89.50	125
1600 4' long, 1,000 watt		6.70	1.194		37	69.50		106.50	148
1800 5' long, 935 watt		5.70	1.404		46.50	81.50		128	177
2000 6' long, 1,500 watt		5	1.600		51.50	93		144.50	201
2400 8' long, 2,000 watt		4	2		62.50	116		178.50	249
2600 9' long, 1,680 watt		3.60	2.222		85	129		214	294
2800 10' long, 1,875 watt		3.30	2.424		154	141		295	390

23 83 Radiant Heating Units

23 83 33 – Electric Radiant Heaters

23 83 33.10 Electric Heating

	23 83 33.10 Electric Heating	Crew	Daily Output	Labor-Hours	Unit	Material	2018 Bare Costs Labor	Equipment	Total	Total Incl O&P
2950	Wall heaters with fan, 120 to 277 volt									
3170	1,000 watt	1 Elec	6	1.333	Ea.	93	77.50		170.50	222
3180	1,250 watt		5	1.600		100	93		193	254
3190	1,500 watt		4	2		100	116		216	290
3230	2,500 watt		3.50	2.286		232	133		365	460
3250	4,000 watt		2.70	2.963		365	172		537	665
3600	Thermostats, integral		16	.500		27.50	29		56.50	75
3800	Line voltage, 1 pole		8	1		15.70	58		73.70	107
3810	2 pole		8	1	↓	21	58		79	113
9990	Minimum labor/equipment charge	↓	.4	2	Job		116		116	180

23 84 Humidity Control Equipment

23 84 13 – Humidifiers

23 84 13.10 Humidifier Units

	23 84 13.10 Humidifier Units	Crew	Daily Output	Labor-Hours	Unit	Material	2018 Bare Costs Labor	Equipment	Total	Total Incl O&P
0010	**HUMIDIFIER UNITS**									
0520	Steam, room or duct, filter, regulators, auto. controls, 220 V									
0560	22 lb./hr.	Q-5	5	3.200	Ea.	3,000	181		3,181	3,575
0580	33 lb./hr.		4	4		3,100	227		3,327	3,750
0620	100 lb./hr.		3	5.333	↓	4,525	300		4,825	5,450
9000	Minimum labor/equipment charge	↓	3.50	4.571	Job		259		259	405

23 91 Prefabricated Equipment Supports

23 91 10 – Prefabricated Curbs, Pads and Stands

23 91 10.10 Prefabricated Pads and Stands

	23 91 10.10 Prefabricated Pads and Stands	Crew	Daily Output	Labor-Hours	Unit	Material	2018 Bare Costs Labor	Equipment	Total	Total Incl O&P
0010	**PREFABRICATED PADS AND STANDS**									
6000	Pad, fiberglass reinforced concrete with polystyrene foam core									
6090	24" x 36"	1 Shee	12	.667	Ea.	39	40		79	106
6280	36" x 36"	Q-9	8	2		56	108		164	232
6300	36" x 40"		7	2.286		61.50	123		184.50	262
6320	36" x 48"	↓	7	2.286	↓	74.50	123		197.50	276

Division Notes

	CREW	DAILY OUTPUT	LABOR-HOURS	UNIT	BARE COSTS				TOTAL INCL O&P
					MAT.	LABOR	EQUIP.	TOTAL	

Estimating Tips

- When estimating material costs for special systems, it is always prudent to obtain manufacturers' quotations for equipment prices and special installation requirements that may affect the total cost.

- For cost modifications for elevated tray installation, add the percentages to labor according to the height of the installation and only to the quantities exceeding the different height levels, not to the total tray quantities. Refer to 26 01 02.20 for labor adjustment factors.

- Do not overlook the costs for equipment used in the installation.

Reference Numbers

Reference numbers are shown at the beginning of some major classifications. These numbers refer to related items in the Reference Section. The reference information may be an estimating procedure, an alternate pricing method, or technical information.

Note: Not all subdivisions listed here necessarily appear. ■

25 05 28.39 Surface Raceways for Integrated Automation	Crew	Daily Output	Labor-Hours	Unit	Material	2018 Bare Costs Labor	Equipment	Total	Total Incl O&P
0010 **SURFACE RACEWAYS FOR INTEGRATED AUTOMATION**									
1000 Din rail, G type, 15 mm, steel, galvanized, 3', for termination blocks	1 Elec	48	.167	Ea.	10.65	9.70		20.35	26.50
1010 32 mm, steel, galvanized, 3', for termination blocks		48	.167		16.35	9.70		26.05	33
1020 35 mm, steel, galvanized, 3', for termination blocks		48	.167		14.55	9.70		24.25	31
1030 Din rail, rail mounted, 35 mm, aluminum, 3', for termination blocks		48	.167		20.50	9.70		30.20	37.50
1040 35 mm, aluminum, 6', for termination blocks		48	.167		25.50	9.70		35.20	43.50
2000 Terminal block, Din mounted, 35 mm, medium screw, connector end bracket		48	.167		2.19	9.70		11.89	17.40
2010 35 mm, large screw, connector end bracket		48	.167		3.43	9.70		13.13	18.75
2020 Large screw, connector end bracket		48	.167		4.21	9.70		13.91	19.65
3000 Terminal block, feed through, 35 mm, Din mounted, 600V, 20A		48	.167		1.04	9.70		10.74	16.15
3010 35 mm, Din mounted, 600V, 30A		48	.167		1.29	9.70		10.99	16.40
3020 Terminal block, fuse holder, LED indicator, Din mounted, 110V, 6A		48	.167		7.55	9.70		17.25	23.50
3030 Din mounted, 220V, 6A		48	.167		7.70	9.70		17.40	23.50
3040 Din mounted, 300V, 6A		48	.167		7.10	9.70		16.80	23

25 36 Integrated Automation Instrumentation & Terminal Devices

25 36 19 – Integrated Automation Current Sensors

25 36 19.10 Integrated Automation Current Sensors	Crew	Daily Output	Labor-Hours	Unit	Material	2018 Bare Costs Labor	Equipment	Total	Total Incl O&P
0010 **INTEGRATED AUTOMATION CURRENT SENSORS**									
1000 Current limiter inrush, 100 to 480 VAC, 10A, power supply	1 Elec	48	.167	Ea.	80	9.70		89.70	103

476

For customer support on your Commercial Renovation Costs with RSMeans data, call 800.448.8182.

Estimating Tips
26 05 00 Common Work Results for Electrical

- Conduit should be taken off in three main categories—power distribution, branch power, and branch lighting—so the estimator can concentrate on systems and components, therefore making it easier to ensure all items have been accounted for.

- For cost modifications for elevated conduit installation, add the percentages to labor according to the height of installation and only to the quantities exceeding the different height levels, not to the total conduit quantities. Refer to 26 01 02.20 for labor adjustment factors.

- Remember that aluminum wiring of equal ampacity is larger in diameter than copper and may require larger conduit.

- If more than three wires at a time are being pulled, deduct percentages from the labor hours of that grouping of wires.

- When taking off grounding systems, identify separately the type and size of wire, and list each unique type of ground connection.

- The estimator should take the weights of materials into consideration when completing a takeoff. Topics to consider include: How will the materials be supported? What methods of support are available? How high will the support structure have to reach? Will the final support structure be able to withstand the total burden? Is the support material included or separate from the fixture, equipment, and material specified?

- Do not overlook the costs for equipment used in the installation. If scaffolding or highlifts are available in the field, contractors may use them in lieu of the proposed ladders and rolling staging.

26 20 00 Low-Voltage Electrical Transmission

- Supports and concrete pads may be shown on drawings for the larger equipment, or the support system may be only a piece of plywood for the back of a panelboard. In either case, they must be included in the costs.

26 40 00 Electrical and Cathodic Protection

- When taking off cathodic protection systems, identify the type and size of cable, and list each unique type of anode connection.

26 50 00 Lighting

- Fixtures should be taken off room by room using the fixture schedule, specifications, and the ceiling plan. For large concentrations of lighting fixtures in the same area, deduct the percentages from labor hours.

Reference Numbers

Reference numbers are shown at the beginning of some major classifications. These numbers refer to related items in the Reference Section. The reference information may be an estimating procedure, an alternate pricing method, or technical information.

Note: Not all subdivisions listed here necessarily appear. ■

26 01 02.20 Labor Adjustment Factors	Crew	Daily Output	Labor-Hours	Unit	Material	2018 Bare Costs Labor	Equipment	Total	Total Incl O&P
0010 **LABOR ADJUSTMENT FACTORS** (For Div. 26, 27 and 28) R260519-90									
0100 Subtract from labor for Economy of Scale for Wire									
0110 4-5 wires						25%			
0120 6-10 wires						30%			
0130 11-15 wires						35%			
0140 over 15 wires						40%			
0150 Labor adjustment factors (For Div. 26, 27, 28 and 48)									
0200 Labor factors: The below are reasonable suggestions, but									
0210 each project must be evaluated for its own peculiarities, and									
0220 the adjustments be increased or decreased depending on the									
0230 severity of the special conditions.									
1000 Add to labor for elevated installation (above floor level)									
1010 10' to 14.5' high						10%			
1020 15' to 19.5' high						20%			
1030 20' to 24.5' high						25%			
1040 25' to 29.5' high						35%			
1050 30' to 34.5' high						40%			
1060 35' to 39.5' high						50%			
1070 40' and higher						55%			
2000 Add to labor for crawl space									
2010 3' high						40%			
2020 4' high						30%			
3000 Add to labor for multi-story building									
3100 For new construction (No elevator available)									
3110 Add for floors 3 thru 10						5%			
3120 Add for floors 11 thru 15						10%			
3130 Add for floors 16 thru 20						15%			
3140 Add for floors 21 thru 30						20%			
3150 Add for floors 31 and up						30%			
3200 For existing structure (Elevator available)									
3210 Add for work on floor 3 and above						2%			
4000 Add to labor for working in existing occupied buildings									
4010 Hospital						35%			
4020 Office building						25%			
4030 School						20%			
4040 Factory or warehouse						15%			
4050 Multi-dwelling						15%			
5000 Add to labor, miscellaneous									
5010 Cramped shaft						35%			
5020 Congested area						15%			
5030 Excessive heat or cold						30%			
6000 Labor factors: the above are reasonable suggestions, but									
6100 each project should be evaluated for its own peculiarities									
6200 Other factors to be considered are:									
6210 Movement of material and equipment through finished areas						10%			
6220 Equipment room min security direct access w/authorization						15%			
6230 Attic space						25%			
6240 No service road						25%			
6250 Poor unloading/storage area, no hydraulic lifts or jacks						20%			
6260 Congested site area/heavy traffic						20%			
7000 Correctional facilities (no compounding division 1 adjustment factors)									
7010 Minimum security w/facilities escort						30%			
7020 Medium security w/facilities and correctional officer escort						40%			

26 01 Operation and Maintenance of Electrical Systems

26 01 02 – Labor Adjustment

26 01 02.20 Labor Adjustment Factors	Crew	Daily Output	Labor-Hours	Unit	Material	2018 Bare Costs Labor	Equipment	Total	Total Incl O&P
7030	Max security w/facilities & correctional officer escort (no inmate contact)						50%			

26 01 50 – Operation and Maintenance of Lighting

26 01 50.81 Luminaire Replacement

		Crew	Daily Output	Labor-Hours	Unit	Material	Labor	Equipment	Total	Total Incl O&P
0010	**LUMINAIRE REPLACEMENT**									
0962	Remove and install new replacement lens, 2' x 2' troffer	1 Elec	16	.500	Ea.	15.90	29		44.90	62.50
0964	2' x 4' troffer		16	.500		14.90	29		43.90	61.50
3200	Remove and replace (reinstall), lighting fixture	↓	4	2	↓		116		116	180

26 05 Common Work Results for Electrical

26 05 05 – Selective Demolition for Electrical

26 05 05.10 Electrical Demolition

			Crew	Daily Output	Labor-Hours	Unit	Material	Labor	Equipment	Total	Total Incl O&P
0010	**ELECTRICAL DEMOLITION**	R260105-30									
0020	Electrical demolition, conduit to 10' high, incl. fittings & hangers										
0100	Rigid galvanized steel, 1/2" to 1" diameter	R024119-10	1 Elec	242	.033	L.F.		1.92		1.92	2.97
0120	1-1/4" to 2"		"	200	.040			2.33		2.33	3.60
0140	2-1/2" to 3-1/2"		2 Elec	302	.053			3.08		3.08	4.76
0200	Electric metallic tubing (EMT), 1/2" to 1"		1 Elec	394	.020			1.18		1.18	1.83
0220	1-1/4" to 1-1/2"			326	.025			1.43		1.43	2.21
0240	2" to 3"		↓	236	.034	↓		1.97		1.97	3.05
0400	Wiremold raceway, including fittings & hangers										
0420	No. 3000		1 Elec	250	.032	L.F.		1.86		1.86	2.88
0440	No. 4000		"	217	.037	"		2.15		2.15	3.31
0500	Channels, steel, including fittings & hangers										
0520	3/4" x 1-1/2"		1 Elec	308	.026	L.F.		1.51		1.51	2.33
0540	1-1/2" x 1-1/2"			269	.030			1.73		1.73	2.67
0560	1-1/2" x 1-7/8"		↓	229	.035			2.03		2.03	3.14
0600	Copper bus duct, indoor, 3 phase										
0610	Including hangers & supports										
0620	225 amp		2 Elec	135	.119	L.F.		6.90		6.90	10.65
0640	400 amp			106	.151			8.80		8.80	13.55
0660	600 amp			86	.186			10.85		10.85	16.75
0680	1,000 amp			60	.267			15.50		15.50	24
0700	1,600 amp			40	.400			23.50		23.50	36
0720	3,000 amp		↓	10	1.600	↓		93		93	144
0800	Plug-in switches, 600 V 3 ph., incl. disconnecting										
0820	wire, conduit terminations, 30 amp		1 Elec	15.50	.516	Ea.		30		30	46.50
0840	60 amp			13.90	.576			33.50		33.50	51.50
0850	100 amp			10.40	.769			45		45	69
0860	200 amp		↓	6.20	1.290			75		75	116
0880	400 amp		2 Elec	5.40	2.963			172		172	266
0900	600 amp			3.40	4.706			274		274	425
0920	800 amp			2.60	6.154			360		360	555
0940	1,200 amp			2	8			465		465	720
0960	1,600 amp		↓	1.70	9.412			550		550	845
1010	Safety switches, 250 or 600 V, incl. disconnection										
1050	of wire & conduit terminations										
1100	30 amp		1 Elec	12.30	.650	Ea.		38		38	58.50
1120	60 amp			8.80	.909			53		53	81.50
1140	100 amp			7.30	1.096			64		64	98.50
1160	200 amp		↓	5	1.600	↓		93		93	144

For customer support on your Commercial Renovation Costs with RSMeans data, call 800.448.8182.

479

26 05 05.10 Electrical Demolition	Crew	Daily Output	Labor-Hours	Unit	Material	2018 Bare Costs Labor	Equipment	Total	Total Incl O&P
1210	Panel boards, incl. removal of all breakers,									
1220	conduit terminations & wire connections									
1230	3 wire, 120/240 V, 100A, to 20 circuits	1 Elec	2.60	3.077	Ea.		179		179	277
1240	200 amps, to 42 circuits	2 Elec	2.60	6.154			360		360	555
1241	225 amps, to 42 circuits		5.60	2.857			166		166	257
1250	400 amps, to 42 circuits	↓	2.20	7.273			425		425	655
1260	4 wire, 120/208 V, 125A, to 20 circuits	1 Elec	2.40	3.333			194		194	300
1270	200 amps, to 42 circuits	2 Elec	2.40	6.667	↓		390		390	600
1300	Transformer, dry type, 1 phase, incl. removal of									
1320	supports, wire & conduit terminations									
1340	1 kVA	1 Elec	7.70	1.039	Ea.		60.50		60.50	93.50
1360	5 kVA		4.70	1.702			99		99	153
1380	10 kVA	↓	3.60	2.222			129		129	200
1400	37.5 kVA	2 Elec	3	5.333			310		310	480
1420	75 kVA	"	2.50	6.400	↓		370		370	575
1440	3 phase to 600 V, primary									
1460	3 kVA	1 Elec	3.87	2.067	Ea.		120		120	186
1480	15 kVA	2 Elec	3.67	4.360			254		254	390
1490	25 kVA		3.51	4.558			265		265	410
1500	30 kVA	↓	3.42	4.678			272		272	420
1530	112.5 kVA	R-3	2.90	6.897			400	44.50	444.50	670
1560	500 kVA		1.40	14.286			830	92.50	922.50	1,375
1570	750 kVA	↓	1.10	18.182	↓		1,050	118	1,168	1,750
1600	Pull boxes & cabinets, sheet metal, incl. removal									
1620	of supports and conduit terminations									
1640	6" x 6" x 4"	1 Elec	31.10	.257	Ea.		14.95		14.95	23
1660	12" x 12" x 4"		23.30	.343			20		20	31
1720	Junction boxes, 4" sq. & oct.		80	.100			5.80		5.80	9
1740	Handy box		107	.075			4.35		4.35	6.70
1760	Switch box		107	.075			4.35		4.35	6.70
1780	Receptacle & switch plates	↓	257	.031	↓		1.81		1.81	2.80
1800	Wire, THW-THWN-THHN, removed from									
1810	in place conduit, to 10' high									
1830	#14	1 Elec	65	.123	C.L.F.		7.15		7.15	11.05
1840	#12		55	.145			8.45		8.45	13.10
1850	#10	↓	45.50	.176			10.25		10.25	15.80
1880	#4	2 Elec	53	.302			17.55		17.55	27
1890	#3		50	.320			18.60		18.60	29
1910	1/0		33.20	.482			28		28	43.50
1920	2/0		29.20	.548			32		32	49.50
1930	3/0		25	.640			37.50		37.50	57.50
1980	400 kcmil		17	.941			55		55	84.50
1990	500 kcmil	↓	16.20	.988	↓		57.50		57.50	89
2000	Interior fluorescent fixtures, incl. supports									
2010	& whips, to 10' high									
2100	Recessed drop-in 2' x 2', 2 lamp	2 Elec	35	.457	Ea.		26.50		26.50	41
2110	2' x 2', 4 lamp		30	.533			31		31	48
2120	2' x 4', 2 lamp		33	.485			28		28	43.50
2140	2' x 4', 4 lamp		30	.533			31		31	48
2160	4' x 4', 4 lamp	↓	20	.800	↓		46.50		46.50	72
2180	Surface mount, acrylic lens & hinged frame									
2200	1' x 4', 2 lamp	2 Elec	44	.364	Ea.		21		21	32.50
2210	6" x 4', 2 lamp	↓	44	.364			21		21	32.50

26 05 05 – Selective Demolition for Electrical

26 05 05.10 Electrical Demolition	Crew	Daily Output	Labor-Hours	Unit	Material	2018 Bare Costs Labor	Equipment	Total	Total Incl O&P
2220	2' x 2', 2 lamp	2 Elec	44	.364	Ea.		21		21	32.50
2260	2' x 4', 4 lamp		33	.485			28		28	43.50
2280	4' x 4', 4 lamp		23	.696			40.50		40.50	62.50
2281	4' x 4', 6 lamp		23	.696			40.50		40.50	62.50
2300	Strip fixtures, surface mount									
2320	4' long, 1 lamp	2 Elec	53	.302	Ea.		17.55		17.55	27
2340	4' long, 2 lamp		50	.320			18.60		18.60	29
2360	8' long, 1 lamp		42	.381			22		22	34.50
2380	8' long, 2 lamp		40	.400			23.50		23.50	36
2400	Pendant mount, industrial, incl. removal									
2410	of chain or rod hangers, to 10' high									
2420	4' long, 2 lamp	2 Elec	35	.457	Ea.		26.50		26.50	41
2421	4' long, 4 lamp		35	.457			26.50		26.50	41
2440	8' long, 2 lamp		27	.593			34.50		34.50	53.50
2460	Interior incandescent, surface, ceiling									
2470	or wall mount, to 12' high									
2480	Metal cylinder type, 75 Watt	2 Elec	62	.258	Ea.		15		15	23
2500	150 Watt	"	62	.258	"		15		15	23
2520	Metal halide, high bay									
2540	400 Watt	2 Elec	15	1.067	Ea.		62		62	96
2560	1,000 Watt		12	1.333			77.50		77.50	120
2580	150 Watt, low bay		20	.800			46.50		46.50	72
2600	Exterior fixtures, incandescent, wall mount									
2620	100 Watt	2 Elec	50	.320	Ea.		18.60		18.60	29
2640	Quartz, 500 Watt		33	.485			28		28	43.50
2660	1,500 Watt		27	.593			34.50		34.50	53.50
2680	Wall pack, mercury vapor									
2700	175 Watt	2 Elec	25	.640	Ea.		37.50		37.50	57.50
2720	250 Watt	"	25	.640	"		37.50		37.50	57.50
9000	Minimum labor/equipment charge	1 Elec	4	2	Job		116		116	180
9900	Add to labor for higher elevated installation									
9905	10' to 14.5' high, add						10%			
9910	15' to 20' high, add						20%			
9920	20' to 25' high, add						25%			
9930	25' to 30' high, add						35%			
9940	30' to 35' high, add						40%			
9950	35' to 40' high, add						50%			
9960	Over 40' high, add						55%			

26 05 19 – Low-Voltage Electrical Power Conductors and Cables

26 05 19.20 Armored Cable

26 05 19.20 Armored Cable	Crew	Daily Output	Labor-Hours	Unit	Material	2018 Bare Costs Labor	Equipment	Total	Total Incl O&P
0010	**ARMORED CABLE**									
0050	600 volt, copper (BX), #14, 2 conductor, solid	1 Elec	2.40	3.333	C.L.F.	41.50	194		235.50	345
0100	3 conductor, solid		2.20	3.636		64.50	212		276.50	395
0150	#12, 2 conductor, solid		2.30	3.478		41.50	202		243.50	360
0200	3 conductor, solid		2	4		71	233		304	440
0250	#10, 2 conductor, solid		2	4		80.50	233		313.50	450
0300	3 conductor, solid		1.60	5		110	291		401	570
0340	#8, 2 conductor, stranded		1.50	5.333		244	310		554	750
0350	3 conductor, stranded		1.30	6.154		244	360		604	825
9010	600 volt, copper (MC) steel clad, #14, 2 wire	R-1A	8.16	1.961		42.50	103		145.50	206
9020	3 wire		7.21	2.219		75.50	116		191.50	263
9030	4 wire		6.69	2.392		184	125		309	395

481

26 05 19.20 Armored Cable

		Crew	Daily Output	Labor-Hours	Unit	Material	2018 Bare Costs Labor	Equipment	Total	Total Incl O&P
9040	#12, 2 wire	R-1A	7.21	2.219	C.L.F.	54.50	116		170.50	240
9050	3 wire		6.69	2.392		70.50	125		195.50	271
9070	#10, 2 wire		6.04	2.649		103	139		242	325
9080	3 wire		5.65	2.832		133	148		281	375
9100	#8, 2 wire, stranded		3.94	4.061		158	213		371	505
9110	3 wire, stranded		3.40	4.706		193	246		439	590
9900	Minimum labor/equipment charge	R-1C	4	8	Job		420	32.50	452.50	680

26 05 19.50 Mineral Insulated Cable

		Crew	Daily Output	Labor-Hours	Unit	Material	2018 Bare Costs Labor	Equipment	Total	Total Incl O&P
0010	**MINERAL INSULATED CABLE** 600 volt									
0100	1 conductor, #12	1 Elec	1.60	5	C.L.F.	350	291		641	835
0200	#10		1.60	5		575	291		866	1,075
0400	#8		1.50	5.333		640	310		950	1,175
0500	#6		1.40	5.714		760	335		1,095	1,350
0600	#4	2 Elec	2.40	6.667		1,025	390		1,415	1,725
0800	#2		2.20	7.273		1,450	425		1,875	2,250
0900	#1		2.10	7.619		1,675	445		2,120	2,500
1000	1/0		2	8		1,950	465		2,415	2,875
1100	2/0		1.90	8.421		2,325	490		2,815	3,325
1200	3/0		1.80	8.889		2,800	515		3,315	3,875
1400	4/0		1.60	10		3,225	580		3,805	4,450
1410	250 kcmil	3 Elec	2.40	10		3,625	580		4,205	4,900
1420	350 kcmil		1.95	12.308		4,150	715		4,865	5,675
1430	500 kcmil		1.95	12.308		5,350	715		6,065	6,975
1500	2 conductor, #12	1 Elec	1.40	5.714		930	335		1,265	1,550
1600	#10		1.20	6.667		1,150	390		1,540	1,850
1800	#8		1.10	7.273		1,225	425		1,650	2,000
2000	#6		1.05	7.619		1,800	445		2,245	2,650
2100	#4	2 Elec	2	8		2,450	465		2,915	3,400
2200	3 conductor, #12	1 Elec	1.20	6.667		1,175	390		1,565	1,875
2400	#10		1.10	7.273		1,325	425		1,750	2,100
2600	#8		1.05	7.619		1,650	445		2,095	2,500
2800	#6		1	8		2,175	465		2,640	3,125
3000	#4	2 Elec	1.80	8.889		2,725	515		3,240	3,775
3100	4 conductor, #12	1 Elec	1.20	6.667		1,225	390		1,615	1,950
3200	#10		1.10	7.273		1,475	425		1,900	2,275
3400	#8		1	8		1,925	465		2,390	2,850
3600	#6		.90	8.889		2,500	515		3,015	3,550
3620	7 conductor, #12		1.10	7.273		1,575	425		2,000	2,375
3640	#10		1	8		2,050	465		2,515	2,975
3800	Terminations, 600 volt, 1 conductor, #12		8	1	Ea.	19.65	58		77.65	112
5500	2 conductor, #12		6.70	1.194		19.65	69.50		89.15	129
5600	#10		6.40	1.250		29.50	73		102.50	145
5800	#8		6.20	1.290		29.50	75		104.50	149
6000	#6		5.70	1.404		29.50	81.50		111	159
6200	#4		5.30	1.509		65.50	88		153.50	208
6400	3 conductor, #12		5.70	1.404		29.50	81.50		111	159
6500	#10		5.50	1.455		29.50	84.50		114	164
6600	#8		5.20	1.538		29.50	89.50		119	171
6800	#6		4.80	1.667		29.50	97		126.50	183
7200	#4		4.60	1.739		65.50	101		166.50	228
7400	4 conductor, #12		4.60	1.739		33	101		134	192
7500	#10		4.40	1.818		33	106		139	199

26 05 19 – Low-Voltage Electrical Power Conductors and Cables

26 05 19.50 Mineral Insulated Cable

		Crew	Daily Output	Labor-Hours	Unit	Material	2018 Bare Costs Labor	2018 Bare Costs Equipment	Total	Total Incl O&P
7600	#8	1 Elec	4.20	1.905	Ea.	33	111		144	207
8400	#6		4	2		69	116		185	256
8500	7 conductor, #12		3.50	2.286		33	133		166	241
8600	#10		3	2.667		71	155		226	320
8800	Crimping tool, plier type					73.50			73.50	81
9000	Stripping tool					278			278	305
9200	Hand vise					83			83	91
9500	Minimum labor/equipment charge	1 Elec	4	2	Job		116		116	180

26 05 19.55 Non-Metallic Sheathed Cable

		Crew	Daily Output	Labor-Hours	Unit	Material	2018 Bare Costs Labor	2018 Bare Costs Equipment	Total	Total Incl O&P
0010	**NON-METALLIC SHEATHED CABLE** 600 volt									
0100	Copper with ground wire (Romex)									
0150	#14, 2 conductor	1 Elec	2.70	2.963	C.L.F.	18.35	172		190.35	286
0200	3 conductor		2.40	3.333		25.50	194		219.50	330
0220	4 conductor		2.20	3.636		39.50	212		251.50	370
0250	#12, 2 conductor		2.50	3.200		24	186		210	315
0300	3 conductor		2.20	3.636		40	212		252	370
0320	4 conductor		2	4		60	233		293	425
0350	#10, 2 conductor		2.20	3.636		41	212		253	370
0400	3 conductor		1.80	4.444		62.50	259		321.50	470
0430	#8, 2 conductor		1.60	5		75.50	291		366.50	535
0450	3 conductor		1.50	5.333		117	310		427	610
0500	#6, 3 conductor		1.40	5.714		197	335		532	730
0550	SE type SER aluminum cable, 3 RHW and									
0600	1 bare neutral, 3 #8 & 1 #8	1 Elec	1.60	5	C.L.F.	56.50	291		347.50	510
0650	3 #6 & 1 #6	"	1.40	5.714		67	335		402	590
0700	3 #4 & 1 #6	2 Elec	2.40	6.667		71.50	390		461.50	680
0750	3 #2 & 1 #4		2.20	7.273		124	425		549	790
0800	3 #1/0 & 1 #2		2	8		182	465		647	920
0850	3 #2/0 & 1 #1		1.80	8.889		212	515		727	1,025
0900	3 #4/0 & 1 #2/0		1.60	10		295	580		875	1,225
6500	Service entrance cap for copper SEU									
6600	100 amp	1 Elec	12	.667	Ea.	8.10	39		47.10	69
6700	150 amp		10	.800		10.90	46.50		57.40	84
6800	200 amp		8	1		17.80	58		75.80	110
9000	Minimum labor/equipment charge		4	2	Job		116		116	180

26 05 19.90 Wire

			Crew	Daily Output	Labor-Hours	Unit	Material	2018 Bare Costs Labor	2018 Bare Costs Equipment	Total	Total Incl O&P
0010	**WIRE**, normal installation conditions in wireway, conduit, cable tray										
0020	600 volt, copper type THW, solid, #14	R260533-22	1 Elec	13	.615	C.L.F.	5.90	36		41.90	62
0030	#12			11	.727		9.65	42.50		52.15	76
0040	#10			10	.800		15.75	46.50		62.25	89.50
0050	Stranded, #14			13	.615		8.60	36		44.60	65
0100	#12			11	.727		13.20	42.50		55.70	80
0120	#10			10	.800		21	46.50		67.50	95
0140	#8			8	1		34.50	58		92.50	128
0160	#6			6.50	1.231		43.50	71.50		115	159
0180	#4		2 Elec	10.60	1.509		69.50	88		157.50	212
0200	#3			10	1.600		105	93		198	259
0220	#2			9	1.778		102	103		205	272
0240	#1			8	2		136	116		252	330
0260	1/0			6.60	2.424		216	141		357	455
0280	2/0			5.80	2.759		205	161		366	475
0300	3/0			5	3.200		235	186		421	545

For customer support on your Commercial Renovation Costs with RSMeans data, call 800.448.8182.

483

26 05 19 – Low-Voltage Electrical Power Conductors and Cables

26 05 19.90 Wire		Crew	Daily Output	Labor-Hours	Unit	Material	2018 Bare Costs Labor	Equipment	Total	Total Incl O&P
0350	4/0	2 Elec	4.40	3.636	C.L.F.	435	212		647	800
0400	250 kcmil	3 Elec	6	4		350	233		583	745
0420	300 kcmil		5.70	4.211		380	245		625	795
0450	350 kcmil		5.40	4.444		465	259		724	910
0480	400 kcmil		5.10	4.706		540	274		814	1,025
0490	500 kcmil		4.80	5		690	291		981	1,200
0510	750 kcmil		3.30	7.273		1,325	425		1,750	2,100
0540	600 volt, aluminum type THHN, stranded, #6	1 Elec	8	1		24	58		82	117
0560	#4	2 Elec	13	1.231		30	71.50		101.50	144
0580	#2		10.60	1.509		41	88		129	181
0600	#1		9	1.778		60	103		163	226
0620	1/0		8	2		72	116		188	260
0640	2/0		7.20	2.222		86	129		215	295
0680	3/0		6.60	2.424		107	141		248	335
0700	4/0		6.20	2.581		119	150		269	365
0720	250 kcmil	3 Elec	8.70	2.759		144	161		305	405
0740	300 kcmil		8.10	2.963		199	172		371	485
0760	350 kcmil		7.50	3.200		201	186		387	510
0780	400 kcmil		6.90	3.478		229	202		431	565
0800	500 kcmil		6	4		258	233		491	645
0850	600 kcmil		5.70	4.211		325	245		570	740
0880	700 kcmil		5.10	4.706		400	274		674	865
9000	Minimum labor/equipment charge	1 Elec	4	2	Job		116		116	180

26 05 26 – Grounding and Bonding for Electrical Systems

26 05 26.80 Grounding		Crew	Daily Output	Labor-Hours	Unit	Material	2018 Bare Costs Labor	Equipment	Total	Total Incl O&P
0010	**GROUNDING**									
0030	Rod, copper clad, 8' long, 1/2" diameter	1 Elec	5.50	1.455	Ea.	19.90	84.50		104.40	153
0040	5/8" diameter		5.50	1.455		21.50	84.50		106	155
0050	3/4" diameter		5.30	1.509		35	88		123	175
0080	10' long, 1/2" diameter		4.80	1.667		22.50	97		119.50	175
0090	5/8" diameter		4.60	1.739		24	101		125	182
0100	3/4" diameter		4.40	1.818		42	106		148	210
0130	15' long, 3/4" diameter		4	2		49	116		165	234
0260	Wire ground bare armored, #8-1 conductor		2	4	C.L.F.	69.50	233		302.50	435
0280	#4-1 conductor		1.60	5		134	291		425	595
0390	Bare copper wire, stranded, #8		11	.727		43	42.50		85.50	113
0400	#6		10	.800		41	46.50		87.50	117
0600	#2	2 Elec	10	1.600		89	93		182	242
0800	3/0		6.60	2.424		315	141		456	565
1000	4/0		5.70	2.807		440	163		603	735
1200	250 kcmil	3 Elec	7.20	3.333		445	194		639	790
1800	Water pipe ground clamps, heavy duty									
2000	Bronze, 1/2" to 1" diameter	1 Elec	8	1	Ea.	29.50	58		87.50	123
2100	1-1/4" to 2" diameter		8	1		28.50	58		86.50	122
2200	2-1/2" to 3" diameter		6	1.333		68	77.50		145.50	195
2400	Grounding, exothermic welding									
2500	Exothermic welding reusable mold, cable to cable, parallel, vertical	1 Elec	8	1	Ea.	345	58		403	470
2505	Splice single		8	1		294	58		352	415
2510	Splice single to double		8	1		345	58		403	470
2520	Cable to cable, termination, Tee		8	1		340	58		398	465
2530	Cable to rod, termination, Tee		8	1		268	58		326	385
2535	Cable to rebar over, termination, Tee		8	1		86	58		144	185

26 05 Common Work Results for Electrical

26 05 26 – Grounding and Bonding for Electrical Systems

26 05 26.80 Grounding

		Crew	Daily Output	Labor-Hours	Unit	Material	2018 Bare Costs Labor	2018 Bare Costs Equipment	Total	Total Incl O&P
2540	Cable to rod, termination, 90 Deg	1 Elec	8	1	Ea.	335	58		393	460
2550	Cable to verticle flat steel		8	1		345	58		403	465
2555	Cable to 45 Deg		8	1		350	58		408	475
2730	Exothermic weld, 4/0 wire to 1" ground rod		7	1.143		9.55	66.50		76.05	114
2800	Brazed connections, #6 wire		12	.667		16.15	39		55.15	78
3000	#2 wire		10	.800		21.50	46.50		68	96
3100	3/0 wire		8	1		32	58		90	126
3200	4/0 wire		7	1.143		36.50	66.50		103	144
3400	250 kcmil wire		5	1.600		43	93		136	191
3600	500 kcmil wire		4	2		53	116		169	238
7000	Exothermic welding kit, multi vertical		8	1		600	58		658	750
9000	Minimum labor/equipment charge		4	2	Job		116		116	180

26 05 29 – Hangers and Supports for Electrical Systems

26 05 29.30 Fittings and Channel Support

		Crew	Daily Output	Labor-Hours	Unit	Material	2018 Bare Costs Labor	2018 Bare Costs Equipment	Total	Total Incl O&P
0010	**FITTINGS & CHANNEL SUPPORT**									
0020	Rooftop channel support									
0200	2-7/8" L x 1-5/8" W, 12 ga. pre galv., dbl. base	1 Elec	46	.174	Ea.	4.07	10.10		14.17	20
0210	2-7/8" L x 1-5/8" W, 12 ga. hot dip galv., dbl. base		46	.174		2.37	10.10		12.47	18.25
0220	2-7/16" L x 1-5/8" W, 12 ga. pre galv., sngl base		48	.167		1.95	9.70		11.65	17.15
0230	2-7/16" L x 1-5/8" W, 12 ga. hot dip galv., sngl base		48	.167		6.95	9.70		16.65	22.50
0240	13/16" L x 1-5/8" W, 14 ga. pre galv., sngl base		56	.143		1.66	8.30		9.96	14.70
0250	13/16" L x 1-5/8" W, 14 ga. hot dip galv., sngl base		56	.143		6.80	8.30		15.10	20.50
0260	1-5/8" L x 1-5/8" W, 12 ga. pre galv., sngl base		54	.148		1.80	8.60		10.40	15.30
0270	1-5/8" L x 1-5/8" W, 12 ga. hot dip galv., sngl base		54	.148		1.90	8.60		10.50	15.40
0280	1-5/8" x 28", 12 ga. pre galv., dbl. H block base		42	.190		4.12	11.10		15.22	21.50
0290	1-5/8" x 36", 12 ga. pre galv., dbl. H block base		40	.200		4.36	11.65		16.01	23
0300	1-5/8" x 42", 12 ga. pre galv., dbl. H block base		34	.235		4.90	13.70		18.60	26.50
0310	1-5/8" x 50", 12 ga. pre galv., dbl. H block base		38	.211		5.20	12.25		17.45	24.50
0320	1-5/8" x 60", 12 ga. pre galv., dbl. H block base		36	.222		5.75	12.95		18.70	26.50
0330	13/16" x 8" H, threaded rod pre galv., H block base		35	.229		2.77	13.30		16.07	23.50
0340	13/16" x 12" H, threaded rod pre galv., H block base		34.60	.231		3.46	13.45		16.91	25
0350	1-5/8" x 16" H, threaded rod dbl. H block base		34	.235		5.15	13.70		18.85	26.50
0360	1-5/8" x 12" H, pre galv., dbl. H block base		34.60	.231		9.90	13.45		23.35	32
0370	1-5/8" x 24" H, pre galv., dbl. H block base		34	.235		10.45	13.70		24.15	32.50
0380	3-1/4" x 24" H, pre galv., dbl. H block base		32	.250		23.50	14.55		38.05	48.50
0390	3-1/4" x 36" H, pre galv., dbl. H block base		30	.267		26	15.50		41.50	52.50

26 05 33 – Raceway and Boxes for Electrical Systems

26 05 33.13 Conduit

		Crew	Daily Output	Labor-Hours	Unit	Material	2018 Bare Costs Labor	2018 Bare Costs Equipment	Total	Total Incl O&P
0010	**CONDUIT** To 10' high, includes 2 terminations, 2 elbows,	R260533-22								
0020	11 beam clamps, and 11 couplings per 100 L.F.									
1161	Field bends, 45° to 90°, 1/2" diameter	1 Elec	53	.151	Ea.		8.80		8.80	13.55
1162	3/4" diameter		47	.170			9.90		9.90	15.30
1163	1" diameter		44	.182			10.60		10.60	16.35
1164	1-1/4" diameter		23	.348			20		20	31.50
1165	1-1/2" diameter		21	.381			22		22	34.50
1166	2" diameter		16	.500			29		29	45
1691	See note on line 26 05 33.13 9995									
1750	Rigid galvanized steel, 1/2" diameter	1 Elec	90	.089	L.F.	2.31	5.15		7.46	10.55
1770	3/4" diameter		80	.100		4.57	5.80		10.37	14.05
1800	1" diameter		65	.123		6.75	7.15		13.90	18.50
1830	1-1/4" diameter		60	.133		4.74	7.75		12.49	17.20
1850	1-1/2" diameter		55	.145		7.90	8.45		16.35	22

26 05 33 – Raceway and Boxes for Electrical Systems

26 05 33.13 Conduit		Crew	Daily Output	Labor-Hours	Unit	Material	2018 Bare Costs Labor	Equipment	Total	Total Incl O&P
1870	2" diameter	1 Elec	45	.178	L.F.	9.95	10.35		20.30	27
1991	Field bends, 45° to 90°, 1/2" diameter		44	.182	Ea.		10.60		10.60	16.35
1992	3/4" diameter		40	.200			11.65		11.65	18
1993	1" diameter		36	.222			12.95		12.95	20
1994	1-1/4" diameter		19	.421			24.50		24.50	38
1995	1-1/2" diameter		18	.444			26		26	40
1996	2" diameter		13	.615	↓		36		36	55.50
2500	Steel, intermediate conduit (IMC), 1/2" diameter		100	.080	L.F.	1.69	4.66		6.35	9.05
2530	3/4" diameter		90	.089		2.05	5.15		7.20	10.25
2550	1" diameter		70	.114		2.58	6.65		9.23	13.10
2570	1-1/4" diameter		65	.123		3.58	7.15		10.73	15
2600	1-1/2" diameter		60	.133		4.76	7.75		12.51	17.25
2630	2" diameter		50	.160		6.15	9.30		15.45	21
2650	2-1/2" diameter	↓	40	.200		9.15	11.65		20.80	28
2670	3" diameter	2 Elec	60	.267		12.10	15.50		27.60	37.50
2700	3-1/2" diameter		54	.296		16.90	17.25		34.15	45
2730	4" diameter	↓	50	.320	↓	11.70	18.60		30.30	42
2731	Field bends, 45° to 90°, 1/2" diameter	1 Elec	44	.182	Ea.		10.60		10.60	16.35
2732	3/4" diameter		40	.200			11.65		11.65	18
2733	1" diameter		36	.222			12.95		12.95	20
2734	1-1/4" diameter		19	.421			24.50		24.50	38
2735	1-1/2" diameter		18	.444			26		26	40
2736	2" diameter		13	.615	↓		36		36	55.50
5000	Electric metallic tubing (EMT), 1/2" diameter		170	.047	L.F.	.80	2.74		3.54	5.10
5020	3/4" diameter		130	.062		1.13	3.58		4.71	6.80
5040	1" diameter		115	.070		1.84	4.05		5.89	8.30
5060	1-1/4" diameter		100	.080		2.99	4.66		7.65	10.50
5080	1-1/2" diameter		90	.089		2.82	5.15		7.97	11.10
5100	2" diameter		80	.100		3.85	5.80		9.65	13.25
5120	2-1/2" diameter	↓	60	.133		4.43	7.75		12.18	16.85
5140	3" diameter	2 Elec	100	.160		5.55	9.30		14.85	20.50
5160	3-1/2" diameter		90	.178		6.95	10.35		17.30	23.50
5180	4" diameter	↓	80	.200	↓	12.10	11.65		23.75	31.50
5200	Field bends, 45° to 90°, 1/2" diameter	1 Elec	89	.090	Ea.		5.25		5.25	8.10
5220	3/4" diameter		80	.100			5.80		5.80	9
5240	1" diameter		73	.110			6.40		6.40	9.85
5260	1-1/4" diameter		38	.211			12.25		12.25	18.95
5280	1-1/2" diameter		36	.222			12.95		12.95	20
5300	2" diameter		26	.308			17.90		17.90	27.50
5320	Offsets, 1/2" diameter		65	.123			7.15		7.15	11.05
5340	3/4" diameter		62	.129			7.50		7.50	11.60
5360	1" diameter		53	.151			8.80		8.80	13.55
5380	1-1/4" diameter		30	.267			15.50		15.50	24
5400	1-1/2" diameter	↓	28	.286	↓		16.65		16.65	25.50
9990	Minimum labor/equipment charge		4	2	Job		116		116	180

26 05 33.16 Boxes for Electrical Systems

		Crew	Daily Output	Labor-Hours	Unit	Material	Labor	Equipment	Total	Total Incl O&P
0010	**BOXES FOR ELECTRICAL SYSTEMS**									
0020	Pressed steel, octagon, 4"	1 Elec	20	.400	Ea.	2.86	23.50		26.36	39
0060	Covers, blank		64	.125		.88	7.30		8.18	12.20
0100	Extension rings		40	.200		3.68	11.65		15.33	22
0150	Square, 4"		20	.400		5.25	23.50		28.75	42
0200	Extension rings	↓	40	.200	↓	3.79	11.65		15.44	22

26 05 33 – Raceway and Boxes for Electrical Systems

26 05 33.16 Boxes for Electrical Systems	Crew	Daily Output	Labor-Hours	Unit	Material	2018 Bare Costs Labor	Equipment	Total	Total Incl O&P
0250 Covers, blank	1 Elec	64	.125	Ea.	.67	7.30		7.97	12
0300 Plaster rings		64	.125		1.54	7.30		8.84	12.95
0652 Switchbox		24	.333		4.61	19.40		24.01	35
1100 Concrete, floor, 1 gang		5.30	1.509	↓	94.50	88		182.50	240
9000 Minimum labor/equipment charge		4	2	Job		116		116	180

26 05 33.17 Outlet Boxes, Plastic

	Crew	Daily Output	Labor-Hours	Unit	Material	2018 Bare Costs Labor	Equipment	Total	Total Incl O&P
0010 **OUTLET BOXES, PLASTIC**									
0050 4" diameter, round with 2 mounting nails	1 Elec	25	.320	Ea.	2.73	18.60		21.33	32
0100 Bar hanger mounted		25	.320		5.30	18.60		23.90	35
0200 4", square with 2 mounting nails		25	.320		5.30	18.60		23.90	35
0300 Plaster ring		64	.125		2.02	7.30		9.32	13.45
0400 Switch box with 2 mounting nails, 1 gang		30	.267		3.03	15.50		18.53	27.50
0500 2 gang		25	.320		3.70	18.60		22.30	33
0600 3 gang		20	.400		4.92	23.50		28.42	41.50
0700 Old work box		30	.267	↓	3.91	15.50		19.41	28.50
9000 Minimum labor/equipment charge		4	2	Job		116		116	180

26 05 33.18 Pull Boxes

	Crew	Daily Output	Labor-Hours	Unit	Material	2018 Bare Costs Labor	Equipment	Total	Total Incl O&P
0010 **PULL BOXES**									
0100 Steel, pull box, NEMA 1, type SC, 6" W x 6" H x 4" D	1 Elec	8	1	Ea.	9.75	58		67.75	101
0200 8" W x 8" H x 4" D		8	1		12.55	58		70.55	104
0300 10" W x 12" H x 6" D		5.30	1.509		28.50	88		116.50	167
0400 16" W x 20" H x 8" D		4	2		82	116		198	270
0500 20" W x 24" H x 8" D		3.20	2.500		96	146		242	330
0600 24" W x 36" H x 8" D		2.70	2.963		151	172		323	435
0650 Pull box, hinged, NEMA 1, 6" W x 6" H x 4" D		8	1		13	58		71	104
0800 12" W x 16" H x 6" D		4.70	1.702		51.50	99		150.50	210
1000 20" W x 20" H x 6" D		3.60	2.222		121	129		250	335
1200 20" W x 20" H x 8" D		3.20	2.500		151	146		297	390
1400 24" W x 36" H x 8" D		2.70	2.963		237	172		409	525
1600 24" W x 42" H x 8" D	↓	2	4	↓	355	233		588	750

26 05 33.23 Wireway

	Crew	Daily Output	Labor-Hours	Unit	Material	2018 Bare Costs Labor	Equipment	Total	Total Incl O&P
0010 **WIREWAY** to 10' high									
0020 For higher elevations, see Section 26 05 36.40									
0100 NEMA 1, screw cover w/fittings and supports, 2-1/2" x 2-1/2"	1 Elec	45	.178	L.F.	12.20	10.35		22.55	29.50
0200 4" x 4"	"	40	.200		13.15	11.65		24.80	32.50
0400 6" x 6"	2 Elec	60	.267		21	15.50		36.50	47
0600 8" x 8"	"	40	.400		28	23.50		51.50	66.50
4475 NEMA 3R, screw cover w/fittings and supports, 4" x 4"	1 Elec	36	.222		16.30	12.95		29.25	38
4480 6" x 6"	2 Elec	55	.291		18.10	16.95		35.05	46
4485 8" x 8"		36	.444		32	26		58	75
4490 12" x 12"	↓	18	.889	↓	57.50	51.50		109	143

26 05 33.95 Cutting and Drilling

	Crew	Daily Output	Labor-Hours	Unit	Material	2018 Bare Costs Labor	Equipment	Total	Total Incl O&P
0010 **CUTTING AND DRILLING**									
0100 Hole drilling to 10' high, concrete wall									
0110 8" thick, 1/2" pipe size	R-31	12	.667	Ea.	.24	39	3.96	43.20	64.50
0120 3/4" pipe size		12	.667		.24	39	3.96	43.20	64.50
0130 1" pipe size		9.50	.842		.39	49	5	54.39	81.50
0140 1-1/4" pipe size		9.50	.842		.39	49	5	54.39	81.50
0150 1-1/2" pipe size		9.50	.842		.39	49	5	54.39	81.50
0160 2" pipe size		4.40	1.818		.52	106	10.80	117.32	175
0170 2-1/2" pipe size		4.40	1.818		.52	106	10.80	117.32	175
0180 3" pipe size	↓	4.40	1.818	↓	.52	106	10.80	117.32	175

26 05 33.95 Cutting and Drilling	Crew	Daily Output	Labor-Hours	Unit	Material	2018 Bare Costs Labor	Equipment	Total	Total Incl O&P
0190	3-1/2" pipe size	R-31	3.30	2.424	Ea.	.67	141	14.40	156.07	235
0200	4" pipe size		3.30	2.424		.67	141	14.40	156.07	235
0500	12" thick, 1/2" pipe size		9.40	.851		.37	49.50	5.05	54.92	82.50
0520	3/4" pipe size		9.40	.851		.37	49.50	5.05	54.92	82.50
0540	1" pipe size		7.30	1.096		.58	64	6.50	71.08	106
0560	1-1/4" pipe size		7.30	1.096		.58	64	6.50	71.08	106
0570	1-1/2" pipe size		7.30	1.096		.58	64	6.50	71.08	106
0580	2" pipe size		3.60	2.222		.78	129	13.20	142.98	215
0590	2-1/2" pipe size		3.60	2.222		.78	129	13.20	142.98	215
0600	3" pipe size		3.60	2.222		.78	129	13.20	142.98	215
0610	3-1/2" pipe size		2.80	2.857		1.01	166	16.95	183.96	277
0630	4" pipe size		2.50	3.200		1.01	186	19	206.01	310
0650	16" thick, 1/2" pipe size		7.60	1.053		.49	61.50	6.25	68.24	102
0670	3/4" pipe size		7	1.143		.49	66.50	6.80	73.79	111
0690	1" pipe size		6	1.333		.77	77.50	7.90	86.17	130
0710	1-1/4" pipe size		5.50	1.455		.77	84.50	8.65	93.92	141
0730	1-1/2" pipe size		5.50	1.455		.77	84.50	8.65	93.92	141
0750	2" pipe size		3	2.667		1.04	155	15.85	171.89	259
0770	2-1/2" pipe size		2.70	2.963		1.04	172	17.60	190.64	287
0790	3" pipe size		2.50	3.200		1.04	186	19	206.04	310
0810	3-1/2" pipe size		2.30	3.478		1.34	202	20.50	223.84	340
0830	4" pipe size		2	4		1.34	233	24	258.34	385
0850	20" thick, 1/2" pipe size		6.40	1.250		.61	73	7.45	81.06	121
0870	3/4" pipe size		6	1.333		.61	77.50	7.90	86.01	129
0890	1" pipe size		5	1.600		.96	93	9.50	103.46	156
0910	1-1/4" pipe size		4.80	1.667		.96	97	9.90	107.86	162
0930	1-1/2" pipe size		4.60	1.739		.96	101	10.35	112.31	168
0950	2" pipe size		2.70	2.963		1.30	172	17.60	190.90	287
0970	2-1/2" pipe size		2.40	3.333		1.30	194	19.80	215.10	325
0990	3" pipe size		2.20	3.636		1.30	212	21.50	234.80	350
1010	3-1/2" pipe size		2	4		1.68	233	24	258.68	390
1030	4" pipe size		1.70	4.706		1.68	274	28	303.68	455
1050	24" thick, 1/2" pipe size		5.50	1.455		.73	84.50	8.65	93.88	141
1070	3/4" pipe size		5.10	1.569		.73	91.50	9.30	101.53	152
1090	1" pipe size		4.30	1.860		1.16	108	11.05	120.21	180
1110	1-1/4" pipe size		4	2		1.16	116	11.90	129.06	194
1130	1-1/2" pipe size		4	2		1.16	116	11.90	129.06	194
1150	2" pipe size		2.40	3.333		1.56	194	19.80	215.36	325
1170	2-1/2" pipe size		2.20	3.636		1.56	212	21.50	235.06	350
1190	3" pipe size		2	4		1.56	233	24	258.56	390
1210	3-1/2" pipe size		1.80	4.444		2.01	259	26.50	287.51	430
1230	4" pipe size		1.50	5.333		2.01	310	31.50	343.51	515
1500	Brick wall, 8" thick, 1/2" pipe size		18	.444		.24	26	2.64	28.88	43
1520	3/4" pipe size		18	.444		.24	26	2.64	28.88	43
1540	1" pipe size		13.30	.602		.39	35	3.57	38.96	58.50
1560	1-1/4" pipe size		13.30	.602		.39	35	3.57	38.96	58.50
1580	1-1/2" pipe size		13.30	.602		.39	35	3.57	38.96	58.50
1600	2" pipe size		5.70	1.404		.52	81.50	8.35	90.37	136
1620	2-1/2" pipe size		5.70	1.404		.52	81.50	8.35	90.37	136
1640	3" pipe size		5.70	1.404		.52	81.50	8.35	90.37	136
1660	3-1/2" pipe size		4.40	1.818		.67	106	10.80	117.47	176
1680	4" pipe size		4	2		.67	116	11.90	128.57	194
1700	12" thick, 1/2" pipe size		14.50	.552		.37	32	3.28	35.65	53.50

26 05 33.95 Cutting and Drilling	Crew	Daily Output	Labor-Hours	Unit	Material	2018 Bare Costs Labor	Equipment	Total	Total Incl O&P
1720	3/4" pipe size	R-31	14.50	.552	Ea.	.37	32	3.28	35.65	53.50
1740	1" pipe size		11	.727		.58	42.50	4.32	47.40	71
1760	1-1/4" pipe size		11	.727		.58	42.50	4.32	47.40	71
1780	1-1/2" pipe size		11	.727		.58	42.50	4.32	47.40	71
1800	2" pipe size		5	1.600		.78	93	9.50	103.28	155
1820	2-1/2" pipe size		5	1.600		.78	93	9.50	103.28	155
1840	3" pipe size		5	1.600		.78	93	9.50	103.28	155
1860	3-1/2" pipe size		3.80	2.105		1.01	123	12.50	136.51	204
1880	4" pipe size		3.30	2.424		1.01	141	14.40	156.41	235
1900	16" thick, 1/2" pipe size		12.30	.650		.49	38	3.86	42.35	63.50
1920	3/4" pipe size		12.30	.650		.49	38	3.86	42.35	63.50
1940	1" pipe size		9.30	.860		.77	50	5.10	55.87	84
1960	1-1/4" pipe size		9.30	.860		.77	50	5.10	55.87	84
1980	1-1/2" pipe size		9.30	.860		.77	50	5.10	55.87	84
2000	2" pipe size		4.40	1.818		1.04	106	10.80	117.84	176
2010	2-1/2" pipe size		4.40	1.818		1.04	106	10.80	117.84	176
2030	3" pipe size		4.40	1.818		1.04	106	10.80	117.84	176
2050	3-1/2" pipe size		3.30	2.424		1.34	141	14.40	156.74	235
2070	4" pipe size		3	2.667		1.34	155	15.85	172.19	259
2090	20" thick, 1/2" pipe size		10.70	.748		.61	43.50	4.44	48.55	72.50
2110	3/4" pipe size		10.70	.748		.61	43.50	4.44	48.55	72.50
2130	1" pipe size		8	1		.96	58	5.95	64.91	97.50
2150	1-1/4" pipe size		8	1		.96	58	5.95	64.91	97.50
2170	1-1/2" pipe size		8	1		.96	58	5.95	64.91	97.50
2190	2" pipe size		4	2		1.30	116	11.90	129.20	194
2210	2-1/2" pipe size		4	2		1.30	116	11.90	129.20	194
2230	3" pipe size		4	2		1.30	116	11.90	129.20	194
2250	3-1/2" pipe size		3	2.667		1.68	155	15.85	172.53	259
2270	4" pipe size		2.70	2.963		1.68	172	17.60	191.28	287
2290	24" thick, 1/2" pipe size		9.40	.851		.73	49.50	5.05	55.28	83
2310	3/4" pipe size		9.40	.851		.73	49.50	5.05	55.28	83
2330	1" pipe size		7.10	1.127		1.16	65.50	6.70	73.36	110
2350	1-1/4" pipe size		7.10	1.127		1.16	65.50	6.70	73.36	110
2370	1-1/2" pipe size		7.10	1.127		1.16	65.50	6.70	73.36	110
2390	2" pipe size		3.60	2.222		1.56	129	13.20	143.76	216
2410	2-1/2" pipe size		3.60	2.222		1.56	129	13.20	143.76	216
2430	3" pipe size		3.60	2.222		1.56	129	13.20	143.76	216
2450	3-1/2" pipe size		2.80	2.857		2.01	166	16.95	184.96	278
2470	4" pipe size	▼	2.50	3.200	▼	2.01	186	19	207.01	310
3000	Knockouts to 8' high, metal boxes & enclosures									
3020	With hole saw, 1/2" pipe size	1 Elec	53	.151	Ea.		8.80		8.80	13.55
3040	3/4" pipe size		47	.170			9.90		9.90	15.30
3050	1" pipe size		40	.200			11.65		11.65	18
3060	1-1/4" pipe size		36	.222			12.95		12.95	20
3070	1-1/2" pipe size		32	.250			14.55		14.55	22.50
3080	2" pipe size		27	.296			17.25		17.25	26.50
3090	2-1/2" pipe size		20	.400			23.50		23.50	36
4010	3" pipe size		16	.500			29		29	45
4030	3-1/2" pipe size		13	.615			36		36	55.50
4050	4" pipe size	▼	11	.727	▼		42.50		42.50	65.50

26 05 39 – Underfloor Raceways for Electrical Systems

26 05 39.30 Conduit In Concrete Slab

	26 05 39.30 Conduit In Concrete Slab	Crew	Daily Output	Labor-Hours	Unit	Material	2018 Bare Costs Labor	Equipment	Total	Total Incl O&P
0010	**CONDUIT IN CONCRETE SLAB** Including terminations,									
0020	fittings and supports									
3230	PVC, schedule 40, 1/2" diameter	1 Elec	270	.030	L.F.	.49	1.72		2.21	3.19
3250	3/4" diameter		230	.035		.56	2.02		2.58	3.75
3270	1" diameter		200	.040		.77	2.33		3.10	4.44
3300	1-1/4" diameter		170	.047		1.01	2.74		3.75	5.35
3330	1-1/2" diameter		140	.057		1.21	3.33		4.54	6.50
3350	2" diameter		120	.067		1.52	3.88		5.40	7.65
4350	Rigid galvanized steel, 1/2" diameter		200	.040		1.99	2.33		4.32	5.80
4400	3/4" diameter		170	.047		4.27	2.74		7.01	8.95
4450	1" diameter		130	.062		6.60	3.58		10.18	12.80
4500	1-1/4" diameter		110	.073		4.21	4.23		8.44	11.20
4600	1-1/2" diameter		100	.080		7.30	4.66		11.96	15.25
4800	2" diameter		90	.089		9.05	5.15		14.20	17.95
9000	Minimum labor/equipment charge		4	2	Job		116		116	180

26 05 39.40 Conduit In Trench

	26 05 39.40 Conduit In Trench	Crew	Daily Output	Labor-Hours	Unit	Material	Labor	Equipment	Total	Total Incl O&P
0010	**CONDUIT IN TRENCH** Includes terminations and fittings									
0020	Does not include excavation or backfill, see Section 31 23 16									
0200	Rigid galvanized steel, 2" diameter	1 Elec	150	.053	L.F.	8.70	3.10		11.80	14.40
0400	2-1/2" diameter	"	100	.080		10.70	4.66		15.36	18.95
0600	3" diameter	2 Elec	160	.100		12.55	5.80		18.35	23
0800	3-1/2" diameter		140	.114		16.55	6.65		23.20	28.50
1000	4" diameter		100	.160		18.90	9.30		28.20	35.50
1200	5" diameter		80	.200		36.50	11.65		48.15	58
1400	6" diameter		60	.267		51.50	15.50		67	80.50
9000	Minimum labor/equipment charge	1 Elec	4	2	Job		116		116	180

26 05 80 – Wiring Connections

26 05 80.10 Motor Connections

		Crew	Daily Output	Labor-Hours	Unit	Material	Labor	Equipment	Total	Total Incl O&P
0010	**MOTOR CONNECTIONS**									
0020	Flexible conduit and fittings, 115 volt, 1 phase, up to 1 HP motor	1 Elec	8	1	Ea.	4.87	58		62.87	95.50
9000	Minimum labor/equipment charge	"	4	2	Job		116		116	180

26 05 90 – Residential Applications

26 05 90.10 Residential Wiring

		Crew	Daily Output	Labor-Hours	Unit	Material	Labor	Equipment	Total	Total Incl O&P
0010	**RESIDENTIAL WIRING**									
0020	20' avg. runs and #14/2 wiring incl. unless otherwise noted									
1000	Service & panel, includes 24' SE-AL cable, service eye, meter,									
1010	Socket, panel board, main bkr., ground rod, 15 or 20 amp									
1020	1-pole circuit breakers, and misc. hardware									
1100	100 amp, with 10 branch breakers	1 Elec	1.19	6.723	Ea.	325	390		715	965
1110	With PVC conduit and wire		.92	8.696		360	505		865	1,175
1120	With RGS conduit and wire		.73	10.959		560	640		1,200	1,600
1150	150 amp, with 14 branch breakers		1.03	7.767		850	450		1,300	1,625
1170	With PVC conduit and wire		.82	9.756		920	570		1,490	1,875
1180	With RGS conduit and wire		.67	11.940		1,175	695		1,870	2,375
1200	200 amp, with 18 branch breakers	2 Elec	1.80	8.889		1,000	515		1,515	1,900
1220	With PVC conduit and wire		1.46	10.959		1,075	640		1,715	2,150
1230	With RGS conduit and wire		1.24	12.903		1,400	750		2,150	2,700
1800	Lightning surge suppressor	1 Elec	32	.250		76	14.55		90.55	107
2000	Switch devices									
2100	Single pole, 15 amp, ivory, with a 1-gang box, cover plate,									
2110	Type NM (Romex) cable	1 Elec	17.10	.468	Ea.	15.50	27		42.50	59

490

For customer support on your Commercial Renovation Costs with RSMeans data, call 800.448.8182.

26 05 90 – Residential Applications

26 05 90.10 Residential Wiring	Crew	Daily Output	Labor-Hours	Unit	Material	2018 Bare Costs Labor	Equipment	Total	Total Incl O&P
2120	Type MC cable	1 Elec	14.30	.559	Ea.	25	32.50		57.50	78
2130	EMT & wire		5.71	1.401		36	81.50		117.50	166
2150	3-way, #14/3, type NM cable		14.55	.550		9.80	32		41.80	60.50
2170	Type MC cable		12.31	.650		22.50	38		60.50	83.50
2180	EMT & wire		5	1.600		30	93		123	177
2200	4-way, #14/3, type NM cable		14.55	.550		18.45	32		50.45	70
2220	Type MC cable		12.31	.650		31.50	38		69.50	93
2230	EMT & wire		5	1.600		38.50	93		131.50	187
2250	S.P., 20 amp, #12/2, type NM cable		13.33	.600		11.10	35		46.10	66
2270	Type MC cable		11.43	.700		19.65	40.50		60.15	84.50
2280	EMT & wire		4.85	1.649		33.50	96		129.50	185
2290	S.P. rotary dimmer, 600 W, no wiring		17	.471		30	27.50		57.50	75.50
2300	S.P. rotary dimmer, 600 W, type NM cable		14.55	.550		33.50	32		65.50	86.50
2320	Type MC cable		12.31	.650		43.50	38		81.50	106
2330	EMT & wire		5	1.600		55	93		148	205
2350	3-way rotary dimmer, type NM cable		13.33	.600		21.50	35		56.50	77.50
2370	Type MC cable		11.43	.700		31	40.50		71.50	97
2380	EMT & wire	▼	4.85	1.649	▼	43	96		139	195
2400	Interval timer wall switch, 20 amp, 1-30 min., #12/2									
2410	Type NM cable	1 Elec	14.55	.550	Ea.	59.50	32		91.50	115
2420	Type MC cable		12.31	.650		65.50	38		103.50	131
2430	EMT & wire	▼	5	1.600	▼	82	93		175	234
2500	Decorator style									
2510	S.P., 15 amp, type NM cable	1 Elec	17.10	.468	Ea.	19.35	27		46.35	63.50
2520	Type MC cable		14.30	.559		29	32.50		61.50	82.50
2530	EMT & wire		5.71	1.401		39.50	81.50		121	170
2550	3-way, #14/3, type NM cable		14.55	.550		13.70	32		45.70	64.50
2570	Type MC cable		12.31	.650		26.50	38		64.50	87.50
2580	EMT & wire		5	1.600		33.50	93		126.50	181
2600	4-way, #14/3, type NM cable		14.55	.550		22.50	32		54.50	74
2620	Type MC cable		12.31	.650		35	38		73	97
2630	EMT & wire		5	1.600		42.50	93		135.50	191
2650	S.P., 20 amp, #12/2, type NM cable		13.33	.600		14.95	35		49.95	70.50
2670	Type MC cable		11.43	.700		23.50	40.50		64	89
2680	EMT & wire		4.85	1.649		37.50	96		133.50	190
2700	S.P., slide dimmer, type NM cable		17.10	.468		30.50	27		57.50	75.50
2720	Type MC cable		14.30	.559		40	32.50		72.50	94.50
2730	EMT & wire		5.71	1.401		52	81.50		133.50	183
2750	S.P., touch dimmer, type NM cable		17.10	.468		46.50	27		73.50	93.50
2770	Type MC cable		14.30	.559		56	32.50		88.50	113
2780	EMT & wire		5.71	1.401		68	81.50		149.50	201
2800	3-way touch dimmer, type NM cable		13.33	.600		49.50	35		84.50	108
2820	Type MC cable		11.43	.700		59	40.50		99.50	128
2830	EMT & wire	▼	4.85	1.649	▼	71	96		167	226
3000	Combination devices									
3100	S.P. switch/15 amp recpt., ivory, 1-gang box, plate									
3110	Type NM cable	1 Elec	11.43	.700	Ea.	20.50	40.50		61	85.50
3120	Type MC cable		10	.800		30	46.50		76.50	105
3130	EMT & wire		4.40	1.818		42	106		148	209
3150	S.P. switch/pilot light, type NM cable		11.43	.700		21	40.50		61.50	86
3170	Type MC cable		10	.800		30.50	46.50		77	106
3180	EMT & wire		4.43	1.806		42.50	105		147.50	209
3190	2-S.P. switches, 2-#14/2, no wiring	▼	14	.571	▼	12.85	33.50		46.35	65.50

491

26 05 90.10 Residential Wiring	Crew	Daily Output	Labor-Hours	Unit	Material	2018 Bare Costs Labor	Equipment	Total	Total Incl O&P
3200	2-S.P. switches, 2-#14/2, type NM cables	1 Elec	10	.800	Ea.	22	46.50		68.50	96.50
3220	Type MC cable		8.89	.900		36.50	52.50		89	121
3230	EMT & wire		4.10	1.951		43.50	114		157.50	223
3250	3-way switch/15 amp recpt., #14/3, type NM cable		10	.800		26.50	46.50		73	102
3270	Type MC cable		8.89	.900		39.50	52.50		92	125
3280	EMT & wire		4.10	1.951		46.50	114		160.50	227
3300	2-3 way switches, 2-#14/3, type NM cables		8.89	.900		34.50	52.50		87	119
3320	Type MC cable		8	1		55.50	58		113.50	151
3330	EMT & wire		4	2		53	116		169	239
3350	S.P. switch/20 amp recpt., #12/2, type NM cable		10	.800		28	46.50		74.50	103
3370	Type MC cable		8.89	.900		34	52.50		86.50	119
3380	EMT & wire		4.10	1.951		50.50	114		164.50	231
3400	Decorator style									
3410	S.P. switch/15 amp recpt., type NM cable	1 Elec	11.43	.700	Ea.	24	40.50		64.50	89.50
3420	Type MC cable		10	.800		34	46.50		80.50	109
3430	EMT & wire		4.40	1.818		45.50	106		151.50	214
3450	S.P. switch/pilot light, type NM cable		11.43	.700		25	40.50		65.50	90.50
3470	Type MC cable		10	.800		34.50	46.50		81	110
3480	EMT & wire		4.40	1.818		46.50	106		152.50	214
3500	2-S.P. switches, 2-#14/2, type NM cables		10	.800		26	46.50		72.50	101
3520	Type MC cable		8.89	.900		40	52.50		92.50	125
3530	EMT & wire		4.10	1.951		47.50	114		161.50	227
3550	3-way/15 amp recpt., #14/3, type NM cable		10	.800		30.50	46.50		77	106
3570	Type MC cable		8.89	.900		43.50	52.50		96	129
3580	EMT & wire		4.10	1.951		50.50	114		164.50	231
3650	2-3 way switches, 2-#14/3, type NM cables		8.89	.900		38.50	52.50		91	124
3670	Type MC cable		8	1		59	58		117	155
3680	EMT & wire		4	2		57	116		173	243
3700	S.P. switch/20 amp recpt., #12/2, type NM cable		10	.800		32	46.50		78.50	107
3720	Type MC cable		8.89	.900		38	52.50		90.50	123
3730	EMT & wire		4.10	1.951		54.50	114		168.50	235
4000	Receptacle devices									
4010	Duplex outlet, 15 amp recpt., ivory, 1-gang box, plate									
4015	Type NM cable	1 Elec	14.55	.550	Ea.	8.40	32		40.40	58.50
4020	Type MC cable		12.31	.650		17.95	38		55.95	78.50
4030	EMT & wire		5.33	1.501		28.50	87.50		116	167
4050	With #12/2, type NM cable		12.31	.650		9.50	38		47.50	69
4070	Type MC cable		10.67	.750		18.05	43.50		61.55	87.50
4080	EMT & wire		4.71	1.699		32	99		131	189
4100	20 amp recpt., #12/2, type NM cable		12.31	.650		17.25	38		55.25	77.50
4120	Type MC cable		10.67	.750		26	43.50		69.50	96
4130	EMT & wire		4.71	1.699		40	99		139	197
4140	For GFI see Section 26 05 90.10 line 4300 below									
4150	Decorator style, 15 amp recpt., type NM cable	1 Elec	14.55	.550	Ea.	12.25	32		44.25	63
4170	Type MC cable		12.31	.650		22	38		60	82.50
4180	EMT & wire		5.33	1.501		32.50	87.50		120	171
4200	With #12/2, type NM cable		12.31	.650		13.35	38		51.35	73
4220	Type MC cable		10.67	.750		22	43.50		65.50	91.50
4230	EMT & wire		4.71	1.699		36	99		135	193
4250	20 amp recpt., #12/2, type NM cable		12.31	.650		21	38		59	82
4270	Type MC cable		10.67	.750		29.50	43.50		73	100
4280	EMT & wire		4.71	1.699		44	99		143	201
4300	GFI, 15 amp recpt., type NM cable		12.31	.650		21	38		59	81.50

26 05 90 – Residential Applications

26 05 90.10 Residential Wiring	Crew	Daily Output	Labor-Hours	Unit	Material	2018 Bare Costs Labor	Equipment	Total	Total Incl O&P
4320	Type MC cable	1 Elec	10.67	.750	Ea.	30.50	43.50		74	101
4330	EMT & wire		4.71	1.699		41	99		140	198
4350	GFI with #12/2, type NM cable		10.67	.750		22	43.50		65.50	91.50
4370	Type MC cable		9.20	.870		30.50	50.50		81	112
4380	EMT & wire		4.21	1.900		44.50	111		155.50	220
4400	20 amp recpt., #12/2, type NM cable		10.67	.750		48.50	43.50		92	121
4420	Type MC cable		9.20	.870		57	50.50		107.50	141
4430	EMT & wire		4.21	1.900		71	111		182	249
4500	Weather-proof cover for above receptacles, add	↓	32	.250	↓	2.10	14.55		16.65	25
4550	Air conditioner outlet, 20 amp-240 volt recpt.									
4560	30' of #12/2, 2 pole circuit breaker									
4570	Type NM cable	1 Elec	10	.800	Ea.	58	46.50		104.50	136
4580	Type MC cable		9	.889		68.50	51.50		120	156
4590	EMT & wire		4	2		81	116		197	270
4600	Decorator style, type NM cable		10	.800		63	46.50		109.50	141
4620	Type MC cable		9	.889		73	51.50		124.50	161
4630	EMT & wire	↓	4	2	↓	86	116		202	275
4650	Dryer outlet, 30 amp-240 volt recpt., 20' of #10/3									
4660	2 pole circuit breaker									
4670	Type NM cable	1 Elec	6.41	1.248	Ea.	53.50	72.50		126	171
4680	Type MC cable		5.71	1.401		62.50	81.50		144	195
4690	EMT & wire	↓	3.48	2.299	↓	75	134		209	289
4700	Range outlet, 50 amp-240 volt recpt., 30' of #8/3									
4710	Type NM cable	1 Elec	4.21	1.900	Ea.	82	111		193	261
4720	Type MC cable		4	2		128	116		244	320
4730	EMT & wire		2.96	2.703		106	157		263	360
4750	Central vacuum outlet, type NM cable		6.40	1.250		50	73		123	167
4770	Type MC cable		5.71	1.401		64.50	81.50		146	197
4780	EMT & wire	↓	3.48	2.299	↓	83	134		217	298
4800	30 amp-110 volt locking recpt., #10/2 circ. bkr.									
4810	Type NM cable	1 Elec	6.20	1.290	Ea.	58	75		133	180
4820	Type MC cable		5.40	1.481		76	86		162	217
4830	EMT & wire	↓	3.20	2.500	↓	93.50	146		239.50	330
4900	Low voltage outlets									
4910	Telephone recpt., 20' of 4/C phone wire	1 Elec	26	.308	Ea.	8	17.90		25.90	36.50
4920	TV recpt., 20' of RG59U coax wire, F type connector	"	16	.500	"	15.55	29		44.55	62
4950	Door bell chime, transformer, 2 buttons, 60' of bellwire									
4970	Economy model	1 Elec	11.50	.696	Ea.	52	40.50		92.50	120
4980	Custom model		11.50	.696		94.50	40.50		135	167
4990	Luxury model, 3 buttons	↓	9.50	.842	↓	185	49		234	280
6000	Lighting outlets									
6050	Wire only (for fixture), type NM cable	1 Elec	32	.250	Ea.	5.75	14.55		20.30	29
6070	Type MC cable		24	.333		11.70	19.40		31.10	43
6080	EMT & wire		10	.800		21.50	46.50		68	96
6100	Box (4"), and wire (for fixture), type NM cable		25	.320		14.55	18.60		33.15	45
6120	Type MC cable		20	.400		20.50	23.50		44	58.50
6130	EMT & wire	↓	11	.727	↓	30.50	42.50		73	99
6200	Fixtures (use with line 6050 or 6100 above)									
6210	Canopy style, economy grade	1 Elec	40	.200	Ea.	23	11.65		34.65	43.50
6220	Custom grade		40	.200		53.50	11.65		65.15	77
6250	Dining room chandelier, economy grade		19	.421		72	24.50		96.50	117
6270	Luxury grade		15	.533		955	31		986	1,100
6310	Kitchen fixture (fluorescent), economy grade		30	.267	↓	71.50	15.50		87	103

For customer support on your Commercial Renovation Costs with RSMeans data, call 800.448.8182.

493

26 05 90.10 Residential Wiring	Crew	Daily Output	Labor-Hours	Unit	Material	2018 Bare Costs Labor	Equipment	Total	Total Incl O&P
6320	Custom grade	1 Elec	25	.320	Ea.	137	18.60		155.60	180
6350	Outdoor, wall mounted, economy grade		30	.267		28.50	15.50		44	55.50
6360	Custom grade		30	.267		121	15.50		136.50	157
6370	Luxury grade		25	.320		248	18.60		266.60	300
6410	Outdoor PAR floodlights, 1 lamp, 150 watt		20	.400		26	23.50		49.50	64.50
6420	2 lamp, 150 watt each		20	.400		45.50	23.50		69	86
6425	Motion sensing, 2 lamp, 150 watt each		20	.400		88.50	23.50		112	134
6430	For infrared security sensor, add		32	.250		99.50	14.55		114.05	133
6450	Outdoor, quartz-halogen, 300 watt flood		20	.400		33	23.50		56.50	72.50
6600	Recessed downlight, round, pre-wired, 50 or 75 watt trim		30	.267		62.50	15.50		78	92.50
6610	With shower light trim		30	.267		86	15.50		101.50	119
6620	With wall washer trim		28	.286		79.50	16.65		96.15	113
6630	With eye-ball trim		28	.286		66.50	16.65		83.15	98.50
6700	Porcelain lamp holder		40	.200		2.92	11.65		14.57	21
6710	With pull switch		40	.200		11.30	11.65		22.95	30.50
6750	Fluorescent strip, 2-20 watt tube, wrap around diffuser, 24"		24	.333		46	19.40		65.40	81
6760	1-34 watt tube, 48"		24	.333		105	19.40		124.40	146
6770	2-34 watt tubes, 48"		20	.400		122	23.50		145.50	170
6800	Bathroom heat lamp, 1-250 watt		28	.286		31	16.65		47.65	60
6810	2-250 watt lamps		28	.286		73	16.65		89.65	106
6820	For timer switch, see Section 26 05 90.10 line 2400									
6900	Outdoor post lamp, incl. post, fixture, 35' of #14/2									
6910	Type NM cable	1 Elec	3.50	2.286	Ea.	300	133		433	535
6920	Photo-eye, add		27	.296		27.50	17.25		44.75	56.50
6950	Clock dial time switch, 24 hr., w/enclosure, type NM cable		11.43	.700		68.50	40.50		109	138
6970	Type MC cable		11	.727		78	42.50		120.50	152
6980	EMT & wire		4.85	1.649		88.50	96		184.50	246
7000	Alarm systems									
7050	Smoke detectors, box, #14/3, type NM cable	1 Elec	14.55	.550	Ea.	33.50	32		65.50	86.50
7070	Type MC cable		12.31	.650		44	38		82	107
7080	EMT & wire		5	1.600		51	93		144	200
7090	For relay output to security system, add					10.75			10.75	11.85
8000	Residential equipment									
8050	Disposal hook-up, incl. switch, outlet box, 3' of flex									
8060	20 amp-1 pole circ. bkr., and 25' of #12/2									
8070	Type NM cable	1 Elec	10	.800	Ea.	28	46.50		74.50	103
8080	Type MC cable		8	1		37.50	58		95.50	131
8090	EMT & wire		5	1.600		54.50	93		147.50	204
8100	Trash compactor or dishwasher hook-up, incl. outlet box,									
8110	3' of flex, 15 amp-1 pole circ. bkr., and 25' of #14/2									
8120	Type NM cable	1 Elec	10	.800	Ea.	15.50	46.50		62	89
8130	Type MC cable		8	1		26.50	58		84.50	119
8140	EMT & wire		5	1.600		41	93		134	189
8150	Hot water sink dispenser hook-up, use line 8100									
8200	Vent/exhaust fan hook-up, type NM cable	1 Elec	32	.250	Ea.	5.75	14.55		20.30	29
8220	Type MC cable		24	.333		11.70	19.40		31.10	43
8230	EMT & wire		10	.800		21.50	46.50		68	96
8250	Bathroom vent fan, 50 CFM (use with above hook-up)									
8260	Economy model	1 Elec	15	.533	Ea.	19.40	31		50.40	69.50
8270	Low noise model		15	.533		44	31		75	96.50
8280	Custom model		12	.667		121	39		160	193
8300	Bathroom or kitchen vent fan, 110 CFM									
8310	Economy model	1 Elec	15	.533	Ea.	68	31		99	123

26 05 90 – Residential Applications

26 05 90.10 Residential Wiring	Crew	Daily Output	Labor-Hours	Unit	Material	2018 Bare Costs Labor	2018 Bare Costs Equipment	Total	Total Incl O&P
8320 Low noise model	1 Elec	15	.533	Ea.	92	31		123	149
8350 Paddle fan, variable speed (w/o lights)									
8360 Economy model (AC motor)	1 Elec	10	.800	Ea.	110	46.50		156.50	193
8362 With light kit		10	.800		149	46.50		195.50	236
8370 Custom model (AC motor)		10	.800		264	46.50		310.50	365
8372 With light kit		10	.800		305	46.50		351.50	405
8380 Luxury model (DC motor)		8	1		315	58		373	440
8382 With light kit		8	1		355	58		413	480
8390 Remote speed switch for above, add	↓	12	.667	↓	33.50	39		72.50	97
8500 Whole house exhaust fan, ceiling mount, 36", variable speed									
8510 Remote switch, incl. shutters, 20 amp-1 pole circ. bkr.									
8520 30' of #12/2, type NM cable	1 Elec	4	2	Ea.	1,225	116		1,341	1,525
8530 Type MC cable		3.50	2.286		1,250	133		1,383	1,575
8540 EMT & wire	↓	3	2.667	↓	1,250	155		1,405	1,625
8600 Whirlpool tub hook-up, incl. timer switch, outlet box									
8610 3' of flex, 20 amp-1 pole GFI circ. bkr.									
8620 30' of #12/2, type NM cable	1 Elec	5	1.600	Ea.	125	93		218	282
8630 Type MC cable		4.20	1.905		131	111		242	315
8640 EMT & wire	↓	3.40	2.353	↓	145	137		282	370
8650 Hot water heater hook-up, incl. 1-2 pole circ. bkr., box;									
8660 3' of flex, 20' of #10/2, type NM cable	1 Elec	5	1.600	Ea.	28.50	93		121.50	176
8670 Type MC cable		4.20	1.905		41	111		152	216
8680 EMT & wire	↓	3.40	2.353	↓	47.50	137		184.50	265
9000 Heating/air conditioning									
9050 Furnace/boiler hook-up, incl. firestat, local on-off switch									
9060 Emergency switch, and 40' of type NM cable	1 Elec	4	2	Ea.	52	116		168	237
9070 Type MC cable		3.50	2.286		66	133		199	278
9080 EMT & wire	↓	1.50	5.333	↓	89	310		399	580
9100 Air conditioner hook up, incl. local 60 amp disc. switch									
9110 3' sealtite, 40 amp, 2 pole circuit breaker									
9130 40' of #8/2, type NM cable	1 Elec	3.50	2.286	Ea.	134	133		267	350
9140 Type MC cable		3	2.667		201	155		356	460
9150 EMT & wire	↓	1.30	6.154	↓	182	360		542	755
9200 Heat pump hook-up, 1-40 & 1-100 amp 2 pole circ. bkr.									
9210 Local disconnect switch, 3' sealtite									
9220 40' of #8/2 & 30' of #3/2									
9230 Type NM cable	1 Elec	1.30	6.154	Ea.	475	360		835	1,075
9240 Type MC cable		1.08	7.407		525	430		955	1,250
9250 EMT & wire	↓	.94	8.511	↓	490	495		985	1,300
9500 Thermostat hook-up, using low voltage wire									
9520 Heating only, 25' of #18-3	1 Elec	24	.333	Ea.	5.70	19.40		25.10	36.50
9530 Heating/cooling, 25' of #18-4	"	20	.400	"	7.15	23.50		30.65	44

26 22 Low-Voltage Transformers

26 22 13 – Low-Voltage Distribution Transformers

26 22 13.10 Transformer, Dry-Type

26 22 13.10 Transformer, Dry-Type		Crew	Daily Output	Labor-Hours	Unit	Material	2018 Bare Costs Labor	Equipment	Total	Total Incl O&P
0010	**TRANSFORMER, DRY-TYPE**									
0050	Single phase, 240/480 volt primary, 120/240 volt secondary									
0100	1 kVA	1 Elec	2	4	Ea.	380	233		613	780
0300	2 kVA		1.60	5		590	291		881	1,100
0500	3 kVA		1.40	5.714		735	335		1,070	1,325
0700	5 kVA		1.20	6.667		1,000	390		1,390	1,700
0900	7.5 kVA	2 Elec	2.20	7.273		1,375	425		1,800	2,175
1100	10 kVA		1.60	10		1,575	580		2,155	2,625
1300	15 kVA		1.20	13.333		2,125	775		2,900	3,525
2300	3 phase, 480 volt primary, 120/208 volt secondary									
2310	Ventilated, 3 kVA	1 Elec	1	8	Ea.	1,125	465		1,590	1,975
2700	6 kVA		.80	10		1,275	580		1,855	2,300
2900	9 kVA		.70	11.429		1,275	665		1,940	2,425
3100	15 kVA	2 Elec	1.10	14.545		1,625	845		2,470	3,075
3300	30 kVA		.90	17.778		1,600	1,025		2,625	3,375
3500	45 kVA		.80	20		1,900	1,175		3,075	3,900
9000	Minimum labor/equipment charge	1 Elec	1	8	Job		465		465	720
9300	Energy efficient transformer 3 ph									
9400	15 kVA, 480VAC delta, 208Y/120VAC	R-3	1.50	13.333	Ea.	1,975	775	86.50	2,836.50	3,475
9405	30 kVA, 480VAC delta, 208Y/120VAC		1.50	13.333		2,525	775	86.50	3,386.50	4,075
9410	45 kVA, 480VAC delta, 208Y/120VAC		1.50	13.333		3,050	775	86.50	3,911.50	4,650
9415	75 kVA, 480VAC delta, 208Y/120VAC		1.25	16		4,500	930	104	5,534	6,525
9420	112 kVA, 480VAC delta, 208Y/120VAC		1.25	16		6,425	930	104	7,459	8,650
9425	150 kVA, 480VAC delta, 208Y/120VAC		1.25	16		10,700	930	104	11,734	13,400
9430	225 kVA, 480VAC delta, 208Y/120VAC		1.25	16		10,700	930	104	11,734	13,400
9435	300 kVA, 480VAC delta, 208Y/120VAC		1.25	16		15,900	930	104	16,934	19,100

26 24 Switchboards and Panelboards

26 24 16 – Panelboards

26 24 16.20 Panelboard and Load Center Circuit Breakers

26 24 16.20 Panelboard and Load Center Circuit Breakers		Crew	Daily Output	Labor-Hours	Unit	Material	2018 Bare Costs Labor	Equipment	Total	Total Incl O&P
0010	**PANELBOARD AND LOAD CENTER CIRCUIT BREAKERS**									
0050	Bolt-on, 10,000 amp I.C., 120 volt, 1 pole									
0100	15-50 amp	1 Elec	10	.800	Ea.	19.20	46.50		65.70	93
0200	60 amp		8	1		19.20	58		77.20	111
0300	70 amp		8	1		28	58		86	121
0350	240 volt, 2 pole									
0400	15-50 amp	1 Elec	8	1	Ea.	51	58		109	147
0500	60 amp		7.50	1.067		34.50	62		96.50	134
0600	80-100 amp		5	1.600		116	93		209	271
0700	3 pole, 15-60 amp		6.20	1.290		140	75		215	270
0800	70 amp		5	1.600		180	93		273	340
0900	80-100 amp		3.60	2.222		195	129		324	415
1000	22,000 amp I.C., 240 volt, 2 pole, 70-225 amp		2.70	2.963		555	172		727	880
1100	3 pole, 70-225 amp		2.30	3.478		275	202		477	615
1200	14,000 amp I.C., 277 volts, 1 pole, 15-30 amp		8	1		36	58		94	130
1300	22,000 amp I.C., 480 volts, 2 pole, 70-225 amp		2.70	2.963		505	172		677	820
1400	3 pole, 70-225 amp		2.30	3.478		770	202		972	1,150
2060	Plug-in tandem, 120/240 V, 2-15 A, 1 pole		11	.727		21	42.50		63.50	88.50
2070	1-15 A & 1-20 A		11	.727		17.60	42.50		60.10	85
2080	2-20 A		11	.727		21	42.50		63.50	88.50
2082	Arc fault circuit interrupter, 120/240 V, 1-15 A & 1-20 A, 1 pole		11	.727		56	42.50		98.50	127

26 24 Switchboards and Panelboards

26 24 16 – Panelboards

26 24 16.20 Panelboard and Load Center Circuit Breakers	Crew	Daily Output	Labor-Hours	Unit	Material	2018 Bare Costs Labor	2018 Bare Costs Equipment	Total	Total Incl O&P
5110	NEMA 1 enclosure only, 600V, 3 p, 14k AIC, 100A	1 Elec	2.20	3.636	Ea.	179	212		391	520
9000	Minimum labor/equipment charge	↓	3	2.667	Job		155		155	240

26 24 16.30 Panelboards Commercial Applications

		Crew	Daily Output	Labor-Hours	Unit	Material	2018 Bare Costs Labor	2018 Bare Costs Equipment	Total	Total Incl O&P
0010	**PANELBOARDS COMMERCIAL APPLICATIONS**									
0050	NQOD, w/20 amp 1 pole bolt-on circuit breakers									
0100	3 wire, 120/240 volts, 100 amp main lugs									
0150	10 circuits	1 Elec	1	8	Ea.	925	465		1,390	1,750
0200	14 circuits		.88	9.091		1,025	530		1,555	1,950
0250	18 circuits		.75	10.667		1,125	620		1,745	2,175
0300	20 circuits		.65	12.308		1,250	715		1,965	2,475
0600	4 wire, 120/208 volts, 100 amp main lugs, 12 circuits		1	8		995	465		1,460	1,825
0650	16 circuits		.75	10.667		1,150	620		1,770	2,200
0700	20 circuits		.65	12.308		1,325	715		2,040	2,550
0750	24 circuits		.60	13.333		1,275	775		2,050	2,600
0800	30 circuits		.53	15.094		1,625	880		2,505	3,150
0850	225 amp main lugs, 32 circuits	2 Elec	.90	17.778		1,825	1,025		2,850	3,625
0900	34 circuits		.84	19.048		1,875	1,100		2,975	3,750
0950	36 circuits		.80	20		1,925	1,175		3,100	3,900
1000	42 circuits		.68	23.529		2,150	1,375		3,525	4,475
1010	400 amp main lugs, 42 circs	↓	.68	23.529	↓	2,150	1,375		3,525	4,475
1200	NEHB, w/20 amp, 1 pole bolt-on circuit breakers									
1250	4 wire, 277/480 volts, 100 amp main lugs, 12 circuits	1 Elec	.88	9.091	Ea.	1,350	530		1,880	2,325
1300	20 circuits	"	.60	13.333		2,025	775		2,800	3,425
1350	225 amp main lugs, 24 circuits	2 Elec	.90	17.778		2,300	1,025		3,325	4,125
1400	30 circuits		.80	20		2,750	1,175		3,925	4,825
1448	32 circuits		4.90	3.265		3,200	190		3,390	3,800
1450	36 circuits	↓	.72	22.222	↓	3,200	1,300		4,500	5,500
1600	NQOD panel, w/20 amp, 1 pole, circuit breakers									
2000	4 wire, 120/208 volts with main circuit breaker									
2050	100 amp main, 24 circuits	1 Elec	.47	17.021	Ea.	1,925	990		2,915	3,625
2100	30 circuits	"	.40	20		2,075	1,175		3,250	4,100
2200	225 amp main, 32 circuits	2 Elec	.72	22.222		3,475	1,300		4,775	5,800
2250	42 circuits		.56	28.571		3,575	1,675		5,250	6,500
2300	400 amp main, 42 circuits		.48	33.333		5,075	1,950		7,025	8,600
2350	600 amp main, 42 circuits	↓	.40	40	↓	7,500	2,325		9,825	11,900
2400	NEHB, with 20 amp, 1 pole circuit breaker									
2450	4 wire, 277/480 volts with main circuit breaker									
2500	100 amp main, 24 circuits	1 Elec	.42	19.048	Ea.	2,650	1,100		3,750	4,600
2550	30 circuits	"	.38	21.053		3,100	1,225		4,325	5,300
2600	225 amp main, 30 circuits	2 Elec	.72	22.222		3,875	1,300		5,175	6,275
2650	42 circuits	"	.56	28.571	↓	4,775	1,675		6,450	7,825
9000	Minimum labor/equipment charge	1 Elec	1	8	Job		465		465	720

26 27 13 – Electricity Metering

26 27 13.10 Meter Centers and Sockets	Crew	Daily Output	Labor-Hours	Unit	Material	2018 Bare Costs Labor	Equipment	Total	Total Incl O&P
0010	**METER CENTERS AND SOCKETS**								
0100	Sockets, single position, 4 terminal, 100 amp	1 Elec	3.20	2.500	Ea.	45	146		191	275
0200	150 amp		2.30	3.478		62	202		264	385
0300	200 amp		1.90	4.211		96	245		341	485
0400	Transformer rated, 20 amp		3.20	2.500		162	146		308	405
0500	Double position, 4 terminal, 100 amp		2.80	2.857		212	166		378	490
0600	150 amp		2.10	3.810		249	222		471	615
0700	200 amp		1.70	4.706		515	274		789	990
1100	Meter centers and sockets, three phase, single pos, 7 terminal, 100 amp		2.80	2.857		105	166		271	375
1200	200 amp		2.10	3.810		214	222		436	575
1400	400 amp		1.70	4.706		810	274		1,084	1,325
9000	Minimum labor/equipment charge		3	2.667	Job		155		155	240

26 27 16 – Electrical Cabinets and Enclosures

26 27 16.10 Cabinets

26 27 16.10 Cabinets	Crew	Daily Output	Labor-Hours	Unit	Material	2018 Bare Costs Labor	Equipment	Total	Total Incl O&P
0010	**CABINETS**								
7000	Cabinets, current transformer								
7050	Single door, 24" H x 24" W x 10" D	1 Elec	1.60	5	Ea.	165	291		456	630
7100	30" H x 24" W x 10" D		1.30	6.154		179	360		539	750
7150	36" H x 24" W x 10" D		1.10	7.273		380	425		805	1,075
7200	30" H x 30" W x 10" D		1	8		246	465		711	990
7250	36" H x 30" W x 10" D		.90	8.889		310	515		825	1,150
7300	36" H x 36" W x 10" D		.80	10		315	580		895	1,250
7500	Double door, 48" H x 36" W x 10" D		.60	13.333		575	775		1,350	1,825
7550	24" H x 24" W x 12" D		1	8		179	465		644	915
9990	Minimum labor/equipment charge		2	4	Job		233		233	360

26 27 23 – Indoor Service Poles

26 27 23.40 Surface Raceway

26 27 23.40 Surface Raceway	Crew	Daily Output	Labor-Hours	Unit	Material	2018 Bare Costs Labor	Equipment	Total	Total Incl O&P
0010	**SURFACE RACEWAY**								
0090	Metal, straight section								
0100	No. 500	1 Elec	100	.080	L.F.	1.07	4.66		5.73	8.40
0110	No. 700		100	.080		1.19	4.66		5.85	8.50
0400	No. 1500, small pancake		90	.089		2.20	5.15		7.35	10.40
0600	No. 2000, base & cover, blank		90	.089		2.24	5.15		7.39	10.45
0800	No. 3000, base & cover, blank		75	.107		4.28	6.20		10.48	14.30
1000	No. 4000, base & cover, blank		65	.123		6.95	7.15		14.10	18.70
1200	No. 6000, base & cover, blank		50	.160		11.75	9.30		21.05	27.50
2400	Fittings, elbows, No. 500		40	.200	Ea.	1.94	11.65		13.59	20
2800	Elbow cover, No. 2000		40	.200		3.64	11.65		15.29	22
2880	Tee, No. 500		42	.190		3.74	11.10		14.84	21
2900	No. 2000		27	.296		12.10	17.25		29.35	40
3000	Switch box, No. 500		16	.500		12.50	29		41.50	59
3400	Telephone outlet, No. 1500		16	.500		14.50	29		43.50	61
3600	Junction box, No. 1500		16	.500		10.30	29		39.30	56.50
3800	Plugmold wired sections, No. 2000								
4000	1 circuit, 6 outlets, 3' long	1 Elec	8	1	Ea.	38	58		96	132
4100	2 circuits, 8 outlets, 6' long	"	5.30	1.509	"	62.50	88		150.50	205
9300	Non-metallic, straight section								
9310	7/16" x 7/8", base & cover, blank	1 Elec	160	.050	L.F.	1.72	2.91		4.63	6.40
9320	Base & cover w/adhesive		160	.050		1.45	2.91		4.36	6.10
9340	7/16" x 1-5/16", base & cover, blank		145	.055		1.88	3.21		5.09	7.05
9350	Base & cover w/adhesive		145	.055		2.21	3.21		5.42	7.40
9370	11/16" x 2-1/4", base & cover, blank		130	.062		2.66	3.58		6.24	8.50

26 27 Low-Voltage Distribution Equipment

26 27 23 – Indoor Service Poles

26 27 23.40 Surface Raceway

		Crew	Daily Output	Labor-Hours	Unit	Material	2018 Bare Costs Labor	Equipment	Total	Total Incl O&P
9380	Base & cover w/adhesive	1 Elec	130	.062	L.F.	3.07	3.58		6.65	8.95
9385	1-11/16" x 5-1/4", two compartment base & cover w/screws		80	.100	↓	9.90	5.80		15.70	19.90
9400	Fittings, elbows, 7/16" x 7/8"		50	.160	Ea.	1.92	9.30		11.22	16.50
9410	7/16" x 1-5/16"		45	.178		1.99	10.35		12.34	18.20
9420	11/16" x 2-1/4"		40	.200		2.15	11.65		13.80	20.50
9425	1-11/16" x 5-1/4"		28	.286		11.85	16.65		28.50	38.50
9430	Tees, 7/16" x 7/8"		35	.229		2.48	13.30		15.78	23
9440	7/16" x 1-5/16"		32	.250		2.54	14.55		17.09	25.50
9450	11/16" x 2-1/4"		30	.267		2.65	15.50		18.15	27
9455	1-11/16" x 5-1/4"		24	.333		18.85	19.40		38.25	50.50
9460	Cover clip, 7/16" x 7/8"		80	.100		.50	5.80		6.30	9.55
9470	7/16" x 1-5/16"		72	.111		.45	6.45		6.90	10.50
9480	11/16" x 2-1/4"		64	.125		.78	7.30		8.08	12.10
9484	1-11/16" x 5-1/4"		42	.190		2.77	11.10		13.87	20
9486	Wire clip, 1-11/16" x 5-1/4"		68	.118		.47	6.85		7.32	11.10
9490	Blank end, 7/16" x 7/8"		50	.160		.72	9.30		10.02	15.20
9500	7/16" x 1-5/16"		45	.178		.80	10.35		11.15	16.90
9510	11/16" x 2-1/4"		40	.200		1.20	11.65		12.85	19.30
9515	1-11/16" x 5-1/4"		38	.211		5.55	12.25		17.80	25
9520	Round fixture box, 5-1/2" diam. x 1"		25	.320		11.30	18.60		29.90	41.50
9530	Device box, 1 gang		30	.267		5.10	15.50		20.60	29.50
9540	2 gang		25	.320	↓	7.65	18.60		26.25	37.50
9990	Minimum labor/equipment charge	↓	5	1.600	Job		93		93	144

26 27 26 – Wiring Devices

26 27 26.20 Wiring Devices Elements

		Crew	Daily Output	Labor-Hours	Unit	Material	2018 Bare Costs Labor	Equipment	Total	Total Incl O&P
0010	**WIRING DEVICES ELEMENTS**									
0200	Toggle switch, quiet type, single pole, 15 amp	1 Elec	40	.200	Ea.	.49	11.65		12.14	18.55
0500	20 amp		27	.296		2.91	17.25		20.16	29.50
0550	Rocker, 15 amp		40	.200		2.31	11.65		13.96	20.50
0560	20 amp		27	.296		6.45	17.25		23.70	33.50
0600	3 way, 15 amp		23	.348		1.52	20		21.52	33
0850	Rocker, 15 amp		23	.348		5.60	20		25.60	37.50
0860	20 amp		18	.444		8.55	26		34.55	49.50
0900	4 way, 15 amp		15	.533		10.70	31		41.70	60
1030	Rocker, 15 amp		15	.533		8.75	31		39.75	57.50
1040	20 amp		11	.727		17.50	42.50		60	85
1650	Dimmer switch, 120 volt, incandescent, 600 watt, 1 pole G		16	.500		22.50	29		51.50	70
2460	Receptacle, duplex, 120 volt, grounded, 15 amp		40	.200		1.30	11.65		12.95	19.45
2470	20 amp		27	.296		9.05	17.25		26.30	36.50
2480	Ground fault interrupting, 15 amp		27	.296		13.75	17.25		31	41.50
2482	20 amp		27	.296		40	17.25		57.25	70.50
2490	Dryer, 30 amp		15	.533		4.44	31		35.44	53
2500	Range, 50 amp		11	.727		10.60	42.50		53.10	77
2600	Wall plates, stainless steel, 1 gang		80	.100		2.56	5.80		8.36	11.80
2800	2 gang		53	.151		4.33	8.80		13.13	18.30
3200	Lampholder, keyless		26	.308		16.85	17.90		34.75	46
3400	Pullchain with receptacle		22	.364		20.50	21		41.50	55
4870	Receptacle box assembly, cast aluminum, 60A, 4P5W, 3P, 120/208V, NEMA PR4		22	.364		590	21		611	680
4875	100A, 4P5W, 3P, 347/600V, NEMA PR4		22	.364		795	21		816	910
4880	200A, 4P5W, 3P, 347/600V, NEMA PR4		22	.364		1,700	21		1,721	1,900
8100	Pin/sleeve, 20A, 480V, DSN1, male inlet, 3 pole, 3 phase, 4 wire		16	.500		43.50	29		72.50	93
8110	20A, 125V, DSN20, male inlet, 3 pole, 3 phase, 3 wire		16	.500	↓	35	29		64	83.50

26 27 26 – Wiring Devices

26 27 26.20 Wiring Devices Elements

	Crew	Daily Output	Labor-Hours	Unit	Material	2018 Bare Costs Labor	Equipment	Total	Total Incl O&P
8120	60A, 480V, DS60, male inlet, 3 pole, 3 phase, 3 wire	1 Elec	16	.500	Ea.	289	29		318	365
8130	Pin/sleeve, 20A, 480V, DSN10, female RCPT, 3P, 3 phase, 4 wire		16	.500		78.50	29		107.50	132
8140	30A, 480V, DS30, female RCPT, 3P, 3 phase, 4 wire		16	.500		310	29		339	385
8500	Wiring device terminal strip, 2 pole, 12-2AWG, 300VAC, CSA, 600VAC, 10A		36	.222		3.16	12.95		16.11	23.50
8505	4 pole, 12-2AWG, 300VAC, CSA, 600VAC, 10A		36	.222		4.46	12.95		17.41	25
8510	6 pole, 12-2AWG, 300VAC, CSA, 600VAC, 10A		36	.222		5.05	12.95		18	25.50
8515	8 pole, 12-2AWG, 300VAC, CSA, 600VAC, 10A		30	.267		5.65	15.50		21.15	30.50
8520	10 pole, 12-2AWG, 300VAC, CSA, 600VAC, 10A		30	.267		6.25	15.50		21.75	31
8525	12 pole, 12-2AWG, 300VAC, CSA, 600VAC, 10A		30	.267		6.90	15.50		22.40	31.50
8530	Wiring device terminal strip jumper, 10A		36	.222		.73	12.95		13.68	21
8535	Wiring device terminal strip, 2 pole, 12-22AWG, 300VAC, CSA, 600VAC, 20A		34	.235		3.42	13.70		17.12	25
8540	4 pole, 12-22AWG, 300VAC, CSA, 600VAC, 20A		34	.235		4.70	13.70		18.40	26
8545	6 pole, 12-22AWG, 300VAC, CSA, 600VAC, 20A		34	.235		5.30	13.70		19	27
8550	8 pole, 12-22AWG, 300VAC, CSA, 600VAC, 20A		30	.267		6	15.50		21.50	30.50
8555	10 pole, 12-22AWG, 300VAC, CSA, 600VAC, 20A		30	.267		6.60	15.50		22.10	31.50
8560	12 pole, 12-22AWG, 300VAC, CSA, 600VAC, 20A		30	.267		7.25	15.50		22.75	32
8565	Wiring device terminal strip jumper, 20A		34	.235		.89	13.70		14.59	22
8570	Wiring device terminal shorting type no cover, 4 pole, 6-10AWG, 600VAC, 50A		30	.267		16.55	15.50		32.05	42.50
8575	6 pole, 6-10AWG, 600VAC, 50A		30	.267		21.50	15.50		37	47.50
8580	8 pole, 6-10AWG, 600VAC, 50A		30	.267		24.50	15.50		40	51
8585	12 pole, 6-10AWG, 600VAC, 50A		30	.267		27	15.50		42.50	54
8590	Wiring device terminal 2 shorting pin 1/2 cover, 2 pole, 8-4AWG, 600V, 45A		30	.267		17.15	15.50		32.65	43
8595	4 pole, 8-4AWG, 600V, 45A		30	.267		21.50	15.50		37	47.50
8600	6 pole, 8-4AWG, 600V, 45A		30	.267		26.50	15.50		42	53.50
8605	8 pole, 8-4AWG, 600V, 45A		30	.267		38.50	15.50		54	66.50
8610	12 pole, 8-4AWG, 600V, 45A		30	.267		46	15.50		61.50	74.50
8615	Wiring device terminal 2 shorting pin w/cover, 4 pole, 8-4AWG, 600V, 45A		30	.267		29	15.50		44.50	56
8620	6 pole, 8-4AWG, 600V, 45A		30	.267		34.50	15.50		50	62
8625	8 pole, 8-4AWG, 600V, 45A		30	.267		46.50	15.50		62	75.50
8630	12 pole, 8-4AWG, 600V, 45A		30	.267		55.50	15.50		71	85
9000	Minimum labor/equipment charge		4	2	Job		116		116	180

26 27 33 – Power Distribution Units

26 27 33.10 Power Distribution Unit Cabinet

	Crew	Daily Output	Labor-Hours	Unit	Material	2018 Bare Costs Labor	Equipment	Total	Total Incl O&P
0010	**POWER DISTRIBUTION UNIT CABINET**									
0100	Power distribution unit, single cabinet, 50 kVA output, 480/208 input	1 Elec	2	4	Ea.	9,000	233		9,233	10,300
0110	75 kVA output, 480/208 input		2	4		9,775	233		10,008	11,200
0120	100 kVA output, 480/208 input		2	4		10,700	233		10,933	12,100
0130	125 kVA output, 480/208 input		1.50	5.333		11,700	310		12,010	13,400
0140	150 kVA output, 480/208 input		1.50	5.333		14,200	310		14,510	16,100
0150	200 kVA output, 480/208 input		1	8		15,700	465		16,165	17,900
0160	225 kVA output, 480/208 input		1	8		16,600	465		17,065	18,900

26 27 73 – Door Chimes

26 27 73.10 Doorbell System

	Crew	Daily Output	Labor-Hours	Unit	Material	2018 Bare Costs Labor	Equipment	Total	Total Incl O&P
0010	**DOORBELL SYSTEM**, incl. transformer, button & signal									
0100	6" bell	1 Elec	4	2	Ea.	139	116		255	335
0200	Buzzer		4	2		109	116		225	300
1000	Door chimes, 2 notes		16	.500		30	29		59	78
1020	with ambient light		12	.667		120	39		159	192
1100	Tube type, 3 tube system		12	.667		226	39		265	310
1180	4 tube system		10	.800		480	46.50		526.50	600
1900	For transformer & button, add		5	1.600		14.20	93		107.20	160
3000	For push button only		24	.333		.94	19.40		20.34	31

26 27 Low-Voltage Distribution Equipment

26 27 73 – Door Chimes

26 27 73.10 Doorbell System	Crew	Daily Output	Labor-Hours	Unit	Material	2018 Bare Costs Labor	Equipment	Total	Total Incl O&P
3200	Bell transformer	1 Elec	16	.500	Ea.	19.35	29		48.35	66.50
9000	Minimum labor/equipment charge		4	2	Job		116		116	180

26 28 Low-Voltage Circuit Protective Devices

26 28 16 – Enclosed Switches and Circuit Breakers

26 28 16.10 Circuit Breakers

		Crew	Daily Output	Labor-Hours	Unit	Material	2018 Bare Costs Labor	Equipment	Total	Total Incl O&P
0010	**CIRCUIT BREAKERS** (in enclosure)									
0100	Enclosed (NEMA 1), 600 volt, 3 pole, 30 amp	1 Elec	3.20	2.500	Ea.	520	146		666	795
0200	60 amp		2.80	2.857		635	166		801	955
0400	100 amp		2.30	3.478		730	202		932	1,125
0500	200 amp		1.50	5.333		1,525	310		1,835	2,150
0600	225 amp		1.50	5.333		1,675	310		1,985	2,325
0700	400 amp	2 Elec	1.60	10.		2,875	580		3,455	4,075
9000	Minimum labor/equipment charge	1 Elec	4	2	Job		116		116	180

26 28 16.13 Circuit Breakers

		Crew	Daily Output	Labor-Hours	Unit	Material	2018 Bare Costs Labor	Equipment	Total	Total Incl O&P
0010	**CIRCUIT BREAKERS**									
0100	Circuit breaker current limiter, 225 to 400, ampere, 1 pole	1 Elec	48	.167	Ea.	770	9.70		779.70	865
0200	Circuit breaker current limiter, 200 to 500, ampere, 1 pole, thermal magnet		46	.174		760	10.10		770.10	850
0300	300 to 500, ampere, 1 pole, thermal magnet		46	.174		1,200	10.10		1,210.10	1,325
0400	Circuit breaker current limiter, 600 to 1000, ampere, 1 pole, tri pac		46	.174		1,650	10.10		1,660.10	1,825

26 28 16.20 Safety Switches

		Crew	Daily Output	Labor-Hours	Unit	Material	2018 Bare Costs Labor	Equipment	Total	Total Incl O&P
0010	**SAFETY SWITCHES**									
0100	General duty 240 volt, 3 pole NEMA 1, fusible, 30 amp	1 Elec	3.20	2.500	Ea.	55.50	146		201.50	286
0200	60 amp		2.30	3.478		94	202		296	420
0300	100 amp		1.90	4.211		159	245		404	555
0400	200 amp		1.30	6.154		345	360		705	935
0500	400 amp	2 Elec	1.80	8.889		905	515		1,420	1,800
1100	Heavy duty, 600 volt, 3 pole NEMA 1 nonfused									
1110	30 amp	1 Elec	3.20	2.500	Ea.	81.50	146		227.50	315
1500	60 amp		2.30	3.478		138	202		340	465
1700	100 amp		1.90	4.211		214	245		459	615
1900	200 amp		1.30	6.154		330	360		690	915
2100	400 amp	2 Elec	1.80	8.889		725	515		1,240	1,600
2300	600 amp		1.20	13.333		1,325	775		2,100	2,650
2500	800 amp		.94	17.021		2,250	990		3,240	4,000
2700	1,200 amp		.80	20		3,500	1,175		4,675	5,650
4350	600 volt, 3 pole, fusible, 30 amp	1 Elec	3.20	2.500		145	146		291	385
4380	60 amp		2.30	3.478		176	202		378	510
4400	100 amp		1.90	4.211		315	245		560	725
4420	200 amp		1.30	6.154		450	360		810	1,050
4440	400 amp	2 Elec	1.80	8.889		1,250	515		1,765	2,175
4450	600 amp		1.20	13.333		1,975	775		2,750	3,375
4460	800 amp		.94	17.021		3,950	990		4,940	5,875
4480	1,200 amp		.80	20		5,400	1,175		6,575	7,750
5510	Heavy duty, 600 volt, 3 pole 3 ph. NEMA 3R fusible, 30 amp	1 Elec	3.10	2.581		245	150		395	500
5520	60 amp		2.20	3.636		310	212		522	665
5530	100 amp		1.80	4.444		435	259		694	880
5540	200 amp		1.20	6.667		630	390		1,020	1,300
5550	400 amp	2 Elec	1.60	10		1,525	580		2,105	2,575
5700	600 volt, 3 pole NEMA 3R nonfused									

For customer support on your Commercial Renovation Costs with RSMeans data, call 800.448.8182.

501

26 28 Low-Voltage Circuit Protective Devices

26 28 16 – Enclosed Switches and Circuit Breakers

26 28 16.20 Safety Switches		Crew	Daily Output	Labor-Hours	Unit	Material	2018 Bare Costs Labor	Equipment	Total	Total Incl O&P
5710	30 amp	1 Elec	3.10	2.581	Ea.	132	150		282	380
5900	60 amp		2.20	3.636		236	212		448	585
6100	100 amp		1.80	4.444		330	259		589	760
6300	200 amp		1.20	6.667		420	390		810	1,050
6500	400 amp	2 Elec	1.60	10		970	580		1,550	1,975
6700	600 amp	"	1	16		2,050	930		2,980	3,700
6900	600 volt, 6 pole NEMA 3R nonfused, 30 amp	1 Elec	2.70	2.963		830	172		1,002	1,175
7100	60 amp		2	4		1,175	233		1,408	1,625
7300	100 amp		1.50	5.333		1,475	310		1,785	2,100
7500	200 amp		1.20	6.667		3,300	390		3,690	4,225
9990	Minimum labor/equipment charge		3	2.667	Job		155		155	240

26 28 16.40 Time Switches

	26 28 16.40 Time Switches	Crew	Daily Output	Labor-Hours	Unit	Material	Labor	Equipment	Total	Total Incl O&P
0010	**TIME SWITCHES**									
0100	Single pole, single throw, 24 hour dial	1 Elec	4	2	Ea.	125	116		241	315
0200	24 hour dial with reserve power		3.60	2.222		565	129		694	820
0300	Astronomic dial		3.60	2.222		240	129		369	465
0400	Astronomic dial with reserve power		3.30	2.424		665	141		806	950
0500	7 day calendar dial		3.30	2.424		140	141		281	370
0600	7 day calendar dial with reserve power		3.20	2.500		253	146		399	505
0700	Photo cell 2,000 watt		8	1		27	58		85	120
1080	Load management device, 4 loads		2	4		1,150	233		1,383	1,600
1100	8 loads		1	8		2,325	465		2,790	3,275
9000	Minimum labor/equipment charge		3.50	2.286	Job		133		133	205

26 28 16.50 Meter Socket Entry Hub

	26 28 16.50 Meter Socket Entry Hub	Crew	Daily Output	Labor-Hours	Unit	Material	Labor	Equipment	Total	Total Incl O&P
0010	**METER SOCKET ENTRY HUB**									
0100	Meter socket entry hub closing cap, 3/4 to 4 inch	1 Elec	40	.200	Ea.	1.94	11.65		13.59	20
0110	Meter socket entry hub, 3/4 conduit		40	.200		15.40	11.65		27.05	35
0120	1 inch conduit		40	.200		15.35	11.65		27	35
0130	1-1/4 inch conduit		40	.200		15.60	11.65		27.25	35
0140	1-1/2 inch conduit		40	.200		15.65	11.65		27.30	35.50
0150	2 inch conduit		40	.200		19.10	11.65		30.75	39
0160	2-1/2 inch conduit		40	.200		50	11.65		61.65	73.50
0170	3 inch conduit		40	.200		90	11.65		101.65	117

26 29 Low-Voltage Controllers

26 29 13 – Enclosed Controllers

26 29 13.40 Relays

	26 29 13.40 Relays	Crew	Daily Output	Labor-Hours	Unit	Material	Labor	Equipment	Total	Total Incl O&P
0010	**RELAYS** Enclosed (NEMA 1)									
0050	600 volt AC, 1 pole, 12 amp	1 Elec	5.30	1.509	Ea.	86	88		174	231
0100	2 pole, 12 amp		5	1.600		86	93		179	239
0200	4 pole, 10 amp		4.50	1.778		115	103		218	286

26 29 23 – Variable-Frequency Motor Controllers

26 29 23.10 Variable Frequency Drives/Adj. Frequency Drives

	26 29 23.10 Variable Frequency Drives/Adj. Frequency Drives		Crew	Daily Output	Labor-Hours	Unit	Material	Labor	Equipment	Total	Total Incl O&P
0010	**VARIABLE FREQUENCY DRIVES/ADJ. FREQUENCY DRIVES**										
0100	Enclosed (NEMA 1), 460 volt, for 3 HP motor size	G	1 Elec	.80	10	Ea.	1,950	580		2,530	3,050
0110	5 HP motor size	G		.80	10		2,225	580		2,805	3,350
0120	7.5 HP motor size	G		.67	11.940		2,700	695		3,395	4,050
0130	10 HP motor size	G		.67	11.940		2,975	695		3,670	4,350
0140	15 HP motor size	G	2 Elec	.89	17.978		3,825	1,050		4,875	5,825
0150	20 HP motor size	G		.89	17.978		4,525	1,050		5,575	6,600

For customer support on your Commercial Renovation Costs with RSMeans data, call 800.448.8182.

26 29 Low-Voltage Controllers

26 29 23 – Variable-Frequency Motor Controllers

26 29 23.10 Variable Frequency Drives/Adj. Frequency Drives		Crew	Daily Output	Labor-Hours	Unit	Material	2018 Bare Costs Labor	Equipment	Total	Total Incl O&P
0160	25 HP motor size	G	2 Elec	.67	23.881	Ea.	5,550	1,400		6,950	8,275
0170	30 HP motor size	G		.67	23.881		6,950	1,400		8,350	9,800
0180	40 HP motor size	G		.67	23.881		8,025	1,400		9,425	11,000
0190	50 HP motor size	G	↓	.53	30.189		10,300	1,750		12,050	14,000
0200	60 HP motor size	G	R-3	.56	35.714		12,400	2,075	231	14,706	17,200
0210	75 HP motor size	G		.56	35.714		14,400	2,075	231	16,706	19,300
0220	100 HP motor size	G		.50	40		18,500	2,325	259	21,084	24,300
0230	125 HP motor size	G		.50	40		20,100	2,325	259	22,684	26,000
0240	150 HP motor size	G		.50	40		24,300	2,325	259	26,884	30,600
0250	200 HP motor size	G	↓	.42	47.619		27,700	2,750	310	30,760	35,100
1100	Custom-engineered, 460 volt, for 3 HP motor size	G	1 Elec	.56	14.286		3,050	830		3,880	4,625
1110	5 HP motor size	G		.56	14.286		3,050	830		3,880	4,625
1120	7.5 HP motor size	G		.47	17.021		3,200	990		4,190	5,025
1130	10 HP motor size	G	↓	.47	17.021		3,350	990		4,340	5,200
1140	15 HP motor size	G	2 Elec	.62	25.806		4,175	1,500		5,675	6,925
1150	20 HP motor size	G		.62	25.806		4,700	1,500		6,200	7,500
1160	25 HP motor size	G		.47	34.043		5,500	1,975		7,475	9,100
1170	30 HP motor size	G		.47	34.043		6,875	1,975		8,850	10,600
1180	40 HP motor size	G		.47	34.043		8,125	1,975		10,100	12,000
1190	50 HP motor size	G	↓	.37	43.243		9,325	2,525		11,850	14,200
1200	60 HP motor size	G	R-3	.39	51.282		14,100	2,975	330	17,405	20,500
1210	75 HP motor size	G		.39	51.282		15,000	2,975	330	18,305	21,500
1220	100 HP motor size	G		.35	57.143		16,400	3,325	370	20,095	23,600
1230	125 HP motor size	G		.35	57.143		17,500	3,325	370	21,195	24,900
1240	150 HP motor size	G		.35	57.143		20,100	3,325	370	23,795	27,700
1250	200 HP motor size	G	↓	.29	68.966	↓	26,200	4,000	445	30,645	35,500
2000	For complex & special design systems to meet specific										
2010	requirements, obtain quote from vendor.										

26 32 Packaged Generator Assemblies

26 32 13 – Engine Generators

26 32 13.16 Gas-Engine-Driven Generator Sets

26 32 13.16 Gas-Engine-Driven Generator Sets		Crew	Daily Output	Labor-Hours	Unit	Material	2018 Bare Costs Labor	Equipment	Total	Total Incl O&P
0010	**GAS-ENGINE-DRIVEN GENERATOR SETS**										
0020	Gas or gasoline operated, includes battery,										
0050	charger, & muffler										
0200	7.5 kW	R-3	.83	24.096	Ea.	8,550	1,400	156	10,106	11,800
0300	11.5 kW		.71	28.169		12,100	1,625	183	13,908	16,000
0400	20 kW		.63	31.746		14,300	1,850	206	16,356	18,800
0500	35 kW	↓	.55	36.364	↓	17,000	2,100	236	19,336	22,200

For customer support on your Commercial Renovation Costs with RSMeans data, call 800.448.8182.

503

26 33 Battery Equipment

26 33 19 – Battery Units

26 33 19.10 Battery Units

		Crew	Daily Output	Labor-Hours	Unit	Material	2018 Bare Costs Labor	Equipment	Total	Total Incl O&P
0010	**BATTERY UNITS**									
0500	Salt water battery, 2.1 kWA, 24V, 750W, 30A, wired in crate, IP22 rated	1 Elec	16.25	.492	Ea.	995	28.50		1,023.50	1,150
0510	2.2 kWh, 48V, 800W, 17A, wired in crate, IP22 rated		16.25	.492		1,200	28.50		1,228.50	1,375
0520	25.9 kWh, 48V, 11,700W, 240A, wired in crate, IP2X rated		16.25	.492		15,000	28.50		15,028.50	16,500
0700	Nickel iron nife battery, deep cycle, 100Ah, 12V, 10 cells		14.25	.561		1,100	32.50		1,132.50	1,250
0710	24V, 10 cells		14.25	.561		2,200	32.50		2,232.50	2,450
0720	48V, 10 cells		14.25	.561		4,375	32.50		4,407.50	4,875
0730	Nickel iron nife battery, deep cycle, 300Ah, 12V, 10 cells		12.25	.653		3,175	38		3,213	3,550
0740	24V, 10 cells		12.25	.653		6,350	38		6,388	7,025
0750	48V, 10 cells		12.25	.653		12,700	38		12,738	14,100

26 51 Interior Lighting

26 51 13 – Interior Lighting Fixtures, Lamps, and Ballasts

26 51 13.50 Interior Lighting Fixtures

		Crew	Daily Output	Labor-Hours	Unit	Material	2018 Bare Costs Labor	Equipment	Total	Total Incl O&P
0010	**INTERIOR LIGHTING FIXTURES** Including lamps, mounting									
0030	hardware and connections									
0100	Fluorescent, C.W. lamps, troffer, recess mounted in grid, RS									
0130	Grid ceiling mount									
0200	Acrylic lens, 1' W x 4' L, two 40 watt	1 Elec	5.70	1.404	Ea.	51.50	81.50		133	183
0300	2' W x 2' L, two U40 watt		5.70	1.404		55.50	81.50		137	187
0600	2' W x 4' L, four 40 watt		4.70	1.702		62	99		161	222
0910	Acrylic lens, 1' W x 4' L, two 32 watt T8 [G]		5.70	1.404		61.50	81.50		143	194
0930	2' W x 2' L, two U32 watt T8 [G]		5.70	1.404		96	81.50		177.50	231
0940	2' W x 4' L, two 32 watt T8 [G]		5.30	1.509		79	88		167	223
0950	2' W x 4' L, three 32 watt T8 [G]		5	1.600		70	93		163	221
0960	2' W x 4' L, four 32 watt T8 [G]		4.70	1.702		74.50	99		173.50	235
1000	Surface mounted, RS									
1030	Acrylic lens with hinged & latched door frame									
1100	1' W x 4' L, two 40 watt	1 Elec	7	1.143	Ea.	65.50	66.50		132	176
1200	2' W x 2' L, two U40 watt		7	1.143		70.50	66.50		137	181
1500	2' W x 4' L, four 40 watt		5.30	1.509		85	88		173	230
1501	2' W x 4' L, six 40 watt T8		5.20	1.538		85	89.50		174.50	232
2100	Strip fixture									
2130	Surface mounted									
2200	4' long, one 40 watt, RS	1 Elec	8.50	.941	Ea.	31	55		86	119
2300	4' long, two 40 watt, RS	"	8	1		43	58		101	138
2600	8' long, one 75 watt, SL	2 Elec	13.40	1.194		55	69.50		124.50	168
2700	8' long, two 75 watt, SL	"	12.40	1.290		66.50	75		141.50	189
3000	Strip, pendent mounted, industrial, white porcelain enamel									
3100	4' long, two 40 watt, RS	1 Elec	5.70	1.404	Ea.	53	81.50		134.50	184
3200	4' long, two 60 watt, HO	"	5	1.600		83	93		176	236
3300	8' long, two 75 watt, SL	2 Elec	8.80	1.818		98.50	106		204.50	271
4220	Metal halide, integral ballast, ceiling, recess mounted									
4230	prismatic glass lens, floating door									
4240	2' W x 2' L, 250 watt	1 Elec	3.20	2.500	Ea.	255	146		401	505
4250	2' W x 2' L, 400 watt	2 Elec	5.80	2.759		365	161		526	655
4260	Surface mounted, 2' W x 2' L, 250 watt	1 Elec	2.70	2.963		345	172		517	645
4270	400 watt	2 Elec	4.80	3.333		405	194		599	750
4280	High bay, aluminum reflector,									
4290	Single unit, 400 watt	2 Elec	4.60	3.478	Ea.	415	202		617	770
4300	Single unit, 1,000 watt		4	4		595	233		828	1,025

504

26 51 13 – Interior Lighting Fixtures, Lamps, and Ballasts

26 51 13.50 Interior Lighting Fixtures

		Crew	Daily Output	Labor-Hours	Unit	Material	2018 Bare Costs Labor	Equipment	Total	Total Incl O&P
4310	Twin unit, 400 watt	2 Elec	3.20	5	Ea.	825	291		1,116	1,350
4320	Low bay, aluminum reflector, 250W DX lamp	1 Elec	3.20	2.500	↓	365	146		511	625
4340	High pressure sodium integral ballast ceiling, recess mounted									
4350	prismatic glass lens, floating door									
4360	2' W x 2' L, 150 watt lamp	1 Elec	3.20	2.500	Ea.	390	146		536	650
4370	2' W x 2' L, 400 watt lamp	2 Elec	5.80	2.759		465	161		626	760
4380	Surface mounted, 2' W x 2' L, 150 watt lamp	1 Elec	2.70	2.963		475	172		647	785
4390	400 watt lamp	2 Elec	4.80	3.333	↓	530	194		724	885
4400	High bay, aluminum reflector,									
4410	Single unit, 400 watt lamp	2 Elec	4.60	3.478	Ea.	380	202		582	735
4430	Single unit, 1,000 watt lamp	"	4	4		550	233		783	965
4440	Low bay, aluminum reflector, 150 watt lamp	1 Elec	3.20	2.500	↓	330	146		476	585
4450	Incandescent, high hat can, round alzak reflector, prewired									
4470	100 watt	1 Elec	8	1	Ea.	64.50	58		122.50	161
4480	150 watt		8	1		105	58		163	205
4500	300 watt	↓	6.70	1.194	↓	242	69.50		311.50	375
4600	Square glass lens with metal trim, prewired									
4630	100 watt	1 Elec	6.70	1.194	Ea.	55.50	69.50		125	168
4680	150 watt		6.70	1.194		98	69.50		167.50	215
4700	200 watt		6.70	1.194		98	69.50		167.50	215
6010	Vapor tight, incandescent, ceiling mounted, 200 watt		6.20	1.290		78.50	75		153.50	202
6100	Fluorescent, surface mounted, 2 lamps, 4' L, RS, 40 watt		3.20	2.500		116	146		262	350
6850	Vandalproof, surface mounted, fluorescent, two 32 watt T8 [G]		3.20	2.500		252	146		398	505
6860	Incandescent, one 150 watt		8	1		90	58		148	189
6900	Mirror light, fluorescent, RS, acrylic enclosure, two 40 watt		8	1		116	58		174	218
6910	One 40 watt		8	1		96	58		154	196
6920	One 20 watt	↓	12	.667	↓	82.50	39		121.50	151
7500	Ballast replacement, by weight of ballast, to 15' high									
7520	Indoor fluorescent, less than 2 lb.	1 Elec	10	.800	Ea.	23.50	46.50		70	98
7540	Two 40W, watt reducer, 2 to 5 lb.		9.40	.851		71	49.50		120.50	155
7560	Two F96 slimline, over 5 lb.		8	1		110	58		168	211
7580	Vaportite ballast, less than 2 lb.		9.40	.851		23.50	49.50		73	103
7600	2 lb. to 5 lb.		8.90	.899		71	52.50		123.50	160
7620	Over 5 lb.		7.60	1.053		110	61.50		171.50	216
7630	Electronic ballast for two tubes		8	1		43.50	58		101.50	138
7640	Dimmable ballast one-lamp [G]		8	1		108	58		166	209
7650	Dimmable ballast two-lamp [G]		7.60	1.053		108	61.50		169.50	213
7660	Dimmable ballast three-lamp		6.90	1.159	↓	130	67.50		197.50	247
9000	Minimum labor/equipment charge	↓	3	2.667	Job		155		155	240

26 51 13.70 Residential Fixtures

		Crew	Daily Output	Labor-Hours	Unit	Material	2018 Bare Costs Labor	Equipment	Total	Total Incl O&P
0010	**RESIDENTIAL FIXTURES**									
0400	Fluorescent, interior, surface, circline, 32 watt & 40 watt	1 Elec	20	.400	Ea.	155	23.50		178.50	207
0500	2' x 2', two U-tube 32 watt T8		8	1		129	58		187	232
0700	Shallow under cabinet, two 20 watt		16	.500		71.50	29		100.50	124
0900	Wall mounted, 4' L, two 32 watt T8, with baffle		10	.800		133	46.50		179.50	218
2000	Incandescent, exterior lantern, wall mounted, 60 watt		16	.500		60.50	29		89.50	112
2100	Post light, 150 W, with 7' post		4	2		270	116		386	475
2500	Lamp holder, weatherproof with 150 W PAR		16	.500		33.50	29		62.50	81.50
2550	With reflector and guard		12	.667		59.50	39		98.50	126
2600	Interior pendent, globe with shade, 150 W		20	.400	↓	201	23.50		224.50	257
9000	Minimum labor/equipment charge		4	2	Job		116		116	180

For customer support on your Commercial Renovation Costs with RSMeans data, call 800.448.8182.

505

26 51 Interior Lighting

26 51 19 – LED Interior Lighting

26 51 19.10 LED Interior Lighting	Crew	Daily Output	Labor-Hours	Unit	Material	2018 Bare Costs Labor	Equipment	Total	Total Incl O&P
0010 **LED INTERIOR LIGHTING**									
7000 Interior LED fixts, tape rope kit, 120V, 3' strip, 4W, w/5' cord, 3,000K	1 Elec	24	.333	Ea.	8.45	19.40		27.85	39.50
7010 5,000K		24	.333		12.05	19.40		31.45	43.50
7020 Interior LED fixts, tape rope kit, 120V, 6' strip, 8W, w/5' cord, 3,000K		23	.348		15.40	20		35.40	48.50
7030 5,000K		23	.348		15.40	20		35.40	48.50
7040 Interior LED fixts, tape rope kit, 120V, 13' strip, 15W, w/5' cord, 3,000K		22	.364		28.50	21		49.50	63.50
7050 5,000K		22	.364		28.50	21		49.50	63.50
7060 Interior LED fixts, tape rope kit, 120V, 20' strip, 22W, w/5' cord, 3,000K		21	.381		41	22		63	80
7070 5,000K		21	.381		41	22		63	80
7080 Interior LED fixts, tape rope kit, 120V, 33' strip, 37W, w/5' cord, 3,000K		20	.400		60.50	23.50		84	103
7090 5,000K		20	.400		60.50	23.50		84	103
7100 Interior LED fixts, tape rope kit, 120V, 50' strip, 55W, w/5' cord, 3,000K		19	.421		92	24.50		116.50	139
7110 5,000K		19	.421		131	24.50		155.50	183
7120 Interior LED tape light, starstrand, dimmable, 12V, super star, 2", 3,000K		25	.320		11.30	18.60		29.90	41.50
7130 Ultra star, 2", 3,000K		25	.320		14.10	18.60		32.70	44.50
7140 Rainbow star, 2", 3,000K		25	.320		11.30	18.60		29.90	41.50
7150 Interior LED tape light, starstrand, dimmable, 24V, elite star, 4", 35K		25	.320		9.85	18.60		28.45	40
7160 12", 35K		25	.320		22.50	18.60		41.10	54
7170 60", 35K		23	.348		98.50	20		118.50	141
7180 120", 35K		21	.381		139	22		161	188
7190 240", 35K		18	.444		273	26		299	340
7200 Interior LED tape lighting channel, 36", AL flex, starstrand, elite star	↓	48	.167	↓	21	9.70		30.70	38

26 52 Safety Lighting

26 52 13 – Emergency and Exit Lighting

26 52 13.10 Emergency Lighting and Battery Units

	Crew	Daily Output	Labor-Hours	Unit	Material	2018 Bare Costs Labor	Equipment	Total	Total Incl O&P
0010 **EMERGENCY LIGHTING AND BATTERY UNITS**									
0300 Emergency light units, battery operated									
0350 Twin sealed beam light, 25 W, 6 V each									
0500 Lead battery operated	1 Elec	4	2	Ea.	144	116		260	340
0700 Nickel cadmium battery operated		4	2	"	345	116		461	555
9000 Minimum labor/equipment charge	↓	4	2	Job		116		116	180

26 52 13.16 Exit Signs

	Crew	Daily Output	Labor-Hours	Unit	Material	2018 Bare Costs Labor	Equipment	Total	Total Incl O&P
0010 **EXIT SIGNS**									
0080 Exit light ceiling or wall mount, incandescent, single face	1 Elec	8	1	Ea.	70	58		128	167
0100 Double face	"	6.70	1.194		50	69.50		119.50	162
1500 Exit sign, 12 V, 1 face, remote end mounted (Type 602A1)	R-19	18	1.111		69.50	65		134.50	177
1780 With emergency battery, explosion proof	"	7.70	2.597	↓	4,375	151		4,526	5,050
9000 Minimum labor/equipment charge	1 Elec	4	2	Job		116		116	180

For customer support on your Commercial Renovation Costs with RSMeans data, call 800.448.8182.

26 55 Special Purpose Lighting

26 55 33 – Hazard Warning Lighting

26 55 33.10 Warning Beacons	Crew	Daily Output	Labor-Hours	Unit	Material	2018 Bare Costs Labor	Equipment	Total	Total Incl O&P
0010 **WARNING BEACONS**									
0015　Surface mount with colored or clear lens									
0100　　Rotating beacon									
0110　　　120V, 40 watt halogen	1 Elec	3.50	2.286	Ea.	106	133		239	320
0120　　　24V, 20 watt halogen	"	3.50	2.286	"	228	133		361	455
0200　　Steady beacon									
0210　　　120V, 40 watt halogen	1 Elec	3.50	2.286	Ea.	108	133		241	325
0220　　　24V, 20 watt		3.50	2.286		112	133		245	330
0230　　　12V DC, incandescent		3.50	2.286		103	133		236	320
0300　　Flashing beacon									
0310　　　120V, 40 watt halogen	1 Elec	3.50	2.286	Ea.	106	133		239	320
0320　　　24V, 20 watt halogen		3.50	2.286		106	133		239	320
0410　　　12V DC with two 6V lantern batteries		7	1.143		107	66.50		173.50	221

26 55 59 – Display Lighting

26 55 59.10 Track Lighting	Crew	Daily Output	Labor-Hours	Unit	Material	2018 Bare Costs Labor	Equipment	Total	Total Incl O&P
0010 **TRACK LIGHTING**									
0080　Track, 1 circuit, 4' section	1 Elec	6.70	1.194	Ea.	42.50	69.50		112	154
0100　　8' section	2 Elec	10.60	1.509		70.50	88		158.50	214
0300　3 circuits, 4' section	1 Elec	6.70	1.194		94	69.50		163.50	211
0400　　8' section	2 Elec	10.60	1.509		133	88		221	282
9000　Minimum labor/equipment charge	1 Elec	3	2.667	Job		155		155	240

26 55 63 – Detention Lighting

26 55 63.10 Detention Lighting Fixtures	Crew	Daily Output	Labor-Hours	Unit	Material	2018 Bare Costs Labor	Equipment	Total	Total Incl O&P
0010 **DETENTION LIGHTING FIXTURES**									
1000　Surface mounted, cold rolled steel, 14 ga., 1' x 4'	1 Elec	5.40	1.481	Ea.	750	86		836	960
1010　　2' x 2'		5.40	1.481		805	86		891	1,025
1020　　2' x 4'		5	1.600		950	93		1,043	1,200
1100　　12 ga., 1' x 4'		5.40	1.481		985	86		1,071	1,200
1110　　2' x 2'		5.40	1.481		805	86		891	1,025
1120　　2' x 4'		5	1.600		1,050	93		1,143	1,300
3000　Fluorescent vandal resistant light fixture, high abuse troffer		5.40	1.481		220	86		306	375
3010　　50" L x 8" W x 4" H		5.40	1.481		203	86		289	355
3020　　32W		5.40	1.481		278	86		364	440
3030　　T8/32W		5.40	1.481		218	86		304	375
3040　　No lamp		5.60	1.429		335	83		418	495
3050　　Lamp included		5.40	1.481		385	86		471	560
3060　　16 ga. cold rolled steel		5.40	1.481		242	86		328	400
3070　16 ga. cold rolled steel, ceiling/wall mount		5.40	1.481		284	86		370	445
3080　　w/lamp, white		5.40	1.481		216	86		302	370
3090　　w/o/lamp, bronze		5.60	1.429		221	83		304	370
3100　　w/lamp 6" L x 4-1/2" W x 8-1/2" H		5.40	1.481		56	86		142	195
3110　　w/lamp 5-7/8" W x 4-1/2" D x 8-1/2" H clear		5.40	1.481		51.50	86		137.50	190
3120　　w/lamp white		5.40	1.481		375	86		461	545
3130　Tall wallpack, w/lamp, bronze		5.40	1.481		111	86		197	255
3140　Tall wallpack, w/lamp, white		5.40	1.481		115	86		201	260

For customer support on your Commercial Renovation Costs with RSMeans data, call 800.448.8182.

507

26 56 Exterior Lighting

26 56 19 – LED Exterior Lighting

26 56 19.55 Roadway LED Luminaire		Crew	Daily Output	Labor-Hours	Unit	Material	2018 Bare Costs Labor	2018 Bare Costs Equipment	Total	Total Incl O&P
0010	**ROADWAY LED LUMINAIRE**									
2000	Exterior fixtures, LED roadway, type 2, 50W, neutral slipfitter bronze	2 Elec	5.20	3.077	Ea.	545	179		724	875
2010	3, 50W, neutral slipfitter bronze		5.20	3.077		545	179		724	875
2020	4, 50W, neutral slipfitter bronze		5.20	3.077		545	179		724	875
2030	Exterior fixtures, LED roadway, type 2, 78W, neutral slipfitter bronze		5.20	3.077		545	179		724	875
2040	3, 78W, neutral slipfitter bronze		5.20	3.077		550	179		729	880
2050	4, 78W, neutral slipfitter bronze		5.20	3.077		545	179		724	875
2060	Exterior fixtures, LED roadway, type 2, 105W, neutral slipfitter bronze		5.20	3.077		645	179		824	985
2070	3, 105W, neutral slipfitter bronze		5.20	3.077		645	179		824	985
2080	4, 105W, neutral slipfitter bronze		5.20	3.077		645	179		824	985
2090	Exterior fixtures, LED roadway, type 2, 125W, neutral slipfitter bronze		5.20	3.077		645	179		824	985

26 56 23 – Area Lighting

26 56 23.10 Exterior Fixtures

26 56 23.10 Exterior Fixtures		Crew	Daily Output	Labor-Hours	Unit	Material	2018 Bare Costs Labor	2018 Bare Costs Equipment	Total	Total Incl O&P
0010	**EXTERIOR FIXTURES** With lamps									
0200	Wall mounted, incandescent, 100 watt	1 Elec	8	1	Ea.	40.50	58		98.50	135
0400	Quartz, 500 watt		5.30	1.509		61.50	88		149.50	204
1100	Wall pack, low pressure sodium, 35 watt		4	2		191	116		307	390
1150	55 watt		4	2		227	116		343	430
1160	High pressure sodium, 70 watt		4	2		187	116		303	385
1170	150 watt		4	2		205	116		321	405
1175	High pressure sodium, 250 watt		4	2		205	116		321	405
1180	Metal halide, 175 watt		4	2		210	116		326	410
1190	250 watt		4	2		236	116		352	440
1195	400 watt		4	2		415	116		531	635
1250	Induction lamp, 40 watt		4	2		470	116		586	695
1260	80 watt		4	2		560	116		676	795
1278	LED, poly lens, 26 watt		4	2		315	116		431	530
1280	110 watt		4	2		775	116		891	1,025
1500	LED, glass lens, 13 watt		4	2		315	116		431	530

26 56 33 – Walkway Lighting

26 56 33.10 Walkway Luminaire

26 56 33.10 Walkway Luminaire		Crew	Daily Output	Labor-Hours	Unit	Material	2018 Bare Costs Labor	2018 Bare Costs Equipment	Total	Total Incl O&P
0010	**WALKWAY LUMINAIRE**									
9000	Minimum labor/equipment charge	1 Elec	3.75	2.133	Job		124		124	192

26 56 33.55 Walkway LED Luminaire

26 56 33.55 Walkway LED Luminaire		Crew	Daily Output	Labor-Hours	Unit	Material	2018 Bare Costs Labor	2018 Bare Costs Equipment	Total	Total Incl O&P
0010	**WALKWAY LED LUMINAIRE**									
0600	LED bollard pole, 6" diam. x 36" H, cast alum, surf. mtd., 8W, 6,000K, 120V	1 Elec	8.50	.941	Ea.	380	55		435	505
0610	6" diam. x 42" H, cast alum, surf. mtd., 12W, 5,100K,120V		8.50	.941		505	55		560	640
0620	6" diam. x 42" H, cast alum, surf. mtd., 18W, 5,100K,120V		8.50	.941		515	55		570	650
0630	6" diam. x 42" H, cast alum, surf. mtd., 24W, 5,100K,120V		8.50	.941		500	55		555	635
0640	9.5" square x 42" H, cast resin, surf. mtd., A19, 9W, 120V		8.50	.941		625	55		680	775
0650	9.75" diam. x 42" H, cast resin, side mount, dome, 9W, 120V		8.50	.941		685	55		740	835
0660	9.75" diam. x 43" H, cast resin, burial base, dome, 9W, 120V		8.50	.941		685	55		740	835
0670	8" diam. x 36" H, concrete security base, w/rebar, 8W, HID		8	1		1,025	58		1,083	1,225
0680	10" diam. x 49" H, concrete base, w/Sch 40 Steel pipe, 8W		6.75	1.185		1,150	69		1,219	1,375
0690	12" diam. x 48" H, concrete base, w/Sch 40 Steel pipe, 8W		6.75	1.185		1,325	69		1,394	1,550
0700	8" diam. x 36" H, concrete, cast aluminum dome, 90 degree, 4.1W		7	1.143		1,025	66.50		1,091.50	1,225
0710	8" diam. x 36" H, concrete, cast aluminum dome, 180 degree, 4.1W		7	1.143		1,250	66.50		1,316.50	1,475
0720	8" diam. x 48" H, concrete, cast aluminum dome, 360 degree, 4.1W		7	1.143		1,550	66.50		1,616.50	1,800

26 56 Exterior Lighting

26 56 36 – Flood Lighting

26 56 36.20 Floodlights	Crew	Daily Output	Labor-Hours	Unit	Material	2018 Bare Costs Labor	2018 Bare Costs Equipment	Total	Total Incl O&P
0010 **FLOODLIGHTS** with ballast and lamp,									
1290 floor mtd, mount with swivel bracket									
1400 Pole mounted, pole not included									
2250 Low pressure sodium, 55 watt	1 Elec	2.70	2.963	Ea.	425	172		597	730
2270 90 watt	"	2	4	"	575	233		808	995

26 61 Lighting Systems and Accessories

26 61 23 – Lamps Applications

26 61 23.10 Lamps	Crew	Daily Output	Labor-Hours	Unit	Material	2018 Bare Costs Labor	2018 Bare Costs Equipment	Total	Total Incl O&P
0010 **LAMPS**									
0080 Fluorescent, rapid start, cool white, 2' long, 20 watt	1 Elec	1	8	C	305	465		770	1,050
0100 4' long, 40 watt		.90	8.889		256	515		771	1,075
0170 4' long, 34 watt energy saver [G]		.90	8.889		820	515		1,335	1,700
0176 2' long, T8, 17 watt energy saver [G]		1	8		335	465		800	1,100
0178 3' long, T8, 25 watt energy saver [G]		.90	8.889		365	515		880	1,200
0180 4' long, T8, 32 watt energy saver [G]		.90	8.889		223	515		738	1,050
0560 Twin tube compact lamp [G]		.90	8.889		450	515		965	1,300
0570 Double twin tube compact lamp [G]		.80	10		985	580		1,565	1,975
0600 Mercury vapor, mogul base, deluxe white, 100 watt		.30	26.667		5,325	1,550		6,875	8,250
0800 400 watt		.30	26.667		5,200	1,550		6,750	8,125
1000 Metal halide, mogul base, 175 watt		.30	26.667		1,075	1,550		2,625	3,600
1100 250 watt		.30	26.667		1,725	1,550		3,275	4,300
1350 High pressure sodium, 70 watt		.30	26.667		1,650	1,550		3,200	4,200
1370 150 watt		.30	26.667		1,675	1,550		3,225	4,250
1500 Low pressure sodium, 35 watt		.30	26.667		16,500	1,550		18,050	20,500
1600 90 watt		.30	26.667		6,500	1,550		8,050	9,550
1800 Incandescent, interior, A21, 100 watt		1.60	5		2,800	291		3,091	3,525
1900 A21, 150 watt		1.60	5		16,600	291		16,891	18,800
2300 R30, 75 watt		1.30	6.154		615	360		975	1,225
2500 Exterior, PAR 38, 75 watt		1.30	6.154		1,900	360		2,260	2,625
2600 PAR 38, 150 watt		1.30	6.154		2,200	360		2,560	2,950
9000 Minimum labor/equipment charge		4	2	Job		116		116	180

Division Notes

	CREW	DAILY OUTPUT	LABOR-HOURS	UNIT	BARE COSTS				TOTAL INCL O&P
					MAT.	LABOR	EQUIP.	TOTAL	

Estimating Tips

27 20 00 Data Communications
27 30 00 Voice Communications
27 40 00 Audio-Video Communications

- When estimating material costs for special systems, it is always prudent to obtain manufacturers' quotations for equipment prices and special installation requirements that may affect the total cost.

- For cost modifications for elevated tray installation, add the percentages to labor according to the height of the installation and only to the quantities exceeding the different height levels, not to the total tray quantities. Refer to 26 01 02.20 for labor adjustment factors.

- Do not overlook the costs for equipment used in the installation. If scissor lifts and boom lifts are available in the field, contractors may use them in lieu of the proposed ladders and rolling staging.

Reference Numbers

Reference numbers are shown at the beginning of some major classifications. These numbers refer to related items in the Reference Section. The reference information may be an estimating procedure, an alternate pricing method, or technical information.

Note: Not all subdivisions listed here necessarily appear. ■

27 13 23 – Communications Optical Fiber Backbone Cabling

27 13 23.13 Communications Optical Fiber	Crew	Daily Output	Labor-Hours	Unit	Material	2018 Bare Costs Labor	Equipment	Total	Total Incl O&P
0010	**COMMUNICATIONS OPTICAL FIBER**								
0040	Specialized tools & techniques cause installation costs to vary.								
0070	Fiber optic, cable, bulk simplex, single mode	1 Elec	8	1	C.L.F.	22.50	58		80.50	115
0080	Multi mode		8	1		29.50	58		87.50	123
0090	4 strand, single mode		7.34	1.090		38	63.50		101.50	140
0095	Multi mode		7.34	1.090		50.50	63.50		114	154
0100	12 strand, single mode		6.67	1.199		72.50	70		142.50	188
0105	Multi mode		6.67	1.199		96.50	70		166.50	214
0150	Jumper				Ea.	33			33	36.50
0200	Pigtail					29			29	32
0300	Connector	1 Elec	24	.333		26	19.40		45.40	58.50
0350	Finger splice		32	.250		38	14.55		52.55	64.50
0400	Transceiver (low cost bi-directional)		8	1		430	58		488	565
0450	Rack housing, 4 rack spaces, 12 panels (144 fibers)		2	4		560	233		793	975
0500	Patch panel, 12 ports		6	1.333		300	77.50		377.50	450
1000	Cable, 62.5 microns, direct burial, 4 fiber	R-15	1200	.040	L.F.	.91	2.29	.24	3.44	4.80
1020	Indoor, 2 fiber	R-19	1000	.020		.42	1.17		1.59	2.26
1040	Outdoor, aerial/duct	"	1670	.012		.65	.70		1.35	1.80
1060	50 microns, direct burial, 8 fiber	R-22	4000	.009		1.27	.50		1.77	2.17
1080	12 fiber		4000	.009		2.08	.50		2.58	3.06
1100	Indoor, 12 fiber		759	.049		1.98	2.62		4.60	6.20
1120	Connectors, 62.5 micron cable, transmission	R-19	40	.500	Ea.	14.35	29		43.35	61
1140	Cable splice		40	.500		17.50	29		46.50	64.50
1160	125 micron cable, transmission		16	1.250		15.10	73		88.10	130
1180	Receiver, 1.2 mile range		20	1		240	58.50		298.50	355
1200	1.9 mile range		20	1		226	58.50		284.50	340
1220	6.2 mile range		5	4		282	233		515	670
1240	Transmitter, 1.2 mile range		20	1		264	58.50		322.50	380
1260	1.9 mile range		20	1		299	58.50		357.50	420
1280	6.2 mile range		5	4		380	233		613	780
1300	Modem, 1.2 mile range		5	4		173	233		406	550
1320	6.2 mile range		5	4		315	233		548	710
1340	1.9 mile range, 12 channel		5	4		2,000	233		2,233	2,550
1360	Repeater, 1.2 mile range		10	2		370	117		487	590
1380	1.9 mile range		10	2		475	117		592	705
1400	6.2 mile range		5	4		920	233		1,153	1,350
1420	1.2 mile range, digital		5	4		440	233		673	840
1510	1/4", black, non-metallic, flexible, tube, liquid tight		8	2.500	C.L.F.	149	146		295	390
1520	3/8"		8	2.500		101	146		247	335
1530	1/2"		7.34	2.725		110	159		269	365
1540	3/4"		7.34	2.725		198	159		357	465
1550	1"		7.34	2.725		278	159		437	550
1560	1-1/4"		6.67	2.999		515	175		690	835
1570	1-1/2"		6.67	2.999		585	175		760	910
1580	2"		6.35	3.150		880	184		1,064	1,250
2040	Fiber optic cable, 48 strand, single mode, steel armor, material		1.67	11.976	M.L.F.	5,050	700		5,750	6,625

27 15 Communications Horizontal Cabling

27 15 01 – Communications Horizontal Cabling Applications

27 15 01.23 Audio-Video Communications Horizontal Cabling	Crew	Daily Output	Labor-Hours	Unit	Material	2018 Bare Costs Labor	Equipment	Total	Total Incl O&P
0010 **AUDIO-VIDEO COMMUNICATIONS HORIZONTAL CABLING**									
3000 S-video cable, MD4, 1M Pro interconnects, 3'	1 Elec	38	.211	Ea.	33	12.25		45.25	55.50
3010 6'		38	.211		37	12.25		49.25	59.50
3020 9'		38	.211		46	12.25		58.25	69.50
3030 16'		36	.222		49.50	12.95		62.45	74.50
3040 22'		36	.222		58	12.95		70.95	84
3300 HDMI cable, coaxial, HDMI/HDMI, Non plenum, 3', 24 kt terminations		40	.200		14.35	11.65		26	34
3310 6', 24 kt terminations		40	.200		16.95	11.65		28.60	36.50
3320 HDMI cable, coaxial, HDMI/HDMI metal, Non plenum, 3', 24 kt terminations		40	.200		16.40	11.65		28.05	36
3330 6', 24 kt terminations		40	.200		19.05	11.65		30.70	39
3340 15', 24 kt terminations		38	.211		26.50	12.25		38.75	48
3350 25', 24 kt terminations		38	.211		34.50	12.25		46.75	57

27 15 10 – Special Communications Cabling

27 15 10.23 Sound and Video Cables and Fittings

	Crew	Daily Output	Labor-Hours	Unit	Material	2018 Bare Costs Labor	Equipment	Total	Total Incl O&P
0010 **SOUND AND VIDEO CABLES & FITTINGS**									
1250 Nonshielded, #22-2 conductor	1 Elec	10	.800	C.L.F.	10.40	46.50		56.90	83.50

27 15 13 – Communications Copper Horizontal Cabling

27 15 13.13 Communication Cables and Fittings

	Crew	Daily Output	Labor-Hours	Unit	Material	2018 Bare Costs Labor	Equipment	Total	Total Incl O&P
0010 **COMMUNICATION CABLES AND FITTINGS**									
5000 High performance unshielded twisted pair (UTP)									
7000 Category 5, #24, 4 pair solid, PVC jacket	1 Elec	7	1.143	C.L.F.	10.20	66.50		76.70	114
7100 4 pair solid, plenum		7	1.143		14.65	66.50		81.15	119
7200 4 pair stranded, PVC jacket		7	1.143		21	66.50		87.50	126
7210 Category 5e, #24, 4 pair solid, PVC jacket		7	1.143		15.20	66.50		81.70	120
7212 4 pair solid, plenum		7	1.143		21.50	66.50		88	127
7214 4 pair stranded, PVC jacket		7	1.143		23.50	66.50		90	129
7240 Category 6, #24, 4 pair solid, PVC jacket		7	1.143		18.35	66.50		84.85	123
7242 4 pair solid, plenum		7	1.143		13.75	66.50		80.25	118
7244 4 pair stranded, PVC jacket		7	1.143		14.05	66.50		80.55	118
7300 Connector, RJ45, category 5		80	.100	Ea.	.75	5.80		6.55	9.85
7302 Shielded RJ45, category 5		72	.111		1.75	6.45		8.20	11.95
7310 Jack, UTP RJ45, category 3		72	.111		.45	6.45		6.90	10.50
7312 Category 5		65	.123		5.50	7.15		12.65	17.10
7314 Category 5e		65	.123		4.98	7.15		12.13	16.55
7316 Category 6		65	.123		3.23	7.15		10.38	14.60
7402 RJ45, green		16.50	.485		6	28		34	50
7408 RJ45, orange		16.50	.485		6.40	28		34.40	50.50
7420 5M, RJ45, 568B, cord set		16.50	.485		32.50	28		60.50	79
7422 3M, RJ45, 568B, cord set		16.50	.485		28	28		56	74.50
7424 1M, RJ45, 568B, cord set		16.50	.485		24.50	28		52.50	70.50
7450 M12 TO RJ45, 2M ultra-loch D-mode, cord set		15.50	.516		33.50	30		63.50	83.50

For customer support on your Commercial Renovation Costs with RSMeans data, call 800.448.8182.

513

27 32 Voice Communications Terminal Equipment

27 32 26 – Ring-Down Emergency Telephones

27 32 26.10 Emergency Phone Stations

		Crew	Daily Output	Labor-Hours	Unit	Material	2018 Bare Costs Labor	2018 Bare Costs Equipment	Total	Total Incl O&P
0010	**EMERGENCY PHONE STATIONS**									
0100	Hands free emergency speaker phone w/LED strobe	1 Elec	7.22	1.108	Ea.	850	64.50		914.50	1,025
0110	Wall/pole mounted emergency phone station w/LED strobe	"	4.60	1.739		1,525	101		1,626	1,825
2000	Call station, 30", pole mounted emergency phone, 120V/ VoIP, blue beacon	2 Elec	5	3.200		1,100	186		1,286	1,500
2010	Call station, 36"		5	3.200		1,575	186		1,761	2,025
2020	Call station, 5'-8"		4.80	3.333		2,600	194		2,794	3,150
2030	Call station, 9'		4.80	3.333		2,700	194		2,894	3,275

27 41 Audio-Video Systems

27 41 33 – Master Antenna Television Systems

27 41 33.10 TV Systems

		Crew	Daily Output	Labor-Hours	Unit	Material	2018 Bare Costs Labor	2018 Bare Costs Equipment	Total	Total Incl O&P
0010	**TV SYSTEMS,** not including rough-in wires, cables & conduits									
5000	Antenna, small	1 Elec	6	1.333	Ea.	55	77.50		132.50	181
5100	Large	"	4	2	"	231	116		347	435

27 51 Distributed Audio-Video Communications Systems

27 51 16 – Public Address Systems

27 51 16.10 Public Address System

		Crew	Daily Output	Labor-Hours	Unit	Material	2018 Bare Costs Labor	2018 Bare Costs Equipment	Total	Total Incl O&P
0010	**PUBLIC ADDRESS SYSTEM**									
0100	Conventional, office	1 Elec	5.33	1.501	Speaker	159	87.50		246.50	310
0200	Industrial		2.70	2.963	"	305	172		477	605
9000	Minimum labor/equipment charge		3.50	2.286	Job		133		133	205

27 51 19 – Sound Masking Systems

27 51 19.10 Sound System

		Crew	Daily Output	Labor-Hours	Unit	Material	2018 Bare Costs Labor	2018 Bare Costs Equipment	Total	Total Incl O&P
0010	**SOUND SYSTEM,** not including rough-in wires, cables & conduits									
2000	Intercom, 30 station capacity, master station	2 Elec	2	8	Ea.	2,300	465		2,765	3,250
2020	10 station capacity	"	4	4		1,250	233		1,483	1,725
3600	House telephone, talking station	1 Elec	1.60	5		555	291		846	1,050
3800	Press to talk, release to listen	"	5.30	1.509		129	88		217	278
4000	System-on button					77			77	85
4200	Door release	1 Elec	4	2		138	116		254	330
4400	Combination speaker and microphone		8	1		235	58		293	350
4600	Termination box		3.20	2.500		74	146		220	305
4800	Amplifier or power supply		5.30	1.509		850	88		938	1,075
5000	Vestibule door unit		16	.500	Name	156	29		185	216
5200	Strip cabinet		27	.296	Ea.	294	17.25		311.25	350
5400	Directory		16	.500	"	139	29		168	197
9000	Minimum labor/equipment charge		3.50	2.286	Job		133		133	205

Estimating Tips

- When estimating material costs for electronic safety and security systems, it is always prudent to obtain manufacturers' quotations for equipment prices and special installation requirements that may affect the total cost.

- Fire alarm systems consist of control panels, annunciator panels, batteries with rack, charger, and fire alarm actuating and indicating devices. Some fire alarm systems include speakers, telephone lines, door closer controls, and other components. Be careful not to overlook the costs related to installation for these items. Also be aware of costs for integrated automation instrumentation and terminal devices, control equipment, control wiring, and programming. Insurance underwriters may have specific requirements for the type of materials to be installed or design requirements based on the hazard to be protected. Local jurisdictions may have requirements not covered by code. It is advisable to be aware of any special conditions.

- Security equipment includes items such as CCTV, access control, and other detection and identification systems to perform alert and alarm functions. Be sure to consider the costs related to installation for this security equipment, such as for integrated automation instrumentation and terminal devices, control equipment, control wiring, and programming.

Reference Numbers

Reference numbers are shown at the beginning of some major classifications. These numbers refer to related items in the Reference Section. The reference information may be an estimating procedure, an alternate pricing method, or technical information.

Note: Not all subdivisions listed here necessarily appear. ■

28 15 Access Control Hardware Devices

28 15 11 – Integrated Credential Readers and Field Entry Management

28 15 11.11 Standard Card Readers

	Crew	Daily Output	Labor-Hours	Unit	Material	2018 Bare Costs Labor	Equipment	Total	Total Incl O&P
0010 **STANDARD CARD READERS**									
0015 Card key access									
0020 Computerized system, processor, proximity reader and cards									
0030 Does not include door hardware, lockset or wiring									
0040 Card key system for 1 door				Ea.	1,425			1,425	1,550
0060 Card key system for 2 doors					2,025			2,025	2,225
0080 Card key system for 4 doors					2,500			2,500	2,750
0100 Processor for card key access system					940			940	1,025
0160 Magnetic lock for electric access, 600 pound holding force					190			190	209
0170 Magnetic lock for electric access, 1200 pound holding force					190			190	209
0200 Proximity card reader					158			158	173

28 15 11.15 Biometric Identity Devices

	Crew	Daily Output	Labor-Hours	Unit	Material	2018 Bare Costs Labor	Equipment	Total	Total Incl O&P
0010 **BIOMETRIC IDENTITY DEVICES**									
0220 Hand geometry scanner, mem of 512 users, excl. striker/power	1 Elec	3	2.667	Ea.	2,300	155		2,455	2,775
0230 Memory upgrade for, adds 9,700 user profiles		8	1		330	58		388	455
0240 Adds 32,500 user profiles		8	1		665	58		723	820
0250 Prison type, memory of 256 users, excl. striker/power		3	2.667		2,700	155		2,855	3,225
0260 Memory upgrade for, adds 3,300 user profiles		8	1		315	58		373	440
0270 Adds 9,700 user profiles		8	1		500	58		558	640
0280 Adds 27,900 user profiles		8	1		680	58		738	835
0290 All weather, mem of 512 users, excl. striker/power		3	2.667		4,075	155		4,230	4,725
0300 Facial & fingerprint scanner, combination unit, excl. striker/power		3	2.667		895	155		1,050	1,225
0310 Access for, for initial setup, excl. striker/power		3	2.667		1,200	155		1,355	1,575

28 18 Security Access Detection Equipment

28 18 11 – Security Access Metal Detectors

28 18 11.13 Security Access Metal Detectors

	Crew	Daily Output	Labor-Hours	Unit	Material	2018 Bare Costs Labor	Equipment	Total	Total Incl O&P
0010 **SECURITY ACCESS METAL DETECTORS**									
0240 Metal detector, hand-held, wand type, unit only				Ea.	129			129	138
0250 Metal detector, walk through portal type, single zone	1 Elec	2	4		3,175	233		3,408	3,825
0260 Multi-zone	"	2	4		4,625	233		4,858	5,425

28 18 13 – Security Access X-Ray Equipment

28 18 13.16 Security Access X-Ray Equipment

	Crew	Daily Output	Labor-Hours	Unit	Material	2018 Bare Costs Labor	Equipment	Total	Total Incl O&P
0010 **SECURITY ACCESS X-RAY EQUIPMENT**									
0290 X-ray machine, desk top, for mail/small packages/letters	1 Elec	4	2	Ea.	3,475	116		3,591	4,000
0300 Conveyor type, incl. monitor		2	4		16,000	233		16,233	18,000
0310 Includes additional features		2	4		28,000	233		28,233	31,200
0320 X-ray machine, large unit, for airports, incl. monitor	2 Elec	1	16		40,000	930		40,930	45,500
0330 Full console	"	.50	32		68,500	1,850		70,350	78,500

28 18 15 – Security Access Explosive Detection Equipment

28 18 15.23 Security Access Explosive Detection Equipment

	Crew	Daily Output	Labor-Hours	Unit	Material	2018 Bare Costs Labor	Equipment	Total	Total Incl O&P
0010 **SECURITY ACCESS EXPLOSIVE DETECTION EQUIPMENT**									
0270 Explosives detector, walk through portal type	1 Elec	2	4	Ea.	44,200	233		44,433	49,000
0280 Hand-held, battery operated				"				25,500	28,100

28 23 Video Management System

28 23 23 – Video Surveillance Systems Infrastructure

28 23 23.50 Video Surveillance Equipments

28 23 23.50 Video Surveillance Equipments	Crew	Daily Output	Labor-Hours	Unit	Material	2018 Bare Costs Labor	2018 Bare Costs Equipment	Total	Total Incl O&P
0010 **VIDEO SURVEILLANCE EQUIPMENTS**									
0200 Video cameras, wireless, hidden in exit signs, clocks, etc., incl. receiver	1 Elec	3	2.667	Ea.	187	155		342	445
0210 Accessories for video recorder, single camera		3	2.667		183	155		338	440
0220 For multiple cameras		3	2.667		1,750	155		1,905	2,175
0230 Video cameras, wireless, for under vehicle searching, complete	↓	2	4	↓	16,000	233		16,233	18,000

28 31 Intrusion Detection

28 31 16 – Intrusion Detection Systems Infrastructure

28 31 16.50 Intrusion Detection

	Crew	Daily Output	Labor-Hours	Unit	Material	2018 Bare Costs Labor	2018 Bare Costs Equipment	Total	Total Incl O&P
0010 **INTRUSION DETECTION**, not including wires & conduits									
0100 Burglar alarm, battery operated, mechanical trigger	1 Elec	4	2	Ea.	280	116		396	490
0200 Electrical trigger		4	2		335	116		451	550
0400 For outside key control, add		8	1		87	58		145	186
0600 For remote signaling circuitry, add		8	1		138	58		196	242
0800 Card reader, flush type, standard		2.70	2.963		810	172		982	1,150
1000 Multi-code		2.70	2.963		1,200	172		1,372	1,600
1200 Door switches, hinge switch		5.30	1.509		62.50	88		150.50	205
1400 Magnetic switch		5.30	1.509		100	88		188	246
1600 Exit control locks, horn alarm		4	2		219	116		335	420
1800 Flashing light alarm		4	2		247	116		363	450
2000 Indicating panels, 1 channel		2.70	2.963		271	172		443	565
2200 10 channel	2 Elec	3.20	5		1,075	291		1,366	1,625
2400 20 channel		2	8		2,500	465		2,965	3,450
2600 40 channel	↓	1.14	14.035		4,500	815		5,315	6,200
2800 Ultrasonic motion detector, 12 V	1 Elec	2.30	3.470		194	202		396	530
3000 Infrared photoelectric detector		4	2		146	116		262	340
3200 Passive infrared detector		4	2		236	116		352	440
3400 Glass break alarm switch		8	1		99.50	58		157.50	199
3420 Switchmats, 30" x 5'		5.30	1.509		105	88		193	252
3440 30" x 25'		4	2		205	116		321	405
3460 Police connect panel		4	2		272	116		388	480
3480 Telephone dialer		5.30	1.509		390	88		478	560
3500 Alarm bell		4	2		105	116		221	296
3520 Siren		4	2		147	116		263	340
3540 Microwave detector, 10' to 200'		2	4		585	233		818	1,000
3560 10' to 350'	↓	2	4	↓	1,775	233		2,008	2,300

28 46 Fire Detection and Alarm

28 46 11 – Fire Sensors and Detectors

28 46 11.21 Carbon-Monoxide Detection Sensors

	Crew	Daily Output	Labor-Hours	Unit	Material	2018 Bare Costs Labor	2018 Bare Costs Equipment	Total	Total Incl O&P
0010 **CARBON-MONOXIDE DETECTION SENSORS**									
8400 Smoke and carbon monoxide alarm battery operated photoelectric low profile	1 Elec	24	.333	Ea.	54.50	19.40		73.90	89.50
8410 low profile photoelectric battery powered		24	.333		28.50	19.40		47.90	61.50
8420 photoelectric low profile sealed lithium		24	.333		52	19.40		71.40	87
8430 Photoelectric low profile sealed lithium smoke and CO with voice combo		24	.333		38	19.40		57.40	72
8500 Carbon monoxide sensor, wall mount 1Mod 1 relay output smoke & heat		24	.333		490	19.40		509.40	570
8700 Carbon monoxide detector, battery operated, wall mounted		16	.500		52	29		81	102
8710 Hardwired, wall and ceiling mounted		8	1		99.50	58		157.50	199
8720 Duct mounted	↓	8	1		345	58		403	465

28 46 Fire Detection and Alarm

28 46 11 – Fire Sensors and Detectors

28 46 11.21 Carbon-Monoxide Detection Sensors

		Crew	Daily Output	Labor-Hours	Unit	Material	2018 Bare Costs Labor	Equipment	Total	Total Incl O&P
8730	Continuous air monitoring system, PC based, incl. remote sensors, min.	2 Elec	.25	64	Ea.	31,200	3,725		34,925	40,100
8740	Maximum	"	.10	160	↓	49,900	9,300		59,200	69,500

28 46 11.27 Other Sensors

		Crew	Daily Output	Labor-Hours	Unit	Material	2018 Bare Costs Labor	Equipment	Total	Total Incl O&P
0010	**OTHER SENSORS**									
5200	Smoke detector, ceiling type	1 Elec	6.20	1.290	Ea.	120	75		195	248
5240	Smoke detector addressable type		6	1.333		224	77.50		301.50	365
5400	Duct type		3.20	2.500		330	146		476	590
5420	Duct addressable type		3.20	2.500		515	146		661	790
8300	Smoke alarm with integrated strobe light 120 V 16 DB 60 fpm flash rate		16	.500		103	29		132	158
8310	Photoelectric smoke detector with strobe 120 V 90 DB ceiling mount		12	.667		158	39		197	234
8320	120 V, 90 DB wall mount	↓	12	.667	↓	166	39		205	242

28 46 11.50 Fire and Heat Detectors

		Crew	Daily Output	Labor-Hours	Unit	Material	2018 Bare Costs Labor	Equipment	Total	Total Incl O&P
0010	**FIRE & HEAT DETECTORS**									
5000	Detector, rate of rise	1 Elec	8	1	Ea.	51.50	58		109.50	147
5100	Fixed temp fire alarm		7	1.143		51.50	66.50		118	160
5200	125 V SRF MT-135F		7.25	1.103		28	64		92	130
5300	MT-195F	↓	7.25	1.103		31.50	64		95.50	134

28 46 21 – Fire Alarm

28 46 21.50 Alarm Panels and Devices

		Crew	Daily Output	Labor-Hours	Unit	Material	2018 Bare Costs Labor	Equipment	Total	Total Incl O&P
0010	**ALARM PANELS AND DEVICES**, not including wires & conduits									
2200	Intercom remote station	1 Elec	8	1	Ea.	71	58		129	168
2400	Intercom outlet		8	1		53	58		111	149
2600	Sound system, intercom handset		8	1		475	58		533	615
3590	Signal device, beacon		8	1		199	58		257	310
3594	2 zone	↓	1.35	5.926		310	345		655	875
3600	4 zone	2 Elec	2	8		380	465		845	1,125
3610	3 zone		2	8		610	465		1,075	1,400
3800	8 zone		1	16		960	930		1,890	2,500
3810	5 zone		1.50	10.667		770	620		1,390	1,800
3900	10 zone		1.25	12.800		945	745		1,690	2,200
4000	12 zone	↓	.67	23.988		2,100	1,400		3,500	4,450
4020	Alarm device, tamper, flow	1 Elec	8	1		240	58		298	355
4025	Fire alarm, loop expander card		16	.500		740	29		769	855
4200	Battery and rack		4	2		415	116		531	635
4400	Automatic charger		8	1		590	58		648	740
4600	Signal bell		8	1		124	58		182	227
4610	Fire alarm signal bell 10" red 20-24 V P		8	1		136	58		194	239
4800	Trouble buzzer or manual station		8	1		84	58		142	183
5425	Duct smoke and heat detector 2 wire		8	1		107	58		165	207
5430	Fire alarm duct detector controller		3	2.667		235	155		390	500
5435	Fire alarm duct detector sensor kit		8	1		75	58		133	173
5440	Remote test station for smoke detector duct type		5.30	1.509		65	88		153	208
5460	Remote fire alarm indicator light		5.30	1.509		18.55	88		106.55	157
5600	Strobe and horn		5.30	1.509		154	88		242	305
5610	Strobe and horn (ADA type)		5.30	1.509		154	88		242	305
5620	Visual alarm (ADA type)		6.70	1.194		104	69.50		173.50	221
5800	Fire alarm horn		6.70	1.194		61.50	69.50		131	175
6000	Door holder, electro-magnetic		4	2		104	116		220	294
6200	Combination holder and closer		3.20	2.500		124	146		270	360
6600	Drill switch		8	1		375	58		433	500
6800	Master box		2.70	2.963		6,475	172		6,647	7,400
7000	Break glass station	↓	8	1	↓	57.50	58		115.50	154

28 46 Fire Detection and Alarm

28 46 21 – Fire Alarm

28 46 21.50 Alarm Panels and Devices	Crew	Daily Output	Labor-Hours	Unit	Material	2018 Bare Costs Labor	Equipment	Total	Total Incl O&P
7800	Remote annunciator, 8 zone lamp	1 Elec	1.80	4.444	Ea.	213	259		472	635
8000	12 zone lamp	2 Elec	2.60	6.154		415	360		775	1,025
8200	16 zone lamp	"	2.20	7.273		375	425		800	1,075

28 47 Mass Notification

28 47 12 – Notification Systems

28 47 12.10 Mass Notification System

		Crew	Daily Output	Labor-Hours	Unit	Material	2018 Bare Costs Labor	Equipment	Total	Total Incl O&P
0010	**MASS NOTIFICATION SYSTEM**									
0100	Wireless command center, 10,000 devices	2 Elec	1.33	12.030	Ea.	2,275	700		2,975	3,575
0200	Option, email notification					1,775			1,775	1,950
0210	Remote device supervision & monitor					2,525			2,525	2,800
0300	Antenna VHF or UHF, for medium range	1 Elec	4	2		87	116		203	276
0310	For high-power transmitter		2	4		1,100	233		1,333	1,550
0400	Transmitter, 25 watt		4	2		1,725	116		1,841	2,075
0410	40 watt		2.66	3.008		2,000	175		2,175	2,475
0420	100 watt		1.33	6.015		6,475	350		6,825	7,675
0500	Wireless receiver/control module for speaker		8	1		279	58		337	395
0600	Desktop paging controller, stand alone		4	2		835	116		951	1,100
0700	Auxiliary level inputs		8.36	.957		92	55.50		147.50	187

Division Notes

	CREW	DAILY OUTPUT	LABOR-HOURS	UNIT	BARE COSTS				TOTAL INCL O&P
					MAT.	LABOR	EQUIP.	TOTAL	

Estimating Tips
31 05 00 Common Work Results for Earthwork

- Estimating the actual cost of performing earthwork requires careful consideration of the variables involved. This includes items such as type of soil, whether water will be encountered, dewatering, whether banks need bracing, disposal of excavated earth, and length of haul to fill or spoil sites, etc. If the project has large quantities of cut or fill, consider raising or lowering the site to reduce costs, while paying close attention to the effect on site drainage and utilities.

- If the project has large quantities of fill, creating a borrow pit on the site can significantly lower the costs.

- It is very important to consider what time of year the project is scheduled for completion. Bad weather can create large cost overruns from dewatering, site repair, and lost productivity from cold weather.

Reference Numbers

Reference numbers are shown at the beginning of some major classifications. These numbers refer to related items in the Reference Section. The reference information may be an estimating procedure, an alternate pricing method, or technical information.

Note: Not all subdivisions listed here necessarily appear. ■

Did you know?

RSMeans data is available through our online application with 24/7 access:

- Search for unit prices by keyword
- Leverage the most up-to-date data
- Build and export estimates

Try it free for 30 days!
www.rsmeans.com/2018freetrial

31 05 Common Work Results for Earthwork

31 05 13 – Soils for Earthwork

31 05 13.10 Borrow

		Daily Output	Labor-Hours	Unit	Material	2018 Bare Costs Labor	Equipment	Total	Total Incl O&P
0010	**BORROW**								
0020	Spread, 200 HP dozer, no compaction, 2 mile RT haul								
0200	Common borrow	B-15	600	.047	C.Y.	12.75	2.21	3.93	18.89	22
0700	Screened loam		600	.047		28	2.21	3.93	34.14	38.50
0800	Topsoil, weed free		600	.047		25	2.21	3.93	31.14	35.50
0900	For 5 mile haul, add	B-34B	200	.040			1.84	2.71	4.55	5.95

31 05 16 – Aggregates for Earthwork

31 05 16.10 Borrow

		Crew	Daily Output	Labor-Hours	Unit	Material	2018 Bare Costs Labor	Equipment	Total	Total Incl O&P
0010	**BORROW**									
0020	Spread, with 200 HP dozer, no compaction, 2 mile RT haul									
0100	Bank run gravel	B-15	600	.047	L.C.Y.	18.50	2.21	3.93	24.64	28.50
0300	Crushed stone (1.40 tons per C.Y.), 1-1/2"		600	.047		28	2.21	3.93	34.14	38.50
0320	3/4"		600	.047		28	2.21	3.93	34.14	38.50
0340	1/2"		600	.047		29.50	2.21	3.93	35.64	40.50
0360	3/8"		600	.047		32.50	2.21	3.93	38.64	44
0400	Sand, washed, concrete		600	.047		36	2.21	3.93	42.14	47.50
0500	Dead or bank sand		600	.047		18.15	2.21	3.93	24.29	28
0600	Select structural fill		600	.047		20	2.21	3.93	26.14	30
0900	For 5 mile haul, add	B-34B	200	.040			1.84	2.71	4.55	5.95

31 06 Schedules for Earthwork

31 06 60 – Schedules for Special Foundations and Load Bearing Elements

31 06 60.14 Piling Special Costs

		Crew	Daily Output	Labor-Hours	Unit	Material	2018 Bare Costs Labor	Equipment	Total	Total Incl O&P
0010	**PILING SPECIAL COSTS**									
0011	Piling special costs, pile caps, see Section 03 30 53.40									
0500	Cutoffs, concrete piles, plain	1 Pile	5.50	1.455	Ea.		74.50		74.50	121
0600	With steel thin shell, add		38	.211			10.80		10.80	17.55
0700	Steel pile or "H" piles		19	.421			21.50		21.50	35
0800	Wood piles		38	.211			10.80		10.80	17.55
1000	Testing, any type piles, test load is twice the design load									
1050	50 ton design load, 100 ton test				Ea.				14,000	15,500
1100	100 ton design load, 200 ton test								20,000	22,000
1200	200 ton design load, 400 ton test								28,000	31,000

31 06 60.15 Mobilization

		Crew	Daily Output	Labor-Hours	Unit	Material	2018 Bare Costs Labor	Equipment	Total	Total Incl O&P
0010	**MOBILIZATION**									
0020	Set up & remove, air compressor, 600 CFM	A-5	3.30	5.455	Ea.		220	14.30	234.30	370
0100	1,200 CFM	"	2.20	8.182			330	21.50	351.50	560
0200	Crane, with pile leads and pile hammer, 75 ton	B-19	.60	107			5,600	3,225	8,825	12,500
0300	150 ton	"	.36	178			9,325	5,350	14,675	20,900

31 14 Earth Stripping and Stockpiling

31 14 13 – Soil Stripping and Stockpiling

31 14 13.23 Topsoil Stripping and Stockpiling

		Crew	Daily Output	Labor-Hours	Unit	Material	2018 Bare Costs Labor	Equipment	Total	Total Incl O&P
0010	**TOPSOIL STRIPPING AND STOCKPILING**									
1500	Loam or topsoil, remove/stockpile on site									
1510	By hand, 6" deep, 50' haul, less than 100 S.Y.	B-1	100	.240	S.Y.		9.70		9.70	15.80
1520	By skid steer, 6" deep, 100' haul, 101-500 S.Y.	B-62	500	.048			2.10	.35	2.45	3.75
1530	100' haul, 501-900 S.Y.	"	900	.027			1.16	.19	1.35	2.08
1540	200' haul, 901-1,100 S.Y.	B-63	1000	.040			1.69	.17	1.86	2.91
1550	By dozer, 200' haul, 1,101-4,000 S.Y.	B-10B	4000	.003	↓		.15	.32	.47	.58

31 23 Excavation and Fill

31 23 16 – Excavation

31 23 16.13 Excavating, Trench

		Crew	Daily Output	Labor-Hours	Unit	Material	2018 Bare Costs Labor	Equipment	Total	Total Incl O&P
0010	**EXCAVATING, TRENCH**									
0011	Or continuous footing									
0020	Common earth with no sheeting or dewatering included									
1400	By hand with pick and shovel 2' to 6' deep, light soil	1 Clab	8	1	B.C.Y.		40		40	64.50
1500	Heavy soil	"	4	2	"		79.50		79.50	129
1700	For tamping backfilled trenches, air tamp, add	A-1G	100	.080	E.C.Y.		3.19	.52	3.71	5.75
1900	Vibrating plate, add	B-18	180	.133	"		5.40	.23	5.63	9
2100	Trim sides and bottom for concrete pours, common earth		1500	.016	S.F.		.65	.03	.68	1.08
2300	Hardpan	↓	600	.040	"		1.62	.07	1.69	2.70
9000	Minimum labor/equipment charge	1 Clab	4	2	Job		79.50		79.50	129

31 23 16.14 Excavating, Utility Trench

		Crew	Daily Output	Labor-Hours	Unit	Material	2018 Bare Costs Labor	Equipment	Total	Total Incl O&P
0010	**EXCAVATING, UTILITY TRENCH**									
0011	Common earth									
0050	Trenching with chain trencher, 12 HP, operator walking									
0100	4" wide trench, 12" deep	B-53	800	.010	L.F.		.51	.08	.59	.90
0150	18" deep		750	.011			.55	.08	.63	.95
0200	24" deep		700	.011			.59	.09	.68	1.03
0300	6" wide trench, 12" deep		650	.012			.63	.10	.73	1.10
0350	18" deep		600	.013			.68	.10	.78	1.19
0400	24" deep		550	.015			.75	.11	.86	1.30
0450	36" deep		450	.018			.91	.14	1.05	1.59
0600	8" wide trench, 12" deep		475	.017			.86	.13	.99	1.50
0650	18" deep		400	.020			1.03	.15	1.18	1.79
0700	24" deep		350	.023			1.17	.18	1.35	2.04
0750	36" deep		300	.027	↓		1.37	.21	1.58	2.39
0900	Minimum labor/equipment charge	↓	2	4	Job		205	31	236	360
1000	Backfill by hand including compaction, add									
1050	4" wide trench, 12" deep	A-1G	800	.010	L.F.		.40	.07	.47	.72
1100	18" deep		530	.015			.60	.10	.70	1.09
1150	24" deep		400	.020			.80	.13	.93	1.43
1300	6" wide trench, 12" deep		540	.015			.59	.10	.69	1.07
1350	18" deep		405	.020			.79	.13	.92	1.42
1400	24" deep		270	.030			1.18	.19	1.37	2.13
1450	36" deep		180	.044			1.77	.29	2.06	3.20
1600	8" wide trench, 12" deep		400	.020			.80	.13	.93	1.43
1650	18" deep		265	.030			1.20	.20	1.40	2.17
1700	24" deep		200	.040			1.59	.26	1.85	2.88
1750	36" deep	↓	135	.059	↓		2.36	.39	2.75	4.25
2000	Chain trencher, 40 HP operator riding									
2050	6" wide trench and backfill, 12" deep	B-54	1200	.007	L.F.		.34	.28	.62	.85

31 23 16.14 Excavating, Utility Trench

		Crew	Daily Output	Labor-Hours	Unit	Material	2018 Bare Costs Labor	Equipment	Total	Total Incl O&P
2100	18" deep	B-54	1000	.008	L.F.		.41	.34	.75	1.02
2150	24" deep		975	.008			.42	.35	.77	1.04
2200	36" deep		900	.009			.46	.37	.83	1.13
2250	48" deep		750	.011			.55	.45	1	1.35
2300	60" deep		650	.012			.63	.52	1.15	1.57
2400	8" wide trench and backfill, 12" deep		1000	.008			.41	.34	.75	1.02
2450	18" deep		950	.008			.43	.35	.78	1.07
2500	24" deep		900	.009			.46	.37	.83	1.13
2550	36" deep		800	.010			.51	.42	.93	1.27
2600	48" deep		650	.012			.63	.52	1.15	1.57
2700	12" wide trench and backfill, 12" deep		975	.008			.42	.35	.77	1.04
2750	18" deep		860	.009			.48	.39	.87	1.18
2800	24" deep		800	.010			.51	.42	.93	1.27
2850	36" deep		725	.011			.57	.46	1.03	1.40
3000	16" wide trench and backfill, 12" deep		835	.010			.49	.40	.89	1.22
3050	18" deep		750	.011			.55	.45	1	1.35
3100	24" deep	▼	700	.011	▼		.59	.48	1.07	1.46
3200	Compaction with vibratory plate, add								35%	35%
5100	Hand excavate and trim for pipe bells after trench excavation									
5200	8" pipe	1 Clab	155	.052	L.F.		2.06		2.06	3.34
5300	18" pipe	"	130	.062	"		2.45		2.45	3.98
9000	Minimum labor/equipment charge	A-1G	4	2	Job		79.50	13	92.50	143

31 23 16.16 Structural Excavation for Minor Structures

		Crew	Daily Output	Labor-Hours	Unit	Material	2018 Bare Costs Labor	Equipment	Total	Total Incl O&P
0010	**STRUCTURAL EXCAVATION FOR MINOR STRUCTURES**									
0015	Hand, pits to 6' deep, sandy soil	1 Clab	8	1	B.C.Y.		40		40	64.50
0100	Heavy soil or clay		4	2			79.50		79.50	129
0300	Pits 6' to 12' deep, sandy soil		5	1.600			64		64	104
0500	Heavy soil or clay		3	2.667			106		106	173
0700	Pits 12' to 18' deep, sandy soil		4	2			79.50		79.50	129
0900	Heavy soil or clay		2	4			159		159	259
1100	Hand loading trucks from stock pile, sandy soil		12	.667			26.50		26.50	43
1300	Heavy soil or clay	▼	8	1	▼		40		40	64.50
1500	For wet or muck hand excavation, add to above								50%	50%
6000	Machine excavation, for spread and mat footings, elevator pits,									
6001	and small building foundations									
6030	Common earth, hydraulic backhoe, 1/2 C.Y. bucket	B-12E	55	.291	B.C.Y.		13.95	7.90	21.85	31
6035	3/4 C.Y. bucket	B-12F	90	.178			8.55	7.45	16	22
6040	1 C.Y. bucket	B-12A	108	.148			7.10	6.85	13.95	18.90
6050	1-1/2 C.Y. bucket	B-12B	144	.111			5.35	6.20	11.55	15.35
6060	2 C.Y. bucket	B-12C	200	.080			3.84	5.25	9.09	11.95
6070	Sand and gravel, 3/4 C.Y. bucket	B-12F	100	.160			7.70	6.70	14.40	19.60
6080	1 C.Y. bucket	B-12A	120	.133			6.40	6.20	12.60	17
6090	1-1/2 C.Y. bucket	B-12B	160	.100			4.80	5.60	10.40	13.80
6100	2 C.Y. bucket	B-12C	220	.073			3.49	4.78	8.27	10.80
6110	Clay, till, or blasted rock, 3/4 C.Y. bucket	B-12F	80	.200			9.60	8.35	17.95	24.50
6120	1 C.Y. bucket	B-12A	95	.168			8.10	7.80	15.90	21.50
6130	1-1/2 C.Y. bucket	B-12B	130	.123			5.90	6.85	12.75	17
6140	2 C.Y. bucket	B-12C	175	.091			4.39	6	10.39	13.60
6230	Sandy clay & loam, hydraulic backhoe, 1/2 C.Y. bucket	B-12E	60	.267			12.80	7.25	20.05	28.50
6235	3/4 C.Y. bucket	B-12F	98	.163			7.85	6.85	14.70	20
6240	1 C.Y. bucket	B-12A	116	.138			6.60	6.40	13	17.60
6250	1-1/2 C.Y. bucket	B-12B	156	.103	▼		4.92	5.75	10.67	14.15

524

For customer support on your Commercial Renovation Costs with RSMeans data, call 800.448.8182.

31 23 Excavation and Fill

31 23 16 – Excavation

31 23 16.16 Structural Excavation for Minor Structures

	Crew	Daily Output	Labor-Hours	Unit	Material	2018 Bare Costs Labor	2018 Bare Costs Equipment	Total	Total Incl O&P
9000 Minimum labor/equipment charge	1 Clab	4	2	Job		79.50		79.50	129

31 23 16.42 Excavating, Bulk Bank Measure

		Crew	Daily Output	Labor-Hours	Unit	Material	Labor	Equipment	Total	Total Incl O&P
0010	**EXCAVATING, BULK BANK MEASURE**	R312316-40								
0011	Common earth piled									
0020	For loading onto trucks, add								15%	15%
0200	Excavator, hydraulic, crawler mtd., 1 C.Y. cap. = 100 C.Y./hr.	R312316-45 B-12A	800	.020	B.C.Y.		.96	.93	1.89	2.55
1200	Front end loader, track mtd., 1-1/2 C.Y. cap. = 70 C.Y./hr.	B-10N	560	.021			1.05	1.04	2.09	2.81
1500	Wheel mounted, 3/4 C.Y. cap. = 45 C.Y./hr.	B-10R	360	.033	↓		1.64	.81	2.45	3.49
5000	Excavating, bulk bank measure, sandy clay & loam piled									
5020	For loading onto trucks, add								15%	15%
5100	Excavator, hydraulic, crawler mtd., 1 C.Y. cap. = 120 C.Y./hr.	B-12A	960	.017	B.C.Y.		.80	.77	1.57	2.13
9000	Minimum labor/equipment charge	B-10L	2	6	Job		295	232	527	725

31 23 16.46 Excavating, Bulk, Dozer

		Crew	Daily Output	Labor-Hours	Unit	Material	Labor	Equipment	Total	Total Incl O&P
0010	**EXCAVATING, BULK, DOZER**									
0011	Open site									
2000	80 HP, 50' haul, sand & gravel	B-10L	460	.026	B.C.Y.		1.28	1.01	2.29	3.15
2200	150' haul, sand & gravel		230	.052			2.56	2.02	4.58	6.30
2400	300' haul, sand & gravel	↓	120	.100			4.91	3.87	8.78	12.05
3000	105 HP, 50' haul, sand & gravel	B-10W	700	.017			.84	.87	1.71	2.30
3200	150' haul, sand & gravel		310	.039			1.90	1.96	3.86	5.20
3300	300' haul, sand & gravel	↓	140	.086			4.21	4.34	8.55	11.50
4000	200 HP, 50' haul, sand & gravel	B-10B	1400	.009			.42	.91	1.33	1.67
4200	150' haul, sand & gravel		595	.020			.99	2.14	3.13	3.93
4400	300' haul, sand & gravel	↓	310	.039	↓		1.90	4.11	6.01	7.55

31 23 23 – Fill

31 23 23.13 Backfill

		Crew	Daily Output	Labor-Hours	Unit	Material	Labor	Equipment	Total	Total Incl O&P
0010	**BACKFILL**	R312323-30								
0015	By hand, no compaction, light soil	1 Clab	14	.571	L.C.Y.		23		23	37
0100	Heavy soil	↓	11	.727	"		29		29	47
0300	Compaction in 6" layers, hand tamp, add to above		20.60	.388	E.C.Y.		15.50		15.50	25
0400	Roller compaction operator walking, add	B-10A	100	.120			5.90	1.80	7.70	11.35
0500	Air tamp, add	B-9D	190	.211			8.45	1.42	9.87	15.30
0600	Vibrating plate, add	A-1D	60	.133			5.30	.53	5.83	9.25
0800	Compaction in 12" layers, hand tamp, add to above	1 Clab	34	.235			9.40		9.40	15.20
0900	Roller compaction operator walking, add	B-10A	150	.080			3.93	1.20	5.13	7.55
1000	Air tamp, add	B-9	285	.140			5.65	.82	6.47	10.05
1100	Vibrating plate, add	A-1E	90	.089	↓		3.54	.45	3.99	6.25
1200	Trench, dozer, no compaction, 60 HP	B-10L	425	.028	L.C.Y.		1.39	1.09	2.48	3.41
1300	Dozer backfilling, bulk, up to 300' haul, no compaction	B-10B	1200	.010	"		.49	1.06	1.55	1.95
1400	Air tamped, add	B-11B	80	.200	E.C.Y.		9.10	3.56	12.66	18.45
1900	Dozer backfilling, trench, up to 300' haul, no compaction	B-10B	900	.013	L.C.Y.		.65	1.41	2.06	2.60
2000	Air tamped, add	B-11B	80	.200	E.C.Y.		9.10	3.56	12.66	18.45
2350	Spreading in 8" layers, small dozer	B-10B	1060	.011	L.C.Y.		.56	1.20	1.76	2.20
2450	Compacting with vibrating plate, 8" lifts	A-1D	73	.110	E.C.Y.		4.37	.44	4.81	7.60

31 23 23.14 Backfill, Structural

		Crew	Daily Output	Labor-Hours	Unit	Material	Labor	Equipment	Total	Total Incl O&P
0010	**BACKFILL, STRUCTURAL**									
0011	Dozer or F.E. loader									
0020	From existing stockpile, no compaction									
1000	55 HP wheeled loader, 50' haul, common earth	B-11C	200	.080	L.C.Y.		3.74	1.56	5.30	7.70
2000	80 HP, 50' haul, sand & gravel	B-10L	1100	.011			.54	.42	.96	1.31
2010	Sandy clay & loam	↓	1070	.011	↓		.55	.43	.98	1.36

For customer support on your Commercial Renovation Costs with RSMeans data, call 800.448.8182.

525

31 23 23.14 Backfill, Structural

		Crew	Daily Output	Labor-Hours	Unit	Material	2018 Bare Costs Labor	2018 Bare Costs Equipment	Total	Total Incl O&P
2020	Common earth	B-10L	975	.012	L.C.Y.		.60	.48	1.08	1.48
2040	Clay		850	.014			.69	.55	1.24	1.70
2400	300' haul, sand & gravel		370	.032			1.59	1.26	2.85	3.91
2410	Sandy clay & loam		360	.033			1.64	1.29	2.93	4.02
2420	Common earth		330	.036			1.79	1.41	3.20	4.39
2440	Clay		290	.041			2.03	1.60	3.63	4.99
3000	105 HP, 50' haul, sand & gravel	B-10W	1350	.009			.44	.45	.89	1.19
3010	Sandy clay & loam		1325	.009			.45	.46	.91	1.21
3020	Common earth		1225	.010			.48	.50	.98	1.32
3040	Clay		1100	.011			.54	.55	1.09	1.46
3300	300' haul, sand & gravel		465	.026			1.27	1.31	2.58	3.46
3310	Sandy clay & loam		455	.026			1.30	1.34	2.64	3.53
3320	Common earth		415	.029			1.42	1.47	2.89	3.87
3340	Clay		370	.032			1.59	1.64	3.23	4.34

31 23 23.16 Fill By Borrow and Utility Bedding

		Crew	Daily Output	Labor-Hours	Unit	Material	2018 Bare Costs Labor	2018 Bare Costs Equipment	Total	Total Incl O&P
0010	**FILL BY BORROW AND UTILITY BEDDING**									
0049	Utility bedding, for pipe & conduit, not incl. compaction									
0050	Crushed or screened bank run gravel	B-6	150	.160	L.C.Y.	21	7	2.08	30.08	36.50
0100	Crushed stone 3/4" to 1/2"		150	.160		28	7	2.08	37.08	44
0200	Sand, dead or bank		150	.160		18.15	7	2.08	27.23	33.50
0500	Compacting bedding in trench	A-1D	90	.089	E.C.Y.		3.54	.35	3.89	6.15
0600	If material source exceeds 2 miles, add for extra mileage.									
0610	See Section 31 23 23.20 for hauling mileage add.									

31 23 23.17 General Fill

		Crew	Daily Output	Labor-Hours	Unit	Material	2018 Bare Costs Labor	2018 Bare Costs Equipment	Total	Total Incl O&P
0010	**GENERAL FILL**									
0011	Spread dumped material, no compaction									
0020	By dozer	B-10B	1000	.012	L.C.Y.		.59	1.27	1.86	2.34
0100	By hand	1 Clab	12	.667	"		26.50		26.50	43
9000	Minimum labor/equipment charge	"	4	2	Job		79.50		79.50	129

31 23 23.20 Hauling

		Crew	Daily Output	Labor-Hours	Unit	Material	2018 Bare Costs Labor	2018 Bare Costs Equipment	Total	Total Incl O&P
0010	**HAULING**									
0011	Excavated or borrow, loose cubic yards									
0012	no loading equipment, including hauling, waiting, loading/dumping									
0013	time per cycle (wait, load, travel, unload or dump & return)									
0014	8 C.Y. truck, 15 MPH avg., cycle 0.5 miles, 10 min. wait/ld./uld.	B-34A	320	.025	L.C.Y.		1.15	1.06	2.21	3.01
0016	cycle 1 mile		272	.029			1.35	1.24	2.59	3.54
0018	cycle 2 miles		208	.038			1.77	1.63	3.40	4.63
0020	cycle 4 miles		144	.056			2.56	2.35	4.91	6.70
0022	cycle 6 miles		112	.071			3.29	3.02	6.31	8.65
0024	cycle 8 miles		88	.091			4.18	3.85	8.03	10.95
0026	20 MPH avg., cycle 0.5 mile		336	.024			1.10	1.01	2.11	2.87
0028	cycle 1 mile		296	.027			1.24	1.14	2.38	3.26
0030	cycle 2 miles		240	.033			1.53	1.41	2.94	4.01
0032	cycle 4 miles		176	.045			2.09	1.92	4.01	5.50
0034	cycle 6 miles		136	.059			2.71	2.49	5.20	7.10
0036	cycle 8 miles		112	.071			3.29	3.02	6.31	8.65
0044	25 MPH avg., cycle 4 miles		192	.042			1.92	1.76	3.68	5
0046	cycle 6 miles		160	.050			2.30	2.12	4.42	6.05
0048	cycle 8 miles		128	.063			2.88	2.65	5.53	7.55
0050	30 MPH avg., cycle 4 miles		216	.037			1.70	1.57	3.27	4.46
0052	cycle 6 miles		176	.045			2.09	1.92	4.01	5.50
0054	cycle 8 miles		144	.056			2.56	2.35	4.91	6.70

31 23 23 — Fill

31 23 23.20 Hauling		Crew	Daily Output	Labor-Hours	Unit	Material	2018 Bare Costs Labor	Equipment	Total	Total Incl O&P
0114	15 MPH avg., cycle 0.5 mile, 15 min. wait/ld./uld.	B-34A	224	.036	L.C.Y.		1.64	1.51	3.15	4.30
0116	cycle 1 mile		200	.040			1.84	1.69	3.53	4.82
0118	cycle 2 miles		168	.048			2.19	2.02	4.21	5.75
0120	cycle 4 miles		120	.067			3.07	2.82	5.89	8.05
0122	cycle 6 miles		96	.083			3.83	3.53	7.36	10.05
0124	cycle 8 miles		80	.100			4.60	4.23	8.83	12.05
0126	20 MPH avg., cycle 0.5 mile		232	.034			1.59	1.46	3.05	4.16
0128	cycle 1 mile		208	.038			1.77	1.63	3.40	4.63
0130	cycle 2 miles		184	.043			2	1.84	3.84	5.25
0132	cycle 4 miles		144	.056			2.56	2.35	4.91	6.70
0134	cycle 6 miles		112	.071			3.29	3.02	6.31	8.65
0136	cycle 8 miles		96	.083			3.83	3.53	7.36	10.05
0144	25 MPH avg., cycle 4 miles		152	.053			2.42	2.23	4.65	6.35
0146	cycle 6 miles		128	.063			2.88	2.65	5.53	7.55
0148	cycle 8 miles		112	.071			3.29	3.02	6.31	8.65
0150	30 MPH avg., cycle 4 miles		168	.048			2.19	2.02	4.21	5.75
0152	cycle 6 miles		144	.056			2.56	2.35	4.91	6.70
0154	cycle 8 miles		120	.067			3.07	2.82	5.89	8.05
0214	15 MPH avg., cycle 0.5 mile, 20 min. wait/ld./uld.		176	.045			2.09	1.92	4.01	5.50
0216	cycle 1 mile		160	.050			2.30	2.12	4.42	6.05
0218	cycle 2 miles		136	.059			2.71	2.49	5.20	7.10
0220	cycle 4 miles		104	.077			3.54	3.26	6.80	9.30
0222	cycle 6 miles		88	.091			4.18	3.85	8.03	10.95
0224	cycle 8 miles		72	.111			5.10	4.70	9.80	13.35
0226	20 MPH avg., cycle 0.5 mile		176	.045			2.09	1.92	4.01	5.50
0228	cycle 1 mile		168	.048			2.19	2.02	4.21	5.75
0230	cycle 2 miles		144	.056			2.56	2.35	4.91	6.70
0232	cycle 4 miles		120	.067			3.07	2.82	5.89	8.05
0234	cycle 6 miles		96	.083			3.83	3.53	7.36	10.05
0236	cycle 8 miles		88	.091			4.18	3.85	8.03	10.95
0244	25 MPH avg., cycle 4 miles		128	.063			2.88	2.65	5.53	7.55
0246	cycle 6 miles		112	.071			3.29	3.02	6.31	8.65
0248	cycle 8 miles		96	.083			3.83	3.53	7.36	10.05
0250	30 MPH avg., cycle 4 miles		136	.059			2.71	2.49	5.20	7.10
0252	cycle 6 miles		120	.067			3.07	2.82	5.89	8.05
0254	cycle 8 miles		104	.077			3.54	3.26	6.80	9.30
0314	15 MPH avg., cycle 0.5 mile, 25 min. wait/ld./uld.		144	.056			2.56	2.35	4.91	6.70
0316	cycle 1 mile		128	.063			2.88	2.65	5.53	7.55
0318	cycle 2 miles		112	.071			3.29	3.02	6.31	8.65
0320	cycle 4 miles		96	.083			3.83	3.53	7.36	10.05
0322	cycle 6 miles		80	.100			4.60	4.23	8.83	12.05
0324	cycle 8 miles		64	.125			5.75	5.30	11.05	15.05
0326	20 MPH avg., cycle 0.5 mile		144	.056			2.56	2.35	4.91	6.70
0328	cycle 1 mile		136	.059			2.71	2.49	5.20	7.10
0330	cycle 2 miles		120	.067			3.07	2.82	5.89	8.05
0332	cycle 4 miles		104	.077			3.54	3.26	6.80	9.30
0334	cycle 6 miles		88	.091			4.18	3.85	8.03	10.95
0336	cycle 8 miles		80	.100			4.60	4.23	8.83	12.05
0344	25 MPH avg., cycle 4 miles		112	.071			3.29	3.02	6.31	8.65
0346	cycle 6 miles		96	.083			3.83	3.53	7.36	10.05
0348	cycle 8 miles		88	.091			4.18	3.85	8.03	10.95
0350	30 MPH avg., cycle 4 miles		112	.071			3.29	3.02	6.31	8.65
0352	cycle 6 miles		104	.077			3.54	3.26	6.80	9.30

31 23 23.20 Hauling		Crew	Daily Output	Labor-Hours	Unit	Material	2018 Bare Costs Labor	Equipment	Total	Total Incl O&P
0354	cycle 8 miles	B-34A	96	.083	L.C.Y.		3.83	3.53	7.36	10.05
0414	15 MPH avg., cycle 0.5 mile, 30 min. wait/ld./uld.		120	.067			3.07	2.82	5.89	8.05
0416	cycle 1 mile		112	.071			3.29	3.02	6.31	8.65
0418	cycle 2 miles		96	.083			3.83	3.53	7.36	10.05
0420	cycle 4 miles		80	.100			4.60	4.23	8.83	12.05
0422	cycle 6 miles		72	.111			5.10	4.70	9.80	13.35
0424	cycle 8 miles		64	.125			5.75	5.30	11.05	15.05
0426	20 MPH avg., cycle 0.5 mile		120	.067			3.07	2.82	5.89	8.05
0428	cycle 1 mile		112	.071			3.29	3.02	6.31	8.65
0430	cycle 2 miles		104	.077			3.54	3.26	6.80	9.30
0432	cycle 4 miles		88	.091			4.18	3.85	8.03	10.95
0434	cycle 6 miles		80	.100			4.60	4.23	8.83	12.05
0436	cycle 8 miles		72	.111			5.10	4.70	9.80	13.35
0444	25 MPH avg., cycle 4 miles		96	.083			3.83	3.53	7.36	10.05
0446	cycle 6 miles		88	.091			4.18	3.85	8.03	10.95
0448	cycle 8 miles		80	.100			4.60	4.23	8.83	12.05
0450	30 MPH avg., cycle 4 miles		96	.083			3.83	3.53	7.36	10.05
0452	cycle 6 miles		88	.091			4.18	3.85	8.03	10.95
0454	cycle 8 miles		80	.100			4.60	4.23	8.83	12.05
0514	15 MPH avg., cycle 0.5 mile, 35 min. wait/ld./uld.		104	.077			3.54	3.26	6.80	9.30
0516	cycle 1 mile		96	.083			3.83	3.53	7.36	10.05
0518	cycle 2 miles		88	.091			4.18	3.85	8.03	10.95
0520	cycle 4 miles		72	.111			5.10	4.70	9.80	13.35
0522	cycle 6 miles		64	.125			5.75	5.30	11.05	15.05
0524	cycle 8 miles		56	.143			6.55	6.05	12.60	17.20
0526	20 MPH avg., cycle 0.5 mile		104	.077			3.54	3.26	6.80	9.30
0528	cycle 1 mile		96	.083			3.83	3.53	7.36	10.05
0530	cycle 2 miles		96	.083			3.83	3.53	7.36	10.05
0532	cycle 4 miles		80	.100			4.60	4.23	8.83	12.05
0534	cycle 6 miles		72	.111			5.10	4.70	9.80	13.35
0536	cycle 8 miles		64	.125			5.75	5.30	11.05	15.05
0544	25 MPH avg., cycle 4 miles		88	.091			4.18	3.85	8.03	10.95
0546	cycle 6 miles		80	.100			4.60	4.23	8.83	12.05
0548	cycle 8 miles		72	.111			5.10	4.70	9.80	13.35
0550	30 MPH avg., cycle 4 miles		88	.091			4.18	3.85	8.03	10.95
0552	cycle 6 miles		80	.100			4.60	4.23	8.83	12.05
0554	cycle 8 miles		72	.111			5.10	4.70	9.80	13.35
1014	12 C.Y. truck, cycle 0.5 mile, 15 MPH avg., 15 min. wait/ld./uld.	B-34B	336	.024			1.10	1.62	2.72	3.54
1016	cycle 1 mile		300	.027			1.23	1.81	3.04	3.96
1018	cycle 2 miles		252	.032			1.46	2.15	3.61	4.72
1020	cycle 4 miles		180	.044			2.04	3.02	5.06	6.60
1022	cycle 6 miles		144	.056			2.56	3.77	6.33	8.25
1024	cycle 8 miles		120	.067			3.07	4.52	7.59	9.90
1025	cycle 10 miles		96	.083			3.83	5.65	9.48	12.35
1026	20 MPH avg., cycle 0.5 mile		348	.023			1.06	1.56	2.62	3.42
1028	cycle 1 mile		312	.026			1.18	1.74	2.92	3.80
1030	cycle 2 miles		276	.029			1.33	1.97	3.30	4.30
1032	cycle 4 miles		216	.037			1.70	2.51	4.21	5.50
1034	cycle 6 miles		168	.048			2.19	3.23	5.42	7.05
1036	cycle 8 miles		144	.056			2.56	3.77	6.33	8.25
1038	cycle 10 miles		120	.067			3.07	4.52	7.59	9.90
1040	25 MPH avg., cycle 4 miles		228	.035			1.61	2.38	3.99	5.20
1042	cycle 6 miles		192	.042			1.92	2.83	4.75	6.20

31 23 23.20 Hauling		Crew	Daily Output	Labor-Hours	Unit	Material	2018 Bare Costs Labor	Equipment	Total	Total Incl O&P
1044	cycle 8 miles	B-34B	168	.048	L.C.Y.		2.19	3.23	5.42	7.05
1046	cycle 10 miles		144	.056			2.56	3.77	6.33	8.25
1050	30 MPH avg., cycle 4 miles		252	.032			1.46	2.15	3.61	4.72
1052	cycle 6 miles		216	.037			1.70	2.51	4.21	5.50
1054	cycle 8 miles		180	.044			2.04	3.02	5.06	6.60
1056	cycle 10 miles		156	.051			2.36	3.48	5.84	7.60
1060	35 MPH avg., cycle 4 miles		264	.030			1.39	2.06	3.45	4.50
1062	cycle 6 miles		228	.035			1.61	2.38	3.99	5.20
1064	cycle 8 miles		204	.039			1.80	2.66	4.46	5.85
1066	cycle 10 miles		180	.044			2.04	3.02	5.06	6.60
1068	cycle 20 miles		120	.067			3.07	4.52	7.59	9.90
1069	cycle 30 miles		84	.095			4.38	6.45	10.83	14.15
1070	cycle 40 miles		72	.111			5.10	7.55	12.65	16.50
1072	40 MPH avg., cycle 6 miles		240	.033			1.53	2.26	3.79	4.95
1074	cycle 8 miles		216	.037			1.70	2.51	4.21	5.50
1076	cycle 10 miles		192	.042			1.92	2.83	4.75	6.20
1078	cycle 20 miles		120	.067			3.07	4.52	7.59	9.90
1080	cycle 30 miles		96	.083			3.83	5.65	9.48	12.35
1082	cycle 40 miles		72	.111			5.10	7.55	12.65	16.50
1084	cycle 50 miles		60	.133			6.15	9.05	15.20	19.80
1094	45 MPH avg., cycle 8 miles		216	.037			1.70	2.51	4.21	5.50
1096	cycle 10 miles		204	.039			1.80	2.66	4.46	5.85
1098	cycle 20 miles		132	.061			2.79	4.11	6.90	9
1100	cycle 30 miles		108	.074			3.41	5.05	8.46	11
1102	cycle 40 miles		84	.095			4.38	6.45	10.83	14.15
1104	cycle 50 miles		72	.111			5.10	7.55	12.65	16.50
1106	50 MPH avg., cycle 10 miles		216	.037			1.70	2.51	4.21	5.50
1108	cycle 20 miles		144	.056			2.56	3.77	6.33	8.25
1110	cycle 30 miles		108	.074			3.41	5.05	8.46	11
1112	cycle 40 miles		84	.095			4.38	6.45	10.83	14.15
1114	cycle 50 miles		72	.111			5.10	7.55	12.65	16.50
1214	15 MPH avg., cycle 0.5 mile, 20 min. wait/ld./uld.		264	.030			1.39	2.06	3.45	4.50
1216	cycle 1 mile		240	.033			1.53	2.26	3.79	4.95
1218	cycle 2 miles		204	.039			1.80	2.66	4.46	5.85
1220	cycle 4 miles		156	.051			2.36	3.48	5.84	7.60
1222	cycle 6 miles		132	.061			2.79	4.11	6.90	9
1224	cycle 8 miles		108	.074			3.41	5.05	8.46	11
1225	cycle 10 miles		96	.083			3.83	5.65	9.48	12.35
1226	20 MPH avg., cycle 0.5 mile		264	.030			1.39	2.06	3.45	4.50
1228	cycle 1 mile		252	.032			1.46	2.15	3.61	4.72
1230	cycle 2 miles		216	.037			1.70	2.51	4.21	5.50
1232	cycle 4 miles		180	.044			2.04	3.02	5.06	6.60
1234	cycle 6 miles		144	.056			2.56	3.77	6.33	8.25
1236	cycle 8 miles		132	.061			2.79	4.11	6.90	9
1238	cycle 10 miles		108	.074			3.41	5.05	8.46	11
1240	25 MPH avg., cycle 4 miles		192	.042			1.92	2.83	4.75	6.20
1242	cycle 6 miles		168	.048			2.19	3.23	5.42	7.05
1244	cycle 8 miles		144	.056			2.56	3.77	6.33	8.25
1246	cycle 10 miles		132	.061			2.79	4.11	6.90	9
1250	30 MPH avg., cycle 4 miles		204	.039			1.80	2.66	4.46	5.85
1252	cycle 6 miles		180	.044			2.04	3.02	5.06	6.60
1254	cycle 8 miles		156	.051			2.36	3.48	5.84	7.60
1256	cycle 10 miles		144	.056			2.56	3.77	6.33	8.25

31 23 23.20 Hauling	Crew	Daily Output	Labor-Hours	Unit	Material	2018 Bare Costs Labor	2018 Bare Costs Equipment	Total	Total Incl O&P
1260	35 MPH avg., cycle 4 miles	B-34B	216	.037	L.C.Y.		1.70	2.51	4.21	5.50
1262	cycle 6 miles		192	.042			1.92	2.83	4.75	6.20
1264	cycle 8 miles		168	.048			2.19	3.23	5.42	7.05
1266	cycle 10 miles		156	.051			2.36	3.48	5.84	7.60
1268	cycle 20 miles		108	.074			3.41	5.05	8.46	11
1269	cycle 30 miles		72	.111			5.10	7.55	12.65	16.50
1270	cycle 40 miles		60	.133			6.15	9.05	15.20	19.80
1272	40 MPH avg., cycle 6 miles		192	.042			1.92	2.83	4.75	6.20
1274	cycle 8 miles		180	.044			2.04	3.02	5.06	6.60
1276	cycle 10 miles		156	.051			2.36	3.48	5.84	7.60
1278	cycle 20 miles		108	.074			3.41	5.05	8.46	11
1280	cycle 30 miles		84	.095			4.38	6.45	10.83	14.15
1282	cycle 40 miles		72	.111			5.10	7.55	12.65	16.50
1284	cycle 50 miles		60	.133			6.15	9.05	15.20	19.80
1294	45 MPH avg., cycle 8 miles		180	.044			2.04	3.02	5.06	6.60
1296	cycle 10 miles		168	.048			2.19	3.23	5.42	7.05
1298	cycle 20 miles		120	.067			3.07	4.52	7.59	9.90
1300	cycle 30 miles		96	.083			3.83	5.65	9.48	12.35
1302	cycle 40 miles		72	.111			5.10	7.55	12.65	16.50
1304	cycle 50 miles		60	.133			6.15	9.05	15.20	19.80
1306	50 MPH avg., cycle 10 miles		180	.044			2.04	3.02	5.06	6.60
1308	cycle 20 miles		132	.061			2.79	4.11	6.90	9
1310	cycle 30 miles		96	.083			3.83	5.65	9.48	12.35
1312	cycle 40 miles		84	.095			4.38	6.45	10.83	14.15
1314	cycle 50 miles		72	.111			5.10	7.55	12.65	16.50
1414	15 MPH avg., cycle 0.5 mile, 25 min. wait/ld./uld.		204	.039			1.80	2.66	4.46	5.85
1416	cycle 1 mile		192	.042			1.92	2.83	4.75	6.20
1418	cycle 2 miles		168	.048			2.19	3.23	5.42	7.05
1420	cycle 4 miles		132	.061			2.79	4.11	6.90	9
1422	cycle 6 miles		120	.067			3.07	4.52	7.59	9.90
1424	cycle 8 miles		96	.083			3.83	5.65	9.48	12.35
1425	cycle 10 miles		84	.095			4.38	6.45	10.83	14.15
1426	20 MPH avg., cycle 0.5 mile		216	.037			1.70	2.51	4.21	5.50
1428	cycle 1 mile		204	.039			1.80	2.66	4.46	5.85
1430	cycle 2 miles		180	.044			2.04	3.02	5.06	6.60
1432	cycle 4 miles		156	.051			2.36	3.48	5.84	7.60
1434	cycle 6 miles		132	.061			2.79	4.11	6.90	9
1436	cycle 8 miles		120	.067			3.07	4.52	7.59	9.90
1438	cycle 10 miles		96	.083			3.83	5.65	9.48	12.35
1440	25 MPH avg., cycle 4 miles		168	.048			2.19	3.23	5.42	7.05
1442	cycle 6 miles		144	.056			2.56	3.77	6.33	8.25
1444	cycle 8 miles		132	.061			2.79	4.11	6.90	9
1446	cycle 10 miles		108	.074			3.41	5.05	8.46	11
1450	30 MPH avg., cycle 4 miles		168	.048			2.19	3.23	5.42	7.05
1452	cycle 6 miles		156	.051			2.36	3.48	5.84	7.60
1454	cycle 8 miles		132	.061			2.79	4.11	6.90	9
1456	cycle 10 miles		120	.067			3.07	4.52	7.59	9.90
1460	35 MPH avg., cycle 4 miles		180	.044			2.04	3.02	5.06	6.60
1462	cycle 6 miles		156	.051			2.36	3.48	5.84	7.60
1464	cycle 8 miles		144	.056			2.56	3.77	6.33	8.25
1466	cycle 10 miles		132	.061			2.79	4.11	6.90	9
1468	cycle 20 miles		96	.083			3.83	5.65	9.48	12.35
1469	cycle 30 miles		72	.111			5.10	7.55	12.65	16.50

31 23 23 – Fill

31 23 23.20 Hauling		Crew	Daily Output	Labor-Hours	Unit	Material	2018 Bare Costs Labor	2018 Bare Costs Equipment	Total	Total Incl O&P
1470	cycle 40 miles	B-34B	60	.133	L.C.Y.		6.15	9.05	15.20	19.80
1472	40 MPH avg., cycle 6 miles		168	.048			2.19	3.23	5.42	7.05
1474	cycle 8 miles		156	.051			2.36	3.48	5.84	7.60
1476	cycle 10 miles		144	.056			2.56	3.77	6.33	8.25
1478	cycle 20 miles		96	.083			3.83	5.65	9.48	12.35
1480	cycle 30 miles		84	.095			4.38	6.45	10.83	14.15
1482	cycle 40 miles		60	.133			6.15	9.05	15.20	19.80
1484	cycle 50 miles		60	.133			6.15	9.05	15.20	19.80
1494	45 MPH avg., cycle 8 miles		156	.051			2.36	3.48	5.84	7.60
1496	cycle 10 miles		144	.056			2.56	3.77	6.33	8.25
1498	cycle 20 miles		108	.074			3.41	5.05	8.46	11
1500	cycle 30 miles		84	.095			4.38	6.45	10.83	14.15
1502	cycle 40 miles		72	.111			5.10	7.55	12.65	16.50
1504	cycle 50 miles		60	.133			6.15	9.05	15.20	19.80
1506	50 MPH avg., cycle 10 miles		156	.051			2.36	3.48	5.84	7.60
1508	cycle 20 miles		120	.067			3.07	4.52	7.59	9.90
1510	cycle 30 miles		96	.083			3.83	5.65	9.48	12.35
1512	cycle 40 miles		72	.111			5.10	7.55	12.65	16.50
1514	cycle 50 miles		60	.133			6.15	9.05	15.20	19.80
1614	15 MPH avg., cycle 0.5 mile, 30 min. wait/ld./uld.		180	.044			2.04	3.02	5.06	6.60
1616	cycle 1 mile		168	.048			2.19	3.23	5.42	7.05
1618	cycle 2 miles		144	.056			2.56	3.77	6.33	8.25
1620	cycle 4 miles		120	.067			3.07	4.52	7.59	9.90
1622	cycle 6 miles		108	.074			3.41	5.05	8.46	11
1624	cycle 8 miles		84	.095			4.38	6.45	10.83	14.15
1625	cycle 10 miles		84	.095			4.38	6.45	10.83	14.15
1626	20 MPH avg., cycle 0.5 mile		180	.044			2.04	3.02	5.06	6.60
1628	cycle 1 mile		168	.048			2.19	3.23	5.42	7.05
1630	cycle 2 miles		156	.051			2.36	3.48	5.84	7.60
1632	cycle 4 miles		132	.061			2.79	4.11	6.90	9
1634	cycle 6 miles		120	.067			3.07	4.52	7.59	9.90
1636	cycle 8 miles		108	.074			3.41	5.05	8.46	11
1638	cycle 10 miles		96	.083			3.83	5.65	9.48	12.35
1640	25 MPH avg., cycle 4 miles		144	.056			2.56	3.77	6.33	8.25
1642	cycle 6 miles		132	.061			2.79	4.11	6.90	9
1644	cycle 8 miles		108	.074			3.41	5.05	8.46	11
1646	cycle 10 miles		108	.074			3.41	5.05	8.46	11
1650	30 MPH avg., cycle 4 miles		144	.056			2.56	3.77	6.33	8.25
1652	cycle 6 miles		132	.061			2.79	4.11	6.90	9
1654	cycle 8 miles		120	.067			3.07	4.52	7.59	9.90
1656	cycle 10 miles		108	.074			3.41	5.05	8.46	11
1660	35 MPH avg., cycle 4 miles		156	.051			2.36	3.48	5.84	7.60
1662	cycle 6 miles		144	.056			2.56	3.77	6.33	8.25
1664	cycle 8 miles		132	.061			2.79	4.11	6.90	9
1666	cycle 10 miles		120	.067			3.07	4.52	7.59	9.90
1668	cycle 20 miles		84	.095			4.38	6.45	10.83	14.15
1669	cycle 30 miles		72	.111			5.10	7.55	12.65	16.50
1670	cycle 40 miles		60	.133			6.15	9.05	15.20	19.80
1672	40 MPH avg., cycle 6 miles		144	.056			2.56	3.77	6.33	8.25
1674	cycle 8 miles		132	.061			2.79	4.11	6.90	9
1676	cycle 10 miles		120	.067			3.07	4.52	7.59	9.90
1678	cycle 20 miles		96	.083			3.83	5.65	9.48	12.35
1680	cycle 30 miles		72	.111			5.10	7.55	12.65	16.50

31 23 23.20 Hauling		Crew	Daily Output	Labor-Hours	Unit	Material	2018 Bare Costs Labor	2018 Bare Costs Equipment	Total	Total Incl O&P
1682	cycle 40 miles	B-34B	60	.133	L.C.Y.		6.15	9.05	15.20	19.80
1684	cycle 50 miles		48	.167			7.65	11.30	18.95	25
1694	45 MPH avg., cycle 8 miles		144	.056			2.56	3.77	6.33	8.25
1696	cycle 10 miles		132	.061			2.79	4.11	6.90	9
1698	cycle 20 miles		96	.083			3.83	5.65	9.48	12.35
1700	cycle 30 miles		84	.095			4.38	6.45	10.83	14.15
1702	cycle 40 miles		60	.133			6.15	9.05	15.20	19.80
1704	cycle 50 miles		60	.133			6.15	9.05	15.20	19.80
1706	50 MPH avg., cycle 10 miles		132	.061			2.79	4.11	6.90	9
1708	cycle 20 miles		108	.074			3.41	5.05	8.46	11
1710	cycle 30 miles		84	.095			4.38	6.45	10.83	14.15
1712	cycle 40 miles		72	.111			5.10	7.55	12.65	16.50
1714	cycle 50 miles		60	.133			6.15	9.05	15.20	19.80
2000	Hauling, 8 C.Y. truck, small project cost per hour	B-34A	8	1	Hr.		46	42.50	88.50	121
2100	12 C.Y. truck	B-34B	8	1			46	68	114	149
2150	16.5 C.Y. truck	B-34C	8	1			46	76	122	158
2175	18 C.Y. 8 wheel truck	B-34I	8	1			46	88.50	134.50	171
2200	20 C.Y. truck	B-34D	8	1			46	78	124	160
2300	Grading at dump, or embankment if required, by dozer	B-10B	1000	.012	L.C.Y.		.59	1.27	1.86	2.34
2310	Spotter at fill or cut, if required	1 Clab	8	1	Hr.		40		40	64.50
9014	18 C.Y. truck, 8 wheels,15 min. wait/ld./uld.,15 MPH, cycle 0.5 mi.	B-34I	504	.016	L.C.Y.		.73	1.40	2.13	2.71
9016	cycle 1 mile		450	.018			.82	1.57	2.39	3.04
9018	cycle 2 miles		378	.021			.97	1.87	2.84	3.61
9020	cycle 4 miles		270	.030			1.36	2.61	3.97	5.05
9022	cycle 6 miles		216	.037			1.70	3.27	4.97	6.35
9024	cycle 8 miles		180	.044			2.04	3.92	5.96	7.60
9025	cycle 10 miles		144	.056			2.56	4.90	7.46	9.50
9026	20 MPH avg., cycle 0.5 mile		522	.015			.71	1.35	2.06	2.62
9028	cycle 1 mile		468	.017			.79	1.51	2.30	2.92
9030	cycle 2 miles		414	.019			.89	1.71	2.60	3.31
9032	cycle 4 miles		324	.025			1.14	2.18	3.32	4.22
9034	cycle 6 miles		252	.032			1.46	2.80	4.26	5.45
9036	cycle 8 miles		216	.037			1.70	3.27	4.97	6.35
9038	cycle 10 miles		180	.044			2.04	3.92	5.96	7.60
9040	25 MPH avg., cycle 4 miles		342	.023			1.08	2.06	3.14	4
9042	cycle 6 miles		288	.028			1.28	2.45	3.73	4.75
9044	cycle 8 miles		252	.032			1.46	2.80	4.26	5.45
9046	cycle 10 miles		216	.037			1.70	3.27	4.97	6.35
9050	30 MPH avg., cycle 4 miles		378	.021			.97	1.87	2.84	3.61
9052	cycle 6 miles		324	.025			1.14	2.18	3.32	4.22
9054	cycle 8 miles		270	.030			1.36	2.61	3.97	5.05
9056	cycle 10 miles		234	.034			1.57	3.02	4.59	5.85
9060	35 MPH avg., cycle 4 miles		396	.020			.93	1.78	2.71	3.45
9062	cycle 6 miles		342	.023			1.08	2.06	3.14	4
9064	cycle 8 miles		288	.028			1.28	2.45	3.73	4.75
9066	cycle 10 miles		270	.030			1.36	2.61	3.97	5.05
9068	cycle 20 miles		162	.049			2.27	4.36	6.63	8.45
9070	cycle 30 miles		126	.063			2.92	5.60	8.52	10.85
9072	cycle 40 miles		90	.089			4.09	7.85	11.94	15.20
9074	40 MPH avg., cycle 6 miles		360	.022			1.02	1.96	2.98	3.80
9076	cycle 8 miles		324	.025			1.14	2.18	3.32	4.22
9078	cycle 10 miles		288	.028			1.28	2.45	3.73	4.75
9080	cycle 20 miles		180	.044			2.04	3.92	5.96	7.60

31 23 23 – Fill

31 23 23.20 Hauling		Crew	Daily Output	Labor-Hours	Unit	Material	2018 Bare Costs Labor	2018 Bare Costs Equipment	Total	Total Incl O&P
9082	cycle 30 miles	B-34I	144	.056	L.C.Y.		2.56	4.90	7.46	9.50
9084	cycle 40 miles		108	.074			3.41	6.55	9.96	12.65
9086	cycle 50 miles		90	.089			4.09	7.85	11.94	15.20
9094	45 MPH avg., cycle 8 miles		324	.025			1.14	2.18	3.32	4.22
9096	cycle 10 miles		306	.026			1.20	2.31	3.51	4.47
9098	cycle 20 miles		198	.040			1.86	3.57	5.43	6.90
9100	cycle 30 miles		144	.056			2.56	4.90	7.46	9.50
9102	cycle 40 miles		126	.063			2.92	5.60	8.52	10.85
9104	cycle 50 miles		108	.074			3.41	6.55	9.96	12.65
9106	50 MPH avg., cycle 10 miles		324	.025			1.14	2.18	3.32	4.22
9108	cycle 20 miles		216	.037			1.70	3.27	4.97	6.35
9110	cycle 30 miles		162	.049			2.27	4.36	6.63	8.45
9112	cycle 40 miles		126	.063			2.92	5.60	8.52	10.85
9114	cycle 50 miles		108	.074			3.41	6.55	9.96	12.65
9214	20 min. wait/ld./uld.,15 MPH, cycle 0.5 mi.		396	.020			.93	1.78	2.71	3.45
9216	cycle 1 mile		360	.022			1.02	1.96	2.98	3.80
9218	cycle 2 miles		306	.026			1.20	2.31	3.51	4.47
9220	cycle 4 miles		234	.034			1.57	3.02	4.59	5.85
9222	cycle 6 miles		198	.040			1.86	3.57	5.43	6.90
9224	cycle 8 miles		162	.049			2.27	4.36	6.63	8.45
9225	cycle 10 miles		144	.056			2.56	4.90	7.46	9.50
9226	20 MPH avg., cycle 0.5 mile		396	.020			.93	1.78	2.71	3.45
9228	cycle 1 mile		378	.021			.97	1.87	2.84	3.61
9230	cycle 2 miles		324	.025			1.14	2.18	3.32	4.22
9232	cycle 4 miles		270	.030			1.36	2.61	3.97	5.05
9234	cycle 6 miles		216	.037			1.70	3.27	4.97	6.35
9236	cycle 8 miles		198	.040			1.86	3.57	5.43	6.90
9238	cycle 10 miles		162	.049			2.27	4.36	6.63	8.45
9240	25 MPH avg., cycle 4 miles		288	.028			1.28	2.45	3.73	4.75
9242	cycle 6 miles		252	.032			1.46	2.80	4.26	5.45
9244	cycle 8 miles		216	.037			1.70	3.27	4.97	6.35
9246	cycle 10 miles		198	.040			1.86	3.57	5.43	6.90
9250	30 MPH avg., cycle 4 miles		306	.026			1.20	2.31	3.51	4.47
9252	cycle 6 miles		270	.030			1.36	2.61	3.97	5.05
9254	cycle 8 miles		234	.034			1.57	3.02	4.59	5.85
9256	cycle 10 miles		216	.037			1.70	3.27	4.97	6.35
9260	35 MPH avg., cycle 4 miles		324	.025			1.14	2.18	3.32	4.22
9262	cycle 6 miles		288	.028			1.28	2.45	3.73	4.75
9264	cycle 8 miles		252	.032			1.46	2.80	4.26	5.45
9266	cycle 10 miles		234	.034			1.57	3.02	4.59	5.85
9268	cycle 20 miles		162	.049			2.27	4.36	6.63	8.45
9270	cycle 30 miles		108	.074			3.41	6.55	9.96	12.65
9272	cycle 40 miles		90	.089			4.09	7.85	11.94	15.20
9274	40 MPH avg., cycle 6 miles		288	.028			1.28	2.45	3.73	4.75
9276	cycle 8 miles		270	.030			1.36	2.61	3.97	5.05
9278	cycle 10 miles		234	.034			1.57	3.02	4.59	5.85
9280	cycle 20 miles		162	.049			2.27	4.36	6.63	8.45
9282	cycle 30 miles		126	.063			2.92	5.60	8.52	10.85
9284	cycle 40 miles		108	.074			3.41	6.55	9.96	12.65
9286	cycle 50 miles		90	.089			4.09	7.85	11.94	15.20
9294	45 MPH avg., cycle 8 miles		270	.030			1.36	2.61	3.97	5.05
9296	cycle 10 miles		252	.032			1.46	2.80	4.26	5.45
9298	cycle 20 miles		180	.044			2.04	3.92	5.96	7.60

31 23 23 – Fill

31 23 23.20 Hauling		Crew	Daily Output	Labor-Hours	Unit	Material	2018 Bare Costs Labor	2018 Bare Costs Equipment	Total	Total Incl O&P
9300	cycle 30 miles	B-34I	144	.056	L.C.Y.		2.56	4.90	7.46	9.50
9302	cycle 40 miles		108	.074			3.41	6.55	9.96	12.65
9304	cycle 50 miles		90	.089			4.09	7.85	11.94	15.20
9306	50 MPH avg., cycle 10 miles		270	.030			1.36	2.61	3.97	5.05
9308	cycle 20 miles		198	.040			1.86	3.57	5.43	6.90
9310	cycle 30 miles		144	.056			2.56	4.90	7.46	9.50
9312	cycle 40 miles		126	.063			2.92	5.60	8.52	10.85
9314	cycle 50 miles		108	.074			3.41	6.55	9.96	12.65
9414	25 min. wait/ld./uld.,15 MPH, cycle 0.5 mi.		306	.026			1.20	2.31	3.51	4.47
9416	cycle 1 mile		288	.028			1.28	2.45	3.73	4.75
9418	cycle 2 miles		252	.032			1.46	2.80	4.26	5.45
9420	cycle 4 miles		198	.040			1.86	3.57	5.43	6.90
9422	cycle 6 miles		180	.044			2.04	3.92	5.96	7.60
9424	cycle 8 miles		144	.056			2.56	4.90	7.46	9.50
9425	cycle 10 miles		126	.063			2.92	5.60	8.52	10.85
9426	20 MPH avg., cycle 0.5 mile		324	.025			1.14	2.18	3.32	4.22
9428	cycle 1 mile		306	.026			1.20	2.31	3.51	4.47
9430	cycle 2 miles		270	.030			1.36	2.61	3.97	5.05
9432	cycle 4 miles		234	.034			1.57	3.02	4.59	5.85
9434	cycle 6 miles		198	.040			1.86	3.57	5.43	6.90
9436	cycle 8 miles		180	.044			2.04	3.92	5.96	7.60
9438	cycle 10 miles		144	.056			2.56	4.90	7.46	9.50
9440	25 MPH avg., cycle 4 miles		252	.032			1.46	2.80	4.26	5.45
9442	cycle 6 miles		216	.037			1.70	3.27	4.97	6.35
9444	cycle 8 miles		198	.040			1.86	3.57	5.43	6.90
9446	cycle 10 miles		180	.044			2.04	3.92	5.96	7.60
9450	30 MPH avg., cycle 4 miles		252	.032			1.46	2.80	4.26	5.45
9452	cycle 6 miles		234	.034			1.57	3.02	4.59	5.85
9454	cycle 8 miles		198	.040			1.86	3.57	5.43	6.90
9456	cycle 10 miles		180	.044			2.04	3.92	5.96	7.60
9460	35 MPH avg., cycle 4 miles		270	.030			1.36	2.61	3.97	5.05
9462	cycle 6 miles		234	.034			1.57	3.02	4.59	5.85
9464	cycle 8 miles		216	.037			1.70	3.27	4.97	6.35
9466	cycle 10 miles		198	.040			1.86	3.57	5.43	6.90
9468	cycle 20 miles		144	.056			2.56	4.90	7.46	9.50
9470	cycle 30 miles		108	.074			3.41	6.55	9.96	12.65
9472	cycle 40 miles		90	.089			4.09	7.85	11.94	15.20
9474	40 MPH avg., cycle 6 miles		252	.032			1.46	2.80	4.26	5.45
9476	cycle 8 miles		234	.034			1.57	3.02	4.59	5.85
9478	cycle 10 miles		216	.037			1.70	3.27	4.97	6.35
9480	cycle 20 miles		144	.056			2.56	4.90	7.46	9.50
9482	cycle 30 miles		126	.063			2.92	5.60	8.52	10.85
9484	cycle 40 miles		90	.089			4.09	7.85	11.94	15.20
9486	cycle 50 miles		90	.089			4.09	7.85	11.94	15.20
9494	45 MPH avg., cycle 8 miles		234	.034			1.57	3.02	4.59	5.85
9496	cycle 10 miles		216	.037			1.70	3.27	4.97	6.35
9498	cycle 20 miles		162	.049			2.27	4.36	6.63	8.45
9500	cycle 30 miles		126	.063			2.92	5.60	8.52	10.85
9502	cycle 40 miles		108	.074			3.41	6.55	9.96	12.65
9504	cycle 50 miles		90	.089			4.09	7.85	11.94	15.20
9506	50 MPH avg., cycle 10 miles		234	.034			1.57	3.02	4.59	5.85
9508	cycle 20 miles		180	.044			2.04	3.92	5.96	7.60
9510	cycle 30 miles		144	.056			2.56	4.90	7.46	9.50

31 23 Excavation and Fill

31 23 23 – Fill

31 23 23.20 Hauling	Crew	Daily Output	Labor-Hours	Unit	Material	2018 Bare Costs Labor	Equipment	Total	Total Incl O&P
9512	cycle 40 miles	B-34I	108	.074	L.C.Y.		3.41	6.55	9.96	12.65
9514	cycle 50 miles		90	.089			4.09	7.85	11.94	15.20
9614	30 min. wait/ld./uld.,15 MPH, cycle 0.5 mi.		270	.030			1.36	2.61	3.97	5.05
9616	cycle 1 mile		252	.032			1.46	2.80	4.26	5.45
9618	cycle 2 miles		216	.037			1.70	3.27	4.97	6.35
9620	cycle 4 miles		180	.044			2.04	3.92	5.96	7.60
9622	cycle 6 miles		162	.049			2.27	4.36	6.63	8.45
9624	cycle 8 miles		126	.063			2.92	5.60	8.52	10.85
9625	cycle 10 miles		126	.063			2.92	5.60	8.52	10.85
9626	20 MPH avg., cycle 0.5 mile		270	.030			1.36	2.61	3.97	5.05
9628	cycle 1 mile		252	.032			1.46	2.80	4.26	5.45
9630	cycle 2 miles		234	.034			1.57	3.02	4.59	5.85
9632	cycle 4 miles		198	.040			1.86	3.57	5.43	6.90
9634	cycle 6 miles		180	.044			2.04	3.92	5.96	7.60
9636	cycle 8 miles		162	.049			2.27	4.36	6.63	8.45
9638	cycle 10 miles		144	.056			2.56	4.90	7.46	9.50
9640	25 MPH avg., cycle 4 miles		216	.037			1.70	3.27	4.97	6.35
9642	cycle 6 miles		198	.040			1.86	3.57	5.43	6.90
9644	cycle 8 miles		180	.044			2.04	3.92	5.96	7.60
9646	cycle 10 miles		162	.049			2.27	4.36	6.63	8.45
9650	30 MPH avg., cycle 4 miles		216	.037			1.70	3.27	4.97	6.35
9652	cycle 6 miles		198	.040			1.86	3.57	5.43	6.90
9654	cycle 8 miles		180	.044			2.04	3.92	5.96	7.60
9656	cycle 10 miles		162	.049			2.27	4.36	6.63	8.45
9660	35 MPH avg., cycle 4 miles		234	.034			1.57	3.02	4.59	5.85
9662	cycle 6 miles		216	.037			1.70	3.27	4.97	6.35
9664	cycle 8 miles		198	.040			1.86	3.57	5.43	6.90
9666	cycle 10 miles		180	.044			2.04	3.92	5.96	7.60
9668	cycle 20 miles		126	.063			2.92	5.60	8.52	10.85
9670	cycle 30 miles		108	.074			3.41	6.55	9.96	12.65
9672	cycle 40 miles		90	.089			4.09	7.85	11.94	15.20
9674	40 MPH avg., cycle 6 miles		216	.037			1.70	3.27	4.97	6.35
9676	cycle 8 miles		198	.040			1.86	3.57	5.43	6.90
9678	cycle 10 miles		180	.044			2.04	3.92	5.96	7.60
9680	cycle 20 miles		144	.056			2.56	4.90	7.46	9.50
9682	cycle 30 miles		108	.074			3.41	6.55	9.96	12.65
9684	cycle 40 miles		90	.089			4.09	7.85	11.94	15.20
9686	cycle 50 miles		72	.111			5.10	9.80	14.90	19
9694	45 MPH avg., cycle 8 miles		216	.037			1.70	3.27	4.97	6.35
9696	cycle 10 miles		198	.040			1.86	3.57	5.43	6.90
9698	cycle 20 miles		144	.056			2.56	4.90	7.46	9.50
9700	cycle 30 miles		126	.063			2.92	5.60	8.52	10.85
9702	cycle 40 miles		108	.074			3.41	6.55	9.96	12.65
9704	cycle 50 miles		90	.089			4.09	7.85	11.94	15.20
9706	50 MPH avg., cycle 10 miles		198	.040			1.86	3.57	5.43	6.90
9708	cycle 20 miles		162	.049			2.27	4.36	6.63	8.45
9710	cycle 30 miles		126	.063			2.92	5.60	8.52	10.85
9712	cycle 40 miles		108	.074			3.41	6.55	9.96	12.65
9714	cycle 50 miles		90	.089			4.09	7.85	11.94	15.20

For customer support on your Commercial Renovation Costs with RSMeans data, call 800.448.8182.

535

31 23 23.23 Compaction

	Crew	Daily Output	Labor-Hours	Unit	Material	2018 Bare Costs Labor	Equipment	Total	Total Incl O&P
0010 **COMPACTION**									
5000 Riding, vibrating roller, 6" lifts, 2 passes	B-10Y	3000	.004	E.C.Y.		.20	.19	.39	.52
5020 3 passes		2300	.005			.26	.25	.51	.69
5040 4 passes		1900	.006			.31	.30	.61	.82
5050 8" lifts, 2 passes		4100	.003			.14	.14	.28	.38
5060 12" lifts, 2 passes		5200	.002			.11	.11	.22	.30
5080 3 passes		3500	.003			.17	.16	.33	.45
5100 4 passes		2600	.005			.23	.22	.45	.60
5600 Sheepsfoot or wobbly wheel roller, 6" lifts, 2 passes	B-10G	2400	.005			.25	.55	.80	.99
5620 3 passes		1735	.007			.34	.76	1.10	1.37
5640 4 passes		1300	.009			.45	1.01	1.46	1.83
5680 12" lifts, 2 passes		5200	.002			.11	.25	.36	.46
5700 3 passes		3500	.003			.17	.38	.55	.68
5720 4 passes		2600	.005			.23	.51	.74	.92
7000 Walk behind, vibrating plate 18" wide, 6" lifts, 2 passes	A-1D	200	.040			1.59	.16	1.75	2.76
7020 3 passes		185	.043			1.72	.17	1.89	2.99
7040 4 passes		140	.057			2.28	.23	2.51	3.95
7200 12" lifts, 2 passes, 21" wide	A-1E	560	.014			.57	.07	.64	1
7220 3 passes		375	.021			.85	.11	.96	1.50
7240 4 passes		280	.029			1.14	.15	1.29	2.01
7500 Vibrating roller 24" wide, 6" lifts, 2 passes	B-10A	420	.029			1.40	.43	1.83	2.70
7520 3 passes		280	.043			2.11	.64	2.75	4.06
7540 4 passes		210	.057			2.81	.86	3.67	5.40
7600 12" lifts, 2 passes		840	.014			.70	.21	.91	1.36
7620 3 passes		560	.021			1.05	.32	1.37	2.02
7640 4 passes		420	.029			1.40	.43	1.83	2.70
8000 Rammer tamper, 6" to 11", 4" lifts, 2 passes	A-1F	130	.062			2.45	.35	2.80	4.37
8050 3 passes		97	.082			3.29	.47	3.76	5.85
8100 4 passes		65	.123			4.90	.71	5.61	8.75
8200 8" lifts, 2 passes		260	.031			1.23	.18	1.41	2.18
8250 3 passes		195	.041			1.64	.24	1.88	2.91
8300 4 passes		130	.062			2.45	.35	2.80	4.37
8400 13" to 18", 4" lifts, 2 passes	A-1G	390	.021			.82	.13	.95	1.48
8450 3 passes		290	.028			1.10	.18	1.28	1.99
8500 4 passes		195	.041			1.64	.27	1.91	2.94
8600 8" lifts, 2 passes		780	.010			.41	.07	.48	.73
8650 3 passes		585	.014			.55	.09	.64	.99
8700 4 passes		390	.021			.82	.13	.95	1.48
9000 Water, 3,000 gal. truck, 3 mile haul	B-45	1888	.008		1.19	.42	.43	2.04	2.45
9010 6 mile haul		1444	.011		1.19	.55	.56	2.30	2.80
9020 12 mile haul		1000	.016		1.19	.80	.80	2.79	3.46
9030 6,000 gal. wagon, 3 mile haul	B-59	2000	.004		1.19	.18	.22	1.59	1.85
9040 6 mile haul	"	1600	.005		1.19	.23	.27	1.69	1.98

31 23 23.24 Compaction, Structural

	Crew	Daily Output	Labor-Hours	Unit	Material	2018 Bare Costs Labor	Equipment	Total	Total Incl O&P
0010 **COMPACTION, STRUCTURAL** R312323-30									
0020 Steel wheel tandem roller, 5 tons	B-10E	8	1.500	Hr.		73.50	18.90	92.40	138
0050 Air tamp, 6" to 8" lifts, common fill	B-9	250	.160	E.C.Y.		6.45	.94	7.39	11.50
0060 Select fill	"	300	.133			5.35	.78	6.13	9.55
0600 Vibratory plate, 8" lifts, common fill	A-1D	200	.040			1.59	.16	1.75	2.76
0700 Select fill	"	216	.037			1.48	.15	1.63	2.56
9000 Minimum labor/equipment charge	1 Clab	4	2	Job		79.50		79.50	129

536

31 25 Erosion and Sedimentation Controls

31 25 14 – Stabilization Measures for Erosion and Sedimentation Control

31 25 14.16 Rolled Erosion Control Mats and Blankets

		Crew	Daily Output	Labor-Hours	Unit	Material	2018 Bare Costs Labor	Equipment	Total	Total Incl O&P
0010	**ROLLED EROSION CONTROL MATS AND BLANKETS**									
0020	Jute mesh, 100 S.Y. per roll, 4' wide, stapled	G B-80A	2400	.010	S.Y.	.91	.40	.10	1.41	1.76
0070	Paper biodegradable mesh	G B-1	2500	.010		.18	.39		.57	.83
0100	Plastic netting, stapled, 2" x 1" mesh, 20 mil	G "	2500	.010		.29	.39		.68	.95
0120	Revegetation mat, webbed	G 2 Clab	1000	.016		2.91	.64		3.55	4.24
0200	Polypropylene mesh, stapled, 6.5 oz./S.Y.	G B-1	2500	.010		1.63	.39		2.02	2.42
0300	Tobacco netting, or jute mesh #2, stapled	G "	2500	.010		.25	.39		.64	.91
1000	Silt fence, install and maintain, remove	G B-62	1300	.018	L.F.	.46	.81	.13	1.40	1.95
1100	Allow 10% per month for maintenance; 6-month max life									
1200	Place and remove hay bales, staked	G A-2	3	8	Ton	288	330	63	681	920
1250	Place and remove hay bales, staked (alt. pricing)	G "	2500	.010	L.F.	4.78	.40	.08	5.26	5.95

31 31 Soil Treatment

31 31 16 – Termite Control

31 31 16.13 Chemical Termite Control

		Crew	Daily Output	Labor-Hours	Unit	Material	2018 Bare Costs Labor	Equipment	Total	Total Incl O&P
0010	**CHEMICAL TERMITE CONTROL**									
0020	Slab and walls, residential	1 Skwk	1200	.007	SF Flr.	.32	.35		.67	.91
0030	SS mesh, no chemicals, avg 1,400 S.F. home, min	G	1000	.008		.32	.42		.74	1.02
0100	Commercial, minimum		2496	.003		.33	.17		.50	.63
0200	Maximum		1645	.005		.50	.25		.75	.95
0390	Minimum labor/equipment charge		4	2	Job		105		105	169
0400	Insecticides for termite control, minimum		14.20	.563	Gal.	69	29.50		98.50	123
0500	Maximum		11	.727	"	118	38		156	192

31 41 Shoring

31 41 13 – Timber Shoring

31 41 13.10 Building Shoring

		Crew	Daily Output	Labor-Hours	Unit	Material	2018 Bare Costs Labor	Equipment	Total	Total Incl O&P
0010	**BUILDING SHORING**									
0020	Shoring, existing building, with timber, no salvage allowance	B-51	2.20	21.818	M.B.F.	890	895	86	1,871	2,525
1000	On cribbing with 35 ton screw jacks, per box and jack		3.60	13.333	Jack	65	545	52.50	662.50	1,025
1090	Minimum labor/equipment charge		2	24	Ea.		985	94.50	1,079.50	1,700
1100	Masonry openings in walls, see Section 02 41 19.16									

31 41 16 – Sheet Piling

31 41 16.10 Sheet Piling Systems

		Crew	Daily Output	Labor-Hours	Unit	Material	2018 Bare Costs Labor	Equipment	Total	Total Incl O&P
0010	**SHEET PILING SYSTEMS**									
0020	Sheet piling, 50,000 psi steel, not incl. wales, 22 psf, left in place	B-40	10.81	5.920	Ton	1,575	310	330	2,215	2,600
0100	Drive, extract & salvage [R314116-40]		6	10.667	"	515	560	595	1,670	2,125
1200	15' deep excavation, 22 psf, left in place		983	.065	S.F.	18.40	3.41	3.62	25.43	29.50
1300	Drive, extract & salvage		545	.117	"	5.80	6.15	6.55	18.50	23.50
2100	Rent steel sheet piling and wales, first month				Ton	310			310	340
2200	Per added month				"	31			31	34
3900	Wood, solid sheeting, incl. wales, braces and spacers,									
3910	drive, extract & salvage, 8' deep excavation	B-31	330	.121	S.F.	1.84	5.15	.67	7.66	11.10
4520	Left in place, 8' deep, 55 S.F./hr.		440	.091	"	3.31	3.86	.50	7.67	10.45
4990	Minimum labor/equipment charge		2	20	Job		850	110	960	1,500
5000	For treated lumber add cost of treatment to lumber									

31 43 Concrete Raising

31 43 13 – Pressure Grouting

31 43 13.13 Concrete Pressure Grouting	Crew	Daily Output	Labor-Hours	Unit	Material	2018 Bare Costs Labor	Equipment	Total	Total Incl O&P
0010 **CONCRETE PRESSURE GROUTING**									
0020 Grouting, pressure, cement & sand, 1:1 mix, minimum	B-61	124	.323	Bag	19.20	13.70	2.59	35.49	46
0100 Maximum		51	.784	"	19.20	33.50	6.30	59	82
0200 Cement and sand, 1:1 mix, minimum		250	.160	C.F.	38.50	6.80	1.29	46.59	55
0300 Maximum		100	.400		57.50	17	3.22	77.72	94.50
0400 Epoxy cement grout, minimum		137	.292		755	12.40	2.35	769.75	855
0500 Maximum		57	.702		755	30	5.65	790.65	885
0700 Alternate pricing method: (Add for materials)									
0710 5 person crew and equipment	B-61	1	40	Day		1,700	320	2,020	3,100

31 46 Needle Beams

31 46 13 – Cantilever Needle Beams

31 46 13.10 Needle Beams

	Crew	Daily Output	Labor-Hours	Unit	Material	2018 Bare Costs Labor	Equipment	Total	Total Incl O&P
0010 **NEEDLE BEAMS**									
0011 Incl. wood shoring 10' x 10' opening									
0400 Block, concrete, 8" thick	B-9	7.10	5.634	Ea.	53	227	33	313	465
0420 12" thick		6.70	5.970		64.50	240	35	339.50	500
0800 Brick, 4" thick with 8" backup block		5.70	7.018		64.50	282	41	387.50	575
1000 Brick, solid, 8" thick		6.20	6.452		53	260	37.50	350.50	520
1040 12" thick		4.90	8.163		64.50	330	48	442.50	660
1080 16" thick		4.50	8.889		87	360	52	499	735
2000 Add for additional floors of shoring	B-1	6	4		53	162		215	320
9000 Minimum labor/equipment charge	"	2	12	Job		485		485	790

31 48 Underpinning

31 48 13 – Underpinning Piers

31 48 13.10 Underpinning Foundations

	Crew	Daily Output	Labor-Hours	Unit	Material	2018 Bare Costs Labor	Equipment	Total	Total Incl O&P
0010 **UNDERPINNING FOUNDATIONS**									
0011 Including excavation,									
0020 forming, reinforcing, concrete and equipment									
0100 5' to 16' below grade, 100 to 500 C.Y.	B-52	2.30	24.348	C.Y.	335	1,125	262	1,722	2,475
0200 Over 500 C.Y.		2.50	22.400		300	1,050	241	1,591	2,275
0400 16' to 25' below grade, 100 to 500 C.Y.		2	28		370	1,300	300	1,970	2,825
0500 Over 500 C.Y.		2.10	26.667		350	1,225	287	1,862	2,700
0700 26' to 40' below grade, 100 to 500 C.Y.		1.60	35		405	1,625	375	2,405	3,475
0800 Over 500 C.Y.		1.80	31.111		370	1,450	335	2,155	3,100
0900 For under 50 C.Y., add					10%	40%			

31 62 Driven Piles

31 62 13 – Concrete Piles

31 62 13.23 Prestressed Concrete Piles

31 62 13.23 Prestressed Concrete Piles	Crew	Daily Output	Labor-Hours	Unit	Material	2018 Bare Costs Labor	Equipment	Total	Total Incl O&P
0010 **PRESTRESSED CONCRETE PILES**, 200 piles									
0020 Unless specified otherwise, not incl. pile caps or mobilization									
3100 Precast, prestressed, 40' long, 10" thick, square	B-19	700	.091	V.L.F.	14.30	4.79	2.75	21.84	26.50
3200 12" thick, square		680	.094		21.50	4.93	2.84	29.27	34.50
3400 14" thick, square	↓	600	.107	↓	24	5.60	3.21	32.81	39

31 62 16 – Steel Piles

31 62 16.13 Steel Piles

	Crew	Daily Output	Labor-Hours	Unit	Material	2018 Bare Costs Labor	Equipment	Total	Total Incl O&P
0010 **STEEL PILES**									
0100 Step tapered, round, concrete filled									
0110 8" tip, 12" butt, 60 ton capacity, 30' depth	B-19	760	.084	V.L.F.	18.45	4.41	2.54	25.40	30.50
0120 60' depth with extension		740	.086		35	4.53	2.61	42.14	48.50
0130 80' depth with extensions	↓	700	.091		54	4.79	2.75	61.54	70
0250 "H" Sections, 50' long, HP8 x 36	↓	640	.100	↓	17.05	5.25	3.01	25.31	30.50

31 62 19 – Timber Piles

31 62 19.10 Wood Piles

	Crew	Daily Output	Labor-Hours	Unit	Material	2018 Bare Costs Labor	Equipment	Total	Total Incl O&P
0010 **WOOD PILES**									
0011 Friction or end bearing, not including									
0050 mobilization or demobilization									
0100 Untreated piles, up to 30' long, 12" butts, 8" points	B-19	625	.102	V.L.F.	11.45	5.35	3.09	19.89	24.50
0200 30' to 39' long, 12" butts, 8" points	"	700	.091	"	11.45	4.79	2.75	18.99	23.50
0800 Treated piles, 12 lb./C.F.,									
0810 friction or end bearing, ASTM class B									
1000 Up to 30' long, 12" butts, 8" points	B-19	625	.102	V.L.F.	13.35	5.35	3.09	21.79	26.50
1100 30' to 39' long, 12" butts, 8" points		700	.091		14.50	4.79	2.75	22.04	26.50
2700 Mobilization for 10,000 L.F. pile job, add		3300	.019			1.02	.58	1.60	2.27
2800 25,000 L.F. pile job, add	↓	8500	.008	↓		.39	.23	.62	.88

31 62 23 – Composite Piles

31 62 23.13 Concrete-Filled Steel Piles

	Crew	Daily Output	Labor-Hours	Unit	Material	2018 Bare Costs Labor	Equipment	Total	Total Incl O&P
0010 **CONCRETE-FILLED STEEL PILES** no mobilization or demobilization									
2600 Pipe piles, 50' L, 8" diam., 29 lb./L.F., no concrete	B-19	500	.128	V.L.F.	21.50	6.70	3.86	32.06	38.50
2700 Concrete filled		460	.139		24	7.30	4.19	35.49	43
2900 10" diameter, 34 lb./L.F., no concrete		500	.128		24	6.70	3.86	34.56	41.50
3000 Concrete filled		450	.142		28.50	7.45	4.29	40.24	48
3200 12" diameter, 44 lb./L.F., no concrete		475	.135		29.50	7.05	4.06	40.61	48.50
3300 Concrete filled		415	.154		35	8.10	4.65	47.75	56.50
3500 14" diameter, 46 lb./L.F., no concrete		430	.149		31.50	7.80	4.48	43.78	52.50
3600 Concrete filled	↓	355	.180	↓	40	9.45	5.45	54.90	65

31 63 Bored Piles

31 63 26 – Drilled Caissons

31 63 26.13 Fixed End Caisson Piles

	Crew	Daily Output	Labor-Hours	Unit	Material	2018 Bare Costs Labor	Equipment	Total	Total Incl O&P
0010 **FIXED END CAISSON PILES**									
0015 Including excavation, concrete, 50 lb. reinforcing									
0020 per C.Y., not incl. mobilization, boulder removal, disposal									
0100 Open style, machine drilled, to 50' deep, in stable ground, no R316326-60									
0110 casings or ground water, 18" diam., 0.065 C.Y./L.F.	B-43	200	.240	V.L.F.	9.40	10.65	12.20	32.25	41
0200 24" diameter, 0.116 C.Y./L.F.		190	.253	"	16.80	11.20	12.85	40.85	50.50
0210 4' bell diameter , add		20	2.400	Ea.	268	106	122	496	600
0500 48" diameter, 0.465 C.Y./L.F.	↓	100	.480	V.L.F.	67.50	21.50	24.50	113.50	135

For customer support on your Commercial Renovation Costs with RSMeans data, call 800.448.8182.

31 63 Bored Piles

31 63 26 – Drilled Caissons

31 63 26.13 Fixed End Caisson Piles

		Crew	Daily Output	Labor-Hours	Unit	Material	2018 Bare Costs Labor	2018 Bare Costs Equipment	Total	Total Incl O&P
0510	9' bell diameter, add	B-43	2	24	Ea.	1,575	1,075	1,225	3,875	4,775
1200	Open style, machine drilled, to 50' deep, in wet ground, pulled									
1220	casing and pumping									
1400	24" diameter, 0.116 C.Y./L.F.	B-48	125	.448	V.L.F.	16.80	20.50	22	59.30	75.50
1410	4' bell diameter, add	"	19.80	2.828	Ea.	268	128	139	535	650
1700	48" diameter, 0.465 C.Y./L.F.	B-49	55	1.600	V.L.F.	67.50	76.50	61	205	263
1710	9' bell diameter, add	"	3.30	26.667	Ea.	1,575	1,275	1,025	3,875	4,875
2300	Open style, machine drilled, to 50' deep, in soft rocks and									
2320	medium hard shales									
2500	24" diameter, 0.116 C.Y./L.F.	B-49	30	2.933	V.L.F.	16.80	140	112	268.80	365
2510	4' bell diameter, add		10.90	8.073	Ea.	268	385	305	958	1,250
2800	48" diameter, 0.465 C.Y./L.F.		10	8.800	V.L.F.	67.50	420	335	822.50	1,125
2810	9' bell diameter, add		1.10	80	Ea.	1,175	3,825	3,050	8,050	10,800
3600	For rock excavation, sockets, add, minimum		120	.733	C.F.		35	28	63	86.50
3650	Average		95	.926			44	35.50	79.50	110
3700	Maximum		48	1.833			87.50	70	157.50	217
3900	For 50' to 100' deep, add				V.L.F.				7%	7%
4000	For 100' to 150' deep, add								25%	25%
4100	For 150' to 200' deep, add								30%	30%
4200	For casings left in place, add				Lb.	1.36			1.36	1.50
4300	For other than 50 lb. reinf. per C.Y., add or deduct				"	1.26			1.26	1.39
4400	For steel I-beam cores, add	B-49	8.30	10.602	Ton	2,150	505	405	3,060	3,625
4500	Load and haul excess excavation, 2 miles	B-34B	178	.045	L.C.Y.		2.07	3.05	5.12	6.65
4600	For mobilization, 50 mile radius, rig to 36"	B-43	2	24	Ea.		1,075	1,225	2,300	3,050
4650	Rig to 84"	B-48	1.75	32			1,450	1,575	3,025	4,050
4700	For low headroom, add								50%	50%
4750	For difficult access, add								25%	25%

31 63 29 – Drilled Concrete Piers and Shafts

31 63 29.13 Uncased Drilled Concrete Piers

		Crew	Daily Output	Labor-Hours	Unit	Material	2018 Bare Costs Labor	2018 Bare Costs Equipment	Total	Total Incl O&P
0010	**UNCASED DRILLED CONCRETE PIERS**									
0020	Unless specified otherwise, not incl. pile caps or mobilization									
0800	Cast in place friction pile, 50' long, fluted,									
0810	tapered steel, 4,000 psi concrete, no reinforcing									
0900	12" diameter, 7 ga.	B-19	600	.107	V.L.F.	29.50	5.60	3.21	38.31	45
1200	18" diameter, 7 ga.	"	480	.133	"	44.50	7	4.02	55.52	64
1300	End bearing, fluted, constant diameter,									
1320	4,000 psi concrete, no reinforcing									
1340	12" diameter, 7 ga.	B-19	600	.107	V.L.F.	31	5.60	3.21	39.81	46.50
1400	18" diameter, 7 ga.	"	480	.133	"	49.50	7	4.02	60.52	70

Estimating Tips

32 01 00 Operations and Maintenance of Exterior Improvements

- Recycling of asphalt pavement is becoming very popular and is an alternative to removal and replacement. It can be a good value engineering proposal if removed pavement can be recycled, either at the project site or at another site that is reasonably close to the project site. Sections on repair of flexible and rigid pavement are included.

32 10 00 Bases, Ballasts, and Paving

- When estimating paving, keep in mind the project schedule. Also note that prices for asphalt and concrete are generally higher in the cold seasons. Lines for pavement markings, including tactile warning systems and fence lines, are included.

32 90 00 Planting

- The timing of planting and guarantee specifications often dictate the costs for establishing tree and shrub growth and a stand of grass or ground cover. Establish the work performance schedule to coincide with the local planting season. Maintenance and growth guarantees can add from 20%–100% to the total landscaping cost and can be contractually cumbersome. The cost to replace trees and shrubs can be as high as 5% of the total cost, depending on the planting zone, soil conditions, and time of year.

Reference Numbers

Reference numbers are shown at the beginning of some major classifications. These numbers refer to related items in the Reference Section. The reference information may be an estimating procedure, an alternate pricing method, or technical information.

Note: Not all subdivisions listed here necessarily appear. ∎

Did you know?

RSMeans data is available through our online application with 24/7 access:

- Search for unit prices by keyword
- Leverage the most up-to-date data
- Build and export estimates

Try it free for 30 days!
www.rsmeans.com/2018freetrial

32 01 13 – Flexible Paving Surface Treatment

32 01 13.61 Slurry Seal (Latex Modified)	Crew	Daily Output	Labor-Hours	Unit	Material	2018 Bare Costs Labor	Equipment	Total	Total Incl O&P
0010 **SLURRY SEAL (LATEX MODIFIED)**									
3600 Waterproofing, membrane, tar and fabric, small area	B-63	233	.172	S.Y.	14.20	7.25	.75	22.20	28
3640 Large area		1435	.028		13.10	1.17	.12	14.39	16.40
3680 Preformed rubberized asphalt, small area		100	.400		19.35	16.85	1.74	37.94	50.50
3720 Large area		367	.109		17.65	4.59	.47	22.71	27.50
3780 Rubberized asphalt (latex) seal	B-45	5000	.003		1.24	.16	.16	1.56	1.79
9000 Minimum labor/equipment charge	1 Clab	2	4	Job		159		159	259

32 01 13.62 Asphalt Surface Treatment

	Crew	Daily Output	Labor-Hours	Unit	Material	Labor	Equipment	Total	Total Incl O&P
0010 **ASPHALT SURFACE TREATMENT**									
3000 Pavement overlay, polypropylene									
3040 6 oz./S.Y., ideal conditions	B-63	10000	.004	S.Y.	.99	.17	.02	1.18	1.38
3080 Adverse conditions		1000	.040		1.33	1.69	.17	3.19	4.37
3120 4 oz./S.Y., ideal conditions		10000	.004		.89	.17	.02	1.08	1.27
3160 Adverse conditions		1000	.040		1.15	1.69	.17	3.01	4.17
3200 Tack coat, emulsion, 0.05 gal./S.Y., 1,000 S.Y.	B-45	2500	.006		.33	.32	.32	.97	1.22
3240 10,000 S.Y.		10000	.002		.26	.08	.08	.42	.51
3270 0.10 gal./S.Y., 1,000 S.Y.		2500	.006		.62	.32	.32	1.26	1.54
3275 10,000 S.Y.		10000	.002		.50	.08	.08	.66	.76
3280 0.15 gal./S.Y., 1,000 S.Y.		2500	.006		.91	.32	.32	1.55	1.86
3320 10,000 S.Y.		10000	.002		.73	.08	.08	.89	1.02

32 01 13.64 Sand Seal

	Crew	Daily Output	Labor-Hours	Unit	Material	Labor	Equipment	Total	Total Incl O&P
0010 **SAND SEAL**									
2080 Sand sealing, sharp sand, asphalt emulsion, small area	B-91	10000	.006	S.Y.	1.49	.31	.22	2.02	2.37
2120 Roadway or large area	"	18000	.004	"	1.28	.17	.12	1.57	1.82
3000 Sealing random cracks, min. 1/2" wide, to 1-1/2", 1,000 L.F.	B-77	2800	.014	L.F.	1.56	.59	.18	2.33	2.86
3040 10,000 L.F.		4000	.010	"	1.08	.41	.12	1.61	2
3080 Alternate method, 1,000 L.F.		200	.200	Gal.	46.50	8.25	2.46	57.21	67
3120 10,000 L.F.		325	.123	"	37.50	5.05	1.51	44.06	51
3200 Multi-cracks (flooding), 1 coat, small area	B-92	460	.070	S.Y.	2.43	2.81	1.15	6.39	8.50
3240 Large area		2850	.011		2.16	.45	.19	2.80	3.31
3280 2 coat, small area		230	.139		14.70	5.60	2.30	22.60	28
3320 Large area		1425	.022		13.35	.91	.37	14.63	16.55
3360 Alternate method, small area		115	.278	Gal.	15.85	11.25	4.59	31.69	41
3400 Large area		715	.045	"	14.90	1.81	.74	17.45	20

32 01 13.66 Fog Seal

	Crew	Daily Output	Labor-Hours	Unit	Material	Labor	Equipment	Total	Total Incl O&P
0010 **FOG SEAL**									
0012 Sealcoating, 2 coat coal tar pitch emulsion over 10,000 S.Y.	B-45	5000	.003	S.Y.	.85	.16	.16	1.17	1.37
0030 1,000 to 10,000 S.Y.	"	3000	.005		.85	.27	.27	1.39	1.65
0100 Under 1,000 S.Y.	B-1	1050	.023		.85	.93		1.78	2.44
0300 Petroleum resistant, over 10,000 S.Y.	B-45	5000	.003		1.32	.16	.16	1.64	1.88
0320 1,000 to 10,000 S.Y.	"	3000	.005		1.32	.27	.27	1.86	2.16
0400 Under 1,000 S.Y.	B-1	1050	.023		1.32	.93		2.25	2.95
0600 Non-skid pavement renewal, over 10,000 S.Y.	B-45	5000	.003		1.38	.16	.16	1.70	1.95
0620 1,000 to 10,000 S.Y.	"	3000	.005		1.38	.27	.27	1.92	2.23
0700 Under 1,000 S.Y.	B-1	1050	.023		1.38	.93		2.31	3.02
0800 Prepare and clean surface for above	A-2	8545	.003			.12	.02	.14	.21
1000 Hand seal asphalt curbing	B-1	4420	.005	L.F.	.63	.22		.85	1.05
1900 Asphalt surface treatment, single course, small area									
1901 0.30 gal./S.Y. asphalt material, 20 lb./S.Y. aggregate	B-91	5000	.013	S.Y.	1.32	.61	.44	2.37	2.92
1910 Roadway or large area		10000	.006		1.21	.31	.22	1.74	2.07
1950 Asphalt surface treatment, dbl. course for small area		3000	.021		2.98	1.02	.74	4.74	5.70
1960 Roadway or large area		6000	.011		2.68	.51	.37	3.56	4.17

32 01 Operation and Maintenance of Exterior Improvements

32 01 13 – Flexible Paving Surface Treatment

32 01 13.66 Fog Seal

		Crew	Daily Output	Labor-Hours	Unit	Material	2018 Bare Costs Labor	2018 Bare Costs Equipment	Total	Total Incl O&P
1980	Asphalt surface treatment, single course, for shoulders	B-91	7500	.009	S.Y.	1.47	.41	.29	2.17	2.59

32 01 13.68 Slurry Seal

		Crew	Daily Output	Labor-Hours	Unit	Material	Labor	Equipment	Total	Total Incl O&P
0010	**SLURRY SEAL**									
0100	Slurry seal, type I, 8 lb. agg./S.Y., 1 coat, small or irregular area	B-90	2800	.023	S.Y.	2.09	1.02	.78	3.89	4.79
0150	Roadway or large area		10000	.006		2.09	.28	.22	2.59	3
0200	Type II, 12 lb. aggregate/S.Y., 2 coats, small or irregular area		2000	.032		4.18	1.42	1.09	6.69	8.10
0250	Roadway or large area		8000	.008		4.18	.36	.27	4.81	5.45
0300	Type III, 20 lb. aggregate/S.Y., 2 coats, small or irregular area		1800	.036		4.93	1.58	1.21	7.72	9.35
0350	Roadway or large area		6000	.011		4.93	.47	.36	5.76	6.60
0400	Slurry seal, thermoplastic coal-tar, type I, small or irregular area		2400	.027		4.20	1.19	.91	6.30	7.55
0450	Roadway or large area		8000	.008		4.20	.36	.27	4.83	5.50
0500	Type II, small or irregular area		2400	.027		5.35	1.19	.91	7.45	8.80
0550	Roadway or large area		7800	.008		5.35	.37	.28	6	6.80

32 01 16 – Flexible Paving Rehabilitation

32 01 16.71 Cold Milling Asphalt Paving

		Crew	Daily Output	Labor-Hours	Unit	Material	Labor	Equipment	Total	Total Incl O&P
0010	**COLD MILLING ASPHALT PAVING**									
5200	Cold planing & cleaning, 1" to 3" asphalt pavmt., over 25,000 S.Y.	B-71	6000	.009	S.Y.		.43	1.07	1.50	1.87
5280	5,000 S.Y. to 10,000 S.Y.	"	4000	.014	"		.65	1.61	2.26	2.80
5300	Asphalt pavement removal from conc. base, no haul									
5320	Rip, load & sweep 1" to 3"	B-70	8000	.007	S.Y.		.32	.22	.54	.76
5330	3" to 6" deep	"	5000	.011			.52	.35	.87	1.22
5340	Profile grooving, asphalt pavement load & sweep, 1" deep	B-71	12500	.004			.21	.52	.73	.90
5350	3" deep		9000	.006			.29	.72	1.01	1.25
5360	6" deep		5000	.011			.52	1.29	1.81	2.25

32 01 16.73 In Place Cold Reused Asphalt Paving

		Crew	Daily Output	Labor-Hours	Unit	Material	Labor	Equipment	Total	Total Incl O&P
0010	**IN PLACE COLD REUSED ASPHALT PAVING**									
5000	Reclamation, pulverizing and blending with existing base									
5040	Aggregate base, 4" thick pavement, over 15,000 S.Y.	B-73	2400	.027	S.Y.		1.30	2.09	3.39	4.37
5080	5,000 S.Y. to 15,000 S.Y.		2200	.029			1.42	2.28	3.70	4.77
5120	8" thick pavement, over 15,000 S.Y.		2200	.029			1.42	2.28	3.70	4.77
5160	5,000 S.Y. to 15,000 S.Y.		2000	.032			1.56	2.51	4.07	5.25

32 01 16.74 In Place Hot Reused Asphalt Paving

		Crew	Daily Output	Labor-Hours	Unit	Material	Labor	Equipment	Total	Total Incl O&P
0010	**IN PLACE HOT REUSED ASPHALT PAVING**									
5500	Recycle asphalt pavement at site									
5520	Remove, rejuvenate and spread 4" deep [G]	B-72	2500	.026	S.Y.	4.38	1.20	4.40	9.98	11.60
5521	6" deep [G]	"	2000	.032	"	6.40	1.51	5.50	13.41	15.50

32 01 17 – Flexible Paving Repair

32 01 17.10 Repair of Asphalt Pavement Holes

		Crew	Daily Output	Labor-Hours	Unit	Material	Labor	Equipment	Total	Total Incl O&P
0010	**REPAIR OF ASPHALT PAVEMENT HOLES** (cold patch)									
0100	Flexible pavement repair holes, roadway, light traffic, 1 C.F. size each	B-37A	24	1	Ea.	9.20	41.50	13.25	63.95	91.50
0150	Group of two, 1 C.F. size each		16	1.500	Set	18.40	62	19.85	100.25	142
0200	Group of three, 1 C.F. size each		12	2	"	27.50	83	26.50	137	194
0300	Medium traffic, 1 C.F. size each	B-37B	24	1.333	Ea.	9.20	54.50	13.25	76.95	113
0350	Group of two, 1 C.F. size each		16	2	Set	18.40	82	19.85	120.25	175
0400	Group of three, 1 C.F. size each		12	2.667	"	27.50	109	26.50	163	237
0500	Highway/heavy traffic, 1 C.F. size each	B-37C	24	1.333	Ea.	9.20	56	21	86.20	124
0550	Group of two, 1 C.F. size each		16	2	Set	18.40	84.50	31.50	134.40	191
0600	Group of three, 1 C.F. size each		12	2.667	"	27.50	112	42	181.50	259
0700	Add police officer and car for traffic control				Hr.	43.50			43.50	48
1000	Flexible pavement repair holes, parking lot, bag material, 1 C.F. size each	B-37D	18	.889	Ea.	51.50	37.50	6.40	95.40	125
1010	Economy bag material, 1 C.F. size each		18	.889	"	37	37.50	6.40	80.90	109

For customer support on your Commercial Renovation Costs with RSMeans data, call 800.448.8182.

543

32 01 17.10 Repair of Asphalt Pavement Holes

	Crew	Daily Output	Labor-Hours	Unit	Material	2018 Bare Costs Labor	Equipment	Total	Total Incl O&P
1100	Group of two, 1 C.F. size each	B-37D	12	1.333	Set	103	56	9.60	168.60	216
1110	Group of two, economy bag, 1 C.F. size each		12	1.333		74.50	56	9.60	140.10	184
1200	Group of three, 1 C.F. size each		10	1.600		155	67.50	11.50	234	293
1210	Economy bag, group of three, 1 C.F. size each		10	1.600		112	67.50	11.50	191	245
1300	Flexible pavement repair holes, parking lot, 1 C.F. single hole	A-3A	4	2	Ea.	51.50	103	31.50	186	254
1310	Economy material, 1 C.F. single hole		4	2	"	37	103	31.50	171.50	238
1400	Flexible pavement repair holes, parking lot, 1 C.F. four holes		2	4	Set	207	205	63.50	475.50	625
1410	Economy material, 1 C.F. four holes		2	4	"	149	205	63.50	417.50	560
1500	Flexible pavement repair holes, large parking lot, bulk matl, 1 C.F. size	B-37A	32	.750	Ea.	9.20	31	9.90	50.10	71

32 01 17.20 Repair of Asphalt Pavement Patches

	Crew	Daily Output	Labor-Hours	Unit	Material	2018 Bare Costs Labor	Equipment	Total	Total Incl O&P
0010	**REPAIR OF ASPHALT PAVEMENT PATCHES**									
0100	Flexible pavement patches, roadway, light traffic, sawcut 10-25 S.F.	B-89	12	1.333	Ea.		64	35	99	141
0150	Sawcut 26-60 S.F.		10	1.600			76.50	42	118.50	168
0200	Sawcut 61-100 S.F.		8	2			96	52.50	148.50	210
0210	Sawcut groups of small size patches		24	.667			32	17.50	49.50	70.50
0220	Large size patches		16	1			48	26.50	74.50	105
0300	Flexible pavement patches, roadway, light traffic, digout 10-25 S.F.	B-6	16	1.500			65.50	19.55	85.05	127
0350	Digout 26-60 S.F.		12	2			87.50	26	113.50	169
0400	Digout 61-100 S.F.		8	3			131	39	170	253
0450	Add 8 C.Y. truck, small project debris haulaway	B-34A	8	1	Hr.		46	42.50	88.50	121
0460	Add 12 C.Y. truck, small project debris haulaway	B-34B	8	1			46	68	114	149
0480	Add flagger for non-intersection medium traffic	1 Clab	8	1			40		40	64.50
0490	Add flasher truck for intersection medium traffic or heavy traffic	A-2B	8	1			44.50	23.50	68	97.50
0500	Flexible pavement patches, roadway, repave, cold, 15 S.F., 4" D	B-37	8	6	Ea.	43.50	253	18.90	315.40	480
0510	6" depth		8	6		66	253	18.90	337.90	505
0520	Repave, cold, 20 S.F., 4" depth		8	6		58	253	18.90	329.90	495
0530	6" depth		8	6		88	253	18.90	359.90	530
0540	Repave, cold, 25 S.F., 4" depth		8	6		72.50	253	18.90	344.40	510
0550	6" depth		8	6		110	253	18.90	381.90	550
0600	Repave, cold, 30 S.F., 4" depth		8	6		87	253	18.90	358.90	525
0610	6" depth		8	6		132	253	18.90	403.90	575
0640	Repave, cold, 40 S.F., 4" depth		8	6		116	253	18.90	387.90	560
0650	6" depth		8	6		175	253	18.90	446.90	625
0680	Repave, cold, 50 S.F., 4" depth		8	6		145	253	18.90	416.90	590
0690	6" depth		8	6		219	253	18.90	490.90	670
0720	Repave, cold, 60 S.F., 4" depth		8	6		174	253	18.90	445.90	620
0730	6" depth		8	6		263	253	18.90	534.90	720
0800	Repave, cold, 70 S.F., 4" depth		7	6.857		203	289	21.50	513.50	710
0810	6" depth		7	6.857		305	289	21.50	615.50	830
0820	Repave, cold, 75 S.F., 4" depth		7	6.857		216	289	21.50	526.50	725
0830	6" depth		7	6.857		330	289	21.50	640.50	850
0900	Repave, cold, 80 S.F., 4" depth		6	8		232	335	25	592	830
0910	6" depth		6	8		350	335	25	710	960
0940	Repave, cold, 90 S.F., 4" depth		6	8		260	335	25	620	860
0950	6" depth		6	8		395	335	25	755	1,000
0980	Repave, cold, 100 S.F., 4" depth		6	8		290	335	25	650	895
0990	6" depth		6	8		440	335	25	800	1,050
1000	Add flasher truck for paving operations in medium or heavy traffic	A-2B	8	1	Hr.		44.50	23.50	68	97.50
1100	Prime coat for repair 15-40 S.F., 25% overspray	B-37A	48	.500	Ea.	8.25	20.50	6.60	35.35	50
1150	41-60 S.F.		40	.600		16.50	25	7.95	49.45	67
1175	61-80 S.F.		32	.750		22.50	31	9.90	63.40	85.50
1200	81-100 S.F.		32	.750		27.50	31	9.90	68.40	91

32 01 17 – Flexible Paving Repair

32 01 17.20 Repair of Asphalt Pavement Patches

		Crew	Daily Output	Labor-Hours	Unit	Material	2018 Bare Costs Labor	Equipment	Total	Total Incl O&P
1210	Groups of patches w/25% overspray	B-37A	3600	.007	S.F.	.27	.28	.09	.64	.85
1300	Flexible pavement repair patches, street repave, hot, 60 S.F., 4" D	B-37	8	6	Ea.	96	253	18.90	367.90	535
1310	6" depth		8	6		145	253	18.90	416.90	590
1320	Repave, hot, 70 S.F., 4" depth		7	6.857		112	289	21.50	422.50	610
1330	6" depth		7	6.857		170	289	21.50	480.50	675
1340	Repave, hot, 75 S.F., 4" depth		7	6.857		119	289	21.50	429.50	620
1350	6" depth		7	6.857		182	289	21.50	492.50	690
1360	Repave, hot, 80 S.F., 4" depth		6	8		128	335	25	488	715
1370	6" depth		6	8		194	335	25	554	785
1380	Repave, hot, 90 S.F., 4" depth		6	8		144	335	25	504	730
1390	6" depth		6	8		218	335	25	578	815
1400	Repave, hot, 100 S.F., 4" depth		6	8		160	335	25	520	750
1410	6" depth		6	8		242	335	25	602	840
1420	Pave hot groups of patches, 4" depth		900	.053	S.F.	1.59	2.24	.17	4	5.55
1430	6" depth		900	.053	"	2.42	2.24	.17	4.83	6.50
1500	Add 8 C.Y. truck for hot asphalt paving operations	B-34A	1	8	Day		370	340	710	960
1550	Add flasher truck for hot paving operations in med/heavy traffic	A-2B	8	1	Hr.		44.50	23.50	68	97.50
2000	Add police officer and car for traffic control				"	43.50			43.50	48

32 01 17.61 Sealing Cracks In Asphalt Paving

		Crew	Daily Output	Labor-Hours	Unit	Material	2018 Bare Costs Labor	Equipment	Total	Total Incl O&P
0010	**SEALING CRACKS IN ASPHALT PAVING**									
0100	Sealing cracks in asphalt paving, 1/8" wide x 1/2" depth, slow set	B-37A	2000	.012	L.F.	.18	.50	.16	.84	1.16
0110	1/4" wide x 1/2" depth		2000	.012		.20	.50	.16	.86	1.19
0130	3/8" wide x 1/2" depth		2000	.012		.23	.50	.16	.89	1.22
0140	1/2" wide x 1/2" depth		1800	.013		.26	.55	.18	.99	1.36
0150	3/4" wide x 1/2" depth		1600	.015		.31	.62	.20	1.13	1.56
0160	1" wido x 1/2" depth		1600	.015		.37	.62	.20	1.19	1.63
0165	1/8" wide x 1" depth, rapid set	B-37F	2000	.016		.18	.66	.24	1.08	1.53
0170	1/4" wide x 1" depth		2000	.016		.35	.66	.24	1.25	1.72
0175	3/8" wide x 1" depth		1800	.010		.53	.73	.27	1.53	2.06
0180	1/2" wide x 1" depth		1800	.018		.71	.73	.27	1.71	2.26
0185	5/8" wide x 1" depth		1800	.018		.89	.73	.27	1.89	2.45
0190	3/4" wide x 1" depth		1600	.020		1.06	.82	.30	2.18	2.83
0195	1" wide x 1" depth		1600	.020		1.42	.82	.30	2.54	3.22
0200	1/8" wide x 2" depth, rapid set		2000	.016		.35	.66	.24	1.25	1.72
0210	1/4" wide x 2" depth		2000	.016		.71	.66	.24	1.61	2.11
0220	3/8" wide x 2" depth		1800	.018		1.06	.73	.27	2.06	2.65
0230	1/2" wide x 2" depth		1800	.018		1.42	.73	.27	2.42	3.04
0240	5/8" wide x 2" depth		1800	.018		1.77	.73	.27	2.77	3.43
0250	3/4" wide x 2" depth		1600	.020		2.13	.82	.30	3.25	4
0260	1" wide x 2" depth		1600	.020		2.84	.82	.30	3.96	4.78
0300	Add flagger for non-intersection medium traffic	1 Clab	8	1	Hr.		40		40	64.50
0400	Add flasher truck for intersection medium traffic or heavy traffic	A-2B	8	1	"		44.50	23.50	68	97.50

32 01 29 – Rigid Paving Repair

32 01 29.61 Partial Depth Patching of Rigid Pavement

		Crew	Daily Output	Labor-Hours	Unit	Material	2018 Bare Costs Labor	Equipment	Total	Total Incl O&P
0010	**PARTIAL DEPTH PATCHING OF RIGID PAVEMENT**									
0100	Rigid pavement repair, roadway, light traffic, 25% pitting 15 S.F.	B-37F	16	2	Ea.	24	82	30.50	136.50	193
0110	Pitting 20 S.F.		16	2		32	82	30.50	144.50	202
0120	Pitting 25 S.F.		16	2		40	82	30.50	152.50	211
0130	Pitting 30 S.F.		16	2		48	82	30.50	160.50	220
0140	Pitting 40 S.F.		16	2		64.50	82	30.50	177	237
0150	Pitting 50 S.F.		12	2.667		80.50	109	40.50	230	310
0160	Pitting 60 S.F.		12	2.667		96.50	109	40.50	246	330

545

32 01 29.61 Partial Depth Patching of Rigid Pavement

		Crew	Daily Output	Labor-Hours	Unit	Material	2018 Bare Costs Labor	2018 Bare Costs Equipment	Total	Total Incl O&P
0170	Pitting 70 S.F.	B-37F	12	2.667	Ea.	113	109	40.50	262.50	345
0180	Pitting 75 S.F.		12	2.667		121	109	40.50	270.50	355
0190	Pitting 80 S.F.		8	4		129	164	61	354	475
0200	Pitting 90 S.F.		8	4		145	164	61	370	490
0210	Pitting 100 S.F.		8	4		161	164	61	386	510
0300	Roadway, light traffic, 50% pitting 15 S.F.		12	2.667		48	109	40.50	197.50	275
0310	Pitting 20 S.F.		12	2.667		64.50	109	40.50	214	292
0320	Pitting 25 S.F.		12	2.667		80.50	109	40.50	230	310
0330	Pitting 30 S.F.		12	2.667		96.50	109	40.50	246	330
0340	Pitting 40 S.F.		12	2.667		129	109	40.50	278.50	365
0350	Pitting 50 S.F.		8	4		161	164	61	386	510
0360	Pitting 60 S.F.		8	4		193	164	61	418	545
0370	Pitting 70 S.F.		8	4		225	164	61	450	580
0380	Pitting 75 S.F.		8	4		241	164	61	466	600
0390	Pitting 80 S.F.		6	5.333		257	219	81	557	725
0400	Pitting 90 S.F.		6	5.333		289	219	81	589	765
0410	Pitting 100 S.F.		6	5.333		320	219	81	620	800
1000	Rigid pavement repair, light traffic, surface patch, 2" deep, 15 S.F.		8	4		96.50	164	61	321.50	440
1010	Surface patch, 2" deep, 20 S.F.		8	4		129	164	61	354	475
1020	Surface patch, 2" deep, 25 S.F.		8	4		161	164	61	386	510
1030	Surface patch, 2" deep, 30 S.F.		8	4		193	164	61	418	545
1040	Surface patch, 2" deep, 40 S.F.		8	4		257	164	61	482	615
1050	Surface patch, 2" deep, 50 S.F.		6	5.333		320	219	81	620	800
1060	Surface patch, 2" deep, 60 S.F.		6	5.333		385	219	81	685	870
1070	Surface patch, 2" deep, 70 S.F.		6	5.333		450	219	81	750	940
1080	Surface patch, 2" deep, 75 S.F.		6	5.333		480	219	81	780	975
1090	Surface patch, 2" deep, 80 S.F.		5	6.400		515	262	97	874	1,100
1100	Surface patch, 2" deep, 90 S.F.		5	6.400		580	262	97	939	1,175
1200	Surface patch, 2" deep, 100 S.F.		5	6.400		645	262	97	1,004	1,225
2000	Add flagger for non-intersection medium traffic	1 Clab	8	1	Hr.		40		40	64.50
2100	Add flasher truck for intersection medium traffic or heavy traffic	A-2B	8	1			44.50	23.50	68	97.50
2200	Add police officer and car for traffic control					43.50			43.50	48

32 01 29.70 Full Depth Patching of Rigid Pavement

		Crew	Daily Output	Labor-Hours	Unit	Material	2018 Bare Costs Labor	2018 Bare Costs Equipment	Total	Total Incl O&P
0010	**FULL DEPTH PATCHING OF RIGID PAVEMENT**									
0015	Pavement preparation includes sawcut, remove pavement and replace									
0020	6" of subbase and one layer of reinforcement									
0030	Trucking and haulaway of debris excluded									
0100	Rigid pavement replace, light traffic, prep, 15 S.F., 6" depth	B-37E	16	3.500	Ea.	20.50	157	52.50	230	330
0110	Replacement preparation 20 S.F., 6" depth		16	3.500		27.50	157	52.50	237	340
0115	25 S.F., 6" depth		16	3.500		34.50	157	52.50	244	345
0120	30 S.F., 6" depth		14	4		41.50	179	60	280.50	400
0125	35 S.F., 6" depth		14	4		48	179	60	287	405
0130	40 S.F., 6" depth		14	4		55	179	60	294	415
0135	45 S.F., 6" depth		14	4		62	179	60	301	420
0140	50 S.F., 6" depth		12	4.667		69	209	70	348	490
0145	55 S.F., 6" depth		12	4.667		75.50	209	70	354.50	495
0150	60 S.F., 6" depth		10	5.600		82.50	251	84.50	418	585
0155	65 S.F., 6" depth		10	5.600		89.50	251	84.50	425	590
0160	70 S.F., 6" depth		10	5.600		96.50	251	84.50	432	600
0165	75 S.F., 6" depth		10	5.600		103	251	84.50	438.50	605
0170	80 S.F., 6" depth		8	7		110	315	105	530	740
0175	85 S.F., 6" depth		8	7		117	315	105	537	750

32 01 29 – Rigid Paving Repair

32 01 29.70 Full Depth Patching of Rigid Pavement	Crew	Daily Output	Labor-Hours	Unit	Material	2018 Bare Costs Labor	Equipment	Total	Total Incl O&P
0180	90 S.F., 6" depth	B-37E	8	7	Ea.	124	315	105	544	755
0185	95 S.F., 6" depth		8	7		131	315	105	551	765
0190	100 S.F., 6" depth	↓	8	7	↓	138	315	105	558	770
0200	Pavement preparation for 8" rigid paving same as 6"									
0290	Pavement preparation for 9" rigid paving same as 10"									
0300	Rigid pavement replace, light traffic, prep, 15 S.F., 10" depth	B-37E	16	3.500	Ea.	33.50	157	52.50	243	345
0302	Pavement preparation 10" includes sawcut, remove pavement and									
0304	replace 6" of subbase and two layers of reinforcement									
0306	Trucking and haulaway of debris excluded									
0310	Replacement preparation 20 S.F., 10" depth	B-37E	16	3.500	Ea.	45	157	52.50	254.50	360
0315	25 S.F., 10" depth		16	3.500		56	157	52.50	265.50	370
0320	30 S.F., 10" depth		14	4		67	179	60	306	425
0325	35 S.F., 10" depth		14	4		52.50	179	60	291.50	410
0330	40 S.F., 10" depth		14	4		89.50	179	60	328.50	450
0335	45 S.F., 10" depth		12	4.667		101	209	70	380	525
0340	50 S.F., 10" depth		12	4.667		112	209	70	391	535
0345	55 S.F., 10" depth		12	4.667		123	209	70	402	545
0350	60 S.F., 10" depth		10	5.600		134	251	84.50	469.50	640
0355	65 S.F., 10" depth		10	5.600		146	251	84.50	481.50	655
0360	70 S.F., 10" depth		10	5.600		157	251	84.50	492.50	665
0365	75 S.F., 10" depth		8	7		168	315	105	588	805
0370	80 S.F., 10" depth		8	7		179	315	105	599	820
0375	85 S.F., 10" depth		8	7		190	315	105	610	830
0380	90 S.F., 10" depth		8	7		202	315	105	622	845
0385	95 S.F., 10" depth		8	7		213	315	105	633	855
0390	100 S.F., 10" depth	↓	8	7	↓	224	315	105	644	865
0500	Add 8 C.Y. truck, small project debris haulaway	B-34A	8	1	Hr.		46	42.50	88.50	121
0550	Add 12 C.Y. truck, small project debris haulaway	B-34B	8	1			46	68	114	149
0600	Add flagger for non-intersection medium traffic	1 Clab	8	1			40		40	64.50
0650	Add flasher truck for intersection medium traffic or heavy traffic	A-2B	8	1			44.50	23.50	68	97.50
0700	Add police officer and car for traffic control				↓	43.50			43.50	48
0990	Concrete will be replaced using quick set mix with aggregate									
1000	Rigid pavement replace, light traffic, repour, 15 S.F., 6" depth	B-37F	18	1.778	Ea.	167	73	27	267	330
1010	20 S.F., 6" depth		18	1.778		222	73	27	322	395
1015	25 S.F., 6" depth		18	1.778		278	73	27	378	455
1020	30 S.F., 6" depth		16	2		335	82	30.50	447.50	530
1025	35 S.F., 6" depth		16	2		390	82	30.50	502.50	595
1030	40 S.F., 6" depth		16	2		445	82	30.50	557.50	655
1035	45 S.F., 6" depth		16	2		500	82	30.50	612.50	715
1040	50 S.F., 6" depth		16	2		555	82	30.50	667.50	775
1045	55 S.F., 6" depth		12	2.667		610	109	40.50	759.50	890
1050	60 S.F., 6" depth		12	2.667		665	109	40.50	814.50	955
1055	65 S.F., 6" depth		12	2.667		725	109	40.50	874.50	1,025
1060	70 S.F., 6" depth		12	2.667		780	109	40.50	929.50	1,075
1065	75 S.F., 6" depth		10	3.200		835	131	48.50	1,014.50	1,175
1070	80 S.F., 6" depth		10	3.200		890	131	48.50	1,069.50	1,250
1075	85 S.F., 6" depth		8	4		945	164	61	1,170	1,375
1080	90 S.F., 6" depth		8	4		1,000	164	61	1,225	1,425
1085	95 S.F., 6" depth		8	4		1,050	164	61	1,275	1,475
1090	100 S.F., 6" depth	↓	8	4	↓	1,100	164	61	1,325	1,550
1099	Concrete will be replaced using 4,500 psi concrete ready mix									
1100	Rigid pavement replace, light traffic, repour, 15 S.F., 6" depth	A-2	12	2	Ea.	345	83	15.70	443.70	530
1110	20-50 S.F., 6" depth	↓	12	2	↓	345	83	15.70	443.70	530

32 01 29 – Rigid Paving Repair

32 01 29.70 Full Depth Patching of Rigid Pavement

		Crew	Daily Output	Labor-Hours	Unit	Material	2018 Bare Costs Labor	Equipment	Total	Total Incl O&P
1120	55-65 S.F., 6" depth	A-2	10	2.400	Ea.	385	99.50	18.85	503.35	600
1130	70-80 S.F., 6" depth		10	2.400		415	99.50	18.85	533.35	635
1140	85-100 S.F., 6" depth		8	3		485	124	23.50	632.50	760
1200	Repour 15-40 S.F., 8" depth		12	2		345	83	15.70	443.70	530
1210	45-50 S.F., 8" depth		12	2		385	83	15.70	483.70	570
1220	55-60 S.F., 8" depth		10	2.400		415	99.50	18.85	533.35	635
1230	65-80 S.F., 8" depth		10	2.400		485	99.50	18.85	603.35	715
1240	85-90 S.F., 8" depth		8	3		525	124	23.50	672.50	800
1250	95-100 S.F., 8" depth		8	3		555	124	23.50	702.50	835
1300	Repour 15-30 S.F., 10" depth		12	2		345	83	15.70	443.70	530
1310	35-40 S.F., 10" depth		12	2		385	83	15.70	483.70	570
1320	45 S.F., 10" depth		12	2		415	83	15.70	513.70	605
1330	50-65 S.F., 10" depth		10	2.400		485	99.50	18.85	603.35	715
1340	70 S.F., 10" depth		10	2.400		525	99.50	18.85	643.35	755
1350	75-80 S.F., 10" depth		10	2.400		555	99.50	18.85	673.35	790
1360	85-95 S.F., 10" depth		8	3		625	124	23.50	772.50	915
1370	100 S.F., 10" depth		8	3		665	124	23.50	812.50	955

32 06 Schedules for Exterior Improvements

32 06 10 – Schedules for Bases, Ballasts, and Paving

32 06 10.10 Sidewalks, Driveways and Patios

		Crew	Daily Output	Labor-Hours	Unit	Material	2018 Bare Costs Labor	Equipment	Total	Total Incl O&P
0010	**SIDEWALKS, DRIVEWAYS AND PATIOS** No base									
0020	Asphaltic concrete, 2" thick	B-37	720	.067	S.Y.	6.85	2.81	.21	9.87	12.30
0100	2-1/2" thick	"	660	.073	"	8.70	3.06	.23	11.99	14.80
0110	Bedding for brick or stone, mortar, 1" thick	D-1	300	.053	S.F.	.78	2.39		3.17	4.75
0120	2" thick	"	200	.080		1.94	3.58		5.52	8
0130	Sand, 2" thick	B-18	8000	.003		.31	.12	.01	.44	.56
0140	4" thick	"	4000	.006		.63	.24	.01	.88	1.09
0300	Concrete, 3,000 psi, CIP, 6 x 6 - W1.4 x W1.4 mesh,									
0310	broomed finish, no base, 4" thick	B-24	600	.040	S.F.	2.01	1.84		3.85	5.15
0350	5" thick		545	.044		2.68	2.03		4.71	6.20
0400	6" thick		510	.047		3.13	2.17		5.30	6.95
0450	For bank run gravel base, 4" thick, add	B-18	2500	.010		.43	.39	.02	.84	1.12
0520	8" thick, add	"	1600	.015		.87	.61	.03	1.51	1.98
0550	Exposed aggregate finish, add to above, minimum	B-24	1875	.013		.13	.59		.72	1.09
0600	Maximum	"	455	.053		.43	2.43		2.86	4.38
0950	Concrete tree grate, 5' square	B-6	25	.960	Ea.	435	42	12.50	489.50	560
0955	Tree well & cover, concrete, 3' square		25	.960		241	42	12.50	295.50	345
0960	Cast iron tree grate with frame, 2 piece, round, 5' diameter		25	.960		1,100	42	12.50	1,154.50	1,275
0980	Square, 5' side		25	.960		1,100	42	12.50	1,154.50	1,275
1700	Redwood, prefabricated, 4' x 4' sections	2 Carp	316	.051	S.F.	4.77	2.57		7.34	9.40
1750	Redwood planks, 1" thick, on sleepers	"	240	.067	"	4.77	3.38		8.15	10.75
2250	Stone dust, 4" thick	B-62	900	.027	S.Y.	3.97	1.16	.19	5.32	6.45
9000	Minimum labor/equipment charge	D-1	2	8	Job		360		360	585

32 06 10.20 Steps

		Crew	Daily Output	Labor-Hours	Unit	Material	2018 Bare Costs Labor	Equipment	Total	Total Incl O&P
0010	**STEPS**									
0011	Incl. excav., borrow & concrete base as required									
0100	Brick steps	B-24	35	.686	LF Riser	17.25	31.50		48.75	70
0200	Railroad ties	2 Clab	25	.640		3.55	25.50		29.05	45.50
0300	Bluestone treads, 12" x 2" or 12" x 1-1/2"	B-24	30	.800		41.50	37		78.50	105
0490	Minimum labor/equipment charge	D-1	2	8	Job		360		360	585

32 06 Schedules for Exterior Improvements

32 06 10 – Schedules for Bases, Ballasts, and Paving

32 06 10.20 Steps	Crew	Daily Output	Labor-Hours	Unit	Material	2018 Bare Costs Labor	2018 Bare Costs Equipment	Total	Total Incl O&P
0500 Concrete, cast in place, see Section 03 30 53.40									
0600 Precast concrete, see Section 03 41 23.50									

32 11 Base Courses

32 11 23 – Aggregate Base Courses

32 11 23.23 Base Course Drainage Layers

		Crew	Daily Output	Labor-Hours	Unit	Material	2018 Bare Costs Labor	2018 Bare Costs Equipment	Total	Total Incl O&P
0010	**BASE COURSE DRAINAGE LAYERS**									
0011	For roadways and large areas									
0050	Crushed 3/4" stone base, compacted, 3" deep	B-36C	5200	.008	S.Y.	2.71	.38	.74	3.83	4.41
0100	6" deep		5000	.008		5.40	.40	.77	6.57	7.45
0200	9" deep		4600	.009		8.15	.43	.84	9.42	10.55
0300	12" deep		4200	.010		10.85	.47	.92	12.24	13.65
0301	Crushed 1-1/2" stone base, compacted to 4" deep	B-36B	6000	.011		4.52	.51	.75	5.78	6.60
0302	6" deep		5400	.012		6.80	.57	.83	8.20	9.25
0303	8" deep		4500	.014		9.05	.68	1	10.73	12.15
0304	12" deep		3800	.017		13.55	.81	1.19	15.55	17.50
0310	Minimum labor/equipment charge	B-36	1	40	Job		1,825	1,550	3,375	4,625
0350	Bank run gravel, spread and compacted									
0370	6" deep	B-32	6000	.005	S.Y.	3.60	.27	.36	4.23	4.78
0390	9" deep		4900	.007		5.40	.33	.44	6.17	6.95
0400	12" deep		4200	.008		7.20	.38	.51	8.09	9.10
6900	For small and irregular areas, add						50%	50%		
6990	Minimum labor/equipment charge	1 Clab	3	2.667	Job		106		106	173
7000	Prepare and roll sub-base, small areas to 2,500 S.Y.	B-32A	1500	.016	S.Y.		.79	.87	1.66	2.21
8000	Large areas over 2,500 S.Y.	"	3500	.007			.34	.37	.71	.95
8050	For roadways	B-32	4000	.008			.40	.54	.94	1.23
9000	Minimum labor/equipment charge	1 Clab	4	2	Job		79.50		79.50	129

32 12 Flexible Paving

32 12 16 – Asphalt Paving

32 12 16.13 Plant-Mix Asphalt Paving

		Crew	Daily Output	Labor-Hours	Unit	Material	2018 Bare Costs Labor	2018 Bare Costs Equipment	Total	Total Incl O&P
0010	**PLANT-MIX ASPHALT PAVING**									
0080	Binder course, 1-1/2" thick	B-25	7725	.011	S.Y.	5.30	.50	.35	6.15	7.05
0120	2" thick		6345	.014		7.05	.61	.42	8.08	9.20
0130	2-1/2" thick		5620	.016		8.80	.69	.48	9.97	11.35
0160	3" thick		4905	.018		10.60	.79	.55	11.94	13.50
0170	3-1/2" thick		4520	.019		12.35	.85	.60	13.80	15.65
0300	Wearing course, 1" thick	B-25B	10575	.009		3.50	.41	.28	4.19	4.80
0340	1-1/2" thick		7725	.012		5.90	.56	.38	6.84	7.75
0380	2" thick		6345	.015		7.90	.68	.46	9.04	10.30
0420	2-1/2" thick		5480	.018		9.75	.78	.53	11.06	12.55
0460	3" thick		4900	.020		11.60	.87	.60	13.07	14.85
0470	3-1/2" thick		4520	.021		13.65	.95	.65	15.25	17.25
0480	4" thick		4140	.023		15.60	1.04	.71	17.35	19.60
0500	Open graded friction course	B-25C	5000	.010		2.42	.43	.47	3.32	3.87
0800	Alternate method of figuring paving costs									
0810	Binder course, 1-1/2" thick	B-25	630	.140	Ton	65	6.10	4.27	75.37	86
0811	2" thick		690	.128		65	5.60	3.90	74.50	85
0812	3" thick		800	.110		65	4.82	3.37	73.19	83

For customer support on your Commercial Renovation Costs with RSMeans data, call 800.448.8182.

549

32 12 Flexible Paving

32 12 16 – Asphalt Paving

32 12 16.13 Plant-Mix Asphalt Paving

		Crew	Daily Output	Labor-Hours	Unit	Material	2018 Bare Costs Labor	2018 Bare Costs Equipment	Total	Total Incl O&P
0813	4" thick	B-25	900	.098	Ton	65	4.28	2.99	72.27	81.50
0850	Wearing course, 1" thick	B-25B	575	.167		71	7.45	5.10	83.55	95.50
0851	1-1/2" thick		630	.152		71	6.80	4.64	82.44	94
0852	2" thick		690	.139		71	6.20	4.24	81.44	92.50
0853	2-1/2" thick		765	.125		71	5.60	3.82	80.42	91
0854	3" thick	↓	800	.120	↓	71	5.35	3.66	80.01	90.50
1000	Pavement replacement over trench, 2" thick	B-37	90	.533	S.Y.	7.30	22.50	1.68	31.48	46
1050	4" thick		70	.686		14.40	29	2.16	45.56	64.50
1080	6" thick	↓	55	.873		23	36.50	2.75	62.25	87.50
3000	Prime coat, emulsion, 0.30 gal./S.Y., 1000 S.Y.	B-45	2500	.006		1.98	.32	.32	2.62	3.04
3100	Tack coat, emulsion, 0.10 gal./S.Y., 1000 S.Y.	"	2500	.006	↓	.66	.32	.32	1.30	1.59

32 12 16.14 Asphaltic Concrete Paving

		Crew	Daily Output	Labor-Hours	Unit	Material	2018 Bare Costs Labor	2018 Bare Costs Equipment	Total	Total Incl O&P
0011	**ASPHALTIC CONCRETE PAVING**, parking lots & driveways									
0015	No asphalt hauling included									
0020	6" stone base, 2" binder course, 1" topping	B-25C	9000	.005	S.F.	1.82	.24	.26	2.32	2.67
0025	2" binder course, 2" topping		9000	.005		2.25	.24	.26	2.75	3.15
0030	3" binder course, 2" topping		9000	.005		2.65	.24	.26	3.15	3.58
0035	4" binder course, 2" topping		9000	.005		3.04	.24	.26	3.54	4.01
0040	1-1/2" binder course, 1" topping		9000	.005		1.63	.24	.26	2.13	2.46
0042	3" binder course, 1" topping		9000	.005		2.22	.24	.26	2.72	3.11
0045	3" binder course, 3" topping		9000	.005		3.08	.24	.26	3.58	4.06
0050	4" binder course, 3" topping		9000	.005		3.47	.24	.26	3.97	4.49
0055	4" binder course, 4" topping		9000	.005		3.90	.24	.26	4.40	4.95
0300	Binder course, 1-1/2" thick		35000	.001		.59	.06	.07	.72	.82
0400	2" thick		25000	.002		.76	.09	.09	.94	1.08
0500	3" thick		15000	.003		1.18	.14	.16	1.48	1.70
0600	4" thick		10800	.004		1.54	.20	.22	1.96	2.26
0800	Sand finish course, 3/4" thick		41000	.001		.31	.05	.06	.42	.48
0900	1" thick	↓	34000	.001		.38	.06	.07	.51	.60
1000	Fill pot holes, hot mix, 2" thick	B-16	4200	.008		.81	.32	.13	1.26	1.55
1100	4" thick		3500	.009		1.18	.38	.16	1.72	2.09
1120	6" thick	↓	3100	.010		1.59	.43	.18	2.20	2.63
1140	Cold patch, 2" thick	B-51	3000	.016		.92	.66	.06	1.64	2.14
1160	4" thick		2700	.018		1.75	.73	.07	2.55	3.18
1180	6" thick	↓	1900	.025	↓	2.72	1.03	.10	3.85	4.78
3000	Prime coat, emulsion, 0.30 gal./S.Y., 1000 S.Y.	B-45	2500	.006	S.Y.	1.98	.32	.32	2.62	3.04
3100	Tack coat, emulsion, 0.10 gal./S.Y., 1000 S.Y.	"	2500	.006	"	.66	.32	.32	1.30	1.59

32 13 Rigid Paving

32 13 13 – Concrete Paving

32 13 13.25 Concrete Pavement, Highways

		Crew	Daily Output	Labor-Hours	Unit	Material	2018 Bare Costs Labor	2018 Bare Costs Equipment	Total	Total Incl O&P
0010	**CONCRETE PAVEMENT, HIGHWAYS**									
0015	Including joints, finishing and curing									
0020	Fixed form, 12' pass, unreinforced, 6" thick	B-26	3000	.029	S.Y.	25.50	1.31	1	27.81	31
0100	8" thick		2750	.032		35	1.43	1.09	37.52	42
0400	12" thick		1800	.049		50.50	2.18	1.67	54.35	61
0410	Conc. pavement, w/jt., fnsh.& curing, fix form, 24' pass, unreinforced, 6"T		6000	.015		24.50	.65	.50	25.65	28.50
0430	8" thick		5500	.016		33.50	.71	.55	34.76	38.50
0460	12" thick	↓	3600	.024		48.50	1.09	.83	50.42	56
0520	Welded wire fabric, sheets for rigid paving 2.33 lb./S.Y.	2 Rodm	389	.041		1.36	2.25		3.61	5.10
0530	Reinforcing steel for rigid paving 12 lb./S.Y.		666	.024	↓	6.05	1.31		7.36	8.75

32 13 Rigid Paving

32 13 13 – Concrete Paving

32 13 13.25 Concrete Pavement, Highways

		Crew	Daily Output	Labor-Hours	Unit	Material	2018 Bare Costs Labor	Equipment	Total	Total Incl O&P
0540	Reinforcing steel for rigid paving 18 lb./S.Y.	2 Rodm	444	.036	S.Y.	9.05	1.97		11.02	13.10
0620	Slip form, 12' pass, unreinforced, 6" thick	B-26A	5600	.016		25	.70	.55	26.25	29
0624	8" thick		5300	.017		34	.74	.59	35.33	39.50
0630	12" thick		3470	.025		49.50	1.13	.90	51.53	57.50
0632	15" thick		2890	.030		62.50	1.36	1.08	64.94	72
0640	Slip form, 24' pass, unreinforced, 6" thick		11200	.008		24.50	.35	.28	25.13	28
0644	8" thick		10600	.008		33	.37	.29	33.66	37
0650	12" thick		6940	.013		48	.57	.45	49.02	54.50
0652	15" thick	▼	5780	.015		60.50	.68	.54	61.72	68
0700	Finishing, broom finish small areas	2 Cefi	120	.133			6.35		6.35	10.05
1000	Curing, with sprayed membrane by hand	2 Clab	1500	.011	▼	1.09	.43		1.52	1.89

32 14 Unit Paving

32 14 13 – Precast Concrete Unit Paving

32 14 13.18 Precast Concrete Plantable Pavers

		Crew	Daily Output	Labor-Hours	Unit	Material	2018 Bare Costs Labor	Equipment	Total	Total Incl O&P
0010	**PRECAST CONCRETE PLANTABLE PAVERS** (50% grass)									
0300	3/4" crushed stone base for plantable pavers, 6" depth	B-62	1000	.024	S.Y.	4.22	1.05	.17	5.44	6.50
0400	8" depth		900	.027		5.60	1.16	.19	6.95	8.30
0500	10" depth		800	.030		7.05	1.31	.22	8.58	10.10
0600	12" depth	▼	700	.034	▼	8.40	1.50	.25	10.15	11.90
0700	Hydro seeding plantable pavers	B-81A	20	.800	M.S.F.	11.40	33.50	17.95	62.85	87
0800	Apply fertilizer and seed to plantable pavers	1 Clab	8	1	"	41.50	40		81.50	110

32 14 16 – Brick Unit Paving

32 14 16.10 Brick Paving

		Crew	Daily Output	Labor-Hours	Unit	Material	2018 Bare Costs Labor	Equipment	Total	Total Incl O&P
0010	**BRICK PAVING**									
0012	4" x 8" x 1-1/2", without joints (4.5 bricks/S.F.)	D-1	110	.145	S.F.	2.57	6.50		9.07	13.45
0100	Grouted, 3/8" joint (3.9 bricks/S.F.)		90	.178		2.00	7.95		10.03	15.30
0200	4" x 8" x 2-1/4", without joints (4.5 bricks/S.F.)		110	.145		2.40	6.50		8.90	13.30
0300	Grouted, 3/8" joint (3.9 bricks/S.F.)		90	.178		2.08	7.95		10.03	15.30
0455	Pervious brick paving, 4" x 8" x 3-1/4", without joints (4.5 bricks/S.F.)	▼	110	.145		3.60	6.50		10.10	14.60
0500	Bedding, asphalt, 3/4" thick	B-25	5130	.017		.65	.75	.52	1.92	2.50
0540	Course washed sand bed, 1" thick	B-18	5000	.005		.35	.19	.01	.55	.72
0580	Mortar, 1" thick	D-1	300	.053		.65	2.39		3.04	4.61
0620	2" thick		200	.080		1.29	3.58		4.87	7.25
1500	Brick on 1" thick sand bed laid flat, 4.5/S.F.		100	.160		2.85	7.15		10	14.85
2000	Brick pavers, laid on edge, 7.2/S.F.		70	.229		4.46	10.25		14.71	21.50
2500	For 4" thick concrete bed and joints, add	▼	595	.027		1.42	1.20		2.62	3.53
2800	For steam cleaning, add	A-1H	950	.008	▼	.11	.34	.08	.53	.75
9000	Minimum labor/equipment charge	1 Bric	2	4	Job		201		201	330

32 14 23 – Asphalt Unit Paving

32 14 23.10 Asphalt Blocks

		Crew	Daily Output	Labor-Hours	Unit	Material	2018 Bare Costs Labor	Equipment	Total	Total Incl O&P
0010	**ASPHALT BLOCKS**									
0020	Rectangular, 6" x 12" x 1-1/4", w/bed & neopr. adhesive	D-1	135	.119	S.F.	8.55	5.30		13.85	18.05
0100	3" thick		130	.123		12	5.50		17.50	22
0300	Hexagonal tile, 8" wide, 1-1/4" thick		135	.119		8.55	5.30		13.85	18.05
0400	2" thick		130	.123		12	5.50		17.50	22
0500	Square, 8" x 8", 1-1/4" thick		135	.119		8.55	5.30		13.85	18.05
0600	2" thick	▼	130	.123	▼	12	5.50		17.50	22
9000	Minimum labor/equipment charge	1 Bric	2	4	Job		201		201	330

For customer support on your Commercial Renovation Costs with RSMeans data, call 800.448.8182.

551

32 14 Unit Paving

32 14 40 – Stone Paving

32 14 40.10 Stone Pavers	Crew	Daily Output	Labor-Hours	Unit	Material	2018 Bare Costs Labor	Equipment	Total	Total Incl O&P
0010 STONE PAVERS									
1300 Slate, natural cleft, irregular, 3/4" thick	D-1	92	.174	S.F.	9.45	7.80		17.25	23
1350 Random rectangular, gauged, 1/2" thick		105	.152		20.50	6.85		27.35	33.50
1400 Random rectangular, butt joint, gauged, 1/4" thick	↓	150	.107		22	4.78		26.78	32.50
1450 For sand rubbed finish, add					9.45			9.45	10.40
1550 Granite blocks, 3-1/2" x 3-1/2" x 3-1/2"	D-1	92	.174		19.30	7.80		27.10	33.50
1600 4" to 12" long, 3" to 5" wide, 3" to 5" thick	"	98	.163	↓	16.10	7.30		23.40	29.50

32 16 Curbs, Gutters, Sidewalks, and Driveways

32 16 13 – Curbs and Gutters

32 16 13.13 Cast-in-Place Concrete Curbs and Gutters

	Crew	Daily Output	Labor-Hours	Unit	Material	2018 Bare Costs Labor	Equipment	Total	Total Incl O&P
0010 CAST-IN-PLACE CONCRETE CURBS AND GUTTERS									
0290 Forms only, no concrete									
0300 Concrete, wood forms, 6" x 18", straight	C-2	500	.096	L.F.	2.93	4.73		7.66	10.85
0400 6" x 18", radius	"	200	.240		3.05	11.80		14.85	22.50
0404 Concrete, wood forms, 6" x 18", straight & concrete	C-2A	500	.096		6.30	4.68		10.98	14.50
0406 6" x 18", radius	"	200	.240		6.45	11.70		18.15	26
0415 Machine formed, 6" x 18", straight	B-69A	2000	.024		3.68	1.05	.59	5.32	6.40
0416 6" x 18", radius	"	900	.053	↓	3.69	2.33	1.30	7.32	9.25

32 16 13.23 Precast Concrete Curbs and Gutters

	Crew	Daily Output	Labor-Hours	Unit	Material	2018 Bare Costs Labor	Equipment	Total	Total Incl O&P
0010 PRECAST CONCRETE CURBS AND GUTTERS									
0550 Precast, 6" x 18", straight	B-29	700	.080	L.F.	9.35	3.50	1.24	14.09	17.20
0600 6" x 18", radius	"	325	.172	"	10.55	7.55	2.68	20.78	26.50

32 16 13.33 Asphalt Curbs

	Crew	Daily Output	Labor-Hours	Unit	Material	2018 Bare Costs Labor	Equipment	Total	Total Incl O&P
0010 ASPHALT CURBS									
0012 Curbs, asphaltic, machine formed, 8" wide, 6" high, 40 L.F./ton	B-27	1000	.032	L.F.	1.63	1.29	.32	3.24	4.25
0100 8" wide, 8" high, 30 L.F./ton		900	.036		2.18	1.43	.35	3.96	5.10
0150 Asphaltic berm, 12" W, 3" to 6" H, 35 L.F./ton, before pavement	↓	700	.046		.04	1.84	.45	2.33	3.54
0200 12" W, 1-1/2" to 4" H, 60 L.F./ton, laid with pavement	B-2	1050	.038	↓	.02	1.53		1.55	2.52

32 16 13.43 Stone Curbs

	Crew	Daily Output	Labor-Hours	Unit	Material	2018 Bare Costs Labor	Equipment	Total	Total Incl O&P
0010 STONE CURBS									
1000 Granite, split face, straight, 5" x 16"	D-13	275	.175	L.F.	15.20	8.40	1.32	24.92	31.50
1100 6" x 18"	"	250	.192		19.95	9.20	1.46	30.61	38.50
1300 Radius curbing, 6" x 18", over 10' radius	B-29	260	.215	↓	24.50	9.40	3.34	37.24	46
1400 Corners, 2' radius	"	80	.700	Ea.	82	30.50	10.85	123.35	151
1600 Edging, 4-1/2" x 12", straight	D-13	300	.160	L.F.	7.60	7.70	1.21	16.51	22
1800 Curb inlets (guttermouth) straight	B-29	41	1.366	Ea.	182	59.50	21	262.50	320
2000 Indian granite (Belgian block)									
2100 Jumbo, 10-1/2" x 7-1/2" x 4", grey	D-1	150	.107	L.F.	8.35	4.78		13.13	17
2150 Pink		150	.107		8.40	4.78		13.18	17.05
2200 Regular, 9" x 4-1/2" x 4-1/2", grey		160	.100		4.45	4.48		8.93	12.20
2250 Pink		160	.100		5.85	4.48		10.33	13.75
2300 Cubes, 4" x 4" x 4", grey		175	.091		3.62	4.10		7.72	10.70
2350 Pink		175	.091		3.72	4.10		7.82	10.80
2400 6" x 6" x 6", pink	↓	155	.103	↓	12.70	4.62		17.32	21.50
2500 Alternate pricing method for Indian granite									
2550 Jumbo, 10-1/2" x 7-1/2" x 4" (30 lb.), grey				Ton	475			475	525
2600 Pink					490			490	540
2650 Regular, 9" x 4-1/2" x 4-1/2" (20 lb.), grey					315			315	345
2700 Pink					410			410	450

32 16 Curbs, Gutters, Sidewalks, and Driveways

32 16 13 – Curbs and Gutters

32 16 13.43 Stone Curbs		Crew	Daily Output	Labor-Hours	Unit	Material	2018 Bare Costs Labor	Equipment	Total	Total Incl O&P
2750	Cubes, 4" x 4" x 4" (5 lb.), grey				Ton	440			440	485
2800	Pink					480			480	525
2850	6" x 6" x 6" (25 lb.), pink					495			495	545
2900	For pallets, add					22			22	24

32 17 Paving Specialties

32 17 13 – Parking Bumpers

32 17 13.13 Metal Parking Bumpers

		Crew	Daily Output	Labor-Hours	Unit	Material	Labor	Equipment	Total	Total Incl O&P
0010	**METAL PARKING BUMPERS**									
0015	Bumper rails for garages, 12 ga. rail, 6" wide, with steel									
0020	posts 12'-6" OC, minimum	E-4	190	.168	L.F.	20.50	9.30	.52	30.32	39
0030	Average		165	.194		26	10.70	.60	37.30	47.50
0100	Maximum		140	.229		31	12.60	.70	44.30	56.50
1300	Pipe bollards, conc. filled/paint, 8' L x 4' D hole, 6" diam.	B-6	20	1.200	Ea.	610	52.50	15.60	678.10	770
1400	8" diam.		15	1.600		690	70	21	781	895
1500	12" diam.		12	2		975	87.50	26	1,088.50	1,250
8900	Security bollards, SS, lighted, hyd., incl. controls, group of 3	L-7	.06	509		39,200	24,800		64,000	83,000
8910	Group of 5	"	.04	683		65,500	33,200		98,700	125,500
9000	Minimum labor/equipment charge	E-4	2	16	Job		880	49.50	929.50	1,550

32 17 13.16 Plastic Parking Bumpers

			Crew	Daily Output	Labor-Hours	Unit	Material	Labor	Equipment	Total	Total Incl O&P
0010	**PLASTIC PARKING BUMPERS**										
1600	Bollards, recycled plastic, 5" x 5" x 5' L	G	B-6	20	1.200	Ea.	115	52.50	15.60	183.10	227
1610	Wheel stops, recycled plastic, yellow, 4" x 6" x 6' L	G		20	1.200		51	52.50	15.60	119.10	157
1620	Speed bumps, recycled plastic, yellow, 3" H x 10" W x 6' L	G		20	1.200		144	52.50	15.60	212.10	259
1630	3" H x 10" W x 9' L	G		20	1.200		217	52.50	15.60	285.10	340

32 17 13.19 Precast Concrete Parking Bumpers

		Crew	Daily Output	Labor-Hours	Unit	Material	Labor	Equipment	Total	Total Incl O&P
0010	**PRECAST CONCRETE PARKING BUMPERS**									
1000	Wheel stops, precast concrete incl. dowels, 6" x 10" x 6'-0"	B-2	120	.333	Ea.	54	13.40		67.40	81.50
1100	8" x 13" x 6'-0"	"	120	.333		62.50	13.40		75.90	91
1540	Bollards, precast concrete, free standing, round, 15" diam. x 36" H	B-6	14	1.714		239	75	22.50	336.50	410
1550	16" diam. x 30" H		14	1.714		300	75	22.50	397.50	480
1560	18" diam. x 34" H		14	1.714		365	75	22.50	462.50	545
1570	18" diam. x 72" H		14	1.714		615	75	22.50	712.50	820
1572	26" diam. x 18" H		12	2		375	87.50	26	488.50	580
1574	26" diam. x 30" H		12	2		630	87.50	26	743.50	865
1578	26" diam. x 50" H		12	2		835	87.50	26	948.50	1,100
1580	Square, 22" x 32" H		12	2		435	87.50	26	548.50	650
1585	22" x 41" H		12	2		640	87.50	26	753.50	875
1590	Hexagon, 26" x 41" H		10	2.400		600	105	31.50	736.50	865

32 17 13.26 Wood Parking Bumpers

		Crew	Daily Output	Labor-Hours	Unit	Material	Labor	Equipment	Total	Total Incl O&P
0010	**WOOD PARKING BUMPERS**									
0020	Parking barriers, timber w/saddles, treated type									
0100	4" x 4" for cars	B-2	520	.077	L.F.	3.07	3.10		6.17	8.45
0200	6" x 6" for trucks	"	520	.077	"	6.45	3.10		9.55	12.15

32 17 23 – Pavement Markings

32 17 23.13 Painted Pavement Markings

		Crew	Daily Output	Labor-Hours	Unit	Material	Labor	Equipment	Total	Total Incl O&P
0010	**PAINTED PAVEMENT MARKINGS**									
0020	Acrylic waterborne, white or yellow, 4" wide, less than 3,000 L.F.	B-78	20000	.002	L.F.	.10	.10	.03	.23	.30
0200	6" wide, less than 3,000 L.F.		11000	.004		.15	.18	.05	.38	.51
0500	8" wide, less than 3,000 L.F.		10000	.005		.20	.20	.05	.45	.60

553

32 17 Paving Specialties

32 17 23 – Pavement Markings

32 17 23.13 Painted Pavement Markings

		Crew	Daily Output	Labor-Hours	Unit	Material	2018 Bare Costs Labor	Equipment	Total	Total Incl O&P
0600	12" wide, less than 3,000 L.F.	B-78	4000	.012	L.F.	.30	.49	.13	.92	1.27
0620	Arrows or gore lines		2300	.021	S.F.	.20	.85	.22	1.27	1.84
0640	Temporary paint, white or yellow, less than 3,000 L.F.		15000	.003	L.F.	.12	.13	.03	.28	.38
0660	Removal	1 Clab	300	.027			1.06		1.06	1.73
0680	Temporary tape	2 Clab	1500	.011		.50	.43		.93	1.24
0710	Thermoplastic, white or yellow, 4" wide, less than 6,000 L.F.	B-79	15000	.003		.35	.11	.08	.54	.66
0730	6" wide, less than 6,000 L.F.		14000	.003		.53	.12	.09	.74	.87
0740	8" wide, less than 6,000 L.F.		12000	.003		.70	.14	.10	.94	1.10
0750	12" wide, less than 6,000 L.F.		6000	.007		1.03	.27	.21	1.51	1.81
0760	Arrows		660	.061	S.F.	.57	2.50	1.87	4.94	6.75
0770	Gore lines		2500	.016		.57	.66	.49	1.72	2.24
0780	Letters		660	.061		.57	2.50	1.87	4.94	6.75

32 17 23.14 Pavement Parking Markings

		Crew	Daily Output	Labor-Hours	Unit	Material	2018 Bare Costs Labor	Equipment	Total	Total Incl O&P
0010	**PAVEMENT PARKING MARKINGS**									
0790	Layout of pavement marking	A-2	25000	.001	L.F.		.04	.01	.05	.07
0800	Lines on pvmt., parking stall, paint, white, 4" wide	B-78B	400	.045	Stall	4.82	1.85	.87	7.54	9.25
0825	Parking stall, small quantities	2 Pord	80	.200		9.65	8.50		18.15	24.50
0830	Lines on pvmt., parking stall, thermoplastic, white, 4" wide	B-79	300	.133		16	5.50	4.12	25.62	31
1000	Street letters and numbers	B-78B	1600	.011	S.F.	.72	.46	.22	1.40	1.78

32 31 Fences and Gates

32 31 13 – Chain Link Fences and Gates

32 31 13.20 Fence, Chain Link Industrial

		Crew	Daily Output	Labor-Hours	Unit	Material	2018 Bare Costs Labor	Equipment	Total	Total Incl O&P
0010	**FENCE, CHAIN LINK INDUSTRIAL**									
0011	Schedule 40, including concrete									
0020	3 strands barb wire, 2" post @ 10' OC, set in concrete, 6' H									
0200	9 ga. wire, galv. steel, in concrete	B-80C	240	.100	L.F.	19.80	4.14	.82	24.76	29.50
0248	Fence, add for vinyl coated fabric				S.F.	.68			.68	.75
0300	Aluminized steel	B-80C	240	.100	L.F.	22	4.14	.82	26.96	31.50
0301	Fence, wrought iron		240	.100		30	4.14	.82	34.96	40.50
0500	6 ga. wire, galv. steel		240	.100		25	4.14	.82	29.96	35
0600	Aluminized steel		240	.100		30.50	4.14	.82	35.46	41
0800	6 ga. wire, 6' high but omit barbed wire, galv. steel		250	.096		20	3.97	.78	24.75	29.50
0900	Aluminized steel, in concrete		250	.096		24	3.97	.78	28.75	34
0920	8' H, 6 ga. wire, 2-1/2" line post, galv. steel, in concrete		180	.133		32	5.50	1.09	38.59	45
0940	Aluminized steel, in concrete		180	.133		39	5.50	1.09	45.59	53
1100	Add for corner posts, 3" diam., galv. steel, in concrete		40	.600	Ea.	88.50	25	4.90	118.40	143
1200	Aluminized steel, in concrete		40	.600		88.50	25	4.90	118.40	143
1300	Add for braces, galv. steel		80	.300		39	12.40	2.45	53.85	65
1350	Aluminized steel		80	.300		50	12.40	2.45	64.85	77.50
1400	Gate for 6' high fence, 1-5/8" frame, 3' wide, galv. steel		10	2.400		208	99.50	19.60	327.10	410
1500	Aluminized steel, in concrete		10	2.400		209	99.50	19.60	328.10	415
2000	5'-0" high fence, 9 ga., no barbed wire, 2" line post, in concrete									
2010	10' OC, 1-5/8" top rail, in concrete									
2100	Galvanized steel, in concrete	B-80C	300	.080	L.F.	21	3.31	.65	24.96	29
2200	Aluminized steel, in concrete		300	.080	"	19.05	3.31	.65	23.01	27
2400	Gate, 4' wide, 5' high, 2" frame, galv. steel, in concrete		10	2.400	Ea.	219	99.50	19.60	338.10	425
2500	Aluminized steel, in concrete		10	2.400	"	197	99.50	19.60	316.10	400
3100	Overhead slide gate, chain link, 6' high, to 18' wide, in concrete		38	.632	L.F.	97	26	5.15	128.15	155
3110	Cantilever type, in concrete	B-80	48	.667		142	29.50	12.75	184.25	218
3120	8' high, in concrete		24	1.333		168	59	25.50	252.50	305

554

32 31 Fences and Gates

32 31 13 – Chain Link Fences and Gates

32 31 13.20 Fence, Chain Link Industrial

		Crew	Daily Output	Labor-Hours	Unit	Material	2018 Bare Costs Labor	Equipment	Total	Total Incl O&P
3130	10' high, in concrete	B-80	18	1.778	L.F.	206	79	34	319	390
9000	Minimum labor/equipment charge	↓	2	16	Job		710	305	1,015	1,475

32 31 13.25 Fence, Chain Link Residential

		Crew	Daily Output	Labor-Hours	Unit	Material	2018 Bare Costs Labor	Equipment	Total	Total Incl O&P
0010	**FENCE, CHAIN LINK RESIDENTIAL**									
0011	Schedule 20, 11 ga. wire, 1-5/8" post									
3300	Residential, 11 ga. wire, 1-5/8" line post @ 10' OC									
3310	1-3/8" top rail									
3350	4' high, galvanized steel	B-80C	475	.051	L.F.	11.80	2.09	.41	14.30	16.85
3400	Aluminized		475	.051	"	15.05	2.09	.41	17.55	20.50
3600	Gate, 3' wide, 1-3/8" frame, galv. steel		10	2.400	Ea.	95.50	99.50	19.60	214.60	288
3700	Aluminized		10	2.400	"	119	99.50	19.60	238.10	315
3900	3' high, galvanized steel		620	.039	L.F.	112	1.60	.32	113.92	126
4000	Aluminized		620	.039	"	112	1.60	.32	113.92	126
4200	Gate, 3' wide, 1-3/8" frame, galv. steel		12	2	Ea.	92	83	16.30	191.30	253
4300	Aluminized	↓	12	2	"	102	83	16.30	201.30	264
9000	Minimum labor/equipment charge	B-1	2	12	Job		485		485	790

32 31 13.26 Tennis Court Fences and Gates

		Crew	Daily Output	Labor-Hours	Unit	Material	2018 Bare Costs Labor	Equipment	Total	Total Incl O&P
0010	**TENNIS COURT FENCES AND GATES**									
3150	Tennis courts, 11 ga. wire, 1-3/4" mesh, 2-1/2" line posts									
3170	1-5/8" top rail, 3" corner and gate posts									
3190	10' high, galvanized steel	B-80	190	.168	L.F.	26.50	7.45	3.22	37.17	44.50
3210	Vinyl covered 9 ga. wire	"	190	.168	"	36	7.45	3.22	46.67	55.50

32 31 13.40 Fence, Fabric and Accessories

		Crew	Daily Output	Labor-Hours	Unit	Material	2018 Bare Costs Labor	Equipment	Total	Total Incl O&P
0010	**FENCE, FABRIC & ACCESSORIES**									
8000	Components only, fabric, galvanized, 4' high	B-80C	585	.041	L.F.	3.55	1.70	.33	5.58	7.05
8040	6' high		430	.056	"	5.35	2.31	.46	8.12	10.10
8200	Posts, line post, 4' high		34	.706	Ea.	14.75	29	5.75	49.50	70
8240	6' high		26	.923	"	21.50	38	7.55	67.05	94
8400	Top rails, 1-3/8" OD		1000	.024	L.F.	1.64	.99	.20	2.83	3.63
8440	1-5/8" OD	↓	1000	.024	"	2.72	.99	.20	3.91	4.82

32 31 13.80 Residential Chain Link Gate

		Crew	Daily Output	Labor-Hours	Unit	Material	2018 Bare Costs Labor	Equipment	Total	Total Incl O&P
0010	**RESIDENTIAL CHAIN LINK GATE**									
0110	Residential 4' gate, single incl. hardware and concrete	B-80C	10	2.400	Ea.	123	99.50	19.60	242.10	320
0120	5'		10	2.400		133	99.50	19.60	252.10	330
0130	6'		10	2.400		143	99.50	19.60	262.10	340
0510	Residential 4' gate, double incl. hardware and concrete		10	2.400		219	99.50	19.60	338.10	425
0520	5'		10	2.400		234	99.50	19.60	353.10	440
0530	6'	↓	10	2.400	↓	273	99.50	19.60	392.10	485

32 31 13.82 Internal Chain Link Gate

		Crew	Daily Output	Labor-Hours	Unit	Material	2018 Bare Costs Labor	Equipment	Total	Total Incl O&P
0010	**INTERNAL CHAIN LINK GATE**									
0110	Internal 6' gate, single incl. post flange, hardware and concrete	B-80C	10	2.400	Ea.	269	99.50	19.60	388.10	480
0120	8'		10	2.400		305	99.50	19.60	424.10	520
0130	10'		10	2.400		420	99.50	19.60	539.10	650
0510	Internal 6' gate, double incl. post flange, hardware and concrete		10	2.400		490	99.50	19.60	609.10	725
0520	8'		10	2.400		565	99.50	19.60	684.10	805
0530	10'	↓	10	2.400	↓	710	99.50	19.60	829.10	970

32 31 13.84 Industrial Chain Link Gate

		Crew	Daily Output	Labor-Hours	Unit	Material	2018 Bare Costs Labor	Equipment	Total	Total Incl O&P
0010	**INDUSTRIAL CHAIN LINK GATE**									
0110	Industrial 8' gate, single incl. hardware and concrete	B-80C	10	2.400	Ea.	460	99.50	19.60	579.10	690
0120	10'		10	2.400		520	99.50	19.60	639.10	760
0510	Industrial 8' gate, double incl. hardware and concrete		10	2.400		715	99.50	19.60	834.10	970

32 31 Fences and Gates

32 31 13 – Chain Link Fences and Gates

32 31 13.84 Industrial Chain Link Gate		Crew	Daily Output	Labor-Hours	Unit	Material	2018 Bare Costs Labor	Equipment	Total	Total Incl O&P
0520	10'	B-80C	10	2.400	Ea.	815	99.50	19.60	934.10	1,075

32 31 13.88 Chain Link Transom

0010	**CHAIN LINK TRANSOM**									
0110	Add for, single transom, 3' wide, incl. components & hardware	B-80C	10	2.400	Ea.	114	99.50	19.60	233.10	310
0120	Add for, double transom, 6' wide, incl. components & hardware	"	10	2.400	"	122	99.50	19.60	241.10	315

32 31 19 – Decorative Metal Fences and Gates

32 31 19.10 Decorative Fence

0010	**DECORATIVE FENCE**									
5300	Tubular picket, steel, 6' sections, 1-9/16" posts, 4' high	B-80C	300	.080	L.F.	38	3.31	.65	41.96	48
5400	2" posts, 5' high		240	.100		42	4.14	.82	46.96	53.50
5600	2" posts, 6' high		200	.120		50.50	4.97	.98	56.45	64.50
5700	Staggered picket, 1-9/16" posts, 4' high		300	.080		38	3.31	.65	41.96	48
5800	2" posts, 5' high		240	.100		43	4.14	.82	47.96	55
5900	2" posts, 6' high		200	.120		50.50	4.97	.98	56.45	64.50
6200	Gates, 4' high, 3' wide	B-1	10	2.400	Ea.	345	97.50		442.50	540
6300	5' high, 3' wide		10	2.400		231	97.50		328.50	415
6400	6' high, 3' wide		10	2.400		281	97.50		378.50	470
6500	4' wide		10	2.400		330	97.50		427.50	525

32 31 23 – Plastic Fences and Gates

32 31 23.20 Fence, Recycled Plastic

0010	**FENCE, RECYCLED PLASTIC**										
9015	Fence rail, made from recycled plastic, various colors, 2 rail	G	B-1	150	.160	L.F.	10.05	6.50		16.55	21.50
9018	3 rail	G		150	.160		12.80	6.50		19.30	24.50
9020	4 rail	G		150	.160		15.55	6.50		22.05	27.50
9030	Fence pole, made from recycled plastic, various colors, 7'	G		96	.250	Ea.	50	10.15		60.15	71.50
9040	Stockade fence, made from recycled plastic, various colors, 4' high	G	B-80C	160	.150	L.F.	33.50	6.20	1.22	40.92	48.50
9050	6' high	G		160	.150	"	36	6.20	1.22	43.42	51
9060	6' pole	G		96	.250	Ea.	38	10.35	2.04	50.39	61
9070	9' pole	G		96	.250	"	43.50	10.35	2.04	55.89	67
9080	Picket fence, made from recycled plastic, various colors, 3' high	G		160	.150	L.F.	22.50	6.20	1.22	29.92	36
9090	4' high	G		160	.150	"	33	6.20	1.22	40.42	47.50
9100	3' high gate	G		8	3	Ea.	22	124	24.50	170.50	253
9110	4' high gate	G		8	3		29	124	24.50	177.50	260
9120	5' high pole	G		96	.250		36	10.35	2.04	48.39	59
9130	6' high pole	G		96	.250		42	10.35	2.04	54.39	65
9140	Pole cap only	G					5.25			5.25	5.80
9150	Keeper pins only	G					.41			.41	.45

32 31 26 – Wire Fences and Gates

32 31 26.10 Fences, Misc. Metal

0010	**FENCES, MISC. METAL**									
0012	Chicken wire, posts @ 4', 1" mesh, 4' high	B-80C	410	.059	L.F.	3.50	2.42	.48	6.40	8.30
0100	2" mesh, 6' high		350	.069		3.93	2.84	.56	7.33	9.55
0200	Galv. steel, 12 ga., 2" x 4" mesh, posts 5' OC, 3' high		300	.080		2.70	3.31	.65	6.66	9.05
0300	5' high		300	.080		3.28	3.31	.65	7.24	9.70
0400	14 ga., 1" x 2" mesh, 3' high		300	.080		3.32	3.31	.65	7.28	9.70
0500	5' high		300	.080		4.44	3.31	.65	8.40	10.95
1000	Kennel fencing, 1-1/2" mesh, 6' long, 3'-6" wide, 6'-2" high	2 Clab	4	4	Ea.	495	159		654	805
1050	12' long		4	4		695	159		854	1,025
1200	Top covers, 1-1/2" mesh, 6' long		15	1.067		132	42.50		174.50	214
1250	12' long		12	1.333		185	53		238	291
4490	Security fence, prison grade, set in concrete, 10' high	B-80	25	1.280	L.F.	61.50	57	24.50	143	186

32 31 Fences and Gates

32 31 26 – Wire Fences and Gates

32 31 26.10 Fences, Misc. Metal

		Crew	Daily Output	Labor-Hours	Unit	Material	2018 Bare Costs Labor	Equipment	Total	Total Incl O&P
4492	Security fence, prison grade, barbed wire, set in concrete, 10' high	B-80	22	1.455	L.F.	62	64.50	28	154.50	203
4494	Security fence, prison grade, razor wire, set in concrete, 10' high		18	1.778		63	79	34	176	234
4500	Security fence, prison grade, set in concrete, 12' high		25	1.280		61.50	57	24.50	143	186
4600	16' high		20	1.600		79	71	30.50	180.50	235
4990	Security fence, prison grade, set in concrete, 10' high		25	1.280		61.50	57	24.50	143	186

32 31 29 – Wood Fences and Gates

32 31 29.20 Fence, Wood Rail

		Crew	Daily Output	Labor-Hours	Unit	Material	2018 Bare Costs Labor	Equipment	Total	Total Incl O&P
0010	**FENCE, WOOD RAIL**									
0012	Picket, No. 2 cedar, Gothic, 2 rail, 3' high	B-1	160	.150	L.F.	8.10	6.10		14.20	18.75
0050	Gate, 3'-6" wide	B-80C	9	2.667	Ea.	78	110	22	210	289
0600	Open rail, rustic, No. 1 cedar, 2 rail, 3' high		160	.150	L.F.	6.05	6.20	1.22	13.47	18.05
0650	Gate, 3' wide		9	2.667	Ea.	83	110	22	215	294
1200	Stockade, No. 2 cedar, treated wood rails, 6' high		160	.150	L.F.	9.10	6.20	1.22	16.52	21.50
1250	Gate, 3' wide		9	2.667	Ea.	97	110	22	229	310
3300	Board, shadow box, 1" x 6", treated pine, 6' high		160	.150	L.F.	12.65	6.20	1.22	20.07	25.50
3400	No. 1 cedar, 6' high		150	.160		25	6.60	1.31	32.91	39.50
3900	Basket weave, No. 1 cedar, 6' high		160	.150		34	6.20	1.22	41.42	49
3950	Gate, 3'-6" wide	B-1	8	3	Ea.	234	122		356	455
4000	Treated pine, 6' high		150	.160	L.F.	16	6.50		22.50	28
4200	Gate, 3'-6" wide		9	2.667	Ea.	178	108		286	370
8000	Posts only, 4' high		40	.600		12.10	24.50		36.60	53
8040	6' high		30	.800		17.80	32.50		50.30	72
9000	Minimum labor/equipment charge	1 Clab	2	4	Job		159		159	259

32 32 Retaining Walls

32 32 29 – Timber Retaining Walls

32 32 29.10 Landscape Timber Retaining Walls

		Crew	Daily Output	Labor-Hours	Unit	Material	2018 Bare Costs Labor	Equipment	Total	Total Incl O&P
0010	**LANDSCAPE TIMBER RETAINING WALLS**									
0100	Treated timbers, 6" x 6"	1 Clab	265	.030	L.F.	2.47	1.20		3.67	4.67
0110	6" x 8'	"	200	.040	"	5.40	1.59		6.99	8.50
0120	Drilling holes in timbers for fastening, 1/2"	1 Carp	450	.018	Inch		.90		.90	1.46
0130	5/8"	"	450	.018	"		.90		.90	1.46
0140	Reinforcing rods for fastening, 1/2"	1 Clab	312	.026	L.F.	.35	1.02		1.37	2.05
0150	5/8"	"	312	.026	"	.55	1.02		1.57	2.26
0160	Reinforcing fabric	2 Clab	2500	.006	S.Y.	2.13	.26		2.39	2.75
0170	Gravel backfill		28	.571	C.Y.	16.80	23		39.80	55.50
0180	Perforated pipe, 4" diameter with silt sock		1200	.013	L.F.	1	.53		1.53	1.96
0190	Galvanized 60d common nails	1 Clab	625	.013	Ea.	.18	.51		.69	1.03
0200	20d common nails	"	3800	.002	"	.04	.08		.12	.18

32 32 53 – Stone Retaining Walls

32 32 53.10 Retaining Walls, Stone

		Crew	Daily Output	Labor-Hours	Unit	Material	2018 Bare Costs Labor	Equipment	Total	Total Incl O&P
0010	**RETAINING WALLS, STONE**									
0015	Including excavation, concrete footing and									
0020	stone 3' below grade. Price is exposed face area.									
0200	Decorative random stone, to 6' high, 1'-6" thick, dry set	D-1	35	.457	S.F.	69	20.50		89.50	109
0300	Mortar set		40	.400		71	17.90		88.90	107
0500	Cut stone, to 6' high, 1'-6" thick, dry set		35	.457		71.50	20.50		92	112
0600	Mortar set		40	.400		72	17.90		89.90	108
0800	Random stone, 6' to 10' high, 2' thick, dry set		45	.356		77.50	15.95		93.45	112
0900	Mortar set		50	.320		81	14.35		95.35	113

32 32 Retaining Walls

32 32 53 – Stone Retaining Walls

32 32 53.10 Retaining Walls, Stone	Crew	Daily Output	Labor-Hours	Unit	Material	2018 Bare Costs Labor	Equipment	Total	Total Incl O&P
1100	Cut stone, 6' to 10' high, 2' thick, dry set	D-1	45	.356	S.F.	78.50	15.95		94.45	112
1200	Mortar set		50	.320	▼	81.50	14.35		95.85	113
9000	Minimum labor/equipment charge	▼	2	8	Job		360		360	585

32 33 Site Furnishings

32 33 33 – Site Manufactured Planters

32 33 33.10 Planters

		Crew	Daily Output	Labor-Hours	Unit	Material	Labor	Equipment	Total	Total Incl O&P
0010	**PLANTERS**									
0012	Concrete, sandblasted, precast, 48" diameter, 24" high	2 Clab	15	1.067	Ea.	660	42.50		702.50	795
0100	Fluted, precast, 7' diameter, 36" high		10	1.600		1,575	64		1,639	1,850
0300	Fiberglass, circular, 36" diameter, 24" high		15	1.067		690	42.50		732.50	830
0400	60" diameter, 24" high	▼	10	1.600	▼	1,150	64		1,214	1,350
9000	Minimum labor/equipment charge	1 Clab	2	4	Job		159		159	259

32 33 43 – Site Seating and Tables

32 33 43.13 Site Seating

		Crew	Daily Output	Labor-Hours	Unit	Material	Labor	Equipment	Total	Total Incl O&P
0010	**SITE SEATING**									
0012	Seating, benches, park, precast conc., w/backs, wood rails, 4' long	2 Clab	5	3.200	Ea.	635	128		763	905
0100	8' long		4	4		1,075	159		1,234	1,425
0300	Fiberglass, without back, one piece, 4' long		10	1.600		1,775	64		1,839	2,050
0400	8' long		7	2.286		2,150	91		2,241	2,500
0500	Steel barstock pedestals w/backs, 2" x 3" wood rails, 4' long		10	1.600		1,350	64		1,414	1,600
0510	8' long		7	2.286		1,725	91		1,816	2,050
0515	Powder coated steel, 4" x 4" plastic slats, 6' long		8	2		545	79.50		624.50	730
0520	3" x 8" wood plank, 4' long		10	1.600		1,500	64		1,564	1,750
0530	8' long		7	2.286		1,500	91		1,591	1,800
0540	Backless, 4" x 4" wood plank, 4' square		10	1.600		1,250	64		1,314	1,475
0550	8' long		7	2.286		1,100	91		1,191	1,375
0560	Powder coated steel, with back and 2 anti-vagrant dividers, 6' long		8	2		1,175	79.50		1,254.50	1,425
0600	Aluminum pedestals, with backs, aluminum slats, 8' long		8	2		470	79.50		549.50	645
0610	15' long		5	3.200		1,000	128		1,128	1,325
0620	Portable, aluminum slats, 8' long		8	2		455	79.50		534.50	630
0630	15' long		5	3.200		460	128		588	710
0800	Cast iron pedestals, back & arms, wood slats, 4' long		8	2		435	79.50		514.50	610
0820	8' long		5	3.200		1,100	128		1,228	1,400
0840	Backless, wood slats, 4' long		8	2		585	79.50		664.50	775
0860	8' long		5	3.200		1,075	128		1,203	1,400
1700	Steel frame, fir seat, 10' long		10	1.600		395	64		459	540
2000	Benches, park, with back, galv. stl. frame, 4" x 4" plastic slats, 6' long		7	2.286	▼	545	91		636	750
9000	Minimum labor/equipment charge	▼	2	8	Job		320		320	520

32 84 Planting Irrigation

32 84 23 – Underground Sprinklers

32 84 23.10 Sprinkler Irrigation System	Crew	Daily Output	Labor-Hours	Unit	Material	2018 Bare Costs Labor	Equipment	Total	Total Incl O&P
0010 **SPRINKLER IRRIGATION SYSTEM**									
0011 For lawns									
0800 Residential system, custom, 1" supply	B-20	2000	.012	S.F.	.25	.54		.79	1.15
0900 1-1/2" supply	"	1800	.013	"	.48	.60		1.08	1.49

32 91 Planting Preparation

32 91 13 – Soil Preparation

32 91 13.23 Structural Soil Mixing

	Crew	Daily Output	Labor-Hours	Unit	Material	2018 Bare Costs Labor	Equipment	Total	Total Incl O&P
0010 **STRUCTURAL SOIL MIXING**									
0100 Rake topsoil, site material, harley rock rake, ideal	B-6	33	.727	M.S.F.		32	9.45	41.45	61.50
0200 Adverse	"	7	3.429			150	44.50	194.50	289
0300 Screened loam, york rake and finish, ideal	B-62	24	1			43.50	7.25	50.75	78
0400 Adverse	"	20	1.200			52.50	8.70	61.20	93.50
1000 Remove topsoil & stock pile on site, 75 HP dozer, 6" deep, 50' haul	B-10L	30	.400			19.65	15.50	35.15	48
1050 300' haul		6.10	1.967			96.50	76	172.50	238
1100 12" deep, 50' haul		15.50	.774			38	30	68	93.50
1150 300' haul		3.10	3.871			190	150	340	465
1200 200 HP dozer, 6" deep, 50' haul	B-10B	125	.096			4.72	10.20	14.92	18.70
1250 300' haul		30.70	.391			19.20	41.50	60.70	76
1300 12" deep, 50' haul		62	.194			9.50	20.50	30	37.50
1350 300' haul		15.40	.779			38.50	82.50	121	152
1400 Alternate method, 75 HP dozer, 50' haul	B-10L	860	.014	C.Y.		.69	.54	1.23	1.68
1450 300' haul	"	114	.105			5.15	4.07	9.22	12.70
1500 200 HP dozer, 50' haul	B-10B	2660	.005			.22	.48	.70	.88
1600 300' haul	"	570	.021			1.03	2.23	3.26	4.10
1800 Rolling topsoil, hand push roller	1 Clab	3200	.003	S.F.		.10		.10	.16
1850 Tractor drawn roller	B-66	10666	.001	"		.04	.02	.06	.09
2000 Root raking and loading, residential, no boulders	B-6	53.30	.450	M.S.F.		19.65	5.85	25.50	38
2100 With boulders		32	.750			33	9.75	42.75	63.50
2200 Municipal, no boulders		200	.120			5.25	1.56	6.81	10.10
2300 With boulders		120	.200			8.75	2.60	11.35	16.85
2400 Large commercial, no boulders	B-10B	400	.030			1.47	3.18	4.65	5.85
2500 With boulders	"	240	.050			2.46	5.30	7.76	9.75
3000 Scarify subsoil, residential, skid steer loader w/scarifiers, 50 HP	B-66	32	.250			12.85	7.70	20.55	28.50
3050 Municipal, skid steer loader w/scarifiers, 50 HP	"	120	.067			3.42	2.06	5.48	7.65
3100 Large commercial, 75 HP, dozer w/scarifier	B-10L	240	.050			2.46	1.94	4.40	6.05
3500 Screen topsoil from stockpile, vibrating screen, wet material (organic)	B-10P	200	.060	C.Y.		2.95	6	8.95	11.30
3550 Dry material	"	300	.040			1.96	4.01	5.97	7.55
3600 Mixing with conditioners, manure and peat	B-10R	550	.022			1.07	.53	1.60	2.28
3650 Mobilization add for 2 days or less operation	B-34K	3	2.667	Job		123	278	401	500
3800 Spread conditioned topsoil, 6" deep, by hand	B-1	360	.067	S.Y.	5.60	2.70		8.30	10.55
3850 300 HP dozer	B-10M	27	.444	M.S.F.	605	22	67.50	694.50	775
4000 Spread soil conditioners, alum. sulfate, 1#/S.Y., hand push spreader	1 Clab	17500	.001	S.Y.	21	.02		21.02	23.50
4050 Tractor spreader	B-66	700	.011	M.S.F.	2,350	.59	.35	2,350.94	2,600
4100 Fertilizer, 0.2#/S.Y., push spreader	1 Clab	17500	.001	S.Y.	.07	.02		.09	.11
4150 Tractor spreader	B-66	700	.011	M.S.F.	8.20	.59	.35	9.14	10.35
4200 Ground limestone, 1#/S.Y., push spreader	1 Clab	17500	.001	S.Y.	.12	.02		.14	.16
4250 Tractor spreader	B-66	700	.011	M.S.F.	13.35	.59	.35	14.29	15.95
4400 Manure, 18#/S.Y., push spreader	1 Clab	2500	.003	S.Y.	8.10	.13		8.23	9.10
4450 Tractor spreader	B-66	280	.029	M.S.F.	900	1.47	.88	902.35	995
4500 Perlite, 1" deep, push spreader	1 Clab	17500	.001	S.Y.	10.25	.02		10.27	11.30

For customer support on your Commercial Renovation Costs with RSMeans data, call 800.448.8182.

559

32 91 13.23 Structural Soil Mixing

		Crew	Daily Output	Labor-Hours	Unit	Material	2018 Bare Costs Labor	2018 Bare Costs Equipment	Total	Total Incl O&P
4550	Tractor spreader	B-66	700	.011	M.S.F.	1,150	.59	.35	1,150.94	1,250
4600	Vermiculite, push spreader	1 Clab	17500	.001	S.Y.	8.35	.02		8.37	9.25
4650	Tractor spreader	B-66	700	.011	M.S.F.	930	.59	.35	930.94	1,025
5000	Spread topsoil, skid steer loader and hand dress	B-62	270	.089	C.Y.	25	3.88	.64	29.52	34.50
5100	Articulated loader and hand dress	B-100	320	.038		25	1.84	3.06	29.90	34
5200	Articulated loader and 75 HP dozer	B-10M	500	.024		25	1.18	3.66	29.84	33.50
5300	Road grader and hand dress	B-11L	1000	.016		25	.75	.64	26.39	29.50
6000	Tilling topsoil, 20 HP tractor, disk harrow, 2" deep	B-66	450	.018	M.S.F.		.91	.55	1.46	2.04
6050	4" deep		360	.022			1.14	.69	1.83	2.55
6100	6" deep	↓	270	.030	↓		1.52	.92	2.44	3.41
6150	26" rototiller, 2" deep	A-1J	1250	.006	S.Y.		.26	.05	.31	.47
6200	4" deep	↓	1000	.008			.32	.06	.38	.59
6250	6" deep	↓	750	.011	↓		.43	.08	.51	.78

32 91 13.26 Planting Beds

		Crew	Daily Output	Labor-Hours	Unit	Material	2018 Bare Costs Labor	2018 Bare Costs Equipment	Total	Total Incl O&P
0010	**PLANTING BEDS**									
0100	Backfill planting pit, by hand, on site topsoil	2 Clab	18	.889	C.Y.		35.50		35.50	57.50
0200	Prepared planting mix, by hand	"	24	.667			26.50		26.50	43
0300	Skid steer loader, on site topsoil	B-62	340	.071			3.08	.51	3.59	5.50
0400	Prepared planting mix	"	410	.059			2.56	.42	2.98	4.57
1000	Excavate planting pit, by hand, sandy soil	2 Clab	16	1			40		40	64.50
1100	Heavy soil or clay	"	8	2			79.50		79.50	129
1200	1/2 C.Y. backhoe, sandy soil	B-11C	150	.107			4.99	2.08	7.07	10.25
1300	Heavy soil or clay	"	115	.139			6.50	2.72	9.22	13.40
2000	Mix planting soil, incl. loam, manure, peat, by hand	2 Clab	60	.267		44	10.65		54.65	66
2100	Skid steer loader	B-62	150	.160	↓	44	7	1.16	52.16	61
3000	Pile sod, skid steer loader	"	2800	.009	S.Y.		.37	.06	.43	.67
3100	By hand	2 Clab	400	.040			1.59		1.59	2.59
4000	Remove sod, F.E. loader	B-10S	2000	.006			.29	.17	.46	.66
4100	Sod cutter	B-12K	3200	.005			.24	.39	.63	.81
4200	By hand	2 Clab	240	.067	↓		2.66		2.66	4.31

32 91 19.13 Topsoil Placement and Grading

		Crew	Daily Output	Labor-Hours	Unit	Material	2018 Bare Costs Labor	2018 Bare Costs Equipment	Total	Total Incl O&P
0010	**TOPSOIL PLACEMENT AND GRADING**									
0700	Furnish and place, truck dumped, screened, 4" deep	B-10S	1300	.009	S.Y.	3.47	.45	.26	4.18	4.83
0800	6" deep	↓	820	.015	"	4.44	.72	.42	5.58	6.50
0810	Minimum labor/equipment charge	↓	2	6	Ea.		295	171	466	660
0900	Fine grading and seeding, incl. lime, fertilizer & seed,									
1000	With equipment	B-14	1000	.048	S.Y.	.46	2.02	.31	2.79	4.11
2000	Minimum labor/equipment charge	1 Clab	4	2	Job		79.50		79.50	129

560

For customer support on your Commercial Renovation Costs with RSMeans data, call 800.448.8182.

32 92 Turf and Grasses

32 92 19 – Seeding

32 92 19.13 Mechanical Seeding

32 92 19.13 Mechanical Seeding	Crew	Daily Output	Labor-Hours	Unit	Material	2018 Bare Costs Labor	Equipment	Total	Total Incl O&P
0010	**MECHANICAL SEEDING**								
0020	Mechanical seeding, 215 lb./acre	B-66	1.50	5.333	Acre	560	274	165	999	1,225
0100	44 lb./M.S.Y.	"	2500	.003	S.Y.	.20	.16	.10	.46	.59
0101	44 lb./M.S.Y.	1 Clab	13950	.001	S.F.	.02	.02		.04	.06
0300	Fine grading and seeding incl. lime, fertilizer & seed,									
0310	with equipment	B-14	1000	.048	S.Y.	.45	2.02	.31	2.78	4.10
0400	Fertilizer hand push spreader, 35 lb./M.S.F.	1 Clab	200	.040	M.S.F.	9.85	1.59		11.44	13.45
0600	Limestone hand push spreader, 50 lb./M.S.F.		180	.044		4.65	1.77		6.42	8
0800	Grass seed hand push spreader, 4.5 lb./M.S.F.		180	.044		22	1.77		23.77	27.50
1000	Hydro or air seeding for large areas, incl. seed and fertilizer	B-81	8900	.003	S.Y.	.50	.13	.06	.69	.82
1100	With wood fiber mulch added	"	8900	.003	"	2	.13	.06	2.19	2.47
1300	Seed only, over 100 lb., field seed, minimum				Lb.	1.85			1.85	2.04
1400	Maximum					1.75			1.75	1.93
1500	Lawn seed, minimum					1.45			1.45	1.60
1600	Maximum					2.50			2.50	2.75
1800	Aerial operations, seeding only, field seed	B-58	50	.480	Acre	600	21	63.50	684.50	765
1900	Lawn seed		50	.480		470	21	63.50	554.50	625
2100	Seed and liquid fertilizer, field seed		50	.480		675	21	63.50	759.50	850
2200	Lawn seed		50	.480		545	21	63.50	629.50	705
9000	Minimum labor/equipment charge	1 Clab	4	2	Job		79.50		79.50	129

32 92 23 – Sodding

32 92 23.10 Sodding Systems

32 92 23.10 Sodding Systems	Crew	Daily Output	Labor-Hours	Unit	Material	2018 Bare Costs Labor	Equipment	Total	Total Incl O&P
0010	**SODDING SYSTEMS**								
0020	Sodding, 1" deep, bluegrass sod, on level ground, over 8 M.S.F.	B-63	22	1.818	M.S.F.	246	76.50	7.90	330.40	405
0200	4 M.S.F.		17	2.353		240	99	10.25	349.25	435
0300	1,000 S.F.		13.50	2.963		290	125	12.90	427.90	535
0500	Sloped ground, over 8 M.S.F.		6	6.667		246	281	29	556	760
0600	4 M.S.F.		5	8		240	335	35	610	850
0700	1,000 S.F.		4	10		290	420	43.50	753.50	1,050
1000	Bent grass sod, on level ground, over 6 M.S.F.		20	2		257	84.50	8.70	350.20	430
1100	3 M.S.F.		18	2.222		271	93.50	9.65	374.15	460
1200	Sodding 1,000 S.F. or less		14	2.857		297	120	12.45	429.45	535
1500	Sloped ground, over 6 M.S.F.		15	2.667		257	112	11.60	380.60	475
1600	3 M.S.F.		13.50	2.963		271	125	12.90	408.90	515
1700	1,000 S.F.		12	3.333		297	140	14.50	451.50	570

32 93 Plants

32 93 13 – Ground Covers

32 93 13.10 Ground Cover Plants

32 93 13.10 Ground Cover Plants	Crew	Daily Output	Labor-Hours	Unit	Material	2018 Bare Costs Labor	Equipment	Total	Total Incl O&P
0010	**GROUND COVER PLANTS**								
0012	Plants, pachysandra, in prepared beds	B-1	15	1.600	C	79.50	65		144.50	193
0200	Vinca minor, 1 yr., bare root, in prepared beds		12	2	"	70	81		151	209
0600	Stone chips, in 50 lb. bags, Georgia marble		520	.046	Bag	4.80	1.87		6.67	8.35
0700	Onyx gemstone		260	.092		16	3.74		19.74	23.50
0800	Quartz		260	.092		16	3.74		19.74	23.50
0900	Pea gravel, truckload lots		28	.857	Ton	30	34.50		64.50	90

32 93 Plants

32 93 33 – Shrubs

32 93 33.10 Shrubs and Trees	Crew	Daily Output	Labor-Hours	Unit	Material	2018 Bare Costs Labor	2018 Bare Costs Equipment	Total	Total Incl O&P
0010 SHRUBS AND TREES									
0011 Evergreen, in prepared beds, B&B									
0100 Arborvitae pyramidal, 4'-5'	B-17	30	1.067	Ea.	103	47	21.50	171.50	213
0150 Globe, 12"-15"	B-1	96	.250		26	10.15		36.15	45
0300 Cedar, blue, 8'-10'	B-17	18	1.778		239	78.50	36	353.50	430
0500 Hemlock, Canadian, 2-1/2'-3'	B-1	36	.667		32.50	27		59.50	79.50
0550 Holly, Savannah, 8'-10' H		9.68	2.479		300	100		400	495
0600 Juniper, andorra, 18"-24"		80	.300		50	12.15		62.15	75
0620 Wiltoni, 15"-18"		80	.300		26.50	12.15		38.65	49.50
0640 Skyrocket, 4-1/2'-5'	B-17	55	.582		110	26	11.85	147.85	176
0660 Blue pfitzer, 2'-2-1/2'	B-1	44	.545		39.50	22		61.50	79.50
0680 Ketleerie, 2-1/2'-3'		50	.480		56	19.45		75.45	93
0700 Pine, black, 2-1/2'-3'		50	.480		62	19.45		81.45	100
0720 Mugo, 18"-24"		60	.400		56	16.20		72.20	88.50
0740 White, 4'-5'	B-17	75	.427		53	18.90	8.70	80.60	98.50
0800 Spruce, blue, 18"-24"	B-1	60	.400		69	16.20		85.20	103
0840 Norway, 4'-5'	B-17	75	.427		86	18.90	8.70	113.60	135
0900 Yew, denisforma, 12"-15"	B-1	60	.400		36.50	16.20		52.70	67
1000 Capitata, 18"-24"		30	.800		34	32.50		66.50	90
1100 Hicksi, 2'-2-1/2'		30	.800		103	32.50		135.50	167

32 93 33.20 Shrubs	Crew	Daily Output	Labor-Hours	Unit	Material	2018 Bare Costs Labor	2018 Bare Costs Equipment	Total	Total Incl O&P
0010 SHRUBS									
0011 Broadleaf Evergreen, planted in prepared beds									
0100 Andromeda, 15"-18", cont	B-1	96	.250	Ea.	33.50	10.15		43.65	53.50
0200 Azalea, 15"-18", cont		96	.250		30.50	10.15		40.65	50.50
0300 Barberry, 9"-12", cont		130	.185		18.70	7.50		26.20	32.50
0400 Boxwood, 15"-18", B&B		96	.250		44.50	10.15		54.65	65.50
0500 Euonymus, emerald gaiety, 12"-15", cont		115	.209		24.50	8.45		32.95	41
0600 Holly, 15"-18", B&B		96	.250		40	10.15		50.15	60.50
0900 Mount laurel, 18"-24", B&B		80	.300		72.50	12.15		84.65	100
1000 Paxistema, 9"-12" H		130	.185		22	7.50		29.50	36
1100 Rhododendron, 18"-24", cont		48	.500		38.50	20.50		59	75.50
1200 Rosemary, 1 gal. cont		600	.040		18.55	1.62		20.17	23
2000 Deciduous, planted in prepared beds, amelanchier, 2'-3', B&B		57	.421		123	17.05		140.05	163
2100 Azalea, 15"-18", B&B		96	.250		30.50	10.15		40.65	50
2300 Bayberry, 2'-3', B&B		57	.421		30	17.05		47.05	60.50
2600 Cotoneaster, 15"-18", B&B		80	.300		27	12.15		39.15	50
2800 Dogwood, 3'-4', B&B	B-17	40	.800		33	35.50	16.25	84.75	111
2900 Euonymus, alatus compacta, 15"-18", cont	B-1	80	.300		27	12.15		39.15	49.50
3200 Forsythia, 2'-3', cont	"	60	.400		17.80	16.20		34	46
3300 Hibiscus, 3'-4', B&B	B-17	75	.427		42	18.90	8.70	69.60	86
3400 Honeysuckle, 3'-4', B&B	B-1	60	.400		27.50	16.20		43.70	57
3500 Hydrangea, 2'-3', B&B	"	57	.421		31	17.05		48.05	61.50
3600 Lilac, 3'-4', B&B	B-17	40	.800		27.50	35.50	16.25	79.25	105
3900 Privet, bare root, 18"-24"	B-1	80	.300		15.15	12.15		27.30	36.50
4100 Quince, 2'-3', B&B	"	57	.421		29	17.05		46.05	59.50
4200 Russian olive, 3'-4', B&B	B-17	75	.427		25	18.90	8.70	52.60	67.50
4400 Spirea, 3'-4', B&B	B-1	70	.343		20.50	13.90		34.40	45
4500 Viburnum, 3'-4', B&B	B-17	40	.800		26	35.50	16.25	77.75	103

32 93 Plants

32 93 43 – Trees

32 93 43.20 Trees		Crew	Daily Output	Labor-Hours	Unit	Material	2018 Bare Costs Labor	2018 Bare Costs Equipment	Total	Total Incl O&P
0010	**TREES**									
0011	Deciduous, in prep. beds, balled & burlapped (B&B)									
0100	Ash, 2" caliper [G]	B-17	8	4	Ea.	200	177	81.50	458.50	595
0200	Beech, 5'-6' [G]		50	.640		210	28.50	13	251.50	291
0300	Birch, 6'-8', 3 stems [G]		20	1.600		167	71	32.50	270.50	335
0500	Crabapple, 6'-8' [G]		20	1.600		139	71	32.50	242.50	305
0600	Dogwood, 4'-5' [G]		40	.800		140	35.50	16.25	191.75	229
0700	Eastern redbud, 4'-5' [G]		40	.800		148	35.50	16.25	199.75	238
0800	Elm, 8'-10' [G]		20	1.600		325	71	32.50	428.50	510
0900	Ginkgo, 6'-7' [G]		24	1.333		148	59	27	234	288
1000	Hawthorn, 8'-10', 1" caliper [G]		20	1.600		161	71	32.50	264.50	325
1100	Honeylocust, 10'-12', 1-1/2" caliper [G]		10	3.200		206	142	65	413	525
1300	Larch, 8' [G]		32	1		128	44.50	20.50	193	234
1400	Linden, 8'-10', 1" caliper [G]		20	1.600		144	71	32.50	247.50	310
1500	Magnolia, 4'-5' [G]		20	1.600		115	71	32.50	218.50	277
1600	Maple, red, 8'-10', 1-1/2" caliper [G]		10	3.200		199	142	65	406	520
1700	Mountain ash, 8'-10', 1" caliper [G]		16	2		177	88.50	40.50	306	380
1800	Oak, 2-1/2"-3" caliper [G]		6	5.333		320	236	108	664	855
2100	Planetree, 9'-11', 1-1/4" caliper [G]		10	3.200		266	142	65	473	590
2200	Plum, 6'-8', 1" caliper [G]		20	1.600		79	71	32.50	182.50	237
2300	Poplar, 9'-11', 1-1/4" caliper [G]		10	3.200		100	142	65	307	410
2500	Sumac, 2'-3' [G]		75	.427		44.50	18.90	8.70	72.10	89
2700	Tulip, 5'-6' [G]		40	.800		46	35.50	16.25	97.75	125
2800	Willow, 6'-8', 1" caliper [G]		20	1.600		97	71	32.50	200.50	257
9000	Minimum labor/equipment charge	1 Clab	4	2	Job		79.50		79.50	129

32 94 Planting Accessories

32 94 13 – Landscape Edging

32 94 13.20 Edging		Crew	Daily Output	Labor-Hours	Unit	Material	2018 Bare Costs Labor	2018 Bare Costs Equipment	Total	Total Incl O&P
0010	**EDGING**									
0050	Aluminum alloy, including stakes, 1/8" x 4", mill finish	B-1	390	.062	L.F.	2.21	2.49		4.70	6.50
0051	Black paint		390	.062		2.56	2.49		5.05	6.85
0052	Black anodized		390	.062		2.96	2.49		5.45	7.30
0100	Brick, set horizontally, 1-1/2 bricks per L.F.	D-1	370	.043		1.39	1.94		3.33	4.69
0150	Set vertically, 3 bricks per L.F.	"	135	.119		3.54	5.30		8.84	12.55
0200	Corrugated aluminum, roll, 4" wide	1 Carp	650	.012		2.19	.62		2.81	3.42
0250	6" wide	"	550	.015		2.74	.74		3.48	4.21
0600	Railroad ties, 6" x 8"	2 Carp	170	.094		2.36	4.77		7.13	10.35
0650	7" x 9"		136	.118		2.62	5.95		8.57	12.60
0750	Redwood 2" x 4"		330	.048		2.26	2.46		4.72	6.50
0800	Steel edge strips, incl. stakes, 1/4" x 5"	B-1	390	.062		4.54	2.49		7.03	9.05
0850	3/16" x 4"	"	390	.062		3.59	2.49		6.08	8
9000	Minimum labor/equipment charge	1 Carp	4	2	Job		101		101	165

32 94 50 – Tree Guying

32 94 50.10 Tree Guying Systems		Crew	Daily Output	Labor-Hours	Unit	Material	2018 Bare Costs Labor	2018 Bare Costs Equipment	Total	Total Incl O&P
0010	**TREE GUYING SYSTEMS**									
0015	Tree guying including stakes, guy wire and wrap									
0100	Less than 3" caliper, 2 stakes	2 Clab	35	.457	Ea.	13.10	18.20		31.30	44
0200	3" to 4" caliper, 3 stakes	"	21	.762	"	19.85	30.50		50.35	71.50
1000	Including arrowhead anchor, cable, turnbuckles and wrap									

32 94 Planting Accessories

32 94 50 – Tree Guying

32 94 50.10 Tree Guying Systems	Crew	Daily Output	Labor-Hours	Unit	Material	2018 Bare Costs Labor	Equipment	Total	Total Incl O&P
1100	Less than 3" caliper, 3" anchors	2 Clab	20	.800	Ea.	21	32		53	75.50
1200	3" to 6" caliper, 4" anchors		15	1.067		31.50	42.50		74	104
1300	6" caliper, 6" anchors		12	1.333		23.50	53		76.50	113
1400	8" caliper, 8" anchors		9	1.778		116	71		187	243

32 96 Transplanting

32 96 23 – Plant and Bulb Transplanting

32 96 23.23 Planting

		Crew	Daily Output	Labor-Hours	Unit	Material	Labor	Equipment	Total	Total Incl O&P
0010	**PLANTING**									
0012	Moving shrubs on site, 12" ball	B-62	28	.857	Ea.		37.50	6.20	43.70	67
0100	24" ball	"	22	1.091	"		47.50	7.90	55.40	85

32 96 23.43 Moving Trees

		Crew	Daily Output	Labor-Hours	Unit	Material	Labor	Equipment	Total	Total Incl O&P
0010	**MOVING TREES**, On site									
0300	Moving trees on site, 36" ball	B-6	3.75	6.400	Ea.		279	83.50	362.50	540
0400	60" ball	"	1	24	"		1,050	310	1,360	2,025

32 96 43 – Tree Transplanting

32 96 43.20 Tree Removal

		Crew	Daily Output	Labor-Hours	Unit	Material	Labor	Equipment	Total	Total Incl O&P
0010	**TREE REMOVAL**									
0100	Dig & lace, shrubs, broadleaf evergreen, 18"-24" high	B-1	55	.436	Ea.		17.70		17.70	28.50
0200	2'-3'	"	35	.686			28		28	45
0300	3'-4'	B-6	30	.800			35	10.40	45.40	67.50
0400	4'-5'	"	20	1.200			52.50	15.60	68.10	101
1000	Deciduous, 12"-15"	B-1	110	.218			8.85		8.85	14.35
1100	18"-24"		65	.369			14.95		14.95	24.50
1200	2'-3'		55	.436			17.70		17.70	28.50
1300	3'-4'	B-6	50	.480			21	6.25	27.25	40.50
2000	Evergreen, 18"-24"	B-1	55	.436			17.70		17.70	28.50
2100	2'-0" to 2'-6"		50	.480			19.45		19.45	31.50
2200	2'-6" to 3'-0"		35	.686			28		28	45
2300	3'-0" to 3'-6"		20	1.200			48.50		48.50	79
3000	Trees, deciduous, small, 2'-3'		55	.436			17.70		17.70	28.50
3100	3'-4'	B-6	50	.480			21	6.25	27.25	40.50
3200	4'-5'		35	.686			30	8.95	38.95	58
3300	5'-6'		30	.800			35	10.40	45.40	67.50
4000	Shade, 5'-6'		50	.480			21	6.25	27.25	40.50
4100	6'-8'		35	.686			30	8.95	38.95	58
4200	8'-10'		25	.960			42	12.50	54.50	81.50
4300	2" caliper		12	2			87.50	26	113.50	169
5000	Evergreen, 4'-5'		35	.686			30	8.95	38.95	58
5100	5'-6'		25	.960			42	12.50	54.50	81.50
5200	6'-7'		19	1.263			55	16.45	71.45	107
5300	7'-8'		15	1.600			70	21	91	135
5400	8'-10'		11	2.182			95.50	28.50	124	184

Estimating Tips

33 10 00 Water Utilities
33 30 00 Sanitary Sewerage Utilities
33 40 00 Storm Drainage Utilities

- Never assume that the water, sewer, and drainage lines will go in at the early stages of the project. Consider the site access needs before dividing the site in half with open trenches, loose pipe, and machinery obstructions. Always inspect the site to establish that the site drawings are complete. Check off all existing utilities on your drawings as you locate them. Be especially careful with underground utilities because appurtenances are sometimes buried during regrading or repaving operations. If you find any discrepancies, mark up the site plan for further research. Differing site conditions can be very costly if discovered later in the project.

- See also Section 33 01 00 for restoration of pipe where removal/replacement may be undesirable. Use of new types of piping materials can reduce the overall project cost. Owners/design engineers should consider the installing contractor as a valuable source of current information on utility products and local conditions that could lead to significant cost savings.

Reference Numbers

Reference numbers are shown at the beginning of some major classifications. These numbers refer to related items in the Reference Section. The reference information may be an estimating procedure, an alternate pricing method, or technical information.

Note: Not all subdivisions listed here necessarily appear. ■

Division 33 Utilities

33 01 30.11 Television Inspection of Sewers

33 01 30.11 Television Inspection of Sewers	Crew	Daily Output	Labor-Hours	Unit	Material	2018 Bare Costs Labor	Equipment	Total	Total Incl O&P
0010 **TELEVISION INSPECTION OF SEWERS**									
0100 Pipe internal cleaning & inspection, cleaning, pressure pipe systems									
0120 Pig method, lengths 1000' to 10,000'									
0140 4" diameter thru 24" diameter, minimum				L.F.				3.60	4.14
0160 Maximum				"				18	21
6000 Sewage/sanitary systems									
6100 Power rodder with header & cutters									
6110 Mobilization charge, minimum				Total				695	800
6120 Mobilization charge, maximum				"				9,125	10,600
6140 Cleaning 4"-12" diameter				L.F.				3.39	3.90
6190 14"-24" diameter								3.98	4.59
6240 30" diameter								5.80	6.65
6250 36" diameter								6.75	7.75
6260 48" diameter								7.70	8.85
6270 60" diameter								8.70	9.95
6280 72" diameter								9.65	11.10
9000 Inspection, television camera with video									
9060 up to 500 linear feet				Total				715	820

33 01 30.72 Cured-In-Place Pipe Lining

33 01 30.72 Cured-In-Place Pipe Lining	Crew	Daily Output	Labor-Hours	Unit	Material	2018 Bare Costs Labor	Equipment	Total	Total Incl O&P
0010 **CURED-IN-PLACE PIPE LINING**									
0011 With cement incl. bypass & cleaning									
0020 Less than 10,000 L.F., urban, 6" to 10"	C-17E	130	.615	L.F.	9.55	32.50	.79	42.84	64
0050 10" to 12"		125	.640		11.75	34	.82	46.57	68.50
0070 12" to 16"		115	.696		12.05	36.50	.89	49.44	73
0100 16" to 20"		95	.842		14.15	44.50	1.08	59.73	88
0200 24" to 36"		90	.889		15.25	47	1.14	63.39	93.50
0300 48" to 72"		80	1		24.50	53	1.28	78.78	113
0500 Rural, 6" to 10"		180	.444		9.55	23.50	.57	33.62	49
0550 10" to 12"		175	.457		11.75	24	.59	36.34	52.50
0570 12" to 16"		160	.500		12.05	26.50	.64	39.19	56.50
0600 16" to 20"		135	.593		14.15	31.50	.76	46.41	67
0700 24" to 36"		125	.640		15.25	34	.82	50.07	72
0800 48" to 72"		100	.800		24.50	42	1.02	67.52	95.50
1000 Greater than 10,000 L.F., urban, 6" to 10"		160	.500		9.55	26.50	.64	36.69	53.50
1050 10" to 12"		155	.516		11.75	27	.66	39.41	57.50
1070 12" to 16"		140	.571		12.05	30	.73	42.78	62.50
1100 16" to 20"		120	.667		14.15	35	.85	50	73
1200 24" to 36"		115	.696		15.25	36.50	.89	52.64	77
1300 48" to 72"		95	.842		24.50	44.50	1.08	70.08	99
1500 Rural, 6" to 10"		215	.372		9.55	19.65	.48	29.68	42.50
1550 10" to 12"		210	.381		11.75	20	.49	32.24	46
1570 12" to 16"		185	.432		12.05	23	.55	35.60	51
1600 16" to 20"		150	.533		14.15	28	.68	42.83	62
1700 24" to 36"		140	.571		15.25	30	.73	45.98	66
1800 48" to 72"		120	.667		24.50	35	.85	60.35	84
2000 Cured in place pipe, non-pressure, flexible felt resin, 400' runs									
2100 6" diameter				L.F.				25.50	28
2200 8" diameter								26.50	29
2300 10" diameter								29	32
2400 12" diameter								34.50	38
2500 15" diameter								59	65
2600 18" diameter								82	90

33 01 Operation and Maintenance of Utilities

33 01 30 – Operation and Maintenance of Sewer Utilities

33 01 30.72 Cured-In-Place Pipe Lining	Crew	Daily Output	Labor-Hours	Unit	Material	2018 Bare Costs Labor	Equipment	Total	Total Incl O&P
2700	21" diameter				L.F.				105	115
2800	24" diameter								177	195
2900	30" diameter								195	215
3000	36" diameter								205	225
3100	48" diameter								218	240

33 05 Common Work Results for Utilities

33 05 07 – Trenchless Installation of Utility Piping

33 05 07.23 Utility Boring and Jacking

		Crew	Daily Output	Labor-Hours	Unit	Material	Labor	Equipment	Total	Total Incl O&P
0010	**UTILITY BORING AND JACKING**									
0011	Casing only, 100' minimum,									
0020	not incl. jacking pits or dewatering									
0100	Roadwork, 1/2" thick wall, 24" diameter casing	B-42	20	3.200	L.F.	117	144	57	318	425
0200	36" diameter		16	4		215	180	71	466	610
0300	48" diameter		15	4.267		297	192	76	565	720
0500	Railroad work, 24" diameter		15	4.267		117	192	76	385	520
0600	36" diameter		14	4.571		215	206	81.50	502.50	660
0700	48" diameter		12	5.333		297	240	95	632	820
0900	For ledge, add								20%	20%
1000	Small diameter boring, 3", sandy soil	B-82	900	.018		21	.81	.09	21.90	24.50
1040	Rocky soil	"	500	.032		21	1.46	.16	22.62	25.50

33 05 07.36 Microtunneling

		Crew	Daily Output	Labor-Hours	Unit	Material	Labor	Equipment	Total	Total Incl O&P
0010	**MICROTUNNELING**									
0011	Not including excavation, backfill, shoring,									
0020	or dewatering, average 50'/day, slurry method									
0100	24" to 48" outside diameter, minimum				L.F.				965	965
0110	Adverse conditions, add				"				500	500
1000	Rent microtunneling machine, average monthly lease				Month				97,500	107,000
1010	Operating technician				Day				630	705
1100	Mobilization and demobilization, minimum				Job				41,200	45,900
1110	Maximum				"				445,500	490,500

33 05 61 – Concrete Manholes

33 05 61.10 Storm Drainage Manholes, Frames and Covers

		Crew	Daily Output	Labor-Hours	Unit	Material	Labor	Equipment	Total	Total Incl O&P
0010	**STORM DRAINAGE MANHOLES, FRAMES & COVERS**									
0020	Excludes footing, excavation, backfill (See line items for frame & cover)									
0050	Brick, 4' inside diameter, 4' deep	D-1	1	16	Ea.	590	715		1,305	1,825
0100	6' deep		.70	22.857		835	1,025		1,860	2,600
0150	8' deep		.50	32		1,075	1,425		2,500	3,525
0200	For depths over 8', add		4	4	V.L.F.	88	179		267	390
1110	Precast, 4' ID, 4' deep	B-22	4.10	7.317	Ea.	845	345	47.50	1,237.50	1,525
1120	6' deep		3	10		1,050	470	65	1,585	1,975
1130	8' deep		2	15		1,200	705	97	2,002	2,550
1140	For depths over 8', add		16	1.875	V.L.F.	128	88	12.15	228.15	296

33 05 63 – Concrete Vaults and Chambers

33 05 63.13 Precast Concrete Utility Structures

		Crew	Daily Output	Labor-Hours	Unit	Material	Labor	Equipment	Total	Total Incl O&P
0010	**PRECAST CONCRETE UTILITY STRUCTURES**, 6" thick									
0050	5' x 10' x 6' high, ID	B-13	2	28	Ea.	1,800	1,225	295	3,320	4,275
0350	Hand hole, precast concrete, 1-1/2" thick									
0400	1'-0" x 2'-0" x 1'-9", ID, light duty	B-1	4	6	Ea.	430	243		673	865
0450	4'-6" x 3'-2" x 2'-0", OD, heavy duty	B-6	3	8	"	1,525	350	104	1,979	2,350

33 05 Common Work Results for Utilities

33 05 97 – Identification and Signage for Utilities

33 05 97.05 Utility Connection	Crew	Daily Output	Labor-Hours	Unit	Material	2018 Bare Costs Labor	Equipment	Total	Total Incl O&P
0010 **UTILITY CONNECTION**									
0020　Water, sanitary, stormwater, gas, single connection	B-14	1	48	Ea.	2,925	2,025	310	5,260	6,825
0030　Telecommunication	"	3	16	"	385	675	104	1,164	1,600

33 05 97.10 Utility Accessories

0010 **UTILITY ACCESSORIES** R312316-40									
0400　Underground tape, detectable, reinforced, alum. foil core, 2"	1 Clab	150	.053	C.L.F.	9	2.13		11.13	13.35
0500　　6"		140	.057	"	36	2.28		38.28	43
9000　Minimum labor/equipment charge		4	2	Job		79.50		79.50	129

33 11 Groundwater Sources

33 11 13 – Potable Water Supply Wells

33 11 13.10 Wells and Accessories

	Crew	Daily Output	Labor-Hours	Unit	Material	Labor	Equipment	Total	Total Incl O&P
0010 **WELLS & ACCESSORIES** R221113-50									
0011　Domestic									
0100　Drilled, 4" to 6" diameter	B-23	120	.333	L.F.		13.40	22.50	35.90	46.50
0200　　8" diameter	"	95.20	.420	"		16.90	28	44.90	58.50
1400　Remove & reset pump, minimum	B-21	4	7	Ea.		325	32.50	357.50	555
1420　　Maximum	"	2	14	"		650	65	715	1,125
1500　Pumps, installed in wells to 100' deep, 4" submersible									
1510　　1/2 HP	Q-1	3.22	4.969	Ea.	700	278		978	1,200
1520　　3/4 HP		2.66	6.015		845	335		1,180	1,450
1600　　1 HP		2.29	6.987		910	390		1,300	1,600
1700　　1-1/2 HP	Q-22	1.60	10		1,800	560	310	2,670	3,175
1800　　2 HP		1.33	12.030		1,725	675	375	2,775	3,350
1900　　3 HP		1.14	14.035		2,225	785	435	3,445	4,150
2000　　5 HP		1.14	14.035		3,050	785	435	4,270	5,050
2050　Remove and install motor only, 4 HP		1.14	14.035		1,300	785	435	2,520	3,125
5000　Wells to 180' deep, 4" submersible, 1 HP	B-21	1.10	25.455		2,375	1,175	118	3,668	4,650
5500　　2 HP		1.10	25.455		3,275	1,175	118	4,568	5,625
6000　　3 HP		1	28		4,425	1,300	130	5,855	7,100
7000　　5 HP		.90	31.111		6,050	1,450	144	7,644	9,125
9000　Minimum labor/equipment charge		1.80	15.556	Job		720	72	792	1,225

33 14 Water Utility Transmission and Distribution

33 14 13 – Public Water Utility Distribution Piping

33 14 13.15 Water Supply, Ductile Iron Pipe

	Crew	Daily Output	Labor-Hours	Unit	Material	Labor	Equipment	Total	Total Incl O&P
0010 **WATER SUPPLY, DUCTILE IRON PIPE** R331113-80									
0020　Not including excavation or backfill									
2000　Pipe, class 50 water piping, 18' lengths									
2020　　Mechanical joint, 4" diameter	B-21A	200	.200	L.F.	30.50	10	1.82	42.32	51.50
3000　　Push-on joint, 4" diameter		400	.100		21	4.99	.91	26.90	32
3020　　　6" diameter		333.33	.120		21.50	6	1.09	28.59	34.50
8000　Piping, fittings, mechanical joint, AWWA C110									
8006　　90° bend, 4" diameter	B-20A	16	2	Ea.	170	97		267	340
8020　　　6" diameter		12.80	2.500		255	121		376	470
8200　　Wye or tee, 4" diameter		10.67	2.999		380	145		525	645
8220　　　6" diameter		8.53	3.751		570	182		752	915
8398　　45° bend, 4" diameter		16	2		211	97		308	385

33 14 Water Utility Transmission and Distribution

33 14 13 – Public Water Utility Distribution Piping

33 14 13.15 Water Supply, Ductile Iron Pipe

		Crew	Daily Output	Labor-Hours	Unit	Material	2018 Bare Costs Labor	Equipment	Total	Total Incl O&P
8400	6" diameter	B-20A	12.80	2.500	Ea.	253	121		374	470
8405	8" diameter		10.67	2.999		365	145		510	630
8450	Decreaser, 6" x 4" diameter		14.22	2.250		232	109		341	430
8460	8" x 6" diameter		11.64	2.749		350	133		483	595
8550	Piping, butterfly valves, cast iron									
8560	4" diameter	B-20	6	4	Ea.	405	179		584	735
9600	Steel sleeve with tap, 4" diameter		3	8		440	355		795	1,075
9620	6" diameter		2	12		510	535		1,045	1,425

33 14 17 – Site Water Utility Service Laterals

33 14 17.15 Tapping, Crosses and Sleeves

		Crew	Daily Output	Labor-Hours	Unit	Material	2018 Bare Costs Labor	Equipment	Total	Total Incl O&P
0010	**TAPPING, CROSSES AND SLEEVES**									
4000	Drill and tap pressurized main (labor only)									
4100	6" main, 1" to 2" service	Q-1	3	5.333	Ea.		298		298	465
4150	8" main, 1" to 2" service	"	2.75	5.818	"		325		325	505
4500	Tap and insert gate valve									
4600	8" main, 4" branch	B-21	3.20	8.750	Ea.		405	40.50	445.50	700
4650	6" branch		2.70	10.370			480	48	528	830
4800	12" main, 6" branch		2.35	11.915			550	55	605	950

33 31 Sanitary Sewerage Piping

33 31 11 – Public Sanitary Sewerage Gravity Piping

33 31 11.15 Sewage Collection, Concrete Pipe

		Crew	Daily Output	Labor-Hours	Unit	Material	2018 Bare Costs Labor	Equipment	Total	Total Incl O&P
0010	**SEWAGE COLLECTION, CONCRETE PIPE**									
0020	See Section 33 41 13.60 for sewage/drainage collection, concrete pipe									

33 31 11.25 Sewage Collection, Polyvinyl Chloride Pipe

		Crew	Daily Output	Labor-Hours	Unit	Material	2018 Bare Costs Labor	Equipment	Total	Total Incl O&P
0010	**SEWAGE COLLECTION, POLYVINYL CHLORIDE PIPE**									
0020	Not including excavation or backfill									
2000	20' lengths, SDR 35, B&S, 4" diameter	B-20	375	.064	L.F.	1.64	2.86		4.50	6.45
2040	6" diameter		350	.069		3.49	3.06		6.55	8.80
2080	13' lengths, SDR 35, B&S, 8" diameter		335	.072		6.50	3.20		9.70	12.35
2120	10" diameter	B-21	330	.085		11.35	3.93	.39	15.67	19.25
4000	Piping, DWV PVC, no exc./bkfill., 10' L, Sch 40, 4" diameter	B-20	375	.064		3.74	2.86		6.60	8.75
4010	6" diameter		350	.069		8.10	3.06		11.16	13.85
4020	8" diameter		335	.072		12.75	3.20		15.95	19.20

33 34 Onsite Wastewater Disposal

33 34 13 – Septic Tanks

33 34 13.13 Concrete Septic Tanks

		Crew	Daily Output	Labor-Hours	Unit	Material	2018 Bare Costs Labor	Equipment	Total	Total Incl O&P
0010	**CONCRETE SEPTIC TANKS**									
0011	Not including excavation or piping									
0015	Septic tanks, precast, 1,000 gallon	B-21	8	3.500	Ea.	1,025	162	16.20	1,203.20	1,425
0020	1,250 gallon		8	3.500		1,200	162	16.20	1,378.20	1,600
0060	1,500 gallon		7	4		1,600	185	18.50	1,803.50	2,100
0100	2,000 gallon		5	5.600		2,275	259	26	2,560	2,950
0140	2,500 gallon		5	5.600		2,425	259	26	2,710	3,125
0180	4,000 gallon		4	7		5,950	325	32.50	6,307.50	7,100
0220	5,000 gallon, 4 piece	B-13	3	18.667		7,600	815	197	8,612	9,900
0300	15,000 gallon, 4 piece	B-13B	1.70	32.941		21,500	1,450	585	23,535	26,600

33 34 13 – Septic Tanks

33 34 13.13 Concrete Septic Tanks

33 34 13.13 Concrete Septic Tanks		Crew	Daily Output	Labor-Hours	Unit	Material	2018 Bare Costs Labor	Equipment	Total	Total Incl O&P
0400	25,000 gallon, 4 piece	B-13B	1.10	50.909	Ea.	42,400	2,225	905	45,530	51,000
0500	40,000 gallon, 4 piece	↓	.80	70		52,500	3,050	1,250	56,800	64,000
0520	50,000 gallon, 5 piece	B-13C	.60	93.333		60,500	4,075	3,125	67,700	76,500
0640	75,000 gallon, cast in place	C-14C	.25	448		73,500	21,500	103	95,103	115,500
0660	100,000 gallon	"	.15	747		91,000	35,900	172	127,072	158,000
1150	Leaching field chambers, 13' x 3'-7" x 1'-4", standard	B-13	16	3.500		500	153	37	690	835
1200	Heavy duty, 8' x 4' x 1'-6"		14	4		284	175	42	501	640
1300	13' x 3'-9" x 1'-6"		12	4.667		1,250	204	49	1,503	1,750
1350	20' x 4' x 1'-6"	↓	5	11.200		1,200	490	118	1,808	2,225
1420	Leaching pit, precast concrete, 6' diameter, 3' deep	B-21	4.70	5.957		815	276	27.50	1,118.50	1,375
1600	Leaching pit, 6'-6" diameter, 6' deep		5	5.600		1,025	259	26	1,310	1,575
1620	8' deep		4	7		1,200	325	32.50	1,557.50	1,875
1700	8' diameter, H-20 load, 6' deep		4	7		1,500	325	32.50	1,857.50	2,200
1720	8' deep		3	9.333		2,550	430	43	3,023	3,550
2000	Velocity reducing pit, precast conc., 6' diameter, 3' deep	↓	4.70	5.957	↓	1,625	276	27.50	1,928.50	2,275

33 34 13.33 Polyethylene Septic Tanks

33 34 13.33 Polyethylene Septic Tanks		Crew	Daily Output	Labor-Hours	Unit	Material	2018 Bare Costs Labor	Equipment	Total	Total Incl O&P
0010	**POLYETHYLENE SEPTIC TANKS**									
0015	High density polyethylene, 1,000 gallon	B-21	8	3.500	Ea.	1,325	162	16.20	1,503.20	1,725
0020	1,250 gallon		8	3.500		1,200	162	16.20	1,378.20	1,575
0025	1,500 gallon	↓	7	4	↓	1,350	185	18.50	1,553.50	1,800

33 34 16 – Septic Tank Effluent Filters

33 34 16.13 Septic Tank Gravity Effluent Filters

33 34 16.13 Septic Tank Gravity Effluent Filters		Crew	Daily Output	Labor-Hours	Unit	Material	2018 Bare Costs Labor	Equipment	Total	Total Incl O&P
0010	**SEPTIC TANK GRAVITY EFFLUENT FILTERS**									
3000	Effluent filter, 4" diameter	1 Skwk	8	1	Ea.	39.50	52.50		92	128
3020	6" diameter		7	1.143		55	60		115	157
3030	8" diameter		7	1.143		250	60		310	370
3040	8" diameter, very fine		7	1.143		495	60		555	640
3050	10" diameter, very fine		6	1.333		237	70		307	370
3060	10" diameter		6	1.333		275	70		345	415
3080	12" diameter		6	1.333		650	70		720	825
3090	15" diameter	↓	5	1.600	↓	1,075	84		1,159	1,300

33 34 51 – Drainage Field Systems

33 34 51.10 Drainage Field Excavation and Fill

33 34 51.10 Drainage Field Excavation and Fill		Crew	Daily Output	Labor-Hours	Unit	Material	2018 Bare Costs Labor	Equipment	Total	Total Incl O&P
0010	**DRAINAGE FIELD EXCAVATION AND FILL**									
2200	Septic tank & drainage field excavation with 3/4 C.Y. backhoe	B-12F	145	.110	C.Y.		5.30	4.62	9.92	13.55
2400	4' trench for disposal field, 3/4 C.Y. backhoe	"	335	.048	L.F.		2.29	2	4.29	5.85
2600	Gravel fill, run of bank	B-6	150	.160	C.Y.	16.80	7	2.08	25.88	32
2800	Crushed stone, 3/4"	"	150	.160	"	36	7	2.08	45.08	53

33 34 51.13 Utility Septic Tank Tile Drainage Field

33 34 51.13 Utility Septic Tank Tile Drainage Field		Crew	Daily Output	Labor-Hours	Unit	Material	2018 Bare Costs Labor	Equipment	Total	Total Incl O&P
0010	**UTILITY SEPTIC TANK TILE DRAINAGE FIELD**									
0015	Distribution box, concrete, 5 outlets	2 Clab	20	.800	Ea.	92.50	32		124.50	154
0020	7 outlets		16	1		92.50	40		132.50	167
0025	9 outlets		8	2		550	79.50		629.50	735
0115	Distribution boxes, HDPE, 5 outlets		20	.800		75	32		107	135
0117	6 outlets		15	1.067		75	42.50		117.50	152
0118	7 outlets		15	1.067		75	42.50		117.50	152
0120	8 outlets	↓	10	1.600		79	64		143	191
0240	Distribution boxes, outlet flow leveler	1 Clab	50	.160		2.22	6.40		8.62	12.80
0300	Precast concrete, galley, 4' x 4' x 4'	B-21	16	1.750	↓	236	81	8.10	325.10	400
0350	HDPE infiltration chamber 12" H x 15" W	2 Clab	300	.053	L.F.	6.95	2.13		9.08	11.10
0351	12" H x 15" W end cap	1 Clab	32	.250	Ea.	18.65	9.95		28.60	36.50

570

For customer support on your Commercial Renovation Costs with RSMeans data, call 800.448.8182.

33 34 Onsite Wastewater Disposal

33 34 51 – Drainage Field Systems

33 34 51.13 Utility Septic Tank Tile Drainage Field

		Crew	Daily Output	Labor-Hours	Unit	Material	2018 Bare Costs Labor	Equipment	Total	Total Incl O&P
0355	chamber 12" H x 22" W	2 Clab	300	.053	L.F.	6.45	2.13		8.58	10.55
0356	12" H x 22" W end cap	1 Clab	32	.250	Ea.	16.80	9.95		26.75	34.50
0360	chamber 13" H x 34" W	2 Clab	300	.053	L.F.	14.15	2.13		16.28	19.05
0361	13" H x 34" W end cap	1 Clab	32	.250	Ea.	50.50	9.95		60.45	71.50
0365	chamber 16" H x 34" W	2 Clab	300	.053	L.F.	19.05	2.13		21.18	24.50
0366	16" H x 34" W end cap	1 Clab	32	.250	Ea.	15.70	9.95		25.65	33.50
0370	chamber 8" H x 16" W	2 Clab	300	.053	L.F.	9.85	2.13		11.98	14.25
0371	8" H x 16" W end cap	1 Clab	32	.250	Ea.	11	9.95		20.95	28.50

33 42 Stormwater Conveyance

33 42 11 – Stormwater Gravity Piping

33 42 11.40 Piping, Storm Drainage, Corrugated Metal

		Crew	Daily Output	Labor-Hours	Unit	Material	2018 Bare Costs Labor	Equipment	Total	Total Incl O&P
0010	**PIPING, STORM DRAINAGE, CORRUGATED METAL**　R221113-50									
0020	Not including excavation or backfill									
2000	Corrugated metal pipe, galvanized									
2020	Bituminous coated with paved invert, 20' lengths									
2040	8" diameter, 16 ga.	B-14	330	.145	L.F.	8.15	6.10	.95	15.20	19.95
2080	12" diameter, 16 ga.	"	210	.229	"	10.55	9.60	1.49	21.64	29

33 42 11.60 Sewage/Drainage Collection, Concrete Pipe

		Crew	Daily Output	Labor-Hours	Unit	Material	2018 Bare Costs Labor	Equipment	Total	Total Incl O&P
0010	**SEWAGE/DRAINAGE COLLECTION, CONCRETE PIPE**									
0020	Not including excavation or backfill									
1000	Non-reinforced pipe, extra strength, B&S or T&G joints									
1010	6" diameter	B-14	265.04	.181	L.F.	7.55	7.60	1.18	16.33	22
1020	8" diameter	↓	224	.214	↓	8.30	9	1.40	18.70	25
1040	12" diameter	↓	200	.240	↓	10.75	10.10	1.56	22.41	30
2000	Reinforced culvert, class 3, no gaskets									
2010	12" diameter	B-14	150	.320	L.F.	12.70	13.45	2.08	28.23	37.50
2020	15" diameter	"	150	.320	"	16.85	13.45	2.08	32.38	42.50

33 42 33 – Stormwater Curbside Drains and Inlets

33 42 33.13 Catch Basins

		Crew	Daily Output	Labor-Hours	Unit	Material	2018 Bare Costs Labor	Equipment	Total	Total Incl O&P
0010	**CATCH BASINS**									
0011	Not including footing & excavation									
1600	Frames & grates, C.I., 24" square, 500 lb.	B-6	7.80	3.077	Ea.	355	134	40	529	650
1700	26" D shape, 600 lb.	"	7	3.429	"	620	150	44.50	814.50	970
3320	Frames and covers, existing, raised for paving, 2", including									
3340	row of brick, concrete collar, up to 12" wide frame	B-6	18	1.333	Ea.	51.50	58	17.35	126.85	169

33 42 33.15 Remove and Replace Catch Basin Cover

		Crew	Daily Output	Labor-Hours	Unit	Material	2018 Bare Costs Labor	Equipment	Total	Total Incl O&P
0010	**REMOVE AND REPLACE CATCH BASIN COVER**									
0023	Remove catch basin cover	1 Clab	80	.100	Ea.		3.99		3.99	6.45
0033	Replace catch basin cover	"	80	.100	"		3.99		3.99	6.45

For customer support on your Commercial Renovation Costs with RSMeans data, call 800.448.8182.

571

33 52 13.16 Gasoline Piping	Crew	Daily Output	Labor-Hours	Unit	Material	2018 Bare Costs Labor	Equipment	Total	Total Incl O&P
0010 **GASOLINE PIPING**									
0020 Primary containment pipe, fiberglass-reinforced									
0030 Plastic pipe 15' & 30' lengths									
0040 2" diameter	Q-6	425	.056	L.F.	6.45	3.32		9.77	12.25
0050 3" diameter		400	.060		10	3.53		13.53	16.50
0060 4" diameter	↓	375	.064	↓	12.95	3.76		16.71	20
0100 Fittings									
0110 Elbows, 90° & 45°, bell ends, 2"	Q-6	24	1	Ea.	41.50	59		100.50	137
0120 3" diameter		22	1.091		51.50	64		115.50	157
0130 4" diameter		20	1.200		66	70.50		136.50	183
0200 Tees, bell ends, 2"		21	1.143		55.50	67		122.50	166
0210 3" diameter		18	1.333		60.50	78.50		139	189
0220 4" diameter		15	1.600		79.50	94		173.50	234
0230 Flanges bell ends, 2"		24	1		31.50	59		90.50	127
0240 3" diameter		22	1.091		37	64		101	141
0250 4" diameter		20	1.200		42.50	70.50		113	157
0260 Sleeve couplings, 2"		21	1.143		11.80	67		78.80	118
0270 3" diameter		18	1.333		16.80	78.50		95.30	140
0280 4" diameter		15	1.600		21.50	94		115.50	170
0290 Threaded adapters, 2"		21	1.143		16.95	67		83.95	124
0300 3" diameter		18	1.333		32	78.50		110.50	158
0310 4" diameter		15	1.600		35.50	94		129.50	186
0320 Reducers, 2"		27	.889		25.50	52.50		78	110
0330 3" diameter		22	1.091		26	64		90	129
0340 4" diameter	↓	20	1.200	↓	34.50	70.50		105	148
1010 Gas station product line for secondary containment (double wall)									
1100 Fiberglass reinforced plastic pipe 25' lengths									
1120 Pipe, plain end, 3" diameter	Q-6	375	.064	L.F.	25.50	3.76		29.26	34
1130 4" diameter		350	.069		30.50	4.03		34.53	40
1140 5" diameter		325	.074		34	4.34		38.34	44.50
1150 6" diameter	↓	300	.080	↓	37	4.70		41.70	48
1200 Fittings									
1230 Elbows, 90° & 45°, 3" diameter	Q-6	18	1.333	Ea.	135	78.50		213.50	270
1240 4" diameter		16	1.500		158	88		246	310
1250 5" diameter		14	1.714		173	101		274	350
1260 6" diameter		12	2		192	118		310	395
1270 Tees, 3" diameter		15	1.600		160	94		254	320
1280 4" diameter		12	2		191	118		309	395
1290 5" diameter		9	2.667		299	157		456	575
1300 6" diameter		6	4		355	235		590	755
1310 Couplings, 3" diameter		18	1.333		51.50	78.50		130	179
1320 4" diameter		16	1.500		113	88		201	261
1330 5" diameter		14	1.714		202	101		303	380
1340 6" diameter		12	2		299	118		417	515
1350 Cross-over nipples, 3" diameter		18	1.333		9.90	78.50		88.40	133
1360 4" diameter		16	1.500		12.05	88		100.05	150
1370 5" diameter		14	1.714		15	101		116	174
1380 6" diameter		12	2		18.05	118		136.05	203
1400 Telescoping, reducers, concentric 4" x 3"		18	1.333		43.50	78.50		122	170
1410 5" x 4"		17	1.412		90	83		173	228
1420 6" x 5"	↓	16	1.500	↓	221	88		309	380

33 52 Hydrocarbon Transmission and Distribution

33 52 16 – Gas Hydrocarbon Piping

33 52 16.20 Piping, Gas Service and Distribution, P.E.	Crew	Daily Output	Labor-Hours	Unit	Material	2018 Bare Costs Labor	Equipment	Total	Total Incl O&P
0010 **PIPING, GAS SERVICE AND DISTRIBUTION, POLYETHYLENE**									
0020 Not including excavation or backfill									
1000 60 psi coils, compression coupling @ 100', 1/2" diameter, SDR 11	B-20A	608	.053	L.F.	.50	2.55		3.05	4.58
1010 1" diameter, SDR 11		544	.059		1.16	2.85		4.01	5.80
1040 1-1/4" diameter, SDR 11		544	.059		1.64	2.85		4.49	6.30
1100 2" diameter, SDR 11		488	.066		2.55	3.17		5.72	7.85
1160 3" diameter, SDR 11		408	.078		5.35	3.80		9.15	11.90
1500 60 psi 40' joints with coupling, 3" diameter, SDR 11	B-21A	408	.098		7.30	4.90	.89	13.09	16.75
1540 4" diameter, SDR 11		352	.114		13	5.65	1.03	19.68	24.50
1600 6" diameter, SDR 11		328	.122		31	6.10	1.11	38.21	45
1640 8" diameter, SDR 11		272	.147		44.50	7.35	1.34	53.19	62
9000 Minimum labor/equipment charge	B-20	2	12	Job		535		535	870

33 71 Electrical Utility Transmission and Distribution

33 71 19 – Electrical Underground Ducts and Manholes

33 71 19.17 Electric and Telephone Underground

	Crew	Daily Output	Labor-Hours	Unit	Material	2018 Bare Costs Labor	Equipment	Total	Total Incl O&P
0010 **ELECTRIC AND TELEPHONE UNDERGROUND**									
0011 Not including excavation									
0200 backfill and cast in place concrete									
4200 Underground duct, banks ready for concrete fill, min. of 7.5"									
4400 between conduits, center to center									
4600 2 @ 2" diameter	2 Elec	240	.067	L.F.	2.18	3.88		6.06	8.40
4800 4 @ 2" diameter		120	.133		4.36	7.75		12.11	16.80
5600 4 @ 4" diameter		80	.200		7.25	11.65		18.90	26
6200 Rigid galvanized steel, 2 @ 2" diameter		180	.089		17.50	5.15		22.65	27.50
6400 4 @ 2" diameter		90	.178		35	10.35		45.35	54.50
7400 4 @ 4" diameter		34	.471		74.50	27.50		102	125
9990 Minimum labor/equipment charge	1 Elec	3.50	2.286	Job		133		133	205

For customer support on your Commercial Renovation Costs with RSMeans data, call 800.448.8182.

573

Division Notes

	CREW	DAILY OUTPUT	LABOR-HOURS	UNIT	BARE COSTS				TOTAL INCL O&P
					MAT.	LABOR	EQUIP.	TOTAL	

Division Notes

Estimating Tips
34 11 00 Rail Tracks
This subdivision includes items that may involve either repair of existing or construction of new railroad tracks. Additional preparation work, such as the roadbed earthwork, would be found in Division 31. Additional new construction siding and turnouts are found in Subdivision 34 72. Maintenance of railroads is found under 34 01 23 Operation and Maintenance of Railways.

34 40 00 Traffic Signals
This subdivision includes traffic signal systems. Other traffic control devices such as traffic signs are found in Subdivision 10 14 53 Traffic Signage.

34 70 00 Vehicle Barriers
This subdivision includes security vehicle barriers, guide and guard rails, crash barriers, and delineators. The actual maintenance and construction of concrete and asphalt pavement are found in Division 32.

Reference Numbers
Reference numbers are shown at the beginning of some major classifications. These numbers refer to related items in the Reference Section. The reference information may be an estimating procedure, an alternate pricing method, or technical information.

Note: Not all subdivisions listed here necessarily appear. ■

34 71 13.26 Vehicle Guide Rails	Crew	Daily Output	Labor-Hours	Unit	Material	2018 Bare Costs Labor	2018 Bare Costs Equipment	Total	Total Incl O&P
0010 **VEHICLE GUIDE RAILS**									
0012 Corrugated stl., galv. stl. posts, 6'-3" OC	B-80	850	.038	L.F.	26	1.67	.72	28.39	32
0200 End sections, galvanized, flared		50	.640	Ea.	100	28.50	12.25	140.75	169
0300 Wrap around end		50	.640	"	143	28.50	12.25	183.75	216
0400 Timber guide rail, 4" x 8" with 6" x 8" wood posts, treated	▼	960	.033	L.F.	12.90	1.48	.64	15.02	17.20

Estimating Tips

Products such as conveyors, material handling cranes and hoists, as well as other items specified in this division, require trained installers. The general contractor may not have any choice as to who will perform the installation or when it will be performed. Long lead times are often required for these products, making early decisions in purchasing and scheduling necessary. The installation of this type of equipment may require the embedment of mounting hardware during construction of floors, structural walls, or interior walls/partitions. Electrical connections will require coordination with the electrical contractor.

Reference Numbers

Reference numbers are shown at the beginning of some major classifications. These numbers refer to related items in the Reference Section. The reference information may be an estimating procedure, an alternate pricing method, or technical information.

Note: Not all subdivisions listed here necessarily appear. ■

41 22 13.10 Crane Rail

41 22 13.10 Crane Rail	Crew	Daily Output	Labor-Hours	Unit	Material	2018 Bare Costs Labor	Equipment	Total	Total Incl O&P
0010 **CRANE RAIL**									
0020 Box beam bridge, no equipment included	E-4	3400	.009	Lb.	1.35	.52	.03	1.90	2.40
0200 Running track only, 104 lb. per yard		5600	.006	"	.67	.31	.02	1	1.30
0210 Running track only, 104 lb. per yard, 20' piece		160	.200	L.F.	23.50	11.05	.62	35.17	45.50

41 22 13.13 Bridge Cranes

41 22 13.13 Bridge Cranes	Crew	Daily Output	Labor-Hours	Unit	Material	2018 Bare Costs Labor	Equipment	Total	Total Incl O&P
0010 **BRIDGE CRANES**									
0100 1 girder, 20' span, 3 ton	M-3	1	34	Ea.	26,700	2,100	143	28,943	32,700
0125 5 ton		1	34		29,300	2,100	143	31,543	35,600
0150 7.5 ton		1	34		34,700	2,100	143	36,943	41,500
0175 10 ton		.80	42.500		46,200	2,625	179	49,004	55,500
0200 15 ton		.80	42.500		59,500	2,625	179	62,304	70,000
0225 30' span, 3 ton		1	34		27,700	2,100	143	29,943	33,900
0250 5 ton		1	34		30,600	2,100	143	32,843	37,000
0275 7.5 ton		1	34		36,500	2,100	143	38,743	43,500
0300 10 ton		.80	42.500		47,900	2,625	179	50,704	57,000
0325 15 ton		.80	42.500		62,500	2,625	179	65,304	73,000
0350 2 girder, 40' span, 3 ton	M-4	.50	72		45,300	4,375	365	50,040	57,000
0375 5 ton		.50	72		47,600	4,375	365	52,340	59,500
0400 7.5 ton		.50	72		52,000	4,375	365	56,740	64,500
0425 10 ton		.40	90		61,500	5,475	460	67,435	76,500
0450 15 ton		.40	90		83,500	5,475	460	89,435	101,000
0475 25 ton		.30	120		98,500	7,300	610	106,410	120,000
0500 50' span, 3 ton		.50	72		52,000	4,375	365	56,740	64,000
0525 5 ton		.50	72		54,000	4,375	365	58,740	66,500
0550 7.5 ton		.50	72		58,000	4,375	365	62,740	70,500
0575 10 ton		.40	90		67,000	5,475	460	72,935	83,000
0600 15 ton		.40	90		87,500	5,475	460	93,435	105,500
0625 25 ton		.30	120		103,000	7,300	610	110,910	125,500

41 22 13.19 Jib Cranes

41 22 13.19 Jib Cranes	Crew	Daily Output	Labor-Hours	Unit	Material	2018 Bare Costs Labor	Equipment	Total	Total Incl O&P
0010 **JIB CRANES**									
0020 Jib crane, wall cantilever, 500 lb. capacity, 8' span	2 Mill	1	16	Ea.	1,425	845		2,270	2,900
0040 12' span		1	16		1,550	845		2,395	3,050
0060 16' span		1	16		1,725	845		2,570	3,225
0080 20' span		1	16		2,200	845		3,045	3,725
0100 1,000 lb. capacity, 8' span		1	16		1,525	845		2,370	3,000
0120 12' span		1	16		1,650	845		2,495	3,150
0130 16' span		1	16		2,200	845		3,045	3,750
0150 20' span		1	16		2,575	845		3,420	4,150

Estimating Tips

- When estimating costs for the installation of electrical power generation equipment, factors to review include access to the job site, access and setting up at the installation site, required connections, uncrating pads, anchors, leveling, final assembly of the components, and temporary protection from physical damage, such as environmental exposure.

- Be aware of the costs of equipment supports, concrete pads, and vibration isolators. Cross-reference them against other trades' specifications. Also, review site and structural drawings for items that must be included in the estimates.

- It is important to include items that are not documented in the plans and specifications but must be priced. These items include, but are not limited to, testing, dust protection, roof penetration, core drilling concrete floors and walls, patching, cleanup, and final adjustments. Add a contingency or allowance for utility company fees for power hookups, if needed.

- The project size and scope of electrical power generation equipment will have a significant impact on cost. The intent of RSMeans cost data is to provide a benchmark cost so that owners, engineers, and electrical contractors will have a comfortable number with which to start a project. Additionally, there are many websites available to use for research and to obtain a vendor's quote to finalize costs.

Reference Numbers

Reference numbers are shown at the beginning of some major classifications. These numbers refer to related items in the Reference Section. The reference information may be an estimating procedure, an alternate pricing method, or technical information.

Note: Not all subdivisions listed here necessarily appear. ■

48 13 13 – Hydroelectric Power Plant Water Turbines

48 13 13.10 Water Turbines and Components	Crew	Daily Output	Labor-Hours	Unit	Material	2018 Bare Costs Labor	Equipment	Total	Total Incl O&P
0010 **WATER TURBINES & COMPONENTS**									
0500 Water turbine, 100-500 watt, 3P, 48VDC, 54"W x 24"H, 260 lbs.	2 Elec	2	8	Ea.	9,875	465		10,340	11,600
0600 Water turbine, 1500-2000 watt, 3P, 130-200VAC to DC, 95 lbs.	2 Elec	3	5.333	Ea.	3,900	310		4,210	4,775

Assemblies Section

Table of Contents

Table of Contents

RSMeans data: Assemblies— How They Work

Assemblies estimating provides a fast and reasonably accurate way to develop construction costs. An assembly is the grouping of individual work items—with appropriate quantities— to provide a cost for a major construction component in a convenient unit of measure.

An assemblies estimate is often used during early stages of design development to compare

the cost impact of various design alternatives on total building cost.

Assemblies estimates are also used as an efficient tool to verify construction estimates.

Assemblies estimates do not require a completed design or detailed drawings. Instead, they are based on the general size of the structure and

other known parameters of the project. The degree of accuracy of an assemblies estimate is generally within +/- 15%.

Most assemblies consist of three major elements: a graphic, the system components, and the cost data itself. The **Graphic** is a visual representation showing the typical appearance of the assembly

① Unique 12-character Identifier

Our assemblies are identified by a **unique 12-character identifier**. The assemblies are numbered using UNIFORMAT II, ASTM Standard E1557. The first 5 characters represent this system to Level 3. The last 7 characters represent further breakdown in order to arrange items in understandable groups of similar tasks. Line numbers are consistent across all our publications, so a line number in any assemblies data set will always refer to the same item.

② Narrative Descriptions

Our assemblies descriptions appear in two formats: narrative and table. **Narrative descriptions** are shown in a hierarchical structure to make them readable. In order to read a complete description, read up through the indents to the top of the section. Include everything that is above and to the left that is not contradicted by information below.

A10 Foundations

A1010 Standard Foundations

This page illustrates and describes a spread footing system including concrete, forms, reinforcing and anchor bolts. Lines within System Components give the unit price and total price on a cost each basis for this system. Prices for spread footing systems are on Line Items A1010 230 1200 thru 2300. Both material quantities and labor costs have been adjusted for the system listed.

Factors: To adjust for job conditions other than normal working situations use Lines A1010 230 2900 thru 4000.

Example: You are to install the system in an existing occupied building. Access to the site and protection to the building are mandatory. Go to Lines A1010 230 3600 and 3800 and apply these percentages to the appropriate MAT. and INST. costs.

System Components	QUANTITY	UNIT	COST EACH MAT.	COST EACH INST.	COST EACH TOTAL
Interior column footing, 3' square 1' thick, 2000 psi concrete, Including forms, reinforcing, and anchor bolts.					
Concrete, 2000 psi	.330	C.Y.	41.91		41.91
Placing concrete	.330	C.Y.		19.65	19.65
Forms, footing, 4 uses	12.000	SFCA	9	72	81
Reinforcing	11.000	Lb.	5.83	7.37	13.20
Anchor bolts, 3/4" diameter	2.000	Ea.	3.12	9.94	13.06
TOTAL			59.86	108.96	168.82

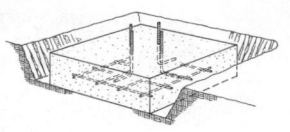

A1010 230	Spread Footing	COST EACH MAT.	COST EACH INST.	COST EACH TOTAL
1200	3' square x 1' thick, 2000 psi concrete	60	109	169
1300	3000 psi concrete	62	109	171
1400	4000 psi concrete	64	109	173
1600	For alternate footing systems:			
1700	4' square x 1' thick, 2000 psi concrete	100	155	255
1800	3000 psi concrete	104	155	259
1900	4000 psi concrete	108	155	263
2100	5' square x 1'-3" thick, 2000 psi concrete	189	254	443
2200	3000 psi concrete	196	254	450
2300	4000 psi concrete	204	254	458
2600				
2700				
2900	Cut & patch to match existing construction, add, minimum	2%	3%	
3000	Maximum	5%	9%	
3100	Dust protection, add, minimum	1%	2%	
3200	Maximum	4%	11%	
3300	Equipment usage curtailment, add, minimum	1%	1%	
3400	Maximum	3%	10%	
3500	Material handling & storage limitation, add, minimum	1%	1%	
3600	Maximum	6%	7%	
3700	Protection of existing work, add, minimum	2%	2%	
3800	Maximum	5%	7%	
3900	Shift work requirements, add, minimum		5%	
4000	Maximum		30%	

For supplemental customizable square foot estimating forms, visit: **www.RSMeans.com/2018books**

in question. It is frequently accompanied by additional explanatory technical information describing the class of items. The **System Components** is a listing of the individual tasks that make up the assembly, including the quantity and unit of measure for each item, along with the cost of material and installation. The **Assemblies**

Data below lists prices for other similar systems with dimensional and/or size variations.

All of our assemblies costs represent the cost for the installing contractor. An allowance for profit has been added to all material, labor, and equipment rental costs. A markup for labor burdens, including workers' compensation, fixed

overhead, and business overhead, is included with installation costs.

The information in RSMeans cost data represents a "national average" cost. This data should be modified to the project location using the **City Cost Indexes** or **Location Factors** tables found in the Reference Section.

A10 Foundations

A1010 Standard Foundations

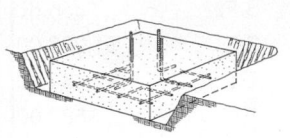

This page illustrates and describes a spread footing system including concrete, forms, reinforcing and anchor bolts. Lines within System Components give the unit price and total price on a cost each basis for this system. Prices for spread footing systems are on Line Items A1010 230 1200 thru 2300. Both material quantities and labor costs have been adjusted for the system listed.

Factors: To adjust for job conditions other than normal working situations use Lines A1010 230 2900 thru 4000.

Example: You are to install the system in an existing occupied building. Access to the site and protection to the building are mandatory. Go to Lines A1010 230 3600 and 3800 and apply these percentages to the appropriate MAT. and INST. costs.

③ Unit of Measure

All RSMeans data: Assemblies include a typical **Unit of Measure** used for estimating that item. For instance, for continuous footings or foundation walls the unit is linear feet (L.F.). For spread footings the unit is each (Ea.). The estimator needs to take special care that the unit in the data matches the unit in the takeoff. Abbreviations and unit conversions can be found in the Reference Section.

③ Unit of Measure

④ System Components

System components are listed separately to detail what is included in the development of the total system price.

System Components ④	QUANTITY	UNIT	COST EACH MAT.	COST EACH INST.	COST EACH TOTAL
Interior column footing, 3' square 1' thick, 2000 psi concrete, including forms, reinforcing, and anchor bolts.					
Concrete, 2000 psi	.330	C.Y.	41.91		41.91
Placing concrete	.330	C.Y.		19.65	19.65
Forms, footing, 4 uses	12.000	SFCA	9	72	81
Reinforcing	11.000	Lb.	5.83	7.37	13.20
Anchor bolts, 3/4" diameter	2.000	Ea.	3.12	9.94	13.06
TOTAL			59.86	108.96	168.82

A1010 230	Spread Footing	COST EACH MAT.	COST EACH INST.	COST EACH TOTAL
1200	3' square x 1' thick, 2000 psi concrete	60	109	169
1300	3000 psi concrete	62	109	171
1400	4000 psi concrete	64	109	173
1600	For alternate footing systems:			
1700	4' square x 1' thick, 2000 psi concrete	100	155	255
1800	3000 psi concrete	104	155	259
1900	4000 psi concrete	108	155	263
2100	5' square x 1'-3" thick, 2000 psi concrete	189	254	443
2200	3000 psi concrete	196	254	450
2300	4000 psi concrete	204	254	458
2600				
2700				
2900	Cut & patch to match existing construction, add, minimum	2%	3%	
3000	Maximum	5%	9%	
3100	Dust protection, add, minimum	1%	2%	
3200	Maximum	4%	11%	
3300	Equipment usage curtailment, add, minimum	1%	1%	
3400	Maximum	3%	10%	
3500	Material handling & storage limitation, add, minimum	1%	1%	
3600	Maximum	6%	7%	
3700	Protection of existing work, add, minimum	2%	2%	
3800	Maximum	5%	7%	
3900	Shift work requirements, add, minimum		5%	
4000	Maximum		30%	

Sample Estimate

This sample demonstrates the elements of an estimate, including a tally of the RSMeans data lines. Published assemblies costs include all markups for labor burden and profit for the installing contractor. This estimate adds a summary of the markups applied by a general contractor on the installing contractor's work. These figures represent the total cost to the owner. The location factor with RSMeans data is applied at the bottom of the estimate to adjust the cost of the work to a specific location.

Project Name:	Interior Fit-out, ABC Office				
Location:	Anywhere, USA		Date: 1/1/2018		R+R
Assembly Number	Description	Qty.	Unit		Subtotal
❶ C1010 124 1200	Wood partition, 2 x 4 @ 16" OC w/5/8" FR gypsum board	560.000	S.F.		$2,990.40
C1020 114 1800	Metal door & frame, flush hollow core, 3'-0" x 7'-0"	2.000	Ea.		$2,560.00
C3010 230 0080	Painting, brushwork, primer & 2 coats	1,120.000	S.F.		$1,500.80
C3020 410 0140	Carpet, tufted, nylon, roll goods, 12' wide, 26 oz	240.000	S.F.		$856.80
C3030 210 6000	Acoustic ceilings, 24" x 48" tile, tee grid suspension	200.000	S.F.		$1,286.00
D5020 125 0560	Receptacles incl plate, box, conduit, wire, 20 A duplex	8.000	Ea.		$2,384.00
D5020 125 0720	Light switch incl plate, box, conduit, wire, 20 A single pole	2.000	Ea.		$582.00
D5020 210 0560	Fluorescent fixtures, recess mounted, 20 per 1000 SF	200.000	S.F.		$2,276.00
	Assembly Subtotal				**$14,436.00**
	Sales Tax @ ❷		5 %		$ 360.90
	General Requirements @ ❸		7 %		$ 1,010.52
	Subtotal A				**$15,807.42**
	GC Overhead @ ❹		5 %		$ 790.37
	Subtotal B				**$16,597.79**
	GC Profit @ ❺		5 %		$ 829.89
	Subtotal C				**$17,427.68**
	Adjusted by Location Factor ❻		115.2		$ 20,076.69
	Architects Fee @ ❼		8 %		$ 1,606.14
	Contingency @ ❽		15 %		$ 3,011.50
	Project Total Cost				**$ 24,694.33**

This estimate is based on an interactive spreadsheet. You are free to download it and adjust it to your methodology. A copy of this spreadsheet is available at **www.RSMeans.com/2018books.**

1 Work Performed

The body of the estimate shows the RSMeans data selected, including line numbers, a brief description of each item, its takeoff quantity and unit, and the total installed cost, including the installing contractor's overhead and profit.

2 Sales Tax

If the work is subject to state or local sales taxes, the amount must be added to the estimate. In a conceptual estimate it can be assumed that one half of the total represents material costs. Therefore, apply the sales tax rate to 50% of the assembly subtotal.

3 General Requirements

This item covers project-wide needs provided by the general contractor. These items vary by project but may include temporary facilities and utilities, security, testing, project cleanup, etc. In assemblies estimates a percentage is used—typically between 5% and 15% of project cost.

4 General Contractor Overhead

This entry represents the general contractor's markup on all work to cover project administration costs.

5 General Contractor Profit

This entry represents the GC's profit on all work performed. The value included here can vary widely by project and is influenced by the GC's perception of the project's financial risk and market conditions.

6 Location Factor

RSMeans published data are based on national average costs. If necessary, adjust the total cost of the project using a location factor from the "Location Factor" table or the "City Cost Indexes" table found in the Reference Section. Use location factors if the work is general, covering the work of multiple trades. If the work is by a single trade (e.g., masonry) use the more specific data found in the City Cost Indexes.

To adjust costs by location factors, multiply the base cost by the factor and divide by 100.

7 Architect's Fee

If appropriate, add the design cost to the project estimate. These fees vary based on project complexity and size. Typical design and engineering fees can be found in the Reference Section.

8 Contingency

A factor for contingency may be added to any estimate to represent the cost of unknowns that may occur between the time that the estimate is performed and the time the project is constructed. The amount of the allowance will depend on the stage of design at which the estimate is done, and the contractor's assessment of the risk involved.

Did you know?

RSMeans data is available through our online application with 24/7 access:

- Search for unit prices by keyword
- Leverage the most up-to-date data
- Build and export estimates

Try it free for 30 days!
www.rsmeans.com/2018freetrial

A1010 Standard Foundations

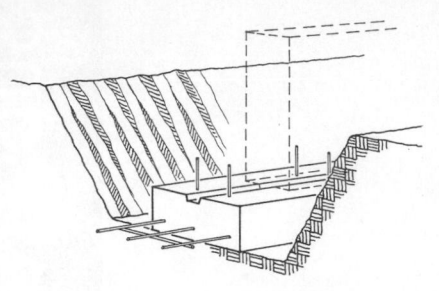

This page illustrates and describes a strip footing system including concrete, forms, reinforcing, keyway and dowels. Lines within System Components give the unit price and total price per linear foot for this system. Prices for strip footing systems are on Line Items A1010 120 1400 thru 2500. Both material quantities and labor costs have been adjusted for the system listed.

Factors: To adjust for job conditions other than normal working situations use Lines A1010 120 2900 thru 4000.

Example: You are to install this footing, and due to a lack of accessibility, only hand tools can be used. Material handling is also a problem. Go to Lines A1010 120 3400 and 3600 and apply these percentages to the appropriate MAT. and INST. costs.

System Components			COST PER L.F.		
	QUANTITY	UNIT	MAT.	INST.	TOTAL
Strip footing, 2'-0" wide x 1'-0" thick, 2000 psi concrete including forms					
Reinforcing, keyway, and dowels.					
Concrete, 2000 psi	.074	C.Y.	9.40		9.40
Placing concrete	.074	C.Y.		2.03	2.03
Forms, footing, 4 uses	2.000	S.F.	4.72	10.30	15.02
Reinforcing	3.170	Lb.	1.68	2.12	3.80
Keyway, 2" x 4", 4 uses	1.000	L.F.	.24	1.24	1.48
Dowels, #4 bars, 2' long, 24" O.C.	.500	Ea.	.39	1.46	1.85
TOTAL			16.43	17.15	33.58

A1010 120	Strip Footing	COST PER L.F.		
		MAT.	INST.	TOTAL
1400	2'-0" wide x 1' thick, 2000 psi concrete	16.45	17.15	33.60
1500	3000 psi concrete	16.85	17.15	34
1600	4000 psi concrete	17.40	17.15	34.55
1800	For alternate footing systems:			
1900	2'-6" wide x 1' thick, 2000 psi concrete	19.30	18.25	37.55
2000	3000 psi concrete	19.85	18.25	38.10
2100	4000 psi concrete	20.50	18.25	38.75
2300	3'-0" wide x 1' thick, 2000 psi concrete	22	19.25	41.25
2400	3000 psi concrete	22.50	19.25	41.75
2500	4000 psi concrete	23.50	19.25	42.75
2700				
2800				
2900	Cut & patch to match existing construction, add, minimum	2%	3%	
3000	Maximum	5%	9%	
3100	Dust protection, add, minimum	1%	2%	
3200	Maximum	4%	11%	
3300	Equipment usage curtailment, add, minimum	1%	1%	
3400	Maximum	3%	10%	
3500	Material handling & storage limitation, add, minimum	1%	1%	
3600	Maximum	6%	7%	
3700	Protection of existing work, add, minimum	2%	2%	
3800	Maximum	5%	7%	
3900	Shift work requirements, add, minimum		5%	
4000	Maximum		30%	

A10 Foundations

A1010 Standard Foundations

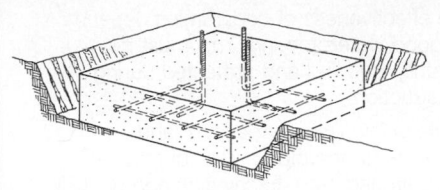

This page illustrates and describes a spread footing system including concrete, forms, reinforcing and anchor bolts. Lines within System Components give the unit price and total price on a cost each basis for this system. Prices for spread footing systems are on Line Items A1010 230 1200 thru 2300. Both material quantities and labor costs have been adjusted for the system listed.

Factors: To adjust for job conditions other than normal working situations use Lines A1010 230 2900 thru 4000.

Example: You are to install the system in an existing occupied building. Access to the site and protection to the building are mandatory. Go to Lines A1010 230 3600 and 3800 and apply these percentages to the appropriate MAT. and INST. costs.

System Components	QUANTITY	UNIT	COST EACH MAT.	COST EACH INST.	COST EACH TOTAL
Interior column footing, 3' square 1' thick, 2000 psi concrete, Including forms, reinforcing, and anchor bolts.					
Concrete, 2000 psi	.330	C.Y.	41.91		41.91
Placing concrete	.330	C.Y.		19.65	19.65
Forms, footing, 4 uses	12.000	SFCA	9	72	81
Reinforcing	11.000	Lb.	5.83	7.37	13.20
Anchor bolts, 3/4" diameter	2.000	Ea.	3.12	9.94	13.06
TOTAL			59.86	108.96	168.82

A1010 230	Spread Footing	COST EACH MAT.	COST EACH INST.	COST EACH TOTAL
1200	3' square x 1' thick, 2000 psi concrete	60	109	109
1300	3000 psi concrete	62	109	171
1400	4000 psi concrete	64	109	173
1600	For alternate footing systems:			
1700	4' square x 1' thick, 2000 psi concrete	100	155	255
1800	3000 psi concrete	104	155	259
1900	4000 psi concrete	108	155	263
2100	5' square x 1'-3" thick, 2000 psi concrete	189	254	443
2200	3000 psi concrete	196	254	450
2300	4000 psi concrete	204	254	458
2600				
2700				
2900	Cut & patch to match existing construction, add, minimum	2%	3%	
3000	Maximum	5%	9%	
3100	Dust protection, add, minimum	1%	2%	
3200	Maximum	4%	11%	
3300	Equipment usage curtailment, add, minimum	1%	1%	
3400	Maximum	3%	10%	
3500	Material handling & storage limitation, add, minimum	1%	1%	
3600	Maximum	6%	7%	
3700	Protection of existing work, add, minimum	2%	2%	
3800	Maximum	5%	7%	
3900	Shift work requirements, add, minimum		5%	
4000	Maximum		30%	

A10 Foundations

A1010 Standard Foundations

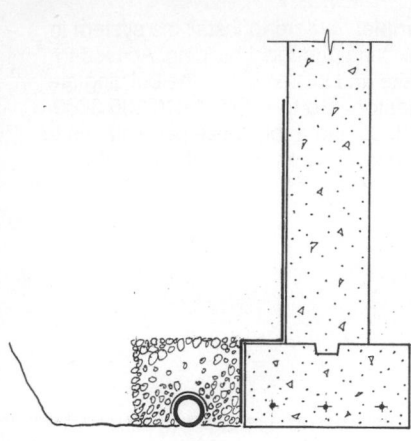

General: Footing drains can be placed either inside or outside of foundation walls depending upon the source of water to be intercepted. If the source of subsurface water is principally from grade or a subsurface stream above the bottom of the footing, outside drains should be used. For high water tables, use inside drains or both inside and outside.

The effectiveness of underdrains depends on good waterproofing. This must be carefully installed and protected during construction.

Costs below include the labor and materials for the pipe and 6" of crushed stone around pipe. Excavation and backfill are not included.

System Components	QUANTITY	UNIT	COST PER L.F.		
			MAT.	INST.	TOTAL
Foundation underdrain, outside, PVC 4" diameter.					
PVC pipe 4" diam. S.D.R. 35	1.000	L.F.	1.80	4.63	6.43
Crushed stone 3/4" to 1/2"	.070	C.Y.	2.14	.94	3.08
TOTAL			3.94	5.57	9.51

A1010 310	Foundation Underdrain	COST PER L.F.		
		MAT.	INST.	TOTAL
1000	Foundation underdrain, outside only, PVC, 4" diameter	3.94	5.55	9.49
1100	6" diameter	6.60	6.15	12.75
1400	Perforated HDPE, 6" diameter	4.82	2.37	7.19
1450	8" diameter	7.70	2.96	10.66
1500	12" diameter	12.40	7.25	19.65
1600	Corrugated metal, 16 ga. asphalt coated, 6" diameter	10.05	5.80	15.85
1650	8" diameter	12.45	6.15	18.60
1700	10" diameter	15.30	7.95	23.25
3000	Outside and inside, PVC, 4" diameter	7.85	11.15	19
3100	6" diameter	13.15	12.35	25.50
3400	Perforated HDPE, 6" diameter	9.65	4.73	14.38
3450	8" diameter	15.40	5.90	21.30
3500	12" diameter	25	14.55	39.55
3600	Corrugated metal, 16 ga., asphalt coated, 6" diameter	20	11.55	31.55
3650	8" diameter	25	12.35	37.35
3700	10" diameter	30.50	15.90	46.40
4700	Cut & patch to match existing construction, add, minimum	2%	3%	
4800	Maximum	5%	9%	
4900	Dust protection, add, minimum	1%	2%	
5000	Maximum	4%	11%	
5100	Equipment usage curtailment, add, minimum	1%	1%	
5200	Maximum	3%	10%	
5300	Material handling & storage limitation, add, minimum	1%	1%	
5400	Maximum	6%	7%	
5500	Protection of existing work, add, minimum	2%	2%	
5600	Maximum	5%	7%	
5700	Shift work requirements, add, minimum		5%	
5800	Maximum		30%	
5900	Temporary shoring and bracing, add, minimum	2%	5%	
6000	Maximum	5%	12%	

A1030 Slab on Grade

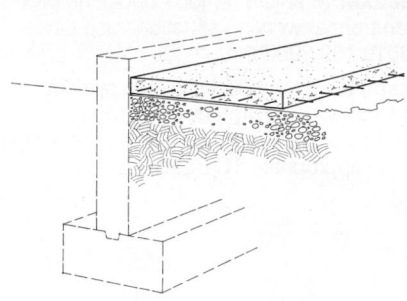

This page illustrates and describes a slab on grade system including slab, bank run gravel, bulkhead forms, placing concrete, welded wire fabric, vapor barrier, steel trowel finish and curing paper. Lines within System Components give the unit price and total price per square foot for this system. Prices for slab on grade systems are on Line Items A1030 110 1400 thru 2600. Both material quantities and labor costs have been adjusted for the system listed.

Factors: To adjust for job conditions other than normal working situations use Lines A1030 110 2900 thru 4000.

Example: You are to install the system at a site where protection of the existing building is required. Go to Line A1030 110 3800 and apply these percentages to the appropriate MAT. and INST. costs.

System Components	QUANTITY	UNIT	COST PER S.F. MAT.	COST PER S.F. INST.	COST PER S.F. TOTAL
Ground slab, 4″ thick, 3000 psi concrete, 4″ granular base, vapor barrier					
Welded wire fabric, screed and steel trowel finish.					
Concrete, 4″ thick, 3000 psi concrete	.012	C.Y.	1.60		1.60
Bank run gravel, 4″ deep	.074	C.Y.	.25	.09	.34
Polyethylene vapor barrier, 10 mil.	.011	C.S.F.	.11	.20	.31
Bulkhead forms, expansion material	.100	L.F.	.03	.42	.45
Welded wire fabric, 6 x 6 - #10/10	.011	C.S.F.	.18	.44	.62
Place concrete	.012	C.Y.		.36	.36
Screed & steel trowel finish	1.000	S.F.		1.03	1.03
TOTAL			2.17	2.54	4.71

A1030 110	Interior Slab on Grade	COST PER S.F. MAT.	COST PER S.F. INST.	COST PER S.F. TOTAL
1400	4″ thick slab, 3000 PSI concrete, 4″ deep bank run gravel	2.17	2.54	4.71
1500	6″ deep bank run gravel	2.36	2.54	4.90
1600	12″ deep bank run gravel	2.79	2.58	5.37
1700				
1800				
1900				
2000	For alternate slab systems:			
2100	5″ thick slab, 3000 psi concrete, 6″ deep bank run gravel	2.76	2.63	5.39
2200	12″ deep bank run gravel	3.19	2.67	5.86
2300				
2400				
2500	6″ thick slab, 3000 psi concrete, 6″ deep bank run gravel	3.29	2.74	6.03
2600	12″ deep bank run gravel	3.72	2.78	6.50
2700				
2900	Cut & patch to match existing construction, add, minimum	2%	3%	
3000	Maximum	5%	9%	
3100	Dust protection, add, minimum	1%	2%	
3200	Maximum	4%	11%	
3300	Equipment usage curtailment, add, minimum	1%	1%	
3400	Maximum	3%	10%	
3500	Material handling & storage limitation, add, minimum	1%	1%	
3600	Maximum	6%	7%	
3700	Protection of existing work, add, minimum	2%	2%	
3800	Maximum	5%	7%	
3900	Shift work requirements, add, minimum		5%	
4000	Maximum		30%	

A2010 Basement Excavation

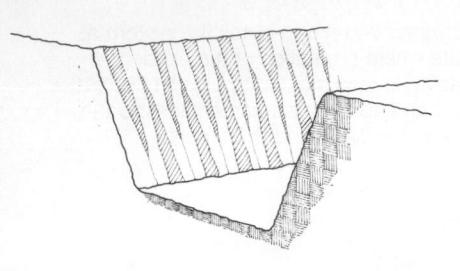

This page illustrates and describes continuous footing and trench excavation systems including a wheel mounted backhoe, operator, equipment rental, fuel, oil, mobilization, no hauling or backfill. Lines within System Components give the unit price and total price per linear foot for this system. Prices for alternate continuous footing and trench excavation systems are on Line Items A2010 130 1200 thru 2700. Both material quantities and labor costs have been adjusted for the system listed.

Factors: To adjust for job conditions other than normal working situations use Lines A2010 130 2900 thru 4000.

Example: You are to install the system and supply dust protection. Go to Line A2010 130 3000 and apply this percentage to the appropriate TOTAL costs.

System Components			COST PER L.F.		
	QUANTITY	UNIT	EQUIP.	LABOR	TOTAL
Continuous footing or trench excav. w/ 3/4 C.Y. wheel mntd backhoe Including operator, equipment rental, fuel, oil and mobilization. Trench Is 4' wide at bottom, 3' deep w/sides sloped 1 to 2 and 200' long. Costs Are based on a production rate of 240 C.Y./day. No hauling/backfill incl.					
Equipment operator	4.000	Hr.		340	340
3/4 C.Y. backhoe, wheel mounted	.500	Day	93.50		93.50
Operating expense (fuel, oil)	4.000	Hr.	78.32		78.32
Mobilization	1.000	Ea.	1,000	1,950	2,950
TOTAL			1,171.82	2,290	3,461.82
COST PER L.F.		L.F.	5.86	11.45	17.31

A2010 130	Excavation, Footings or Trench	COST PER L.F.		
		EQUIP.	LABOR	TOTAL
1200	For alternate trench sizes, 4' bottom with sloped sides:			
1300	2' deep, 50' long	6.80	19.20	26
1400	100' long	3.56	9.60	13.16
1500	300' long	3.88	7.65	11.53
1600	4' deep, 50' long	7.10	19.20	26.30
1700	100' long	3.86	9.60	13.46
1800	300' long	1.97	4.48	6.45
1900	6' deep, 50' long	7.55	19.20	26.75
2000	100' long	4.07	10.90	14.97
2100	300' long	3.07	6.60	9.67
2200	8' deep, 50' long	8.10	19.20	27.30
2300	100' long	5.75	13	18.75
2400	300' long	4.22	6.60	10.82
2500	10' deep, 50' long	9.80	22.50	32.30
2600	100' long	7.25	16	23.25
2700	300' long	5.60	11.55	17.15
2800				
2900	Dust protection, add, minimum		2%	2%
3000	Maximum		11%	11%
3100	Equipment usage curtailment, add, minimum		1%	1%
3200	Maximum		10%	10%
3300	Material handling & storage limitation, add, minimum		1%	1%
3400	Maximum		7%	7%
3500	Protection of existing work, add, minimum		2%	2%
3600	Maximum		7%	7%
3700	Shift work requirements, add, minimum		5%	5%
3800	Maximum		30%	30%
3900	Temporary shoring and bracing, add, minimum		5%	5%
4000	Maximum		12%	12%

A2010 Basement Excavation

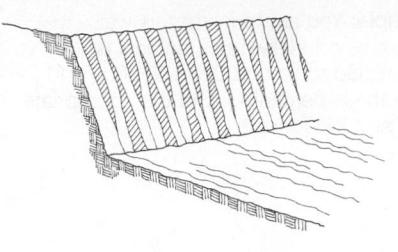

This page illustrates and describes foundation excavation systems including a backhoe-loader, operator, equipment rental, fuel, oil, mobilization, hauling material, no backfilling. Lines within System Components give the unit price and total price per cubic yard for this system. Prices for foundation excavation systems are on Line Items A2010 140 1600 thru 2200. Both material quantities and labor costs have been adjusted for the system listed.

Factors: To adjust for job conditions other than normal working situations use Lines A2010 140 2900 thru 4000.

Example: You are to install the system with the use of temporary shoring and bracing. Go to Line A2010 140 3900 and apply these percentages to the appropriate TOTAL costs.

System Components			COST PER C.Y.		
	QUANTITY	UNIT	EQUIP.	LABOR	TOTAL
Foundation excav w/ 3/4 C.Y. backhoe-loader, incl. operator, equip Rental, fuel, oil, and mobilization. Hauling of excavated material is Included. Prices based on one day production of 360 C.Y. in medium soil Without backfilling.					
Equipment operator	8.000	Hr.		680	680
Backhoe-loader, 3/4 C.Y.	1.000	Day	187		187
Operating expense (fuel, oil)	8.000	Hr.	156.64		156.64
Hauling, 12 C.Y. trucks, 1 mile round trip	360.000	L.C.Y.	853.20	846	1,699.20
Mobilization	1.000	Ea.	233	620	853
TOTAL			1,429.84	2,146	3,575.84
COST PER C.Y.		C.Y.	3.96	5.94	9.93

A2010 140	Excavation, Foundation	COST PER C.Y.		
		EQUIP.	LABOR	TOTAL
1600	100 C.Y.	5.15	12.40	17.55
1700	200 C.Y.	4.50	7.35	11.85
1800	300 C.Y.	4.14	6.30	10.44
1850	360 C.Y.	3.96	5.97	9.93
1900	400 C.Y.	3.89	5.75	9.64
2000	500 C.Y.	3.80	5.50	9.30
2100	600 C.Y.	3.69	5.25	8.94
2200	700 C.Y.	3.64	5.10	8.74
2300				
2400				
2900	Dust protection, add, minimum		2%	2%
3000	Maximum		11%	11%
3100	Equipment usage curtailment, add, minimum		1%	1%
3200	Maximum		10%	10%
3300	Material handling & storage limitation, add, minimum		1%	1%
3400	Maximum		7%	7%
3500	Protection of existing work, add, minimum		2%	2%
3600	Maximum		7%	7%
3700	Shift work requirements, add, minimum		5%	5%
3800	Maximum		30%	30%
3900	Temporary shoring and bracing, add, minimum		5%	5%
4000	Maximum		12%	12%

595

A2020 Basement Walls

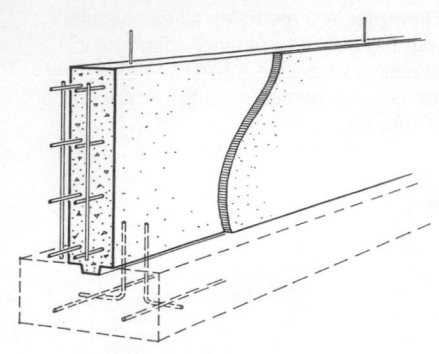

This page illustrates and describes a concrete wall system including concrete, placing concrete, forms, reinforcing, insulation, waterproofing and anchor bolts. Lines within System Components give the unit price and total price per linear foot for this system. Prices for concrete wall systems are on Line Items A2020 120 1400 thru 2600. Both material quantities and labor costs have been adjusted for the system listed.

Factors: To adjust for job conditions other than normal working situations use Lines A2020 120 2900 thru 4000.

Example: You are to install this wall system where delivery of material is difficult. Go to Line A2020 120 3600 and apply these percentages to the appropriate MAT. and INST. costs.

System Components	QUANTITY	UNIT	COST PER L.F.		
			MAT.	INST.	TOTAL
Cast in place concrete foundation wall, 8" thick, 3' high, 2500 psi					
Concrete including forms, reinforcing, waterproofing, and anchor bolts.					
Concrete, 2500 psi, 8" thick, 3' high	.070	C.Y.	9.10		9.10
Forms, wall, 4 uses	6.000	S.F.	6.84	45.60	52.44
Reinforcing	6.000	Lb.	3.18	2.82	6
Placing concrete	.070	C.Y.		3.80	3.80
Waterproofing	3.000	S.F.	1.47	3.75	5.22
Rigid insulaton, 1" polystyrene	3.000	S.F.	.90	2.91	3.81
Anchor bolts, 1/2" diameter, 4' O.C.	.250	Ea.	.44	1.25	1.69
TOTAL			21.93	60.13	82.06

A2020 120	Concrete Wall	COST PER L.F.		
		MAT.	INST.	TOTAL
1400	8" thick, 2500 psi concrete, 3' high	22	60	82
1500	4' high	30	80	110
1600	6' high	44.50	120	164.50
1700	8' high	59.50	159	218.50
1800	3500 psi concrete, 4' high	30.50	80	110.50
1900	6' high	46	120	166
2000	8' high	61	159	220
2100	12" thick, 2500 psi concrete, 4' high	38.50	84.50	123
2200	6' high	56	125	181
2300	8' high	74.50	166	240.50
2400	3500 psi concrete, 4' high	39.50	84.50	124
2500	8' high	76.50	166	242.50
2600	10' high	95	207	302
2700				
2900	Cut & patch to match existing construction, add, minimum	2%	3%	
3000	Maximum	5%	9%	
3100	Dust protection, add, minimum	1%	2%	
3200	Maximum	4%	11%	
3300	Equipment usage curtailment, add, minimum	1%	1%	
3400	Maximum	3%	10%	
3500	Material handling & storage limitation, add, minimum	1%	1%	
3600	Maximum	6%	7%	
3700	Protection of existing work, add, minimum	2%	2%	
3800	Maximum	5%	7%	
3900	Shift work requirements, add, minimum		5%	
4000	Maximum		30%	

A2020 Basement Walls

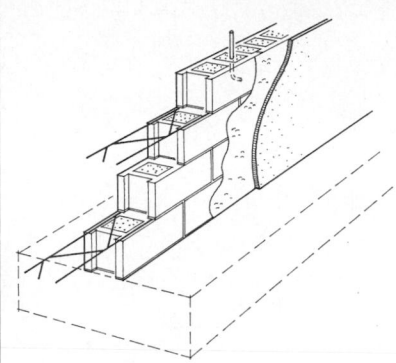

This page illustrates and describes a concrete block wall system including concrete block, masonry reinforcing, parging, waterproofing insulation and anchor bolts. Lines within System Components give the unit price and total price per linear foot for this system. Prices for concrete block wall systems are on Line Items A2020 140 1200 thru 2600. Both material quantities and labor costs have been adjusted for the system listed.

Factors: To adjust for job conditions other than normal working situations use Lines A2020 140 2900 thru 4000.

Example: You are to install the system to match an existing foundation wall. Go to Line A2020 140 3000 and apply these percentages to the appropriate MAT. and INST. costs.

System Components	QUANTITY	UNIT	COST PER L.F.		
			MAT.	INST.	TOTAL
Concrete block, 8″ thick, masonry reinforcing, parged and Waterproofed, insulation and anchor bolts, wall 2′-8″ high.					
Concrete block, 8″ x 8″ x 16″	2.670	S.F.	9.13	18.82	27.95
Masonry reinforcing	2.000	L.F.	.48	.44	.92
Parging	.193	S.Y.	.93	2.84	3.77
Waterproofing	2.670	S.F.	1.31	3.34	4.65
Insulation, 1″ rigid polystyrene	2.670	S.F.	.80	2.59	3.39
Anchor bolts, 1/2″ diameter, 4′ O.C.	.250	Ea.	.44	1.25	1.69
TOTAL			13.09	29.28	42.37

A2020 140	Concrete Block Wall	COST PER L.F.		
		MAT.	INST.	TOTAL
1200	8″ thick block, 2′-8″ high	13.10	29	42.10
1300	4′ high	19.40	43	62.40
1400	6′ high	29	64.50	93.50
1500	8′ high	38.50	85.50	124
1600	Grouted solid, 4′ high	25	58	83
1700	6′ high	37.50	86	123.50
1800	8′ high	50	114	164
2100	12″ thick block, 4′ high	27	62	89
2200	6′ high	40	92.50	132.50
2300	8′ high	53.50	123	176.50
2400	Grouted solid, 4′ high	36	77.50	113.50
2500	6′ high	54	116	170
2600	8′ high	72	154	226
2700				
2900	Cut & patch to match existing construction, add, minimum	2%	3%	
3000	Maximum	5%	9%	
3100	Dust protection, add, minimum	1%	2%	
3200	Maximum	4%	11%	
3300	Equipment usage curtailment, add, minimum	1%	1%	
3400	Maximum	3%	10%	
3500	Material handling & storage limitation, add, minimum	1%	1%	
3600	Maximum	6%	7%	
3700	Protection of existing work, add, minimum	2%	2%	
3800	Maximum	5%	7%	
3900	Shift work requirements, add, minimum		5%	
4000	Maximum		30%	

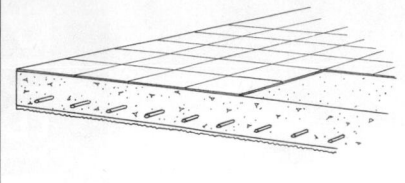

This page illustrates and describes a reinforced concrete slab system including concrete, placing concrete, formwork, reinforcing, steel trowel finish, V.A. floor tile and acoustical spray ceiling finish. Lines within System Components give the unit price per square foot for this system. Prices for reinforced concrete slab systems are on Line Items B1010 225 1400 thru 1800. Both material quantities and labor costs have been adjusted for the system listed.

Factors: To adjust for job conditions other than normal working situations use Lines B1010 225 2700 thru 4000.

Example: You are to install the system to match an existing floor system. Go to Line B1010 225 2800 and apply these percentages to the appropriate MAT. and INST. costs.

System Components	QUANTITY	UNIT	COST PER S.F.		
			MAT.	INST.	TOTAL
Flat slab system, reinforced concrete with vinyl tile floor, sprayed					
Acoustical ceiling finish, not including columns.					
Concrete, 4000 psi, 6" thick	.020	C.Y.	2.80		2.80
Placing concrete	.020	C.Y.		.78	.78
Formwork, 4 use	1.000	S.F.	1.38	6.85	8.23
Edge form	.120	S.F.	.07	1.33	1.40
Reinforcing steel	3.000	Lb.	1.59	2.01	3.60
Steel trowel finish	1.000	S.F.		1.36	1.36
Vinyl floor tile, 1/8" thick, premium colors/patterns	1.000	S.F.	7.70	1.18	8.88
Acoustical spray ceiling finish, solvent based	1.000	S.F.	.31	.19	.50
TOTAL			13.85	13.70	27.55

B1010 225	Floor - Ceiling, Concrete Slab	COST PER S.F.		
		MAT.	INST.	TOTAL
1400	Concrete, 4000 psi, 6" thick	13.85	13.70	27.55
1500	7" thick	14.35	14.25	28.60
1600	8" thick	15.10	15.05	30.15
1700	9" thick	15.75	15.60	31.35
1800	10" thick	16.50	16.40	32.90
1900				
2000				
2100				
2200				
2300				
2400				
2500				
2700	Cut & patch to match existing construction, add, minimum	2%	3%	
2800	Maximum	5%	9%	
2900	Dust protection, add, minimum	1%	2%	
3000	Maximum	4%	11%	
3100	Equipment usage curtailment, add, minimum	1%	1%	
3200	Maximum	3%	10%	
3300	Material handling & storage limitation, add, minimum	1%	1%	
3400	Maximum	6%	7%	
3500	Protection of existing work, add, minimum	2%	2%	
3600	Maximum	5%	7%	
3700	Shift work requirements, add, minimum		5%	
3800	Maximum		30%	
3900	Temporary shoring and bracing, add, minimum	2%	5%	
4000	Maximum	5%	12%	

B1010 Floor Construction

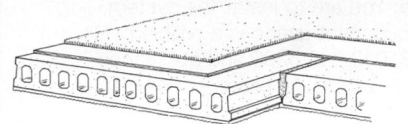

This page illustrates and describes a hollow core prestressed concrete panel system including hollow core slab, grout, carpet, carpet padding, and sprayed ceiling. Lines within System Components give the unit price and total price per square foot for this system. Prices for hollow core prestressed concrete panel systems are on Line Items B1010 232 1200 thru 1600. Both material and labor costs have been adjusted for the system listed.

Factors: To adjust for job conditions other than normal working situations use Lines B1010 232 2700 thru 3800.

Example: You are to install the system where dust control is a major concern. Go to Line B1010 232 2800 and apply these percentages to the appropriate MAT. and INST. costs.

System Components	QUANTITY	UNIT	COST PER S.F.		
			MAT.	INST.	TOTAL
Precast hollow core plank with carpeted floors, padding And sprayed textured ceiling.					
Hollow core plank, 4" thick with grout topping	1.000	S.F.	8.80	3.66	12.46
Nylon carpet, 26 oz. medium traffic	.110	S.Y.	4.79	.87	5.66
Carpet padding, 20 oz./sq. yd.	.110	S.Y.	.57	.43	1
Sprayed texture ceiling	1.000	S.F.	.04	.78	.82
TOTAL			14.20	5.74	19.94

B1010 232	Floor - Ceiling, Conc. Panel	COST PER S.F.		
		MAT.	INST.	TOTAL
1200	Hollow core concrete plank, with grout, 4" thick	14.20	5.75	19.95
1300	6" thick	15.25	5.20	20.45
1400	8" thick	16.40	4.82	21.22
1500	10" thick	16.80	4.52	21.32
1600	12" thick	17.20	4.27	21.47
1700				
1800				
1900				
2000				
2100				
2200				
2300				
2400				
2500				
2700	Dust protection, add, minimum	1%	2%	
2800	Maximum	4%	11%	
2900	Equipment usage curtailment, add, minimum	1%	1%	
3000	Maximum	3%	10%	
3100	Material handling & storage limitation, add, minimum	1%	1%	
3200	Maximum	6%	7%	
3300	Protection of existing work, add, minimum	2%	2%	
3400	Maximum	5%	7%	
3500	Shift work requirements, add, minimum		5%	
3600	Maximum		30%	
3700	Temporary shoring and bracing, add, minimum	2%	5%	
3800	Maximum	5%	12%	

B1010 Floor Construction

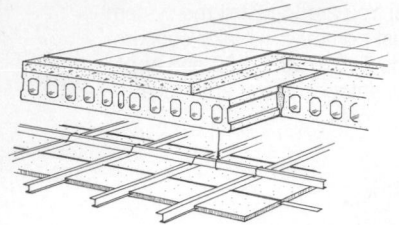

This page illustrates and describes a hollow core plank system including precast hollow core plank, concrete topping, V.A. tile, concealed suspension system and ceiling tile. Lines within System Components give the unit price and total price per square foot for this system. Prices for hollow core plank systems are on Line Items B1010 233 1400 thru 1800. Both material quantities and labor costs have been adjusted for the system listed.

Factors: To adjust for job conditions other than normal working situations use Lines B1010 233 2900 thru 4000.

Example: You are to install the system where equipment usage is a problem. Go to Line B1010 233 3200 and apply these percentages to the appropriate MAT. and INST. costs.

System Components			COST PER S.F.		
	QUANTITY	UNIT	MAT.	INST.	TOTAL
Precast hollow core plank with 2″ topping, vinyl composition floor tile					
And suspended acoustical tile ceiling.					
Hollow core concrete plank, 4″ thick	1.000	S.F.	8.80	3.66	12.46
Finishing floors, granolithic topping, 1:1:1-1/2 mix, 2″ thick	1.000	S.F.	1.60	6.04	7.64
Vinyl composition tile, .08″ thick	1.000	S.F.	1.71	1.18	2.89
Concealed Z bar suspension system, 12″ module	1.000	S.F.	1.02	1.27	2.29
Ceiling tile, mineral fiber, 3/4″ thick	1.000	S.F.	3.36	2.19	5.55
TOTAL			16.49	14.34	30.83

B1010 233	Floor - Ceiling, Conc. Plank	COST PER S.F.		
		MAT.	INST.	TOTAL
1400	Hollow core concrete plank, 4″ thick	16.50	14.35	30.85
1500	6″ thick	17.55	13.85	31.40
1600	8″ thick	18.70	13.45	32.15
1700	10″ thick	19.10	13.10	32.20
1800	12″ thick	19.50	12.90	32.40
1900				
2000				
2100				
2200				
2300				
2400				
2500				
2600				
2700				
2900	Dust protection, add, minimum	1%	2%	
3000	Maximum	4%	11%	
3100	Equipment usage curtailment, add, minimum	1%	1%	
3200	Maximum	3%	10%	
3300	Material handling & storage limitation, add, minimum	1%	1%	
3400	Maximum	6%	7%	
3500	Protection of existing work, add, minimum	2%	2%	
3600	Maximum	5%	7%	
3700	Shift work requirements, add, minimum		5%	
3800	Maximum		30%	
3900	Temporary shoring and bracing, add, minimum	2%	5%	
4000	Maximum	5%	12%	

B1010 Floor Construction

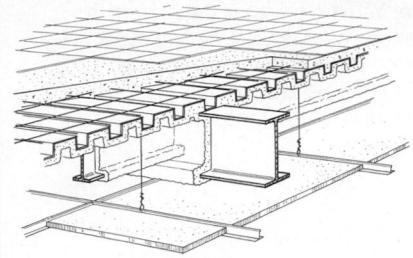

This page illustrates and describes a structural steel w/metal decking and concrete system including steel beams, steel decking, shear studs, concrete, placing concrete, edge form, steel trowel finish, curing, wire fabric, fireproofing, beams and decking, tile floor, and suspended ceiling. Lines within System Components give the unit price and total price per square foot for this system. Prices for alternate structural steel w/metal decking and concrete systems are on Line Items B1010 243 1900 thru 2400. Both material quantities and labor costs have been adjusted for the system listed.

Factors: To adjust for job conditions other than normal working situations use Lines B1010 243 2900 thru 4000.

Example: You are to install the system where material handling and storage are a problem. Go to Line B1010 243 3400 and apply these percentages to the appropriate MAT. and INST. costs.

System Components			COST PER S.F.		
	QUANTITY	UNIT	MAT.	INST.	TOTAL
Composite structural beams, 20 ga. 3″ deep steel decking, shear studs, 3000 psi, concrete, placing concrete, edge forms, steel trowel finish, Welded wire fabric fireproofing, tile floor and suspended ceiling.					
Steel deck, 20 gage, 3″ deep, galvanized	1.000	S.F.	3	1.04	4.04
Structural steel framing	.004	Ton	11.80	2.89	14.69
Shear studs, 3/4″	.250	Ea.	.14	.46	.60
Concrete, 3000 psi, 5″ thick	.015	C.Y.	2		2
Place concrete	.015	C.Y.		.58	.58
Edge form	.140	L.F.	.03	.70	.73
Steel trowel finish	1.000	S.F.		1.36	1.36
Welded wire fabric 6 x 6 - #10/10	.011	C.S.F.	.18	.44	.62
Fireproofing, sprayed	1.880	S.F.	1.17	2.20	3.37
Vinyl composition floor tile	1.000	S.F.	1.34	1.18	2.52
Suspended acoustical ceiling	1.000	S.F.	4.17	1.73	5.90
TOTAL			23.83	12.58	36.41

B1010 243	Floor - Ceiling, Struc. Steel	COST PER S.F.		
		MAT.	INST.	TOTAL
1900	Composite deck, galvanized, 3″ deep, 22 ga.	23.50	12.50	36
1950	20 ga.	24	12.55	36.55
2000	18 ga.	24.50	12.60	37.10
2100	16 ga.	26	12.65	38.65
2200	Composite deck, galvanized, 2″ deep, 22 ga.	23.50	12.35	35.85
2300	20 ga.	23.50	12.40	35.90
2400	16 ga.	25	12.50	37.50
2500				
2900	Dust protection, add, minimum	1%	2%	
3000	Maximum	4%	11%	
3100	Equipment usage curtailment, add, minimum	1%	1%	
3200	Maximum	3%	10%	
3300	Material handling & storage limitation, add, minimum	1%	1%	
3400	Maximum	6%	7%	
3500	Protection of existing work, add, minimum	2%	2%	
3600	Maximum	5%	7%	
3700	Shift work requirements, add, minimum		5%	
3800	Maximum		30%	
3900	Temporary shoring and bracing, add, minimum	2%	5%	
4000	Maximum	5%	12%	

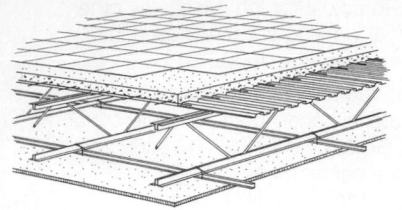

This page illustrates and describes an open web joist and steel slab-form system including open web steel joist, slab form, concrete, placing concrete, wire fabric, steel trowel finish, tile floor and plasterboard. Lines within System Components give the unit price and total price per square foot for this system. Prices for open web joists and steel slab-form systems are on Line Items B1010 245 1800 thru 2200. Both material quantities and labor costs have been adjusted for the systems listed.

Factors: To adjust for job conditions other than normal working situations use Lines B1010 245 2900 thru 4000.

Example: You are to install the system in a congested commercial area and most work will be done at night. Go to Line B1010 245 3800 and apply this percentage to the appropriate INST. cost.

System Components			COST PER S.F.		
	QUANTITY	UNIT	MAT.	INST.	TOTAL
Open web steel joists, 24″ O.C., slab form, 3000 psi concrete, welded					
Wire fabric, steel trowel finish, floor tile, 5/8″ drywall ceiling.					
Open web steel joists, 12″ deep, 5.2#/L.F., 24″ O.C.	2.630	Lb.	2.81	1.23	4.04
Slab form, 28 gage, 9/16″ deep, galvanized	1.000	S.F.	1.75	.78	2.53
Concrete, 3000 psi, 2-1/2″ thick	.008	C.Y.	1.06		1.06
Placing concrete	.008	C.Y.		.31	.31
Welded wire fabric 6 x 6 - #10/10	.011	C.S.F.	.18	.44	.62
Steel trowel finish	1.000	S.F.		1.36	1.36
Vinyl composition floor tile	1.000	S.F.	1.34	1.18	2.52
Ceiling furring, 3/4″ channels, 24″ O.C.	1.000	S.F.	.26	1.48	1.74
Gypsum drywall, 5/8″ thick, finished	1.000	S.F.	.44	1.72	2.16
Paint ceiling	1.000	S.F.	.22	.72	.94
TOTAL			8.06	9.22	17.28

B1010 245	Floor - Ceiling, Steel Joists	COST PER S.F.		
		MAT.	INST.	TOTAL
1800	Open web joists, 12″ deep, 5.2#/L.F.	8.05	9.20	17.25
1900	16″ deep, 6.6#/L.F.	8.80	9.50	18.30
2000	20″ deep, 8.4# L.F.	9.75	9.95	19.70
2100	24″ deep, 11.5# L.F.	11.40	10.70	22.10
2200	26″ deep, 12.8# L.F.	12.10	11	23.10
2300				
2400				
2500				
2600				
2700				
2900	Dust protection, add, minimum	1%	2%	
3000	Maximum	4%	11%	
3100	Equipment usage curtailment, add, minimum	1%	1%	
3200	Maximum	3%	10%	
3300	Material handling & storage limitation, add, minimum	1%	1%	
3400	Maximum	6%	7%	
3500	Protection of existing work, add, minimum	2%	2%	
3600	Maximum	5%	7%	
3700	Shift work requirements, add, minimum		5%	
3800	Maximum		30%	
3900	Temporary shoring and bracing, add, minimum	2%	5%	
4000	Maximum	5%	12%	

B1010 Floor Construction

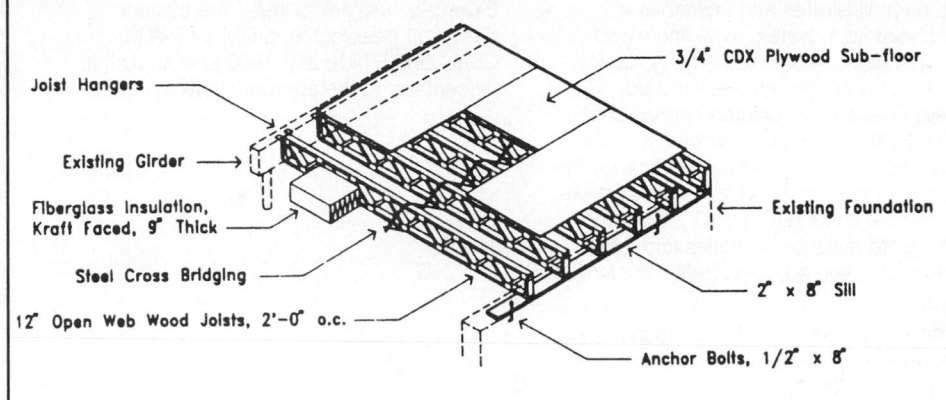

Joist Hangers

Existing Girder →

Fiberglass Insulation,
Kraft Faced, 9" Thick

Steel Cross Bridging

12" Open Web Wood Joists, 2'-0" o.c.

3/4" CDX Plywood Sub-floor

← Existing Foundation

← 2" x 8" Sill

Anchor Bolts, 1/2" x 8"

Open Web Wood Joists

This page illustrates and describes floor framing systems including sills, joist hangers, joists, bridging, subfloor and insulation. Lines within systems components give the unit price and total price per square foot for this system. Prices for alternate joist types are on line items B1010 262 0100 through 0700. Both material quantities and labor costs have been adjusted for the system listed.

Factors: To adjust for job conditions other than normal working situations use lines B1010 262 2700 through 4000.

Example: You are to install the system where delivery of material is difficult. Go to line B1010 262 3400 and apply these percentages to the appropriate MAT. and INST. costs.

System Components			COST PER S.F.		
	QUANTITY	UNIT	MAT.	INST.	TOTAL
Open web wood joists, sill, butt to girder, 5/8″ subfloor, insul.					
Anchor bolt, hook type, with nut and washer, 1/2″ x 8″ long	.250	Ea.	.04	.12	.16
Joist & beam hanger, 18 ga. galvanized	.750	Ea.	.11	.30	.41
Bridging, steel, compression, for 2″ x 12″ joists	.010	C.Pr.	.01	.03	.04
Framing, open web wood joists, 12″ deep	.001	M.L.F.	3.98	1.50	5.48
Framing, sills, 2″ x 8″	.001	M.B.F.	.73	1.98	2.71
Sub-floor, plywood, CDX, 3/4″ thick	1.000	S.F.	.84	.98	1.82
Insulation, fiberglass, kraft face, 9″ thick x 23″ wide	1.000	S.F.	.80	.49	1.29
TOTAL			6.51	5.40	11.91

B1010 262	Open Web Wood Joists	COST PER S.F.		
		MAT.	INST.	TOTAL
0100	12″ open web joists, sill, butt to girder, 5/8″ sub-floor	6.50	5.40	11.90
0200	14″ joists	6.85	5.50	12.35
0300	16″ joists	6.80	5.60	12.40
0400	18″ joists	7.20	5.70	12.90
0450	Wood struc." I " joists, 24″ O.C., to 24′ span, 50 PSF LL, butt to girder	4.72	5	9.72
0500	55 PSF LL	5.05	5.10	10.15
0600	to 30′ span, 45 PSF LL	5.70	4.92	10.62
0700	55 PSF LL	7.65	5	12.65
2700	Cut & patch to match existing construction, add, minimum	2%	3%	
2800	Maximum	5%	9%	
2900	Dust protection, add, minimum	1%	2%	
3000	Maximum	4%	11%	
3100	Equipment usage curtailment, add, minimum	1%	1%	
3200	Maximum	3%	10%	
3300	Material handling & storage limitation, add, minimum	1%	1%	
3400	Maximum	6%	7%	
3500	Protection of existing work, add, minimum	2%	2%	
3600	Maximum	5%	7%	
3700	Shift work requirements, add, minimum		5%	
3800	Maximum		30%	
3900	Temporary shoring and bracing, add, minimum	2%	5%	
4000	Maximum	5%	12%	

This page illustrates and describes a wood joist floor system including wood joist, oak floor, sub-floor, bridging, sand and finish floor, furring, plasterboard, taped, finished and painted ceiling. Lines within System Components give the unit price and total price per square foot for this system. Prices for wood joist floor systems are on Line Items B1010 263 1700 thru 2300. Both material quantities and labor costs have been adjusted for the system listed.

Factors: To adjust for job conditions other than normal working situations use Lines B1010 263 2700 thru 4000.

Example: You are to install the system during off peak hours, 6 P.M. to 2 A.M. Go to Line B1010 263 3800 and apply this percentage to the appropriate INST. costs.

System Components			COST PER S.F.		
	QUANTITY	UNIT	MAT.	INST.	TOTAL
Wood joists, 2″ x 8″, 16″ O.C.,oak floor (sanded & finished),1/2″ sub					
Floor, 1″x 3″ bridging, furring, 5/8″ drywall, taped, finished and painted.					
Wood joists, 2″ x 8″, 16″ O.C.	1.000	L.F.	.98	1.20	2.18
Subfloor plywood CDX 1/2″	1.000	SF Flr.	.68	.88	1.56
Bridging 1″ x 3″	.150	Pr.	.11	.76	.87
Oak flooring, No. 1 common	1.000	S.F.	4.65	3.87	8.52
Sand and finish floor	1.000	S.F.	.23	1.75	1.98
Wood furring, 1″ x 3″, 16″ O.C.	1.000	L.F.	.47	1.88	2.35
Gypsum drywall, 5/8″ thick	1.000	S.F.	.39	.73	1.12
Tape and finishing	1.000	S.F.	.05	.66	.71
Paint ceiling	1.000	S.F.	.22	.72	.94
TOTAL			7.78	12.45	20.23

B1010 263	Floor-Ceiling, Wood Joists	COST PER S.F.		
		MAT.	INST.	TOTAL
1700	16″ on center, 2″ x 6″ joists	7.50	12.30	19.80
1750	2″ x 8″	7.80	12.45	20.25
1800	2″ x 10″	8.35	12.70	21.05
1900	2″ x 12″	8.75	12.75	21.50
2000	2″ x 14″	9.30	12.95	22.25
2100	12″ on center, 2″ x 10″ joists	8.75	13.10	21.85
2200	2″ x 12″	9.25	13.15	22.40
2300	2″ x 14″	9.90	13.40	23.30
2400				
2500				
2700	Cut & patch to match existing construction, add, minimum	2%	3%	
2800	Maximum	5%	9%	
2900	Dust protection, add, minimum	1%	2%	
3000	Maximum	4%	11%	
3100	Equipment usage curtailment, add, minimum	1%	1%	
3200	Maximum	3%	10%	
3300	Material handling & storage limitation, add, minimum	1%	1%	
3400	Maximum	6%	7%	
3500	Protection of existing work, add, minimum	2%	2%	
3600	Maximum	5%	7%	
3700	Shift work requirements, add, minimum		5%	
3800	Maximum		30%	
3900	Temporary shoring and bracing, add, minimum	2%	5%	
4000	Maximum	5%	12%	

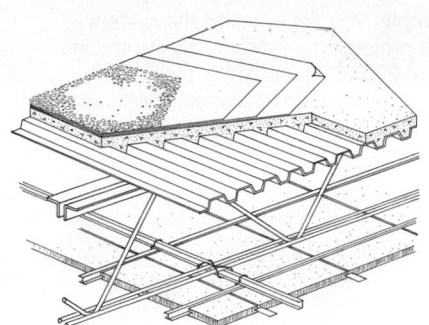

This page illustrates and describes a flat roof bar joist system including asphalt and gravel roof, lightweight concrete, metal decking, open web joists, and suspended acoustic ceiling. Lines within System Components give the unit price and total price per square foot for this system. Prices for flat roof bar joist systems are on Line Items B1020 106 1200 thru 1600. Both material quantities and labor costs have been adjusted for the system listed.

Factors: To adjust for job conditions other than normal working situations use Lines B1020 106 2700 thru 4000.

Example: You are to install the system where material handling and storage present some problem. Go to Line B1020 106 3300 and apply these percentages to the appropriate MAT. and INST. costs.

System Components	QUANTITY	UNIT	MAT.	INST.	TOTAL
Three ply asphalt and gravel roof on lightweight concrete, metal decking on Web joist 4' O.C., with suspended acoustic ceiling.					
Open web joist, 10" deep, 4' O.C.	1.250	Lb.	1.34	.59	1.93
Metal decking, 1-1/2" deep, 22 Ga., galvanized	1.000	S.F.	2.62	.69	3.31
Gravel roofing, 3 ply asphalt	.010	C.S.F.	1.48	2.12	3.60
Perlite roof fill, 3" thick	1.000	S.F.	2.34	.65	2.99
Suspended ceiling, 3/4" mineral fiber on "Z" bar suspension	1.000	S.F.	2.66	4.39	7.05
TOTAL			10.44	8.44	18.88

B1020 106	Steel Joist Roof & Ceiling	MAT.	INST.	TOTAL
1200	Open web bar joist, 10" deep, 5#/L.F.	10.45	8.45	18.90
1300	16" deep, 5#/L.F.	10.85	8.65	19.50
1400	18" deep, 6.6#/L.F.	11.25	8.80	20.05
1500	20" deep, 9.6#/L.F.	11.65	9	20.65
1600	24" deep, 11.52#/L.F.	12.20	9.20	21.40
1700				
1800				
1900				
2000				
2100				
2200				
2300				
2400				
2500				
2700	Cut & patch to match existing construction, add, minimum	2%	3%	
2800	Maximum	5%	9%	
2900	Dust protection, add, minimum	1%	2%	
3000	Maximum	4%	11%	
3100	Equipment usage curtailment, add, minimum	1%	1%	
3200	Maximum	3%	10%	
3300	Material handling & storage limitation, add, minimum	1%	1%	
3400	Maximum	6%	7%	
3500	Protection of existing work, add, minimum	2%	2%	
3600	Maximum	5%	7%	
3700	Shift work requirements, add, minimum		5%	
3800	Maximum		30%	
3900	Temporary shoring and bracing, add, minimum	2%	5%	
4000	Maximum	5%	12%	

607

B1020 Roof Construction

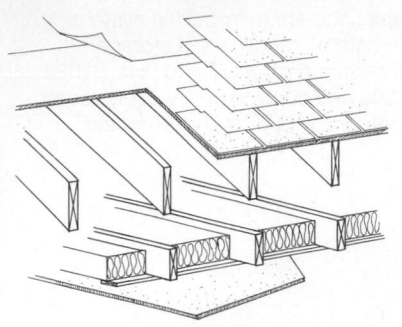

This page illustrates and describes a wood frame roof system including rafters, ceiling joists, sheathing, building paper, asphalt shingles, roof trim, furring, insulation, plaster and paint. Lines within System Components give the unit price and total price per square foot for this system. Prices for wood frame roof systems are on Line Items B1020 202 1800 thru 2700. Both material quantities and labor costs have been adjusted for the system listed.

Factors: To adjust for job conditions other than normal working situations use Lines B1020 202 3300 thru 4000.

Example: You are to install the system while protecting existing work. Go to Line B1020 202 3800 and apply these percentages to the appropriate MAT. and INST. costs.

System Components	QUANTITY	UNIT	COST PER S.F. MAT.	COST PER S.F. INST.	COST PER S.F. TOTAL
Wood frame roof system, 4 in 12 pitch, including rafters, sheathing, Shingles, insulation, drywall, thin coat plaster, and painting.					
Rafters, 2" x 6", 16" O.C., 4 in 12 pitch	1.080	L.F.	.77	1.43	2.20
Ceiling joists, 2" x 6", 16 O.C.	1.000	L.F.	.71	1.05	1.76
Sheathing, 1/2" CDX	1.080	S.F.	.73	1.02	1.75
Building paper, 15# felt	.011	C.S.F.	.06	.20	.26
Asphalt, standard strip shingles, inorganic, class A, 30 year	.011	C.S.F.	1.16	1.38	2.54
Roof trim	.100	L.F.	.18	.29	.47
Furring, 1" x 3", 16" O.C.	1.000	L.F.	.47	1.88	2.35
Fiberglass insulation, 6" batts	1.000	S.F.	.68	.49	1.17
Gypsum board, 1/2" thick	1.000	S.F.	.37	.73	1.10
Thin coat plaster	1.000	S.F.	.12	.82	.94
Paint, roller, 2 coats	1.000	S.F.	.22	.72	.94
TOTAL			5.47	10.01	15.48

B1020 202	Wood Frame Roof & Ceiling	COST PER S.F. MAT.	COST PER S.F. INST.	COST PER S.F. TOTAL
1800	Rafters 16" O.C., 2" x 6"	5.45	10	15.45
1900	2" x 8"	5.75	10.10	15.85
2000	2" x 10"	6.35	10.85	17.20
2100	2" x 12"	6.85	11.05	17.90
2200	Rafters 24" O.C., 2" x 6"	5.10	9.40	14.50
2300	2" x 8"	5.30	9.45	14.75
2400	2" x 10"	5.75	10	15.75
2500	2" x 12"	6.10	10.15	16.25
2600	Roof pitch, 6 in 12, add	3%	10%	
2700	8 in 12, add	5%	12%	
2800				
2900				
3000				
3100				
3300	Cut & patch to match existing construction, add, minimum	2%	3%	
3400	Maximum	5%	9%	
3500	Material handling & storage limitation, add, minimum	1%	1%	
3600	Maximum	6%	7%	
3700	Protection of existing work, add, minimum	2%	2%	
3800	Maximum	5%	7%	
3900	Shift work requirements, add, minimum		5%	
4000	Maximum		30%	

B10 Superstructure

B1020 Roof Construction

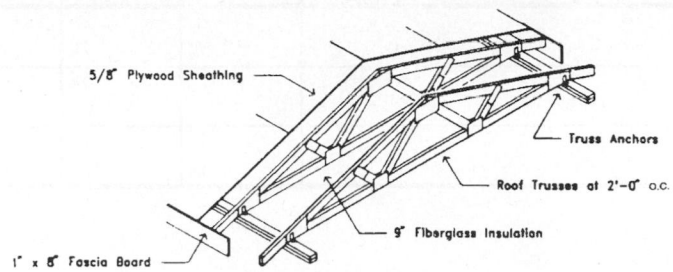

5/8" Plywood Sheathing

Truss Anchors

Roof Trusses at 2'-0" o.c.

9" Fiberglass Insulation

1" x 8" Fascia Board

Wood Fabricated Roof Truss System

This page illustrates and describes a sloped roof truss framing system including trusses, truss anchors, sheathing, fascia and insulation. Lines B1020 210 1500 through 1800 are for a flat roof framing system including flat roof trusses, truss anchors, sheathing, fascia and insulation. Lines within system components give the unit price and total price per square foot for this system. Prices for alternate sizes and systems are given in line items B1020 210 0100 through 1800. Both material quantities and labor costs have been adjusted for the system listed.

Factors: To adjust for job conditions other than normal working situations use lines B1020 210 2700 through 4000.

Example: You are to install the system where crane placement and movement will be difficult. Go to line B1020 210 3200 and apply these percentages to the appropriate MAT. and INST. costs.

System Components	QUANTITY	UNIT	COST PER S.F.		
			MAT.	INST.	TOTAL
Roof truss, 2' O.C., 4/12 slope, 12' span, 5/8 sheath., fascia, insul.					
Timber connectors, rafter anchors, galv., 1 1/2" x 5 1/4"	.084	Ea.	.04	.38	.42
Plywood sheathing, CDX, 5/8" thick	1.054	S.F.	.89	1.06	1.95
Truss, 2' O.C., metal plate connectors, 12' span	.042	Ea.	1.68	2.04	3.72
Molding, fascia trim, 1" x 8"	.167	L.F.	.30	.48	.78
Insulation, fiberglass, kraft face, 9" thick x 23" wide	1.000	S.F.	.80	.49	1.29
TOTAL			3.71	4.45	8.16

B1020 210	Roof Truss, Wood	COST PER S.F.		
		MAT.	INST.	TOTAL
0100	Roof truss, 2'O.C., 4/12p, 1'ovhg, 12'span, 5/8"sheath, fascia, insul.	3.71	4.45	8.16
0200	20' span	3.94	3.68	7.62
0300	24' span	3.70	3.43	7.13
0400	26' span	5.10	4.54	9.64
0500	28' span	3.57	3.36	6.93
0600	30' span	3.70	3.32	7.02
0700	32' span	3.84	3.26	7.10
0800	34' span	3.61	3.23	6.84
0890	60' span	5	3.28	8.28
0900	8/12 pitch, 20' span	4.33	3.95	8.28
1000	24' span	4.37	3.70	8.07
1100	26' span	4.25	3.62	7.87
1200	28' span	4.41	3.61	8.02
1300	32' span	4.38	3.54	7.92
1400	36' span	4.62	3.47	8.09
1500	Flat roof frame, fab.strl.joists, 2' O.C., 15'-24' span, 50 PSF LL	4.04	3.02	7.06
1600	55 PSF LL	4.39	3.09	7.48
1700	24'-30'span, 45 PSF LL	5	2.88	7.88
1800	55 PSF LL	6.90	2.97	9.87
1900				
2700	Cut & patch to match existing construction, add, minimum	2%	3%	
2800	Maximum	5%	9%	
2900	Dust protection, add, minimum	1%	2%	
3000	Maximum	4%	11%	
3100	Equipment usage curtailment, add, minimum	1%	1%	
3200	Maximum	3%	10%	
3300	Material handling & storage limitation, add, minimum	1%	1%	
3400	Maximum	6%	7%	

B1020 Roof Construction

B1020 210	Roof Truss, Wood	COST PER S.F.		
		MAT.	INST.	TOTAL
3500	Protection of existing work, add, minimum	2%	2%	
3600	Maximum	5%	7%	
3700	Shift work requirements, add, minimum		5%	
3800	Maximum		30%	
3900	Temporary shoring and bracing, add, minimum	2%	5%	
4000	Maximum	5%	12%	

B2010 Exterior Walls

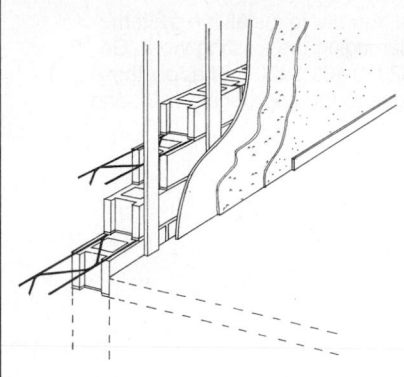

This page illustrates and describes a masonry concrete block system including concrete block wall, pointed, reinforcing, waterproofing, gypsum plaster on gypsum lath, and furring. Lines within System Components give the unit price and total price per square foot for this system. Prices for masonry concrete block wall systems are on Line Items B2010 108 1400 thru 2200. Both material quantities and labor costs have been adjusted for the system listed.

Factors: To adjust for job conditions other than normal working situations use Lines B2010 108 2700 thru 4000.

Example: You are to install the system and match existing construction at several locations. Go to Line B2010 108 2800 and apply these percentages to the appropriate MAT. and INST. costs.

System Components	QUANTITY	UNIT	COST PER S.F. MAT.	COST PER S.F. INST.	COST PER S.F. TOTAL
Concrete block, 8″ thick, reinforced every 2 courses, waterproofing gypsum					
Plaster over gypsum lath on 1″ x 3″ furring, interior painting & baseboard.					
Concrete block, 8″ x 8″ x 16″, reinforced	1.000	S.F.	4.96	7.20	12.16
Silicone waterproofing, 2 coats	1.000	S.F.	.87	.21	1.08
Bituminous coating, 1/16″ thick	1.000	S.F.	.40	1.25	1.65
On concrete	1.000	L.F.	.51	2.53	3.04
Gypsum lath, 3/8″ thick	.110	S.Y.	.38	.80	1.18
Gypsum plaster, 2 coats	.110	S.Y.	.46	3.07	3.53
Painting, 2 coats	1.000	S.F.	.14	.56	.70
Baseboard wood, 9/16″ x 2-5/8″	.100	L.F.	.13	.27	.40
Paint baseboard, primer + 1 coat enamel	.100	L.F.	.02	.07	.09
TOTAL			**7.87**	**15.96**	**23.83**

B2010 108	Masonry Wall, Concrete Block	COST PER S.F. MAT.	COST PER S.F. INST.	COST PER S.F. TOTAL
1400	8″ thick block, regular	7.85	15.95	23.80
1500	Fluted 2 sides	8.95	18.75	27.70
1600	Deep grooved	7.45	18.75	26.20
1700	Slump block	9.75	16.30	26.05
1800	Split rib	8.60	18.75	27.35
1900				
2000	12″ thick block, regular	7.60	19.05	26.65
2100	Slump block	17.35	17.75	35.10
2200	Split rib	9.55	21.50	31.05
2300				
2700	Cut & patch to match existing construction, add, minimum	2%	3%	
2800	Maximum	5%	9%	
2900	Dust protection, add, minimum	1%	2%	
3000	Maximum	4%	11%	
3100	Equipment usage curtailment, add, minimum	1%	1%	
3200	Maximum	3%	10%	
3300	Material handling & storage limitation, add, minimum	1%	1%	
3400	Maximum	6%	7%	
3500	Protection of existing work, add, minimum	2%	2%	
3600	Maximum	5%	7%	
3700	Shift work requirements, add, minimum		5%	
3800	Maximum		30%	
3900	Temporary shoring and bracing, add, minimum	2%	5%	
4000	Maximum	5%	12%	

B2010 Exterior Walls

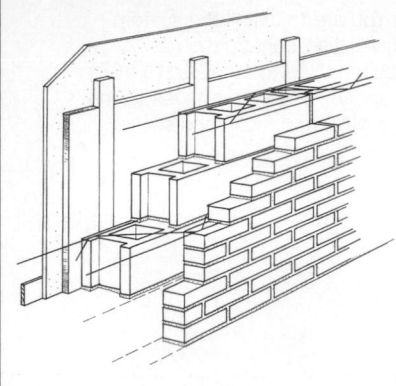

This page illustrates and describes a masonry wall, brick-stone system including brick, concrete block, durawall, insulation, plasterboard, taped and finished, furring, baseboard and painting interior. Lines within System Components give the unit price and total price per square foot for this system. Prices for masonry wall, brick-stone systems are on Line Item B2010 131 1400 thru 2500. Both material quantities and labor costs have been adjusted for the system listed.

Factors: To adjust for job conditions other than normal working situations use Lines B2010 131 3100 thru 4200.

Example: You are to install the system without damaging the existing work. Go to Line B2010 131 3900 and apply these percentages to the appropriate MAT. and INST. costs.

System Components	QUANTITY	UNIT	COST PER S.F.		
			MAT.	INST.	TOTAL
Face brick, 4″thick, concrete block back-up, reinforce every second course,					
3/4″insulation, furring, 1/2″drywall, taped, finish, and painted, baseboard					
Face brick, 4″ brick	1.000	S.F.	4.47	13.60	18.07
Concrete back-up block, reinforced 8″ thick	1.000	S.F.	3.17	7.60	10.77
3/4″ rigid polystyrene insulation	1.000	S.F.	.63	.82	1.45
Furring, 1″ x 3″, wood, 16″ O.C.	1.000	L.F.	.51	1.33	1.84
Drywall, 1/2″ thick	1.000	S.F.	.37	.66	1.03
Taping & finishing	1.000	S.F.	.05	.66	.71
Painting, 2 coats	1.000	S.F.	.22	.72	.94
Baseboard, wood, 9/16″ x 2-5/8″	.100	L.F.	.13	.27	.40
Paint baseboard, primer + 1 coat enamel	.100	L.F.	.02	.07	.09
TOTAL			9.57	25.73	35.30

B2010 131	Masonry Wall, Brick - Stone	COST PER S.F.		
		MAT.	INST.	TOTAL
1400	Face brick, standard, red, 4″ x 2-2/3″ x 8″ (6.75/S.F.)	9.55	25.50	35.05
1500	Norman, 4″ x 2-2/3″ x 12″ (4.5 per S.F.)	12.15	21.50	33.65
1600	Roman, 4″ x 2″ x 12″ (6.0 per S.F.)	13.10	24	37.10
1700	Engineer, 4″ x 3-1/5″ x 8″ (5.63 per S.F.)	9.40	23.50	32.90
1800	S.C.R., 6″ x 2-2/3″ x 12″ (4.5 per S.F.)	12.05	22	34.05
1900	Jumbo, 6″ x 4″ x 12″ (3.0 per S.F.)	11.10	19.05	30.15
2000	Norwegian, 6″ x 3-1/5″ x 12″ (3.75 per S.F.)	11.35	20	31.35
2100				
2200				
2300	Stone, veneer, fieldstone, 6″ thick	12.55	22	34.55
2400	Marble, 2″ thick	58.50	37	95.50
2500	Limestone, 2″ thick	37	23	60
3100	Cut & patch to match existing construction, add, minimum	2%	3%	
3200	Maximum	5%	9%	
3300	Dust protection, add, minimum	1%	2%	
3400	Maximum	4%	11%	
3500	Equipment usage curtailment, add, minimum	1%	1%	
3600	Maximum	3%	10%	
3700	Material handling & storage limitation, add, minimum	1%	1%	
3800	Maximum	6%	7%	
3900	Protection of existing work, add, minimum	2%	2%	
4000	Maximum	5%	7%	
4100	Shift work requirements, add, minimum		5%	
4200	Maximum		30%	

B2010 Exterior Walls

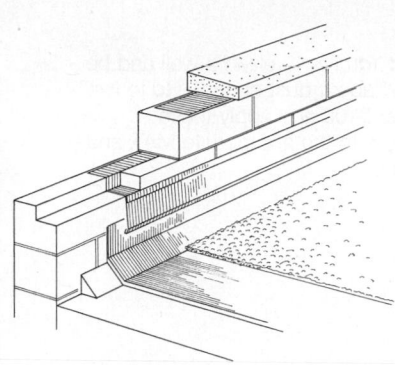

This page illustrates and describes a parapet wall system including wall, coping, flashing and cant strip. Lines within System Components give the unit price and cost per lineal foot for this system. Prices for parapet wall systems are on Lines B2010 142 1300 thru 2400. Both material quantities and labor costs have been adjusted for the system listed.

Factors: To adjust for job conditions other than normal working situations, use Lines B2010 142 2900 thru 4000.

Example: You are to install the system without damaging the adjacent property. Go to Line B2010 142 3600 and apply these percentages to the appropriate MAT. and INST. costs.

System Components	QUANTITY	UNIT	COST PER L.F. MAT.	COST PER L.F. INST.	COST PER L.F. TOTAL
Concrete block parapet, incl. reinf., coping, flashing & cant strip, 2'high					
8" concrete block	2.000	S.F.	9.92	14.40	24.32
Masonry reinforcing	1.000	L.F.	.24	.22	.46
Coping, precast	1.000	L.F.	27	16.70	43.70
Roof cant	1.000	L.F.	.55	1.93	2.48
16 oz.	1.000	L.F.	7.15	2.93	10.08
Cap flashing	1.000	L.F.	8.90	5.45	14.35
TOTAL			53.76	41.63	95.39

B2010 142	Parapet Wall	COST PER L.F. MAT.	COST PER L.F. INST.	COST PER L.F. TOTAL
1300	Concrete block, 8" thick	54	41.50	95.50
1400	12" thick	65.50	54.50	120
1500	Split rib block, 8" thick	55	47	102
1600	12" thick	63	55.50	118.50
1700	Brick, common brick, 8" wall	64	71	135
1800	12" wall	74	90	164
1900	4" brick, 4" backup block	57	68.50	125.50
2000	8" backup block	59	70.50	129.50
2100	Stucco on masonry	54	46.50	100.50
2200	On wood frame	22.50	38	60.50
2300	Wood, T 1-11 siding	33	45.50	78.50
2400	Boards, 1" x 6" cedar	40	50.50	90.50
2500				
2600				
2900	Dust protection, add, minimum	1%	2%	
3000	Maximum	4%	11%	
3100	Equipment usage curtailment, add, minimum	1%	1%	
3200	Maximum	3%	10%	
3300	Material handling & storage limitation, add, minimum	1%	1%	
3400	Maximum	6%	7%	
3500	Protection of existing work, add, minimum	2%	2%	
3600	Maximum	5%	7%	
3700	Shift work requirements, add, minimum		5%	
3800	Maximum		30%	
3900	Temporary shoring and bracing, add, minimum	2%	5%	
4000	Maximum	5%	12%	

This page illustrates and describes a masonry cleaning and restoration system, including staging, cleaning, repointing. Lines within System Components give the unit price and cost per square foot for this system. Prices for systems are on Lines B2010 143 1300 thru 2500. Both material quantities and labor costs have been adjusted for the system listed.

Factors: To adjust for conditions other than normal working situations, use Lines B2010 143 2900 thru 4200.

Example: You are to clean a wall and be concerned about dust control. Go to line B2010 143 3100 and apply these percentages to the appropriate MAT. and INST. costs.

System Components			COST PER S.F.		
	QUANTITY	UNIT	MAT.	INST.	TOTAL
Repoint existing building, brick, common bond, high pressure cleaning, Water only, soft old mortar.					
Scaffolding,steel tubular,bldg ext wall face, labor only, 1 to 5 stories	.010	C.S.F.		2.47	2.47
Cleaning masonry, high pressure wash, water only, avg soil, bio staining	1.100	S.F.		1.73	1.73
Repoint, brick common bond	1.000	S.F.	.61	6.85	7.46
TOTAL			.61	11.05	11.66

B2010 143	Masonry Restoration - Cleaning	COST PER S.F.		
		MAT.	INST.	TOTAL
1300	Brick, common bond	.61	11.05	11.66
1400	Flemish bond	.64	11.50	12.14
1500	English bond	.64	12.20	12.84
1700	Stone, 2' x 2' blocks	.82	8.30	9.12
1800	2' x 4' blocks	.62	7.25	7.87
2000	Add to above prices for alternate cleaning systems:			
2100	Chemical brush and wash	.05	.73	.78
2200	High pressure chemical and water	.08	.59	.67
2300	Sandblasting, wet system	.45	1.69	2.14
2400	Dry system	.42	.99	1.41
2500	Steam cleaning		.49	.49
2600				
2900	Cut & patch to match existing construction, add, minimum	2%	3%	
3000	Maximum	5%	9%	
3100	Dust protection, add, minimum	1%	2%	
3200	Maximum	4%	11%	
3300	Equipment usage curtailment, add, minimum	1%	1%	
3400	Maximum	3%	10%	
3500	Material handling & storage limitation, add, minimum	1%	1%	
3600	Maximum	6%	7%	
3900	Protection of existing work, add, minimum	2%	2%	
4000	Maximum	5%	7%	
4100	Shift work requirements, add, minimum		5%	
4200	Maximum		30%	

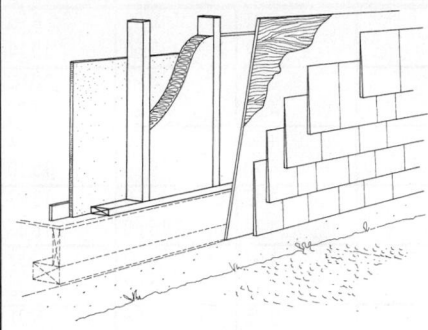

This page illustrates and describes a wood frame exterior wall system including wood studs, sheathing, felt, insulation, plasterboard, taped and finished, baseboard and painted interior. Lines within System Components give the unit price and total price per square foot for this system. Prices for wood frame exterior wall systems are on Line Items B2010 150 1600 thru 2700. Both material quantities and labor costs have been adjusted for the system listed.

Factors: To adjust for job conditions other than normal working situations use Lines B2010 150 3100 thru 4200.

Example: You are to install the system with need for complete temporary bracing. Go to Line B2010 150 4200 and apply these percentages to the appropriate MAT. and INST. costs.

System Components	QUANTITY	UNIT	COST PER S.F. MAT.	COST PER S.F. INST.	COST PER S.F. TOTAL
Wood stud wall, cedar shingle siding, building paper, plywood sheathing, Insulation, 5/8" drywall, taped, finished and painted, baseboard.					
2" x 4" wood studs, 16" O.C.	.100	L.F.	.48	1.32	1.80
1/2" CDX sheathing	1.000	S.F.	.68	.94	1.62
18" No. 1 red cedar shingles, 7-1/2" exposure	.008	C.S.F.	1.63	2.34	3.97
15# felt paper	.010	C.S.F.	.06	.18	.24
3-1/2" fiberglass insulation	1.000	S.F.	.52	.49	1.01
5/8" drywall	1.000	S.F.	.39	.66	1.05
Drywall finishing adder	1.000	S.F.	.05	.66	.71
Baseboard trim, stock pine, 9/16" x 3-1/2", painted	.100	L.F.	.13	.27	.40
Paint baseboard, primer + 1 coat enamel	.100	L.F.	.02	.07	.09
Paint, 2 coats, interior	1.000	S.F.	.22	.72	.94
TOTAL			4.18	7.65	11.83

B2010 150	Wood Frame Exterior Wall	COST PER S.F. MAT.	COST PER S.F. INST.	COST PER S.F. TOTAL
1600	Cedar shingle siding, painted	4.18	7.65	11.83
1700	Aluminum siding, horizontal clapboard	5.30	7.85	13.15
1800	Cedar bevel siding, 1/2" x 6", vertical , painted	7.40	7.55	14.95
1900	Redwood siding 1" x 4" to 1" x 6" vertical, T.&G.	7.85	8.30	16.15
2000	Board and batten	7.80	6.90	14.70
2100	Ship lap siding	7.75	7.30	15.05
2200	Plywood, grooved (T1-11) fir	4.06	7.25	11.31
2300	Redwood	5.40	7.25	12.65
2400	Southern yellow pine	4.13	7.25	11.38
2500	Masonry on stud wall, stucco, wire and plaster	3.56	11.55	15.11
2600	Stone veneer	10	15.05	25.05
2700	Brick veneer, brick $275 per M	7	18.90	25.90
3100	Cut & patch to match existing construction, add, minimum	2%	3%	
3200	Maximum	5%	9%	
3300	Dust protection, add, minimum	1%	2%	
3400	Maximum	4%	11%	
3500	Material handling & storage limitation, add, minimum	1%	1%	
3600	Maximum	6%	7%	
3700	Protection of existing work, add, minimum	2%	2%	
3800	Maximum	5%	7%	
3900	Shift work requirements, add, minimum		5%	
4000	Maximum		30%	
4100	Temporary shoring and bracing, add, minimum	2%	5%	
4200	Maximum	5%	12%	

B2010 Exterior Walls

B2010 190	Selective Price Sheet	COST PER S.F.		
		MAT.	INST.	TOTAL
0100	Exterior surface, masonry, concrete block, standard 4" thick	2.01	6.95	8.96
0200	6" thick	2.64	7.50	10.14
0300	8" thick	2.82	8	10.82
0400	12" thick	4.69	10.30	14.99
0500	Split rib, 4" thick	4.47	8.70	13.17
0600	8" thick	5.70	10	15.70
0700	Brick running bond, standard size, 6.75/S.F.	4.47	13.60	18.07
0800	Buff, 6.75/S.F.	4.73	13.60	18.33
0900	Stucco, on frame	1.02	6.30	7.32
1000	On masonry	.42	4.90	5.32
1100	Metal, aluminum, horizontal, plain	2.75	2.56	5.31
1200	Insulated	3.17	2.56	5.73
1300	Vertical, plain	2.31	2.56	4.87
1400	Insulated	2.41	2.67	5.08
1500	Wood, beveled siding, "A" grade cedar, 1/2" x 6"	4.86	2.23	7.09
1600	1/2" x 8"	8.25	2	10.25
1700	Shingles, 16" #1 red, 7-1/2" exposure	2.16	3.20	5.36
1800	18" perfections, 7-1/2" exposure	2.35	2.69	5.04
1900	Handsplit, 10" exposure	3.30	2.63	5.93
2000	White cedar, 7-1/2" exposure	1.55	3.30	4.85
2100	Vertical, board & batten, redwood	5.30	2.99	8.29
2200	White pine	2.73	2	4.73
2300	T.&G. boards, redwood, 1" x 4"	8.25	2.99	11.24
2400	1' x 8"	8.70	2	10.70
2500				
2600	Interior surface, drywall, taped & finished, standard, 1/2"	.43	1.36	1.79
2700	5/8" thick	.44	1.36	1.80
2800	Fire resistant, 1/2" thick	.46	1.36	1.82
2900	5/8" thick	.45	1.36	1.81
3000	Moisture resistant, 1/2" thick	.51	1.36	1.87
3100	5/8" thick	.54	1.36	1.90
3200	Core board, 1" thick	.85	2.74	3.59
3300	Plaster, gypsum, 2 coats	.46	3.09	3.55
3400	3 coats	.65	3.74	4.39
3500	Perlite or vermiculite, 2 coats	.73	3.57	4.30
3600	3 coats	.78	4.44	5.22
3700	Gypsum lath, standard, 3/8" thick	.39	.81	1.20
3800	1/2" thick	.31	.86	1.17
3900	Fire resistant, 3/8" thick	.31	.98	1.29
4000	1/2" thick	.35	1.06	1.41
4100	Metal lath, diamond, 2.5 lb.	.51	.81	1.32
4200	Rib, 3.4 lb.	.58	.98	1.56
4300	Framing metal studs including top and bottom			
4400	Runners, walls 10' high			
4500	24" O.C., non load bearing 20 gauge, 2-1/2" wide	.33	1.10	1.43
4600	3-5/8" wide	.38	1.12	1.50
4700	4" wide	.46	1.12	1.58
4800	6" wide	.55	1.13	1.68
4900	Load bearing 18 gauge, 2-1/2" wide	.71	1.23	1.94
5000	3-5/8" wide	.84	1.26	2.10
5100	4" wide	.94	1.60	2.54
5200	6" wide	1.12	1.31	2.43
5300	16" O.C., non load bearing 20 gauge, 2-1/2" wide	.45	1.69	2.14
5400	3-5/8" wide	.52	1.71	2.23
5500	4" wide	.63	1.73	2.36
5600	6" wide	.74	1.76	2.50
5700	Load bearing 18 gauge, 2-1/2" wide	.96	1.71	2.67
5800	3-5/8" wide	1.14	1.74	2.88

B2010 Exterior Walls

B2010 190	Selective Price Sheet	COST PER S.F.		
		MAT.	INST.	TOTAL
5900	4" wide	1.18	1.78	2.96
6000	6" wide	1.51	1.81	3.32
6100	Framing wood studs incl. double top plate and			
6200	Single bottom plate, walls 10' high			
6300	24" O.C., 2" x 4"	.37	1.06	1.43
6400	2" x 6"	.57	1.15	1.72
6500	16" O.C., 2" x 4"	.48	1.32	1.80
6600	2" x 6"	.75	1.47	2.22
6700	Sheathing, boards, 1" x 6"	1.86	2.03	3.89
6800	1" x 8"	2.28	1.72	4
6900	Plywood, 3/8" thick	.66	1.10	1.76
7000	1/2" thick	.68	1.17	1.85
7100	5/8" thick	.84	1.25	2.09
7200	3/4" thick	1.02	1.35	2.37
7300	Wood fiber, 5/8" thick	.77	1.10	1.87
7400	Gypsum weatherproof, 1/2" thick	.51	1.17	1.68
7500	Insulation, fiberglass batts, 3-1/2" thick, R13	.45	.94	1.39
7600	6" thick, R19	.56	1.10	1.66
7700	Poured 4" thick, fiberglass wool, R4/inch	.68	3.29	3.97
7800	Mineral wool, R3/inch	.54	3.29	3.83
7900	Polystyrene, R4/inch	1.56	3.29	4.85
8000	Perlite or vermiculite, R2.7/inch	5.80	3.29	9.09
8100	Rigid, fiberglass, R4.3/inch, 1" thick	.59	.66	1.25
8200	R8.7/inch, 2" thick	1.18	.74	1.92

B2020 Exterior Windows

B2020 108			Vinyl Clad Windows				
	MATERIAL	TYPE	GLAZING	SIZE	DETAIL	COST PER UNIT		
						MAT.	INST.	TOTAL
1000	Vinyl clad	casement	insul. glass	1'-4" x 4'-0"		445	152	597
1010				2'-0" x 3'-0"		375	152	527
1020				2'-0" x 4'-0"		425	159	584
1030				2'-0" x 5'-0"		480	169	649
1040				2'-0" x 6'-0"		505	169	674
1050				2'-4" x 4'-0"		480	169	649
1060				2'-6" x 5'-0"		605	164	769
1070				3'-0" x 5'-0"		825	169	994
1080				4'-0" x 3'-0"		900	169	1,069
1090				4'-0" x 4'-0"		780	169	949
1100				4'-8" x 4'-0"		855	169	1,024
1110				4'-8" x 5'-0"		965	196	1,161
1120				4'-8" x 6'-0"		1,075	196	1,271
1130				6'-0" x 4'-0"		990	196	1,186
1140				6'-0" x 5'-0"		1,075	196	1,271
3000		double-hung	insul. glass	2'-0" x 4'-0"		475	146	621
3050				2'-0" x 5'-0"		520	146	666
3100				2'-4" x 4'-0"		495	152	647
3150				2'-4" x 4'-8"		475	152	627
3200				2'-4" x 6'-0"		520	152	672
3250				2'-6" x 4'-0"		475	152	627
3300				2'-8" x 4'-0"		690	196	886
3350				2'-8" x 5'-0"		555	169	724
3375				2'-8" x 6'-0"		545	169	714
3400				3'-0 x 3'-6"		435	152	587
3450				3'-0 x 4'-0"		500	159	659
3500				3'-0 x 4'-8"		550	159	709
3550				3'-0 x 5'-0"		565	169	734
3600				3'-0 x 6'-0"		500	169	669
3700				4'-0 x 5'-0"		695	180	875
3800				4'-0 x 6'-0"		860	180	1,040

B2020 Exterior Windows

This page illustrates and describes an aluminum window system including double hung aluminum window, exterior and interior trim, hardware and insulating glass. Lines within System Components give the unit price and total price on a cost each basis for this system. Prices for aluminum window systems are on Line Items B2020 110 1000 thru 2400. Both material quantities and labor costs have been adjusted for the system listed.

Factors: To adjust for job conditions other than normal working situations use Lines B2020 110 3100 thru 4000.

Example: You are to install the system and cut and patch to match existing construction. Go to Line B2020 110 3200 and apply these percentages to the appropriate MAT. and INST. costs.

System Components	QUANTITY	UNIT	COST EACH		
			MAT.	INST.	TOTAL
Double hung aluminum window, 2'-4" x 2'-6" exterior and interior trim, Hardware glazed with insulating glass.					
Double hung aluminum window, 2'-4" x 2' 6"	1.000	Ea.	265.27	43.43	308.70
Interior Trim	1.000	Set	35.50	50.50	86
Hardware	1.000	Set	6	27.50	33.50
TOTAL			306.77	121.43	428.20

B2020 110	Windows - Aluminum	COST EACH		
		MAT.	INST.	TOTAL
1000	Aluminum, double hung, 2'-4" x 2'-6"	305	121	426
1100	2'-8" x 4'-6"	595	183	778
1200	3'- 4" x 5'-6"	925	274	1,199
1300	Casement, 3'-6" x 2'-4"	415	139	554
1400	4'-6" x 2'-4"	525	172	697
1500	5'-6" x 2'-4"	675	233	908
1600	Projected window, 2'-1" x 3'-0"	298	125	423
1700	3'-5" x 3'-0"	470	170	640
1800	4'-0" x 4'-0"	745	257	1,002
1900	Horizontal sliding 3'-0" x 3'-0"	262	145	407
2000	3'-6" x 4'-0"	390	198	588
2100	5'-0" x 6'-0"	825	360	1,185
2200	Picture window, 3'-8" x 3'-1"	284	162	446
2300	4'-4" x 4'-5"	460	236	696
2400	5'-8" x 4'-9"	670	340	1,010
2500				
2600				
2700				
2800				
2900				
3100	Cut & patch to match existing construction, add, minimum	2%	3%	
3200	Maximum	5%	9%	
3300	Dust protection, add, minimum	1%	2%	
3400	Maximum	4%	11%	
3500	Material handling & storage limitation, add, minimum	1%	1%	
3600	Maximum	6%	7%	
3700	Protection of existing work, add, minimum	2%	2%	
3800	Maximum	5%	7%	
3900	Shift work requirements, add, minimum		5%	
4000	Maximum		30%	

B2020 Exterior Windows

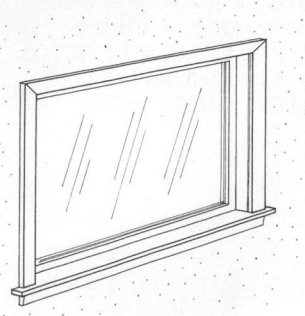

This page illustrates and describes a wood window system including wood picture window, exterior and interior trim, hardware and insulating glass. Lines within System Components give the unit price and total price on a cost each basis for this system. Prices for wood window systems are on Line Items B2020 112 1200 thru 2900. Both material quantities and labor costs have been adjusted for the system listed.

Factors: To adjust for job conditions other than normal working situations use Lines B2020 112 3100 thru 4000.

Example: You are to install the above system where dust control is vital. Go to Line B2020 112 3400 and apply these percentages to the appropriate MAT. and INST. costs.

System Components	QUANTITY	UNIT	COST EACH MAT.	COST EACH INST.	COST EACH TOTAL
Wood picture window 4'-0" x 4'-6", exterior and interior trim, hardware, Glazed with insulating glass.					
4'-0" x 4'-6" wood picture window with insulating glass	1.000	Ea.	585	120	705
Oak interior trim	1.000	Set	85	110	195
Hardware	1.000	Set	6	27.50	33.50
TOTAL			676	257.50	933.50

B2020 112	Windows - Wood	MAT.	INST.	TOTAL
1200	Picture window, 4'-0" x 4'-6"	675	258	933
1300	5'-0" x 4'-0"	755	258	1,013
1400	6'-0" x 4'-6"	795	270	1,065
1500	Bow, bay window, 8'-0" x 5'-0", standard	2,650	270	2,920
1600	Deluxe	1,975	270	2,245
1700	Bow, bay window, 12'-0" x 6'-0" standard	3,325	355	3,680
1800	Deluxe	2,725	355	3,080
1900	Double hung, 3'-0" x 4'-0"	375	167	542
2000	4'-0" x 4'-6"	490	220	710
2100	Casement 2'-0" x 3'-0"	340	144	484
2200	2 leaf, 4'-0" x 4'-0"	900	240	1,140
2300	3 leaf, 6'-0" x 6'-0"	1,550	385	1,935
2400	Awning, 2'-10" x 1'-10"	370	144	514
2500	3'-6" x 2'-4"	430	167	597
2600	4'-0" x 3'-0"	635	220	855
2700	Horizontal sliding 3'-0" x 2'-0"	390	144	534
2800	4'-0" x 3'-6"	470	167	637
2900	6'-0" x 5'-0"	695	220	915
3100	Cut & patch to match existing construction, add, minimum	2%	3%	
3200	Maximum	5%	9%	
3300	Dust protection, add, minimum	1%	2%	
3400	Maximum	4%	11%	
3500	Material handling & storage limitation, add, minimum	1%	1%	
3600	Maximum	6%	7%	
3700	Protection of existing work, add, minimum	2%	2%	
3800	Maximum	5%	7%	
3900	Shift work requirements, add, minimum		5%	
4000	Maximum		30%	

B2020 Exterior Windows

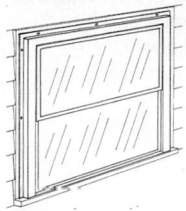

This page illustrates and describes storm window and door systems based on a cost each price. Prices for storm window and door systems are on Line Items B2020 116 0400 thru 2400. Both material quantities and labor costs have been adjusted for the system listed.

Factors: To adjust for job conditions other than normal working situations use Lines B2020 116 3100 thru 4000 and apply these percentages to the appropriate MAT and INST. costs.

Example: You are to install the system and protect all existing construction. Go to Line B2020 116 3800 and apply these percentages to the appropriate MAT. and INST. costs.

B2020 116	Storm Windows & Doors	COST EACH		
		MAT.	INST.	TOTAL
0350	Storm Window and door systems, single glazed			
0400	Window, custom aluminum anodized, 2'-0" x 3'-5"	109	44	153
0500	2'-6" x 5'-0"	132	47	179
0600	4'-0" x 6'-0"	259	52.50	311.50
0700	White painted aluminum, 2'-0" x 3'-5"	127	44	171
0800	2'-6" x 5'-0"	193	47	240
0900	4'-0" x 6'-0"	320	52.50	372.50
1000	Average quality aluminum, anodized, 2'-0" x 3'-5"	107	44	151
1100	2'-6" x 5'-0"	136	47	183
1200	4'-0" x 6'-0"	149	52.50	201.50
1300	White painted aluminum, 2'-0" x 3'-5"	105	44	149
1400	2'-6" x 5'-0"	110	47	157
1500	4'-0" x 6'-0"	127	52.50	179.50
1600	Mill finish, 2'-0" x 3'-5"	99	44	143
1700	2'-6" x 5'-0"	105	47	152
1800	4'-0" x 6'-0"	127	52.50	179.50
1900				
2000	Door, aluminum anodized 3'-0" x 6'-8"	210	94	304
2100	White painted aluminum, 3'-0" x 6'-8"	345	94	439
2200	Mill finish, 3'-0" x 6'-8"	297	94	391
2300				
2400	Wood, storm and screen, painted	360	146	506
2500				
2600				
2700				
2800				
3100	Cut & patch to match existing construction, add, minimum	2%	3%	
3200	Maximum	5%	9%	
3300	Dust protection, add, minimum	1%	2%	
3400	Maximum	4%	11%	
3500	Material handling & storage limitation, add, minimum	1%	1%	
3600	Maximum	6%	7%	
3700	Protection of existing work, add, minimum	2%	2%	
3800	Maximum	5%	7%	
3900	Shift work requirements, add, minimum		5%	
4000	Maximum		30%	

B2020 Exterior Windows

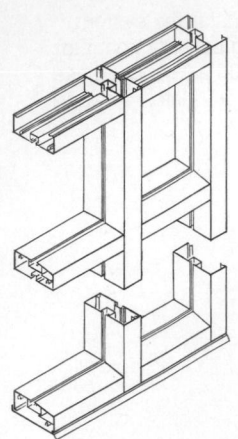

This page illustrates and describes a window wall system including aluminum tube framing, caulking, and glass. Lines within System Components give the unit price and total price per square foot for this system. Prices for window wall systems are on Line Items B2020 124 1300 thru 2600. Both material quantities and labor costs have been adjusted for the system listed.

Factors: To adjust for job conditions other than normal working situations use Lines B2020 124 3100 thru 4000.

Example: You are to install the system with need for complete dust protection. Go to Line B2020 124 3400 and apply these percentages to the appropriate MAT. and INST. costs.

System Components	QUANTITY	UNIT	COST PER S.F.		
			MAT.	INST.	TOTAL
Window wall, including aluminum header, sill, mullions, caulking And glass					
Header, mill finish, 1-3/4" x 4-1/2" deep	.167	L.F.	3.76	2.52	6.28
Sill, mill finish, 1-3/4" x 4-1/2" deep	.167	L.F.	3.59	2.46	6.05
Vertical mullion, 1-3/4" x 4-1/2" deep, 6' O.C.	.191	L.F.	9.13	6.14	15.27
Caulking	.381	L.F.	.10	.87	.97
Glass, 1/4" plate	1.000	S.F.	10.90	10.45	21.35
TOTAL			27.48	22.44	49.92

B2020 124	Aluminum Frame, Window Wall	COST PER S.F.		
		MAT.	INST.	TOTAL
1300	Mill finish, 1-3/4" x 4-1/2" deep, 1/4" plate glass	27.50	22.50	50
1400	2" x 4-1/2" deep, insulating glass	24.50	26	50.50
1500	Thermo-break, 2-1/4" x 4-1/2" deep, insulating glass	25.50	26.50	52
1600	Bronze finish, 1-3/4" x 4-1/2" deep, 1/4" plate glass	30.50	24.50	55
1700	2" x 4-1/2" deep, insulating glass	26.50	28	54.50
1800	Thermo-break, 2-1/4" x 4-1/2" deep, insulating glass	28	28.50	56.50
2000	Black finish, 1-3/4" x 4-1/2" deep, 1/4" plate glass	32	25.50	57.50
2100	2" x 4-1/2" deep, insulating glass	28	29	57
2200	Thermo-break, 2-1/4" x 4-1/2" deep, insulating glass	29.50	29.50	59
2300				
2400	Stainless steel, 1-3/4" x 4-1/2" deep, 1/4" plate glass	40	31	71
2500	2" x 4-1/2" deep, insulating glass	34	35	69
2600	Thermo-break, 2-1/4" x 4-1/2" deep, insulating glass	36	35.50	71.50
2700				
3100	Cut & patch to match existing construction, add, minimum	2%	3%	
3200	Maximum	5%	9%	
3300	Dust protection, add, minimum	1%	2%	
3400	Maximum	4%	11%	
3500	Material handling, & storage limitation, add, minimum	1%	1%	
3600	Maximum	6%	7%	
3700	Protection of existing work, add, minimum	2%	2%	
3800	Maximum	5%	7%	
3900	Shift work requirements, add, minimum		5%	
4000	Maximum		30%	

B2030 Exterior Doors

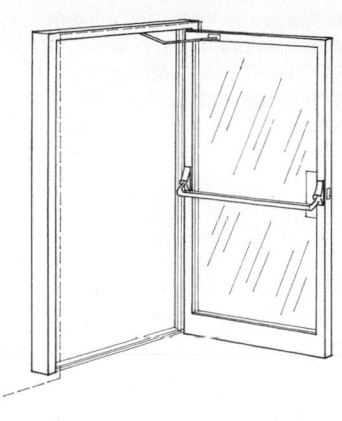

This page illustrates and describes a commercial metal door system, including a single aluminum and glass door, narrow stiles, jamb, hardware weatherstripping, panic hardware and closer. Lines within System Components give the unit price and total price on a cost each basis for this system. Prices for commercial metal door systems are on Line Items B2030 125 1200 thru 2500. Both material quantities and labor costs have been adjusted for the system listed.

Factors: To adjust for job conditions other than normal working situations, use Lines B2030 125 3100 thru 4000.

Example: You are to install the system and cut and patch to match existing construction. Go to Line B2030 125 3200 and apply these percentages to the appropriate MAT. and INST. costs.

System Components	QUANTITY	UNIT	COST EACH		
			MAT.	INST.	TOTAL
Single aluminum and glass door, 3'-0"x7'-0", with narrow stiles, ext. jamb.					
Weatherstripping, 1/2" tempered insul. glass, panic hardware, and closer.					
Aluminum door, 3'-0" x 7'-0" x 1-3/4", narrow stiles	1.000	Ea.	1,050	745	1,795
Tempered insulating glass, 1/2" thick	20.000	S.F.	580	460	1,040
Panic hardware	1.000	Set	900	110	1,010
Surface mount regular arm	1.000	Ea.	183	101	284
TOTAL			2,713	1,416	4,129

B2030 125	Doors, Metal - Commercial	COST EACH		
		MAT.	INST.	TOTAL
1200	Single aluminum and glass, 3'-0" x 7'-0"	2,725	1,425	4,150
1300	With transom, 3'-0" x 10'-0"	3,375	1,675	5,050
1400	Anodized aluminum and glass, 3'-0" x 7'-0"	3,075	1,675	4,750
1500	With transom, 3'-0" x 10'-0"	3,875	1,975	5,850
1600	Steel, deluxe, hollow metal 3'-0" x 7'-0"	1,900	550	2,450
1700	With transom 3'-0" x 10'-0"	2,325	625	2,950
1800	Fire door, "A" label, 3'-0" x 7'-0"	2,275	550	2,825
1900	Double, aluminum and glass, 6'-0" x 7'-0"	4,050	2,425	6,475
2000	With transom, 6'-0" x 10'-0"	4,875	3,000	7,875
2100	Anodized aluminum and glass 6'-0" x 7'-0"	4,500	2,850	7,350
2200	With transom, 6'-0" x 10'-0"	5,425	3,475	8,900
2300	Steel, deluxe, hollow metal, 6'-0" x 7'-0"	3,125	955	4,080
2400	With transom, 6'-0" x 10'-0"	3,975	1,100	5,075
2500	Fire door, "A" label, 6'-0" x 7'-0"	3,800	945	4,745
2800				
2900				
3100	Cut & patch to match existing construction, add, minimum	2%	3%	
3200	Maximum	5%	9%	
3300	Dust protection, add, minimum	1%	2%	
3400	Maximum	4%	11%	
3500	Material handling & storage limitation, add, minimum	1%	1%	
3600	Maximum	6%	7%	
3700	Protection of existing work, add, minimum	2%	2%	
3800	Maximum	5%	7%	
3900	Shift work requirements, add, minimum		5%	
4000	Maximum		30%	

623

B2030 Exterior Doors

This page illustrates and describes sliding door systems including a sliding door, frame, interior and exterior trim with exterior staining. Lines within System Components give the unit price and total price on a cost each basis for this system. Prices for sliding door systems are on Line Items B2030 150 1000 thru 2400. Both material quantities and labor costs have been adjusted for the system listed.

Factors: To adjust for job conditions other than normal working situations use Lines B2030 150 2700 thru 4000.

Example: You are to install the system with temporary shoring and bracing. Go to Line B2030 150 3900 and apply these percentages to the appropriate MAT. and INST. costs.

System Components	QUANTITY	UNIT	COST EACH		
			MAT.	INST.	TOTAL
Sliding wood door, 6'-0" x 6'-8", with wood frame, interior and exterior					
Trim and exterior staining.					
Sliding wood door, standard, 6'-0" x 6'-8", insulated glass	1.000	Ea.	1,325	330	1,655
Interior & exterior trim	1.000	Set	31.20	52.60	83.80
Stain door	1.000	Ea.	2.59	32	34.59
Stain trim	1.000	Ea.	3	13.60	16.60
TOTAL			1,361.79	428.20	1,789.99

B2030 150	Doors, Sliding - Patio	COST EACH		
		MAT.	INST.	TOTAL
1000	Wood, standard, 6'-0" x 6'-8", insulated glass	1,350	430	1,780
1100	8'-0" x 6'-8"	1,725	545	2,270
1200	12'-0" x 6'-8"	2,775	660	3,435
1300	Vinyl coated, 6'-0" x 6'-8"	1,625	430	2,055
1400	8'-0" x 6'-8"	2,075	545	2,620
1500	12'-0" x 6'-8"	3,450	660	4,110
1700	Aluminum, standard, 6'-0" x 6'-8", insulated glass	965	460	1,425
1800	8'-0" x 6'-8"	1,700	585	2,285
1900	12'-0" x 6'-8"	1,800	695	2,495
2000	Anodized, 6'-0" x 6'-8"	1,750	460	2,210
2100	8'-0" x 6'-8"	1,950	585	2,535
2200	12'-0" x 6'-8"	3,375	695	4,070
2300				
2400	Deduct for single glazing	112		112
2500				
2700	Cut & patch to match existing construction, add, minimum	2%	3%	
2800	Maximum	5%	9%	
2900	Dust protection, add, minimum	1%	2%	
3000	Maximum	4%	11%	
3100	Equipment usage curtailment, add, minimum	1%	1%	
3200	Maximum	3%	10%	
3300	Material handling & storage limitation, add, minimum	1%	1%	
3400	Maximum	6%	7%	
3500	Protection of existing, work, add, minimum	2%	2%	
3600	Maximum	5%	7%	
3700	Shift work requirements, add, minimum		5%	
3800	Maximum		30%	
3900	Temporary shoring and bracing, add, minimum	2%	5%	
4000	Maximum	5%	12%	

B2030 Exterior Doors

This page illustrates and describes residential door systems including a door, frame, trim, hardware, weatherstripping, stained and finished. Lines within System Components give the unit price and total price on a cost each basis for this system. Prices for residential door systems are on Line Items B2030 240 1500 thru 2200. Both material quantities and labor costs have been adjusted for the system listed.

Factors: To adjust for job conditions other than normal working situations use Lines B2030 240 3100 thru 4000.

Example: You are to install the system with a material handling and storage limitation. Go to Line B2030 240 3600 and apply these percentages to the appropriate MAT. and INST. costs.

System Components	QUANTITY	UNIT	COST EACH MAT.	COST EACH INST.	COST EACH TOTAL
Single solid wood colonial door, 3' x 6'-8" with wood frame and trim,					
Stained and finished, including hardware and weatherstripping.					
Solid wood colonial door, fir 3' x 6'-8" x 1-3/4", hinges	1.000	Ea.	615	88	703
Hinges, ball bearing	1.000	Ea.	39		39
Exterior frame, with trim	1.000	Set	182.75	59.67	242.42
Interior trim	1.000	Set	26.50	55	81.50
Average quality	1.000	Set	57	47	104
Sill, oak 8" deep	3.500	Ea.	82.25	46.03	128.28
Weatherstripping	1.000	Set	27.50	86.50	114
Stained and finished	1.000	Ea.	5.55	68.50	74.05
TOTAL			1,035.55	450.70	1,486.25

B2030 240	Doors, Residential - Exterior	COST EACH MAT.	COST EACH INST.	COST EACH TOTAL
1400	Single doors, 3' x 6'-8"			
1500	Solid colonial door, fir	1,025	450	1,475
1600	Hollow metal exterior door, plain	900	420	1,320
1700	Solid core wood door, plain	585	430	1,015
1800				
1900	Double doors, 6' x 6'-8"			
2000	Solid colonial double doors, fir	1,750	600	2,350
2100	Hollow metal exterior doors, plain	1,500	560	2,060
2200	Hollow core wood doors, plain	860	580	1,440
2300				
2800				
2900				
3100	Cut & patch to match existing construction, add, minimum	2%	3%	
3200	Maximum	5%	9%	
3300	Dust protection, add, minimum	1%	2%	
3400	Maximum	4%	11%	
3500	Material handling & storage limitation, add, minimum	1%	1%	
3600	Maximum	6%	7%	
3700	Protection of existing work, add, minimum	2%	2%	
3800	Maximum	5%	7%	
3900	Shift work requirements, add, minimum		5%	
4000	Maximum		30%	

B2030 Exterior Doors

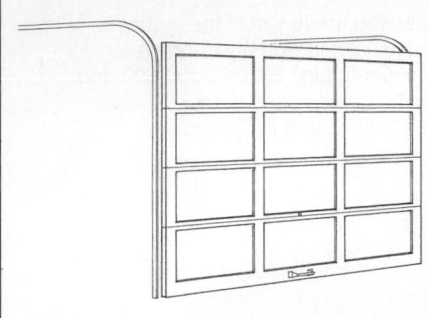

This page illustrates and describes overhead door systems including an overhead door, track, hardware, trim, and electric door opener. Lines within System Components give the unit price and total price on a cost each basis for this system. Prices for overhead door systems are on Line Items B2030 410 1200 thru 2300. Both material quantities and labor costs have been adjusted for the system listed.

Factors: To adjust for job conditions other than normal working situations use Lines B2030 410 3100 thru 4000.

Example: You are to install the system and match existing construction. Go to Line B2030 410 3200 and apply these percentages to the appropriate MAT. and INST. costs.

System Components	QUANTITY	UNIT	COST EACH MAT.	COST EACH INST.	COST EACH TOTAL
Wood overhead door, commercial sectional, including track, door, hardware					
And trim, electrically operated.					
Commercial, heavy duty wood door, 8' x 8' x 1-3/4" thick	1.000	Ea.	1,350	660	2,010
Frame, 2 x 8, pressure treated	1.000	Set	29.52	83.28	112.80
Wood trim	1.000	Set	30.96	65.76	96.72
Painting, two coats	1.000	Ea.	23.04	218.88	241.92
Electric trolley operator	1.000	Ea.	1,225	330	1,555
TOTAL			2,658.52	1,357.92	4,016.44

B2030 410	Doors, Overhead	COST EACH MAT.	COST EACH INST.	COST EACH TOTAL
1200	Commercial, wood, 1-3/4" thick, 8' x 8'	2,650	1,350	4,000
1300	12' x 12'	3,950	1,900	5,850
1400	14' x 14'	5,400	2,250	7,650
1500	Fiberglass & aluminum 12' x 12'	5,050	1,950	7,000
1600	20' x 20'	9,750	4,925	14,675
1700	Residential, wood, 9' x 7'	1,200	515	1,715
1800	16' x 7'	1,950	765	2,715
1900	Hardboard faced, 9' x 7'	855	515	1,370
2000	16' x 7'	1,525	765	2,290
2100	Fiberglass & aluminum, 9' x 7'	1,200	610	1,810
2200	16' x 7'	1,900	765	2,665
2300	For residential electric opener, add	695	82.50	777.50
2400				
2500				
2600				
2700				
2800				
2900				
3100	Cut & patch to match existing construction, add, minumum	2%	3%	
3200	Maximum	5%	9%	
3300	Dust protection, add, minimum	1%	2%	
3400	Maximum	4%	11%	
3500	Material handling & storage limitation, add, minimum	1%	1%	
3600	Maximum	6%	7%	
3700	Protection of existing work, add, minimum	2%	2%	
3800	Maximum	5%	7%	
3900	Shift work requirements, add, minimum		5%	
4000	Maximum		30%	

B2030 Exterior Doors

B2030 810	Selective Price Sheet	COST EACH		
		MAT.	INST.	TOTAL
0100	Door closer, rack and pinion	221	110	331
0200	Backcheck and adjustable power	221	110	331
0300	Regular, hinge face mount, all sizes, regular arm	221	110	331
0400	Hold open arm	106	110	216
0500	Top jamb mount, all sizes, regular arm	221	110	331
0600	Hold open arm	106	110	216
0700	Stop face mount, all sizes, regular arm	221	110	331
0800	Hold open arm	106	110	216
0900	Fusible link, hinge face mount, all sizes, regular arm	269	110	379
1000	Hold open arm	154	110	264
1100	Top jamb mount, all sizes, regular arm	269	110	379
1200	Hold open arm	154	110	264
1300	Stop face mount, all sizes, regular arm	269	110	379
1400	Hold open arm	154	110	264
1500				
1600				
1700	Door stops			
1800				
1900	Holder & bumper, floor or wall	39.50	20.50	60
2000	Wall bumper	13.95	20.50	34.45
2100	Floor bumper	3.65	20.50	24.15
2200	Plunger type, door mounted	34	20.50	54.50
2300	Hinges, full mortise, material only, per pair			
2400	Low frequency, 4-1/2" x 4-1/2", steel base, USP	25.50		25.50
2500	Brass base, US10	60.50		60.50
2600	Stainless steel base, US32	84.50		84.50
2700	Average frequency, 4-1/2" x 4-1/2", steel base, USP	43.50		43.50
2800	Brass base, US10	70		70
2900	Stainless steel base, US32	85		85
3000	High frequency, 4-1/2" x 4-1/2", steel base, USP	73.50		73.50
3100	Brass base, US10	83.50		83.50
3200	Stainless steel base, US32	115		115
3300	Kick plate			
3400				
3500	6" high, for 3'-0" door, aluminum	34.50	44	78.50
3600	Bronze	66	44	110
3700	Panic device			
3800				
3900	For rim locks, single door, exit	900	110	1,010
4000	Outside key and pull	1,475	132	1,607
4100	Bar and vertical rod, exit only	1,075	132	1,207
4200	Outside key and pull	1,375	165	1,540
4300	Lockset			
4400				
4500	Heavy duty, cylindrical, passage doors	139	55	194
4600	Classroom	580	82.50	662.50
4700	Bedroom, bathroom, and inner office doors	173	55	228
4800	Apartment, office, and corridor doors	340	66	406
4900	Standard duty, cylindrical, exit doors	167	66	233
5100	Passage doors	88	55	143
5200	Public restroom, classroom, & office doors	231	82.50	313.50
5300	Deadlock, mortise, heavy duty	200	73	273
5400	Double cylinder	200	73	273
5500	Entrance lock, cylinder, deadlocking latch	199	73	272
5600	Deadbolt	214	82.50	296.50
5700	Commercial, mortise, wrought knob, keyed, minimum	212	82.50	294.50
5800	Maximum	580	94	674
5900	Cast knob, keyed, minimum	291	73	364

B2030 810	Selective Price Sheet	COST EACH		
		MAT.	INST.	TOTAL
6000	Maximum	590	73	663
6100	Push-pull			
6200				
6300	Aluminum	15.95	55	70.95
6400	Bronze	29	55	84
6500	Door pull, designer style, minimum	28	55	83
6600	Maximum	270	120	390
6700	Threshold			
6800				
6900	3'-0" long door saddles, aluminum, minimum	11.45	13.70	25.15
7000	Maximum	55	55	110
7100	Bronze, minimum	50	10.95	60.95
7200	Maximum	178	55	233
7300	Rubber, 1/2" thick, 5-1/2" wide	45	33	78
7400	2-3/4" wide	54	33	87
7500	Weatherstripping, per set			
7600				
7700	Doors, wood frame, interlocking for 3' x 7' door, zinc	52.50	219	271.50
7800	Bronze	66.50	219	285.50
7900	Wood frame, spring type for 3' x 7' door, bronze	27.50	86.50	114
8000	Metal frame, spring type for 3' x 7' door, bronze	54	219	273
8100	For stainless steel, spring type, add	133%		
8200				
8300	Metal frame, extruded sections, 3' x 7' door, aluminum	31	219	250
8400	Bronze	90.50	219	309.50

B3010   Roof Coverings

B3010 160	Selective Price Sheet	COST PER S.F.		
		MAT.	INST.	TOTAL
0100	Roofing, built-up, asphalt roll roof, 3 ply organic/mineral surface	.78	1.73	2.51
0200	3 plies glass fiber felt type iv, 1 ply mineral surface	1.35	1.87	3.22
0300	Cold applied, 3 ply		.61	.61
0400	Coal tar pitch, 4 ply tarred felt	2.19	2.22	4.41
0500	Mopped, 3 ply glass fiber	1.81	2.46	4.27
0600	4 ply organic felt	2.19	2.22	4.41
0700	Elastomeric, hypalon, neoprene unreinforced	1.59	4.11	5.70
0800	Polyester reinforced	1.72	4.86	6.58
0900	Neoprene, 5 coats 60 mils	5.30	14.35	19.65
1000	Over 10,000 S.F.	5.30	7.45	12.75
1600				
1700	Asphalt, strip, 210-235#/sq.	.83	1.14	1.97
1800	235-240#/sq.	1.05	1.25	2.30
1900	Class A laminated	1.21	1.39	2.60
2000	Class C laminated	1.21	1.57	2.78
2100	Slate, buckingham, 3/16" thick	6.10	3.60	9.70
2200	Black, 1/4" thick	6.10	3.60	9.70
2300	Wood, shingles, 16" no. 1, 5" exp.	3.25	2.63	5.88
2400	Red cedar, 18" perfections	2.78	2.39	5.17
2500	Shakes, 24", 10" exposure	3.30	2.63	5.93
2600	18", 8-1/2" exposure	3	3.30	6.30
2700	Insulation, ceiling batts, fiberglass, 3-1/2" thick, R13	.37	.49	.86
2800	6" thick, R19	.48	.49	.97
2900	9" thick, R30	.80	.57	1.37
3000	12" thick, R38	1.16	.57	1.73
3100	Mineral fiber, 3-1/2" thick, R13	.85	.41	1.26
3200	6" thick, R19	1.33	.41	1.74
3300	Roof deck, fiberboard, 1" thick, R2.78	.76	.80	1.56
3400	Mineral fiber, 2" thick, R5.26	1.39	.80	2.19
3500	Perlite boards, 3/4" thick, R2.08	.54	.80	1.34
3600	2" thick, R5.26	1.31	.89	2.20
3700	Polystyrene extruded, 15 PSI, 1" thick, R5.26	.73	.52	1.25
3800	2" thick, R10	.91	.59	1.50
3900	40 PSI, 1" thick, R5	1.08	.52	1.60
4000	Tapered for drainage	1.08	.54	1.62
4300	Ceiling, plaster, gypsum, 2 coats	.46	3.57	4.03
4400	3 coats	.65	4.21	4.86
4500	Perlite or vermiculite, 2 coats	.73	4.15	4.88
4600	3 coats	.78	5.20	5.98
4700	Gypsum lath, plain 3/8" thick	.39	.81	1.20
4800	1/2" thick	.31	.86	1.17
4900	Firestop, 3/8" thick	.31	.98	1.29
5000	1/2" thick	.35	1.06	1.41
5100	Metal lath, rib, 2.75 lb.	.44	.92	1.36
5200	3.40 lb.	.58	.98	1.56
5300	Diamond, 2.50 lb.	.51	.92	1.43
5400	3.40 lb.	.55	1.15	1.70
5500	Drywall, taped and finished standard, 1/2" thick	.43	1.72	2.15
5600	5/8" thick	.44	1.72	2.16
5700	Fire resistant, 1/2" thick	.46	1.72	2.18
5800	5/8" thick	.45	1.72	2.17
5900	Water resist., 1/2" thick	.51	1.72	2.23
6000	5/8" thick	.54	1.72	2.26
6100	Finish, instead of taping			
6200	For thin coat plaster, add	.12	.82	.94
6300	Finish textured spray, add	.04	.78	.82
6500	Tile, stapled glued, mineral fiber plastic coated, 5/8" thick	2.57	2.19	4.76
6600	3/4" thick	3.36	2.19	5.55

629

B3010 Roof Coverings

B3010 160	Selective Price Sheet	COST PER S.F.		
		MAT.	INST.	TOTAL
6700	Wood fiber, 1/2" thick	1.32	1.65	2.97
6800	3/4" thick	1.28	1.65	2.93
6900	Suspended, fiberglass film faced, 5/8" thick	1.39	1.05	2.44
7000	3" thick	3.85	1.46	5.31
7100	Mineral fiber 5/8" thick, standard face	.96	.98	1.94
7200	Aluminum faced	4.30	1.10	5.40
7500	Ceiling suspension systems, for tile, "T" bar, class "A", 2' x 4' grid	.87	.82	1.69
7600	2' x 2' grid	1.12	1.01	2.13
7700	Concealed "Z" bar, 12" module	1.02	1.27	2.29
7800				
7900	Plaster/drywall, 3/4" channels, steel furring, 16" O.C.	.39	2.14	2.53
8000	24" O.C.	.26	1.48	1.74
8100	1-1/2" channels, 16" O.C.	.52	2.39	2.91
8200	24" O.C.	.35	1.59	1.94
8300	Ceiling framing, 2" x 4" studs, 16" O.C.	.35	1.10	1.45
8400	24" O.C.	.23	.73	.96

B3020 Roof Openings

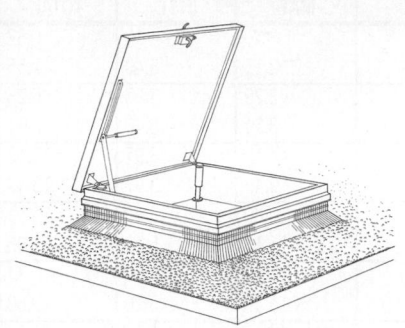

This page illustrates and describes a roof hatch system. Lines within System Components give the unit price and total cost each for this system. Prices for systems are on Lines B3020 230 1400 thru 2400. Both material quantities and labor costs have been adjusted for the system listed.

Factors: To adjust for job conditions other than normal working situtations use Lines B3020 230 2900 thru 4000.

Example: You are to install the system and cut and patch to match existing construction. Use line B3020 230 3000 and apply these percentages to the appropriate MAT. and INST. costs.

System Components	QUANTITY	UNIT	COST EACH MAT.	COST EACH INST.	COST EACH TOTAL
Roof hatch, 2'-6" x 3'-0", aluminum, curb and cover included.					
Through steel construction.					
Roof hatch, 2'-6" x 3'-0", aluminum	1.000	Ea.	975	254	1,229
Cutout decking	11.000	L.F.	4.51	27.06	31.57
Frame opening	44.000	L.F.	39.16	394.24	433.40
Flashing, 16 oz. copper	16.000	S.F.	142.40	87.20	229.60
Cant strip, 4 x 4, mineral fiber	12.000	L.F.	6.60	23.16	29.76
TOTAL			1,167.67	785.66	1,953.33

B3020 230	Roof Hatches, Skylights	COST EACH MAT.	COST EACH INST.	COST EACH TOTAL
1400	Roof hatch, aluminum, curb and cover, 2'-6" x 3'-0"	1,175	785	1,960
1500	2'-6" x 4'-6"	1,450	960	2,410
1600	2'-6" x 8'-0"	2,600	1,375	3,975
1700	Skylight, plexiglass dome, with curb mounting, 30" x 32"	725	745	1,470
1800	30" x 45"	785	930	1,715
1900	40" x 45"	930	1,250	2,180
2000	Smoke hatch, unlabeled, 2'-6" x 3'-0"	1,400	850	2,250
2100	2'-6" x 4'-6"	1,750	1,025	2,775
2200	2'-6" x 8'-0"	2,850	1,400	4,250
2300				
2400	For steel ladder, add per vertical linear foot	40	37	77
2450				
2600				
2700				
2900	Cut & patch to match existing construction, add, minimum	2%	3%	
3000	Maximum	5%	9%	
3100	Dust protection, add, minimum	1%	2%	
3200	Maximum	4%	11%	
3300	Equipment usage curtailment, add, minimum	1%	1%	
3400	Maximum	3%	10%	
3500	Material handling & storage limitation, add, minimum	1%	1%	
3600	Maximum	6%	7%	
3700	Protection of existing work, add, minimum	2%	2%	
3800	Maximum	5%	7%	
3900	Shift work requirements, add, minimum		5%	
4000	Maximum		30%	

B3020 240	Selective Price Sheet	COST PER UNIT		
		MAT.	INST.	TOTAL
0100	Downspouts per L.F., aluminum, enameled .024" thick, 2" x 3"	2.33	4.19	6.52
0200	3" x 4"	2.15	5.40	7.55
0300	Round .025" thick, 3" diam.	2.29	3.97	6.26
0400	4" diam.	3.51	5.40	8.91
0500	Copper, round 16 oz. stock, 2" diam.	9.15	3.97	13.12
0600	3" diam.	9.45	3.97	13.42
0700	4" diam.	10.80	5.20	16
0800	5" diam.	16.15	5.80	21.95
0900	Rectangular, 2" x 3"	12.40	3.97	16.37
1000	3" x 4"	15.40	5.20	20.60
1100	Lead coated copper, round, 2" diam.	25	3.97	28.97
1200	3" diam.	25.50	3.97	29.47
1300	4" diam.	26.50	5.20	31.70
1400	5" diam.	26	5.80	31.80
1500	Rectangular, 2" x 3"	27	3.97	30.97
1600	3" x 4"	36.50	5.20	41.70
1700	Steel galvanized, round 28 gauge, 3" diam.	2.28	3.97	6.25
1800	4" diam.	2.29	5.20	7.49
1900	5" diam.	3.88	5.80	9.68
2000	6" diam.	3.83	7.20	11.03
2100	Rectangular, 2" x 3"	4.21	3.97	8.18
2200	3" x 4"	4.75	5.20	9.95
2300	Elbows, aluminum, round, 3" diam.	3.41	7.55	10.96
2400	4" diam.	7.20	7.55	14.75
2500	Rectangular, 2" x 3"	1.02	7.55	8.57
2600	3" x 4"	3.41	7.55	10.96
2700	Copper, round 16 oz., 2" diam.	10.40	7.55	17.95
2800	3" diam.	10	7.55	17.55
2900	4" diam.	15.30	7.55	22.85
3100	Rectangular, 2" x 3"	10.15	7.55	17.70
3200	3" x 4"	16.65	7.55	24.20
3300	Drip edge per L.F., aluminum, 5" wide	.64	1.65	2.29
3400	8" wide	1.66	1.65	3.31
3500	28" wide	8.60	6.60	15.20
3600				
3700	Steel galvanized, 5" wide	.67	1.65	2.32
3800	8" wide	.91	1.65	2.56
3900				
4000				
4100				
4200				
4300	Flashing 12" wide per S.F., aluminum, mill finish, .013" thick	.91	4.32	5.23
4400	.019" thick	1.54	4.32	5.86
4500	.040" thick	2.52	4.32	6.84
4600	.050" thick	3.03	4.32	7.35
4700	Copper, mill finish, 16 oz.	8.90	5.45	14.35
4800	20 oz.	11.80	5.70	17.50
4900	24 oz.	16.25	5.95	22.20
5000	32 oz.	21	6.25	27.25
5100	Lead, 2.5 lb./S.F., 12" wide	6.75	4.64	11.39
5200	Over 12" wide	4.44	4.64	9.08
5900	Polyvinyl chloride, black, .010" thick	.29	2.20	2.49
6000	.020" thick	.29	2.20	2.49
6100	.030" thick	.36	2.20	2.56
6200	.056" thick	.94	2.20	3.14
6300	Steel, galvanized, 20 gauge	1.40	4.82	6.22
6400	30 gauge	.66	3.92	4.58
6500	Stainless, 32 gauge, .010" thick	3.69	4.04	7.73

B3020 Roof Openings

B3020 240	Selective Price Sheet	COST PER UNIT		
		MAT.	INST.	TOTAL
6600	28 gauge, .015" thick	5.15	4.04	9.19
6700	26 gauge, .018" thick	4.95	4.04	8.99
6800	24 gauge, .025" thick	5.50	4.04	9.54
6900	Gutters per L.F., aluminum, 5" box, .027" thick	3.18	6.25	9.43
7000	.032" thick	3.88	6.05	9.93
7100	Copper, half round, 4" wide	9.35	6.05	15.40
7200	6" wide	12.40	6.40	18.80
7300	Steel, 26 gauge galvanized, 5" wide	2.72	6.05	8.77
7400	6" wide	2.75	6.05	8.80
7500	Wood, treated hem-fir, 3" x 4"	12.10	6.60	18.70
7600	4" x 5"	24	6.60	30.60
7700	Reglet per L.F., aluminum, .025" thick	1.73	2.93	4.66
7800	Copper, 10 oz.	7.15	2.93	10.08
7900	Steel, galvanized, 24 gauge	1.35	2.93	4.28
8000	Stainless, .020" thick	4.24	2.93	7.17
8100	Counter flashing 12" wide per L.F., aluminum, .032" thick	2.38	5.05	7.43
8200	Aluminum, .025" thick	2.38	5.05	7.43
8300	Steel, galvanized, 24 gauge	1.38	5.05	6.43
8400	Stainless, .020" thick	6.95	5.05	12

For customer support on your Commercial Renovation Costs with RSMeans data, call 800.448.8182.

C1010 Partitions

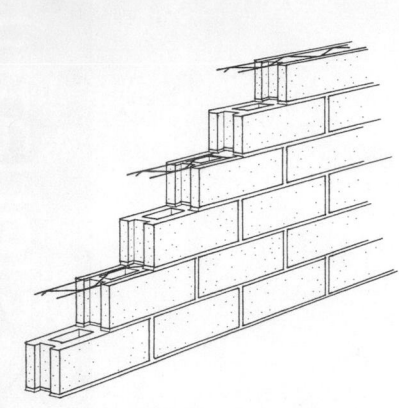

This page illustrates and describes a concrete block wall system including concrete block, horizontal reinforcing alternate courses, mortar, and tooled joints both sides. Lines within System Components give the unit price and total price per square foot for this system. Prices for concrete block wall systems are on Line Items C1010 106 0800 thru 2100. Both material quantities and labor costs have been adjusted for the system listed.

Factors: To adjust for job conditions other than normal working situations use Lines C1010 106 2900 thru 4000.

Example: You are to install the system and protect all existing work. Go to Line C1010 106 3600 and apply these percentages to the appropriate MAT. and INST. costs.

System Components	QUANTITY	UNIT	COST PER S.F.		
			MAT.	INST.	TOTAL
Concrete block partition, including horizontal reinforcing every second Course, mortar, tooled joints both sides.					
8" x 16" concrete block, normal weight, 4" thick	1.000	S.F.	2.01	6.95	8.96
Horizontal reinforcing every second course	.750	L.F.	.18	.17	.35
TOTAL			2.19	7.12	9.31

C1010 106	Partitions, Concrete Block	COST PER S.F.		
		MAT.	INST.	TOTAL
0800	8" x 16" concrete block, normal weight, 4" thick	2.19	7.10	9.29
0900	6" thick	2.82	7.65	10.47
1000	8" thick	3	8.15	11.15
1100	10" thick	3.54	8.45	11.99
1200	12" thick	4.87	10.45	15.32
1300	8" x 16" concrete block, lightweight, 4" thick	2.32	6.95	9.27
1400	6" thick	3.21	7.45	10.66
1500	8" thick	3.87	7.95	11.82
1600	10" thick	4.66	8.25	12.91
1700	12" thick	4.88	10.15	15.03
1800	8" x 16" glazed concrete block, 4" thick	12.45	8.85	21.30
1900	8" thick	14.05	9.80	23.85
2000	Structural facing tile, 6T series, glazed 2 sides, 4" thick	19.35	15.50	34.85
2100	6" thick	23	16.35	39.35
2200				
2300				
2400				
2500				
2600				
2900	Cut & patch to match existing construction, add, minimum	2%	3%	
3000	Maximum	5%	9%	
3100	Dust protection, add, minimum	1%	2%	
3200	Maximum	4%	11%	
3300	Material handling & storage limitation, add, minimum	1%	1%	
3400	Maximum	6%	7%	
3500	Protection of existing work, add, minimum	2%	2%	
3600	Maximum	5%	7%	
3700	Shift work requirements, add, minimum		5%	
3810	Maximum		30%	
3900	Temporary shoring and bracing, add, minimum	2%	5%	
4000	Maximum	5%	12%	

C1010 Partitions

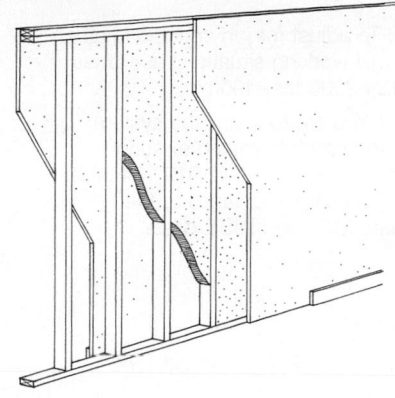

This page illustrates and describes a wood stud partition system including wood studs with plates, gypsum plasterboard – taped and finished, insulation, baseboard and painting. Lines within System Components give the unit price and total price per square foot for this system. Prices for wood stud partition systems are on Line Items C1010 132 1300 thru 2700. Both material quantities and labor costs have been adjusted for the system listed.

Factors: To adjust for job conditions other than normal working situations use Lines C1010 132 2900 thru 4000.

Example: You are to install the system where material handling and storage present a serious problem. Go to Line C1010 132 3400 and apply these percentages to the appropriate MAT. and INST. costs.

System Components	QUANTITY	UNIT	COST PER S.F. MAT.	COST PER S.F. INST.	COST PER S.F. TOTAL
Wood stud wall, 2"x4", 16" O.C., dbl. top plate, sngl bot. plate, 5/8" dwl. taped, finished and painted on 2 faces, insulation, baseboard, wall 8' high.					
Wood studs, 2" x 4", 16" O.C., 8' high	1.000	S.F.	.52	1.64	2.16
Gypsum drywall, 5/8" thick	2.000	S.F.	.78	1.32	2.10
Taping and finishing	2.000	S.F.	.10	1.32	1.42
Insulation, 3-1/2" fiberglass batts	1.000	S.F.	.52	.49	1.01
Baseboard	.200	L.F.	.32	.69	1.01
Paint baseboard, primer + 2 coats	.200	L.F.	.06	.40	.46
Painting, roller, 2 coats	2.000	S.F.	.44	1.44	1.88
TOTAL			2.74	7.30	10.04

C1010 132	Partitions, Wood Stud	COST PER S.F. MAT.	COST PER S.F. INST.	COST PER S.F. TOTAL
1300	2" x 3" studs, 8' high, 16" O.C.	2.71	7.20	9.91
1400	24" O.C.	2.60	6.90	9.50
1500	10' high, 16" O.C.	2.68	6.90	9.58
1600	24" O.C.	2.57	6.65	9.22
1650	2" x 4" studs, 8' high, 16" O.C.	2.74	7.30	10.04
1700	24" O.C.	2.62	7	9.62
1800	10' high, 16" O.C.	2.70	7	9.70
1900	24" O.C.	2.59	6.70	9.29
2000	12' high, 16" O.C.	2.66	7	9.66
2100	24" O.C.	2.55	6.70	9.25
2200	2" x 6" studs, 8' high, 16" O.C.	3.01	7.50	10.51
2300	24" O.C.	2.84	7.10	9.94
2400	10' high, 16" O.C.	2.97	7.15	10.12
2500	24" O.C.	2.79	6.80	9.59
2600	12' high, 16" O.C.	2.90	7.15	10.05
2700	24" O.C.	2.73	6.85	9.58
2900	Cut & patch to match existing construction, add, minimum	2%	3%	
3000	Maximum	5%	9%	
3100	Dust protection, add, minimum	1%	2%	
3200	Maximum	4%	11%	
3300	Material handling & storage limitation, add, minimum	1%	1%	
3400	Maximum	6%	7%	
3500	Protection of existing work, add, minimum	2%	2%	
3600	Maximum	5%	7%	
3700	Shift work requirements, add, minimum		5%	
3800	Maximum		30%	
3900	Temporary shoring and bracing, add, minimum	2%	5%	
4000	Maximum	5%	12%	

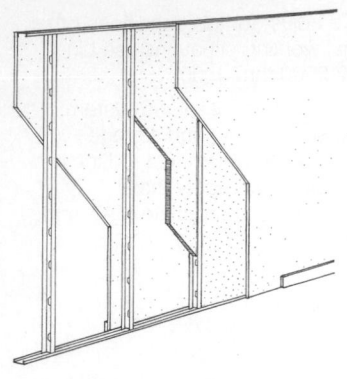

This page illustrates and describes a non-load bearing metal stud partition system including metal studs with runners, gypsum plasterboard, taped and finished, insulation, baseboard and painting. Lines within System Components give the unit price and total price per square foot for this system. Prices for non-load bearing metal stud partition systems are on Line Items C1010 134 1300 thru 2300. Both material quantities and labor costs have been adjusted for the system listed.

Factors: To adjust for job conditions other than normal working situations use Lines C1010 134 2900 thru 4000.

Example: You are to install the system and cut and patch to match existing construction. Go to Line C1010 134 3000 and apply these percentages to the appropriate MAT. and INST. costs.

System Components	QUANTITY	UNIT	COST PER S.F. MAT.	COST PER S.F. INST.	COST PER S.F. TOTAL
Non-load bearing metal studs, including top & bottom runners, 5/8" drywall, Taped, finished and painted 2 faces, insulation, painted baseboard.					
Metal studs, 25 ga., 3-5/8" wide, 24" O.C.	1.000	S.F.	.31	.89	1.20
Gypsum drywall, 5/8" thick	2.000	S.F.	.78	1.32	2.10
Taping & finishing	2.000	S.F.	.10	1.32	1.42
Insulation, 3-1/2" fiberglass batts	1.000	S.F.	.52	.49	1.01
Baseboard	.200	L.F.	.26	.55	.81
Paint baseboard, primer + 2 coats	.200	L.F.	.04	.32	.36
Painting, roller work, 2 coats	2.000	S.F.	.44	1.44	1.88
TOTAL			2.45	6.33	8.78

C1010 134	Partitions, Metal Stud, NLB	COST PER S.F. MAT.	COST PER S.F. INST.	COST PER S.F. TOTAL
1300	Non-load bearing, 25 ga., 24" O.C., 2-1/2" wide	2.41	6.30	8.71
1350	3-5/8" wide	2.45	6.35	8.80
1400	6" wide	2.55	6.35	8.90
1500	16" O.C., 2-1/2" wide	2.48	6.55	9.03
1600	3-5/8" wide	2.53	6.55	9.08
1700	6" wide	2.65	6.60	9.25
1800	20 ga., 24" O.C., 2-1/2" wide	2.47	6.55	9.02
1900	3-5/8" wide	2.60	6.55	9.15
2000	6" wide	2.69	6.55	9.24
2100	16" O.C., 2-1/2" wide	2.55	6.80	9.35
2200	3-5/8" wide	2.72	6.85	9.57
2300	6" wide	2.83	6.85	9.68
2400				
2500				
2600				
2700				
2900	Cut & patch to match existing construction, add, minimum	2%	3%	
3000	Maximum	5%	9%	
3100	Dust protection, add, minimum	1%	2%	
3200	Maximum	4%	11%	
3300	Material handling & storage limitation, add, minimum	1%	1%	
3400	Maximum	6%	7%	
3500	Protection of existing work, add, minimum	2%	2%	
3600	Maximum	5%	7%	
3700	Shift work requirements, add, minimum		5%	
3800	Maximum		30%	
3900	Temporary shoring and bracing, add, minimum	2%	5%	
4000	Maximum	5%	12%	

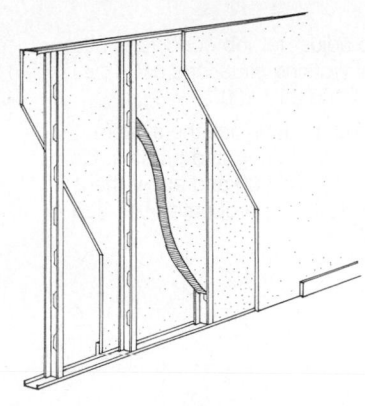

This page illustrates and describes a drywall system including gypsum plasterboard, taped and finished, metal studs with runners, insulation, baseboard and painting. Lines within System Components give the unit price and total price per square foot for this system. Prices for drywall systems are on Line Items C1010 136 1200 thru 1900. Both material quantities and labor costs have been adjusted for the system listed.

Factors: To adjust for job conditions other than normal working situations use Lines C1010 136 2900 thru 4000.

Example: You are to install the system and control dust in the work area. Go to Line C1010 136 3100 and apply these percentages to the appropriate MAT. and INST. costs.

System Components	QUANTITY	UNIT	COST PER S.F. MAT.	COST PER S.F. INST.	COST PER S.F. TOTAL
Gypsum drywall, taped, finished and painted 2 faces, galvanized metal studs					
Including top & bottom runners, insulation, painted baseboard, wall 10'high					
Gypsum drywall, 5/8″ thick, standard	2.000	S.F.	.78	1.32	2.10
Taping and finishing	2.000	S.F.	.10	1.32	1.42
Metal studs, 20 ga., 3-5/8″ wide, 24″ O.C.	1.000	S.F.	.38	1.12	1.50
Insulation, 3-1/2″ fiberglass batts	1.000	S.F.	.52	.49	1.01
Baseboard	.200	L.F.	.26	.55	.81
Paint baseboard, primer + 2 coats	.200	L.F.	.03	.14	.17
Painting, roller 2 coats	2.000	S.F.	.32	1.36	1.68
TOTAL			2.39	6.30	8.69

C1010 136	Partitions, Drywall	COST PER S.F. MAT.	COST PER S.F. INST.	COST PER S.F. TOTAL
1200	Gypsum drywall, 5/8″ thick, standard	2.39	6.30	8.69
1300	Fire resistant	2.34	6.05	8.39
1400	Water resistant	2.50	6.05	8.55
1500	1/2″ thick, standard	2.28	6.05	8.33
1600	Fire resistant	2.36	6.05	8.41
1700	Water resistant	2.46	6.05	8.51
1800	3/8″ thick, vinyl faced, standard	2.66	7.65	10.31
1900	5/8″ thick, vinyl faced, fire resistant	3.28	7.65	10.93
2000				
2100				
2200				
2300				
2400				
2500				
2600				
2700				
2900	Cut & patch to match existing construction, add, minimum	2%	3%	
3000	Maximum	5%	9%	
3100	Dust protection, add, minimum	1%	2%	
3200	Maximum	4%	11%	
3300	Material handling & storage limitation, add, minimum	1%	1%	
3400	Maximum	6%	7%	
3500	Protection of existing work, add, minimum	2%	2%	
3600	Maximum	5%	7%	
3700	Shift work requirements, add, minimum		5%	
3800	Maximum		30%	
3900	Temporary shoring and bracing, add, minimum	2%	5%	
4000	Maximum	5%	12%	

For customer support on your Commercial Renovation Costs with RSMeans data, call 800.448.8182.

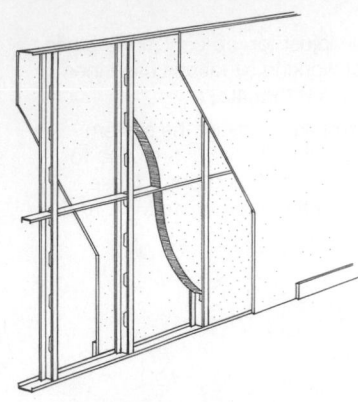

This page illustrates and describes a load bearing metal stud wall system including metal studs, sheetrock–taped and finished, insulation, baseboard and painting. Lines within System Components give the unit price and total price per square foot for this system. Prices for load bearing metal stud wall systems are on Line Items C1010 138 1500 thru 2500. Both material quantities and labor costs have been adjusted for the system listed.

Factors: To adjust for job conditions other than normal working situations use Lines C1010 138 2900 thru 4000.

Example: You are to install the system using temporary shoring and bracing. Go to Line C1010 138 3900 and apply these percentages to the appropriate MAT. and INST. costs.

System Components	QUANTITY	UNIT	COST PER S.F.		
			MAT.	INST.	TOTAL
Load bearing, 18 ga., 3-5/8″, galvanized metal studs, 24″ O.C.,including					
Top and bottom runners, 1/2″ drywall, taped, finished and painted 2					
Faces, 3″ insulation, and painted baseboard, wall 10′ high.					
Metal studs, 24″ OC, 18 ga., 3-5/8″ wide, galvanized	1.000	S.F.	.84	1.26	2.10
Gypsum drywall, 1/2″ thick	2.000	S.F.	.74	1.32	2.06
Taping and finishing	2.000	S.F.	.10	1.32	1.42
Insulation, 3-1/2″ fiberglass batts	1.000	S.F.	.52	.49	1.01
Baseboard	.200	L.F.	.26	.55	.81
Paint baseboard, primer + 2 coats	.200	L.F.	.04	.32	.36
Painting, roller 2 coats	2.000	S.F.	.44	1.44	1.88
TOTAL			2.94	6.70	9.64

C1010 138	Partitions, Metal Stud, LB	COST PER S.F.		
		MAT.	INST.	TOTAL
1500	Load bearing, 18 ga., 24″ O.C., 2-1/2″ wide	2.81	6.65	9.46
1550	3-5/8″ wide	2.94	6.70	9.64
1600	6″ wide	3.22	6.75	9.97
1700	16″ O.C. 2-1/2″ wide	2.98	7	9.98
1800	3-5/8″ wide	3.14	7	10.14
1900	6″ wide	3.49	7.05	10.54
2000	16 ga., 24″ O.C., 2-1/2″ wide	2.92	6.85	9.77
2100	3-5/8″ wide	3.08	6.85	9.93
2200	6″ wide	3.40	6.95	10.35
2300	16″ O.C., 2-1/2″ wide	3.13	7.20	10.33
2400	3-5/8″ wide	3.33	7.25	10.58
2500	6″ wide	3.72	7.30	11.02
2600				
2700				
2900	Cut & patch to match existing construction, add, minimum	2%	3%	
3000	Maximum	5%	9%	
3100	Dust protection, add, minumum	1%	2%	
3200	Maximum	4%	11%	
3300	Material handling & storage limitation, add, minimum	1%	1%	
3400	Maximum	6%	7%	
3500	Protection of existing work, add, minimum	2%	2%	
3600	Maximum	5%	7%	
3700	Shift work requirements, add, minimum		5%	
3800	Maximum		30%	
3900	Temporary shoring and bracing, add, minimum	2%	5%	
4000	Maximum	5%	12%	

C1010 Partitions

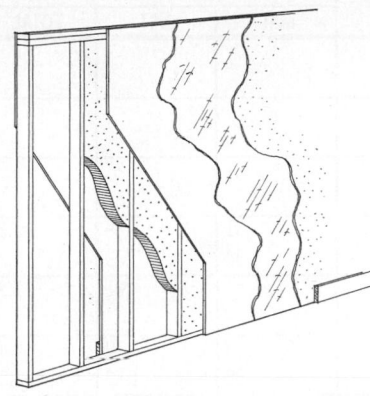

This page illustrates and describes a plaster and lath system including gypsum plaster, gypsum lath, wood studs with plates, insulation, baseboard and painting. Lines within System Components give the unit price and price per square foot for this system. Prices for plaster and lath systems are on Line Items C1010 148 1400 thru 2500. Both material quantities and labor costs have been adjusted for the system listed.

Factors: To adjust for job conditions other than normal working situations use Lines C1010 148 2900 thru 4000.

Example: You are to install the system during evening hours only. Go to Line C1010 148 3800 and apply this percentage to the appropriate INST. costs.

System Components	QUANTITY	UNIT	COST PER S.F. MAT.	COST PER S.F. INST.	COST PER S.F. TOTAL
Gypsum plaster, 2 coats, over 3/8" lath, 2 faces, 2" x 4" wood Stud partition, 24"O.C. incl. double top plate, single bottom plate, 3-1/2" Insulation, baseboard and painting.					
Gypsum plaster, 2 coats	.220	S.Y.	.91	6.14	7.05
Lath, gypsum, 3/8" thick	.220	S.Y.	.76	1.61	2.37
Wood studs, 2" x 4", 24" O.C.	1.000	S.F.	.37	1.06	1.43
Insulation, 3-1/2" fiberglass batts	1.000	S.F.	.52	.49	1.01
Baseboard, 9/16" x 3-1/2"	.200	L.F.	.26	.55	.81
Paint baseboard, primer + 2 coats	.200	L.F.	.03	.14	.17
Paint, 2 coats	2.000	S.F.	.44	1.44	1.88
TOTAL			3.29	11.43	14.72

C1010 148	Partitions, Plaster & Lath	COST PER S.F. MAT.	COST PER S.F. INST.	COST PER S.F. TOTAL
1400	Gypsum plaster, 2 coats	3.29	11.40	14.69
1500	3 coats	3.68	12.70	16.38
1600	Perlite plaster, 2 coats	3.82	12.35	16.17
1700	3 coats	3.93	14.10	18.03
1800				
1900				
2000	For alternate lath systems:			
2100	Gypsum lath, 1/2" thick	3.67	12.45	16.12
2200	Foil back, 3/8" thick	3.84	12.55	16.39
2300	1/2" thick	3.89	12.70	16.59
2400	Metal lath, 2.5 Lb. diamond	4.07	12.35	16.42
2500	3.4 Lb. diamond	4.16	12.55	16.71
2600				
2700				
2900	Cut & patch to match existing construction, add, minimum	2%	3%	
3000	Maximum	5%	9%	
3100	Dust protection, add, minimum	1%	2%	
3200	Maximum	4%	11%	
3300	Material handling & storage limitation, add, minimum	1%	1%	
3400	Maximum	6%	7%	
3500	Protection of existing work, add, minimum	2%	2%	
3600	Maximum	5%	7%	
3700	Shift work requirements, add, minimum		5%	
3800	Maximum		30%	
3900	Temporary shoring and bracing, add, minimum	2%	5%	
4000	Maximum	5%	12%	

C1010 170	Selective Price Sheet	COST PER S.F.		
		MAT.	INST.	TOTAL
0100	Studs			
0200				
0300	24" O.C. metal, 10' high wall, including			
0400	Top and bottom runners			
0500	Non load bearing, galvanized 25 Ga., 1-5/8" wide	.21	.87	1.08
0600	2-1/2" wide	.27	.88	1.15
0700	3-5/8" wide	.31	.89	1.20
0800	4" wide	.34	.89	1.23
0900	6" wide	.41	.91	1.32
1000				
1100	Galvanized 20 Ga., 2-1/2" wide	.33	1.10	1.43
1200	3-5/8" wide	.38	1.12	1.50
1300	4" wide	.46	1.12	1.58
1400	6" wide	.55	1.13	1.68
1500	Load bearing, painted 18 Ga., 2-1/2" wide	.71	1.23	1.94
1600	3-5/8" wide	.84	1.26	2.10
1700	4" wide	.75	1.28	2.03
1800	6" wide	1.12	1.31	2.43
1900	Galvanized 18 Ga., 2-1/2" wide	.71	1.23	1.94
2000	3-5/8" wide	.84	1.26	2.10
2100	4" wide	.75	1.28	2.03
2200	6" wide	1.12	1.31	2.43
2300	Galvanized 16 Ga., 2-1/2" wide	.82	1.40	2.22
2400	3-5/8" wide	.98	1.43	2.41
2500	4" wide	1.03	1.47	2.50
2600	6" wide	1.30	1.50	2.80
2700				
2800				
2900	24" O.C. wood, 10' high wall, including			
3000	Double top plate and shoe			
3100	2" x 4"	.43	1.24	1.67
3200	2" x 6"	.66	1.33	1.99
3300				
3400				
3500	Furring, 24" O.C. 10' high wall, metal, 3/4" channels	.26	1.48	1.74
3600	1-1/2" channels	.35	1.59	1.94
3700	Wood, on wood, 1" x 2" strips	.15	.60	.75
3800	1" x 3" strips	.24	.60	.84
3900	On masonry, 1" x 2" strips	.16	.67	.83
4000	1" x 3" strips	.26	.67	.93
4100	On concrete, 1" x 2" strips	.16	1.27	1.43
4200	1" x 3" strips	.26	1.27	1.53
4300	Studs			
4400				
4500	16" O.C., metal, 10' high wall, including			
4600	Top and bottom runners			
4700	Non load bearing, galvanized 25 Ga., 1-5/8" wide	.28	1.33	1.61
4800	2-1/2" wide	.37	1.34	1.71
4900	3-5/8" wide	.42	1.37	1.79
5000	4" wide	.47	1.39	1.86
5100	6" wide	.55	1.40	1.95
5200				
5300	Galvanized 20 Ga., 2-1/2" wide	.45	1.69	2.14
5400	3-5/8" wide	.52	1.71	2.23
5500	4" wide	.63	1.73	2.36
5600	6" wide	.74	1.76	2.50
5700	Load bearing, painted 18 Ga., 2-1/2" wide	.96	1.71	2.67
5800	3-5/8" wide	1.14	1.74	2.88

C1010 Partitions

C1010 170	Selective Price Sheet	COST PER S.F.		
		MAT.	INST.	TOTAL
5900	4" wide	1.18	1.78	2.96
6000	6" wide	1.51	1.81	3.32
6100	Galvanized 18 Ga., 2-1/2" wide	.96	1.71	2.67
6200	3-5/8" wide	1.14	1.74	2.88
6300	4" wide	1.18	1.78	2.96
6400	6" wide	1.51	1.81	3.32
6500	Galvanized 16 Ga., 2-1/2" wide	1.13	1.94	3.07
6600	3-5/8" wide	1.35	2	3.35
6700	4" wide	1.41	2.05	3.46
6800	6" wide	1.78	2.05	3.83
6900				
7000				
7100	16" O.C., wood, 10' high wall, including			
7200	Double top plate and shoe			
7300	2" x 4"	.53	1.43	1.96
7400	2" x 6"	.66	1.29	1.95
7500				
7600				
7700	Furring, 16" O.C. 10' high wall, metal, 3/4" channels	.39	2.14	2.53
7800	1-1/2" channels	.52	2.39	2.91
7900	Wood, on wood, 1" x 2" strips	.22	.90	1.12
8000	1" x 3" strips	.35	.90	1.25
8100	On masonry, 1" x 2" strips	.24	1	1.24
8200	1" x 3" strips	.38	1	1.38
8300	On concrete, 1" x 2" strips	.24	1.90	2.14
8400	1" x 3" strips	.38	1.90	2.28

C1010 Partitions

C1010 180	Selective Price Sheet, Drywall & Plaster	COST PER S.F.		
		MAT.	INST.	TOTAL
0100	Lath, gypsum perforated			
0200				
0300	Lath, gypsum, perforated, regular, 3/8" thick	.39	.81	1.20
0400	1/2" thick	.31	.98	1.29
0500	Fire resistant, 3/8" thick	.31	.98	1.29
0600	1/2" thick	.35	1.06	1.41
0700	Foil back, 3/8" thick	.40	.92	1.32
0800	1/2" thick	.42	.98	1.40
0900				
1000	Metal lath			
1100	Metal lath, diamond, painted, 2.5 lb.	.51	.81	1.32
1200	3.4 lb.	.55	.92	1.47
1300	Rib painted, 2.75 lb	.44	.92	1.36
1400	3.40 lb	.58	.98	1.56
1500				
1600				
1700	Plaster, gypsum, 2 coats	.46	3.09	3.55
1800	3 coats	.65	3.74	4.39
1900	Perlite/vermiculite, 2 coats	.73	3.57	4.30
2000	3 coats	.78	4.44	5.22
2100	Bondcrete, 1 coat	.53	1.63	2.16
2200				
2500				
2600				
2700	Drywall, standard, 3/8" thick, no finish included	.40	.66	1.06
2800	1/2" thick, no finish included	.37	.66	1.03
2900	Taped and finished	.43	1.36	1.79
3000	5/8" thick, no finish included	.39	.66	1.05
3100	Taped and finished	.44	1.36	1.80
3200	Fire resistant, 1/2" thick, no finish included	.41	.66	1.07
3300	Taped and finished	.46	1.36	1.82
3400	5/8" thick, no finish included	.40	.66	1.06
3500	Taped and finished	.45	1.36	1.81
3600	Water resistant, 1/2" thick, no finish included	.46	.66	1.12
3700	Taped and finished	.51	1.36	1.87
3800	5/8" thick, no finish included	.48	.66	1.14
3900	Taped and finished	.54	1.36	1.90
4000	Finish, instead of taping			
4100	For thin coat plaster, add	.12	.82	.94
4200	Finish, textured spray, add	.04	.78	.82

C1010 Partitions

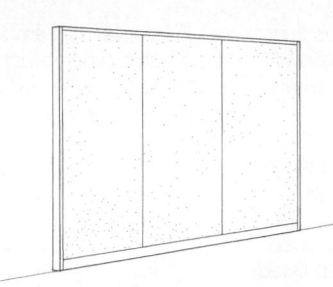

This page illustrates and describes a movable office partition system including demountable office partitions of various styles and sizes based on a cost per square foot basis. Prices for movable office partition systems are on Line Items C1010 210 0300 thru 3600.

C1010 210	Partitions, Movable Office	COST PER S.F.		
		MAT.	INST.	TOTAL
0250	Office partition, demountable, no deduction for door opening, add for doors			
0300	Air wall, cork finish, semi acoustic, 1-5/8" thick, minimum	42.50	4.05	46.55
0400	Maximum	47	6.95	53.95
0500	Acoustic, 2" thick, minimum	40	4.32	44.32
0600	Maximum	61.50	5.85	67.35
0700				
0800				
0900	In-plant modular office system, w/prehung steel door			
1000	3" thick honeycomb core panels			
1100	12' x 12', 2 wall	10.60	.54	11.14
1200	4 wall	25	1.64	26.64
1300				
1400				
1500	Gypsum, demountable, 3" to 3-3/4" thick x 9' high, vinyl clad	7.35	3.05	10.40
1600	Fabric clad	18.30	3.33	21.63
1700	1.75 system, vinyl clad hardboard, paper honeycomb core panel			
1800	1-3/4" to 2-1/2" thick x 9' high	12.30	3.05	15.35
1900	Unitized gypsum panel, 2" to 2-1/2" thick x 9' high, vinyl clad	15.85	3.05	18.90
2000	Fabric clad	26	3.33	29.33
2100				
2200				
2300	Unitized mineral fiber panel system, 2-1/4" thick x 9' high			
2400	Vinyl clad mineral fiber	15.75	3.05	18.80
2500	Fabric clad mineral fiber	23.50	3.33	26.83
2600				
2700	Movable steel walls, modular system			
2800	Unitized panels, 48" wide x 9' high			
2900	Baked enamel, pre-finished	17.85	2.44	20.29
3000	Fabric clad	26	2.61	28.61
3100	For acoustical partitions, add, unsealed	2.60		2.60
3200	Sealed	12.10		12.10
3300				
3400				
3500	Note: For door prices, see Div. 08			
3600	For door hardware prices, see Div. 08			
3700				
3800				
3900				
4000				

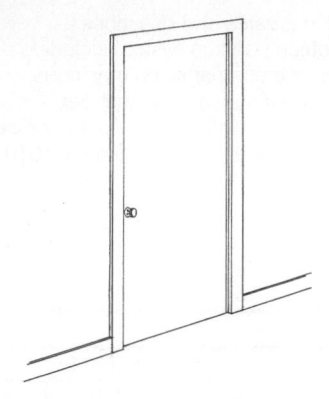

This page illustrates and describes flush interior door systems including hollow core door, jamb, header and trim with hardware. Lines within System Components give the unit price and total price on a cost each basis for this system. Prices for flush interior door systems are on Line items C1020 106 1000 thru 2400. Both material quantities and labor costs have been adjusted for the system listed.

Factors: To adjust for job conditions other than normal working situations use Lines C1020 106 2700 thru 3800.

Example: You are to install the system in an area where dust protection is vital. Go to Line C1020 106 3000 and apply these percentages to the appropriate MAT. and INST. costs.

System Components	QUANTITY	UNIT	COST EACH		
			MAT.	INST.	TOTAL
Single hollow core door, include jamb, header, trim and hardware, painted.					
Hollow core Lauan, 1-3/8" thick, 2'-0" x 6'-8", painted	1.000	Ea.	66	73	139
Paint door and frame, 1 coat	1.000	Ea.	2.59	32	34.59
Wood jamb, 4-9/16" deep	1.000	Set	84.80	56.16	140.96
Trim, casing	1.000	Set	49.92	84.16	134.08
Hardware, hinges	1.000	Set	25.50		25.50
Hardware, lockset	1.000	Set	41	41	82
TOTAL			269.81	286.32	556.13

C1020 106	Doors, Interior Flush, Wood	COST EACH		
		MAT.	INST.	TOTAL
1000	Lauan (Mahogany) hollow core, 1-3/8" x 2'-0" x 6'-8"	270	286	556
1100	2'-6" x 6'-8"	283	291	574
1200	2'-8" x 6'-8"	288	294	582
1300	3'-0" x 6'-8"	298	305	603
1500	Birch, hollow core, 1-3/8" x 2'-0" x 6'-8"	280	286	566
1600	2'-6" x 6'-8"	295	291	586
1700	2'-8" x 6'-8"	300	294	594
1800	3'-0" x 6'-8"	310	305	615
1900	Solid core, pre-hung, 1-3/8" x 2'-6" x 6'-8"	535	295	830
2000	2'-8" x 6'-8"	440	299	739
2100	3'-0" x 6'-8"	570	310	880
2200				
2400	For metal frame instead of wood, add	50%	20%	
2600				
2700	Cut & patch to match existing construction, add, minimum	2%	3%	
2800	Maximum	5%	9%	
2900	Dust protection, add, minimum	1%	2%	
3000	Maximum	4%	11%	
3100	Equipment usage curtailment, add, minimum	1%	1%	
3200	Maximum	3%	10%	
3300	Material handling & storage limitation, add, minimum	1%	1%	
3400	Maximum	6%	7%	
3500	Protection of existing work, add, minimum	2%	2%	
3600	Maximum	5%	7%	
3700	Shift work requirements, add, minimum		5%	
3800	Maximum		30%	
3900				

This page illustrates and describes interior, solid and louvered door systems including a pine panel door, wood jambs, header, and trim with hardware. Lines within System Components give the unit price and total price on a cost each basis for this system. Prices for interior, solid and louvered systems are on Line Items C1020 108 1100 thru 2400. Both material quantities and labor costs have been adjusted for the system listed.

Factors: To adjust for job conditions other than normal working situations use Lines C1020 108 2900 thru 4000.

Example: You are to install the system during night hours only. Go to Line C1020 108 4000 and apply these percentages to the appropriate INST. costs.

System Components	QUANTITY	UNIT	COST EACH MAT.	COST EACH INST.	COST EACH TOTAL
Single interior door, including jamb, header, trim and hardware, painted.					
Solid pine panel door, 1-3/8" thick, 2'-0" x 6'-8"	1.000	Ea.	245	73	318
Paint door and frame, 1 coat	1.000	Ea.	2.59	32	34.59
Wooden jamb, 4-5/8" deep	1.000	Set	84.80	56.16	140.96
Trim, casing	1.000	Set	49.92	84.16	134.08
Hardware, hinges	1.000	Set	25.50		25.50
Hardware, lockset	1.000	Set	41	41	82
TOTAL			448.81	286.32	735.13

C1020 108	Doors, Interior Solid & Louvered	COST EACH MAT.	COST EACH INST.	COST EACH TOTAL
1000				
1100	Solid pine, painted raised panel, 1-3/8" x 2'-0" x 6'-8"	450	286	736
1200	2'-6" x 6'-8"	485	291	776
1300	2'-8" x 6'-8"	495	294	789
1400	3'-0" x 6'-8"	515	305	820
1500				
1600	Louvered pine, painted 1'-6" x 6'-8"	365	283	648
1700	2'-0" x 6'-8"	390	291	681
1800	2'-6" x 6'-8"	430	294	724
1900	3'-0" x 6'-8"	455	305	760
2200	For prehung door, deduct	5%	30%	
2300				
2400	For metal frame instead of wood, add	50%	20%	
2500				
2900	Cut & patch to match existing construction, add, minimum	2%	3%	
3000	Maximum	5%	9%	
3100	Dust protection, add, minimum	1%	2%	
3200	Maximum	4%	11%	
3300	Equipment usage curtailment, add, minimum	1%	1%	
3400	Maximum	3%	10%	
3500	Material handling & storage limitation, add, minimum	1%	1%	
3600	Maximum	6%	7%	
3700	Protection of existing work, add, minimum	2%	2%	
3800	Maximum	5%	7%	
3900	Shift work requirements, add, minimum		5%	
4000	Maximum		30%	

C1020 Interior Doors

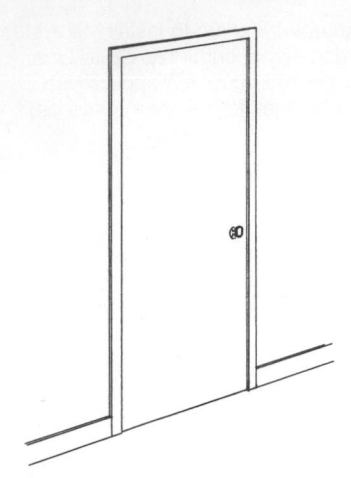

This page illustrates and describes interior metal door systems including a metal door, metal frame and hardware. Lines within System Components give the unit price and total price on a cost each for this system. Prices for interior metal door systems are on Line Items C1020 110 1000 thru 2000. Both material quantities and labor costs have been adjusted for the system listed.

Factors: To adjust for job conditions other than normal working situations use Lines C1020 110 2900 thru 4000.

Example: You are to install the system while protecting existing construction. Go to Line C1020 110 3700 and apply these percentages to the appropriate MAT. and INST. costs.

System Components	QUANTITY	UNIT	COST EACH MAT.	COST EACH INST.	COST EACH TOTAL
Single metal door, including frame and hardware.					
Hollow metal door, 1-3/8" thick, 2'-6" x 6'-8", painted	1.000	Ea.	350	88	438
Metal frame, 6-1/4" deep	1.000	Set	187	82.50	269.50
Paint door and frame, 1 coat	1.000	Ea.	2.59	32	34.59
Hardware, hinges	1.000	Set	78.50		78.50
Hardware, passage lockset	1.000	Set	41	41	82
TOTAL			659.09	243.50	902.59

C1020 110	Doors, Interior Flush, Metal	COST EACH MAT.	COST EACH INST.	COST EACH TOTAL
1000	Hollow metal doors, 1-3/8" thick, 2'-6" x 6'-8"	660	244	904
1100	1-3/8" thick, 2'-8" x 6'-8"	665	244	909
1200	3'-0" x 7'-0"	655	255	910
1300				
1400	Interior fire door, 1-3/8" thick, 2'-6" x 6'-8"	1,200	244	1,444
1500	2'-8" x 6'-8"	1,025	244	1,269
1600	3'-0" x 7'-0"	1,100	255	1,355
1700				
1800	Add to fire doors:			
1900	Baked enamel finish	30%	15%	
2000	Galvanizing	20%		
2200				
2300				
2400				
2900	Cut & patch to match existing construction, add, minimum	2%	3%	
3000	Maximum	5%	9%	
3100	Dust protection, add, minimum	1%	2%	
3200	Maximum	4%	11%	
3300	Equipment usage curtailment, add, minimum	1%	1%	
3400	Maximum	3%	10%	
3500	Material handling & storage limitation, add, minimum	1%	1%	
3600	Maximum	6%	7%	
3700	Protection of existing work, add, minimum	2%	2%	
3800	Maximum	5%	7%	
3900	Shift work requirements, add, minimum		5%	
4000	Maximum		30%	

C1020 Interior Doors

This page illustrates and describes an interior closet door system including an interior closet door, painted, with trim and hardware. Prices for interior closet door systems are on Line Items C1020 112 0500 thru 2200. Both material quantities and labor costs have been adjusted for the system listed.

Factors: To adjust for job conditions other than normal working situations use Lines C1020 112 2900 thru 4000.

Example: You are to install the system and match the existing construction. Go to Line C1020 112 2900 and apply these percentages to the appropriate MAT. and INST. costs.

C1020 112	Doors, Closet	COST PER SET		
		MAT.	INST.	TOTAL
0350	Interior closet door painted, including frame, trim and hardware, prehung.			
0400	Bi-fold doors			
0500	Pine paneled, 3'-0" x 6'-8"	425	237	662
0600	6'-0" x 6'-8"	620	320	940
0700	Birch, hollow core, 3'-0" x 6'-8"	205	258	463
0800	6'-0" x 6'-8"	305	345	650
0900	Lauan, hollow core, 3'-0" x 6'-8"	189	237	426
1000	6'-0" x 6'-8"	276	320	596
1100	Louvered pine, 3'-0" x 6'-8"	370	237	607
1200	6'-0" x 6'-8"	470	320	790
1300				
1400	Sliding, bi-passing closet doors			
1500	Pine paneled, 4'-0" x 6'-8"	655	253	908
1600	6'-0" x 6'-8"	1,050	320	1,370
1700	Birch, hollow core, 4'-0" x 6'-8"	395	253	648
1800	6'-0" x 6'-8"	500	320	820
1900	Lauan, hollow core, 4'-0" x 6'-8"	297	253	550
2000	6'-0" x 6'-8"	315	320	635
2100	Louvered pine, 4'-0" x 6'-8"	645	253	898
2200	6'-0" x 6'-8"	920	320	1,240
2300				
2400				
2500				
2600				
2700				
2800				
2900	Cut & patch to match existing construction, add, minimum	2%	3%	
3000	Maximum	5%	9%	
3100	Dust protection, add, minimum	1%	2%	
3200	Maximum	4%	11%	
3300	Equipment usage curtailment, add, minimum	1%	1%	
3400	Maximum	3%	10%	
3500	Material handling & storage limitation, add, minimum	1%	1%	
3600	Maximum	6%	7%	
3700	Protection of existing work, add, minimum	2%	2%	
3800	Maximum	5%	7%	
3900	Shift work requirements, add, minimum		5%	
4000	Maximum		30%	

C2010 Stair Construction

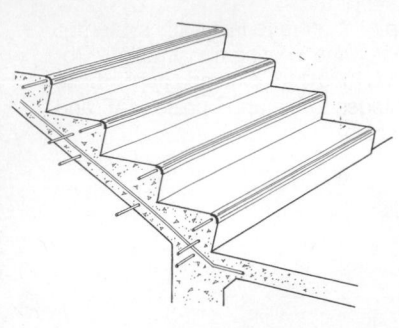

This page illustrates and describes a stair system based on a cost per flight price. Prices for various stair systems are on Line Items C2010 130 0700 thru 3200. Both material quantities and labor costs have been adjusted for the system listed.

Factors: To adjust for job conditions other than normal working situations use Lines C2010 130 3500 thru 4200.

Example: You are to install the system during evenings only. Go to Line C2010 130 4200 and apply this percentage to the appropriate MAT. and INST. costs.

System Components	QUANTITY	UNIT	COST PER FLIGHT		
			MAT.	INST.	TOTAL
Below are various stair systems based on cost per flight of stairs, no side Walls. Stairs are 4'-0" wide, railings are included unless otherwise noted.					

C2010 130	Stairs	COST PER FLIGHT		
		MAT.	INST.	TOTAL
0700	Concrete, cast in place, no nosings, no railings, 12 risers	320	2,225	2,545
0800	24 risers	640	4,450	5,090
0900	Add for 1 intermediate landing	170	615	785
1000	Concrete, cast in place, with nosings, no railings, 12 risers	1,225	2,475	3,700
1100	24 risers	2,475	4,900	7,375
1200	Add for 1 intermediate landing	246	635	881
1300	Steel, grating tread, safety nosing, 12 risers	5,275	1,175	6,450
1400	24 risers	10,600	2,350	12,950
1500	Add for intermediate landing	1,300	310	1,610
1600	Steel, metal pan tread for concrete in-fill, 12 risers	8,125	1,175	9,300
1700	24 risers	16,200	2,350	18,550
1800	Add for intermediate landing	1,300	310	1,610
1900	Spiral, industrial, 4'-6" diameter, 12 risers	6,475	835	7,310
2000	24 risers	13,000	1,650	14,650
2100	Wood, box stairs, oak treads, 12 risers	2,550	750	3,300
2200	24 risers	5,075	1,500	6,575
2300	Add for 1 intermediate landing	214	168	382
2400	Wood, basement stairs, no risers, 12 steps	845	305	1,150
2500	24 steps	1,700	610	2,310
2600	Add for 1 intermediate landing	47.50	32	79.50
2700	Wood, open, rough sawn cedar, 12 steps	2,850	605	3,455
2800	24 steps	5,725	1,200	6,925
2900	Add for 1 intermediate landing	24	33	57
3000	Wood, residential, oak treads, 12 risers	2,775	3,300	6,075
3100	24 risers	5,550	6,600	12,150
3200	Add for 1 intermediate landing	190	152	342
3500	Dust protection, add, minimum	1%	2%	
3600	Maximum	4%	11%	
3700	Material handling & storage limitation, add, minimum	1%	1%	
3800	Maximum	6%	7%	
3900	Protection of existing work, add, minimum	2%	2%	
4000	Maximum	5%	7%	
4100	Shift work requirements, add, minimum		5%	
4200	Maximum		30%	

C3010 210	Selective Price Sheet	COST PER S.F.		
		MAT.	INST.	TOTAL
0100	Painting, on plaster or drywall, brushwork, primer and 1 ct.	.16	.80	.96
0200	Primer and 2 ct.	.23	1.28	1.51
0300	Rollerwork, primer and 1 ct.	.16	.68	.84
0400	Primer and 2 ct.	.23	1.09	1.32
0500	Woodwork incl. puttying, brushwork, primer and 1 ct.	.14	1.21	1.35
0600	Primer and 2 ct.	.22	1.61	1.83
0700	Wood trim to 6" wide, enamel, primer and 1 ct.	.15	.68	.83
0800	Primer and 2 ct.	.23	.87	1.10
0900	Cabinets and casework, enamel, primer and 1 ct.	.15	1.37	1.52
1000	Primer and 2 ct.	.23	1.68	1.91
1100	On masonry or concrete, latex, brushwork, primer and 1 ct.	.32	1.14	1.46
1200	Primer and 2 ct.	.42	1.63	2.05
1300	For block filler, add	.20	1.40	1.60
1400				
1500	Varnish, wood trim, sealer 1 ct., sanding, puttying, quality work	.14	2.47	2.61
1600	Medium work	.11	1.92	2.03
1700	Without sanding	.17	.29	.46
1800				
1900	Wall coverings, wall paper, at low price per double roll, avg workmanship	.67	.85	1.52
2000	At medium price per double roll, average workmanship	1.31	1.02	2.33
2100	At high price per double roll, quality workmanship	2.32	1.26	3.58
2200				
2300	Grass cloths with lining paper	1.52	1.37	2.89
2400	Premium texture/color	3.47	1.56	5.03
2500	Vinyl, fabric backed, light weight	1.18	.85	2.03
2600	Medium weight	1	1.14	2.14
2700	Heavy weight	1.54	1.26	2.80
2800				
2900	Cork tiles, 12" x 12", 3/16" thick	4.69	2.28	6.97
3000	5/16" thick	3.66	2.33	5.99
3100	Granular surface, 12" x 36", 1/2" thick	1.43	1.42	2.85
3200	1" thick	1.83	1.48	3.31
3300	Aluminum foil	1.13	1.99	3.12
3400				
3500	Tile, ceramic, adhesive set, 4-1/4" x 4-1/4"	2.87	5.60	8.47
3600	6" x 6"	3.93	6.05	9.98
3700	Decorated, 4-1/4" x 4-1/4", minimum	4.05	3.93	7.98
3800	Color group 4	57.50	5.90	63.40
3900	For epoxy grout, add	.45	1.33	1.78
4000	Pregrouted sheets	6.20	4.43	10.63
4100	Glass mosaics, 3/4" tile on 12" sheets, minimum	22	14.55	36.55
4200	Color group 8	46	16.60	62.60
4300	Metal, tile pattern, 4' x 4' sheet, 24 ga., nailed			
4400	Stainless steel	31	2.57	33.57
4500	Aluminized steel	21	2.57	23.57
4600				
4700	Brick, interior veneer, 4" face brick, running bond, minimum	4.39	13.95	18.34
4800	Maximum	4.39	13.95	18.34
4900	Simulated, urethane pieces, set in mastic	9.45	4.39	13.84
5000	Fiberglass panels	11	3.29	14.29
5100	Wall coating, on drywall, thin coat, plain	.12	.82	.94
5200	Stipple	.12	.82	.94
5300	Textured spray	.04	.78	.82
5400				
5500	Paneling not incl. furring or trim, hardboard, tempered, 1/8" thick	.50	2.63	3.13
5600	1/4" thick	.70	2.63	3.33
5700	Plastic faced, 1/8" thick	.68	2.63	3.31
5800	1/4" thick	.92	2.63	3.55

C3010 Wall Finishes

C3010 210	Selective Price Sheet	COST PER S.F.		
		MAT.	INST.	TOTAL
5900	Woodgrained, 1/4" thick, minimum	.72	2.63	3.35
6000	Maximum	1.56	3.10	4.66
6100	Plywood, 4' x 8' shts. 1/4" thick, prefin., birch faced, min.	1.62	2.63	4.25
6200	Maximum	1.21	3.76	4.97
6300	Walnut, minimum	2.72	2.63	5.35
6400	Maximum	2.71	3.29	6
6500	Mahogany, African	2.70	3.29	5.99
6600	Philippine	.72	2.63	3.35
6700	Chestnut	5.80	3.51	9.31
6800	Pecan	2.75	3.29	6.04
6900	Rosewood	3.41	4.12	7.53
7000	Teak	3.43	3.29	6.72
7100	Aromatic cedar, plywood	2.48	3.29	5.77
7200	Particle board	1.27	3.29	4.56
7300	Wood board, 3/4" thick, knotty pine	2.26	4.39	6.65
7400	Rough sawn cedar	3.63	4.39	8.02
7500	Redwood, clear	5.45	4.39	9.84
7600	Aromatic cedar	2.66	4.79	7.45

For customer support on your Commercial Renovation Costs with RSMeans data, call 800.448.8182.

C30 Interior Finishes

C3020 Floor Finishes

C3020 430	Selective Price Sheet	COST PER S.F.		
		MAT.	INST.	TOTAL
0100	Flooring, carpet, acrylic, 26 oz. light traffic	2.66	.88	3.54
0200	35 oz. heavy traffic	5.95	.88	6.83
0300	Nylon anti-static, 15 oz. light traffic	2.50	.88	3.38
0400	22 oz. medium traffic	4.44	.88	5.32
0500	26 oz. heavy traffic	6.45	.94	7.39
0600	28 oz. heavy traffic	7.20	.94	8.14
0700	Tile, foamed back, needle punch	4.52	1.04	5.56
0800	Tufted loop	3.62	1.04	4.66
0900	Wool, 36 oz. medium traffic	12	.94	12.94
1000	42 oz. heavy traffic	13.30	.94	14.24
1100	Composition, epoxy, with colored chips, minimum	3.39	4.84	8.23
1200	Maximum	4.88	6.65	11.53
1300	Trowelled, minimum	3.93	5.85	9.78
1400	Maximum	6.50	6.80	13.30
1500	Terrazzo, 1/4" thick, chemical resistant, minimum	7	7.15	14.15
1600	Maximum	10.65	9.55	20.20
1700	Resilient, asphalt tile, 1/8" thick	1.57	1.48	3.05
1800	Conductive flooring, rubber, 1/8" thick	7.75	1.88	9.63
1900	Cork tile 1/8" thick, standard finish	5.75	1.88	7.63
2000	Urethane finish	5.80	1.88	7.68
2100	PVC sheet goods for gyms, 1/4" thick	7.55	7.40	14.95
2200	3/8" thick	11.05	9.85	20.90
2300	Vinyl composition 12" x 12" tile, plain, 1/16" thick	1.34	1.18	2.52
2400	1/8" thick	1.95	1.18	3.13
2500	Vinyl tile, 12" x 12" x 1/8" thick, minimum	7.70	1.18	8.88
2600	Premium colors/pattern	6.85	1.18	8.03
2700	Vinyl sheet goods, backed, .080" thick, plain pattern/colors	4.64	2.57	7.21
2900	Slate, random rectangular, 1/4" thick	24.50	7.80	32.30
3000	1/2" thick	22.50	11.15	33.65
3100	Natural cleft, irregular, 3/4" thick	10.40	12.70	23.10
3200	For sand rubbed finish, add	10.40		10.40
3300	Terrazzo, cast in place, bonded 1-3/4" thick, gray cement	3.93	19.05	22.98
3400	White cement	4.58	19.05	23.63
3500	Not bonded 3" thick, gray cement	4.84	20.50	25.34
3600	White cement	5.90	20.50	26.40
3700	Precast, 12" x 12" x 1" thick	27	19.50	46.50
3800	1-1/4" thick	28	19.50	47.50
3900	16" x 16" x 1-1/4" thick	30.50	23.50	54
4000	1-1/2" thick	28	26	54
4100	Marble travertine, standard, 12" x 12" x 3/4" thick	9.65	17.70	27.35
4200				
4300	Tile, ceramic, natural clay, thin set	4.74	5.80	10.54
4400	Porcelain, thin set	6.90	5.60	12.50
4500	Specialty, decorator finish	14	5.80	19.80
4600				
4700	Quarry, red, mud set, 4" x 4" x 1/2" thick	9.60	8.85	18.45
4800	6" x 6" x 1/2" thick	9.10	7.60	16.70
4900	Brown, imported, 6" x 6" x 3/4" thick	7.90	8.85	16.75
5000	8" x 8" x 1" thick	10.50	9.65	20.15
5100	Slate, Vermont, thin set, 6" x 6" x 1/4" thick	8.45	5.90	14.35
5200				
5300	Wood, maple strip, 25/32" x 2-1/4", finished, select	5.55	4.16	9.71
5400	2nd and better	5.45	4.16	9.61
5500	Oak, 25/32" x 2-1/4" finished, clear	3.93	4.16	8.09
5600	No. 1 common	4.82	4.16	8.98
5700	Parquet, standard, 5/16", finished, minimum	6.15	4.41	10.56
5800	Maximum	11	6.90	17.90
5900	Custom, finished, plain pattern	18.85	6.60	25.45

C3020 430	Selective Price Sheet	COST PER S.F.		
		MAT.	INST.	TOTAL
6000	Intricate pattern	27.50	13.15	40.65
6100	Prefinished, oak, 2-1/4" wide	5.75	3.87	9.62
6200	Ranch plank	8	4.54	12.54
6300	Sleepers on concrete, treated, 24" O.C., 1" x 2"	.26	.44	.70
6400	1" x 3"	.46	.56	1.02
6500	2" x 4"	.61	.78	1.39
6600	2" x 6"	.79	.89	1.68
6700	Refinish old floors, soft wood	.23	1.29	1.52
6800	Hard wood	.23	3.98	4.21
6900	Subflooring, plywood, CDX, 1/2" thick	.68	.88	1.56
7000	5/8" thick	.84	.98	1.82
7100	3/4" thick	1.02	1.05	2.07
7200				
7300	1" x 10" boards, S4S, laid regular	2.33	1.20	3.53
7400	Laid diagonal	2.33	1.46	3.79
7500	1" x 8" boards, S4S, laid regular	2.28	1.32	3.60
7600	Laid diagonal	2.28	1.55	3.83
7700	Underlayment, plywood, underlayment grade, 3/8" thick	1.14	.88	2.02
7800	1/2" thick	1.35	.91	2.26
7900	5/8" thick	1.50	.94	2.44
8000	3/4" thick	1.62	1.01	2.63
8100	Particle board, 3/8" thick	.45	.88	1.33
8200	1/2" thick	.47	.91	1.38
8300	5/8" thick	.61	.94	1.55
8400	3/4" thick	.74	1.01	1.75

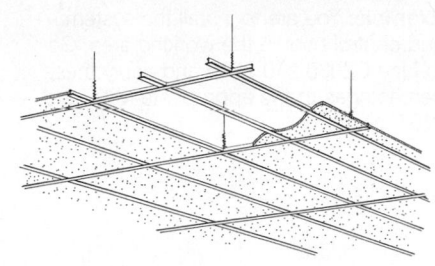

This page illustrates suspended acoustical board systems including acoustic ceiling board, hangers, and T bar suspension. Lines within System Components give the unit price and total price per square foot for this system. Prices for suspended acoustical board systems are on Line Items C3030 210 1000 thru 2400. Both material quantities and labor costs have been adjusted for the system listed.

Factors: To adjust for job conditions other than normal working situations use Lines C3030 210 2900 thru 4000.

Example: You are to install the system and protect existing construction. Go to Line C3030 210 3800 and apply these percentages to the appropriate MAT. and INST. costs.

System Components			COST PER S.F.		
	QUANTITY	UNIT	MAT.	INST.	TOTAL
Suspended acoustical ceiling board installed on exposed grid system.					
Fiberglass boards, film faced, 2' x 4', 5/8" thick	1.000	S.F.	1.39	1.05	2.44
Hangers, #12 wire	1.000	S.F.	.03	.63	.66
T bar suspension system, 2' x 4' grid	1.000	S.F.	.87	.82	1.69
TOTAL			2.29	2.50	4.79

C3030 210	Ceiling, Suspended Acoustical	COST PER S.F.		
		MAT.	INST.	TOTAL
1000	2' x 4' grid, fiberglass board, film faced, 5/8" thick	2.29	2.50	4.79
1100	Mineral fiber board, aluminum faced	5.20	2.46	7.66
1200	Standard faced	1.86	2.43	4.29
1300	Plastic faced	3.78	3.10	6.88
1400	Fiberglass, film faced, 3" thick, R11	4.75	2.91	7.66
1500	Grass cloth faced, 3/4" thick	4.19	2.77	6.96
1600	1" thick	4.87	2.81	7.68
1700	1-1/2" thick, nubby face	3.90	2.84	6.74
2200				
2300				
2400	Add for 2' x 2' grid system	.22	.20	.42
2500				
2600				
2700				
2900	Cut & patch to match existing construction, add, minimum	2%	3%	
3000	Maximum	5%	9%	
3100	Dust protection, add, minimum	1%	2%	
3200	Maximum	4%	11%	
3300	Equipment usage curtailment, add, minimum	1%	1%	
3400	Maximum	3%	10%	
3500	Material handling & storage limitation, add, minimum	1%	1%	
3600	Maximum	6%	7%	
3700	Protection of existing work, add, minimum	2%	2%	
3800	Maximum	5%	7%	
3900	Shift work requirements, add, minimum		5%	
4000	Maximum		30%	

C3030 Ceiling Finishes

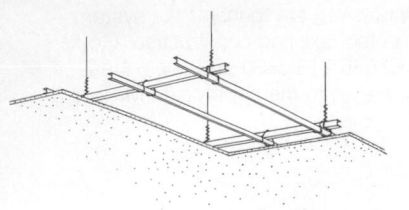

This page illustrates and describes suspended gypsum board systems including gypsum board, metal furring, taping, finished and painted. Lines within System Components give the unit price and total price per square foot for this system. Prices for suspended gypsum board systems are on Line Items C3030 220 1400 thru 1700. Both material quantities and labor costs have been adjusted for the system listed.

Factors: To adjust for job conditions other than normal working situations use Lines C3030 220 2900 thru 4000.

Example: You are to install the system and control dust in the working area. Go to Line C3030 220 3200 and apply these percentages to the appropriate MAT. and INST. costs.

System Components			COST PER S.F.		
	QUANTITY	UNIT	MAT.	INST.	TOTAL
Suspended ceiling gypsum board,4' x 8' x 5/8" thick,					
On metal furring, taped, finished and painted.					
Gypsum drywall, 4' x 8', 5/8" thick, screwed	1.000	S.F.	.39	.73	1.12
Main runners, 1-1/2" C.R.C., 4' O.C.	.500	S.F.	.18	.80	.98
25 ga., channels, 2' O.C.	1.000	S.F.	.26	1.48	1.74
Taped and finished	1.000	S.F.	.05	.66	.71
Paint, 2 coats, roller work	1.000	S.F.	.22	.72	.94
TOTAL			1.10	4.39	5.49

C3030 220	Ceilings, Suspended Gypsum Board	COST PER S.F.		
		MAT.	INST.	TOTAL
1400	4' x 8' x 5/8", gypsum drywall, 2 coats paint	1.10	4.39	5.49
1500	Thin coat plaster, 2 coats paint	.95	3.83	4.78
1600	Spray-on sand finish, no paint	.87	3.79	4.66
1700	12" x 12" x 3/4" acoustical wood fiber tile	2.11	4.66	6.77
1800				
1900				
2000				
2100				
2200				
2300				
2400				
2500				
2600				
2700				
2900	Cut & patch to match existing construction, add, minimum	2%	3%	
3000	Maximum	5%	9%	
3100	Dust protection, add, minimum	1%	2%	
3200	Maximum	4%	11%	
3300	Equipment usage curtailment, add, minimum	1%	1%	
3400	Maximum	3%	10%	
3500	Material handling & storage limitation, add, minimum	1%	1%	
3600	Maximum	6%	7%	
3700	Protection of existing work, add, minimum	2%	2%	
3800	Maximum	5%	7%	
3900	Shift work requirements, add, minimum		5%	
4000	Maximum		30%	

C3030 Ceiling Finishes

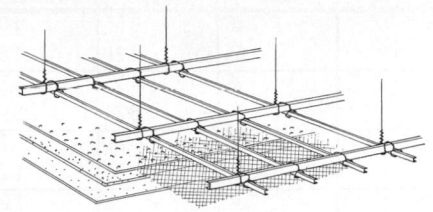

This page illustrates and describes suspended plaster and lath systems including gypsum plaster, lath, furring and runners with ceiling painted. Lines within System Components give the unit price and total price per square foot for this system. Prices for suspended plaster and lath systems are on Line Items C3030 230 1200 thru 2300. Both material quantities and labor costs have been adjusted for the system listed.

Factors: To adjust for job conditions other than normal working situations use Lines C3030 230 2900 thru 4000.

Example: You are to install the system to match existing construction. Go to Line C3030 230 2900 and apply these percentages to the appropriate MAT. and INST. costs.

System Components	QUANTITY	UNIT	COST PER S.F.		
			MAT.	INST.	TOTAL
Gypsum plaster, 3 coats, on 3.4# rib lath, on 3/4" C.R.C. furring on 1-1/2" main runners, ceiling painted.					
Gypsum plaster, 3 coats	.110	S.Y.	.65	4.17	4.82
3.4# rib lath	.110	S.Y.	.57	.97	1.54
Main runners, 1-1/2" C.R.C. 24" O.C.	.333	S.F.	.35	1.59	1.94
Furring 3/4" C.R.C. 16" OC	1.000	S.F.	.39	2.14	2.53
Painting, 2 coats, roller work	1.000	S.F.	.22	.72	.94
TOTAL			2.18	9.59	11.77

C3030 230	Ceiling, Suspended Plaster	COST PER S.F.		
		MAT.	INST.	TOTAL
1200	Gypsum plaster, 3 coats, on 3.4# rib lath	2.18	9.60	11.78
1300	On 2.5# diamond lath	2.11	9.50	11.61
1400	On 3/8" gypsum lath	1.99	9.50	11 49
1500	2 coats, on 3.4# rib lath	1.91	9.60	11.51
1600	On 2.5# diamond lath	1.92	8.85	10.77
1700	On 3/8" gypsum lath	1.80	8.85	10.65
1800	Perlite plaster, 3 coats, on 3.4# rib lath	2.31	10.55	12.86
1900	On 2.5# diamond lath	2.24	10.50	12.74
2000	On 3/8" gypsum lath	2.12	10.50	12.62
2100	2 coats, on 3.4# rib lath	2.25	9.55	11.80
2200	On 2.5# diamond lath	2.18	9.45	11.63
2300	On 3/8" gypsum lath	2.06	9.45	11.51
2400				
2500				
2600				
2700				
2900	Cut & patch to match existing construction, add, minimum	2%	3%	
3000	Maximum	5%	9%	
3100	Dust protection, add, minimum	1%	2%	
3200	Maximum	4%	11%	
3300	Equipment usage curtailment, add, minimum	1%	1%	
3400	Maximum	3%	10%	
3500	Material handling & storage limitation, add, minimum	1%	1%	
3600	Maximum	6%	7%	
3700	Protection of existing work, add, minimum	2%	2%	
3800	Maximum	5%	7%	
3900	Shift work requirements, add, minimum		5%	
4000	Maximum		30%	

C3030 Ceiling Finishes

C3030 250	Selective Price Sheet	COST PER S.F.		
		MAT.	INST.	TOTAL
0100	Ceiling, plaster, gypsum, 2 coats	.46	3.57	4.03
0200	3 coats	.65	4.21	4.86
0300	Perlite or vermiculite, 2 coats	.73	4.15	4.88
0400	3 coats	.78	5.20	5.98
0500	Gypsum lath, plain, 3/8" thick	.39	.81	1.20
0600	1/2" thick	.31	.86	1.17
0700	Firestop, 3/8" thick	.31	.98	1.29
0800	1/2" thick	.35	1.06	1.41
0900	Metal lath, rib, 2.75 lb.	.44	.92	1.36
1000	3.40 lb.	.58	.98	1.56
1100	Diamond, 2.50 lb.	.51	.92	1.43
1200	3.40 lb.	.55	1.15	1.70
1500	Drywall, standard, 1/2" thick, no finish included	.37	.73	1.10
1600	Taped and finished	.43	1.72	2.15
1700	5/8" thick, no finish included	.39	.73	1.12
1800	Taped and finished	.44	1.72	2.16
1900	Fire resistant, 1/2" thick, no finish included	.41	.73	1.14
2000	Taped and finished	.46	1.72	2.18
2100	5/8" thick, no finish included	.40	.73	1.13
2200	Taped and finished	.45	1.72	2.17
2300	Water resistant, 1/2" thick, no finish included	.46	.73	1.19
2400	Taped and finished	.51	1.72	2.23
2500	5/8" thick, no finish included	.48	.73	1.21
2600	Taped and finished	.54	1.72	2.26
2700	Finish, instead of taping			
2800	For thin coat plaster, add	.12	.82	.94
2900	Finish, textured spray, add	.04	.78	.82
3000				
3100				
3200				
3300	Tile, stapled or glued, mineral fiber plastic coated, 5/8" thick	2.57	2.19	4.76
3400	3/4" thick	3.36	2.19	5.55
3500	Wood fiber, 1/2" thick	1.32	1.65	2.97
3600	3/4" thick	1.28	1.65	2.93
3700	Suspended, fiberglass, film faced, 5/8" thick	1.39	1.05	2.44
3800	3" thick	3.85	1.46	5.31
3900	Mineral fiber, 5/8" thick, standard	.96	.98	1.94
4000	Aluminum	4.30	1.10	5.40
4100	Wood fiber, reveal edge, 1" thick			
4200	3" thick			
4300	Framing, metal furring, 3/4" channels, 12" O.C.	.44	2.95	3.39
4400	16" O.C.	.39	2.14	2.53
4500	24" O.C.	.26	1.48	1.74
4600				
4700	1-1/2" channels, 12" O.C.	.58	3.27	3.85
4800	16" O.C.	.52	2.39	2.91
4900	24" O.C.	.35	1.59	1.94
5000				
5100				
5200				
5300	Ceiling suspension systems, for tile,			
5400	Concealed "Z" bar, 12" module	1.02	1.27	2.29
5500	Class A, "T" bar 2'-0" x 4'-0" grid	.87	.82	1.69
5600	2'-0" x 2'-0" grid	1.12	1.01	2.13
5700	Carrier channels for lighting fixtures, add	.24	1.43	1.67
5800				
5900				
6000				

C30 Interior Finishes

C3030 Ceiling Finishes

C3030 250	Selective Price Sheet	COST PER S.F.		
		MAT.	INST.	TOTAL
6100	Wood, furring 1" x 3", on wood, 12" O.C.	.47	1.88	2.35
6200	16" O.C..	.35	1.41	1.76
6300	24" O.C.	.24	.94	1.18
6400				
6500	On concrete, 12" O.C.	.51	3.14	3.65
6600	16" O.C.	.38	2.36	2.74
6700	24" O.C.	.26	1.57	1.83
6800				
6900	Joists, 2" x 4", 12" O.C.	.59	1.35	1.94
7000	16" O.C.	.50	1.13	1.63
7100	24" O.C.	.40	.92	1.32
7200	32" O.C.	.31	.70	1.01
7300	2" x 6", 12" O.C.	.93	1.37	2.30
7400	16" O.C.	.77	1.14	1.91
7500	24" O.C.	.61	.90	1.51
7600	32" O.C.	.45	.67	1.12

D Services

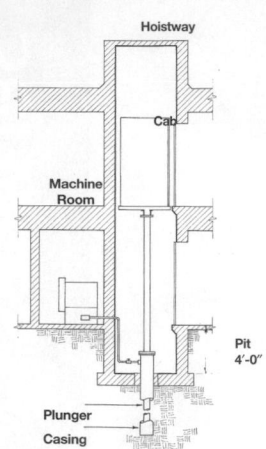

Hoistway

Cab

Machine Room

Pit 4'-0"

Plunger

Casing

This page illustrates and describes oil hydraulic elevator systems. Prices for various oil hydraulic elevator systems are on Line Items D1010 130 0400 thru 2000. Both material quantities and labor costs have been adjusted for the system listed.

Factors: To adjust for job conditions other than normal working situations use Lines D1010 130 3100 thru 4000.

Example: You are to install the system with limited equipment usage. Go to Line D1010 130 3400 and apply this percentage to the appropriate TOTAL costs.

D1010 130	Oil Hydraulic Elevators	COST EACH		
		MAT.	INST.	TOTAL
0310	Oil-hydraulic elevator systems			
0320	Including piston and piston shaft			
0330				
0400	1500 Lb. passenger, 2 floors	47,200	20,300	67,500
0500	3 floors	58,000	31,300	89,300
0600	4 floors	82,500	44,100	126,600
0700	5 floors	93,000	55,000	148,000
0800	6 floors	103,000	65,500	168,500
0900				
1000	2500 Lb. passenger, 2 floors	50,500	20,300	70,800
1100	3 floors	61,500	31,300	92,800
1200	4 floors	81,500	44,100	125,600
1300	5 floors	92,000	55,000	147,000
1400	6 floors	106,500	65,500	172,000
1500				
1600	4000 Lb. passenger, 2 floors	57,500	20,300	77,800
1700	3 floors	69,500	31,300	100,800
1800	4 floors	93,000	44,100	137,100
1900	5 floors	103,000	55,000	158,000
2000	6 floors	113,000	65,500	178,500
2100				
2200				
2300				
2400				
2500				
2600				
2700				
2800				
2900				
3000				
3100	Dust protection, add, minimum	1%	2%	
3200	Maximum	4%	11%	
3300	Equipment usage curtailment, add, minimum	1%	1%	
3400	Maximum	3%	10%	
3500	Material handling & storage limitation, add, minimum	1%	1%	
3600	Maximum	6%	7%	
3700	Protection of existing work, add, minimum	2%	2%	
3800	Maximum	5%	7%	
3900	Shift work requirements, add, minimum		5%	
4000	Maximum		30%	

D2010 Plumbing Fixtures

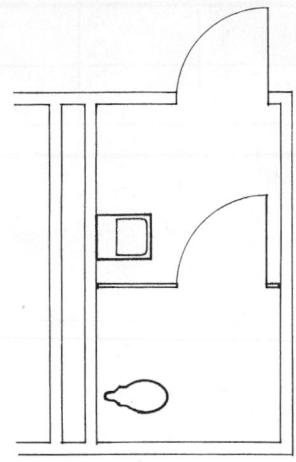

This page illustrates and describes a women's public restroom system including a water closet, lavatory, accessories, and service piping. Lines within System Components give the unit price and total price on a cost each basis for this system. Prices for women's public restroom systems are on Line Items D2010 956 1300 thru 1700. Both material quantities and labor costs have been adjusted for the system listed.

Factors: To adjust for job conditions other than normal working situations use Lines D2010 956 2900 thru 4000.

Example: You are to install the system and protect surrounding area from dust. Go to Line D2010 956 3100 and apply these percentages to the MAT. and INST. costs.

System Components	QUANTITY	UNIT	COST EACH MAT.	COST EACH INST.	COST EACH TOTAL
Public women's restroom incl. water closet, lavatory, accessories and Necessary service piping to install this system in one wall.					
Water closet, wall mounted, one piece	1.000	Ea.	1,075	240	1,315
Rough-in waste and vent for water closet	1.000	Set	1,300	545	1,845
Lavatory, 20″ x 18″ P.E. cast iron with accessories	1.000	Ea.	270	174	444
Rough-in waste and vent for lavatory	1.000	Set	515	840	1,355
Toilet partition, painted metal between walls, floor mounted	1.000	Ea.	430	219	649
For handicap unit, add	1.000	Ea.	360		360
Grab bar, 36″ long	1.000	Ea.	77	66	143
Mirror, 18″ x 24″, with stainless steel shelf	1.000	Ea.	208	33	241
Napkin/tampon dispenser, recessed	1.000	Ea.	680	44	724
Soap Dispenser, chrome, surface mounted, liquid	1.000	Ea.	56	33	89
Toilet tissue dispenser, surface mounted, stainless steel	1.000	Ea.	19.85	22	41.85
Towel dispenser, surface mounted, stainless steel	1.000	Ea.	47	41	88
TOTAL			5,037.85	2,257	7,294.85

D2010 956	Plumbing - Public Restroom	COST EACH MAT.	COST EACH INST.	COST EACH TOTAL
1200				
1300	Public women's restroom, one water closet, one lavatory	5,050	2,250	7,300
1400	Two water closets, two lavatories	9,425	4,300	13,725
1500				
1600	For each additional water closet over 2, add	3,075	995	4,070
1700	For each additional lavatory over 2, add	1,050	1,075	2,125
1800				
1900				
2400	NOTE: PLUMBING APPROXIMATIONS			
2500	WATER CONTROL: water meter, backflow preventer,			
2600	Shock absorbers, vacuum breakers, mixer....10 to 15% of fixtures			
2700	PIPE AND FITTINGS: .... 30 to 60% of fixtures			
2800				
2900	Cut & patch to match existing construction, add, minimum	2%	3%	
3000	Maximum	5%	9%	
3100	Dust protection, add, minimum	1%	2%	
3200	Maximum	4%	11%	
3300	Equipment usage curtailment, add, minimum	1%	1%	
3400	Maximum	3%	10%	
3500	Material handling & storage limitation, add, minimum	1%	1%	

D2010 Plumbing Fixtures

D2010 956	Plumbing - Public Restroom	COST EACH		
		MAT.	INST.	TOTAL
3600	Maximum	6%	7%	
3700	Protection of existing work, add, minimum	2%	2%	
3800	Maximum	5%	7%	
3900	Shift work requirements, add, minimum		5%	
4000	Maximum		30%	

D2010 Plumbing Fixtures

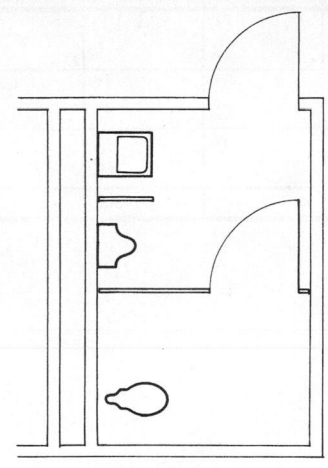

This page illustrates and describes a men's public restroom system including a water closet, urinal, lavatory, accessories and service piping. Lines within System Components give the unit price and total price on a cost each basis for this system. Prices for men's public restroom systems are on Line Items D2010 957 1700 thru 2200. Both material quantities and labor costs have been adjusted for the system listed.

Factors: To adjust for job conditions other than normal working situations use Lines D2010 957 2900 thru 4000.

Example: You are to install the system and match existing construction. Go to Line D2010 957 3000 and apply these percentages to the appropriate MAT. and INST. costs.

System Components	QUANTITY	UNIT	COST EACH MAT.	COST EACH INST.	COST EACH TOTAL
Public men's restroom incl. water closet, urinal, lavatory, accessories, And necessary service piping to install this system in one wall.					
Water closet, wall mounted, one piece	1.000	Ea.	1,075	240	1,315
Rough-in waste & vent for water closet	1.000	Set	1,300	545	1,845
Urinal, wall hung	1.000	Ea.	310	465	775
Rough-in waste & vent for urinal	1.000	Set	740	490	1,230
Lavatory, 20" x 18", P.E. cast iron with accessories	1.000	Ea.	270	174	444
Rough-in waste & vent for lavatory	1.000	Set	515	840	1,355
Partition, painted mtl., between walls, floor mntd.	1.000	Ea.	430	219	649
For handicap unit, add	1.000	Ea.	360		360
Grab bars	1.000	Ea.	77	66	143
Urinal screen, powder coated steel, wall mounted	1.000	Ea.	187	132	319
Mirror, 18" x 24" with stainless steel shelf	1.000	Ea.	208	33	241
Soap dispenser, surface mounted, liquid	1.000	Ea.	56	33	89
Toilet tissue dispenser, surface mounted, stainless steel	1.000	Ea.	19.85	22	41.85
Towel dispenser, surface mounted, stainless steel	1.000	Ea.	47	41	88
TOTAL			5,594.85	3,300	8,894.85

D2010 957	Plumbing - Public Restroom	COST EACH MAT.	COST EACH INST.	COST EACH TOTAL
1600				
1700	Public men's restroom, one water closet, one urinal, one lavatory	5,600	3,300	8,900
1800	Two water closets, two urinals, two lavatories	11,200	6,425	17,625
1900				
2000	For each additional water closet over 2, add	3,075	995	4,070
2100	For each additional urinal over 2, add	1,225	1,075	2,300
2200	For each additional lavatory over 2, add	1,050	1,075	2,125
2300				
2400	NOTE: PLUMBING APPROXIMATIONS			
2500	WATER CONTROL: water meter, backflow preventer,			
2600	Shock absorbers, vacuum breakers, mixer....10 to 15% of fixtures			
2700	PIPE AND FITTINGS: .... 30 to 60% of fixtures			
2800				
2900	Cut & patch to match existing construction, add, minimum	2%	3%	
3000	Maximum	5%	9%	
3100	Dust protection, add, minimum	1%	2%	
3200	Maximum	4%	11%	
3300	Equipment usage curtailment, add, minimum	1%	1%	

D2010 Plumbing Fixtures

D2010 957	Plumbing - Public Restroom	COST EACH		
		MAT.	INST.	TOTAL
3400	Maximum	3%	10%	
3500	Material handling, & storage limitation, add, minimum	1%	1%	
3600	Maximum	6%	7%	
3700	Protection of existing work, add, minimum	2%	2%	
3800	Maximum	5%	7%	
3900	Shift work requirements, add, minimum		5%	
4000	Maximum		30%	

D2010 Plumbing Fixtures

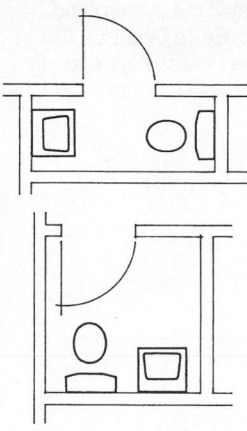

This page illustrates and describes a two fixture lavatory system including a water closet, lavatory, accessories and all service piping. Lines within System Components give the unit price and total price on a cost each basis for this system. Prices for two fixture lavatory systems are on Line Items D2010 958 1800 and 1900. Both material quantities and labor costs have been adjusted for the system listed.

Factors: To adjust for job conditions other than normal working situations use Lines D2010 958 2900 thru 4000.

Example: You are to install the system while controlling dust in the work area. Go to Line D2010 958 3200 and apply these percentages to the appropriate MAT. and INST. costs.

System Components	QUANTITY	UNIT	COST EACH		
			MAT.	INST.	TOTAL
Two fixture bathroom incl. water closet, lavatory, accessories and Necessary service piping to install this system in 2 walls.					
Water closet, floor mounted, 2 piece, close coupled	1.000	Ea.	253	263	516
Rough in waste & vent for water closet	1.000	Set	395	455	850
Lavatory, 20" x 18", P.E. cast iron with accessories	1.000	Ea.	270	174	444
Rough in waste & vent for lavatory	1.000	Set	515	840	1,355
Additional service piping					
1/2" copper pipe with sweat solder joints	10.000	L.F.	36.80	95.50	132.30
2" black schedule 40 steel pipe with threaded couplings	12.000	L.F.	157.80	264	421.80
4" cast iron soil pipe with lead and oakum joints	7.000	L.F.	269.50	178.50	448
Accessories					
Toilet tissue dispenser, chrome, single roll	1.000	Ea.	19.85	22	41.85
18" long stainless steel towel bar	1.000	Ea.	47	28.50	75.50
Medicine cabinet with mirror, 20" x 16", unlighted	1.000	Ea.	109	47	156
TOTAL			2,072.95	2,367.50	4,440.45

D2010 958	Plumbing - Two Fixture Bathroom	COST EACH		
		MAT.	INST.	TOTAL
1800	Water closet, lavavatory & accessories, in 2 walls with all service piping	2,075	2,375	4,450
1900	Above system installed in 1 wall	1,900	2,050	3,950
2000				
2400	NOTE: PLUMBING APPROXIMATIONS			
2500	WATER CONTROL: water meter, backflow preventer,			
2600	Shock absorbers, vacuum breakers, mixer....10 to 15% of fixtures			
2700	PIPE AND FITTINGS: .... 30 to 60% of fixtures			
2800				
2900	Cut & patch to match existing construction, add, minimum	2%	3%	
3000	Maximum	5%	9%	
3100	Dust protection, add, minimum	1%	2%	
3200	Maximum	4%	11%	
3300	Equipment usage curtailment, add, minimum	1%	1%	
3400	Maximum	3%	10%	
3500	Material handling & storage limitation, add, minimum	1%	1%	
3600	Maximum	6%	7%	
3700	Protection of existing work, add, minimum	2%	2%	
3800	Maximum	5%	7%	
3900	Shift work requirements, add, minimum		5%	
4000	Maximum		30%	

D2010 Plumbing Fixtures

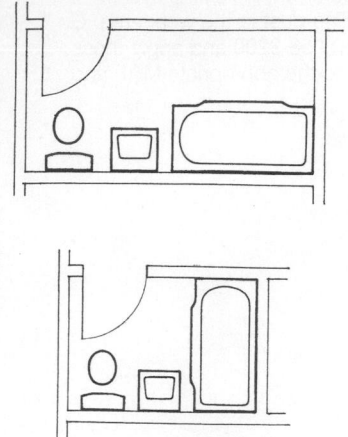

This page illustrates and describes a three fixture bathroom system including a water closet, tub, lavatory, accessories and service piping. Lines within System Components give the unit price and total price on a cost each basis for this system. Prices for a three fixture bathroom system is on Line Item D2010 959 1700. Both material quantities and labor costs have been adjusted for the system listed.

Factors: To adjust for job conditions other than normal working situations use Lines D2010 959 2900 thru 4000.

Example: You are to install the system and protect all existing work. Go to Line D2010 959 3800 and apply these percentages to the appropriate MAT. and INST. costs.

System Components	QUANTITY	UNIT	COST EACH MAT.	COST EACH INST.	COST EACH TOTAL
Three fixture bathroom incl. water closet, bathtub, lavatory, accessories,					
And necessary service piping to install this system in 1 wall.					
Water closet, tank vit china flr mtd close cpld, 2 pc. w/seat supply & stop	1.000	Ea.	253	263	516
For rough in, supply, waste and vent	.930	Set	367.35	423.15	790.50
Bath tub, P.E., cast iron, 5' long, with accessories	1.000	Ea.	1,300	315	1,615
Rough-in, supply, waste and vent, for all above tubs, add	.950	Set	446.50	641.25	1,087.75
Lavatory, 20" x 18" P.E. cast iron with accessories	1.000	Ea.	270	174	444
Rough-in, supply, waste and vent for above lavatories	1.000	Set	515	840	1,355
Accessories					
Toilet tissue dispenser, surface mounted, SS, single roll	1.000	Ea.	19.85	22	41.85
Towel bar, stainless steel, 18" long	2.000	Ea.	94	57	151
Medicine cabinets with mirror, sst. frame, 16" x 22", unlighted	1.000	Ea.	109	47	156
TOTAL			3,374.70	2,782.40	6,157.10

D2010 959	Plumbing - Three Fixture Bathroom	COST EACH MAT.	COST EACH INST.	COST EACH TOTAL
1600				
1700	Water closet, lavatory, bathtub, in 1 wall with all service piping	3,375	2,775	6,150
2400	NOTE: PLUMBING APPROXIMATIONS			
2500	WATER CONTROL: water meter, backflow preventer,			
2600	Shock absorbers, vacuum breakers, mixer....10 to 15% of fixtures			
2700	PIPE AND FITTINGS: .... 30 to 60% of fixtures			
2800				
2900	Cut & patch to match existing construction, add, minimum	2%	3%	
3000	Maximum	5%	9%	
3100	Dust protection, add, minimum	1%	2%	
3200	Maximum	4%	11%	
3300	Equipment usage curtailment, add, minimum	1%	1%	
3400	Maximum	3%	10%	
3500	Material handling & storage limitation, add, minimum	1%	1%	
3600	Maximum	6%	7%	
3700	Protection of existing work, add, minimum	2%	2%	
3800	Maximum	5%	7%	
3900	Shift work requirements, add, minimum		5%	
4000	Maximum		30%	

D20 Plumbing

D2010 Plumbing Fixtures

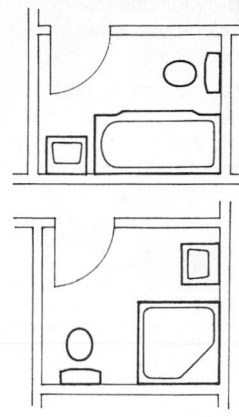

This page illustrates and describes a three fixture bathroom system including a water closet, tub, lavatory, accessories, and service piping. Lines within System Components give the unit price and total price on a cost each basis for this system. Prices for a three fixture bathroom system are on Line Item D2010 960 1900 thru 2000. Both material quantities and labor costs have been adjusted for the system listed.

Factors: To adjust for job conditions other than normal working situations use Lines D2010 960 2900 thru 4000.

Example: You are to install the system and protect the surrounding area from dust. Go to Line D2010 960 3100 and apply these percentages to the appropriate MAT. and INST. costs.

System Components	QUANTITY	UNIT	COST EACH		
			MAT.	INST.	TOTAL
Three fixture bathroom incl. water closet, bathtub, lavatory, accessories,					
And necessary service piping to install this system in 2 walls.					
Water closet, floor mounted, 2 piece, close coupled	1.000	Ea.	253	263	516
Rough-in waste & vent for water closet	1.000	Set	395	455	850
Bathtub, P.E. cast iron, 5' long with accessories	1.000	Ea.	1,300	315	1,615
Rough-in waste & vent for bathtub	1.000	Set	470	675	1,145
Lavatory, 20" x 18" P.E. cast iron with accessories	1.000	Ea.	270	174	444
Rough-in waste & vent for lavatory	1.000	Set	515	840	1,355
Additional service piping					
1-1/4" copper DWV type tubing, sweat solder joints	6.000	L.F.	76.50	77.40	153.90
2" black schedule 40 steel pipe with threaded couplings	12.000	L.F.	157.80	264	421.80
Accessories					
Toilet tissue dispenser, chrome, single roll	1.000	Ea.	19.85	22	41.85
18" long stainless steel towel bar	2.000	Ea.	94	57	151
Medicine cabinet with mirror, 20" x 16", unlighted	1.000	Ea.	109	47	156
TOTAL			3,660.15	3,189.40	6,849.55

D2010 960	Plumbing - Three Fixture Bathroom	COST EACH		
		MAT.	INST.	TOTAL
1900	Water closet, lavatory, bathtub, in 2 walls with all service piping	3,650	3,200	6,850
2000	Above system with corner contour tub, P.E. cast iron	5,425	3,200	8,625
2300				
2400	NOTE: PLUMBING APPROXIMATIONS			
2500	WATER CONTROL: water meter, backflow preventer,			
2600	Shock absorbers, vacuum breakers, mixer....10 to 15% of fixtures			
2700	PIPE AND FITTINGS: .... 30 to 60% of fixtures			
2800				
2900	Cut & patch to match existing construction, add, minimum	2%	3%	
3000	Maximum	5%	9%	
3100	Dust protection, add, minimum	1%	2%	
3200	Maximum	4%	11%	
3300	Equipment usage curtailment, add, minimum	1%	1%	
3400	Maximum	3%	10%	
3500	Material handling & storage limitation, add, minimum	1%	1%	
3600	Maximum	6%	7%	
3700	Protection of existing work, add, minimum	2%	2%	
3800	Maximum	5%	7%	
3900	Shift work requirements, add, minimum		5%	
4000	Maximum		30%	

D2010 Plumbing Fixtures

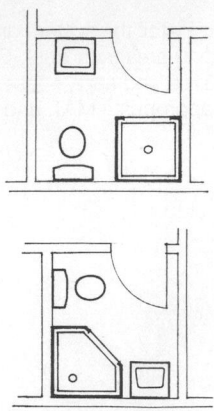

This page illustrates and describes a three fixture bathroom system including a water closet, shower, lavatory, accessories, and service piping, Lines within System Components give the unit price and total price on a cost each basis for this system. Prices for three fixture bathroom systems are on Line Items D2010 961 2100 thru 2200. Both material quantities and labor costs have been adjusted for the system listed.

Factors: To adjust for job conditions other than normal working situations use Lines D2010 961 2900 thru 4000.

Example: You are to install the system and protect existing construction. Go to Line D2010 961 3700 and apply these percentages to the appropriate MAT. and INST. costs.

System Components	QUANTITY	UNIT	COST EACH		
			MAT.	INST.	TOTAL
Three fixture bathroom incl. water closet, shower, lavatory, accessories,					
And necessary service piping to install this system in 2 walls.					
Water closet, floor mounted, 2 piece, close coupled	1.000	Ea.	253	263	516
Rough-in waste & vent for water closet	1.000	Set	395	455	850
32″ shower, enameled stall, molded stone receptor	1.000	Ea.	1,275	279	1,554
Rough-in waste & vent for shower	1.000	Set	410	680	1,090
Lavatory, 20″ x 18″ P.E. cast iron with accessories	1.000	Ea.	270	174	444
Rough-in waste & vent for lavatory	1.000	Set	515	840	1,355
Additional service piping					
1-1/4″ Copper DWV type tubing, sweat solder joints	4.000	L.F.	51	51.60	102.60
2″ black schedule 40 steel pipe with threaded couplings	6.000	L.F.	78.90	132	210.90
4″ cast iron soil with lead and oakum joints	7.000	L.F.	269.50	178.50	448
Accessories					
Toilet tissue dispenser, chrome, single roll	1.000	Ea.	19.85	22	41.85
18″ long stainless steel towel bar	2.000	Ea.	94	57	151
Medicine cabinet with mirror, 20″ x 16″, unlighted	1.000	Ea.	109	47	156
TOTAL			3,740.25	3,179.10	6,919.35

D2010 961	Plumbing - Three Fixture Bathroom	COST EACH		
		MAT.	INST.	TOTAL
2100	Water closet, 32″ shower, lavavatory, in 2 walls with all service piping	3,750	3,175	6,925
2200	Above system with 36″ corner angle shower	4,400	3,200	7,600
2300				
2400	NOTE: PLUMBING APPROXIMATIONS			
2500	WATER CONTROL: water meter, backflow preventer,			
2600	Shock absorbers, vacuum breakers, mixer....10 to 15% of fixtures			
2700	PIPE AND FITTINGS: .... 30 to 60% of fixtures			
2800				
2900	Cut & patch to match existing construction, add, minimum	2%	3%	
3000	Maximum	5%	9%	
3100	Dust protection, add, minimum	1%	2%	
3200	Maximum	4%	11%	
3300	Equipment usage curtailment, add, minimum	1%	1%	
3400	Maximum	3%	10%	
3500	Material handling & storage limitation, add, minimum	1%	1%	
3600	Maximum	6%	7%	
3700	Protection of existing work, add, minimum	2%	2%	
3800	Maximum	5%	7%	
3900	Shift work requirements, add, minimum		5%	
4000	Maximum		30%	

D2010 Plumbing Fixtures

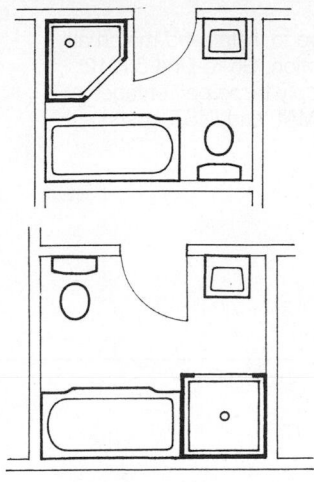

This page illustrates and describes a four fixture bathroom system including a water closet, shower, bathtub, lavatory, accessories, and service piping. Lines within System Components give the unit price and total price on a cost each basis for this system. Prices for four fixture bathroom systems are on Line Items D2010 962 1800 thru 1900. Both material quantities and labor costs have been adjusted for the system listed.

Factors: To adjust for job conditions other than normal working situations use Lines D2010 962 2900 thru 4000.

Example: You are to install the system during weekends and evenings. Go to Line D2010 962 4000 and apply these percentages to the appropriate MAT. and INST. costs.

System Components	QUANTITY	UNIT	COST EACH		
			MAT.	INST.	TOTAL
Four fixture bathroom incl. water closet, shower, bathtub, lavatory					
Accessories and necessary service piping to install this system in 2 walls.					
Water closet, floor mounted, 2 piece, close coupled	1.000	Ea.	253	263	516
Rough-in waste & vent for water closet	1.000	Set	395	455	850
32" shower, enameled steel stall, molded stone receptor	1.000	Ea.	1,275	279	1,554
Rough-in waste & vent for shower	1.000	Set	410	680	1,090
Bathtub, P.E. cast iron, 5' long with accessories	1.000	Ea.	1,300	315	1,615
Rough-in waste & vent for bathtub	1.000	Set	470	675	1,145
Lavatory, 20" x 18" P.E. cast iron with accessories	1.000	Ea.	270	174	444
Rough-in waste & vent for lavatory	1.000	Set	515	840	1,355
Accessories					
Toilet tissue dispenser, chrome, single roll	1.000	Ea.	19.85	22	41.85
18" long stainless steel towel bar	2.000	Ea.	94	57	151
Medicine cabinet with mirror, 20" x 16" unlighted	1.000	Ea.	109	47	156
TOTAL			5,110.85	3,807	8,917.85

D2010 962	Plumbing - Four Fixture Bathroom	COST EACH		
		MAT.	INST.	TOTAL
1800	Water closet, 32" shower, bathtub, lavavatory, in 2 walls	5,100	3,800	8,900
1900	Above system with 36" corner angle shower, plumbing in 3 walls	5,950	4,175	10,125
2000				
2300				
2400	NOTE: PLUMBING APPROXIMATIONS			
2500	WATER CONTROL: water meter, backflow preventer,			
2600	Shock absorbers, vacuum breakers, mixer....10 to 15% of fixtures			
2700	PIPE AND FITTINGS: .... 30 to 60% of fixtures			
2800				
2900	Cut & patch to match existing construction, add, minimum	2%	3%	
3000	Maximum	5%	9%	
3100	Dust protection, add, minimum	1%	2%	
3200	Maximum	4%	11%	
3300	Equipment usage curtailment, add, minimum	1%	1%	
3400	Maximum	3%	10%	
3500	Material handling & storage limitation, add, minimum	1%	1%	
3600	Maximum	6%	7%	
3700	Protection of existing work, add, minimum	2%	2%	
3800	Maximum	5%	7%	
3900	Shift work requirements, add, minimum		5%	
4000	Maximum		30%	

671

D2010 Plumbing Fixtures

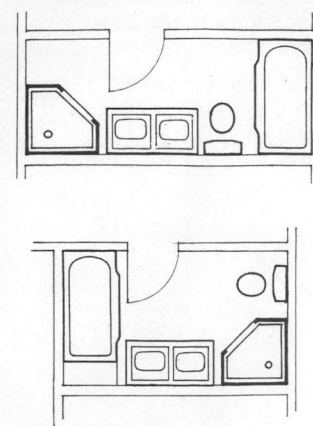

This page illustrates and describes a five fixture bathroom system including a water closet, shower, bathtub, lavatories, accessories, and service piping. Lines within System Components give the unit price and total price on a cost each basis for this system. Prices for five fixture bathroom systems are on Line Items D2010 963 1800 thru 1900. Both material quantities and labor costs have been adjusted for the system listed.

Factors: To adjust for job conditions other than normal working situations use Lines D2010 963 2900 thru 4000.

Example: You are to install and match any existing construction. Go to Line D2010 963 2900 and apply these percentages to the appropriate MAT. and INST. costs.

System Components	QUANTITY	UNIT	COST EACH		
			MAT.	INST.	TOTAL
Five fixture bathroom incl. water closet, shower, bathtub, 2 lavatories,					
Accessories and necessary service piping to install this system in 1 wall.					
Water closet, floor mounted, 2 piece, close coupled	1.000	Ea.	253	263	516
Rough-in waste & vent for water closet	1.000	Set	395	455	850
36" corner angle shower, enameled steel stall, molded stone receptor	1.000	Ea.	1,925	290	2,215
Rough-in waste & vent for shower	1.000	Set	410	680	1,090
Bathtub, P.E. cast iron, 5' long with accessories	1.000	Ea.	1,300	315	1,615
Rough-in waste & vent for bathtub	1.000	Set	470	675	1,145
Countertop, plastic laminate, with backsplash	2.000	Ea.	182	88	270
Vanity base, 2 doors, 24" wide	2.000	Ea.	385	120	505
Lavatories, 20" x 18" cabinet mntd, PECI	2.000	Ea.	660	436	1,096
Rough-in waste & vent for lavatories	1.600	Set	419.20	968	1,387.20
Accessories					
Toilet tissue dispenser, chrome, single roll	1.000	Ea.	19.85	22	41.85
18" long stainless steel towel bars	2.000	Ea.	94	57	151
Medicine cabinet with mirror, 20" x 16", unlighted	2.000	Ea.	218	94	312
TOTAL			6,731.05	4,463	11,194.05

D2010 963	Plumbing - Five Fixture Bathroom	COST EACH		
		MAT.	INST.	TOTAL
1800	Water closet, shower, bathtub, 2 lavatories, on 1 wall	6,725	4,475	11,200
1900	Above system installed in 2 walls with all necessary service piping	6,825	4,625	11,450
2000				
2100				
2200				
2300				
2400	NOTE: PLUMBING APPROXIMATIONS			
2500	WATER CONTROL: water meter, backflow preventer,			
2600	Shock absorbers, vacuum breakers, mixer....10 to 15% of fixtures			
2700	PIPE AND FITTINGS: .... 30 to 60% of fixtures			
2800				
2900	Cut & patch to match existing construction, add, minimum	2%	3%	
3000	Maximum	5%	9%	
3100	Dust protection, add, minimum	1%	2%	
3200	Maximum	4%	11%	
3300	Equipment usage curtailment, add, minimum	1%	1%	
3400	Maximum	3%	10%	
3500	Material handling & storage limitation, add, minimum	1%	1%	

D2010 Plumbing Fixtures

D2010 963	Plumbing - Five Fixture Bathroom	COST EACH		
		MAT.	INST.	TOTAL
3600	Maximum	6%	7%	
3700	Protection of existing work, add, minimum	2%	2%	
3800	Maximum	5%	7%	
3900	Shift work requirements, add, minimum		5%	
4000	Maximum		30%	

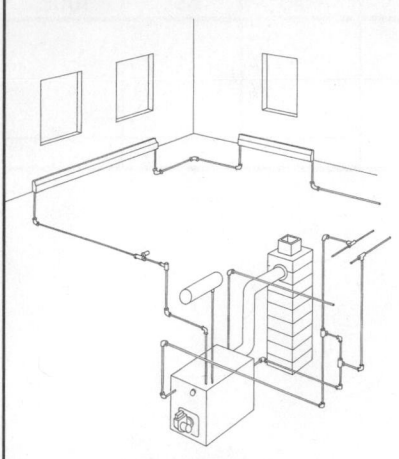

This page illustrates and describes an oil fired hot water baseboard system including an oil fired boiler, fin tube radiation and all fittings and piping. Lines within System Components give the unit price and total price per square foot for this system. Prices for oil fired hot water baseboard systems are on Line Items D3010 540 1600 thru 2100. Both material quantities and labor costs have been adjusted for the system listed.

Factors: To adjust for job conditions other than normal working situations use Lines D3010 540 2900 thru 4000.

Example: You are to install the system while protecting all existing work. Go to Line D3010 540 3700 and apply these percentages to the appropriate MAT. and INST. costs.

System Components	QUANTITY	UNIT	COST PER S.F.		
			MAT.	INST.	TOTAL
Oil fired hot water baseboard system including boiler, fin tube radiation And all necessary fittings and piping.					
Area to 800 S.F.					
Boiler, cast iron, w/oil piping, 97 MBH	1.000	Ea.	2,593.75	1,718.75	4,312.50
Copper piping	130.000	L.F.	1,638	1,735.50	3,373.50
Fin tube radiation	68.000	L.F.	588.20	1,666	2,254.20
Circulator	1.000	Ea.	710	232	942
Oil tank	1.000	Ea.	565	282	847
Expansion tank, ASME	1.000	Ea.	915	101	1,016
TOTAL			7,009.95	5,735.25	12,745.20
COST PER S.F.		S.F.	8.76	7.17	15.93

D3010 540	Heating - Oil Fired Hot Water	COST PER S.F.		
		MAT.	INST.	TOTAL
1600	Cast iron boiler, area to 800 S.F.	8.75	7.16	15.91
1700	To 1000 S.F.	7.70	6.15	13.85
1800	To 1200 S.F.	6.65	5.60	12.25
1900	To 1600 S.F.	7.90	5.55	13.45
2000	To 2000 S.F.	6.60	4.96	11.56
2100	To 3000 S.F.	4.86	4.17	9.03
2200				
2300				
2700				
2800				
2900	Cut & patch to match existing construction, add, minimum	2%	3%	
3000	Maximum	5%	9%	
3100	Dust protection, add, minimum	1%	2%	
3200	Maximum	4%	11%	
3300	Equipment usage curtailment, add, minimum	1%	1%	
3400	Maximum	3%	10%	
3500	Material handling & storage limitation, add, minimum	1%	1%	
3600	Maximum	6%	7%	
3700	Protection of existing work, add, minimum	2%	2%	
3800	Maximum	5%	7%	
3900	Shift work requirements, add, minimum		5%	
4000	Maximum		30%	

D3010 Energy Supply

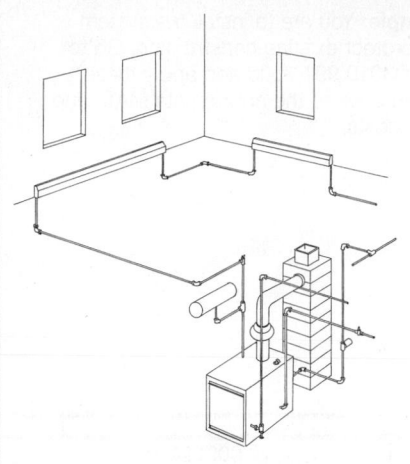

This page illustrates and describes a gas fired hot water baseboard system including a gas fired boiler, fin tube radiation, fittings and piping. Lines within System Components give the unit price and total price per square foot for this system. Prices for gas fired hot water baseboard systems are on Line Items D3010 550 1400 thru 1900. Both material quantities and labor costs have been adjusted for the system listed.

Factors: To adjust for job conditions other than normal working situations use Lines D3010 550 2900 thru 4000.

Example: You are to install the system with minimal equipment usage. Go to Line D3010 550 3400 and apply these percentages to the appropriate MAT. and INST. costs.

System Components	QUANTITY	UNIT	COST PER S.F. MAT.	COST PER S.F. INST.	COST PER S.F. TOTAL
Gas fired hot water baseboard system including boiler, fin tube radiation,					
All necessary fittings and piping.					
Area to 800 S.F.					
Cast iron boiler, insulating jacket, gas piping, 80 MBH	1.000	Ea.	2,656.25	2,562.50	5,218.75
Copper piping	130.000	L.F.	1,638	1,735.50	3,373.50
Fin tube radiation	68.000	L.F.	588.20	1,666	2,254.20
Circulator, flange connection	1.000	Ea.	710	232	942
Expansion tank, ASME	1.000	Ea.	915	101	1,016
TOTAL			6,507.45	6,297	12,804.45
COST PER S.F.		S.F.	8.13	7.87	16

D3010 550	Heating - Gas Fired Hot Water	COST PER S.F. MAT.	COST PER S.F. INST.	COST PER S.F. TOTAL
1400	Cast iron boiler, area to 800 S.F.	8.13	7.88	16.01
1500	To 1000 S.F.	6.85	6.90	13.75
1600	To 1200 S.F.	5.95	6.20	12.15
1700	To 1600 S.F.	7.35	6.05	13.40
1800	To 2000 S.F.	6.45	5.40	11.85
1900	To 3000 S.F.	4.98	4.73	9.71
2000				
2100				
2900	Cut & patch to match existing construction, add, minimum	2%	3%	
3000	Maximum	5%	9%	
3100	Dust protection, add, minimum	1%	2%	
3200	Maximum	4%	11%	
3300	Equipment usage curtailment, add, minimum	1%	1%	
3400	Maximum	3%	10%	
3500	Material handling & storage limitation, add, minimum	1%	1%	
3600	Maximum	6%	7%	
3700	Protection of existing work, add, minimum	2%	2%	
3800	Maximum	5%	7%	
3900	Shift work requirements, add, minimum		5%	
4000	Maximum		30%	

D3010 Energy Supply

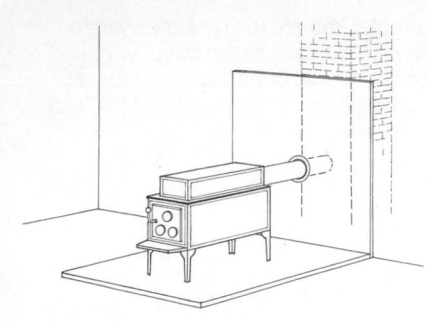

This page illustrates and describes a wood burning stove system including a free standing stove, preformed hearth, masonry chimney, and necessary piping and fittings. Lines within System Components give the unit price and total price on a cost each basis for this system. Prices for wood burning stove systems are on Line Items D3010 991 1400 thru 2000. Both material quantities and labor costs have been adjusted for the system listed.

Factors: To adjust for job conditions other than normal working situations use Lines D3010 991 2900 thru 4000.

Example: You are to install the system and protect existing construction. Go to Line D3010 991 3500 and apply these percentages to the appropriate MAT., and INST. costs.

System Components	QUANTITY	UNIT	COST EACH MAT.	COST EACH INST.	COST EACH TOTAL
Cast iron, free standing, wood burning stove with preformed hearth, masonry Chimney, and all necessary piping and fittings to install in chimney.					
Cast iron wood burning stove, stove pipe	1.000	Ea.	1,475	1,025	2,500
Wall panel or hearth, non-combustible	1.000	Ea.	1,045	185.25	1,230.25
16" x 16" brick chimney, 8" x 8" flue	20.000	V.L.F.	700	1,612.50	2,312.50
Concrete in place, chimney foundations	.500	C.Y.	94	135.44	229.44
TOTAL			3,314	2,958.19	6,272.19

D3010 991	Wood Burning Stoves	COST EACH MAT.	COST EACH INST.	COST EACH TOTAL
1400 1500	Cast iron, free standing with preformed hearth	3,325	2,950	6,275
1600 1700	Installed in existing fireplace	2,025	1,125	3,150
1800 1900	Installed with insulated metal chimney System including ceiling package,			
2000 2100	Metal chimney to 10 L.F.	3,850	1,575	5,425
2200 2300				
2400 2500				
2600 2700				
2900 3000	Dust protection, add, minimum Maximum	1% 4%	2% 11%	
3100 3200	Equipment usage curtailment, add, minimum Maximum	1% 3%	1% 10%	
3300 3400	Material handling & storage limitation, add, minimum Maximum	1% 6%	1% 7%	
3500 3600	Protection of existing work, add, minimum Maximum	2% 5%	2% 7%	
3700 3800	Shift work requirements, add, minimum Maximum		5% 30%	
3900 4000	Temporary shoring and bracing, add, minimum Maximum	2% 5%	5% 12%	

D3010 Energy Supply

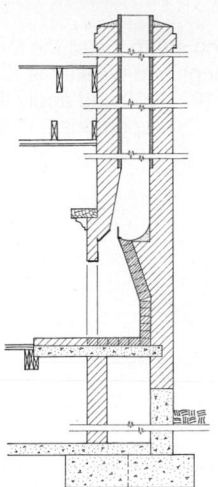

This page illustrates and describes masonry fireplace systems including a brick fireplace, footing, foundation, hearth, firebox, chimney and flue. Lines within System Components give the unit price and total price on a cost each basis for this system. Prices for masonry fireplace systems are on Line Items D3010 992 1200 thru 1500. Both material quantities and labor costs have been adjusted for the system listed.

Factors: To adjust for job conditions other than normal working situations use Lines D3010 992 2700 thru 4000.

Example: You are to install the system with some temporary shoring and bracing. Go to Line D3010 992 3900 and apply these percentages to the appropriate MAT. and INST. costs.

System Components	QUANTITY	UNIT	COST EACH MAT.	COST EACH INST.	COST EACH TOTAL
Brick masonry fireplace, including footing, foundation, hearth, firebox,					
Chimney and flue, chimney 12' above firebox.					
Footing, 4' x 7' x 12" thick, 3000 psi concrete	1.040	C.Y.	212.16	323.44	535.60
Foundation, 12" concrete block	180.000	S.F.	945	2,106	3,051
Fireplace, brick faced, 6'-0" wide x 5'-0" high	1.000	Ea.	645	2,925	3,570
Hearth	1.000	Ea.	234	585	819
Chimney, 20" x 20", one 12" x 12" flue, 12 V.L.F.	12.000	V.L.F.	630	1,026	1,656
Mantle, wood	7.000	L.F.	61.95	128.10	190.05
TOTAL			2,728.11	7,093.54	9,821.65

D3010 992	Masonry Fireplace	COST EACH MAT.	COST EACH INST.	COST EACH TOTAL
1200	Chimney, 20" x 20", one 12" x 12" flue, 12 V.L.F.	2,725	7,100	9,825
1300	18 V.L.F.	3,050	7,600	10,650
1400	24 V.L.F.	3,350	8,125	11,475
1500	Fieldstone face instead of brick	3,025	7,100	10,125
1600				
1700				
1800				
1900				
2000				
2100				
2200				
2300				
2700	Cut & patch to match existing construction, add, minimum	2%	3%	
2800	Maximum	5%	9%	
2900	Dust protection, add, minimum	1%	2%	
3000	Maximum	4%	11%	
3100	Equipment usage curtailment, add, minimum	1%	1%	
3200	Maximum	3%	10%	
3300	Material handling & storage limitation, add, minimum	1%	1%	
3400	Maximum	6%	7%	
3500	Protection of existing work, add, minimum	2%	2%	
3600	Maximum	5%	7%	
3700	Shift work requirements, add, minimum		5%	
3800	Maximum		30%	
3900	Temporary shoring and bracing, add, minimum	2%	5%	
4000	Maximum	5%	12%	

677

D3020 Heat Generating Systems

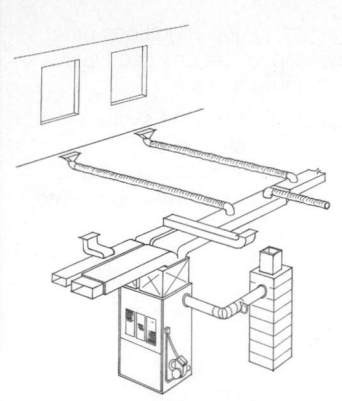

This page illustrates and describes an oil fired forced air system including an oil fired furnace, ductwork, registers and hookups. Lines within System Components give the unit price and total price per square foot for this system. Prices for oil fired forced air systems are on Line Items D3020 122 1600 thru 2800. Both material quantities and labor costs have been adjusted for the system listed.

Factors: To adjust for job conditions other than normal working situations use Lines D3020 122 3100 thru 4200.

Example: You are to install the system during evenings and weekends. Go to Line D3020 122 4200 and apply this percentage to the appropriate INST. cost.

System Components	QUANTITY	UNIT	COST PER S.F. MAT.	COST PER S.F. INST.	COST PER S.F. TOTAL
Oil fired hot air heating system including furnace, ductwork, registers And all necessary hookups.					
Area to 800 S.F., heat only					
Furnace, oil, atomizing gun type burner, w/oil piping	1.000	Ea.	3,875	468.75	4,343.75
Oil tank, steel, 275 gallon	1.000	Ea.	565	282	847
Duct, galvanized, steel	312.000	Lb.	196.56	2,808	3,004.56
Insulation, blanket type, duct work	270.000	S.F.	1,347.30	4,171.50	5,518.80
Flexible duct, 6" diameter, insulated	100.000	L.F.	342	520	862
Registers, baseboard, gravity, 12" x 6"	8.000	Ea.	71.20	264	335.20
Return, damper, 18" x 36"	1.000	Ea.	109	75.50	184.50
TOTAL			6,506.06	8,589.75	15,095.81
COST PER S.F.		S.F.	8.13	10.74	18.87

D3020 122	Heating-Cooling, Oil, Forced Air	COST PER S.F. MAT.	COST PER S.F. INST.	COST PER S.F. TOTAL
1600	Oil fired, area to 800 S.F.	8.13	10.75	18.88
1700	To 1000 S.F.	6.55	8.65	15.20
1800	To 1200 S.F.	5.60	7.50	13.10
1900	To 1600 S.F.	4.42	6.05	10.47
2000	To 2000 S.F.	4.51	8.25	12.76
2100	To 3000 S.F.	3.42	6.55	9.97
2200	For combined heating and cooling systems:			
2300	Oil fired, heating and cooling, area to 800 S.F.	10.80	12	22.80
2400	To 1000 S.F.	8.65	9.75	18.40
2500	To 1200 S.F.	7.35	8.40	15.75
2600	To 1600 S.F.	5.90	6.75	12.65
2700	To 2000 S.F.	5.75	8.85	14.60
2800	To 3000 S.F.	4.27	7	11.27
3100	Cut & patch to match existing construction, add, minimum	2%	3%	
3200	Maximum	5%	9%	
3300	Dust protection, add, minimum	1%	2%	
3400	Maximum	4%	11%	
3500	Equipment usage curtailment, add, minimum	1%	1%	
3600	Maximum	3%	10%	
3700	Material handling & storage limitation, add, minimum	1%	1%	
3800	Maximum	6%	7%	
3900	Protection of existing work, add, minimum	2%	2%	
4000	Maximum	5%	7%	
4100	Shift work requirements, add, minimum		5%	
4200	Maximum		30%	

D3020 Heat Generating Systems

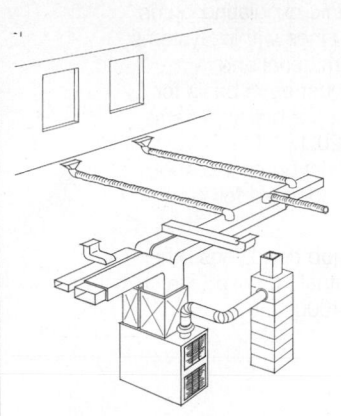

This page illustrates and describes a gas fired forced air system including a gas fired furnace, ductwork, registers and hookups. Lines within System Components give the unit price and total price per square foot for this system. Prices for gas fired forced air systems are on Line Items D3020 124 1400 thru 2600. Both material quantities and labor costs have been adjusted for the system listed.

Factors: To adjust for job conditions other than normal working situations use Lines D3020 124 2900 thru 4000.

Example: You are to install the system with material handling and storage limitations. Go to Line D3020 124 3500 and apply these percentages to the appropriate MAT. and INST. costs.

System Components			COST PER S.F.		
	QUANTITY	UNIT	MAT.	INST.	TOTAL
Gas fired hot air heating system including furnace, ductwork, registers					
And all necessary hookups.					
Area to 800 S.F., heat only					
Furnace, gas, AGA certified, direct drive, w/gas piping, 44 MBH	1.000	Ea.	887.50	425	1,312.50
Duct, galvanized steel	312.000	Lb.	196.56	2,808	3,004.56
Insulation, blanket type, ductwork	270.000	S.F.	1,347.30	4,171.50	5,518.80
Flexible duct, 6" diameter, insulated	100.000	L.F.	342	520	862
Registers, baseboard, gravity, 12" x 6"	8.000	Ea.	71.20	264	335.20
Return, damper, 18" x 36"	1.000	Ea.	109	75.50	184.50
TOTAL			2,953.56	8,264	11,217.56
COST PER S.F.		S.F.	3.69	10.33	14.02

D3020 124	Heating-Cooling, Gas, Forced Air	COST PER S.F.		
		MAT.	INST.	TOTAL
1400	Gas fired, area to 800 S.F.	3.69	10.34	14.03
1500	To 1000 S.F.	2.98	8.35	11.33
1600	To 1200 S.F.	2.63	7.25	9.88
1700	To 1600 S.F.	2.47	6.55	9.02
1800	To 2000 S.F.	2.76	8.10	10.86
1900	To 3000 S.F.	2.17	6.45	8.62
2000	For combined heating and cooling systems:			
2100	Gas fired, heating and cooling, area to 800 S.F.	6.35	11.60	17.95
2200	To 1000 S.F.	5.10	9.40	14.50
2300	To 1200 S.F.	4.39	8.10	12.49
2400	To 1600 S.F.	3.94	7.25	11.19
2500	To 2000 S.F.	3.98	8.75	12.73
2600	To 3000 S.F.	3.02	6.90	9.92
2900	Cut & patch to match existing construction, add, minimum	2%	3%	
3000	Maximum	5%	9%	
3100	Dust protection, add, minimum	1%	2%	
3200	Maximum	4%	11%	
3300	Equipment usage curtailment, add, minimum	1%	1%	
3400	Maximum	3%	10%	
3500	Material handling & storage limitation, add, minimum	1%	1%	
3600	Maximum	6%	7%	
3700	Protection of existing work, add, minimum	2%	2%	
3800	Maximum	5%	7%	
3900	Shift work requirements, add, minimum		5%	
4000	Maximum		30%	

D3020 Heat Generating Systems

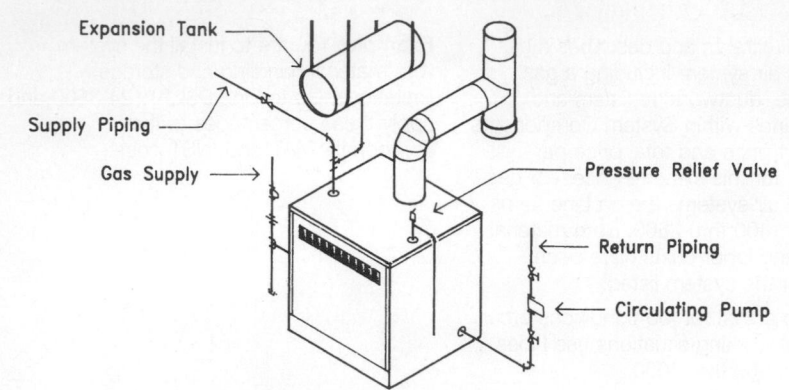

Expansion Tank

Supply Piping

Gas Supply

Pressure Relief Valve

Return Piping

Circulating Pump

Cast Iron Boiler, Hot Water, Gas Fired

This page illustrates and describes boilers including expansion tank, circulating pump and all service piping. Lines within Systems Components give the material and installation price on a cost each basis for the components. Prices for boiler systems are on Line Items D3020 130 1010 thru D3020 138 1040. Material quantities and labor costs have been adjusted for the system listed.

Factors: To adjust for job conditions other than normal working situations use Line D3020 138 2700 thru 4000.

System Components	QUANTITY	UNIT	COST EACH		
			MAT.	INST.	TOTAL
Cast iron boiler, including piping, chimney and accessories					
Boilers, gas fired, std controls, CI, insulated, HW, gross output 100 MBH	1.000	Ea.	2,700	2,200	4,900
Pipe, black steel, Sch 40, threaded, W/coupling & hangers, 10' OC, 3/4" dia.	20.000	L.F.	103.28	285.75	389.03
Pipe, black steel, Sch 40, threaded, W/coupling & hangers, 10' OC, 1" dia.	20.000	L.F.	103.63	313.90	417.53
Elbow, 90°, black, straight, 3/4" dia.	9.000	Ea.	43.65	499.50	543.15
Elbow, 90°, black, straight, 1" dia.	6.000	Ea.	50.70	357	407.70
Tee, black, straight, 3/4" dia.	2.000	Ea.	15.50	172	187.50
Tee, black, straight, 1" dia.	2.000	Ea.	26.50	193	219.50
Tee, black, reducing, 1" dia.	2.000	Ea.	42	193	235
Pipe cap, black, 3/4" dia.	1.000	Ea.	5.50	24	29.50
Union, black with brass seat, 3/4" dia.	2.000	Ea.	43	119	162
Union, black with brass seat, 1" dia.	2.000	Ea.	56	129	185
Pipe nipples, black, 3/4" dia.	5.000	Ea.	11.48	31.75	43.23
Pipe nipples, black, 1" dia.	3.000	Ea.	7.23	21.90	29.13
Valves, bronze, gate, N.R.S., threaded, class 150, 3/4" size	2.000	Ea.	192	77	269
Valves, bronze, gate, N.R.S., threaded, class 150, 1" size	2.000	Ea.	240	81	321
Gas cock, brass, 3/4" size	1.000	Ea.	17.40	35	52.40
Thermometer, stem type, 9" case, 8" stem, 3/4" NPT	2.000	Ea.	402	56	458
Tank, steel, liquid expansion, ASME, painted, 15 gallon capacity	1.000	Ea.	790	83	873
Pump, circulating, bronze, flange connection, 3/4" to 1-1/2" size, 1/8 HP	1.000	Ea.	1,225	232	1,457
Vent chimney, all fuel, pressure tight, double wall, SS, 6" dia.	20.000	L.F.	1,220	450	1,670
Vent chimney, elbow, 90° fixed, 6" dia.	2.000	Ea.	910	90	1,000
Vent chimney, Tee, 6" dia.	2.000	Ea.	532	113	645
Vent chimney, ventilated roof thimble, 6" dia.	1.000	Ea.	278	52	330
Vent chimney, flat roof flashing, 6" dia.	1.000	Ea.	119	45	164
Vent chimney, stack cap, 6" diameter	1.000	Ea.	325	29.50	354.50
Insulation, fiberglass pipe covering, 1" wall, 1" IPS	20.000	L.F.	21.80	118	139.80
TOTAL			9,480.67	6,001.30	15,481.97

D3020 130	Boiler, Cast Iron, Hot Water, Gas	COST EACH		
		MAT.	INST.	TOTAL
1000	For alternate boilers:			
1010	Boiler, cast iron, gas, hot water, 100 MBH	9,475	6,000	15,475
1020	200 MBH	10,900	6,875	17,775
1030	320 MBH	10,700	7,475	18,175
1040	440 MBH	14,900	9,675	24,575
1050	544 MBH	24,600	15,100	39,700
1060	765 MBH	29,400	16,200	45,600
1070	1088 MBH	31,400	17,300	48,700

D3020 Heat Generating Systems

D3020 130	Boiler, Cast Iron, Hot Water, Gas	COST EACH		
		MAT.	INST.	TOTAL
1080	1530 MBH	37,600	23,000	60,600
1090	2312 MBH	44,200	27,000	71,200

D3020 134	Boiler, Cast Iron, Steam, Gas	COST EACH		
		MAT.	INST.	TOTAL
1010	Boiler, cast iron, gas, steam, 100 MBH	7,775	5,825	13,600
1020	200 MBH	18,200	11,000	29,200
1030	320 MBH	20,000	12,300	32,300
1040	544 MBH	28,400	19,200	47,600
1050	765 MBH	31,800	20,000	51,800
1060	1275 MBH	37,300	22,000	59,300
1070	2675 MBH	51,500	32,100	83,600

D3020 136	Boiler, Cast Iron, Hot Water, Gas/Oil	COST EACH		
		MAT.	INST.	TOTAL
1010	Boiler, cast iron, gas & oil, hot water, 584 MBH	30,100	16,700	46,800
1020	876 MBH	38,000	18,000	56,000
1030	1168 MBH	47,800	20,300	68,100
1040	1460 MBH	57,500	24,500	82,000
1050	2044 MBH	62,500	25,300	87,800
1060	2628 MBH	74,500	30,500	105,000

D3020 138	Boiler, Cast Iron, Steam, Gas/Oil	COST EACH		
		MAT.	INST.	TOTAL
1010	Boiler, cast iron, gas & oil, steam, 810 MBH	34,700	21,800	56,500
1020	1360 MBH	39,900	23,200	63,100
1030	2040 MBH	48,600	29,800	78,400
1040	2700 MBH	52,000	33,200	85,200
2700	Cut & patch to match existing construction, add, minimum	2%	3%	
2800	Maximum	5%	9%	
2900	Dust protection, add, minimum	1%	2%	
3000	Maximum	4%	11%	
3100	Equipment usage curtailment, add, minimum	1%	1%	
3200	Maximum	3%	10%	
3300	Material handling & storage limitation, add, minimum	1%	1%	
3400	Maximum	6%	7%	
3500	Protection of existing work, add, minimum	2%	2%	
3600	Maximum	5%	7%	
3700	Shift work requirements, add, minimum		5%	
3800	Maximum		30%	
3900	Temporary shoring and bracing, add, minimum	2%	5%	
4000	Maximum	5%	12%	

For customer support on your Commercial Renovation Costs with RSMeans data, call 800.448.8182.

D3020 Heat Generating Systems

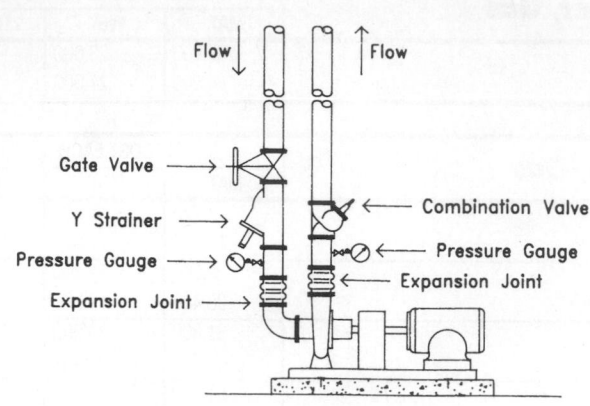

Flow ↓ Flow ↑

Gate Valve →
Y Strainer →
Pressure Gauge →
Expansion Joint →

← Combination Valve
← Pressure Gauge
← Expansion Joint

Base Mounted End—Suction Pump

This page illustrates and describes end suction base mounted circulating pumps, including strainer and piping. Lines within system components give the material and installation price on a cost each basis for the components. Prices for circulating pumps are on Line items D3020 330 1010 thru 1050. Material quantities and labor costs have been adjusted for the system listed.

Factors: To adjust for job conditions other than normal working situations uses Lines D3020 330 2700 thru 4000.

System Components	QUANTITY	UNIT	COST EACH		
			MAT.	INST.	TOTAL
Circulator pump, end suction, including pipe, Valves, fittings and guages.					
Pump, circulating, CI, base mounted, 2-1/2" size, 3 HP, to 150 GPM	1.000	Ea.	9,675	775	10,450
Pipe, black steel, Sch. 40, on yoke & roll hangers, 10' O.C., 2-1/2" dia	12.000	L.F.	199.20	370.68	569.88
Elbow, 90°, weld joint, steel, 2-1/2" pipe size	1.000	Ea.	30.50	182.15	212.65
Flange, weld neck, 150 LB, 2-1/2" pipe size	8.000	Ea.	280	728.56	1,008.56
Valve, iron body, gate, 125 lb., N.R.S., flanged, 2-1/2" size	1.000	Ea.	665	279	944
Strainer, Y type, iron body, flanged, 125 lb., 2-1/2" pipe size	1.000	Ea.	174	282	456
Multipurpose valve, CI body, 2" size	1.000	Ea.	555	98	653
Expansion joint, flanged spool, 6" F to F, 2-1/2" dia	2.000	Ea.	588	228	816
T-O-L, weld joint, socket, 1/4" pipe size, nozzle	2.000	Ea.	16.40	126.66	143.06
T-O-L, weld joint, socket, 1/2" pipe size, nozzle	2.000	Ea.	16.40	132.92	149.32
Control gauges, pressure or vacuum, 3-1/2" diameter dial	2.000	Ea.	41	49	90
Pressure/temperature relief plug, 316 SS, 3/4" OD, 7-1/2" insertion	2.000	Ea.	109	49	158
Insulation, fiberglass pipe covering, 1-1/2" wall, 2-1/2" IPS	12.000	L.F.	37.80	86.40	124.20
Water flow	1.000	Ea.	1,300	1,175	2,475
Pump balancing	1.000	Ea.		345	345
TOTAL			13,687.30	4,907.37	18,594.67

D3020 330	Circulating Pump Systems, End Suction	COST EACH		
		MAT.	INST.	TOTAL
1000	For alternate pump sizes:			
1010	Pump, base mtd with motor, end-suction, 2-1/2" size, 3 HP, to 150 GPM	13,700	4,925	18,625
1020	3" size, 5 HP, to 225 GPM	16,500	5,550	22,050
1030	4" size, 7-1/2 HP, to 350 GPM	19,300	6,600	25,900
1040	5" size, 15 HP, to 1000 GPM	28,100	9,750	37,850
1050	6" size, 25 HP, to 1550 GPM	37,200	11,800	49,000
2700	Cut & patch to match existing construction, add, minimum	2%	3%	
2800	Maximum	5%	9%	
2900	Dust protection, add, minimum	1%	2%	
3000	Maximum	4%	11%	
3100	Equipment usage curtailment, add, minimum	1%	1%	
3200	Maximum	3%	10%	
3300	Material handling & storage limitation, add, minimum	1%	1%	
3400	Maximum	6%	7%	
3500	Protection of existing work, add, minimum	2%	2%	
3600	Maximum	5%	7%	
3700	Shift work requirements, add, minimum		5%	
3800	Maximum		30%	

D3020 Heat Generating Systems

D3020 330	Circulating Pump Systems, End Suction	COST EACH		
		MAT.	INST.	TOTAL
3900	Temporary shoring and bracing, add, minimum	2%	5%	
4000	Maximum	5%	12%	

D3040 Distribution Systems

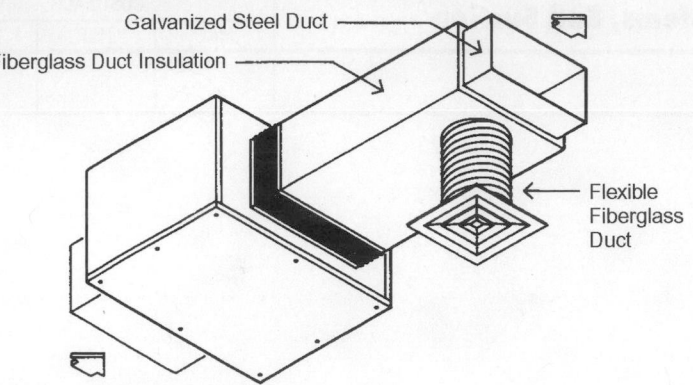

Galvanized Steel Duct

Fiberglass Duct Insulation

Flexible Fiberglass Duct

Horizontal Fan Coil Air Conditioning System

This page illustrates and describes fan coil air conditioners including duct work, duct installation, piping and diffusers. Lines within Systems Components give the material and installation price on a cost each basis for the components. Prices for fan coil A/C unit systems are on Line Items D3040 118 1010 thru D3040 126 1120. Material quantities and labor costs have been adjusted for the system listed.

Factors: To adjust for job conditions other than normal working situations use Lines D3040 126 2700 thru 4000.

System Components	QUANTITY	UNIT	COST EACH		
			MAT.	INST.	TOTAL
Fan coil A/C system, horizontal, with housing, controls, 2 pipe, 1/2 ton					
Fan coil A/C, horizontal housing, filters, chilled water, 1/2 ton cooling	1.000	Ea.	1,450	176	1,626
Pipe, black steel, Sch 40, threaded, W/cplgs & hangers, 10' OC, 3/4" dia.	20.000	L.F.	98.69	273.05	371.74
Elbow, 90°, black, straight, 3/4" dia.	6.000	Ea.	29.10	333	362.10
Tee, black, straight, 3/4" dia.	2.000	Ea.	15.50	172	187.50
Union, black with brass seat, 3/4" dia.	2.000	Ea.	43	119	162
Pipe nipples, 3/4" diam	3.000	Ea.	6.89	19.05	25.94
Valves, bronze, gate, N.R.S., threaded, class 150, 3/4" size	2.000	Ea.	192	77	269
Circuit setter, balance valve, bronze body, threaded, 3/4" pipe size	1.000	Ea.	95	39	134
Insulation, fiberglass pipe covering, 1" wall, 3/4" IPS	20.000	L.F.	20.20	113	133.20
Ductwork, 12" x 8" fabricated, galvanized steel, 12 LF	55.000	Lb.	34.65	495	529.65
Insulation, ductwork, blanket type, fiberglass, 1" thk, 1-1/2 LB density	40.000	S.F.	199.60	618	817.60
Diffusers, aluminum, OB damper, ceiling, perf, 24"x24" panel size, 6"x6"	2.000	Ea.	348	94	442
Ductwork, flexible, fiberglass fabric, insulated, 1"thk, PE jacket, 6" dia	16.000	L.F.	54.72	83.20	137.92
Round volume control damper 6" dia.	2.000	Ea.	74	69	143
Fan coil unit balancing	1.000	Ea.		98	98
Re-heat coil balancing	1.000	Ea.		145	145
Diffuser/register, high, balancing	2.000	Ea.		272	272
TOTAL			2,661.35	3,195.30	5,856.65

D3040 118	Fan Coil A/C Unit, Two Pipe	COST EACH		
		MAT.	INST.	TOTAL
1000	For alternate A/C units:			
1010	Fan coil A/C system, cabinet mounted, controls, 2 pipe, 1/2 Ton	1,100	1,450	2,550
1020	1 Ton	1,550	1,625	3,175
1030	1-1/2 Ton	1,550	1,650	3,200
1040	2 Ton	2,675	2,025	4,700
1050	3 Ton	3,500	2,100	5,600

D3040 120	Fan Coil A/C Unit, Two Pipe, Electric Heat	COST EACH		
		MAT.	INST.	TOTAL
1010	Fan coil A/C system, cabinet mntd, elect. ht, controls, 2 pipe, 1/2 Ton	2,025	1,450	3,475
1020	1 Ton	2,400	1,625	4,025
1030	1-1/2 Ton	2,825	1,650	4,475
1040	2 Ton	4,275	2,050	6,325
1050	3 Ton	7,025	2,100	9,125

D3040 Distribution Systems

D3040 122	Fan Coil A/C Unit, Four Pipe	COST EACH		
		MAT.	INST.	TOTAL
1010	Fan coil A/C system, cabinet mounted, controls, 4 pipe, 1/2 Ton	1,850	2,825	4,675
1020	1 Ton	2,400	3,025	5,425
1030	1-1/2 Ton	2,425	3,050	5,475
1040	2 Ton	3,225	3,150	6,375
1050	3 Ton	4,500	3,400	7,900

D3040 124	Fan Coil A/C, Horizontal, Duct Mount, 2 Pipe	COST EACH		
		MAT.	INST.	TOTAL
1010	Fan coil A/C system, horizontal w/housing, controls, 2 pipe, 1/2 Ton	2,650	3,200	5,850
1020	1 Ton	3,575	5,075	8,650
1030	1-1/2 Ton	4,475	6,825	11,300
1040	2 Ton	5,850	8,475	14,325
1050	3 Ton	7,475	11,800	19,275
1060	3-1/2 Ton	7,775	12,100	19,875
1070	4 Ton	8,200	14,100	22,300
1080	5 Ton	9,950	18,100	28,050
1090	6 Ton	10,100	19,300	29,400
1100	7 Ton	10,200	19,800	30,000
1110	8 Ton	10,200	20,100	30,300
1120	10 Ton	10,900	20,600	31,500

D3040 126	Fan Coil A/C, Horiz., Duct Mount, 2 Pipe, Elec. Ht.	COST EACH		
		MAT.	INST.	TOTAL
1010	Fan coil A/C system, horiz. hsng, elect. ht, ctrls, 2 pipe, 1/2 Ton	2,925	3,200	6,125
1020	1 Ton	3,950	5,075	9,025
1030	1-1/2 Ton	4,825	6,825	11,650
1040	2 Ton	6,425	8,475	14,900
1050	3 Ton	10,100	12,000	22,100
1060	3-1/2 Ton	11,200	12,100	23,300
1070	4 Ton	12,600	14,100	26,700
1080	5 Ton	15,500	18,100	33,600
1090	6 Ton	15,700	19,300	35,000
1100	7 Ton	16,100	20,000	36,100
1110	8 Ton	16,400	20,700	37,100
1120	10 Ton	17,600	19,800	37,400
2700	Cut & patch to match existing construction, add, minimum	2%	3%	
2800	Maximum	5%	9%	
2900	Dust protection, add, minimum	1%	2%	
3000	Maximum	4%	11%	
3100	Equipment usage curtailment, add, minimum	1%	1%	
3200	Maximum	3%	10%	
3300	Material handling & storage limitation, add, minimum	1%	1%	
3400	Maximum	6%	7%	
3500	Protection of existing work, add, minimum	2%	2%	
3600	Maximum	5%	7%	
3700	Shift work requirements, add, minimum		5%	
3800	Maximum		30%	
3900	Temporary shoring and bracing, add, minimum	2%	5%	
4000	Maximum	5%	12%	

For customer support on your Commercial Renovation Costs with RSMeans data, call 800.448.8182.

D3040 Distribution Systems

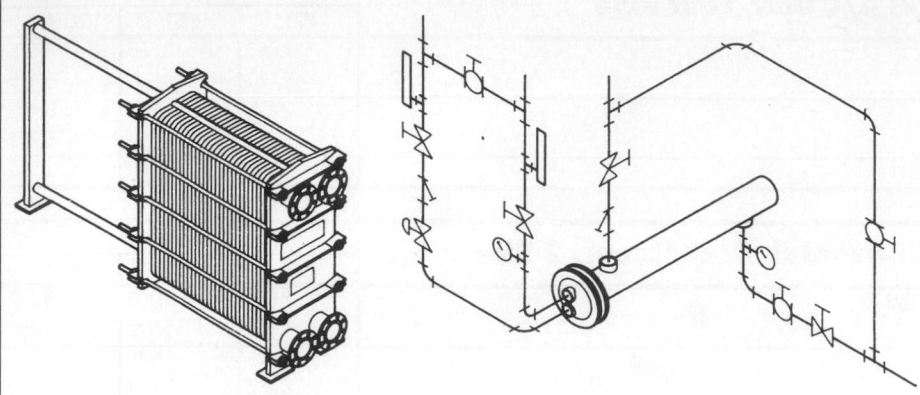

Plate Heat Exchanger   Shell and Tube Heat Exchanger

This page illustrates and describes plate and shell and tube type heat exchangers including pressure gauge and exchanger control system. Lines within Systems Components give the material and installation price on a cost each basis for the components. Prices for heat exchanger systems are on Line Items D3040 610 1010 thru D3040 620 1040. Material quantities and labor costs have been adjusted for the system listed.

Factors To adjust for job conditions other then normal working situations use Lines D3040 620 2700 thru 4000.

System Components	QUANTITY	UNIT	COST EACH		
			MAT.	INST.	TOTAL
Shell and tube type heat exchanger					
with related valves and piping.					
Heat exchanger, 4 pass, 3/4" O.D. copper tubes, by steam at 10 psi, 40 GPM	1.000	Ea.	6,100	355	6,455
Pipe, black steel, Sch 40, threaded, w/couplings & hangers, 10' OC, 2" dia.	40.000	L.F.	664.08	1,111	1,775.08
Elbow, 90°, straight, 2" dia.	16.000	Ea.	504	1,240	1,744
Reducer, black steel, concentric, 2" dia.	2.000	Ea.	71	133	204
Valves, bronze, gate, rising stem, threaded, class 150, 2" size	6.000	Ea.	1,602	423	2,025
Valves, bronze, globe, class 150, rising stem, threaded, 2" size	2.000	Ea.	1,210	141	1,351
Strainers, Y type, bronze body, screwed, 125 lb, 2" pipe size	2.000	Ea.	380	121	501
Valve, electric motor actuated, brass, 2 way, screwed, 1-1/2" pipe size	1.000	Ea.	900	59.50	959.50
Union, black with brass seat, 2" dia.	8.000	Ea.	468	656	1,124
Tee, black, straight, 2" dia.	9.000	Ea.	360	1,143	1,503
Tee, black, reducing run and outlet, 2" dia.	4.000	Ea.	232	508	740
Pipe nipple, black, 2" dia.	21.000	Ea.	6.58	11	17.58
Thermometers, stem type, 9" case, 8" stem, 3/4" NPT	2.000	Ea.	402	56	458
Gauges, pressure or vacuum, 3-1/2" diameter dial	2.000	Ea.	41	49	90
Insulation, fiberglass pipe covering, 1" wall, 2" IPS	40.000	L.F.	67.20	260	327.20
Heat exchanger control system	1.000	Ea.	3,125	2,475	5,600
Coil balancing	1.000	Ea.		145	145
TOTAL			16,132.86	8,886.50	25,019.36

D3040 610	Heat Exchanger, Plate Type	COST EACH		
		MAT.	INST.	TOTAL
1000	For alternate heat exchangers:			
1010	Plate heat exchanger, 400 GPM	62,000	17,900	79,900
1020	800 GPM	100,000	22,800	122,800
1030	1200 GPM	150,500	30,800	181,300
1040	1800 GPM	202,000	38,300	240,300

D3040 620	Heat Exchanger, Shell & Tube	COST EACH		
		MAT.	INST.	TOTAL
1010	Shell & tube heat exchanger, 40 GPM	16,100	8,875	24,975
1020	96 GPM	26,900	15,800	42,700
1030	240 GPM	49,900	21,400	71,300
1040	600 GPM	98,000	30,000	128,000
2700	Cut & patch to match existing construction, add, minimum	2%	3%	
2800	Maximum	5%	9%	
2900	Dust protection, add, minimum	1%	2%	
3000	Maximum	4%	11%	

D3040 Distribution Systems

D3040 620	Heat Exchanger, Shell & Tube	COST EACH		
		MAT.	INST.	TOTAL
3100	Equipment usage curtailment, add, minimum	1%	1%	
3200	Maximum	3%	10%	
3300	Material handling & storage limitation, add, minimum	1%	1%	
3400	Maximum	6%	7%	
3500	Protection of existing work, add, minimum	2%	2%	
3600	Maximum	5%	7%	
3700	Shift work requirements, add, minimum		5%	
3800	Maximum		30%	
3900	Temporary shoring and bracing, add, minimum	2%	5%	
4000	Maximum	5%	12%	

D3050 Terminal & Package Units

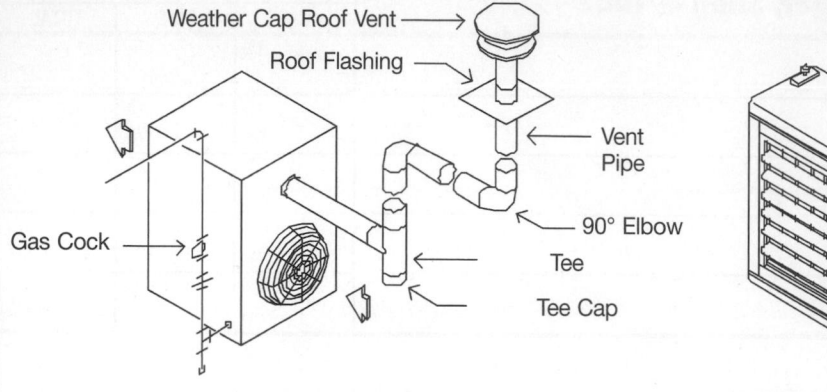

Gas Unit Heater

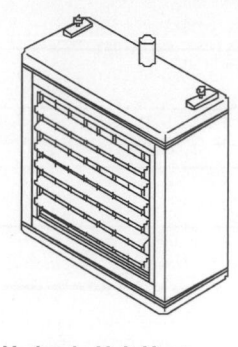

Hydronic Unit Heater

This page illustrates and describes unit heaters including piping and vents. Lines within Systems Components give the material and installation price on a cost each basis for this system. Prices for unit heater systems are on Line items D3050 120 1010 thru D3050 140 1040. Material quantities and labor costs have been adjusted for the system listed.

Factors: To adjust for job conditions other than normal working situations use Lines D3050 180 2700-4000.

System Components	QUANTITY	UNIT	COST EACH MAT.	COST EACH INST.	COST EACH TOTAL
Unit heaters complete with piping, thermostats and chimney as required					
Space heater, propeller fan, 20 MBH output	1.000	Ea.	1,725	166	1,891
Pipe, black steel, Sch 40, threaded, w/coupling & hangers, 10' OC, 3/4" dia	20.000	L.F.	96.39	266.70	363.09
Elbow, 90°, black steel, straight, 1/2" dia.	3.000	Ea.	12.09	154.50	166.59
Elbow, 90°, black steel, straight, 3/4" dia.	3.000	Ea.	14.55	166.50	181.05
Tee, black steel, reducing, 3/4" dia.	1.000	Ea.	14.65	86	100.65
Union, black with brass seat, 3/4" dia.	1.000	Ea.	21.50	59.50	81
Pipe nipples, black, 1/2" dia	2.000	Ea.	8.40	24.60	33
Pipe nipples, black, 3/4" dia	2.000	Ea.	4.59	12.70	17.29
Pipe cap, black, 3/4" dia.	1.000	Ea.	5.50	24	29.50
Gas cock, brass, 3/4" size	1.000	Ea.	17.40	35	52.40
Thermostat, 1 set back, electric, timed	1.000	Ea.	84.50	98	182.50
Wiring, thermostat hook-up,25' of #18-3	1.000	Ea.	6.25	30	36.25
Vent chimney, prefab metal, U.L. listed, gas, double wall, galv. st, 4" dia	12.000	L.F.	108	239.40	347.40
Vent chimney, gas, double wall, galv. steel, elbow 90°, 4" dia.	2.000	Ea.	51	80	131
Vent chimney, gas, double wall, galv. steel, Tee, 4" dia.	1.000	Ea.	36	52	88
Vent chimney, gas, double wall, galv. steel, T cap, 4" dia.	1.000	Ea.	2.41	32.50	34.91
Vent chimney, gas, double wall, galv. steel, roof flashing, 4" dia.	1.000	Ea.	8.70	40	48.70
Vent chimney, gas, double wall, galv. steel, top, 4" dia.	1.000	Ea.	17.95	31	48.95
TOTAL			2,234.88	1,598.40	3,833.28

D3050 120	Unit Heaters, Gas		MAT.	INST.	TOTAL
1000	For alternate unit heaters:				
1010	Space heater, suspended, gas fired, propeller fan, 20 MBH		2,225	1,600	3,825
1020	60 MBH		2,550	1,650	4,200
1030	100 MBH		2,950	1,750	4,700
1040	160 MBH		3,475	1,925	5,400
1050	200 MBH		4,300	2,150	6,450
1060	280 MBH		5,125	2,275	7,400
1070	320 MBH		6,575	2,600	9,175

D3050 130	Unit Heaters, Hydronic		MAT.	INST.	TOTAL
1010	Space heater, suspended, horiz. mount, HW, prop. fan, 20 MBH		2,300	1,700	4,000
1020	60 MBH		3,075	1,925	5,000
1030	100 MBH		3,700	2,100	5,800
1040	150 MBH		4,200	2,325	6,525

D3050 Terminal & Package Units

D3050 130	Unit Heaters, Hydronic	COST EACH		
		MAT.	INST.	TOTAL
1050	200 MBH	4,800	2,600	7,400
1060	300 MBH	5,775	3,075	8,850

D3050 140	Cabinet Unit Heaters, Hydronic	COST EACH		
		MAT.	INST.	TOTAL
1010	Unit heater, cabinet type, horizontal blower, hot water, 20 MBH	2,850	1,675	4,525
1020	60 MBH	4,125	1,850	5,975
1030	100 MBH	4,650	2,025	6,675
1040	120 MBH	4,750	2,100	6,850
2700	Cut & patch to match existing construction, add, minimum	2%	3%	
2800	Maximum	5%	9%	
2900	Dust protection, add, minimum	1%	2%	
3000	Maximum	4%	11%	
3100	Equipment usage curtailment, add, minimum	1%	1%	
3200	Maximum	3%	10%	
3300	Material handling & storage limitation, add, minimum	1%	1%	
3400	Maximum	6%	7%	
3500	Protection of existing work, add, minimum	2%	2%	
3600	Maximum	5%	7%	
3700	Shift work requirements, add, minimum		5%	
3800	Maximum		30%	
3900	Temporary shoring and bracing, add, minimum	2%	5%	
4000	Maximum	5%	12%	

D3050 Terminal & Package Units

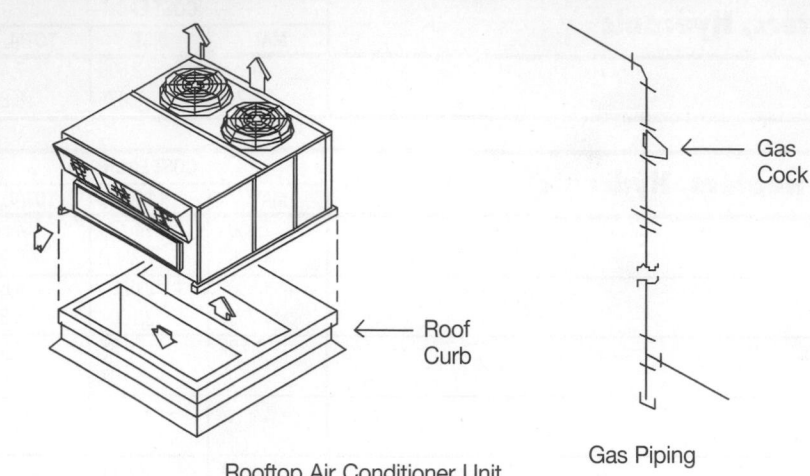

Rooftop Air Conditioner Unit

Gas Cock ←

Roof Curb ←

Gas Piping

This page illustrates and describes packaged air conditioners including rooftop DX unit, pipes and fittings. Lines within Systems Components give the material and installation price on a cost each basis for the components. Prices for A/C systems are on Line Items D3050 175 1010 thru D3050 180 1060. Material quantities and labor costs have been adjusted for the system listed.

Factors: To adjust for job conditions other than normal working situations use Lines D3050 175 2700 thru 4000.

System Components	QUANTITY	UNIT	COST EACH MAT.	COST EACH INST.	COST EACH TOTAL
Packaged rooftop DX unit, gas, with related piping and valves.					
Roof top A/C, curb, economizer, sgl zone, elec cool, gas ht, 5 ton, 112 MBH	1.000	Ea.	4,800	2,525	7,325
Pipe, black steel, Sch 40, threaded, w/cplgs & hangers, 10' OC, 1" dia	20.000	L.F.	101.22	306.60	407.82
Elbow, 90°, black, straight, 3/4" dia.	3.000	Ea.	14.55	166.50	181.05
Elbow, 90°, black, straight, 1" dia.	3.000	Ea.	25.35	178.50	203.85
Tee, black, reducing, 1" dia.	1.000	Ea.	21	96.50	117.50
Union, black with brass seat, 1" dia.	1.000	Ea.	28	64.50	92.50
Pipe nipple, black, 3/4" dia	2.000	Ea.	9.18	25.40	34.58
Pipe nipple, black, 1" dia	2.000	Ea.	4.82	14.60	19.42
Cap, black, 1" dia.	1.000	Ea.	6.65	26	32.65
Gas cock, brass, 1" size	1.000	Ea.	34.50	40.50	75
Control system, pneumatic, Rooftop A/C unit	1.000	Ea.	3,225	2,775	6,000
Rooftop unit heat/cool balancing	1.000	Ea.		525	525
TOTAL			8,270.27	6,744.10	15,014.37

D3050 175	Rooftop Air Conditioner, Const. Volume	COST EACH MAT.	COST EACH INST.	COST EACH TOTAL
1000	For alternate A/C systems:			
1010	A/C, Rooftop, DX cool, gas heat, curb, economizer, fltrs, 5 Ton	8,275	6,750	15,025
1020	7-1/2 Ton	9,875	7,025	16,900
1030	12-1/2 Ton	15,000	7,700	22,700
1040	18 Ton	18,900	8,425	27,325
1050	25 Ton	38,800	9,600	48,400
1060	40 Ton	52,000	12,800	64,800

D3050 180	Rooftop Air Conditioner, Variable Air Volume	COST EACH MAT.	COST EACH INST.	COST EACH TOTAL
1010	A/C, Rooftop, DX cool, gas heat, curb, ecmizr, fltrs, VAV, 12-1/2 Ton	18,300	8,225	26,525
1020	18 Ton	23,100	9,050	32,150
1030	25 Ton	43,800	10,400	54,200
1040	40 Ton	60,500	14,100	74,600
1050	60 Ton	85,000	19,200	104,200
1060	80 Ton	109,500	24,000	133,500
2700	Cut & patch to match existing construction, add, minimum	2%	3%	
2800	Maximum	5%	9%	
2900	Dust protection, add, minimum	1%	2%	
3000	Maximum	4%	11%	
3100	Equipment usage curtailment, add, minimum	1%	1%	
3200	Maximum	3%	10%	

D3050 Terminal & Package Units

D3050 180	Rooftop Air Conditioner, Variable Air Volume	COST EACH		
		MAT.	INST.	TOTAL
3300	Material handling & storage limitation, add, minimum	1%	1%	
3400	Maximum	6%	7%	
3500	Protection of existing work, add, minimum	2%	2%	
3600	Maximum	5%	7%	
3700	Shift work requirements, add, minimum		5%	
3800	Maximum		30%	
3900	Temporary shoring and bracing, add, minimum	2%	5%	
4000	Maximum	5%	12%	

D3050 Terminal & Package Units

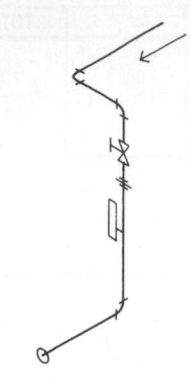
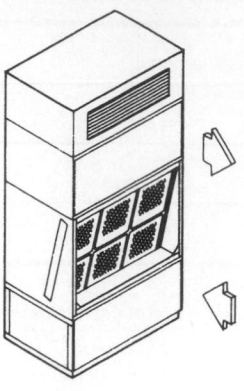
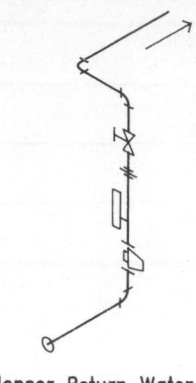

Condenser Supply Water Piping

Condenser Return Water Piping

Self—Contained Air Conditioner Unit, Water Cooled

This page illustrates and describes packaged electrical air conditioning units including multizone control and related piping. Lines within Systems Components give the material and installation price on a cost each basis for the components. Prices for packaged electrical A/C unit systems are on Line Items D3050 210 1010 thru D3050 225 1060. Material quantities and labor costs have been adjusted for the system listed.

Factors: To adjust for job conditions other than normal working situations use Lines D3050 225 2700 thru 4000.

System Components	QUANTITY	UNIT	COST EACH		
			MAT.	INST.	TOTAL
Packaged water cooled electric air conditioning unit with related piping and valves.					
Self-contained, water cooled, elect. heat, not inc tower, 5 ton, const vol	1.000	Ea.	8,975	1,825	10,800
Pipe, black steel, Sch 40, threaded, w/cplg & hangers, 10' OC, 1-1/4" dia.	20.000	L.F.	121.48	336.48	457.96
Elbow, 90°, black, straight, 1-1/4" dia.	6.000	Ea.	83.70	381	464.70
Tee, black, straight, 1-1/4" dia.	2.000	Ea.	43	199	242
Tee, black, reducing, 1-1/4" dia.	2.000	Ea.	75	199	274
Thermometers, stem type, 9" case, 8" stem, 3/4" NPT	2.000	Ea.	402	56	458
Union, black with brass seat, 1-1/4" dia.	2.000	Ea.	81	133	214
Pipe nipple, black, 1-1/4" dia	3.000	Ea.	8.48	23.48	31.96
Valves, bronze, gate, N.R.S., threaded, class 150, 1-1/4" size	2.000	Ea.	326	103	429
Circuit setter, bal valve, bronze body, threaded, 1-1/4" pipe size	1.000	Ea.	180	52.50	232.50
Insulation, fiberglass pipe covering, 1" wall, 1-1/4" IPS	20.000	L.F.	23.60	124	147.60
Control system, pneumatic, A/C Unit with heat	1.000	Ea.	3,225	2,775	6,000
Re-heat coil balancing	1.000	Ea.		145	145
Rooftop unit heat/cool balancing	1.000	Ea.		525	525
TOTAL			13,544.26	6,877.46	20,421.72

D3050 210	AC Unit, Package, Elec. Ht., Water Cooled		COST EACH		
			MAT.	INST.	TOTAL
1000	For alternate A/C systems:				
1010	A/C, Self contained, single pkg., water cooled, elect. heat, 5 Ton		13,500	6,875	20,375
1020	10 Ton		20,800	8,475	29,275
1030	20 Ton		27,900	10,700	38,600
1040	30 Ton		48,400	11,700	60,100
1050	40 Ton		61,500	13,000	74,500
1060	50 Ton		82,500	15,500	98,000

D3050 215	AC Unit, Package, Elec. Ht., Water Cooled, VAV		COST EACH		
			MAT.	INST.	TOTAL
1010	A/C, Self contained, single pkg., water cooled, elect. ht, VAV, 10 Ton		23,800	8,475	32,275
1020	20 Ton		32,800	10,700	43,500
1030	30 Ton		56,000	11,700	67,700
1040	40 Ton		70,500	13,000	83,500
1050	50 Ton		94,000	15,500	109,500
1060	60 Ton		105,500	18,800	124,300

D3050 Terminal & Package Units

D3050 220	AC Unit, Package, Hot Water Coil, Water Cooled	COST EACH		
		MAT.	INST.	TOTAL
1010	A/C, Self contn'd, single pkg., water cool, H/W ht, const. vol, 5 Ton	11,900	9,550	21,450
1020	10 Ton	17,200	11,400	28,600
1030	20 Ton	42,500	14,200	56,700
1040	30 Ton	55,000	18,000	73,000
1050	40 Ton	62,500	21,900	84,400
1060	50 Ton	74,000	26,800	100,800

D3050 225	AC Unit, Package, HW Coil, Water Cooled, VAV	COST EACH		
		MAT.	INST.	TOTAL
1010	A/C, Self contn'd, single pkg., water cool, H/W ht, VAV, 10 Ton	20,200	10,700	30,900
1020	20 Ton	28,800	13,800	42,600
1030	30 Ton	48,900	17,200	66,100
1040	40 Ton	64,000	18,800	82,800
1050	50 Ton	84,000	22,100	106,100
1060	60 Ton	97,000	23,900	120,900
2700	Cut & patch to match existing construction, add, minimum	2%	3%	
2800	Maximum	5%	9%	
2900	Dust protection, add, minimum	1%	2%	
3000	Maximum	4%	11%	
3100	Equipment usage curtailment, add, minimum	1%	1%	
3200	Maximum	3%	10%	
3300	Material handling & storage limitation, add, minimum	1%	1%	
3400	Maximum	6%	7%	
3500	Protection of existing work, add, minimum	2%	2%	
3600	Maximum	5%	7%	
3700	Shift work requirements, add, minimum		5%	
3800	Maximum		30%	
3900	Temporary shoring and bracing, add, minimum	2%	5%	
4000	Maximum	5%	12%	

D3050 Terminal & Package Units

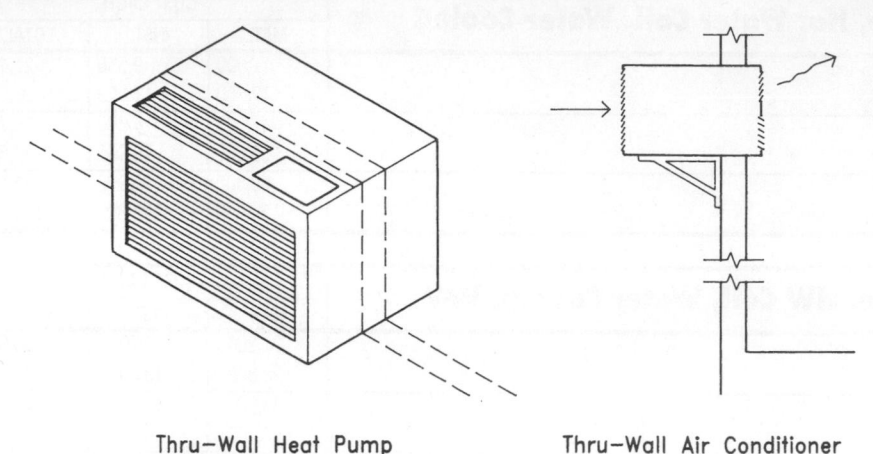

Thru—Wall Heat Pump

Thru—Wall Air Conditioner

This page illustrates and describes thru-wall air conditioning and heat pump units. Lines within Systems Components give the material and installation price on a cost each basis for the components. Prices for thru-wall unit systems are on Line Items D3050 255 1010 thru D3050 260 1050. Material quantities and labor costs have been adjusted for the system listed.

Factors: To adjust for job conditions other than normal working conditions use Lines D3050 260 2700 thru 4000.

System Components	QUANTITY	UNIT	COST EACH		
			MAT.	INST.	TOTAL
Electric thru-wall air conditioning unit.					
A/C unit, thru-wall, electric heat., cabinet, louver, 1/2 ton	1.000	Ea.	790	235	1,025
TOTAL			790	235	1,025

D3050 255	Thru-Wall A/C Unit	COST EACH		
		MAT.	INST.	TOTAL
1000	For alternate A/C units:			
1010	A/C Unit, thru-the-wall, supplemental elect. heat, cabinet, louver, 1/2 Ton	790	235	1,025
1020	3/4 Ton	1,025	282	1,307
1030	1 Ton	1,600	355	1,955
1040	1-1/2 Ton	2,575	590	3,165
1050	2 Ton	2,625	745	3,370

D3050 260	Thru-Wall Heat Pump	COST EACH		
		MAT.	INST.	TOTAL
1010	Heat pump, thru-the-wall, cabinet, louver, 1/2 Ton	3,600	176	3,776
1020	3/4 Ton	3,800	235	4,035
1030	1 Ton	3,525	355	3,880
1040	Supplemental. elect. heat, 1-1/2 Ton	3,550	910	4,460
1050	2 Ton	3,525	940	4,465
2700	Cut & patch to match existing construction, add, minimum	2%	3%	
2800	Maximum	5%	9%	
2900	Dust protection, add, minimum	1%	2%	
3000	Maximum	4%	11%	
3100	Equipment usage curtailment, add, minimum	1%	1%	
3200	Maximum	3%	10%	
3300	Material handling & storage limitation, add, minimum	1%	1%	
3400	Maximum	6%	7%	
3500	Protection of existing work, add, minimum	2%	2%	
3600	Maximum	5%	7%	
3700	Shift work requirements, add, minimum		5%	
3800	Maximum		30%	
3900	Temporary shoring and bracing, add, minimum	2%	5%	
4000	Maximum	5%	12%	

D4010 Sprinklers

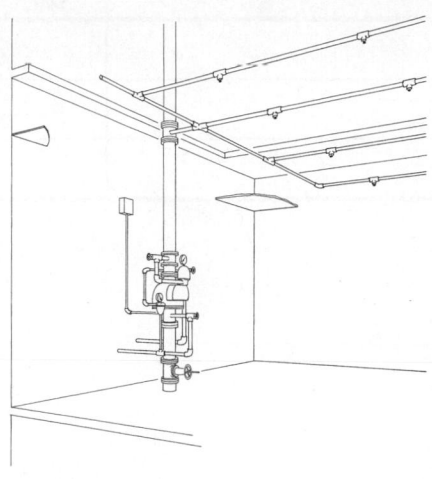

This page illustrates and describes a dry type fire sprinkler system. Lines within System Components give the unit cost of a system for a 2,000 square foot building. Lines D4010 320 1700 thru 2400 give the square foot costs for alternate systems. Both material quantities and labor costs have been adjusted for the system listed.

Factors: To adjust for conditions other than normal working conditions, use Lines D4010 320 2900 thru 4000.

System Components	QUANTITY	UNIT	COST EACH		
			MAT.	INST.	TOTAL
Pre-action fire sprinkler system, ordinary hazard, open area to 2000 S.F.					
On one floor.					
Air compressor	1.000	Ea.	1,250	575	1,825
4" OS & Y valve	1.000	Ea.	730	465	1,195
Pipe riser 4" diameter	10.000	L.F.	185.50	392.60	578.10
Sprinkler head supply piping, 1"	163.000	L.F.	361.50	1,095	1,456.50
Sprinkler head supply piping, 1-1/4"	163.000	L.F.	310.75	860.75	1,171.50
Sprinkler head supply piping, 2-1/2"	163.000	L.F.	670.70	1,064	1,734.70
Pipe fittings, 1"	8.000	Ea.	106	772	878
Pipe fittings, 1-1/4"	8.000	Ea.	172	796	968
Pipe fittings, 2-1/2"	4.000	Ea.	392	620	1,012
Pipe fittings, 4"	2.000	Ea.	700	700	1,400
Detectors	2.000	Ea.	1,600	93	1,693
Sprinkler heads	16.000	Ea.	183.20	744	927.20
Fire department connection	1.000	Ea.	835	269	1,104
TOTAL			7,496.65	8,446.35	15,943

D4010 320	Fire Sprinkler Systems, Dry	COST PER S.F.		
		MAT.	INST.	TOTAL
1700	Ordinary hazard, one floor, area to 2000 S.F./floor	3.77	4.23	8
1800	For each additional floor, add per floor	1.55	3.53	5.08
1900	Area to 3200 S.F./ floor	2.94	4.21	7.15
2000	For each additional floor, add per floor	1.56	3.77	5.33
2100	Area to 5000 S.F./ floor	3.20	4.28	7.48
2200	For each additional floor, add per floor	1.99	3.98	5.97
2300	Area to 8000 S.F./ floor	2.39	3.59	5.98
2400	For each additional floor, add per floor	1.54	3.40	4.94
2500				
2600				
2700				
2800				
2900	Cut & patch to match existing construction, add, minimum	2%	3%	
3000	Maximum	5%	9%	
3100	Dust protection, add, minimum	1%	2%	
3200	Maximum	4%	11%	
3300	Equipment usage curtailment, add, minimum	1%	1%	
3400	Maximum	3%	10%	

695

D4010 Sprinklers

D4010 320	Fire Sprinkler Systems, Dry	COST PER S.F.		
		MAT.	INST.	TOTAL
3500	Material handling & storage limitation, add, minimum	1%	1%	
3600	Maximum	6%	7%	
3700	Protection of existing work, add, minimum	2%	2%	
3800	Maximum	5%	7%	
3900	Shift work requirements, add, minimum		5%	
4000	Maximum		30%	

D4010 Sprinklers

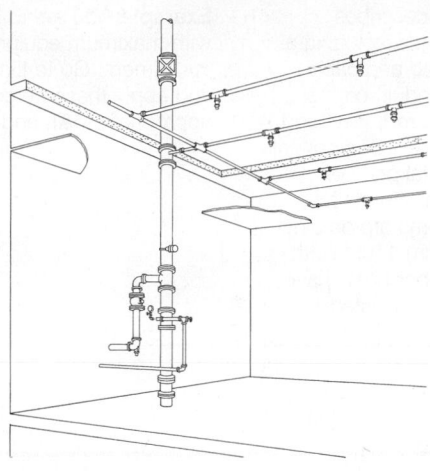

This page illustrates and describes a wet type fire sprinkler system. Lines within System Components give the unit cost of a system for a 2,000 square foot building. Lines D4010 420 1900 thru 2600 give the square foot costs for alternate systems. Both material quantities and labor costs have been adjusted for the system listed.

Factors: To adjust for conditions other than normal working conditions, use Lines D4010 420 2900 thru 4000.

System Components	QUANTITY	UNIT	COST EACH MAT.	COST EACH INST.	COST EACH TOTAL
Wet pipe fire sprinkler system, ordinary hazard, open area to 2000 S.F.					
On one floor.					
4" OS&Y valve	1.000	Ea.	730	465	1,195
Wet pipe alarm valve	1.000	Ea.	2,100	675	2,775
Water motor alarm	1.000	Ea.	490	187	677
3" check valve	1.000	Ea.	415	450	865
Pipe riser, 4" diameter	10.000	L.F.	185.50	392.60	578.10
Water gauges and trim	1.000	Set	3,400	1,350	4,750
Electric fire horn	1.000	Ea.	67.50	107	174.50
Sprinkler head supply piping, 1" diameter	168.000	L.F.	361.50	1,095	1,456.50
Sprinkler head supply piping, 1-1/2" diameter	168.000	L.F.	343.75	957	1,300.75
Sprinkler head supply piping, 2-1/2" diameter	168.000	L.F.	670.70	1,064	1,734.70
Pipe fittings, 1"	8.000	Ea.	106	772	878
Pipe fittings, 1-1/2"	8.000	Ea.	212	856	1,068
Pipe fittings, 2-1/2"	4.000	Ea.	392	620	1,012
Pipe fittings, 4"	2.000	Ea.	700	700	1,400
Sprinkler heads	16.000	Ea.	183.20	744	927.20
Fire department connection	1.000	Ea.	835	269	1,104
TOTAL			11,192.15	10,703.60	21,895.75

D4010 420	Fire Sprinkler Systems, Wet	COST PER S.F. MAT.	COST PER S.F. INST.	COST PER S.F. TOTAL
1900	Ordinary hazard, one floor, area to 2000 S.F./floor	5.60	5.35	10.95
2000	For each additional floor, add per floor	1.58	3.61	5.19
2100	Area to 3200 S.F./floor	4.07	4.86	8.93
2200	For each additional floor, add per floor	1.56	3.77	5.33
2300	Area to 5000 S.F./ floor	3.60	4.68	8.28
2400	For each additional floor, add per floor	1.99	3.98	5.97
2500	Area to 8000 S.F./ floor	2.54	3.83	6.37
2600	For each additional floor, add per floor	1.67	3.61	5.28
2700				
2900	Cut & patch to match existing construction, add minimum	2%	3%	
3000	Maximum	5%	9%	
3100	Dust protection, add, minimum	1%	2%	
3200	Maximum	4%	11%	
3300	Equipment usage curtailment, add, minimum	1%	1%	
3400	Maximum	3%	10%	
3500	Material handling & storage limitation, add, minimum	1%	1%	
3600	Maximum	6%	7%	

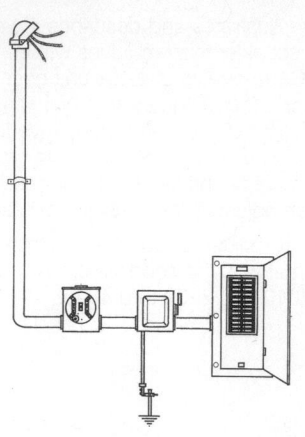

This page illustrates and describes commercial service systems including a meter socket, service head and cable, entrance switch, steel conduit, copper wire, panel board, ground rod, wire, and conduit. Lines within System Components give the unit price and total price on a cost each basis for this system. Prices for commercial service systems are on Line Items D5010 210 1500 thru 1700. Both material quantities and labor costs have been adjusted for the system listed.

Factors: To adjust for job conditions other than normal working situations use Lines D5010 210 2900 thru 4000.

Example: You are to install the system with maximum equipment usage curtailment. Go to Line D5010 210 3400 and apply these percentages to the appropriate MAT. and INST. costs.

System Components	QUANTITY	UNIT	COST EACH		
			MAT.	INST.	TOTAL
Commercial electric service including service breakers, metering 120/208 Volt, 3 phase, 4 wire, feeder, and panel board.					
100 Amp Service					
Meter socket	1.000	Ea.	49.50	225	274.50
Service head	.200	C.L.F.	27.20	144	171.20
Service entrance switch	1.000	Ea.	175	380	555
Rigid steel conduit	20.000	L.F.	78.80	221	299.80
600 volt copper wire #3	1.000	C.L.F.	115	144	259
Panel board, 24 circuits, 20 Amp breakers	1.000	Ea.	1,400	1,200	2,600
Ground rod	1.000	Ea.	22	131	153
TOTAL			1,867.50	2,445	4,312.50

D5010 210	Commercial Service - 3 Phase	COST EACH		
		MAT.	INST.	TOTAL
1500	120/208 Volt, 3 phase, 4 wire service, 60 Amp	1,400	1,900	3,300
1550	100 Amp	1,875	2,450	4,325
1600	200 Amp	3,675	4,300	7,975
1700	400 Amp	6,550	5,375	11,925
1800				
1900				
2000				
2100				
2200				
2300				
2400				
2500				
2900	Cut & patch to match existing construction, add, minimum	2%	3%	
3000	Maximum	5%	9%	
3100	Dust protection, add, minimum	1%	2%	
3200	Maximum	4%	11%	
3300	Equipment usage curtailment, add, minimum	1%	1%	
3400	Maximum	3%	10%	
3500	Material handling & storage limitation, add, minimum	1%	1%	
3600	Maximum	6%	7%	
3700	Protection of existing work, add, minimum	2%	2%	
3800	Maximum	5%	7%	
3900	Shift work requirements, add, minimum		5%	
4000	Maximum		30%	

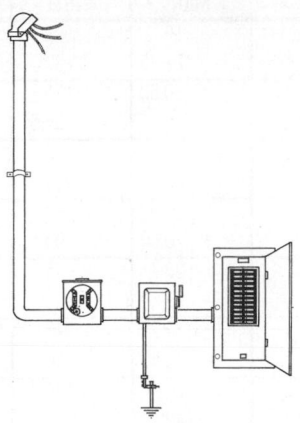

This page illustrates and describes a residential, single phase system including a weather cap, service entrance cable, meter socket, entrance switch, ground rod, ground cable, EMT, and panelboard. Lines within System Components give the unit price and total price on a cost each basis for this system. Prices for a residential, single phase system are also given. Both material quantities and labor costs have been adjusted for the system listed.

Factors: To adjust for job conditions other than normal working situations use Lines D5010 220 2900 thru 4000.

Example: You are to install the system with a minimum equipment usage curtailment. Go to Line D5010 220 3300 and apply these percentages to the appropriate MAT. and INST. costs.

System Components	QUANTITY	UNIT	COST EACH MAT.	COST EACH INST.	COST EACH TOTAL
100 Amp Service, single phase					
Weathercap	1.000	Ea.	8.90	60	68.90
Service entrance cable	.200	C.L.F.	27.20	144	171.20
Meter socket	1.000	Ea.	49.50	225	274.50
Entrance disconnect switch	1.000	Ea.	175	380	555
Ground rod, with clamp	1.000	Ea.	24.50	150	174.50
Ground cable	.100	C.L.F.	14.70	45	59.70
Panelboard, 12 circuit	1.000	Ea.	159	600	759
TOTAL			458.80	1,604	2,062.80
200 Amp Service , single phase					
Weathercap	1.000	Ea.	19.55	90	109.55
Service entrance cable	.200	C.L.F.	65	205	270
Meter socket	1.000	Ea.	105	380	485
Entrance disconnect switch	1.000	Ea.	380	555	935
Ground rod, with clamp	1.000	Ea.	46.50	163	209.50
Ground cable	.100	C.L.F.	22.60	24.80	47.40
3/4" EMT	10.000	L.F.	12.40	55.50	67.90
Panelboard, 24 circuit	1.000	Ea.	280	995	1,275
TOTAL			931.05	2,468.30	3,399.35

D5010 220	Residential Service - Single Phase	COST EACH MAT.	COST EACH INST.	COST EACH TOTAL
2800				
2810	100 Amp Service, single phase	460	1,600	2,060
2820	200 Amp Service, single phase	930	2,475	3,405
2900	Cut & patch to match existing construction, add, minimum	2%	3%	
3000	Maximum	5%	9%	
3100	Dust protection, add, minimum	1%	2%	
3200	Maximum	4%	11%	
3300	Equipment usage curtailment, add, minimum	1%	1%	
3400	Maximum	3%	10%	
3500	Material handling & storage limitation, add, minimum	1%	1%	
3600	Maximum	6%	7%	
3700	Protection of existing work, add, minimum	2%	2%	
3800	Maximum	5%	7%	
3900	Shift work requirements, add, minimum		5%	
4000	Maximum		30%	

D5020 Lighting and Branch Wiring

D5020 180	Selective Price Sheet	COST EACH		
		MAT.	INST.	TOTAL
0100	Using non-metallic sheathed, cable, air conditioning receptacle	19	72	91
0200	Disposal wiring	21.50	80	101.50
0300	Dryer circuit	30.50	131	161.50
0400	Duplex receptacle	19	55.50	74.50
0500	Fire alarm or smoke detector	99	72	171
0600	Furnace circuit & switch	27.50	120	147.50
0700	Ground fault receptacle	56.50	90	146.50
0800	Heater circuit	20.50	90	110.50
0900	Lighting wiring	23	45	68
1000	Range circuit	77	180	257
1100	Switches single pole	12.20	45	57.20
1200	3-way	16.50	60	76.50
1300	Water heater circuit	21	144	165
1400	Weatherproof receptacle	140	120	260
1500	Using BX cable, air conditioning receptacle	28.50	86.50	115
1600	Disposal wiring	30	96	126
1700	Dryer circuit	40.50	156	196.50
1800	Duplex receptacle	28.50	66.50	95
1900	Fire alarm or smoke detector	99	72	171
2000	Furnace circuit & switch	38	144	182
2100	Ground fault receptacle	66	109	175
2200	Heater circuit	27	109	136
2300	Lighting wiring	29	54	83
2400	Range circuit	114	218	332
2500	Switches, single pole	21.50	54	75.50
2600	3-way	23.50	72	95.50
2700	Water heater circuit	34.50	171	205.50
2800	Weatherproof receptacle	147	144	291
2900	Using EMT conduit, air conditioning receptacle	44	107	151
3000	Disposal wiring	47.50	120	167.50
3100	Dryer circuit	54	194	248
3200	Duplex receptacle	44	82.50	126.50
3300	Fire alarm or smoke detector	119	107	226
3400	Furnace circuit & switch	50	180	230
3500	Ground fault receptacle	83	133	216
3600	Heater circuit	40.50	133	173.50
3700	Lighting wiring	38.50	67	105.50
3800	Range circuit	95	266	361
3900	Switches, single pole	37	67	104
4000	3-way	35	90	125
4100	Water heater circuit	42	212	254
4200	Weatherproof receptacle	162	180	342
4300	Using aluminum conduit, air conditioning receptacle	41.50	144	185.50
4400	Disposal wiring	45	160	205
4500	Dryer circuit	58	257	315
4600	Duplex receptacle	39.50	111	150.50
4700	Fire alarm or smoke detector	122	144	266
4800	Furnace circuit & switch	51.50	240	291.50
4900	Ground fault receptacle	87	180	267
5000	Heater circuit	40.50	180	220.50
5100	Lighting wiring	43.50	90	133.50
5200	Range circuit	107	360	467
5300	Switches, single pole	42.50	90	132.50
5400	3-way	40.50	120	160.50
5500	Water heater circuit	48	288	336
5600	Weatherproof receptacle	174	240	414
5700	Using galvanized steel conduit	43.50	153	196.50
5800	Disposal wiring	43.50	171	214.50

D5020 Lighting and Branch Wiring

D5020 180	Selective Price Sheet	COST EACH		
		MAT.	INST.	TOTAL
5900	Dryer circuit	110	277	387
6000	Duplex receptacle	43.50	118	161.50
6100	Fire alarm or smoke detector	146	153	299
6200	Furnace circuit & switch	75	257	332
6300	Ground fault receptacle	89.50	189	278.50
6400	Heater circuit	66.50	189	255.50
6500	Lighting wiring	68.50	96	164.50
6600	Range circuit	109	380	489
6700	Switches, single pole	45	96	141
6800	3-way	91	124	215
6900	Water heater circuit	49	300	349
7000	Weatherproof receptacle	171	257	428
7100				
7200				
7300				
7400				
7500				
7600				
7700				
7800				
7900				
8000				
8100				
8200				
8300				
8400				

For customer support on your Commercial Renovation Costs with RSMeans data, call 800.448.8182.

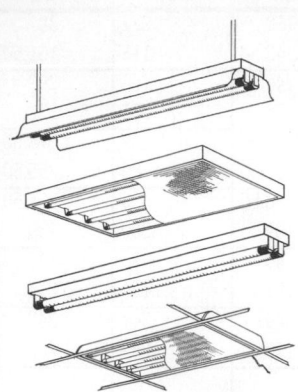

This page illustrates and describes fluorescent lighting systems including a fixture, lamp, outlet box and wiring. Lines within System Components give the unit price and total price on a cost each basis for this system. Prices for fluorescent lighting systems are on Line Items D5020 248 1200 thru 1500. Both material quantities and labor costs have been adjusted for the system listed.

Factors: To adjust for job conditions other than normal working situations use Lines D5020 248 2900 thru 4000.

Example: You are to install the system during evening hours. Go to Line D5020 248 3900 and apply this percentage to the appropriate INST. cost.

System Components			COST EACH		
	QUANTITY	UNIT	MAT.	INST.	TOTAL
Fluorescent lighting, including fixture, lamp, outlet box and wiring.					
Recessed lighting fixture, on suspended system	1.000	Ea.	68.50	153	221.50
Outlet box	1.000	Ea.	5.80	40	45.80
#12 wire	.660	C.L.F.	6.96	43.23	50.19
Conduit, EMT, 1/2" conduit	20.000	L.F.	17.40	84.60	102
TOTAL			98.66	320.83	419.49

D5020 248	Lighting, Fluorescent	COST EACH		
		MAT.	INST.	TOTAL
1200	Recessed lighting fixture, on suspended system	98.50	320	418.50
1300	Surface mounted, 2' x 4', acrylic prismatic diffuser	123	335	458
1400	Strip fixture, 8' long, two 8' lamps	102	315	417
1500	Pendant mounted, industrial, 8' long, with reflectors	137	365	502
1600				
1700				
1800				
1900				
2000				
2100				
2200				
2300				
2900	Cut & patch to match existing construction, add, minimum	2%	3%	
3000	Maximum	5%	9%	
3100	Dust protection, add, minimum	1%	2%	
3200	Maximum	4%	11%	
3300	Equipment usage curtailment, add, minimum	1%	1%	
3400	Maximum	3%	10%	
3500	Material handling & storage limitation, add, minimum	1%	1%	
3600	Maximum	6%	7%	
3700	Protection of existing work, add, minimum	2%	2%	
3800	Maximum	5%	7%	
3900	Shift work requirements, add, minimum		5%	
4000	Maximum		30%	

D5020 Lighting and Branch Wiring

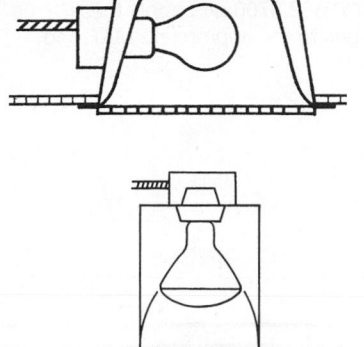

This page illustrates and describes incandescent lighting systems including a fixture, lamp, outlet box, conduit and wiring. Lines within System Components give the unit price and total price on a cost each basis for this system. Prices for an alternate incandescent lighting system are also given. Both material quantities and labor costs have been adjusted for the system listed.

Factors: To adjust for conditions other than normal working situations use Lines D5020 250 2900 thru 4000.

Example: You are to install the system and cut and match existing construction. Go to Line D5020 250 3000 and apply this percentage to the appropriate INST. costs.

System Components			COST EACH		
	QUANTITY	UNIT	MAT.	INST.	TOTAL
Incandescent light fixture, including lamp, outlet box, conduit and wiring.					
Recessed wide reflector with flat glass lens	1.000	Ea.	108	107	215
Outlet box	1.000	Ea.	5.80	40	45.80
Armored cable, 3 wire	.200	C.L.F.	14.20	65	79.20
TOTAL			128	212	340
Recessed flood light fixture, including lamp, outlet box, conduit & wire					
Recessed, R-40 flood lamp with refractor	1.000	Ea.	115	90	205
150 watt R-40 flood lamp	.010	Ea.	6.80	5.55	12.35
Outlet box	1.000	Ea.	5.80	40	45.80
Outlet box cover	1.000	Ea.	.74	11.25	11.99
Romex, 12-2 with ground	.200	C.L.F.	5.30	65	70.30
Conduit, 1/2" EMT	20.000	L.F.	17.40	84.60	102
TOTAL			151.04	296.40	447.44

D5020 250	Lighting, Incandescent	COST EACH		
		MAT.	INST.	TOTAL
2810	Incandescent light fixture, including lamp, outlet box, conduit and wiring.	128	212	340
2820	Recessed flood light fixture, including lamp, outlet box, conduit & wire	151	296	447
2900	Cut & patch to match existing construction, add, minimum	2%	3%	
3000	Maximum	5%	9%	
3100	Dust protection, add, minimum	1%	2%	
3200	Maximum	4%	11%	
3300	Equipment usage curtailment, add, minimum	1%	1%	
3400	Maximum	3%	10%	
3500	Material handling & storage limitation, add, minimum	1%	1%	
3600	Maximum	6%	7%	
3700	Protection of existing work, add, minimum	2%	2%	
3800	Maximum	5%	7%	
3900	Shift work requirements, add, minimum		5%	
4000	Maximum		30%	

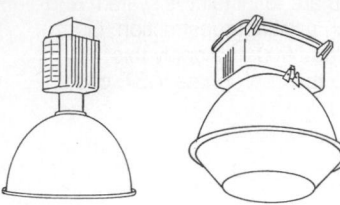

This page illustrates and describes high intensity lighting systems including a lamp, EMT conduit, EMT "T" fitting with cover, and wire. Lines within System Components give the unit price and total price on a cost each basis for this system. Prices for high intensity lighting systems are on Line Items D5020 252 1100 thru 1800. Both material quantities and labor costs have been adjusted for the system listed.

Factors: To adjust for job conditions other than normal working situations use Lines D5020 252 2900 thru 4000.

Example: You are to install the system and protect existing construction. Go to Line D5020 252 3700 and apply these percentages to the appropriate MAT. and INST. costs.

System Components	QUANTITY	UNIT	COST EACH		
			MAT.	INST.	TOTAL
High intensity lighting system consisting of 400 watt metal halide fixture					
And lamp with 1/2″ EMT conduit and fittings using #12 wire, high bay.					
Single unit, 400 watt	1.000	Ea.	455	315	770
Electric metallic tubing (EMT), 1/2″ diameter	30.000	Ea.	26.10	126.90	153
1/2″ EMT "T" fitting with cover	1.000	Ea.	5.80	45	50.80
Wire, 600 volt, type THWN-THHN, copper, solid, #12	.600	Ea.	6.33	39.30	45.63
TOTAL			493.23	526.20	1,019.43

D5020 252	Lighting, High Intensity	COST EACH		
		MAT.	INST.	TOTAL
1100	High bay: 400 watt, metal halide fixture and lamp	495	525	1,020
1200	High pressure sodium and lamp	460	525	985
1400	1000 watt, metal halide	695	570	1,265
1500	High pressure sodium	645	570	1,215
1700	Low bay: 250 watt, metal halide	440	435	875
1800	150 watt, high pressure sodium	400	435	835
1900				
2000				
2100				
2200				
2300				
2400				
2500				
2600				
2700				
2900	Cut & patch to match existing construction, add, minimum	2%	3%	
3000	Maximum	5%	9%	
3100	Dust protection, add, minimum	1%	2%	
3200	Maximum	4%	11%	
3300	Equipment usage curtailment, add, minimum	1%	1%	
3400	Maximum	3%	10%	
3500	Material handling & storage limitation, add, minimum	1%	1%	
3600	Maximum	6%	7%	
3700	Protection of existing work, add, minimum	2%	2%	
3800	Maximum	5%	7%	
3900	Shift work requirements, add, minimum		5%	
4000	Maximum		30%	

D5090   Other Electrical Systems

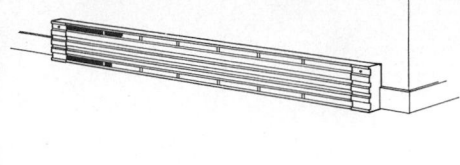

This page illustrates and describes baseboard heat systems including a thermostat, outlet box, breaker, and feed. Lines within System Components give the unit price and total price on a cost each basis for this system. Prices for baseboard heat systems are on Line Items D5090 530 1500 thru 1800. Both material quantities and labor costs have been adjusted for the system listed.

Factors: To adjust for job conditions other than normal working situations use Lines D5090 530 2900 thru 4000.

Example: You are to install the system during evenings and weekends only. Go to Line D5090 530 4000 and apply this percentage to the appropriate INST. cost.

System Components	QUANTITY	UNIT	COST EACH		
			MAT.	INST.	TOTAL
Baseboard heat including thermostat, outlet box, breaker and feed.					
Electric baseboard heater, 4' long	1.000	Ea.	41	107	148
Thermostat, integral	1.000	Ea.	30	45	75
Romex, 12-3 with ground	.400	C.L.F.	17.60	144	161.60
Panel board breaker, 15-50 amp	1.000	Ea.	56.50	90	146.50
Outlet box	1.000	Ea.	5.80	40	45.80
TOTAL			150.90	426	576.90

D5090 530	Heat, Baseboard	COST EACH		
		MAT.	INST.	TOTAL
1400	Heat, electric baseboard			
1500	2' long	140	410	550
1550	4' long	151	425	576
1600	6' long	166	465	631
1700	8' long	179	500	679
1800	10' long	280	535	815
1900				
2000				
2100				
2200				
2300				
2400				
2500				
2600				
2700				
2900	Cut & patch to match existing construction, add, minimum	2%	3%	
3000	Maximum	5%	9%	
3100	Dust protection, add, minimum	1%	2%	
3200	Maximum	4%	11%	
3300	Equipment usage curtailment, add, minimum	1%	1%	
3400	Maximum	3%	10%	
3500	Material handling & storage limitation, add, minimum	1%	1%	
3600	Maximum	6%	7%	
3700	Protection of existing work, add, minimum	2%	2%	
3800	Maximum	5%	7%	
3900	Shift work requirements, add, minimum		5%	
4000	Maximum		30%	

This page illustrates and describes kitchen systems including top and bottom cabinets, custom laminated plastic top, single bowl sink, and appliances. Lines within System Components give the unit price and total price on a cost each basis for this system. Prices for kitchen systems are on Line Items E1090 310 1400 thru 1600. Both material quantities and labor costs have been adjusted for the system listed.

Factors: To adjust for job conditions other than normal working situations use Lines E1090 310 2900 thru 4000.

Example: You are to install the system and protect the work area from dust. Go to Line E1090 310 3200 and apply these percentages to the appropriate MAT. and INST. costs.

System Components	QUANTITY	UNIT	COST EACH MAT.	COST EACH INST.	COST EACH TOTAL
Kitchen cabinets including wall and base cabinets, custom laminated					
Plastic top,sink & appliances,no plumbing or electrical rough-in included.					
Prefinished wood cabinets, average quality, wall and base	20.000	L.F.	4,440	880	5,320
Custom laminated plastic counter top	20.000	L.F.	380	440	820
Stainless steel sink, 22" x 25"	1.000	Ea.	750	249	999
Faucet, top mount	1.000	Ea.	91.50	77.50	169
Dishwasher, built-in	1.000	Ea.	325	375	700
Compactor, built-in	1.000	Ea.	760	132	892
Range hood, vented, 30" wide	1.000	Ea.	103	279	382
TOTAL			6,849.50	2,432.50	9,282

E1090 310	Kitchens	COST EACH MAT.	COST EACH INST.	COST EACH TOTAL
1400	Prefinished wood cabinets, average quality	6,850	2,425	9,275
1500	Prefinished wood cabinets, high quality	12,700	2,775	15,475
1600	Custom cabinets, built in place, high quality	17,000	3,200	20,200
1700				
1800				
1900				
2000				
2100				
2200	NOTE: No plumbing or electric rough-ins are included in the above			
2300	Prices, for plumbing see Division 22, for electric see Division 26.			
2400				
2500				
2600				
2700				
2900	Cut & patch to match existing construction, add, minimum	2%	3%	
3000	Maximum	5%	9%	
3100	Dust protection, add, minimum	1%	2%	
3200	Maximum	4%	11%	
3300	Equipment usage curtailment, add, minimum	1%	1%	
3400	Maximum	3%	10%	
3500	Material handling & storage limitation, add, minimum	1%	1%	
3600	Maximum	6%	7%	
3700	Protection of existing work, add, minimum	2%	2%	
3800	Maximum	5%	7%	
3900	Shift work requirements, add, minimum		5%	
4000	Maximum		30%	

E1090 315	Selective Price Sheet	COST EACH		
		MAT.	INST.	TOTAL
0100	Cabinets standard wood, base, one drawer one door, 12" wide	330	53	383
0200	15" wide	345	55	400
0300	18" wide	375	56.50	431.50
0400	21" wide	390	58	448
0500	24" wide	455	59	514
0600				
0700	Two drawers two doors, 27" wide	490	60	550
0800	30" wide	535	61.50	596.50
0900	33" wide	555	63	618
1000	36" wide	575	65	640
1100	42" wide	610	66.50	676.50
1200	48" wide	650	69.50	719.50
1300	Drawer base (4 drawers), 12" wide	350	53	403
1400	15" wide	355	55	410
1500	18" wide	390	56.50	446.50
1600	24" wide	430	59	489
1700	Sink or range base, 30" wide	445	61.50	506.50
1800	33" wide	475	63	538
1900	36" wide	500	65	565
2000	42" wide	525	66.50	591.50
2100	Corner base, 36" wide	820	73	893
2200	Lazy Susan with revolving door	1,050	80	1,130
2300	Cabinets standard wood, wall two doors, 12" high, 30" wide	300	53	353
2400	36" wide	83.50	55	138.50
2500	15" high, 30" high	305	55	360
2600	36" wide	365	58	423
2700	24" high, 30" wide	410	56.50	466.50
2800	36" wide	450	58	508
2900	30" high, 30" wide	450	68	518
3000	36" wide	515	70	585
3100	42" wide	555	71	626
3200	48" wide	625	71.50	696.50
3300	One door, 30" high, 12" wide	287	60	347
3400	15" wide	300	61.50	361.50
3500	18" wide	330	63	393
3600	24" wide	385	65	450
3700	Corner, 30" high, 24" wide	440	73	513
3800	36" wide	535	80	615
3900	Broom, 84" high, 24" deep, 18" wide	820	132	952
4000	Oven, 84" high, 24" deep, 27" wide	1,250	165	1,415
4100	Valance board, 4' long	79.50	13.30	92.80
4200	6" long	119	19.90	138.90
4300	Counter tops, laminated plastic, stock 25" wide w/backsplash, min.	19	22	41
4400	Maximum	43	26.50	69.50
4500	Custom, 7/8" thick, no splash	38	22	60
4600	Cove splash	45	22	67
4700	1-1/4" thick, no splash	42.50	23.50	66
4800	Square splash	47.50	23.50	71
4900	Post formed	13.85	22	35.85
5000				
5100	Maple laminated 1-1/2" thick, no splash	97	23.50	120.50
5200	Square splash	116	23.50	139.50
5300				
5400				
5500	Appliances, range, free standing, minimum	515	104	619
5600	Maximum	2,525	259	2,784
5700	Built-in, minimum	880	120	1,000
5800	Maximum	1,925	660	2,585

E1090   Other Equipment

E1090 315	Selective Price Sheet	COST EACH		
		MAT.	INST.	TOTAL
5900	Counter top range 4 burner, minimum	360	120	480
6000	Maximum	1,925	240	2,165
6100	Compactor, built-in, minimum	760	132	892
6200	Maximum	1,275	219	1,494
6300	Dishwasher, built-in, minimum	325	375	700
6400	Maximum	500	745	1,245
6500	Garbage disposer, sink-pipe, minimum	116	149	265
6600	Maximum	224	149	373
6700	Range hood, 30″ wide, 2 speed, minimum	103	279	382
6800	Maximum	1,000	465	1,465
6900	Refrigerator, no frost, 12 cu. ft.	485	104	589
7000	20 cu. ft.	590	173	763
7100	Plumb. not incl. rough-ins, sinks porc. C.I., single bowl, 21″ x 24″	335	249	584
7200	21″ x 30″	725	249	974
7300	Double bowl, 20″ x 32″	415	290	705
7400				
7500	Stainless steel, single bowl, 19″ x 18″	675	249	924
7600	22″ x 25″	750	249	999
7700				
7800				
7900				
8000				
8100				
8200				
8300				
8400				

For customer support on your Commercial Renovation Costs with RSMeans data, call 800.448.8182.

G1030 Site Earthwork

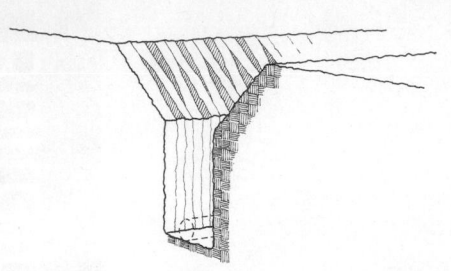

This page illustrates and describes utility trench excavation with backfill systems including a trench, backfill, excavated material, utility pipe or ductwork not included. Lines within System Components give the unit price and total price on a linear foot basis for this system. Prices for utility trench excavation with backfill systems are on Line Items G1030 810 1100 thru 2500. Both material

quantities and labor costs have been adjusted for the system listed.

Factors: To adjust for job conditions other than normal working situations use Lines G1030 810 3100 thru 4000.

Example: You are to install the system and protect existing construction. Go to Line G1030 810 3600 and apply these percentages to the appropriate TOTAL cost.

System Components			COST PER L.F.		
	QUANTITY	UNIT	MAT.	INST.	TOTAL
Cont. utility trench excav. with 3/8 C.Y. wheel mtd. backhoe, Backfilling with excav. mat. and comp. in 12″ lifts. Trench is 2′ wide by 4′ dp. Cost of utility piping or ductwork is not incl. Cost based on an Excavation production rate of 150 C.Y. daily no hauling included.					
Machine excavate trench, 2′ wide by 4′ deep	.296	C.Y.	.68	2.35	3.03
Dozer backfill with excavated material	.296	C.Y.	.36	.65	1.01
Compact in 12″ lifts, vibrating plate	.296	C.Y.	.07	2.59	2.66
TOTAL			1.11	5.59	6.70

G1030 810	Excavation, Utility Trench	COST PER L.F.		
		MAT.	INST.	TOTAL
1100	2′ wide x 2′ deep	.56	2.81	3.37
1200	3′ deep	.84	4.19	5.05
1250	4′ deep	1.11	5.60	6.70
1300	With sloping sides, 2′ wide x 5′ deep	3.12	15.75	18.85
1400	6′ deep	4.15	21	25
1500	7′ deep	5.35	27	32.50
1600	8′ deep	6.65	33.50	40
1700	9′ deep	8.10	41	49
1800	10′ deep	9.70	49	58.50
2000	For hauling excavated material up to 2 miles & backfilling w/ gravel			
2100	Gravel, 2′ wide by 2′ deep, add	3.40	3.79	7.19
2200	4′ deep, add	6.80	7.55	14.35
2300	6′ deep, add	25.50	28	53.50
2400	8′ deep, add	41	45.50	86.50
2500	10′ deep, add	59.50	66	125.50
2600				
2700				
2800	For shallow, hand excavated trenches, no backfill, to 6′ dp, light soil		64.50	64.50
2900	Heavy soil		129	129
3000				
3100	Equipment usage curtailment, add, minimum	1%	1%	
3200	Maximum	3%	10%	
3300	Material handling & storage limitation, add, minimum	1%	1%	
3400	Maximum	6%	7%	
3500	Protection of existing work, add, minimum	2%	2%	
3600	Maximum	5%	7%	
3700	Shift work requirements, add, minimum		5%	
3800	Maximum		30%	
3900	Temporary shoring and bracing, add, minimum	2%	5%	
4000	Maximum	5%	12%	

G20 Site Improvements

G2010 Roadways

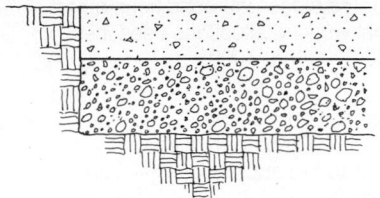

This page illustrates and describes driveway systems including concrete slab, gravel base, broom finish, compaction, and joists. Lines within System Components give the unit price and total price on a cost each basis for this system. Prices for driveway systems are on Line Items G2010 240 1200 thru 2700. Both material quantities and labor costs have been adjusted for the system listed.

Factors: To adjust for job conditions other than normal working situations use Lines G2010 240 2900 thru 4000.

Example: You are to install the system and match existing construction. Go to Line G2010 240 3000 and apply these percentages to the appropriate MAT. and INST. costs.

System Components	QUANTITY	UNIT	COST EACH		
			MAT.	INST.	TOTAL
Complete driveway, 10' x 30', 4" concrete slab, 3000 psi, on 6" crushed					
Stone base, concrete broom finished and cured with 10' x 10' joints.					
Grade and compact subgrade	33.000	S.Y.		72.93	72.93
Place and compact 6" crushed stone base	33.000	S.Y.	196.35	48.84	245.19
Place and remove edgeforms (4 use)	80.000	L.F.	25.60	332	357.60
Concrete for slab, 3000 psi	4.000	C.Y.	532		532
Place concrete, 4" thick slab, direct chute	4.000	C.Y.		118.04	118.04
Screed, float and broom finish slab	4.000	C.Y.		279	279
Spray on membrane curing compound	3.000	S.F.	40.05	32.70	72.75
Saw cut 3" deep joints	60.000	L.F.	9	102.60	111.60
TOTAL			803	986.11	1,789.11

G2010 240	Driveways	COST EACH		
		MAT.	INST.	TOTAL
1200	Concrete: 10' x 30' with 4" concrete on 6" crushed stone	805	985	1,790
1300	With 6" concrete	1,075	1,050	2,125
1400	20' x 30' with 4" concrete	1,600	1,875	3,475
1500	With 6" concrete	2,150	2,000	4,150
1600	10' wide, for each additional 10' length over 30', 4" thick, add	257	315	572
1700	6" thick, add	355	340	695
1800	20' wide, for each additional 10' length over 30', 4" thick, add	510	600	1,110
1900	6" thick, add	675	640	1,315
2000	Asphalt: 10' x 30' with 2" binder, 1" topping on 6" cr. st., sealed	635	1,125	1,760
2100	3" binder, 1" topping	765	1,125	1,890
2200	20' x 30' with 2" binder, 1" topping	1,275	1,375	2,650
2300	3" binder, 1" topping	1,525	1,400	2,925
2400	10' wide for each add'l. 10' length over 30', 2" b + 1" t, add	212	86.50	298.50
2500	3" binder, 1" topping	255	91	346
2600	20' wide for each add'l. 10' length over 30', 2" b + 1" t, add	425	173	598
2700	3" binder, 1" topping	510	182	692
2800				
2900	Cut & patch to match existing construction, add, minimum	2%	3%	
3000	Maximum	5%	9%	
3100	Dust protection, add, minimum	1%	2%	
3200	Maximum	4%	11%	
3300	Equipment usage curtailment, add, minimum	1%	1%	
3400	Maximum	3%	10%	
3500	Material handling & storage limitation, add, minimum	1%	1%	
3600	Maximum	6%	7%	
3700	Protection of existing work, add, minimum	2%	2%	
3800	Maximum	5%	7%	
3900	Shift work requirements, add, minimum		5%	
4000	Maximum		30%	

For customer support on your Commercial Renovation Costs with RSMeans data, call 800.448.8182.

G2020 Parking Lots

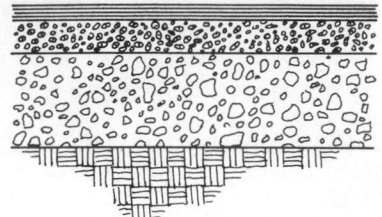

This page illustrates and describes asphalt parking lot systems including asphalt binder, topping, crushed stone base, painted parking stripes and concrete parking blocks. Lines within System Components give the unit price and total price per square yard for this system. Prices for asphalt parking lot systems are on Line Items G2020 230 1400 thru 2500. Both material quantities and labor costs have been adjusted for the system listed.

Factors: To adjust for job conditions other than normal working situations use Lines G2020 230 2900 thru 4000.

Example: You are to install the system and match existing construction. Go to Line G2020 230 2900 and apply these percentages to the appropriate MAT. and INST. costs.

System Components			COST PER S.Y.		
	QUANTITY	UNIT	MAT.	INST.	TOTAL
Parking lot consisting of 2″ asphalt binder and 1″ topping on 6″					
Crushed stone base with painted parking stripes and concrete parking blocks.					
Fine grade and compact subgrade	1.000	S.Y.		2.21	2.21
6″ crushed stone base, stone	.320	Ton	5.95	1.48	7.43
2″ asphalt binder	1.000	S.Y.	7.75	1.45	9.20
1″ asphalt topping	1.000	S.Y.	3.85	.95	4.80
Paint parking stripes	.500	L.F.	.06	.10	.16
6″ x 10″ x 6′ precast concrete parking blocks	.020	Ea.	1.19	.44	1.63
Mobilization of equipment	.005	Ea.		4.27	4.27
TOTAL			18.80	10.90	29.70

G2020 230	Parking Lots, Asphalt	COST PER S.Y.		
		MAT.	INST.	TOTAL
1400	2″ binder plus 1″ topping on 6″ crushed stone	18.80	10.90	29.70
1500	9″ crushed stone	22	11.05	33.05
1600	12″ crushed stone	25	11.20	36.20
1700	On bank run gravel, 6″ deep	16.80	10.25	27.05
1800	9″ deep	18.80	10.40	29.20
1900	12″ deep	21	10.60	31.60
2000	3″ binder plus 1″ topping on 6″ crushed stone	22.50	11.30	33.80
2100	9″ deep crushed stone	25.50	11.40	36.90
2200	12″ deep crushed stone	28.50	11.60	40.10
2300	On bank run gravel, 6″ deep	20.50	10.65	31.15
2400	9″ deep	22.50	10.85	33.35
2500	12″ deep	24.50	11	35.50
2600				
2700				
2900	Cut & patch to match existing construction, add, minimum	2%	3%	
3000	Maximum	5%	9%	
3100	Dust protection, add, minimum	1%	2%	
3200	Maximum	4%	11%	
3300	Equipment usage curtailment, add, minimum	1%	1%	
3400	Maximum	3%	10%	
3500	Material handling & storage limitation, add, minimum	1%	1%	
3600	Maximum	6%	7%	
3700	Protection of existing work, add, minimum	2%	2%	
3800	Maximum	5%	7%	
3900	Shift work requirements, add, minimum		5%	
4000	Maximum		30%	

G2020 Parking Lots

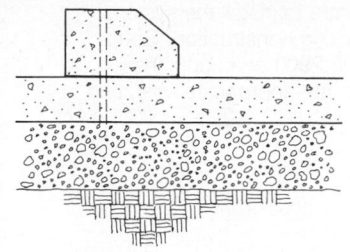

This page illustrates and describes concrete parking lot systems including a concrete slab, unreinforced, compacted gravel base, broom finished, cured, joints, painted parking stripes, and parking blocks. Lines within System Components give the unit price and total price per square yard for this system. Prices for concrete parking lot systems are on Line Items G2020 240 1500 thru 2500. All materials have been adjusted according to the system listed.

Factors: To adjust for job conditions other than normal working situations use Lines G2020 240 2900 thru 4000.

Example: You are to install the system and protect existing construction. Go to Line G2020 240 3800 and apply these percentages to the appropriate MAT. and INST. costs.

System Components			COST PER S.Y.		
	QUANTITY	UNIT	MAT.	INST.	TOTAL
Parking lot, 4" slab, 3000 psi concrete on 6" crushed stone base					
With 12' x 12' joints, painted parking strips, conc. parking blocks.					
Grade and compact subgrade	1.000	S.Y.		2.21	2.21
6" crushed stone base, stone	.320	Ton	5.95	1.48	7.43
Place and remove edge forms, 4 use	.250	L.F.	.08	1.04	1.12
Concrete for slab, 3000 psi	1.000	S.F.	14.63		14.63
Place concrete, 4" thick slab, direct chute	1.000	S.F.		3.25	3.25
Spray on membrane curing compound	1.000	S.F.	1.20	.69	1.89
Saw cut 3" deep joints	2.250	L.F.	.34	3.85	4.19
Paint parking stripes	.500	L.F.	.06	.10	.16
6" x 10" x 6' precast concrete parking blocks	.020	Ea.	1.19	.44	1.63
TOTAL			23.45	13.06	36.51

G2020 240	Parking Lots, Concrete	COST PER S.Y.		
		MAT.	INST.	TOTAL
1500	4" concrete slab, 3000 psi on 3" crushed stone	20.50	13	33.50
1550	6" crushed stone	23.50	13.05	36.55
1600	9" crushed stone	26.50	13.15	39.65
1700	On bank run gravel, 6" deep	21.50	12.40	33.90
1800	9" deep	23.50	12.60	36.10
1900	On compacted subgrade only	17.50	11.60	29.10
2000	6" concrete slab, 3000 psi on 3" crushed stone	27.50	14.60	42.10
2100	6" crushed stone	30.50	14.60	45.10
2200	9" crushed stone	33.50	14.75	48.25
2300	On bank run gravel, 6" deep	28.50	13.95	42.45
2400	9" deep	30.50	14.15	44.65
2500	On compacted subgrade only	30.50	14.60	45.10
2600				
2700				
2900	Cut & patch to match existing construction, add, minimum	2%	3%	
3000	Maximum	5%	9%	
3100	Dust protection, add, minimum	1%	2%	
3200	Maximum	4%	11%	
3300	Equipment usage curtailment, add, minimum	1%	1%	
3400	Maximum	3%	10%	
3500	Material handling & storage limitation, add, minimum	1%	1%	
3600	Maximum	6%	7%	
3700	Protection of existing work, add, minimum	2%	2%	
3800	Maximum	5%	7%	
3900	Shift work requirements, add, minimum		5%	
4000	Maximum		30%	

715

G2030 Pedestrian Paving

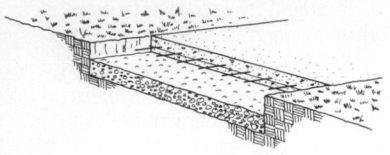

This page illustrates and describes sidewalk systems including concrete, welded wire and broom finish. Lines within System Components give unit price and total price per square foot for this system. Prices for sidewalk systems are on Line Items G2030 105 1700 thru 1900. Both material quantities and labor costs have been adjusted for the system listed.

Factors: To adjust for job conditions other than normal working situations use Lines G2030 105 2900 thru 4000.

Example: You are to install the system and match existing construction. Go to Line G2030 105 2900 and apply these percentages to the appropriate MAT. and INST. costs.

System Components	QUANTITY	UNIT	COST PER S.F. MAT.	COST PER S.F. INST.	COST PER S.F. TOTAL
4″ thick concrete sidewalk with welded wire fabric					
3000 psi air entrained concrete, broom finish.					
Gravel fill, 4″ deep	.012	C.Y.	.25	.09	.34
Compact fill	.012	C.Y.		.03	.03
Hand grade	1.000	S.F.		2.32	2.32
Edge form	.250	L.F.	.08	1.04	1.12
Welded wire fabric	.011	S.F.	.18	.44	.62
Concrete, 3000 psi air entrained	.012	C.Y.	1.60		1.60
Place concrete	.012	C.Y.		.36	.36
Broom finish	1.000	S.F.		.93	.93
TOTAL			2.11	5.21	7.32

G2030 105	Sidewalks	COST PER S.F. MAT.	COST PER S.F. INST.	COST PER S.F. TOTAL
1700	Asphalt (bituminous), 2″ thick	1.08	2.97	4.05
1750	4″ thick	2.11	5.20	7.31
1800	Brick, on sand, bed, 4.5 brick per S.F.	3.14	11.70	14.84
1900	Flagstone, slate, 1″ thick, rectangular	17.45	12.70	30.15
2000				
2100				
2200				
2300				
2400				
2500				
2600				
2700				
2900	Cut & patch to match existing construction, add, minimum	2%	3%	
3000	Maximum	5%	9%	
3100	Dust protection, add, minimum	1%	2%	
3200	Maximum	4%	11%	
3300	Equipment usage curtailment, add, minimum	1%	1%	
3400	Maximum	3%	10%	
3500	Material handling & storage limitation, add, minimum	1%	1%	
3600	Maximum	6%	7%	
3700	Protection of existing work, add, minimum	2%	2%	
3800	Maximum	5%	7%	
3900	Shift work requirements, add, minimum		5%	
4000	Maximum		30%	

G2050 Landscaping

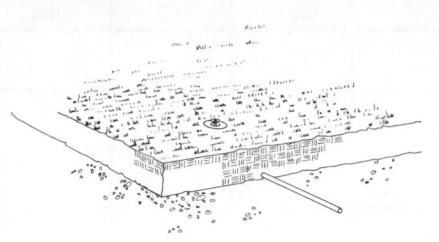

This page describes landscaping—lawn establishment systems including loam, lime, fertilizer, side and top mulching. Lines within system components give the unit price and total price per square yard for this system. Prices for landscaping—lawn establishment systems are on Line Items G2050 420 1000 thru 1200. Both material quantities and labor costs have been adjusted for the system listed.

Factors: To adjust for job conditions other than normal working situations use Lines G2050 420 2900 thru 4000.

Example: You are to install the system and provide dust protection. Go to Line G2050 420 3200 and apply these percentages to the appropriate MAT. and INST. costs.

System Components	QUANTITY	UNIT	COST PER S.Y. MAT.	COST PER S.Y. INST.	COST PER S.Y. TOTAL
Establishing lawns with loam, lime, fertilizer, seed and top mulching					
On rough graded areas.					
Furnish and place loam 4" deep	.110	C.Y.	3.82	1.01	4.83
Fine grade, lime, fertilize and seed	1.000	S.Y.	.51	3.60	4.11
Hay mulch, 1 bale/M.S.F.	1.000	S.Y.	.28	.63	.91
Rolling with hand roller	1.000	S.Y.		1.14	1.14
TOTAL			4.61	6.38	10.99

G2050 420	Landscaping - Lawn Establishment	COST PER S.Y. MAT.	COST PER S.Y. INST.	COST PER S.Y. TOTAL
1000	Establishing lawns with loam, lime, fertilizer, seed and hay mulch	4.61	6.40	11.01
1100	Above system with jute mesh in place of hay mulch	5.35	6.50	11.85
1200	Above system with sod in place of seed	6.40	2.26	8.66
1300				
1400				
1500				
1600				
1700				
1800				
1900				
2000				
2100				
2200				
2300				
2400				
2500				
2600				
2700				
2900	Cut & patch to match existing construction, add, minimum	2%	3%	
3000	Maximum	5%	9%	
3100	Dust protection, add, minimum	1%	2%	
3200	Maximum	4%	11%	
3300	Equipment usage curtailment, add, minimum	1%	1%	
3400	Maximum	3%	10%	
3500	Material handling & storage limitation, add, minimum	1%	1%	
3600	Maximum	6%	7%	
3700	Protection of existing work, add, minimum	2%	2%	
3800	Maximum	5%	7%	
3900	Shift work requirements, add, minimum		5%	
4000	Maximum		30%	

G2050 Landscaping

G2050 510	Selective Price Sheet	COST PER UNIT		
		MAT.	INST.	TOTAL
0100	Shrubs and trees, Evergreen, in prepared beds			
0200				
0300	Arborvitae pyramidal, 4'-5'	113	100	213
0400	Globe, 12"-15"	28.50	16.45	44.95
0500	Cedar, blue, 8'-10'	263	166	429
0600	Hemlock, Canadian, 2-1/2'-3'	35.50	44	79.50
0700	Juniper, andora, 18"-24"	55	19.75	74.75
0800	Wiltoni, 15"-18"	29.50	19.75	49.25
0900	Skyrocket, 4-1/2'-5'	121	54.50	175.50
1000	Blue pfitzer, 2'-2-1/2'	43.50	36	79.50
1100	Ketleerie, 2-1/2'-3'	61.50	31.50	93
1200	Pine, black, 2-1/2'-3'	68.50	31.50	100
1300	Mugo, 18"-24"	62	26.50	88.50
1400	White, 4'-5'	58.50	40	98.50
1500	Spruce, blue, 18"-24"	76	26.50	102.50
1600	Norway, 4'-5'	94.50	40	134.50
1700	Yew, denisforma, 12"-15"	40.50	26.50	67
1800	Capitata, 18"-24"	37.50	52.50	90
1900	Hicksi, 2'-2-1/2'	114	52.50	166.50
2000				
2100	Trees, Deciduous, in prepared beds			
2200				
2300	Beech, 5'-6'	231	60	291
2400	Dogwood, 4'-5'	154	75	229
2500	Elm, 8'-10'	360	150	510
2600	Magnolia, 4'-5'	127	150	277
2700	Maple, red, 8'-10', 1-1/2" caliper	219	299	518
2800	Oak, 2-1/2"-3" caliper	355	500	855
2900	Willow, 6'-8', 1" caliper	107	150	257
3000				
3100	Shrubs, Broadleaf Evergreen, in prepared beds			
3200				
3300	Andromeda, 15"-18", container	37	16.45	53.45
3400	Azalea, 15"-18", container	34	16.45	50.45
3500	Barberry, 9"-12", container	20.50	12.15	32.65
3600	Boxwood, 15"-18", B & B	49	16.45	65.45
3700	Euonymus, emerald gaiety, 12"-15", container	27	13.75	40.75
3800	Holly, 15"-18", B & B	44	16.45	60.45
3900	Mount laurel, 18"-24", B & B	80	19.75	99.75
4000	Privet, 18"-24", B & B	24	12.15	36.15
4100	Rhododendron, 18"-24", container	42.50	33	75.50
4200	Rosemary, 1 gal. container	20.50	2.63	23.13
4300	Deciduous, amalanchier, 2'-3', B & B	135	27.50	162.50
4400	Azalea, 15"-18", B & B	33.50	16.45	49.95
4500	Bayberry, 2'-3', B & B	33	27.50	60.50
4600	Cotoneaster, 15"-18", B & B	30	19.75	49.75
4700	Dogwood, 3'-4', B & B	36.50	75	111.50
4800	Euonymus, alatus compacta, 15"-18", container	29.50	19.75	49.25
4900	Forsythia, 2'-3', container	19.55	26.50	46.05
5000	Honeysuckle, 3'-4', B & B	30.50	26.50	57
5100	Hydrangea, 2'-3', B & B	34	27.50	61.50
5200	Lilac, 3'-4', B & B	30	75	105
5300	Quince, 2'-3', B & B	32	27.50	59.50
5400				
5500	Ground Cover			
5600				
5700	Plants, Pachysandra, prepared beds, per hundred	87.50	105	192.50
5800	Vinca minor or English ivy, per hundred	77	132	209

G2050 Landscaping

G2050 510	Selective Price Sheet	COST PER UNIT		
		MAT.	INST.	TOTAL
5900	Plant bed prep. 18″ dp., mach., per square foot	2.69	.47	3.16
6000	By hand, per square foot	2.69	3.34	6.03
6100	Stone chips, Georgia marble, 50# bags, per bag	5.30	3.04	8.34
6200	Onyx gemstone, per bag	17.60	6.05	23.65
6300	Quartz, per bag	17.60	6.05	23.65
6400	Pea gravel, truck load lots, per cubic yard	33.50	56.50	90
6500	Mulch, polyethylene mulch, per square yard	.58	.52	1.10
6600	Wood chips, 2″ deep, per square yard	1.74	2.35	4.09
6700	Peat moss, 1″ deep, per square yard	5.40	.58	5.98
6800	Erosion control per square yard, Jute mesh stapled	1	.76	1.76
6900	Plastic netting, stapled	.32	.63	.95
7000	Polypropylene mesh, stapled	1.79	.63	2.42
7100	Tobacco netting, stapled	.28	.63	.91
7200	Lawns per square yard, seeding incl. fine grade, limestone, fert. and seed	.51	3.60	4.11
7300	Sodding, incl. fine grade, level ground	2.44	1.20	3.64
7500	Edging, per linear foot, Redwood, untreated, 1″ x 4″	1.28	2.63	3.91
7600	2″ x 4″	2.49	3.99	6.48
7700	Stl. edge strips, 1/4″ x 5″ inc. stakes	4.91	3.99	8.90
7800	3/16″ x 4″	4.16	3.99	8.15
7900	Brick edging, set on edge	3.06	8.65	11.71
8000	Set flat	1.53	3.16	4.69
8100				
8200				
8300				
8400				

For customer support on your Commercial Renovation Costs with RSMeans data, call 800.448.8182.

G3020 Sanitary Sewer

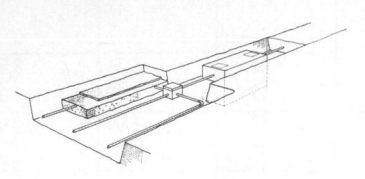

This page illustrates and describes drainage and utilities – septic system including a tank, distribution box, excavation, piping, crushed stone and backfill. Lines within System Components give the unit price and total price on a cost each basis for this system. Prices for drainage and utilities – septic systems are on Line Items G3020 610 1600 thru 2000. Both material quantities and labor costs have been adjusted for the system listed.

Factors: To adjust for job conditions other than normal working situations use Lines G3020 610 3100 thru 4000.

Example: You are to install the system with a material handling limitation. Go to Line G3020 610 3500 and apply these percentages to the appropriate MAT. and INST. costs.

System Components	QUANTITY	UNIT	COST EACH MAT.	COST EACH INST.	COST EACH TOTAL
Septic system including tank, distribution box, excavation, piping, crushed Stone and backfill for a 1000 S.F. leaching field.					
Precast concrete septic tank, 1000 gal. capacity	1.000	Ea.	1,150	278.80	1,428.80
Concrete distribution box	1.000	Ea.	102	64.50	166.50
Sch. 40 sewer pipe, PVC, 4″ diameter	25.000	L.F.	306	787.10	1,093.10
Sch. 40 sewer pipe fittings, PVC, 4″ diameter	8.000	Ea.	172	676	848
Excavation for septic tank	120.000	C.Y.		243.90	243.90
Excavation for disposal field	120.000	C.Y.		591.86	591.86
Crushed stone backfill	76.000	C.Y.	1,406	1,025.24	2,431.24
Backfill with excavated material	26.000	C.Y.		50.70	50.70
Asphalt felt paper, #15	6.000	Sq.	35.40	106.80	142.20
TOTAL			3,171.40	3,824.90	6,996.30

G3020 610	Septic Systems	COST EACH MAT.	COST EACH INST.	COST EACH TOTAL
1600	1000 gal. tank with 1000 S.F. field	3,175	3,825	7,000
1700	1000 gal. tank with 2000 S.F. field	4,400	5,625	10,025
1800	With leaching pits	3,175	2,250	5,425
1900	2000 gal. tank with 2000 S.F. field	5,750	5,800	11,550
2000	With leaching pits	4,625	2,900	7,525
2100				
2200				
2300				
2400				
2500				
2600				
2700				
2800				
2900				
3100	Dust protection, add, minimum	1%	2%	
3200	Maximum	4%	11%	
3300	Equipment usage curtailment, add, minimum	1%	1%	
3400	Maximum	3%	10%	
3500	Material handling & storage limitation, add, minimum	1%	1%	
3600	Maximum	6%	7%	
3700	Protection of existing work, add, minimum	2%	2%	
3800	Maximum	5%	7%	
3900	Shift work requirements, add, minimum		5%	
4000	Maximum		30%	

G3060 Fuel Distribution

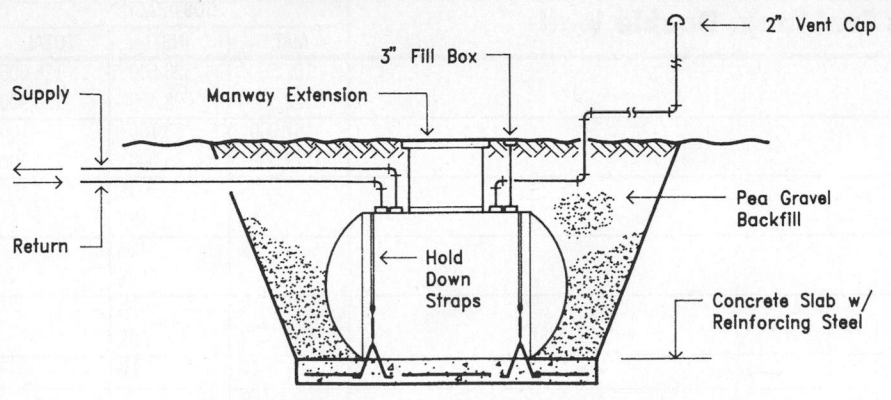

2" Vent Cap

3" Fill Box

Manway Extension

Supply

Return

Hold Down Straps

Pea Gravel Backfill

Concrete Slab w/ Reinforcing Steel

Fiberglass Underground Fuel Storage Tank, Single Wall

This page illustrates and describes fiberglass tanks including hold down slab, reinforcing steel and peastone gravel. Lines within Systems Components give the material and installation price on a cost each basis for the components. Prices for fiberglass tank systems are on Line Items G3060 310 1010 thru G3060 310 1100. Material quantities and labor costs have been adjusted for the system listed.

Factors: To adjust for job conditions other than normal working situations use Lines G3060 310 2700 thru 4000.

System Components	QUANTITY	UNIT	COST EACH		
			MAT.	INST.	TOTAL
Double wall underground fiberglass storage tanks including, excavation, hold down slab, backfill, manway extension and related piping.					
Excavating trench, no sheeting or dewatering, 6'-10 ' D, 3/4 CY hyd backhoe	18.000	Ea.		156.96	156.96
Corrosion resistance wrap & coat, small diam. pipe, 1" diam., add	120.000	Ea.	121.20		121.20
Fiberglass-reinforced elbows 90 & 45 degree, 2 inch	7.000	Ea.	318.50	640.50	959
Fiberglass-reinforced threaded adapters, 2 inch	2.000	Ea.	37.30	210	247.30
Remote tank gauging system, 30', 5" pointer travel	1.000	Ea.	6,175	390	6,565
Corrosion resistance wrap & coat, small diam. pipe. 4 " diam. add	30.000	Ea.	81.60		81.60
Pea gravel	9.000	Ea.	1,098	508.50	1,606.50
Reinforcing in hold down pad, #3 to #7	120.000	Ea.	63.60	80.40	144
Tank leak detector system, 8 channel, external monitoring	1.000	Ea.	1,300		1,300
Tubbing copper, type L, 3/4" diam.	120.000	Ea.	592.80	1,224	1,816.80
Elbow 90 degree ,wrought cu x cu 3/4 " diam.	4.000	Ea.	11.08	162	173.08
Pipe, black steel welded, sch 40, 3" diam.	30.000	Ea.	561	1,020.30	1,581.30
Elbow 90 degree, steel welded butt joint, 3 " diam.	3.000	Ea.	96	624.90	720.90
Fiberglass-reinforced plastic pipe, 2 inch	156.000	Ea.	1,107.60	803.40	1,911
Fiberglass-reinforced plastic pipe, 3 inch	36.000	Ea.	1,008	210.60	1,218.60
Fiberglass-reinforced elbows 90 & 45 degree, 3 inch	3.000	Ea.	444	366	810
Tank leak detection system, probes, well monitoring, liquid phase detection	2.000	Ea.	1,450		1,450
Double wall fiberglass annular space	1.000	Ea.	410		410
Flange, weld neck, 150 lb, 3 " pipe size	1.000	Ea.	38.50	104.15	142.65
Vent protector/breather, 2" diam.	1.000	Ea.	50	24.50	74.50
Corrosion resistance wrap & coat, small diam. pipe, 2" diam., add	36.000	Ea.	48.96		48.96
Concrete hold down pad, 6' x 5' x 8" thick	30.000	Ea.	105.60	51.90	157.50
600 gallon capacity, underground storage tank, double wall fiberglass	1.000	Ea.	9,150	585	9,735
Tank, for hold-downs 500-4,000 gal., add	1.000	Ea.	550	187	737
Foot valve, single poppet, 3/4" diam.	1.000	Ea.	102	43.50	145.50
Fuel fill box, locking inner cover, 3" diam.	1.000	Ea.	126	157	283
Fuel oil specialties, valve,ball chk, globe type fusible, 3/4" diam.	2.000	Ea.	171	78	249
TOTAL			25,217.74	7,628.61	32,846.35

G3060 310	Fiberglass Fuel Tank, Double Wall	COST EACH		
		MAT.	INST.	TOTAL
1010	Storage tank, fuel, underground, double wall fiberglass, 600 Gal.	25,200	7,625	32,825
1020	2500 Gal.	39,100	10,500	49,600
1030	4000 Gal.	46,500	13,000	59,500
1040	6000 Gal.	54,500	14,600	69,100
1050	8000 Gal.	63,500	17,700	81,200
1060	10,000 Gal.	72,000	20,400	92,400

G3060 Fuel Distribution

G3060 310	Fiberglass Fuel Tank, Double Wall	COST EACH		
		MAT.	INST.	TOTAL
1070	15,000 Gal.	102,500	23,500	126,000
1080	20,000 Gal.	122,500	28,400	150,900
1090	25,000 Gal.	157,000	33,100	190,100
1100	30,000 Gal.	179,500	35,000	214,500
2700	Cut & patch to match existing construction, add, minimum	2%	3%	
2800	Maximum	5%	9%	
2900	Dust protection, add, minimum	1%	2%	
3000	Maximum	4%	11%	
3100	Equipment usage curtailment, add, minimum	1%	1%	
3200	Maximum	3%	10%	
3300	Material handling & storage limitation, add, minimum	1%	1%	
3400	Maximum	6%	7%	
3500	Protection of existing work, add, minimum	2%	2%	
3600	Maximum	5%	7%	
3700	Shift work requirements, add, minimum		5%	
3800	Maximum		30%	
3900	Temporary shoring and bracing, add, minimum	2%	5%	
4000	Maximum	5%	12%	

G3060 Fuel Distribution

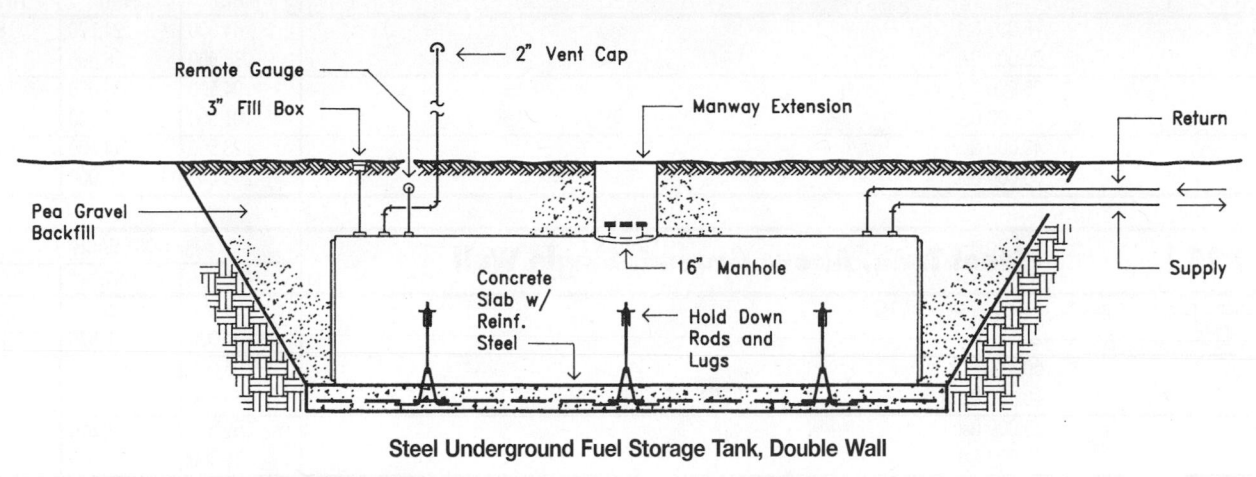

Steel Underground Fuel Storage Tank, Double Wall

System Components	QUANTITY	UNIT	COST EACH		
			MAT.	INST.	TOTAL
Double wall underground steel storage tank including, excavation, hold down slab, backfill, manway extension and related piping.					
Fiberglass-reinforced plastic pipe, 2 inch	156.000	Ea.	1,107.60	803.40	1,911
Fiberglass-reinforced plastic pipe, 3 inch	36.000	Ea.	1,008	210.60	1,218.60
Fiberglass-reinforced elbows 90 & 45 degree, 3 inch	3.000	Ea.	444	366	810
Tank leak detection system, probes, well monitoring, liquid phase detection	2.000	Ea.	1,450		1,450
Flange, weld neck, 150 lb, 3" pipe size	1.000	Ea.	38.50	104.15	142.65
Vent protector/breather, 2" diam.	1.000	Ea.	50	24.50	74.50
Concrete hold down pad, 6' x 5' x 8" thick	30.000	Ea.	105.60	51.90	157.50
Tanks, for hold-downs, 500 2,000 gal, add	1.000	Ea	203	187	390
Double wall steel tank annular space	1.000	Ea.	330		330
Foot valve, single poppet, 3/4" diam.	1.000	Ea.	102	43.50	145.50
Fuel fill box, locking inner cover, 3" diam.	1.000	Ea.	126	157	283
Fuel oil specialties, valve, ball chk, globe type fusible, 3/4" diam.	2.000	Ea.	171	78	249
Pea gravel	9.000	Ea.	1,098	508.50	1,606.50
Reinforcing in hold down pad, #3 to #7	120.000	Ea.	63.60	80.40	144
Tank, for manways, add	1.000	Ea.	1,025		1,025
Tank leak detector system, 8 channel, external monitoring	1.000	Ea.	1,300		1,300
Tubbing copper, type L, 3/4" diam.	120.000	Ea.	592.80	1,224	1,816.80
Elbow 90 degree, wrought cu x cu 3/4" diam.	8.000	Ea.	22.16	324	346.16
Pipe, black steel welded, sch 40, 3" diam.	30.000	Ea.	561	1,020.30	1,581.30
Elbow 90 degree, steel welded butt joint, 3" diam.	3.000	Ea.	96	624.90	720.90
Excavating trench, no sheeting or dewatering, 6'-10' D, 3/4 CY hyd backhoe	18.000	Ea.		156.96	156.96
Fiberglass-reinforced elbows 90 & 45 degree, 2 inch	7.000	Ea.	318.50	640.50	959
Fiberglass-reinforced threaded adapters, 2 inch	2.000	Ea.	37.30	210	247.30
500 gallon capacity, underground storage tank, double wall steel	1.000	Ea.	3,300	590	3,890
Remote tank gauging system, 30', 5" pointer travel	1.000	Ea.	6,175	390	6,565
TOTAL			19,725.06	7,795.61	27,520.67

G3060 320	Steel Fuel Tank, Double Wall	COST EACH		
		MAT.	INST.	TOTAL
1010	Storage tank, fuel, underground, double wall steel, 500 Gal.	19,700	7,775	27,475
1020	2000 Gal.	25,700	10,200	35,900
1030	4000 Gal.	35,000	13,100	48,100
1040	6000 Gal.	41,100	14,500	55,600
1050	8000 Gal.	45,900	18,000	63,900
1060	10,000 Gal.	52,500	20,700	73,200

G3060 Fuel Distribution

G3060 320	Steel Fuel Tank, Double Wall	COST EACH		
		MAT.	INST.	TOTAL
1070	15,000 Gal.	64,000	24,100	88,100
1080	20,000 Gal.	88,500	28,800	117,300
1090	25,000 Gal.	94,500	33,000	127,500
1100	30,000 Gal.	101,500	35,200	136,700
1110	40,000 Gal.	198,500	44,100	242,600
1120	50,000 Gal.	240,500	53,000	293,500

G3060 325	Steel Tank, Above Ground, Single Wall	COST EACH		
		MAT.	INST.	TOTAL
1010	Storage tank, fuel, above ground, single wall steel, 550 Gal.	15,300	5,450	20,750
1020	2000 Gal.	24,000	5,575	29,575
1030	5000 Gal.	35,000	8,525	43,525
1040	10,000 Gal.	56,500	9,100	65,600
1050	15,000 Gal.	69,500	9,350	78,850
1060	20,000 Gal.	86,500	9,650	96,150
1070	25,000 Gal.	99,000	9,900	108,900
1080	30,000 Gal.	116,000	10,400	126,400

G3060 330	Steel Tank, Above Ground, Double Wall	COST EACH		
		MAT.	INST.	TOTAL
1010	Storage tank, fuel, above ground, double wall steel, 500 Gal.	14,000	5,500	19,500
1020	2000 Gal.	18,400	5,650	24,050
1030	4000 Gal.	26,300	5,750	32,050
1040	6000 Gal.	29,500	8,850	38,350
1050	8000 Gal.	32,700	9,100	41,800
1060	10,000 Gal.	47,200	9,250	56,450
1070	15,000 Gal.	58,000	9,600	67,600
1080	20,000 Gal.	61,500	9,900	71,400
1090	25,000 Gal.	73,000	10,200	83,200
1100	30,000 Gal.	85,000	10,600	95,600
2700	Cut & patch to match existing construction, add, minimum	2%	3%	
2800	Maximum	5%	9%	
2900	Dust protection, add, minimum	1%	2%	
3000	Maximum	4%	11%	
3100	Equipment usage curtailment, add, minimum	1%	1%	
3200	Maximum	3%	10%	
3300	Material handling & storage limitation, add, minimum	1%	1%	
3400	Maximum	6%	7%	
3500	Protection of existing work, add, minimum	2%	2%	
3600	Maximum	5%	7%	
3700	Shift work requirements, add, minimum		5%	
3800	Maximum		30%	
3900	Temporary shoring and bracing, add, minimum	2%	5%	
4000	Maximum	5%	12%	

G40 Site Electrical Utilities

G4020 Site Lighting

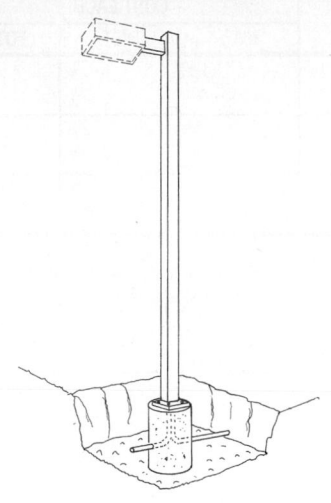

This page illustrates and describes light poles for parking or walkway area lighting. Included are aluminum or steel light poles, single or multiple fixture bracket arms, excavation, concrete footing, backfill and compaction. Lines within system components give the unit price and total price on a cost per each basis. Prices for systems are shown on lines G4020 210 0200 through 1440. Both material quantities and labor costs have been adjusted for the system listed.

Factors: To adjust for job conditions other than normal working situations use lines G4020 210 2700 through 3600.

Example: You are to install the system where you must be careful not to damage existing walks or parking areas. Go to line G4020 2110 3300 and apply these percentages to the appropriate MAT. and INST. costs.

System Components	QUANTITY	UNIT	COST EACH MAT.	COST EACH INST.	COST EACH TOTAL
Light poles, aluminum, 20' high, 1 arm bracket					
Aluminum light pole, 20', no concrete base	1.000	Ea.	1,150	669	1,819
Bracket arm for aluminum light pole	1.000	Ea.	148	90	238
Excavation by hand, pits to 6' deep, heavy soil or clay	2.368	C.Y.		305.47	305.47
Footing, concrete incl forms, reinforcing, spread, under 1 C.Y.	.465	C.Y.	94.86	144.62	239.48
Backfill by hand	1.903	C.Y.		89.44	89.44
Compaction vibrating plate	1.903	C.Y.		11.89	11.89
TOTAL			1,392.86	1,310.42	2,703.28

G4020 210	Light Pole (Installed)	COST EACH MAT.	COST EACH INST.	COST EACH TOTAL
0200	Light pole, aluminum, 20' high, 1 arm bracket	1,400	1,300	2,700
0240	2 arm brackets	1,550	1,300	2,850
0280	3 arm brackets	1,700	1,350	3,050
0320	4 arm brackets	1,850	1,375	3,225
0360	30' high, 1 arm bracket	2,450	1,650	4,100
0400	2 arm brackets	2,600	1,650	4,250
0440	3 arm brackets	2,750	1,700	4,450
0480	4 arm brackets	2,900	1,700	4,600
0680	40' high, 1 arm bracket	3,000	2,225	5,225
0720	2 arm brackets	3,150	2,225	5,375
0760	3 arm brackets	3,275	2,250	5,525
0800	4 arm brackets	3,425	2,275	5,700
0840	Steel, 20' high, 1 arm bracket	1,500	1,375	2,875
0880	2 arm brackets	1,600	1,375	2,975
0920	3 arm brackets	1,525	1,425	2,950
0960	4 arm brackets	1,625	1,425	3,050
1000	30' high, 1 arm bracket	1,950	1,775	3,725
1040	2 arm brackets	2,050	1,775	3,825
1080	3 arm brackets	1,975	1,825	3,800
1120	4 arm brackets	2,075	1,825	3,900
1320	40' high, 1 arm bracket	2,550	2,375	4,925
1360	2 arm brackets	2,650	2,375	5,025
1400	3 arm brackets	2,575	2,400	4,975
1440	4 arm brackets	2,675	2,400	5,075
2700	Cut & patch to match existing construction, add, minimum	2%	3%	
2800	Maximum	5%	9%	
2900	Equipment usage curtailment, add, minimum	1%	1%	
3000	Maximum	3%	10%	

725

G4020 Site Lighting

G4020 210	Light Pole (Installed)	COST EACH		
		MAT.	INST.	TOTAL
3100	Material handling & storage limitation, add, minimum	1%	1%	
3200	Maximum	6%	7%	
3300	Protection of existing work, add, minimum	2%	2%	
3400	Maximum	5%	7%	
3500	Shift work requirements, add, minimum		5%	
3600	Maximum		30%	

All the reference information is in one section, making it easy to find what you need to know . . . and easy to use the data set on a daily basis. This section is visually identified by a vertical black bar on the page edges.

In this Reference Section, we've included Equipment Rental Costs, a listing of rental and operating costs; Crew Listings, a full listing of all crews and equipment, and their costs; Historical Cost Indexes for cost comparisons over time; City Cost Indexes and Location Factors for adjusting costs to the region you are in; Reference Tables, where you will find explanations, estimating information and procedures, or technical data; Change Orders, information on pricing changes to contract documents; and an explanation of all the Abbreviations in the data set.

Table of Contents

Estimating Tips

- This section contains the average costs to rent and operate hundreds of pieces of construction equipment. This is useful information when estimating the time and material requirements of any particular operation in order to establish a unit or total cost. Bare equipment costs shown on a unit cost line include not only rental, but also operating costs for equipment under normal use.

Rental Costs

- Equipment rental rates are obtained from the following industry sources throughout North America: contractors, suppliers, dealers, manufacturers, and distributors.

- Rental rates vary throughout the country, with larger cities generally having lower rates. Lease plans for new equipment are available for periods in excess of six months, with a percentage of payments applying toward purchase.

- Monthly rental rates vary from 2% to 5% of the purchase price of the equipment depending on the anticipated life of the equipment and its wearing parts.

- Weekly rental rates are about 1/3 of the monthly rates, and daily rental rates are about 1/3 of the weekly rate.

- Rental rates can also be treated as reimbursement costs for contractor-owned equipment. Owned equipment costs include depreciation, loan payments, interest, taxes, insurance, storage, and major repairs.

Operating Costs

- The operating costs include parts and labor for routine servicing, such as the repair and replacement of pumps, filters, and worn lines. Normal operating expendables, such as fuel, lubricants, tires, and electricity (where applicable), are also included.

- Extraordinary operating expendables with highly variable wear patterns, such as diamond bits and blades, are excluded. These costs can be found as material costs in the Unit Price section.

- The hourly operating costs listed do not include the operator's wages.

Equipment Cost/Day

- Any power equipment required by a crew is shown in the Crew Listings with a daily cost.

- This daily cost of equipment needed by a crew includes both the rental cost and the operating cost and is based on dividing the weekly rental rate by 5 (number of working days in the week), and then adding the hourly operating cost times 8 (the number of hours in a day). This "Equipment Cost/Day" is shown in the far right column of the Equipment Rental section.

- If equipment is needed for only one or two days, it is best to develop your own cost by including components for daily rent and hourly operating costs. This is important when the listed Crew for a task does not contain the equipment needed, such as a crane for lifting mechanical heating/cooling equipment up onto a roof.

- If the quantity of work is less than the crew's Daily Output shown for a Unit Price line item that includes a bare unit equipment cost, the recommendation is to estimate one day's rental cost and operating cost for equipment shown in the Crew Listing for that line item.

Mobilization, Demobilization Costs

- The cost to move construction equipment from an equipment yard or rental company to the job site and back again is not included in equipment rental costs listed in the Reference Section, nor in the bare equipment cost of any unit price line item, nor in any equipment costs shown in the Crew Listings.

- Mobilization (to the site) and demobilization (from the site) costs can be found in the Unit Price section.

- If a piece of equipment is already at the job site, it is not appropriate to utilize mobilization, demobilization costs again in an estimate. ■

01 54 33 | Equipment Rental

			UNIT	HOURLY OPER. COST	RENT PER DAY	RENT PER WEEK	RENT PER MONTH	EQUIPMENT COST/DAY	
10	0010	**CONCRETE EQUIPMENT RENTAL** without operators	R015433 -10							10
	0200	Bucket, concrete lightweight, 1/2 C.Y.	Ea.	.85	24.50	73	219	21.40	
	0300	1 C.Y.		.95	28.50	85	255	24.60	
	0400	1-1/2 C.Y.		1.20	38.50	115	345	32.60	
	0500	2 C.Y.		1.30	46.50	140	420	38.40	
	0580	8 C.Y.		6.35	263	790	2,375	208.80	
	0600	Cart, concrete, self-propelled, operator walking, 10 C.F.		2.85	58.50	175	525	57.80	
	0700	Operator riding, 18 C.F.		4.80	98.50	295	885	97.40	
	0800	Conveyer for concrete, portable, gas, 16" wide, 26' long		10.60	130	390	1,175	162.80	
	0900	46' long		11.00	155	465	1,400	181	
	1000	56' long		11.15	163	490	1,475	187.20	
	1100	Core drill, electric, 2-1/2 H.P., 1" to 8" bit diameter		1.56	58.50	175	525	47.50	
	1150	11 H.P., 8" to 18" cores		5.40	115	345	1,025	112.20	
	1200	Finisher, concrete floor, gas, riding trowel, 96" wide		9.65	148	445	1,325	166.20	
	1300	Gas, walk-behind, 3 blade, 36" trowel		2.15	23	69	207	31	
	1400	4 blade, 48" trowel		3.05	28	84	252	41.20	
	1500	Float, hand-operated (Bull float), 48" wide		.08	13.65	41	123	8.85	
	1570	Curb builder, 14 H.P., gas, single screw		13.90	275	825	2,475	276.20	
	1590	Double screw		14.80	330	995	2,975	317.40	
	1600	Floor grinder, concrete and terrazzo, electric, 22" path		2.95	183	550	1,650	133.60	
	1700	Edger, concrete, electric, 7" path		1.12	56.50	170	510	42.95	
	1750	Vacuum pick-up system for floor grinders, wet/dry		1.62	90	270	810	66.95	
	1800	Mixer, powered, mortar and concrete, gas, 6 C.F., 18 H.P.		7.40	123	370	1,100	133.20	
	1900	10 C.F., 25 H.P.		9.00	148	445	1,325	161	
	2000	16 C.F.		9.35	172	515	1,550	177.80	
	2100	Concrete, stationary, tilt drum, 2 C.Y.		7.40	242	725	2,175	204.20	
	2120	Pump, concrete, truck mounted, 4" line, 80' boom		29.20	1,075	3,240	9,725	881.60	
	2140	5" line, 110' boom		36.60	1,375	4,115	12,300	1,116	
	2160	Mud jack, 50 C.F. per hr.		6.45	128	385	1,150	128.60	
	2180	225 C.F. per hr.		8.80	147	440	1,325	158.40	
	2190	Shotcrete pump rig, 12 C.Y./hr.		13.95	225	675	2,025	246.60	
	2200	35 C.Y./hr.		15.65	242	725	2,175	270.20	
	2600	Saw, concrete, manual, gas, 18 H.P.		5.40	46.50	140	420	71.20	
	2650	Self-propelled, gas, 30 H.P.		7.55	70	210	630	102.40	
	2675	V-groove crack chaser, manual, gas, 6 H.P.		1.80	18.35	55	165	25.40	
	2700	Vibrators, concrete, electric, 60 cycle, 2 H.P.		.47	9	27	81	9.15	
	2800	3 H.P.		.60	11.65	35	105	11.80	
	2900	Gas engine, 5 H.P.		1.50	16.35	49	147	21.80	
	3000	8 H.P.		2.00	16	48	144	25.60	
	3050	Vibrating screed, gas engine, 8 H.P.		2.82	88	264	790	75.35	
	3120	Concrete transit mixer, 6 x 4, 250 H.P., 8 C.Y., rear discharge		48.75	600	1,805	5,425	751	
	3200	Front discharge		56.60	735	2,205	6,625	893.80	
	3300	6 x 6, 285 H.P., 12 C.Y., rear discharge		55.65	700	2,095	6,275	864.20	
	3400	Front discharge		57.95	735	2,210	6,625	905.60	
20	0010	**EARTHWORK EQUIPMENT RENTAL** without operators	R015433 -10							20
	0040	Aggregate spreader, push type, 8' to 12' wide	Ea.	2.60	26.50	80	240	36.80	
	0045	Tailgate type, 8' wide		2.55	33.50	100	300	40.40	
	0055	Earth auger, truck mounted, for fence & sign posts, utility poles		12.60	455	1,365	4,100	373.80	
	0060	For borings and monitoring wells		42.60	695	2,090	6,275	758.80	
	0070	Portable, trailer mounted		2.30	33	99	297	38.20	
	0075	Truck mounted, for caissons, water wells		85.75	2,925	8,790	26,400	2,444	
	0080	Horizontal boring machine, 12" to 36" diameter, 45 H.P.		22.70	197	590	1,775	299.60	
	0090	12" to 48" diameter, 65 H.P.		30.60	340	1,020	3,050	448.80	
	0095	Auger, for fence posts, gas engine, hand held		.45	6.35	19	57	7.40	
	0100	Excavator, diesel hydraulic, crawler mounted, 1/2 C.Y. cap.		21.30	440	1,325	3,975	435.40	
	0120	5/8 C.Y. capacity		28.60	580	1,740	5,225	576.80	
	0140	3/4 C.Y. capacity		31.95	690	2,070	6,200	669.60	
	0150	1 C.Y. capacity		39.90	705	2,115	6,350	742.20	

01 54 33 | Equipment Rental

		UNIT	HOURLY OPER. COST	RENT PER DAY	RENT PER WEEK	RENT PER MONTH	EQUIPMENT COST/DAY	
20								**20**
0200	1-1/2 C.Y. capacity	Ea.	47.45	855	2,570	7,700	893.60	
0300	2 C.Y. capacity		54.15	1,025	3,095	9,275	1,052	
0320	2-1/2 C.Y. capacity		78.90	1,300	3,900	11,700	1,411	
0325	3-1/2 C.Y. capacity		113.65	2,150	6,430	19,300	2,195	
0330	4-1/2 C.Y. capacity		143.70	2,650	7,925	23,800	2,735	
0335	6 C.Y. capacity		181.40	3,400	10,180	30,500	3,487	
0340	7 C.Y. capacity		175.20	3,125	9,365	28,100	3,275	
0342	Excavator attachments, bucket thumbs		3.35	252	755	2,275	177.80	
0345	Grapples		3.05	220	660	1,975	156.40	
0346	Hydraulic hammer for boom mounting, 4000 ft lb.		13.20	365	1,095	3,275	324.60	
0347	5000 ft lb.		15.50	450	1,350	4,050	394	
0348	8000 ft lb.		22.85	655	1,970	5,900	576.80	
0349	12,000 ft lb.		24.90	785	2,350	7,050	669.20	
0350	Gradall type, truck mounted, 3 ton @ 15' radius, 5/8 C.Y.		44.95	850	2,550	7,650	869.60	
0370	1 C.Y. capacity		60.20	1,275	3,805	11,400	1,243	
0400	Backhoe-loader, 40 to 45 H.P., 5/8 C.Y. capacity		12.25	248	745	2,225	247	
0450	45 H.P. to 60 H.P., 3/4 C.Y. capacity		17.80	283	850	2,550	312.40	
0460	80 H.P., 1-1/4 C.Y. capacity		19.50	380	1,145	3,425	385	
0470	112 H.P., 1-1/2 C.Y. capacity		31.90	605	1,820	5,450	619.20	
0482	Backhoe-loader attachment, compactor, 20,000 lb.		6.25	148	445	1,325	139	
0485	Hydraulic hammer, 750 ft lb.		3.55	102	305	915	89.40	
0486	Hydraulic hammer, 1200 ft lb.		6.55	208	625	1,875	177.40	
0500	Brush chipper, gas engine, 6" cutter head, 35 H.P.		9.15	110	330	990	139.20	
0550	Diesel engine, 12" cutter head, 130 H.P.		24.00	335	1,005	3,025	393	
0600	15" cutter head, 165 H.P.		26.05	400	1,200	3,600	448.40	
0750	Bucket, clamshell, general purpose, 3/8 C.Y.		1.40	40	120	360	35.20	
0800	1/2 C.Y.		1.50	48.50	145	435	41	
0850	3/4 C.Y.		1.65	56.50	170	510	47.20	
0900	1 C.Y.		1.70	61.50	185	555	50.60	
0950	1-1/2 C.Y.		2.75	85	255	765	73	
1000	2 C.Y.		2.90	93.50	280	840	79.20	
1010	Bucket, dragline, medium duty, 1/2 C.Y.		.80	24.50	73	219	21	
1020	3/4 C.Y.		.80	25.50	77	231	21.80	
1030	1 C.Y.		.85	27	81	243	23	
1040	1-1/2 C.Y.		1.30	41.50	125	375	35.40	
1050	2 C.Y.		1.35	45	135	405	37.80	
1070	3 C.Y.		2.10	65	195	585	55.80	
1200	Compactor, manually guided 2-drum vibratory smooth roller, 7.5 H.P.		7.20	203	610	1,825	179.60	
1250	Rammer/tamper, gas, 8"		2.25	46.50	140	420	46	
1260	15"		2.50	53.50	160	480	52	
1300	Vibratory plate, gas, 18" plate, 3000 lb. blow		2.15	24.50	73	219	31.80	
1350	21" plate, 5000 lb. blow		2.60	33	99	297	40.60	
1370	Curb builder/extruder, 14 H.P., gas, single screw		13.90	275	825	2,475	276.20	
1390	Double screw		14.80	330	995	2,975	317.40	
1500	Disc harrow attachment, for tractor		.47	78.50	235	705	50.75	
1810	Feller buncher, shearing & accumulating trees, 100 H.P.		40.55	810	2,430	7,300	810.40	
1860	Grader, self-propelled, 25,000 lb.		32.75	745	2,235	6,700	709	
1910	30,000 lb.		32.35	640	1,925	5,775	643.80	
1920	40,000 lb.		54.75	1,250	3,765	11,300	1,191	
1930	55,000 lb.		67.80	1,650	4,975	14,900	1,537	
1950	Hammer, pavement breaker, self-propelled, diesel, 1000 to 1250 lb.		28.40	460	1,380	4,150	503.20	
2000	1300 to 1500 lb.		42.60	920	2,760	8,275	892.80	
2050	Pile driving hammer, steam or air, 4150 ft lb. @ 225 bpm		11.80	565	1,695	5,075	433.40	
2100	8750 ft lb. @ 145 bpm		14.30	805	2,410	7,225	596.40	
2150	15,000 ft lb. @ 60 bpm		14.65	845	2,530	7,600	623.20	
2200	24,450 ft lb. @ 111 bpm		15.65	935	2,800	8,400	685.20	
2250	Leads, 60' high for pile driving hammers up to 20,000 ft lb.		3.60	84.50	253	760	79.40	
2300	90' high for hammers over 20,000 ft lb.		5.35	148	444	1,325	131.60	

01 54 33 | Equipment Rental

		UNIT	HOURLY OPER. COST	RENT PER DAY	RENT PER WEEK	RENT PER MONTH	EQUIPMENT COST/DAY	
2350	Diesel type hammer, 22,400 ft lb.	Ea.	18.25	470	1,405	4,225	427	20
2400	41,300 ft lb.		26.50	600	1,800	5,400	572	
2450	141,000 ft lb.		41.80	945	2,840	8,525	902.40	
2500	Vib. elec. hammer/extractor, 200 kW diesel generator, 34 H.P.		40.35	695	2,080	6,250	738.80	
2550	80 H.P.		70.10	1,000	3,000	9,000	1,161	
2600	150 H.P.		129.40	1,925	5,780	17,300	2,191	
2800	Log chipper, up to 22" diameter, 600 H.P.		46.10	660	1,980	5,950	764.80	
2850	Logger, for skidding & stacking logs, 150 H.P.		44.05	830	2,485	7,450	849.40	
2860	Mulcher, diesel powered, trailer mounted		17.10	220	660	1,975	268.80	
2900	Rake, spring tooth, with tractor		14.54	360	1,075	3,225	331.30	
3000	Roller, vibratory, tandem, smooth drum, 20 H.P.		7.65	150	450	1,350	151.20	
3050	35 H.P.		10.05	252	755	2,275	231.40	
3100	Towed type vibratory compactor, smooth drum, 50 H.P.		25.25	370	1,105	3,325	423	
3150	Sheepsfoot, 50 H.P.		25.35	370	1,115	3,350	425.80	
3170	Landfill compactor, 220 H.P.		71.35	1,575	4,710	14,100	1,513	
3200	Pneumatic tire roller, 80 H.P.		12.90	390	1,175	3,525	338.20	
3250	120 H.P.		19.45	645	1,930	5,800	541.60	
3300	Sheepsfoot vibratory roller, 240 H.P.		62.05	1,375	4,100	12,300	1,316	
3320	340 H.P.		83.10	2,025	6,075	18,200	1,880	
3350	Smooth drum vibratory roller, 75 H.P.		23.30	650	1,950	5,850	576.40	
3400	125 H.P.		27.60	730	2,195	6,575	659.80	
3410	Rotary mower, brush, 60", with tractor		19.00	340	1,025	3,075	357	
3420	Rototiller, walk-behind, gas, 5 H.P.		2.21	76.50	229	685	63.50	
3422	8 H.P.		2.80	87	261	785	74.60	
3440	Scrapers, towed type, 7 C.Y. capacity		6.30	122	365	1,100	123.40	
3450	10 C.Y. capacity		7.10	165	495	1,475	155.80	
3500	15 C.Y. capacity		7.60	192	575	1,725	175.80	
3525	Self-propelled, single engine, 14 C.Y. capacity		131.25	2,400	7,200	21,600	2,490	
3550	Dual engine, 21 C.Y. capacity		137.25	2,275	6,850	20,600	2,468	
3600	31 C.Y. capacity		184.55	3,500	10,500	31,500	3,576	
3640	44 C.Y. capacity		225.40	4,425	13,270	39,800	4,457	
3650	Elevating type, single engine, 11 C.Y. capacity		61.05	1,100	3,335	10,000	1,155	
3700	22 C.Y. capacity		114.65	2,275	6,835	20,500	2,284	
3710	Screening plant, 110 H.P. w/5' x 10' screen		20.55	385	1,160	3,475	396.40	
3720	5' x 16' screen		25.96	495	1,480	4,450	503.70	
3850	Shovel, crawler-mounted, front-loading, 7 C.Y. capacity		204.00	3,750	11,240	33,700	3,880	
3855	12 C.Y. capacity		332.00	5,200	15,567	46,700	5,769	
3860	Shovel/backhoe bucket, 1/2 C.Y.		2.70	71.50	215	645	64.60	
3870	3/4 C.Y.		2.75	78.50	235	705	69	
3880	1 C.Y.		2.85	88.50	265	795	75.80	
3890	1-1/2 C.Y.		3.00	102	305	915	85	
3910	3 C.Y.		3.35	137	410	1,225	108.80	
3950	Stump chipper, 18" deep, 30 H.P.		6.93	214	643	1,925	184.05	
4110	Dozer, crawler, torque converter, diesel 80 H.P.		25.05	440	1,320	3,950	464.40	
4150	105 H.P.		34.25	555	1,670	5,000	608	
4200	140 H.P.		42.20	845	2,540	7,625	845.60	
4260	200 H.P.		61.65	1,300	3,900	11,700	1,273	
4310	300 H.P.		82.45	1,950	5,845	17,500	1,829	
4360	410 H.P.		109.35	2,350	7,080	21,200	2,291	
4370	500 H.P.		137.25	2,875	8,600	25,800	2,818	
4380	700 H.P.		229.75	5,275	15,805	47,400	4,999	
4400	Loader, crawler, torque conv., diesel, 1-1/2 C.Y., 80 H.P.		29.60	570	1,715	5,150	579.80	
4450	1-1/2 to 1-3/4 C.Y., 95 H.P.		30.55	670	2,005	6,025	645.40	
4510	1-3/4 to 2-1/4 C.Y., 130 H.P.		47.50	1,000	2,995	8,975	979	
4530	2-1/2 to 3-1/4 C.Y., 190 H.P.		57.55	1,250	3,720	11,200	1,204	
4560	3-1/2 to 5 C.Y., 275 H.P.		71.55	1,475	4,435	13,300	1,459	
4610	Front end loader, 4WD, articulated frame, diesel, 1 to 1-1/4 C.Y., 70 H.P.		16.10	270	810	2,425	290.80	
4620	1-1/2 to 1-3/4 C.Y., 95 H.P.		19.35	315	940	2,825	342.80	

		UNIT	HOURLY OPER. COST	RENT PER DAY	RENT PER WEEK	RENT PER MONTH	EQUIPMENT COST/DAY	
20	4650	1-3/4 to 2 C.Y., 130 H.P.	Ea.	20.50	380	1,140	3,425	392	**20**
	4710	2-1/2 to 3-1/2 C.Y., 145 H.P.		29.15	485	1,450	4,350	523.20	
	4730	3 to 4-1/2 C.Y., 185 H.P.		32.20	530	1,585	4,750	574.60	
	4760	5-1/4 to 5-3/4 C.Y., 270 H.P.		53.15	930	2,785	8,350	982.20	
	4810	7 to 9 C.Y., 475 H.P.		90.60	1,750	5,275	15,800	1,780	
	4870	9 to 11 C.Y., 620 H.P.		131.70	2,625	7,880	23,600	2,630	
	4880	Skid-steer loader, wheeled, 10 C.F., 30 H.P. gas		9.50	163	490	1,475	174	
	4890	1 C.Y., 78 H.P., diesel		18.35	365	1,090	3,275	364.80	
	4892	Skid-steer attachment, auger		.82	136	408	1,225	88.15	
	4893	Backhoe		.74	123	369	1,100	79.70	
	4894	Broom		.71	118	355	1,075	76.70	
	4895	Forks		.16	27	81	243	17.50	
	4896	Grapple		.69	115	346	1,050	74.70	
	4897	Concrete hammer		1.05	175	526	1,575	113.60	
	4898	Tree spade		.60	100	300	900	64.80	
	4899	Trencher		.69	115	344	1,025	74.30	
	4900	Trencher, chain, boom type, gas, operator walking, 12 H.P.		4.10	48.50	145	435	61.80	
	4910	Operator riding, 40 H.P.		16.30	345	1,030	3,100	336.40	
	5000	Wheel type, diesel, 4' deep, 12" wide		69.35	900	2,700	8,100	1,095	
	5100	6' deep, 20" wide		87.50	2,100	6,285	18,900	1,957	
	5150	Chain type, diesel, 5' deep, 8" wide		16.30	345	1,030	3,100	336.40	
	5200	Diesel, 8' deep, 16" wide		95.25	2,150	6,430	19,300	2,048	
	5202	Rock trencher, wheel type, 6" wide x 18" deep		43.65	925	2,775	8,325	904.20	
	5206	Chain type, 18" wide x 7' deep		106.10	3,050	9,160	27,500	2,681	
	5210	Tree spade, self-propelled		14.24	390	1,167	3,500	347.30	
	5250	Truck, dump, 2-axle, 12 ton, 8 C.Y. payload, 220 H.P.		23.95	245	735	2,200	338.60	
	5300	Three axle dump, 16 ton, 12 C.Y. payload, 400 H.P.		41.60	350	1,050	3,150	542.80	
	5310	Four axle dump, 25 ton, 18 C.Y. payload, 450 H.P.		50.00	510	1,530	4,600	706	
	5350	Dump trailer only, rear dump, 16-1/2 C.Y.		5.75	145	435	1,300	133	
	5400	20 C.Y.		6.20	163	490	1,475	147.60	
	5450	Flatbed, single axle, 1-1/2 ton rating		18.30	70	210	630	188.40	
	5500	3 ton rating		22.25	100	300	900	238	
	5550	Off highway rear dump, 25 ton capacity		61.80	1,425	4,245	12,700	1,343	
	5600	35 ton capacity		66.10	1,525	4,595	13,800	1,448	
	5610	50 ton capacity		81.20	1,675	5,060	15,200	1,662	
	5620	65 ton capacity		84.95	1,950	5,820	17,500	1,844	
	5630	100 ton capacity		119.60	2,850	8,560	25,700	2,669	
	6000	Vibratory plow, 25 H.P., walking		6.80	61.50	185	555	91.40	
40	0010	**GENERAL EQUIPMENT RENTAL** without operators							**40**
	0020	Aerial lift, scissor type, to 20' high, 1200 lb. capacity, electric	R015433 -10	Ea.	3.40	51.50	155	465	58.20	
	0030	To 30' high, 1200 lb. capacity		3.85	68.50	205	615	71.80	
	0040	Over 30' high, 1500 lb. capacity		5.15	122	365	1,100	114.20	
	0070	Articulating boom, to 45' high, 500 lb. capacity, diesel	R015433 -15	9.70	273	820	2,450	241.60	
	0075	To 60' high, 500 lb. capacity		13.70	470	1,410	4,225	391.60	
	0080	To 80' high, 500 lb. capacity		16.10	560	1,685	5,050	465.80	
	0085	To 125' high, 500 lb. capacity		18.40	780	2,335	7,000	614.20	
	0100	Telescoping boom to 40' high, 500 lb. capacity, diesel		11.60	315	950	2,850	282.80	
	0105	To 45' high, 500 lb. capacity		12.40	320	958	2,875	290.80	
	0110	To 60' high, 500 lb. capacity		16.20	540	1,625	4,875	454.60	
	0115	To 80' high, 500 lb. capacity		21.70	625	1,875	5,625	548.60	
	0120	To 100' high, 500 lb. capacity		28.80	805	2,420	7,250	714.40	
	0125	To 120' high, 500 lb. capacity		29.25	845	2,530	7,600	740	
	0195	Air compressor, portable, 6.5 CFM, electric		.91	13	39	117	15.10	
	0196	Gasoline		.66	19.65	59	177	17.10	
	0200	Towed type, gas engine, 60 CFM		9.30	51.50	155	465	105.40	
	0300	160 CFM		10.65	53.50	160	480	117.20	
	0400	Diesel engine, rotary screw, 250 CFM		12.05	118	355	1,075	167.40	
	0500	365 CFM		15.80	142	425	1,275	211.40	

For customer support on your Commercial Renovation Costs with RSMeans data, call 800.448.8182.

733

01 54 33 | Equipment Rental

		UNIT	HOURLY OPER. COST	RENT PER DAY	RENT PER WEEK	RENT PER MONTH	EQUIPMENT COST/DAY	
40	0550	450 CFM	Ea.	19.70	177	530	1,600	263.60	40
	0600	600 CFM		33.90	243	730	2,200	417.20	
	0700	750 CFM		34.10	252	755	2,275	423.80	
	0930	Air tools, breaker, pavement, 60 lb.		.55	10.35	31	93	10.60	
	0940	80 lb.		.55	10.65	32	96	10.80	
	0950	Drills, hand (jackhammer), 65 lb.		.65	17.65	53	159	15.80	
	0960	Track or wagon, swing boom, 4" drifter		54.80	925	2,775	8,325	993.40	
	0970	5" drifter		63.45	1,100	3,325	9,975	1,173	
	0975	Track mounted quarry drill, 6" diameter drill		104.15	1,650	4,945	14,800	1,822	
	0980	Dust control per drill		1.04	24.50	74	222	23.10	
	0990	Hammer, chipping, 12 lb.		.60	27	81	243	21	
	1000	Hose, air with couplings, 50' long, 3/4" diameter		.07	11.35	34	102	7.35	
	1100	1" diameter		.08	13	39	117	8.45	
	1200	1-1/2" diameter		.21	35	105	315	22.70	
	1300	2" diameter		.24	40	120	360	25.90	
	1400	2-1/2" diameter		.35	58.50	175	525	37.80	
	1410	3" diameter		.40	66.50	200	600	43.20	
	1450	Drill, steel, 7/8" x 2'		.08	13.65	41	123	8.85	
	1460	7/8" x 6'		.11	17.65	53	159	11.50	
	1520	Moil points		.03	4.67	14	42	3.05	
	1525	Pneumatic nailer w/accessories		.48	32	96	288	23.05	
	1530	Sheeting driver for 60 lb. breaker		.04	7.35	22	66	4.70	
	1540	For 90 lb. breaker		.15	9.65	29	87	7	
	1550	Spade, 25 lb.		.50	7.35	22	66	8.40	
	1560	Tamper, single, 35 lb.		.59	39.50	118	355	28.30	
	1570	Triple, 140 lb.		.89	59	177	530	42.50	
	1580	Wrenches, impact, air powered, up to 3/4" bolt		.45	13	39	117	11.40	
	1590	Up to 1-1/4" bolt		.55	23.50	71	213	18.60	
	1600	Barricades, barrels, reflectorized, 1 to 99 barrels		.03	5.35	16	48	3.45	
	1610	100 to 200 barrels		.02	4.13	12.40	37	2.65	
	1620	Barrels with flashers, 1 to 99 barrels		.04	6	18	54	3.90	
	1630	100 to 200 barrels		.03	4.80	14.40	43	3.10	
	1640	Barrels with steady burn type C lights		.05	8	24	72	5.20	
	1650	Illuminated board, trailer mounted, with generator		3.30	133	400	1,200	106.40	
	1670	Portable barricade, stock, with flashers, 1 to 6 units		.04	6	18	54	3.90	
	1680	25 to 50 units		.03	5.60	16.80	50.50	3.60	
	1685	Butt fusion machine, wheeled, 1.5 H.P. electric, 2" - 8" diameter pipe		2.63	167	500	1,500	121.05	
	1690	Tracked, 20 H.P. diesel, 4" - 12" diameter pipe		11.21	560	1,680	5,050	425.70	
	1695	83 H.P. diesel, 8" - 24" diameter pipe		49.46	2,525	7,560	22,700	1,908	
	1700	Carts, brick, gas engine, 1000 lb. capacity		2.95	61.50	185	555	60.60	
	1800	1500 lb., 7-1/2' lift		3.00	65	195	585	63	
	1822	Dehumidifier, medium, 6 lb./hr., 150 CFM		1.16	72.50	218	655	52.90	
	1824	Large, 18 lb./hr., 600 CFM		2.20	138	413	1,250	100.20	
	1830	Distributor, asphalt, trailer mounted, 2000 gal., 38 H.P. diesel		10.75	350	1,050	3,150	296	
	1840	3000 gal., 38 H.P. diesel		12.35	380	1,140	3,425	326.80	
	1850	Drill, rotary hammer, electric		1.12	27	81	243	25.15	
	1860	Carbide bit, 1-1/2" diameter, add to electric rotary hammer		.03	5	15	45	3.25	
	1865	Rotary, crawler, 250 H.P.		136.15	2,225	6,690	20,100	2,427	
	1870	Emulsion sprayer, 65 gal., 5 H.P. gas engine		2.77	103	309	925	83.95	
	1880	200 gal., 5 H.P. engine		7.30	172	515	1,550	161.40	
	1900	Floor auto-scrubbing machine, walk-behind, 28" path		5.41	350	1,055	3,175	254.30	
	1930	Floodlight, mercury vapor, or quartz, on tripod, 1000 watt		.46	22	66	198	16.90	
	1940	2000 watt		.63	27.50	82	246	21.45	
	1950	Floodlights, trailer mounted with generator, 1 - 300 watt light		3.60	76.50	230	690	74.80	
	1960	2 - 1000 watt lights		4.50	102	305	915	97	
	2000	4 - 300 watt lights		4.25	96.50	290	870	92	
	2005	Foam spray rig, incl. box trailer, compressor, generator, proportioner		23.78	515	1,545	4,625	499.25	
	2015	Forklift, pneumatic tire, rough terr, straight mast, 5000 lb, 12' lift, gas		19.00	212	635	1,900	279	

01 54 33 | Equipment Rental

		UNIT	HOURLY OPER. COST	RENT PER DAY	RENT PER WEEK	RENT PER MONTH	EQUIPMENT COST/DAY	
2025	8000 lb., 12' lift	Ea.	22.75	283	850	2,550	352	40
2030	5000 lb., 12' lift, diesel		15.70	237	710	2,125	267.60	
2035	8000 lb., 12' lift, diesel		16.75	268	805	2,425	295	
2045	All terrain, telescoping boom, diesel, 5000 lb., 10' reach, 19' lift		17.25	325	980	2,950	334	
2055	6600 lb., 29' reach, 42' lift		21.10	380	1,140	3,425	396.80	
2065	10,000 lb., 31' reach, 45' lift		23.65	490	1,475	4,425	484.20	
2070	Cushion tire, smooth floor, gas, 5000 lb. capacity		8.25	76.50	230	690	112	
2075	8000 lb. capacity		11.40	95	285	855	148.20	
2085	Diesel, 5000 lb. capacity		7.75	83.50	250	750	112	
2090	12,000 lb. capacity		12.05	130	390	1,175	174.40	
2095	20,000 lb. capacity		17.00	165	495	1,475	235	
2100	Generator, electric, gas engine, 1.5 kW to 3 kW		2.70	11.35	34	102	28.40	
2200	5 kW		3.35	14.35	43	129	35.40	
2300	10 kW		6.25	35	105	315	71	
2400	25 kW		7.60	86.50	260	780	112.80	
2500	Diesel engine, 20 kW		9.20	76.50	230	690	119.60	
2600	50 kW		15.85	100	300	900	186.80	
2700	100 kW		28.20	137	410	1,225	307.60	
2800	250 kW		56.05	260	780	2,350	604.40	
2850	Hammer, hydraulic, for mounting on boom, to 500 ft lb.		2.85	86.50	260	780	74.80	
2860	1000 ft lb.		4.70	133	400	1,200	117.60	
2900	Heaters, space, oil or electric, 50 MBH		1.47	8	24	72	16.55	
3000	100 MBH		2.73	11.35	34	102	28.65	
3100	300 MBH		7.84	40	120	360	86.70	
3150	500 MBH		12.75	45	135	405	129	
3200	Hose, water, suction with coupling, 20' long, 2" diameter		.02	3	9	27	1.95	
3210	3" diameter		.03	4.33	13	39	2.85	
3220	4" diameter		.03	5	15	45	3.25	
3230	6" diameter		.11	17.65	53	159	11.50	
3240	8" diameter		.28	46.50	140	420	30.25	
3250	Discharge hose with coupling, 50' long, 2" diameter		.01	1.33	4	12	.90	
3260	3" diameter		.01	2.33	7	21	1.50	
3270	4" diameter		.02	3.67	11	33	2.35	
3280	6" diameter		.06	9.35	28	84	6.10	
3290	8" diameter		.24	40	120	360	25.90	
3295	Insulation blower		.83	6	18	54	10.25	
3300	Ladders, extension type, 16' to 36' long		.18	30	90	270	19.45	
3400	40' to 60' long		.67	112	335	1,000	72.35	
3405	Lance for cutting concrete		2.23	58.50	176	530	53.05	
3407	Lawn mower, rotary, 22", 5 H.P.		1.15	25	75	225	24.20	
3408	48" self-propelled		2.86	90	270	810	76.90	
3410	Level, electronic, automatic, with tripod and leveling rod		1.05	70	210	630	50.40	
3430	Laser type, for pipe and sewer line and grade		2.13	142	425	1,275	102.05	
3440	Rotating beam for interior control		.90	60	180	540	43.20	
3460	Builder's optical transit, with tripod and rod		.10	16.35	49	147	10.60	
3500	Light towers, towable, with diesel generator, 2000 watt		4.25	96.50	290	870	92	
3600	4000 watt		4.50	102	305	915	97	
3700	Mixer, powered, plaster and mortar, 6 C.F., 7 H.P.		2.05	20.50	62	186	28.80	
3800	10 C.F., 9 H.P.		2.20	33.50	100	300	37.60	
3850	Nailer, pneumatic		.48	32	96	288	23.05	
3900	Paint sprayers complete, 8 CFM		.94	62.50	188	565	45.10	
4000	17 CFM		1.69	112	337	1,000	80.90	
4020	Pavers, bituminous, rubber tires, 8' wide, 50 H.P., diesel		31.55	550	1,645	4,925	581.40	
4030	10' wide, 150 H.P.		96.50	1,875	5,655	17,000	1,903	
4050	Crawler, 8' wide, 100 H.P., diesel		87.60	2,025	6,105	18,300	1,922	
4060	10' wide, 150 H.P.		104.50	2,350	7,015	21,000	2,239	
4070	Concrete paver, 12' to 24' wide, 250 H.P.		87.45	1,625	4,875	14,600	1,675	
4080	Placer-spreader-trimmer, 24' wide, 300 H.P.		117.20	2,375	7,115	21,300	2,361	

	01 54 33	Equipment Rental	UNIT	HOURLY OPER. COST	RENT PER DAY	RENT PER WEEK	RENT PER MONTH	EQUIPMENT COST/DAY	
40	4100	Pump, centrifugal gas pump, 1-1/2" diameter, 65 GPM	Ea.	3.90	53.50	160	480	63.20	**40**
	4200	2" diameter, 130 GPM		5.00	63.50	190	570	78	
	4300	3" diameter, 250 GPM		5.15	63.50	190	570	79.20	
	4400	6" diameter, 1500 GPM		22.30	197	590	1,775	296.40	
	4500	Submersible electric pump, 1-1/4" diameter, 55 GPM		.41	17.65	53	159	13.90	
	4600	1-1/2" diameter, 83 GPM		.45	20.50	61	183	15.80	
	4700	2" diameter, 120 GPM		1.65	25.50	76	228	28.40	
	4800	3" diameter, 300 GPM		2.94	45	135	405	50.50	
	4900	4" diameter, 560 GPM		14.70	167	500	1,500	217.60	
	5000	6" diameter, 1590 GPM		21.94	218	655	1,975	306.50	
	5100	Diaphragm pump, gas, single, 1-1/2" diameter		1.12	54.50	164	490	41.75	
	5200	2" diameter		4.00	68.50	205	615	73	
	5300	3" diameter		4.05	68.50	205	615	73.40	
	5400	Double, 4" diameter		5.85	113	340	1,025	114.80	
	5450	Pressure washer 5 GPM, 3000 psi		3.95	53.50	160	480	63.60	
	5460	7 GPM, 3000 psi		4.90	63.50	190	570	77.20	
	5500	Trash pump, self-priming, gas, 2" diameter		3.80	23.50	70	210	44.40	
	5600	Diesel, 4" diameter		6.95	95	285	855	112.60	
	5650	Diesel, 6" diameter		16.90	167	500	1,500	235.20	
	5655	Grout pump		19.50	275	825	2,475	321	
	5700	Salamanders, L.P. gas fired, 100,000 BTU		2.93	14	42	126	31.85	
	5705	50,000 BTU		1.67	11.35	34	102	20.15	
	5720	Sandblaster, portable, open top, 3 C.F. capacity		.60	27	81	243	21	
	5730	6 C.F. capacity		1.00	40	120	360	32	
	5740	Accessories for above		.14	22.50	68	204	14.70	
	5750	Sander, floor		.77	17.65	53	159	16.75	
	5760	Edger		.52	15	45	135	13.15	
	5800	Saw, chain, gas engine, 18" long		1.80	22.50	67	201	27.80	
	5900	Hydraulic powered, 36" long		.80	66.50	200	600	46.40	
	5950	60" long		.80	68.50	205	615	47.40	
	6000	Masonry, table mounted, 14" diameter, 5 H.P.		1.32	56.50	170	510	44.55	
	6050	Portable cut-off, 8 H.P.		1.85	33.50	100	300	34.80	
	6100	Circular, hand held, electric, 7-1/4" diameter		.23	5	15	45	4.85	
	6200	12" diameter		.23	8	24	72	6.65	
	6250	Wall saw, w/hydraulic power, 10 H.P.		3.30	33.50	100	300	46.40	
	6275	Shot blaster, walk-behind, 20" wide		4.85	293	880	2,650	214.80	
	6280	Sidewalk broom, walk-behind		2.39	85	255	765	70.10	
	6300	Steam cleaner, 100 gallons per hour		3.35	80	240	720	74.80	
	6310	200 gallons per hour		4.40	96.50	290	870	93.20	
	6340	Tar kettle/pot, 400 gallons		15.15	76.50	230	690	167.20	
	6350	Torch, cutting, acetylene-oxygen, 150' hose, excludes gases		.45	15	45	135	12.60	
	6360	Hourly operating cost includes tips and gas		21.00				168	
	6410	Toilet, portable chemical		.13	22	66	198	14.25	
	6420	Recycle flush type		.16	27	81	243	17.50	
	6430	Toilet, fresh water flush, garden hose,		.19	32.50	97	291	20.90	
	6440	Hoisted, non-flush, for high rise		.16	26.50	79	237	17.10	
	6465	Tractor, farm with attachment		17.80	340	1,025	3,075	347.40	
	6480	Trailers, platform, flush deck, 2 axle, 3 ton capacity		1.60	21	63	189	25.40	
	6500	25 ton capacity		6.25	138	415	1,250	133	
	6600	40 ton capacity		8.00	193	580	1,750	180	
	6700	3 axle, 50 ton capacity		8.65	215	645	1,925	198.20	
	6800	75 ton capacity		10.90	285	855	2,575	258.20	
	6810	Trailer mounted cable reel for high voltage line work		5.79	276	827	2,475	211.70	
	6820	Trailer mounted cable tensioning rig		11.48	545	1,640	4,925	419.85	
	6830	Cable pulling rig		72.98	3,075	9,210	27,600	2,426	
	6850	Portable cable/wire puller, 8000 lb. max pulling capacity		3.72	167	502	1,500	130.15	
	6900	Water tank trailer, engine driven discharge, 5000 gallons		7.20	150	450	1,350	147.60	
	6925	10,000 gallons		9.70	207	620	1,850	201.60	

		UNIT	HOURLY OPER. COST	RENT PER DAY	RENT PER WEEK	RENT PER MONTH	EQUIPMENT COST/DAY	
40	6950	Water truck, off highway, 6000 gallons	Ea.	70.16	805	2,420	7,250	1,045	**40**
	7010	Tram car for high voltage line work, powered, 2 conductor		6.85	150	449	1,350	144.60	
	7020	Transit (builder's level) with tripod		.10	16.35	49	147	10.60	
	7030	Trench box, 3000 lb., 6' x 8'		.56	93.50	280	840	60.50	
	7040	7200 lb., 6' x 20'		.75	125	375	1,125	81	
	7050	8000 lb., 8' x 16'		1.08	180	540	1,625	116.65	
	7060	9500 lb., 8' x 20'		1.21	201	603	1,800	130.30	
	7065	11,000 lb., 8' x 24'		1.27	211	633	1,900	136.75	
	7070	12,000 lb., 10' x 20'		1.50	251	752	2,250	162.40	
	7100	Truck, pickup, 3/4 ton, 2 wheel drive		9.90	60	180	540	115.20	
	7200	4 wheel drive		10.20	75	225	675	126.60	
	7250	Crew carrier, 9 passenger		14.00	90	270	810	166	
	7290	Flat bed truck, 20,000 lb. GVW		14.90	130	390	1,175	197.20	
	7300	Tractor, 4 x 2, 220 H.P.		21.00	203	610	1,825	290	
	7410	330 H.P.		30.80	280	840	2,525	414.40	
	7500	6 x 4, 380 H.P.		35.15	325	975	2,925	476.20	
	7600	450 H.P.		43.30	395	1,185	3,550	583.40	
	7610	Tractor, with A frame, boom and winch, 225 H.P.		24.10	282	845	2,525	361.80	
	7620	Vacuum truck, hazardous material, 2500 gallons		12.85	305	910	2,725	284.80	
	7625	5000 gallons		13.11	425	1,270	3,800	358.90	
	7650	Vacuum, HEPA, 16 gallon, wet/dry		.90	18	54	162	18	
	7655	55 gallon, wet/dry		.81	27	81	243	22.70	
	7660	Water tank, portable		.74	123	370	1,100	79.90	
	7690	Sewer/catch basin vacuum, 14 C.Y., 1500 gallons		17.59	635	1,910	5,725	522.70	
	7700	Welder, electric, 200 amp		3.99	16.35	49	147	41.70	
	7800	300 amp		5.90	20	60	180	59.20	
	7900	Gas engine, 200 amp		9.10	24.50	74	222	87.60	
	8000	300 amp		10.35	26	78	234	98.40	
	8100	Wheelbarrow, any size		.06	10.65	32	96	6.90	
	8200	Wrecking ball, 4000 lb.		2.45	71.50	215	645	62.60	
50	0010	**HIGHWAY EQUIPMENT RENTAL** without operators	Ea.						**50**
	0050	Asphalt batch plant, portable drum mixer, 100 ton/hr.		85.49	1,500	4,505	13,500	1,585	
	0060	200 ton/hr.		97.81	1,600	4,800	14,400	1,742	
	0070	300 ton/hr.		116.21	1,875	5,625	16,900	2,055	
	0100	Backhoe attachment, long stick, up to 185 H.P., 10.5' long		.37	24.50	73	219	17.55	
	0140	Up to 250 H.P., 12' long		.41	27	81	243	19.50	
	0180	Over 250 H.P., 15' long		.56	37	111	335	26.70	
	0200	Special dipper arm, up to 100 H.P., 32' long		1.14	75.50	227	680	54.50	
	0240	Over 100 H.P., 33' long		1.42	94.50	284	850	68.15	
	0280	Catch basin/sewer cleaning truck, 3 ton, 9 C.Y., 1000 gal.		35.10	405	1,210	3,625	522.80	
	0300	Concrete batch plant, portable, electric, 200 C.Y./hr.		24.34	545	1,630	4,900	520.70	
	0520	Grader/dozer attachment, ripper/scarifier, rear mounted, up to 135 H.P.		3.15	61.50	185	555	62.20	
	0540	Up to 180 H.P.		4.10	91.50	275	825	87.80	
	0580	Up to 250 H.P.		5.70	145	435	1,300	132.60	
	0700	Pvmt. removal bucket, for hyd. excavator, up to 90 H.P.		2.10	56.50	170	510	50.80	
	0740	Up to 200 H.P.		2.25	71.50	215	645	61	
	0780	Over 200 H.P.		2.45	88.50	265	795	72.60	
	0900	Aggregate spreader, self-propelled, 187 H.P.		50.00	730	2,185	6,550	837	
	1000	Chemical spreader, 3 C.Y.		3.15	45	135	405	52.20	
	1900	Hammermill, traveling, 250 H.P.		68.23	2,200	6,620	19,900	1,870	
	2000	Horizontal borer, 3" diameter, 13 H.P. gas driven		5.50	56.50	170	510	78	
	2150	Horizontal directional drill, 20,000 lb. thrust, 78 H.P. diesel		27.50	680	2,045	6,125	629	
	2160	30,000 lb. thrust, 115 H.P.		33.65	1,050	3,135	9,400	896.20	
	2170	50,000 lb. thrust, 170 H.P.		48.35	1,325	4,005	12,000	1,188	
	2190	Mud trailer for HDD, 1500 gallons, 175 H.P., gas		24.10	158	475	1,425	287.80	
	2200	Hydromulcher, diesel, 3000 gallon, for truck mounting		16.35	253	760	2,275	282.80	
	2300	Gas, 600 gallon		7.40	103	310	930	121.20	
	2400	Joint & crack cleaner, walk behind, 25 H.P.		3.10	51.50	155	465	55.80	

R015433 -10

01 54 33 | Equipment Rental

		UNIT	HOURLY OPER. COST	RENT PER DAY	RENT PER WEEK	RENT PER MONTH	EQUIPMENT COST/DAY	
50	2500	Filler, trailer mounted, 400 gallons, 20 H.P.	Ea.	8.40	218	655	1,975	198.20	**50**
	3000	Paint striper, self-propelled, 40 gallon, 22 H.P.		6.75	162	485	1,450	151	
	3100	120 gallon, 120 H.P.		18.90	405	1,220	3,650	395.20	
	3200	Post drivers, 6" I-Beam frame, for truck mounting		12.45	390	1,175	3,525	334.60	
	3400	Road sweeper, self-propelled, 8' wide, 90 H.P.		35.95	670	2,005	6,025	688.60	
	3450	Road sweeper, vacuum assisted, 4 C.Y., 220 gallons		56.05	655	1,960	5,875	840.40	
	4000	Road mixer, self-propelled, 130 H.P.		45.95	800	2,405	7,225	848.60	
	4100	310 H.P.		75.55	2,150	6,425	19,300	1,889	
	4220	Cold mix paver, incl. pug mill and bitumen tank, 165 H.P.		94.60	2,300	6,915	20,700	2,140	
	4240	Pavement brush, towed		3.40	96.50	290	870	85.20	
	4250	Paver, asphalt, wheel or crawler, 130 H.P., diesel		94.25	2,275	6,845	20,500	2,123	
	4300	Paver, road widener, gas, 1' to 6', 67 H.P.		46.65	940	2,825	8,475	938.20	
	4400	Diesel, 2' to 14', 88 H.P.		56.75	1,125	3,355	10,100	1,125	
	4600	Slipform pavers, curb and gutter, 2 track, 75 H.P.		56.30	1,200	3,615	10,800	1,173	
	4700	4 track, 165 H.P.		36.95	825	2,470	7,400	789.60	
	4800	Median barrier, 215 H.P.		57.45	1,275	3,805	11,400	1,221	
	4901	Trailer, low bed, 75 ton capacity		11.05	268	805	2,425	249.40	
	5000	Road planer, walk behind, 10" cutting width, 10 H.P.		2.50	33.50	100	300	40	
	5100	Self-propelled, 12" cutting width, 64 H.P.		8.00	115	345	1,025	133	
	5120	Traffic line remover, metal ball blaster, truck mounted, 115 H.P.		46.70	800	2,395	7,175	852.60	
	5140	Grinder, truck mounted, 115 H.P.		51.05	850	2,555	7,675	919.40	
	5160	Walk-behind, 11 H.P.		3.55	55	165	495	61.40	
	5200	Pavement profiler, 4' to 6' wide, 450 H.P.		218.90	3,450	10,350	31,100	3,821	
	5300	8' to 10' wide, 750 H.P.		336.50	4,550	13,635	40,900	5,419	
	5400	Roadway plate, steel, 1" x 8' x 20'		.09	14.35	43	129	9.30	
	5600	Stabilizer, self-propelled, 150 H.P.		41.50	680	2,045	6,125	741	
	5700	310 H.P.		77.65	1,825	5,485	16,500	1,718	
	5800	Striper, truck mounted, 120 gallon paint, 460 H.P.		48.50	505	1,510	4,525	690	
	5900	Thermal paint heating kettle, 115 gallons		7.73	26.50	80	240	77.85	
	6000	Tar kettle, 330 gallon, trailer mounted		11.65	60	180	540	129.20	
	7000	Tunnel locomotive, diesel, 8 to 12 ton		29.90	600	1,800	5,400	599.20	
	7005	Electric, 10 ton		28.40	685	2,060	6,175	639.20	
	7010	Muck cars, 1/2 C.Y. capacity		2.25	26	78	234	33.60	
	7020	1 C.Y. capacity		2.45	33.50	100	300	39.60	
	7030	2 C.Y. capacity		2.60	38.50	115	345	43.80	
	7040	Side dump, 2 C.Y. capacity		2.80	46.50	140	420	50.40	
	7050	3 C.Y. capacity		3.80	51.50	155	465	61.40	
	7060	5 C.Y. capacity		5.50	66.50	200	600	84	
	7100	Ventilating blower for tunnel, 7-1/2 H.P.		2.16	51.50	155	465	48.30	
	7110	10 H.P.		2.39	53.50	160	480	51.10	
	7120	20 H.P.		3.56	69.50	208	625	70.10	
	7140	40 H.P.		6.02	80	240	720	96.15	
	7160	60 H.P.		8.75	98.50	295	885	129	
	7175	75 H.P.		10.31	153	460	1,375	174.50	
	7180	200 H.P.		20.73	305	920	2,750	349.85	
	7800	Windrow loader, elevating		54.10	1,350	4,045	12,100	1,242	
60	0010	**LIFTING AND HOISTING EQUIPMENT RENTAL** without operators							**60**
	0150	Crane, flatbed mounted, 3 ton capacity	Ea.	14.10	205	615	1,850	235.80	
	0200	Crane, climbing, 106' jib, 6000 lb. capacity, 410 fpm		41.13	1,750	5,260	15,800	1,381	
	0300	101' jib, 10,250 lb. capacity, 270 fpm		48.18	2,225	6,670	20,000	1,719	
	0500	Tower, static, 130' high, 106' jib, 6200 lb. capacity at 400 fpm		45.23	2,025	6,080	18,200	1,578	
	0520	Mini crawler spider crane, up to 24" wide, 1990 lb. lifting capacity		12.54	755	2,265	6,800	553.30	
	0525	Up to 30" wide, 6450 lb. lifting capacity		14.57	840	2,520	7,550	620.55	
	0530	Up to 52" wide, 6680 lb. lifting capacity		23.17	1,325	3,960	11,900	977.35	
	0535	Up to 55" wide, 8920 lb. lifting capacity		25.87	1,500	4,500	13,500	1,107	
	0540	Up to 66" wide, 13,350 lb. lifting capacity		35.03	2,050	6,120	18,400	1,504	
	0600	Crawler mounted, lattice boom, 1/2 C.Y., 15 tons at 12' radius		37.10	885	2,660	7,975	828.80	
	0700	3/4 C.Y., 20 tons at 12' radius		49.46	1,100	3,320	9,950	1,060	

R015433 -10

R312316 -45

For customer support on your Commercial Renovation Costs with RSMeans data, call 800.448.8182.

01 54 33 | Equipment Rental

		UNIT	HOURLY OPER. COST	RENT PER DAY	RENT PER WEEK	RENT PER MONTH	EQUIPMENT COST/DAY
0800	1 C.Y., 25 tons at 12' radius	Ea.	65.95	1,375	4,100	12,300	1,348
0900	1-1/2 C.Y., 40 tons at 12' radius		66.45	1,400	4,190	12,600	1,370
1000	2 C.Y., 50 tons at 12' radius		88.90	2,050	6,145	18,400	1,940
1100	3 C.Y., 75 tons at 12' radius		76.00	1,825	5,500	16,500	1,708
1200	100 ton capacity, 60' boom		86.05	1,975	5,920	17,800	1,872
1300	165 ton capacity, 60' boom		105.00	2,325	6,970	20,900	2,234
1400	200 ton capacity, 70' boom		140.85	3,125	9,385	28,200	3,004
1500	350 ton capacity, 80' boom		182.50	4,125	12,375	37,100	3,935
1600	Truck mounted, lattice boom, 6 x 4, 20 tons at 10' radius		37.84	1,300	3,900	11,700	1,083
1700	25 tons at 10' radius		40.92	1,425	4,240	12,700	1,175
1800	8 x 4, 30 tons at 10' radius		44.29	1,500	4,520	13,600	1,258
1900	40 tons at 12' radius		47.09	1,575	4,720	14,200	1,321
2000	60 tons at 15' radius		52.56	1,675	5,000	15,000	1,420
2050	82 tons at 15' radius		58.46	1,775	5,340	16,000	1,536
2100	90 tons at 15' radius		65.33	1,950	5,820	17,500	1,687
2200	115 tons at 15' radius		73.60	2,175	6,500	19,500	1,889
2300	150 tons at 18' radius		80.95	2,275	6,845	20,500	2,017
2350	165 tons at 18' radius		85.95	2,425	7,260	21,800	2,140
2400	Truck mounted, hydraulic, 12 ton capacity		31.00	415	1,240	3,725	496
2500	25 ton capacity		37.45	485	1,455	4,375	590.60
2550	33 ton capacity		50.70	890	2,675	8,025	940.60
2560	40 ton capacity		51.00	905	2,710	8,125	950
2600	55 ton capacity		56.60	900	2,705	8,125	993.80
2700	80 ton capacity		77.85	1,500	4,475	13,400	1,518
2720	100 ton capacity		76.80	1,525	4,595	13,800	1,533
2740	120 ton capacity		101.90	1,825	5,460	16,400	1,907
2760	150 ton capacity		107.85	1,975	5,960	17,900	2,055
2800	Self-propelled, 4 x 4, with telescoping boom, 5 ton		15.00	232	695	2,075	259
2900	12-1/2 ton capacity		20.45	335	1,000	3,000	363.60
3000	15 ton capacity		33.95	520	1,560	4,675	583.60
3050	20 ton capacity		25.40	615	1,840	5,525	571.20
3100	25 ton capacity		37.30	615	1,850	5,550	668.40
3150	40 ton capacity		44.10	635	1,910	5,725	734.80
3200	Derricks, guy, 20 ton capacity, 60' boom, 75' mast		23.07	430	1,288	3,875	442.15
3300	100' boom, 115' mast		36.55	735	2,210	6,625	734.40
3400	Stiffleg, 20 ton capacity, 70' boom, 37' mast		25.74	555	1,670	5,000	539.90
3500	100' boom, 47' mast		39.84	895	2,680	8,050	854.70
3550	Helicopter, small, lift to 1250 lb. maximum, w/pilot		97.10	3,475	10,400	31,200	2,857
3600	Hoists, chain type, overhead, manual, 3/4 ton		.15	.33	1	3	1.40
3900	10 ton		.80	6	18	54	10
4000	Hoist and tower, 5000 lb. cap., portable electric, 40' high		5.19	247	742	2,225	189.90
4100	For each added 10' section, add		.12	19.35	58	174	12.55
4200	Hoist and single tubular tower, 5000 lb. electric, 100' high		7.03	345	1,036	3,100	263.45
4300	For each added 6'-6" section, add		.20	33.50	101	305	21.80
4400	Hoist and double tubular tower, 5000 lb., 100' high		7.56	380	1,141	3,425	288.70
4500	For each added 6'-6" section, add		.22	37	111	335	23.95
4550	Hoist and tower, mast type, 6000 lb., 100' high		8.14	395	1,183	3,550	301.70
4570	For each added 10' section, add		.14	22.50	68	204	14.70
4600	Hoist and tower, personnel, electric, 2000 lb., 100' @ 125 fpm		17.23	1,050	3,150	9,450	767.85
4700	3000 lb., 100' @ 200 fpm		19.70	1,200	3,570	10,700	871.60
4800	3000 lb., 150' @ 300 fpm		21.85	1,325	4,000	12,000	974.80
4900	4000 lb., 100' @ 300 fpm		22.62	1,350	4,080	12,200	996.95
5000	6000 lb., 100' @ 275 fpm	▼	24.32	1,425	4,270	12,800	1,049
5100	For added heights up to 500', add	L.F.	.01	1.67	5	15	1.10
5200	Jacks, hydraulic, 20 ton	Ea.	.05	2	6	18	1.60
5500	100 ton		.40	12	36	108	10.40
6100	Jacks, hydraulic, climbing w/50' jackrods, control console, 30 ton cap.		2.13	142	426	1,275	102.25
6150	For each added 10' jackrod section, add	▼	.05	3.33	10	30	2.40

01 54 33 | Equipment Rental

			UNIT	HOURLY OPER. COST	RENT PER DAY	RENT PER WEEK	RENT PER MONTH	EQUIPMENT COST/DAY	
60	6300	50 ton capacity	Ea.	3.43	228	685	2,050	164.45	60
	6350	For each added 10' jackrod section, add		.06	4	12	36	2.90	
	6500	125 ton capacity		8.95	595	1,790	5,375	429.60	
	6550	For each added 10' jackrod section, add		.61	40.50	121	365	29.10	
	6600	Cable jack, 10 ton capacity with 200' cable		1.79	119	357	1,075	85.70	
	6650	For each added 50' of cable, add		.22	14.35	43	129	10.35	
70	0010	**WELLPOINT EQUIPMENT RENTAL** without operators	R015433 -10						70
	0020	Based on 2 months rental							
	0100	Combination jetting & wellpoint pump, 60 H.P. diesel	Ea.	15.83	350	1,057	3,175	338.05	
	0200	High pressure gas jet pump, 200 H.P., 300 psi	"	34.42	300	903	2,700	455.95	
	0300	Discharge pipe, 8" diameter	L.F.	.01	.57	1.71	5.15	.40	
	0350	12" diameter		.01	.84	2.53	7.60	.60	
	0400	Header pipe, flows up to 150 GPM, 4" diameter		.01	.52	1.56	4.68	.40	
	0500	400 GPM, 6" diameter		.01	.61	1.83	5.50	.45	
	0600	800 GPM, 8" diameter		.01	.84	2.53	7.60	.60	
	0700	1500 GPM, 10" diameter		.01	.89	2.66	8	.60	
	0800	2500 GPM, 12" diameter		.03	1.68	5.03	15.10	1.25	
	0900	4500 GPM, 16" diameter		.03	2.15	6.44	19.30	1.55	
	0950	For quick coupling aluminum and plastic pipe, add		.03	2.22	6.67	20	1.55	
	1100	Wellpoint, 25' long, with fittings & riser pipe, 1-1/2" or 2" diameter	Ea.	.07	4.44	13.31	40	3.20	
	1200	Wellpoint pump, diesel powered, 4" suction, 20 H.P.		7.07	203	609	1,825	178.35	
	1300	6" suction, 30 H.P.		9.51	252	756	2,275	227.30	
	1400	8" suction, 40 H.P.		12.87	345	1,036	3,100	310.15	
	1500	10" suction, 75 H.P.		19.01	405	1,211	3,625	394.30	
	1600	12" suction, 100 H.P.		27.56	645	1,930	5,800	606.50	
	1700	12" suction, 175 H.P.		39.50	710	2,130	6,400	742	
80	0010	**MARINE EQUIPMENT RENTAL** without operators	R015433 -10						80
	0200	Barge, 400 ton, 30' wide x 90' long	Ea.	18.05	1,150	3,455	10,400	835.40	
	0240	800 ton, 45' wide x 90' long		21.95	1,425	4,240	12,700	1,024	
	2000	Tugboat, diesel, 100 H.P.		28.80	228	685	2,050	367.40	
	2040	250 H.P.		54.40	410	1,225	3,675	680.20	
	2080	380 H.P.		122.45	1,225	3,685	11,100	1,717	
	3000	Small work boat, gas, 16-foot, 50 H.P.		12.50	63.50	190	570	138	
	4000	Large, diesel, 48-foot, 200 H.P.		74.90	1,300	3,930	11,800	1,385	

Crew No.	Bare Costs Hr.	Daily	Incl. Subs O&P Hr.	Daily	Cost Per Labor-Hour Bare Costs	Incl. O&P
Crew A-1	Hr.	Daily	Hr.	Daily	Bare Costs	Incl. O&P
1 Building Laborer	$39.85	$318.80	$64.70	$517.60	$39.85	$64.70
1 Concrete Saw, Gas Manual		71.20		78.32	8.90	9.79
8 L.H., Daily Totals		$390.00		$595.92	$48.75	$74.49

Crew No.						
Crew A-1A	Hr.	Daily	Hr.	Daily	Bare Costs	Incl. O&P
1 Skilled Worker	$52.35	$418.80	$84.35	$674.80	$52.35	$84.35
1 Shot Blaster, 20"		214.80		236.28	26.85	29.54
8 L.H., Daily Totals		$633.60		$911.08	$79.20	$113.89

Crew No.						
Crew A-1B	Hr.	Daily	Hr.	Daily	Bare Costs	Incl. O&P
1 Building Laborer	$39.85	$318.80	$64.70	$517.60	$39.85	$64.70
1 Concrete Saw		102.40		112.64	12.80	14.08
8 L.H., Daily Totals		$421.20		$630.24	$52.65	$78.78

Crew No.						
Crew A-1C	Hr.	Daily	Hr.	Daily	Bare Costs	Incl. O&P
1 Building Laborer	$39.85	$318.80	$64.70	$517.60	$39.85	$64.70
1 Chain Saw, Gas, 18"		27.80		30.58	3.48	3.82
8 L.H., Daily Totals		$346.60		$548.18	$43.33	$68.52

Crew No.						
Crew A-1D	Hr.	Daily	Hr.	Daily	Bare Costs	Incl. O&P
1 Building Laborer	$39.85	$318.80	$64.70	$517.60	$39.85	$64.70
1 Vibrating Plate, Gas, 18"		31.80		34.98	3.98	4.37
8 L.H., Daily Totals		$350.60		$552.58	$43.83	$69.07

Crew No.						
Crew A-1E	Hr.	Daily	Hr.	Daily	Bare Costs	Incl. O&P
1 Building Laborer	$39.85	$318.80	$64.70	$517.60	$39.85	$64.70
1 Vibrating Plate, Gas, 21"		40.60		44.66	5.08	5.58
8 L.H., Daily Totals		$359.40		$562.26	$44.92	$70.28

Crew No.						
Crew A-1F	Hr.	Daily	Hr.	Daily	Bare Costs	Incl. O&P
1 Building Laborer	$39.85	$318.80	$64.70	$517.60	$39.85	$64.70
1 Rammer/Tamper, Gas, 8"		46.00		50.60	5.75	6.33
8 L.H., Daily Totals		$364.80		$568.20	$45.60	$71.03

Crew No.						
Crew A-1G	Hr.	Daily	Hr.	Daily	Bare Costs	Incl. O&P
1 Building Laborer	$39.85	$318.80	$64.70	$517.60	$39.85	$64.70
1 Rammer/Tamper, Gas, 15"		52.00		57.20	6.50	7.15
8 L.H., Daily Totals		$370.80		$574.80	$46.35	$71.85

Crew No.						
Crew A-1H	Hr.	Daily	Hr.	Daily	Bare Costs	Incl. O&P
1 Building Laborer	$39.85	$318.80	$64.70	$517.60	$39.85	$64.70
1 Exterior Steam Cleaner		74.80		82.28	9.35	10.29
8 L.H., Daily Totals		$393.60		$599.88	$49.20	$74.98

Crew No.						
Crew A-1J	Hr.	Daily	Hr.	Daily	Bare Costs	Incl. O&P
1 Building Laborer	$39.85	$318.80	$64.70	$517.60	$39.85	$64.70
1 Cultivator, Walk-Behind, 5 H.P.		63.50		69.85	7.94	8.73
8 L.H., Daily Totals		$382.30		$587.45	$47.79	$73.43

Crew No.						
Crew A-1K	Hr.	Daily	Hr.	Daily	Bare Costs	Incl. O&P
1 Building Laborer	$39.85	$318.80	$64.70	$517.60	$39.85	$64.70
1 Cultivator, Walk-Behind, 8 H.P.		74.60		82.06	9.32	10.26
8 L.H., Daily Totals		$393.40		$599.66	$49.17	$74.96

Crew No.						
Crew A-1M	Hr.	Daily	Hr.	Daily	Bare Costs	Incl. O&P
1 Building Laborer	$39.85	$318.80	$64.70	$517.60	$39.85	$64.70
1 Snow Blower, Walk-Behind		70.10		77.11	8.76	9.64
8 L.H., Daily Totals		$388.90		$594.71	$48.61	$74.34

Crew No.	Bare Costs Hr.	Daily	Incl. Subs O&P Hr.	Daily	Cost Per Labor-Hour Bare Costs	Incl. O&P
Crew A-2	Hr.	Daily	Hr.	Daily	Bare Costs	Incl. O&P
2 Laborers	$39.85	$637.60	$64.70	$1035.20	$41.40	$66.95
1 Truck Driver (light)	44.50	356.00	71.45	571.60		
1 Flatbed Truck, Gas, 1.5 Ton		188.40		207.24	7.85	8.63
24 L.H., Daily Totals		$1182.00		$1814.04	$49.25	$75.58

Crew No.						
Crew A-2A	Hr.	Daily	Hr.	Daily	Bare Costs	Incl. O&P
2 Laborers	$39.85	$637.60	$64.70	$1035.20	$41.40	$66.95
1 Truck Driver (light)	44.50	356.00	71.45	571.60		
1 Flatbed Truck, Gas, 1.5 Ton		188.40		207.24		
1 Concrete Saw		102.40		112.64	12.12	13.33
24 L.H., Daily Totals		$1284.40		$1926.68	$53.52	$80.28

Crew No.						
Crew A-2B	Hr.	Daily	Hr.	Daily	Bare Costs	Incl. O&P
1 Truck Driver (light)	$44.50	$356.00	$71.45	$571.60	$44.50	$71.45
1 Flatbed Truck, Gas, 1.5 Ton		188.40		207.24	23.55	25.91
8 L.H., Daily Totals		$544.40		$778.84	$68.05	$97.36

Crew No.						
Crew A-3A	Hr.	Daily	Hr.	Daily	Bare Costs	Incl. O&P
1 Equip. Oper. (light)	$51.30	$410.40	$80.95	$647.60	$51.30	$80.95
1 Pickup Truck, 4x4, 3/4 Ton		126.60		139.26	15.82	17.41
8 L.H., Daily Totals		$537.00		$786.86	$67.13	$98.36

Crew No.						
Crew A-3B	Hr.	Daily	Hr.	Daily	Bare Costs	Incl. O&P
1 Equip. Oper. (medium)	$53.75	$430.00	$84.80	$678.40	$49.88	$79.35
1 Truck Driver (heavy)	46.00	368.00	73.90	591.20		
1 Dump Truck, 12 C.Y., 400 H.P.		542.80		597.08		
1 F.E. Loader, W.M., 2.5 C.Y.		523.20		575.52	66.63	73.29
16 L.H., Daily Totals		$1864.00		$2442.20	$116.50	$152.64

Crew No.						
Crew A-3C	Hr.	Daily	Hr.	Daily	Bare Costs	Incl. O&P
1 Equip. Oper. (light)	$51.30	$410.40	$80.95	$647.60	$51.30	$80.95
1 Loader, Skid Steer, 78 H.P.		364.80		401.28	45.60	50.16
8 L.H., Daily Totals		$775.20		$1048.88	$96.90	$131.11

Crew No.						
Crew A-3D	Hr.	Daily	Hr.	Daily	Bare Costs	Incl. O&P
1 Truck Driver (light)	$44.50	$356.00	$71.45	$571.60	$44.50	$71.45
1 Pickup Truck, 4x4, 3/4 Ton		126.60		139.26		
1 Flatbed Trailer, 25 Ton		133.00		146.30	32.45	35.70
8 L.H., Daily Totals		$615.60		$857.16	$76.95	$107.15

Crew No.						
Crew A-3E	Hr.	Daily	Hr.	Daily	Bare Costs	Incl. O&P
1 Equip. Oper. (crane)	$56.10	$448.80	$88.55	$708.40	$51.05	$81.22
1 Truck Driver (heavy)	46.00	368.00	73.90	591.20		
1 Pickup Truck, 4x4, 3/4 Ton		126.60		139.26	7.91	8.70
16 L.H., Daily Totals		$943.40		$1438.86	$58.96	$89.93

Crew No.						
Crew A-3F	Hr.	Daily	Hr.	Daily	Bare Costs	Incl. O&P
1 Equip. Oper. (crane)	$56.10	$448.80	$88.55	$708.40	$51.05	$81.22
1 Truck Driver (heavy)	46.00	368.00	73.90	591.20		
1 Pickup Truck, 4x4, 3/4 Ton		126.60		139.26		
1 Truck Tractor, 6x4, 380 H.P.		476.20		523.82		
1 Lowbed Trailer, 75 Ton		249.40		274.34	53.26	58.59
16 L.H., Daily Totals		$1669.00		$2237.02	$104.31	$139.81

741

Crew No.	Bare Costs		Incl. Subs O&P		Cost Per Labor-Hour	
Crew A-3G	Hr.	Daily	Hr.	Daily	Bare Costs	Incl. O&P
1 Equip. Oper. (crane)	$56.10	$448.80	$88.55	$708.40	$51.05	$81.22
1 Truck Driver (heavy)	46.00	368.00	73.90	591.20		
1 Pickup Truck, 4x4, 3/4 Ton		126.60		139.26		
1 Truck Tractor, 6x4, 450 H.P.		583.40		641.74		
1 Lowbed Trailer, 75 Ton		249.40		274.34	59.96	65.96
16 L.H., Daily Totals		$1776.20		$2354.94	$111.01	$147.18
Crew A-3H	Hr.	Daily	Hr.	Daily	Bare Costs	Incl. O&P
1 Equip. Oper. (crane)	$56.10	$448.80	$88.55	$708.40	$56.10	$88.55
1 Hyd. Crane, 12 Ton (Daily)		628.60		691.46	78.58	86.43
8 L.H., Daily Totals		$1077.40		$1399.86	$134.68	$174.98
Crew A-3I	Hr.	Daily	Hr.	Daily	Bare Costs	Incl. O&P
1 Equip. Oper. (crane)	$56.10	$448.80	$88.55	$708.40	$56.10	$88.55
1 Hyd. Crane, 25 Ton (Daily)		759.40		835.34	94.92	104.42
8 L.H., Daily Totals		$1208.20		$1543.74	$151.03	$192.97
Crew A-3J	Hr.	Daily	Hr.	Daily	Bare Costs	Incl. O&P
1 Equip. Oper. (crane)	$56.10	$448.80	$88.55	$708.40	$56.10	$88.55
1 Hyd. Crane, 40 Ton (Daily)		1313.00		1444.30	164.13	180.54
8 L.H., Daily Totals		$1761.80		$2152.70	$220.22	$269.09
Crew A-3K	Hr.	Daily	Hr.	Daily	Bare Costs	Incl. O&P
1 Equip. Oper. (crane)	$56.10	$448.80	$88.55	$708.40	$52.35	$82.63
1 Equip. Oper. (oiler)	48.60	388.80	76.70	613.60		
1 Hyd. Crane, 55 Ton (Daily)		1353.00		1488.30		
1 P/U Truck, 3/4 Ton (Daily)		139.20		153.12	93.26	102.59
16 L.H., Daily Totals		$2329.80		$2963.42	$145.61	$185.21
Crew A-3L	Hr.	Daily	Hr.	Daily	Bare Costs	Incl. O&P
1 Equip. Oper. (crane)	$56.10	$448.80	$88.55	$708.40	$52.35	$82.63
1 Equip. Oper. (oiler)	48.60	388.80	76.70	613.60		
1 Hyd. Crane, 80 Ton (Daily)		2113.00		2324.30		
1 P/U Truck, 3/4 Ton (Daily)		139.20		153.12	140.76	154.84
16 L.H., Daily Totals		$3089.80		$3799.42	$193.11	$237.46
Crew A-3M	Hr.	Daily	Hr.	Daily	Bare Costs	Incl. O&P
1 Equip. Oper. (crane)	$56.10	$448.80	$88.55	$708.40	$52.35	$82.63
1 Equip. Oper. (oiler)	48.60	388.80	76.70	613.60		
1 Hyd. Crane, 100 Ton (Daily)		2144.00		2358.40		
1 P/U Truck, 3/4 Ton (Daily)		139.20		153.12	142.70	156.97
16 L.H., Daily Totals		$3120.80		$3833.52	$195.05	$239.60
Crew A-3N	Hr.	Daily	Hr.	Daily	Bare Costs	Incl. O&P
1 Equip. Oper. (crane)	$56.10	$448.80	$88.55	$708.40	$56.10	$88.55
1 Tower Crane (monthly)		1180.00		1298.00	147.50	162.25
8 L.H., Daily Totals		$1628.80		$2006.40	$203.60	$250.80
Crew A-3P	Hr.	Daily	Hr.	Daily	Bare Costs	Incl. O&P
1 Equip. Oper. (light)	$51.30	$410.40	$80.95	$647.60	$51.30	$80.95
1 A.T. Forklift, 31' reach, 45' lift		484.20		532.62	60.52	66.58
8 L.H., Daily Totals		$894.60		$1180.22	$111.83	$147.53
Crew A-3Q	Hr.	Daily	Hr.	Daily	Bare Costs	Incl. O&P
1 Equip. Oper. (light)	$51.30	$410.40	$80.95	$647.60	$51.30	$80.95
1 Pickup Truck, 4x4, 3/4 Ton		126.60		139.26		
1 Flatbed Trailer, 3 Ton		25.40		27.94	19.00	20.90
8 L.H., Daily Totals		$562.40		$814.80	$70.30	$101.85

Crew No.	Bare Costs		Incl. Subs O&P		Cost Per Labor-Hour	
Crew A-3R	Hr.	Daily	Hr.	Daily	Bare Costs	Incl. O&P
1 Equip. Oper. (light)	$51.30	$410.40	$80.95	$647.60	$51.30	$80.95
1 Forklift, Smooth Floor, 8,000 Lb.		148.20		163.02	18.52	20.38
8 L.H., Daily Totals		$558.60		$810.62	$69.83	$101.33
Crew A-4	Hr.	Daily	Hr.	Daily	Bare Costs	Incl. O&P
2 Carpenters	$50.70	$811.20	$82.30	$1316.80	$47.98	$77.63
1 Painter, Ordinary	42.55	340.40	68.30	546.40		
24 L.H., Daily Totals		$1151.60		$1863.20	$47.98	$77.63
Crew A-5	Hr.	Daily	Hr.	Daily	Bare Costs	Incl. O&P
2 Laborers	$39.85	$637.60	$64.70	$1035.20	$40.37	$65.45
.25 Truck Driver (light)	44.50	89.00	71.45	142.90		
.25 Flatbed Truck, Gas, 1.5 Ton		47.10		51.81	2.62	2.88
18 L.H., Daily Totals		$773.70		$1229.91	$42.98	$68.33
Crew A-6	Hr.	Daily	Hr.	Daily	Bare Costs	Incl. O&P
1 Instrument Man	$52.35	$418.80	$84.35	$674.80	$50.42	$81.33
1 Rodman/Chainman	48.50	388.00	78.30	626.40		
1 Level, Electronic		50.40		55.44	3.15	3.46
16 L.H., Daily Totals		$857.20		$1356.64	$53.58	$84.79
Crew A-7	Hr.	Daily	Hr.	Daily	Bare Costs	Incl. O&P
1 Chief of Party	$63.65	$509.20	$99.60	$796.80	$54.83	$87.42
1 Instrument Man	52.35	418.80	84.35	674.80		
1 Rodman/Chainman	48.50	388.00	78.30	626.40		
1 Level, Electronic		50.40		55.44	2.10	2.31
24 L.H., Daily Totals		$1366.40		$2153.44	$56.93	$89.73
Crew A-8	Hr.	Daily	Hr.	Daily	Bare Costs	Incl. O&P
1 Chief of Party	$63.65	$509.20	$99.60	$796.80	$53.25	$85.14
1 Instrument Man	52.35	418.80	84.35	674.80		
2 Rodmen/Chainmen	48.50	776.00	78.30	1252.80		
1 Level, Electronic		50.40		55.44	1.58	1.73
32 L.H., Daily Totals		$1754.40		$2779.84	$54.83	$86.87
Crew A-9	Hr.	Daily	Hr.	Daily	Bare Costs	Incl. O&P
1 Asbestos Foreman	$56.80	$454.40	$91.00	$728.00	$56.36	$90.30
7 Asbestos Workers	56.30	3152.80	90.20	5051.20		
64 L.H., Daily Totals		$3607.20		$5779.20	$56.36	$90.30
Crew A-10A	Hr.	Daily	Hr.	Daily	Bare Costs	Incl. O&P
1 Asbestos Foreman	$56.80	$454.40	$91.00	$728.00	$56.47	$90.47
2 Asbestos Workers	56.30	900.80	90.20	1443.20		
24 L.H., Daily Totals		$1355.20		$2171.20	$56.47	$90.47
Crew A-10B	Hr.	Daily	Hr.	Daily	Bare Costs	Incl. O&P
1 Asbestos Foreman	$56.80	$454.40	$91.00	$728.00	$56.42	$90.40
3 Asbestos Workers	56.30	1351.20	90.20	2164.80		
32 L.H., Daily Totals		$1805.60		$2892.80	$56.42	$90.40
Crew A-10C	Hr.	Daily	Hr.	Daily	Bare Costs	Incl. O&P
3 Asbestos Workers	$56.30	$1351.20	$90.20	$2164.80	$56.30	$90.20
1 Flatbed Truck, Gas, 1.5 Ton		188.40		207.24	7.85	8.63
24 L.H., Daily Totals		$1539.60		$2372.04	$64.15	$98.83

Crew No.	Bare Costs Hr.	Bare Costs Daily	Incl. Subs O&P Hr.	Incl. Subs O&P Daily	Cost Per Labor-Hour Bare Costs	Cost Per Labor-Hour Incl. O&P
Crew A-10D						
2 Asbestos Workers	$56.30	$900.80	$90.20	$1443.20	$54.33	$86.41
1 Equip. Oper. (crane)	56.10	448.80	88.55	708.40		
1 Equip. Oper. (oiler)	48.60	388.80	76.70	613.60		
1 Hydraulic Crane, 33 Ton		940.60		1034.66	29.39	32.33
32 L.H., Daily Totals		$2679.00		$3799.86	$83.72	$118.75
Crew A-11						
1 Asbestos Foreman	$56.80	$454.40	$91.00	$728.00	$56.36	$90.30
7 Asbestos Workers	56.30	3152.80	90.20	5051.20		
2 Chip. Hammers, 12 Lb., Elec.		42.00		46.20	.66	.72
64 L.H., Daily Totals		$3649.20		$5825.40	$57.02	$91.02
Crew A-12						
1 Asbestos Foreman	$56.80	$454.40	$91.00	$728.00	$56.36	$90.30
7 Asbestos Workers	56.30	3152.80	90.20	5051.20		
1 Trk-Mtd Vac, 14 CY, 1500 Gal.		522.70		574.97		
1 Flatbed Truck, 20,000 GVW		197.20		216.92	11.25	12.37
64 L.H., Daily Totals		$4327.10		$6571.09	$67.61	$102.67
Crew A-13						
1 Equip. Oper. (light)	$51.30	$410.40	$80.95	$647.60	$51.30	$80.95
1 Trk-Mtd Vac, 14 CY, 1500 Gal.		522.70		574.97		
1 Flatbed Truck, 20,000 GVW		197.20		216.92	89.99	98.99
8 L.H., Daily Totals		$1130.30		$1439.49	$141.29	$179.94
Crew B-1						
1 Labor Foreman (outside)	$41.85	$334.80	$67.90	$543.20	$40.52	$65.77
2 Laborers	39.85	637.60	64.70	1035.20		
24 L.H., Daily Totals		$972.40		$1578.40	$40.52	$65.77
Crew B-1A						
1 Labor Foreman (outside)	$41.85	$334.80	$67.90	$543.20	$40.52	$65.77
2 Laborers	39.85	637.60	64.70	1035.20		
2 Cutting Torches		25.20		27.72		
2 Sets of Gases		336.00		369.60	15.05	16.56
24 L.H., Daily Totals		$1333.60		$1975.72	$55.57	$82.32
Crew B-1B						
1 Labor Foreman (outside)	$41.85	$334.80	$67.90	$543.20	$44.41	$71.46
2 Laborers	39.85	637.60	64.70	1035.20		
1 Equip. Oper. (crane)	56.10	448.80	88.55	708.40		
2 Cutting Torches		25.20		27.72		
2 Sets of Gases		336.00		369.60		
1 Hyd. Crane, 12 Ton		496.00		545.60	26.79	29.47
32 L.H., Daily Totals		$2278.40		$3229.72	$71.20	$100.93
Crew B-1C						
1 Labor Foreman (outside)	$41.85	$334.80	$67.90	$543.20	$40.52	$65.77
2 Laborers	39.85	637.60	64.70	1035.20		
1 Telescoping Boom Lift, to 60'		454.60		500.06	18.94	20.84
24 L.H., Daily Totals		$1427.00		$2078.46	$59.46	$86.60
Crew B-1D						
2 Laborers	$39.85	$637.60	$64.70	$1035.20	$39.85	$64.70
1 Small Work Boat, Gas, 50 H.P.		138.00		151.80		
1 Pressure Washer, 7 GPM		77.20		84.92	13.45	14.80
16 L.H., Daily Totals		$852.80		$1271.92	$53.30	$79.50

Crew No.	Bare Costs Hr.	Bare Costs Daily	Incl. Subs O&P Hr.	Incl. Subs O&P Daily	Cost Per Labor-Hour Bare Costs	Cost Per Labor-Hour Incl. O&P
Crew B-1E						
1 Labor Foreman (outside)	$41.85	$334.80	$67.90	$543.20	$40.35	$65.50
3 Laborers	39.85	956.40	64.70	1552.80		
1 Work Boat, Diesel, 200 H.P.		1385.00		1523.50		
2 Pressure Washers, 7 GPM		154.40		169.84	48.11	52.92
32 L.H., Daily Totals		$2830.60		$3789.34	$88.46	$118.42
Crew B-1F						
2 Skilled Workers	$52.35	$837.60	$84.35	$1349.60	$48.18	$77.80
1 Laborer	39.85	318.80	64.70	517.60		
1 Small Work Boat, Gas, 50 H.P.		138.00		151.80		
1 Pressure Washer, 7 GPM		77.20		84.92	8.97	9.86
24 L.H., Daily Totals		$1371.60		$2103.92	$57.15	$87.66
Crew B-1G						
2 Laborers	$39.85	$637.60	$64.70	$1035.20	$39.85	$64.70
1 Small Work Boat, Gas, 50 H.P.		138.00		151.80	8.63	9.49
16 L.H., Daily Totals		$775.60		$1187.00	$48.48	$74.19
Crew B-1H						
2 Skilled Workers	$52.35	$837.60	$84.35	$1349.60	$48.18	$77.80
1 Laborer	39.85	318.80	64.70	517.60		
1 Small Work Boat, Gas, 50 H.P.		138.00		151.80	5.75	6.33
24 L.H., Daily Totals		$1294.40		$2019.00	$53.93	$84.13
Crew B-1J						
1 Labor Foreman (inside)	$40.35	$322.80	$65.50	$524.00	$40.10	$65.10
1 Laborer	39.85	318.80	64.70	517.60		
16 L.H., Daily Totals		$641.60		$1041.60	$40.10	$65.10
Crew B-1K						
1 Carpenter Foreman (inside)	$51.20	$409.60	$83.10	$664.80	$50.95	$82.70
1 Carpenter	50.70	405.60	82.30	658.40		
16 L.H., Daily Totals		$815.20		$1323.20	$50.95	$82.70
Crew B-2						
1 Labor Foreman (outside)	$41.85	$334.80	$67.90	$543.20	$40.25	$65.34
4 Laborers	39.85	1275.20	64.70	2070.40		
40 L.H., Daily Totals		$1610.00		$2613.60	$40.25	$65.34
Crew B-2A						
1 Labor Foreman (outside)	$41.85	$334.80	$67.90	$543.20	$40.52	$65.77
2 Laborers	39.85	637.60	64.70	1035.20		
1 Telescoping Boom Lift, to 60'		454.60		500.06	18.94	20.84
24 L.H., Daily Totals		$1427.00		$2078.46	$59.46	$86.60
Crew B-3						
1 Labor Foreman (outside)	$41.85	$334.80	$67.90	$543.20	$44.55	$71.65
2 Laborers	39.85	637.60	64.70	1035.20		
1 Equip. Oper. (medium)	53.75	430.00	84.80	678.40		
2 Truck Drivers (heavy)	46.00	736.00	73.90	1182.40		
1 Crawler Loader, 3 C.Y.		1204.00		1324.40		
2 Dump Trucks, 12 C.Y., 400 H.P.		1085.60		1194.16	47.70	52.47
48 L.H., Daily Totals		$4428.00		$5957.76	$92.25	$124.12
Crew B-3A						
4 Laborers	$39.85	$1275.20	$64.70	$2070.40	$42.63	$68.72
1 Equip. Oper. (medium)	53.75	430.00	84.80	678.40		
1 Hyd. Excavator, 1.5 C.Y.		893.60		982.96	22.34	24.57
40 L.H., Daily Totals		$2598.80		$3731.76	$64.97	$93.29

Crew No.	Bare Costs Hr.	Daily	Incl. Subs O&P Hr.	Daily	Cost Per Labor-Hour Bare Costs	Incl. O&P
Crew B-3B	Hr.	Daily	Hr.	Daily	Bare Costs	Incl. O&P
2 Laborers	$39.85	$637.60	$64.70	$1035.20	$44.86	$72.03
1 Equip. Oper. (medium)	53.75	430.00	84.80	678.40		
1 Truck Driver (heavy)	46.00	368.00	73.90	591.20		
1 Backhoe Loader, 80 H.P.		385.00		423.50		
1 Dump Truck, 12 C.Y., 400 H.P.		542.80		597.08	28.99	31.89
32 L.H., Daily Totals		$2363.40		$3325.38	$73.86	$103.92
Crew B-3C	Hr.	Daily	Hr.	Daily	Bare Costs	Incl. O&P
3 Laborers	$39.85	$956.40	$64.70	$1552.80	$43.33	$69.72
1 Equip. Oper. (medium)	53.75	430.00	84.80	678.40		
1 Crawler Loader, 4 C.Y.		1459.00		1604.90	45.59	50.15
32 L.H., Daily Totals		$2845.40		$3836.10	$88.92	$119.88
Crew B-4	Hr.	Daily	Hr.	Daily	Bare Costs	Incl. O&P
1 Labor Foreman (outside)	$41.85	$334.80	$67.90	$543.20	$41.21	$66.77
4 Laborers	39.85	1275.20	64.70	2070.40		
1 Truck Driver (heavy)	46.00	368.00	73.90	591.20		
1 Truck Tractor, 220 H.P.		290.00		319.00		
1 Flatbed Trailer, 40 Ton		180.00		198.00	9.79	10.77
48 L.H., Daily Totals		$2448.00		$3721.80	$51.00	$77.54
Crew B-5	Hr.	Daily	Hr.	Daily	Bare Costs	Incl. O&P
1 Labor Foreman (outside)	$41.85	$334.80	$67.90	$543.20	$44.11	$70.90
4 Laborers	39.85	1275.20	64.70	2070.40		
2 Equip. Oper. (medium)	53.75	860.00	84.80	1356.80		
1 Air Compressor, 250 cfm		167.40		184.14		
2 Breakers, Pavement, 60 lb.		21.20		23.32		
2 -50' Air Hoses, 1.5"		45.40		49.94		
1 Crawler Loader, 3 C.Y.		1204.00		1324.40	25.68	28.25
56 L.H., Daily Totals		$3908.00		$5552.20	$69.79	$99.15
Crew B-5D	Hr.	Daily	Hr.	Daily	Bare Costs	Incl. O&P
1 Labor Foreman (outside)	$41.85	$334.80	$67.90	$543.20	$44.34	$71.28
4 Laborers	39.85	1275.20	64.70	2070.40		
2 Equip. Oper. (medium)	53.75	860.00	84.80	1356.80		
1 Truck Driver (heavy)	46.00	368.00	73.90	591.20		
1 Air Compressor, 250 cfm		167.40		184.14		
2 Breakers, Pavement, 60 lb.		21.20		23.32		
2 -50' Air Hoses, 1.5"		45.40		49.94		
1 Crawler Loader, 3 C.Y.		1204.00		1324.40		
1 Dump Truck, 12 C.Y., 400 H.P.		542.80		597.08	30.95	34.05
64 L.H., Daily Totals		$4818.80		$6740.48	$75.29	$105.32
Crew B-6	Hr.	Daily	Hr.	Daily	Bare Costs	Incl. O&P
2 Laborers	$39.85	$637.60	$64.70	$1035.20	$43.67	$70.12
1 Equip. Oper. (light)	51.30	410.40	80.95	647.60		
1 Backhoe Loader, 48 H.P.		312.40		343.64	13.02	14.32
24 L.H., Daily Totals		$1360.40		$2026.44	$56.68	$84.44
Crew B-6B	Hr.	Daily	Hr.	Daily	Bare Costs	Incl. O&P
2 Labor Foremen (outside)	$41.85	$669.60	$67.90	$1086.40	$40.52	$65.77
4 Laborers	39.85	1275.20	64.70	2070.40		
1 S.P. Crane, 4x4, 5 Ton		259.00		284.90		
1 Flatbed Truck, Gas, 1.5 Ton		188.40		207.24		
1 Butt Fusion Mach., 4"-12" diam.		425.70		468.27	18.19	20.01
48 L.H., Daily Totals		$2817.90		$4117.21	$58.71	$85.78

Crew No.	Bare Costs Hr.	Daily	Incl. Subs O&P Hr.	Daily	Cost Per Labor-Hour Bare Costs	Incl. O&P
Crew B-6C	Hr.	Daily	Hr.	Daily	Bare Costs	Incl. O&P
2 Labor Foremen (outside)	$41.85	$669.60	$67.90	$1086.40	$40.52	$65.77
4 Laborers	39.85	1275.20	64.70	2070.40		
1 S.P. Crane, 4x4, 12 Ton		363.60		399.96		
1 Flatbed Truck, Gas, 3 Ton		238.00		261.80		
1 Butt Fusion Mach., 8"-24" diam.		1908.00		2098.80	52.28	57.51
48 L.H., Daily Totals		$4454.40		$5917.36	$92.80	$123.28
Crew B-7	Hr.	Daily	Hr.	Daily	Bare Costs	Incl. O&P
1 Labor Foreman (outside)	$41.85	$334.80	$67.90	$543.20	$42.50	$68.58
4 Laborers	39.85	1275.20	64.70	2070.40		
1 Equip. Oper. (medium)	53.75	430.00	84.80	678.40		
1 Brush Chipper, 12", 130 H.P.		393.00		432.30		
1 Crawler Loader, 3 C.Y.		1204.00		1324.40		
2 Chain Saws, Gas, 36" Long		92.80		102.08	35.20	38.72
48 L.H., Daily Totals		$3729.80		$5150.78	$77.70	$107.31
Crew B-7A	Hr.	Daily	Hr.	Daily	Bare Costs	Incl. O&P
2 Laborers	$39.85	$637.60	$64.70	$1035.20	$43.67	$70.12
1 Equip. Oper. (light)	51.30	410.40	80.95	647.60		
1 Rake w/Tractor		331.30		364.43		
2 Chain Saws, Gas, 18"		55.60		61.16	16.12	17.73
24 L.H., Daily Totals		$1434.90		$2108.39	$59.79	$87.85
Crew B-7B	Hr.	Daily	Hr.	Daily	Bare Costs	Incl. O&P
1 Labor Foreman (outside)	$41.85	$334.80	$67.90	$543.20	$43.00	$69.34
4 Laborers	39.85	1275.20	64.70	2070.40		
1 Equip. Oper. (medium)	53.75	430.00	84.80	678.40		
1 Truck Driver (heavy)	46.00	368.00	73.90	591.20		
1 Brush Chipper, 12", 130 H.P.		393.00		432.30		
1 Crawler Loader, 3 C.Y.		1204.00		1324.40		
2 Chain Saws, Gas, 36" Long		92.80		102.08		
1 Dump Truck, 8 C.Y., 220 H.P.		338.60		372.46	36.22	39.84
56 L.H., Daily Totals		$4436.40		$6114.44	$79.22	$109.19
Crew B-7C	Hr.	Daily	Hr.	Daily	Bare Costs	Incl. O&P
1 Labor Foreman (outside)	$41.85	$334.80	$67.90	$543.20	$43.00	$69.34
4 Laborers	39.85	1275.20	64.70	2070.40		
1 Equip. Oper. (medium)	53.75	430.00	84.80	678.40		
1 Truck Driver (heavy)	46.00	368.00	73.90	591.20		
1 Brush Chipper, 12", 130 H.P.		393.00		432.30		
1 Crawler Loader, 3 C.Y.		1204.00		1324.40		
2 Chain Saws, Gas, 36" Long		92.80		102.08		
1 Dump Truck, 12 C.Y., 400 H.P.		542.80		597.08	39.87	43.85
56 L.H., Daily Totals		$4640.60		$6339.06	$82.87	$113.20
Crew B-8	Hr.	Daily	Hr.	Daily	Bare Costs	Incl. O&P
1 Labor Foreman (outside)	$41.85	$334.80	$67.90	$543.20	$46.21	$73.92
2 Laborers	39.85	637.60	64.70	1035.20		
2 Equip. Oper. (medium)	53.75	860.00	84.80	1356.80		
1 Equip. Oper. (oiler)	48.60	388.80	76.70	613.60		
2 Truck Drivers (heavy)	46.00	736.00	73.90	1182.40		
1 Hyd. Crane, 25 Ton		590.60		649.66		
1 Crawler Loader, 3 C.Y.		1204.00		1324.40		
2 Dump Trucks, 12 C.Y., 400 H.P.		1085.60		1194.16	45.00	49.50
64 L.H., Daily Totals		$5837.40		$7899.42	$91.21	$123.43

For customer support on your Commercial Renovation Costs with RSMeans data, call 800.448.8182.

Crews - Renovation

Crew No.	Bare Costs		Incl. Subs O&P		Cost Per Labor-Hour	

Crew B-9	Hr.	Daily	Hr.	Daily	Bare Costs	Incl. O&P
1 Labor Foreman (outside)	$41.85	$334.80	$67.90	$543.20	$40.25	$65.34
4 Laborers	39.85	1275.20	64.70	2070.40		
1 Air Compressor, 250 cfm		167.40		184.14		
2 Breakers, Pavement, 60 lb.		21.20		23.32		
2 -50' Air Hoses, 1.5"		45.40		49.94	5.85	6.43
40 L.H., Daily Totals		$1844.00		$2871.00	$46.10	$71.78

Crew B-9A	Hr.	Daily	Hr.	Daily	Bare Costs	Incl. O&P
2 Laborers	$39.85	$637.60	$64.70	$1035.20	$41.90	$67.77
1 Truck Driver (heavy)	46.00	368.00	73.90	591.20		
1 Water Tank Trailer, 5000 Gal.		147.60		162.36		
1 Truck Tractor, 220 H.P.		290.00		319.00		
2 -50' Discharge Hoses, 3"		3.00		3.30	18.36	20.19
24 L.H., Daily Totals		$1446.20		$2111.06	$60.26	$87.96

Crew B-9B	Hr.	Daily	Hr.	Daily	Bare Costs	Incl. O&P
2 Laborers	$39.85	$637.60	$64.70	$1035.20	$41.90	$67.77
1 Truck Driver (heavy)	46.00	368.00	73.90	591.20		
2 -50' Discharge Hoses, 3"		3.00		3.30		
1 Water Tank Trailer, 5000 Gal.		147.60		162.36		
1 Truck Tractor, 220 H.P.		290.00		319.00		
1 Pressure Washer		63.60		69.96	21.01	23.11
24 L.H., Daily Totals		$1509.80		$2181.02	$62.91	$90.88

Crew B-9D	Hr.	Daily	Hr.	Daily	Bare Costs	Incl. O&P
1 Labor Foreman (outside)	$41.85	$334.80	$67.90	$543.20	$40.25	$65.34
4 Common Laborers	39.85	1275.20	64.70	2070.40		
1 Air Compressor, 250 cfm		167.40		184.14		
2 -50' Air Hoses, 1.5"		45.40		49.94		
2 Air Powered Tampers		56.60		62.26	6.74	7.41
40 L.H., Daily Totals		$1879.40		$2909.94	$46.98	$72.75

Crew B-10	Hr.	Daily	Hr.	Daily	Bare Costs	Incl. O&P
1 Equip. Oper. (medium)	$53.75	$430.00	$84.80	$678.40	$49.12	$78.10
.5 Laborer	39.85	159.40	64.70	258.80		
12 L.H., Daily Totals		$589.40		$937.20	$49.12	$78.10

Crew B-10A	Hr.	Daily	Hr.	Daily	Bare Costs	Incl. O&P
1 Equip. Oper. (medium)	$53.75	$430.00	$84.80	$678.40	$49.12	$78.10
.5 Laborer	39.85	159.40	64.70	258.80		
1 Roller, 2-Drum, W.B., 7.5 H.P.		179.60		197.56	14.97	16.46
12 L.H., Daily Totals		$769.00		$1134.76	$64.08	$94.56

Crew B-10B	Hr.	Daily	Hr.	Daily	Bare Costs	Incl. O&P
1 Equip. Oper. (medium)	$53.75	$430.00	$84.80	$678.40	$49.12	$78.10
.5 Laborer	39.85	159.40	64.70	258.80		
1 Dozer, 200 H.P.		1273.00		1400.30	106.08	116.69
12 L.H., Daily Totals		$1862.40		$2337.50	$155.20	$194.79

Crew B-10C	Hr.	Daily	Hr.	Daily	Bare Costs	Incl. O&P
1 Equip. Oper. (medium)	$53.75	$430.00	$84.80	$678.40	$49.12	$78.10
.5 Laborer	39.85	159.40	64.70	258.80		
1 Dozer, 200 H.P.		1273.00		1400.30		
1 Vibratory Roller, Towed, 23 Ton		423.00		465.30	141.33	155.47
12 L.H., Daily Totals		$2285.40		$2802.80	$190.45	$233.57

Crew B-10D	Hr.	Daily	Hr.	Daily	Bare Costs	Incl. O&P
1 Equip. Oper. (medium)	$53.75	$430.00	$84.80	$678.40	$49.12	$78.10
.5 Laborer	39.85	159.40	64.70	258.80		
1 Dozer, 200 H.P.		1273.00		1400.30		
1 Sheepsft. Roller, Towed		425.80		468.38	141.57	155.72
12 L.H., Daily Totals		$2288.20		$2805.88	$190.68	$233.82

Crew B-10E	Hr.	Daily	Hr.	Daily	Bare Costs	Incl. O&P
1 Equip. Oper. (medium)	$53.75	$430.00	$84.80	$678.40	$49.12	$78.10
.5 Laborer	39.85	159.40	64.70	258.80		
1 Tandem Roller, 5 Ton		151.20		166.32	12.60	13.86
12 L.H., Daily Totals		$740.60		$1103.52	$61.72	$91.96

Crew B-10F	Hr.	Daily	Hr.	Daily	Bare Costs	Incl. O&P
1 Equip. Oper. (medium)	$53.75	$430.00	$84.80	$678.40	$49.12	$78.10
.5 Laborer	39.85	159.40	64.70	258.80		
1 Tandem Roller, 10 Ton		231.40		254.54	19.28	21.21
12 L.H., Daily Totals		$820.80		$1191.74	$68.40	$99.31

Crew B-10G	Hr.	Daily	Hr.	Daily	Bare Costs	Incl. O&P
1 Equip. Oper. (medium)	$53.75	$430.00	$84.80	$678.40	$49.12	$78.10
.5 Laborer	39.85	159.40	64.70	258.80		
1 Sheepsfoot Roller, 240 H.P.		1316.00		1447.60	109.67	120.63
12 L.H., Daily Totals		$1905.40		$2384.80	$158.78	$198.73

Crew B-10H	Hr.	Daily	Hr.	Daily	Bare Costs	Incl. O&P
1 Equip. Oper. (medium)	$53.75	$430.00	$84.80	$678.40	$49.12	$78.10
.5 Laborer	39.85	159.40	64.70	258.80		
1 Diaphragm Water Pump, 2"		73.00		80.30		
1 -20' Suction Hose, 2"		1.95		2.15		
2 -50' Discharge Hoses, 2"		1.80		1.98	6.40	7.04
12 L.H., Daily Totals		$666.15		$1021.63	$55.51	$85.14

Crew B-10I	Hr.	Daily	Hr.	Daily	Bare Costs	Incl. O&P
1 Equip. Oper. (medium)	$53.75	$430.00	$84.80	$678.40	$49.12	$78.10
.5 Laborer	39.85	159.40	64.70	258.80		
1 Diaphragm Water Pump, 4"		114.80		126.28		
1 -20' Suction Hose, 4"		3.25		3.58		
2 -50' Discharge Hoses, 4"		4.70		5.17	10.23	11.25
12 L.H., Daily Totals		$712.15		$1072.22	$59.35	$89.35

Crew B-10J	Hr.	Daily	Hr.	Daily	Bare Costs	Incl. O&P
1 Equip. Oper. (medium)	$53.75	$430.00	$84.80	$678.40	$49.12	$78.10
.5 Laborer	39.85	159.40	64.70	258.80		
1 Centrifugal Water Pump, 3"		79.20		87.12		
1 -20' Suction Hose, 3"		2.85		3.13		
2 -50' Discharge Hoses, 3"		3.00		3.30	7.09	7.80
12 L.H., Daily Totals		$674.45		$1030.76	$56.20	$85.90

Crew B-10K	Hr.	Daily	Hr.	Daily	Bare Costs	Incl. O&P
1 Equip. Oper. (medium)	$53.75	$430.00	$84.80	$678.40	$49.12	$78.10
.5 Laborer	39.85	159.40	64.70	258.80		
1 Centr. Water Pump, 6"		296.40		326.04		
1 -20' Suction Hose, 6"		11.50		12.65		
2 -50' Discharge Hoses, 6"		12.20		13.42	26.68	29.34
12 L.H., Daily Totals		$909.50		$1289.31	$75.79	$107.44

| Crew No. | Bare Costs | | Incl. Subs O&P | | Cost Per Labor-Hour | |

Crew B-10L

Crew B-10L	Hr.	Daily	Hr.	Daily	Bare Costs	Incl. O&P
1 Equip. Oper. (medium)	$53.75	$430.00	$84.80	$678.40	$49.12	$78.10
.5 Laborer	39.85	159.40	64.70	258.80		
1 Dozer, 80 H.P.		464.40		510.84	38.70	42.57
12 L.H., Daily Totals		$1053.80		$1448.04	$87.82	$120.67

Crew B-10M

Crew B-10M	Hr.	Daily	Hr.	Daily	Bare Costs	Incl. O&P
1 Equip. Oper. (medium)	$53.75	$430.00	$84.80	$678.40	$49.12	$78.10
.5 Laborer	39.85	159.40	64.70	258.80		
1 Dozer, 300 H.P.		1829.00		2011.90	152.42	167.66
12 L.H., Daily Totals		$2418.40		$2949.10	$201.53	$245.76

Crew B-10N

Crew B-10N	Hr.	Daily	Hr.	Daily	Bare Costs	Incl. O&P
1 Equip. Oper. (medium)	$53.75	$430.00	$84.80	$678.40	$49.12	$78.10
.5 Laborer	39.85	159.40	64.70	258.80		
1 F.E. Loader, T.M., 1.5 C.Y.		579.80		637.78	48.32	53.15
12 L.H., Daily Totals		$1169.20		$1574.98	$97.43	$131.25

Crew B-10O

Crew B-10O	Hr.	Daily	Hr.	Daily	Bare Costs	Incl. O&P
1 Equip. Oper. (medium)	$53.75	$430.00	$84.80	$678.40	$49.12	$78.10
.5 Laborer	39.85	159.40	64.70	258.80		
1 F.E. Loader, T.M., 2.25 C.Y.		979.00		1076.90	81.58	89.74
12 L.H., Daily Totals		$1568.40		$2014.10	$130.70	$167.84

Crew B-10P

Crew B-10P	Hr.	Daily	Hr.	Daily	Bare Costs	Incl. O&P
1 Equip. Oper. (medium)	$53.75	$430.00	$84.80	$678.40	$49.12	$78.10
.5 Laborer	39.85	159.40	64.70	258.80		
1 Crawler Loader, 3 C.Y.		1204.00		1324.40	100.33	110.37
12 L.H., Daily Totals		$1793.40		$2261.60	$149.45	$188.47

Crew B-10Q

Crew B-10Q	Hr.	Daily	Hr.	Daily	Bare Costs	Incl. O&P
1 Equip. Oper. (medium)	$53.75	$430.00	$84.80	$678.40	$49.12	$78.10
.5 Laborer	39.85	159.40	64.70	258.80		
1 Crawler Loader, 4 C.Y.		1459.00		1604.90	121.58	133.74
12 L.H., Daily Totals		$2048.40		$2542.10	$170.70	$211.84

Crew B-10R

Crew B-10R	Hr.	Daily	Hr.	Daily	Bare Costs	Incl. O&P
1 Equip. Oper. (medium)	$53.75	$430.00	$84.80	$678.40	$49.12	$78.10
.5 Laborer	39.85	159.40	64.70	258.80		
1 F.E. Loader, W.M., 1 C.Y.		290.80		319.88	24.23	26.66
12 L.H., Daily Totals		$880.20		$1257.08	$73.35	$104.76

Crew B-10S

Crew B-10S	Hr.	Daily	Hr.	Daily	Bare Costs	Incl. O&P
1 Equip. Oper. (medium)	$53.75	$430.00	$84.80	$678.40	$49.12	$78.10
.5 Laborer	39.85	159.40	64.70	258.80		
1 F.E. Loader, W.M., 1.5 C.Y.		342.80		377.08	28.57	31.42
12 L.H., Daily Totals		$932.20		$1314.28	$77.68	$109.52

Crew B-10T

Crew B-10T	Hr.	Daily	Hr.	Daily	Bare Costs	Incl. O&P
1 Equip. Oper. (medium)	$53.75	$430.00	$84.80	$678.40	$49.12	$78.10
.5 Laborer	39.85	159.40	64.70	258.80		
1 F.E. Loader, W.M., 2.5 C.Y.		523.20		575.52	43.60	47.96
12 L.H., Daily Totals		$1112.60		$1512.72	$92.72	$126.06

Crew B-10U

Crew B-10U	Hr.	Daily	Hr.	Daily	Bare Costs	Incl. O&P
1 Equip. Oper. (medium)	$53.75	$430.00	$84.80	$678.40	$49.12	$78.10
.5 Laborer	39.85	159.40	64.70	258.80		
1 F.E. Loader, W.M., 5.5 C.Y.		982.20		1080.42	81.85	90.03
12 L.H., Daily Totals		$1571.60		$2017.62	$130.97	$168.13

Crew B-10V

Crew B-10V	Hr.	Daily	Hr.	Daily	Bare Costs	Incl. O&P
1 Equip. Oper. (medium)	$53.75	$430.00	$84.80	$678.40	$49.12	$78.10
.5 Laborer	39.85	159.40	64.70	258.80		
1 Dozer, 700 H.P.		4999.00		5498.90	416.58	458.24
12 L.H., Daily Totals		$5588.40		$6436.10	$465.70	$536.34

Crew B-10W

Crew B-10W	Hr.	Daily	Hr.	Daily	Bare Costs	Incl. O&P
1 Equip. Oper. (medium)	$53.75	$430.00	$84.80	$678.40	$49.12	$78.10
.5 Laborer	39.85	159.40	64.70	258.80		
1 Dozer, 105 H.P.		608.00		668.80	50.67	55.73
12 L.H., Daily Totals		$1197.40		$1606.00	$99.78	$133.83

Crew B-10X

Crew B-10X	Hr.	Daily	Hr.	Daily	Bare Costs	Incl. O&P
1 Equip. Oper. (medium)	$53.75	$430.00	$84.80	$678.40	$49.12	$78.10
.5 Laborer	39.85	159.40	64.70	258.80		
1 Dozer, 410 H.P.		2291.00		2520.10	190.92	210.01
12 L.H., Daily Totals		$2880.40		$3457.30	$240.03	$288.11

Crew B-10Y

Crew B-10Y	Hr.	Daily	Hr.	Daily	Bare Costs	Incl. O&P
1 Equip. Oper. (medium)	$53.75	$430.00	$84.80	$678.40	$49.12	$78.10
.5 Laborer	39.85	159.40	64.70	258.80		
1 Vibr. Roller, Towed, 12 Ton		576.40		634.04	48.03	52.84
12 L.H., Daily Totals		$1165.80		$1571.24	$97.15	$130.94

Crew B-11A

Crew B-11A	Hr.	Daily	Hr.	Daily	Bare Costs	Incl. O&P
1 Equipment Oper. (med.)	$53.75	$430.00	$84.80	$678.40	$46.80	$74.75
1 Laborer	39.85	318.80	64.70	517.60		
1 Dozer, 200 H.P.		1273.00		1400.30	79.56	87.52
16 L.H., Daily Totals		$2021.80		$2596.30	$126.36	$162.27

Crew B-11B

Crew B-11B	Hr.	Daily	Hr.	Daily	Bare Costs	Incl. O&P
1 Equipment Oper. (light)	$51.30	$410.40	$80.95	$647.60	$45.58	$72.83
1 Laborer	39.85	318.80	64.70	517.60		
1 Air Powered Tamper		28.30		31.13		
1 Air Compressor, 365 cfm		211.40		232.54		
2 -50' Air Hoses, 1.5"		45.40		49.94	17.82	19.60
16 L.H., Daily Totals		$1014.30		$1478.81	$63.39	$92.43

Crew B-11C

Crew B-11C	Hr.	Daily	Hr.	Daily	Bare Costs	Incl. O&P
1 Equipment Oper. (med.)	$53.75	$430.00	$84.80	$678.40	$46.80	$74.75
1 Laborer	39.85	318.80	64.70	517.60		
1 Backhoe Loader, 48 H.P.		312.40		343.64	19.52	21.48
16 L.H., Daily Totals		$1061.20		$1539.64	$66.33	$96.23

Crew B-11K

Crew B-11K	Hr.	Daily	Hr.	Daily	Bare Costs	Incl. O&P
1 Equipment Oper. (med.)	$53.75	$430.00	$84.80	$678.40	$46.80	$74.75
1 Laborer	39.85	318.80	64.70	517.60		
1 Trencher, Chain Type, 8' D		2048.00		2252.80	128.00	140.80
16 L.H., Daily Totals		$2796.80		$3448.80	$174.80	$215.55

Crew B-11L

Crew B-11L	Hr.	Daily	Hr.	Daily	Bare Costs	Incl. O&P
1 Equipment Oper. (med.)	$53.75	$430.00	$84.80	$678.40	$46.80	$74.75
1 Laborer	39.85	318.80	64.70	517.60		
1 Grader, 30,000 Lbs.		643.80		708.18	40.24	44.26
16 L.H., Daily Totals		$1392.60		$1904.18	$87.04	$119.01

Crew B-11M

Crew B-11M	Hr.	Daily	Hr.	Daily	Bare Costs	Incl. O&P
1 Equipment Oper. (med.)	$53.75	$430.00	$84.80	$678.40	$46.80	$74.75
1 Laborer	39.85	318.80	64.70	517.60		
1 Backhoe Loader, 80 H.P.		385.00		423.50	24.06	26.47
16 L.H., Daily Totals		$1133.80		$1619.50	$70.86	$101.22

Crew No.	Bare Costs		Incl. Subs O&P		Cost Per Labor-Hour	
	Hr.	Daily	Hr.	Daily	Bare Costs	Incl. O&P
Crew B-11S	Hr.	Daily	Hr.	Daily	Bare Costs	Incl. O&P
1 Equipment Operator (med.)	$53.75	$430.00	$84.80	$678.40	$49.12	$78.10
.5 Laborer	39.85	159.40	64.70	258.80		
1 Dozer, 300 H.P.		1829.00		2011.90		
1 Ripper, Beam & 1 Shank		87.80		96.58	159.73	175.71
12 L.H., Daily Totals		$2506.20		$3045.68	$208.85	$253.81
Crew B-11T	Hr.	Daily	Hr.	Daily	Bare Costs	Incl. O&P
1 Equipment Operator (med.)	$53.75	$430.00	$84.80	$678.40	$49.12	$78.10
.5 Laborer	39.85	159.40	64.70	258.80		
1 Dozer, 410 H.P.		2291.00		2520.10		
1 Ripper, Beam & 2 Shanks		132.60		145.86	201.97	222.16
12 L.H., Daily Totals		$3013.00		$3603.16	$251.08	$300.26
Crew B-11W	Hr.	Daily	Hr.	Daily	Bare Costs	Incl. O&P
1 Equipment Operator (med.)	$53.75	$430.00	$84.80	$678.40	$46.13	$74.04
1 Common Laborer	39.85	318.80	64.70	517.60		
10 Truck Drivers (heavy)	46.00	3680.00	73.90	5912.00		
1 Dozer, 200 H.P.		1273.00		1400.30		
1 Vibratory Roller, Towed, 23 Ton		423.00		465.30		
10 Dump Trucks, 8 C.Y., 220 H.P.		3386.00		3724.60	52.94	58.23
96 L.H., Daily Totals		$9510.80		$12698.20	$99.07	$132.27
Crew B-11Y	Hr.	Daily	Hr.	Daily	Bare Costs	Incl. O&P
1 Labor Foreman (outside)	$41.85	$334.80	$67.90	$543.20	$44.71	$71.76
5 Common Laborers	39.85	1594.00	64.70	2588.00		
3 Equipment Operators (med.)	53.75	1290.00	84.80	2035.20		
1 Dozer, 80 H.P.		464.40		510.84		
2 Rollers, 2-Drums, W.B., 7.5 H.P.		359.20		395.12		
4 Vibrating Plates, Gas, 21"		162.40		178.64	13.69	15.06
72 L.H., Daily Totals		$4204.80		$6251.00	$58.40	$86.82
Crew B-12A	Hr.	Daily	Hr.	Daily	Bare Costs	Incl. O&P
1 Equip. Oper. (crane)	$56.10	$448.80	$88.55	$708.40	$47.98	$76.63
1 Laborer	39.85	318.80	64.70	517.60		
1 Hyd. Excavator, 1 C.Y.		742.20		816.42	46.39	51.03
16 L.H., Daily Totals		$1509.80		$2042.42	$94.36	$127.65
Crew B-12B	Hr.	Daily	Hr.	Daily	Bare Costs	Incl. O&P
1 Equip. Oper. (crane)	$56.10	$448.80	$88.55	$708.40	$47.98	$76.63
1 Laborer	39.85	318.80	64.70	517.60		
1 Hyd. Excavator, 1.5 C.Y.		893.60		982.96	55.85	61.44
16 L.H., Daily Totals		$1661.20		$2208.96	$103.83	$138.06
Crew B-12C	Hr.	Daily	Hr.	Daily	Bare Costs	Incl. O&P
1 Equip. Oper. (crane)	$56.10	$448.80	$88.55	$708.40	$47.98	$76.63
1 Laborer	39.85	318.80	64.70	517.60		
1 Hyd. Excavator, 2 C.Y.		1052.00		1157.20	65.75	72.33
16 L.H., Daily Totals		$1819.60		$2383.20	$113.72	$148.95
Crew B-12D	Hr.	Daily	Hr.	Daily	Bare Costs	Incl. O&P
1 Equip. Oper. (crane)	$56.10	$448.80	$88.55	$708.40	$47.98	$76.63
1 Laborer	39.85	318.80	64.70	517.60		
1 Hyd. Excavator, 3.5 C.Y.		2195.00		2414.50	137.19	150.91
16 L.H., Daily Totals		$2962.60		$3640.50	$185.16	$227.53
Crew B-12E	Hr.	Daily	Hr.	Daily	Bare Costs	Incl. O&P
1 Equip. Oper. (crane)	$56.10	$448.80	$88.55	$708.40	$47.98	$76.63
1 Laborer	39.85	318.80	64.70	517.60		
1 Hyd. Excavator, .5 C.Y.		435.40		478.94	27.21	29.93
16 L.H., Daily Totals		$1203.00		$1704.94	$75.19	$106.56

Crew No.	Bare Costs		Incl. Subs O&P		Cost Per Labor-Hour	
Crew B-12F	Hr.	Daily	Hr.	Daily	Bare Costs	Incl. O&P
1 Equip. Oper. (crane)	$56.10	$448.80	$88.55	$708.40	$47.98	$76.63
1 Laborer	39.85	318.80	64.70	517.60		
1 Hyd. Excavator, .75 C.Y.		669.60		736.56	41.85	46.03
16 L.H., Daily Totals		$1437.20		$1962.56	$89.83	$122.66
Crew B-12G	Hr.	Daily	Hr.	Daily	Bare Costs	Incl. O&P
1 Equip. Oper. (crane)	$56.10	$448.80	$88.55	$708.40	$47.98	$76.63
1 Laborer	39.85	318.80	64.70	517.60		
1 Crawler Crane, 15 Ton		828.80		911.68		
1 Clamshell Bucket, .5 C.Y.		41.00		45.10	54.36	59.80
16 L.H., Daily Totals		$1637.40		$2182.78	$102.34	$136.42
Crew B-12H	Hr.	Daily	Hr.	Daily	Bare Costs	Incl. O&P
1 Equip. Oper. (crane)	$56.10	$448.80	$88.55	$708.40	$47.98	$76.63
1 Laborer	39.85	318.80	64.70	517.60		
1 Crawler Crane, 25 Ton		1348.00		1482.80		
1 Clamshell Bucket, 1 C.Y.		50.60		55.66	87.41	96.15
16 L.H., Daily Totals		$2166.20		$2764.46	$135.39	$172.78
Crew B-12I	Hr.	Daily	Hr.	Daily	Bare Costs	Incl. O&P
1 Equip. Oper. (crane)	$56.10	$448.80	$88.55	$708.40	$47.98	$76.63
1 Laborer	39.85	318.80	64.70	517.60		
1 Crawler Crane, 20 Ton		1060.00		1166.00		
1 Dragline Bucket, .75 C.Y.		21.80		23.98	67.61	74.37
16 L.H., Daily Totals		$1849.40		$2415.98	$115.59	$151.00
Crew B-12J	Hr.	Daily	Hr.	Daily	Bare Costs	Incl. O&P
1 Equip. Oper. (crane)	$56.10	$448.80	$88.55	$708.40	$47.98	$76.63
1 Laborer	39.85	318.80	64.70	517.60		
1 Gradall, 5/8 C.Y.		869.60		956.56	54.35	59.78
16 L.H., Daily Totals		$1637.20		$2182.56	$102.33	$136.41
Crew B-12K	Hr.	Daily	Hr.	Daily	Bare Costs	Incl. O&P
1 Equip. Oper. (crane)	$56.10	$448.80	$88.55	$708.40	$47.98	$76.63
1 Laborer	39.85	318.80	64.70	517.60		
1 Gradall, 3 Ton, 1 C.Y.		1243.00		1367.30	77.69	85.46
16 L.H., Daily Totals		$2010.60		$2593.30	$125.66	$162.08
Crew B-12L	Hr.	Daily	Hr.	Daily	Bare Costs	Incl. O&P
1 Equip. Oper. (crane)	$56.10	$448.80	$88.55	$708.40	$47.98	$76.63
1 Laborer	39.85	318.80	64.70	517.60		
1 Crawler Crane, 15 Ton		828.80		911.68		
1 F.E. Attachment, .5 C.Y.		64.60		71.06	55.84	61.42
16 L.H., Daily Totals		$1661.00		$2208.74	$103.81	$138.05
Crew B-12M	Hr.	Daily	Hr.	Daily	Bare Costs	Incl. O&P
1 Equip. Oper. (crane)	$56.10	$448.80	$88.55	$708.40	$47.98	$76.63
1 Laborer	39.85	318.80	64.70	517.60		
1 Crawler Crane, 20 Ton		1060.00		1166.00		
1 F.E. Attachment, .75 C.Y.		69.00		75.90	70.56	77.62
16 L.H., Daily Totals		$1896.60		$2467.90	$118.54	$154.24
Crew B-12N	Hr.	Daily	Hr.	Daily	Bare Costs	Incl. O&P
1 Equip. Oper. (crane)	$56.10	$448.80	$88.55	$708.40	$47.98	$76.63
1 Laborer	39.85	318.80	64.70	517.60		
1 Crawler Crane, 25 Ton		1348.00		1482.80		
1 F.E. Attachment, 1 C.Y.		75.80		83.38	88.99	97.89
16 L.H., Daily Totals		$2191.40		$2792.18	$136.96	$174.51

For customer support on your Commercial Renovation Costs with RSMeans data, call 800.448.8182.

Crew No.	Bare Costs		Incl. Subs O&P		Cost Per Labor-Hour	

Crew B-12O	Hr.	Daily	Hr.	Daily	Bare Costs	Incl. O&P
1 Equip. Oper. (crane)	$56.10	$448.80	$88.55	$708.40	$47.98	$76.63
1 Laborer	39.85	318.80	64.70	517.60		
1 Crawler Crane, 40 Ton		1370.00		1507.00		
1 F.E. Attachment, 1.5 C.Y.		85.00		93.50	90.94	100.03
16 L.H., Daily Totals		$2222.60		$2826.50	$138.91	$176.66

Crew B-12P	Hr.	Daily	Hr.	Daily	Bare Costs	Incl. O&P
1 Equip. Oper. (crane)	$56.10	$448.80	$88.55	$708.40	$47.98	$76.63
1 Laborer	39.85	318.80	64.70	517.60		
1 Crawler Crane, 40 Ton		1370.00		1507.00		
1 Dragline Bucket, 1.5 C.Y.		35.40		38.94	87.84	96.62
16 L.H., Daily Totals		$2173.00		$2771.94	$135.81	$173.25

Crew B-12Q	Hr.	Daily	Hr.	Daily	Bare Costs	Incl. O&P
1 Equip. Oper. (crane)	$56.10	$448.80	$88.55	$708.40	$47.98	$76.63
1 Laborer	39.85	318.80	64.70	517.60		
1 Hyd. Excavator, 5/8 C.Y.		576.80		634.48	36.05	39.66
16 L.H., Daily Totals		$1344.40		$1860.48	$84.03	$116.28

Crew B-12S	Hr.	Daily	Hr.	Daily	Bare Costs	Incl. O&P
1 Equip. Oper. (crane)	$56.10	$448.80	$88.55	$708.40	$47.98	$76.63
1 Laborer	39.85	318.80	64.70	517.60		
1 Hyd. Excavator, 2.5 C.Y.		1411.00		1552.10	88.19	97.01
16 L.H., Daily Totals		$2178.60		$2778.10	$136.16	$173.63

Crew B-12T	Hr.	Daily	Hr.	Daily	Bare Costs	Incl. O&P
1 Equip. Oper. (crane)	$56.10	$448.80	$88.55	$708.40	$47.98	$76.63
1 Laborer	39.85	318.80	64.70	517.60		
1 Crawler Crane, 75 Ton		1708.00		1878.80		
1 F.E. Attachment, 3 C.Y.		108.80		119.68	113.55	124.91
16 L.H., Daily Totals		$2584.40		$3224.48	$161.53	$201.53

Crew B-12V	Hr.	Daily	Hr.	Daily	Bare Costs	Incl. O&P
1 Equip. Oper. (crane)	$56.10	$448.80	$88.55	$708.40	$47.98	$76.63
1 Laborer	39.85	318.80	64.70	517.60		
1 Crawler Crane, 75 Ton		1708.00		1878.80		
1 Dragline Bucket, 3 C.Y.		55.80		61.38	110.24	121.26
16 L.H., Daily Totals		$2531.40		$3166.18	$158.21	$197.89

Crew B-12Y	Hr.	Daily	Hr.	Daily	Bare Costs	Incl. O&P
1 Equip. Oper. (crane)	$56.10	$448.80	$88.55	$708.40	$45.27	$72.65
2 Laborers	39.85	637.60	64.70	1035.20		
1 Hyd. Excavator, 3.5 C.Y.		2195.00		2414.50	91.46	100.60
24 L.H., Daily Totals		$3281.40		$4158.10	$136.72	$173.25

Crew B-12Z	Hr.	Daily	Hr.	Daily	Bare Costs	Incl. O&P
1 Equip. Oper. (crane)	$56.10	$448.80	$88.55	$708.40	$45.27	$72.65
2 Laborers	39.85	637.60	64.70	1035.20		
1 Hyd. Excavator, 2.5 C.Y.		1411.00		1552.10	58.79	64.67
24 L.H., Daily Totals		$2497.40		$3295.70	$104.06	$137.32

Crew B-13	Hr.	Daily	Hr.	Daily	Bare Costs	Incl. O&P
1 Labor Foreman (outside)	$41.85	$334.80	$67.90	$543.20	$43.71	$70.28
4 Laborers	39.85	1275.20	64.70	2070.40		
1 Equip. Oper. (crane)	56.10	448.80	88.55	708.40		
1 Equip. Oper. (oiler)	48.60	388.80	76.70	613.60		
1 Hyd. Crane, 25 Ton		590.60		649.66	10.55	11.60
56 L.H., Daily Totals		$3038.20		$4585.26	$54.25	$81.88

Crew B-13A	Hr.	Daily	Hr.	Daily	Bare Costs	Incl. O&P
1 Labor Foreman (outside)	$41.85	$334.80	$67.90	$543.20	$45.86	$73.53
2 Laborers	39.85	637.60	64.70	1035.20		
2 Equipment Operators (med.)	53.75	860.00	84.80	1356.80		
2 Truck Drivers (heavy)	46.00	736.00	73.90	1182.40		
1 Crawler Crane, 75 Ton		1708.00		1878.80		
1 Crawler Loader, 4 C.Y.		1459.00		1604.90		
2 Dump Trucks, 8 C.Y., 220 H.P.		677.20		744.92	68.65	75.51
56 L.H., Daily Totals		$6412.60		$8346.22	$114.51	$149.04

Crew B-13B	Hr.	Daily	Hr.	Daily	Bare Costs	Incl. O&P
1 Labor Foreman (outside)	$41.85	$334.80	$67.90	$543.20	$43.71	$70.28
4 Laborers	39.85	1275.20	64.70	2070.40		
1 Equip. Oper. (crane)	56.10	448.80	88.55	708.40		
1 Equip. Oper. (oiler)	48.60	388.80	76.70	613.60		
1 Hyd. Crane, 55 Ton		993.80		1093.18	17.75	19.52
56 L.H., Daily Totals		$3441.40		$5028.78	$61.45	$89.80

Crew B-13C	Hr.	Daily	Hr.	Daily	Bare Costs	Incl. O&P
1 Labor Foreman (outside)	$41.85	$334.80	$67.90	$543.20	$43.71	$70.28
4 Laborers	39.85	1275.20	64.70	2070.40		
1 Equip. Oper. (crane)	56.10	448.80	88.55	708.40		
1 Equip. Oper. (oiler)	48.60	388.80	76.70	613.60		
1 Crawler Crane, 100 Ton		1872.00		2059.20	33.43	36.77
56 L.H., Daily Totals		$4319.60		$5994.80	$77.14	$107.05

Crew B-13D	Hr.	Daily	Hr.	Daily	Bare Costs	Incl. O&P
1 Laborer	$39.85	$318.80	$64.70	$517.60	$47.98	$76.63
1 Equip. Oper. (crane)	56.10	448.80	88.55	708.40		
1 Hyd. Excavator, 1 C.Y.		742.20		816.42		
1 Trench Box		81.00		89.10	51.45	56.59
16 L.H., Daily Totals		$1590.80		$2131.52	$99.42	$133.22

Crew B-13E	Hr.	Daily	Hr.	Daily	Bare Costs	Incl. O&P
1 Laborer	$39.85	$318.80	$64.70	$517.60	$47.98	$76.63
1 Equip. Oper. (crane)	56.10	448.80	88.55	708.40		
1 Hyd. Excavator, 1.5 C.Y.		893.60		982.96		
1 Trench Box		81.00		89.10	60.91	67.00
16 L.H., Daily Totals		$1742.20		$2298.06	$108.89	$143.63

Crew B-13F	Hr.	Daily	Hr.	Daily	Bare Costs	Incl. O&P
1 Laborer	$39.85	$318.80	$64.70	$517.60	$47.98	$76.63
1 Equip. Oper. (crane)	56.10	448.80	88.55	708.40		
1 Hyd. Excavator, 3.5 C.Y.		2195.00		2414.50		
1 Trench Box		81.00		89.10	142.25	156.47
16 L.H., Daily Totals		$3043.60		$3729.60	$190.22	$233.10

Crew B-13G	Hr.	Daily	Hr.	Daily	Bare Costs	Incl. O&P
1 Laborer	$39.85	$318.80	$64.70	$517.60	$47.98	$76.63
1 Equip. Oper. (crane)	56.10	448.80	88.55	708.40		
1 Hyd. Excavator, .75 C.Y.		669.60		736.56		
1 Trench Box		81.00		89.10	46.91	51.60
16 L.H., Daily Totals		$1518.20		$2051.66	$94.89	$128.23

Crew B-13H	Hr.	Daily	Hr.	Daily	Bare Costs	Incl. O&P
1 Laborer	$39.85	$318.80	$64.70	$517.60	$47.98	$76.63
1 Equip. Oper. (crane)	56.10	448.80	88.55	708.40		
1 Gradall, 5/8 C.Y.		869.60		956.56		
1 Trench Box		81.00		89.10	59.41	65.35
16 L.H., Daily Totals		$1718.20		$2271.66	$107.39	$141.98

For customer support on your Commercial Renovation Costs with RSMeans data, call 800.448.8182.

Crews - Renovation

Crew No.	Bare Costs Hr.	Bare Costs Daily	Incl. Subs O&P Hr.	Incl. Subs O&P Daily	Cost Per Labor-Hour Bare Costs	Cost Per Labor-Hour Incl. O&P
Crew B-13I	Hr.	Daily	Hr.	Daily	Bare Costs	Incl. O&P
1 Laborer	$39.85	$318.80	$64.70	$517.60	$47.98	$76.63
1 Equip. Oper. (crane)	56.10	448.80	88.55	708.40		
1 Gradall, 3 Ton, 1 C.Y.		1243.00		1367.30		
1 Trench Box		81.00		89.10	82.75	91.03
16 L.H., Daily Totals		$2091.60		$2682.40	$130.72	$167.65

Crew No.	Hr.	Daily	Hr.	Daily	Bare Costs	Incl. O&P
Crew B-13J	Hr.	Daily	Hr.	Daily	Bare Costs	Incl. O&P
1 Laborer	$39.85	$318.80	$64.70	$517.60	$47.98	$76.63
1 Equip. Oper. (crane)	56.10	448.80	88.55	708.40		
1 Hyd. Excavator, 2.5 C.Y.		1411.00		1552.10		
1 Trench Box		81.00		89.10	93.25	102.58
16 L.H., Daily Totals		$2259.60		$2867.20	$141.22	$179.20

Crew B-13K	Hr.	Daily	Hr.	Daily	Bare Costs	Incl. O&P
2 Equip. Opers. (crane)	$56.10	$897.60	$88.55	$1416.80	$56.10	$88.55
1 Hyd. Excavator, .75 C.Y.		669.60		736.56		
1 Hyd. Hammer, 4000 ft-lb		324.60		357.06		
1 Hyd. Excavator, .75 C.Y.		669.60		736.56	103.99	114.39
16 L.H., Daily Totals		$2561.40		$3246.98	$160.09	$202.94

Crew B-13L	Hr.	Daily	Hr.	Daily	Bare Costs	Incl. O&P
2 Equip. Opers. (crane)	$56.10	$897.60	$88.55	$1416.80	$56.10	$88.55
1 Hyd. Excavator, 1.5 C.Y.		893.60		982.96		
1 Hyd. Hammer, 5000 ft-lb		394.00		433.40		
1 Hyd. Excavator, .75 C.Y.		669.60		736.56	122.33	134.56
16 L.H., Daily Totals		$2854.80		$3569.72	$178.43	$223.11

Crew B-13M	Hr.	Daily	Hr.	Daily	Bare Costs	Incl. O&P
2 Equip. Opers. (crane)	$56.10	$897.60	$88.55	$1416.80	$56.10	$88.55
1 Hyd. Excavator, 2.5 C.Y.		1411.00		1552.10		
1 Hyd. Hammer, 8000 ft-lb		576.80		634.48		
1 Hyd. Excavator, 1.5 C.Y.		893.60		982.96	180.09	198.10
16 L.H., Daily Totals		$3779.00		$4586.34	$236.19	$286.65

Crew B-13N	Hr.	Daily	Hr.	Daily	Bare Costs	Incl. O&P
2 Equip. Opers. (crane)	$56.10	$897.60	$88.55	$1416.80	$56.10	$88.55
1 Hyd. Excavator, 3.5 C.Y.		2195.00		2414.50		
1 Hyd. Hammer, 12,000 ft-lb		669.20		736.12		
1 Hyd. Excavator, 1.5 C.Y.		893.60		982.96	234.86	258.35
16 L.H., Daily Totals		$4655.40		$5550.38	$290.96	$346.90

Crew B-14	Hr.	Daily	Hr.	Daily	Bare Costs	Incl. O&P
1 Labor Foreman (outside)	$41.85	$334.80	$67.90	$543.20	$42.09	$67.94
4 Laborers	39.85	1275.20	64.70	2070.40		
1 Equip. Oper. (light)	51.30	410.40	80.95	647.60		
1 Backhoe Loader, 48 H.P.		312.40		343.64	6.51	7.16
48 L.H., Daily Totals		$2332.80		$3604.84	$48.60	$75.10

Crew B-14A	Hr.	Daily	Hr.	Daily	Bare Costs	Incl. O&P
1 Equip. Oper. (crane)	$56.10	$448.80	$88.55	$708.40	$50.68	$80.60
.5 Laborer	39.85	159.40	64.70	258.80		
1 Hyd. Excavator, 4.5 C.Y.		2735.00		3008.50	227.92	250.71
12 L.H., Daily Totals		$3343.20		$3975.70	$278.60	$331.31

Crew B-14B	Hr.	Daily	Hr.	Daily	Bare Costs	Incl. O&P
1 Equip. Oper. (crane)	$56.10	$448.80	$88.55	$708.40	$50.68	$80.60
.5 Laborer	39.85	159.40	64.70	258.80		
1 Hyd. Excavator, 6 C.Y.		3487.00		3835.70	290.58	319.64
12 L.H., Daily Totals		$4095.20		$4802.90	$341.27	$400.24

Crew B-14C	Hr.	Daily	Hr.	Daily	Bare Costs	Incl. O&P
1 Equip. Oper. (crane)	$56.10	$448.80	$88.55	$708.40	$50.68	$80.60
.5 Laborer	39.85	159.40	64.70	258.80		
1 Hyd. Excavator, 7 C.Y.		3275.00		3602.50	272.92	300.21
12 L.H., Daily Totals		$3883.20		$4569.70	$323.60	$380.81

Crew B-14F	Hr.	Daily	Hr.	Daily	Bare Costs	Incl. O&P
1 Equip. Oper. (crane)	$56.10	$448.80	$88.55	$708.40	$50.68	$80.60
.5 Laborer	39.85	159.40	64.70	258.80		
1 Hyd. Shovel, 7 C.Y.		3880.00		4268.00	323.33	355.67
12 L.H., Daily Totals		$4488.20		$5235.20	$374.02	$436.27

Crew B-14G	Hr.	Daily	Hr.	Daily	Bare Costs	Incl. O&P
1 Equip. Oper. (crane)	$56.10	$448.80	$88.55	$708.40	$50.68	$80.60
.5 Laborer	39.85	159.40	64.70	258.80		
1 Hyd. Shovel, 12 C.Y.		5769.00		6345.90	480.75	528.83
12 L.H., Daily Totals		$6377.20		$7313.10	$531.43	$609.42

Crew B-14J	Hr.	Daily	Hr.	Daily	Bare Costs	Incl. O&P
1 Equip. Oper. (medium)	$53.75	$430.00	$84.80	$678.40	$49.12	$78.10
.5 Laborer	39.85	159.40	64.70	258.80		
1 F.E. Loader, 8 C.Y.		1780.00		1958.00	148.33	163.17
12 L.H., Daily Totals		$2369.40		$2895.20	$197.45	$241.27

Crew B-14K	Hr.	Daily	Hr.	Daily	Bare Costs	Incl. O&P
1 Equip. Oper. (medium)	$53.75	$430.00	$84.80	$678.40	$49.12	$78.10
.5 Laborer	39.85	159.40	64.70	258.80		
1 F.E. Loader, 10 C.Y.		2630.00		2893.00	219.17	241.08
12 L.H., Daily Totals		$3219.40		$3830.20	$268.28	$319.18

Crew D-15	Hr.	Daily	Hr.	Daily	Bare Costs	Incl. O&P
1 Equipment Oper. (med.)	$53.75	$430.00	$84.80	$678.40	$47.34	$75.70
.5 Laborer	39.85	159.40	64.70	258.80		
2 Truck Drivers (heavy)	46.00	736.00	73.90	1182.40		
2 Dump Trucks, 12 C.Y., 400 H.P.		1085.60		1194.16		
1 Dozer, 200 H.P.		1273.00		1400.30	84.24	92.66
28 L.H., Daily Totals		$3684.00		$4714.06	$131.57	$168.36

Crew B-16	Hr.	Daily	Hr.	Daily	Bare Costs	Incl. O&P
1 Labor Foreman (outside)	$41.85	$334.80	$67.90	$543.20	$41.89	$67.80
2 Laborers	39.85	637.60	64.70	1035.20		
1 Truck Driver (heavy)	46.00	368.00	73.90	591.20		
1 Dump Truck, 12 C.Y., 400 H.P.		542.80		597.08	16.96	18.66
32 L.H., Daily Totals		$1883.20		$2766.68	$58.85	$86.46

Crew B-17	Hr.	Daily	Hr.	Daily	Bare Costs	Incl. O&P
2 Laborers	$39.85	$637.60	$64.70	$1035.20	$44.25	$71.06
1 Equip. Oper. (light)	51.30	410.40	80.95	647.60		
1 Truck Driver (heavy)	46.00	368.00	73.90	591.20		
1 Backhoe Loader, 48 H.P.		312.40		343.64		
1 Dump Truck, 8 C.Y., 220 H.P.		338.60		372.46	20.34	22.38
32 L.H., Daily Totals		$2067.00		$2990.10	$64.59	$93.44

Crew B-17A	Hr.	Daily	Hr.	Daily	Bare Costs	Incl. O&P
2 Labor Foremen (outside)	$41.85	$669.60	$67.90	$1086.40	$42.95	$69.59
6 Laborers	39.85	1912.80	64.70	3105.60		
1 Skilled Worker Foreman (out)	54.35	434.80	87.55	700.40		
1 Skilled Worker	52.35	418.80	84.35	674.80		
80 L.H., Daily Totals		$3436.00		$5567.20	$42.95	$69.59

For customer support on your Commercial Renovation Costs with RSMeans data, call 800.448.8182.

Crews - Renovation

Crew B-17B	Hr.	Daily	Hr.	Daily	Bare Costs	Incl. O&P
2 Laborers	$39.85	$637.60	$64.70	$1035.20	$44.25	$71.06
1 Equip. Oper. (light)	51.30	410.40	80.95	647.60		
1 Truck Driver (heavy)	46.00	368.00	73.90	591.20		
1 Backhoe Loader, 48 H.P.		312.40		343.64		
1 Dump Truck, 12 C.Y., 400 H.P.		542.80		597.08	26.73	29.40
32 L.H., Daily Totals		$2271.20		$3214.72	$70.97	$100.46

Crew B-18	Hr.	Daily	Hr.	Daily	Bare Costs	Incl. O&P
1 Labor Foreman (outside)	$41.85	$334.80	$67.90	$543.20	$40.52	$65.77
2 Laborers	39.85	637.60	64.70	1035.20		
1 Vibrating Plate, Gas, 21"		40.60		44.66	1.69	1.86
24 L.H., Daily Totals		$1013.00		$1623.06	$42.21	$67.63

Crew B-19	Hr.	Daily	Hr.	Daily	Bare Costs	Incl. O&P
1 Pile Driver Foreman (outside)	$53.30	$426.40	$86.70	$693.60	$52.41	$84.29
4 Pile Drivers	51.30	1641.60	83.45	2670.40		
2 Equip. Oper. (crane)	56.10	897.60	88.55	1416.80		
1 Equip. Oper. (oiler)	48.60	388.80	76.70	613.60		
1 Crawler Crane, 40 Ton		1370.00		1507.00		
1 Lead, 90' High		131.60		144.76		
1 Hammer, Diesel, 22k ft-lb		427.00		469.70	30.13	33.15
64 L.H., Daily Totals		$5283.00		$7515.86	$82.55	$117.44

Crew B-19A	Hr.	Daily	Hr.	Daily	Bare Costs	Incl. O&P
1 Pile Driver Foreman (outside)	$53.30	$426.40	$86.70	$693.60	$52.41	$84.29
4 Pile Drivers	51.30	1641.60	83.45	2670.40		
2 Equip. Oper. (crane)	56.10	897.60	88.55	1416.80		
1 Equip. Oper. (oiler)	48.60	388.80	76.70	613.60		
1 Crawler Crane, 75 Ton		1708.00		1878.80		
1 Lead, 90' High		131.60		144.76		
1 Hammer, Diesel, 41k ft-lb		572.00		629.20	37.68	41.45
64 L.H., Daily Totals		$5766.00		$8047.16	$90.09	$125.74

Crew B-19B	Hr.	Daily	Hr.	Daily	Bare Costs	Incl. O&P
1 Pile Driver Foreman (outside)	$53.30	$426.40	$86.70	$693.60	$52.41	$84.29
4 Pile Drivers	51.30	1641.60	83.45	2670.40		
2 Equip. Oper. (crane)	56.10	897.60	88.55	1416.80		
1 Equip. Oper. (oiler)	48.60	388.80	76.70	613.60		
1 Crawler Crane, 40 Ton		1370.00		1507.00		
1 Lead, 90' High		131.60		144.76		
1 Hammer, Diesel, 22k ft-lb		427.00		469.70		
1 Barge, 400 Ton		835.40		918.94	43.19	47.51
64 L.H., Daily Totals		$6118.40		$8434.80	$95.60	$131.79

Crew B-19C	Hr.	Daily	Hr.	Daily	Bare Costs	Incl. O&P
1 Pile Driver Foreman (outside)	$53.30	$426.40	$86.70	$693.60	$52.41	$84.29
4 Pile Drivers	51.30	1641.60	83.45	2670.40		
2 Equip. Oper. (crane)	56.10	897.60	88.55	1416.80		
1 Equip. Oper. (oiler)	48.60	388.80	76.70	613.60		
1 Crawler Crane, 75 Ton		1708.00		1878.80		
1 Lead, 90' High		131.60		144.76		
1 Hammer, Diesel, 41k ft-lb		572.00		629.20		
1 Barge, 400 Ton		835.40		918.94	50.73	55.81
64 L.H., Daily Totals		$6601.40		$8966.10	$103.15	$140.10

Crew B-20	Hr.	Daily	Hr.	Daily	Bare Costs	Incl. O&P
1 Labor Foreman (outside)	$41.85	$334.80	$67.90	$543.20	$44.68	$72.32
1 Skilled Worker	52.35	418.80	84.35	674.80		
1 Laborer	39.85	318.80	64.70	517.60		
24 L.H., Daily Totals		$1072.40		$1735.60	$44.68	$72.32

Crew B-20A	Hr.	Daily	Hr.	Daily	Bare Costs	Incl. O&P
1 Labor Foreman (outside)	$41.85	$334.80	$67.90	$543.20	$48.39	$76.66
1 Laborer	39.85	318.80	64.70	517.60		
1 Plumber	62.15	497.20	96.70	773.60		
1 Plumber Apprentice	49.70	397.60	77.35	618.80		
32 L.H., Daily Totals		$1548.40		$2453.20	$48.39	$76.66

Crew B-21	Hr.	Daily	Hr.	Daily	Bare Costs	Incl. O&P
1 Labor Foreman (outside)	$41.85	$334.80	$67.90	$543.20	$46.31	$74.64
1 Skilled Worker	52.35	418.80	84.35	674.80		
1 Laborer	39.85	318.80	64.70	517.60		
.5 Equip. Oper. (crane)	56.10	224.40	88.55	354.20		
.5 S.P. Crane, 4x4, 5 Ton		129.50		142.45	4.63	5.09
28 L.H., Daily Totals		$1426.30		$2232.25	$50.94	$79.72

Crew B-21A	Hr.	Daily	Hr.	Daily	Bare Costs	Incl. O&P
1 Labor Foreman (outside)	$41.85	$334.80	$67.90	$543.20	$49.93	$79.04
1 Laborer	39.85	318.80	64.70	517.60		
1 Plumber	62.15	497.20	96.70	773.60		
1 Plumber Apprentice	49.70	397.60	77.35	618.80		
1 Equip. Oper. (crane)	56.10	448.80	88.55	708.40		
1 S.P. Crane, 4x4, 12 Ton		363.60		399.96	9.09	10.00
40 L.H., Daily Totals		$2360.80		$3561.56	$59.02	$89.04

Crew B-21B	Hr.	Daily	Hr.	Daily	Bare Costs	Incl. O&P
1 Labor Foreman (outside)	$41.85	$334.80	$67.90	$543.20	$43.50	$70.11
3 Laborers	39.85	956.40	64.70	1552.80		
1 Equip. Oper. (crane)	56.10	448.80	88.55	708.40		
1 Hyd. Crane, 12 Ton		496.00		545.60	12.40	13.64
40 L.H., Daily Totals		$2236.00		$3350.00	$55.90	$83.75

Crew B-21C	Hr.	Daily	Hr.	Daily	Bare Costs	Incl. O&P
1 Labor Foreman (outside)	$41.85	$334.80	$67.90	$543.20	$43.71	$70.28
4 Laborers	39.85	1275.20	64.70	2070.40		
1 Equip. Oper. (crane)	56.10	448.80	88.55	708.40		
1 Equip. Oper. (oiler)	48.60	388.80	76.70	613.60		
2 Cutting Torches		25.20		27.72		
2 Sets of Gases		336.00		369.60		
1 Lattice Boom Crane, 90 Ton		1687.00		1855.70	36.58	40.23
56 L.H., Daily Totals		$4495.80		$6188.62	$80.28	$110.51

Crew B-22	Hr.	Daily	Hr.	Daily	Bare Costs	Incl. O&P
1 Labor Foreman (outside)	$41.85	$334.80	$67.90	$543.20	$46.97	$75.56
1 Skilled Worker	52.35	418.80	84.35	674.80		
1 Laborer	39.85	318.80	64.70	517.60		
.75 Equip. Oper. (crane)	56.10	336.60	88.55	531.30		
.75 S.P. Crane, 4x4, 5 Ton		194.25		213.68	6.47	7.12
30 L.H., Daily Totals		$1603.25		$2480.57	$53.44	$82.69

Crew B-22A	Hr.	Daily	Hr.	Daily	Bare Costs	Incl. O&P
1 Labor Foreman (outside)	$41.85	$334.80	$67.90	$543.20	$46.00	$74.04
1 Skilled Worker	52.35	418.80	84.35	674.80		
2 Laborers	39.85	637.60	64.70	1035.20		
1 Equipment Operator, Crane	56.10	448.80	88.55	708.40		
1 S.P. Crane, 4x4, 5 Ton		259.00		284.90		
1 Butt Fusion Mach., 4"-12" diam.		425.70		468.27	17.12	18.83
40 L.H., Daily Totals		$2524.70		$3714.77	$63.12	$92.87

For customer support on your Commercial Renovation Costs with RSMeans data, call 800.448.8182.

Crew No.	Bare Costs		Incl. Subs O&P		Cost Per Labor-Hour	
Crew B-22B	Hr.	Daily	Hr.	Daily	Bare Costs	Incl. O&P
1 Labor Foreman (outside)	$41.85	$334.80	$67.90	$543.20	$46.00	$74.04
1 Skilled Worker	52.35	418.80	84.35	674.80		
2 Laborers	39.85	637.60	64.70	1035.20		
1 Equip. Oper. (crane)	56.10	448.80	88.55	708.40		
1 S.P. Crane, 4x4, 5 Ton		259.00		284.90		
1 Butt Fusion Mach., 8"-24" diam.		1908.00		2098.80	54.17	59.59
40 L.H., Daily Totals		$4007.00		$5345.30	$100.18	$133.63

Crew No.	Bare Costs		Incl. Subs O&P		Cost Per Labor-Hour	
Crew B-22C	Hr.	Daily	Hr.	Daily	Bare Costs	Incl. O&P
1 Skilled Worker	$52.35	$418.80	$84.35	$674.80	$46.10	$74.53
1 Laborer	39.85	318.80	64.70	517.60		
1 Butt Fusion Mach., 2"-8" diam.		121.05		133.16	7.57	8.32
16 L.H., Daily Totals		$858.65		$1325.56	$53.67	$82.85

Crew No.	Bare Costs		Incl. Subs O&P		Cost Per Labor-Hour	
Crew B-23	Hr.	Daily	Hr.	Daily	Bare Costs	Incl. O&P
1 Labor Foreman (outside)	$41.85	$334.80	$67.90	$543.20	$40.25	$65.34
4 Laborers	39.85	1275.20	64.70	2070.40		
1 Drill Rig, Truck-Mounted		2444.00		2688.40		
1 Flatbed Truck, Gas, 3 Ton		238.00		261.80	67.05	73.75
40 L.H., Daily Totals		$4292.00		$5563.80	$107.30	$139.10

Crew No.	Bare Costs		Incl. Subs O&P		Cost Per Labor-Hour	
Crew B-23A	Hr.	Daily	Hr.	Daily	Bare Costs	Incl. O&P
1 Labor Foreman (outside)	$41.85	$334.80	$67.90	$543.20	$45.15	$72.47
1 Laborer	39.85	318.80	64.70	517.60		
1 Equip. Oper. (medium)	53.75	430.00	84.80	678.40		
1 Drill Rig, Truck-Mounted		2444.00		2688.40		
1 Pickup Truck, 3/4 Ton		115.20		126.72	106.63	117.30
24 L.H., Daily Totals		$3642.80		$4554.32	$151.78	$189.76

Crew No.	Bare Costs		Incl. Subs O&P		Cost Per Labor-Hour	
Crew B-23B	Hr.	Daily	Hr.	Daily	Bare Costs	Incl. O&P
1 Labor Foreman (outside)	$41.85	$334.80	$67.90	$543.20	$45.15	$72.47
1 Laborer	39.85	318.80	64.70	517.60		
1 Equip. Oper. (medium)	53.75	430.00	84.80	678.40		
1 Drill Rig, Truck-Mounted		2444.00		2688.40		
1 Pickup Truck, 3/4 Ton		115.20		126.72		
1 Centr. Water Pump, 6"		296.40		326.04	118.98	130.88
24 L.H., Daily Totals		$3939.20		$4880.36	$164.13	$203.35

Crew No.	Bare Costs		Incl. Subs O&P		Cost Per Labor-Hour	
Crew B-24	Hr.	Daily	Hr.	Daily	Bare Costs	Incl. O&P
1 Cement Finisher	$47.55	$380.40	$75.20	$601.60	$46.03	$74.07
1 Laborer	39.85	318.80	64.70	517.60		
1 Carpenter	50.70	405.60	82.30	658.40		
24 L.H., Daily Totals		$1104.80		$1777.60	$46.03	$74.07

Crew No.	Bare Costs		Incl. Subs O&P		Cost Per Labor-Hour	
Crew B-25	Hr.	Daily	Hr.	Daily	Bare Costs	Incl. O&P
1 Labor Foreman (outside)	$41.85	$334.80	$67.90	$543.20	$43.82	$70.47
7 Laborers	39.85	2231.60	64.70	3623.20		
3 Equip. Oper. (medium)	53.75	1290.00	84.80	2035.20		
1 Asphalt Paver, 130 H.P.		2123.00		2335.30		
1 Tandem Roller, 10 Ton		231.40		254.54		
1 Roller, Pneum. Whl., 12 Ton		338.20		372.02	30.60	33.66
88 L.H., Daily Totals		$6549.00		$9163.46	$74.42	$104.13

Crew No.	Bare Costs		Incl. Subs O&P		Cost Per Labor-Hour	
Crew B-25B	Hr.	Daily	Hr.	Daily	Bare Costs	Incl. O&P
1 Labor Foreman (outside)	$41.85	$334.80	$67.90	$543.20	$44.65	$71.67
7 Laborers	39.85	2231.60	64.70	3623.20		
4 Equip. Oper. (medium)	53.75	1720.00	84.80	2713.60		
1 Asphalt Paver, 130 H.P.		2123.00		2335.30		
2 Tandem Rollers, 10 Ton		462.80		509.08		
1 Roller, Pneum. Whl., 12 Ton		338.20		372.02	30.46	33.50
96 L.H., Daily Totals		$7210.40		$10096.40	$75.11	$105.17

Crew No.	Bare Costs		Incl. Subs O&P		Cost Per Labor-Hour	
Crew B-25C	Hr.	Daily	Hr.	Daily	Bare Costs	Incl. O&P
1 Labor Foreman (outside)	$41.85	$334.80	$67.90	$543.20	$44.82	$71.93
3 Laborers	39.85	956.40	64.70	1552.80		
2 Equip. Oper. (medium)	53.75	860.00	84.80	1356.80		
1 Asphalt Paver, 130 H.P.		2123.00		2335.30		
1 Tandem Roller, 10 Ton		231.40		254.54	49.05	53.95
48 L.H., Daily Totals		$4505.60		$6042.64	$93.87	$125.89

Crew No.	Bare Costs		Incl. Subs O&P		Cost Per Labor-Hour	
Crew B-25D	Hr.	Daily	Hr.	Daily	Bare Costs	Incl. O&P
1 Labor Foreman (outside)	$41.85	$334.80	$67.90	$543.20	$45.02	$72.23
3 Laborers	39.85	956.40	64.70	1552.80		
2.125 Equip. Oper. (medium)	53.75	913.75	84.80	1441.60		
.125 Truck Driver (heavy)	46.00	46.00	73.90	73.90		
.125 Truck Tractor, 6x4, 380 H.P.		59.52		65.48		
.125 Dist. Tanker, 3000 Gallon		40.85		44.94		
1 Asphalt Paver, 130 H.P.		2123.00		2335.30		
1 Tandem Roller, 10 Ton		231.40		254.54	49.10	54.01
50 L.H., Daily Totals		$4705.73		$6311.75	$94.11	$126.24

Crew No.	Bare Costs		Incl. Subs O&P		Cost Per Labor-Hour	
Crew B-25E	Hr.	Daily	Hr.	Daily	Bare Costs	Incl. O&P
1 Labor Foreman (outside)	$41.85	$334.80	$67.90	$543.20	$45.21	$72.50
3 Laborers	39.85	956.40	64.70	1552.80		
2.250 Equip. Oper. (medium)	53.75	967.50	84.80	1526.40		
.25 Truck Driver (heavy)	46.00	92.00	73.90	147.80		
.25 Truck Tractor, 6x4, 380 H.P.		119.05		130.96		
.25 Dist. Tanker, 3000 Gallon		81.70		89.87		
1 Asphalt Paver, 130 H.P.		2123.00		2335.30		
1 Tandem Roller, 10 Ton		231.40		254.54	49.14	54.05
52 L.H., Daily Totals		$4905.85		$6580.86	$94.34	$126.56

Crew No.	Bare Costs		Incl. Subs O&P		Cost Per Labor-Hour	
Crew B-26	Hr.	Daily	Hr.	Daily	Bare Costs	Incl. O&P
1 Labor Foreman (outside)	$41.85	$334.80	$67.90	$543.20	$44.60	$71.65
6 Laborers	39.85	1912.80	64.70	3105.60		
2 Equip. Oper. (medium)	53.75	860.00	84.80	1356.80		
1 Rodman (reinf.)	54.65	437.20	87.30	698.40		
1 Cement Finisher	47.55	380.40	75.20	601.60		
1 Grader, 30,000 Lbs.		643.80		708.18		
1 Paving Mach. & Equip.		2361.00		2597.10	34.15	37.56
88 L.H., Daily Totals		$6930.00		$9610.88	$78.75	$109.21

Crew No.	Bare Costs		Incl. Subs O&P		Cost Per Labor-Hour	
Crew B-26A	Hr.	Daily	Hr.	Daily	Bare Costs	Incl. O&P
1 Labor Foreman (outside)	$41.85	$334.80	$67.90	$543.20	$44.60	$71.65
6 Laborers	39.85	1912.80	64.70	3105.60		
2 Equip. Oper. (medium)	53.75	860.00	84.80	1356.80		
1 Rodman (reinf.)	54.65	437.20	87.30	698.40		
1 Cement Finisher	47.55	380.40	75.20	601.60		
1 Grader, 30,000 Lbs.		643.80		708.18		
1 Paving Mach. & Equip.		2361.00		2597.10		
1 Concrete Saw		102.40		112.64	35.31	38.84
88 L.H., Daily Totals		$7032.40		$9723.52	$79.91	$110.49

Crew No.	Bare Costs		Incl. Subs O&P		Cost Per Labor-Hour	
Crew B-26B	Hr.	Daily	Hr.	Daily	Bare Costs	Incl. O&P
1 Labor Foreman (outside)	$41.85	$334.80	$67.90	$543.20	$45.37	$72.75
6 Laborers	39.85	1912.80	64.70	3105.60		
3 Equip. Oper. (medium)	53.75	1290.00	84.80	2035.20		
1 Rodman (reinf.)	54.65	437.20	87.30	698.40		
1 Cement Finisher	47.55	380.40	75.20	601.60		
1 Grader, 30,000 Lbs.		643.80		708.18		
1 Paving Mach. & Equip.		2361.00		2597.10		
1 Concrete Pump, 110' Boom		1116.00		1227.60	42.92	47.22
96 L.H., Daily Totals		$8476.00		$11516.88	$88.29	$119.97

Crew No.	Hr.	Bare Costs Daily	Hr.	Incl. Subs O&P Daily	Cost Per Labor-Hour Bare Costs	Incl. O&P
Crew B-26C	Hr.	Daily	Hr.	Daily	Bare Costs	Incl. O&P
1 Labor Foreman (outside)	$41.85	$334.80	$67.90	$543.20	$43.69	$70.34
6 Laborers	39.85	1912.80	64.70	3105.60		
1 Equip. Oper. (medium)	53.75	430.00	84.80	678.40		
1 Rodman (reinf.)	54.65	437.20	87.30	698.40		
1 Cement Finisher	47.55	380.40	75.20	601.60		
1 Paving Mach. & Equip.		2361.00		2597.10		
1 Concrete Saw		102.40		112.64	30.79	33.87
80 L.H., Daily Totals		$5958.60		$8336.94	$74.48	$104.21
Crew B-27	Hr.	Daily	Hr.	Daily	Bare Costs	Incl. O&P
1 Labor Foreman (outside)	$41.85	$334.80	$67.90	$543.20	$40.35	$65.50
3 Laborers	39.85	956.40	64.70	1552.80		
1 Berm Machine		317.40		349.14	9.92	10.91
32 L.H., Daily Totals		$1608.60		$2445.14	$50.27	$76.41
Crew B-28	Hr.	Daily	Hr.	Daily	Bare Costs	Incl. O&P
2 Carpenters	$50.70	$811.20	$82.30	$1316.80	$47.08	$76.43
1 Laborer	39.85	318.80	64.70	517.60		
24 L.H., Daily Totals		$1130.00		$1834.40	$47.08	$76.43
Crew B-29	Hr.	Daily	Hr.	Daily	Bare Costs	Incl. O&P
1 Labor Foreman (outside)	$41.85	$334.80	$67.90	$543.20	$43.71	$70.28
4 Laborers	39.85	1275.20	64.70	2070.40		
1 Equip. Oper. (crane)	56.10	448.80	88.55	708.40		
1 Equip. Oper. (oiler)	48.60	388.80	76.70	613.60		
1 Gradall, 5/8 C.Y.		869.60		956.56	15.53	17.08
56 L.H., Daily Totals		$3317.20		$4892.16	$59.24	$87.36
Crew B-30	Hr.	Daily	Hr.	Daily	Bare Costs	Incl. O&P
1 Equip. Oper. (medium)	$53.75	$430.00	$84.80	$678.40	$48.58	$77.53
2 Truck Drivers (heavy)	46.00	736.00	73.90	1182.40		
1 Hyd. Excavator, 1.5 C.Y.		893.60		982.96		
2 Dump Trucks, 12 C.Y., 400 H.P.		1085.60		1194.16	82.47	90.71
24 L.H., Daily Totals		$3145.20		$4037.92	$131.05	$168.25
Crew B-31	Hr.	Daily	Hr.	Daily	Bare Costs	Incl. O&P
1 Labor Foreman (outside)	$41.85	$334.80	$67.90	$543.20	$42.42	$68.86
3 Laborers	39.85	956.40	64.70	1552.80		
1 Carpenter	50.70	405.60	82.30	658.40		
1 Air Compressor, 250 cfm		167.40		184.14		
1 Sheeting Driver		7.00		7.70		
2 -50' Air Hoses, 1.5"		45.40		49.94	5.50	6.04
40 L.H., Daily Totals		$1916.60		$2996.18	$47.91	$74.90
Crew B-32	Hr.	Daily	Hr.	Daily	Bare Costs	Incl. O&P
1 Laborer	$39.85	$318.80	$64.70	$517.60	$50.27	$79.78
3 Equip. Oper. (medium)	53.75	1290.00	84.80	2035.20		
1 Grader, 30,000 Lbs.		643.80		708.18		
1 Tandem Roller, 10 Ton		231.40		254.54		
1 Dozer, 200 H.P.		1273.00		1400.30	67.13	73.84
32 L.H., Daily Totals		$3757.00		$4915.82	$117.41	$153.62
Crew B-32A	Hr.	Daily	Hr.	Daily	Bare Costs	Incl. O&P
1 Laborer	$39.85	$318.80	$64.70	$517.60	$49.12	$78.10
2 Equip. Oper. (medium)	53.75	860.00	84.80	1356.80		
1 Grader, 30,000 Lbs.		643.80		708.18		
1 Roller, Vibratory, 25 Ton		659.80		725.78	54.32	59.75
24 L.H., Daily Totals		$2482.40		$3308.36	$103.43	$137.85

Crew No.	Hr.	Bare Costs Daily	Hr.	Incl. Subs O&P Daily	Cost Per Labor-Hour Bare Costs	Incl. O&P
Crew B-32B	Hr.	Daily	Hr.	Daily	Bare Costs	Incl. O&P
1 Laborer	$39.85	$318.80	$64.70	$517.60	$49.12	$78.10
2 Equip. Oper. (medium)	53.75	860.00	84.80	1356.80		
1 Dozer, 200 H.P.		1273.00		1400.30		
1 Roller, Vibratory, 25 Ton		659.80		725.78	80.53	88.59
24 L.H., Daily Totals		$3111.60		$4000.48	$129.65	$166.69
Crew B-32C	Hr.	Daily	Hr.	Daily	Bare Costs	Incl. O&P
1 Labor Foreman (outside)	$41.85	$334.80	$67.90	$543.20	$47.13	$75.28
2 Laborers	39.85	637.60	64.70	1035.20		
3 Equip. Oper. (medium)	53.75	1290.00	84.80	2035.20		
1 Grader, 30,000 Lbs.		643.80		708.18		
1 Tandem Roller, 10 Ton		231.40		254.54		
1 Dozer, 200 H.P.		1273.00		1400.30	44.75	49.23
48 L.H., Daily Totals		$4410.60		$5976.62	$91.89	$124.51
Crew B-33A	Hr.	Daily	Hr.	Daily	Bare Costs	Incl. O&P
1 Equip. Oper. (medium)	$53.75	$430.00	$84.80	$678.40	$49.78	$79.06
.5 Laborer	39.85	159.40	64.70	258.80		
.25 Equip. Oper. (medium)	53.75	107.50	84.80	169.60		
1 Scraper, Towed, 7 C.Y.		123.40		135.74		
1.25 Dozers, 300 H.P.		2286.25		2514.88	172.12	189.33
14 L.H., Daily Totals		$3106.55		$3757.42	$221.90	$268.39
Crew B-33B	Hr.	Daily	Hr.	Daily	Bare Costs	Incl. O&P
1 Equip. Oper. (medium)	$53.75	$430.00	$84.80	$678.40	$49.78	$79.06
.5 Laborer	39.85	159.40	64.70	258.80		
.25 Equip. Oper. (medium)	53.75	107.50	84.80	169.60		
1 Scraper, Towed, 10 C.Y.		155.80		171.38		
1.25 Dozers, 300 H.P.		2286.25		2514.88	174.43	191.88
14 L.H., Daily Totals		$3138.95		$3793.05	$224.21	$270.93
Crew B-33C	Hr.	Daily	Hr.	Daily	Bare Costs	Incl. O&P
1 Equip. Oper. (medium)	$53.75	$430.00	$84.80	$678.40	$49.78	$79.06
.5 Laborer	39.85	159.40	64.70	258.80		
.25 Equip. Oper. (medium)	53.75	107.50	84.80	169.60		
1 Scraper, Towed, 15 C.Y.		175.80		193.38		
1.25 Dozers, 300 H.P.		2286.25		2514.88	175.86	193.45
14 L.H., Daily Totals		$3158.95		$3815.05	$225.64	$272.50
Crew B-33D	Hr.	Daily	Hr.	Daily	Bare Costs	Incl. O&P
1 Equip. Oper. (medium)	$53.75	$430.00	$84.80	$678.40	$49.78	$79.06
.5 Laborer	39.85	159.40	64.70	258.80		
.25 Equip. Oper. (medium)	53.75	107.50	84.80	169.60		
1 S.P. Scraper, 14 C.Y.		2490.00		2739.00		
.25 Dozer, 300 H.P.		457.25		502.98	210.52	231.57
14 L.H., Daily Totals		$3644.15		$4348.77	$260.30	$310.63
Crew B-33E	Hr.	Daily	Hr.	Daily	Bare Costs	Incl. O&P
1 Equip. Oper. (medium)	$53.75	$430.00	$84.80	$678.40	$49.78	$79.06
.5 Laborer	39.85	159.40	64.70	258.80		
.25 Equip. Oper. (medium)	53.75	107.50	84.80	169.60		
1 S.P. Scraper, 21 C.Y.		2468.00		2714.80		
.25 Dozer, 300 H.P.		457.25		502.98	208.95	229.84
14 L.H., Daily Totals		$3622.15		$4324.57	$258.73	$308.90

Crew No.	Bare Costs		Incl. Subs O&P		Cost Per Labor-Hour	

Crew B-33F

	Hr.	Daily	Hr.	Daily	Bare Costs	Incl. O&P
1 Equip. Oper. (medium)	$53.75	$430.00	$84.80	$678.40	$49.78	$79.06
.5 Laborer	39.85	159.40	64.70	258.80		
.25 Equip. Oper. (medium)	53.75	107.50	84.80	169.60		
1 Elev. Scraper, 11 C.Y.		1155.00		1270.50		
.25 Dozer, 300 H.P.		457.25		502.98	115.16	126.68
14 L.H., Daily Totals		$2309.15		$2880.28	$164.94	$205.73

Crew B-33G

	Hr.	Daily	Hr.	Daily	Bare Costs	Incl. O&P
1 Equip. Oper. (medium)	$53.75	$430.00	$84.80	$678.40	$49.78	$79.06
.5 Laborer	39.85	159.40	64.70	258.80		
.25 Equip. Oper. (medium)	53.75	107.50	84.80	169.60		
1 Elev. Scraper, 22 C.Y.		2284.00		2512.40		
.25 Dozer, 300 H.P.		457.25		502.98	195.80	215.38
14 L.H., Daily Totals		$3438.15		$4122.18	$245.58	$294.44

Crew B-33K

	Hr.	Daily	Hr.	Daily	Bare Costs	Incl. O&P
1 Equipment Operator (med.)	$53.75	$430.00	$84.80	$678.40	$49.78	$79.06
.25 Equipment Operator (med.)	53.75	107.50	84.80	169.60		
.5 Laborer	39.85	159.40	64.70	258.80		
1 S.P. Scraper, 31 C.Y.		3576.00		3933.60		
.25 Dozer, 410 H.P.		572.75		630.02	296.34	325.97
14 L.H., Daily Totals		$4845.65		$5670.43	$346.12	$405.03

Crew B-34A

	Hr.	Daily	Hr.	Daily	Bare Costs	Incl. O&P
1 Truck Driver (heavy)	$46.00	$368.00	$73.90	$591.20	$46.00	$73.90
1 Dump Truck, 8 C.Y., 220 H.P.		338.60		372.46	42.33	46.56
8 L.H., Daily Totals		$706.60		$963.66	$88.33	$120.46

Crew B-34B

	Hr.	Daily	Hr.	Daily	Bare Costs	Incl. O&P
1 Truck Driver (heavy)	$46.00	$368.00	$73.90	$591.20	$46.00	$73.90
1 Dump Truck, 12 C.Y., 400 H.P.		542.80		597.08	67.85	74.64
8 L.H., Daily Totals		$910.80		$1188.28	$113.85	$148.54

Crew B-34C

	Hr.	Daily	Hr.	Daily	Bare Costs	Incl. O&P
1 Truck Driver (heavy)	$46.00	$368.00	$73.90	$591.20	$46.00	$73.90
1 Truck Tractor, 6x4, 380 H.P.		476.20		523.82		
1 Dump Trailer, 16.5 C.Y.		133.00		146.30	76.15	83.77
8 L.H., Daily Totals		$977.20		$1261.32	$122.15	$157.66

Crew B-34D

	Hr.	Daily	Hr.	Daily	Bare Costs	Incl. O&P
1 Truck Driver (heavy)	$46.00	$368.00	$73.90	$591.20	$46.00	$73.90
1 Truck Tractor, 6x4, 380 H.P.		476.20		523.82		
1 Dump Trailer, 20 C.Y.		147.60		162.36	77.97	85.77
8 L.H., Daily Totals		$991.80		$1277.38	$123.97	$159.67

Crew B-34E

	Hr.	Daily	Hr.	Daily	Bare Costs	Incl. O&P
1 Truck Driver (heavy)	$46.00	$368.00	$73.90	$591.20	$46.00	$73.90
1 Dump Truck, Off Hwy., 25 Ton		1343.00		1477.30	167.88	184.66
8 L.H., Daily Totals		$1711.00		$2068.50	$213.88	$258.56

Crew B-34F

	Hr.	Daily	Hr.	Daily	Bare Costs	Incl. O&P
1 Truck Driver (heavy)	$46.00	$368.00	$73.90	$591.20	$46.00	$73.90
1 Dump Truck, Off Hwy., 35 Ton		1448.00		1592.80	181.00	199.10
8 L.H., Daily Totals		$1816.00		$2184.00	$227.00	$273.00

Crew B-34G

	Hr.	Daily	Hr.	Daily	Bare Costs	Incl. O&P
1 Truck Driver (heavy)	$46.00	$368.00	$73.90	$591.20	$46.00	$73.90
1 Dump Truck, Off Hwy., 50 Ton		1662.00		1828.20	207.75	228.53
8 L.H., Daily Totals		$2030.00		$2419.40	$253.75	$302.43

Crew B-34H

	Hr.	Daily	Hr.	Daily	Bare Costs	Incl. O&P
1 Truck Driver (heavy)	$46.00	$368.00	$73.90	$591.20	$46.00	$73.90
1 Dump Truck, Off Hwy., 65 Ton		1844.00		2028.40	230.50	253.55
8 L.H., Daily Totals		$2212.00		$2619.60	$276.50	$327.45

Crew B-34I

	Hr.	Daily	Hr.	Daily	Bare Costs	Incl. O&P
1 Truck Driver (heavy)	$46.00	$368.00	$73.90	$591.20	$46.00	$73.90
1 Dump Truck, 18 C.Y., 450 H.P.		706.00		776.60	88.25	97.08
8 L.H., Daily Totals		$1074.00		$1367.80	$134.25	$170.97

Crew B-34J

	Hr.	Daily	Hr.	Daily	Bare Costs	Incl. O&P
1 Truck Driver (heavy)	$46.00	$368.00	$73.90	$591.20	$46.00	$73.90
1 Dump Truck, Off Hwy., 100 Ton		2669.00		2935.90	333.63	366.99
8 L.H., Daily Totals		$3037.00		$3527.10	$379.63	$440.89

Crew B-34K

	Hr.	Daily	Hr.	Daily	Bare Costs	Incl. O&P
1 Truck Driver (heavy)	$46.00	$368.00	$73.90	$591.20	$46.00	$73.90
1 Truck Tractor, 6x4, 450 H.P.		583.40		641.74		
1 Lowbed Trailer, 75 Ton		249.40		274.34	104.10	114.51
8 L.H., Daily Totals		$1200.80		$1507.28	$150.10	$188.41

Crew B-34L

	Hr.	Daily	Hr.	Daily	Bare Costs	Incl. O&P
1 Equip. Oper. (light)	$51.30	$410.40	$80.95	$647.60	$51.30	$80.95
1 Flatbed Truck, Gas, 1.5 Ton		188.40		207.24	23.55	25.91
8 L.H., Daily Totals		$598.80		$854.84	$74.85	$106.86

Crew B-34M

	Hr.	Daily	Hr.	Daily	Bare Costs	Incl. O&P
1 Equip. Oper. (light)	$51.30	$410.40	$80.95	$647.60	$51.30	$80.95
1 Flatbed Truck, Gas, 3 Ton		238.00		261.80	29.75	32.73
8 L.H., Daily Totals		$648.40		$909.40	$81.05	$113.68

Crew B-34N

	Hr.	Daily	Hr.	Daily	Bare Costs	Incl. O&P
1 Truck Driver (heavy)	$46.00	$368.00	$73.90	$591.20	$49.88	$79.35
1 Equip. Oper. (medium)	53.75	430.00	84.80	678.40		
1 Truck Tractor, 6x4, 380 H.P.		476.20		523.82		
1 Flatbed Trailer, 40 Ton		180.00		198.00	41.01	45.11
16 L.H., Daily Totals		$1454.20		$1991.42	$90.89	$124.46

Crew B-34P

	Hr.	Daily	Hr.	Daily	Bare Costs	Incl. O&P
1 Pipe Fitter	$63.00	$504.00	$98.05	$784.40	$53.75	$84.77
1 Truck Driver (light)	44.50	356.00	71.45	571.60		
1 Equip. Oper. (medium)	53.75	430.00	84.80	678.40		
1 Flatbed Truck, Gas, 3 Ton		238.00		261.80		
1 Backhoe Loader, 48 H.P.		312.40		343.64	22.93	25.23
24 L.H., Daily Totals		$1840.40		$2639.84	$76.68	$109.99

Crew B-34Q

	Hr.	Daily	Hr.	Daily	Bare Costs	Incl. O&P
1 Pipe Fitter	$63.00	$504.00	$98.05	$784.40	$54.53	$86.02
1 Truck Driver (light)	44.50	356.00	71.45	571.60		
1 Equip. Oper. (crane)	56.10	448.80	88.55	708.40		
1 Flatbed Trailer, 25 Ton		133.00		146.30		
1 Dump Truck, 8 C.Y., 220 H.P.		338.60		372.46		
1 Hyd. Crane, 25 Ton		590.60		649.66	44.26	48.68
24 L.H., Daily Totals		$2371.00		$3232.82	$98.79	$134.70

Crew No.	Bare Costs Hr.	Daily	Incl. Subs O&P Hr.	Daily	Cost Per Labor-Hour Bare Costs	Incl. O&P
Crew B-34R	Hr.	Daily	Hr.	Daily	Bare Costs	Incl. O&P
1 Pipe Fitter	$63.00	$504.00	$98.05	$784.40	$54.53	$86.02
1 Truck Driver (light)	44.50	356.00	71.45	571.60		
1 Equip. Oper. (crane)	56.10	448.80	88.55	708.40		
1 Flatbed Trailer, 25 Ton		133.00		146.30		
1 Dump Truck, 8 C.Y., 220 H.P.		338.60		372.46		
1 Hyd. Crane, 25 Ton		590.60		649.66		
1 Hyd. Excavator, 1 C.Y.		742.20		816.42	75.18	82.70
24 L.H., Daily Totals		$3113.20		$4049.24	$129.72	$168.72
Crew B-34S	Hr.	Daily	Hr.	Daily	Bare Costs	Incl. O&P
2 Pipe Fitters	$63.00	$1008.00	$98.05	$1568.80	$57.02	$89.64
1 Truck Driver (heavy)	46.00	368.00	73.90	591.20		
1 Equip. Oper. (crane)	56.10	448.80	88.55	708.40		
1 Flatbed Trailer, 40 Ton		180.00		198.00		
1 Truck Tractor, 6x4, 380 H.P.		476.20		523.82		
1 Hyd. Crane, 80 Ton		1518.00		1669.80		
1 Hyd. Excavator, 2 C.Y.		1052.00		1157.20	100.82	110.90
32 L.H., Daily Totals		$5051.00		$6417.22	$157.84	$200.54
Crew B-34T	Hr.	Daily	Hr.	Daily	Bare Costs	Incl. O&P
2 Pipe Fitters	$63.00	$1008.00	$98.05	$1568.80	$57.02	$89.64
1 Truck Driver (heavy)	46.00	368.00	73.90	591.20		
1 Equip. Oper. (crane)	56.10	448.80	88.55	708.40		
1 Flatbed Trailer, 40 Ton		180.00		198.00		
1 Truck Tractor, 6x4, 380 H.P.		476.20		523.82		
1 Hyd. Crane, 80 Ton		1518.00		1669.80	67.94	74.74
32 L.H., Daily Totals		$3999.00		$5260.02	$124.97	$164.38
Crew B-34U	Hr.	Daily	Hr.	Daily	Bare Costs	Incl. O&P
1 Truck Driver (heavy)	$46.00	$368.00	$73.90	$591.20	$48.65	$77.42
1 Equip. Oper. (light)	51.30	410.40	80.95	647.60		
1 Truck Tractor, 220 H.P.		290.00		319.00		
1 Flatbed Trailer, 25 Ton		133.00		146.30	26.44	29.08
16 L.H., Daily Totals		$1201.40		$1704.10	$75.09	$106.51
Crew B-34V	Hr.	Daily	Hr.	Daily	Bare Costs	Incl. O&P
1 Truck Driver (heavy)	$46.00	$368.00	$73.90	$591.20	$51.13	$81.13
1 Equip. Oper. (crane)	56.10	448.80	88.55	708.40		
1 Equip. Oper. (light)	51.30	410.40	80.95	647.60		
1 Truck Tractor, 6x4, 450 H.P.		583.40		641.74		
1 Equipment Trailer, 50 Ton		198.20		218.02		
1 Pickup Truck, 4x4, 3/4 Ton		126.60		139.26	37.84	41.63
24 L.H., Daily Totals		$2135.40		$2946.22	$88.97	$122.76
Crew B-34W	Hr.	Daily	Hr.	Daily	Bare Costs	Incl. O&P
5 Truck Drivers (heavy)	$46.00	$1840.00	$73.90	$2956.00	$48.71	$77.79
2 Equip. Opers. (crane)	56.10	897.60	88.55	1416.80		
1 Equip. Oper. (mechanic)	56.30	450.40	88.85	710.80		
1 Laborer	39.85	318.80	64.70	517.60		
4 Truck Tractors, 6x4, 380 H.P.		1904.80		2095.28		
2 Equipment Trailers, 50 Ton		396.40		436.04		
2 Flatbed Trailers, 40 Ton		360.00		396.00		
1 Pickup Truck, 4x4, 3/4 Ton		126.60		139.26		
1 S.P. Crane, 4x4, 20 Ton		571.20		628.32	46.65	51.32
72 L.H., Daily Totals		$6865.80		$9296.10	$95.36	$129.11

Crew No.	Bare Costs Hr.	Daily	Incl. Subs O&P Hr.	Daily	Cost Per Labor-Hour Bare Costs	Incl. O&P
Crew B-35	Hr.	Daily	Hr.	Daily	Bare Costs	Incl. O&P
1 Labor Foreman (outside)	$41.85	$334.80	$67.90	$543.20	$50.15	$79.82
1 Skilled Worker	52.35	418.80	84.35	674.80		
1 Welder (plumber)	62.15	497.20	96.70	773.60		
1 Laborer	39.85	318.80	64.70	517.60		
1 Equip. Oper. (crane)	56.10	448.80	88.55	708.40		
1 Equip. Oper. (oiler)	48.60	388.80	76.70	613.60		
1 Welder, Electric, 300 amp		59.20		65.12		
1 Hyd. Excavator, .75 C.Y.		669.60		736.56	15.18	16.70
48 L.H., Daily Totals		$3136.00		$4632.88	$65.33	$96.52
Crew B-35A	Hr.	Daily	Hr.	Daily	Bare Costs	Incl. O&P
1 Labor Foreman (outside)	$41.85	$334.80	$67.90	$543.20	$48.68	$77.66
2 Laborers	39.85	637.60	64.70	1035.20		
1 Skilled Worker	52.35	418.80	84.35	674.80		
1 Welder (plumber)	62.15	497.20	96.70	773.60		
1 Equip. Oper. (crane)	56.10	448.80	88.55	708.40		
1 Equip. Oper. (oiler)	48.60	388.80	76.70	613.60		
1 Welder, Gas Engine, 300 amp		98.40		108.24		
1 Crawler Crane, 75 Ton		1708.00		1878.80	32.26	35.48
56 L.H., Daily Totals		$4532.40		$6335.84	$80.94	$113.14
Crew B-36	Hr.	Daily	Hr.	Daily	Bare Costs	Incl. O&P
1 Labor Foreman (outside)	$41.85	$334.80	$67.90	$543.20	$45.81	$73.38
2 Laborers	39.85	637.60	64.70	1035.20		
2 Equip. Oper. (medium)	53.75	860.00	84.80	1356.80		
1 Dozer, 200 H.P.		1273.00		1400.30		
1 Aggregate Spreader		36.80		40.48		
1 Tandem Roller, 10 Ton		231.40		254.54	38.53	42.38
40 L.H., Daily Totals		$3373.60		$4630.52	$84.34	$115.76
Crew B-36A	Hr.	Daily	Hr.	Daily	Bare Costs	Incl. O&P
1 Labor Foreman (outside)	$41.85	$334.80	$67.90	$543.20	$48.08	$76.64
2 Laborers	39.85	637.60	64.70	1035.20		
4 Equip. Oper. (medium)	53.75	1720.00	84.80	2713.60		
1 Dozer, 200 H.P.		1273.00		1400.30		
1 Aggregate Spreader		36.80		40.48		
1 Tandem Roller, 10 Ton		231.40		254.54		
1 Roller, Pneum. Whl., 12 Ton		338.20		372.02	33.56	36.92
56 L.H., Daily Totals		$4571.80		$6359.34	$81.64	$113.56
Crew B-36B	Hr.	Daily	Hr.	Daily	Bare Costs	Incl. O&P
1 Labor Foreman (outside)	$41.85	$334.80	$67.90	$543.20	$47.82	$76.30
2 Laborers	39.85	637.60	64.70	1035.20		
4 Equip. Oper. (medium)	53.75	1720.00	84.80	2713.60		
1 Truck Driver (heavy)	46.00	368.00	73.90	591.20		
1 Grader, 30,000 Lbs.		643.80		708.18		
1 F.E. Loader, Crl, 1.5 C.Y.		645.40		709.94		
1 Dozer, 300 H.P.		1829.00		2011.90		
1 Roller, Vibratory, 25 Ton		659.80		725.78		
1 Truck Tractor, 6x4, 450 H.P.		583.40		641.74		
1 Water Tank Trailer, 5000 Gal.		147.60		162.36	70.45	77.50
64 L.H., Daily Totals		$7569.40		$9843.10	$118.27	$153.80

754

Crew No.		Bare Costs		Incl. Subs O&P		Cost Per Labor-Hour	

Crew B-36C

	Hr.	Daily	Hr.	Daily	Bare Costs	Incl. O&P
1 Labor Foreman (outside)	$41.85	$334.80	$67.90	$543.20	$49.82	$79.24
3 Equip. Oper. (medium)	53.75	1290.00	84.80	2035.20		
1 Truck Driver (heavy)	46.00	368.00	73.90	591.20		
1 Grader, 30,000 Lbs.		643.80		708.18		
1 Dozer, 300 H.P.		1829.00		2011.90		
1 Roller, Vibratory, 25 Ton		659.80		725.78		
1 Truck Tractor, 6x4, 450 H.P.		583.40		641.74		
1 Water Tank Trailer, 5000 Gal.		147.60		162.36	96.59	106.25
40 L.H., Daily Totals		$5856.40		$7419.56	$146.41	$185.49

Crew B-37

	Hr.	Daily	Hr.	Daily	Bare Costs	Incl. O&P
1 Labor Foreman (outside)	$41.85	$334.80	$67.90	$543.20	$42.09	$67.94
4 Laborers	39.85	1275.20	64.70	2070.40		
1 Equip. Oper. (light)	51.30	410.40	80.95	647.60		
1 Tandem Roller, 5 Ton		151.20		166.32	3.15	3.46
48 L.H., Daily Totals		$2171.60		$3427.52	$45.24	$71.41

Crew B-37A

	Hr.	Daily	Hr.	Daily	Bare Costs	Incl. O&P
2 Laborers	$39.85	$637.60	$64.70	$1035.20	$41.40	$66.95
1 Truck Driver (light)	44.50	356.00	71.45	571.60		
1 Flatbed Truck, Gas, 1.5 Ton		188.40		207.24		
1 Tar Kettle, T.M.		129.20		142.12	13.23	14.56
24 L.H., Daily Totals		$1311.20		$1956.16	$54.63	$81.51

Crew B-37B

	Hr.	Daily	Hr.	Daily	Bare Costs	Incl. O&P
3 Laborers	$39.85	$956.40	$64.70	$1552.80	$41.01	$66.39
1 Truck Driver (light)	44.50	356.00	71.45	571.60		
1 Flatbed Truck, Gas, 1.5 Ton		188.40		207.24		
1 Tar Kettle, T.M.		129.20		142.12	9.93	10.92
32 L.H., Daily Totals		$1630.00		$2473.76	$50.94	$77.31

Crew B-37C

	Hr.	Daily	Hr.	Daily	Bare Costs	Incl. O&P
2 Laborers	$39.85	$637.60	$64.70	$1035.20	$42.17	$68.08
2 Truck Drivers (light)	44.50	712.00	71.45	1143.20		
2 Flatbed Trucks, Gas, 1.5 Ton		376.80		414.48		
1 Tar Kettle, T.M.		129.20		142.12	15.81	17.39
32 L.H., Daily Totals		$1855.60		$2735.00	$57.99	$85.47

Crew B-37D

	Hr.	Daily	Hr.	Daily	Bare Costs	Incl. O&P
1 Laborer	$39.85	$318.80	$64.70	$517.60	$42.17	$68.08
1 Truck Driver (light)	44.50	356.00	71.45	571.60		
1 Pickup Truck, 3/4 Ton		115.20		126.72	7.20	7.92
16 L.H., Daily Totals		$790.00		$1215.92	$49.38	$76.00

Crew B-37E

	Hr.	Daily	Hr.	Daily	Bare Costs	Incl. O&P
3 Laborers	$39.85	$956.40	$64.70	$1552.80	$44.80	$71.82
1 Equip. Oper. (light)	51.30	410.40	80.95	647.60		
1 Equip. Oper. (medium)	53.75	430.00	84.80	678.40		
2 Truck Drivers (light)	44.50	712.00	71.45	1143.20		
4 Barrels w/Flasher		15.60		17.16		
1 Concrete Saw		102.40		112.64		
1 Rotary Hammer Drill		25.15		27.66		
1 Hammer Drill Bit		3.25		3.58		
1 Loader, Skid Steer, 30 H.P.		174.00		191.40		
1 Conc. Hammer Attach.		113.60		124.96		
1 Vibrating Plate, Gas, 18"		31.80		34.98		
2 Flatbed Trucks, Gas, 1.5 Ton		376.80		414.48	15.05	16.55
56 L.H., Daily Totals		$3351.40		$4948.86	$59.85	$88.37

Crew B-37F

	Hr.	Daily	Hr.	Daily	Bare Costs	Incl. O&P
3 Laborers	$39.85	$956.40	$64.70	$1552.80	$41.01	$66.39
1 Truck Driver (light)	44.50	356.00	71.45	571.60		
4 Barrels w/Flasher		15.60		17.16		
1 Concrete Mixer, 10 C.F.		161.00		177.10		
1 Air Compressor, 60 cfm		105.40		115.94		
1 -50' Air Hose, 3/4"		7.35		8.09		
1 Spade (Chipper)		8.40		9.24		
1 Flatbed Truck, Gas, 1.5 Ton		188.40		207.24	15.19	16.71
32 L.H., Daily Totals		$1798.55		$2659.17	$56.20	$83.10

Crew B-37G

	Hr.	Daily	Hr.	Daily	Bare Costs	Incl. O&P
1 Labor Foreman (outside)	$41.85	$334.80	$67.90	$543.20	$42.09	$67.94
4 Laborers	39.85	1275.20	64.70	2070.40		
1 Equip. Oper. (light)	51.30	410.40	80.95	647.60		
1 Berm Machine		317.40		349.14		
1 Tandem Roller, 5 Ton		151.20		166.32	9.76	10.74
48 L.H., Daily Totals		$2489.00		$3776.66	$51.85	$78.68

Crew B-37H

	Hr.	Daily	Hr.	Daily	Bare Costs	Incl. O&P
1 Labor Foreman (outside)	$41.85	$334.80	$67.90	$543.20	$42.09	$67.94
4 Laborers	39.85	1275.20	64.70	2070.40		
1 Equip. Oper. (light)	51.30	410.40	80.95	647.60		
1 Tandem Roller, 5 Ton		151.20		166.32		
1 Flatbed Truck, Gas, 1.5 Ton		188.40		207.24		
1 Tar Kettle, T.M.		129.20		142.12	9.77	10.74
48 L.H., Daily Totals		$2489.20		$3776.88	$51.86	$78.69

Crew B-37I

	Hr.	Daily	Hr.	Daily	Bare Costs	Incl. O&P
3 Laborers	$39.85	$956.40	$64.70	$1552.80	$44.80	$71.82
1 Equip. Oper. (light)	51.30	410.40	80.95	647.60		
1 Equip. Oper. (medium)	53.75	430.00	84.80	678.40		
2 Truck Drivers (light)	44.50	712.00	71.45	1143.20		
4 Barrels w/Flasher		15.60		17.16		
1 Concrete Saw		102.40		112.64		
1 Rotary Hammer Drill		25.15		27.66		
1 Hammer Drill Bit		3.25		3.58		
1 Air Compressor, 60 cfm		105.40		115.94		
1 -50' Air Hose, 3/4"		7.35		8.09		
1 Spade (Chipper)		8.40		9.24		
1 Loader, Skid Steer, 30 H.P.		174.00		191.40		
1 Conc. Hammer Attach.		113.60		124.96		
1 Concrete Mixer, 10 C.F.		161.00		177.10		
1 Vibrating Plate, Gas, 18"		31.80		34.98		
2 Flatbed Trucks, Gas, 1.5 Ton		376.80		414.48	20.08	22.09
56 L.H., Daily Totals		$3633.55		$5259.23	$64.88	$93.91

Crew B-37J

	Hr.	Daily	Hr.	Daily	Bare Costs	Incl. O&P
1 Labor Foreman (outside)	$41.85	$334.80	$67.90	$543.20	$42.09	$67.94
4 Laborers	39.85	1275.20	64.70	2070.40		
1 Equip. Oper. (light)	51.30	410.40	80.95	647.60		
1 Air Compressor, 60 cfm		105.40		115.94		
1 -50' Air Hose, 3/4"		7.35		8.09		
2 Concrete Mixers, 10 C.F.		322.00		354.20		
2 Flatbed Trucks, Gas, 1.5 Ton		376.80		414.48		
1 Shot Blaster, 20"		214.80		236.28	21.38	23.52
48 L.H., Daily Totals		$3046.75		$4390.19	$63.47	$91.46

Crew No.	Bare Costs		Incl. Subs O&P		Cost Per Labor-Hour	
Crew B-37K	Hr.	Daily	Hr.	Daily	Bare Costs	Incl. O&P
1 Labor Foreman (outside)	$41.85	$334.80	$67.90	$543.20	$42.09	$67.94
4 Laborers	39.85	1275.20	64.70	2070.40		
1 Equip. Oper. (light)	51.30	410.40	80.95	647.60		
1 Air Compressor, 60 cfm		105.40		115.94		
1 -50' Air Hose, 3/4"		7.35		8.09		
2 Flatbed Trucks, Gas, 1.5 Ton		376.80		414.48		
1 Shot Blaster, 20"		214.80		236.28	14.67	16.14
48 L.H., Daily Totals		$2724.75		$4035.99	$56.77	$84.08

Crew B-38	Hr.	Daily	Hr.	Daily	Bare Costs	Incl. O&P
1 Labor Foreman (outside)	$41.85	$334.80	$67.90	$543.20	$45.32	$72.61
2 Laborers	39.85	637.60	64.70	1035.20		
1 Equip. Oper. (light)	51.30	410.40	80.95	647.60		
1 Equip. Oper. (medium)	53.75	430.00	84.80	678.40		
1 Backhoe Loader, 48 H.P.		312.40		343.64		
1 Hyd. Hammer (1200 lb.)		177.40		195.14		
1 F.E. Loader, W.M., 4 C.Y.		574.60		632.06		
1 Pvmt. Rem. Bucket		61.00		67.10	28.14	30.95
40 L.H., Daily Totals		$2938.20		$4142.34	$73.45	$103.56

Crew B-39	Hr.	Daily	Hr.	Daily	Bare Costs	Incl. O&P
1 Labor Foreman (outside)	$41.85	$334.80	$67.90	$543.20	$42.09	$67.94
4 Laborers	39.85	1275.20	64.70	2070.40		
1 Equip. Oper. (light)	51.30	410.40	80.95	647.60		
1 Air Compressor, 250 cfm		167.40		184.14		
2 Breakers, Pavement, 60 lb.		21.20		23.32		
2 -50' Air Hoses, 1.5"		45.40		49.94	4.88	5.36
48 L.H., Daily Totals		$2254.40		$3518.60	$46.97	$73.30

Crew B-40	Hr.	Daily	Hr.	Daily	Bare Costs	Incl. O&P
1 Pile Driver Foreman (outside)	$53.30	$426.40	$86.70	$693.60	$52.41	$84.29
4 Pile Drivers	51.30	1641.60	83.45	2670.40		
2 Equip. Oper. (crane)	56.10	897.60	88.55	1416.80		
1 Equip. Oper. (oiler)	48.60	388.80	76.70	613.60		
1 Crawler Crane, 40 Ton		1370.00		1507.00		
1 Vibratory Hammer & Gen.		2191.00		2410.10	55.64	61.20
64 L.H., Daily Totals		$6915.40		$9311.50	$108.05	$145.49

Crew B-40B	Hr.	Daily	Hr.	Daily	Bare Costs	Incl. O&P
1 Labor Foreman (outside)	$41.85	$334.80	$67.90	$543.20	$44.35	$71.21
3 Laborers	39.85	956.40	64.70	1552.80		
1 Equip. Oper. (crane)	56.10	448.80	88.55	708.40		
1 Equip. Oper. (oiler)	48.60	388.80	76.70	613.60		
1 Lattice Boom Crane, 40 Ton		1321.00		1453.10	27.52	30.27
48 L.H., Daily Totals		$3449.80		$4871.10	$71.87	$101.48

Crew B-41	Hr.	Daily	Hr.	Daily	Bare Costs	Incl. O&P
1 Labor Foreman (outside)	$41.85	$334.80	$67.90	$543.20	$41.35	$66.91
4 Laborers	39.85	1275.20	64.70	2070.40		
.25 Equip. Oper. (crane)	56.10	112.20	88.55	177.10		
.25 Equip. Oper. (oiler)	48.60	97.20	76.70	153.40		
.25 Crawler Crane, 40 Ton		342.50		376.75	7.78	8.56
44 L.H., Daily Totals		$2161.90		$3320.85	$49.13	$75.47

Crew No.	Bare Costs		Incl. Subs O&P		Cost Per Labor-Hour	
Crew B-42	Hr.	Daily	Hr.	Daily	Bare Costs	Incl. O&P
1 Labor Foreman (outside)	$41.85	$334.80	$67.90	$543.20	$45.08	$73.14
4 Laborers	39.85	1275.20	64.70	2070.40		
1 Equip. Oper. (crane)	56.10	448.80	88.55	708.40		
1 Equip. Oper. (oiler)	48.60	388.80	76.70	613.60		
1 Welder	54.65	437.20	93.20	745.60		
1 Hyd. Crane, 25 Ton		590.60		649.66		
1 Welder, Gas Engine, 300 amp		98.40		108.24		
1 Horz. Boring Csg. Mch.		448.80		493.68	17.78	19.56
64 L.H., Daily Totals		$4022.60		$5932.78	$62.85	$92.70

Crew B-43	Hr.	Daily	Hr.	Daily	Bare Costs	Incl. O&P
1 Labor Foreman (outside)	$41.85	$334.80	$67.90	$543.20	$44.35	$71.21
3 Laborers	39.85	956.40	64.70	1552.80		
1 Equip. Oper. (crane)	56.10	448.80	88.55	708.40		
1 Equip. Oper. (oiler)	48.60	388.80	76.70	613.60		
1 Drill Rig, Truck-Mounted		2444.00		2688.40	50.92	56.01
48 L.H., Daily Totals		$4572.80		$6106.40	$95.27	$127.22

Crew B-44	Hr.	Daily	Hr.	Daily	Bare Costs	Incl. O&P
1 Pile Driver Foreman (outside)	$53.30	$426.40	$86.70	$693.60	$51.32	$82.79
4 Pile Drivers	51.30	1641.60	83.45	2670.40		
2 Equip. Oper. (crane)	56.10	897.60	88.55	1416.80		
1 Laborer	39.85	318.80	64.70	517.60		
1 Crawler Crane, 40 Ton		1370.00		1507.00		
1 Lead, 60' High		79.40		87.34		
1 Hammer, Diesel, 15K ft.-lbs.		623.20		685.52	32.38	35.62
64 L.H., Daily Totals		$5357.00		$7578.26	$83.70	$118.41

Crew B-45	Hr.	Daily	Hr.	Daily	Bare Costs	Incl. O&P
1 Equip. Oper. (medium)	$53.75	$430.00	$84.80	$678.40	$49.88	$79.35
1 Truck Driver (heavy)	46.00	368.00	73.90	591.20		
1 Dist. Tanker, 3000 Gallon		326.80		359.48		
1 Truck Tractor, 6x4, 380 H.P.		476.20		523.82	50.19	55.21
16 L.H., Daily Totals		$1601.00		$2152.90	$100.06	$134.56

Crew B-46	Hr.	Daily	Hr.	Daily	Bare Costs	Incl. O&P
1 Pile Driver Foreman (outside)	$53.30	$426.40	$86.70	$693.60	$45.91	$74.62
2 Pile Drivers	51.30	820.80	83.45	1335.20		
3 Laborers	39.85	956.40	64.70	1552.80		
1 Chain Saw, Gas, 36" Long		46.40		51.04	.97	1.06
48 L.H., Daily Totals		$2250.00		$3632.64	$46.88	$75.68

Crew B-47	Hr.	Daily	Hr.	Daily	Bare Costs	Incl. O&P
1 Blast Foreman (outside)	$41.85	$334.80	$67.90	$543.20	$44.33	$71.18
1 Driller	39.85	318.80	64.70	517.60		
1 Equip. Oper. (light)	51.30	410.40	80.95	647.60		
1 Air Track Drill, 4"		993.40		1092.74		
1 Air Compressor, 600 cfm		417.20		458.92		
2 -50' Air Hoses, 3"		86.40		95.04	62.38	68.61
24 L.H., Daily Totals		$2561.00		$3355.10	$106.71	$139.80

Crew B-47A	Hr.	Daily	Hr.	Daily	Bare Costs	Incl. O&P
1 Drilling Foreman (outside)	$41.85	$334.80	$67.90	$543.20	$48.85	$77.72
1 Equip. Oper. (heavy)	56.10	448.80	88.55	708.40		
1 Equip. Oper. (oiler)	48.60	388.80	76.70	613.60		
1 Air Track Drill, 5"		1173.00		1290.30	48.88	53.76
24 L.H., Daily Totals		$2345.40		$3155.50	$97.72	$131.48

Crews - Renovation

Crew B-47C	Hr.	Daily	Hr.	Daily	Bare Costs	Incl. O&P
1 Laborer	$39.85	$318.80	$64.70	$517.60	$45.58	$72.83
1 Equip. Oper. (light)	51.30	410.40	80.95	647.60		
1 Air Compressor, 750 cfm		423.80		466.18		
2 -50' Air Hoses, 3"		86.40		95.04		
1 Air Track Drill, 4"		993.40		1092.74	93.97	103.37
16 L.H., Daily Totals		$2232.80		$2819.16	$139.55	$176.20

Crew B-47E	Hr.	Daily	Hr.	Daily	Bare Costs	Incl. O&P
1 Labor Foreman (outside)	$41.85	$334.80	$67.90	$543.20	$40.35	$65.50
3 Laborers	39.85	956.40	64.70	1552.80		
1 Flatbed Truck, Gas, 3 Ton		238.00		261.80	7.44	8.18
32 L.H., Daily Totals		$1529.20		$2357.80	$47.79	$73.68

Crew B-47G	Hr.	Daily	Hr.	Daily	Bare Costs	Incl. O&P
1 Labor Foreman (outside)	$41.85	$334.80	$67.90	$543.20	$43.21	$69.56
2 Laborers	39.85	637.60	64.70	1035.20		
1 Equip. Oper. (light)	51.30	410.40	80.95	647.60		
1 Air Track Drill, 4"		993.40		1092.74		
1 Air Compressor, 600 cfm		417.20		458.92		
2 -50' Air Hoses, 3"		86.40		95.04		
1 Gunite Pump Rig		321.00		353.10	56.81	62.49
32 L.H., Daily Totals		$3200.80		$4225.80	$100.03	$132.06

Crew B-47H	Hr.	Daily	Hr.	Daily	Bare Costs	Incl. O&P
1 Skilled Worker Foreman (out)	$54.35	$434.80	$87.55	$700.40	$52.85	$85.15
3 Skilled Workers	52.35	1256.40	84.35	2024.40		
1 Flatbed Truck, Gas, 3 Ton		238.00		261.80	7.44	8.18
32 L.H., Daily Totals		$1929.20		$2986.60	$60.29	$93.33

Crew B-48	Hr.	Daily	Hr.	Daily	Bare Costs	Incl. O&P
1 Labor Foreman (outside)	$41.85	$334.80	$67.90	$543.20	$45.34	$72.60
3 Laborers	39.85	956.40	64.70	1552.80		
1 Equip. Oper. (crane)	56.10	448.80	88.55	708.40		
1 Equip. Oper. (oiler)	48.60	388.80	76.70	613.60		
1 Equip. Oper. (light)	51.30	410.40	80.95	647.60		
1 Centr. Water Pump, 6"		296.40		326.04		
1 -20' Suction Hose, 6"		11.50		12.65		
1 -50' Discharge Hose, 6"		6.10		6.71		
1 Drill Rig, Truck-Mounted		2444.00		2688.40	49.25	54.17
56 L.H., Daily Totals		$5297.20		$7099.40	$94.59	$126.78

Crew B-49	Hr.	Daily	Hr.	Daily	Bare Costs	Incl. O&P
1 Labor Foreman (outside)	$41.85	$334.80	$67.90	$543.20	$47.70	$76.40
3 Laborers	39.85	956.40	64.70	1552.80		
2 Equip. Oper. (crane)	56.10	897.60	88.55	1416.80		
2 Equip. Oper. (oilers)	48.60	777.60	76.70	1227.20		
1 Equip. Oper. (light)	51.30	410.40	80.95	647.60		
2 Pile Drivers	51.30	820.80	83.45	1335.20		
1 Hyd. Crane, 25 Ton		590.60		649.66		
1 Centr. Water Pump, 6"		296.40		326.04		
1 -20' Suction Hose, 6"		11.50		12.65		
1 -50' Discharge Hose, 6"		6.10		6.71		
1 Drill Rig, Truck-Mounted		2444.00		2688.40	38.05	41.86
88 L.H., Daily Totals		$7546.20		$10406.26	$85.75	$118.25

Crew B-50	Hr.	Daily	Hr.	Daily	Bare Costs	Incl. O&P
2 Pile Driver Foremen (outside)	$53.30	$852.80	$86.70	$1387.20	$49.63	$80.14
6 Pile Drivers	51.30	2462.40	83.45	4005.60		
2 Equip. Oper. (crane)	56.10	897.60	88.55	1416.80		
1 Equip. Oper. (oiler)	48.60	388.80	76.70	613.60		
3 Laborers	39.85	956.40	64.70	1552.80		
1 Crawler Crane, 40 Ton		1370.00		1507.00		
1 Lead, 60' High		79.40		87.34		
1 Hammer, Diesel, 15K ft.-lbs.		623.20		685.52		
1 Air Compressor, 600 cfm		417.20		458.92		
2 -50' Air Hoses, 3"		86.40		95.04		
1 Chain Saw, Gas, 36" Long		46.40		51.04	23.42	25.76
112 L.H., Daily Totals		$8180.60		$11860.86	$73.04	$105.90

Crew B-51	Hr.	Daily	Hr.	Daily	Bare Costs	Incl. O&P
1 Labor Foreman (outside)	$41.85	$334.80	$67.90	$543.20	$40.96	$66.36
4 Laborers	39.85	1275.20	64.70	2070.40		
1 Truck Driver (light)	44.50	356.00	71.45	571.60		
1 Flatbed Truck, Gas, 1.5 Ton		188.40		207.24	3.92	4.32
48 L.H., Daily Totals		$2154.40		$3392.44	$44.88	$70.68

Crew B-52	Hr.	Daily	Hr.	Daily	Bare Costs	Incl. O&P
1 Carpenter Foreman (outside)	$52.70	$421.60	$85.55	$684.40	$46.39	$74.74
1 Carpenter	50.70	405.60	82.30	658.40		
3 Laborers	39.85	956.40	64.70	1552.80		
1 Cement Finisher	47.55	380.40	75.20	601.60		
.5 Rodman (reinf.)	54.65	218.60	87.30	349.20		
.5 Equip. Oper. (medium)	53.75	215.00	84.80	339.20		
.5 Crawler Loader, 3 C.Y.		602.00		662.20	10.75	11.82
56 L.H., Daily Totals		$3199.60		$4847.80	$57.14	$86.57

Crew B-53	Hr.	Daily	Hr.	Daily	Bare Costs	Incl. O&P
1 Equip. Oper. (light)	$51.30	$410.40	$80.95	$647.60	$51.30	$80.95
1 Trencher, Chain, 12 H.P.		61.80		67.98	7.72	8.50
8 L.H., Daily Totals		$472.20		$715.58	$59.02	$89.45

Crew B-54	Hr.	Daily	Hr.	Daily	Bare Costs	Incl. O&P
1 Equip. Oper. (light)	$51.30	$410.40	$80.95	$647.60	$51.30	$80.95
1 Trencher, Chain, 40 H.P.		336.40		370.04	42.05	46.26
8 L.H., Daily Totals		$746.80		$1017.64	$93.35	$127.21

Crew B-54A	Hr.	Daily	Hr.	Daily	Bare Costs	Incl. O&P
.17 Labor Foreman (outside)	$41.85	$56.92	$67.90	$92.34	$52.02	$82.34
1 Equipment Operator (med.)	53.75	430.00	84.80	678.40		
1 Wheel Trencher, 67 H.P.		1095.00		1204.50	116.99	128.69
9.36 L.H., Daily Totals		$1581.92		$1975.24	$169.01	$211.03

Crew B-54B	Hr.	Daily	Hr.	Daily	Bare Costs	Incl. O&P
.25 Labor Foreman (outside)	$41.85	$83.70	$67.90	$135.80	$51.37	$81.42
1 Equipment Operator (med.)	53.75	430.00	84.80	678.40		
1 Wheel Trencher, 150 H.P.		1957.00		2152.70	195.70	215.27
10 L.H., Daily Totals		$2470.70		$2966.90	$247.07	$296.69

Crew B-54D	Hr.	Daily	Hr.	Daily	Bare Costs	Incl. O&P
1 Laborer	$39.85	$318.80	$64.70	$517.60	$46.80	$74.75
1 Equipment Operator (med.)	53.75	430.00	84.80	678.40		
1 Rock Trencher, 6" Width		904.20		994.62	56.51	62.16
16 L.H., Daily Totals		$1653.00		$2190.62	$103.31	$136.91

For customer support on your Commercial Renovation Costs with RSMeans data, call 800.448.8182.

757

Crew No.	Bare Costs		Incl. Subs O&P		Cost Per Labor-Hour	
					Bare Costs	Incl. O&P
Crew B-54E	Hr.	Daily	Hr.	Daily		
1 Laborer	$39.85	$318.80	$64.70	$517.60	$46.80	$74.75
1 Equipment Operator (med.)	53.75	430.00	84.80	678.40		
1 Rock Trencher, 18" Width		2681.00		2949.10	167.56	184.32
16 L.H., Daily Totals		$3429.80		$4145.10	$214.36	$259.07
Crew B-55	Hr.	Daily	Hr.	Daily	Bare Costs	Incl. O&P
2 Laborers	$39.85	$637.60	$64.70	$1035.20	$41.40	$66.95
1 Truck Driver (light)	44.50	356.00	71.45	571.60		
1 Truck-Mounted Earth Auger		758.80		834.68		
1 Flatbed Truck, Gas, 3 Ton		238.00		261.80	41.53	45.69
24 L.H., Daily Totals		$1990.40		$2703.28	$82.93	$112.64
Crew B-56	Hr.	Daily	Hr.	Daily	Bare Costs	Incl. O&P
1 Laborer	$39.85	$318.80	$64.70	$517.60	$45.58	$72.83
1 Equip. Oper. (light)	51.30	410.40	80.95	647.60		
1 Air Track Drill, 4"		993.40		1092.74		
1 Air Compressor, 600 cfm		417.20		458.92		
1 -50' Air Hose, 3"		43.20		47.52	90.86	99.95
16 L.H., Daily Totals		$2183.00		$2764.38	$136.44	$172.77
Crew B-57	Hr.	Daily	Hr.	Daily	Bare Costs	Incl. O&P
1 Labor Foreman (outside)	$41.85	$334.80	$67.90	$543.20	$46.26	$73.92
2 Laborers	39.85	637.60	64.70	1035.20		
1 Equip. Oper. (crane)	56.10	448.80	88.55	708.40		
1 Equip. Oper. (light)	51.30	410.40	80.95	647.60		
1 Equip. Oper. (oiler)	48.60	388.80	76.70	613.60		
1 Crawler Crane, 25 Ton		1348.00		1482.80		
1 Clamshell Bucket, 1 C.Y.		50.60		55.66		
1 Centr. Water Pump, 6"		296.40		326.04		
1 -20' Suction Hose, 6"		11.50		12.65		
20 -50' Discharge Hoses, 6"		122.00		134.20	38.09	41.90
48 L.H., Daily Totals		$4048.90		$5559.35	$84.35	$115.82
Crew B-58	Hr.	Daily	Hr.	Daily	Bare Costs	Incl. O&P
2 Laborers	$39.85	$637.60	$64.70	$1035.20	$43.67	$70.12
1 Equip. Oper. (light)	51.30	410.40	80.95	647.60		
1 Backhoe Loader, 48 H.P.		312.40		343.64		
1 Small Helicopter, w/ Pilot		2857.00		3142.70	132.06	145.26
24 L.H., Daily Totals		$4217.40		$5169.14	$175.72	$215.38
Crew B-59	Hr.	Daily	Hr.	Daily	Bare Costs	Incl. O&P
1 Truck Driver (heavy)	$46.00	$368.00	$73.90	$591.20	$46.00	$73.90
1 Truck Tractor, 220 H.P.		290.00		319.00		
1 Water Tank Trailer, 5000 Gal.		147.60		162.36	54.70	60.17
8 L.H., Daily Totals		$805.60		$1072.56	$100.70	$134.07
Crew B-59A	Hr.	Daily	Hr.	Daily	Bare Costs	Incl. O&P
2 Laborers	$39.85	$637.60	$64.70	$1035.20	$41.90	$67.77
1 Truck Driver (heavy)	46.00	368.00	73.90	591.20		
1 Water Tank Trailer, 5000 Gal.		147.60		162.36		
1 Truck Tractor, 220 H.P.		290.00		319.00	18.23	20.06
24 L.H., Daily Totals		$1443.20		$2107.76	$60.13	$87.82

Crew No.	Bare Costs		Incl. Subs O&P		Cost Per Labor-Hour	
					Bare Costs	Incl. O&P
Crew B-60	Hr.	Daily	Hr.	Daily		
1 Labor Foreman (outside)	$41.85	$334.80	$67.90	$543.20	$46.98	$74.92
2 Laborers	39.85	637.60	64.70	1035.20		
1 Equip. Oper. (crane)	56.10	448.80	88.55	708.40		
2 Equip. Oper. (light)	51.30	820.80	80.95	1295.20		
1 Equip. Oper. (oiler)	48.60	388.80	76.70	613.60		
1 Crawler Crane, 40 Ton		1370.00		1507.00		
1 Lead, 60' High		79.40		87.34		
1 Hammer, Diesel, 15K ft.-lbs.		623.20		685.52		
1 Backhoe Loader, 48 H.P.		312.40		343.64	42.59	46.85
56 L.H., Daily Totals		$5015.80		$6819.10	$89.57	$121.77
Crew B-61	Hr.	Daily	Hr.	Daily	Bare Costs	Incl. O&P
1 Labor Foreman (outside)	$41.85	$334.80	$67.90	$543.20	$42.54	$68.59
3 Laborers	39.85	956.40	64.70	1552.80		
1 Equip. Oper. (light)	51.30	410.40	80.95	647.60		
1 Cement Mixer, 2 C.Y.		204.20		224.62		
1 Air Compressor, 160 cfm		117.20		128.92	8.04	8.84
40 L.H., Daily Totals		$2023.00		$3097.14	$50.58	$77.43
Crew B-62	Hr.	Daily	Hr.	Daily	Bare Costs	Incl. O&P
2 Laborers	$39.85	$637.60	$64.70	$1035.20	$43.67	$70.12
1 Equip. Oper. (light)	51.30	410.40	80.95	647.60		
1 Loader, Skid Steer, 30 H.P.		174.00		191.40	7.25	7.97
24 L.H., Daily Totals		$1222.00		$1874.20	$50.92	$78.09
Crew B-62A	Hr.	Daily	Hr.	Daily	Bare Costs	Incl. O&P
2 Laborers	$39.85	$637.60	$64.70	$1035.20	$43.67	$70.12
1 Equip. Oper. (light)	51.30	410.40	80.95	647.60		
1 Loader, Skid Steer, 30 H.P.		174.00		191.40		
1 Trencher Attachment		74.30		81.73	10.35	11.38
24 L.H., Daily Totals		$1296.30		$1955.93	$54.01	$81.50
Crew B-63	Hr.	Daily	Hr.	Daily	Bare Costs	Incl. O&P
4 Laborers	$39.85	$1275.20	$64.70	$2070.40	$42.14	$67.95
1 Equip. Oper. (light)	51.30	410.40	80.95	647.60		
1 Loader, Skid Steer, 30 H.P.		174.00		191.40	4.35	4.79
40 L.H., Daily Totals		$1859.60		$2909.40	$46.49	$72.73
Crew B-63B	Hr.	Daily	Hr.	Daily	Bare Costs	Incl. O&P
1 Labor Foreman (inside)	$40.35	$322.80	$65.50	$524.00	$42.84	$68.96
2 Laborers	39.85	637.60	64.70	1035.20		
1 Equip. Oper. (light)	51.30	410.40	80.95	647.60		
1 Loader, Skid Steer, 78 H.P.		364.80		401.28	11.40	12.54
32 L.H., Daily Totals		$1735.60		$2608.08	$54.24	$81.50
Crew B-64	Hr.	Daily	Hr.	Daily	Bare Costs	Incl. O&P
1 Laborer	$39.85	$318.80	$64.70	$517.60	$42.17	$68.08
1 Truck Driver (light)	44.50	356.00	71.45	571.60		
1 Power Mulcher (small)		139.20		153.12		
1 Flatbed Truck, Gas, 1.5 Ton		188.40		207.24	20.48	22.52
16 L.H., Daily Totals		$1002.40		$1449.56	$62.65	$90.60
Crew B-65	Hr.	Daily	Hr.	Daily	Bare Costs	Incl. O&P
1 Laborer	$39.85	$318.80	$64.70	$517.60	$42.17	$68.08
1 Truck Driver (light)	44.50	356.00	71.45	571.60		
1 Power Mulcher (Large)		268.80		295.68		
1 Flatbed Truck, Gas, 1.5 Ton		188.40		207.24	28.57	31.43
16 L.H., Daily Totals		$1132.00		$1592.12	$70.75	$99.51

Crews - Renovation

Crew B-66

Crew B-66	Bare Costs Hr.	Daily	Incl. Subs O&P Hr.	Daily	Cost Per Labor-Hour Bare Costs	Incl. O&P
1 Equip. Oper. (light)	$51.30	$410.40	$80.95	$647.60	$51.30	$80.95
1 Loader-Backhoe, 40 H.P.		247.00		271.70	30.88	33.96
8 L.H., Daily Totals		$657.40		$919.30	$82.17	$114.91

Crew B-67

Crew B-67	Hr.	Daily	Hr.	Daily	Bare Costs	Incl. O&P
1 Millwright	$52.75	$422.00	$82.25	$658.00	$52.02	$81.60
1 Equip. Oper. (light)	51.30	410.40	80.95	647.60		
1 R.T. Forklift, 5,000 Lb., diesel		267.60		294.36	16.73	18.40
16 L.H., Daily Totals		$1100.00		$1599.96	$68.75	$100.00

Crew B-67B

Crew B-67B	Hr.	Daily	Hr.	Daily	Bare Costs	Incl. O&P
1 Millwright Foreman (inside)	$53.25	$426.00	$83.00	$664.00	$53.00	$82.63
1 Millwright	52.75	422.00	82.25	658.00		
16 L.H., Daily Totals		$848.00		$1322.00	$53.00	$82.63

Crew B-68

Crew B-68	Hr.	Daily	Hr.	Daily	Bare Costs	Incl. O&P
2 Millwrights	$52.75	$844.00	$82.25	$1316.00	$52.27	$81.82
1 Equip. Oper. (light)	51.30	410.40	80.95	647.60		
1 R.T. Forklift, 5,000 Lb., diesel		267.60		294.36	11.15	12.27
24 L.H., Daily Totals		$1522.00		$2257.96	$63.42	$94.08

Crew B-68A

Crew B-68A	Hr.	Daily	Hr.	Daily	Bare Costs	Incl. O&P
1 Millwright Foreman (inside)	$53.25	$426.00	$83.00	$664.00	$52.92	$82.50
2 Millwrights	52.75	844.00	82.25	1316.00		
1 Forklift, Smooth Floor, 8,000 Lb.		148.20		163.02	6.17	6.79
24 L.H., Daily Totals		$1418.20		$2143.02	$59.09	$89.29

Crew B-68B

Crew B-68B	Hr.	Daily	Hr.	Daily	Bare Costs	Incl. O&P
1 Millwright Foreman (inside)	$53.25	$426.00	$83.00	$664.00	$57.06	$88.67
2 Millwrights	52.75	844.00	82.25	1316.00		
2 Electricians	58.20	931.20	89.90	1438.40		
2 Plumbers	62.15	994.40	96.70	1547.20		
1 R.T. Forklift, 5,000 Lb., gas		279.00		306.90	4.98	5.48
56 L.H., Daily Totals		$3474.60		$5272.50	$62.05	$94.15

Crew B-68C

Crew B-68C	Hr.	Daily	Hr.	Daily	Bare Costs	Incl. O&P
1 Millwright Foreman (inside)	$53.25	$426.00	$83.00	$664.00	$56.59	$87.96
1 Millwright	52.75	422.00	82.25	658.00		
1 Electrician	58.20	465.60	89.90	719.20		
1 Plumber	62.15	497.20	96.70	773.60		
1 R.T. Forklift, 5,000 Lb., gas		279.00		306.90	8.72	9.59
32 L.H., Daily Totals		$2089.80		$3121.70	$65.31	$97.55

Crew B-68D

Crew B-68D	Hr.	Daily	Hr.	Daily	Bare Costs	Incl. O&P
1 Labor Foreman (inside)	$40.35	$322.80	$65.50	$524.00	$43.83	$70.38
1 Laborer	39.85	318.80	64.70	517.60		
1 Equip. Oper. (light)	51.30	410.40	80.95	647.60		
1 R.T. Forklift, 5,000 Lb., gas		279.00		306.90	11.63	12.79
24 L.H., Daily Totals		$1331.00		$1996.10	$55.46	$83.17

Crew B-68E

Crew B-68E	Hr.	Daily	Hr.	Daily	Bare Costs	Incl. O&P
1 Struc. Steel Foreman (inside)	$55.15	$441.20	$94.05	$752.40	$54.75	$93.37
3 Struc. Steel Workers	54.65	1311.60	93.20	2236.80		
1 Welder	54.65	437.20	93.20	745.60		
1 Forklift, Smooth Floor, 8,000 Lb.		148.20		163.02	3.71	4.08
40 L.H., Daily Totals		$2338.20		$3897.82	$58.45	$97.45

Crew B-68F

Crew B-68F	Hr.	Daily	Hr.	Daily	Bare Costs	Incl. O&P
1 Skilled Worker Foreman (out)	$54.35	$434.80	$87.55	$700.40	$53.02	$85.42
2 Skilled Workers	52.35	837.60	84.35	1349.60		
1 R.T. Forklift, 5,000 Lb., gas		279.00		306.90	11.63	12.79
24 L.H., Daily Totals		$1551.40		$2356.90	$64.64	$98.20

Crew B-68G

Crew B-68G	Hr.	Daily	Hr.	Daily	Bare Costs	Incl. O&P
2 Structural Steel Workers	$54.65	$874.40	$93.20	$1491.20	$54.65	$93.20
1 R.T. Forklift, 5,000 Lb., gas		279.00		306.90	17.44	19.18
16 L.H., Daily Totals		$1153.40		$1798.10	$72.09	$112.38

Crew B-69

Crew B-69	Hr.	Daily	Hr.	Daily	Bare Costs	Incl. O&P
1 Labor Foreman (outside)	$41.85	$334.80	$67.90	$543.20	$44.35	$71.21
3 Laborers	39.85	956.40	64.70	1552.80		
1 Equip. Oper. (crane)	56.10	448.80	88.55	708.40		
1 Equip. Oper. (oiler)	48.60	388.80	76.70	613.60		
1 Hyd. Crane, 80 Ton		1518.00		1669.80	31.63	34.79
48 L.H., Daily Totals		$3646.80		$5087.80	$75.97	$106.00

Crew B-69A

Crew B-69A	Hr.	Daily	Hr.	Daily	Bare Costs	Incl. O&P
1 Labor Foreman (outside)	$41.85	$334.80	$67.90	$543.20	$43.78	$70.33
3 Laborers	39.85	956.40	64.70	1552.80		
1 Equip. Oper. (medium)	53.75	430.00	84.80	678.40		
1 Concrete Finisher	47.55	380.40	75.20	601.60		
1 Curb/Gutter Paver, 2-Track		1173.00		1290.30	24.44	26.88
48 L.H., Daily Totals		$3274.60		$4666.30	$68.22	$97.21

Crew B-69B

Crew B-69B	Hr.	Daily	Hr.	Daily	Bare Costs	Incl. O&P
1 Labor Foreman (outside)	$41.85	$334.80	$67.90	$543.20	$43.78	$70.33
3 Laborers	39.85	956.40	64.70	1552.80		
1 Equip. Oper. (medium)	53.75	430.00	84.80	678.40		
1 Cement Finisher	47.55	380.40	75.20	601.60		
1 Curb/Gutter Paver, 4-Track		789.60		868.56	16.45	18.09
48 L.H., Daily Totals		$2891.20		$4244.56	$60.23	$88.43

Crew B-70

Crew B-70	Hr.	Daily	Hr.	Daily	Bare Costs	Incl. O&P
1 Labor Foreman (outside)	$41.85	$334.80	$67.90	$543.20	$46.09	$73.77
3 Laborers	39.85	956.40	64.70	1552.80		
3 Equip. Oper. (medium)	53.75	1290.00	84.80	2035.20		
1 Grader, 30,000 Lbs.		643.80		708.18		
1 Ripper, Beam & 1 Shank		87.80		96.58		
1 Road Sweeper, S.P., 8' wide		688.60		757.46		
1 F.E. Loader, W.M., 1.5 C.Y.		342.80		377.08	31.48	34.63
56 L.H., Daily Totals		$4344.20		$6070.50	$77.58	$108.40

Crew B-71

Crew B-71	Hr.	Daily	Hr.	Daily	Bare Costs	Incl. O&P
1 Labor Foreman (outside)	$41.85	$334.80	$67.90	$543.20	$46.09	$73.77
3 Laborers	39.85	956.40	64.70	1552.80		
3 Equip. Oper. (medium)	53.75	1290.00	84.80	2035.20		
1 Pvmt. Profiler, 750 H.P.		5419.00		5960.90		
1 Road Sweeper, S.P., 8' wide		688.60		757.46		
1 F.E. Loader, W.M., 1.5 C.Y.		342.80		377.08	115.19	126.70
56 L.H., Daily Totals		$9031.60		$11226.64	$161.28	$200.48

Crew No.	Bare Costs		Incl. Subs O&P		Cost Per Labor-Hour	
Crew B-72	Hr.	Daily	Hr.	Daily	Bare Costs	Incl. O&P
1 Labor Foreman (outside)	$41.85	$334.80	$67.90	$543.20	$47.05	$75.15
3 Laborers	39.85	956.40	64.70	1552.80		
4 Equip. Oper. (medium)	53.75	1720.00	84.80	2713.60		
1 Pvmt. Profiler, 750 H.P.		5419.00		5960.90		
1 Hammermill, 250 H.P.		1870.00		2057.00		
1 Windrow Loader		1242.00		1366.20		
1 Mix Paver, 165 H.P.		2140.00		2354.00		
1 Roller, Pneum. Whl., 12 Ton		338.20		372.02	172.02	189.22
64 L.H., Daily Totals		$14020.40		$16919.72	$219.07	$264.37

Crew B-73	Hr.	Daily	Hr.	Daily	Bare Costs	Incl. O&P
1 Labor Foreman (outside)	$41.85	$334.80	$67.90	$543.20	$48.79	$77.66
2 Laborers	39.85	637.60	64.70	1035.20		
5 Equip. Oper. (medium)	53.75	2150.00	84.80	3392.00		
1 Road Mixer, 310 H.P.		1889.00		2077.90		
1 Tandem Roller, 10 Ton		231.40		254.54		
1 Hammermill, 250 H.P.		1870.00		2057.00		
1 Grader, 30,000 Lbs.		643.80		708.18		
.5 F.E. Loader, W.M., 1.5 C.Y.		171.40		188.54		
.5 Truck Tractor, 220 H.P.		145.00		159.50		
.5 Water Tank Trailer, 5000 Gal.		73.80		81.18	78.51	86.36
64 L.H., Daily Totals		$8146.80		$10497.24	$127.29	$164.02

Crew B-74	Hr.	Daily	Hr.	Daily	Bare Costs	Incl. O&P
1 Labor Foreman (outside)	$41.85	$334.80	$67.90	$543.20	$48.59	$77.45
1 Laborer	39.85	318.80	64.70	517.60		
4 Equip. Oper. (medium)	53.75	1720.00	84.80	2713.60		
2 Truck Drivers (heavy)	46.00	736.00	73.90	1182.40		
1 Grader, 30,000 Lbs.		643.80		708.18		
1 Ripper, Beam & 1 Shank		87.80		96.58		
2 Stabilizers, 310 H.P.		3436.00		3779.60		
1 Flatbed Truck, Gas, 3 Ton		238.00		261.80		
1 Chem. Spreader, Towed		52.20		57.42		
1 Roller, Vibratory, 25 Ton		659.80		725.78		
1 Water Tank Trailer, 5000 Gal.		147.60		162.36		
1 Truck Tractor, 220 H.P.		290.00		319.00	86.80	95.48
64 L.H., Daily Totals		$8664.80		$11067.52	$135.39	$172.93

Crew B-75	Hr.	Daily	Hr.	Daily	Bare Costs	Incl. O&P
1 Labor Foreman (outside)	$41.85	$334.80	$67.90	$543.20	$48.96	$77.96
1 Laborer	39.85	318.80	64.70	517.60		
4 Equip. Oper. (medium)	53.75	1720.00	84.80	2713.60		
1 Truck Driver (heavy)	46.00	368.00	73.90	591.20		
1 Grader, 30,000 Lbs.		643.80		708.18		
1 Ripper, Beam & 1 Shank		87.80		96.58		
2 Stabilizers, 310 H.P.		3436.00		3779.60		
1 Dist. Tanker, 3000 Gallon		326.80		359.48		
1 Truck Tractor, 6x4, 380 H.P.		476.20		523.82		
1 Roller, Vibratory, 25 Ton		659.80		725.78	100.54	110.60
56 L.H., Daily Totals		$8372.00		$10559.04	$149.50	$188.55

Crew B-76	Hr.	Daily	Hr.	Daily	Bare Costs	Incl. O&P
1 Dock Builder Foreman (outside)	$53.30	$426.40	$86.70	$693.60	$52.29	$84.19
5 Dock Builders	51.30	2052.00	83.45	3338.00		
2 Equip. Oper. (crane)	56.10	897.60	88.55	1416.80		
1 Equip. Oper. (oiler)	48.60	388.80	76.70	613.60		
1 Crawler Crane, 50 Ton		1940.00		2134.00		
1 Barge, 400 Ton		835.40		918.94		
1 Hammer, Diesel, 15K ft.-lbs.		623.20		685.52		
1 Lead, 60' High		79.40		87.34		
1 Air Compressor, 600 cfm		417.20		458.92		
2 -50' Air Hoses, 3"		86.40		95.04	55.30	60.83
72 L.H., Daily Totals		$7746.40		$10441.76	$107.59	$145.02

Crew B-76A	Hr.	Daily	Hr.	Daily	Bare Costs	Incl. O&P
1 Labor Foreman (outside)	$41.85	$334.80	$67.90	$543.20	$43.23	$69.58
5 Laborers	39.85	1594.00	64.70	2588.00		
1 Equip. Oper. (crane)	56.10	448.80	88.55	708.40		
1 Equip. Oper. (oiler)	48.60	388.80	76.70	613.60		
1 Crawler Crane, 50 Ton		1940.00		2134.00		
1 Barge, 400 Ton		835.40		918.94	43.37	47.70
64 L.H., Daily Totals		$5541.80		$7506.14	$86.59	$117.28

Crew B-77	Hr.	Daily	Hr.	Daily	Bare Costs	Incl. O&P
1 Labor Foreman (outside)	$41.85	$334.80	$67.90	$543.20	$41.18	$66.69
3 Laborers	39.85	956.40	64.70	1552.80		
1 Truck Driver (light)	44.50	356.00	71.45	571.60		
1 Crack Cleaner, 25 H.P.		55.80		61.38		
1 Crack Filler, Trailer Mtd.		198.20		218.02		
1 Flatbed Truck, Gas, 3 Ton		238.00		261.80	12.30	13.53
40 L.H., Daily Totals		$2139.20		$3208.80	$53.48	$80.22

Crew B-78	Hr.	Daily	Hr.	Daily	Bare Costs	Incl. O&P
1 Labor Foreman (outside)	$41.85	$334.80	$67.90	$543.20	$40.96	$66.36
4 Laborers	39.85	1275.20	64.70	2070.40		
1 Truck Driver (light)	44.50	356.00	71.45	571.60		
1 Paint Striper, S.P., 40 Gallon		151.00		166.10		
1 Flatbed Truck, Gas, 3 Ton		238.00		261.80		
1 Pickup Truck, 3/4 Ton		115.20		126.72	10.50	11.55
48 L.H., Daily Totals		$2470.20		$3739.82	$51.46	$77.91

Crew B-78A	Hr.	Daily	Hr.	Daily	Bare Costs	Incl. O&P
1 Equip. Oper. (light)	$51.30	$410.40	$80.95	$647.60	$51.30	$80.95
1 Line Rem. (Metal Balls) 115 H.P.		852.60		937.86	106.58	117.23
8 L.H., Daily Totals		$1263.00		$1585.46	$157.88	$198.18

Crew B-78B	Hr.	Daily	Hr.	Daily	Bare Costs	Incl. O&P
2 Laborers	$39.85	$637.60	$64.70	$1035.20	$41.12	$66.51
.25 Equip. Oper. (light)	51.30	102.60	80.95	161.90		
1 Pickup Truck, 3/4 Ton		115.20		126.72		
1 Line Rem.,11 H.P.,Walk Behind		61.40		67.54		
.25 Road Sweeper, S.P., 8' wide		172.15		189.37	19.38	21.31
18 L.H., Daily Totals		$1088.95		$1580.72	$60.50	$87.82

Crew B-78C	Hr.	Daily	Hr.	Daily	Bare Costs	Incl. O&P
1 Labor Foreman (outside)	$41.85	$334.80	$67.90	$543.20	$40.96	$66.36
4 Laborers	39.85	1275.20	64.70	2070.40		
1 Truck Driver (light)	44.50	356.00	71.45	571.60		
1 Paint Striper, T.M., 120 Gal.		690.00		759.00		
1 Flatbed Truck, Gas, 3 Ton		238.00		261.80		
1 Pickup Truck, 3/4 Ton		115.20		126.72	21.73	23.91
48 L.H., Daily Totals		$3009.20		$4332.72	$62.69	$90.27

Crew No.	Bare Costs		Incl. Subs O&P		Cost Per Labor-Hour	

Crew B-78D	Hr.	Daily	Hr.	Daily	Bare Costs	Incl. O&P
2 Labor Foremen (outside)	$41.85	$669.60	$67.90	$1086.40	$40.72	$66.02
7 Laborers	39.85	2231.60	64.70	3623.20		
1 Truck Driver (light)	44.50	356.00	71.45	571.60		
1 Paint Striper, T.M., 120 Gal.		690.00		759.00		
1 Flatbed Truck, Gas, 3 Ton		238.00		261.80		
3 Pickup Trucks, 3/4 Ton		345.60		380.16		
1 Air Compressor, 60 cfm		105.40		115.94		
1 -50' Air Hose, 3/4"		7.35		8.09		
1 Breaker, Pavement, 60 lb.		10.60		11.66	17.46	19.21
80 L.H., Daily Totals		$4654.15		$6817.85	$58.18	$85.22

Crew B-78E	Hr.	Daily	Hr.	Daily	Bare Costs	Incl. O&P
2 Labor Foremen (outside)	$41.85	$669.60	$67.90	$1086.40	$40.57	$65.80
9 Laborers	39.85	2869.20	64.70	4658.40		
1 Truck Driver (light)	44.50	356.00	71.45	571.60		
1 Paint Striper, T.M., 120 Gal.		690.00		759.00		
1 Flatbed Truck, Gas, 3 Ton		238.00		261.80		
4 Pickup Trucks, 3/4 Ton		460.80		506.88		
2 Air Compressors, 60 cfm		210.80		231.88		
2 -50' Air Hoses, 3/4"		14.70		16.17		
2 Breakers, Pavement, 60 lb.		21.20		23.32	17.04	18.74
96 L.H., Daily Totals		$5530.30		$8115.45	$57.61	$84.54

Crew B-78F	Hr.	Daily	Hr.	Daily	Bare Costs	Incl. O&P
2 Labor Foremen (outside)	$41.85	$669.60	$67.90	$1086.40	$40.47	$65.64
11 Laborers	39.85	3506.80	64.70	5693.60		
1 Truck Driver (light)	44.50	356.00	71.45	571.60		
1 Paint Striper, T.M., 120 Gal.		690.00		759.00		
1 Flatbed Truck, Gas, 3 Ton		238.00		261.80		
7 Pickup Trucks, 3/4 Ton		806.40		887.04		
3 Air Compressors, 60 cfm		316.20		347.82		
3 -50' Air Hoses, 3/4"		22.05		24.25		
3 Breakers, Pavement, 60 lb.		31.80		34.98	18.79	20.67
112 L.H., Daily Totals		$6636.85		$9666.50	$59.26	$86.31

Crew B-79	Hr.	Daily	Hr.	Daily	Bare Costs	Incl. O&P
1 Labor Foreman (outside)	$41.85	$334.80	$67.90	$543.20	$41.18	$66.69
3 Laborers	39.85	956.40	64.70	1552.80		
1 Truck Driver (light)	44.50	356.00	71.45	571.60		
1 Paint Striper, T.M., 120 Gal.		690.00		759.00		
1 Heating Kettle, 115 Gallon		77.85		85.64		
1 Flatbed Truck, Gas, 3 Ton		238.00		261.80		
2 Pickup Trucks, 3/4 Ton		230.40		253.44	30.91	34.00
40 L.H., Daily Totals		$2883.45		$4027.47	$72.09	$100.69

Crew B-79A	Hr.	Daily	Hr.	Daily	Bare Costs	Incl. O&P
1.5 Equip. Oper. (light)	$51.30	$615.60	$80.95	$971.40	$51.30	$80.95
.5 Line Remov. (Grinder) 115 H.P.		459.70		505.67		
1 Line Rem. (Metal Balls) 115 H.P.		852.60		937.86	109.36	120.29
12 L.H., Daily Totals		$1927.90		$2414.93	$160.66	$201.24

Crew B-79B	Hr.	Daily	Hr.	Daily	Bare Costs	Incl. O&P
1 Laborer	$39.85	$318.80	$64.70	$517.60	$39.85	$64.70
1 Set of Gases		168.00		184.80	21.00	23.10
8 L.H., Daily Totals		$486.80		$702.40	$60.85	$87.80

Crew B-79C	Hr.	Daily	Hr.	Daily	Bare Costs	Incl. O&P
1 Labor Foreman (outside)	$41.85	$334.80	$67.90	$543.20	$40.80	$66.12
5 Laborers	39.85	1594.00	64.70	2588.00		
1 Truck Driver (light)	44.50	356.00	71.45	571.60		
1 Paint Striper, T.M., 120 Gal.		690.00		759.00		
1 Heating Kettle, 115 Gallon		77.85		85.64		
1 Flatbed Truck, Gas, 3 Ton		238.00		261.80		
3 Pickup Trucks, 3/4 Ton		345.60		380.16		
1 Air Compressor, 60 cfm		105.40		115.94		
1 -50' Air Hose, 3/4"		7.35		8.09		
1 Breaker, Pavement, 60 lb.		10.60		11.66	26.34	28.97
56 L.H., Daily Totals		$3759.60		$5325.08	$67.14	$95.09

Crew B-79D	Hr.	Daily	Hr.	Daily	Bare Costs	Incl. O&P
2 Labor Foremen (outside)	$41.85	$669.60	$67.90	$1086.40	$40.93	$66.34
5 Laborers	39.85	1594.00	64.70	2588.00		
1 Truck Driver (light)	44.50	356.00	71.45	571.60		
1 Paint Striper, T.M., 120 Gal.		690.00		759.00		
1 Heating Kettle, 115 Gallon		77.85		85.64		
1 Flatbed Truck, Gas, 3 Ton		238.00		261.80		
4 Pickup Trucks, 3/4 Ton		460.80		506.88		
1 Air Compressor, 60 cfm		105.40		115.94		
1 -50' Air Hose, 3/4"		7.35		8.09		
1 Breaker, Pavement, 60 lb.		10.60		11.66	24.84	27.33
64 L.H., Daily Totals		$4209.60		$5995.00	$65.78	$93.67

Crew B-79E	Hr.	Daily	Hr.	Daily	Bare Costs	Incl. O&P
2 Labor Foremen (outside)	$41.85	$669.60	$67.90	$1086.40	$40.72	$66.02
7 Laborers	39.85	2231.60	64.70	3623.20		
1 Truck Driver (light)	44.50	356.00	71.45	571.60		
1 Paint Striper, T.M., 120 Gal.		690.00		759.00		
1 Heating Kettle, 115 Gallon		77.85		85.64		
1 Flatbed Truck, Gas, 3 Ton		238.00		261.80		
5 Pickup Trucks, 3/4 Ton		576.00		633.60		
2 Air Compressors, 60 cfm		210.80		231.88		
2 -50' Air Hoses, 3/4"		14.70		16.17		
2 Breakers, Pavement, 60 lb.		21.20		23.32	22.86	25.14
80 L.H., Daily Totals		$5085.75		$7292.60	$63.57	$91.16

Crew B-80	Hr.	Daily	Hr.	Daily	Bare Costs	Incl. O&P
1 Labor Foreman (outside)	$41.85	$334.80	$67.90	$543.20	$44.38	$71.25
1 Laborer	39.85	318.80	64.70	517.60		
1 Truck Driver (light)	44.50	356.00	71.45	571.60		
1 Equip. Oper. (light)	51.30	410.40	80.95	647.60		
1 Flatbed Truck, Gas, 3 Ton		238.00		261.80		
1 Earth Auger, Truck-Mtd.		373.80		411.18	19.12	21.03
32 L.H., Daily Totals		$2031.80		$2952.98	$63.49	$92.28

Crew B-80A	Hr.	Daily	Hr.	Daily	Bare Costs	Incl. O&P
3 Laborers	$39.85	$956.40	$64.70	$1552.80	$39.85	$64.70
1 Flatbed Truck, Gas, 3 Ton		238.00		261.80	9.92	10.91
24 L.H., Daily Totals		$1194.40		$1814.60	$49.77	$75.61

Crew B-80B	Hr.	Daily	Hr.	Daily	Bare Costs	Incl. O&P
3 Laborers	$39.85	$956.40	$64.70	$1552.80	$42.71	$68.76
1 Equip. Oper. (light)	51.30	410.40	80.95	647.60		
1 Crane, Flatbed Mounted, 3 Ton		235.80		259.38	7.37	8.11
32 L.H., Daily Totals		$1602.60		$2459.78	$50.08	$76.87

Crew No.	Bare Costs		Incl. Subs O&P		Cost Per Labor-Hour	
Crew B-80C	Hr.	Daily	Hr.	Daily	Bare Costs	Incl. O&P
2 Laborers	$39.85	$637.60	$64.70	$1035.20	$41.40	$66.95
1 Truck Driver (light)	44.50	356.00	71.45	571.60		
1 Flatbed Truck, Gas, 1.5 Ton		188.40		207.24		
1 Manual Fence Post Auger, Gas		7.40		8.14	8.16	8.97
24 L.H., Daily Totals		$1189.40		$1822.18	$49.56	$75.92
Crew B-81	Hr.	Daily	Hr.	Daily	Bare Costs	Incl. O&P
1 Laborer	$39.85	$318.80	$64.70	$517.60	$46.53	$74.47
1 Equip. Oper. (medium)	53.75	430.00	84.80	678.40		
1 Truck Driver (heavy)	46.00	368.00	73.90	591.20		
1 Hydromulcher, T.M., 3000 Gal.		282.80		311.08		
1 Truck Tractor, 220 H.P.		290.00		319.00	23.87	26.25
24 L.H., Daily Totals		$1689.60		$2417.28	$70.40	$100.72
Crew B-81A	Hr.	Daily	Hr.	Daily	Bare Costs	Incl. O&P
1 Laborer	$39.85	$318.80	$64.70	$517.60	$42.17	$68.08
1 Truck Driver (light)	44.50	356.00	71.45	571.60		
1 Hydromulcher, T.M., 600 Gal.		121.20		133.32		
1 Flatbed Truck, Gas, 3 Ton		238.00		261.80	22.45	24.70
16 L.H., Daily Totals		$1034.00		$1484.32	$64.63	$92.77
Crew B-82	Hr.	Daily	Hr.	Daily	Bare Costs	Incl. O&P
1 Laborer	$39.85	$318.80	$64.70	$517.60	$45.58	$72.83
1 Equip. Oper. (light)	51.30	410.40	80.95	647.60		
1 Horiz. Borer, 6 H.P.		78.00		85.80	4.88	5.36
16 L.H., Daily Totals		$807.20		$1251.00	$50.45	$78.19
Crew B-82A	Hr.	Daily	Hr.	Daily	Bare Costs	Incl. O&P
2 Laborers	$39.85	$637.60	$64.70	$1035.20	$45.58	$72.83
2 Equip. Opers. (light)	51.30	820.80	80.95	1295.20		
2 Dump Trucks, 8 C.Y., 220 H.P.		677.20		744.92		
1 Flatbed Trailer, 25 Ton		133.00		146.30		
1 Horiz. Dir. Drill, 20k lb. Thrust		629.00		691.90		
1 Mud Trailer for HDD, 1500 Gal.		287.80		316.58		
1 Pickup Truck, 4x4, 3/4 Ton		126.60		139.26		
1 Flatbed Trailer, 3 Ton		25.40		27.94		
1 Loader, Skid Steer, 78 H.P.		364.80		401.28	70.12	77.13
32 L.H., Daily Totals		$3702.20		$4798.58	$115.69	$149.96
Crew B-82B	Hr.	Daily	Hr.	Daily	Bare Costs	Incl. O&P
2 Laborers	$39.85	$637.60	$64.70	$1035.20	$45.58	$72.83
2 Equip. Opers. (light)	51.30	820.80	80.95	1295.20		
2 Dump Trucks, 8 C.Y., 220 H.P.		677.20		744.92		
1 Flatbed Trailer, 25 Ton		133.00		146.30		
1 Horiz. Dir. Drill, 30k lb. Thrust		896.20		985.82		
1 Mud Trailer for HDD, 1500 Gal.		287.80		316.58		
1 Pickup Truck, 4x4, 3/4 Ton		126.60		139.26		
1 Flatbed Trailer, 3 Ton		25.40		27.94		
1 Loader, Skid Steer, 78 H.P.		364.80		401.28	78.47	86.32
32 L.H., Daily Totals		$3969.40		$5092.50	$124.04	$159.14

Crew No.	Bare Costs		Incl. Subs O&P		Cost Per Labor-Hour	
Crew B-82C	Hr.	Daily	Hr.	Daily	Bare Costs	Incl. O&P
2 Laborers	$39.85	$637.60	$64.70	$1035.20	$45.58	$72.83
2 Equip. Opers. (light)	51.30	820.80	80.95	1295.20		
2 Dump Trucks, 8 C.Y., 220 H.P.		677.20		744.92		
1 Flatbed Trailer, 25 Ton		133.00		146.30		
1 Horiz. Dir. Drill, 50k lb. Thrust		1188.00		1306.80		
1 Mud Trailer for HDD, 1500 Gal.		287.80		316.58		
1 Pickup Truck, 4x4, 3/4 Ton		126.60		139.26		
1 Flatbed Trailer, 3 Ton		25.40		27.94		
1 Loader, Skid Steer, 78 H.P.		364.80		401.28	87.59	96.35
32 L.H., Daily Totals		$4261.20		$5413.48	$133.16	$169.17
Crew B-82D	Hr.	Daily	Hr.	Daily	Bare Costs	Incl. O&P
1 Equip. Oper. (light)	$51.30	$410.40	$80.95	$647.60	$51.30	$80.95
1 Mud Trailer for HDD, 1500 Gal.		287.80		316.58	35.98	39.57
8 L.H., Daily Totals		$698.20		$964.18	$87.28	$120.52
Crew B-83	Hr.	Daily	Hr.	Daily	Bare Costs	Incl. O&P
1 Tugboat Captain	$53.75	$430.00	$84.80	$678.40	$46.80	$74.75
1 Tugboat Hand	39.85	318.80	64.70	517.60		
1 Tugboat, 250 H.P.		680.20		748.22	42.51	46.76
16 L.H., Daily Totals		$1429.00		$1944.22	$89.31	$121.51
Crew B-84	Hr.	Daily	Hr.	Daily	Bare Costs	Incl. O&P
1 Equip. Oper. (medium)	$53.75	$430.00	$84.80	$678.40	$53.75	$84.80
1 Rotary Mower/Tractor		357.00		392.70	44.63	49.09
8 L.H., Daily Totals		$787.00		$1071.10	$98.38	$133.89
Crew B-85	Hr.	Daily	Hr.	Daily	Bare Costs	Incl. O&P
3 Laborers	$39.85	$956.40	$64.70	$1552.80	$43.86	$70.56
1 Equip. Oper. (medium)	53.75	430.00	84.80	678.40		
1 Truck Driver (heavy)	46.00	368.00	73.90	591.20		
1 Telescoping Boom Lift, to 80'		548.60		603.46		
1 Brush Chipper, 12", 130 H.P.		393.00		432.30		
1 Pruning Saw, Rotary		6.65		7.32	23.71	26.08
40 L.H., Daily Totals		$2702.65		$3865.47	$67.57	$96.64
Crew B-86	Hr.	Daily	Hr.	Daily	Bare Costs	Incl. O&P
1 Equip. Oper. (medium)	$53.75	$430.00	$84.80	$678.40	$53.75	$84.80
1 Stump Chipper, S.P.		184.05		202.46	23.01	25.31
8 L.H., Daily Totals		$614.05		$880.86	$76.76	$110.11
Crew B-86A	Hr.	Daily	Hr.	Daily	Bare Costs	Incl. O&P
1 Equip. Oper. (medium)	$53.75	$430.00	$84.80	$678.40	$53.75	$84.80
1 Grader, 30,000 Lbs.		643.80		708.18	80.47	88.52
8 L.H., Daily Totals		$1073.80		$1386.58	$134.22	$173.32
Crew B-86B	Hr.	Daily	Hr.	Daily	Bare Costs	Incl. O&P
1 Equip. Oper. (medium)	$53.75	$430.00	$84.80	$678.40	$53.75	$84.80
1 Dozer, 200 H.P.		1273.00		1400.30	159.13	175.04
8 L.H., Daily Totals		$1703.00		$2078.70	$212.88	$259.84
Crew B-87	Hr.	Daily	Hr.	Daily	Bare Costs	Incl. O&P
1 Laborer	$39.85	$318.80	$64.70	$517.60	$50.97	$80.78
4 Equip. Oper. (medium)	53.75	1720.00	84.80	2713.60		
2 Feller Bunchers, 100 H.P.		1620.80		1782.88		
1 Log Chipper, 22" Tree		764.80		841.28		
1 Dozer, 105 H.P.		608.00		668.80		
1 Chain Saw, Gas, 36" Long		46.40		51.04	76.00	83.60
40 L.H., Daily Totals		$5078.80		$6575.20	$126.97	$164.38

Crews - Renovation

Crew No.	Bare Costs		Incl. Subs O&P		Cost Per Labor-Hour	
Crew B-88	Hr.	Daily	Hr.	Daily	Bare Costs	Incl. O&P
1 Laborer	$39.85	$318.80	$64.70	$517.60	$51.76	$81.93
6 Equip. Oper. (medium)	53.75	2580.00	84.80	4070.40		
2 Feller Bunchers, 100 H.P.		1620.80		1782.88		
1 Log Chipper, 22" Tree		764.80		841.28		
2 Log Skidders, 50 H.P.		1698.80		1868.68		
1 Dozer, 105 H.P.		608.00		668.80		
1 Chain Saw, Gas, 36" Long		46.40		51.04	84.62	93.08
56 L.H., Daily Totals		$7637.60		$9800.68	$136.39	$175.01

Crew B-89	Hr.	Daily	Hr.	Daily	Bare Costs	Incl. O&P
1 Equip. Oper. (light)	$51.30	$410.40	$80.95	$647.60	$47.90	$76.20
1 Truck Driver (light)	44.50	356.00	71.45	571.60		
1 Flatbed Truck, Gas, 3 Ton		238.00		261.80		
1 Concrete Saw		102.40		112.64		
1 Water Tank, 65 Gal.		79.90		87.89	26.27	28.90
16 L.H., Daily Totals		$1186.70		$1681.53	$74.17	$105.10

Crew B-89A	Hr.	Daily	Hr.	Daily	Bare Costs	Incl. O&P
1 Skilled Worker	$52.35	$418.80	$84.35	$674.80	$46.10	$74.53
1 Laborer	39.85	318.80	64.70	517.60		
1 Core Drill (Large)		112.20		123.42	7.01	7.71
16 L.H., Daily Totals		$849.80		$1315.82	$53.11	$82.24

Crew B-89B	Hr.	Daily	Hr.	Daily	Bare Costs	Incl. O&P
1 Equip. Oper. (light)	$51.30	$410.40	$80.95	$647.60	$47.90	$76.20
1 Truck Driver (light)	44.50	356.00	71.45	571.60		
1 Wall Saw, Hydraulic, 10 H.P.		46.40		51.04		
1 Generator, Diesel, 100 kW		307.60		338.36		
1 Water Tank, 65 Gal.		79.90		87.89		
1 Flatbed Truck, Gas, 3 Ton		238.00		261.80	41.99	46.19
16 L.H., Daily Totals		$1438.30		$1958.29	$89.89	$122.39

Crew B-90	Hr.	Daily	Hr.	Daily	Bare Costs	Incl. O&P
1 Labor Foreman (outside)	$41.85	$334.80	$67.90	$543.20	$44.50	$71.46
3 Laborers	39.85	956.40	64.70	1552.80		
2 Equip. Oper. (light)	51.30	820.80	80.95	1295.20		
2 Truck Drivers (heavy)	46.00	736.00	73.90	1182.40		
1 Road Mixer, 310 H.P.		1889.00		2077.90		
1 Dist. Truck, 2000 Gal.		296.00		325.60	34.14	37.55
64 L.H., Daily Totals		$5033.00		$6977.10	$78.64	$109.02

Crew B-90A	Hr.	Daily	Hr.	Daily	Bare Costs	Incl. O&P
1 Labor Foreman (outside)	$41.85	$334.80	$67.90	$543.20	$48.08	$76.64
2 Laborers	39.85	637.60	64.70	1035.20		
4 Equip. Oper. (medium)	53.75	1720.00	84.80	2713.60		
2 Graders, 30,000 Lbs.		1287.60		1416.36		
1 Tandem Roller, 10 Ton		231.40		254.54		
1 Roller, Pneum. Whl., 12 Ton		338.20		372.02	33.16	36.48
56 L.H., Daily Totals		$4549.60		$6334.92	$81.24	$113.12

Crew B-90B	Hr.	Daily	Hr.	Daily	Bare Costs	Incl. O&P
1 Labor Foreman (outside)	$41.85	$334.80	$67.90	$543.20	$47.13	$75.28
2 Laborers	39.85	637.60	64.70	1035.20		
3 Equip. Oper. (medium)	53.75	1290.00	84.80	2035.20		
1 Roller, Pneum. Whl., 12 Ton		338.20		372.02		
1 Road Mixer, 310 H.P.		1889.00		2077.90	46.40	51.04
48 L.H., Daily Totals		$4489.60		$6063.52	$93.53	$126.32

Crew B-90C	Hr.	Daily	Hr.	Daily	Bare Costs	Incl. O&P
1 Labor Foreman (outside)	$41.85	$334.80	$67.90	$543.20	$45.50	$72.98
4 Laborers	39.85	1275.20	64.70	2070.40		
3 Equip. Oper. (medium)	53.75	1290.00	84.80	2035.20		
3 Truck Drivers (heavy)	46.00	1104.00	73.90	1773.60		
3 Road Mixers, 310 H.P.		5667.00		6233.70	64.40	70.84
88 L.H., Daily Totals		$9671.00		$12656.10	$109.90	$143.82

Crew B-90D	Hr.	Daily	Hr.	Daily	Bare Costs	Incl. O&P
1 Labor Foreman (outside)	$41.85	$334.80	$67.90	$543.20	$44.63	$71.71
6 Laborers	39.85	1912.80	64.70	3105.60		
3 Equip. Oper. (medium)	53.75	1290.00	84.80	2035.20		
3 Truck Drivers (heavy)	46.00	1104.00	73.90	1773.60		
3 Road Mixers, 310 H.P.		5667.00		6233.70	54.49	59.94
104 L.H., Daily Totals		$10308.60		$13691.30	$99.12	$131.65

Crew B-90E	Hr.	Daily	Hr.	Daily	Bare Costs	Incl. O&P
1 Labor Foreman (outside)	$41.85	$334.80	$67.90	$543.20	$45.39	$72.78
4 Laborers	39.85	1275.20	64.70	2070.40		
3 Equip. Oper. (medium)	53.75	1290.00	84.80	2035.20		
1 Truck Driver (heavy)	46.00	368.00	73.90	591.20		
1 Road Mixer, 310 H.P.		1889.00		2077.90	26.24	28.86
72 L.H., Daily Totals		$5157.00		$7317.90	$71.63	$101.64

Crew B-91	Hr.	Daily	Hr.	Daily	Bare Costs	Incl. O&P
1 Labor Foreman (outside)	$41.85	$334.80	$67.90	$543.20	$47.82	$76.30
2 Laborers	39.85	637.60	64.70	1035.20		
4 Equip. Oper. (medium)	53.75	1720.00	84.80	2713.60		
1 Truck Driver (heavy)	46.00	368.00	73.90	591.20		
1 Dist. Tanker, 3000 Gallon		326.80		359.48		
1 Truck Tractor, 6x4, 380 H.P.		476.20		523.82		
1 Aggreg. Spreader, S.P.		837.00		920.70		
1 Roller, Pneum. Whl., 12 Ton		338.20		372.02		
1 Tandem Roller, 10 Ton		231.40		254.54	34.52	37.98
64 L.H., Daily Totals		$5270.00		$7313.76	$82.34	$114.28

Crew B-91B	Hr.	Daily	Hr.	Daily	Bare Costs	Incl. O&P
1 Laborer	$39.85	$318.80	$64.70	$517.60	$46.80	$74.75
1 Equipment Oper. (med.)	53.75	430.00	84.80	678.40		
1 Road Sweeper, Vac. Assist.		840.40		924.44	52.52	57.78
16 L.H., Daily Totals		$1589.20		$2120.44	$99.33	$132.53

Crew B-91C	Hr.	Daily	Hr.	Daily	Bare Costs	Incl. O&P
1 Laborer	$39.85	$318.80	$64.70	$517.60	$42.17	$68.08
1 Truck Driver (light)	44.50	356.00	71.45	571.60		
1 Catch Basin Cleaning Truck		522.80		575.08	32.67	35.94
16 L.H., Daily Totals		$1197.60		$1664.28	$74.85	$104.02

Crew B-91D	Hr.	Daily	Hr.	Daily	Bare Costs	Incl. O&P
1 Labor Foreman (outside)	$41.85	$334.80	$67.90	$543.20	$46.30	$74.09
5 Laborers	39.85	1594.00	64.70	2588.00		
5 Equip. Oper. (medium)	53.75	2150.00	84.80	3392.00		
2 Truck Drivers (heavy)	46.00	736.00	73.90	1182.40		
1 Aggreg. Spreader, S.P.		837.00		920.70		
2 Truck Tractors, 6x4, 380 H.P.		952.40		1047.64		
2 Dist. Tankers, 3000 Gallon		653.60		718.96		
2 Pavement Brushes, Towed		170.40		187.44		
2 Rollers Pneum. Whl., 12 Ton		676.40		744.04	31.63	34.80
104 L.H., Daily Totals		$8104.60		$11324.38	$77.93	$108.89

For customer support on your Commercial Renovation Costs with RSMeans data, call 800.448.8182.

Crew No.	Bare Costs		Incl. Subs O&P		Cost Per Labor-Hour	
	Hr.	Daily	Hr.	Daily	Bare Costs	Incl. O&P
Crew B-92	Hr.	Daily	Hr.	Daily	Bare Costs	Incl. O&P
1 Labor Foreman (outside)	$41.85	$334.80	$67.90	$543.20	$40.35	$65.50
3 Laborers	39.85	956.40	64.70	1552.80		
1 Crack Cleaner, 25 H.P.		55.80		61.38		
1 Air Compressor, 60 cfm		105.40		115.94		
1 Tar Kettle, T.M.		129.20		142.12		
1 Flatbed Truck, Gas, 3 Ton		238.00		261.80	16.51	18.16
32 L.H., Daily Totals		$1819.60		$2677.24	$56.86	$83.66
Crew B-93	Hr.	Daily	Hr.	Daily	Bare Costs	Incl. O&P
1 Equip. Oper. (medium)	$53.75	$430.00	$84.80	$678.40	$53.75	$84.80
1 Feller Buncher, 100 H.P.		810.40		891.44	101.30	111.43
8 L.H., Daily Totals		$1240.40		$1569.84	$155.05	$196.23
Crew B-94A	Hr.	Daily	Hr.	Daily	Bare Costs	Incl. O&P
1 Laborer	$39.85	$318.80	$64.70	$517.60	$39.85	$64.70
1 Diaphragm Water Pump, 2"		73.00		80.30		
1 -20' Suction Hose, 2"		1.95		2.15		
2 -50' Discharge Hoses, 2"		1.80		1.98	9.59	10.55
8 L.H., Daily Totals		$395.55		$602.02	$49.44	$75.25
Crew B-94B	Hr.	Daily	Hr.	Daily	Bare Costs	Incl. O&P
1 Laborer	$39.85	$318.80	$64.70	$517.60	$39.85	$64.70
1 Diaphragm Water Pump, 4"		114.80		126.28		
1 -20' Suction Hose, 4"		3.25		3.58		
2 -50' Discharge Hoses, 4"		4.70		5.17	15.34	16.88
8 L.H., Daily Totals		$441.55		$652.63	$55.19	$81.58
Crew B-94C	Hr.	Daily	Hr.	Daily	Bare Costs	Incl. O&P
1 Laborer	$39.85	$318.80	$64.70	$517.60	$39.85	$64.70
1 Centrifugal Water Pump, 3"		79.20		87.12		
1 -20' Suction Hose, 3"		2.85		3.13		
2 -50' Discharge Hoses, 3"		3.00		3.30	10.63	11.69
8 L.H., Daily Totals		$403.85		$611.15	$50.48	$76.39
Crew B-94D	Hr.	Daily	Hr.	Daily	Bare Costs	Incl. O&P
1 Laborer	$39.85	$318.80	$64.70	$517.60	$39.85	$64.70
1 Centr. Water Pump, 6"		296.40		326.04		
1 -20' Suction Hose, 6"		11.50		12.65		
2 -50' Discharge Hoses, 6"		12.20		13.42	40.01	44.01
8 L.H., Daily Totals		$638.90		$869.71	$79.86	$108.71
Crew C-1	Hr.	Daily	Hr.	Daily	Bare Costs	Incl. O&P
3 Carpenters	$50.70	$1216.80	$82.30	$1975.20	$47.99	$77.90
1 Laborer	39.85	318.80	64.70	517.60		
32 L.H., Daily Totals		$1535.60		$2492.80	$47.99	$77.90
Crew C-2	Hr.	Daily	Hr.	Daily	Bare Costs	Incl. O&P
1 Carpenter Foreman (outside)	$52.70	$421.60	$85.55	$684.40	$49.23	$79.91
4 Carpenters	50.70	1622.40	82.30	2633.60		
1 Laborer	39.85	318.80	64.70	517.60		
48 L.H., Daily Totals		$2362.80		$3835.60	$49.23	$79.91
Crew C-2A	Hr.	Daily	Hr.	Daily	Bare Costs	Incl. O&P
1 Carpenter Foreman (outside)	$52.70	$421.60	$85.55	$684.40	$48.70	$78.72
3 Carpenters	50.70	1216.80	82.30	1975.20		
1 Cement Finisher	47.55	380.40	75.20	601.60		
1 Laborer	39.85	318.80	64.70	517.60		
48 L.H., Daily Totals		$2337.60		$3778.80	$48.70	$78.72

Crew No.	Bare Costs		Incl. Subs O&P		Cost Per Labor-Hour	
Crew C-3	Hr.	Daily	Hr.	Daily	Bare Costs	Incl. O&P
1 Rodman Foreman (outside)	$56.65	$453.20	$90.45	$723.60	$50.78	$81.25
4 Rodmen (reinf.)	54.65	1748.80	87.30	2793.60		
1 Equip. Oper. (light)	51.30	410.40	80.95	647.60		
2 Laborers	39.85	637.60	64.70	1035.20		
3 Stressing Equipment		31.20		34.32		
.5 Grouting Equipment		79.20		87.12	1.73	1.90
64 L.H., Daily Totals		$3360.40		$5321.44	$52.51	$83.15
Crew C-4	Hr.	Daily	Hr.	Daily	Bare Costs	Incl. O&P
1 Rodman Foreman (outside)	$56.65	$453.20	$90.45	$723.60	$55.15	$88.09
3 Rodmen (reinf.)	54.65	1311.60	87.30	2095.20		
3 Stressing Equipment		31.20		34.32	.97	1.07
32 L.H., Daily Totals		$1796.00		$2853.12	$56.13	$89.16
Crew C-4A	Hr.	Daily	Hr.	Daily	Bare Costs	Incl. O&P
2 Rodmen (reinf.)	$54.65	$874.40	$87.30	$1396.80	$54.65	$87.30
4 Stressing Equipment		41.60		45.76	2.60	2.86
16 L.H., Daily Totals		$916.00		$1442.56	$57.25	$90.16
Crew C-5	Hr.	Daily	Hr.	Daily	Bare Costs	Incl. O&P
1 Rodman Foreman (outside)	$56.65	$453.20	$90.45	$723.60	$54.28	$86.41
4 Rodmen (reinf.)	54.65	1748.80	87.30	2793.60		
1 Equip. Oper. (crane)	56.10	448.80	88.55	708.40		
1 Equip. Oper. (oiler)	48.60	388.80	76.70	613.60		
1 Hyd. Crane, 25 Ton		590.60		649.66	10.55	11.60
56 L.H., Daily Totals		$3630.20		$5488.86	$64.83	$98.02
Crew C-6	Hr.	Daily	Hr.	Daily	Bare Costs	Incl. O&P
1 Labor Foreman (outside)	$41.85	$334.80	$67.90	$543.20	$41.47	$66.98
4 Laborers	39.85	1275.20	64.70	2070.40		
1 Cement Finisher	47.55	380.40	75.20	601.60		
2 Gas Engine Vibrators		51.20		56.32	1.07	1.17
48 L.H., Daily Totals		$2041.60		$3271.52	$42.53	$68.16
Crew C-7	Hr.	Daily	Hr.	Daily	Bare Costs	Incl. O&P
1 Labor Foreman (outside)	$41.85	$334.80	$67.90	$543.20	$43.44	$69.79
5 Laborers	39.85	1594.00	64.70	2588.00		
1 Cement Finisher	47.55	380.40	75.20	601.60		
1 Equip. Oper. (medium)	53.75	430.00	84.80	678.40		
1 Equip. Oper. (oiler)	48.60	388.80	76.70	613.60		
2 Gas Engine Vibrators		51.20		56.32		
1 Concrete Bucket, 1 C.Y.		24.60		27.06		
1 Hyd. Crane, 55 Ton		993.80		1093.18	14.86	16.34
72 L.H., Daily Totals		$4197.60		$6201.36	$58.30	$86.13
Crew C-8	Hr.	Daily	Hr.	Daily	Bare Costs	Incl. O&P
1 Labor Foreman (outside)	$41.85	$334.80	$67.90	$543.20	$44.32	$71.03
3 Laborers	39.85	956.40	64.70	1552.80		
2 Cement Finishers	47.55	760.80	75.20	1203.20		
1 Equip. Oper. (medium)	53.75	430.00	84.80	678.40		
1 Concrete Pump (Small)		881.60		969.76	15.74	17.32
56 L.H., Daily Totals		$3363.60		$4947.36	$60.06	$88.35
Crew C-8A	Hr.	Daily	Hr.	Daily	Bare Costs	Incl. O&P
1 Labor Foreman (outside)	$41.85	$334.80	$67.90	$543.20	$42.75	$68.73
3 Laborers	39.85	956.40	64.70	1552.80		
2 Cement Finishers	47.55	760.80	75.20	1203.20		
48 L.H., Daily Totals		$2052.00		$3299.20	$42.75	$68.73

Crew No.	Bare Costs Hr.	Daily	Incl. Subs O&P Hr.	Daily	Cost Per Labor-Hour Bare Costs	Incl. O&P
Crew C-8B	**Hr.**	**Daily**	**Hr.**	**Daily**	**Bare Costs**	**Incl. O&P**
1 Labor Foreman (outside)	$41.85	$334.80	$67.90	$543.20	$43.03	$69.36
3 Laborers	39.85	956.40	64.70	1552.80		
1 Equip. Oper. (medium)	53.75	430.00	84.80	678.40		
1 Vibrating Power Screed		75.35		82.89		
1 Roller, Vibratory, 25 Ton		659.80		725.78		
1 Dozer, 200 H.P.		1273.00		1400.30	50.20	55.22
40 L.H., Daily Totals		$3729.35		$4983.36	$93.23	$124.58
Crew C-8C	**Hr.**	**Daily**	**Hr.**	**Daily**	**Bare Costs**	**Incl. O&P**
1 Labor Foreman (outside)	$41.85	$334.80	$67.90	$543.20	$43.78	$70.33
3 Laborers	39.85	956.40	64.70	1552.80		
1 Cement Finisher	47.55	380.40	75.20	601.60		
1 Equip. Oper. (medium)	53.75	430.00	84.80	678.40		
1 Shotcrete Rig, 12 C.Y./hr		246.60		271.26		
1 Air Compressor, 160 cfm		117.20		128.92		
4 -50' Air Hoses, 1"		33.80		37.18		
4 -50' Air Hoses, 2"		103.60		113.96	10.44	11.49
48 L.H., Daily Totals		$2602.80		$3927.32	$54.23	$81.82
Crew C-8D	**Hr.**	**Daily**	**Hr.**	**Daily**	**Bare Costs**	**Incl. O&P**
1 Labor Foreman (outside)	$41.85	$334.80	$67.90	$543.20	$45.14	$72.19
1 Laborer	39.85	318.80	64.70	517.60		
1 Cement Finisher	47.55	380.40	75.20	601.60		
1 Equipment Oper. (light)	51.30	410.40	80.95	647.60		
1 Air Compressor, 250 cfm		167.40		184.14		
2 -50' Air Hoses, 1"		16.90		18.59	5.76	6.34
32 L.H., Daily Totals		$1628.70		$2512.73	$50.90	$78.52
Crew C-8E	**Hr.**	**Daily**	**Hr.**	**Daily**	**Bare Costs**	**Incl. O&P**
1 Labor Foreman (outside)	$41.85	$334.80	$67.90	$543.20	$43.38	$69.69
3 Laborers	39.85	956.40	64.70	1552.80		
1 Cement Finisher	47.55	380.40	75.20	601.60		
1 Equipment Oper. (light)	51.30	410.40	80.95	647.60		
1 Shotcrete Rig, 35 C.Y./hr		270.20		297.22		
1 Air Compressor, 250 cfm		167.40		184.14		
4 -50' Air Hoses, 1"		33.80		37.18		
4 -50' Air Hoses, 2"		103.60		113.96	11.98	13.18
48 L.H., Daily Totals		$2657.00		$3977.70	$55.35	$82.87
Crew C-10	**Hr.**	**Daily**	**Hr.**	**Daily**	**Bare Costs**	**Incl. O&P**
1 Laborer	$39.85	$318.80	$64.70	$517.60	$44.98	$71.70
2 Cement Finishers	47.55	760.80	75.20	1203.20		
24 L.H., Daily Totals		$1079.60		$1720.80	$44.98	$71.70
Crew C-10B	**Hr.**	**Daily**	**Hr.**	**Daily**	**Bare Costs**	**Incl. O&P**
3 Laborers	$39.85	$956.40	$64.70	$1552.80	$42.93	$68.90
2 Cement Finishers	47.55	760.80	75.20	1203.20		
1 Concrete Mixer, 10 C.F.		161.00		177.10		
2 Trowels, 48" Walk-Behind		82.40		90.64	6.09	6.69
40 L.H., Daily Totals		$1960.60		$3023.74	$49.02	$75.59
Crew C-10C	**Hr.**	**Daily**	**Hr.**	**Daily**	**Bare Costs**	**Incl. O&P**
1 Laborer	$39.85	$318.80	$64.70	$517.60	$44.98	$71.70
2 Cement Finishers	47.55	760.80	75.20	1203.20		
1 Trowel, 48" Walk-Behind		41.20		45.32	1.72	1.89
24 L.H., Daily Totals		$1120.80		$1766.12	$46.70	$73.59

Crew No.	Bare Costs Hr.	Daily	Incl. Subs O&P Hr.	Daily	Cost Per Labor-Hour Bare Costs	Incl. O&P
Crew C-10D	**Hr.**	**Daily**	**Hr.**	**Daily**	**Bare Costs**	**Incl. O&P**
1 Laborer	$39.85	$318.80	$64.70	$517.60	$44.98	$71.70
2 Cement Finishers	47.55	760.80	75.20	1203.20		
1 Vibrating Power Screed		75.35		82.89		
1 Trowel, 48" Walk-Behind		41.20		45.32	4.86	5.34
24 L.H., Daily Totals		$1196.15		$1849.01	$49.84	$77.04
Crew C-10E	**Hr.**	**Daily**	**Hr.**	**Daily**	**Bare Costs**	**Incl. O&P**
1 Laborer	$39.85	$318.80	$64.70	$517.60	$44.98	$71.70
2 Cement Finishers	47.55	760.80	75.20	1203.20		
1 Vibrating Power Screed		75.35		82.89		
1 Cement Trowel, 96" Ride-On		166.20		182.82	10.06	11.07
24 L.H., Daily Totals		$1321.15		$1986.51	$55.05	$82.77
Crew C-10F	**Hr.**	**Daily**	**Hr.**	**Daily**	**Bare Costs**	**Incl. O&P**
1 Laborer	$39.85	$318.80	$64.70	$517.60	$44.98	$71.70
2 Cement Finishers	47.55	760.80	75.20	1203.20		
1 Telescoping Boom Lift, to 60'		454.60		500.06	18.94	20.84
24 L.H., Daily Totals		$1534.20		$2220.86	$63.92	$92.54
Crew C-11	**Hr.**	**Daily**	**Hr.**	**Daily**	**Bare Costs**	**Incl. O&P**
1 Struc. Steel Foreman (outside)	$56.65	$453.20	$96.60	$772.80	$54.36	$91.23
6 Struc. Steel Workers	54.65	2623.20	93.20	4473.60		
1 Equip. Oper. (crane)	56.10	448.80	88.55	708.40		
1 Equip. Oper. (oiler)	48.60	388.80	76.70	613.60		
1 Lattice Boom Crane, 150 Ton		2017.00		2218.70	28.01	30.82
72 L.H., Daily Totals		$5931.00		$8787.10	$82.38	$122.04
Crew C-12	**Hr.**	**Daily**	**Hr.**	**Daily**	**Bare Costs**	**Incl. O&P**
1 Carpenter Foreman (outside)	$52.70	$421.60	$85.55	$684.40	$50.13	$80.95
3 Carpenters	50.70	1216.80	82.30	1975.20		
1 Laborer	39.85	318.80	64.70	517.60		
1 Equip. Oper. (crane)	56.10	448.80	88.55	708.40		
1 Hyd. Crane, 12 Ton		496.00		545.60	10.33	11.37
48 L.H., Daily Totals		$2902.00		$4431.20	$60.46	$92.32
Crew C-13	**Hr.**	**Daily**	**Hr.**	**Daily**	**Bare Costs**	**Incl. O&P**
1 Struc. Steel Worker	$54.65	$437.20	$93.20	$745.60	$53.33	$89.57
1 Welder	54.65	437.20	93.20	745.60		
1 Carpenter	50.70	405.60	82.30	658.40		
1 Welder, Gas Engine, 300 amp		98.40		108.24	4.10	4.51
24 L.H., Daily Totals		$1378.40		$2257.84	$57.43	$94.08
Crew C-14	**Hr.**	**Daily**	**Hr.**	**Daily**	**Bare Costs**	**Incl. O&P**
1 Carpenter Foreman (outside)	$52.70	$421.60	$85.55	$684.40	$49.11	$78.93
5 Carpenters	50.70	2028.00	82.30	3292.00		
4 Laborers	39.85	1275.20	64.70	2070.40		
4 Rodmen (reinf.)	54.65	1748.80	87.30	2793.60		
2 Cement Finishers	47.55	760.80	75.20	1203.20		
1 Equip. Oper. (crane)	56.10	448.80	88.55	708.40		
1 Equip. Oper. (oiler)	48.60	388.80	76.70	613.60		
1 Hyd. Crane, 80 Ton		1518.00		1669.80	10.54	11.60
144 L.H., Daily Totals		$8590.00		$13035.40	$59.65	$90.52

For customer support on your Commercial Renovation Costs with RSMeans data, call 800.448.8182.

Crew No.	Bare Costs Hr.	Daily	Incl. Subs O&P Hr.	Daily	Cost Per Labor-Hour Bare Costs	Incl. O&P
Crew C-14A	Hr.	Daily	Hr.	Daily	Bare Costs	Incl. O&P
1 Carpenter Foreman (outside)	$52.70	$421.60	$85.55	$684.40	$50.54	$81.64
16 Carpenters	50.70	6489.60	82.30	10534.40		
4 Rodmen (reinf.)	54.65	1748.80	87.30	2793.60		
2 Laborers	39.85	637.60	64.70	1035.20		
1 Cement Finisher	47.55	380.40	75.20	601.60		
1 Equip. Oper. (medium)	53.75	430.00	84.80	678.40		
1 Gas Engine Vibrator		25.60		28.16		
1 Concrete Pump (Small)		881.60		969.76	4.54	4.99
200 L.H., Daily Totals		$11015.20		$17325.52	$55.08	$86.63
Crew C-14B	Hr.	Daily	Hr.	Daily	Bare Costs	Incl. O&P
1 Carpenter Foreman (outside)	$52.70	$421.60	$85.55	$684.40	$50.42	$81.39
16 Carpenters	50.70	6489.60	82.30	10534.40		
4 Rodmen (reinf.)	54.65	1748.80	87.30	2793.60		
2 Laborers	39.85	637.60	64.70	1035.20		
2 Cement Finishers	47.55	760.80	75.20	1203.20		
1 Equip. Oper. (medium)	53.75	430.00	84.80	678.40		
1 Gas Engine Vibrator		25.60		28.16		
1 Concrete Pump (Small)		881.60		969.76	4.36	4.80
208 L.H., Daily Totals		$11395.60		$17927.12	$54.79	$86.19
Crew C-14C	Hr.	Daily	Hr.	Daily	Bare Costs	Incl. O&P
1 Carpenter Foreman (outside)	$52.70	$421.60	$85.55	$684.40	$48.08	$77.71
6 Carpenters	50.70	2433.60	82.30	3950.40		
2 Rodmen (reinf.)	54.65	874.40	87.30	1396.80		
4 Laborers	39.85	1275.20	64.70	2070.40		
1 Cement Finisher	47.55	380.40	75.20	601.60		
1 Gas Engine Vibrator		25.60		28.16	.23	.25
112 L.H., Daily Totals		$5410.80		$8731.76	$48.31	$77.96
Crew C-14D	Hr.	Daily	Hr.	Daily	Bare Costs	Incl. O&P
1 Carpenter Foreman (outside)	$52.70	$421.60	$85.55	$684.40	$50.22	$81.24
18 Carpenters	50.70	7300.80	82.30	11851.20		
2 Rodmen (reinf.)	54.65	874.40	87.30	1396.80		
2 Laborers	39.85	637.60	64.70	1035.20		
1 Cement Finisher	47.55	380.40	75.20	601.60		
1 Equip. Oper. (medium)	53.75	430.00	84.80	678.40		
1 Gas Engine Vibrator		25.60		28.16		
1 Concrete Pump (Small)		881.60		969.76	4.54	4.99
200 L.H., Daily Totals		$10952.00		$17245.52	$54.76	$86.23
Crew C-14E	Hr.	Daily	Hr.	Daily	Bare Costs	Incl. O&P
1 Carpenter Foreman (outside)	$52.70	$421.60	$85.55	$684.40	$49.07	$78.97
2 Carpenters	50.70	811.20	82.30	1316.80		
4 Rodmen (reinf.)	54.65	1748.80	87.30	2793.60		
3 Laborers	39.85	956.40	64.70	1552.80		
1 Cement Finisher	47.55	380.40	75.20	601.60		
1 Gas Engine Vibrator		25.60		28.16	.29	.32
88 L.H., Daily Totals		$4344.00		$6977.36	$49.36	$79.29
Crew C-14F	Hr.	Daily	Hr.	Daily	Bare Costs	Incl. O&P
1 Labor Foreman (outside)	$41.85	$334.80	$67.90	$543.20	$45.21	$72.06
2 Laborers	39.85	637.60	64.70	1035.20		
6 Cement Finishers	47.55	2282.40	75.20	3609.60		
1 Gas Engine Vibrator		25.60		28.16	.36	.39
72 L.H., Daily Totals		$3280.40		$5216.16	$45.56	$72.45

Crew No.	Bare Costs Hr.	Daily	Incl. Subs O&P Hr.	Daily	Cost Per Labor-Hour Bare Costs	Incl. O&P
Crew C-14G	Hr.	Daily	Hr.	Daily	Bare Costs	Incl. O&P
1 Labor Foreman (outside)	$41.85	$334.80	$67.90	$543.20	$44.54	$71.16
2 Laborers	39.85	637.60	64.70	1035.20		
4 Cement Finishers	47.55	1521.60	75.20	2406.40		
1 Gas Engine Vibrator		25.60		28.16	.46	.50
56 L.H., Daily Totals		$2519.60		$4012.96	$44.99	$71.66
Crew C-14H	Hr.	Daily	Hr.	Daily	Bare Costs	Incl. O&P
1 Carpenter Foreman (outside)	$52.70	$421.60	$85.55	$684.40	$49.36	$79.56
2 Carpenters	50.70	811.20	82.30	1316.80		
1 Rodman (reinf.)	54.65	437.20	87.30	698.40		
1 Laborer	39.85	318.80	64.70	517.60		
1 Cement Finisher	47.55	380.40	75.20	601.60		
1 Gas Engine Vibrator		25.60		28.16	.53	.59
48 L.H., Daily Totals		$2394.80		$3846.96	$49.89	$80.14
Crew C-14L	Hr.	Daily	Hr.	Daily	Bare Costs	Incl. O&P
1 Carpenter Foreman (outside)	$52.70	$421.60	$85.55	$684.40	$46.99	$76.11
6 Carpenters	50.70	2433.60	82.30	3950.40		
4 Laborers	39.85	1275.20	64.70	2070.40		
1 Cement Finisher	47.55	380.40	75.20	601.60		
1 Gas Engine Vibrator		25.60		28.16	.27	.29
96 L.H., Daily Totals		$4536.40		$7334.96	$47.25	$76.41
Crew C-14M	Hr.	Daily	Hr.	Daily	Bare Costs	Incl. O&P
1 Carpenter Foreman (outside)	$52.70	$421.60	$85.55	$684.40	$48.72	$78.36
2 Carpenters	50.70	811.20	82.30	1316.80		
1 Rodman (reinf.)	54.65	437.20	87.30	698.40		
2 Laborers	39.85	637.60	64.70	1035.20		
1 Cement Finisher	47.55	380.40	75.20	601.60		
1 Equip. Oper. (medium)	53.75	430.00	84.80	678.40		
1 Gas Engine Vibrator		25.60		28.16		
1 Concrete Pump (Small)		881.60		969.76	14.18	15.59
64 L.H., Daily Totals		$4025.20		$6012.72	$62.89	$93.95
Crew C-15	Hr.	Daily	Hr.	Daily	Bare Costs	Incl. O&P
1 Carpenter Foreman (outside)	$52.70	$421.60	$85.55	$684.40	$47.04	$75.77
2 Carpenters	50.70	811.20	82.30	1316.80		
3 Laborers	39.85	956.40	64.70	1552.80		
2 Cement Finishers	47.55	760.80	75.20	1203.20		
1 Rodman (reinf.)	54.65	437.20	87.30	698.40		
72 L.H., Daily Totals		$3387.20		$5455.60	$47.04	$75.77
Crew C-16	Hr.	Daily	Hr.	Daily	Bare Costs	Incl. O&P
1 Labor Foreman (outside)	$41.85	$334.80	$67.90	$543.20	$44.32	$71.03
3 Laborers	39.85	956.40	64.70	1552.80		
2 Cement Finishers	47.55	760.80	75.20	1203.20		
1 Equip. Oper. (medium)	53.75	430.00	84.80	678.40		
1 Gunite Pump Rig		321.00		353.10		
2 -50' Air Hoses, 3/4"		14.70		16.17		
2 -50' Air Hoses, 2"		51.80		56.98	6.92	7.61
56 L.H., Daily Totals		$2869.50		$4403.85	$51.24	$78.64
Crew C-16A	Hr.	Daily	Hr.	Daily	Bare Costs	Incl. O&P
1 Laborer	$39.85	$318.80	$64.70	$517.60	$47.17	$74.97
2 Cement Finishers	47.55	760.80	75.20	1203.20		
1 Equip. Oper. (medium)	53.75	430.00	84.80	678.40		
1 Gunite Pump Rig		321.00		353.10		
2 -50' Air Hoses, 3/4"		14.70		16.17		
2 -50' Air Hoses, 2"		51.80		56.98		
1 Telescoping Boom Lift, to 60'		454.60		500.06	26.32	28.95
32 L.H., Daily Totals		$2351.70		$3325.51	$73.49	$103.92

766

Crew No.	Bare Costs		Incl. Subs O&P		Cost Per Labor-Hour	
Crew C-17	Hr.	Daily	Hr.	Daily	Bare Costs	Incl. O&P
2 Skilled Worker Foremen (out)	$54.35	$869.60	$87.55	$1400.80	$52.75	$84.99
8 Skilled Workers	52.35	3350.40	84.35	5398.40		
80 L.H., Daily Totals		$4220.00		$6799.20	$52.75	$84.99
Crew C-17A	Hr.	Daily	Hr.	Daily	Bare Costs	Incl. O&P
2 Skilled Worker Foremen (out)	$54.35	$869.60	$87.55	$1400.80	$52.79	$85.03
8 Skilled Workers	52.35	3350.40	84.35	5398.40		
.125 Equip. Oper. (crane)	56.10	56.10	88.55	88.55		
.125 Hyd. Crane, 80 Ton		189.75		208.72	2.34	2.58
81 L.H., Daily Totals		$4465.85		$7096.48	$55.13	$87.61
Crew C-17B	Hr.	Daily	Hr.	Daily	Bare Costs	Incl. O&P
2 Skilled Worker Foremen (out)	$54.35	$869.60	$87.55	$1400.80	$52.83	$85.08
8 Skilled Workers	52.35	3350.40	84.35	5398.40		
.25 Equip. Oper. (crane)	56.10	112.20	88.55	177.10		
.25 Hyd. Crane, 80 Ton		379.50		417.45		
.25 Trowel, 48" Walk-Behind		10.30		11.33	4.75	5.23
82 L.H., Daily Totals		$4722.00		$7405.08	$57.59	$90.31
Crew C-17C	Hr.	Daily	Hr.	Daily	Bare Costs	Incl. O&P
2 Skilled Worker Foremen (out)	$54.35	$869.60	$87.55	$1400.80	$52.87	$85.12
8 Skilled Workers	52.35	3350.40	84.35	5398.40		
.375 Equip. Oper. (crane)	56.10	168.30	88.55	265.65		
.375 Hyd. Crane, 80 Ton		569.25		626.17	6.86	7.54
83 L.H., Daily Totals		$4957.55		$7691.02	$59.73	$92.66
Crew C-17D	Hr.	Daily	Hr.	Daily	Bare Costs	Incl. O&P
2 Skilled Worker Foremen (out)	$54.35	$869.60	$87.55	$1400.80	$52.91	$85.16
8 Skilled Workers	52.35	3350.40	84.35	5398.40		
.5 Equip. Oper. (crane)	56.10	224.40	88.55	354.20		
.5 Hyd. Crane, 80 Ton		759.00		834.90	9.04	9.94
84 L.H., Daily Totals		$5203.40		$7988.30	$61.95	$95.10
Crew C-17E	Hr.	Daily	Hr.	Daily	Bare Costs	Incl. O&P
2 Skilled Worker Foremen (out)	$54.35	$869.60	$87.55	$1400.80	$52.75	$84.99
8 Skilled Workers	52.35	3350.40	84.35	5398.40		
1 Hyd. Jack with Rods		102.25		112.47	1.28	1.41
80 L.H., Daily Totals		$4322.25		$6911.68	$54.03	$86.40
Crew C-18	Hr.	Daily	Hr.	Daily	Bare Costs	Incl. O&P
.125 Labor Foreman (outside)	$41.85	$41.85	$67.90	$67.90	$40.07	$65.06
1 Laborer	39.85	318.80	64.70	517.60		
1 Concrete Cart, 10 C.F.		57.80		63.58	6.42	7.06
9 L.H., Daily Totals		$418.45		$649.08	$46.49	$72.12
Crew C-19	Hr.	Daily	Hr.	Daily	Bare Costs	Incl. O&P
.125 Labor Foreman (outside)	$41.85	$41.85	$67.90	$67.90	$40.07	$65.06
1 Laborer	39.85	318.80	64.70	517.60		
1 Concrete Cart, 18 C.F.		97.40		107.14	10.82	11.90
9 L.H., Daily Totals		$458.05		$692.64	$50.89	$76.96
Crew C-20	Hr.	Daily	Hr.	Daily	Bare Costs	Incl. O&P
1 Labor Foreman (outside)	$41.85	$334.80	$67.90	$543.20	$42.80	$68.92
5 Laborers	39.85	1594.00	64.70	2588.00		
1 Cement Finisher	47.55	380.40	75.20	601.60		
1 Equip. Oper. (medium)	53.75	430.00	84.80	678.40		
2 Gas Engine Vibrators		51.20		56.32		
1 Concrete Pump (Small)		881.60		969.76	14.57	16.03
64 L.H., Daily Totals		$3672.00		$5437.28	$57.38	$84.96

Crew No.	Bare Costs		Incl. Subs O&P		Cost Per Labor-Hour	
Crew C-21	Hr.	Daily	Hr.	Daily	Bare Costs	Incl. O&P
1 Labor Foreman (outside)	$41.85	$334.80	$67.90	$543.20	$42.80	$68.92
5 Laborers	39.85	1594.00	64.70	2588.00		
1 Cement Finisher	47.55	380.40	75.20	601.60		
1 Equip. Oper. (medium)	53.75	430.00	84.80	678.40		
2 Gas Engine Vibrators		51.20		56.32		
1 Concrete Conveyer		187.20		205.92	3.73	4.10
64 L.H., Daily Totals		$2977.60		$4673.44	$46.52	$73.02
Crew C-22	Hr.	Daily	Hr.	Daily	Bare Costs	Incl. O&P
1 Rodman Foreman (outside)	$56.65	$453.20	$90.45	$723.60	$54.92	$87.68
4 Rodmen (reinf.)	54.65	1748.80	87.30	2793.60		
.125 Equip. Oper. (crane)	56.10	56.10	88.55	88.55		
.125 Equip. Oper. (oiler)	48.60	48.60	76.70	76.70		
.125 Hyd. Crane, 25 Ton		73.83		81.21	1.76	1.93
42 L.H., Daily Totals		$2380.53		$3763.66	$56.68	$89.61
Crew C-23	Hr.	Daily	Hr.	Daily	Bare Costs	Incl. O&P
2 Skilled Worker Foremen (out)	$54.35	$869.60	$87.55	$1400.80	$52.75	$84.64
6 Skilled Workers	52.35	2512.80	84.35	4048.80		
1 Equip. Oper. (crane)	56.10	448.80	88.55	708.40		
1 Equip. Oper. (oiler)	48.60	388.80	76.70	613.60		
1 Lattice Boom Crane, 90 Ton		1687.00		1855.70	21.09	23.20
80 L.H., Daily Totals		$5907.00		$8627.30	$73.84	$107.84
Crew C-24	Hr.	Daily	Hr.	Daily	Bare Costs	Incl. O&P
2 Skilled Worker Foremen (out)	$54.35	$869.60	$87.55	$1400.80	$52.75	$84.64
6 Skilled Workers	52.35	2512.80	84.35	4048.80		
1 Equip. Oper. (crane)	56.10	448.80	88.55	708.40		
1 Equip. Oper. (oiler)	48.60	388.80	76.70	613.60		
1 Lattice Boom Crane, 150 Ton		2017.00		2218.70	25.21	27.73
80 L.H., Daily Totals		$6237.00		$8990.30	$77.96	$112.38
Crew C-25	Hr.	Daily	Hr.	Daily	Bare Costs	Incl. O&P
2 Rodmen (reinf.)	$54.65	$874.40	$87.30	$1396.80	$43.75	$72.92
2 Rodmen Helpers	32.85	525.60	58.55	936.80		
32 L.H., Daily Totals		$1400.00		$2333.60	$43.75	$72.92
Crew C-27	Hr.	Daily	Hr.	Daily	Bare Costs	Incl. O&P
2 Cement Finishers	$47.55	$760.80	$75.20	$1203.20	$47.55	$75.20
1 Concrete Saw		102.40		112.64	6.40	7.04
16 L.H., Daily Totals		$863.20		$1315.84	$53.95	$82.24
Crew C-28	Hr.	Daily	Hr.	Daily	Bare Costs	Incl. O&P
1 Cement Finisher	$47.55	$380.40	$75.20	$601.60	$47.55	$75.20
1 Portable Air Compressor, Gas		17.10		18.81	2.14	2.35
8 L.H., Daily Totals		$397.50		$620.41	$49.69	$77.55
Crew C-29	Hr.	Daily	Hr.	Daily	Bare Costs	Incl. O&P
1 Laborer	$39.85	$318.80	$64.70	$517.60	$39.85	$64.70
1 Pressure Washer		63.60		69.96	7.95	8.74
8 L.H., Daily Totals		$382.40		$587.56	$47.80	$73.44
Crew C-30	Hr.	Daily	Hr.	Daily	Bare Costs	Incl. O&P
1 Laborer	$39.85	$318.80	$64.70	$517.60	$39.85	$64.70
1 Concrete Mixer, 10 C.F.		161.00		177.10	20.13	22.14
8 L.H., Daily Totals		$479.80		$694.70	$59.98	$86.84

Crews - Renovation

Crew No.	Bare Costs		Incl. Subs O&P		Cost Per Labor-Hour	
Crew C-31	Hr.	Daily	Hr.	Daily	Bare Costs	Incl. O&P
1 Cement Finisher	$47.55	$380.40	$75.20	$601.60	$47.55	$75.20
1 Grout Pump		321.00		353.10	40.13	44.14
8 L.H., Daily Totals		$701.40		$954.70	$87.67	$119.34
Crew C-32	Hr.	Daily	Hr.	Daily	Bare Costs	Incl. O&P
1 Cement Finisher	$47.55	$380.40	$75.20	$601.60	$43.70	$69.95
1 Laborer	39.85	318.80	64.70	517.60		
1 Crack Chaser Saw, Gas, 6 H.P.		25.40		27.94		
1 Vacuum Pick-Up System		66.95		73.64	5.77	6.35
16 L.H., Daily Totals		$791.55		$1220.79	$49.47	$76.30
Crew D-1	Hr.	Daily	Hr.	Daily	Bare Costs	Incl. O&P
1 Bricklayer	$50.30	$402.40	$82.05	$656.40	$44.80	$73.08
1 Bricklayer Helper	39.30	314.40	64.10	512.80		
16 L.H., Daily Totals		$716.80		$1169.20	$44.80	$73.08
Crew D-2	Hr.	Daily	Hr.	Daily	Bare Costs	Incl. O&P
3 Bricklayers	$50.30	$1207.20	$82.05	$1969.20	$46.34	$75.55
2 Bricklayer Helpers	39.30	628.80	64.10	1025.60		
.5 Carpenter	50.70	202.80	82.30	329.20		
44 L.H., Daily Totals		$2038.80		$3324.00	$46.34	$75.55
Crew D-3	Hr.	Daily	Hr.	Daily	Bare Costs	Incl. O&P
3 Bricklayers	$50.30	$1207.20	$82.05	$1969.20	$46.13	$75.22
2 Bricklayer Helpers	39.30	628.80	64.10	1025.60		
.25 Carpenter	50.70	101.40	82.30	164.60		
42 L.H., Daily Totals		$1937.40		$3159.40	$46.13	$75.22
Crew D-4	Hr.	Daily	Hr.	Daily	Bare Costs	Incl. O&P
1 Bricklayer	$50.30	$402.40	$82.05	$656.40	$45.05	$72.80
2 Bricklayer Helpers	39.30	628.80	64.10	1025.60		
1 Equip. Oper. (light)	51.30	410.40	80.95	647.60		
1 Grout Pump, 50 C.F./hr.		128.60		141.46	4.02	4.42
32 L.H., Daily Totals		$1570.20		$2471.06	$49.07	$77.22
Crew D-5	Hr.	Daily	Hr.	Daily	Bare Costs	Incl. O&P
1 Bricklayer	50.30	402.40	82.05	656.40	50.30	82.05
8 L.H., Daily Totals		$402.40		$656.40	$50.30	$82.05
Crew D-6	Hr.	Daily	Hr.	Daily	Bare Costs	Incl. O&P
3 Bricklayers	$50.30	$1207.20	$82.05	$1969.20	$45.04	$73.44
3 Bricklayer Helpers	39.30	943.20	64.10	1538.40		
.25 Carpenter	50.70	101.40	82.30	164.60		
50 L.H., Daily Totals		$2251.80		$3672.20	$45.04	$73.44
Crew D-7	Hr.	Daily	Hr.	Daily	Bare Costs	Incl. O&P
1 Tile Layer	$46.85	$374.80	$74.00	$592.00	$42.02	$66.40
1 Tile Layer Helper	37.20	297.60	58.80	470.40		
16 L.H., Daily Totals		$672.40		$1062.40	$42.02	$66.40
Crew D-8	Hr.	Daily	Hr.	Daily	Bare Costs	Incl. O&P
3 Bricklayers	$50.30	$1207.20	$82.05	$1969.20	$45.90	$74.87
2 Bricklayer Helpers	39.30	628.80	64.10	1025.60		
40 L.H., Daily Totals		$1836.00		$2994.80	$45.90	$74.87
Crew D-9	Hr.	Daily	Hr.	Daily	Bare Costs	Incl. O&P
3 Bricklayers	$50.30	$1207.20	$82.05	$1969.20	$44.80	$73.08
3 Bricklayer Helpers	39.30	943.20	64.10	1538.40		
48 L.H., Daily Totals		$2150.40		$3507.60	$44.80	$73.08

Crew No.	Bare Costs		Incl. Subs O&P		Cost Per Labor-Hour	
Crew D-10	Hr.	Daily	Hr.	Daily	Bare Costs	Incl. O&P
1 Bricklayer Foreman (outside)	$52.30	$418.40	$85.30	$682.40	$49.50	$80.00
1 Bricklayer	50.30	402.40	82.05	656.40		
1 Bricklayer Helper	39.30	314.40	64.10	512.80		
1 Equip. Oper. (crane)	56.10	448.80	88.55	708.40		
1 S.P. Crane, 4x4, 12 Ton		363.60		399.96	11.36	12.50
32 L.H., Daily Totals		$1947.60		$2959.96	$60.86	$92.50
Crew D-11	Hr.	Daily	Hr.	Daily	Bare Costs	Incl. O&P
1 Bricklayer Foreman (outside)	$52.30	$418.40	$85.30	$682.40	$47.30	$77.15
1 Bricklayer	50.30	402.40	82.05	656.40		
1 Bricklayer Helper	39.30	314.40	64.10	512.80		
24 L.H., Daily Totals		$1135.20		$1851.60	$47.30	$77.15
Crew D-12	Hr.	Daily	Hr.	Daily	Bare Costs	Incl. O&P
1 Bricklayer Foreman (outside)	$52.30	$418.40	$85.30	$682.40	$45.30	$73.89
1 Bricklayer	50.30	402.40	82.05	656.40		
2 Bricklayer Helpers	39.30	628.80	64.10	1025.60		
32 L.H., Daily Totals		$1449.60		$2364.40	$45.30	$73.89
Crew D-13	Hr.	Daily	Hr.	Daily	Bare Costs	Incl. O&P
1 Bricklayer Foreman (outside)	$52.30	$418.40	$85.30	$682.40	$48.00	$77.73
1 Bricklayer	50.30	402.40	82.05	656.40		
2 Bricklayer Helpers	39.30	628.80	64.10	1025.60		
1 Carpenter	50.70	405.60	82.30	658.40		
1 Equip. Oper. (crane)	56.10	448.80	88.55	708.40		
1 S.P. Crane, 4x4, 12 Ton		363.60		399.96	7.58	8.33
48 L.H., Daily Totals		$2667.60		$4131.16	$55.58	$86.07
Crew D-14	Hr.	Daily	Hr.	Daily	Bare Costs	Incl. O&P
3 Bricklayers	$50.30	$1207.20	$82.05	$1969.20	$47.55	$77.56
1 Bricklayer Helper	39.30	314.40	64.10	512.80		
32 L.H., Daily Totals		$1521.60		$2482.00	$47.55	$77.56
Crew E-1	Hr.	Daily	Hr.	Daily	Bare Costs	Incl. O&P
1 Welder Foreman (outside)	$56.65	$453.20	$96.60	$772.80	$54.20	$90.25
1 Welder	54.65	437.20	93.20	745.60		
1 Equip. Oper. (light)	51.30	410.40	80.95	647.60		
1 Welder, Gas Engine, 300 amp		98.40		108.24	4.10	4.51
24 L.H., Daily Totals		$1399.20		$2274.24	$58.30	$94.76
Crew E-2	Hr.	Daily	Hr.	Daily	Bare Costs	Incl. O&P
1 Struc. Steel Foreman (outside)	$56.65	$453.20	$96.60	$772.80	$54.28	$90.66
4 Struc. Steel Workers	54.65	1748.80	93.20	2982.40		
1 Equip. Oper. (crane)	56.10	448.80	88.55	708.40		
1 Equip. Oper. (oiler)	48.60	388.80	76.70	613.60		
1 Lattice Boom Crane, 90 Ton		1687.00		1855.70	30.13	33.14
56 L.H., Daily Totals		$4726.60		$6932.90	$84.40	$123.80
Crew E-3	Hr.	Daily	Hr.	Daily	Bare Costs	Incl. O&P
1 Struc. Steel Foreman (outside)	$56.65	$453.20	$96.60	$772.80	$55.32	$94.33
1 Struc. Steel Worker	54.65	437.20	93.20	745.60		
1 Welder	54.65	437.20	93.20	745.60		
1 Welder, Gas Engine, 300 amp		98.40		108.24	4.10	4.51
24 L.H., Daily Totals		$1426.00		$2372.24	$59.42	$98.84

Crews - Renovation

Crew No.		Bare Costs	Incl. Subs O&P		Cost Per Labor-Hour	
Crew E-3A	Hr.	Daily	Hr.	Daily	Bare Costs	Incl. O&P
1 Struc. Steel Foreman (outside)	$56.65	$453.20	$96.60	$772.80	$55.32	$94.33
1 Struc. Steel Worker	54.65	437.20	93.20	745.60		
1 Welder	54.65	437.20	93.20	745.60		
1 Welder, Gas Engine, 300 amp		98.40		108.24		
1 Telescoping Boom Lift, to 40'		282.80		311.08	15.88	17.47
24 L.H., Daily Totals		$1708.80		$2683.32	$71.20	$111.81
Crew E-4	Hr.	Daily	Hr.	Daily	Bare Costs	Incl. O&P
1 Struc. Steel Foreman (outside)	$56.65	$453.20	$96.60	$772.80	$55.15	$94.05
3 Struc. Steel Workers	54.65	1311.60	93.20	2236.80		
1 Welder, Gas Engine, 300 amp		98.40		108.24	3.08	3.38
32 L.H., Daily Totals		$1863.20		$3117.84	$58.23	$97.43
Crew E-5	Hr.	Daily	Hr.	Daily	Bare Costs	Incl. O&P
2 Struc. Steel Foremen (outside)	$56.65	$906.40	$96.60	$1545.60	$54.59	$91.77
5 Struc. Steel Workers	54.65	2186.00	93.20	3728.00		
1 Equip. Oper. (crane)	56.10	448.80	88.55	708.40		
1 Welder	54.65	437.20	93.20	745.60		
1 Equip. Oper. (oiler)	48.60	388.80	76.70	613.60		
1 Lattice Boom Crane, 90 Ton		1687.00		1855.70		
1 Welder, Gas Engine, 300 amp		98.40		108.24	22.32	24.55
80 L.H., Daily Totals		$6152.60		$9305.14	$76.91	$116.31
Crew E-6	Hr.	Daily	Hr.	Daily	Bare Costs	Incl. O&P
3 Struc. Steel Foremen (outside)	$56.65	$1359.60	$96.60	$2318.40	$54.53	$91.75
9 Struc. Steel Workers	54.65	3934.80	93.20	6710.40		
1 Equip. Oper. (crane)	56.10	448.80	88.55	708.40		
1 Welder	54.65	437.20	93.20	745.60		
1 Equip. Oper. (oiler)	48.60	388.80	76.70	613.60		
1 Equip. Oper. (light)	51.30	410.40	80.95	647.60		
1 Lattice Boom Crane, 90 Ton		1687.00		1855.70		
1 Welder, Gas Engine, 300 amp		98.40		108.24		
1 Air Compressor, 160 cfm		117.20		128.92		
2 Impact Wrenches		37.20		40.92	15.15	16.67
128 L.H., Daily Totals		$8919.40		$13877.78	$69.68	$108.42
Crew E-7	Hr.	Daily	Hr.	Daily	Bare Costs	Incl. O&P
1 Struc. Steel Foreman (outside)	$56.65	$453.20	$96.60	$772.80	$54.59	$91.77
4 Struc. Steel Workers	54.65	1748.80	93.20	2982.40		
1 Equip. Oper. (crane)	56.10	448.80	88.55	708.40		
1 Equip. Oper. (oiler)	48.60	388.80	76.70	613.60		
1 Welder Foreman (outside)	56.65	453.20	96.60	772.80		
2 Welders	54.65	874.40	93.20	1491.20		
1 Lattice Boom Crane, 90 Ton		1687.00		1855.70		
2 Welders, Gas Engine, 300 amp		196.80		216.48	23.55	25.90
80 L.H., Daily Totals		$6251.00		$9413.38	$78.14	$117.67
Crew E-8	Hr.	Daily	Hr.	Daily	Bare Costs	Incl. O&P
1 Struc. Steel Foreman (outside)	$56.65	$453.20	$96.60	$772.80	$54.35	$91.15
4 Struc. Steel Workers	54.65	1748.80	93.20	2982.40		
1 Welder Foreman (outside)	56.65	453.20	96.60	772.80		
4 Welders	54.65	1748.80	93.20	2982.40		
1 Equip. Oper. (crane)	56.10	448.80	88.55	708.40		
1 Equip. Oper. (oiler)	48.60	388.80	76.70	613.60		
1 Equip. Oper. (light)	51.30	410.40	80.95	647.60		
1 Lattice Boom Crane, 90 Ton		1687.00		1855.70		
4 Welders, Gas Engine, 300 amp		393.60		432.96	20.01	22.01
104 L.H., Daily Totals		$7732.60		$11768.66	$74.35	$113.16

Crew No.		Bare Costs	Incl. Subs O&P		Cost Per Labor-Hour	
Crew E-9	Hr.	Daily	Hr.	Daily	Bare Costs	Incl. O&P
2 Struc. Steel Foremen (outside)	$56.65	$906.40	$96.60	$1545.60	$54.53	$91.75
5 Struc. Steel Workers	54.65	2186.00	93.20	3728.00		
1 Welder Foreman (outside)	56.65	453.20	96.60	772.80		
5 Welders	54.65	2186.00	93.20	3728.00		
1 Equip. Oper. (crane)	56.10	448.80	88.55	708.40		
1 Equip. Oper. (oiler)	48.60	388.80	76.70	613.60		
1 Equip. Oper. (light)	51.30	410.40	80.95	647.60		
1 Lattice Boom Crane, 90 Ton		1687.00		1855.70		
5 Welders, Gas Engine, 300 amp		492.00		541.20	17.02	18.73
128 L.H., Daily Totals		$9158.60		$14140.90	$71.55	$110.48
Crew E-10	Hr.	Daily	Hr.	Daily	Bare Costs	Incl. O&P
1 Welder Foreman (outside)	$56.65	$453.20	$96.60	$772.80	$55.65	$94.90
1 Welder	54.65	437.20	93.20	745.60		
1 Welder, Gas Engine, 300 amp		98.40		108.24		
1 Flatbed Truck, Gas, 3 Ton		238.00		261.80	21.02	23.13
16 L.H., Daily Totals		$1226.80		$1888.44	$76.67	$118.03
Crew E-11	Hr.	Daily	Hr.	Daily	Bare Costs	Incl. O&P
2 Painters, Struc. Steel	$43.50	$696.00	$76.80	$1228.80	$44.54	$74.81
1 Building Laborer	39.85	318.80	64.70	517.60		
1 Equip. Oper. (light)	51.30	410.40	80.95	647.60		
1 Air Compressor, 250 cfm		167.40		184.14		
1 Sandblaster, Portable, 3 C.F.		21.00		23.10		
1 Set Sand Blasting Accessories		14.70		16.17	6.35	6.98
32 L.H., Daily Totals		$1628.30		$2617.41	$50.88	$81.79
Crew E-11A	Hr.	Daily	Hr.	Daily	Bare Costs	Incl. O&P
2 Painters, Struc. Steel	$43.50	$696.00	$76.80	$1228.80	$44.54	$74.81
1 Building Laborer	39.85	318.80	64.70	517.60		
1 Equip. Oper. (light)	51.30	410.40	80.95	647.60		
1 Air Compressor, 250 cfm		167.40		184.14		
1 Sandblaster, Portable, 3 C.F.		21.00		23.10		
1 Set Sand Blasting Accessories		14.70		16.17		
1 Telescoping Boom Lift, to 60'		454.60		500.06	20.55	22.61
32 L.H., Daily Totals		$2082.90		$3117.47	$65.09	$97.42
Crew E-11B	Hr.	Daily	Hr.	Daily	Bare Costs	Incl. O&P
2 Painters, Struc. Steel	$43.50	$696.00	$76.80	$1228.80	$42.28	$72.77
1 Building Laborer	39.85	318.80	64.70	517.60		
2 Paint Sprayer, 8 C.F.M.		90.20		99.22		
1 Telescoping Boom Lift, to 60'		454.60		500.06	22.70	24.97
24 L.H., Daily Totals		$1559.60		$2345.68	$64.98	$97.74
Crew E-12	Hr.	Daily	Hr.	Daily	Bare Costs	Incl. O&P
1 Welder Foreman (outside)	$56.65	$453.20	$96.60	$772.80	$53.98	$88.78
1 Equip. Oper. (light)	51.30	410.40	80.95	647.60		
1 Welder, Gas Engine, 300 amp		98.40		108.24	6.15	6.76
16 L.H., Daily Totals		$962.00		$1528.64	$60.13	$95.54
Crew E-13	Hr.	Daily	Hr.	Daily	Bare Costs	Incl. O&P
1 Welder Foreman (outside)	$56.65	$453.20	$96.60	$772.80	$54.87	$91.38
.5 Equip. Oper. (light)	51.30	205.20	80.95	323.80		
1 Welder, Gas Engine, 300 amp		98.40		108.24	8.20	9.02
12 L.H., Daily Totals		$756.80		$1204.84	$63.07	$100.40
Crew E-14	Hr.	Daily	Hr.	Daily	Bare Costs	Incl. O&P
1 Welder Foreman (outside)	$56.65	$453.20	$96.60	$772.80	$56.65	$96.60
1 Welder, Gas Engine, 300 amp		98.40		108.24	12.30	13.53
8 L.H., Daily Totals		$551.60		$881.04	$68.95	$110.13

Crew No.	Bare Costs Hr.	Daily	Incl. Subs O&P Hr.	Daily	Cost Per Labor-Hour Bare Costs	Incl. O&P
Crew E-16	Hr.	Daily	Hr.	Daily	Bare Costs	Incl. O&P
1 Welder Foreman (outside)	$56.65	$453.20	$96.60	$772.80	$55.65	$94.90
1 Welder	54.65	437.20	93.20	745.60		
1 Welder, Gas Engine, 300 amp		98.40		108.24	6.15	6.76
16 L.H., Daily Totals		$988.80		$1626.64	$61.80	$101.67

Crew E-17	Hr.	Daily	Hr.	Daily	Bare Costs	Incl. O&P
1 Struc. Steel Foreman (outside)	$56.65	$453.20	$96.60	$772.80	$55.65	$94.90
1 Structural Steel Worker	54.65	437.20	93.20	745.60		
16 L.H., Daily Totals		$890.40		$1518.40	$55.65	$94.90

Crew E-18	Hr.	Daily	Hr.	Daily	Bare Costs	Incl. O&P
1 Struc. Steel Foreman (outside)	$56.65	$453.20	$96.60	$772.80	$54.87	$92.20
3 Structural Steel Workers	54.65	1311.60	93.20	2236.80		
1 Equipment Operator (med.)	53.75	430.00	84.80	678.40		
1 Lattice Boom Crane, 20 Ton		1083.00		1191.30	27.07	29.78
40 L.H., Daily Totals		$3277.80		$4879.30	$81.94	$121.98

Crew E-19	Hr.	Daily	Hr.	Daily	Bare Costs	Incl. O&P
1 Struc. Steel Foreman (outside)	$56.65	$453.20	$96.60	$772.80	$54.20	$90.25
1 Structural Steel Worker	54.65	437.20	93.20	745.60		
1 Equip. Oper. (light)	51.30	410.40	80.95	647.60		
1 Lattice Boom Crane, 20 Ton		1083.00		1191.30	45.13	49.64
24 L.H., Daily Totals		$2383.80		$3357.30	$99.33	$139.89

Crew E-20	Hr.	Daily	Hr.	Daily	Bare Costs	Incl. O&P
1 Struc. Steel Foreman (outside)	$56.65	$453.20	$96.60	$772.80	$54.33	$90.98
5 Structural Steel Workers	54.65	2186.00	93.20	3728.00		
1 Equip. Oper. (crane)	56.10	448.80	88.55	708.40		
1 Equip. Oper. (oiler)	48.60	388.80	76.70	613.60		
1 Lattice Boom Crane, 40 Ton		1321.00		1453.10	20.64	22.70
64 L.H., Daily Totals		$4797.80		$7275.90	$74.97	$113.69

Crew E-22	Hr.	Daily	Hr.	Daily	Bare Costs	Incl. O&P
1 Skilled Worker Foreman (out)	$54.35	$434.80	$87.55	$700.40	$53.02	$85.42
2 Skilled Workers	52.35	837.60	84.35	1349.60		
24 L.H., Daily Totals		$1272.40		$2050.00	$53.02	$85.42

Crew E-24	Hr.	Daily	Hr.	Daily	Bare Costs	Incl. O&P
3 Structural Steel Workers	$54.65	$1311.60	$93.20	$2236.80	$54.42	$91.10
1 Equipment Operator (med.)	53.75	430.00	84.80	678.40		
1 Hyd. Crane, 25 Ton		590.60		649.66	18.46	20.30
32 L.H., Daily Totals		$2332.20		$3564.86	$72.88	$111.40

Crew E-25	Hr.	Daily	Hr.	Daily	Bare Costs	Incl. O&P
1 Welder Foreman (outside)	$56.65	$453.20	$96.60	$772.80	$56.65	$96.60
1 Cutting Torch		12.60		13.86	1.58	1.73
8 L.H., Daily Totals		$465.80		$786.66	$58.23	$98.33

Crew E-26	Hr.	Daily	Hr.	Daily	Bare Costs	Incl. O&P
1 Struc. Steel Foreman (outside)	$56.65	$453.20	$96.60	$772.80	$56.01	$94.19
1 Struc. Steel Worker	54.65	437.20	93.20	745.60		
1 Welder	54.65	437.20	93.20	745.60		
.25 Electrician	58.20	116.40	89.90	179.80		
.25 Plumber	62.15	124.30	96.70	193.40		
1 Welder, Gas Engine, 300 amp		98.40		108.24	3.51	3.87
28 L.H., Daily Totals		$1666.70		$2745.44	$59.52	$98.05

Crew E-27	Hr.	Daily	Hr.	Daily	Bare Costs	Incl. O&P
1 Struc. Steel Foreman (outside)	$56.65	$453.20	$96.60	$772.80	$54.33	$90.98
5 Struc. Steel Workers	54.65	2186.00	93.20	3728.00		
1 Equip. Oper. (crane)	56.10	448.80	88.55	708.40		
1 Equip. Oper. (oiler)	48.60	388.80	76.70	613.60		
1 Hyd. Crane, 12 Ton		496.00		545.60		
1 Hyd. Crane, 80 Ton		1518.00		1669.80	31.47	34.62
64 L.H., Daily Totals		$5490.80		$8038.20	$85.79	$125.60

Crew F-3	Hr.	Daily	Hr.	Daily	Bare Costs	Incl. O&P
4 Carpenters	$50.70	$1622.40	$82.30	$2633.60	$51.78	$83.55
1 Equip. Oper. (crane)	56.10	448.80	88.55	708.40		
1 Hyd. Crane, 12 Ton		496.00		545.60	12.40	13.64
40 L.H., Daily Totals		$2567.20		$3887.60	$64.18	$97.19

Crew F-4	Hr.	Daily	Hr.	Daily	Bare Costs	Incl. O&P
4 Carpenters	$50.70	$1622.40	$82.30	$2633.60	$51.25	$82.41
1 Equip. Oper. (crane)	56.10	448.80	88.55	708.40		
1 Equip. Oper. (oiler)	48.60	388.80	76.70	613.60		
1 Hyd. Crane, 55 Ton		993.80		1093.18	20.70	22.77
48 L.H., Daily Totals		$3453.80		$5048.78	$71.95	$105.18

Crew F-5	Hr.	Daily	Hr.	Daily	Bare Costs	Incl. O&P
1 Carpenter Foreman (outside)	$52.70	$421.60	$85.55	$684.40	$51.20	$83.11
3 Carpenters	50.70	1216.80	82.30	1975.20		
32 L.H., Daily Totals		$1638.40		$2659.60	$51.20	$83.11

Crew F-6	Hr.	Daily	Hr.	Daily	Bare Costs	Incl. O&P
2 Carpenters	$50.70	$811.20	$82.30	$1316.80	$47.44	$76.51
2 Building Laborers	39.85	637.60	64.70	1035.20		
1 Equip. Oper. (crane)	56.10	448.80	88.55	708.40		
1 Hyd. Crane, 12 Ton		496.00		545.60	12.40	13.64
40 L.H., Daily Totals		$2393.60		$3606.00	$59.84	$90.15

Crew F-7	Hr.	Daily	Hr.	Daily	Bare Costs	Incl. O&P
2 Carpenters	$50.70	$811.20	$82.30	$1316.80	$45.27	$73.50
2 Building Laborers	39.85	637.60	64.70	1035.20		
32 L.H., Daily Totals		$1448.80		$2352.00	$45.27	$73.50

Crew G-1	Hr.	Daily	Hr.	Daily	Bare Costs	Incl. O&P
1 Roofer Foreman (outside)	$45.95	$367.60	$81.95	$655.60	$41.06	$73.21
4 Roofers Composition	43.95	1406.40	78.35	2507.20		
2 Roofer Helpers	32.85	525.60	58.55	936.80		
1 Application Equipment		181.00		199.10		
1 Tar Kettle/Pot		167.20		183.92		
1 Crew Truck		166.00		182.60	9.18	10.10
56 L.H., Daily Totals		$2813.80		$4665.22	$50.25	$83.31

Crew G-2	Hr.	Daily	Hr.	Daily	Bare Costs	Incl. O&P
1 Plasterer	$46.45	$371.60	$74.30	$594.40	$42.02	$67.53
1 Plasterer Helper	39.75	318.00	63.60	508.80		
1 Building Laborer	39.85	318.80	64.70	517.60		
1 Grout Pump, 50 C.F./hr.		128.60		141.46	5.36	5.89
24 L.H., Daily Totals		$1137.00		$1762.26	$47.38	$73.43

For customer support on your Commercial Renovation Costs with RSMeans data, call 800.448.8182.

Crew No.	Bare Costs		Incl. Subs O&P		Cost Per Labor-Hour	

Crew G-2A

Crew G-2A	Hr.	Daily	Hr.	Daily	Bare Costs	Incl. O&P
1 Roofer Composition	$43.95	$351.60	$78.35	$626.80	$38.88	$67.20
1 Roofer Helper	32.85	262.80	58.55	468.40		
1 Building Laborer	39.85	318.80	64.70	517.60		
1 Foam Spray Rig, Trailer-Mtd.		499.25		549.17		
1 Pickup Truck, 3/4 Ton		115.20		126.72	25.60	28.16
24 L.H., Daily Totals		$1547.65		$2288.70	$64.49	$95.36

Crew G-3	Hr.	Daily	Hr.	Daily	Bare Costs	Incl. O&P
2 Sheet Metal Workers	$59.80	$956.80	$94.25	$1508.00	$49.83	$79.47
2 Building Laborers	39.85	637.60	64.70	1035.20		
32 L.H., Daily Totals		$1594.40		$2543.20	$49.83	$79.47

Crew G-4	Hr.	Daily	Hr.	Daily	Bare Costs	Incl. O&P
1 Labor Foreman (outside)	$41.85	$334.80	$67.90	$543.20	$40.52	$65.77
2 Building Laborers	39.85	637.60	64.70	1035.20		
1 Flatbed Truck, Gas, 1.5 Ton		188.40		207.24		
1 Air Compressor, 160 cfm		117.20		128.92	12.73	14.01
24 L.H., Daily Totals		$1278.00		$1914.56	$53.25	$79.77

Crew G-5	Hr.	Daily	Hr.	Daily	Bare Costs	Incl. O&P
1 Roofer Foreman (outside)	$45.95	$367.60	$81.95	$655.60	$39.91	$71.15
2 Roofers Composition	43.95	703.20	78.35	1253.60		
2 Roofer Helpers	32.85	525.60	58.55	936.80		
1 Application Equipment		181.00		199.10	4.53	4.98
40 L.H., Daily Totals		$1777.40		$3045.10	$44.44	$76.13

Crew G-6A	Hr.	Daily	Hr.	Daily	Bare Costs	Incl. O&P
2 Roofers Composition	$43.95	$703.20	$78.35	$1253.60	$43.95	$78.35
1 Small Compressor, Electric		15.10		16.61		
2 Pneumatic Nailers		46.10		50.71	3.83	4.21
16 L.H., Daily Totals		$764.40		$1320.92	$47.77	$82.56

Crew G 7	Hr.	Daily	Hr	Daily	Bare Costs	Incl. O&P
1 Carpenter	$50.70	$405.60	$82.30	$658.40	$50.70	$82.30
1 Small Compressor, Electric		15.10		16.61		
1 Pneumatic Nailer		23.05		25.36	4.77	5.25
8 L.H., Daily Totals		$443.75		$700.37	$55.47	$87.55

Crew H-1	Hr.	Daily	Hr.	Daily	Bare Costs	Incl. O&P
2 Glaziers	$48.50	$776.00	$78.30	$1252.80	$51.58	$85.75
2 Struc. Steel Workers	54.65	874.40	93.20	1491.20		
32 L.H., Daily Totals		$1650.40		$2744.00	$51.58	$85.75

Crew H-2	Hr.	Daily	Hr.	Daily	Bare Costs	Incl. O&P
2 Glaziers	$48.50	$776.00	$78.30	$1252.80	$45.62	$73.77
1 Building Laborer	39.85	318.80	64.70	517.60		
24 L.H., Daily Totals		$1094.80		$1770.40	$45.62	$73.77

Crew H-3	Hr.	Daily	Hr.	Daily	Bare Costs	Incl. O&P
1 Glazier	$48.50	$388.00	$78.30	$626.40	$43.15	$70.20
1 Helper	37.80	302.40	62.10	496.80		
16 L.H., Daily Totals		$690.40		$1123.20	$43.15	$70.20

Crew H-4	Hr.	Daily	Hr.	Daily	Bare Costs	Incl. O&P
1 Carpenter	$50.70	$405.60	$82.30	$658.40	$47.04	$75.74
1 Carpenter Helper	37.80	302.40	62.10	496.80		
.5 Electrician	58.20	232.80	89.90	359.60		
20 L.H., Daily Totals		$940.80		$1514.80	$47.04	$75.74

Crew J-1	Hr.	Daily	Hr.	Daily	Bare Costs	Incl. O&P
3 Plasterers	$46.45	$1114.80	$74.30	$1783.20	$43.77	$70.02
2 Plasterer Helpers	39.75	636.00	63.60	1017.60		
1 Mixing Machine, 6 C.F.		133.20		146.52	3.33	3.66
40 L.H., Daily Totals		$1884.00		$2947.32	$47.10	$73.68

Crew J-2	Hr.	Daily	Hr.	Daily	Bare Costs	Incl. O&P
3 Plasterers	$46.45	$1114.80	$74.30	$1783.20	$44.67	$71.28
2 Plasterer Helpers	39.75	636.00	63.60	1017.60		
1 Lather	49.15	393.20	77.55	620.40		
1 Mixing Machine, 6 C.F.		133.20		146.52	2.77	3.05
48 L.H., Daily Totals		$2277.20		$3567.72	$47.44	$74.33

Crew J-3	Hr.	Daily	Hr.	Daily	Bare Costs	Incl. O&P
1 Terrazzo Worker	$47.00	$376.00	$74.25	$594.00	$43.05	$68.03
1 Terrazzo Helper	39.10	312.80	61.80	494.40		
1 Floor Grinder, 22" Path		133.60		146.96		
1 Terrazzo Mixer		177.80		195.58	19.46	21.41
16 L.H., Daily Totals		$1000.20		$1430.94	$62.51	$89.43

Crew J-4	Hr.	Daily	Hr.	Daily	Bare Costs	Incl. O&P
2 Cement Finishers	$47.55	$760.80	$75.20	$1203.20	$44.98	$71.70
1 Laborer	39.85	318.80	64.70	517.60		
1 Floor Grinder, 22" Path		133.60		146.96		
1 Floor Edger, 7" Path		42.95		47.24		
1 Vacuum Pick-Up System		66.95		73.64	10.15	11.16
24 L.H., Daily Totals		$1323.10		$1988.65	$55.13	$82.86

Crew J-4A	Hr.	Daily	Hr.	Daily	Bare Costs	Incl. O&P
2 Cement Finishers	$47.55	$760.80	$75.20	$1203.20	$43.70	$69.95
2 Laborers	39.85	637.60	64.70	1035.20		
1 Floor Grinder, 22" Path		133.60		146.96		
1 Floor Edger, 7" Path		42.95		47.24		
1 Vacuum Pick-Up System		66.95		73.64		
1 Floor Auto Scrubber		254.30		279.73	15.50	17.11
32 L.H., Daily Totals		$1896.20		$2785.98	$59.26	$87.06

Crew J-4B	Hr.	Daily	Hr.	Daily	Bare Costs	Incl. O&P
1 Laborer	$39.85	$318.80	$64.70	$517.60	$39.85	$64.70
1 Floor Auto Scrubber		254.30		279.73	31.79	34.97
8 L.H., Daily Totals		$573.10		$797.33	$71.64	$99.67

Crew J-6	Hr.	Daily	Hr.	Daily	Bare Costs	Incl. O&P
2 Painters	$42.55	$680.80	$68.30	$1092.80	$44.06	$70.56
1 Building Laborer	39.85	318.80	64.70	517.60		
1 Equip. Oper. (light)	51.30	410.40	80.95	647.60		
1 Air Compressor, 250 cfm		167.40		184.14		
1 Sandblaster, Portable, 3 C.F.		21.00		23.10		
1 Set Sand Blasting Accessories		14.70		16.17	6.35	6.98
32 L.H., Daily Totals		$1613.10		$2481.41	$50.41	$77.54

Crew J-7	Hr.	Daily	Hr.	Daily	Bare Costs	Incl. O&P
2 Painters	$42.55	$680.80	$68.30	$1092.80	$42.55	$68.30
1 Floor Belt Sander		16.75		18.43		
1 Floor Sanding Edger		13.15		14.47	1.87	2.06
16 L.H., Daily Totals		$710.70		$1125.69	$44.42	$70.36

Crew No.	Bare Costs		Incl. Subs O&P		Cost Per Labor-Hour	
Crew K-1	Hr.	Daily	Hr.	Daily	Bare Costs	Incl. O&P
1 Carpenter	$50.70	$405.60	$82.30	$658.40	$47.60	$76.88
1 Truck Driver (light)	44.50	356.00	71.45	571.60		
1 Flatbed Truck, Gas, 3 Ton		238.00		261.80	14.88	16.36
16 L.H., Daily Totals		$999.60		$1491.80	$62.48	$93.24
Crew K-2	Hr.	Daily	Hr.	Daily	Bare Costs	Incl. O&P
1 Struc. Steel Foreman (outside)	$56.65	$453.20	$96.60	$772.80	$51.93	$87.08
1 Struc. Steel Worker	54.65	437.20	93.20	745.60		
1 Truck Driver (light)	44.50	356.00	71.45	571.60		
1 Flatbed Truck, Gas, 3 Ton		238.00		261.80	9.92	10.91
24 L.H., Daily Totals		$1484.40		$2351.80	$61.85	$97.99
Crew L-1	Hr.	Daily	Hr.	Daily	Bare Costs	Incl. O&P
1 Electrician	$58.20	$465.60	$89.90	$719.20	$60.17	$93.30
1 Plumber	62.15	497.20	96.70	773.60		
16 L.H., Daily Totals		$962.80		$1492.80	$60.17	$93.30
Crew L-2	Hr.	Daily	Hr.	Daily	Bare Costs	Incl. O&P
1 Carpenter	$50.70	$405.60	$82.30	$658.40	$44.25	$72.20
1 Carpenter Helper	37.80	302.40	62.10	496.80		
16 L.H., Daily Totals		$708.00		$1155.20	$44.25	$72.20
Crew L-3	Hr.	Daily	Hr.	Daily	Bare Costs	Incl. O&P
1 Carpenter	$50.70	$405.60	$82.30	$658.40	$54.85	$87.19
.5 Electrician	58.20	232.80	89.90	359.60		
.5 Sheet Metal Worker	59.80	239.20	94.25	377.00		
16 L.H., Daily Totals		$877.60		$1395.00	$54.85	$87.19
Crew L-3A	Hr.	Daily	Hr.	Daily	Bare Costs	Incl. O&P
1 Carpenter Foreman (outside)	$52.70	$421.60	$85.55	$684.40	$55.07	$88.45
.5 Sheet Metal Worker	59.80	239.20	94.25	377.00		
12 L.H., Daily Totals		$660.80		$1061.40	$55.07	$88.45
Crew L-4	Hr.	Daily	Hr.	Daily	Bare Costs	Incl. O&P
2 Skilled Workers	$52.35	$837.60	$84.35	$1349.60	$47.50	$76.93
1 Helper	37.80	302.40	62.10	496.80		
24 L.H., Daily Totals		$1140.00		$1846.40	$47.50	$76.93
Crew L-5	Hr.	Daily	Hr.	Daily	Bare Costs	Incl. O&P
1 Struc. Steel Foreman (outside)	$56.65	$453.20	$96.60	$772.80	$55.14	$93.02
5 Struc. Steel Workers	54.65	2186.00	93.20	3728.00		
1 Equip. Oper. (crane)	56.10	448.80	88.55	708.40		
1 Hyd. Crane, 25 Ton		590.60		649.66	10.55	11.60
56 L.H., Daily Totals		$3678.60		$5858.86	$65.69	$104.62
Crew L-5A	Hr.	Daily	Hr.	Daily	Bare Costs	Incl. O&P
1 Struc. Steel Foreman (outside)	$56.65	$453.20	$96.60	$772.80	$55.51	$92.89
2 Structural Steel Workers	54.65	874.40	93.20	1491.20		
1 Equip. Oper. (crane)	56.10	448.80	88.55	708.40		
1 S.P. Crane, 4x4, 25 Ton		668.40		735.24	20.89	22.98
32 L.H., Daily Totals		$2444.80		$3707.64	$76.40	$115.86

Crew No.	Bare Costs		Incl. Subs O&P		Cost Per Labor-Hour	
Crew L-5B	Hr.	Daily	Hr.	Daily	Bare Costs	Incl. O&P
1 Struc. Steel Foreman (outside)	$56.65	$453.20	$96.60	$772.80	$57.01	$91.57
2 Structural Steel Workers	54.65	874.40	93.20	1491.20		
2 Electricians	58.20	931.20	89.90	1438.40		
2 Steamfitters/Pipefitters	63.00	1008.00	98.05	1568.80		
1 Equip. Oper. (crane)	56.10	448.80	88.55	708.40		
1 Equip. Oper. (oiler)	48.60	388.80	76.70	613.60		
1 Hyd. Crane, 80 Ton		1518.00		1669.80	21.08	23.19
72 L.H., Daily Totals		$5622.40		$8263.00	$78.09	$114.76
Crew L-6	Hr.	Daily	Hr.	Daily	Bare Costs	Incl. O&P
1 Plumber	$62.15	$497.20	$96.70	$773.60	$60.83	$94.43
.5 Electrician	58.20	232.80	89.90	359.60		
12 L.H., Daily Totals		$730.00		$1133.20	$60.83	$94.43
Crew L-7	Hr.	Daily	Hr.	Daily	Bare Costs	Incl. O&P
2 Carpenters	$50.70	$811.20	$82.30	$1316.80	$48.67	$78.36
1 Building Laborer	39.85	318.80	64.70	517.60		
.5 Electrician	58.20	232.80	89.90	359.60		
28 L.H., Daily Totals		$1362.80		$2194.00	$48.67	$78.36
Crew L-8	Hr.	Daily	Hr.	Daily	Bare Costs	Incl. O&P
2 Carpenters	$50.70	$811.20	$82.30	$1316.80	$52.99	$85.18
.5 Plumber	62.15	248.60	96.70	386.80		
20 L.H., Daily Totals		$1059.80		$1703.60	$52.99	$85.18
Crew L-9	Hr.	Daily	Hr.	Daily	Bare Costs	Incl. O&P
1 Labor Foreman (inside)	$40.35	$322.80	$65.50	$524.00	$45.29	$74.01
2 Building Laborers	39.85	637.60	64.70	1035.20		
1 Struc. Steel Worker	54.65	437.20	93.20	745.60		
.5 Electrician	58.20	232.80	89.90	359.60		
36 L.H., Daily Totals		$1630.40		$2664.40	$45.29	$74.01
Crew L-10	Hr.	Daily	Hr.	Daily	Bare Costs	Incl. O&P
1 Struc. Steel Foreman (outside)	$56.65	$453.20	$96.60	$772.80	$55.80	$92.78
1 Structural Steel Worker	54.65	437.20	93.20	745.60		
1 Equip. Oper. (crane)	56.10	448.80	88.55	708.40		
1 Hyd. Crane, 12 Ton		496.00		545.60	20.67	22.73
24 L.H., Daily Totals		$1835.20		$2772.40	$76.47	$115.52
Crew L-11	Hr.	Daily	Hr.	Daily	Bare Costs	Incl. O&P
2 Wreckers	$39.85	$637.60	$66.65	$1066.40	$46.77	$75.70
1 Equip. Oper. (crane)	56.10	448.80	88.55	708.40		
1 Equip. Oper. (light)	51.30	410.40	80.95	647.60		
1 Hyd. Excavator, 2.5 C.Y.		1411.00		1552.10		
1 Loader, Skid Steer, 78 H.P.		364.80		401.28	55.49	61.04
32 L.H., Daily Totals		$3272.60		$4375.78	$102.27	$136.74
Crew M-1	Hr.	Daily	Hr.	Daily	Bare Costs	Incl. O&P
3 Elevator Constructors	$82.20	$1972.80	$126.60	$3038.40	$78.09	$120.26
1 Elevator Apprentice	65.75	526.00	101.25	810.00		
5 Hand Tools		50.00		55.00	1.56	1.72
32 L.H., Daily Totals		$2548.80		$3903.40	$79.65	$121.98

For customer support on your Commercial Renovation Costs with RSMeans data, call 800.448.8182.

Crew No.	Bare Costs Hr.	Daily	Incl. Subs O&P Hr.	Daily	Cost Per Labor-Hour Bare Costs	Incl. O&P
Crew M-3	Hr.	Daily	Hr.	Daily	Bare Costs	Incl. O&P
1 Electrician Foreman (outside)	$60.20	$481.60	$93.00	$744.00	$61.51	$95.71
1 Common Laborer	39.85	318.80	64.70	517.60		
.25 Equipment Operator (med.)	53.75	107.50	84.80	169.60		
1 Elevator Constructor	82.20	657.60	126.60	1012.80		
1 Elevator Apprentice	65.75	526.00	101.25	810.00		
.25 S.P. Crane, 4x4, 20 Ton		142.80		157.08	4.20	4.62
34 L.H., Daily Totals		$2234.30		$3411.08	$65.71	$100.33

Crew M-4	Hr.	Daily	Hr.	Daily	Bare Costs	Incl. O&P
1 Electrician Foreman (outside)	$60.20	$481.60	$93.00	$744.00	$60.93	$94.86
1 Common Laborer	39.85	318.80	64.70	517.60		
.25 Equipment Operator, Crane	56.10	112.20	88.55	177.10		
.25 Equip. Oper. (oiler)	48.60	97.20	76.70	153.40		
1 Elevator Constructor	82.20	657.60	126.60	1012.80		
1 Elevator Apprentice	65.75	526.00	101.25	810.00		
.25 S.P. Crane, 4x4, 40 Ton		183.70		202.07	5.10	5.61
36 L.H., Daily Totals		$2377.10		$3616.97	$66.03	$100.47

Crew Q-1	Hr.	Daily	Hr.	Daily	Bare Costs	Incl. O&P
1 Plumber	$62.15	$497.20	$96.70	$773.60	$55.92	$87.03
1 Plumber Apprentice	49.70	397.60	77.35	618.80		
16 L.H., Daily Totals		$894.80		$1392.40	$55.92	$87.03

Crew Q-1A	Hr.	Daily	Hr.	Daily	Bare Costs	Incl. O&P
.25 Plumber Foreman (outside)	$64.15	$128.30	$99.80	$199.60	$62.55	$97.32
1 Plumber	62.15	497.20	96.70	773.60		
10 L.H., Daily Totals		$625.50		$973.20	$62.55	$97.32

Crew Q-1C	Hr.	Daily	Hr.	Daily	Bare Costs	Incl. O&P
1 Plumber	$62.15	$497.20	$96.70	$773.60	$55.20	$86.28
1 Plumber Apprentice	49.70	397.60	77.35	618.80		
1 Equip. Oper. (medium)	53.75	430.00	84.80	678.40		
1 Trencher, Chain Type, 8' D		2048.00		2252.80	85.33	93.87
24 L.H., Daily Totals		$3372.80		$4323.60	$140.53	$180.15

Crew Q-2	Hr.	Daily	Hr.	Daily	Bare Costs	Incl. O&P
2 Plumbers	$62.15	$994.40	$96.70	$1547.20	$58.00	$90.25
1 Plumber Apprentice	49.70	397.60	77.35	618.80		
24 L.H., Daily Totals		$1392.00		$2166.00	$58.00	$90.25

Crew Q-3	Hr.	Daily	Hr.	Daily	Bare Costs	Incl. O&P
1 Plumber Foreman (inside)	$62.65	$501.20	$97.50	$780.00	$59.16	$92.06
2 Plumbers	62.15	994.40	96.70	1547.20		
1 Plumber Apprentice	49.70	397.60	77.35	618.80		
32 L.H., Daily Totals		$1893.20		$2946.00	$59.16	$92.06

Crew Q-4	Hr.	Daily	Hr.	Daily	Bare Costs	Incl. O&P
1 Plumber Foreman (inside)	$62.65	$501.20	$97.50	$780.00	$59.16	$92.06
1 Plumber	62.15	497.20	96.70	773.60		
1 Welder (plumber)	62.15	497.20	96.70	773.60		
1 Plumber Apprentice	49.70	397.60	77.35	618.80		
1 Welder, Electric, 300 amp		59.20		65.12	1.85	2.04
32 L.H., Daily Totals		$1952.40		$3011.12	$61.01	$94.10

Crew Q-5	Hr.	Daily	Hr.	Daily	Bare Costs	Incl. O&P
1 Steamfitter	$63.00	$504.00	$98.05	$784.40	$56.70	$88.22
1 Steamfitter Apprentice	50.40	403.20	78.40	627.20		
16 L.H., Daily Totals		$907.20		$1411.60	$56.70	$88.22

Crew Q-6	Hr.	Daily	Hr.	Daily	Bare Costs	Incl. O&P
2 Steamfitters	$63.00	$1008.00	$98.05	$1568.80	$58.80	$91.50
1 Steamfitter Apprentice	50.40	403.20	78.40	627.20		
24 L.H., Daily Totals		$1411.20		$2196.00	$58.80	$91.50

Crew Q-7	Hr.	Daily	Hr.	Daily	Bare Costs	Incl. O&P
1 Steamfitter Foreman (inside)	$63.50	$508.00	$98.80	$790.40	$59.98	$93.33
2 Steamfitters	63.00	1008.00	98.05	1568.80		
1 Steamfitter Apprentice	50.40	403.20	78.40	627.20		
32 L.H., Daily Totals		$1919.20		$2986.40	$59.98	$93.33

Crew Q-8	Hr.	Daily	Hr.	Daily	Bare Costs	Incl. O&P
1 Steamfitter Foreman (inside)	$63.50	$508.00	$98.80	$790.40	$59.98	$93.33
1 Steamfitter	63.00	504.00	98.05	784.40		
1 Welder (steamfitter)	63.00	504.00	98.05	784.40		
1 Steamfitter Apprentice	50.40	403.20	78.40	627.20		
1 Welder, Electric, 300 amp		59.20		65.12	1.85	2.04
32 L.H., Daily Totals		$1978.40		$3051.52	$61.83	$95.36

Crew Q-9	Hr.	Daily	Hr.	Daily	Bare Costs	Incl. O&P
1 Sheet Metal Worker	$59.80	$478.40	$94.25	$754.00	$53.83	$84.83
1 Sheet Metal Apprentice	47.85	382.80	75.40	603.20		
16 L.H., Daily Totals		$861.20		$1357.20	$53.83	$84.83

Crew Q-10	Hr.	Daily	Hr.	Daily	Bare Costs	Incl. O&P
2 Sheet Metal Workers	$59.80	$956.80	$94.25	$1508.00	$55.82	$87.97
1 Sheet Metal Apprentice	47.85	382.80	75.40	603.20		
24 L.H., Daily Totals		$1339.60		$2111.20	$55.82	$87.97

Crew Q-11	Hr.	Daily	Hr.	Daily	Bare Costs	Incl. O&P
1 Sheet Metal Foreman (inside)	$60.30	$482.40	$95.05	$760.40	$56.94	$89.74
2 Sheet Metal Workers	59.80	956.80	94.25	1508.00		
1 Sheet Metal Apprentice	47.85	382.80	75.40	603.20		
32 L.H., Daily Totals		$1822.00		$2871.60	$56.94	$89.74

Crew Q-12	Hr.	Daily	Hr.	Daily	Bare Costs	Incl. O&P
1 Sprinkler Installer	$59.90	$479.20	$93.45	$747.60	$53.90	$84.08
1 Sprinkler Apprentice	47.90	383.20	74.70	597.60		
16 L.H., Daily Totals		$862.40		$1345.20	$53.90	$84.08

Crew Q-13	Hr.	Daily	Hr.	Daily	Bare Costs	Incl. O&P
1 Sprinkler Foreman (inside)	$60.40	$483.20	$94.20	$753.60	$57.02	$88.95
2 Sprinkler Installers	59.90	958.40	93.45	1495.20		
1 Sprinkler Apprentice	47.90	383.20	74.70	597.60		
32 L.H., Daily Totals		$1824.80		$2846.40	$57.02	$88.95

Crew Q-14	Hr.	Daily	Hr.	Daily	Bare Costs	Incl. O&P
1 Asbestos Worker	$56.30	$450.40	$90.20	$721.60	$50.67	$81.17
1 Asbestos Apprentice	45.05	360.40	72.15	577.20		
16 L.H., Daily Totals		$810.80		$1298.80	$50.67	$81.17

Crew Q-15	Hr.	Daily	Hr.	Daily	Bare Costs	Incl. O&P
1 Plumber	$62.15	$497.20	$96.70	$773.60	$55.92	$87.03
1 Plumber Apprentice	49.70	397.60	77.35	618.80		
1 Welder, Electric, 300 amp		59.20		65.12	3.70	4.07
16 L.H., Daily Totals		$954.00		$1457.52	$59.63	$91.09

773

Crew Q-16

	Bare Costs Hr.	Daily	Incl. Subs O&P Hr.	Daily	Bare Costs	Incl. O&P
2 Plumbers	$62.15	$994.40	$96.70	$1547.20	$58.00	$90.25
1 Plumber Apprentice	49.70	397.60	77.35	618.80		
1 Welder, Electric, 300 amp		59.20		65.12	2.47	2.71
24 L.H., Daily Totals		$1451.20		$2231.12	$60.47	$92.96

Crew Q-17

	Bare Costs Hr.	Daily	Incl. Subs O&P Hr.	Daily	Bare Costs	Incl. O&P
1 Steamfitter	$63.00	$504.00	$98.05	$784.40	$56.70	$88.22
1 Steamfitter Apprentice	50.40	403.20	78.40	627.20		
1 Welder, Electric, 300 amp		59.20		65.12	3.70	4.07
16 L.H., Daily Totals		$966.40		$1476.72	$60.40	$92.30

Crew Q-17A

	Bare Costs Hr.	Daily	Incl. Subs O&P Hr.	Daily	Bare Costs	Incl. O&P
1 Steamfitter	$63.00	$504.00	$98.05	$784.40	$56.50	$88.33
1 Steamfitter Apprentice	50.40	403.20	78.40	627.20		
1 Equip. Oper. (crane)	56.10	448.80	88.55	708.40		
1 Hyd. Crane, 12 Ton		496.00		545.60		
1 Welder, Electric, 300 amp		59.20		65.12	23.13	25.45
24 L.H., Daily Totals		$1911.20		$2730.72	$79.63	$113.78

Crew Q-18

	Bare Costs Hr.	Daily	Incl. Subs O&P Hr.	Daily	Bare Costs	Incl. O&P
2 Steamfitters	$63.00	$1008.00	$98.05	$1568.80	$58.80	$91.50
1 Steamfitter Apprentice	50.40	403.20	78.40	627.20		
1 Welder, Electric, 300 amp		59.20		65.12	2.47	2.71
24 L.H., Daily Totals		$1470.40		$2261.12	$61.27	$94.21

Crew Q-19

	Bare Costs Hr.	Daily	Incl. Subs O&P Hr.	Daily	Bare Costs	Incl. O&P
1 Steamfitter	$63.00	$504.00	$98.05	$784.40	$57.20	$88.78
1 Steamfitter Apprentice	50.40	403.20	78.40	627.20		
1 Electrician	58.20	465.60	89.90	719.20		
24 L.H., Daily Totals		$1372.80		$2130.80	$57.20	$88.78

Crew Q-20

	Bare Costs Hr.	Daily	Incl. Subs O&P Hr.	Daily	Bare Costs	Incl. O&P
1 Sheet Metal Worker	$59.80	$478.40	$94.25	$754.00	$54.70	$85.84
1 Sheet Metal Apprentice	47.85	382.80	75.40	603.20		
.5 Electrician	58.20	232.80	89.90	359.60		
20 L.H., Daily Totals		$1094.00		$1716.80	$54.70	$85.84

Crew Q-21

	Bare Costs Hr.	Daily	Incl. Subs O&P Hr.	Daily	Bare Costs	Incl. O&P
2 Steamfitters	$63.00	$1008.00	$98.05	$1568.80	$58.65	$91.10
1 Steamfitter Apprentice	50.40	403.20	78.40	627.20		
1 Electrician	58.20	465.60	89.90	719.20		
32 L.H., Daily Totals		$1876.80		$2915.20	$58.65	$91.10

Crew Q-22

	Bare Costs Hr.	Daily	Incl. Subs O&P Hr.	Daily	Bare Costs	Incl. O&P
1 Plumber	$62.15	$497.20	$96.70	$773.60	$55.92	$87.03
1 Plumber Apprentice	49.70	397.60	77.35	618.80		
1 Hyd. Crane, 12 Ton		496.00		545.60	31.00	34.10
16 L.H., Daily Totals		$1390.80		$1938.00	$86.92	$121.13

Crew Q-22A

	Bare Costs Hr.	Daily	Incl. Subs O&P Hr.	Daily	Bare Costs	Incl. O&P
1 Plumber	$62.15	$497.20	$96.70	$773.60	$51.95	$81.83
1 Plumber Apprentice	49.70	397.60	77.35	618.80		
1 Laborer	39.85	318.80	64.70	517.60		
1 Equip. Oper. (crane)	56.10	448.80	88.55	708.40		
1 Hyd. Crane, 12 Ton		496.00		545.60	15.50	17.05
32 L.H., Daily Totals		$2158.40		$3164.00	$67.45	$98.88

Crew Q-23

	Bare Costs Hr.	Daily	Incl. Subs O&P Hr.	Daily	Bare Costs	Incl. O&P
1 Plumber Foreman (outside)	$64.15	$513.20	$99.80	$798.40	$60.02	$93.77
1 Plumber	62.15	497.20	96.70	773.60		
1 Equip. Oper. (medium)	53.75	430.00	84.80	678.40		
1 Lattice Boom Crane, 20 Ton		1083.00		1191.30	45.13	49.64
24 L.H., Daily Totals		$2523.40		$3441.70	$105.14	$143.40

Crew R-1

	Bare Costs Hr.	Daily	Incl. Subs O&P Hr.	Daily	Bare Costs	Incl. O&P
1 Electrician Foreman	$58.70	$469.60	$90.70	$725.60	$54.40	$84.03
3 Electricians	58.20	1396.80	89.90	2157.60		
2 Electrician Apprentices	46.55	744.80	71.90	1150.40		
48 L.H., Daily Totals		$2611.20		$4033.60	$54.40	$84.03

Crew R-1A

	Bare Costs Hr.	Daily	Incl. Subs O&P Hr.	Daily	Bare Costs	Incl. O&P
1 Electrician	$58.20	$465.60	$89.90	$719.20	$52.38	$80.90
1 Electrician Apprentice	46.55	372.40	71.90	575.20		
16 L.H., Daily Totals		$838.00		$1294.40	$52.38	$80.90

Crew R-1B

	Bare Costs Hr.	Daily	Incl. Subs O&P Hr.	Daily	Bare Costs	Incl. O&P
1 Electrician	$58.20	$465.60	$89.90	$719.20	$50.43	$77.90
2 Electrician Apprentices	46.55	744.80	71.90	1150.40		
24 L.H., Daily Totals		$1210.40		$1869.60	$50.43	$77.90

Crew R-1C

	Bare Costs Hr.	Daily	Incl. Subs O&P Hr.	Daily	Bare Costs	Incl. O&P
2 Electricians	$58.20	$931.20	$89.90	$1438.40	$52.38	$80.90
2 Electrician Apprentices	46.55	744.80	71.90	1150.40		
1 Portable Cable Puller, 8000 lb.		130.15		143.16	4.07	4.47
32 L.H., Daily Totals		$1806.15		$2731.97	$56.44	$85.37

Crew R-2

	Bare Costs Hr.	Daily	Incl. Subs O&P Hr.	Daily	Bare Costs	Incl. O&P
1 Electrician Foreman	$58.70	$469.60	$90.70	$725.60	$54.64	$84.68
3 Electricians	58.20	1396.80	89.90	2157.60		
2 Electrician Apprentices	46.55	744.80	71.90	1150.40		
1 Equip. Oper. (crane)	56.10	448.80	88.55	708.40		
1 S.P. Crane, 4x4, 5 Ton		259.00		284.90	4.63	5.09
56 L.H., Daily Totals		$3319.00		$5026.90	$59.27	$89.77

Crew R-3

	Bare Costs Hr.	Daily	Incl. Subs O&P Hr.	Daily	Bare Costs	Incl. O&P
1 Electrician Foreman	$58.70	$469.60	$90.70	$725.60	$57.98	$89.95
1 Electrician	58.20	465.60	89.90	719.20		
.5 Equip. Oper. (crane)	56.10	224.40	88.55	354.20		
.5 S.P. Crane, 4x4, 5 Ton		129.50		142.45	6.47	7.12
20 L.H., Daily Totals		$1289.10		$1941.45	$64.45	$97.07

Crew R-4

	Bare Costs Hr.	Daily	Incl. Subs O&P Hr.	Daily	Bare Costs	Incl. O&P
1 Struc. Steel Foreman (outside)	$56.65	$453.20	$96.60	$772.80	$55.76	$93.22
3 Struc. Steel Workers	54.65	1311.60	93.20	2236.80		
1 Electrician	58.20	465.60	89.90	719.20		
1 Welder, Gas Engine, 300 amp		98.40		108.24	2.46	2.71
40 L.H., Daily Totals		$2328.80		$3837.04	$58.22	$95.93

Crew No.	Bare Costs		Incl. Subs O&P		Cost Per Labor-Hour	

Crew R-5	Hr.	Daily	Hr.	Daily	Bare Costs	Incl. O&P
1 Electrician Foreman	$58.70	$469.60	$90.70	$725.60	$50.83	$79.86
4 Electrician Linemen	58.20	1862.40	89.90	2876.80		
2 Electrician Operators	58.20	931.20	89.90	1438.40		
4 Electrician Groundmen	37.80	1209.60	62.10	1987.20		
1 Crew Truck		166.00		182.60		
1 Flatbed Truck, 20,000 GVW		197.20		216.92		
1 Pickup Truck, 3/4 Ton		115.20		126.72		
.2 Hyd. Crane, 55 Ton		198.76		218.64		
.2 Hyd. Crane, 12 Ton		99.20		109.12		
.2 Earth Auger, Truck-Mtd.		74.76		82.24		
1 Tractor w/Winch		361.80		397.98	13.78	15.16
88 L.H., Daily Totals		$5685.72		$8362.21	$64.61	$95.03

Crew R-6	Hr.	Daily	Hr.	Daily	Bare Costs	Incl. O&P
1 Electrician Foreman	$58.70	$469.60	$90.70	$725.60	$50.83	$79.86
4 Electrician Linemen	58.20	1862.40	89.90	2876.80		
2 Electrician Operators	58.20	931.20	89.90	1438.40		
4 Electrician Groundmen	37.80	1209.60	62.10	1987.20		
1 Crew Truck		166.00		182.60		
1 Flatbed Truck, 20,000 GVW		197.20		216.92		
1 Pickup Truck, 3/4 Ton		115.20		126.72		
.2 Hyd. Crane, 55 Ton		198.76		218.64		
.2 Hyd. Crane, 12 Ton		99.20		109.12		
.2 Earth Auger, Truck-Mtd.		74.76		82.24		
1 Tractor w/Winch		361.80		397.98		
3 Cable Trailers		635.10		698.61		
.5 Tensioning Rig		209.93		230.92		
.5 Cable Pulling Rig		1213.00		1334.30	37.17	40.89
88 L.H., Daily Totals		$7743.74		$10626.04	$88.00	$120.75

Crew R-7	Hr.	Daily	Hr.	Daily	Bare Costs	Incl. O&P
1 Electrician Foreman	$58.70	$469.60	$90.70	$725.60	$41.28	$66.87
5 Electrician Groundmen	37.80	1512.00	62.10	2484.00		
1 Crew Truck		166.00		182.60	3.46	3.80
48 L.H., Daily Totals		$2147.60		$3392.20	$44.74	$70.67

Crew R-8	Hr.	Daily	Hr.	Daily	Bare Costs	Incl. O&P
1 Electrician Foreman	$58.70	$469.60	$90.70	$725.60	$51.48	$80.77
3 Electrician Linemen	58.20	1396.80	89.90	2157.60		
2 Electrician Groundmen	37.80	604.80	62.10	993.60		
1 Pickup Truck, 3/4 Ton		115.20		126.72		
1 Crew Truck		166.00		182.60	5.86	6.44
48 L.H., Daily Totals		$2752.40		$4186.12	$57.34	$87.21

Crew R-9	Hr.	Daily	Hr.	Daily	Bare Costs	Incl. O&P
1 Electrician Foreman	$58.70	$469.60	$90.70	$725.60	$48.06	$76.10
1 Electrician Lineman	58.20	465.60	89.90	719.20		
2 Electrician Operators	58.20	931.20	89.90	1438.40		
4 Electrician Groundmen	37.80	1209.60	62.10	1987.20		
1 Pickup Truck, 3/4 Ton		115.20		126.72		
1 Crew Truck		166.00		182.60	4.39	4.83
64 L.H., Daily Totals		$3357.20		$5179.72	$52.46	$80.93

Crew R-10	Hr.	Daily	Hr.	Daily	Bare Costs	Incl. O&P
1 Electrician Foreman	$58.70	$469.60	$90.70	$725.60	$54.88	$85.40
4 Electrician Linemen	58.20	1862.40	89.90	2876.80		
1 Electrician Groundman	37.80	302.40	62.10	496.80		
1 Crew Truck		166.00		182.60		
3 Tram Cars		433.80		477.18	12.50	13.75
48 L.H., Daily Totals		$3234.20		$4758.98	$67.38	$99.15

Crew R-11	Hr.	Daily	Hr.	Daily	Bare Costs	Incl. O&P
1 Electrician Foreman	$58.70	$469.60	$90.70	$725.60	$55.35	$86.22
4 Electricians	58.20	1862.40	89.90	2876.80		
1 Equip. Oper. (crane)	56.10	448.80	88.55	708.40		
1 Common Laborer	39.85	318.80	64.70	517.60		
1 Crew Truck		166.00		182.60		
1 Hyd. Crane, 12 Ton		496.00		545.60	11.82	13.00
56 L.H., Daily Totals		$3761.60		$5556.60	$67.17	$99.22

Crew R-12	Hr.	Daily	Hr.	Daily	Bare Costs	Incl. O&P
1 Carpenter Foreman (inside)	$51.20	$409.60	$83.10	$664.80	$47.44	$77.19
4 Carpenters	50.70	1622.40	82.30	2633.60		
4 Common Laborers	39.85	1275.20	64.70	2070.40		
1 Equip. Oper. (medium)	53.75	430.00	84.80	678.40		
1 Steel Worker	54.65	437.20	93.20	745.60		
1 Dozer, 200 H.P.		1273.00		1400.30		
1 Pickup Truck, 3/4 Ton		115.20		126.72	15.78	17.35
88 L.H., Daily Totals		$5562.60		$8319.82	$63.21	$94.54

Crew R-13	Hr.	Daily	Hr.	Daily	Bare Costs	Incl. O&P
1 Electrician Foreman	$58.70	$469.60	$90.70	$725.60	$56.37	$87.47
3 Electricians	58.20	1396.80	89.90	2157.60		
.25 Equip. Oper. (crane)	56.10	112.20	88.55	177.10		
1 Equipment Oiler	48.60	388.80	76.70	613.60		
.25 Hydraulic Crane, 33 Ton		235.15		258.67	5.60	6.16
42 L.H., Daily Totals		$2602.55		$3932.57	$61.97	$93.63

Crew R-15	Hr.	Daily	Hr.	Daily	Bare Costs	Incl. O&P
1 Electrician Foreman	$58.70	$469.60	$90.70	$725.60	$57.13	$88.54
4 Electricians	58.20	1862.40	89.90	2876.80		
1 Equipment Oper. (light)	51.30	410.40	80.95	647.60		
1 Telescoping Boom Lift, to 40'		282.80		311.08	5.89	6.48
48 L.H., Daily Totals		$3025.20		$4561.08	$63.02	$95.02

Crew R-18	Hr.	Daily	Hr	Daily	Bare Costs	Incl. O&P
.25 Electrician Foreman	$58.70	$117.40	$90.70	$181.40	$51.07	$78.88
1 Electrician	58.20	465.60	89.90	719.20		
2 Electrician Apprentices	46.55	744.80	71.90	1150.40		
26 L.H., Daily Totals		$1327.80		$2051.00	$51.07	$78.88

Crew R-19	Hr.	Daily	Hr.	Daily	Bare Costs	Incl. O&P
.5 Electrician Foreman	$58.70	$234.80	$90.70	$362.80	$58.30	$90.06
2 Electricians	58.20	931.20	89.90	1438.40		
20 L.H., Daily Totals		$1166.00		$1801.20	$58.30	$90.06

Crew R-21	Hr.	Daily	Hr.	Daily	Bare Costs	Incl. O&P
1 Electrician Foreman	$58.70	$469.60	$90.70	$725.60	$58.21	$89.97
3 Electricians	58.20	1396.80	89.90	2157.60		
.1 Equip. Oper. (medium)	53.75	43.00	84.80	67.84		
.1 S.P. Crane, 4x4, 25 Ton		66.84		73.52	2.04	2.24
32.8 L.H., Daily Totals		$1976.24		$3024.56	$60.25	$92.21

Crew R-22	Hr.	Daily	Hr.	Daily	Bare Costs	Incl. O&P
.66 Electrician Foreman	$58.70	$309.94	$90.70	$478.90	$53.27	$82.29
2 Electricians	58.20	931.20	89.90	1438.40		
2 Electrician Apprentices	46.55	744.80	71.90	1150.40		
37.28 L.H., Daily Totals		$1985.94		$3067.70	$53.27	$82.29

For customer support on your Commercial Renovation Costs with RSMeans data, call 800.448.8182.

Crews - Renovation

Crew No.	Bare Costs		Incl. Subs O & P		Cost Per Labor-Hour	
Crew R-30	Hr.	Daily	Hr.	Daily	Bare Costs	Incl. O&P
.25 Electrician Foreman (outside)	$60.20	$120.40	$93.00	$186.00	$47.06	$74.63
1 Electrician	58.20	465.60	89.90	719.20		
2 Laborers (Semi-Skilled)	39.85	637.60	64.70	1035.20		
26 L.H., Daily Totals		$1223.60		$1940.40	$47.06	$74.63
Crew R-31	Hr.	Daily	Hr.	Daily	Bare Costs	Incl. O&P
1 Electrician	$58.20	$465.60	$89.90	$719.20	$58.20	$89.90
1 Core Drill, Electric, 2.5 H.P.		47.50		52.25	5.94	6.53
8 L.H., Daily Totals		$513.10		$771.45	$64.14	$96.43

Historical Cost Indexes

The table below lists both the RSMeans® historical cost index based on Jan. 1, 1993 = 100 as well as the computed value of an index based on Jan. 1, 2018 costs. Since the Jan. 1, 2018 figure is estimated, space is left to write in the actual index figures as they become available through the quarterly *RSMeans Construction Cost Indexes*.

To compute the actual index based on Jan. 1, 2018 = 100, divide the historical cost index for a particular year by the actual Jan. 1, 2018 construction cost index. Space has been left to advance the index figures as the year progresses.

Year	Historical Cost Index Jan. 1, 1993 = 100		Current Index Based on Jan. 1, 2018 = 100		Year	Historical Cost Index Jan. 1, 1993 = 100	Current Index Based on Jan. 1, 2018 = 100		Year	Historical Cost Index Jan. 1, 1993 = 100	Current Index Based on Jan. 1, 2018 = 100	
	Est.	Actual	Est.	Actual		Actual	Est.	Actual		Actual	Est.	Actual
Oct 2018*					July 2003	132.0	61.2		July 1985	82.6	38.3	
July 2018*					2002	128.7	59.6		1984	82.0	38.0	
April 2018*					2001	125.1	58.0		1983	80.2	37.1	
Jan 2018*	215.8		100.0	100.0	2000	120.9	56.0		1982	76.1	35.3	
July 2017		213.6	99.0		1999	117.6	54.5		1981	70.0	32.4	
2016		207.3	96.1		1998	115.1	53.3		1980	62.9	29.1	
2015		206.2	95.6		1997	112.8	52.3		1979	57.8	26.8	
2014		204.9	94.9		1996	110.2	51.1		1978	53.5	24.8	
2013		201.2	93.2		1995	107.6	49.9		1977	49.5	22.9	
2012		194.6	90.2		1994	104.4	48.4		1976	46.9	21.7	
2011		191.2	88.6		1993	101.7	47.1		1975	44.8	20.8	
2010		183.5	85.0		1992	99.4	46.1		1974	41.4	19.2	
2009		180.1	83.5		1991	96.8	44.9		1973	37.7	17.5	
2008		180.4	83.6		1990	94.3	43.7		1972	34.8	16.1	
2007		169.4	78.5		1989	92.1	42.7		1971	32.1	14.9	
2006		162.0	75.1		1988	89.9	41.6		1970	28.7	13.3	
2005		151.6	70.3		1987	87.7	40.6		1969	26.9	12.5	
2004		143.7	66.6		1986	84.2	39.0		1968	24.9	11.5	

Adjustments to Costs

The "Historical Cost Index" can be used to convert national average building costs at a particular time to the approximate building costs for some other time.

Example:

Estimate and compare construction costs for different years in the same city.

To estimate the national average construction cost of a building in 1970, knowing that it cost $900,000 in 2018:

INDEX in 1970 = 28.7

INDEX in 2018 = 215.8

Note: The city cost indexes for Canada can be used to convert U.S. national averages to local costs in Canadian dollars.

Example:

To estimate and compare the cost of a building in Toronto, ON in 2018 with the known cost of $600,000 (US$) in New York, NY in 2018:

INDEX Toronto = 110.8

INDEX New York = 134.6

$$\frac{\text{INDEX Toronto}}{\text{INDEX New York}} \times \text{Cost New York} = \text{Cost Toronto}$$

$$\frac{110.8}{134.6} \times \$600,000 = .823 \times \$600,000 = \$493,908$$

The construction cost of the building in Toronto is $493,908 (CN$).

Time Adjustment Using the Historical Cost Indexes:

$$\frac{\text{Index for Year A}}{\text{Index for Year B}} \times \text{Cost in Year B} = \text{Cost in Year A}$$

$$\frac{\text{INDEX 1970}}{\text{INDEX 2018}} \times \text{Cost 2018} = \text{Cost 1970}$$

$$\frac{28.7}{215.8} \times \$900,000 = .133 \times \$900,000 = \$119,694$$

The construction cost of the building in 1970 was $119,694.

*Historical Cost Index updates and other resources are provided on the following website:
http://info.thegordiangroup.com/RSMeans.html

How to Use the City Cost Indexes

What you should know before you begin

RSMeans City Cost Indexes (CCI) are an extremely useful tool for when you want to compare costs from city to city and region to region.

This publication contains average construction cost indexes for 731 U.S. and Canadian cities covering over 930 three-digit zip code locations, as listed directly under each city.

Keep in mind that a City Cost Index number is a percentage ratio of a specific city's cost to the national average cost of the same item at a stated time period.

In other words, these index figures represent relative construction factors (or, if you prefer, multipliers) for material and installation costs, as well as the weighted average for Total In Place costs for each CSI MasterFormat division. Installation costs include both labor and equipment rental costs. When estimating equipment rental rates only for a specific location, use 01 54 33 EQUIPMENT RENTAL COSTS in the Reference Section.

The 30 City Average Index is the average of 30 major U.S. cities and serves as a national average.

Index figures for both material and installation are based on the 30 major city average of 100 and represent the cost relationship as of July 1, 2017. The index for each division is computed from representative material and labor quantities for that division. The weighted average for each city is a weighted total of the components listed above it. It does not include relative productivity between trades or cities.

As changes occur in local material prices, labor rates, and equipment rental rates (including fuel costs), the impact of these changes should be accurately measured by the change in the City Cost Index for each particular city (as compared to the 30 city average).

Therefore, if you know (or have estimated) building costs in one city today, you can easily convert those costs to expected building costs in another city.

In addition, by using the Historical Cost Index, you can easily convert national average building costs at a particular time to the approximate building costs for some other time. The City Cost Indexes can then be applied to calculate the costs for a particular city.

Quick calculations

Location Adjustment Using the City Cost Indexes:

$$\frac{\text{Index for City A}}{\text{Index for City B}} \times \text{Cost in City B} = \text{Cost in City A}$$

Time Adjustment for the National Average Using the Historical Cost Index:

$$\frac{\text{Index for Year A}}{\text{Index for Year B}} \times \text{Cost in Year B} = \text{Cost in Year A}$$

Adjustment from the National Average:

$$\frac{\text{Index for City A}}{100} \times \text{National Average Cost} = \text{Cost in City A}$$

Since each of the other RSMeans data sets contains many different items, any *one* item multiplied by the particular city index may give incorrect results. However, the larger the number of items compiled, the closer the results should be to actual costs for that particular city.

The City Cost Indexes for Canadian cities are calculated using Canadian material and equipment prices and labor rates in Canadian dollars. Therefore, indexes for Canadian cities can be used to convert U.S. national average prices to local costs in Canadian dollars.

How to use this section

1. Compare costs from city to city.

In using the RSMeans Indexes, remember that an index number is not a fixed number but a ratio: It's a percentage ratio of a building component's cost at any stated time to the national average cost of that same component at the same time period. Put in the form of an equation:

$$\frac{\text{Specific City Cost}}{\text{National Average Cost}} \times 100 = \text{City Index Number}$$

Therefore, when making cost comparisons between cities, do not subtract one city's index number from the index number of another city and read the result as a percentage difference. Instead, divide one city's index number by that of the other city. The resulting number may then be used as a multiplier to calculate cost differences from city to city.

The formula used to find cost differences between cities for the purpose of comparison is as follows:

$$\frac{\text{City A Index}}{\text{City B Index}} \times \text{City B Cost (Known)} = \text{City A Cost (Unknown)}$$

In addition, you can use RSMeans CCI to calculate and compare costs division by division between cities using the same basic formula. (Just be sure that you're comparing similar divisions.)

2. Compare a specific city's construction costs with the national average.

When you're studying construction location feasibility, it's advisable to compare a prospective project's cost index with an index of the national average cost.

For example, divide the weighted average index of construction costs of a specific city by that of the 30 City Average, which = 100.

$$\frac{\text{City Index}}{100} = \% \text{ of National Average}$$

As a result, you get a ratio that indicates the relative cost of construction in that city in comparison with the national average.

3. Convert U.S. national average to actual costs in Canadian City.

$$\frac{\text{Index for Canadian City}}{100} \times \text{National Average Cost} = \text{Cost in Canadian City in \$ CAN}$$

4. Adjust construction cost data based on a national average.

When you use a source of construction cost data which is based on a national average (such as RSMeans cost data), it is necessary to adjust those costs to a specific location.

$$\frac{\text{City Index}}{100} \times \frac{\text{Cost Based on}}{\text{National Average Costs}} = \frac{\text{City Cost}}{\text{(Unknown)}}$$

5. When applying the City Cost Indexes to demolition projects, use the appropriate division installation index. For example, for removal of existing doors and windows, use the Division 8 (Openings) index.

What you might like to know about how we developed the Indexes

The information presented in the CCI is organized according to the Construction Specifications Institute (CSI) MasterFormat 2014 classification system.

To create a reliable index, RSMeans researched the building type most often constructed in the United States and Canada. Because it was concluded that no one type of building completely represented the building construction industry, nine different types of buildings were combined to create a composite model.

The exact material, labor, and equipment quantities are based on detailed analyses of these nine building types, and then each quantity is weighted in proportion to expected usage. These various material items, labor hours, and equipment rental rates are thus combined to form a composite building representing as closely as possible the actual usage of materials, labor, and equipment in the North American building construction industry.

The following structures were chosen to make up that composite model:

1. Factory, 1 story
2. Office, 2–4 stories
3. Store, Retail
4. Town Hall, 2–3 stories
5. High School, 2–3 stories
6. Hospital, 4–8 stories
7. Garage, Parking
8. Apartment, 1–3 stories
9. Hotel/Motel, 2–3 stories

For the purposes of ensuring the timeliness of the data, the components of the index for the composite model have been streamlined. They currently consist of:

- specific quantities of 66 commonly used construction materials;
- specific labor-hours for 21 building construction trades; and
- specific days of equipment rental for 6 types of construction equipment (normally used to install the 66 material items by the 21 trades.) Fuel costs and routine maintenance costs are included in the equipment cost.

Material and equipment price quotations are gathered quarterly from cities in the United States and Canada. These prices and the latest negotiated labor wage rates for 21 different building trades are used to compile the quarterly update of the City Cost Index.

The 30 major U.S. cities used to calculate the national average are:

Atlanta, GA	Memphis, TN
Baltimore, MD	Milwaukee, WI
Boston, MA	Minneapolis, MN
Buffalo, NY	Nashville, TN
Chicago, IL	New Orleans, LA
Cincinnati, OH	New York, NY
Cleveland, OH	Philadelphia, PA
Columbus, OH	Phoenix, AZ
Dallas, TX	Pittsburgh, PA
Denver, CO	St. Louis, MO
Detroit, MI	San Antonio, TX
Houston, TX	San Diego, CA
Indianapolis, IN	San Francisco, CA
Kansas City, MO	Seattle, WA
Los Angeles, CA	Washington, DC

What the CCI does not indicate

The weighted average for each city is a total of the divisional components weighted to reflect typical usage. It does not include the productivity variations between trades or cities.

In addition, the CCI does not take into consideration factors such as the following:

- managerial efficiency
- competitive conditions
- automation
- restrictive union practices
- unique local requirements
- regional variations due to specific building codes

City Cost Indexes

DIVISION		UNITED STATES 30 CITY AVERAGE			ANNISTON 362			BIRMINGHAM 350-352			BUTLER 369			DECATUR 356			DOTHAN 363		
		MAT.	INST.	TOTAL	MAT.	INST.	TOTAL	MAT.	INST.	TOTAL	MAT.	INST.	TOTAL	MAT.	INST.	TOTAL	MAT.	INST.	TOTAL
015433	CONTRACTOR EQUIPMENT		100.0	100.0		104.4	104.4		104.5	104.5		101.9	101.9		104.4	104.4		101.9	101.9
0241, 31 - 34	SITE & INFRASTRUCTURE, DEMOLITION	100.0	100.0	100.0	86.8	92.4	90.7	93.0	92.8	92.8	99.8	87.9	91.5	86.4	91.9	90.3	97.5	88.2	91.0
0310	Concrete Forming & Accessories	100.0	100.0	100.0	88.7	66.5	69.6	92.5	67.5	70.9	85.3	67.1	69.6	95.0	62.6	67.1	93.9	68.2	71.7
0320	Concrete Reinforcing	100.0	100.0	100.0	87.2	72.0	79.5	95.5	71.9	83.6	92.3	72.2	82.2	89.4	68.9	79.1	92.3	71.6	81.9
0330	Cast-in-Place Concrete	100.0	100.0	100.0	91.6	67.5	82.5	102.4	68.4	89.5	89.3	67.9	81.2	100.5	67.4	87.9	89.3	67.6	81.1
03	CONCRETE	100.0	100.0	100.0	92.3	69.6	81.9	92.6	70.3	82.4	93.2	70.0	82.6	92.4	67.2	80.9	92.5	70.3	82.3
04	MASONRY	100.0	100.0	100.0	92.9	70.6	79.1	90.4	68.0	76.5	97.6	70.6	80.9	88.3	70.1	77.0	99.0	60.0	74.8
05	METALS	100.0	100.0	100.0	92.1	95.2	93.1	92.2	95.4	93.2	91.1	96.3	92.7	94.2	94.2	94.2	91.2	95.7	92.6
06	WOOD, PLASTICS & COMPOSITES	100.0	100.0	100.0	88.9	67.1	76.8	92.2	67.1	78.2	83.7	67.1	74.5	102.1	61.5	79.5	95.3	68.6	80.5
07	THERMAL & MOISTURE PROTECTION	100.0	100.0	100.0	95.4	65.2	82.6	98.3	67.3	85.2	95.4	67.8	83.8	97.3	67.0	84.5	95.4	64.9	82.5
08	OPENINGS	100.0	100.0	100.0	95.8	68.8	89.6	101.0	68.8	93.6	95.8	68.8	89.6	107.7	64.8	97.9	95.8	69.6	89.8
0920	Plaster & Gypsum Board	100.0	100.0	100.0	88.0	66.7	73.7	87.9	66.7	73.7	85.0	66.7	72.7	93.9	60.9	71.8	95.2	68.3	77.1
0950, 0980	Ceilings & Acoustic Treatment	100.0	100.0	100.0	77.6	66.7	70.2	82.6	66.7	71.9	77.6	66.7	70.2	82.9	60.9	68.1	77.6	68.3	71.3
0960	Flooring	100.0	100.0	100.0	80.5	82.5	81.0	92.3	75.9	87.8	84.9	82.5	84.2	87.0	82.5	85.7	90.0	84.1	88.3
0970, 0990	Wall Finishes & Painting/Coating	100.0	100.0	100.0	86.2	62.6	70.9	87.8	60.0	71.6	86.2	48.9	64.5	83.5	62.6	71.3	86.2	80.5	82.9
09	FINISHES	100.0	100.0	100.0	79.2	69.2	73.7	87.3	68.2	76.9	81.9	68.0	74.4	83.4	66.0	74.0	84.6	72.5	78.1
COVERS	DIVS. 10 - 14, 25, 28, 41, 43, 44, 46	100.0	100.0	100.0	100.0	71.6	93.7	100.0	84.9	96.6	100.0	78.5	95.2	100.0	70.8	93.5	100.0	73.0	94.0
21, 22, 23	FIRE SUPPRESSION, PLUMBING & HVAC	100.0	100.0	100.0	100.9	51.3	79.7	100.0	63.1	84.2	98.2	66.9	84.8	100.0	65.9	85.5	98.2	65.3	84.1
26, 27, 3370	ELECTRICAL, COMMUNICATIONS & UTIL.	100.0	100.0	100.0	96.4	60.2	77.6	98.1	61.5	79.1	98.5	67.4	82.3	93.9	66.0	79.4	97.2	76.1	86.3
MF2016	WEIGHTED AVERAGE	100.0	100.0	100.0	94.5	68.5	83.1	95.9	71.4	85.2	94.8	72.8	85.2	96.0	71.3	85.2	94.9	72.9	85.3

ALABAMA

DIVISION		EVERGREEN 364			GADSDEN 359			HUNTSVILLE 357-358			JASPER 355			MOBILE 365-366			MONTGOMERY 360-361		
		MAT.	INST.	TOTAL	MAT.	INST.	TOTAL	MAT.	INST.	TOTAL	MAT.	INST.	TOTAL	MAT.	INST.	TOTAL	MAT.	INST.	TOTAL
015433	CONTRACTOR EQUIPMENT		101.9	101.9		104.4	104.4		104.4	104.4		104.4	104.4		101.9	101.9		101.9	101.9
0241, 31 - 34	SITE & INFRASTRUCTURE, DEMOLITION	100.2	87.9	91.6	91.9	92.4	92.3	86.1	92.3	90.4	91.8	92.5	92.3	92.9	89.0	90.1	91.3	88.2	89.1
0310	Concrete Forming & Accessories	82.1	68.0	69.9	87.2	66.9	69.9	95.0	64.3	68.5	92.3	67.2	70.7	93.0	68.0	71.4	94.6	67.0	70.8
0320	Concrete Reinforcing	92.3	72.2	82.2	95.0	71.9	83.3	89.4	71.1	80.2	89.4	72.1	80.7	89.8	71.6	80.6	97.6	71.5	84.5
0330	Cast-in-Place Concrete	89.3	67.6	81.1	100.6	67.5	88.0	97.9	66.6	86.0	111.5	68.3	95.1	93.7	67.6	83.7	90.6	67.7	81.9
03	CONCRETE	93.5	70.3	82.9	96.8	69.7	84.4	91.2	68.1	80.6	100.5	70.2	86.6	89.1	70.2	80.5	87.4	69.8	79.4
04	MASONRY	97.6	70.1	80.6	86.8	66.4	74.2	89.6	65.0	74.4	84.5	71.7	76.6	95.8	60.0	73.6	93.1	60.5	72.9
05	METALS	91.2	96.2	92.7	92.2	95.5	93.2	94.2	94.9	94.4	92.1	95.5	93.2	93.1	95.5	93.8	92.2	95.0	93.1
06	WOOD, PLASTICS & COMPOSITES	80.3	68.6	73.8	92.9	67.1	78.5	102.1	64.3	81.1	99.2	67.1	81.4	93.9	68.6	79.9	92.3	67.1	78.3
07	THERMAL & MOISTURE PROTECTION	95.4	67.8	83.7	97.5	66.6	84.4	97.2	65.7	83.9	97.5	63.7	83.2	95.0	64.9	82.3	93.6	65.0	81.5
08	OPENINGS	95.8	69.6	89.8	104.2	68.8	96.0	107.4	67.0	98.2	104.1	68.8	96.0	98.7	69.6	92.0	88.8	68.8	90.9
0920	Plaster & Gypsum Board	84.4	68.3	73.6	86.7	66.7	73.3	93.9	63.8	73.7	90.7	66.7	74.6	92.1	68.3	76.1	88.8	66.7	74.0
0950, 0980	Ceilings & Acoustic Treatment	77.6	68.3	71.3	80.3	66.7	71.1	84.8	63.8	70.7	80.3	66.7	71.1	82.9	68.3	73.0	84.2	66.7	72.4
0960	Flooring	82.9	82.5	82.8	83.3	77.5	81.7	87.0	77.5	84.3	85.3	82.5	84.5	89.5	65.1	82.8	87.8	65.1	81.5
0970, 0990	Wall Finishes & Painting/Coating	86.2	48.9	64.4	83.5	60.0	69.8	83.5	62.7	71.4	83.5	60.0	69.8	89.4	48.9	65.8	86.5	60.0	71.1
09	FINISHES	81.3	68.8	74.5	81.3	68.1	74.1	83.8	66.4	74.3	82.3	69.5	75.3	84.7	65.2	74.1	85.7	65.7	74.8
COVERS	DIVS. 10 - 14, 25, 28, 41, 43, 44, 46	100.0	71.4	93.6	100.0	84.4	96.5	100.0	83.7	96.4	100.0	72.8	93.9	100.0	85.4	96.8	100.0	84.6	96.6
21, 22, 23	FIRE SUPPRESSION, PLUMBING & HVAC	98.2	66.7	84.7	102.1	62.9	85.3	100.0	62.2	83.8	102.1	63.0	85.4	99.8	66.7	85.6	99.9	65.5	85.2
26, 27, 3370	ELECTRICAL, COMMUNICATIONS & UTIL.	95.8	67.4	81.0	93.8	61.5	77.0	94.8	66.0	79.8	93.6	60.2	76.2	99.1	58.0	77.7	99.6	76.1	87.4
MF2016	WEIGHTED AVERAGE	94.5	72.6	84.9	96.1	71.0	85.1	96.0	70.7	84.9	96.5	71.2	85.5	95.4	70.2	84.4	94.8	72.3	85.0

DIVISION		ALABAMA PHENIX CITY 368			SELMA 367			TUSCALOOSA 354			ALASKA ANCHORAGE 995-996			FAIRBANKS 997			JUNEAU 998		
		MAT.	INST.	TOTAL	MAT.	INST.	TOTAL	MAT.	INST.	TOTAL	MAT.	INST.	TOTAL	MAT.	INST.	TOTAL	MAT.	INST.	TOTAL
015433	CONTRACTOR EQUIPMENT		101.9	101.9		101.9	101.9		104.4	104.4		115.3	115.3		115.3	115.3		115.3	115.3
0241, 31 - 34	SITE & INFRASTRUCTURE, DEMOLITION	103.9	88.2	92.9	97.4	88.2	90.9	86.6	92.5	90.7	127.0	130.4	129.4	126.5	130.4	129.2	147.3	130.4	135.5
0310	Concrete Forming & Accessories	88.7	67.2	70.1	86.4	67.1	69.8	94.9	67.2	71.0	125.4	119.5	120.3	130.1	118.0	119.7	131.0	119.5	121.1
0320	Concrete Reinforcing	92.3	65.9	78.9	92.3	72.1	82.1	89.4	71.9	80.6	158.3	118.8	138.4	152.5	118.8	135.5	137.9	118.8	128.3
0330	Cast-in-Place Concrete	89.3	67.8	81.2	89.3	67.8	81.1	102.0	67.8	89.0	108.1	119.1	112.3	117.0	119.4	117.9	127.8	119.1	124.5
03	CONCRETE	96.2	68.9	83.7	92.0	70.0	81.9	93.1	69.9	82.5	117.6	118.4	118.0	110.2	117.9	113.7	123.5	118.4	121.2
04	MASONRY	97.6	70.6	80.8	100.9	70.6	82.1	88.5	66.8	75.1	175.5	121.0	141.7	185.3	121.0	145.4	165.8	121.0	138.0
05	METALS	91.1	93.6	91.9	91.1	95.8	92.6	93.5	95.6	94.1	120.0	104.9	115.4	121.2	105.0	116.2	120.0	104.9	113.3
06	WOOD, PLASTICS & COMPOSITES	88.6	67.0	76.6	85.7	67.1	75.4	102.1	67.1	82.6	117.8	117.9	117.9	129.1	115.8	121.7	124.7	117.9	120.9
07	THERMAL & MOISTURE PROTECTION	95.8	67.8	84.0	95.3	67.8	83.7	97.3	66.8	84.4	169.2	117.9	147.5	180.8	116.9	153.8	183.4	117.9	155.7
08	OPENINGS	95.8	67.1	89.2	95.8	68.8	89.6	107.4	68.8	98.6	128.9	117.3	126.3	134.8	116.4	130.6	129.7	117.3	126.8
0920	Plaster & Gypsum Board	89.3	66.6	74.1	87.3	66.7	73.5	93.9	66.7	75.6	143.2	118.3	126.5	166.9	116.1	132.8	147.9	118.3	128.0
0950, 0980	Ceilings & Acoustic Treatment	77.6	66.6	70.2	77.6	66.7	70.2	84.8	66.7	72.6	131.3	118.3	122.5	130.9	116.1	120.9	143.6	118.3	126.5
0960	Flooring	86.8	82.5	85.6	85.4	82.5	84.6	87.0	77.5	84.3	119.0	125.1	120.7	117.5	125.1	119.6	125.7	125.1	125.5
0970, 0990	Wall Finishes & Painting/Coating	86.2	80.5	82.9	86.2	60.0	70.9	83.5	60.0	69.8	103.4	115.3	110.3	106.4	119.9	114.3	108.9	115.3	112.6
09	FINISHES	83.5	71.4	76.9	82.1	69.2	75.1	83.8	68.2	75.3	125.6	120.4	122.8	126.1	119.7	122.6	128.4	120.4	124.1
COVERS	DIVS. 10 - 14, 25, 28, 41, 43, 44, 46	100.0	84.6	96.6	100.0	72.1	93.8	100.0	84.6	96.6	100.0	111.1	102.5	100.0	110.9	102.4	100.0	111.1	102.5
21, 22, 23	FIRE SUPPRESSION, PLUMBING & HVAC	98.2	65.4	84.2	98.2	65.5	84.2	100.0	68.3	86.5	100.8	106.4	103.2	100.6	110.1	104.7	100.6	106.4	103.1
26, 27, 3370	ELECTRICAL, COMMUNICATIONS & UTIL.	97.9	67.7	82.2	97.0	76.1	86.1	94.3	61.5	77.3	114.4	111.5	112.9	118.2	111.5	114.7	103.1	111.5	107.5
MF2016	WEIGHTED AVERAGE	95.4	72.7	85.4	94.6	73.6	85.4	96.0	72.3	85.6	118.9	114.8	117.1	120.1	115.3	118.0	118.8	114.8	117.0

DIVISION		ALASKA KETCHIKAN 999			ARIZONA CHAMBERS 865			ARIZONA FLAGSTAFF 860			ARIZONA GLOBE 855			ARIZONA KINGMAN 864			ARIZONA MESA/TEMPE 852		
		MAT.	INST.	TOTAL	MAT.	INST.	TOTAL	MAT.	INST.	TOTAL	MAT.	INST.	TOTAL	MAT.	INST.	TOTAL	MAT.	INST.	TOTAL
015433	CONTRACTOR EQUIPMENT		115.3	115.3		90.7	90.7		90.7	90.7		92.0	92.0		90.7	90.7		92.0	92.0
0241, 31 - 34	SITE & INFRASTRUCTURE, DEMOLITION	183.9	130.4	146.4	72.9	94.9	88.3	92.6	94.9	94.2	105.2	95.9	98.7	72.9	94.8	88.3	94.5	95.9	95.5
0310	Concrete Forming & Accessories	121.3	119.5	119.7	97.2	68.7	72.6	102.4	68.7	73.3	93.4	68.7	72.1	95.5	68.6	72.3	96.5	68.8	72.6
0320	Concrete Reinforcing	113.8	118.8	116.3	100.0	81.8	90.9	99.8	81.8	90.8	110.6	81.9	96.1	100.2	81.8	90.9	111.3	81.9	96.5
0330	Cast-in-Place Concrete	236.9	119.1	192.1	89.7	68.4	81.6	89.8	68.4	81.6	82.7	68.1	77.2	89.4	68.4	81.4	83.4	68.1	77.6
03	CONCRETE	187.2	118.4	155.7	92.0	70.8	82.3	111.6	70.8	93.0	103.7	70.8	88.6	91.7	70.8	82.1	94.6	70.8	83.8
04	MASONRY	192.8	121.0	148.2	95.7	59.3	73.1	95.8	58.7	72.8	103.6	59.2	76.0	95.7	59.3	73.1	103.7	59.2	76.1
05	METALS	121.3	104.9	116.3	101.6	73.5	92.9	102.1	73.5	93.3	98.7	74.4	91.3	102.2	73.4	93.4	99.0	74.5	91.5
06	WOOD, PLASTICS & COMPOSITES	119.9	117.9	118.8	100.1	70.5	83.6	105.8	70.5	86.2	97.0	70.7	82.4	95.3	70.5	81.5	100.3	70.7	83.8
07	THERMAL & MOISTURE PROTECTION	185.6	117.9	157.0	93.6	70.1	83.6	95.4	69.8	84.6	99.9	68.8	86.7	93.5	69.9	83.5	99.7	68.5	86.5
08	OPENINGS	130.4	117.3	127.4	105.9	68.4	97.3	106.0	73.0	98.4	96.4	68.8	90.1	106.1	70.8	98.0	96.5	70.9	90.6
0920	Plaster & Gypsum Board	155.6	118.3	130.6	88.6	69.9	76.0	91.8	69.9	77.1	89.0	69.9	76.2	81.2	69.9	73.6	91.3	69.9	76.9
0950, 0980	Ceilings & Acoustic Treatment	124.2	118.3	120.2	105.7	69.9	81.6	106.6	69.9	81.9	95.0	69.9	78.1	106.6	69.9	81.9	95.0	69.9	78.1
0960	Flooring	117.5	125.1	119.6	91.1	60.1	82.5	93.5	60.1	84.3	102.7	60.1	90.9	89.8	60.1	81.6	104.6	60.1	92.3
0970, 0990	Wall Finishes & Painting/Coating	106.4	115.3	111.6	88.8	64.0	74.3	88.8	64.0	74.3	94.1	64.0	76.5	88.8	64.0	74.3	94.1	64.0	76.5
09	FINISHES	127.1	120.4	123.5	90.7	66.4	77.5	93.9	66.4	78.9	97.7	66.6	80.8	89.5	66.4	76.9	97.5	66.6	80.7
COVERS	DIVS. 10 - 14, 25, 28, 41, 43, 44, 46	100.0	111.1	102.5	100.0	84.1	96.5	100.0	84.0	96.4	100.0	84.4	96.5	100.0	84.0	96.4	100.0	84.4	96.5
21, 22, 23	FIRE SUPPRESSION, PLUMBING & HVAC	98.7	106.4	102.0	98.0	77.2	89.1	100.2	78.2	90.8	97.0	77.3	88.6	98.0	78.0	89.4	100.1	77.4	90.4
26, 27, 3370	ELECTRICAL, COMMUNICATIONS & UTIL.	118.2	111.5	114.7	103.2	86.2	94.4	102.2	63.1	81.9	99.0	66.4	82.0	103.2	66.4	84.0	95.9	66.4	80.5
MF2016	WEIGHTED AVERAGE	130.8	114.8	123.8	97.9	75.0	87.9	101.7	72.1	88.8	99.2	72.4	87.4	97.9	72.5	86.8	98.2	72.5	87.0

DIVISION		ARIZONA PHOENIX 850, 853			ARIZONA PRESCOTT 863			ARIZONA SHOW LOW 859			ARIZONA TUCSON 856 - 857			ARKANSAS BATESVILLE 725			ARKANSAS CAMDEN 717		
		MAT.	INST.	TOTAL	MAT.	INST.	TOTAL	MAT.	INST.	TOTAL	MAT.	INST.	TOTAL	MAT.	INST.	TOTAL	MAT.	INST.	TOTAL
015433	CONTRACTOR EQUIPMENT		93.0	93.0		90.7	90.7		92.0	92.0		92.0	92.0		92.2	92.2		92.2	92.2
0241, 31 - 34	SITE & INFRASTRUCTURE, DEMOLITION	95.1	94.5	94.7	79.8	94.8	90.3	107.5	95.9	99.4	90.1	95.9	94.2	73.6	90.1	85.2	82.5	90.1	87.8
0310	Concrete Forming & Accessories	101.4	72.9	76.9	98.7	68.7	72.8	100.5	68.8	73.1	96.9	68.8	72.7	83.8	57.4	61.1	81.0	57.4	60.7
0320	Concrete Reinforcing	109.2	79.3	94.2	99.8	81.8	90.8	111.3	81.9	96.5	92.4	81.8	87.1	90.9	65.5	78.1	94.9	65.2	79.9
0330	Cast-in-Place Concrete	83.0	70.7	78.3	89.7	68.3	81.6	82.8	68.1	77.2	85.9	68.1	79.1	68.9	75.8	71.5	80.4	75.8	78.7
03	CONCRETE	97.2	73.3	86.2	97.4	70.8	85.2	106.1	70.8	89.9	93.2	70.8	83.0	73.3	66.1	70.0	79.5	66.0	73.4
04	MASONRY	96.1	62.9	75.5	95.8	59.3	73.1	103.6	59.2	76.1	90.2	58.6	70.6	92.6	43.2	62.0	109.8	37.7	65.1
05	METALS	100.5	75.7	92.9	102.1	73.4	93.3	98.5	74.4	91.1	99.9	74.5	92.0	99.1	73.7	91.3	100.9	73.3	92.4
06	WOOD, PLASTICS & COMPOSITES	103.4	74.2	87.1	101.4	70.7	84.3	104.6	70.7	85.7	100.6	70.7	83.9	93.1	58.2	73.7	93.0	58.2	73.6
07	THERMAL & MOISTURE PROTECTION	100.9	70.4	88.0	94.1	70.2	84.0	100.1	68.6	86.8	100.8	68.3	87.1	103.1	59.0	84.4	96.4	57.5	79.9
08	OPENINGS	97.7	74.4	92.4	106.0	68.8	97.5	95.7	70.9	90.0	92.9	73.1	88.4	99.7	54.2	89.2	102.6	56.1	91.9
0920	Plaster & Gypsum Board	99.2	73.3	81.8	88.7	70.1	76.2	93.3	69.9	77.6	96.2	69.9	78.5	82.0	57.5	65.5	85.1	57.5	66.5
0950, 0980	Ceilings & Acoustic Treatment	105.6	73.3	83.8	104.9	70.1	81.4	95.0	69.9	78.1	95.8	69.9	78.4	86.7	57.5	67.0	82.1	57.5	65.5
0960	Flooring	106.7	60.1	93.0	92.0	60.1	83.2	106.6	60.1	93.7	95.7	60.1	85.8	84.5	54.2	76.1	92.0	36.3	76.6
0970, 0990	Wall Finishes & Painting/Coating	98.9	66.2	79.8	88.8	64.0	74.3	94.1	64.0	76.5	94.8	64.0	76.8	90.5	36.0	58.7	93.1	50.5	68.3
09	FINISHES	102.2	69.8	84.6	91.4	66.5	77.9	99.7	66.6	81.7	95.5	66.6	79.8	78.4	53.9	65.1	80.9	52.3	65.3
COVERS	DIVS. 10 - 14, 25, 28, 41, 43, 44, 46	100.0	86.5	97.0	100.0	84.1	96.5	100.0	84.4	96.5	100.0	84.4	96.5	100.0	67.2	92.7	100.0	67.2	92.7
21, 22, 23	FIRE SUPPRESSION, PLUMBING & HVAC	99.9	80.2	91.5	100.2	77.2	90.4	97.0	77.3	88.6	100.0	77.5	90.4	97.1	53.2	78.3	97.0	57.9	80.3
26, 27, 3370	ELECTRICAL, COMMUNICATIONS & UTIL.	101.4	63.1	81.5	101.9	66.4	83.4	96.2	66.4	80.7	98.1	61.0	78.8	95.5	62.6	78.4	94.4	59.0	76.0
MF2016	WEIGHTED AVERAGE	99.5	74.1	88.4	99.3	72.2	87.5	99.4	72.5	87.6	97.1	71.8	86.0	92.6	60.8	78.7	95.0	60.5	79.9

DIVISION		ARKANSAS FAYETTEVILLE 727			ARKANSAS FORT SMITH 729			ARKANSAS HARRISON 726			ARKANSAS HOT SPRINGS 719			ARKANSAS JONESBORO 724			ARKANSAS LITTLE ROCK 720 - 722		
		MAT.	INST.	TOTAL	MAT.	INST.	TOTAL	MAT.	INST.	TOTAL	MAT.	INST.	TOTAL	MAT.	INST.	TOTAL	MAT.	INST.	TOTAL
015433	CONTRACTOR EQUIPMENT		92.2	92.2		92.2	92.2		92.2	92.2		92.2	92.2		116.6	116.6		92.2	92.2
0241, 31 - 34	SITE & INFRASTRUCTURE, DEMOLITION	73.0	90.1	85.0	78.1	90.1	86.5	78.5	90.1	86.6	85.8	90.1	88.8	97.6	108.2	105.0	86.4	90.3	89.1
0310	Concrete Forming & Accessories	79.4	57.6	60.6	97.9	57.9	63.5	88.1	57.3	61.5	78.6	57.6	60.5	87.2	57.6	61.7	92.2	59.5	64.0
0320	Concrete Reinforcing	91.0	62.8	76.7	92.1	65.2	78.6	90.6	65.6	78.0	93.1	65.5	79.2	87.8	65.7	76.6	97.4	68.4	82.7
0330	Cast-in-Place Concrete	68.9	75.9	71.6	78.8	76.1	77.7	76.4	75.8	76.2	82.3	75.9	79.9	75.1	77.2	75.9	76.3	76.2	76.3
03	CONCRETE	73.0	65.7	69.7	80.1	66.3	73.8	79.7	66.0	73.4	82.8	66.2	75.2	77.3	67.7	72.9	81.0	67.6	74.9
04	MASONRY	84.0	42.7	58.4	90.9	63.4	73.8	93.0	44.3	62.8	82.3	42.9	57.9	85.2	40.8	57.6	86.7	67.2	74.6
05	METALS	99.1	72.9	91.0	101.4	75.2	93.3	100.2	73.6	92.1	100.9	73.8	92.6	95.6	89.1	93.6	101.6	76.8	94.0
06	WOOD, PLASTICS & COMPOSITES	89.7	58.2	72.2	109.4	58.2	80.9	98.9	58.2	76.3	90.4	58.2	72.5	97.3	58.6	75.7	93.9	59.7	74.9
07	THERMAL & MOISTURE PROTECTION	103.9	58.9	84.8	104.2	64.7	87.5	103.4	59.3	84.7	96.7	58.9	80.7	108.7	58.3	87.4	98.2	65.9	84.5
08	OPENINGS	99.7	56.1	89.7	101.7	56.6	91.4	100.5	53.1	89.6	102.6	56.0	91.9	105.4	57.2	94.3	95.9	58.2	87.2
0920	Plaster & Gypsum Board	81.4	57.5	65.3	88.3	57.5	67.6	87.3	57.5	67.3	83.8	57.5	66.1	96.1	57.5	70.1	95.4	59.0	71.0
0950, 0980	Ceilings & Acoustic Treatment	86.7	57.5	67.0	88.4	57.5	67.5	88.4	57.5	67.5	82.1	57.5	65.5	90.7	57.5	68.3	90.5	59.0	69.3
0960	Flooring	81.7	53.1	73.8	91.0	75.0	86.6	86.7	54.2	77.7	90.9	53.1	80.4	61.2	48.0	57.5	90.9	83.9	88.9
0970, 0990	Wall Finishes & Painting/Coating	90.5	48.2	65.8	90.5	58.0	67.3	90.5	31.1	55.9	93.1	50.5	68.3	80.1	44.1	59.1	88.8	50.5	66.5
09	FINISHES	77.5	54.9	65.2	81.8	60.1	70.0	80.5	53.3	65.7	80.7	55.3	66.9	76.2	53.7	64.0	87.8	62.8	74.2
COVERS	DIVS. 10 - 14, 25, 28, 41, 43, 44, 46	100.0	67.2	92.7	100.0	79.3	95.4	100.0	68.3	92.9	100.0	67.2	92.7	100.0	66.3	92.5	100.0	79.5	95.4
21, 22, 23	FIRE SUPPRESSION, PLUMBING & HVAC	97.2	62.3	82.3	100.3	49.2	78.4	97.1	49.4	76.7	97.0	53.5	78.4	100.6	53.2	80.4	99.9	53.6	80.1
26, 27, 3370	ELECTRICAL, COMMUNICATIONS & UTIL.	90.0	54.7	71.6	93.0	59.0	75.4	94.2	54.7	73.6	96.3	68.8	82.0	98.8	62.8	80.1	99.7	68.8	83.7
MF2016	WEIGHTED AVERAGE	91.5	61.6	78.5	95.0	63.1	81.0	93.9	58.9	78.6	94.3	62.0	80.1	94.5	63.7	81.0	95.3	66.6	82.7

ARKANSAS / CALIFORNIA

DIVISION		PINE BLUFF 716			RUSSELLVILLE 728			TEXARKANA 718			WEST MEMPHIS 723			ALHAMBRA 917-918			ANAHEIM 928		
		MAT.	INST.	TOTAL	MAT.	INST.	TOTAL	MAT.	INST.	TOTAL	MAT.	INST.	TOTAL	MAT.	INST.	TOTAL	MAT.	INST.	TOTAL
015433	CONTRACTOR EQUIPMENT		92.2	92.2		92.2	92.2		93.5	93.5		116.6	116.6		95.8	95.8		101.0	101.0
0241, 31 - 34	SITE & INFRASTRUCTURE, DEMOLITION	87.8	90.2	89.4	74.9	90.1	85.5	98.9	92.3	94.3	104.6	108.5	107.3	101.0	109.1	106.7	102.0	109.9	107.5
0310	Concrete Forming & Accessories	78.4	58.2	61.0	84.4	57.3	61.0	84.1	57.5	61.1	92.6	57.9	62.7	116.4	134.6	132.1	104.9	138.5	133.9
0320	Concrete Reinforcing	94.8	68.1	81.3	91.5	65.4	78.4	94.4	65.5	79.8	87.8	63.9	75.7	104.8	128.4	116.7	94.8	128.1	111.6
0330	Cast-in-Place Concrete	82.3	76.2	80.0	72.2	75.8	73.6	89.7	75.9	84.5	78.7	77.3	78.2	82.8	129.0	100.4	87.9	132.7	104.9
03	CONCRETE	83.5	67.0	75.9	76.1	65.9	71.5	81.9	66.1	74.7	83.8	67.5	76.4	94.6	130.3	110.9	95.6	133.4	112.9
04	MASONRY	116.6	63.4	83.6	90.2	41.8	60.2	96.5	42.2	62.8	73.9	40.6	53.3	112.0	141.0	130.0	78.0	136.0	114.0
05	METALS	101.7	76.4	93.9	99.1	73.4	91.2	93.8	73.6	87.6	94.6	88.7	92.8	86.5	109.6	93.6	104.9	111.5	106.9
06	WOOD, PLASTICS & COMPOSITES	89.9	58.2	72.3	94.4	58.2	74.2	97.4	58.2	75.6	102.9	58.6	78.2	99.6	130.3	116.7	99.7	135.3	119.5
07	THERMAL & MOISTURE PROTECTION	96.7	64.6	83.1	104.1	58.6	84.8	97.4	58.7	81.0	109.2	58.2	87.6	98.9	130.1	112.1	106.2	134.4	118.1
08	OPENINGS	103.8	57.4	93.2	99.7	55.3	89.5	108.9	55.3	96.6	102.8	56.0	92.1	91.2	131.2	100.4	107.1	134.0	113.3
0920	Plaster & Gypsum Board	83.5	57.5	66.0	82.0	57.5	65.5	86.5	57.5	67.0	98.4	57.5	70.9	96.2	131.4	119.9	105.7	136.3	126.2
0950, 0980	Ceilings & Acoustic Treatment	82.1	57.5	65.5	86.7	57.5	67.0	85.5	57.5	66.6	88.8	57.5	67.7	110.3	131.4	124.5	111.1	136.3	128.1
0960	Flooring	90.6	75.0	86.3	84.1	54.2	75.8	92.8	45.8	79.8	63.4	48.0	59.2	102.9	116.0	106.5	96.4	118.6	103.3
0970, 0990	Wall Finishes & Painting/Coating	93.1	50.8	68.4	90.5	33.2	57.1	93.1	41.7	63.1	80.1	50.5	62.9	108.0	120.9	115.5	89.3	118.5	106.4
09	FINISHES	80.6	60.1	69.5	78.5	53.6	65.0	82.9	53.1	66.7	77.5	54.6	65.1	102.3	129.3	117.0	97.4	132.8	116.7
COVERS	DIVS. 10 - 14, 25, 28, 41, 43, 44, 46	100.0	79.3	95.4	100.0	67.2	92.7	100.0	67.2	92.7	100.0	68.1	92.9	100.0	119.7	104.4	100.0	120.9	104.6
21, 22, 23	FIRE SUPPRESSION, PLUMBING & HVAC	100.2	53.5	80.2	97.2	52.9	78.2	100.0	56.7	81.6	97.5	65.4	83.8	96.9	125.5	109.1	100.0	133.2	114.2
26, 27, 3370	ELECTRICAL, COMMUNICATIONS & UTIL.	94.6	64.4	78.9	93.0	54.7	73.1	96.2	56.4	75.5	100.3	64.6	81.7	117.8	126.8	122.5	91.1	108.6	100.2
MF2016	WEIGHTED AVERAGE	96.9	65.0	82.9	92.7	59.4	78.1	95.7	60.7	80.3	94.2	66.6	82.1	98.0	125.9	110.2	99.1	125.9	110.8

CALIFORNIA

DIVISION		BAKERSFIELD 932-933			BERKELEY 947			EUREKA 955			FRESNO 936-938			INGLEWOOD 903-905			LONG BEACH 906-908		
		MAT.	INST.	TOTAL	MAT.	INST.	TOTAL	MAT.	INST.	TOTAL	MAT.	INST.	TOTAL	MAT.	INST.	TOTAL	MAT.	INST.	TOTAL
015433	CONTRACTOR EQUIPMENT		98.9	98.9		102.7	102.7		98.7	98.7		98.9	98.9		97.8	97.8		97.8	97.8
0241, 31 - 34	SITE & INFRASTRUCTURE, DEMOLITION	99.6	107.6	105.2	107.4	109.4	108.8	113.2	106.2	108.3	104.1	107.0	106.1	91.7	105.8	101.6	99.1	105.9	103.8
0310	Concrete Forming & Accessories	104.7	134.4	130.3	115.2	165.0	158.2	113.6	150.3	145.2	102.1	149.9	143.3	108.3	134.9	131.2	103.3	135.0	130.6
0320	Concrete Reinforcing	90.0	128.0	109.2	90.6	130.1	110.5	102.9	129.5	116.3	75.5	129.1	102.5	94.7	128.4	111.7	93.9	128.4	111.3
0330	Cast-in-Place Concrete	89.3	131.6	105.4	117.4	133.0	123.3	95.4	129.0	108.2	96.9	128.8	109.0	80.8	131.6	100.1	92.0	131.7	107.1
03	CONCRETE	88.1	131.1	107.8	102.9	145.7	122.5	105.9	137.6	120.4	92.1	137.3	112.8	88.9	131.4	108.3	98.3	131.5	113.4
04	MASONRY	92.0	135.4	118.9	118.6	155.8	141.7	103.0	150.3	132.4	96.7	145.1	126.7	74.2	141.1	115.7	83.2	141.1	119.1
05	METALS	102.6	110.5	105.1	106.6	114.8	109.1	104.7	112.8	107.2	103.0	111.9	105.8	96.3	111.4	100.9	96.2	111.5	100.9
06	WOOD, PLASTICS & COMPOSITES	96.1	130.7	115.4	110.2	170.1	143.5	113.7	153.2	135.7	103.1	153.2	130.9	100.0	130.7	117.1	94.2	130.7	114.5
07	THERMAL & MOISTURE PROTECTION	102.6	124.5	111.9	102.7	153.7	124.3	104.5	145.8	125.1	94.0	133.0	110.5	103.4	130.2	114.7	103.7	131.8	115.6
08	OPENINGS	95.1	130.0	103.1	96.2	158.5	110.5	106.3	140.2	114.1	96.7	142.4	107.2	91.0	131.4	100.3	91.0	131.4	100.3
0920	Plaster & Gypsum Board	95.3	131.4	119.5	107.3	171.5	150.4	110.7	154.5	140.1	92.8	154.5	134.2	103.7	131.4	122.3	100.1	131.4	121.1
0950, 0980	Ceilings & Acoustic Treatment	97.9	131.4	120.5	100.5	171.5	148.4	114.5	154.5	141.5	95.3	154.5	135.2	109.2	131.4	124.2	109.2	131.4	124.2
0960	Flooring	100.9	116.0	105.1	116.9	136.3	122.3	101.2	130.6	109.4	103.5	114.8	106.6	109.0	116.0	110.9	105.7	116.0	108.6
0970, 0990	Wall Finishes & Painting/Coating	91.8	108.5	101.5	108.5	165.7	141.9	90.7	127.2	112.0	103.6	120.9	113.7	107.3	120.9	115.3	107.3	120.9	115.3
09	FINISHES	94.9	128.3	113.1	106.8	161.6	136.6	102.3	146.0	126.0	94.3	142.7	120.6	106.1	129.5	118.8	105.1	129.5	118.4
COVERS	DIVS. 10 - 14, 25, 28, 41, 43, 44, 46	100.0	112.1	102.7	100.0	131.3	107.0	100.0	128.1	106.2	100.0	128.0	106.2	100.0	120.6	104.6	100.0	120.6	104.6
21, 22, 23	FIRE SUPPRESSION, PLUMBING & HVAC	100.0	123.0	109.9	97.1	161.4	124.6	96.9	122.8	108.0	100.2	128.4	112.3	96.6	122.9	107.8	96.6	125.6	109.0
26, 27, 3370	ELECTRICAL, COMMUNICATIONS & UTIL.	106.3	105.1	105.6	102.5	159.0	131.9	97.6	120.7	109.6	96.5	108.5	102.7	100.5	126.8	114.2	100.2	126.8	114.1
MF2016	WEIGHTED AVERAGE	98.3	121.3	108.3	102.5	148.9	122.9	102.4	130.0	114.4	98.2	128.2	111.3	95.5	125.5	108.6	97.1	126.1	109.8

CALIFORNIA

DIVISION		LOS ANGELES 900-902			MARYSVILLE 959			MODESTO 953			MOJAVE 935			OAKLAND 946			OXNARD 930		
		MAT.	INST.	TOTAL	MAT.	INST.	TOTAL	MAT.	INST.	TOTAL	MAT.	INST.	TOTAL	MAT.	INST.	TOTAL	MAT.	INST.	TOTAL
015433	CONTRACTOR EQUIPMENT		103.6	103.6		98.7	98.7		98.7	98.7		98.9	98.9		102.7	102.7		97.8	97.8
0241, 31 - 34	SITE & INFRASTRUCTURE, DEMOLITION	97.7	111.1	107.1	109.7	106.3	107.3	104.3	106.5	105.8	97.3	107.6	104.5	112.3	109.4	110.3	104.3	105.7	105.3
0310	Concrete Forming & Accessories	106.3	138.9	134.4	103.8	150.0	143.7	100.3	149.9	143.1	113.8	134.3	131.5	104.4	165.0	156.7	105.3	138.6	134.0
0320	Concrete Reinforcing	95.4	128.6	112.1	102.9	128.9	116.0	106.6	128.8	117.8	90.4	128.0	109.4	92.7	130.1	111.6	88.8	128.1	108.6
0330	Cast-in-Place Concrete	83.4	132.6	102.1	106.6	128.9	115.1	95.5	128.8	108.1	83.8	131.6	102.0	111.3	133.0	119.6	97.6	132.0	110.7
03	CONCRETE	95.8	133.7	113.1	106.7	137.4	120.7	97.9	137.3	115.9	84.1	131.1	105.6	102.9	145.7	122.5	92.2	133.2	110.9
04	MASONRY	89.0	142.4	122.1	104.0	140.4	126.5	101.3	140.4	125.5	94.8	135.4	120.0	126.1	155.8	144.5	98.3	134.6	120.6
05	METALS	102.8	113.6	106.2	104.2	112.1	106.6	101.0	111.9	104.3	100.6	110.4	103.6	101.9	114.8	105.9	98.3	111.2	102.3
06	WOOD, PLASTICS & COMPOSITES	103.9	135.7	121.6	101.2	153.2	130.1	96.9	153.2	128.2	102.2	130.7	118.1	98.8	170.1	138.5	97.4	135.4	118.5
07	THERMAL & MOISTURE PROTECTION	99.5	134.1	114.1	109.4	135.1	120.3	108.9	132.4	118.9	99.9	122.2	109.3	101.4	153.7	123.6	103.2	132.5	115.6
08	OPENINGS	100.8	134.9	108.6	105.5	143.1	114.1	104.2	143.6	113.3	91.4	130.0	100.3	96.2	158.5	110.5	94.1	134.0	103.3
0920	Plaster & Gypsum Board	99.3	136.3	124.1	103.3	154.5	137.7	105.7	154.5	138.5	102.6	131.4	121.9	101.9	171.5	148.7	96.6	136.3	123.2
0950, 0980	Ceilings & Acoustic Treatment	116.2	136.3	129.7	113.6	154.5	141.2	111.1	154.5	140.4	96.7	131.4	120.1	103.3	171.5	149.3	99.2	136.3	124.2
0960	Flooring	106.9	118.6	110.1	96.9	121.2	103.6	97.3	121.0	103.9	104.5	116.0	107.7	109.7	136.3	117.1	96.7	118.6	102.7
0970, 0990	Wall Finishes & Painting/Coating	103.9	120.9	113.8	90.7	134.7	116.4	90.7	131.9	114.7	88.6	107.0	99.3	108.5	165.7	141.9	98.6	115.4	104.2
09	FINISHES	107.1	133.1	121.3	99.3	145.2	124.3	98.8	144.9	123.9	94.6	128.0	112.8	104.8	161.6	135.7	92.1	132.5	114.1
COVERS	DIVS. 10 - 14, 25, 28, 41, 43, 44, 46	100.0	118.7	104.2	100.0	128.1	106.2	100.0	128.0	106.2	100.0	112.1	102.7	100.0	131.3	107.0	100.0	121.1	104.7
21, 22, 23	FIRE SUPPRESSION, PLUMBING & HVAC	100.0	130.1	112.9	96.9	125.4	109.1	100.0	124.1	110.3	96.9	122.4	107.8	100.2	161.4	126.4	100.0	133.2	114.2
26, 27, 3370	ELECTRICAL, COMMUNICATIONS & UTIL.	98.9	130.6	115.4	94.5	118.3	106.9	96.8	107.3	102.3	95.0	107.9	101.7	101.7	159.0	131.5	100.4	117.2	109.1
MF2016	WEIGHTED AVERAGE	99.9	129.3	112.8	101.6	128.8	113.5	100.4	126.9	112.0	95.2	121.4	106.7	102.5	148.9	122.8	97.6	126.5	110.2

CALIFORNIA

DIVISION		PALM SPRINGS 922			PALO ALTO 943			PASADENA 910 - 912			REDDING 960			RICHMOND 948			RIVERSIDE 925		
		MAT.	INST.	TOTAL	MAT.	INST.	TOTAL	MAT.	INST.	TOTAL	MAT.	INST.	TOTAL	MAT.	INST.	TOTAL	MAT.	INST.	TOTAL
015433	CONTRACTOR EQUIPMENT		99.8	99.8		102.7	102.7		95.8	95.8		98.7	98.7		102.7	102.7		99.8	99.8
0241, 31 - 34	SITE & INFRASTRUCTURE, DEMOLITION	93.4	107.9	103.5	103.7	109.4	107.7	97.7	109.1	105.7	133.9	106.3	114.5	111.6	109.4	110.1	100.5	107.9	105.7
0310	Concrete Forming & Accessories	101.7	134.9	130.3	102.6	164.7	156.1	105.4	134.5	130.5	104.3	149.7	143.5	117.8	164.6	158.2	105.4	138.5	133.9
0320	Concrete Reinforcing	108.3	128.2	118.3	90.6	129.6	110.3	105.8	128.4	117.2	127.7	128.9	128.3	90.6	129.6	110.3	105.2	128.1	116.7
0330	Cast-in-Place Concrete	83.8	132.6	102.4	99.3	132.9	112.1	78.6	128.9	97.7	115.6	128.9	120.6	114.2	132.8	121.3	91.2	132.7	106.9
03	CONCRETE	90.1	131.7	109.1	92.6	145.5	116.8	90.4	130.2	108.6	117.1	137.2	126.3	104.8	145.4	123.4	96.0	133.4	113.1
04	MASONRY	76.0	138.5	114.8	100.7	148.3	130.2	97.5	141.0	124.5	132.5	140.4	137.4	118.4	148.3	136.9	77.0	135.7	113.4
05	METALS	105.5	111.5	107.3	99.4	114.0	103.9	86.6	109.5	93.6	106.4	112.1	108.2	99.5	113.8	103.9	104.9	111.5	106.9
06	WOOD, PLASTICS & COMPOSITES	94.7	130.6	114.7	96.4	169.9	137.3	86.5	130.3	110.9	111.3	152.7	134.3	113.4	170.1	145.0	99.7	135.3	119.5
07	THERMAL & MOISTURE PROTECTION	105.5	133.6	117.4	100.6	152.2	122.4	98.5	130.1	111.9	130.6	135.1	132.5	101.3	149.3	121.6	106.4	133.4	117.8
08	OPENINGS	102.8	131.4	109.3	96.2	157.4	110.3	91.2	131.2	100.4	119.0	142.9	124.5	96.2	157.5	110.3	105.7	134.0	112.2
0920	Plaster & Gypsum Board	100.9	131.4	121.4	100.3	171.3	148.0	90.3	131.4	117.9	104.9	154.1	137.9	108.5	171.5	150.8	104.9	136.3	126.0
0950, 0980	Ceilings & Acoustic Treatment	107.7	131.4	123.7	101.4	171.3	148.5	110.3	131.4	124.5	141.6	154.1	150.0	101.4	171.5	148.7	116.3	136.3	129.7
0960	Flooring	99.4	122.3	105.7	108.2	136.3	116.0	96.4	116.0	101.8	91.7	124.9	100.9	119.5	136.3	124.2	101.0	118.6	105.9
0970, 0990	Wall Finishes & Painting/Coating	87.8	118.5	105.7	105.8	165.7	141.9	108.0	120.9	115.9	100.4	134.7	120.4	108.5	165.7	141.9	87.8	118.5	105.7
09	FINISHES	96.0	130.3	114.7	103.2	161.4	134.9	99.3	129.3	115.6	104.6	145.6	126.9	108.3	161.6	137.3	99.1	132.8	117.4
COVERS	DIVS. 10 - 14, 25, 28, 41, 43, 44, 46	100.0	120.4	104.5	100.0	131.3	107.0	100.0	119.7	104.4	100.0	128.0	106.2	100.0	131.2	106.9	100.0	120.9	104.6
21, 22, 23	FIRE SUPPRESSION, PLUMBING & HVAC	96.9	125.3	109.1	97.1	155.6	122.1	96.9	122.8	108.0	100.5	125.4	111.1	97.1	154.1	121.5	100.0	133.2	114.2
26, 27, 3370	ELECTRICAL, COMMUNICATIONS & UTIL.	94.1	110.8	102.8	101.6	169.0	136.6	114.7	126.8	121.0	99.0	118.3	109.1	102.2	136.0	119.8	90.9	108.0	99.8
MF2016	WEIGHTED AVERAGE	97.2	123.9	108.9	98.5	148.0	120.2	96.0	125.3	108.8	109.1	128.8	117.7	101.7	143.0	119.8	99.0	125.6	110.7

CALIFORNIA

DIVISION		SACRAMENTO 942, 956 - 958			SALINAS 939			SAN BERNARDINO 923 - 924			SAN DIEGO 919 - 921			SAN FRANCISCO 940 - 941			SAN JOSE 951		
		MAT.	INST.	TOTAL	MAT.	INST.	TOTAL	MAT.	INST.	TOTAL	MAT.	INST.	TOTAL	MAT.	INST.	TOTAL	MAT.	INST.	TOTAL
015433	CONTRACTOR EQUIPMENT		101.4	101.4		98.9	98.9		99.8	99.8		100.8	100.8		109.7	109.7		100.5	100.5
0241, 31 - 34	SITE & INFRASTRUCTURE, DEMOLITION	93.6	115.5	108.9	118.0	107.1	110.4	79.5	107.9	99.4	106.2	106.5	106.4	113.9	113.0	113.3	135.4	101.4	111.6
0310	Concrete Forming & Accessories	102.5	152.8	145.9	108.4	153.1	147.0	109.2	138.4	134.4	105.3	126.0	123.1	108.4	165.8	157.9	106.2	164.9	156.8
0320	Concrete Reinforcing	86.2	129.3	107.9	89.2	129.4	109.5	105.2	128.3	116.8	102.4	128.3	115.4	105.6	130.3	118.1	94.2	129.7	112.1
0330	Cast-in-Place Concrete	94.2	129.9	107.7	96.4	129.3	108.9	63.0	132.6	89.5	90.0	119.2	101.1	124.3	133.3	127.7	110.9	132.2	119.0
03	CONCRETE	93.9	138.7	114.4	101.1	139.0	118.4	71.0	133.4	99.5	100.6	123.2	110.9	114.2	146.7	129.1	103.9	145.7	123.0
04	MASONRY	101.6	146.1	129.2	94.9	146.5	126.9	83.3	138.5	117.5	89.8	132.5	116.3	135.9	156.5	148.7	131.2	149.8	142.7
05	METALS	97.6	108.5	101.0	103.2	113.2	106.3	104.9	111.6	107.0	102.8	113.0	106.0	107.4	122.2	112.0	99.3	119.1	105.4
06	WOOD, PLASTICS & COMPOSITES	92.2	156.5	128.0	101.8	156.3	132.1	103.2	135.3	121.1	97.9	122.6	111.6	101.1	170.1	139.5	109.7	169.9	143.2
07	THERMAL & MOISTURE PROTECTION	110.6	139.1	122.7	100.5	143.3	118.6	104.5	135.1	117.5	103.2	118.2	109.8	106.5	156.0	127.4	105.4	151.6	125.0
08	OPENINGS	109.6	145.4	117.8	95.3	150.9	108.1	102.8	134.6	110.1	100.1	125.8	106.0	102.2	158.5	115.2	95.7	158.4	110.1
0920	Plaster & Gypsum Board	97.6	157.7	137.9	97.6	157.7	138.0	106.3	136.3	126.4	96.2	123.0	114.2	99.9	171.5	148.0	101.8	171.5	148.6
0950, 0980	Ceilings & Acoustic Treatment	101.4	157.7	139.3	96.7	157.7	137.8	111.1	136.3	128.1	120.8	123.0	122.3	110.6	171.5	151.7	111.2	171.5	151.9
0960	Flooring	108.4	125.7	113.2	98.7	137.9	109.6	102.9	118.6	107.3	106.2	118.6	109.6	110.2	137.9	117.9	90.7	136.3	103.4
0970, 0990	Wall Finishes & Painting/Coating	105.2	134.7	122.4	89.4	165.7	133.9	87.8	120.9	107.1	105.6	120.9	114.5	106.4	175.4	146.7	91.0	165.7	134.6
09	FINISHES	102.1	148.0	127.1	94.0	153.3	126.3	97.4	132.9	116.7	107.0	123.7	116.1	106.4	162.9	137.2	97.8	161.4	132.4
COVERS	DIVS. 10 - 14, 25, 28, 41, 43, 44, 46	100.0	128.9	106.4	100.0	128.4	106.3	100.0	118.1	104.0	100.0	114.5	103.2	100.0	131.4	107.0	100.0	130.8	106.8
21, 22, 23	FIRE SUPPRESSION, PLUMBING & HVAC	100.1	128.1	112.1	96.9	134.5	113.0	96.9	129.9	111.1	99.9	126.5	111.3	100.0	189.7	138.4	100.0	166.0	128.3
26, 27, 3370	ELECTRICAL, COMMUNICATIONS & UTIL.	97.4	122.0	110.2	96.0	128.7	113.0	94.1	111.6	103.2	101.3	103.1	102.2	101.8	174.7	139.7	99.7	170.0	136.3
MF2016	WEIGHTED AVERAGE	100.0	131.6	113.8	98.8	134.8	114.6	94.8	125.7	108.4	101.0	119.5	109.1	106.3	158.5	129.1	102.4	150.4	123.4

CALIFORNIA

DIVISION		SAN LUIS OBISPO 934			SAN MATEO 944			SAN RAFAEL 949			SANTA ANA 926 - 927			SANTA BARBARA 931			SANTA CRUZ 950		
		MAT.	INST.	TOTAL	MAT.	INST.	TOTAL	MAT.	INST.	TOTAL	MAT.	INST.	TOTAL	MAT.	INST.	TOTAL	MAT.	INST.	TOTAL
015433	CONTRACTOR EQUIPMENT		98.9	98.9		102.7	102.7		101.8	101.8		99.8	99.8		98.9	98.9		100.5	100.5
0241, 31 - 34	SITE & INFRASTRUCTURE, DEMOLITION	110.0	107.7	108.4	109.5	109.4	109.4	106.3	115.2	112.5	91.8	107.9	103.0	104.3	107.7	106.7	135.0	101.3	111.4
0310	Concrete Forming & Accessories	115.5	136.8	133.9	108.4	164.9	157.1	112.9	164.9	157.7	109.4	134.7	131.2	106.1	138.6	134.1	106.2	153.5	147.0
0320	Concrete Reinforcing	90.4	128.1	109.4	90.6	129.8	110.4	91.4	129.9	110.8	108.9	128.2	118.6	88.8	128.1	108.6	116.2	129.4	122.9
0330	Cast-in-Place Concrete	103.6	131.8	114.3	110.6	132.9	119.1	128.6	132.0	129.9	80.5	132.5	100.2	97.2	131.9	110.4	110.2	131.1	118.1
03	CONCRETE	99.5	132.3	114.4	101.6	145.6	121.7	122.8	145.1	133.0	87.7	131.6	107.8	92.0	133.1	110.8	106.4	140.1	121.8
04	MASONRY	96.5	135.6	120.7	118.1	151.6	138.8	96.9	152.5	131.4	73.1	138.8	113.8	95.1	135.6	120.2	134.9	146.7	142.2
05	METALS	101.3	111.0	104.3	99.3	114.3	103.9	100.6	110.8	103.7	105.0	111.4	106.9	98.9	111.2	102.7	106.5	117.8	110.0
06	WOOD, PLASTICS & COMPOSITES	104.3	133.2	120.4	103.6	170.1	140.6	99.4	169.9	138.6	105.1	130.6	119.3	97.4	135.4	118.5	109.7	156.4	135.7
07	THERMAL & MOISTURE PROTECTION	100.7	131.7	113.8	101.0	152.4	122.7	105.3	151.9	125.0	105.8	133.7	117.6	100.1	132.6	113.9	105.0	145.8	122.2
08	OPENINGS	93.4	131.4	102.1	96.2	157.5	110.3	107.1	158.0	118.7	102.1	131.4	108.8	95.1	134.4	104.1	97.0	150.9	109.3
0920	Plaster & Gypsum Board	103.5	134.0	124.0	105.2	171.5	149.7	106.9	171.5	150.3	108.0	131.4	123.7	96.6	136.3	123.2	110.4	157.7	142.1
0950, 0980	Ceilings & Acoustic Treatment	96.7	134.0	121.9	101.4	171.5	148.7	108.2	171.5	150.9	111.1	131.4	124.8	99.2	136.3	124.2	111.9	157.7	142.8
0960	Flooring	105.4	116.0	108.4	112.3	136.3	118.9	124.7	130.6	126.3	103.5	116.0	107.0	97.8	118.6	103.6	94.7	137.9	106.7
0970, 0990	Wall Finishes & Painting/Coating	88.6	110.9	101.6	105.8	165.7	141.9	104.6	164.8	139.7	87.8	118.5	105.7	88.6	115.4	104.2	91.2	165.7	134.7
09	FINISHES	96.0	129.9	114.4	105.5	161.6	136.0	108.9	160.3	136.9	98.9	129.2	115.4	92.6	132.3	114.2	100.4	153.4	129.2
COVERS	DIVS. 10 - 14, 25, 28, 41, 43, 44, 46	100.0	127.6	106.1	100.0	131.3	107.0	100.0	130.6	106.8	100.0	120.4	104.5	100.0	118.3	104.1	100.0	128.7	106.4
21, 22, 23	FIRE SUPPRESSION, PLUMBING & HVAC	96.9	127.2	109.9	97.1	157.3	122.9	97.1	185.0	134.7	96.9	122.0	107.6	100.0	133.2	114.2	100.0	134.9	115.0
26, 27, 3370	ELECTRICAL, COMMUNICATIONS & UTIL.	95.0	112.5	104.1	101.6	161.1	132.6	98.8	123.8	111.8	94.1	110.0	102.4	93.9	114.8	104.8	98.9	128.7	114.4
MF2016	WEIGHTED AVERAGE	98.1	124.4	109.6	100.8	147.7	121.3	103.8	148.4	123.3	96.8	122.9	108.3	96.9	126.3	109.8	104.3	135.1	117.8

For customer support on your Commercial Renovation Costs with RSMeans data, call 800.448.8182.

DIVISION		SANTA ROSA 954 MAT.	INST.	TOTAL	STOCKTON 952 MAT.	INST.	TOTAL	SUSANVILLE 961 MAT.	INST.	TOTAL	VALLEJO 945 MAT.	INST.	TOTAL	VAN NUYS 913 - 916 MAT.	INST.	TOTAL	ALAMOSA 811 MAT.	INST.	TOTAL
								CALIFORNIA									COLORADO		
015433	CONTRACTOR EQUIPMENT		99.3	99.3		98.7	98.7		98.7	98.7		101.8	101.8		95.8	95.8		92.8	92.8
0241, 31 - 34	SITE & INFRASTRUCTURE, DEMOLITION	106.1	106.5	106.4	104.0	106.5	105.8	142.3	106.2	117.0	94.8	115.3	109.1	116.1	109.1	111.2	143.5	88.4	104.9
0310	Concrete Forming & Accessories	102.6	164.2	155.7	104.2	152.2	145.6	105.4	141.9	136.9	103.6	163.8	155.5	112.1	134.5	131.4	102.5	63.8	69.1
0320	Concrete Reinforcing	103.8	130.0	117.0	106.6	128.8	117.8	127.7	128.8	128.3	92.5	129.8	111.3	105.8	128.4	117.2	107.9	67.8	87.7
0330	Cast-in-Place Concrete	104.7	130.5	114.5	93.0	128.8	106.6	105.1	128.8	114.1	102.5	131.1	113.4	82.9	128.9	100.4	100.0	73.6	89.9
03	CONCRETE	106.6	144.6	124.0	96.9	138.3	115.9	119.5	133.7	126.0	98.7	144.2	119.5	104.7	130.2	116.4	110.6	68.6	91.4
04	MASONRY	101.8	153.3	133.7	101.2	140.4	125.5	131.1	140.4	136.8	73.7	151.4	121.9	112.0	141.0	130.0	130.4	60.9	87.3
05	METALS	105.3	115.2	108.3	101.2	111.9	104.5	105.2	111.9	107.3	100.5	110.1	103.5	85.7	109.5	93.0	103.7	78.2	95.9
06	WOOD, PLASTICS & COMPOSITES	96.1	169.6	137.0	102.3	156.3	132.3	113.0	142.3	129.3	89.3	169.9	134.1	94.6	130.3	114.5	97.8	64.1	79.0
07	THERMAL & MOISTURE PROTECTION	106.3	152.0	125.6	109.4	134.0	119.8	132.3	131.9	132.1	103.3	151.8	123.8	99.7	130.1	112.6	104.3	69.0	89.4
08	OPENINGS	103.7	158.3	116.3	104.2	143.2	113.2	119.9	137.1	123.9	108.9	156.3	119.8	91.0	131.2	100.3	96.4	67.0	89.6
0920	Plaster & Gypsum Board	102.7	171.5	148.9	105.7	157.7	140.6	105.7	143.4	131.0	101.6	171.5	148.6	94.3	131.4	119.2	78.0	63.0	67.9
0950, 0980	Ceilings & Acoustic Treatment	111.1	171.5	151.8	119.7	157.7	145.3	134.1	143.4	140.3	110.1	171.5	151.5	107.7	131.4	123.7	102.3	63.0	75.8
0960	Flooring	99.9	128.7	107.9	97.3	121.0	103.9	92.1	124.9	101.2	118.0	136.3	123.1	99.8	116.0	104.3	109.3	73.0	99.2
0970, 0990	Wall Finishes & Painting/Coating	87.8	155.6	127.4	90.7	128.3	112.6	100.4	134.7	120.4	105.6	164.8	140.1	108.0	120.9	115.5	103.1	61.9	79.1
09	FINISHES	98.0	158.5	130.9	100.5	146.3	125.4	104.4	139.5	123.5	105.4	161.0	135.6	101.7	129.2	116.7	99.6	64.8	80.7
COVERS	DIVS. 10 - 14, 25, 28, 41, 43, 44, 46	100.0	129.7	106.6	100.0	128.4	106.3	100.0	126.9	106.0	100.0	130.2	106.7	100.0	119.7	104.4	100.0	84.4	96.5
21, 22, 23	FIRE SUPPRESSION, PLUMBING & HVAC	96.9	182.2	133.4	100.0	124.1	110.3	97.4	125.4	109.4	100.2	143.2	118.5	96.9	122.8	108.0	96.9	65.5	83.5
26, 27, 3370	ELECTRICAL, COMMUNICATIONS & UTIL.	94.4	124.4	110.0	96.8	110.8	104.1	99.4	118.3	109.2	95.0	126.8	111.5	114.7	126.8	121.0	97.7	65.1	80.8
MF2016	WEIGHTED AVERAGE	101.1	147.3	121.3	100.5	127.7	112.4	108.8	127.0	116.8	99.5	139.6	117.0	99.1	125.3	110.6	103.3	69.0	88.3

DIVISION		BOULDER 803 MAT.	INST.	TOTAL	COLORADO SPRINGS 808 - 809 MAT.	INST.	TOTAL	DENVER 800 - 802 MAT.	INST.	TOTAL	DURANGO 813 MAT.	INST.	TOTAL	FORT COLLINS 805 MAT.	INST.	TOTAL	FORT MORGAN 807 MAT.	INST.	TOTAL
								COLORADO											
015433	CONTRACTOR EQUIPMENT		94.7	94.7		92.6	92.6		99.1	99.1		92.8	92.8		94.7	94.7		94.7	94.7
0241, 31 - 34	SITE & INFRASTRUCTURE, DEMOLITION	97.3	95.7	96.2	99.1	90.7	93.2	104.2	102.4	102.9	136.6	88.4	102.8	109.6	95.6	99.8	99.5	95.4	96.6
0310	Concrete Forming & Accessories	103.0	74.9	78.8	93.8	60.3	64.9	102.0	63.8	69.1	108.6	63.6	69.8	100.6	69.9	74.1	103.5	53.3	60.2
0320	Concrete Reinforcing	97.7	67.9	82.7	96.9	67.9	82.3	96.9	68.0	82.3	107.9	67.8	87.7	97.8	67.9	82.7	97.9	67.9	82.8
0330	Cast-in-Place Concrete	106.0	74.2	93.9	108.9	73.3	95.4	113.6	72.7	98.0	115.0	73.5	99.2	119.7	72.8	101.9	104.0	72.7	92.1
03	CONCRETE	103.2	73.8	89.7	106.8	66.9	88.6	109.4	68.3	90.6	112.8	68.4	92.5	114.2	71.0	94.5	101.6	63.5	84.2
04	MASONRY	97.3	64.7	77.1	98.9	62.7	76.4	100.1	65.1	78.4	117.1	60.9	82.2	115.2	62.2	82.3	112.5	62.2	81.3
05	METALS	98.1	77.1	91.7	101.3	77.4	93.9	103.8	78.4	96.0	103.7	78.0	95.8	99.4	77.1	92.5	97.8	76.9	91.4
06	WOOD, PLASTICS & COMPOSITES	105.7	77.7	90.1	96.1	59.1	75.5	103.0	63.6	81.1	106.7	64.1	83.0	103.3	72.2	86.0	105.7	49.9	74.6
07	THERMAL & MOISTURE PROTECTION	98.9	73.5	88.1	99.6	70.7	87.4	97.4	71.6	86.5	104.3	69.0	89.4	99.2	71.8	87.6	98.8	69.3	86.3
08	OPENINGS	98.3	74.5	92.8	102.6	64.2	93.8	105.3	66.7	96.4	103.4	67.0	95.0	98.2	71.5	92.1	98.2	59.1	89.2
0920	Plaster & Gypsum Board	116.2	77.5	90.2	100.0	58.2	71.9	111.9	63.0	79.1	90.8	63.0	72.2	110.6	71.8	84.6	116.2	48.9	71.0
0950, 0980	Ceilings & Acoustic Treatment	90.2	77.5	81.6	97.9	58.2	71.1	102.5	63.0	75.9	102.3	63.0	75.8	90.2	71.8	77.8	90.2	48.9	62.4
0960	Flooring	111.8	73.0	101.0	102.8	73.0	94.5	109.0	73.0	99.0	114.7	73.0	103.1	107.7	73.0	98.1	112.3	73.0	101.4
0970, 0990	Wall Finishes & Painting/Coating	101.5	61.9	78.4	101.2	61.9	78.3	108.2	61.9	81.2	103.1	61.9	79.1	101.5	61.9	78.4	101.5	61.9	78.4
09	FINISHES	102.8	73.4	86.8	99.4	61.8	79.0	104.8	64.6	82.9	102.1	64.8	81.8	101.4	69.5	84.0	102.9	56.3	77.6
COVERS	DIVS. 10 - 14, 25, 28, 41, 43, 44, 46	100.0	85.8	96.8	100.0	83.2	96.3	100.0	83.5	96.3	100.0	84.4	96.5	100.0	84.4	96.5	100.0	81.9	96.0
21, 22, 23	FIRE SUPPRESSION, PLUMBING & HVAC	96.9	73.3	86.8	100.1	80.6	91.8	99.9	75.5	89.5	96.9	57.3	80.0	100.0	72.0	88.0	96.9	74.8	87.4
26, 27, 3370	ELECTRICAL, COMMUNICATIONS & UTIL.	95.2	79.5	87.1	98.2	75.8	86.5	99.8	79.6	89.3	97.2	53.1	74.3	95.2	79.5	87.1	95.5	79.4	87.2
MF2016	WEIGHTED AVERAGE	98.8	75.9	88.8	101.0	73.2	88.8	102.8	74.6	90.5	103.7	65.5	87.0	102.1	74.3	89.9	99.4	71.2	87.0

DIVISION		GLENWOOD SPRINGS 816 MAT.	INST.	TOTAL	GOLDEN 804 MAT.	INST.	TOTAL	GRAND JUNCTION 815 MAT.	INST.	TOTAL	GREELEY 806 MAT.	INST.	TOTAL	MONTROSE 814 MAT.	INST.	TOTAL	PUEBLO 810 MAT.	INST.	TOTAL
								COLORADO											
015433	CONTRACTOR EQUIPMENT		96.0	96.0		94.7	94.7		96.0	96.0		94.7	94.7		94.4	94.4		92.8	92.8
0241, 31 - 34	SITE & INFRASTRUCTURE, DEMOLITION	153.2	96.4	113.4	110.6	95.7	100.2	135.9	96.6	108.4	95.9	95.6	95.7	146.3	92.4	108.5	127.3	88.4	100.0
0310	Concrete Forming & Accessories	99.5	53.2	59.6	96.1	59.6	64.6	107.5	74.3	78.9	98.5	73.9	77.3	99.0	53.5	59.8	104.7	60.3	66.4
0320	Concrete Reinforcing	106.7	67.8	87.1	97.9	67.9	82.8	107.0	67.9	87.3	97.7	67.9	82.7	106.5	67.8	87.0	103.1	67.9	85.3
0330	Cast-in-Place Concrete	99.9	72.4	89.5	104.1	74.1	92.7	110.7	73.9	96.7	100.0	72.8	89.7	100.0	73.0	89.7	99.2	73.6	89.5
03	CONCRETE	115.7	63.4	91.8	112.7	66.8	91.7	109.3	73.5	92.9	98.3	72.8	86.6	106.8	63.7	87.1	99.5	67.0	84.7
04	MASONRY	101.7	62.1	77.2	115.6	64.7	84.0	137.4	65.4	92.7	109.0	62.2	80.0	110.2	60.9	79.6	97.9	62.0	75.7
05	METALS	103.4	77.7	95.5	98.0	77.0	91.6	105.1	78.1	96.8	99.3	77.1	92.5	102.5	78.0	95.0	106.8	78.3	98.0
06	WOOD, PLASTICS & COMPOSITES	93.1	50.0	69.2	98.3	57.1	75.4	104.4	77.0	89.2	97.6	77.6	87.8	94.4	50.2	69.8	100.3	59.4	77.5
07	THERMAL & MOISTURE PROTECTION	104.2	67.9	88.8	99.9	71.2	87.8	103.3	72.2	90.1	98.5	72.4	87.4	104.4	67.5	88.8	102.8	69.1	88.5
08	OPENINGS	102.4	59.2	92.5	98.2	63.1	90.2	103.1	74.1	96.5	98.2	74.5	92.8	103.5	59.3	93.4	98.2	64.4	90.5
0920	Plaster & Gypsum Board	120.4	48.9	72.4	108.3	56.4	73.4	132.9	76.7	95.1	109.0	77.4	87.8	77.3	48.9	58.2	82.0	58.2	66.0
0950, 0980	Ceilings & Acoustic Treatment	101.5	48.9	66.0	90.2	56.4	67.4	101.5	76.7	84.7	90.2	77.4	81.6	102.3	48.9	66.3	110.0	58.2	75.1
0960	Flooring	108.4	73.0	98.6	105.2	73.0	96.3	114.0	73.0	102.6	106.7	73.0	97.2	112.0	73.0	101.1	110.6	73.0	100.1
0970, 0990	Wall Finishes & Painting/Coating	103.1	61.9	79.1	101.5	61.9	78.4	103.1	61.9	79.1	101.5	61.9	78.4	103.1	61.9	79.1	103.1	61.9	79.1
09	FINISHES	105.3	56.5	78.8	101.0	61.2	79.4	106.9	73.0	88.5	100.0	72.7	85.2	100.3	56.6	76.5	100.2	62.0	79.4
COVERS	DIVS. 10 - 14, 25, 28, 41, 43, 44, 46	100.0	82.2	96.0	100.0	83.5	96.3	100.0	86.0	96.9	100.0	85.0	96.7	100.0	82.5	96.1	100.0	83.9	96.4
21, 22, 23	FIRE SUPPRESSION, PLUMBING & HVAC	96.9	66.6	83.9	96.9	73.3	86.8	99.9	76.4	89.9	100.0	72.0	88.0	96.9	77.1	88.4	99.9	66.6	85.7
26, 27, 3370	ELECTRICAL, COMMUNICATIONS & UTIL.	94.7	53.1	73.1	95.5	79.5	87.2	96.9	53.1	74.1	95.2	79.5	87.1	96.9	53.1	74.1	97.7	66.2	81.3
MF2016	WEIGHTED AVERAGE	103.5	65.9	87.0	101.0	72.6	88.6	105.5	73.0	91.3	99.3	75.1	88.7	102.4	67.8	87.3	101.3	68.8	87.1

DIVISION		COLORADO SALIDA 812			CONNECTICUT BRIDGEPORT 066			BRISTOL 060			HARTFORD 061			MERIDEN 064			NEW BRITAIN 060		
		MAT.	INST.	TOTAL	MAT.	INST.	TOTAL	MAT.	INST.	TOTAL	MAT.	INST.	TOTAL	MAT.	INST.	TOTAL	MAT.	INST.	TOTAL
015433	CONTRACTOR EQUIPMENT		94.4	94.4		98.8	98.8		98.8	98.8		98.8	98.8		99.3	99.3		98.8	98.8
0241, 31 - 34	SITE & INFRASTRUCTURE, DEMOLITION	136.3	92.4	105.5	105.9	103.2	104.0	105.0	103.2	103.7	100.3	103.2	102.3	102.9	103.9	103.6	105.2	103.2	103.8
0310	Concrete Forming & Accessories	107.4	53.3	60.8	102.2	118.7	116.5	102.2	118.6	116.3	102.1	118.6	116.3	101.9	118.5	116.3	102.6	118.6	116.4
0320	Concrete Reinforcing	106.2	67.8	86.8	117.4	142.1	129.8	117.4	142.1	129.8	112.6	142.1	127.5	117.4	142.1	129.8	117.4	142.1	129.8
0330	Cast-in-Place Concrete	114.5	72.9	98.7	98.7	129.3	110.3	92.5	129.2	106.4	96.9	129.2	109.2	89.0	129.2	104.3	94.0	129.2	107.4
03	CONCRETE	107.8	63.6	87.6	96.2	125.6	109.7	93.3	125.5	108.0	93.5	125.5	108.2	91.6	125.5	107.1	94.0	125.5	108.4
04	MASONRY	138.9	60.9	90.5	113.9	130.1	124.0	105.1	130.1	120.6	104.4	130.1	120.3	104.8	130.1	120.5	107.5	130.1	121.5
05	METALS	102.2	77.8	94.7	96.2	116.4	102.4	96.2	116.2	102.4	101.2	116.3	105.8	93.6	116.2	100.6	92.7	116.3	100.0
06	WOOD, PLASTICS & COMPOSITES	101.3	50.2	72.9	110.3	116.7	113.9	110.3	116.7	113.9	96.7	116.7	107.8	110.3	116.7	113.9	110.3	116.7	113.9
07	THERMAL & MOISTURE PROTECTION	103.2	67.5	88.1	100.1	124.8	110.6	100.3	121.4	109.2	105.4	121.4	112.1	100.3	121.3	109.2	100.3	121.6	109.3
08	OPENINGS	96.5	59.3	87.9	95.5	122.4	101.7	95.5	122.4	101.7	94.0	122.4	100.5	97.8	122.4	103.4	95.5	122.4	101.7
0920	Plaster & Gypsum Board	77.6	48.9	58.3	115.6	116.8	116.4	115.6	116.8	116.4	103.9	116.8	112.5	116.8	116.8	116.8	115.6	116.8	116.4
0950, 0980	Ceilings & Acoustic Treatment	102.3	48.9	66.3	95.3	116.8	109.8	95.3	116.8	109.8	93.2	116.8	109.1	98.5	116.8	110.8	95.3	116.8	109.8
0960	Flooring	117.6	73.0	105.2	87.3	131.6	99.6	87.3	124.4	97.6	90.2	131.6	101.7	87.3	124.4	97.6	87.3	124.4	97.6
0970, 0990	Wall Finishes & Painting/Coating	103.1	61.9	79.1	92.3	128.8	113.6	92.3	128.8	113.6	94.4	128.8	114.5	92.3	128.8	113.6	92.3	128.8	113.6
09	FINISHES	101.0	56.6	76.6	91.4	121.4	107.7	91.4	120.2	107.0	93.6	121.4	108.7	92.2	120.2	107.4	91.4	120.2	107.0
COVERS	DIVS. 10 - 14, 25, 28, 41, 43, 44, 46	100.0	82.6	96.1	100.0	113.2	102.9	100.0	113.2	102.9	100.0	113.2	102.9	100.0	113.2	102.9	100.0	113.2	102.9
21, 22, 23	FIRE SUPPRESSION, PLUMBING & HVAC	96.9	62.9	82.3	100.1	118.8	108.1	100.1	118.7	108.1	100.0	118.8	108.0	97.0	118.7	106.3	100.1	118.7	108.1
26, 27, 3370	ELECTRICAL, COMMUNICATIONS & UTIL.	97.1	60.8	78.2	99.0	108.3	103.8	99.0	106.3	102.8	98.5	110.5	104.8	98.9	105.7	102.5	99.0	106.3	102.8
MF2016	WEIGHTED AVERAGE	103.0	65.8	86.7	98.5	118.4	107.2	97.7	117.9	106.5	98.4	118.6	107.3	96.6	117.8	105.9	97.4	117.9	106.3

DIVISION		CONNECTICUT NEW HAVEN 065			NEW LONDON 063			NORWALK 068			STAMFORD 069			WATERBURY 067			WILLIMANTIC 062		
		MAT.	INST.	TOTAL	MAT.	INST.	TOTAL	MAT.	INST.	TOTAL	MAT.	INST.	TOTAL	MAT.	INST.	TOTAL	MAT.	INST.	TOTAL
015433	CONTRACTOR EQUIPMENT		99.3	99.3		99.3	99.3		98.8	98.8		98.8	98.8		98.8	98.8		98.8	98.8
0241, 31 - 34	SITE & INFRASTRUCTURE, DEMOLITION	105.1	103.7	104.1	97.4	103.7	101.8	105.6	103.1	103.9	106.3	103.2	104.1	105.7	103.2	103.9	105.6	102.9	103.7
0310	Concrete Forming & Accessories	101.9	118.6	116.3	101.9	118.6	116.3	102.2	118.7	116.4	102.2	119.1	116.7	102.2	118.6	116.4	102.2	118.5	116.2
0320	Concrete Reinforcing	117.4	142.1	129.8	92.0	142.1	117.3	117.4	142.1	129.8	117.4	142.2	129.9	117.4	142.1	129.8	117.4	142.1	129.8
0330	Cast-in-Place Concrete	95.6	129.2	108.4	81.5	129.2	99.6	97.1	129.3	109.3	98.7	129.4	110.4	97.1	129.2	110.3	92.2	129.2	106.2
03	CONCRETE	106.6	125.5	115.3	82.3	125.5	102.1	95.4	125.6	109.2	96.2	125.8	109.8	96.2	125.6	109.6	93.1	125.5	107.9
04	MASONRY	105.4	130.1	120.7	103.7	130.1	120.1	104.8	130.1	120.5	105.6	130.1	120.8	105.6	130.1	120.8	105.0	130.1	120.6
05	METALS	92.9	116.3	100.1	92.7	116.3	99.9	96.2	116.4	102.4	96.2	116.6	102.5	96.2	116.3	102.4	96.0	116.2	102.2
06	WOOD, PLASTICS & COMPOSITES	110.3	116.7	113.9	110.3	116.7	113.9	110.3	116.7	113.9	110.3	116.7	113.9	110.3	116.7	113.9	110.3	116.7	113.9
07	THERMAL & MOISTURE PROTECTION	100.4	121.7	109.4	100.2	121.4	109.1	100.3	124.8	110.7	100.3	124.8	110.7	100.3	121.8	109.4	100.5	121.0	109.2
08	OPENINGS	95.5	122.4	101.7	97.9	122.4	103.5	95.5	122.4	101.7	95.5	122.4	101.7	95.5	122.4	101.7	97.9	122.4	103.5
0920	Plaster & Gypsum Board	115.6	116.8	116.4	115.6	116.8	116.4	115.6	116.8	116.4	115.6	116.8	116.4	115.6	116.8	116.4	115.6	116.8	116.4
0950, 0980	Ceilings & Acoustic Treatment	95.3	116.8	109.8	93.4	116.8	109.2	95.3	116.8	109.8	95.3	116.8	109.8	95.3	116.8	109.8	93.4	116.8	109.2
0960	Flooring	87.3	131.6	99.6	87.3	131.6	99.6	87.3	124.4	97.6	87.3	131.6	99.6	87.3	129.1	98.9	87.3	128.0	98.6
0970, 0990	Wall Finishes & Painting/Coating	92.3	128.8	113.6	92.3	128.8	113.6	92.3	128.8	113.6	92.3	128.8	113.6	92.3	128.8	113.6	92.3	128.8	113.6
09	FINISHES	91.4	121.4	107.7	90.6	121.4	107.3	91.4	120.2	107.1	91.5	121.4	107.7	91.3	121.0	107.4	91.1	120.8	107.3
COVERS	DIVS. 10 - 14, 25, 28, 41, 43, 44, 46	100.0	113.2	102.9	100.0	113.2	102.9	100.0	113.2	102.9	100.0	113.4	103.0	100.0	113.2	102.9	100.0	113.2	102.9
21, 22, 23	FIRE SUPPRESSION, PLUMBING & HVAC	100.1	118.8	108.1	97.0	118.7	106.3	100.1	118.8	108.1	100.1	118.8	108.1	100.1	118.8	108.1	100.1	118.6	108.0
26, 27, 3370	ELECTRICAL, COMMUNICATIONS & UTIL.	98.9	112.0	105.7	96.0	109.6	103.1	99.0	110.7	105.1	99.0	158.5	130.0	98.5	111.4	105.2	99.0	110.5	105.0
MF2016	WEIGHTED AVERAGE	98.9	118.9	107.6	94.7	118.5	105.1	98.0	118.6	107.0	98.1	125.5	110.1	98.1	118.7	107.1	97.9	118.5	106.9

DIVISION		D.C. WASHINGTON 200 - 205			DELAWARE DOVER 199			NEWARK 197			WILMINGTON 198			FLORIDA DAYTONA BEACH 321			FORT LAUDERDALE 333		
		MAT.	INST.	TOTAL	MAT.	INST.	TOTAL	MAT.	INST.	TOTAL	MAT.	INST.	TOTAL	MAT.	INST.	TOTAL	MAT.	INST.	TOTAL
015433	CONTRACTOR EQUIPMENT		106.4	106.4		122.9	122.9		122.9	122.9		123.1	123.1		101.9	101.9		94.8	94.8
0241, 31 - 34	SITE & INFRASTRUCTURE, DEMOLITION	104.0	96.5	98.7	104.1	114.8	111.6	103.8	114.8	111.5	105.0	115.2	112.1	102.4	88.7	92.8	94.2	76.1	81.5
0310	Concrete Forming & Accessories	99.2	72.3	76.0	95.9	102.3	101.4	98.3	102.3	101.8	96.7	102.3	101.5	97.3	63.6	68.3	95.8	56.5	61.9
0320	Concrete Reinforcing	100.6	87.1	93.8	105.3	114.3	109.9	100.8	114.3	107.6	105.3	114.3	109.9	92.6	65.9	79.2	92.3	57.5	74.7
0330	Cast-in-Place Concrete	109.9	81.0	98.9	101.0	107.7	103.5	88.3	107.7	95.7	98.2	107.7	101.8	88.8	65.8	80.0	88.6	62.7	78.7
03	CONCRETE	107.2	79.0	94.3	96.2	107.4	101.3	91.2	107.4	98.6	95.0	107.4	100.7	86.4	66.5	77.3	87.4	60.7	75.2
04	MASONRY	101.8	82.1	89.6	101.0	100.5	100.7	103.1	100.5	101.5	101.3	100.5	100.8	89.7	59.6	71.0	94.7	63.5	75.4
05	METALS	101.9	96.6	100.3	102.0	124.3	108.9	103.5	124.3	109.9	102.0	124.3	108.9	96.0	90.9	94.5	94.8	87.8	92.7
06	WOOD, PLASTICS & COMPOSITES	99.7	70.2	83.3	91.3	100.3	96.3	96.8	100.3	98.8	87.3	100.3	94.5	100.5	62.9	79.6	90.3	56.4	71.4
07	THERMAL & MOISTURE PROTECTION	99.8	85.6	93.8	100.7	113.2	106.0	105.0	113.2	108.5	100.1	113.2	105.6	97.2	65.8	83.9	98.4	62.2	83.1
08	OPENINGS	99.8	74.6	94.0	90.6	110.5	95.2	91.5	110.5	95.9	88.7	110.5	93.7	95.3	62.0	87.7	96.2	56.3	87.0
0920	Plaster & Gypsum Board	106.8	69.3	81.6	101.9	100.3	100.8	98.9	100.3	99.8	101.3	100.3	100.6	88.7	62.4	71.0	97.7	55.7	69.5
0950, 0980	Ceilings & Acoustic Treatment	99.2	69.3	79.0	94.2	100.3	98.3	96.1	100.3	98.9	91.6	100.3	97.5	83.2	62.4	69.2	87.2	55.7	65.9
0960	Flooring	97.4	79.0	92.3	87.5	111.5	94.1	86.5	111.5	93.4	88.0	111.5	94.5	96.8	58.0	86.0	94.1	78.5	89.8
0970, 0990	Wall Finishes & Painting/Coating	103.1	74.9	86.7	88.0	110.6	101.2	86.7	110.6	100.6	83.9	110.6	99.5	94.7	62.4	75.8	87.8	63.1	73.3
09	FINISHES	96.3	72.8	83.5	92.2	104.0	98.6	88.9	104.0	97.1	91.7	104.0	98.4	88.2	61.9	73.9	88.7	60.9	73.6
COVERS	DIVS. 10 - 14, 25, 28, 41, 43, 44, 46	100.0	95.9	99.1	100.0	106.2	101.4	100.0	106.2	101.4	100.0	106.2	101.4	100.0	82.6	96.1	100.0	81.6	95.9
21, 22, 23	FIRE SUPPRESSION, PLUMBING & HVAC	100.0	89.9	95.7	100.0	122.0	109.4	100.2	122.0	109.5	100.0	122.0	109.4	99.9	76.4	89.8	99.9	58.9	82.4
26, 27, 3370	ELECTRICAL, COMMUNICATIONS & UTIL.	100.6	102.8	101.7	97.3	112.7	105.3	98.9	112.7	106.1	97.3	112.7	105.3	92.7	62.2	76.8	94.6	69.4	81.4
MF2016	WEIGHTED AVERAGE	101.1	87.6	95.2	98.0	112.4	104.3	97.9	112.4	104.3	97.6	112.5	104.1	94.9	71.0	84.4	95.1	66.0	82.4

FLORIDA

DIVISION		FORT MYERS 339, 341			GAINESVILLE 326, 344			JACKSONVILLE 320, 322			LAKELAND 338			MELBOURNE 329			MIAMI 330 - 332, 340		
		MAT.	INST.	TOTAL	MAT.	INST.	TOTAL	MAT.	INST.	TOTAL	MAT.	INST.	TOTAL	MAT.	INST.	TOTAL	MAT.	INST.	TOTAL
015433	CONTRACTOR EQUIPMENT		101.9	101.9		101.9	101.9		101.9	101.9		101.9	101.9		101.9	101.9		94.8	94.8
0241, 31 - 34	SITE & INFRASTRUCTURE, DEMOLITION	105.9	88.2	93.5	110.0	88.5	95.0	102.4	88.5	92.7	107.8	88.7	94.4	109.1	88.9	94.9	95.5	76.1	81.9
0310	Concrete Forming & Accessories	91.7	56.5	61.3	92.5	56.6	61.5	97.1	56.6	62.1	88.0	62.5	66.0	93.7	63.8	67.9	100.0	56.2	62.3
0320	Concrete Reinforcing	93.3	56.0	74.5	98.2	58.4	78.1	92.6	58.2	75.3	95.6	76.5	86.0	93.6	65.9	79.7	99.1	56.9	77.8
0330	Cast-in-Place Concrete	92.4	62.6	81.1	101.9	64.0	87.5	89.6	63.9	79.9	94.5	65.3	83.4	107.0	65.9	91.4	85.6	62.5	76.8
03	CONCRETE	88.0	60.4	75.4	97.0	61.4	80.7	86.8	61.3	75.2	89.6	67.6	79.6	97.5	66.6	83.4	85.8	60.5	74.2
04	MASONRY	88.4	63.5	73.0	103.0	56.6	74.2	89.4	54.1	67.6	104.5	58.8	76.2	87.5	59.6	70.2	95.0	54.4	69.8
05	METALS	96.8	87.2	93.8	94.9	87.9	92.8	94.6	87.7	92.5	96.7	94.3	95.9	104.1	91.0	100.1	95.2	86.8	92.6
06	WOOD, PLASTICS & COMPOSITES	87.2	56.4	70.0	94.4	55.5	72.8	100.5	55.5	75.5	82.4	62.1	71.1	96.1	62.9	77.6	96.2	56.4	74.0
07	THERMAL & MOISTURE PROTECTION	98.2	62.5	83.1	97.6	60.9	82.1	97.5	60.2	81.7	98.1	62.6	83.0	97.7	64.6	83.7	98.2	59.7	81.9
08	OPENINGS	97.4	56.0	87.9	94.9	56.2	86.0	95.3	56.2	86.3	97.3	64.0	89.7	94.5	64.6	87.6	98.8	56.3	89.0
0920	Plaster & Gypsum Board	93.8	55.7	68.2	85.5	54.8	64.9	88.7	54.8	65.9	90.5	61.6	71.1	85.5	62.4	70.0	87.7	55.7	66.2
0950, 0980	Ceilings & Acoustic Treatment	82.7	55.7	64.5	77.6	54.8	62.2	83.2	54.8	64.0	82.7	61.6	68.5	82.3	62.4	68.9	86.8	55.7	65.8
0960	Flooring	91.1	76.7	87.1	94.2	58.0	84.2	96.8	58.0	86.0	88.9	58.0	80.3	94.4	58.0	84.3	93.0	58.0	83.3
0970, 0990	Wall Finishes & Painting/Coating	91.7	63.0	75.0	94.7	62.4	75.8	94.7	63.0	76.2	91.7	63.0	75.0	94.7	83.5	88.2	85.8	63.1	72.5
09	FINISHES	88.4	60.5	73.2	86.6	56.7	70.4	88.2	56.8	71.1	87.4	61.2	73.2	87.4	64.2	74.8	87.5	56.7	70.7
COVERS	DIVS. 10 - 14, 25, 28, 41, 43, 44, 46	100.0	78.4	95.2	100.0	80.8	95.7	100.0	79.0	95.3	100.0	80.5	95.7	100.0	82.6	96.1	100.0	81.6	95.9
21, 22, 23	FIRE SUPPRESSION, PLUMBING & HVAC	98.2	57.0	80.6	98.7	61.9	82.9	99.9	61.9	83.6	98.2	59.5	81.7	99.9	76.4	89.8	99.9	57.3	81.7
26, 27, 3370	ELECTRICAL, COMMUNICATIONS & UTIL.	96.4	62.9	79.0	92.9	59.0	75.3	92.4	64.1	77.7	94.8	63.4	78.5	93.7	61.0	76.7	98.2	73.3	85.2
MF2016	WEIGHTED AVERAGE	95.4	65.4	82.3	96.4	65.0	82.7	94.7	65.3	81.8	96.2	67.8	83.7	97.6	71.2	86.0	95.6	64.6	82.0

FLORIDA

DIVISION		ORLANDO 327 - 328, 347			PANAMA CITY 324			PENSACOLA 325			SARASOTA 342			ST. PETERSBURG 337			TALLAHASSEE 323		
		MAT.	INST.	TOTAL	MAT.	INST.	TOTAL	MAT.	INST.	TOTAL	MAT.	INST.	TOTAL	MAT.	INST.	TOTAL	MAT.	INST.	TOTAL
015433	CONTRACTOR EQUIPMENT		101.9	101.9		101.9	101.9		101.9	101.9		101.9	101.9		101.9	101.9		101.9	101.9
0241, 31 - 34	SITE & INFRASTRUCTURE, DEMOLITION	99.7	88.7	92.0	113.1	88.0	95.5	113.8	88.5	96.0	113.3	88.7	96.1	109.4	88.3	94.6	101.7	88.5	92.5
0310	Concrete Forming & Accessories	100.1	63.3	68.4	96.4	61.4	66.2	94.5	64.7	68.8	95.4	62.5	67.0	95.2	60.3	65.2	97.1	56.6	62.2
0320	Concrete Reinforcing	96.2	65.9	80.9	96.7	69.1	82.8	99.2	68.6	83.7	93.3	76.5	84.8	95.6	76.5	86.0	99.8	58.4	78.9
0330	Cast-in-Place Concrete	105.9	65.7	90.7	94.2	59.9	81.2	116.4	65.1	96.9	103.2	65.3	88.8	95.5	62.1	82.8	91.4	64.0	81.0
03	CONCRETE	94.1	66.3	81.4	95.1	64.0	80.9	104.9	67.2	87.7	95.2	67.6	82.6	91.1	65.5	79.4	88.3	61.4	76.0
04	MASONRY	95.3	59.6	73.2	94.1	58.6	72.1	114.3	58.7	79.8	92.0	58.8	71.4	144.1	53.2	87.7	91.7	56.6	69.9
05	METALS	95.2	90.6	93.8	95.8	92.1	94.6	96.8	91.4	95.1	98.7	94.3	97.4	97.6	93.2	96.3	92.3	87.9	91.0
06	WOOD, PLASTICS & COMPOSITES	89.8	62.9	74.9	99.3	65.4	80.4	97.3	65.4	79.5	98.1	62.1	78.1	91.7	62.1	75.2	96.0	55.5	73.5
07	THERMAL & MOISTURE PROTECTION	100.0	65.8	85.5	97.8	59.9	81.8	97.7	63.0	83.0	98.2	62.6	83.1	98.3	60.0	82.1	96.9	60.8	81.7
08	OPENINGS	98.1	62.0	89.8	93.3	64.0	86.6	93.3	64.0	86.6	99.5	64.0	91.3	96.1	63.6	88.6	100.1	56.2	90.0
0920	Plaster & Gypsum Board	94.7	62.4	73.0	87.8	64.9	72.4	95.5	64.9	75.0	94.7	61.6	72.5	96.1	61.6	72.9	93.3	54.8	67.4
0950, 0980	Ceilings & Acoustic Treatment	89.4	62.4	71.2	82.3	64.9	70.6	82.3	64.9	70.6	87.3	61.6	69.9	84.6	61.6	69.1	87.9	54.8	65.6
0960	Flooring	91.3	58.0	82.1	96.3	76.7	90.9	92.2	58.0	82.7	101.4	58.0	89.4	93.0	58.0	83.3	92.9	58.0	83.2
0970, 0990	Wall Finishes & Painting/Coating	93.4	63.0	75.7	94.7	60.6	74.8	94.7	62.4	75.8	96.3	63.0	76.9	91.7	63.6	75.3	91.9	63.0	75.1
09	FINISHES	90.4	61.9	74.9	88.8	64.3	75.5	88.4	63.1	74.6	94.3	61.2	76.3	89.9	59.9	73.6	89.8	56.8	71.8
COVERS	DIVS. 10 - 14, 25, 28, 41, 43, 44, 46	100.0	82.6	96.1	100.0	78.0	95.1	100.0	81.1	95.8	100.0	80.5	95.7	100.0	78.6	95.2	100.0	78.6	95.2
21, 22, 23	FIRE SUPPRESSION, PLUMBING & HVAC	100.0	57.8	81.9	99.9	51.9	79.4	99.9	63.0	84.1	99.9	59.5	82.6	99.9	56.6	81.4	99.9	67.8	86.2
26, 27, 3370	ELECTRICAL, COMMUNICATIONS & UTIL.	96.6	62.7	78.9	91.6	59.0	74.6	94.8	53.5	73.3	95.7	63.9	79.2	94.8	64.0	78.8	96.7	59.0	77.1
MF2016	WEIGHTED AVERAGE	96.8	67.0	83.8	96.2	65.1	82.5	98.8	67.1	84.9	98.1	67.9	84.9	99.0	65.9	84.5	95.6	66.2	82.7

DIVISION		FLORIDA TAMPA 335 - 336, 346			WEST PALM BEACH 334, 349			GEORGIA ALBANY 317, 398			ATHENS 306			ATLANTA 300 - 303, 399			AUGUSTA 308 - 309		
		MAT.	INST.	TOTAL	MAT.	INST.	TOTAL	MAT.	INST.	TOTAL	MAT.	INST.	TOTAL	MAT.	INST.	TOTAL	MAT.	INST.	TOTAL
015433	CONTRACTOR EQUIPMENT		101.9	101.9		94.8	94.8		95.7	95.7		94.5	94.5		96.4	96.4		94.5	94.5
0241, 31 - 34	SITE & INFRASTRUCTURE, DEMOLITION	109.9	88.7	95.1	91.1	76.1	80.6	104.3	79.0	86.6	102.6	95.1	97.3	99.4	95.3	96.5	95.6	95.4	95.5
0310	Concrete Forming & Accessories	98.3	62.4	67.4	99.4	56.3	62.2	94.0	68.9	72.4	91.4	45.1	51.5	96.0	71.7	75.1	92.5	72.9	75.6
0320	Concrete Reinforcing	92.3	76.5	84.3	94.9	56.9	75.7	88.1	71.9	79.9	102.0	65.6	83.6	101.2	72.0	86.5	102.4	68.1	85.1
0330	Cast-in-Place Concrete	93.4	65.3	82.7	84.2	62.5	76.0	88.3	69.0	81.0	109.3	69.9	94.3	112.7	71.4	97.0	103.3	70.2	90.7
03	CONCRETE	89.8	67.6	79.7	84.4	60.5	73.5	86.4	71.2	79.5	104.9	58.6	83.7	107.1	72.4	91.2	97.3	71.8	85.6
04	MASONRY	95.7	58.8	72.8	94.3	54.4	69.5	92.2	69.2	77.9	78.5	79.4	79.1	91.9	69.4	77.9	92.2	69.3	78.0
05	METALS	96.7	94.3	96.0	93.9	86.8	91.7	97.8	80.9	89.1	93.1	80.0	89.1	93.9	85.1	91.2	92.8	81.3	89.3
06	WOOD, PLASTICS & COMPOSITES	95.7	62.1	77.0	95.3	56.4	73.6	85.9	69.6	76.8	97.0	37.4	63.9	99.2	73.1	84.7	98.5	75.3	85.6
07	THERMAL & MOISTURE PROTECTION	98.6	62.6	83.3	98.1	61.3	82.5	97.3	68.9	85.3	97.9	70.2	86.2	99.4	72.4	88.0	97.7	71.7	86.7
08	OPENINGS	97.4	64.0	89.7	95.7	56.3	86.7	87.7	71.6	84.0	93.1	52.0	83.6	97.6	73.5	92.1	93.1	73.8	88.7
0920	Plaster & Gypsum Board	98.4	61.6	73.7	102.1	55.7	70.9	103.6	69.3	80.5	93.6	36.1	55.0	96.0	72.5	80.2	94.9	75.0	81.6
0950, 0980	Ceilings & Acoustic Treatment	87.2	61.6	69.9	82.7	55.7	64.5	83.8	69.3	74.0	98.8	36.1	56.6	91.6	72.5	78.7	99.8	75.0	83.1
0960	Flooring	94.1	58.0	84.1	95.9	58.0	85.4	96.9	67.4	88.7	95.8	86.6	93.3	98.8	67.4	90.1	96.0	67.4	88.1
0970, 0990	Wall Finishes & Painting/Coating	91.7	63.0	75.0	87.8	62.4	72.9	89.1	95.9	93.1	97.5	95.9	96.6	101.3	95.9	98.2	97.5	80.4	87.5
09	FINISHES	91.1	61.2	74.9	88.7	56.7	71.3	92.2	71.1	80.7	96.2	56.0	74.4	96.3	73.2	83.7	96.0	72.8	83.4
COVERS	DIVS. 10 - 14, 25, 28, 41, 43, 44, 46	100.0	80.5	95.7	100.0	81.6	95.9	100.0	84.9	96.6	100.0	81.2	95.8	100.0	86.0	96.9	100.0	85.3	96.7
21, 22, 23	FIRE SUPPRESSION, PLUMBING & HVAC	99.9	59.5	82.6	98.2	56.8	80.5	99.9	71.1	87.6	97.0	68.5	84.8	100.0	70.1	87.2	100.1	69.2	86.9
26, 27, 3370	ELECTRICAL, COMMUNICATIONS & UTIL.	94.5	62.2	77.7	95.6	69.4	81.9	95.0	63.6	78.7	99.1	67.5	82.7	98.5	71.4	84.4	99.7	70.5	84.5
MF2016	WEIGHTED AVERAGE	96.6	67.6	83.9	94.2	64.0	81.0	95.0	73.4	85.5	96.6	69.1	84.5	98.7	75.0	88.3	96.8	74.2	86.9

For customer support on your Commercial Renovation Costs with RSMeans data, call 800.448.8182.

GEORGIA

DIVISION		COLUMBUS 318 - 319			DALTON 307			GAINESVILLE 305			MACON 310 - 312			SAVANNAH 313 - 314			STATESBORO 304		
		MAT.	INST.	TOTAL	MAT.	INST.	TOTAL	MAT.	INST.	TOTAL	MAT.	INST.	TOTAL	MAT.	INST.	TOTAL	MAT.	INST.	TOTAL
015433	CONTRACTOR EQUIPMENT		95.7	95.7		110.4	110.4		94.5	94.5		105.7	105.7		96.6	96.6		97.6	97.6
0241, 31 - 34	SITE & INFRASTRUCTURE, DEMOLITION	104.2	79.1	86.6	102.0	100.6	101.0	102.5	95.0	97.2	106.0	94.5	97.9	104.0	80.6	87.6	103.2	81.3	87.8
0310	Concrete Forming & Accessories	93.9	69.0	72.4	84.4	66.7	69.1	94.9	42.5	49.7	93.6	69.1	72.5	96.1	71.8	75.2	79.2	53.1	56.7
0320	Concrete Reinforcing	87.9	71.9	79.9	101.4	62.7	81.9	101.8	65.5	83.4	89.1	72.0	80.4	94.7	68.2	81.3	101.0	36.7	68.5
0330	Cast-in-Place Concrete	88.0	69.1	80.8	106.2	68.5	91.9	114.9	69.7	97.7	86.8	70.4	80.6	91.7	68.9	83.1	109.1	68.6	93.7
03	CONCRETE	86.2	71.3	79.4	104.1	68.5	87.8	106.7	57.3	84.1	85.8	71.8	79.4	87.7	71.9	80.4	104.5	57.8	83.2
04	MASONRY	92.2	69.2	77.9	79.6	78.7	79.1	87.0	79.4	82.3	104.5	69.3	82.7	87.2	69.2	76.0	81.6	79.4	80.2
05	METALS	97.4	98.3	97.7	94.1	94.7	94.3	92.4	79.4	88.4	93.1	98.4	94.7	94.2	96.4	94.9	97.6	84.9	93.7
06	WOOD, PLASTICS & COMPOSITES	85.9	69.6	76.8	79.7	68.3	73.3	100.8	34.7	64.0	92.6	69.6	79.8	90.3	73.6	81.0	73.6	49.3	60.1
07	THERMAL & MOISTURE PROTECTION	97.2	70.5	85.9	100.0	72.4	88.3	97.9	69.9	86.0	95.7	72.7	85.9	95.3	70.1	84.6	98.6	69.5	86.3
08	OPENINGS	87.7	71.6	84.0	94.1	68.3	88.2	93.1	50.5	83.3	87.5	71.6	83.8	94.4	72.9	89.4	95.0	50.5	84.8
0920	Plaster & Gypsum Board	103.6	69.3	80.5	82.4	67.8	72.6	95.9	33.3	53.9	109.5	69.3	82.5	98.9	73.4	81.8	82.3	48.4	59.6
0950, 0980	Ceilings & Acoustic Treatment	83.8	69.3	74.0	113.1	67.8	82.6	98.8	33.3	54.7	79.0	69.3	72.4	90.3	73.4	78.9	109.4	48.4	68.3
0960	Flooring	96.9	67.4	88.7	96.6	86.6	93.8	97.4	86.6	94.4	76.4	67.4	73.9	92.5	67.4	85.6	113.6	86.6	106.1
0970, 0990	Wall Finishes & Painting/Coating	89.1	95.9	93.1	87.6	73.2	79.2	97.5	95.9	96.6	91.1	95.9	93.9	88.3	87.9	88.1	95.3	73.2	82.2
09	FINISHES	92.1	71.1	80.7	105.3	71.6	87.0	96.9	54.4	73.8	81.9	71.1	76.0	92.1	72.6	81.5	108.8	60.6	82.6
COVERS	DIVS. 10 - 14, 25, 28, 41, 43, 44, 46	100.0	85.0	96.7	100.0	28.8	84.2	100.0	36.8	85.9	100.0	85.0	96.7	100.0	85.0	96.7	100.0	42.9	87.3
21, 22, 23	FIRE SUPPRESSION, PLUMBING & HVAC	100.0	67.4	86.1	97.1	63.2	82.6	97.0	68.3	84.7	100.0	70.5	87.4	100.1	68.3	86.5	97.6	69.1	85.4
26, 27, 3370	ELECTRICAL, COMMUNICATIONS & UTIL.	95.2	71.5	82.9	107.6	65.7	85.8	99.1	71.4	84.7	93.9	66.2	79.5	98.7	72.4	85.0	99.3	55.7	76.6
MF2016	WEIGHTED AVERAGE	94.9	73.8	85.7	98.3	72.2	86.9	97.2	67.6	84.2	93.8	75.1	85.6	95.4	74.4	86.2	98.7	66.3	84.5

DIVISION		GEORGIA VALDOSTA 316			WAYCROSS 315			HAWAII HILO 967			HONOLULU 968			STATES & POSS., GUAM 969			IDAHO BOISE 836 - 837		
		MAT.	INST.	TOTAL	MAT.	INST.	TOTAL	MAT.	INST.	TOTAL	MAT.	INST.	TOTAL	MAT.	INST.	TOTAL	MAT.	INST.	TOTAL
015433	CONTRACTOR EQUIPMENT		95.7	95.7		95.7	95.7		99.3	99.3		99.3	99.3		165.6	165.6		97.4	97.4
0241, 31 - 34	SITE & INFRASTRUCTURE, DEMOLITION	113.8	79.0	89.4	110.6	80.3	89.4	156.3	106.9	121.7	164.8	106.9	124.2	201.3	102.9	132.3	88.9	96.8	94.3
0310	Concrete Forming & Accessories	84.4	42.8	48.5	86.4	42.8	48.5	107.6	122.5	120.4	118.4	122.5	121.9	110.0	53.9	61.7	98.7	81.3	83.7
0320	Concrete Reinforcing	90.0	61.4	75.5	90.0	61.6	75.7	133.0	113.9	123.4	154.7	113.9	134.1	243.0	28.2	134.7	101.6	77.6	89.5
0330	Cast-in-Place Concrete	86.5	68.9	79.8	97.7	68.7	86.7	194.4	122.0	166.9	156.3	122.0	143.3	168.5	100.5	142.7	90.5	84.8	88.3
03	CONCRETE	91.1	57.5	75.7	94.1	67.7	82.0	151.7	119.8	137.1	142.6	119.8	132.2	153.7	66.6	113.9	96.1	82.0	89.7
04	MASONRY	97.6	79.4	86.3	98.5	79.4	86.7	153.6	121.4	133.6	140.0	121.4	128.5	217.7	36.2	105.2	125.6	89.0	102.9
05	METALS	97.0	94.1	96.1	96.0	89.7	94.1	112.9	102.3	109.6	125.9	102.3	118.7	145.3	77.1	124.4	108.1	82.1	100.1
06	WOOD, PLASTICS & COMPOSITES	74.1	34.5	52.0	75.7	66.4	70.5	114.9	122.4	119.1	129.4	122.4	125.5	126.7	56.3	87.5	92.4	79.9	85.5
07	THERMAL & MOISTURE PROTECTION	97.5	60.3	82.1	97.2	71.1	86.2	130.8	119.0	125.8	148.7	119.0	136.1	153.9	60.9	114.5	94.3	86.3	90.9
08	OPENINGS	84.3	49.2	76.2	84.6	64.2	79.9	117.0	118.9	117.5	131.5	118.9	128.6	121.6	45.9	104.3	97.1	74.3	91.9
0920	Plaster & Gypsum Board	95.7	33.2	53.7	95.7	66.0	75.8	108.1	122.9	118.0	151.3	122.9	132.2	216.4	44.6	101.0	92.4	79.4	83.7
0950, 0980	Ceilings & Acoustic Treatment	81.3	33.2	48.8	79.4	66.0	70.3	131.5	122.9	125.7	140.9	122.9	128.8	246.8	44.6	110.5	102.4	79.4	86.9
0960	Flooring	90.9	88.2	90.2	92.0	86.6	90.5	107.3	136.7	115.5	125.7	136.7	128.8	126.6	41.8	103.1	93.8	81.4	90.4
0970, 0990	Wall Finishes & Painting/Coating	89.1	95.9	93.1	89.1	73.2	79.8	95.6	137.2	119.9	103.4	137.2	123.1	100.4	33.0	61.1	93.6	41.3	63.1
09	FINISHES	89.8	54.6	70.6	89.3	70.6	79.2	108.7	127.1	118.7	124.8	127.1	126.1	178.9	50.4	109.0	92.7	77.2	84.3
COVERS	DIVS. 10 - 14, 25, 28, 41, 43, 44, 46	100.0	80.6	95.7	100.0	50.7	89.0	100.0	112.1	102.7	100.0	112.1	102.7	100.0	68.5	93.0	100.0	90.4	97.9
21, 22, 23	FIRE SUPPRESSION, PLUMBING & HVAC	100.0	70.6	87.4	98.1	65.4	84.1	100.5	109.4	104.3	100.5	109.4	104.3	103.5	34.9	74.2	100.1	74.9	89.3
26, 27, 3370	ELECTRICAL, COMMUNICATIONS & UTIL.	93.5	54.6	73.2	97.3	55.7	75.6	105.3	121.7	113.9	106.9	121.7	114.6	151.8	37.9	92.6	97.9	67.8	82.2
MF2016	WEIGHTED AVERAGE	95.1	67.3	82.9	95.3	69.5	83.8	116.9	116.0	116.5	121.1	116.0	118.9	139.1	53.4	101.6	100.5	79.8	91.4

DIVISION		IDAHO COEUR D'ALENE 838			IDAHO FALLS 834			LEWISTON 835			POCATELLO 832			TWIN FALLS 833			ILLINOIS BLOOMINGTON 617		
		MAT.	INST.	TOTAL	MAT.	INST.	TOTAL	MAT.	INST.	TOTAL	MAT.	INST.	TOTAL	MAT.	INST.	TOTAL	MAT.	INST.	TOTAL
015433	CONTRACTOR EQUIPMENT		92.2	92.2		97.4	97.4		92.2	92.2		97.4	97.4		97.4	97.4		105.2	105.2
0241, 31 - 34	SITE & INFRASTRUCTURE, DEMOLITION	87.9	90.9	90.0	86.5	96.8	93.7	94.7	91.8	92.7	89.9	96.8	94.7	97.0	96.6	96.7	95.7	100.5	99.1
0310	Concrete Forming & Accessories	108.9	79.6	83.6	92.8	80.1	81.8	113.9	81.6	86.0	98.8	80.9	83.4	100.0	79.9	82.7	83.4	119.9	114.9
0320	Concrete Reinforcing	110.0	99.2	104.6	103.8	77.6	90.6	110.0	99.3	104.6	102.1	77.6	89.7	104.1	77.6	90.7	96.1	105.6	100.9
0330	Cast-in-Place Concrete	97.8	82.6	92.0	86.2	83.2	85.0	101.7	86.0	95.7	93.0	84.7	89.9	95.5	83.1	90.8	97.7	117.5	105.2
03	CONCRETE	102.6	84.2	94.2	88.3	80.9	84.9	106.0	86.2	97.0	95.3	81.8	89.1	102.9	80.8	92.8	93.2	117.2	104.2
04	MASONRY	127.3	85.4	101.3	120.6	86.3	99.4	127.7	87.7	102.9	123.1	89.0	102.0	125.9	86.4	101.4	115.2	122.8	119.9
05	METALS	101.6	89.5	97.9	116.4	81.9	105.8	101.0	89.8	97.6	116.5	81.8	105.8	116.5	81.7	105.8	99.9	125.4	107.7
06	WOOD, PLASTICS & COMPOSITES	97.5	78.4	86.8	86.7	79.9	82.9	102.8	79.4	89.8	92.4	79.9	85.5	93.5	79.9	85.9	86.7	118.5	104.4
07	THERMAL & MOISTURE PROTECTION	147.3	83.8	120.4	93.5	76.4	86.3	147.6	85.5	121.4	94.0	77.2	86.9	94.9	83.3	90.0	96.9	115.8	104.9
08	OPENINGS	114.3	76.9	105.7	100.1	74.3	94.1	108.0	80.9	101.8	97.8	69.2	91.2	100.9	64.9	92.6	94.5	124.8	101.5
0920	Plaster & Gypsum Board	166.3	77.9	107.0	77.5	79.4	78.8	168.0	79.0	108.2	79.4	79.4	79.4	81.0	79.4	79.9	92.8	119.1	110.4
0950, 0980	Ceilings & Acoustic Treatment	136.6	77.9	97.1	103.2	79.4	87.1	136.6	79.0	97.8	110.0	79.4	89.4	105.7	79.4	88.0	89.2	119.1	109.3
0960	Flooring	132.6	81.4	118.4	93.4	81.4	90.1	136.1	81.4	120.9	97.1	81.4	92.7	98.3	81.4	93.7	84.6	126.3	96.7
0970, 0990	Wall Finishes & Painting/Coating	112.8	75.9	91.3	93.6	41.3	63.1	112.8	75.9	91.3	93.5	41.3	63.1	93.6	41.3	63.1	85.6	140.9	117.8
09	FINISHES	155.8	79.2	114.1	90.3	76.5	82.8	157.2	80.4	115.4	93.4	77.2	84.6	93.9	76.5	84.5	86.9	123.7	106.9
COVERS	DIVS. 10 - 14, 25, 28, 41, 43, 44, 46	100.0	95.0	98.9	100.0	89.5	97.7	100.0	95.9	99.1	100.0	90.4	97.9	100.0	89.5	97.7	100.0	107.6	101.7
21, 22, 23	FIRE SUPPRESSION, PLUMBING & HVAC	99.6	83.2	92.6	100.9	72.3	88.7	100.9	88.1	95.4	99.9	74.8	89.2	99.9	70.1	87.2	96.9	107.5	101.4
26, 27, 3370	ELECTRICAL, COMMUNICATIONS & UTIL.	90.2	86.1	88.1	89.9	68.2	78.6	88.4	82.6	85.4	95.4	69.7	82.0	91.3	68.2	79.3	95.6	95.4	95.5
MF2016	WEIGHTED AVERAGE	108.1	84.7	97.8	100.0	78.5	90.6	108.2	86.3	98.6	101.5	79.5	91.9	102.8	77.8	91.8	96.8	113.0	103.9

ILLINOIS

DIVISION		CARBONDALE 629			CENTRALIA 628			CHAMPAIGN 618 - 619			CHICAGO 606 - 608			DECATUR 625			EAST ST. LOUIS 620 - 622		
		MAT.	INST.	TOTAL	MAT.	INST.	TOTAL	MAT.	INST.	TOTAL	MAT.	INST.	TOTAL	MAT.	INST.	TOTAL	MAT.	INST.	TOTAL
015433	CONTRACTOR EQUIPMENT		114.0	114.0		114.0	114.0		106.1	106.1		100.2	100.2		106.1	106.1		114.0	114.0
0241, 31 - 34	SITE & INFRASTRUCTURE, DEMOLITION	100.6	101.6	101.3	100.9	102.3	101.9	104.7	101.6	102.5	101.9	104.4	103.6	97.6	101.8	100.5	102.9	102.4	102.5
0310	Concrete Forming & Accessories	90.5	106.1	104.0	92.1	110.8	108.3	89.6	117.8	113.9	99.8	160.4	152.0	92.6	118.2	114.7	88.0	111.9	108.6
0320	Concrete Reinforcing	98.3	102.6	100.5	98.3	102.8	100.6	96.1	100.9	98.6	104.7	150.8	128.0	96.3	101.2	98.8	98.2	103.1	100.7
0330	Cast-in-Place Concrete	90.8	103.6	95.7	91.3	117.2	101.1	113.3	112.9	113.2	123.6	153.6	135.0	99.2	116.5	105.8	92.8	118.0	102.4
03	CONCRETE	83.1	106.3	93.7	83.5	113.2	97.1	105.5	113.8	109.3	109.9	155.7	130.8	93.5	115.2	103.4	84.6	114.0	98.0
04	MASONRY	78.1	109.9	97.8	78.1	112.4	99.4	139.8	121.5	128.5	102.6	164.9	141.2	73.8	122.8	104.2	78.3	117.2	102.4
05	METALS	97.8	130.5	107.9	97.9	132.9	108.6	99.9	121.7	106.6	97.9	146.9	113.0	101.6	121.2	107.6	99.0	133.3	109.5
06	WOOD, PLASTICS & COMPOSITES	90.9	102.7	97.5	93.3	108.3	101.6	93.2	117.1	106.5	104.3	160.0	135.3	91.2	117.1	105.6	88.4	108.9	99.8
07	THERMAL & MOISTURE PROTECTION	93.9	101.4	97.1	93.9	108.0	99.9	97.7	114.6	104.9	96.7	150.2	119.3	99.6	113.2	105.4	94.0	107.6	99.7
08	OPENINGS	89.4	115.7	95.4	89.4	118.8	96.1	95.1	119.4	100.7	101.9	170.3	117.6	100.5	119.5	104.9	89.4	119.1	96.2
0920	Plaster & Gypsum Board	97.3	102.9	101.1	98.3	108.6	105.2	94.7	117.6	110.1	103.3	161.7	142.5	99.7	117.6	111.7	96.3	109.2	105.0
0950, 0980	Ceilings & Acoustic Treatment	87.5	102.9	97.9	87.5	108.6	101.7	89.2	117.6	108.3	102.2	161.7	142.3	93.4	117.6	109.7	87.5	109.2	102.1
0960	Flooring	107.2	116.4	109.8	108.0	112.1	109.2	87.8	117.3	96.0	94.4	163.4	113.5	96.1	117.8	102.2	106.2	112.1	107.9
0970, 0990	Wall Finishes & Painting/Coating	94.5	104.8	100.5	94.5	115.2	106.6	85.6	117.6	104.2	101.4	162.5	137.0	86.5	111.6	101.1	94.5	115.2	106.6
09	FINISHES	91.6	106.6	99.7	92.0	111.1	102.4	88.8	118.3	104.8	99.3	162.1	133.4	91.4	118.1	105.9	91.2	112.4	102.7
COVERS	DIVS. 10 - 14, 25, 28, 41, 43, 44, 46	100.0	103.4	100.8	100.0	104.2	100.9	100.0	106.6	101.5	100.0	127.1	106.0	100.0	107.1	101.6	100.0	104.6	101.0
21, 22, 23	FIRE SUPPRESSION, PLUMBING & HVAC	96.9	105.8	100.7	96.9	94.1	95.7	96.9	105.1	100.4	100.0	138.4	116.4	100.1	98.2	99.3	100.0	97.7	99.0
26, 27, 3370	ELECTRICAL, COMMUNICATIONS & UTIL.	94.9	107.4	101.4	96.2	107.4	102.0	98.7	95.4	97.0	96.4	135.5	116.7	98.6	92.6	95.5	95.8	100.5	98.2
MF2016	WEIGHTED AVERAGE	93.2	108.8	100.0	93.4	108.8	100.1	100.3	110.6	104.8	100.8	145.8	120.5	97.3	109.0	102.4	94.4	109.4	100.9

ILLINOIS

DIVISION		EFFINGHAM 624			GALESBURG 614			JOLIET 604			KANKAKEE 609			LA SALLE 613			NORTH SUBURBAN 600 - 603		
		MAT.	INST.	TOTAL	MAT.	INST.	TOTAL	MAT.	INST.	TOTAL	MAT.	INST.	TOTAL	MAT.	INST.	TOTAL	MAT.	INST.	TOTAL
015433	CONTRACTOR EQUIPMENT		106.1	106.1		105.2	105.2		96.6	96.6		96.6	96.6		105.2	105.2		96.6	96.6
0241, 31 - 34	SITE & INFRASTRUCTURE, DEMOLITION	101.3	100.6	100.8	98.1	100.1	99.5	97.6	102.8	101.3	91.6	102.0	98.9	97.5	101.1	100.0	96.9	102.6	100.9
0310	Concrete Forming & Accessories	97.1	116.3	113.7	89.5	118.9	114.9	99.1	158.3	150.1	92.7	141.6	134.9	103.1	122.5	119.8	98.4	153.7	146.1
0320	Concrete Reinforcing	99.5	92.1	95.8	95.6	102.7	99.2	104.7	141.0	123.0	105.2	134.9	120.4	95.8	133.8	115.0	104.7	141.1	125.6
0330	Cast-in-Place Concrete	98.8	111.0	103.4	100.6	106.8	103.0	112.4	152.0	127.4	104.8	139.2	117.8	100.5	126.2	110.3	112.5	150.1	126.7
03	CONCRETE	94.3	110.8	101.8	96.0	112.6	103.6	102.8	152.2	125.4	96.7	139.1	116.1	96.8	126.5	110.4	102.8	150.4	124.6
04	MASONRY	82.0	115.2	102.6	115.4	120.6	118.6	101.8	163.4	140.0	98.1	147.8	128.9	115.4	145.7	134.2	98.6	159.1	136.2
05	METALS	98.7	114.1	103.4	99.9	122.9	107.0	95.9	138.4	108.9	95.9	133.9	107.5	100.0	143.1	113.2	97.0	140.1	110.2
06	WOOD, PLASTICS & COMPOSITES	93.5	117.1	106.6	93.0	118.6	107.3	102.9	157.3	133.2	96.1	139.8	120.4	107.7	119.5	114.3	101.3	152.6	129.8
07	THERMAL & MOISTURE PROTECTION	99.1	110.2	103.8	97.1	110.0	102.6	101.1	149.2	121.5	100.2	138.3	116.3	97.3	124.0	108.6	101.4	144.3	119.6
08	OPENINGS	95.3	116.0	100.0	94.5	117.3	99.7	99.5	165.7	114.7	92.7	153.2	106.6	94.5	133.5	103.5	99.6	163.8	114.3
0920	Plaster & Gypsum Board	99.6	117.6	111.7	94.7	119.3	111.2	96.1	159.0	138.3	94.1	141.0	125.6	101.7	120.1	114.1	99.1	154.1	136.0
0950, 0980	Ceilings & Acoustic Treatment	87.5	117.6	107.8	89.2	119.3	109.4	101.8	159.0	140.3	101.8	141.0	128.2	89.2	120.1	110.1	101.8	154.1	137.1
0960	Flooring	97.2	110.9	101.0	87.6	126.3	98.3	92.4	158.3	110.7	89.3	159.3	108.7	94.2	125.7	102.9	92.9	158.3	111.0
0970, 0990	Wall Finishes & Painting/Coating	86.5	109.3	99.8	85.6	99.9	93.9	93.5	170.4	138.4	93.5	140.9	121.2	85.6	139.1	116.8	95.5	159.5	132.9
09	FINISHES	90.6	115.7	104.3	88.2	119.2	105.1	94.7	160.6	130.6	93.2	145.2	121.4	91.1	123.5	108.7	95.4	156.0	128.3
COVERS	DIVS. 10 - 14, 25, 28, 41, 43, 44, 46	100.0	78.1	95.1	100.0	102.4	100.5	100.0	126.9	106.0	100.0	116.9	103.8	100.0	103.1	100.7	100.0	120.6	104.6
21, 22, 23	FIRE SUPPRESSION, PLUMBING & HVAC	97.0	104.2	100.1	96.9	105.2	100.5	100.0	135.4	115.2	96.9	129.7	111.0	96.9	123.8	108.4	99.9	133.2	114.2
26, 27, 3370	ELECTRICAL, COMMUNICATIONS & UTIL.	96.5	107.3	102.2	96.5	86.4	91.2	95.6	127.2	112.0	91.0	126.4	109.5	93.7	126.5	110.8	95.5	132.0	114.5
MF2016	WEIGHTED AVERAGE	95.8	108.7	101.5	97.4	108.9	102.5	98.8	142.0	117.7	95.6	133.5	112.2	97.6	126.5	110.3	98.9	140.6	117.1

ILLINOIS

DIVISION		PEORIA 615 - 616			QUINCY 623			ROCK ISLAND 612			ROCKFORD 610 - 611			SOUTH SUBURBAN 605			SPRINGFIELD 626 - 627		
		MAT.	INST.	TOTAL	MAT.	INST.	TOTAL	MAT.	INST.	TOTAL	MAT.	INST.	TOTAL	MAT.	INST.	TOTAL	MAT.	INST.	TOTAL
015433	CONTRACTOR EQUIPMENT		105.2	105.2		106.1	106.1		105.2	105.2		105.2	105.2		96.6	96.6		106.1	106.1
0241, 31 - 34	SITE & INFRASTRUCTURE, DEMOLITION	98.4	100.6	99.9	100.2	101.1	100.8	96.2	99.6	98.6	97.9	102.0	100.8	96.9	102.3	100.7	101.2	101.6	101.5
0310	Concrete Forming & Accessories	92.3	120.3	116.4	95.0	114.5	111.8	91.0	102.7	101.1	96.5	132.2	127.3	98.4	153.7	146.1	93.3	117.8	114.4
0320	Concrete Reinforcing	93.2	103.0	98.1	99.1	84.5	91.7	95.6	96.2	95.9	88.2	128.6	108.6	104.7	146.1	125.6	99.6	101.2	100.4
0330	Cast-in-Place Concrete	97.6	120.0	106.1	99.0	101.6	100.0	98.4	102.1	99.8	99.9	131.2	111.8	112.5	150.1	126.7	94.1	110.7	100.4
03	CONCRETE	93.3	117.8	104.5	93.9	105.5	99.2	94.0	101.5	97.4	94.0	131.7	111.2	102.8	150.4	124.6	91.7	113.1	101.4
04	MASONRY	114.5	123.6	120.2	105.4	111.0	108.9	115.2	101.9	106.9	89.1	139.1	120.1	98.6	159.1	136.2	84.6	121.6	107.5
05	METALS	102.6	124.2	109.3	98.7	112.4	102.9	99.9	118.7	105.7	102.6	140.4	114.2	97.0	140.1	110.2	99.1	121.3	105.9
06	WOOD, PLASTICS & COMPOSITES	100.5	118.6	110.6	91.1	117.1	105.5	94.6	102.2	98.8	100.5	129.6	116.7	101.3	152.6	129.8	92.2	117.1	106.0
07	THERMAL & MOISTURE PROTECTION	97.8	116.6	105.8	99.1	105.4	101.8	97.1	100.7	98.6	100.2	132.0	113.7	101.4	144.3	119.6	101.6	114.6	107.1
08	OPENINGS	100.6	124.1	106.0	96.0	113.6	100.0	94.5	106.7	97.3	100.6	140.6	109.7	99.6	163.8	114.3	98.8	118.8	103.4
0920	Plaster & Gypsum Board	98.5	119.3	112.4	98.3	117.6	111.3	94.7	102.4	99.9	98.5	130.5	120.0	99.1	154.1	136.0	100.8	117.6	112.1
0950, 0980	Ceilings & Acoustic Treatment	94.3	119.3	111.1	87.5	117.6	107.8	89.2	102.4	98.1	94.3	130.5	118.7	101.8	154.1	137.1	94.7	117.6	110.2
0960	Flooring	91.1	126.3	100.9	96.1	114.1	101.1	88.7	98.2	91.4	91.1	125.7	100.7	92.9	158.3	111.0	99.3	117.8	104.5
0970, 0990	Wall Finishes & Painting/Coating	85.6	140.9	117.8	86.5	115.2	103.2	85.6	99.9	93.9	85.6	149.2	122.7	95.5	159.5	132.9	84.7	115.2	102.5
09	FINISHES	90.8	123.9	108.8	90.1	115.7	104.0	88.4	101.4	95.5	90.8	132.6	113.6	95.4	156.0	128.3	96.9	118.1	108.4
COVERS	DIVS. 10 - 14, 25, 28, 41, 43, 44, 46	100.0	107.8	101.7	100.0	79.3	95.4	100.0	100.7	100.1	100.0	116.0	103.6	100.0	120.6	104.6	100.0	106.6	101.5
21, 22, 23	FIRE SUPPRESSION, PLUMBING & HVAC	100.0	102.7	101.1	97.0	101.5	98.9	96.9	100.1	98.3	100.1	118.6	108.0	99.9	133.2	114.2	100.0	103.0	101.3
26, 27, 3370	ELECTRICAL, COMMUNICATIONS & UTIL.	97.5	96.6	97.0	94.1	86.8	90.3	89.0	93.8	91.5	97.7	132.6	115.9	95.5	132.0	114.5	101.1	86.8	93.7
MF2016	WEIGHTED AVERAGE	99.3	112.2	104.9	96.7	103.9	99.8	96.4	101.9	98.8	98.2	128.3	111.4	98.9	140.6	117.1	97.8	108.8	102.6

INDIANA

DIVISION		ANDERSON 460			BLOOMINGTON 474			COLUMBUS 472			EVANSVILLE 476 - 477			FORT WAYNE 467 - 468			GARY 463 - 464		
		MAT.	INST.	TOTAL	MAT.	INST.	TOTAL	MAT.	INST.	TOTAL	MAT.	INST.	TOTAL	MAT.	INST.	TOTAL	MAT.	INST.	TOTAL
015433	CONTRACTOR EQUIPMENT		96.7	96.7		83.4	83.4		83.4	83.4		113.9	113.9		96.7	96.7		96.7	96.7
0241, 31 - 34	SITE & INFRASTRUCTURE, DEMOLITION	97.9	93.4	94.7	87.8	92.1	90.8	84.4	91.9	89.7	93.4	121.8	113.3	98.9	93.2	94.9	98.5	97.1	97.5
0310	Concrete Forming & Accessories	94.6	81.2	83.0	99.0	81.6	84.0	93.2	79.1	81.0	92.6	82.3	83.7	92.7	74.8	77.3	94.7	114.8	112.1
0320	Concrete Reinforcing	90.1	85.7	87.8	87.1	85.3	86.2	87.5	85.4	86.4	95.4	78.6	86.9	90.1	77.1	83.5	90.1	113.1	101.7
0330	Cast-in-Place Concrete	100.6	78.7	92.3	99.1	77.8	91.0	98.7	75.7	89.9	94.7	87.6	92.0	106.9	75.0	94.8	105.1	116.3	109.4
03	CONCRETE	89.8	81.6	86.0	97.8	80.5	89.9	97.1	78.6	88.6	97.9	83.6	91.4	92.6	76.0	85.0	91.9	114.7	102.3
04	MASONRY	88.8	77.2	81.6	91.1	73.5	80.2	91.0	74.6	80.8	86.5	80.6	82.8	91.8	74.5	81.0	90.2	111.7	103.5
05	METALS	96.3	89.9	94.3	98.2	74.8	91.0	98.2	74.2	90.9	91.5	83.2	88.9	96.3	86.6	93.3	96.3	106.3	99.4
06	WOOD, PLASTICS & COMPOSITES	96.5	81.3	88.0	111.0	82.0	94.9	106.1	78.8	90.9	91.8	81.4	86.0	96.3	74.4	84.1	93.8	113.1	104.5
07	THERMAL & MOISTURE PROTECTION	110.1	77.9	96.5	96.6	78.9	89.1	96.0	78.5	88.6	100.9	84.8	94.1	109.8	74.0	94.7	108.5	108.2	108.4
08	OPENINGS	93.5	79.7	90.3	100.3	80.1	95.6	96.4	78.3	92.3	94.2	78.3	90.5	93.5	72.0	88.5	93.5	116.6	98.8
0920	Plaster & Gypsum Board	105.8	81.1	89.2	99.0	82.3	87.8	96.4	78.9	84.7	94.8	80.5	85.2	105.1	74.0	84.2	98.9	113.7	108.8
0950, 0980	Ceilings & Acoustic Treatment	89.7	81.1	83.9	79.8	82.3	81.5	79.8	78.9	79.2	83.6	80.5	81.5	89.7	74.0	79.1	89.7	113.7	105.9
0960	Flooring	94.0	79.3	89.9	98.5	83.7	94.4	93.3	83.7	90.6	93.0	75.5	88.1	94.0	73.2	88.2	94.0	116.8	100.3
0970, 0990	Wall Finishes & Painting/Coating	93.1	68.8	78.9	86.0	79.8	82.4	86.0	79.8	82.4	91.1	88.2	89.4	93.1	73.4	81.6	93.1	123.7	110.9
09	FINISHES	91.2	79.7	85.0	90.6	81.9	85.9	88.6	80.0	83.9	89.2	81.5	85.0	91.0	74.4	82.0	90.2	116.2	104.3
COVERS	DIVS. 10 - 14, 25, 28, 41, 43, 44, 46	100.0	91.0	98.0	100.0	87.5	97.2	100.0	87.2	97.1	100.0	95.1	98.9	100.0	90.5	97.9	100.0	107.2	101.6
21, 22, 23	FIRE SUPPRESSION, PLUMBING & HVAC	100.0	80.2	91.5	99.7	79.3	91.0	96.6	80.1	89.6	100.0	79.3	91.2	100.0	73.3	88.6	100.0	107.1	103.0
26, 27, 3370	ELECTRICAL, COMMUNICATIONS & UTIL.	87.9	87.8	87.9	99.9	88.1	93.8	99.2	87.7	93.2	96.0	86.0	90.8	88.6	76.5	82.3	99.4	114.1	107.0
MF2016	WEIGHTED AVERAGE	95.1	83.3	89.9	97.9	81.3	90.6	96.3	80.8	89.5	95.5	85.5	91.1	95.7	77.7	87.8	96.5	110.4	102.6

INDIANA

DIVISION		INDIANAPOLIS 461 - 462			KOKOMO 469			LAFAYETTE 479			LAWRENCEBURG 470			MUNCIE 473			NEW ALBANY 471		
		MAT.	INST.	TOTAL	MAT.	INST.	TOTAL	MAT.	INST.	TOTAL	MAT.	INST.	TOTAL	MAT.	INST.	TOTAL	MAT.	INST.	TOTAL
015433	CONTRACTOR EQUIPMENT		84.3	84.3		96.7	96.7		83.4	83.4		103.6	103.6		95.3	95.3		93.1	93.1
0241, 31 - 34	SITE & INFRASTRUCTURE, DEMOLITION	99.0	90.4	93.0	94.3	93.3	93.6	85.3	92.1	90.0	83.1	107.0	100.3	87.7	92.6	91.1	80.1	94.4	90.1
0310	Concrete Forming & Accessories	99.3	84.6	86.6	97.7	77.5	80.2	90.9	82.5	83.6	89.8	78.1	79.7	90.7	80.6	82.0	88.3	78.8	80.1
0320	Concrete Reinforcing	91.1	85.7	88.4	81.4	85.5	83.5	87.1	85.6	86.3	86.4	77.5	81.9	96.4	85.6	90.9	87.7	81.4	84.5
0330	Cast-in-Place Concrete	98.8	86.0	93.9	99.6	82.5	93.1	99.2	81.7	92.6	92.8	75.0	86.0	104.2	77.8	94.2	95.8	74.7	87.8
03	CONCRETE	95.7	84.7	90.7	86.9	81.2	84.3	97.3	82.3	90.4	90.5	77.5	84.6	96.2	81.1	89.3	95.8	78.1	87.7
04	MASONRY	90.5	79.2	83.5	88.5	75.2	80.3	96.5	77.1	84.5	75.7	73.3	74.2	92.9	77.3	83.2	82.3	69.5	74.4
05	METALS	97.1	75.3	90.4	92.7	89.6	91.8	96.6	75.1	90.0	93.3	85.3	90.8	100.0	89.9	96.9	95.2	81.8	91.1
06	WOOD, PLASTICS & COMPOSITES	98.8	85.1	91.2	99.5	76.2	86.5	103.3	83.0	92.0	90.1	77.9	83.3	105.1	80.8	91.6	92.0	80.1	85.4
07	THERMAL & MOISTURE PROTECTION	101.3	81.2	92.8	109.0	77.7	95.8	95.9	81.0	89.6	101.7	77.7	91.5	99.1	79.3	90.7	88.0	72.9	81.6
08	OPENINGS	99.7	81.8	95.6	89.0	76.9	86.2	94.9	80.6	91.6	96.1	74.7	91.2	93.7	79.4	90.4	93.7	78.8	90.3
0920	Plaster & Gypsum Board	94.7	85.0	88.2	110.4	75.8	87.2	94.0	83.3	86.8	73.6	77.9	76.5	94.8	81.1	85.6	92.5	80.0	84.1
0950, 0980	Ceilings & Acoustic Treatment	90.9	85.0	86.9	89.7	75.8	80.4	76.4	83.3	81.0	87.0	77.9	80.9	79.8	81.1	80.7	83.6	80.0	81.2
0060	Flooring	96.2	83.7	92.7	98.2	87.4	95.2	92.2	83.7	89.8	68.4	83.7	72.6	92.3	79.3	88.7	90.4	57.1	81.2
0970, 0990	Wall Finishes & Painting/Coating	98.6	79.8	87.6	93.1	71.0	80.2	86.0	85.1	85.4	86.8	74.5	79.6	86.0	68.8	76.0	91.1	68.2	77.8
09	FINISHES	94.6	84.1	88.9	93.0	78.4	85.0	87.3	83.1	85.0	79.2	79.0	79.1	87.8	79.1	83.1	88.3	73.7	80.4
COVERS	DIVS. 10 - 14, 25, 28, 41, 43, 44, 46	100.0	91.9	98.2	100.0	87.8	97.3	100.0	90.2	97.8	100.0	85.7	96.8	100.0	89.9	97.8	100.0	85.2	96.7
21, 22, 23	FIRE SUPPRESSION, PLUMBING & HVAC	99.9	80.8	91.7	96.9	80.2	89.8	96.6	79.4	89.3	97.6	75.9	88.3	99.7	80.1	91.3	96.9	77.0	88.4
26, 27, 3370	ELECTRICAL, COMMUNICATIONS & UTIL.	102.2	87.8	94.7	92.2	81.5	86.7	98.6	83.0	90.5	93.9	73.7	83.4	91.8	78.2	84.7	94.5	77.1	85.5
MF2016	WEIGHTED AVERAGE	98.3	83.2	91.7	93.5	81.7	88.3	96.0	81.6	89.7	92.8	79.6	87.0	96.3	81.7	89.9	94.0	78.0	87.0

INDIANA / IOWA

DIVISION		SOUTH BEND 465 - 466			TERRE HAUTE 478			WASHINGTON 475			BURLINGTON 526			CARROLL 514			CEDAR RAPIDS 522 - 524		
		MAT.	INST.	TOTAL	MAT.	INST.	TOTAL	MAT.	INST.	TOTAL	MAT.	INST.	TOTAL	MAT.	INST.	TOTAL	MAT.	INST.	TOTAL
015433	CONTRACTOR EQUIPMENT		108.4	108.4		113.9	113.9		113.9	113.9		101.5	101.5		101.5	101.5		98.2	98.2
0241, 31 - 34	SITE & INFRASTRUCTURE, DEMOLITION	97.4	93.9	94.9	95.2	122.2	114.1	95.0	122.4	114.2	99.5	96.6	97.5	88.3	97.4	94.7	100.7	96.1	97.5
0310	Concrete Forming & Accessories	94.5	79.6	81.6	93.4	80.8	82.6	94.3	82.5	84.1	93.0	82.5	83.9	81.3	80.6	80.7	98.7	87.9	89.4
0320	Concrete Reinforcing	89.4	86.2	87.8	95.4	85.7	90.5	88.2	85.1	86.7	91.7	86.1	88.9	92.4	82.9	87.6	92.4	82.7	87.5
0330	Cast-in-Place Concrete	99.1	81.5	92.4	91.7	82.4	88.2	99.8	87.8	95.3	107.6	57.4	88.5	107.6	58.9	89.1	107.9	85.3	99.3
03	CONCRETE	88.3	83.0	85.9	100.8	82.4	92.4	106.6	84.9	96.7	94.6	75.2	85.7	93.5	74.3	84.7	94.7	86.7	91.0
04	MASONRY	93.4	77.2	83.4	94.2	76.4	83.2	86.7	80.3	82.8	100.4	75.0	84.6	102.1	74.2	84.8	106.2	83.4	92.0
05	METALS	96.3	105.9	99.2	92.2	86.2	90.3	86.7	86.1	86.5	90.2	95.0	91.7	90.2	94.0	91.4	92.7	95.0	93.4
06	WOOD, PLASTICS & COMPOSITES	94.3	78.6	85.6	93.8	80.4	86.3	94.2	81.3	87.0	89.0	81.4	84.8	76.4	84.4	80.9	95.3	87.6	91.1
07	THERMAL & MOISTURE PROTECTION	101.2	82.0	93.1	101.0	83.0	93.4	100.9	84.5	93.9	104.2	78.5	93.3	104.5	73.4	91.4	105.2	83.5	96.0
08	OPENINGS	95.2	78.5	91.4	94.7	79.2	91.1	91.5	80.1	88.9	93.5	84.9	91.5	97.9	82.6	94.4	98.4	85.4	95.4
0920	Plaster & Gypsum Board	95.1	78.3	83.9	94.8	79.4	84.5	95.1	80.4	85.2	109.4	81.0	90.3	104.8	84.1	90.9	113.9	87.7	96.3
0950, 0980	Ceilings & Acoustic Treatment	91.6	78.3	82.7	83.6	79.4	80.8	79.4	80.4	80.1	98.0	81.0	86.5	98.0	84.1	88.7	100.6	87.7	91.9
0960	Flooring	91.6	91.1	91.5	93.0	80.0	89.4	93.9	81.1	90.3	94.1	73.0	88.3	88.2	84.1	87.1	107.9	89.1	102.7
0970, 0990	Wall Finishes & Painting/Coating	86.9	86.3	86.5	91.1	83.4	86.6	91.1	88.2	89.4	94.8	89.1	91.5	94.8	76.1	83.9	96.4	74.0	83.3
09	FINISHES	91.1	82.2	86.2	89.2	80.8	84.6	88.9	82.6	85.5	94.1	81.6	87.3	90.2	81.6	85.5	99.4	86.8	92.6
COVERS	DIVS. 10 - 14, 25, 28, 41, 43, 44, 46	100.0	92.1	98.2	100.0	92.4	98.3	100.0	95.1	98.9	100.0	91.7	98.1	100.0	64.9	92.2	100.0	93.7	98.6
21, 22, 23	FIRE SUPPRESSION, PLUMBING & HVAC	99.9	78.1	90.6	100.0	79.7	91.3	96.9	80.9	90.0	97.1	82.6	90.9	97.1	73.8	87.2	100.2	84.5	93.5
26, 27, 3370	ELECTRICAL, COMMUNICATIONS & UTIL.	100.1	88.6	94.1	94.3	88.1	91.1	94.8	86.0	90.2	100.8	76.3	88.0	101.4	81.2	90.9	98.4	79.6	88.6
MF2016	WEIGHTED AVERAGE	96.3	85.0	91.4	96.3	85.4	91.5	94.8	86.5	91.1	96.0	82.3	90.0	95.8	79.9	88.8	98.2	86.5	93.1

IOWA

DIVISION		COUNCIL BLUFFS 515 MAT.	INST.	TOTAL	CRESTON 508 MAT.	INST.	TOTAL	DAVENPORT 527-528 MAT.	INST.	TOTAL	DECORAH 521 MAT.	INST.	TOTAL	DES MOINES 500-503, 509 MAT.	INST.	TOTAL	DUBUQUE 520 MAT.	INST.	TOTAL
015433	CONTRACTOR EQUIPMENT		97.5	97.5		101.5	101.5		101.5	101.5		101.5	101.5		103.4	103.4		97.0	97.0
0241, 31 - 34	SITE & INFRASTRUCTURE, DEMOLITION	103.5	92.5	95.8	98.0	98.3	98.2	99.8	99.6	99.7	98.1	96.3	96.8	104.4	101.2	102.2	98.5	93.2	94.8
0310	Concrete Forming & Accessories	80.8	76.2	76.8	78.5	83.7	83.0	98.2	100.5	100.2	90.7	74.7	76.9	96.4	86.0	87.4	82.0	85.4	85.0
0320	Concrete Reinforcing	94.3	80.0	87.1	91.5	82.4	86.9	92.4	108.9	100.7	91.7	71.0	81.3	96.0	83.4	89.6	91.1	82.5	86.8
0330	Cast-in-Place Concrete	112.1	81.9	100.7	108.8	63.4	91.6	103.9	100.3	102.5	104.7	79.7	95.2	92.6	93.3	92.9	105.6	87.2	98.6
03	CONCRETE	96.9	79.8	89.1	93.9	77.2	86.3	92.7	102.2	97.1	92.6	76.7	85.3	87.5	89.0	88.2	91.6	86.2	89.2
04	MASONRY	107.2	77.5	88.8	101.9	72.4	83.6	102.8	96.1	98.7	122.3	69.7	89.7	89.0	86.0	87.1	107.1	71.8	85.2
05	METALS	97.7	93.1	96.3	94.5	93.5	94.2	92.7	106.3	96.9	90.3	88.6	89.8	100.5	100.7	100.6	91.2	94.0	92.1
06	WOOD, PLASTICS & COMPOSITES	75.3	75.6	75.5	72.1	84.4	78.9	95.3	99.9	97.9	86.1	74.3	79.5	91.4	84.4	87.5	76.9	86.2	82.0
07	THERMAL & MOISTURE PROTECTION	104.6	76.9	92.9	107.5	80.1	95.9	104.6	95.7	100.9	104.4	72.7	91.0	99.1	86.7	93.9	104.9	80.7	94.6
08	OPENINGS	97.4	78.7	93.1	105.0	84.4	100.3	98.4	100.5	98.9	96.5	77.4	92.1	96.9	87.5	94.7	97.5	86.6	95.0
0920	Plaster & Gypsum Board	104.8	75.3	85.0	102.5	84.1	90.2	113.9	100.0	104.6	108.1	73.7	85.0	94.5	84.1	87.5	104.8	86.2	92.3
0950, 0980	Ceilings & Acoustic Treatment	98.0	75.3	82.7	89.8	84.1	86.0	100.6	100.0	100.2	98.0	73.7	81.6	91.5	84.1	86.5	98.0	86.2	90.0
0960	Flooring	86.8	89.1	87.4	78.6	73.0	77.1	96.6	94.5	96.0	93.5	73.0	87.8	87.5	94.1	89.3	98.8	73.0	91.7
0970, 0990	Wall Finishes & Painting/Coating	90.9	74.0	81.0	84.3	77.7	80.4	94.8	94.6	94.7	94.8	85.0	89.1	85.8	85.0	85.3	95.6	82.0	87.7
09	FINISHES	90.8	78.2	84.0	83.4	81.5	82.4	95.9	98.8	97.5	93.7	75.6	83.9	89.5	87.0	88.1	94.7	82.9	88.3
COVERS	DIVS. 10 - 14, 25, 28, 41, 43, 44, 46	100.0	90.6	97.9	100.0	67.5	92.8	100.0	97.4	99.4	100.0	81.1	95.8	100.0	93.6	98.6	100.0	92.5	98.3
21, 22, 23	FIRE SUPPRESSION, PLUMBING & HVAC	100.2	74.8	89.3	96.9	81.7	90.4	100.2	97.8	99.2	97.1	73.1	86.9	99.9	83.9	93.0	100.2	78.7	91.0
26, 27, 3370	ELECTRICAL, COMMUNICATIONS & UTIL.	103.6	82.8	92.8	93.3	81.2	87.0	96.6	92.5	94.4	98.4	49.9	73.2	104.3	83.8	93.6	102.2	77.1	89.1
MF2016	WEIGHTED AVERAGE	99.0	81.1	91.2	96.2	82.1	90.0	97.3	98.6	97.9	96.8	74.0	86.8	97.2	88.6	93.5	97.3	82.8	91.0

IOWA

DIVISION		FORT DODGE 505 MAT.	INST.	TOTAL	MASON CITY 504 MAT.	INST.	TOTAL	OTTUMWA 525 MAT.	INST.	TOTAL	SHENANDOAH 516 MAT.	INST.	TOTAL	SIBLEY 512 MAT.	INST.	TOTAL	SIOUX CITY 510 - 511 MAT.	INST.	TOTAL
015433	CONTRACTOR EQUIPMENT		101.5	101.5		101.5	101.5		97.0	97.0		97.5	97.5		101.5	101.5		101.5	101.5
0241, 31 - 34	SITE & INFRASTRUCTURE, DEMOLITION	107.1	95.1	98.7	107.1	96.2	99.5	99.0	90.9	93.3	102.1	92.9	95.7	109.7	96.3	100.3	111.4	97.5	101.6
0310	Concrete Forming & Accessories	79.1	75.2	75.8	83.1	74.6	75.7	88.7	74.9	76.8	82.3	57.9	61.3	82.7	38.9	44.9	98.7	75.7	78.9
0320	Concrete Reinforcing	91.5	69.0	80.2	91.4	82.4	86.9	91.7	86.1	88.9	94.3	69.4	81.7	94.3	68.9	81.5	92.4	79.6	85.9
0330	Cast-in-Place Concrete	102.2	43.5	79.9	102.2	74.7	91.7	108.3	66.7	92.5	108.5	79.3	97.4	106.3	58.0	87.9	106.9	73.3	94.2
03	CONCRETE	89.5	64.2	77.9	89.8	77.0	83.9	94.2	75.0	85.4	94.4	68.8	82.7	93.4	52.7	74.8	94.0	76.5	86.0
04	MASONRY	100.8	51.7	70.3	114.2	68.4	85.8	103.6	55.1	73.5	106.8	73.2	86.0	126.0	52.0	80.1	100.0	69.6	81.1
05	METALS	94.6	87.6	92.4	94.6	93.4	94.2	90.2	94.8	91.6	96.7	88.1	94.1	90.4	87.1	89.4	92.7	92.5	92.6
06	WOOD, PLASTICS & COMPOSITES	72.5	84.4	79.1	76.3	74.3	75.2	83.5	81.2	82.2	76.9	52.8	63.5	77.6	35.7	54.3	95.3	75.7	84.4
07	THERMAL & MOISTURE PROTECTION	106.8	67.0	90.0	106.3	73.3	92.3	105.0	69.5	90.0	103.8	67.4	88.4	104.1	54.6	83.2	104.6	73.0	91.3
08	OPENINGS	99.0	70.3	92.4	91.4	80.5	88.9	97.9	81.7	94.2	89.1	59.0	82.2	94.6	43.3	82.8	98.4	77.1	93.5
0920	Plaster & Gypsum Board	102.5	84.1	90.2	102.5	73.7	83.2	105.7	81.0	89.1	104.8	51.9	69.2	104.8	34.1	57.3	113.9	75.2	87.9
0950, 0980	Ceilings & Acoustic Treatment	89.8	84.1	86.0	89.8	73.7	79.0	98.0	81.0	86.5	98.0	51.9	66.9	98.0	34.1	54.9	100.6	75.2	83.5
0960	Flooring	79.8	73.0	77.9	81.7	73.0	79.3	102.0	73.0	93.9	87.5	78.1	84.9	89.1	73.0	84.7	96.6	75.7	90.8
0970, 0990	Wall Finishes & Painting/Coating	84.3	85.0	84.7	84.3	85.0	84.7	95.6	85.0	89.4	90.9	63.7	75.0	94.8	59.8	74.4	94.8	69.5	80.0
09	FINISHES	85.3	77.1	80.8	85.9	75.3	80.2	95.9	76.0	85.1	90.9	60.9	74.6	93.4	45.6	67.4	97.4	74.9	85.2
COVERS	DIVS. 10 - 14, 25, 28, 41, 43, 44, 46	100.0	84.1	96.5	100.0	88.5	97.4	100.0	84.5	96.5	100.0	61.7	91.5	100.0	75.9	94.6	100.0	90.6	97.9
21, 22, 23	FIRE SUPPRESSION, PLUMBING & HVAC	96.9	70.6	85.7	96.9	79.5	89.5	97.1	72.7	86.7	97.1	87.1	92.8	97.1	66.3	84.0	100.2	79.5	91.4
26, 27, 3370	ELECTRICAL, COMMUNICATIONS & UTIL.	99.3	73.2	85.7	98.4	49.9	73.2	100.6	74.9	87.2	98.4	79.8	88.7	98.4	49.9	73.2	98.4	72.6	85.0
MF2016	WEIGHTED AVERAGE	95.9	73.0	85.9	95.8	76.0	87.1	96.7	76.2	87.7	96.4	76.5	87.7	97.1	61.0	81.3	97.9	79.2	89.7

IOWA / KANSAS

DIVISION		SPENCER 513 MAT.	INST.	TOTAL	WATERLOO 506 - 507 MAT.	INST.	TOTAL	BELLEVILLE 669 MAT.	INST.	TOTAL	COLBY 677 MAT.	INST.	TOTAL	DODGE CITY 678 MAT.	INST.	TOTAL	EMPORIA 668 MAT.	INST.	TOTAL
015433	CONTRACTOR EQUIPMENT		101.5	101.5		101.5	101.5		107.1	107.1		107.1	107.1		107.1	107.1		105.2	105.2
0241, 31 - 34	SITE & INFRASTRUCTURE, DEMOLITION	109.8	95.0	99.5	112.5	97.4	101.9	110.7	97.2	101.3	109.2	97.9	101.3	111.7	96.8	101.3	102.5	94.7	97.0
0310	Concrete Forming & Accessories	88.7	38.7	45.6	94.3	71.6	74.7	93.9	56.2	61.4	99.6	63.6	68.6	92.9	63.3	67.3	85.1	68.2	70.5
0320	Concrete Reinforcing	94.3	68.9	81.5	92.1	82.9	87.5	96.8	63.4	80.0	104.6	63.5	83.9	102.0	63.3	82.5	95.5	63.9	79.6
0330	Cast-in-Place Concrete	106.3	68.7	92.0	109.4	83.4	99.5	120.9	87.7	108.3	115.6	91.6	106.5	117.7	91.3	107.6	116.9	91.7	107.3
03	CONCRETE	93.8	56.3	76.7	95.3	78.7	87.7	111.5	70.0	92.5	108.9	74.7	93.3	110.3	74.4	93.9	104.1	76.9	91.6
04	MASONRY	126.0	52.0	80.1	101.5	77.7	86.7	92.0	55.2	69.2	102.2	61.5	77.0	112.5	59.9	79.9	97.9	68.9	79.9
05	METALS	90.4	87.0	89.3	97.0	94.7	96.3	98.7	86.4	95.0	99.1	87.3	95.5	100.6	86.0	96.1	98.5	87.9	95.2
06	WOOD, PLASTICS & COMPOSITES	83.4	35.7	56.9	89.3	67.4	77.1	98.2	53.6	73.4	107.0	60.1	80.9	99.0	60.1	77.4	89.6	65.9	76.4
07	THERMAL & MOISTURE PROTECTION	105.1	55.2	84.0	106.5	79.9	95.3	93.7	63.5	80.9	95.4	66.9	83.3	95.3	66.1	83.0	91.8	78.4	86.1
08	OPENINGS	105.9	43.3	91.5	91.7	76.7	88.3	98.2	55.9	88.5	103.4	63.3	94.2	103.4	59.5	93.3	96.1	65.2	89.0
0920	Plaster & Gypsum Board	105.7	34.1	57.6	111.0	66.7	81.2	95.4	52.4	66.5	100.6	59.1	72.7	95.0	59.1	70.9	92.7	65.1	74.2
0950, 0980	Ceilings & Acoustic Treatment	98.0	34.1	54.9	92.4	66.7	75.0	84.0	52.4	62.7	81.5	59.1	66.4	81.5	59.1	66.4	84.0	65.1	71.3
0960	Flooring	91.9	73.0	86.7	86.7	83.4	85.8	88.5	70.8	83.6	87.4	70.8	82.8	83.6	70.8	80.1	83.8	70.8	80.2
0970, 0990	Wall Finishes & Painting/Coating	94.8	59.8	74.4	84.3	85.0	84.7	84.5	57.6	68.8	91.7	57.6	71.8	91.7	57.6	71.8	84.5	57.6	68.8
09	FINISHES	94.3	44.4	67.2	89.4	74.3	81.2	86.6	57.8	71.0	86.3	63.3	73.8	84.5	63.3	72.9	83.8	66.7	74.5
COVERS	DIVS. 10 - 14, 25, 28, 41, 43, 44, 46	100.0	74.6	94.4	100.0	91.0	98.0	100.0	85.3	96.7	100.0	88.2	97.4	100.0	88.2	97.4	100.0	88.8	97.5
21, 22, 23	FIRE SUPPRESSION, PLUMBING & HVAC	97.1	70.7	85.9	100.0	83.1	92.8	96.9	71.7	86.1	96.9	72.5	86.5	100.0	72.5	88.2	96.9	75.9	87.9
26, 27, 3370	ELECTRICAL, COMMUNICATIONS & UTIL.	100.1	49.9	74.0	95.1	65.0	79.5	102.2	66.2	83.5	99.6	71.7	85.1	96.8	71.7	83.7	99.5	70.0	84.2
MF2016	WEIGHTED AVERAGE	98.7	62.1	82.7	97.2	80.2	89.7	99.1	69.9	86.3	99.8	73.5	88.3	100.9	72.9	88.7	97.4	75.8	87.9

For customer support on your Commercial Renovation Costs with RSMeans data, call 800.448.8182.

City Cost Indexes

		KANSAS																
		FORT SCOTT			HAYS			HUTCHINSON			INDEPENDENCE			KANSAS CITY			LIBERAL		
	DIVISION	667			676			675			673			660 - 662			679		
		MAT.	INST.	TOTAL	MAT.	INST.	TOTAL	MAT.	INST.	TOTAL	MAT.	INST.	TOTAL	MAT.	INST.	TOTAL	MAT.	INST.	TOTAL
015433	CONTRACTOR EQUIPMENT		106.1	106.1		107.1	107.1		107.1	107.1		107.1	107.1		103.6	103.6		107.1	107.1
0241, 31 - 34	SITE & INFRASTRUCTURE, DEMOLITION	99.3	94.7	96.1	114.1	97.4	102.4	93.1	97.9	96.5	112.8	97.9	102.4	93.3	94.9	94.4	113.8	97.5	102.4
0310	Concrete Forming & Accessories	101.8	83.7	86.2	97.1	61.1	66.1	87.6	58.6	62.6	108.3	70.4	75.6	98.5	98.7	98.7	93.4	60.7	65.2
0320	Concrete Reinforcing	94.9	101.7	98.3	102.0	63.4	82.5	102.0	63.3	82.5	101.4	64.9	83.0	92.0	104.4	98.3	103.5	63.3	83.2
0330	Cast-in-Place Concrete	108.4	86.8	100.2	91.2	88.0	90.0	84.5	91.3	87.1	118.2	91.5	108.0	92.8	99.9	95.5	91.2	87.6	89.8
03	CONCRETE	99.3	88.6	94.4	100.4	72.3	87.6	83.7	72.3	78.5	111.4	77.9	96.1	91.4	100.6	95.6	102.3	71.9	88.4
04	MASONRY	99.3	55.9	72.4	111.6	55.3	76.6	102.1	59.9	75.9	99.3	64.6	77.8	99.8	98.2	98.8	110.3	53.6	75.1
05	METALS	98.4	100.3	99.0	98.7	87.0	95.1	98.5	86.0	94.7	98.4	86.5	94.8	106.0	106.3	106.1	99.0	85.8	94.9
06	WOOD, PLASTICS & COMPOSITES	108.2	90.6	98.4	103.9	60.1	79.5	94.1	53.8	71.7	117.3	69.0	90.4	104.1	98.6	101.0	99.6	60.1	77.6
07	THERMAL & MOISTURE PROTECTION	92.8	75.2	85.4	95.7	64.2	82.4	94.2	65.4	82.0	95.4	77.8	88.0	92.5	100.6	95.9	95.9	63.2	82.0
08	OPENINGS	96.1	87.5	94.1	103.3	59.5	93.3	103.3	56.0	92.4	101.1	64.0	92.6	97.4	98.0	97.6	103.4	59.5	93.3
0920	Plaster & Gypsum Board	97.7	90.4	92.8	97.9	59.1	71.8	94.0	52.6	66.2	107.5	68.2	81.1	91.4	98.6	96.2	95.6	59.1	71.1
0950, 0980	Ceilings & Acoustic Treatment	84.0	90.4	88.3	81.5	59.1	66.4	81.5	52.6	62.0	81.5	68.2	72.6	84.0	98.6	93.9	81.5	59.1	66.4
0960	Flooring	97.9	70.7	90.3	86.2	70.8	81.9	80.8	70.8	78.0	91.6	70.7	85.8	78.8	99.7	84.6	83.9	69.7	79.9
0970, 0990	Wall Finishes & Painting/Coating	86.1	81.0	83.1	91.7	57.6	71.8	91.7	57.6	71.8	91.7	57.6	71.8	91.5	102.4	97.8	91.7	57.6	71.8
09	FINISHES	89.0	82.2	85.3	86.0	61.7	72.8	82.0	59.6	69.8	88.6	69.1	78.0	84.1	99.3	92.4	85.3	61.4	72.3
COVERS	DIVS. 10 - 14, 25, 28, 41, 43, 44, 46	100.0	88.5	97.5	100.0	86.1	96.9	100.0	87.5	97.2	100.0	89.2	97.6	100.0	95.9	99.1	100.0	86.1	96.9
21, 22, 23	FIRE SUPPRESSION, PLUMBING & HVAC	96.9	69.0	85.0	96.9	69.3	85.1	96.9	70.4	85.6	96.9	72.9	86.6	99.9	98.8	99.4	96.9	67.9	84.5
26, 27, 3370	ELECTRICAL, COMMUNICATIONS & UTIL.	98.8	71.7	84.7	98.6	71.7	84.6	94.1	64.4	78.6	96.1	72.3	83.7	104.4	98.4	101.3	96.8	71.7	83.7
MF2016	WEIGHTED AVERAGE	97.3	79.1	89.3	99.1	71.3	86.9	95.0	70.5	84.3	99.6	75.6	89.1	98.4	99.3	98.8	99.0	70.6	86.6

		KANSAS									KENTUCKY								
		SALINA			TOPEKA			WICHITA			ASHLAND			BOWLING GREEN			CAMPTON		
	DIVISION	674			664 - 666			670 - 672			411 - 412			421 - 422			413 - 414		
		MAT.	INST.	TOTAL	MAT.	INST.	TOTAL	MAT.	INST.	TOTAL	MAT.	INST.	TOTAL	MAT.	INST.	TOTAL	MAT.	INST.	TOTAL
015433	CONTRACTOR EQUIPMENT		107.1	107.1		105.2	105.2		107.1	107.1		100.3	100.3		93.1	93.1		99.5	99.5
0241, 31 - 34	SITE & INFRASTRUCTURE, DEMOLITION	102.0	97.4	98.8	96.9	94.5	95.2	98.0	98.6	98.4	115.9	82.6	92.6	80.2	94.2	90.0	89.3	95.5	93.6
0310	Concrete Forming & Accessories	89.4	56.3	60.9	97.2	69.0	72.8	95.7	56.1	61.6	85.8	95.9	94.5	84.7	79.2	79.9	87.7	83.2	83.8
0320	Concrete Reinforcing	101.4	63.4	82.2	91.6	90.6	91.1	99.5	101.3	100.4	89.2	94.7	92.0	86.5	79.1	82.8	87.3	95.2	91.3
0330	Cast-in-Place Concrete	102.1	88.0	96.8	97.5	89.9	94.6	95.5	76.3	88.2	87.2	97.1	91.0	86.5	72.5	81.2	96.5	70.6	86.7
03	CONCRETE	97.4	70.2	85.0	93.2	81.2	87.7	93.2	72.5	83.7	91.7	97.2	94.2	90.1	77.2	84.2	93.7	81.2	88.0
04	MASONRY	127.8	55.3	82.8	93.5	67.0	77.1	99.3	51.5	69.7	93.3	94.3	93.9	95.4	69.8	79.5	92.1	57.1	70.4
05	METALS	100.5	87.0	96.3	102.6	97.7	101.1	102.6	99.3	101.6	94.3	109.5	99.0	95.9	84.4	92.4	95.2	90.8	93.8
06	WOOD, PLASTICS & COMPOSITES	95.6	53.6	72.2	101.0	68.1	82.7	104.0	53.8	76.1	74.4	95.5	86.2	86.9	79.8	83.0	85.5	90.6	88.3
07	THERMAL & MOISTURE PROTECTION	94.8	63.5	81.6	96.5	77.7	88.6	94.1	60.6	79.9	92.0	91.7	91.9	88.0	79.2	84.2	101.0	70.9	88.3
08	OPENINGS	103.3	55.9	92.4	103.3	74.0	96.6	105.7	64.7	96.2	93.0	94.4	93.3	93.7	81.2	90.8	95.0	88.7	93.6
0920	Plaster & Gypsum Board	94.0	52.4	66.1	100.8	67.3	78.3	93.7	52.6	66.1	61.6	95.6	84.4	88.2	79.7	82.5	88.2	89.9	89.4
0950, 0980	Ceilings & Acoustic Treatment	81.5	52.4	61.9	88.7	67.3	74.3	85.3	52.6	63.3	78.9	95.6	90.2	83.6	79.7	81.0	83.6	89.9	87.9
0960	Flooring	82.2	70.8	79.0	89.0	70.0	84.4	92.6	70.8	86.6	73.6	85.9	77.0	88.3	64.4	81.7	90.4	65.7	83.5
0970, 0990	Wall Finishes & Painting/Coating	91.7	57.6	71.8	88.8	70.8	78.3	92.0	57.6	71.9	93.2	91.7	92.3	91.1	89.2	90.0	91.1	56.3	70.8
09	FINISHES	83.2	57.8	69.4	92.1	68.5	79.3	90.3	57.5	72.4	76.5	93.9	85.9	87.0	77.3	81.7	87.8	78.1	82.5
COVERS	DIVS. 10 - 14, 25, 28, 41, 43, 44, 46	100.0	86.1	96.9	100.0	83.3	96.3	100.0	85.4	96.8	100.0	91.7	98.1	100.0	87.4	97.2	100.0	49.5	88.8
21, 22, 23	FIRE SUPPRESSION, PLUMBING & HVAC	100.0	70.3	87.3	100.0	76.0	89.8	99.8	69.8	87.0	96.7	87.6	92.8	100.0	80.2	91.5	96.9	76.8	88.3
26, 27, 3370	ELECTRICAL, COMMUNICATIONS & UTIL.	96.5	74.1	84.8	103.2	73.7	87.8	100.2	74.1	86.6	92.3	93.2	92.8	94.8	78.9	86.5	92.4	93.1	92.8
MF2016	WEIGHTED AVERAGE	99.6	70.8	87.0	99.1	78.1	89.9	99.2	72.1	87.3	93.5	93.4	93.5	94.6	79.9	88.2	94.7	80.2	88.3

		KENTUCKY																	
		CORBIN			COVINGTON			ELIZABETHTOWN			FRANKFORT			HAZARD			HENDERSON		
	DIVISION	407 - 409			410			427			406			417 - 418			424		
		MAT.	INST.	TOTAL	MAT.	INST.	TOTAL	MAT.	INST.	TOTAL	MAT.	INST.	TOTAL	MAT.	INST.	TOTAL	MAT.	INST.	TOTAL
015433	CONTRACTOR EQUIPMENT		99.5	99.5		103.6	103.6		93.1	93.1		99.5	99.5		99.5	99.5		113.9	113.9
0241, 31 - 34	SITE & INFRASTRUCTURE, DEMOLITION	95.1	96.0	95.7	84.6	107.6	100.7	74.6	93.9	88.1	93.0	96.4	95.4	87.0	96.6	93.7	83.1	121.3	109.9
0310	Concrete Forming & Accessories	83.2	74.4	75.6	83.3	73.1	74.5	79.6	75.4	75.9	94.0	76.0	78.4	84.3	83.8	83.9	90.7	80.5	81.9
0320	Concrete Reinforcing	90.9	62.1	76.4	86.0	81.1	83.5	86.9	82.7	84.8	95.8	80.5	88.1	87.7	62.3	74.9	86.6	77.8	82.2
0330	Cast-in-Place Concrete	91.8	75.2	85.5	92.3	76.3	86.2	78.2	69.9	75.0	90.9	77.4	85.8	92.8	72.9	85.2	76.5	88.8	81.2
03	CONCRETE	87.1	73.1	80.7	92.1	76.5	84.9	82.3	75.2	79.1	89.3	77.7	84.0	90.6	76.6	84.2	87.3	83.2	85.4
04	MASONRY	90.0	62.6	73.0	106.9	75.1	87.1	79.2	64.4	70.0	85.5	73.3	77.9	90.8	59.2	71.2	99.0	82.1	88.5
05	METALS	93.9	78.4	89.1	93.2	89.9	92.2	95.1	85.1	92.0	96.6	86.2	93.4	95.2	78.8	90.2	86.4	84.6	85.9
06	WOOD, PLASTICS & COMPOSITES	72.1	75.0	73.7	83.7	70.2	76.1	82.4	77.5	79.7	89.6	75.0	81.5	82.7	90.6	87.1	89.5	79.4	83.9
07	THERMAL & MOISTURE PROTECTION	105.1	72.2	91.2	101.9	75.1	90.5	87.4	71.1	80.5	102.6	75.7	91.2	100.9	72.7	89.0	100.2	80.7	91.9
08	OPENINGS	89.4	62.5	83.2	97.0	74.1	91.7	93.7	75.6	89.6	95.3	77.2	91.2	95.4	80.7	92.0	91.9	79.8	89.1
0920	Plaster & Gypsum Board	93.1	73.9	80.2	70.8	70.0	70.2	87.5	77.3	80.6	92.5	73.9	80.0	87.5	89.9	89.1	91.5	78.4	82.7
0950, 0980	Ceilings & Acoustic Treatment	80.1	73.9	75.9	86.2	70.0	75.2	83.6	77.3	79.4	88.7	73.9	78.7	83.6	89.9	87.9	79.4	78.4	78.7
0960	Flooring	85.0	65.7	79.6	65.9	85.9	71.4	85.5	78.0	83.5	90.5	65.7	83.6	88.5	65.7	82.2	92.0	80.4	88.8
0970, 0990	Wall Finishes & Painting/Coating	90.9	64.8	75.7	86.8	74.4	79.5	91.1	70.1	78.9	93.2	71.2	80.4	91.1	56.3	70.8	91.1	90.7	90.9
09	FINISHES	83.2	71.9	77.0	78.1	75.0	76.4	85.7	75.0	79.9	89.9	73.6	81.0	87.0	78.6	82.4	87.1	81.8	84.2
COVERS	DIVS. 10 - 14, 25, 28, 41, 43, 44, 46	100.0	90.7	97.9	100.0	89.3	97.6	100.0	72.8	94.0	100.0	59.0	90.9	100.0	50.2	88.9	100.0	58.0	90.6
21, 22, 23	FIRE SUPPRESSION, PLUMBING & HVAC	97.0	76.4	88.2	97.7	79.0	89.7	97.0	77.1	88.5	100.0	80.2	91.5	96.9	77.6	88.7	97.0	79.0	89.3
26, 27, 3370	ELECTRICAL, COMMUNICATIONS & UTIL.	92.6	93.1	92.9	95.8	76.5	85.8	92.2	78.9	85.3	101.1	75.5	87.8	92.4	93.1	92.8	94.3	75.8	84.7
MF2016	WEIGHTED AVERAGE	92.7	77.7	86.2	94.7	80.5	88.5	91.5	77.2	85.3	96.0	78.5	88.3	94.1	78.7	87.4	92.4	83.0	88.3

For customer support on your Commercial Renovation Costs with RSMeans data, call 800.448.8182.

791

KENTUCKY

DIVISION		LEXINGTON 403 - 405			LOUISVILLE 400 - 402			OWENSBORO 423			PADUCAH 420			PIKEVILLE 415 - 416			SOMERSET 425 - 426		
		MAT.	INST.	TOTAL	MAT.	INST.	TOTAL	MAT.	INST.	TOTAL	MAT.	INST.	TOTAL	MAT.	INST.	TOTAL	MAT.	INST.	TOTAL
015433	CONTRACTOR EQUIPMENT		99.5	99.5		93.1	93.1		113.9	113.9		113.9	113.9		100.3	100.3		99.5	99.5
0241, 31 - 34	SITE & INFRASTRUCTURE, DEMOLITION	97.2	97.8	97.6	91.6	94.2	93.4	93.3	122.0	113.4	85.9	121.2	110.6	127.4	81.8	95.5	79.6	96.0	91.1
0310	Concrete Forming & Accessories	94.2	76.7	79.1	94.7	78.4	80.7	89.2	80.5	81.7	87.2	82.3	82.9	94.5	78.1	80.3	85.3	74.8	76.3
0320	Concrete Reinforcing	99.8	83.0	91.3	95.8	82.9	89.3	86.6	79.8	83.1	87.1	81.0	84.0	89.7	94.5	92.1	86.9	62.3	74.5
0330	Cast-in-Place Concrete	93.9	80.8	88.9	90.9	71.7	83.6	89.2	88.3	88.9	81.5	84.5	82.6	95.9	92.3	94.5	76.5	92.1	82.4
03	CONCRETE	89.9	79.6	85.2	89.3	77.3	83.8	99.1	83.4	91.9	91.7	83.1	87.8	105.0	87.4	97.0	77.5	79.1	78.2
04	MASONRY	88.2	73.3	79.0	86.5	69.4	75.9	91.3	82.0	85.5	94.1	78.8	84.6	90.8	86.8	88.3	85.7	65.4	73.1
05	METALS	96.4	86.9	93.5	97.5	85.5	93.8	87.9	86.5	87.5	84.9	86.7	85.4	94.2	108.0	98.5	95.1	79.0	90.2
06	WOOD, PLASTICS & COMPOSITES	86.0	75.0	79.8	89.7	79.8	84.2	87.5	79.4	83.0	85.3	82.2	83.6	83.3	77.0	79.8	83.2	75.0	78.6
07	THERMAL & MOISTURE PROTECTION	105.3	76.5	93.1	102.4	74.2	90.5	100.9	80.1	92.1	100.3	84.7	93.7	92.8	82.6	88.5	100.2	73.2	88.8
08	OPENINGS	89.6	77.1	86.8	85.9	76.9	83.9	91.9	80.2	89.2	91.2	82.6	89.2	93.6	76.6	89.7	94.4	69.3	88.7
0920	Plaster & Gypsum Board	102.4	73.9	83.3	92.8	79.7	84.0	90.2	78.4	82.3	89.2	81.3	83.9	64.5	76.6	72.6	87.5	73.9	78.4
0950, 0980	Ceilings & Acoustic Treatment	83.5	73.9	77.0	88.7	79.7	82.6	79.4	78.4	78.7	79.4	81.3	80.7	78.9	76.6	77.3	83.6	73.9	77.1
0960	Flooring	89.9	65.7	83.2	89.8	64.4	82.7	91.3	64.4	83.9	90.3	80.4	87.5	77.9	65.7	74.5	88.9	65.7	82.4
0970, 0990	Wall Finishes & Painting/Coating	90.9	80.9	85.0	92.5	70.1	79.5	91.1	90.7	90.9	91.1	76.5	82.6	93.2	69.9	79.9	91.1	70.1	78.9
09	FINISHES	86.7	74.6	80.1	89.6	74.9	81.6	87.3	78.2	82.3	86.5	81.3	83.7	79.0	74.6	76.6	86.4	72.5	78.8
COVERS	DIVS. 10 - 14, 25, 28, 41, 43, 44, 46	100.0	91.8	98.2	100.0	89.3	97.6	100.0	96.9	99.3	100.0	88.1	97.3	100.0	48.3	88.5	100.0	90.7	97.9
21, 22, 23	FIRE SUPPRESSION, PLUMBING & HVAC	100.1	79.0	91.1	100.0	78.7	90.9	100.0	80.3	91.6	97.0	79.8	89.7	96.7	82.8	90.8	97.0	75.9	88.0
26, 27, 3370	ELECTRICAL, COMMUNICATIONS & UTIL.	95.2	78.9	86.7	101.1	78.9	89.5	94.3	75.8	84.7	96.5	78.1	87.0	95.1	66.7	80.3	92.8	93.1	93.0
MF2016	WEIGHTED AVERAGE	94.8	80.3	88.5	95.1	79.1	88.1	94.7	84.3	90.2	92.6	84.5	89.1	95.9	81.4	89.6	91.9	79.2	86.4

LOUISIANA

DIVISION		ALEXANDRIA 713 - 714			BATON ROUGE 707 - 708			HAMMOND 704			LAFAYETTE 705			LAKE CHARLES 706			MONROE 712		
		MAT.	INST.	TOTAL	MAT.	INST.	TOTAL	MAT.	INST.	TOTAL	MAT.	INST.	TOTAL	MAT.	INST.	TOTAL	MAT.	INST.	TOTAL
015433	CONTRACTOR EQUIPMENT		93.5	93.5		91.4	91.4		91.9	91.9		91.9	91.9		91.4	91.4		93.5	93.5
0241, 31 - 34	SITE & INFRASTRUCTURE, DEMOLITION	105.3	92.4	96.3	102.3	91.1	94.5	98.9	90.4	93.0	99.9	92.7	94.9	100.7	91.8	94.5	105.3	92.3	96.2
0310	Concrete Forming & Accessories	80.2	62.9	65.3	95.5	74.7	77.5	77.9	57.3	60.1	95.0	70.9	74.2	95.8	71.0	74.4	79.8	62.2	64.7
0320	Concrete Reinforcing	96.3	65.1	80.5	91.7	55.6	73.5	93.0	53.7	73.2	94.3	54.0	74.0	94.3	54.3	74.1	95.1	65.0	80.0
0330	Cast-in-Place Concrete	93.8	68.2	84.0	93.5	77.2	87.3	90.1	65.7	80.8	89.7	69.5	82.0	94.4	69.5	84.9	93.8	67.4	83.7
03	CONCRETE	87.1	65.8	77.3	88.8	72.5	81.4	86.9	60.4	74.8	87.8	67.9	78.7	90.1	68.0	80.0	86.9	65.2	77.0
04	MASONRY	114.2	62.9	82.4	87.1	64.9	73.4	91.7	54.7	68.8	91.7	65.3	75.4	91.1	69.7	77.8	108.7	61.6	79.5
05	METALS	92.5	75.1	87.2	96.6	70.4	88.5	87.6	69.1	81.9	86.9	69.8	81.7	86.9	70.2	81.8	92.5	75.0	87.1
06	WOOD, PLASTICS & COMPOSITES	92.9	61.6	75.5	102.1	76.5	87.9	85.6	56.8	69.6	105.4	71.3	86.4	103.6	71.3	85.6	92.3	61.6	75.2
07	THERMAL & MOISTURE PROTECTION	97.9	67.4	85.0	95.3	69.9	84.5	97.1	63.7	83.0	97.6	69.5	85.7	97.3	70.5	86.0	97.9	66.6	84.6
08	OPENINGS	110.5	63.8	99.8	99.7	70.4	93.0	95.5	57.0	86.6	99.1	64.9	91.3	99.1	64.9	91.3	110.5	62.2	99.4
0920	Plaster & Gypsum Board	84.0	60.9	68.5	99.3	76.3	83.9	101.2	60.9	70.9	109.0	70.9	83.4	109.0	70.9	83.4	83.7	60.9	68.5
0950, 0980	Ceilings & Acoustic Treatment	83.0	60.9	68.1	94.8	76.3	82.3	101.6	56.0	70.9	99.9	70.9	80.3	100.9	70.9	80.7	83.0	60.9	68.1
0960	Flooring	90.9	64.9	83.7	86.9	60.4	79.6	88.1	58.4	79.8	96.8	64.3	87.8	96.8	67.4	88.6	90.5	58.2	81.5
0970, 0990	Wall Finishes & Painting/Coating	93.1	62.5	75.2	89.4	62.5	73.7	94.9	62.5	76.0	94.9	62.5	76.0	94.9	62.5	76.0	93.1	62.5	75.2
09	FINISHES	82.0	62.8	71.6	89.7	71.2	79.6	91.4	57.6	73.0	94.8	69.0	80.8	95.0	69.7	81.2	81.8	61.1	70.6
COVERS	DIVS. 10 - 14, 25, 28, 41, 43, 44, 46	100.0	80.6	95.7	100.0	87.2	97.2	100.0	80.4	95.6	100.0	86.8	97.1	100.0	86.9	97.1	100.0	80.1	95.6
21, 22, 23	FIRE SUPPRESSION, PLUMBING & HVAC	100.2	65.4	85.3	100.0	66.4	85.6	97.1	63.0	82.5	100.2	66.6	85.8	100.2	66.6	85.8	100.2	64.6	85.0
26, 27, 3370	ELECTRICAL, COMMUNICATIONS & UTIL.	93.2	64.9	78.5	97.9	59.8	78.1	93.7	70.5	81.7	94.8	67.4	80.5	94.4	69.6	81.5	94.8	60.5	77.0
MF2016	WEIGHTED AVERAGE	96.9	68.2	84.3	96.3	70.1	84.8	93.2	65.1	80.9	94.9	70.1	84.0	95.1	70.9	84.5	96.7	66.9	83.7

		LOUISIANA									MAINE								
DIVISION		NEW ORLEANS 700 - 701			SHREVEPORT 710 - 711			THIBODAUX 703			AUGUSTA 043			BANGOR 044			BATH 045		
		MAT.	INST.	TOTAL	MAT.	INST.	TOTAL	MAT.	INST.	TOTAL	MAT.	INST.	TOTAL	MAT.	INST.	TOTAL	MAT.	INST.	TOTAL
015433	CONTRACTOR EQUIPMENT		87.5	87.5		93.5	93.5		91.9	91.9		98.8	98.8		98.8	98.8		98.8	98.8
0241, 31 - 34	SITE & INFRASTRUCTURE, DEMOLITION	99.9	93.5	95.4	107.6	92.3	96.9	101.2	92.4	95.1	90.4	99.3	96.6	93.6	100.7	98.6	91.6	99.3	97.0
0310	Concrete Forming & Accessories	97.3	69.3	73.1	94.8	62.8	67.2	89.4	67.2	70.3	101.1	77.7	80.9	93.8	79.5	81.5	89.7	77.9	79.5
0320	Concrete Reinforcing	96.2	55.0	75.4	96.3	53.8	74.9	93.0	67.5	80.1	102.3	79.0	90.5	92.1	80.8	86.4	91.1	80.2	85.6
0330	Cast-in-Place Concrete	91.1	69.3	82.8	96.7	67.5	85.6	96.8	67.4	85.6	86.7	112.4	96.5	70.0	113.9	86.7	70.0	112.5	86.1
03	CONCRETE	93.1	66.7	81.0	90.5	63.6	78.2	91.7	67.6	80.7	88.1	90.5	89.2	81.7	92.1	86.5	81.7	90.8	85.9
04	MASONRY	96.9	60.6	74.4	98.1	58.9	73.8	114.7	53.9	77.0	102.7	66.9	80.5	118.3	95.5	104.1	125.5	86.9	101.5
05	METALS	96.8	62.7	86.3	96.6	70.8	88.7	87.6	68.9	81.8	101.1	89.5	97.6	91.0	92.7	91.5	89.4	91.2	89.9
06	WOOD, PLASTICS & COMPOSITES	103.7	71.4	85.7	99.1	62.4	78.6	92.8	68.7	79.4	98.6	76.5	86.3	97.9	76.5	86.0	92.5	76.5	83.6
07	THERMAL & MOISTURE PROTECTION	97.9	68.4	85.4	95.6	66.0	83.1	97.0	65.1	83.5	106.2	70.1	90.9	102.8	84.7	95.1	102.8	76.3	91.5
08	OPENINGS	103.1	65.3	94.5	105.4	59.0	94.8	100.0	59.4	90.7	100.6	80.3	96.0	98.0	80.3	94.0	98.0	80.3	93.9
0920	Plaster & Gypsum Board	98.6	70.9	80.0	93.5	61.7	72.2	102.5	68.2	79.5	108.2	75.4	86.2	112.7	75.4	87.7	108.5	75.4	86.3
0950, 0980	Ceilings & Acoustic Treatment	105.8	70.9	82.3	89.7	61.7	70.8	101.6	68.2	79.1	104.0	75.4	84.7	86.6	75.4	79.1	85.6	75.4	78.8
0960	Flooring	103.8	60.4	91.8	94.9	58.2	84.7	94.1	40.3	79.2	88.5	48.4	77.4	85.0	104.4	90.4	83.3	48.4	73.6
0970, 0990	Wall Finishes & Painting/Coating	99.0	62.5	77.7	85.6	62.5	72.1	96.1	62.5	76.5	90.5	93.0	91.9	90.7	93.0	92.0	90.7	93.0	92.0
09	FINISHES	101.3	67.1	82.7	88.7	61.7	74.0	91.6	61.6	76.3	93.1	72.7	82.0	87.4	84.8	86.0	86.1	72.7	78.8
COVERS	DIVS. 10 - 14, 25, 28, 41, 43, 44, 46	100.0	85.6	96.8	100.0	85.7	96.8	100.0	85.2	96.7	100.0	100.9	100.2	100.0	104.0	100.9	100.0	100.9	100.2
21, 22, 23	FIRE SUPPRESSION, PLUMBING & HVAC	100.1	64.0	84.7	99.9	64.8	84.9	97.1	64.5	83.2	100.0	71.9	88.0	100.2	73.0	88.5	97.1	72.0	86.3
26, 27, 3370	ELECTRICAL, COMMUNICATIONS & UTIL.	98.4	70.5	83.9	100.2	64.5	81.6	92.5	70.5	81.1	101.5	77.6	89.1	99.7	73.2	86.0	97.8	77.6	87.3
MF2016	WEIGHTED AVERAGE	98.7	68.4	85.5	97.9	66.7	84.2	95.6	67.4	83.3	98.4	79.8	90.3	95.8	85.0	91.1	94.8	82.2	89.3

City Cost Indexes

MAINE

DIVISION		HOULTON 047 MAT.	INST.	TOTAL	KITTERY 039 MAT.	INST.	TOTAL	LEWISTON 042 MAT.	INST.	TOTAL	MACHIAS 046 MAT.	INST.	TOTAL	PORTLAND 040-041 MAT.	INST.	TOTAL	ROCKLAND 048 MAT.	INST.	TOTAL
015433	CONTRACTOR EQUIPMENT		98.8	98.8		98.8	98.8		98.8	98.8		98.8	98.8		98.8	98.8		98.8	98.8
0241, 31-34	SITE & INFRASTRUCTURE, DEMOLITION	93.4	99.3	97.5	82.9	99.3	94.4	91.3	100.7	97.9	92.8	99.3	97.3	91.2	100.7	97.8	89.3	99.3	96.3
0310	Concrete Forming & Accessories	97.6	77.7	80.4	87.8	78.2	79.6	99.4	79.6	82.3	94.7	77.7	80.0	101.1	79.6	82.5	95.7	77.7	80.2
0320	Concrete Reinforcing	92.1	78.7	85.3	85.3	80.3	82.8	113.1	80.8	96.8	92.1	78.7	85.3	102.3	80.8	91.4	92.1	78.7	85.3
0330	Cast-in-Place Concrete	70.0	112.4	86.1	73.3	112.6	88.2	71.4	113.9	87.6	70.0	112.4	86.1	86.7	113.9	97.0	71.5	112.4	87.0
03	CONCRETE	82.6	90.4	86.2	78.8	91.0	84.4	82.1	92.1	86.7	82.1	90.4	85.9	87.4	92.1	89.6	79.8	90.4	84.7
04	MASONRY	100.2	62.4	76.8	115.6	86.9	97.8	100.8	95.5	97.5	100.2	62.4	76.8	100.1	95.5	99.5	93.7	62.4	74.3
05	METALS	89.6	89.1	89.5	85.8	91.5	87.6	94.3	92.7	93.8	89.6	89.1	89.5	101.2	92.7	98.6	89.5	89.1	89.4
06	WOOD, PLASTICS & COMPOSITES	101.8	76.5	87.7	95.8	76.5	85.0	103.6	76.5	88.5	98.8	76.5	86.4	97.3	76.5	85.7	99.6	76.5	86.7
07	THERMAL & MOISTURE PROTECTION	102.9	69.9	88.9	102.4	75.4	91.0	102.6	84.7	95.0	102.8	69.1	88.6	106.5	84.7	97.2	102.5	69.1	88.4
08	OPENINGS	98.1	80.3	94.0	97.4	84.0	94.3	101.3	80.3	96.5	98.1	80.3	94.0	99.9	80.3	95.4	98.0	80.3	94.0
0920	Plaster & Gypsum Board	115.0	75.4	88.4	104.8	75.4	85.1	118.0	75.4	89.4	113.4	75.4	87.9	108.6	75.4	86.3	113.4	75.4	87.9
0950, 0980	Ceilings & Acoustic Treatment	85.6	75.4	78.8	99.5	75.4	83.3	95.1	75.4	81.8	85.6	75.4	78.8	102.2	75.4	84.1	85.6	75.4	78.8
0960	Flooring	86.2	45.2	74.9	85.4	50.4	75.7	87.8	104.4	92.4	85.4	45.2	74.2	89.2	104.4	93.4	85.8	45.2	74.5
0970, 0990	Wall Finishes & Painting/Coating	90.7	93.0	92.0	84.8	106.8	97.6	90.7	93.0	92.0	90.7	93.0	92.0	93.2	93.0	93.1	90.7	93.0	92.0
09	FINISHES	88.0	72.1	79.3	87.8	74.5	80.6	90.3	84.8	87.3	87.4	72.1	79.1	93.7	84.8	88.8	87.2	72.1	79.0
COVERS	DIVS. 10-14, 25, 28, 41, 43, 44, 46	100.0	100.9	100.2	100.0	100.9	100.2	100.0	104.1	100.9	100.0	100.9	100.2	100.0	104.1	100.9	100.0	100.9	100.2
21, 22, 23	FIRE SUPPRESSION, PLUMBING & HVAC	97.1	71.9	86.3	97.0	82.6	90.9	100.2	73.0	88.5	97.1	71.9	86.3	100.0	73.0	88.5	97.1	71.9	86.3
26, 27, 3370	ELECTRICAL, COMMUNICATIONS & UTIL.	101.7	77.6	89.2	96.0	77.6	86.4	101.7	77.6	89.2	101.7	77.6	89.2	103.8	77.6	90.2	101.6	77.6	89.1
MF2016	WEIGHTED AVERAGE	94.4	79.2	87.8	93.0	84.9	89.5	96.3	85.6	91.6	94.2	79.2	87.7	98.7	85.6	93.0	93.5	79.2	87.2

MAINE / MARYLAND

DIVISION		WATERVILLE 049 MAT.	INST.	TOTAL	ANNAPOLIS 214 MAT.	INST.	TOTAL	BALTIMORE 210-212 MAT.	INST.	TOTAL	COLLEGE PARK 207-208 MAT.	INST.	TOTAL	CUMBERLAND 215 MAT.	INST.	TOTAL	EASTON 216 MAT.	INST.	TOTAL
015433	CONTRACTOR EQUIPMENT		98.8	98.8		105.7	105.7		105.6	105.6		111.0	111.0		105.7	105.7		105.7	105.7
0241, 31-34	SITE & INFRASTRUCTURE, DEMOLITION	93.3	99.3	97.5	105.3	94.1	97.5	102.5	98.2	99.4	101.3	98.6	99.4	95.9	95.3	95.5	103.1	92.4	95.6
0310	Concrete Forming & Accessories	89.2	77.7	79.3	97.4	75.7	78.7	98.8	78.3	81.2	85.6	73.8	75.4	90.9	85.8	86.5	89.1	74.1	76.2
0320	Concrete Reinforcing	92.1	79.0	85.5	104.8	90.0	97.3	100.6	90.1	95.3	100.6	89.1	94.8	91.6	85.8	88.7	90.9	87.7	89.3
0330	Cast-in-Place Concrete	70.0	112.4	86.1	110.5	74.5	96.8	109.9	76.3	97.1	116.0	78.2	101.7	94.8	87.3	91.9	105.3	68.1	91.1
03	CONCRETE	83.1	90.5	86.5	98.2	79.3	89.5	107.2	80.7	95.1	106.9	79.7	94.4	89.6	87.6	88.7	97.6	75.8	87.6
04	MASONRY	110.8	66.9	83.6	99.5	70.6	81.6	101.8	70.6	82.5	115.6	76.5	91.4	101.0	87.9	92.9	115.2	61.1	81.6
05	METALS	89.6	89.5	89.6	102.1	105.0	103.0	101.7	99.9	101.1	88.5	107.3	94.3	98.1	105.0	100.2	98.4	101.0	99.2
06	WOOD, PLASTICS & COMPOSITES	92.0	76.5	83.4	94.2	77.4	84.8	101.2	80.7	89.8	83.4	71.7	76.9	90.2	84.9	87.2	88.3	80.9	84.1
07	THERMAL & MOISTURE PROTECTION	102.9	70.1	89.0	103.1	79.8	93.2	101.1	80.8	92.5	100.6	82.9	93.1	99.9	85.6	93.8	100.1	75.8	89.8
08	OPENINGS	98.1	80.3	94.0	97.3	83.0	94.0	99.8	84.8	96.4	95.2	79.6	91.6	100.1	87.8	97.3	98.3	84.3	95.1
0920	Plaster & Gypsum Board	108.5	75.4	86.3	100.1	77.0	84.6	104.3	80.1	88.1	98.7	71.0	80.1	105.7	84.7	91.6	105.7	80.6	88.9
0950, 0980	Ceilings & Acoustic Treatment	85.6	75.4	78.8	91.7	77.0	81.8	100.1	80.1	86.6	101.6	71.0	81.0	103.0	84.7	90.7	103.0	80.6	87.9
0960	Flooring	83.0	40.4	73.4	86.2	80.0	84.5	99.3	80.0	93.9	89.3	80.0	86.7	87.9	96.3	90.2	87.0	80.0	85.1
0970, 0990	Wall Finishes & Painting/Coating	90.7	93.0	92.0	87.3	76.5	81.0	102.9	76.5	87.5	104.1	74.3	86.7	97.3	88.5	92.2	97.3	74.3	83.9
09	FINISHES	86.2	72.7	78.8	86.1	76.2	80.7	99.8	78.2	88.1	92.8	74.1	82.6	94.1	88.0	90.8	94.3	75.7	84.2
COVERS	DIVS. 10-14, 25, 28, 41, 43, 44, 46	100.0	100.7	100.2	100.0	87.9	97.3	100.0	89.1	97.6	100.0	84.0	96.4	100.0	92.5	98.3	100.0	70.9	93.5
21, 22, 23	FIRE SUPPRESSION, PLUMBING & HVAC	97.1	71.9	86.3	100.1	82.1	92.4	100.1	82.2	92.4	97.0	86.1	92.3	96.8	74.9	87.5	96.8	75.8	87.8
26, 27, 3370	ELECTRICAL, COMMUNICATIONS & UTIL.	101.7	77.6	89.1	100.1	90.9	95.3	100.7	90.9	95.6	100.2	98.6	99.4	98.1	80.6	89.0	97.6	64.1	80.2
MF2016	WEIGHTED AVERAGE	94.7	79.8	88.2	98.9	84.2	92.4	101.4	84.7	94.1	97.9	86.8	93.0	96.9	86.3	92.3	98.6	76.6	88.9

MARYLAND / MASSACHUSETTS

DIVISION		ELKTON 219 MAT.	INST.	TOTAL	HAGERSTOWN 217 MAT.	INST.	TOTAL	SALISBURY 218 MAT.	INST.	TOTAL	SILVER SPRING 209 MAT.	INST.	TOTAL	WALDORF 206 MAT.	INST.	TOTAL	BOSTON 020-022, 024 MAT.	INST.	TOTAL
015433	CONTRACTOR EQUIPMENT		105.7	105.7		105.7	105.7		105.7	105.7		102.4	102.4		102.4	102.4		103.0	103.0
0241, 31-34	SITE & INFRASTRUCTURE, DEMOLITION	89.8	93.9	92.6	94.0	95.2	94.8	103.1	92.4	95.6	89.7	90.0	89.9	96.1	90.1	91.9	95.5	102.1	100.1
0310	Concrete Forming & Accessories	94.8	90.7	91.3	90.0	82.5	83.5	102.8	52.6	59.5	94.1	73.1	75.9	101.3	70.5	74.8	101.6	139.0	133.8
0320	Concrete Reinforcing	90.9	113.2	102.2	91.6	85.8	88.7	90.9	67.3	79.0	99.4	89.0	94.1	100.0	89.0	94.4	101.1	147.6	124.6
0330	Cast-in-Place Concrete	85.3	74.0	81.0	90.3	87.3	89.2	105.3	66.1	90.4	118.8	78.6	103.5	133.1	78.6	112.4	96.0	139.2	112.4
03	CONCRETE	82.3	89.9	85.8	86.0	86.0	86.0	98.4	61.8	81.7	104.8	79.3	93.1	115.7	78.1	98.5	100.9	139.9	118.7
04	MASONRY	99.6	69.9	81.2	106.6	87.9	95.0	114.9	57.5	79.3	115.1	76.9	91.4	98.5	76.9	85.1	112.0	148.2	134.5
05	METALS	98.4	113.6	103.1	98.3	104.9	100.3	98.4	93.0	96.8	92.9	103.1	96.1	92.9	103.1	96.1	101.1	133.3	111.0
06	WOOD, PLASTICS & COMPOSITES	95.3	98.1	96.9	89.5	80.4	84.4	105.3	53.8	76.6	89.9	70.9	79.3	96.8	67.5	80.5	101.9	139.5	122.8
07	THERMAL & MOISTURE PROTECTION	99.6	81.6	92.0	100.3	86.7	94.5	100.3	71.1	87.9	104.3	88.0	97.4	104.8	87.6	97.5	106.0	137.6	119.3
08	OPENINGS	98.3	100.6	98.9	98.3	82.5	94.7	98.6	63.7	90.6	87.4	79.1	85.5	88.1	77.3	85.6	98.7	145.2	109.3
0920	Plaster & Gypsum Board	109.3	98.4	102.0	105.7	80.1	88.5	116.2	52.8	73.6	104.3	71.0	81.9	107.3	67.5	80.6	108.8	140.4	130.0
0950, 0980	Ceilings & Acoustic Treatment	103.0	98.4	99.9	103.9	80.1	87.9	103.0	52.8	69.2	110.1	71.0	83.7	110.1	67.5	81.4	109.4	140.4	130.3
0960	Flooring	89.5	80.0	86.9	87.5	96.3	89.9	93.4	80.0	89.7	95.4	80.0	91.2	99.2	80.0	93.8	95.3	163.7	114.3
0970, 0990	Wall Finishes & Painting/Coating	97.3	74.3	83.9	97.3	74.3	83.9	97.3	74.3	83.9	111.7	74.3	89.9	111.7	74.3	89.9	97.1	158.1	132.7
09	FINISHES	94.7	88.1	91.1	94.0	83.8	88.5	97.7	58.8	76.5	93.8	73.6	82.8	95.6	71.6	82.5	102.7	146.3	126.4
COVERS	DIVS. 10-14, 25, 28, 41, 43, 44, 46	100.0	58.1	90.7	100.0	92.0	98.2	100.0	79.7	95.5	100.0	84.4	96.5	100.0	84.0	96.4	100.0	116.7	103.7
21, 22, 23	FIRE SUPPRESSION, PLUMBING & HVAC	96.8	80.5	89.8	99.9	86.5	94.2	96.8	73.7	86.9	97.0	85.9	92.3	97.0	85.8	92.2	100.1	126.1	111.2
26, 27, 3370	ELECTRICAL, COMMUNICATIONS & UTIL.	99.3	89.1	94.0	97.9	80.6	88.9	96.4	63.5	79.3	97.5	98.6	98.1	95.1	98.6	96.9	101.4	129.8	116.2
MF2016	WEIGHTED AVERAGE	95.8	87.4	92.1	97.2	87.7	93.0	99.0	69.8	86.2	97.2	85.8	92.2	97.9	85.2	92.3	101.2	133.1	115.2

For customer support on your Commercial Renovation Costs with RSMeans data, call 800.448.8182.

MASSACHUSETTS

DIVISION		BROCKTON 023			BUZZARDS BAY 025			FALL RIVER 027			FITCHBURG 014			FRAMINGHAM 017			GREENFIELD 013		
		MAT.	INST.	TOTAL	MAT.	INST.	TOTAL	MAT.	INST.	TOTAL	MAT.	INST.	TOTAL	MAT.	INST.	TOTAL	MAT.	INST.	TOTAL
015433	CONTRACTOR EQUIPMENT		101.4	101.4		101.4	101.4		102.3	102.3		98.8	98.8		100.7	100.7		98.8	98.8
0241, 31 - 34	SITE & INFRASTRUCTURE, DEMOLITION	94.0	103.8	100.9	84.2	103.8	97.9	93.0	103.9	100.6	85.3	103.7	98.2	82.0	103.6	97.1	88.9	102.5	98.4
0310	Concrete Forming & Accessories	98.6	125.5	121.8	96.3	124.9	120.9	98.6	125.1	121.4	92.9	125.1	120.7	100.3	125.5	122.0	91.3	109.4	106.9
0320	Concrete Reinforcing	102.0	147.3	124.9	81.8	124.4	103.3	102.0	124.5	113.4	84.6	147.0	116.1	84.6	147.1	116.1	88.3	120.1	104.3
0330	Cast-in-Place Concrete	89.4	138.6	108.1	74.3	138.4	98.6	86.5	138.9	106.4	78.7	138.5	101.4	78.7	138.7	101.5	80.9	122.3	96.6
03	CONCRETE	92.2	133.3	111.0	78.1	129.0	101.4	90.8	129.2	108.4	76.7	132.9	102.3	79.2	133.3	103.9	80.0	115.4	96.2
04	MASONRY	103.0	144.1	128.5	96.0	142.3	124.7	103.7	142.3	127.6	102.0	140.7	126.0	108.7	140.8	128.6	106.9	122.8	116.8
05	METALS	97.0	128.7	106.8	91.9	118.8	100.1	97.0	119.2	103.8	94.0	125.0	103.5	94.1	128.3	104.6	96.5	111.5	101.1
06	WOOD, PLASTICS & COMPOSITES	101.4	122.2	113.0	98.3	122.2	111.6	101.4	122.4	113.1	99.7	122.2	112.2	105.9	122.0	114.8	97.8	107.1	103.0
07	THERMAL & MOISTURE PROTECTION	100.1	132.9	114.0	99.0	129.5	111.9	100.0	127.8	111.8	100.8	126.7	111.7	100.9	131.5	113.8	100.8	110.2	104.8
08	OPENINGS	98.8	133.2	106.7	94.6	122.4	101.0	98.8	122.4	104.2	100.1	133.2	107.7	90.9	133.1	100.6	100.5	113.1	103.4
0920	Plaster & Gypsum Board	92.1	122.4	112.5	87.9	122.4	111.1	92.1	122.4	112.5	108.3	122.4	117.8	111.0	122.4	118.6	109.3	106.9	107.7
0950, 0980	Ceilings & Acoustic Treatment	108.7	122.4	117.9	93.3	122.4	112.9	108.7	122.4	117.9	93.1	122.4	112.9	93.1	122.4	112.9	101.6	106.9	105.2
0960	Flooring	90.8	163.7	111.0	88.7	163.7	109.5	89.9	163.7	110.4	86.6	163.7	108.0	88.3	163.7	109.2	85.8	140.4	101.0
0970, 0990	Wall Finishes & Painting/Coating	90.5	137.4	117.9	90.5	137.4	117.9	90.5	137.4	117.9	87.8	137.4	116.7	88.7	137.4	117.1	87.8	114.3	103.3
09	FINISHES	92.8	133.9	115.2	88.0	133.9	112.9	92.6	134.1	115.1	87.5	133.9	112.7	88.2	133.7	113.0	89.4	115.7	103.7
COVERS	DIVS. 10 - 14, 25, 28, 41, 43, 44, 46	100.0	115.1	103.4	100.0	115.1	103.4	100.0	115.7	103.5	100.0	109.7	102.2	100.0	114.7	103.3	100.0	105.3	101.2
21, 22, 23	FIRE SUPPRESSION, PLUMBING & HVAC	100.3	109.7	104.3	97.2	109.2	102.3	100.3	109.3	104.1	97.4	110.4	102.9	97.4	123.4	108.5	97.4	102.4	99.5
26, 27, 3370	ELECTRICAL, COMMUNICATIONS & UTIL.	100.1	103.2	101.7	97.5	103.2	100.4	100.0	103.2	101.6	101.4	102.5	101.9	98.1	129.3	114.3	101.4	98.2	99.7
MF2016	WEIGHTED AVERAGE	97.9	121.8	108.4	92.8	119.5	104.5	97.7	119.6	107.3	94.4	120.8	105.9	93.7	127.9	108.7	95.7	109.0	101.5

MASSACHUSETTS

DIVISION		HYANNIS 026			LAWRENCE 019			LOWELL 018			NEW BEDFORD 027			PITTSFIELD 012			SPRINGFIELD 010 - 011		
		MAT.	INST.	TOTAL	MAT.	INST.	TOTAL	MAT.	INST.	TOTAL	MAT.	INST.	TOTAL	MAT.	INST.	TOTAL	MAT.	INST.	TOTAL
015433	CONTRACTOR EQUIPMENT		101.4	101.4		101.4	101.4		98.8	98.8		102.3	102.3		98.8	98.8		98.8	98.8
0241, 31 - 34	SITE & INFRASTRUCTURE, DEMOLITION	90.3	103.8	99.8	94.1	103.8	100.9	93.0	103.8	100.5	91.4	103.9	100.2	94.0	102.2	99.7	93.5	102.5	99.8
0310	Concrete Forming & Accessories	90.9	124.9	120.2	101.5	125.8	122.4	98.0	125.7	121.9	98.6	125.2	121.5	98.0	107.8	106.4	98.2	109.4	107.9
0320	Concrete Reinforcing	81.8	124.4	103.3	104.7	139.0	122.0	105.6	138.7	122.3	102.0	124.5	113.4	87.7	120.8	104.4	105.6	120.1	112.9
0330	Cast-in-Place Concrete	81.6	138.4	103.2	91.0	138.8	109.2	82.8	138.7	104.0	76.2	138.9	100.0	90.3	119.5	101.4	86.1	122.3	99.9
03	CONCRETE	84.0	129.0	104.6	92.4	132.0	110.5	84.7	131.7	106.2	86.0	129.3	105.8	86.0	113.8	98.7	86.3	115.4	99.6
04	MASONRY	102.0	142.3	127.0	114.6	144.9	133.4	100.9	140.7	125.6	101.9	142.3	126.9	101.5	113.8	109.2	101.2	122.8	114.6
05	METALS	93.4	118.8	101.2	96.9	125.7	105.7	96.9	122.2	104.6	97.0	119.3	103.9	96.7	111.5	101.2	99.5	111.5	103.2
06	WOOD, PLASTICS & COMPOSITES	92.1	122.2	108.8	106.1	122.2	115.0	105.6	122.2	114.8	101.4	122.4	113.1	105.6	107.5	106.7	105.6	107.1	106.4
07	THERMAL & MOISTURE PROTECTION	99.5	129.5	112.2	101.5	133.1	114.8	101.2	131.9	114.2	99.9	127.8	111.7	101.3	106.9	103.7	101.2	110.2	105.0
08	OPENINGS	95.2	122.4	101.4	94.8	130.9	103.1	101.6	130.9	108.3	98.8	127.1	105.3	101.6	113.6	104.3	101.6	113.1	104.2
0920	Plaster & Gypsum Board	83.7	122.4	109.7	113.6	122.4	119.5	113.6	122.4	119.5	92.1	122.4	112.5	113.6	107.3	109.4	113.6	106.9	109.1
0950, 0980	Ceilings & Acoustic Treatment	101.9	122.4	115.7	103.5	122.4	116.3	103.5	122.4	116.3	108.7	122.4	117.9	103.5	107.3	106.1	103.5	106.9	105.8
0960	Flooring	86.0	163.7	107.6	88.8	163.7	109.6	88.8	163.7	109.6	89.9	163.7	110.4	89.2	132.9	101.4	88.2	140.4	102.7
0970, 0990	Wall Finishes & Painting/Coating	90.5	137.4	117.9	87.9	137.4	116.8	87.8	137.4	116.7	90.5	137.4	117.9	87.8	114.3	103.3	89.0	114.3	103.7
09	FINISHES	88.6	133.9	113.2	91.4	133.9	114.5	91.4	133.9	114.5	92.4	134.1	115.1	91.5	113.2	103.3	91.3	115.7	104.6
COVERS	DIVS. 10 - 14, 25, 28, 41, 43, 44, 46	100.0	115.1	103.4	100.0	115.2	103.4	100.0	115.2	103.4	100.0	115.7	103.5	100.0	103.7	100.8	100.0	105.3	101.2
21, 22, 23	FIRE SUPPRESSION, PLUMBING & HVAC	100.3	109.2	104.1	100.1	125.9	111.1	100.1	125.8	111.1	100.3	109.3	104.1	100.1	98.1	99.2	100.1	102.4	101.1
26, 27, 3370	ELECTRICAL, COMMUNICATIONS & UTIL.	97.9	103.2	100.6	100.5	129.8	115.7	100.1	129.6	114.0	100.9	103.2	102.1	100.9	99.2	99.5	101.0	98.2	99.5
MF2016	WEIGHTED AVERAGE	95.1	119.5	105.8	98.0	128.5	111.4	97.1	127.1	110.3	97.0	119.8	107.0	97.3	106.5	101.3	97.8	109.0	102.7

DIVISION		MASSACHUSETTS WORCESTER 015 - 016			MICHIGAN ANN ARBOR 481			BATTLE CREEK 490			BAY CITY 487			DEARBORN 481			DETROIT 482		
		MAT.	INST.	TOTAL	MAT.	INST.	TOTAL	MAT.	INST.	TOTAL	MAT.	INST.	TOTAL	MAT.	INST.	TOTAL	MAT.	INST.	TOTAL
015433	CONTRACTOR EQUIPMENT		98.8	98.8		112.1	112.1		98.9	98.9		112.1	112.1		112.1	112.1		96.3	96.3
0241, 31 - 34	SITE & INFRASTRUCTURE, DEMOLITION	93.4	103.7	100.6	82.2	96.7	92.3	92.1	85.8	87.7	73.6	96.3	89.5	81.9	96.8	92.3	97.3	100.7	99.7
0310	Concrete Forming & Accessories	98.6	125.1	121.5	96.4	106.3	105.0	95.2	80.9	82.9	96.5	82.6	84.5	96.3	107.0	105.5	99.3	107.0	105.9
0320	Concrete Reinforcing	105.6	147.0	126.5	98.4	104.5	101.5	88.6	80.9	84.7	98.4	103.7	101.1	98.4	104.6	101.5	99.3	106.5	102.9
0330	Cast-in-Place Concrete	85.6	138.5	105.7	88.4	101.5	93.3	85.3	95.8	89.3	84.6	87.7	85.7	86.4	102.3	92.5	103.2	101.9	102.7
03	CONCRETE	86.1	132.9	107.5	89.2	105.2	96.5	84.8	85.8	85.3	87.4	89.6	88.4	88.3	105.8	96.3	101.3	104.5	102.8
04	MASONRY	100.7	140.7	125.5	97.8	100.5	99.5	97.3	81.7	87.7	97.4	81.0	87.2	97.7	102.4	100.6	96.1	96.3	96.2
05	METALS	99.5	125.0	107.3	97.2	117.5	103.4	99.2	84.1	94.6	97.8	115.4	103.2	97.2	117.7	103.5	98.5	96.3	97.8
06	WOOD, PLASTICS & COMPOSITES	106.0	122.2	115.0	90.9	107.9	100.4	89.2	79.2	83.6	90.9	82.1	86.0	90.9	107.9	100.4	95.7	107.8	102.4
07	THERMAL & MOISTURE PROTECTION	101.2	126.7	112.0	106.1	102.7	104.6	97.5	81.9	90.9	103.3	84.5	95.4	104.3	106.2	105.1	103.5	106.7	104.9
08	OPENINGS	101.6	133.2	108.8	96.6	103.3	98.1	91.7	77.3	88.4	96.6	85.1	93.9	96.6	102.7	98.0	98.3	103.7	99.5
0920	Plaster & Gypsum Board	113.6	122.4	119.5	105.4	107.8	107.0	98.0	75.2	82.7	105.4	81.3	89.2	105.4	107.8	107.0	100.6	107.8	105.4
0950, 0980	Ceilings & Acoustic Treatment	103.5	122.4	116.3	88.9	107.8	101.6	82.4	75.2	77.5	89.8	81.3	84.0	88.9	107.8	101.6	96.9	107.8	104.3
0960	Flooring	88.8	163.7	109.6	90.1	113.4	96.6	83.8	74.5	84.4	90.1	81.3	87.7	89.6	107.4	94.6	96.7	107.4	100.3
0970, 0990	Wall Finishes & Painting/Coating	87.8	137.4	116.7	82.0	99.9	92.4	86.7	79.7	82.6	82.0	82.7	82.4	82.0	98.1	91.4	92.3	98.9	96.7
09	FINISHES	91.4	133.9	114.5	90.3	107.3	99.5	85.7	79.0	82.0	90.1	81.7	85.5	90.2	106.5	99.0	97.4	106.6	102.4
COVERS	DIVS. 10 - 14, 25, 28, 41, 43, 44, 46	100.0	109.7	102.2	100.0	97.1	99.4	100.0	97.6	99.5	100.0	91.2	98.0	100.0	97.6	99.5	100.0	102.2	100.5
21, 22, 23	FIRE SUPPRESSION, PLUMBING & HVAC	100.1	110.4	104.5	100.2	95.2	98.0	100.2	86.0	94.1	100.2	81.9	92.4	100.2	104.2	101.9	100.0	104.2	101.8
26, 27, 3370	ELECTRICAL, COMMUNICATIONS & UTIL.	101.0	102.5	101.7	96.3	100.9	98.7	93.5	82.1	87.6	95.5	87.1	91.1	96.3	101.5	99.0	99.0	101.4	100.3
MF2016	WEIGHTED AVERAGE	97.7	120.8	107.8	96.2	102.4	98.9	94.8	83.7	89.9	95.7	88.3	92.5	96.1	104.7	99.8	99.2	103.0	100.9

MICHIGAN

DIVISION		FLINT 484 - 485			GAYLORD 497			GRAND RAPIDS 493, 495			IRON MOUNTAIN 498 - 499			JACKSON 492			KALAMAZOO 491		
		MAT.	INST.	TOTAL	MAT.	INST.	TOTAL	MAT.	INST.	TOTAL	MAT.	INST.	TOTAL	MAT.	INST.	TOTAL	MAT.	INST.	TOTAL
015433	CONTRACTOR EQUIPMENT		112.1	112.1		106.8	106.8		98.9	98.9		93.6	93.6		106.8	106.8		98.9	98.9
0241, 31 - 34	SITE & INFRASTRUCTURE, DEMOLITION	71.5	96.2	88.8	86.6	82.6	83.8	91.0	85.7	87.3	95.1	91.6	92.7	109.1	85.1	92.3	92.4	85.8	87.7
0310	Concrete Forming & Accessories	99.2	85.8	87.6	93.9	73.2	76.0	93.8	79.4	81.4	85.7	79.8	80.6	90.5	84.1	85.0	95.2	80.8	82.8
0320	Concrete Reinforcing	98.4	104.1	101.3	82.6	89.0	85.8	93.0	80.7	86.8	82.5	85.7	84.1	80.2	104.0	92.2	88.6	78.7	83.6
0330	Cast-in-Place Concrete	89.0	90.2	89.4	85.1	82.2	84.0	89.5	94.0	91.2	100.7	70.8	89.3	84.9	92.1	87.6	87.0	95.8	90.3
03	CONCRETE	89.7	91.9	90.7	81.7	81.0	81.4	88.6	84.5	86.7	90.1	78.4	84.8	77.0	91.6	83.7	87.9	85.4	86.7
04	MASONRY	97.9	89.6	92.7	108.0	75.7	88.0	90.2	77.2	82.2	93.2	80.8	85.5	87.5	87.5	87.5	95.8	81.7	87.1
05	METALS	97.2	116.1	103.0	100.6	113.0	104.4	96.3	83.6	92.4	100.0	91.8	97.5	100.8	114.1	104.9	99.2	83.2	94.3
06	WOOD, PLASTICS & COMPOSITES	94.3	84.5	88.9	82.6	71.9	76.6	93.2	77.6	84.5	78.2	79.8	79.1	81.4	82.1	81.8	89.2	79.2	83.6
07	THERMAL & MOISTURE PROTECTION	103.7	88.7	97.3	96.2	70.8	85.4	98.8	74.1	88.3	100.0	76.4	90.0	95.4	89.0	92.7	97.5	81.9	90.9
08	OPENINGS	96.6	86.3	94.2	92.1	80.0	89.3	101.3	77.0	95.8	99.0	71.4	92.7	91.2	88.5	90.6	91.7	77.2	88.4
0920	Plaster & Gypsum Board	107.4	83.8	91.5	97.8	70.1	79.2	101.4	73.6	82.7	52.7	79.8	70.9	96.2	80.6	85.7	98.0	75.2	82.7
0950, 0980	Ceilings & Acoustic Treatment	88.9	83.8	85.4	81.5	70.1	73.8	95.1	73.6	80.6	80.5	79.8	80.0	81.5	80.6	80.9	82.4	75.2	77.5
0960	Flooring	90.1	90.9	90.3	81.8	90.5	84.2	90.3	80.5	87.6	100.0	92.4	97.9	80.5	83.1	81.2	88.3	74.5	84.4
0970, 0990	Wall Finishes & Painting/Coating	82.0	81.9	81.9	83.1	80.9	81.9	87.9	80.4	83.5	100.4	67.9	81.5	83.1	98.1	91.9	86.7	79.7	82.6
09	FINISHES	89.8	85.7	87.5	85.2	76.5	80.5	91.3	79.1	84.7	85.6	80.9	83.1	86.1	84.6	85.3	85.7	79.0	82.0
COVERS	DIVS. 10 - 14, 25, 28, 41, 43, 44, 46	100.0	92.5	98.3	100.0	80.5	95.7	100.0	97.3	99.4	100.0	89.0	97.6	100.0	93.7	98.6	100.0	97.6	99.5
21, 22, 23	FIRE SUPPRESSION, PLUMBING & HVAC	100.2	86.8	94.5	97.3	75.3	87.9	100.0	82.9	92.7	97.2	82.1	90.7	97.3	86.9	92.8	100.2	81.5	92.2
26, 27, 3370	ELECTRICAL, COMMUNICATIONS & UTIL.	96.3	94.1	95.2	91.7	75.9	83.5	98.5	87.6	92.8	97.4	81.3	89.0	95.2	99.8	97.6	93.4	78.3	85.5
MF2016	WEIGHTED AVERAGE	95.9	92.3	94.4	94.1	80.7	88.2	96.4	82.9	90.5	96.0	82.4	90.0	93.4	91.8	92.7	95.1	82.1	89.4

MICHIGAN / MINNESOTA

DIVISION		LANSING 488 - 489			MUSKEGON 494			ROYAL OAK 480, 483			SAGINAW 486			TRAVERSE CITY 496			BEMIDJI 566		
		MAT.	INST.	TOTAL	MAT.	INST.	TOTAL	MAT.	INST.	TOTAL	MAT.	INST.	TOTAL	MAT.	INST.	TOTAL	MAT.	INST.	TOTAL
015433	CONTRACTOR EQUIPMENT		112.1	112.1		98.9	98.9		91.7	91.7		112.1	112.1		93.6	93.6		100.3	100.3
0241, 31 - 34	SITE & INFRASTRUCTURE, DEMOLITION	94.1	96.5	95.8	90.1	85.7	87.0	86.1	95.8	92.9	74.6	96.2	89.8	80.8	91.1	88.0	96.1	99.6	98.6
0310	Concrete Forming & Accessories	93.3	78.4	80.5	95.5	79.2	81.5	91.7	106.6	104.6	96.4	83.4	85.2	85.7	72.9	74.7	85.0	85.2	85.1
0320	Concrete Reinforcing	101.2	103.8	102.5	89.2	80.6	84.9	89.2	90.0	89.6	98.4	103.6	101.1	83.7	77.9	80.8	94.4	98.7	96.6
0330	Cast-in-Place Concrete	99.3	89.0	95.4	85.0	93.9	88.4	77.3	102.0	86.7	87.3	87.6	87.4	78.5	78.5	78.5	102.1	107.1	104.0
03	CONCRETE	92.1	88.2	90.3	83.2	84.3	83.7	76.7	101.5	88.0	88.7	89.9	89.3	73.9	76.6	75.1	89.9	96.4	92.9
04	MASONRY	96.9	88.1	91.5	94.4	77.2	83.8	91.5	100.8	97.3	99.3	81.0	87.9	91.4	76.9	82.4	100.1	100.9	100.6
05	METALS	96.3	115.5	102.2	96.9	83.4	92.8	100.4	91.8	97.8	97.2	115.3	102.8	99.9	88.6	96.4	93.5	116.4	100.6
06	WOOD, PLASTICS & COMPOSITES	89.9	74.6	81.4	86.0	77.6	81.3	86.7	107.9	98.5	87.3	83.5	85.2	78.2	72.6	75.1	66.8	80.7	74.5
07	THERMAL & MOISTURE PROTECTION	101.7	86.1	95.1	96.4	74.4	87.1	101.9	102.6	102.2	104.6	84.8	96.2	99.0	68.4	86.0	106.1	91.9	100.1
08	OPENINGS	100.4	80.6	95.9	91.0	77.6	87.9	96.4	103.2	98.0	94.6	85.9	92.6	99.0	66.1	91.5	98.8	101.8	99.5
0920	Plaster & Gypsum Board	99.3	73.6	82.0	77.1	73.6	74.7	103.1	107.8	106.2	105.4	82.6	90.1	52.7	72.5	66.0	108.9	80.5	89.9
0950, 0980	Ceilings & Acoustic Treatment	88.7	73.6	78.5	83.2	73.6	76.7	88.1	107.8	101.4	88.9	82.6	84.7	80.5	72.5	75.1	125.3	80.5	95.1
0960	Flooring	91.0	83.1	89.5	87.0	76.4	84.1	87.2	107.4	92.8	90.1	81.3	87.7	100.0	90.5	97.4	90.6	90.4	90.6
0970, 0990	Wall Finishes & Painting/Coating	89.6	80.4	84.2	84.9	79.1	81.5	83.6	98.1	92.0	82.0	82.7	82.4	100.4	41.2	65.9	90.9	108.5	101.1
09	FINISHES	89.4	78.3	83.4	82.0	77.5	79.6	89.2	106.2	98.4	90.0	82.5	85.9	84.6	72.7	78.1	97.8	88.1	92.5
COVERS	DIVS. 10 - 14, 25, 28, 41, 43, 44, 46	100.0	91.7	98.2	100.0	97.3	99.4	100.0	102.4	100.5	100.0	91.4	98.1	100.0	87.2	97.2	100.0	95.3	99.0
21, 22, 23	FIRE SUPPRESSION, PLUMBING & HVAC	100.0	87.1	94.5	99.9	82.9	92.7	97.3	99.9	98.4	100.2	81.4	92.2	97.2	79.0	89.4	97.3	82.8	91.1
26, 27, 3370	ELECTRICAL, COMMUNICATIONS & UTIL.	97.9	92.5	95.1	93.8	77.0	85.1	98.3	101.4	99.9	94.7	89.2	91.8	93.2	75.9	84.2	104.4	104.2	104.3
MF2016	WEIGHTED AVERAGE	97.0	90.1	94.0	93.6	81.2	88.2	94.3	100.5	97.0	95.6	88.7	92.6	92.9	78.4	86.6	97.0	96.0	96.6

MINNESOTA

DIVISION		BRAINERD 564			DETROIT LAKES 565			DULUTH 556 - 558			MANKATO 560			MINNEAPOLIS 553 - 555			ROCHESTER 559		
		MAT.	INST.	TOTAL	MAT.	INST.	TOTAL	MAT.	INST.	TOTAL	MAT.	INST.	TOTAL	MAT.	INST.	TOTAL	MAT.	INST.	TOTAL
015433	CONTRACTOR EQUIPMENT		103.0	103.0		100.3	100.3		103.4	103.4		103.0	103.0		107.9	107.9		103.4	103.4
0241, 31 - 34	SITE & INFRASTRUCTURE, DEMOLITION	98.6	104.6	102.8	94.3	99.8	98.2	99.8	102.9	102.0	95.2	104.3	101.6	97.9	108.3	105.2	98.8	102.6	101.4
0310	Concrete Forming & Accessories	85.9	85.7	85.7	82.1	84.9	84.5	99.5	96.7	97.1	93.6	95.3	95.1	98.1	114.7	112.4	98.1	98.9	98.8
0320	Concrete Reinforcing	93.2	98.6	95.9	94.4	98.5	96.5	95.2	99.0	97.2	93.0	106.0	99.6	94.9	106.5	100.8	92.9	106.2	99.6
0330	Cast-in-Place Concrete	111.0	111.6	111.2	99.2	110.0	103.3	102.2	101.6	102.0	102.3	99.2	101.1	101.5	115.9	106.9	99.7	99.9	99.8
03	CONCRETE	93.8	98.2	95.8	87.6	97.3	92.0	95.1	99.9	97.3	89.4	99.6	94.1	98.0	114.3	105.5	91.5	101.6	96.2
04	MASONRY	125.1	106.4	113.5	124.5	103.7	111.6	99.2	110.1	106.0	112.9	104.3	107.6	105.4	118.5	113.5	98.8	107.8	104.4
05	METALS	94.7	116.0	101.2	93.5	115.7	100.3	101.2	118.8	106.6	94.5	119.7	102.3	101.2	124.0	108.2	100.9	122.0	107.4
06	WOOD, PLASTICS & COMPOSITES	83.6	77.8	80.4	64.1	78.0	71.8	93.1	93.6	93.3	92.5	93.3	93.0	95.7	112.1	104.8	95.7	97.4	96.6
07	THERMAL & MOISTURE PROTECTION	104.4	101.8	103.3	105.9	100.6	103.7	105.1	104.0	104.6	104.8	92.8	99.7	103.4	116.7	109.0	110.0	93.5	103.0
08	OPENINGS	85.8	100.2	89.1	98.8	100.3	99.1	105.8	106.0	105.8	90.3	112.2	95.3	101.3	122.6	106.2	100.7	113.5	103.6
0920	Plaster & Gypsum Board	94.8	77.8	83.3	107.9	77.8	87.7	92.7	93.9	93.5	99.4	93.7	95.6	100.8	112.7	108.8	103.6	97.8	99.7
0950, 0980	Ceilings & Acoustic Treatment	57.8	77.8	71.3	125.3	77.8	93.3	94.3	93.9	94.0	57.8	93.7	82.0	99.6	112.7	108.4	96.3	97.8	97.3
0960	Flooring	89.5	90.4	89.7	89.4	90.4	89.7	92.7	122.4	100.9	91.4	86.5	90.0	100.8	116.9	105.3	93.6	86.5	91.6
0970, 0990	Wall Finishes & Painting/Coating	85.3	108.5	98.8	90.9	87.2	88.7	85.5	109.2	99.3	96.2	102.6	100.0	100.9	129.5	117.6	85.7	102.6	95.6
09	FINISHES	81.8	88.0	85.2	97.1	85.5	90.8	90.5	102.2	96.8	83.6	94.7	89.6	99.8	116.7	109.0	91.4	97.5	94.7
COVERS	DIVS. 10 - 14, 25, 28, 41, 43, 44, 46	100.0	96.9	99.3	100.0	96.7	99.3	100.0	97.9	99.5	100.0	97.1	99.4	100.0	105.6	101.2	100.0	98.1	99.6
21, 22, 23	FIRE SUPPRESSION, PLUMBING & HVAC	96.5	86.3	92.1	97.3	85.3	92.2	100.0	96.3	98.4	96.5	86.4	92.2	99.9	108.4	103.6	100.0	93.9	97.4
26, 27, 3370	ELECTRICAL, COMMUNICATIONS & UTIL.	102.1	99.5	100.8	104.2	70.3	86.6	103.1	99.5	101.2	108.5	92.0	99.9	101.2	104.4	102.9	101.6	92.0	96.6
MF2016	WEIGHTED AVERAGE	95.9	97.5	96.6	97.8	92.0	95.3	99.8	102.7	101.1	96.0	98.0	96.9	100.4	113.0	105.9	98.8	100.9	99.7

DIVISION		MINNESOTA																MISSISSIPPI		
		SAINT PAUL 550 - 551			ST. CLOUD 563			THIEF RIVER FALLS 567			WILLMAR 562			WINDOM 561			BILOXI 395			
		MAT.	INST.	TOTAL	MAT.	INST.	TOTAL	MAT.	INST.	TOTAL	MAT.	INST.	TOTAL	MAT.	INST.	TOTAL	MAT.	INST.	TOTAL	
015433	CONTRACTOR EQUIPMENT		103.4	103.4		103.0	103.0		100.3	100.3		103.0	103.0		103.0	103.0		102.5	102.5	
0241, 31 - 34	SITE & INFRASTRUCTURE, DEMOLITION	96.0	103.6	101.3	93.7	105.6	102.1	95.1	99.4	98.1	93.2	104.6	101.2	87.0	103.9	98.8	102.9	89.7	93.6	
0310	Concrete Forming & Accessories	97.1	114.4	112.0	83.3	113.7	109.5	85.6	84.2	84.4	83.1	89.8	88.9	87.3	85.8	86.0	93.4	66.9	70.5	
0320	Concrete Reinforcing	95.5	106.5	101.1	93.2	105.8	99.6	94.8	98.4	96.6	92.8	105.6	99.3	92.8	105.0	99.0	83.1	50.1	66.4	
0330	Cast-in-Place Concrete	101.2	114.4	106.2	98.0	114.1	104.1	101.2	83.5	94.5	99.5	82.2	92.9	86.1	87.3	86.5	114.5	67.1	96.5	
03	CONCRETE	94.6	113.7	103.3	85.5	113.1	98.1	88.7	87.8	88.3	85.5	91.2	88.1	76.7	91.1	83.3	98.1	65.7	83.3	
04	MASONRY	97.7	118.6	110.7	109.0	108.7	108.8	100.1	98.4	99.0	113.0	102.9	106.7	124.2	85.2	100.0	97.5	61.5	75.2	
05	METALS	101.2	123.5	108.1	95.3	120.3	103.0	93.7	114.9	100.2	94.4	119.2	102.1	94.4	117.9	101.6	89.4	84.8	88.0	
06	WOOD, PLASTICS & COMPOSITES	97.1	111.8	105.3	81.2	111.8	98.2	67.7	80.7	75.0	80.9	83.7	82.5	84.9	83.7	84.2	102.6	67.6	83.1	
07	THERMAL & MOISTURE PROTECTION	106.0	116.3	110.4	104.5	106.0	105.1	106.9	87.4	98.6	104.3	97.9	101.6	104.3	83.6	95.5	94.8	63.9	81.7	
08	OPENINGS	100.3	121.5	105.2	90.7	121.5	97.8	98.8	101.8	99.5	87.8	106.0	91.9	91.3	106.0	94.7	100.8	57.8	90.9	
0920	Plaster & Gypsum Board	102.0	112.7	109.1	94.8	112.7	106.8	108.6	80.5	89.7	94.8	83.8	87.4	94.8	83.8	87.4	97.0	67.2	77.0	
0950, 0980	Ceilings & Acoustic Treatment	95.8	112.7	107.2	57.8	112.7	94.8	125.3	80.5	95.1	57.8	83.8	75.4	57.8	83.8	75.4	90.2	67.2	74.7	
0960	Flooring	87.7	116.9	95.8	86.3	90.4	87.4	90.3	90.4	90.3	87.6	90.4	88.4	90.2	90.4	90.2	88.6	57.5	80.0	
0970, 0990	Wall Finishes & Painting/Coating	88.1	123.2	108.6	96.2	123.2	112.0	90.9	87.2	88.7	90.9	87.2	88.7	90.9	102.6	97.7	81.9	49.0	62.7	
09	FINISHES	91.5	115.8	104.7	81.2	111.2	97.5	97.6	85.7	91.1	81.2	89.1	85.5	81.5	88.4	85.2	85.7	63.6	73.7	
COVERS	DIVS. 10 - 14, 25, 28, 41, 43, 44, 46	100.0	102.8	100.6	100.0	102.6	100.6	100.0	95.2	98.9	100.0	97.5	99.4	100.0	94.3	98.7	100.0	72.7	93.9	
21, 22, 23	FIRE SUPPRESSION, PLUMBING & HVAC	99.9	111.3	104.8	99.6	108.2	103.2	97.3	82.4	91.0	96.5	101.9	98.8	96.5	82.3	90.4	99.9	57.6	81.8	
26, 27, 3370	ELECTRICAL, COMMUNICATIONS & UTIL.	101.6	113.2	107.6	102.1	113.2	107.9	101.6	70.2	85.3	102.1	79.7	90.5	108.5	92.0	99.9	100.0	56.7	77.5	
MF2016	WEIGHTED AVERAGE	98.9	114.1	105.6	95.2	111.3	102.3	96.6	89.2	93.4	94.3	97.4	95.6	94.6	92.3	93.6	96.7	65.5	83.1	

DIVISION		MISSISSIPPI																	
		CLARKSDALE 386			COLUMBUS 397			GREENVILLE 387			GREENWOOD 389			JACKSON 390 - 392			LAUREL 394		
		MAT.	INST.	TOTAL	MAT.	INST.	TOTAL	MAT.	INST.	TOTAL	MAT.	INST.	TOTAL	MAT.	INST.	TOTAL	MAT.	INST.	TOTAL
015433	CONTRACTOR EQUIPMENT		102.5	102.5		102.5	102.5		102.5	102.5		102.5	102.5		102.5	102.5		102.5	102.5
0241, 31 - 34	SITE & INFRASTRUCTURE, DEMOLITION	97.3	88.3	91.0	101.7	89.2	92.9	102.8	89.5	93.5	100.1	88.1	91.7	98.0	89.5	92.1	106.6	88.4	93.9
0310	Concrete Forming & Accessories	84.9	45.7	51.1	82.3	47.6	52.4	81.7	64.5	66.9	94.0	45.5	52.2	90.7	65.6	69.1	82.4	61.8	64.6
0320	Concrete Reinforcing	100.7	64.8	82.6	89.0	35.9	62.2	101.2	51.5	76.1	100.7	44.0	72.1	92.2	51.1	71.5	89.6	32.9	61.0
0330	Cast-in-Place Concrete	102.2	50.8	82.7	116.8	56.1	93.7	105.3	57.2	87.0	109.9	50.4	87.3	99.8	65.7	86.8	114.2	52.1	90.6
03	CONCRETE	93.7	52.9	75.1	99.5	50.8	77.2	99.1	61.5	81.9	99.9	49.1	76.7	92.4	64.8	79.8	102.0	55.2	80.6
04	MASONRY	92.5	41.0	60.6	124.2	47.2	76.4	138.0	69.1	95.3	93.2	40.9	60.8	103.3	59.2	75.9	120.1	43.8	72.8
05	METALS	89.3	87.2	88.6	86.5	79.3	84.3	90.4	85.7	88.9	89.3	77.4	85.7	95.7	85.2	92.5	86.6	76.0	83.3
06	WOOD, PLASTICS & COMPOSITES	85.2	46.4	63.6	88.6	47.3	65.6	82.0	65.8	73.0	97.6	46.4	69.1	97.2	67.0	80.4	89.7	67.6	77.4
07	THERMAL & MOISTURE PROTECTION	95.8	49.5	76.2	94.8	54.5	77.7	96.3	63.8	82.5	96.2	51.8	77.4	93.1	62.4	80.1	94.9	55.5	78.2
08	OPENINGS	96.7	49.8	85.9	100.4	43.3	87.3	96.4	57.1	87.4	96.7	43.6	84.5	101.2	57.8	91.3	97.3	54.4	87.4
0920	Plaster & Gypsum Board	87.5	45.5	59.3	88.0	46.4	60.0	87.2	65.3	72.5	98.1	45.5	62.7	85.8	66.6	72.9	87.6	67.2	74.0
0950, 0980	Ceilings & Acoustic Treatment	84.7	45.5	58.2	84.9	46.4	58.9	88.5	65.3	72.9	84.7	45.5	58.2	90.8	66.6	74.5	84.9	67.2	72.9
0960	Flooring	95.7	47.6	82.3	82.9	53.4	74.7	94.0	47.6	81.1	101.4	47.6	86.5	85.9	57.5	78.1	81.6	47.6	72.1
0970, 0990	Wall Finishes & Painting/Coating	92.1	49.0	67.0	81.9	49.0	62.7	92.1	49.0	67.0	92.1	49.0	67.0	82.6	49.0	63.0	81.9	49.0	62.7
09	FINISHES	88.7	46.0	65.5	81.8	48.4	63.6	89.4	60.0	73.4	92.2	46.0	67.1	85.7	62.7	73.2	81.9	59.2	69.5
COVERS	DIVS. 10 - 14, 25, 28, 41, 43, 44, 46	100.0	49.8	88.8	100.0	50.8	89.1	100.0	71.8	93.7	100.0	49.8	88.8	100.0	71.9	93.7	100.0	35.4	85.6
21, 22, 23	FIRE SUPPRESSION, PLUMBING & HVAC	98.6	53.3	79.2	98.2	54.9	79.7	100.0	58.9	82.4	98.6	53.5	79.3	99.9	58.9	82.4	98.2	43.4	74.8
26, 27, 3370	ELECTRICAL, COMMUNICATIONS & UTIL.	95.6	43.9	68.7	97.5	46.6	71.0	95.6	56.7	75.4	95.6	41.2	67.3	101.5	56.7	78.2	99.0	56.7	77.0
MF2016	WEIGHTED AVERAGE	94.7	55.2	77.4	96.6	55.8	78.7	98.3	65.6	83.9	96.0	53.2	77.3	97.3	65.3	83.3	96.7	56.3	79.0

DIVISION		MISSISSIPPI									MISSOURI								
		MCCOMB 396			MERIDIAN 393			TUPELO 388			BOWLING GREEN 633			CAPE GIRARDEAU 637			CHILLICOTHE 646		
		MAT.	INST.	TOTAL	MAT.	INST.	TOTAL	MAT.	INST.	TOTAL	MAT.	INST.	TOTAL	MAT.	INST.	TOTAL	MAT.	INST.	TOTAL
015433	CONTRACTOR EQUIPMENT		102.5	102.5		102.5	102.5		102.5	102.5		110.5	110.5		110.5	110.5		104.8	104.8
0241, 31 - 34	SITE & INFRASTRUCTURE, DEMOLITION	94.8	88.2	90.2	98.8	89.7	92.4	94.8	88.2	90.2	89.1	94.9	93.1	90.7	94.8	93.6	105.8	92.9	96.8
0310	Concrete Forming & Accessories	82.3	47.0	51.9	79.6	65.3	67.3	82.2	47.4	52.2	93.6	91.5	91.8	86.4	78.3	79.4	82.9	96.1	94.2
0320	Concrete Reinforcing	90.1	34.5	62.1	89.0	51.1	69.9	98.5	44.1	71.0	106.8	96.7	101.7	108.1	78.4	93.1	98.0	101.3	99.7
0330	Cast-in-Place Concrete	101.3	50.3	81.9	108.4	66.7	92.6	102.2	71.0	90.3	91.4	98.5	94.1	90.4	88.1	89.6	94.8	89.0	92.6
03	CONCRETE	89.0	48.1	70.3	93.5	65.0	80.5	93.2	57.0	76.7	92.8	96.4	94.5	92.0	83.7	88.2	98.6	95.3	97.1
04	MASONRY	125.4	40.4	72.7	97.1	60.8	74.6	126.7	43.6	75.2	113.0	93.7	101.0	109.9	79.8	91.2	103.0	92.1	96.3
05	METALS	86.7	75.3	83.2	87.6	85.2	86.8	89.2	77.6	85.7	97.5	119.0	104.1	98.6	110.2	102.2	92.4	110.9	98.1
06	WOOD, PLASTICS & COMPOSITES	88.6	48.9	66.6	86.1	65.8	74.8	82.5	47.3	62.9	99.6	91.5	95.1	92.6	75.3	83.0	93.0	97.5	95.5
07	THERMAL & MOISTURE PROTECTION	94.3	51.9	76.3	94.5	63.0	81.1	95.8	53.1	77.7	100.1	99.3	99.8	99.6	85.4	93.6	92.5	94.5	93.3
08	OPENINGS	100.4	43.3	87.3	100.1	57.1	90.2	96.6	47.0	85.3	102.9	98.4	101.9	102.9	74.4	96.4	92.1	97.6	93.3
0920	Plaster & Gypsum Board	88.0	48.0	61.2	88.0	65.3	72.8	87.2	46.4	59.8	103.9	91.6	95.6	103.2	75.0	84.3	100.6	97.3	98.4
0950, 0980	Ceilings & Acoustic Treatment	84.9	48.0	60.1	86.8	65.3	72.3	84.7	46.4	58.8	91.4	91.6	91.5	91.4	75.0	80.3	91.1	97.3	95.3
0960	Flooring	82.9	47.6	73.1	81.5	57.5	74.9	94.2	47.6	81.3	95.6	98.5	96.4	92.2	87.2	90.8	94.7	104.7	97.5
0970, 0990	Wall Finishes & Painting/Coating	81.9	49.0	62.7	81.9	49.0	62.7	81.9	47.3	66.0	96.0	105.8	101.8	96.0	70.1	80.9	95.5	94.2	94.8
09	FINISHES	81.2	47.4	62.8	81.4	62.4	71.1	88.2	47.0	65.8	97.7	94.0	95.7	96.5	77.7	86.3	97.8	97.8	97.8
COVERS	DIVS. 10 - 14, 25, 28, 41, 43, 44, 46	100.0	52.8	89.5	100.0	72.3	93.8	100.0	50.8	89.1	100.0	82.7	96.2	100.0	92.8	98.4	100.0	83.4	96.3
21, 22, 23	FIRE SUPPRESSION, PLUMBING & HVAC	98.2	52.8	78.8	99.9	59.8	82.7	98.7	54.5	79.8	96.9	95.9	96.4	99.9	99.4	99.7	97.0	100.9	98.7
26, 27, 3370	ELECTRICAL, COMMUNICATIONS & UTIL.	96.1	46.8	70.5	99.0	58.7	78.1	95.3	46.7	70.0	101.8	79.6	90.3	101.7	101.0	101.3	95.9	77.9	86.5
MF2016	WEIGHTED AVERAGE	95.0	53.7	76.9	95.1	65.9	82.3	96.2	55.9	78.6	98.7	95.0	97.1	99.1	91.5	95.8	96.5	95.1	95.9

For customer support on your Commercial Renovation Costs with RSMeans data, call 800.448.8182.

MISSOURI

DIVISION		COLUMBIA 652			FLAT RIVER 636			HANNIBAL 634			HARRISONVILLE 647			JEFFERSON CITY 650 - 651			JOPLIN 648		
		MAT.	INST.	TOTAL	MAT.	INST.	TOTAL	MAT.	INST.	TOTAL	MAT.	INST.	TOTAL	MAT.	INST.	TOTAL	MAT.	INST.	TOTAL
015433	CONTRACTOR EQUIPMENT		114.0	114.0		110.5	110.5		110.5	110.5		104.8	104.8		114.0	114.0		108.3	108.3
0241, 31 - 34	SITE & INFRASTRUCTURE, DEMOLITION	95.1	99.0	97.8	91.8	94.7	93.9	87.0	94.9	92.5	97.1	94.2	95.1	94.7	99.0	97.7	106.3	97.8	100.4
0310	Concrete Forming & Accessories	83.2	81.6	81.8	99.9	86.0	87.9	91.9	83.2	84.4	80.4	102.0	99.0	94.4	81.6	83.3	93.4	74.5	77.1
0320	Concrete Reinforcing	87.7	96.2	92.0	108.1	103.5	105.8	106.2	96.7	101.4	97.6	106.8	102.3	94.7	96.2	95.5	101.2	86.9	94.0
0330	Cast-in-Place Concrete	83.5	86.5	84.6	94.4	94.3	94.3	86.5	98.4	91.0	97.1	103.1	99.4	88.4	86.5	87.7	102.7	78.1	93.3
03	CONCRETE	80.7	87.8	83.9	95.7	93.7	94.8	89.2	92.7	90.8	94.6	103.9	98.9	87.4	87.8	87.6	97.4	79.2	89.1
04	MASONRY	144.4	87.5	109.1	110.2	78.5	90.5	104.7	93.7	97.9	97.4	102.3	100.4	102.0	87.5	93.0	96.3	81.3	87.0
05	METALS	102.6	118.1	107.4	97.4	120.9	104.6	97.5	118.6	104.0	92.9	114.6	99.5	102.2	118.1	107.1	95.6	101.0	97.2
06	WOOD, PLASTICS & COMPOSITES	85.6	79.1	81.8	107.7	85.2	95.2	97.9	80.8	88.4	89.9	101.6	96.4	97.8	79.1	87.4	103.5	73.0	86.5
07	THERMAL & MOISTURE PROTECTION	93.6	87.7	91.1	100.3	93.1	97.3	100.0	96.2	98.4	91.7	104.2	97.0	100.8	87.7	95.2	91.6	83.1	88.0
08	OPENINGS	99.6	84.1	96.0	102.9	97.0	101.6	102.9	85.0	98.8	92.1	104.5	95.0	97.6	84.1	94.5	93.2	78.0	89.7
0920	Plaster & Gypsum Board	93.1	78.6	83.3	109.8	85.1	93.2	103.6	80.6	88.1	96.2	101.5	99.8	96.8	78.6	84.6	107.0	72.1	83.6
0950, 0980	Ceilings & Acoustic Treatment	93.0	78.6	83.3	91.4	85.1	87.2	91.4	80.6	84.1	91.1	101.5	98.1	92.5	78.6	83.1	91.9	72.1	78.6
0960	Flooring	88.5	75.5	84.9	98.9	87.2	95.6	94.9	100.4	96.5	90.3	106.5	94.8	93.9	75.5	88.8	119.6	75.5	107.4
0970, 0990	Wall Finishes & Painting/Coating	92.4	81.7	86.2	96.0	75.0	83.8	96.0	98.4	97.4	99.8	107.6	104.3	86.0	81.7	83.5	95.1	79.3	85.9
09	FINISHES	85.9	79.9	82.7	99.7	83.9	91.1	97.3	86.7	91.6	95.5	103.4	99.3	91.2	79.9	85.1	104.7	75.4	88.8
COVERS	DIVS. 10 - 14, 25, 28, 41, 43, 44, 46	100.0	96.6	99.3	100.0	93.5	98.5	100.0	81.6	95.9	100.0	85.8	96.8	100.0	96.6	99.3	100.0	82.2	96.0
21, 22, 23	FIRE SUPPRESSION, PLUMBING & HVAC	99.9	99.9	99.9	96.9	99.7	97.8	96.9	99.9	98.2	96.9	102.7	99.4	99.9	100.0	100.0	100.1	71.8	88.0
26, 27, 3370	ELECTRICAL, COMMUNICATIONS & UTIL.	95.4	83.8	89.4	106.3	101.0	103.6	100.5	79.6	89.6	102.1	101.5	101.8	100.6	83.8	91.8	93.9	69.7	81.3
MF2016	WEIGHTED AVERAGE	98.1	92.5	95.7	99.6	95.7	97.9	97.6	93.6	95.9	95.9	102.8	98.9	97.8	92.5	95.5	97.8	79.6	89.8

MISSOURI

DIVISION		KANSAS CITY 640 - 641			KIRKSVILLE 635			POPLAR BLUFF 639			ROLLA 654 - 655			SEDALIA 653			SIKESTON 638		
		MAT.	INST.	TOTAL	MAT.	INST.	TOTAL	MAT.	INST.	TOTAL	MAT.	INST.	TOTAL	MAT.	INST.	TOTAL	MAT.	INST.	TOTAL
015433	CONTRACTOR EQUIPMENT		105.7	105.7		100.2	100.2		102.6	102.6		114.0	114.0		103.6	103.6		102.6	102.6
0241, 31 - 34	SITE & INFRASTRUCTURE, DEMOLITION	98.5	98.4	98.4	90.9	89.6	90.0	78.0	93.5	88.8	94.0	99.4	97.8	93.7	94.1	94.0	81.4	94.1	90.3
0310	Concrete Forming & Accessories	93.3	102.0	100.8	84.6	80.8	81.3	84.8	79.9	80.6	90.2	96.5	95.7	88.0	80.7	81.7	85.9	76.7	78.0
0320	Concrete Reinforcing	96.1	96.9	96.5	107.0	85.9	96.3	110.2	78.3	94.1	88.1	96.4	92.3	86.9	106.2	96.6	109.5	78.3	93.8
0330	Cast-in-Place Concrete	100.6	104.0	101.9	94.3	86.4	91.3	72.6	86.4	77.8	85.4	98.0	90.2	89.2	85.3	87.7	77.5	86.5	80.9
03	CONCRETE	97.4	102.4	99.7	107.7	84.9	97.3	82.9	83.1	83.0	82.3	98.5	89.7	96.0	88.0	92.3	86.6	81.7	84.4
04	MASONRY	103.4	102.3	102.7	116.8	87.6	98.7	108.2	76.1	88.3	117.4	88.2	99.3	123.8	84.1	99.2	107.9	76.1	88.2
05	METALS	102.6	109.1	104.6	97.2	102.8	98.9	97.8	99.5	98.3	102.1	118.7	107.2	100.9	111.9	104.3	98.2	99.7	98.7
06	WOOD, PLASTICS & COMPOSITES	98.5	101.5	100.2	85.3	78.8	81.7	84.4	79.9	81.9	92.9	98.1	95.8	86.0	79.1	82.1	85.9	75.4	80.1
07	THERMAL & MOISTURE PROTECTION	91.8	104.7	97.3	106.8	89.2	99.3	104.9	83.4	95.8	93.8	95.8	94.7	99.7	90.6	95.9	105.1	84.5	96.4
08	OPENINGS	100.1	101.9	100.5	108.2	82.2	102.2	109.3	76.9	101.9	99.6	94.6	98.4	104.5	88.7	100.9	109.3	74.4	101.3
0920	Plaster & Gypsum Board	102.5	101.5	101.8	98.9	78.6	85.3	99.3	79.7	86.1	95.1	98.2	97.2	88.8	78.6	81.9	100.9	75.0	83.5
0950, 0980	Ceilings & Acoustic Treatment	95.1	101.5	99.4	89.7	78.6	82.2	91.4	79.7	83.5	93.0	98.2	96.5	93.0	78.6	83.3	91.4	75.0	80.3
0960	Flooring	96.5	106.5	99.3	74.1	75.5	74.5	87.2	87.2	87.2	92.1	75.5	87.5	71.8	75.5	72.8	87.7	87.2	87.6
0970, 0990	Wall Finishes & Painting/Coating	98.4	107.6	103.7	92.0	81.9	86.1	91.2	70.1	78.9	92.4	94.3	93.5	92.4	81.9	86.3	91.2	70.1	78.9
09	FINISHES	99.1	103.4	101.4	96.4	79.8	87.4	96.2	79.6	87.2	87.4	92.7	90.3	84.7	79.1	81.7	96.8	77.3	86.2
COVERS	DIVS. 10 - 14, 25, 28, 41, 43, 44, 46	100.0	100.3	100.1	100.0	81.4	95.9	100.0	91.9	98.2	100.0	100.3	100.1	100.0	91.4	98.1	100.0	91.6	98.1
21, 22, 23	FIRE SUPPRESSION, PLUMBING & HVAC	99.9	102.6	101.1	96.9	99.3	97.9	96.9	96.8	96.8	96.9	100.4	98.4	96.8	97.4	97.0	96.9	96.8	96.9
26, 27, 3370	ELECTRICAL, COMMUNICATIONS & UTIL.	103.8	101.5	102.6	100.7	77.9	88.8	100.9	100.9	100.9	94.0	83.8	88.7	95.0	101.5	98.4	100.0	100.9	100.5
MF2016	WEIGHTED AVERAGE	100.3	102.7	101.4	101.1	88.5	95.6	97.5	89.7	94.1	96.2	96.9	96.5	98.6	93.2	96.2	98.0	89.1	94.1

MISSOURI / MONTANA

DIVISION		SPRINGFIELD 656 - 658			ST. JOSEPH 644 - 645			ST. LOUIS 630 - 631			BILLINGS 590 - 591			BUTTE 597			GREAT FALLS 594		
		MAT.	INST.	TOTAL	MAT.	INST.	TOTAL	MAT.	INST.	TOTAL	MAT.	INST.	TOTAL	MAT.	INST.	TOTAL	MAT.	INST.	TOTAL
015433	CONTRACTOR EQUIPMENT		106.1	106.1		104.8	104.8		111.1	111.1		100.6	100.6		100.3	100.3		100.3	100.3
0241, 31 - 34	SITE & INFRASTRUCTURE, DEMOLITION	96.4	96.6	96.5	100.8	92.0	94.6	96.0	99.4	98.4	90.7	97.6	95.6	95.9	96.4	96.3	99.5	97.4	98.0
0310	Concrete Forming & Accessories	96.1	76.6	79.3	92.2	90.2	90.5	97.1	104.1	103.2	94.7	67.7	71.4	82.8	67.5	69.6	94.6	66.9	70.7
0320	Concrete Reinforcing	84.6	95.8	90.2	94.9	106.4	100.7	98.8	104.9	101.9	95.4	80.9	88.1	103.5	80.8	92.1	95.4	80.8	88.1
0330	Cast-in-Place Concrete	90.7	77.5	85.7	95.4	100.1	97.1	102.1	102.8	102.3	108.9	71.3	94.7	119.9	71.1	101.4	126.7	71.2	105.6
03	CONCRETE	92.2	81.5	87.3	93.6	97.4	95.4	98.6	105.0	101.5	92.4	72.2	83.1	95.9	72.0	85.0	100.6	71.8	87.4
04	MASONRY	94.1	81.7	86.4	99.3	91.8	94.7	94.1	111.1	104.7	127.7	75.4	95.3	123.6	75.4	93.7	127.9	75.4	95.3
05	METALS	107.5	104.0	106.4	98.9	113.4	103.3	103.4	121.5	109.0	105.2	88.1	99.9	99.4	87.9	95.9	102.7	88.0	98.2
06	WOOD, PLASTICS & COMPOSITES	93.4	75.3	83.4	103.5	89.1	95.5	99.9	102.6	101.4	86.5	64.5	74.3	74.6	64.5	69.0	87.6	63.6	74.3
07	THERMAL & MOISTURE PROTECTION	97.9	78.4	89.6	92.1	91.6	91.9	97.9	107.1	101.8	107.6	71.1	92.2	107.4	71.0	92.0	108.0	70.6	92.2
08	OPENINGS	107.0	86.8	102.4	97.3	97.6	97.4	101.5	107.4	102.8	96.8	66.4	89.8	95.1	66.4	88.5	98.0	65.9	90.6
0920	Plaster & Gypsum Board	96.1	74.7	81.7	108.8	88.6	95.3	108.2	103.0	104.7	113.2	63.9	80.1	112.3	63.9	79.8	121.9	62.9	82.3
0950, 0980	Ceilings & Acoustic Treatment	93.0	74.7	80.7	99.6	88.6	92.2	93.1	103.0	99.8	91.9	63.9	73.0	97.8	63.9	74.9	99.5	62.9	74.9
0960	Flooring	91.1	75.5	86.8	99.9	104.7	101.2	98.0	100.4	98.7	93.7	79.7	89.8	91.1	79.7	87.9	97.9	79.7	92.9
0970, 0990	Wall Finishes & Painting/Coating	86.9	103.9	96.8	95.5	107.6	102.5	97.0	108.5	103.7	95.2	70.8	81.0	94.1	70.8	80.5	94.1	70.8	80.5
09	FINISHES	89.6	79.6	84.1	101.5	94.5	97.7	100.1	103.7	102.1	91.6	69.6	79.7	91.6	69.6	79.7	95.6	69.1	81.2
COVERS	DIVS. 10 - 14, 25, 28, 41, 43, 44, 46	100.0	94.0	98.7	100.0	97.4	99.4	100.0	102.2	100.5	100.0	92.2	98.3	100.0	92.2	98.3	100.0	92.1	98.2
21, 22, 23	FIRE SUPPRESSION, PLUMBING & HVAC	100.0	72.3	88.1	100.1	89.0	95.4	99.9	105.8	102.4	100.1	74.4	89.1	100.2	70.8	87.6	100.2	70.8	87.6
26, 27, 3370	ELECTRICAL, COMMUNICATIONS & UTIL.	99.1	71.7	84.8	102.1	77.9	89.5	104.3	101.0	102.6	98.6	72.0	84.8	105.4	71.2	87.6	98.0	71.2	84.1
MF2016	WEIGHTED AVERAGE	99.6	81.7	91.7	98.9	92.8	96.2	100.5	106.2	103.0	100.0	76.4	89.6	99.8	75.3	89.1	101.3	75.3	89.9

797

MONTANA

DIVISION		HAVRE 595 MAT.	INST.	TOTAL	HELENA 596 MAT.	INST.	TOTAL	KALISPELL 599 MAT.	INST.	TOTAL	MILES CITY 593 MAT.	INST.	TOTAL	MISSOULA 598 MAT.	INST.	TOTAL	WOLF POINT 592 MAT.	INST.	TOTAL
015433	CONTRACTOR EQUIPMENT		100.3	100.3		100.3	100.3		100.3	100.3		100.3	100.3		100.3	100.3		100.3	100.3
0241, 31 - 34	SITE & INFRASTRUCTURE, DEMOLITION	102.8	96.5	98.4	90.0	96.4	94.5	86.6	96.3	93.4	92.5	96.4	95.2	79.6	96.2	91.2	108.8	96.5	100.2
0310	Concrete Forming & Accessories	76.4	67.4	68.6	97.5	67.4	71.6	85.7	67.5	70.0	93.2	67.5	71.0	85.7	67.4	69.9	86.4	66.8	69.5
0320	Concrete Reinforcing	104.3	80.8	92.5	110.2	79.6	94.8	106.2	85.6	95.8	104.0	80.8	92.3	105.2	85.6	95.3	105.4	80.1	92.7
0330	Cast-in-Place Concrete	129.1	71.1	107.1	93.7	71.1	85.1	104.1	71.1	91.6	114.0	71.1	97.7	88.4	71.1	81.8	127.7	71.1	106.2
03	CONCRETE	103.3	72.0	89.0	89.2	71.8	81.2	86.4	72.9	80.2	93.0	72.0	83.4	76.0	72.8	74.5	106.4	71.6	90.5
04	MASONRY	124.6	75.4	94.1	115.4	75.4	90.6	122.6	75.8	93.6	129.4	75.4	96.1	149.3	75.8	103.7	131.0	75.4	96.5
05	METALS	95.7	87.8	93.3	101.4	87.5	97.1	95.5	89.6	93.7	94.8	87.9	92.7	96.1	89.4	94.0	94.9	87.6	92.7
06	WOOD, PLASTICS & COMPOSITES	67.1	64.5	65.7	90.0	64.5	75.8	77.5	64.5	70.3	85.0	64.5	73.6	77.5	64.5	70.3	77.3	63.6	69.7
07	THERMAL & MOISTURE PROTECTION	107.8	66.2	90.2	102.4	71.0	89.1	107.0	73.2	92.7	107.3	66.3	90.0	106.6	70.6	91.3	108.3	66.2	90.5
08	OPENINGS	95.6	66.4	88.9	99.5	66.2	91.9	95.6	67.5	89.1	95.1	66.4	88.5	95.1	67.5	88.8	95.1	65.8	88.4
0920	Plaster & Gypsum Board	108.0	63.9	78.4	106.4	63.9	77.8	112.3	63.9	79.8	121.1	63.9	82.7	112.3	63.9	79.8	115.9	62.9	80.3
0950, 0980	Ceilings & Acoustic Treatment	97.8	63.9	74.9	96.8	63.9	74.6	97.8	63.9	74.9	96.1	63.9	74.4	97.8	63.9	74.9	96.1	62.9	73.7
0960	Flooring	88.6	79.7	86.1	98.9	79.7	93.6	92.8	79.7	89.2	97.7	79.7	92.8	92.8	79.7	89.2	94.3	79.7	90.2
0970, 0990	Wall Finishes & Painting/Coating	94.1	70.8	80.5	94.4	70.8	80.6	94.1	70.8	80.5	94.1	70.8	80.5	94.1	56.0	71.9	94.1	70.8	80.5
09	FINISHES	90.9	69.6	79.3	97.2	69.6	82.2	91.6	69.6	79.7	94.4	69.6	80.9	91.1	68.1	78.6	93.8	69.1	80.4
COVERS	DIVS. 10 - 14, 25, 28, 41, 43, 44, 46	100.0	92.2	98.3	100.0	92.2	98.3	100.0	92.2	98.3	100.0	92.2	98.3	100.0	92.2	98.3	100.0	92.1	98.2
21, 22, 23	FIRE SUPPRESSION, PLUMBING & HVAC	97.1	70.8	85.8	100.2	70.8	87.6	97.1	68.8	85.0	97.1	74.4	87.4	100.2	68.8	86.8	97.1	74.4	87.4
26, 27, 3370	ELECTRICAL, COMMUNICATIONS & UTIL.	98.0	71.2	84.1	104.6	71.2	87.2	102.2	68.7	84.8	98.0	76.6	86.8	103.2	67.7	84.8	98.0	76.6	86.8
MF2016	WEIGHTED AVERAGE	98.9	75.2	88.5	99.6	75.3	88.9	96.8	75.0	87.2	97.8	76.7	88.6	97.4	74.5	87.4	99.9	76.5	89.7

NEBRASKA

DIVISION		ALLIANCE 693 MAT.	INST.	TOTAL	COLUMBUS 686 MAT.	INST.	TOTAL	GRAND ISLAND 688 MAT.	INST.	TOTAL	HASTINGS 689 MAT.	INST.	TOTAL	LINCOLN 683 - 685 MAT.	INST.	TOTAL	MCCOOK 690 MAT.	INST.	TOTAL
015433	CONTRACTOR EQUIPMENT		98.2	98.2		105.2	105.2		105.2	105.2		105.2	105.2		105.2	105.2		105.2	105.2
0241, 31 - 34	SITE & INFRASTRUCTURE, DEMOLITION	99.4	101.2	100.6	100.8	95.2	96.9	105.6	95.2	98.3	104.4	95.2	97.9	91.1	95.2	94.0	102.7	95.2	97.4
0310	Concrete Forming & Accessories	86.3	55.8	60.0	95.2	74.8	77.6	94.8	71.2	74.4	98.0	74.3	77.6	93.7	76.4	78.8	91.6	55.9	60.9
0320	Concrete Reinforcing	111.2	88.0	99.5	97.2	86.9	92.0	96.6	76.3	86.4	96.6	73.4	84.9	95.9	76.5	86.2	103.4	69.7	86.4
0330	Cast-in-Place Concrete	107.8	77.3	96.2	110.5	80.9	99.2	117.0	80.3	103.0	117.0	62.2	96.2	91.8	80.4	87.5	116.6	58.6	94.6
03	CONCRETE	115.6	69.7	94.6	100.3	80.2	91.1	105.0	76.5	92.0	105.2	71.2	89.7	88.5	78.9	84.1	105.1	61.0	85.0
04	MASONRY	110.3	75.1	88.5	116.5	78.6	93.0	109.1	75.1	88.0	118.6	75.1	91.6	96.9	78.6	85.6	105.5	75.1	86.6
05	METALS	104.0	84.1	97.9	98.3	98.4	98.3	100.1	94.2	98.3	101.0	92.8	98.5	102.1	94.5	99.8	98.5	90.8	96.2
06	WOOD, PLASTICS & COMPOSITES	86.6	50.1	66.3	99.2	74.1	85.2	98.5	69.4	82.3	102.0	74.1	86.5	98.3	76.3	86.1	95.2	50.1	70.1
07	THERMAL & MOISTURE PROTECTION	102.5	65.9	87.0	101.7	80.5	92.7	101.8	79.1	92.2	101.9	78.7	92.1	98.2	80.9	90.9	97.5	76.1	88.4
08	OPENINGS	94.1	59.7	86.2	93.8	74.2	89.3	93.8	69.7	88.3	93.8	70.9	88.6	101.3	70.3	94.1	94.6	55.1	85.5
0920	Plaster & Gypsum Board	82.7	48.9	60.0	94.6	73.5	80.4	93.9	68.7	77.0	95.6	73.5	80.7	102.7	75.8	84.6	94.7	48.9	63.9
0950, 0980	Ceilings & Acoustic Treatment	93.2	48.9	63.3	85.9	73.5	77.5	85.9	68.7	74.3	85.9	73.5	77.5	95.1	75.8	82.1	89.2	48.9	62.0
0960	Flooring	89.7	83.4	88.0	82.4	89.7	84.5	82.2	83.4	82.5	83.5	83.4	83.5	93.7	89.7	92.6	88.5	83.4	87.1
0970, 0990	Wall Finishes & Painting/Coating	150.3	52.9	93.5	73.3	61.6	66.5	73.3	65.3	68.7	73.3	61.6	66.5	91.4	81.3	85.5	85.6	46.1	62.6
09	FINISHES	91.4	59.5	74.1	83.6	76.0	79.4	83.7	72.3	77.5	84.3	74.7	79.1	94.2	79.4	86.2	89.1	58.8	72.6
COVERS	DIVS. 10 - 14, 25, 28, 41, 43, 44, 46	100.0	60.9	91.3	100.0	84.1	96.5	100.0	89.0	97.5	100.0	84.1	96.5	100.0	89.7	97.7	100.0	60.3	91.2
21, 22, 23	FIRE SUPPRESSION, PLUMBING & HVAC	97.1	73.4	86.9	97.0	76.3	88.2	100.1	80.6	91.8	97.0	76.0	88.0	100.0	80.6	91.7	96.9	76.1	88.0
26, 27, 3370	ELECTRICAL, COMMUNICATIONS & UTIL.	93.5	65.6	79.0	93.7	80.2	86.7	92.4	65.7	78.5	91.8	81.4	86.4	101.4	65.7	82.8	95.5	65.6	80.0
MF2016	WEIGHTED AVERAGE	100.4	71.9	87.9	97.3	81.4	90.3	98.5	78.4	89.7	98.5	79.0	89.9	98.3	80.2	90.4	98.0	71.5	86.4

NEBRASKA / NEVADA

DIVISION		NORFOLK 687 MAT.	INST.	TOTAL	NORTH PLATTE 691 MAT.	INST.	TOTAL	OMAHA 680 - 681 MAT.	INST.	TOTAL	VALENTINE 692 MAT.	INST.	TOTAL	CARSON CITY 897 MAT.	INST.	TOTAL	ELKO 898 MAT.	INST.	TOTAL
015433	CONTRACTOR EQUIPMENT		94.4	94.4		105.2	105.2		94.4	94.4		97.9	97.9		97.4	97.4		97.4	97.4
0241, 31 - 34	SITE & INFRASTRUCTURE, DEMOLITION	83.6	94.3	91.1	103.9	95.2	97.8	90.9	94.3	93.3	87.3	100.3	96.4	89.2	97.8	95.2	71.8	96.1	88.8
0310	Concrete Forming & Accessories	81.9	75.5	76.4	94.1	70.3	73.6	93.3	76.1	78.5	82.7	55.3	59.1	105.0	79.7	83.2	110.2	77.4	82.0
0320	Concrete Reinforcing	97.3	64.6	80.8	102.8	73.3	87.9	95.9	76.5	86.1	103.4	64.2	83.6	99.0	113.3	106.2	106.2	91.7	98.9
0330	Cast-in-Place Concrete	111.1	61.5	92.2	116.6	63.3	96.3	92.1	78.2	86.8	102.8	56.3	85.1	102.5	82.7	94.9	99.2	74.1	89.7
03	CONCRETE	98.9	69.3	85.4	105.2	69.7	89.0	88.6	77.4	83.5	102.5	58.3	82.3	96.7	86.7	92.1	94.4	79.0	87.4
04	MASONRY	123.3	78.6	95.5	93.6	75.1	82.1	97.9	78.6	85.9	105.4	75.0	86.6	115.7	72.4	88.8	122.3	67.3	88.2
05	METALS	101.9	79.4	95.0	97.7	92.2	96.0	102.1	84.3	96.7	110.3	78.2	100.4	105.7	94.4	102.2	109.1	86.6	102.2
06	WOOD, PLASTICS & COMPOSITES	82.5	75.8	78.8	97.1	69.4	81.7	94.5	75.8	84.1	81.0	49.7	63.6	92.3	77.8	84.2	101.7	77.8	88.4
07	THERMAL & MOISTURE PROTECTION	101.6	79.2	92.1	97.4	78.2	89.3	98.1	80.3	90.6	98.3	75.0	88.4	105.7	80.4	95.0	102.3	71.8	89.4
08	OPENINGS	95.4	69.7	89.5	93.9	68.4	88.0	99.3	76.4	94.0	96.3	55.2	86.9	98.7	80.0	94.4	101.5	75.3	95.5
0920	Plaster & Gypsum Board	94.7	75.8	82.0	94.7	68.7	77.2	101.0	75.8	84.0	96.1	48.9	64.4	99.3	77.3	84.5	101.3	77.3	85.2
0950, 0980	Ceilings & Acoustic Treatment	99.8	75.8	83.6	89.2	68.7	75.4	97.2	75.8	82.8	105.6	48.9	67.4	101.4	77.3	85.1	100.1	77.3	84.7
0960	Flooring	102.0	89.7	98.6	89.6	83.4	87.9	94.3	89.7	93.0	114.6	83.4	105.9	96.0	71.0	89.0	101.0	71.0	92.6
0970, 0990	Wall Finishes & Painting/Coating	127.6	61.6	89.1	85.6	61.6	71.6	101.4	75.4	86.2	150.8	63.8	100.0	93.9	80.4	86.0	93.4	80.4	85.8
09	FINISHES	99.9	76.9	87.4	89.4	71.9	79.9	91.6	78.4	86.5	109.6	60.4	82.8	96.1	77.5	86.0	92.8	76.5	84.0
COVERS	DIVS. 10 - 14, 25, 28, 41, 43, 44, 46	100.0	86.1	96.9	100.0	62.4	91.6	100.0	88.7	97.5	100.0	58.3	90.7	100.0	95.2	98.9	100.0	90.5	97.9
21, 22, 23	FIRE SUPPRESSION, PLUMBING & HVAC	96.7	76.7	88.1	100.0	75.6	89.6	100.0	76.9	90.1	96.5	76.2	87.8	100.1	80.1	91.6	98.6	77.7	89.7
26, 27, 3370	ELECTRICAL, COMMUNICATIONS & UTIL.	92.7	83.9	88.1	93.8	65.6	79.2	102.7	83.9	92.9	91.2	66.1	78.2	103.2	93.2	98.0	100.4	93.2	96.7
MF2016	WEIGHTED AVERAGE	98.8	78.6	90.0	97.8	75.2	87.9	98.4	80.8	90.7	100.5	70.5	87.4	101.0	84.9	93.9	100.6	81.2	92.1

City Cost Indexes

NEVADA / NEW HAMPSHIRE

DIVISION		ELY 893 MAT.	INST.	TOTAL	LAS VEGAS 889-891 MAT.	INST.	TOTAL	RENO 894-895 MAT.	INST.	TOTAL	CHARLESTON 036 MAT.	INST.	TOTAL	CLAREMONT 037 MAT.	INST.	TOTAL	CONCORD 032-033 MAT.	INST.	TOTAL
015433	CONTRACTOR EQUIPMENT		97.4	97.4		97.4	97.4		97.4	97.4		98.8	98.8		98.8	98.8		98.8	98.8
0241, 31 - 34	SITE & INFRASTRUCTURE, DEMOLITION	77.9	97.6	91.7	81.2	100.1	94.4	77.6	97.8	91.8	84.9	99.8	95.3	78.8	99.8	93.5	90.2	101.2	97.9
0310	Concrete Forming & Accessories	103.4	99.8	100.3	104.3	108.8	108.1	99.7	79.8	82.5	85.5	78.7	79.6	91.2	78.7	80.4	95.9	93.5	93.9
0320	Concrete Reinforcing	105.0	91.9	98.4	96.5	120.6	108.7	99.2	119.4	109.4	85.3	83.3	84.3	85.3	83.3	84.3	99.6	83.7	91.6
0330	Cast-in-Place Concrete	106.5	100.1	104.1	103.1	106.7	104.5	112.7	82.7	101.3	87.7	104.5	94.1	80.6	104.5	89.7	103.4	115.5	108.0
03	CONCRETE	102.6	98.3	100.6	97.9	109.5	103.2	102.5	87.8	95.8	87.3	88.7	88.0	80.2	88.7	84.1	94.6	99.3	96.7
04	MASONRY	127.9	78.1	97.0	114.8	99.2	105.1	121.5	72.4	91.0	95.6	77.4	84.3	95.9	77.4	84.4	104.0	100.1	101.6
05	METALS	109.1	89.7	103.1	117.6	101.6	112.7	110.8	96.4	106.4	93.0	89.3	91.9	93.0	89.3	91.9	99.6	90.6	96.9
06	WOOD, PLASTICS & COMPOSITES	92.6	101.0	97.3	90.7	106.9	99.7	87.0	77.8	81.9	94.2	83.2	88.1	100.0	83.2	90.7	94.9	92.7	93.7
07	THERMAL & MOISTURE PROTECTION	102.9	95.8	99.9	115.7	100.3	109.2	102.4	80.4	93.1	101.9	97.6	100.1	101.7	97.6	100.0	106.1	108.3	107.0
08	OPENINGS	101.4	88.2	98.4	100.4	111.9	103.0	99.2	81.4	95.1	97.3	82.9	94.0	98.4	82.9	94.8	95.7	88.1	94.0
0920	Plaster & Gypsum Board	96.7	101.1	99.7	91.9	107.2	102.1	87.0	77.3	80.5	104.8	82.4	89.7	106.1	82.4	90.2	107.3	92.1	97.1
0950, 0980	Ceilings & Acoustic Treatment	100.1	101.1	100.8	107.8	107.2	107.4	105.2	77.3	86.4	99.5	82.4	87.9	99.5	82.4	87.9	98.6	92.1	94.2
0960	Flooring	98.4	71.0	90.8	90.1	102.2	93.5	95.0	71.0	88.4	83.9	106.7	90.2	86.4	106.7	92.0	92.5	106.7	96.4
0970, 0990	Wall Finishes & Painting/Coating	93.4	117.1	107.2	95.9	114.9	107.0	93.4	80.4	85.8	84.8	94.1	90.2	84.8	94.1	90.2	88.8	94.1	91.9
09	FINISHES	92.0	96.9	94.7	90.5	108.3	100.1	90.5	77.5	83.4	86.6	85.4	85.9	87.1	85.4	86.2	90.6	95.7	93.3
COVERS	DIVS. 10 - 14, 25, 28, 41, 43, 44, 46	100.0	72.7	93.9	100.0	103.9	100.9	100.0	95.2	98.9	100.0	85.5	96.8	100.0	85.5	96.8	100.0	104.8	101.1
21, 22, 23	FIRE SUPPRESSION, PLUMBING & HVAC	98.6	99.5	99.0	100.3	100.7	100.5	100.2	80.1	91.6	97.0	60.1	81.2	97.0	60.1	81.2	100.0	85.2	93.7
26, 27, 3370	ELECTRICAL, COMMUNICATIONS & UTIL.	100.7	98.2	99.4	104.9	113.8	109.5	101.1	93.2	97.0	97.2	49.1	72.2	97.2	49.1	72.2	98.0	79.5	88.4
MF2016	WEIGHTED AVERAGE	101.9	94.1	98.5	103.1	105.2	104.0	101.8	85.3	94.6	94.3	76.3	86.4	93.5	76.3	86.0	97.9	92.4	95.5

NEW HAMPSHIRE / NEW JERSEY

DIVISION		KEENE 034 MAT.	INST.	TOTAL	LITTLETON 035 MAT.	INST.	TOTAL	MANCHESTER 031 MAT.	INST.	TOTAL	NASHUA 030 MAT.	INST.	TOTAL	PORTSMOUTH 038 MAT.	INST.	TOTAL	ATLANTIC CITY 082, 084 MAT.	INST.	TOTAL
015433	CONTRACTOR EQUIPMENT		98.8	98.8		98.8	98.8		98.8	98.8		98.8	98.8		98.8	98.8		96.3	96.3
0241, 31 - 34	SITE & INFRASTRUCTURE, DEMOLITION	91.9	100.1	97.6	79.0	100.1	93.7	92.1	101.2	98.5	93.4	101.0	98.7	86.9	101.6	97.2	89.7	104.6	100.2
0310	Concrete Forming & Accessories	89.8	80.3	81.6	102.0	80.3	83.3	95.9	93.7	94.0	97.8	93.6	94.2	86.9	93.5	92.6	110.1	143.9	139.3
0320	Concrete Reinforcing	85.3	83.4	84.3	86.0	83.4	84.7	99.6	83.7	91.6	106.5	83.7	95.0	85.3	83.7	84.5	75.8	142.1	109.2
0330	Cast-in-Place Concrete	88.1	106.5	95.1	79.0	106.6	89.5	103.4	115.6	108.0	83.4	115.4	95.5	79.1	116.0	93.1	77.9	142.7	102.5
03	CONCRETE	87.0	90.2	88.4	79.7	90.2	84.5	94.5	99.4	96.8	87.0	99.3	92.6	79.8	99.5	88.8	82.1	141.5	109.2
04	MASONRY	98.4	80.7	87.4	108.5	80.7	91.2	100.7	100.1	100.4	100.7	100.0	100.3	96.0	100.3	98.7	111.3	140.6	129.5
05	METALS	93.7	89.7	92.5	93.8	89.7	92.5	101.2	90.8	98.0	99.3	90.3	96.6	95.3	91.5	94.2	96.5	115.7	102.4
06	WOOD, PLASTICS & COMPOSITES	98.4	83.2	89.9	109.2	83.2	94.7	95.0	92.7	93.7	107.6	92.7	99.3	95.4	92.7	93.9	118.0	144.6	132.8
07	THERMAL & MOISTURE PROTECTION	102.4	99.1	101.0	101.8	99.1	100.7	106.3	108.3	107.2	102.8	108.3	105.1	102.4	108.3	104.9	101.0	137.4	116.4
08	OPENINGS	96.0	86.5	93.8	99.3	82.9	95.5	96.7	91.8	95.5	100.6	91.1	98.4	100.9	81.2	96.4	96.8	140.3	100.6
0920	Plaster & Gypsum Board	105.1	82.4	89.8	118.9	82.4	94.4	107.3	92.1	97.1	114.5	92.1	99.5	104.8	92.1	96.3	116.0	145.4	135.7
0950, 0980	Ceilings & Acoustic Treatment	99.5	82.4	87.9	99.5	82.4	87.9	99.6	92.1	94.5	111.5	92.1	98.4	100.4	92.1	94.8	91.4	145.4	127.8
0960	Flooring	86.0	106.7	91.8	96.7	105.6	99.2	89.8	106.7	94.5	89.7	106.7	94.4	84.1	106.7	90.4	92.5	168.1	113.4
0970, 0990	Wall Finishes & Painting/Coating	84.8	108.1	98.4	84.8	94.1	90.2	90.8	108.1	100.9	84.8	106.9	97.7	84.8	106.9	97.7	83.0	145.9	119.6
09	FINISHES	88.4	87.8	88.0	91.9	86.0	88.7	91.9	97.2	94.8	93.1	97.1	95.3	87.4	97.1	92.7	89.7	149.9	122.4
COVERS	DIVS. 10 - 14, 25, 28, 41, 43, 44, 46	100.0	93.2	98.5	100.0	98.5	99.7	100.0	104.8	101.1	100.0	104.8	101.1	100.0	104.8	101.1	100.0	119.2	104.3
21, 22, 23	FIRE SUPPRESSION, PLUMBING & HVAC	97.0	63.3	82.6	97.0	69.8	85.4	100.1	85.2	93.7	100.1	85.2	93.7	100.1	85.2	93.7	99.8	131.7	113.4
26, 27, 3370	ELECTRICAL, COMMUNICATIONS & UTIL.	97.2	58.6	77.2	98.2	51.9	74.1	98.1	79.5	88.4	98.1	79.4	89.0	97.7	79.5	88.2	93.2	144.6	119.9
MF2016	WEIGHTED AVERAGE	94.8	79.6	88.1	94.9	79.8	88.3	98.3	92.7	95.9	95.5	92.6	95.5	95.1	92.4	93.9	95.7	134.7	112.8

NEW JERSEY

DIVISION		CAMDEN 081 MAT.	INST.	TOTAL	DOVER 078 MAT.	INST.	TOTAL	ELIZABETH 072 MAT.	INST.	TOTAL	HACKENSACK 076 MAT.	INST.	TOTAL	JERSEY CITY 073 MAT.	INST.	TOTAL	LONG BRANCH 077 MAT.	INST.	TOTAL
015433	CONTRACTOR EQUIPMENT		96.3	96.3		98.8	98.8		98.8	98.8		98.8	98.8		96.3	96.3		95.9	95.9
0241, 31 - 34	SITE & INFRASTRUCTURE, DEMOLITION	90.9	104.6	100.5	96.5	106.3	103.4	100.3	106.3	104.5	97.5	106.3	103.6	88.7	106.2	100.9	92.5	105.9	101.9
0310	Concrete Forming & Accessories	101.3	143.8	137.9	99.4	146.5	140.0	111.6	146.5	141.7	99.4	146.4	139.9	103.7	146.4	140.6	104.2	145.6	139.9
0320	Concrete Reinforcing	99.5	132.6	116.2	78.1	150.5	114.6	78.1	150.5	114.6	78.1	150.5	114.6	101.3	150.5	126.1	78.1	150.5	114.6
0330	Cast-in-Place Concrete	75.6	140.8	100.3	85.2	136.1	104.5	73.2	136.1	97.1	83.2	136.1	103.3	66.4	136.1	92.9	73.8	140.8	99.3
03	CONCRETE	82.6	139.1	108.4	84.2	142.1	110.7	80.8	142.1	108.8	82.7	142.1	109.9	78.6	141.9	107.5	82.3	143.1	110.1
04	MASONRY	101.4	140.6	125.7	93.5	143.0	124.2	111.1	143.0	130.9	97.6	143.0	125.8	89.3	143.0	122.6	103.2	141.1	126.7
05	METALS	102.3	112.3	105.3	94.3	123.7	103.4	95.9	123.7	104.4	94.4	123.7	103.4	100.2	120.6	106.4	94.4	120.4	102.4
06	WOOD, PLASTICS & COMPOSITES	106.7	144.6	127.8	102.4	146.2	126.7	117.3	146.2	133.3	102.4	146.2	126.7	103.2	146.2	127.1	104.2	146.1	127.5
07	THERMAL & MOISTURE PROTECTION	100.8	138.4	116.8	108.7	136.4	120.4	108.9	136.4	120.5	108.4	135.6	119.9	108.1	136.4	120.1	108.3	135.2	119.7
08	OPENINGS	99.1	138.1	108.0	104.2	144.2	113.4	102.6	144.2	112.1	102.0	144.2	111.7	100.7	144.2	110.7	96.6	144.2	107.6
0920	Plaster & Gypsum Board	111.5	145.4	134.3	107.2	147.0	133.9	114.4	147.0	136.3	107.2	147.0	133.9	110.2	147.0	134.9	109.1	147.0	134.6
0950, 0980	Ceilings & Acoustic Treatment	101.0	145.4	130.9	87.8	147.0	127.7	89.7	147.0	128.3	87.8	147.0	127.7	97.3	147.0	130.8	87.8	147.0	127.7
0960	Flooring	88.6	168.1	113.4	81.8	191.9	112.4	86.9	191.9	116.0	81.8	191.9	116.0	82.9	191.9	113.1	83.1	187.0	117.3
0970, 0990	Wall Finishes & Painting/Coating	83.0	145.9	119.6	84.0	145.9	120.1	84.0	145.9	120.1	84.0	145.9	120.1	84.1	145.9	120.1	84.1	145.9	120.1
09	FINISHES	89.7	149.9	122.4	85.4	155.4	123.5	88.8	155.4	125.0	85.3	155.4	123.4	87.8	155.4	124.6	86.3	154.2	123.2
COVERS	DIVS. 10 - 14, 25, 28, 41, 43, 44, 46	100.0	119.2	104.3	100.0	131.8	107.1	100.0	131.8	107.1	100.0	131.8	107.1	100.0	131.8	107.1	100.0	119.4	104.3
21, 22, 23	FIRE SUPPRESSION, PLUMBING & HVAC	100.0	131.6	113.5	99.9	137.5	116.0	100.1	137.5	116.1	99.9	137.5	116.0	100.1	137.5	116.1	99.9	136.5	115.6
26, 27, 3370	ELECTRICAL, COMMUNICATIONS & UTIL.	97.7	139.5	119.4	94.8	142.9	119.8	95.4	142.9	120.1	94.8	140.9	118.8	99.5	140.9	121.0	94.4	137.2	116.7
MF2016	WEIGHTED AVERAGE	97.0	133.3	112.9	95.7	138.2	114.3	96.8	138.2	114.9	95.5	137.9	114.0	95.9	137.6	114.1	95.0	136.2	113.1

NEW JERSEY

DIVISION		NEW BRUNSWICK 088-089			NEWARK 070-071			PATERSON 074-075			POINT PLEASANT 087			SUMMIT 079			TRENTON 085-086		
		MAT.	INST.	TOTAL	MAT.	INST.	TOTAL	MAT.	INST.	TOTAL	MAT.	INST.	TOTAL	MAT.	INST.	TOTAL	MAT.	INST.	TOTAL
015433	CONTRACTOR EQUIPMENT		95.9	95.9		98.8	98.8		98.8	98.8		95.9	95.9		98.8	98.8		95.9	95.9
0241, 31 - 34	SITE & INFRASTRUCTURE, DEMOLITION	102.2	106.0	104.9	100.9	106.3	104.7	99.4	106.3	104.2	103.7	105.9	105.3	98.1	106.3	103.9	89.0	105.9	100.9
0310	Concrete Forming & Accessories	104.3	146.4	140.6	104.4	146.5	140.7	101.4	146.4	140.2	98.8	145.5	139.1	102.2	146.5	140.4	100.5	145.3	139.1
0320	Concrete Reinforcing	76.7	150.5	113.9	99.7	150.5	125.3	101.3	150.5	126.1	76.7	150.5	113.9	78.1	150.5	114.6	100.7	119.9	110.4
0330	Cast-in-Place Concrete	96.3	141.9	113.6	93.9	136.1	109.9	84.7	136.1	104.2	96.3	142.9	114.0	70.7	136.1	95.6	94.8	140.6	112.2
03	CONCRETE	97.5	143.8	118.7	91.6	142.1	114.7	87.1	142.1	112.2	97.2	143.8	118.5	78.2	142.1	107.4	91.9	137.6	112.8
04	MASONRY	109.7	143.0	130.4	96.6	143.0	125.4	94.0	143.0	124.4	98.0	141.1	124.7	96.6	143.0	124.7	101.9	141.1	126.2
05	METALS	96.6	120.5	103.9	102.0	123.7	108.7	95.5	123.7	104.2	96.6	120.3	103.9	94.3	123.7	103.4	102.0	109.1	104.2
06	WOOD, PLASTICS & COMPOSITES	111.5	146.1	130.8	100.0	146.2	125.7	105.0	146.2	127.9	104.1	146.1	127.5	106.1	146.2	128.4	93.5	146.1	122.8
07	THERMAL & MOISTURE PROTECTION	101.3	134.8	115.5	109.8	136.4	121.1	108.7	135.6	120.1	101.3	137.6	116.7	109.1	136.4	120.6	101.7	138.6	117.3
08	OPENINGS	92.0	144.2	104.0	100.6	144.2	110.6	107.3	144.2	115.8	93.8	144.2	105.3	108.3	144.2	116.5	98.5	136.7	107.3
0920	Plaster & Gypsum Board	113.3	147.0	136.0	102.5	147.0	132.4	110.2	147.0	134.9	108.4	147.0	134.3	109.1	147.0	134.6	103.4	147.0	132.7
0950, 0980	Ceilings & Acoustic Treatment	91.4	147.0	128.9	98.8	147.0	131.3	97.3	147.0	130.8	91.4	147.0	128.9	87.8	147.0	127.7	97.9	147.0	131.0
0960	Flooring	90.0	191.9	118.3	87.4	191.9	116.3	82.9	191.9	113.1	87.5	168.1	109.8	83.1	191.9	113.3	91.9	187.0	118.2
0970, 0990	Wall Finishes & Painting/Coating	83.0	145.9	119.6	85.4	145.9	120.6	84.0	145.9	120.1	83.0	145.9	119.6	84.0	145.9	120.1	86.6	145.9	121.2
09	FINISHES	89.6	155.3	125.4	90.7	155.4	125.9	88.0	155.4	124.6	88.2	150.9	122.3	86.4	155.4	123.9	92.4	154.2	126.0
COVERS	DIVS. 10 - 14, 25, 28, 41, 43, 44, 46	100.0	131.7	107.0	100.0	131.8	107.1	100.0	131.8	107.1	100.0	116.4	103.7	100.0	131.8	107.1	100.0	119.4	104.3
21, 22, 23	FIRE SUPPRESSION, PLUMBING & HVAC	99.8	137.5	115.9	100.0	137.5	116.1	100.1	137.5	116.1	99.8	136.5	115.5	99.9	137.5	116.0	100.2	136.2	115.6
26, 27, 3370	ELECTRICAL, COMMUNICATIONS & UTIL.	93.9	141.2	118.5	103.1	140.9	122.8	99.5	142.9	122.1	93.2	137.2	116.1	95.4	142.9	120.1	101.2	135.9	119.2
MF2016	WEIGHTED AVERAGE	97.4	137.8	115.1	99.1	137.9	116.1	97.4	138.2	115.3	96.8	135.8	113.9	95.8	138.2	114.3	98.5	133.9	114.0

NEW JERSEY / NEW MEXICO

DIVISION		NEW JERSEY VINELAND 080, 083			ALBUQUERQUE 870 - 872			CARRIZOZO 883			CLOVIS 881			FARMINGTON 874			GALLUP 873		
		MAT.	INST.	TOTAL	MAT.	INST.	TOTAL	MAT.	INST.	TOTAL	MAT.	INST.	TOTAL	MAT.	INST.	TOTAL	MAT.	INST.	TOTAL
015433	CONTRACTOR EQUIPMENT		96.3	96.3		111.0	111.0		111.0	111.0		111.0	111.0		111.0	111.0		111.0	111.0
0241, 31 - 34	SITE & INFRASTRUCTURE, DEMOLITION	93.9	104.7	101.4	92.3	102.4	99.4	112.0	102.4	105.3	98.8	102.4	101.3	98.8	102.4	101.3	108.5	102.4	104.2
0310	Concrete Forming & Accessories	96.0	144.0	137.4	99.0	64.1	68.9	97.1	64.1	68.6	97.1	64.0	68.5	99.1	64.1	68.9	99.1	64.1	68.9
0320	Concrete Reinforcing	75.8	135.4	105.9	97.0	71.0	83.9	109.7	71.0	90.2	110.9	71.0	90.8	106.1	71.0	88.4	101.5	71.0	86.1
0330	Cast-in-Place Concrete	84.0	140.9	105.6	93.6	70.1	84.7	94.7	70.1	85.3	94.6	70.0	85.3	94.5	70.1	85.2	88.9	70.1	81.8
03	CONCRETE	86.3	139.8	110.8	93.3	68.8	82.1	114.3	68.8	93.5	102.6	68.7	87.1	96.7	68.8	83.9	102.9	68.8	87.3
04	MASONRY	98.6	141.1	125.0	107.3	60.0	78.0	106.9	60.0	77.8	106.9	60.0	77.8	116.9	60.0	81.6	101.6	60.0	75.8
05	METALS	96.5	113.9	101.8	109.2	91.0	103.6	105.2	91.0	100.8	104.8	90.8	100.5	106.8	91.0	101.9	105.9	91.0	101.3
06	WOOD, PLASTICS & COMPOSITES	101.0	144.6	125.2	94.7	64.6	77.9	92.4	64.6	77.0	92.4	64.6	77.0	94.8	64.6	78.0	94.8	64.6	78.0
07	THERMAL & MOISTURE PROTECTION	100.8	137.6	116.4	96.9	70.8	85.8	99.5	70.8	87.3	98.2	70.8	86.6	97.1	70.8	86.0	98.2	70.8	86.6
08	OPENINGS	93.3	139.1	103.8	98.5	65.8	91.0	96.6	65.8	89.5	96.7	65.8	89.7	100.8	65.8	92.8	100.9	65.8	92.8
0920	Plaster & Gypsum Board	106.8	145.4	132.7	100.6	63.4	75.6	77.6	63.4	68.1	77.6	63.4	68.1	92.7	63.4	73.0	92.7	63.4	73.0
0950, 0980	Ceilings & Acoustic Treatment	91.4	145.4	127.8	99.3	63.4	75.1	102.3	63.4	76.1	102.3	63.4	76.1	97.1	63.4	74.4	97.1	63.4	74.4
0960	Flooring	86.6	168.1	109.2	87.7	69.5	82.7	98.5	69.5	90.4	98.5	69.5	90.4	89.3	69.5	83.8	89.3	69.5	83.8
0970, 0990	Wall Finishes & Painting/Coating	83.0	145.9	119.6	95.3	52.8	70.5	93.6	52.8	69.8	93.6	52.8	69.8	89.9	52.8	68.3	89.9	52.8	68.3
09	FINISHES	87.0	150.0	121.3	89.4	63.6	75.4	93.9	63.6	77.4	92.6	63.6	76.8	88.3	63.6	74.9	89.6	63.6	75.4
COVERS	DIVS. 10 - 14, 25, 28, 41, 43, 44, 46	100.0	119.4	104.3	100.0	84.0	96.5	100.0	84.0	96.5	100.0	84.0	96.5	100.0	84.0	96.5	100.0	84.0	96.5
21, 22, 23	FIRE SUPPRESSION, PLUMBING & HVAC	99.8	131.9	113.5	100.2	66.2	86.9	98.1	69.2	85.7	98.1	68.8	85.6	100.0	69.2	86.8	98.1	69.2	85.7
26, 27, 3370	ELECTRICAL, COMMUNICATIONS & UTIL.	93.2	144.6	119.9	88.1	88.5	88.3	91.7	88.5	90.0	89.4	88.5	88.9	86.2	88.5	87.4	85.6	88.5	87.1
MF2016	WEIGHTED AVERAGE	95.0	134.4	112.2	98.5	75.1	88.3	101.1	75.1	89.7	98.9	75.0	88.4	99.1	75.1	88.6	98.9	75.1	88.5

NEW MEXICO

DIVISION		LAS CRUCES 880			LAS VEGAS 877			ROSWELL 882			SANTA FE 875			SOCORRO 878			TRUTH/CONSEQUENCES 879		
		MAT.	INST.	TOTAL	MAT.	INST.	TOTAL	MAT.	INST.	TOTAL	MAT.	INST.	TOTAL	MAT.	INST.	TOTAL	MAT.	INST.	TOTAL
015433	CONTRACTOR EQUIPMENT		85.7	85.7		111.0	111.0		111.0	111.0		111.0	111.0		111.0	111.0		85.7	85.7
0241, 31 - 34	SITE & INFRASTRUCTURE, DEMOLITION	99.0	80.8	86.3	97.9	102.4	101.1	101.0	102.4	102.0	102.2	102.4	102.3	94.4	102.4	100.0	114.9	80.9	91.0
0310	Concrete Forming & Accessories	93.8	63.0	67.2	99.1	64.1	68.9	97.1	64.1	68.6	97.8	64.1	68.7	99.1	64.1	68.9	96.8	63.0	67.6
0320	Concrete Reinforcing	106.4	70.9	88.5	103.3	71.0	87.0	110.9	71.0	90.8	96.5	71.0	83.7	105.3	71.0	88.0	98.4	70.9	84.5
0330	Cast-in-Place Concrete	89.4	62.6	79.2	91.9	70.1	83.6	94.6	70.1	85.3	97.5	70.1	87.1	90.0	70.1	82.5	98.6	62.6	85.0
03	CONCRETE	81.3	65.3	74.0	94.3	68.8	82.6	103.4	68.8	87.6	93.5	68.8	82.2	93.1	68.8	82.0	87.1	65.3	77.1
04	MASONRY	102.2	59.6	75.8	101.9	60.0	75.9	118.3	60.0	82.1	94.9	60.0	73.2	101.8	60.0	75.9	99.0	59.6	74.6
05	METALS	103.6	83.0	97.2	105.6	91.0	101.1	106.1	91.0	101.4	102.7	91.0	99.1	105.9	91.0	101.3	105.5	83.0	98.6
06	WOOD, PLASTICS & COMPOSITES	81.4	63.5	71.4	94.8	64.6	78.0	92.4	64.6	77.0	93.9	64.6	77.6	94.8	64.6	78.0	85.9	63.5	73.5
07	THERMAL & MOISTURE PROTECTION	85.7	65.9	77.3	96.7	70.8	85.7	98.4	70.8	86.7	99.2	70.8	87.2	96.6	70.8	85.7	85.6	65.9	77.2
08	OPENINGS	92.3	65.2	86.1	97.3	65.8	90.1	96.6	65.8	89.5	100.0	65.8	92.2	97.1	65.8	89.9	90.7	65.2	84.9
0920	Plaster & Gypsum Board	75.9	63.4	67.5	92.7	63.4	73.0	77.6	63.4	68.1	107.8	63.4	78.0	92.7	63.4	73.0	94.1	63.4	73.5
0950, 0980	Ceilings & Acoustic Treatment	88.0	63.4	71.4	97.1	63.4	74.4	102.3	63.4	76.1	97.7	63.4	74.6	97.1	63.4	74.4	85.9	63.4	70.7
0960	Flooring	129.6	69.5	112.9	89.3	69.5	83.8	98.5	69.5	90.4	97.0	69.5	89.4	89.3	69.5	83.8	118.0	69.5	104.5
0970, 0990	Wall Finishes & Painting/Coating	82.8	52.8	65.3	89.9	52.8	68.3	93.6	52.8	69.8	95.9	52.8	70.8	89.9	52.8	68.3	82.3	52.8	65.1
09	FINISHES	102.6	62.7	80.9	88.2	63.6	74.8	92.7	63.6	76.9	96.6	63.6	78.6	88.1	63.6	74.8	100.0	62.8	79.8
COVERS	DIVS. 10 - 14, 25, 28, 41, 43, 44, 46	100.0	81.5	95.9	100.0	84.0	96.5	100.0	84.0	96.5	100.0	84.0	96.5	100.0	84.0	96.5	100.0	81.5	95.9
21, 22, 23	FIRE SUPPRESSION, PLUMBING & HVAC	100.4	68.9	86.9	98.1	69.2	85.7	99.9	69.2	86.8	100.1	69.2	86.9	98.1	69.2	85.7	98.1	68.9	85.6
26, 27, 3370	ELECTRICAL, COMMUNICATIONS & UTIL.	91.4	88.5	89.9	87.7	88.5	88.1	90.8	88.5	89.6	99.8	88.5	93.9	86.0	88.5	87.3	89.6	88.5	89.0
MF2016	WEIGHTED AVERAGE	96.4	71.7	85.6	97.2	75.1	87.5	100.3	75.1	89.3	99.1	75.1	88.6	96.8	75.1	87.3	96.6	71.8	85.7

For customer support on your Commercial Renovation Costs with RSMeans data, call 800.448.8182.

		NEW MEXICO			NEW YORK															
	DIVISION	TUCUMCARI			ALBANY			BINGHAMTON			BRONX			BROOKLYN			BUFFALO		
		884			120 - 122			137 - 139			104			112			140 - 142		
		MAT.	INST.	TOTAL	MAT.	INST.	TOTAL	MAT.	INST.	TOTAL	MAT.	INST.	TOTAL	MAT.	INST.	TOTAL	MAT.	INST.	TOTAL
015433	CONTRACTOR EQUIPMENT		111.0	111.0		115.8	115.8		121.0	121.0		106.9	106.9		112.3	112.3		98.9	98.9
0241, 31 - 34	SITE & INFRASTRUCTURE, DEMOLITION	98.4	102.4	101.2	83.9	104.7	98.5	96.2	93.2	94.1	92.6	113.7	107.4	120.9	125.5	124.1	99.6	101.6	101.0
0310	Concrete Forming & Accessories	97.1	64.0	68.5	99.3	105.9	105.0	100.6	90.7	92.0	99.1	193.1	180.2	106.7	182.5	172.1	102.3	120.3	117.8
0320	Concrete Reinforcing	108.7	71.0	89.7	103.7	119.5	111.7	100.4	104.2	102.3	96.4	172.4	134.8	101.8	222.6	162.8	108.6	115.3	111.9
0330	Cast-in-Place Concrete	94.6	70.0	85.3	79.7	116.1	93.5	105.1	107.6	106.1	86.4	173.4	119.5	107.2	171.7	131.7	108.3	125.3	114.8
03	CONCRETE	101.9	68.7	86.7	84.1	112.9	97.3	93.4	101.3	97.0	89.4	181.7	131.6	104.6	184.0	140.9	105.1	120.6	112.2
04	MASONRY	118.3	60.0	82.1	90.5	116.0	106.3	106.8	104.1	105.1	90.1	182.8	147.6	117.2	182.8	157.9	113.5	123.6	119.7
05	METALS	104.8	90.8	100.5	101.2	128.8	109.7	94.2	134.4	106.6	90.8	170.3	115.2	102.1	167.5	122.2	99.4	108.2	102.1
06	WOOD, PLASTICS & COMPOSITES	92.4	64.6	77.0	94.4	102.3	98.8	105.6	86.6	95.0	98.0	197.8	153.6	107.2	183.4	149.6	100.9	119.7	111.4
07	THERMAL & MOISTURE PROTECTION	98.2	70.8	86.6	106.6	110.3	108.2	107.1	95.0	102.0	104.4	168.9	131.7	107.3	166.6	132.4	104.8	113.3	108.4
08	OPENINGS	96.5	65.8	89.5	95.4	104.5	97.5	91.9	92.8	92.1	92.8	200.7	117.6	89.4	190.6	112.7	99.8	114.0	103.1
0920	Plaster & Gypsum Board	77.6	63.4	68.1	102.5	102.3	102.3	107.2	85.9	92.9	96.7	200.3	166.3	103.4	185.8	158.7	105.4	120.1	115.3
0950, 0980	Ceilings & Acoustic Treatment	102.3	63.4	76.1	99.6	102.3	101.4	94.1	85.9	88.6	85.2	200.3	162.8	89.9	185.8	154.5	97.3	120.1	112.7
0960	Flooring	98.5	69.5	90.4	90.6	111.1	96.3	99.4	99.2	99.4	92.6	189.9	119.6	106.3	189.9	129.5	95.2	118.5	101.7
0970, 0990	Wall Finishes & Painting/Coating	93.6	52.8	69.8	93.8	102.3	98.8	87.4	99.6	94.5	103.8	159.9	136.5	114.2	159.9	140.8	97.8	115.8	108.3
09	FINISHES	92.5	63.6	76.8	90.3	105.9	98.8	91.4	92.2	91.8	92.6	190.5	145.8	104.3	181.9	146.5	100.1	120.1	111.0
COVERS	DIVS. 10 - 14, 25, 28, 41, 43, 44, 46	100.0	84.0	96.5	100.0	102.0	100.4	100.0	98.8	99.7	100.0	137.9	108.4	100.0	135.5	107.9	100.0	108.1	101.8
21, 22, 23	FIRE SUPPRESSION, PLUMBING & HVAC	98.1	68.8	85.6	100.2	109.0	103.9	100.6	99.6	100.2	100.2	170.4	130.2	99.7	170.2	129.9	100.0	103.4	101.5
26, 27, 3370	ELECTRICAL, COMMUNICATIONS & UTIL.	91.7	88.5	90.0	95.5	110.9	103.5	99.4	100.8	100.1	94.2	181.5	139.6	99.1	181.5	141.9	98.9	106.0	102.6
MF2016	WEIGHTED AVERAGE	99.5	75.0	88.8	95.8	111.2	102.5	97.2	101.7	99.2	94.7	173.4	129.2	101.6	172.5	132.7	101.2	111.6	105.8

		NEW YORK																	
	DIVISION	ELMIRA			FAR ROCKAWAY			FLUSHING			GLENS FALLS			HICKSVILLE			JAMAICA		
		148 - 149			116			113			128			115, 117, 118			114		
		MAT.	INST.	TOTAL	MAT.	INST.	TOTAL	MAT.	INST.	TOTAL	MAT.	INST.	TOTAL	MAT.	INST.	TOTAL	MAT.	INST.	TOTAL
015433	CONTRACTOR EQUIPMENT		123.0	123.0		112.3	112.3		112.3	112.3		115.8	115.8		112.3	112.3		112.3	112.3
0241, 31 - 34	SITE & INFRASTRUCTURE, DEMOLITION	102.2	92.8	95.6	124.2	125.5	125.1	124.2	125.5	125.1	74.3	104.2	95.3	113.8	124.6	121.3	118.4	125.5	123.4
0310	Concrete Forming & Accessories	83.5	96.0	94.3	93.1	193.0	179.2	97.0	193.0	179.8	84.8	96.3	94.8	89.6	162.1	152.1	97.0	193.0	179.8
0320	Concrete Reinforcing	108.1	108.3	108.2	101.8	222.7	162.8	103.5	222.7	163.6	99.3	111.3	105.4	101.8	222.5	162.7	101.8	222.7	162.8
0330	Cast-in-Place Concrete	98.3	106.7	101.5	115.9	171.8	137.1	115.9	171.8	137.1	77.0	110.8	89.8	98.5	168.2	125.0	107.2	171.8	131.7
03	CONCRETE	89.5	104.3	96.2	110.7	188.8	146.4	111.2	188.8	146.7	78.3	105.3	90.6	96.6	173.4	131.7	104.0	188.8	142.7
04	MASONRY	105.1	105.1	105.1	122.0	182.8	159.7	115.9	182.8	157.4	96.6	108.1	103.7	111.7	178.0	152.8	119.9	182.8	158.9
05	METALS	95.2	137.2	108.1	102.1	167.5	122.2	102.1	167.5	122.2	94.6	125.1	104.0	103.6	164.4	122.3	102.1	167.5	122.2
06	WOOD, PLASTICS & COMPOSITES	84.7	94.3	90.0	91.1	197.5	150.3	95.6	197.5	152.3	87.2	92.9	90.4	87.8	159.5	127.7	95.6	197.5	152.3
07	THERMAL & MOISTURE PROTECTION	103.2	94.6	100.0	107.2	168.2	133.0	107.2	168.2	133.0	99.7	105.3	102.1	106.9	161.4	130.0	107.0	168.2	132.9
08	OPENINGS	98.3	97.3	98.0	88.1	198.4	113.5	88.1	198.4	113.5	90.4	98.6	92.3	88.5	177.4	100.9	88.1	190.4	113.5
0920	Plaster & Gypsum Board	101.8	94.1	96.6	92.5	200.3	164.9	94.8	200.3	165.6	95.0	92.6	93.4	92.1	161.2	138.5	94.8	200.3	165.6
0950, 0980	Ceilings & Acoustic Treatment	98.5	94.1	95.5	79.7	200.3	161.0	79.7	200.3	161.0	85.6	92.6	90.3	78.7	161.2	134.4	79.7	200.3	161.0
0960	Flooring	86.5	102.4	90.9	101.1	189.9	125.7	102.6	189.9	126.9	82.3	111.1	90.3	100.0	176.4	121.2	102.6	189.9	126.9
0970, 0990	Wall Finishes & Painting/Coating	94.8	91.5	92.9	114.2	159.9	140.8	114.2	159.9	140.8	89.8	97.3	94.2	114.2	159.9	140.8	114.2	159.9	140.8
09	FINISHES	92.0	96.5	94.5	99.6	190.2	148.9	100.4	190.2	149.3	82.4	98.2	91.0	98.2	164.1	134.1	100.0	190.2	149.1
COVERS	DIVS. 10 - 14, 25, 28, 41, 43, 44, 46	100.0	97.9	99.5	100.0	137.1	108.3	100.0	137.1	108.3	100.0	98.1	99.6	100.0	124.9	105.5	100.0	137.1	108.3
21, 22, 23	FIRE SUPPRESSION, PLUMBING & HVAC	97.1	96.5	96.9	96.6	170.2	128.1	96.6	170.2	128.1	97.1	104.4	100.3	99.7	159.5	125.3	96.6	170.2	128.1
26, 27, 3370	ELECTRICAL, COMMUNICATIONS & UTIL.	97.7	103.8	100.9	105.9	181.5	145.2	105.9	181.5	145.2	90.7	110.9	101.2	98.6	142.4	121.4	97.6	181.5	141.2
MF2016	WEIGHTED AVERAGE	96.5	103.0	99.3	100.2	174.8	133.9	101.9	174.8	133.8	91.4	106.4	98.0	99.6	158.9	125.6	100.2	174.8	132.9

		NEW YORK																	
	DIVISION	JAMESTOWN			KINGSTON			LONG ISLAND CITY			MONTICELLO			MOUNT VERNON			NEW ROCHELLE		
		147			124			111			127			105			108		
		MAT.	INST.	TOTAL	MAT.	INST.	TOTAL	MAT.	INST.	TOTAL	MAT.	INST.	TOTAL	MAT.	INST.	TOTAL	MAT.	INST.	TOTAL
015433	CONTRACTOR EQUIPMENT		91.9	91.9		112.3	112.3		112.3	112.3		112.3	112.3		106.9	106.9		106.9	106.9
0241, 31 - 34	SITE & INFRASTRUCTURE, DEMOLITION	103.7	93.5	96.6	145.6	120.7	128.1	122.0	125.5	124.5	140.6	121.2	127.0	97.5	109.9	106.2	97.3	109.9	106.1
0310	Concrete Forming & Accessories	83.6	89.1	88.3	86.2	128.1	122.4	101.3	193.0	180.4	93.3	128.1	123.3	89.4	140.1	133.1	104.5	140.1	135.2
0320	Concrete Reinforcing	108.3	114.5	111.4	99.8	156.7	128.5	101.8	222.7	162.8	99.0	156.7	128.1	95.2	170.9	133.4	95.3	170.9	133.4
0330	Cast-in-Place Concrete	102.0	85.3	95.7	103.9	145.2	119.6	110.6	171.8	133.8	97.3	145.2	115.5	96.5	150.7	117.1	96.5	150.7	117.1
03	CONCRETE	92.4	92.3	92.4	98.3	138.6	116.7	107.0	188.8	144.4	93.6	138.6	114.2	98.7	149.4	121.9	98.1	149.4	121.5
04	MASONRY	114.6	101.4	106.4	111.3	150.7	135.7	114.0	182.8	156.7	104.2	150.7	133.0	95.9	154.7	132.4	95.9	154.7	132.4
05	METALS	92.7	102.6	95.7	102.9	134.2	112.5	102.1	167.5	122.2	102.9	134.2	112.5	90.6	162.5	112.7	90.8	162.4	112.8
06	WOOD, PLASTICS & COMPOSITES	83.3	86.1	84.8	88.4	122.0	107.1	101.4	197.5	154.9	95.2	122.0	110.1	88.6	134.9	114.3	105.4	134.9	121.8
07	THERMAL & MOISTURE PROTECTION	102.7	94.0	99.0	123.4	143.5	131.9	107.2	168.2	133.0	123.1	143.5	131.7	105.4	146.6	122.8	105.4	146.6	122.9
08	OPENINGS	98.1	94.8	97.3	92.6	140.8	103.7	88.1	198.4	113.4	88.3	140.8	100.3	92.8	165.9	109.6	92.9	165.9	109.6
0920	Plaster & Gypsum Board	88.9	85.6	86.7	97.2	122.8	114.4	99.4	200.3	167.2	97.9	122.8	114.6	91.7	135.6	121.2	104.9	135.6	125.5
0950, 0980	Ceilings & Acoustic Treatment	95.1	85.6	88.7	76.3	122.8	107.6	79.7	200.3	161.0	76.3	122.8	107.6	83.5	135.6	118.6	83.5	135.6	118.6
0960	Flooring	89.0	102.4	92.7	98.7	162.9	116.5	104.3	189.9	128.0	101.1	162.9	118.2	84.6	173.4	109.3	92.0	173.4	114.6
0970, 0990	Wall Finishes & Painting/Coating	96.4	95.1	95.6	118.7	129.0	124.7	114.2	159.9	140.8	118.7	129.0	124.7	102.2	159.9	135.9	102.2	159.9	135.9
09	FINISHES	90.6	91.4	91.0	96.0	134.0	116.7	101.3	190.2	149.7	96.4	134.0	116.9	89.6	147.5	121.1	93.5	147.5	122.8
COVERS	DIVS. 10 - 14, 25, 28, 41, 43, 44, 46	100.0	96.6	99.3	100.0	115.9	103.5	100.0	137.1	108.3	100.0	115.9	103.5	100.0	119.9	104.4	100.0	116.8	103.7
21, 22, 23	FIRE SUPPRESSION, PLUMBING & HVAC	97.0	89.0	93.6	97.1	125.7	109.3	99.7	170.2	129.9	97.1	121.8	107.6	97.2	140.6	115.8	97.2	140.6	115.8
26, 27, 3370	ELECTRICAL, COMMUNICATIONS & UTIL.	96.7	93.8	95.2	92.0	121.0	107.1	98.1	181.5	141.4	92.0	121.0	107.1	92.4	149.6	122.2	92.4	149.6	122.2
MF2016	WEIGHTED AVERAGE	96.6	93.9	95.4	100.0	131.6	113.8	101.2	174.8	133.5	98.5	130.8	112.7	95.1	145.6	117.2	95.5	145.5	117.4

NEW YORK

DIVISION		NEW YORK 100 - 102			NIAGARA FALLS 143			PLATTSBURGH 129			POUGHKEEPSIE 125 - 126			QUEENS 110			RIVERHEAD 119		
		MAT.	INST.	TOTAL	MAT.	INST.	TOTAL	MAT.	INST.	TOTAL	MAT.	INST.	TOTAL	MAT.	INST.	TOTAL	MAT.	INST.	TOTAL
015433	CONTRACTOR EQUIPMENT		104.9	104.9		91.9	91.9		96.6	96.6		112.3	112.3		112.3	112.3		112.3	112.3
0241, 31 - 34	SITE & INFRASTRUCTURE, DEMOLITION	101.4	111.7	108.6	106.0	95.0	98.3	113.2	101.7	105.2	141.7	121.1	127.3	117.1	125.5	123.0	115.0	124.7	121.8
0310	Concrete Forming & Accessories	108.5	196.3	184.2	83.5	118.6	113.8	90.5	95.8	95.1	86.2	175.7	163.3	89.9	193.0	178.8	94.2	162.2	152.8
0320	Concrete Reinforcing	102.0	236.2	169.7	106.8	115.6	111.2	103.8	115.6	109.8	99.8	156.8	128.5	103.5	222.7	163.6	103.7	222.5	163.6
0330	Cast-in-Place Concrete	99.9	181.8	131.0	105.5	130.7	115.1	94.2	106.0	98.7	100.7	142.7	116.6	101.8	171.8	128.4	100.1	168.6	126.1
03	CONCRETE	103.3	195.8	145.6	94.6	121.5	106.9	90.9	102.5	96.2	95.8	159.3	124.8	99.8	188.8	140.5	97.2	173.6	132.1
04	MASONRY	103.5	189.8	157.0	121.8	130.5	127.2	90.9	101.5	97.5	103.8	145.5	129.7	108.1	182.8	154.4	117.5	178.4	155.3
05	METALS	102.0	169.5	122.8	95.3	104.3	98.0	98.8	101.6	99.7	102.9	135.3	112.9	102.1	167.5	122.2	104.1	164.5	122.7
06	WOOD, PLASTICS & COMPOSITES	102.4	197.9	155.5	83.2	113.4	100.0	93.6	92.4	93.0	88.4	188.0	143.9	87.9	197.5	148.9	92.7	159.5	129.9
07	THERMAL & MOISTURE PROTECTION	107.4	173.4	135.3	102.8	115.4	108.1	117.8	102.0	111.1	123.4	148.4	133.9	106.8	168.2	132.8	107.8	161.6	130.6
08	OPENINGS	95.8	201.1	119.9	98.1	110.0	100.9	98.1	99.5	98.5	92.7	177.3	112.1	88.1	198.4	113.4	88.5	177.4	108.9
0920	Plaster & Gypsum Board	102.2	200.3	168.1	88.9	113.7	105.6	114.8	91.7	99.3	97.2	190.6	159.9	92.1	200.3	164.8	93.3	161.2	138.9
0950, 0980	Ceilings & Acoustic Treatment	100.5	200.3	167.8	95.1	113.7	107.7	104.7	91.7	95.9	76.3	190.6	153.3	79.7	200.3	161.0	79.6	161.2	134.6
0960	Flooring	93.9	189.9	120.5	89.0	118.5	97.2	104.6	111.1	106.4	98.7	167.7	117.8	100.0	189.9	125.0	101.1	176.4	122.0
0970, 0990	Wall Finishes & Painting/Coating	100.5	164.8	138.0	96.4	116.5	108.1	114.4	99.5	105.7	118.7	129.0	124.7	114.2	159.9	140.8	114.2	159.9	140.8
09	FINISHES	97.5	192.8	149.3	90.7	118.6	105.9	94.7	98.2	96.6	95.8	172.5	137.5	98.6	190.2	148.4	98.8	163.9	134.2
COVERS	DIVS. 10 - 14, 25, 28, 41, 43, 44, 46	100.0	143.3	109.6	100.0	107.0	101.6	100.0	96.6	99.2	100.0	121.4	104.8	100.0	137.1	108.3	100.0	125.0	105.6
21, 22, 23	FIRE SUPPRESSION, PLUMBING & HVAC	100.1	174.1	131.8	97.0	107.7	101.6	97.1	104.5	100.3	97.1	125.3	109.1	99.7	170.2	129.9	99.9	159.6	125.5
26, 27, 3370	ELECTRICAL, COMMUNICATIONS & UTIL.	100.9	188.1	146.3	95.4	104.4	100.1	88.8	96.9	93.0	92.0	128.0	110.8	98.6	181.5	141.7	100.1	142.4	122.1
MF2016	WEIGHTED AVERAGE	100.6	178.2	134.6	97.6	111.9	103.9	96.6	100.9	98.5	99.2	142.4	118.1	99.6	174.8	132.6	100.4	159.0	126.1

NEW YORK

DIVISION		ROCHESTER 144 - 146			SCHENECTADY 123			STATEN ISLAND 103			SUFFERN 109			SYRACUSE 130 - 132			UTICA 133 - 135		
		MAT.	INST.	TOTAL	MAT.	INST.	TOTAL	MAT.	INST.	TOTAL	MAT.	INST.	TOTAL	MAT.	INST.	TOTAL	MAT.	INST.	TOTAL
015433	CONTRACTOR EQUIPMENT		119.8	119.8		115.8	115.8		106.9	106.9		106.9	106.9		115.8	115.8		115.8	115.8
0241, 31 - 34	SITE & INFRASTRUCTURE, DEMOLITION	94.1	108.8	104.4	84.6	104.7	98.7	101.5	113.7	110.1	94.5	109.2	104.8	94.8	103.5	100.9	73.2	103.1	94.1
0310	Concrete Forming & Accessories	99.8	101.5	101.3	101.1	105.9	105.3	88.9	182.9	169.9	97.8	147.6	140.7	99.7	93.4	94.2	100.8	90.5	91.9
0320	Concrete Reinforcing	110.7	108.4	109.5	98.3	119.5	109.0	96.4	222.7	160.1	95.3	156.9	126.4	101.4	104.2	102.8	101.4	103.5	102.5
0330	Cast-in-Place Concrete	97.7	107.7	101.5	92.9	116.1	101.7	96.5	173.5	125.8	93.6	146.5	113.7	97.6	107.7	101.4	89.3	106.3	95.8
03	CONCRETE	94.3	106.1	99.7	91.2	112.9	101.2	100.5	184.9	139.1	95.3	147.9	119.3	96.0	101.6	98.6	94.0	99.7	96.6
04	MASONRY	99.6	107.8	104.7	93.3	116.0	107.4	102.0	182.8	152.1	95.4	148.3	128.2	99.3	105.3	103.0	90.4	103.7	98.6
05	METALS	101.4	123.2	108.1	98.7	128.8	108.0	88.9	170.5	113.9	88.9	135.0	103.1	97.7	120.4	104.7	95.8	120.0	103.2
06	WOOD, PLASTICS & COMPOSITES	98.0	100.0	99.1	105.3	102.3	103.6	87.1	183.7	140.9	97.8	149.4	126.5	101.9	90.1	95.4	101.9	86.7	93.4
07	THERMAL & MOISTURE PROTECTION	102.5	103.6	102.9	101.2	110.3	105.1	104.8	167.3	131.3	105.3	145.0	122.1	101.7	98.8	100.5	90.1	98.7	93.7
08	OPENINGS	102.3	101.6	102.1	96.1	104.5	98.0	92.8	192.9	115.8	92.9	155.9	107.3	94.2	92.8	93.9	96.9	90.8	95.5
0920	Plaster & Gypsum Board	105.9	100.1	102.0	104.3	102.3	103.0	91.8	185.8	154.9	96.0	150.5	132.6	98.5	89.8	92.7	98.5	86.2	90.3
0950, 0980	Ceilings & Acoustic Treatment	97.0	100.1	99.1	92.4	102.3	99.1	85.2	185.8	153.0	83.5	150.5	128.7	94.1	89.8	91.2	94.1	86.2	88.8
0960	Flooring	90.1	111.0	95.9	89.5	111.1	95.5	88.1	189.9	116.3	88.1	181.4	114.0	88.9	97.6	91.3	86.9	97.7	89.9
0970, 0990	Wall Finishes & Painting/Coating	94.7	102.9	99.5	89.8	102.3	97.1	103.8	159.9	136.5	102.2	132.7	120.0	91.2	103.2	98.2	84.8	103.2	95.5
09	FINISHES	95.2	103.5	99.7	87.8	105.9	97.6	91.3	182.1	140.7	90.9	153.2	124.8	90.5	94.5	92.7	88.9	92.3	90.7
COVERS	DIVS. 10 - 14, 25, 28, 41, 43, 44, 46	100.0	102.0	100.5	100.0	102.0	100.4	100.0	136.3	108.1	100.0	119.5	104.3	100.0	99.0	99.8	100.0	93.5	98.6
21, 22, 23	FIRE SUPPRESSION, PLUMBING & HVAC	100.1	91.8	96.5	100.2	109.0	104.0	100.2	170.4	130.2	97.2	128.8	110.7	100.2	96.7	98.7	100.2	96.4	98.6
26, 27, 3370	ELECTRICAL, COMMUNICATIONS & UTIL.	102.7	94.9	98.6	94.7	110.9	103.1	94.2	181.5	139.6	99.0	121.0	110.5	99.4	106.7	103.2	97.5	106.7	102.3
MF2016	WEIGHTED AVERAGE	99.5	102.7	100.9	96.1	111.2	102.7	96.4	172.2	129.6	95.1	136.0	113.0	97.6	102.0	99.5	95.7	100.8	97.9

DIVISION		NEW YORK WATERTOWN 136			NEW YORK WHITE PLAINS 106			NEW YORK YONKERS 107			NORTH CAROLINA ASHEVILLE 287 - 288			NORTH CAROLINA CHARLOTTE 281 - 282			NORTH CAROLINA DURHAM 277		
		MAT.	INST.	TOTAL	MAT.	INST.	TOTAL	MAT.	INST.	TOTAL	MAT.	INST.	TOTAL	MAT.	INST.	TOTAL	MAT.	INST.	TOTAL
015433	CONTRACTOR EQUIPMENT		115.8	115.8		106.9	106.9		106.9	106.9		101.9	101.9		101.9	101.9		107.3	107.3
0241, 31 - 34	SITE & INFRASTRUCTURE, DEMOLITION	80.9	103.6	96.8	92.4	109.9	104.7	99.2	110.0	106.7	97.4	81.6	86.3	99.1	81.8	87.0	101.5	90.7	93.9
0310	Concrete Forming & Accessories	85.8	97.5	95.9	102.8	140.1	135.0	103.0	151.3	144.7	91.1	63.4	67.2	96.1	62.9	67.5	95.2	63.4	67.7
0320	Concrete Reinforcing	102.1	104.2	103.2	95.3	170.9	133.4	99.1	171.0	135.4	94.9	68.0	81.3	99.2	67.0	82.9	98.2	67.0	82.5
0330	Cast-in-Place Concrete	104.0	110.2	106.3	85.7	150.7	110.4	95.9	150.9	116.8	109.3	71.3	94.9	111.5	70.6	95.9	98.2	71.3	88.0
03	CONCRETE	106.4	104.4	105.5	89.1	149.4	116.7	98.1	154.6	124.0	99.5	68.6	85.4	99.2	68.0	84.9	94.2	68.4	82.4
04	MASONRY	91.6	109.3	102.6	94.8	154.7	131.9	98.6	154.7	133.4	90.1	63.7	73.7	93.9	62.3	74.3	91.3	62.3	72.4
05	METALS	95.8	120.4	103.4	90.5	162.5	112.6	98.6	163.0	118.4	97.8	91.7	95.9	98.6	91.4	96.4	113.7	91.3	106.8
06	WOOD, PLASTICS & COMPOSITES	84.4	94.2	89.9	103.1	134.9	120.8	103.0	149.4	128.8	90.7	61.8	74.7	91.2	61.8	74.8	94.9	61.8	76.5
07	THERMAL & MOISTURE PROTECTION	90.4	102.0	95.3	105.1	146.6	122.7	105.4	149.5	124.1	101.0	65.3	85.9	95.4	64.6	82.4	107.4	65.3	89.6
08	OPENINGS	96.9	97.3	97.0	92.9	165.9	109.6	96.5	173.7	114.2	93.7	62.9	86.6	98.6	62.7	90.3	100.8	62.7	92.0
0920	Plaster & Gypsum Board	89.7	94.0	92.6	98.9	135.6	123.6	102.3	150.5	134.7	106.6	60.6	75.7	101.6	60.6	74.1	98.7	60.6	73.1
0950, 0980	Ceilings & Acoustic Treatment	94.1	94.0	94.0	83.5	135.6	118.6	99.7	150.5	133.9	83.8	60.6	68.2	87.8	60.6	69.5	86.3	60.6	69.0
0960	Flooring	80.0	97.7	84.9	90.5	173.4	113.5	90.0	189.9	117.7	92.8	63.5	84.6	92.1	63.5	84.1	96.8	63.5	87.6
0970, 0990	Wall Finishes & Painting/Coating	84.8	97.2	92.0	102.2	159.9	135.9	102.2	159.9	135.9	104.5	59.7	78.4	97.0	59.7	75.3	98.8	59.7	76.0
09	FINISHES	86.2	97.0	92.1	91.7	147.5	122.0	95.7	159.3	130.3	90.1	62.7	75.2	90.0	62.4	75.0	89.7	62.7	75.0
COVERS	DIVS. 10 - 14, 25, 28, 41, 43, 44, 46	100.0	100.5	100.1	100.0	119.9	104.4	100.0	129.1	106.5	100.0	85.2	96.7	100.0	84.8	96.6	100.0	85.2	96.7
21, 22, 23	FIRE SUPPRESSION, PLUMBING & HVAC	100.2	91.4	96.5	100.4	140.6	117.6	100.4	140.8	117.7	100.5	62.6	84.3	99.9	63.1	84.1	100.5	62.6	84.3
26, 27, 3370	ELECTRICAL, COMMUNICATIONS & UTIL.	99.3	95.2	97.2	92.4	149.6	122.2	99.1	164.7	133.2	100.5	61.1	80.0	99.7	62.3	80.3	98.0	60.6	78.5
MF2016	WEIGHTED AVERAGE	97.3	100.7	98.8	94.6	145.6	116.7	98.9	150.8	121.6	97.6	68.3	84.8	98.1	68.3	85.0	100.2	68.9	86.5

City Cost Indexes

DIVISION		NORTH CAROLINA																	
		ELIZABETH CITY 279			FAYETTEVILLE 283			GASTONIA 280			GREENSBORO 270, 272 - 274			HICKORY 286			KINSTON 285		
		MAT.	INST.	TOTAL	MAT.	INST.	TOTAL	MAT.	INST.	TOTAL	MAT.	INST.	TOTAL	MAT.	INST.	TOTAL	MAT.	INST.	TOTAL
015433	CONTRACTOR EQUIPMENT		111.5	111.5		107.3	107.3		101.9	101.9		107.3	107.3		107.3	107.3		107.3	107.3
0241, 31 - 34	SITE & INFRASTRUCTURE, DEMOLITION	105.8	92.5	96.5	96.8	90.6	92.5	97.3	81.9	86.5	101.3	90.8	94.0	96.4	89.4	91.5	95.2	89.3	91.1
0310	Concrete Forming & Accessories	81.6	63.6	66.0	90.8	62.8	66.7	97.3	63.6	68.2	95.0	63.3	67.7	87.7	63.4	66.7	84.1	62.6	65.6
0320	Concrete Reinforcing	96.2	71.9	84.0	98.6	67.0	82.6	95.3	67.0	81.0	97.1	67.0	81.9	94.9	67.0	80.8	94.4	66.9	80.5
0330	Cast-in-Place Concrete	98.4	73.0	88.8	114.7	70.5	97.9	106.9	71.4	93.4	97.5	71.3	87.6	109.3	71.3	94.9	105.5	70.4	92.2
03	CONCRETE	94.4	70.0	83.2	101.4	67.9	86.1	98.0	68.6	84.5	93.6	68.4	82.1	99.3	68.4	85.2	96.1	67.8	83.2
04	MASONRY	98.8	62.3	76.2	93.2	62.3	74.1	94.5	63.7	75.4	82.5	63.7	70.9	78.9	63.7	69.5	85.8	62.3	71.2
05	METALS	100.0	94.0	98.2	117.8	91.3	109.7	98.3	91.6	96.3	106.2	91.3	101.7	97.8	91.3	95.8	96.6	91.1	94.9
06	WOOD, PLASTICS & COMPOSITES	80.1	61.8	70.3	90.0	61.8	74.3	98.5	61.8	78.1	94.6	61.8	76.4	85.9	61.8	72.5	83.0	61.8	71.2
07	THERMAL & MOISTURE PROTECTION	106.6	64.5	88.8	100.4	64.6	85.3	101.2	65.3	86.0	107.1	65.3	89.4	101.3	65.3	86.1	101.2	64.6	85.7
08	OPENINGS	97.6	64.1	89.9	93.8	62.7	86.7	97.5	62.7	89.5	100.8	62.7	91.5	93.8	62.7	86.6	93.9	62.7	86.7
0920	Plaster & Gypsum Board	92.3	60.6	71.0	110.8	60.6	77.1	112.7	60.6	77.7	100.2	60.6	73.6	106.6	60.6	75.7	106.4	60.6	75.7
0950, 0980	Ceilings & Acoustic Treatment	86.3	60.6	69.0	85.5	60.6	68.8	87.3	60.6	69.3	86.3	60.6	69.0	83.8	60.6	68.2	87.3	60.6	69.3
0960	Flooring	89.2	81.6	87.1	93.0	63.5	84.8	96.1	81.6	92.1	96.8	63.5	87.6	92.7	81.6	89.6	90.0	81.6	87.7
0970, 0990	Wall Finishes & Painting/Coating	98.8	59.7	76.0	104.5	59.7	78.4	104.5	59.7	78.4	98.8	59.7	76.0	104.5	59.7	78.4	104.5	59.7	78.4
09	FINISHES	86.9	66.5	75.8	91.1	62.4	75.5	92.6	66.4	78.3	90.0	62.7	75.1	90.3	66.4	77.3	89.9	66.0	76.9
COVERS	DIVS. 10 - 14, 25, 28, 41, 43, 44, 46	100.0	82.7	96.1	100.0	84.7	96.6	100.0	85.2	96.7	100.0	82.4	96.1	100.0	85.2	96.7	100.0	84.7	96.6
21, 22, 23	FIRE SUPPRESSION, PLUMBING & HVAC	97.3	60.5	81.6	100.2	61.9	83.9	100.5	61.5	83.8	100.4	62.6	84.2	97.4	61.5	82.0	97.4	60.3	81.5
26, 27, 3370	ELECTRICAL, COMMUNICATIONS & UTIL.	97.7	68.2	82.4	100.3	60.6	79.6	100.0	62.3	80.4	97.1	61.1	78.4	98.4	62.3	79.6	98.2	58.9	77.8
MF2016	WEIGHTED AVERAGE	97.2	70.4	85.5	101.3	68.4	86.9	98.4	68.7	85.4	98.6	68.9	85.6	96.1	69.3	84.4	95.8	68.2	83.7

DIVISION		NORTH CAROLINA															NORTH DAKOTA		
		MURPHY 289			RALEIGH 275 - 276			ROCKY MOUNT 278			WILMINGTON 284			WINSTON-SALEM 271			BISMARCK 585		
		MAT.	INST.	TOTAL	MAT.	INST.	TOTAL	MAT.	INST.	TOTAL	MAT.	INST.	TOTAL	MAT.	INST.	TOTAL	MAT.	INST.	TOTAL
015433	CONTRACTOR EQUIPMENT		101.9	101.9		107.3	107.3		107.3	107.3		101.9	101.9		107.3	107.3		100.3	100.3
0241, 31 - 34	SITE & INFRASTRUCTURE, DEMOLITION	98.7	80.1	85.6	101.2	90.7	93.8	103.9	90.7	94.7	98.5	81.5	86.6	101.7	90.8	94.1	102.1	98.6	99.6
0310	Concrete Forming & Accessories	97.9	62.7	67.5	95.5	62.8	67.3	87.5	62.9	66.3	92.5	62.8	66.9	96.8	63.4	68.0	102.8	74.9	78.7
0320	Concrete Reinforcing	94.4	66.3	80.2	99.2	67.0	82.9	96.2	67.0	81.5	95.6	67.0	81.2	97.1	67.0	81.9	93.8	94.8	94.3
0330	Cast-in-Place Concrete	113.1	70.5	96.9	101.1	70.5	89.5	96.3	70.6	86.5	108.9	70.5	94.3	99.9	71.3	89.1	102.9	85.8	96.4
03	CONCRETE	102.4	67.7	86.6	92.5	67.9	81.3	95.1	68.0	82.7	99.4	67.9	85.0	94.9	68.4	82.8	92.5	82.8	88.1
04	MASONRY	82.0	62.3	69.8	80.9	62.3	69.4	76.1	62.3	67.6	79.4	62.3	68.8	82.8	63.7	71.0	103.6	83.3	91.0
05	METALS	95.6	91.1	94.2	98.7	91.3	96.5	99.2	91.4	96.8	97.3	91.3	95.5	103.4	91.3	99.7	103.2	93.7	100.3
06	WOOD, PLASTICS & COMPOSITES	99.1	61.7	78.3	92.7	61.8	75.5	86.6	61.8	72.8	92.5	61.8	75.5	94.6	61.8	76.4	97.2	70.7	82.5
07	THERMAL & MOISTURE PROTECTION	101.2	64.6	85.7	101.8	64.6	86.1	107.1	64.6	89.1	101.0	64.6	85.6	107.1	65.3	89.4	105.0	84.7	96.4
08	OPENINGS	93.7	62.4	86.5	99.1	62.7	90.7	96.8	62.7	89.0	93.9	62.7	86.7	100.8	62.7	92.0	106.5	81.0	100.7
0920	Plaster & Gypsum Board	111.9	60.5	77.4	92.6	60.6	71.1	94.2	60.6	71.7	108.7	60.6	76.4	100.2	60.6	73.6	102.8	70.3	81.0
0950, 0980	Ceilings & Acoustic Treatment	83.8	60.6	68.1	83.3	60.6	68.0	84.6	60.6	68.4	85.5	60.6	68.8	86.3	60.6	69.0	110.2	70.3	83.3
0960	Flooring	96.4	81.0	92.3	90.4	63.5	82.9	92.6	81.6	89.6	93.5	63.5	85.2	96.8	63.5	87.6	86.0	52.6	76.7
0970, 0990	Wall Finishes & Painting/Coating	104.5	59.7	78.4	93.8	59.7	73.9	98.8	59.7	76.0	104.5	59.7	78.4	98.8	59.7	76.0	91.2	60.1	73.1
09	FINISHES	92.1	65.9	77.9	88.4	62.4	74.2	88.0	66.0	76.0	91.0	62.4	75.4	90.0	62.7	75.1	96.0	69.6	81.7
COVERS	DIVS. 10 - 14, 25, 28, 41, 43, 44, 46	100.0	84.7	96.6	100.0	84.7	96.6	100.0	84.7	96.6	100.0	84.7	96.6	100.0	85.2	96.7	100.0	90.9	98.0
21, 22, 23	FIRE SUPPRESSION, PLUMBING & HVAC	97.4	60.7	81.7	99.9	61.9	83.7	97.3	60.6	81.6	100.5	61.9	84.0	100.4	62.6	84.2	100.1	76.9	90.2
26, 27, 3370	ELECTRICAL, COMMUNICATIONS & UTIL.	101.4	58.9	79.3	100.5	59.5	79.1	99.5	60.6	79.3	100.8	58.9	79.0	97.1	58.9	77.3	99.5	76.7	87.6
MF2016	WEIGHTED AVERAGE	96.9	67.5	84.0	96.9	68.3	84.4	96.3	68.6	84.2	97.2	67.5	84.2	98.4	68.7	85.4	100.3	81.4	92.0

DIVISION		NORTH DAKOTA																	
		DEVILS LAKE 583			DICKINSON 586			FARGO 580 - 581			GRAND FORKS 582			JAMESTOWN 584			MINOT 587		
		MAT.	INST.	TOTAL	MAT.	INST.	TOTAL	MAT.	INST.	TOTAL	MAT.	INST.	TOTAL	MAT.	INST.	TOTAL	MAT.	INST.	TOTAL
015433	CONTRACTOR EQUIPMENT		100.3	100.3		100.3	100.3		100.3	100.3		100.3	100.3		100.3	100.3		100.3	100.3
0241, 31 - 34	SITE & INFRASTRUCTURE, DEMOLITION	106.9	98.6	101.1	114.4	98.6	103.3	97.6	98.6	98.3	110.4	98.6	102.1	105.9	98.6	100.8	108.0	98.6	101.4
0310	Concrete Forming & Accessories	99.3	74.8	78.1	89.1	74.8	76.8	96.9	75.0	78.0	92.8	74.8	77.3	90.7	74.7	76.9	88.8	74.8	76.8
0320	Concrete Reinforcing	95.3	94.7	95.0	96.2	94.8	95.5	95.4	95.1	95.3	93.9	94.7	94.3	95.9	95.1	95.5	97.2	94.8	96.0
0330	Cast-in-Place Concrete	123.1	85.8	108.9	111.4	85.8	101.6	100.8	85.8	95.1	111.3	85.8	101.6	121.5	85.7	107.9	111.3	85.8	101.6
03	CONCRETE	101.6	82.7	93.0	100.3	82.8	92.3	93.4	82.9	88.6	97.9	82.8	91.0	100.2	82.8	92.2	96.6	82.8	90.3
04	MASONRY	120.5	82.6	97.0	122.1	83.3	98.0	107.1	89.0	95.9	114.1	82.6	94.6	133.8	89.0	106.0	113.2	83.3	94.7
05	METALS	98.9	93.5	97.3	98.8	93.6	97.2	101.2	94.2	99.1	98.9	93.5	97.2	98.9	93.9	97.4	99.2	93.7	97.5
06	WOOD, PLASTICS & COMPOSITES	92.5	70.7	80.5	80.7	70.7	75.1	92.5	70.7	80.4	84.7	70.7	76.9	82.4	70.7	75.9	80.4	70.7	75.0
07	THERMAL & MOISTURE PROTECTION	105.7	84.5	96.7	106.3	84.7	97.1	105.9	86.7	97.8	105.9	84.5	96.7	105.5	86.7	97.6	105.7	84.7	96.8
08	OPENINGS	100.2	81.0	95.8	100.1	81.0	95.7	101.4	81.0	96.7	98.7	81.0	94.7	100.1	81.0	95.7	98.9	81.0	94.8
0920	Plaster & Gypsum Board	121.8	70.3	87.2	112.6	70.3	84.2	101.0	70.3	80.4	113.9	70.3	84.6	113.6	70.3	84.5	112.6	70.3	84.2
0950, 0980	Ceilings & Acoustic Treatment	108.4	70.3	82.7	108.4	70.3	82.7	99.9	70.3	79.9	108.4	70.3	82.7	108.4	70.3	82.7	108.4	70.3	82.7
0960	Flooring	92.6	52.6	81.5	86.1	52.6	76.8	96.1	52.6	84.1	88.0	52.6	78.2	86.8	52.6	77.3	85.8	52.6	76.6
0970, 0990	Wall Finishes & Painting/Coating	89.6	58.9	71.7	89.6	58.9	71.7	90.8	70.4	78.9	89.6	69.0	77.6	89.6	58.9	71.7	89.6	60.1	72.4
09	FINISHES	96.2	69.4	81.6	93.8	69.4	80.5	96.9	70.6	82.6	94.0	70.5	81.2	93.2	69.4	80.2	93.0	69.5	80.2
COVERS	DIVS. 10 - 14, 25, 28, 41, 43, 44, 46	100.0	90.8	98.0	100.0	90.9	98.0	100.0	90.9	98.0	100.0	90.9	98.0	100.0	90.8	98.0	100.0	90.9	98.0
21, 22, 23	FIRE SUPPRESSION, PLUMBING & HVAC	97.2	81.5	90.5	97.2	75.2	87.8	100.1	77.0	90.2	100.3	74.4	89.2	97.2	72.8	86.8	100.3	75.0	89.5
26, 27, 3370	ELECTRICAL, COMMUNICATIONS & UTIL.	97.4	48.0	71.7	105.8	71.7	88.0	103.5	73.3	87.8	100.8	63.0	81.1	97.4	48.0	71.7	103.8	76.7	89.7
MF2016	WEIGHTED AVERAGE	100.1	78.2	90.5	100.8	80.3	91.8	100.1	81.8	92.1	100.1	79.0	90.8	100.2	77.1	90.1	100.0	81.0	91.7

NORTH DAKOTA / OHIO

DIVISION		WILLISTON 588 MAT.	INST.	TOTAL	AKRON 442-443 MAT.	INST.	TOTAL	ATHENS 457 MAT.	INST.	TOTAL	CANTON 446-447 MAT.	INST.	TOTAL	CHILLICOTHE 456 MAT.	INST.	TOTAL	CINCINNATI 451-452 MAT.	INST.	TOTAL
015433	CONTRACTOR EQUIPMENT		100.3	100.3		91.6	91.6		87.4	87.4		91.6	91.6		98.5	98.5		96.0	96.0
0241, 31 - 34	SITE & INFRASTRUCTURE, DEMOLITION	108.4	96.3	99.9	97.4	99.2	98.7	115.8	88.9	97.0	97.6	99.0	98.6	101.6	99.8	100.4	96.9	99.1	98.4
0310	Concrete Forming & Accessories	94.5	74.5	77.3	100.5	86.0	88.0	92.6	87.6	88.3	100.5	76.9	80.2	95.0	80.2	82.2	98.6	82.5	84.7
0320	Concrete Reinforcing	98.1	94.7	96.4	103.4	90.2	96.8	89.1	83.5	86.3	103.4	78.6	90.9	86.0	78.3	82.1	91.2	79.4	85.2
0330	Cast-in-Place Concrete	111.3	85.7	101.6	98.9	90.3	95.6	109.4	86.2	100.6	99.8	88.5	95.5	99.3	83.4	93.3	94.8	76.6	87.9
03	CONCRETE	97.8	82.6	90.9	98.3	87.6	93.4	101.1	85.9	94.1	98.7	80.9	90.6	95.5	81.4	89.1	94.6	80.1	88.0
04	MASONRY	107.4	83.3	92.5	93.5	91.6	92.3	80.1	81.2	80.8	94.1	79.1	84.8	87.3	91.2	89.7	90.5	78.0	82.7
05	METALS	99.0	93.3	97.3	96.8	80.1	91.7	103.2	77.5	95.3	96.8	74.8	90.0	95.2	85.7	92.3	97.5	83.8	93.3
06	WOOD, PLASTICS & COMPOSITES	86.2	70.7	77.6	107.2	85.0	94.8	86.4	90.0	88.8	107.5	75.2	89.6	99.7	77.0	87.1	99.4	83.9	90.7
07	THERMAL & MOISTURE PROTECTION	105.9	84.7	96.9	103.3	92.7	98.8	98.5	87.5	93.9	104.4	87.6	97.3	100.4	85.8	94.3	98.6	81.5	91.4
08	OPENINGS	100.2	81.0	95.8	109.6	85.3	104.0	97.1	84.9	94.3	103.3	73.2	96.4	88.7	74.4	85.4	96.6	80.1	92.8
0920	Plaster & Gypsum Board	113.9	70.3	84.6	103.3	84.5	90.7	95.6	90.2	92.0	104.3	74.4	84.2	98.3	76.7	83.8	97.8	83.8	88.4
0950, 0980	Ceilings & Acoustic Treatment	108.4	70.3	82.7	93.4	84.5	87.4	108.9	90.2	96.3	93.4	74.4	80.6	102.5	76.7	85.1	96.2	83.8	87.8
0960	Flooring	88.9	52.6	78.8	95.1	86.2	92.6	120.1	78.2	108.5	95.2	73.1	89.1	98.1	78.2	92.6	99.4	80.8	94.3
0970, 0990	Wall Finishes & Painting/Coating	89.6	58.9	71.7	98.2	96.3	96.3	100.7	88.8	93.7	98.2	72.2	83.0	97.9	80.1	87.5	97.5	74.6	84.1
09	FINISHES	94.2	69.4	80.7	97.4	86.6	91.5	101.5	86.5	93.3	97.6	75.0	85.3	98.4	79.3	88.0	97.7	81.6	88.9
COVERS	DIVS. 10 - 14, 25, 28, 41, 43, 44, 46	100.0	90.9	98.0	100.0	93.6	98.6	100.0	88.0	97.3	100.0	91.6	98.1	100.0	87.9	97.3	100.0	89.9	97.5
21, 22, 23	FIRE SUPPRESSION, PLUMBING & HVAC	97.2	75.2	87.8	100.0	89.0	95.3	96.9	76.3	88.1	100.0	78.0	90.6	97.4	94.6	96.2	99.9	77.1	90.2
26, 27, 3370	ELECTRICAL, COMMUNICATIONS & UTIL.	101.1	72.6	86.3	98.3	85.0	91.4	100.8	90.6	95.5	97.6	85.6	91.4	100.1	76.5	87.8	99.1	74.8	86.5
MF2016	WEIGHTED AVERAGE	99.3	80.2	90.9	99.7	88.2	94.7	99.1	83.7	92.4	99.1	81.0	91.1	96.2	86.1	91.7	97.7	80.9	90.3

OHIO

DIVISION		CLEVELAND 441 MAT.	INST.	TOTAL	COLUMBUS 430-432 MAT.	INST.	TOTAL	DAYTON 453-454 MAT.	INST.	TOTAL	HAMILTON 450 MAT.	INST.	TOTAL	LIMA 458 MAT.	INST.	TOTAL	LORAIN 440 MAT.	INST.	TOTAL
015433	CONTRACTOR EQUIPMENT		92.7	92.7		91.4	91.4		91.5	91.5		98.5	98.5		90.8	90.8		91.6	91.6
0241, 31 - 34	SITE & INFRASTRUCTURE, DEMOLITION	96.0	97.5	97.1	101.7	91.9	94.8	97.1	99.2	98.5	97.0	99.5	98.8	109.1	89.0	95.0	96.8	99.5	98.7
0310	Concrete Forming & Accessories	100.1	90.7	92.0	97.7	77.4	80.1	96.8	73.2	76.5	96.8	71.9	75.3	92.6	79.2	81.1	100.6	77.7	80.8
0320	Concrete Reinforcing	103.9	92.0	97.9	96.2	80.3	88.2	91.2	81.5	86.3	91.2	79.4	85.2	89.1	81.6	85.3	103.4	90.5	96.9
0330	Cast-in-Place Concrete	95.8	99.3	97.1	96.9	82.2	91.3	85.1	78.8	82.7	91.2	77.2	85.9	100.5	88.0	95.8	94.1	93.7	94.0
03	CONCRETE	98.5	93.4	96.2	96.2	79.6	88.6	86.9	76.6	82.2	89.8	75.7	83.4	94.0	82.7	88.8	96.0	85.2	91.1
04	MASONRY	100.0	99.2	99.5	91.9	86.5	88.5	86.3	77.2	80.6	86.7	79.4	82.2	108.7	78.6	90.0	90.2	94.6	92.9
05	METALS	98.4	84.9	94.3	97.5	80.8	92.4	96.7	77.8	90.9	96.8	86.9	93.7	103.2	80.7	96.3	97.4	81.5	92.5
06	WOOD, PLASTICS & COMPOSITES	101.1	88.3	93.9	99.4	76.1	86.4	103.5	71.7	85.8	102.1	69.0	83.7	86.4	78.7	82.1	107.2	73.4	88.4
07	THERMAL & MOISTURE PROTECTION	100.2	98.4	99.5	100.4	85.8	94.2	104.8	79.1	93.9	100.5	80.5	92.1	98.0	85.0	92.5	104.3	94.2	100.0
08	OPENINGS	97.9	87.5	95.5	99.7	74.4	93.9	95.8	72.8	90.5	93.5	71.5	88.4	97.1	75.9	92.2	103.3	78.9	97.7
0920	Plaster & Gypsum Board	103.2	87.7	92.8	97.2	75.5	82.6	100.1	71.3	80.8	100.1	68.5	78.9	95.6	78.0	83.8	103.3	72.6	82.7
0950, 0980	Ceilings & Acoustic Treatment	89.0	87.7	88.1	90.9	75.5	80.5	104.3	71.3	82.0	103.3	68.5	79.8	108.0	78.0	87.7	93.4	72.6	79.4
0960	Flooring	96.5	92.6	95.4	92.8	78.2	88.8	101.8	71.9	93.5	99.2	80.8	94.1	119.0	77.9	107.6	95.2	90.8	94.0
0970, 0990	Wall Finishes & Painting/Coating	101.1	95.0	97.5	100.4	80.1	88.5	97.9	71.6	82.6	97.9	71.0	82.2	100.7	71.9	83.9	98.2	95.0	96.3
09	FINISHES	97.8	91.2	94.2	93.9	77.3	84.8	99.9	72.3	84.8	98.9	72.8	84.7	100.5	78.0	88.3	97.4	81.0	88.5
COVERS	DIVS. 10 - 14, 25, 28, 41, 43, 44, 46	100.0	98.0	99.6	100.0	89.0	97.6	100.0	87.2	97.1	100.0	87.4	97.2	100.0	87.1	97.1	100.0	94.5	98.8
21, 22, 23	FIRE SUPPRESSION, PLUMBING & HVAC	100.0	93.0	97.0	99.9	85.7	93.9	100.8	75.7	90.1	100.5	77.5	90.7	96.9	83.1	91.0	100.0	90.0	95.7
26, 27, 3370	ELECTRICAL, COMMUNICATIONS & UTIL.	98.0	95.7	96.8	99.0	84.7	91.6	97.8	76.4	86.6	98.0	76.7	86.9	101.1	76.4	88.3	97.7	80.8	89.0
MF2016	WEIGHTED AVERAGE	98.9	93.4	96.5	98.1	83.3	91.6	96.7	77.9	88.5	96.6	79.4	89.1	99.4	81.1	91.4	98.6	86.9	93.5

OHIO

DIVISION		MANSFIELD 448-449 MAT.	INST.	TOTAL	MARION 433 MAT.	INST.	TOTAL	SPRINGFIELD 455 MAT.	INST.	TOTAL	STEUBENVILLE 439 MAT.	INST.	TOTAL	TOLEDO 434-436 MAT.	INST.	TOTAL	YOUNGSTOWN 444-445 MAT.	INST.	TOTAL
015433	CONTRACTOR EQUIPMENT		91.6	91.6		91.2	91.2		91.5	91.5		95.1	95.1		95.1	95.1		91.6	91.6
0241, 31 - 34	SITE & INFRASTRUCTURE, DEMOLITION	93.4	99.1	97.4	96.3	95.0	95.4	97.4	99.1	98.6	141.4	103.5	114.8	99.9	95.5	96.8	97.3	99.0	98.5
0310	Concrete Forming & Accessories	89.8	76.3	78.2	95.6	77.0	79.5	96.8	73.3	76.5	96.8	82.3	84.3	99.2	88.1	89.6	100.5	79.3	82.2
0320	Concrete Reinforcing	94.1	79.3	86.7	88.6	80.5	84.5	91.2	81.5	86.3	86.2	96.8	91.5	96.2	84.5	90.3	103.4	86.4	94.8
0330	Cast-in-Place Concrete	91.6	83.0	88.3	84.0	81.1	82.9	87.4	78.7	84.1	91.2	88.6	90.2	92.0	91.7	91.9	97.9	88.6	94.4
03	CONCRETE	90.6	78.9	85.2	83.4	78.9	81.4	88.0	76.6	82.8	87.2	86.5	86.9	91.0	88.7	89.9	97.8	83.4	91.2
04	MASONRY	92.6	91.7	92.0	93.5	89.2	90.8	86.5	76.9	80.5	80.8	91.3	87.3	99.9	90.7	94.2	93.8	87.3	89.7
05	METALS	97.6	76.0	91.0	96.5	78.9	91.1	96.7	77.9	90.9	93.1	82.2	89.7	97.3	85.9	93.8	96.8	78.2	91.1
06	WOOD, PLASTICS & COMPOSITES	94.9	73.4	83.0	96.6	76.1	85.2	105.0	71.7	86.5	92.1	80.0	85.4	100.5	88.0	93.5	107.2	78.2	91.1
07	THERMAL & MOISTURE PROTECTION	102.5	90.4	97.4	96.9	89.1	93.6	104.6	78.9	93.8	108.7	89.1	100.4	97.0	91.9	94.9	104.5	85.8	96.6
08	OPENINGS	103.7	73.2	96.7	93.4	71.5	88.4	94.0	72.8	89.1	93.7	82.0	91.0	96.2	85.1	93.6	103.3	78.6	97.6
0920	Plaster & Gypsum Board	96.9	72.6	80.6	98.5	75.5	83.1	100.1	71.3	80.8	97.6	79.1	85.1	100.5	87.7	91.9	103.3	77.5	86.0
0950, 0980	Ceilings & Acoustic Treatment	94.3	72.6	79.7	97.1	75.5	82.5	104.3	71.3	82.0	94.2	79.1	84.0	97.1	87.7	90.8	93.4	77.5	82.7
0960	Flooring	89.8	92.2	90.4	92.2	92.2	92.2	101.8	71.9	93.5	119.4	92.8	112.0	92.5	88.8	91.5	95.2	83.4	91.9
0970, 0990	Wall Finishes & Painting/Coating	98.2	76.5	85.5	104.9	76.5	88.3	97.9	71.6	82.6	118.5	88.8	101.2	104.9	87.3	94.6	98.2	82.0	88.7
09	FINISHES	94.8	78.6	86.0	95.3	79.5	86.7	99.9	72.2	84.8	112.9	84.3	97.4	96.0	88.0	91.6	97.5	79.9	87.9
COVERS	DIVS. 10 - 14, 25, 28, 41, 43, 44, 46	100.0	91.7	98.2	100.0	86.1	96.9	100.0	87.1	97.1	100.0	89.0	97.5	100.0	94.8	98.8	100.0	91.8	98.2
21, 22, 23	FIRE SUPPRESSION, PLUMBING & HVAC	96.9	89.0	93.5	96.9	85.1	91.9	100.8	81.1	92.4	97.3	93.3	95.6	100.0	93.5	97.2	100.0	83.8	93.1
26, 27, 3370	ELECTRICAL, COMMUNICATIONS & UTIL.	95.6	70.6	82.6	93.5	70.6	81.6	97.8	84.7	91.0	88.3	107.1	98.1	99.1	109.2	104.4	97.7	77.1	87.0
MF2016	WEIGHTED AVERAGE	96.8	82.8	90.7	94.4	81.6	88.8	96.7	80.2	89.5	96.2	91.8	94.3	97.5	93.1	95.6	98.9	83.3	92.1

DIVISION		OHIO			OKLAHOMA												
		ZANESVILLE			ARDMORE			CLINTON			DURANT			ENID			GUYMON	
		437 - 438			734			736			747			737			739	
		MAT.	INST.	TOTAL	MAT.	INST.	TOTAL	MAT.	INST.	TOTAL	MAT.	INST.	TOTAL	MAT.	INST.	TOTAL	MAT.	INST.	TOTAL
015433	CONTRACTOR EQUIPMENT		91.2	91.2		82.8	82.8		81.8	81.8		81.8	81.8		81.8	81.8		81.8	81.8
0241, 31 - 34	SITE & INFRASTRUCTURE, DEMOLITION	99.4	95.0	96.3	101.6	97.4	98.6	103.1	95.8	98.0	95.5	93.4	94.0	104.7	95.8	98.5	107.6	95.5	99.1
0310	Concrete Forming & Accessories	92.7	76.9	79.1	91.6	57.8	62.4	90.2	57.8	62.2	84.2	57.4	61.1	93.7	57.9	62.8	96.9	57.6	63.0
0320	Concrete Reinforcing	88.1	93.0	90.6	81.1	66.5	73.7	81.6	66.5	74.0	92.1	65.6	78.7	81.0	66.5	73.7	81.6	62.7	72.0
0330	Cast-in-Place Concrete	88.5	80.9	85.6	103.5	73.9	92.2	100.0	73.9	90.1	92.0	73.8	85.1	100.0	74.0	90.1	100.0	73.6	90.0
03	CONCRETE	87.0	81.1	84.3	89.8	64.7	78.3	89.1	64.7	78.0	85.7	64.3	75.9	89.6	64.8	78.3	92.1	63.9	79.2
04	MASONRY	91.2	77.7	82.8	95.8	57.9	72.3	119.7	57.9	81.4	88.7	64.8	73.8	101.5	57.9	74.5	98.0	56.6	72.4
05	METALS	98.0	83.6	93.6	102.0	61.3	89.5	102.1	61.3	89.6	91.3	60.9	82.0	103.6	61.4	90.7	102.6	59.1	89.3
06	WOOD, PLASTICS & COMPOSITES	92.4	76.1	83.3	104.1	57.0	77.9	103.1	57.0	77.5	91.5	57.0	72.3	106.5	57.0	79.0	110.0	57.0	80.5
07	THERMAL & MOISTURE PROTECTION	97.0	82.8	91.0	105.3	65.2	88.3	105.4	65.2	88.4	99.3	66.2	85.2	105.5	65.2	88.5	105.9	64.8	88.5
08	OPENINGS	93.4	78.1	89.9	103.2	57.0	92.6	103.2	57.0	92.6	96.9	56.7	87.7	104.4	57.3	93.6	103.3	56.1	92.5
0920	Plaster & Gypsum Board	95.2	75.5	82.0	88.9	56.4	67.0	88.5	56.4	66.9	77.9	56.4	63.4	89.2	56.4	67.1	89.2	56.4	67.1
0950, 0980	Ceilings & Acoustic Treatment	97.1	75.5	82.5	86.0	56.4	66.0	86.0	56.4	66.0	82.3	56.4	64.8	86.0	56.4	66.0	86.0	56.4	66.0
0960	Flooring	90.3	78.2	87.0	88.5	60.6	80.7	87.4	60.6	79.9	90.8	54.6	80.8	89.1	74.8	85.1	90.6	60.6	82.3
0970, 0990	Wall Finishes & Painting/Coating	104.9	80.1	90.4	84.0	45.1	61.3	84.0	45.1	61.3	90.9	45.1	64.2	84.0	45.1	61.3	84.0	44.2	60.8
09	FINISHES	94.5	77.0	85.0	81.8	55.9	67.7	81.6	55.9	67.7	82.5	54.8	67.4	82.2	58.8	69.5	83.1	57.2	69.0
COVERS	DIVS. 10 - 14, 25, 28, 41, 43, 44, 46	100.0	82.8	96.0	100.0	79.1	95.3	100.0	79.1	95.3	100.0	79.0	95.3	100.0	79.1	95.3	100.0	79.0	95.3
21, 22, 23	FIRE SUPPRESSION, PLUMBING & HVAC	96.9	90.9	94.3	97.2	67.8	84.7	97.2	67.8	84.7	97.2	67.8	84.6	100.3	67.9	86.4	97.2	65.6	83.7
26, 27, 3370	ELECTRICAL, COMMUNICATIONS & UTIL.	93.6	74.0	83.4	93.9	71.2	82.1	94.8	71.2	82.5	96.2	62.1	78.4	94.8	71.2	82.5	96.4	57.5	76.2
MF2016	WEIGHTED AVERAGE	95.0	82.5	89.5	96.7	66.7	83.6	97.9	66.6	84.2	93.2	65.6	81.1	98.3	67.0	84.6	97.8	63.8	82.9

DIVISION		OKLAHOMA																
		LAWTON			MCALESTER			MIAMI			MUSKOGEE			OKLAHOMA CITY			PONCA CITY	
		735			745			743			744			730 - 731			746	
		MAT.	INST.	TOTAL	MAT.	INST.	TOTAL	MAT.	INST.	TOTAL	MAT.	INST.	TOTAL	MAT.	INST.	TOTAL	MAT.	INST.	TOTAL
015433	CONTRACTOR EQUIPMENT		82.8	82.8		81.8	81.8		93.5	93.5		93.5	93.5		83.1	83.1		81.8	81.8
0241, 31 - 34	SITE & INFRASTRUCTURE, DEMOLITION	100.6	97.4	98.3	88.7	95.5	93.5	90.0	92.8	91.9	90.3	92.7	92.0	99.0	97.9	98.2	96.0	95.8	95.9
0310	Concrete Forming & Accessories	96.9	57.9	63.3	82.3	43.2	48.6	95.1	57.6	62.7	99.5	57.4	63.2	94.3	57.9	62.9	90.6	57.7	62.3
0320	Concrete Reinforcing	81.2	66.5	73.8	91.8	65.2	78.4	90.3	66.5	78.3	91.2	64.9	77.9	89.7	66.5	78.0	91.2	66.5	78.7
0330	Cast-in-Place Concrete	96.7	74.0	88.1	80.7	73.5	78.0	84.6	74.9	80.9	85.6	74.8	81.5	96.7	74.0	88.1	94.5	73.9	86.7
03	CONCRETE	86.0	64.8	76.3	76.8	57.7	68.1	81.1	66.0	74.2	82.7	65.6	74.9	88.8	64.8	77.9	87.8	64.7	77.2
04	MASONRY	97.8	57.9	73.1	106.4	57.8	76.3	91.6	58.0	70.7	108.3	51.3	73.0	100.3	57.9	74.0	84.4	57.9	68.0
05	METALS	107.5	61.4	93.3	91.2	60.0	81.6	91.2	76.7	86.7	92.7	75.3	87.4	100.4	61.4	88.4	91.2	61.2	82.0
06	WOOD, PLASTICS & COMPOSITES	109.0	57.0	80.1	89.1	37.9	60.6	103.7	57.1	77.8	107.9	57.1	79.7	94.9	57.0	73.8	99.4	57.0	75.8
07	THERMAL & MOISTURE PROTECTION	105.3	65.2	88.3	98.9	62.1	83.3	99.3	65.2	84.8	99.6	62.9	84.1	96.2	65.5	83.2	99.4	65.5	85.1
08	OPENINGS	106.2	57.3	95.0	96.9	46.1	85.2	96.9	57.1	87.7	98.1	56.8	88.6	101.5	57.3	91.4	96.9	57.0	87.7
0920	Plaster & Gypsum Board	90.7	56.4	67.7	76.9	36.7	49.9	83.8	56.4	65.4	85.8	56.4	66.0	96.0	56.4	69.4	82.8	56.4	65.1
0950, 0980	Ceilings & Acoustic Treatment	92.8	56.4	68.2	82.3	36.7	51.6	82.3	56.4	64.8	90.9	56.4	67.6	92.6	56.4	68.2	82.3	56.4	64.8
0960	Flooring	90.9	74.8	86.5	89.7	60.6	81.6	97.0	51.6	85.2	99.4	35.1	81.6	88.2	74.8	84.5	93.0	60.6	84.7
0970, 0990	Wall Finishes & Painting/Coating	84.0	45.1	61.3	90.9	44.2	63.7	90.9	44.2	63.7	90.9	44.2	63.7	87.1	45.1	62.6	90.9	45.1	64.2
09	FINISHES	83.9	58.8	70.3	81.5	44.5	61.4	84.6	54.8	68.3	87.3	51.3	67.7	86.5	58.8	71.4	84.2	56.9	69.4
COVERS	DIVS. 10 - 14, 25, 28, 41, 43, 44, 46	100.0	79.1	95.3	100.0	77.0	94.9	100.0	79.5	95.4	100.0	79.5	95.4	100.0	79.1	95.3	100.0	79.1	95.3
21, 22, 23	FIRE SUPPRESSION, PLUMBING & HVAC	100.3	67.9	86.4	97.2	62.4	82.3	97.2	62.5	82.4	100.3	62.4	84.1	100.1	67.9	86.3	97.2	62.5	82.4
26, 27, 3370	ELECTRICAL, COMMUNICATIONS & UTIL.	96.4	69.4	82.4	94.7	64.5	79.0	96.0	64.6	79.7	94.3	70.9	82.1	101.7	71.2	85.8	94.3	67.6	80.4
MF2016	WEIGHTED AVERAGE	98.7	66.9	84.8	92.6	61.1	78.8	92.9	65.8	81.0	95.1	65.3	82.1	97.7	67.2	84.4	93.3	65.1	81.0

DIVISION		OKLAHOMA												OREGON				
		POTEAU			SHAWNEE			TULSA			WOODWARD			BEND			EUGENE	
		749			748			740 - 741			738			977			974	
		MAT.	INST.	TOTAL	MAT.	INST.	TOTAL	MAT.	INST.	TOTAL	MAT.	INST.	TOTAL	MAT.	INST.	TOTAL	MAT.	INST.	TOTAL
015433	CONTRACTOR EQUIPMENT		92.2	92.2		81.8	81.8		93.5	93.5		81.8	81.8		98.9	98.9		98.9	98.9
0241, 31 - 34	SITE & INFRASTRUCTURE, DEMOLITION	77.2	88.1	84.9	99.1	95.8	96.8	96.7	91.8	93.3	103.5	95.8	98.1	109.4	102.5	104.6	99.4	102.6	101.6
0310	Concrete Forming & Accessories	88.6	57.3	61.6	84.1	57.7	61.3	99.6	57.6	63.4	90.3	45.5	51.6	108.8	96.2	97.9	105.3	96.3	97.5
0320	Concrete Reinforcing	92.2	66.5	79.2	91.2	62.8	76.8	91.4	66.5	78.9	81.0	66.5	73.7	100.0	109.8	104.9	104.4	109.8	107.1
0330	Cast-in-Place Concrete	84.6	74.8	80.9	97.6	73.9	88.6	93.2	75.0	86.3	100.0	73.9	90.1	105.4	101.3	103.9	102.1	101.4	101.8
03	CONCRETE	83.2	65.8	75.3	89.4	64.0	77.8	87.9	66.1	77.9	89.4	59.2	75.6	103.8	99.9	102.0	95.6	99.9	97.6
04	MASONRY	91.9	58.0	70.9	107.8	57.9	76.8	92.5	58.0	71.1	91.0	57.9	70.5	104.0	100.8	102.0	101.0	100.8	100.9
05	METALS	91.2	76.4	86.7	91.1	59.9	81.5	95.8	76.8	89.9	102.2	61.2	89.6	96.3	94.5	95.7	97.9	94.6	96.3
06	WOOD, PLASTICS & COMPOSITES	96.1	57.1	74.4	91.3	57.0	72.2	107.1	57.1	79.3	103.3	40.5	68.3	101.8	95.1	98.1	97.8	95.1	96.3
07	THERMAL & MOISTURE PROTECTION	99.4	65.2	84.9	99.4	64.9	84.8	99.6	66.1	85.4	105.5	63.6	87.8	115.4	99.4	108.6	114.5	102.5	109.4
08	OPENINGS	96.9	57.1	87.7	96.9	56.1	87.5	99.8	57.1	90.0	103.2	47.9	90.5	101.2	98.9	100.6	101.4	98.9	100.9
0920	Plaster & Gypsum Board	80.9	56.4	64.4	77.9	56.4	63.4	85.8	56.4	66.0	88.5	39.4	55.5	115.2	94.8	101.5	113.2	94.8	100.8
0950, 0980	Ceilings & Acoustic Treatment	82.3	56.4	64.8	82.3	56.4	64.8	90.9	56.4	67.6	86.0	39.4	54.6	90.8	94.8	93.5	91.7	94.8	93.8
0960	Flooring	93.3	54.6	82.6	90.8	60.6	82.4	98.3	62.5	88.4	87.4	57.7	79.1	98.4	105.4	100.3	96.6	105.4	99.1
0970, 0990	Wall Finishes & Painting/Coating	90.9	45.1	64.2	90.9	44.2	63.7	90.9	44.2	63.7	84.0	45.1	61.3	90.9	72.8	80.3	90.9	72.8	80.3
09	FINISHES	82.3	54.9	67.4	82.7	55.8	68.1	87.1	56.3	70.4	81.8	45.6	62.0	96.1	95.1	95.6	94.5	95.1	94.8
COVERS	DIVS. 10 - 14, 25, 28, 41, 43, 44, 46	100.0	79.3	95.4	100.0	79.1	95.3	100.0	79.5	95.4	100.0	77.3	94.9	100.0	99.7	99.9	100.0	99.7	99.9
21, 22, 23	FIRE SUPPRESSION, PLUMBING & HVAC	97.2	62.4	82.4	97.2	67.8	84.6	100.3	64.3	84.9	97.2	67.8	84.7	97.0	102.2	99.2	100.1	102.2	101.0
26, 27, 3370	ELECTRICAL, COMMUNICATIONS & UTIL.	94.4	64.6	78.9	96.2	71.2	83.2	96.2	64.6	79.8	96.3	71.2	83.2	101.0	94.5	97.6	99.6	94.5	97.0
MF2016	WEIGHTED AVERAGE	92.5	65.4	80.6	94.7	66.3	82.3	96.0	66.4	83.0	96.7	63.8	82.3	100.0	98.7	99.4	99.1	98.8	99.0

OREGON

DIVISION		KLAMATH FALLS 976			MEDFORD 975			PENDLETON 978			PORTLAND 970 - 972			SALEM 973			VALE 979		
		MAT.	INST.	TOTAL	MAT.	INST.	TOTAL	MAT.	INST.	TOTAL	MAT.	INST.	TOTAL	MAT.	INST.	TOTAL	MAT.	INST.	TOTAL
015433	CONTRACTOR EQUIPMENT		98.9	98.9		98.9	98.9		96.4	96.4		98.9	98.9		98.9	98.9		96.4	96.4
0241, 31 - 34	SITE & INFRASTRUCTURE, DEMOLITION	113.4	102.5	105.8	107.1	102.5	103.9	106.8	95.5	98.9	102.1	102.6	102.4	93.9	102.5	99.9	94.0	95.4	95.0
0310	Concrete Forming & Accessories	101.7	95.9	96.7	100.8	95.9	96.5	102.2	96.2	97.1	106.4	96.3	97.7	105.9	96.2	97.5	108.7	95.1	97.0
0320	Concrete Reinforcing	100.0	109.7	104.9	101.7	109.7	105.8	99.2	109.8	104.5	105.2	109.8	107.5	113.1	109.8	111.4	96.7	109.6	103.2
0330	Cast-in-Place Concrete	105.4	101.2	103.8	105.4	101.2	103.8	106.2	100.2	103.9	104.9	101.4	103.6	96.4	101.3	98.3	83.6	102.2	90.6
03	CONCRETE	106.4	99.7	103.4	101.4	99.7	100.6	88.6	99.5	93.6	97.1	99.9	98.4	93.3	99.9	96.3	74.1	99.6	85.8
04	MASONRY	118.1	100.8	107.4	98.0	100.8	99.7	108.5	100.9	103.8	102.7	100.8	101.5	107.2	100.8	103.2	106.6	100.9	103.1
05	METALS	96.3	94.3	95.7	96.6	94.2	95.8	103.0	95.0	100.5	98.1	94.6	97.0	103.7	94.5	100.9	102.9	93.9	100.1
06	WOOD, PLASTICS & COMPOSITES	92.9	95.1	94.1	91.8	95.1	93.6	94.6	95.2	94.9	98.7	95.1	96.7	92.5	95.1	93.9	102.9	95.2	98.6
07	THERMAL & MOISTURE PROTECTION	115.6	95.2	107.0	115.2	95.2	106.8	108.0	95.8	102.9	114.4	99.4	108.1	110.5	99.4	105.8	107.4	93.4	101.5
08	OPENINGS	101.2	98.9	100.7	104.3	98.9	103.0	97.7	99.0	98.0	99.1	98.9	99.1	102.7	98.9	101.8	97.7	89.1	95.7
0920	Plaster & Gypsum Board	110.0	94.8	99.8	109.4	94.8	99.6	96.7	94.8	95.4	113.1	94.8	100.8	110.9	94.8	100.1	102.9	94.8	97.5
0950, 0980	Ceilings & Acoustic Treatment	98.4	94.8	96.0	105.1	94.8	98.2	65.0	94.8	85.1	93.6	94.8	94.4	99.6	94.8	96.4	65.0	94.8	85.1
0960	Flooring	95.2	105.4	98.1	94.6	105.4	97.6	66.1	105.4	77.0	94.5	105.4	97.5	98.0	105.4	100.0	68.2	105.4	78.5
0970, 0990	Wall Finishes & Painting/Coating	90.9	68.2	77.7	90.9	68.2	77.7	83.5	75.0	78.6	90.7	75.0	81.6	90.6	75.0	81.5	83.5	72.8	77.2
09	FINISHES	96.4	94.6	95.4	99.4	94.6	95.5	68.1	95.4	83.0	94.2	95.3	94.8	96.1	95.3	95.6	68.7	95.1	83.1
COVERS	DIVS. 10 - 14, 25, 28, 41, 43, 44, 46	100.0	99.6	99.9	100.0	99.6	99.9	100.0	99.9	100.0	100.0	99.7	99.9	100.0	99.7	99.9	100.0	99.9	100.0
21, 22, 23	FIRE SUPPRESSION, PLUMBING & HVAC	97.0	106.1	100.9	100.1	102.2	101.0	98.9	108.7	103.1	100.1	106.2	102.7	100.0	102.2	101.0	98.9	73.5	88.1
26, 27, 3370	ELECTRICAL, COMMUNICATIONS & UTIL.	99.7	77.5	88.2	103.1	77.5	89.8	92.2	96.3	94.3	99.8	103.5	101.7	107.2	94.5	100.6	92.2	67.7	79.4
MF2016	WEIGHTED AVERAGE	101.0	96.9	99.2	100.6	96.1	98.6	96.0	99.7	97.6	99.4	100.8	100.0	100.9	98.7	100.0	93.8	87.7	91.1

PENNSYLVANIA

DIVISION		ALLENTOWN 181			ALTOONA 166			BEDFORD 155			BRADFORD 167			BUTLER 160			CHAMBERSBURG 172		
		MAT.	INST.	TOTAL	MAT.	INST.	TOTAL	MAT.	INST.	TOTAL	MAT.	INST.	TOTAL	MAT.	INST.	TOTAL	MAT.	INST.	TOTAL
015433	CONTRACTOR EQUIPMENT		115.8	115.8		115.8	115.8		114.1	114.1		115.8	115.8		115.8	115.8		115.0	115.0
0241, 31 - 34	SITE & INFRASTRUCTURE, DEMOLITION	93.5	102.2	99.6	96.3	102.1	100.3	105.9	99.9	101.7	92.5	101.0	98.4	87.6	102.8	98.3	91.3	99.4	97.0
0310	Concrete Forming & Accessories	99.1	109.4	108.0	84.2	83.2	83.3	82.6	79.9	80.3	86.3	101.0	99.0	85.6	88.9	88.4	86.9	80.0	81.0
0320	Concrete Reinforcing	101.4	117.0	109.3	98.3	104.2	101.2	96.4	102.0	99.2	100.4	104.0	102.2	99.0	120.4	109.8	99.0	111.9	105.5
0330	Cast-in-Place Concrete	88.5	105.2	94.8	98.6	91.3	95.8	111.9	86.3	102.2	94.1	95.3	94.6	87.1	99.6	91.8	91.3	97.7	93.7
03	CONCRETE	90.8	110.1	99.6	86.8	91.2	88.8	103.3	87.6	96.1	92.7	100.6	96.3	78.9	99.5	88.3	94.5	93.3	93.9
04	MASONRY	94.3	97.1	96.0	97.8	87.0	91.1	105.7	83.7	92.0	95.2	82.1	87.1	100.0	96.7	97.9	100.0	82.3	89.0
05	METALS	98.0	123.2	105.7	92.0	115.6	99.3	97.2	114.0	102.3	95.8	115.4	101.8	91.7	123.4	101.5	95.5	117.8	102.4
06	WOOD, PLASTICS & COMPOSITES	101.3	111.6	107.0	80.5	81.3	80.9	84.0	79.1	81.3	86.9	106.1	97.6	81.9	85.6	84.0	88.2	79.2	83.2
07	THERMAL & MOISTURE PROTECTION	101.7	112.1	106.1	100.7	92.8	97.4	99.6	89.0	95.1	101.5	91.5	97.3	100.2	96.3	98.5	95.7	86.1	91.6
08	OPENINGS	94.2	109.9	97.8	87.9	85.6	87.4	96.3	84.5	93.6	94.1	99.4	95.3	87.9	96.1	89.8	91.4	84.7	89.9
0920	Plaster & Gypsum Board	96.8	111.8	106.9	88.5	80.7	83.2	100.5	78.5	85.8	89.6	106.2	100.7	88.5	85.2	86.3	104.3	78.5	87.0
0950, 0980	Ceilings & Acoustic Treatment	86.4	111.8	103.5	89.0	80.7	83.4	98.8	78.5	85.2	89.1	106.2	100.6	89.9	85.2	86.7	88.5	78.5	81.8
0960	Flooring	88.9	100.5	92.1	82.8	104.6	88.9	94.7	82.4	91.3	83.2	104.6	89.1	83.7	111.3	91.4	87.3	82.4	85.9
0970, 0990	Wall Finishes & Painting/Coating	91.2	106.5	100.1	86.8	111.0	100.9	96.7	111.0	105.1	91.2	105.3	99.4	86.8	111.0	100.9	88.6	105.3	98.3
09	FINISHES	88.8	108.0	99.2	86.5	89.2	88.0	99.4	83.3	90.6	86.6	102.9	95.5	86.5	94.4	90.8	86.5	82.7	84.5
COVERS	DIVS. 10 - 14, 25, 28, 41, 43, 44, 46	100.0	103.2	100.7	100.0	97.9	99.5	100.0	96.4	99.2	100.0	100.1	100.0	100.0	100.0	100.0	100.0	95.6	99.0
21, 22, 23	FIRE SUPPRESSION, PLUMBING & HVAC	100.2	112.0	105.3	99.7	85.6	93.7	96.9	82.3	90.6	97.2	94.4	96.0	96.7	97.1	96.8	97.2	88.1	93.3
26, 27, 3370	ELECTRICAL, COMMUNICATIONS & UTIL.	98.7	98.0	98.3	89.5	110.6	100.5	95.6	110.6	103.4	92.8	110.6	102.1	90.1	109.1	100.0	89.2	86.3	87.7
MF2016	WEIGHTED AVERAGE	96.5	107.7	101.4	93.2	95.1	94.0	98.7	92.2	95.8	94.8	100.2	97.2	91.4	101.5	95.8	94.4	90.8	92.8

PENNSYLVANIA

DIVISION		DOYLESTOWN 189			DUBOIS 158			ERIE 164 - 165			GREENSBURG 156			HARRISBURG 170 - 171			HAZLETON 182		
		MAT.	INST.	TOTAL	MAT.	INST.	TOTAL	MAT.	INST.	TOTAL	MAT.	INST.	TOTAL	MAT.	INST.	TOTAL	MAT.	INST.	TOTAL
015433	CONTRACTOR EQUIPMENT		93.5	93.5		114.1	114.1		115.8	115.8		114.1	114.1		115.0	115.0		115.8	115.8
0241, 31 - 34	SITE & INFRASTRUCTURE, DEMOLITION	107.5	89.2	94.7	110.8	100.0	103.2	93.2	102.4	99.6	101.8	101.9	101.9	91.4	100.3	97.6	86.9	101.5	97.1
0310	Concrete Forming & Accessories	83.3	126.4	120.5	82.1	80.7	80.9	98.4	87.0	88.5	88.8	92.8	92.2	98.6	87.5	89.0	80.9	86.7	85.9
0320	Concrete Reinforcing	98.1	124.0	111.2	95.7	104.1	99.9	100.4	105.2	102.8	95.7	120.1	108.0	108.4	114.5	111.5	98.4	113.0	105.8
0330	Cast-in-Place Concrete	83.6	133.8	102.6	107.9	94.7	102.9	96.9	93.3	95.5	103.8	99.2	102.0	89.8	100.9	94.0	83.6	99.1	89.5
03	CONCRETE	86.5	127.8	105.4	105.1	91.1	98.7	85.8	93.8	89.4	98.2	101.0	99.5	89.9	98.3	93.7	83.7	97.0	89.8
04	MASONRY	97.7	126.7	115.7	106.2	84.1	92.5	86.9	90.6	89.2	116.1	92.9	101.7	94.2	87.4	90.0	107.4	91.5	97.6
05	METALS	95.5	114.2	101.3	97.2	114.2	102.4	92.2	116.2	99.6	97.1	121.7	104.6	102.0	121.2	107.9	97.7	119.9	104.5
06	WOOD, PLASTICS & COMPOSITES	82.6	126.6	107.1	83.1	79.1	80.9	84.0	84.5	90.2	90.1	91.3	90.7	99.8	86.4	92.3	81.3	83.8	82.7
07	THERMAL & MOISTURE PROTECTION	99.2	127.8	111.3	99.9	91.2	96.2	101.2	92.1	97.4	99.5	95.8	97.9	98.6	103.9	100.8	101.0	88.0	100.9
08	OPENINGS	96.5	129.8	104.2	96.3	84.5	93.6	88.1	89.0	88.3	96.3	99.2	97.0	97.8	89.3	95.8	94.7	93.1	94.4
0920	Plaster & Gypsum Board	87.5	127.3	114.2	99.5	78.5	85.4	96.8	83.9	88.2	101.6	91.0	94.5	108.2	86.0	93.3	87.9	83.3	84.8
0950, 0980	Ceilings & Acoustic Treatment	85.6	127.3	113.7	98.8	78.5	85.2	86.4	83.9	84.7	98.0	91.0	93.3	97.0	86.0	89.6	87.4	83.3	84.6
0960	Flooring	73.8	147.3	94.2	94.5	104.6	97.3	89.6	104.6	93.7	98.8	82.4	94.2	89.8	95.6	91.4	80.4	94.4	84.3
0970, 0990	Wall Finishes & Painting/Coating	90.8	145.9	122.9	96.7	111.0	105.1	96.3	91.8	93.7	96.7	111.0	105.1	90.6	88.6	89.4	91.2	108.8	101.5
09	FINISHES	80.4	132.4	108.7	99.6	87.5	93.0	89.8	90.0	89.9	100.3	92.5	96.0	92.0	88.3	90.0	84.8	90.0	87.6
COVERS	DIVS. 10 - 14, 25, 28, 41, 43, 44, 46	100.0	113.0	102.8	100.0	96.9	99.3	100.0	100.4	100.1	100.0	100.4	100.1	100.0	97.7	99.5	100.0	96.8	99.3
21, 22, 23	FIRE SUPPRESSION, PLUMBING & HVAC	96.7	127.8	110.0	96.9	83.0	91.0	99.7	100.5	100.1	96.9	90.9	94.3	100.2	92.0	96.7	97.2	95.7	96.5
26, 27, 3370	ELECTRICAL, COMMUNICATIONS & UTIL.	92.3	127.0	110.3	96.2	110.6	103.7	91.1	92.4	91.8	96.2	110.6	103.7	95.9	88.7	92.2	93.6	91.7	92.6
MF2016	WEIGHTED AVERAGE	94.0	123.5	106.9	99.1	93.5	96.7	93.1	96.8	94.7	98.6	99.9	99.2	97.3	95.2	96.3	94.4	96.8	95.4

PENNSYLVANIA

DIVISION		INDIANA 157			JOHNSTOWN 159			KITTANNING 162			LANCASTER 175-176			LEHIGH VALLEY 180			MONTROSE 188		
		MAT.	INST.	TOTAL	MAT.	INST.	TOTAL	MAT.	INST.	TOTAL	MAT.	INST.	TOTAL	MAT.	INST.	TOTAL	MAT.	INST.	TOTAL
015433	CONTRACTOR EQUIPMENT		114.1	114.1		114.1	114.1		115.8	115.8		115.0	115.0		115.8	115.8		115.8	115.8
0241, 31 - 34	SITE & INFRASTRUCTURE, DEMOLITION	99.9	100.9	100.6	106.4	101.1	102.7	90.2	102.4	98.8	83.1	100.4	95.2	90.8	103.0	99.4	89.5	102.2	98.4
0310	Concrete Forming & Accessories	83.2	89.9	89.0	82.1	83.1	83.0	85.6	93.9	92.7	89.0	87.7	87.9	92.5	127.5	122.7	81.8	90.6	89.4
0320	Concrete Reinforcing	95.0	120.1	107.7	96.4	119.9	108.2	99.0	120.2	109.7	98.6	114.6	106.7	98.4	111.3	104.9	103.1	119.9	111.6
0330	Cast-in-Place Concrete	101.8	97.7	100.2	112.9	91.2	104.6	90.5	98.0	93.3	77.6	100.9	86.5	90.5	109.4	97.7	88.7	97.8	92.2
03	CONCRETE	95.7	99.1	97.3	104.3	93.7	99.5	81.4	101.1	90.4	83.1	98.4	90.1	89.9	118.8	103.1	88.6	99.6	93.6
04	MASONRY	102.2	92.4	96.1	103.1	87.3	93.3	102.6	92.3	96.2	106.1	88.9	95.4	94.3	109.9	104.0	94.3	95.5	95.0
05	METALS	97.2	121.2	104.6	97.2	119.8	104.1	91.8	122.4	101.2	95.5	121.4	103.4	97.6	120.9	104.8	95.9	123.0	104.2
06	WOOD, PLASTICS & COMPOSITES	84.7	88.8	87.0	83.1	81.2	82.0	81.9	93.6	88.4	90.9	86.4	88.4	93.2	130.3	113.8	82.0	87.3	85.0
07	THERMAL & MOISTURE PROTECTION	99.4	95.4	97.7	99.6	91.4	96.2	100.3	96.0	98.5	95.2	104.4	99.1	101.4	113.8	106.7	101.1	90.6	96.6
08	OPENINGS	96.3	93.5	95.7	96.3	89.3	94.7	87.9	100.5	90.8	91.4	89.3	90.9	94.7	118.9	100.3	91.4	94.4	92.1
0920	Plaster & Gypsum Board	100.8	88.4	92.5	99.3	80.7	86.8	88.5	93.4	91.8	106.2	86.0	92.6	90.5	131.0	117.7	88.3	86.9	87.4
0950, 0980	Ceilings & Acoustic Treatment	98.8	88.4	91.8	98.0	80.7	86.3	89.9	93.4	92.2	88.5	86.0	86.8	87.4	131.0	116.8	89.1	86.9	87.6
0960	Flooring	95.5	104.6	98.0	94.5	82.4	91.1	83.7	104.6	89.5	88.3	101.1	91.8	85.8	99.5	89.6	81.1	104.6	87.6
0970, 0990	Wall Finishes & Painting/Coating	96.7	111.0	105.1	96.7	111.0	105.1	86.8	111.0	100.9	88.6	88.6	88.6	91.2	62.5	74.4	91.2	108.8	101.5
09	FINISHES	99.2	94.3	96.5	99.0	85.5	91.6	86.6	97.2	92.4	86.5	89.4	88.1	87.0	116.8	103.2	85.6	94.2	90.3
COVERS	DIVS. 10 - 14, 25, 28, 41, 43, 44, 46	100.0	99.9	99.9	100.0	97.9	99.5	100.0	100.2	100.0	100.0	97.9	99.5	100.0	102.9	100.7	100.0	98.4	99.7
21, 22, 23	FIRE SUPPRESSION, PLUMBING & HVAC	96.9	83.3	91.1	96.9	82.1	90.6	96.7	93.3	95.2	97.2	92.3	95.1	97.2	113.0	103.9	97.2	98.6	97.8
26, 27, 3370	ELECTRICAL, COMMUNICATIONS & UTIL.	96.2	110.6	103.7	96.2	110.6	103.7	89.5	110.6	100.5	90.4	94.7	92.6	93.6	136.6	116.0	92.8	96.2	94.6
MF2016	WEIGHTED AVERAGE	97.4	97.8	97.6	98.7	94.7	96.9	91.9	101.2	95.9	93.2	96.4	94.6	94.9	117.4	104.7	93.8	99.5	96.3

PENNSYLVANIA

DIVISION		NEW CASTLE 161			NORRISTOWN 194			OIL CITY 163			PHILADELPHIA 190-191			PITTSBURGH 150-152			POTTSVILLE 179		
		MAT.	INST.	TOTAL	MAT.	INST.	TOTAL	MAT.	INST.	TOTAL	MAT.	INST.	TOTAL	MAT.	INST.	TOTAL	MAT.	INST.	TOTAL
015433	CONTRACTOR EQUIPMENT		115.8	115.8		100.1	100.1		115.8	115.8		100.3	100.3		102.6	102.6		115.0	115.0
0241, 31 - 34	SITE & INFRASTRUCTURE, DEMOLITION	88.0	102.7	98.3	97.0	100.2	99.2	86.6	101.2	96.8	98.8	100.6	100.1	106.0	100.4	102.1	86.1	100.6	96.3
0310	Concrete Forming & Accessories	85.6	94.6	93.4	84.1	126.7	120.8	85.6	93.4	92.3	101.5	141.9	136.4	99.0	98.9	99.0	80.6	88.9	87.8
0320	Concrete Reinforcing	97.8	101.9	99.8	99.5	126.6	113.2	99.0	97.7	98.3	103.3	126.7	115.1	96.9	120.6	108.8	97.8	114.5	106.2
0330	Cast-in-Place Concrete	87.9	99.2	92.2	83.4	134.1	102.6	85.3	97.6	90.0	91.2	137.7	108.9	111.3	99.9	107.0	82.7	101.3	89.8
03	CONCRETE	79.3	98.7	88.2	85.9	128.5	105.4	77.8	96.9	86.5	98.9	136.6	116.2	104.8	103.0	104.0	86.5	99.1	92.2
04	MASONRY	99.6	96.6	97.7	110.2	127.3	120.8	98.9	88.4	92.4	95.7	130.8	117.5	99.0	101.0	100.2	99.4	91.3	94.4
05	METALS	91.8	116.3	99.3	99.2	114.8	104.0	91.8	114.7	98.8	101.9	115.1	106.0	98.6	107.9	101.5	95.7	121.4	103.6
06	WOOD, PLASTICS & COMPOSITES	81.9	94.0	88.6	82.0	126.6	106.8	81.9	93.6	88.4	96.3	144.5	123.1	102.6	99.0	100.6	81.0	85.6	83.6
07	THERMAL & MOISTURE PROTECTION	100.2	95.7	98.3	104.6	128.2	114.6	100.1	93.2	97.2	101.5	136.9	116.5	99.8	99.1	99.5	95.3	103.1	98.6
08	OPENINGS	87.9	93.0	89.1	86.9	129.6	96.7	87.9	95.8	89.7	97.1	140.3	107.0	98.7	103.5	99.8	91.5	95.0	92.3
0920	Plaster & Gypsum Board	88.5	93.7	92.0	86.5	127.3	113.9	88.5	93.4	91.8	98.0	145.5	129.9	98.0	98.8	98.5	101.0	85.2	90.4
0950, 0980	Ceilings & Acoustic Treatment	89.9	93.7	92.5	91.0	127.3	115.4	89.9	93.4	92.2	101.3	145.5	131.1	92.6	98.8	96.8	88.5	85.2	86.3
0960	Flooring	83.7	111.3	91.4	81.8	147.3	99.9	83.7	106.1	89.9	96.7	147.3	110.8	104.0	107.7	105.0	84.3	101.1	89.0
0970, 0990	Wall Finishes & Painting/Coating	86.8	111.0	100.9	87.9	145.9	121.7	86.8	111.0	100.9	94.6	150.2	127.0	99.8	111.0	106.3	88.6	108.8	100.4
09	FINISHES	86.5	99.1	93.3	84.4	132.5	110.6	86.4	97.4	92.4	100.4	144.5	124.4	101.1	101.5	101.3	84.8	92.3	88.9
COVERS	DIVS. 10 - 14, 25, 28, 41, 43, 44, 46	100.0	100.8	100.2	100.0	113.2	102.9	100.0	100.1	100.0	100.0	120.0	104.4	100.0	101.8	100.4	100.0	95.2	98.9
21, 22, 23	FIRE SUPPRESSION, PLUMBING & HVAC	96.7	100.9	98.5	96.9	126.8	109.7	96.7	96.4	96.6	100.1	136.7	115.7	99.9	98.9	99.5	97.2	97.6	97.4
26, 27, 3370	ELECTRICAL, COMMUNICATIONS & UTIL.	90.1	100.7	95.6	93.7	150.0	123.0	91.7	109.1	100.8	100.1	155.2	128.7	98.5	111.6	105.3	88.8	91.0	89.9
MF2016	WEIGHTED AVERAGE	91.5	100.9	95.6	94.5	127.6	109.0	91.3	99.6	95.0	99.7	134.6	115.0	100.3	103.0	101.5	93.0	97.9	95.1

PENNSYLVANIA

DIVISION		READING 195-196			SCRANTON 184-185			STATE COLLEGE 168			STROUDSBURG 183			SUNBURY 178			UNIONTOWN 154		
		MAT.	INST.	TOTAL	MAT.	INST.	TOTAL	MAT.	INST.	TOTAL	MAT.	INST.	TOTAL	MAT.	INST.	TOTAL	MAT.	INST.	TOTAL
015433	CONTRACTOR EQUIPMENT		122.9	122.9		115.8	115.8		115.0	115.0		115.8	115.8		115.8	115.8		114.1	114.1
0241, 31 - 34	SITE & INFRASTRUCTURE, DEMOLITION	101.2	113.7	110.0	94.0	102.2	99.8	84.4	100.6	95.8	88.5	103.0	98.7	98.7	101.6	100.7	100.4	101.7	101.3
0310	Concrete Forming & Accessories	100.1	90.2	91.6	99.2	90.8	92.0	84.8	83.4	83.6	87.2	95.5	94.3	92.2	86.1	86.9	76.5	92.6	90.4
0320	Concrete Reinforcing	100.8	143.2	122.2	101.4	120.1	110.8	99.6	104.2	102.0	101.7	120.2	111.0	100.6	114.4	107.6	95.7	120.1	108.0
0330	Cast-in-Place Concrete	74.7	102.2	85.1	92.4	98.1	94.5	89.1	91.4	90.0	87.0	104.8	93.8	90.4	99.7	94.0	101.8	98.9	100.7
03	CONCRETE	84.8	105.0	94.0	92.6	99.8	95.9	92.8	91.3	92.1	87.4	104.2	95.1	89.9	97.2	93.3	95.3	100.7	97.8
04	MASONRY	99.2	95.4	96.8	94.6	97.9	96.6	100.3	87.5	92.3	92.4	109.7	103.1	99.9	84.4	90.3	117.8	92.8	102.3
05	METALS	99.5	132.9	109.7	99.0	123.3	107.1	95.6	115.8	101.8	97.7	123.6	105.7	95.4	121.1	103.3	96.9	121.2	104.4
06	WOOD, PLASTICS & COMPOSITES	99.4	86.4	92.0	101.3	87.3	93.5	88.7	81.3	84.5	87.7	87.3	87.5	88.9	85.6	87.1	77.6	91.2	85.2
07	THERMAL & MOISTURE PROTECTION	104.9	107.9	106.2	101.6	95.1	98.8	100.8	101.5	101.1	101.2	99.0	100.3	96.2	96.3	96.2	99.3	95.8	97.6
08	OPENINGS	91.2	102.2	93.8	94.2	94.4	94.2	91.2	85.6	89.9	94.7	97.3	95.3	91.6	92.2	91.7	96.3	98.7	96.8
0920	Plaster & Gypsum Board	96.3	86.0	89.4	98.5	86.9	90.7	90.4	80.7	83.9	89.2	86.9	87.6	100.2	85.2	90.1	97.7	91.0	93.2
0950, 0980	Ceilings & Acoustic Treatment	84.0	86.0	85.3	94.1	86.9	89.2	86.5	80.7	82.6	85.6	86.9	86.5	85.1	85.2	85.1	98.0	91.0	93.3
0960	Flooring	86.5	103.5	91.2	88.9	98.2	91.5	86.3	103.5	91.1	83.8	99.5	88.1	85.1	104.6	90.5	91.3	82.4	88.8
0970, 0990	Wall Finishes & Painting/Coating	86.7	106.5	98.2	91.2	113.2	104.0	91.2	111.0	102.8	91.2	108.8	101.5	88.6	108.8	100.4	96.7	108.9	103.8
09	FINISHES	86.1	92.9	89.8	90.5	93.8	92.3	86.2	89.1	87.8	85.7	96.5	91.6	85.7	91.2	88.7	97.3	92.2	94.6
COVERS	DIVS. 10 - 14, 25, 28, 41, 43, 44, 46	100.0	99.8	99.9	100.0	98.6	99.7	100.0	97.1	99.3	100.0	102.6	100.6	100.0	95.2	98.9	100.0	100.4	100.1
21, 22, 23	FIRE SUPPRESSION, PLUMBING & HVAC	100.2	111.1	104.8	100.2	98.8	99.6	97.2	91.6	94.8	97.2	104.8	100.4	97.2	89.6	93.9	96.9	85.6	92.1
26, 27, 3370	ELECTRICAL, COMMUNICATIONS & UTIL.	99.8	94.7	97.1	98.7	96.2	97.4	92.0	110.6	101.7	93.6	142.9	119.2	89.1	93.2	91.2	93.3	110.6	102.3
MF2016	WEIGHTED AVERAGE	96.1	105.2	100.1	97.2	100.0	98.4	94.4	96.5	95.3	94.3	110.4	101.3	93.9	95.1	94.4	97.6	98.6	98.1

For customer support on your Commercial Renovation Costs with RSMeans data, call 800.448.8182.

807

PENNSYLVANIA

DIVISION		WASHINGTON 153			WELLSBORO 169			WESTCHESTER 193			WILKES-BARRE 186 - 187			WILLIAMSPORT 177			YORK 173 - 174		
		MAT.	INST.	TOTAL	MAT.	INST.	TOTAL	MAT.	INST.	TOTAL	MAT.	INST.	TOTAL	MAT.	INST.	TOTAL	MAT.	INST.	TOTAL
015433	CONTRACTOR EQUIPMENT		114.1	114.1		115.8	115.8		100.1	100.1		115.8	115.8		115.8	115.8		115.0	115.0
0241, 31 - 34	SITE & INFRASTRUCTURE, DEMOLITION	100.5	102.0	101.6	96.0	101.6	99.9	102.9	101.1	101.6	86.5	102.2	97.5	89.3	101.7	98.0	86.9	100.4	96.3
0310	Concrete Forming & Accessories	83.3	94.6	93.1	85.8	85.9	85.9	90.8	127.3	122.3	89.8	89.4	89.4	88.8	86.8	87.1	84.0	87.7	87.1
0320	Concrete Reinforcing	95.7	120.4	108.2	99.6	119.9	109.8	98.5	143.9	121.4	100.4	120.1	110.3	99.8	114.5	107.2	100.6	114.6	107.6
0330	Cast-in-Place Concrete	101.8	99.3	100.8	93.3	92.7	93.1	92.3	135.2	108.6	83.6	97.9	89.0	76.5	93.6	83.0	83.1	100.9	89.9
03	CONCRETE	95.8	101.9	98.6	95.0	95.7	95.3	93.4	132.2	111.1	84.5	99.0	91.1	78.5	95.5	86.2	87.6	98.4	92.5
04	MASONRY	102.2	97.4	99.2	101.4	84.9	91.2	104.9	129.4	120.1	107.7	97.5	101.4	91.4	88.0	89.3	101.0	88.9	93.5
05	METALS	96.9	122.2	104.7	95.7	122.2	103.8	99.2	121.5	106.1	95.8	123.2	104.2	95.5	120.8	103.2	97.1	121.3	104.6
06	WOOD, PLASTICS & COMPOSITES	84.8	93.6	89.7	86.3	85.6	85.9	88.6	126.6	109.7	90.0	85.6	87.6	85.5	86.4	86.0	84.3	86.4	85.5
07	THERMAL & MOISTURE PROTECTION	99.4	97.7	98.6	101.8	88.8	96.3	105.0	124.7	113.3	101.0	94.7	98.4	95.6	90.8	93.6	95.3	104.4	99.2
08	OPENINGS	96.2	100.5	97.2	94.1	93.1	93.9	86.9	134.5	97.8	91.4	93.5	91.9	91.6	92.7	91.8	91.4	89.3	90.9
0920	Plaster & Gypsum Board	100.6	93.4	95.7	89.0	85.2	86.4	87.1	127.3	114.1	89.9	85.2	86.7	101.0	86.0	90.9	101.3	86.0	91.0
0950, 0980	Ceilings & Acoustic Treatment	98.0	93.4	94.9	86.5	85.2	85.6	91.0	127.3	115.4	89.1	85.2	86.4	88.5	86.0	86.8	87.6	86.0	86.5
0960	Flooring	95.6	82.4	91.9	82.8	101.1	87.9	84.6	147.3	102.0	84.6	98.2	88.4	84.0	91.7	86.2	85.6	101.1	89.9
0970, 0990	Wall Finishes & Painting/Coating	96.7	108.9	103.8	91.2	108.8	101.5	87.9	142.1	119.5	91.2	106.5	100.1	88.6	106.5	99.0	88.6	88.6	88.6
09	FINISHES	99.1	93.6	96.1	86.2	91.0	88.8	85.9	132.7	111.3	86.7	92.0	89.5	85.4	88.9	87.3	85.1	89.4	87.5
COVERS	DIVS. 10 - 14, 25, 28, 41, 43, 44, 46	100.0	100.7	100.2	100.0	92.9	98.4	100.0	111.2	102.5	100.0	98.2	99.6	100.0	97.5	99.5	100.0	97.8	99.5
21, 22, 23	FIRE SUPPRESSION, PLUMBING & HVAC	96.9	97.0	97.0	97.2	89.0	93.7	96.9	128.1	110.2	97.2	98.6	97.8	97.2	91.5	94.7	100.2	92.3	96.8
26, 27, 3370	ELECTRICAL, COMMUNICATIONS & UTIL.	95.6	110.6	103.4	92.8	93.6	93.2	93.6	109.4	101.9	93.6	91.7	92.6	89.6	75.9	82.4	90.4	81.2	85.6
MF2016	WEIGHTED AVERAGE	97.3	102.2	99.4	95.5	94.7	95.1	95.5	123.7	107.8	94.1	98.9	96.2	91.9	92.8	92.3	94.4	94.5	94.4

DIVISION		PUERTO RICO SAN JUAN 009			RHODE ISLAND NEWPORT 028			RHODE ISLAND PROVIDENCE 029			SOUTH CAROLINA AIKEN 298			SOUTH CAROLINA BEAUFORT 299			SOUTH CAROLINA CHARLESTON 294		
		MAT.	INST.	TOTAL	MAT.	INST.	TOTAL	MAT.	INST.	TOTAL	MAT.	INST.	TOTAL	MAT.	INST.	TOTAL	MAT.	INST.	TOTAL
015433	CONTRACTOR EQUIPMENT		87.1	87.1		101.1	101.1		101.1	101.1		106.8	106.8		106.8	106.8		106.8	106.8
0241, 31 - 34	SITE & INFRASTRUCTURE, DEMOLITION	131.9	89.9	102.5	88.7	102.8	98.6	90.6	102.8	99.1	128.2	89.6	101.2	123.3	89.0	99.3	108.2	89.3	95.0
0310	Concrete Forming & Accessories	91.0	21.7	31.3	98.6	119.9	117.0	96.9	119.9	117.1	95.7	65.0	69.3	94.7	39.8	47.4	93.9	66.0	69.0
0320	Concrete Reinforcing	189.9	18.4	103.4	102.0	114.6	108.4	104.8	114.6	109.7	95.8	62.6	79.0	94.8	26.6	60.4	94.6	66.1	80.2
0330	Cast-in-Place Concrete	95.0	32.8	71.4	72.7	118.7	90.2	93.2	118.7	102.9	90.9	67.6	82.1	90.9	67.2	81.9	106.7	67.6	91.8
03	CONCRETE	96.0	26.0	64.0	84.3	118.1	99.8	90.6	118.1	103.2	104.7	67.2	87.6	102.0	49.6	78.0	97.6	67.8	84.0
04	MASONRY	88.8	20.8	46.6	97.3	123.6	113.6	99.3	123.6	114.4	78.0	66.7	71.0	92.2	66.7	76.4	93.6	66.7	76.9
05	METALS	116.6	38.7	92.7	97.0	111.8	101.5	101.2	111.8	104.4	97.0	91.0	95.1	97.0	81.0	92.1	98.9	92.1	96.8
06	WOOD, PLASTICS & COMPOSITES	102.4	21.0	57.1	101.3	119.0	111.2	101.1	119.0	111.1	95.0	66.8	79.3	93.7	33.9	60.4	92.6	66.8	78.2
07	THERMAL & MOISTURE PROTECTION	130.6	25.9	86.3	99.6	117.7	107.2	103.1	117.7	109.3	99.1	66.3	85.2	98.8	58.1	81.5	97.9	65.9	84.3
08	OPENINGS	154.0	19.4	123.1	98.8	118.8	103.4	98.7	118.8	103.3	95.3	63.9	88.1	95.4	39.6	82.6	99.2	64.8	91.3
0920	Plaster & Gypsum Board	168.8	18.9	68.2	91.4	119.2	110.0	103.3	119.2	113.9	99.3	65.8	76.8	102.6	32.0	55.2	104.1	65.8	78.4
0950, 0980	Ceilings & Acoustic Treatment	250.5	18.9	94.4	96.7	119.2	111.8	101.6	119.2	113.4	86.4	65.8	72.5	89.0	32.0	50.5	89.0	65.8	73.3
0960	Flooring	203.1	21.3	152.6	89.9	127.6	100.3	87.3	127.6	98.5	95.9	94.9	95.6	97.3	80.1	92.6	97.0	81.6	92.7
0970, 0990	Wall Finishes & Painting/Coating	188.0	22.6	91.5	90.5	116.3	105.6	87.0	116.3	104.1	94.9	69.0	79.8	94.9	59.5	74.2	94.9	69.0	79.8
09	FINISHES	196.9	21.8	101.7	90.1	121.5	107.2	91.0	121.5	107.6	91.7	71.1	80.5	92.6	48.1	68.4	90.8	68.9	78.8
COVERS	DIVS. 10 - 14, 25, 28, 41, 43, 44, 46	100.0	19.8	82.1	100.0	107.5	101.7	100.0	107.5	101.7	100.0	72.1	93.8	100.0	76.9	94.9	100.0	72.0	93.8
21, 22, 23	FIRE SUPPRESSION, PLUMBING & HVAC	103.9	18.2	67.2	100.3	108.7	103.9	100.1	108.7	103.8	97.4	57.1	80.2	97.4	57.0	80.1	100.5	59.4	82.9
26, 27, 3370	ELECTRICAL, COMMUNICATIONS & UTIL.	124.8	17.5	69.0	100.9	96.6	98.6	101.1	96.6	98.7	100.9	65.3	82.4	104.6	32.6	67.2	103.0	60.1	80.7
MF2016	WEIGHTED AVERAGE	120.7	27.7	79.9	96.3	112.0	103.2	98.1	112.0	104.2	98.0	69.2	85.4	98.7	56.8	80.3	98.9	68.8	85.7

DIVISION		SOUTH CAROLINA COLUMBIA 290 - 292			SOUTH CAROLINA FLORENCE 295			SOUTH CAROLINA GREENVILLE 296			SOUTH CAROLINA ROCK HILL 297			SOUTH CAROLINA SPARTANBURG 293			SOUTH DAKOTA ABERDEEN 574		
		MAT.	INST.	TOTAL	MAT.	INST.	TOTAL	MAT.	INST.	TOTAL	MAT.	INST.	TOTAL	MAT.	INST.	TOTAL	MAT.	INST.	TOTAL
015433	CONTRACTOR EQUIPMENT		106.8	106.8		106.8	106.8		106.8	106.8		106.8	106.8		106.8	106.8		100.3	100.3
0241, 31 - 34	SITE & INFRASTRUCTURE, DEMOLITION	106.9	89.3	94.6	117.4	89.1	97.6	112.4	89.4	96.3	110.2	88.7	95.1	112.2	89.4	96.3	102.0	98.2	99.3
0310	Concrete Forming & Accessories	95.3	65.0	69.2	82.1	64.9	67.3	93.4	65.0	68.9	91.5	64.7	68.4	96.2	65.0	69.3	94.1	46.1	52.7
0320	Concrete Reinforcing	99.9	66.1	82.8	94.2	66.1	80.0	94.1	66.1	80.0	95.0	66.1	80.4	94.1	66.1	80.0	96.7	72.5	84.5
0330	Cast-in-Place Concrete	107.9	67.6	92.6	90.9	67.4	82.0	90.8	67.6	82.0	90.8	67.5	82.0	90.8	67.6	82.0	112.5	78.9	99.8
03	CONCRETE	95.8	67.8	83.0	96.5	67.6	83.3	95.1	67.8	82.6	93.0	67.6	81.4	95.3	67.8	82.7	99.3	63.6	83.0
04	MASONRY	87.3	66.7	74.5	78.0	66.7	71.0	75.6	66.7	70.1	99.6	66.7	79.2	78.0	66.7	71.0	113.6	73.4	88.7
05	METALS	96.1	92.1	94.9	97.8	91.5	96.0	97.8	92.1	96.0	97.0	91.8	95.4	97.8	92.1	96.0	100.9	83.0	95.4
06	WOOD, PLASTICS & COMPOSITES	93.1	66.8	78.5	79.9	66.8	72.7	92.2	66.8	78.1	90.9	66.8	77.5	96.1	66.8	79.8	92.4	34.7	60.3
07	THERMAL & MOISTURE PROTECTION	94.2	65.8	82.2	98.2	65.9	84.5	98.1	65.9	84.5	98.0	61.4	82.5	98.1	65.9	84.5	99.8	74.9	89.3
08	OPENINGS	95.4	64.8	88.4	95.4	64.8	88.4	95.3	64.8	88.3	95.4	64.8	88.3	95.4	64.8	88.3	95.7	41.4	83.2
0920	Plaster & Gypsum Board	94.8	65.8	75.3	93.6	65.8	74.9	98.4	65.8	76.5	98.0	65.8	76.4	101.0	65.8	77.4	112.0	33.3	59.1
0950, 0980	Ceilings & Acoustic Treatment	89.7	65.8	73.6	87.3	65.8	72.8	86.4	65.8	72.5	86.4	65.8	72.5	86.4	65.8	72.5	94.2	33.3	53.1
0960	Flooring	87.1	80.2	85.2	88.4	80.1	86.1	94.7	80.1	90.6	93.6	80.1	89.9	96.1	80.1	91.6	92.3	46.3	79.6
0970, 0990	Wall Finishes & Painting/Coating	89.0	69.0	77.3	94.9	69.0	79.8	94.9	69.0	79.8	94.9	69.0	79.8	94.9	69.0	79.8	90.4	34.9	58.0
09	FINISHES	86.8	68.6	76.9	87.6	68.6	77.3	89.6	68.6	78.2	88.9	68.6	77.9	90.3	68.6	78.5	91.4	43.0	65.1
COVERS	DIVS. 10 - 14, 25, 28, 41, 43, 44, 46	100.0	72.1	93.8	100.0	72.1	93.8	100.0	72.1	93.8	100.0	72.0	93.8	100.0	72.1	93.8	100.0	79.6	95.5
21, 22, 23	FIRE SUPPRESSION, PLUMBING & HVAC	100.1	58.7	82.4	100.5	58.7	82.7	100.5	58.7	82.7	97.4	57.1	80.2	100.5	58.7	82.7	100.2	54.8	80.8
26, 27, 3370	ELECTRICAL, COMMUNICATIONS & UTIL.	103.1	63.0	82.3	100.8	63.0	81.1	103.0	61.8	81.6	103.0	61.8	81.6	103.0	61.8	81.6	101.4	65.9	82.9
MF2016	WEIGHTED AVERAGE	96.9	69.1	84.7	97.0	69.0	84.8	97.1	68.9	84.8	97.1	68.3	84.5	97.3	68.9	84.9	99.7	64.5	84.3

SOUTH DAKOTA

	DIVISION	MITCHELL 573			MOBRIDGE 576			PIERRE 575			RAPID CITY 577			SIOUX FALLS 570 - 571			WATERTOWN 572		
		MAT.	INST.	TOTAL	MAT.	INST.	TOTAL	MAT.	INST.	TOTAL	MAT.	INST.	TOTAL	MAT.	INST.	TOTAL	MAT.	INST.	TOTAL
015433	CONTRACTOR EQUIPMENT		100.3	100.3		100.3	100.3		100.3	100.3		100.3	100.3		101.3	101.3		100.3	100.3
0241, 31 - 34	SITE & INFRASTRUCTURE, DEMOLITION	98.8	98.1	98.3	98.8	98.2	98.3	100.4	97.6	98.4	100.3	97.6	98.4	93.4	100.0	98.0	98.7	98.1	98.3
0310	Concrete Forming & Accessories	93.3	46.2	52.7	84.4	46.6	51.8	96.5	47.5	54.2	101.5	59.1	64.9	100.7	64.5	69.5	81.2	44.9	49.9
0320	Concrete Reinforcing	96.2	72.7	84.3	98.6	72.7	85.6	98.6	97.0	97.8	90.4	97.2	93.8	96.8	97.3	97.0	93.5	41.1	67.1
0330	Cast-in-Place Concrete	109.4	54.8	88.6	109.4	79.0	97.8	104.2	79.8	94.9	108.5	80.1	97.7	92.5	80.8	88.1	109.4	57.9	89.8
03	CONCRETE	97.0	55.3	77.9	96.7	63.9	81.7	93.4	68.6	82.0	96.3	74.0	86.1	87.8	76.8	82.8	95.9	50.4	75.1
04	MASONRY	101.4	77.9	86.8	110.1	75.9	88.9	112.1	75.7	89.6	110.2	77.2	89.7	92.5	77.9	83.4	137.5	76.9	99.9
05	METALS	99.9	83.0	94.7	100.0	83.5	94.9	103.2	90.0	99.1	102.7	90.8	99.1	102.9	92.9	99.8	99.9	90.1	91.4
06	WOOD, PLASTICS & COMPOSITES	91.3	35.2	60.1	81.0	34.9	55.4	99.4	35.5	63.8	96.0	50.2	70.5	98.2	57.2	75.4	77.5	34.7	53.7
07	THERMAL & MOISTURE PROTECTION	99.6	76.2	89.7	99.7	78.8	90.8	101.9	75.7	90.8	100.3	80.4	91.9	98.4	81.9	91.4	99.4	75.7	89.4
08	OPENINGS	94.8	41.6	82.6	97.3	41.0	84.4	100.5	59.3	91.0	99.4	67.5	92.1	102.0	71.3	95.0	94.8	34.0	80.8
0920	Plaster & Gypsum Board	110.4	33.8	58.9	105.1	33.5	57.0	102.2	34.0	56.4	111.4	49.2	69.6	104.1	56.4	72.1	102.8	33.3	56.1
0950, 0980	Ceilings & Acoustic Treatment	89.9	33.8	52.1	94.2	33.5	53.3	94.2	34.0	53.6	95.9	49.2	64.4	89.7	56.4	67.2	89.9	33.3	51.7
0960	Flooring	91.9	46.3	79.3	87.5	49.1	76.9	93.3	33.0	76.6	91.7	76.8	87.6	95.8	77.8	90.8	86.1	46.3	75.1
0970, 0990	Wall Finishes & Painting/Coating	90.4	38.2	60.4	90.4	39.2	60.5	95.1	74.0	80.3	90.4	98.0	94.8	97.4	98.0	99.7	90.4	34.9	58.0
09	FINISHES	90.0	43.1	64.5	88.8	43.6	64.2	94.6	47.1	68.8	91.4	64.7	76.9	95.0	69.0	80.9	87.2	42.5	62.9
COVERS	DIVS. 10 - 14, 25, 28, 41, 43, 44, 46	100.0	79.6	95.5	100.0	79.5	95.4	100.0	81.4	95.9	100.0	83.0	96.2	100.0	83.8	96.4	100.0	44.8	87.7
21, 22, 23	FIRE SUPPRESSION, PLUMBING & HVAC	97.1	51.0	77.4	97.1	70.8	85.8	100.1	79.9	91.5	100.2	80.0	91.5	100.1	71.1	87.7	97.1	53.8	78.6
26, 27, 3370	ELECTRICAL, COMMUNICATIONS & UTIL.	99.7	65.9	82.1	101.4	42.0	70.5	104.1	50.4	76.1	97.9	50.4	73.2	102.5	65.9	83.5	98.9	46.1	71.4
MF2016	WEIGHTED AVERAGE	97.5	63.1	82.4	98.2	65.0	83.7	100.4	70.5	87.3	99.5	74.4	88.5	98.5	76.4	88.8	98.7	57.6	80.7

TENNESSEE

	DIVISION	CHATTANOOGA 373 - 374			COLUMBIA 384			COOKEVILLE 385			JACKSON 383			JOHNSON CITY 376			KNOXVILLE 377 - 379		
		MAT.	INST.	TOTAL	MAT.	INST.	TOTAL	MAT.	INST.	TOTAL	MAT.	INST.	TOTAL	MAT.	INST.	TOTAL	MAT.	INST.	TOTAL
015433	CONTRACTOR EQUIPMENT		107.3	107.3		101.9	101.9		101.9	101.9		108.2	108.2		101.0	101.0		101.0	101.0
0241, 31 - 34	SITE & INFRASTRUCTURE, DEMOLITION	102.4	99.4	99.6	88.0	88.5	88.4	93.5	85.3	87.7	96.6	98.7	98.1	108.8	84.7	91.9	88.9	87.6	88.0
0310	Concrete Forming & Accessories	96.0	59.9	64.9	81.9	64.7	67.0	82.0	32.9	39.7	88.8	42.6	49.0	82.5	60.0	63.1	94.2	64.5	68.6
0320	Concrete Reinforcing	96.9	67.1	81.9	88.5	64.7	76.5	88.5	64.2	76.2	88.4	66.0	77.1	97.5	63.6	80.4	96.9	63.7	80.2
0330	Cast-in-Place Concrete	94.9	64.4	83.4	91.1	62.3	80.2	103.3	40.9	79.6	100.8	69.2	88.8	76.4	59.6	70.0	89.2	65.9	80.4
03	CONCRETE	93.2	64.5	80.1	92.1	65.6	80.0	102.2	43.7	75.4	93.0	58.2	77.1	103.2	62.2	84.5	90.8	66.5	79.7
04	MASONRY	99.5	58.7	74.2	116.4	56.4	79.2	111.3	39.8	67.0	116.7	42.7	70.8	113.5	42.1	69.2	77.9	52.8	62.4
05	METALS	96.9	88.9	94.5	90.2	89.6	90.0	90.3	87.6	89.5	92.6	89.6	91.7	94.1	86.7	91.9	97.4	87.2	94.3
06	WOOD, PLASTICS & COMPOSITES	104.1	58.9	78.9	72.3	66.4	69.0	72.5	30.8	49.3	88.5	41.5	62.3	77.4	65.6	70.8	90.4	65.6	76.6
07	THERMAL & MOISTURE PROTECTION	99.4	64.1	84.4	93.8	62.5	80.5	94.3	50.8	76.9	96.2	56.2	79.3	95.1	54.8	78.1	92.8	63.2	80.2
08	OPENINGS	101.2	60.4	91.8	92.4	55.5	83.9	92.4	36.7	79.6	99.6	45.8	87.2	97.5	62.6	89.5	94.7	58.0	86.3
0920	Plaster & Gypsum Board	78.9	58.3	65.1	85.4	66.0	72.3	85.4	29.4	47.8	87.3	40.4	55.8	99.1	65.2	76.3	105.9	65.2	78.5
0950, 0980	Ceilings & Acoustic Treatment	108.2	58.3	74.6	79.6	66.0	70.4	79.6	29.4	45.7	89.0	40.4	56.2	104.5	65.2	78.0	105.4	65.2	78.3
0960	Flooring	100.0	58.4	88.5	79.9	57.0	73.6	80.0	52.9	72.5	79.5	57.3	73.3	94.0	46.4	80.8	99.5	52.5	86.5
0970, 0990	Wall Finishes & Painting/Coating	96.9	62.2	76.7	83.0	59.0	69.0	83.0	59.0	69.0	84.6	59.0	69.7	93.7	61.4	74.9	93.7	60.2	74.2
09	FINISHES	96.6	59.4	76.3	84.6	62.7	72.7	85.0	37.9	59.4	84.9	45.5	63.5	100.0	58.0	77.2	93.6	61.8	76.3
COVERS	DIVS. 10 - 14, 25, 28, 41, 43, 44, 46	100.0	70.5	93.4	100.0	69.5	93.2	100.0	37.2	86.0	100.0	63.7	91.9	100.0	78.8	95.3	100.0	82.6	96.1
21, 22, 23	FIRE SUPPRESSION, PLUMBING & HVAC	100.2	59.7	82.9	98.0	75.9	88.6	98.0	69.8	86.0	100.1	61.1	83.4	99.9	57.3	81.7	99.9	63.0	84.1
26, 27, 3370	ELECTRICAL, COMMUNICATIONS & UTIL.	100.1	87.5	93.6	92.9	53.4	72.3	94.4	63.2	78.2	99.0	54.4	75.8	91.1	43.1	66.1	96.3	58.0	76.4
MF2016	WEIGHTED AVERAGE	98.6	70.3	86.2	94.2	68.4	82.9	95.6	57.7	79.0	97.1	60.6	81.1	98.8	63.0	82.0	95.3	66.2	82.5

TENNESSEE / TEXAS

	DIVISION	TENNESSEE — MCKENZIE 382			TENNESSEE — MEMPHIS 375, 380 - 381			TENNESSEE — NASHVILLE 370 - 372			TEXAS — ABILENE 795 - 796			TEXAS — AMARILLO 790 - 791			TEXAS — AUSTIN 786 - 787		
		MAT.	INST.	TOTAL	MAT.	INST.	TOTAL	MAT.	INST.	TOTAL	MAT.	INST.	TOTAL	MAT.	INST.	TOTAL	MAT.	INST.	TOTAL
015433	CONTRACTOR EQUIPMENT		101.9	101.9		100.8	100.8		105.2	105.2		93.5	93.5		93.5	93.5		92.5	92.5
0241, 31 - 34	SITE & INFRASTRUCTURE, DEMOLITION	93.2	85.6	87.8	89.6	94.3	92.9	96.7	97.6	97.3	98.2	92.2	94.0	98.4	91.8	93.5	96.7	90.8	92.6
0310	Concrete Forming & Accessories	89.7	35.4	42.9	96.0	64.8	69.1	94.5	65.9	69.8	98.9	61.9	67.0	97.6	54.0	60.0	94.4	53.6	59.2
0320	Concrete Reinforcing	88.6	66.1	77.2	104.1	68.1	85.9	103.4	64.9	84.0	90.4	51.2	70.6	96.8	51.1	73.8	101.1	48.7	74.6
0330	Cast-in-Place Concrete	101.1	54.1	83.2	92.7	77.6	86.9	87.3	65.7	79.1	87.4	68.2	80.1	80.8	68.1	76.0	93.6	68.2	84.0
03	CONCRETE	100.7	49.7	77.4	95.6	70.9	84.3	94.0	67.0	81.6	82.8	63.0	73.7	82.3	59.4	71.8	88.6	58.7	74.9
04	MASONRY	115.4	47.0	73.0	98.0	57.8	73.1	84.6	57.8	68.0	95.1	63.2	75.3	94.5	62.7	74.8	97.3	63.2	76.1
05	METALS	90.3	88.9	89.9	99.8	84.4	95.1	96.9	85.9	93.5	107.8	70.7	96.4	102.7	70.5	92.8	97.2	69.0	88.5
06	WOOD, PLASTICS & COMPOSITES	80.9	33.1	54.3	98.0	65.1	79.7	91.2	66.7	77.6	104.6	63.7	81.8	103.2	53.3	75.4	93.1	52.4	70.5
07	THERMAL & MOISTURE PROTECTION	94.2	53.6	77.0	94.3	67.2	82.8	95.6	64.3	82.4	99.6	66.2	85.5	97.8	64.4	83.7	96.6	66.0	83.7
08	OPENINGS	92.4	38.7	80.1	97.7	61.8	89.4	96.0	63.4	88.5	102.2	59.1	92.3	103.4	53.3	91.9	102.8	50.6	90.8
0920	Plaster & Gypsum Board	88.3	31.8	50.4	90.4	64.4	72.9	91.6	66.0	74.4	84.9	63.1	70.2	94.1	52.4	66.1	88.6	51.6	63.7
0950, 0980	Ceilings & Acoustic Treatment	79.6	31.8	47.4	98.1	64.4	75.4	102.6	66.0	77.9	91.7	63.1	72.4	95.1	52.4	66.3	94.3	51.6	65.5
0960	Flooring	82.8	57.0	75.7	102.7	58.4	90.4	95.1	58.4	84.9	92.9	71.5	87.0	93.9	67.4	86.6	87.7	67.4	82.1
0970, 0990	Wall Finishes & Painting/Coating	83.0	46.2	61.5	91.3	59.6	72.8	100.7	70.3	83.0	91.0	54.0	69.4	86.6	54.0	67.6	92.4	46.2	65.5
09	FINISHES	86.2	39.3	60.7	98.1	62.5	78.8	98.0	64.6	79.8	83.7	63.1	72.5	90.6	56.1	71.8	89.6	54.7	70.6
COVERS	DIVS. 10 - 14, 25, 28, 41, 43, 44, 46	100.0	24.7	83.3	100.0	82.8	96.2	100.0	83.2	96.3	100.0	80.6	95.7	100.0	75.4	94.5	100.0	79.1	95.3
21, 22, 23	FIRE SUPPRESSION, PLUMBING & HVAC	98.0	61.0	82.2	100.0	71.3	87.7	100.0	76.6	90.0	100.3	49.3	78.5	99.9	53.2	79.9	100.0	59.1	82.5
26, 27, 3370	ELECTRICAL, COMMUNICATIONS & UTIL.	94.2	58.5	75.6	99.1	65.7	81.7	96.1	63.2	79.0	98.5	56.1	76.5	101.5	62.2	81.1	95.2	59.4	76.6
MF2016	WEIGHTED AVERAGE	95.8	56.8	78.7	98.4	70.7	86.3	96.7	71.7	85.8	97.7	62.6	82.3	97.6	62.2	82.1	96.7	62.7	81.8

For customer support on your Commercial Renovation Costs with RSMeans data, call 800.448.8182.

809

City Cost Indexes

TEXAS

DIVISION		BEAUMONT 776 - 777			BROWNWOOD 768			BRYAN 778			CHILDRESS 792			CORPUS CHRISTI 783 - 784			DALLAS 752 - 753		
		MAT.	INST.	TOTAL	MAT.	INST.	TOTAL	MAT.	INST.	TOTAL	MAT.	INST.	TOTAL	MAT.	INST.	TOTAL	MAT.	INST.	TOTAL
015433	CONTRACTOR EQUIPMENT		97.5	97.5		93.5	93.5		97.5	97.5		93.5	93.5		103.8	103.8		107.7	107.7
0241, 31 - 34	SITE & INFRASTRUCTURE, DEMOLITION	87.6	96.3	93.7	106.2	92.0	96.3	79.0	96.5	91.2	109.4	90.4	96.1	142.0	87.1	103.6	108.7	97.1	100.6
0310	Concrete Forming & Accessories	103.2	57.2	63.5	98.2	61.5	66.5	82.1	60.9	63.8	97.3	61.4	66.3	98.6	54.5	60.6	94.7	62.6	67.0
0320	Concrete Reinforcing	96.9	64.0	80.3	81.3	51.1	66.1	99.3	52.8	75.9	90.6	51.1	70.7	89.2	48.6	68.7	84.7	51.3	67.8
0330	Cast-in-Place Concrete	88.0	68.3	80.5	107.7	61.5	90.1	70.5	62.3	67.4	89.7	61.5	79.0	114.2	70.4	97.6	108.7	70.4	94.1
03	CONCRETE	90.9	63.2	78.3	94.5	60.5	78.9	76.0	60.9	69.1	90.5	60.4	76.7	96.5	61.0	80.3	99.7	64.9	83.8
04	MASONRY	102.5	64.3	78.8	128.3	63.2	87.9	142.8	63.2	93.4	99.2	62.7	76.6	85.5	63.3	71.7	100.6	61.8	76.3
05	METALS	93.5	77.2	88.5	95.5	70.3	87.8	93.3	72.9	87.0	105.2	70.2	94.5	93.5	84.7	90.8	97.6	83.5	93.3
06	WOOD, PLASTICS & COMPOSITES	114.0	56.6	82.0	104.7	63.7	81.8	79.1	62.1	69.7	104.0	63.7	81.6	122.5	53.9	84.3	95.3	64.0	77.9
07	THERMAL & MOISTURE PROTECTION	102.0	66.8	87.1	94.7	64.7	82.0	93.9	66.3	82.2	100.3	63.1	84.6	102.4	65.2	86.6	92.6	67.6	82.0
08	OPENINGS	94.9	57.8	86.4	98.0	59.1	89.1	95.3	58.2	86.8	97.4	59.1	88.6	103.9	51.6	91.9	101.2	59.3	91.6
0920	Plaster & Gypsum Board	97.1	55.9	69.4	85.3	63.1	70.4	85.0	61.5	69.3	84.5	63.1	70.1	96.1	52.9	67.1	93.8	63.1	73.2
0950, 0980	Ceilings & Acoustic Treatment	98.5	55.9	69.8	78.6	63.1	68.1	90.5	61.5	71.0	90.0	63.1	71.8	95.6	52.9	66.8	97.3	63.1	74.2
0960	Flooring	121.1	78.9	109.4	77.9	57.7	72.3	90.5	70.8	85.0	91.4	57.7	82.0	104.4	71.6	95.3	97.3	69.1	89.5
0970, 0990	Wall Finishes & Painting/Coating	88.7	56.3	69.8	89.7	54.0	68.9	86.8	60.0	71.2	91.0	54.0	69.4	104.9	46.2	70.7	102.2	54.0	74.1
09	FINISHES	94.7	60.6	76.2	76.5	60.3	67.7	81.8	62.3	71.2	84.1	60.3	71.1	98.2	56.5	75.5	97.4	62.8	78.6
COVERS	DIVS. 10 - 14, 25, 28, 41, 43, 44, 46	100.0	80.9	95.8	100.0	76.4	94.8	100.0	80.1	95.6	100.0	76.5	94.8	100.0	81.0	95.8	100.0	81.5	95.9
21, 22, 23	FIRE SUPPRESSION, PLUMBING & HVAC	100.2	64.3	84.8	97.1	47.3	75.8	97.1	63.8	82.8	97.2	53.2	78.4	100.2	50.7	79.0	99.9	60.6	83.1
26, 27, 3370	ELECTRICAL, COMMUNICATIONS & UTIL.	97.0	66.9	81.4	96.5	48.7	71.6	95.2	66.9	80.5	98.5	58.9	77.9	92.8	57.0	74.2	97.8	61.9	79.2
MF2016	WEIGHTED AVERAGE	96.5	68.0	84.0	96.9	60.2	80.8	94.0	67.3	82.3	97.6	62.7	82.3	98.7	62.4	82.8	99.2	67.6	85.4

TEXAS

DIVISION		DEL RIO 788			DENTON 762			EASTLAND 764			EL PASO 798 - 799, 885			FORT WORTH 760 - 761			GALVESTON 775		
		MAT.	INST.	TOTAL	MAT.	INST.	TOTAL	MAT.	INST.	TOTAL	MAT.	INST.	TOTAL	MAT.	INST.	TOTAL	MAT.	INST.	TOTAL
015433	CONTRACTOR EQUIPMENT		92.5	92.5		103.4	103.4		93.5	93.5		93.6	93.6		93.5	93.5		110.9	110.9
0241, 31 - 34	SITE & INFRASTRUCTURE, DEMOLITION	122.7	90.1	99.9	105.2	83.9	90.3	109.2	90.1	95.8	95.7	92.6	93.5	101.6	92.9	95.5	103.0	95.1	97.4
0310	Concrete Forming & Accessories	95.3	53.5	59.2	105.4	61.6	67.6	98.9	61.3	66.4	97.9	58.2	63.7	95.2	62.1	66.7	91.7	58.4	63.0
0320	Concrete Reinforcing	89.8	48.5	69.0	82.6	51.1	66.7	81.5	51.1	66.2	96.7	51.3	73.8	87.5	51.2	69.2	98.8	61.6	80.0
0330	Cast-in-Place Concrete	123.0	61.3	99.6	83.3	62.6	75.4	113.8	61.4	93.9	81.6	68.0	76.4	99.1	68.3	87.4	93.6	63.4	82.1
03	CONCRETE	116.3	56.3	88.8	74.2	62.0	68.6	99.1	60.3	81.4	82.7	61.2	72.9	89.6	63.1	77.5	92.1	62.8	78.7
04	MASONRY	100.5	63.1	77.3	137.5	61.8	90.6	96.9	63.2	76.0	85.9	64.4	72.6	95.9	61.8	74.7	98.6	63.3	76.7
05	METALS	93.3	68.2	85.6	95.2	86.0	92.3	95.3	70.1	87.6	102.7	70.1	92.7	97.1	70.9	89.1	94.9	92.9	94.3
06	WOOD, PLASTICS & COMPOSITES	100.9	52.8	74.1	116.1	63.8	87.0	111.8	63.7	85.0	88.7	58.5	71.9	95.1	63.7	77.6	96.5	58.5	75.3
07	THERMAL & MOISTURE PROTECTION	98.9	65.1	84.6	92.7	65.2	81.1	95.1	64.7	82.2	95.6	65.6	82.9	91.2	66.6	80.8	93.1	66.9	82.0
08	OPENINGS	99.2	50.1	87.9	115.6	58.7	102.5	70.6	59.1	67.9	98.6	54.0	88.3	99.7	59.1	90.4	99.5	58.3	90.0
0920	Plaster & Gypsum Board	92.4	51.9	65.2	89.9	63.1	71.9	85.3	61.3	70.4	92.3	57.7	69.1	95.5	63.1	73.7	91.6	57.7	68.8
0950, 0980	Ceilings & Acoustic Treatment	92.0	51.9	65.0	84.6	63.1	70.1	78.6	63.1	68.1	93.4	57.7	69.4	87.8	63.1	71.1	94.8	57.7	69.8
0960	Flooring	89.9	57.7	81.0	73.7	57.7	69.3	98.6	57.7	87.3	91.1	73.7	86.3	94.8	68.1	87.4	106.4	70.8	96.5
0970, 0990	Wall Finishes & Painting/Coating	92.7	44.5	64.6	99.7	51.1	71.4	91.1	54.0	69.5	95.5	49.9	68.9	92.0	54.1	69.9	97.5	56.1	73.4
09	FINISHES	90.8	52.8	70.1	75.7	60.1	67.2	83.3	60.3	70.8	89.3	60.0	73.4	89.0	62.4	74.6	91.6	59.7	74.3
COVERS	DIVS. 10 - 14, 25, 28, 41, 43, 44, 46	100.0	76.7	94.8	100.0	75.9	94.6	100.0	74.7	94.4	100.0	75.9	94.6	100.0	80.6	95.7	100.0	81.4	95.9
21, 22, 23	FIRE SUPPRESSION, PLUMBING & HVAC	97.0	61.2	81.7	97.1	58.1	80.4	97.1	47.2	75.8	99.9	66.3	85.5	100.0	57.7	81.9	97.0	63.9	82.8
26, 27, 3370	ELECTRICAL, COMMUNICATIONS & UTIL.	94.8	63.2	78.3	98.8	61.9	79.6	96.4	61.9	78.4	97.5	52.3	74.0	97.8	61.9	79.1	96.7	67.0	81.3
MF2016	WEIGHTED AVERAGE	99.5	62.9	83.4	96.8	65.3	83.0	93.6	61.8	79.7	96.0	64.7	82.3	96.6	65.1	82.8	96.2	69.0	84.3

TEXAS

DIVISION		GIDDINGS 789			GREENVILLE 754			HOUSTON 770 - 772			HUNTSVILLE 773			LAREDO 780			LONGVIEW 756		
		MAT.	INST.	TOTAL	MAT.	INST.	TOTAL	MAT.	INST.	TOTAL	MAT.	INST.	TOTAL	MAT.	INST.	TOTAL	MAT.	INST.	TOTAL
015433	CONTRACTOR EQUIPMENT		92.5	92.5		104.3	104.3		102.5	102.5		97.5	97.5		92.5	92.5		95.1	95.1
0241, 31 - 34	SITE & INFRASTRUCTURE, DEMOLITION	107.8	90.3	95.6	99.6	87.1	90.9	101.9	94.7	96.8	94.2	96.2	95.6	101.3	90.8	93.9	99.4	93.8	95.5
0310	Concrete Forming & Accessories	92.7	53.5	58.9	86.6	59.8	63.5	97.1	58.6	63.9	89.3	56.6	61.1	95.3	54.6	60.2	83.1	61.5	64.5
0320	Concrete Reinforcing	90.3	48.6	69.3	85.2	51.1	68.0	98.6	52.9	75.5	99.6	52.7	75.9	89.8	48.6	69.0	84.1	51.1	67.4
0330	Cast-in-Place Concrete	104.2	61.5	88.0	108.1	62.7	90.9	86.9	64.5	78.4	96.9	62.2	83.7	87.9	68.2	80.4	125.8	61.5	101.4
03	CONCRETE	93.6	56.4	76.6	93.4	61.1	78.6	91.3	61.0	77.4	98.4	58.9	80.3	88.0	59.2	74.8	110.5	60.4	87.6
04	MASONRY	107.9	63.2	80.2	155.5	61.8	97.3	89.3	63.3	76.4	141.0	63.2	92.8	93.8	63.2	74.8	151.7	61.7	95.9
05	METALS	92.7	68.7	85.4	95.1	84.6	91.9	97.6	78.4	91.7	93.2	72.6	86.9	95.8	69.1	87.6	88.3	69.3	82.5
06	WOOD, PLASTICS & COMPOSITES	100.0	52.8	73.7	85.5	61.3	72.0	101.9	58.7	77.9	87.9	56.6	70.5	100.9	53.8	74.7	80.3	63.8	71.1
07	THERMAL & MOISTURE PROTECTION	99.4	65.2	84.9	92.6	64.3	80.6	95.1	68.1	83.7	95.1	65.7	82.6	98.2	66.1	84.6	94.2	63.7	81.3
08	OPENINGS	98.3	50.8	87.4	97.8	57.8	88.6	102.5	56.3	91.9	95.3	55.1	86.1	99.2	51.4	88.2	88.4	58.7	81.6
0920	Plaster & Gypsum Board	91.4	51.9	64.9	85.2	60.3	68.5	90.8	57.7	68.6	89.0	55.9	66.7	93.2	52.9	66.2	84.2	63.1	70.0
0950, 0980	Ceilings & Acoustic Treatment	92.0	51.9	65.0	92.8	60.3	70.9	91.7	57.7	68.8	90.5	55.9	67.2	95.4	52.9	66.8	88.5	63.1	71.4
0960	Flooring	89.2	57.7	80.4	91.1	57.7	81.8	107.5	71.6	97.5	95.0	57.7	84.7	89.8	67.4	83.6	96.4	57.7	85.7
0970, 0990	Wall Finishes & Painting/Coating	92.7	46.2	65.6	96.3	54.0	71.6	95.4	58.3	73.8	86.8	58.3	70.1	92.7	46.2	65.6	87.0	51.1	66.0
09	FINISHES	89.4	53.0	69.6	91.9	58.9	74.0	97.3	60.6	77.3	84.6	56.6	69.4	89.9	55.5	71.2	97.3	60.0	77.0
COVERS	DIVS. 10 - 14, 25, 28, 41, 43, 44, 46	100.0	73.2	94.0	100.0	76.7	94.8	100.0	81.9	96.0	100.0	72.6	93.9	100.0	78.9	95.3	100.0	76.7	94.8
21, 22, 23	FIRE SUPPRESSION, PLUMBING & HVAC	97.1	62.6	82.3	96.9	58.9	80.7	100.1	65.2	85.2	97.1	63.7	82.8	100.1	61.3	83.5	96.9	57.5	80.0
26, 27, 3370	ELECTRICAL, COMMUNICATIONS & UTIL.	92.0	58.7	74.7	94.7	61.9	77.7	98.7	68.5	83.0	95.2	62.4	78.2	94.9	60.3	76.9	95.1	52.8	73.1
MF2016	WEIGHTED AVERAGE	96.1	62.6	81.4	98.6	65.2	84.0	98.3	68.0	85.0	97.4	65.1	83.3	96.1	63.5	81.8	98.9	62.8	83.1

TEXAS

DIVISION		LUBBOCK 793-794 MAT.	INST.	TOTAL	LUFKIN 759 MAT.	INST.	TOTAL	MCALLEN 785 MAT.	INST.	TOTAL	MCKINNEY 750 MAT.	INST.	TOTAL	MIDLAND 797 MAT.	INST.	TOTAL	ODESSA 797 MAT.	INST.	TOTAL
015433	CONTRACTOR EQUIPMENT		106.2	106.2		95.1	95.1		103.9	103.9		104.3	104.3		106.2	106.2		93.5	93.5
0241, 31-34	SITE & INFRASTRUCTURE, DEMOLITION	123.4	91.3	100.9	94.0	95.4	95.0	146.7	87.2	105.0	95.9	87.1	89.8	126.6	90.4	101.2	98.5	92.9	94.6
0310	Concrete Forming & Accessories	97.7	54.4	60.3	86.0	56.7	60.7	99.5	53.6	59.9	85.8	59.8	63.3	101.5	61.9	67.4	98.9	62.0	67.1
0320	Concrete Reinforcing	91.7	51.2	71.3	85.6	61.5	73.4	89.3	48.6	68.8	85.2	51.1	68.0	92.8	51.2	71.8	90.4	51.2	70.6
0330	Cast-in-Place Concrete	87.6	69.3	80.6	112.5	61.4	93.1	124.1	62.7	100.8	101.4	62.7	86.7	93.1	69.3	84.0	87.4	68.2	80.1
03	CONCRETE	81.7	61.1	72.2	101.8	59.9	82.7	104.2	57.9	83.0	88.4	61.1	75.9	85.7	64.4	76.0	82.8	63.1	73.8
04	MASONRY	94.5	62.8	74.8	116.7	63.1	83.4	100.1	63.3	77.3	166.9	61.8	101.7	111.1	62.8	81.1	95.1	62.7	75.0
05	METALS	111.7	86.5	104.1	95.2	72.4	88.2	93.2	84.5	90.5	95.0	84.6	91.8	109.9	86.4	102.7	107.2	70.8	96.0
06	WOOD, PLASTICS & COMPOSITES	103.8	53.4	75.8	87.6	56.8	70.4	121.2	52.9	83.2	84.6	61.3	71.6	108.5	63.8	83.6	104.6	63.7	81.8
07	THERMAL & MOISTURE PROTECTION	88.9	65.3	78.9	94.0	62.8	80.8	102.1	66.0	86.8	92.4	64.3	80.5	89.2	66.4	79.6	99.6	65.5	85.2
08	OPENINGS	109.8	53.4	96.9	67.2	57.7	65.0	103.1	50.1	90.9	97.8	57.7	88.6	111.3	59.2	99.3	102.2	59.1	92.3
0920	Plaster & Gypsum Board	85.1	52.4	63.1	83.2	55.9	64.8	96.8	51.9	66.7	85.2	60.3	68.5	87.0	63.1	70.9	84.9	63.1	70.2
0950, 0980	Ceilings & Acoustic Treatment	92.6	52.4	65.5	84.2	55.9	65.1	95.4	51.9	66.1	92.8	60.3	70.9	90.9	63.1	72.1	91.7	63.1	72.4
0960	Flooring	86.9	70.7	82.4	129.3	57.7	109.4	103.9	76.9	96.4	90.7	57.7	81.5	88.1	67.4	82.4	92.9	67.4	85.9
0970, 0990	Wall Finishes & Painting/Coating	101.4	54.0	73.8	87.0	54.0	67.7	104.9	44.5	69.7	96.3	54.0	71.6	101.4	54.0	73.8	91.0	54.0	69.4
09	FINISHES	86.1	56.9	70.2	106.1	56.6	79.2	98.5	52.8	75.8	91.6	58.9	73.8	86.8	62.3	73.5	83.7	62.2	72.0
COVERS	DIVS. 10-14, 25, 28, 41, 43, 44, 46	100.0	79.8	95.5	100.0	73.0	94.0	100.0	79.0	95.3	100.0	76.7	94.8	100.0	76.8	94.8	100.0	76.5	94.8
21, 22, 23	FIRE SUPPRESSION, PLUMBING & HVAC	99.7	49.3	78.2	96.9	60.4	81.3	97.1	50.7	77.2	96.9	58.9	80.7	96.6	49.4	76.4	100.3	55.3	81.0
26, 27, 3370	ELECTRICAL, COMMUNICATIONS & UTIL.	97.2	62.2	79.0	96.3	64.8	79.9	92.6	33.4	61.8	94.8	61.9	77.7	97.2	62.2	79.0	98.6	62.2	79.6
MF2016	WEIGHTED AVERAGE	99.3	63.4	83.6	95.7	64.9	82.2	99.6	58.6	81.7	98.4	65.2	83.9	100.0	64.8	84.6	97.6	64.5	83.1

TEXAS

DIVISION		PALESTINE 758 MAT.	INST.	TOTAL	SAN ANGELO 769 MAT.	INST.	TOTAL	SAN ANTONIO 781-782 MAT.	INST.	TOTAL	TEMPLE 765 MAT.	INST.	TOTAL	TEXARKANA 755 MAT.	INST.	TOTAL	TYLER 757 MAT.	INST.	TOTAL
015433	CONTRACTOR EQUIPMENT		95.1	95.1		93.5	93.5		93.3	93.3		93.5	93.5		95.1	95.1		95.1	95.1
0241, 31-34	SITE & INFRASTRUCTURE, DEMOLITION	99.5	94.1	95.7	102.4	92.5	95.4	101.1	93.5	95.8	90.3	92.3	91.7	88.3	96.6	94.1	98.6	93.9	95.3
0310	Concrete Forming & Accessories	77.6	61.6	63.8	98.5	53.8	59.9	93.6	54.2	59.6	101.9	53.5	60.2	92.7	62.2	66.4	87.6	61.7	65.3
0320	Concrete Reinforcing	83.5	51.1	67.1	81.2	51.2	66.1	96.0	48.6	72.1	81.4	48.6	64.8	83.3	51.2	67.1	84.1	51.1	67.5
0330	Cast-in-Place Concrete	102.8	61.5	87.1	101.6	68.2	88.9	86.2	69.6	79.9	83.2	61.4	74.9	103.6	68.3	90.2	123.5	61.6	100.0
03	CONCRETE	102.6	60.5	83.4	90.1	59.3	76.0	90.4	59.3	76.2	76.6	56.4	67.4	94.1	63.1	79.9	109.8	60.5	87.3
04	MASONRY	110.8	61.7	80.4	124.7	63.2	86.6	93.5	63.3	74.8	136.5	63.2	91.0	171.1	61.7	103.3	161.3	61.7	99.6
05	METALS	95.0	69.4	87.1	95.7	70.6	88.0	97.6	66.7	88.1	95.4	68.9	87.2	88.2	70.0	82.6	94.9	69.5	87.1
06	WOOD, PLASTICS & COMPOSITES	79.3	63.8	70.7	104.9	52.8	75.9	97.6	53.0	72.8	114.5	52.8	80.2	90.6	63.8	75.7	89.0	63.8	75.0
07	THERMAL & MOISTURE PROTECTION	94.6	64.8	82.0	94.5	65.7	82.3	93.0	67.5	82.2	94.1	64.7	81.7	93.7	66.1	82.0	94.4	64.8	81.9
08	OPENINGS	67.2	59.1	65.3	98.0	53.1	87.7	101.6	50.9	90.0	67.2	52.5	63.8	88.3	59.1	81.6	67.1	59.1	65.3
0920	Plaster & Gypsum Board	80.9	63.1	68.9	85.3	51.9	62.9	91.3	51.9	64.9	85.3	51.9	62.9	87.8	63.1	71.2	83.2	63.1	69.7
0950, 0980	Ceilings & Acoustic Treatment	84.2	63.1	70.0	78.6	51.9	60.6	94.5	51.9	65.8	78.6	51.9	60.6	88.5	63.1	71.4	84.2	63.1	70.0
0960	Flooring	120.2	57.7	102.8	78.0	57.7	72.3	101.6	71.6	93.3	100.2	57.7	88.4	104.8	67.4	94.4	131.5	57.7	111.0
0970, 0990	Wall Finishes & Painting/Coating	87.0	54.0	67.7	89.7	54.0	68.9	91.5	46.2	65.1	91.1	46.2	64.9	87.0	54.0	67.7	87.0	54.0	67.7
09	FINISHES	103.6	60.3	80.1	76.3	53.9	64.1	99.3	55.9	75.7	82.5	53.0	66.5	99.7	62.3	79.4	107.1	60.3	81.7
COVERS	DIVS. 10-14, 25, 28, 41, 43, 44, 46	100.0	76.7	94.8	100.0	79.2	95.4	100.0	79.2	95.4	100.0	75.3	94.5	100.0	80.9	95.7	100.0	80.8	95.7
21, 22, 23	FIRE SUPPRESSION, PLUMBING & HVAC	96.9	57.6	80.1	97.1	49.3	76.7	100.1	64.0	84.6	97.1	53.6	78.5	96.9	56.7	79.7	96.9	60.5	81.3
26, 27, 3370	ELECTRICAL, COMMUNICATIONS & UTIL.	92.6	51.1	71.0	100.4	57.6	78.2	97.3	62.0	79.0	97.6	54.9	75.4	96.2	57.6	76.1	95.1	56.7	75.1
MF2016	WEIGHTED AVERAGE	95.0	62.7	80.9	96.4	60.7	80.8	97.8	64.4	83.2	91.9	60.4	78.1	97.9	64.4	83.2	98.9	64.2	83.7

TEXAS / UTAH

DIVISION		VICTORIA 779 MAT.	INST.	TOTAL	WACO 766-767 MAT.	INST.	TOTAL	WAXAHACHIE 751 MAT.	INST.	TOTAL	WHARTON 774 MAT.	INST.	TOTAL	WICHITA FALLS 763 MAT.	INST.	TOTAL	LOGAN 843 MAT.	INST.	TOTAL
015433	CONTRACTOR EQUIPMENT		109.5	109.5		93.5	93.5		104.3	104.3		110.9	110.9		93.5	93.5		96.6	96.6
0241, 31-34	SITE & INFRASTRUCTURE, DEMOLITION	107.8	92.2	96.9	99.1	92.7	94.6	97.7	87.4	90.5	112.8	94.5	100.0	99.9	92.9	95.0	101.4	93.9	96.1
0310	Concrete Forming & Accessories	91.0	53.6	58.8	100.4	62.0	67.3	85.8	61.6	65.0	86.3	55.5	59.7	100.4	62.0	67.3	102.2	67.1	71.9
0320	Concrete Reinforcing	94.9	48.6	71.5	81.1	48.7	64.7	85.2	51.1	68.0	98.7	52.2	75.2	81.1	51.2	66.0	102.1	86.6	94.3
0330	Cast-in-Place Concrete	105.1	63.3	89.2	90.1	68.2	81.8	107.0	62.9	90.2	107.9	62.3	90.6	96.2	68.3	85.6	86.3	75.2	82.1
03	CONCRETE	99.4	58.3	80.6	83.0	62.6	73.7	92.4	62.0	78.5	103.5	59.4	83.3	85.9	63.1	75.5	103.3	73.8	89.8
04	MASONRY	116.5	63.3	83.5	94.9	63.2	75.2	156.0	61.9	97.6	99.8	63.2	77.1	95.4	62.7	75.1	108.4	67.9	83.3
05	METALS	93.4	87.8	91.7	97.7	69.4	89.0	95.1	84.9	92.0	94.9	88.8	93.0	97.6	70.8	89.4	107.9	84.0	100.6
06	WOOD, PLASTICS & COMPOSITES	99.9	52.9	73.7	112.8	63.7	85.4	84.6	63.9	73.1	90.0	55.2	70.6	112.8	63.7	85.4	84.2	67.1	74.6
07	THERMAL & MOISTURE PROTECTION	96.9	64.1	83.0	95.0	66.9	83.1	92.5	64.5	80.6	93.5	66.2	81.9	95.0	65.5	82.5	96.5	71.6	85.9
08	OPENINGS	99.3	52.1	88.4	78.7	58.5	74.1	97.8	59.2	89.0	99.4	54.2	89.0	78.7	59.1	74.2	92.4	66.9	86.5
0920	Plaster & Gypsum Board	88.5	51.9	63.9	85.9	63.1	70.5	85.5	63.1	70.4	87.3	54.3	65.1	85.9	63.1	70.5	77.8	66.2	70.0
0950, 0980	Ceilings & Acoustic Treatment	95.6	51.9	66.2	81.2	63.1	69.0	94.5	63.1	73.3	94.8	54.3	67.5	81.2	63.1	69.0	103.2	66.2	78.2
0960	Flooring	104.9	57.7	91.8	99.5	67.4	90.6	90.7	57.7	81.5	103.4	70.4	94.2	100.3	76.0	93.6	98.5	59.3	87.6
0970, 0990	Wall Finishes & Painting/Coating	97.7	58.3	74.7	91.1	54.0	69.5	96.3	54.0	71.6	97.5	58.3	74.6	93.5	54.0	70.4	93.6	56.7	72.1
09	FINISHES	89.3	54.4	70.3	83.3	62.2	71.9	92.2	60.5	74.9	90.8	57.9	72.9	83.8	64.0	73.0	93.4	64.0	77.4
COVERS	DIVS. 10-14, 25, 28, 41, 43, 44, 46	100.0	72.5	93.9	100.0	80.6	95.7	100.0	77.0	94.9	100.0	72.8	93.9	100.0	76.5	94.8	100.0	85.0	96.7
21, 22, 23	FIRE SUPPRESSION, PLUMBING & HVAC	97.1	63.6	82.8	100.2	59.1	82.6	96.9	52.5	77.9	97.0	61.0	81.6	100.2	52.7	79.9	99.9	70.0	87.1
26, 27, 3370	ELECTRICAL, COMMUNICATIONS & UTIL.	101.4	57.0	78.3	100.5	55.6	77.2	94.8	61.9	77.7	100.4	62.5	80.7	102.3	56.1	78.3	96.4	69.2	82.3
MF2016	WEIGHTED AVERAGE	98.3	64.9	83.7	93.6	64.4	80.8	98.5	64.3	83.5	98.2	66.2	84.2	94.2	63.3	80.7	100.2	72.9	88.3

For customer support on your Commercial Renovation Costs with RSMeans data, call 800.448.8182.

811

UTAH / VERMONT

		OGDEN 842, 844			PRICE 845			PROVO 846 - 847			SALT LAKE CITY 840 - 841			BELLOWS FALLS 051			BENNINGTON 052		
DIVISION		MAT.	INST.	TOTAL	MAT.	INST.	TOTAL	MAT.	INST.	TOTAL	MAT.	INST.	TOTAL	MAT.	INST.	TOTAL	MAT.	INST.	TOTAL
015433	CONTRACTOR EQUIPMENT		96.6	96.6		95.6	95.6		95.6	95.6		96.6	96.6		98.8	98.8		98.8	98.8
0241, 31 - 34	SITE & INFRASTRUCTURE, DEMOLITION	88.8	93.9	92.4	98.8	92.1	94.1	97.5	92.2	93.8	88.5	93.8	92.2	92.2	101.5	98.7	91.5	101.5	98.5
0310	Concrete Forming & Accessories	102.2	67.1	71.9	104.6	66.9	72.1	103.9	67.1	72.1	104.5	67.1	72.2	93.8	106.3	104.6	91.6	106.2	104.1
0320	Concrete Reinforcing	101.7	86.6	94.1	109.4	86.6	97.9	110.4	86.6	98.4	103.9	86.6	95.2	80.9	81.4	81.1	80.9	81.4	81.1
0330	Cast-in-Place Concrete	87.6	75.2	82.9	86.4	75.2	82.1	86.4	75.2	82.2	95.7	75.2	88.0	90.7	115.5	100.1	90.7	115.5	100.1
03	CONCRETE	92.9	73.8	84.2	104.4	73.7	90.4	102.9	73.8	89.6	112.0	73.8	94.5	85.2	104.7	94.2	85.1	104.6	94.0
04	MASONRY	102.9	67.9	81.2	114.4	67.9	85.6	114.5	67.9	85.6	116.9	67.9	86.5	104.2	89.8	95.3	115.7	89.8	99.7
05	METALS	108.4	84.0	100.9	105.0	83.8	98.5	106.0	84.0	99.3	112.1	84.0	103.4	93.4	89.7	92.3	93.4	89.6	92.2
06	WOOD, PLASTICS & COMPOSITES	84.2	67.1	74.6	87.2	67.1	76.0	85.5	67.1	75.2	85.9	67.1	75.4	108.3	111.9	110.3	105.6	111.9	109.1
07	THERMAL & MOISTURE PROTECTION	95.3	71.6	85.3	98.3	71.6	87.0	98.4	71.6	87.0	102.8	71.6	89.6	98.0	92.6	95.7	98.0	92.6	95.7
08	OPENINGS	92.4	66.9	86.5	96.3	63.1	88.7	96.4	66.9	89.6	94.3	66.9	88.0	100.4	100.8	100.5	100.4	100.8	100.5
0920	Plaster & Gypsum Board	77.8	66.2	70.0	80.5	66.2	70.9	78.1	66.2	70.1	90.8	66.2	74.3	105.6	111.9	109.8	103.9	111.9	109.3
0950, 0980	Ceilings & Acoustic Treatment	103.2	66.2	78.2	103.2	66.2	78.2	103.2	66.2	78.2	95.5	66.2	75.8	93.8	111.9	106.0	93.8	111.9	106.0
0960	Flooring	96.4	59.3	86.1	99.7	59.3	88.5	99.4	59.3	88.2	100.4	59.3	89.0	91.1	104.5	94.8	90.4	104.5	94.3
0970, 0990	Wall Finishes & Painting/Coating	93.6	56.7	72.1	93.6	56.7	72.1	93.6	56.7	72.1	96.7	56.7	73.3	86.2	106.1	97.8	86.2	106.1	97.8
09	FINISHES	91.5	64.0	76.6	94.6	64.0	78.0	94.0	64.0	77.7	93.7	64.0	77.6	88.1	107.0	98.4	87.6	107.0	98.2
COVERS	DIVS. 10 - 14, 25, 28, 41, 43, 44, 46	100.0	85.0	96.7	100.0	85.0	96.7	100.0	85.0	96.7	100.0	85.0	96.7	100.0	95.7	99.0	100.0	95.6	99.0
21, 22, 23	FIRE SUPPRESSION, PLUMBING & HVAC	99.9	70.0	87.1	98.3	65.7	84.4	99.9	70.0	87.1	100.1	70.0	87.2	97.2	90.8	94.5	97.2	90.8	94.5
26, 27, 3370	ELECTRICAL, COMMUNICATIONS & UTIL.	96.7	69.2	82.4	101.6	69.2	84.7	97.0	69.2	82.5	99.3	69.2	83.6	105.2	84.0	94.1	105.1	57.5	80.4
MF2016	WEIGHTED AVERAGE	98.2	72.9	87.2	100.8	71.7	88.1	100.6	72.8	88.5	102.8	72.9	89.7	96.0	95.3	95.7	96.5	91.6	94.3

VERMONT

		BRATTLEBORO 053			BURLINGTON 054			GUILDHALL 059			MONTPELIER 056			RUTLAND 057			ST. JOHNSBURY 058		
DIVISION		MAT.	INST.	TOTAL	MAT.	INST.	TOTAL	MAT.	INST.	TOTAL	MAT.	INST.	TOTAL	MAT.	INST.	TOTAL	MAT.	INST.	TOTAL
015433	CONTRACTOR EQUIPMENT		98.8	98.8		98.8	98.8		98.8	98.8		98.8	98.8		98.8	98.8		98.8	98.8
0241, 31 - 34	SITE & INFRASTRUCTURE, DEMOLITION	93.1	101.5	99.0	96.9	101.4	100.0	91.1	96.9	95.2	95.4	101.4	99.6	95.6	101.4	99.6	91.2	100.5	97.7
0310	Concrete Forming & Accessories	94.0	106.3	104.6	97.3	82.6	84.6	92.0	99.4	98.4	97.3	105.3	104.2	94.3	82.6	84.2	90.7	99.8	98.6
0320	Concrete Reinforcing	80.0	81.4	80.7	99.4	81.2	90.2	81.6	81.2	81.4	99.4	81.3	90.2	101.1	81.2	91.1	80.0	81.3	80.7
0330	Cast-in-Place Concrete	93.5	115.5	101.9	108.2	114.7	110.7	87.9	106.3	94.9	108.1	114.7	110.6	88.8	114.7	98.6	87.9	106.5	94.9
03	CONCRETE	87.2	104.7	95.2	96.9	93.7	95.4	82.9	98.4	90.0	96.8	104.0	100.1	87.7	93.7	90.4	82.6	98.7	90.0
04	MASONRY	115.2	89.8	99.5	113.8	89.0	98.4	115.6	74.5	90.1	110.1	89.0	97.0	94.8	89.0	91.2	144.6	74.5	101.2
05	METALS	93.4	89.8	92.3	101.1	88.8	97.4	93.4	88.6	91.9	99.1	89.0	96.0	99.2	88.8	96.0	93.4	89.0	92.1
06	WOOD, PLASTICS & COMPOSITES	108.7	111.9	110.5	100.6	81.5	89.9	105.1	111.9	108.9	97.3	111.9	105.5	108.8	81.5	93.6	100.5	111.9	106.9
07	THERMAL & MOISTURE PROTECTION	98.1	92.6	95.8	105.1	88.9	98.2	97.9	85.6	92.7	105.0	92.2	99.6	98.2	88.9	94.3	97.8	85.6	92.6
08	OPENINGS	100.4	101.0	100.5	102.9	80.2	97.7	100.4	97.1	99.6	103.7	97.1	102.2	103.6	80.2	98.2	100.4	97.1	99.6
0920	Plaster & Gypsum Board	105.6	111.9	109.8	105.9	80.6	88.9	111.5	111.9	111.8	105.5	111.9	109.8	106.0	80.6	88.9	112.8	111.9	112.2
0950, 0980	Ceilings & Acoustic Treatment	93.8	111.9	106.0	100.8	80.6	87.1	93.8	111.9	106.0	101.0	111.9	108.3	98.4	80.6	86.4	93.8	111.9	106.0
0960	Flooring	91.2	104.5	94.9	97.3	104.5	99.3	94.0	104.5	96.9	98.4	104.5	100.1	91.1	104.5	94.8	97.1	104.5	99.2
0970, 0990	Wall Finishes & Painting/Coating	86.2	106.1	97.8	95.8	92.4	93.8	86.2	92.4	89.8	92.2	92.4	92.3	86.2	92.4	89.8	86.2	92.4	89.8
09	FINISHES	88.2	107.0	98.4	95.8	87.2	91.1	89.7	101.6	96.2	95.7	105.3	100.9	89.2	87.2	88.1	90.8	101.6	96.7
COVERS	DIVS. 10 - 14, 25, 28, 41, 43, 44, 46	100.0	95.7	99.0	100.0	92.0	98.2	100.0	90.4	97.9	100.0	95.3	99.0	100.0	92.0	98.2	100.0	90.4	97.9
21, 22, 23	FIRE SUPPRESSION, PLUMBING & HVAC	97.2	90.8	94.5	100.0	68.2	86.4	97.2	60.7	81.6	96.9	68.2	84.7	100.3	68.2	86.6	97.2	60.7	81.6
26, 27, 3370	ELECTRICAL, COMMUNICATIONS & UTIL.	105.1	84.0	94.1	105.0	55.6	79.3	105.2	57.5	80.4	104.3	55.6	79.0	105.2	55.6	79.4	105.2	57.5	80.4
MF2016	WEIGHTED AVERAGE	96.9	95.3	96.2	101.0	80.9	92.2	96.4	81.1	89.7	99.8	85.9	93.7	98.1	80.9	90.6	97.8	81.5	90.7

VERMONT / VIRGINIA

		WHITE RIVER JCT. 050			ALEXANDRIA 223			ARLINGTON 222			BRISTOL 242			CHARLOTTESVILLE 229			CULPEPER 227		
DIVISION		MAT.	INST.	TOTAL	MAT.	INST.	TOTAL	MAT.	INST.	TOTAL	MAT.	INST.	TOTAL	MAT.	INST.	TOTAL	MAT.	INST.	TOTAL
015433	CONTRACTOR EQUIPMENT		98.8	98.8		108.6	108.6		107.3	107.3		107.3	107.3		111.5	111.5		107.3	107.3
0241, 31 - 34	SITE & INFRASTRUCTURE, DEMOLITION	95.7	100.6	99.1	115.4	92.9	99.6	125.6	90.6	101.1	109.8	89.9	95.9	114.2	92.4	99.0	113.1	90.5	97.3
0310	Concrete Forming & Accessories	88.9	100.3	98.8	90.8	71.5	74.1	89.8	71.4	73.9	86.0	66.9	69.5	84.5	48.5	53.5	81.7	71.0	72.5
0320	Concrete Reinforcing	80.9	81.3	81.1	85.2	82.8	84.0	96.2	80.6	88.3	96.2	70.0	83.0	95.5	70.3	82.8	96.2	80.5	88.3
0330	Cast-in-Place Concrete	93.5	107.3	98.8	107.7	78.3	96.5	104.8	78.3	94.7	104.3	46.4	82.3	108.5	78.8	97.2	107.3	78.1	96.2
03	CONCRETE	88.7	99.2	93.5	100.0	77.3	89.6	104.5	76.9	91.9	100.8	62.1	83.1	101.5	65.0	84.8	98.5	76.6	88.5
04	MASONRY	129.7	76.0	96.4	91.2	73.7	80.4	116.3	73.7	86.3	95.3	47.3	65.5	120.2	56.9	80.9	107.2	73.7	86.4
05	METALS	93.4	89.0	92.1	103.4	100.5	102.5	102.0	99.6	101.3	100.8	93.8	98.7	101.1	95.8	99.5	101.2	98.8	100.4
06	WOOD, PLASTICS & COMPOSITES	102.6	111.9	107.8	94.6	69.7	80.7	91.3	69.7	79.2	84.5	73.2	78.2	83.1	41.6	60.0	81.9	69.7	75.1
07	THERMAL & MOISTURE PROTECTION	98.2	86.2	93.1	102.6	81.4	93.6	104.7	80.7	94.5	104.0	60.7	85.7	103.6	69.4	89.1	103.8	80.7	94.0
08	OPENINGS	100.4	97.1	99.6	98.5	74.1	92.9	96.6	73.6	91.3	99.6	67.1	92.2	97.8	55.8	88.2	98.1	73.6	92.5
0920	Plaster & Gypsum Board	102.6	111.9	108.8	103.9	68.7	80.3	100.6	68.7	79.2	97.1	72.3	80.4	97.1	39.1	58.2	97.3	68.7	78.1
0950, 0980	Ceilings & Acoustic Treatment	93.8	111.9	106.0	94.5	68.7	77.1	92.8	68.7	76.5	91.9	72.3	78.7	91.9	39.1	56.3	92.8	68.7	76.5
0960	Flooring	89.3	104.5	93.5	96.3	78.7	91.4	94.7	78.7	90.3	91.2	58.9	82.3	89.7	58.9	81.2	89.7	78.7	86.7
0970, 0990	Wall Finishes & Painting/Coating	86.2	92.4	89.8	114.7	75.0	91.5	114.7	75.0	91.5	101.5	56.6	75.3	101.5	60.0	77.3	114.7	60.0	82.8
09	FINISHES	87.5	102.0	95.4	94.4	72.4	82.4	94.3	72.4	82.4	91.0	65.3	77.1	90.6	49.5	68.3	91.3	70.7	80.1
COVERS	DIVS. 10 - 14, 25, 28, 41, 43, 44, 46	100.0	90.9	98.0	100.0	89.2	97.6	100.0	86.7	97.0	100.0	77.6	95.0	100.0	82.1	96.0	100.0	86.7	97.0
21, 22, 23	FIRE SUPPRESSION, PLUMBING & HVAC	97.2	61.4	81.9	100.4	85.3	93.9	100.4	85.2	93.9	97.3	47.7	76.1	97.3	70.4	85.8	97.3	70.6	85.9
26, 27, 3370	ELECTRICAL, COMMUNICATIONS & UTIL.	105.2	55.6	79.4	98.0	99.6	98.8	95.7	101.2	98.6	97.6	38.4	66.8	97.6	72.2	84.4	99.9	96.9	98.4
MF2016	WEIGHTED AVERAGE	97.7	81.6	90.7	99.8	84.7	93.2	100.7	84.5	93.7	98.6	60.5	82.0	99.8	69.3	86.5	99.1	80.5	91.0

City Cost Indexes

VIRGINIA

DIVISION		FAIRFAX 220 - 221			FARMVILLE 239			FREDERICKSBURG 224 - 225			GRUNDY 246			HARRISONBURG 228			LYNCHBURG 245		
		MAT.	INST.	TOTAL	MAT.	INST.	TOTAL	MAT.	INST.	TOTAL	MAT.	INST.	TOTAL	MAT.	INST.	TOTAL	MAT.	INST.	TOTAL
015433	CONTRACTOR EQUIPMENT		107.3	107.3		111.5	111.5		107.3	107.3		107.3	107.3		107.3	107.3		107.3	107.3
0241, 31 - 34	SITE & INFRASTRUCTURE, DEMOLITION	124.3	90.6	100.7	108.4	91.4	96.5	112.7	90.4	97.1	107.6	89.2	94.7	121.3	89.0	98.6	108.5	90.6	96.0
0310	Concrete Forming & Accessories	84.5	71.3	73.1	95.5	45.6	52.5	84.5	68.5	70.7	89.1	38.7	45.6	80.7	39.4	45.1	86.1	73.0	74.8
0320	Concrete Reinforcing	96.2	80.6	88.3	94.9	69.9	82.3	96.9	82.7	89.8	94.9	45.4	69.9	96.2	59.1	77.5	95.5	70.8	83.1
0330	Cast-in-Place Concrete	104.8	78.2	94.7	106.5	88.0	99.5	106.4	78.1	95.6	104.3	52.0	84.5	104.7	55.0	85.8	104.3	77.6	94.2
03	CONCRETE	104.2	76.8	91.7	97.2	66.6	83.2	98.3	75.9	88.0	99.4	46.6	75.3	102.0	50.6	78.5	99.3	75.7	88.5
04	MASONRY	106.2	73.7	86.0	104.3	52.1	71.9	106.3	69.6	83.6	96.1	55.7	71.1	104.2	60.3	77.0	111.7	65.8	83.2
05	METALS	101.3	99.5	100.7	98.0	91.8	96.1	101.2	100.3	100.9	100.8	76.1	93.2	101.1	88.8	97.3	101.0	95.9	99.5
06	WOOD, PLASTICS & COMPOSITES	84.5	69.7	76.2	93.8	40.2	64.0	84.5	67.7	75.1	87.1	31.9	56.4	81.0	35.5	55.7	84.5	74.6	79.0
07	THERMAL & MOISTURE PROTECTION	104.4	76.4	92.6	104.2	56.0	83.8	103.8	78.7	93.1	103.9	48.3	80.4	104.2	64.2	87.3	103.8	72.6	90.6
08	OPENINGS	96.6	73.6	91.3	97.6	45.3	85.6	97.8	72.0	91.9	99.6	31.6	84.0	98.1	45.3	86.0	98.1	67.9	91.2
0920	Plaster & Gypsum Board	97.3	68.7	78.1	106.4	37.7	60.3	97.3	66.7	76.7	97.1	29.9	52.0	97.1	33.6	54.5	97.1	73.7	81.4
0950, 0980	Ceilings & Acoustic Treatment	92.8	68.7	76.5	87.6	37.7	53.9	92.8	66.7	75.2	91.9	29.9	50.1	91.9	33.6	52.6	91.9	73.7	79.7
0960	Flooring	91.5	78.7	88.0	92.0	53.4	81.3	91.5	76.5	87.4	92.5	29.8	75.1	89.5	78.7	86.5	91.2	69.1	85.1
0970, 0990	Wall Finishes & Painting/Coating	114.7	75.0	91.5	99.7	59.8	76.4	114.7	59.6	82.6	101.5	32.7	61.3	114.7	50.5	77.3	101.5	56.6	75.3
09	FINISHES	92.9	72.4	81.7	90.5	47.3	67.0	91.8	68.3	79.0	91.2	35.0	60.7	91.7	46.2	67.0	90.8	70.6	79.9
COVERS	DIVS. 10 - 14, 25, 28, 41, 43, 44, 46	100.0	86.7	97.0	100.0	77.7	95.0	100.0	81.5	95.9	100.0	75.8	94.6	100.0	62.5	91.6	100.0	81.2	95.8
21, 22, 23	FIRE SUPPRESSION, PLUMBING & HVAC	97.3	85.2	92.1	97.4	46.9	75.8	97.3	83.2	91.3	97.3	67.4	84.5	97.3	68.1	84.8	97.3	70.7	85.9
26, 27, 3370	ELECTRICAL, COMMUNICATIONS & UTIL.	98.7	96.9	97.8	92.2	45.1	67.7	95.9	96.9	96.4	97.6	41.7	68.5	97.8	91.2	94.4	98.6	71.6	84.6
MF2016	WEIGHTED AVERAGE	100.0	83.8	92.9	97.4	58.6	80.4	98.7	82.2	91.5	98.5	56.0	79.9	99.5	67.3	85.4	99.2	75.2	88.7

VIRGINIA

DIVISION		NEWPORT NEWS 236			NORFOLK 233 - 235			PETERSBURG 238			PORTSMOUTH 237			PULASKI 243			RICHMOND 230 - 232		
		MAT.	INST.	TOTAL	MAT.	INST.	TOTAL	MAT.	INST.	TOTAL	MAT.	INST.	TOTAL	MAT.	INST.	TOTAL	MAT.	INST.	TOTAL
015433	CONTRACTOR EQUIPMENT		111.6	111.6		112.2	112.2		111.5	111.5		111.5	111.5		107.3	107.3		111.5	111.5
0241, 31 - 34	SITE & INFRASTRUCTURE, DEMOLITION	107.3	92.4	96.9	105.7	93.5	97.1	111.0	92.4	97.9	105.9	91.7	95.9	106.9	89.7	94.8	102.2	92.4	95.3
0310	Concrete Forming & Accessories	94.8	62.5	67.0	100.4	62.6	67.8	88.8	63.7	67.2	85.0	49.0	53.9	89.1	41.4	48.0	95.2	82.3	84.1
0320	Concrete Reinforcing	94.7	67.5	81.0	102.5	67.5	84.8	94.3	70.9	82.5	94.3	66.8	80.5	94.9	58.2	76.4	105.2	70.9	87.9
0330	Cast-in-Place Concrete	103.5	78.0	93.8	107.5	79.1	96.7	109.9	79.3	98.3	102.5	62.6	87.3	104.3	86.4	97.5	98.4	79.3	91.1
03	CONCRETE	94.5	70.5	83.6	96.5	71.0	84.8	99.7	72.1	87.1	93.4	59.0	77.7	99.4	62.2	82.4	92.2	80.6	86.9
04	MASONRY	98.6	64.5	77.4	96.6	64.5	76.7	112.9	64.5	82.9	104.6	49.4	70.4	90.8	56.6	69.6	97.3	64.5	77.0
05	METALS	100.3	95.1	98.7	102.0	95.2	99.9	98.0	96.8	97.7	99.3	93.6	97.5	100.9	87.4	96.7	102.0	97.0	100.5
06	WOOD, PLASTICS & COMPOSITES	92.8	61.4	75.3	96.1	61.4	76.8	85.2	62.4	72.5	82.0	46.9	62.5	87.1	33.5	57.3	95.9	87.4	91.2
07	THERMAL & MOISTURE PROTECTION	104.1	71.6	90.4	104.0	71.8	88.9	104.1	73.6	91.2	104.2	62.1	86.4	103.9	59.0	84.4	102.5	76.4	91.4
08	OPENINGS	98.0	60.3	89.3	96.2	61.6	88.2	97.3	67.3	90.4	98.0	49.9	87.0	99.7	38.3	85.6	100.2	81.1	95.8
0920	Plaster & Gypsum Board	107.4	59.5	75.2	100.3	59.5	72.9	100.7	60.5	73.7	101.2	44.6	63.2	97.1	31.5	53.1	101.4	86.2	91.2
0950, 0980	Ceilings & Acoustic Treatment	91.8	59.5	70.0	95.4	59.5	71.2	88.4	60.5	69.6	91.8	44.6	60.0	91.9	31.5	51.2	91.4	86.2	87.9
0960	Flooring	92.0	69.1	85.7	87.7	69.1	82.5	88.0	74.3	84.2	85.0	58.9	77.8	92.5	58.9	83.2	87.3	74.3	83.7
0970, 0990	Wall Finishes & Painting/Coating	99.7	60.0	76.5	97.4	60.0	75.6	99.7	60.0	76.5	99.7	60.0	76.5	101.5	50.5	71.8	92.7	60.0	73.6
09	FINISHES	91.3	63.0	75.9	90.7	63.0	75.6	88.9	64.6	75.7	88.4	50.5	67.8	91.2	42.9	64.9	91.0	79.4	84.7
COVERS	DIVS. 10 - 14, 25, 28, 41, 43, 44, 46	100.0	80.5	95.7	100.0	80.5	95.7	100.0	83.9	96.4	100.0	66.5	92.6	100.0	76.0	94.7	100.0	86.7	97.0
21, 22, 23	FIRE SUPPRESSION, PLUMBING & HVAC	100.5	64.0	84.9	100.1	67.3	86.1	97.4	69.8	85.6	100.5	64.3	85.1	97.3	66.2	84.0	100.1	69.8	87.1
26, 27, 3370	ELECTRICAL, COMMUNICATIONS & UTIL.	94.7	63.7	78.6	97.9	63.8	80.2	94.8	72.2	83.1	93.1	63.8	77.9	97.6	56.7	76.3	98.3	72.2	84.7
MF2016	WEIGHTED AVERAGE	98.2	70.4	86.0	98.5	71.3	86.6	98.3	73.9	87.6	97.7	64.3	83.1	98.2	62.7	82.7	98.5	77.9	89.5

VIRGINIA / WASHINGTON

DIVISION		ROANOKE 240 - 241			STAUNTON 244			WINCHESTER 226			CLARKSTON 994			EVERETT 982			OLYMPIA 985		
		MAT.	INST.	TOTAL	MAT.	INST.	TOTAL	MAT.	INST.	TOTAL	MAT.	INST.	TOTAL	MAT.	INST.	TOTAL	MAT.	INST.	TOTAL
015433	CONTRACTOR EQUIPMENT		107.2	107.2		111.5	111.5		107.3	107.3		91.5	91.5		102.2	102.2		102.2	102.2
0241, 31 - 34	SITE & INFRASTRUCTURE, DEMOLITION	107.5	90.6	95.7	110.7	91.3	97.1	119.9	90.5	99.3	107.5	89.6	94.9	92.9	110.9	105.5	93.0	110.9	105.5
0310	Concrete Forming & Accessories	95.2	73.1	76.1	88.8	50.2	55.5	82.9	69.5	71.4	111.4	67.0	73.1	116.2	102.6	104.5	105.3	102.4	102.8
0320	Concrete Reinforcing	95.9	70.8	83.2	95.5	70.0	82.7	95.6	82.8	89.1	112.0	98.3	105.1	108.9	111.8	110.4	113.8	111.8	112.8
0330	Cast-in-Place Concrete	118.5	87.8	106.8	108.5	88.6	100.9	104.7	67.1	90.5	87.8	84.1	86.4	98.6	109.1	102.6	90.8	109.0	97.7
03	CONCRETE	103.9	79.3	92.6	100.8	68.9	86.2	101.5	72.6	88.3	95.4	78.7	87.8	90.1	105.9	97.3	86.8	105.9	95.5
04	MASONRY	97.4	65.8	77.8	107.0	55.6	75.1	101.6	72.7	83.7	100.8	84.8	90.9	112.7	103.6	107.1	107.1	103.6	104.9
05	METALS	103.2	96.0	101.0	101.1	91.6	98.2	101.2	100.1	100.9	91.3	87.6	90.1	103.3	96.7	101.3	103.0	96.3	101.0
06	WOOD, PLASTICS & COMPOSITES	95.4	74.6	83.8	87.1	45.5	64.0	83.1	67.9	74.6	107.4	61.2	81.7	112.0	101.6	106.2	97.0	101.6	99.6
07	THERMAL & MOISTURE PROTECTION	103.7	76.1	92.0	103.5	57.5	84.1	104.3	78.3	93.3	158.3	81.7	125.8	111.9	105.8	109.3	111.6	101.9	107.5
08	OPENINGS	98.5	67.9	91.5	98.1	48.1	86.6	99.7	71.8	93.3	118.1	68.3	106.7	106.1	104.0	105.6	110.5	104.3	109.0
0920	Plaster & Gypsum Board	103.9	73.7	83.6	97.1	43.2	60.9	97.3	66.9	76.9	149.2	60.1	89.3	112.7	101.8	105.4	107.0	101.8	103.5
0950, 0980	Ceilings & Acoustic Treatment	94.5	73.7	80.5	91.9	43.2	59.1	91.9	66.9	75.3	108.8	60.1	75.9	105.7	101.8	103.1	105.7	101.8	103.1
0960	Flooring	96.3	69.1	88.8	92.1	35.0	76.2	91.0	78.7	87.6	86.7	81.9	85.3	105.2	101.5	104.2	93.1	101.5	95.8
0970, 0990	Wall Finishes & Painting/Coating	101.5	56.6	75.3	101.5	32.1	61.0	114.7	81.1	95.1	78.6	75.4	76.7	91.5	95.7	93.9	87.1	95.7	92.1
09	FINISHES	93.4	70.6	81.0	91.1	44.5	65.8	92.3	71.8	81.1	105.5	69.0	85.8	103.3	101.3	102.2	94.0	101.3	98.0
COVERS	DIVS. 10 - 14, 25, 28, 41, 43, 44, 46	100.0	81.1	95.8	100.0	78.8	95.3	100.0	86.1	96.9	100.0	92.9	98.4	100.0	100.0	100.0	100.0	102.5	100.6
21, 22, 23	FIRE SUPPRESSION, PLUMBING & HVAC	100.4	66.4	85.9	97.3	60.1	81.4	97.3	84.7	91.9	97.7	83.9	91.8	100.2	101.8	100.9	100.1	101.8	100.8
26, 27, 3370	ELECTRICAL, COMMUNICATIONS & UTIL.	97.6	57.8	76.9	96.6	72.2	83.9	96.3	96.9	96.6	86.6	98.0	92.6	108.2	98.4	103.1	106.5	98.3	102.2
MF2016	WEIGHTED AVERAGE	100.3	73.0	88.3	99.0	65.8	84.5	99.3	82.9	92.2	100.6	83.4	93.1	102.0	102.4	102.2	100.7	102.3	101.4

For customer support on your Commercial Renovation Costs with RSMeans data, call 800.448.8182.

WASHINGTON

DIVISION	RICHLAND 993			SEATTLE 980-981, 987			SPOKANE 990-992			TACOMA 983-984			VANCOUVER 986			WENATCHEE 988		
	MAT.	INST.	TOTAL	MAT.	INST.	TOTAL	MAT.	INST.	TOTAL	MAT.	INST.	TOTAL	MAT.	INST.	TOTAL	MAT.	INST.	TOTAL
015433 CONTRACTOR EQUIPMENT		91.5	91.5		103.7	103.7		91.5	91.5		102.2	102.2		98.4	98.4		102.2	102.2
0241, 31 - 34 SITE & INFRASTRUCTURE, DEMOLITION	109.4	89.7	95.6	99.3	110.1	106.9	108.8	89.7	95.4	96.1	110.9	106.5	106.0	97.8	100.3	105.3	108.3	107.4
0310 Concrete Forming & Accessories	111.5	80.0	84.4	112.3	102.9	104.2	116.8	79.6	84.7	106.5	102.5	103.0	107.1	94.7	96.4	108.4	78.9	82.9
0320 Concrete Reinforcing	107.4	98.4	102.9	114.0	111.9	112.9	108.1	98.3	103.2	107.4	111.8	109.6	108.5	110.9	109.7	108.5	98.4	103.4
0330 Cast-in-Place Concrete	88.0	85.5	87.0	102.5	108.8	104.9	91.4	85.3	89.1	101.3	109.0	104.3	113.2	100.7	108.4	103.5	93.8	99.8
03 CONCRETE	95.0	85.1	90.5	98.2	106.2	101.9	97.1	84.8	91.5	91.7	105.8	98.2	101.1	99.4	100.3	98.8	87.4	93.6
04 MASONRY	101.5	86.8	92.4	119.1	103.6	109.5	102.0	86.8	92.6	112.2	103.6	106.9	111.7	87.8	96.9	115.6	84.4	96.3
05 METALS	91.7	88.2	90.6	103.7	99.9	102.5	94.0	87.8	92.1	105.1	96.3	102.4	102.7	96.4	100.8	102.5	87.9	98.0
06 WOOD, PLASTICS & COMPOSITES	107.6	77.0	90.6	107.3	101.6	104.1	116.2	77.0	94.4	101.5	101.6	101.6	93.0	93.9	93.5	103.2	76.8	88.5
07 THERMAL & MOISTURE PROTECTION	159.4	85.7	128.2	107.1	105.2	106.3	155.6	87.1	126.6	111.6	102.0	107.6	112.0	97.3	105.8	111.1	87.9	101.3
08 OPENINGS	116.1	75.9	106.9	106.5	104.9	106.1	116.7	79.5	108.2	106.9	104.3	106.3	103.2	98.7	102.2	106.4	76.7	99.5
0920 Plaster & Gypsum Board	149.2	76.3	100.2	108.6	101.8	104.0	139.6	76.3	97.1	108.8	101.8	104.1	106.9	94.1	98.3	111.7	76.3	87.9
0950, 0980 Ceilings & Acoustic Treatment	115.5	76.3	89.1	112.0	101.8	105.1	110.7	76.3	87.5	109.1	101.8	104.2	104.2	94.1	97.4	101.3	76.3	84.4
0960 Flooring	87.0	81.9	85.6	103.6	101.5	103.0	86.0	81.9	84.9	97.9	101.5	98.9	103.6	83.6	98.1	100.9	81.9	95.6
0970, 0990 Wall Finishes & Painting/Coating	78.6	79.9	79.3	105.4	95.7	99.7	78.7	79.9	79.4	91.5	95.7	93.9	94.0	75.2	83.0	91.5	74.1	81.4
09 FINISHES	107.4	79.3	92.1	106.9	101.3	103.9	105.0	79.3	91.0	101.4	101.3	101.3	99.2	90.1	94.2	102.1	77.9	88.9
COVERS DIVS. 10 - 14, 25, 28, 41, 43, 44, 46	100.0	95.3	99.0	100.0	102.6	100.6	100.0	95.2	98.9	100.0	102.5	100.6	100.0	98.0	99.6	100.0	94.0	98.7
21, 22, 23 FIRE SUPPRESSION, PLUMBING & HVAC	100.8	112.6	105.9	100.0	113.5	105.8	100.8	86.6	94.7	100.2	100.8	100.9	100.3	101.5	100.8	97.1	94.2	95.8
26, 27, 3370 ELECTRICAL, COMMUNICATIONS & UTIL.	84.6	98.0	91.6	107.4	113.1	110.4	83.0	81.4	82.1	108.0	98.3	103.0	113.8	101.6	107.5	108.8	93.3	100.7
MF2016 WEIGHTED AVERAGE	101.1	92.7	97.4	103.6	107.3	105.2	101.4	85.0	94.2	102.4	102.3	102.3	103.3	97.2	100.6	102.6	89.5	96.9

WASHINGTON / WEST VIRGINIA

DIVISION	YAKIMA 989			BECKLEY 258 - 259			BLUEFIELD 247 - 248			BUCKHANNON 262			CHARLESTON 250 - 253			CLARKSBURG 263 - 264		
	MAT.	INST.	TOTAL	MAT.	INST.	TOTAL	MAT.	INST.	TOTAL	MAT.	INST.	TOTAL	MAT.	INST.	TOTAL	MAT.	INST.	TOTAL
015433 CONTRACTOR EQUIPMENT		102.2	102.2		107.3	107.3		107.3	107.3		107.3	107.3		107.3	107.3		107.3	107.3
0241, 31 - 34 SITE & INFRASTRUCTURE, DEMOLITION	98.6	109.4	106.2	101.5	91.8	94.7	101.5	91.8	94.7	107.7	91.7	96.5	100.7	92.8	95.2	108.5	91.7	96.7
0310 Concrete Forming & Accessories	107.0	98.0	99.2	86.9	90.2	89.7	86.0	89.9	89.3	85.5	86.4	86.3	97.7	91.3	92.2	83.0	86.4	85.9
0320 Concrete Reinforcing	108.0	98.5	103.2	91.8	89.4	90.6	94.5	70.8	82.6	95.1	94.3	94.7	99.3	89.7	94.4	95.1	97.3	96.2
0330 Cast-in-Place Concrete	108.2	84.6	99.3	104.4	93.1	100.1	102.0	92.9	98.6	101.7	93.4	98.5	101.0	94.5	98.5	111.5	91.6	103.9
03 CONCRETE	96.2	92.9	94.7	95.9	92.0	94.1	95.4	88.7	92.3	98.3	91.3	95.1	94.5	93.1	93.8	102.4	91.2	97.3
04 MASONRY	105.1	84.5	92.3	93.0	88.8	90.4	92.5	88.8	90.2	104.5	86.5	93.3	88.7	94.2	92.1	108.4	86.5	94.8
05 METALS	103.2	88.7	98.7	102.8	104.5	103.3	101.1	97.9	100.1	101.3	106.1	102.8	101.4	104.9	102.5	101.3	107.1	103.1
06 WOOD, PLASTICS & COMPOSITES	101.9	101.6	101.8	85.1	90.5	88.1	86.3	90.5	88.6	85.6	85.9	85.8	92.3	90.5	91.3	82.3	85.9	84.3
07 THERMAL & MOISTURE PROTECTION	111.7	88.9	102.1	106.5	88.6	98.9	103.7	88.6	97.3	104.0	87.6	97.1	103.1	90.6	97.8	103.9	87.3	96.9
08 OPENINGS	106.3	101.3	105.2	96.3	85.6	93.8	100.1	81.3	95.7	100.1	84.1	96.4	94.4	85.6	92.4	100.1	84.8	96.6
0920 Plaster & Gypsum Board	108.4	101.8	103.9	98.9	90.1	93.0	96.4	90.1	92.1	96.8	85.4	89.1	99.9	90.1	93.3	94.1	85.4	88.3
0950, 0980 Ceilings & Acoustic Treatment	103.2	101.8	102.2	84.1	90.1	88.1	90.2	90.1	90.1	91.9	85.4	87.5	95.2	90.1	91.7	91.9	85.4	87.5
0960 Flooring	98.8	81.9	94.1	91.6	100.6	94.1	88.9	100.6	92.2	88.7	96.5	90.8	94.0	100.6	95.8	87.6	96.5	90.1
0970, 0990 Wall Finishes & Painting/Coating	91.5	79.9	84.7	92.6	91.7	92.1	101.5	91.7	95.8	101.5	87.9	93.6	88.2	91.7	90.3	101.5	87.9	93.6
09 FINISHES	100.6	93.2	96.6	88.2	92.6	90.6	89.3	92.6	91.1	90.1	88.3	89.1	93.4	93.2	93.3	89.4	88.3	88.8
COVERS DIVS. 10 - 14, 25, 28, 41, 43, 44, 46	100.0	99.3	99.8	100.0	93.8	98.6	100.0	93.8	98.6	100.0	92.4	98.3	100.0	94.1	98.7	100.0	92.4	98.3
21, 22, 23 FIRE SUPPRESSION, PLUMBING & HVAC	100.2	111.8	105.1	97.5	91.4	94.9	97.3	87.1	92.9	97.3	91.2	94.7	100.1	92.8	97.0	97.3	91.1	94.7
26, 27, 3370 ELECTRICAL, COMMUNICATIONS & UTIL.	111.0	98.1	104.3	94.6	87.0	90.7	96.6	87.0	91.6	98.0	91.9	94.8	99.7	88.8	94.0	98.0	91.9	94.8
MF2016 WEIGHTED AVERAGE	102.5	98.2	100.6	97.3	91.8	94.8	97.5	89.6	94.0	98.8	91.5	95.6	97.9	93.3	95.9	99.4	91.6	96.0

WEST VIRGINIA

DIVISION	GASSAWAY 266			HUNTINGTON 255 - 257			LEWISBURG 249			MARTINSBURG 254			MORGANTOWN 265			PARKERSBURG 261		
	MAT.	INST.	TOTAL	MAT.	INST.	TOTAL	MAT.	INST.	TOTAL	MAT.	INST.	TOTAL	MAT.	INST.	TOTAL	MAT.	INST.	TOTAL
015433 CONTRACTOR EQUIPMENT		107.3	107.3		107.3	107.3		107.3	107.3		107.3	107.3		107.3	107.3		107.3	107.3
0241, 31 - 34 SITE & INFRASTRUCTURE, DEMOLITION	105.2	91.7	95.7	106.2	92.8	96.8	117.2	91.8	99.4	105.5	92.5	96.4	102.5	92.6	95.5	110.7	92.7	98.1
0310 Concrete Forming & Accessories	84.9	86.3	86.1	98.6	91.0	92.1	83.3	89.5	88.6	87.0	77.7	79.0	83.3	86.5	86.1	87.6	86.8	86.9
0320 Concrete Reinforcing	95.1	94.1	94.6	93.1	93.6	93.3	95.1	70.6	82.7	91.8	94.2	93.0	95.1	97.3	96.2	94.5	94.3	94.4
0330 Cast-in-Place Concrete	106.5	92.4	101.1	113.9	98.4	108.0	102.1	92.8	98.6	109.4	86.9	100.8	101.7	93.4	98.5	104.0	93.5	100.0
03 CONCRETE	98.7	90.9	95.1	101.2	95.0	98.3	105.0	88.4	97.4	99.6	84.8	92.8	95.1	91.9	93.6	100.4	91.5	96.3
04 MASONRY	109.4	84.8	94.1	91.9	96.7	94.9	95.5	85.9	89.6	94.6	87.6	90.2	126.6	86.5	101.7	82.0	88.8	86.2
05 METALS	101.2	105.8	102.6	105.3	106.6	105.7	101.2	97.3	100.0	103.2	99.5	102.1	101.3	107.2	103.1	102.0	106.1	103.2
06 WOOD, PLASTICS & COMPOSITES	84.9	85.9	85.4	95.8	90.0	92.6	82.6	90.5	87.0	85.1	75.3	79.6	82.6	85.9	84.4	85.7	85.4	85.5
07 THERMAL & MOISTURE PROTECTION	103.7	87.0	96.6	106.6	91.7	100.3	104.6	87.8	97.5	106.7	83.0	96.7	103.8	87.6	96.9	103.8	88.4	97.3
08 OPENINGS	98.2	84.1	95.0	95.6	86.2	93.4	100.1	81.3	95.8	98.2	68.9	91.4	101.4	84.8	97.6	99.0	83.9	95.5
0920 Plaster & Gypsum Board	96.1	85.4	88.9	106.2	89.6	95.1	94.1	90.1	91.4	99.3	74.4	82.6	94.1	85.4	88.3	97.1	84.9	88.9
0950, 0980 Ceilings & Acoustic Treatment	91.9	85.4	87.5	85.9	89.6	88.4	91.9	90.1	90.7	85.9	74.4	78.2	91.9	85.4	87.5	91.9	84.9	87.2
0960 Flooring	88.4	100.6	91.8	99.0	109.2	101.8	87.7	100.6	91.3	91.6	96.6	93.0	87.7	96.5	90.2	91.6	97.2	93.2
0970, 0990 Wall Finishes & Painting/Coating	101.5	91.7	95.8	92.6	91.7	92.1	101.5	67.4	81.6	92.6	87.5	89.7	101.5	87.9	93.6	101.5	87.5	93.3
09 FINISHES	89.7	89.6	89.6	91.8	94.6	93.3	90.4	89.9	90.2	88.8	81.6	84.9	89.0	88.3	88.6	91.1	88.6	89.8
COVERS DIVS. 10 - 14, 25, 28, 41, 43, 44, 46	100.0	92.4	98.3	100.0	93.9	98.6	100.0	67.7	92.8	100.0	84.0	96.4	100.0	86.7	97.0	100.0	93.0	98.4
21, 22, 23 FIRE SUPPRESSION, PLUMBING & HVAC	97.3	86.5	92.7	100.6	89.8	96.0	97.3	91.2	94.7	97.5	84.0	91.7	97.3	91.2	94.7	100.3	89.2	95.6
26, 27, 3370 ELECTRICAL, COMMUNICATIONS & UTIL.	98.0	88.8	93.2	98.2	91.0	94.4	94.2	87.0	90.5	100.2	76.0	87.6	98.1	91.9	94.9	98.1	78.3	87.8
MF2016 WEIGHTED AVERAGE	98.8	89.9	94.9	99.7	93.8	97.1	99.1	88.9	94.6	98.8	84.4	92.5	99.4	91.6	96.0	98.8	89.5	94.8

City Cost Indexes

		WEST VIRGINIA									WISCONSIN								
DIVISION		PETERSBURG			ROMNEY			WHEELING			BELOIT			EAU CLAIRE			GREEN BAY		
		268			267			260			535			547			541 - 543		
		MAT.	INST.	TOTAL	MAT.	INST.	TOTAL	MAT.	INST.	TOTAL	MAT.	INST.	TOTAL	MAT.	INST.	TOTAL	MAT.	INST.	TOTAL
015433	CONTRACTOR EQUIPMENT		107.3	107.3		107.3	107.3		107.3	107.3		101.9	101.9		102.7	102.7		100.3	100.3
0241, 31 - 34	SITE & INFRASTRUCTURE, DEMOLITION	101.7	92.5	95.3	104.6	92.6	96.2	111.4	92.6	98.2	95.8	106.6	103.4	97.0	104.3	102.1	100.3	99.8	99.9
0310	Concrete Forming & Accessories	86.4	77.8	79.0	82.5	78.2	78.8	89.1	86.8	87.1	96.4	92.3	92.9	95.0	105.8	104.4	103.6	105.6	105.3
0320	Concrete Reinforcing	94.5	94.1	94.3	95.1	94.3	94.7	93.9	97.3	95.6	95.5	131.6	113.7	91.4	110.6	101.1	89.4	106.3	97.9
0330	Cast-in-Place Concrete	101.7	87.0	96.1	106.5	87.1	99.1	104.0	96.6	101.2	106.4	99.1	103.6	100.2	99.8	100.0	103.6	103.8	103.7
03	CONCRETE	95.2	85.2	90.6	98.6	85.4	92.6	100.4	93.1	97.1	96.1	101.7	98.7	91.8	104.5	97.6	94.7	105.1	99.5
04	MASONRY	99.5	86.5	91.4	98.5	87.6	90.7	107.6	87.0	94.8	100.3	102.3	101.6	92.0	98.7	96.1	123.3	99.7	108.6
05	METALS	101.4	105.4	102.6	101.4	105.9	102.8	102.1	107.3	103.7	96.5	110.6	100.8	96.3	104.5	98.8	98.9	103.9	100.4
06	WOOD, PLASTICS & COMPOSITES	86.5	75.3	80.3	81.8	75.3	78.2	87.1	85.9	86.4	93.5	88.8	90.9	98.3	107.8	103.6	103.0	107.8	105.7
07	THERMAL & MOISTURE PROTECTION	103.8	82.9	94.9	103.9	83.9	95.4	104.2	88.8	97.7	100.7	94.0	97.9	103.5	98.9	101.6	105.8	100.8	103.7
08	OPENINGS	101.4	78.2	96.1	101.3	78.2	96.0	99.8	84.8	96.4	96.3	106.1	98.5	101.8	107.2	103.1	98.2	107.0	100.2
0920	Plaster & Gypsum Board	96.8	74.4	81.8	93.8	74.4	80.8	97.1	85.4	89.2	97.3	88.9	91.7	110.1	108.4	109.0	106.9	108.4	107.9
0950, 0980	Ceilings & Acoustic Treatment	91.9	74.4	80.1	91.9	74.4	80.1	91.9	85.4	87.5	86.3	88.9	88.1	90.3	108.4	102.5	84.6	108.4	100.7
0960	Flooring	89.5	96.5	91.4	87.5	98.5	90.6	92.5	96.5	93.6	91.3	125.2	100.7	80.8	112.6	89.6	96.5	121.4	103.4
0970, 0990	Wall Finishes & Painting/Coating	101.5	87.9	93.6	101.5	87.9	93.6	101.5	87.9	93.6	96.8	104.5	101.3	84.9	79.9	82.0	94.0	83.6	88.0
09	FINISHES	89.8	82.0	85.6	89.1	82.4	85.5	91.4	88.5	89.8	92.0	94.0	95.8	87.5	105.0	97.0	91.9	107.0	100.1
COVERS	DIVS. 10 - 14, 25, 28, 41, 43, 44, 46	100.0	55.1	90.0	100.0	91.2	98.0	100.0	87.8	97.3	100.0	94.1	98.7	100.0	96.4	99.2	100.0	96.3	99.2
21, 22, 23	FIRE SUPPRESSION, PLUMBING & HVAC	97.3	86.7	92.8	97.3	86.6	92.7	100.4	91.4	96.6	100.1	95.8	98.3	100.1	87.3	94.7	100.4	85.6	94.1
26, 27, 3370	ELECTRICAL, COMMUNICATIONS & UTIL.	101.1	76.0	88.0	100.4	76.0	87.7	95.4	91.9	93.6	99.3	85.2	91.9	104.0	84.8	94.0	98.6	81.3	89.6
MF2016	WEIGHTED AVERAGE	98.5	85.0	92.6	98.6	86.4	93.3	100.0	91.9	96.5	97.7	98.6	98.1	97.6	97.3	97.5	99.6	96.5	98.2

		WISCONSIN																	
DIVISION		KENOSHA			LA CROSSE			LANCASTER			MADISON			MILWAUKEE			NEW RICHMOND		
		531			546			538			537			530, 532			540		
		MAT.	INST.	TOTAL	MAT.	INST.	TOTAL	MAT.	INST.	TOTAL	MAT.	INST.	TOTAL	MAT.	INST.	TOTAL	MAT.	INST.	TOTAL
015433	CONTRACTOR EQUIPMENT		99.8	99.8		102.7	102.7		101.9	101.9		101.9	101.9		89.4	89.4		103.0	103.0
0241, 31 - 34	SITE & INFRASTRUCTURE, DEMOLITION	101.3	104.1	103.3	90.7	104.3	100.2	94.9	106.4	103.0	93.9	107.3	103.3	93.2	95.3	94.6	95.9	104.4	101.9
0310	Concrete Forming & Accessories	104.3	111.2	110.3	82.6	105.7	102.5	95.8	96.3	96.2	100.5	95.7	96.3	100.8	112.4	110.8	90.5	91.3	92.8
0320	Concrete Reinforcing	95.3	113.9	104.7	91.1	103.0	97.1	97.0	102.8	99.9	95.7	103.2	99.5	102.9	114.2	108.6	88.4	110.2	99.4
0330	Cast-in-Place Concrete	115.7	108.3	112.9	90.1	101.6	94.5	105.8	98.7	103.1	102.4	104.8	103.3	93.9	110.9	100.4	104.3	79.8	95.0
03	CONCRETE	101.0	110.4	105.3	83.4	103.8	92.7	95.9	98.4	97.0	93.7	100.3	96.7	95.7	111.2	102.8	89.8	91.9	90.7
04	MASONRY	97.8	110.7	105.8	91.1	98.7	95.8	100.3	95.9	97.6	95.6	99.6	98.1	105.8	114.7	111.3	118.1	95.7	104.2
05	METALS	97.4	105.5	99.9	96.2	102.5	98.1	93.9	99.3	95.5	98.3	100.6	99.0	97.9	97.4	97.8	96.6	103.5	98.7
06	WOOD, PLASTICS & COMPOSITES	97.3	110.8	104.8	84.3	107.8	97.4	92.8	95.7	94.4	93.9	93.4	93.6	98.8	111.3	105.7	87.8	91.7	90.0
07	THERMAL & MOISTURE PROTECTION	100.9	107.2	103.6	103.0	98.7	101.1	100.5	93.2	97.4	98.3	101.2	99.5	102.8	109.9	105.8	104.6	88.3	97.7
08	OPENINGS	90.9	110.8	95.5	101.8	100.2	101.4	92.3	93.5	92.5	102.3	98.3	101.3	101.0	111.1	103.3	87.9	91.8	88.8
0920	Plaster & Gypsum Board	87.3	111.6	103.4	105.0	108.4	107.3	96.5	96.1	96.2	104.3	93.6	97.1	100.3	111.6	107.9	95.6	92.0	93.2
0950, 0980	Ceilings & Acoustic Treatment	86.3	111.6	103.4	89.5	108.4	102.2	82.9	96.1	91.8	89.6	93.6	92.3	96.5	111.6	106.7	57.0	92.0	80.6
0960	Flooring	108.3	121.4	112.0	74.8	118.0	86.8	90.8	105.3	94.8	89.5	113.7	96.2	96.6	117.4	102.3	80.6	100.8	95.2
0970, 0990	Wall Finishes & Painting/Coating	108.0	118.8	114.3	84.9	81.2	82.8	96.8	102.1	99.9	90.1	104.5	98.5	106.8	123.4	116.5	96.2	85.4	89.9
09	FINISHES	96.8	114.3	106.3	84.3	106.3	96.3	91.1	99.0	95.4	91.7	99.6	96.0	100.5	114.7	108.2	82.4	95.7	89.7
COVERS	DIVS. 10 - 14, 25, 28, 41, 43, 44, 46	100.0	101.5	100.3	100.0	96.4	99.2	100.0	86.7	97.0	100.0	99.7	99.9	100.0	102.9	100.6	100.0	88.9	97.5
21, 22, 23	FIRE SUPPRESSION, PLUMBING & HVAC	100.3	97.0	98.9	100.1	87.3	94.6	97.0	87.1	92.8	100.0	96.7	98.6	99.9	105.5	102.3	96.5	87.4	92.6
26, 27, 3370	ELECTRICAL, COMMUNICATIONS & UTIL.	99.6	97.3	98.4	104.3	84.8	94.2	99.0	84.8	91.6	100.0	94.1	96.9	99.1	101.5	100.4	102.1	84.8	93.1
MF2016	WEIGHTED AVERAGE	98.4	105.0	101.3	95.9	96.9	96.4	96.0	93.9	95.1	98.0	98.9	98.4	99.3	106.4	102.5	95.6	92.7	94.3

		WISCONSIN																	
DIVISION		OSHKOSH			PORTAGE			RACINE			RHINELANDER			SUPERIOR			WAUSAU		
		549			539			534			545			548			544		
		MAT.	INST.	TOTAL	MAT.	INST.	TOTAL	MAT.	INST.	TOTAL	MAT.	INST.	TOTAL	MAT.	INST.	TOTAL	MAT.	INST.	TOTAL
015433	CONTRACTOR EQUIPMENT		100.3	100.3		101.9	101.9		101.9	101.9		100.3	100.3		103.0	103.0		100.3	100.3
0241, 31 - 34	SITE & INFRASTRUCTURE, DEMOLITION	92.1	99.4	97.2	85.8	106.2	100.1	95.6	108.1	104.3	104.1	99.5	100.9	92.6	104.1	100.7	87.9	100.7	96.9
0310	Concrete Forming & Accessories	87.1	89.6	89.3	88.5	91.5	91.1	96.8	111.3	109.3	84.9	90.3	89.6	88.6	89.6	89.5	86.6	105.0	102.4
0320	Concrete Reinforcing	89.6	98.5	94.1	97.1	102.9	100.0	95.5	114.0	104.8	89.7	96.8	93.3	88.4	106.9	97.8	89.7	102.8	96.3
0330	Cast-in-Place Concrete	96.1	94.2	95.4	90.8	99.3	94.0	104.3	108.1	105.8	108.9	96.9	104.4	98.3	101.9	99.6	89.6	95.8	91.9
03	CONCRETE	85.1	93.2	88.8	83.9	96.4	89.6	95.2	110.4	102.1	96.6	94.2	95.5	84.6	97.3	90.4	80.3	101.4	90.0
04	MASONRY	105.5	97.3	100.4	99.1	97.0	97.8	100.2	110.7	106.7	123.9	94.7	105.8	117.4	100.8	107.1	105.0	94.8	98.7
05	METALS	96.8	99.4	97.6	94.6	99.7	96.2	98.2	105.5	100.4	96.6	99.1	97.4	97.6	103.8	99.5	96.5	101.8	98.1
06	WOOD, PLASTICS & COMPOSITES	83.9	88.9	86.7	83.7	88.8	86.5	93.7	110.8	103.2	81.6	88.9	85.7	86.3	87.5	87.0	83.4	107.8	97.0
07	THERMAL & MOISTURE PROTECTION	104.8	82.2	95.2	99.9	88.9	95.3	100.8	106.8	103.3	105.7	82.7	96.0	104.2	89.4	98.0	104.6	84.2	96.0
08	OPENINGS	94.6	89.4	93.4	92.4	95.4	93.1	96.3	110.8	99.6	94.6	88.2	93.2	87.4	93.4	88.8	94.8	100.6	96.2
0920	Plaster & Gypsum Board	94.4	88.9	90.7	89.6	88.9	89.1	97.3	111.6	106.9	94.4	88.9	90.7	95.4	87.7	90.2	94.4	108.4	103.8
0950, 0980	Ceilings & Acoustic Treatment	84.6	88.9	87.5	85.4	88.9	87.8	86.3	111.6	103.4	84.6	88.9	87.5	57.8	87.7	78.0	84.6	108.4	100.7
0960	Flooring	87.6	121.4	97.0	87.2	114.8	94.9	91.3	121.4	99.6	87.0	111.7	93.9	90.7	127.2	100.8	87.5	111.7	94.2
0970, 0990	Wall Finishes & Painting/Coating	90.9	94.9	93.3	96.8	102.1	99.9	96.8	121.2	111.0	90.9	83.6	86.6	85.3	110.6	100.1	90.9	83.6	86.6
09	FINISHES	86.7	95.2	91.3	89.0	96.6	93.1	92.0	114.6	104.3	87.5	92.7	90.3	81.8	98.6	90.9	86.4	105.1	96.5
COVERS	DIVS. 10 - 14, 25, 28, 41, 43, 44, 46	100.0	86.4	97.0	100.0	87.4	97.2	100.0	101.5	100.3	100.0	87.6	97.2	100.0	87.2	97.2	100.0	96.3	99.2
21, 22, 23	FIRE SUPPRESSION, PLUMBING & HVAC	97.3	80.9	90.3	97.0	96.5	96.8	100.1	97.1	98.8	97.3	86.4	92.6	96.5	91.5	94.4	97.3	86.8	92.8
26, 27, 3370	ELECTRICAL, COMMUNICATIONS & UTIL.	102.6	76.3	88.9	102.7	94.1	98.2	99.1	103.5	101.4	102.0	77.6	89.3	106.9	98.7	102.6	103.8	80.2	91.5
MF2016	WEIGHTED AVERAGE	95.7	89.2	92.9	94.4	96.6	95.4	97.8	106.2	101.5	98.3	90.1	94.7	95.4	97.1	96.1	95.0	94.5	94.8

For customer support on your Commercial Renovation Costs with RSMeans data, call 800.448.8182.

815

WYOMING

DIVISION		CASPER 826 MAT.	INST.	TOTAL	CHEYENNE 820 MAT.	INST.	TOTAL	NEWCASTLE 827 MAT.	INST.	TOTAL	RAWLINS 823 MAT.	INST.	TOTAL	RIVERTON 825 MAT.	INST.	TOTAL	ROCK SPRINGS 829 - 831 MAT.	INST.	TOTAL
015433	CONTRACTOR EQUIPMENT		97.4	97.4		97.4	97.4		97.4	97.4		97.4	97.4		97.4	97.4		97.4	97.4
0241, 31 - 34	SITE & INFRASTRUCTURE, DEMOLITION	100.6	94.7	96.5	96.1	94.7	95.1	88.0	94.2	92.4	102.6	94.2	96.7	95.8	94.1	94.6	92.2	94.2	93.6
0310	Concrete Forming & Accessories	100.5	55.2	61.4	101.0	68.4	72.9	92.2	73.8	76.4	96.3	74.1	77.1	91.3	62.9	66.8	98.3	63.9	68.7
0320	Concrete Reinforcing	111.9	85.2	98.4	107.4	85.3	96.2	115.3	85.0	100.0	115.0	85.0	99.9	116.1	85.0	100.4	116.1	83.2	99.5
0330	Cast-in-Place Concrete	104.4	77.5	94.2	98.6	77.6	90.6	99.6	76.5	90.8	99.6	76.7	90.9	99.6	76.5	90.8	99.6	76.5	90.8
03	CONCRETE	100.5	68.9	86.1	98.0	74.9	87.5	98.3	77.0	88.6	111.3	77.1	95.7	106.2	72.0	90.6	98.8	72.2	86.6
04	MASONRY	103.9	66.1	80.5	107.3	67.3	82.5	103.9	61.8	77.8	103.9	61.8	77.8	103.9	61.8	77.8	168.0	57.5	99.5
05	METALS	103.6	81.5	96.8	106.1	81.8	98.6	102.3	81.2	95.8	102.3	81.4	95.9	102.4	81.1	95.9	103.2	80.3	96.1
06	WOOD, PLASTICS & COMPOSITES	94.7	49.8	69.7	91.0	67.7	78.0	82.2	76.2	78.8	85.8	76.2	80.4	81.2	61.5	70.3	90.5	62.8	75.1
07	THERMAL & MOISTURE PROTECTION	102.5	67.1	87.5	97.1	69.4	85.4	98.7	66.3	85.0	100.3	77.9	90.8	99.7	64.8	84.9	98.8	67.9	85.7
08	OPENINGS	103.1	61.3	93.5	104.0	71.2	96.5	108.1	75.9	100.7	107.7	75.9	100.4	107.9	67.8	98.7	108.4	68.1	99.2
0920	Plaster & Gypsum Board	103.4	48.5	66.5	91.7	66.8	75.0	88.5	75.5	79.8	88.9	75.5	79.9	88.5	60.5	69.7	100.0	61.8	74.4
0950, 0980	Ceilings & Acoustic Treatment	103.1	48.5	66.3	98.7	66.8	77.2	100.6	75.5	83.7	100.6	75.5	83.7	100.6	60.5	73.6	100.6	61.8	74.5
0960	Flooring	105.8	73.6	96.9	105.0	73.6	96.3	98.7	47.1	84.4	101.6	47.1	86.5	98.2	63.7	88.7	103.9	47.1	88.1
0970, 0990	Wall Finishes & Painting/Coating	91.5	61.0	73.7	97.4	61.0	76.2	94.1	61.0	74.8	94.1	61.0	74.8	94.1	61.0	74.8	94.1	61.0	74.8
09	FINISHES	99.0	57.8	76.6	95.8	68.3	80.8	91.0	67.6	78.3	93.3	67.6	79.3	91.7	62.3	75.7	94.2	59.7	75.4
COVERS	DIVS. 10 - 14, 25, 28, 41, 43, 44, 46	100.0	84.5	96.6	100.0	86.5	97.0	100.0	94.5	98.8	100.0	94.5	98.8	100.0	80.5	95.7	100.0	84.7	96.6
21, 22, 23	FIRE SUPPRESSION, PLUMBING & HVAC	100.1	72.6	88.3	100.0	72.6	88.3	98.3	71.8	87.0	98.3	71.8	87.0	98.3	71.7	87.0	99.9	71.7	87.9
26, 27, 3370	ELECTRICAL, COMMUNICATIONS & UTIL.	101.9	63.7	82.0	104.0	64.6	83.5	102.7	63.3	82.2	102.7	63.3	82.2	102.7	62.4	81.8	100.8	80.9	90.5
MF2016	WEIGHTED AVERAGE	101.4	70.3	87.8	101.3	73.5	89.2	99.9	73.1	88.2	102.1	73.5	89.6	101.2	70.7	87.8	103.9	72.7	90.2

WYOMING / CANADA

DIVISION		SHERIDAN 828 MAT.	INST.	TOTAL	WHEATLAND 822 MAT.	INST.	TOTAL	WORLAND 824 MAT.	INST.	TOTAL	YELLOWSTONE NAT'L PA 821 MAT.	INST.	TOTAL	BARRIE, ONTARIO MAT.	INST.	TOTAL	BATHURST, NEW BRUNSWICK MAT.	INST.	TOTAL
015433	CONTRACTOR EQUIPMENT		97.4	97.4		97.4	97.4		97.4	97.4		97.4	97.4		103.0	103.0		102.8	102.8
0241, 31 - 34	SITE & INFRASTRUCTURE, DEMOLITION	96.2	94.7	95.1	93.0	94.2	93.8	90.2	94.2	93.0	90.3	94.2	93.0	119.2	99.9	105.7	107.0	95.7	99.1
0310	Concrete Forming & Accessories	99.0	63.7	68.6	94.1	50.6	56.6	94.1	63.9	68.0	94.2	63.9	68.1	124.2	86.3	91.5	103.6	60.6	66.5
0320	Concrete Reinforcing	116.1	85.0	100.4	115.3	84.4	99.8	116.1	85.0	100.4	118.1	85.0	101.4	175.3	89.7	132.1	137.7	59.4	98.2
0330	Cast-in-Place Concrete	103.0	77.5	93.3	103.9	76.4	93.5	99.6	76.5	90.8	99.6	76.5	90.8	158.9	86.5	131.4	117.7	59.1	95.4
03	CONCRETE	106.3	72.7	91.0	103.2	66.3	86.3	98.5	72.4	86.6	98.8	72.5	86.8	139.8	87.3	115.8	113.6	60.8	89.5
04	MASONRY	104.2	63.3	78.9	104.3	53.4	72.7	103.9	61.8	77.8	104.0	61.8	77.8	167.9	93.1	121.5	162.5	59.9	98.9
05	METALS	105.9	81.3	98.3	102.2	80.3	95.5	102.4	81.1	95.9	103.0	80.6	96.1	110.3	93.5	105.2	113.7	75.0	101.8
06	WOOD, PLASTICS & COMPOSITES	92.2	61.5	75.1	83.9	45.0	62.2	83.9	62.8	72.2	83.9	62.8	72.2	118.0	84.7	99.5	98.4	60.6	77.4
07	THERMAL & MOISTURE PROTECTION	100.0	67.6	86.3	99.0	60.3	82.6	98.8	66.3	85.0	98.2	66.3	84.7	113.8	90.3	103.9	109.4	61.3	89.0
08	OPENINGS	108.6	67.8	99.2	106.6	58.6	95.6	108.3	68.5	99.1	101.3	68.1	93.7	92.4	85.0	90.7	86.4	53.8	78.9
0920	Plaster & Gypsum Board	112.5	60.5	77.6	88.5	43.5	58.3	88.5	61.8	70.6	88.7	61.8	70.7	151.5	84.2	106.3	126.6	59.5	81.5
0950, 0980	Ceilings & Acoustic Treatment	103.5	60.5	74.5	100.6	43.5	62.1	100.6	61.8	74.5	101.5	61.8	74.7	95.3	84.2	87.9	112.5	59.5	76.7
0960	Flooring	102.5	63.7	91.7	100.3	46.3	85.3	100.3	47.1	85.5	100.3	47.1	85.5	118.1	92.3	110.9	98.6	43.8	83.4
0970, 0990	Wall Finishes & Painting/Coating	96.3	61.0	75.7	94.1	61.0	74.8	94.1	61.0	74.8	94.1	61.0	74.8	104.4	87.6	94.6	109.9	50.1	75.0
09	FINISHES	98.6	63.0	79.2	91.8	49.0	68.5	91.5	59.7	74.2	91.7	59.8	74.4	109.8	87.7	97.8	104.7	56.5	78.5
COVERS	DIVS. 10 - 14, 25, 28, 41, 43, 44, 46	100.0	85.8	96.8	100.0	83.5	96.3	100.0	79.3	95.4	100.0	79.4	95.4	139.2	67.6	123.3	131.1	60.2	115.4
21, 22, 23	FIRE SUPPRESSION, PLUMBING & HVAC	98.3	72.6	87.3	98.3	71.7	86.9	98.3	71.7	87.0	98.3	71.7	87.0	104.5	97.5	101.5	104.6	67.5	88.7
26, 27, 3370	ELECTRICAL, COMMUNICATIONS & UTIL.	105.3	62.4	83.0	102.7	80.9	91.4	102.7	80.9	91.4	101.6	90.9	96.1	116.1	87.6	101.3	112.0	59.0	84.4
MF2016	WEIGHTED AVERAGE	102.8	71.5	89.1	100.6	69.2	86.9	100.1	73.0	88.2	99.4	74.4	88.4	117.0	91.0	105.6	111.3	65.0	91.0

CANADA

DIVISION		BRANDON, MANITOBA MAT.	INST.	TOTAL	BRANTFORD, ONTARIO MAT.	INST.	TOTAL	BRIDGEWATER, NOVA SCOTIA MAT.	INST.	TOTAL	CALGARY, ALBERTA MAT.	INST.	TOTAL	CAP-DE-LA-MADELEINE, QUEBEC MAT.	INST.	TOTAL	CHARLESBOURG, QUEBEC MAT.	INST.	TOTAL
015433	CONTRACTOR EQUIPMENT		105.8	105.8		103.1	103.1		102.6	102.6		128.2	128.2		103.3	103.3		103.3	103.3
0241, 31 - 34	SITE & INFRASTRUCTURE, DEMOLITION	134.2	98.8	109.4	119.6	100.3	106.1	103.7	97.5	99.3	123.1	120.8	121.5	99.2	98.9	99.0	99.2	98.9	99.0
0310	Concrete Forming & Accessories	143.5	69.4	79.6	123.5	93.5	97.6	96.8	71.2	74.7	123.2	99.2	102.5	129.2	83.0	89.4	129.2	83.0	89.4
0320	Concrete Reinforcing	192.0	56.3	123.5	169.0	88.3	128.3	144.0	49.2	96.2	135.2	84.5	109.6	144.0	74.8	109.1	144.0	74.8	109.1
0330	Cast-in-Place Concrete	118.3	74.1	101.5	135.8	107.5	125.0	139.5	70.4	113.3	143.2	108.3	129.9	109.6	92.2	103.0	109.6	92.2	103.0
03	CONCRETE	127.0	69.7	100.8	127.0	97.6	113.6	125.3	68.0	99.1	129.2	100.5	116.1	111.7	85.2	99.6	111.7	85.2	99.6
04	MASONRY	221.4	63.3	123.3	168.8	97.4	124.5	164.2	68.4	104.8	200.4	90.6	132.3	164.9	79.6	112.0	164.9	79.6	112.0
05	METALS	133.5	80.4	117.2	112.4	94.3	106.8	111.7	77.6	101.2	129.5	104.6	121.9	110.8	86.9	103.5	110.8	86.9	103.5
06	WOOD, PLASTICS & COMPOSITES	150.9	70.0	105.9	119.8	92.0	104.3	90.7	70.6	79.5	96.4	98.8	97.8	130.5	82.8	103.9	130.5	82.8	103.9
07	THERMAL & MOISTURE PROTECTION	131.5	72.1	106.4	118.6	96.0	109.0	113.2	70.9	95.3	129.9	97.8	116.3	111.9	88.3	101.9	111.9	88.3	101.9
08	OPENINGS	102.2	63.0	93.2	90.3	91.1	90.5	84.5	64.4	79.9	83.5	87.8	84.4	91.6	76.0	88.0	91.6	76.0	88.0
0920	Plaster & Gypsum Board	113.1	68.8	83.3	116.6	91.7	99.9	122.4	69.8	87.1	121.5	97.9	105.6	144.8	82.2	102.8	144.8	82.2	102.8
0950, 0980	Ceilings & Acoustic Treatment	119.2	68.8	85.2	103.2	91.7	95.5	103.2	69.8	80.6	142.1	97.9	112.3	103.2	82.2	89.0	103.2	82.2	89.0
0960	Flooring	130.3	65.4	112.3	112.1	92.3	106.6	94.7	62.3	85.7	117.9	88.6	109.8	112.1	90.9	106.2	112.1	90.9	106.2
0970, 0990	Wall Finishes & Painting/Coating	117.0	56.5	81.8	109.0	96.2	101.5	109.0	62.0	81.6	117.1	111.2	113.7	109.0	87.4	96.4	109.0	87.4	96.4
09	FINISHES	118.4	67.6	90.8	106.3	93.8	99.5	101.1	68.8	83.6	122.9	99.1	109.6	109.4	85.2	96.3	109.4	85.2	96.3
COVERS	DIVS. 10 - 14, 25, 28, 41, 43, 44, 46	131.1	62.7	115.9	131.1	69.7	117.5	131.1	63.3	116.0	131.1	95.5	123.2	131.1	78.8	119.5	131.1	78.8	119.5
21, 22, 23	FIRE SUPPRESSION, PLUMBING & HVAC	104.7	82.2	95.1	104.6	100.4	102.8	104.6	82.8	95.3	105.5	93.3	100.3	105.1	87.7	97.6	105.1	87.7	97.6
26, 27, 3370	ELECTRICAL, COMMUNICATIONS & UTIL.	111.1	66.6	88.0	111.4	87.1	98.7	116.0	62.0	87.9	107.3	99.1	103.1	110.4	68.9	88.8	110.4	68.9	88.8
MF2016	WEIGHTED AVERAGE	123.6	73.8	101.8	114.3	94.9	105.8	112.3	73.4	95.3	119.5	98.8	110.4	111.7	83.6	99.4	111.7	83.6	99.4

City Cost Indexes

CANADA

DIVISION		CHARLOTTETOWN, PRINCE EDWARD ISLAND			CHICOUTIMI, QUEBEC			CORNER BROOK, NEWFOUNDLAND			CORNWALL, ONTARIO			DALHOUSIE, NEW BRUNSWICK			DARTMOUTH, NOVA SCOTIA		
		MAT.	INST.	TOTAL	MAT.	INST.	TOTAL	MAT.	INST.	TOTAL	MAT.	INST.	TOTAL	MAT.	INST.	TOTAL	MAT.	INST.	TOTAL
015433	CONTRACTOR EQUIPMENT		120.7	120.7		103.4	103.4		104.0	104.0		103.1	103.1		102.6	102.6		102.9	102.9
0241, 31 - 34	SITE & INFRASTRUCTURE, DEMOLITION	133.6	108.5	116.0	103.9	98.6	100.2	138.7	96.1	108.8	117.7	99.7	105.1	102.8	95.5	97.7	124.9	97.7	105.9
0310	Concrete Forming & Accessories	118.4	55.4	64.1	131.9	91.0	96.7	120.7	79.3	85.0	121.3	86.5	91.3	102.4	60.8	66.5	110.5	71.2	76.6
0320	Concrete Reinforcing	152.2	48.0	99.6	103.9	94.2	99.0	176.3	50.4	112.8	169.0	88.0	128.2	147.3	59.4	102.9	184.1	49.2	116.0
0330	Cast-in-Place Concrete	149.0	59.1	114.9	110.8	97.0	105.6	137.7	67.0	110.8	122.1	97.9	112.9	115.9	59.1	94.3	132.9	70.4	109.2
03	CONCRETE	134.7	57.3	99.3	104.6	93.8	99.7	159.1	70.7	118.7	120.4	91.1	107.0	118.1	60.9	91.9	141.6	68.1	108.0
04	MASONRY	177.7	57.2	103.0	164.8	90.3	118.6	217.1	77.5	130.5	167.6	89.3	119.0	164.5	59.9	99.6	232.1	68.4	130.6
05	METALS	137.2	81.7	120.2	113.4	92.4	107.0	133.9	77.1	116.5	112.2	93.0	106.3	106.9	75.1	97.2	134.6	77.8	117.1
06	WOOD, PLASTICS & COMPOSITES	99.9	54.8	74.8	131.2	91.4	109.0	129.3	85.4	104.9	118.1	85.5	100.0	97.5	60.6	77.0	117.5	70.6	91.4
07	THERMAL & MOISTURE PROTECTION	128.5	59.1	99.1	109.8	98.0	104.8	136.2	69.2	107.9	118.4	90.5	106.6	116.2	61.3	92.9	133.2	71.0	106.9
08	OPENINGS	85.1	48.0	76.6	90.5	77.8	87.5	108.7	70.0	99.8	91.6	84.7	90.0	87.3	53.8	79.6	92.6	64.4	86.1
0920	Plaster & Gypsum Board	119.5	53.0	74.9	144.8	91.0	108.7	147.0	84.8	105.3	173.2	85.1	114.1	130.5	59.5	82.8	142.6	69.8	93.7
0950, 0980	Ceilings & Acoustic Treatment	128.0	53.0	77.4	111.7	91.0	97.8	120.0	84.8	96.3	105.7	85.1	91.8	106.5	59.5	74.8	126.9	69.8	88.4
0960	Flooring	106.3	58.4	93.0	114.3	90.9	107.8	112.4	52.4	95.7	112.1	90.9	106.2	100.6	67.1	91.3	107.3	62.3	94.8
0970, 0990	Wall Finishes & Painting/Coating	117.5	41.7	73.3	109.9	105.2	107.1	116.9	59.0	83.1	109.0	89.6	97.7	112.5	50.1	76.1	116.9	62.0	84.9
09	FINISHES	116.7	54.8	83.0	111.5	93.4	101.7	118.1	73.7	94.0	114.4	87.8	99.9	105.9	61.1	81.6	116.0	68.8	90.3
COVERS	DIVS. 10 - 14, 25, 28, 41, 43, 44, 46	131.1	60.4	115.4	131.1	82.5	120.3	131.1	63.1	116.0	131.1	67.3	116.9	131.1	60.1	115.3	131.1	63.3	116.1
21, 22, 23	FIRE SUPPRESSION, PLUMBING & HVAC	104.9	61.4	86.3	104.6	81.5	94.7	104.7	68.7	89.3	105.1	98.3	102.2	104.7	67.5	88.8	104.7	82.8	95.4
26, 27, 3370	ELECTRICAL, COMMUNICATIONS & UTIL.	114.3	50.1	80.9	108.7	85.6	96.7	108.8	54.4	80.5	111.7	88.0	99.4	113.9	55.8	83.6	113.0	62.0	86.5
MF2016	WEIGHTED AVERAGE	120.9	62.8	95.5	111.1	88.9	101.4	128.0	71.5	103.3	114.2	91.3	104.2	111.3	65.2	91.1	124.6	73.5	102.2

CANADA

DIVISION		EDMONTON, ALBERTA			FORT MCMURRAY, ALBERTA			FREDERICTON, NEW BRUNSWICK			GATINEAU, QUEBEC			GRANBY, QUEBEC			HALIFAX, NOVA SCOTIA		
		MAT.	INST.	TOTAL	MAT.	INST.	TOTAL	MAT.	INST.	TOTAL	MAT.	INST.	TOTAL	MAT.	INST.	TOTAL	MAT.	INST.	TOTAL
015433	CONTRACTOR EQUIPMENT		128.5	128.5		105.5	105.5		120.7	120.7		103.3	103.3		103.3	103.3		118.3	118.3
0241, 31 - 34	SITE & INFRASTRUCTURE, DEMOLITION	129.3	121.0	123.5	124.8	101.0	108.1	113.6	110.3	111.3	99.0	98.9	99.0	99.5	98.9	99.1	107.8	109.8	109.2
0310	Concrete Forming & Accessories	119.7	99.2	102.0	121.4	91.7	95.7	120.2	61.4	69.5	129.2	82.9	89.3	129.2	82.9	89.3	117.4	85.0	89.4
0320	Concrete Reinforcing	135.7	84.5	109.9	156.5	84.3	120.1	151.1	59.6	105.0	152.2	74.8	113.1	152.2	74.7	113.1	152.2	72.3	111.9
0330	Cast-in-Place Concrete	147.7	108.4	132.7	180.3	103.5	151.1	116.8	59.5	95.0	108.1	92.2	102.0	111.7	92.2	104.3	106.9	86.0	98.9
03	CONCRETE	131.2	100.6	117.2	146.3	94.7	122.7	119.4	62.1	93.2	112.1	85.2	99.8	113.8	85.1	100.7	114.0	84.3	100.4
04	MASONRY	196.0	90.6	130.7	209.5	87.3	133.7	182.5	61.2	107.3	164.8	79.6	112.0	165.1	79.6	112.1	180.1	87.1	122.4
05	METALS	136.2	105.0	126.6	139.9	92.8	125.4	134.2	87.1	119.7	110.8	86.8	103.4	111.0	86.7	103.6	138.4	98.1	126.0
06	WOOD, PLASTICS & COMPOSITES	96.8	98.8	97.9	114.2	90.9	101.2	109.4	60.9	82.4	130.5	82.8	103.9	130.5	82.8	103.9	100.5	84.0	91.3
07	THERMAL & MOISTURE PROTECTION	129.7	97.8	116.2	127.1	93.9	113.0	123.7	61.3	97.3	111.9	88.3	101.9	111.9	86.7	101.2	129.0	86.1	110.8
08	OPENINGS	81.8	87.8	83.2	91.6	83.4	89.7	86.1	52.8	78.5	91.6	71.4	87.0	91.6	71.4	87.0	88.6	76.8	85.9
0920	Plaster & Gypsum Board	116.7	97.9	104.1	117.4	90.3	99.2	123.5	59.5	80.5	114.5	82.2	92.8	114.5	82.2	92.8	118.1	83.3	94.7
0950, 0080	Ceilings & Acoustic Treatment	137.2	97.9	110.7	110.8	90.3	97.0	133.9	59.5	83.7	103.2	82.2	89.0	103.2	82.2	89.0	128.5	83.3	97.9
0960	Flooring	114.0	88.6	107.0	112.1	88.6	105.6	109.5	70.0	98.6	112.1	90.9	106.2	112.1	90.9	106.2	99.6	83.8	95.2
0970, 0990	Wall Finishes & Painting/Coating	108.3	111.2	110.0	109.1	92.3	99.3	114.8	63.8	85.1	109.0	87.4	96.4	109.0	87.4	96.4	112.8	91.7	100.5
09	FINISHES	115.6	99.1	106.6	109.2	91.4	99.5	118.5	63.4	88.5	105.3	85.2	94.4	105.3	85.2	94.4	110.4	85.9	97.1
COVERS	DIVS. 10 - 14, 25, 28, 41, 43, 44, 46	131.1	95.5	123.2	131.1	92.3	122.5	131.1	60.7	115.5	131.1	78.8	119.5	131.1	78.8	119.5	131.1	68.7	117.2
21, 22, 23	FIRE SUPPRESSION, PLUMBING & HVAC	105.5	93.3	100.3	105.1	97.1	101.7	105.0	76.7	92.9	105.1	87.7	97.6	104.6	87.7	97.4	105.3	82.4	95.5
26, 27, 3370	ELECTRICAL, COMMUNICATIONS & UTIL.	110.3	99.1	104.5	105.6	81.8	93.3	111.9	73.6	92.0	110.4	68.9	88.8	111.0	68.9	89.1	109.4	87.0	97.7
MF2016	WEIGHTED AVERAGE	120.4	98.8	110.9	123.4	91.9	109.7	118.2	72.5	98.2	111.4	83.4	99.1	111.6	83.3	99.2	117.4	87.2	104.2

CANADA

DIVISION		HAMILTON, ONTARIO			HULL, QUEBEC			JOLIETTE, QUEBEC			KAMLOOPS, BRITISH COLUMBIA			KINGSTON, ONTARIO			KITCHENER, ONTARIO		
		MAT.	INST.	TOTAL	MAT.	INST.	TOTAL	MAT.	INST.	TOTAL	MAT.	INST.	TOTAL	MAT.	INST.	TOTAL	MAT.	INST.	TOTAL
015433	CONTRACTOR EQUIPMENT		118.4	118.4		103.3	103.3		103.3	103.3		106.7	106.7		105.4	105.4		105.3	105.3
0241, 31 - 34	SITE & INFRASTRUCTURE, DEMOLITION	112.1	113.0	112.7	99.0	98.9	99.0	99.7	98.9	99.2	122.5	102.7	108.6	117.7	103.7	107.9	97.2	105.0	102.7
0310	Concrete Forming & Accessories	122.8	95.7	99.5	129.2	82.9	89.3	129.2	83.0	89.4	121.4	86.3	91.1	121.4	86.5	91.3	113.1	88.6	92.0
0320	Concrete Reinforcing	150.1	98.8	124.2	152.2	74.8	113.1	144.0	74.8	109.1	112.8	78.3	95.4	169.0	88.0	128.2	104.2	98.7	101.4
0330	Cast-in-Place Concrete	118.5	103.5	112.8	108.1	92.2	102.1	112.6	92.2	104.9	97.5	96.8	97.2	122.1	97.9	112.9	113.4	99.5	108.1
03	CONCRETE	116.3	99.6	108.7	112.1	85.2	99.8	113.2	85.2	100.4	121.2	88.9	106.4	122.3	91.1	108.0	100.5	94.4	97.7
04	MASONRY	170.9	101.7	128.0	164.8	79.6	112.0	165.2	79.6	112.1	172.2	87.6	119.7	174.7	89.3	121.8	145.6	99.0	116.7
05	METALS	131.7	106.4	123.9	111.0	86.8	103.6	111.0	86.9	103.6	112.8	89.2	105.6	113.5	92.9	107.2	121.0	97.5	113.8
06	WOOD, PLASTICS & COMPOSITES	99.9	94.7	97.0	130.5	82.8	103.9	130.5	82.8	103.9	102.5	84.5	92.5	118.1	85.6	100.0	107.6	86.4	95.8
07	THERMAL & MOISTURE PROTECTION	122.0	101.6	113.4	111.9	88.3	101.9	111.9	88.3	101.9	128.2	85.8	110.3	118.4	91.6	107.1	115.0	98.8	108.2
08	OPENINGS	85.6	93.8	87.5	91.6	71.4	87.0	91.6	76.0	88.0	88.4	82.5	87.1	91.6	84.4	90.0	83.1	87.4	84.1
0920	Plaster & Gypsum Board	121.1	94.1	103.0	114.5	82.2	92.8	144.8	82.2	102.8	102.2	83.7	89.8	176.2	85.2	115.1	105.6	86.0	92.4
0950, 0980	Ceilings & Acoustic Treatment	130.0	94.1	105.8	103.2	82.2	89.0	103.2	82.2	89.0	103.2	83.7	90.0	118.5	85.2	96.1	105.1	86.0	92.2
0960	Flooring	109.7	97.8	106.4	112.1	90.9	106.2	112.1	90.9	106.2	111.0	52.0	94.6	112.1	90.9	106.2	100.1	97.8	99.5
0970, 0990	Wall Finishes & Painting/Coating	102.9	105.4	104.3	109.0	87.4	96.4	109.0	87.4	96.4	109.0	79.9	92.0	109.0	82.8	93.7	103.1	95.1	98.4
09	FINISHES	113.5	97.1	104.6	105.3	85.2	94.4	109.4	85.2	96.3	105.8	79.6	91.5	117.3	87.1	100.9	99.1	90.5	94.4
COVERS	DIVS. 10 - 14, 25, 28, 41, 43, 44, 46	131.1	93.3	122.7	131.1	78.8	119.5	131.1	78.8	119.5	131.1	87.9	121.5	131.1	67.3	116.9	131.1	90.7	122.2
21, 22, 23	FIRE SUPPRESSION, PLUMBING & HVAC	105.5	92.2	99.8	104.6	87.7	97.4	104.6	87.7	97.4	104.6	91.2	98.9	105.1	98.4	102.2	104.8	90.9	98.8
26, 27, 3370	ELECTRICAL, COMMUNICATIONS & UTIL.	108.4	102.7	105.4	112.3	68.9	89.7	111.0	68.9	89.1	114.3	78.3	95.6	111.7	86.8	98.7	110.8	100.1	105.2
MF2016	WEIGHTED AVERAGE	116.0	99.6	108.8	111.5	83.4	99.2	111.9	83.6	99.5	114.1	87.2	102.3	115.3	91.4	104.8	109.1	95.1	103.0

For customer support on your Commercial Renovation Costs with RSMeans data, call 800.448.8182.

City Cost Indexes

CANADA

DIVISION		LAVAL, QUEBEC			LETHBRIDGE, ALBERTA			LLOYDMINSTER, ALBERTA			LONDON, ONTARIO			MEDICINE HAT, ALBERTA			MONCTON, NEW BRUNSWICK		
		MAT.	INST.	TOTAL	MAT.	INST.	TOTAL	MAT.	INST.	TOTAL	MAT.	INST.	TOTAL	MAT.	INST.	TOTAL	MAT.	INST.	TOTAL
015433	CONTRACTOR EQUIPMENT		103.3	103.3		105.5	105.5		105.5	105.5		119.8	119.8		105.5	105.5		102.8	102.8
0241, 31 - 34	SITE & INFRASTRUCTURE, DEMOLITION	99.5	98.9	99.1	117.2	101.6	106.3	117.1	101.0	105.8	104.6	112.2	109.9	115.8	101.0	105.5	106.4	96.0	99.1
0310	Concrete Forming & Accessories	129.5	82.9	89.3	122.8	91.7	96.0	120.9	82.0	87.3	121.9	90.0	94.4	122.7	81.9	87.5	103.6	63.2	68.8
0320	Concrete Reinforcing	152.2	74.8	113.1	156.5	84.3	120.1	156.5	84.3	120.1	150.1	97.6	123.6	156.5	84.3	120.1	137.7	71.0	104.1
0330	Cast-in-Place Concrete	111.7	92.2	104.3	135.2	103.5	123.2	125.5	99.8	115.7	117.1	101.6	111.2	125.5	99.8	115.7	113.3	69.7	96.7
03	CONCRETE	113.8	85.2	100.7	125.1	94.8	111.2	120.3	89.1	106.0	115.6	96.2	106.7	120.4	89.0	106.1	111.5	68.0	91.6
04	MASONRY	165.1	79.6	112.1	183.2	87.3	123.7	164.1	80.9	112.5	179.6	99.9	130.2	164.1	80.9	112.5	162.1	61.4	99.7
05	METALS	110.9	86.8	103.5	134.6	92.8	121.8	112.8	92.7	106.6	131.7	107.7	124.3	113.0	92.6	106.7	113.7	85.9	105.2
06	WOOD, PLASTICS & COMPOSITES	130.6	82.8	104.0	117.6	90.9	102.7	114.2	81.2	95.8	106.2	87.7	95.9	117.6	81.2	97.3	98.4	62.6	78.4
07	THERMAL & MOISTURE PROTECTION	112.4	88.3	102.2	124.2	93.9	111.4	120.7	89.3	107.4	124.0	99.5	113.6	127.4	89.3	111.3	113.9	65.5	93.4
08	OPENINGS	91.6	71.4	87.0	91.6	83.4	89.7	91.6	78.0	88.5	80.4	88.8	82.4	91.6	78.0	88.5	86.4	61.5	80.7
0920	Plaster & Gypsum Board	114.8	82.2	92.9	106.7	90.3	95.7	102.7	80.3	87.6	123.0	86.9	98.8	105.0	80.3	88.4	126.6	61.5	82.8
0950, 0980	Ceilings & Acoustic Treatment	103.2	82.2	89.0	110.8	90.3	97.0	103.2	80.3	87.8	138.5	86.9	103.7	103.2	80.3	87.8	112.5	61.5	78.1
0960	Flooring	112.1	90.9	106.2	112.1	88.6	105.6	112.1	88.6	105.6	106.2	97.8	103.9	112.1	88.6	105.6	98.6	68.7	90.3
0970, 0990	Wall Finishes & Painting/Coating	109.0	87.4	96.4	108.9	100.8	104.2	109.1	78.5	91.3	109.3	102.0	105.0	108.9	78.5	91.2	109.9	51.4	75.8
09	FINISHES	105.3	85.2	94.4	106.8	92.4	99.0	104.9	82.6	92.7	116.2	92.3	103.2	105.0	82.6	92.8	104.7	62.9	82.0
COVERS	DIVS. 10 - 14, 25, 28, 41, 43, 44, 46	131.1	78.8	119.5	131.1	92.3	122.5	131.1	89.0	121.8	131.1	92.1	122.5	131.1	89.0	121.8	131.1	61.8	115.7
21, 22, 23	FIRE SUPPRESSION, PLUMBING & HVAC	104.8	87.7	97.5	104.9	93.8	100.2	105.1	93.7	100.2	105.5	89.4	98.6	104.6	90.5	98.6	104.6	69.9	89.8
26, 27, 3370	ELECTRICAL, COMMUNICATIONS & UTIL.	109.3	68.9	88.3	107.1	81.8	94.0	104.8	81.8	92.9	105.3	100.1	102.6	104.8	81.8	92.9	115.8	60.8	87.2
MF2016	WEIGHTED AVERAGE	111.5	83.4	99.2	118.3	91.4	106.5	112.7	88.1	101.9	115.6	97.0	107.5	112.9	87.4	101.7	111.5	69.3	93.0

CANADA

DIVISION		MONTREAL, QUEBEC			MOOSE JAW, SASKATCHEWAN			NEW GLASGOW, NOVA SCOTIA			NEWCASTLE, NEW BRUNSWICK			NORTH BAY, ONTARIO			OSHAWA, ONTARIO		
		MAT.	INST.	TOTAL	MAT.	INST.	TOTAL	MAT.	INST.	TOTAL	MAT.	INST.	TOTAL	MAT.	INST.	TOTAL	MAT.	INST.	TOTAL
015433	CONTRACTOR EQUIPMENT		123.5	123.5		101.7	101.7		102.9	102.9		102.8	102.8		103.5	103.5		105.3	105.3
0241, 31 - 34	SITE & INFRASTRUCTURE, DEMOLITION	114.0	113.5	113.6	117.1	95.1	101.7	119.1	97.7	104.1	107.0	95.7	99.1	134.5	99.6	110.0	108.3	103.9	105.2
0310	Concrete Forming & Accessories	124.4	91.6	96.1	106.3	57.2	64.0	110.5	71.2	76.6	103.6	60.8	66.7	144.3	84.0	92.3	118.5	88.7	92.8
0320	Concrete Reinforcing	126.8	94.3	110.4	110.4	63.4	86.7	176.3	49.2	112.2	137.7	59.4	98.2	207.8	87.6	147.1	165.2	93.6	129.1
0330	Cast-in-Place Concrete	130.9	99.1	118.8	121.7	67.0	100.9	132.9	70.4	109.2	117.7	59.1	95.4	128.0	84.7	111.5	131.1	87.2	114.4
03	CONCRETE	119.3	95.5	108.4	105.8	62.7	86.1	140.6	68.1	107.4	113.6	61.0	89.5	142.4	85.3	116.3	122.7	89.4	107.5
04	MASONRY	176.1	90.3	122.9	162.8	58.8	98.3	216.7	68.4	124.7	162.5	59.9	98.9	223.0	85.4	137.7	148.6	91.9	113.4
05	METALS	139.5	104.4	128.7	109.8	76.7	99.6	131.8	77.8	115.2	113.7	75.2	101.9	132.7	92.8	120.4	111.5	96.2	106.8
06	WOOD, PLASTICS & COMPOSITES	105.6	91.9	98.0	100.1	55.7	75.4	117.5	70.6	91.4	98.4	60.6	77.4	153.9	84.2	115.1	114.1	87.4	99.3
07	THERMAL & MOISTURE PROTECTION	119.2	98.2	110.3	111.3	63.0	90.9	133.2	71.0	106.9	113.9	61.3	91.6	139.8	86.8	117.4	116.0	91.1	105.4
08	OPENINGS	84.8	79.9	83.7	87.6	54.0	79.9	92.6	64.4	86.1	86.4	53.8	78.9	100.6	82.3	96.4	88.5	88.9	88.6
0920	Plaster & Gypsum Board	125.4	91.0	102.3	99.6	54.5	69.3	140.9	69.8	93.1	126.6	59.5	81.5	134.7	83.7	100.4	108.8	87.1	94.2
0950, 0980	Ceilings & Acoustic Treatment	140.4	91.0	107.1	103.2	54.5	70.3	119.2	69.8	85.9	112.5	59.5	76.7	119.2	83.7	95.3	101.7	87.1	91.8
0960	Flooring	110.0	93.1	105.3	102.4	56.7	89.7	107.3	62.3	94.8	98.6	67.1	89.9	130.3	90.9	119.4	102.9	100.3	102.2
0970, 0990	Wall Finishes & Painting/Coating	104.5	105.2	104.9	109.0	63.8	82.6	116.9	62.0	84.9	109.9	50.1	75.0	116.9	88.9	100.6	103.1	109.5	106.9
09	FINISHES	116.1	94.2	104.2	101.3	57.5	77.5	114.2	68.8	89.5	104.7	61.1	81.0	121.1	86.1	102.1	100.3	92.9	96.3
COVERS	DIVS. 10 - 14, 25, 28, 41, 43, 44, 46	131.1	83.7	120.6	131.1	59.8	115.3	131.1	63.3	116.1	131.1	60.2	115.4	131.1	66.1	116.7	131.1	90.4	122.1
21, 22, 23	FIRE SUPPRESSION, PLUMBING & HVAC	105.6	81.6	95.4	105.1	73.3	91.5	104.7	82.8	95.4	104.6	67.5	88.7	104.7	96.3	101.1	104.8	101.9	103.5
26, 27, 3370	ELECTRICAL, COMMUNICATIONS & UTIL.	108.4	85.7	96.6	113.3	59.5	85.3	109.1	62.0	84.6	111.4	59.0	84.1	109.4	87.9	98.2	111.9	90.3	100.7
MF2016	WEIGHTED AVERAGE	118.0	91.7	106.5	110.1	66.7	91.1	122.6	73.5	101.1	111.3	65.7	91.3	125.6	89.2	109.7	111.6	94.7	104.2

CANADA

DIVISION		OTTAWA, ONTARIO			OWEN SOUND, ONTARIO			PETERBOROUGH, ONTARIO			PORTAGE LA PRAIRIE, MANITOBA			PRINCE ALBERT, SASKATCHEWAN			PRINCE GEORGE, BRITISH COLUMBIA		
		MAT.	INST.	TOTAL	MAT.	INST.	TOTAL	MAT.	INST.	TOTAL	MAT.	INST.	TOTAL	MAT.	INST.	TOTAL	MAT.	INST.	TOTAL
015433	CONTRACTOR EQUIPMENT		121.6	121.6		103.0	103.0		103.1	103.1		105.4	105.4		101.7	101.7		106.7	106.7
0241, 31 - 34	SITE & INFRASTRUCTURE, DEMOLITION	108.4	114.4	112.6	119.2	99.7	105.6	119.6	99.7	105.7	118.3	98.5	104.5	112.6	95.3	100.5	126.0	102.7	109.7
0310	Concrete Forming & Accessories	122.2	90.0	94.4	124.2	82.5	88.2	123.5	85.0	90.3	122.9	68.9	76.3	106.3	57.0	63.8	111.6	81.3	85.4
0320	Concrete Reinforcing	130.3	97.6	113.8	175.3	89.7	132.1	169.0	88.1	128.2	156.5	56.3	105.9	115.3	63.3	89.1	112.8	78.3	95.4
0330	Cast-in-Place Concrete	126.9	104.4	118.4	158.9	80.6	129.2	135.8	86.2	116.9	125.5	73.4	105.7	110.3	66.9	93.8	122.2	96.8	112.5
03	CONCRETE	117.8	97.3	108.4	139.8	83.6	114.1	127.0	86.4	108.4	114.2	69.3	93.7	101.0	62.6	83.4	132.3	86.7	111.4
04	MASONRY	172.3	102.5	129.0	167.9	90.8	120.1	168.8	91.8	121.0	167.3	62.3	102.2	161.9	58.8	98.0	174.2	87.6	120.5
05	METALS	132.4	109.0	125.2	110.3	93.3	105.1	112.4	93.1	106.4	113.0	80.2	102.9	109.8	76.5	99.6	112.8	89.2	105.6
06	WOOD, PLASTICS & COMPOSITES	106.4	86.5	95.4	118.0	80.9	97.4	119.8	82.8	99.2	117.5	70.0	91.1	100.1	55.7	75.4	102.5	77.7	88.7
07	THERMAL & MOISTURE PROTECTION	130.7	100.6	118.0	113.8	87.6	102.7	118.6	92.4	107.5	111.8	71.6	94.8	111.2	61.9	90.3	122.0	85.0	106.4
08	OPENINGS	89.1	88.1	88.9	92.4	81.6	89.9	90.3	84.0	88.9	91.6	63.0	85.1	86.5	54.0	79.1	88.4	78.7	86.2
0920	Plaster & Gypsum Board	129.7	85.7	100.2	151.5	80.4	103.8	116.6	82.4	93.6	104.6	68.8	80.6	99.6	54.5	69.3	102.2	76.6	85.0
0950, 0980	Ceilings & Acoustic Treatment	134.2	85.7	101.5	95.3	80.4	85.3	103.2	82.4	89.1	103.2	68.8	80.0	103.2	54.5	70.3	103.7	76.6	85.3
0960	Flooring	102.3	93.2	99.8	118.1	92.3	110.9	112.1	90.9	106.2	112.1	65.4	99.2	102.4	56.7	89.7	107.7	71.2	97.6
0970, 0990	Wall Finishes & Painting/Coating	110.0	96.7	102.2	104.4	87.6	94.6	109.0	91.1	98.6	109.1	56.5	78.4	109.0	54.4	77.1	109.0	79.9	92.0
09	FINISHES	116.1	90.8	102.4	109.8	84.9	96.3	106.3	86.7	95.6	105.0	67.3	84.5	101.3	56.5	76.9	104.8	78.8	90.7
COVERS	DIVS. 10 - 14, 25, 28, 41, 43, 44, 46	131.1	90.7	122.1	139.2	66.4	123.0	131.1	67.5	117.0	131.1	62.3	115.8	131.1	59.8	115.3	131.1	87.2	121.4
21, 22, 23	FIRE SUPPRESSION, PLUMBING & HVAC	105.6	91.1	99.4	104.5	96.2	101.0	104.6	99.8	102.6	104.6	81.7	94.8	105.1	66.0	88.4	104.6	91.2	98.9
26, 27, 3370	ELECTRICAL, COMMUNICATIONS & UTIL.	108.5	100.9	104.5	117.6	86.7	101.5	111.4	87.5	99.0	113.0	57.5	84.1	113.3	59.5	85.3	111.4	78.3	94.2
MF2016	WEIGHTED AVERAGE	117.2	98.0	108.8	117.1	89.2	104.9	114.3	91.1	104.1	112.7	72.2	94.9	109.2	65.0	89.8	115.1	86.6	102.6

For customer support on your Commercial Renovation Costs with RSMeans data, call 800.448.8182.

City Cost Indexes

CANADA

DIVISION		QUEBEC CITY, QUEBEC			RED DEER, ALBERTA			REGINA, SASKATCHEWAN			RIMOUSKI, QUEBEC			ROUYN-NORANDA, QUEBEC			SAINT HYACINTHE, QUEBEC		
		MAT.	INST.	TOTAL	MAT.	INST.	TOTAL	MAT.	INST.	TOTAL	MAT.	INST.	TOTAL	MAT.	INST.	TOTAL	MAT.	INST.	TOTAL
015433	CONTRACTOR EQUIPMENT		124.2	124.2		105.5	105.5		124.2	124.2		103.3	103.3		103.3	103.3		103.3	103.3
0241, 31 - 34	SITE & INFRASTRUCTURE, DEMOLITION	115.4	113.3	113.9	115.8	101.0	105.5	127.4	115.0	118.7	99.3	98.5	98.8	99.0	98.9	99.0	99.5	98.9	99.1
0310	Concrete Forming & Accessories	122.5	91.7	95.9	136.9	81.9	89.5	124.8	95.6	99.6	129.2	91.0	96.3	129.2	82.9	89.3	129.2	82.9	89.3
0320	Concrete Reinforcing	129.3	94.3	111.7	156.5	84.3	120.1	147.1	91.7	119.1	108.6	94.2	101.3	152.2	74.8	113.1	152.2	74.8	113.1
0330	Cast-in-Place Concrete	132.7	98.9	119.9	125.5	99.8	115.7	144.3	100.7	127.7	113.7	97.0	107.4	108.1	92.2	102.1	111.7	92.2	104.3
03	CONCRETE	120.4	95.6	109.0	121.3	89.0	106.6	129.7	97.4	114.9	109.2	93.7	102.1	112.1	85.2	99.8	113.8	85.2	100.7
04	MASONRY	170.9	90.3	120.9	164.1	80.9	112.5	194.2	91.1	130.3	164.5	90.3	118.5	164.8	79.6	112.0	165.0	79.6	112.1
05	METALS	139.5	106.1	129.3	113.0	92.6	106.7	143.2	103.9	131.1	110.5	92.3	105.0	111.0	86.8	103.6	111.0	86.8	103.6
06	WOOD, PLASTICS & COMPOSITES	111.7	91.8	100.6	117.6	81.2	97.3	109.4	96.2	102.0	130.5	91.4	108.7	130.5	82.8	103.9	130.5	82.8	103.9
07	THERMAL & MOISTURE PROTECTION	115.8	98.4	108.4	137.7	89.3	117.2	132.7	90.0	114.7	111.9	98.0	106.0	111.9	88.3	101.9	112.2	88.3	102.1
08	OPENINGS	88.5	87.4	88.2	91.6	78.0	88.5	86.9	83.4	86.1	91.2	77.8	84.1	91.6	71.4	87.0	91.6	71.4	87.0
0920	Plaster & Gypsum Board	128.6	91.0	103.4	105.0	80.3	88.4	134.8	95.6	108.5	144.6	91.0	108.6	114.3	82.2	92.8	114.3	82.2	92.8
0950, 0980	Ceilings & Acoustic Treatment	137.2	91.0	106.1	103.2	80.3	87.8	141.5	95.6	110.6	102.3	91.0	94.7	102.3	82.2	88.8	102.3	82.2	88.8
0960	Flooring	110.8	93.1	105.9	114.5	88.6	107.3	113.4	97.8	109.1	113.4	90.9	107.2	112.1	90.9	106.2	112.1	90.9	106.2
0970, 0990	Wall Finishes & Painting/Coating	122.1	105.2	112.2	108.9	78.5	91.2	115.1	93.1	102.2	112.1	105.2	108.1	109.0	87.4	96.4	109.0	87.4	96.4
09	FINISHES	117.5	94.1	104.8	105.8	82.6	93.2	125.2	96.7	109.7	109.9	93.4	100.9	105.1	85.2	94.3	105.1	85.2	94.3
COVERS	DIVS. 10 - 14, 25, 28, 41, 43, 44, 46	131.1	83.4	120.5	131.1	89.0	121.8	131.1	71.8	117.9	131.1	82.5	120.3	131.1	78.8	119.5	131.1	78.8	119.5
21, 22, 23	FIRE SUPPRESSION, PLUMBING & HVAC	105.5	81.6	95.3	104.6	90.5	98.6	105.2	90.5	98.9	104.6	81.5	94.7	104.6	87.7	97.4	100.8	87.7	95.2
26, 27, 3370	ELECTRICAL, COMMUNICATIONS & UTIL.	109.6	85.7	97.2	104.8	81.8	92.9	111.4	96.6	103.7	111.0	85.6	97.8	111.0	68.9	89.1	111.6	68.9	89.4
MF2016	WEIGHTED AVERAGE	118.5	92.2	107.0	113.3	87.4	102.0	122.8	95.4	110.8	111.3	88.9	101.5	111.4	83.4	99.1	110.8	83.4	98.8

CANADA

DIVISION		SAINT JOHN, NEW BRUNSWICK			SARNIA, ONTARIO			SASKATOON, SASKATCHEWAN			SAULT STE MARIE, ONTARIO			SHERBROOKE, QUEBEC			SOREL, QUEBEC		
		MAT.	INST.	TOTAL	MAT.	INST.	TOTAL	MAT.	INST.	TOTAL	MAT.	INST.	TOTAL	MAT.	INST.	TOTAL	MAT.	INST.	TOTAL
015433	CONTRACTOR EQUIPMENT		102.8	102.8		103.1	103.1		101.5	101.5		103.1	103.1		103.3	103.3		103.3	103.3
0241, 31 - 34	SITE & INFRASTRUCTURE, DEMOLITION	107.1	97.4	100.3	118.2	99.8	105.3	114.1	97.9	102.7	107.9	99.4	101.9	99.5	98.9	99.1	99.7	98.9	99.2
0310	Concrete Forming & Accessories	123.5	66.3	74.1	122.2	91.9	96.1	106.0	95.1	96.6	111.8	87.9	91.2	129.2	82.9	89.3	129.2	83.0	89.4
0320	Concrete Reinforcing	137.7	71.6	104.4	120.7	89.5	104.9	117.9	91.6	104.6	108.4	88.2	98.2	152.2	74.8	113.1	144.0	74.8	109.1
0330	Cast-in-Place Concrete	115.9	71.4	99.0	125.3	99.4	115.5	119.1	98.1	111.1	112.6	84.8	102.0	111.7	92.2	104.3	112.6	92.2	104.9
03	CONCRETE	114.1	70.1	94.0	115.8	94.3	105.9	106.2	95.5	101.3	101.5	87.3	95.0	113.8	85.2	100.7	113.2	85.2	100.4
04	MASONRY	183.9	71.5	114.2	180.5	94.4	127.1	173.5	91.1	122.4	165.5	93.9	121.1	165.1	79.6	112.1	165.2	79.6	112.1
05	METALS	113.6	87.3	105.5	112.3	93.5	106.5	107.8	90.6	102.5	111.5	95.2	106.5	110.8	86.8	103.4	111.0	86.9	103.6
06	WOOD, PLASTICS & COMPOSITES	120.4	64.3	89.2	118.7	90.9	103.3	97.6	95.7	96.6	106.6	88.7	96.7	130.5	82.8	103.9	130.5	82.8	103.9
07	THERMAL & MOISTURE PROTECTION	114.2	72.4	96.5	118.7	96.0	109.1	112.3	89.2	102.5	117.4	91.2	106.3	111.9	88.3	101.9	111.9	88.3	101.9
08	OPENINGS	86.3	61.3	80.6	92.9	87.8	91.7	87.1	83.2	86.2	84.7	87.1	85.2	91.6	71.4	87.0	91.6	76.0	88.0
0920	Plaster & Gypsum Board	139.3	63.3	88.2	139.9	90.7	106.8	115.6	95.6	102.2	107.8	88.4	94.8	114.3	82.2	92.8	144.6	82.2	102.7
0050, 0080	Ceilings & Acoustic Treatment	117.0	63.3	81.0	107.4	90.7	96.1	123.5	95.6	104.7	103.2	88.4	93.2	102.3	82.2	88.8	102.3	82.2	88.8
0960	Flooring	110.1	68.7	98.6	112.1	99.5	108.6	105.0	97.8	103.0	105.2	96.4	102.7	112.1	90.9	106.2	112.1	90.9	106.2
0970, 0990	Wall Finishes & Painting/Coating	109.9	84.2	94.9	109.0	103.0	105.5	112.5	93.1	101.2	109.0	95.4	101.1	109.0	87.4	96.4	109.0	87.4	96.4
09	FINISHES	111.0	68.1	87.7	110.3	94.8	101.9	109.3	96.3	102.3	102.3	90.5	95.9	105.1	85.2	94.3	109.2	85.2	96.2
COVERS	DIVS. 10 - 14, 25, 28, 41, 43, 44, 46	131.1	63.0	116.0	131.1	68.8	117.3	131.1	71.0	117.8	131.1	89.5	121.9	131.1	78.8	119.5	131.1	78.8	119.5
21, 22, 23	FIRE SUPPRESSION, PLUMBING & HVAC	104.6	77.9	93.2	104.6	105.8	105.1	105.0	90.3	98.7	104.6	94.0	100.1	105.1	87.7	97.6	104.6	87.7	97.4
26, 27, 3370	ELECTRICAL, COMMUNICATIONS & UTIL.	118.5	86.3	101.7	114.3	89.7	101.5	112.6	96.6	104.3	113.0	87.9	100.0	111.0	68.9	89.1	111.0	68.9	89.1
MF2016	WEIGHTED AVERAGE	113.8	77.0	97.7	114.3	95.5	106.0	110.8	92.5	102.7	109.6	91.7	101.8	111.7	83.4	99.3	111.9	83.6	99.5

CANADA

DIVISION		ST CATHARINES, ONTARIO			ST JEROME, QUEBEC			ST JOHNS, NEWFOUNDLAND			SUDBURY, ONTARIO			SUMMERSIDE, PRINCE EDWARD ISLAND			SYDNEY, NOVA SCOTIA		
		MAT.	INST.	TOTAL	MAT.	INST.	TOTAL	MAT.	INST.	TOTAL	MAT.	INST.	TOTAL	MAT.	INST.	TOTAL	MAT.	INST.	TOTAL
015433	CONTRACTOR EQUIPMENT		103.1	103.1		103.3	103.3		121.7	121.7		103.1	103.1		102.9	102.9		102.9	102.9
0241, 31 - 34	SITE & INFRASTRUCTURE, DEMOLITION	97.4	101.4	100.2	99.0	98.9	99.0	121.6	111.7	114.7	97.5	101.0	99.9	129.3	95.0	105.3	114.8	97.7	102.9
0310	Concrete Forming & Accessories	111.1	94.9	97.1	129.2	82.9	89.3	122.8	88.0	92.8	107.1	89.9	92.3	110.8	55.1	62.7	110.5	71.2	76.6
0320	Concrete Reinforcing	105.1	98.7	101.9	152.2	74.8	113.1	161.2	85.0	122.8	106.0	102.5	104.3	173.8	48.0	110.3	176.3	49.2	112.2
0330	Cast-in-Place Concrete	108.3	101.4	105.6	108.1	92.2	102.1	140.6	101.1	125.6	109.1	98.0	104.9	126.2	57.7	100.1	102.5	70.4	90.4
03	CONCRETE	98.1	97.9	98.0	112.1	85.2	99.8	132.7	93.0	114.5	98.4	95.0	96.8	148.5	55.9	106.2	126.2	68.1	99.6
04	MASONRY	145.2	101.3	118.0	164.8	79.6	112.0	184.6	93.7	128.2	145.3	97.6	115.7	215.9	57.2	117.5	214.2	68.4	123.8
05	METALS	111.4	97.4	107.1	111.0	86.8	103.6	135.7	101.0	125.1	110.8	97.0	106.6	131.8	71.0	113.1	131.8	77.8	115.2
06	WOOD, PLASTICS & COMPOSITES	105.3	94.2	99.1	130.5	82.8	103.9	102.1	85.8	93.0	101.5	88.7	94.4	117.9	54.3	82.5	117.5	70.6	91.4
07	THERMAL & MOISTURE PROTECTION	115.0	103.0	109.9	111.9	88.3	101.9	139.2	98.0	121.8	114.5	97.5	107.3	132.6	60.1	101.9	133.2	71.0	106.9
08	OPENINGS	82.6	92.1	84.8	91.6	71.4	87.0	85.7	77.0	83.7	83.3	87.6	84.3	104.6	47.8	91.5	92.6	64.4	86.1
0920	Plaster & Gypsum Board	97.0	94.0	95.0	114.3	82.2	92.8	139.6	84.8	102.8	98.8	88.4	91.8	141.5	53.0	82.1	140.9	69.8	93.1
0950, 0980	Ceilings & Acoustic Treatment	101.7	94.0	96.5	102.3	82.2	88.8	129.6	84.8	99.4	96.6	88.4	91.1	119.2	53.0	74.6	119.2	69.8	85.9
0960	Flooring	98.9	94.5	97.7	112.1	90.9	106.2	111.1	54.2	95.3	97.1	96.4	96.9	107.4	58.4	93.8	107.3	62.3	94.8
0970, 0990	Wall Finishes & Painting/Coating	103.1	105.4	104.4	109.0	87.4	96.4	112.5	103.1	107.0	103.1	96.1	99.0	116.9	41.7	73.0	116.9	62.0	84.9
09	FINISHES	96.9	96.0	96.4	105.1	85.2	94.3	119.6	83.3	99.8	95.6	91.5	93.4	115.3	54.5	82.2	114.2	68.8	89.5
COVERS	DIVS. 10 - 14, 25, 28, 41, 43, 44, 46	131.1	70.5	117.7	131.1	78.8	119.5	131.1	69.7	117.5	131.1	90.8	122.2	131.1	59.4	115.2	131.1	63.3	116.1
21, 22, 23	FIRE SUPPRESSION, PLUMBING & HVAC	104.8	90.8	98.8	104.6	87.7	97.4	105.3	86.3	97.2	104.5	90.2	98.4	104.7	61.3	86.2	104.7	82.8	95.4
26, 27, 3370	ELECTRICAL, COMMUNICATIONS & UTIL.	112.4	100.0	106.4	111.7	68.9	89.4	109.3	85.9	97.1	112.5	101.3	106.7	108.2	50.1	78.0	109.1	62.0	84.6
MF2016	WEIGHTED AVERAGE	107.1	96.1	102.3	111.4	83.4	99.1	120.7	90.3	107.4	106.9	94.9	101.6	125.1	60.5	96.8	120.6	73.5	100.0

819

		CANADA																
DIVISION		THUNDER BAY, ONTARIO			TIMMINS, ONTARIO			TORONTO, ONTARIO			TROIS RIVIERES, QUEBEC			TRURO, NOVA SCOTIA			VANCOUVER, BRITISH COLUMBIA		
		MAT.	INST.	TOTAL	MAT.	INST.	TOTAL	MAT.	INST.	TOTAL	MAT.	INST.	TOTAL	MAT.	INST.	TOTAL	MAT.	INST.	TOTAL
015433	CONTRACTOR EQUIPMENT		103.1	103.1		103.1	103.1		121.8	121.8		103.7	103.7		102.6	102.6		137.0	137.0
0241, 31 - 34	SITE & INFRASTRUCTURE, DEMOLITION	102.2	101.3	101.6	119.6	99.3	105.4	119.5	115.3	116.6	115.2	99.2	104.0	103.9	97.5	99.4	114.7	124.5	121.6
0310	Concrete Forming & Accessories	118.4	93.2	96.6	123.5	84.0	89.4	122.8	101.0	104.0	151.9	83.1	92.5	96.8	71.2	74.7	122.6	88.8	93.5
0320	Concrete Reinforcing	94.0	98.2	96.1	169.0	87.6	127.9	150.1	100.6	125.1	176.3	74.8	125.1	144.0	49.2	96.2	123.7	93.7	108.6
0330	Cast-in-Place Concrete	119.2	100.6	112.1	135.8	84.6	116.3	116.9	113.6	115.6	106.2	92.3	100.9	141.0	70.4	114.2	125.3	93.5	113.2
03	CONCRETE	105.4	96.7	101.4	127.0	85.3	107.9	115.6	105.9	111.1	128.5	85.3	108.7	126.0	68.0	99.5	118.7	92.7	106.8
04	MASONRY	145.8	101.5	118.3	168.8	85.4	117.1	170.9	108.4	132.1	218.4	79.6	132.3	164.3	68.4	104.8	170.9	88.0	119.5
05	METALS	111.4	96.4	106.8	112.4	92.7	106.3	132.3	109.7	125.3	131.0	87.1	117.5	111.7	77.6	101.2	139.8	111.4	131.1
06	WOOD, PLASTICS & COMPOSITES	114.1	91.5	101.6	119.8	84.1	99.9	105.5	98.5	101.6	168.1	82.8	120.7	90.7	70.6	79.5	100.9	86.8	93.1
07	THERMAL & MOISTURE PROTECTION	115.3	100.2	108.9	118.6	86.8	105.1	133.0	108.0	122.4	131.7	88.3	113.3	113.2	70.9	95.3	133.0	88.3	114.1
08	OPENINGS	81.9	90.0	83.8	90.3	82.3	88.5	84.1	98.0	87.3	102.2	76.0	96.2	84.5	64.4	79.9	83.0	87.2	84.0
0920	Plaster & Gypsum Board	124.9	91.3	102.3	116.6	83.7	94.5	119.4	98.1	105.1	169.3	82.2	110.8	122.4	69.8	87.1	114.2	85.4	94.8
0950, 0980	Ceilings & Acoustic Treatment	96.6	91.3	93.0	103.2	83.7	90.0	128.1	98.1	107.9	118.3	82.2	94.0	103.2	69.8	80.6	135.5	85.4	101.7
0960	Flooring	102.9	101.2	102.4	112.1	90.9	106.2	104.5	103.7	104.3	130.3	90.9	119.4	94.7	62.3	85.7	119.7	91.9	112.0
0970, 0990	Wall Finishes & Painting/Coating	103.1	97.1	99.6	109.0	88.9	97.3	100.8	109.5	105.9	116.9	87.4	99.7	109.0	62.0	81.6	110.8	89.2	98.2
09	FINISHES	101.2	95.0	97.9	106.3	86.1	95.3	110.6	102.3	106.1	124.8	85.2	103.3	101.1	68.8	83.6	119.1	89.3	102.9
COVERS	DIVS. 10 - 14, 25, 28, 41, 43, 44, 46	131.1	70.7	117.7	131.1	66.0	116.7	131.1	95.5	123.2	131.1	78.8	119.5	131.1	63.3	116.0	131.1	92.0	122.4
21, 22, 23	FIRE SUPPRESSION, PLUMBING & HVAC	104.8	91.0	98.9	104.6	96.3	101.1	105.4	99.0	102.7	104.7	87.7	97.5	104.6	82.8	95.3	105.5	78.6	94.0
26, 27, 3370	ELECTRICAL, COMMUNICATIONS & UTIL.	110.8	99.5	104.9	113.0	87.9	100.0	106.2	103.1	104.6	109.4	68.9	88.3	110.7	62.0	85.4	107.1	83.0	94.6
MF2016	WEIGHTED AVERAGE	108.3	95.4	102.7	114.4	89.2	103.4	115.9	104.2	110.8	123.3	83.6	105.9	111.9	73.4	95.1	118.0	91.2	106.3

		CANADA																
DIVISION		VICTORIA, BRITISH COLUMBIA			WHITEHORSE, YUKON			WINDSOR, ONTARIO			WINNIPEG, MANITOBA			YARMOUTH, NOVA SCOTIA			YELLOWKNIFE, NWT		
		MAT.	INST.	TOTAL	MAT.	INST.	TOTAL	MAT.	INST.	TOTAL	MAT.	INST.	TOTAL	MAT.	INST.	TOTAL	MAT.	INST.	TOTAL
015433	CONTRACTOR EQUIPMENT		109.7	109.7		130.0	130.0		103.1	103.1		123.9	123.9		102.9	102.9		129.9	129.9
0241, 31 - 34	SITE & INFRASTRUCTURE, DEMOLITION	125.5	106.8	112.4	146.2	116.0	125.0	93.8	101.3	99.0	117.4	111.2	113.1	118.9	97.7	104.1	154.5	122.1	131.8
0310	Concrete Forming & Accessories	110.8	87.6	90.8	133.6	59.2	69.5	118.4	91.4	95.1	125.4	66.7	74.8	110.5	71.2	76.6	135.9	79.3	87.1
0320	Concrete Reinforcing	115.3	93.5	104.3	159.8	65.3	112.1	103.0	97.5	100.2	142.1	61.3	101.3	176.3	49.2	112.2	156.6	67.9	111.9
0330	Cast-in-Place Concrete	123.4	90.8	111.0	179.5	73.1	139.1	110.9	102.1	107.6	144.8	73.4	117.6	131.5	70.4	108.3	177.1	90.8	144.3
03	CONCRETE	134.7	90.0	114.2	155.6	67.0	115.1	99.5	96.4	98.1	129.3	69.6	102.0	139.9	68.1	107.1	154.2	82.6	121.5
04	MASONRY	176.5	88.0	121.6	250.8	60.4	132.8	145.4	100.5	117.5	185.3	66.2	111.4	216.6	68.4	124.7	244.4	71.8	137.4
05	METALS	109.0	92.3	103.9	148.5	92.5	131.3	111.4	96.9	106.9	143.2	90.0	126.8	131.8	77.8	115.2	147.4	96.0	131.6
06	WOOD, PLASTICS & COMPOSITES	101.3	86.2	92.9	123.4	57.7	86.8	114.1	89.8	100.6	107.8	67.3	85.2	117.5	70.6	91.4	129.7	80.6	102.4
07	THERMAL & MOISTURE PROTECTION	122.2	89.3	108.3	142.8	64.9	109.9	115.1	99.6	108.5	125.3	71.0	102.3	133.2	71.0	106.9	144.1	81.4	117.5
08	OPENINGS	88.6	82.6	87.2	104.4	55.6	93.2	81.7	89.5	83.5	83.9	61.0	78.7	92.6	64.4	86.1	96.5	68.8	90.1
0920	Plaster & Gypsum Board	108.8	85.4	93.1	167.4	55.6	92.3	109.8	89.5	96.2	121.9	65.4	84.0	140.9	69.8	93.1	172.4	79.2	109.8
0950, 0980	Ceilings & Acoustic Treatment	105.6	85.4	92.0	161.4	55.6	90.1	96.6	89.5	91.8	143.8	65.4	91.0	119.2	69.8	85.9	155.1	79.2	103.9
0960	Flooring	110.4	71.2	99.5	125.4	58.3	106.8	102.9	98.5	101.7	109.0	70.4	98.3	107.3	62.3	94.8	123.4	87.6	113.5
0970, 0990	Wall Finishes & Painting/Coating	112.5	89.2	98.9	119.7	55.6	82.3	103.1	98.1	100.2	107.7	54.4	76.6	116.9	62.0	84.9	118.9	79.3	95.8
09	FINISHES	108.0	85.3	95.6	142.3	58.4	96.6	98.8	93.4	95.9	119.8	66.4	90.8	114.2	68.8	89.5	138.6	80.3	106.9
COVERS	DIVS. 10 - 14, 25, 28, 41, 43, 44, 46	131.1	67.4	116.9	131.1	62.6	115.9	131.1	70.1	117.6	131.1	64.7	116.4	131.1	63.3	116.1	131.1	66.1	116.7
21, 22, 23	FIRE SUPPRESSION, PLUMBING & HVAC	104.7	76.6	92.7	105.3	74.6	92.1	104.8	90.9	98.8	105.5	65.5	88.4	104.7	82.8	95.4	105.4	92.1	99.7
26, 27, 3370	ELECTRICAL, COMMUNICATIONS & UTIL.	112.8	82.4	97.0	132.0	60.3	94.7	115.6	101.0	108.0	112.8	65.1	88.0	109.1	62.0	84.6	121.5	81.8	100.8
MF2016	WEIGHTED AVERAGE	115.3	85.7	102.3	135.9	71.3	107.6	107.6	95.2	102.2	121.3	72.0	99.7	122.5	73.5	101.1	133.4	86.2	112.7

Costs shown in RSMeans cost data publications are based on national averages for materials and installation. To adjust these costs to a specific location, simply multiply the base cost by the factor and divide by 100 for that city. The data is arranged alphabetically by state and postal zip code numbers. For a city not listed, use the factor for a nearby city with similar economic characteristics.

STATE/ZIP	CITY	MAT.	INST.	TOTAL
ALABAMA				
350-352	Birmingham	95.9	71.4	85.2
354	Tuscaloosa	96.0	72.3	85.6
355	Jasper	96.5	71.2	85.5
356	Decatur	96.0	71.3	85.2
357-358	Huntsville	96.0	70.7	84.9
359	Gadsden	96.1	71.0	85.1
360-361	Montgomery	94.8	72.3	85.0
362	Anniston	94.5	68.5	83.1
363	Dothan	94.9	72.9	85.3
364	Evergreen	94.5	72.6	84.9
365-366	Mobile	95.4	70.2	84.4
367	Selma	94.6	73.6	85.4
368	Phenix City	95.4	72.7	85.4
369	Butler	94.8	72.8	85.2
ALASKA				
995-996	Anchorage	118.9	114.8	117.1
997	Fairbanks	120.1	115.3	118.0
998	Juneau	118.8	114.8	117.0
999	Ketchikan	130.8	114.8	123.8
ARIZONA				
850,853	Phoenix	99.5	74.1	88.4
851,852	Mesa/Tempe	98.2	72.5	87.0
855	Globe	99.2	72.4	87.4
856-857	Tucson	97.1	71.8	86.0
859	Show Low	99.4	72.5	87.6
860	Flagstaff	101.7	72.1	88.8
863	Prescott	99.3	72.2	87.5
864	Kingman	97.9	72.5	86.8
865	Chambers	97.9	75.0	87.9
ARKANSAS				
716	Pine Bluff	96.9	65.0	82.9
717	Camden	95.0	60.5	79.9
718	Texarkana	95.7	60.7	80.3
719	Hot Springs	94.3	62.0	80.1
720-722	Little Rock	95.3	66.6	82.7
723	West Memphis	94.2	66.6	82.1
724	Jonesboro	94.5	63.7	81.0
725	Batesville	92.6	60.8	78.7
726	Harrison	93.9	58.9	78.6
727	Fayetteville	91.5	61.6	78.5
728	Russellville	92.7	59.4	78.1
729	Fort Smith	95.0	63.1	81.0
CALIFORNIA				
900-902	Los Angeles	99.9	129.3	112.8
903-905	Inglewood	95.5	125.5	108.6
906-908	Long Beach	97.1	126.1	109.8
910-912	Pasadena	96.0	125.3	108.8
913-916	Van Nuys	99.1	125.3	110.6
917-918	Alhambra	98.0	125.9	110.2
919-921	San Diego	101.0	119.5	109.1
922	Palm Springs	97.2	123.9	108.9
923-924	San Bernardino	94.8	125.7	108.4
925	Riverside	99.0	125.6	110.7
926-927	Santa Ana	96.8	122.9	108.3
928	Anaheim	99.1	125.9	110.8
930	Oxnard	97.6	126.5	110.2
931	Santa Barbara	96.9	126.3	109.8
932-933	Bakersfield	98.3	121.3	108.3
934	San Luis Obispo	98.1	124.4	109.6
935	Mojave	95.2	124.1	106.7
936-938	Fresno	98.2	128.2	111.3
939	Salinas	98.8	134.8	114.6
940-941	San Francisco	106.3	158.5	129.1
942,956-958	Sacramento	100.0	131.6	113.8
943	Palo Alto	98.5	148.0	120.2
944	San Mateo	100.8	147.7	121.3
945	Vallejo	99.5	139.6	117.0
946	Oakland	102.5	148.9	122.8
947	Berkeley	102.5	148.9	122.8
948	Richmond	101.7	143.0	119.8
949	San Rafael	103.8	148.4	123.3
950	Santa Cruz	104.3	135.1	117.8

STATE/ZIP	CITY	MAT.	INST.	TOTAL
CALIFORNIA (CONT'D)				
951	San Jose	102.4	150.4	123.4
952	Stockton	100.5	127.7	112.4
953	Modesto	100.4	126.9	112.0
954	Santa Rosa	101.1	147.3	121.3
955	Eureka	102.4	130.0	114.4
959	Marysville	101.6	128.8	113.5
960	Redding	109.1	128.8	117.7
961	Susanville	108.8	127.0	116.8
COLORADO				
800-802	Denver	102.8	74.6	90.5
803	Boulder	98.8	75.9	88.8
804	Golden	101.0	72.6	88.6
805	Fort Collins	102.1	74.3	89.9
806	Greeley	99.3	75.1	88.7
807	Fort Morgan	99.4	71.2	87.0
808-809	Colorado Springs	101.0	73.2	88.8
810	Pueblo	101.3	68.8	87.1
811	Alamosa	103.3	69.0	88.3
812	Salida	103.0	65.8	86.7
813	Durango	103.7	65.5	87.0
814	Montrose	102.4	67.8	87.3
815	Grand Junction	105.5	73.0	91.3
816	Glenwood Springs	103.5	65.9	87.0
CONNECTICUT				
060	New Britain	97.4	117.9	106.3
061	Hartford	98.4	118.6	107.3
062	Willimantic	97.9	118.5	106.9
063	New London	94.7	118.5	105.1
064	Meriden	96.6	117.8	105.9
065	New Haven	98.9	118.9	107.6
066	Bridgeport	98.5	118.4	107.2
067	Waterbury	98.1	118.7	107.1
068	Norwalk	98.0	118.6	107.0
069	Stamford	98.1	125.5	110.1
D.C.				
200-205	Washington	101.1	87.6	95.2
DELAWARE				
197	Newark	97.9	112.4	104.3
198	Wilmington	97.6	112.5	104.1
199	Dover	98.0	112.4	104.3
FLORIDA				
320,322	Jacksonville	94.7	65.3	81.8
321	Daytona Beach	94.9	71.0	84.4
323	Tallahassee	95.6	66.2	82.7
324	Panama City	96.2	65.1	82.5
325	Pensacola	98.8	67.1	84.9
326,344	Gainesville	96.4	65.0	82.7
327-328,347	Orlando	96.8	67.0	83.8
329	Melbourne	97.6	71.2	86.0
330-332,340	Miami	95.6	64.6	82.0
333	Fort Lauderdale	95.1	66.0	82.4
334,349	West Palm Beach	94.2	64.0	81.0
335-336,346	Tampa	96.6	67.6	83.9
337	St. Petersburg	99.0	65.9	84.5
338	Lakeland	96.2	67.8	83.7
339,341	Fort Myers	95.4	65.4	82.3
342	Sarasota	98.1	67.9	84.9
GEORGIA				
300-303,399	Atlanta	98.7	75.0	88.3
304	Statesboro	98.7	66.3	84.5
305	Gainesville	97.2	67.6	84.2
306	Athens	96.6	69.1	84.5
307	Dalton	98.3	72.2	86.9
308-309	Augusta	96.8	74.2	86.9
310-312	Macon	93.8	75.1	85.6
313-314	Savannah	95.4	74.4	86.2
315	Waycross	95.3	69.5	84.0
316	Valdosta	95.1	67.3	82.9
317,398	Albany	95.0	73.4	85.5
318-319	Columbus	94.9	73.8	85.7

STATE/ZIP	CITY	MAT.	INST.	TOTAL
HAWAII				
967	Hilo	116.9	116.0	116.5
968	Honolulu	121.1	116.0	118.9
STATES & POSS.				
969	Guam	139.1	53.4	101.6
IDAHO				
832	Pocatello	101.5	79.5	91.9
833	Twin Falls	102.8	77.8	91.8
834	Idaho Falls	100.0	78.5	90.6
835	Lewiston	108.2	86.3	98.6
836-837	Boise	100.5	79.8	91.4
838	Coeur d'Alene	108.1	84.7	97.8
ILLINOIS				
600-603	North Suburban	98.9	140.6	117.1
604	Joliet	98.8	142.0	117.7
605	South Suburban	98.9	140.6	117.1
606-608	Chicago	100.8	145.8	120.5
609	Kankakee	95.6	133.5	112.2
610-611	Rockford	98.2	128.3	111.4
612	Rock Island	96.4	101.9	98.8
613	La Salle	97.6	126.5	110.3
614	Galesburg	97.4	108.9	102.5
615-616	Peoria	99.3	112.2	104.9
617	Bloomington	96.8	113.0	103.9
618-619	Champaign	100.3	110.6	104.8
620-622	East St. Louis	94.4	109.4	100.9
623	Quincy	96.7	103.9	99.8
624	Effingham	95.8	108.7	101.5
625	Decatur	97.3	109.0	102.4
626-627	Springfield	97.8	108.8	102.6
628	Centralia	93.4	108.8	100.1
629	Carbondale	93.2	108.8	100.0
INDIANA				
460	Anderson	95.1	83.3	89.9
461-462	Indianapolis	98.3	83.2	91.7
463-464	Gary	96.5	110.4	102.6
465-466	South Bend	96.3	85.0	91.4
467-468	Fort Wayne	95.7	77.7	87.8
469	Kokomo	93.5	81.7	88.3
470	Lawrenceburg	92.8	79.6	87.0
471	New Albany	94.0	78.0	87.0
472	Columbus	96.3	80.8	89.5
473	Muncie	96.3	81.7	89.9
474	Bloomington	97.9	81.3	90.6
475	Washington	94.8	86.5	91.1
476-477	Evansville	95.5	85.5	91.1
478	Terre Haute	96.3	85.4	91.5
479	Lafayette	96.0	81.6	89.7
IOWA				
500-503,509	Des Moines	97.2	88.6	93.5
504	Mason City	95.8	76.0	87.1
505	Fort Dodge	95.9	73.0	85.9
506-507	Waterloo	97.2	80.2	89.7
508	Creston	96.2	82.1	90.0
510-511	Sioux City	97.9	79.2	89.7
512	Sibley	97.1	61.0	81.3
513	Spencer	98.7	62.1	82.7
514	Carroll	95.8	79.9	88.8
515	Council Bluffs	99.0	81.1	91.2
516	Shenandoah	96.4	76.5	87.7
520	Dubuque	97.3	82.8	91.0
521	Decorah	96.8	74.0	86.8
522-524	Cedar Rapids	98.2	86.5	93.1
525	Ottumwa	96.7	76.2	87.7
526	Burlington	96.0	82.3	90.0
527-528	Davenport	97.3	98.6	97.9
KANSAS				
660-662	Kansas City	98.4	99.3	98.8
664-666	Topeka	99.1	78.1	89.9
667	Fort Scott	97.3	79.1	89.3
668	Emporia	97.4	75.8	87.9
669	Belleville	99.1	69.9	86.3
670-672	Wichita	99.2	72.1	87.3
673	Independence	99.6	75.6	89.1
674	Salina	99.6	70.8	87.0
675	Hutchinson	95.0	70.5	84.3
676	Hays	99.1	71.3	86.9
677	Colby	99.8	73.5	88.3

STATE/ZIP	CITY	MAT.	INST.	TOTAL
KANSAS (CONT'D)				
678	Dodge City	100.9	72.9	88.7
679	Liberal	99.0	70.6	86.6
KENTUCKY				
400-402	Louisville	95.1	79.1	88.1
403-405	Lexington	94.8	80.3	88.5
406	Frankfort	96.0	78.5	88.3
407-409	Corbin	92.7	77.7	86.2
410	Covington	94.7	80.5	88.5
411-412	Ashland	93.5	93.4	93.5
413-414	Campton	94.7	80.2	88.3
415-416	Pikeville	95.9	81.4	89.6
417-418	Hazard	94.1	78.7	87.4
420	Paducah	92.6	84.5	89.1
421-422	Bowling Green	94.6	79.9	88.2
423	Owensboro	94.7	84.3	90.2
424	Henderson	92.4	83.0	88.3
425-426	Somerset	91.9	79.2	86.4
427	Elizabethtown	91.5	77.2	85.3
LOUISIANA				
700-701	New Orleans	98.7	68.4	85.5
703	Thibodaux	95.6	67.4	83.3
704	Hammond	93.2	65.1	80.9
705	Lafayette	94.9	70.1	84.0
706	Lake Charles	95.1	70.9	84.5
707-708	Baton Rouge	96.3	70.1	84.8
710-711	Shreveport	97.9	66.7	84.2
712	Monroe	96.7	66.9	83.7
713-714	Alexandria	96.9	68.2	84.3
MAINE				
039	Kittery	93.0	84.9	89.5
040-041	Portland	98.7	85.6	93.0
042	Lewiston	96.3	85.6	91.6
043	Augusta	98.4	79.8	90.3
044	Bangor	95.8	85.0	91.1
045	Bath	94.8	82.2	89.3
046	Machias	94.2	79.2	87.7
047	Houlton	94.4	79.2	87.8
048	Rockland	93.5	79.2	87.2
049	Waterville	94.7	79.8	88.2
MARYLAND				
206	Waldorf	97.9	85.2	92.3
207-208	College Park	97.9	86.8	93.0
209	Silver Spring	97.2	85.8	92.2
210-212	Baltimore	101.4	84.7	94.1
214	Annapolis	98.9	84.2	92.4
215	Cumberland	96.9	86.3	92.3
216	Easton	98.6	76.6	88.9
217	Hagerstown	97.2	87.7	93.0
218	Salisbury	99.0	69.8	86.2
219	Elkton	95.8	87.4	92.1
MASSACHUSETTS				
010-011	Springfield	97.8	109.0	102.7
012	Pittsfield	97.3	106.5	101.3
013	Greenfield	95.7	109.0	101.5
014	Fitchburg	94.4	120.8	105.9
015-016	Worcester	97.7	120.8	107.8
017	Framingham	93.7	127.9	108.7
018	Lowell	97.1	127.1	110.3
019	Lawrence	98.0	128.5	111.4
020-022, 024	Boston	101.2	133.1	115.2
023	Brockton	97.9	121.8	108.4
025	Buzzards Bay	92.8	119.5	104.5
026	Hyannis	95.1	119.5	105.8
027	New Bedford	97.0	119.8	107.0
MICHIGAN				
480,483	Royal Oak	94.3	100.5	97.0
481	Ann Arbor	96.2	102.4	98.9
482	Detroit	99.2	103.0	100.9
484-485	Flint	95.9	92.3	94.4
486	Saginaw	95.6	88.7	92.6
487	Bay City	95.7	88.3	92.5
488-489	Lansing	97.0	90.1	94.0
490	Battle Creek	94.8	83.7	89.9
491	Kalamazoo	95.1	82.1	89.4
492	Jackson	93.4	91.8	92.7
493,495	Grand Rapids	96.4	82.9	90.5
494	Muskegon	93.6	81.2	88.2

Location Factors - Commercial

STATE/ZIP	CITY	MAT.	INST.	TOTAL
MICHIGAN (CONT'D)				
496	Traverse City	92.9	78.4	86.6
497	Gaylord	94.1	80.7	88.2
498-499	Iron Mountain	96.0	82.4	90.0
MINNESOTA				
550-551	Saint Paul	98.9	114.1	105.6
553-555	Minneapolis	100.4	113.0	105.9
556-558	Duluth	99.8	102.7	101.1
559	Rochester	98.8	100.9	99.7
560	Mankato	96.0	98.0	96.9
561	Windom	94.6	92.3	93.6
562	Willmar	94.3	97.4	95.6
563	St. Cloud	95.2	111.3	102.3
564	Brainerd	95.9	97.5	96.6
565	Detroit Lakes	97.8	92.0	95.3
566	Bemidji	97.0	96.0	96.6
567	Thief River Falls	96.6	89.2	93.4
MISSISSIPPI				
386	Clarksdale	94.7	55.2	77.4
387	Greenville	98.3	65.6	83.9
388	Tupelo	96.2	55.9	78.6
389	Greenwood	96.0	53.2	77.3
390-392	Jackson	97.3	65.3	83.3
393	Meridian	95.1	65.9	82.3
394	Laurel	96.7	56.3	79.0
395	Biloxi	96.7	65.5	83.1
396	McComb	95.0	53.7	76.9
397	Columbus	96.6	55.8	78.7
MISSOURI				
630-631	St. Louis	100.5	106.2	103.0
633	Bowling Green	98.7	95.0	97.1
634	Hannibal	97.6	93.6	95.9
635	Kirksville	101.1	88.5	95.6
636	Flat River	99.6	95.7	97.9
637	Cape Girardeau	99.1	91.5	95.8
638	Sikeston	98.0	89.1	94.1
639	Poplar Bluff	97.5	89.7	94.1
640-641	Kansas City	100.3	102.7	101.4
644-645	St. Joseph	98.9	92.8	96.2
646	Chillicothe	96.5	95.1	95.9
647	Harrisonville	95.9	102.8	98.9
648	Joplin	97.8	79.6	89.8
650-651	Jefferson City	97.8	92.5	95.5
652	Columbia	98.1	92.5	95.7
653	Sedalia	98.6	93.2	96.2
654-655	Rolla	96.2	96.9	96.5
656-658	Springfield	99.6	81.7	91.7
MONTANA				
590-591	Billings	100.0	76.4	89.6
592	Wolf Point	99.9	76.5	89.7
593	Miles City	97.8	76.7	88.6
594	Great Falls	101.3	75.3	89.9
595	Havre	98.9	75.2	88.5
596	Helena	99.6	75.3	88.9
597	Butte	99.8	75.3	89.1
598	Missoula	97.4	74.5	87.4
599	Kalispell	96.8	75.0	87.2
NEBRASKA				
680-681	Omaha	98.4	80.8	90.7
683-685	Lincoln	98.3	80.2	90.4
686	Columbus	97.3	81.4	90.3
687	Norfolk	98.8	78.6	90.0
688	Grand Island	98.5	78.4	89.7
689	Hastings	98.5	79.0	89.9
690	McCook	98.0	71.5	86.4
691	North Platte	97.8	75.2	87.9
692	Valentine	100.5	70.5	87.4
693	Alliance	100.4	71.9	87.9
NEVADA				
889-891	Las Vegas	103.1	105.2	104.0
893	Ely	101.9	94.1	98.5
894-895	Reno	101.8	85.3	94.6
897	Carson City	101.0	84.9	93.9
898	Elko	100.6	81.2	92.1
NEW HAMPSHIRE				
030	Nashua	97.7	92.6	95.5
031	Manchester	98.3	92.7	95.9

STATE/ZIP	CITY	MAT.	INST.	TOTAL
NEW HAMPSHIRE (CONT'D)				
032-033	Concord	97.9	92.4	95.5
034	Keene	94.8	79.6	88.1
035	Littleton	94.9	79.8	88.3
036	Charleston	94.3	76.3	86.4
037	Claremont	93.5	76.3	86.0
038	Portsmouth	95.1	92.4	93.9
NEW JERSEY				
070-071	Newark	99.1	137.9	116.1
072	Elizabeth	96.8	138.2	114.9
073	Jersey City	95.9	137.6	114.1
074-075	Paterson	97.4	138.2	115.3
076	Hackensack	95.5	137.9	114.0
077	Long Branch	95.0	136.2	113.1
078	Dover	95.7	138.2	114.3
079	Summit	95.8	138.2	114.3
080,083	Vineland	95.0	134.4	112.2
081	Camden	97.0	133.3	112.9
082,084	Atlantic City	95.7	134.7	112.8
085-086	Trenton	98.5	133.9	114.0
087	Point Pleasant	96.8	135.8	113.9
088-089	New Brunswick	97.4	137.8	115.1
NEW MEXICO				
870-872	Albuquerque	98.5	75.1	88.3
873	Gallup	98.9	75.1	88.5
874	Farmington	99.1	75.1	88.6
875	Santa Fe	99.1	75.1	88.6
877	Las Vegas	97.2	75.1	87.5
878	Socorro	96.8	75.1	87.3
879	Truth/Consequences	96.6	71.8	85.7
880	Las Cruces	96.4	71.7	85.6
881	Clovis	98.9	75.0	88.4
882	Roswell	100.3	75.1	89.3
883	Carrizozo	101.1	75.1	89.7
884	Tucumcari	99.5	75.0	88.8
NEW YORK				
100-102	New York	100.6	178.2	134.6
103	Staten Island	96.4	172.2	129.6
104	Bronx	94.7	173.4	129.2
105	Mount Vernon	95.1	145.6	117.2
106	White Plains	94.6	145.6	116.9
107	Yonkers	98.9	150.8	121.6
108	New Rochelle	95.5	145.5	117.4
109	Suffern	95.1	136.0	113.0
110	Queens	99.6	174.8	132.6
111	Long Island City	101.2	174.8	133.5
112	Brooklyn	101.6	172.5	132.7
113	Flushing	101.9	174.8	133.8
114	Jamaica	100.2	174.8	132.9
115,117,118	Hicksville	99.6	158.9	125.6
116	Far Rockaway	102.0	174.8	133.9
119	Riverhead	100.4	159.0	126.1
120-122	Albany	95.8	111.2	102.5
123	Schenectady	96.1	111.2	102.7
124	Kingston	100.0	131.6	113.8
125-126	Poughkeepsie	99.2	142.4	118.1
127	Monticello	98.5	130.8	112.7
128	Glens Falls	91.4	106.4	98.0
129	Plattsburgh	96.6	100.9	98.5
130-132	Syracuse	97.6	102.0	99.5
133-135	Utica	95.7	100.8	97.9
136	Watertown	97.3	100.7	98.8
137-139	Binghamton	97.2	101.7	99.2
140-142	Buffalo	102.1	111.6	105.8
143	Niagara Falls	97.6	111.9	103.9
144-146	Rochester	99.5	102.7	100.9
147	Jamestown	96.6	93.9	95.4
148-149	Elmira	96.5	103.0	99.3
NORTH CAROLINA				
270,272-274	Greensboro	98.6	68.9	85.6
271	Winston-Salem	98.4	68.7	85.4
275-276	Raleigh	96.9	68.3	84.4
277	Durham	100.2	68.9	86.5
278	Rocky Mount	96.3	68.6	84.2
279	Elizabeth City	97.2	70.4	85.5
280	Gastonia	98.4	68.7	85.4
281-282	Charlotte	98.1	68.3	85.0
283	Fayetteville	101.3	68.4	86.9
284	Wilmington	97.2	67.5	84.2
285	Kinston	95.8	68.2	83.7

823

STATE/ZIP	CITY	MAT.	INST.	TOTAL
NORTH CAROLINA (CONT'D)				
286	Hickory	96.1	69.3	84.4
287-288	Asheville	97.6	68.3	84.8
289	Murphy	96.9	67.5	84.0
NORTH DAKOTA				
580-581	Fargo	100.1	81.8	92.1
582	Grand Forks	100.1	79.0	90.8
583	Devils Lake	100.1	78.2	90.5
584	Jamestown	100.2	77.1	90.1
585	Bismarck	100.3	81.4	92.0
586	Dickinson	100.8	80.3	91.8
587	Minot	100.0	81.0	91.7
588	Williston	99.3	80.2	90.9
OHIO				
430-432	Columbus	98.1	83.3	91.6
433	Marion	94.4	81.6	88.8
434-436	Toledo	97.5	93.1	95.6
437-438	Zanesville	95.0	82.5	89.5
439	Steubenville	96.2	91.8	94.3
440	Lorain	98.6	86.9	93.5
441	Cleveland	98.9	93.4	96.5
442-443	Akron	99.7	88.2	94.7
444-445	Youngstown	98.9	83.3	92.1
446-447	Canton	99.1	81.0	91.1
448-449	Mansfield	96.8	82.8	90.7
450	Hamilton	96.6	79.4	89.1
451-452	Cincinnati	97.7	80.9	90.3
453-454	Dayton	96.7	77.9	88.5
455	Springfield	96.7	80.2	89.5
456	Chillicothe	96.2	86.1	91.7
457	Athens	99.1	83.7	92.4
458	Lima	99.4	81.1	91.4
OKLAHOMA				
730-731	Oklahoma City	97.7	67.2	84.4
734	Ardmore	96.7	66.7	83.6
735	Lawton	98.7	66.9	84.8
736	Clinton	97.9	66.6	84.2
737	Enid	98.3	67.0	84.6
738	Woodward	96.7	63.8	82.3
739	Guymon	97.8	63.8	82.9
740-741	Tulsa	96.0	66.4	83.0
743	Miami	92.9	65.8	81.0
744	Muskogee	95.1	65.3	82.1
745	McAlester	92.6	61.1	78.8
746	Ponca City	93.3	65.1	81.0
747	Durant	93.2	65.6	81.1
748	Shawnee	94.7	66.3	82.3
749	Poteau	92.5	65.4	80.6
OREGON				
970-972	Portland	99.4	100.8	100.0
973	Salem	100.9	98.7	100.0
974	Eugene	99.1	98.8	99.0
975	Medford	100.6	96.1	98.6
976	Klamath Falls	101.0	96.9	99.2
977	Bend	100.0	98.7	99.4
978	Pendleton	96.0	99.7	97.6
979	Vale	93.8	87.7	91.1
PENNSYLVANIA				
150-152	Pittsburgh	100.3	103.0	101.5
153	Washington	97.3	102.2	99.4
154	Uniontown	97.6	98.6	98.1
155	Bedford	98.7	92.2	95.8
156	Greensburg	98.6	99.9	99.2
157	Indiana	97.4	97.8	97.6
158	Dubois	99.1	93.5	96.7
159	Johnstown	98.7	94.7	96.9
160	Butler	91.4	101.5	95.8
161	New Castle	91.5	100.9	95.6
162	Kittanning	91.9	101.2	95.9
163	Oil City	91.3	99.6	95.0
164-165	Erie	93.1	96.8	94.7
166	Altoona	93.2	95.1	94.0
167	Bradford	94.8	100.2	97.2
168	State College	94.4	96.5	95.3
169	Wellsboro	95.5	94.7	95.1
170-171	Harrisburg	97.3	95.2	96.3
172	Chambersburg	94.4	90.8	92.8
173-174	York	94.4	94.5	94.4
175-176	Lancaster	93.2	96.4	94.6

STATE/ZIP	CITY	MAT.	INST.	TOTAL
PENNSYLVANIA (CONT'D)				
177	Williamsport	91.9	92.8	92.3
178	Sunbury	93.9	95.1	94.4
179	Pottsville	93.0	97.9	95.1
180	Lehigh Valley	94.9	117.4	104.7
181	Allentown	96.5	107.7	101.4
182	Hazleton	94.4	96.8	95.4
183	Stroudsburg	94.3	110.4	101.3
184-185	Scranton	97.2	100.0	98.4
186-187	Wilkes-Barre	94.1	98.9	96.2
188	Montrose	93.8	99.5	96.3
189	Doylestown	94.0	123.5	106.9
190-191	Philadelphia	99.7	134.6	115.0
193	Westchester	95.5	123.7	107.8
194	Norristown	94.5	127.6	109.0
195-196	Reading	96.1	105.2	100.1
PUERTO RICO				
009	San Juan	120.7	27.7	79.9
RHODE ISLAND				
028	Newport	96.3	112.0	103.2
029	Providence	98.1	112.0	104.2
SOUTH CAROLINA				
290-292	Columbia	96.9	69.1	84.7
293	Spartanburg	97.3	68.9	84.9
294	Charleston	98.9	68.8	85.7
295	Florence	97.0	69.0	84.8
296	Greenville	97.1	68.9	84.8
297	Rock Hill	97.1	68.3	84.5
298	Aiken	98.0	69.2	85.4
299	Beaufort	98.7	56.8	80.3
SOUTH DAKOTA				
570-571	Sioux Falls	98.5	76.4	88.8
572	Watertown	98.7	57.6	80.7
573	Mitchell	97.5	63.1	82.4
574	Aberdeen	99.7	64.5	84.3
575	Pierre	100.4	70.5	87.3
576	Mobridge	98.2	65.0	83.7
577	Rapid City	99.5	74.4	88.5
TENNESSEE				
370-372	Nashville	96.7	71.7	85.8
373-374	Chattanooga	98.6	70.3	86.2
375,380-381	Memphis	98.4	70.7	86.3
376	Johnson City	98.8	60.3	82.0
377-379	Knoxville	95.3	66.2	82.5
382	McKenzie	95.8	56.8	78.7
383	Jackson	97.1	60.6	81.1
384	Columbia	94.2	68.4	82.9
385	Cookeville	95.6	57.7	79.0
TEXAS				
750	McKinney	98.4	65.2	83.9
751	Waxahackie	98.5	64.3	83.5
752-753	Dallas	99.2	67.6	85.4
754	Greenville	98.6	65.2	84.0
755	Texarkana	97.9	64.4	83.2
756	Longview	98.9	62.8	83.1
757	Tyler	98.9	64.2	83.7
758	Palestine	95.0	62.7	80.9
759	Lufkin	95.7	64.9	82.2
760-761	Fort Worth	96.6	65.1	82.8
762	Denton	96.8	65.3	83.0
763	Wichita Falls	94.2	63.3	80.7
764	Eastland	93.6	61.8	79.7
765	Temple	91.9	60.4	78.1
766-767	Waco	93.6	64.4	80.8
768	Brownwood	96.9	60.2	80.8
769	San Angelo	96.4	60.7	80.8
770-772	Houston	98.3	68.0	85.0
773	Huntsville	97.4	65.1	83.3
774	Wharton	98.2	66.2	84.2
775	Galveston	96.2	69.0	84.3
776-777	Beaumont	96.5	68.0	84.0
778	Bryan	94.0	67.3	82.3
779	Victoria	98.3	64.9	83.7
780	Laredo	96.1	63.5	81.8
781-782	San Antonio	97.8	64.4	83.2
783-784	Corpus Christi	98.7	62.4	82.8
785	McAllen	99.6	58.6	81.7
786-787	Austin	96.7	62.7	81.8

Location Factors - Commercial

STATE/ZIP	CITY	MAT.	INST.	TOTAL
TEXAS (CONT'D)				
788	Del Rio	99.5	62.9	83.4
789	Giddings	96.1	62.6	81.4
790-791	Amarillo	97.6	62.2	82.1
792	Childress	97.6	62.7	82.3
793-794	Lubbock	99.3	63.4	83.6
795-796	Abilene	97.7	62.6	82.3
797	Midland	100.0	64.8	84.6
798-799,885	El Paso	96.0	64.7	82.3
UTAH				
840-841	Salt Lake City	102.8	72.9	89.7
842,844	Ogden	98.2	72.9	87.2
843	Logan	100.2	72.9	88.3
845	Price	100.8	71.7	88.1
846-847	Provo	100.6	72.8	88.5
VERMONT				
050	White River Jct.	97.7	81.6	90.7
051	Bellows Falls	96.0	95.3	95.7
052	Bennington	96.5	91.6	94.3
053	Brattleboro	96.9	95.3	96.2
054	Burlington	101.0	80.9	92.2
056	Montpelier	99.8	85.9	93.7
057	Rutland	98.1	80.9	90.6
058	St. Johnsbury	97.8	81.5	90.7
059	Guildhall	96.4	81.1	89.7
VIRGINIA				
220-221	Fairfax	100.0	83.8	92.9
222	Arlington	100.7	84.5	93.7
223	Alexandria	99.8	84.7	93.2
224-225	Fredericksburg	98.7	82.2	91.5
226	Winchester	99.3	82.9	92.2
227	Culpeper	99.1	80.5	91.0
228	Harrisonburg	99.5	67.3	85.4
229	Charlottesville	99.8	69.3	86.5
230-232	Richmond	98.5	77.9	89.5
233-235	Norfolk	98.5	71.3	86.6
236	Newport News	98.2	70.4	86.0
237	Portsmouth	97.7	64.3	83.1
238	Petersburg	98.3	73.9	87.6
239	Farmville	97.4	58.6	80.4
240-241	Roanoke	100.3	73.0	88.3
242	Bristol	98.6	60.5	82.0
243	Pulaski	98.2	62.7	82.7
244	Staunton	99.0	65.8	84.5
245	Lynchburg	99.2	75.2	88.7
246	Grundy	98.5	56.0	79.9
WASHINGTON				
980-981,987	Seattle	103.6	107.3	105.2
982	Everett	102.0	102.4	102.2
983-984	Tacoma	102.4	102.3	102.3
985	Olympia	100.7	102.3	101.4
986	Vancouver	103.3	97.2	100.6
988	Wenatchee	102.6	89.5	96.9
989	Yakima	102.5	98.2	100.6
990-992	Spokane	101.4	85.0	94.2
993	Richland	101.1	92.7	97.4
994	Clarkston	100.6	83.4	93.1
WEST VIRGINIA				
247-248	Bluefield	97.5	89.6	94.0
249	Lewisburg	99.1	88.9	94.6
250-253	Charleston	97.9	93.3	95.9
254	Martinsburg	98.8	84.4	92.5
255-257	Huntington	99.7	93.8	97.1
258-259	Beckley	97.3	91.8	94.8
260	Wheeling	100.0	91.9	96.5
261	Parkersburg	98.8	89.5	94.8
262	Buckhannon	98.8	91.5	95.6
263-264	Clarksburg	99.4	91.6	96.0
265	Morgantown	99.4	91.6	96.0
266	Gassaway	98.8	89.9	94.9
267	Romney	98.6	86.4	93.3
268	Petersburg	98.5	85.0	92.6
WISCONSIN				
530,532	Milwaukee	99.3	106.6	102.5
531	Kenosha	98.4	105.0	101.3
534	Racine	97.8	106.2	101.5
535	Beloit	97.7	98.6	98.1
537	Madison	98.0	98.9	98.4

STATE/ZIP	CITY	MAT.	INST.	TOTAL
WISCONSIN (CONT'D)				
538	Lancaster	96.0	93.9	95.1
539	Portage	94.4	96.6	95.4
540	New Richmond	95.6	92.7	94.3
541-543	Green Bay	99.6	96.5	98.2
544	Wausau	95.0	94.5	94.8
545	Rhinelander	98.3	90.1	94.7
546	La Crosse	95.9	96.9	96.4
547	Eau Claire	97.6	97.3	97.5
548	Superior	95.4	97.1	96.1
549	Oshkosh	95.7	89.2	92.9
WYOMING				
820	Cheyenne	101.3	73.5	89.2
821	Yellowstone Nat'l Park	99.4	74.4	88.4
822	Wheatland	100.6	69.2	86.9
823	Rawlins	102.1	73.5	89.6
824	Worland	100.1	73.0	88.2
825	Riverton	101.2	70.7	87.8
826	Casper	101.4	70.3	87.8
827	Newcastle	99.9	73.1	88.2
828	Sheridan	102.8	71.5	89.1
829-831	Rock Springs	103.9	72.7	90.2
CANADIAN FACTORS (reflect Canadian currency)				
ALBERTA				
	Calgary	119.5	98.8	110.4
	Edmonton	120.4	98.8	110.9
	Fort McMurray	123.4	91.9	109.7
	Lethbridge	118.3	91.4	106.5
	Lloydminster	112.7	88.1	101.9
	Medicine Hat	112.9	87.4	101.7
	Red Deer	113.3	87.4	102.0
BRITISH COLUMBIA				
	Kamloops	114.1	87.2	102.3
	Prince George	115.1	86.6	102.6
	Vancouver	118.0	91.2	106.3
	Victoria	115.3	85.7	102.3
MANITOBA				
	Brandon	123.6	73.8	101.8
	Portage la Prairie	112.7	72.2	94.9
	Winnipeg	121.3	72.0	99.7
NEW BRUNSWICK				
	Bathurst	111.3	65.0	91.0
	Dalhousie	111.3	65.2	91.1
	Fredericton	118.2	72.5	98.2
	Moncton	111.5	69.3	93.0
	Newcastle	111.3	65.7	91.3
	St. John	113.8	77.0	97.7
NEWFOUNDLAND				
	Corner Brook	128.0	71.5	103.3
	St. Johns	120.7	90.3	107.4
NORTHWEST TERRITORIES				
	Yellowknife	133.4	86.2	112.7
NOVA SCOTIA				
	Bridgewater	112.3	73.4	95.3
	Dartmouth	124.6	73.5	102.2
	Halifax	117.4	87.2	104.2
	New Glasgow	122.6	73.5	101.1
	Sydney	120.6	73.5	100.0
	Truro	111.9	73.4	95.1
	Yarmouth	122.5	73.5	101.1
ONTARIO				
	Barrie	117.0	91.0	105.6
	Brantford	114.3	94.9	105.8
	Cornwall	114.2	91.3	104.2
	Hamilton	116.0	99.6	108.8
	Kingston	115.3	91.4	104.8
	Kitchener	109.1	95.1	103.0
	London	115.6	97.0	107.5
	North Bay	125.6	89.2	109.7
	Oshawa	111.6	94.7	104.2
	Ottawa	117.2	98.0	108.8
	Owen Sound	117.1	89.2	104.9
	Peterborough	114.3	91.1	104.1
	Sarnia	114.3	95.5	106.0
ONTARIO (CONT'D)				

STATE/ZIP	CITY	MAT.	INST.	TOTAL
	Sault Ste. Marie	109.6	91.7	101.8
	St. Catharines	107.1	96.1	102.3
	Sudbury	106.9	94.9	101.6
	Thunder Bay	108.3	95.4	102.7
	Timmins	114.4	89.2	103.4
	Toronto	115.9	104.2	110.8
	Windsor	107.6	95.2	102.2
PRINCE EDWARD ISLAND				
	Charlottetown	120.9	62.8	95.5
	Summerside	125.1	60.5	96.8
QUEBEC				
	Cap-de-la-Madeleine	111.7	83.6	99.4
	Charlesbourg	111.7	83.6	99.4
	Chicoutimi	111.1	88.9	101.4
	Gatineau	111.4	83.4	99.1
	Granby	111.6	83.3	99.2
	Hull	111.5	83.4	99.2
	Joliette	111.9	83.6	99.5
	Laval	111.5	83.4	99.2
	Montreal	118.0	91.7	106.5
	Quebec City	118.5	92.2	107.0
	Rimouski	111.3	88.9	101.5
	Rouyn-Noranda	111.4	83.4	99.1
	Saint-Hyacinthe	110.8	83.4	98.8
	Sherbrooke	111.7	83.4	99.3
	Sorel	111.9	83.6	99.5
	Saint-Jerome	111.4	83.4	99.1
	Trois-Rivieres	123.3	83.6	105.9
SASKATCHEWAN				
	Moose Jaw	110.1	66.7	91.1
	Prince Albert	109.2	65.0	89.8
	Regina	122.8	95.4	110.8
	Saskatoon	110.8	92.5	102.7
YUKON				
	Whitehorse	135.9	71.3	107.6

R011105-05   Tips for Accurate Estimating

1. Use pre-printed or columnar forms for orderly sequence of dimensions and locations and for recording telephone quotations.

2. Use only the front side of each paper or form except for certain pre-printed summary forms.

3. Be consistent in listing dimensions: For example, length x width x height. This helps in rechecking to ensure that, the total length of partitions is appropriate for the building area.

4. Use printed (rather than measured) dimensions where given.

5. Add up multiple printed dimensions for a single entry where possible.

6. Measure all other dimensions carefully.

7. Use each set of dimensions to calculate multiple related quantities.

8. Convert foot and inch measurements to decimal feet when listing. Memorize decimal equivalents to .01 parts of a foot (1/8″ equals approximately .01′).

9. Do not "round off" quantities until the final summary.

10. Mark drawings with different colors as items are taken off.

11. Keep similar items together, different items separate.

12. Identify location and drawing numbers to aid in future checking for completeness.

13. Measure or list everything on the drawings or mentioned in the specifications.

14. It may be necessary to list items not called for to make the job complete.

15. Be alert for: Notes on plans such as N.T.S. (not to scale); changes in scale throughout the drawings; reduced size drawings; discrepancies between the specifications and the drawings.

16. Develop a consistent pattern of performing an estimate. For example:
    a. Start the quantity takeoff at the lower floor and move to the next higher floor.
    b. Proceed from the main section of the building to the wings.
    c. Proceed from south to north or vice versa, clockwise or counterclockwise.
    d. Take off floor plan quantities first, elevations next, then detail drawings.

17. List all gross dimensions that can be either used again for different quantities, or used as a rough check of other quantities for verification (exterior perimeter, gross floor area, individual floor areas, etc.).

18. Utilize design symmetry or repetition (repetitive floors, repetitive wings, symmetrical design around a center line, similar room layouts, etc.). Note: Extreme caution is needed here so as not to omit or duplicate an area.

19. Do not convert units until the final total is obtained. For instance, when estimating concrete work, keep all units to the nearest cubic foot, then summarize and convert to cubic yards.

20. When figuring alternatives, it is best to total all items involved in the basic system, then total all items involved in the alternates. Therefore you work with positive numbers in all cases. When adds and deducts are used, it is often confusing whether to add or subtract a portion of an item; especially on a complicated or involved alternate.

827

R011105-50   Metric Conversion Factors

Description: This table is primarily for converting customary U.S. units in the left hand column to SI metric units in the right hand column. In addition, conversion factors for some commonly encountered Canadian and non-SI metric units are included.

If You Know		Multiply By		To Find
Length	Inches	x	25.4[a]	=	Millimeters
	Feet	x	0.3048[a]	=	Meters
	Yards	x	0.9144[a]	=	Meters
	Miles (statute)	x	1.609	=	Kilometers
Area	Square inches	x	645.2	=	Square millimeters
	Square feet	x	0.0929	=	Square meters
	Square yards	x	0.8361	=	Square meters
Volume (Capacity)	Cubic inches	x	16,387	=	Cubic millimeters
	Cubic feet	x	0.02832	=	Cubic meters
	Cubic yards	x	0.7646	=	Cubic meters
	Gallons (U.S. liquids)[b]	x	0.003785	=	Cubic meters[c]
	Gallons (Canadian liquid)[b]	x	0.004546	=	Cubic meters[c]
	Ounces (U.S. liquid)[b]	x	29.57	=	Milliliters[c, d]
	Quarts (U.S. liquid)[b]	x	0.9464	=	Liters[c, d]
	Gallons (U.S. liquid)[b]	x	3.785	=	Liters[c, d]
Force	Kilograms force[d]	x	9.807	=	Newtons
	Pounds force	x	4.448	=	Newtons
	Pounds force	x	0.4536	=	Kilograms force[d]
	Kips	x	4448	=	Newtons
	Kips	x	453.6	=	Kilograms force[d]
Pressure, Stress, Strength (Force per unit area)	Kilograms force per square centimeter[d]	x	0.09807	=	Megapascals
	Pounds force per square inch (psi)	x	0.006895	=	Megapascals
	Kips per square inch	x	6.895	=	Megapascals
	Pounds force per square inch (psi)	x	0.07031	=	Kilograms force per square centimeter[d]
	Pounds force per square foot	x	47.88[a]	=	Pascals
	Pounds force per square foot	x	4.882[a]	=	Kilograms force per square meter[d]
Flow	Cubic feet per minute	x	0.4719	=	Liters per second
	Gallons per minute	x	0.0631	=	Liters per second
	Gallons per hour	x	1.05	=	Milliliters per second
Bending Moment Or Torque	Inch-pounds force	x	0.01152	=	Meter-kilograms force[d]
	Inch-pounds force	x	0.1130	=	Newton-meters
	Foot-pounds force	x	0.1383	=	Meter-kilograms force[d]
	Foot-pounds force	x	1.356	=	Newton-meters
	Meter-kilograms force[d]	x	9.807	=	Newton-meters
Mass	Ounces (avoirdupois)	x	28.35	=	Grams
	Pounds (avoirdupois)	x	0.4536	=	Kilograms
	Tons (metric)	x	1000	=	Kilograms
	Tons, short (2000 pounds)	x	907.2	=	Kilograms
	Tons, short (2000 pounds)	x	0.9072	=	Megagrams[e]
Mass per Unit Volume	Pounds mass per cubic foot	x	16.02	=	Kilograms per cubic meter
	Pounds mass per cubic yard	x	0.5933	=	Kilograms per cubic meter
	Pounds mass per gallon (U.S. liquid)[b]	x	119.8	=	Kilograms per cubic meter
	Pounds mass per gallon (Canadian liquid)[b]	x	99.78	=	Kilograms per cubic meter
Temperature	Degrees Fahrenheit	(F-32)/1.8		=	Degrees Celsius
	Degrees Fahrenheit	(F+459.67)/1.8		=	Degrees Kelvin
	Degrees Celsius	C+273.15		=	Degrees Kelvin

[a]The factor given is exact
[b]One U.S. gallon = 0.8327 Canadian gallon
[c]1 liter = 1000 milliliters = 1000 cubic centimeters
 1 cubic decimeter = 0.001 cubic meter

[d]Metric but not SI unit
[e]Called "tonne" in England and
 "metric ton" in other metric countries

R011105-60   Weights and Measures

Measures of Length
1 Mile = 1760 Yards = 5280 Feet
1 Yard = 3 Feet = 36 inches
1 Foot = 12 Inches
1 Mil = 0.001 Inch
1 Fathom = 2 Yards = 6 Feet
1 Rod = 5.5 Yards = 16.5 Feet
1 Hand = 4 Inches
1 Span = 9 Inches
1 Micro-inch = One Millionth Inch or 0.000001 Inch
1 Micron = One Millionth Meter + 0.00003937 Inch

Surveyor's Measure
1 Mile = 8 Furlongs = 80 Chains
1 Furlong = 10 Chains = 220 Yards
1 Chain = 4 Rods = 22 Yards = 66 Feet = 100 Links
1 Link = 7.92 Inches

Square Measure
1 Square Mile = 640 Acres = 6400 Square Chains
1 Acre = 10 Square Chains = 4840 Square Yards =
    43,560 Sq. Ft.
1 Square Chain = 16 Square Rods = 484 Square Yards =
    4356 Sq. Ft.
1 Square Rod = 30.25 Square Yards = 272.25 Square Feet = 625 Square
    Lines
1 Square Yard = 9 Square Feet
1 Square Foot = 144 Square Inches
An Acre equals a Square 208.7 Feet per Side

Cubic Measure
1 Cubic Yard = 27 Cubic Feet
1 Cubic Foot = 1728 Cubic Inches
1 Cord of Wood = 4 x 4 x 8 Feet = 128 Cubic Feet
1 Perch of Masonry = 16½ x 1½ x 1 Foot = 24.75 Cubic Feet

Avoirdupois or Commercial Weight
1 Gross or Long Ton = 2240 Pounds
1 Net or Short Ton = 2000 Pounds
1 Pound = 16 Ounces = 7000 Grains
1 Ounce = 16 Drachms = 437.5 Grains
1 Stone = 14 Pounds

Power
1 British Thermal Unit per Hour = 0.2931 Watts
1 Ton (Refrigeration) = 3.517 Kilowatts
1 Horsepower (Boiler) = 9.81 Kilowatts
1 Horsepower (550 ft-lb/s) = 0.746 Kilowatts

Shipping Measure
For Measuring Internal Capacity of a Vessel:
    1 Register Ton = 100 Cubic Feet

For Measurement of Cargo:
    Approximately 40 Cubic Feet of Merchandise is considered a Shipping
    Ton, unless that bulk would weigh more than 2000 Pounds, in which case
    Freight Charge may be based upon weight.

40 Cubic Feet = 32.143 U.S. Bushels = 31.16 Imp. Bushels

Liquid Measure
1 Imperial Gallon = 1.2009 U.S. Gallon = 277.42 Cu. In.
1 Cubic Foot = 7.48 U.S. Gallons

R011110-10   Architectural Fees

Tabulated below are typical percentage fees by project size, for good professional architectural service. Fees may vary from those listed depending upon degree of design difficulty and economic conditions in any particular area.

Rates can be interpolated horizontally and vertically. Various portions of the same project requiring different rates should be adjusted proportionately. For alterations, add 50% to the fee for the first $500,000 of project cost and add 25% to the fee for project cost over $500,000.

Architectural fees tabulated below include Structural, Mechanical and Electrical Engineering Fees. They do not include the fees for special consultants such as kitchen planning, security, acoustical, interior design, etc.

Civil Engineering fees are included in the Architectural fee for project sites requiring minimal design such as city sites. However, separate Civil Engineering fees must be added when utility connections require design, drainage calculations are needed, stepped foundations are required, or provisions are required to protect adjacent wetlands.

Building Types	Total Project Size in Thousands of Dollars						
	100	250	500	1,000	5,000	10,000	50,000
Factories, garages, warehouses, repetitive housing	9.0%	8.0%	7.0%	6.2%	5.3%	4.9%	4.5%
Apartments, banks, schools, libraries, offices, municipal buildings	12.2	12.3	9.2	8.0	7.0	6.6	6.2
Churches, hospitals, homes, laboratories, museums, research	15.0	13.6	12.7	11.9	9.5	8.8	8.0
Memorials, monumental work, decorative furnishings	—	16.0	14.5	13.1	10.0	9.0	8.3

R011110-30   Engineering Fees

Typical **Structural Engineering Fees** based on type of construction and total project size. These fees are included in Architectural Fees.

Type of Construction	Total Project Size (in thousands of dollars)			
	$500	$500-$1,000	$1,000-$5,000	Over $5000
Industrial buildings, factories & warehouses	Technical payroll times 2.0 to 2.5	1.60%	1.25%	1.00%
Hotels, apartments, offices, dormitories, hospitals, public buildings, food stores		2.00%	1.70%	1.20%
Museums, banks, churches and cathedrals		2.00%	1.75%	1.25%
Thin shells, prestressed concrete, earthquake resistive		2.00%	1.75%	1.50%
Parking ramps, auditoriums, stadiums, convention halls, hangars & boiler houses		2.50%	2.00%	1.75%
Special buildings, major alterations, underpinning & future expansion	↓	Add to above 0.5%	Add to above 0.5%	Add to above 0.5%

For complex reinforced concrete or unusually complicated structures, add 20% to 50%.

Typical **Mechanical and Electrical Engineering Fees** are based on the size of the subcontract. The fee structure for both is shown below. These fees are included in Architectural Fees.

Type of Construction	Subcontract Size							
	$25,000	$50,000	$100,000	$225,000	$350,000	$500,000	$750,000	$1,000,000
Simple structures	6.4%	5.7%	4.8%	4.5%	4.4%	4.3%	4.2%	4.1%
Intermediate structures	8.0	7.3	6.5	5.6	5.1	5.0	4.9	4.8
Complex structures	10.1	9.0	9.0	8.0	7.5	7.5	7.0	7.0

For renovations, add 15% to 25% to applicable fee.

R012153-10   Repair and Remodeling

Cost figures are based on new construction utilizing the most cost-effective combination of labor, equipment and material with the work scheduled in proper sequence to allow the various trades to accomplish their work in an efficient manner.

The costs for repair and remodeling work must be modified due to the following factors that may be present in any given repair and remodeling project.

1. Equipment usage curtailment due to the physical limitations of the project, with only hand-operated equipment being used.

2. Increased requirement for shoring and bracing to hold up the building while structural changes are being made and to allow for temporary storage of construction materials on above-grade floors.

3. Material handling becomes more costly due to having to move within the confines of an enclosed building. For multi-story construction, low capacity elevators and stairwells may be the only access to the upper floors.

4. Large amount of cutting and patching and attempting to match the existing construction is required. It is often more economical to remove entire walls rather than create many new door and window openings. This sort of trade-off has to be carefully analyzed.

5. Cost of protection of completed work is increased since the usual sequence of construction usually cannot be accomplished.

6. Economies of scale usually associated with new construction may not be present. If small quantities of components must be custom fabricated due to job requirements, unit costs will naturally increase. Also, if only small work areas are available at a given time, job scheduling between trades becomes difficult and subcontractor quotations may reflect the excessive start-up and shut-down phases of the job.

7. Work may have to be done on other than normal shifts and may have to be done around an existing production facility which has to stay in production during the course of the repair and remodeling.

8. Dust and noise protection of adjoining non-construction areas can involve substantial special protection and alter usual construction methods.

9. Job may be delayed due to unexpected conditions discovered during demolition or removal. These delays ultimately increase construction costs.

10. Piping and ductwork runs may not be as simple as for new construction. Wiring may have to be snaked through walls and floors.

11. Matching "existing construction" may be impossible because materials may no longer be manufactured. Substitutions may be expensive.

12. Weather protection of existing structure requires additional temporary structures to protect building at openings.

13. On small projects, because of local conditions, it may be necessary to pay a tradesman for a minimum of four hours for a task that is completed in one hour.

All of the above areas can contribute to increased costs for a repair and remodeling project. Each of the above factors should be considered in the planning, bidding and construction stage in order to minimize the increased costs associated with repair and remodeling jobs.

R012153-60   Security Factors

Contractors entering, working in, and exiting secure facilities often lose productive time during a normal workday. The recommended allowances in this section are intended to provide for the loss of productivity by increasing labor costs. Note that different costs are associated with searches upon entry only and searches upon entry and exit. Time spent in a queue is unpredictable and not part of these allowances. Contractors should plan ahead for this situation.

Security checkpoints are designed to reflect the level of security required to gain access or egress. An extreme example is when contractors, along with any materials, tools, equipment, and vehicles, must be physically searched and have all materials, tools, equipment, and vehicles inventoried and documented prior to both entry and exit.

Physical searches without going through the documentation process represent the next level and take up less time.

Electronic searches—passing through a detector or x-ray machine with no documentation of materials, tools, equipment, and vehicles—take less time than physical searches.

Visual searches of materials, tools, equipment, and vehicles represent the next level of security.

Finally, access by means of an ID card or displayed sticker takes the least amount of time.

Another consideration is if the searches described above are performed each and every day, or if they are performed only on the first day with access granted by ID card or displayed sticker for the remainder of the project. The figures for this situation have been calculated to represent the initial check-in as described and subsequent entry by ID card or displayed sticker for up to 20 days on site. For the situation described above, where the time period is beyond 20 days, the impact on labor cost is negligible.

There are situations where tradespeople must be accompanied by an escort and observed during the work day. The loss of freedom of movement will slow down productivity for the tradesperson. Costs for the observer have not been included. Those costs are normally born by the owner.

R012157-20   Construction Time Requirements

The table below lists the construction durations for various building types along with their respective project sizes and project values. Design time runs 25% to 40% of construction time.

Building Type	Size S.F.	Project Value	Construction Duration
Industrial/Warehouse	100,000	$8,000,000	14 months
	500,000	$32,000,000	19 months
	1,000,000	$75,000,000	21 months
Offices/Retail	50,000	$7,000,000	15 months
	250,000	$28,000,000	23 months
	500,000	$58,000,000	34 months
Institutional/Hospitals/Laboratory	200,000	$45,000,000	31 months
	500,000	$110,000,000	52 months
	750,000	$160,000,000	55 months
	1,000,000	$210,000,000	60 months

R012909-80   Sales Tax by State

State sales tax on materials is tabulated below (5 states have no sales tax). Many states allow local jurisdictions, such as a county or city, to levy additional sales tax.

Some projects may be sales tax exempt, particularly those constructed with public funds.

State	Tax (%)	State	Tax (%)	State	Tax (%)	State	Tax (%)
Alabama	4	Illinois	6.25	Montana	0	Rhode Island	7
Alaska	0	Indiana	7	Nebraska	5.5	South Carolina	6
Arizona	5.6	Iowa	6	Nevada	6.85	South Dakota	5
Arkansas	6.5	Kansas	6.5	New Hampshire	0	Tennessee	7
California	7.25	Kentucky	6	New Jersey	7	Texas	6.25
Colorado	2.9	Louisiana	4	New Mexico	5.125	Utah	5.95
Connecticut	6.35	Maine	5.5	New York	4	Vermont	6
Delaware	0	Maryland	6	North Carolina	4.75	Virginia	5.3
District of Columbia	5.75	Massachusetts	6.25	North Dakota	5	Washington	6.5
Florida	6	Michigan	6	Ohio	5.75	West Virginia	6
Georgia	4	Minnesota	6.875	Oklahoma	4.5	Wisconsin	5
Hawaii	4	Mississippi	7	Oregon	0	Wyoming	4
Idaho	6	Missouri	4.225	Pennsylvania	6	Average	5.11 %

Sales Tax by Province (Canada)

GST - a value-added tax, which the government imposes on most goods and services provided in or imported into Canada. PST - a retail sales tax, which five of the provinces impose on the prices of most goods and some

services. QST - a value-added tax, similar to the federal GST, which Quebec imposes. HST - Three provinces have combined their retail sales taxes with the federal GST into one harmonized tax.

Province	PST (%)	QST (%)	GST(%)	HST(%)
Alberta	0	0	5	0
British Columbia	7	0	5	0
Manitoba	8	0	5	0
New Brunswick	0	0	0	15
Newfoundland	0	0	0	15
Northwest Territories	0	0	5	0
Nova Scotia	0	0	0	15
Ontario	0	0	0	13
Prince Edward Island	0	0	0	15
Quebec	0	9.975	5	0
Saskatchewan	6	0	5	0
Yukon	0	0	5	0

General Requirements   R0129 Payment Procedures

R012909-85   Unemployment Taxes and Social Security Taxes

State unemployment tax rates vary not only from state to state, but also with the experience rating of the contractor. The federal unemployment tax rate is 6.0% of the first $7,000 of wages. This is reduced by a credit of up to 5.4% for timely payment to the state. The minimum federal unemployment tax is 0.6% after all credits.

Social security (FICA) for 2018 is estimated at time of publication to be 7.65% of wages up to $127,200.

R012909-86   Unemployment Tax by State

Information is from the U.S. Department of Labor, state unemployment tax rates.

State	Tax (%)	State	Tax (%)	State	Tax (%)	State	Tax (%)
Alabama	6.74	Illinois	7.75	Montana	6.12	Rhode Island	9.79
Alaska	5.4	Indiana	7.474	Nebraska	5.4	South Carolina	5.46
Arizona	8.91	Iowa	8	Nevada	5.4	South Dakota	9.5
Arkansas	6.0	Kansas	7.6	New Hampshire	7.5	Tennessee	10.0
California	6.2	Kentucky	10.0	New Jersey	5.8	Texas	7.5
Colorado	8.9	Louisiana	6.2	New Mexico	5.4	Utah	7.2
Connecticut	6.8	Maine	5.4	New York	8.5	Vermont	8.4
Delaware	8.0	Maryland	7.50	North Carolina	5.76	Virginia	6.27
District of Columbia	7	Massachusetts	11.13	North Dakota	10.72	Washington	5.7
Florida	5.4	Michigan	10.3	Ohio	8.7	West Virginia	7.5
Georgia	5.4	Minnesota	9.0	Oklahoma	5.5	Wisconsin	12.0
Hawaii	5.6	Mississippi	5.4	Oregon	5.4	Wyoming	8.8
Idaho	5.4	Missouri	9.75	Pennsylvania	10.89	Median	7.47%

General Requirements   R0131 Project Management & Coordination

R013113-40   Builder's Risk Insurance

Builder's risk insurance is insurance on a building during construction. Premiums are paid by the owner or the contractor. Blasting, collapse and underground insurance would raise total insurance costs.

R013113-50   General Contractor's Overhead

There are two distinct types of overhead on a construction project: Project overhead and main office overhead. Project overhead includes those costs at a construction site not directly associated with the installation of construction materials. Examples of project overhead costs include the following:

1. Superintendent
2. Construction office and storage trailers
3. Temporary sanitary facilities
4. Temporary utilities
5. Security fencing
6. Photographs
7. Cleanup
8. Performance and payment bonds

The above project overhead items are also referred to as general requirements and therefore are estimated in Division 1. Division 1 is the first division listed in the CSI MasterFormat but it is usually the last division estimated. The sum of the costs in Divisions 1 through 49 is referred to as the sum of the direct costs.

All construction projects also include indirect costs. The primary components of indirect costs are the contractor's main office overhead and profit. The amount of the main office overhead expense varies depending on the following:

1. Owner's compensation
2. Project managers' and estimators' wages
3. Clerical support wages
4. Office rent and utilities
5. Corporate legal and accounting costs
6. Advertising
7. Automobile expenses
8. Association dues
9. Travel and entertainment expenses

These costs are usually calculated as a percentage of annual sales volume. This percentage can range from 35% for a small contractor doing less than $500,000 to 5% for a large contractor with sales in excess of $100 million.

R013113-55   Installing Contractor's Overhead

Installing contractors (subcontractors) also incur costs for general requirements and main office overhead.

Included within the total incl. overhead and profit costs is a percent mark-up for overhead that includes:

1. Compensation and benefits for office staff and project managers
2. Office rent, utilities, business equipment, and maintenance
3. Corporate legal and accounting costs

4. Advertising
5. Vehicle expenses (for office staff and project managers)
6. Association dues
7. Travel, entertainment
8. Insurance
9. Small tools and equipment

R013113-60   Workers' Compensation Insurance Rates by Trade

The table below tabulates the national averages for workers' compensation insurance rates by trade and type of building. The average "Insurance Rate" is multiplied by the "% of Building Cost" for each trade. This produces the "Workers' Compensation" cost by % of total labor cost, to be added for each trade by building type to determine the weighted average workers' compensation rate for the building types analyzed.

Trade	Insurance Rate (% Labor Cost) Range		Average	% of Building Cost Office Bldgs.	Schools & Apts.	Mfg.	Workers' Compensation Office Bldgs.	Schools & Apts.	Mfg.
Excavation, Grading, etc.	2.7 % to	20.1%	8.5%	4.8%	4.9%	4.5%	0.41%	0.42%	0.38%
Piles & Foundations	5.3 to	29.8	13.4	7.1	5.2	8.7	0.95	0.70	1.17
Concrete	4.1 to	28.0	11.8	5.0	14.8	3.7	0.59	1.75	0.44
Masonry	3.9 to	49.3	13.8	6.9	7.5	1.9	0.95	1.04	0.26
Structural Steel	5.3 to	59.1	21.2	10.7	3.9	17.6	2.27	0.83	3.73
Miscellaneous & Ornamental Metals	3.3 to	24.4	10.6	2.8	4.0	3.6	0.30	0.42	0.38
Carpentry & Millwork	4.4 to	32.4	13.0	3.7	4.0	0.5	0.48	0.52	0.07
Metal or Composition Siding	5.5 to	107.2	19.0	2.3	0.3	4.3	0.44	0.06	0.82
Roofing	5.5 to	120.3	29.0	2.3	2.6	3.1	0.67	0.75	0.90
Doors & Hardware	3.2 to	32.4	11.0	0.9	1.4	0.4	0.10	0.15	0.04
Sash & Glazing	4.7 to	25.5	12.1	3.5	4.0	1.0	0.42	0.48	0.12
Lath & Plaster	3.0 to	31.6	10.7	3.3	6.9	0.8	0.35	0.74	0.09
Tile, Marble & Floors	2.7 to	18.3	8.7	2.6	3.0	0.5	0.23	0.26	0.04
Acoustical Ceilings	2.4 to	46.3	8.5	2.4	0.2	0.3	0.20	0.02	0.03
Painting	3.3 to	38.8	11.2	1.5	1.6	1.6	0.17	0.18	0.18
Interior Partitions	4.4 to	32.4	13.0	3.9	4.3	4.4	0.51	0.56	0.57
Miscellaneous Items	2.3 to	97.7	11.2	5.2	3.7	9.7	0.58	0.42	1.09
Elevators	1.3 to	13.7	4.7	2.1	1.1	2.2	0.10	0.05	0.10
Sprinklers	2.0 to	15.5	6.7	0.5	—	2.0	0.03	—	0.13
Plumbing	1.7 to	14.0	6.3	4.9	7.2	5.2	0.31	0.45	0.33
Heat., Vent., Air Conditioning	3.3 to	17.8	8.3	13.5	11.0	12.9	1.12	0.91	1.07
Electrical	1.9 to	11.6	5.2	10.1	8.4	11.1	0.53	0.44	0.58
Total	1.3 % to	120.3%	—	100.0%	100.0%	100.0%	11.71%	11.15%	12.52%

Overall Weighted Average   11.79%

Workers' Compensation Insurance Rates by States

The table below lists the weighted average Workers' Compensation base rate for each state with a factor comparing this with the national average of 11.8%.

State	Weighted Average	Factor	State	Weighted Average	Factor	State	Weighted Average	Factor
Alabama	15.0%	127	Kentucky	10.4%	88	North Dakota	6.2%	53
Alaska	10.4	88	Louisiana	18.7	158	Ohio	7.2	61
Arizona	9.6	81	Maine	10.4	88	Oklahoma	8.9	75
Arkansas	7.0	59	Maryland	11.3	96	Oregon	9.3	79
California	22.2	188	Massachusetts	11.2	95	Pennsylvania	21.1	179
Colorado	7.5	64	Michigan	8.2	69	Rhode Island	13.7	116
Connecticut	17.5	148	Minnesota	16.9	143	South Carolina	16.5	140
Delaware	13.9	118	Mississippi	11.8	100	South Dakota	11.8	100
District of Columbia	9.1	77	Missouri	12.4	105	Tennessee	8.6	73
Florida	11.1	94	Montana	8.8	75	Texas	6.6	56
Georgia	31.9	270	Nebraska	13.5	114	Utah	7.4	63
Hawaii	8.5	72	Nevada	7.5	64	Vermont	10.9	92
Idaho	9.4	80	New Hampshire	12.0	102	Virginia	6.9	58
Illinois	21.1	179	New Jersey	14.8	125	Washington	9.1	77
Indiana	4.1	35	New Mexico	13.3	113	West Virginia	4.5	38
Iowa	13.7	116	New York	19.2	163	Wisconsin	12.2	103
Kansas	6.5	55	North Carolina	15.8	134	Wyoming	5.6	47

Weighted Average for U.S. is   11.8% of payroll = 100%

The weighted average skilled worker rate for 35 trades is 11.8%. For bidding purposes, apply the full value of Workers' Compensation directly to total labor costs, or if labor is 38%, materials 42% and overhead and profit 20% of total cost, carry 38/80 x 11.8% = 6.0% of cost (before overhead and profit) into overhead. Rates vary not only from state to state but also with the experience rating of the contractor.

Rates are the most current available at the time of publication.

835

For customer support on your Commercial Renovation Costs with RSMeans data, call 800.448.8182.

R013113-80   Performance Bond

This table shows the cost of a Performance Bond for a construction job scheduled to be completed in 12 months. Add 1% of the premium cost per month for jobs requiring more than 12 months to complete. The rates are "standard" rates offered to contractors that the bonding company considers financially sound and capable of doing the work. Preferred rates are offered by some bonding companies based upon financial strength of the contractor. Actual rates vary from contractor to contractor and from bonding company to bonding company. Contractors should prequalify through a bonding agency before submitting a bid on a contract that requires a bond.

Contract Amount		Building Construction Class B Projects			Highways & Bridges					
					Class A New Construction			Class A-1 Highway Resurfacing		
First $ 100,000 bid		$25.00 per M			$15.00 per M			$9.40 per M		
Next 400,000 bid		$ 2,500	plus $15.00	per M	$ 1,500	plus $10.00	per M	$ 940	plus $7.20	per M
Next 2,000,000 bid		8,500	plus 10.00	per M	5,500	plus 7.00	per M	3,820	plus 5.00	per M
Next 2,500,000 bid		28,500	plus 7.50	per M	19,500	plus 5.50	per M	15,820	plus 4.50	per M
Next 2,500,000 bid		47,250	plus 7.00	per M	33,250	plus 5.00	per M	28,320	plus 4.50	per M
Over 7,500,000 bid		64,750	plus 6.00	per M	45,750	plus 4.50	per M	39,570	plus 4.00	per M

R015423-10   Steel Tubular Scaffolding

On new construction, tubular scaffolding is efficient up to 60' high or five stories. Above this it is usually better to use a hung scaffolding if construction permits. Swing scaffolding operations may interfere with tenants. In this case, the tubular is more practical at all heights.

In repairing or cleaning the front of an existing building the cost of tubular scaffolding per S.F. of building front increases as the height increases above the first tier. The first tier cost is relatively high due to leveling and alignment.

The minimum efficient crew for erecting and dismantling is three workers. They can set up and remove 18 frame sections per day up to 5 stories high. For 6 to 12 stories high, a crew of four is most efficient. Use two or more on top and two on the bottom for handing up or hoisting. They can also set up and remove 18 frame sections per day. At 7' horizontal spacing, this will run about 800 S.F. per day of erecting and dismantling. Time for placing and removing planks must be added to the above. A crew of three can place and remove 72 planks per day up to 5 stories. For over 5 stories, a crew of four can place and remove 80 planks per day.

The table below shows the number of pieces required to erect tubular steel scaffolding for 1000 S.F. of building frontage. This area is made up of a scaffolding system that is 12 frames (11 bays) long by 2 frames high.

For jobs under twenty-five frames, add 50% to rental cost. Rental rates will be lower for jobs over three months duration. Large quantities for long periods can reduce rental rates by 20%.

Description of Component	Number of Pieces for 1000 S.F. of Building Front	Unit
5' Wide Standard Frame, 6'-4" High	24	Ea.
Leveling Jack & Plate	24	
Cross Brace	44	
Side Arm Bracket, 21"	12	
Guardrail Post	12	
Guardrail, 7' section	22	
Stairway Section	2	
Stairway Starter Bar	1	
Stairway Inside Handrail	2	
Stairway Outside Handrail	2	
Walk-Thru Frame Guardrail	2	

Scaffolding is often used as falsework over 15' high during construction of cast-in-place concrete beams and slabs. Two foot wide scaffolding is generally used for heavy beam construction. The span between frames depends upon the load to be carried with a maximum span of 5'.

Heavy duty shoring frames with a capacity of 10,000#/leg can be spaced up to 10' O.C. depending upon form support design and loading.

Scaffolding used as horizontal shoring requires less than half the material required with conventional shoring.

On new construction, erection is done by carpenters.

Rolling towers supporting horizontal shores can reduce labor and speed the job. For maintenance work, catwalks with spans up to 70' can be supported by the rolling towers.

R015423-20   Pump Staging

Pump staging is generally not available for rent. The table below shows the number of pieces required to erect pump staging for 2400 S.F. of building frontage. This area is made up of a pump jack system that is 3 poles (2 bays) wide by 2 poles high.

Item	Number of Pieces for 2400 S.F. of Building Front	Unit
Aluminum pole section, 24' long	6	Ea.
Aluminum splice joint, 6' long	3	
Aluminum foldable brace	3	
Aluminum pump jack	3	
Aluminum support for workbench/back safety rail	3	
Aluminum scaffold plank/workbench, 14" wide x 24' long	4	
Safety net, 22' long	2	
Aluminum plank end safety rail	2	

The cost in place for this 2400 S.F. will depend on how many uses are realized during the life of the equipment.

R015433-10   Contractor Equipment

Rental Rates shown elsewhere in the data set pertain to late model high quality machines in excellent working condition, rented from equipment dealers. Rental rates from contractors may be substantially lower than the rental rates from equipment dealers depending upon economic conditions; for older, less productive machines, reduce rates by a maximum of 15%. Any overtime must be added to the base rates. For shift work, rates are lower. Usual rule of thumb is 150% of one shift rate for two shifts; 200% for three shifts.

For periods of less than one week, operated equipment is usually more economical to rent than renting bare equipment and hiring an operator.

Costs to move equipment to a job site (mobilization) or from a job site (demobilization) are not included in rental rates, nor in any Equipment costs on any Unit Price line items or crew listings. These costs can be found elsewhere. If a piece of equipment is already at a job site, it is not appropriate to utilize mob/demob costs in an estimate again.

Rental rates vary throughout the country with larger cities generally having lower rates. Lease plans for new equipment are available for periods in excess of six months with a percentage of payments applying toward purchase.

Rental rates can also be treated as reimbursement costs for contractor-owned equipment. Owned equipment costs include depreciation, loan payments, interest, taxes, insurance, storage, and major repairs.

Monthly rental rates vary from 2% to 5% of the cost of the equipment depending on the anticipated life of the equipment and its wearing parts. Weekly rates are about 1/3 the monthly rates and daily rental rates are about 1/3 the weekly rates.

The hourly operating costs for each piece of equipment include costs to the user such as fuel, oil, lubrication, normal expendables for the equipment, and a percentage of the mechanic's wages chargeable to maintenance. The hourly operating costs listed do not include the operator's wages.

The daily cost for equipment used in the standard crews is figured by dividing the weekly rate by five, then adding eight times the hourly operating cost to give the total daily equipment cost, not including the operator. This figure is in the right hand column of the Equipment listings under Equipment Cost/Day.

Pile Driving rates shown for the pile hammer and extractor do not include leads, cranes, boilers or compressors. Vibratory pile driving requires an added field specialist during set-up and pile driving operation for the electric model. The hydraulic model requires a field specialist for set-up only. Up to 125 reuses of sheet piling are possible using vibratory drivers. For normal conditions, crane capacity for hammer type and size is as follows.

Crane Capacity	Hammer Type and Size		
	Air or Steam	Diesel	Vibratory
25 ton	to 8,750 ft.-lb.		70 H.P.
40 ton	15,000 ft.-lb.	to 32,000 ft.-lb.	170 H.P.
60 ton	25,000 ft.-lb.		300 H.P.
100 ton		112,000 ft.-lb.	

Cranes should be specified for the job by size, building and site characteristics, availability, performance characteristics, and duration of time required.

Backhoes & Shovels rent for about the same as equivalent size cranes but maintenance and operating expenses are higher. The crane operator's rate must be adjusted for high boom heights. Average adjustments: for 150' boom add 2% per hour; over 185', add 4% per hour; over 210', add 6% per hour; over 250', add 8% per hour and over 295', add 12% per hour.

Tower Cranes of the climbing or static type have jibs from 50' to 200' and capacities at maximum reach range from 4,000 to 14,000 pounds. Lifting capacities increase up to maximum load as the hook radius decreases.

Typical rental rates, based on purchase price, are about 2% to 3% per month.

Erection and dismantling run between 500 and 2000 labor hours. Climbing operation takes 10 labor hours per 20' climb. Crane dead time is about 5 hours per 40' climb. If crane is bolted to side of the building add cost of ties and extra mast sections. Climbing cranes have from 80' to 180' of mast while static cranes have 80' to 800' of mast.

Truck Cranes can be converted to tower cranes by using tower attachments. Mast heights over 400' have been used.

A single 100' high material **Hoist and Tower** can be erected and dismantled in about 400 labor hours; a double 100' high hoist and tower in about 600 labor hours. Erection times for additional heights are 3 and 4 labor hours

per vertical foot respectively up to 150', and 4 to 5 labor hours per vertical foot over 150' high. A 40' high portable Buck hoist takes about 160 labor hours to erect and dismantle. Additional heights take 2 labor hours per vertical foot to 80' and 3 labor hours per vertical foot for the next 100'. Most material hoists do not meet local code requirements for carrying personnel.

A 150' high **Personnel Hoist** requires about 500 to 800 labor hours to erect and dismantle. Budget erection time at 5 labor hours per vertical foot for all trades. Local code requirements or labor scarcity requiring overtime can add up to 50% to any of the above erection costs.

Earthmoving Equipment: The selection of earthmoving equipment depends upon the type and quantity of material, moisture content, haul distance, haul road, time available, and equipment available. Short haul cut and fill operations may require dozers only, while another operation may require excavators, a fleet of trucks, and spreading and compaction equipment. Stockpiled material and granular material are easily excavated with front end loaders. Scrapers are most economically used with hauls between 300' and 1-1/2 miles if adequate haul roads can be maintained. Shovels are often used for blasted rock and any material where a vertical face of 8' or more can be excavated. Special conditions may dictate the use of draglines, clamshells, or backhoes. Spreading and compaction equipment must be matched to the soil characteristics, the compaction required and the rate the fill is being supplied.

R015436-50   Mobilization

Costs to move rented construction equipment to a job site from an equipment dealer's or contractor's yard (mobilization) or off the job site (demobilization) are not included in the rental or operating rates, nor in the equipment cost on a unit price line or in a crew listing. These costs can be found consolidated in the Mobilization section of the data and elsewhere in particular site work sections. If a piece of equipment is already on the job site, it is not appropriate to include mob/demob costs in a new estimate that requires use of that equipment. The following table identifies approximate sizes of rented construction equipment that would be hauled on a towed trailer. Because this listing is not all-encompassing, the user can infer as to what size trailer might be required for a piece of equipment not listed.

3-ton Trailer	20-ton Trailer	40-ton Trailer	50-ton Trailer
20 H.P. Excavator	110 H.P. Excavator	200 H.P. Excavator	270 H.P. Excavator
50 H.P. Skid Steer	165 H.P. Dozer	300 H.P. Dozer	Small Crawler Crane
35 H.P. Roller	150 H.P. Roller	400 H.P. Scraper	500 H.P. Scraper
40 H.P. Trencher	Backhoe	450 H.P. Art. Dump Truck	500 H.P. Art. Dump Truck

839

R024119-10 Demolition Defined

Whole Building Demolition - Demolition of the whole building with no concern for any particular building element, component, or material type being demolished. This type of demolition is accomplished with large pieces of construction equipment that break up the structure, load it into trucks and haul it to a disposal site, but disposal or dump fees are not included. Demolition of below-grade foundation elements, such as footings, foundation walls, grade beams, slabs on grade, etc., is not included. Certain mechanical equipment containing flammable liquids or ozone-depleting refrigerants, electric lighting elements, communication equipment components, and other building elements may contain hazardous waste, and must be removed, either selectively or carefully, as hazardous waste before the building can be demolished.

Foundation Demolition - Demolition of below-grade foundation footings, foundation walls, grade beams, and slabs on grade. This type of demolition is accomplished by hand or pneumatic hand tools, and does not include saw cutting, or handling, loading, hauling, or disposal of the debris.

Gutting - Removal of building interior finishes and electrical/mechanical systems down to the load-bearing and sub-floor elements of the rough building frame, with no concern for any particular building element, component, or material type being demolished. This type of demolition is accomplished by hand or pneumatic hand tools, and includes loading into trucks, but not hauling, disposal or dump fees, scaffolding, or shoring. Certain mechanical equipment containing flammable liquids or ozone-depleting refrigerants, electric lighting elements, communication equipment components, and other building elements may contain hazardous waste, and must be removed, either selectively or carefully, as hazardous waste, before the building is gutted.

Selective Demolition - Demolition of a selected building element, component, or finish, with some concern for surrounding or adjacent elements, components, or finishes (see the first Subdivision (s) at the beginning of appropriate Divisions). This type of demolition is accomplished by hand or pneumatic hand tools, and does not include handling, loading, storing, hauling, or disposal of the debris, scaffolding, or shoring. "Gutting" methods may be used in order to save time, but damage that is caused to surrounding or adjacent elements, components, or finishes may have to be repaired at a later time.

Careful Removal - Removal of a piece of service equipment, building element or component, or material type, with great concern for both the removed item and surrounding or adjacent elements, components or finishes. The purpose of careful removal may be to protect the removed item for later re-use, preserve a higher salvage value of the removed item, or replace an item while taking care to protect surrounding or adjacent elements, components, connections, or finishes from cosmetic and/or structural damage. An approximation of the time required to perform this type of removal is 1/3 to 1/2 the time it would take to install a new item of like kind (see Reference Number R220105-10). This type of removal is accomplished by hand or pneumatic hand tools, and does not include loading, hauling, or storing the removed item, scaffolding, shoring, or lifting equipment.

Cutout Demolition - Demolition of a small quantity of floor, wall, roof, or other assembly, with concern for the appearance and structural integrity of the surrounding materials. This type of demolition is accomplished by hand or pneumatic hand tools, and does not include saw cutting, handling, loading, hauling, or disposal of debris, scaffolding, or shoring.

Rubbish Handling - Work activities that involve handling, loading or hauling of debris. Generally, the cost of rubbish handling must be added to the cost of all types of demolition, with the exception of whole building demolition.

Minor Site Demolition - Demolition of site elements outside the footprint of a building. This type of demolition is accomplished by hand or pneumatic hand tools, or with larger pieces of construction equipment, and may include loading a removed item onto a truck (check the Crew for equipment used). It does not include saw cutting, hauling or disposal of debris, and, sometimes, handling or loading.

R024119-20 Dumpsters

Dumpster rental costs on construction sites are presented in two ways.

The cost per week rental includes the delivery of the dumpster; its pulling or emptying once per week, and its final removal. The assumption is made that the dumpster contractor could choose to empty a dumpster by simply bringing in an empty unit and removing the full one. These costs also include the disposal of the materials in the dumpster.

The Alternate Pricing can be used when actual planned conditions are not approximated by the weekly numbers. For example, these lines can be used when a dumpster is needed for 4 weeks and will need to be emptied 2 or 3 times per week. Conversely the Alternate Pricing lines can be used when a dumpster will be rented for several weeks or months but needs to be emptied only a few times over this period.

R024119-30 Rubbish Handling Chutes

To correctly estimate the cost of rubbish handling chute systems, the individual components must be priced separately. First choose the size of the system; a 30-inch diameter chute is quite common, but the sizes range from 18 to 36 inches in diameter. The 30-inch chute comes in a standard weight and two thinner weights. The thinner weight chutes are sometimes chosen for cost savings, but they are more easily damaged.

There are several types of major chute pieces that make up the chute system. The first component to consider is the top chute section (top intake hopper) where the material is dropped into the chute at the highest point. After determining the top chute, the intermediate chute pieces called the regular chute sections are priced. Next, the number of chute control door sections (intermediate intake hoppers) must be determined. In the more complex systems, a chute control door section is provided at each floor level. The last major component to consider is bolt down frames; these are usually provided at every other floor level.

There are a number of accessories to consider for safe operation and control. There are covers for the top chute and the chute door sections. The top chute can have a trough that allows for better loading of the chute. For the safest operation, a chute warning light system can be added that will warn the other chute intake locations not to load while another is being used. There are dust control devices that spray a water mist to keep down the dust as the debris is loaded into a Dumpster. There are special breakaway cords that are used to prevent damage to the chute if the dumpster is removed without disconnecting from the chute. There are chute liners that can be installed to protect the chute structure from physical damage from rough abrasive materials. Warning signs can be posted at each floor level that is provided with a chute control door section.

In summary, a complete rubbish handling chute system will include one top section, several intermediate regular sections, several intermediate control door (intake hopper) sections and bolt down frames at every other floor level starting with the top floor. If so desired, the system can also include covers and a light warning system for a safer operation. The bottom of the chute should always be above the Dumpster and should be tied off with a breakaway cord to the Dumpster.

Existing Conditions — R0265 Underground Storage Tank Removal

R026510-20 Underground Storage Tank Removal

Underground storage tank removal can be divided into two categories: non-leaking and leaking. Prior to removing an underground storage tank, tests should be made, with the proper authorities present, to determine whether a tank has been leaking or the surrounding soil has been contaminated.

To safely remove liquid underground storage tanks:
1. Excavate to the top of the tank.
2. Disconnect all piping.
3. Open all tank vents and access ports.
4. Remove all liquids and/or sludge.
5. Purge the tank with an inert gas.
6. Provide access to the inside of the tank and clean out the interior using proper personal protective equipment (PPE).
7. Excavate soil surrounding the tank using proper PPE for on-site personnel.
8. Pull and properly dispose of the tank.
9. Clean up the site of all contaminated material.
10. Install new tanks or close the excavation.

Existing Conditions — R0282 Asbestos Remediation

R028213-20 Asbestos Removal Process

Asbestos removal is accomplished by a specialty contractor who understands the federal and state regulations regarding the handling and disposal of the material. The process of asbestos removal is divided into many individual steps. An accurate estimate can be calculated only after all the steps have been priced.

The steps are generally as follows:
1. Obtain an asbestos abatement plan from an industrial hygienist.
2. Monitor the air quality in and around the removal area and along the path of travel between the removal area and transport area. This establishes the background contamination.
3. Construct a two part decontamination chamber at entrance to removal area.
4. Install a HEPA filter to create a negative pressure in the removal area.
5. Install wall, floor and ceiling protection as required by the plan, usually 2 layers of fireproof 6 mil polyethylene.
6. Industrial hygienist visually inspects work area to verify compliance with plan.
7. Provide temporary supports for conduit and piping affected by the removal process.
8. Proceed with asbestos removal and bagging process. Monitor air quality as described in Step #2. Discontinue operations when contaminate levels exceed applicable standards.
9. Document the legal disposal of materials in accordance with EPA standards.
10. Thoroughly clean removal area including all ledges, crevices and surfaces.
11. Post abatement inspection by industrial hygienist to verify plan compliance.
12. Provide a certificate from a licensed industrial hygienist attesting that contaminate levels are within acceptable standards before returning area to regular use.

841

R028319-60   Lead Paint Remediation Methods

Lead paint remediation can be accomplished by the following methods.

1. Abrasive blast
2. Chemical stripping
3. Power tool cleaning with vacuum collection system
4. Encapsulation
5. Remove and replace
6. Enclosure

Each of these methods has strengths and weaknesses depending on the specific circumstances of the project. The following is an overview of each method.

1. **Abrasive blasting** is usually accomplished with sand or recyclable metallic blast. Before work can begin, the area must be contained to ensure the blast material with lead does not escape to the atmosphere. The use of vacuum blast greatly reduces the containment requirements. Lead abatement equipment that may be associated with this work includes a negative air machine. In addition, it is necessary to have an industrial hygienist monitor the project on a continual basis. When the work is complete, the spent blast sand with lead must be disposed of as a hazardous material. If metallic shot was used, the lead is separated from the shot and disposed of as hazardous material. Worker protection includes disposable clothing and respiratory protection.

2. **Chemical stripping** requires strong chemicals to be applied to the surface to remove the lead paint. Before the work can begin, the area under/adjacent to the work area must be covered to catch the chemical and removed lead. After the chemical is applied to the painted surface it is usually covered with paper. The chemical is left in place for the specified period, then the paper with lead paint is pulled or scraped off. The process may require several chemical applications. The paper with chemicals and lead paint adhered to it, plus the containment and loose scrapings collected by a HEPA (High Efficiency Particulate Air Filter) vac, must be disposed of as hazardous material. The chemical stripping process usually requires a neutralizing agent and several wash downs after the paint is removed. Worker protection includes a neoprene or other compatible protective clothing and respiratory

protection with face shield. An industrial hygienist is required intermittently during the process.

3. **Power tool cleaning** is accomplished using shrouded needle blasting guns. The shrouding with different end configurations is held up against the surface to be cleaned. The area is blasted with hardened needles and the shroud captures the lead with a HEPA vac and deposits it in a holding tank. An industrial hygienist monitors the project. Protective clothing and a respirator are required until air samples prove otherwise. When the work is complete the lead must be disposed of as hazardous material.

4. **Encapsulation** is a method that leaves the well bonded lead paint in place after the peeling paint has been removed. Before the work can begin, the area under/adjacent to the work must be covered to catch the scrapings. The scraped surface is then washed with a detergent and rinsed. The prepared surface is covered with approximately 10 mils of paint. A reinforcing fabric can also be embedded in the paint covering. The scraped paint and containment must be disposed of as hazardous material. Workers must wear protective clothing and respirators.

5. **Removing and replacing** are effective ways to remove lead paint from windows, gypsum walls and concrete masonry surfaces. The painted materials are removed and new materials are installed. Workers should wear a respirator and tyvek suit. The demolished materials must be disposed of as hazardous waste if it fails the TCLP (Toxicity Characteristic Leachate Process) test.

6. **Enclosure** is the process that permanently seals lead painted materials in place. This process has many applications such as covering lead painted drywall with new drywall, covering exterior construction with tyvek paper then residing, or covering lead painted structural members with aluminum or plastic. The seams on all enclosing materials must be securely sealed. An industrial hygienist monitors the project, and protective clothing and a respirator are required until air samples prove otherwise.

All the processes require clearance monitoring and wipe testing as required by the hygienist.

R031113-10   Wall Form Materials

Aluminum Forms

Approximate weight is 3 lbs. per S.F.C.A. Standard widths are available from 4″ to 36″ with 36″ most common. Standard lengths of 2′, 4′, 6′ to 8′ are available. Forms are lightweight and fewer ties are needed with the wider widths. The form face is either smooth or textured.

Metal Framed Plywood Forms

Manufacturers claim over 75 reuses of plywood and over 300 reuses of steel frames. Many specials such as corners, fillers, pilasters, etc. are available. Monthly rental is generally about 15% of purchase price for first month and 9% per month thereafter with 90% of rental applied to purchase for the first month and decreasing percentages thereafter. Aluminum framed forms cost 25% to 30% more than steel framed.

After the first month, extra days may be prorated from the monthly charge. Rental rates do not include ties, accessories, cleaning, loss of hardware or freight in and out. Approximate weight is 5 lbs. per S.F. for steel; 3 lbs. per S.F. for aluminum.

Forms can be rented with option to buy.

Plywood Forms, Job Fabricated

There are two types of plywood used for concrete forms.

1. Exterior plyform which is completely waterproof. This is face oiled to facilitate stripping. Ten reuses can be expected with this type with 25 reuses possible.
2. An overlaid type consists of a resin fiber fused to exterior plyform. No oiling is required except to facilitate cleaning. This is available in both high density (HDO) and medium density overlaid (MDO). Using HDO, 50 reuses can be expected with 200 possible.

Plyform is available in 5/8″ and 3/4″ thickness. High density overlaid is available in 3/8″, 1/2″, 5/8″ and 3/4″ thickness.

5/8″ thick is sufficient for most building forms, while 3/4″ is best on heavy construction.

Plywood Forms, Modular, Prefabricated

There are many plywood forming systems without frames. Most of these are manufactured from 1-1/8″ (HDO) plywood and have some hardware attached. These are used principally for foundation walls 8′ or less high. With care and maintenance, 100 reuses can be attained with decreasing quality of surface finish.

Steel Forms

Approximate weight is 6-1/2 lbs. per S.F.C.A. including accessories. Standard widths are available from 2″ to 24″, with 24″ most common. Standard lengths are from 2′ to 8′, with 4′ the most common. Forms are easily ganged into modular units.

Forms are usually leased for 15% of the purchase price per month prorated daily over 30 days.

Rental may be applied to sale price, and usually rental forms are bought. With careful handling and cleaning 200 to 400 reuses are possible.

Straight wall gang forms up to 12′ x 20′ or 8′ x 30′ can be fabricated. These crane handled forms usually lease for approx. 9% per month.

Individual job analysis is available from the manufacturer at no charge.

R031113-40   Forms for Reinforced Concrete

Design Economy

Avoid many sizes in proportioning beams and columns.

From story to story avoid changing column dimensions. Gain strength by adding steel or using a richer mix. If a change in size of column is necessary, vary one dimension only to minimize form alterations. Keep beams and columns the same width.

From floor to floor in a multi-story building vary beam depth, not width, as that will leave the slab panel form unchanged. It is cheaper to vary the strength of a beam from floor to floor by means of a steel area than by 2″ changes in either width or depth.

Cost Factors

Material includes the cost of lumber, cost of rent for metal pans or forms if used, nails, form ties, form oil, bolts and accessories.

Labor includes the cost of carpenters to make up, erect, remove and repair, plus common labor to clean and move. Having carpenters remove forms minimizes repairs.

Improper alignment and condition of forms will increase finishing cost. When forms are heavily oiled, concrete surfaces must be neutralized before finishing. Special curing compounds will cause spillages to spall off in first frost. Gang forming methods will reduce costs on large projects.

Materials Used

Boards are seldom used unless their architectural finish is required. Generally, steel, fiberglass and plywood are used for contact surfaces. Labor on plywood is 10% less than with boards. The plywood is backed up with

2 x 4′s at 12″ to 32″ O.C. Walers are generally 2 - 2 x 4′s. Column forms are held together with steel yokes or bands. Shoring is with adjustable shoring or scaffolding for high ceilings.

Reuse

Floor and column forms can be reused four or possibly five times without excessive repair. Remember to allow for 10% waste on each reuse.

When modular sized wall forms are made, up to twenty uses can be expected with exterior plyform.

When forms are reused, the cost to erect, strip, clean and move will not be affected. 10% replacement of lumber should be included and about one hour of carpenter time for repairs on each reuse per 100 S.F.

The reuse cost for certain accessory items normally rented on a monthly basis will be lower than the cost for the first use.

After the fifth use, new material required plus time needed for repair prevent the form cost from dropping further; it may go up. Much depends on care in stripping, the number of special bays, changes in beam or column sizes and other factors.

Costs for multiple use of formwork may be developed as follows:

2 Uses	3 Uses	4 Uses
$\dfrac{(\text{1st Use} + \text{Reuse})}{2}$ = avg. cost/2 uses	$\dfrac{(\text{1st Use} + \text{2 Reuses})}{3}$ = avg. cost/3 uses	$\dfrac{(\text{1st use} + \text{3 Reuses})}{4}$ = avg. cost/4 uses

843

R031113-60 Formwork Labor-Hours

Item	Unit	Hours Required			Total Hours	Multiple Use		
		Fabricate	Erect & Strip	Clean & Move	1 Use	2 Use	3 Use	4 Use
Beam and Girder, interior beams, 12" wide	100 S.F.	6.4	8.3	1.3	16.0	13.3	12.4	12.0
Hung from steel beams		5.8	7.7	1.3	14.8	12.4	11.6	11.2
Beam sides only, 36" high		5.8	7.2	1.3	14.3	11.9	11.1	10.7
Beam bottoms only, 24" wide		6.6	13.0	1.3	20.9	18.1	17.2	16.7
Box out for openings		9.9	10.0	1.1	21.0	16.6	15.1	14.3
Buttress forms, to 8' high		6.0	6.5	1.2	13.7	11.2	10.4	10.0
Centering, steel, 3/4" rib lath			1.0		1.0			
3/8" rib lath or slab form	▼		0.9		0.9			
Chamfer strip or keyway	100 L.F.		1.5		1.5	1.5	1.5	1.5
Columns, fiber tube 8" diameter			20.6		20.6			
12"			21.3		21.3			
16"			22.9		22.9			
20"			23.7		23.7			
24"			24.6		24.6			
30"	▼		25.6		25.6			
Columns, round steel, 12" diameter			22.0		22.0	22.0	22.0	22.0
16"			25.6		25.6	25.6	25.6	25.6
20"			30.5		30.5	30.5	30.5	30.5
24"	▼		37.7		37.7	37.7	37.7	37.7
Columns, plywood 8" x 8"	100 S.F.	7.0	11.0	1.2	19.2	16.2	15.2	14.7
12" x 12"		6.0	10.5	1.2	17.7	15.2	14.4	14.0
16" x 16"		5.9	10.0	1.2	17.1	14.7	13.8	13.4
24" x 24"		5.8	9.8	1.2	16.8	14.4	13.6	13.2
Columns, steel framed plywood 8" x 8"			10.0	1.0	11.0	11.0	11.0	11.0
12" x 12"			9.3	1.0	10.3	10.3	10.3	10.3
16" x 16"			8.5	1.0	9.5	9.5	9.5	9.5
24" x 24"			7.8	1.0	8.8	8.8	8.8	8.8
Drop head forms, plywood		9.0	12.5	1.5	23.0	19.0	17.7	17.0
Coping forms		8.5	15.0	1.5	25.0	21.3	20.0	19.4
Culvert, box			14.5	4.3	18.8	18.8	18.8	18.8
Curb forms, 6" to 12" high, on grade		5.0	8.5	1.2	14.7	12.7	12.1	11.7
On elevated slabs	▼	6.0	10.8	1.2	18.0	15.5	14.7	14.3
Edge forms to 6" high, on grade	100 L.F.	2.0	3.5	0.6	6.1	5.6	5.4	5.3
7" to 12" high	100 S.F.	2.5	5.0	1.0	8.5	7.8	7.5	7.4
Equipment foundations		10.0	18.0	2.0	30.0	25.5	24.0	23.3
Flat slabs, including drops		3.5	6.0	1.2	10.7	9.5	9.0	8.8
Hung from steel		3.0	5.5	1.2	9.7	8.7	8.4	8.2
Closed deck for domes		3.0	5.8	1.2	10.0	9.0	8.7	8.5
Open deck for pans		2.2	5.3	1.0	8.5	7.9	7.7	7.6
Footings, continuous, 12" high		3.5	3.5	1.5	8.5	7.3	6.8	6.6
Spread, 12" high		4.7	4.2	1.6	10.5	8.7	8.0	7.7
Pile caps, square or rectangular		4.5	5.0	1.5	11.0	9.3	8.7	8.4
Grade beams, 24" deep		2.5	5.3	1.2	9.0	8.3	8.0	7.9
Lintel or Sill forms		8.0	17.0	2.0	27.0	23.5	22.3	21.8
Spandrel beams, 12" wide		9.0	11.2	1.3	21.5	17.5	16.2	15.5
Stairs			25.0	4.0	29.0	29.0	29.0	29.0
Trench forms in floor		4.5	14.0	1.5	20.0	18.3	17.7	17.4
Walls, Plywood, at grade, to 8' high		5.0	6.5	1.5	13.0	11.0	9.7	9.5
8' to 16'		7.5	8.0	1.5	17.0	13.8	12.7	12.1
16' to 20'		9.0	10.0	1.5	20.5	16.5	15.2	14.5
Foundation walls, to 8' high		4.5	6.5	1.0	12.0	10.3	9.7	9.4
8' to 16' high		5.5	7.5	1.0	14.0	11.8	11.0	10.6
Retaining wall to 12' high, battered		6.0	8.5	1.5	16.0	13.5	12.7	12.3
Radial walls to 12' high, smooth		8.0	9.5	2.0	19.5	16.0	14.8	14.3
2' chords		7.0	8.0	1.5	16.5	13.5	12.5	12.0
Prefabricated modular, to 8' high		—	4.3	1.0	5.3	5.3	5.3	5.3
Steel, to 8' high		—	6.8	1.2	8.0	8.0	8.0	8.0
8' to 16' high		—	9.1	1.5	10.6	10.3	10.2	10.2
Steel framed plywood to 8' high		—	6.8	1.2	8.0	7.5	7.3	7.2
8' to 16' high	▼	—	9.3	1.2	10.5	9.5	9.2	9.0

R032110-10   Reinforcing Steel Weights and Measures

Bar Designation No.**	Nominal Weight Lb./Ft.	U.S. Customary Units			SI Units			
		Nominal Dimensions*			Nominal Dimensions*			
		Diameter in.	Cross Sectional Area, in.^{2}	Perimeter in.	Nominal Weight kg/m	Diameter mm	Cross Sectional Area, cm^2	Perimeter mm
3	.376	.375	.11	1.178	.560	9.52	.71	29.9
4	.668	.500	.20	1.571	.994	12.70	1.29	39.9
5	1.043	.625	.31	1.963	1.552	15.88	2.00	49.9
6	1.502	.750	.44	2.356	2.235	19.05	2.84	59.8
7	2.044	.875	.60	2.749	3.042	22.22	3.87	69.8
8	2.670	1.000	.79	3.142	3.973	25.40	5.10	79.8
9	3.400	1.128	1.00	3.544	5.059	28.65	6.45	90.0
10	4.303	1.270	1.27	3.990	6.403	32.26	8.19	101.4
11	5.313	1.410	1.56	4.430	7.906	35.81	10.06	112.5
14	7.650	1.693	2.25	5.320	11.384	43.00	14.52	135.1
18	13.600	2.257	4.00	7.090	20.238	57.33	25.81	180.1

* The nominal dimensions of a deformed bar are equivalent to those of a plain round bar having the same weight per foot as the deformed bar.

** Bar numbers are based on the number of eighths of an inch included in the nominal diameter of the bars.

R032110-80   Shop-Fabricated Reinforcing Steel

The material prices for reinforcing, shown in the unit cost sections of the data set, are for 50 tons or more of shop-fabricated reinforcing steel and include:

1. Mill base price of reinforcing steel
2. Mill grade/size/length extras
3. Mill delivery to the fabrication shop
4. Shop storage and handling
5. Shop drafting/detailing
6. Shop shearing and bending
7. Shop listing
8. Shop delivery to the job site

Both material and installation costs can be considerably higher for small jobs consisting primarily of smaller bars, while material costs may be slightly lower for larger jobs.

R032205-30　Common Stock Styles of Welded Wire Fabric

This table provides some of the basic specifications, sizes, and weights of welded wire fabric used for reinforcing concrete.

New Designation Spacing — Cross Sectional Area (in.) — (Sq. in. 100)		Old Designation Spacing — Wire Gauge (in.) — (AS & W)		Steel Area per Foot				Approximate Weight per 100 S.F.	
				Longitudinal		Transverse			
				in.	cm	in.	cm	lbs	kg
Rolls	6 x 6 — W1.4 x W1.4	6 x 6 — 10 x 10		.028	.071	.028	.071	21	9.53
	6 x 6 — W2.0 x W2.0	6 x 6 — 8 x 8	1	.040	.102	.040	.102	29	13.15
	6 x 6 — W2.9 x W2.9	6 x 6 — 6 x 6		.058	.147	.058	.147	42	19.05
	6 x 6 — W4.0 x W4.0	6 x 6 — 4 x 4		.080	.203	.080	.203	58	26.91
	4 x 4 — W1.4 x W1.4	4 x 4 — 10 x 10		.042	.107	.042	.107	31	14.06
	4 x 4 — W2.0 x W2.0	4 x 4 — 8 x 8	1	.060	.152	.060	.152	43	19.50
	4 x 4 — W2.9 x W2.9	4 x 4 — 6 x 6		.087	.227	.087	.227	62	28.12
	4 x 4 — W4.0 x W4.0	4 x 4 — 4 x 4		.120	.305	.120	.305	85	38.56
Sheets	6 x 6 — W2.9 x W2.9	6 x 6 — 6 x 6		.058	.147	.058	.147	42	19.05
	6 x 6 — W4.0 x W4.0	6 x 6 — 4 x 4		.080	.203	.080	.203	58	26.31
	6 x 6 — W5.5 x W5.5	6 x 6 — 2 x 2	2	.110	.279	.110	.279	80	36.29
	4 x 4 — W1.4 x W1.4	4 x 4 — 4 x 4		.120	.305	.120	.305	85	38.56

NOTES: 1. Exact W—number size for 8 gauge is W2.1
　　　　 2. Exact W—number size for 2 gauge is W5.4

The above table was compiled with the following excerpts from the WRI Manual of Standard Practices, 7th Edition, Copyright 2006. Reproduced with permission of the Wire Reinforcement Institute, Inc.:

1. Chapter 3, page 7, Table 1 Common Styles of Metric Wire Reinforcement (WWR) With Equivalent US Customary Units
2. Chapter 6, page 19, Table 5 Customary Units
3. Chapter 6, Page 23, Table 7 Customary Units (in.) Welded Plain Wire Reinforcement
4. Chapter 6, Page 25 Table 8 Wire Size Comparison
5. Chapter 9, Page 30, Table 9 Weight of Longitudinal Wires Weight (Mass) Estimating Tables
6. Chapter 9, Page 31, Table 9M Weight of Longitudinal Wires Weight (Mass) Estimating Tables
7. Chapter 9, Page 32, Table 10 Weight of Transverse Wires Based on 62″ lengths of transverse wire (60″ width plus 1″ overhand each side)
8. Chapter 9, Page 33, Table 10M Weight of Transverse Wires

R033105-10 Proportionate Quantities

The tables below show both quantities per S.F. of floor areas as well as form and reinforcing quantities per C.Y. Unusual structural requirements would increase the ratios below. High strength reinforcing would reduce the steel weights. Figures are for 3000 psi concrete and 60,000 psi reinforcing unless specified otherwise.

Type of Construction	Live Load	Span	Per S.F. of Floor Area				Per C.Y. of Concrete		
			Concrete	Forms	Reinf.	Pans	Forms	Reinf.	Pans
Flat Plate	50 psf	15 Ft.	.46 C.F.	1.06 S.F.	1.71 lb.		62 S.F.	101 lb.	
		20	.63	1.02	2.40		44	104	
		25	.79	1.02	3.03		35	104	
	100	15	.46	1.04	2.14		61	126	
		20	.71	1.02	2.72		39	104	
		25	.83	1.01	3.47		33	113	
Flat Plate (waffle construction) 20" domes	50	20	.43	1.00	2.10	.84 S.F.	63	135	53 S.F.
		25	.52	1.00	2.90	.89	52	150	46
		30	.64	1.00	3.70	.87	42	155	37
	100	20	.51	1.00	2.30	.84	53	125	45
		25	.64	1.00	3.20	.83	42	135	35
		30	.76	1.00	4.40	.81	36	160	29
Waffle Construction 30" domes	50	25	.69	1.06	1.83	.68	42	72	40
		30	.74	1.06	2.39	.69	39	87	39
		35	.86	1.05	2.71	.69	33	85	39
		40	.78	1.00	4.80	.68	35	165	40
Flat Slab (two way with drop panels)	50	20	.62	1.03	2.34		45	102	
		25	.77	1.03	2.99		36	105	
		30	.95	1.03	4.09		29	116	
	100	20	.64	1.03	2.83		43	119	
		25	.79	1.03	3.88		35	133	
		30	.96	1.03	4.66		29	131	
	200	20	.73	1.03	3.03		38	112	
		25	.86	1.03	4.23		32	133	
		30	1.06	1.03	5.30		26	135	
One Way Joists 20" Pans	50	15	.36	1.04	1.40	.93	78	105	70
		20	.42	1.05	1.80	.94	67	120	60
		25	.47	1.05	2.60	.94	60	150	54
	100	15	.38	1.07	1.90	.93	77	140	66
		20	.44	1.08	2.40	.94	67	150	58
		25	.52	1.07	3.50	.94	55	185	49
One Way Joists 8" x 16" filler blocks	50	15	.34	1.06	1.80	.81 Ea.	84	145	64 Ea.
		20	.40	1.08	2.20	.82	73	145	55
		25	.46	1.07	3.20	.83	63	190	49
	100	15	.39	1.07	1.90	.81	74	130	56
		20	.46	1.09	2.80	.82	64	160	48
		25	.53	1.10	3.60	.83	56	190	42
One Way Beam & Slab	50	15	.42	1.30	1.73		84	111	
		20	.51	1.28	2.61		68	138	
		25	.64	1.25	2.78		53	117	
	100	15	.42	1.30	1.90		84	122	
		20	.54	1.35	2.69		68	154	
		25	.69	1.37	3.93		54	145	
	200	15	.44	1.31	2.24		80	137	
		20	.58	1.40	3.30		65	163	
		25	.69	1.42	4.89		53	183	
Two Way Beam & Slab	100	15	.47	1.20	2.26		69	130	
		20	.63	1.29	3.06		55	131	
		25	.83	1.33	3.79		43	123	
	200	15	.49	1.25	2.70		41	149	
		20	.66	1.32	4.04		54	165	
		25	.88	1.32	6.08		41	187	

R033105-10 Proportionate Quantities (cont.)

4000 psi Concrete and 60,000 psi Reinforcing—Form and Reinforcing Quantities per C.Y.

Item	Size	Forms	Reinforcing	Minimum	Maximum
	10" x 10"	130 S.F.C.A.	#5 to #11	220 lbs.	875 lbs.
	12" x 12"	108	#6 to #14	200	955
	14" x 14"	92	#7 to #14	190	900
	16" x 16"	81	#6 to #14	187	1082
	18" x 18"	72	#6 to #14	170	906
	20" x 20"	65	#7 to #18	150	1080
Columns	22" x 22"	59	#8 to #18	153	902
(square tied)	24" x 24"	54	#8 to #18	164	884
	26" x 26"	50	#9 to #18	169	994
	28" x 28"	46	#9 to #18	147	864
	30" x 30"	43	#10 to #18	146	983
	32" x 32"	40	#10 to #18	175	866
	34" x 34"	38	#10 to #18	157	772
	36" x 36"	36	#10 to #18	175	852
	38" x 38"	34	#10 to #18	158	765
	40" x 40"	32	#10 to #18	143	692

Item	Size	Form	Spiral	Reinforcing	Minimum	Maximum
	12" diameter	34.5 L.F.	190 lbs.	#4 to #11	165 lbs.	1505 lb.
		34.5	190	#14 & #18	—	1100
	14"	25	170	#4 to #11	150	970
		25	170	#14 & #18	800	1000
	16"	19	160	#4 to #11	160	950
		19	160	#14 & #18	605	1080
	18"	15	150	#4 to #11	160	915
		15	150	#14 & #18	480	1075
	20"	12	130	#4 to #11	155	865
		12	130	#14 & #18	385	1020
	22"	10	125	#4 to #11	165	775
		10	125	#14 & #18	320	995
	24"	9	120	#4 to #11	195	800
		9	120	#14 & #18	290	1150
Columns	26"	7.3	100	#4 to #11	200	729
(spirally reinforced)		7.3	100	#14 & #18	235	1035
	28"	6.3	95	#4 to #11	175	700
		6.3	95	#14 & #18	200	1075
	30"	5.5	90	#4 to #11	180	670
		5.5	90	#14 & #18	175	1015
	32"	4.8	85	#4 to #11	185	615
		4.8	85	#14 & #18	155	955
	34"	4.3	80	#4 to #11	180	600
		4.3	80	#14 & #18	170	855
	36"	3.8	75	#4 to #11	165	570
		3.8	75	#14 & #18	155	865
	40"	3.0	70	#4 to #11	165	500
		3.0	70	#14 & #18	145	765

R033105-10 Proportionate Quantities (cont.)

3000 psi Concrete and 60,000 psi Reinforcing—Form and Reinforcing Quantities per C.Y.						
Item	Type	Loading	Height	C.Y./L.F.	Forms/C.Y.	Reinf./C.Y.
Retaining Walls	Cantilever	Level Backfill	4 Ft.	0.2 C.Y.	49 S.F.	35 lbs.
			8	0.5	42	45
			12	0.8	35	70
			16	1.1	32	85
			20	1.6	28	105
		Highway Surcharge	4	0.3	41	35
			8	0.5	36	55
			12	0.8	33	90
			16	1.2	30	120
			20	1.7	27	155
		Railroad Surcharge	4	0.4	28	45
			8	0.8	25	65
			12	1.3	22	90
			16	1.9	20	100
			20	2.6	18	120
	Gravity, with Vertical Face	Level Backfill	4	0.4	37	None
			7	0.6	27	
			10	1.2	20	
		Sloping Backfill	4	0.3	31	
			7	0.8	21	↓
			10	1.6	15	

		Live Load in Kips per Linear Foot						
	Span	Under 1 Kip		2 to 3 Kips		4 to 5 Kips		6 to 7 Kips	
		Forms	Reinf.	Forms	Reinf.	Forms	Reinf.	Forms	Reinf.
Beams	10 Ft.	—	—	90 S.F.	170 #	85 S.F.	175 #	75 S.F.	185 #
	16	130 S.F.	165 #	85	180	75	180	65	225
	20	110	170	75	185	62	200	51	200
	26	90	170	65	215	62	215	—	—
	30	85	175	60	200	—	—	—	—

Item	Size	Type	Forms per C.Y.	Reinforcing per C.Y.
Spread Footings	Under 1 C.Y.	1,000 psf soil	24 S.F.	44 lbs.
		5,000	24	42
		10,000	24	52
	1 C.Y. to 5 C.Y.	1,000	14	49
		5,000	14	50
		10,000	14	50
	Over 5 C.Y.	1,000	9	54
		5,000	9	52
		10,000	9	56
Pile Caps (30 Ton Concrete Piles)	Under 5 C.Y.	shallow caps	20	65
		medium	20	50
		deep	20	40
	5 C.Y. to 10 C.Y.	shallow	14	55
		medium	15	45
		deep	15	40
	10 C.Y. to 20 C.Y.	shallow	11	60
		medium	11	45
		deep	12	35
	Over 20 C.Y.	shallow	9	60
		medium	9	45
		deep	10	40

R033105-10   Proportionate Quantities (cont.)

		3000 psi Concrete and 60,000 psi Reinforcing — Form and Reinforcing Quantities per C.Y.					
Item	Size	Pile Spacing	50 T Pile	100 T Pile	50 T Pile	100 T Pile	
Pile Caps (Steel H Piles)	Under 5 C.Y.	24" O.C.	24 S.F.	24 S.F.	75 lbs.	90 lbs.	
		30"	25	25	80	100	
		36"	24	24	80	110	
	5 C.Y. to 10 C.Y.	24"	15	15	80	110	
		30"	15	15	85	110	
		36"	15	15	75	90	
	Over 10 C.Y.	24"	13	13	85	90	
		30"	11	11	85	95	
		36"	10	10	85	90	

		8" Thick		10" Thick		12" Thick		15" Thick	
	Height	Forms	Reinf.	Forms	Reinf.	Forms	Reinf.	Forms	Reinf.
Basement Walls	7 Ft.	81 S.F.	44 lbs.	65 S.F.	45 lbs.	54 S.F.	44 lbs.	41 S.F.	43 lbs.
	8		44		45		44		43
	9		46		45		44		43
	10		57		45		44		43
	12		83		50		52		43
	14		116		65		64		51
	16				86		90		65
	18						106		70

R033105-20   Materials for One C.Y. of Concrete

This is an approximate method of figuring quantities of cement, sand and coarse aggregate for a field mix with waste allowance included.

With crushed gravel as coarse aggregate, to determine barrels of cement required, divide 10 by total mix; that is, for 1:2:4 mix, 10 divided by 7 = 1-3/7 barrels.

If the coarse aggregate is crushed stone, use 10-1/2 instead of 10 as given for gravel.

To determine tons of sand required, multiply barrels of cement by parts of sand and then by 0.2; that is, for the 1:2:4 mix, as above, 1-3/7 x 2 x .2 = .57 tons.

Tons of crushed gravel are in the same ratio to tons of sand as parts in the mix, or 4/2 x .57 = 1.14 tons.

1 bag cement = 94#	1 C.Y. sand or crushed gravel = 2700#	1 C.Y. crushed stone = 2575#
4 bags = 1 barrel	1 ton sand or crushed gravel = 20 C.F.	1 ton crushed stone = 21 C.F.

Average carload of cement is 692 bags; of sand or gravel is 56 tons.

Do not stack stored cement over 10 bags high.

R033105-70   Placing Ready-Mixed Concrete

For ground pours allow for 5% waste when figuring quantities.

Prices in the front of the data set assume normal deliveries. If deliveries are made before 8 A.M. or after 5 P.M. or on Saturday afternoons add 30%. Negotiated discounts for large volumes are not included in prices in front of the data set.

For the lower floors without truck access, concrete may be wheeled in rubber-tired buggies, conveyer handled, crane handled or pumped. Pumping is economical if there is top steel. Conveyers are more efficient for thick slabs.

At higher floors the rubber-tired buggies may be hoisted by a hoisting tower and wheeled to the location. Placement by a conveyer is limited to three floors and is best for high-volume pours. Pumped concrete is best when the building has no crane access. Concrete may be pumped directly as high as thirty-six stories using special pumping techniques. Normal maximum height is about fifteen stories.

The best pumping aggregate is screened and graded bank gravel rather than crushed stone.

Pumping downward is more difficult than pumping upward. The horizontal distance from pump to pour may increase preparation time prior to pour. Placing by cranes, either mobile, climbing or tower types, continues as the most efficient method for high-rise concrete buildings.

R033105-85   Lift Slabs

The cost advantage of the lift slab method is due to placing all concrete, reinforcing steel, inserts and electrical conduit at ground level and in reduction of formwork. Minimum economical project size is about 30,000 S.F. Slabs may be tilted for parking garage ramps.

It is now used in all types of buildings and has gone up to 22 stories high in apartment buildings. Current trend is to use post-tensioned flat plate slabs with spans from 22' to 35'. Cylindrical void forms are used when deep slabs are required. One pound of prestressing steel is about equal to seven pounds of conventional reinforcing.

To be considered cured for stressing and lifting, a slab must have attained 75% of design strength. Seven days are usually sufficient with four to five days possible if high early strength cement is used. Slabs can be stacked using two coats of a non-bonding agent to insure that slabs do not stick to each other. Lifting is done by companies specializing in this work. Lift rate is 5' to 15' per hour with an average of 10' per hour. Total areas up to 33,000 S.F. have been lifted at one time. 24 to 36 jacking columns are common. Most economical bay sizes are 24' to 28' with four to fourteen stories most efficient. Continuous design reduces reinforcing steel cost. Use of post-tensioned slabs allows larger bay sizes.

R033543-10   Polished Concrete Floors

A polished concrete floor has a glossy mirror-like appearance and is created by grinding the concrete floor with finer and finer diamond grits, similar to sanding wood, until the desired level of reflective clarity and sheen are achieved. The technical term for this type of polished concrete is bonded abrasive polished concrete. The basic piece of equipment used in the polishing process is a walk-behind planetary grinder for working large floor areas. This grinder drives diamond-impregnated abrasive discs, which progress from coarse- to fine-grit discs.

The process begins with the use of very coarse diamond segments or discs bonded in a metallic matrix. These segments are coarse enough to allow the removal of pits, blemishes, stains, and light coatings from the floor surface in preparation for final smoothing. The condition of the original concrete surface will dictate the grit coarseness of the initial grinding step which will generally end up being a three- to four-step process using ever finer grits. The purpose of this initial grinding step is to remove surface coatings and blemishes and to cut down into the cream for very fine aggregate exposure, or deeper into the fine aggregate layer just below the cream layer, or even deeper into the coarse aggregate layer. These initial grinding steps will progress up to the 100/120 grit. If wet grinding is done, a waste slurry is produced that must be removed between grit changes and disposed of properly. If dry grinding is done, a high performance vacuum will pick up the dust during grinding and collect it in bags which must be disposed of properly.

The process continues with honing the floor in a series of steps that progresses from 100-grit to 400-grit diamond abrasive discs embedded in a plastic or resin matrix. At some point during, or just prior to, the honing step, one or two coats of stain or dye can be sprayed onto the surface to give color to the concrete, and two coats of densifier/hardener must be applied to the floor surface and allowed to dry. This sprayed-on densifier/hardener will penetrate about 1/8" into the concrete to make the surface harder, denser and more abrasion-resistant.

The process ends with polishing the floor surface in a series of steps that progresses from resin-impregnated 800-grit (medium polish) to 1500-grit (high polish) to 3000-grit (very high polish), depending on the desired level of reflective clarity and sheen.

The Concrete Polishing Association of America (CPAA) has defined the flooring options available when processing concrete to a desired finish. The first category is aggregate exposure, the grinding of a concrete surface with bonded abrasives, in as many abrasive grits necessary, to achieve one of the following classes:

© Concrete Polishing Association of America "Glossary."

A. Cream – very little surface cut depth; little aggregate exposure

B. Fine aggregate (salt and pepper) – surface cut depth of 1/16"; fine aggregate exposure with little or no medium aggregate exposure at random locations

C. Medium aggregate – surface cut depth of 1/8"; medium aggregate exposure with little or no large aggregate exposure at random locations

D. Large aggregate – surface cut depth of 1/4"; large aggregate exposure with little or no fine aggregate exposure at random locations

The second CPAA defined category is reflective clarity and sheen, the polishing of a concrete surface with the minimum number of bonded abrasives as indicated to achieve one of the following levels:

1. Ground – flat appearance with none to very slight diffused reflection; none to very low reflective sheen; using a minimum total of 4 grit levels up 100-grit

2. Honed – matte appearance with or without slight diffused reflection; low to medium reflective sheen; using a minimum total of 5 grit levels up to 400-grit

3. Semi-polished – objects being reflected are not quite sharp and crisp but can be easily identified; medium to high reflective sheen; using a minimum total of 6 grit levels up to 800-grit

4. Highly-polished – objects being reflected are sharp and crisp as would be seen in a mirror-like reflection; high to highest reflective sheen; using a minimum total of up to 8 grit levels up to 1500-grit or 3000-grit

The CPAA defines reflective clarity as the degree of sharpness and crispness of the reflection of overhead objects when viewed 5' above and perpendicular to the floor surface. Reflective sheen is the degree of gloss reflected from a surface when viewed at least 20' from and at an angle to the floor surface. These terms are relatively subjective. The final outcome depends on the internal makeup and surface condition of the original concrete floor, the experience of the floor polishing crew, and the expectations of the owner. Before the grinding, honing, and polishing work commences on the main floor area, it might be beneficial to do a mock-up panel in the same floor but in an out of the way place to demonstrate the sequence of steps with increasingly fine abrasive grits and to demonstrate the final reflective clarity and reflective sheen. This mock-up panel will be within the area of, and part of, the final work.

851

R034105-30 Prestressed Precast Concrete Structural Units

Type	Location	Depth	Span in Ft.		Live Load Lb. per S.F.
Double Tee 8' to 10'	Floor	28" to 34"	60 to 80		50 to 80
	Roof	12" to 24"	30 to 50		40
	Wall	Width 8'	Up to 55' high		Wind
Multiple Tee 8'	Roof	8" to 12"	15 to 40		40
	Floor	8" to 12"	15 to 30		100
Plank or	Roof or Floor		Roof	Floor	
		4"	13	12	40 for Roof
		6"	22	18	
		8"	26	25	
		10"	33	29	100 for Floor
		12"	42	32	
Single Tee 8' to 10'	Roof	28"	40		
		32"	80		
		36"	100		40
		48"	120		
AASHO Girder	Bridges	Type 4	100		
		5	110		Highway
		6	125		
Box Beam 4'	Bridges	15"	40		
		27"	to		Highway
		33"	100		

The majority of precast projects today utilize double tees rather than single tees because of speed and ease of installation. As a result casting beds at manufacturing plants are normally formed for double tees. Single tee projects will therefore require an initial set up charge to be spread over the individual single tee costs.

For floors, a 2" to 3" topping is field cast over the shapes. For roofs, insulating concrete or rigid insulation is placed over the shapes.

Member lengths up to 40' are standard haul, 40' to 60' require special permits and lengths over 60' must be escorted. Excessive width and/or length can add up to 100% on hauling costs.

Large heavy members may require two cranes for lifting which would increase erection costs by about 45%. An eight man crew can install 12 to 20 double tees, or 45 to 70 quad tees or planks per day.

Grouting of connections must also be included.

Several system buildings utilizing precast members are available. Heights can go up to 22 stories for apartment buildings. The optimum design ratio is 3 S.F. of surface to 1 S.F. of floor area.

R035216-10 Lightweight Concrete

Lightweight aggregate concrete is usually purchased ready mixed, but it can also be field mixed.

Vermiculite or Perlite comes in bags of 4 C.F. under various trade names. The weight is about 8 lbs. per C.F. For insulating roof fill use 1:6 mix. For a structural deck use 1:4 mix over gypsum boards, steeltex, steel centering, etc., supported by closely spaced joists or bulb trees. For structural slabs use 1:3:2 vermiculite sand concrete over steeltex, metal lath, steel centering, etc., on joists spaced 2'-0" O.C. for maximum L.L. of 80 P.S.F. Use same mix

for slab base fill over steel flooring or regular reinforced concrete slab when tile, terrazzo or other finish is to be laid over.

For slabs on grade use 1:3:2 mix when tile, etc., finish is to be laid over. If radiant heating units are installed use a 1:6 mix for a base. After coils are in place, cover with a regular granolithic finish (mix 1:3:2) to a minimum depth of 1-1/2" over top of units.

Reinforce all slabs with 6 x 6 or 10 x 10 welded wire mesh.

Masonry | R0421 Clay Unit Masonry

R042110-20 Common and Face Brick

Common building brick manufactured according to ASTM C62 and facing brick manufactured according to ASTM C216 are the two standard bricks available for general building use.

Building brick is made in three grades: SW, where high resistance to damage caused by cyclic freezing is required; MW, where moderate resistance to cyclic freezing is needed; and NW, where little resistance to cyclic freezing is needed. Facing brick is made in only the two grades SW and MW. Additionally, facing brick is available in three types: FBS, for general use; FBX, for general use where a higher degree of precision and lower permissible variation in size than FBS are needed; and FBA, for general use to produce characteristic architectural effects resulting from non-uniformity in size and texture of the units.

In figuring the material cost of brickwork, an allowance of 25% mortar waste and 3% brick breakage was included. If bricks are delivered palletized

with 280 to 300 per pallet, or packaged, allow only 1-1/2% for breakage. Packaged or palletized delivery is practical when a job is big enough to have a crane or other equipment available to handle a package of brick. This is so on all industrial work but not always true on small commercial buildings.

The use of buff and gray face is increasing, and there is a continuing trend to the Norman, Roman, Jumbo and SCR brick.

Common red clay brick for backup is not used that often. Concrete block is the most usual backup material with occasional use of sand lime or cement brick. Building brick is commonly used in solid walls for strength and as a fire stop.

Brick panels built on the ground and then crane erected to the upper floors have proven to be economical. This allows the work to be done under cover and without scaffolding.

R042110-50 Brick, Block & Mortar Quantities

Running Bond						For Other Bonds Standard Size Add to S.F. Quantities in Table to Left		
Number of Brick per S.F. of Wall - Single Wythe with 3/8" Joints				C.F. of Mortar per M Bricks, Waste Included				
Type Brick	Nominal Size (incl. mortar) L H W	Modular Coursing	Number of Brick per S.F.	3/8" Joint	1/2" Joint	Bond Type	Description	Factor
Standard	8 x 2-2/3 x 4	3C=8"	6.75	8.1	10.3	Common	full header every fifth course	+20%
Economy	8 x 4 x 4	1C=4"	4.50	9.1	11.6		full header every sixth course	+16.7%
Engineer	8 x 3-1/5 x 4	5C=16"	5.63	8.5	10.8	English	full header every second course	+50%
Fire	9 x 2-1/2 x 4-1/2	2C=5"	6.40	550 # Fireclay	—	Flemish	alternate headers every course	+33.3%
Jumbo	12 x 4 x 6 or 8	1C=4"	3.00	22.5	29.2		every sixth course	+5.6%
Norman	12 x 2-2/3 x 4	3C=8"	4.50	11.2	14.3	Header = W x H exposed		+100%
Norwegian	12 x 3-1/5 x 4	5C=16"	3.75	11.7	14.9	Rowlock = H x W exposed		+100%
Roman	12 x 2 x 4	2C=4"	6.00	10.7	13.7	Rowlock stretcher = L x W exposed		+33.3%
SCR	12 x 2-2/3 x 6	3C=8"	4.50	21.8	28.0	Soldier = H x L exposed		—
Utility	12 x 4 x 4	1C=4"	3.00	12.3	15.7	Sailor = W x L exposed		-33.3%

Concrete Blocks Nominal Size		Approximate Weight per S.F.		Blocks per 100 S.F.	Mortar per M block, waste included	
		Standard	Lightweight		Partitions	Back up
2"	x 8" x 16"	20 PSF	15 PSF	113	27 C.F.	36 C.F.
4"		30	20		41	51
6"		42	30		56	66
8"		55	38		72	82
10"		70	47		87	97
12"		85	55		102	112

Brick & Mortar Quantities
©Brick Industry Association. 2009 Feb. Technical Notes on
Brick Construction 10:
  Dimensioning and Estimating Brick Masonry. Reston (VA): BIA. Table 1
  Modular Brick Sizes and Table 4 Quantity Estimates for Brick Masonry.

R050521-20   Welded Structural Steel

Usual weight reductions with welded design run 10% to 20% compared with bolted or riveted connections. This amounts to about the same total cost compared with bolted structures since field welding is more expensive than bolts. For normal spans of 18' to 24' figure 6 to 7 connections per ton.

Trusses — For welded trusses add 4% to weight of main members for connections. Up to 15% less steel can be expected in a welded truss compared to one that is shop bolted. Cost of erection is the same whether shop bolted or welded.

General — Typical electrodes for structural steel welding are E6010, E6011, E60T and E70T. Typical buildings vary between 2# to 8# of weld rod per

ton of steel. Buildings utilizing continuous design require about three times as much welding as conventional welded structures. In estimating field erection by welding, it is best to use the average linear feet of weld per ton to arrive at the welding cost per ton. The type, size and position of the weld will have a direct bearing on the cost per linear foot. A typical field welder will deposit 1.8# to 2# of weld rod per hour manually. Using semiautomatic methods can increase production by as much as 50% to 75%.

Metals

R0512 Structural Steel Framing

R051223-10   Structural Steel

The bare material prices for structural steel, shown in the unit cost sections of the data set, are for 100 tons of shop-fabricated structural steel and include:

1. Mill base price of structural steel
2. Mill scrap/grade/size/length extras
3. Mill delivery to a metals service center (warehouse)
4. Service center storage and handling
5. Service center delivery to a fabrication shop
6. Shop storage and handling
7. Shop drafting/detailing
8. Shop fabrication
9. Shop coat of primer paint
10. Shop listing
11. Shop delivery to the job site

In unit cost sections of the data set that contain items for field fabrication of steel components, the bare material cost of steel includes:

1. Mill base price of structural steel
2. Mill scrap/grade/size/length extras
3. Mill delivery to a metals service center (warehouse)
4. Service center storage and handling
5. Service center delivery to the job site

R051223-15   Structural Steel Estimating for Repair and Remodeling Projects

The correct approach to estimating structural steel is dependent upon the amount of steel required for the particular project. If the project is a sizable addition to a building, the data can be used directly from the data set. This is not the case however if the project requires a small amount of steel, for instance to reinforce existing roof or floor structural systems. To better understand this, please refer to the unit price line for a W16x31 beam with bolted connections. Assume your project requires the reinforcement of the structural members of the roof system of a 3 story building required for the installation a new piece of HVAC equipment. The project will need 4 pieces of W16x31 that are 30' long. After pricing this 120 L.F. job using the unit prices for a W16x31, an analysis will reveal that the price is wholly inadequate.

The first problem is apparent if you examine the amount of steel that can be installed per day. The unit price line indicates 900 linear feet per day can be installed by a 5 man crew with an 90 ton crane. This productivity is correct for new construction but certainly is not for repair and remodeling work. Installation of new structural steel considers that each member is installed from the foundation to the roof of the structure in a planned and systematic manner with a crane having unrestricted access to all parts of the project. Additionally each connection is planned and detailed with full field access for fit-up and final bolting. The erection is planned and progresses such that interferences and conflicts with other structural members are minimized, if not completely eliminated. All of these assumptions are clearly not the case with a repair and remodeling job, and a significant decrease in the stated productivity will be observed.

A crane will certainly be needed to lift the members into the general area of the project but in most cases will not be able to place the beams into

their final position. An opening in the existing roof may not be large enough to permit the beams to pass through and it may be necessary to bring them into the building through existing windows or doors. Moving the beams to the actual area where they will be installed may involve hand labor and the use of dollies. Finally, hoists and/or jacks may be needed for final positioning.

The connection of new members to existing can often be accomplished by field bolting with accurate field measurements and good planning but in many cases access to both sides of existing members is not possible and field welding becomes the only alternative. In addition to the cost of the actual welding, protection of existing finishes, systems and structure and fire protection must be considered.

New beams can never be installed tight to the existing decks or floors which they must support and the use of shims and tack welding becomes necessary. Additionally, further planning and cost are involved in assuring that existing loads are minimized during the installation and shimming process.

It is apparent that installation of structural steel as part of a repair and remodeling project involves more than simply installing the members and estimating the cost in the same manner as new construction. The best procedure for estimating the total cost are adequate planning and coordination of each process and activity that will be needed. Unit costs for the materials, labor and equipment can then be attached to each needed activity and a final, complete price can be determined.

R053100-10 Decking Descriptions

General - All Deck Products

A steel deck is made by cold forming structural grade sheet steel into a repeating pattern of parallel ribs. The strength and stiffness of the panels are the result of the ribs and the material properties of the steel. Deck lengths can be varied to suit job conditions, but because of shipping considerations, are usually less than 40 feet. Standard deck width varies with the product used but full sheets are usually 12″, 18″, 24″, 30″, or 36″. The deck is typically furnished in a standard width with the ends cut square. Any cutting for width, such as at openings or for angular fit, is done at the job site.

The deck is typically attached to the building frame with arc puddle welds, self-drilling screws, or powder or pneumatically driven pins. Sheet to sheet fastening is done with screws, button punching (crimping), or welds.

Composite Floor Deck

After installation and adequate fastening, a floor deck serves several purposes. It (a) acts as a working platform, (b) stabilizes the frame, (c) serves as a concrete form for the slab, and (d) reinforces the slab to carry the design loads applied during the life of the building. Composite decks are distinguished by the presence of shear connector devices as part of the deck. These devices are designed to mechanically lock the concrete and deck together so that the concrete and the deck work together to carry subsequent floor loads. These shear connector devices can be rolled-in embossments, lugs, holes, or wires welded to the panels. The deck profile can also be used to interlock concrete and steel.

Composite deck finishes are either galvanized (zinc coated) or phosphatized/painted. Galvanized deck has a zinc coating on both the top and bottom surfaces. The phosphatized/painted deck has a bare (phosphatized) top surface that will come into contact with the concrete. This bare top surface can be expected to develop rust before the concrete is placed. The bottom side of the deck has a primer coat of paint.

A composite floor deck is normally installed so the panel ends do not overlap on the supporting beams. Shear lugs or panel profile shapes often prevent a tight metal to metal fit if the panel ends overlap; the air gap caused by overlapping will prevent proper fusion with the structural steel supports when the panel end laps are shear stud welded.

Adequate end bearing of the deck must be obtained as shown on the drawings. If bearing is actually less in the field than shown on the drawings, further investigation is required.

Roof Deck

A roof deck is not designed to act compositely with other materials. A roof deck acts alone in transferring horizontal and vertical loads into the building frame. Roof deck rib openings are usually narrower than floor deck rib openings. This provides adequate support of the rigid thermal insulation board.

A roof deck is typically installed to endlap approximately 2″ over supports. However, it can be butted (or lapped more than 2″) to solve field fit problems. Since designers frequently use the installed deck system as part of the horizontal bracing system (the deck as a diaphragm), any fastening substitution or change should be approved by the designer. Continuous perimeter support of the deck is necessary to limit edge deflection in the finished roof and may be required for diaphragm shear transfer.

Standard roof deck finishes are galvanized or primer painted. The standard factory applied paint for roof decks is a primer paint and is not intended to weather for extended periods of time. Field painting or touching up of abrasions and deterioration of the primer coat or other protective finishes is the responsibility of the contractor.

Cellular Deck

A cellular deck is made by attaching a bottom steel sheet to a roof deck or composite floor deck panel. A cellular deck can be used in the same manner as a floor deck. Electrical, telephone, and data wires are easily run through the chase created between the deck panel and the bottom sheet.

When used as part of the electrical distribution system, the cellular deck must be installed so that the ribs line up and create a smooth cell transition at abutting ends. The joint that occurs at butting cell ends must be taped or otherwise sealed to prevent wet concrete from seeping into the cell. Cell interiors must be free of welding burrs, or other sharp intrusions, to prevent damage to wires.

When used as a roof deck, the bottom flat plate is usually left exposed to view. Care must be maintained during erection to keep good alignment and prevent damage.

A cellular deck is sometimes used with the flat plate on the top side to provide a flat working surface. Installation of the deck for this purpose requires special methods for attachment to the frame because the flat plate, now on the top, can prevent direct access to the deck material that is bearing on the structural steel. It may be advisable to treat the flat top surface to prevent slipping.

A cellular deck is always furnished galvanized or painted over galvanized.

Form Deck

A form deck can be any floor or roof deck product used as a concrete form. Connections to the frame are by the same methods used to anchor floor and roof decks. Welding washers are recommended when welding a deck that is less than 20 gauge thickness.

A form deck is furnished galvanized, prime painted, or uncoated. Galvanized deck must be used for those roof deck systems where a form deck is used to carry a lightweight insulating concrete fill.

R061110-30   Lumber Product Material Prices

The price of forest products fluctuates widely from location to location and from season to season depending upon economic conditions. The bare material prices in the unit cost sections of the data set show the National Average material prices in effect Jan. 1 of this data year. It must be noted that lumber prices in general may change significantly during the year.

Availability of certain items depends upon geographic location and must be checked prior to firm-price bidding.

R061636-20   Plywood

There are two types of plywood used in construction: interior, which is moisture-resistant but not waterproofed, and exterior, which is waterproofed.

The grade of the exterior surface of the plywood sheets is designated by the first letter: A, for smooth surface with patches allowed; B, for solid surface with patches and plugs allowed; C, which may be surface plugged or may have knot holes up to 1″ wide; and D, which is used only for interior type plywood and may have knot holes up to 2-1/2″ wide. "Structural Grade" is specifically designed for engineered applications such as box beams. All CC & DD grades have roof and floor spans marked on them.

Underlayment-grade plywood runs from 1/4″ to 1-1/4″ thick. Thicknesses 5/8″ and over have optional tongue and groove joints which eliminate the need for blocking the edges. Underlayment 19/32″ and over may be referred to as Sturd-i-Floor.

The price of plywood can fluctuate widely due to geographic and economic conditions.

Typical uses for various plywood grades are as follows:

AA-AD Interior — cupboards, shelving, paneling, furniture

BB Plyform — concrete form plywood

CDX — wall and roof sheathing

Structural — box beams, girders, stressed skin panels

AA-AC Exterior — fences, signs, siding, soffits, etc.

Underlayment — base for resilient floor coverings

Overlaid HDO — high density for concrete forms & highway signs

Overlaid MDO — medium density for painting, siding, soffits & signs

303 Siding — exterior siding, textured, striated, embossed, etc.

R075113-20   Built-Up Roofing

Asphalt is available in kegs of 100 lbs. each; coal tar pitch in 560 lb. kegs. Prepared roofing felts are available in a wide range of sizes, weights and characteristics. However, the most commonly used are #15 (432 S.F. per roll, 13 lbs. per square) and #30 (216 S.F. per roll, 27 lbs. per square).

Inter-ply bitumen varies from 24 lbs. per sq. (asphalt) to 30 lbs. per sq. (coal tar) per ply, MF4@ 25%. Flood coat bitumen also varies from 60 lbs. per sq. (asphalt) to 75 lbs. per sq. (coal tar), MF4@ 25%. Expendable equipment (mops, brooms, screeds, etc.) runs about 16% of the bitumen cost. For new, inexperienced crews this factor may be much higher.

A rigid insulation board is typically applied in two layers. The first is mechanically attached to nailable decks or spot or solid mopped to non-nailable decks; the second layer is then spot or solid mopped to the first layer. Membrane application follows the insulation, except in protected membrane roofs, where the membrane goes down first and the insulation on top, followed with ballast (stone or concrete pavers). Insulation and related labor costs are NOT included in prices for built-up roofing.

Thermal & Moist. Protec. R0752 Modified Bituminous Membrane Roofing

Reference Tables

R075213-30 Modified Bitumen Roofing

The cost of modified bitumen roofing is highly dependent on the type of installation that is planned. Installation is based on the type of modifier used in the bitumen. The two most popular modifiers are atactic polypropylene (APP) and styrene butadiene styrene (SBS). The modifiers are added to heated bitumen during the manufacturing process to change its characteristics. A polyethylene, polyester or fiberglass reinforcing sheet is then sandwiched between layers of this bitumen. When completed, the result is a pre-assembled, built-up roof that has increased elasticity and weatherability. Some manufacturers include a surfacing material such as ceramic or mineral granules, metal particles or sand.

The preferred method of adhering SBS-modified bitumen roofing to the substrate is with hot-mopped asphalt (much the same as built-up roofing). This installation method requires a tar kettle/pot to heat the asphalt, as well as the labor, tools and equipment necessary to distribute and spread the hot asphalt.

The alternative method for applying APP and SBS modified bitumen is as follows. A skilled installer uses a torch to melt a small pool of bitumen off the membrane. This pool must form across the entire roll for proper adhesion. The installer must unroll the roofing at a pace slow enough to melt the bitumen, but fast enough to prevent damage to the rest of the membrane.

Modified bitumen roofing provides the advantages of both built-up and single-ply roofing. Labor costs are reduced over those of built-up roofing because only a single ply is necessary. The elasticity of single-ply roofing is attained with the reinforcing sheet and polymer modifiers. Modifieds have some self-healing characteristics and because of their multi-layer construction, they offer the reliability and safety of built-up roofing.

R078413-30 Firestopping

Firestopping is the sealing of structural, mechanical, electrical and other penetrations through fire-rated assemblies. The basic components of firestop systems are safing insulation and firestop sealant on both sides of wall penetrations and the top side of floor penetrations.

Pipe penetrations are assumed to be through concrete, grout, or joint compound and can be sleeved or unsleeved. Costs for the penetrations and sleeves are not included. An annular space of 1″ is assumed. Escutcheons are not included.

A metallic pipe is assumed to be copper, aluminum, cast iron or similar metallic material. An insulated metallic pipe is assumed to be covered with a thermal insulating jacket of varying thickness and materials.

A non-metallic pipe is assumed to be PVC, CPVC, FR Polypropylene or similar plastic piping material. Intumescent firestop sealants or wrap strips are included. Collars on both sides of wall penetrations and a sheet metal plate on the underside of floor penetrations are included.

Ductwork is assumed to be sheet metal, stainless steel or similar metallic material. Duct penetrations are assumed to be through concrete, grout or joint compound. Costs for penetrations and sleeves are not included. An annular space of 1/2″ is assumed.

Multi-trade openings include costs for sheet metal forms, firestop mortar, wrap strips, collars and sealants as necessary.

Structural penetrations joints are assumed to be 1/2″ or less. CMU walls are assumed to be within 1-1/2″ of the metal deck. Drywall walls are assumed to be tight to the underside of metal decking.

Metal panel, glass or curtain wall systems include a spandrel area of 5′ filled with mineral wool foil-faced insulation. Fasteners and stiffeners are included.

R081313-20   Steel Door Selection Guide

Standard steel doors are classified into four levels, as recommended by the Steel Door Institute in the chart below. Each of the four levels offers a range of construction models and designs to meet architectural requirements for preference and appearance, including full flush, seamless, and stile & rail. Recommended minimum gauge requirements are also included.

For complete standard steel door construction specifications and available sizes, refer to the Steel Door Institute Technical Data Series, ANSI A250.8-98 (SDI-100), and ANSI A250.4-94 Test Procedure and Acceptance Criteria for Physical Endurance of Steel Door and Hardware Reinforcements.

Level		Model	Construction	For Full Flush or Seamless		
				Min. Gauge	Thickness (in)	Thickness (mm)
I	Standard Duty	1	Full Flush			
		2	Seamless	20	0.032	0.8
II	Heavy Duty	1	Full Flush			
		2	Seamless	18	0.042	1.0
III	Extra Heavy Duty	1	Full Flush			
		2	Seamless			
		3	*Stile & Rail	16	0.053	1.3
IV	Maximum Duty	1	Full Flush			
		2	Seamless	14	0.067	1.6

*Stiles & rails are 16 gauge; flush panels, when specified, are 18 gauge

R085123-10   Steel Sash

An ironworker crew will erect 25 S.F. or 1.3 sash unit per hour, whichever is less.

A mechanic will point 30 L.F. per hour.

A painter will paint 90 S.F. per coat per hour.

A Glazier production depends on light size.

Allow 1 lb. special steel sash putty per 16″ x 20″ light.

R085313-20   Replacement Windows

Replacement windows are typically measured per United Inch.

United Inches are calculated by rounding the width and height of the window opening up to the nearest inch, then adding the two figures.

The labor cost for replacement windows includes removal of sash, existing sash balance or weights, parting bead where necessary and installation of new window.

Debris hauling and dump fees are not included.

R087110-10   Hardware Finishes

This table describes hardware finishes used throughout the industry. It also shows the base metal and the respective symbols in the three predominate systems of identification. Many of these are used in pricing descriptions in Division Eight.

US″	BMHA*	CDN	Base	Description
US P	600	CP	Steel	Primed for Painting
US 1B	601	C1B	Steel	Bright Black Japanned
US 2C	602	C2C	Steel	Zinc Plated
US 2G	603	C2G	Steel	Zinc Plated
US 3	605	C3	Brass	Bright Brass, Clear Coated
US 4	606	C4	Brass	Satin Brass, Clear Coated
US 5	609	C5	Brass	Satin Brass, Blackened, Satin Relieved, Clear Coated
US 7	610	C7	Brass	Satin Brass, Blackened, Bright Relieved, Clear Coated
US 9	611	C9	Bronze	Bright Bronze, Clear Coated
US 10	612	C10	Bronze	Satin Bronze, Clear Coated
US 10A	641	C10A	Steel	Antiqued Bronze, Oiled and Lacquered
US 10B	613	C10B	Bronze	Antiqued Bronze, Oiled
US 11	616	C11	Bronze	Satin Bronze, Blackened, Satin Relieved, Clear Coated
US 14	618	C14	Brass/Bronze	Bright Nickel Plated, Clear Coated
US 15	619	C15	Brass/Bronze	Satin Nickel, Clear Coated
US 15A	620	C15A	Brass/Bronze	Satin Nickel Plated, Blackened, Satin Relieved, Clear Coated
US 17A	621	C17A	Brass/Bronze	Nickel Plated, Blackened, Relieved, Clear Coated
US 19	622	C19	Brass/Bronze	Flat Black Coated
US 20	623	C20	Brass/Bronze	Statuary Bronze, Light
US 20A	624	C20A	Brass/Bronze	Statuary Bronze, Dark
US 26	625	C26	Brass/Bronze	Bright Chromium
US 26D	626	C26D	Brass/Bronze	Satin Chromium
US 20	627	C27	Aluminum	Satin Aluminum Clear
US 28	628	C28	Aluminum	Anodized Dull Aluminum
US 32	629	C32	Stainless Steel	Bright Stainless Steel
US 32D	630	C32D	Stainless Steel	Stainless Steel
US 3	632	C3	Steel	Bright Brass Plated, Clear Coated
US 4	633	C4	Steel	Satin Brass, Clear Coated
US 7	636	C7	Steel	Satin Brass Plated, Blackened, Bright Relieved, Clear Coated
US 9	637	C9	Steel	Bright Bronze Plated, Clear Coated
US 5	638	C5	Steel	Satin Brass Plated, Blackened, Bright Relieved, Clear Coated
US 10	639	C10	Steel	Satin Bronze Plated, Clear Coated
US 10B	640	C10B	Steel	Antique Bronze, Oiled
US 10A	641	C10A	Steel	Antiqued Bronze, Oiled and Lacquered
US 11	643	C11	Steel	Satin Bronze Plated, Blackened, Bright Relieved, Clear Coated
US 14	645	C14	Steel	Bright Nickel Plated, Clear Coated
US 15	646	C15	Steel	Satin Nickel Plated, Clear Coated
US 15A	647	C15A	Steel	Nickel Plated, Blackened, Bright Relieved, Clear Coated
US 17A	648	C17A	Steel	Nickel Plated, Blackened, Relieved, Clear Coated
US 20	649	C20	Steel	Statuary Bronze, Light
US 20A	650	C20A	Steel	Statuary Bronze, Dark
US 26	651	C26	Steel	Bright Chromium Plated
US 26D	652	C26D	Steel	Satin Chromium Plated

* - BMHA Builders Hardware Manufacturing Association
″ - US Equivalent
^ - Canadian Equivalent
Japanning is imitating Asian lacquer work

859

R088110-10 Glazing Productivity

Some glass sizes are estimated by the "united inch" (height + width). The table below shows the number of lights glazed in an eight-hour period by the crew size indicated, for glass up to 1/4″ thick. Square or nearly square lights are more economical on a S.F. basis. Long slender lights will have a high S.F. installation cost. For insulated glass reduce production by 33%. For 1/2″ float glass reduce production by 50%. Production time for glazing with two glaziers per day averages: 1/4″ float glass 120 S.F.; 1/2″ float glass 55 S.F.; 1/2″ insulated glass 95 S.F.; 3/4″ insulated glass 75 S.F.

Glazing Method	United Inches per Light							
	40″	60″	80″	100″	135″	165″	200″	240″
Number of Men in Crew	1	1	1	1	2	3	3	4
Industrial sash, putty	60	45	24	15	18	—	—	—
With stops, putty bed	50	36	21	12	16	8	4	3
Wood stops, rubber	40	27	15	9	11	6	3	2
Metal stops, rubber	30	24	14	9	9	6	3	2
Structural glass	10	7	4	3	—	—	—	—
Corrugated glass	12	9	7	4	4	4	3	—
Storefronts	16	15	13	11	7	6	4	4
Skylights, putty glass	60	36	21	12	16	—	—	—
Thiokol set	15	15	11	9	9	6	3	2
Vinyl set, snap on	18	18	13	12	12	7	5	4
Maximum area per light	2.8 S.F.	6.3 S.F.	11.1 S.F.	17.4 S.F.	31.6 S.F.	47 S.F.	69 S.F.	100 S.F.

R092000-50   Lath, Plaster and Gypsum Board

Gypsum board lath is available in 3/8″ thick x 16″ wide x 4′ long sheets as a base material for multi-layer plaster applications. It is also available as a base for either multi-layer or veneer plaster applications in 1/2″ and 5/8″ thick–4′ wide x 8′, 10′ or 12′ long sheets. Fasteners are screws or blued ring shank nails for wood framing and screws for metal framing.

Metal lath is available in diamond mesh patterns with flat or self-furring profiles. Paper backing is available for applications where excessive plaster waste needs to be avoided. A slotted mesh ribbed lath should be used in areas where the span between structural supports is greater than normal. Most metal lath comes in 27″ x 96″ sheets. Diamond mesh weighs 1.75, 2.5 or 3.4 pounds per square yard, slotted mesh lath weighs 2.75 or 3.4 pounds per square yard. Metal lath can be nailed, screwed or tied in place.

Many **accessories** are available. Corner beads, flat reinforcing strips, casing beads, control and expansion joints, furring brackets and channels are some examples. Note that accessories are not included in plaster or stucco line items.

Plaster is defined as a material or combination of materials that when mixed with a suitable amount of water, forms a plastic mass or paste. When applied to a surface, the paste adheres to it and subsequently hardens, preserving in a rigid state the form or texture imposed during the period of elasticity.

Gypsum plaster is made from ground calcined gypsum. It is mixed with aggregates and water for use as a base coat plaster.

Vermiculite plaster is a fire-retardant plaster covering used on steel beams, concrete slabs and other heavy construction materials. Vermiculite is a group name for certain clay minerals, hydrous silicates or aluminum, magnesium and iron that have been expanded by heat.

Perlite plaster is a plaster using perlite as an aggregate instead of sand. Perlite is a volcanic glass that has been expanded by heat.

Gauging plaster is a mix of gypsum plaster and lime putty that when applied produces a quick drying finish coat.

Veneer plaster is a one or two component gypsum plaster used as a thin finish coat over special gypsum board.

Keenes cement is a white cementitious material manufactured from gypsum that has been burned at a high temperature and ground to a fine powder. Alum is added to accelerate the set. The resulting plaster is hard and strong and accepts and maintains a high polish, hence it is used as a finishing plaster.

Stucco is a Portland cement based plaster used primarily as an exterior finish.

Plaster is used on both interior and exterior surfaces. Generally it is applied in multiple-coat systems. A three-coat system uses the terms scratch, brown and finish to identify each coat. A two-coat system uses base and finish to describe each coat. Each type of plaster and application system has attributes that are chosen by the designer to best fit the intended use.

Gypsum Plaster Quantities for 100 S.Y.	2 Coat, 5/8″ Thick		3 Coat, 3/4″ Thick		
	Base	Finish	Scratch	Brown	Finish
	1:3 Mix	2:1 Mix	1:2 Mix	1:3 Mix	2:1 Mix
Gypsum plaster	1,300 lb.		1,350 lb.	650 lb.	
Sand	1.75 C.Y.		1.85 C.Y.	1.35 C.Y.	
Finish hydrated lime		340 lb.			340 lb.
Gauging plaster		170 lb.			170 lb.

Vermiculite or Perlite Plaster Quantities for 100 S.Y.	2 Coat, 5/8″ Thick		3 Coat, 3/4″ Thick		
	Base	Finish	Scratch	Brown	Finish
Gypsum plaster	1,250 lb.		1,450 lb.	800 lb.	
Vermiculite or perlite	7.8 bags		8.0 bags	3.3 bags	
Finish hydrated lime		340 lb.			340 lb.
Gauging plaster		170 lb.			170 lb.

Stucco–Three-Coat System Quantities for 100 S.Y.	On Wood Frame	On Masonry
Portland cement	29 bags	21 bags
Sand	2.6 C.Y.	2.0 C.Y.
Hydrated lime	180 lb.	120 lb.

861

R092910-10 Levels of Gypsum Board Finish

In the past, contract documents often used phrases such as "industry standard" and "workmanlike finish" to specify the expected quality of gypsum board wall and ceiling installations. The vagueness of these descriptions led to unacceptable work and disputes.

In order to resolve this problem, four major trade associations concerned with the manufacture, erection, finish and decoration of gypsum board wall and ceiling systems developed an industry-wide *Recommended Levels of Gypsum Board Finish*.

The finish of gypsum board walls and ceilings for specific final decoration is dependent on a number of factors. A primary consideration is the location of the surface and the degree of decorative treatment desired. Painted and unpainted surfaces in warehouses and other areas where appearance is normally not critical may simply require the taping of wallboard joints and 'spotting' of fastener heads. Blemish-free, smooth, monolithic surfaces often intended for painted and decorated walls and ceilings in habitated structures, ranging from single-family dwellings through monumental buildings, require additional finishing prior to the application of the final decoration.

Other factors to be considered in determining the level of finish of the gypsum board surface are (1) the type of angle of surface illumination (both natural and artificial lighting), and (2) the paint and method of application or the type and finish of wallcovering specified as the final decoration. Critical lighting conditions, gloss paints, and thin wall coverings require a higher level of gypsum board finish than heavily textured surfaces which are subsequently painted or surfaces which are to be decorated with heavy grade wall coverings.

The following descriptions were developed by the Association of the Wall and Ceiling Industries-International (AWCI), Ceiling & Interior Systems Construction Association (CISCA), Gypsum Association (GA), and Painting and Decorating Contractors of America (PDCA) as a guide.

Level 0: Used in temporary construction or wherever the final decoration has not been determined. Unfinished. No taping, finishing or corner beads are required. Also could be used where non-predecorated panels will be used in demountable-type partitions that are to be painted as a final finish.

Level 1: Frequently used in plenum areas above ceilings, in attics, in areas where the assembly would generally be concealed, or in building service corridors and other areas not normally open to public view. Some degree of sound and smoke control is provided; in some geographic areas, this level is referred to as "fire-taping," although this level of finish does not typically meet fire-resistant assembly requirements. Where a fire resistance rating is required for the gypsum board assembly, details of construction should be in accordance with reports of fire tests of assemblies that have met the requirements of the fire rating acceptable.

All joints and interior angles shall have tape embedded in joint compound. Accessories are optional at specifier discretion in corridors and other areas with pedestrian traffic. Tape and fastener heads need not be covered with joint compound. Surface shall be free of excess joint compound. Tool marks and ridges are acceptable.

Level 2: It may be specified for standard gypsum board surfaces in garages, warehouse storage, or other similar areas where surface appearance is not of primary importance.

All joints and interior angles shall have tape embedded in joint compound and shall be immediately wiped with a joint knife or trowel, leaving a thin coating of joint compound over all joints and interior angles. Fastener heads and accessories shall be covered with a coat of joint compound. Surface shall be free of excess joint compound. Tool marks and ridges are acceptable.

Level 3: Typically used in areas receiving heavy texture (spray or hand applied) finishes before final painting, or where commercial-grade (heavy duty) wall coverings are to be applied as the final decoration. This level of finish should not be used where smooth painted surfaces or where lighter weight wall coverings are specified. The prepared surface shall be coated with a drywall primer prior to the application of final finishes.

All joints and interior angles shall have tape embedded in joint compound and shall be immediately wiped with a joint knife or trowel, leaving a thin coating of joint compound over all joints and interior angles. One additional coat of joint compound shall be applied over all joints and interior angles. Fastener heads and accessories shall be covered with two separate coats of joint compound. All joint compounds shall be smooth and free of tool marks and ridges. The prepared surface shall be covered with a drywall primer prior to the application of the final decoration.

Level 4: This level should be used where residential grade (light duty) wall coverings, flat paints, or light textures are to be applied. The prepared surface shall be coated with a drywall primer prior to the application of final finishes. Release agents for wall coverings are specifically formulated to minimize damage if coverings are subsequently removed.

The weight, texture, and sheen level of the wall covering material selected should be taken into consideration when specifying wall coverings over this level of drywall treatment. Joints and fasteners must be sufficiently concealed if the wall covering material is lightweight, contains limited pattern, has a glossy finish, or has any combination of these features. In critical lighting areas, flat paints applied over light textures tend to reduce joint photographing. Gloss, semi-gloss, and enamel paints are not recommended over this level of finish.

All joints and interior angles shall have tape embedded in joint compound and shall be immediately wiped with a joint knife or trowel, leaving a thin coating of joint compound over all joints and interior angles. In addition, two separate coats of joint compound shall be applied over all flat joints and one separate coat of joint compound applied over interior angles. Fastener heads and accessories shall be covered with three separate coats of joint compound. All joint compounds shall be smooth and free of tool marks and ridges. The prepared surface shall be covered with a drywall primer like Sheetrock first coat prior to the application of the final decoration.

Level 5: The highest quality finish is the most effective method to provide a uniform surface and minimize the possibility of joint photographing and of fasteners showing through the final decoration. This level of finish is required where gloss, semi-gloss, or enamel is specified; when flat joints are specified over an untextured surface; or where critical lighting conditions occur. The prepared surface shall be coated with a drywall primer prior to the application of the final decoration.

All joints and interior angles shall have tape embedded in joint compound and be immediately wiped with a joint knife or trowel, leaving a thin coating of joint compound over all joints and interior angles. Two separate coats of joint compound shall be applied over all flat joints and one separate coat of joint compound applied over interior angles. Fastener heads and accessories shall be covered with three separate coats of joint compound.

A thin skim coat of joint compound shall be trowel applied to the entire surface. Excess compound is immediately troweled off, leaving a film or skim coating of compound completely covering the paper. As an alternative to a skim coat, a material manufactured especially for this purpose may be applied such as Sheetrock Tuff-Hide primer surfacer. The surface must be smooth and free of tool marks and ridges. The prepared surface shall be covered with a drywall primer prior to the application of the final decoration.

862

For customer support on your Commercial Renovation Costs with RSMeans data, call 800.448.8182.

R142000-10 Freight Elevators

Capacities run from 2,000 lbs. to over 100,000 lbs. with 3,000 lbs. to 10,000 lbs. most common. Travel speeds are generally lower and the control is less intricate than on passenger elevators. Costs in the unit price sections are for hydraulic and geared elevators.

R142000-20 Elevator Selective Costs See R142000-40 for cost development.

A. Base Unit	Passenger		Freight		Hospital	
	Hydraulic	Electric	Hydraulic	Electric	Hydraulic	Electric
Capacity	1,500 lb.	2,000 lb.	2,000 lb.	4,000 lb.	4,000 lb.	4,000 lb.
Speed	100 F.P.M.	200 F.P.M.	100 F.P.M.	200 F.P.M.	100 F.P.M.	200 F.P.M.
#Stops/Travel Ft.	2/12	4/40	2/20	4/40	2/20	4/40
Push Button Oper.	Yes	Yes	Yes	Yes	Yes	Yes
Telephone Box & Wire	"	"	"	"	"	"
Emergency Lighting	"	"	No	No	"	"
Cab	Plastic Lam. Walls	Plastic Lam. Walls	Painted Steel	Painted Steel	Plastic Lam. Walls	Plastic Lam. Walls
Cove Lighting	Yes	Yes	No	No	Yes	Yes
Floor	V.C.T.	V.C.T.	Wood w/Safety Treads	Wood w/Safety Treads	V.C.T.	V.C.T.
Doors, & Speedside Slide	Yes	Yes	Yes	Yes	Yes	Yes
Gates, Manual	No	No	No	No	No	No
Signals, Lighted Buttons	Car and Hall	Car and Hall	Car and Hall	Car and Hall	Car and Hall	Car and Hall
O.H. Geared Machine	N.A.	Yes	N.A.	Yes	N.A.	Yes
Variable Voltage Contr.	"	"	N.A.	"	"	"
Emergency Alarm	Yes	"	Yes	"	Yes	"
Class "A" Loading	N.A.	N.A.	"	"	N.A.	N.A.

R142000-30 Passenger Elevators

Electric elevators are used generally but hydraulic elevators can be used for lifts up to 70' and where large capacities are required. Hydraulic speeds are limited to 200 F.P.M. but cars are self leveling at the stops. On low rises, hydraulic installation runs about 15% less than standard electric types but on higher rises this installation cost advantage is reduced. Maintenance of hydraulic elevators is about the same as the electric type but the underground portion is not included in the maintenance contract.

In electric elevators there are several control systems available, the choice of which will be based upon elevator use, size, speed and cost criteria. The two types of drives are geared for low speeds and gearless for 450 F.P.M. and over.

The tables on the preceding pages illustrate typical installed costs of the various types of elevators available.

R142000-40 Elevator Cost Development

To price a new car or truck from the factory, you must start with the manufacturer's basic model, then add or exchange optional equipment and features. The same is true for pricing elevators.

Requirement: One-passenger elevator, five-story hydraulic, 2,500 lb. capacity, 12' floor to floor, speed 150 F.P.M., emergency power switching and maintenance contract.

Example:

Description	Adjustment
A. Base Elevator: Hydraulic Passenger, 1500 lb. Capacity, 100 fpm, 2 Stops, Standard Finish	1 Ea.
B. Capacity Adjustment (2,500 lb.)	1 Ea.
C. Excess Travel Adjustment: 48' Total Travel (4 x 12') minus 12' Base Unit Travel =	36 V.L.F.
D. Stops Adjustment: 5 Total Stops minus 2 Stops (Base Unit) =	3 Stops
E. Speed Adjustment (150 F.P.M.)	1 Ea.
F. Options:	
1. Intercom Service	1 Ea.
2. Emergency Power Switching, Automatic	1 Ea.
3. Stainless Steel Entrance Doors	5 Ea.
4. Maintenance Contract (12 Months)	1 Ea.
5. Position Indicator for main floor level (none indicated in Base Unit)	1 Ea.

R220102-20   Labor Adjustment Factors

Labor Adjustment Factors are provided for Divisions 21, 22, and 23 to assist the mechanical estimator account for the various complexities and special conditions of any particular project. While a single percentage has been entered on each line of Division 22 01 02.20, it should be understood that these are just suggested midpoints of ranges of values commonly used by mechanical estimators. They may be increased or decreased depending on the severity of the special conditions.

The group for "existing occupied buildings" has been the subject of requests for explanation. Actually there are two stages to this group: buildings that are existing and "finished" but unoccupied, and those that also are occupied. Buildings that are "finished" may result in higher labor costs due to the

workers having to be more careful not to damage finished walls, ceilings, floors, etc. and may necessitate special protective coverings and barriers. Also corridor bends and doorways may not accommodate long pieces of pipe or larger pieces of equipment. Work above an already hung ceiling can be very time consuming. The addition of occupants may force the work to be done on premium time (nights and/or weekends), eliminate the possible use of some preferred tools such as pneumatic drivers, powder charged drivers, etc. The estimator should evaluate the access to the work area and just how the work is going to be accomplished to arrive at an increase in labor costs over "normal" new construction productivity.

R220105-10   Demolition (Selective vs. Removal for Replacement)

Demolition can be divided into two basic categories.

One type of demolition involves the removal of material with no concern for its replacement. The labor-hours to estimate this work are found under "Selective Demolition" in the Fire Protection, Plumbing and HVAC Divisions. It is selective in that individual items or all the material installed as a system or trade grouping such as plumbing or heating systems are removed. This may be accomplished by the easiest way possible, such as sawing, torch cutting, or sledge hammering as well as simple unbolting.

The second type of demolition is the removal of some items for repair or replacement. This removal may involve careful draining, opening of unions,

disconnecting and tagging electrical connections, capping pipes/ducts to prevent entry of debris or leakage of the material contained as well as transporting the item away from its in-place location to a truck/dumpster. An approximation of the time required to accomplish this type of demolition is to use half of the time indicated as necessary to install a new unit. For example: installation of a new pump might be listed as requiring 6 labor-hours so if we had to estimate the removal of the old pump we would allow an additional 3 hours for a total of 9 hours. That is, the complete replacement of a defective pump with a new pump would be estimated to take 9 labor-hours.

R221113-50   Pipe Material Considerations

1. Malleable fittings should be used for gas service.
2. Malleable fittings are used where there are stresses/strains due to expansion and vibration.
3. Cast fittings may be broken as an aid to disassembling heating lines frozen by long use, temperature and minerals.
4. A cast iron pipe is extensively used for underground and submerged service.
5. Type M (light wall) copper tubing is available in hard temper only and is used for nonpressure and less severe applications than K and L.

6. Type L (medium wall) copper tubing, available hard or soft for interior service.
7. Type K (heavy wall) copper tubing, available in hard or soft temper for use where conditions are severe. For underground and interior service.
8. Hard drawn tubing requires fewer hangers or supports but should not be bent. Silver brazed fittings are recommended, but soft solder is normally used.
9. Type DMV (very light wall) copper tubing designed for drainage, waste and vent plus other non-critical pressure services.

Domestic/Imported Pipe and Fittings Costs

The prices shown in this publication for steel/cast iron pipe and steel, cast iron, and malleable iron fittings are based on domestic production sold at the normal trade discounts. The above listed items of foreign manufacture may be available at prices 1/3 to 1/2 of those shown. Some imported items after minor machining or finishing operations are being sold as domestic to further complicate the system.

Caution: Most pipe prices in this data set also include a coupling and pipe hangers which for the larger sizes can add significantly to the per foot cost and should be taken into account when comparing "book cost" with the quoted supplier's cost.

R260105-30   Electrical Demolition (Removal for Replacement)

The purpose of this reference number is to provide a guide to users for electrical "removal for replacement" by applying the rule of thumb: 1/3 of new installation time (typical range from 20% to 50%) for removal. Remember to use reasonable judgment when applying the suggested percentage factor. For example:

Contractors have been requested to remove an existing fluorescent lighting fixture and replace with a new fixture utilizing energy saver lamps and electronic ballast:

In order to fully understand the extent of the project, contractors should visit the job site and estimate the time to perform the renovation work in accordance with applicable national, state and local regulation and codes.

The contractor may need to add extra labor hours to his estimate if he discovers unknown concealed conditions such as: contaminated asbestos ceiling, broken acoustical ceiling tile and need to repair, patch and touch-up paint all the damaged or disturbed areas, tasks normally assigned to general contractors. In addition, the owner could request that the contractors salvage the materials removed and turn over the materials to the owner or dispose of the materials to a reclamation station. The normal removal item is 0.5 labor hour for a lighting fixture and 1.5 labor-hours for new installation time. Revise the estimate times from 2 labor-hours work up to a minimum 4 labor-hours work for just fluorescent lighting fixture.

For removal of large concentrations of lighting fixtures in the same area, apply an "economy of scale" to reduce estimating labor hours.

R260519-90   Wire

Wire quantities are taken off by either measuring each cable run or by extending the conduit and raceway quantities times the number of conductors in the raceway. Ten percent should be added for waste and tie-ins. Keep in mind that the unit of measure of wire is C.L.F. not L.F. as in raceways so the formula would read:

$$\frac{\text{(L.F. Raceway x No. of Conductors) x 1.10}}{100} = \text{C.L.F.}$$

Price per C.L.F. of wire includes:
1. Setting up wire coils or spools on racks
2. Attaching wire to pull in means
3. Measuring and cutting wire
4. Pulling wire into a raceway
5. Identifying and tagging

Price does not include:
1. Connections to breakers, panelboards, or equipment
2. Splices

Job Conditions: Productivity is based on new construction to a height of 15′ using rolling staging in an unobstructed area. Material staging is assumed to be within 100′ of work being performed.

Economy of Scale: If more than three wires at a time are being pulled, deduct the following percentages from the labor of that grouping:

4-5 wires	25%
6-10 wires	30%
11-15 wires	35%
over 15	40%

If a wire pull is less than 100′ in length and is interrupted several times by boxes, lighting outlets, etc., it may be necessary to add the following lengths to each wire being pulled:

Junction box to junction box	2 L.F.
Lighting panel to junction box	6 L.F.
Distribution panel to sub panel	8 L.F.
Switchboard to distribution panel	12 L.F.
Switchboard to motor control center	20 L.F.
Switchboard to cable tray	40 L.F.

Measure of Drops and Riser: It is important when taking off wire quantities to include the wire for drops to electrical equipment. If heights of electrical equipment are not clearly stated, use the following guide:

	Bottom A.F.F.	Top A.F.F.	Inside Cabinet
Safety switch to 100A	5′	6′	2′
Safety switch 400 to 600A	4′	6′	3′
100A panel 12 to 30 circuit	4′	6′	3′
42 circuit panel	3′	6′	4′
Switch box	3′	3′6″	1′
Switchgear	0′	8′	8′
Motor control centers	0′	8′	8′
Transformers - wall mount	4′	8′	2′
Transformers - floor mount	0′	12′	4′

R260533-22 Conductors in Conduit

The table below lists the maximum number of conductors for various sized conduit using THW, TW or THWN insulations.

Copper Wire Size	1/2"			3/4"			1"			1-1/4"			1-1/2"			2"			2-1/2"			3"		3-1/2"		4"	
	TW	THW	THWN	TW	THW	THWN	TW	THW	THWN	TW	THW	THWN	TW	THW	THWN	TW	THW	THWN	TW	THW	THWN	THW	THWN	THW	THWN	THW	THWN
#14	9	6	13	15	10	24	25	16	39	44	29	69	60	40	94	99	65	154	142	93		143		192			
#12	7	4	10	12	8	18	19	13	29	35	24	51	47	32	70	78	53	114	111	76	164	117		157			
#10	5	4	6	9	6	11	15	11	18	26	19	32	36	26	44	60	43	73	85	61	104	95	160	127		163	
#8	2	1	3	4	3	5	7	5	9	12	10	16	17	13	22	28	22	36	40	32	51	49	79	66	106	85	136
#6		1	1		2	4		4	6		7	11		10	15		16	26		23	37	36	57	48	76	62	98
#4		1	1		1	2		3	4		5	7		7	9		12	16		17	22	27	35	36	47	47	60
#3		1	1		1	1		2	3		4	6		6	8		10	13		15	19	23	29	31	39	40	51
#2		1	1		1	1		2	3		4	5		5	7		9	11		13	16	20	25	27	33	34	43
#1					1	1		1	1		3	3		4	5		6	8		9	12	14	18	19	25	25	32
1/0					1	1		1	1		2	3		3	4		5	7		8	10	12	15	16	21	21	27
2/0					1	1		1	1		1	2		3	3		5	6		7	8	10	13	14	17	18	22
3/0					1	1		1	1		1	1		2	3		4	5		6	7	9	11	12	14	15	18
4/0						1		1	1		1	1		1	2		3	4		5	6	7	9	10	12	13	15
250 kcmil								1	1		1	1		1	1		2	3		4	4	6	7	8	10	10	12
300								1	1		1	1		1	1		2	3		3	4	5	6	7	8	9	11
350									1		1	1		1	1		1	2		3	3	4	5	6	7	8	9
400											1	1		1	1		1	1		2	3	4	5	5	6	7	8
500											1	1		1	1		1	1		1	2	3	4	4	5	6	7
600														1			1	1		1	1	3	3	4	4	5	5
700																	1	1		1	1	2	3	3	4	4	5
750																	1	1		1	1	2	2	3	3	4	4

Reprinted with permission from NFPA 70-2014, *National Electrical Code®*, Copyright © 2013, National Fire Protection Association, Quincy, MA. This reprinted material is not the complete and official position of the NFPA on the referenced subject, which is represented solely by the standard in its entirety.

R312316-40   Excavating

The selection of equipment used for structural excavation and bulk excavation or for grading is determined by the following factors.
1. Quantity of material
2. Type of material
3. Depth or height of cut
4. Length of haul
5. Condition of haul road
6. Accessibility of site
7. Moisture content and dewatering requirements
8. Availability of excavating and hauling equipment

Some additional costs must be allowed for hand trimming the sides and bottom of concrete pours and other excavation below the general excavation.

Number of B.C.Y. per truck = 1.5 C.Y. bucket x 8 passes = 12 loose C.Y.

$$= 12 \times \frac{100}{118} = 10.2 \text{ B.C.Y. per truck}$$

Truck Haul Cycle:

Load truck, 8 passes	=	4 minutes
Haul distance, 1 mile	=	9 minutes
Dump time	=	2 minutes
Return, 1 mile	=	7 minutes
Spot under machine	=	1 minute
		23 minute cycle

Add the mobilization and demobilization costs to the total excavation costs. When equipment is rented for more than three days, there is often no mobilization charge by the equipment dealer. On larger jobs outside of urban areas, scrapers can move earth economically provided a dump site or fill area and adequate haul roads are available. Excavation within sheeting bracing or cofferdam bracing is usually done with a clamshell and production

When planning excavation and fill, the following should also be considered.
1. Swell factor
2. Compaction factor
3. Moisture content
4. Density requirements

A typical example for scheduling and estimating the cost of excavation of a 15′ deep basement on a dry site when the material must be hauled off the site is outlined below.

Assumptions:
1. Swell factor, 18%
2. No mobilization or demobilization
3. Allowance included for idle time and moving on job
4. No dewatering, sheeting, or bracing
5. No truck spotter or hand trimming

Fleet Haul Production per day in B.C.Y.

$$4 \text{ trucks} \times \frac{50 \text{ min. hour}}{23 \text{ min. haul cycle}} \times 8 \text{ hrs.} \times 10.2 \text{ B.C.Y.}$$

$$= 4 \times 2.2 \times 8 \times 10.2 = 718 \text{ B.C.Y./day}$$

is low, since the clamshell may have to be guided by hand between the bracing. When excavating or filling an area enclosed with a wellpoint system, add 10% to 15% to the cost to allow for restricted access. When estimating earth excavation quantities for structures, allow work space outside the building footprint for construction of the foundation and a slope of 1:1 unless sheeting is used.

867

R312316-45 Excavating Equipment

The table below lists theoretical hourly production in C.Y./hr. bank measure for some typical excavation equipment. Figures assume 50 minute hours, 83% job efficiency, 100% operator efficiency, 90° swing and properly sized hauling units, which must be modified for adverse digging and loading conditions. Actual production costs in the front of the data set average about 50% of the theoretical values listed here.

Equipment	Soil Type	B.C.Y. Weight	% Swell	1 C.Y.	1-1/2 C.Y.	2 C.Y.	2-1/2 C.Y.	3 C.Y.	3-1/2 C.Y.	4 C.Y.
Hydraulic Excavator	Moist loam, sandy clay	3400 lb.	40%	165	195	200	275	330	385	440
"Backhoe"	Sand and gravel	3100	18	140	170	225	240	285	330	380
15' Deep Cut	Common earth	2800	30	150	180	230	250	300	350	400
	Clay, hard, dense	3000	33	120	140	190	200	240	260	320
	Moist loam, sandy clay	3400	40	170 (6.0)	245 (7.0)	295 (7.8)	335 (8.4)	385 (8.8)	435 (9.1)	475 (9.4)
Power Shovel	Sand and gravel	3100	18	165 (6.0)	225 (7.0)	275 (7.8)	325 (8.4)	375 (8.8)	420 (9.1)	460 (9.4)
Optimum Cut (Ft.)	Common earth	2800	30	145 (7.8)	200 (9.2)	250 (10.2)	295 (11.2)	335 (12.1)	375 (13.0)	425 (13.8)
	Clay, hard, dense	3000	33	120 (9.0)	175 (10.7)	220 (12.2)	255 (13.3)	300 (14.2)	335 (15.1)	375 (16.0)
	Moist loam, sandy clay	3400	40	130 (6.6)	180 (7.4)	220 (8.0)	250 (8.5)	290 (9.0)	325 (9.5)	385 (10.0)
Drag Line	Sand and gravel	3100	18	130 (6.6)	175 (7.4)	210 (8.0)	245 (8.5)	280 (9.0)	315 (9.5)	375 (10.0)
Optimum Cut (Ft.)	Common earth	2800	30	110 (8.0)	160 (9.0)	190 (9.9)	220 (10.5)	250 (11.0)	280 (11.5)	310 (12.0)
	Clay, hard, dense	3000	33	90 (9.3)	130 (10.7)	160 (11.8)	190 (12.3)	225 (12.8)	250 (13.3)	280 (12.0)

				Wheel Loaders				Track Loaders		
				3 C.Y.	4 C.Y.	6 C.Y.	8 C.Y.	2-1/4 C.Y.	3 C.Y.	4 C.Y.
	Moist loam, sandy clay	3400	40	260	340	510	690	135	180	250
Loading Tractors	Sand and gravel	3100	18	245	320	480	650	130	170	235
	Common earth	2800	30	230	300	460	620	120	155	220
	Clay, hard, dense	3000	33	200	270	415	560	110	145	200
	Rock, well-blasted	4000	50	180	245	380	520	100	130	180

R312323-30 Compacting Backfill

Compaction of fill in embankments, around structures, in trenches, and under slabs is important to control settlement. Factors affecting compaction are:
1. Soil gradation
2. Moisture content
3. Equipment used
4. Depth of fill per lift
5. Density required

Production Rate:

$$\frac{1.75' \text{ plate width x 50 F.P.M. x 50 min./hr. x .67' lift}}{27 \text{ C.F. per C.Y.}} = 108.5 \text{ C.Y./hr.}$$

Production Rate for 4 Passes:

$$\frac{108.5 \text{ C.Y.}}{4 \text{ passes}} = 27.125 \text{ C.Y./hr. x 8 hrs.} = 217 \text{ C.Y./day}$$

Example:

Compact granular fill around a building foundation using a 21" wide x 24" vibratory plate in 8" lifts. Operator moves at 50 F.P.M. working a 50 minute hour to develop 95% Modified Proctor Density with 4 passes.

Earthwork | R3141 Shoring

R314116-40   Wood Sheet Piling

Wood sheet piling may be used for depths to 20' where there is no ground water. If moderate ground water is encountered Tongue & Groove sheeting will help to keep it out. When considerable ground water is present, steel sheeting must be used.

For estimating purposes on trench excavation, sizes are as follows:

Depth	Sheeting	Wales	Braces	B.F. per S.F.
To 8'	3 x 12's	6 x 8's, 2 line	6 x 8's, @ 10'	4.0 @ 8'
8' x 12'	3 x 12's	10 x 10's, 2 line	10 x 10's, @ 9'	5.0 average
12' to 20'	3 x 12's	12 x 12's, 3 line	12 x 12's, @ 8'	7.0 average

Sheeting to be toed in at least 2' depending upon soil conditions. A five person crew with an air compressor and sheeting driver can drive and brace 440 SF/day at 8' deep, 360 SF/day at 12' deep, and 320 SF/day at 16' deep.

For normal soils, piling can be pulled in 1/3 the time to install. Pulling difficulty increases with the time in the ground. Production can be increased by high pressure jetting.

Earthwork | R3163 Drilled Caissons

R316326-60   Caissons

The three principal types of caissons are:

(1) **Belled Caissons,** which except for shallow depths and poor soil conditions, are generally recommended. They provide more bearing than shaft area. Because of its conical shape, no horizontal reinforcement of the bell is required.

(2) **Straight Shaft Caissons** are used where relatively light loads are to be supported by caissons that rest on high value bearing strata. While the shaft is larger in diameter than for belled types this is more than offset by the savings in time and labor.

(3) **Keyed Caissons** are used when extremely heavy loads are to be carried. A keyed or socketed caisson transfers its load into rock by a combination of end-bearing and shear reinforcing of the shaft. The most economical shaft often consists of a steel casing, a steel wide flange core and concrete. Allowable compressive stresses of .225 f'c for concrete, 16,000 psi for the wide flange core, and 9,000 psi for the steel casing are commonly used. The usual range of shaft diameter is 18" to 84". The number of sizes specified for any one project should be limited due to the problems of casing and auger storage. When hand work is to be performed, shaft diameters should not be less than 32". When inspection of borings is required a minimum shaft diameter of 30" is recommended. Concrete caissons are intended to be poured against earth excavation so permanent forms, which add to cost, should not be used if the excavation is clean and the earth is sufficiently impervious to prevent excessive loss of concrete.

Soil Conditions for Belling		
Good	Requires Handwork	Not Recommended
Clay	Hard Shale	Silt
Sandy Clay	Limestone	Sand
Silty Clay	Sandstone	Gravel
Clayey Silt	Weathered Mica	Igneous Rock
Hard-pan		
Soft Shale		
Decomposed Rock		

Exterior Improvements | R3292 Turf & Grasses

R329219-50   Seeding

The type of grass is determined by light, shade and moisture content of soil plus intended use. Fertilizer should be disked 4" before seeding. For steep slopes disk five tons of mulch and lay two tons of hay or straw on surface per acre after seeding. Surface mulch can be staked, lightly disked or tar emulsion sprayed. Material for mulch can be wood chips, peat moss, partially rotted hay or straw, wood fibers and sprayed emulsions. Hemp seed blankets with fertilizer are also available. For spring seeding, watering is necessary. Late fall seeding may have to be reseeded in the spring. Hydraulic seeding, power mulching, and aerial seeding can be used on large areas.

Reference Tables

R331113-80   Piping Designations

There are several systems currently in use to describe pipe and fittings. The following paragraphs will help to identify and clarify classifications of piping systems used for water distribution.

Piping may be classified by schedule. Piping schedules include 5S, 10S, 10, 20, 30, Standard, 40, 60, Extra Strong, 80, 100, 120, 140, 160 and Double Extra Strong. These schedules are dependent upon the pipe wall thickness. The wall thickness of a particular schedule may vary with pipe size.

Ductile iron pipe for water distribution is classified by Pressure Classes such as Class 150, 200, 250, 300 and 350. These classes are actually the rated water working pressure of the pipe in pounds per square inch (psi). The pipe in these pressure classes is designed to withstand the rated water working pressure plus a surge allowance of 100 psi.

The American Water Works Association (AWWA) provides standards for various types of **plastic pipe.** C-900 is the specification for polyvinyl chloride (PVC) piping used for water distribution in sizes ranging from 4″ through 12″. C-901 is the specification for polyethylene (PE) pressure pipe, tubing and fittings used for water distribution in sizes ranging from 1/2″ through 3″. C-905 is the specification for PVC piping sizes 14″ and greater.

PVC pressure-rated pipe is identified using the standard dimensional ratio (SDR) method. This method is defined by the American Society for Testing and Materials (ASTM) Standard D 2241. This pipe is available in SDR numbers 64, 41, 32.5, 26, 21, 17, and 13.5. A pipe with an SDR of 64 will have the thinnest wall while a pipe with an SDR of 13.5 will have the thickest wall. When the pressure rating (PR) of a pipe is given in psi, it is based on a line supplying water at 73 degrees F.

The National Sanitation Foundation (NSF) seal of approval is applied to products that can be used with potable water. These products have been tested to ANSI/NSF Standard 14.

Valves and strainers are classified by American National Standards Institute (ANSI) Classes. These Classes are 125, 150, 200, 250, 300, 400, 600, 900, 1500 and 2500. Within each class there is an operating pressure range dependent upon temperature. Design parameters should be compared to the appropriate material dependent, pressure-temperature rating chart for accurate valve selection.

Change Orders

Change Order Considerations

A change order is a written document usually prepared by the design professional and signed by the owner, the architect/engineer, and the contractor. A change order states the agreement of the parties to: an addition, deletion, or revision in the work; an adjustment in the contract sum, if any; or an adjustment in the contract time, if any. Change orders, or "extras", in the construction process occur after execution of the construction contract and impact architects/engineers, contractors, and owners.

Change orders that are properly recognized and managed can ensure orderly, professional, and profitable progress for everyone involved in the project. There are many causes for change orders and change order requests. In all cases, change orders or change order requests should be addressed promptly and in a precise and prescribed manner. The following paragraphs include information regarding change order pricing and procedures.

The Causes of Change Orders

Reasons for issuing change orders include:

- Unforeseen field conditions that require a change in the work
- Correction of design discrepancies, errors, or omissions in the contract documents
- Owner-requested changes, either by design criteria, scope of work, or project objectives
- Completion date changes for reasons unrelated to the construction process
- Changes in building code interpretations, or other public authority requirements that require a change in the work
- Changes in availability of existing or new materials and products

Procedures

Properly written contract documents must include the correct change order procedures for all parties—owners, design professionals, and contractors—to follow in order to avoid costly delays and litigation.

Being "in the right" is not always a sufficient or acceptable defense. The contract provisions requiring notification and documentation must be adhered to within a defined or reasonable time frame.

The appropriate method of handling change orders is by a written proposal and acceptance by all parties involved. Prior to starting work on a project, all parties should identify their authorized agents who may sign and accept change orders, as well as any limits placed on their authority.

Time may be a critical factor when the need for a change arises. For such cases, the contractor might be directed to proceed on a "time and materials" basis, rather than wait for all paperwork to be processed—a delay that could impede progress. In this situation, the contractor must still follow the prescribed change order procedures including, but not limited to, notification and documentation.

Lack of documentation can be very costly, especially if legal judgments are to be made, and if certain field personnel are no longer available. For time and material change orders, the contractor should keep accurate daily records of all labor and material allocated to the change.

Owners or awarding authorities who do considerable and continual building construction (such as the federal government) realize the inevitability of change orders for numerous reasons, both predictable and unpredictable. As a result, the federal government, the American Institute of Architects (AIA), the Engineers Joint Contract Documents Committee (EJCDC), and other contractor, legal, and technical organizations have developed standards and procedures to be followed by all parties to achieve contract continuance and timely completion, while being financially fair to all concerned.

Pricing Change Orders

When pricing change orders, regardless of their cause, the most significant factor is when the change occurs. The need for a change may be perceived in the field or requested by the architect/engineer *before* any of the actual installation has begun, or may evolve or appear *during* construction when the item of work in question is partially installed. In the latter cases, the original sequence of construction is disrupted, along with all contiguous and supporting systems. Change orders cause the greatest impact when they occur *after* the installation has been completed and must be uncovered, or even replaced. Post-completion changes may be caused by necessary design changes, product failure, or changes in the owner's requirements that are not discovered until the building or the systems begin to function.

Specified procedures of notification and record keeping must be adhered to and enforced regardless of the stage of construction: *before, during,* or *after* installation. Some bidding documents anticipate change orders by requiring that unit prices including overhead and profit percentages—for additional as well as deductible changes—be listed. Generally these unit prices do not fully take into account the ripple effect, or impact on other trades, and should be used for general guidance only.

When pricing change orders, it is important to classify the time frame in which the change occurs. There are two basic time frames for change orders: *pre-installation change orders*, which occur before the start of construction, and *post-installation change orders*, which involve reworking after the original installation. Change orders that occur between these stages may be priced according to the extent of work completed using a combination of techniques developed for pricing *pre-* and *post-installation* changes.

Factors To Consider When Pricing Change Orders

As an estimator begins to prepare a change order, the following questions should be reviewed to determine their impact on the final price.

General

- *Is the change order work* pre-installation *or* post-installation?

  Change order work costs vary according to how much of the installation has been completed. Once workers have the project scoped in their minds, even though they have not started, it can be difficult to refocus. Consequently they may spend more than the normal amount of time understanding the change. Also, modifications to work in place, such as trimming or refitting, usually take more time than was initially estimated. The greater the amount of work in place, the more reluctant workers are to change it. Psychologically they may resent the change and as a result the rework takes longer than normal. Post-installation change order estimates must include demolition of existing work as required to accomplish the change. If the work is performed at a later time, additional obstacles, such as building finishes, may be present which must be protected. Regardless of whether the change occurs

For customer support on your Commercial Renovation Costs with RSMeans data, call 800.448.8182.

871

pre-installation or post-installation, attempt to isolate the identifiable factors and price them separately. For example, add shipping costs that may be required pre-installation or any demolition required post-installation. Then analyze the potential impact on productivity of psychological and/or learning curve factors and adjust the output rates accordingly. One approach is to break down the typical workday into segments and quantify the impact on each segment.

Change Order Installation Efficiency

The labor-hours expressed (for new construction) are based on average installation time, using an efficiency level. For change order situations, adjustments to this efficiency level should reflect the daily labor-hour allocation for that particular occurrence.

- *Will the change substantially delay the original completion date?*

A significant change in the project may cause the original completion date to be extended. The extended schedule may subject the contractor to new wage rates dictated by relevant labor contracts. Project supervision and other project overhead must also be extended beyond the original completion date. The schedule extension may also put installation into a new weather season. For example, underground piping scheduled for October installation was delayed until January. As a result, frost penetrated the trench area, thereby changing the degree of difficulty of the task. Changes and delays may have a ripple effect throughout the project. This effect must be analyzed and negotiated with the owner.

- *What is the net effect of a deduct change order?*

In most cases, change orders resulting in a deduction or credit reflect only bare costs. The contractor may retain the overhead and profit based on the original bid.

Materials

- *Will you have to pay more or less for the new material, required by the change order, than you paid for the original purchase?*

The same material prices or discounts will usually apply to materials purchased for change orders as new construction. In some instances, however, the contractor may forfeit the advantages of competitive pricing for change orders. Consider the following example:

A contractor purchased over $20,000 worth of fan coil units for an installation and obtained the maximum discount. Some time later it was determined the project required an additional matching unit. The contractor has to purchase this unit from the original supplier to ensure a match. The supplier at this time may not discount the unit because of the small quantity, and he is no longer in a competitive situation. The impact of quantity on purchase can add between 0% and 25% to material prices and/or subcontractor quotes.

- *If materials have been ordered or delivered to the job site, will they be subject to a cancellation charge or restocking fee?*

Check with the supplier to determine if ordered materials are subject to a cancellation charge. Delivered materials not used as a result of a change order may be subject to a restocking fee if returned to the supplier. Common restocking charges run between 20% and 40%. Also, delivery charges to return the goods to the supplier must be added.

Labor

- *How efficient is the existing crew at the actual installation?*

Is the same crew that performed the initial work going to do the change order? Possibly the change consists of the installation of a unit identical to one already installed; therefore, the change should take less time. Be sure to consider this potential productivity increase and modify the productivity rates accordingly.

- *If the crew size is increased, what impact will that have on supervision requirements?*

Under most bargaining agreements or management practices, there is a point at which a working foreman is replaced by a nonworking foreman. This replacement increases project overhead by adding a nonproductive worker. If additional workers are added to accelerate the project or to perform changes while maintaining the schedule, be sure to add additional supervision time if warranted. Calculate the hours involved and the additional cost directly if possible.

- *What are the other impacts of increased crew size?*

The larger the crew, the greater the potential for productivity to decrease. Some of the factors that cause this productivity loss are: overcrowding (producing restrictive conditions in the working space) and possibly a shortage of any special tools and equipment required. Such factors affect not only the crew working on the elements directly involved in the change order, but other crews whose movements may also be hampered. As the crew increases, check its basic composition for changes by the addition or deletion of apprentices or nonworking foreman, and quantify the potential effects of equipment shortages or other logistical factors.

- *As new crews, unfamiliar with the project, are brought onto the site, how long will it take them to become oriented to the project requirements?*

The orientation time for a new crew to become 100% effective varies with the site and type of project. Orientation is easiest at a new construction site and most difficult at existing, very restrictive renovation sites. The type of work also affects orientation time. When all elements of the work are exposed, such as concrete or masonry work, orientation is decreased. When the work is concealed or less visible, such as existing electrical systems, orientation takes longer. Usually orientation can be accomplished in one day or less. Costs for added orientation should be itemized and added to the total estimated cost.

- *How much actual production can be gained by working overtime?*

Short term overtime can be used effectively to accomplish more work in a day. However, as overtime is scheduled to run beyond several weeks, studies have shown marked decreases in output. The following chart shows the effect of long term overtime on worker efficiency. If the anticipated change requires extended overtime to keep the job on schedule, these factors can be used as a guide to predict the impact on time and cost. Add project overhead, particularly supervision, that may also be incurred.

Days per Week	Hours per Day	Production Efficiency					Payroll Cost Factors	
		1st Week	2nd Week	3rd Week	4th Week	Average 4 Weeks	@ 1-1/2 Times	@ 2 Times
5	8	100%	100%	100%	100%	100%	100%	100%
	9	100	100	95	90	96.25	105.6	111.1
	10	100	95	90	85	92.50	110.0	120.0
	11	95	90	75	65	81.25	113.6	127.3
	12	90	85	70	60	76.25	116.7	133.3
6	8	100	100	95	90	96.25	108.3	116.7
	9	100	95	90	85	92.50	113.0	125.9
	10	95	90	85	80	87.50	116.7	133.3
	11	95	85	70	65	78.75	119.7	139.4
	12	90	80	65	60	73.75	122.2	144.4
7	8	100	95	85	75	88.75	114.3	128.6
	9	95	90	80	70	83.75	118.3	136.5
	10	90	85	75	65	78.75	121.4	142.9
	11	85	80	65	60	72.50	124.0	148.1
	12	85	75	60	55	68.75	126.2	152.4

Effects of Overtime

Caution: Under many labor agreements, Sundays and holidays are paid at a higher premium than the normal overtime rate.

The use of long-term overtime is counterproductive on almost any construction job; that is, the longer the period of overtime, the lower the actual production rate. Numerous studies have been conducted, and while they have resulted in slightly different numbers, all reach the same conclusion. The figure above tabulates the effects of overtime work on efficiency.

As illustrated, there can be a difference between the *actual* payroll cost per hour and the *effective* cost per hour for overtime work. This is due to the reduced production efficiency with the increase in weekly hours beyond 40. This difference between actual and effective cost results from overtime work over a prolonged period. Short-term overtime work does not result in as great a reduction in efficiency and, in such cases, effective cost may not vary significantly from the actual payroll cost. As the total hours per week are increased on a regular basis, more time is lost due to fatigue, lowered morale, and an increased accident rate.

As an example, assume a project where workers are working 6 days a week, 10 hours per day. From the figure above (based on productivity studies), the average effective productive hours over a 4-week period are:

$$0.875 \times 60 = 52.5$$

Depending upon the locale and day of week, overtime hours may be paid at time and a half or double time. For time and a half, the overall (average) *actual* payroll cost (including regular and overtime hours) is determined as follows:

$$\frac{40 \text{ reg. hrs.} + (20 \text{ overtime hrs.} \times 1.5)}{60 \text{ hrs.}} = 1.167$$

Based on 60 hours, the payroll cost per hour will be 116.7% of the normal rate at 40 hours per week. However, because the effective production (efficiency) for 60 hours is reduced to the equivalent of 52.5 hours, the effective cost of overtime is calculated as follows:

For time and a half:

$$\frac{40 \text{ reg. hrs.} + (20 \text{ overtime hrs.} \times 1.5)}{52.5 \text{ hrs.}} = 1.33$$

The installed cost will be 133% of the normal rate (for labor).

Thus, when figuring overtime, the actual cost per unit of work will be higher than the apparent overtime payroll dollar increase, due to the reduced productivity of the longer work week. These efficiency calculations are true only for those cost factors determined by hours worked. Costs that are applied weekly or monthly, such as equipment rentals, will not be similarly affected.

Equipment

- *What equipment is required to complete the change order?*

Change orders may require extending the rental period of equipment already on the job site, or the addition of special equipment brought in to accomplish the change work. In either case, the additional rental charges and operator labor charges must be added.

Summary

The preceding considerations and others you deem appropriate should be analyzed and applied to a change order estimate. The impact of each should be quantified and listed on the estimate to form an audit trail.

Change orders that are properly identified, documented, and managed help to ensure the orderly, professional, and profitable progress of the work. They also minimize potential claims or disputes at the end of the project.

Back by customer demand!

You asked and we listened. For customer convenience and estimating ease, we have made the 2018 Project Costs available for download at **www.RSMeans.com/2018books**. You will also find sample estimates, an RSMeans data overview video and book registration form to receive quarterly data updates throughout 2018.

Estimating Tips

- The cost figures available in the download were derived from hundreds of projects contained in the RSMeans database of completed construction projects. They include the contractor's overhead and profit. The figures have been adjusted to January of the current year.

- These projects were located throughout the U.S. and reflect a tremendous variation in square foot (S.F.) costs. This is due to differences, not only in labor and material costs, but also in individual owners' requirements. For instance, a bank in a large city would have different features than one in a rural area. This is true of all the different types of buildings analyzed. Therefore, caution should be exercised when using these Project Costs. For example, for courthouses, costs in the database are local courthouse costs and will not apply to the larger, more elaborate federal courthouses.

- None of the figures "go with" any others. All individual cost items were computed and tabulated separately. Thus, the sum of the median figures for plumbing, HVAC, and electrical will not normally total up to the total mechanical and electrical costs arrived at by separate analysis and tabulation of the projects.

- Each building was analyzed as to total and component costs and percentages. The figures were arranged in ascending order with the results tabulated as shown. The 1/4 column shows that 25% of the projects had lower costs and 75% had higher. The 3/4 column shows that 75% of the projects had lower costs and 25% had higher. The median column shows that 50% of the projects had lower costs and 50% had higher.

- Project Costs are useful in the conceptual stage when no details are available. As soon as details become available in the project design, the square foot approach should be discontinued and the project priced as to its particular components. When more precision is required, or for estimating the replacement cost of specific buildings, the current edition of *RSMeans Square Foot Costs* should be used.

- In using the figures in this section, it is recommended that the median column be used for preliminary figures if no additional information is available. The median figures, when multiplied by the total city construction cost index figures (see City Cost Indexes) and then multiplied by the project size modifier at the end of this section, should present a fairly accurate base figure, which would then have to be adjusted in view of the estimator's experience, local economic conditions, code requirements, and the owner's particular requirements. There is no need to factor the percentage figures, as these should remain constant from city to city.

- The editors of this data would greatly appreciate receiving cost figures on one or more of your recent projects, which would then be included in the averages for next year. All cost figures received will be kept confidential, except that they will be averaged with other similar projects to arrive at square foot cost figures for next year.

See the website above for details and the discount available for submitting one or more of your projects.

Did you know?

RSMeans data is available through our online application with 24/7 access:

- Search for unit prices by keyword
- Leverage the most up-to-date data
- Build and export estimates

Try it free for 30 days!
www.rsmeans.com/2018freetrial

50 17 00	Project Costs		UNIT	UNIT COSTS			% OF TOTAL		
				1/4	MEDIAN	3/4	1/4	MEDIAN	3/4
01	**0000**	**Auto Sales with Repair**	S.F.						**01**
	0100	Architectural		98.50	110	119	58%	64%	67%
	0200	Plumbing		8.25	8.65	11.55	4.84%	5.20%	6.80%
	0300	Mechanical		11.05	14.80	16.35	6.40%	8.70%	10.15%
	0400	Electrical		17	21	26.50	9.05%	11.70%	15.90%
	0500	Total Project Costs		165	173	177			
02	**0000**	**Banking Institutions**	S.F.						**02**
	0100	Architectural		149	183	222	59%	65%	69%
	0200	Plumbing		6	8.35	11.60	2.12%	3.39%	4.19%
	0300	Mechanical		11.90	16.45	19.40	4.41%	5.10%	10.75%
	0400	Electrical		29	35	54	10.45%	13.05%	15.90%
	0500	Total Project Costs		247	278	340			
03	**0000**	**Court House**	S.F.						**03**
	0100	Architectural		78.50	78.50	78.50	54.50%	54.50%	54.50%
	0200	Plumbing		2.96	2.96	2.96	2.07%	2.07%	2.07%
	0300	Mechanical		18.55	18.55	18.55	12.95%	12.95%	12.95%
	0400	Electrical		24	24	24	16.60%	16.60%	16.60%
	0500	Total Project Costs		143	143	143			
04	**0000**	**Data Centers**	S.F.						**04**
	0100	Architectural		177	177	177	68%	68%	68%
	0200	Plumbing		9.70	9.70	9.70	3.71%	3.71%	3.71%
	0300	Mechanical		24.50	24.50	24.50	9.45%	9.45%	9.45%
	0400	Electrical		23.50	23.50	23.50	9%	9%	9%
	0500	Total Project Costs		261	261	261			
05	**0000**	**Detention Centers**	S.F.						**05**
	0100	Architectural		164	174	184	52%	53%	60.50%
	0200	Plumbing		17.35	21	25.50	5.15%	7.10%	7.25%
	0300	Mechanical		22	31.50	37.50	7.55%	9.50%	13.80%
	0400	Electrical		36	42.50	55.50	10.90%	14.85%	17.95%
	0500	Total Project Costs		278	293	345			
06	**0000**	**Fire Stations**	S.F.						**06**
	0100	Architectural		97	120	174	49%	55.50%	63%
	0200	Plumbing		9.45	13.30	15.75	4.86%	5.80%	6.30%
	0300	Mechanical		12.80	18.65	24	5.45%	8.25%	9.70%
	0400	Electrical		21.50	27.50	32.50	8.35%	12.80%	14.70%
	0500	Total Project Costs		195	222	294			
07	**0000**	**Gymnasium**	S.F.						**07**
	0100	Architectural		82	108	108	57%	64.50%	64.50%
	0200	Plumbing		2.01	6.60	6.60	1.58%	3.48%	3.48%
	0300	Mechanical		3.09	28	28	2.42%	14.65%	14.65%
	0400	Electrical		10.10	19.60	19.60	7.95%	10.35%	10.35%
	0500	Total Project Costs		128	189	189			
08	**0000**	**Hospitals**	S.F.						**08**
	0100	Architectural		99.50	164	178	43%	47.50%	48%
	0200	Plumbing		7.30	13.95	30	6%	7.45%	7.65%
	0300	Mechanical		48.50	54.50	71	14.20%	17.95%	23.50%
	0400	Electrical		22	44	57.50	10.95%	13.75%	16.85%
	0500	Total Project Costs		233	345	375			
09	**0000**	**Industrial Buildings**	S.F.						**09**
	0100	Architectural		42	66.50	216	46%	54%	56.50%
	0200	Plumbing		1.62	6.10	12.30	2%	3.06%	6.30%
	0300	Mechanical		4.47	8.50	40.50	4.77%	5.55%	14.80%
	0400	Electrical		6.80	7.80	65	7.85%	13.55%	16.20%
	0500	Total Project Costs		74.50	97	400			
10	**0000**	**Medical Clinics & Offices**	S.F.						**10**
	0100	Architectural		79.50	118	152	48.50%	55.50%	62.50%
	0200	Plumbing		8.10	12.20	19.65	4.44%	6.40%	8.05%
	0300	Mechanical		13.70	25	42.50	8.30%	11.05%	17.25%
	0400	Electrical		18.60	25	36	8.55%	12.10%	14.65%
	0500	Total Project Costs		156	208	272			

50 17 00 | Project Costs

			UNIT	UNIT COSTS			% OF TOTAL		
				1/4	MEDIAN	3/4	1/4	MEDIAN	3/4
11	0000	**Mixed Use**	S.F.							11
	0100	Architectural		85	120	182	45.50%	52.50%	70%
	0200	Plumbing		5.70	8.70	11	2.84%	3.44%	4.18%
	0300	Mechanical		14.05	36.50	65	8.05%	13.75%	20%
	0400	Electrical		14.85	36.50	49	8.30%	12.95%	15.80%
	0500	Total Project Costs		179	201	320			
12	0000	**Multi-Family Housing**	S.F.							12
	0100	Architectural		70.50	106	158	56.50%	61.50%	66.50%
	0200	Plumbing		5.90	10.20	13.90	4.81%	6.30%	7.40%
	0300	Mechanical		6.60	8.80	35.50	4.92%	6.90%	11.20%
	0400	Electrical		9.20	14.35	17.40	6.20%	8.45%	10.25%
	0500	Total Project Costs		104	194	233			
13	0000	**Nursing Home & Assisted Living**	S.F.							13
	0100	Architectural		66.50	87	116	51.50%	56.50%	69%
	0200	Plumbing		6.75	10.45	11.70	6.05%	7.40%	8.80%
	0300	Mechanical		5.90	7.55	17.05	4.04%	6.70%	9.55%
	0400	Electrical		9.75	12.80	21.50	7%	10.05%	12.20%
	0500	Total Project Costs		114	135	179			
14	0000	**Office Buildings**	S.F.							14
	0100	Architectural		90.50	120	168	56.50%	62%	71.50%
	0200	Plumbing		4.73	7.05	11.45	2.62%	3.48%	5.85%
	0300	Mechanical		10.25	15.80	23.50	5.50%	8.20%	11.10%
	0400	Electrical		11.85	20	31.50	7.75%	10%	12.70%
	0500	Total Project Costs		150	186	262			
15	0000	**Parking Garage**	S.F.							15
	0100	Architectural		29.50	36	37.50	70%	79%	88%
	0200	Plumbing		.97	1.01	1.90	2.05%	2.70%	2.83%
	0300	Mechanical		.75	1.16	4.39	2.11%	3.62%	3.81%
	0400	Electrical		2.58	2.83	5.90	5.30%	6.35%	7.95%
	0500	Total Project Costs		36	43.50	47			
16	0000	**Parking Garage/Mixed Use**	S.F.							16
	0100	Architectural		95.50	104	106	61%	62%	65.50%
	0200	Plumbing		3.06	4.01	6.15	2.47%	2.72%	3.66%
	0300	Mechanical		13.10	14.75	21	7.80%	13.10%	13.60%
	0400	Electrical		13.70	19.70	20.50	8.20%	12.65%	18.15%
	0500	Total Project Costs		156	162	168			
17	0000	**Police Stations**	S.F.							17
	0100	Architectural		108	121	152	49%	56.50%	61%
	0200	Plumbing		14.25	17.10	17.20	5.05%	5.55%	9.05%
	0300	Mechanical		32.50	45	46.50	13%	14.55%	16.55%
	0400	Electrical		24.50	26.50	28	9.15%	12.10%	14%
	0500	Total Project Costs		202	248	282			
18	0000	**Police/Fire**	S.F.							18
	0100	Architectural		105	105	320	55.50%	66%	68%
	0200	Plumbing		8.40	8.70	32	5.45%	5.50%	5.55%
	0300	Mechanical		12.85	20.50	73.50	8.35%	12.70%	12.80%
	0400	Electrical		14.65	18.70	84	9.50%	11.75%	14.55%
	0500	Total Project Costs		154	159	580			
19	0000	**Public Assembly Buildings**	S.F.							19
	0100	Architectural		119	148	200	58%	61.50%	64%
	0200	Plumbing		5.65	7.40	11.45	2.60%	3.32%	4.01%
	0300	Mechanical		14.95	23	33	6.90%	9.15%	12.75%
	0400	Electrical		19.70	24.50	38.50	8.65%	10.75%	13%
	0500	Total Project Costs		188	242	335			
20	0000	**Recreational**	S.F.							20
	0100	Architectural		103	162	219	53.50%	59%	66%
	0200	Plumbing		8.40	15.30	20	3.12%	4.76%	7.40%
	0300	Mechanical		12.65	19.25	30.50	5.35%	7.90%	12.75%
	0400	Electrical		14.15	24.50	33	7.35%	8.50%	10.65%
	0500	Total Project Costs		183	272	375			

50 17 00	Project Costs		UNIT COSTS			% OF TOTAL			
		UNIT	1/4	MEDIAN	3/4	1/4	MEDIAN	3/4	
21	0000 **Restaurants**	S.F.							**21**
	0100 Architectural		50.50	179	231	57.50%	60.50%	78%	
	0200 Plumbing		9.45	29.50	37	7.35%	7.75%	8.95%	
	0300 Mechanical		7.25	18.30	44.50	6.50%	10.80%	18.30%	
	0400 Electrical		11.65	21.50	48.50	6.75%	8.45%	12.45%	
	0500 Total Project Costs	↓	65	283	400				
22	0000 **Retail**	S.F.							**22**
	0100 Architectural		52	57.50	104	60%	60%	64.50%	
	0200 Plumbing		5.35	8.05	10.15	5.05%	6.70%	9%	
	0300 Mechanical		4.88	7	8.60	5.05%	6.15%	6.60%	
	0400 Electrical		6.85	10.95	17.60	7.90%	10.15%	12.25%	
	0500 Total Project Costs	↓	87	106	173				
23	0000 **Schools**	S.F.							**23**
	0100 Architectural		88.50	114	152	52.50%	57%	61.50%	
	0200 Plumbing		7.10	9.85	14.40	3.75%	4.82%	7%	
	0300 Mechanical		16.85	23	35.50	9.50%	12.25%	14.55%	
	0400 Electrical		16.65	23	29	9.45%	11.40%	13.05%	
	0500 Total Project Costs	↓	151	206	272				
24	0000 **University, College & Private School Classroom & Admin Buildings**	S.F.							**24**
	0100 Architectural		115	142	179	50.50%	55%	59.50%	
	0200 Plumbing		6.55	10.15	14.35	2.74%	4.30%	6.35%	
	0300 Mechanical		24.50	35.50	43	10.10%	12.15%	14.70%	
	0400 Electrical		18.55	26	31.50	7.65%	9.50%	11.55%	
	0500 Total Project Costs	↓	191	264	350				
25	0000 **University, College & Private School Dormitories**	S.F.							**25**
	0100 Architectural		75	132	140	54.50%	65%	68.50%	
	0200 Plumbing		9.95	14.05	21	6.45%	7.30%	9.15%	
	0300 Mechanical		4.46	18.95	30	4.13%	9%	12.05%	
	0400 Electrical		5.30	18.40	28	4.75%	7.35%	12.30%	
	0500 Total Project Costs	↓	111	211	249				
26	0000 **University, College & Private School Science, Eng. & Lab Buildings**	S.F.							**26**
	0100 Architectural		129	137	178	50.50%	56.50%	58%	
	0200 Plumbing		8.90	9.30	24.50	3.29%	3.95%	8.40%	
	0300 Mechanical		40.50	64	64	13.20%	21.50%	23.50%	
	0400 Electrical		26	30	35.50	8.95%	9.70%	12.85%	
	0500 Total Project Costs	↓	265	271	295				
27	0000 **University, College & Private School Student Union Buildings**	S.F.							**27**
	0100 Architectural		69.50	102	102	51%	59.50%	59.50%	
	0200 Plumbing		5	23	23	4.27%	11.45%	11.45%	
	0300 Mechanical		11.25	29.50	29.50	9.60%	14.55%	14.55%	
	0400 Electrical		15.40	25.50	25.50	12.80%	13.15%	13.15%	
	0500 Total Project Costs	↓	117	201	201				
28	0000 **Warehouses**	S.F.							**28**
	0100 Architectural		44	68.50	162	61.50%	67.50%	72%	
	0200 Plumbing		2.28	4.88	9.40	2.82%	3.72%	5%	
	0300 Mechanical		2.70	15.40	24	4.56%	8.20%	10.70%	
	0400 Electrical		4.91	18.45	30.50	7.50%	10.10%	18.30%	
	0500 Total Project Costs	↓	65.50	116	226				

Square Foot Project Size Modifier

One factor that affects the S.F. cost of a particular building is the size. In general, for buildings built to the same specifications in the same locality, the larger building will have the lower S.F. cost. This is due mainly to the decreasing contribution of the exterior walls plus the economy of scale usually achievable in larger buildings. The Area Conversion Scale shown below will give a factor to convert costs for the typical size building to an adjusted cost for the particular project.

The Square Foot Base Size lists the median costs, most typical project size in our accumulated data, and the range in size of the projects.

The Size Factor for your project is determined by dividing your project area in S.F. by the typical project size for the particular Building Type. With this factor, enter the Area Conversion Scale at the appropriate Size Factor and determine the appropriate cost multiplier for your building size.

Example: Determine the cost per S.F. for a 152,600 S.F. Multi-family housing.

$$\frac{\text{Proposed building area} = 152,600 \text{ S.F.}}{\text{Typical size from below} = 76,300 \text{ S.F.}} = 2.00$$

Enter Area Conversion scale at 2.0, intersect curve, read horizontally the appropriate cost multiplier of .94. Size adjusted cost becomes .94 x $194.00 = $182.36 based on national average costs.

Note: For Size Factors less than .50, the Cost Multiplier is 1.1
For Size Factors greater than 3.5, the Cost Multiplier is .90

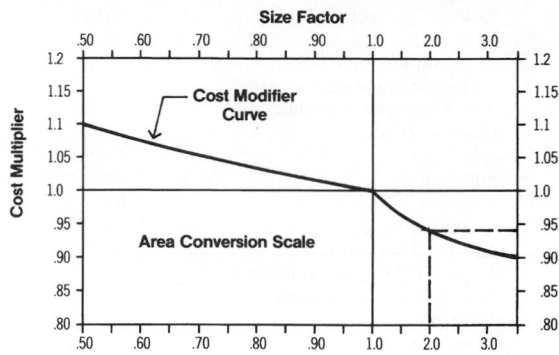

System	Median Cost (Total Project Costs)	Typical Size Gross S.F. (Median of Projects)	Typical Range (Low – High) (Projects)
Auto Sales with Repair	$173.00	25,300	8,200 – 28,700
Banking Institutions	278.00	26,300	3,300 – 28,700
Detention Centers	293.00	42,000	12,300 – 183,300
Fire Stations	222.00	14,600	6,300 – 29,600
Hospitals	345.00	137,500	54,700 – 410,300
Industrial Buildings	$97.00	16,900	5,100 – 200,600
Medical Clinics & Offices	208.00	5,500	2,600 – 327,000
Mixed Use	201.00	49,900	14,400 – 49,900
Multi-Family Housing	194.00	76,300	12,500 – 1,161,500
Nursing Home & Assisted Living	135.00	16,200	1,500 – 242,600
Office Buildings	186.00	10,000	1,100 – 930,000
Parking Garage	43.50	174,600	99,900 – 287,000
Parking Garage/Mixed Use	162.00	5,300	5,300 – 318,000
Police Stations	248.00	15,400	15,400 – 31,600
Public Assembly Buildings	242.00	30,500	2,200 – 235,300
Recreational	272.00	2,300	1,500 – 223,800
Restaurants	283.00	6,100	5,500 – 42,000
Retail	106.00	28,700	5,800 – 61,000
Schools	206.00	30,000	5,500 – 410,800
University, College & Private School Classroom & Admin Buildings	264.00	89,200	9,400 – 196,200
University, College & Private School Dormitories	211.00	50,800	1,500 – 126,900
University, College & Private School Science, Eng. & Lab Buildings	271.00	39,800	36,000 – 117,600
Warehouses	116.00	2,100	600 – 303,800

Abbreviations

A	Area Square Feet; Ampere	Brk., brk	Brick	Csc	Cosecant
AAFES	Army and Air Force Exchange Service	brkt	Bracket	C.S.F.	Hundred Square Feet
		Brs.	Brass	CSI	Construction Specifications Institute
ABS	Acrylonitrile Butadiene Stryrene; Asbestos Bonded Steel	Brz.	Bronze		
		Bsn.	Basin	CT	Current Transformer
A.C., AC	Alternating Current; Air-Conditioning; Asbestos Cement; Plywood Grade A & C	Btr.	Better	CTS	Copper Tube Size
		BTU	British Thermal Unit	Cu	Copper, Cubic
		BTUH	BTU per Hour	Cu. Ft.	Cubic Foot
		Bu.	Bushels	cw	Continuous Wave
ACI	American Concrete Institute	BUR	Built-up Roofing	C.W.	Cool White; Cold Water
ACR	Air Conditioning Refrigeration	BX	Interlocked Armored Cable	Cwt.	100 Pounds
ADA	Americans with Disabilities Act	°C	Degree Centigrade	C.W.X.	Cool White Deluxe
AD	Plywood, Grade A & D	c	Conductivity, Copper Sweat	C.Y.	Cubic Yard (27 cubic feet)
Addit.	Additional	C	Hundred; Centigrade	C.Y./Hr.	Cubic Yard per Hour
Adh.	Adhesive	C/C	Center to Center, Cedar on Cedar	Cyl.	Cylinder
Adj.	Adjustable	C-C	Center to Center	d	Penny (nail size)
af	Audio-frequency	Cab	Cabinet	D	Deep; Depth; Discharge
AFFF	Aqueous Film Forming Foam	Cair.	Air Tool Laborer	Dis., Disch.	Discharge
AFUE	Annual Fuel Utilization Efficiency	Cal.	Caliper	Db	Decibel
AGA	American Gas Association	Calc	Calculated	Dbl.	Double
Agg.	Aggregate	Cap.	Capacity	DC	Direct Current
A.H., Ah	Ampere Hours	Carp.	Carpenter	DDC	Direct Digital Control
A hr.	Ampere-hour	C.B.	Circuit Breaker	Demob.	Demobilization
A.H.U., AHU	Air Handling Unit	C.C.A.	Chromate Copper Arsenate	d.f.t.	Dry Film Thickness
A.I.A.	American Institute of Architects	C.C.F.	Hundred Cubic Feet	d.f.u.	Drainage Fixture Units
AIC	Ampere Interrupting Capacity	cd	Candela	D.H.	Double Hung
Allow.	Allowance	cd/sf	Candela per Square Foot	DHW	Domestic Hot Water
alt., alt	Alternate	CD	Grade of Plywood Face & Back	DI	Ductile Iron
Alum.	Aluminum	CDX	Plywood, Grade C & D, exterior glue	Diag.	Diagonal
a.m.	Ante Meridiem			Diam., Dia	Diameter
Amp.	Ampere	Cefi.	Cement Finisher	Distrib.	Distribution
Anod.	Anodized	Cem.	Cement	Div.	Division
ANSI	American National Standards Institute	CF	Hundred Feet	Dk.	Deck
		C.F.	Cubic Feet	D.L.	Dead Load; Diesel
APA	American Plywood Association	CFM	Cubic Feet per Minute	DLH	Deep Long Span Bar Joist
Approx.	Approximate	CFRP	Carbon Fiber Reinforced Plastic	dlx	Deluxe
Apt.	Apartment	c.g.	Center of Gravity	Do.	Ditto
Asb.	Asbestos	CHW	Chilled Water; Commercial Hot Water	DOP	Dioctyl Phthalate Penetration Test (Air Filters)
A.S.B.C.	American Standard Building Code				
Asbe.	Asbestos Worker	C.I., CI	Cast Iron	Dp., dp	Depth
ASCE	American Society of Civil Engineers	C.I.P., CIP	Cast in Place	D.P.S.T.	Double Pole, Single Throw
A.S.H.R.A.E.	American Society of Heating, Refrig. & AC Engineers	Circ.	Circuit	Dr.	Drive
		C.L.	Carload Lot	DR	Dimension Ratio
ASME	American Society of Mechanical Engineers	CL	Chain Link	Drink.	Drinking
		Clab.	Common Laborer	D.S.	Double Strength
ASTM	American Society for Testing and Materials	Clam	Common Maintenance Laborer	D.S.A.	Double Strength A Grade
		C.L.F.	Hundred Linear Feet	D.S.B.	Double Strength B Grade
Attchmt.	Attachment	CLF	Current Limiting Fuse	Dty.	Duty
Avg., Ave.	Average	CLP	Cross Linked Polyethylene	DWV	Drain Waste Vent
AWG	American Wire Gauge	cm	Centimeter	DX	Deluxe White, Direct Expansion
AWWA	American Water Works Assoc.	CMP	Corr. Metal Pipe	dyn	Dyne
Bbl.	Barrel	CMU	Concrete Masonry Unit	e	Eccentricity
B&B, BB	Grade B and Better; Balled & Burlapped	CN	Change Notice	E	Equipment Only; East; Emissivity
		Col.	Column	Ea.	Each
B&S	Bell and Spigot	CO₂	Carbon Dioxide	EB	Encased Burial
B.&W.	Black and White	Comb.	Combination	Econ.	Economy
b.c.c.	Body-centered Cubic	comm.	Commercial, Communication	E.C.Y	Embankment Cubic Yards
B.C.Y.	Bank Cubic Yards	Compr.	Compressor	EDP	Electronic Data Processing
BE	Bevel End	Conc.	Concrete	EIFS	Exterior Insulation Finish System
B.F.	Board Feet	Cont., cont	Continuous; Continued, Container	E.D.R.	Equiv. Direct Radiation
Bg. cem.	Bag of Cement	Corkbd.	Cork Board	Eq.	Equation
BHP	Boiler Horsepower; Brake Horsepower	Corr.	Corrugated	EL	Elevation
		Cos	Cosine	Elec.	Electrician; Electrical
B.I.	Black Iron	Cot	Cotangent	Elev.	Elevator; Elevating
bidir.	bidirectional	Cov.	Cover	EMT	Electrical Metallic Conduit; Thin Wall Conduit
Bit., Bitum.	Bituminous	C/P	Cedar on Paneling		
Bit., Conc.	Bituminous Concrete	CPA	Control Point Adjustment	Eng.	Engine, Engineered
Bk.	Backed	Cplg.	Coupling	EPDM	Ethylene Propylene Diene Monomer
Bkrs.	Breakers	CPM	Critical Path Method		
Bldg., bldg	Building	CPVC	Chlorinated Polyvinyl Chloride	EPS	Expanded Polystyrene
Blk.	Block	C.Pr.	Hundred Pair	Eqhv.	Equip. Oper., Heavy
Bm.	Beam	CRC	Cold Rolled Channel	Eqlt.	Equip. Oper., Light
Boil.	Boilermaker	Creos.	Creosote	Eqmd.	Equip. Oper., Medium
bpm	Blows per Minute	Crpt.	Carpet & Linoleum Layer	Eqmm.	Equip. Oper., Master Mechanic
BR	Bedroom	CRT	Cathode-ray Tube	Eqol.	Equip. Oper., Oilers
Brg., brng.	Bearing	CS	Carbon Steel, Constant Shear Bar Joist	Equip.	Equipment
Brhe.	Bricklayer Helper			ERW	Electric Resistance Welded
Bric.	Bricklayer				

Abbreviations

E.S.	Energy Saver	H	High Henry	Lath.	Lather
Est.	Estimated	HC	High Capacity	Lav.	Lavatory
esu	Electrostatic Units	H.D., HD	Heavy Duty; High Density	lb.; #	Pound
E.W.	Each Way	H.D.O.	High Density Overlaid	L.B., LB	Load Bearing; L Conduit Body
EWT	Entering Water Temperature	HDPE	High Density Polyethylene Plastic	L. & E.	Labor & Equipment
Excav.	Excavation	Hdr.	Header	lb./hr.	Pounds per Hour
excl	Excluding	Hdwe.	Hardware	lb./L.F.	Pounds per Linear Foot
Exp., exp	Expansion, Exposure	H.I.D., HID	High Intensity Discharge	lbf/sq.in.	Pound-force per Square Inch
Ext., ext	Exterior; Extension	Help.	Helper Average	L.C.L.	Less than Carload Lot
Extru.	Extrusion	HEPA	High Efficiency Particulate Air	L.C.Y.	Loose Cubic Yard
f.	Fiber Stress		Filter	Ld.	Load
F	Fahrenheit; Female; Fill	Hg	Mercury	LE	Lead Equivalent
Fab., fab	Fabricated; Fabric	HIC	High Interrupting Capacity	LED	Light Emitting Diode
FBGS	Fiberglass	HM	Hollow Metal	L.F.	Linear Foot
F.C.	Footcandles	HMWPE	High Molecular Weight	L.F. Hdr	Linear Feet of Header
f.c.c.	Face-centered Cubic		Polyethylene	L.F. Nose	Linear Foot of Stair Nosing
f'c.	Compressive Stress in Concrete;	HO	High Output	L.F. Rsr	Linear Foot of Stair Riser
	Extreme Compressive Stress	Horiz.	Horizontal	Lg.	Long; Length; Large
F.E.	Front End	H.P., HP	Horsepower; High Pressure	L & H	Light and Heat
FEP	Fluorinated Ethylene Propylene	H.P.F.	High Power Factor	LH	Long Span Bar Joist
	(Teflon)	Hr.	Hour	L.H.	Labor Hours
F.G.	Flat Grain	Hrs./Day	Hours per Day	L.L., LL	Live Load
F.H.A.	Federal Housing Administration	HSC	High Short Circuit	L.L.D.	Lamp Lumen Depreciation
Fig.	Figure	Ht.	Height	lm	Lumen
Fin.	Finished	Htg.	Heating	lm/sf	Lumen per Square Foot
FIPS	Female Iron Pipe Size	Htrs.	Heaters	lm/W	Lumen per Watt
Fixt.	Fixture	HVAC	Heating, Ventilation & Air-	LOA	Length Over All
FJP	Finger jointed and primed		Conditioning	log	Logarithm
Fl. Oz.	Fluid Ounces	Hvy.	Heavy	L-O-L	Lateralolet
Flr.	Floor	HW	Hot Water	long.	Longitude
Flrs.	Floors	Hyd.; Hydr.	Hydraulic	L.P., LP	Liquefied Petroleum; Low Pressure
FM	Frequency Modulation;	Hz	Hertz (cycles)	L.P.F.	Low Power Factor
	Factory Mutual	I.	Moment of Inertia	LR	Long Radius
Fmg.	Framing	IBC	International Building Code	L.S.	Lump Sum
FM/UL	Factory Mutual/Underwriters Labs	I.C.	Interrupting Capacity	Lt.	Light
Fdn.	Foundation	ID	Inside Diameter	Lt. Ga.	Light Gauge
FNPT	Female National Pipe Thread	I.D.	Inside Dimension; Identification	L.T.L.	Less than Truckload Lot
Fori.	Foreman, Inside	I.F.	Inside Frosted	Lt. Wt.	Lightweight
Foro.	Foreman, Outside	I.M.C.	Intermediate Metal Conduit	L.V.	Low Voltage
Fount.	Fountain	In.	Inch	M	Thousand; Material; Male;
fpm	Feet per Minute	Incan.	Incandescent		Light Wall Copper Tubing
FPT	Female Pipe Thread	Incl.	Included; Including	M²CA	Meters Squared Contact Area
Fr	Frame	Int.	Interior	m/hr.; M.H.	Man-hour
F.R.	Fire Rating	Inst.	Installation	mA	Milliampere
FRK	Foil Reinforced Kraft	Insul., insul	Insulation/Insulated	Mach.	Machine
FSK	Foil/Scrim/Kraft	I.P.	Iron Pipe	Mag. Str.	Magnetic Starter
FRP	Fiberglass Reinforced Plastic	I.P.S., IPS	Iron Pipe Size	Maint.	Maintenance
FS	Forged Steel	IPT	Iron Pipe Threaded	Marb.	Marble Setter
FSC	Cast Body; Cast Switch Box	I.W.	Indirect Waste	Mat; Mat'l.	Material
Ft., ft	Foot; Feet	J	Joule	Max.	Maximum
Ftng.	Fitting	J.I.C.	Joint Industrial Council	MBF	Thousand Board Feet
Ftg.	Footing	K	Thousand; Thousand Pounds;	MBH	Thousand BTU's per hr.
Ft lb.	Foot Pound		Heavy Wall Copper Tubing, Kelvin	MC	Metal Clad Cable
Furn.	Furniture	K.A.H.	Thousand Amp. Hours	MCC	Motor Control Center
FVNR	Full Voltage Non-Reversing	kcmil	Thousand Circular Mils	M.C.F.	Thousand Cubic Feet
FVR	Full Voltage Reversing	KD	Knock Down	MCFM	Thousand Cubic Feet per Minute
FXM	Female by Male	K.D.A.T.	Kiln Dried After Treatment	M.C.M.	Thousand Circular Mils
Fy.	Minimum Yield Stress of Steel	kg	Kilogram	MCP	Motor Circuit Protector
g	Gram	kG	Kilogauss	MD	Medium Duty
G	Gauss	kgf	Kilogram Force	MDF	Medium-density fibreboard
Ga.	Gauge	kHz	Kilohertz	M.D.O.	Medium Density Overlaid
Gal., gal.	Gallon	Kip	1000 Pounds	Med.	Medium
Galv., galv	Galvanized	KJ	Kilojoule	MF	Thousand Feet
GC/MS	Gas Chromatograph/Mass	K.L.	Effective Length Factor	M.F.B.M.	Thousand Feet Board Measure
	Spectrometer	K.L.F.	Kips per Linear Foot	Mfg.	Manufacturing
Gen.	General	Km	Kilometer	Mfrs.	Manufacturers
GFI	Ground Fault Interrupter	KO	Knock Out	mg	Milligram
GFRC	Glass Fiber Reinforced Concrete	K.S.F.	Kips per Square Foot	MGD	Million Gallons per Day
Glaz.	Glazier	K.S.I.	Kips per Square Inch	MGPH	Million Gallons per Hour
GPD	Gallons per Day	kV	Kilovolt	MH, M.H.	Manhole; Metal Halide; Man-Hour
gpf	Gallon per Flush	kVA	Kilovolt Ampere	MHz	Megahertz
GPH	Gallons per Hour	kVAR	Kilovar (Reactance)	Mi.	Mile
gpm, GPM	Gallons per Minute	KW	Kilowatt	MI	Malleable Iron; Mineral Insulated
GR	Grade	KWh	Kilowatt-hour	MIPS	Male Iron Pipe Size
Gran.	Granular	L	Labor Only; Length; Long;	mj	Mechanical Joint
Grnd.	Ground		Medium Wall Copper Tubing	m	Meter
GVW	Gross Vehicle Weight	Lab.	Labor	mm	Millimeter
GWB	Gypsum Wall Board	lat	Latitude	Mill.	Millwright
				Min., min.	Minimum, Minute

Abbreviations

Misc.	Miscellaneous	PCM	Phase Contrast Microscopy	SBS	Styrene Butadiere Styrene
ml	Milliliter, Mainline	PDCA	Painting and Decorating	SC	Screw Cover
M.L.F.	Thousand Linear Feet		Contractors of America	SCFM	Standard Cubic Feet per Minute
Mo.	Month	P.E., PE	Professional Engineer;	Scaf.	Scaffold
Mobil.	Mobilization		Porcelain Enamel;	Sch., Sched.	Schedule
Mog.	Mogul Base		Polyethylene; Plain End	S.C.R.	Modular Brick
MPH	Miles per Hour	P.E.C.I.	Porcelain Enamel on Cast Iron	S.D.	Sound Deadening
MPT	Male Pipe Thread	Perf.	Perforated	SDR	Standard Dimension Ratio
MRGWB	Moisture Resistant Gypsum	PEX	Cross Linked Polyethylene	S.E.	Surfaced Edge
	Wallboard	Ph.	Phase	Sel.	Select
MRT	Mile Round Trip	P.I.	Pressure Injected	SER, SEU	Service Entrance Cable
ms	Millisecond	Pile.	Pile Driver	S.F.	Square Foot
M.S.F.	Thousand Square Feet	Pkg.	Package	S.F.C.A.	Square Foot Contact Area
Mstz.	Mosaic & Terrazzo Worker	Pl.	Plate	S.F. Flr.	Square Foot of Floor
M.S.Y.	Thousand Square Yards	Plah.	Plasterer Helper	S.F.G.	Square Foot of Ground
Mtd., mtd., mtd	Mounted	Plas.	Plasterer	S.F. Hor.	Square Foot Horizontal
Mthe.	Mosaic & Terrazzo Helper	plf	Pounds Per Linear Foot	SFR	Square Feet of Radiation
Mtng.	Mounting	Pluh.	Plumber Helper	S.F. Shlf.	Square Foot of Shelf
Mult.	Multi; Multiply	Plum.	Plumber	S4S	Surface 4 Sides
MUTCD	Manual on Uniform Traffic Control	Ply.	Plywood	Shee.	Sheet Metal Worker
	Devices	p.m.	Post Meridiem	Sin.	Sine
M.V.A.	Million Volt Amperes	Pntd.	Painted	Skwk.	Skilled Worker
M.V.A.R.	Million Volt Amperes Reactance	Pord.	Painter, Ordinary	SL	Saran Lined
MV	Megavolt	pp	Pages	S.L.	Slimline
MW	Megawatt	PP, PPL	Polypropylene	Sldr.	Solder
MXM	Male by Male	P.P.M.	Parts per Million	SLH	Super Long Span Bar Joist
MYD	Thousand Yards	Pr.	Pair	S.N.	Solid Neutral
N	Natural; North	P.E.S.B.	Pre-engineered Steel Building	SO	Stranded with oil resistant inside
nA	Nanoampere	Prefab.	Prefabricated		insulation
NA	Not Available; Not Applicable	Prefin.	Prefinished	S-O-L	Socketolet
N.B.C.	National Building Code	Prop.	Propelled	sp	Standpipe
NC	Normally Closed	PSF, psf	Pounds per Square Foot	S.P.	Static Pressure; Single Pole; Self-
NEMA	National Electrical Manufacturers	PSI, psi	Pounds per Square Inch		Propelled
	Assoc.	PSIG	Pounds per Square Inch Gauge	Spri.	Sprinkler Installer
NEHB	Bolted Circuit Breaker to 600V.	PSP	Plastic Sewer Pipe	spwg	Static Pressure Water Gauge
NFPA	National Fire Protection Association	Pspr.	Painter, Spray	S.P.D.T.	Single Pole, Double Throw
NLB	Non-Load-Bearing	Psst.	Painter, Structural Steel	SPF	Spruce Pine Fir; Sprayed
NM	Non-Metallic Cable	P.T.	Potential Transformer		Polyurethane Foam
nm	Nanometer	P. & T.	Pressure & Temperature	S.P.S.T.	Single Pole, Single Throw
No.	Number	Ptd.	Painted	SPT	Standard Pipe Thread
NO	Normally Open	Ptns.	Partitions	Sq.	Square; 100 Square Feet
N.O.C.	Not Otherwise Classified	Pu	Ultimate Load	Sq. Hd.	Square Head
Nose.	Nosing	PVC	Polyvinyl Chloride	Sq. In.	Square Inch
NPT	National Pipe Thread	Pvmt.	Pavement	S.S.	Single Strength; Stainless Steel
NQOD	Combination Plug-on/Bolt on	PRV	Pressure Relief Valve	S.S.B.	Single Strength B Grade
	Circuit Breaker to 240V.	Pwr.	Power	sst, ss	Stainless Steel
N.R.C., NRC	Noise Reduction Coefficient/	Q	Quantity Heat Flow	Sswk.	Structural Steel Worker
	Nuclear Regulator Commission	Qt.	Quart	Sswl.	Structural Steel Welder
N.R.S.	Non Rising Stem	Quan., Qty.	Quantity	St.; Stl.	Steel
ns	Nanosecond	Q.C.	Quick Coupling	STC	Sound Transmission Coefficient
NTP	Notice to Proceed	r	Radius of Gyration	Std.	Standard
nW	Nanowatt	R	Resistance	Stg.	Staging
OB	Opposing Blade	R.C.P.	Reinforced Concrete Pipe	STK	Select Tight Knot
OC	On Center	Rect.	Rectangle	STP	Standard Temperature & Pressure
OD	Outside Diameter	recpt.	Receptacle	Stpi.	Steamfitter, Pipefitter
O.D.	Outside Dimension	Reg.	Regular	Str.	Strength; Starter; Straight
ODS	Overhead Distribution System	Reinf.	Reinforced	Strd.	Stranded
O.G.	Ogee	Req'd.	Required	Struct.	Structural
O.H.	Overhead	Res.	Resistant	Sty.	Story
O&P	Overhead and Profit	Resi.	Residential	Subj.	Subject
Oper.	Operator	RF	Radio Frequency	Subs.	Subcontractors
Opng.	Opening	RFID	Radio-frequency Identification	Surf.	Surface
Orna.	Ornamental	Rgh.	Rough	Sw.	Switch
OSB	Oriented Strand Board	RGS	Rigid Galvanized Steel	Swbd.	Switchboard
OS&Y	Outside Screw and Yoke	RHW	Rubber, Heat & Water Resistant;	S.Y.	Square Yard
OSHA	Occupational Safety and Health		Residential Hot Water	Syn.	Synthetic
	Act	rms	Root Mean Square	S.Y.P.	Southern Yellow Pine
Ovhd.	Overhead	Rnd.	Round	Sys.	System
OWG	Oil, Water or Gas	Rodm.	Rodman	t.	Thickness
Oz.	Ounce	Rofc.	Roofer, Composition	T	Temperature; Ton
P.	Pole; Applied Load; Projection	Rofp.	Roofer, Precast	Tan	Tangent
p.	Page	Rohe.	Roofer Helpers (Composition)	T.C.	Terra Cotta
Pape.	Paperhanger	Rots.	Roofer, Tile & Slate	T & C	Threaded and Coupled
P.A.P.R.	Powered Air Purifying Respirator	R.O.W.	Right of Way	T.D.	Temperature Difference
PAR	Parabolic Reflector	RPM	Revolutions per Minute	TDD	Telecommunications Device for
P.B., PB	Push Button	R.S.	Rapid Start		the Deaf
Pc., Pcs.	Piece, Pieces	Rsr	Riser	T.E.M.	Transmission Electron Microscopy
P.C.	Portland Cement; Power Connector	RT	Round Trip	temp	Temperature, Tempered, Temporary
P.C.F.	Pounds per Cubic Foot	S.	Suction; Single Entrance; South	TFFN	Nylon Jacketed Wire

882

Abbreviations

TFE	Tetrafluoroethylene (Teflon)	U.L., UL	Underwriters Laboratory	w/	With
T. & G.	Tongue & Groove;	Uld.	Unloading	W.C., WC	Water Column; Water Closet
	Tar & Gravel	Unfin.	Unfinished	W.F.	Wide Flange
Th., Thk.	Thick	UPS	Uninterruptible Power Supply	W.G.	Water Gauge
Thn.	Thin	URD	Underground Residential	Wldg.	Welding
Thrded	Threaded		Distribution	W. Mile	Wire Mile
Tilf.	Tile Layer, Floor	US	United States	W-O-L	Weldolet
Tilh.	Tile Layer, Helper	USGBC	U.S. Green Building Council	W.R.	Water Resistant
THHN	Nylom Jacketed Wire	USP	United States Primed	Wrck.	Wrecker
THW.	Insulated Strand Wire	UTMCD	Uniform Traffic Manual For Control	WSFU	Water Supply Fixture Unit
THWN	Nylon Jacketed Wire		Devices	W.S.P.	Water, Steam, Petroleum
T.L., TL	Truckload	UTP	Unshielded Twisted Pair	WT., Wt.	Weight
T.M.	Track Mounted	V	Volt	WWF	Welded Wire Fabric
Tot.	Total	VA	Volt Amperes	XFER	Transfer
T-O-L	Threadolet	VAT	Vinyl Asbestos Tile	XFMR	Transformer
tmpd	Tempered	V.C.T.	Vinyl Composition Tile	XHD	Extra Heavy Duty
TPO	Thermoplastic Polyolefin	VAV	Variable Air Volume	XHHW	Cross-Linked Polyethylene Wire
T.S.	Trigger Start	VC	Veneer Core	XLPE	Insulation
Tr.	Trade	VDC	Volts Direct Current	XLP	Cross-linked Polyethylene
Transf.	Transformer	Vent.	Ventilation	Xport	Transport
Trhv.	Truck Driver, Heavy	Vert.	Vertical	Y	Wye
Trlr	Trailer	V.F.	Vinyl Faced	yd	Yard
Trlt.	Truck Driver, Light	V.G.	Vertical Grain	yr	Year
TTY	Teletypewriter	VHF	Very High Frequency	Δ	Delta
TV	Television	VHO	Very High Output	%	Percent
T.W.	Thermoplastic Water Resistant	Vib.	Vibrating	~	Approximately
	Wire	VLF	Vertical Linear Foot	Ø	Phase; diameter
UCI	Uniform Construction Index	VOC	Volatile Organic Compound	@	At
UF	Underground Feeder	Vol.	Volume	#	Pound; Number
UGND	Underground Feeder	VRP	Vinyl Reinforced Polyester	<	Less Than
UHF	Ultra High Frequency	W	Wire; Watt; Wide; West	>	Greater Than
U.I.	United Inch			Z	Zone

Index

886

Index

887

Index

893

Index

For customer support on your Commercial Renovation Costs with RSMeans data, call 800.448.8182.

Index

907

Index

909

X

Y

Z

For customer support on your Commercial Renovation Costs with RSMeans data, call 800.448.8182.

Division Notes

	CREW	DAILY OUTPUT	LABOR-HOURS	UNIT	BARE COSTS				TOTAL INCL O&P
					MAT.	LABOR	EQUIP.	TOTAL	

Division Notes

	CREW	DAILY OUTPUT	LABOR-HOURS	UNIT	BARE COSTS				TOTAL INCL O&P
					MAT.	LABOR	EQUIP.	TOTAL	

Division Notes

		CREW	DAILY OUTPUT	LABOR-HOURS	UNIT	BARE COSTS				TOTAL INCL O&P
						MAT.	LABOR	EQUIP.	TOTAL	

Division Notes

	CREW	DAILY OUTPUT	LABOR-HOURS	UNIT	BARE COSTS				TOTAL INCL O&P
					MAT.	LABOR	EQUIP.	TOTAL	

Division Notes

	CREW	DAILY OUTPUT	LABOR-HOURS	UNIT	BARE COSTS				TOTAL INCL O&P
					MAT.	LABOR	EQUIP.	TOTAL	

Division Notes

		CREW	DAILY OUTPUT	LABOR-HOURS	UNIT	BARE COSTS				TOTAL INCL O&P
						MAT.	LABOR	EQUIP.	TOTAL	
					CREW	DAILY OUTPUT	LABOR-HOURS	UNIT	MAT.	LABOR	EQUIP.	TOTAL	TOTAL INCL O&P

Division Notes

		CREW	DAILY OUTPUT	LABOR-HOURS	UNIT	BARE COSTS				TOTAL INCL O&P
						MAT.	LABOR	EQUIP.	TOTAL	

Other Data & Services

A tradition of excellence in construction cost information and services since 1942

Table of Contents
Annual Cost Guides
Online Estimating Solution
Seminars and Professional Development

For more information visit our website at www.RSMeans.com

Unit prices according to the latest MasterFormat®

Cost Data Selection Guide

The following table provides definitive information on the content of each cost data publication. The number of lines of data provided in each unit price or assemblies division, as well as the number of crews, is listed for each data set. The presence of other elements such as reference tables, square foot models, equipment rental costs, historical cost indexes, and city cost indexes, is also indicated. You can use the table to help select the RSMeans data set that has the quantity and type of information you most need in your work.

Unit Cost Divisions	Building Construction	Mechanical	Electrical	Commercial Renovation	Square Foot	Site Work Landsc.	Green Building	Interior	Concrete Masonry	Open Shop	Heavy Construction	Light Commercial	Facilities Construction	Plumbing	Residential
1	590	411	428	530	0	524	200	331	473	589	527	273	1065	421	178
2	777	280	86	733	0	991	207	399	214	776	733	481	1220	287	274
3	1688	340	230	1081	0	1480	986	354	2034	1688	1690	482	1788	316	389
4	960	21	0	920	0	725	180	615	1158	928	615	533	1175	0	447
5	1901	158	155	1093	0	852	1799	1106	729	1901	1037	979	1918	204	746
6	2453	18	18	2111	0	110	589	1528	281	2449	123	2141	2125	22	2661
7	1596	215	128	1634	0	580	763	532	523	1593	26	1329	1697	227	1049
8	2140	80	3	2733	0	255	1140	1813	105	2142	0	2328	2966	0	1552
9	2107	86	45	1931	0	309	455	2193	412	2050	15	1756	2356	54	1521
10	1090	17	10	685	0	234	32	899	136	1090	34	589	1181	237	224
11	1097	201	166	541	0	135	56	925	29	1064	0	231	1117	164	110
12	548	0	2	299	0	219	147	1551	14	515	0	273	1574	23	217
13	744	149	158	253	0	366	125	254	78	720	267	109	760	115	104
14	273	36	0	223	0	0	0	257	0	273	0	12	293	16	6
21	130	0	41	37	0	0	0	296	0	130	0	121	668	688	259
22	1165	7557	160	1226	0	1572	1063	849	20	1154	1681	875	7505	9414	719
23	1194	6995	581	938	0	157	898	787	38	1177	110	886	5235	1917	480
25	0	0	14	14	0	0	0	0	0	0	0	0	0	0	0
26	1512	491	10456	1293	0	811	644	1159	55	1438	600	1360	10237	399	636
27	94	0	447	101	0	0	0	71	0	94	39	67	388	0	56
28	143	79	223	124	0	0	28	97	0	127	0	70	209	57	41
31	1510	733	610	806	0	3265	288	7	1217	1455	3276	604	1569	660	613
32	836	49	8	905	0	4472	353	405	314	808	1889	440	1751	142	487
33	1246	1076	534	252	0	3021	38	0	237	523	3058	128	1698	2085	154
34	107	0	47	4	0	190	0	0	31	62	212	0	136	0	0
35	18	0	0	0	0	327	0	0	0	18	442	0	84	0	0
41	62	0	0	33	0	8	0	22	0	61	31	0	68	14	0
44	75	79	0	0	0	0	0	0	0	0	0	0	75	75	0
46	23	16	0	0	0	274	261	0	0	23	264	0	33	33	0
48	10	0	38	2	0	0	23	0	0	10	17	10	23	0	10
Totals	26089	19087	14588	20502	0	20877	10275	16450	8098	24858	16686	16077	50914	17570	12933

Assem Div	Building Construction	Mechanical	Electrical	Commercial Renovation	Square Foot	Site Work Landscape	Assemblies	Green Building	Interior	Concrete Masonry	Heavy Construction	Light Commercial	Facilities Construction	Plumbing	Asm Div	Residential
A		15	0	188	164	577	598	0	0	536	571	154	24	0	1	378
B		0	0	848	2554	0	5661	56	329	1976	368	2094	174	0	2	211
C		0	0	647	954	0	1334	0	1642	146	0	844	255	0	3	588
D		1057	941	712	1859	72	2538	330	825	0	0	1345	1105	1088	4	851
E		0	0	86	261	0	301	0	5	0	0	258	5	0	5	391
F		0	0	0	114	0	143	0	0	0	0	114	0	0	6	357
G		527	447	318	312	3377	792	0	0	534	1349	205	293	677	7	307
															8	760
															9	80
															10	0
															11	0
															12	0
Totals		1599	1388	2799	6218	4026	11367	386	2801	3192	2288	5014	1856	1765		3923

Reference Section	Building Construction Costs	Mechanical	Electrical	Commercial Renovation	Square Foot	Site Work Landscape	Assem.	Green Building	Interior	Concrete Masonry	Open Shop	Heavy Construction	Light Commercial	Facilities Construction	Plumbing	Resi.
Reference Tables	yes	yes	yes	yes	no	yes	yes	yes	yes	yes	yes	yes	yes	yes	yes	yes
Models					111			25					50			28
Crews	578	578	578	556		578		578	578	578	555	578	555	556	578	555
Equipment Rental Costs	yes	yes	yes	yes		yes		yes	yes	yes	yes	yes	yes	yes	yes	yes
Historical Cost Indexes	yes	yes	yes	yes	yes	yes	yes	yes	yes	yes	yes	yes	yes	yes	yes	no
City Cost Indexes	yes	yes	yes	yes	yes	yes	yes	yes	yes	yes	yes	yes	yes	yes	yes	yes

Our Online Estimating Solution

Competitive Cost Estimates Made Easy

Our online estimating solution is a web-based service that provides accurate and up-to-date cost information to help you build competitive estimates or budgets in less time.

Quick, intuitive, easy to use and automatically updated, you'll gain instant access to hundreds of thousands of material, labor, and equipment costs from RSMeans' comprehensive database, delivering the information you need to build budgets and competitive estimates every time.

With our online estimating solutions, you can perform quick searches to locate specific costs and adjust costs to reflect prices in your geographic area. Tag and store your favorites for fast access to your frequently used line items and assemblies and clone estimates to save time. System notifications will alert you as updated data becomes available. This data is automatically updated throughout the year.

Our visual, interactive estimating features help you create, manage, save and share estimates with ease! You'll enjoy increased flexibility with customizable advanced reports. Easily edit custom report templates and import your company logo onto your estimates.

	Core	Advanced	Complete
Unit Prices	✓	✓	✓
Assemblies	⊘	✓	✓
Sq Foot Models	⊘	⊘	✓
Editable Sq Foot Models	⊘	⊘	✓
Editable Assembly Components	⊘	✓	✓
Custom Cost Data	⊘	✓	✓
User Defined Components	⊘	✓	✓
Advanced Reporting & Customization	⊘	✓	✓
Union Labor Type	✓	✓	✓

Continue to check our website at www.RSMeans.com for more product offerings.

Estimate with Precision

Find everything you need to develop complete, accurate estimates.

- Verified costs for construction materials
- Equipment rental costs
- Crew sizing, labor hours and labor rates
- Localized costs for U.S. and Canada

Save Time & Increase Efficiency

Make cost estimating and calculating faster and easier than ever with secure, online estimating tools.

- Quickly locate costs in the searchable database
- Create estimates in minutes with RSMeans cost lines
- Tag and store favorites for fast access to frequently used items

Improve Planning & Decision-Making

Back your estimates with complete, accurate and up-to-date cost data for informed business decisions.

- Verify construction costs from third parties
- Check validity of subcontractor proposals
- Evaluate material and assembly alternatives

Increase Profits

Use our online estimating solution to estimate projects quickly and accurately, so you can gain an edge over your competition.

- Create accurate and competitive bids
- Minimize the risk of cost overruns
- Reduce variability

Access the data online

Search for unit prices by keyword   Leverage the most up-to-date data   Build and export estimates

Try it free for 30 days   www.rsmeans.com/2018freetrial

2018 Seminar Schedule   ☎ 877-620-6245

Note: call for exact dates, locations, and details as some cities are subject to change.

Location	Dates	Location	Dates
Seattle, WA	January and August	San Francisco, CA	June
Dallas/Ft. Worth, TX	January	Bethesda, MD	June
Austin, TX	February	El Segundo, CA	August
Anchorage, AK	March and September	Dallas, TX	September
Las Vegas, NV	March	Raleigh, NC	October
New Orleans, LA	March	Salt Lake City, UT	October
Washington, DC	April and September	Baltimore, MD	November
Phoenix, AZ	April	Orlando, FL	November
Toronto	May	San Diego, CA	December
Denver, CO	May	San Antonio, TX	December

Gordian also offers a suite of online RSMeans data self-paced offerings. Check our website www.RSMeans.com for more information.

Self-Paced Professional Development Courses

Training on how to use RSMeans data and estimating tools, as well as Professional Development courses on industry topics are now offered in a convenient self-paced format. These courses are on-demand and allow more flexibility to learn around your busy schedule, while saving the cost of travel and time.

Current course offerings include:
Facilities Construction Estimating—our best-selling live class now available as an on-demand training course! Let the subject matter experts of construction estimating—the RSMeans Engineering Staff—walk you through the basics and much more of estimating for renovation and facilities construction.

RSMeansOnline.com Training—learn the ins and outs of the flagship delivery method of RSMeans data!

The Construction Process—how much do you and your team really know about the ins and outs of the "contract-side" of a construction project? This self-paced course will clarify best practices for items such as schedules, change orders, and project closeout.

These self-paced training courses can be completed over the course of 45 days and are comprised of multiple lessons with documentation, video presentation, software simulation, assessment quizzes and certificate of completion.

Site Work Estimating with RSMeans data

This new one-day program focuses directly on site work costs, a unique portion of most construction projects that often is the wild card in determining whether you have developed a good estimate or not. Accurately scoping, quantifying, and pricing site preparation, underground utility work, and improvements to exterior site elements are often the most difficult estimating tasks on any project. The program takes the participant from preparing a never-developed site through underground utility installation, pad preparation, paving and sidewalks, and landscaping. Attendees will use the full array of site work cost data through the RSMeans online program and participate in exercises to strengthen their estimating skills.

Some of what you'll learn:
- Evaluation of site work and understanding site scope of work.
- Site work estimating topics including: site clearing, grading, excavation, disposal and trucking of materials, erosion control devices, backfill and compaction, underground utilities, paving, sidewalks, fences & gates, and seeding & planting.
- Unit price site work estimates—Correct use of RSMeans site work cost data to develop a cost estimate.
- Using and modifying assemblies—Save valuable time when estimating site work activities using custom assemblies.

Who should attend: Engineers, contractors, estimators, project managers, owner's representatives, and others who are concerned with the proper preparation and/or evaluation of site work estimates.

Please bring a laptop with ability to access the internet.

Professional Development

Training for our Online Estimating Solution

Construction estimating is vital to the decision-making process at each state of every project. Our online solution works the way you do. It's systematic, flexible and intuitive. In this one-day class you will see how you can estimate any phase of any project faster and better.

Some of what you'll learn:
- Customizing our online estimating solution
- Making the most of RSMeans "Circle Reference" numbers
- How to integrate your cost data
- Generating reports, exporting estimates to MS Excel, sharing, collaborating and more

Also offered as a self-paced or on-site training program!

Maintenance & Repair Estimating for Facilities

This two-day course teaches attendees how to plan, budget, and estimate the cost of ongoing and preventive maintenance and repair for existing buildings and grounds.

Some of what you'll learn:
- The most financially favorable maintenance, repair, and replacement scheduling and estimating
- Auditing and value engineering facilities
- Preventive planning and facilities upgrading
- Determining both in-house and contract-out service costs
- Annual, asset-protecting M&R plan

Who should attend: facility managers, maintenance supervisors, buildings and grounds superintendents, plant managers, planners, estimators, and others involved in facilities planning and budgeting.

Facilities Construction Estimating

In this two-day course, professionals working in facilities management can get help with their daily challenges to establish budgets for all phases of a project.

Some of what you'll learn:
- Determining the full scope of a project
- Identifying the scope of risks and opportunities
- Creative solutions to estimating issues
- Organizing estimates for presentation and discussion
- Special techniques for repair/remodel and maintenance projects
- Negotiating project change orders

Who should attend: facility managers, engineers, contractors, facility tradespeople, planners, and project managers.

Practical Project Management for Construction Professionals

In this two-day course, acquire the essential knowledge and develop the skills to effectively and efficiently execute the day-to-day responsibilities of the construction project manager.

Some of what you'll learn:
- General conditions of the construction contract
- Contract modifications: change orders and construction change directives
- Negotiations with subcontractors and vendors
- Effective writing: notification and communications
- Dispute resolution: claims and liens

Who should attend: architects, engineers, owners' representatives, and project managers.

Construction Cost Estimating: Concepts and Practice

This one-day introductory course to improve estimating skills and effectiveness starts with the details of interpreting bid documents and ends with the summary of the estimate and bid submission.

Some of what you'll learn:
- Using the plans and specifications to create estimates
- The takeoff process—deriving all tasks with correct quantities
- Developing pricing using various sources; how subcontractor pricing fits in
- Summarizing the estimate to arrive at the final number
- Formulas for area and cubic measure, adding waste and adjusting productivity to specific projects
- Evaluating subcontractors' proposals and prices
- Adding insurance and bonds
- Understanding how labor costs are calculated
- Submitting bids and proposals

Who should attend: project managers, architects, engineers, owners' representatives, contractors, and anyone who's responsible for budgeting or estimating construction projects.

Mechanical & Electrical Estimating

This two-day course teaches attendees how to prepare more accurate and complete mechanical/electrical estimates, avoid the pitfalls of omission and double-counting, and understand the composition and rationale within the RSMeans mechanical/electrical database.

Some of what you'll learn:
- The unique way mechanical and electrical systems are interrelated
- M&E estimates—conceptual, planning, budgeting, and bidding stages
- Order of magnitude, square foot, assemblies, and unit price estimating
- Comparative cost analysis of equipment and design alternatives

Who should attend: architects, engineers, facilities managers, mechanical and electrical contractors, and others who need a highly reliable method for developing, understanding, and evaluating mechanical and electrical contracts.

Unit Price Estimating

This interactive two-day seminar teaches attendees how to interpret project information and process it into final, detailed estimates with the greatest accuracy level.

The most important credential an estimator can take to the job is the ability to visualize construction and estimate accurately.

Some of what you'll learn:

- Interpreting the design in terms of cost
- The most detailed, time-tested methodology for accurate pricing
- Key cost drivers—material, labor, equipment, staging, and subcontracts
- Understanding direct and indirect costs for accurate job cost accounting and change order management

Who should attend: corporate and government estimators and purchasers, architects, engineers, and others who need to produce accurate project estimates.

Life Cycle Cost Estimating for Facility Asset Managers

Life Cycle Cost Estimating will take the attendee through choosing the correct RSMeans database to use and then correctly applying RSMeans data to their specific life cycle application. Conceptual estimating through RSMeans new building models, conceptual estimating of major existing building projects through RSMeans renovation models, pricing specific renovation elements, estimating repair, replacement and preventive maintenance costs today and forward up to 30 years will be covered.

Some of what you'll learn:
- Cost implications of managing assets
- Planning projects and initial & life cycle costs
- How to use RSMeans data online

Who should attend: facilities owners and managers and anyone involved in the financial side of the decision making process in the planning, design, procurement, and operation of facility real assets.

Please bring a laptop with ability to access the internet.

Training for our CD Estimating Solution

This one-day course helps users become more familiar with the functionality of the CD. Each menu, icon, screen, and function found in the program is explained in depth. Time is devoted to hands-on estimating exercises.

Some of what you'll learn:
- Searching the database using all navigation methods
- Exporting RSMeans data to your preferred spreadsheet format
- Viewing crews, assembly components, and much more
- Automatically regionalizing the database

This training session requires you to bring a laptop computer to class.

When you register for this course you will receive an outline for your laptop requirements.

Also offered as a self-paced or on-site training program!

Assessing Scope of Work for Facilities Construction Estimating

This two-day practical training program addresses the vital importance of understanding the scope of projects in order to produce accurate cost estimates for facility repair and remodeling.

Some of what you'll learn:
- Discussions of site visits, plans/specs, record drawings of facilities, and site specific lists
- Review of CSI divisions, including means, methods, materials, and the challenges of scoping each topic
- Exercises in scope identification and scope writing for accurate estimating of projects
- Hands-on exercises that require scope, take-off, and pricing

Who should attend: corporate and government estimators, planners, facility managers, and others who need to produce accurate project estimates.

Facilities Estimating Using the CD

This two-day class combines hands-on skill-building with best estimating practices and real-life problems. You will learn key concepts, tips, pointers, and guidelines to save time and avoid cost oversights and errors.

Some of what you'll learn:
- Estimating process concepts
- Customizing and adapting RSMeans cost data
- Establishing scope of work to account for all known variables
- Budget estimating: when, why, and how
- Site visits: what to look for and what you can't afford to overlook
- How to estimate repair and remodeling variables

This training session requires you to bring a laptop computer to class.

Who should attend: facility managers, architects, engineers, contractors, facility tradespeople, planners, project managers, and anyone involved with JOC, SABRE, or IDIQ.

Building Systems and the Construction Process

This one-day course was written to assist novices and those outside the industry in obtaining a solid understanding of the construction process - from both a building systems and construction administration approach.

Some of what you'll learn:
- Various systems used and how components come together to create a building
- Start with foundation and end with the physical systems of the structure such as HVAC and Electrical
- Focus on the process from start of design through project closeout

This training session requires you to bring a laptop computer to class.

Who should attend: building professionals or novices to help make the crossover to the construction industry; suited for anyone responsible for providing high level oversight on construction projects.

Professional Development

Registration Information

Register early and save up to $100!

Register 30 days before the start date of a seminar and save $100 off your total fee. Note: This discount can be applied only once per order. It cannot be applied to team discount registrations or any other special offer.

How to register

By Phone
Register by phone at 877-620-6245

Online
Register online at
www.RSMeans.com/products/seminars.aspx

Note: Purchase Orders or Credits Cards are required to register.

Two-day seminar registration fee - $1,045.

One-Day Construction Cost Estimating or Building Systems and the Construction Process - $630.

Government pricing

All federal government employees save off the regular seminar price. Other promotional discounts cannot be combined with the government discount.

Team discount program

For over five attendee registrations. Call for pricing: 781-422-5115

Refund policy

Cancellations will be accepted up to ten business days prior to the seminar start. There are no refunds for cancellations received later than ten working days prior to the first day of the seminar. A $150 processing fee will be applied for all cancellations. Written notice of the cancellation is required. Substitutions can be made at any time before the session starts. No-shows are subject to the full seminar fee.

Note: Pricing subject to change.

AACE approved courses

Many seminars described and offered here have been approved for 14 hours (1.4 recertification credits) of credit by the AACE International Certification Board toward meeting the continuing education requirements for recertification as a Certified Cost Engineer/Certified Cost Consultant.

AIA Continuing Education

We are registered with the AIA Continuing Education System (AIA/CES) and are committed to developing quality learning activities in accordance with the CES criteria. Many seminars meet the AIA/CES criteria for Quality Level 2. AIA members may receive 14 learning units (LUs) for each two-day RSMeans course.

Daily course schedule

The first day of each seminar session begins at 8:30 a.m. and ends at 4:30 p.m. The second day begins at 8:00 a.m. and ends at 4:00 p.m. Participants are urged to bring a hand-held calculator since many actual problems will be worked out in each session.

Continental breakfast

Your registration includes the cost of a continental breakfast and a morning and afternoon refreshment break. These informal segments allow you to discuss topics of mutual interest with other seminar attendees. (You are free to make your own lunch and dinner arrangements.)

Hotel/transportation arrangements

We arrange to hold a block of rooms at most host hotels. To take advantage of special group rates when making your reservation, be sure to mention that you are attending the RSMeans Institute data seminar. You are, of course, free to stay at the lodging place of your choice. (Hotel reservations and transportation arrangements should be made directly by seminar attendees.)

Important

Class sizes are limited, so please register as soon as possible.

Professional Development